3rd European Conference on Unsaturated Soils (E-UNSAT 2016)

E3S Web of Conferences Volume 9 (2016)

Paris, France
12 - 14 September 2016

Part 1 of 2

Editors:

P. Delage
Y.-J. Cui
S. Ghabezloo
J.-M. Pereira
A.-M. Tang

ISBN: 978-1-5108-3018-9

Printed from e-media with permission by:

Curran Associates, Inc.
57 Morehouse Lane
Red Hook, NY 12571

Some format issues inherent in the e-media version may also appear in this print version.

Printed by Curran Associates, Inc. (2016)

For additional information, please contact EDP Sciences – Web of Conferences at the address below.

EDP Sciences – Web of Conferences
17, Avenue du Hoggar
Parc d'Activité de Courtabœuf
BP 112
F-91944 Les Ulis Cedex A
France

Phone: +33 (0) 1 69 18 75 75
Fax: +33 (0) 1 69 28 84 91

contact-edps@webofconferences.org

Additional copies of this publication are available from:

Curran Associates, Inc.
57 Morehouse Lane
Red Hook, NY 12571 USA
Phone: 845-758-0400
Fax: 845-758-2633
Email: curran@proceedings.com
Web: www.proceedings.com

TABLE OF CONTENTS

PART 1

PART 2

E3S Web of Conferences
Open access conference proceedings
in Environment, Energy & Earth Sciences
A flexible, fast and efficient
publication service,
for an open access recording
of your event
and a wide dissemination of
your proceedings
www.e3s-conferences.org
edp open
Open Access Publishing
from EDP Sciences

ISSN (electronic edition): 2267-1242

Preface

Geotechnical research into the field of Unsaturated soils started in the 50's with some seminal contributions including pioneering papers by Croney, Coleman and Bishop. An important further step occurred with the venue of the UNSAT'95 - First International Conference on Unsaturated Soils held in Paris in 1995, following a series of successful Conferences on Swelling soils with the last one held in Dallas in 1992. This series of conferences sponsored by the TC6/TC106 Technical Committee of the International Society of Soil Mechanics and Geotechnical Engineering (ISSMGE) was afterwards followed by a series of successful Regional Conferences that started in Singapore in 2000 with the Asia-Pacific series, followed by the European one, started in Durham (UK) in 2008 and, finally, the Pan-American series started in Cartagena de las Indias (Colombia) in 2013.

In the meantime, research into the behaviour of unsaturated soils expanded in a spectacular manner in terms of laboratory/in-situ investigations and conceptual/numerical modelling. Researches were aimed at solving practical engineering problems based on sound analyses of coupled multi-physics phenomena that became less mysterious thanks to the sound and deep investigations carried out all over the world by a constantly increasing number of research groups.

In this context, the E-UNSAT2016 Paris Conference attracted quite a good response to the call for papers with 166 accepted papers from all over the world, quite a large number for a regional Unsat Conference. 18 Distinguished Lecturers (Keynote, Plenary and Invited Lecturers) also provide insights into the more advanced achievements in the field, playing an important role in the interest and scientific outcome of the Conference. The topics that drew particular interest are *Advances in experimental methods*, *Water retention and transport properties* and *Mechanical behaviour*, with also significant contributions on *Numerical and Constitutive modelling*. The practical applications that have been dealt with are *Dams and dykes*, *Soil-Vegetation-Atmosphere interaction*, *Slope stability* and *Geo-environmental applications* (a session organised together with the ISSMGE TC215).

As expected and confirmed by the large number and the high quality of the papers received, the E-UNSAT2016 Paris Conference fostered fruitful presentations and discussions on up to date ideas, concepts and achievements in Unsaturated Soils, from many researchers from Europe and other continents.

Prof. Pierre DELAGE
Chairman of E-UNSAT2016 Conference
Ecole des Ponts ParisTech, France

Committees and Acknowledgements to the reviewers

Organising committee

- Pierre Delage (Conference chair)
- Yu-Jun Cui (Conference co-chair)
- Siavash Ghabezloo
- Jean-Michel Pereira
- Anh-Minh Tang
- Séverine Beaunier

Advisory committee

- E. Alonso (*Spain*)
- C. Augarde (*UK*)
- M. Bardanis (*Greece*)
- O. Buzzi (*Australia*)
- R. Charlier (*Belgium*)
- B. Caicedo (*Colombia*)
- R. Cardoso (*Portugal*)
- F. Casini (*Italy*)
- T.M.P De Campos (*Brazil*)
- F. D'Onza (*Italy*)
- D. Fredlund (*Canada*)
- D. Gallipoli (*France*)
- K. Garbulewski (*Poland*)
- B. Gatmiri (*Iran*)
- A. Gens (*Spain*)
- G. Habibagahi (*Iran*)
- S. Houston (*USA*)
- L. Hoyos (*USA*)
- E. Imre (Hungary)
- C. Jommi (*Netherlands*)
- N. Khalili (*Australia*)
- L. Laloui (*Switzerland*)
- C. Mancuso (*Italy*)
- D. Manzanal (*Argentina*)
- F. Marinho (*Brazil*)
- D. Masin (*Czech Republic*)
- F. Masrouri (*France*)
- G. Miller (*USA*)
- C. Ng (*Hong Kong*)
- T. Nishimura (*Japan*)
- L. Oldecop (*Argentina*)
- H. Rahardjo (*Singapore*)
- E. Romero (*Spain*)
- A. Russell (*Australia*)
- M. Sanchez (*USA*)
- T. Schanz (*Germany*)
- D. Sheng (*Australia*)
- S. Springman (*Switzerland*)
- H.W. Sun (*Hong Kong*)
- D. Toll (*UK*)
- S. Tripathy (*UK*)
- S. Vanapalli (*Canada*)
- J. Vaunat (*Spain*)
- S. Wheeler (*UK*)
- T. Zhang (*China*)

Careful reviewing is an essential process to ensure the high quality of the papers accepted at the Conference and published in the Proceedings. The Organising Committee is most grateful to the following reviewers for their kind and helpful contribution to the success of the Conference.

Alonso Eduardo, Ashayeri Iman, Augarde Charles, Baille Wiebke, Bicalho Katia, Biglari Mahnoosh, Buzzi Olivier, Caicedo Bernardo, Cardoso Rafaela, Chen Yonggui, Cordão-Neto Manoel, Cuisinier Olivier, Dangla Patrick, Deng Yongfeng, Fabbri Antonin, Feia Sadok, Fleureau Jean-Marie, Fredlund Delwyn, Gallipoli Domenico, Gens Antonio, Ghayoomi Majid, Habibagahi Ghassem, Heyerdahl Håkon, Hong Zhenshun, Khelifa Abbeche, Le Pense Solenn, Lloret Antonio, Manzanal Diego, Mašin David, Mendes João, Morvan Mathilde, Nikooee Ehsan, Nishimura Tomoyoshi, Rahardjo Harianto, Roux Jean-Noël, Russo Giacomo, Salager Simon, Seyedi Darius, Sheng Daichao, Stefanou Ioannis, Sun Wenjing, Tang Chao-Sheng, Teng Jidong, Tripathy Snehasis, Vanapalli Sai, Villar María Victoria, Wang Qiong, Zeng Ling-Ling, Zha Fusheng, Zhou Annan.

Supported by

École des Ponts
ParisTech

SIMSG ISSMGE

TC 106 - Unsaturated Soils

CFMS

Comité français de mécanique des sols et de géotechnique

LABEX MMCD

ÉCOLE DOCTORALE UNIVERSITÉ PARIS-EST
Sciences, Ingénierie et Environnement

edp sciences

E3S Web of Conferences

Statement of Peer review

In submitting conference proceedings to E3S Web of Conferences, I certify to the Publisher that I adhere to the **Policy on Publishing Integrity** of the journal in order to safeguard good scientific practice in publishing.

1. All articles have been subjected to peer review administered by the proceedings editors.
2. Reviews have been conducted by expert referees, who have been requested to provide unbiased and constructive comments aimed, whenever possible, at improving the work.
3. Proceedings editors have taken all reasonable steps to ensure the quality of the materials they publish and their decision to accept or reject a paper for publication has been based only on the merits of the work and the relevance to the journal.

Title, date and place of the conference

3rd European Conference on Unsaturated Soils
12-14 September 2016
Ecole des Ponts, Marne La Vallée

Proceedings editor(s):

P. Delage, Y.J. Cui, S. Ghabezloo, J.M. Pereira, A.M Tang

Date and editor's signature

DOI: 10.1051/e3sconf/20160901001

Unsaturated railway track-bed materials

Yu-Jun Cui

Ecole des Ponts ParisTech, U.R. Navier/CERMES, 6 – 8 av. Blaise Pascal, Cité Descartes, Champs – sur – Marne, 77455 Marne – la – Vallée cedex 2, France

Abstract. Railway track-bed materials are mostly in unsaturated state, and their hydro-mechanical properties depend strongly on their water contents or suctions. In France, problematic soils such as collapsible loess and swelling marl can be origin of instability problems for new lines for high speed trains, while the hydro-mechanical behaviour of interlayer soil formed mainly by interpenetration of ballast and subgrade soil is of concern for conventional lines. In the latter case, the main challenge relies in the large variability of the interlayer soils owing to the variability of natural subgrade soils involved in the railway network. In this paper, a study on interlayer soils is presented, that covers a large spectrum of aspects: geotechnical/geophysical site characterisation, laboratory investigation into the hydraulic behaviour and mechanical behaviour as well as the mud pumping/interlayer creation phenomena, and field monitoring. The combined effect of water content and fines content were emphasized in the laboratory investigation, whilst interaction between atmosphere and track was focused on in the field monitoring. The results show clearly that it is of paramount importance to consider the unsaturated aspect of track-bed materials when analysing the overall behaviour of tracks, in particular when clay is involved in fine fraction.

1 Introduction

In France, most railway tracks are ballasted and the whole network is composed of new lines for high speed trains and conventional lines for other categories. In the case of new lines, a sub-layer of gravel is placed between ballast and subgrade, and sometimes a capping layer is put with treated soil between the sub-layer and subgrade (see Figure 1). This design has been proven to be successful when dealing with sub-grades of "noble soils". But when special soils such as expansive soils and collapsible soils are involved, additional measures should be taken in order to ensure the stability of tracks under the combined effect of water infiltration and cyclic loading. Indeed, in the case of northern line (Paris – Lille), a subgrade of loess is involved. Under the effects of rainfall and dynamic loading, track instability problems were identified, leading to an extensive study to identify the main instability mechanisms (collapse upon wetting, fatigue under large number of loading cycles and liquefaction in the saturated sate) [1-6]. In the case of Mediterranean Line (Paris – Marseille), a subgrade of marl was encountered and its swelling due to the coupled effect of water infiltration and excavation-related unloading was revealed to be one of the main mechanisms for the significant track heave identified [7].

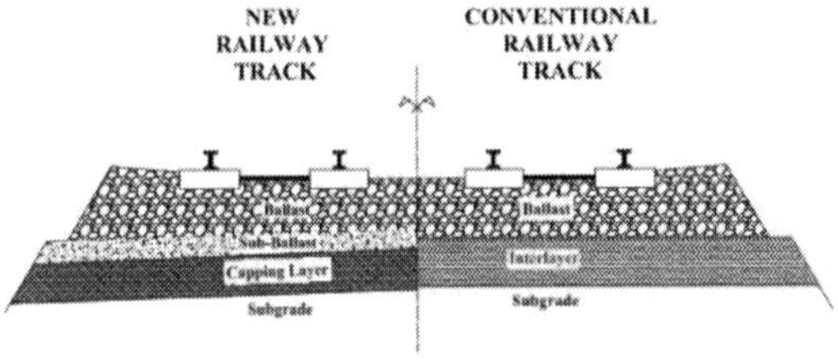

Figure 1. Comparison between new and conventional railway tracks.

As opposed to the new lines, the conventional lines have their ballast directly placed on the subgrade during their construction. Under the effect of train circulation over time, an interlayer was created mainly by the interpenetration of ballast and subgrade soils (see Figure 1). This layer is found to be well "compacted" by train circulation over long time: the in-situ dry density can be as high as 2.4 Mg/m^3 [8]. Thereby, it constitutes a good bearing layer in the sub-structure. This is why it has been decided to keep this layer in the French program of conventional track renewal. Note that this track renewal program is important because it almost involves the whole French network: the conventional lines represent 94% of the whole network of about 30 000 km [9]. Figure 2 shows a picture of soil profile taken when during sampling at the site of Sénissiat, France, after removing the ballast layer. An interlayer composed of coarse grains and fine particles can be clearly identified. When approaching the subgrade layer, the fraction of coarse grains seems to decrease, so that a transition layer can be defined that differentiated from interlayer by a lower coarse grain fraction.

Figure 3 shows a comparison of grain size distribution between sub-ballast and the interlayer soil taken at the site of Sénissiat. It appears that the main difference between the sub-ballast and the interlayer soil lies in the fines content: there is no grain smaller than 0.1 mm in sub-ballast whereas the fraction of grains smaller than 0.1 mm can reach 20% in the interlayer soil. Due to the presence of fines in the interlayer, the hydro-mechanical behaviour of the interlayer soil can be greatly affected by the sensitivity of the fines to changes in water content. Furthermore, as the nature and the amount of fines can be quite different from one site to another, the variability of materials constituting the interlayer is also an important issue to be addressed.

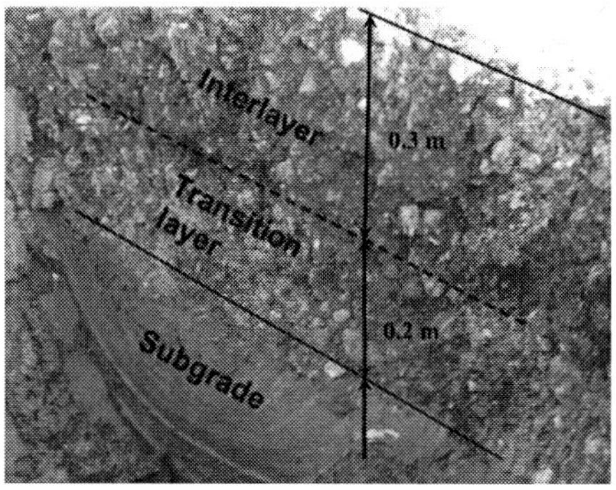

Figure 2. Interlayer at the site of Sénissiat.

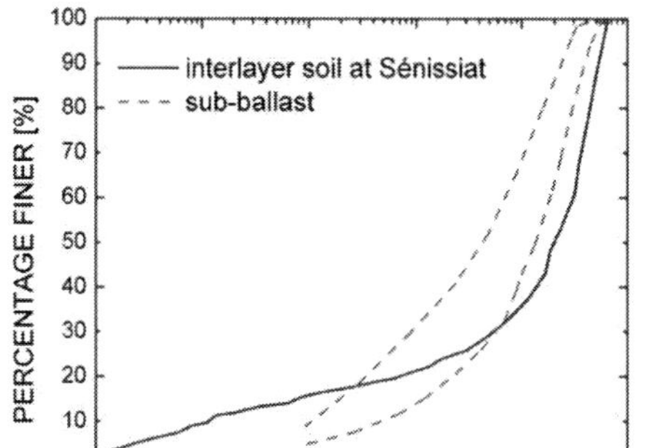

Figure 3. Comparison of grain size distribution between sub-ballast and interlayer soil.

In this paper, the hydro-mechanical behaviour of interlayer materials is addressed through laboratory testing, physical modelling and field monitoring. A large scale infiltration cell equipped with tensiometer and TDR was developed to study the unsaturated hydraulic behaviour, and a large scale triaxial cell was employed to investigate the effects of water content and fines content under cyclic loading conditions. A physical model was developed to study the interlayer creation and the mud pumping phenomena. The field monitoring was conducted to investigate the interaction between the atmosphere and track.

2 Site selection and prospection

A site on a conventional line was needed to conduct the study on the hydro-mechanical behaviour of interlayer materials. To make the site representative, different criteria should be applied. For instance, for the selection of Vierzon site [10], it was imposed that the track components as rail and sleepers should not have more than ten years since their last renewal. In addition, the maximum service speed should be comprised between 200 and 220 km/h. A site with tracks in an alignment was also required to obtain the same solicitation at each side of each track. After selection of the site, geophysical and geotechnical prospections were conducted in order to investigate the site characteristics such as the Rayleigh wave velocity that is related to the critical velocity of train and the profile of soils of different geotechnical properties.

Different geophysical and geotechnical methods have been used in railway auscultations. The commonly used geophysical techniques are the multichannel analysis of surface waves (MASW), the ground penetrating radar or down-hole tests. MASW is used to analyse the propagation wave velocities in soil and the thicknesses of different layers of soil [11]. Ground penetrating radar (GPR) defines the thicknesses of different layers of substructure [12-13]. Down-hole seismic test is suitable for estimating the shear wave velocities of different track-bed materials [14]. On the other hand, most of the geophysical methods present some difficulties in track-bed investigations. For instance, the MASW is not easy to apply on track surface with ballast because common geophone pins are not adapted to be installed into large grains as ballast due to contact failure. Other geophysical methods, as GPR, are more suitable for qualitative auscultation of track-bed layers and for estimation of the thicknesses of track layers. For the geotechnical techniques, the mostly used ones are the dynamic penetration test (as PANDA tests) and the dynamic plate loading test (as the light weight deflectometer, LWD) [15-16]. They can help define the quality of a track-bed and the required thickness for different layers. The LWD test estimates the dynamic modulus of a soil, and is used for quality control. The PANDA test now becomes a common technique in France [17] since it allows prospecting the stiffness of existing tracks. The PANDA test measures the tip resistance of different materials. The elastic modulus of materials can be further estimated from their tip resistances using empirical equations.

The geo-endoscopic test uses a little video camera (wired to a data logger with a soft cable) to observe the soil. The camera is introduced into the hole of a previously performed PANDA test (15 mm of diameter). The results allow a qualitative characterization of soil, and even the estimation of the soil grain size distribution.

DOI: 10.1051/e3sconf/20160901001

The geo-endoscopic test is largely employed test in track prospections since the images and video obtained from this test allow the determination of different materials constituting the investigated soil layers: the thickness, the soil nature and even the state of saturation. This information can help better analyse the PANDA test results. In practice, PANDA and geo-endoscopic tests are often coupled for a same auscultation point.

Figure 4 shows the situations of prospection tests at the Vierzon site. Figure 5 shows different materials observed by geo-endoscope and Figure 6 gives an example of PANDA test results [10].

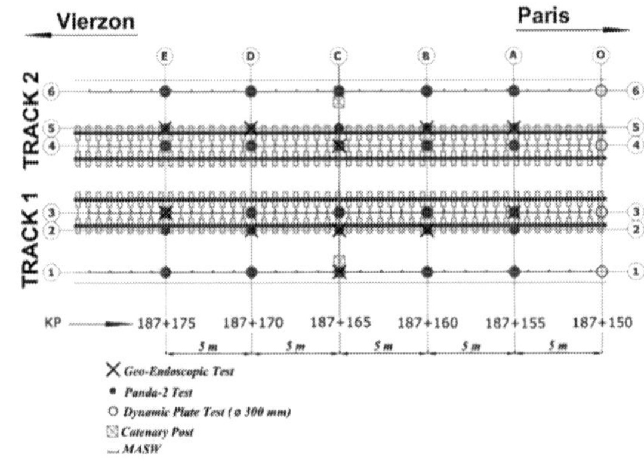

Figure 4. Prospection tests performed on the 'Vierzon experimental site'.

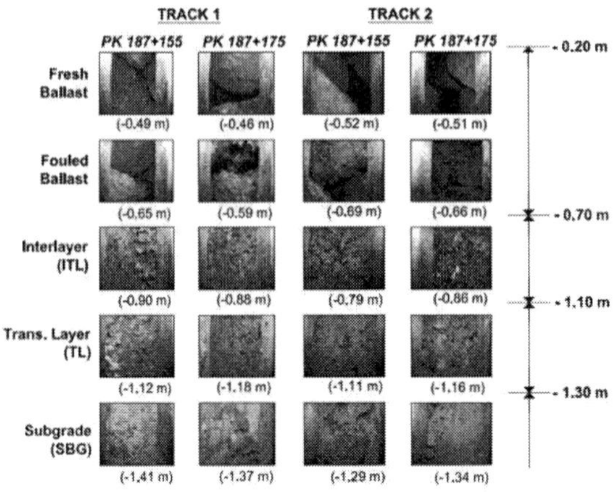

Figure 5. Different materials observed by Geo-endoscopic tests.

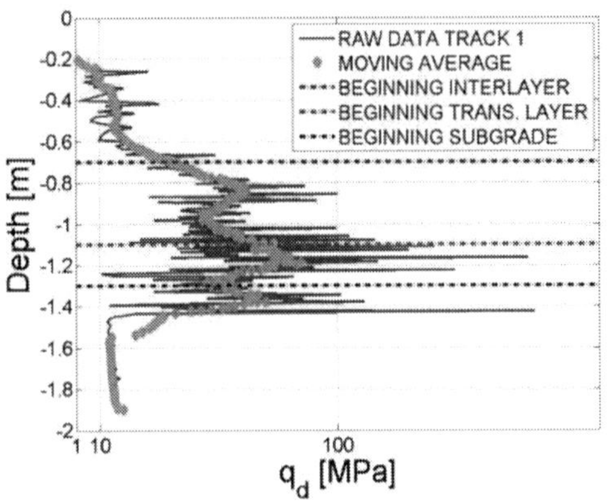

Figure 6. PANDA test result on track 1 and moving average (of 8 records).

All the test results allow the soil profile at Vierzon site to be well defined (see Figure 7), with 50 cm ballast (fresh and fouled), 40 cm interlayer (ITL), about 20 cm transition layer (TL) overlying the SBG. The water table is at 1.20 m depth from the rail level, near the interface between TL and SBG. Thus, the interlayer material is expected to be unsaturated.

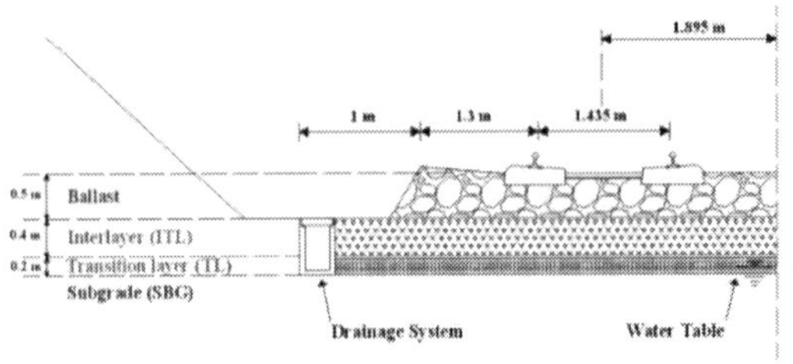

Figure 7. Profile of Vierzon site defined by the prospection tests.

3 Laboratory investigation

The laboratory investigation results are presented in three parts: i) hydraulic behaviour of interlayer soil, ii) mechanical behaviour of interlayer soil, and iii) physical modelling of the interlayer creation and mud pumping phenomena. Emphasis was put on the role of suction generated in the part of fine particles in the global hydraulic behaviour, while the roles of both suction/water content and fines content were stressed for the overall mechanical behaviour.

3.1 Hydraulic behaviour

The effect of fines content on the hydraulic behavior of interlayer soil was studied through the comparison between the fines (< 2mm) and ITL_{10} that is an interlayer soil with addition of a quantity of sub-grade representing 10% interlayer soil by dry mass [18].

The fines content was determined using a small-scale infiltration column of 50 mm in diameter and 200 mm in height, while ITL_{10} was tested in a large-scale infiltration column (300 mm in diameter and 600 mm in height) equipped with five TDRs for measuring water content and five tensiometers for measuring pore-water pressure [19]. The dry unit mass and water content of fines were taken equal to those of the fine particles contained in the sample of interlayer soil. Unlike the large-scale column where both suction and water content were monitored, the small-scale column has only suction monitored. To obtain the water content changes during infiltration, the soil-water retention curve (SWRC) was needed. This was done separately using an oedometer cell equipped with a high-capacity tensiometer (see Le et al. [20] for more details).

The instantaneous profile method [21-25] was applied for the determination of hydraulic conductivity for each soil. Note that this method is based on the generalized Darcy's law. The hydraulic gradient is determined by considering the slope of suction isochrones and the water volume passing through a given section between times t and $t+dt$ is used for calculating the water flux.

Figure 8 depicts the comparison of SWRC between ITL_{10} and fines. The curves of the two soils start from almost the same point - around 97% degree of saturation and 1.7 kPa suction. Two drying curves were determined for fines, while one drying curve and two wetting curves were determined for ITL_{10}. It is observed that the two drying curves of ITL_{10} are almost the same, so are the two wetting curves of fines, suggesting a good repeatability of the tests. A limited hysteresis was also identified for fines. Moreover, from 3 kPa suction, the curves of fines start to separate from those of ITL_{10}. The curves of drying path for the two soils are almost parallel from 10 kPa suction. The curves of ITL_{10} stop at 71 kPa while the curves of fines stop at 389 kPa due to the different capacities of the tensiometers used for the two soils. The gap between two curves is about 10% degree of saturation at the end of the curves for ITL_{10}.

In Figure 9, the hydraulic conductivity of ITL_{10} and fines is plotted versus suction, together with the fitting curves using van Genuchten's and Brook-Corey's models. It can be observed that in unsaturated state the wetting and drying curves of the interlayer soil are quite close to those of fines, suggesting that the hydraulic conductivity of the interlayer soil is mainly governed by the hydraulic conductivity of the fines contained in it. In other words, in unsaturated state, water transfer in the interlayer soil takes place mainly through the network of pores between fine particles, coarse elements like ballast behaving as inert materials. By contrast, in saturated state, a value of 1.67×10^{-5} m/s was obtained for ITL_{10}, higher than the value for fines (2.6×10^{-6} m/s). This difference is considered as being significant because as opposed to the determination of unsaturated hydraulic conductivity, the determination of saturated hydraulic conductivity can be deemed accurate. The higher value for ITL_{10} can be explained as follows: the macro-pores in ITL_{10} are larger than those in fines and water flow in saturated state took place mainly through macro-pores. Thereby, the water flow mechanism in saturated state is different from that in unsaturated state.

From a practical point of view, Figure 9 shows that when determining the unsaturated hydraulic conductivity of interlayer soils, it is not necessary to use large-scale experimental devices to match the soil grain size; smaller devices can be used to determine their hydraulic conductivity by testing the fine particles only, provided that equivalent dry density is accounted for. This is however not valid for saturated state.

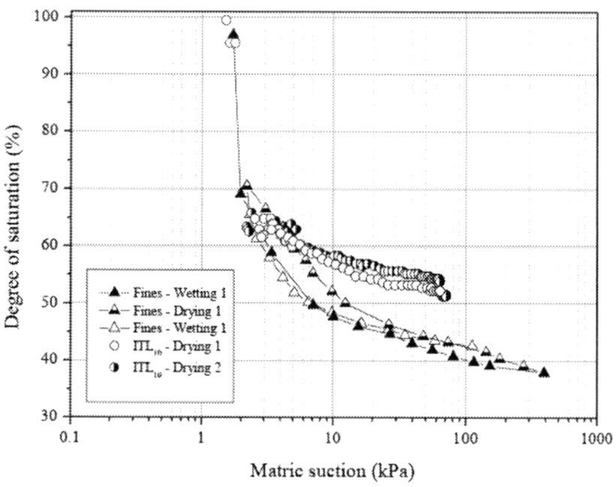

Figure 8. Comparison of SWRC between ITL_{10} and fines.

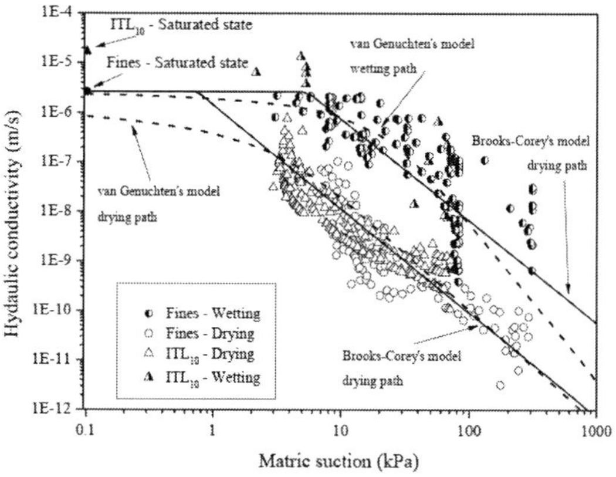

Figure 9. Comparison of hydraulic conductivity between ITL_{10} and fines.

On the other hand, Figure 9 also shows an uncommon phenomenon of hysteresis in hydraulic conductivity for ITL_{10} and fines. Indeed, it is observed that the curve of wetting path lies above the curve of drying path. The same phenomenon was observed by Wayllace and Lu [26]. The following interpretations can be attempted. Due to the different kinetic between the fast liquid transfer in wetting and the long vapour transfer in drying, the time needed for drying was much longer than that for wetting. This phenomenon was also reported by Toker et al. [27] and discussed by Munoz-Castelblanco et al. [5]. A higher hydraulic conductivity can be expected in that case for wetting path. In this study, the calculation of hydraulic conductivity was performed based on the suction evolution given by tensiometers. Assuming that in the compacted soils, both macro-pores and micro-pores exist. During wetting, the macro-pores were filled with water more quickly than micro-pores. Moreover, in the micro-pores, there were always air bubbles preventing the total saturation. By contrast, during drying, all pores (micro and macro) participated in the evaporation process. As a result, when water filled the macro-pores, the tensiometers immediately gave the suction changes corresponding to the water flow through the macro-pores, even though the suction in micro-pores would be higher. On the contrary, when water evaporates during drying, the tensiometers gave the suction changes that involve both macro- and micro-pores, in a much slower fashion. In other words, the suction measured by the tensiometers was probably under-estimated for wetting paths. Côté and Roy [28] also reported that one re-saturating stage is not enough to fully saturate a soil sample because of the air bubbles trapped in micro-pores. This can also explain the uncommon hysteresis mentioned before.

3.2 Mechanical behaviour

The mechanical behaviour has been investigated using a large scale cyclic triaxial cell that can host a sample of 300 mm diameter and 600 mm height [29]. Emphasis was put on the effects of water content and fines content on the resilient modulus and permanent deformation [30]. Based on the experimental results, a constitutive model was developed, taking into account both the effect of stress level and the effect of degree of saturation [9].

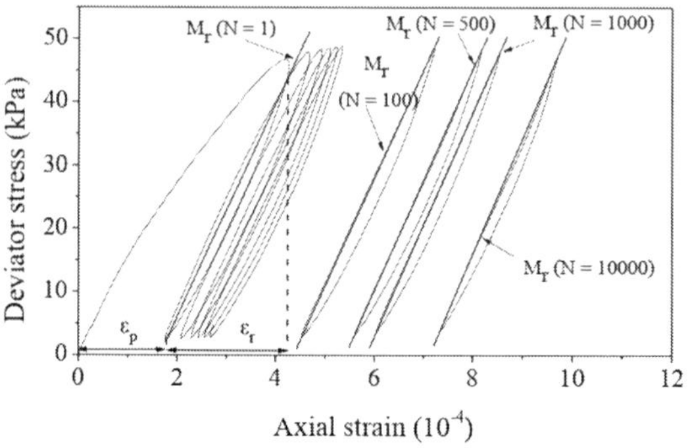

Figure 10. Deviator stress versus axial strain for test ITL5w6 - Determination of resilient modulus.

Figure 10 presents typical curves of deviator stress versus axial strain for the first cycles in a test on the interlayer soil with addition of 5% subgrade soil and at 6% water content - ITL5w6 [31]. It also shows the definition of permanent axial strain (ε_p), the resilient axial strain (ε_r) through the first cycle. The resilient modulus (M_r) corresponds to the secant slope of curves (see Figure 10).

Three water contents (4%, 6% and 12%) and four fines contents were considered (0% - natural ITL soil ITL_0, 5% - addition of 5% subgrade soil ITL_5, 10% - addition of 10% subgrade soil ITL_{10}, as well as -10% - removal of 10% fines from the natural ITL soil ITL_{-10}). Note that the water content of 12% corresponds to the saturated state, and only one test was carried out on the soil with 5% fine addition. The end-level resilient modulus values (after 30 000 cycles for each deviator stress level) are plotted versus deviator stress in Figure 11 for ITL_{10}, ITL_0 and ITL_{-10}. The value of ITL_{-10} at 4% water content is the highest; the values of 12% water content are lower than those of 4% water content but higher than those of 6% water content. The relative positions of these curves show that the water content effect is not the same for the three materials. It appears that the increase of water content brings a negative effect to the resilient modulus of the soil that has the highest fines content (ITL_{10}). However, for ITL_0 and ITL_{-10}, the resilient modulus decreased when water content changed from 4% to 6% but increased when water content was increased to 12%. Note that only a decreasing trend of resilient modulus with growing saturation level for granular material was observed by Lekarp at al. [32]. More studies are required to clarify this issue.

Figure 12 presents the results according to the water content: 4% (Figure 12a), 6% (Figure 12b) and 12% (Figure 12a). The effect of fines content can be observed clearly. In the case of w = 4%, the results of ITL_{-10} and ITL_{10} are almost identical for the deviator stress up to 100 kPa. But much lower values are observed for ITL_0 for the deviator stress up to 140 kPa. Beyond 140 kPa deviator stress, the values become almost the same for ITL_{-10} and ITL_0. At 6% water content, the resilient modulus of ITL_0 is also the smallest, showing that when decreasing the fines content (from ITL_{10} to ITL_0), a reduction of resilient modulus is produced. However, when the fines content is decreased to a level as low as ITL_{-10}, the resilient modulus starts to increase. Under the saturation conditions (w = 12% in Figure 12c), it is observed that the greater the fines content, the lower the resilient modulus.

 DOI: 10.1051/e3sconf/20160901001

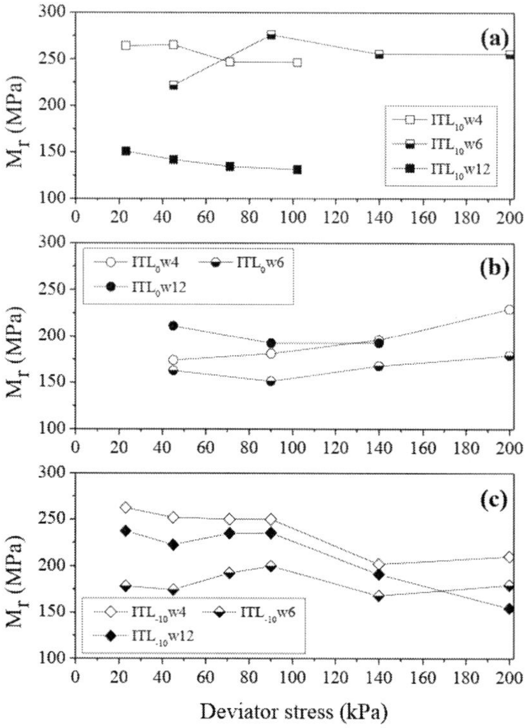

Figure 11. End-stage resilient modulus versus deviator stress - Effect of water content. a) ITL_{10}; b) ITL_0; c) ITL_{-10}.

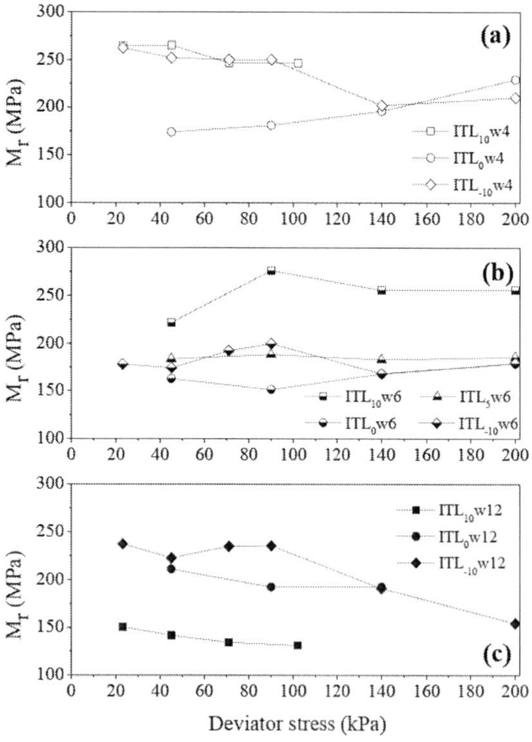

Figure 12. End-stage resilient modulus versus deviator stress - Effect of fines content. a) $w = 4\%$; b) $w = 6\%$ and c) $w = 12\%$.

Figure 13 presents the permanent axial strains obtained at the end of each loading step (same number of cycles) versus the corresponding maximum deviator stress Δq_{max} for each soil at three different water contents, allowing analyzing the effects of stress level and water content on the permanent axial strain. For ITL_{-10} (Figure 13a), the curves are similar up to 90 kPa deviator stress. Beyond 90 kPa, the permanent axial strain of $ITL_{-10}w4C$ is smaller than those of $ITL_{-10}w6C$ and $ITL_{-10}w12C$. Note that failure occurred for the nearly saturated specimen (w = 12%) when Δq_{max} = 200 kPa. For ITL_0 (Figure 13b), three curves are well separated. At all stress levels, the higher the water content, the larger the permanent axial strain. Failure was also observed for the nearly saturated specimen at Δq_{max} = 200 kPa. For ITL_{10} (Figure 13c), the curve of the specimens at w = 4 and 6% are similar, while for the nearly saturated specimen ($ITL_{10}w12C$), the permanent axial strain is much larger. The effect of water content depends not only on the soil nature but also on the variation range of water content. Changing water content from 4% to 6% does not significantly affect the permanent axial strain of ITL_{10} but induces significant changes in the case of lower fines content (ITL_{-10} and ITL_0). On the contrary, increasing water content from 6% to 12% induces a significant change in the permanent axial strain for ITL_{10}, but not for ITL_{-10} when the stress is lower than 140 kPa. At a given stress level, increasing water content results in an increase in permanent axial strain.

Figure 14 shows the permanent axial strains obtained at the end of each loading step versus the corresponding q_{max} for each water content value, allowing analyzing the effect of fines content on the permanent axial strain. In the case of w = 4% and 6% (Figures 14a and14b, respectively), the permanent axial strains of $ITL_{-10}w4C$ and $ITL_{-10}w6C$ are significantly higher than the others, suggesting that the lower the fines content, the larger the permanent axial strain. In nearly saturated state (Figure 14c), the effect of fines content is inversed: the soil having the highest fines content (ITL_{10}) exhibits the largest permanent axial strain [30]. Uthus et al. [33] also observed that the effect of water content on the permanent axial strain depends on the fines content. This phenomenon can be explained as follows: in unsaturated state, water is mainly trapped by fine particles. At the same density and water content, if the fines content is higher, the soil suction must be higher. As a result, the soil is mechanically more resistant. On the other hand, fine particles are well known to be very sensitive to changes in water content, and when the water content is 12%, the soil becomes saturated and its suction approaches zero, leading to a significant decrease in fine particles strength and further the overall mechanical behavior of soil. Similar observations were reported by Fortunato et al. [34], Seif El Dine et al. [35] and Huang et al. [36].

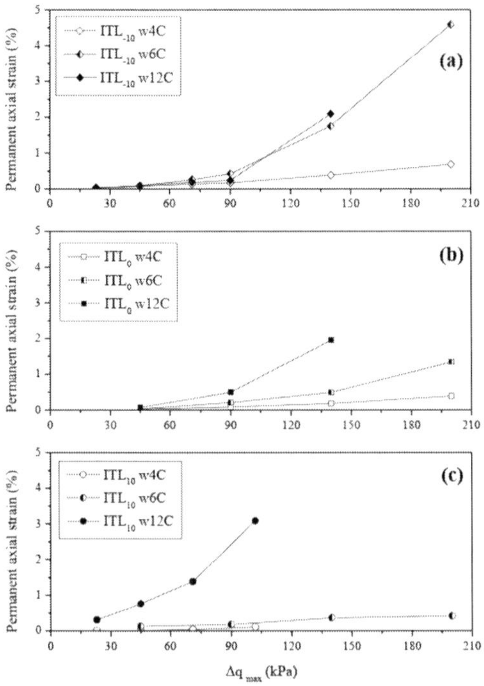

Figure 13. Effect of water content on the end-stage permanent axial strain for ITL_{-10}, ITL_0 and ITL_{10}.

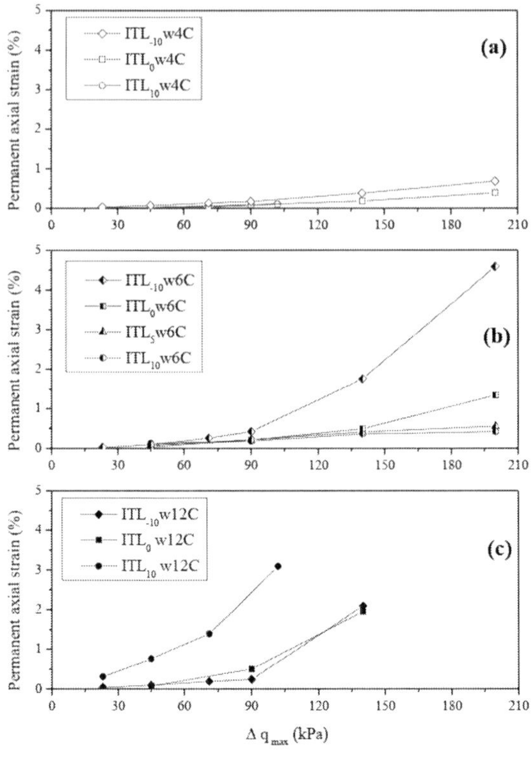

Figure 14. Effect of fines content on the end-stage permanent axial strain at various water contents.

On the whole, at a given water content in dry condition, adding more fines has a positive effect on the mechanical behavior of interlayer soil (the permanent axial strain is reduced), while in nearly saturated condition, adding more fines boosts the axial strain. This means that in the railway context, during the assessment and exploitation of the interlayer soil, the effect of water content and fines content must be taken into account simultaneously. The interlayer soil containing a larger quantity of fine particles have to be protected from water infiltration in order to avoid any increase of water content that would decrease its mechanical performance.

Gidel et al. [37] studied the permanent strain behaviour of two Unbound Granular Materials UGMs (0/20 limestone UGM and 0/10 UGM obtained from a micro-granite) following various stress paths with a $\Delta q/\Delta p$ ratio comprised between 0 and 3. Based on the results obtained, they developed a constitutive model relating permanent axial strain to the maximum applied cyclic stress and the number of cycles. The general form of this model is as follows:

$$\varepsilon_1^{\,p} = g(\Delta p_{\max}, \Delta q_{\max}).f(N) \tag{1}$$

In this model, the effects of number of cycles and maximum applied stress are separated, making easier the determination of these functions. For the function *f*(*N*), the model proposed by Hornych et al. [36] was adopted (Eq. 2). This model was verified and validated using the experimental results for UGMs.

$$f(N) = \varepsilon_1^{\,p^*} = \varepsilon_1^{\,p}(N) - \varepsilon_1^{\,p}(100) = A\left(1 - \left(\frac{N}{100}\right)^{-B}\right)$$

(for $N > 100$ cycles) (2)

where $\varepsilon_1^{\,p^*}$ is the permanent strain after the first 100 cycles, A and B are regression parameters. A depends on the stress level and the material; it represents the maximum permanent strain when the number of cycles tends to infinity. B controls the evolution of permanent strain with the increase of cycle number. According to Gidel et al. [37], this relationship is valid only for $N >$ 100 cycles because the first one hundred loading cycles correspond to the bedding down of the material.

For the function $g(\Delta p_{max}, \Delta q_{max})$, Gidel et al. [37] showed that the permanent axial strain increases with the increase of mean stress p and this axial strain increase is highly dependent on the stress path defined by $\Delta q_{max}/\Delta p_{max}$. The relationship between the permanent axial strain and the maximum applied stress in the case of $q > 0$ is then given as follows:

$$g(\Delta p_{\max}, \Delta q_{\max}) = \varepsilon_1^{\,p0}\left(\frac{l_{\max}}{p_a}\right)^{n} \frac{1}{\left(m + \frac{s}{\Delta p_{\max}} - \frac{\Delta q_{\max}}{\Delta p_{\max}}\right)} \tag{3}$$

where $\varepsilon_1^{\,p0}$, m, n, s are parameters; p_a =100 kPa and

$$l_{\max} = \sqrt{\Delta p_{\max}^2 + \Delta q_{\max}^2}$$

The influence of water content on the mechanical behaviour of granular materials under cyclic loading has been observed experimentally in several studies [39-42]. The results obtained in this study on the interlayer soil are in agreement with these works. To extend the model of Gidel et al. [37] to the effect of water content or degree of saturation, a function relating the permanent axial strain to the water content and the applied deviator stress, $t(w, \Delta q_{max})$, was defined, as follows [29, 9]:

$$t(w, \Delta q_{\max}) = \varepsilon_1^{p0}.(w+a).\left(\frac{\Delta q_{\max}}{p_a}\right)^{\alpha} \quad (4)$$

where ε_1^{p0}, a and α are parameters and p_a =100 kPa.

In order to account for the stress level, number of cycles and water content of soil, Eq. 4 is extended as follows:

$$\varepsilon_1^p = t(w, \Delta q_{\max}).f(N) \quad (5)$$

where $f(N)$ is calculated using Eq. 2 and $t(w, \Delta q_{max})$ is calculated using Eq. 4.

To validate the model, two tests on the natural interlayer soil of Sénissiat at two water contents, w = 4% (TC1) and w = 6% (TC2) are used for determining the model parameters. These parameters are then used to simulate a test on the saturated sample (w = 4%, TC3). The parameters determined are: ε_1^{p0} = 9.14×10-4, a = -3.36; α = 2.54; A = 0.76; B = 0.17.

Figure 15a shows that the proposed model fits well with the experimental results. For test TC1, in the last loading stage, the model value is close to the measured value at the end of the test; but for test TC2, larger difference between the simulation and the experimental results can be observed in the last loading stage. Figure 15b presents a zoom of the results for the first three loading stages. It is observed that the simulation of test TC1 is better than that of test TC2. This is related to the final value of permanent axial strain at the end of each loading stage. Indeed, it was observed that stabilisation of permanent axial strain occurred for test TC1 but no stabilisation for TC2. Better simulation results can be expected if stabilised permanent axial strains are accounted for in the determination of parameters.

Figure 16 presents the comparison between the experimental results and the simulation for test TC3 (saturated state). It can be observed that the parameters determined from tests TC1 (w = 4%) and TC2 (w = 6%) enable a satisfactory simulation of test TC3 (w = 12%) for the first two loading stages. Significant differences between the simulation and the results are however observed for the third loading stage that corresponds to large deformations (ε_1^p > 150 × 10^{-4}). This can be explained by the "shakedown theory" [43-44]. Actually, there are three categories of material response under repeated loading: plastic shakedown - category A (steady deformation behaviour), plastic creep - category B (failure at large number of cycles), and incremental collapse - category C (failure at low number of cycles). The critical stress levels can be defined according to the accumulated permanent axial strain. Following AFNOR [45], the plastic creep limit is reached when ε_1^p (5000) - ε_1^p (3000) > 4 × 10^{-4} for UGMs, where ε_1^p (5000), ε_1^p (3000) are the accumulated permanent axial strains at 5000 cycles and 3000 cycles, respectively. For the interlayer soil, this value is 1 × 10^{-4} for TC1, 4 × 10^{-4} for TC2 in the last loading stage, and 5.6 × 10^{-4} for TC3, in the third loading stage. This shows that tests TC1 and TC2 are of categories A and B respectively at a deviator stress Δq_{max} = 200 kPa, while test TC3 at Δq_{max} = 140 kPa entre category C. This means that the proposed model can only fit well with the permanent deformation of categories A and B.

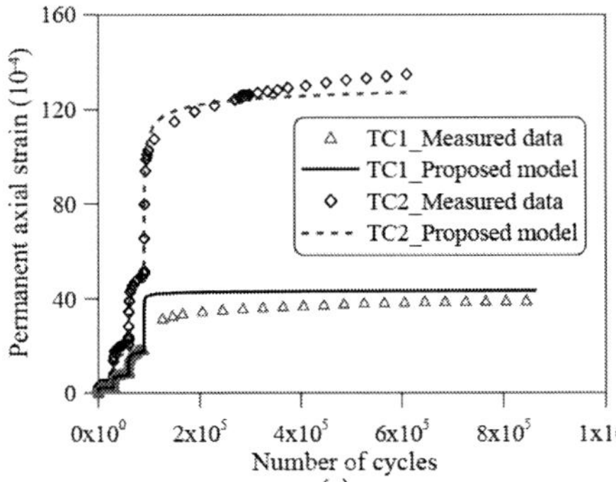

(a)

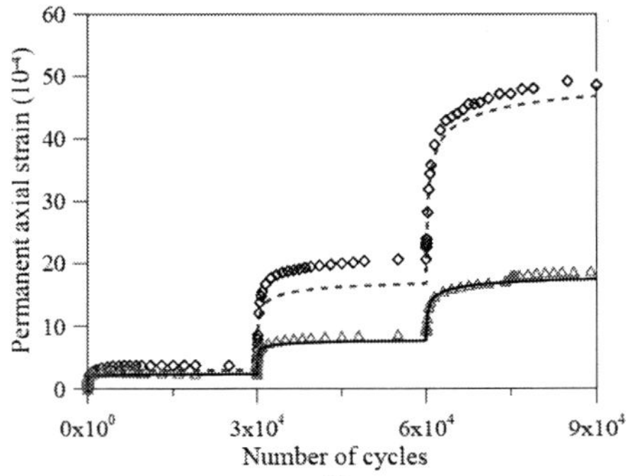

(b)

Figure 15. Determination of parameters by fitting the experimental results from tests TC1 (w = 4%) and TC2 (w = 6%).

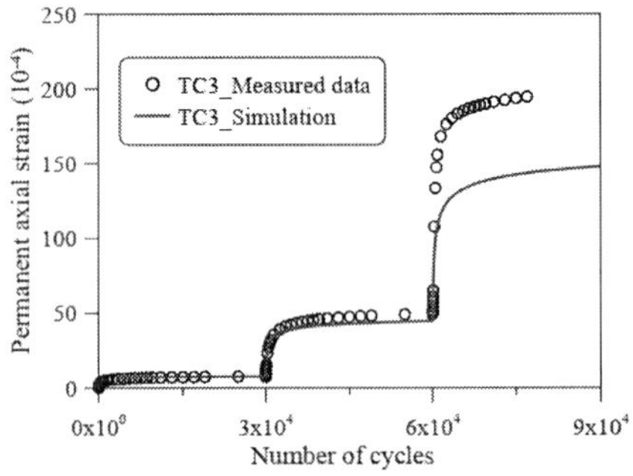

Figure 16. Verification of the proposed model – Simulation of test TC3 (saturated state).

3.3 Physical modelling of interlayer creation of mud-pumping phenomena

Mud pumping which is characterized by the fast upward migration of subgrade fine particles through the ballast voids has been known to be the worst degradation phenomenon for the railway substructure. This phenomenon was reported by several authors [39, 46-52]. Basically, the phenomena of interlayer creation and mud pumping are both related to the migration of fine particles and the interaction between ballast and subgrade. Thus, a good knowledge on the mechanism related to the migration of fine particles is crucial for better understanding the interlayer creation and the mud-pumping, and also for further proposing efficient and economic methods for railway track maintenance.

A physical model was developed for this purpose [53-54]. Figure 17 show it with a 3D view (Figure 17a), a side view (Figure 17b) and a view of cross section A-A (Figure 17c). The cylindrical cell has an internal diameter of 550 mm, a wall thickness of 20 mm and a height of 600 mm. The wall was made of Poly (methyl methacrylate) - PMMA which is a transparent thermoplastic allowing the observation of sample from outside. A digital camera connected to a computer, allowing the visual monitoring of the ballast/sub-soil interface. Three time-domain refectory probes (TDR1 to TDR3) embedded in the sub-soil (h = 120, 160, 200 mm) allowed the volumetric water content to be monitored. Three tensiometers (T1 to T3) [55-58] installed in couple with TDRs at different heights (h = 120, 160, 200 mm) were used for suction monitoring. One pressure sensor was installed at the bottom of the apparatus (h = 0 mm) for measuring the positive pore water pressure under saturated conditions. A hydraulic actuator with the integrated displacement and force sensors was used for monotonic or cyclic loadings.

A silt was used as subgrade soil. It was compacted in five layers of 40 mm thick each and one layer of 20 mm thick, making a total of 220 mm thickness for the whole subgrade layer. For the dry unit mass, three values were considered: 1.4, 1.5 and 1.6 Mg/m^3. The TDR probes were placed between the compaction layers. A 160 mm ballast layer was placed on the sub-soil and the surface of ballast layer was arranged in order to be horizontal for ensuring a good ballast/piston contact. The whole apparatus was put under the hydraulic actuator. Finally, the tensiometers were installed.

All the tests started with the sub-soil in unsaturated state (w = 16 %) and with a pre-loading stage: monotonic loading from 0 to 100 kPa at a rate of 2 kN/min; low-frequency cyclic loading from 30 to 100 kPa; 0.1 Hz cyclic loading for 20 cycles; 1 Hz cyclic loading for 50 cycles and 2 Hz cyclic loading for 100 cycles. Afterwards, a 5 Hz cyclic loading for 500 000 cycles was applied. The value of 5 Hz frequency represents the train circulation at 100 km/h and the applied stress was chosen according to the stress distribution in the conventional railway tracks in France.

In order to study the effect of water content or degree of saturation, once the 500 000 cycles ended, the sub-soil was saturated from the bottom under a hydraulic head of 12 kPa. After saturation, the water level was maintained at 20 mm above the ballast/sub-soil interface in order to ensure the saturated state of the sub-soil. Before the loading under saturated condition, a pressure sensor was installed at h = 0 mm. Monotonic loading at the same rate as in the unsaturated case was then applied followed by the 5 Hz cyclic loading. The test ended when fine particles were observed on the surface of the ballast layer or when the number of cycles reached 500 000.

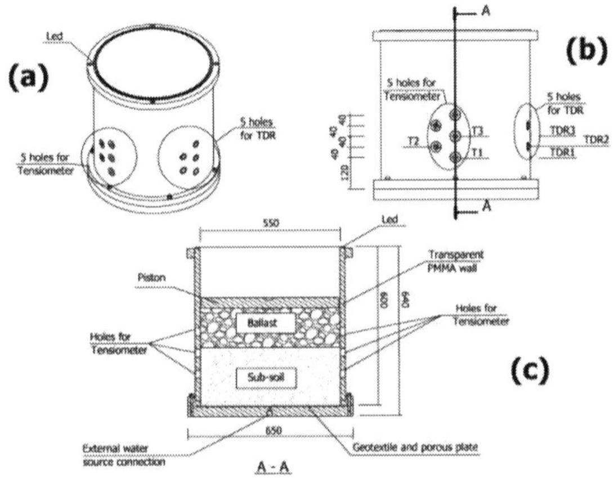

Figure 17. Schematic view of the apparatus developed.

The photographs taken by the digital camera before and after loading for 1.4 Mg/m^3 allowed the visualization of the movement of ballast particles and the evolution of the sub-soil surface (Figure 18). A limited ballast movement can be seen and the ballast/sub-soil interface did not change significantly. The same phenomenon was observed for the two other densities, 1.5 and 1.6 Mg/m^3.

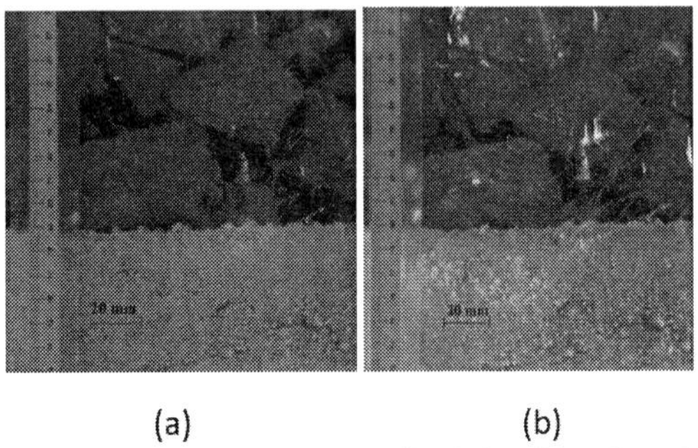

(a) (b)

Figure 18: Photographs of the ballast/sub-soil interface: a) before monotonic loading and b) after 500 000 cycles 5 Hz loading (w = 16% and $\rho_{d\text{-}initial}$ = 1.4 Mg/m^3).

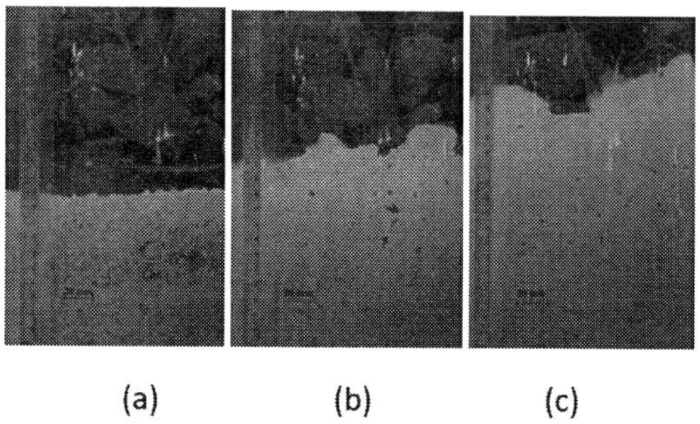

Figure 19: Photographs showing the evolution of the ballast/sub-soil interface ($\rho_{d\text{-}initial}$ = 1.4 Mg/m^3).

The photographs taken at the ballast/sub-soil interface at three moments (after the saturation, after the monotonic loading and after the cyclic loading) are presented in Figure 19. It can be observed that the fine particles were pumped up significantly especially during the cyclic loading. In the test at 1.5 Mg/m^3 dry density, also significant migration of fine particles was identified. However, this migration was found limited in the test at 1.6 Mg/m^3.

During the loadings under saturated condition, five reference vertical sections were chosen and the interface evolutions at these sections were analyzed and presented in Figure 20. In general, the interface evolutions followed the same trend and the pumping levels of fine particles are uniform across the section. This is confirmed by the visual observation after the tests. Two key moments (start of monotonic and cyclic loadings) can be clearly identified by the distinguished changes of pumping level.

In order to verify the nature of the migration of fine particles, a comparison was made between the volume of ballast particles penetrated into the sub-soil V1 and the volume of ballast layer voids filled by the pumped up fine particles V2:

$$V_1 = \frac{h_1 \times A}{1+e}$$
$$V_2 = \frac{h_2 \times A}{1+e} e \qquad (6)$$

where h1 is the settlement of ballast layer, A is the sample cross section and e is the void ratio of ballast layer, h_2 is the pumping up level of fine particles in the ballast layer.

Assuming that the volume of fine particles migrated due to the ballast particles penetration and the fine particles filling the voids of the ballast layer are equal, i.e., $V_1 = V_2$, the following expression can be obtained:

$$h_2 = \frac{h_1}{e} \qquad (7)$$

Indraratna et al. [59] reported that the void ratio of ballast e can vary from 0.74 (compacted) to 0.95 (un-compacted). A value of 0.75 was proposed by Tutumluer et al. [60] and Huang et al. [36]. In this study, as the sample underwent 500 000 cycles of 5 Hz loading, the ballast layer can be considered as compacted. From Eq. 7, the theoretical pumping level of fine particles is calculated using $e = 0.74$ and plotted also in Figure 20. It can be seen clearly that the theoretical pumping level for

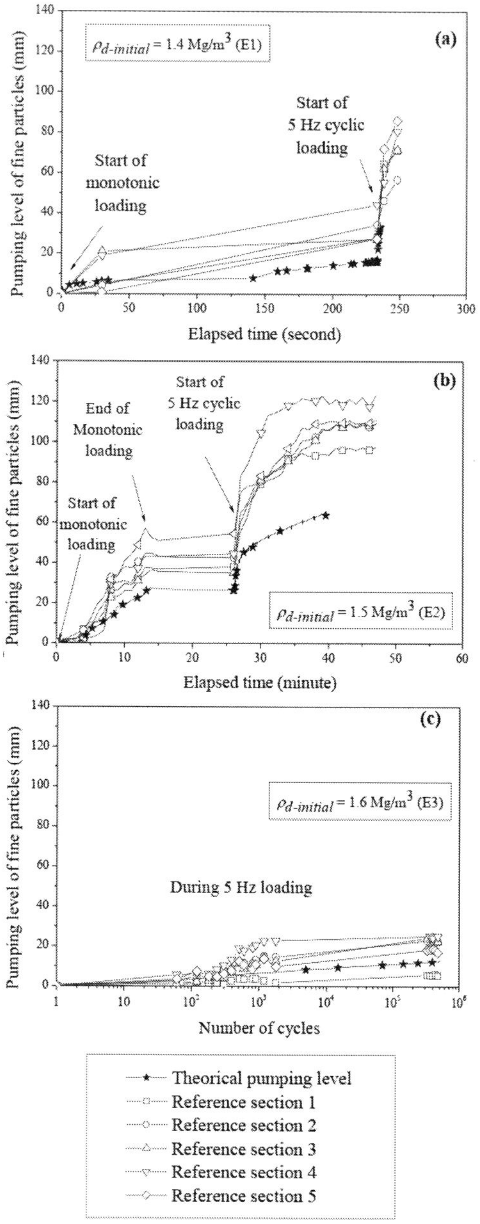

Figure 20. Pumping level of fines particles.

tests at 1.4 and 1.5 Mg/m^3 are much smaller than the real pumping level deduced from the photographs. This suggests that fine particles moved up not only because of the ballast penetration, but also the dissipation of excess pore water pressure. This corresponds to the phenomenon of mud pumping. By contrast, in the case of test at 1.6 Mg/m^3, the theoretical curve stays in the middle of those of reference sections. This suggests that the migration of fine particles was solely due to the ballast penetration. This corresponds to the phenomenon of interlayer creation. Thus, the stiffness of subgrade plays an essential role in the creation of interlayer or in the phenomenon of mud pumping. This is consistent with a case study conducted by Duong et al. [61].

 DOI: 10.1051/e3sconf/20160901001

In reality, most sub-soils are not always saturated. This explains why the interlayer creation is more common than the mud pumping.

4 Field investigation

To investigate the behaviour of railway tracks, it is of paramount importance to conduct field monitoring with real train loadings, real composition of tracks as well as real climatic conditions. The field monitoring includes several parts: i) acceleration due to wheel-rail interactions, ii) surface acceleration/velocity and displacement under train loadings, iii) acceleration/velocity and displacement in the track-bed under train loadings, iv) interaction between the atmosphere and track-bed. In this section, an example of field monitoring focusing on the atmosphere/track-bed interaction is presented [62].

The experimental site is at Moulin Blanc, located in the region of Nord - Pas de Calais, France, in the middle of the 'Douai – Valenciennes' line that was constructed in the late XIXth century. This site is in a cutting zone, and was instrumented at PK 230+400 of the line number 262000 of the French railway network. Different trains are involved: High velocity (TGV), Intercity (TER) and Freight. The train velocity at the site is limited to 110 km/h. Different soils were identified at this site, in particular the interlayer soil of blackish colour with a thickness varying from 10 cm to 57 cm, and the subgrade soil consisting of either sandy or clayey silt. The thickness of the ballast layer is about 50 cm, with about 30 cm fresh ballast and 20 cm fouled ballast.

The field instrumentation was done at PK 230+400 of the line. Based on the depth of interlayer, it was decided to install the suction/temperature probes at -0.20, -0.30 and -0.50 m depths under the ballast layer in two zones A and B, at 20 m distance from each other (see Figure 21a). Two other suction/temperature probes were installed in a zone out of tracks (zone C) and at -0.30 and -0.50 m depths. A temperature sensor was installed at the platform surface in zone C. A total of 8 suction probes and 9 temperature sensors were installed. A weather station was installed on the top of the excavation slope to monitor solar radiation, air temperature, air humidity and rainfall (see Figure 21b). Two piezometers of 10 m depth each were installed near zone A and zone B respectively for the water table monitoring (see Figure 21b).

Four piezo-electric accelerometers ICP/PCB-601A12 were also used, with a measurement range of ±10 g, a frequency ranging from 0.47 Hz to 4000 Hz and a resonant frequency of 16000 Hz. Two were implemented in zone A at -0.10 and -0.40 m depths respectively, and two others in zone B at -0.10 and -0.40 m depths respectively (see Figure 21). Moreover, strain gauges were glued on the rail for the measurement of rail strain under the passages of trains.

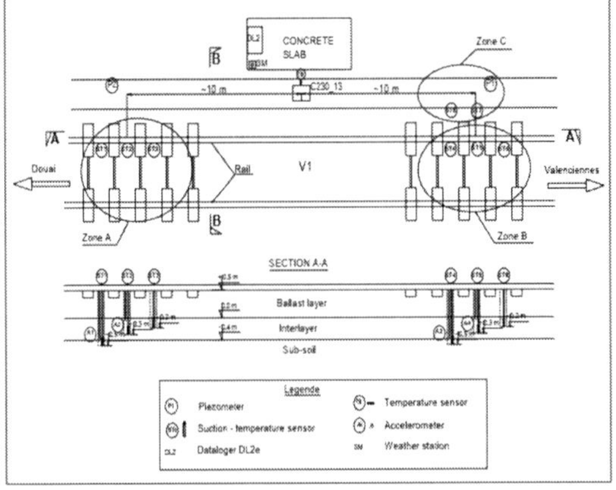

(a)

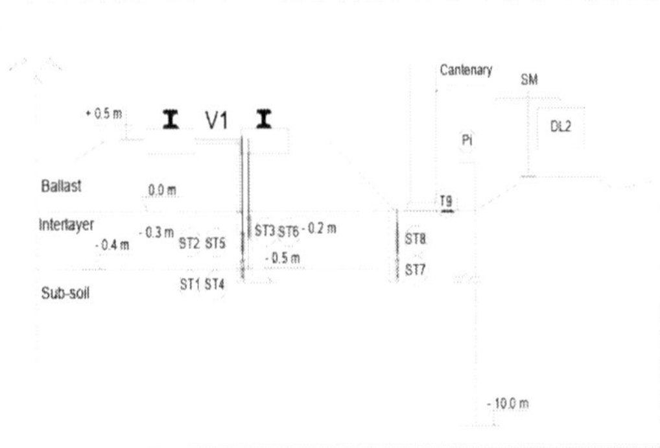

(b)

Figure 21. Site instrumentation. (a) Longitudinal section; (b) Transversal section.

Figure 22 depicts the evolutions of the air parameters measured by the weather station, including air temperature (Figure 22a), solar radiation (Figure 22b), wind speed (Figure 22c), relative humidity (Figure 22d) and precipitation (Figure 22e). Note that due to a technical problem, the monitoring of relative humidity was interrupted in the beginning and in the middle of the recording period. It appears that the solar radiation directly conditions the air temperature: in winter the solar radiation is low giving rise to low air temperatures, whereas in summer the solar radiation is higher generating higher temperatures. For the wind speed, it appears that there are few differences from one season to another, except the exceptional wind recoded in early November 2009 at a speed close to 12 m/s. The variations of relative humidity are more pronounced in the period from March to July 2010 than in the period from September to December 2009. This is consistent with the precipitation. Indeed, there were more strong rainfalls in the second period that in the first (see Figure 22e). Note that all data presented in Figure 22 correspond to a recording frequency of 30 min.

The air data presented in Figure 22 can be used to determine the potential evaporation rate using the model of Penman-Monteith [63]:

$$ET_0 = \frac{0.408\Delta(R_n - G) + \gamma \frac{900}{T+273} u_2 (e_s - e_a)}{\Delta + \gamma(1 + 0.34u_2)} \quad (8)$$

where
ET_0 is the potential evaporation (mm.day^{-1});
R_n is the net solar radiation (MJ.m^{-2}.day^{-1});
G is the soil heat flux (MJ.m^{-2}.day^{-1});
T is the daily average temperature at 2 m above the ground surface (°C);
u_2 is the wind speed at 2 m above the ground surface (m.s^{-1});
e_s is the saturated vapour pressure (kPa);
e_a is the actual vapour pressure that can be deduced from the air relative humidity using Kelvin's law (kPa);
Δ is the slope of the curve of saturated vapour pressure versus temperature (kPa.°C^{-1});
γ is the psychometric constant (kPa .°C^{-1}).

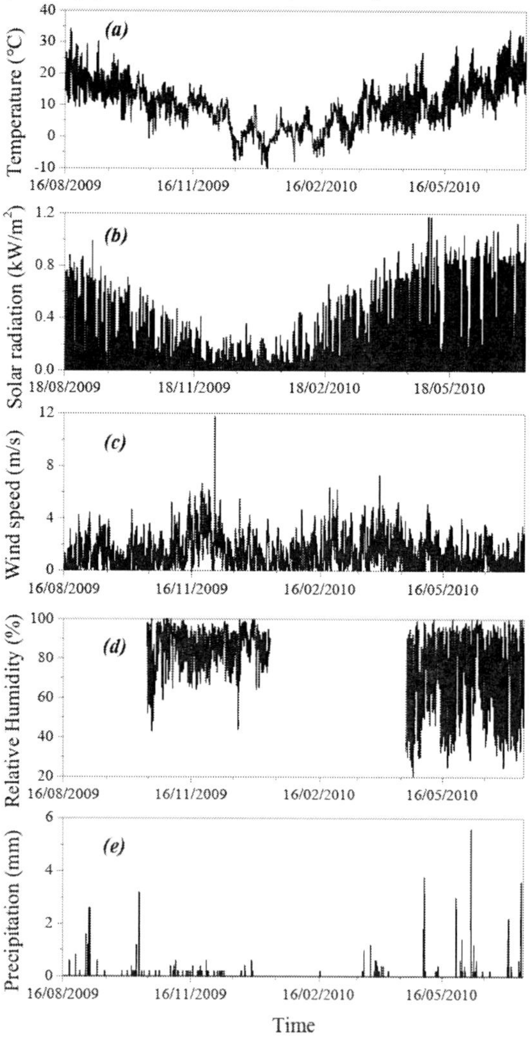

Figure 22. Air data. (a) Temperature; (b) Solar radiation; (c) Wind speed; (d) Relative humidity; (e) Precipitation.

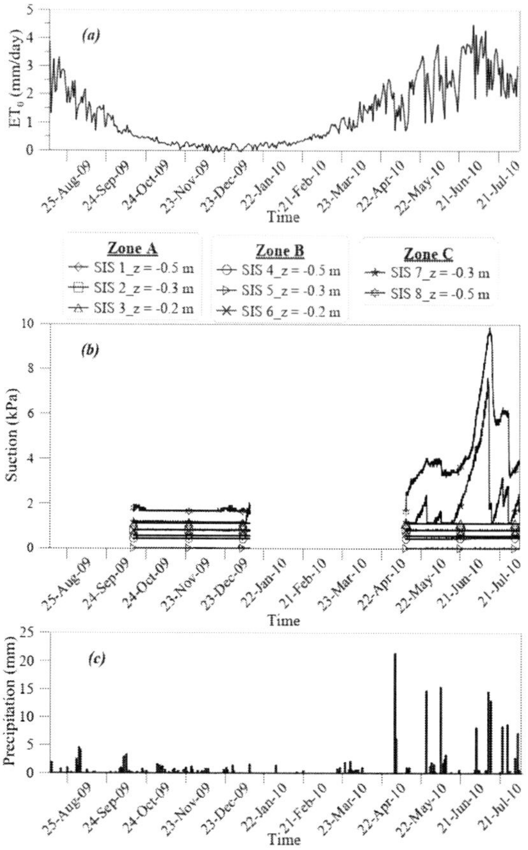

Figure 23. Potential evaporation (a), suction (b) and daily rainfall (c).

As mentioned before, the measurements of air relative humidity were not complete in the monitoring period. Thereby, the data recorded in a Méteo France weather station in Cambrai (close to the Moulin Blanc site) were used in the calculation. Figure 23a shows the variations of ET_0 over the monitoring period. It appears clearly that the value of ET_0 is large in summer and small even equal to zero in winter, in agreement with the variations of solar radiation and air temperature shown in Figure 22.

Figure 23b shows the variations of suction in the three zones and at different depths. Again, because of the technical recording problem mentioned previously, the data are not complete and only the period from September to December 2009 and the period from April to July 2010 are covered. From the limited available data, it appears that in the relatively cold season (from September to December 2009), the suction values remained stable and small in all the three zones. By contrast, in the hot season (from April to July 2010), the values in zone A and zone B still remained stable but showed large variations in zone C. The difference between zone A/zone B and zone C is to be attributed to the effect of ballast layer.

Figure 23c shows the daily precipitation in the monitoring period. The difference between this figure and Figure 22e is that the former corresponds to a frequency of every day (precipitation amount for each day) whilst the latter corresponds to a frequency of every 30 min (precipitation amount for every 30 min). It can be observed that adopting a different frequency significantly changes the precipitation distribution feature: more details can be found with a high frequency (30 min) while the precipitation values are more averaged in the case of low frequency (one day). It appears in Figure 23c that the strong rainfall events occurred in the period from April to July 2010. This suggests that the large variations of suction identified in zone C in this period (see Figure 22b) are not related to the rainfalls, but to the high water evaporation rate. In other words, at the site of Moulin Blanc in the monitoring period the evaporation rate was larger than the water infiltration rate in the zone out of tracks – the soil was being dried over time. This explains why no water table was detected by the two piezometers installed till 10 m depth. Further examination shows that under the ballast layer (zones A and B) suction did not change, indicating the effective protection of the ballast layer against water evaporation on one hand and water infiltration on the other hand.

5 Conclusion

The hydro-mechanical behaviour of unsaturated railway track-bed materials were investigated through laboratory testing, physical modelling as well as field monitoring. The results obtained allow the following conclusions to be drawn.

The Geo-endoscopic tests allow the identification of different materials constituting the platform, their natures and thicknesses. A good correlation of thicknesses is obtained between the geo-endoscopic tests and the PANDA penetrometer tests. This site prospection is essential to well define the soil profile, in particular the interlayer position.

Hysteresis exists for both the soil water retention curve and the hydraulic conductivity changes with suction. The wetting process was found to be much faster than the drying process, and the hydraulic conductivity during wetting is always higher than that during drying.

The water retention curves of ITL_{10} and fines are different, illustrating an obvious effect of soil texture. On the contrary, in terms of unsaturated hydraulic conductivity, a good agreement was identified between the results of two soils, regardless of the drying or wetting paths. This suggests that water transfer in the unsaturated interlayer soil takes place mainly through the network of pores between fine particles, coarse elements like ballast behaving as inert materials. On the contrary, in saturated state, a higher value was obtained for ITL_{10}, suggesting that in this case the hydraulic conductivity is mainly governed by the water flow through macro-pores. Thereby, the water flow mechanism in saturated state is different from that in unsaturated state. From a practical point of view, this finding shows that to determine the unsaturated hydraulic conductivity of interlayer soils, a device as small as the small-scale infiltration cell can be employed by testing the fine particles only, provided that equivalent dry density is taken into account. However, this is not valid for the determination of saturated hydraulic conductivity.

The effects of water content and of fines content on the resilient modulus are linked. Indeed, under unsaturated conditions, due to the suction effect, the soil having high fines content showed higher resilient modulus. On the contrary, when the soil approached the saturated conditions, the fine particles provided a negative effect.

In terms of permanent axial strain, it was also observed that in dry condition (w = 4% and 6%), adding fines decreases the permanent axial strain due to the suction effect. On the contrary, in nearly saturated condition (w = 12%), the higher the fines content the higher the axial permanent strain. This is because that the interlayer soil is weakened mechanically due to the sensitivity of fine particles to water content changes in nearly saturated condition. From a practical point of view, the increase of fines content on the mechanical behavior of interlayer is acceptable in case of lower water content (dry state). But in case of high water content (nearly saturated state), the interlayer with high fines content needs to be protected from water infiltration. Otherwise, significant decrease in mechanical performance can be expected.

In order to predict the permanent axial strain, a constitutive model accounting for the effects of stress level, number of cycles and water content of soil was developed based on the results of cyclic triaxial tests with multi-stage loading procedure and the model proposed by Gidel et al. [37]. The parameters were first determined by fitting the tests with 4% and 6% water contents. The parameters determined were then used to simulate the test corresponding to the saturated state (w = 12%). The simulation was in good agreement with the experimental results, indicating the performance of the proposed model for the prediction of permanent deformation within categories A and B. Further study is however necessary to better describe the deformation behaviour for category C.

The presence of water was found to be the most crucial factor for the migration of fine particles. In the unsaturated state, both the ballast and sub-soil settlements occurred, but without the migration of fine particles. Under the near saturated state, the ballast/sub-surface interface moved up, and the pumping level depends on the sub-soil dry unit mass. In the case of $\rho_{d\text{-}initial}$ = 1.4 Mg/m^3 and 1.5 Mg/m^3, excess pore water pressure dissipation took place, bringing fine particles upward. This corresponds to the mud pumping phenomenon. In the case of $\rho_{d\text{-}initial}$ = 1.6 Mg/m^3, there was just the interpenetration of ballast and sub-soil, resulting in a mixture layer namely interlayer.

In the field conditions, the suction under the ballast layer remained low and unchanged over long time. However, significant changes can take place in the zone where the soil ground is directly exposed to the atmosphere without being covered by ballast. These changes are mainly due to water evaporation. Comparison between the potential evaporation rate and the precipitation at the site of Moulin Blanc showed that the site was in deficiency in water supply. This explained the absence of water table till the depth of the installed piezometers (10 m).

Acknowledgements

The author would like to thank all his co-workers from Ecole des Ponts ParisTech – ENPC (Trong Vinh Duong, Viet Nam Trinh, Francesco Lamas-Lopez, Anh Minh Tang, Jean-Claude Dupla, Jean Canou) and from the French Railway Company - SNCF (Nicolas Calon, Alain Robinet, Sofia Costa D'Aguiar, Lucie Lambert, Olivier Schoen).

References

1. Y.J. Cui, A.M. Tang, D. Marcial, J.M. Terpereau, G. Marchadier, X. Boulay. Geotech. Test. J. **30**(3), 227–233 (2007)
2. C. Yang, Y.J. Cui, J.M. Pereira, M.S. Huang. Comput. Geotech. **35**, 853–859 (2008)
3. J.P. Karam, Y.J. Cui, A.M. Tang, J.M. Terpereau, G. Marchadier. Soils Found. **49**(3), 421–429 (2009)
4. J.A. Munoz-Castelblanco, J.M. Pereira, P. Delage, Y.J. Cui. Géotechnique **62**(2), 95–106 (2011)
5. J.A. Munoz-Castelblanco, J.M. Pereira, P. Delage, Y.J. Cui. Geotech. Test. J. **35**(1), 11–17 (2012a)
6. J.A. Munoz-Castelblanco, P. Delage, J.M. Pereira, Y.J. Cui. Géotechnique **62**(7), 595–604 (2012b)
7. A.M. Tang, Y.J. Cui, V.N. Trinh, Y. Szerman, G. Marchadier. Eng. Geol. **106**(1–2), 68–77 (2009)
8. V.N. Trinh, A.M. Tang, Y.J. Cui, J. Canou, J.C. Dupla, N. Calon, L. Lambert, A. Robinet, O. Schoen. R. F. G. **134–135**, 65–74 (2011)
9. Y.J. Cui, T.V. Duong, A.M. Tang, J.C. Dupla, N. Calon, A. Robinet. J. Zhejiang Univ. Sci. A **14**(4), 244–255 (2013).
10. F. Lamas-Lopez, Y.J. Cui, N. Calon, S. Costa D'Aguiar. Soils Found. **56**(2), 240–250 (2016)
11. G. Degrande, L. Schillemans. J. Sound Vibration **247**(1), 131–144 (2001).
12. L.J. Su, C. Rujikiatkamjorn, B. Indraratna. (2010). Geotech. Test. J. **33**(5) (2010)
13. T.R. Sussmann, E.T. Selig, J.P. Hyslip. NDT & E INT, 157–167 (2003)
14. J. Hunter, B. Benjumea, J. Harris, R. Miller, S. Pullan, R. Burns, R. Good. Soil Dyn. Earthq. Eng. **22**(9-12), 931–941 (2002)
15. Z. Tompai. Period. Polytech-Civ. **52**, 97–102 (2008)
16. P.K. Woodward, J. Kennedy, O. Laghrouche, D.P. Connolly, G. Medero. Transportation Geotechnics **1**(4), 214–224 (2014)
17. V. Alves Fernandes, F. Lopez-Caballero, S. Costa d'Aguiar. Comput. Geotech. **55**, 267–276 (2014)
18. T.V. Duong, Y.J. Cui, A.M. Tang, J.C. Dupla, N. Calon. Can. Geotech. J. **51**(7), 735–746 (2014a)
19. T.V. Duong, V.N. Trinh, Y.J. Cui, A.M. Tang, N. Calon 2013. Geotech. Test. J. **36**(1), 54–63 (2013a)
20. T.T. Le, Y.J. Cui, J.J. Muñoz, P. Delage, A.M. Tang, X.L. Li. Front. Archit. Civ. Eng. China **5**(2), 160–170 (2011)
21. D.E. Daniel. Soil Sci. Am. J. **46**(6), 1125–1129 (1982)
22. P. Delage, Y.J. Cui. 2001. Article C302. Ed. *Techniques Ingénieur* (2001)
23. L.B. Bruckler, P. Angulo-Jaramillo, R. Ruy. Soil Sci. Am. J. **66**, 384–395 (2002)
24. Y.J. Cui, A.M. Tang, C. Loiseau, P. Delage. Phy. Chem. Earth PT A/B/C, 33, S462–S471 (2008a)
25. W.M. Ye, Y.J. Cui, L.X. Qian, B. Chen. Eng. Geol. **108**(3), 169– 176 (2009)
26. A. Wayllace, N. Lu. Geotech. Test. J.**35**(1), 1–15 (2011)
27. N. Toker, J. Germaine, K. Sjoblomt, P. Culligan. Géotechnique **54**(3), 179–186 (2004)
28. J. Côté, M. Roy. 1998. *Rapport de l'études et recherches en transports du Québec*, 177p (1998)
29. V.N. Trinh, A.M. Tang, Y.J. Cui, J. Canou, J.C Dupla, N. Calon, L. Lambert, A. Robinet, O. Schoen. Soils Found. **52**(3), 511–523 (2012)
30. T.V. Duong, A.M. Tang, Y.J. Cui, V.N. Trinh, J.C. Dupla, N. Calon, J. Canou, A. Robinet. Soils Found. **53**(6), 868–878 (2013b)
31. T.V. Duong, Y.J. Cui, A.M. Tang, J.C. Dupla, J. Canou, N. Calon, A. Robinet. Acta Geotech. 51–59 (2016)
32. F. Lekarp, U. Isacsson, A. Dawson. J. Transp. Eng. **126**(1), 66–75 (2000)
33. L. Uthus, I. Hoff, I. Horvli. *Proceeding of the 7th International Conference on the Bearing Capacity of Roads, Railways and Airfields*. Trondheim, Norway, 1–13 (2005)
34. E. Fortunato, A. Pinelo, M. Matos Fernandes. Soils Found. **50**(1), 55–62 (2010)
35. S. Seif El Dine, J.C. Dupla, R. Frank, J. Canou, Y. Kazan. Can. Geotech. J. **47**(4), 425–438 (2010)
36. H. Huang, E. Tutumluer, W. Dombrow. T.R.B. 2117, 93–101 (2009)
37. G. Gidel, P. Hornych, J.J. Chauvin, D. Breysse, A. Denis. Bulletin de Liaison des Laboratoires des Ponts et Chaussées 233, 5–21 (2001)
38. P. Hornych, J.F. Corté, J.L. Paute. Bulletin de Liaison des Laboratoires des Ponts et Chaussées **184**, 77-84 (1993).
39. E.T. Selig, J.M. Waters. Thomas Telford (1994)
40. G. Gidel, D. Breysse A. Denis. Revue française de génie civil 6 (5), 789–799 (2002)
41. J. Ekblad. Road Mater. Pavement **7**(3), 369–404 (2006)
42. S. Werkmeister, R. Numrich, A.R. Dawson, F. Wellner. Transportation Res. Rec. **1837**, 61–70 (2003)

43. S. Werkmeister, A.R. Dawson, F. Wellner. J. Transp. Eng. **130**(5), 665–674 (2004)
44. R. García-Rojo, H.J. Herrmann, H.J. Granul. Matter **7**(2), 109–118.
45. AFNOR - EN 13286-7 - Unbound and hydraulically bound mixtures - Part 7: *Cyclic load triaxial test for unbound mixtures* (2004)
46. D.J. Ayres. Geotext. Geomembranes **3**(2-3), 129–142 (1986)
47. G.P. Raymond. Geotext. Geomembranes **17**(4), 213–230 (1999)
48. T. Sussmann, K. Maser, D. Kutrubes, F. Heyns, E. Selig. *Symposium on the Application of Geophysics to Engineering and Environmental Problems* (2001)
49. B. Burns, G.S. Ghataora, P. Sharley. *Proceedings of the First International Conference on Railway Foundations*, University of Birmingham, UK, 385–393 (2006)
50. G.S. Ghataora, B. Burns, M.P.N. Burrow, H.T. Evdorides. *Proceedings of the First International Conference on Railway Foundations*, University of Birmingham, UK, 355–366 (2006)
51. E.S. Aw. *PhD Dissertation*, Massachusetts Institute of Technology, USA (2007)
52. B. Indraratna, W. Salim, C. Rujikiatkamjorn. CRC Press (2011)
53. T.V. Duong, Y.J. Cui, A.M. Tang, J.C. Dupla, N. Calon, J. Canou, A. Robinet. Eng. Geol. **171**, 45–58 (2014b)
54. T.V. Duong, Y.J. Cui, A.M. Tang, J.C. Dupla, J. Canou, N. Calon, A. Robinet, B. Chabot, E. De Laure Geotech. Test. J. **37**(5), 1–12 (2014c)
55. A. Ridley, K. Dineen, J. Burland, P. Vaughan. Géotechnique **53** (2), 241–254 (2003)
56. Y.J. Cui, A.M. Tang, A. Mantho, E. De Laure. Geotech. Test. J. **31** (1), 95–100 (2008b)
57. D.G Toll, S.D.N. Lourenço, J. Mendes. Eng. Geol. **165**, 29–37 (2012)
58. S.D.N. Lourenço, D. Gallipoli, D.G. Toll, C.E. Augarde, F.D. Evans. Can. Geotech. J. 48(2), 327–335 (2011)
59. B. Indraratna, D. Ionescu, D. Christie, R. Chowdhurry. Australian Geomechanics Journal **12**, 48 – 61 (1997).
60. E. Tutumluer, W. Dombrow, H. Huang. *AREMA 2008 Annual Conference & Exposition September* 21-24, Salt Lake City, USA (2008)
61. T.V. Duong, Y.J. Cui, A.M. Tang, N. Calon, A. Robinet. Bull. Eng. Geol. Environ. **74**(1), 259–270 (2015)
62. Y.J. Cui, F. Lamas-Lopez, V.N. Trinh, N. Calon, S. Costa D'Aguiar, J.C. Dupla, A.M. Tang, J. Canou, A. Robinet. Transportation Geotechnics **1**(3), 91–105 (2014)
63. R.G. Allen, L.S. Pereira, D. Raes, M. Smith. F.A.O. 56 (1998)

Investigation on chemical effects on GMZ bentonite used as buffer materials

Wei-Min Ye

[1] *Department of Geotechnical Engineering, College of civil engineering, Tongji University, Shanghai 200092, China*
[2] *Key Laboratory of Geotechnical and Underground Engineering of Ministry of Education, Tongji University, Shanghai 200092, China*

Abstract. GMZ bentonite has been selected as a buffer/backfill material for the design of Chinese HLW repository due to its suitable properties. Investigations have been conducted on this material to have a better understanding of its engineering behavior. In this paper, a review of the studies on the chemical effects in GMZ bentonite is presented. The adsorption of radionuclides Ni(II), Cr(III) and Eu(III) etc. on GMZ bentonite is investigated and the influences of external chemical agents are assessed. Some modifications on GMZ bentonite have been implemented to improve its adsorption properties. In the meantime, the influences of NaOH solutions with variations in concentration and pH values on hydraulic conductivity, mechanical behavior and mineralogy of GMZ bentonite are investigated with consideration of temperature. Effects of salt solutions are also assessed in different studies using solutions of Na^+ and Ca^{2+}, which are the predominant cations in the groundwater of Beishan area. It is observed that both cations influence the buffering behavior of bentonite at a different degree. Further investigations should be done for better understanding the influences of coupling factors and time on the chemical behavior of material.

1 Introduction

Deep geological storage is a widely accepted technology for disposal of high-level radioactive wastes. Bentonite or bentonite-based materials has been found to be an appropriate material for surrounding the habitual canisters in hard rocks where the waste is placed ([1-2]) due to its particular micro-porous structure, low hydraulic conductivity, high swelling and self-sealing capacity and good sorption properties ([3-8]).

Until now, different kinds of bentonite have been selected as the best candidate to act as a buffer/backfill material in HLW repositories. FEBEX bentonite, obtained from deposits located in Almería, has been selected for such purpose by the Spanish Agency for Radioactive Waste Management (ENRESA) ([9]). MX-80 bentonite, extracted from Wyoming, has also been chosen as a buffer material in many disposal concepts in countries such as Finland ([10]), France ([11]), Sweden ([12]) and Germany ([13]). The Kunigel V1 and FoCa bentonites have been selected by Japan and France as candidates for HLW disposal ([14-15]).

During the long-term operation of the repository, the engineered barrier is characterized by coupled thermal-hydraulic-mechanical (THM) behavior involving interactive processes such as the decay heat of vitrified waste, the infiltration of groundwater and the stress generation due to the ground pressure, the thermal loading and swelling pressure of the buffer material will take place ([16]). In order to understand the behavior of the canister/barrier/host rock system, the coupled THM processes within the compacted bentonite need to be modelled ([16-19]). For this reason, a number of theoretical proposals concerning THM analysis have been reported ([20-25]).

Meanwhile, chemical reactions inducing changes of buffer/backfill property of the artificial barrier could happen. Moreover, the degradation of the concrete in the repository could induce a notable rise of the groundwater pH, which means an exposure of the bentonite to alkaline solutions ([26]). Then, bentonite properties such as radionuclides adsorption, hydraulic conductivity and swelling behavior may be affected by these external factors, results in an incorrect performance of the material for our expected engineering purposes in the isolation of HLW from the environment.

A great number of experimental studies dealing with the chemical effects on behavior of clayey soils are available in literature ([27-31]). Common view can be reached that effects of salt solutions on the expansion of clays have been always attributed to the effects of the diffuse double layer (DDL) ([31-32]). The behavior of clay particles is strongly influenced by the net inter-particle repulsive minus attractive stress (R - A) ([33-35]). The thickness of the DDL and the net repulsive stress decreases as the concentration of the solution in the pores increases, which leads to true effective stress increases, resulting in contraction of the aggregates and shrinkage of the sample ([33]).

[a] ye_tju@tongji.edu.cn

The Chinese program of radioactive waste disposal in a deep geological media was launched in middle 1980s. According to the Chinese long-term plan for the implementation of such repositories, it will be put into operation by the middle of the 21st Century ([36]). Currently, a preliminary concept of a multi-barrier HLW repository, which consists of a shaft-tunnel structure and located in saturated zone in granites ([37]) , is adopted.

Gao-Miao-Zi (GMZ) bentonite has been selected for the construction of the engineered barrier ([4,38]). Investigations have been conducted on GMZ bentonite to understand and characterize its engineering behaviors ([4-5,36,38-39]). In the meantime, Beishan in Gansu province has been chosen as the preferred construction site for Chinese repository ([4]). Previous in-situ investigation results show that the average total dissolved solids (TDS) of the groundwater in Beishan is 3-12g/L, up to 80 g/L ([40]). Obviously, investigation on behaviour of GMZ bentonite with consideration of the influence of pore fluid concentration is of great importance for Chinese HLW disposal program.

Investigations on mineralogy and chemical composition, mechanical properties, hydraulic behavior, swelling behavior, thermal conductivity, microstructure and volume change behavior were performed on GMZ bentonite ([4-5,36,38-39,41-45]. among others). China-Mock-Up ([46]) was performed to assess the behavior of GMZ bentonite under coupled THM conditions

In this paper, an exhaustive review of the accumulated knowledge from literature about the chemical effects on the behavior of a GMZ bentonite-based buffer/backfill material in a HLW disposal is presented. Achievements in certain chemical-related factors affecting the properties of GMZ bentonite, such as adsorption capacity, hydraulic conductivity, and mechanical behavior, among others are presented. Problems that should be investigated are suggested for GMZ bentonite in the future.

2 GMZ bentonite

GMZ bentonite is a montmorillonite-dominant clay originated in the later Jurassic period ([41]). It is extracted from a vast deposit in the Inner Mongolia Chinese autonomous region, about 300 km northwest from Beijing. The deposit covers an area of about 72 km^2, with 120 million tons of Na-bentonite from a total of 160 million tons of clay materials (Fig. 1) ([5]). The characteristic bedded ores have a waxy appearance and a soapy texture. Its formation was caused by the mineralization due to an interaction with ground water and weathering of the firstly formed continental volcanic sediments ([41]).

Investigations showed that GMZ bentonite is composed of 75.4% montmorillonite 11.7% quartz, 7.3% cristobalite, 4.3% feldspar, 0.5% calcite and 0.8% kaolinite ([42]). X-ray fluorescence spectrometry results show that the composition of GMZ bentonite is SiO_2 67.43%, Al_2O_3 14.20%, TFe_2O_3 2.40%, Na_2O 1.75%, CaO 1.13%, K_2O 0.73%, FeO 0.29%, TiO_2 0.12%, MgO 0.10%, MnO 0.02% and P_2O_5 0.02% ([4,42]).

Figure 1. Open pit of GMZ bentonite deposit ([5]).

Table 1. Mineralogical composition of GMZ bentonite compared to the other three, in % ([9,42,47-49]).

Bentonite	GMZ	MX-80	FEBEX	Kunigel-V1
Montmorillonite	75.4	65-82	92±3	47
Quarz	11.7	4-12	2±1	0.6
Cristobalite	7.3	-	-	-
Feldspar	4.3	5-8	-	-
Plagioclase	-	8.2±2.7	2±1	4
Kaolinite	0.8	-	-	-
Chalcedony	-	-	-	37

The high content of montmorillonite results in a high cation exchange capacity (CEC = 77.30 meq/100 g), a large plasticity index (Ip= 275), and large specific surface area (S = 570m^2/g). The major exchangeable cations are Na^+ (43.36 meq/100 g), Ca^{2+} (29.14 meq/100 g), Mg^{2+} (12.33 meq/100 g) and K^+ (2.51 meq/100 g) ([42]). The mineral composition and some physical properties for the most used bentonites can be found in Table 1 and Table 2, respectively.

Table 2. Some physical properties of GMZ bentonite compared to the other three ([9,42,47-49]).

Sample	G_s	w_L (%)	w_P (%)	I_p	S (m^2/g)
GMZ	2.66	313	38	275	570
Mx-80	2.76	520	42	478	562
FEBEX	2.7	102 ± 4	53 ± 3	42 ± 7	725
Kunigel-V1	2.79	415	32	383	687

With these characteristics and especially for the GMZ bentonite in the Chinese case, bentonites are ideal to be used as the buffer/backfill materials, which have suitable thermal, hydraulic, mechanical and physico-chemical properties.

3 Chemical effects on the adsorption of radionuclides

As backfill materials for engineered barrier, bentonite should has good adsorption capacity for preventing the escaping of the radionuclides, which possibly leaked from the canister, to the environment. Up to now, many contributions have made to investigate the chemical effects on the capacity of GMZ bentonite for the adsorption radionuclides ([50-61]).

The sorption of an important radionuclide such as Ni(II) on GMZ bentonite was studied under the effects of pH, ionic strength, foreign ions, humic acid and temperature ([58]). The results indicated that the sorption process of Ni(II) quickly achieved the equilibration. Its kinetics can be described reasonably well by the following pseudo-second-order model ([62]):

$$\frac{t}{q_t} = \frac{1}{2kq_e^2} + \frac{1}{q_e}t \quad (1)$$

Where, k (g mg^{-1} h^{-1}) is the pseudo-second-order rate constant of sorption, qt (mg g^{-1} of dry mass) is the amount of Ni(II) sorbed on the surface of the adsorbent at time t (h), and qe (mg g−1 of dry mass) is the equilibrium sorption capacity.

The sorption of Ni(II) was found to be strongly dependent on pH. It increased with an increase in pH values lower than 8.5, and then maintained a high level at values above 8.5. At a low pH, the sorption was affected by the foreign cations, but not for high pH. The same behavior was observed in the case of ionic strength. However, the presence of different foreign anions had no obvious influence on the sorption. The sorption process was dominated in a different way depending on the pH. For low values, it was mainly dominated by outer-sphere complexation and exchange with Na^+/H^+; for high values, it was controlled by inner-sphere surface complexation. Humic acid (HA) also affected the process, increasing it at low pH and decreasing it at high pH.

Lagmuir, Freundlich and DR models were used to simulate the temperature-dependent sorption isotherms of Ni(II) and thermodynamic parameters were calculated. The results showed that the sorption of Ni(II) on GMZ bentonite was endothermic and spontaneous.

Other studies were also performed on the capacity of GMZ bentonite for the adsorption of Cr(III), which can also be found in HLW. Influences of ionic strength, pH, temperature and adsorbent dose variations on the process were studied ([54]).

The results showed that the adsorption capacity of bentonite increased with increasing pH from 2 to 7. There was no significant change for higher pH values. NaCl solutions were used to assess the influence of ionic strength, which was considerable for pH under 7. The adsorption of Cr(III) was also increased by increasing temperature (Fig. 2), suggesting that is an endothermic process and agreeing with other studies on different bentonites ([63]). It was also observed that the bentonite amount was also an important influencing factor, since its increase produced as well an increase in the removal of Cr(III).

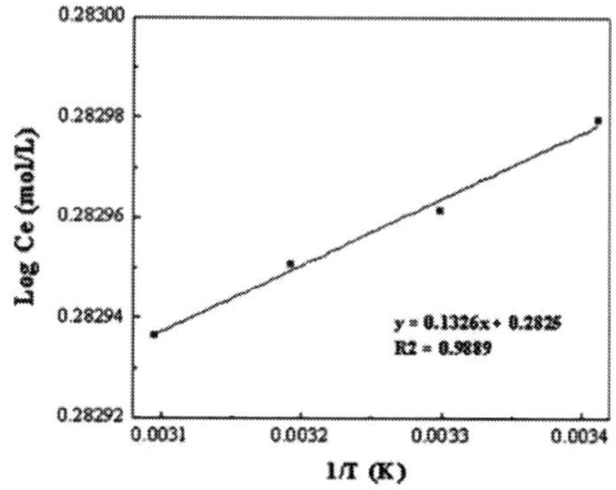

Figure 2. Effect of temperature on the adsorption of Cr(III) onto GMZ bentonite ([54]).

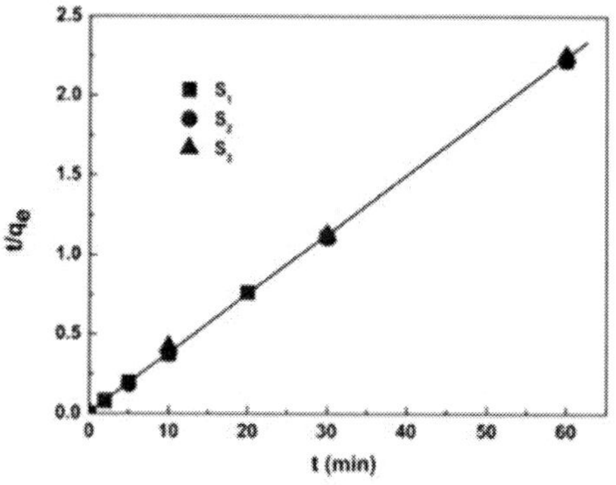

Figure 3. Absorption isotherms ([55]).

Adsorption of Eu(III) on di(2-thylhexly) phosphoric acid-immobilized magnetic GMZ bentonite was also studied under the influences of temperature, pH, contact time and concentration of metal ion ([55]). In this case, adsorption could also be described by the Langmuir isotherm and fitted the pseudo-second-order model described by equation (1). Fig. 3 shows the adsorption isotherms of Eu(III) on the modified bentonite.

The adsorption capacity was relatively low at high pH values, possibly explained by the competitive adsorption between Eu(III) and H^+. With an increase in iron oxide particles, the adsorption capacity was decreased. Otherwise, increasing temperature induced an increase in the mobility of Eu(III) ions, which permitted the adsorption to be increased. The thermodynamic parameters indicated that the process was exothermic and spontaneous.

Efforts also have been made to improve the radionuclide adsorption capacity by using modified magnetic GMZ bentonite instead of the unmodified one ([64]). La^{3+}, Eu^{3+} and Yb^{3+} adsorption was evaluated and the improvement of the method was confirmed. The influence of pH and metal ions was as well studied, obtaining similar results to other experiments ([55]). The method was strongly recommended due to the easy removal of the adsorbent and its excellent efficiency.

The use of rare earth elements (REEs) as substitutes for radionuclides such as 238U, 213Th, 83Sr and 234Ra due to their similar chemical properties and ionic radius is an important issue that has also been investigated ([53]). La(III) was used as a nonradioactive chemical homologue for the commonly used lanthanides and actinides, and its adsorption on GMZ bentonite was studied. Influences of pH, bentonite content, contact time and ionic strength were assessed.

The results showed that the equilibrium on the adsorption of La(III) on GMZ bentonite was quickly achieved and that the process followed the pseudo-second-order model. Equation (1) was also valid in this case.

The effectivity of the removal of La(III) from aqueous solution by the adsorption onto GMZ bentonite was demonstrated in the equilibrium batch experiment. Other experiments proved the improvement of the removal technique using 8-hydroxyquinoline to modify the bentonite ([65]).

As the adsorption was considerably fast, it was deduced that the main process involved was the chemical adsorption rather than the physical one. It also followed Langmuir isotherm. The influence of pH was notable, since there was a gradual increase of La(III) adsorption with a pH value between 2 and 8. Then, it maintained a high level without more significant variation with increasing pH. The maximum removal of La(III) was observed at pH 8 and reached the 96%. Surface complexation was the main process for the control of adsorption in this case due to this strong dependence on pH ([53]).

Increasing bentonite content also induced an increase in the adsorption of La(III). It was influenced by ionic strength as well, since an increase in the concentration of KCl produced a reduction in the removal of La(III), probably as a result of competitive adsorption between K^+ and La(III). The results also showed an exchange between La^{3+} and Na^+, K^+, Ca^{2+} and Mg^{2+}.

4 Chemical effects on hydraulic properties

4.1 Influences of alkaline solutions

The groundwater of Beishan area may eventually reach high pH values as a result of the degradation of concrete in the HLW repository. Studies have conducted on GMZ bentonite using alkaline solutions (such as NaOH) in pHs and temperatures ([39]). The mineralogy of GMZ bentonite specimens was checked before and after the tests by X-ray diffraction (XDR) explorations.

The saturated hydraulic conductivity was obtained by means of Darcy's law. At different temperatures, the measured hydraulic conductivity experimented an increase with increasing pH of NaOH solutions (Fig. 4). The test results also revealed that, at a given pH, there was a significant increase in the saturated hydraulic conductivity with increasing temperature and its rate also depends on temperature.

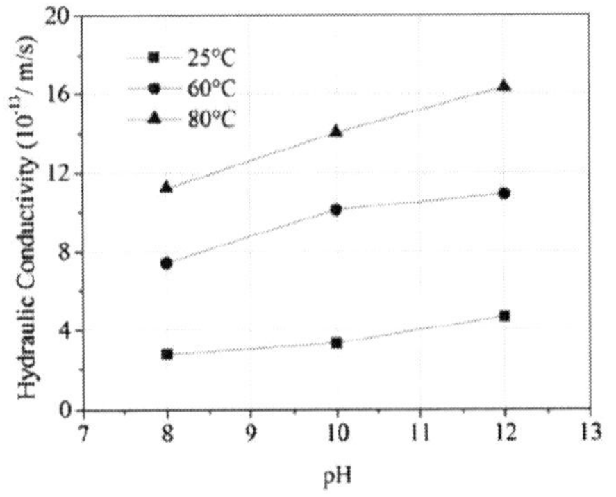

Figure 4. Relationships between the hydraulic conductivity and pH value of NaOH solutions at different temperatures ([39]).

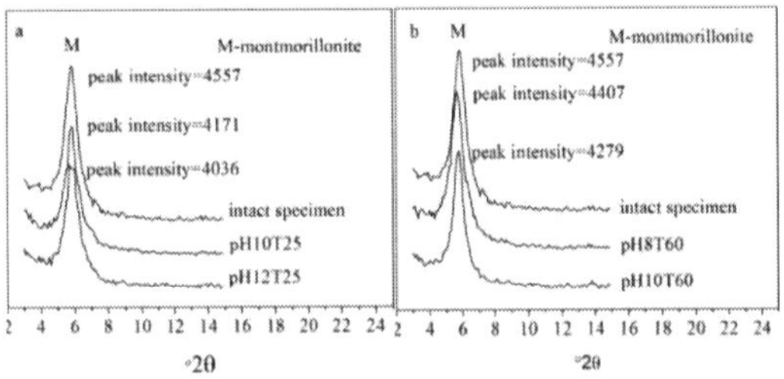

Figure 5. X-ray diffraction profiles of compacted GMZ01 bentonite specimens ([39]).

XRD results showed a decrease of montmorillonite content in GMZ bentonite after the infiltration of NaOH solution as a result of its dissolution (Fig. 5). Increasing pH induced an increase in the amount of montmorillonite dissolved, in a rate controlled by temperature. The increase in the hydraulic conductivity was explained by the decrease of the dry density of compacted bentonite as a result of the dissolution of montmorillonite. The reason for the effect of temperature on hydraulic conductivity could be the increase of effective cross section area of the pore space and the evolution of the viscosity of the injected alkaline solution.

4.2 Influences of saline solutions

Literatures show that influence of interactions between bentonite and pore fluid could be drastic to the performance of the bentonite barrier for groundwater with higher salinities ([66]). Beishan area groundwater mainly contains two predominant ions: Na^+ and Ca^{2+} ([40]). Therefore, studies have also been conducted on the behavior of GMZ bentonite against saline solutions. Different concentrations of NaCl solutions were used in some laboratory tests to assess its influences in the hydraulic conductivity of GMZ bentonite ([43]). For a void ratio lower than 0.55, salinity had no significant influence on the saturated hydraulic conductivity. Otherwise, it increased with salinity for higher void ratio values. It was deduced that, at large pressures, bentonite tended to acquire similar fabric and structure. Therefore, the increase on the hydraulic conductivity as a consequence of increasing saline solution could be prevented with high values of confining stresses.

DOI: 10.1051/e3sconf/20160902001

The increase in the hydraulic conductivity was explained by the reduction of the bentonite swelling capacity caused by the NaCl solutions and by the diffuse double layer (DDL) theory. The increase in saline concentration induce a decrease in the thickness of the diffuse layer ([67,34]), which results in the increase of the effective porosity of GMZ bentonite ([43]).
It was also discovered by other tests that the increase on the infiltrating solutions concentration induced an increase on the saturated hydraulic conductivity of bentonite as well ([68]). The results showed that Na^+ had more influence on the hydraulic conductivity than that of Ca^{2+} (Fig. 6), due to the less clogging of macro-pores induced by the NaCl solution compared to the CaCl2 one ([69,31]).

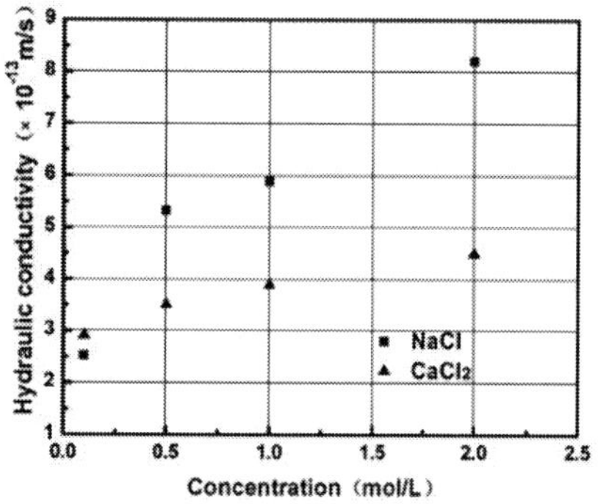

Figure 6. Comparison between the effects of the ion type on the hydraulic conductivity of GMZ bentonite ([68]).

5 Chemical effects on mechanical properties

5.1 Swelling pressure

5.1.1 Influences of alkaline solutions

Some studies investigated the influences of alkaline solutions on the swelling pressure of GMZ bentonite. A series of tests using NaOH solutions at different pHs and temperatures were conducted ([39]). At the beginning of the infiltration, the swelling pressure rapidly increased until reaching a peak, and then a decline was noticed before the full saturation of the specimen. After reaching the complete saturation at constant volume, the swelling pressure increased until reaching a constant value, defined as the equilibrium swelling pressure. The shape of the swelling pressure evolution curve was characterized by these two peaks (Fig. 7) and was in accordance with results achieved in similar studies using GMZ bentonite ([70,68]) and FEBEX bentonite ([71]).

Nonetheless, pH and temperature were observed to affect this "double-peak" shape. Increasing pH and temperature slightly eased off this shape. At a given temperature, increasing pH of NaOH solution induced a decrease in the swelling pressure, and its rate was controlled basically by pH for high temperatures. Similar results were found in different studies using Azraq Green clay ([72]) and MX-80 bentonite ([73]). Also, as in some other studies ([74]), it was noticed that the increase in temperature for a constant pH value caused a decrease in the swelling pressure as well (also appreciable in Fig. 7).

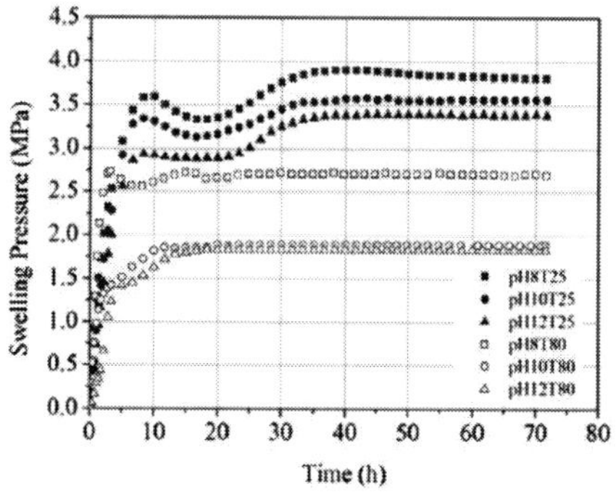

Figure 7. Swelling pressure evolution curves of GMZ bentonite ([39]).

The relationship between the montmorillonite content of GMZ bentonite and its swelling properties is very important and was the answer to some results in the tests. The decreasing of the swelling pressure due to the increase of the pH of NaOH solution was attributed to the dissolution of montmorillonite, which is the main responsible for the swelling behavior of GMZ bentonite ([39]).

5.1.2 Influences of saline solutions

Some studies also used salts to evaluate its effects on the swelling pressure of GMZ bentonite. De-ionized water and different concentrations of NaCl and $CaCl_2$ solutions were tested on compacted GMZ bentonite ([68]).
Results revealed that salinity had a considerable influence in the swelling properties of bentonite. Increasing solutions concentration induced a decrease in the swelling pressure. There was a difference in the effect between both solutions, resulting in a bigger impact on the swelling pressure induced by the $CaCl_2$ one (Fig. 8). It was also observed that the swelling pressure reached stability faster as the concentration of infiltrating solutions was higher.

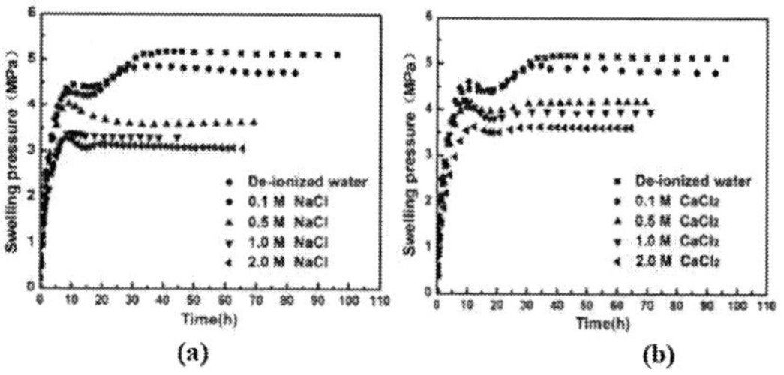

Figure 8. Influence of the concentration of (a) NaCl solutions and (b) CaCl2 solutions on the swelling pressure of GMZ bentonite ([68]).

It is generally accepted that there are two mechanisms of swelling of bentonite when it is exposed to water solutions: the crystalline swelling and the diffuse double-layer swelling ([75-76]). In Fig. 9, the stage I corresponds to crystalline swelling (first peak in Fig. 8), followed by a collapse of the soil skeleton induced by swelling of aggregates (between stage I and II). Stage III represents the second peak of Fig. 8 caused by the diffuse double-layer repulsion, which becomes the governing swelling mechanism. The DDL theory is used to explain the reduction of the swelling pressure with increasing salt concentration, since it causes a decrease in the DDL thickness and, hence, a reduction in the repulsive forces between clay particles ([67,34]).

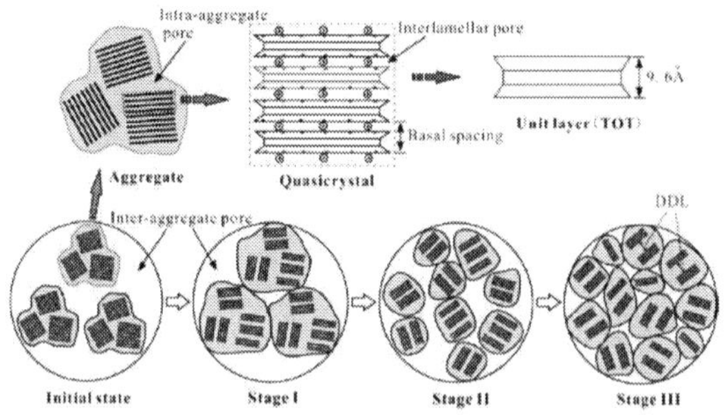

Figure 9. Conceptual diagrams of constant-volume swelling process and microstructure of compacted GMZ bentonite ([68,77-78]).

Some other studies also evaluated the influence of saline solutions on the swelling behavior of GMZ bentonite. Different concentrations of NaCl and Na_2SO_4 solutions were used on both GMZ Na- and Ca-bentonites for a series of constant-volume tests in which the dry density of bentonite was also varied ([79]).

The results showed differences between the behaviors of Na- and Ca-bentonite. The swelling pressure of Na-bentonite had a linear decrease with increasing saline solutions. In the case of Ca-bentonite, the swelling pressure first increased linearly with the concentration of solutions until a certain value, when the swelling pressure had a linear decrease.

5.2 Volume changes

The influence of saline solutions on the volume change behavior of GMZ bentonite was also investigated. Some swelling tests under different vertical loads were conducted using NaCl solutions to evaluate its effect on the swelling strain and compressibility of GMZ bentonite ([43]).

The results revealed that swelling strain of bentonite was increased by increasing concentration of saline solutions. This process was dominated by the dissipation of matric suction and the increase of diffuse ion layer repulsion ([43]). The compression index (C_c^*) is a parameter which describes how the void ratio varies as a function of the effective stress. It was observed to increase first and then turned to decrease while the void ratio decreased or the vertical stress increased for different concentrations of NaCl solution. If the concentration was increased, then C_c^* experimented a decrease. The secondary consolidation coefficient (C_α), which refers to the part of the compression that is recorded after pore water pressure cease to change, increased linearly with C_c^*. A bi-linear relationship between C_α and the swelling index (C_s^*) was established during unloading.

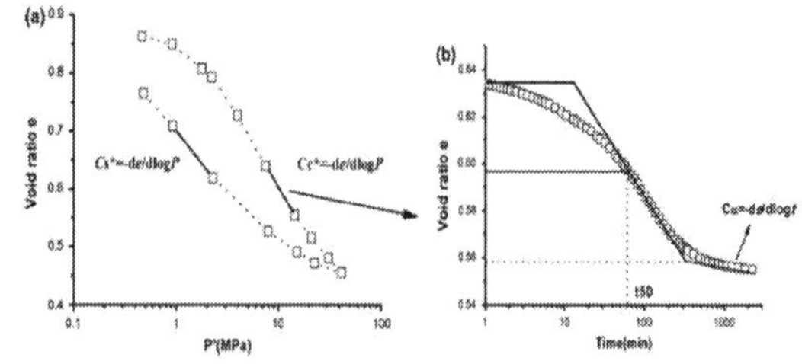

Figure 10. Determination of the parameters C_c^*, C_s^* and C_α ([43]).

The influence of cations on the swelling strain of GMZ bentonite was studied as well in a series of tests using different concentrations of NaCl and $CaCl_2$ solutions ([44]). Results revealed that the swelling strain decreased in a form of power function with increasing concentration of saline solutions. At a given salt concentration, the bentonite specimen tested with NaCl solutions had a higher swelling strain than that tested with the $CaCl_2$ ones.

During the injection of the saline solutions, it was noticed that the primary swell coefficient of bentonite decreased with increasing concentration of the solutions, and the secondary swell in contrast was more slowly. Inhibition of salt solution on the swelling strain of bentonite mainly happened during the secondary swell. DDL theory also explains this phenomena, since the diffusion layer thickness is notably reduced due to the increasing salt concentration, resulting in a reduction of the repulsion between the clay particles ([67,34]) and, hence, a reduction in swelling strain ([80-81,31]). The difference between the influence of Na^+ and Ca^{2+} were explained by the larger basal space of Na-bentonite than that of Ca-bentonite. With $CaCl_2$ infiltrations, some Na^+ is replaced by Ca^{2+} transforming part of the bentonite into Ca-bentonite ([82]), which has less basal space than Na-bentonite and, consequently, less extensive swelling ([44]).

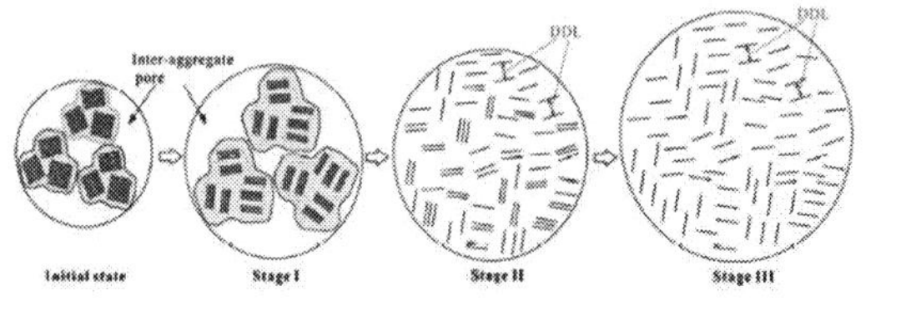

Figure 11. Conceptual diagram of micro-structural swelling process of compacted GMZ01 bentonite ([44,77-78]).

After swelling tests under constant load, when GMZ bentonite specimens were totally saturated with saline solutions and stable, compression tests were conducted applying the conventional step loading method ([43]). It was found that there was an almost linear relationship between the odometer modulus (Eodo) and the vertical effective stress (P'). As the vertical pressure increased or the void ratio decreased upon loading, the consolidation coefficient (C_v) was observed to first decrease and then turn to increase. C_v incremented with the increase of NaCl concentration, regardless the specimen was under loading or unloading.

Another important issue is the cyclically change of the chemistry of the groundwater during a long-term operation of a repository as a result of the fluctuations and temperature among other factors. To study how this cycles can influence the swelling behavior of GMZ bentonite, some studies have been conducted using different concentrations of $CaCl_2$ solutions on a series of one-dimensional free swelling tests with cyclic infiltrations of this saline solutions ([45]).

The results revealed that the total swelling strains of bentonite had an increase with increasing concentration of the solutions. It happened after more than one cycle of salinization-desalinization, since in the first salinization process the trend of the swelling strain was inversed. At a given concentration of salt solution, and especially for the highest ones, the total vertical displacement on bentonite happened basically during the first salinization-desalinization cycle. The total swelling strains measured over the first salinization decreased as the concentration of salts increased, which means that the concentration of $CaCl_2$ had a notable restricting effect on the swelling behavior of bentonite ([45]).

In the first desalinization, the relative swelling strain increased abruptly with an increase in the concentration of the saline solutions. In contrast to the tests with low concentrations of salt solutions, in those for high concentrations obvious relative swelling strains were noticed over desalinization and contraction strains were observed over salinization processes. It was also noticed that as the number of salinization-desalinization cycles increased, the swelling strain of the specimens was incremented as well (Fig. 12). When exposed to salinization-desalinization cycle, the swelling strain and its associated contraction strain were almost reversible from the third salinization process.

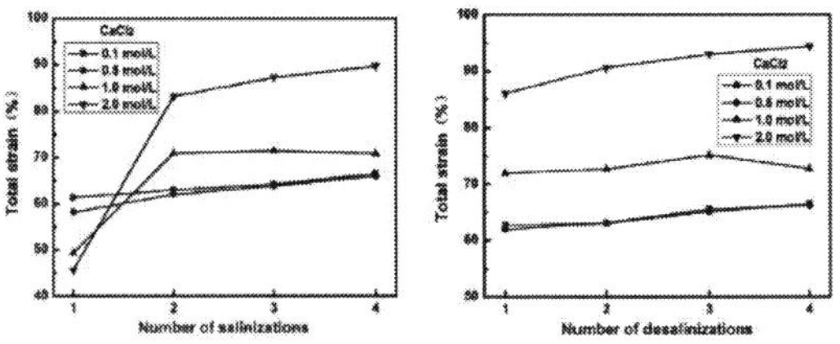

Figure 12. Effect of the number of salinizations and desalinizations on the total strain of GMZ bentonite ([45]).

6 Conclusions

The sorption of Ni(II) on GMZ bentonite is endothermic and spontaneous. It increases with an increase in pH values lower than 8.5, and then maintain a high level at values above 8.5. It is also dependent on foreign cations at a low pH, and independent at high pH values. The presence of humic acid enhances the sorption at low pH values, and reduces it at high pH values.

Removal of Cr(III) by means of adsorption onto GMZ bentonite is increased with increasing pH until a value of 7. Temperature and bentonite concentration also enhance the adsorption process, which is strongly dependent on ionic strength as well. Adsorption of Eu(III) can be improved using di(2-thylhexly) phosphoric acid-immobilized magnetic GMZ bentonite. For high pH values, the adsorption is low and it can be increased by increasing temperature. The adsorption of La(III) gradually increases with pH between 2 and 8. The adsorption process is probably dominated by surface complexation. It is also increased with an increment of bentonite content and decreases with increasing KCl concentration.

The hydraulic conductivity of GMZ bentonite increases with the pH of NaOH solutions at a rate which depends on the temperature. The infiltration of NaOH solution in GMZ bentonite produces a "double peak" structure in the swelling pressure evolution curve, in which the swelling pressure first increase rapidly until reaching a peak, and then, there is an intermediate phase with a decline until reaching another peak at full saturation and constant volume. The increase of pH and temperature induces a decrease in swelling pressure. For a given temperature, NaOH presence in GMZ bentonite induces a decrease of montmorillonite content. The increase of pH and/or temperature also produces a decrease of montmorillonite. The swelling pressure is observed to decrease with decreasing montmorillonite content.

The increase of NaCl and $CaCl_2$ saline solutions also increase the hydraulic conductivity of bentonite. For high concentrations of solution, the hydraulic conductivity is more influenced by the presence of the Na-based salt.

An increase of the concentration of NaCl and $CaCl_2$ solutions also produce a decrease in the swelling pressure, which is higher with the $CaCl_2$ one. Stability is faster reached as much high is the concentration of salt solutions. Swelling strain of GMZ bentonite decrease as the concentration of NaCl solution increases. For given concentrations, the swelling strain of the bentonite with infiltrations of NaCl solutions is higher than that with $CaCl_2$. An increase of the concentration of the solutions induce a decrease on the compression index C_c^* and an increase on the consolidation coefficient C_V.

There are differences in the influence of saline solutions on the swelling behavior between Na-bentonite and Ca-bentonite.

The increasing number of salinization-desalinization cycles induce an increase in the volume change of GMZ bentonite with $CaCl_2$ solutions.

Considering the complexity of the operation conditions and long run of the repository, further investigation is needed to totally understand the influences of coupling factors and time on chemical behavior of the material in the HLW repository.

Acknowledgements

The author would like to thank his colleagues and students: Isaac Moya-Carreras, Yong-Gui Chen, Bao Chen and Yu-Jun Cui, Chun-Ming Zhu, Feng Zhang and Yong He etc. for their contributions to this work.

The author is also grateful to China Atomic Energy Authority (Project [2011]1051) and the National Natural

Science Foundation of China (Projects No. 41527801, 40802069, 41030748) for the financial supports.

References

1. Pusch, R. 1983. KBS report 83-46.
2. Chapman, N.A., McKinley, I.G. 1989. John Wiley & Sons.
3. Romero, E., Gens, A., Lloret, A., 1999. Engineering Geology. 54,117–127
4. Ye, W.M., Cui, Y.J., Qian, L.X., Chen, B. 2009. Engineering Geology 108, 169-176.
5. Ye, W.M., Chen, Y.G., Chen, B., Wang, Q., Wang, J. 2010. Engineering Geology 116, 12-20.
6. Zhang, H.Y., Cui, S.L., Zhang, M., Jia, L.Y. 2012. Nuclear Engineering and Design 242, 115-123.
7. Li, X.L., Bastiaens, P., Van Marcke P., Verstricht, J., Chen, G.J., Weetjens, E., Sillen, X. 2010. Journal of Rock Mechanics and Geotechnical Engineering 2(2), 103-110.
8. Tripathy S., Thomas H. R. and Bag R., 2015. Journal of Hazardous, Toxic, and Radioactive Waste. 10.1061/(ASCE)HZ.2153-5515.0000272.
9. Lloret, A., Villar, M.V. 2007. Physysics and Chemistry of the Earth 32, 701-715.
10. Dixon, D.A. 2000. POSIVA Report 2000-04. Posiva Oy, Helsinki, Finland.
11. Hurel, C., Marmier, N. 2010. Journal of Radio-analytical Nuclear Chemistry 284, 225-230.
12. Nakashima, Y. 2006. Clay Minerals 41, 659-668.
13. Herbert, H.-J., Moog, H.C. 1999. Engineering Geology 54, 55-65.
14. Imbert, C., Villar, M.V. 2006. Applied Clay Science 32, 197-209.
15. Tang, A.M., Cui, Y.J., Le, T.T. 2008. Applied Clay Science 41, 181-189.
16. Kanno, T., Fujita, T., Takeuchi, S., Ishikawa, H., Hara, K., Nakano, M. 1999. International Journal for Numerical and Analytical Methods in Geomechanics 23, 1281-1307.
17. Nguyen, T.S., Selvadurai, A.P.S., Armand, G. 2005. International Journal of Rock Mechanics and Mining Sciences 42, 639-651.
18. Gens, A., Olivella, S. 2006. Chinese Journal of Rock Mechanics and Engineering 25, 670-680.
19. Gens, A., Sánchez, M., Guimarães, L.D.N., Alonso, E.E., Lloret, A., Olivella, S., Villar, M.V., Huertas, F. 2009. Geotechnique 59, 377-399.
20. Olivella, S., Carrera, J., Gens, A., Alonso, E.E. 1994. Transport in Porous Media 15, 271-293.
21. Gawin, D., Baggio, P., Schrefler, B.A. 1995. International Journal for Numerical Methods in Fluids 20, 969-987.
22. Thomas, H.R., He, Y. 1995. Geotechnique 45, 677-689.
23. Liu, X.Y., Zhang, C.Y., Liu, Q.S. 2008. Advanced Materials Research 33-37, 639-644.
24. Åkesson, M., Jacinto, A.C., Gatabin, C., Sanchez, M., Ledesma, A. 2009. Geotechnique 59, 307-318.
25. Chen, Y.F., Zhou, C.B., Jing, L.R. 2009. Computers and Geotechnics 36, 1308-1329.
26. Fernández, R., Cuevas, J., Sánchez, L. 2006. Applied Geochemistry 21(6), 977-992.
27. Tripathy, S., Sridharan, A., Schanz, T., 2004. Can. Geotech. J. 41, 437–450.
28. Rao M, Shivananda P. 2005. Can Geotech J 42:307–315
29. Sánchez, L., Cuevas, J., Ramírez, S., de León, D.R., Fernández, R., de la Villa, R. V., Leguey, S. 2006. Applied Clay Science 33(2), 125-141.
30. Rao, S.M., Thyagaraj, T., Thomas, H.R., 2006. Geotechnique 56 (10), 707–713
31. Castellanos, E., Villar, M.V., Romero, E., Lloret, A., Gens, A. 2008. Physics and Chemistry of the Earth 33, S516–S526.
32. Siddiqua, S., Blatz, J., Siemens, G., 2011. Can. Geotech. J. 48, 199–213
33. Barbour, S.L., Fredlund, D.G. 1989. Canadian Geotechnical Journal 26, 551–562
34. Mitchell, J.K. 1993. John Wiley and sons, New York.
35. Barbour, S.L., Yang N. 1993. Can Geotech J 30:920–934
36. Ye, W.M., Borrell, N.C., Zhu, J.Y., Chen, B., Chen, Y.G. 2014a. Engineering Geology 169, 41-49.
37. CAEA (China Atomic Energy Authority) 2006. Ministry of environment protection & ministry of science and technology (in Chinese).
38. Wang, J., Chen, W.M., Su, R., Guo, Y.H., Jin., Y.X. 2006. Chinese Journal of Rock Mechanics and Engineering 25, 801-812 (in Chinese).
39. Ye, W.M., Zheng, Z.J., Chen, B., Chen, Y.G., Cui, Y.J., Wang, J. 2014b. Applied Clay Science 101, 192-198.
40. Guo, Y.H., Yang, T.X., Liu, F.S. 2001. Uranium Geology 17 (3), 184-189.
41. Liu, Y.M., Wen, Z.J. 2003. Mineral Rocks 23, 42-45 (in Chinese).
42. Wen, Z.J. 2005. Acta Petrologica et Mineralogica 24, 583-586 (in Chinese).
43. Ye, W.M., Zhang, F., Chen, B., Chen, Y.G., Wang, Q., Cui, Y.J. 2014c. Environmental Earth Sciences 72, 2621-2630.
44. Ye, W.M., Zhu, C.M., Chen, Y.G., Chen, B., Cui, Y.J., Wang, J. 2015. Environmental Earth Sciences 74, 793-802.
45. Zhu, C.M., Ye, W.M., Chen, Y.G., Chen, B., Cui, Y.J. 2015. Engineering Geology 184, 104-110.
46. Wang, J., Liu, Y.M., Zhao, X.G. 2011. BRIUG PEBS Internal Deliverable DB-2.1.
47. Madsen, F.T. 1998. Clay Minerals 33, 109-129.
48. Montes, H.G., Duplay, J., Martinez, L., Geraud, Y., Rousset-Tournier, B. 2003. Applied Clay Science 23, 309–321.
49. Nakashima, Y. 2004. Journal of Nuclear Science and Technology 41, 981-992.
50. Li, J.X., Hu, J., Sheng, G.D., Zhao, G.X., Huang, Q. 2009. Colloids and Surfaces A: Physicochemical and Engineering Aspects 349, 195-201.
51. Wang, S.W., Dong, Y.H., He, M.L., Chen, L., Yu, X.J. 2009a. Applied Clay Science 43, 164-171.
52. Wang, S.W., Hu, J., Li, J.X., Dong, Y.H. 2009b. Journal of Hazardous Materials 167, 44-51.

53. Chen, Y.G., Zhu, C.M., Sun, Y.H., Duan, H.Y., Ye, W.M., Wu, D.B. 2012a. Journal of Radioanalytical Nuclear Chemistry 292, 1339-1347.
54. Chen, Y.G., He, Y., Ye, W.M., Lin, C.H., Zhang, X.F., Ye. B. 2012b. Environmental Earth Sciences 67, 1261-1268.
55. Chen, Y.G., Zhu, B.H., Wu, D.B., Wang, Q.G., Yang, Y.H., Ye, W.M, Guo, J.F. 2012c. Chemical Engineering Journal 181-182, 387-396.
56. Chen, Y.G., He, Y., Ye, W.M., Sui, W.H., Xiao, M.M. 2013. Transactions of Nonferrous Metals Society of China 23, 3482-3489.
57. Chen, Y.G., He, Y., Ye, W.M., Jia, L.Y. 2015. Journal of Industrial and Engineering Chemistry 26, 335-339.
58. Yang, S.T., Li, J.X., Lu, Y., Chen, Y.X., Wang, X.K. 2009. Applied Radiation and Isotopes 67, 1600-1608.
59. Pan, D.Q., Fan, Q.H., Li, P., Liu, P.L., Wu, W.S. 2011. Chemical Engineering Journal 172, 898-905.
60. Zhao, D.L., Chen, S.H., Yang, S.B., Yang, X., Yang, S.T. 2011. Chemical Engineering Journal 166, 1010-1016.
61. Wu, T., Wang, H., Zheng, Q., Zhao, Y.L., Van Loon, L.R. 2014. Applied Clay Science 101, 136-140.
62. Ho, Y.S., McKay, G. 1998. Process Safety and Environmental Protection 76, 183-191.
63. Tahir S.S., Naseem R. 2007. Separation and Purification Technology 53, 312-321.
64. Wu, D.B., Zhu, C.M., Chen, Y.G., Zhu, B.H., Yang, Y.H., Wang, Q.G., Ye, W.M. 2012. Applied Clay Science 62-63, 87-93.
65. Chen, Y.G., Huang, R.Q., Zhu, C.M., Wu, D.B., Sun, Y.H., He, Y., Ye, W.M. 2014. Journal of Radioanalytical Nuclear Chemistry 299, 665-674.
66. Rao, S.M., Thyagaraj, T., 2007. Appl. Clay Sci. 38 (1), 113–129
67. Yong, R.N., Warkentin, B.P. 1975. Elsevier, Amsterdam.
68. Zhu, C.M., Ye, W.M., Chen, Y.G., Chen, B., Cui, Y.J. 2013. Engineering Geology 166, 74-80.
69. Pusch, R. 2001. SKB Technical Report TR-01-07, Stockholm.
70. Ye, W.M., Wan, M., Chen, B., Chen, Y.G., Cui, Y.J., Wang, J. 2013. Environment Earth Science 68(1), 281–288.
71. Villar, M.V., Lloret, A. 2008. Applied Clay Science 39, 38-49.
72. Abdullah, W.S., Alshibli, K.A., Al-Zou'bi, M.S. 1999. Applied Clay Science 15, 447-462.
73. Herbert, H.-J., Kasbohm, J., Sprenger, H., Fernández, A.M., Christian, R. 2008. Physics and Chemistry of the Earth 33(supplement1), S327–S342.
74. Lingnau, B.E., Graham, J., Yarechewski, D., Tanaka, N., Gray, M.N. 1996. Engineering Geology 41(1-4), 103-115.
75. Madsen, F.T., Müller-VonMoos, M. 1989. Applied Clay Science 4, 143–156.
76. Savage, D. 2005. SKI Report 54.
77. Villar, M.V. 2002. Publicación Técnica ENRESA 01/2002, Madrid (44 pp).
78. Suzuki, S., Prayongphan, S., Ichikawa, Y., Chae, B. 2005. Applied Clay Science 29, 89–98.
79. Sun, D.A., Zhang, L., Li, J., Zhang, B.C. 2015. Applied Clay Science 105-106, 207-216.
80. Karnland, O. 1997. SKB technical report 97-31. Swedish Nuclear Fuel and Waste Management Co, Stockholm.
81. Mata, C. 2003. Ph.D. Thesis. Universitat Politecnica de Catalunya, Barcelona.
82. Montes, H.G., Geraud, Y. 2004. Colloids and Surfaces A: Physicochemical and Engineering Aspects 235, 17–23.

Determination of the relationship among capillary pressure, saturation and interfacial area: a pore unit assembly approach

E. Nikooee[1,2], T. Sweijen[1] and S. Majid Hassanizadeh[1a]

[1]*Department of Earth Sciences, Utrecht University, Utrecht, The Netherlands*
[2]*Department of Civil and Environmental Engineering, School of Civil and Material Engineering, Shiraz University, Shiraz, Iran*

Abstract. Three state variables namely, degree of saturation (S^w), capillary pressure (P_c) and specific air-water interfacial area (a^{aw}) are indispensable for modelling coupled processes relevant to unsaturated soils mechanics, agriculture, and contaminant hydrology. They play a key role in simulating various phenomena and the determination of various parameters and physical characteristics such as the unsaturated soil shear strength, field capacity, wilting point, air and water diffusivity and the rate of dissolution of contaminants. The determination of soil water retention curve (S^w-P_c) as well as the specific interfacial area (a^{aw}) using available experimental techniques is a challenging and time consuming task. Therefore, a numerical technique that employs basic soil properties to obtain these variables is of much value and high practical and theoretical importance. In the current study, the porous network extracted from a discrete element model (the so-called pore unit-assembly) has been used to directly model the drying and wetting processes inside a granular soil packing and to obtain the values of P_c, S^w and a^{aw}. The results of the simulations are in good agreement with the experimental data, which points to the efficacy and adequacy of the introduced algorithms and involved assumptions for this purpose.

1 Introduction

The soil water retention curve is a fundamental relationship in unsaturated soil mechanics, based on which different hydraulic and mechanical parameters are defined. Knowledge of this curve is essential for determining the contribution of matric suction to the shear strength in unsaturated soils and obtaining different hydraulic parameters such as water diffusivity. It is a key part of models used for simulating water flow and solute transport as well as soil mechanical behaviour [1].

The soil water retention curve, however, is not unique and for different hydraulic paths (drying, wetting, and scanning), different equations need to be introduced to capture variation of capillary pressure versus degree of saturation. This phenomenon has been referred to as hydraulic hysteresis [2-6]. There is also hysteresis in mechanical properties such as effective stress parameter [7].

The hydraulic (and associated mechanical) hysteresis can be attributed to the fact that at different hydraulic paths and the same degree of saturation, the water is distributed differently inside the medium [8]. In other words, at the same saturation, different macroscopically measured capillary pressure values are found due to different distributions of water phase (manifested in different air-water interface configurations). Therefore, in addition to saturation, a macroscale variable is required to account for many possible distributions of water at a given saturation. In other words, a new state variable is needed for hydro-mechanical models. This is proposed to be the specific interfacial area, defined as the amount of air-water interfacial area per total volume of an REV. The inclusion of this variable in coupled hydro-mechanical models can allow for modeling hysteresis. That is, the hysteretic capillary pressure-saturation curves should be replaced by a unique capillary pressure-saturation-interfacial area surface [7-10].

The measurement of capillary pressure and saturation is already a time consuming task. For the investigation of the relationship among capillary pressure, saturation and interfacial area, one needs also to deal with the measurement of specific interfacial area. It adds further challenge and difficulty and is often done through sophisticated experimental methods such as tracer technique (e.g. [11]) and x-ray micro-tomography (e.g. [12]). Another approach for the study of role of interfacial area has been the use of pore-scale computational models such as lattice Boltzmann [13, 14], SPH [15], and pore-network modelling [16]. These models, however, often suffer from serious difficulties to be used for practical purposes in geotechnical analysis. They need either the soil porous network or are highly computationally demanding, and moreover, they cannot easily and directly make use of the soil basic properties such as porosity and grain size distribution.

The purpose of this study is to present a method to use a grain scale model (DEM) rather than a pore-based

[a] Corresponding author: S.M.Hassanizadeh@uu.nl

model. The merit of such an approach is that the grain size distribution and porosity can be directly used when constructing the model of the granular packing and then the porous network would be obtained from the packing. Next, wetting and drying are simulated therein and the required state variables (air-water interfacial area, capillary pressure and saturation) are obtained.

In what follows, the required steps, namely, packing construction, extraction of porous network, and invasion criteria for wetting and drying are described and simulations versus experimental data are presented. Finally, concluding remarks and future perspectives are presented.

2 Model description

2.1 Model assumptions and considerations

In order to start with a medium where less microstructural and mineralogical complexities and heterogeneities are present to affect a^{aw}-S^{w}-P_c relationship, this study is focused on modelling a packing consisted of glass beads. Therefore, at this stage, our focus is mainly on the coarse-grained material. We consider rigid grains and disregard possible mineralogical effects (such as water absorption by the grains and reaction with the grains). Other assumptions are that grains are modelled as spheres and the contact angle is assumed to be equal to 0° for drying and wetting. The contact angle hysteresis will be added in future investigations.

2.2 Construction of packing

For constructing the packing, we use the open source discrete element model YADE-DEM [17]. Discrete element technique allows for adjusting particle packing while the grains are allowed to move and interact. We can thus start with grains at arbitrary locations and let them move and reach a stable configuration under their own weight as well as external pressure. We take the model domain to be a box, with dimensions 0.02m x 0.02m x 0.02m, within which we randomly position a cloud of 4000 particles. Grain sizes are chosen from a particle size distribution for the glass bead packing used in drying-wetting experiments of [18]. Values of mechanical parameters of the particles used in our simulations are reported in Table 1. The values of mechanical parameters of the domain boundaries were the same as those for the particles. In this study, Hertz-Mindlin contact law has been used.

The initial porosity of the cloud of particles is larger than 0.99. To compact the cloud of particles into a loose packing, we apply a confining pressure of one kPa on all boundaries of the box. The model is then run until the cloud of particles is compacted and a force equilibrium is reached, which corresponds to an unbalanced force smaller than 10^{-6} N. The packing in equilibrium has a porosity value of 0.34.

The next step is to simulate an oedometer test. For this purpose, we fix all boundaries except for the top boundary. A confining stress of 1MPa is applied to the top boundary to simulate compaction of the packing. Further details on the utilized discrete element approach can be found in [17].

Table 1. Input micromechanics parameters for glass beads (Chung and Ooi, 2008) [19]

Property	Value	Unit
Density	2530	kg/m³
Young's Modulus	41	GPa
Poisson's ratio	0.22	-
Friction angle	0.1*	°
Number of glass beads	4000	-

* Different value selected to match the porosity value of 0.34 reported by [18]

2.3 Extraction of pore network of the packing

The pore structure of the packing needs to be extracted in order to simulate the wetting and drying processes. The pore network of a granular packing consists of pore bodies and throats. Pore bodies constitute the greatest portion of total void volume.They are connected by throats which are considered to be without any volume (either as 2d surface elements or 3d but with negligible volume). There are various pore network extraction techniques originally introduced for extraction of the pore network from three-dimensional images of porous media [20]. Here we use a regular tessellation technique (see [21] and [22]). The triangulation results in the division of pore space into several tetrahedra, each containing one pore unit. Vertices of each tetrahedron are located at the centre of the four particles embracing a pore unit (Figure 1). The scalene triangle formed between two neighbouring tetrahedron is considered as the pore throat connecting the two.

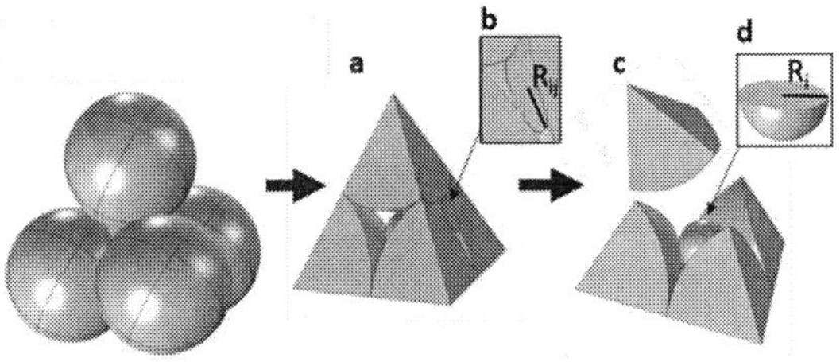

Figure 1. (a) One pore unit embraced by four particles; (b) Inscribed circle in a pore throat, with radius R_{ij} (c, d) Inscribed sphere in a pore body with radius R_i.

2.4 Invasion algorithms

Following Yuan et al. [23], we have used the Mayer-Stowe-Princen (MSP) [24-26] to determine throat entry pressure surpassing which the air invades a pore throat and connected pore units during drainage. For imbibition, we assume that the critical curvature at which water fills a pore unit is that of the inscribed sphere of the pore unit. The capillary pressure corresponding to the radius R_i of the inscribed sphere in pore unit is given by:

$$P_{ci} = \frac{2\gamma}{R_i} \tag{1}$$

where γ is the interfacial tension between water and air (72 dynes/cm^2). However, we need to account for cooperative filling during imbibition; the more the number of air-saturated neighbouring pores, the smaller the capillary pressure at which imbibition occurs. To account for this, we follow the method of Jerauld and Salter (1990) [27] to modify equation (1):

$$P_{ci} = \frac{2\gamma}{R_i N_{air}} \tag{2}$$

where N_{air} is the number of air-saturated neighbouring pore units.

The radius $_{of}$ the inscribed sphere is found by using the Cayley-Menger determinant, which is a method for finding the radius of an inscribed sphere for four touching or non-touching particles. For more information, the reader is referred to MacKay (1973) and Michelucci and Foufou (2004) [28,29].

2.5 Interfacial area determination

To determine the air-water interfacial area, we calculate the area of main menisci, also-called "terminal menisci", of interfaces inside a pore throat and pore unit. We neglect the water films on the solid surfaces and the liquid bridges that may be present between two touching particles. The position of an air-water interface (and thus, the amount of interfacial area) depends on the capillary pressure. Let's assume that an interface moves from pore unit i towards pore throat ij during drainage, and the reverse occurs during imbibition. Thus, capillary pressure varies between a minimum value of $P_{ci}=2\gamma/R_i$ and a maximum value of $P_{cij}=\gamma/R_{ij}$. Then, the area of the meniscus varies between a maximum and a minimum value at these locations, respectively. The minimum value of the the air-water interfacial area is simply equal to the area of the scalene triangle that forms the pore throat ij. The maximum value of the the air-water interfacial area is simply equal to the area of the meniscus formed on the inscribed sphere of pore body i. It has an area ($A^{aw}_{ij@i}$) defined by:

$$A^{aw}_{ij@i} = 4\pi r_i^2 \frac{V_{ij\cap i}}{V_i} \tag{3}$$

where $V_{ij\cap i}$ is the volume of a portion of pore unit i that is assigned to pore throat ij and V_i is the volume of pore unit i. In between these two extremes, we assume that the air-water interfacial area changes proportional to $(\gamma/P_c)^2$.

During imbibition, it may happen that an interface remains hanging on the inscribed sphere of a pore unit, as the inscribed sphere has a much larger radius than the pore throat. When the capillary pressure is smaller than the entry pressure of pore throat ij but larger than the capillary pressure corresponding to the inscribed sphere, then we assume that the air-water interface remains hanging on the narrowest opening of pore throat ij. When the capillary pressure is lower than that of the inscribed sphere, spontaneous imbibition occurs which fills the pore unit. This will create new interfaces in other pore throats. Finally, the interfaces that are located next to a residual phase are assumed to be located exactly at the narrowest opening of a pore throat

3 Results and discussion

Figure 2 shows the pore unit and throat size distributions of the pore unit assembly extracted from the modelled packing. Figure 3 illustrates the capillary pressure-saturation curve of sample with porosity of 0.34 as compared to the experimental data. The agreement between the simulated and experimental data is not fully satisfactory. Although in most parts of the drying and imbibition curves a good agreement is seen, the experimental residual saturation is not fully captured. This can be attributed to phenomena which have not been accounted for in our model such as film flow or the presence of liquid bridges, and their coalescence.

The variation of air-water specific interfacial area versus saturation has been depicted in Figure 4, where experimental data are also plotted. Here also, we find discrepancy in estimating values of air-water interface at low degrees of saturations.

Furthermore, the variation of air-solid specific interfacial area, a^{as}, versus saturation are presented in Figure 5; as it can be seen, less hysteresis is present in this relationship as compared to the air-water specific interfacial area versus degree of saturation. a^{as} versus saturation shows almost a linear relationship with degree of saturation. This is in agreement with the results of Culligan et al. (2004) [18].

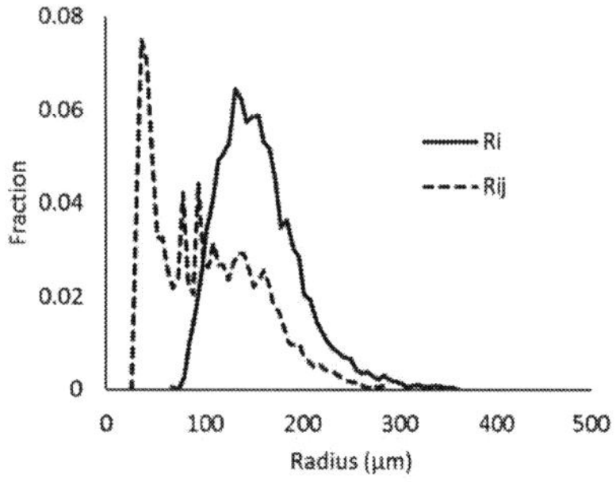

Figure 2. Pore throat radii (Rij) distribution and the distribution of inscribed spheres in pore units (Ri).

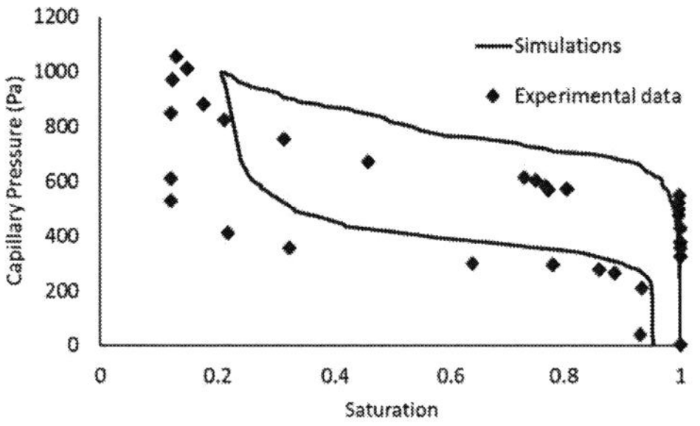

Figure 3. Capillary pressure-saturation curve (simulation versus experiments), experimental data from [18].

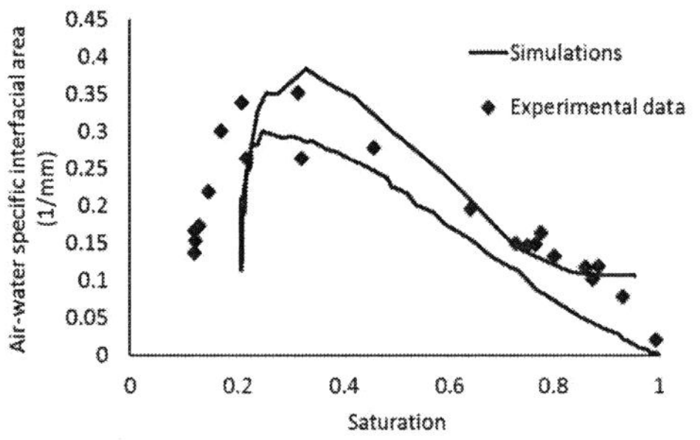

Figure 4. Air-water specific interfacial area versus saturation (simulation versus experiments), experimental data from [18].

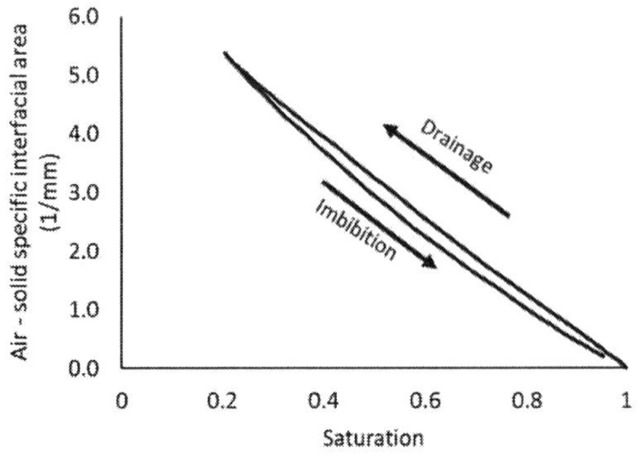

Figure 5. Simulated air-solid specific interfacial area versus saturation

4 Concluding remarks

In this study, a method for the determination of interfacial areas in an assembly of pore units and throats was presented. The porous network was extracted from granular packing by regular tessellation. Our results were of reasonable accuracy as compared to the experimental data. However, further comparison with experimental data of other packings and granular materials (e.g. sand) and additional validation is necessary to assess the scope of application and the accuracy of the proposed method. Inclusion of further complexities to the model such as including contact angle hysteresis, as well as required local rules (relationships) to account for film flow, trapping and snap off would enhance the capabilities of the model for a better estimation of the values of air-water interfacial area and its variation with capillary pressure and saturation.

Acknowledgement

The first and third authors would like to thank the European Research Council for ERC advanced research funding under the European Union's Seventh Framework Programme (FP/2007-2013) / ERC Grant Agreement no. 341225 which has supported them during this study. The second author gratefully acknowledges financial support from the Technology Foundation STW, the technological branch of the Netherlands Organisation of Scientific Research, NWO, and the Dutch ministry of Economic Affairs under contract no. 12538, entitled Interfacial effects in ionized media.

References

1. M. Tuller, D. Or, Retention of water in soil and the soil water characteristic curve. Encyclopedia of Soils in the Environment, **4**, 278-289 (2004).

2. D. B. Jaynes, Comparison of soil-water hysteresis models. Journal of Hydrology, 75(1), 287-299 (1984).
3. S. J., Wheeler, R. S. Sharma, M. S. R. Buisson, Coupling of hydraulic hysteresis and stress–strain behaviour in unsaturated soils. Géotechnique, 53(1), 41-54 (2003).
4. R. Tamagnini, An extended Cam-clay model for unsaturated soils with hydraulic hysteresis. Géotechnique, 54(3), 223-228 (2004).
5. A. Khosravi, J.S. McCartney, Impact of hydraulic hysteresis on the small-strain shear modulus of low plasticity soils. Journal of Geotechnical and Geoenvironmental Engineering, 138(11), 1326-1333 (2012).
6. C. Yang, D., Sheng, J.P. Carter, Effect of hydraulic hysteresis on seepage analysis for unsaturated soils. Computers and Geotechnics, 41, 36-56. ISO 690, (2012).
7. E. Nikooee, G. Habibagahi, S.M, Hassanizadeh, A. Ghahramani, Effective stress in unsaturated soils: A thermodynamic approach based on the interfacial energy and hydromechanical coupling. Transport in porous media,96 (2), 369-396 (2013).
8. N. K. Karadimitriou, Two-phase flow experimental studies in micro-models, PhD Dissertation, Depart of Earth Sciences, Utrecht University, 145 (2013).
9. S.M. Hassanizadeh, W.G. Gray, Mechanics and thermodynamics of multiphase flow in porous media including interphase boundaries. Advances in water resources, 13(4), 169-186 (1990).
10. W. G. Gray, S.M. Hassanizadeh, Thermodynamic basis of capillary pressure in porous media. Water Resources Research, 29(10), 3389-3405 (1993).
11. H. Kim, P.S.C. Rao, M. D. Annable, Gaseous tracer technique for estimating air–water interfacial areas and interface mobility. Soil Science Society of America Journal, 63(6), 1554-1560 (1999).
12. M. L. Brusseau, S. Peng, G. Schnaar, A. Murao, Measuring air-water interfacial areas with X-ray microtomography and interfacial partitioning tracer tests. Environmental science & technology, 41(6), 1956-1961 (2007).
13. M. L. Porter, M. G. Schaap, D. Wildenschild, Lattice-Boltzmann simulations of the capillary pressure–saturation–interfacial area relationship for porous media. Advances in Water Resources, 32(11), 1632-1640 (2009).
14. S. A. Galindo-Torres, A. Scheuermann, L. Li, Boundary effects on the Soil Water Characteristic Curves obtained from lattice Boltzmann simulations. Computers and Geotechnics, 71, 136-146 (2016).
15. R. Sivanesapillai, N. Falkner, A. Hartmaier, H. Steeb. A CSF-SPH method for simulating drainage and imbibition at pore-scale resolution while tracking interfacial areas. Advances in Water Resources (2015).
16. V. Joekar-Niasar, S. M. Hassanizadeh. Pore-network modeling of wicking: a two-phase flow approach. Wicking in Porous Materials: Traditional and Modern Modeling Approaches, 237-262 (2012).
17. V. Šmilauer, Cohesive particle model using the discrete element method on the yade platform (Doctoral dissertation, Université de Grenoble; and Czech Technical University in Prague (2010).
18. K.A. Culligan, D. Wildenschild, B. Christensen, W. G., Gray, M.L. Rivers, A. F. Tompson, Interfacial area measurements for unsaturated flow through a porous medium. Water Resources Research, 40(12) (2004).
19. Y. C. Chung, J. Y. Ooi, A study of influence of gravity on bulk behaviour of particulate solid. Particuology, 6(6), 467-474 (2008).
20. H. Dong, Micro-CT imaging and pore network extraction (Doctoral dissertation, Department of Earth Science and Engineering, Imperial College London) (2008).
21. B. Chareyre, A. Cortis, E. Catalano, E. Barthélemy, Pore-scale modeling of viscous flow and induced forces in dense sphere packings. Transport in porous media, 94(2), 595-615 (2012).
22. T. Sweijen, E. Nikooee, S. Majid Hassanizadeh, B. Chareyre, The effects of swelling and porosity change on capillarity: DEM coupled with a pore-unit assembly method, Transport in Porous Media (in press).
23. C. Yuan, B. Chareyre, F. Darve, Pore-scale simulations of drainage in granular materials: finite size effects and the representative elementary volume. Advances in Water Resources (2015), doi: 10.1016/j.advwatres.2015.11.018.
24. M. Prodanovic, S.L. Bryant. A level set method for determining critical curvatures for drainage and imbibition. J Colloid Interface Sc., 304(2), 442–58 (2006).
25. R.P. Mayer, R.A. Stowe. Mercury porosimetry breakthrough pressure for penetration between packed spheres. J Colloid Sci. 20(8), 893–911 (1965).
26. H. M. Princen, Capillary phenomena in assemblies of parallel cylinders: I. Capillary rise between two cylinders. J Colloid Interface Science, 30(1), 69–75 (1969).
27. G. R. Jerauld, S.J. Salter, The effect of pore-structure on hysteresis in relative permeability and capillary pressure: pore-level modeling. Transport in Porous Media, 5(2), 103-151 (1990).
28. A. MacKay, To find the largest sphere which can be inscribed between four others. Acta Crystallographica Section A: Crystal Physics, Diffraction, Theoretical and General Crystallography, 29(3), 308-309 (1973).
29. D. Michelucci, S. Foufou, Using cayley-menger determinants for geometric constraint solving. Paper presented at the Proceedings of the Ninth ACM Symposium on Solid Modeling and Applications, 285-290 (2004).

Transient boundary conditions in the frame of THM-processes at nuclear waste repositories

Tom Schanz[1]

[1]*Chair of Foundation Engineering, Soil and Rock Mechanics, Ruhr-Universität Bochum, Germany*

Abstract. In nuclear waste repositories, initially unsaturated buffer is subjected to constant heat emitted by waste canister in conjunction with peripheral hydration through water from host rock. The transient hydration process can be potraied as transformation of initial heterogeneity towards homogeneity as final stage. In this context, this paper addresses the key issue of hydro mechanical behaviour of compacted buffer in context of clay microstructure and its evolution under repository relevant loading paths and material heterogeneity. This paper also introduces a unique column experiment facility available at Ruhr Universität Bochum, Germany. The facility has been designed as a forerunner of field scale testing program to simulate the transient hydration process of compacted buffer as per German reference disposal concept. The device is unique in terms of having proficiency to capture the transient material response under various possible repository relevant loading paths with higher precision level by monitor the key parameters like temperature, total suction, water content and axial & radial swelling pressure at three different sections along the length of compacted soil sample. In general, a larger spectrum of loading paths/scenarios, which may arise in the nuclear repository, can be covered precisely with existing device.

1 Introduction

In recent years, considerable research has been carried out to evaluate the behavior of compacted clay based buffer material in the frame of coupled thermo-hydro-mechanical (THM) processes at nuclear waste repositories. In this context, initially unsaturated buffer is subjected to radioactive decay heat emitted by waste canister in conjunction with peripheral hydration through water from host rock. The process of hydration triggers swelling of inter layers space in clay particles and transform the material from initial heterogeneous to homogeneous one as final state. This shift of heterogeneity towards homogeneity simultaneously changes the local volume constraint conditions. The challenge is to anticipate the material behavior under such transient boundary conditions in the frame of THM-processes.

The inherent heterogeneity in compacted clays is characterized by its two distinct continua of pore space namely macropores and micropores, whereas the global heterogeneity in compacted buffer bricks in disposal pits is attributed by technological voids. The technological voids are referred to as the macro-pores related to different interfaces involving the buffer material [1-2]. Dual porosity model appears to be well suited to depict this characteristic of compacted clays. The macroscopic volume change due clay-water interaction is associated with evolution of microstructure and its interaction with macrostructure. Thus, to anticipate the complex hydro-mechanical behaviour of compacted buffer in nuclear waste repository, the influence of initial heterogeneity and applied boundary conditions on microstructure evolution must be considered.

Based on the inputs from phenomenological investigations on microstructure evolution, various dual porosity models have been reported such as volume change behavior due to hydration [3-5], coupled THM behaviour in frame of nuclear waste repository [6], water retention behaviour [7]. Enhancement of existing dual porosity models requires incorporating the inputs regarding clay microstructure and its evolution from phenomenological investigations identical to the actual one. The errors associated with element tests due to scale effect, steady state boundary conditions and lack of generating repository relevant loading scenarios can be eliminated in phenomenological experiments.

This paper addresses the key issue of hydro mechanical behaviour of compacted buffer in context of clay microstructure and its evolution under repository relevant loading paths and material heterogeneity. This paper also introduces a unique column experiment facility available at Ruhr Universität Bochum, Germany. The facility has been designed as a forerunner of field scale testing program to simulate the transient hydration process of compacted buffer as per German reference disposal concept.

2 Dual porosity models: theoretical background

The studies pertinent to clay microstructure reveal that as compacted clay samples have different structural levels, which can be scrutinized by its pore size distribution

Corresponding author: tom.schanz@rub.de

curve. The pore size distribution curve based on MIP data revealed about two or more overlapping but distinct continua [3, 8-10]. Later these continua or structural levels were termed as micro and macrostructure. Clay microstructure level comprises the aggregation of clay particles or quasi-crystals. Thus the intra aggregate pores include inter layers and inter particles pores. Where as the pores in between aggregates are inter aggregate or macropores. A typical PSD curve for compacted bentonite is shown in Figure 1.

In dual porosity model the overall medium is assumed to have two interacting sub domain named as micro (m) and macropores (M). These sub domains are two totally distinct environments due to associated physicochemcial forces, hence show peculiar response and sensitivity to applied boundary conditions. For an example, the change in macroporosity is dominant during compaction, whereas the change in microporosity is dominant during hydration (Figure 1). The measurable volume change on macrolevel during clay water interaction is the upscaled microstructure response. Therefore, the material behavior during hydration is governed by initial heterogeneity in terms of micro to macroporosity fraction, evolution of microstructure under applied loading paths and its interaction with macrostructure. The micro to macroporosity fraction is attributed by various factores like initial physical state (powder/pellets), initial water content and dry density, fine content for clay based mixtures, sample preparation method, soil deformation history in term of wetting and drying cycles, clay mineralogy, geological (host rock convergence) and construction features (technical voids).

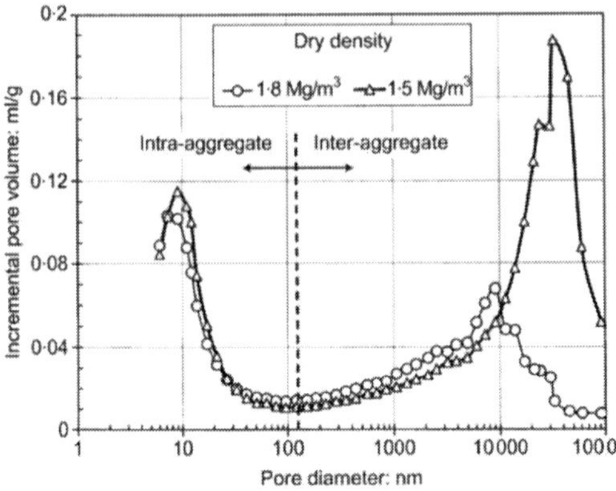

Figure 1. Distribution of incremental pore volume for two compacted bentonite samples at different dry densities. Mercury intrusion porosimetry test [9].

The microstructure after compaction depends on initial water content for a given dry density. The aggregates form, while mixing the clay powder with water and the aggregates themselves associate to form assemblies of aggregates. The aggregates are more prone to deformation of the wet side of optimum, thus filling the larger inter-aggregate voids and creating a single class of pores. Contrary, the association of aggregates remains more stable at the dry side of optimum water content, thus, the larger inter-aggregate pores exist additionally to the pores inside the aggregate. Figure 2 shows the PSDs of as compacted sample and saturated sample having same void ratio [12]. Distinct bimodal PSD features can be seen for as compacted sample at optimum moisture content, where as a weak bimodal PSD features are evident in PSD of saturated sample. In other words, the PSD is leading from a heterogeneous PSD to a more homogeneous one. Considering the initial physical state of material, the PSD of bentonite pellets mixed with powdered bentonite is shown in Figure 3. The other influencing factors for initial micro to macroporosity fraction are not discussed here in detail, as the paper mainly addresses the evolution of microstructure under transient boundary conditions.

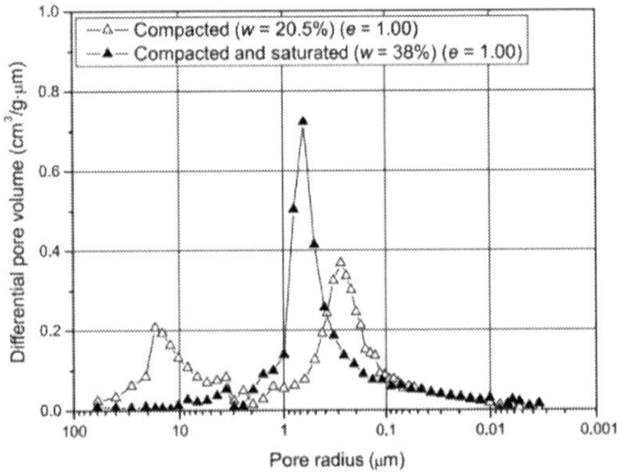

Figure 2. Comparison of PSDs of saturated compacted sample and samples compacted at optimum water content having same void ratio [11].

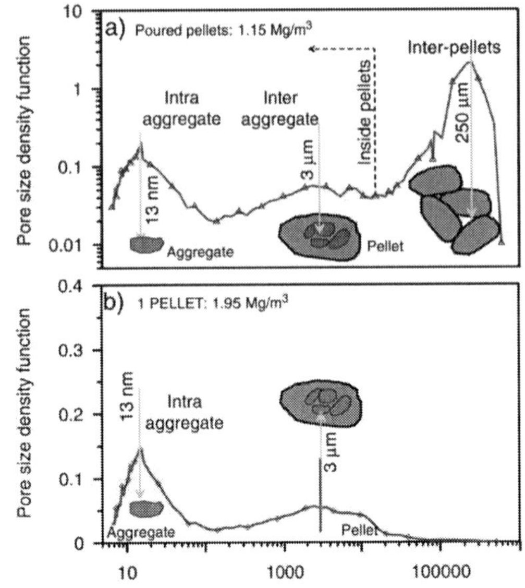

Figure 3. Pores size distribution of the bentonite pellets (a) Sampled poured into the sample holder. (b) High density pellets [12].

3 Microstructure evolution

Within the frame of coupled THM processes at nuclear waste repository, the augmentation of microporosity and its transformation as macroscopic swelling potential are highly senssitive to locally evolved mechanical constraints / transient boundary conditions and controlled by initial heterogeneity (e_m/e_M) and applied global boundary conditions. During hydration, the initial heterogeneity converts into homogeneity, the process can be portraied as a transient shift of one continua (microporosity e_m) towards the other continua (macroporosity e_M). The rate of this shift depends on the applied boundary conditions. Considerable research works have been reported to address the issue of volume constraints on the material response either in from of swelling pressure or in from of water retention behaviour [5, 13-23].

Pusch and Yong (2003) distinguished three boundary conditions that are believed to cause different rates of microstrucure evolution, such as clay exposed to water vapour, non-pressurized liquid water and pressurized water. For the clay hydrated with water vapour, the water molecules migrate into the open channels and adsorbed on exposed mineral surface. The water molecules then migrate into the elementary sheets that have higher hydration potential. As a result, swelling occurs uniformly for all the clay particles in an aggregate.

For the clay exposed to non-pressurized liquid water, water is absorbed by capillary forces in the open channels or macro pores. Then the water migrates into finer void and further into the elementary sheets. This results in expanding of clay particles near the boundary of the aggregate and closes the open channels or macro pores. Thus in this case, the hydration process is then controlled by diffusion process in its later stages.

For the clay exposed to pressurized water, water is pressed into the large channels and move quickly. The penetrating water displaces air and compresses the unsaturated matrix. The hydration rate is faster than those of first and second conditions (i.e., water exposed to vapor and non pressurized water). However, when the large channels become closed by the expansion of the clay aggregates, the hydration process follows the same procedure as the second procedure. This occurs in case of the heavily compacted bentonite. Whereas in case of compacted bentonite with available technical voids, all the large channels surrounding the compacted bricks are initially filled with water following the Darcy's advective flow regime and later bentonite bricks get hydrated through these water filled channels. With the progressive hydration process, these wider channels are filled due to expansion of compacted bentonite bricks due to swelling.

Agus and Schanz (2005) determine the PSD of the compacted bentonite sand mixture at three different states; namely, as-prepared, oven-dried, and swollen states under free swell condition. The PSD curve of the as-compacted, oven-dried, and swollen specimens was reported as bimodal indicating the presence of micro- and macropores in the specimens (Figure 4). Clay clusters enlarged and separated each other without being disintegrated when the specimen was wetted under unconfined conditions. During unconfined wetting, water was absorbed in the micro- and macro-pores which led to an increase in the volume of both pores. The similar sensitivity of the intra-aggregate pores to hydraulic and mechanical paths was exemplified by Romero et al. (2011) in Figure 5. The as-compacted sample is characterized by a clear bimodal distribution with dominant pore modes around 13 μm (macropores) and 60 nm (micropores). Loading the sample at vertical stress of 0.6 MPa shifts the size of dominant macropores towards a slightly lower value, while the micropore volume and its dominant pore size are not affected. On the contrary, the two saturation paths affect both inter-aggregate samples display a similar fabric dominated by the single peak at around 1.2 μm, which is originated by both expansion of the micropores and reduction of macropores.

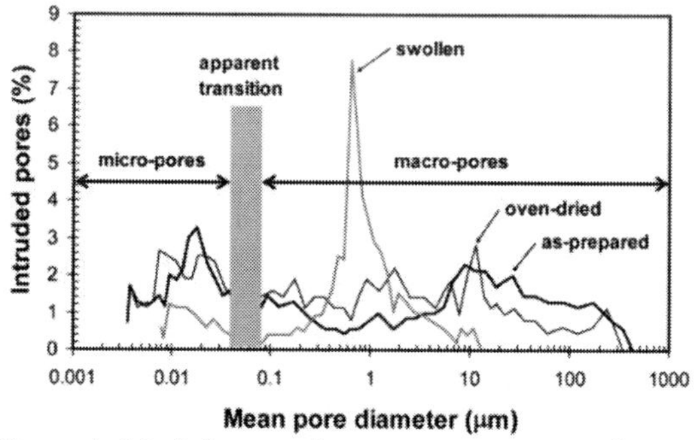

Figure 4. Intruded pore volume versus mean pore diameter plotted from the MIP data [15].

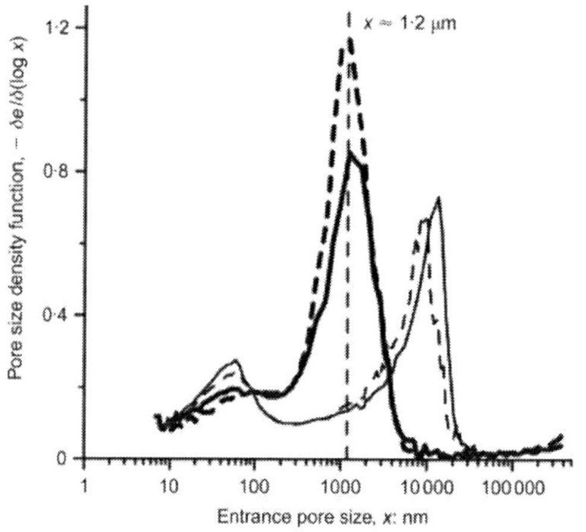

Figure 5. PSD of compacted Boom clay subjected to different hydro mechanical paths, as compacted sample (continuous line), vertical loading (dash line), constant volume (bold cont. line) and free swell condition (bold dash line) [23].

Considering the effect of different hydraulic path in terms of wetting and drying cycles, Seiphoori et al. (2014) observed an irreversible modification of pore size distribution curve, when the material approached a fully saturated state during the first wetting, and a new hydraulic domain was consequently created (Figure 6). The samples E and B show completely distinct PSD, although having the same degree of saturation, whereas the sample D, E and G show similar PSD having different degree of saturation. This permanent structure modification seems to occur after completion of first wetting path (D) and will remains permanent during the subsequent wetting and drying cycles. Considering the material in as compacted state with bimodal PSD and the state after first wetting with uni-modal PSD will modify the water retention capacity.

The changes in PSD domains in terms of micro to macroporosity fraction are highly sensitive to the involved physico-chemical processes as a function of applied loading paths and sequences, especially in the frame of couple THM processes at nuclear waste repository. Hence the phenomenological experiments provide additional key inputs related to understand the complex material behaviour.

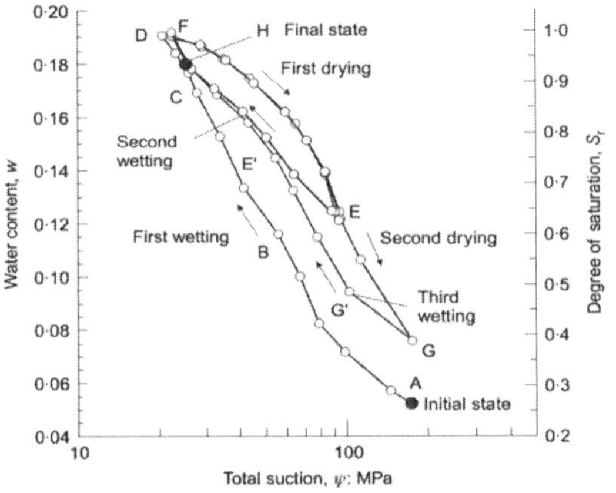

(a) Water retention behaviour and cyclic variation in total suction.

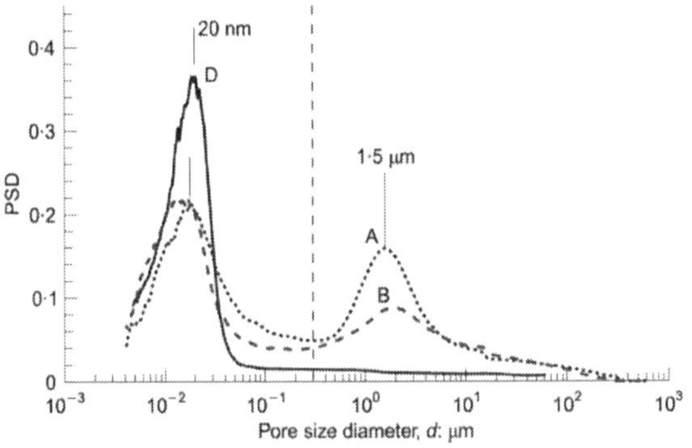

(b) PSD of as compacted sample (A), sample with higher degree of saturation (B) and at saturation stage (D).

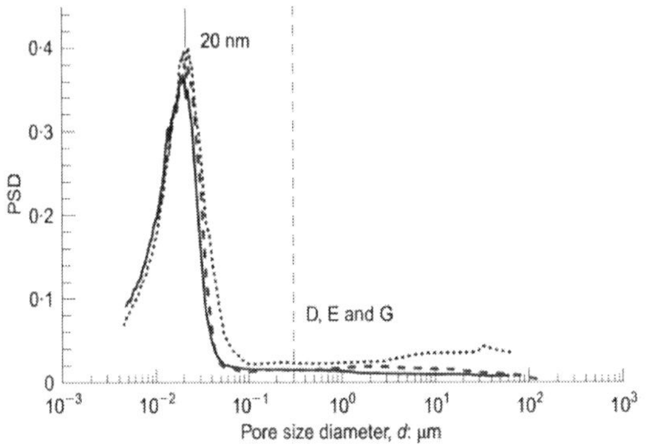

(c) PSD of fully saturated sample (D), end of first drying (E) and end of second drying (G).

Figure 6. Microstructure evolution for compacted MX-80 granular bentonite during different phases of wetting and drying [21].

4 Column experiment setup at RUB

As compared to field scale tests, the small scale phenomenological tests are more economical and have precise control over applied boundary conditions. Column type experiment devices have been extensively used for such proto type tests [24-30]. In column experimental studies, the transient hydration process in compacted buffer is simulated under repository relevant loading paths and the material response is observed by monitoring the key THM parameters like temperature, relative humidity and swelling pressure etc.

A unique column experiment facility is designed at Ruhr Universität Bochum as a forerunner of field scale test. The device is unique in terms of having proficiency to capture the transient material response under various possible repository relevant loading paths with higher precision level by monitor the key parameters like temperature, soil total suction, water content and axial & radial swelling pressure at three different sections along the total length of 30 cm. In general, a larger spectrum of loading paths / scenarios, which may arise in the nuclear repository, can be covered precisely with this devise.

4.1 Description of Column Testing Device

The entire device (Figure 7) can be assembled with five main components namely, (a) boundary control units (top & bottom), (b) PTFE rings for sample (middle), (c) monitoring sensors, (d) outer confining cell, and (e) rigid frame structure.

4.1.1 Top and bottom boundary control units

These units are encapsulated in 10 cm thick PTFE rings to ensure the one dimensional heat flow along the length of sample. Temperature is controlled by external units for heating and cooling separately. Silicon oil is used as a heat transport medium, which is pumped continuously in a closed circuit from external control unit to a small stainless steel (SS) chamber located at top and bottom of soil sample. For the hydration, water is supplied externally via. a SS pipe and SS porous stone, this porous stone is kept in contact with soil sample. The chamber, pipe and porous stone are designed and installed as a single unit to insure the thermal equilibrium in between the externally supplied water and the heating or cooling chambers.

The target temperature for heating (maximum 80°C) and cooling (10°C) can be controlled with precise rate using external units. The temperature of circulating Silicon oil is monitored constantly and adjusted according to target temperature. Both heating and cooling control units have three Pt100 sensors to perform the above task, which are installed in porous stone, inside the chamber and in Silicon oil reservoirs of external control units. The overall dimensions of top and bottom boundary control units are 35 cm as outer diameter and 20 cm of height.

4.1.2 PTFE rings for sample (middle section)

The middle section consists 03 PTFE rings having thermal conductivity as low as 0.23 W/mK. The rings have inner diameter as 15 cm, outer diameter as 35 cm and height as 10 cm. Each compacted clay sample (total

03) has the identical dimensions (diameter 15 cm and height 10cm) to fit into corresponding PTFE ring. The individual sample is compacted under uni-axial vertical pressure of 20-30 MPa in specially designed compaction mould with target initial dry density in the range of 1.8 to 2 g/cc for 50:50 bentonite sand mixture with 8-9% initial water content.

After compaction, the mould having compacted sample is placed over the bottom most PTFE ring. The sample is extruded from mould and pushed directly into the ring using a hydraulic jack. The same procedure is followed to push other two samples (middle and top) over each other, the provision of peripheral O-ring in between these rings ensure the air and water tight joint. The tightening screws made up by PTFE are further deployed to make the proper connection in between rings. Once all the sample rings are stacked over each other, the compacted soil sample has overall height of 30cm with 15 cm diameter. The top boundary control unit is placed over the stacked sample rings with O-ring and PTFE tightening screws. After assembling the sample rings in between top and bottom boundary control units, the confining structure is placed around the assembled column testing units to make sure constant volume condition. Finally the monitoring sensors are installed in each sample ring.

4.1.3 Monitoring sensors

To monitor the transient material behavior, each sample ring is having four sensor installation points perpendicular to each other at the same horizontal level as shown in Figure 8a. Thus we have individual sensor to monitor four key parameters as temperature, relative humidity, water content and radial swelling pressure at the same level. Each sensor is installed with its specially designed PTFE adaptor. These PTFE adaptors with O-rings are designed to ensure the air and water tight joints with minimum disturbance to sample condition. The detailed description of individual sensor is not discussed here. Additionally, the axial swelling pressure will be measured at top and bottom using load cells.

To capture the effect of transient boundary conditions during hydration, the swelling pressure will be measured by total pressure transducers in between two sample interfaces. These total pressure transducers will be embedded in the center of compacted sample between first and second; second and third sample rings. To get the transient profile of total suction along the sample length with Kelvin's equation using transient RH profiles and temperature profiles, one must be ensure about the proper thermal equilibrium between location of measuring temperature and relative humidity. The schematic view of soil sample ring with sensor installation locations are shown in Figure 8.

5 Discussion

Transient hydration process of compacted buffer material under coupled THM processes at nuclear waste repository indicates the transformation of its initial heterogeneity into homogeneity as final state. This process is identical to the process of getting more stable lower energy state by dissipating the initial high energy. In transient hydration process, the initial higher energy can be correlated with initial higher total suction, which reduces slowly with elapsed time.

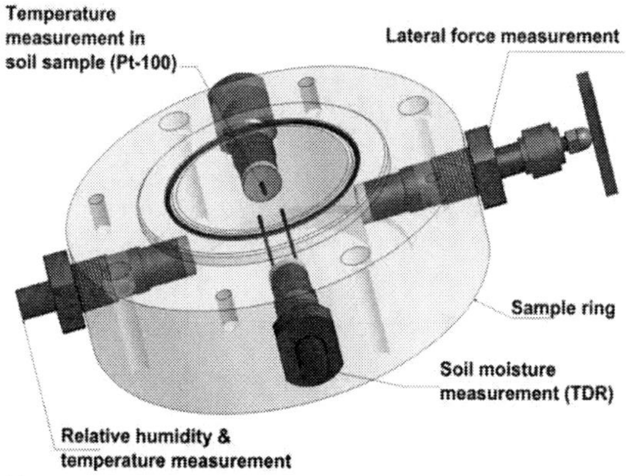

(a) PTFE sample ring with installed sensors.

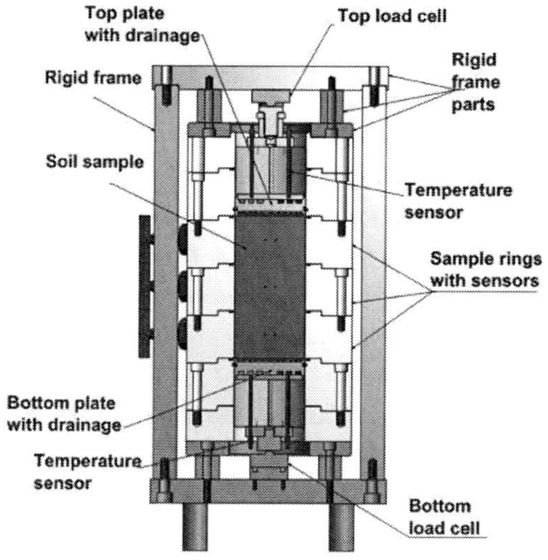

(b) Overall schematic view

Figure 7. Schematic view of column testing device at RUB, (a) individual PTFE ring, and (b) overall column device.

The micro structure evolution drives the homogenization process, which involves the shift of initial bi-modal pore size distribution into uni-modal nature. The changes in PSD domains in terms of micro to macroporosity fraction are highly sensitive to the involved physico-chemical processes as a function of applied loading paths, especially in the frame of couple THM processes at nuclear waste repository. Hence the phenomenological experiments provide additional key inputs related to understand the complex soil water retention and volume change behaviour. The SWRC perform in laboratory under unconfined condition using axis translation and vapour equilibrium technique are not directly applicable for phenomenological investigations due to difference initial soil state, different fabric arrangement and transient boundary conditions. Similarly the swelling pressure test results under steady state boundary conditions like constant volume or free swell conditions will differ from the swelling pressure under transient boundary conditions in small scale phenomenological investigations. This ambiguity in material behaviour under transient boundary conditions is a major challenge for researchers, which can be removed by defining a suitable state parameter, which essentially describe the path dependent microstructure evolution and its interaction with macrostructure.

References

[1]. Martin, P. L., Barcala, J. M. & Huertas, F. (2006). Large-scale and long-term coupled thermo-hydro-mechanic experiments with bentonite: the FEBEX mock-up test. J. of Iberian Geol. 32 (2), 259–282.

[2]. Wang, Q., Tang, A. M., Cui, Y., J., Delage, P., Barnichon, J. D. & Ye, W. M. (2013). The effects of technological voids on the hydro-mechanical behaviour of compacted bentonite–sand mixture. Soils and Foundations. 53(2), 232-245.

[3]. Gens, A. & Alonso, E. E. (1992). A framework for the behaviour of unsaturated expansive clays. Can. Geotech. J. 29 (6), 1013–1032.

[4]. Alonso, E. E., Vaunat, J. & Gens, A. (1999). Modelling the mechanical behaviour of expansive clays. Engng Geol. 54 (1), 173–183.

[5]. Cui, Y. J., Loiseau, C. & Delage, P. (2002). Microstructure changes of a confined swelling soil due to suction controlled hydration. Proceedings of the 3rd Intl. Conf. on Unsat. Soils, Brazil, 2, 593–598.

[6]. Sánchez, M., Gens, A., Guimarães, L. & Olivella, S. (2005). A double structure generalized plasticity model for expansive materials. Int. J. for Num. and Anal. Metd. in Geomech. 29, 751–787.

[7]. Tuller, R. & Or, D. (2003). Hydraulic functions for swelling soils: pore scale considerations. J. Hydrol. 272 (1), 50-71.

[8]. Romero, E. (2013). A microstructural inside into compacted clayey soils and their hydraulic properties. Engg. Geol. 165, 3-19.

[9]. Lloret, A., Villar, M. V., Sanchez, M., Gens, A., Pintado, X. & Alonso, E. (2003). Mechanical behaviour of heavily compacted bentonite under high suction changes. Géotechnique. 53 (1), 27–40.

[10]. Delage, P., Marcial, D., Cui, Y. & Ruiz, X. (2006). Ageing effects in a compacted bentonite: a microstructure approach. Géotechnique. 56 (5), 291–304.

[11]. Hoffmann, C., Alonso, E. E. & Romero, E. (2007). Hydro-mechanical behaviour of bentonite pellet mixtures. Phys. Chem. Earth. 32 (8), 832–849.

[12]. Li, X. & Zhang, L. M. (2009). Characterization of dual structure pore size distribution of soil. Can. Geotech. J. 46, 129-141.

[13]. Pusch, R. (2001). Experimental study of the effect of high porewater salinity on the physical properties of a natural smectitic clay, SKB tech. report 01-07.

[14]. Pusch, R. & Yong, R. (2003). Water saturation and retention of hydrophilic clay buffer—microstructural aspects. Appl. Clay Sci. 23 (1), 61–68.

[15]. Agus, S. S. & Schanz, T. (2005). Effect of shrinking and swelling on microstructures and fabric of a compacted bentonite-sand mixture. In Proceedings of the international conference on problematic soils, North Cyprus. 2, 543–550.

[16]. Likos, W. J. & Lu, N. (2006). Pore-scale analysis of bulk volume change from crystalline swelling in Na^+ and Ca^{+2} smectite. Clays and Clay Minerals. 54, 516-529.

[17]. Lloret, A. & Villar, M. (2007). Advances on the knowledge of the thermo-hydro-mechanical behaviour of heavily compacted 'FEBEX' bentonite. Phys. Chem. Earth, Parts A/B/C 32 (8), 701–715.

[18]. Likos, W. J. & Wayllace, A. (2010). Porosity evolution of free and confined bentonites during interlayer hydration. Clays and Clay Minerals. 58 (3), 399–414.

[19]. Gens, A., Valleján, B., Sánchez, M., Imbert, C., Villar, M. V. & Van Geet, M. (2011). Hydromechanical behaviour of a heterogeneous compacted soil: experimental observations and modelling. Geotechnique. 61 (5), 367–386.

[20]. Romero, E. (2001). Controlled–suction techniques. In Proceedings of the 4th simpósio Brasileiro de solos não saturados, Brazil, 535–542.

[21]. Seiphoori, A., Ferrari, A. & Laloui, L. (2014). Water retention behaviour and microstructural evolution of MX-80 bentonite during wetting and drying cycles. Géotechnique. 64 (9), 721–734.

[22]. Manca, D., Ferrari, A. & Laloui, L. (2016). Fabric evolution and the related swelling behaviour of a sand/bentonite mixture upon hydro-chemo-mechanical loadings. Géotechnique. 66 (1), 41–57.

[23]. Romero, E., Della Vecchia, G. & Jommi, C. (2011). An insight into the water retention properties of compacted clayey soils. Géotechnique. 61 (4), 313–328.

[24]. Mohamed, A., Yong, R. & Kjartanson, B. (1992). Temperature and moisture distributions in a clay buffer material due to thermal gradients. In: MRS Proceedings. 294, 417-425.

[25]. Cuevas, J., Villar, M., Fernández, A., Gomez, P., Martín, P. (1997). Pore waters extracted from compacted bentonite subjected to simultaneous heating and hydration. Applied Geochemistry. 12 (4). 473-481.

[26]. Gatabin, C. & Billaud, P. (2005). Bentonite THM mock-up experiments: Sensors data report. CEA, Rapport NT-DPC/SCCME 05-300-A. CEA, Paris.

[27]. Villar, M., Gomez-Espina, R. & Martin, P. (2006). Band 1081, Informes técnicos Ciemat.

[28]. Tripathy, S., Thomas, H., & Bag, R. (2015). Waste, 10.1061/ (ASCE) HZ.2153-5515.0000272 , D4015002.

[29]. Ye, W. M., Wan, M., Chen, B., Chen, Y. G., Cui, Y. J. & Wang, J. (2012). Temperature effects on the unsaturated permeability of the densely compacted GMZ01 bentonite under confined conditions. Engg. Geology. 126, 1-7.

[30]. Schanz, T., Nguyen-Tuan, L. & Datcheva, M. (2013). A column experiment to study the thermo-hydro-mechanical behaviour of expansive soils. Rock Mech. and Rock Engg. 46, 1287-1301.

 DOI: 10.1051/e3sconf/20160903002

Challenges in treating earthen construction materials as unsaturated soils

Charles E. Augarde[1,a], Christopher T.S. Beckett[2], Jonathan C. Smith[3], and Andrew J. Corbin[1]

[1] *School of Engineering and Computing Sciences, Durham University, Durham, UK*
[2] *Civil, Environmental and Mining Engineering, University of Western Australia, Australia*
[3] *Cundall, Regent Centre, Newcastle upon Tyne, UK*

Abstract. Earthen construction is a loosely defined term covering both the materials and methods for creating structural components from mixtures of subsoil, often with the addition of chemical or mechanical stabilisers. There is evidence of Man creating earthen structures for thousands of years, and there are many world heritage sites containing earthen structures, some of which present issues in terms of conservation. In some parts of the world there is a growing market for new-build earthen structures and a key issue here is the lack of design codes. Since these materials are composed mainly of particulates and water it is natural to regard them as geotechnical in nature, where friction and the presence of water have a key influence on material properties, however until very recently this was not the case, with earthen construction materials regarded as weak concrete or masonry. In this paper we examine these opposing views and discuss the issues associated with regarding these materials as unsaturated soils. The paper is illustrated with outcomes from research at Durham University carried out over the past ten years.

1 Introduction

Earthen construction, meaning the construction of structural components (walls, arches, domes and lintels) from subsoil, is receiving increased interest, despite it being something Man has done for thousands of years, and the oft-repeated "fact" that one third of the world's population live in structures at least partially made of "earth". Interest in the developed world is being driven by the potential green credentials of these materials, i.e. low embodied energy as compared to fired masonry or concrete blockwork, and heat/humidity buffering which could reduce in-life energy costs. There are also drivers associated with the preservation of heritage structures made of soil, of which there are many famous examples [1]. Specific techniques of earthen construction are unit-based (e.g. Adobe, compressed earth blocks) and insitu (rammed earth, Cob). In all cases the base materials are selected mixtures of gravel, sand and clay, to which stabilisers (chemical or mechanical) are sometimes added (e.g. cement or fibres respectively) [1, 2, 3].

If the interest in earthen construction (EC from here) is to lead to environmental benefits, it is clear that the specification, design and construction using these materials will have to change. At present, EC is seen as a niche building option in Europe, beloved by architects but neither widely available nor cheap. In Western Australia, Canada and California, however, there are mature EC markets for small to medium-size buildings, due in part to the climates of these places, but also because of the activity of local contractors and consultants, notably David Easton in California, Meror Krayenhoff in Canada and Steven Dobson in Australia, who came together (with the first three authors of this paper) at the First International Conference on Rammed Earth Construction in 2015 in Perth, Australia [4]. However, even in the countries where there is measurable activity, expansion of the market is prevented by the virtual absence of design codes, unlike the situation for steel, concrete, masonry and timber which all have their own well-established codes, certainly in the developed world. In contrast, much earthen construction is designed either from a craft point of view or, if engineering is involved, the material is assumed to be homogeneous, continuous and much like a weak concrete or masonry.

While some consider that the lack of design codes stems in part from the concrete and masonry industries, perceiving a threat to their products, the absence is mainly due to a lack of quality, rigorous scientific investigation and the fact that, until 5-10 years ago, it was led by structural engineers who ignored both the particulate nature of, and the importance of water to, these materials, both of which are regarded as key in our understanding of geotechnical materials. It is for this reason that this paper is in this conference, because better understanding of the behaviour of EC materials may be achieved by regarding them as *manufactured* unsaturated soils. To explore this idea we consider the various ways that EC materials are viewed and survey the relatively few existing links to other areas of geotechnical research, highlighting some challenges; challenges which should be regarded not as barriers but the starting point for future research.

[a] Corresponding author: charles.augarde@dur.ac.uk

2 The two views of earthen construction materials

Specific guidance for the design and construction of structural components using various EC materials can be found in the landmark book by Houben and Guillard [2] and the more recent UK publication by Walker *et al.* [3]. Only New Zealand has anything approaching a Eurocode for earthen construction. What characterises all these sources of guidance is the structural engineering approach taken to the material, with very little acknowledgement of the particulate nature of the source materials, and the effect of water. Strength is not predicted using a scientific approach; rather, rules often rely on previous experience with similar materials. Structural members are sized in a similar way to the approach used with masonry in Europe, and it is the structure that is designed, e.g. effective lengths and slendernesses are checked rather than states of stress at material points. The structural approach is also evident in the small research literature on EC materials. Most testing reported on EC materials is closely related to concrete testing, e.g. unconfined compression, fracture tests [5] and insitu testing of structural components [6]

An alternative is to regard EC materials as unsaturated soils, where suction plays a major role in shear strength, but where one is studying a structural component rather than a body of soil. It appears that the first publication to make the link, between suction and the strength of earthen construction materials, is the 2007 conference paper of Gelard *et al.* [7], which came at the same time as the group at Durham was carrying out the experimental tests which later appeared in [8]. The latter tests were constant water content, unconfined compression tests on unstabilised rammed earth samples where the suctions were measured during testing using high capacity tensiometers (developed by Toll and co-workers). A clear link was shown in [8] between suction and shear strength, although at much lower suctions than usually experienced with EC materials in the field after drying. Later work, making use of filter paper measurements of much higher suctions, served to confirm this approach (e.g. the results shown in Fig. 1 from [9]). Recently, Gallipoli *et al.* [10] also put forward the case for a geotechnical approach to these materials.

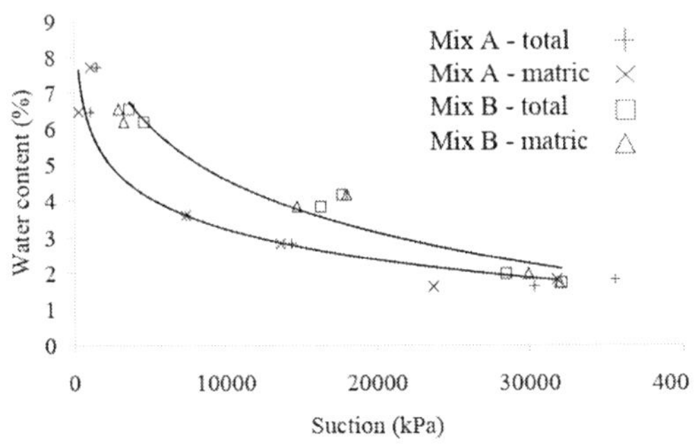

Figure 1: Water retention curves for various samples of rammed earth at varying drying levels (from [9]).

3 The challenges

On the face of it, it sounds entirely reasonable to adopt the geotechnical approach described above, however there are a number of challenges to be addressed and our assessment is that in fact a mixture of the two approaches (geotechnical and structural) is necessary. While there are some parallels with current geotechnical research which might be extended to EC materials, these are often inappropriate.

Taking rammed earth as an example, we can define this material in the following way: cross-anisotropic (due to the means of production), very low degree of saturation, high dry density, brittle, possibly containing chemical bonding or reinforcement and having a very large particle size distribution (clay, sand and gravel). In use the material is placed so that a large surface area is subjected to wetting and drying, a very different situation as compared to standard "buried" geotechnics.

3.1 Constitutive modelling

The development of constitutive models is one of the main activities undertaken by the geotechnical research community, and this is particularly the case in unsaturated soils. It is worth considering why constitutive models are needed. For instance, a large body of work in unsaturated soils has been aimed at understanding and predicting the long term behaviour of nuclear waste containment systems including bentonite. Primarily the interest is hydraulic, i.e. will leakage occur? Considering EC materials, strength and durability are the main concerns, rather than movements and permeability so one might question if a full-functioning constitutive model is actually needed for EC materials.

A common reason for developing a constitutive model is for use in numerical modelling using finite elements and there have been some moves in this direction by researchers in EC materials. Nowamooz & Chazallon [11] developed a finite element model of a rammed earth wall using a non-linear elastic – perfectly plastic model with a Drucker-Prager yield criterion. The effect of suction was included assuming a linear increase in tensile strength similar to the Barcelona Basic Model. More recently Gerard *et al.* [12] presented results from tests on an unsaturated Belgian clayey silt with similar properties to a typical RE mix (without a gravel fraction), results of which were then used to validate a constitutive model based on a generalized effective stress (Bishop stress) approach with values suggested for the effective stress parameter χ. The attempt to fit an EC material into one of the two established methods for dealing with effective stress in unsaturated soils is an interesting development, also seen in [13].

Other, perhaps more satisfying possibilities for a constitutive modelling framework for EC materials are those proposed for compacted unsaturated soils, compaction being a key ingredient in the production of many EC materials. In these models it is recognised that microstructural changes (i.e. void and particle size distribution) should be taken into account. In particular, many geotechnical studies have shown that compacted clay forms aggregated structures which behave like large particles inside which water is trapped, being separate to water held in the pendular regime between the particles (both aggregates and larger solid particles). A recent example of such a modelling framework can be found in [14] in which the microstructure is quantified by the ratio of microvoids (i.e. voids within the clay aggregates) to total void ratio, a state parameter on which the stress and suction behaviour can be based. Another possible starting point is the work of Koliji [15]. These particular models (as are many others for compacted soils in geotechnics) are developed for clays and silty-clays and therefore consideration would be needed to extend this to EC materials which contain a much wider particle size range. In addition, the gap grading inherent in EC materials due to manufacture is a feature unlikely to be present to such a marked degree in the natural soils routinely tested in geotechnics.

A much greater challenge is to consider how to extend these models to accommodate the very brittle behaviour and fracture common in EC materials. Fracture mechanics here (as for other materials) is concerned with fracture initiation and propagation and, of the little scientific work published on this, most have adopted linear elastic fracture mechanics principles as a start (e.g. the work presented in Brune *et al.* [16] on Roman mortars). It is here that it seems necessary to work with tests established for rocks and bonded materials such as concrete, and to investigate the constitutive models used for these materials. Continuum elasto-plasticity deals poorly with discontinuities whether they are fractures or shear bands (to cite one other area of geotechnics).

Some ideas on appropriate tests for fracture testing for EC materials are explored in [17] in which the development and use of a wedge splitting device for earthen construction materials is described and demonstrated. Fig. 2 shows the device and Fig. 3 shows an example of a fractured stabilised RE specimen obtained with this rig. With this device one can obtain reliable and repeatable fracture energies for these materials, data which can inform a constitutive model for fracture. Adopting tests like this is pragmatic, similar to the use of Brazilian tests for tensile strengths.

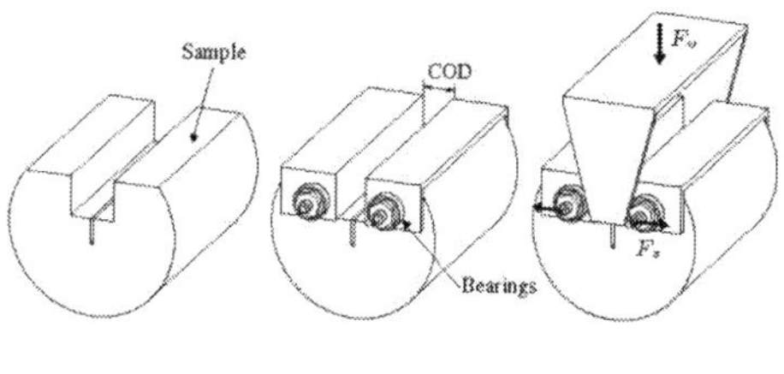

Figure 2: Wedge splitting device for obtaining the fracture energy of earthen construction materials (from [17]).

However, it seems unlikely that a single, holistic constitutive model can be developed to bridge the gap from low-suction to very brittle behaviour and thus be able to model the full range of wetting and drying cycles likely to be experienced by EC materials in the field. As with other engineering problems in which fracture has to be introduced, it may be necessary to change the numerical modelling paradigm to allow discontinuity, e.g. to use XFEM, and there is recent work of just this nature with EC materials in mind [18].

3.2 Investigating the microstructure

As indicated above, it is clear that microstructure is important in determining the mechanical and hydraulic properties of unsaturated soils and this must also be the case with EC materials. Microstructural investigations of soils have traditionally used mercury intrusion porosimetry (MIP) to determine void size distributions (or rather pore entry diameters which are something different), and more recently x-ray computed tomography (XRCT) has appeared to have become mature enough to become useful to geotechnical engineers.

A PhD thesis [19] recently completed at Durham contains a new body of research assessing the various means of investigating the microstructures of EC materials (rammed earth in particular) with an emphasis on XRCT. In [19], as a precursor to the use of XRCT on EC materials, a study is presented to assess the way that XRCT has been used to date in geotechnics, where three key journals (Géotechnique, Géotechnique Letters and Granular Matter), and two recent conference proceedings were surveyed (UNSAT2014 and IS-Cambridge 2014). A total of 40 papers were found this way and from each the following information was obtained (where provided):

- material analysed;
- sample size;
- voxel size, or resolution;
- XRCT scan descriptions;
- results presented.

There is insufficient space here to cite all the articles consulted (full details are in [19]) however some overall statistics are presented which are of interest. Thirty papers provide information regarding the material scanned (Figure 4a shows the range of particle sizes). It is clear that the majority study particle sizes (actually object sizes) between 0.1mm and 2mm, i.e. sands. This is clearly at odds with many natural soils and certainly with all EC materials. Twenty-seven papers provide

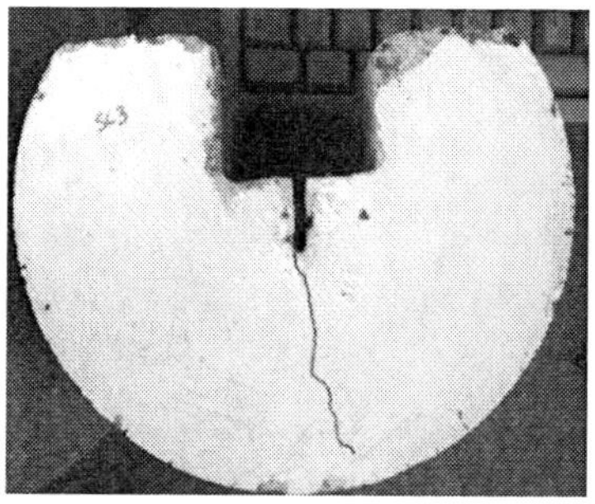

Figure 3: An example of a fractured stabilised RE specimen using the fracture in [17].

 DOI: 10.1051/e3sconf/20160903002

information regarding the size and shape of the (cylindrical) samples scanned. Five of these investigations took smaller cores from larger samples on which other laboratory experiments had already been performed. Figure 4b shows the diameter of the samples scanned in each of these 27 papers from which it can be seen that a large proportion of investigations only scanned small samples (<12mm) whilst a significant minority, 5 investigations, performed scans on much larger (e.g. oedometer size) samples. It was surprising to find that the 38mm diameter triaxial cylinder was only used in a single surveyed publication when one might expect XRCT to be a useful tool in tandem with standard triaxial tests. While no discussion of the choices of sample size can be found in any of the papers surveyed it is presumed that the XRCT machines performing the scans required small sample sizes to obtain the required resolution, or to fit within the scanning chamber. This survey is, of course, by no means the full picture of XRCT use in geotechnics but the trends are clear; the soils scanned tend to be those which will "scan well" and there are questions about the viability of current XRCT scanners for investigation of soils with varying particle sizes.

There is a conflict in XRCT scanning between wishing to obtain the highest resolution and the largest area of coverage. One can rarely achieve both, and with a compacted material with a range of particle sizes (e.g. a rammed earth mix) one cannot see right "down to the clay". Instead a pragmatic approach must be adopted where sample size is chosen to balance the capabilities of the XRCT machine and the desire for representative samples. A very small sample will scan well but is unrepresentative of a mix where there could be large sand and gravel particles present, which is the case with EC materials such as rammed earth.

It would appear then that microstructures of EC materials require a two-stage process of investigation: MIP to determine the microstructure up to the micron level of the clay aggregates, and XRCT above that. There are clear challenges in sampling for the former (i.e. how representative of the clay aggregates in a large same of an EC material is a given sample of a size suitable for MIP). It could be that the microstructure at the sand/gravel level is not significant and the main source of strength lies in the clay aggregates, however this remains speculation without further experimental work.

3.3 Additives and materials

EC materials can contain both chemical and mechanical additives, e.g. cement and fibres respectively. Clay soils have long been stabilised by the addition of lime or cement and there is a mature research literature on this topic, examples including the many papers by Consoli and co-workers (e.g. [20]) where, interestingly, extensive

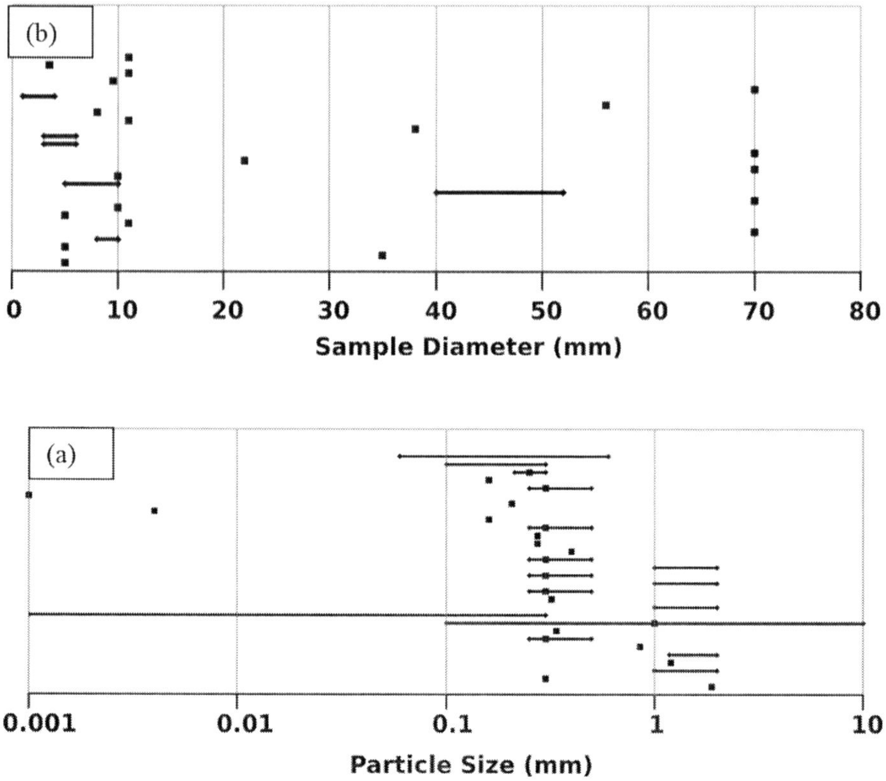

Figure 4: Results from the literature survey on XRCT use in geotechnics

DOI: 10.1051/e3sconf/20160903002

use is made of unconfined compression and Brazilian tests, as found in the "structural" literature on EC materials. Constitutive models have also been developed for these materials, e.g. [21]. Modern stabilised rammed earth contains between 1% and 15% cement by mass so at the higher end the contribution to strength from cementation will dwarf anything from suction and the material could be regarded as a weak concrete. At the lower end, findings for cement stabilised clays may have some applicability but the addition of compaction in the case of the EC materials may make it inappropriate. Contractors, particularly in Australia, seem to have concerns that the presence of clay in their rammed earth mixes will inhibit the cement hydration, and this tends to lead to a choice of mix with very low levels of clay.

There is also a widespread belief in the EC building community that the type of clay is of prime importance, i.e. an avoidance of any expansive materials. This can further reduce the environmental benefit as clays may have to be imported rather than being won locally. In fact, recent studies have shown that there is negligible deterioration in the properties of rammed earth mixes which contain up to 20% of their clay fraction as expansive. Figure 5 shows results of filter paper tests on

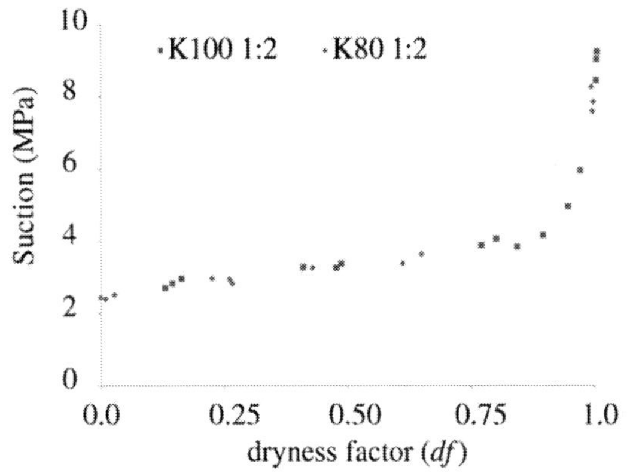

Figure 5: Results for filter paper tests on sand/clay mixtures. K100 is pure kaolin. K80 is 80% kaolin/20% bentonite [22].

two samples of a clay-sand mix (in proportion 1:2, similar to a rammed earth mix without gravel) in which one has a clay fraction entirely composed of kaolin and the other has a clay mix of 80% kaolin and 20% bentonite. The plot may be unusual to geotechnical engineers as suction is plotted against "dryness factor", *df* which is a measure of the ratio of the water content at optimum (at the point of compaction), *df* = 0 and final equilibrated dry state after a number of days, *df* = 1. The two materials follow a similar path and other mechanical tests also show similar agreement [22]. This is an indication that for these mixed soil materials the expansive clay behaviour is inhibited in some way.

Fracture inhibition in brittle materials usually means reinforcement in tension, and fibre-reinforcement is a key feature of many earthen construction materials, e.g. the straw in adobe bricks and cob is a form of tensile

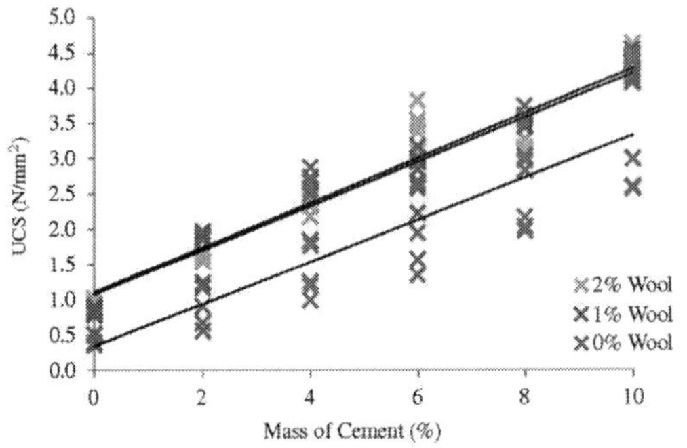

Figure 6: UCS results for stabilised rammed earth samples with varying amounts of cement stabiliser and wool fibre reinforcement [17].

reinforcement. It is an intriguing question to consider what role these fibres play in terms of water storage and distribution, and the nature of the bond between the fibres and the surrounding material. A recent review of the use of fibres in EC materials can be found in [23].

Investigations have been carried out at Durham on the properties of fibre-reinforced mixes at the macro scale and also the fibre/earth bond itself. Corbin & Augarde [17] demonstrated the major change in fracture behaviour between un-reinforced and reinforced stabilised RE and also the increase in unconfined compressive strength (UCS) with wool reinforcement (an example plot is given in Fig. 6). Investigations of the fibre-EC material bond behaviour are presented in [24] including the development of a new test rig to carry out pull out tests on samples of earthen materials. In this study pull-out loads were measured using a jute fibre embedded in both stabilised and unstabilised rammed earth mixes. Water content, fibre embedment length and dry density were all varied. Two example results plots are presented here in Figs. 7 and 8.

In Fig. 7, peak pull-out forces are plotted for a large number of tests where interface failure (i.e. loss of bond between fibre and soil) occurred. The majority of results follow a trend of major increases in pull-out force for the lowest water contents and less marked differences for higher water contents. The link between water content and strength might be seen as counterintuitive when one considers shrinkage would be greater the drier the soil gets, and hence one might conclude the interface strength should decrease as soil shrinks away from the main fibre axis. While this might be the case, other studies [17] suggest that the main bonding occurs between the soil and "microfibres" extending outwards from the main fibre axis, rather than the main fibre axis alone, and these bonds could be less influenced by shrinkage.

Also clear from Fig. 8, and in many of the results in [24], is the presence of peak and residual strengths, potentially unsafe for design when the materials are unstabilised. This behaviour is thought to be associated with dilation of the soil increasing bond strength initially, followed by frictional failure. What is also clear is that some stabilisers serve to add ductility (cement) while others make the material less robust (lime). Clearly, single fibre studies have to be scaled up to the macro-material case but these findings are interesting for revealing mechanisms of failure. The behaviour of fibre-reinforced soil has of course been studied by the geotechnical engineering community for a number of years (e.g. [25, 26]) however these studies are usually for natural soils at much lower compaction levels and higher saturations than the conditions found in earthen construction.

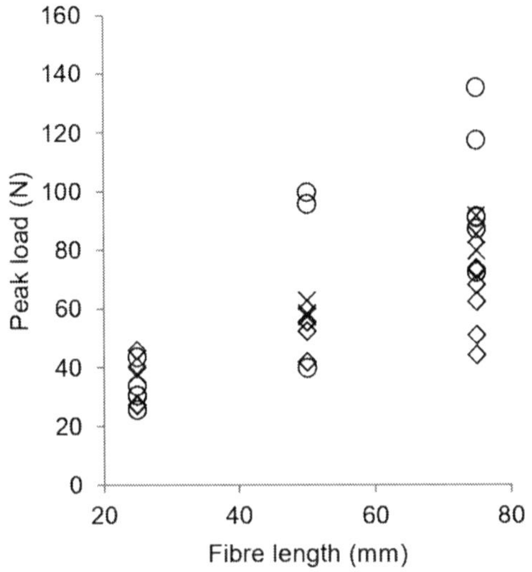

Figure 7: Results from pull-out tests of fibres embedded in rammed earth samples. Peak loads are shown for varying water contents: 3% wc (circles), 7% wc (crosses), 11% wc (diamonds) [24].

4 The future

This paper has outlined links to existing research in geotechnics and also identified gaps which need to be filled if we are to develop a scientific understanding of EC materials. With a robust and validated constitutive framework for EC materials, taking aspects of modelling from unsaturated soil mechanics and from brittle bonded material modelling, e.g. for weak concrete, it is likely that design codes could follow.

The ultimate goal of earthen construction material research, from an engineering viewpoint, is to improve material modelling and structural design. However, the earthen construction industry is small and practitioners are generally unfamiliar with geotechnical terminology or analysis concepts. Rather, they will defer to familiar procedures and materials to reduce risk. The most significant challenge facing the treatment of earthen construction materials as unsaturated soils is the power and familiarity of existing, simple empirical techniques and the need to convince industry (and researchers) of the benefits that can be achieved by adopting new practices.

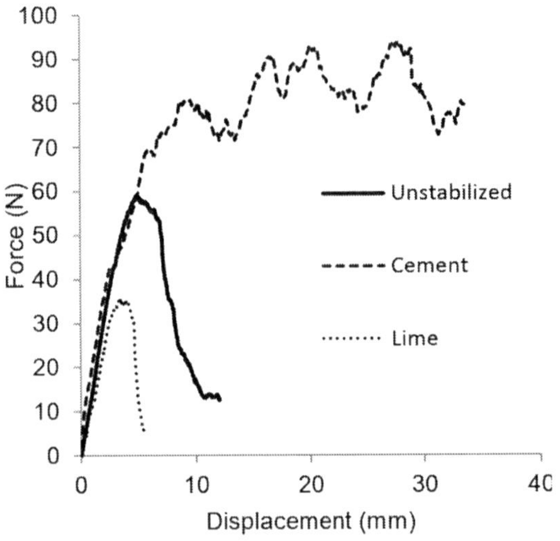

Figure 8: Results from pull-out tests of fibres embedded in rammed earth samples. Force/displacement results for 50 mm fibres with and without stabilisers (9%) [24].

Several core unsaturated soil mechanics concepts are already familiar to and indeed at the heart of earthen construction, for example:

- the need to compact material to achieve a high density and sufficient strength;
- the need to correctly identify the right water content to aid compaction;
- the need to control material grading to ensure sufficient compactability;
- the need to limit clay content to reduce shrinkage;
- that material strength increases on drying;
- that materials must breathe to avoid water build-up and possible failure.

To address the challenge of industry uptake, these concepts must be 're-packaged' in an unsaturated soils framework and the ability of that framework to predict their effects demonstrated. For example, Fig. 9 shows the unconfined compressive strengths of two rammed earth soil mixes equilibrated to high and low suction conditions. Soil water retention curves for the two soils are shown in Figure 10, derived from combined filter paper (suctions <10 MPa) and vapour pressure results (suctions >10MPa). Material properties are given in Table 1. 100mm cube specimens were compacted at the optimum water content (OWC) and equilibrated under a range of humidity and temperature conditions using an environmental chamber. Equilibration suctions under given conditions were calculated using the well-known Kelvin equation. Humidities between 30% and 90% and temperatures between 15°C and 40°C were investigated: here, only "high suction" (30% humidity, 40°C, 168.3 MPa suction) and "low suction" (90% humidity, 15°C, 14.0 MPa suction) will be discussed.

Figure 9 shows that strength almost doubled between the high and low suction conditions for both soils. It is unlikely that a practitioner would have the time or the facilities to derive full water retention curves for a given soil, for example Figure 10, or determine a full yield surface for their material, for example as in [12]. However, a change in strength of the magnitude shown in Figure 9 would be critical to a project's feasibility. For example, NZS 4297 (the New Zealand rammed earth construction standard) requires a minimum compressive strength of 1.3 MPa for a soil to be considered suitable for construction; if a proposed soil were tested under inappropriate conditions, it might either be rejected unnecessarily, increasing costs as alternative soils must be sourced, or approved inappropriately which could potentially lead to catastrophic failure. Such outcomes cannot be discovered using traditional testing or analysis methods, for example those based on concrete or masonry. For an industry based on risk minimisation, an appreciation of the implications of unsaturated soil behaviour is clearly a benefit.

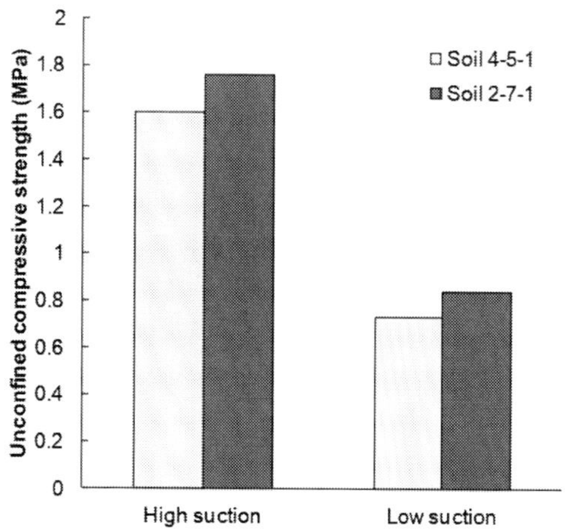

Figure 9: Soil 4-5-1 and 2-7-1 UCS under high and low equilibration suction conditions

Table 1. Soil 4-5-1 and 2-7-1 material properties. MPT: Modified Proctor Test (BS1377); $\rho_{d_{max}}$: maximum dry density. Percentages and water contents by mass.

Soil	Silty clay (%)	Sand (%)	Gravel (%)	OWC (%)	MPT $\rho_{d_{max}}$ (kg/m^3)
4-5-1	40	50	10	12.0	1940
2-7-1	20	70	10	12.0	1960

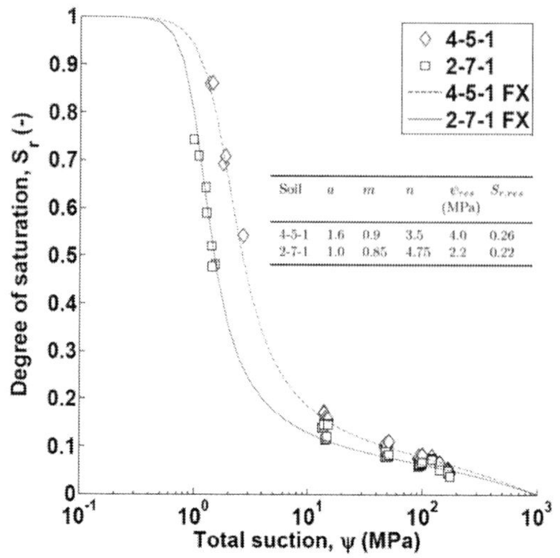

Figure 10: Soil water retention curves for soils 4-5-1 and 2-7-1 and approximations using the Fredlund and Xing [27] model (parameter values given in inset table).

5 Conclusions

This paper sets out the case for geotechnical engineers interested in unsaturated soil mechanics to consider applying their ideas to earthen construction materials. It identifies where existing work might be of use and also identifies some challenges where discussion and research is required. It is to be hoped that the paper will inspire younger researchers to take a look at these fascinating materials.

It seems clear that engineers need to find an approach which makes use of a range of test procedures from geotechnics and structures and combines this with unsaturated soil mechanics principles. A useful goal would be a EC mix design procedure similar to that used for concrete, which begins with the selection of a target compressive strength, a link to the water cement ratio to deliver that strength followed by inclusion of steps to ensure the right workability. It is a procedure that is robust, within the limit state framework it occupies, and a similar approach for EC materials could allow one to specify a manufactured unsaturated soil.

Finally, another area of potential interest linked to EC, and one perhaps closer to traditional geotechnical engineering, is the exploitation of suction as a source of strength in unsaturated soils; this is effectively what one is doing when creating EC materials and could be employed on lower suction situations such as mass retaining walls, foundations and, in particular, temporary works. In fact this is one part of a recently commenced EC-funded research network "Training Engineers and Researchers to Rethink geotechnical Engineering for a low carbon future" (TERRE) led by Strathclyde University [28].

DOI: 10.1051/e3sconf/20160903002

Acknowledgements

The work described in this paper draws from experiences supervising the research projects of a number of PhD and MEng students at Durham over the past 5 years. The assistance provided by the Durham XRCT Facility, which was funded in part by the UK EPSRC (grants EP/K036084/1 & EP/K024698/1), is gratefully acknowledged.

References

1. P.A. Jaquin, C.E. Augarde, *Earth Building: History, Science & Conservation,* IHS BRE Press: Bracknell, (2012).
2. H. Houben, H. Guillard, *Earth construction: a comprehensive guide.* London: ITDG Publishing (1994).
3. P. Walker, R. Keable, J. Martin, V. Maniatidis, *Rammed earth design and construction guidelines.* Watford: BRE Bookshop, (2005).
4. D. Ciancio, C.T.S. Beckett, C.E. Augarde, P.A. Jaquin, ICE Proc. Constr. Mats, to appear (2016).
5. S. Lenci, F. Clementi, T. Sadowski, Eng. Frac. Mech. **87**:62-72 (2012).
6. I. Lombillo, L. Villegas, E. Fodde, C. Thomas, Constr. Bldg Mats **51**:451-460, (2014).
7. D. Gelard, L. Fontaine, S. Maximilien, C. Olagnon, J-P. Laurent, H. Houben, H. Van Damme, *Proc. Int. Symp. Earth. Struct., Bangalore*, 22-24 August (2007).
8. P.A. Jaquin, C.E. Augarde, D.G. Toll, D. Gallipoli, Géotechnique, **59**(5):487–490 (2009).
9. P.A. Jaquin, C.E. Augarde, L. Legrand in *Proc. 1st EUNSAT,* Durham, UK, 417-422 (2008).
10. D. Gallipoli, A.W. Bruno, C. Perlot, N. Salmon, in *Unsaturated Soils: Research and Applications*, 55-62 (2014).
11. H. Nowamooz, C. Chazallon Constr. Bldg Mats, **25**:2112-2121 (2011).
12. P. Gerard. M. Mahdad, A.R. McCormack, B. Francois, Constr. Bldg Mats **95**:437–447(2015).
13. C.T.S. Beckett, J.C. Smith, D. Ciancio, C.E. Augarde, Géotechnique Letters, **5**:254-260, (2015).
14. E.E. Alonso, N.M. Pinyol, A. Gens, Géotechnique, **63**(6):463-478 (2013).
15. A. Koliji, PhD Thesis, EPFL, Switzerland, (2008).
16. Brune, P.F., Ingraffea, A.R., Jackson, M.D. & Perucchio, R. Eng. Frac. Mech., 102: 65-76, (2013).
17. A. Corbin, C.E. Augarde, Proc. Mat. Sci **3**:1675-1680 (2014).
18. L. Miccoli, D.V. Oliveira, R.A. Silva, U. Müller, L. Schueremans, Mat. Struct., **48**:3443–3456 (2015).
19. J.C. Smith, PhD Thesis, Durham University, (2015).
20. N.C. Consoli, L. Da Silva Lopes, L. Jr, B.S. Consoli, L. Festugato, Géotechnique **64**:165-170 (2014).
21. V. Robin, A.A. Javadi, O. Cuisinier, F. Masrouri, Comp. Geotech, **66**:189-202 (2015).
22. J.C. Smith, C.E. Augarde in *Proc. First International Conference on Rammed Earth Construction (ICREC)*, 145-149, (2015).
23. S.M. Hejazi, M. Sheikhazdeh, S.M. Abtahi, A. Zadhoush, Constr. Bldg Mats **30**:100-116 (2012).
24. D. Readle, S. Coghlan, J.C. Smith, A.J. Corbin, C.E. Augarde. ICE Proc. Constr. Mats, to appear (2016).
25. J.G. Zornberg, Géotechnique **52**:593-604 (2002).
26. A. Diambra, E. Ibraim, Géotechnique, **65**:296-308, (2015).
27. TERRE, http://cordis.europa.eu/project/rcn/198308_en.html
28. D.G. Fredlund, A. Xing, Can. Geo. J **31**:521-532 (1994).

 DOI: 10.1051/e3sconf/20160903003

Modelling suction instabilities in soils at varying degrees of saturation

Giuseppe Buscarnera[1,a] and Constance Mihalache[1]

[1]*Northwestern University, Evanston, IL 60208, United States*

Abstract. Wetting paths imparted by the natural environment and/or human activities affect the state of soils in the near-surface, promoting transitions across different regimes of saturation. This paper discusses a set of techniques aimed at quantifying the role of hydrologic processes on the hydro-mechanical stability of soil specimens subjected to saturation events. Emphasis is given to the mechanical conditions leading to coupled flow/deformation instabilities. For this purpose, energy balance arguments for three-phase systems are used to derive second-order work expressions applicable to various regimes of saturation. Controllability analyses are then performed to relate such work input with constitutive singularities that reflect the loss of strength under coupled and/or uncoupled hydro-mechanical forcing. A suction-dependent plastic model is finally used to track the evolution of stability conditions in samples subjected to wetting, thus quantifying the growth of the potential for coupled failure modes upon increasing degree of saturation. These findings are eventually linked with the properties of the field equations that govern pore pressure transients, thus disclosing a conceptual link between the onset of coupled hydro-mechanical failures and the evolution of suction with time. Such results point out that mathematical instabilities caused by a non-linear suction dependent behaviour play an important role in the advanced constitutive and/or numerical tools that are commonly used for the analysis of geomechanical problems in the unsaturated zone, and further stress that the relation between suction transients and soil deformations is a key factor for the interpretation of runaway failures caused by intense saturation events.

1 Introduction

Humanity faces unprecedented challenges related with population growth, increasing energy demand, and degradation of ecosystems [1]. In this context, significant threats come from the deterioration of the climate to a point that vast communities are exposed to growing risks of extreme events [2]. To address such crises, a key role is played by geotechnologies [3], i.e. interventions on near-surface and/or underground deposits for the purpose of environmental remediation (e.g., disposal of CO_2 and hazardous wastes) and risk management (e.g., protection of cities and/or coastal areas from sea level rise and other geohazards). Research on the multi-physics of geological systems, as well as on the couplings between mechanical, hydrologic, chemical and biological agents in the natural environment, is therefore a new frontier to be explored to devise innovative technological solutions able to face the abovementioned challenges [4].

From a scientific viewpoint, intriguing problems derive from the inherent complexity of soils and rocks. Geomaterials are indeed porous and heterogeneous media hosting multiple fluids which establish physico-chemical interactions with the solids. Any alteration of their in-situ state affects the strength and deformation properties, thus playing a role in applications such as the underground storage of hazardous products, infrastructure management and geohazard forecasting. In the latter case, the literature offers numerous examples of near-surface instabilities caused by environmental agents. A recent example is the catastrophic event that took place at Oso (USA) in 2014 [5]. This case involved a sudden failure of unconsolidated glacial deposits with average inclination lower than 20°. The mobilized masses crossed at high velocity a 1-km floodplain, claiming the lives of 43 people and suggesting that liquefaction mediated by transient flow played a role in the unstable propagation of the mass [6], thus stressing the need to quantify the processes that may lead to such kinds of events. Dramatic evidences of destructive near-surface failures also derive from other geological settings, such as unsaturated residual soils, volcanic sediments and airborne deposits [7-9], i.e. materials in which water-sensitivity generates strong hydro-mechanical couplings leading to solid-fluid transitions upon saturation [10].

It can thus be argued that, unless the multi-physical processes controlling soil and rock failure are understood, limited progress can be achieved in assessing the spatial and/or temporal distribution of geologic hazards. In this paper particular emphasis is given to a specific form of multi-physical interactions, which involve changes of the saturation conditions within the pores, thus impacting the loss of unsaturated soil strength, as well as the alteration of suction transients in deformable ground. Specifically, the goal is to stress the role of solid-fluid interactions in failure events in shallow unsaturated slopes, as well as to discuss the role that a loss of mathematical stability may have in the constitutive/numerical tools that are widely used for geomechanical analyses in the unsaturated zone.

[a] Corresponding author: g-buscarnera@northwestern.edu

DOI: 10.1051/e3sconf/20160903003

2 Second-order energy input in partially saturated soils

Previous works by Nova and coworkers [11, 12] have elucidated the connection between second-order energy input to a solid and the singularity of the principal minors of the constitutive tensor. Such relation has been pivotal to correlate non-positive values of second-order work to the loss of control of loading variables, and hence to the failure of soil specimens. In hydro-mechanical contexts characterized by multiple fluids, the incremental response of a porous medium is controlled by both mechanical and hydraulic agents (e.g., pressure changes, fluid injection or extraction). As a result, additional control variables must be added to the constitutive law, requiring an adaptation of both second-order work input and control theory. Such an extension for unsaturated soils with two continuous fluid phases (which, in usual environmental contexts, coincide with water and air) has been discussed by Buscarnera and Nova [13] and Buscarnera and di Prisco [14], and it is rooted in the following expression of second-order work:

$$d^2W = \dot{\sigma}^*_{ij}\dot{\varepsilon}_{ij} - n\dot{s}\dot{S}_r + n(1-S_r)\dot{u}_a\dot{\rho}_a / \rho_a \qquad (1)$$

where $\dot{\sigma}^*_{ij} = \dot{\sigma}_{ij} - S_r\dot{u}_w\delta_{ij} - (1-S_r)\dot{u}_a\delta_{ij}$ is a stress measure work-conjugate with the strains, $\dot{\varepsilon}_{ij}$, $\dot{s}$ the suction rate, n the porosity, S_r the degree of saturation, and ρ_a the air density. In most applications, only the first two terms of the second-order work are considered, thus implicitly assuming passive air phase (i.e., constant atmospheric pressure). Nevertheless, cases of relevance for failure events in the near-surface can be envisaged when, upon a substantial increase in degree of saturation, the air loses continuity because of air-entrapment, possibly causing undrained conditions for the air phase and simultaneous growth of both air and water pressures (Fig. 1).

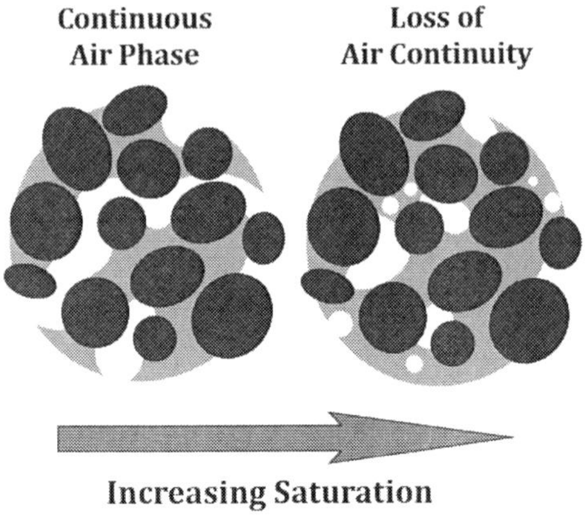

Figure 1. Schematics indicating transition from a continuous air phase characterized by drained conditions (i.e. $\dot{u}_a = 0$) to states characterized by air entrapment and growth of air pressure due to lack of drainage (i.e. $\dot{m}_a = 0$)

In the following, Equation (1) will be used to study coupled failure mechanisms under unsaturated conditions that may imply a simultaneous loss of control of variables associated with the two pore fluids. For this purpose, to express the second-order energy input as a function of variables commonly used in laboratory experiments, the expression of d^2W is rearranged as follows:

$$d^2W = \dot{\sigma}_{ij}\dot{\varepsilon}_{ij} + \dot{u}_w\dot{e}_w / (1+e) + n(1-S_r)\dot{u}_a(\dot{m}_a / m_a) \qquad (2)$$

where σ_{ij} is the total stress, $e_w = eS_r$ is the water ratio, and m_a is the mass of air in the porous medium. This choice enables expression of the second-order energy input in terms of variables that are directly imposed during soil testing (i.e., total stress, σ_{ij}, suction, s, and mass of air and/or water, i.e. e_w and/or m_a), thus offering a convenient approach to link the energetics of hydro-mechanical perturbations to the constitutive description of the soil.

3 Controllability analyses

The second-order work input can be linked to constitutive laws to compute failure indices related with given hydro-mechanical controls (a strategy that is usually referred to as *controllability analysis* [12]). Here, various drainage conditions will be considered, with the goal to assess the role of the pore fluids on the onset of failures involving either uncontrolled water pressures or sharp variations of the pressures of both water and air.

3.1 Controllability conditions

Let us consider a constitutive law for a three-phasic soil written in incremental form. To define a simple tangent constitutive operator compatible with the second-order energy input, let us restrict the analysis to stress variables reflecting axisymmetric stress conditions, as follows:

$$\begin{Bmatrix} \dot{p}^* \\ \dot{q} \\ n\dot{s} \\ n\dot{u}_a \end{Bmatrix} = \begin{bmatrix} D_{PP} & D_{PQ} & D_{PH} & 0 \\ D_{QP} & D_{QQ} & D_{QH} & 0 \\ D_{HP} & D_{HQ} & D_{HH} & 0 \\ 0 & 0 & 0 & D_{GG} \end{bmatrix} \begin{Bmatrix} \dot{\varepsilon}_v \\ \dot{\varepsilon}_s \\ -\dot{S}_r \\ (1-S_r)\frac{\dot{\rho}_a}{\rho_a} \end{Bmatrix} \qquad (3)$$

where $\dot{p}^* = \dot{\sigma}^*_{ij}\delta_{ij}/3$ is a mean skeleton stress increment working for the volumetric strain rate, $\dot{\varepsilon}_v$, while $\dot{q}$ is the deviatoric stress increment, working for the shear strain rate, $\dot{\varepsilon}_s$. The constitutive matrix in (3) is built under the simplification that the pressure-density relation for the air is uncoupled from the other hydro-mechanical variables, thus enabling the use of basic state equations for perfect gases to define the term D_{GG}. The other diagonal terms reflect either bulk or shear stiffness (D_{PP} and D_{QQ}) or the water retention behaviour (D_{HH}). Cross couplings are also included in form of hydrologic effects on the stress-strain response (D_{PH} and D_{QH}) or to reflect the strain-dependence of the retention curve (D_{HP} and D_{HQ}). While the latter form of coupling can be included with no alterations of the general strategy of analysis [15], for the sake of analytical simplicity here the study will be restricted to uncoupled water retention responses (i.e., $D_{HP} = D_{HQ} = 0$), thus focusing solely on the suction dependence of the solid skeleton (i.e., $D_{PH} = D_{QH} \neq 0$).

The detection of constitutive singularities involves the analysis of specific loading scenarios. In particular, for a given choice of control variables, $\dot{\varphi}$ (i.e., variables that are directly imposed) and response variables, $\dot{\psi}$ (i.e., variables work-conjugate to $\dot{\varphi}$ and obtained as an outcome of the soil response) it is possible to state:

$$\dot{\varphi} = \mathbf{X}\dot{\psi} \tag{4}$$

where $\mathbf{X}$ is a control matrix compatible with the second-order work input. It thus follows that a loss of $\dot{\varphi}$ control leading to undefined values of $\dot{\psi}$ is associated with the following singularity condition [12, 16]:

$$\det \mathbf{X} = 0 \tag{5}$$

To facilitate the selection of the control variables, it is convenient to perform a linear transformation replacing the variables in (3) with the two following sets of work-conjugate variables:

$$\dot{\xi} = \begin{Bmatrix} \dot{p} \\ \dot{q} \\ \dot{u}_w \\ \dot{u}_a \end{Bmatrix} \text{ and } \dot{\eta} = \begin{Bmatrix} \dot{\varepsilon}_v \\ \dot{\varepsilon}_s \\ -e_w/(1+e) \\ n(1-S_r)\dot{m}_a/m_a \end{Bmatrix} \tag{6}$$

which are consistent with the expression of d^2W given in (2). This choice enables rewriting of the constitutive laws in (3) as follows:

$$\dot{\xi} = \mathbf{\Lambda}\dot{\eta} = \mathbf{T}_\sigma \mathbf{D} \mathbf{T}_\sigma^{\mathrm{T}} \dot{\eta} \tag{7}$$

where the following transformation is used:

$$\dot{\xi} = \mathbf{T}_\sigma \begin{Bmatrix} \dot{p}^* \\ \dot{q} \\ n\dot{s} \\ n\dot{u}_a \end{Bmatrix} = \begin{bmatrix} 1 & 0 & -S_r/n & 1/n \\ 0 & 1 & 0 & 0 \\ 0 & 0 & -1/n & 1/n \\ 0 & 0 & 0 & 1/n \end{bmatrix} \begin{Bmatrix} \dot{p}^* \\ \dot{q} \\ n\dot{s} \\ n\dot{u}_a \end{Bmatrix} \tag{8}$$

Hereafter, three control scenarios will be used to assess the impact of hydro-mechanical couplings on the constitutive singularities: (i) stress-control combined with drained conditions for both fluids (i.e., $\dot{\varphi} = [\dot{p}, \dot{q}, \dot{u}_w, \dot{u}_a]$); (ii) stress-control plus undrained conditions for water and drained conditions for air (i.e. $\dot{\varphi} = [\dot{p}, \dot{q}, \dot{e}_w/(1+e), \dot{u}_a]$); (iii) stress control combined with undrained conditions for both fluids ($\dot{\varphi} = [\dot{p}, \dot{q}, \dot{e}_w/(1+e), n(1-S_r)\dot{m}_a/\dot{m}_a]$). Scenario (i) leads to the following singularity condition:

$$\det \mathbf{X}_S = \frac{D_{GG}D_{HH}}{n^4}\left(D_{PP}D_{QQ} - D_{PQ}D_{QP}\right) = 0 \tag{9}$$

which resembles the failure index that was obtained by di Prisco et al. [17] for shear failure in saturated ground and by Buscarnera and di Prisco [18] for unsaturated slopes at constant suction. By contrast, the scenario (ii) produces the following singularity condition:

$$\det \mathbf{X}_W = \frac{D_{GG}}{n^2}\left(D^*_{PP}D_{QQ} - D_{PQ}D^*_{QP}\right) = 0 \tag{10}$$

in which

$$D^*_{PP} = D_{PP} - \frac{S_r}{n}\left(D_{PH} - \frac{S_r}{n}D_{HH}\right)$$
$$D^*_{QP} = D_{QP} - \frac{S_r}{n}D_{QH} \tag{11}$$

are corrective terms related with the coupling effects. The expression in (10) resembles that found by Buscarnera and di Prisco [10] for failure under water-undrained paths and/or water injection under the hypothesis of passive air phase. Finally, scenario (iii) provides the following loss of controllability condition:

$$\det \mathbf{X}_{WA} = \left(D^*_{PP}D_{QQ} - D_{PQ}D^*_{QP}\right) + \frac{1}{n^2}D_{GG}D_{QQ} = 0 \tag{12}$$

in which the potential for air pressurization contributes to the singularity of $\mathbf{X}$ via a term affected by the air-phase compressibility (D_{GG}) and by the shearing response of the skeleton (D_{QQ}). In the following, such indices will be computed with a constitutive model for unsaturated soils and will be used to assess the potential for coupled failure in volcanic soils involved in flowslide events [9].

3.2 Constitutive Models

An elastoplastic constitutive law based on that proposed Buscarnera and Nova [19] will be used for the following analyses. This model includes versatile yield surface and plastic potential expressions proposed by Lagioia et al. [20] to reproduce non-associated flow and a saturation-dependent hardening rule based on earlier propositions by Jommi and di Prisco [21],

$$\dot{p}_s = \frac{1}{B_p} p_s(\dot{\varepsilon}^p_v + \xi_s \dot{\varepsilon}^p_s) - r_{sw} p_s \dot{S}_r \tag{13}$$

where p_s is the internal variable associated with the size of the yield surface, ε^p_v and ε^p_s are volumetric and shear components of the plastic strain, respectively, and B_p, ξ_s and r_{sw} are constitutive parameters. In particular, the parameter r_{sw} controls the contraction of the yield surface that would occur during saturation (i.e., when the suction increment is negative), which plays a crucial role on soil instabilities induced by saturation [15]. To define the soil response for stress states inside the yield surface, the model is completed by an uncoupled hypoelastic law based on the following tangent bulk and shear moduli:

$$K = K_0\left(\frac{p'}{p_0}\right)^n \text{ and } G = G_0\left(\frac{p'}{p_0}\right)^n \tag{14}$$

which describe power laws linking the soil stiffness to the mean skeleton stress, $p' = p - S_r u_w - (1-S_r)u_a$, and where K_0 and G_0 are reference values of bulk and shear moduli, respectively, while p_0 is a reference pressure.

Finally, to link the skeleton response to hydrologic processes, the model is integrated by a water retention curve based on the well-known expression proposed by Van Genuchten:

$$S_r = \frac{1}{\left[1+\left(a_{vg}s\right)^{n_{vg}}\right]^{m_{vg}}} \tag{15}$$

as well as by the ideal gas law, which under isothermal conditions can be arranged as follows to relate changes in air density to the air pressure rate:

$$\frac{\dot{\rho}_a}{\rho_a} = \frac{\dot{u}_a}{u_a} \tag{16}$$

3.3 Calibration Strategy

To illustrate the use of controllability indices, this section makes reference to a class of volcanic soils involved in flowslide events that took place in the Campania region of Italy [9, 22]. While the site-specific properties of such soils vary across the region, some common features can be identified for the soils of the area, which include the suction-sensitivity of both compressibility and shear strength and their high potential for liquefaction.

To study their behaviour, here a baseline calibration is obtained based on data reported by [23] and [24] for soils retrieved from a specific site of the region (Table I). This calibration will be later applied to reinterpret flow failure processes detected at the laboratory scale on a similar pyroclastic soil from another site [22].

Figure 2 illustrates the calibration of the retention curve parameters with reference to data reported by [23] and [25]. The range of retention curve variability reported by [23] for different classes of Campanian soil and stress conditions is also illustrated in the figure.

The parameters controlling the shearing response of the Campanian ashes have been assessed on the basis of suction-controlled triaxial compression tests conducted at varying levels of suction and net confinement (Fig. 3). The good agreement between model predictions and data corroborates the validity of the average skeleton stress for the simulation of shear failure in Campanian soils.

To assess the non-associativity of the plastic flow, the liquefaction resistance of saturated samples subjected to undrained shear has been considered [26], identifying a set of parameters capturing accurately the lower bound of the instability line emerging from tests (Fig. 4).

Table 1. Baseline model parameters for Campanian ash.

Description	Symbol	Value
Elastic Constants	K_0	5000 *kPa*
	G_0	3500 *kPa*
	p_0	100 *kPa*
	n	0.5
Logarithmic Plastic Compressibility	B_p	0.032
Yield Surface Shape Parameters	a_f	0.99
	m_f	1.50
	M_{cf}	1.10
	M_{ef}	0.88
Plastic Potential Shape Parameters	a_g	0.50
	m_g	1.50
	M_{cg}	1.35
	M_{eg}	1.08
Suction-Hardening Parameter	r_{sw}	3.5
Dilatancy Parameter	ξ_s	-0.05
SWCC Shape Parameters	a_{vg}	0.17
	m_{vg}	0.31
	n_{vg}	1.3

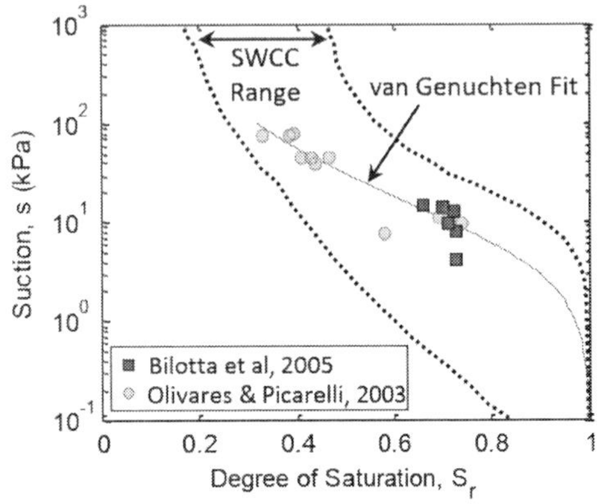

Figure 2. Calibration of the water retention curve (SWCC) of Campanian ashes (data after [23,25]).

Finally, to have a first-order assessment of the ability of the model to reproduce plastic strains due to wetting, the suction-hardening characteristics have been calibrated to mimic the compression behaviour of Campanian soils under saturated and unsaturated conditions, as well as the occurrence of wetting-induced compaction in oedometric experiments conducted at varying levels of vertical net stress (Fig. 5). A satisfactory agreement between model predictions and data was found, corroborating the choice made for the suction-hardening law in Equation (13) and its applicability for the study of non-linear phenomena promoted by saturation paths.

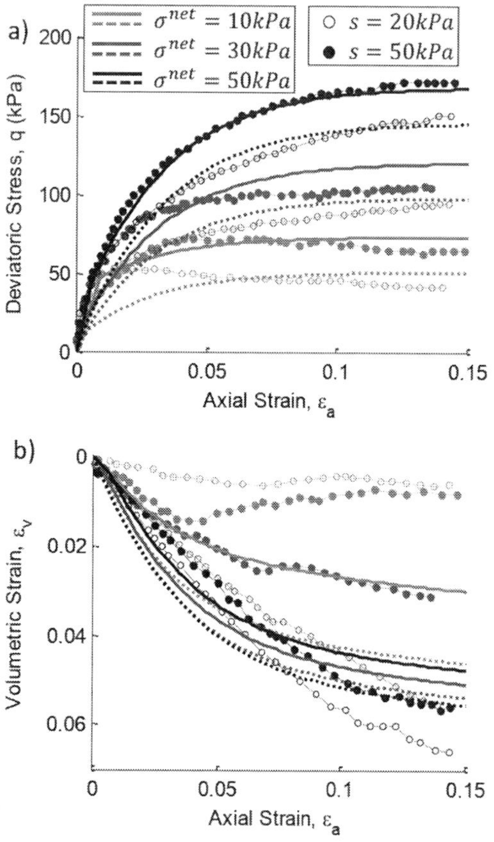

Figure 3. Simulated shear and volumetric responses at different levels of suction and net confinements (data after [23]).

3.4 Numerical Assessment of Unstable States

The set of parameters calibrated above provides a starting point to inspect the evolution of the state of stability of ashy soils during wetting. An example is the unstable behaviour observed upon suction reduction in pyroclastic soils under constant stresses (Fig. 6). Similar experiments have been reported by Olivares and Damiano [22] on soil samples from a neighbouring area. Such tests were based on an initial suction of 82kPa gradually reduced at a rate of 1.4kPa/hr to ensure drained conditions. The applied loads remained unchanged, thereby imposing a constant stress deviator under decreasing mean skeleton stress. Axial and volumetric strains were tracked throughout the test, along with the degree of saturation. As wetting progressed, a sharp increase in rate of soil deformation was observed in correspondence with a critical suction value close to saturated conditions, which was indicative of unstable behaviour and, ultimately, failure.

The material tested by [22] exhibits different hydraulic retention behaviour from that of the baseline calibration, but similar wetting paths can be simulated using the parameters in Table 1, with stress conditions as shown in Fig. 7b. The resultant volumetric compaction observed during these simulations is presented in Fig. 7a, along with points at which the controllability criteria given in (9), (10) and (12) are violated.

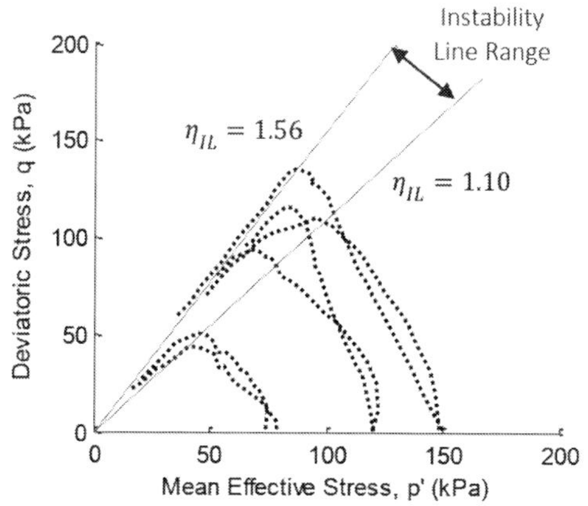

Figure 4. Computed instability lines from undrained triaxial compression tests on saturated ashy soil (data after [22]).

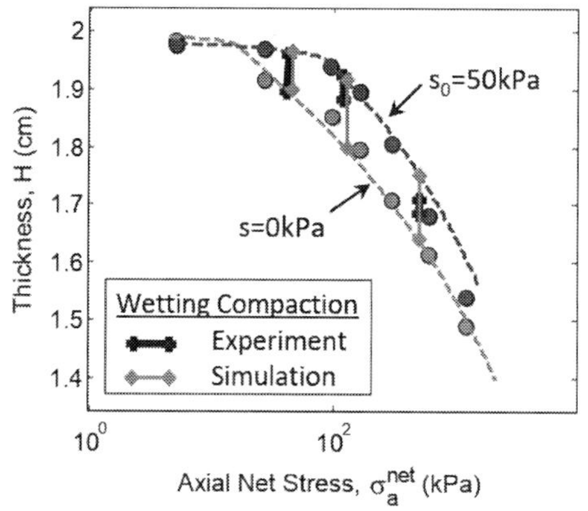

Figure 5. Compression behaviour of saturated and unsaturated samples and simulations of wetting-collapse (data after [24]).

It can be noted that the controllability indices computed with these parameters suggest that different hypotheses about the dominating control conditions at the onset of failure yield different times of potential instability. In particular, the first index to vanish upon saturation is found to be that associated with unstable pressurization of both water and air phases (det$\mathbf{X}_{WA}$), shortly followed by that associated with a pressurization of the water phase (det$\mathbf{X}_W$) under constant air pressure, and finally by the index associated with constant suction failure (det$\mathbf{X}_S$). Most importantly, it must be noted that the saturation levels where these indices vanish are in proximity of states at which spontaneous undrained response and sharp acceleration of the soil deformation characteristics were found [22], thus corroborating the connection between vanishing controllability indices and unstable couplings between fluid flow and deformations. Such links will be explored in greater detail in the next section in light of the field equations that control suction transients in deformable unsaturated ground.

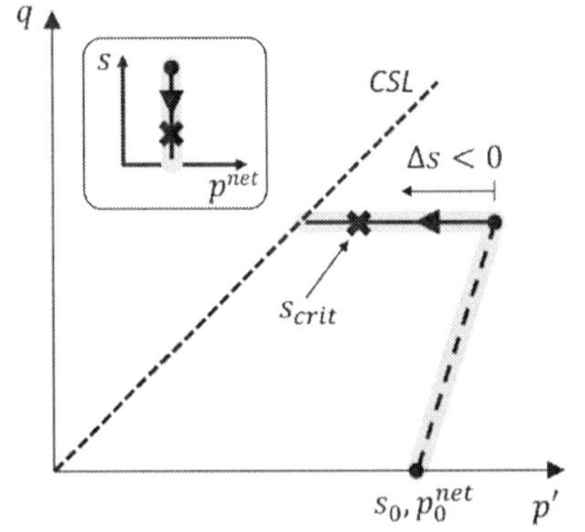

Figure 6. Schematic of constant shear stress path under suction-controlled wetting, with critical suction (s_{crit}) attained prior to full saturation.

4 Implications for the stability of suction transients

The analyses discussed above point out the important role of hydro-mechanical couplings on an activation of unstable processes. Most notably, they stress the fact that such failures imply deformation-induced transient alterations of the pressure regime of both fluids. Hence, it is of interest to inspect the implications that constitutive instabilities of unsaturated soils have on the equations that are commonly used to track the spatial and temporal variations of the pressure regimes of deformable deposits. To achieve this goal, an idealized one-dimensional flow condition is considered (Fig. 8) to derive a mass balance equation. For the sake of simplicity, only pore water flow is considered, and, in agreement with previous analyses, axisymmetric stress states are used to model triaxial compression tests characterized by transient saturation.

These hypotheses rely on simplifying assumptions similar to those used by di Prisco [27] to study pressure transients in saturated soils. In particular, it is assumed that water flow takes place only along the vertical axis of the sample, thus ignoring flow along horizontal planes. In addition, the state of stress, strain and suction is assumed to be uniform until the onset of transient effects, total stress control is hypothesized to be a good representation of the local constitutive response, and the presence of shear stresses at the bases of the sample is neglected.

For vertical flow, the mass balance equation can be written as follows:

$$\frac{\partial q_z}{\partial z} = -\frac{1}{1+e}\frac{\partial e_w}{\partial t} \tag{17}$$

where q_z is the Darcian velocity of water along the vertical axis and the storage of water volume over time is expressed in terms of water ratio, e_w.

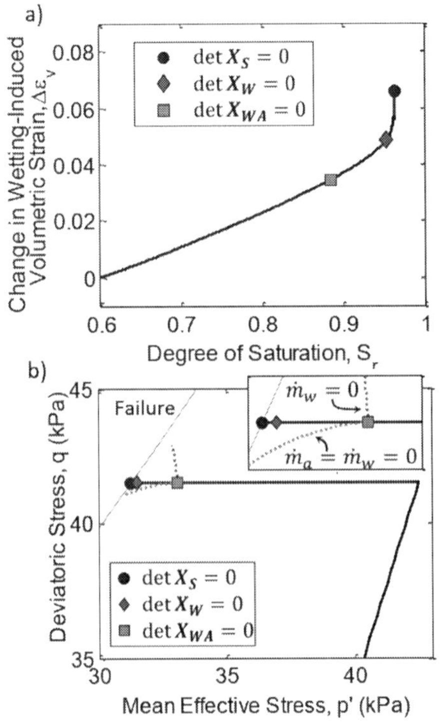

Figure 7. Computed volumetric strains during wetting with points of loss of controllability for suction control (det$\mathbf{X_S}$), constrained water mass (det$\mathbf{X_W}$), and constrained air and water mass (det$\mathbf{X_{WA}}$) (a) and consequences of switching to water-undrained shearing ($\dot{m}_w = 0$) or shearing under simultaneous undrained conditions for air and water ($\dot{m}_a = \dot{m}_w = 0$) (b).

By introducing the Darcy's law in the following form

$$q_z = -k_w \frac{\partial}{\partial z}\left(\frac{u_w}{\rho_w g} + z\right) \tag{18}$$

where k_w is the suction-dependent permeability for water flow, ρ_w is the mass density of water and g is the gravitational acceleration, Equation (17) leads to:

$$\frac{k_w}{\rho_w g}\frac{\partial^2 u_w}{\partial z^2} + \frac{\partial k_w}{\partial u_w}\frac{\partial u_w}{\partial z}\frac{\partial}{\partial z}\left(\frac{u_w}{\rho_w g} + z\right) = \frac{1}{1+e}\frac{\partial e_w}{\partial t} \tag{19}$$

To write the mass balance equation (19) in terms of water pressures it is possible to use the constitutive equations of a deformable unsaturated soil. For this purpose, relations expressed in terms of tangent plastic compliance are adopted:

$$\begin{Bmatrix} \dot{\varepsilon}_v \\ \dot{\varepsilon}_s \\ -\dot{S}_r \end{Bmatrix} = \begin{bmatrix} C_{PP} & C_{PQ} & C_{PH} \\ C_{QP} & C_{QQ} & C_{QH} \\ 0 & 0 & C_{HH} \end{bmatrix} \begin{Bmatrix} \dot{p}^* \\ \dot{q} \\ n\dot{s} \end{Bmatrix} \tag{20}$$

where hypotheses similar to those adopted for the matrix in (3) are used by neglecting the compliance terms linked with the strain-dependence of the retention curve (i.e., $C_{HP} = C_{HQ} = 0$). It is worth noting that, since only water pressure transients are considered, the effect of the air pressure is neglected in the constitutive law. This implies a hypothesis of passive air (constant air pressure) which implies negligible values for the last term in Equations (1) and (2). In other words, (20) is consistent with a second-order work expression accounting only for soil deformations and suction changes [14].

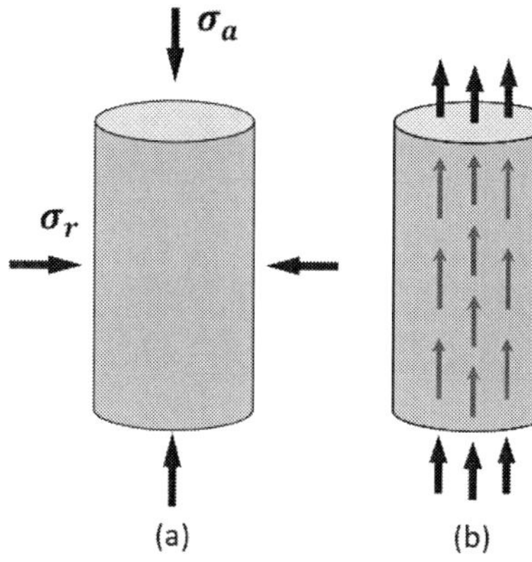

Figure 8. Schematic describing axisymmetric stress conditions in a triaxial compression test (a) and one-dimensional water flow (b).

By using a transformation of variables similar to that used to pass from the variables in (3) to those in (6), it is possible to rewrite (20) in terms of increments of water ratio, as follows:

$$\begin{Bmatrix} \dot{\varepsilon}_v \\ \dot{\varepsilon}_s \\ \dfrac{-\dot{e}_w}{1+e} \end{Bmatrix} = \begin{bmatrix} C_{PP} & C_{PQ} & S_r C_{PP} + nC_{PH} \\ C_{QP} & C_{QQ} & S_r C_{QP} + nC_{QH} \\ S_r C_{PP} & S_r C_{PQ} & C^*_{HH} \end{bmatrix} \begin{Bmatrix} \dot{p}^{net} \\ \dot{q} \\ \dot{s} \end{Bmatrix} \tag{21}$$

where C^*_{HH} is a modified compliance term defined as:

$$C^*_{HH} = n^2 C_{HH} + S_r^2 C_{PP} + nS_r C_{PH} \tag{22}$$

which reflects the role of isotropic compliance and hydro-mechanical coupling. It is worth noting that the variables in (21) are compatible with the following form of d^2W

$$d^2W = \dot{p}^{net}\dot{\varepsilon}_v + \dot{q}\dot{\varepsilon}_s - \dot{s}\dot{e}_w / (1+e) \tag{23}$$

which was shown by [14] to be equivalent to (1) under constant air pressure and axisymmetric stress conditions.

By using (21) to express the rate of e_w in terms of stress and suction increments, and by assuming that under constant air pressure $\dot{s} \approx -\dot{u}_w$, it is possible to rearrange (19) as follows:

$$\frac{\partial u_w}{\partial t} - A\frac{\partial^2 u_w}{\partial z^2} + b = 0 \tag{24}$$

where

$$A = \frac{k_w}{\rho_w g}\left(C^*_{HH}\right)^{-1} \tag{25a}$$

$$b = -S_r\left(C^*_{HH}\right)^{-1}\left[C_{PP}\frac{\partial p^{net}}{\partial t} + C_{PQ}\frac{\partial q}{\partial t} + \zeta\left(u_w, \frac{\partial u_w}{\partial z}\right)\right] \tag{25b}$$

$$\zeta\left(u_w, \frac{\partial u_w}{\partial z}\right) = \frac{1}{S_r}\frac{\partial k_w}{\partial u_w}\frac{\partial u_w}{\partial z}\frac{\partial}{\partial z}\left(\frac{u_w}{\rho_w g} + z\right) \tag{25c}$$

The differential form in Equation (24) is convenient to study the ill-posedness of the temporal evolution of a pressure field controlled by diffusive effects [28], and it has been used in early works by Rice [29] and di Prisco [27] to study pore pressure instabilities in fully saturated geomaterials. In particular, while the function b reflects forcing associated with external loading and advective effects linked with inhomogeneous permeability, the sign of the function A in (25a) is related with the spontaneous temporal evolution of the pore pressure field under lack of external forcing (i.e., homogeneous equation). In other words, the sign of A is associated with the parabolicity of the differential form (24), implying that a self-stabilizing diffusive evolution of the excess pore water pressures (and, hence, of suction transients) is possible only when A is positive, i.e. under the condition

$$C^*_{HH} = n^2 C_{HH} + S_r^2 C_{PP} + nS_r C_{PH} > 0 \tag{26}$$

Violations of the condition above suggests the possibility of "backwards" diffusion processes, according to which small perturbations of the pore pressure field can quickly grow into singularities [27,29]. Since (26) is a principal minor of the constitutive matrix in (21), it is interesting to explore the connection of this loss of well-posedness of the transient water mass balance equation in terms of controllability. Indeed, by writing the loss of control conditions as a function of compliance terms, it is readily apparent that the loss of control for a loading scenario characterized by $\dot{\varphi} = \left[\dot{p}, \dot{q}, \dot{e}_w / (1+e)\right]$ provides:

$$\det \mathbf{X}_W = 0 \iff C^*_{HH} = 0 \tag{27}$$

implying that a loss of parabolicity of the water pressure field coincides with the loss of control for water-undrained loading. This finding suggests an additional interpretation of coupled failure modes at the material point level, as it shows that loss of controllability of pore pressures may imply alterations of the differential form of the field equations, thus impacting the robustness of the numerical techniques used for coupled flow/deformation analysis, as well as the interpretation of measurements of time-varying values of suction and/or degree of saturation in deformable unsaturated ground.

5 Conclusions

This paper has discussed a set of analytical techniques to detect coupled failure events in unsaturated soil samples, i.e. states at which a simultaneous singularity of hydraulic (e.g., air and/or water pressure) and mechanical variables (e.g., axial and/or volumetric strains) is possible. For this purpose, energy statements defining a second-order work input in three-phase media have been linked to a set of constitutive relations for partially saturated soils, with the purpose to detect singularities of the tangent constitutive operator. A range of hydro-mechanical control conditions has been considered, encompassing constant suction and stress control (drained loading), constant water content under stress control (i.e., water-undrained conditions) and simultaneous lack of drainage for both the water and the air phase under imposed total stresses. It has been shown that the choice of control conditions impacts considerably the singularity of the control matrix, implying vanishing values of determinants that are met at different states during a saturation process. Such possibility was explored in detail by making reference to a particular volcanic soil involved in past rainfall-induced flowslides. In particular, after defining a baseline calibration of the parameters of a hydro-mechanical model reproducing first-order features of the selected soil, a constant shear wetting path was performed to interpret the unstable behaviour witnessed in similar tests reported in the literature. Conditions for loss of control were monitored during the path and points of controllability violation were identified. Most notably, such points of potential instability were detected to occur in proximity of the state at which failure accompanied by an increase of the rates of deformation and pore pressure growth were measured in laboratory experiments. While the unstable mode corresponding to simultaneous air and water-undrained conditions was shown to be the first to be activated upon wetting, it was also shown that water-undrained instability involving unstable pressurization of the water phase was activated shortly afterwards. Both failure modes were found to anticipate the onset of shear-failure at constant suction, suggesting that coupled controllability indices provide advantages to identify failure modes involving simultaneous changes of deformation and pore pressures. To further corroborate such findings, it was shown that loss of controllability conditions can also be associated with the loss of well-posedness of the field equations that govern the mass balance of water flow and, hence, the stability of suction transients in deformable unsaturated ground. As a result, the tools discussed in this paper find multiple applications not only in the detection of failure mechanisms relevant for geohazard initiation, but also to assess the robustness of the analytical/numerical techniques that are routinely used to study coupled flow/deformation problems in unsaturated ground, as well as to conceive regularization strategies mitigating the loss of computational stability in presence of highly non-linear constitutive relations.

References

1. J. E. Cohen, Science, **302**, 1172-1175 (2003)
2. J. F. Mitchell, J. Lowe, R. A. Wood, M. Vellinga, Phyl. Trans. Roy. Soc., **364**, 2117-2133 (2006)
3. J. C. Santamarina, G. C. Cho, KSCE J. of Civ. Eng., **15**, 607-610 (2011)
4. A. Gens, Geotechnique, 60, 3-74, 2010.
5. J. Wartman, D. R. Montgomery, S. A. Anderson, J. R. Keaton, J. Benoît, J., dela Chapelle, R. Gilbert, Geomorphology, **253**, 275-288 (2016)
6. R. M. Iverson, D. L. George, Geotechnique, **66**, 175-187, (2016)
7. D. Fuchu, C. F. Lee, W. Sijing, Eng. Geol., **51**, 279-290, (1999)
8. D. Zhang, G. Wang, C. Luo, J. Chen, Y. Zhou, Landslides, **61**, 55-60 (2009)
9. G. Sorbino, M. V. Nicotera, Eng. Geol., **165**, 105-132 (2013)
10. G. Buscarnera, C. di Prisco, Geotechnique, **63**, 801-816 (2013)
11. R. Nova, J. Mech. Behav. Mat., **5**, 193-202 (1994)
12. S. Imposimato, R. Nova, Mech. Coh. Fri. Mater.. **3** 65-87 (1998)
13. G. Buscarnera, R. Nova, Int. J. Numer. Anal. Meth. Geomech. **35** 179-200 (2011)
14. G. Buscarnera, C. di Prisco, Int. J. Numer. Anal. Meth. Geomech. **36**, 36-49 (2012)
15. G. Buscarnera, Acta Geotech., **9**, 313-327 (2014)
16. G. Buscarnera, G. Dattola, C. di Prisco, Int. J. Solids Structures, **48**, 1867-1878 (2011)
17. C. di Prisco, R. Matiotti, R. Nova, Géotechnique, **45**, 479-496 (1995)
18. G. Buscarnera, C. di Prisco, Geotechnique Letters **1**, 85-90 (2011)
19. G. Buscarnera, R. Nova, Int. J. Numer. Anal. Meth. Geomech. **33** 1055-1086 (2009)
20. R. Lagioia, A. M. Puzrin, D. M. Potts, Computers and Geotechnics, **19**, 171-191 (1996)
21. C. Jommi, C. di Prisco, Proc. Conf. on Role of Fluids in Geotech. Eng., Mondovi, **1**, 167-188 (1994)
22. L. Olivares, E. Damiano, J. Geotech. Geoenviron. Eng, **133**, 51-62 (2007)
23. E. Bilotta, L. Cascini, V. Foresta, G. Sorbino, Geotech. Geol. Eng., **23**, 365-402 (2005)
24. E. Bilotta, V. Foresta, G. Migliaro, *Unsaturated Soils*, 849-860 (2006)
25. L. Olivares, L. Picarelli, Géotechnique, **53**, 283-287 (2003)
26. G. Buscarnera, A. J. Whittle, J. Geotech. Geoenviron. Eng, **139**, 420-432 (2013)
27. C. di Prisco, Computers and Geotechnics, **18**, 225-244 (1996)
28. O. A. Ladyzhenskaia, V. A., Solonnikov, N. N. Uraltseva, **23**, American Math. Soc. (1988)
29. J. R. Rice J. Geophys. Res, **80**, 1531-1536 (1975)

Inputs from in situ experiments to the understanding of the unsaturated behaviour of Callovo-Oxfordian claystone

Gilles Armand[1,a], Hippolyte Djizanne[1] , Jad Zghondi[1], Rémi de La Vaissière[1], Jean Talandier[2] and Nathalie Conil[1]

[1]*Andra, R&D Division, Meuse/Haute-Marne Underground Research Laboratory, 52290, Bure, France*
[2]*Andra, R&D Division, 1-7 rue Jean Monnet, 92290, Chatenay-Malabry, France*

Abstract. The French national radioactive waste management agency (Andra) research program is dedicated to preparing the construction and operation of a deep geological disposal facility for high-level and intermediate-level long-lived radioactive waste (HL, IL-LLW) in the Callovo-Oxfordian claystone (COx). The characterization of the COx thermo-hydro-mechanical (THM) behavior, at different scales of interest, must gradually give relevant data for design and safety calculations. The effects of saturation and desaturation of COx claystone are studied in laboratory conditions (sample scale) and in situ (drift scale), in order to improve knowledge on ventilation effect at gallery wall as galleries will remains open during operational phase (more than 100 years for some specific galleries) in the repository. The Saturation Damaged Zone (SDZ) experiment is outlined and its results are discussed. This experimentation aims to change the relative humidity in an isolated portion of a gallery in order to follow the HM behavior of the surrounding rock mass. Drying and wetting cycles could induce in certain cases cracks and swelling and modify the hydromechanical behavior of the claystone around the gallery (Young modulus, strength, creep…). The long term behavior of the COx claystone at the vicinity of gallery is then studied by performing climatic, hydraulic, geological and geomechanical measurements. Results of the in situ experiment are discussed with respect to the identified process on samples. The discussion given on this paper intends to highlights the inputs from 7 years of an in situ experiment to better understand the unsaturated behavior of the COx claystone.

1 Introduction

Clay formations in their natural state exhibit very favorable conditions for repository of radioactive waste, as they generally have a very low hydraulic conductivity, small molecular diffusion and significant retention capacity for radionuclide. Nevertheless, one concern regarding waste disposal is that, due to the necessary underground excavations and the associated disturbance and damage in the area close to these excavations, the favorable properties of such formations could change and the host rock could lose part of its barrier function and thus negatively influence the performance of a repository. That is why, in order to demonstrate feasibility of a radioactive waste repository in a claystone formation, the French national radioactive waste management agency (Andra) started in 2000 to build the Meuse/Haute-Marne underground research laboratory (URL) at Bure (at nearly 300 km East of Paris). The host formation consists of a Callovo-Oxfordian claystone (COx) laying between 420 m and 550 m depth, which is overlain and underlain by poorly permeable carbonate formations.

First, the main objective of the research was to characterize the confining properties of the clay through in situ hydrogeological tests, chemical measurements and diffusion experiments and to demonstrate that the construction and operation of a geological disposal will not introduce pathways for radionuclides migration [1]. The excavation worksite in the host layer is a scientific experimentation in itself to characterize the impacts of digging, to understand the hydro mechanical behaviour of the claystone and to study the excavated damaged zone (EDZ).

In France, the repository, if it will be allowed, will be operated during at least 100 years after it commissioning. Some of the drift will remain opened over a long period during which ventilation will activate. Then the effect of ventilation on rock mass behaviour around drift has to be understood, in particular to evaluate effect of ventilation over excavation damaged zone (EDZ). One of the experiments, called Saturation Damaged Zone (SDZ), is dedicated to the understanding of change of relative humidity over the drift behaviour and the excavation damaged zone (extent and permeability measurement).

It is known that drying test on clay soil leads to the fracturing of the clay. On claystone, fractures associated with the desaturation of argillaceous medium have been observed on gallery fronts in several underground research laboratories, e.g., in Tournemire [2] and Mont Terri underground research laboratory. Hedan et al [3] demonstrated the ability of the Digital image correlation

[a] Corresponding author: gilles.armand@andra.fr

(DIC) technique, to achieve a full displacement field measurement in order to monitor opening and closure of crack under climatic conditions changes at Tournemire URL. They show that cracks opening/closure are clearly correlated to changes in RH and T. Strains concentrate into narrow zones or bands indicating typical strain localization as also observed in the laboratory on samples. These narrow zones or bands were desiccation cracks (typical aperture 0.5 mm) that were observable from the displacement fields. The strain fields between these desiccation cracks were homogeneous at the spatial resolution used in this study. The hydric strains were anisotropic with larger strains perpendicular to desiccation cracks than those parallel to the same cracks. As observed at the Mont-Terri site, desiccation cracks close in summer and open in winter. However, contrary to the measurements acquired at the Swiss site, the crack apertures of the desiccation cracks were reversible after one year of data acquisition. Such annual climatic changes with high relative humidity in summer are not as severe at the Meuse/Haute-Marne URL due to the fact that accesses to the lab are through shafts with control ventilation.

At the main level of the Meuse/Haute-Marne URL, Armand et al [4] show that excavation works induce an important fracture pattern (shear an extension fractures) whose extent depends on the drift orientation versus the in situ principal stresses. The effect of relative humidity change on the behavior of the excavation induced fractures network is key point of SDZ experiment.

This paper first recalls some of the results of drying test performed on sample in order to investigate the desaturation effect on hydro mechanical behavior of the COx claystone. The major part of this paper is devoted to the description of SDZ experiment and results.

2 Callovo-oxfordian claystone

2.1 Mineralogy

Sediments of the Callovo–Oxfordian unit consist of a dominant clay fraction associated with carbonate, quartz with minor feldspars, and accessory minerals [5]. At the MHM URL main level (-490 m depth), the clay fraction is high (40–60%), and clay minerals consist of illite, ordered illite/smectite mixed layers, kaolinite, tri-octahedral, iron-rich chlorite, and minor biotite.

The Callovo–Oxfordian clay-rich rock porosity lies between 14% and 20% at the MHM URL site and is close to 18% at the URL main level [6], and natural water content ranges between 5% and 8%.

2.2 Microstructure

At the URL level, Callovo-Oxfordian claystone could be considered as a clay matrix (between 40 to 60% clay content) with inclusions of carbonate and tectosilicates grains. Those grains are rigid (regarding to clay matrix), not porous and their sizes are in a range of 0.1 to 100μm. As a consequence, most of the porosity is located in the clay matrix [7], and there is a correlation between clay content and porosity at the scale of Callovo-Oxfordian geological layer. Performed investigations to characterize this porosity indicate that pore network is composed mainly of meso/micro pores and shows a very low connectivity for pores over 40 nm [8]. Using various imaging techniques such as electron microscopy or Xray tomography mineral maps have been obtained [9] highlighting a preferential orientation parallel to the bedding of carbonate and tectosilicates inclusions (Figure 1). Nevertheless, orientation of clay particles and aggregates with the bedding plan is not too marked as it could be found in other indurated clays such as Opalinus Clay in Switzerland. This leads to a slight anisotropy in most properties of the rock especially in terms of solute diffusion or water permeability.

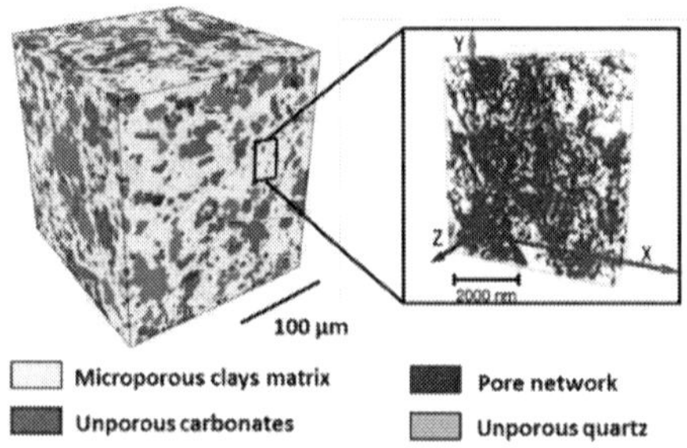

Figure 1. View of the microstruture of COx at different scales (a) clay matrix with carbonate and quartz inclusions [8], (b) pore network obtained by FIB/SEM methods [9].

3 Effect of desaturation/saturation observed on sample

Many authors have studied the effect of drying and wetting cycles of COx claystone samples. Sample are subjected to air with a control Relative Humidity and mass or/and strain changes are measured. Some devices rely on a static atmosphere with saline solution [10-12], other used complex system with a steam generator and a heating channel, in which air is blown through the chamber [13].

Guillon et al [10] have performed static atmosphere drying tests which highlighted the HM coupling and mechanical anisotropy of the samples. Both wetting and drying paths were explored, allowing the potential hysteretic behaviour of the samples. During imbibition, the swelling of the sample is mainly achieved by the interaction of water with swelling clay sheets (illite/smectite interstratified, see [14]). The opening of some cracks can also be responsible for the dilation of the sample.

On the drying path, water is removed from the clay minerals that shrink. As the material progressively dehydrates, water menisci form in the porous space. These menisci strengthen the cohesion between grains and improve the apparent stiffness of the rock [15, 11]. This issue is highlighted by triaxial tests that were performed at the end of the drying tests or in other study. Zhang et al [16] performed micro-indentation and mini-compression tests on the COx claystone samples (11 mm× 22 mm) under different values of relative humidity.

The choice of sample size was justified by the fact that the sample should be large enough with respect to characteristic size of heterogeneities (mineral grains and bedding planes) and reasonably small in order to get uniform distribution of water saturation. They found that the mechanical properties of the claystone are strongly influenced by water saturation and structural anisotropy. The elastic modulus in the normal orientation to bedding planes is smaller than that in the parallel orientation. The deformation at failure state is larger in the perpendicular orientation than that in the parallel one due to progressive compaction of bedding planes. The elastic modulus and failure stress increase with desaturation of material.

At a smaller scale, Wang et al [17] also studied the evolution of the COx claystone under hydric cycles using techniques based on the combination of environmental scanning electron microscope (ESEM) and digital image correlation (DIC) [11], in order to quantify local strain field. The observation is carried out on zones of several hundred micrometers, so the evolution of such material under hydric loadings can be studied at the scale of inclusion-matrix composite. A heterogeneous strain field is evidenced at microscale during wetting. Nonlinear deformation is observed at high relative humidity (RH) which is related not only to damage, but also to the nonlinear swelling of the clay mineral itself, controlled by different local mechanisms depending on relative humidity. This sensitivity to variations in water content leads to enhanced variations of mechanical parameters in response to water saturation.

Among many factors related to desaturation and resaturation process, the suction plays an essential role due to the small size of claystone's pores (Figure 1). At mesoscopic scale, the suction generates not only an overall confining effect but also compaction or swelling of bedding planes. Such effects of water content are related to the modification of microstructure in clayey rocks [7]. Drying and wetting processes modify the distance between clay platelets, leading to modification of the mechanical properties of clay aggregates.

Zhang and Rothfuchs [18] performed creep test on COx claystone. **Figure 2** compares the creep curves of the specimens loaded parallel and perpendicular to the bedding plane. The total deformation perpendicular to the bedding plane is larger than that parallel to the bedding. However, the pure creep strains and the creep rates are not significantly dependent on the loading direction. In the last creep phase at 15 MPa, two initially saturated specimens were dried by exposing them to room air. The desaturation turned off creep rate, and even caused a significant shrinkage of 0.3%, independent of the loading direction to the bedding.

The question is how all those processes observed at sample scale will interact and affect the mechanical behavior of a gallery? Are those processes will just appear at the gallery wall or reach the rock mass deeper? Most of the experiment on desaturation/saturation have been performed on intact rock samples, also in situ at the main level of the MHM URL, Armand et al [4] show that excavation work induced a large fractures network depending of the drift orientation (see Section 4.5)? What will be the effect of desaturation/saturation on the fractured rock at the drift wall? A specific experiment, called Saturation Damaged Zone (SDZ), has been emplaced in 2009 at the main level of the MHM URL to try to better understand what really happened at this scale.

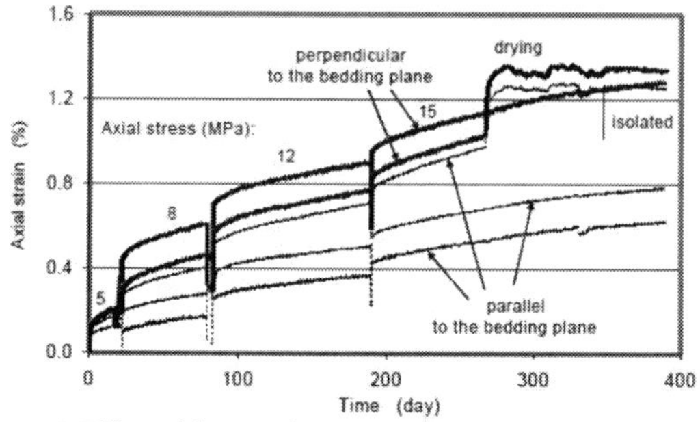

Figure 2. Effect of desaturation on creep test [18].

4 In situ experiment SDZ

4.1 Objectives

The objective of the SDZ experimentation is to study the effect of desaturation and saturation generated by the ventilation of galleries over the excavation damaged zone (EDZ). In this experiment (Figure 3), it is thus issue of:

- Measure changes in water saturation profiles in COx claystone around a drift under controlled ventilation conditions (temperature and relative humidity) over drying / wetting cycles;
- Characterize the EDZ and its development around a drift under controlled ventilation conditions (opening and closing of cracks initially present in COx claystone, appearance of new cracks, cracks connectivity ...);
- Characterize the mechanical evolution of a gallery subject to cycles of drying/wetting (deformation, convergence).

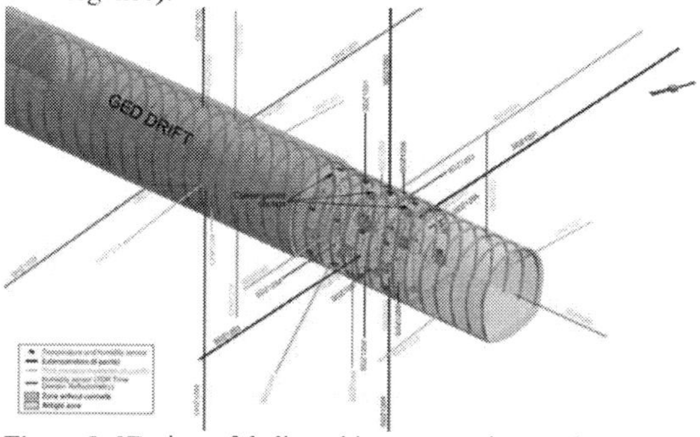

Figure 3. 3D view of dedicated instrumentations to the SDZ experimentation.

4.2 Experimental setup

The area dedicated to the SDZ experimentation (SDZ area) corresponds to the extremity of the GED drift parallel to the minor horizontal stress (σ_h). This area includes an airlock of 2 m long, an experimental area without coating of 7.2 m long (Zone 1) and a sealed area of 5 m long (Zone 2). The positioning of these areas dedicated to SDZ experimentation is shown in Figure 4. The part of the GED drift forward the SDZ area (Zone 1 and Zone 2) is called in the paper Zone 0. The airlock doors are supposed airtight and waterproof.

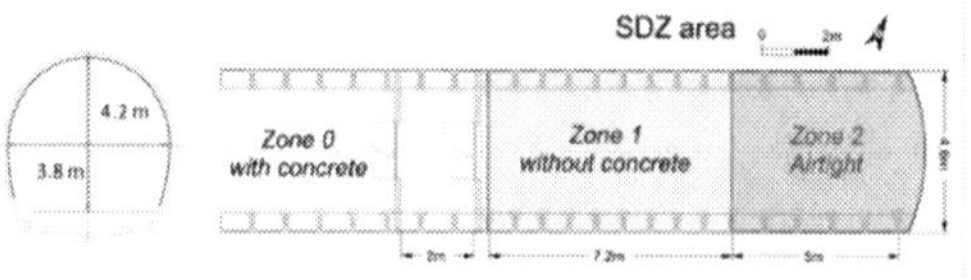

Figure 4. 2D view of SDZ area in the GED drift.

The excavation of the SDZ area took place from 12/01/2008 to 01/23/2009 (2 months). This period includes a stop work between 12/18/2008 and 01/05/2009. The floor of this area was sunk about 2 weeks after the end of the excavation. The ground support of Zones 0, 1 and 2 is ensured by a HA25 radial bolting (resin over their entire length) and sliding arches spaced 1 m in Zone 0 and 0.8 m in the SDZ area. The contact between the arches and the rocks is provided with a cement-filled Bulflex. Compressible slims are used to make the junction between the slab and the side elements of the arches. In the Zones 0 and 2, a wire mesh and fiber shotcrete 20~30 cm thick are set wall. The slab is 43 cm thick concrete. The theoretical final section of the Zones 0, 1 and 2 after excavation is identical, see Figure 4; the excavated diameter is 4.40 m. In Zone 2 assumed waterproof, a geotextile membrane and a heat-sealed plastic entirely cover the concrete. The Zone 1 is not coated and considered the main area of study.

The ventilation system dedicated to the SDZ experimentation was installed between July and August 2009. The regulation of air relative humidity (RH) and temperature is achieved through a cold battery, air heaters and steam humidifier. Several ventilation operation tests were conducted during September and October 2009. Four dates are important in the analysis of this experiment:

• 09/14/2009: closures airlock doors, no ventilation control,
• 09/01/2011: change of ventilation in the GED gallery,
• 10/03/2011: Start control of ventilation at SDZ area (RH = 30%, T = 23°C),
• 07/09/2013: change of relative humidity (RH = 60%, T = 23°C).

4.3 Main instrumentations

To achieve the above-mentioned objectives, several measures were carried out on geological, geomechanical, hydraulic and climatic aims. However, there are two major families of instrumentations: those carried out into the gallery and those carried out in the surrounding rock mass. Figure 3 shows the implementation of both families of instrumentations in the SDZ area.

Into the gallery

• arches sliding measurement by potentiometric displacement sensor,
• apparent cracks evolution measurements by potentiometric displacement sensor,
• hydrogeological test to measure hydraulic conductivities and permeabilities,
• geological survey of the front face and walls during excavation,
• climate sensors (relative humidity, temperature and air velocity),
• convergence-measuring gauge (with studs and rods);

In the surrounding rock mass

• geological survey on cored boreholes and front face (cracks analysis, core water content, mineralogy, calcimetry ...),
• boreholes extensometers (single-point and multipoint) to capture rock displacement,
• boreholes pore pressures and permeability measurements,
• boreholes for TDR measurement of the saturation of the rock mass (water content),
• boreholes equipped with FD probes for measuring volumetric water content.

Thereafter the analysis will focus on the long-term behavior of a controlled ventilation gallery in the MHM URL.

4.4 Ventilation of the SDZ area

The measurements of the temperature and relative humidity at the inlet and at the outlet of Zone 1 are compared to those recorded in the Zone 0 in Figure 5. The four key dates are shown with the curves (dashed lines). This reflects that it is difficult to maintain a strictly constant temperature and relative humidity in the gallery. Seasonal variations are observed on Zone 0 measurements. At the scale of a circular section, a gradient, although small is always observed between ceiling and floor of the gallery.

Before closing airlock doors, many fluctuations were recorded on measurements. After closing the airlock, measured daily variations correspond to the fact that the drift ventilation retransmits external thermal fluctuations meaning a likely connection between the Zone 0 and the SDZ area. The temperature is high because the lighting lamps remained lit in this period. It is also observed on each circular section, of the Zone 1, an increasing thermal gradient from the ground to the roof. This thermal gradient is between 1°C and 1.5°C.

The set of Relative Humidity measurements presents similar behavior over the observation period. Both levels of 30% RH and 60% RH are visible throughout the Zone 1. It is also observed on each circular portion of the Zone 1, a decreasing gradient of moisture from floor to roof. This gradient is of the order of 4% and is also favored by the position of the arrival of moist air. The trend of rising relative humidity corresponds to the gradual re-saturation of air and probably reflects the efficiency of the isolation of the SDZ area from the rest of the GED gallery. Recent years, the ventilation of the SDZ facility loses its efficiency.

 DOI: 10.1051/e3sconf/20160903004

The SDZ area is 12.5 m long and the air flow is not laminar. Despite the position of a sensor inlet and outlet of the cooling unit, it is very difficult to make a precise assessment of the ventilation. Borehole temperature measurements in the first meters of Zone 2 (supposed sealed) teaches us that it is not possible to seal a portion of gallery as there is a connected fractures network.

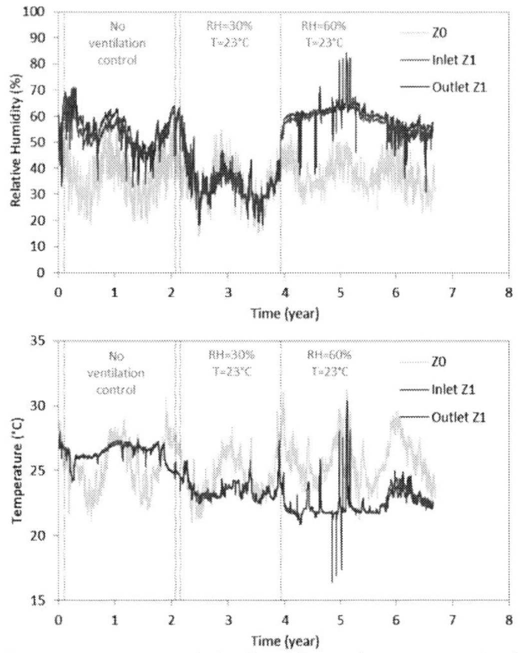

Figure 5. Average relative humidity and temperature in the Zone 0 (Z0) and Zone 1 (inlet Z1 and outlet Z1).

4.5 Excavation induced fracture network

An important characterization investigation of excavation-induced fracture network around openings has been conducted at the main level of Meuse/Haute-Marne URL by structural analysis of the front face and walls, the analysis of cores and resin impregnated cores [19].

Armand et al. [4] described the induced fractures network observed at the main level of the URL. Drift excavations exhibit extensional (mode I) and shear (mode II) fractures that were not observed in drifts at the –445 m level. Spalling is not the prevailing mechanism of failure due to the high-level mean stress compared to the Callovo-Oxfordian claystone strength. Shear failure seems to occur first from the excavation front face. Shear bands have been observed by the resin impregnation method at smaller scale confirming the shear failure mechanism. Armand et al. [20] showed that the fracture pattern and extent depend on the drift orientation versus the state of in situ stress. The maximum extent of fracture zones in different drifts can be summarized as follows:

(1) Drifts parallel to σ_H: the extensional and shear fractures are observed 0.4 and one diameter from the wall, respectively.

(2) At the floor and ceiling, shear and extensional fracture zones are mixed up to 0.15 diameter.

(3) Drifts parallel to σ_h: the extensional and shear fractures are observed up to 0.5 and one diameter respectively, at the roof and the ceiling. At the wall, shear and extensional fracture zones are mixed up to 0.2 diameter.

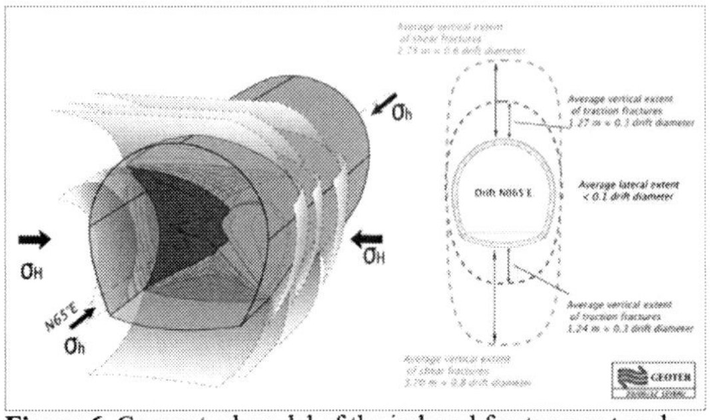

Figure 6. Conceptual model of the induced fractures network around a drift parallel to the horizontal major stress [4].

Figure 6 illustrates the model proposed to represent the fracture networks for drifts parallel to the minor horizontal principal stress, and Table 1 gives the extent of the different types of fracture zones for both orientations. Geological survey on cored borehole in the SDZ area confirms that the excavation induced fractures network follow the model given by Armand et al [4].

The area with unloading fractures (mode I) exhibits the highest hydraulic conductivity up to 10^{-10} m/s. Those high values are due to the fracture transmissivity and not the permeability of the rock matrix. Farther from the wall, the shear fractures exhibit low transmissivity, which did not affect considerably the average hydraulic conductivity. We can distinguish two zones: one with extensional and shear fractures which form a connected fractures network with potentially high transmissivity; one deeper from the gallery wall, where just remains some shear fractures or any (depending gallery orientation) with a slight decreasing hydraulic conductivity gradient to the pristine rock mass.

Table 1. Extent of shear and extensional fractures around drift at the main level of the URL.

Drift Orientation		Extensional fractures extent (*D*)			Shear fractures extent (*D*)		
		Min.	Ave.	Max.	Min.	Ave.	Max.
N65 // σ_h	C	0.2	0.3	0.4	0.5	0.6	0.8
	W	0.1	0.1	0.2	-	-	-
	F	0.2	0.4	0.5	0.8	0.8	1.1
N155 // σ_H	C	-	0.1	0.15	-	-	-
	W	0.01	0.2	0.4	0.7	0.8	1.0
	F	-	0.1	0.15	-	-	-

Note: D: is the drift's diameter;
C: ceiling;
W: wall;
F: floor;
'-' means no extent further the extensional fractures.

The extent of the fractured zone has been followed during time over a period of nearly 10 years for the older gallery at the MHM URL. No significant change has been noticed yet.

5 Main results

5.1 Convergence and deformation of the rock mass

In this section, an analysis of rock deformation and support loading is made on the basis of convergence rods and measurement sections (SMR) constituted of 8 boreholes extensometers including 4 single-point (1.2 m) and 4 multi-points (7 anchors spread over 20 and 30 m). To evaluate the effect of ventilation and coating, a comparison is made between SMR measurements at Zone 0 and that of SDZ area (Zone 1). Those two SMR are distant of 13 m. Single-point and multipoint extensometers are positioned at the same inter arch. The measurements presented are recorded between the head of the extensometer (considered the reference point) and anchors (7). The positive displacement represents an extension (convergence of the walls). The SMR of Zone 0 was set during excavation (1.5 m from the front) and SMR of Zone 1 have been put in place 149 days after the excavation. Arbitrarily, these measurements are compared from a time t = 0 corresponding to a time tracking commencing 450 days after the excavation of the concerned zone; either from 01/25/2010 for Zone 0 and from 04/15/2010 for Zone 1. This means that short term convergence or displacements due to the excavation are not included in this study.

In the SDZ area (Zone 1); convergence rods were installed in pairs (vertical and horizontal). Figure 7 shows the measurements obtained by the convergence of rods at 1.2 and 3.6 m door of the airlock (Zone 1). The most important displacements and convergence speeds are measured on the vertical rods. Note that, convergence speeds are equivalent in Zone 0 and Zone 1. The convergence report (C_v/C_h) measured on all sections of the GED gallery is greater than 4.

A comparison of the displacements measured in the first meter by single-point and multipoint extensometers is done on Figure 8. Positive displacements represent convergence and negative displacements represent shrinkage. In this case (small strain), shrinkage could depict the effect of rock desaturation. It is observed that changes in relative humidity directly affect the rock mass in a field close to the wall. Last Relative Humidity change contributed to accelerating the strain rates. Extension areas are localized in roof and floor. The magnitude of the vertical displacement is thus the most important.

Thereafter, a comparison is done between the multipoint extensometers of the two SMR in Zone 0 and Zone 1. For this purpose, it represents what happens in horizontal (Figure 9) and vertical (Figure 10) borehole.

It may be noted that:

• The increase in relative humidity has contributed to increase the strain rate in in the first meter of the wall (Figure 7, Figure 8) meaning that, creep of COx claystone is dependent of relative humidity. This effect is not preponderant beyond the first meter on vertical multipoint extensometers (Figure 9), certainly given the magnitude of the measured deformations.

• Conversely, climate variations in the SDZ area have no effect on the arch sliding measurements since the postponement of the rock mass applied stresses on the arches remain still too important. The change in the boundary condition of relative humidity at drift wall has negligible impact on the drift convergence.

• The type of coating (shotcrete 20-30 cm thick or any) does not present a significant impact on the hygrometric condition at drift wall. Sprayed concrete does not play a structural effect in this case; it serves to prevent the falling blocks.

• The relative displacements among the different anchors beyond 5 m depth are very low (negligible).

These observations are insufficient to determine, at this stage of the experiment, if the re-saturation of the rock mass generated swelling deformation under in situ confinement.

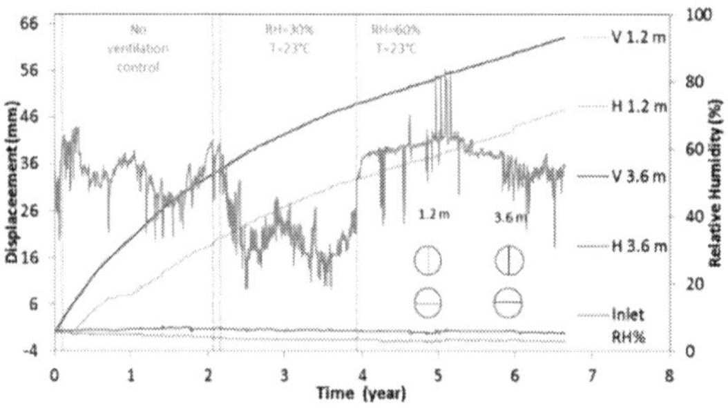

Figure 7. Inlet RH and convergence measurements in zone 1 of the area SDZ (Zone 1).

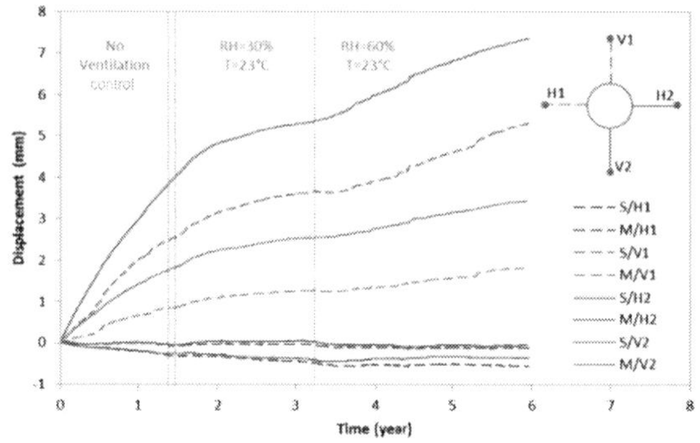

Figure 8. Displacement measurements comparison between single-point (S) and multipoint (M) borehole extensometers in the SDZ area (Zone 1).

DOI: 10.1051/e3sconf/20160903004

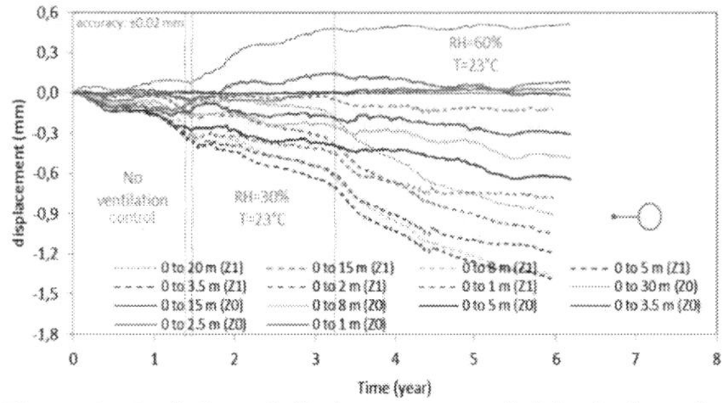

Figure 9. Evolution of displacement recorded by horizontal multipoint extensometers in Zone 0 and Zone 1, starting 450 days after excavation.

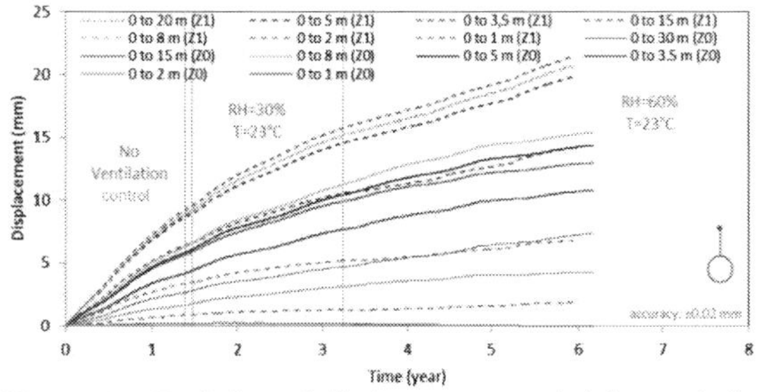

Figure 10. Evolution of displacement recorded by vertical multipoint extensometers in Zone 0 and Zone 1, starting 450 days after excavation.

5.2 Pore pressure and permeability measurements

In this section, the evolution of pore pressure is assessed at 3 m and 6 m (Figure 11) from drift wall through horizontal and vertical borehole. Pore pressure at this level (-490 m) is around 47 bar [20]. These figures show that the temperature fluctuations have a direct impact on pore pressures changes in the rock mass and thus its volumetric strain. Observed peaks on Figure 11 represent hydraulic tests to determine permeability. In long-term, no difference is observed at 3 m from drift wall on measurements between Zone 1 and Zone 2 (sealed area or not) on pore pressure measurements. This means that there is no difference between boundary conditions (bare / coated / waterproof wall). The sealing wall does not preclude the connectivity of the fracture network in the side wall. Ahead of the front, pore pressures at 3 m and 6 m remain high despite the extensive fracturing certainly because of the increase in vertical stress due to excavation. Pore pressures continue to decline over time (diffusive phenomenon).

Permeability measurements were performed (between 2009 and 2014) through gas and/or hydraulic tests. The aims were to assess the EDZ hydraulic conductivity with time and the degree of saturation of COx claystone [21, 22]. The results of these campaigns show that the permeability of the rock into the damaged area does not evolve much post excavation (Figure 12). Note that clay formations in their natural state generally have a very low hydraulic conductivity ($\sim 10^{-13}$ m/s).

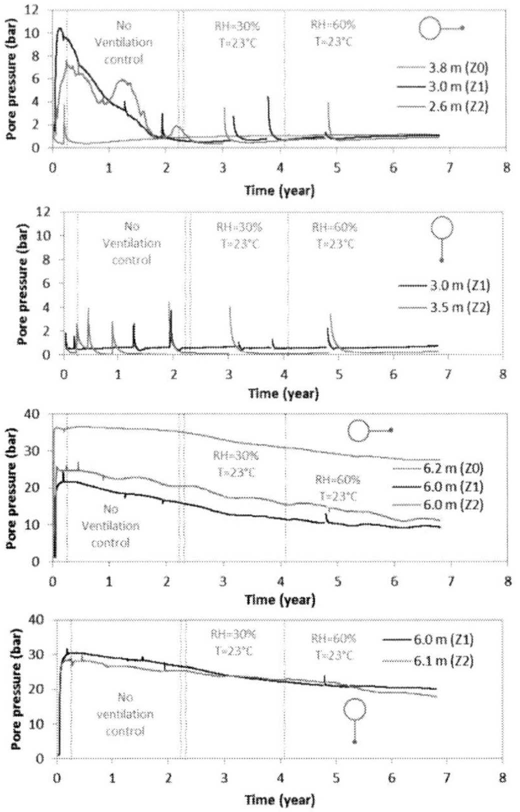

Figure 11. Evolution of borehole pore pressure measurements at 3 m and 6 m deep in the Zone 0, 1 and 2.

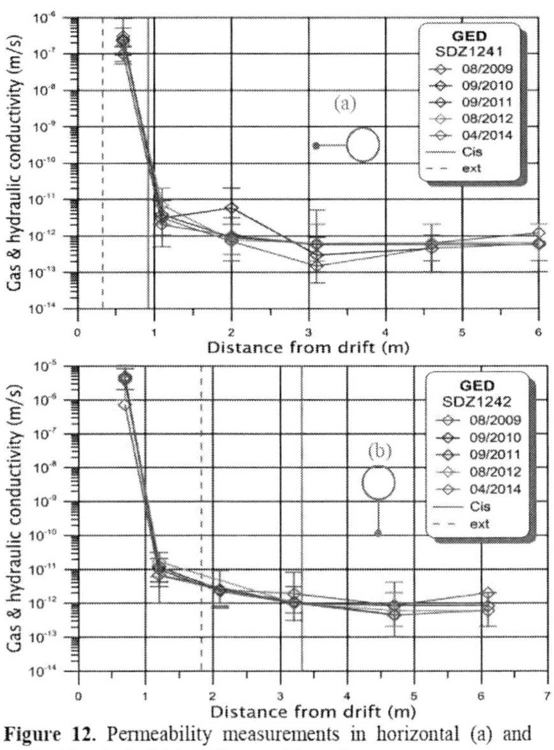

Figure 12. Permeability measurements in horizontal (a) and vertical borehole (b) in SDZ area (Zone 1).

For all boreholes, permeability values in excess of 3.0 m away from the gallery are characteristic of undisturbed COx claystone, or micro-cracks close to the limit with a undisturbed COx claystone. Permeability values characteristic of a micro-fractured COx claystone were observed between 1 and 3 m from wall. Permeability values obtained to the nearest interval of the gallery (~0.6 m) are characteristic of an unsaturated fractured zone (tire testing). Saturated intervals closest to the gallery mostly indicate pressures near or below atmospheric pressure, and the high specific storage coefficients.

5.3 Water content and saturation of the rock mass

In this part, rock mass saturation profile is evaluated in the first meters from the wall through horizontal and vertical boreholes. For this purpose, gravimetric (Figure 14-a) and boreholes water content measurements (Figure 13 and Figure 15) are carried out.

In Figure 14-a, an increase in gravimetric water content measurement is observed in the first meters from the wall, subsequently the water content is between 7 and 8% over 2 and 7 m. Geological survey were carried out between 2008 and 2015 on annual basis. The change in mineralogy at the cores sample scale justifies this value range. This measure is similar to that measured in other galleries of the underground laboratory. An extension of the low water content area with time is not observed. The changes in relative humidity do not affect the rock saturation measurements in the near field of the wall.

The TDR (Time Domain Reflectometry) packer system shown in Figure 13 is composed of a dual shutter completion which defines six measurement points of the rock mass water content. The determination of the water content using TDR is based on the principle that the relative dielectric number of water is much larger than that of other rock constituents (solid, air). The dielectric number can be measured by determining the propagation velocity of an electromagnetic wave along uncoated metallic waveguides. Figure 14-b highlights a desaturation of the wall even before the installation of the controlled ventilation facility in the SDZ area. One of the main difficulties that may arise from using the TDR method is to ensure optimal contact between the probe and the measuring medium (rock mass). The results of these sensors are difficult to interpret after a year of measurement. It seems that complete desaturation of the rock/sensor interface (opening of the interface) lasted about 2 years or not proper functioning. This shows that it is very difficult to make a continuous measurement of the rock mass water content. Desaturation is related to the fracture zone due to the extension. So, the reliable measurement between the two methods is gravimetric.

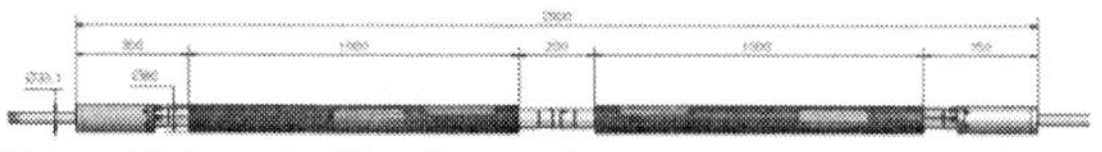

Figure 13. Layout of the TDR packer system

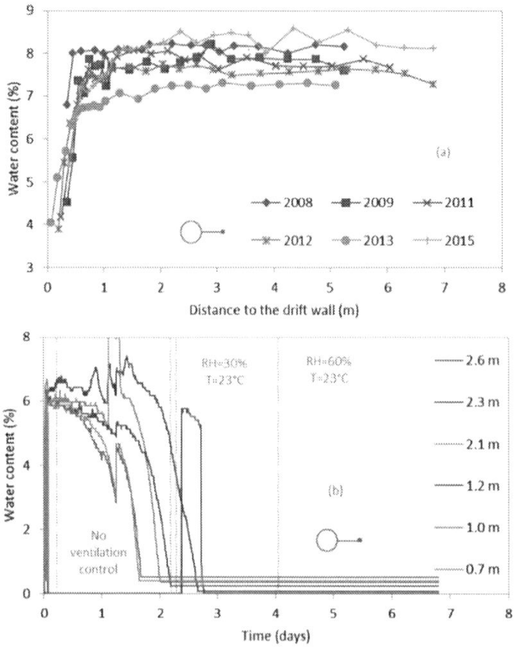

Figure 14. Gravimetric (a) and borehole (b) water content measurements in the SDZ area (Zone 1).

The volumetric water content is measured through FD probe by propagation over 100 mm of an electromagnetic field. FD probe is equipped with a completion, defining six measurement points in the surrounding rock mass (Figure 15). Results are given in Figure 16-a for a horizontal borehole and in Figure 16-b for a vertical borehole. This shows that, the volumetric water content measurements are not sensitive to relative humidity changes at drift wall. This observation is consistent with the gravimetric water content measurement shown on Figure 13-a.

Figure 15. Layout of FD probe (L=1,35 m, Φ=25 mm, 950 g).

DOI: 10.1051/e3sconf/20160903004

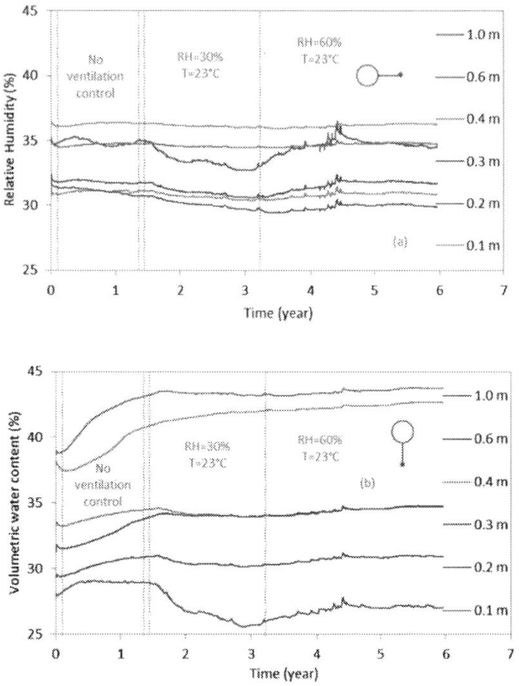

Figure 16. Volumetric water content changes in the rock in SDZ area (Zone 1).

5.4 Selected cracks evolution at drift wall

Crack development measurements are carried out by potentiometric displacement sensors also called fissurometers. These sensors are fixed on two ankles anchored on either side of the selected crack. Followed crack is chosen opportunistically. 2 types of cracks are selected: fractures generated by shearing (S) and extension (E). Figure 17 shows the data acquired from fissurometers since their installation. Positive displacements reflect the closing of the studied crack. It is observed that most fissurometers appears to have identical developments trends: decreasing slightly regardless of the type of cracks and, a very small opening. The closing rate observed at 30% RH is greater than that observed at 60% RH (very low). Most cracks therefore have almost stopped evolving since the last change of relative humidity. Only one tensile fracture continues to close.

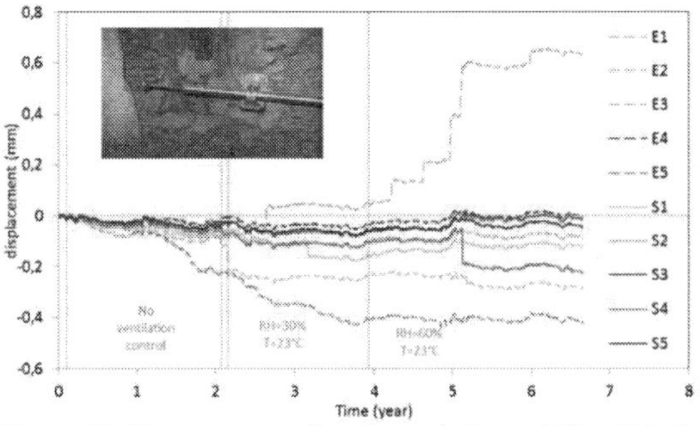

Figure 17. Measurements of cracks evolution at drift wall in the SDZ area (Zone 1).

6 Discussions

From experimental point of view, desaturation experiment on COx claystone at gallery scale remains challenging. In fact, desaturation or water content measurements are difficult in such rigid and low permeable media. The online measurements (TDR, volumetric water content) are not working properly and even failed. The TDR system was based on a packer system in order to get a good contact between the needles and the rock. At the beginning it seems to work but on larger time desaturation occurs or the system downward slide or failed. For the volumetric water content, evolution is somehow coherent with gravimetric water content evolution, but the initial gradient in the rock from the gallery wall is strange and need further analysis. Capacitive and psychrometric measurement hygrometers have been installed in cross section of the Ventilation Experiment (VE) at Mont Terri [23] and performed successfully. At MHM URL, doubt on those values remains. Due to the excavation induced fractures network, it could be possible that the highest water content is not the deepest in the rock wall due to the presence of connected fractures. Geoelectric measurements have been although used in VE experiment at Mont Terri (Mayor et al, 2007), but could not be used easily in the context of SDZ experiment due to the presence of metallic rock bolt necessary for the gallery support. The fact, that saturation degree in field remains uncertain to measure continuously in the gallery wall, makes interpretation of desaturation cycles more difficult. Better systems have to be developed to have more reliable data.

On the side wall, water content gradient is observed in the first meters from the wall and have not evolve since the excavation of the gallery despite low RH in the gallery air. The extent of the lower water content matches with the extent of the induced fracture zone with shear and extensional fractures. Nevertheless, a suction state develops up to a larger distance about 3 m from the wall for horizontal boreholes.

As the wall is highly fractured during the excavation, there is no evidence of desiccation cracks outbreak during the drying period as seen at Tournemire [3]. A slight opening of the excavation induced fractures is measured over 1.5 years under RH equal to 30 % and the increase of RH in the gallery air seems to interrupt this opening. In the surrounding rock mass the increase in RH has contributed to increase the strain rate in in the first meter of the wall, but deeper in rock mass no effect was observed. In the first meter, creep of COx claystone is dependent of relative humidity as demonstrate by Zhang & Rothfuchs [18] on samples.

Strain, permeability and pore pressure measurements give the same insight of a low impact of the desaturation included in the excavation induced fractures. Those results are in accordance with what has been observed during the VE experiment at Mont Terri even if the excavation induced fractures network is less developed at Mont Terri and nearly not existing around VE gallery. Mayor et al. [23] show that only very small shrinkages of the rock have been detected during the desaturation period. Moreover they conclude that the thermal and hydraulic rock characteristics will not be practically affected by the ventilation except in a narrow ring around the drift (less than 30 cm for a gallery of 2.4 m diameter), where the degree of rock saturation can be lower than 98 %.

The excavation induced fractures are a major issue to understand desaturation around the GED. de La Vaissiere et al [24] have performed gas circulation tests at the wall of a drift parallel to the major stress (in this situation the excavation induced fractures network is mainly develop at the wall at the MHM URL, see section 4.5). These tests showed that fractures may be connected to one another and allow gas circulation only in the zone where extensional fractures are observed. We can assume that it is the case also along the GED gallery. Vinsot et al [25] tracked oxidation front defined as the set of deepest points away from the drift walls where oxidation reactions have occurred. The oxidized features encountered mainly consisted of weathered pyrite minerals associated with gypsum, as well as iron oxides and hydroxides and are analyzed on drill core. Finally, over 6.5 years, they found that the oxidation front has propagated deeper into the fractured zone and has developed in the shear zone. Nevertheless, all the oxidized features were encountered at less than 1.8 m from the drift walls, and none were found within the deepest part of the shear zone over the maximum lag time of the study (6.5 years). The evolution of the oxidation front over time implies that gas diffusion was facilitated in the shear fracture when compared to the pristine rock. Those chemical results underline the role of the fracture network which has to be considered in modelling in order to predict the gallery behavior during and after excavation.

Pardoen et al [26] shows that the fractures induced by excavation process in the gallery vicinity could be reproduced by shear banding, with strain localisation properly reproduced by means of a regularisation method. Taking into account intrinsic permeability evolution in the shear band conduct with localisation methods seems to be promising to better reproduced the behavior of GED gallery, with its anisotropy, its long-term evolution and the influence of desaturation.

7 Conclusions and perspectives

An in situ experiment, called SDZ, has been implemented at the main level of the MHM URL in order to characterize the influence of controlled ventilation in an experimental drift. The aim is to analyse the hydro-mechanical behaviour of the rock mass and the excavation damaged zone evolution. The desaturation process, the permeability and the deformation of the surrounding clay rock are most particularly studied during the tests. Relative humidity (of 30% then 60%) has been applied over a period of more than 4 years.

Effect of ventilation with dry air remains small and concentrated in the vicinity of the drift in the zone with shear and extensional fractures. The extent of the lower water content matches with the extent of the extensional fractures. Unfortunately, some of the online water content measurements failed or were uncertain which increases the difficulty of following of saturation field. Better systems have to be developed to have more reliable data. No evidence of desiccation cracks outbreak has been observed, but excavation induced fractures which are widespread at the drift wall show small opening during the driest phase. The opening tendency decreases with RH increase.

The experiment already provided inputs over 7 years of drying phase and will continue with an increase of relative humidity up to 90% to see if swelling occurs and the impact in the connected excavation induced fracture network. This increase of HR will represent the closing of the drift in the nuclear disposal with ventilation stop. In parallel, effort on modelling will be done through benchmark exercise in order to improve constitutive modelling of COx claysone behavior in unsaturated state.

References

1. J. Delay, A. Vinsot, J.-M. Krieguer, H. Rebours, G. Armand, Phys. Chem. Earth J., **32** (1/7), 2-18, (2007).
2. J.-M. Matray, S. Savoye, J. Cabrera, Eng. Geol. ,**90**, 1-16, (2007).
3. S. Hedan, A.-L. Fauchille, V. Valle, J. Cabrera, P. Cosenza, Int. J. Rock Mech. Min. Sci. **68**, 22-35, (2014).
4. G. Armand, F. Leveau, C. Nussbaum, R. de La Vaissiere, A. Noiret, D. Jaeggi, P. Landrein, C. Righini, Rock Mech. Rock Eng. **47**, Issue 1, 21-41 (2014).
5. C. Lerouge, S. Grangeon, E.C. Gaucher, C. Tournassat, P. Agrinier, C. Guerrot, D. Widory, C. Fléhoc, G. Wille, C. Ramboz, A. Vinsot, S. Buschaert,. Geochim. Cosmochim. Acta, 75, 2633-2663, (2011).
6. B. Yven, S. Sammartino, Y. Geraud, F. Homand, F. Villieras, Mémoires de la Société géologique de France **178**, 73-90, (2007).
7. J.-C. Robinet, P. Sardini, M. Siitari-Kauppi, D. Prêt, B. Yven, Sediment. Geol. **321**, 1-10, (2015).
8. J.-C. Robinet, P. Sardini, D. Coelho, J.-C. Parneix, D. Prêt, S. Sammartino, E. Boller, S. Altmann, Water Resour. Res. **48**, (2012).
9. Y. Song, C. Davy, D. Troadec, A.-M. Blanchenet, F. Skoczylas, J. Talandier, J.-C. Robinet, Mar. Petrol. Geol. **65**, 63-82, (2015).
10. T. Guillon, R. Giot, A. Giraud, G. Armand, Acta Geotech. **7**, 313-332, (2012).
11. M. Bornert, F. Valès, H. Gharbi, D. Nguyen Minh, Strain **46**, 33-46, (2010).
12. Q.T. Pham, F. Valès, L. Malinsky, D. Nguyen Minh, H. Gharbi, Phys. Chem. Earth J. **32**, 646-655, (2007).
13. P. Gerard, A. Leonard, J.P. Masekanya, R. Charlier, F. Collin, Int. J. Numer. Anal. Meth. Geomech. **34**, 1297-1320, (2010).
14. H.G. Montes, J. Duplay, L. Martinez, S. Escoffier, D. Rousset, Appl. Clay Sci. **25**, 187-194, (2004).
15. L. Dormieux, J. Sanahuja, S. Maghous, C R Mecanique 334:19-24, (2006).
16. F. Zhang, S.Y. Xie, D.W. Hu, J.F. Shao, B. Gatmiri, Appl. Clay Sci. **69**, 79-86, (2012).

17. L. Wang, M. Bornert, S. Chanchole, Poromechanics V, 1635-1643 (2013).
18. C.L. Zhang, T. Rothfuchs, Appl. Clay Sci., **26** (1-4), 325-336, (2004).
19. G. Armand, Y. Wileveau, J. Morel, M. Cruchaudet, H. Rebours, *Proc. 11th Congress of the International Society for Rock Mechanics*, Riberiro e Sousa, Olalla&Grossmann (eds), 33-36, (2007).
20. G. Armand, A. Noiret, J. Zghondi, D.M. Seyedi, Rock Mech. and Geotech. Eng. J. **5**, 221–230, (2013).
21. P. Bossart, P.M. Meier, A. Moeri, T. Trick, J.-C. Mayor, Eng. Geol., **66**, 19-38, (2002).
22. T. Fierz, M. Piedevache, J. Delay, G. Armand, J. Morel. *Proc. 7th International Symposium on Field Measurements in Geomechanics*, (2007).
23. J.C. Mayor, M. Velasco, J.L. García-Siñeriz, Phys Chem Earth J., **32**(8-14):616-628, (2007)
24. R. de La Vaissière, G. Armand, J. Talandier, J. Hydrol. 521, 141–156 (2015).
25. A. Vinsot, F. Leveau, A. Bouchet, A. Arnould, Clays in Natural and Engineered Barriers for radioactive Waste Confinement, **400**, 37, (2014).
26. B. Pardoen, S. Levasseur, F. Collin, Rock Mech. Rock Eng. **48**, 2, 691-714, (2015).

Cracking in desiccating soils

Alberto Ledesma[1a]

[1] *Universitat Politècnica de Catalunya – UPC-BarcelonaTech , Barcelona, Spain*

Abstract. Soil shrinkage is produced typically under desiccating conditions. Eventually shrinkage may generate cracks in the soil mass, a phenomenon that is being studied by several researchers, because its prediction is far from being a routine in Soil Mechanics. Within this context, Unsaturated Soil Mechanics provides a promising framework to understand the mechanisms involved. In addition to that, physical modelling of desiccating soils constitutes a good tool to explore the nature of this problem. In this paper, a description of the processes involved in desiccating soils is presented first. There is a debate in the Soil Mechanics community about the conditions leading to the initiation of a crack and some of those issues are addressed in the paper. Then, two physical modelling strategies to study desiccating soils are briefly described: a laboratory environmental chamber and a field controlled test. The chamber was able to reproduce drying and wetting cycles by controlling the relative humidity of the environment that affected the moisture conditions of a soil tray. Temperature was a controlled variable as well. The second experiment refers to a field test, exposing a large soil tray to standard environmental actions in the field, that is, cycles of temperature, solar radiation, relative humidity, rain, wind, etc. In this case, actions cannot be controlled, but they can be measured. An effort was devoted to measure internal variables in the soil mass, mainly temperature, suction and water content. Additionally changes of weight were recorded as well. In essence the paper gives a general overview of the mechanisms involved in soil cracking and the physical modelling procedures employed to analyse this phenomenon in practice.

1 Introduction

Shrinkage of soils due to desiccation is a common process in natural dry environments. Eventually, that shrinkage may induce cracking and soil becomes degraded loosing strength, bearing capacity, and increasing permeability. In that context, volume changes (shrinkage) and cracking are phenomena related to the interaction between soil and atmosphere ([1]-[4]). Cracking on the ground surface due to desiccation has been reported quite often, including some giant cracks described by Harris [5]. Recently, cracks observed on Mars surface have been attributed to desiccation processes as well [6].

The analysis of cracking in desiccating soils is a topic with increasing interest in the geotechnical community. Preliminary publications are quite isolated (i.e. [7]), but the publications have increased significantly after the 90's ([8], [9]). Physical modelling was considered by many researchers a convenient tool to understand the processes involved in soil cracking due to desiccation. Laboratory experiments included not only the observation of drying samples, but also the development of environmental controlled chambers to observe the pattern of cracking ([10]-[13]). In recent years, numerical simulations of soil cracking have been presented using different strategies involving finite elements with interfaces or discrete elements ([14]-[16]). The topic, initially covered by soil scientists, is becoming more popular among geotechnical experts and was included in the list of themes in the 2013 Geo-Congress organized by ASCE in San Diego [17].

This paper presents first an analysis of the problem by using an Unsaturated Soil Mechanics approach. This is particularly useful despite the fact that cracks initiate when soil is still saturated. However water pressures in the soil are at that moment negative and therefore suction must be taken into account as most of the soil properties are suction dependent. Some basic laboratory experiments are also described. The second part of the paper includes the description of two physical models that have been used in this case: an environmental chamber, controlling relative humidity and temperature over a monitored soil tray, and a field experiment where a large soil container is subjected to atmospheric actions, i.e., solar radiation, temperature, wind, relative humidity changes, rain, etc. Both tests are important to understand the mechanisms involved in the initiation and propagation of the cracks. The field test has been recently set up and it is described in detail in another paper in this Conference [18]. However, here it is presented as a complement to the laboratory environmental chamber.

[a] Corresponding author: alberto.ledesma@upc.edu

DOI: 10.1051/e3sconf/20160903005

Both experiments are compared and the advantages and disadvantages of each test are briefly discussed.

2 Soil drying

When a soil sample is continuously dried in the laboratory, a reduction of volume (shrinkage) is observed up to a point after which the total volume remains constant. The process is related to the evaporation of pore water and that explains the reduction of volume. A typical shrinkage curve is depicted in figure 1, representing total volume vs water content, starting at point A and drying the sample to points B and finally D. The Atterberg limits have been represented in the horizontal line, from liquid limit (w_L) to plastic limit (w_P) and shrinkage limit (w_S). Line C-D represents the path where total volume is constant and it is assumed that air enters the soil pores. A classical convention in Soil Mechanics considers that soil is unsaturated when water content is below shrinkage limit [19]. Also, between plastic and shrinkage limits, soil is assumed to be saturated, but with water in tension. That may not be precise in some cases, as liquid and plastic limits are defined in an empirical manner, but it is convenient in order to have a framework to understand this drying process. Real experiments provide curved paths, instead of straight lines A-B-C-D. In fact, gradient of line A-C is similar for most soils, as it depends on the specific weight of solid particles and water only. If soil is saturated, the total volume, V_T, is obtained by adding the volume of solid, V_S, and water, V_w:

$$V_T = V_W + V_S = W_S\left[\frac{w}{\gamma_w} + \frac{1}{\gamma_S}\right] \quad (1)$$

where W_S is the weight of solid particles, w is the gravimetric water content, γ_w is the specific unit weight of water and γ_S the specific unit weight of solid particles. Thus for a particular value of solid particles weight, gradient of line A-B is well defined, as shown in figure 2 ([20]). That gradient becomes 1 (that is, 45°), if the volume per 100 g of dry soil is represented in the vertical axis.

It is possible to extend equation (1) for unsaturated conditions and straight lines are obtained depending on the degree of saturation. In that figure, gradients corresponding to unsaturated conditions (S<100%, being S degree of saturation) are depicted as well. This is an indication that curved paths after B suggest a desaturation of the soil sample, so deviation from line A-B indicates air entry.

Although gradient of line A-B (figures 1 and 2) is similar for most soils, the volume change when drying a sample from the liquid limit to the plastic limit is very much soil dependent. High Plasticity Index (PI) values (PI = w_L - w_P where w_L is the liquid limit and w_P is the plastic limit) results in a large volume change when drying and therefore, a tendency to crack because of inhomogeneous deformations. This is consistent with the observations: desiccation cracks are more frequent on high plasticity clayey deposits.

3 Crack initiation

Desiccating a sample in the laboratory as indicated in figure 1 may not produce any crack. If the strain field is homogeneous, a global shrinkage without cracks will develop. However, a crack should be expected to appear when the strain field is not homogeneous. Cracking seems to depend on boundary conditions. Nevertheless, cracks may initiate where there is a defect in the soil mass. There are evidences of both situations reported in the literature. In fact, cracks are consequence of heterogeneity either in the strain field or in the soil mass.

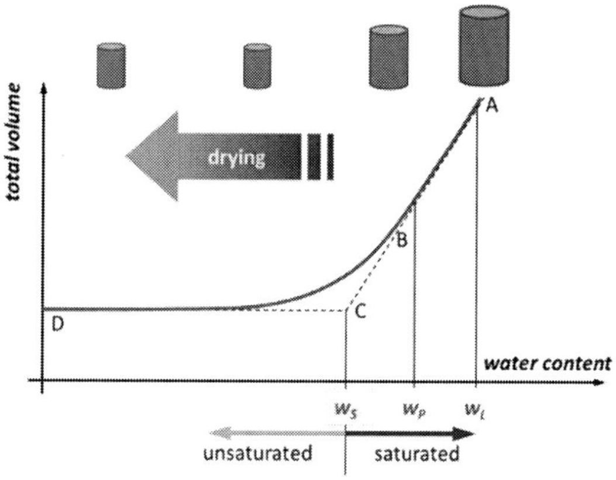

Figure 1. Typical soil shrinkage curve: total volume vs water content.

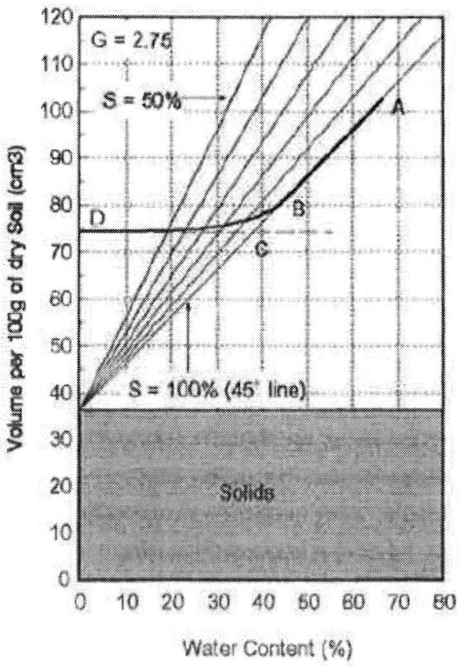

Figure 2. Quantitative shrinkage curve for a soil (from Marinho, [20])

 DOI: 10.1051/e3sconf/20160903005

On the one hand, Shin and Santamarina ([21]) suggested that cracks initiate at the bottom of the defects. They performed experiments by desiccating soil samples (slurry montmorillonite) in plastic containers and identified the initiation of cracks with heterogeneities or defects in the soil mass. Defects are random and difficult to predict in practice. Because of that, when attempting to simulate this process numerically, some researchers have assigned random material properties to each soil element [22]. In this manner, cracks developed first in those elements where tensile strength was lower.

On the other hand, Ávila et al ([23]) showed that in some cases cracks can be predicted as they depend directly on the stress and strain fields. They carried out desiccating experiments with Bogotá clay, CH (Soil Universal classification), w_L=62%, w_P=30% and being the main minerals Kaolinite (55%), Smectite (17%) and Illite (15%). Several soil "I-shaped" trays were prepared by using that clay desiccated under laboratory conditions. Figure 3 presents pictures of 3 experiments starting at different water contents and after different desiccation times. Note that cracks were produced always in the same areas, that is, location of cracks could be reasonably predicted due to the restraint of boundaries in the shrinkage of the samples.

Therefore, the experiments show that cracks are related to heterogeneities either in the soil properties or in the stresses. If the material is homogeneous, then stress/strain field becomes crucial and boundary conditions may play a fundamental role. However, if there is any heterogeneity in the soil mass, defects will be the ideal place for crack initiation. When performing experiments in the laboratory it is important to prepare a homogeneous soil mass (starting from a liquid consistency facilitates that task), in order to minimise the defects within the soil mass.

In a laboratory environment, the initiation of the cracks seems to be controlled by hydro-mechanical conditions, assuming that temperature is constant. Experiments indicate that cracks initiate when the soil is still under saturated conditions, although suction may be developed. That situation corresponds to the zone in the water retention curve close to 100% of saturation, as shown in Figure 4, corresponding to the clay used in the experiments reported in [23]. Therefore, suction plays a role close to the crack initiation, and Unsaturated Soil Mechanics seems to be a convenient approach to analyse this particular problem.

Most recent works dealing with cracking in drying soils consider the tensile strength as failure criterion to define whether a crack initiates or not in a particular point of the soil mass ([10], [15], [16], [22]). This failure criterion, however, has been criticised as too simplistic ([21]). There is some evidence that other mechanisms are involved in the initiation of the crack. In fact when suction appears, soil is compressed in all directions and therefore, when tensile stresses develop in one direction, some compression due to suction is acting in other directions and a shear failure component should be expected [23]. That makes the stress state three dimensional and complicates the analysis.

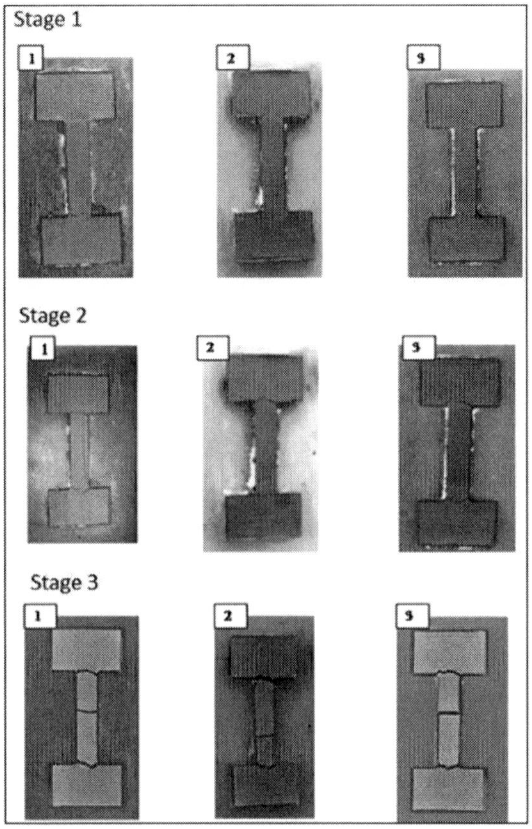

Figure 3. Desiccation of 3 "I-shaped" soil trays starting from different initial water content conditions. First row: pictures after 4 hours, second row: after 4 hours 35 min, third row: after 22 hours. [23]

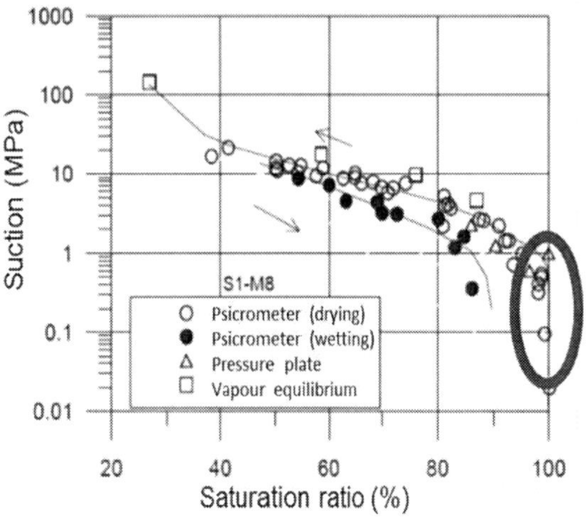

Figure 4. Soil water characteristic curve for the Bogotá clay used in previous experiments. Curves measured in drying and wetting paths using different techniques [23]. The zone marked with a blue thick ellipse corresponds to the conditions where cracking initiation may be expected: close to saturation and low suction values.

DOI: 10.1051/e3sconf/20160903005

Tensile strength criterion can be used for the sake of simplicity if suction effects are taken into account to some extent. Indeed tensile strength depends on water content and therefore, on suction, as shown in Figure 5, from [23]. Increasing water content decreases tensile strength, approaching a very small value when reaching liquid consistency.

A hydro-mechanical (HM) analysis of the desiccation problem is feasible and can be carried out by considering the equilibrium equation, the solid mass balance equation and the water mass balance equation considering unsaturated conditions. Air balance equation is not required if air pressure is assumed to be constant in the soil mass. In addition to that, some further ingredients are required to simulate drying: generalised Darcy's law, unsaturated permeability depending on degree of saturation, water retention curve and tensile strength depending on suction. While the soil is a continuous medium, the conventional coupled HM formulations used for porous media may apply. Therefore, soil drying up to the crack initiation can be simulated numerically by using standard finite element codes solving HM equations in unsaturated conditions [10].

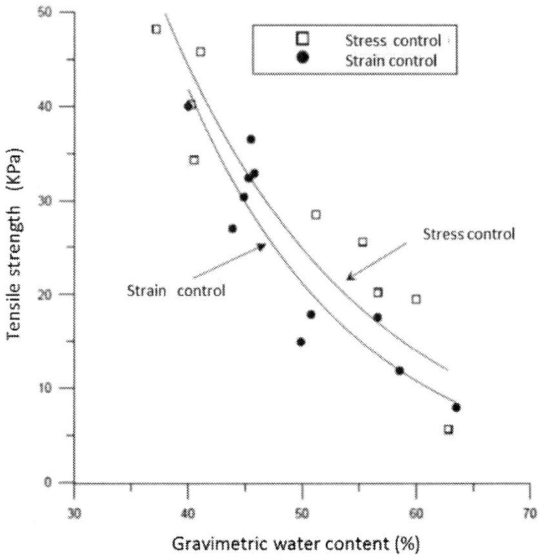

Figure 5. Tensile strength as a function of gravimetric water content for Bogotá clay, measured in stress controlled and strain controlled tests [23].

4 Crack propagation

Propagation of cracks can be simulated by extending the previous coupled HM formulations to take into account crack opening when tensile stress reaches the tensile strength. However, additional difficulties arise, as cracks may constitute a new boundary where evaporation can take place. Figure 6 presents the results of simulating the drying process of a soil tray by using an HM formulation [24]. Soil separates from the tray wall, so soil nodes are released when tensile stress reaches the tensile strength and the suction boundary condition is extended to the released nodes.

It seems appropriate to consider that propagation of cracks is a phenomenon involving release of energy and therefore, a fracture mechanics approach may be convenient to explain that behaviour. There are experimental evidences of that relationship between crack propagation and fracture mechanics including the confirmation of size effect laws ([25]). Linear Elastic Fracture Mechanics was initially used as theoretical framework, but soon, non-elastic theories were proposed, as the cohesive-crack method [26]. In this context, some attempts to measure fracture mechanics properties of soil samples have been reported [27]. However, still most of the works simulating numerically crack propagation in soils use a tensile strength criterion to check whether a crack propagates or not. From a numerical point of view, they use a technique based on initial joints between element faces, that is, a sort of Voronoi-based discretization technique, capable of representing discrete fracture networks ([15], [16]). Discrete element method is particularly appropriate for this type of problems [14]. Adapting the hydraulic problem (unsaturated flow) to the new cracked geometry constitutes a challenge in this case.

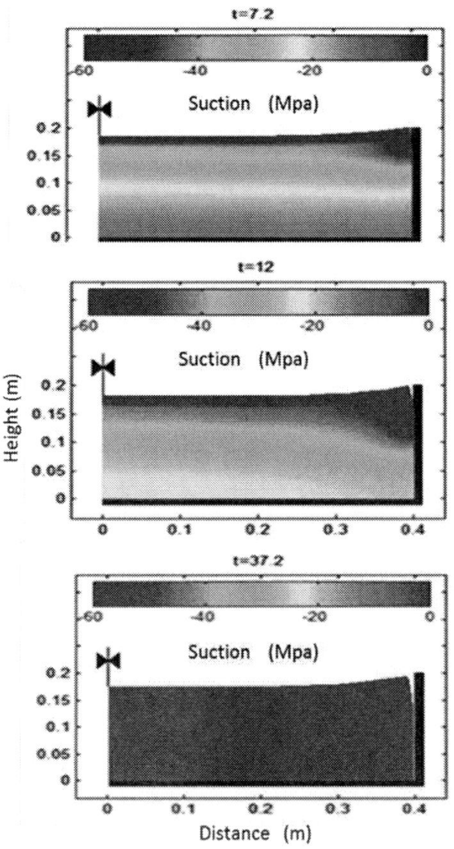

Figure 6. Numerical simulation of a desiccation test at different times (hours) developing a crack in the soil-wall contact ([24]).

5 Physical modelling: Laboratory tests

Cracking patterns are very complex to predict in advance. Therefore, the use of physical models may be convenient to understand the processes involved in practical cases. In the laboratory, environmental chambers have been developed in the last decade to reproduce soil drying at an intermediate scale ([12], [29]). Figure 7 shows one of this chambers developed at UPC [28], where it is possible to control temperature and relative humidity inside the chamber, imposing dry conditions to a soil tray. Those trays are instrumented, so the time evolution of main variables is recorded, including weight of the tray and suction, water content and temperature at particular locations inside the soil sample.

Modern environmental chambers include a system to circulate dry or wet air so cycles of drying and wetting episodes can be imposed to the soil sample. This is to create conditions similar to what it is found in nature. Figure 8 shows two pictures of one experiment lasting for more than two months, where first a drying boundary condition is applied to the sample (80 cm in diameter) and then a wetting is imposed (air relative humidity close to 100%). It should be pointed out that wetting may close some small cracks due to the expansion when reducing suction, but what dominates is the formation of new cracks. This effect is related to the reduction of tensile strength when reducing suction, as suggested in Figure 5.

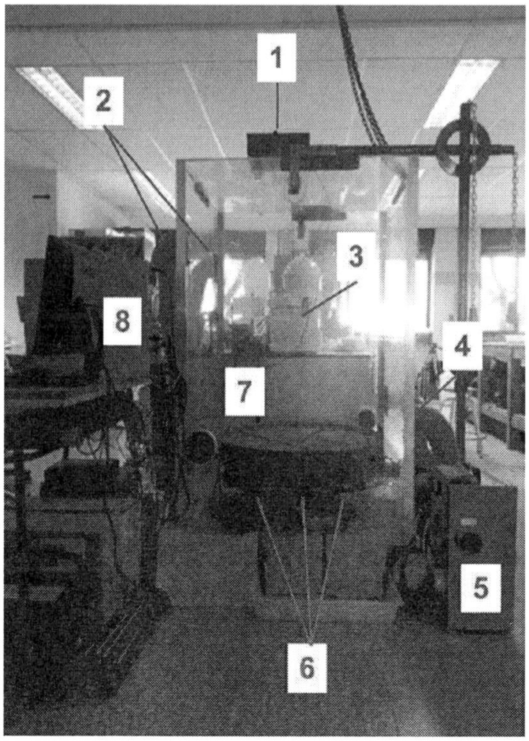

Figure 7. Environmental chamber to study soil desiccation. 1. Digital camera; 2. Dry/Wet air entry; 3. Thermometer; 4. Wet air exit; 5. Dehumidifier; 6. Load cells; 7. Soil specimen; 8. Computer and data logger. ([12], [28]).

In some cases, the vertical cuts at each side of a surficial crack become unstable when wetting the soil. Then small wedges fall down to the crack due to the reduction in apparent cohesion and new cracks appear. This effect can be observed in the lower picture of Figure 8.

These types of experiments have been useful to understand some of the mechanisms involved in desiccating soils. The experience from the environmental chambers confirms that suction gradient between the soil and the air of the chamber is the main driving mechanism producing desiccation in this context. Temperature gradients can be also applied and may have influence on the evaporation of water from the soil. Drying processes are quite fast in general, whereas wetting the sample after drying seems to be more difficult. Indeed there is a hysteretic effect and to saturate the sample once it has been cracked requires imposing a 100% of relative humidity to the environment. This is due to the low permeability of the dry cells between cracks.

Finally, some experiments have shown that sometimes cracks do not start at the soil surface, but at the bottom of the specimen [30]. Differential drying between surface and bottom may lead to curling of the sample as well.

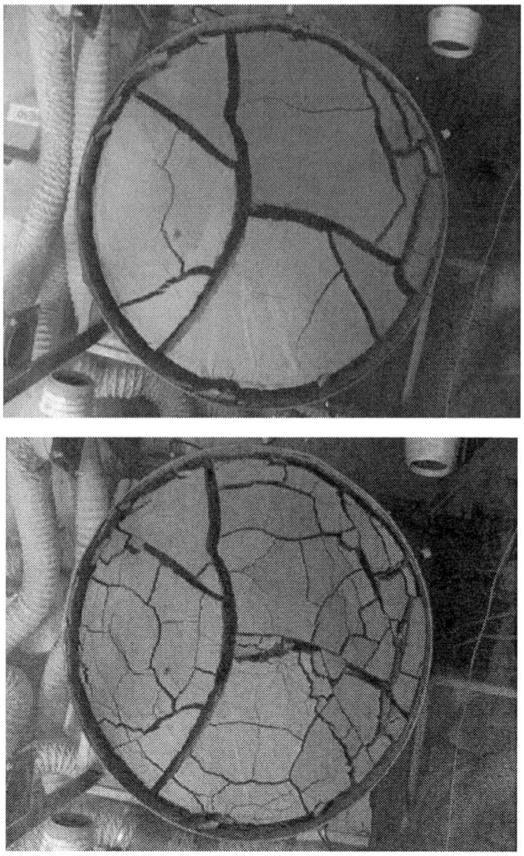

Figure 8. Pattern of cracks in a soil tray after a long period inside the environmental chamber. Above: After 1 month of continuous drying. Below: new cracks develop when wetting again after a drying period. [28].

6 Physical modelling: Field tests

In addition to environmental chambers, physical modelling may refer to field controlled experiments as well. This is a natural extension of the work carried out in the laboratory, with the aim of understanding the processes involved in actual desiccating conditions. It becomes then evident that other variables, not considered so far, are involved in the physics of the test, basically meteorological variables, as solar radiation and wind velocity. From the experience in environmental chambers, there is increasing evidence that close to the soil surface most of the variables exhibit important gradients. Therefore, an effort should be devoted to the measurement close to the soil surface. Note that sensors may constitute a preferential flow path and a particular onset of cracking. There is always a compromise between installation of sensors and maintaining the soil mass as homogeneous as possible.

Figure 9 shows a picture of a field test that has been recently set up at UPC on a site called Agrópolis, near Barcelona airport. Details of the deployment of the experiment are described in a companion paper in this Conference [18]. That Figure shows the external instrumentation installed with the objective of measuring as close to the surface as possible. Remote sensors have been installed as well: temperature infra-red sensor and a reflectometer sensor, trying to obtain information on temperature and water content in a remote manner. Meteorological variables are also provided by a weather station located nearby. The installation replicates a typical soil tray used in the laboratory, but using larger dimensions: 3 m x 3 m x 0.5m depth. Load cells control continuously the weight loss due to drying or any eventual weight gain due to rain.

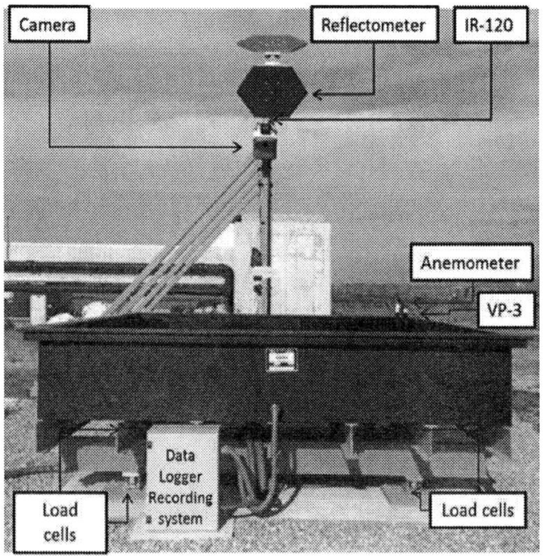

Figure 9. Overview of the field test and main components of the external instrumentation. IR-120 is an infra-red remote temperature sensor. VP-3 is a relative humidity, air temperature and vapour pressure sensors.

The test has been in operation from one year and it is being processed right now. Figure 10 presents a preliminary result concerning temperature values in a vertical line including soil and air above soil surface. It becomes evident that temperature profile is quite complex and that a gradient is developed around soil surface. An interpretation of this type of results is still an ongoing work. It could be stated that in the atmosphere, just above soil surface, there is a "limit layer" where the main gradients in terms of temperature and relative humidity develop. When water evaporates, a layer of wet air above soil surface remains there, unless wind takes away that moisture speeding the drying process. Therefore, wind velocity above soil surface becomes an important factor.

Another factor that seems crucial is solar radiation. In Mediterranean climate, heat flow due to solar radiation has an important influence on the temperature profile and therefore on the capacity to evaporate water. Note that radiation may enter the soil mass and thus it is not just a simple boundary condition on the soil surface. A numerical simulation of heat transport in the soil should take into account all the mechanisms involved, including radiation.

These two factors constitute the main difference between experiments performed in the laboratory and field experiments. Figure 11 presents the comparison between two of those experiments: a test in an environmental chamber was designed to reproduce the same weight loss – time curve measured in the field test shown in Figure 9. In the laboratory, almost a constant relative humidity of 30% was imposed in the environmental chamber, whereas in the field, relative humidity changed a lot in the range 20% - 95%, but with average values above 50%. The lines in Figure 11 show the evolution of the water loss measured in terms of weight change divided by surface area, to compare dimensions of laboratory and field trays.

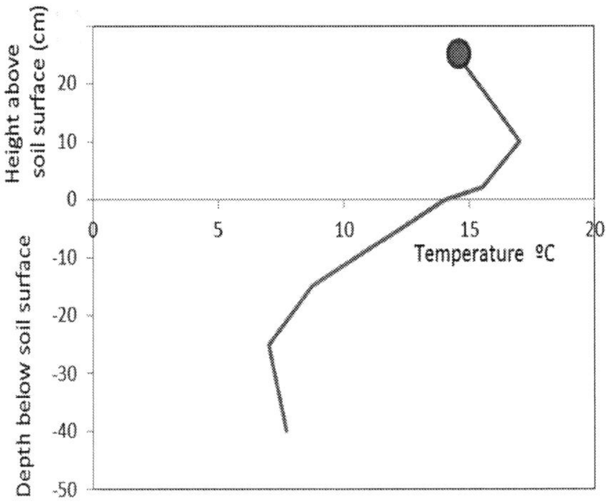

Figure 10. Temperature profile in the soil and above soil surface (maximum temperatures measured for a particular day). Dot corresponds to the temperature data from the Weather station.

 DOI: 10.1051/e3sconf/20160903005

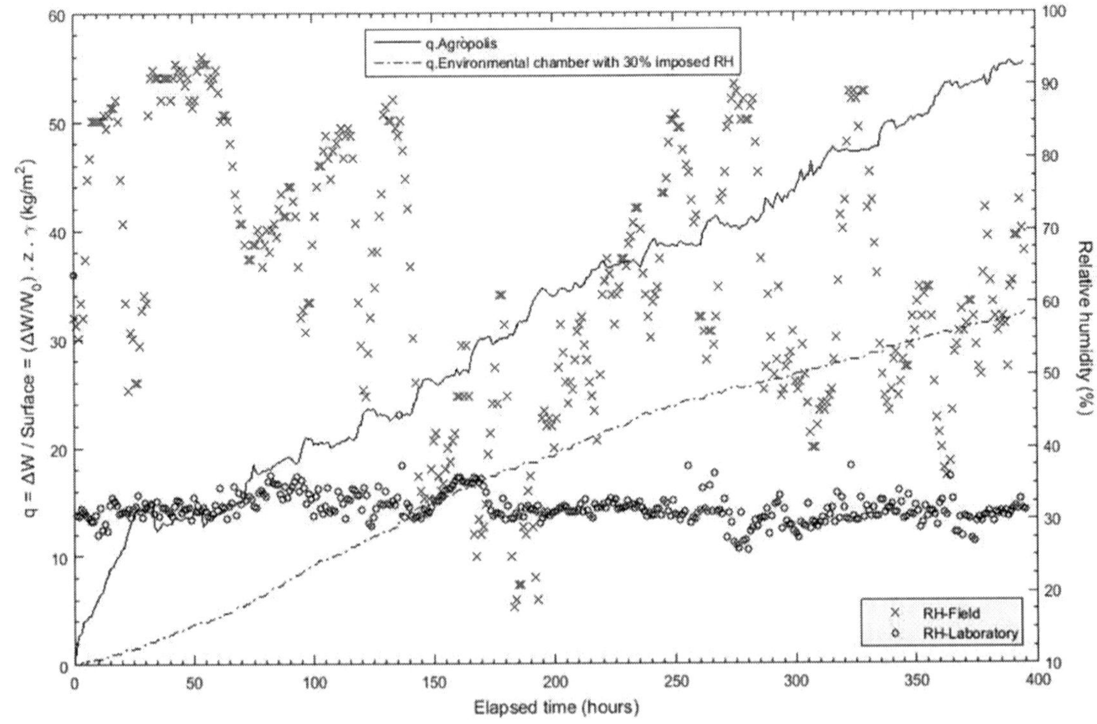

Figure 11. Comparison between laboratory environmental chamber and field test for a period of time. Evolution of relative humidity and loss of water measured as change of weight divided by exposed soil surface.

Analysis of Figure 11 indicates that comparing environmental chamber and field experiment is not straightforward and this is because some of the variables are different in both tests. It can be seen on that Figure that the loss of water per unit area is more important in the field, with relative humilities well above 50%, than in the environmental chamber imposing an extreme 30% of constant relative humidity. The reason for that discrepancy is the contribution of heat transported by radiation and wind velocity that are not reproduced in an environmental chamber. Those effects should be taken into account in the future when developing numerical simulations of cracking in desiccating soils in nature.

7 Conclusions

The paper presents a general perspective of the shrinkage and cracking of desiccating soils describing briefly the mechanisms involved and the methods and techniques available to study this type of problems. The initiation of the crack seems to be controlled by the hydro-mechanical conditions of the soil mass and therefore, an Unsaturated Soil Mechanics approach seems to be a convenient framework. Most of the criteria for crack initiation and propagation are based on tensile strength and future works should improve that, taking into account the effect of suction compressing in other directions. Propagation of cracks seems to be controlled by Fracture Mechanics variables and there is still room for improvement, regarding the numerical simulation of this phenomenon, as the information on Fracture Mechanics properties of unsaturated soils is scarce.

Physical modelling of drying soils is always a valuable tool to understand the processes involved in a large scale experiment. However, environmental chambers are based typically on controlling relative humidity boundary condition, whereas in the field, a more complex boundary condition acts on the soil surface. When comparing laboratory environmental chambers with field experiments it becomes apparent that heat transported by radiation plays an important role at least in Mediterranean areas. Also wind velocity close to the soil surface is fundamental to characterize the boundary condition. The interaction between soil and atmosphere, a topic identified as fundamental by several researchers, is still difficult to characterize and constitutes an important issue for future works.

Acknowledgements

The work described here has been carried out in collaboration with many people, in the context of several research projects during the last 15 years. The author thanks his colleagues, Profs. P.C. Prat and A. Lloret and the PhD students from that period: G. Ávila, R. Rodríguez, M. Sánchez, M.R. Lakshmikantha, H.U. Levatti, A. Cuadrado and J. Cordero.

Financial support from the Spanish Ministry of Economy and Competitiveness (grant BIA2012-36498) is gratefully acknowledged.

References

1. G. Blight, “Interactions between the atmosphere and the earth,” *Géotechnique*, vol. **47**, 4, pp. 715–767, (1997).
2. G. Blight, “Solar heating of the soil and evaporation from a soil surface,” *Géotechnique*, vol. **59**, no. 4, pp. 355–363, Jan. (2009).
3. Y. J. J. Cui, Y. F. F. Lu, P. Delage, and M. Riffard, “Field simulation of in situ water content and temperature changes due to ground atmospheric interactions,” *Géotechnique*, vol. **55**, no. 7, pp. 557–567, (2005).
4. Y.J. Cui, A.N. Ta, S. Hemmati, A. M. Tang, B. Gatmiri, "Experimental and numerical investigation of soil-atmosphere interaction". *Engineering Geology*, vol. **165** , p. 20-28, (2013).
5. R.C. Harris. "Giant desiccation cracks in Arizona". *Arizona Geology*, Arizona Geological Survey, vol. **34**, n. 2 (2004)
6. M.R. El Maarry, J. Kodikara, S. Wijessoriya, W. J. Markiewicz, N. Thomas. "Desiccation mechanism for formation of giant polygons on Earth and intermediate-sized polygons on Mars: Results from a pre-fracture model". *Earth and Planetary Sciences Letters*, **323-324**, p. 19-26, (2012).
7. A. Corte, A. Higashi. "Experimental research on desiccation cracks in soil". US Army Snow Ice and Permafrost Research Establishment. Research Report No 66, Corps of Engineers, Wilmette, Illinois, USA. (1960).
8. A.N. Abu-Hejleh, D. Znidarcic, "Desiccation theory for soft cohesive soils". *ASCE J. of Geotechnical Engineering*, vol. **121**, pp 493-502, (1995).
9. J.M. Konrad, R. Ayad, "Desiccation of a sensitive clay: field experimental observations", *Canadian Geotechnical Journal*, vol. **34**, pp 929-942. (1997).
10. R. Rodríguez, M. Sánchez, A. Ledesma, A. Lloret, "Experimental and numerical analysis of desiccation of a mining waste". *Canadian Geotechnical Journal*, **44**, pp 644-658, (2007).
11. H. Peron, T. Hueckel, L. Laloui, L.B. Hu, "Fundamentals of desiccation cracking of fine-grained soils: experimental characterisation and mechanisms identification", *Canadian Geotechnical Journal*, **46**, pp 1177-1201, (2009).
12. M.R. Lakshmikantha. Experimental and theoretical analysis of cracking in drying soils." PhD thesis, UPC-BarcelonaTech, Spain, (2009), PhD thesis repository: http://hdl.handle.net/10803/352470
13. S. Costa, J. Kodikara, B. Shannon, "Salient factors controlling desiccation cracking of clay in laboratory experiments". *Géotechnique,* vol **63**, pp 18-29 (2013).
14. H. Peron, J.Y. Delenne, L. Laloui, M.S. El Youssoufi, "Discrete element modelling of drying shrinkage and cracking of soils". *Computers and Geotechnics*, vol. **36**, p. 61-69, (2009).
15. M. Sánchez, O.L. Manzoli, L.J.N. Guimaraes, "Modelling 3-D desiccation soil crack networks using a mesh fragmentation technique". *Computers and Geotechnics,*vol. **62**, p. 27-39, (2014).
16. D. Asahina, J.E. Houseworth, J.T. Birkholzer, J. Rutquvist, J.E. Bolander, "Hydro-mechanical model for wetting/drying and fracture development in geomaterials." *Computers and Geosciences*, vol. **65**, p. 13-23. (2014).
17. C.L. Meehan, D. Pradel, M.A. Pando, J.F. Labuz, "Stability and performance of slopes and embankments", Geo-Congress 2013. San Diego (California). ASCE, Geotechnical Special Publication No 231, (2013).
18. J. Cordero, A. Cuadrado, P. Prat, A. Ledesma, "Description of a field test involving cracking in a drying soil". E-UNSAT 2016. This Conference, (2016).
19. R.D. Holtz, W.D. Kovacs. "An Introduction to Geotechnical Engineering", Prentice Hall, Englewood Cliffs, New Jersey, (1981).
20. F.A.M. Marinho. "Shrinkage behaviour of some plastic soils". PhD thesis, University of London, London, UK (1994).
21. H. Shin, J.C. Santamarina. "Desiccation cracks in saturated fine-grained soils: Particle level phenomena and effective stress analysis. *Géotechnique*, vol. **51**, pp. 961-972. (2011).
22. H. Peron, J.Y. Delenne, L. Laloui, M.S. El Youssoufi. "Discrete element modelling of drying shrinkage and cracking of soils". *Computers and Geotechnics*, vol. **36**, pp. 61-69. (2009).
23. G. Ávila, A. Ledesma, A. Lloret. "One-dimensional cracking model in clayey soils". Proc. 18th Int. Conference on Soil Mechanics and Geotech. Engng., Paris (2013).
24. H.U. Levatti. "Estudio experimental y análisis numérico de la desecación en suelos arcillosos". PhD Thesis (in Spanish). UPC-BarcelonaTech, Spain (2015).
25. M.R. Lakshmikantha, P.C. Prat, A. Ledesma. "Experimental evidence of size effect in soil cracking". *Canadian Geotechnical Journal*, vol. **49**, p. 264-284, (2012).
26. A.L. Amarasiri, J.K. Kodikara. "Numerical modeling of desiccation cracking using the cohesive crack method". *ASCE International Journal of Geomechanics*, vol. **13**, n. 3, p. 213-221. (2013).
27. G. Ávila, A. Ledesma, A. Lloret. "Measurement of fracture mechanics parameters for the analysis of cracking in soils". *Proc. 3rd Int. Conference on Unsaturated Soils - UNSAT 2002*. Recife, Brazil. (2002).
28. A.N. Ta. "Étude de l'interaction sol-atmosphère en chambre environnementale". PhD Thesis (in French). École Nationale Ponts et Chaussées- Université Paris Est. (2009).
29. J. Cordero, A. Cuadrado, A. Ledesma, P. Prat. "Patterns of cracking in soils due to drying and wetting cycles". *Proc. Int. Conference on Unsaturated Soils, Research and Application, UNSAT-2014*. pp. 381-387. Sydney (2014).
30. M.R. Lakshmikantha, R. Reig, P.C. Prat, A. Ledesma. "Origin and mechanism of cracks seen at the bottom of a desiccating soil specimen. *ASCE Geo-Congress*, San Diego, USA, p. 790-799. (2013).

Advances in the monitoring of geo-structure subjected to climate loading

Alessandro Tarantino[1,a], Domenico Gallipoli[2], Cristina Jommi[3], Joao Mendes[2], Anastasia Capotosto[1], Alessia Amabile[1], Matteo Pedrotti[1], Annarita Pozzato[1], Vojtěch Beneš[4], and Fabio Bottaro[5], Heiner Denzer[6], and Fabian Boeck[7]

[1]*University of Strathclyde, Department of Civil and Environmental Engineering, G11XJ Glasgow, UK*
[2]*Laboratoire SIAME, Equipe Géomatériaux et Structures du Génie Civil, Université de Pau et des Pays de l'Adour, 64600 Anglet, France*
[3]*Department of Geoscience & Engineering, Delft University of Technology, 2600 GA Delft, Netherlands*
[4]*G Impuls Praha, 170 00 Praha 7-Holešovice, Czech Republic*
[5]*Tecnopenta srl, 35037 Teolo (PD), Italy*
[6]*Pessl Instruments GmbH, Weiz, 8160 Austria*
[7]*UMS GmbH, 81379 Munich, Germany*

Abstract. The paper presents results achieved within the project MAGIC, a project funded by the European Commission under the Marie-Curie Industry Academia Partnerships and Pathways (IAPP) scheme. The project MAGIC aims to advance the state-of-the art in the monitoring of geo-structures subjected to climate loading by filling some of the gaps in current monitoring technologies. The project involves a partnership between academic and industrial partners to boost knowledge transfer and promote the development of 'industrial' instruments and services. The paper presents developments concerning the measurement of pore-water tension (suction in excess of 100 kPa) and the integration of geotechnical and geophysical monitoring.

1 Introduction

Major geotechnical infrastructure (road and railway embankments, dams, flood embankments, excavations, cut slopes) and natural geo-structures such a slopes are exposed to the atmosphere and are strongly affected by climate loading. Monitoring systems are crucial to assess climate hazard to geo-structures, minimise climate vulnerability via early-warning systems, and devise adaptation measures to enhance climate-resilience.

The literature is rich of research contributions documenting approaches to geo-structure monitoring. However, there appears to be a gap in industrial geotechnical monitoring. Companies provide a wide range of traditional instruments and monitoring services but are generally unprepared to devise solutions for climate interaction monitoring.

The project MAGIC 'Monitoring systems to Assess Geotechnical Infrastructure subjected to Climatic hazards', funded by the European Commission under the Marie-Curie Industry-Academia Partnerships and Pathways (IAPP) scheme aims to address this gap. It involves three academic and four industrial partners:

- *University of Strathclyde, UK*
- *Université de Pau et des Pays de l'Adour, France*
- *Delft University of Technology, Netherlands*
- *G Impuls Praha spol. s r.o., Czech Republic*
- *Tecnopenta srl, Italy*
- *Pessl Instruments GmbH, Austria*
- *UMS GmbH, Germany*

Systems for monitoring the ground-atmosphere interaction are routinely devised for agricultural applications. These include most of the sensors that are required for geotechnical installations and can therefore serve as a basis for industrial geotechnical monitoring systems. The partner PESSL is indeed specialised in agricultural monitoring and brings into the project the experience in the installation and operation of sensors under a wide range of soil and climate conditions.

Systems based on local sensors pose two challenges, that is i) where to install the sensors to ensure that the local measurement is representative of the ground/geo-structure as a whole and ii) how to extrapolate the spatial distribution of the variables measured locally. Systems based on local sensors also suffer from another limitation, they are difficult to implement for the case where geotechnical hazard has to be assessed at the regional scale (e.g. flood embankments and shallow landslides at the catchment scale). These issues can be addressed successfully by integrating the geotechnical monitoring with the electrical geophysical monitoring. Preliminary geophysical investigation can be of great help to characterise the geo-structure and optimise location of 'geotechnical' sensors. At the same time, electrical geophysical 'diffuse' monitoring allows mapping the spatial distribution of electrical resistivity and, hence, of water content, which can integrate the more accurate information acquired from local sensors. The partner GIMPULS makes available this expertise to the project.

Sensors currently included in agricultural installations cover most the requirement for geotechnical monitoring

[a] Corresponding author: alessandro.tarantino@strath.ac.uk

with one exception, the measurement of pore-water tension (suction). Conventional tensiometers only cover the range 0-80 kPa, which is generally exceeded during the summer period. Sensors for indirect measurement of suction based on the measurement of electrical or thermal properties of a porous element in contact with the soil are generally suitable for agricultural but not for geotechnical applications. These sensors suffer from slow response and poor accuracy in the low suction range and are therefore not suitable for capturing rapid changes in suction associated with rainwater infiltration or propagation of a water front. High capacity tensiometers (HCT) can potentially address this gap. However, no HCTs are available commercially to date. MAGIC has therefore involved two companies, UMS and TECNOPENTA, to address the problem of shallow and deep measurement of pore-water tension respectively.

The paper therefore presents some of the advances made in the development of instruments for the direct measurement of suction in excess of 100 kPa and the integration of geophysical monitoring with geotechnical monitoring.

2 Shallow measurement of pore-water tension

The high-capacity tensiometer (HCT) for pore-water tension measurement was invented in the early 90's by Ridley and Burland [1]. Since then, a number of prototypes have been developed essentially for research purposes as documented by [2], [3], and [4] with the most recent HCT presented by Mendes and Buzzi in 2013 [5].

Despite a large number of examples of successful measurement of pore-water tension in the laboratory, there are essentially no 'industrial' HCT tensiometers available on the market. This is likely due to the non-conventional design of HCT, the relatively complex experimental procedures required for tensiometer conditioning (including its first saturation), and the relatively limited evidence of HCT successful performance in the field.

This section presents a number of studies aimed at simplifying HCT design and conditioning procedures in order to make HCT more suitable for industrial applications. This study involves cooperation between the Université de Pau et des Pays de l'Adour (UPPA), UMS GmbH, the University of Strathclyde, and PESSL Instruments.

2.1. Design concept

Two main design concepts have been put forward over the years, integral-strain gauge diaphragm tensiometers and 'assembled' tensiometers, i.e. tensiometers obtained by fitting a High Air-Entry Value (HAEV) ceramic disk onto a commercial transducer.

The advantage of integral-strain gauged diaphragms is that connections and, hence, potential entrapment of cavitation nuclei are minimised. However, integral-strain gauged diaphragms are home-made and they may therefore suffer from relatively poor repeatability and are relatively expensive not being mass-produced. In addition, they tend to be temperature sensitive and this makes them suitable for laboratory but not field use.

For industrial applications, the 'assembled' design might be a much more convenient choice. Mendes et al. [6] have investigated the performance of assembled tensiometers based on ceramic flush diaphragm, ceramic cavity diaphragm, and stainless steel flush diaphragm transducers (UPPA HCT design). The performance of all these prototypes appeared to be satisfactory in terms of maximum sustainable tension and comparable to the performance of HCT based on integral strain-gauged diaphragms. This is also consistent with the good performance of other epoxy-assembled transducers as discussed by Delage et al. [4].

Figure 1 shows the three UPPA prototypes based on ceramic pressure transducers. In prototype 4, the HAEV ceramic filter was directly glued onto a ceramic cavity diaphragm transducer. The water reservoir was formed by the transducer cavity having a relatively large volume (~430 mm^3) compared to the majority of HCT prototypes published in the literature. If the maximum sustainable tension is investigated by letting the ceramic filter dry (Figure 2), it can be observed that Prototype 4 exhibits a maximum sustainable tension of around -1800kPa, which is comparable to that of Prototypes 2 and 3 having a small water reservoir.

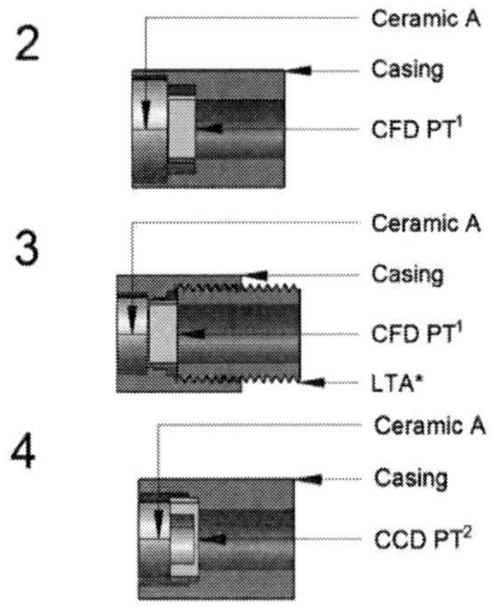

Figure 1. UPPA HCT prototypes based on ceramic sensors [6]

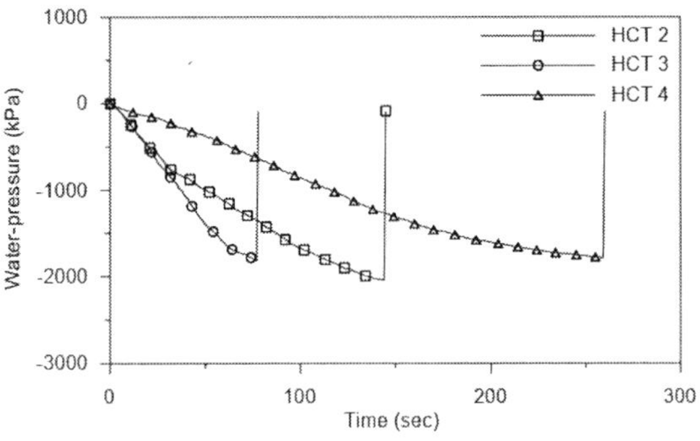

Figure 2. Response upon drying of UPPA tensiometers [6]

It therefore appears that the reservoir size does not have a marked influence on measuring range. This is consistent with the findings of Mendes and Buzzi [5] and Tarantino and Mongiovì [7] suggesting that cavitation initiates in the porous ceramic rather than in the reservoir.

Successful performance can then be achieved by mounting a HAEV ceramic onto a commercial pressure transducer and that there is apparently no need to push the size of the water reservoir to very small values (i.e. advanced machining skills and technologies are not required).

2.2 Conditioning and periodical re-saturation

Initial saturation of the porous ceramic has always been considered a critical step to ensure adequate performance of HCT. Procedures have become more and more sophisticated, involving preliminary removal of adsorbed water via pre-heating or silica gel adsorption, ceramic evacuation in the absence of water to maximise vacuum, and ceramic flooding under vacuum [3]. These procedures are difficult to implement in an industrial laboratory. In addition, accidental desaturation in the field is an event that should be considered in industrial practice. If this happens, it will be very unpractical (and unacceptably time-consuming) sending the HCT back to the manufacturer for reinitiating the initial saturation procedure.

The procedure based on the 'one-shot' initial saturation was therefore investigated by Pedrotti et al. [8] using a tensiometer similar in design to [9]. An air-dried HCT was placed in the saturation chamber and then almost immediately pressurised to 4 MPa. The response of the tensiometer during the first hour of the 'one-shot' saturation is shown in Figure 3. It took around 8 min (from t=2min to t=10min) for the tensiometer to record the external pressure applied. This time lag is clearly associated with the ingress of water into the porous ceramic.

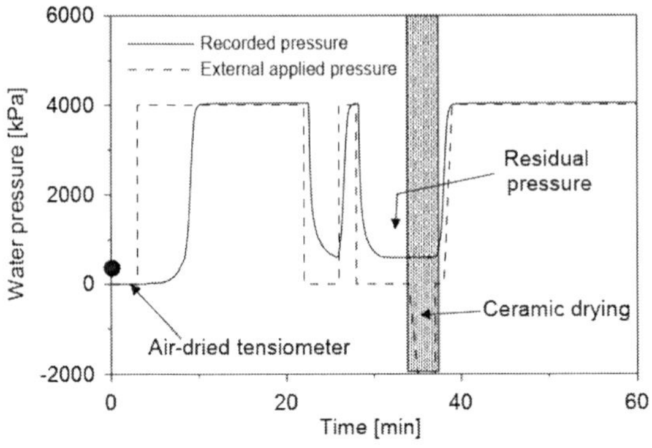

Figure 3. Pressurization-depressurization cycles within the first hour of one-shot saturation (Pedrotti et al. [3]).

The pressure in the saturation chamber was then cycled twice between 4 MPa and 0 MPa (from t=21min to t=34min). When releasing the external pressure, a significant residual pressure of about 600 kPa remained 'locked' in the tensiometer. The unbalance between the external water pressure and the water pressure in the tensiometer reservoir is likely carried by the menisci present in the unsaturated ceramic filter. This unbalance, although associated with an extreme condition, is clear evidence that the (negative) water pressure in the tensiometer water reservoir may differ from the 'external' water pressure applied on the outer side of the ceramic filter. Before re-applying a positive pressure, an attempt was made to generate water tension in the tensiometer by exposing the porous ceramic to air-drying (from t=35min to t=37min). However, no water tension could be recorded.

After the first hour, a constant external pressure of 4MPa was applied to the HCT. During the subsequent 8-day period, this external pressure was periodically released to zero for a few minutes only. Upon each pressure release, the residual excess pore water pressure was recorded. Figure 4 shows the residual excess water pressure versus the time elapsed from the first saturation. The residual excess water pressure recorded upon pressure release attained 0 kPa after 8 days of almost continuous 4MPa pressurization (circle in Figure 4). The absence of a residual excess water pressure was taken as an indicator of adequate saturation of the ceramic filter. By fitting the residual excess water pressure with an exponential function, a time of about 4 days can be estimated as the time required to achieve saturation of the ceramic filter in 'one-shot'.

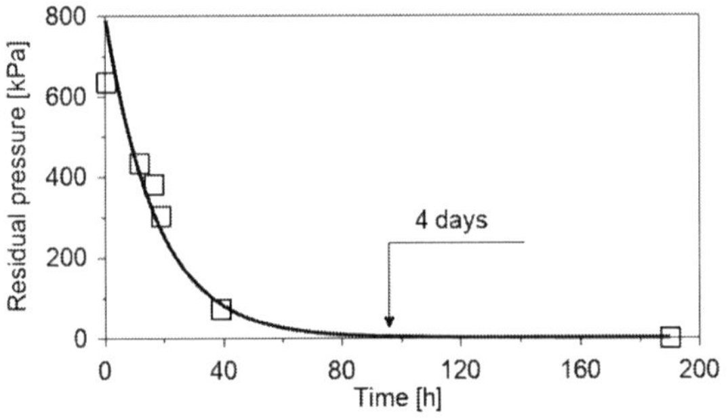

Figure 4. Residual pressure at atmospheric pressure after pressurization (Pedrotti et al. [3])

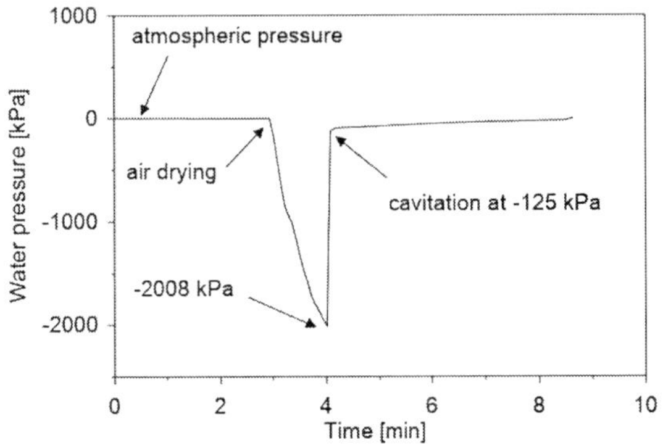

Figure 5. Air-drying cavitation (after ~8 days of pressurisation) (Pedrotti et al. [3]).

After 8 day pressurisation, the tensiometer was indeed able to sustain negative water pressures for the first time as Figure 5. Noticeably, the water pressure recorded was much higher than the nominal air-entry value of the porous ceramic (1.5 MPa). This response is generally accepted as evidence of adequate saturation of the porous ceramic. Following the results shown in Figure 5, new tensiometers have been saturated in one-shot. The performance of other tensiometers saturated using the 'one-shot' procedure is discussed in section 3.

2.3 Laboratory and field performance

Although Toll et al. [10] and Cui et al [11] have presented successful examples of field measurement, evidence of field performance of HCTs still remains scarce. Within the project MAGIC, field measurements are planned in two sites in France and one site in Scotland.

The layout of the station installed in Orist (France) by UPPA and PESSL is shown in Figure 6. It includes a weather station (rain gauge, and sensors for air humidity and temperature, solar radiation, and wind speed), the tensiometer developed at UPPA similar to Prototype 3 in Figure 1 (MAGIC tensiometer), two water content sensors (ERC-5 by Decagon), one porous ceramic suction sensor (MPS6 by Decagon), and one temperature sensor (combined with the MPS6)

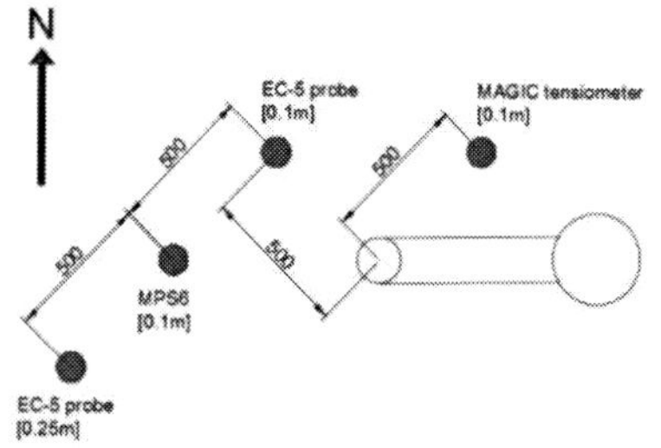

Figure 6. Layout of the system at Orist - Landes (France). EC-5 = capacitive water content sensor, MPS6 = porous ceramic suction sensor, MAGIC = high-capacity tensiometer.

The response of the instruments over 1 month is shown in Figure 7. The MAGIC tensiometer and the water content sensor are subjected to significant daily fluctuations associated with soil temperature changes (sensors were installed very close to ground surface where ground temperature is significantly affected by the daily fluctuation of air temperature).

To remove the effect of daily fluctuation, data were first smoothed by running average. With this correction, the output from the MAGIC tensiometer (thin black curve in Figure 7) still appears to be in phase with the temperature and not very consistent with the changes in water content during the initial dry period (until about 1/1/2016).

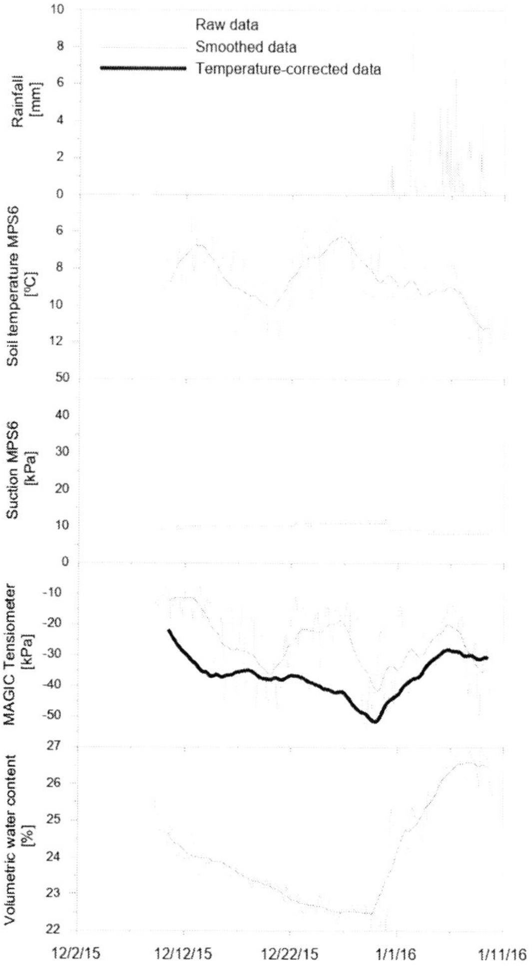

Figure 7. Response of the MAGIC HCT in the field and comparison with other sensors

The temperature effect on MAGIC tensiometer response clearly appears when the MAGIC tensiometer pore-water pressure is plotted against soil temperature (Figure 8). If the MAGIC tensiometer data are corrected using the linear regression in Figure 8, the pore-water pressure (thick black curve Figure 7) now appears to be consistent with the water content data. In fact, during the rainfall event started on 1/1/2016, both pore-water pressure and water content increase.

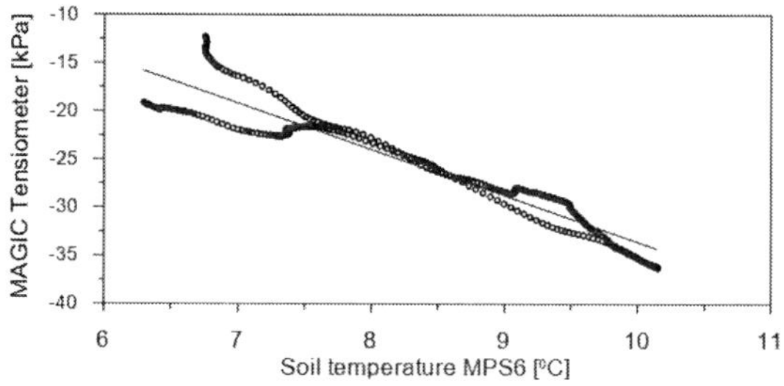

Figure 8. Correlation between MAGIC tensiometer response and soil temperature.

DOI: 10.1051/e3sconf/20160904001

The response of the MAGIC tensiometer therefore appears to be satisfactory if temperature effects are corrected for. This MAGIC tensiometer has been measuring suction uninterruptedly for almost 8 months to date (from August 2015 to March 2016) with peaks in suction reaching about 200 kPa during the summer period. This appears to be one of the longest field measurement performance recorded for a HCT.

Concerning the temperature sensitivity of the MAGIC tensiometer, it is worth noticing that the ceramic pressure sensor itself was temperature-compensated and its sensitivity, measured in the laboratory before installing it in the tensiometer casing, only resulted in 0.16 kPa ^{0}C. The higher temperature sensitivity shown by the tensiometer in the field is therefore likely due to differential thermal deformation of the pressure transducer and the casing. These should be eliminated in future tensiometer designs.

Finally, it is worth noticing that the suction sensor MPS6 (see Figure 7) dod not respond to changes in water content presumably because suction is very close to the limit of the measurement range.

3 Deep measurement of pore-water tension

High-capacity tensiometers (including the one developed within the MAGIC project) are suitable for shallow installations. However, if suction needs to be measured at depths greater than say 1 m, the installation becomes problematic. One option is to drill a borehole and install the HCT at its bottom. However, the contact might be difficult to ensure and a closure should be devised that allows removal of the tensiometer while preventing soil water evaporation during measurement. Another option is the probe locator presented by Toll et al. [10] consisting of a 3 m long PVC cylinder buried into the ground. Openings at various depths on the sides of the probe locator allow the HCTs to be installed in contact with the ground (HCTs are slid down through appropriate guides). However, this installation is relatively costly and complex to implement and it is unlikely that this system is adopted commercially at this stage.

The option pursued by the MAGIC project is to install the HCT on a penetrometer shaft having the same size and connections of a CPT shaft. In this way, the HCT can be pushed down into the ground using the very common CPT push system. This work results form cooperation between the University of Strathclyde and Tecnpoenta srl.

3.1. HCT for the deep installation

The HCT mounted on the penetrometer shaft is shown in Figure 9. It is similar in concept to the one developed by [9] with its size reduced to accommodate it into the 36 mm diameter penetrometer shaft.

The HCT was saturated using the one-shot procedure. The response of two penetrometer HCTs after 5-day pressurisation at 4 MPa is shown in Figure 10. They could both measure pore-water tension successfully with cavitation occurring at around 800 kPa, which is the air-entry value of the porous ceramic used.

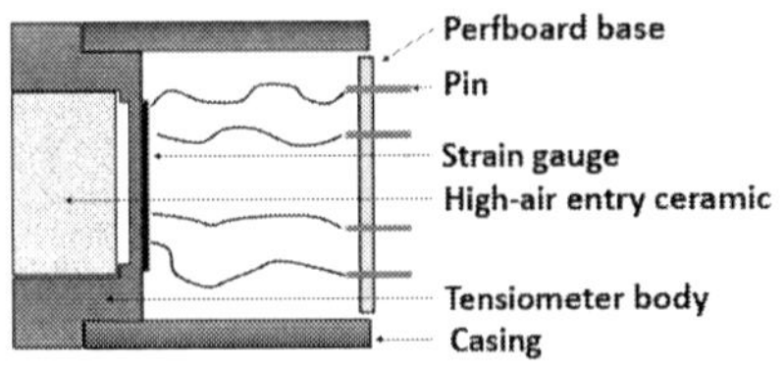

Figure 9. HCT mounted on the penetrometer shaft.

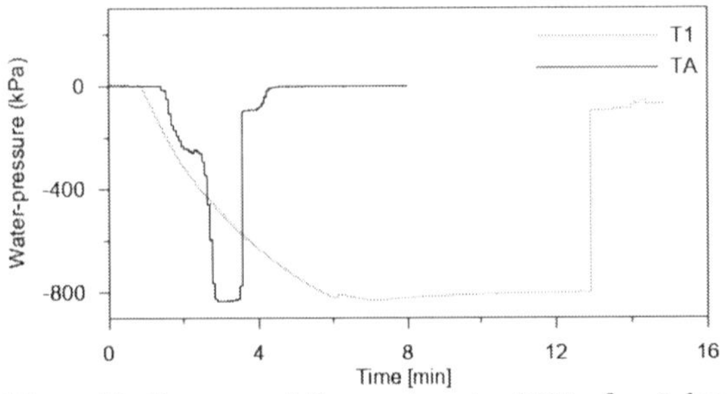

Figure 10. Response of the penetrometer HCT after 5-day pressurisation at 4 MPa.

To test their response, these two penetrometer HCTs were simultaneously installed on a clay sample using the suction measurement box shown in [9]. The HCTs were installed just after the 5-day one-shot pressurisation and another 5-day pressurisation at 4MPa following the cavitation test shown in Figure 10.

The aim of this test was to assess the performance of the two HCTs. They were installed in pair as this is the only way to ensure that pore-water tension measurement is reliable. As shown in Figure 3, the pressure measured by the tensiometer may differ from the external pore-water pressure if menisci are present in the ceramic filter. However, if the two HCTs are measuring the same pore-water tension, this is the best evidence that pore-water tension is measured reliably. It is extremely unlikely that random meniscus configuration in the two HCTs are leading to the same 'false' water tension in the water reservoir of the two tensiometers.

The measurement is presented in Figure 11. It shows fluctuation in suction due to water drops condensing on the lid of the box and dripping on the sample.

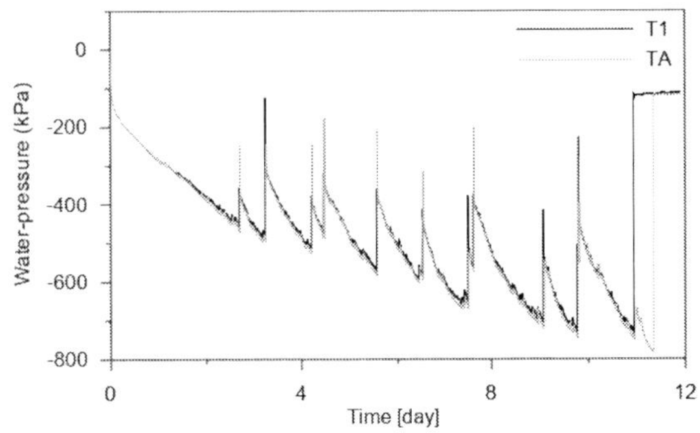

Figure 11. Long-term measurement using two penetrometer HCTs.

The two HCTs were successfully measuring the same pore-water tension despite the fluctuations due to water dripping. This result is noteworthy as the two HCTs were saturated using the one-shot procedure. Measurement duration is also remarkable (~11 days), with cavitation occurring only because the 800 kPa air-entry value of the ceramic filter was approached.

3.2 Prototype penetrometer shaft and laboratory measurements

The prototype penetrometer shaft carrying the HCT is shown in Figure 12. The main issue in the measurement of pore-water tension in this configuration is the contact between the tensiometer ceramic filter and the soil. In surface installations, a soil paste is usually interposed between the tensiometer ceramic filter and the soil [12]. A paste cannot be applied onto the penetrometer HCT and contact can only be ensured by the radial compression of the soil surrounding the penetrometer shaft during penetration.

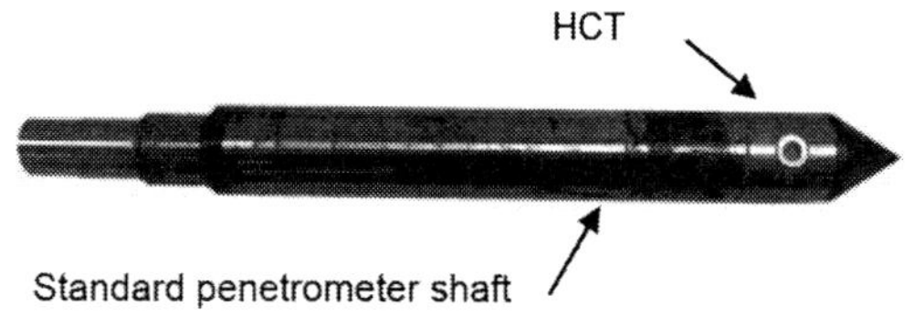

Figure 12. Prototype penetrometer shaft

To test whether adequate contact can be established during penetration, the penetrometer was installed through a soil sample contained into a 100 mm diameter and 150mm height mould as shown in **Figure 13**. A HCT was installed at the top of the sample (through the confining lid) to double check the measurement of the penetrometer HCT.

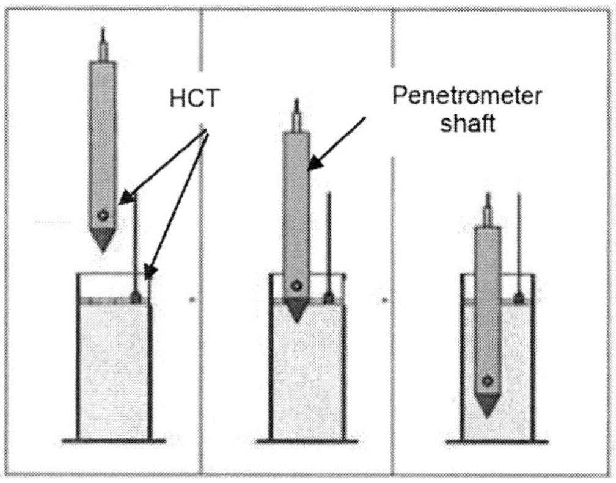

Figure 13. Mock-up scale test to simulate measurement during penetration.

The result on a test performed on a lightly compacted sample prepared with 80% Sand and 20% Kaolin (*w*=40% with respect to the kaolin mass) is shown in Figure 14.

The pore-water pressure measured by the penetrometer HCT decreased during penetration and then increased to an equilibrium value after the shaft penetration was stopped. The top HCT initially recorded an over pressure (due to the soil compression associated with the shaft penetration) and then levelled off to the same value recorded by the penetrometer HCT. This test shows that contact could be established even under unfavourable conditions. In fact, the soils included a high fraction of course-grained material (this makes the contact more difficult to achieve) and the soil was lightly compacted (low radial stresses develops during penetration because of the low density).

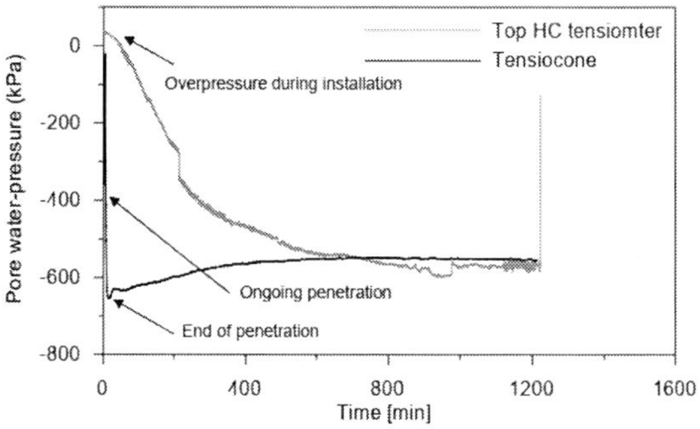

Figure 14. Penetration test in the laboratory

3.3 Field test

Following the tests in the laboratory, a test was carried out in the field. The penetrometer HCT was installed from the crest of a flood embankment with the phreatic surface located at around 7m below the crest. The HCT was therefor expected to measure suction in the range 0-70 kPa as the pore-water pressure is essentially hydrostatic below the asphalt-covered embankment crest. The results of the field test are shown in Figure 15.

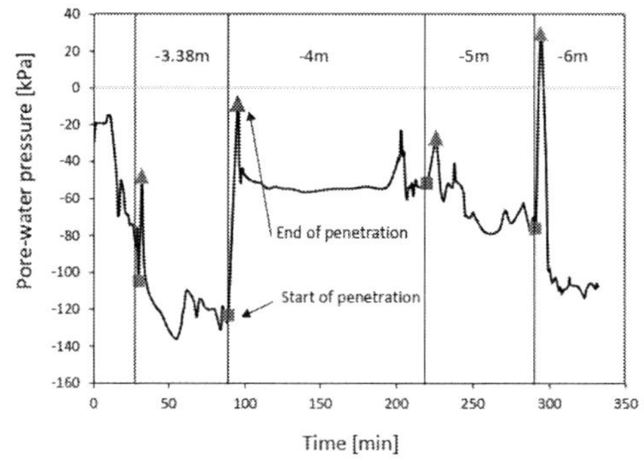

Figure 15. Field test of the penetrometer HCT

It can be observed that equilibrium could be reached after stopping the penetration, indicating that adequate contact could be established between the penetrometer HCT and the ground. However, the pore-water pressure readings were not consistent with a hydrostatic profile and not consistent with the values of suction established by the 7 m deep phreatic surface.

Following these results, possible mechanical effects on pressure reading were inspected in the laboratory. The penetrometer was subjected to an axial load while keeping the penetrometer HCT at zero pressure (by covering the ceramic filter with wet cotton). It can be observed that spurious readings are significant. The compression of the shaft induces deformation in the HCT resulting in false pressure readings. This explains the almost random values recorded during the field penetration test. The major challenge in the design of the penetrometer shaft would therefore be isolating mechanically the HCT from the shaft and this will be addressed in future design of the penetrometer shaft.

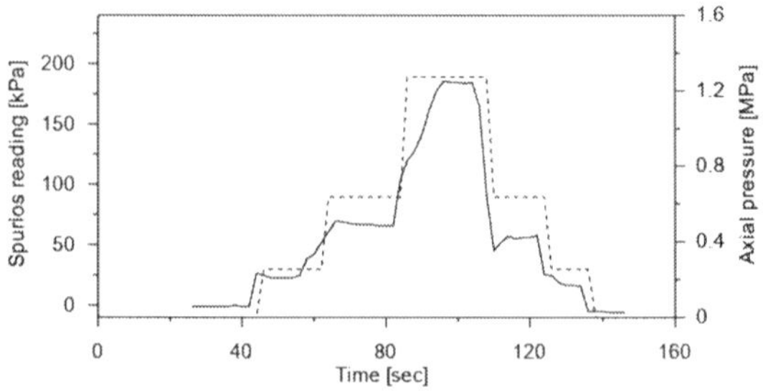

Figure 16. Mechanical effects on HCT tensiometer signal

4 Geophysical monitoring

Geophysical electrical methods can be used to monitor the water regime in geo-structures. Three methods were explored in the MAGIC project, the Electrical Resistivity Tomography (ERT), the Ground Penetrating Radar (GPR), and Slingram electromagnetic method (EM).

The ERT allows mapping the electrical resistivity of the ground. Above the groundwater table, the electrical resistivity is directly associated with the soil (volumetric) water content and the ERT therefore allows mapping the water regime in the vadose zone. The ERT is a relatively time-consuming technique compared to other geophysical electrical methods as it requires the installation of electrodes at the ground surface and relatively large number of scans.

The GPR allows for faster scanning of geo-structures and, similarly to ERT, provides a 2-D mapping of the ground. However, interpretation of GPR data is much more complex than ERT and it generally allows for the detection the phreatic surface/ground water table only.

The advantage of ERT and GPR in geotechnical design is that that the mapping water flow is faster and cheaper than methods based on conventional geotechnical instrumentation built upon local sensors (piezometer cells, tensiometers, and water content sensors). This makes these techniques appealing especially for hazard analysis at the regional scale (e.g. flood embankments and shallow landslides).

The EM is also a fast-scanning method and returns a single apparent resistivity of the ground. Different penetration depths can be achieved by varying either the distance between the transmitter and receiver coils or the frequency of the oscillating primary field.

The EM is the cheapest, fastest, and easiest-to-process monitoring technique and this makes it very suitable to be used as an early warning of flood embankments impending breach, provided suitable procedures are developed to interpret the resistivity data.

4.1 Case studies

Two case studies were considered to explore the potential of geophysical electrical techniques in geotechnical analysis. The first geo-structure is a flood embankment of the Adige River located in S. Florian in the province of Bolzano in Italy (Figure 17). A segment of the embankment was instrumented with piezometer cells to monitor the phreatic surface. The same segment was inspected using ERT, GPR, and EM.

The embankment is made of an inner core of silty sand and an outer 'shell' of sandy gravel (Figure 18). The embankment foundation consists of a sandy silt layer overlying a sandy gravel material. Samples were taken from the embankment core and the upper foundation layer to determine the relationship between electrical resistivity and volumetric water content and therefore allow converting ERT data into water content distribution.

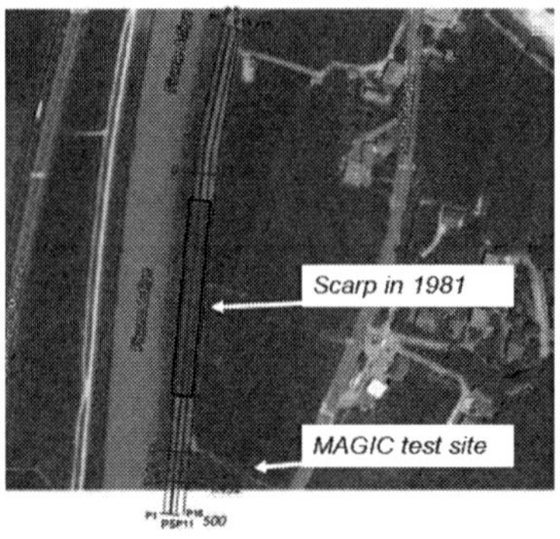

Figure 17. Adige river embankment in S.Florian, Italy

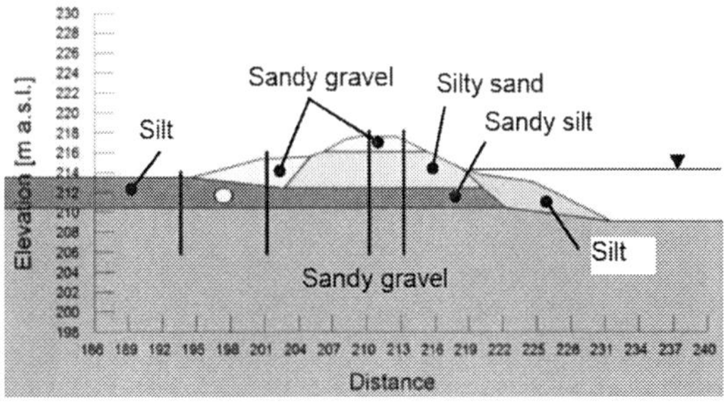

Figure 18. Cross section of the Adige river flood embankment

The second geo-structure is a flood embankment protecting the Leendert de Boerspolder, a small island (polder) in the Kagerplassen in the Netherlands. Flood embankments are made of clay (though mixed with recycled materials used when the water barrier was built well over 500 years ago) and are founded on peat bog.

DOI: 10.1051/e3sconf/20160904001

Water level is almost at the crest on the water side whereas it is about 50-100 cm below ground on the landside.

Figure 19. View of the Leendert de Boerspolder [13].

4.2 Phreatic surface detection using GPR

GPR measurement consists of high-frequency electromagnetic waves radiated into the investigated medium. If the investigated medium shows changes of dielectric permittivity properties, a portion of the wave is reflected back and registered by a receiving antenna. The travel time of the reflected wave, which is controlled by the dielectric permittivity of the medium, indicates the depth of the reflecting interface.

Fully digital dual-trace radar device SIR 20 (GSSI - USA) was used for the measurements. Considering the expected depth of the investigated embankments, antennas of carrier frequency of 900, 400, and 100 MHz were used. This frequency allows for penetration depth ranging between 1 m to 6–8 m. The measurement was taken in a time window of 30–200 ns. Measured data have been processed with the RADAN 6.6 (GSSI – USA) program.

Figure 20 shows GPR reflection data from the Adige river embankment. The thick curve represents the blind prediction of the phreatic surface as inferred from high amplitude reflections. On the other hand, the dashed line represents the phreatic surface obtained from numerical simulation calibrated to the piezometer water level measured in the field. It can be observed that GPR estimation of the phreatic surface agrees reasonably well with the one inferred from local piezometer cells.

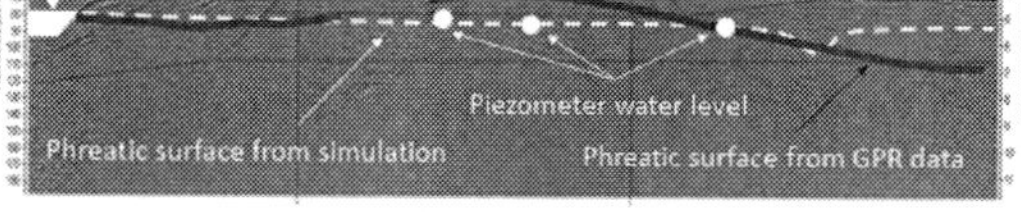

Figure 20.GPR data interpretation for the Adige river embankment, November 2014 (K452).

Figure 21 shows the GPR reflection data from the Leendert de Boerspolder embankment. Again, the thick curves represent the blind prediction of the phreatic surface as inferred from high amplitude reflections. Although no phreatic surface is simulated here, it can be observed that the prediction matches reasonably well the piezometer water level measured in the field on the land side and the water level on the left-hand water side (not shown in the figure), which is close to the embankment crest.

In summary, GPR appears to be able to detect the location of the phreatic surface in both fine-grained and coarse grained materials. It would be expected that the 'sampling' depth of GPR in clays is much lower than the one in granular materials due to dissipative attenuation occurring in clay. It is therefore likely that the phreatic surface down to about 6m detected in the Adige river coarse-grained embankment could not have been detected if the embankment was made of clay.

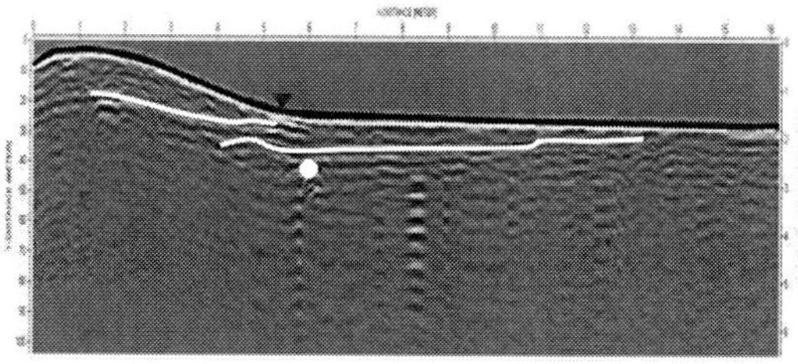

Figure 21. GPR data interpretation for the Leendert de Boerspolder embankment (K234)

4.3 Water content mapping using ERT

Electrical resistivity tomography (ERT) belongs to direct-current geo-electrical resistivity methods. The principle of the method is based on the measurement of resistivity using a large number of electrodes in an array configuration. The electrodes are connected by a special cable that enables the current to be injected through a pair of electrode and the electrical potential to be red by the remaining electrodes in the array. A multiplexer sends the current to different pairs of electrodes through a set sequence. The distribution of electrical resistivity in the ground is obtained by tomographic reconstruction (inverse problem).

For the resistivity tomography measurement, the device ARES (GF Instruments, Czech Republic) has been used. 2-D resistivity sections were compiled using the programme Res2Dinv (GEOTOMO Software - Malaysia). The ERT measurement was carried out using a Schlumberger electrode arrays and the results were processed using the 3rd iteration of least squares method.

Figure 22 shows the ERT model for the Adige river embankment. The darker region at the top of the embankment close to the crest denotes higher resistivity do to the lower content in turn associated with the embankment pavement preventing infiltration of incident rainwater. High resistivity is also observed at the berm on the right-hand side (landside) due to the presence of a drainage system.

The ERT data were collected in three cross-sections along a 27 m long embankment segment in November 2014 (wet season) and June 2015 (dry season). Electrical resistivity data were then converted into volumetric water content using calibration curves developed in the laboratory on samples taken from the lower embankment and upper foundation. The ERT data were finally

 DOI: 10.1051/e3sconf/20160904001

compared to the water content derived from the numerical simulation of water flow in turn 'calibrated' to field piezometer data.

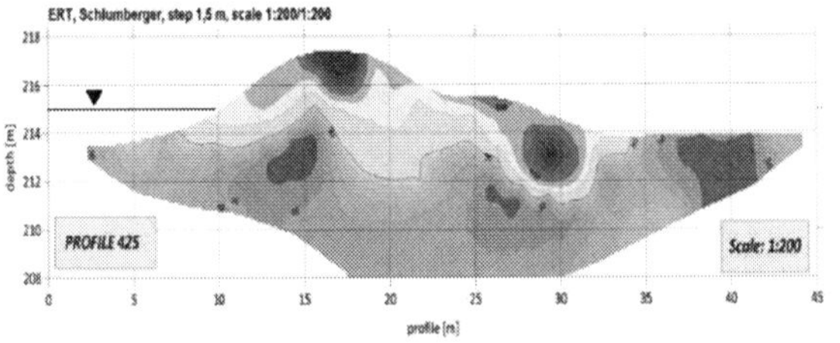

Figure 22. ERT data interpretation for the Adige river embankment, November 2014 (K425).

Figure 23 shows the performance of the ERT prediction against the data from the simulation for the November 2014 measurement. The agreement is very good for the water content in the upper and lower parts of the embankment. Prediction is less accurate in the saturated foundation layer. This can be attributed to the low sensitivity of the calibration curve at high water content and low resistivity.

Figure 24 shows the performance of the ERT prediction against the data from the simulation for the June 2015 measurement. The ERT simulation for the lower embankment is less accurate. The overestimation of the resistivity is likely associated with the drier upper layer that 'shadows' the inner core of the embankment. In fact, the ERT data in the lower embankment appears to become less accurate as the upper layer is approached. Further studies on the inversion method can probably eliminate this bias.

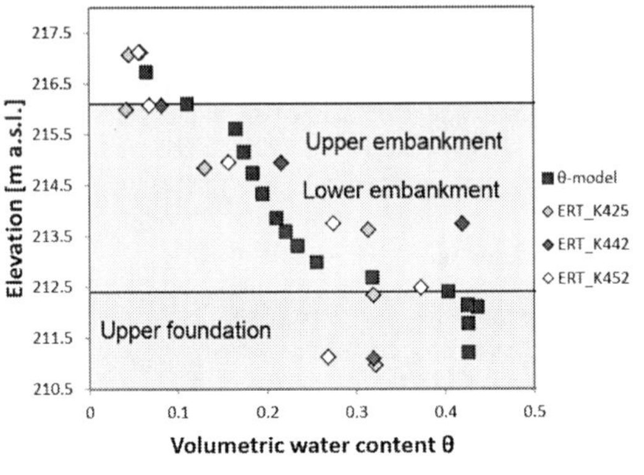

Figure 23. ERT data from Adige river embankment and comparison with simulated water content (November 2014).

Figure 25 show the ERT data from the Leendert de Boerspolder embankment. Contours of equal electrical resistivity close to the ground surface seem to follow the phreatic surface (water level on the left-hand side of the embankment is close to the crest and generates a right-ward water flow). However, it would be difficult to locate the phreatic surface purely on ERT data. If the ERT data are integrated with the tentative location of the phreatic surface as inferred from GPR data (dashed curves), the phreatic surface can be located confidently as shown in Figure 25b.

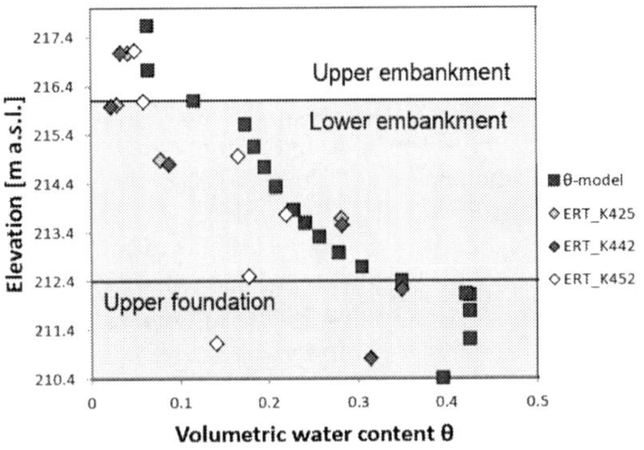

Figure 24. ERT data from Adige river embankment and comparison with simulated water content (June 2015).

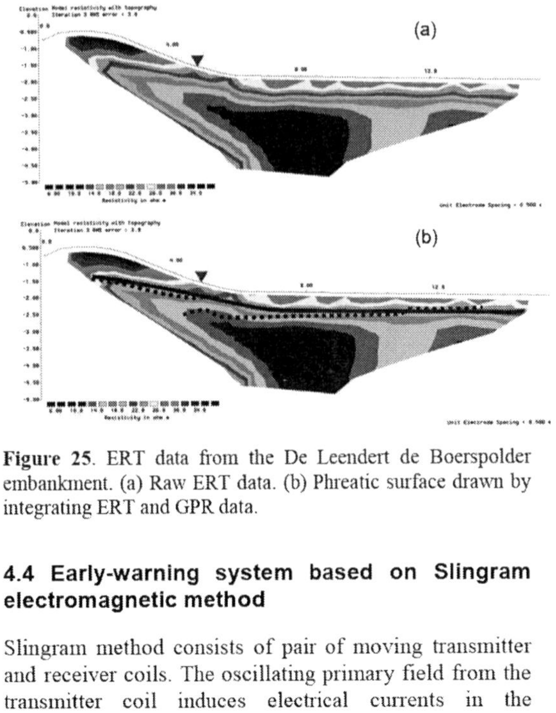

Figure 25. ERT data from the De Leendert de Boerspolder embankment. (a) Raw ERT data. (b) Phreatic surface drawn by integrating ERT and GPR data.

4.4 Early-warning system based on Slingram electromagnetic method

Slingram method consists of pair of moving transmitter and receiver coils. The oscillating primary field from the transmitter coil induces electrical currents in the subsurface that, in turn, generates a secondary magnetic field, which is picked up by the receiver coil. This secondary magnetic field is further divided into in-phase (real) and quadrature (imaginary) components, in turn dependent on the subsurface electrical conductivity.

In the project MAGIC, we used two types of equipment for multi depth EM profiling, the multi-frequency sensor GEM2 (GEOPHEX, USA) and the multi-distance sensor CMD (GF Instruments, Czech Republic). Only data from the GEM2 are discussed herein.

Measurements were carried out by walking with the instrument over the embankment in cross diection (Figure

 DOI: 10.1051/e3sconf/20160904001

26). The higher the distance between the ground surface and the phreatic surface, the higher will be the resistivity of the ground. Figure 27a shows the change in water depth when walking from the left-hand to the right-hand side of the embankment. Water depth is lower in November 2014 (wet season, high river level) than June 2015 (dry season, low river level)). The change in resistivity is shown in Figure 27b and appears to be consistent with water depth. In this figure, raw resistivity data were smoothened and shifted to align the valleys in the water depth and resistivity profiles.

A correlation between equivalent electrical resistivity and water depth was established for the data from the June 2015 survey considering the portion of the crest dashed in Figure 26.

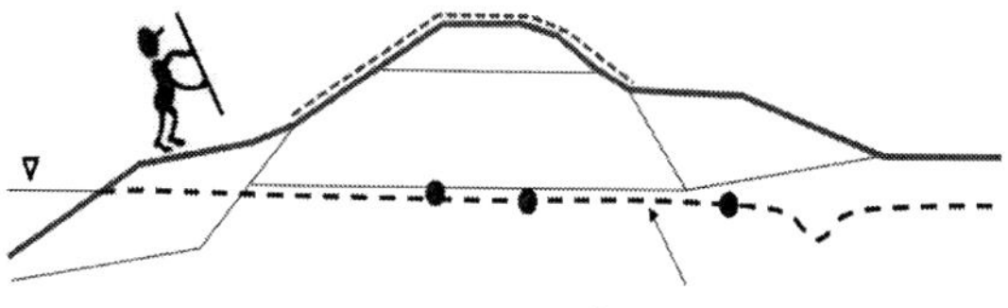

Figure 26. Inspection of the embankment cross-section using GEM2. The portion of the crest dashed indicates the segment where water depth was predicted based on GEM2 data.

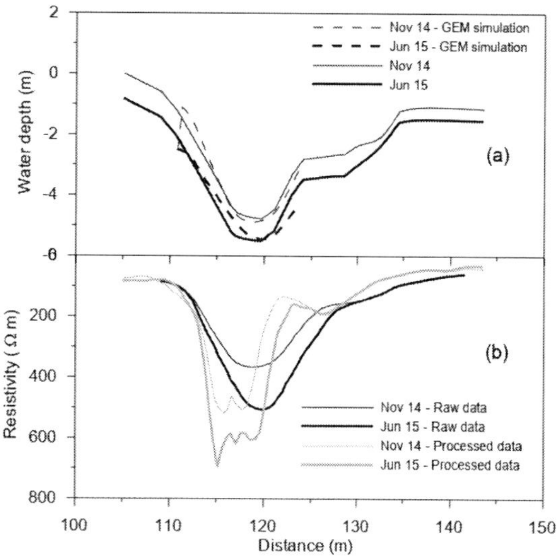

Figure 27. Electrical resistivity from EM profiling of the Adige River embankment associated with the November 2014 and June 2015 survey.

This correlation is shown in (Figure 28) and data were fitted using a straight line. Based on this correlation, the water depth profile was simulated from the EM resistivity data. The results of this simulation are shown in Figure 27a. The simulation of the June 2015 profile is obviously good as the correlation was indeed built upon the June 2015 data. On the other hand, the simulation of the November 2014 water depth profile should be regarded as prediction and it appears to reproduce fairly the field data.

In other words, if one was walking across the embankment in November 2015 during the wet season and was interpreting the electrical resistivity data based on the correlation developed for the June 2015 data (dry season), the raise of the phreatic surface in the embankment could have been properly detected. This shows the potential of the EM to be used as an indicator of water through-flow regime and, hence, of impeding breaching conditions.

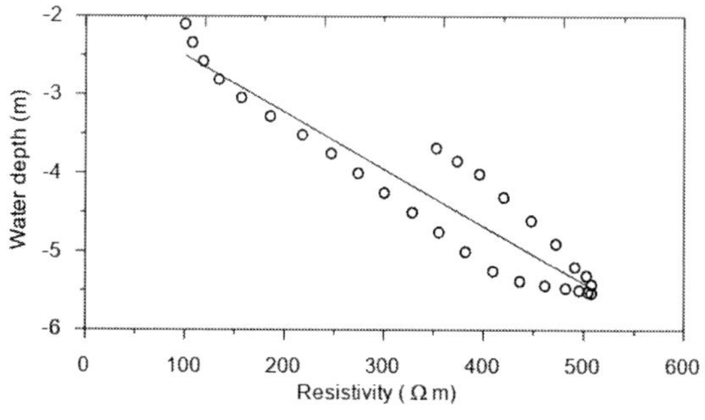

Figure 28. Calibration of GEM2 electrical resistivity against water depth (data relative to the June 2015 survey).

5 Conclusions

The paper has presented some advances in the monitoring of geo-structures exposed to climate loading with the aim of i) filling some of the gaps in current monitoring technologies and ii) developing 'industrial' instruments and approaches, i.e. ready to be implemented in routine geotechnical problems.

One of the gaps in 'industrial' monitoring technologies is the direct measurement of pore-water tension (suction in excess of 100 kPa) via high-capacity tensiometers. We have shown that tensiometer design and experimental procedures can be significantly simplified to facilitate the development of an 'industrial version of the HCT. 'Assembled' tensiometers, based on a high air-entry ceramic fitted onto a commercial pressure transducer, are simpler to manufacture and have been shown to perform to the same standard as the integral-strain gauge tensiometers. Procedures for saturation can also be simplified as the one-shot saturation proved equally effective as procedures based on pre-evacuation of the ceramic filter.

We have also addressed the problem of deep measurement of pore-water tension by incorporating a HCT into a penetrometer shaft. Preliminary tests in the laboratory and in the field have demonstrated that adequate contact can be achieved between the soil and the ceramic filter even in absence of any interposed soil paste. Problems still remain to be solved concerning the mechanical effects on HCT response.

Electrical geophysical methods have also proven to be an extremely valuable tool to assess the water regime in geo-structures. GPR is capable of detecting phreatic surface/groundwater table in geo-structures in a faster and more economical way than traditional piezometer cells. If GPR is used in conjunction with traditional piezometer cells installed in strategic locations, significant time and financial saving can be made.

ERT allows for relative accurate mapping of water content if electrical conductivity versus volumetric water content can be assessed experimentally. Again, if used in conjunction with local sensors, this technique can allow identifying anomalies or discontinuities in the water flow regime that are impossible to detect using local sensors.

Finally, the EM technique can be potentially used as a system for early-warning of impending failure triggered by water flow. For example, this technique can complement visual inspection of operators patrolling flood embankments during the flood event.

Acknowledgements

We wish to thank all who have contributed to the work presented in this paper: Georg von Unold (UMS, Germany), Maura Bellio, Alice Pavanini, and Aurelio Busetto (Tecnopenta, Italy), Vlado Budinski (GImpuls Praha, Czech Republic), Gabriel Gomez (PESSL Instrument, Austria), Brunella Balzano and Arianna Pagano (University of Strathclyde), and Amin Askarinejad (TU Delft). We also wish to thank the Autonomous Province of Bolzano for making available data from the Adige River embankment. The financial contribution of the European Commission to this research through the Marie Curie Industry-Academia Partnership and Pathways Network MAGIC (Monitoring systems to Assess Geotechnical Infrastructure subjected to Climatic hazards) – PIAPP-GA-2012-324426 – is gratefully acknowledged.

References

Ridley and Burland [1] A.M. Ridley, and J.B. Burland. Géotechnique, 43(2): 321-324 (1993).
[2] A. Tarantino. Proc. 3rd Int. Conf. on Unsaturated Soils, Recife, Brasil, 3: 1005-1017 (2004).
[3] F.A.M. Marinho, A. Take and A. Tarantino. Geotechnical and Geological Engineering, 26(6): 615-631 (2008).
[4] P. Delage, E. Romero and & A. Tarantino. *Proc. 1st European Conference on Unsaturated Soils*, Durham, UK: 33-52 (2008).
[5] J. Mendes and O. Buzzi, Can. Geotech. J. 50(5): 550-556 (2013).
[6] J. Mendes, D Gallipoli, F. Boeck, G. von Unold and A. Tarantino. *E-UNSAT 2016, 3rd European Conference on Unsaturated Soils,* Paris, France (2016).
[7] A. Tarantino and L. Mongiovi. Geotechnical and Geological Engineering 19: 189-210 (2001).
[8] M. Pedrotti, F Boeck, and A. Tarantino. *Proc. UNSAT 2014, Unsaturated Soils: Research & Applications,* Sydney, Australia: 1651–1657 (2016)
[9] A. Tarantino and L. Mongiovi. Geotechnique, 53: 137-41 (2003).
[10] D.G. Toll, S.D.N Lourenco, J. Mendes, D. Gallipoli, Evans, F.D., Augarde, C.E., Cui, Y.J., Tang, A.M., Rojas, J.C., Pagano, L., Mancuso, C., Zingariello, C., and A. Tarantino. Quarterly Journal of Engineering Geology and Hydrogeology, 44(1): 23-33 (2013).
[11] Y.J. Cui, A.M. Tang, A.T. Mantho and E. De Laure. Geotechnical Testing Journal 31 (1): 95–100 (2008).
[12] O.M. Oliveira and F.A.M Marinho. Geotechnical Testing Journal, 31(1): 1-5 (2008).
[13] van Veen A.E. Beheerplan Leendert de Boerspolder. Van landbouwpolder naar waterbergingslocatie (http://www.leendertdeboerspolder.stowa.nl/).

Fabric and clay activity in soil water retention behaviour

Cristina Jommi[1,a] and Gabriele Della Vecchia[2]

[1]*Delft University of Technology, Department of Geoscience and Engineering, Stevinweg 1, 2628 CN Delft, The Netherlands*
[2]*Politecnico di Milano, Department of Civil and Environmental Engineering, piazza Leonardo da Vinci 32, 20133 Milano, Italy*

Abstract. Modelling the water retention behaviour requires proper understanding of all the processes which affect the amount of water stored in the pore network, depending on the soil state and the soil history. Traditionally, in many applications a single water content – suction curve is used. This approach limits the applicability of the retention data to practical cases, especially when fine grain soils are dealt with, when the deformability and activity of the clay fraction significantly affect the interaction with water. On the other side, water retention is being recognised more and more as a fundamental information in the description of the mechanical response of the soil, as it provides the key connection to the partial volumetric strains in a deformation process. With reference to the work performed at the Politecnico di Milano in the last years, a contribution on the understanding and modelling the coupled water retention- mechanical response in deformable soils is presented. The contribution aims to: (i) summarise the mechanisms which contribute to water retention; (ii) point out the role played by an evolving fabric and the fluid properties on water retention; and (iii) provide an overview on some of the consequences of evolving water retention properties on the mechanical behaviour.

1 Introduction

Water retention was early recognised as fundamental information to describe the hydraulic response of unsaturated soils [1]. Early studies pointed out the role of the type of clay and of clay content on the retention potential of soils. The role of organic content was also addressed extensively, as most of the models were developed for applications in the field of agronomy [e.g. 2,3]. Conversely, when unsaturated soil mechanics started being developed, most of the attention was devoted to compressibility and strength, in an attempt to describe the mechanical behaviour of unsaturated soils as a unique function of suction. The role of water retention on the mechanical response was initially disregarded, with the exception of the numerical analysis of unsaturated consolidation, which was requiring some information on water storage. Simple non linear functions, fitting available data on water content and suction were used to the scope [e.g. 4].

At the end of the '90s, inherent coupling between the retention mechanisms and the mechanical response started being investigated and included in models for unsaturated soil behaviour [e.g. 5-7]. A comprehensive approach to soil water retention was introduced by Romero and co-workers, who started analysing the role played by fabric on the retention properties of compacted soils [8,9]. Vaunat et al. [10] developed the first fully coupled hydro-mechanical model, which included an explicit description of the soil water retention behaviour.

In the last years, a huge body of research has been devoted to better quantify and model the influence of deformation on the water retention behaviour, especially focussing on the dependence of the air-entry value on the void ratio [11–14]. A more comprehensive insight into the role of the fabric on water storage can be provided by analysing the soil microstructure features by means of experimental techniques. In general, mercury intrusion porosimetry (MIP) is the preferred method to this aim, after Delage introduced its systematic use to analyse the consequences of mechanical loads on the pore size distribution (PSD) [15]. Mercury intrusion porosimetry has a number of limitations, which were recently reviewed by Romero and Simms [16]. Most importantly, MIP provides only a scalar measure of the apparent size of interconnected pores, while the pore network topology depends on the complex 3-D fabric of the different structure levels of the soil. Nonetheless, relevant basic information can be extracted from MIP data, which can be used to account for the soil fabric evolution in modelling the unsaturated soil behaviour.

Multiscale hydro-mechanical coupling is suggested by MIP data. This originates from the interplay behaviour of the different microstructure levels. At the scale of element testing in the laboratory, the hydraulic and the mechanical response depend on the mutual interaction between the different fabric levels, which govern the internal evolution of the soil microstructure. Some of these aspects are discussed hereafter, based on the work performed by the authors in the last years.

[a] Corresponding author: c.jommi@tudelft.nl

DOI: 10.1051/e3sconf/20160904002

2 Water retention in soils

Traditionally, water retention in soils is claimed to be described by a "soil water retention curve", also called sometimes "soil water characteristic curve". Actually, different mechanisms contribute to the current amount of water stored into a given soil, which cannot be fully described by a single curve, nor is a characteristic of the soil. More properly, a water retention domain must be described, which reflects the dependence of the adsorption and storage potential of a soil from the current fabric and from its interaction with a specific fluid.

In the formulation of conceptual water retention models, stored and adsorbed water have been traditionally quantified in different ways in the various fields of application. Either mass based measures like water content, w (mass of water over solid mass), or volumetric measures, as volumetric water content θ_w (volume of water over total soil volume) and degree of saturation S_r (volume of water over volume of voids), have been adopted. Each of the previous measures presents advantages and shortcomings, as in an evolving fabric neither the mass, nor the volume can play the role of a constant reference [17]. The water ratio, e_w (volume of water over solid volume) might be somehow preferred, as it physically represents a measure of the mass of water (to which energy is referred), it can directly substitute the water ratio, $e_w = w\,G_s$, where G_s is the specific density of the soil particles, and it can be equally used as a measure of the degree of saturation, when normalised to the void ratio e, $S_r = e_w / e$.

Romero and Vaunat [9] used the water ratio in their fundamental paper, which was the starting point of many following developments. Their conceptual synthesis, copied in figure 1, clarifies the different exchange and retention mechanism, which dominates the storage response in the different suction ranges.

The multiscale character of water retention is also demonstrated in figure 1: in the lower suction range, a storage mechanism prevails, which depends on void ratio.

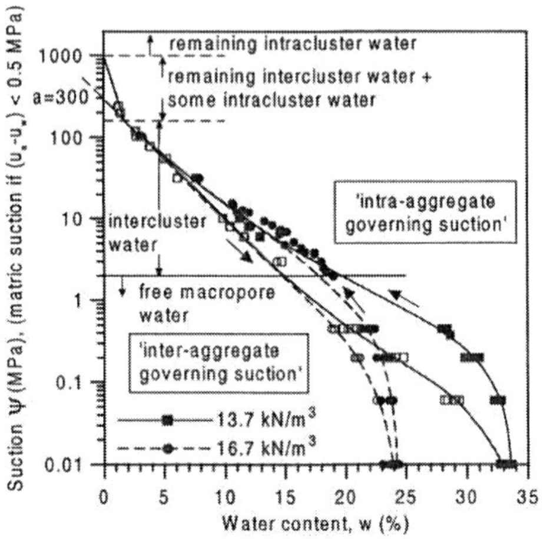

Figure 1. Water retention domain of Boom clay compacted dry of optimum, showing the different levels of fabric in which water can be stored (from Romero and Vaunat [9]).

At low water contents, the influence of initial and current void ratio is less significant, as the suction–water content relationship is mainly dependent on the mineralogical composition of the clay. Also, Romero et al. [8] and Romero and Vaunat [9] evidenced that a link exist between the different physical mechanisms dominating water storage in the different suction intervals and the pore size distribution, PSD, of the soil.

3 Evolution of water retention properties

Experimental evidence, especially on compacted soils, showed that the fabric of a given soil is not unique. It was demonstrated that the size of the aggregates is sensitive to the compaction water content [18], to the chemical composition of the pore fluid [19], and to the hydromechanical path history [20]. Furthermore, the fabric developed during the soil formation is not fixed, but evolves along multiphysics generalised stress path, as discussed in [21-24].

Sensitivity of the different classes of pores to hydraulic and mechanical paths was extensively discussed by Romero et al. [25], with reference to data on compacted Boom clay. In figure 2, the PSD of the as-compacted sample is compared to those of samples subjected to various hydromechanical paths after compaction. One sample was loaded at constant water content, a second one underwent wetting at constant volume, and the last one experienced free saturation at null vertical stress. The as-compacted sample is characterised by a bimodal distribution. Loading the sample at vertical stress of 0.6 MPa shifts the size of dominant macropores towards a slightly lower value, while the micropore volume and its dominant pore size are not affected.

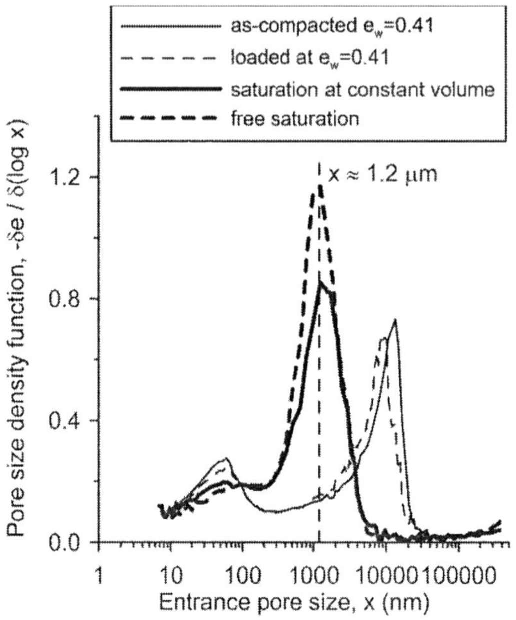

Figure 2. PSD functions of compacted Boom clay, subjected to different hydromechanical paths (from Romero et al. [25]).

On the contrary, the two saturation paths affect both inter-aggregate and intra-aggregate pores. After saturation, the two samples display a similar fabric dominated by a single peak, which is originated by both expansion of the micropores and reduction of macropores. The PSD function of a sample, which was further subjected to a drying path up to a total suction of 100 MPa after wetting at constant volume, showed that the inter-aggregate porosity eventually shrunk significantly, while the intra-aggregate porosity was almost completely recovered. The conclusion from the work was that the aggregate fabric tends to swell and shrink almost reversibly, becoming an inherent feature of the soil. On the contrary, the porosity between the aggregates suffers from irreversible modifications, and it is responsible of the irreversible strains observed at the laboratory test scale.

3.1 Modelling the water retention domain

The previous experimental observation promoted the development of a multiscale approach in modelling the water retention domain, which is based on the separation between intra-aggregate and inter-aggregate water ratios, having different evolution rules. The separation reads:

$$e_w = e_{wm} + e_{wM} \tag{1}$$

where the subscripts m indicate the intra-aggregate and M the inter-aggregate water ratios, respectively. The intra-aggregate pore volume was assumed to evolve reversibly. For water ratios lower than a characteristic value e^*_m, the aggregates are unsaturated, and the inter-aggregate voids are empty. Vice versa, for $e_w > e^*_m$, the aggregates are saturated, and the available water also partly fills the inter-aggregate voids. The delimiting value between the two regions, e^*_m, belongs to the $e_w = e_m$ line, and describes empty inter-aggregate voids and saturated intra-aggregate pore space. The data for $e_w > e^*_m$ are of interest in the description of the interaction between the two porosity levels. As a first approximation, a linear relationship between e_m and e_w was proposed to describe the evolution of the intra-aggregate pore volume:

$$e_m = e^*_m + \beta(e_w - e^*_m) \tag{2}$$

where the parameter β describes the slope of the approximating linear interpolation function and quantifies the swelling tendency of the aggregates.

Based on this assumption, a conceptual water retention model was presented by Romero et al. [25], and further discussed by Della Vecchia et al. [26]. Having assumed that the void ratio does not influence the intra-aggregate retention mechanisms, the intra-aggregate portion of the retention domain is delimited by fixed main wetting (W) and main drying (D) branches, uniquely linking the suction to a measure of water content:

$$e_{wD,W} = \frac{b_{D,W}}{\ln\left(\frac{s_{max}}{s^*_{mD,W}}\right)} e^*_m \left[\frac{b_{D,W} + \ln\left(\frac{s_{max}}{s^*_{mD,W}}\right)}{b_{D,W} + \ln\left(\frac{s}{s^*_{mD,W}}\right)} - 1\right] \tag{3}$$

where the subscripts D and W stand for the drying and the wetting branches, respectively. In the previous equation, e_w is the water ratio on the drying or the wetting branch of the curve, s_{max} is the value of suction corresponding to $e_w = 0$, $s^*_{mD;W}$ are the suction values corresponding to e_{wm}, that is, the smallest value of water ratio corresponding to saturated microvoids and empty macrovoids, on the drying and the wetting branches, respectively. The average slope of the two branches is given by the two parameters $b_{D,W}$.

The portion of the water retention domain that describes the retention mechanism of the inter-aggregate pore space may be scaled in the range $e \geq e_w \geq e_m$ with the expression:

$$e_{wD,W} = e_m + (e - e_m)\left[1 - \frac{\ln\left(1 + \frac{s}{s_{mD,W}}\right)}{2}\right] \cdot \left[\frac{1}{1 + (\alpha_{D,W} s)^{n_{D,W}}}\right]^{m_{D,W}} \tag{4}$$

where the subscripts D and W again indicate the drying and the wetting branches, respectively.

The microscopic void ratio e_m was continuously adjusted following the relationship in Equation (2). In this way, swelling of the aggregates, and therefore, reduction of the inter-aggregate pore volume at constant void ratio by invasion of the aggregates, is explicitly taken into account by the parameter β. Explicit introduction of the intra-aggregate void ratio e_m and of its evolution in the phenomenological retention equations characterises the multiscale interaction at the hydraulic level. The variable s_m is assumed to change following the microscopic portion of the relevant branch of the water retention domain

$$s_{mD,W} = s^*_{mD,W} \exp\left[\frac{b_{D,W} e^*_m \left(b_{D,W} + \ln\left(\frac{s_{max}}{s^*_{mD,W}}\right)\right)}{e_{wmD,W} \ln\left(\frac{s_{max}}{s^*_{mD,W}}\right) + b_{D,W} e^*_m} - b_{D,W}\right] \tag{5}$$

Imposing the two analytical expressions in Equations (3) and (2) to be continuous together with their first derivatives in ($s_{D,W} = s_{mD,W}$, $e_w = e_{wm}$) gives a unique dependence of α on the set of independent parameters, which implicitly describes the dependence of the air entry value, AEV, on the macroscopic void ratio.

In figure 3, experimental data of compacted Boom clay at different void ratios are compared with the model calibration. In spite of the model having been developed specifically on the experimental results from Boom clay, it proved able to reproduce the water retention domain for soils having different activity, as the comparison between data and calibrated curves for Barcelona silt and Febex bentonite, shown in figure 4, demonstrates.

It is worth noting that the model was formulated accounting for the experimental evidence from MIP data, which were partially exploited also on a quantitative basis to describe the microscopic void ratio evolution as a function of the water ratio. The number of parameters is comparable – or less – to that of other formulations developed to account for changes in the air entry value with void ratio. At the same time, as the formulation is physically based, some of these parameters can be inferred from basic soil properties, namely the Atterberg limits or the specific surface. The correlations presented in the papers by Romero et al. [25] and Della Vecchia et al. [27] provide the initial reference for the acceptable range of parameters. The model is able to track fabric changes, measured by the intra-aggregate and inter-aggregate porosities. Following Della Vecchia et al [27], the approach can be conceptually extended to any number of fabric levels.

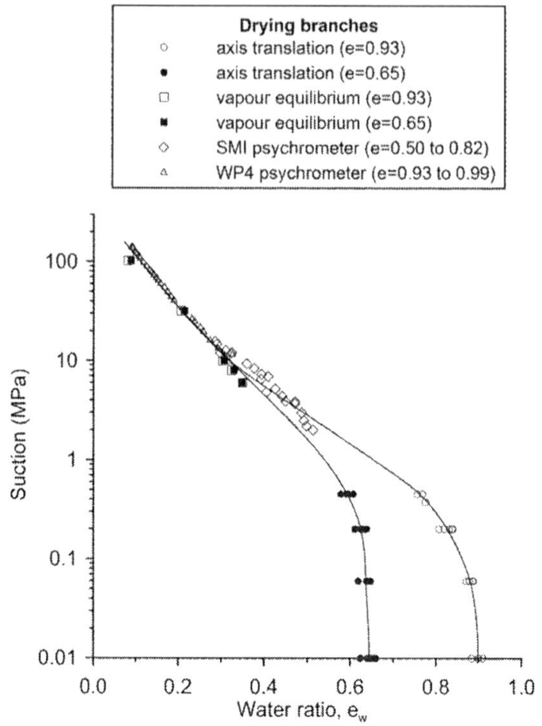

Figure 3. Water retention data and model for Boom clay compacted on the dry side of optimum (from Romero et al. [25]).

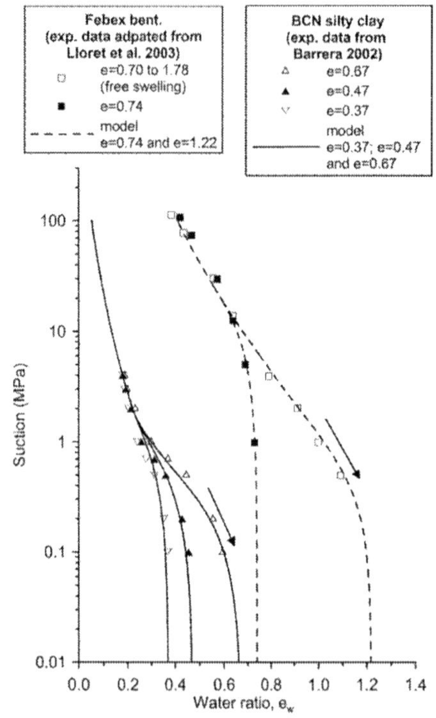

Figure 4. Water retention data and model for the wetting branches of the water retention domain of Barcelona silt and Febex bentonite (from Romero et al. [25]).

3.2 Influence of soil-water chemistry

The activity of the soil is dealt with by the parameter β, which actually is a function of the water chemistry too, for active soils. The influence of the water chemistry on the swelling potential of a given soil can be dealt with by changing β.

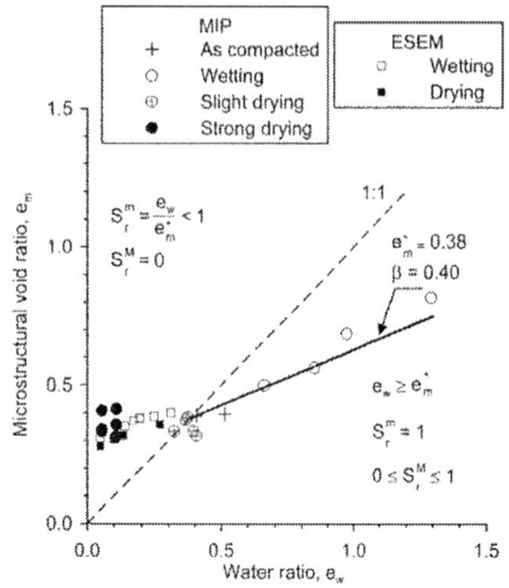

Figure 5. Evolution of microstructural void ratio with water ratio from MIP and ESEM data on compacted Boom clay (from Romero et al. [25]).

The role of β in the description of the microfabric evolution is demonstrated with reference to data on compacted Boom clay in figure 5. For water content in excess of e^*_m , the ratio between the microstructural void ratio, which measures the size of the aggregates, and the water content is measured by β. The linear approximation depicted in figure 5 can be used as a first approximation until saturated conditions are approached.

The parameter β quantifies how much the aggregates swell as a function of water availability. As swelling of the aggregates is due to the physico-chemical interaction of the solid particles with water, the chemical composition of water is playing a role too. Swelling of the aggregates will reduce the amount of macropores for a given total porosity, affecting the shape of the retention domain. An example is shown in figure 6, where the drying branches of the water retention curves of saturated bentonite-enriched sand mixtures, with active and inert bentonite, are compared. In the second case, the swelling potential of the bentonite was reduced by changing the chemical composition of the water by adding lead nitrate. The curves derived from MIP data clearly demonstrate the difference between the amount of water which can be stored in a swelling or non-swelling microstructure for a given suction. In figure 7 data from a wetting-drying path on the same material with active bentonite are simulated using a value of $\beta = 0.85$. Simulating inert bentonite by using a value of $\beta = 0.11$, would predict much lower values of water ratio for the same suction, as demonstrated by the continuous line in the figure.

Musso et al. [19] further exploited the approach in analysing the chemo-hydro-mechanical response of active clay, using MIP data to calibrate a relationship describing the dependence of the microscopic void ratio on the osmotic suction.

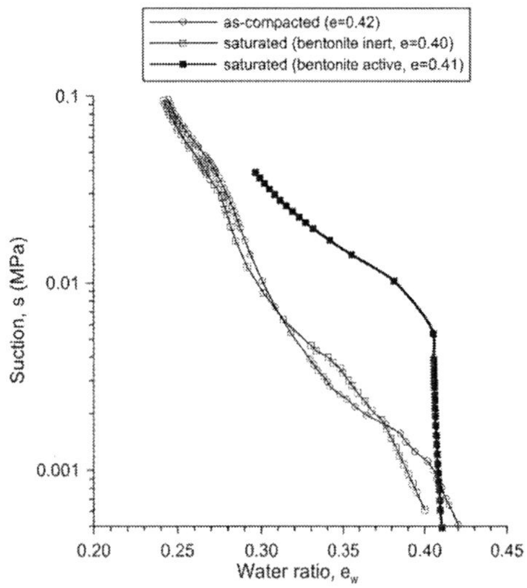

Figure 6. Drying branches of the water retention curves derived from MIP data for saturated bentonite-enriched sand mixtures, with active and inert bentonite, compared to the as-compacted state (from Romero et al. [25]).

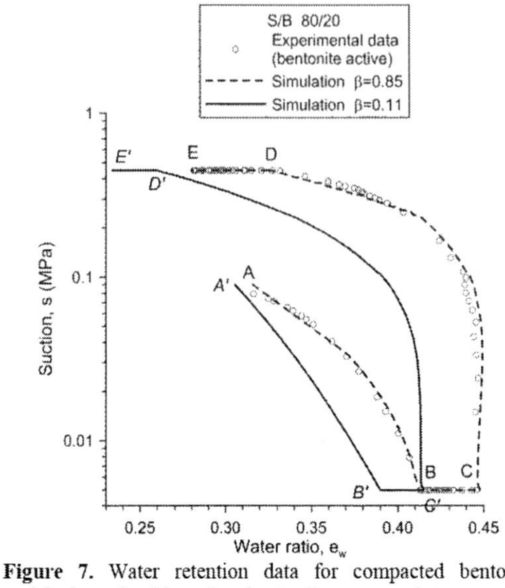

Figure 7. Water retention data for compacted bentonite-enriched sand with active bentonite compared to numerical simulations along the stress path ABCDE (from Romero et al. [25]).

4 Hydro-mechanical coupling

The evolution of the fabric, which governs the water retention potential, equally affects the mechanical response of soil. On the one hand, for a given water content, different suctions are expected for different soil fabrics. On the other hand, changing the size of the aggregates will change the amount and the size of the inter-aggregate pores, which are mostly responsible for the irreversible volumetric strains.

4.1 The role of volumetric strain on the retention domain

Experimental data on water retention typically show that the air-entry value, changes as a function of the void ratio. Advanced models for water retention [e.g. 11] tend to include this feature explicitly in the model equations. An interesting feature of the model recalled in the section 3.1 is that it implicitly predicts the evolution of the air-entry value, AEV, and of the air-occlusion value, AOV, with the total void ratio e and the intra-aggregate void ratio e_m, without adding any dedicated parameter. Model predictions of void ratio effects on the AEV and AOV of a compacted scaly clay are presented in Figure 8 (from [28]), compared with the values proposed in Airò Farulla et al. [29]. Model responses in terms of iso-suction lines is presented in Figure 9 in the plane (w;e), both for main wetting and main drying paths. The model correctly predicts the retention behaviour to be independent from the dry density for high values of suction: for $s > s_m$ the predicted iso-suction curves are vertical. For lower constant suction values, variation in void ratio implies variations in water content.

The difference between the main drying and the main wetting curves for a given suction defines the width of the hysteretic domain, which also changes with void

DOI: 10.1051/e3sconf/20160904002

ratio. To characterise the hydraulic hysteretic behaviour of the material, Airò Farulla et al. [29] proposed a non-dimensional hysteresis index, HI, defined as

$$HI = 1 - \frac{A_w}{A_d} \tag{6}$$

where A_w and A_d are the areas subtended by the main wetting and the main drying curves, respectively, calculated as

$$A = \int_{s_{min}}^{s_{max}} e_w \, ds \tag{7}$$

The change in the hysteresis index as a function of void ratio predicted by the model is presented in Figure 10, together with the values calculated by [29]. The model is able to reproduce the increase of the size of the hysteresis loop with void ratio, again without introducing any dedicated parameter.

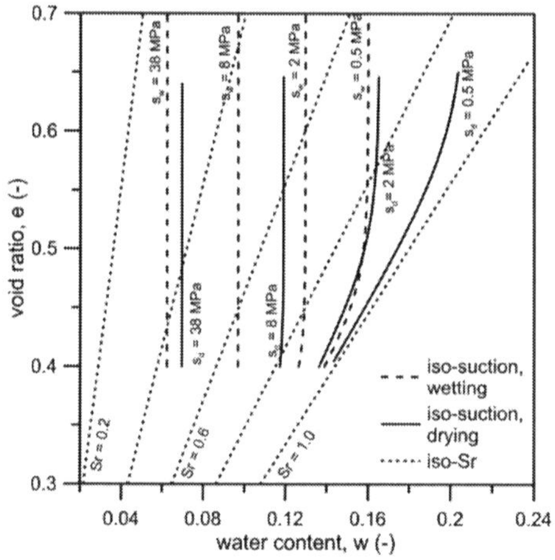

Figure 8. Water content vs. void ratio at constant suction. (from Della Vecchia et al. [28]).

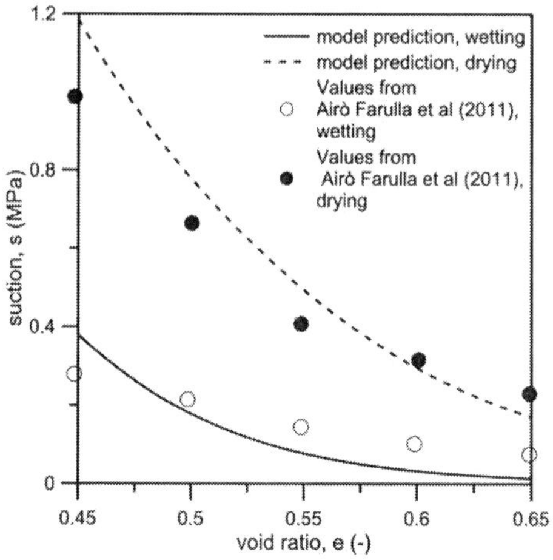

Figure 9. Evolution of AEV and AOV with void ratio: model predictions (from Della Vecchia et al. [28]) and data (from Airò Farulla et al. [29]).

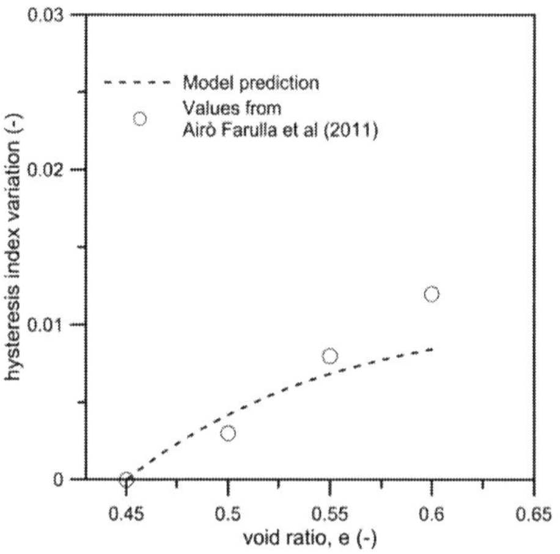

Figure 10. Model prediction of the hysteresis index variation (with respect to the reference void ratio e=0.45) with void ratio. (data from [29]).

4.2 The role of microfabric on the volumetric response

In the description of the mechanical response, different combination of "operative" stresses can be adopted for modelling the mechanical behaviour of unsaturated soils. Modelling the water retention as an integral part of the behaviour requires two stress variables to provide a complete formulation [30, 31]. Recently, Vaunat et al. [32] suggested a unified approach to stress variables, encompassing most of the previous choices, and linking the so-called Bishop's stress to the fabric evolution.

Here a simple volumetric average for the fluid pressures is used in the definition of the average skeleton stress:

$$\sigma'_{ij} = \sigma_{ij} - u_a \, \delta_{ij} + S_r \, (u_a - u_w) \, \delta_{ij} \tag{8}$$

where u_a is the air pressure, u_w is the water pressure and S_r is the degree of saturation (δ_{ij} is the Kronecker delta).

In the previous volumetric average, information on the water retention, depending on microstructural features, is essential in providing the link between the degree of saturation and the suction $s = u_a - u_w$. Besides, fabric evolution due to water content changes influences the stress–strain response of the material as well.

The experimental evidence suggests that the slope of the normal consolidation line is seldom constant with suction, even if expressed in terms of the average skeleton stress [33-36]. Koliji [36] suggested that this is a consequence of the influence of suction on the soil structure, where the term structure was used to indicate the combined effects of fabric, composition and interparticle forces. An increase in the post-yield compressibility in the effective stress plane at increasing suction was observed, which was interpreted as the experimental evidence of a plastic degradation mechanism. Because of suction, the preconsolidation pressure increases, and a state to the right of the virgin compression curve for saturated conditions may be reached by the soil. Increasing the applied effective stress, the soil destructurates, and the compression index increases.

Similar evidence was recognised for compacted Boom clay [33]. Figure 11 shows the evolution of the elastic–plastic volumetric compliance in oedometer conditions $C_{oed} = -\Delta e/\Delta \ln\sigma'_v$ as a function of water ratio. The data were obtained from tests performed on samples prepared at different water contents, statically compacted at the same dry density (ρ_d = 1.37 Mg/m^3), and then subjected to compression in oedometer at constant water content. In the figure, other data from compaction stages at constant water content by Romero [37] are reported for comparison. The dependence of C_{oed} on water content was interpreted as a consequence of the related change of the size of the aggregates. To describe this dependence, a new state variable, D, was introduced by Della Vecchia et al [26,33], which describes the amount of available pore space invaded by the aggregates with respect to the current available pore space:

$$D = \frac{e_m - e_m^*}{e \ - e_m^*} = \beta \frac{e_w - e_m^*}{e \ - e_m^*} \tag{9}$$

The state variable D is a function of the parameters e_m and β, and it represents the mechanical counterpart of the influence of the size of the aggregates on the hydraulic behaviour. The variable D increases with the size of the aggregates, and it takes the value 0 for completely empty macrostructure and β in saturated conditions. While the dependence of the volumetric compliance on the water ratio is evidenced in Figure 11, Figure 12 clearly shows that the state parameter D, which represents a measure of the water ratio dependent fabric, is better able to normalise the elastic–plastic compliance data.

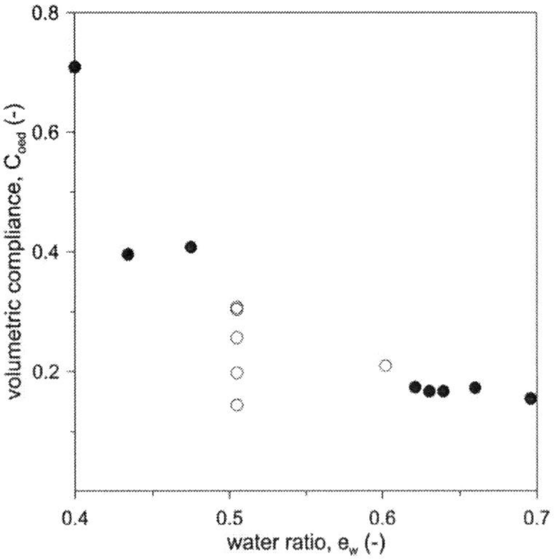

Figure 11. Evolution with water ratio of elastic-plastic logarithmic compliance under oedometer conditions for compacted Boom clay (data from [33] and [37]).

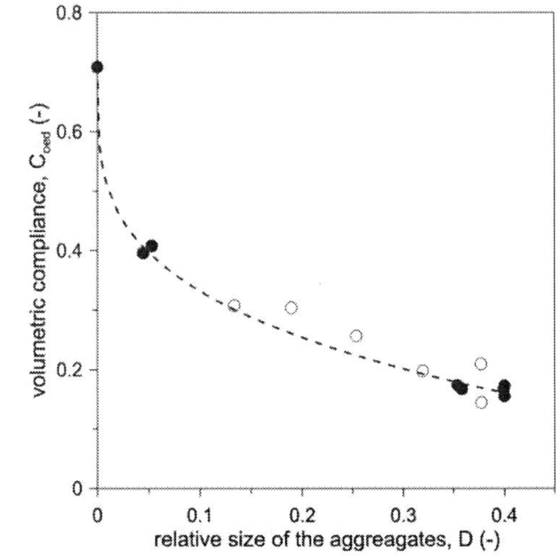

Figure 12. Evolution with the relative size of the aggregates of elastic-plastic logarithmic compliance under oedometer conditions for compacted Boom clay (data from [33] and [37]).

Further insight into the role of mutual interaction between the different fabric levels is provided by analysing the theoretical response of the inter-aggregate e_M and the intra-aggregate e_m voids during a cyclic isotropic wetting–drying test [26], shown in Figure 13. Intra-aggregate void ratio evolution was calculated by Equation (2), whereas e_M was obtained as the difference between the predicted values of total void ratio e and e_m. During the first wetting stage, e_m increases with increasing water content, despite the global collapse of the sample. During the subsequent drying–wetting cycle, a hysteretic response of the aggregate size is predicted. Reversibility of intra-aggregate evolution with water content implies a hysteretic response in terms of suction because of the nature of the adopted water retention model. Increase in e_m with decreasing e leads to a pronounced variation of the inter-aggregate void ratio e_M because of both global collapse of the sample and invasion of the aggregates in the inter-aggregate pores. In Figure 14, this concept is highlighted with reference to the first wetting stage. Reduction of the inter-aggregate void ratio Δe_M, defined as

$$\Delta e_M = \Delta e - \Delta e_m \tag{10}$$

can be interpreted as the sum of the contributions of the total collapse measured on the sample, $\Delta e_{coll} = \Delta e$, and of the invasion of the aggregates in the macrovoids $\Delta e_{inv} = -\Delta e_m$:

$$\Delta e_M = \Delta e_{coll} + \Delta e_{inv} \tag{11}$$

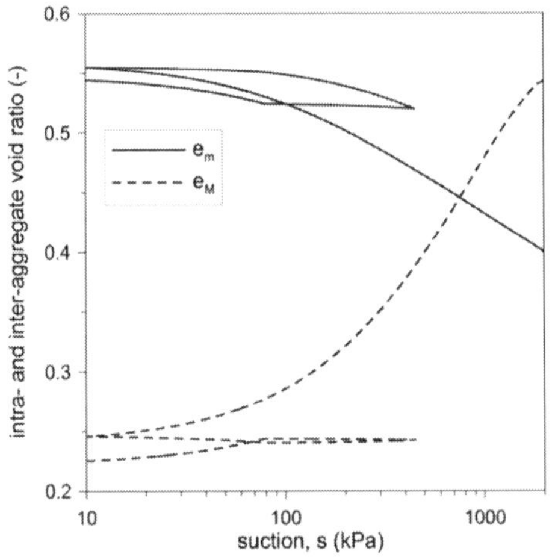

Figure 13. Predicted evolution of intra-aggregate and inter-aggregate void ratio with suction during an isotropic wetting test (from [26]).

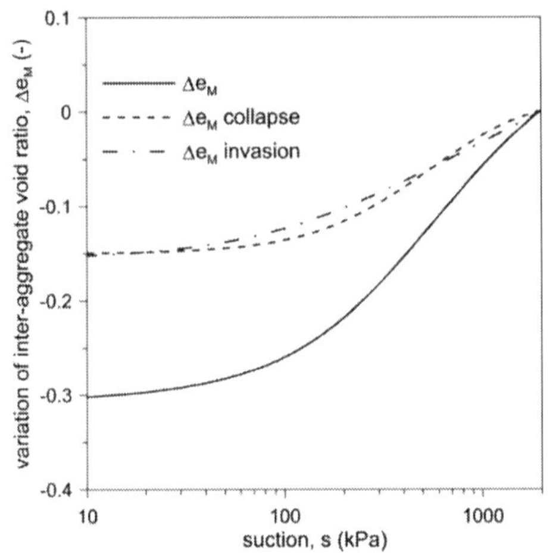

Figure 14. Variation of inter-aggregate void ratio during volumetric collapse (from [26]).

5 Selected applications in the context of earth constructions

The evolution of the fabric, which governs the water retention potential and the mechanical behaviour, is particularly relevant on the behaviour of compacted soils used in earth construction. Although the fabric created during compaction marks the response of the material, continuous interaction with the atmosphere and the groundwater produces irreversible strains, which change the hydro-mechanical response of the soil in the field, with respect to the as-compacted conditions. Two examples are briefly reported in the following. The first one refers to the Po river embankments, where samples retrieved from the field were used to quantify the change in the state of the compacted material during the lifetime of the construction. The second example shows the changes in the fabric estimated for a compacted soil stabilised with a biological technique.

5.1 Data from the Po river embankments

After a major flood occurred in 1952, which caused significant damage in the low Po river floodplain, an extensive investigation was started to characterise the typical soils, mainly clayey sandy silts, used in the construction of the flood protection embankments. One of the aims was to better quantify the optimal compaction conditions and verify the effectiveness in the field of the compaction procedures. Design specifications indicated that compaction should have guaranteed a dry density not lower than 95% optimum Proctor reference, with a tolerance on the water content of ± 2%. Data collected on samples retrieved in the field after compaction revealed that, in spite the design specifications being mostly respected, the actual as-compacted state was dominated by the seasonal variations in the water content of the available soil. During the wet season, higher water content resulted in compaction on the wet side of optimum with lower densities than the optimum. Conversely, in the dry season higher densities were usually obtained, but at the expenses of water content lower than the optimum.

Samples retrieved 250 days after the end of the construction, revealed that, irrespective of the initial as compacted state, the degree of saturation of the soil had increased and that the dry density had decreased, with more relevant changes for the samples compacted in the dry season [38].

Cyclic wetting-drying cycles, typically characterising the interaction of the earth construction with the atmosphere after the end of the construction, were simulated with the coupled hydro-mechanical model [39], to verify whether the fabric changes could be responsible of the evolution of the state of the samples in the field. The simulated as-compacted states are reported in Figure 15, together with the reference experimental compaction curves from [38]. In the figure, the simulated dry densities and water contents after cyclic wetting-drying are reported too. The results of the simulations fall in the same range of the samples collected in situ 250 days after the end of the construction, confirming that water content increases and dry density tends to decrease due to cyclic wetting-drying cycles experienced at the stress levels investigated.

A deeper insight into the coupled volumetric mechanisms may be provided by analysing the results of the numerical simulation of a wetting–drying–wetting cycle in detail. The results of the numerical simulations for the tests run imposing a net vertical stress of $\sigma_v = 10$ kPa are reported in Figures 16 and 17. An irreversible increment of water content is always to be expected as a result of the hysteretic retention mechanisms of compacted soils. Independently of the initial state of the compacted soil, the first wetting stage promotes an irreversible increase in water content. In the following drying and wetting paths the water content cycles in a smaller range. The results for void ratio, collected in Figure 17, highlight that if the soil is compacted at the wet of optimum it may undergo significant shrinkage upon first drying. The swelling strain depends slightly on the applied stress, but it is mostly governed again by the retention properties of the soil. For the soil compacted wet of optimum, the main effect of suction is to act as an external load, hence inducing positive plastic volumetric strains.

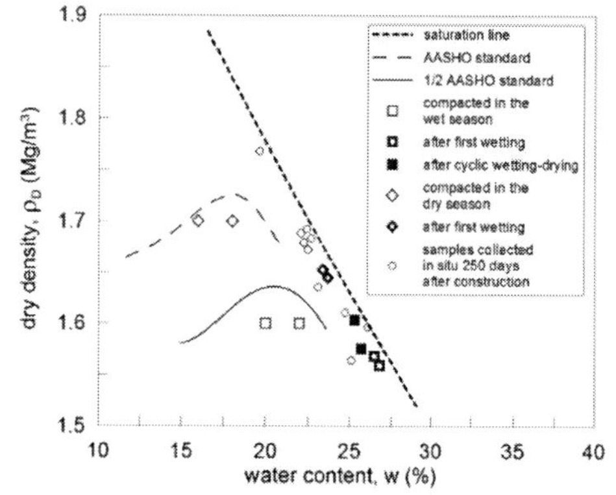

Figure 15. Simulated as-compacted states and states after a single and cyclic wetting stage (from Jommi & Della Vecchia [39]). Experimental data from [38].

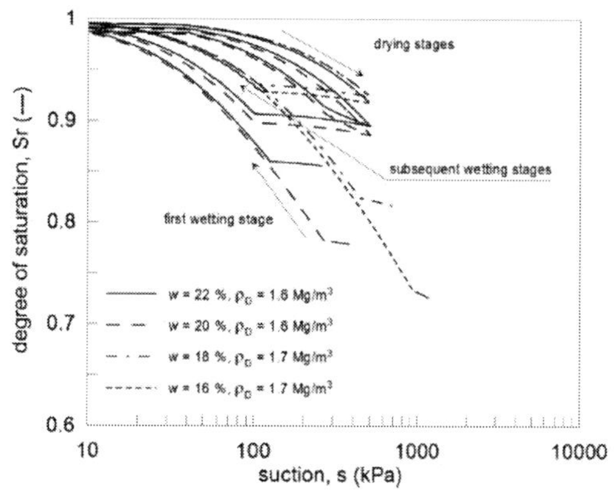

Figure 16. Simulated as-compacted states and states after a single and cyclic wetting stage (from Jommi & Della Vecchia [39]).

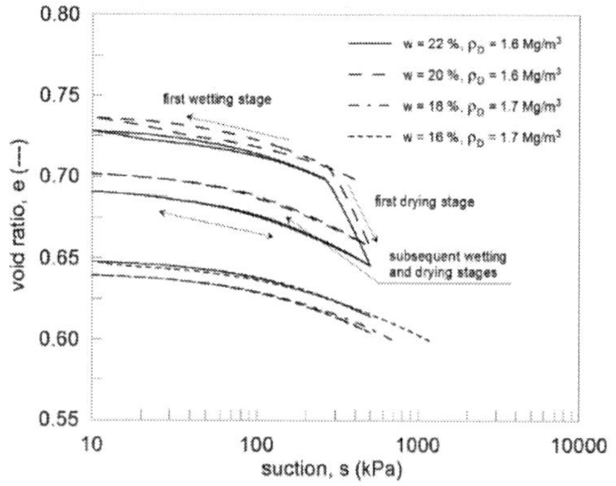

Figure 17. Simulated as-compacted states and states after a single and cyclic wetting stage (from Jommi & Della Vecchia [39]).

Following on the same line, Azizi [40] recently presented the results of a comprehensive investigation aimed at understanding the general consequences of drying-wetting cycles on the coupled hydromechanical response of compacted soils for earth constructions. The results of the study show that significant change in the fabric of the soil occur in the first drying-wetting cycles, which affect to a large extent the water retention properties, as well as the mechanical behaviour although to a less extent.

5.2 Biological improvement of compacted soils

A stabilisation technique, based on a microbiological treatment, was studied for different Iberian soils used in compacted linear earth constructions, resulting in the production of calcite by bio-mineralisation. To evaluate the advantages of the stabilisation procedure at the material level, laboratory tests were run on the soils, treated with different amounts of microorganisms during different treatment time intervals, and compacted afterwards [41]. The experimental data suggested that precipitation of calcium carbonate from bacteria had taken place in the pores which are slightly bigger than the characteristic size of the bacteria themselves. Comparison of hydro-mechanical data on the natural and the treated soils showed that the change in the pore size distribution is reflected consistently on the water retention properties, on the shear strength, and on the compressibility of the compacted samples. In view of the field applications, the data were exploited to develop a simple constitutive model for the stabilised soil, starting from the constitutive properties of its natural untreated counterpart [42].

Based on the results of the microstructure investigation [41], microbiological precipitation of calcite seems to produce two effects at the microscopic level. Firstly, it changes the pore size distribution due to microbiological calcite filling the pores having diameter in the range 3 μm ÷ 30 μm. Secondly, the liquid limit and the plasticity index of the soil decrease as the amount of microorganism increases.

The voids in which calcite precipitates do not seem to be affected by changes in the total void ratio, as the comparison between the relevant pore size distributions suggests (see [41]). Under this hypothesis, precipitation of calcite produces a decrease of the inter-aggregate pore space, with respect to the untreated sample, of an amount Δe^{BIO}:

$$e_M = e - (e_m + \Delta e^{BIO}) \qquad (12)$$

in which the microbiologically affected pore volume may be estimated as:

$$\Delta e^{BIO} = \int_{3\mu m}^{30\mu m} (PSD_{nat} - PSD_{trated})dD \qquad (13)$$

Besides, as the plastic characteristics of the soil are slightly decreased by the presence of calcite, the value of e_m* decreased consistently (Fig. 18, from [42]).

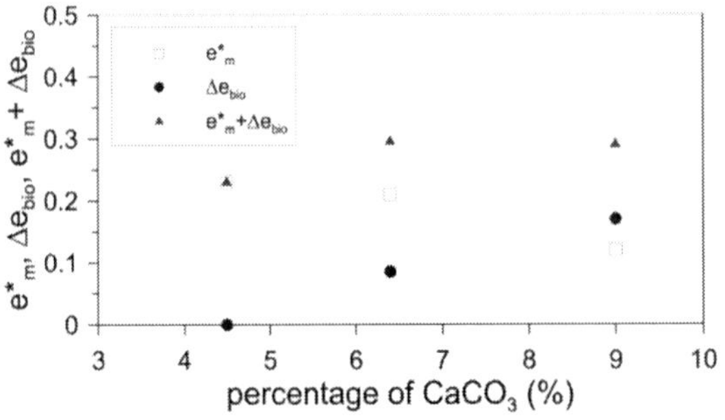

Figure 18. Variation of the microscopic pore space with calcite content at the end of the microbiological treatment (from [42]).

After compaction, specimens of different samples were loaded under vertical stresses representing the as-compacted condition in the field. They were further flooded at constant stress and eventually subjected to a loading-unloading cycle. The results showed that: (i) the amount of collapse induced by wetting was a function of the void ratio at the time of flooding, irrespective of the microbiological treatment; (ii) the biological treatment reduced the plastic compliance after collapse.

The hydro-mechanical oedometer tests were simulated by adopting the hydromechanical coupled framework presented in [43, 26]. The retention curves for the treated and the untreated soil were calibrated on the results of MIP data. The slope of the elastic-plastic compression line λ and the apparent pre-consolidation pressure were calibrated for the different samples on part of the data from the laboratory. All the other parameters were the same for the natural and the treated samples [42]. The comparison between the numerical simulations and the experimental data is reported in Figures 19 and 20. They show that a very good match can be obtained for both samples, adopting the same constitutive coupled hydro-mechanical approach, by suitable calibration of few parameters.

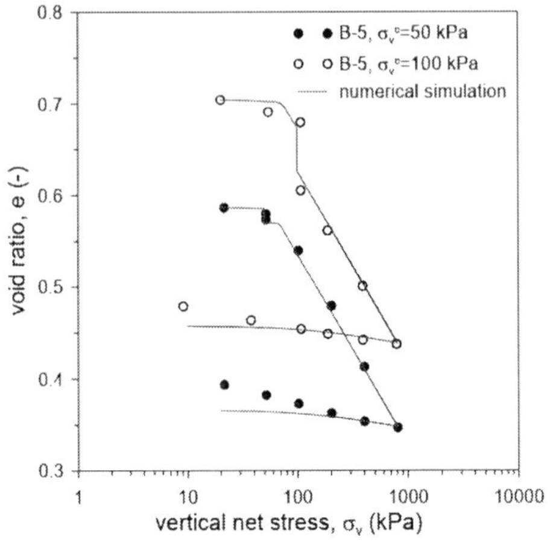

Figure 19. Comparison between experimental data (symbols) and numerical simulations of the oedometer tests on the original compacted soil (from [42]).

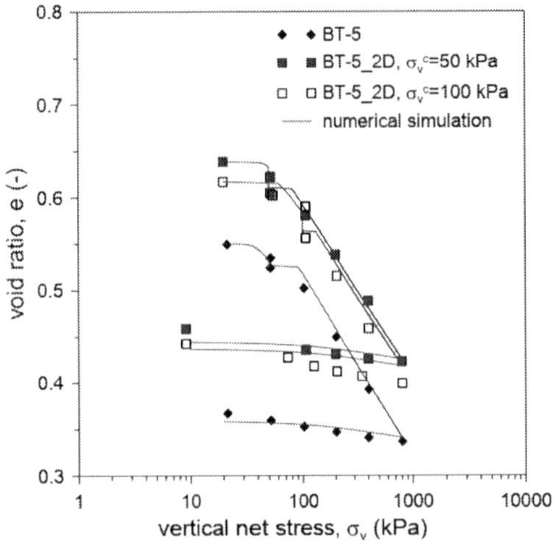

Figure 20. Comparison between experimental data (symbols) and numerical simulations of the oedometer tests on the compacted samples microbiologically (from [42]).

6 Conclusions

The results of the studies presented in this review summarise a number of findings, also coming from other research groups, which are considered to be relevant in the proper interpretation of the coupled hydromechanical behaviour of unsaturated soils. Although the work presented here mostly refers to compacted soils, the general principles equally apply to natural and reconstituted soils.

The key concept in the understanding of the water retention behaviour is that it is the macroscopic evidence of the mutual interaction of different fabric levels with water. This includes the chemical composition of water as far as it reacts with soil particles and aggregates.

The soil formation process gives the footprint of the hydraulic and mechanical response, but the following hydromechanical history affects the fabric of the soil, modifying its storage and deformational response. Water retention is not a characteristic of the soil neither it depends merely on void ratio. Although not complete, the pore size distribution is a better representation of the fabric features which influence the water storage, as well as the mechanical response. MIP can be usefully employed to this aim, as well as to detect relevant changes in the soil fabric during generalised hydromechanical paths.

An underlying assumption to this investigation, as well as all the other works cited, is that the different fabric levels are in thermodynamic equilibrium along any hydromechanical loading path. Removing this hypothesis will provide further insight into the time dependent behaviour of unsaturated soils, and a more general view on the dynamics of coupled process in unsaturated soils.

References

1. W.R. Gardner, Soil Science **85**, 228-232 (1958)
2. S.C. Gupta, W. Larson, Water Res. Res. **15**, 1633-1635 (1979)
3. T. Mayr, N.J. Jarvis, Geoderma **91**, 1-9 (1999)
4. A. Lloret, E.E. Alonso, Géotechnique **30** (4), 449-477 (1980)
5. S. Vanapalli, D. Fredlund, D. Pufhal, Géotechnique **49** (2), 143-159 (1999)
6. K. Kawai, S. Kato, D. Karube, *Proc. 1st Asian Conf. on Unsaturated Soils. Unsaturated Soils for Asia*, 329-334 (2000)
7. C. Ng, Y. Pang, J. Geotech. Geoenv. Eng. **126**(2), 157-166 (2000)
8. E. Romero, A. Gens, A. Lloret, *Eng. Geol.* **54** (1-2), 117-127 (1999)
9. E. Romero, J. Vanuat, *Experimental Evidence and Theoretical Approaches in Unsaturated Soils*, 91-106 (2000)
10. J. Vaunat, E. Romero, C. Jommi, *Experimental Evidence and Theoretical Approaches in Unsaturated Soils* 121-138 (2000)
11. D. Gallipoli, S.J. Wheeler, M. Karstunen, Géotechnique **53**(1), 105-112 (2003)
12. M. Nuth, L. Laloui, Comp. Geot. **35**, 835-844 (2008)
13. A. Tarantino, Géotechnique **59**(9), 751-762 (2008)
14. D. Masin, Int. J. Num. Anal. Meth. Geomech. **34**(1), 73-90 (2010)
15. P. Delage, G. Lefebvre, Can. Geot. J. **21**, 21-35 (1984)
16. E. Romero, P.H. Simms, Geotech. Geol. Eng. **26** (6), 705–722 (2008).
17. D. Masin, Engineering Geology **165**, 73-88 (2013)
18. R. Thom, R. Sivakumar, V. Sivakumar, E. Murray, P.A. MacKinnon, Géotechnique **57**(7), 469-474 (2007)
19. G. Musso, E. Romero, G. Della Vecchia, Géotechnique **63**(3), 206-220 (2013)
20. P. Delage, Géotechnique **60**(5), 353-368 (2010).
21. O. Cuisinier, L. Laloui, Int. J. Num. Anal. Meth. Geomech. **28**(6), 483-499 (2004).
22. A. Koliji, L. Laloui. O. Cuisinier. L. Vulliet, Transp. Porous Media **64**, 261-278 (2006).
23. R. Monroy, L. Zdravkoviv, A. Ridley, Géotechnique **60**(2), 105-119 (2010)
24. Q. Wang, A.M. Tang, Y.J. Cui, P. Delage, B. Gatmiri, Eng. Geol. **124** (4): 59-66 (2012).
25. E. Romero, G. Della Vecchia, C. Jommi, Géotechnique 61(4), 313-328 (2011).
26. G. Della Vecchia, C. Jommi, E. Romero, Int. J. Num. Anal. Meth. Geomech. **37**, 503-535 (2013).
27. G. Della Vecchia, A.C. Dieudonnè, C. Jommi, R. Charlier, Int. J. Num. Anal. Meth. Geomech. **39** (7), 702-723(2015).
28. G. Della Vecchia, C. Airò Farulla, C. Jommi, *Unsaturated Soils: Research and Application, Proceedings of the Second European Conference on Unsaturated Soils* 2, 55-62 (2012)
29. C. Airò Farulla, A. Battiato, A. Ferrari, *Unsaturated Soils. Proceedings of the Fifth International Conference on Unsaturated Soils* 1, 417–422 (2011)
30. A. Gens, *Proc. 1st Int. Conf. Unsat. Soils* (3), 1179-1200 (1996)
31. G.T. Houlsby, Géotechnique **47** (1), 193-196 (1997)
32. J. Vaunat, F. Casini, E. Romero, *submitted to Géotechnique* (2016).
33. G. Della Vecchia, PhD Thesis, Politecnico di Milano (2009)
34. S. Wheeler, V. Sivakumar, Gèotechnique **45**, 35-53 (1995)
35. D. Sun, H. Matsuoka, Y. Xu, Geot. Test. J. **27**(4), 362-370 (2004)
36. A. Koliji, PhD Thesis, EPFL (2008)
37. E. Romero, PhD Thesis, UPC (1999)
38. F. Colleselli, G. Cortellazzo, C. Jommi, R. Lagioia, *XX Italian National Congress of Geotechnics*, 435-443 (1999)
39. C. Jommi, G. Della Vecchia, *Mechanics of Unsaturated Geomaterials*, 353-373 (2010).
40. A. Azizi, PhD Thesis, Politecnico di Milano (2016)
41. L. Morales, E. Romero, C. Jommi, E. Garzón, A. Giménez, Acta Geotechnica (10), 157-171 (2015)
42. G. Della Vecchia, L. Morales, E. Garzon, C. Jommi, E. Romero, *Unsaturated Soils, proceedings of the Fifth International Conference on Unsaturated Soils*, 795-801 (2010)
43. E. Romero, C. Jommi, Water Resources Research **44**, W12412:1–W12412:16 (2008)

DOI: 10.1051/e3sconf/20160904003

Modelling an in-situ ventilation test in the Andra Underground Research Facilities

Frédéric Collin[1,a], Jean Talandier[2] and Benoit Pardoen[3]

[1]*University of Liège, ArGEnCo Department, 9 Allée de la Découverte 4000 Liège, Belgique*
[2]*Andra, Research and Development Division, Rue Jean Monnet 1-7, 92298 Châtenay-Malabry, France*
[3]*Université Laval, Département de Génie Civil et de Génie des Eaux, 1065 Avenue de la Médecine, Québec, Canada*

Abstract. Wastes resulting from the nuclear electricity production have to be isolated from the biosphere for a very long period of time. For this purpose, deep underground repository in weak permeable geological layers is considered as a reliable solution for the nuclear waste storage. It is however well established that during excavation, the underground drilling process engenders cracks and eventually fractures [1] that deteriorate the hydro-mechanical properties of the surrounding host material in the so-called Excavation Damaged Zone (EDZ). The EDZ behaviour is a major issue because it may constitute a preferential flow path for radionuclide migration. Consequently, the characterisation of the material transport properties and of the transfer kinetics that occur around galleries still need to be investigated. The EDZ properties may be also affected by host rock-gallery air interactions. Ventilation induced drying may also provoke additional cracking, which potentially alters the transport properties of the damaged zone. Large-scale air ventilation experiments are performed in Underground Research Laboratories (URL) that have been constructed to check the feasibility of the repository. A numerical modelling of the SDZ air ventilation test (Andra URL) performed in a low permeability rock is proposed in order to both predict the development of the EDZ during excavation and study the air interaction with the host formation during maintenance phases.

1 Introduction

Wastes resulting from the nuclear electricity production have to be isolated from the biosphere for a very long period of time. For this purpose, deep underground repository in low permeable geological layers is considered as a reliable solution for the nuclear waste storage. Networks of galleries and access shafts will be constructed in these geological formations. The design of these underground structures must ensure that the good confining properties of the host rock will not be affected during the different steps of the storage. It is however well established that during excavation, the underground drilling process engenders cracks and eventually fractures [1] that deteriorates the hydro-mechanical properties of the surrounding host material. These modifications take place in a zone called Excavation Damaged Zone (EDZ), located around the galleries, which is affected by important modifications of the material flow characteristics such as permeability increase [2]. Since a low hydraulic conductivity is required to ensure a safe long-term disposal, the EDZ behaviour is a major issue because it may constitute a preferential flow path for radionuclide migration. Consequently, the characterisation of the material transport properties and of the transfer kinetics that occur around galleries still need to be investigated.

The EDZ properties may be also affected by host rock-gallery air interactions. Ventilation during excavation and maintenance phases has a direct influence on relative humidity and temperature of the gallery atmosphere. Ventilation induced drying may also provoke additional cracking, which potentially modifies the transport properties of the damaged zone.

In order to better understand the impact of these processes, large-scale air ventilation experiments are performed in Underground Research Laboratories (URL) that have been constructed to check the feasibility of the repository. The objective of the in-situ experiment (SDZ) performed in the Andra URL (Meuse Haute Marne region in France) is the characterisation of the structure and the size of the excavated damaged zone (EDZ), potentially influenced by the desaturation and possible resaturation due to controlled gallery ventilation conditions.

A numerical modelling of the SDZ air ventilation test performed in a low permeability rock is proposed in order to both predict the development of the EDZ during excavation and study the air interaction with the host formation during maintenance phases. The studied rock is the Callovo-Oxfordian claystone (COx), which is envisaged for deep underground repository of nuclear wastes in France [3]. Numerous modelling of the EDZ behaviour have been realised in the past decades. Theses modelling often include separately hydro-mechanical coupling [4, 5], permeability variation [6], flow transfers

[a] Corresponding author: f.collin@ulg.ac.be

[7], or strain localisation [8]. In the present study, the objective is to predict, in a fully coupled approach, the shear band pattern in the EDZ, the permeability increase within the fractures and the impact of ventilation on the two latter processes. The numerical developments were performed in the non-linear finite element code Lagamine [9].

EDZ evidences and the considered large-scale air ventilation experiment are firstly detailed in section 2. Then, the hydro-mechanical constitutive model is developed in section 3 for an unsaturated porous media. Finally, the gallery excavation and the air ventilation experiment are numerically modelled in the section 4. The first numerical phase is the gallery excavation modelling during which the development of the fractures that compose the EDZ is reproduced by shear banding [8]. Considering the fractured rock as a continuous medium at the macroscale, the intrinsic hydraulic permeability evolution is reproduced through a strain-dependent relation [10]. Succeeding to the excavation, the air ventilation experiment is reproduced in the underground gallery. In this second numerical phase, the exchanges at gallery wall between the claystone and the cavity air are characterised with a non-classical mixed hydraulic boundary condition [11, 12] that involves both water vapour and liquid water transfers.

2 SDZ experiment

The in-situ experiment (SDZ), devoted to the characterisation of the structure and the size of the excavated damaged zone (EDZ), is located in an experimental zone isolated from the GED experimental horizontal gallery by a 2 m long airlock. The test zone contains a first part without covering (with a length of 7.2 m) and, at the end, another 5 m long zone with a concrete covering with a thickness of 20 cm and a thin impervious geotextile (Figure 1).

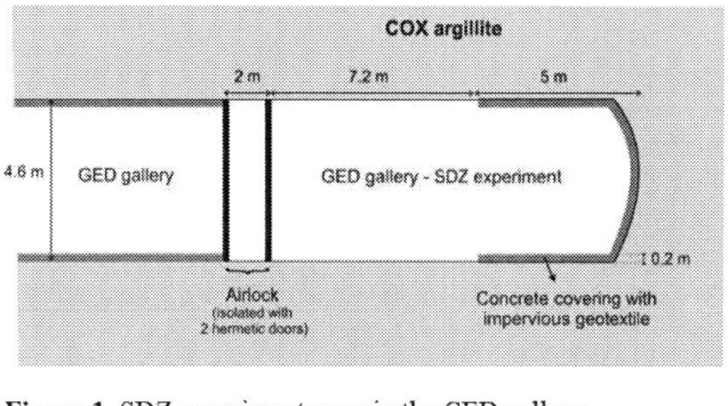

Figure 1. SDZ experiment zone in the GED gallery

A series of drilling in the SDZ zone and in the GED gallery allow for a description of the fracturing around the galleries as well as permeability and water content measurement. During the test, the pore pressure evolution in different directions and radial distances is monitored. Furthermore, a geological survey of the fractured zone as well as measurements of the gallery convergence is also performed by Andra after the gallery excavation. All these measurements allow for a better observation of the coupled hydro mechanical couplings in the EDZ.

2.1 Excavation damaged zone

The EDZ has been carefully investigated in the Andra URL through, for instance, fracture measurements, permeability analysis, and flow transfers. For the Callovo-Oxfordian claystone, induced extension and shear fractures are detected in the proximity of the galleries, with shearing as the principal failure mechanism because of the high in situ stress environment [13]. The SDZ ventilation experiment is located in the GED experimental gallery, which is oriented parallel to the minor horizontal principal stress. Around this gallery, the permeability increase has been highlighted by measurements performed under saturated conditions in boreholes that are drilled in different orientations. Three zones can be defined: an undisturbed zone with a permeability lower than 10^{-19} m^2 far from the gallery, a slightly disturbed zone with a permeability ranging from 10^{-19} m^2 to 10^{-17} m^2, and a highly disturbed zone close to the gallery with a permeability higher than 10^{-17} m^2 (increase higher than 2 orders of magnitude). The extents of the zones are detailed in Fig. 2.

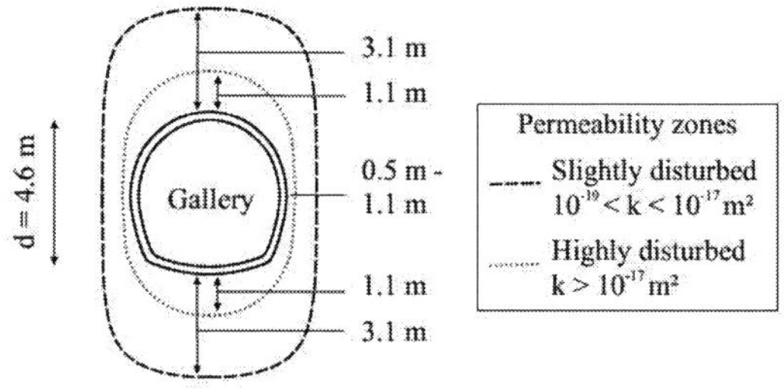

Figure 2. SDZ experiment zone in the GED gallery

2.2 Ventilation experiment

Ventilation can lead to rock desaturation, stress modification, as well as modification of the fracturing structure close to the drifts. The major objective of the SDZ ventilation test is to characterise the influence of controlled ventilation on the hydraulic transfers taking place around the galleries, especially in the EDZ. In the experimental zone isolated from the rest of the laboratory, different levels of hygrometry are applied in order to highlight the effect of drainage and wetting. The ventilation can be sequenced in different phases (Fig. 3): (1) the first phase corresponds to the global laboratory ventilation before the airlock closure, (2) once the airlock is closed the ventilation is stopped in the experimental zone, later a controlled ventilation is imposed with hygrometric conditions of (3) 30 % of relative humidity and then of (4) 60 % of relative humidity. The air temperature T and relative humidity RH evolutions are monitored in the experimental zone in different gallery sections.

The pore water pressure evolution is monitored inside the rock mass, around the experimental zone without covering. After equilibrium of the sensors with the rock mass, the progressive drainage of the rock is monitored

 DOI: 10.1051/e3sconf/20160904003

with a decrease of the pore water pressure. One can observe that the drainage is not important in the far field, where the pore water pressures are close to the initial rock water pressure. It is more important around the gallery where the measurements are about the atmospheric pressure (in a 3 m ring). In the excavation damaged zone, we can expect that the increase of permeability and the drainage lead to rock desaturation. In the long term, the two controlled ventilation phases (phases (3) and (4) in Fig. 3) only have a minor effect on the measured pore water pressure.

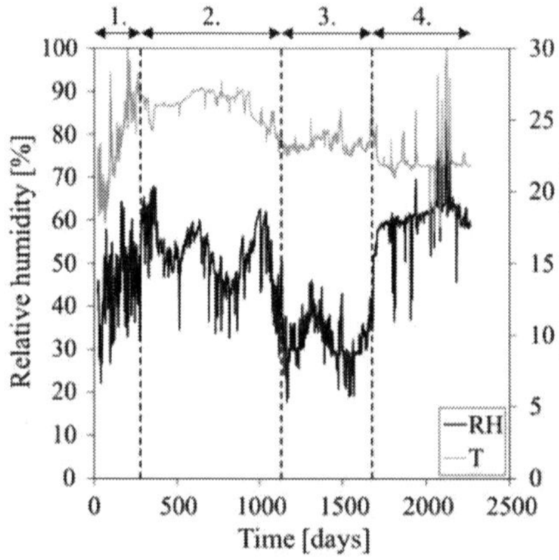

Figure 3. Temperature and relative humidity evolution in the SDZ gallery.

To characterise the desaturation, water content measurements have also been performed on core samples coming from horizontal boreholes drilled in the experimental zone without covering. They indicate, firstly, a strong desaturation close to the gallery wall with 3 % ≤ w ≤ 5 % that can extend up to 0.5 m depth, secondly, a moderated desaturation deeper in the rock, and thirdly, a limited desaturation or no desaturation deeper than 2 m with 6.7 % ≤ w ≤ 8.2 % for the saturated claystone.

3 Framework for an unsaturated media with microstructure

The observations and the measurements of the SDZ experiment clearly evidence the existence of the excavation damaged zone, where modifications of mechanical and flow properties appear. In this work, the numerical modelling of the EDZ aims at representing the complexity of this zone in a fully coupled approach. The goal is to be able to predict the localisation of the deformation in shear bands, which are the precursor of fractures [14]. Inside these localisation bands, the permeability is deeply increased. In fact, the coupling between mechanical deformation, permeability, and hydraulic flow is indubitably of paramount importance to obtain an adequate representation of the EDZ hydro-mechanical behaviour.

The first step is to define a correct and robust modelling of the strain localisation leading finally to the rupture. However, it is well known that classical finite element models suffer of a pathological problem of mesh dependency for strain localisation modelling. This is due to properties of the underlying mathematical problem. There is a need of an internal length scale for a correct modelling of the post peak behaviour. Several approaches have been proposed to introduce such internal length scale in the problem. Among the different regularization techniques, we propose in the following to use the second gradient model [15], which is a particular model of a microstructure medium. The second grade model has been further extended to saturated biphasic medium [16] and unsaturated conditions [8].

A more pronounced permeability increase inside the fractures (shear bands) due to strain localisation effect is expected. Different approaches exist in the literature to predict the evolution of the permeability with the mechanical loading (micromechanics, FE^2 computation, semi-empirical formulation …). A strain-dependent evolution of the material intrinsic permeability will be used in this work, in order to introduce this hydro-mechanical coupling effect.

The second step is the modelling of the gallery air ventilation. The SDZ experiment in the experimental zone without covering will be reproduced after the gallery excavation. The air-rock interaction has also to be characterised to reproduce the exchanges at gallery wall correctly.

In the following numerical modelling, these two steps will be investigated with an elaborated description of the hydro-mechanical coupling, especially within the EDZ. To describe the numerical model, the balance equations are firstly recalled. Then the constitutive equations of the mechanical and the fluid transfer problems are presented, with special emphasis on the different coupling existing between the mechanical and the hydraulic parts.

3.1 Balance of momentum

Let us recall first the governing equations of a monophasic medium with microstructure. This kind of models can be traced back to the pioneering work of the Cosserat brothers [17], Mindlin [18] and Germain [19]. In the framework of microstructure continuum theory, a microkinematic gradient field v_{ij} is introduced to describe strain and rotation at the microscale. With respect to classical continuum mechanics, additional terms are then added in the internal virtual work of a given body [19].

In the local second gradient model used in the following, an assumption is added: the micro kinematic gradient is equal to the macro-deformation gradient F_{ij}

$$v_{ij} = F_{ij} \tag{1}$$

Finally, the principle of virtual work can be rewritten as follows:

$$W_i^* = \int_{\Omega^t} \left(\sigma_{ij}^t \frac{\partial u_i^*}{\partial x_j^t} + \Sigma_{ijk}^t \frac{\partial^2 u_i^*}{\partial x_j^t \partial x_k^t} \right) d\Omega^t \tag{2}$$

where σ^t_{ij} is the Cauchy stress and Σ^t_{ijk} is the double stress dual to the second gradient of the macro displacement.

Assuming that the boundary Ω^t is regular, the external virtual work can be defined as follows:

$$W^*_e = \int_{\Omega^t} \rho g_i u^*_i d\Omega^t + \int_{\Gamma^t_\sigma} \left(\bar{t}_i u^*_i + \bar{T}_i D u^*_i\right) d\Gamma^t \quad (3)$$

where $\bar{t}_i$ is the external (classical) forces per unit area and $\bar{T}_i$ an additional external (double) force per unit area, both applied on a part Γ_σ of the boundary of Ω^t. D denotes the normal derivative of any quantity q, (Dq = $(\delta q/\delta x_k)n_k$).

The previous virtual work equation of second gradient models can of course be used in a finite element code. However, this equation needs the use of C^1 functions for the displacement field as second derivatives of the displacement are involved. In order to avoid such functions in the virtual work principle (Eq. 2), the equalities between v_{ij} and F_{ij} and between v^*_{ij} and v^*_{ij} are introduced through a field of Lagrange multipliers λ_{ij} related to a weak form of the constraint (Eq. 1) (see [15]):

$$\int_{\Omega^t} \left(\sigma^t_{ij} \frac{\partial u^*_i}{\partial x^t_j} + \Sigma^t_{ijk} \frac{\partial v^*_{ij}}{\partial x^t_k} \right) d\Omega^t - \int_{\Omega^t} \lambda^t_{ij} \left(\frac{\partial u^*_i}{\partial x^t_j} - v^*_{ij} \right) d\Omega^t - W^*_e = 0 \quad (4)$$

$$\int_{\Omega^t} \lambda^*_{ij} \left(\frac{\partial u^t_i}{\partial x^t_j} - v^t_{ij} \right) d\Omega^t = 0 \quad (5)$$

3.2 Fluid mass balance equation

The water mass balance equation reads in a weak form:

$$\int_{\Omega^t} \left(\dot{M}^t p^*_w - m^t_i \frac{\partial p^*_w}{\partial x^t_i} \right) d\Omega^t = \int_{\Omega^t} Q^t p^*_w d\Omega^t - \int_{\Gamma^t_q} \bar{q}^t p^*_w d\Gamma^t \quad (6)$$

where p^*_w is the virtual pore water pressure field, Q^t is a sink term and Γ_q is the part of the boundary where the input water mass per unit area $\bar{q}$ is prescribed. M^t and m^t_i are respectively the mass of the water inside the current configuration of the skeleton Ω^t and the mass flow.

The mass flow m^t_i is defined as follows:

$$m^t_i = -\rho^t_w \frac{\kappa_{ij} k^t_{r,w}}{\mu_w} \left(\frac{\partial p^t_w}{\partial x^t_j} + \rho^t_w g_j \right) \quad (7)$$

where κ_{ij} is the anisotropic intrinsic permeability, $k^t_{r,w}$ is water relative permeability and μ_w is the water dynamic viscosity.

The compressible fluid is assumed to respect the following relationship [20]. This predicts an increase of water density as a function of the pore water pressure, defining χ_w as the water bulk modulus:

$$\dot{\rho}^t_w = \frac{\rho^t_w}{\chi_w} \dot{p}^t_w \quad (8)$$

The time derivative of the water mass is obtained directly by considering the influence of compressibility of the fluid phase and the solid grains, variation of water saturation and modification of the porous volume.

3.3 Constitutive law

Geomaterials exhibit a porous structure where an assembly of solid grains forms the solid skeleton and fluids can fill in the porous space between these grains. Generally, two fluids are considered: a wetting and a non-wetting fluid corresponding respectively to a liquid and a gaseous phase. The effect of fluid pressures on the Cauchy total stress field σ_{ij} is defined assuming Biot's postulate of effective stress [21]. This definition accounts for the relative deformability of the solid grains by introducing its compressibility through the Biot's tensor:

$$\sigma_{ij} = \sigma'_{ij} + b_{ij} S_{r,w} p_w \quad (9)$$

where σ'_{ij} is the Biot's effective stress field defined under soil mechanics convention in which compressive stress is positive, b_{ij} is Biot's tensor and p_w is the pore water pressure, negative if suction.

An isotropic transverse linear elastic-perfectly plastic model (with a Van Eekelen yield surface) is assumed to reproduce the mechanical behaviour of the porous media [22]. The elastic and plastic theories are defined hereafter for anisotropic materials, which means that the elastoplastic properties depend on the coordinate axes to which the properties are referred and therefore depend on the orientation.

By separating the elastic and plastic components of the strain tensor, a linear elastic relationship links the elastic strain rate ε^e_{ij} to the effective stress rate σ'_{ij} through the Hooke's law. C^e_{ijkl} is the Hooke anisotropic elastic tensor [23] that must be symmetric because of thermodynamic requirement [24]. In case of orthotropy and cross-anisotropy, the properties are related to the orthotropic (and orthogonal) axes e_1, e_2 and e_3 and the (e_1; e_2; e_3) space represents the orthotropic configuration [25]. For cross-anisotropy, five independent parameters are needed to describe the material elasticity: $E_\perp$; $E_{//}$; $\nu_{////}$; $\nu_{//\perp}$; $G_{//\perp}$, where the subscripts // and $\perp$ indicate the directions parallel and perpendicular to the bedding, respectively. We consider that (e_1; e_2) is the isotropic planes orientation (bedding planes for sedimentary materials) and e_3 is the normal to these planes. The Hooke's law being formulate in the orthotropic axes, a change of the reference system has to be computed to obtain the stress state.

Another important aspect of the anisotropic elasticity is the compressibility of the solid grains skeleton. It is expressed through Biot's symmetric tensor which is defined for anisotropic material as follows [26]:

$$b_{ij} = \delta_{ij} - \frac{C^e_{ijkk}}{3K_s} \quad [10]$$

where δ_{ij} is the Kronecker symbol and K_s is the bulk modulus of the solid phase.

In addition to the anisotropic elastic behaviour, the plastic behaviour and the strength parameters of the material can also be anisotropic and therefore depend on direction and orientation. Various failure criteria for anisotropic materials have been developed [27] and hereafter, we focus on the material cohesion anisotropy which is defined with a second order microstructure fabric tensor a_{ij} describing the cohesion spatial distribution. This tensor is a measure of the material fabric and its eigenvectors correspond to the principal material microstructure axes, i.e. the orthotropic axes, e_1; e_2; e_3. The cohesion corresponds to the projection of this tensor on a generalised unit loading vector l_i that characterises the loading direction relative to the material microstructure orientations [28]. Each component l_i corresponds to the stress resultant acting on facets of normal e_i and is defined as follows:

$$l_i = \sqrt{\frac{\sigma_{i1}^{\#'2}+\sigma_{i2}^{\#'2}+\sigma_{i3}^{\#'2}}{\sigma_{ij}^{\#'2}\sigma_{ij}^{\#'2}}} \qquad [11]$$

where $\sigma_{ij}^{\#'}$ is expressed in reference to the material axes. Therefore, the cohesion specifies the effect of load orientation relative to the material axes [29]:

$$c_0 = a_{ij}\, l_i l_j \qquad [12]$$

4 Results and comparison

When considering deep underground repository of nuclear waste, there is a need for a more accurate modelling of the EDZ and of the gallery air ventilation effect on the flow transfers taking place around the galleries. Therefore, a hydro-mechanical modelling of gallery excavation and of the SDZ ventilation test is realised in two-dimensional plane strain state.

4.1. Numerical model

The considered rock is the Callovo-Oxfordian claystone which is a transversely isotropic material with horizontal bedding planes and an initial anisotropic stress state. For a gallery oriented along the minor horizontal principal stress, the initial stress state and pore water pressure are:

$\sigma_{x,0}$ = 15.6 MPa

$\sigma_{y,0} = \sigma_{z,0}$ = 12 MPa

$p_{w,0}$ = 4.5 MPa

where $p_{w,0}$ is the initial pore water pressure and $\sigma_{x,0}$, $\sigma_{y,0}$, $\sigma_{z,0}$ are the initial total stresses.

The modelling considers the SDZ experimental zone without covering and only one gallery quarter for symmetry reason. The mesh geometry is a gallery with a 2.3 m radius and is discretized with 2520 elements and 9802 nodes. To establish the symmetry, the symmetry x- and y-axes are considered as impervious (no water flow) and the normal displacement, as well as the normal derivative of the radial displacement, cancel along these axes [8]. At the mesh external boundaries, the normal total stress and the pore water pressure are imposed constant.

4.2. Excavation phase

As a first step, the drilling process and the EDZ development are modelled. During the excavation, the major objectives are to characterise the development of fractures induced by rock deconfinement and the modification of the hydraulic properties. The fractures are represented with shear banding and the permeability evolution is addressed with a strain-dependent relation.

The gallery excavation is performed with a convergence-confinement method [30] and decreasing stresses at gallery wall with the excavation front crossing the studied section after one week. The pore water pressure in the gallery is assumed to decrease rapidly to one atmospheric pressure when the excavation front crosses the studied section. It is assumed that the gallery air remains fully saturated with water vapour because the gallery is not ventilated yet. It means that the air relative humidity is of 100 % and the water pressure in the cavity corresponds to the atmospheric pressure.

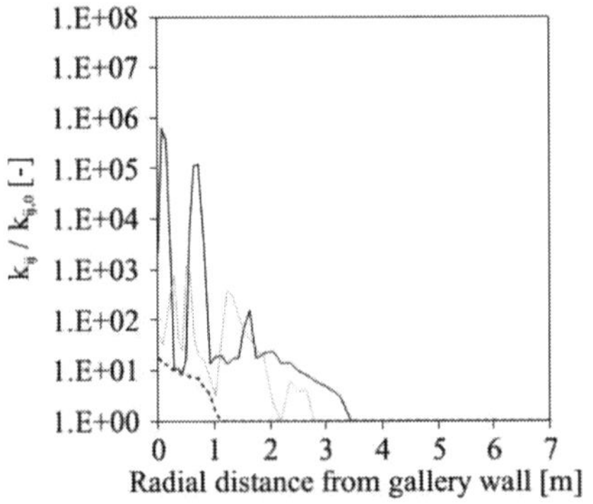

Figure 4. Intrinsic permeability at the end of excavation in vertical (black), oblique (grey) and horizontal (dotted line) direction.

The evolution of the intrinsic hydraulic permeability is reproduced with a strain-dependent formulation. The evolution parameters for the permeability are calibrated based on the shear banding results and on in situ experimental measurements of permeability around the GED gallery (Fig. 2). An appropriate solution considers a formulation mixing the deformation and the plastic indicator (yield index). The following equation uses the equivalent deviatoric strain $\hat{\varepsilon}_{ij}$ and the yield index YI, leading to:

$$\kappa_{ij} = \kappa_{ij,0}\left(1 + \beta\langle YI - YI^{thr}\rangle \hat{\varepsilon}_{ij}^{3}\right) \qquad [13]$$

The results for $\beta = 1 \times 10^{10}$ and $YI^{thr} = 0.95$ illustrated in Fig. 4 show a permeability increase in a quite good agreement with the experimental measurements for all the directions. The permeability evolution during the excavation and the comparison to experimental data are consistent. In the far field, the permeability corresponds to the undamaged rock and is well captured in the modelling. In the near field, the permeability corresponds to the fractured rock (EDZ).

4.3. Ventilation phase

The second part of the numerical modelling consists in applying, after the excavation, an air ventilation in the underground gallery that corresponds to the SDZ ventilation experiment. After the first step, the fracture description as well as the permeability evolution and inhomogeneity in the EDZ have been taken into account. The following challenge is to analyse whether the introduced hydro-mechanical coupling and EDZ description allow reproducing the drainage and desaturation provoked in the surrounding media by the air ventilation. The principal objectives of the SDZ test modelling are to characterise, firstly, the air-rock transfers at gallery wall, and secondly, the influence of a controlled gallery ventilation on the claystone behaviour.

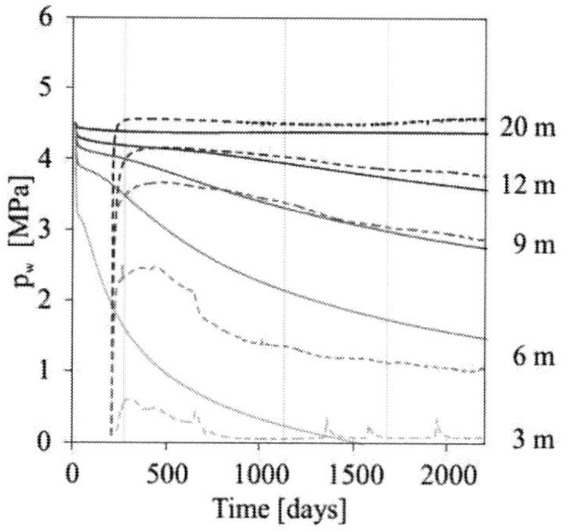

Figure 5. Pore water pressure evolution in horizontal boreholes with comparison to experimental measurements for different distances from gallery wall

After the drilling, the ventilation is progressively applied in the underground gallery with a drying of the air which is numerically reproduced by a ventilation initiation phase. During this initiation phase (between 21 and 50 days of simulation), the water pressure in the gallery air decreases from the atmospheric pressure to a value corresponding to the first measurement of air humidity. Then, the air ventilation is applied in the gallery by means of the hygrometric measurements. During the ventilation, the pore water pressure and the water vapour density in the cavity correspond to the relative humidity of the cavity air by Kelvin's law.

The first result that can be analysed is the claystone progressive drainage which occurs due to the ventilation. In Fig. 5 is illustrated the comparison between numerical results and experimental measurements of pore water pressure evolution in horizontal boreholes drilled around the drift. It is obvious that the equilibrium phase of the measure (increase of p_w) is not reproduced by the numerical modelling which starts with the initial pore water pressure value of the clayey rock, $p_{w,0}$ = 4.5 MPa. Only the drainage is reproduced and has to be compared to experimental values after about 275 days of modelling. A close correspondence is observed between the numerical results and the experimental measurements in the different directions. The pore overpressures observed after the excavation in the vertical direction result of hydro-mechanical couplings mainly related to the anisotropy of the initial stress state. In this direction, the matching with the experimental data is slightly less good which may be related to the lower vertical permeability ($k_{\perp\perp} < k_{////}$) and to the positions of the shear bands that condition the intrinsic permeability evolution.

The vapour mass transfer coefficient used to obtain the results of Fig. 5 is of $\alpha = 10^{-3}$ m/s. This transfer coefficient is introduced through a non-classical mixed hydraulic boundary condition [11, 12] to represent the long-term exchange (evaporation) at gallery wall. It has indeed an influence on the drainage and its value is chosen to obtain a good reproduction of the experimental data. In the long term, the two controlled ventilation phases (phases (3) and (4) in Fig. 3) do not have a visible effect on the numerical results.

5 Conclusions

The presented study is devoted to the numerical modelling of failure process and hydraulic transfer occurring around underground galleries subjected to air ventilation, in a cross-anisotropic argillaceous rock. The fractures induced by the excavation process in the vicinity of the gallery are reproduced by shear banding, and the strain localisation is properly modelled by means of a regularisation method. The evolutions of flow transfers and transport properties have been characterised around the underground drift, especially in the excavation damaged zone. The evolution of intrinsic permeability is addressed by using a strain-dependent relation that involves plastic strain as well as a restricted part of the elastic strain. At the end of excavation, the permeability increase is of several orders of magnitude in the shear bands as measured experimentally.

The flow transfers in the damaged zone and in the undisturbed material are studied through the modelling of a large-scale gallery air ventilation experiment (SDZ). The air-rock interaction at gallery wall involves both water vapour and liquid water transfers that are implemented with a mixed hydraulic boundary condition. Such condition is relevant for the reproduction of the transfers because an instantaneous equilibrium between the air and the rock may not be assumed beforehand. Equilibrium is reached in the long term by evaporation process during the gallery air ventilation. Regarding the different ventilation phases, it has been observed that the progressive rock drainage is mostly affected by the long-term ventilation, not by the constant ventilation phases of the SDZ experiment.

The numerical results which are presented indicate that a strain localisation approach for the EDZ modelling, linked to an intrinsic permeability evolution, allows to predict the hydraulic transfer in the Callovo-Oxfordian claystone.

References

1. MS Diederichs. Rock Fracture and Collapse Under Low Confinement Conditions. Rock Mech Rock Eng **36**(5):339–81 (2003).
2. CF Tsang, F Bernier, C. Davies. Geohydromechanical processes in the Excavation Damaged Zone in crystalline rock, rock salt, and indurated and plastic clays - in the context of radioactive waste disposal. Int J Rock Mech Min Sci **42**(1):109–25 (2005).
3. Andra. Dossier 2005 Argile. Synthesis: Evaluation of the feasibility of a geological repository in an argillaceous formation, Meuse/Haute Marne site. Tech. Rep.; Paris, France; (2005).
4. R Plassart, R Fernandes, A Giraud, D Hoxha, F Laigle. Hydromechanical modelling of an excavation in an underground research laboratory with an elastoviscoplastic behaviour law and regularization by second gradient of dilation. Int J Rock Mech Min Sci **58**:23–33 (2013).
5. Y Jia, HB Bian, G Duveau, K Su, JF Shao. Hydromechanical modelling of shaft excavation in Meuse/Haute-Marne laboratory. Phys Chem Earth **33**:S422–35 (2008).
6. S Levasseur, F Collin, R Charlier, D Kondo. A micro-macro approach of permeability evolution in rocks excavation damaged zones. Comput Geotech **49**:245–52 (2013).
7. R Charlier, F Collin, B Pardoen, J Talandier, JP Radu, P Gerard. An unsaturated hydro-mechanical modelling of two in-situ experiments in Callovo-Oxfordian argillite. Eng Geol **165**:46–63 (2013).
8. B Pardoen, S Levasseur, F Collin. Using Local Second Gradient Model and Shear Strain Localisation to Model the Excavation Damaged Zone in Unsaturated Claystone. Rock Mech Rock Eng **48**(2):691–714. doi:10.1007/s00603-014-0580-2 (2015).
9. F Collin, XL Li, JP Radu, R Charlier. Thermo-hydro-mechanical coupling in clay barriers. Engineering Geology **64**:179-193 (2002).
10. B. Pardoen, J. Talandier, F. Collin. Permeability evolution and water transfer in the excavation damaged zone of a ventilated gallery. Int. J. Rock Mech. Min. Sci., under review (2016)
11. P Gerard, R Charlier, R Chambon, F Collin. Influence of evaporation and seepage on the convergence of a ventilated cavity. Water Resour Res **44**(5):1–16 (2008).
12. TA Ghezzehei, RC Trautz, S Finsterle, PJ Cook, CF Ahlers. Modeling coupled evaporation and seepage in ventilated cavities. Vadose Zone J **3**(3):806–18 (2004).
13. G. Armand, A. Noiret, J. Zghondi, D.M. Seyedi. Short- and long-term behaviors of drifts in the Callovo-Oxfordian claystone at the Meuse/Haute-Marne Underground Research Laboratory. Journal of Rock Mechanics and Geotechnical Engineering Volume 5, Issue 3, June 2013, Pages 221–230
14. J Desrues. Hydro-mechanical coupling and strain localization in saturated porous media. Rev Eur Génie Civ **9**(5-6):619–34 (2005).
15. R Chambon, D Caillerie, NE Hassan. One-dimensional localisation studied with a second grade model. Eur J Mech A-Solid **17**(4):637–56 (1998).
16. F Collin, R Chambon, R Charlier. A finite element method for poro mechanical modelling of geotechnical problems using local second gradient models. Int J Numer Meth Engng **65**(11):1749–72 (2006).
17. E Cosserat, F Cosserat. Théorie des Corps Déformables. Paris: Hermann (1909).
18. RD Mindlin. Second gradient of strain and surface-tension in linear elasticity. Int J Solids Struct **1**:417–38 (1965).
19. P Germain. The method of virtual power in continuum mechanics. Part 2 Microstructure. SIAM J. Appl. Math. **25**, 556–575 (1973).
20. Lewis RW, Schrefler BA. The Finite Element Method in the Static and Dynamic Deformation and Consolidation of Porous Media. Wiley: New York (2000)
21. MA Biot. General theory for three-dimensional consolidation. J. Appl. Phys. **12**, 155–164 (1941).
22. B. Pardoen, D.M. Seyedi , F. Collin Shear banding modelling in cross-anisotropic rocks International Journal of Solids and Structures **72**:63–87 (2015)
23. J Graham, GT Houlsby. Anisotropic elasticity of a natural clay. Géotechnique **33**:165–180 (1983).
24. AEH Love. A Treatise on the Mathematical Theory of Elasticity, Cambridge University Press (1927).
25. B Amadei. Rock anisotropy and the theory of stress measurements. Lecture Notes in Engineering Series, vol. 2. Springer-Verlag, New York (1983).
26. AHD Cheng. Material coefficients of anisotropic poroelasticity. Int. J. Rock Mech. Min. Sci. **34**:199–205 (1997)
27. G Duveau, JF Shao, JP Henry. Assessment of some failure criteria for strongly anisotropic materials. Mech. Cohes-Frict Mat. **3**:1–26 (1998).
28. S Pietruszczak, Z Mroz. On failure criteria for anisotropic cohesive-frictional materials. Int. J. Numer. Anal. Methods Geomech. **25**:509–524 (2001).
29. S Pietruszczak, D Lydzba, JF Shao. Modelling of inherent anisotropy in sedimentary rocks. Int. J. Solids Struct. **39**:637–648 (2002)

30. M Panet, A Guenot. Analysis of convergence behind the face of a tunnel. In:Proceedings of the 3rd International Symposium : Tunnelling 82. Brighton: 1065 Institution of Mining and Metallurgy; 197–204 (1982).

 DOI: 10.1051/e3sconf/20160904004

Indirect determination of soil water content

Philippe Cosenza[1,a]

[1]University of Poitiers, CNRS, UMR 7285 IC2MP, ENSI Poitiers, 86073 Poitiers cedex 09, France

Abstract. This manuscript reviews the recent developments in indirect methods used to measure soil watent content (SWC). It focuses on three types of methodological approaches: (i) high-frequency electromagnetic methodologies (dielectric methods and Ground Penetrating Radar), (ii) low-frequency electromagnetic methods (resistivity method and spectral induced polarization-SIP) and (iii) emerging methods (laboratory and ground-based nuclear magnetic resonance-NMR techniques and methods based on fiber optic temperature sensing systems). For each method the physical principles and the corresponding methodology are briefly described first, followed by the recent methodological advances. These recent advances address the following issues: (a) integration of different types of sensors to design new multi-functional devices, (b) miniaturization of low-cost and easily-installed electromagnetic sensors, (c) fusion and assimilation of data acquired from multiple modalities with different sampling rates and spatial resolutions, (d) physical understanding of spectra and distributed parameters measured by SIP and NMR methods, (e) improvement of geophysical inversion techniques combined with fast data acquisitions (4D monitoring of SWC).

1 Introduction

Soil water content (SWC) is a key parameter to understand the mechanical behavior of unsaturated soils. It controls many mechanical characteristics (e.g., plasticity, strength, desiccation cracking) and plays an important role in the stability of numerous geotechnical systems (engineered earth structures, landslides and artificial slopes such as embankments and cuttings). In recent decades, a variety of innovative methodologies used to measure SWC at different scales has been developed. These methodologies include mainly electrical and electromagnetic probes and geophysical techniques (i.e., non-invasive and non-destructive methods with a rapid spatial coverage and high-resolution of measurements).

This contribution reviews the recent developments in the indirect and geophysical methods used to measure SWC. Contrary to the standard gravimetric method, these approaches are indirect because they measure a physical parameter which is sensitive to the amount of soil water: they relate this physical parameter (or its variation) to SWC (mainly volumetric SWC in the following). In addition, this review focuses on laboratory and "local" field scales (typically 0.1-100 m) relevant in geotechnical engineering; the remote sensing techniques operating on regional scales are not considered.

Three groups of methodological approaches used to estimate SWC are presented and discussed: (i) high-frequency electromagnetic methodologies often referred to dielectric methods, (ii) low-frequency electromagnetic methods (resistivity method and spectral induced polarization) and finally (iii) emerging methods (nuclear magnetic resonance- NMR techniques and methods based on fiber optic temperature sensing systems). For each method the physical principles and the corresponding methodology are briefly described first, followed by the recent methodological advances obtained in the last years.

2 High-frequency electromagnetic methods

2.1 Physical principles and methodologies

High-frequency (HF) electromagnetic methods are based on the measurement of the soil relative dielectric permittivity ε which is a complex parameter given by:

$$\varepsilon(\omega) = \varepsilon'(\omega) - j\left[\varepsilon''(\omega) + \frac{\sigma_{DC}}{\omega\varepsilon_0}\right] \quad (1)$$

where ε' and ε'' are the real and imaginary component, respectively (both adimensionless variables); ω is the angular frequency of the imposed electromagnetic field ($\omega=2\pi f$ with frequency, f expressed in Hertz); j is the imaginary unit ($j^2=-1$); ε_0 (8.854 10^{-12} $N^{-1}.m^{-2}.C^2$) is the vacuum dielectric permittivity and σ_{DC} is the direct current (DC) electrical conductivity (S. m^{-1}).

HF electromagnetic methods take advantage of the differences in real relative permittivity values, ε', between soil phases; the water, air and solid phases exhibit real relative permittivity values of approximately 80, 1 and 3-10, respectively.

[a] Corresponding author: philippe.cosenza@univ-poitiers.fr

HF electromagnetic sensors can be schematically classified in two groups: (a) transmission line methods such as time-domain reflectometry (TDR) and time domain transmission (TDT) and (b) capacitances probes often termed HF impedance measurements.

In the first group, TDR and TDT methods both use the propagation velocity of an electromagnetic wave along a pair (or a parallel group) of metallic rods to obtain the permittivity of the geomaterial in which they are embedded. TDR probes are characterized by the rise time of the step pulse used to generate the electromagnetic propagation; the shorter the time, the higher the frequency content of the signal. Generally, rise times shorter than 300 ps are used (e.g., [1]). These probes generally operate at frequencies above 0.5 GHz (up to 2-3 GHz), i.e., in a frequency range that minimizes the effects associated with the imaginary part of ε. TDR probes are expected to estimate SWC to an accuracy of approximately 2% without soil specific calibration in coarse-grained soils [1].

In the second group, capacitor sensors use a fixed frequency or frequency shift and operate in the frequency range between approximately 5 and 150 MHz. The use of this frequency range leads to two difficulties: (i) the measurement can be affected by dielectric dispersion (i.e., real component ε' is frequency dependent) mainly due to clay minerals (e.g., [2]) and (ii) the measurement cannot fully separate the dielectric response associated ε' and the conduction phenomena associated with σ_{DC} (see equation (1)). In most cases, a soil-specific or site-specific calibration is mandatory. However, compared to TDR probes, the geometry of impedance sensors is more adaptable for short electrodes and borehole applications and thus allows interesting methodological developments, as explained below.

Regarding field applications, HF electromagnetic sensors are often considered as "point" measurements due their limited "support" (i.e., the volume in which the average value of SWC is obtained). Ground Penetrating Radar (GPR), a geophysical method, offers potential to overcome this limited spatial resolution. GPR is based on the transmission and reflection of HF (typically between 10 MHz and 1 GHz) electromagnetic waves within the subsurface. The objective here is to measure following different geometrical arrangements of transmitter and receiver antennas, the velocity of these electromagnetic waves directly linked to the relative permittivity ε. GPR like numerous other geophysical techniques is a high-resolution method (down to a few cm or dm) able to quantify the spatial variations of relative permittivity ε over large areas (easily several hundreds of m^2). It is often argued that GPR is an important technique to reduce the gap between "point" measurements and regional approaches, such as remote sensing techniques.

2.2 Recent methodological advances

2.2.1 Wireless sensing applications

Major technical progress regarding HF electromagnetic sensors will likely be achieved with the use of capacitance sensors and more precisely the family of the TDT sensors. These low-cost sensors that use the oscillation frequency of a ring oscillator to estimate ε, currently show interesting features suitable for wireless sensing applications. All electronics are included in the head of the probe removing the need for long cables and multiplexers (e.g., [3]). This advance has opened the opportunity to develop "smart" SWC monitoring systems in the field.

However, one has to keep in mind that wireless or distributed sensing applications still suffer from the following disadvantages:

(a) These methodologies are not standardized; i.e., several commercial probes with different sizes and operating frequencies exist.

(b) Consequently, sensor calibration is often required and may be fastidious in the field when different types of soils are present.

(c) A SWC measurement is often a "point" measurement at the field scale and the interpolation of data between sensors still remains a geostatistical challenge.

2.2.2 Multi-functional sensing probes

The integration of various types of data in a single sensor is particularly attractive because it (a) minimizes the soil disturbance and (b) validates deterministic modeling tools that couple water flow, solute transport, heat fluxes and from geotechnical perspective, hydro-mechanical processes. Recent works on this topic include the following developments:

- A combined tensiometer-moisture probe: a coiled TDR device that can be wound around a high suction tensiometer [4] (Figure 1). The complex geometry of this TDR probe requires a soil-specific calibration. The main objective here is to obtain the soil water retention curve in situ by measuring both SWC and suction.

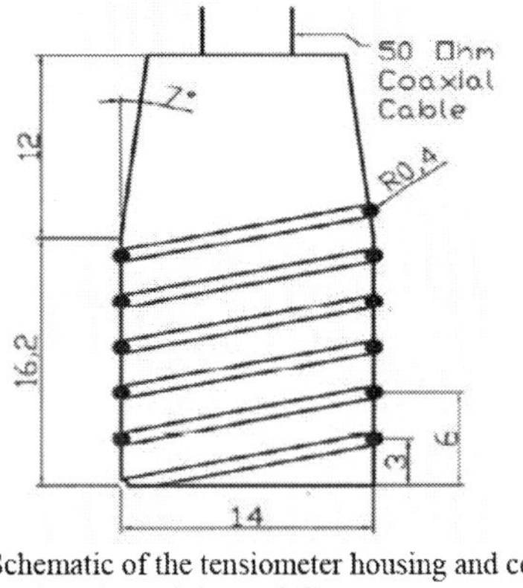

Figure 1. Schematic of the tensiometer housing and coiled TDR probe (dimensions in mm) (From [4]).

- A thermo-TDR probe which combines dual-probe heat pulse (DPHP) with TDR technology (e.g., [5]). This probe estimates both thermal properties and electromagnetic properties (relative permittivity and DC conductivity).

- Combined penetrometer-moisture probes. These tools address the study of soil compaction and the investigation of the heterogeneous hydrological properties of a foot-slope area (e.g., [6]).

2.2.3 Multi-frequency and high-resolution GPR

During the past decade, several advances in the GPR systems have been made. Increases in data-acquisition speeds have made it possible to collect more data in a short time whereas new global positioning systems (GPS) allow higher resolution information for the spatial registration of data. Moreover, multi-channel/multi-frequency systems now provide varying investigation depths and vertical resolving capabilities.

All these developments have stimulated researches to improve in situ SWC monitoring especially through time-lapse or 4D acquisitions. Recent studies (e.g., [7-8]) have shown the abilities of these new techniques to monitor dynamic hydrological processes and specific transient spatial changes in SWC in simple geological contexts (i.e., sandy formations).

However, validation of these techniques in more complex geological contexts still remains. Indeed, a radar image or a radargram acquired in a natural system results from different heterogeneities whose origins are not necessarily hydrological. Moreover, the recent work of [9] performed in more complex geological contexts suggests that the highest frequencies (900 MHz in their case) are preferred to optimally characterize soil moisture conditions.

3 Low-frequency electromagnetic methods

3.1 Physical principles and methodologies

Among low-frequency (LF) electromagnetic methods, the direct-current (DC) resistivity method is likely the most known in geotechnical engineering. From a practical point of view, a DC current (in fact a low-frequency alternating current) is injected into the soil by a pair of electrodes and the value of soil electrical resistance is obtained by measuring the resulting difference in voltage on two other electrodes. The soil resistivity is then calculated from the measured resistance by multiplying the latter by a "geometrical factor" that depends on the arrangements of different electrodes and the geometry of the soil sample in a laboratory experiment.

Soil resistivity is primarily sensitive to the number of electrical charge carriers (ions) in the interstitial solution and thus to the solution content. When smectite-type clay minerals are present, a significant part of these charge carriers are captured in the so-called electrical double layer at the clay minerals surface and contribute to an electrical surface conduction which has to be considered in the interpretation of soil electrical data.

The most popular geometrical configuration in the field is currently 2D electrical resistivity tomography (ERT) (also recently called electrical resistivity imaging, ERI). The main objective here is to obtain a 2D distribution of electrical resistivity in the subsurface. For this purpose, a multi-electrode system consisting of a series of electrodes attached to a resistivity meter is used. A current is injected by two source electrodes and the resultant potential distribution is measured at two or more receiving pairs of electrodes. In the ERT method, this procedure is repeated for as many combination of source and receiver electrodes as desired. Usually this procedure may involve the collection of several hundred (or even more) multi-electrode combinations. This collection of apparent resistivity values is then inverted by appropriate (commercial) inversion software to obtain a resistivity image or an interpreted 2D resistivity section.

It is worth noting that recent case studies have illustrated the ability of 3D ERT methods to monitor soil moisture dynamics in the field (e.g., [10-11]) and in laboratory (e.g.,[12]): this 4D ERT methodology is often referred to a time-lapse ERT method.

3.2 Recent methodological advances

3.2.1 Multi-functional devices

In the laboratory, a new electrical probe has been designed to monitor SWC changes in a triaxial apparatus [13]. This probe is composed of four circular electrodes (1.5 mm diameters) disposed in a square-grid scheme (inter-electrode distances of 6 mm) as presented in Figure 2. A dielectric matrix (Araldite epoxy resin) is used to accommodate the electrodes in order to avoid any electrical current lines between them.

Figure 2. In-house constructed resistivity probe (the four electrodes are separated by distances of 6 mm) (from [13]).

This electrical probe is associated with a specific holding system to ensure a good electrical contact with the triaxial specimen. In the triaxial experiments performed by [13], a suction probe is also fixed on the triaxial sample opposite to the electrical probe.

In the same way, to obtain a multi-functional device, a modified oedometer cell developed by [14] simultaneously measures the electrical resistivity and the settlement of the sample. The oedometer ring, which is composed of Teflon (an electrical insulating material) allows for the coupling of a classical oedometer procedure and an electrical resistivity measurement. The

 DOI: 10.1051/e3sconf/20160904004

two metallic porous plates of the oedometer cell are used as the current and potential electrodes. Although this dual-function device was initially designed to study the relationship between soil resistivity and soil compaction, it can be easily used to monitor SWC of a soil sample if a proper calibration of the measured resistivity is conducted.

3.2.2 Electrical mapping methods

Automatic resistivity profiling (ARP)

To our knowledge, the most rapid geophysical method to map electrical resistivity over large areas is the automatic resistivity profiling (ARP, formerly, multi continuous electrical profiling). The ARP method is currently used in archeology and soil science to map electrical resistivity at different depths of investigation and over large areas (typically a few ha). However, surprisingly, this technique is not so commonly used in civil engineering although some case studies illustrating its capacity to map SWC do exist (e.g., [15-16]).

In practice, ARP is a mobile multi-electrode system pulled by an off-road vehicule and coupled to a GPS device. The example given in Figure 3 is a V-shaped array composed of three receiving/voltage pairs of electrodes (three "dipoles" M_1N_1, M_2N_2 and M_3N_3) and one injection pair of electrodes (AB) inserted into the topsoil. The distance between the injection electrodes and the voltage measurement electrodes is 0.5 m for the M_1N_1 dipole, 1 m for the M_2N_2 dipole and 2 m for the M_3N_3 dipole, which correspond approximately to three different depths of investigation (Figure 3b).

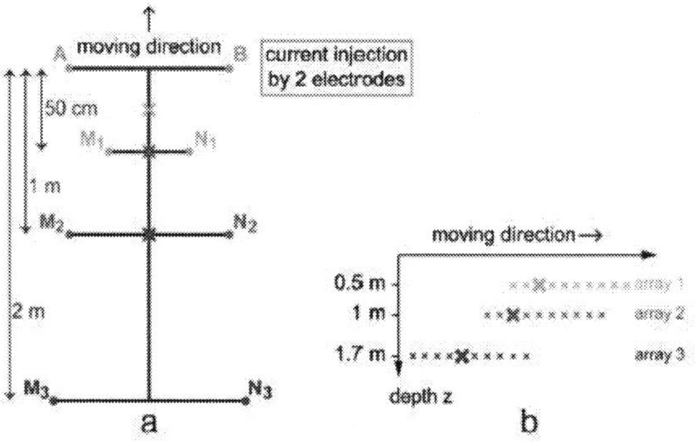

Figure 3. a) Configuration of an ARP device (crosses show the three measurement locations) and b) principle of multi-depth measurements (crosses show the depths of investigation at 0.5, 1 and 1.7 m of the three measurements corresponding to the three electrode arrays) (from [17]).

Capacitive resistivity mapping

The conventional electrical resistivity method has a major disadvantage; direct (galvanic) contact of electrodes with soil is required. This requirement can be problematic in areas with very rigid and/or impenetrable ground cover (i.e., dry and cracked soils, highways, hard rocks and permafrost). Moreover, galvanic measurements on a high resistive surface induce high contact impedances that decrease the accuracy of resistivity measurements. In this context, electrostatic method, often referred to as a capacitive resistivity method is an emerging geophysical technique with the potential to overcome both problems. Indeed following this method, no physical contact between the instrument and the surface of the measured material is required; two poles located in the air, outside the medium under study, produce the electrical field that generates electrical conduction in soil. These transmitter poles constitute an open capacitor whose complex impedance can be calculated from the voltage measured between two (or more) receiver poles.

The recent work of [18] has shown the capability of the electrostatic method to capture the water movement in the unsaturated zone. Moreover, it is worth noting that the electrostatic method can be used in an ARP configuration [19] (Figure 4).

Figure 4. Electrostatic resistivity meter with two depths of investigation for soil mapping at large scale (from [19]).

3.2.3 Spectral induced polarization method

When the soil resistivity measurement is repeated for different frequencies of the injected current (typically in the range of 0.1 Hz to 1 kHz), a spectrum of soil resistivity can be obtained: this methodology refers to the spectral induced polarization (SIP) method or impedance spectroscopy in laboratory applications. In practice, a spectrum of soil complex impedance (or complex resistivity) is measured because a phase shift exists between the injected electric current and measured voltage. The spectrum of this phase shift demonstrates the presence of polarization mechanisms occurring at low frequencies (typically <1 kHz) which are closely related to the pore-scale properties of soil that in part control hydraulic permeability (e.g.,[20]). Moreover, as for conventional electrical resistivity method, the spectrum of the in-phase (or real) component of complex impedance is directly associated with electrical conduction phenomena which in turn are related to the solution/water content.

Finally, this means that SIP spectra allow to estimate (a) SWC but also (b) some microstructural features (e.g., distribution of pore sizes or grain sizes, see Figure 5) which can be related to water permeability or specific surface.

In practice, the main difficulties with obtaining accurate SIP measurements are due to (a) the effects of

 DOI: 10.1051/e3sconf/20160904004

electromagnetic and/or capacitive coupling between the wiring that is used to connect the electronics to electrodes and (b) electrode polarization effect (non-polarizing electrodes have to be specifically designed). Moreover, despite recent theoretical advances, the physical meaning of SIP spectra acquired on a broad frequency bandwidth is still an intensive area of research (e.g.,[21]).

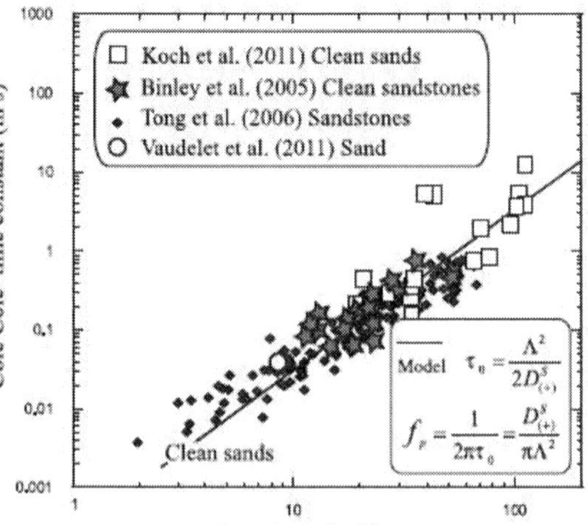

Figure 5. Pore-throat size versus Cole-Cole relaxation time inferred from SIP data obtained from clean sands (from [20]).

4 Emerging and promising methods

4.1 Nuclear magnetic resonance techniques

Nuclear magnetic resonance (NMR) methodology was originally developed in the oil industry. As SIP method, it deserves better recognition in civil engineering because this method is able to determine SWC and may provide some precious microstructural features (e.g., distribution of pore sizes) of geomaterials.

This method is based on the existence of a strong magnetic moment of the proton in the hydrogen nucleus. At thermal equilibrium in a static magnetic field, the volume of interstitial water in a soil sample exhibits a small net magnetic moment that results from the sum of all the magnetic moment associated with each of the protons in the volume. This net magnetic moment is aligned with the static magnetic field (usually Earth's magnetic field in the field) and its modulus, called E_0, is proportional to the number of protons. If the number of protons of the hydrogen nuclei is assumed to be proportional to volumetric SWC, it implies that the modulus E_0 is also proportional to volumetric SWC, which can be then estimated by an *ad hoc* calibration.

In fact, in a classical NMR relaxometry experiment, the moments of protons, initially at thermal equilibrium are perturbed by an energizing pulse tuned to the Larmor frequency which is an intrinsic physical property of a given nucleus. If this pulse is applied and then removed, these moments precess from their thermal equilibrium and then relax back to this same thermal equilibrium. As these moments relax, they emit a measurable magnetic signal which allows for the calculation of two parameters: relaxation times T_1 and T_2 associated with relaxation longitudinal and transversal to the static field, respectively. In the context of geomaterial, the T_2 distribution (with respect to the proportion of the magnetic field relaxing in the transverse direction) is often related to the pore size distribution (e.g., [22]). This explains the empirical relationships between NMR T_2 and hydraulic permeability existing in the petroleum literature. However, it should be noted that the use of NMR is problematic in presence of (a) iron metals associated with large magnetic susceptibilities (ferromagnetic materials) and (b) clay minerals associated with very short T_2.

Regarding laboratory and ground-based NMR methods, recent advances mainly address the following issues:

(a) Understanding the link between NMR measurements and hydraulic parameters of the unsaturated zone i.e., the soil water retention curve. With respect to this objective, recent laboratory investigations suggest that quantitative linear relationships exist between some geometrical features of the T_2 distribution (peak value and area) and the degree of saturation of silts [23]. This observation needs verification for others soils.

(b) Improvement of the in situ technique and inversion procedure to image SWC from field data. With regard to this issue, 3D numerical tools validated by field measurements are now available (e.g., [24]).

4.2 Methods based on fiber optic temperature sensing systems

Among the numerous optical methods used to determine SWC, those based on distributed temperature sensing (DTS) systems using fiber optic cables are growing in surface water hydrology [1, 25]. Their main advantages lie in their high spatial and temporal resolution in comparison with "point" measurements associated with HF electromagnetic sensors. Recent works suggest that DTS technology could provide estimates of SWC of lengths up to 10 km with spatial resolution on the order of 1 cm and sub-daily temporal resolution [26]. In these DTS methods, temperature is considered as a tracer of water movement; the temperature measurement is based on the thermal sensitivity of the relative intensities of backscattered Raman Stokes and anti-Stokes that arise from collisions with electrons in the core of the glass fiber [27].

From a practical point of view, the DTS methods to estimate SWC can be divided in two groups. In the first group, the passive methods simply record temperature variations of soil induced by the diurnal temperature cycle; no external source of energy is required. Its application may be questionable when the thermal response to the amplitude of diurnal cycle is not large enough to allow accurate estimation of SWC. The second group of methods refers to the active methods that couple a DTS system and heat source (or with a heat pulse method).

Recent advances regarding DTS systems mainly concern the active methods for which the works focused on the performances of heated optic cables (accuracy and suitability), the technology required to their insertion into the ground [27-28] and the limitations for long-term field applications [26].

5 Summary and concluding remarks

To summarize, our review highlights the following current trends and areas of research for indirect determination of SWC:

- *Integration* of different types of sensors to design new *multi-functional* devices. These multi-functional devices allow for experimentally addressing the numerous coupled hydro-mechanical problems existing in geomechanics.
- *Miniaturization* of low-cost and easily installed electromagnetic sensors. Recent advances in electronics allow to address the emerging smart SWC monitoring in the field (e.g., wireless sensing applications).
- *Data fusion and assimilation*. This aspect is a consequence of the previous points. Combining different and numerous data from multiple modalities with different sampling rates and spatial resolutions is becoming an emerging domain of research.
- *Physical understanding of spectra or distributions measured in SIP and NMR methods*. The practical objective here is to obtain by a single set of data, both SWC and microstructural features of the studied geomaterial.
- *Improvements of inversion techniques*. Improvements in geophysical inversion techniques combined with fast data acquisitions (thanks to technological advances as well) allows time-lapse and 4D methodologies to monitor SWC both in the laboratory and in the field.

However, it should be emphasized that accuracy of most indirect methods described here is still problematic in saline soils and clay geomaterials for which a material-specific calibration is always required. Regarding the latter (clay soils and clay rocks), the understanding and modeling of their physical parameters with respect to their hydration state remains a challenging area of research.

References

1. Robinson D. A., Campbell C. S., Hopmans J. W., Hornbuckle B. K., Jones, S. B., Knight R., Wendroth O. Vadose Zone J., **7**, 1 (2008)
2. Cosenza P., Tabbagh A. Appl. Clay Sci., **26**, 1 (2004).
3. Qu W., Bogena H. R., Huisman J. A., Vereecken H. Vadose Zone J. **12**, 2 (2013).
4. Toll D. G., Hassan A. A., King J. M., Asquith J. D. *Proc. of the 18th International Conference on Soil Mechanics and Geotechnical Engineering, Paris. France*, pp. 1199-1202 (2013).
5. Zhang B., Han C., Yu X. B. J. Rock Mech. and Geotech. Eng., **7**, 155-162 (2015).
6. Masaoka N., Kosugi K. I., Yamakawa Y., Mizuyama, T., Tsutsumi, D. Vadose Zone J. **11**, 2 (2012).
7. Klenk P., Jaumann S., Roth K.. Hydrol. and Earth Syst. Sci., **19**, pp. 1125-1139 (2015).
8. Mangel A. R., Lytle B. A., Moysey S. M. The Leading Edge. **34**, pp. 190-196 (2015).
9. Steelman C. M., Endres, A. L. J. Hydrol. **436**, pp. 51-66 (2012).
10. Chambers J. E., Gunn D. A., Wilkinson P. B., Meldrum P. I., Haslam E., Holyoake S., Wragg J. Near Surf. Geophys. **12**, pp. 61-72 (2014).
11. Gance J., Malet J. P., Supper R., Sailhac P., Ottowitz, D., Jochum B. J. of Appl. Geophys. in press (2016).
12. Clement R., Moreau S. J. Appl. Geophys. In press (2016).
13. Munoz-Castelblanco JA, Delage, P., Pereira, JM, Cui, YJ. Géotechnique. **62**, pp. 595–604 (2012).
14. Ghorbani, A., Cosenza P., Badrzadeh Y., Ansari A. Eur. J. Environ. Civil Eng. **17**, pp. 84-98 (2013).
15. Besson A., Cousin I., Bourennane H., Nicoullaud B., Pasquier C., Richard G., King, D. Eur. J. of Soil Sci. **61**, pp. 120-132 (2010).
16. Bourennane H., Nicoullaud B., Couturier A., Pasquier C., Mary B., King D. Geoderma. **183**, pp. 32-40 (2012).
17. Buvat S., Thiesson J., Michelin J., Nicoullaud B., Bourennane H., Coquet Y., Tabbagh A. Geoderma. **232**, 317-327 (2014).
18. Niu Q., Wang Y. H., Zhao K. J. of Appl. Geophys. **106**, 119-127 (2014) .
19. Dabas M., Tabbagh A., Flageul S. *3rd Global Workshop on Proximal Soil Sensing, Potsdam, Germany*. pp 27 (2013).
20. Revil A., Florsch N., Camerlynck C. Geophys. J. Intern. **198**, pp. 1016-1033 (2014).
21. Okay G., Leroy P., Ghorbani A., Cosenza P., Camerlynck C., Cabrera J., Florsch N, Revil A. Geophys. **79**, pp. E353-E375 (2014).
22. Tian H., Wei C., Wei H., Yan R., Chen P. Appl. Mag. Reson. **45**, pp. 49-61 (2014).
23. Chen P., Wei C. F., Liu, J., Xue, W. J. *Unsaturated Soil Mechanics-from Theory to Practice: Proceedings of the 6th Asia Pacific Conference on Unsaturated Soils, Guilin, China*. pp 181. CRC Press, (2015).
24. Jiang C., Müller-Petke M., Lin J., Yaramanci U. J. Appl. Geophys. **116**, pp.17-27 (2015).
25. Bakker M., Caljé R., Schaars F., van der Made K. J., de Haas S. Water Resour. Res. **51**, pp. 2760-2772 (2015).
26. Sourbeer, J. J., Loheide, S. P. Hydrol. Process. In press (2015).
27. Tyler S. W., Selker J. S., Hausner M. B., Hatch C. E., Torgersen T., Thodal C. E., Schladow S. Water Resour. Res. **45**, 4 (2009).
28. Coleman, T. I., Parker, B. L., Maldaner, C. H., & Mondanos, M. J. J. Hydrol. **528**, pp. 449-462 (2015).

 DOI: 10.1051/e3sconf/20160904005

Infiltration-induced Slope Instability: a multi-scale approach

Francesca Casini[1,a], Amin Askarinejad[2] and Sarah M. Springman[3]

[1]*Università di Roma Tor Vergata, Dipartimento di Ingegneria Civile e Ingegneria Informatica, 00133 Roma, Italy*
[2]*TU Delft, Faculty of Civil Engineering and Geosciences, Geoengineering section, Delft, The Netherlands*
[3]*ETH Zürich, Institute for Geotechnical Engineering, Zürich, Switzerland*

Abstract. Precipitation, together with erosion and earthquakes, have been recognized as the main triggering factors of shallow landslides. However, there are relatively few well-documented cases where direct relationships could be established between occurrence and features of shallow landslides, the rainfall characteristics (*e.g.* intensity, duration) and water retention curves. A field experiment has been performed on a steep forested slope located on the east-facing banks of the river Rhine in Ruedlingen, northern Switzerland. The aim of the experiments was to study the triggering mechanisms of the landslides induced by rainfall. The pore pressure and the degree of saturation, which are linked through the water retention curve, represent two of the main variables affecting the mechanical behaviour of unsaturated soils, and their relationships to rainfall are complex. The difference in the determination of water retention curves at different scales are analysed in this paper for Ruedlingen soil together with their effects on mechanical behaviour at multi-scale.

1 Introduction

Rainfall induced landslides are one of the major natural hazards in Switzerland [1-2-3-4]. Experience in Hong Kong shows that they occur more frequently when short-duration rainfall is of high intensity [5-6] . The location of the failure is strongly influenced by spatial and temporal variations in seepage induced by pore water pressure [6].

The slope stability depends directly on the distribution of water and of the pore pressure inside its mass. These factors are affected by the hydraulic properties of unsaturated soils, such as the water retention curve and the hydraulic conductivity function [e.g. 7-8].

A landslide triggering experiment was carried out near Ruedlingen in North East Switzerland, on a slope overlooking the river Rhine, in autumn 2008 and spring 2009. This attempted to replicate the effects of a heavy rainfall event from May 2002, in which 100 mm rain fell in 40 minutes [9], causing 42 superficial landslides.

The slope was subjected to extreme rainfall by artificial means in October 2008 over a period of 4 days, with a rainfall intensity of 15 mm/h for the first 2 days, and an intensity of 30 mm/h for the rest. The sprinklers were distributed at constant spacing along the central line of the slope. Some surface movements were detected during this extreme event, although failure did not occur. Subsequently, a range of measures was implemented, such as relocating the distribution of the sprinklers to provide more rainfall to the upper part of the slope, so that a failure mechanism was triggered in March 2009, incorporating about 150 m^3 of debris [1, 4, 10, 11].

It is essential to understand the hydro-mechanical (HM) behaviour of the unsaturated soils subject to rainfall to be able to prevent the landslides and the HM factors affecting the loss of strength. HM behaviour observed in situ and in the laboratory are compared, highlighting common trends and discrepancies in preventing shallow landslides being induced by rainfall.

2 Landslide test site and soil classification

The experimental area is located on an east-facing slope on the banks of the river Rhine

[a] Corresponding author: francesca.casini@uniroma2.it

 DOI: 10.1051/e3sconf/20160904005

(Figure 1a). The altitude is about 350 metres above sea level. The average gradient was determined using a total station theodolite to be 38° with maximum of 43° in the middle of the slope. The surface is slightly concave; the middle longitudinal line is 0.3 to 0.5 m lower than the sides, except that the slope drains marginally to the north at the bottom [4]. According to DPL tests (Dynamic Probing with a "Light" 10 kg weight), the bedrock level is reached between 0.5 m and 5 m. The site is located in the Swiss lowlands and the underlying rock consists mainly of a sequence of Molasse, which is the naturally cemented sediment that was deposited in the foreland basin of the Alps under successive fresh and saltwater regimes.

Three cluster locations (Figure 1b) were instrumented with tensiometers, Decagons, TDRs, piezometers, rain gauges, pressure cells, and deformation probes. The instruments were installed from 15 to 150 cm depth below the surface at each of three cluster locations.

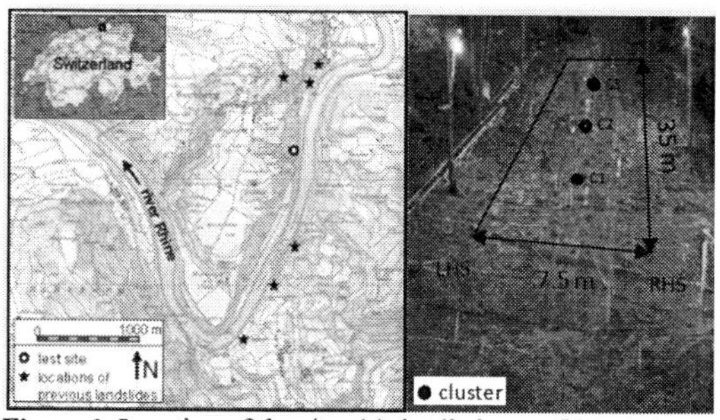

Figure 1. Location of the site: (a) detailed map; (b) test site, geometry and 3 instrument clusters (C1-C3) (after [10]).

Combined sprinkling and dye tracer tests were carried out at the three different locations, revealing a high infiltration capacity of the soil. No overland flow was observed during the sprinkling at any of the locations. In all of the soil profiles, staining of the soil was mainly homogeneous and little preferential flow was visible, although deeper percolation into the Molasse sandstone was revealed by the observation of several stained joints below the subsoil. The experimental slope was originally forested, with covertures of circa 80% to heights between 5-20 m. Shrubs up to 1-5 m high and ground vegetation each cover about 50% of the surface.

In order to understand the in situ triggering experiment, an extensive soil investigation was conducted to understand how the strength is mobilised in the soil and how this may be related to change in water content and void ratio in the slope [10, 12]. The investigation includes tests on both undisturbed and reconstituted soil samples, in saturated and unsaturated conditions. Water retention curves and hydraulic conductivity functions have been obtained in the laboratory at different scales [12] and in situ [8]. Mobilised shear strength has been investigated at multi-scale, with standard and modified shear box apparatuses in saturated and unsaturated conditions [13, 14, 15]. Unsaturated and saturated statically compacted and natural specimens have been tested under triaxial stress conditions to characterise the mechanical behaviour of the soil at different gravimetric water contents in a triaxial stress path apparatuses and to link the mechanical behaviour with the soil-water retention curve obtained under suction-controlled conditions with different void ratios [7, 10].

Moreover, a series of Unsaturated Constant Shear Drained (U-CSD) triaxial tests were performed by [11] to replicate the stress path that a soil element experiences during the rise of pore water pressure in a slope, during and/or after a rainfall event.

The soil can be classified as medium to low plasticity silty sand (ML) according to USCS. Activity of the soil comes from the chloritic-smectitic clay fraction [14]. The activity, I_A, is higher than 1.25 in the upper part of the soil profile and decreases from I_A = 1. 25 to I_A = 0.75 below 1 m depth. The clay fraction increases with depth from 4% at shallow depths to 10% at about 2 m. The silt fraction also increases with depth from 25% to 32%, while the sand fraction decreases from 67% to 56%. The increase of the fine fraction with depth may be due to internal erosion, sedimentological and morphological reasons, and downward transport promoted by infiltrating water [10].

3 Water retention curve: laboratory versus in situ results

The water retention curves (WRCs) under suction controlled conditions have been obtained on specimens compacted at different void ratios [7]. A wetting path has been followed at three void ratios (Figure 2), while results for the drying path have also been obtained for a specimen with a void ratio e = 0.85. Particular attention has been paid to the evolution of the wetting path dependent on the void ratios, because it is the typical path experienced under field conditions during a rainfall event. The evolution of the WRC with void ratio plays a fundamental role in the evaluation of the stability of steep slopes due to the coupling effects of suction and degree of saturation (e.g. [2, 7]).

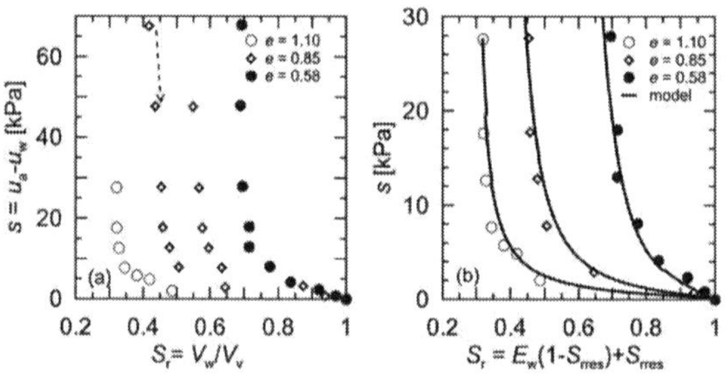

Figure 2. Water retention curves of Ruedlingen silty sand: (a) experimental data at different void ratios along wetting–drying paths; (b) modelling of the wetting paths (after [11]).

The wetting water storage mechanism has been described using a modified van Genuchten [17] model (after [12]) in terms of dimensionless water ratio $E_w = E_w(s, n)$ as follows:

$$E_w = \frac{e_w - e_{wres}}{e - e_{wres}} = \left\{1 + \left[s / P(n)\right]^{1/[1-l(n)]}\right\}^{-l(n)} \tag{1}$$
$$P(n) = P_0 \exp^{b(n_0 - n)};\ l(n) = l_0 \exp^{c(n_0 - n)}$$

where $e_w = V_w/V_s$, V_s is the volume of solids, e_{wres} and S_{rres} are the residual water void and saturation ratios, $P(n)$ and $l(n)$ are functions depending on porosity n, via parameters b and c. The parameters obtained by fitting the experimental WRCs are reported in Table 1 and compared in Figure 2b, in terms of S_r versus s.

Table 1. WRC parameters

n_0	P_0 (kPa)	b	l_0	l
0.524	0.7	9	0.5	-2.5

The model fits the experimental results well over the variation of void ratio and in the range of suction investigated.

In reality, the rainfall was not uniformly distributed, as had been approximated in [1]. The artificial rainfall was applied in situ by 10 equi-spaced sprinklers, located on the middle longitudinal line of the test-field. The lower sprinklers experienced higher hydraulic heads as the water was supplied from the upper part of the slope.

Insitu WRCs and those derived in the laboratory are presented in Figures 3 for a depth of 30 cm in cluster 2. The first wetting branch (W1) shows higher values of suction at the same VWC compared with first drying curve (D1), which is in contrast with the literature data and theory on hysteretic behaviour [18]. This can be attributed to increase in volume during wetting and to the heterogeneity in the flow paths [8]. Similar trend are reported for different depths (see [8]).

The drying branch of WRC for clusters 1 and 2 at different depths are enlarged and shown in Figure 4. The curves obtained for depths of 120 cm and 90 cm in cluster 1 show air entry values of around 2.2 kPa, which is lower than that implied by the laboratory WRC. This can be caused by void ratio differences between soil at these depths and the void ratio of the natural specimen tested in the laboratory (e = 0.8).

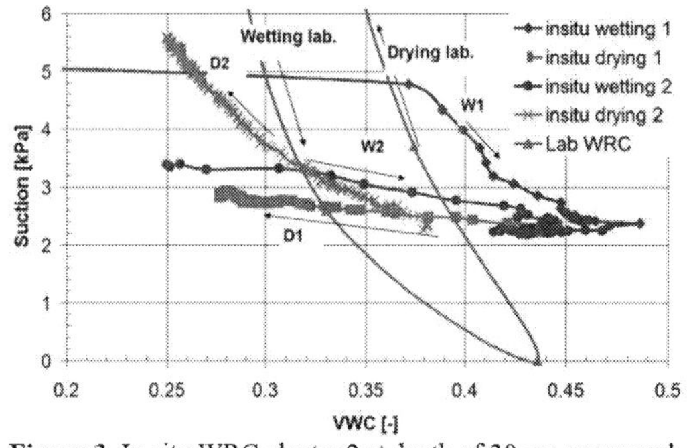

Figure 3. In situ WRC cluster 2 at depth of 30 cm compared with laboratory results (after [8])

Similarly, in cluster 2 (Figure 5) at a depth of 120 cm, the air entry value is comparable to the value at 150 cm (~ 3.0 kPa), while the soil has a lower void ratio at 120 cm compared with that at 150 cm. The trend lines of dry branch of WRC in situ are less steep than in the laboratory.

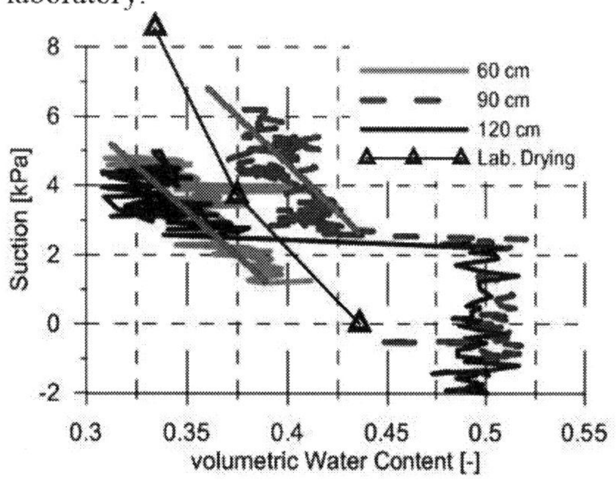

Figure 4. Drying curve of in situ WRC for cluster 1 compared with laboratory results (after [8])

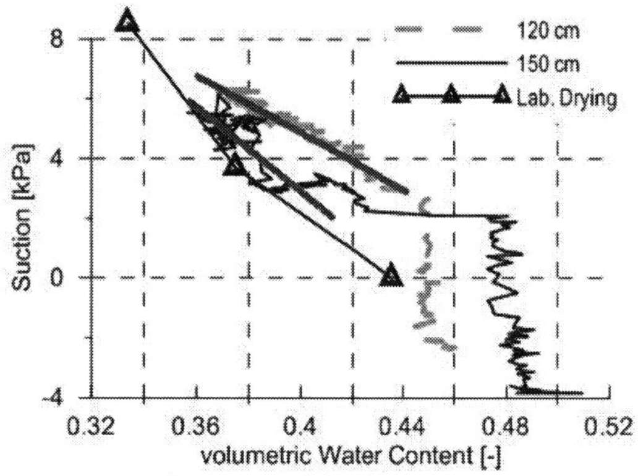

Figure 5. Drying curve of in situ WRC for cluster 2 compared with laboratory results (after [8])

4 Mechanical behaviour: laboratory versus in situ experiments

An extensive soil investigation was conducted to understand how the strength is mobilised in the soil and how this may be related to change in water content and void ratio in the slope (*e.g.* 7; 10; 14; 15). The investigation includes triaxial tests and shear tests on both undisturbed and reconstituted soil samples.

4.1. Triaxial stress path tests

The main objectives of the experimental programme were to characterise the behaviour expected during rainfall infiltration, under unsaturated and saturated conditions, and to investigate the potential for instability of the Ruedlingen soil during shearing to failure. Conventional stress path tests were performed on specimens with three gravimetric water contents, as well as on paths approaching those experienced in the field (more details in [7, 10]).

Standard drained (CIDC) and undrained (CIUC) shear compression tests were performed, together with constant axial load shear compression tests under decreasing isotropic stress (by increasing pore water pressure). In the latter case, both drained (CADCAF) and undrained (CADCAUF) conditions were studied after an initial drained stage (CAD), to analyse the influence of the hydraulic boundary conditions on the eventual failure mechanism of the soil specimens [7, 10].

The results of the constant axial load phase for the same water content and different $\eta = q/p'$ values are reported in Figure 6 in the p'–q and ε_q–ε_v planes, where the mean effective stress is defined as $p'=p-u_a+S_r(u_a-u_w)$ with the suction obtained from equation 1, knowing the water ratio E_w. The total stress was decreased for specimens TX3 and TX4, while maintaining a constant axial load, which promotes a slight increase in volumetric strain (Figures 6a-b). This is likely to be the result of elastic swelling as the stress path moves inside the yield locus.

After an initial contractant phase, a more pronounced increase of the swelling volumetric strain is observed for specimen TX2, which was compressed to $\eta = 0.375$, denoting a dilatant mode of ultimate failure with plastic volume increase. The response of the specimen to an increase in stress obliquity ($\eta \geq 0.75$) during the consolidation phase shows that the magnitude of strain in axial extension is suppressed in favour of greater magnitude of strain in radial extension, as the deviator strain is always positive; whereas it is negative initially for specimen TX2 with $\eta = 0.375$.

The specimens with w = 0.25 (TX6, TX7, and TX8; Figures 6c–d) exhibit a significant increase in volume with increasing shear strain for two of the three obliquities investigated. Plastic volumetric strains increase as the stress path approaches and crosses the critical state line, demonstrating a dilatant mode of ultimate failure. The same behaviour is observed for saturated specimens TX11 and TX12, while specimen TX10, with η = 0.375, exhibits less pronounced decreases of deviatoric stress (Figures 6e–f).

The framework proposed by [19] (after [20]) has been adopted to study the failure mechanisms of loose and dense sand samples. They concluded that the instability can occur under both undrained and drained conditions as long as the stress path leads the stress state into the zone of instability, distinguishing the occurrence of conditional stability (continuously decreasing mean effective stress in drained conditions) from that of run-away failure (under undrained conditions due to the progressive increase of pore-water pressure). The validity of the framework has been extended to unsaturated conditions, with summary data reported in Figure 7. The void ratio at critical state $e_{csl} = v_{csl} - 1$ is evaluated through $v_{csl}=N_M(\chi)-\lambda \ln p'$ with parameter $\chi = S_r$ (more details in [7]).

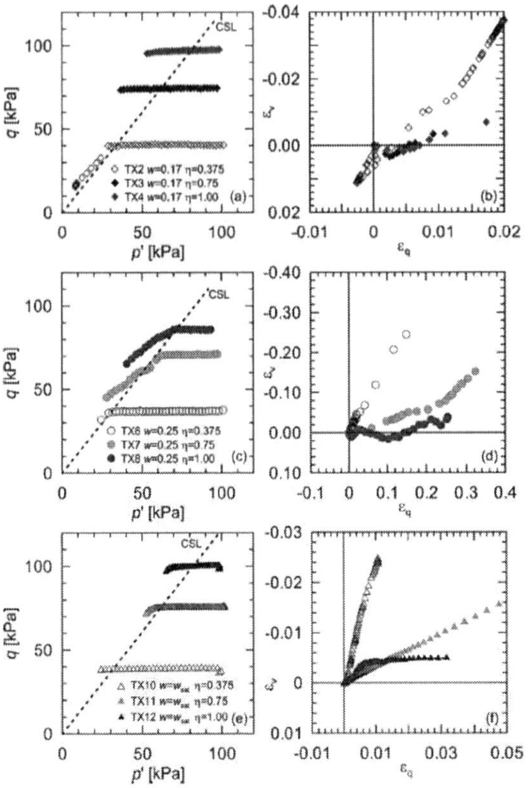

Figure 6. Constant axial load (CAL) phase: (a) p'–q plane w = 0.17; (b) ε_q–ε_v plane w = 0.17; (c) p'-q plane w = 0.25; (d) ε_q–ε_v plane w = 0.25; (e) p'–q plane $w = w_{sat}$; (f) ε_q–ε_v plane $w = w_{sat}$ (after [7]).

The framework is used to examine the influences of the stress state, void ratio, and degree of saturation on the stability behaviour of granular soils in slopes subject to saturation. The point representative of instability in the slopes under different water content and stress ratio at consolidation fits the trend of increasing η_{IL} with decreasing $e_{csl} - e_{IL}$ well. The bonding induced by partial saturation and different void ratios at the end of the consolidation phase explains the higher instability ratio exhibited by the samples with η = 0 at decreasing water content (more details in [8]).

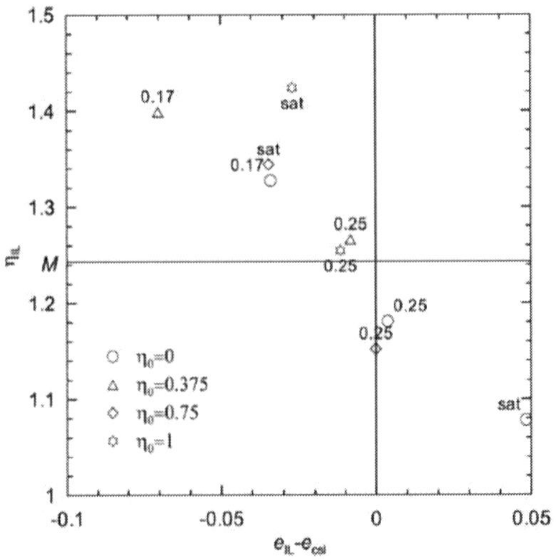

Figure 7. Potential instability line _IL as a function of the distance $e_{IL} - e_{csl}$: from unsaturated to saturated condition (after [7])

4.2. Shear tests

Direct shear tests have been performed on statically compacted specimens of Ruedlingen silty sand with a void ratio e=0.90 and three different gravimetric water contents of w= 0.15, 0.20 and 0.25. The experimental results and the Mohr-Coulomb envelope using the Bishop effective stress and the water retention curve (equation 1) are reported in Figure 8. The friction angle φ' = 31° is obtained from:

$$\tau = (\sigma_n - u_a)\tan\varphi' + S_r(u_a - u_w)\tan\varphi' \quad (3)$$

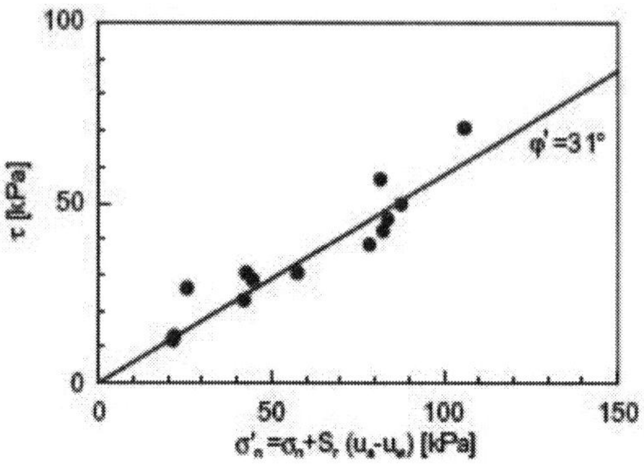

Figure 8. Shear strength envelope (after [15])

4.2.1 Limit Equilibrium Analyses

Based on the experimental results, a simplified slope stability analysis is carried out with shear strength parameters determined under unsaturated conditions for homogeneous ground. Extending the approach developed by [2] to a laterally limited slide, the factor of safety FoS becomes:

$$FoS = \frac{\frac{c^*}{\cos^2\alpha}\left(1+2K\frac{z}{d\cdot\cos(\alpha)}\right)+\gamma z\tan\varphi'\left(1+K\frac{z}{d\cdot\cos^3(\alpha)}\right)}{\gamma z\tan\alpha} \qquad (2)$$

where $c^*=S_r\ (u_a-u_w)\ \tan\varphi'$, α is the inclination of the slope, z the depth of the sliding surface, d the width of the slope, K the earth pressure coefficient. The groundwater table is assumed to lie below the sliding surface and the shear resistance can be fully mobilised to depth z along the lateral sides, according to equation 2.

In the following parametric evaluation of equation 2, a slope inclination of $\alpha = 38°$, $\gamma=\gamma_s\cdot(1+w)/(1+e)$ $\gamma_s=25.8$ kN/m^3, a width of $d = 8$ m, a void ratio $e = 0.9$ and an earth pressure coefficient of $K = 1 - \sin(\varphi')$ ($\varphi'=31°$) were chosen as representative values for the field conditions.

The depth, for which the factor of safety (FoS) reaches unity, is plotted in Figure 9 versus the saturation degree. All points to the left of the curves represent stress states where the shear stress exceeds the shear resistance. Shallow slides are stable (to the right of the curve) due to the apparent cohesion provided by suction and degree of saturation, as represented by c*, whereas very deep slides are unlikely to happen due to the friction mobilised on the base and side planes of the failure mechanism. The difference between drying and wetting conditions originate from different degrees of saturation and unit weight at the same values of suction.

When plotting the critical depth z (FoS=1) versus the suction (Figure 10), the difference between the drying and the wetting branch of the WRC becomes less evident. During a rainfall event, the soil is saturated gradually. When a suction s =2.9 kPa and S_r = 0.54 are exceeded, a sliding surface is formed at a depth of about 1.12 m.

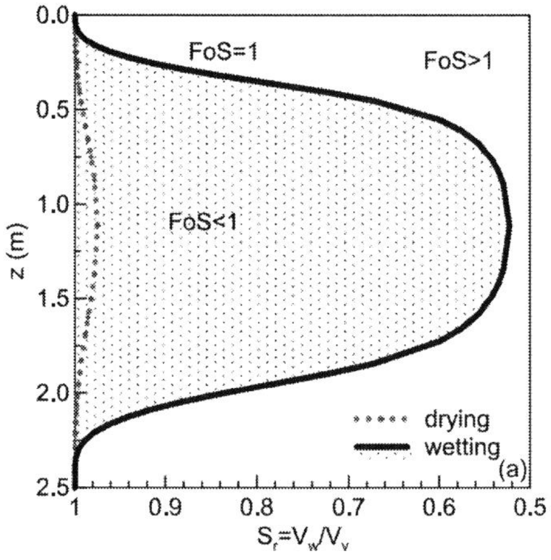

Figure 9 Depth z for FoS = 1 from Equation (2) as a function of saturation degree for conditions at cluster 3 (slope inclination α = 38°, width d = 8 m, void ratio e = 0.9, earth pressure coefficient K = 1 – sin φ', φ'=31°) (after [15]).

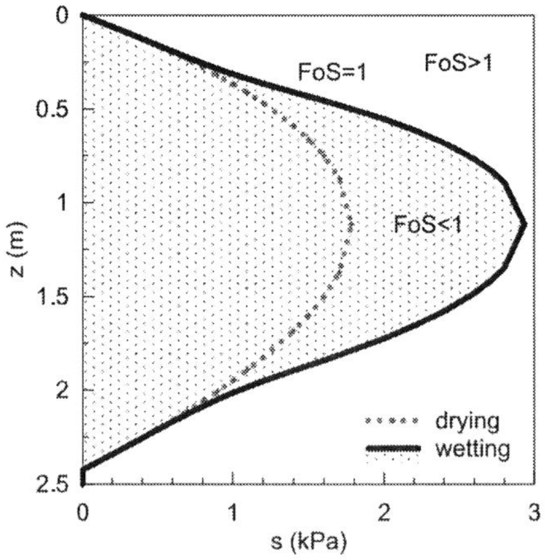

Figure 10 Depth z for FoS = 1 as a function of suction (after [15])

4.2.2. Comparison with in situ behaviour

The evolution of Volumetric Water Content (VWC) during the failure experiment is reported in Figure 11a at different depths in cluster 3 (see Figure 1). The evolution of pore water pressure recorded by piezometers P4 and P5, which were located inside the eventual failure zone, are also reported in Figure 11b. The TDR installed at a depth of 1.2 m, which was the nearest instrument to the eventual slip surface, measured a decrease in water content about 1 hour before the failure (see [1, 21]).

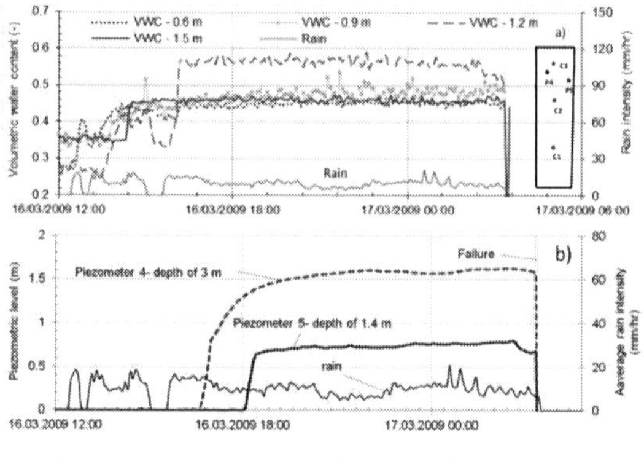

Figure 11 Time evolution during the landslide triggering experiment: a) volumetric water content in cluster 3-C3; b) piezometric level at P4 and P5 (after [1])

The piezometers P4 and P5, which were located inside the eventual failure zone, also showed a decrease in the piezometric level (Figure 11b). Piezometer P4 is installed at a depth of 3 m from the slope surface and reacted faster to the start of rainfall and attained a higher value than P5. This response can be explained by the fact that this piezometer was installed in a higher location where more intense rainfall had been provided to the slope.

The observed decreases in the piezometric levels was more pronounced in P5 (approximately 0.12 m), which was nearer to the failure surface. The decrease time corresponds well to the time when the VWC at 1.2 m also slightly decreased (about 1 hour before the landslide) [1].

These observed decreases in the VWC and piezometric level at the failure surface, before the landslide was triggered, can be attributed to the dilation of the soil at the failure surface, as observed also along the CADCAF stress path in triaxial cell (Figure 6).

The depth of failure and the pore water pressure acting on the slope are fairly well predicted by the limit equilibrium analyses, which account for an apparent cohesion dependent upon suction and saturation degree and have been extended to a lateral limited slide.

Conclusion

A field experiment has been performed on a steep forested slope with the aim to study the triggering mechanisms of landslides induced by rainfall.

The hydro-mechanical behaviour of the silty sand has been characterised in the laboratory and in situ through a wide range of tests conducted under saturated and unsaturated conditions. Since the average slope of the field exceeds the critical state internal friction angle, suctions developing through partial saturation play a fundamental role in stabilising the slope.

Partial agreement was observed comparing the in-situ values with a laboratory-derived WRC. Discrepancies could be attributed either to sampling effects or to drainage paths of different degrees of freedom: primarily one dimensional flow in small specimens in the laboratory, while three-dimensional flow will dominate under insitu conditions.

The soil exhibits dilatant behaviour along CADCAF stress paths, simulating the decreases of mean effective stress at constant axial force. This dilatant behaviour has been also confirmed from the in situ monitoring of volumetric water content and pore water pressure in the hour preceding the landslide.

The depth of the failure has been predicted with a simple Limit Equilibrium Analyse that accounts for a box-shaped failure mechanism and additional shear strength due to suction. The prediction agrees fairly well with the depth of failure observed in situ.

References

1. Askarinejad, A., Casini, F., Bischof, P., Beck, A., Springman, S.M., Italian Geotechnical Journal, **3**, 12 (2012)
2. Springman, S.M., Jommi, C., Teysseire P., Géotechnique **53**, 1 (2003)
3. Lateltin O., Haemmig Ch., Raetzo H., Bonnard C., Journal of the International Consortium on Landslides, **2**, 4 (2005)
4. Springman, S.M., Kienzler, P., Casini, F., Askarinejad, A. In Hamza M., Shaien, M., El-Mossallamy Y. (eds): Proc. 17th International Conference on Soil Mechanics and Geotechnical Engineering, Egypt, 5-9 Oct.. IOS Press. (2009)
5. Brand, E.W., Premchitt, J., Phillipson, H.B. , Relationship between rainfall and landslides in Hong Kong. Proc. 4th Int. Symp. on Landslides 1 (1984)
6. Fannin, R.J., Eliadorani, A., Wilkinson, J.M.T. , Géotechnique **55**, 6 (2005)
7. Casini F, Serri V, Springman S. Canadian Geotechnical Journal; **50**, 1 (2013)
8. Askarinejad, A., Casini, F., Kienzler, P., Springman, S.M., In Unsaturated Soils – Proc. 5th International Conference on Unsaturated Soils, Barcelona, Spain. Vol. **1** (2011)
9. Fischer, C., López, J., Springman, S.M. Intern. Conf. on Land., Hong Kong:. (2003)
10. Casini, F., Jommi, C., Springman, S.M. Granular Matter **12**, 3 (2010)
11. Askarinejad, A., Failure mechanisms in unsaturated silty sand slopes triggered by rainfall DSc thesis, ETH Zurich (2013)
12. Casini, F, Canadian Geotechnical Journal **49**, 8 (2012)
13. Minder, P., Semester thesis, ETH, Zurich (2008)
14. Colombo L., Master Thesis, Politecnico di Milano (2009)
15. Casini, F., Minder, P., Springman, S.M., In: Unsaturated Soils – Proc. 5th International Conference on Unsaturated Soils, Barcelona, Spain. Vol. **1** (2011)
16. Askarinejad, A., Beck, A., Springman, S.M. Canadian Geotechnical Journal. **52**, 6 (2015)
17. van Genuchten MTh. Soil Science Society of American Journal **44** (1980)
18. Mualem, Y., Water Resour.Res., 9 (1973)
19. Chu J, Leroueil S, Leong WK. Canadian Geotechnical Journal **40** (2003)
20. Been K, Jefferies MG. Géotechnique **35**, 2 (1985)
21. Springman, S.M., Askarinejad, A., Casini, F., Friedel, S., Kienzler, P., Teysseire, P., Thielen, A. Acta Geotechnica Slovenica, **1** (2012).

SEALEX in-situ experiments-performance tests of repository seals: experimental observations and modelling

Nadia Mokni[a], Jean-Dominique Barnichon

Institute for Radiation Protection and Nuclear safety (IRSN), France.

ABSTRACT. The paper describes observations and numerical analysis of SEALEX performance tests installed in Tournemire Underground Research Laboratory (URL). One of the objectives of the large scale in-situ tests is to investigate the impact of technological gaps on the long term performance of bentonite based seals. The swelling cores consist of pre-compacted blocks of a natural sodic Wyoming bentonite (MX80 type) mixed with quartz sand in a ratio of 70/30 (in dry mass) with different geometries (monolithic disks or four jointed disks). Several technological gaps exist within the in situ tests: Gaps between the blocks and annular gap with variable width between the bentonite-based core and the host rock. All the tests are extensively instrumented for monitoring the main Hydro-Mechanical (HM) variables. Comparison of the experimental results showed that the presence of technological gaps constituted new hydration sources (annular gaps) and flow paths (gaps between the blocks) that changed the saturation kinetics. A coupled HM formulation that incorporates the relevant processes involved in the problem under consideration has been adopted to analyse the effect of the annular technological gap on dry density homogenization of the bentonite based core as hydration progresses. Technological gaps were demonstrated to have an impact on dry density distribution.

1 Introduction

Highly compacted bentonites are often considered as sealing materials due to their low permeability, high radionuclide retardation capacity and high swelling ability (Pusch, 1979; Yong et al., 1986; Villar, 2008). Under geological disposal conditions bentonite will progressively hydrate and swell filling up the existing technological gaps. Various laboratory and in-situ experiments were conducted in the past to characterize the Hydro-Mechanical (HM) behaviour of these materials ([2], [3], [5], [7-12], [17], [18], [20]). However, little attention has been paid to the role and influence of technological gaps on the HM properties of the sealing materials during the saturation phase and especially on the degree of homogeneity of the bentonite based core at the end of the transient period. Still, the existence of technological gaps is generally neglected in numerical modelling and perfect contact is assumed between the different components of the sealing system.

In this context the SEALEX project was launched to specifically focus on long-term performance of sealing systems. This project relies on a series of in-situ experiments emplaced in IRSN's Underground Research Laboratory (URL) at Tournemire (South France). The SEALEX in-situ experiments aim to (i) quantify the impact of intra-core geometry (technological gaps) on the hydraulic properties of sealing systems, (ii) test the long-term hydraulic performance of sealing systems in normal conditions for different clay core compositions (pure bentonite or bentonite/sand mixtures) and conditionings (pre-compacted blocks or in-situ compacted), and (iii) investigate the concept of robustness by considering altered scenarios, such as an incidental decrease of the swelling pressure (for instance originating from the failure of the confining plugs). Accordingly, the experimental

[a] Corresponding author: nadia.mokni@irsn.fr

program was progressively implemented with six in-situ tests, changing one parameter at a time (core composition, conditioning, geometry…). The installation phase was undertaken from December 2010 to October 2013 and all tests are still currently ongoing. During the operational phases of the in-situ tests, the injected volume of water, swelling pressure at radial and axial directions, pore water pressure and relative humidity changes have been monitored at several positions within the seal.

The present paper presents the installation, analysis and interpretation of the experimental results of three ongoing SEALEX in-situ experiments (named PT-N1, PT-N2 and PT-A1). The analysis specifically focuses on the impact of technological gaps on the long-term performance of clay-based seals. A coupled HM formulation that incorporates the relevant processes involved in the problem under consideration has been adopted to analyse the effect of annular technological gap with non-uniform thickness, located between the bentonite based core and the host rock, on dry density homogenization of the bentonite seal as hydration progresses.

2 Description of SEALEX in-situ tests

2.1 Test site location

All the in-situ tests were conducted in IRSN's Tournemire URL (Figure.1), located in a Mesozoic sedimentary basin on the western edge of the French Causses (S France). The tests are situated in the 180 m thick shale formation (Toarcian-Domerian) that features typical anisotropic sheet textures induced by compaction processes. The mineral composition of the Tournemire shale is relatively homogeneous for the upper Toarcian section. It contains 39-51 wt% of phyllosilicates (smectite, illite, kaolinite, chlorite and mica), 24-36 wt% of quartz, 16-17 wt% of carbonates (bioclasts and diagenetic calcite are predominant and dolomite and siderite are in small proportions) and 4-8 wt% of K-feldspaths. A small quantity of pyrite and organic matter are prevalent in the clay matrix. The shale is characterized by low porosity (8-11 wt%) with a water content of 3.5-4 wt% in the saturation state. Hydraulic permeability measurements give 10^{-21}-10^{-22} m^2 in laboratory and 10^{-18}-10^{-21} m^2 in-situ.

2.2 Tested material

The used bentonite is from Wyoming, USA, with high montmorillonite content (80%). The cation exchange capacity (CEC) is 76 meq/100 g (Wang et al., 2012). It was provided by the Laviosa-MPC company, under the commercial name Gelclay WH2. The major exchangeable cation is Na+ (83%). The liquid limit is 575 %, the plastic limit is 53 % and the unit mass is 2.77 g/cm^3 ([21, [22]). The quartz sand used in the mixture comes from Eure and Loire (France) with a unit mass of 2.65 g/cm^3. It was sieved at 2mm prior to being mixed with bentonite.

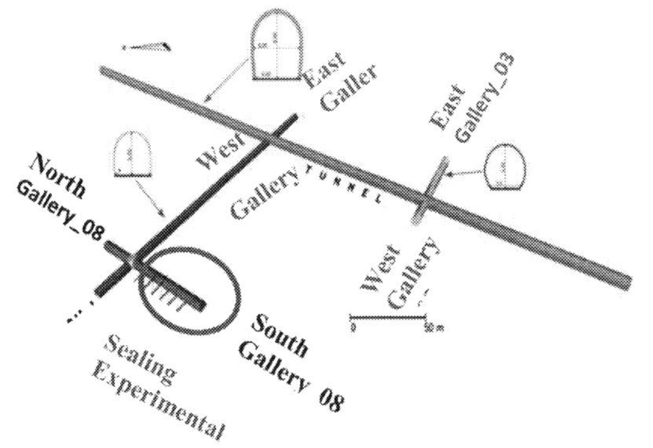

Figure. 1. Location of SEALEX in-situ experiments in Tournemire URL.

For the laboratory and in-situ tests, synthetic water used for hydration has the same chemical composition as the pore water of the Callovo-Oxfordian claystone from ANDRA URL in Meuse Haute la Marne (France) ([22]).

The bentonite blocks were obtained through uniaxial compaction of the mixture at a water content w=11% and dry density ρ_d= 1.97 Mg/m^3. This initial dry density was calculated taking into account the available radial technological gap. So that, the final dry density after filling the annular gap would be 1.67 Mg/m^3.

2.3 Testing geometry and operations

Six SEALEX in-situ tests have been emplaced in horizontal boreholes (dip 0 ± 2°), of 60 cm diameter and 5.4 m length, excavated into a 180 m thick shale formation (Toarcian-Domerian) at a height of 1.50 m above the URL gallery's floor (Figure.1). Each experiment consists of a bentonite-based core mechanically confined at both ends, which represents a generic seal mock-up, except for the artificial saturation system (Figure.2).

SEALEX in-situ tests PT-N1, PT-N2 and PT-A1 have been designed to investigate the impact of intra-core

DOI: 10.1051/e3sconf/20160904006

geometry (technological gaps) on the performance of the clay-based seal. For these tests, the swelling core consists of pre-compacted blocks of a bentonite/sand mixture in a ratio of 70/30 (in dry mass) with different geometries. Two types of technological gaps exist within the experiments: gaps between the clay-based core and the rock (annular gap with non-uniform thickness) (Figure 2) and gaps within the clay core: between the block (PT-N1, PT-A1 and PT-N2) and within the blocks (PT-N2) (Figure.3). Dimensions of the clay core (56 cm diameter, 120 cm in length) have been defined from scoping calculations of the re-saturation kinetics ([4]).

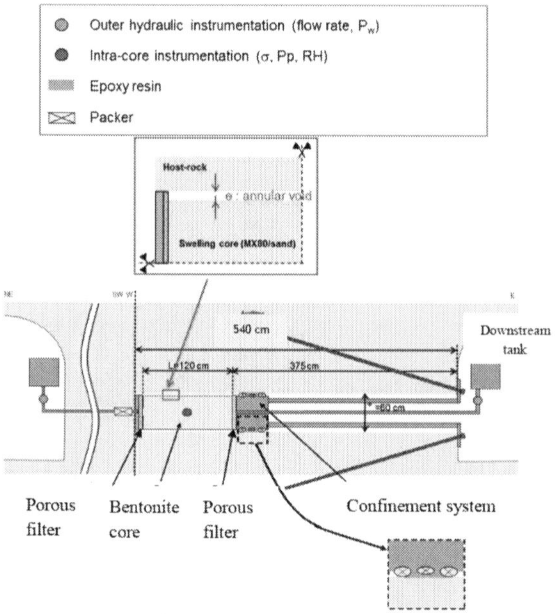

Figure. 2. Layout of SEALEX in-situ tests.

The clay seals in PT-N1 and PT-A1 are made up of 8 monolithic pre-compacted disks (56 cm in diameter and 15 cm thick) (Figure. 3a). For PT-N2 the disks are divided in 4 sections with internal joints (Figure. 3b). The core is confined between two fixed stainless steel lids, namely upstream and downstream lid, which provide both hydraulic and mechanical closure (Figure. 2). The lids consist basically of a stainless steel circular plate and a cylindrical stainless steel tube welded all around its periphery. It has a series of three rubber inflatable cushions (O-ring) all around and passing for the hydration tubings. Once inflated with water, the O-rings close the gap between the plate and the surrounding rock.

A chamber filled up with an injected grout is left between the upstream lid and the flat end of the main borehole. The hydration surfaces at both sides of the lids are based on special geotextile mats fed with water distribution tubes. The reduction in thickness of the mats due to the swelling of the buffer is minimal up to 4 MPa.

The hydration system is equipped with a water distribution panel fed by a weighted stainless steel water tank connected with inflow lines to both hydration surfaces (downstream and upstream). The confining system consists of a support tube inserted into the cylindrical tube of the downstream lid up to rest against the circular plate and of a closure plate (1400 mm in diameter) placed at the outer face of the borehole and secured by four bolts anchored to the rock. For test PT-A1, a second tube is added. The coupling acts as a piston and allows the movement of the plate (in contact with the bentonite core) when the confinement is released.

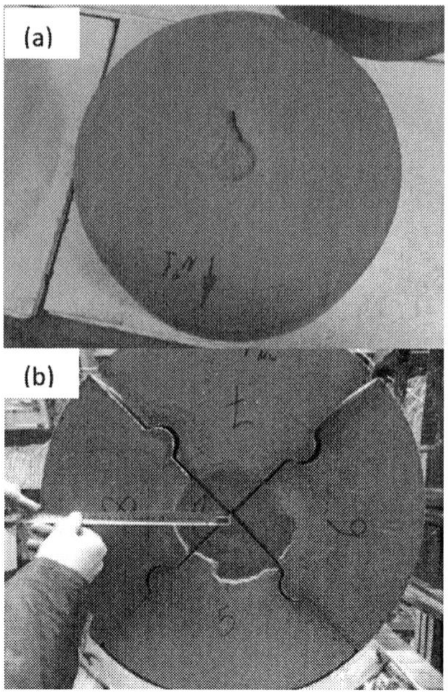

Figure. 3 (a) Bentonite monolithic disks used in tests PT-N1 and PT-A1. (b) Jointed quarters of disks used in PT-N2.

In order to avoid potential flow paths along cables and sensors, and thereby, to limit as much as possible the disturbance of the clay core wireless sensors have been used. Three types of sensors have been installed within the compacted blocks: 5 total pressure cells, 8 pore pressure sensors and 8 relative humidity sensors

[a] Corresponding author: nadia.mokni@irsn.fr

DOI: 10.1051/e3sconf/20160904006

(Figure.4). Three total pressure sensors have been installed on the surface of the core at 60 cm from the downstream saturation face to measure the radial swelling pressure. Two total pressure sensors have been installed at 0 and 120 cm from the downstream saturation face to measure axial swelling pressure. The blocks have been mechanized to house the sensors and the wireless transmitter. The remaining gaps have been then filled up with MX80 bentonite powder to minimize preferential flow paths.

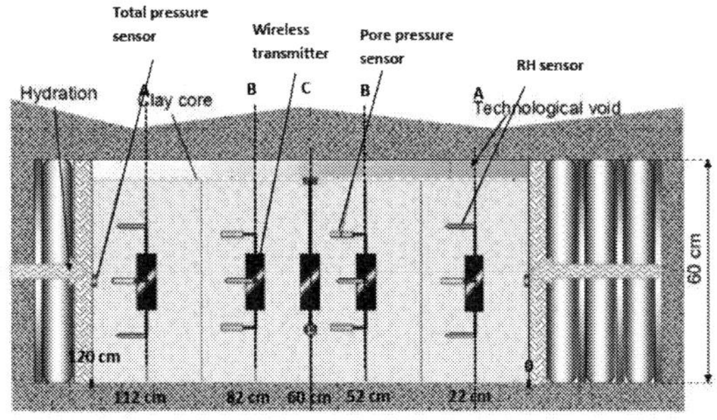

Figure. 4. Distribution of RH, pore pressure, and swelling pressure sensors in the seal.

Hydration has been performed following 2 phases: First, a back pressure of 0.2 MPa was applied in the tank, which allowed filling the gap between the rock and the device in 2 hours. Second, the back pressure was removed, and the upper level of the water in the tank was constantly maintained at 1m above the axis of the borehole (i.e. 1m water head).

3 Results and comparison

3.1 Impact of technological gaps on saturation kinetics

Eight relative humidity sensors were installed in each bentonite core at 22, 52, 82 and 112 cm from the downstream hydration face (Figure. 4). Figure 5 compares RH evolution over time measured at RH-52-1 and RH-52-2 in PT-N1, PT-N2 and PT-A1. The curves show a monotonic increase of RH corresponding to the progress of saturation caused by water drawn from the hydration downstream and upstream ends and the annular technological gap. Independently of the distance to the wetting sources, a faster increase of RH is observed in PT-N2 which is probably the consequence of water initially infiltrating the intra and inter-disks technological gaps.

3.2 Impact of technological gaps on swelling pressure

During hydration radial swelling pressures were monitored using three total pressure cells installed on the surface of the bentonite based core at 60 cm from the downstream saturation face. Figure 6 compares the evolution of the radial swelling pressure at each sensor for PT-N1 and PT-N2. No comparison could be made with PT-A1 due to the lack of data (malfunctioning of the data acquisition system (DAS)). For PT-N1, the swelling pressures increased over time at different rates, depending on the sensor position. Higher swelling rates are observed at the sensor located at bottom right hand side (sensor S03). However, the swelling pressure level off at 400 days and reaches 1.8 MPa after 600 days. Slower swelling rates are observed at S02 and S01, but the swelling pressures increase at constant rates until the interruption of the data acquisition at 800 days (due to a problem in the DAS unit). The lowest swelling pressure is measured at the top of the core (sensor S02), where the annular gap between the bentonite and the rock was wider and therefore the confinement was lower.

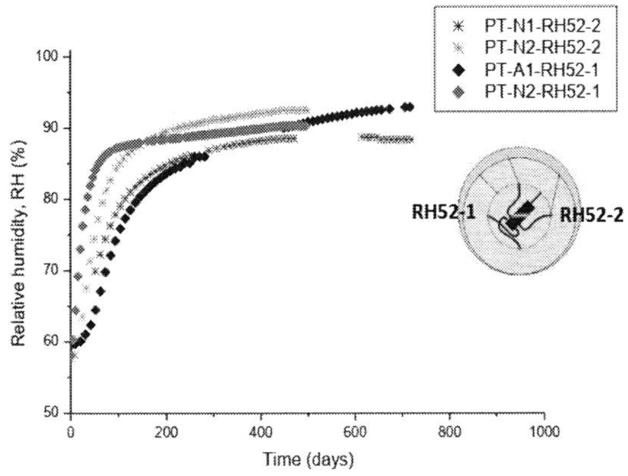

Figure. 5. Comparison of RH evolution in PT-N2, PT-N1 and PT-A1 measured at different distance from the hydrations sources.

In test PT-N2, a significantly lower and rapid stabilization (after 200 days) of the swelling pressure is observed. At S02 located on the top of the clay core, where the annular technological gap is the wider, comparable swelling rates are measured in PT-N1 and PT-

DOI: 10.1051/e3sconf/20160904006

N2 at least during the first 200 days. After this time period, the swelling pressure levels off in PT-N2. The observed lower swelling rates might be attributed to the initial self-sealing of the additional technological gaps existing within and between the clay jointed disks in this in-situ test.

The significant difference between the radial swelling pressures, observed in PT-N1, suggests a heterogeneous structural distribution of the bentonite/sand mixture within the blocks. Saba et al. (2014a, b) used microfocus X-ray tomography (?CT) to examine, in 3D, the internal microstructure of a bentonite/sand (70/30) sample (38 mm diameter, 10 mm height), as compacted at ?d=1.8 $Mg/m^3$and after four months of hydration, in constant volume conditions. Continuity between the sand grains with interconnected inter-grain large pores has been detected at different locations within bentonite/sand mixture. Furthermore, a concentration of larger macropores between either sand grains or bentonite grains has been observed around the sample. Saba et al. (2014b) concluded that this heterogeneity has an impact on the swelling pressure measured on the radial and axial directions.

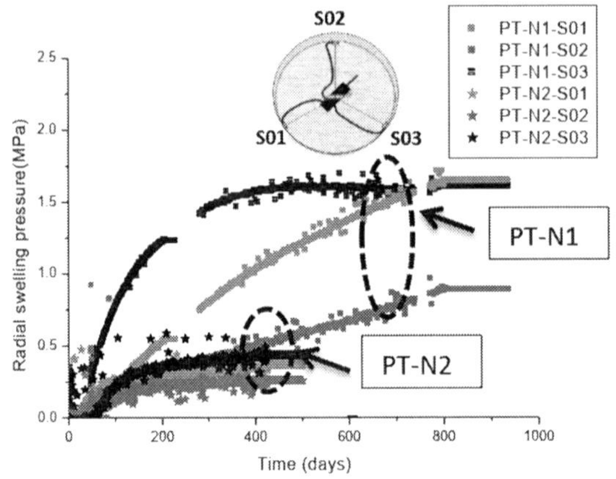

Figure. 6. Evolution of radial swelling pressure in tests PT-N2 and PT-N1 measured in 3 sensors located at 60 cm from the downstream hydration face.

3.3 Impact of technological gaps on dry density distribution

In order to investigate the effect of closure of technological gaps on bentonite dry density homogenisation, a hydro-mechanical coupled analysis of SEALEX in-situ test PT-N1 has been performed. The numerical analysis has been carried out using the finite element code CODE_BRIGHT ([6]).

A key feature in the simulation lies in the appropriate modeling of the bentonite-sand mixture hydro-mechanical behavior. A detailed description of the HM constitutive laws used to describe the behaviour of the bentonite based seal could be found in [13]. To represent the mechanical behaviour of the material, the well-known mechanical elasto-plastic Barcelona Basic Model, (BBM) ([1]) has been used.

The model parameters have been first calibrated against conventional laboratory tests ([13]). Two sets of laboratory experiments have been modelled: the first involves characterising the hydro-mechanical behaviour of the 70/30 MX80 bentonite/sand mixture (infiltration tests, swelling/ compression tests under controlled suction conditions); the second is a 1/10th scale mock-up of the SEALEX experiment ([13]).

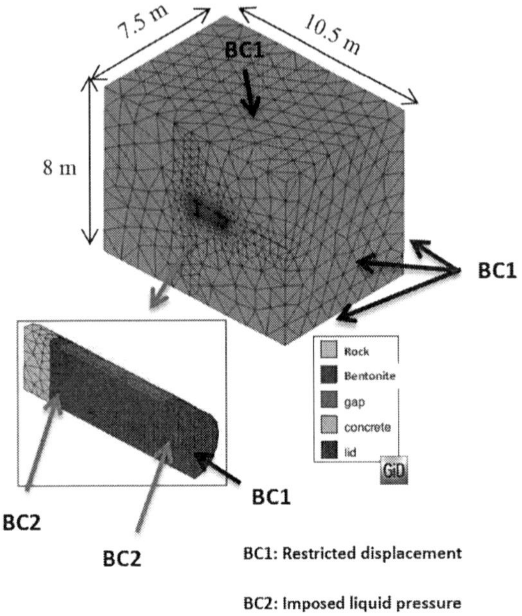

Figure. 7. Model geometry, mesh and boundary conditions.

The HM coupled analysis has been performed using a 3-D finite element model that includes the hydration systems (downstream and upstream lids), the concrete block between the host rock and upstream lid, the bentonite based core, the surrounding host-rock and the annular gap with variable thickness that exists between the bentonite based core and the host rock. Only one half of the problem is analysed because of symmetry (Figure.

[a] Corresponding author: nadia.mokni@irsn.fr

DOI: 10.1051/e3sconf/20160904006

7). A fundamental issue in modeling SEALEX in-situ test lies in properly representing the initially air-filled gap with variable thickness (0 at the bottom of the bentonite based core and 4 cm at the top). As a first approach, this gap is modelled as a material with high porosity and high permeability. A dependence of the retention curve and permeability on porosity has also been taken into account. The mechanical response of the gap is achieved using a nonlinear elastic model with stiffness values evolving with porosity defined as ([13]),

$$\begin{aligned} &if \quad \phi > \phi_0, \quad E = \frac{E_0}{\phi - \phi_0} \\ &else \qquad\quad E = E_{max} \end{aligned} \tag{1}$$

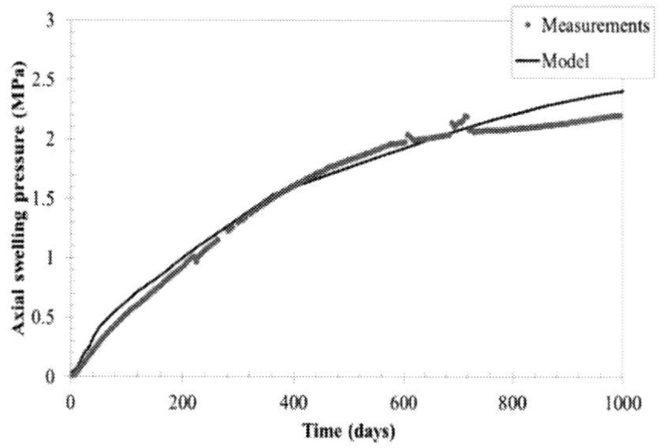

Figure. 8. Evolution of axial swelling pressure at a sensor placed at the downstream saturation face in test PT-N1.

Comparing modelling results with field observations, the model predict reasonably well the evolution of axial swelling pressure (Figure.8), and relative humidity (Figure.9) at different points of the clay core. Changes in dry density for three different times (140, 500 and 1000 Days) are displayed in Figure. 10.

A heterogeneous distribution of dry density is observed. For a given hydration time, low densities prevail at the outer boundary due to the swelling of the bentonite to fill up the non-uniform thickness annular gap, but they increases progressively towards the centre of the core. As hydration progresses, this decrease in density prevails wider into the interior core.

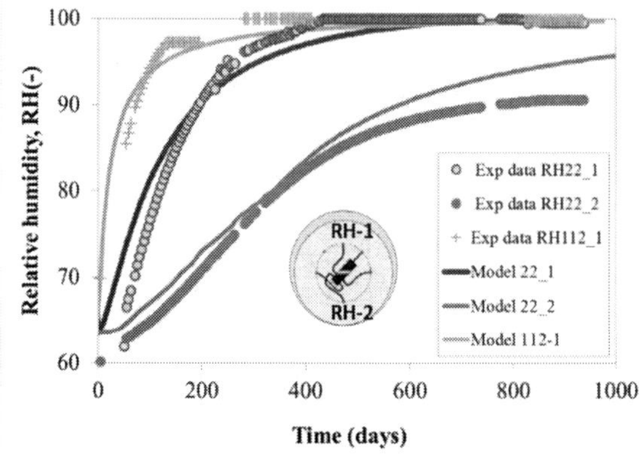

Figure. 9. Evolution of relative humidity at different distance from hydrations sources for SEALEX performance test PT-N1. Experimental and Modelling results.

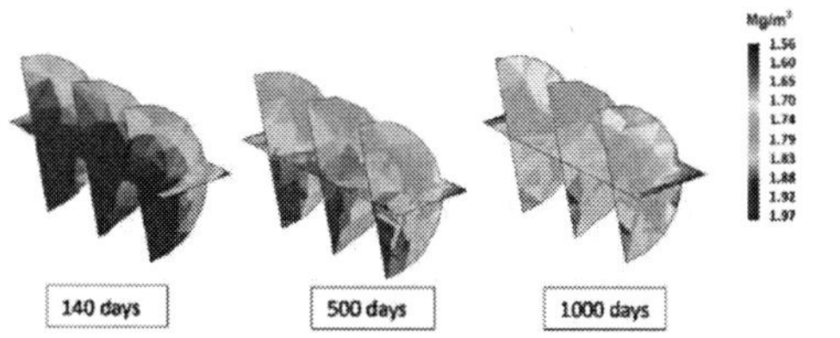

Figure. 10. Contour maps of dry density from Time=140 to 1000 days.

4 Concluding remarks

Technological gaps left between the pre-compacted plug and the host rock may impact the long term performance of the sealing system. For PT-N1 and PT-A1 which have similar clay core geometries, higher saturation rates have been observed at RH sensors located close to the annular technological gap. The latter constituted a new hydration source that allowed together with the installed hydration system the simultaneous radial and axial hydrations of the bentonite core. The presence of additional technological gaps within the clay jointed disks in PT-N2 did have a great impact on the saturations kinetics since it constituted additional infiltration paths. The effect of technological gaps was also evidenced when comparing the swelling pressure kinetics: the evolutions rates of radial swelling pressure were significantly lower in PT-N2. This difference is attributed to the initial self-sealing of the technological gaps existing within the clay disks in PT-N2. Technological gap was demonstrated to have an impact on dry density distribution. Model results demonstrated the non-uniform closure of the gap which induces a heterogeneous distribution of dry density

Acknowledgments

The authors would like to thank AITEMIN (Spain) and ENPC-CERMES (France) our partners in this project.

References

1. Alonso, E. E., Gens, A., and Josa, A. Géotechnique 40(3), 405-430. (1990)
2. Börgesson, L., Chijimatsu M., Fujita, T., Nguyen, T.S., Rutqvist, J., Jing, L.,. International Journal of Rock Mechanics and Mining Sciences 38, 95-104. (2001)
3. Barnichon, J.D., and Deleruyelle, F. EUROSAFE. (2009)
4. Barnichon J.D., Dick P. and Bauer C.,. Qian & Zhou (eds) Taylor & Francis Group, London. ISBN 978-0-415-80444-8, pages 1391-1394. (2012)
5. Delage P., Howat M. D. and Cui Y. J.,. Engineering Geology 50, 31-48. (1998)
6. DIT-UPC, 2000. CODE-BRIGHT. User-s guide. CIMNE, Barcelona.
7. Gens, A., Vallejan, B., Sánchez, M., Imbert, C., Villar, M .V, and Van Geet, M., 2011. Geotechnique 61(5):367-386, (2011).
8. Imbert C. and Villar M.V. Applied Clay Science, 32(3-4):197–209. (2006)
9. Lloret, A., and Villar, M.V., Sanchez, M., Gens, A.. Geotechnique 53 (1), 27-40.(2003)
10. Lloret, A., and Villar, M.V., Sanchez, M., Gens, A.,. Physics and Chemistry of the Earth 32 701-715. (2007)
11. Mayor, J.C., García-Siñeriz, J.L., Alonso, E.E., Alheid, H-J., Blümbling, P.. Final report Publicación técnica ENRESA 02/05. Madrid, 101 pp.(2005)
12. Martino, J.B., Dixon, D.A., Vignal, B., Fujita, T. WM'06 Conference, Febururay 26-March 2, 2006, Tucson, AZ.
13. Mokni, N., Barnichon, J.D. Engineering Geology 205: 81-92.(2016)
14. Pusch, R., 1979. Nuclear technology 45 (2), 153-157 United States.(1979)
15. Saba, S., Delage, P., Lenoir, N., Cui, Y.J., Tang, A.M. and Barnichon, J.D. Engineering Geology 168: 141-148. (2014a)
16. Saba, S., Barnichon, J.D, Cui, Y.J., Tang, A.M. and Delage, P.,. Journal of Rock Mechanics and Geotechnical Engineering 6: 126-132. (2014b)
17. Volckaert, G., Dereeper, B., Put, M., Ortiz, L., Gens, A., Vaunat, J., Villar, M.V., Martín, P.L., Imbert, C., Lassabatère, T., Mouche, E., Cany,F.. Reseal project-Phase I. EUR 19612. Commission of the European Communities, Luxembourg. 273.pp.(2000)
18. Van Geet, M., Bastiaens, W., Volckaert, G., Weetjens, E., Sillen, X., Maes, N., Imbert, C., Billaud, P., Touzé, G., Fillippi, M., Plas, F., Villar, M.V., García-Gutiérrez, M., Mingarro, M., GensA., Vallejan, B.,. RESEAL II. Final report.: EUR 24261, Nuclear Science and Technology, European Commission, Luxembourg, 288 pp.(2009)
19. Villar M.V.. Publicación Técnica ENRESA 01/2002, Madrid, 258 pp.(2008)
20. Villar M.V., Gómez-Espina, R., Campos, R., Barrios, I., Gutiérrez-Nebot, L.. Unsaturated soils: Research and Applications, vol. 1. Springer, Berlin, pp. 137-144.(2012)
21. Wang Q., Tang A.M., Cui Y.J., Delage P. and Gatmiri B.,. Engineering Geology 124: 59-66, doi : 10.1016/j.engeo.2011.10.003.(2012)
22. Wang, Q., Tang, A. M., Cui, Y.J., Barnichon J.D. & Ye W.M.. Engineering Geology, Vol. 162, 79–87.(2013a)
23. Yong, R.N., Boonsinsuk P. and Wong G., 1986. Canadian Geotechnical Journal, 23(2):216–228.(1986)

[a] Corresponding author: nadia.mokni@irsn.fr

Two-phase flow effects on the CO_2 injection pressure evolution and implications for the caprock geomechanical stability

Víctor Vilarrasa[1,a], Jesús Carrera[2] and Sebastià Olivella[3]

[1]*École Polytechnique Fédérale de Lausanne, 1015 Lausanne, Switzerland*
[2]*GHS, Institute of Environmental Assessment and Water Research (IDAEA), CSIC, 08034 Barcelona, Spain*
[3]*Dept Geotechnical Engineering and Geosciences, Technical University of Catalonia (UPC-BarcelonaTech), 08034 Barcelona, Spain*

Abstract. Geologic carbon storage is considered to be one of the main solutions to significantly reduce CO_2 emissions to the atmosphere to mitigate climate change. CO_2 injection in deep geological formations entails a two-phase flow, being CO_2 the non-wetting phase. One of the main concerns of geologic carbon storage is whether the overpressure induced by CO_2 injection may compromise the caprock integrity and faults stability. We numerically investigate the two-phase flow effects that govern the overpressure evolution generated by CO_2 injection and how this overpressure affects the caprock geomechanical stability. We find that fluid pressure increases sharply at the beginning of injection because CO_2 has to displace the brine that fills the pores around the injection well, which reduces the relative permeability. However, overpressure decreases subsequently because once CO_2 fills the pores around the injection well, CO_2 can flow easily due to its low viscosity and because the relative permeability to CO_2 increases. Furthermore, the pressure drop that occurs in the capillary fringe due to two-phase flow interference decreases as the CO_2 plume becomes larger. This overpressure evolution induced by CO_2 injection, which remains practically constant with time after the initial peak, is very beneficial for maintaining caprock stability. Thus, the sealing capacity of the caprock will be maintained, preventing CO_2 leakage to occur across the caprock.

1 Introduction

Despite the known necessity to significantly reduce atmospheric carbon dioxide (CO_2) concentration, CO_2 emissions continue increasing worldwide. The rate of increase is such that climate models predict a rise in the mean temperature of the Earth that may have very negative effects on the environment. Governments are aware of these potential negative effects and have recently reached an agreement at the COP21 meeting in Paris to take the necessary measures to avoid a temperature increase higher than 2 °C with respect to the pre-industrial temperature. According to the International Energy Agency (IEA), to achieve such objective, CO_2 emissions should be reduced by more than half with respect to the current CO_2 emissions by 2050 [1]. Furthermore, the IEA considers that geologic carbon storage should contribute with one fifth of the total reduction, which represents storing around 8 Gt of CO_2 per year in deep geological formations by 2050. Currently, only a few Mt of CO_2 are stored each year, mainly in pilot test sites [2, 3]. Thus, a tremendous amount of work has to be done before geologic carbon storage can become a real solution.

Geologic carbon storage consists in capturing CO_2 from point sources, such as industries or hydrocarbon based power plants, transporting the captured CO_2 to the injection well and finally injecting CO_2 in deep saline formations, where it will be permanently kept away from the atmosphere. CO_2 forms a plume around the injection well, displacing the formation brine laterally [4, 5]. CO_2 is the non-wetting phase and is partially miscible in brine [6]. Brine with dissolved CO_2 is denser than brine without dissolved CO_2, which leads to convection cells that enhance CO_2 dissolution [7-9], leading to safer storage. Furthermore, CO_2 is a very compressible fluid: it is a gas in the atmosphere, but at pressures greater than 7.38 MPa and temperatures above 31.04 °C CO_2 becomes a supercritical fluid. The transition to a supercritical fluid usually occurs at a depth of around 800 m [4]. CO_2 will be stored in sedimentary formations deeper than this depth because the high density of supercritical CO_2 makes the storage efficient in terms of volume. Still, supercritical CO_2 is lighter than the resident brine and thus, it tends to float. For this reason, a low-permeability and high entry pressure rock, known as caprock, overlying the storage formation, is required. This caprock provides a hydrodynamic trap for CO_2 that prevents CO_2 from migrating upwards (Figure 1). Apart from a liquid-like density, supercritical CO_2 has a low gas-like dynamic viscosity, which is around one order of magnitude lower than that of brine. Therefore, CO_2 flows more easily than brine. Furthermore, CO_2 injection generates overpressure, reducing the effective stresses, which induces deformations and brings the stress state closer to failure conditions.

[a] Corresponding author: victor.vilarrasariano@epfl.ch

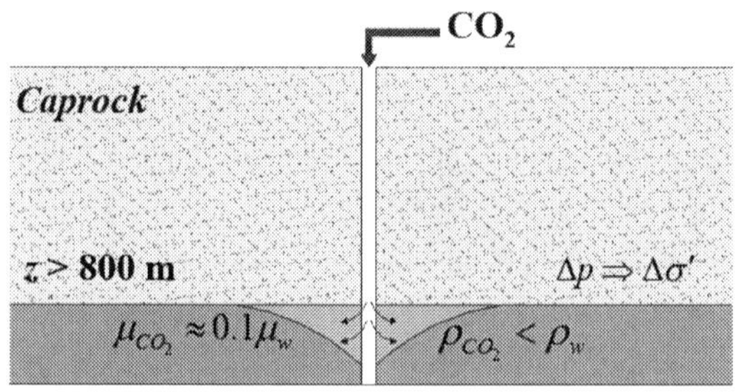

Figure 1. Schematic representation of geologic carbon storage.

One of the main concerns of geologic carbon storage is whether the caprock sealing capacity and faults stability will be maintained [10]. Since CO_2 tends to migrate upwards due to buoyancy, maintaining the caprock sealing capacity is crucial to avoid CO_2 leakage [11]. Nevertheless, sedimentary basins usually present sequences of permeable formations alternated with low permeable rocks that may act as secondary caprocks [12]. In such cases, CO_2 trapping in secondary caprocks would hinder CO_2 from reaching freshwater aquifers or the surface even though leakage occurred across the primary caprock. On the other hand, maintaining fault stability is important to avoid inducing felt seismic events [13]. Even though fault reactivation will not lead to CO_2 leakage [14, 15], it may induce earthquakes that may be felt by the local population. Felt seismicity induced by fluid injection in deep geological formations has led to the closure of several geo-energy related projects (e.g., a geothermal project at Basel, Switzerland [16], the seasonal gas storage project of Castor, Spain [17] and a wastewater disposal project at Guy-Greenbrier, Arkansas [18]). Thus, felt induced seismicity should be avoided to achieve a successful deployment of geological carbon storage.

To predict the geomechanical response of rocks, first, the fluid pressure distribution and evolution needs to be known. Overpressure evolution is well known in single phase flow, presenting a linear buildup with the logarithm of time. However, overpressure becomes more complex both in space and time for two-phase flow, like in CO_2 injection in deep saline formations. Despite this complexity, several studies have aimed at developing analytical solutions of overpressure. Fluid pressure evolution was calculated analytically by Mathias et al. [19] based on the analytical solution of Nordbotten et al. [20] for the CO_2 plume position. Vilarrasa et al. [21] also derived the fluid pressure evolution resulting from the analytical solutions for the CO_2 plume position of Nordbotten et al. [20] and Dentz and Tartakovsky [22], and incorporated a correction in these solutions to account for CO_2 compressibility. In all cases, fluid pressure was predicted to monotonically build up with time as a result of CO_2 injection at a constant mass flow rate. However, a semi-analytical solution that made more realistic assumptions yielded a fluid pressure evolution that peaks shortly after the beginning of injection, but that slightly decreases subsequently [23]. This semi-analytical solution accounts for the buoyancy of CO_2 within the injection well, which leads to (*i*) a CO_2 plume that may not reach the bottom of the storage formation and (*ii*) a CO_2 injection rate that is not uniformly distributed along the whole thickness of the injection well. To validate these analytical and semi-analytical solutions, CO_2 injection in the field is needed.

The experience with overpressure evolution in CO_2 injection sites is quite limited. However, a few pilot tests have already performed CO_2 injection. For example, at Ketzin, Germany, CO_2 pressure underwent a sharp increase at the beginning of injection, but it became practically constant shortly afterwards [24]. Thus, overpressure evolution in the field is more similar to that predicted by the semi-analytical solution of Vilarrasa et al. [23] than to that derived from other analytical solutions. Numerical solutions also predict the sharp increase in injection pressure at the beginning of injection followed by a progressive pressure drop [25-29]. Nevertheless, little attention has been paid to understand the driving mechanisms of overpressure evolution induced by CO_2 injection.

The objective of this paper is to numerically investigate how overpressure evolves when injecting CO_2 in deep saline formations and to analyze how this injection affects to caprock stability. In the following section, we explain the numerical model and capillary properties that we use. Next, we explain how injection pressure evolves when CO_2 is injected through a vertical well. Then, we present the geomechanical implications of overpressure on the caprock integrity. Finally, we summarize and draw the conclusions of this study.

2 Methods

CO_2 injection in deep geological formations is a two-phase flow problem that requires solving the mass conservation of each fluid phase [30]

$$\frac{\partial(\varphi S_\alpha \rho_\alpha)}{\partial t} + \nabla\cdot(\rho_\alpha \mathbf{q}_\alpha) = r_\alpha, \qquad \alpha = c, w \qquad (1)$$

where φ [$L^3\,L^{-3}$] is porosity, S_α [-] is saturation of the α -phase, ρ_α [$M\,L^{-3}$] is density of the α -phase, t [T] is time, $\mathbf{q}_\alpha$ [$L^3\,L^{-2}\,T^{-1}$] is the volumetric flux, r_α [$M\,L^{-3}\,T^{-1}$] is the phase change term, i.e., CO_2 dissolution into water and water evaporation into CO_2, and α is either CO_2-rich phase, c, or aqueous phase, w.

Momentum conservation for the fluid phases is expressed using Darcy's law

$$\mathbf{q}_\alpha = -\frac{k k_{r\alpha}}{\mu_\alpha}\left(\nabla p_\alpha + \rho_\alpha g \nabla z\right), \qquad \alpha = c, w, \qquad (2)$$

where k [L^2] is intrinsic permeability, $k_{r\alpha}$ [-] is the α -phase relative permeability, μ_α [$M\,L^{-1}\,T^{-1}$] is the α -phase viscosity, p_α [$M\,L^{-1}\,T^{-2}$] is the α -phase pressure and g [$L\,T^{-2}$] is gravity.

To account for geomechanics, the momentum balance of the solid phase has to be solved. Neglecting inertial terms, it reduces to the equilibrium of stresses

$$\nabla \cdot \boldsymbol{\sigma} + \mathbf{b} = \mathbf{0}, \quad (3)$$

where $\boldsymbol{\sigma}$ [M L^{-1} T^{-2}] is the stress tensor and $\mathbf{b}$ [M L^{-2} T^{-2}] is the body forces vector. The stress-strain relationship, assuming elasticity, is given by Hooke's law

$$\boldsymbol{\varepsilon} = \frac{\sigma'_m}{3K}\mathbf{I} + \frac{1}{2G}\left(\boldsymbol{\sigma}' - \sigma'_m \mathbf{I}\right), \quad (4)$$

where $\boldsymbol{\varepsilon}$ [-] is the elastic strain tensor, $\boldsymbol{\sigma}'$ [M L^{-1} T^{-2}] is the effective stress tensor, $\sigma'_m = \left(\sigma'_x + \sigma'_y + \sigma'_z\right)/3$ [M L^{-1} T^{-2}] is the mean effective stress, $\mathbf{I}$ [-] is the identity matrix, $K = E/(3(1-2\nu))$ [M L^{-1} T^{-2}] is the bulk modulus, $G = E/(2(1+\nu))$ [M L^{-1} T^{-2}] is the shear modulus, E [M L^{-1} T^{-2}] is the Young's modulus and ν [-] is Poisson ratio.

We model CO_2 injection at a constant mass flow rate through a vertical well. Due to the symmetry of the problem, the model is axisymmetric. We include in the model the storage formation, the caprock and the baserock (Figure 2). The top of the storage formation is at a depth of 1500 m. We assume isothermal conditions, with a constant temperature of the storage formation equal to 60 °C, which corresponds to a surface temperature of 10 °C and a geothermal gradient of 33 °C/km.

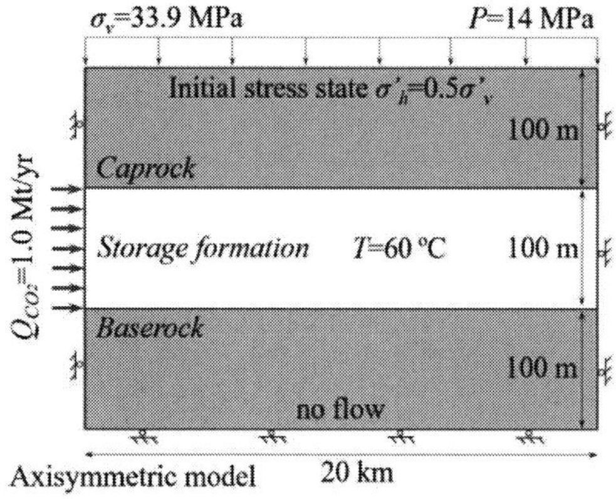

Figure 2. Schematic representation of the numerical model for simulating CO_2 injection in an extensive deep saline formation.

The storage formation is assumed to be a permeable sandstone and the caprock and baserock a low permeable shale. The properties of these rocks are detailed in Table 1. The retention curves of the rocks (Figure 3) follow a van Genuchten model [31]

$$S_{ew} = \frac{S_w - S_{rl}}{S_{ls} - S_{rl}} = \left(1 + \left(\frac{p_c - p_w}{p_0}\right)^{\frac{1}{1-\lambda}}\right)^{\lambda}, \quad (5)$$

where p_0 [M L^{-1} T^{-2}] is the entry pressure, λ [-] is the shape function of the retention curve, S_{ew} [-] is the effective water saturation, S_{rl} [-] is the residual liquid saturation and S_{ls} [-] is the maximum liquid saturation. The relative permeability of the storage formation follows a power law of effective saturation, with a cubic law for water and a power equal to 6 for CO_2, which reproduces the higher multiphase flow interference effects for CO_2 than for water that are usually measured in the laboratory [32] (Figure 4).

Table 1. Material properties of the storage formation, the caprock and baserock.

Property	Storage formation	Caprock and baserock
Permeability, k (m^2)	10^{-13}	10^{-18}
Gas entry pressure, p_0 (MPa)	0.02	5
van Genuchten λ (-)	0.6	0.3
Residual liquid saturation, S_{rl} (-)	0.3	0.4
Maximum liquid saturation, S_{ls} (-)	1.0	1.0
Relative water permeability, k_{rw} (-)	S_{ew}^3	S_{ew}^6
Relative CO_2 permeability, k_{rc} (-)	$(1-S_{ew})^6$	$(1-S_{ew})^6$
Porosity, φ (-)	0.1	0.01
Young's modulus, E (GPa)	10	5
Poisson ratio, ν (-)	0.3	0.3

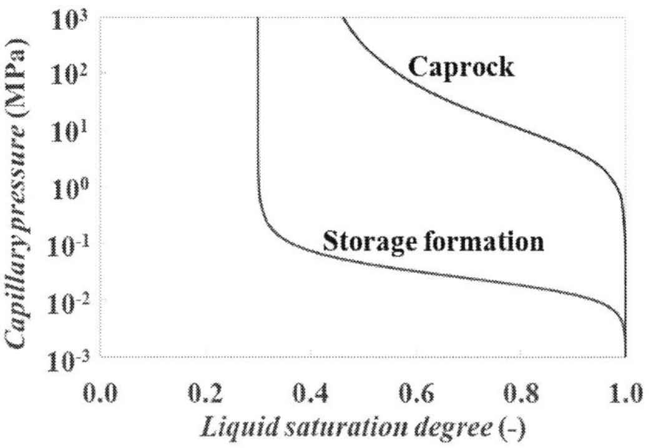

Figure 3. Retention curve of the storage formation and the caprock. The high permeable storage formation is characterized by a low entry pressure and a small slope of the retention curve, indicating that the pores significantly desaturate for small increases in the capillary pressure. In contrast, the low permeable caprock presents a high entry pressure and a high increase in the capillary pressure is needed to desaturate the pores (note the logarithmic scale of the vertical axis).

We inject 1 Mt/yr of CO_2 during 1 year. Since the model extends laterally 20 km, the pressure perturbation cone, which is slightly larger than 10 km after 1 year, does not reach the outer boundary during the injection time. Thus, the outer boundary condition does not influence the results and the storage formation in our model is an infinitely acting aquifer.

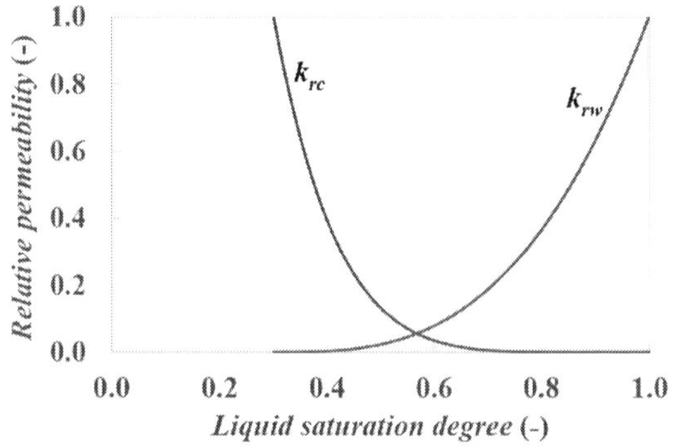

Figure 4. Relative permeability to water and CO_2 of the storage formation.

We simulate CO_2 injection in a deep saline formation using the finite element code CODE_BRIGHT [33, 34], extended for CO_2 injection [25]. CO_2 density is calculated using the cubic Redlich-Kwong equation of state with the parameters proposed for CO_2 by Spycher et al. [6]. CO_2 viscosity is calculated according to the empirical expression developed by Altunin and Sakhabetdinov [35]. The mesh is made of structured quadrilateral elements. Radially, the size of the elements is of a few cm close to the injection well and increases progressively up to 3000 m next to the outer boundary. Vertically, the storage formation is discretized with 5 m thick elements. The element size grows in the caprock and baserock from 5 m at the contact with the storage formation to 25 m far away from it. We performed a mesh sensitivity analysis to ensure that results are not influenced by further refinements.

3 Results

3.1. CO_2 injection pressure evolution

Figure 5 shows the overpressure evolution at the injection well induced by injection of 1 Mt/yr of CO_2. CO_2 injection pressure is characterized by an initial peak followed by a slight decrease in overpressure. The sharp increase at the beginning of injection is caused by the very low relative permeability to CO_2 that occurs as the pores around the injection well start to desaturate (recall Figure 4). Initially, the capillary fringe, where two-phase flow effects are dominant, completely surrounds the injection well (Figure 6a). As long as the capillary fringe surrounds the injection well, the low relative permeability values act as a negative skin effect of the injection well, causing the sharp increase in injection pressure. However, as CO_2 injection continues, the capillary fringe is displaced away from the injection well and CO_2 eventually fills the pores around the injection well, reaching the residual liquid saturation. At this stage, the relative permeability to CO_2 approaches unity and since the viscosity of supercritical CO_2 is one order of magnitude lower than that of brine, CO_2 flows easily through the well-connected pores filled with CO_2 and overpressure starts to drop (Figure 6b). Due to buoyancy, CO_2 tends to migrate upwards and advances preferentially through the top of the storage formation. As the capillary fringe is displaced away from the injection well, the pressure drop that occurs through it progressively decreases (see the magnitude of the arrows in Figure 7). While the retention curve and relative permeability curves have a small effect on the magnitude of the induced overpressure, the retention curve controls the CO_2 distribution within the CO_2 plume and the extent of the CO_2 plume [36]. Overall, overpressure induced by CO_2 injection tends to slightly decrease with time after the initial sharp increase.

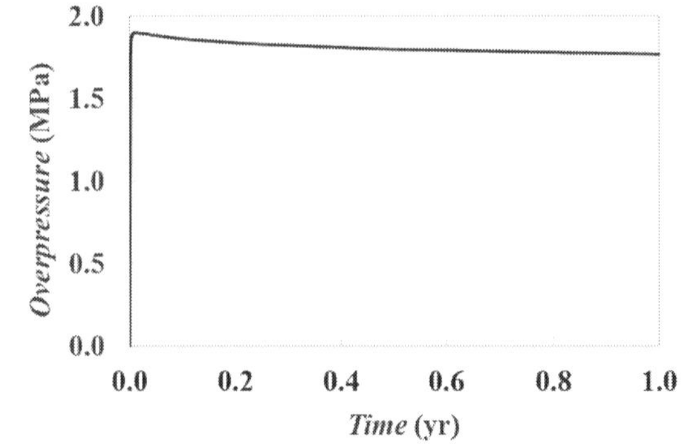

Figure 5. Overpressure evolution as a result of injecting 1 Mt/yr of CO_2.

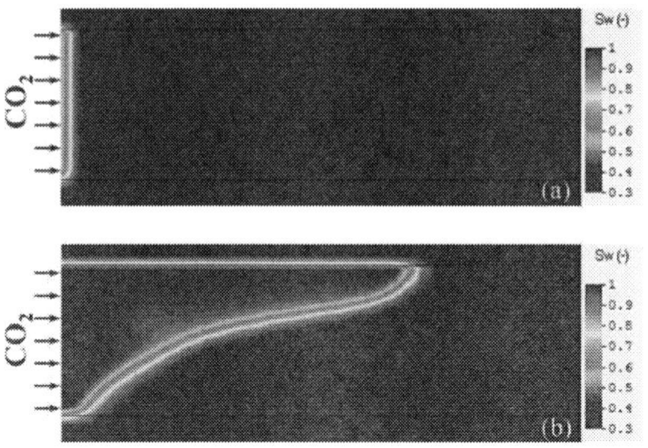

Figure 6. Cross section of an axisymmetric CO_2 plume (a) at the beginning of injection, coinciding with the peak in overpressure caused by the low values of the relative permeability in the capillary fringe and (b) once the CO_2 has filled the pores in the vicinity of the injection well and can flow easily due to its low viscosity, which induces a slight decrease in overpressure as the capillary fringes moves away from the injection well. Note that CO_2 tends to advance preferentially through the top of the storage formation due to buoyancy.

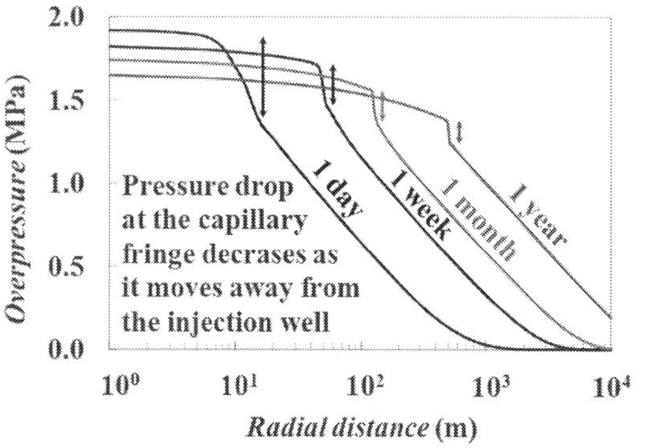

Figure 7. Fluid overpressure at the top of the storage formation as a function of the distance from the injection well for several injection times. Note that the pressure drop at the capillary fringe (indicated with an arrow) decreases as the capillary fringe moves away from the injection well.

3.2 Caprock geomechanical stability

Fluid pressure evolution has a direct effect on the caprock geomechanical stability. Even though the low permeability of the caprock prevents overpressure from propagating quickly into the caprock, the portion of the caprock that is close to the interface with the storage formation undergoes an overpressure similar to that of the storage formation, but delayed. Since overpressure induces a reduction in the effective stresses, the stress state evolves towards failure conditions (Figure 8). The initial sharp increase in injection pressure shifts the Mohr circle towards the failure envelope. However, the horizontal stresses increase in response to overpressure, which causes a decrease in the size of the Mohr circle. As a result, the caprock remains stable, being still some margin before failure conditions are reached. In the long-term, after the peak in overpressure, caprock stability slightly improves as a result of the slight decrease in overpressure. Thus, the most critical situation occurs at the beginning of injection. Consequently, it may be advisable to progressively inject CO_2 at the beginning of injection to avoid inducing a sharp increase in overpressure as the pores around the injection well get filled with CO_2.

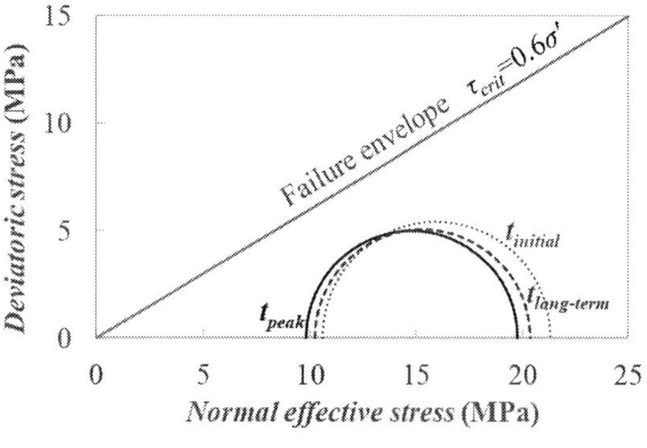

Figure 8. Stability evolution of the lower portion of the caprock next to the injection well as a result of injecting 1 Mt/yr of CO_2.

This overpressure evolution induced by CO_2 injection significantly differs from that generated by water injection. Recently, the feasibility of geologic carbon storage to significantly reduce CO_2 emissions has been put in doubt because it was compared with wastewater disposal [37], which is inducing a large number of large earthquakes (magnitude greater than 4) in the central US [38]. However, the geomechanical response of CO_2 injection will be completely different to that of wastewater disposal, making geological carbon storage a safe option for mitigating climate change [15, 39].

4 Summary and conclusions

CO_2 injection in deep geologic formations implies a two-phase flow in which the relative permeability plays a major role on the fluid pressure evolution. Overpressure induced by CO_2 injection peaks at the beginning of injection caused by the low values of the relative permeability to CO_2 as the pores start to desaturate. However, once the pores surrounding the injection well get filled with CO_2 and the residual liquid saturation is reached, the relative permeability to CO_2 approaches one and since the viscosity of supercritical CO_2 is one order of magnitude lower than that of brine, overpressure reaches a maximum and starts decreasing slightly. Additionally, the pressure drop that occurs in the capillary fringe due to two-phase flow interference decreases as the capillary fringe is displaced away from the injection well. The fact that overpressure does not continuously increase with time is beneficial to maintain the caprock integrity. Thus, the caprock sealing capacity is not altered and CO_2 leakage across the caprock does not take place. The peak in overpressure at the beginning of injection coincides with the least stable situation. To minimize the initial changes on the caprock stability, a good practice would be to progressively inject CO_2, so that the pores around the injection well slowly fill with CO_2 and the peak in injection pressure would be avoided. Overall, CO_2 injection induces an overpressure that can be easily controlled to avoid damaging the caprock or reactivating faults, which makes geologic carbon storage a feasible option to significantly reduce CO_2 emissions to mitigate climate change.

References

1. IEA. *Energy technology perspectives. Scenarios & strategies to 2050. Executive summary*, International Energy Agency (2010).
2. K. Michael, A. Golab, V. Shulakova, J. Ennis-king, G. Allinson, A. Sharma, T. Aiken, Int. J. Greenh. Gas Contr., **4**, 659–667 (2010).
3. J.T. Birkholzer, C. Oldenburg, Q. Zhou, Int. J. Greenh. Gas Contr., **40**, 203–220 (2015).
4. S. Bachu, Environ. Geol., **44**, 277–289 (2003).
5. M.L. Szulczewski, C.W. MacMinn, H.J. Herzog, R. Juanes, Proc. Nat. Acad. Sci., **109**, 5185-5189 (2012).
6. N. Spycher, K. Pruess, J. Ennis-king, Geochimica et Cosmochimica Acta, **67**, 3015–3031 (2003).

7. A. Riaz, M.A. Hesse, H.A. Tchelepi, F.M. Orr, Jr., J. Fluid Mech., **548**, 87–111 (2006).
8. J.J. Hidalgo, J. Carrera, J. Fluid Mech., **640**, 443–454 (2009).
9. G.S.H. Pau, J.B. Bell, K. Pruess, A.S. Almgren, M.J. Lijewski, K. Zhang, Adv. Water Res., **33**, 443–455 (2010).
10. J. Rutqvist, Int. J. Geotech. Geol. Eng., **30**, 525–551 (2012).
11. J.M. Miocic, S. Gilfillan, C. McDermott, R.S. Haszeldine, Energy Procedia, **40**, 320-328 (2013).
12. J.T. Birkholzer, Q. Zhou, C.F. Tsang, Int. J. Greenh. Gas Contr., **3**, 181-194 (2009).
13. F. Cappa, J. Rutqvist, Geophys. Res. Lett., **38**, L17313 (2011).
14. J. Verdon, Environ. Res. Lett., **9**, 064022 (2014).
15. V. Vilarrasa, J. Carrera, Proc. Nat. Acad. Sci., **112**, 5938-5943 (2015).
16. M.O. Häring, U. Schanz, F. Ladner, B.C. Dyer, Geothermics, **37**, 469–95 (2008).
17. S. Cesca, F. Grigoli, S. Heimann, A. Gonzalez, E. Buforn, S. Maghsoudi, E. Blanch, T. Dahm, Geophys. J. Int., **198**, 941-953 (2014).
18. S. Horton, Seismolo. Res. Lett., **83**, 250-260 (2012).
19. S.A. Mathias, P.E. Hardisty, M.R. Trudell, R.W. Zimmerman, Transp. Porous Media, **79**, 265–284 (2009).
20. J.M. Nordbotten, M.A. Celia, S. Bachu, Transp. Porous Media, **58**, 339–360 (2005).
21. V. Vilarrasa, D. Bolster, M. Dentz, S. Olivella, J. Carrera, Transp. Porous Media, **85**, 619–639 (2010).
22. M. Dentz, D.M. Tartakovsky, Transp. Porous Media, **79**, 15–27 (2009).
23. V. Vilarrasa, J. Carrera, D. Bolster, M. Dentz, Transp. Porous Media, **97**, 43-65 (2013).
24. J. Henninges, A. Liebscher, A. Bannach, W. Brandt, S. Hurter, S. Köhler, F. Möller, CO2SINK Group, Energy Procedia, **4**, 6085–6090 (2011).
25. V. Vilarrasa, D. Bolster, S. Olivella, J. Carrera, Int. J. Greenh. Gas Contr., **4**, 910–919 (2010).
26. R.T. Okwen, M.T. Stewart, J.A. Cunningham, Int. J. Greenh. Gas Contr., **5**, 1140-1148 (2011).
27. K.Y. Kim, W.S. Han, J. Oh, T. Kim, J.C. Kim, Transp. Porous Media, **92**, 397-418 (2012).
28. M.J. Martinez, P. Newell, J.E. Bishop, D.Z. Turner, Int. J. Greenh. Gas Contr., **17**, 148-160 (2013).
29. M. Burton, N. Kumar, S.L. Bryant, Energy Procedia, **1**, 3091-3098 (2009).
30. J.Bear, *Dynamics of fluids in porous media* (Elsevier, New York, 1972).
31. R. van Genuchten, Soil Sci. Soc. Am. J., **44**, 892–898 (1980).
32. B. Bennion, S. Bachu, SPE Reserv. Eval. Eng., 11, 487–96 (2008).
33. S. Olivella, J. Carrera, A. Gens, E.E. Alonso, Transp. Porous Media, **15**, 271–293 (1994).
34. S. Olivella, A. Gens, J. Carrera, E.E. Alonso, Eng. Computations, **13**, 87–112 (1996).
35. V.V. Altunin, M.A. Sakhabetdinov, Teploenergetika, **8**, 85–89 (1972).
36. V. Vilarrasa, J. Rutqvist, A.P. Rinaldi, Greenhouse Gases: Sci. & Tech., **5**, 449-461 (2015).
37. M.D. Zoback, S.M. Gorelick, Proc. Nat. Acad. Sci., **109**, 10164–10168 (2012).
38. W.L. Ellsworth, Science, **341**, 1225942 (2013).
39. V. Vilarrasa, J. Carrera, Proc. Nat. Acad. Sci., **112**, E4511-E4511 (2015).

Chemo-physical evolution and microstructure features of lime treated soils

Giacomo Russo[1,a],

[1] *Department of Civil and Mechanical Engineering, University of Cassino and Southern Lazio, Italy*

Abstract. In the paper some results on the effects of chemo-physical evolution of clay-lime-water suspensions on the microstructure of a lime treated kaolin have been presented. A multi-scale investigation on the sedimentation behaviour of clay suspensions under different pore water chemistry has been developed highlighting the chemo-physical mechanisms controlling particle arrangement and the soil fabric formation. The results evidenced the key role of ionic exchange in the short term on the microstructure features of the lime treated soil.

1 Introduction

The addition of lime to a clayey soil has a strong impact on the physical and mechanical properties in the short and long term after treatment [1-7]. Several studies on the basic lime-clay-reaction mechanisms have been widely reported in literature [8-14] allowing a better understanding of chemo-physical evolution of the system and microstructural features induced after the addition of lime [15-17].

When lime is added to the soil water system, either in the form of quicklime (CaO) or as hydrated lime ($Ca(OH)_2$), a change in the pore water chemistry occurs. For quicklime the first reaction taking place is hydration reaction, highly exothermic, leading to the formation of hydrated lime (portlandite) in the soil water system. The dissociation reaction of the hydrated lime leads to an increase of the pH and to a high concentration of calcium ions in the pore water. The ions dissolved in the clay water system tend to counteract the charge on the basal face and lateral face sites of clay particles surface, controlling both the chemo-physical interactions and the arrangement of clay particles. The high pH environment induced by lime addition promotes the dissolution of the alumino-silicate constituents of clay minerals. Cementitious compounds, such as hydrated calcium silicates (C-S-H) or hydrated calcium aluminates (C-A-H), are formed as products of pozzolanic reactions between silica and alumina and lime.

The effects of the chemo-physical evolution of the clay-lime-water systems on microstructure play a key role in the mechanical improvement of lime treated soils. In the paper some results on the effects of chemo-physical evolution of clay-lime-water suspensions on the microstructure of a lime treated kaolin have been presented. A multi-scale investigation on the sedimentation behaviour of clay suspensions under different pore water chemistry has been developed highlighting the chemo-physical mechanisms controlling particle arrangement and the soil fabric formation. Sedimentation behaviour of clay suspension is closely related to the electrochemical interactions acting at particle level [18-24]. Changes in the pore water chemistry induced by addition of lime affect the extent of attraction and repulsion forces among particles, controlling the microstructural features of the sediment formed from suspensions. The sedimentation tests are then a useful tool for understanding the effects of pore water chemistry on the microstructural features of the formaed soil skeleton.

In the paper the results of several sedimentation tests performed on lime treated kaolin suspensions have been reported and discussed. The chemo-physical evolution of the suspensions has also been studied at particle scale level. The electro-kinetic properties of clay particles governing the interaction forces have been inferred from Zeta-potential measurements. Soil fabric evolution and its effects on the physical properties have been investigated by means of Dynamic Light Scattering measurements. A discussion section has been finally devoted to the interpretation of results linking different scale level of the experimental analysis.

2 Experimental procedures

2.1 Soil

Speswhite kaolin was provided by Imerys Minerals UK. The specific gravity G_s is 2.6, and its surface area is 14 m^2/g. The pH value is 4.6. Atterberg limits are respectively w_L = 70% and w_P = 32%, with a plasticity index IP equal to 38%. The mineralogical composition is mainly formed by kaolinite clay minerals with a small amount of quartz and muscovite. No swelling clay phases were detected after the ethylene glycol treatment.

2.2 Sedimentation tests

The sedimentation behaviour of kaolin suspensions has been investigated as a function of lime content and cation valence. Suspensions were prepared in graduated cylinders with mixture of 50 g of dry kaolin at 100% of water content. Different amounts of CaO and KOH were added to reach the required pH and electrolytes concentrations. Additional water was added up to a initial volume of 1000 ml. The pH value was increased by adding KOH. During the test the suspensions heights were monitored as a function of time.

Two typical curves can be identified (Figure 1) from sedimentation tests, depending on the degree of particles association in the suspension. Figure 1a shows the

[a] Corresponding author: giarusso@unicas.it

settling of a flocculated system, with the suspension height progressively reduced over time leaving a clear supernatant fluid. Figure 1b shows a dispersed system maintaining its configuration over time. The height of the suspension in the latter case is not relevantly modified and a partition between the suspension height and a clear supernatant fluid is not well recognisable.

In a flocculated system three stages of the sedimentation process can be identified (Figure 2). During the initial stage (induction period), the electrochemical interactions [22] induce the aggregation from small clusters to aggregates depending on the surface charge distribution of particles. Its duration is linked to the degree of particles or aggregates association [25]. A slight translation of the interface between suspended solids and water can be appreciated in this stage. The clear separation between the suspended solids and the supernatant liquid indicates that sedimentation and self-weight consolidation stages are ongoing. The flocculated soil particles settle under gravity forming a deposit of increasing height which undergoes self-weight consolidation. Settling and self-weight consolidation take place simultaneously, as evidenced in Figure 2 with the time t*. The soil formation line (Figure 2) delimitates the sedimentation and the self-weight consolidation regions and is characterised by a constant value of the void ratio e_m, the soil-formation void ratio, which represents the void ratio characterising the soil skeleton at zero effective stresses [26]. At time t_s, corresponding to the intersection between the soil formation line and the line representing the upper interface of the suspension, an abrupt reduction of settling velocity occurs, indicating the completion of the sedimentation stage [23, 27]. During the subsequent self-weight consolidation stage the sediment reduces its height with a decreasing rate until the final height. During the self-weight consolidation the effective stresses gradually develop due to the mechanical interaction between aggregates [23, 28] until the soil skeleton is in equilibrium under its own weight. The duration t_f of primary consolidation process generally lasts a time interval of two order of magnitudes larger than t_s, time necessary to complete the free settling stage [23, 27, 29].

2.3 Microstructure analyses

The Zeta potential measurements are based on the electrophoretic mobility of particles under an induced electric field. Mobility is then converted to zeta potential applying the Henry equation with the Smoluchovski approximation [30]. Diluted dispersions with 100 mg/l solid content and different concentration of calcium were prepared. The grain size fraction lesser than 1 μm was extracted and used for zeta potential measurement. The pH of dilute dispersions were adjusted between 2 and 12.4 by adding KOH. The measurement was performed at 25 °C in a capillary cell (DTS1061) with Malvern Nano Zetasizer apparatus. For each sample at least 3 measurements were performed assuring the repeatability of the results.

The hydrodynamic size of soil particles was determined using dynamic light scattering (DLS)

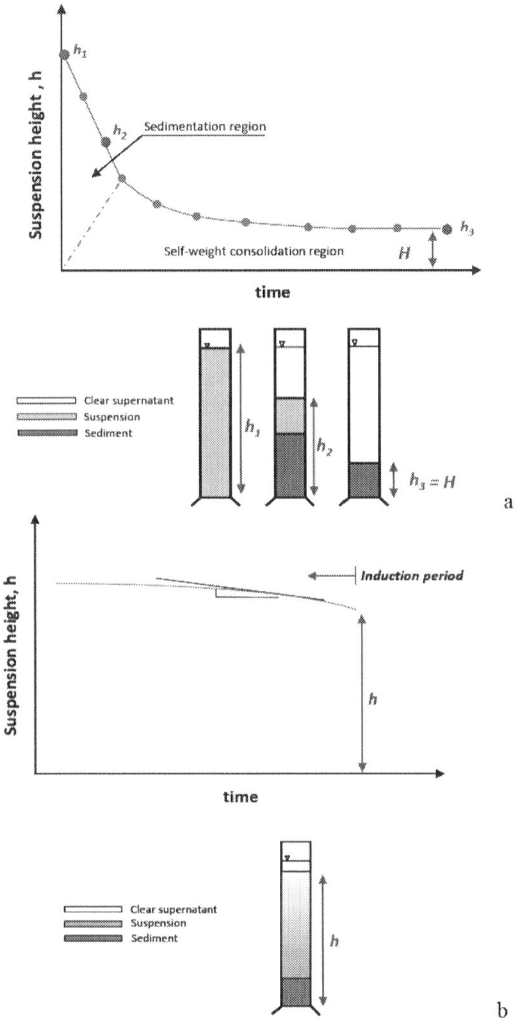

Figure 1. Sedimentation behaviour of suspensions a) flocculated system, b) dispersed system

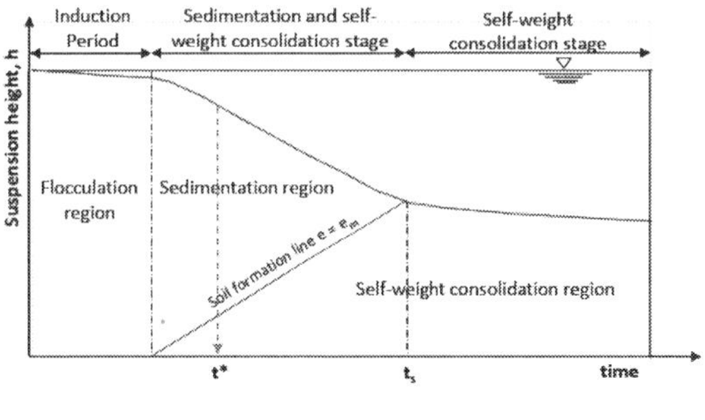

Figure 2. Stages of sedimentation process (after Vitale, 2016)

technique, based on the interaction between a set of particles and an incident radiation. Details of the technique are referred in [31].

The grain size fraction lesser than 1 μm was extracted by dispersing about 10 g of sample in 1 L ultrapure water and centrifugation method. The suspension (100 mg/l) was prepared at different concentrations of $Ca(OH)_2$ and

DOI: 10.1051/e3sconf/20160904008

KOH. Hydrodynamic size measurements were then performed at 25 °C and laser scattering at 173° in a capillary cell with Malvern Nano Zetasizer apparatus.

3 Results

Sedimentation curves of kaolin suspensions prepared at increasing lime percentages are reported in Figure 3. The evolution of the suspension height is shown as a function of time with reference to two different time scales, namely the whole duration of the test (up to 13000 min, Figure 3a) and the initial stage of the test (up to 200 min, Figure 3b). Not treated kaolin suspension is characterised by a relatively long initial stage (induction period) during which the clayey particles show a negligible settling. A small percentage of lime (1% CaO) promotes a more efficient particles association process due to alkaline environment and divalent cations, as showed by shorter induction period. It is noteworthy that no remarkable changes of the final sediment height were recorded (Figure 3a). As shown in Figure 3b, the higher amount of lime for 3% CaO mixes promotes a higher degree of flocculation with respect to 1%CaO mixes. As a consequence, a shorter induction period has been detected (Figure 3b) with an increase of the final sediment height after the self-weight consolidation stage (Figure 3a). No significant effects on the sedimentation behaviour and final suspension height were observed for samples treated with higher lime percentages (namely 5% CaO and 7% CaO). Therefore, 3% CaO can be regarded as the minimum amount of lime required to promote an effective aggregation of particle in the short term.

The influence of cation valence on the sedimentation characteristics of kaolin suspensions is shown in Figure 4. Sample prepared with KOH at pH = 12.4 showed a long induction period (Figure 4a), with particles dispersed in suspension and a not well recognizable sediment at the bottom of the cylinder (Figure 4b).

A further insight in the effects of pore water chemistry on the aggregation of kaolin particles has been highlighted by microstructural analyses results. In Figure 5 the zeta potential measurements of not treated and lime treated kaolin suspensions prepared at increasing pH have been presented. An increase of the zeta potential magnitude with increasing pH has been detected as consequence of deprotonation reactions respectively on the lateral and octahedral basal faces of kaolinite. The addition of lime (calcium hydroxide at increasing concentration) induces a decrease of the zeta potential decreases with increasing pH. In particular, the reduction is sharp for pH higher than 9. Calcium undergoes hydrolysis at higher pH (pH > 10) leading to the formation of monovalent species [33, 34]. According to [14], the observed behaviour can be linked to the formation and specific adsorption of first hydrolysis product of calcium (i.e. $CaOH^+$).

The evolution of the size of aggregated particle structure with time was monitored by Dynamic Light Scattering (DLS) measurements. For kaolin suspension in de-ionized water (pH = 4.6) (Figure 6), the measured average size of aggregations increases with time as a

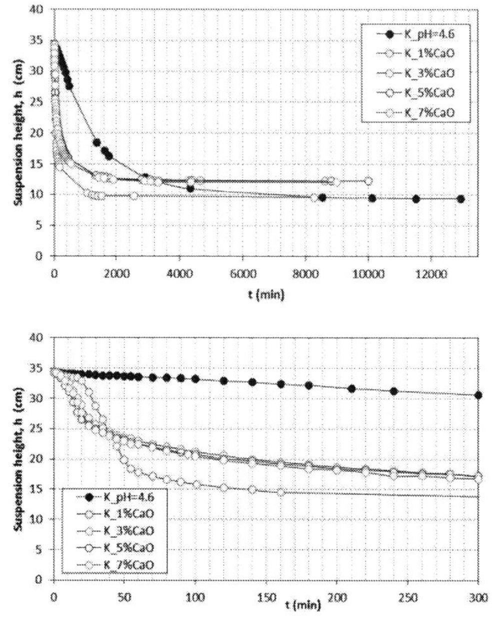

a

b

Figure 3. Sedimentation curves of increasing lime content suspensions: a) whole duration of the test, b) initial stage

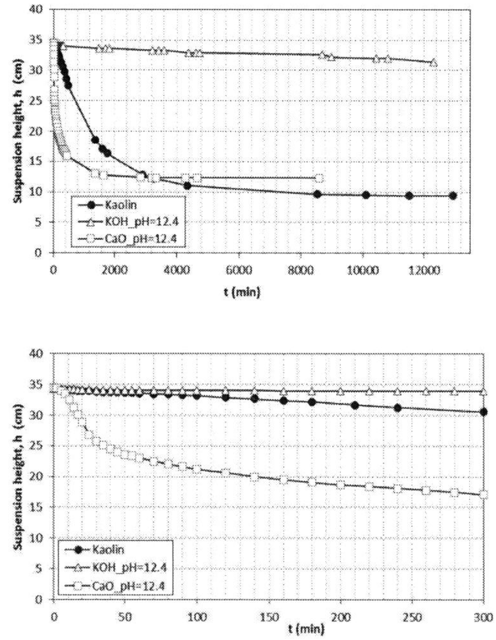

a

b

Figure 4. Sedimentation curves of suspension prepared with different valence ions: a) whole duration of the test, b) initial stage

consequence of positively charged edge and negatively charged basal face mode of particle aggregation referred to as edge-to-face [35]. As the pH increased to 12.4 by adding KOH, the size of aggregates is smaller and remains unchanged over time. The kaolinite particles are not flocculated and their mutual interaction is weak (i.e. dispersed state). The high negative surface charge, as indicated by higher zeta potential (Figure 5), increases the electrostatic repulsive forces preventing particles collisions and aggregations. The presence of calcium ($Ca(OH)_2$) and the alkaline environment (pH=12.4) favour the larger average size of kaolinite aggregates and a strongly flocculated structure. Calcium ions counteract the effect of higher surface charge and repulsive forces, promoting particles collisions and aggregations.

4 Discussion

A pH dependent surface charge behaviour of kaolin suspension was observed from zeta potential measurements. In acidic environment the interaction between particles is the result of Coulombian attractions between positively charged edges and negatively charged faces (edge-to-face arrangement). In alkaline environment the absolute value of zeta potential increases due to the increase of negative charges on the kaolinite edge surfaces (deprotonation reaction). The results of sedimentation test showed a flocculated behaviour of kaolin suspensions in deionized water at pH=4.6, confirmed by DLS results highlighting the increase of the average aggregate size as a consequence of the edge-to-face flocculation. The increase of pH provided by the addition of KOH induces a suspension stability. The repulsive forces between charged particles prevent flocculation, as confirmed by the constant size of kaolinite particles over time detected by DLS measurements.

The addition of lime induces a change in the pore water chemistry, with an increase of the pH and the divalent cation concentration, affecting the sedimentation behaviour of kaolin suspension. A larger decrease of the absolute value of the zeta potential (Figure 5) is the consequence of calcium adsorption at the surface of the negatively charged particles. The repulsive forces are no longer able to counteract the van der Waals attractions [36], and an increase of the flocculated aggregates size is detected, as showed by DLS measurements results.

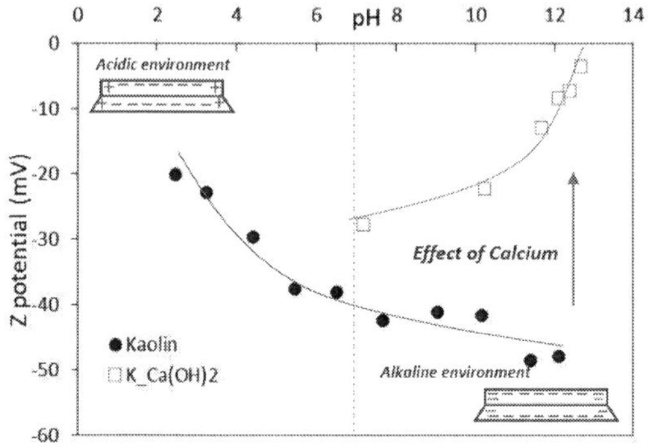

Figure 5. Zeta potential measurements of not treated and lime treated suspensions

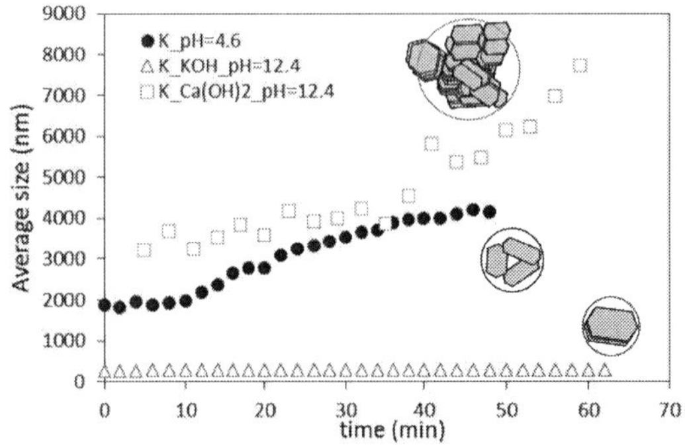

Figure 6. Evolution of aggregates size by Dynamic Light Scattering

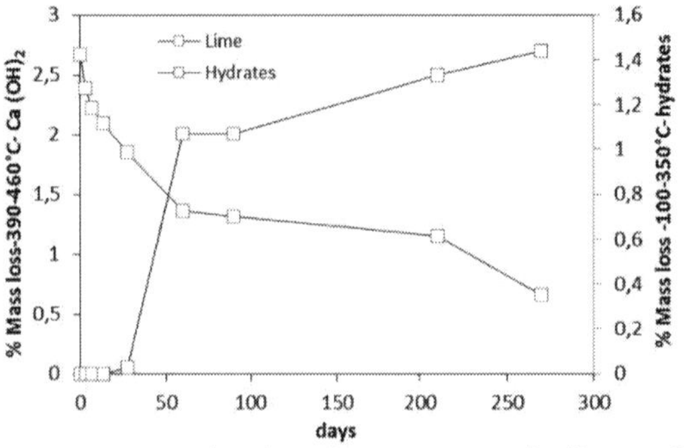

Figure 7. Termogravimetric analyses on 3% CaO kaolin treated samples (after Vitale, 2016)

The aggregation of particles of lime treated suspensions can be considered not affected by the ongoing of pozzolanic reactions with reference to the time intervals analysed during the sedimentation tests (about 9 days). As reported by [31], thermo-gravimetric analyses performed on 3% CaO treated kaolin samples showed a low reactivity in the short term, as showed in Figure 7, where the mass losses of hydrated lime (390°C - 460°C) and cementitious compounds (110-350°C) after the treatment have been represented as function of time. A significant consumption of portlandite and formation of new cementitious phases has been detected only after times longer than 28 days.

The increase of the average dimensions of clay aggregates in saturated lime solution is consistent with the increase of the soil-formation void ratio e_m determined from the sedimentation curves. The soil formation void ratio e_m, measured at the top of the sediments layer corresponds to the soil skeleton void ratio under zero effective stresses. Its value is representative of the soil skeleton void ratio induced only by the pore water chemistry. In order to estimate its value for each suspension, the end of sedimentation stage time t_s and the corresponding thickness of the sediment h_s were evaluated from the sedimentation curves. A lower limit of soil-formation void ratio e_s can be inferred by considering that the total thickness of the solid matter h_z (i.e. the material height) for each of the sedimentation tests is constant throughout the sedimentation/consolidation process [27], that is

$$h_z = \frac{h_0}{1+e_0} = \frac{h_s}{1+e_s} = \text{constant} \quad (1)$$

where h_0 and e_0 are respectively the initial height and the initial void ratio of the suspension. The value of e_m must satisfy the following relationship:

$$e_m \geq e_s = \frac{h_s}{h_z} - 1 \quad (2)$$

Since for each treated sample the time to achieve full consolidation t_f was found to be two order of magnitude larger than t_s [27] the difference between the calculated e_s and the soil-formation void ratio e_m has been considered negligible. Therefore, e_m was assumed as the representative value of void ratio when the soil skeleton is completely formed without any mechanical effect (i.e. zero effective stresses). The estimated values of t_s and h_s and the calculated values of e_s for 1% CaO and 3% CaO mixes have been reported in Table 1.

Table 1. Estimated and calculated values from sedimentation curves

Test	e_0 (-)	h_0 (cm)	t_s (min)	h_s (cm)	e_s (-)	t_f (min)
K_1%CaO	50	34.5	52.5	19.1	27.3	1530
K_3%CaO	50	34.5	25	28	40.4	2833
K_5%CaO	50	34.5	20	27.5	39.7	2886

The final height of the sediments after the soil skeleton formation, at the end of the tests, is affected by the pore water chemistry. The negative charge neutralization induced by the addition of lime causes a reduction of inter-particle repulsions promoting van der Waals attractions between particles. During the consolidation stage, the effective stresses increase inducing the reduction of the soil skeleton volume, due to the self-weight effect and the van der Waals attractions. The contribution of the attractive forces results in lower volume strains during consolidation, and thus higher sediment heights, as showed by lime treated suspensions which experience a consolidation stage characterized by lower settlements due to the higher contribution of van de Waals attractions.

5 Conclusions

In the paper some results on the effects of chemo-physical evolution of clay-lime-water suspensions on the microstructure of a lime treated kaolin have been presented. A multi-scale investigation on the sedimentation behaviour of clay suspensions under different pore water chemistry has been developed highlighting the chemo-physical mechanisms controlling particle arrangement and the soil fabric formation.

The results evidenced the key role of ionic exchange in the short term in terms of microstructure features of the treated soil. No effects of pozzolanic reactions were overlapped to the observed behaviour due to the low reactivity of kaolinite. The arrangement of kaolin particles in suspension is strongly pH dependent, with a flocculated face to edge arrangement at pH 4.6 turning to a dispersed one at higher pH. The addition of calcium ions in suspension induces the formation of larger particles aggregates, formed by face to face stacking of particles grouped face to edge. The sedimentation behaviour in presence of calcium is consistent with the effects of particle arrangement, highlighting a shorter induction period, a faster settling stage and a shorter consolidation stage due to the aggregation induced by the pore water chemistry.

Acknowledgments

The Author is grateful to Dr. Enza Vitale and Dr. Dimitri Deneele for their valuable support during the experimental stage and for the helpful discussions during the interpretation of the results. The microstructural analyses were performed at Institut des Materiaux Jean Rouxel, Université de Nantes, France. A special thank to Prof. Guy Ouvrard for his support to the research.

References

1. G. R. Glenn, R. L. Handy, Highway Research Board, Washington, D.C. (1963)
2. H. Brandl, Proc. 10th Int. Conf. Soil Mech. Found. Eng., Stockholm, 3, 587-594, (1981)
3. F.G. Bell, Engineering Geology, 42, 223-237, (1996)
4. C.D.F. Rogers, S. Glendinning, In Lime Stabilisation, C.D.F. Rogers, S. Glendinning and N. Dixon (eds.), 99-112, Thomas Telford, (1996)
5. J. Locat, H. Tremblay, S. Leroueil, Canadian Geotechnical Journal 33 (4), 654–669, (1996).
6. P.V. Sivapullaiah, A. Sridharan, A.N. Ramesh, Canadian Geotechnical Journal 37:1358-1367, (2000)
7. H. Tremblay, S. Leroueil, J. Locat, Canadian Geotechnical Journal, 38: 567- 579, (2001)
8. J. L. Eades, R. E. Grim, Highway Research Board, Bulletin 262, pp.51-53, (1960)
9. S. Diamond, E.B. Kinter. Presentation at the 44th Annual Meeting, Highway Research Board, 92, Washington, (1965)
10. D.I. Boardman, S. Glendinning, C.D. Rogers, Geotechnique 50, No. 6, 533-543, (2001)
11. N. Maubec, Thése de doctorat, Faculté des sciences et des techniques, Université de Nantes (2010)
12. L. Delavernhe, Thèse de doctorat, Université de Nantes, (2011)
13. E. Pomakhina, D. Deneele, A. C. Gaillot, M. Paris, G. Ouvrard, Cement and Concrete Research, **14**, 626-632, (2012)
14. Y. C. Chemeda, D. Deneele, G. E. Christidis, G. Ouvrard, Applied Clay Science, 107, 1-13, (2015)

15. G. Russo, S. Dal Vecchio, G. Mascolo, In T. Schanz, ed. Experimental Unsaturated Soil Mechanics. Springer Berlin Heidelberg, pp. 49–56, (2007)
16. Z. Metelková, J. Boháč, R. Přikryl, I. Sedlářová, Applied Clay Science 61 (2012) 37–43, (2012)
17. G. Russo, G. Modoni, Géotechnique Letters, 3, pp.93–9, (2013)
18. K. Ma, A.C. Pierre, Clays and Clay Minerals, 40, 586-592, (1992)
19. K. Ma, A.C. Pierre, Clays and Clay Minerals, 47, 4, 522-526, (1999)
20. A.M. Palomino, J. C. Santamarina, Clay and Clay Minerals, Vol. 53, No. 3, 209-222, (2005)
21. A. Kaya, A.H. Oren, Y. Yukselen, Marine Georesources and Geotechnology, 24: 203-218, (2006)
22. G. Imai, Soils and Foundations, Vol. 20, No.2, Japanese Society of Soil Mechanics and Foundation Engineering 61-77, (1980)
23. G. Imai, Soils and Foundations, Vol. 21, No.1, Japanese Society of Soil Mechanics and Foundation Engineering 7-20, (1981)
24. A.S. Michaels, J.C. Bolger, Industrial Engineering Chemistry & Fundamentals, 1, 24-33, (1962)
25. A. M. Palomino, S. E. Burns, J. C. Santamarina, Clays and Clay Minerals, Vol. 56, No. 6, 599-611, (2008)
26. V. Pane, PhD thesis, Department of Civil Engineering, University of Colorado, Boulder, (1985)
27. V. Pane, R. L.Schiffman, Géotechnique 47, No. 2, 273-288, (1997)
28. G. C. Sills, Phil. Trans. R. Soc. Lond., 356, 2515-2534, (1998)
29. K. Been, G.C. Sills, Geotechnique, 31 (4), 519-535, (1981)
30. R.J. Hunter, Academic Press, New York, (1981)
31. E. Vitale, PhD thesis, University of Cassino and Southern Lazio, (2016)
32. E. Vitale, D. Deneele, G. Russo, G. Ouvrard, Applied Clay Science, in press, (2016)
33. P. Mpofu, J. Addai-Mensah, J. Ralston, J. Colloid Interface Sci. 261, 349–359, (2003)
34. P. Mpofu, J. Addai-Mensah, J. Ralston, International Journal of Mineral Processing 75, 155–171, (2005)
35. E. Tombacz, , M. Szekeres, Appl. Clay Sci. 34, 105–124, (2006)
36. H. Van Olphen, John Wiley, (1977).

Challenges in monitoring and managing engineered slopes in a changing climate

Paul N Hughes[1,a], Rosalind Hen-Jones[2], Ross A Stirling[2] , Stephanie Glendinnning[2] , David A Gunn[3] , Jonathon E Chambers[3], Tom A Dijkstra[3], Joel Smethurst[4], and Kristine Flesjo[5]

[1]*School of Engineering and Computing Sciences, Durham University, Durham, UK.*
[2]*School of Civil Engineering and Geosciences, Newcastle University, Newcastle, UK*
[3]*British Geological Survey, Keyworth, UK*
[4]*School of Engineering and the Environment, University of Southampton, UK*
[5] *Norwegian Public Roads Administration, Oslo, Norway*

Abstract. Geotechnical asset owners need to know which parts of their asset network are vulnerable to climate change induced failure in order to optimise future investment. Protecting these vulnerable slopes requires monitoring systems capable of identifying and alerting to asset operators changes in the internal conditions that precede failure. Current monitoring systems are heavily reliant on point sensors which can be difficult to interpret across slope scale. This paper presents challenges to producing such a system and research being carried out to address some of these using electrical resistance tomography (ERT). Experimental results show that whilst it is possible to measure soil water content indirectly via resistivity the relationship between resistivity and water content will change over time for a given slope. If geotechnical parameters such as pore water pressure are to be estimated using this method then ERT systems will require integrating with more conventional geotechnical instrumentation to ensure correct representative information is provided. The paper also presents examples of how such data can be processed and communicated to asset owners for the purposes of asset management.

1 Introduction

There is an urgent need to better understand the medium to long-term implications of climate change on engineered slopes to make pro-active interventions more cost-effective and to better understand the scale of financial and societal consequences of future failures requiring re-active mitigation, repair or replacement. Many components of our transport infrastructure system that supports our population and economy have little or no resilience, so even minor changes in environmental conditions can cause compromises of serviceability or even lead to complete failure resulting in potentially severe consequences for asset owners and users. Engineered slopes (cuttings, embankments, flood defences) are a critical component of this infrastructure and where slopes exhibit excess deformation this causes delays and economic loss to European industry and the general public, while at the same time posing a significant safety hazard. Climate change projections across Europe are varied and there will be important regional differences in how this is manifested and how this will impact on the performance of engineered slopes for transport infrastructure. One of the important regional examples that requires further investigation includes an increased likelihood of summer drought leading to drying/cracking of soil which will in turn lead to increased macro permeability and hence more rapid ingress of water during extreme precipitation) [1,2].

1.1 Climate change context

Projections of future climate change suggest a move toward drier summers and wetter winters in northern Europe and a reduction in annual rainfall in southern Europe [3], with associated changes in ground condition and hence implications for slope stability. It is well understood that increasing ground water decreases soil strength and can lead to swelling of some clay soils [4] and that conversely drying increases soil strength but causes shrinkage and desiccation cracking [5]. These moisture-driven changes have the potential to increase the incidence of failure across a range of earth structures [6,2] affecting road and rail networks. Whilst the magnitude of these impacts is not yet fully understood, engineers and asset managers require reliable and cost effective systems to monitor the condition of these assets and direct maintenance activities at the most vulnerable parts of the network.

Asset owners need to know which parts of their asset network are vulnerable to climate change induced failure in order to optimise future investment. We require models to help us identify vulnerable elements of infrastructure networks, based on prediction of failures,

[a] Corresponding author: paul.hughes2@durham.ac.uk

to be able to provide sound advice on maintenance, remediation and adaptation measures. As an example, approximately one third of the total asset value of the UK transport network is derived from infrastructure slopes [7]. Maintaining this network is costly; it is estimated that in the UK, Network Rail spent £70 million in 2007/2008 on preventative works to stabilize earthworks [8,9].

COST Action TU1202 is a coalition of researchers addressing the challenges of engineered slope infrastructure resilience and adaptation to climate change in Europe. The research network has four focus areas, slope data and monitoring; soil/vegetation/climate interactions; slope numerical modelling and risk assessment. Overcoming the challenges of monitoring and managing slope stability risk in a changing climate requires integration of these approaches. This paper presents some of the work conducted by COST TU1202 towards this integration.

1.2 Monitoring

At-risk engineered slopes are routinely monitored using geotechnical instrumentation that focusses on the measurement of ground water condition and deformation. These instruments are usually augmented by meteorological monitoring in the form of weather stations or simple rain gauges. Rainfall amount is the most widely used parameter for identifying a trigger value for early warning systems for landslides. However, it is recognized that there is a series of processes separating the parameter being measured (rainfall) and the pore water pressure change that will cause the slope to fail and that the measurement of pore water pressure is the most reliable predictor of failure in a slope [10] and therefore the more useful parameter to use when planning preventative maintenance in engineered slopes. Deformation monitoring informs when a slope is moving and provides valuable information about rates and magnitudes of movement which is also useful in planning remediation works.

Traditionally, ground water and deformation monitoring in engineering applications has been performed through the use of piezometers, tensiometers, inclinometers and extensometers [10,11,12] which have increased in sophistication and reduced in price over time. Improvements in digital communications and power management in these instruments have enabled a much greater degree of autonomous operation reducing the demands for manual reading of instruments. However, direct monitoring in this way remains expensive (from both an equipment and human resource perspective) and is only able to provide single point values which may be unreliable and require much effort to resolve a spatially-integrated cross-sectional model.

Monitoring of ground movement over larger areas can be accomplished using aerial reconnaissance and LIDAR and is done regularly by asset owners such as Network Rail in the UK [13] but these surveys provide topographical information only [14] and therefore are not currently capable of capturing potentially rapid changes in subsurface conditions preceding slope failure. Geophysical monitoring using techniques such as Electrical Resistance Tomography (ERT) has the potential to bridge this gap in scale and has been used by numerous researchers to monitor larger areas than traditional instrumentation [15,16].

Protection of at risk slope assets requires risk-based early intervention. However, such interventions require monitoring systems capable of identifying and alerting to asset operators the changes in the internal conditions that precede failure, in real time. The creation of such a system requires:

1. Instrumentation that can sample a sufficiently large area/volume to be representative of at-risk slopes.
2. Automated logging systems that can convert proxy measurements (such as geophysical data) into geotechnical engineering parameters.
3. Communications and data management systems that can collect and integrate multiple data types from multiple sources coupled to robust modelling capable of processing and analyzing environmental and slope data in near real time.
4. Decision support interfaces that can communicate the analysis to asset operators in a timely fashion.

Achieving this capability presents a number of technical and research challenges. Research presented in this paper focusses on one potential method to overcome challenges 1 and 2. Progress towards overcoming challenges 3 and 4 are presented in the discussion.

2. Electrical Resistivity Tomography

Electrical Resistivity Tomography (ERT) is a ground imaging technique that is being increasingly applied to the characterisation and monitoring of the subsurface [17]. Resistivity is particularly sensitive to changes in pore fluid resistivity and saturation as the principal mode of current flow in the subsurface is through electrolytic conduction in the pore fluid; consequently, ERT is widely used in hydro-geophysical investigations [18]. ERT can also be used to distinguish between lithologies of contrasting resistivity, where resistivity can vary due to differing porosities [19] or due to the presence of clay minerals [20,21]. A key advantage of using ERT techniques is the capability to monitor large volumes of the ground using either 2D or 3D arrays (Figure 1).

Although there are an increasing number of studies using three-dimensional ERT (using electrode arrays) as a means of characterising and monitoring unstable slopes [15,22,23,24], relevant geophysical - geotechnical relationships require further validation. As elevated water contents and a corresponding reduction of soil suction are associated with shear failure, their interaction with soil resistivity is key to the development of a slope stability assessment system.

Many studies have investigated the relationship between electrical resistivity and water content for clays

 DOI: 10.1051/e3sconf/20160904009

both in the laboratory [25,26,27] and in the field [28,29]. However, in order to fully resolve these relationships, it is necessary to investigate how they are affected by repeated seasonal cycles, which have been shown to progressively weaken clay fills [30]. This process is likely to be exacerbated by the increasingly frequent and extreme weather events suggested by climate change projections.

Studies on partially-saturated rocks [31] and sands [32,33] observed hysteresis in the electrical response to varying degrees of saturation between imbibition (wetting) and drainage; a study by Muñoz-Castelblanco et al, [34] on a natural unsaturated loess found soil resistivity to be independent of whether a drying or a wetting path was followed. However, there is little research into the effects of repeated seasonal cycles on the resistivity response of volume-sensitive clay soils. Hysteresis in near-surface soils is well-established in the soil water retention curve [35,36,37] whereby at a given water content a decrease in soil suction is observed between the drying and the wetting paths, due to entrapped air.

Figure 1. 3D Resistivity array at Bionics test site.

If ERT is to fulfil its potential in geotechnical monitoring then it is necessary to understand how soil suction and resistivity interact when subjected to seasonally varying water content, in order to be able to interpret geophysical information gathered from electrical resistivity tomography arrays. To this end, an experimental programme integrating field monitoring and multi-scale laboratory tests has been undertaken on a Glacial Till clay material obtained from a full-scale test embankment in Northumberland, United Kingdom, which forms part of the BIONICS field research project [2,6]. The experiments have been designed to establish the relationships between soil water content, resistivity and pore water pressure, with a particular focus on whether these relationships remain constant over time.

2.1 Resistivity as a pore water pressure proxy

BIONICS clay soil was tested to establish its soil water retention (SWR) and electrical resistivity properties. In the SWR phase of the experiment thirteen 38 mm diameter x 8 mm length discs were formed by placing 18.5 g of clay (at 22% water content) into a compaction cell at a strain rate of 0.33 mm/min to achieve a target dry density of approximately 1.6 Mg/m^3 (chosen to represent in-situ conditions at the BIONICS test site). The specimens were then subjected to moisture cycling: drying was achieved by allowing the specimens to air-dry in a temperature-controlled environment (20°C), until their masses corresponded to target water contents at regular intervals between 22% and the residual; wetting was achieved by spraying specimens using a hand-powered water mister. Eight specimens were used for stage 1a (drying), whilst five specimens were reserved for stage 1b (wetting). Following moisture cycling, the specimens were wrapped in plastic film and allowed to homogenise for a further 24 hours. The specimens were then put in a WP4 dew-point potentiometer [38] and their pore pressures recorded before being oven-dried to confirm their water content.

In the resistivity phase of the experiment seventy five cylindrical soil specimens (38 mm diameter by 76 mm length) were prepared using a steel mould filled by tamping of four approximately equal layers to a target density of 1.6mg/m3 to represent field conditions. The drying component of moisture cycling was achieved by the same method as used for the SWR samples wetting was achieved by placing specimens in a "humidity chamber" (an insulated box with two 90 ml/hour mist generators submerged in deionised water, with a grate above to hold the specimens (Figure 2)). The reason for the difference in wetting procedure with respect to SWRC specimens is the larger volume of these specimens such that they required a more intense and prolonged wetting environment to achieve the same moisture content. Resistivity samples were subjected two complete cycles of drying and re-wetting. Specimens were then tested for resistivity using the two-point method, in accordance with BS 1377-3: 1990 [39]. To improve contact resistance at the soil-electrode interface, the disc electrodes were coated with a layer of Nyogel conductive grease.

In addition to the above tests the effects of desiccation cracking on soil resistivity were also investigated using the two-point method. In order to stimulate desiccation cracks, fourteen specimens were prepared with in-built planes of weakness which would be more vulnerable to cracking as shrinkage occurred during drying. This was achieved by tamping after the addition of each of the four layers, creating a deliberately smooth surface, orthogonal to the direction of current flow. For this series of tests, a single stage of drying only was performed.

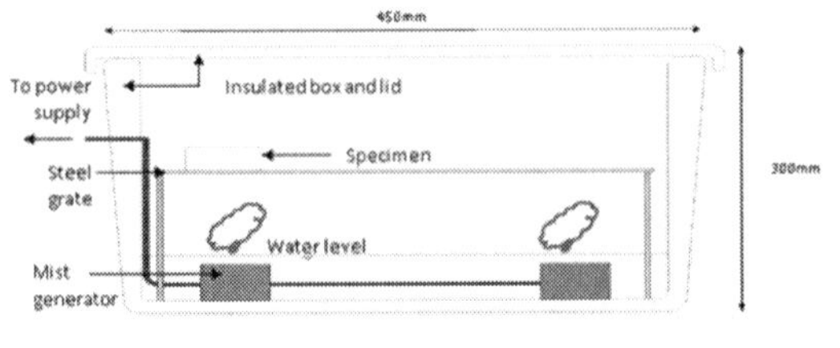

Figure 2. Humidity Chamber

2.2 Pore water pressure / resistivity results

The WP4 dewpoint potentiometer was used to measure suction values, which were subsequently fitted using the van Genuchten [40] expression. Both continuous drying and wetting paths were fitted using the van Genuchten fitting parameters n = 1.54, α = 0.0097 m^{-1} and n = 1.29, α = 0.1001 m^{-1} respectively (where $m = 1-1/n$) (Figure 3). The drying curve rapidly de-saturates from θs = 0.36 at an approximate air entry value of 600 kPa. The drying path is shown to fit well, though the wetting path suffers from a reduced number of data points. However, in the measured suction range, the fitted curve is shown to pass through the majority of points. Traditionally, for the wetting path, re-saturated water content is observed to be reduced from the initial content due to the entrapment of air. The presented curve displays an inferred, elevated re-saturated water content on the wetting path considered as a product of increased porosity as a direct result of the formation of micro-cracking and permanent fabric modification, as described in later in the paper. This trend is predicted due to the extreme drying (desiccation) that the specimens had undergone during the latter stages of drying prior to re-wetting. However, limitations in the dewpoint potentiometer technique do not allow behaviour at very low suctions to be accurately investigated.

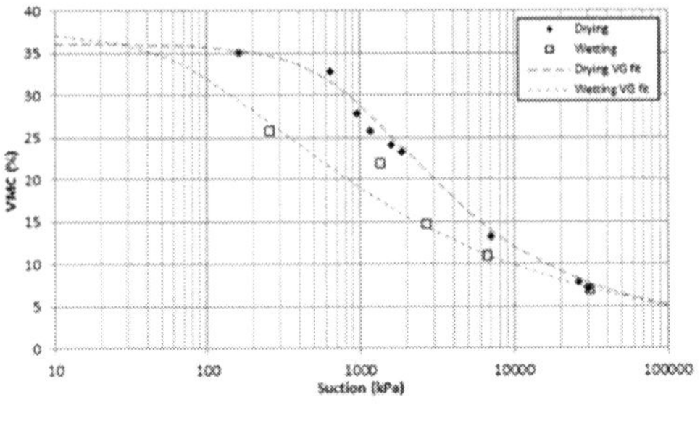

Figure 3. Soil water retention curve

Figure 4 shows the water content - resistivity data are separated into four individual stages (stage 1 drying and wetting; stage 2 re-drying and re-wetting). Little hysteresis can be observed between the stages, demonstrating that the relationship is not significantly affected by whether a drying or a wetting path is followed. However, if the data are divided broadly into stages 1 and 2 (each comprising a full dry-wet cycle), then a hysteretic inverse power relationship is evident. There is a shift of the resistivity - water content path centred at approximately 22% VWC, such that below this point, stage 2 specimens have elevated values of resistivity with respect to stage 1, with the opposite being true beyond this point, as is illustrated by two grey arrows. As before, resistivity error bars of +/- 14% show that this shift is likely to be significant.

Figure 5 shows a drying curve comparing intact specimens and those with built-in horizontal fracture planes (orthogonal to the current flow). As can be seen in the figure, fractured specimens exhibited higher values of resistivity for a given water content, consistent with an increased porosity as air within the voids acts to impede current flow. Error bars of +/- 14% shown on the figure demonstrate that this shift values is significant.

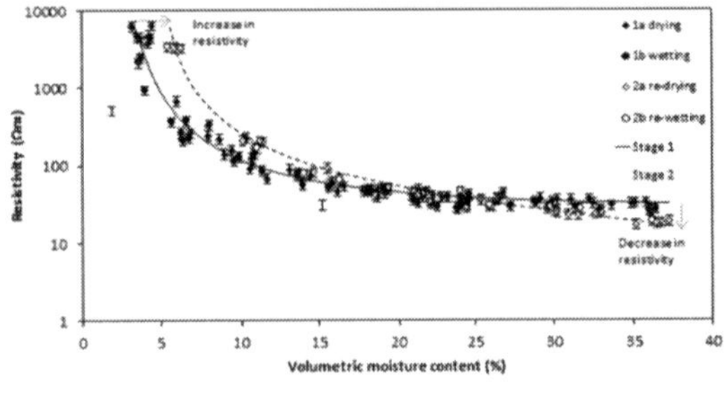

Figure 4. Water content - resistivity relationship separated by seasonal stage

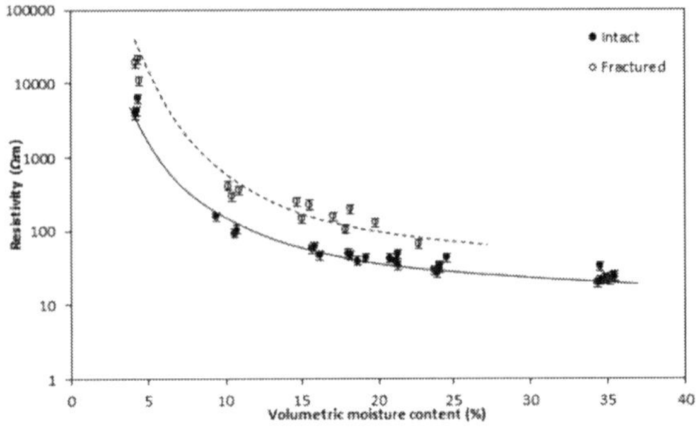

Figure 5. Comparison of water content - resistivity relationship for fractured and intact specimens

2.3 Implications of results on geophysical/geotechnical monitoring

Because of its volume-sensitivity, desiccation cracking is associated with clay subjected to dry-wet cycles. In Figure 5, the water content – resistivity relationship for deliberately-fractured specimens is presented alongside that for intact specimens, showing elevated values of resistivity resulting from cracking, resulting from the insulating nature of air acting to impede current flow. Therefore, it could be considered that for clay subjected to such extremes, there would be an increase in resistivity at a given water content with ongoing seasonal cycling, as fractures develop.

 DOI: 10.1051/e3sconf/20160904009

In un-fractured samples exposed to wetting and drying cycles (figure 4) whilst no hysteresis of the resistivity water content relationship can be observed between the wetting and drying paths within one cycle. However, it can be seen that there is a shift between cycle 1 and cycle 2. Below 22% VWC specimens subjected to more than one dry-wet cycle have increased values of resistivity with respect to those subjected to less than one. This can be attributed to the development of desiccation cracks. The opposite is true at water contents above 22%, a possible explanation for this reversal between the two phases observed at higher water contents is the dissolution of clay ions in the pore water, such that they become further mobilised with ongoing dry-wet cycling, with existing cracks acting as a high conductivity conduit upon filling with water.

The implications of these results are that if resistivity is to be used as a proxy for water content in geotechnical monitoring a static relationship should not be assumed. It should be anticipated that the relationship between resistivity and water content will change over time in response to environmentally driven structural changes within the soil. It may be necessary to co-locate TDR based point sensors within ERT arrays so that the relationship can be updated over time with new calibration curves applied to the monitoring system automatically in order that representative water contents can be reported in near real time.

Whilst water content is useful in predicting slope behaviour accurate prediction requires knowledge of the pore water distribution within the slope. If the water content distribution is known then it may be possible to estimate pwp using an appropriate soil water retention curve. However, given the hysteretic nature of soil water retention behaviour and the non-uniform periods of wetting and drying that can be expected to occur in the field it is likely that values obtained in this way are only a rough estimate of pwp.

3. Further challenges

Geophysical techniques such as ERT provide a potential solution to challenges 1 and 2 listed in section 1.2. The other challenges listed are also being addressed by researchers, stakeholders and instrument manufactures. There are increasingly sophisticated commercial systems that collect and store data, process it into engineering units, and post it onto secure web portals where it can be viewed. Alarms can be set to alert key personnel if certain pre-set trigger levels are exceeded. Standard data formats such as AGS-M, which enable easier sharing of information, are becoming common [41]. These are likely to become more important as assets are monitored over longer periods, giving flexibility in updating hardware and software and interoperability between proprietary systems. Collection and monitoring of more information is part of a technological trend towards 'big data', which is becoming increasingly important across wide areas of the European economy. Data on engineered slopes may be generated during design, construction and operational phases, i.e. the whole life cycle of the asset; geotechnical monitoring information may be part of this data-set. Many large highway and railway infrastructure owners increasingly store information on their assets within large databases, many of which are linked to geographical information systems (GIS). These form a digital representation of the physical and functional characteristics of an asset, and act as a resource for sharing and visualising information and knowledge (e.g. HAGDMS the Highways Agency (now Highways England) Geotechnical data Management System.).

Semi-empirical systems and process-response models continue to be developed [1]. Major transport infrastructure asset managers have progressed investigations into geotechnical risk through the use of morphological and geotechnical characteristics and a variety of research consortia collaborate with asset managers to progress the science underpinning management, maintenance and intervention strategies [1, 2, 8, 9, 14, 41]. Coupling system with near-future weather data (e.g. impending storms) will enable assessing the probability of disruptive slope failure and much can be learned in this context from the flood forecasting community. Prioritisation of sections most at risk will also provide clarity where local monitoring data will be most beneficial and cost-effective. Increasingly large datasets (e.g. Network Rails complete detailed LiDAR coverage of the UK network) creates its own problems in terms of storage, management and processing of large amounts of data, particularly where increasingly sophisticated models require multiple iterations and calculation intensive operations. However, this should not be seen as an impediment as increasing computing power, particularly from cloud-based parallel processing, will make complex operations much more achievable in the future. Traditional monitoring approaches produce periodic reports, which might be attached to an asset within the GIS system. The capability of current systems to hold large data sets is less certain, and may become challenging as the number of sensors and frequency of data logging increases. However, GIS systems that distribute on a fine spatial scale risk information, often in real time (for example, linked to antecedent and forecast rainfall), are becoming more commonplace, and it is plausible that in the future this could include effective integration of regularly updated, multiple datasets (including near real-time weather and asset monitoring data). A good example of this is the Norwegian national system XGEO where data inputs and derived information are available on a 1 km grid at a national scale (Figure 6).

DOI: 10.1051/e3sconf/20160904009

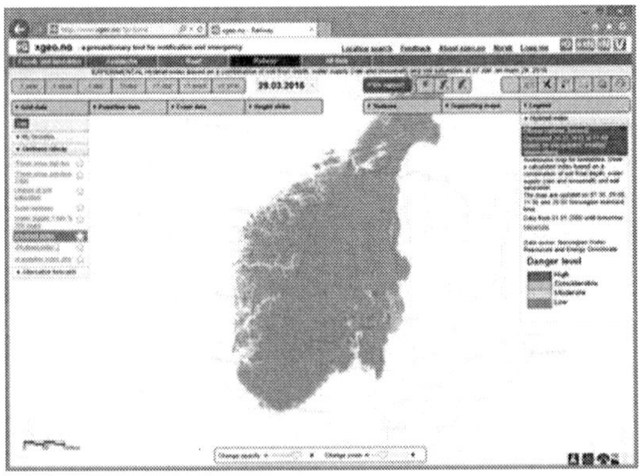

Figure 6. Norwegian XGEO system, an example of (example shows colour coded landslide hazard).

4. Conclusions

There remain significant challenges to be overcome in the development of integrated stability monitoring systems that can provide real time warning of impending failure over large lengths of engineered slope. The aging nature of our infrastructure coupled with the anticipated effects of climate change makes the need for these systems more pressing. However, many of the technologies and techniques required to produce such a system are either in development or are currently deployed. Geophysical monitoring has the potential to provide data from large slope volumes which can be integrated with direct point measurement of water content, pore water pressure and weather data to provide live calibration. Mobile communications and database systems already exist that are capable of transmitting and processing slope data and metrological information which can be presented as slope failure risk via user friendly web pages.

More work is still required to characterise slope material properties and establish geophysical/geotechnical property interrelationships. This characterisation will require an understanding of the way soil geotechnical properties change and deteriorate due to environmental drivers such as wetting and drying, freezing and thawing and vegetation growth. Systems will also need to be coupled with robust numerical modelling capable of taking these deterioration factors into account. Achieving this will require collaboration between researchers and developer across a range of fields of expertise including geotechnics, hydrology, transport planning, electronics, communications software development, climate change and with asset owners to ensure systems met the demands of industry.

Acknowledgements

The authors would like to acknowledge funding provided by EU-COST Action TU1202, the Natural Environment Research Council (GA/12S/007), the BGS British University Funding Initiative (S223), iSMART (EP/K027050/1), and Assessing the Underworld (EP/K021699/1). Tom Dijkstra, Jon Chambers and David Gunn publish with the permission of the Executive Director of the British Geological Survey.

References

1. Dijkstra, T. & Dixon, N. 2010. *QJEG&H*, **43**, 4, 371-385.
2. Glendinning S, Hughes PN, Helm P, Chambers J, Mendes J, Gunn D, Wilkinson P, Uhlemann S (2014). Acta Geotech 9(5):799
3. EEA 2015. European Environment Agency, Denmark.
4. Driscoll R (1983). Géotechnique 33:93–105
5. Stirling RA. 2014. Multiphase modelling of desiccation cracking in compacted soils. PhD Thesis, Newcastle University, UK. Hughes PN, Glendinning S, Mendes J, Parkin G, Toll DG, Gallipoli D, Miller P (2009). Proc Inst Civil Eng Eng Sustain 162(2):67-79
6. Perry J, Pedley M, Reid M (2001). CIRIA, Report C550.
7. Rail Accident and Investigations Branch (RAIB) (2008) Network Rail's management of existing earthworks. Department for Transport, London
8. Dijkstra T, Dixon N, Crosby C, Frost M, Gunn D, Fleming P, Wilks J (2014) Proc Inst Civil Eng-Trans 167(5):269-280.
9. Toll DG, Lourenço SDN, Mendes J, Gallipoli D, Evans FD, Augarde CE, Cui YJ, Tang AM, Rojas JC, Pagano L, Mancuso C, Zingariello C, Tarantino A (2011) QJEG&H 44(1):23-33
10. Dunnicliff J (1988) Geotechnical Instrumentation for Measuring Field Performance. J Wiley&Sons, New York
11. Cui YJ, Tang AM, Mantho A, De Laure E (2008) Geotech Testing Journal 31(1):95-100
12. Birch G, Anderson I (2011). Ground Engineering, May 2011
13. Jaboyedoff M, Demers D, Locat J, Locat A, Locat P, Oppikofer T, Robitaille D, Turme D (2009). Can Geotech J 46(12):1379-1390
14. Chambers JE, Wilkinson PB, Kuras O, Ford JR, Gunn DA, Meldrum PI, Pennington CVL, Weller AL, Hobbs PRN, Ogilvy RD (2011). Geomorphology 125:472-484
15. Springman SM, Askarinejad A, Casini F, Friedel S, Kienzler P, Teysseire P, Thielen A (2012). Acta Slovenica Geotechnica 1:5-29
16. Loke MH, Chambers JE, Rucker DF, Kuras O, Wilkinson PB (2013). J Appl Geophys 95:135-156
17. Binley A, Hubbard SS, Huisman JA, Revil A, Robinson DA, Singha K, Slater LD (2015). Water Resour Res 51
18. Archie GE (1942). Transactions of the AIME 146
19. Shevnin V, Mousatov A, Ryjov A, Delgado-Rodriquez O (2007). Geophys Prospect 55
20. Russell EJF, Barker RD (2010). Near Surf Geophys
21. Friedel S, Thielen A, Springman SM (2006). J Appl Geophys 60(2)

 DOI: 10.1051/e3sconf/20160904009

22. Merritt AJ, Chambers JE, Murphy W, Wilkinson PB, West LJ, Gunn DA, Meldrum PI, Kirkham M, Dixon N (2014). Landslides 11:537-550
23. Perrone A, Lapenna V, Piscitelli S (2014). Earth-Sci Rev.
24. McCarter WJ (1984). Géotechnique 34(2):263-267
25. Fukue M, Minatoa T, Horibe H, Taya N (1999). Eng Geol 54:43–53
26. Kibria G, Hossain M (2012). J Geotech Geoenviron 138(12):1520–1529
27. Michot D, Benderitter Y, Dorigny A, Nicoullaud B, King D, Tabbagh A (2003). Water Resour Res 39(5):1138
28. Chrétien M, Lataste JF, Fabre R, Denis A (2014). Eng Geol 169: 112-123
29. Loveridge FA, Spink TW, O'Brien AS, Briggs KM, Butcher D (2010). QJEH&H 43(4):461-472
30. Roberts JJ, Lin W (1997). Water Res Res 33(4):577–587.
31. Knight R (1991). Geophysics 56:2139–2147
32. Kavian M, Slob EC, Mulder WA (2011). J Geophys Res 116: B08214
33. Muñoz-Castelblanco J, Pereira J, Delage P, Cui YJ (2012). ASTM Geotechnical Testing Journal 35(1):11-17
34. Fredlund DG, Xing A (1994). Can Geotech J 31(4):521–532
35. Šimůnek J, Kodešová R, Gribb M, Van Genuchten MT (1999). Water Resour Res 35(5):1329-1345
36. Fredlund, DG, Sheng D, Zhao J (2011). *Can Geotech J* 48:186-198
37. Decagon Devices (2014) WP4C Dewpoint Potentiometer, operator's manual. Decagon Devices Inc.
38. British Standards Institution (1990) BS 1377-2:, BSI.
39. Van Genuchten MT (1980). Soil Sci Soc Am J 44(5):892-898
40. Richards, D.J., Chandler, R.J. and Lock, A.C. (2003). *Proc Inst of Civil Engineers: Geotechnical Engineering*, 156, (1), 47-55.
41. Sadler, J., Griffin, D., Gilchrist, A., Austin, J., Kit, O., and Heavisides, J. (2015). GeoSRM - An Online Geospatial Safety Risk Model.

 DOI: 10.1051/e3sconf/20160905001

A study on the saturation degree dependency of the seismic behaviour of retaining walls

M.J. Momeni[1] G. Habibagahi[2a] E. Nikooee[3]

[1]*Graduate student, Department of Civil and Environmental Engineering, School of Engineering, Shiraz University, Shiraz, Iran*
[2]*Professor, Department of Civil and Environmental Engineering, School of Engineering, Shiraz University, Shiraz, Iran*
[3] *Assistant Professor, Department of Civil and Environmental Engineering, School of Engineering, Shiraz University, Shiraz, Iran*

Abstract. Retaining walls are important geotechnical structures that are often used in soil slopes and trenches to bring ground surface at appropriate level for the construction of roads, highways and buildings. It is common practice to assume that the soil behind a retaining structure is either fully saturated or completely dry. However, for the case the soil is partially saturated, mechanical behaviour of the soil above the water table is different than that of the dry soil. Thus, it is necessary to investigate the effect of the variations of degree of saturation on lateral pressure behind retaining walls. In this research, the seismic behaviour of unsaturated soils behind a retaining structure is analysed. A finite difference code was employed to conduct the necessary analyses and a series of equivalent linear analyses is performed to reveal the effect of the degree of saturation on the general response of the retaining structures. The required functions for unsaturated zone were defined and implemented in the code. For this purpose, a soil water retention function was employed and the soil shear modulus is assumed to vary with the mean effective stress for both saturated and unsaturated zones which naturally introduces the required hydro-mechanical coupling in unsaturated and saturated zone. The results of the analyses compared to the conventional methods which does not include the unsaturated mechanical properties, indicate that in the unsaturated state, the increase in the effective stress, and hence, the shear modulus considerably affects the seismic forces on the retaining wall.

1 Introduction

Retaining Walls are important geotechnical structures that are often used to create desired change in ground elevation in slopes and trenches. In analysis of the most of lateral pressure problems, such as retaining walls, slope stability, and shallow foundations, it is assumed that the soil is either completely dry or fully saturated. But in reality, soil is partially saturated and has different shear strength as compared to dry soil due to the presence of matric suction.

Matric suction increases contact forces between grains and makes unsaturated soils stronger than dry soils [1]. It has been a common practice by geotechnical practitioners not to consider the partial saturation effect on shear strength which makes the design conservative. However considering the effect of partial saturation can decreases construction costs significantly. Furthermore, the effects of such consideration on the behavior of earth structures on various conditions should be, carefully, taken into account one of which is their seismic behavior. Usually geotechnical engineers deal with risky earth structures, such as unstable slopes. In such cases, a complete analysis of existing stresses that includes properties of partially saturated regions located above water level is critical to identify instability reasons [2].

In order to determine partial saturation effects on stability of earth structures, methodologies should be expanded to analyze failure conditions in partially saturated soils. To date, there has not been much work on this issue. In particular, the seismic behavior of retaining walls taking into account the possible presence of adjacent unsaturated regions have not been thoroughly looked in to. The objective of current research is, therefore, to study the effects of the variation of the degree of saturation of soils on the effective stress, and shear modulus (G) of soil and accordingly the seismic behavior of unsaturated retaining wall.

2 Dynamic analysis methodology

2.1 Equivalent Linear Method

In this study, the Equivalent Linear method (E.L.) was used. E.L. starts a dynamic analysis with specified soil stiffness. In the first stage, FLAC processes the entire earthquake record and identifies the peak shear strain at each Gauss numerical integration point in each element.

[a]Corresponding author: habibg@shirazu.ac.ir

The shear modulus is then modified according to a specified G reduction function and the process is repeated. This iterative procedure continues until the required G modifications are within a specified tolerance [3, 4]. In each iteration, the values of damping ratio and the shear modulus were updated based on the current values of the confining pressure, the effective stress state of the element and the maximum shear strain in that iteration.

2.2 seismic analysis of unsaturated soils

Due to the presence of matric suction, the heterogeneity induced by the variation of the degree of saturation in different elements and different state of the effective stress than that of the fully saturated soil or dry case, careful consideration of the hydromechanical coupling is necessary in the seismic analysis of unsaturated soils. Hereafter, the parameters and variables, and the required steps for such analysis are explained (sections 2.2.1 to 2.2.4).

2.2.1 Formulating the effective stress in unsaturated elements

Effective stress in unsaturated soils was calculated using the following equation (Bishop 1959):

$$\sigma' = \sigma - u_a + \chi(u_a - u_w) \tag{1}$$

Where:
σ denotes total stress, χ is effective stress parameter, the term (u_a - u_w) denotes soil matric suction. u_a specifies air pressure and u_w is pore water pressure.

There are various formulations for the effective stress parameter. As discussed by Nikooee et al. (2013) and Pereira et al. (2010), degree of saturation is not a good candidate for the effective stress parameter, especially in fine grained soils [5, 6]. Recently Lu et al. (2010) suggested the use of effective degree of saturation for the effective stress parameter to account for the fact that the immobile water does not contribute to the mechanical behaviour of unsaturated soils [7]. Furthermore, Oh and Lu (2014) have shown that the use of effective degree of saturation results in a reasonable approximation of suction stress under different hydraulic and mechanical states [8]. Therefore, in this study the effective stress parameter was assumed to be effective degree of saturation.

$$\chi = \frac{S_r - S_{r0}}{1 - S_{r0}} \tag{2}$$

Where S_r is degree of saturation and S_{r0} is the residual degree of saturation.

2.2.2 Calculation of the shear modulus in unsaturated region

During dynamic or seismic excitation, the soil sample tends to soften in response to shear strain cycle. This softening is often expressed as a ratio of maximum shear modulus G_{max}. Equation 3 was therefore used to estimate G/G_{max} [4, 9]. In this equation, plasticity index and effective confining pressure are two main variables. Effective confining pressure was calculated by means of equation 1 for vertical effective stress. The shear modulus ratio was calculated as follows:

$$\frac{G}{G_{max}} = K(\gamma, PI)(\sigma'_m)^{m(\gamma,PI)-m_0} \tag{3}$$

$$K(\gamma, PI) = 0.5\left\{1 + tanh\left[ln\left(\frac{0.000102 + n(PI)}{\gamma}\right)^{0.492}\right]\right\} \tag{4}$$

$$m(\gamma, PI) - m_0 = 0.272\left\{1 - \tanh\left[\ln\left(\frac{0.000556}{\gamma}\right)^{0.4}\right]\right\} \times exp(-0.0145PI^{1.3}) \tag{5}$$

Where the sub-function $n(PI)$ is estimated using equations 6 to 9:

$$n(PI) = 0.0 \qquad for\ PI = 0 \tag{6}$$

$$n(PI) = 3.37 \times 10^{-6}\ PI^{1.404} \qquad for\ 0 < PI < 15 \tag{7}$$

$$n(PI) = 7.00 \times 10^{-7}\ PI^{1.976} \qquad for\ 15 < PI < 70 \tag{8}$$

$$n(PI) = 2.70 \times 10^{-5}\ PI^{1.115} \qquad for\ PI > 70 \tag{9}$$

γ is the maximum shear strain in each zone during earthquake which is multiplied by 0.65 to be converted to equivalent shear strain. In equation 3, mean effective stress (σ'_m) should be in kPa. Maximum shear modulus is a function of confining stress as follows [4]:

$$G_{max} = 21.7 \times P_a \times K_{2,max} \times \left[\frac{\sigma'_m}{P_a}\right]^{0.5} \tag{10}$$

In this equation P_a is the atmospheric pressure, $K_{2,max}$ is a constant which is defined using relative density and σ'_m is the mean effective stress which is calculated using equation 1 in unsaturated case:

$$\sigma'_{xx} = \sigma_{xx} - u_a + \chi(u_a - u_w) \tag{11}$$

$$\sigma'_{yy} = \sigma_{yy} - u_a + \chi(u_a - u_w) \tag{12}$$

$$\sigma'_{zz} = \sigma_{zz} - u_a + \chi(u_a - u_w) \tag{13}$$

$$\sigma'_m = \frac{\sigma'_{yy} + \sigma'_{xx} + \sigma'_{zz}}{3} \tag{14}$$

Parameters and variables used in equations 11 to 14 are as follows:

σ'_{yy}= Vertical effective stress

σ'_{xx} and σ'_{zz}= Horizontal effective stresses

χ = Effective stress parameter

u_a=Air pressure

2.2.3 Determining damping ratio

Damping ratio of soil is also effected by cycle shear strain and should be modified in each dynamic analysis in equivalent linear method. This was done by employing the following damping ratio function (equation 15). Variables in the equation are plasticity index, ratio of shear modulus to maximum shear modulus, and therefore, it depends indirectly also on the mean effective stress:

$$\xi = 0.333 \frac{1+\exp(-0.0145 PI^{1.3})}{2}\left[0.586\left(\frac{G}{G_{max}}\right)^2 - 1.547\frac{G}{G_{max}} + 1\right] \quad (15)$$

2.2.4 Unsaturated shear strength

The shear strength formulation employs Bishop effective stress equation for the effective stress. It reads as follows:

$$\tau = c' + (\sigma_n - u_a) tan\varphi' + \chi (u_a - u_w) tan\varphi' \quad (16)$$

C'= Effective cohesion

σ_n = Total vertical stress on failure plane

φ' = Internal friction angle of soil

3 Statement of problem and modelling

In this study, the finite difference code FLAC was utilized for creating the model and analyzing the unsaturated seismic behavior of the retaining wall. The model was consisted of 3000 elements. A combination of mesh refinement around the retaining wall and coarser elements far from the retaining wall was considered. This way the computational cost can be kept low without considerable loss in precision. (Figure 1).

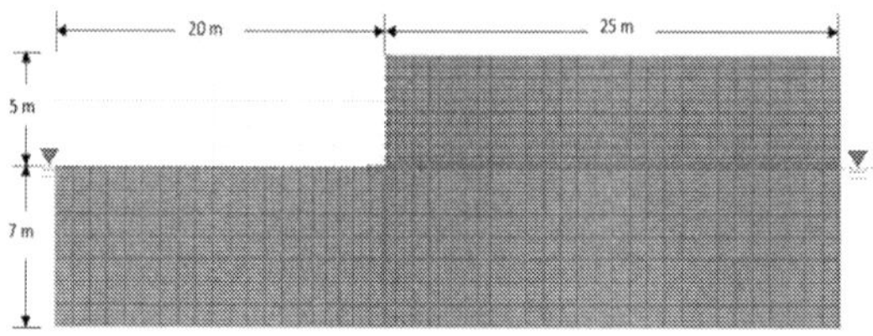

Figure 1. Geometry and dimensions of the model

Elastic beam elements were used to model the concrete retaining wall. Four parameters were required: cross-sectional area (0.5 m^2), mass density, elastic modulus, and second moment of area (0.004 m^4). Table 1 shows properties of retaining wall materials and the soil beneath foundation. For modeling the interaction between the concrete retaining wall and soil, interface elements were used that have four parameters (S, T,k_n, k_s), where, S is a function of the interface friction angle ($\delta = \frac{2}{3}\varphi = 24^0$), φ is the friction angle of backfill soil. T specifies tensile strength and has been set to zero (i.e., cohesionless soil). k_n and k_s are normal and shear stiffness, respectively [10]. As a rule of thumb, k_n and k_s should be set to 10 times the equivalent stiffness of the neighboring zone [2], i.e., $10\, max\left[\frac{K+\frac{4}{3}G_{max}}{\Delta y_{min}}\right]$ where K and G_{max} are the bulk and small strain shear moduli, respectively, and Δy_{min} is the smallest width of the zone (elements) perpendicular to the interface.Water level is in depth of 5 meters from ground surface(Figure 1), above which and behind the wall(due to capillary rise), there is an unsaturated region and the unsaturated properties have been assigned to the elements. Mohr-Coulomb criteria is used to simulate strain-stress behavior of soil. Element are 2D four nodes type. Dimension of elements are chosen so that the element are either rectangular or square shaped. For accurate representation of wave transmission in the numerical models, it is suggested to use element dimensions (ΔL) be less than 1/8 to 1/10 of λ of the maximum entering wave frequency where λ is the wavelength. Based on maximum wavelength considered in this study, maximum allowable dimensions for elements were1.5 m. In order to reduce the boundary effects, the lateral boundaries of model should be in distant from the wall. For this purpose, a distance of 4-5 times of wall height was chosen from each side of the wall.

Table 1. Properties of soil layers and wall

	Density (Kg/m³)	Poison's Ratio	Friction Angle (Degree)	Shear Modulus (kPa)
Concrete Retaining Wall	2300	0.2	-	10700000
Backfill Soil	1600	0.3	35	13000
Foundation Soil	1750	0.26	40	36000

Table 2. Van Genuchten parameters

Van Genuchten parameters	value
a_{vg}	0.75
n_{vg}	1.88
m_{vg}	0.23
w_r (%)	0.2

In current study, the lateral boundaries of the main grid are coupled to the free-field grid by viscous dashpots to simulate a quiet boundary. The lower boundary is fixed (no displacement is allowed in vertical and horizontal direction). The displacement of soil elements adjacent to the boundary of wall is controlled by wall (compatibility condition).

DOI: 10.1051/e3sconf/20160905001

The wall backfill soil is sand with dry density of 1600 kg/m^3 and foundation soil is compacted sand and gravel mixture with dry density of 1750 kg/m^3 (designated as GW based on the unified soil classification, 70% Gravel and 30 % sand). Density and elasticity parameters of the aforementioned soils are reported in Table 1. It is noteworthy that the considered input parameters are obtained from soilvision database. The backfill soil is soil no. 10700 and the soil under foundation is soil no. 18719 in this database. The retention properties of the considered soil types have been reported in table 2.Modeling the retaining wall was performed utilizing the elastic beam element and through a finite difference code. In order to consider interaction between soil and structure, the contact surface between soil and the wall was introduced to the model using boundary elements. Friction angle was considered two third of the internal friction angle of the backfill soil. As common, Rayleigh damping of 0.05 was selected for initial analysis in order to create dynamic model. Two scenarios for modeling the backfill soil were considered. In the first modeling scenario, the backfill soil was considered dry. In the second case, the capillary rise (the unsaturated region behind the retaining wall) was taken into account. As the soil under foundation has been considered saturated in our models, the two modeling scenarios are called sat-dry and sat-unsat hereafter.

The AEROPUETRTO MEXICALI 1979 earthquake record was applied to the bottom boundary of the model as an accelerogram (Figure 2). Maximum earthquake acceleration is reduced (the record is therefore scaled) to the maximum acceleration for Shiraz (Iran) which is 0.3g.

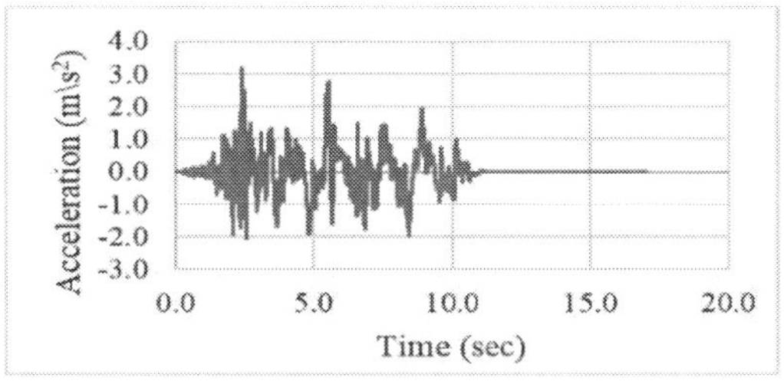

Figure 2. Input accelerogram

4. Results and discussion

4.1 Effect of unsaturated zone on mean effective stress

Figure 3 presents the comparison of mean effective stress σ'_m in dry and unsaturated modeling scenarios considered for the backfill (to recall for the derivation of the mean effective stress, please see equations 11 to 14).

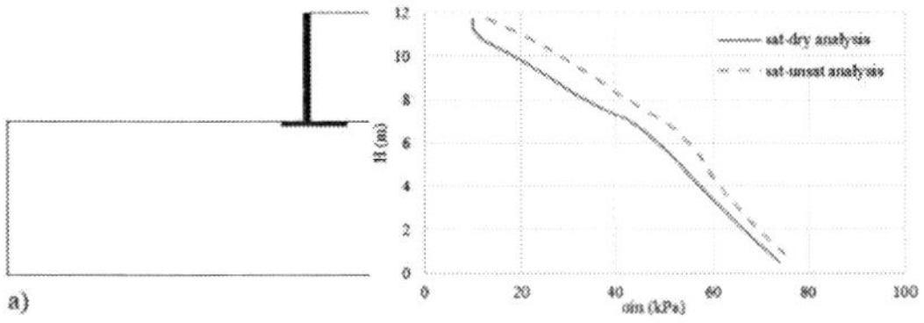

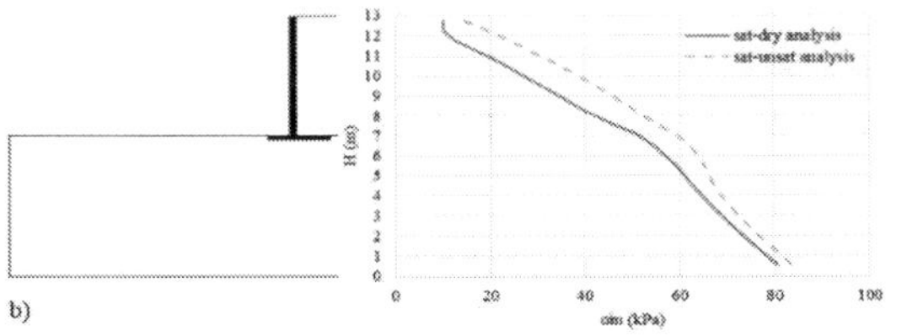

Figure 3. Mean effective stress versus height in dry and unsaturated backfill conditions (a. Wall=5m, b. Wall=6m)

As expected, the mean effective stress will increase with depth since the vertical effective stress increases with depth (equation 14). As it will be discussed in next section, the mean effective stress and its variation with the degree of saturation play an important role in seismic behavior of earth structures.

4.2. Effect of unsaturated zone on shear modulus

In this section, the presence of negative pore water pressure in unsaturated region and its effects on the shear modulus of soil is taken into account. The results are then compared to the condition where the backfill soil is dry (Figure 4).

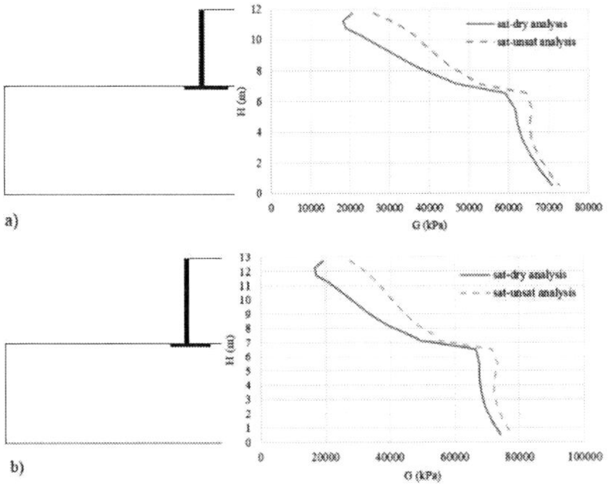

Figure 4. Shear modulus versus height in dry and unsaturated backfill conditions (a. Wall=5m, b. Wall=6m)

As equation 3 and equation 10 indicate, shear modulus and maximum shear modulus have direct relationship with the effective stress level. It could, therefore, be concluded that variation of shear modulus in dry and unsaturated soil is similar to the variation in mean effective stress for these two conditions which is confirmed by Figure 3. Larger shear modulus resulted from presence of suction will inevitably result in larger soil stiffness which has considerable impact on the seismic behavior of the retaining wall.

4.3. Effect of unsaturated zone on resultant force applied on the wall and its point of effect

The lateral earth force from earthquake at each nodal point on the wall was determined and the resultant total lateral force was then calculated. The variation of resultant force during the earthquake is illustrated in Figure 5. The resultant force was calculated using the following formulae:

$$F_{i,j}^{top} = \frac{1}{2}.h_i.\sigma_{i,j}^{top} = \frac{1}{2}.h_i.\sigma_{i,j}$$

$$F_{i,j}^{bottom} = \frac{1}{2}.h_i.\sigma_{i,j}^{bottom} = \frac{1}{2}.h_i.\sigma_{i,j}$$

$$P_j = \sum_i (F_{i,j}^{top} + F_{i,j}^{bottom}) = \sum_i \sigma_{i,j}.h_i \qquad (17)$$

As it is shown in Figure 5, the resultant force applied on the wall is fluctuating during the earthquake in both dry and unsaturated conditions. This is due to variation in horizontal stresses during the earthquake. From this figure it is obvious that the resultant force in unsaturated condition is less than dry condition. The increase in soil shear strength due to matric suction is the main reason for this phenomenon. In both conditions, the maximum resultant force occurs at the same time (t=5.7 sec).

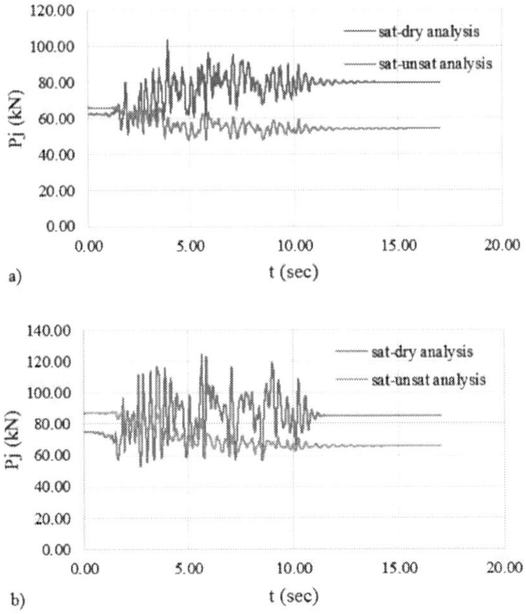

Figure 5. Variation of resultant force applied on the wall during earthquake in dry and unsaturated backfill conditions. (a. Wall=5m, b. Wall=6m)

The point of application of the resultant dynamic forces on the wall (Figure 6), is calculated using the following equation.

$$Y_j = \frac{\sum_i h_i.\sigma_{i,j}.y_i}{\sum_i h_i.\sigma_{i,j}} \qquad (18)$$

Where Y_j is the vertical distance from base of the retaining wall to the point of application of total resultant force in time j, and y_i is the vertical distance from base of the retaining wall to the center of element i.

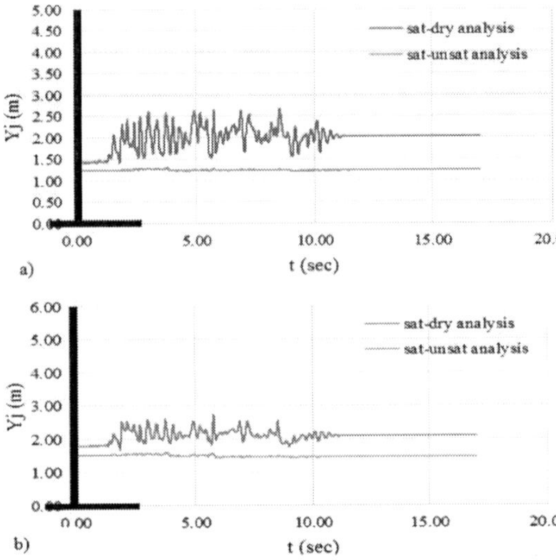

Figure 6. The point of effect of resultant dynamic forces on the wall. (a. Wall=5m, b. Wall=6m)

4.4 Effect of negative pore water pressure on earth pressure coefficient of soil

Horizontal pressure coefficients at any time can be calculated from applied horizontal forces using the following equation. K_{vj} is vertical earthquake coefficient and is assumed to be zero in this study.

$$K_{FLAC} = \frac{2P_{FLAC}}{\gamma_t H^2 (1-k_{vj})} \qquad (19)$$

Furthermore, lateral pressure coefficient under seismic load was calculated using Mononobe-Okabe method. The values of horizontal inertial coefficient, k_h, at the middle of the backfill portion of the structural wedge (expressed in terms of gravitational acceleration) were computed during the FLAC analysis (Figure 7).

In order to compare the results of numerical simulations for dry and unsaturated analyses with those from Mononobe-Okabe equation, they are all plotted on the same graph. These graphs are plotted for walls with heights of 5 and 6 meters, respectively (Figures 8 and 9). In these graphs lateral earth pressure coefficient K obtained from numerical simulation as well as values from Mononobe-Okabe equation have been depicted versus coefficient of horizontal earthquake acceleration.

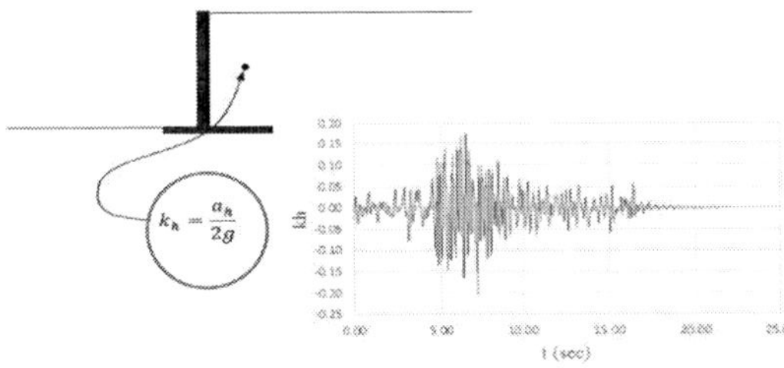

Figure 7. Coefficient of earthquake horizontal acceleration (k_h) at the middle of backfill soil above the heel

The lateral pressure coefficient of soil corresponding to the maximum and minimum points of horizontal acceleration coefficient of earthquake were calculated for both dry and unsaturated conditions. These coefficients are smaller in unsaturated condition as compared to dry condition (Figures 8 and 9). Also by comparing these coefficients to the results from Mononobe-Okabe, we can see again that the lateral active earth pressure coefficient for unsaturated soil condition is somehow smaller than those predicted using Mononobe-Okabe equation. In these figures, KAE denotes active each pressure coefficient, and KPE stands for the passive seismic earth pressure coefficient.

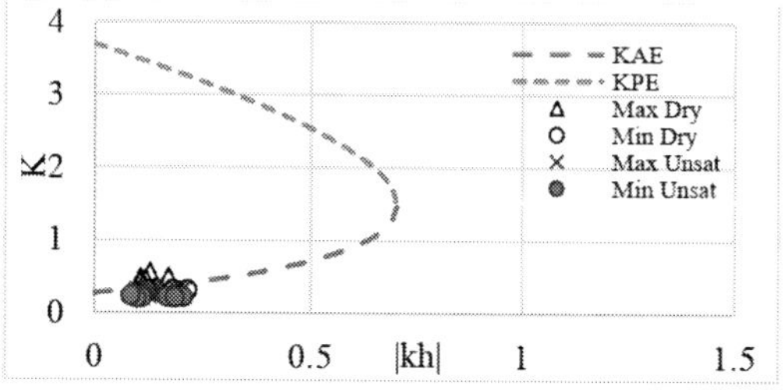

Figure 8. Comparison of horizontal pressure coefficients of soil in dry and unsaturated conditions and the values of Mononobe-Okabe (wall of 5-meter height)

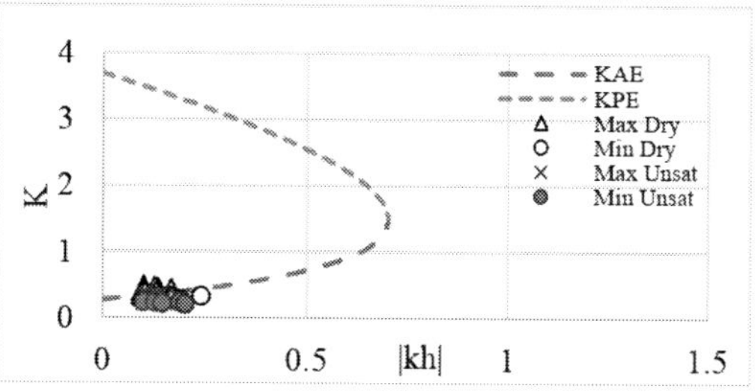

Figure 9. Comparison of horizontal pressure coefficients of soil in dry and unsaturated conditions and the values of Mononobe-Okabe method (wall of 6-meter height)

5. Summary and conclusion

The main results of this study can be summarized as follows:

1. The maximum resultant force in unsaturated condition is smaller and the application point is located lower than the dry condition. Hence, it results in smaller overturning moment and increased stability for the wall.
2. The increase in effective stress from suction in unsaturated region results in an increase in the shear modulus proportional to the change in effective stress.
3. As expected, soil suction increases the shear strength significantly, resulting in smaller lateral force on the wall during earthquake.
4. The lateral pressures calculated from the numerical simulation indicate that the lateral active earth pressure coefficient is somehow smaller than those predicted using Mononobe-Okabe equation which shows that this theory is on the safe side for calculating dynamic forces behind the wall, resulted by earthquake in unsaturated condition.
5. As it was expected, the resultant force appears at an elevation above H/3 in dry condition. However, the results of this study for unsaturated soil condition indicate that application point of the resultant force is at an elevation below H/3 measured from the wall heel.

Acknowledgement

The authors wish to thank R. Behzadpour, Ph.D. student, Department of Civil and Environmental Engineering, Washington State University, USA, for his valuable comments on the manuscript.

References

1. S. A. Stanier, A. Tarantino, Active earth force in 'cohesionless' unsaturated soils using bound theorems of plasticity. Proc. of 5th Int. conf. on Unsaturated Soils, Barcelona, 1081-1086 (2010).
2. T. Vo, A. R. Russell, Slip line theory applied to a retaining wall-unsaturated soil interaction problem. Computers and Geotechnics, **55**, 416-428 (2013).
3. FLAC2d/User's manual, Fifth Edition (FLAC Version 7.0), Itasca Consulting Group Inc., USA, September (2011)
4. S. L. Kramer, *Geotechnical Earthquake Engineering, Prentice Hall*, Inc. (1996).
5. E. Nikooee, G. Habibagahi, S. M. Hassanizadeh, A. Ghahramani, Effective stress in unsaturated soils: A thermodynamic approach based on the interfacial energy and hydromechanical coupling. Transport in porous media, **96(2)**, 369-396 (2013).
6. J. M. Pereira, O. Coussy, E. E. Alonso, J. Vaunt, S. Olivella, Is the degree of saturation a good candidate for Bishop's X parameter? Proceedings of the 5th International Conference on Unsaturated Soils 2. DOI: 10.1201/b10526-141 (2010).
7. N. Lu, J. W. Godt, D. T. Wu, A closed-form equation for effective stress in unsaturated soil. Water Resources Research, **46(5)**, (2010).
8. S. Oh, N. Lu, Uniqueness of the suction stress characteristic curve under different confining stress conditions. Vadose Zone Journal, **13(5)**, (2014).
9. I. Ishibashi and X. Zhang, Unified dynamic shear moduli and damping ratios of sand and clay, Soils and foundations, **33(1)**, 182-191 (1993).
10. R. A. Green and R. M. Ebeling Seismic Analysis of Cantilever Retaining Walls, Phase 1. ERDC/ITL TR-02-3, US Army Corps of Engineers, Engineer Research and Development Center, (2002).

Study on vertical permeability regularity and collapsibility of a large thickness loess foundation by in-situ testing

Yang Kong[1], Xuefeng Huang[2,a], Huaining Ruan[1], Zhende Zhu[1] and Chong Shi[1]

[1]Hohai University, Institute of Geotechnical Engineering, 210098 Nanjing, China
[2]LEU, Department of Architectural Engineering, 401331 Chongqing, China

Abstract. In many regions in China there is a wide distribution of dead-weight collapsible loess that experience frequent disasters resulting in enormous losses to national property. Loess collapsibility is the leading cause of disasters in loess areas. Many international scholars are working collaboratively to avoid property losses and human casualties caused by the collapsible loess. Loess collapsibility is closely related to the vertical permeability regularity, so the vertical permeability regularity should be the first critical factor taken into account in studying loess collapsibility. In Lanzhou city, the in-situ soaking test is carried out on an unsaturated natural loess foundation with a thickness of 36.5m, the test pit diameter is 40m, TDR moisture probes are embedded, and the test lasts 282 days. During the test, the vertical migration of water is observed and the temporal and spatial variations of volumetric water content are considered. In the detailed research of the vertical permeability regularity, the relation between permeability regularity and collapsibility are discussed. For shallow layer soil, soil volumetric water content reaches saturation at wetting front, the water flows downward due to gravity and the infiltration rate is larger. For deep layer soil, the moisture will increase because of the upper soil permeability and its own suction, but, due to soil compression caused by the occurrence of the upper soil collapsibility, the infiltration rate is significantly reduced. The temporal and spatial variations of water volumetric content can be used to determine whether a loess collapsible deformation has occurred. The limit permeability depth in this test is 25m, and the limit permeability depth can also be served as the collapsibility evaluation depth of the unsaturated natural Malan loess foundation.

1 Introduction

Lack of capital and lack of techniques are important reasons that restrict the development of infrastructure construction in developing countries. China's quest for a social harmonious development is always accompanied with a series of geotechnical engineering problems. It is well known that the unsaturated natural loess is widely distributed in northwest China, and the collapsible characteristic of loess is harmful to the construction of large industrial and civil buildings. A large number of industrial and civil buildings that were built on the collapsible loess foundation have been destroyed, the road pavement has cracked and the water conservancy facilities have failed which caused huge property losses when water flowed into collapsible loess foundation (Liu, 1996). It is an urgent need to find an approach to avoid the damage caused by water permeability. Therefore, it is necessary to study the water permeability regularity on the unsaturated natural loess foundation with the dead-weight collapsible characteristic. Permeability is one of the most important engineering properties of unsaturated natural loess, and many engineering properties such as the size of collapsible deformation value and the collapsible deformation coefficient are all closely related to the permeability. The permeability is mainly reflected in the water flowing regularity. So far, the research on water flow regularity of unsaturated natural loess is far from enough (Liu *et al.*, 2004; Huang and Li, 2011).

The experimental method and theoretical calculation of the unsaturated soil permeability coefficient have a certain degree of difficulty (Fredlund, 1993; Chen *et al.*, 2006). The permeability coefficient of unsaturated natural loess is closely related to volumetric water content. The permeability coefficient of unsaturated natural loess can be obtained by direct methods and indirect methods. The indirect methods include the soil water character curve method and the empirical formula method, but its computational accuracy is difficult to grasp. In-situ non-destructive testing of water content is the key testing technique of in-situ soaking test for unsaturated natural loess, which can study the collapsible deformation of unsaturated natural loess and directly get the permeability coefficient.

The TDR method is also called the Domain Reflectometry Time method, which is used to determine the soil dielectric constant and convert it into soil volumetric water content. The theoretical model had been established as early as 1939, which was initially used in

[a] Corresponding author: hxfen60@163.com

the telecommunication industry to find the cable breakpoint. Monitoring of soil water content was firstly proposed by Canadian scientist, Topp, in 1980, and was used to determine the water content of farmland in 1985. Compared with the traditional soil volumetric water content testing techniques, it has the advantages of not destroying the sample, is fast, simple and accurate and can be measured continuously with automatic data collection. Since the 1990s, TDR has been used as the basic equipment for testing soil water content (Friel and Or, 1999; Thomsen *et al.*, 2000; Skierucha, 2001; Manoel *et al.*, 2001; Souza and Matsura, 2003; Giraldi and Iannelli, 2009; Scudiero *et al.*, 2012).

In 1963, the hydraulic structure of a Romanian steel industry was built on the dead-weight collapsible loess foundation, which had a thickness of 24m. The foundation settlement was stable after three months of soaking, and basically eliminated the collapsibility. Soviet construction research institute used the pre-soaking method and the explosion method to encrypt loess at Polos in 1964, after 1 year of soaking, the water content was closed to the natural water content state. Soil tests showed that the loess under a 5 - 6 m depth basically eliminated the collapsibility. As a main research method on collapsible deformation and foundation treatment, the pre-soaking method can reflect the actual soil water flow state and provide the analysis basis for loess permeability regularity research, and these results indoor tests cannot achieve. Since the 1960s, a great number of pre-soaking tests in China have made great contributions to the theoretical research of loess in China and even the world. But studying the permeability regularity of unsaturated natural loess through the large soaking test has rarely been reported. TDR moisture probes are embedded in six artificial exploratory wells in this test to study the loess moisture permeability regularity and the collapsible deformation, so the study has a practical significance.

2 In-situ tests

The test site is located in the Peace town of Lanzhou City, Gansu Province, China. Based on the geological prospecting, the geomorphic unit belongs to Huangshui river the southern bank's class-II terrace, and the geomorphic unit is single. According to local deep wells data, the underground stable water depth in this site is greater than 70m, the total thickness of the soil layer is larger than 38 m, in which collapsible loess's thickness is approximately 36.6m, and the full thickness soil is collapsible loess. According to the geological examination, it is the Malan loess, and its geotechnical engineering properties are listed in Table 1.

Table 1. Properties of soil layer in test site.

Soil layer	Thickness (m)	Buried depth (m)
Cultivated soil (Q_3^{ml})	0.50	0 - 0.50
Silt (Q_3^{fl})	4.50	0.50 - 9.00
Silty clay (Q_3^{fl})	31.50	1.50 - 36.00
Gravel	--	> 38.0m

The test pit shape is circular, with a diameter of 40m, and the diameter length is larger than the thickness of the dead-weight collapsible loess. According to the soil's physical and mechanical indexes of addressed exploration reports and corresponding exploratory wells investigation, there is a homogeneous distribution of soil. There is no adverse geological body, impermeable soil layer or other impurities in the soil of the test site itself or the surroundings.

Six exploratory wells are excavated in both inside and outside of the test pit. Two parallel exploratory wells that have a 5m distance from the center are excavated to study the vertical permeability regularity of water. Four wells that are 0m, 5m, 8m, and 11m distance from the test pit edge are excavated to study the level permeability regularity of water. The six exploratory wells are located in one linear cross-section as is shown in Figure 1.

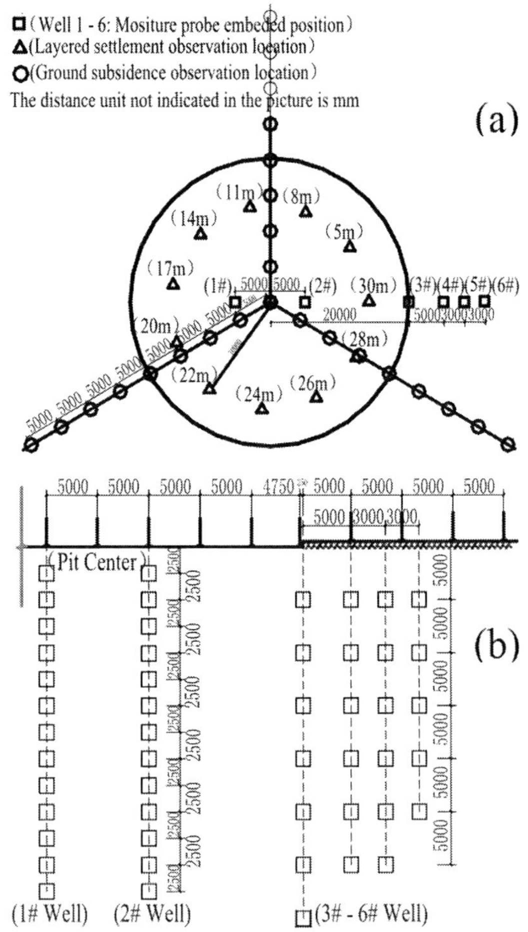

Figure 1. Soaking test arrangement scheme, (a) floor plan and (b) vertical cross-section schematic view.

No.1 and No.2 exploratory wells are embedded with 13 TDR moisture probes, and the distance between each TDR probe is 2.5m. The number of TDR probes that have been embedded in exploratory wells 3 - 6 is respectively 7, 6, 6, and 5. The minimum height TDR probes are embedded at a depth of 2.5m or 5.0m.

In order to obtain the authentic natural loess test data, above all, some certain vertical exploratory wells of a 32.5m length or other lengths and diameter 1.30m in the test pit should be excavated. At a certain depth of the exploratory well, the TDR probe should be embedded at the end of a 1.5m length and diameter 0.35m lateral space that has been excavated by the Luoyang shovel. According to the above method, the TDR moisture probe is buried in the natural soil with the greatest possibility. The TDR moisture probe should be installed with an angle of 45 degrees to reduce the influence caused by poor soil characteristics. The worker should set aside 2m extra length cables to ensure that the collapsible loess will not pull out the instrument. The TDR moisture probe installed method as is shown in Figure 2.

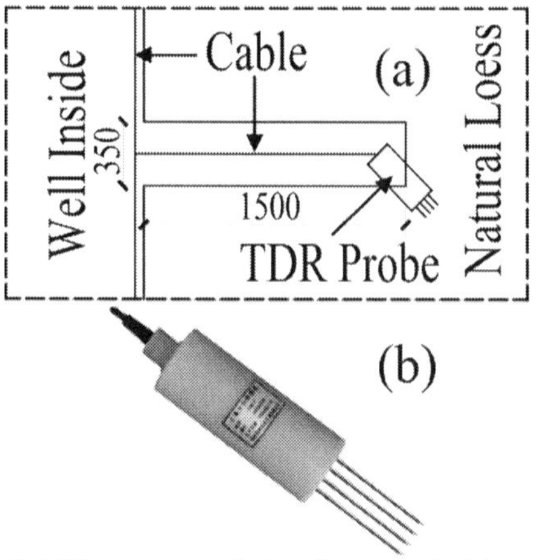

Figure 2. TDR moisture probe installed method, a) installation sketch map and b) a TDR moisture probe.

After the TDR moisture probes are all embedded in a certain well, the exploratory well should be compacted by the rammer, and the compacted density should be slightly larger than the original soil density to prevent the occurrence of embankment collapse. Around the cable and the well surroundings, the isolation water treatment should be done (with 3:7 lime - soil) to prevent downward water penetration along the cable or around the shaft lining.

3 Test results

The in-situ soaking test lasts about 282 days on an unsaturated natural loess foundation with a thickness of 36.5m. The water soaking observation time is 140 days and the water pausing time is 142 days. The soaking test site panorama is shown in Figure 3.

Figure 3. Soaking test site panorama.

According to the above experimental method and the above text depiction, the authors obtained a considerable experimental data of the undisturbed natural loess. The test data measured by the TDR moisture can truly reflect the vertical permeability and the collapsibility of the natural soil.

Because of the limited space, this paper only gives the No.1 exploratory well testing results. The volumetric water content variation curves at different depths and testing stages measured by the TDR moisture probes are shown in Figure 4 (a - m). In these figures, the different curve thickness and colours represent the different testing stages.

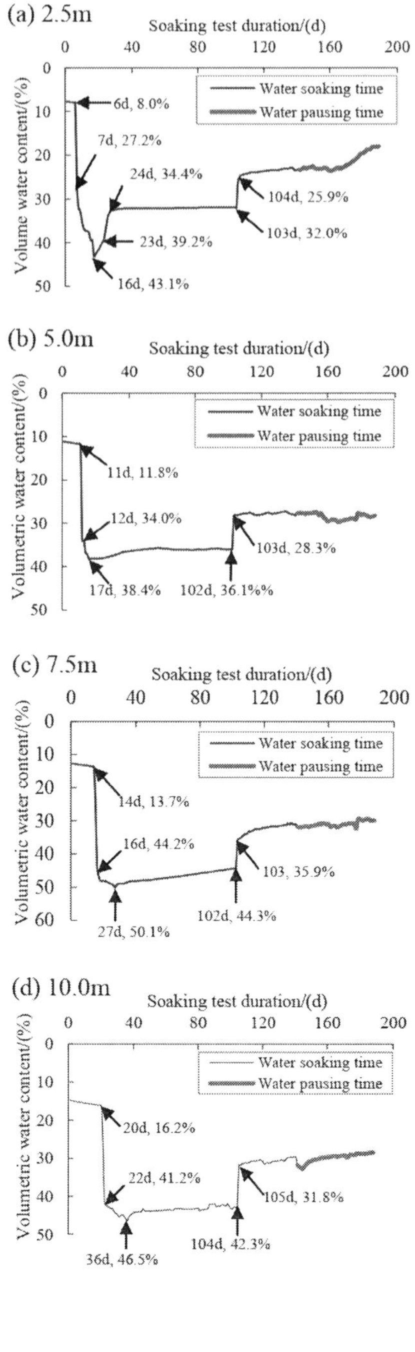

DOI: 10.1051/e3sconf/20160905002

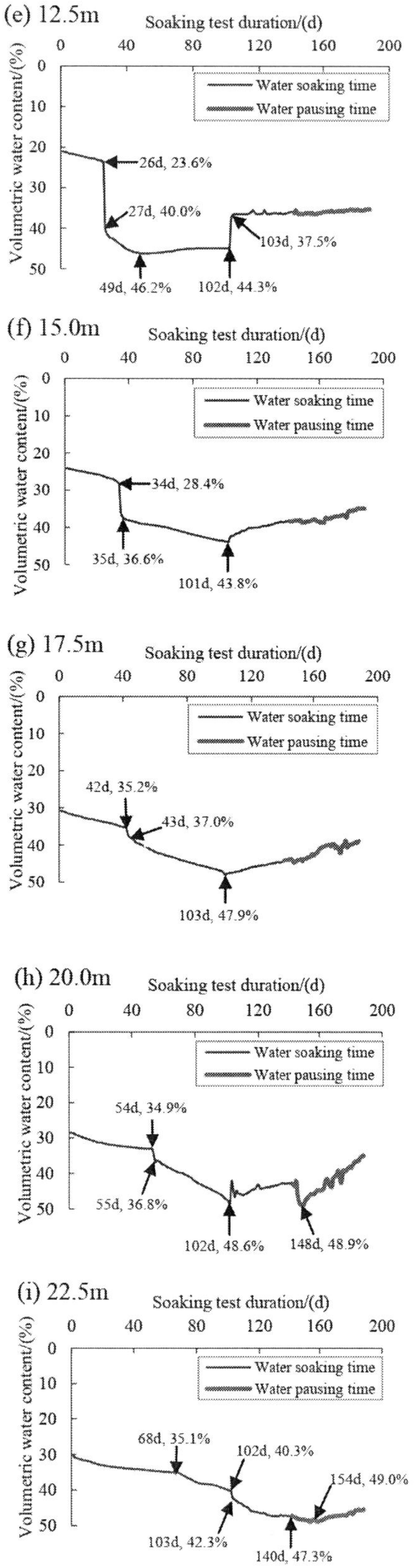

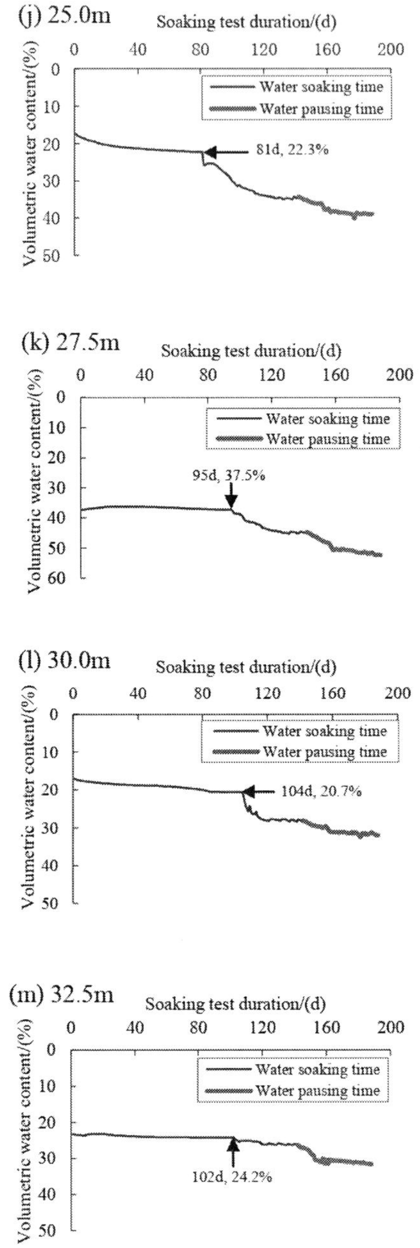

Figure 4. Soil volumetric water content variation curves at different depths and testing stages. Figure 4 (a) - (m) respectively represent depth 2.5 - 32.5 m.

The variation curve of volumetric water content and soaking duration at depth 2.5m in No.1 exploratory well is shown in Figure 4a. As is shown in the figure, the volumetric water content increases steadily to about 8.0% during the first 6 days, then, the volumetric water content has a rapid increase on the 6th - 7th day, and the volumetric water content increases gradually into the peak water content 39.2% on the 16th day (the calculated saturation is larger than 85%). The volumetric water content decreases gradually on the 16th – 23rd day, and then the volumetric water content has a rapid decrease on the 23rd - 24th day. The volumetric water content decreases gradually on the 24th – 103rd day, and the volumetric water content has a rapid decrease on the 103rd - 104th day again. After that, the volumetric water content decreases slowly till the end of testing period. From the depiction of Figure 4a, we know that the water reaches the depth 2.5m on the 7th day, and the loess of depth 2.5m reaches the saturation on the 16th day. There are two sudden decreases in the figure, the 23rd - 24th day and the 103rd - 104th day, it is analyzed that the dead-weight collapsible deformation happened on the 23rd - 24th day, and the collapsible deformation happened on the 103rd - 104th day. The reason for this phenomenon is that there is a wet deformation of loess in the period from 7th – 23rd day, and the loess structural strength decreases gradually. The loess structural strength decreases proportional to the initial dead-weight collapsible pressure till the 23rd - 24th day, so, the soil particle skeleton lost stability and rearranged the particles, led to the stilted particles wedged into the pores of loess, and then followed a dramatically dead-weight collapsible deformation. As the results of the dead-weight collapsible deformation, the loess has a relatively compact and stable structure. From then on, the water weight in the soaking pit is equivalent to the effect of additional load and it is calculated that the influence depth of the additional load has not exceeded 5.0m. There is a slowly decreasing of strength in the surrounding loess (with a new strength and structure) to the collapsible pressure, accompanied a dramatically collapsible deformation on the 103rd - 104th day. It is found that the dead-weight collapsible deformation and collapsible deformation of loess are the results of the displacements of loess structural units and the permeability movements caused by water flowing into the loess. Figure 4a shows that development regularity of volumetric water content at depth 2.5m can be divided into 6 stages: the initial slow increasing stage, the rapid increasing stage, the rapid decreasing stage, the middle slow decreasing stage, the rapid decreasing stage, the final slow decreasing stage.

The variation curve of volumetric water content and soaking duration at 5.0m in No.1 exploratory well is shown in Figure 4b. As is shown in the figure, the volumetric water content increases steadily to about 11.2% during the first 11 days, then, the volumetric water content has a rapid increase on the 11th - 12th day, and the volumetric water content increases gradually to the peak water content of 38.4% on the 17th day. The volumetric water content decreases slowly on the 17th - 102nd day, and then the volumetric water content has a rapid decrease from 36.1% to 28.3% on the 102nd - 103rd day. It is showed that the collapsible deformation happened on the 102nd – 103rd day. Figure 4b - 4e are the variation curves of volumetric water content and soaking duration from the depth 5.0m to depth 12.5m, and these figures have the similar development trend. Figure 4b - 4e show that development regularity of volumetric water content at depth 5.0m - 12.5m can be divided into 5 stages: the initial slow increasing stage, the rapid increasing stage, the middle slow decreasing stage, the rapid decreasing stage, the final slow decreasing stage.

The variation curves of volumetric water content and soaking duration from the depth 15.0m to depth 20.0m in No.1 exploratory well are shown in Figure 4f - 4h, and these figures have the similar variation regularity. As are shown in Figure 4f, the volumetric water content increases rapidly from 28.4% to 36.6% on the 34th - 35th day, and then the volumetric water content increases gradually to the peak water content 48.8% on the 101st day. As is shown in Figure 4i, the volumetric water content increases rapidly from 40.3% to 42.3% on the 102nd – 103rd day, and the volumetric water content increases gradually to the peak water content (saturation) 49.0% on the 154th day (water pausing time). According to the analysis before this depth, the variation regularity is dramatically different from the former. After the rapid increase on the 34th - 35th day, the volumetric water content continues to increase until the peak value, and then maintains a steady decrease. Figure 4g and Figure 4h have the similar development trend as Figure 4f. The volumetric water content had a small increase in water pausing time at depth 20.0m in Figure 4h, it is estimated that the water infiltrated into the TDR moisture probe from cracks at this point, but this does not alter the water content's steady decreasing trend. Figure 4f - 4h show that development regularity of volumetric water content at depth 15.0m - 20.0m can be divided into 4 stages: the initial slow increasing stage, the rapid increasing stage, the middle slow increasing stage, and the final slow decreasing stage.

The variation curves of volumetric water content and soaking duration from the depth 25.0m to depth 32.5m in No.1 exploratory well are shown in Figure 4j - 4m, and these figures have the similar variation regularity. The water content increases continuously from the first day of soaking test, but the variation curves are relatively smooth and it can be divided into three increasing stages. The first stage is the initial stage of the soaking test because the water has not penetrated into the layer, resulting in little change in the volume of water content, almost in a straight line. The second stage is the mid water soaking time and the beginning of the water pausing time. In this stage volumetric water content increases steadily, and the stage water content increment increases significantly. The gradually increasing trend is larger than that of the first phase of increase. The third phase is the volumetric water content stabilization phase of the later water pausing time, but it has always maintained a very slow increasing trend.

In conclusion, loess collapsibility causes structural damages. As a result, the original void is compacted and is accompanied by the volumetric water content decreasing. It is analyzed that the dead-weight collapsible deformation happened on the 23rd - 24th day, and the collapsible deformation happened on the 103rd - 104th day at depth 2.5m according to Figure 4a. Figure 4b to Figure 4e reflect that there is a dead-weight collapsible deformation that happened on about the 102nd - 104th day at depth 5.0m - 12.5m and the extent of dead-collapsible deformation decreases with the increase of depth. It is analyzed that the dead-weight collapsible deformation happened on the 101st - 102nd day at depth 15.0m - 20.0 m according to Figure 4f - 4h. The degree of dead-weight collapsible deformation decreases with the increase of depth. Figure 4i reflects that a slight dead-weight collapsible deformation happened on about the 154th day at depth 22.5 m, and it is remarkable that it belongs to the water pausing time. In short, although, there is a difference in volumetric water content change process and collapsible deformation process due to the variation of test conditions and soil distributions according to Figure 4a - 4i, it points out that there is a close relationship between volumetric water content and collapsible deformation. It can be indirectly determined from the TDR moisture probe measured volumetric water content variation curve whether there is a dead-weight collapsible deformation or collapsible deformation, and even study the deformation variation with time.

At a depth of 22.5m and a depth smaller than 22.5m, figures (Figure 4a - 4i) are all ultimately have the symbol of volumetric water content decreasing before the researchers stop recording the data. The volumetric water content in figures (Figure 4j - 4m) with a depth of 25.0m and a depth greater than 25.0m has a slow increasing in water pausing time according to the trend of the data after stopping recording. Volumetric water content has kept an increasing trend under the depth of 22.5m, and the phenomenon indicates that the soil has not reached the saturation point. This is also reflected in the other side that the water permeability is not static on unsaturated natural loess foundation with large thickness, and, with the increase of the soil depth, there is a permeability limited depth. For the shallow soil, water infiltration rate is faster, but for the deeper soil, water infiltration rate is very slow. This is closely related to the occurrence of collapsible deformation of the upper part of the soil. Collapsible deformation caused the compaction of the upper soil layer, and led to voids becoming smaller and, moreover, blocked the further proliferation and permeability of water. The lower part of the soil volumetric continued to increase on the one hand because of the slow soil water permeability in the upper part, on the other hand because of the effect of soil suction.

4 Conclusions

An in-situ soaking test is carried out on an unsaturated natural loess foundation with a thickness of 36.5m, and TDR moisture probes are embedded. During the test, the vertical migration of water is observed. According to the temporal and spatial variation of volumetric water content, the vertical permeability regularity of a large thickness unsaturated natural loess foundation has been cleared, and it can also be used to determine whether there is an occurred loess collapsible deformation. The limit depth of permeability in this test is 25m, and it served as the collapsible evaluation depth of the unsaturated natural Malan loess foundation, which not only ensures the safety of building foundations, but also reduces construction costs.

Acknowledgements

The authors would like to thank Dr. Yao, Zhihua and M.S. LIU, Changling for their contributions to this test. The authors would also like to thank B.S. Lisa Gott for her kind help to correct and polish the language.

References

1. Liu, Z. D. *Loess mechanics and engineering* (Science and Technology Press of Shanxi, Xian, 1996)
2. Liu, B. J., Xie, Y. L., & Yu, Y. C. In-site testing study on infiltration in unsaturated loess. Chinese Journal of Rock Mechanics and Engineering, 23(24), 4156–4160 (2004)
3. Huang, X. F., Li, J. Test study on influence of immersion on earth pressure and horizontal displacement of high slope of unsaturated intact loess. Chinese Journal of Rock Mechanics and Engineering, 30(3), 635–642 (2011)
4. Fredlund, D. G., H., Rahardjo. *Soils Mechanics for Unsaturated Soils* (John Wiley & Sons, Ins, 1993)
5. Chen, Z. H., Sun, S. G., Fang, X. W., Zhou, H. Q., Xie, Y. Y. Recent advances of the measuring technology for unsaturated soils and special soils. Chinese Journal of Geotechnical Engineering, 28(2), 147-169 (2006)
6. Friel, R., & Or, D. Frequency analysis of time-domain reflectometry (TDR) with application to dielectric spectroscopy of soil constituents. Geophysics, 64(3), 707 (1999)
7. Thomsen, a., Hansen, B., & Schelde, K. Application of TDR to water level measurement. Journal of Hydrology, 236(3-4), 252–258 (2000)
8. Skierucha, W. Accuracy of soil moisture measurement by TDR technique. International Agrophysics, 14, 417–426 (2000)
9. Manoel, C., Vaz, P., & Hopmans, J. W. Simultaneous measurement of soil penetration resistance and water content with a combined penetrometer - TDR moisture probe. Soil Science Society of America Journal, 65, 4–12 (2001)
10. Souza, C. F., & Matsura, E. E. Multi-wire time domain reflectometry (TDR) probe with electrical impedance discontinuities for measuring water content distribution. Agricultural Water Management [AGRIC. WATER MANAGE.]., 59(3), 205–216 (2003)
11. Giraldi, D., & Iannelli, R. Measurements of water content distribution in vertical subsurface flow constructed wetlands using a capacitance probe: benefits and limitations. Desalination, 243(1-3), 182–194 (2009)
12. Scudiero, E., Berti, A., Teatini, P., & Morari, F. Simultaneous monitoring of soil water content and salinity with a low-cost capacitance-resistance probe. Sensors (Switzerland), 12(12), 17588–17607 (2012)

Experimental study on the coefficient of lateral earth pressure in unsaturated soils

Ali Pirjalili[1,a], Aliakbar Golshani[2] and Ali Mirzaii[3]

[1]*M.Sc. Student, Department of Soil and Foundation Engineering, Faculty of Civil and Environmental Engineering, Tarbiat Modares University, Iran*
[2]*Assistant Professor, Department of Soil and Foundation Engineering, Faculty of Civil and Environmental Engineering, Tarbiat Modares University, Iran*
[3]*Assistant Professor, Department of Civil Engineering, Faculty of Engineering, University of Kashan, Iran*

Abstract. Accurate measurement of lateral earth pressure coefficient, *k*, is a key step in any geotechnical investigation. Since conventional approaches for obtaining this value leads to high risk of uncertainty in the results, this paper introduces a new experimental approach for measuring the value of *k*, particularly in unsaturated soils. Details of the new instrument designed for this purpose is provided in the manuscript, which is capable of obtaining *k* value of unsaturated soils in a range of matrix suction or degree of saturation. Finally, some test results are presented that examine the accuracy of the novel apparatus in order to measure the coefficient of lateral earth pressure in unsaturated soils.

1 Introduction

Lateral earth pressure is a crucial issue, which affects obviously on many geotechnical applications analyses such as; underground structures, slope stability and retaining walls. Generally, lateral earth pressure refers to the coefficient of lateral earth pressure, *k*; therefore, obtaining an accurate and reliable values is necessary. Typically, simplifying assumptions have been considered in calculation of this factor by using conventional equations; however, experimental studies provide better estimations of true soil behaviour.

Conventional soil mechanic principles are based on saturated mode of soil. Also, previous studies for evaluating of at-rest lateral earth pressures coefficient, k_0, were based on this suggestion. (Terzaghi 1920; Chi-in 1957; kjellman 1936; Okochi and Tatsuoka 19884; lo and Chu 1990; Chu et al.1992; Mesri and Hayat 1993; Chu and Gun 2004; Northcutt and Wijewickreme 2013).[1-9]

In spite of the fact that most analysis are based on saturated soil mechanic principles, the major part of the earth land are composed of unsaturated soils. As mechanical properties of saturated and unsaturated soils are obviously different, it is necessary to study about *k* value in unsaturated mode.

Because of test procedure complexity, there are a few experimental studies which seeking *k* coefficient in unsaturated soils. Zhang et al. (2009) have performed a series of tests to evaluate k_0 in a recompacted expansive soil by a stress-path triaxial system. They showed that k_0 increasing by increasing in saturation degree of soil[10]. In addition, Sr. Oh et al. (2013) have used a suction-controlled k_0 consolidation triaxial test procedure to evaluate k_0 in two different soils. Results of this study show decreasing in k_0 value by increasing matrix suction. Recently, Li et al. (2014) have carried out a centrifugal test on a clayey soil to examine k_0 and they observed k_0 increasing by saturation degree increase. In this study, it proved that k_0 value in unsaturated soil is less than that obtained in saturated soil.[11, 12]

Nowadays available information about the coefficient of lateral earth pressure behavior in unsaturated soils is limited to some experimental data that mainly focused on k_0, however at-rest condition exists scarcely in geotechnical applications. This paper introduces a suction-controlled approach by using a new device whereby *k* could be evaluated in at-rest state.

2 Apparatus Feature

Fig.1 shows a schematic of test procedure designed to evaluate *k* in a suction-controlled system. To this end, a new device has been designed and built in Tarbiat Modares University for the first time.

Test system consists of four main parts; ring device, loading system, data acquisition, and an automatic volume change. To measure the coefficient of lateral earth pressure, lateral pressure measurement was needed in order to calculate vertical overburden pressure changes into lateral pressure rate. A Ring device was used for measuring lateral pressure. Fig.2 demonstrates its components. As shown, the specimen ring is a flexible

[a] alipirjalili@yahoo.com

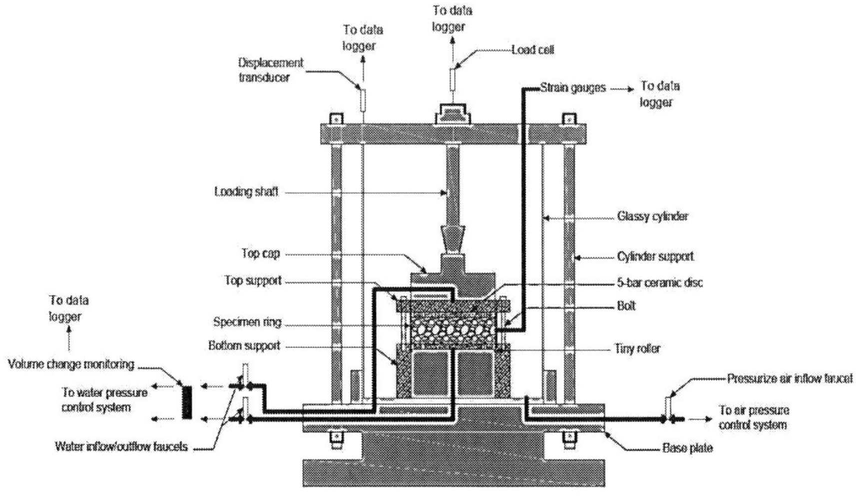

Figure 1. Test Procedure

thin ring by 1mm thickness, 20 mm sample height, and 70 mm inner diameter.

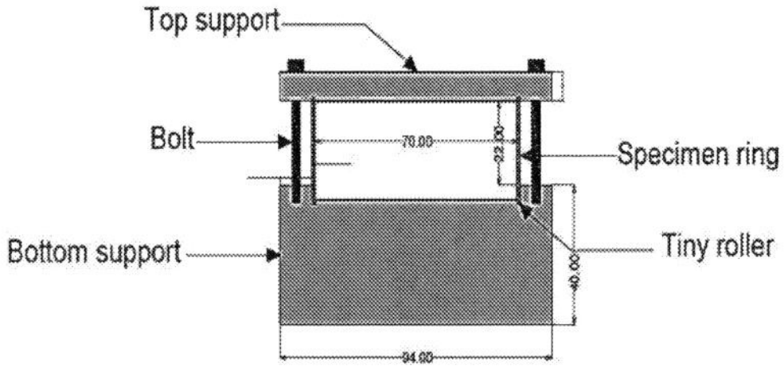

Figure 2. Ring Device Components

The ring device was placed in a triaxial chamber and it used for measuring lateral pressure of soil specimens (Fig. 1). In order to fix the specimen ring in axial direction, two supports have been used and ring leans on some tiny rollers located inside the top and bottom supports. In addition, top and bottom supports were fixed to each other moderately. Three strain gauges installed on ring outer wall in order to monitor lateral strain, as shown in Fig.3.

Figure 3. Ring Device

For interpreting recorded data, strain gauges installed on specimen ring were calibrated by installing specimen ring between two caps on top and bottom of ring, which were tightened to each other moderately. To calibrate specimen ring, several defined pore air pressures were applied inside the ring and strain gauges records allocated to applied pressure.

This process was repeated three times in seven pressure points and a calibration factor was reported for every cycle. Fig.4 shows results of tests performed for calibration procedure. A fit linear relationship was observed between results for every cycle, as shown in Fig.4. Consequently, an average factor from three cycles was reported as calibration factor. It should be mentioned that recorded data in main test were less than 0.2 (in data output unit, as shown in Fig. 4), which data demonstrated sufficient adjacency to each other.

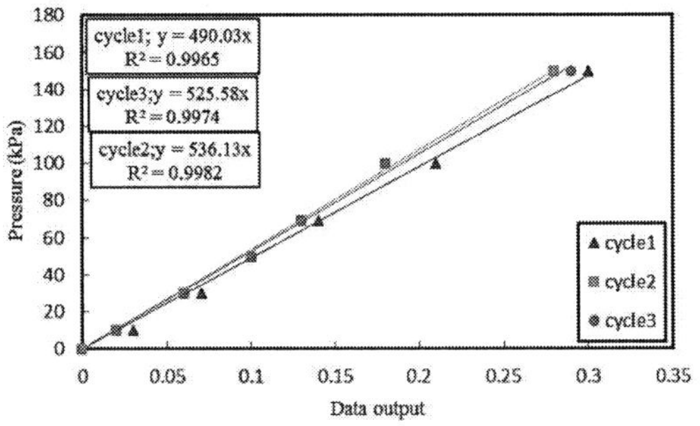

Figure 4. Calibration Results

3 Material and Test procedure

Some validation tests have been performed to check accuracy of proposed approach. In this study, Firouzkouh clay was chosen in order to examine the dependency of k value on matrix suction and initial void ratio in unsaturated soil. The mechanical properties of this soil summarized in Table 1. Other properties of Firouzkouh clay investigated in a series of experimental studies by Mirzaii and Yasrobi in 2012. (2012a, b). [13, 14]

The specimens were compacted in two diverse initial void ratios, 0.71 and 0.92. Moreover, three different matrix suctions were defined to apply on the soil specimens in order to examine dependency of *k* value on initial void ratio and matrix suction in same time.

Table 1. Soil Mechanical Property

Property	amount
Specific Gravity	2.75
Maximum dry unit weight, kN/m^3	18.06
Liquid limit %	30
Plastic index %	9
Percentage < 4.75 mm	100
Percentage < 75 μm	88.7
Percentage < 2 μm	20
Optimum water content %	14.1
Soil classification	CL

The soil specimen was wetted in optimum water content and it was mounted into the triaxial chamber. The coefficient of at-rest lateral earth pressure was measured in 10, 50 and 90 kPa of matrix suction. To achieve that, the pore air pressure was increased gradually from 0 to 300 kPa and after that, the pore water pressure was increased to the defined pressures (290,250 and 210 kPa) in which calculated for desired matrix suctions (10, 50 and 90 kPa). To achieve to the equalization stage, the specimen has been kept under this condition for about 18-24 hrs to wet. By allowing water and air draining in samples, matrix suction were constant during the all tests.

Within wetting stage, water inflow to specimen was observed for all specimens. The specimens' compression during the wetting stage was controlled by an automatic volume change accurately. The wetting stage considered to be finished when the water inflow rate reaches to a steady mode and water inflow volume remained more-or-less constant, as shown in Fig.5.

After equalization stage, the specimen was loaded axially by a static system with a constant rate of 0.5mm per minute, which was slow enough for air and water drainage. In a same with equalization stage, in this step allowing water and air drainage kept the matrix suction constant during the test. In addition, using two porous stone in top and bottom of the sample beside low height of each sample, made the drainage process possible. To monitor constancy of matrix suction, the water and air pressure gauges were observed which were constant during the test.

Loading process has been kept running until the measured *k* attained to a stable condition (this condition normally occurred about 0.4-0.6% specimen radial strain). As radial strains were not enough to reach to active condition, measured *k* amounts belong to a state between active and at-rest condition. During the test, matrix suction were remained constant by allowing water and air exceeding to the soil.

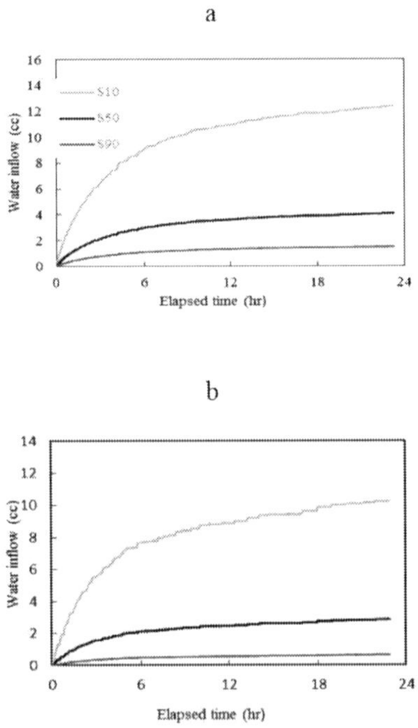

Figure 5. Wetting Stage (*a*: e_0=0.71, *b*: e_0=0.92)

4 Results

The coefficients of lateral earth pressure values versus specimens' vertical strain were measured during the test and they were demonstrated in Fig.6. This figure presents that *k* value reached to a steady value in a range of 0.4-0.6% vertical strain in every soil specimen. Steady *k* value for each test has been summarized in Table 2.

Table 2. Final *k* Amounts

Matrix Suction *kPa*	**Initial void ratio**	
	0.71	0.92
	***k* amount**	
10	0.411	0.481
50	0.391	0.455
90	0.371	0.429

The *k* sensitivity to matrix suction changing is demonstrated in Fig. 7 that measured *k* are plotted versus diverse matrix suctions were applied to the soil for both specimens compacted in 0.71 and 0.92 initial void ratio, respectively. According to this figure, experimental results show that the *k* values decreased by matrix suction increasing obviously.

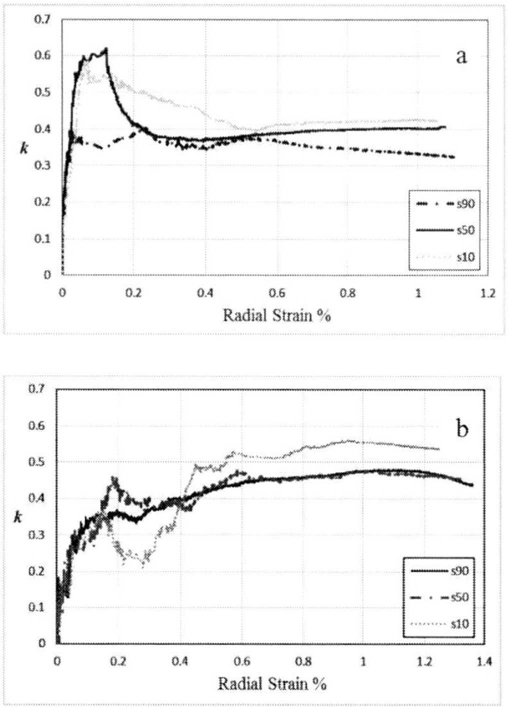

Figure 6. *k* Amount versus Vertical Strain (*a*: e_0=0.71, *b*: e_0=0.92)

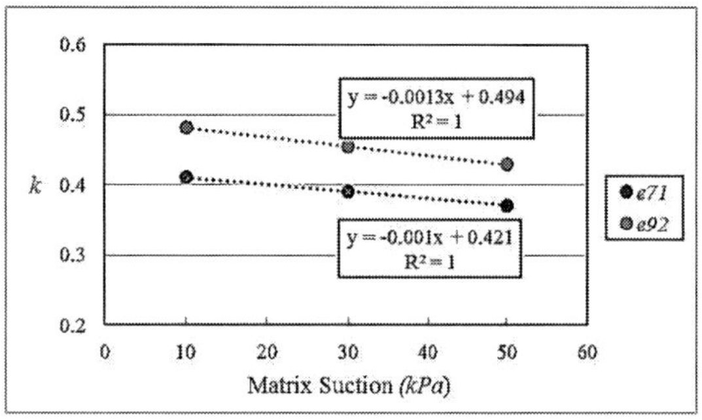

Figure 7. *k* Changes versus a Range of Matrix Suctions

With regards to Fig.7, dependency of *k* values to matrix suction appears to be a linear relationship and this behavior was observed in both specimen by 0.71 and 0.92 of initial void ratio. By caparisoning *k* amount in both soil samples with the initial compaction void ratio of 0.71 and 0.92 (see Fig. 7), it could be concluded that obtained *k* in soil by 0.71 initial void ratio is less than that in specimen compacted in 0.92 void ratio initially. Therefore, *k* value for looser soil is more than that in denser one, which this behavior has also been observed in saturated soils. (Chu and Gun 2004)[8]

5 Conclusion

In order to examine dependency of the coefficient of lateral earth pressure, *k*, on initial void ratio and matrix suction in unsaturated soil, a new method has been introduced in this paper. To achieve this goal, a new ring device has been designed and built in Tarbiat Modares University. In order to validate this new method, some tests have been performed on a clayey soil and results show that this method is reliable enough to measure *k* amount in unsaturated soils. In addition, results have been interpreted in manuscript showed that *k* amount decreased by increasing matrix suction in a linear relationship. This phenomena has been observed in both specimen compacted in 0.71 void ratio initially and 0.92. Furthermore, *k* values in soil specimens by 0.71 initial void ratio are more than that in sample by 0.92 initial void ratio. The measured *k* amounts have been belonged to a state between active and at-rest condition.

References

1. K.v. Terzaghi, Eng. News-Rec., **85**(14), 632-637 (1920)
2. H. Chi-in, China Civil Eng. J., **2**, 002 (1957)
3. W. Kjellman, *Proceedings of the 1st International Conference on Soil Mechanics and Foundation Engineering*. (1936)
4. Y. Okochi, and F. Tatsuoka, Soil and Fund., **24**(3), 52-68 (1984)
5. J. Chu, and S.C.R. Lo, *On the implementation of strain path testing*, University College, Australian Defence Force Academy, University of New South Wales (1990)
6. J. Chu, S.C.R. Lo, and I.K. Lee, J. Geotech. Eng.-ASCE, **118**(2), 191-208 (1992)
7. G. Mesri, and T. Hayat, Can. Geotech. J., **30**(4), 647-666 (1993.)
8. Chu, J. and C. Gan, Geotechnique, **54**(4), 285-288 (2004)
9. S. Northcutt, and D. Wijewickreme, Can. Geotech. J., **50**(5), 457-466 (2013)
10. R. Zhang, J.L. Zheng, and H.P. Yang, *Recent Advancement in Soil Behavior, in Situ Test Methods, Pile Foundations, and Tunneling@ sSelected Papers from the 2009 GeoHunan International Conference*, ASCE (2009)
11. S. Oh, et al., J. Geotech. Geoenviron., 2012.
12. G.E.J. Poinern, *A Laboratory Course in Nanoscience and Nanotechnology*, CRC Press (2014)
13. A. Mirzaii, and S.S. Yasrobi, Transport porous med., **95**(3), 497-505.(2012)
14. A. Mirzaii, and S. Yasrobi, Géotechnique Lett., **2**(October-December), 193-198 (2012)

Bearing capacity of strip footings on unsaturated soils by the slip line theory

Thanh Vo[1,a] and Adrian Russell[1]

[1]*Centre for Infrastructure Engineering and Safety, School of Civil and Environmental Engineering, The University of New South Wales, Sydney, NSW, Australia*

Abstract. The bearing capacity of strip footings on unsaturated soils is studied using slip line theory. The suction profiles considered are non-uniform with depth and correspond to vertical flow of water by infiltration or evaporation. The slip line theory assumes a plastic equilibrium state of a Mohr-Coulomb soil in which suction influences are included using the effective stress concept. This paper shows that the influence of a non-uniform suction profile on bearing capacity is significant. When the contribution of suction to the effective stress can be approximated as a function that varies linearly with depth, the effects of suction on bearing capacity can be represented in dimensionless form in separate charts for smooth and rough footings. Using the charts the bearing capacity can be determined for any combination of friction angle, footing width, surcharge, soil unit weight and linear profiles of cohesion and the contribution of suction to the effective stress.

1 Introduction

Foundations are ubiquitous to geotechnical engineering. Current design tools for estimating ultimate bearing capacity are only applicable when foundation soils are fully saturated or dry. Shallow foundations, however, operate in the upper fewer meters of the ground, often above the water table where soils are variably saturated and may never become fully saturated. As a result the true margins of safety cannot be known for shallow foundations on unsaturated soils using existing tools.

The significant influence that suction can have on bearing capacity has been recognised in the literature. Meyerhof [1] and Vesic [2] recommended suction be treated as an apparent cohesion and shallow foundations be designed based on the highest possible ground water table. Vesic [2] suggested that when the distance between the ground water table and the base of the foundation is less than the foundation width then the soil unit weight should be modified to account for suction above the water level. More recently Oloo et al. [3] and Costa et al. [4] incorporated suction into Terzaghi-type bearing capacity calculations by treating it as an apparent cohesion that was constant with depth. A suction related term, in addition to cohesion, contributed to bearing capacity when scaled by the bearing capacity factor N_c. Xu [5] observed in experiments a suction dependant increase in bearing capacity in unsaturated expansive soils. Jahanandish et al. [6] accounted for this increase in an analysis using zero extension line theory. Again suction was treated as an apparent cohesion constant with depth that contributed to bearing capacity through N_c. It is recognised that suction and apparent cohesion have similar but separate effects on stress equilibrium.

Never before has the effect of a non-uniform suction profile with depth been considered in a bearing capacity analysis of frictional soil. Suction may be highest at very shallow depths where soil interacts with the environment and evaporation may occur. But suction may decay with depth and vanish all together the ground water table. This non-uniformity introduces uncertainties in to how suction may influence bearing capacity and to what extent it may be relied upon during the service life of the footing.

The slip line theory [7-9] is widely used to calculate the ultimate bearing capacity of a strip footing. It involves assuming limit stresses are reached at every point in the soil foundation while an equilibrium stress state is maintained. This avoids use of arbitrarily assumed failure mechanisms. The aim of this paper is to study the influence of steady state suction profiles, set up through evaporation or infiltration, on the bearing capacity of a surface strip footing using the slip line theory.

2 Slip line governing equations

2.1 Effective stress and failure criterion

The governing equations are formulated based on the effective stress concept for unsaturated soils [10]. The effective stress (σ') is defined as:

$$\sigma'=\sigma-u_a+\chi(u_a-u_w) \qquad (1)$$

[a] Corresponding author: thanh.vo@unsw.edu.au

where $\sigma \equiv$ total stress, $u_a \equiv$ pore air pressure, $u_w \equiv$ pore water pressure and $\chi \equiv$ effective stress parameter. Eq. 1 can be rewritten as:

$$\sigma' = \sigma_n + \chi s \qquad (2)$$

where $s = u_a - u_w$ and $\sigma_n = \sigma - u_a \equiv$ net stress, that is the total stress in excess of pore air pressure. When pore air pressure is equal to atmospheric pressure and is taken as the pressure datum, the net stress and total stress become the same. The subscript n may then be removed to simplify notation ($\sigma' = \sigma + \chi s$).

The expression for χ when the hydraulic state of the soil is located on the main drying or the main wetting branch of the soil-water characteristic curve is (Khalili and Khabbaz, [11]):

$$\chi = 1,\ s \leq s_e \qquad (3.1)$$
$$\chi = (s/s_e)^{-0.55},\ s > s_e \qquad (3.2)$$

where $s_e \equiv$ suction value separating saturated from unsaturated states ($s_e = s_{ae} \equiv$ air entry suction and applies when the hydraulic state is on the main drying branch, $s_e = s_{ex} \equiv$ air expulsion suction and applies when the hydraulic state is on the main wetting branch). Eq. 3 is appealing because it is simple and has been validated extensively using experimental data.

The Mohr-Coulomb failure criterion may be written as:

$$\sigma_{xx} = [(1 + \sin\varphi' \cos 2\theta)\sigma_m' - c'\cot\varphi'] - \chi s \qquad (4.1)$$
$$\sigma_{yy} = [(1 - \sin\varphi' \cos 2\theta)\sigma_m' - c'\cot\varphi'] - \chi s \qquad (4.2)$$
$$\sigma_{xy} = \sin\varphi' \sin 2\theta \sigma_m' \qquad (4.3)$$

where $\sigma_{xx}, \sigma_{yy} \equiv$ normal stresses in the x and y directions, respectively, $\sigma_m' \equiv$ effective mean stress, $\sigma_{xy} \equiv$ shear stress, $c' \equiv$ soil cohesion, $\varphi' \equiv$ soil friction angle and $\theta \equiv$ angle between the vertical axis and the major principal stress direction. Here φ' is taken to be a material constant for simplicity (ignoring the slight dependence it may have on soil density, confining pressure, loading history and suction). c' and χs will be set to be zero, constant or vary with depth as will be clearly indicated in the paper.

2.2 Stress equations

The coordinates system used by Sokolovski [9] is adopted in this paper where $x \equiv$ vertical direction and $y \equiv$ horizontal direction (Figure 1). The stress components in Eq. 4 can be differentiated with respect to distance and substituted into the static equilibrium equation. After some simplifications two families of stress characteristic curves are obtained: the η family defined by Eq. 5 and the ξ family defined by Eq. 6:

$$dy = \tan(\theta + \mu)dx \qquad (5.1)$$
$$d\sigma_m' + 2\tan\varphi' \sigma_m' d\theta = C_1 \qquad (5.2)$$
$$dy = \tan(\theta - \mu)dx \qquad (6.1)$$
$$d\sigma_m' - 2\tan\varphi' \sigma_m' d\theta = C_2 \qquad (6.2)$$

where $C_1 = [\gamma_t + \partial(c'\cot\varphi' + \chi s)/\partial x](dx - \tan\varphi' dy)$ and $C_2 = [\gamma_t + \partial(c'\cot\varphi' + \chi s)/\partial x](dx + \tan\varphi' dy)$, $\gamma_t \equiv$ soil unit weight and $\mu = \pi/4 - \varphi'/2$. To arrive at Eqs. 5 and 6, it has been assumed cohesion and suction do not change in the horizontal direction. The term $\partial(c'\cot\varphi' + \chi s)/\partial x$ in Eqs. 5 and 6 indicates that $c'\cot\varphi'$ and χs have similar but separate effects on stress equilibrium.

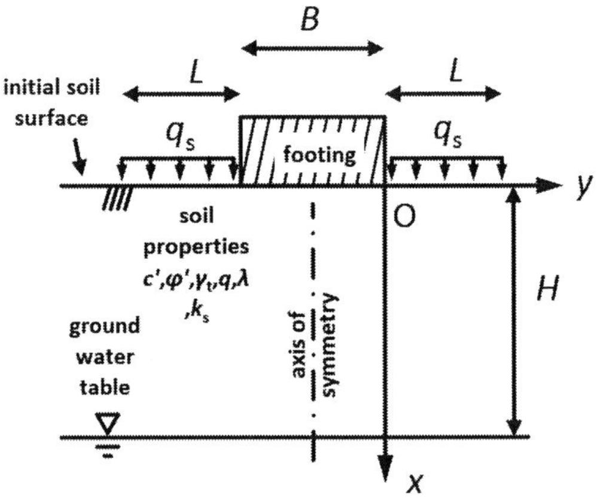

Figure 1. Problem geometry and parameters

2.3 Suction function and water pressure profile

A steady state suction function was derived by Lu and Griffiths [12] by combining Darcy's law [13, 14] for vertical unsaturated flow and Gardner's permeability model [15]. A modified version of this function is used here. The modification ensures that a saturated condition is achieved when $s \leq s_e$, contrary to it being achieved for $s = 0$ in the original formulation. The permeability model used is:

$$k = k_s,\ s \leq s_e \qquad (7.1)$$
$$k = k_s \exp[A_1(1 - s/s_e)],\ s > s_e \qquad (7.2)$$

where $k \equiv$ unsaturated hydraulic conductivity, $k_s \equiv$ saturated hydraulic conductivity and $A_1 \equiv$ a fitting constant ($A_1 = 1$ is adopted in this paper). Eq. 7 is combined with Darcy's law for vertical unsaturated flow to obtain:

$$s = \gamma_w,\ s \leq s_e \qquad (8.1)$$
$$s = s_e - (s_e/A_1)\ln\{(1 + q/k_s)\exp[A_1(1 - \gamma_w \bar{x}/s_e)] - q/k_s\},\ s > s_e \qquad (8.2)$$

in which $\bar{x}$ is the height above a horizontal ground water table (Figure 2). The water table is at a distance H below the soil surface and $\bar{x} = H - x$.

As Eq. 8 is relatively simple with physically intuitive parameters it is adopted in this paper to compute steady state suction profiles typical of sands and silts. The influence of the suction profiles on stress equilibrium can be found by calculating $\partial(c'\cot\varphi'+\chi s)/\partial x$ in Eq. 6. For constant c' it is found that $\partial(c'\cot\varphi'+\chi s)/\partial x=$

$$\partial(\chi s)/\partial x=-\gamma_w,\ s\leq s_e \text{ where } x\geq H-s_e/\gamma_w \quad (9.1)$$

$$\partial(\chi s)/\partial x=C_3/C_4,\ s>s_e \text{ where } x<H-s_e/\gamma_w \quad (9.2)$$

with $C_3=-0.45(1+q/k_s)\gamma_w A_2$ and
$C_4=[A_2(1+q/k_s)-q/k_s]\{1-(1/A_1)\ln[A_2(1+q/k_s)-q/k_s]\}^{0.55}$
in which $A_2=\exp\{A_1[1-\gamma_w(H-x)/s_e]\}$.

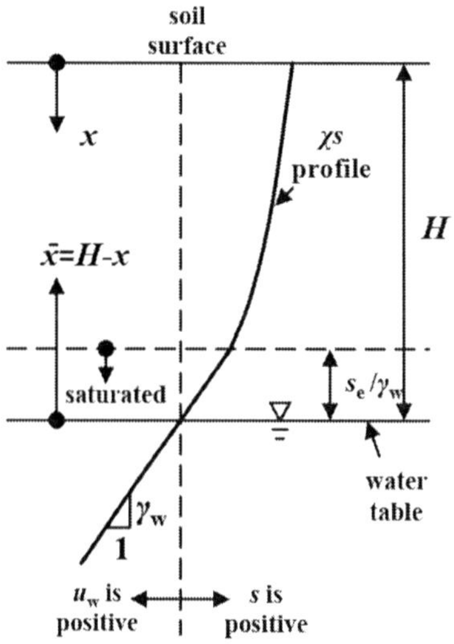

Figure 2. A typical χs profile in soil

3 The strip footing problem

3.1. Problem definition

A typical strip footing is shown in Figure 1. Only the case when the footing load is larger than the surcharge load is considered here which Sokolovski [9] referred to as the heaving problem. The suction profile in the unsaturated region is assumed to be at a steady state. The water pressure in the saturated region is assumed hydrostatic and the effect of water exchange with the unsaturated region is assumed negligible. The unsaturated flow occurs in the vertical direction only and may be either an evaporative (positive) flow for which the soil's hydraulic state is assumed to always be on the main drying curve or infiltrative (negative) flow for which the hydraulic state is assumed to always be on the main wetting curve. The effects of soil deformation and changes to soil structure on s_e are assumed negligible. Because the problem is symmetric about a vertical line (shown in Figure 1) it is possible to work with one-half of the geometry. The right-half of the geometry is referred to in this paper.

3.2 Boundary conditions

At the soil surface away from the footing contact interface the two variables σ_m' and θ can be written as:

$$\sigma_m'=(q_s+\chi s+c'\cot\varphi')/(1+\sin\varphi'\cos 2\theta) \quad (10)$$

$$\theta=\pi/2 \quad (11)$$

The singular point O (shown in Figure 1) is treated as a characteristic curve in the limit extending from point O1 on the soil surface to point O2 at the edge of the footing. At a point M on this curve the variable σ_m' is evaluated as:

$$\sigma_{m'(M)}=\sigma_{m'(O1)}\exp[2\tan\varphi'(\theta_{(O1)}-\theta_{(M)})] \quad (12)$$

At the soil-footing interface:

$$\theta=-\Delta/2-\delta'/2 \quad (13)$$

where $\Delta=\sin^{-1}(\sin(\delta'/\sin\varphi'))$. For rough footings $\delta'=\varphi'$, $\Delta=\pi/2$ and $\theta=-\pi/4-\varphi'/2$. For smooth footings $\delta'=0$, $\Delta=0$ and $\theta=0$. For $0<\delta'<\varphi'$, θ depends on the adopted δ'. Another boundary condition can be derived for the soil-footing interface from the coordinates system (Figure 1). For a smooth footing:

$$x=0 \quad (14)$$

along the entire interface. For $0<\delta'\leq\varphi'$, Eq. 14 only applies to part of the footing width [16, 17].

It has been widely discussed [18] that plastic yielding (according to the assumed Mohr-Coulomb failure criterion) cannot occur at the axis of symmetry (Figure 1) for $0<\delta'\leq\varphi'$. Different approaches have been adopted to circumvent the problem of not having zero shear stress on the axis of symmetry for $0<\delta'\leq\varphi'$. The approach adopted here is similar to Davis and Booker [16] and Martin [17] in assuming that a soil wedge is trapped between the plastic zone and the footing. A stress field is searched so that $\theta=0$ is satisfied on the axis of symmetry. The exact shape and extent of the wedge is governed by the problem parameters. It extends to the edge of the footing or to a certain point on the footing.

3.3 Solution procedures

A standard finite difference scheme is used to discretise the governing equations (Eq. 5 and Eq. 6) in the same way as Sokolovski [9]. Details of numerical implementation for dry soils can be found in references [9, 17]. For unsaturated soils the variables (σ_m',θ,x,y) at each grid point of the finite difference mesh need to be updated with Eq. 9 until convergence. It is also necessary to adopt an adaptive subdivision strategy to control the level of mesh distortion in the fan zone. An adaptive subdivision strategy similar to Martin [17] is used here.

4 Results of slip line analyses

Results are presented in total normal stresses (denoted $\sigma_{xx\_interface}$) on a soil-footing interface. For simplicity a soil unit weight γ_t=17 kN/m^3 is adopted. It will be shown later that γ_t has a significant effect on bearing capacity.

Typical parameters of sands (φ'=40°,45°) and silts (φ'=25°,30°) whilst unsaturated are used to compute the

χs profiles for evaporation (q=1.15×10^{-8} m/s) and infiltration (q=-3.14×10^{-8} m/s), as listed in Table 1.

Table 1. Parameters adopted in analyses.

φ' (°)	s_e (kPa)	k_s (m/s)	q (m/s)
25	15 (infiltration)	1×10^{-7}	-3.14×10^{-8}
25	75 (evaporation)	1×10^{-7}	1.15×10^{-8}
30	15 (infiltration)	1×10^{-7}	-3.14×10^{-8}
30	75 (evaporation)	1×10^{-7}	1.15×10^{-8}
40	2 (infiltration)	3×10^{-4}	-3.14×10^{-8}
40	10 (evaporation)	3×10^{-4}	1.15×10^{-8}
45	2 (infiltration)	3×10^{-4}	-3.14×10^{-8}
45	10 (evaporation)	3×10^{-4}	1.15×10^{-8}

For H=10 m the corresponding χs profiles computed using Eqs. 3 and 8 are shown in Figure 3a for silt and Figure 3b for sand. They show that the χs profiles may be approximated by linear functions.

4.1. Smooth and rough footings

Figure 4a and Figure 4b show distributions of $\sigma_{xx\_interface}$ for a two meters wide smooth strip footing. H=10 m is used. Figure 5a and Figure 5b show distributions of $\sigma_{xx\_interface}$ for a rough footing, B=2 m and H=10 m. $\sigma_{xx\_interface}$ for the rough footing is higher than that for the smooth footing.

The results show that $\sigma_{xx\_interface}$, and thus the bearing capacity, is strongly linked to whether the sand or silt is undergoing evaporation or infiltration. This is due to a significant difference in the χs profiles (Figure 3).

4.2. Influence of footing width

The influence of B is investigated by repeating the analyses in section 4.1 for a typical sand (φ'=40°) and silt (φ'=25°) in steady state evaporation. B=6 m and B=18 m are considered along with B=2 m. The results are shown in Figure 6.

Figure 6 shows that the results for a particular unsaturated soil with a certain φ' value but for different footing widths do not coincide when presented in $\sigma_{xx\_interface}/(\gamma_t B)$~$y/B$, contrary to what is observed for dry soils [19].

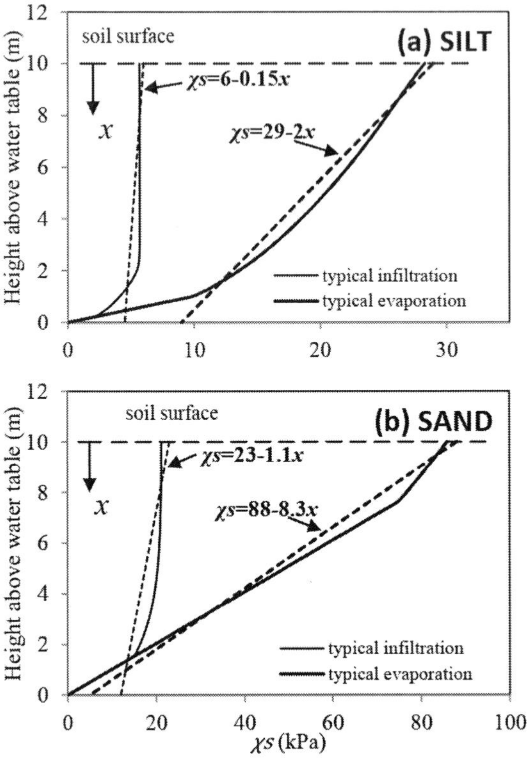

Figure 3. Typical infiltration and evaporation in sand and silt

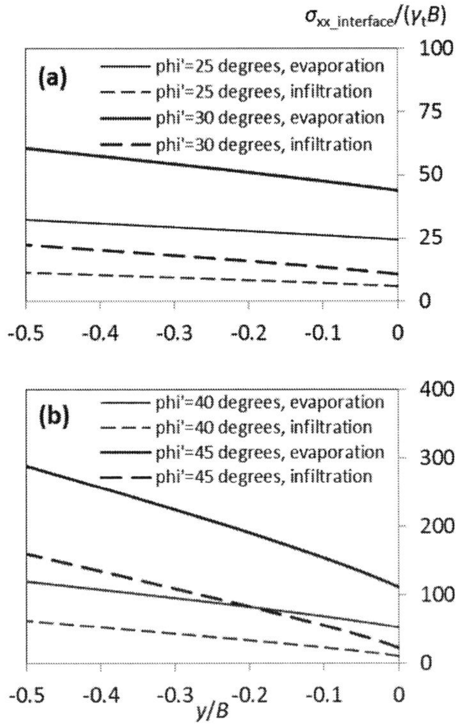

Figure 4. Total normal stress at a smooth soil-footing interface

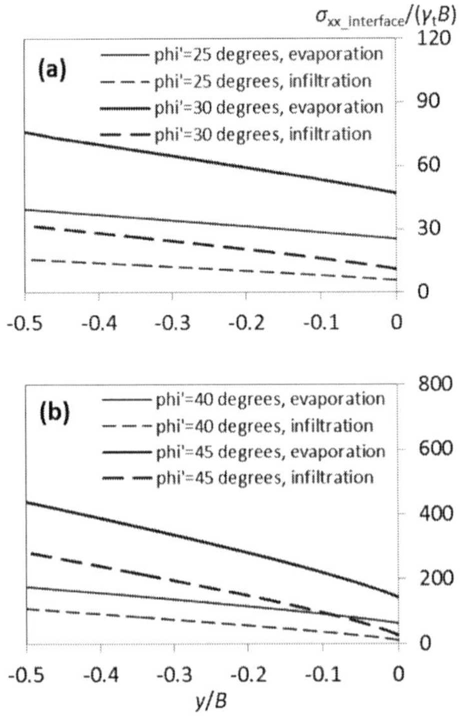

Figure 5. Total normal stress at a rough soil-footing interface

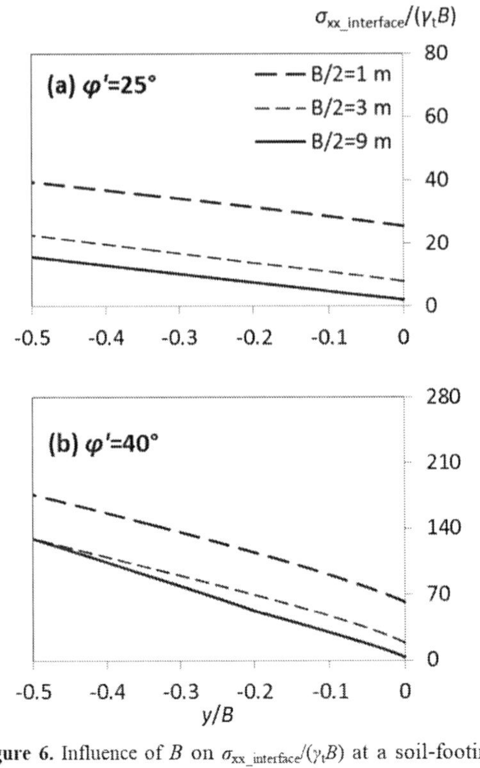

Figure 6. Influence of B on $\sigma_{xx\_interface}/(\gamma_t B)$ at a soil-footing interface

The results for unsaturated silt and evaporation are most divergent. The amount of divergence depends on χs and its variation with depth, which is scale dependent. To understand this it is helpful to refer to the results of Davis and Booker [20] who studied the influence of a linearly varying cohesion profile on the bearing capacity of clays, as cohesion has a similar influence to $\chi s/\cot\varphi'$. Davis and Booker [20] identified $c_0'/(K_c B)$ (where $c_0' \equiv$ soil cohesion at soil surface and $K_c \equiv$ rate of cohesion increase with depth) to be the variable controlling the bearing capacity. This can be shown readily by formulating the governing equations in non-dimensional form [9, 17]. It can be inferred from this that the divergence of the results in Figure 6 is due to χs at the soil surface and $\partial(\chi s)/\partial x$ being different for each case. An alternate presentation is needed to see coincidence of results which is discussed below.

5 Incorporating a suction profile into bearing capacity calculation

The analyses completed above involve χs profiles which are nonlinear with depth. Here it is shown how linear approximations to those profiles may be incorporated into a bearing capacity calculation.

Consider first dry soils in which φ' and γ_t are constant but have a cohesion profile that varies linearly with depth according to $c'=c_0'+K_c x$. It was shown in Martin [17] that the dimensionless ratio:

$$V=q_u/(c_0'+q_s\tan\varphi') \quad (15)$$

is dependent on the dimensionless factor:

$$F=(K_c B+\gamma_t B\tan\varphi')/(c_0'+q_s\tan\varphi') \quad (16)$$

and the values of φ'. To incorporate suction influences it is recognised that $c'\cot\varphi'$ and χs have similar (and independent) effects on the governing equations. If a linear suction profile exists defined by:

$$\chi s=(\chi s)_0+K_{\chi s}x \quad (17)$$

where $(\chi s)_0$ is the value of χs at the ground surface and $K_{\chi s}=\partial(\chi s)/\partial x$ is a constant, along with a linearly varying cohesion profile with depth, and constant φ' and γ_t, then a more general form of the ratio V becomes:

$$V=q_u/(c_0'+(\chi s)_0\tan\varphi'+q_s\tan\varphi') \quad (18)$$

and this must be dependent on a modified F:

$$F=[(K_c+K_{\chi s}\tan\varphi'+\gamma_t\tan\varphi')B]/(c_0'+(\chi s)_0\tan\varphi'+q_s\tan\varphi') \quad (19)$$

along with φ'. The parameters $(\chi s)_0$ and $K_{\chi s}$ may be estimated using linear fits to the suction profiles arrived at using Eqs. 3 and 8 as shown in Figure 3. The errors associated with using a linear approximation for the χs profiles rather than the curved profiles are small.

The unique relationships between V and F for different φ' values are shown graphically in Figure 7a and Figure 7b for smooth and rough footings, respectively.

This figure can be used to determine q_u for any combination of φ', c_0', K_c, B, $(\chi s)_0$, $K_{\chi s}$, q_s and γ_t.
In Figure 7, as $F \to 0$, $V \to N_c$ and as $F \to \infty$, $V \to [N_\gamma/(2\tan\varphi')]F$. These limiting conditions are represented by the straight dash lines.

6 Conclusion

The bearing capacity of a shallow strip footing in unsaturated soils has been investigated by the slip line method. The stress field governing equations are formulated based on the effective stress concept for unsaturated soils. Bearing capacity analyses are carried out for typical sands and silts in steady state evaporation and infiltration.. When the contribution of suction to effective stress varies linearly with depth the influence of suction can be captured by plotting results in dimensionless form. These plots are a useful tool for practitioners to estimate bearing capacity.

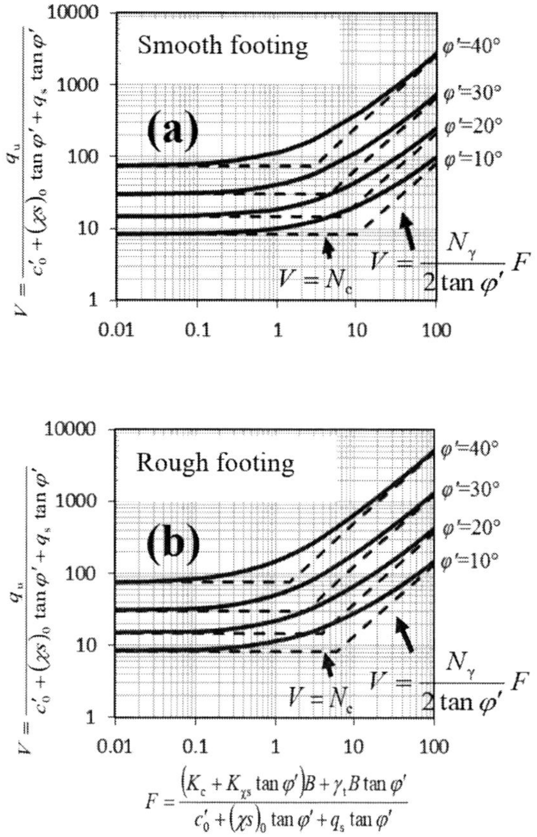

Figure 7. The unique relationship between V and F for different φ' values

References

[1] Meyerhof GG. Influence of roughness of base and ground water conditions on the ultimate bearing capacity of foundations. Géotechnique. 1955;5:227-42.

[2] Vesic A. Analysis of ultimate loads of shallow foundations. Journal of the Soil Mechanics and Foundations Division, ASCE. 1973;99:45-73.

[3] Oloo SY, Fredlund DG, Gan JK-M. Bearing capacity of unpaved roads. Canadian Geotechnical Journal. 1997;34:398-407.

[4] Costa YD, Cintra JC, Zornberg JG. Influence of matric suction on the results of plate load tests performed on a lateritic soil deposit. Geotechnical Testing Journal. 2003;26:219-26.

[5] Xu Y. Bearing capacity of unsaturated expansive soils. Geotechnical and Geological Engineering. 2004;22:611-25.

[6] Jahanandish M, Habibagahi G, Veiskarami M. Bearing capacity factor, Nγ , for unsaturated soils by ZEL method. Acta Geotechnica. 2010;5:177-88.

[7] Prandtl L. Über die Häerte plastischer Köerper. Nachrichten von der Gesellschaft der Wissenschaften zu Göttingen, Mathematisch-Physikalische Klasse. 1920;12:74-85.

[8] Reissner H. Zum erddruck problem. The 1st International Congress on Applied Mechanics. Delft1924. p. 295-311.

[9] Sokolovski WV. Statics of Soil Media. 2nd ed. London: Butterworths, 1954.

[10] Bishop AW. The principle of effective stress. Teknisk Ukeblad. 1959;106:859-63.

[11] Khalili N, Khabbaz MH. A unique relationship for chi for the determination of the shear strength of unsaturated soils. Géotechnique. 1998;48:681-7.

[12] Lu N, Griffiths DV. Profiles of steady-state suction stress in unsaturated soils. Journal of Geotechnical and Geoenvironmental Engineering. 2004;130:1063-76.

[13] Buckingham E. Studies on the movement of soil moisture. Bureau of Soils Bulletin 38. Washington, USA: U.S. Deparment of Agriculture, 1907.

[14] Richards LA. Capillary conduction of liquids through porous mediums. Physics. 1931;1:318.

[15] Gardner WR. Steady state solutions of the unsaturated moisture flow equation with application to evaporation from a water table. Soil Science. 1958;85:228-32.

[16] Davis EH, Booker JR. The bearing capacity of strip footings from the standpoint of plasticity theory. The 1st Australian New Zealand Conference in Geomechanics. Melbourne1971. p. 276-82.

[17] Martin CM. User guide for ABC - analysis of bearing capacity version 1.0. Oxford: University of Oxford, 2004.

[18] Martin CM. Exact bearing capacity calculation using the method of characteristics. In: Barla G, Barla M, editors. The 11th Conference of IACMAG. Turin, Italy2005. p. 441-50.

[19] Kumar J. The variation of Nγ with footing roughness using the method of characteristics. International Journal for Numerical and Analytical Methods in Geomechanics. 2009;33:275-84.

[20] Davis EH, Booker JR. The effect of increasing strength with depth on the bearing capacity of clays. Géotechnique. 1973;23:551-63.

 DOI: 10.1051/e3sconf/20160905005

Numerical analysis of energy piles under different boundary conditions and thermal loading cycles

Ali Khosravi[1,a], Aria Moradshahi[2], John S. McCartney[3], and Mirmohammadreza Kabiri[4]

[1] *Assistant Professor, Civil Engineering Department, Sharif University of Technology, Tehran, Iran*
[2] *Graduate Student, Civil Engineering Department, Sharif University of Technology, Tehran, Iran*
[3] *Associate Professor, Department of Structural Engineering, University of California San Diego, La Jolla, USA*
[4] *Ph.D., Dept. of Civil, Environmental, and Architectural Engineering, Univ. of Colorado Boulder, Boulder, USA*

Abstract. The thermo- mechanical behavior of energy piles has been studied extensively in recent years. In the present study, a numerical model was adapted to study the effect of various parameters (e.g. heating/cooling temperature, head loading condition and soil stiffness) on the thermo-mechanical behavior of an energy pile installed in unsaturated sandstone. The results from the simulations were compared with measurements from a thermal response test on a prototype energy pile installed beneath a 1-story building at the US Air Force Academy (USAFA) in Colorado Springs, CO. A good agreement was achieved between the results obtained from the prototype and the numerical models. A parametric evaluation were also carried out which indicated the significance of the stiffness of the unsaturated sandstone and pile's head loading condition on stress-strain response of the energy pile during heating/cooling cycles.

1 Introduction

In recent years, reinforced concrete piles have been widely used as geothermal heat exchangers to access the relatively constant temperature of the ground for efficient heating and cooling of buildings [1-15]. Deformation in energy piles is a complex process due to interaction between soil and pile and the effects of temperature change on the thermal expansion and contraction of the pile and surrounding subsurface. Deformations in this case may occur due to initial mechanical loading associated with construction of the overlying building, and thermo-elastic expansion and contraction of the reinforced concrete during heating and cooling, as well as settlement or heave of the surrounding subsurface.

Reviewing the literature, it was found that some aspects of soil-pile behavior have gained less attention than their actual impact in energy piles. Considering the fact that a pile foundation may be subjected to different vertical loads, it is essential to consider the effect of overlying structures on thermo-mechanical response of an energy pile. Additionally, strength properties of soil surrounding the pile seems to have an influence on lateral and axial deformations [17-21]. This study seeks to examine these two aspects by means of a 2D finite difference analysis to predict the deformation behavior of an energy pile during thermo-mechanical loading.

First, the model was validated based on the results from a series of comprehensive full-scale in-situ tests, investigating the behavior of eight energy piles installed beneath a one-story building at the US Air Force Academy (USAFA) in Colorado Springs, CO. Pile deformations were then evaluated at different temperatures during active heating and ambient cooling. Finally, a parametric study was carried out to examine the effects of different head conditions and surrounding soil strength characteristics followed by a discussion on obtained results.

2 Case study energy pile

2.1 Pile description and instrumentation

The case pile is an energy pile with a circular section of diameter, D, of 0.61 m and length, L, of 15.6 m which was selected from an energy pile testing research project at the Field Engineering and Readiness Laboratory (FERL) of the US Air Force Academy (USAFA), Colorado Springs, CO (Pile 4 in Fig. 1). This pile was part of a supporting system of eight drilled shaft energy piles which were used to support a one-story building. The Young's modulus and coefficient of thermal expansion of the foundation were reported to be 30 GPa and 13 με/°C respectively [12].

The soil profile of the site together with the geometry of the pile, and embedded instrumentation are presented in Figure 2. As presented in this figure, the top layer of the soil system consists of a layer of medium dense, sandy fill with silt and gravel, having a thickness of 1 m

[a] Corresponding author: khosravi@sharif.edu

and a dry unit weight of 18.4 kN/m^3. The second layer is an approximately 1 m thick medium dense sandy-silty gravel layer having a dry unit weight of 19.2 kN/m^3. Below the second layer is Dawson–Arkose bedrock (sandstone), from a depth of approximately 2 m below the surface to the maximum depth of investigation. The foundation unit weight is 25 kN/m^3. The water table was considered to be at a depth greater than 16 m based on the results of exploration.

Different instrumentation was incorporated into the energy pile to investigate the axial strain and stress behavior of the piles during thermo-mechanical loading. Distribution of axial strain of the pile with depth was measured using a set of twelve Geokon Model 4200 vibrating wire strain gauges (VWSGs) (Fig. 2). A series of ten Geokon model 3810 thermistor strings were also used for monitoring temperature variations in the soil surrounding the energy pile.

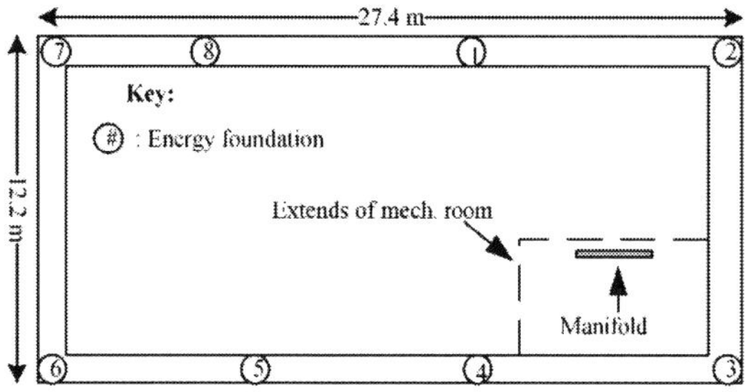

Figure 1. Schematic of the location of the energy piles under the building

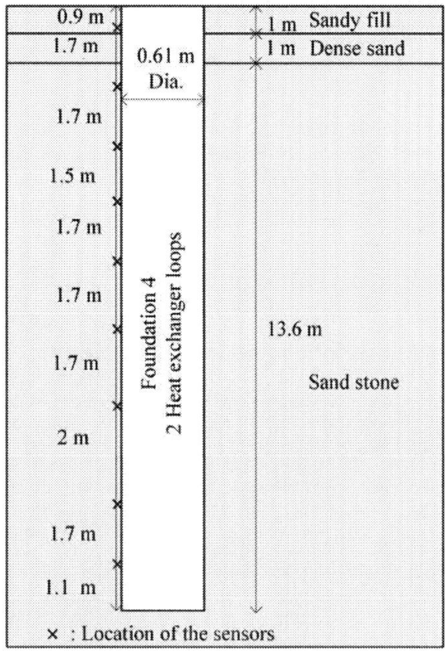

Figure 2. Schematic of soil layers and foundation instrumentation for Foundation 4

Different cycles of heating and cooling were imposed to the foundation using a heating system consisting of a series of loops of HDPE heat exchanger tubing with an average diameter of 20 mm attached to the inside part of the reinforcing cages of the foundations, and a heating unit to heat and circulate a 20 % propylene glycol–water mixture through the heat exchanger loops. More details about the heating system are presented in [16].

The thermal loading process included an initial heating of the foundation for approximately 498 hours, after which fluid circulation in the foundation was stopped and the cooling process began for almost 700-1200 hours. During the heating and cooling cycles, thermo-mechanical behavior of the foundation was obtained for times corresponding to foundation temperature average changes of 6 °C. The profiles of foundation temperature are presented in Figure 3. As shown in this figure, variations in temperature with depth is relatively constant, except for depths below 11 m, where slight changes in the temperature were observed with depth.

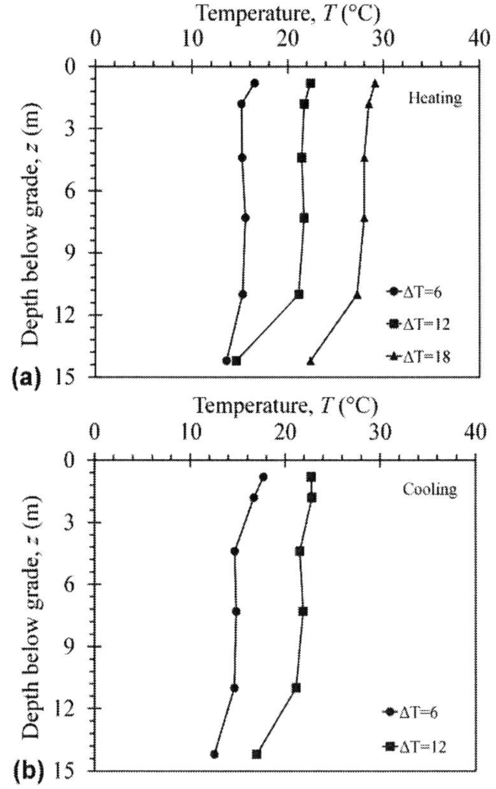

Figure 3. Profiles of foundation temperatures during different cycles of (a) heating; (b) cooling

2.2 Numerical model description

A finite difference method was adopted to simulate an axisymmetric model for energy pile analysis during subsequent cycles of heating and cooling, using the finite difference code FLAC^{2D} version 4. The finite difference method has been reported to be an efficient way to analyze the behavior of energy piles, [8], as it considers complexities arising from boundary conditions, thermo-hydro-mechanic loading, geometry, soil properties and so forth in solving equations that are related to thermo-hydro-mechanical problems. The model and its mesh discretization are presented in Figure 4. As shown in this figure, except a 1 m dense sand overlain by a 1 m sandy fill along the shaft, rest of surrounding soil is unsaturated mostly sandstone. The boundary distances in horizontal and vertical directions, from shaft and pile tip are 10 m and 25 m in the axisymmetric model, respectively. Due to high stress gradients near the pile shaft, smaller mesh dimension were adopted around the pile (25×25 cm within 5 m from the shaft) than those outside this zone. The Mohr-coulomb failure criterion was used for the geomaterial and a fully linear elastic behavior was considered for the energy pile. The properties of the materials used in this study are presented in Tables 1 and 2.

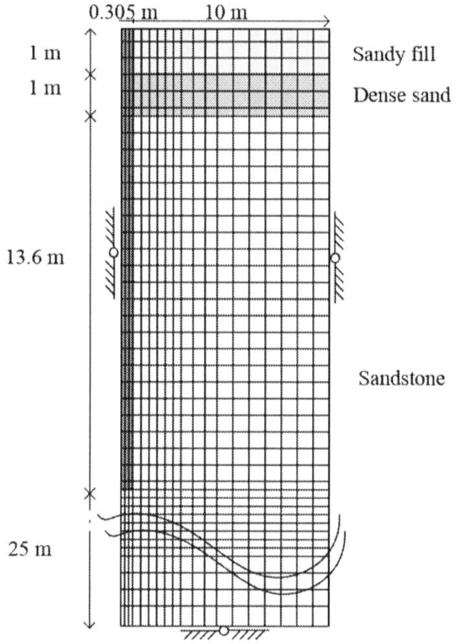

Figure 4. Geometry and 2D mesh discretization of the model

Table 1. Properties of the model

	Sandy fill	Dense sand	Sandstone	Foundation
Bulk Mod. (MPa)	60	90	200	15000
Shear Mod.(MPa)	22.5	33.75	67.55	6500
Thermal conductivity (W/m°C)	1.11	0.8	1.233	1.8
Specific heat extraction (J/kg°C)	800	700	800	900
Coefficient of thermal expansion (×10^{-6} J/°C)	6	5	6	13
Poisson's ratio	0.3	0.25	0.2	0.2

Table 2. Properties of the soil-pile interfaces

	Sandy fill	Dense sand	Sandstone
Normal stiffness (MN/m^3)	140	200	450
Shear stiffness (MN/m^3)	50	120	350
Cohesion (kPa)	100	50	0
Friction (°)	20	30	35

To analyze the thermal behavior of the energy pile, the pile was initially subjected to a vertical mechanical load of about 300 kN due to the dead weight of the building. The energy pile was then subjected to different cycles of temperature changes of ±19°C from an initial temperature of 10°C which is the reported mean ground temperature. Figure 5 shows the thermal loading path considered for the analysis. As shown in this figure, two subsequent cycles of heating and cooling were applied to the pile. During each cycle, the pile temperature was elevated to an average temperature of 29°C from a ground temperature in 500 h and then decreased to the ground temperature in 1200 h.

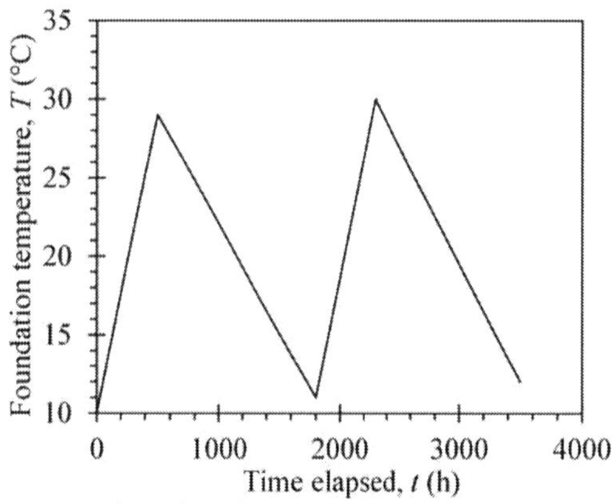

Figure 5. Thermal loading path for two cycles of heating and cooling

3 Results

Figures 6 and 7 demonstrate the results of the analyses in terms of thermal axial strain (Fig. 6a) and stress (Fig. 6b) against depth during heating and cooling cycles, respectively. The filed measurements reported by [12], were also presented in these two figure for comparison purposes. As can be observed, the model provided profiles of stress and strain very similar to in-situ trends at different temperatures. Based on the results presented in these figures, the thermal axial strains became more negative with temperature increase, indicating the expansion of the pile during heating. While during cooling cycles, the thermal axial strains were reduced, showing a contraction trend in the pile. The thermal axial strain profiles of both in-situ and numerical analyses have relatively consistent shapes at different temperatures. As a result of soil–structure interaction, due to the mobilization of shear resistance along sides of the foundation, a nonlinear distribution in thermal strain was observed with depth similar to reports from [2] and [14]. The thermal axial strain profiles showed an initial decrease in magnitude with depth to a depth of about 9 m (the location of the null point), followed by axial strain increase at greater depths. The thermal axial stress experienced an initial increase with depth to a maximum value at a depth of 8-10 (the null point). The stress appeared to decrease at greater depths. The same trends for axial strain and stress with depth were observed during both heating and cooling cycles.

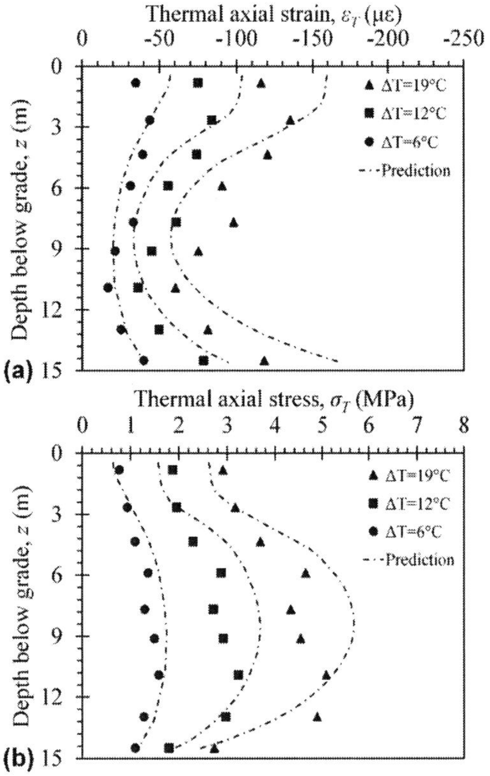

Figure 6. Thermal axial (a) strain; (b) stress; for heating to different temperatures

Thermally induced displacements of the energy pile during heating and cooling cycles are shown in Figure 8. Based on results presented in this figure, for depths above the null point, the pile would experience an upward movement during heating with an uplift in the surface, while the lower part of the pile would experience expansion with a downward movement. During cooling, the behavior was totally different. The energy pile tends to contract above the null point and upper portions of the pile were moving downward while lower parts were moving upward resulting in settlement of the surface. Additionally, as observed in Figure 9, the magnitudes of vertical displacement decreases during the second cycles of heating and cooling.

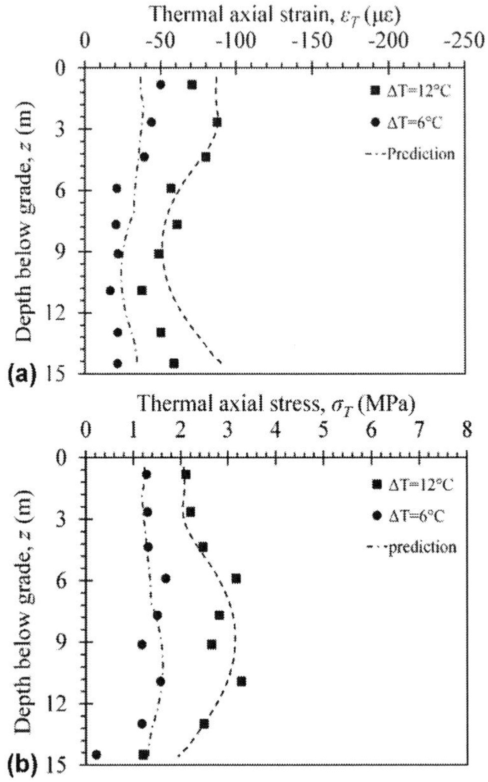

Figure 7. Thermal axial (a) strain; (b) stress; for cooling to different temperatures

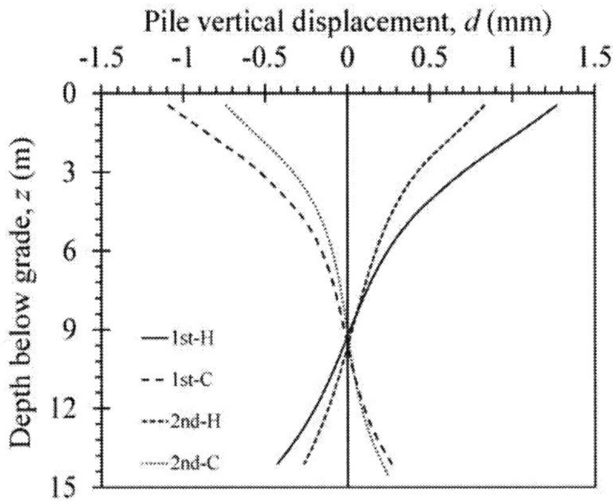

Figure 8. Pile displacement values for heating and cooling cycles

Figure 9 provides the mobilized side shear stresses measured along the soil-pile interface for different cycles of heating and cooling. During the heating cycle, at upper part of the pile lower magnitudes of shear stress were observed while at lower part, higher shear stresses were mobilized. Side shear stresses during the cooling cycle show significantly lower values. The point at which the sign of the mobilized side shear stress changes, shows the location of the null point where the axial thermal stress and strain are maximum and minimum respectively.

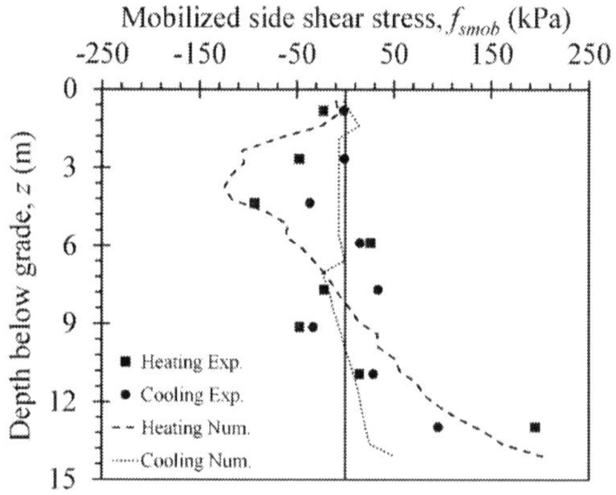

Figure 9. Mobilized side shear stress

4 Parametric evaluation

A parametric evaluation was performed in this study to explore the effect of head-structure stiffness and soil properties on the behavior of an energy pile. The impact of overlying structure was investigated by applying vertical loads, p, with different magnitudes to the pile head, and the stiffness of the surrounding soil was assessed through changing soil's Young modulus, E, of the sandstone layer.

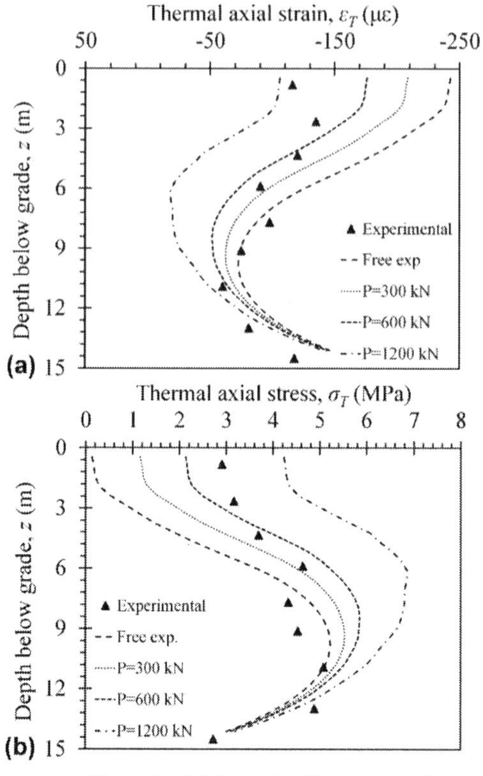

Figure 10. Thermal axial (a) strain; (b) stress; for heating and different weight of the building for heating to 19°C

4.1 Effect of different head loading conditions on pile behavior

The impact of head loading condition on the response of energy piles was investigated in this study by performing an evaluation process of four different values of vertical load. Results of the evaluation process are presented in Figure 10. In this figure, the data points represent the measured in-situ data for a pile without a vertical load and the dashed lines are the predicted stress/strain values.

As presented in this figure, by increasing the vertical load applied to the pile head, a larger thermal axial stress was observed at the top of the pile, while the profiles of strain shifted to the left. The rate of changes in the thermal axial stresses and strains with the load of overlying structure was also different through the depth. For the values of p of 0 to 1200 kN, axial strain at the pile head would range from -250 to -110 με, while no change would happen at the toe where the behavior was mostly controlled by the stiffness of the underlying soil.

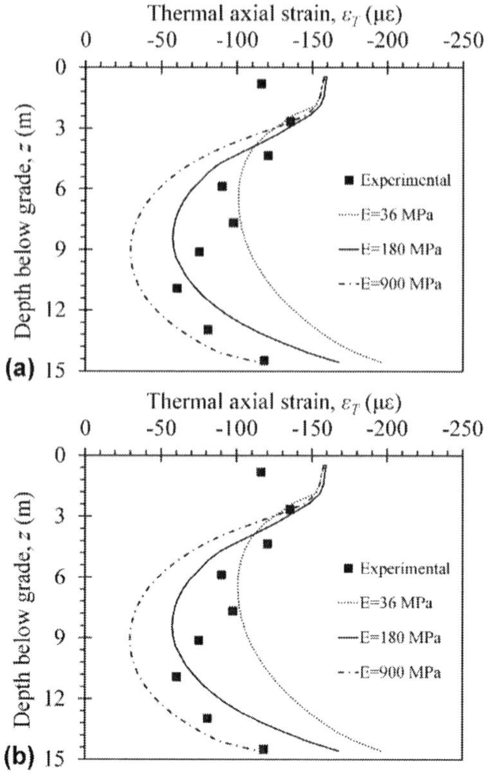

Figure 11. Thermal axial stress and strain profiles for different stiffness of surrounding soil

Based on the results of numerical modeling, the maximum thermal axial stress for the pile that was subjected to p=1200 kN was about 6.8 MPa which was in good agreement with the value of 6.45 MPa reported by [16]. It is clear from this figure that the null point moved upward as the head restraint was increased.

4.2 Effect of surrounding soil stiffness on pile behavior

Results for axial stress and strain are presented in Figure 11 for different values of Young's modulus of the surrounding soil (36 MPa, 180 MPa, and 900 MPa). As observed in this figure, the resistance of the surrounding soil may have a significant influence on the distribution of the thermal axial stress and strain profiles which is almost similar to the effect of overlying structure. As the value of the Young's modulus of the soil increased, higher thermal axial stress and lower thermal axial strain were observed.

5 Conclusion

Energy piles are important elements since they function as load-bearing elements while they are capable of providing thermal energy for buildings. Heating and cooling cycles due to seasonal fluctuations in temperature, weight of the overlying structure, and stiffness of the surrounding soil play important role in affecting the thermo-mechanical behavior of energy piles.

A numerical simulation was conducted using finite difference method to investigate the thermo-mechanical behavior of the energy piles. The study considered different conditions for temperature, head-restraint, and soil stiffness to provide a comprehensive understanding of the factors affecting the pile response. The results of the study proved that, heating and cooling cycles significantly affect the distribution of thermal axial stress and strain. Also during heating cycle, the pile was observed to expand and surface uplift was observed, while during cooling cycle the surface settled due to contraction of the pile. The measured deformations were lower in second cycles of heating and cooling. The study also indicated that the effect of the overlying structure and the surrounding soil stiffness are similar. Increasing the weight of the structure and stiffness of the soil results in higher values for thermal axial stress and lower values for thermal axial strain. The results of the numerical study was completely in agreement with the observed response of a case study energy pile at the US Air Force Academy (USAFA) in Colorado Springs, CO.

Acknowledgement

J. McCartney would like to thank the support from DoD ESTCP project EW-201153 and NSF project CMMI-0928159. The views in this paper are those of the authors alone.

References

1. K.D. Murphy, J.S. McCartney, K.S. Henry, Acta Geotechnica. **10**(2), 179-195 (2015)
2. L. Laloui, M. Nuth, L. Vulliet, Int. J. Nem. Anal. Meth. Geomech. **30**, 763-781 (2006)
3. M.E. Suryatriyastuti, H. Mroueh, S. Burlon, Com. And Geotech. **55**, 378-391 (2014)
4. B.L. Amatya, K. Soga, P.J. Bourn-webb, T. Amis, L. Laloui, Geotech. **62**(6), 503-519 (2012)
5. D. Adam, R. Markiewicz, Geotech. **59**(3), 229-236 (2009)
6. S. O'Connell, S.F. Cassidy, Int. Geoth. Con. *Reykjavik*, 12-18 (2003)
7. H. Brandl, Geotech. **56**(2), 81-122 (2006)
8. M.E. Suryatriyastuti, H. Mroueh, S. Burlon, Renew. And Sust. Ene. Rev. **16**, 3344-3354 (2012)
9. T. Mimouni, L. Laloui, Acta Geotech. **9**(3), 355-366 (2014)
10. L. Laloui, A. Di Donna, *Proceeding of the ICE-Civil. Eng.* **64**(4), 184-191 (2011)
11. C. Knellwolf, H. Peron, L. Laloui, J. Geotech, Geoenviron. Eng. **137**(12), 890-902 (2011)
12. J.S. McCartney, K.D. Murphy, DFI. J. **6**(2), 28-36 (2012)
13. J. Gao, X. Zhang, J. Liu, K. Li, J. Yang, Appl. Ene. **85**(10), 901-910 (2008)
14. P.J. Bourne-Webb, B. Amatya, K. Soga, T. Amis, C. Davidson, P. Payne, Geotech. **59**(3), 237-248 (2009)
15. W. Wang, R. Regueiro, J.S. McCartney, Geotech. and Geolog. Eng. **33**(2), 373-388.
16. K.D. Murphy, J.S. McCartney, Geotech. and Geolog. Eng. **33**(2), 343-356 (2015)
17. C.G. Olgun, T.Y. Ozudogru, S.L. Abdelaziz, A. Senol, Acta Geotech. **10**(5), 553-569 (2014a)
18. C.G. Olgun, J.S. McCartney, J. Deep piles Inst. **8**(2), 58-72 (2014)
19. C.J. Wood, H. Liu, S.B. Riffat, Geotech. **59**(3), 287-290 (2009)
20. O. Ghasemi-Fare, P. Basu, Ene. and Buil. **66**, 470-479 (2013)
21. K.D. Murphy, J.S. McCartney, Geotech. and Geolog. Eng. **33**(2), 343-356 (2015)

Effect of plastic soil on a retaining wall subjected to surcharge loading

Khawla Al-Juari[1,a], Suhail Khattab and Mohamed Al-Shamam
[1]University of Mosul, Civil Engineering Department, Iraq

Abstract. The seasonal variation and climatic changes play a significant role that affects the stresses exerted on a retaining wall, and the state of stresses in the soil mass behind the wall especially for highly expansive soil. These stresses resulted in the wall moving either away or towards the soil. In this study, a laboratory physical model of the retaining wall formed of a box having (950×900×600) mm dimensions with one side representing the wall being developed. After the soil being laid out in the box in specified layers, specified conditions of saturation and normal stresses were applied. The wall is allowed to move horizontally in several distances (0.1, 0.2, 0.3, 0.6, 0.8, 1.0 , 2.0, 3.0 and 4.0) mm, and the stresses being measured, then the vertical loading was released. The main measured variables during the tests are; the active and passive earth pressures, vertical movement of the soil, total suction and time. Results showed that the lateral earth pressure along the depth of the wall largely decreased when wall moved away from the soil. Total suction was slightly affected during wall's movement. At unloading stage, the lateral earth pressure decreased at the upper half of wall height, but increased at the other wall part. Total suction was increased at all depths during this stage.

1 Introduction

Retaining walls mainly used as lateral support to soil and rock. Cantilever retaining wall is one of the common types of retaining structures. Typical cantilevered walls are T-shaped, L-shaped, or reverse L-shaped. The pressures exerted on these structures can be divided into three categories: at rest, active and passive earth pressures. The direction of wall's movement is the important factor that limits the type of the pressure exerted on retaining structures. Active and passive cases generated when the wall moved away and toward the soil respectively. Whereas at rest case generated when the wall was constructed in a rigid condition and no movement occurred.

Generally, the traditional method in the design of the retaining wall with granular material as a backfill was used, due to simplicity in calculations with no chemical interaction between the soil and the wall and between the soil particles themselves. The effect of lateral swelling pressure exerted on the retaining wall when the clay soil is used as a backfill material was also ignored. Where the lateral swelling pressure is equal to seven times the lateral pressure of the overburden soil without clay [12]. The lateral swelling pressure is more than the vertical swelling pressure and vertical applied stresses [1, 3, 8, 15, 18]. Consequently, the damage and distress observed in the retaining wall and walls of the building is mainly due to lateral swelling pressure [5, 8].

Several studies were conducted on laboratory physical model of a retaining wall using granular soil as a backfill material [4, 6, 9, 10, 13, 17, 19, 20]. Whereas, a very limited studies used clayey soil as a backfill material due to the difficulties in dealing with large quantities of the clayey soil. The lateral swelling pressure increased rapidly to a certain depth of the wall and reduced the lateral swelling pressure transmitted to the wall by different approaches [2, 3,7, 12, 14].

This study is, therefore, investigated and considered several factors; the effect of wall's movement toward and far away from the backfill soil on the magnitude of lateral pressure applied on the wall, vertical deformation of the backfill soil and total suction.

2 Experimental work and testing program

2.1 Soil and methodology

According to the aim of the study, disturbed and undisturbed samples of highly swelling potential are obtained from Al-Majmoua quarter, Mosul city.

The physical properties of the soil used are shown in Table 1.

Table 1. Engineering properties of the soil used

Property		Value
Liquid Limit (%)		89
Plastic Limit (%)		39
Plasticity index (%)		50
Shrinkage limit (%)		12
Linear shrinkage (%)		20.75
Soil classification		CH
Specific gravity		2.8
Total soluble salts (%)		5.6
Organic matter (%)		1.8
Sulphate salts		0.45
Gypsum content (%)		5.12
Sand (%)		14
Silt (%)		34
Clay (%)		52
Max. dry unit weight (kN)/m^3	Standard compaction	14.4
Optimum moisture content OMC (%)	Standard compaction	27.4
Max. dry unit weight (kN)/m^3	Modified compaction	16.8
Optimum moisture content (%)	Modified compaction	13.8

[a] Corresponding author: khawla_19752012@yahoo.com

The soil classified as (CH) according to the unified soil classification system. The X-ray diffraction analysis of the soil showed that the major clay minerals are montmorillonite, palygorskite, and kaolinite. Whereas, the minor clay minerals are chlorite and illite. The non clay minerals are calcite and quartz.

The experimental set up was designed and built by the authors at The University of Mosul, and it is formed of a steel structure and consists mainly of a box 900×600 mm in plain and 950 mm in height. Two sides of the box were made of mild steel plate of 8 mm thickness, the third side was made of fiberglass plates with 20 mm thickness to allow observation of the clay deformation during testing. The fourth side of the box was a cantilever movable retaining wall with dimensions 800×450 mm and thickness 8 mm. At the center line of the retaining wall, four circular openings with diameter 50 mm were made. The height of these openings from the top of the retaining wall were 100, 285, 470 and 655 mm which were used for the installation of the pressure cell transducers. At the edges of the retaining wall, special plastic material was used to prevent leakage and evaporation during the test. The setup was designed to enable lateral swelling pressure, the at rest swelling pressure, active and passive earth pressures measurements. Figure (1) presents the layout of the experimental setup used in this investigation.

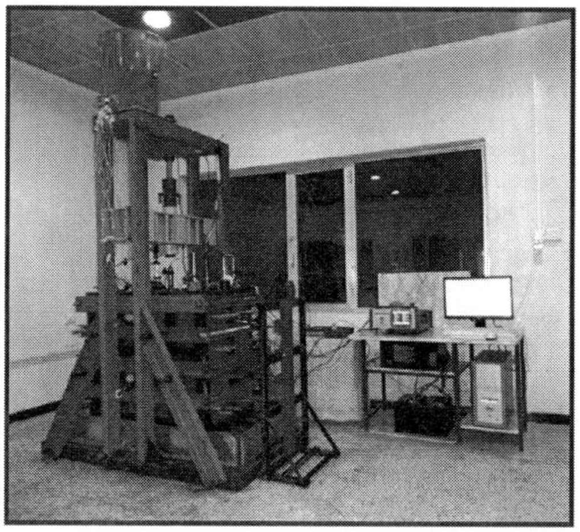

Figure 1. Large scale physical model

2.2 Soil layout in the box

At the bottom of the box, a layer of gravel with 100 mm in thickness, was used to permit a uniformly distribution of water under the bottom surface of the soil. Also, a layer of geogrid was used above the gravel layer to allow the water to pass though it easily.

In order to simulate the plain strain condition, sidewall friction between the soil and box's sides was reduced as low as possible. So, the sides of the box were lubricated with silicon grease. Also, plastic sheets of 70 mm height were placed on the silicon grease to cover overall height of the box's sides.

The soil was placed in layers and compacted in the box using static compaction method. Compaction pressure applied by electrical hydraulic press until reaching a dry unit weight of 14.4 kN/m^3 and optimum moisture content 27.4 %. Where the initial degree of saturation of the soil sample was 81.27%. The compaction was achieved in the box in 14 layers. The thickness of each layer was 50 mm to reach an overall depth of 700 mm. The compaction pressure was kept for two hours to prevent rebound of soil, and insure that the top soil is on the required level. Five holes were bored by a special hand auger with diameter of 20 mm. Three of them were made on the left side of the box at depths of 100, 345 and 590 mm. Also, two holes were made on the back of the box at depths of 100 and 345 mm. These holes were used to install calibrated water probes which have the capacity of 239 kPa. The probes were used for measuring the total suction of the soil. Beside that, the probes were utilized to check the variation of the total suction at different depths and different locations on the same soil levels.

When compaction process completed, a rectangular steel plate of dimensions 786×468 mm was placed at the top of the compacted soil surface, where the center line of the steel plate and soil surface was coincided. Then, all the equipments were installed in appropriate positions. So, three linear variable displacement transducer LVDTs with capacity of 100 mm were fitted on the longitudinal center line surface. Four pressure cells with capacity of 2 N/mm^2 were installed on the retaining wall. Finally, another three LVDTs with capacity of 25 mm were installed on a steel mesh which was placed in front of the retaining wall, and their pointer were localized on the pressure cells to measure the wall's movement.. Figure (2) shows a schematic diagram of the model with their instrumentation devices.

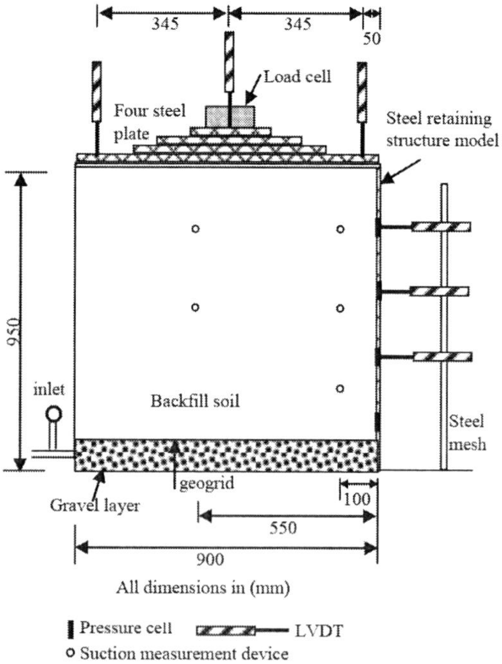

Figure 2. Schematic diagram of the model with their instrumentation devices

2.3 Saturation and load application

Expansive soil could exert a lateral pressure on retaining wall, when suction changes within soil due to rainfall and seasonal change of water table. Thus in this study, the soil was saturated from the top and bottom of the box using a water tank which was placed at the top of hydraulic press frame and filled with water as shown in Figure (1), which was used to provide water on all sides of a rectangular plate and consequently, water penetrate to the soil mass by gravity. The water inlet at the bottom of the box was connected to the water source and water can rise by the capillarity.

After saturation stage, the soil was subjected to the vertical load by a mechanical hydraulic jack. Four plates with load cell and their circular disk were arranged, and placed at the surface of the compacted soil, the center line of mechanically hydraulic jack, load cell and jointed plates were coincided with the center line of the top soil surface. More details of results of the saturation and loading stages can be found in [16].

3 Results and discussion

3.1 Wall's movement stage

The soil sample was subjected to saturation and loading up to 214.3 kPa. The wall remained stable without movement, using four steel tides and guides at different depths. Also, the wall was fixed by seven rivets at the bottom of the wall. The wall's movement represented active case of the retaining wall. The wall was moved horizontally away from the soil at distances 0.1, 0.2, 0.3, 0.6, 0.8, 1.0 , 2.0, 3.0 and 4.0 mm after the rivets at the bottom of the wall were opened, and the movement was controlled by the guides. The movement of the wall was measured by LVDTs which were localized on each pressure cell and monitored by data logger. At each step of wall's movement the following parameters were studied:

1. Effect of wall's movement on the lateral stresses.
2. Effect of wall's movement on the settlement of the backfill soil.
3. Effect of wall's movement on the total suction of the backfill soil.

3.1.1 Effect of wall's movement on the lateral stresses

Figure (3) shows the relationship between lateral pressures and wall's movement. It can be noted, that the lateral stresses decreased with the increase of wall's movement at all depths. These results coincide with findings of [11, 17] when moving the retaining wall away from the backfill sandy soil in horizontal and rotation movement about the base of the wall. Rapid decrease in lateral stresses occurred when the wall's movement range between 0.1 to 1.0 mm, where at 1.0 mm the percent of decrease reached 90, 85, 75 and 55 % at depths 0.1, 0.285, 0.47 and 0.655 m, respectively. Then, the decrease will be slow with the increase of wall's movement. It should be noted, the rate of decrease in lateral stresses, at wall's movement more than 1.0 mm and at depths 0.47 and 0.655 m are more than that at depths 0.1 and 0.285 m.

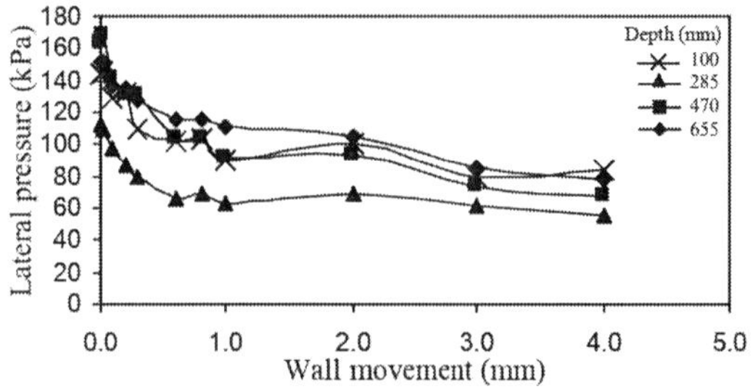

Figure 3. Varaition of lateral pressure with wall movement

To investigate the effect of elapsed time on the behavior of lateral stresses during steps of wall's movement, the wall was left for many hours at each step of wall's movement until the lateral stresses reached an equilibrium state. Figure (4) shows results that illustrate the effect of elapsed time on the distribution of lateral stresses. Generally, the lateral stresses suddenly decreased with high values immediately after wall's movement, then increased within 30 minutes and reached an equilibrium states after hours.

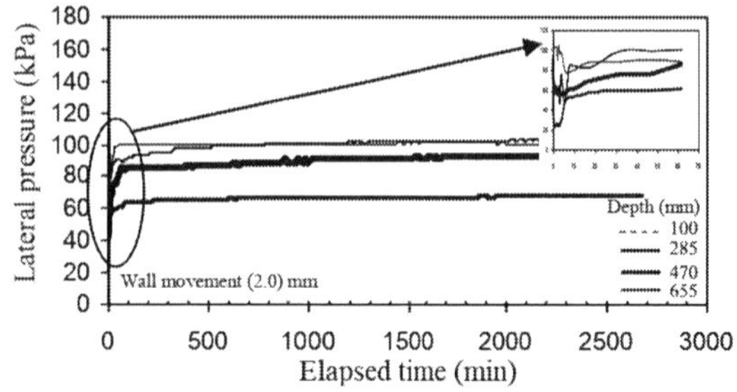

Figure 4. Varaition of lateral pressure with elapsed time

3.1.2 Effect of wall's movement on the settlement of the backfill soil

The settlement of the backfill expansive soil was measured, when the wall's movement was at different steps 0.1, 0.2, 0.3, 0.6, 0.8, 1.0, 2.0, 3.0 and 4.0 mm. The settlement was measured using two LVDTs, one of them fixed directly behind the wall, the other was fixed away from the wall. Figure (5) shows the relationship between the settlement of the backfill expansive soil and wall's movement. The settlement increased with the increase of wall's movement. The settlement of soil when the wall was moved 4.0 mm is equal to 11.04 and 9.15 times the settlement of soil, when the wall was moved 0.1 mm behind the wall and away from the wall, respectively. The percent of increase in settlement increased from (1.04, 1.43) up to (8.79, 13.84) % when the wall was moved from (0.1 to 4) mm at away from and behind the wall, respectively.

It should be noted, the settlement behind the wall was more than that away from the wall at each step of wall's movement. From these results, it can be concluded that the surface of the backfill soil was sloped toward the retaining wall.

DOI: 10.1051/e3sconf/20160905006

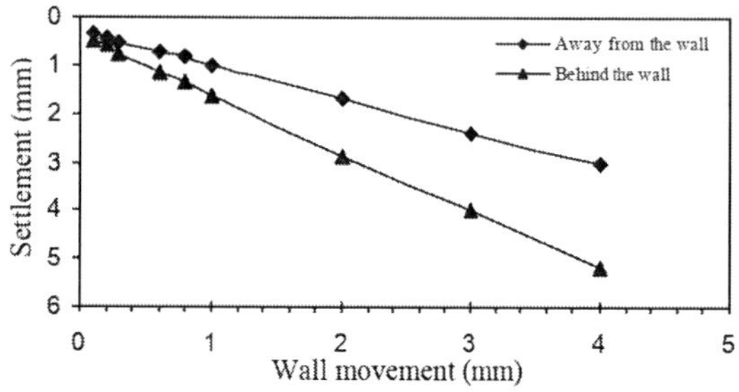

Figure 5. Settlement of the backfill Soil with wall movement

3.1.3 Effect of wall's movement on the total suction of the backfill soil

To study the effect of wall's movement on the total suction of the soil, readings of total suction were obtained from water mark monitor during each step of wall's movement with one hour interval.

Figure (6) shows the variation of total suction with wall's movement at different locations and depths. The total suction increased slightly with wall's movement up to a maximum value of 7.0 kPa at the middle side of the soil sample during the test. Also, the total suction increased at the whole depth. The range of increase in total suction when the wall was moved from (0.1 to 1.0) mm was (0 to 4.0) kPa. Then, when the wall was moved from (1.0 to 4.0) mm the range of increase of the total suction was (1.0 to 3.0) kPa. This implies that the wall's movement has a very limited effect on the total suction of the soil.

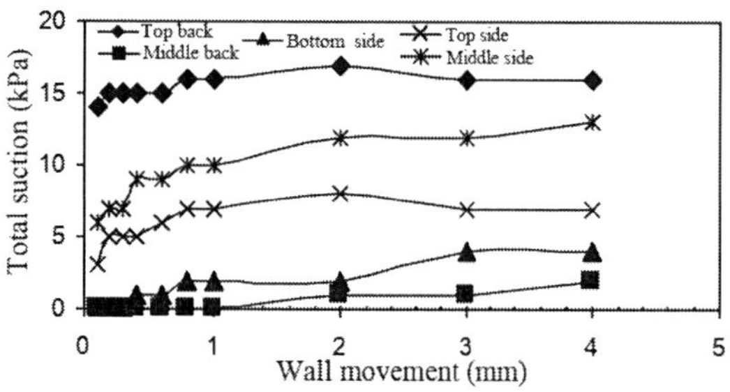

Figure 6. Variation of total suction during wall movement

3.2 Unloading Stage

Table (2) shows the unloading steps of the vertical stresses. During this stage, the wall was left to move freely. Where all guides that limit the movement of the wall were removed. At each step of the vertical stresses release, the following parameters were measured by sensors:

1. Effect of unloading on the lateral stresses.
2. Effect of unloading on the vertical stresses with time.
3. Effect of unloading on the settlement of the backfill soil.
4. Effect of unloading on the total suction of the backfill soil.

Table 2. Steps of unloading stage

Steps	Vertical stress (kPa)	
	From	To
1	214.3	160.7
2	160.7	107.1
3	107.1	53.57
4	53.57	26.78
5	26.78	13.39
6	13.39	0

3.2.1 Effect of unloading on the lateral stresses

Figure (7) shows the results of lateral stresses at different depths due to the unloading stage. The lateral stresses at the depths 0.1 and 0.285 m decreased with the decrease of vertical stresses. However, at depths 0.47 and 0.655 m the lateral stresses increased. This is not identical with the findings of [17]. This behavior is may be attributed to the unloading which caused a decrease in lateral stresses as the soil attempted to move away from the wall and then a decrease in the lateral pressure observed. However, the upper part of the wall moved towards the soil when the soil rebounded, in which this upper part movement of the wall caused a rotation of the bottom part away from the soil. This in turn caused the soil, at the deepest depths of 0.47 and 0.655 m, to be swelled towards the wall. Therefore, an increase of horizontal stresses on the wall and lateral pressure observed.

The decrease in the lateral stresses at depth 0.1 m was more than that at depth 0.285 m at all levels of vertical stresses release. Thus, the slope of the curve as shown in Figure (7) at depth 0.1 m is more than that at depth 0.285 m.

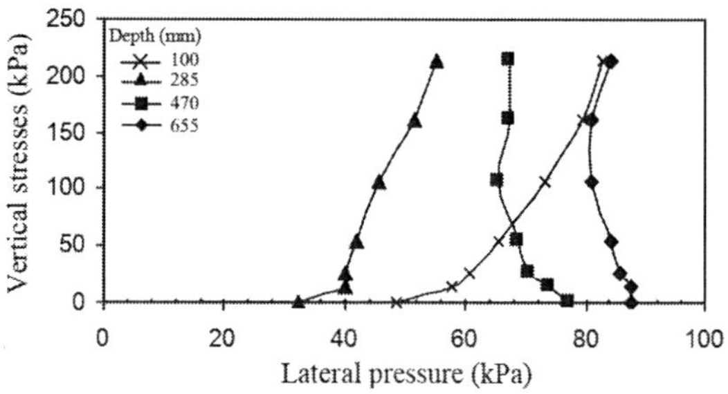

Figure 7. Change in lateral pressure at unloading stage

3.2.2 Effect of unloading on the vertical stresses with time

During the unloading stage the vertical stresses was left for many hours, the vertical stresses reduced to values less than the initial values within several minutes, then increased with elapsed time. Figure (8) shows the variation in vertical stresses with elapsed time. It should be noted, that the vertical stresses suddenly decreased and reached the minimum value, within the time ranged between 2.5 to 5.8 min as shown in Table (3). Then, followed by an increase in the vertical stresses with time until reached an equilibrium state. The rate of increase in vertical stresses decreased with the reduction in vertical stresses. Therefore, the time required to obtain the stability of vertical stresses increased with the decrease of vertical stresses, which ranged between 44.25 to 1945.99 min. Table (3) also shows the amount and percent of return in the vertical stresses with time. It should be noted, the percent of return increased with the reduction in the vertical stresses, which reached two times at stress level 13.39 kPa. The return in the vertical stresses is related to the ability of highly expansive soil to exert uplift pressure.

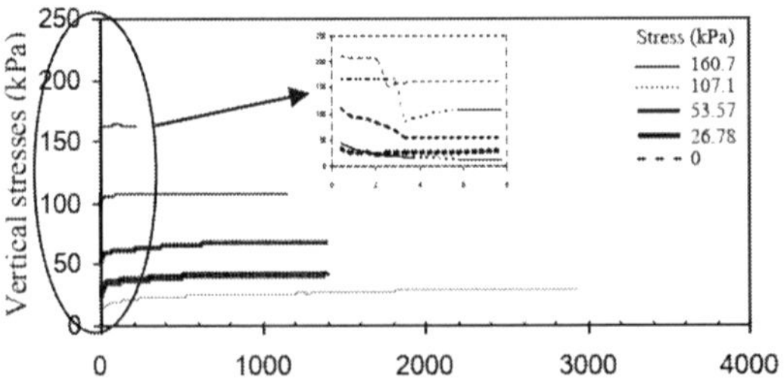

Figure 8. Variation in the vertical stresses with time

Table 3. Variation of the vertical stresses with time

Vertical stress (kPa)	Initial vertical stress (kPa)	Decrease of vertical stress (kPa)	Time (min)	Final vertical stress (kPa)	Return of vertical stress (kPa)	Time (min)	% of return of vertical stress
160	156	58	2.5	163	7	44	4.7
107	85	78	3.3	108	23	838	26
53	51	57	3.3	66	15	598	29
26	22	44	2.1	42	20	1410	93
13	9	33	5.8	28	19	1946	199

3.2.3 Effect of unloading on the settlement of the backfill soil

In this stage, the applied vertical stresses were released in steps as mentioned previously. So, the backfill soil rebounded. Figure (9) shows the decrease in settlement with the release of applied vertical stresses. It should be noted the rebound curves have the same trend. Table (4) shows the amount and percent of decrease in the settlement of the backfill soil behind the wall, which is slightly more than that away from the wall, that is refer to the uniformity in rebound of expansive soil when the vertical stress was released.

The reduction in the settlement was small up to vertical stress 53.57 kPa, then increased especially at stress level 13.39 and 0 kPa. The amount and percent of decrease in settlement at stress level 13.39 kPa are more than twice that at stress level 26.78 kPa.

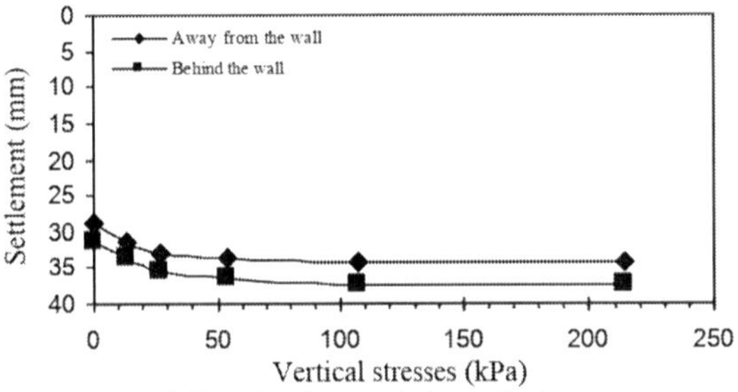

Figure 9. Variation of settlement during unloading

Table 4. Variation of settlement at unloading stage

Vertical stresses (kPa)	Settlement away from the wall (mm)	Decrease in settlement (mm)	Settlement behind the wall (mm)	percent of decrease in settlement
214.3	34.35	-	37.5	-
107.1	34.19	0.4367	37.37	0.3467
53.57	33.59	2.183	36.61	2.3733
26.78	32.95	4.0466	35.66	4.9067
13.39	31.25	8.9965	33.71	10.107
0	28.69	16.448	31.37	16.347

3.2.4 Effect of unloading on the total suction of the backfill soil

The total suction at all depths and locations increased with the reduction of the vertical stresses as shown in Figure (10). The maximum increase in total suction was 20 kPa occurred at the top side of the box. The curves in Figure (10) at the top side and top back of the box have the same trend because of the difference in the total suction between them was small, which is related to the plastic nature of highly expansive soil, that helped the top layer to increase its height uniformly and permit the air entrance to the soil. Thus, the total suction increased on the same values.

At middle layer, the total suction increased with a rate slightly less than that at the top layer. This is because the top layer was more affected by the reduction of the vertical stresses than the middle and bottom layers. The increase in the total suction in the middle back was more than that of the middle side until reaching the same values of 20 kPa when the vertical stresses became 13.39 kPa.

At the bottom layer, the total suction increased with a very slow rate compared to the top and middle layers. The increase in the total suction decreased with the decrease of the vertical stresses to reach 1 kPa when the vertical stresses were reduced from 13.39 to 0 kPa.

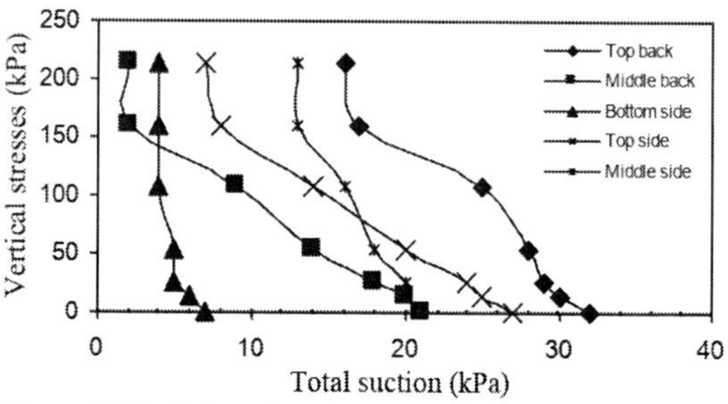

Figure 10. Variation of total suction during unloading stage

4 Conclusions

From the study carried out, the following conclusions can be withdrawn;

1- The lateral stresses largely decreased when the retaining wall was moved away from the backfill soil.
2- The settlement of the backfill soil increased with increasing the wall's movement away from the backfill soil. Also, the settlement behind the wall was more than that away from the wall at each step of wall's movement.
3- The effect of the wall's movement on the total suction was marginal.
4- During the unloading stage, the lateral stresses increased at the middle and bottom layers and decreased at the top layer.
5- The same amount of the settlement was obtained behind and away from the wall during the unloading stage.
6- The total suction increased during unloading, and the increase in the total suction at the top layer of soil sample is more than that at the middle and bottom layers.

References

1- B. Fourie, "Laboratory Evaluation of Lateral Swelling Pressure". J. Geo. E. ASCE. **115**, 1481-1486 (1989).
2- B. Ikizler, M. Aytekin, E. Nas, "Laboratory study of expanded polystyrene (EPS) geofoam used with expansive soils". *Geot and Geom*. **26**, 189-195 (2008).
3- C. Clayton, I. Symons, J. Hiedra-Cobo, "The Pressure of Clay Backfill Against Retaining Structures" Can. Geot. J. **28**, 282-297 (1991).
4- C. Huang, W. Luo, "Behavior of Soil Retaining Walls on Deformable Foundations". E. Geol. **105**, 1-10(2009).
5- E. Marsh, R. Walsh, "Common Causes of Retaining-Wall Distress: Case Study". J. Per. Con. Fac.. **10**, 35-38 (1996).
6- L. Tagle, F. Villalobos, "Experimental Study of the Lateral Earth Pressure On Retaining Structures In Soils Reinforced With Geogrids". *Rev. Ing. Constr.* **26**, 299-320 (2011).
7- L. Wang, S. Liu, B. Zhou, "Experimental Study on the Inclusion of Soilbags in Retaining Walls Constructed in Expansive Soils". *Geot. Geom. J.* 43, 89-96 (2015).
8- M. Ozer, R. Ulusay, N. Isik, "Evaluation of Damage to Light Structures Erected on a Fill Material Rich in Expansive Soil". B. E. Geol. Env. **71**, 21-36 (2012).
9- M. Pinto, T. Cousens, "Effect Of The Foundation Quality On A Geotextile-Reinforced, Brick-Faced Soil Retaining Wall", Geos. Int. **7**, 217-242 (2000).
10- O. Ertugrul, A. Trandafir, "Lateral Earth Pressures on flexible Cantilever Retaining Walls with Deformable Geofoam Inclusions". E. Geo. **158**, 23-33(2013).
11- O. Mohamed, Y. Taha, S. Abd El-Aziz, "Experimental Study on the Effect of Lateral Swelling Pressure of Expansive Soil on Retaining Structure". *J. E. Sci.* **42**, 84-92 (2014).
12- O. Mohamed, Y.Taha, S. Abd El-Aziz, "Field Study of the Distribution of Lateral Swelling Pressure of Expansive Soil on Retaining Structure". *J. E. Sci.* 42, 289-302 (2014).
13- R. Chen, Y. Chiu, " Model Tests Of Geocell Retaining Structures", *Geot. Geom*. **26**, 56–70 (2008).
14- R. Katti, E. Bhangale, K. Moza, "Lateral Pressure in Expansive Soil With and Without a Cohesive Non-Swelling Soil Layer, Application to Earth Pressures on Cross Drainage Structures in Canals and Key Walls in Dams", C. B. Irr. P. **110**, 1-303 (1983).
15- R. Thompson, R. Mckeen,. "Heave Prediction Using Soil Suction a Case History". *S. S App. Geot. E. P. ASCE*. 1-13 (1995).
16- S.A. Khattab, M.K. Al-Shamam, K.A. Al-Juari, "Model Test on Lateral Earth Pressure of highly Expansive soil". Proc. 6th Asia pacific Conf. on Unsat. Guilin, China, 709-716 (2015).
17- T. Dave, S. Dasaka, "Transition of Earth Pressure on Rigid Retaining Walls Subjected to Surcharge Loading". *Int. J. Geo. E.* **6**, 427-435 (2012).
18- Y. Ertekin, "Measurement of Lateral Swell Pressure with Thin Wall Oedometer Technique". MSc. Thesis, Middle East Technical University (1991).
19- Y. Tsukamoto, K. Ishihara, T. Higuchi, H .Aoki, "Influence of Geogrid Reinforcement on Lateral Earth Pressures Against Model Retaining Walls". Geo. Int. **6**,195-218 (1999).
20- Z. Mengxi, Z. Huai, "Model Test on Sand Retaining Wall Reinforced with Denti-Strip Inclusion". Sci. Ch. S. E. Tech. Sci.. **51**, 2269-2279 (2008).

Macro-microscopic one-dimensional compression of wet granular soils by experimental investigation

Vinh-Du Than[1,2,a], Patrick Aimedieu[1], Jean-Michel Pereira[1], Jean-Noël Roux[1], Michel Bornert[1] and Anh-Minh Tang[1]

[1] *Université Paris-Est, Laboratoire Navier (UMR 8205), CNRS, ENPC, IFSTTAR, 77455 Marne-la-Vallée Cedex 2, France*
[2] *The University of Danang, College of Technology, Department of Civil Engineering, Danang City, Vietnam*

Abstract. The presence of liquid menisci between soil particles in wet granular soils plays a key role in the macroscopic behavior. This liquid bridge between particles may also allow the existence of metastable microstructures that lead to mechanical instabilities when the soil in subjected to external loadings. In this work, the authors present an investigation into macroscopic and microscopic behaviors of wet granular soils in very loose state under one-dimensional (1D) compression by using oedometer test combined with the X-ray computed tomography (X-ray CT) observations. The specimen is prepared in the presence of a small amount of a water meniscus by using the controlled pluviation method. The influences of various mechanical parameters are first studied on the compression curve. The microstructure along those compression curves are then analyzed by using X-ray CT to observe the arrangement of microstructure under growing of applied force.

1 Introduction

Wet granular soils are usually found in many natural processes. In the field of geotechnical engineering, the appearance of water menisci between soil particles is the most important factor which can be governed the macroscopic behavior. Hence, the cohesive soils can be stable at very loose state however they are very sensitive to the external force. So far, there are a lot of studies in the macro-micromechanical behavior of wet granular materials, such as: cohesive soils [1,2], wet beads [3], loessic soils [4,5], wet sand [6,7], etc. In general, the behavior of this material under the growing of external load (oedometric or isotropic compression tests) is normally characterized by the irreversible compression and decompression curves.

Various investigations of the microstructure of loose unsaturated granular soils have been carried out by the scanning electron microscopy (SEM) [3,4] and the microfocus X-ray computed tomography (μCT) [7]. The SEM techniques require a cautious preliminary dehydration of the samples, often by freeze-drying. They provide local observations on a small part of millimetric samples. The μCT is a high-resolution non-destructive 3D observation technique that does not need any sample pre-treatment and that gives further 3D information on the whole sample.

Yet, wet granular soils, especially in very loose unsaturated states, have less frequently been investigated by the μCT than dense states. In this work, glass beads with capillary bonds in the pendular state were investigated by the oedometric compression tests and μCT tests for further macro-microscopic investigations of a wet bead sample.

[a] Corresponding author: vinh-du.than@cermes.enpc.fr

2 Material and methods

2.1 Material

The studied material is glass beads, an industrial product, with the spherical beads. The size of diameters from 80 μm to 104 μm is chosen for this study. With these diameters, particle weight looses relevance with respect to capillary forces [8], and they are large enough for the X-ray CT tests. The specific density of this material was equal to 2,460 kg/m^3. The grain size distribution is shown in Figure 1.

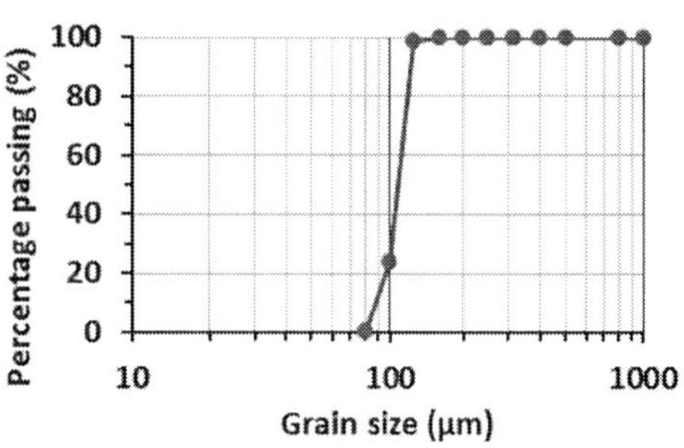

Figure 1. Cumulative grain size distribution of glass beads.

DOI: 10.1051/e3sconf/20160906001

2.2 Specimen preparation

The specimen was prepared as following steps:

- The glass beads were mixed with water at initial selected water content of w_0 = 3.50 %, in order to ensure that the specimens were always maintained in pendular state.
- The oedometric cell (50 mm of internal diameter for oedometer tests -- hereafter called cellϕ50 and 20 mm of internal diameter for X-ray CT tests -- hereafter called cellϕ20) was placed on the support system (see Figure 2). It was then filled with the wetted glass beads by using a controlled pluviation method as follows. The filling was performed by placing the wetted glass beads on a fixed sieve with a 200 μm opening. This size allows passing for maximum of two grains. The sieve was vibrated vertically with a vibration machine. Thanks to the shocks, clusters of glass beads passed sequentially through the sieve to finally fall into the cell. The height of free fall (h_{fall}) from bottom sieve to bottom of the cell was between 80 and 260 mm.
- All specimens were prepared in an air-conditioned room. The preparation of each specimen took approximately 30 minutes. These conditions allow minimizing the water evaporation during the preparation.
- The specimen top surface was then perfected by using a small iron wire to cut the excess soil. This action must to do carefully and gently in order to avoid any perturbation to initial structure. Height of specimen is 20mm (cellϕ50) and 10mm (cellϕ20).
- Finally, all specimens were tested to measure the water content after preparation (w_{ap}).

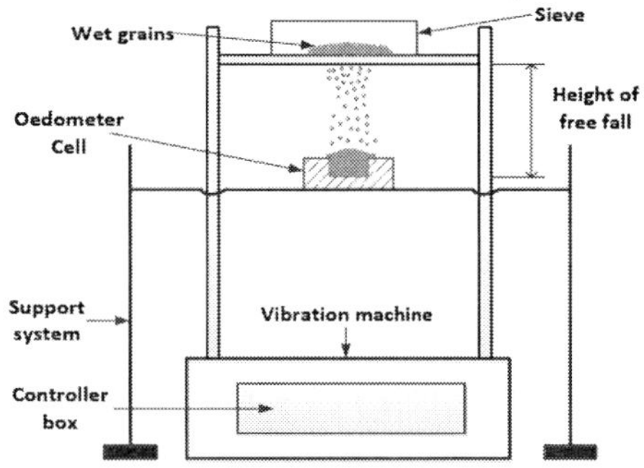

Figure 2. Schematic of system for preparing the specimen.

The initial parameters of samples after specimen preparation with cellϕ50 and cellϕ20, listed in Table 1.

The water content of all specimens after preparation is strongly affected by the height of free fall. Indeed, within same initial conditions, the evaporation of samples is proportional to the height of free fall. Clear evidences are given in Table 1, the maximal evaporation is 54.29 % for highest height of 260 mm and minimal ones is 22.57 % corresponds to smallest height of 80 mm. Although the water content after preparation is remained so small, these specimens are still stable at very loose state. This confirms that the capillary forces between grains play a key role to maintain the stability of specimen's structure. Moreover, the relationship between initial void ratio e_0 and the height of free fall h_{fall} is shown in Figure 3, in which the blue square point is the sample for cellϕ20. Indeed, the higher the heights of free fall, the smaller the initial void ratios.

Table 1. Initial parameters (height of free fall h_{fall}, porosity n, water content after preparation w_{ap} and initial void ratio e_0)

Tests \ Parameters	h_{fall} (mm)	n (%)	w_{ap} (%)	e_0
OD01	80	70.19	2.71	2.35
OD02	80	70.12	2.70	2.35
OD03	120	66.56	2.00	1.99
OD04	120	66.86	2.30	2.02
OD05	150	63.72	2.50	1.76
OD06	150	63.95	2.20	1.77
OD07	260	60.60	1.80	1.54
OD08	260	60.02	1.60	1.50
CT01	80	69.66	2.85	2.29

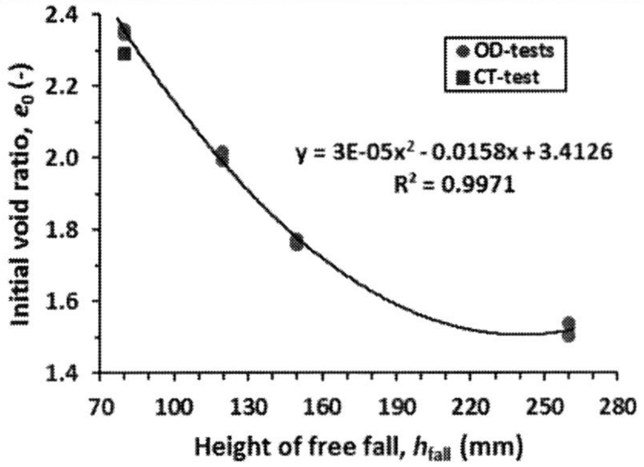

Figure 3. Initial void ratio e_0 versus height of free fall h_{fall}.

2.3 Oedometric compression process

The specimen was carefully placed on the loading platform of the compression system (see Figure 3). The piston is fixed to the load cell which has maximum capacity of 1,000 N. A LVDT (Linear Variable Differential Transformer) was installed to measure the displacement of specimen. The cell was then moved upward with a rate of 0.5 mm/min. All signals of load cell and LVDT were received by the data acquisition system using the LabVIEW software.

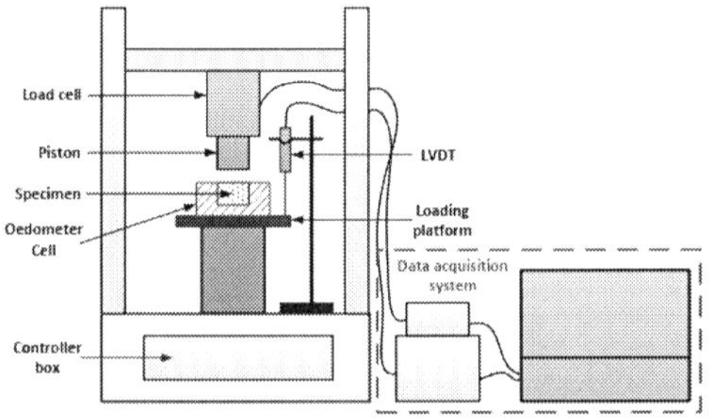

Figure 4. Schematic of oedometric compression system.

Figure 5 illustrates the typical relationship between applied forces and time during the compression process. Based on this curve, the initial vertical stress was specified after removing the noise's effect, especially the moment when the surface of specimen touched the piston. The settlement of specimens was then determined from the displacement of LVDT. The void ratio was finally calculated from the settlement and initial void ratio of specimen. All compacted specimens were also retested in order to measure the final height. This parameter was used to verify the final state of compression curve.

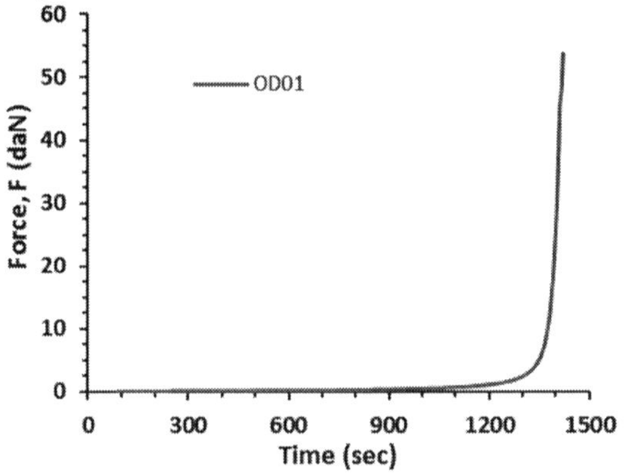

Figure 5. Applied force versus time during compaction process.

2.4 X-ray computed tomography

X-ray computed tomography (X-ray CT) is a non-destructive method that allows to obtain three-dimensional images of various materials (for more information on principles and applications, see [9,10]). It generally works well for porous media.

X-ray CT scans were performed using an UltraTom microtomograph, equipped with a 230 kV micro-focus Hamamatsu L10801 source, associated with a PaxScan Varian 2520 V bidimentional sensor.

The specimen used for those scans was prepared in Sec. 2.2 with cellϕ20. The cell was made of PMMA (PolyMethyl MethAcrylate).

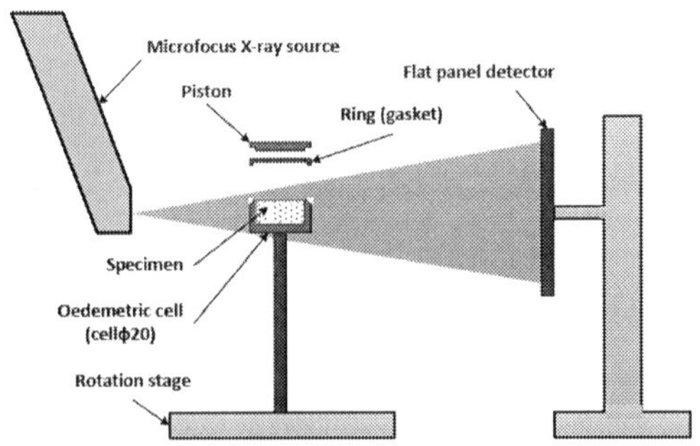

Figure 6. Schematic of the X-ray CT scanner.

Four scans (denoted by S1, S2, S3 and S4) were carried out, corresponding to four points on the compression curve. For each point, the position of the piston was fixed during scantime. Those four states represented different heights of the specimen: h_1 = 10 mm, h_2 = 8.7 mm, h_3 = 7.5 mm and h_4 = 6.4 mm. The corresponding the macroscopic void ratios were respectively calculated e_1 = 2.29, e_2 = 1.87, e_3 = 1.47 and e_4 = 1.11. X-ray source voltage and current were 70kV and 130μA. Each scantime was 80 min long. Voxel size was 13 μm. In the present study, this resolution was applied just to analyze the homogeneity of specimen. Images were reconstructed using the Xact software (made by RX-Solutions). 3D-images are 16 bit-depth, and were converted to 8 bit-depth to reduce memory usage. The image analysis was carried out by using ImageJ [13] and FiJi softwares, which is another version ImageJ for 3D modeling [14].

3 Results

3.1 Macroscopic behavior

All the specimens were carried out with very loose initial structures, from e_0 = 2.35 to e_0 = 1.50. Figure 7 shows the compression curves for different values of initial void ratios. All the tests were duplicated to ensure the repetition. The initial stress at bottom with the weight of sample is 0.246 kPa.

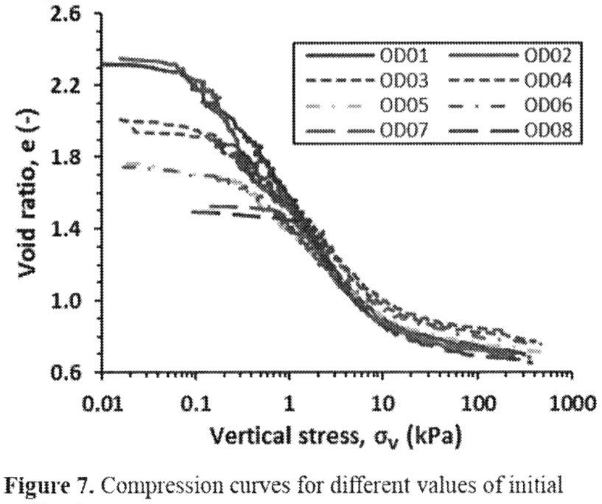

Figure 7. Compression curves for different values of initial void ratios.

Based on the shape of these curves and the range of stress, three stages can be identified. The first stage, under the low stress, void ratio *e* was slightly decreased. The void ratio was then sharply decreased in the intermediate stage. Finally, at third stage, void ratio was gradually decreased when the vertical stress was larger than 10 kPa. Especially, these curves were very close in this stage. In other words, when no further rearrangement is possible, the decrease of void ratio was very small because of the elastic deformation of grains at the contact points. Interestingly, there were only two stages 1 and 2 on the oedometric compression curve in the usual soil mechanics behavior [4,11]. This is a special behavior of the studied material under the 1D compression tests.

3.2 Microscopic behavior

The reconstructed 3D sample volume at initial state is shown in Figure 8. µCT technique allows distinguishing between various components, according to their density and atomic composition, thanks to grey levels. On examining reconstructed pictures, lighter grey levels typically correspond to more absorbing materials, while darker grey levels corresponds to less absorbing materials.

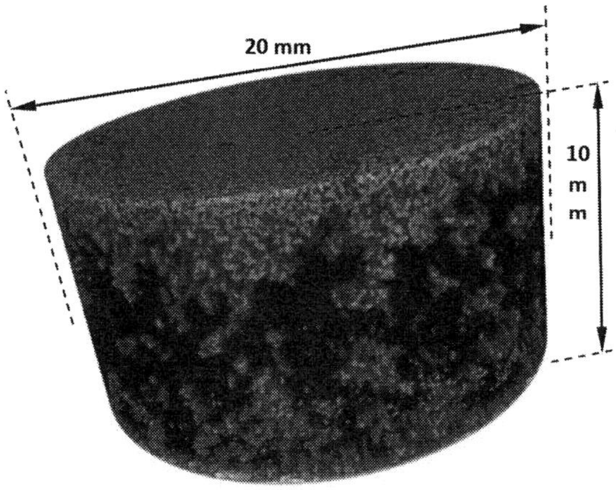

Figure 8. 3D reconstructed image of the sample at initial void ratio (S1).

The density is not homogeneous within the cylindrical sample, especially at the periphery where more voids are observed plus a larger density with less voids at top and bottom of sample. This is due to the specimen preparation, and also to the boundary conditions imposed by the ring. In other words, the friction between the ring and grains and the cohesion of grains resisting displacement the other particles.

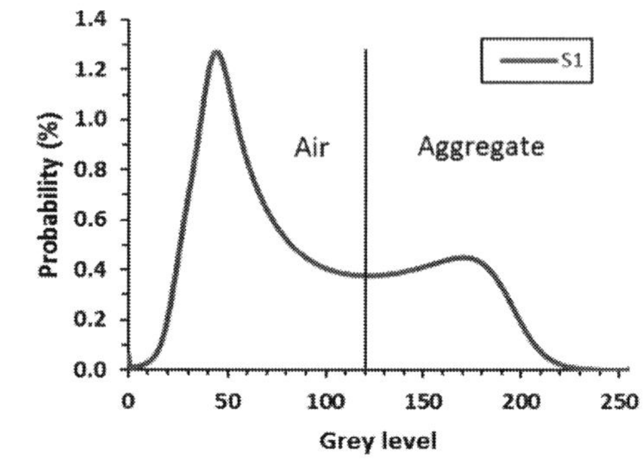

Figure 9. Grey level histogram for the scan S1.

Figure 9 illustrates the representative histograms of grey level for the first scans. Grey levels range from 0 to 255. A threshold, shown on the same figure, was used to delineate o border between air dominated portions of the sample (left of the line) and aggregates. The grey-level threshold value was found in the interval [110:130]. This allowed recalculating the total void ratio, as well as the void ratio distribution during the compaction process. The IsoData method [12] was chosen for the segmentation process. The grey-level threshold values following 128, 122, 130 and 128 were selected for scans S1, S2, S3 and S4, respectively.

Figure 10 shows the segmentation process corresponding to three consecutive stages: a) initial state; b) during the thresholding process and c) segmented image. After thresholding, the aggregates (darker grey levels) and the air (lighter grey levels) were clearly distinguished. The void ratio was then recalculated by dividing the number of black voxels by the total number of voxels of the segmented image.

(a)
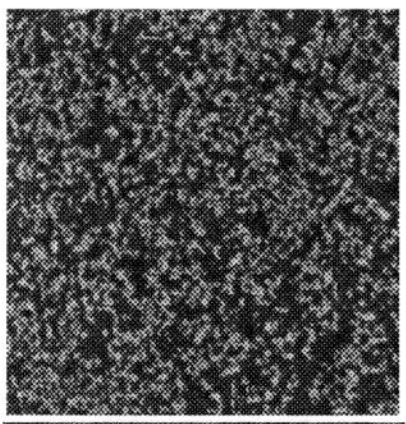
(b)

(c)
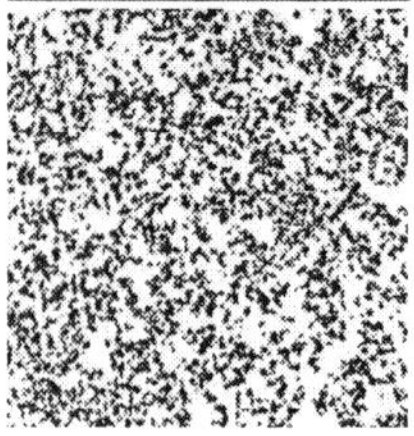

Figure 10. Images at different steps of segmentation: (a) original image, (b) during the thresholding process and (c) segmented image (binary image).

The void ratio distribution according to the height of the specimen during the compression process is illustrated in Figure 11. Ten layers of sample are split corresponding to the same thickness from first layer h_1 to last layer h_{10} (from bottom to top), as described in the inset figure.

At the initial state S1, from bottom to top, the void ratio is largest at sixth layer (e_{max} = 2.35) because of the influence of the wall in specimen preparation process. Meanwhile the sample is denser at the bottom layers and densest at the top layers. This is caused by the end of preparation process when the grains were spread on the surface of the cell, no longer the wall's effect. The lower layers were steadily compacted by the grains above them. This confirms that the void ratio at this region is smallest. For the step 2 (S2), the structure is compacted and rearranged, void ratio in the central and top layers is reduced, especially the central layers. The structure is notably perturbed in step 3 (S3). The void ratio is reduced of 49.80 % compared to its at bottom layers and of 23.25 % at the top layers. At the densest state (S4), a strong decrease of void ratio obtained at the central layers, while this void ratio is very close to it's at the bottom layer (h_1). The average void ratio of four curves (2.31, 1.91, 1.48 and 1.14, corresponding to S1, S2, S3 and S4) are very close to the macroscopic ones, which are given in Sec. 2.4.

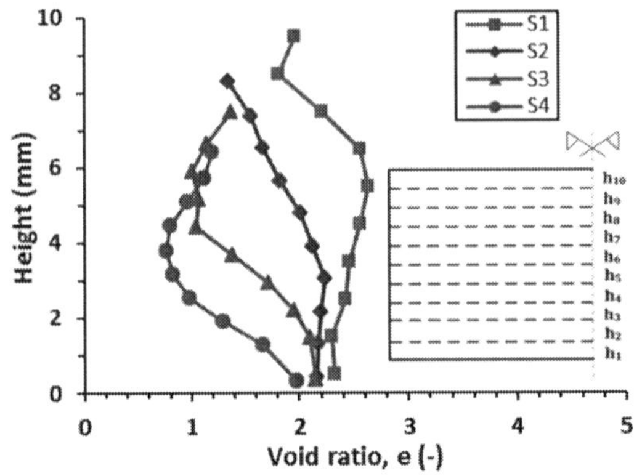

Figure 11. Void ratio distribution according to the height of sample for different compaction steps.

A supplemental evidence for the void ratio distribution during the compaction process is shown in Figure 12.

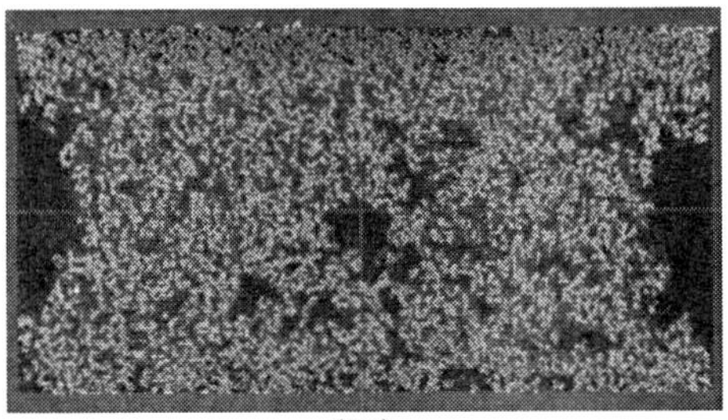

(S1)

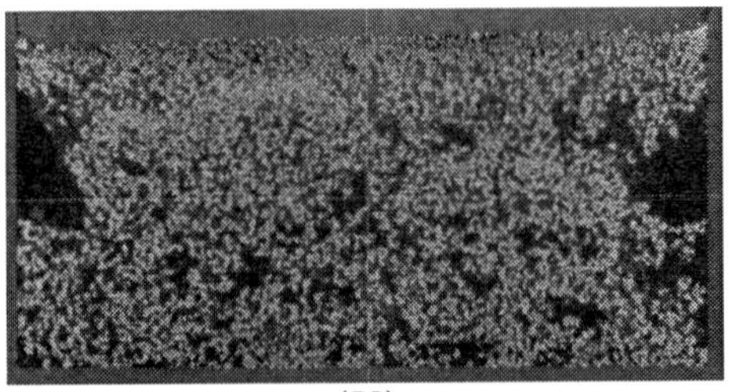

(S2)

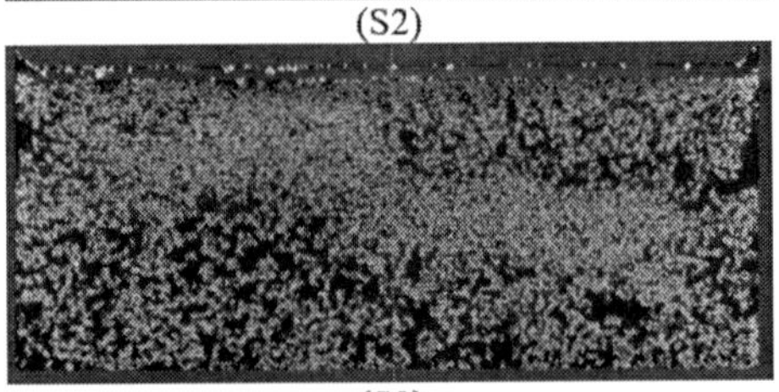

(S3)

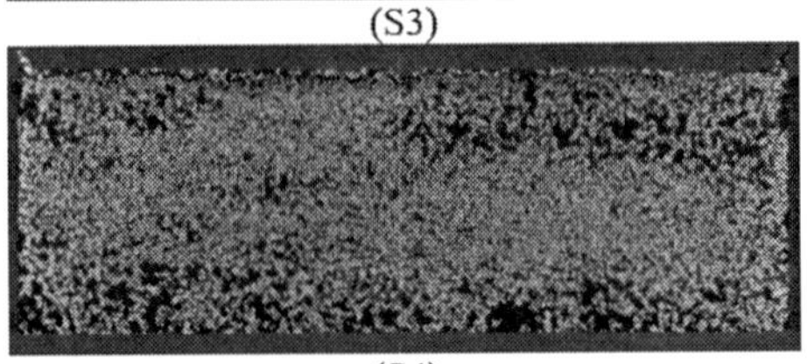

(S4)

Figure 12. Vertical slice at the center of the sample for 4 states.

Figure 12 clearly shows the void ratio distribution during the compaction process along the vertical direction. Under a growing stress, the structure is strongly rearranged in the central layers because the loosest state is obtained in this region while it is slightly affected at the bottom and top layers. When the specimen becomes denser in central, it can be against the settlement of top layers. Moreover, relying on the cohesive forces, the grains at bottom layers can be create arches which strongly support above structures. This explains why it is really difficult to compact this specimen at step 4 although volumetric strain is only 40 %.

In addition, the spatial distribution of the void ratio is investigated as a radial function, considering different concentric cylinders having the same height for each scan. Ten concentric cylinders from the center (smallest radius R_1 = 0.5 mm) to the contour (radius R_{10} of 9.5 mm, see Figure 13-a).

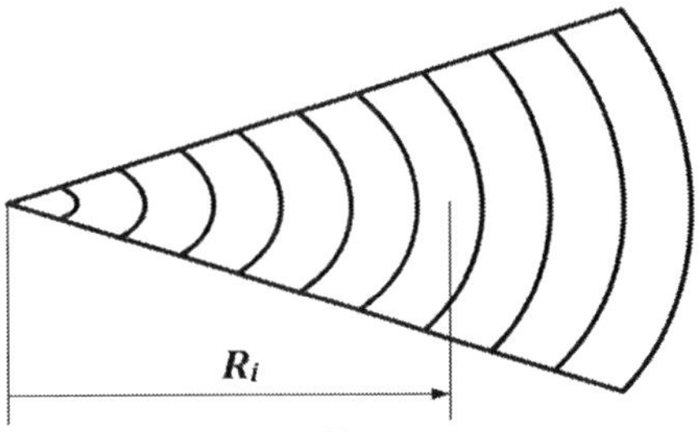

(a)

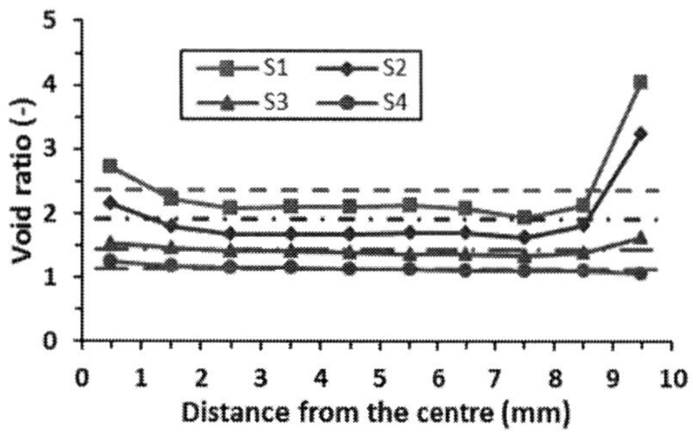

(b)

Figure 13. (a) Concentric cylinders investigated. (b) The change of void ratio respects to the average radii of the selected cylinders.

The calculated void ratios correspond to the cylinders for 4 steps of compaction process, as shown in Figure 13-b. Each curve, illustrates the radial void variation for each of the four steps. The mean distribution of void is rather regular, except for the first and the last point of step 1 and step 2. Void ratios are larger at the center and periphery of the cylinders, especially in steps 1 and 2. Once again, this demonstrates the remarkable effect of the wall which is already mentioned above. Furthermore, during the specimen preparation, the grains firstly fell into the center of the cell. They were then dispersed around until spread to the cell. This leads to the existence of many macro pores in cylinder R_1. Hence, under low stress, the void ratio in this cylinder is larger than the void ratios in the intermediate cylinders (see the first point of two curves S1 and S2).

In spite of the precautions and carefulness taken during the specimen preparation, μCT clearly demonstrated that the sample was heterogeneous with looser zones in the center and the periphery. More complete analyses of the corresponding microstructural changes are being carried out.

4 Conclusions and Perspectives

These results show that a very loose yet stable-state of wet granular soils can be made for tests, even using very small water content. Besides, the oedometric tests exhibit three stages of compression curve of cohesive soils with different values of low initial densities. This is not existent in the usual soil mechanics.

μCT investigations of the microstructure allowed quantifying the degree of homogeneity and the arching effect at the bottom of sample by combining with the image processing.

The research program developed here can be extended into several directions. For example, the pore size distribution during the compaction test, images will be analyzed by using the Volumetric Digital Image Correlation tools. A comparison with simulation results will be mentioned for macro-microscopic behaviors.

Acknowledgments

The first author of the paper is supported by the Scholarship (the Project No 911) of Ministry of Education and Training, the Vietnamese Government to do this work.

References

[1] D. M. Wood, *Soil Behaviour and Critical State Soil Mechanics* (Cambridge University Press, 1990).

[2] J. K. Mitchell and K. Soga, *Fundamentals of Soil Behavior* (Wiley, 1976).

[3] Z. Fournier, D. Geromichalos, S. Herminghaus, M. M. Kohonen, F. Mugele, M. Scheel, M. Schulz, B. Schulz, R. Seemann, and A. Skudelny, J. Phys. Condens. Matter **17**, 477 (2005).

[4] J. Munõz-Castelblanco, P. Delage, J.-M. Pereira, and Y. J. Cui, Géotechnique Lett. **1**, 17 (2011).

[5] M. Jiang, H. Hu, and F. Liu, Can. Geotech. J. **1157**, 1147 (2012).

[6] N. Mitarai and F. Nori, Adv. Phys. **00**, 1 (2006).

[7] J.-F. Bruchon, J.-M. Pereira, M. Vandamme, N. Lenoir, P. Delage, and M. Bornert, Granul. Matter **15**, 783 (2013).

[8] J. C. Santamarina, in *Soil Behav. Soft Gr. Constr.*, edited by C. C. Ladd (2001), pp. 1–32.

[9] J. Banhart, *Advanced Tomographic Methods in Materials Research and Engineering* (Oxford University Press, New York, 2008).

[10] J. Hsied, *Computed Tomography - Principles, Design, Artifacts, and Recent Advances*, Second Edi (SPIE and John Wiley & Sons, Inc., United States of America, 2009).

[11] A. M. Tang, Y. J. Cui, J. Eslami, and P. Défossez, Geoderma **148**, 282 (2009).

[12] S. Ridler, T.W. Calvard, IEEE Trans. Syst. Man Cybern. **8**, 630 (1978).

[13] http://imagej.nih.gov/ij/index.html

[14] http://fiji.sc/Fiji

 DOI: 10.1051/e3sconf/20160906002

Grain scale mechanisms for capillary collapse in a loose unsaturated pyroclastic soil

Sabatino Cuomo[1,a], Mariagiovanna Moscariello[1] and Simon Salager[2]

[1] *University of Salerno, Department of Civil Engineering, Laboratory of Geotechnics, Italy*

[2]*University Grenoble Alpes, INP Grenoble, 3SR Laboratory, France*

Abstract. Soil collapse may occur passing from unsaturated to saturated conditions, thus causing major problems, among which one can mention a poor performance of the structures or the occurrence of landslides turning into flows. The mechanisms of the soil collapse have been studied at macroscopic scale since many years, while few observations at microscopic level are available. In this work, the mechanisms of capillary collapse were investigated for a volcanic (air-fall) pyroclastic soil of Southern Italy, which is characterized by an open metastable structure and is frequently involved into catastrophic rainfall-induced landslides. The experimental investigation was performed through X-ray Computed Tomography, which allows reconstructing 3D images of the specimen from the spatial distribution of the linear attenuation coefficient. The tests were carried out on coarse sand. During the tests, the specimens were loaded by its self-weight without any external load, and the suction was gradually reduced until the specimen collapse occurs. The aims of the experimental program were: i) follow the transformation of the specimen's microstructure; ii) evaluate the variation in terms of water content, porosity and grains spatial distribution; iii) analyse the effect of grain size distribution on the development of capillary forces and mass forces. The experimental evidences outline that, for the coarse pyroclastic sand, the collapse occurs at a very low suction, while it is not mandatory to reach the complete saturation.

1. Introduction

The collapse is commonly defined as the decrease in total volume of a soil resulting from the wetting-induced breakdown of its structure at essentially unchanging total vertical stress [1]. It occurs in granular soils, when capillary forces are of the same order of magnitude as other interparticle forces (i.e. skeleton forces, weight of particles, cementation) and when the microstructure is loose enough.

Several authors [2, 3, 4] studied the collapse that occurs in loose unsaturated soils and concluded that: i) the collapse deformation due to a suction reduction depends mainly upon the density and stress state under which it occurs; ii) the volume change due to a suction decrease mainly depends on the mean net stress and the initial void ratio, and the maximum collapse occurs when the mean net stress equals the initial yielding mean net stress of the unsaturated soil; iii) for unsaturated specimens compacted at the same initial density, the oedometric compression curves obtained for specimens saturated at different mean net stresses are more or less the same; the compression curve for initially unsaturated specimens wetted to zero suction is influenced by the value of initial density.

The interest for the collapse is that it causes various problems in geotechnical engineering, such as poor performance of the structures and the occurrence of rainfall-induced landslides. A significant example is represented by the landslides which involve the pyroclastic deposits of Southern Italy made of unsaturated soils originated from the explosive activity of the Somma-Vesuvius volcanic apparatus [5, 6, 7]. For these soils, an extensive data-set is available on the hydro-mechanical behaviour under both saturated and unsaturateted conditions. Indeed, the collapse of the pyroclastic soils was investigated at the specimen scale through wetting tests in triaxial and oedometric conditions. However, the capillary collapse and the modifications of the microstructure during the wetting process have been less observed. In this paper, the mechanism of capillary collapse was investigated for a pyroclastic soil of Southern Italy through the X-ray Computed Tomography, which is a non-destructive technique that allows visualizing the internal structure of a body through the measurement of density and atomic composition [8]. So far, the applications of the tomography to porous media have been mainly focused on: i) the analysis of porosity and fluid flow; ii) the evaluation of density, water content and volume fractions [9]; iii) the characterization of the microstructure of asphalts and concretes [8]; iv) the characterization of the local deformations within a soil specimen in a mechanical testing device [8].

[a] Corresponding author: scuomo@unisa.it

2. X-ray Computed Tomography

The basic components of X-ray CT scanners are an X-ray source, a detector and a rotation system. Two possible configurations exist [10]: i) the X-ray source and the detector turn around the symmetry axis of the object (common for medical field); ii) the object turns around its symmetry axis while the detector is at a fixed place (common in the industrial field).
The experimental technique used in the present work was proposed by Arthur [11] and requires the acquisition of radiographs from different positions during a slow rotation of the specimen around its vertical symmetry axis.
After the acquisition, the images are processed – aimed to reconstruct the entire body volume – through the application of digital filters to improve the image quality and to reduce the artefacts related to the X-ray CT technique (i.e. misalignment of source, detector and axis of rotation; detector defects; anomalous bright pixels in the radiographs).
The reconstruction procedure provides several contiguous cross-sections, either vertical or horizontal, which allows 3D visualization of the selected features and the quantification of the 3D volumes
After the body reconstruction, the regions occupied by each phase can be recognized by visual interpretation – obtaining qualitative estimates –, or using computational methods to obtain quantitative results [12]. The recognition of the phases which constitutes the object is commonly called "segmentation". Specifically, the segmentation consists into labelling each voxel (volumetric pixel) based on the ratio of its gray value to a reference threshold [13]. This is done referring to the "intensity histogram", which represents the number of voxels as a function of the intensity.
The segmentation is not an immediate process for geomaterials and in particular for unsaturated soils, which are composed by three main phases (grains, air and water). The histogram of an unsaturated soil is schematically composed of three normal distributions and therefore three peaks exist. Moreover, the tomographic images are obtained by absorption contrast, so it is possible to distinguish the phases by intensity or gray values. The low-density phases like air or water have lower intensity than solid grains; thus, the voxels occupied by air are darker than the solid voxels, while water has an intermediate gray level (Fig. 1).
The histograms of the different phases may overlap due to two artefacts: Gaussian Noise (GN) and Partial Volume Effect (PVE).
The GN is homogenously present in any image; while, the PVE is related to the existence of so called "mixels", i.e. voxels in which two or more phases are present. A mixel shows an intermediate grey value in between the two Gaussian curves, and the grey value depends on the percentage of each phase in the voxel. Consequently, a single threshold of the grey value is not sufficient to distinguish each phase properly.
To overcome the difficulties related to the CT images of unsaturated soils, a kind of region growing method is used [12]. The method was developed as one of region-based segmentation methods for digital images [14], and postulates that the neighboring voxel within one region have similar values [15]. The region-growing type segmentation is applied on 3D X-ray 8 bits (0-255 gray level range) reconstructed images. The procedure is composed of four steps: i) partial thresholding; ii) PVE filtering; iii) simultaneous phase growing; iv) filling of interface.
1) The "partial thresholding" consists in seeding the regions where each phase is present. The first step consists into selecting the regions, where the solid phase is certainly present. Then, the peaks of grains and air histograms are chosen. In this way, the grains threshold isolate the 50% upper part of the grains Gaussian curve and the air threshold isolate the 50% lower part of the air Gaussian curve. For the middle Gaussian curve (water), the thresholds are defined to select a range centred at the peak of the water curve and extended 25% to the left (dark) and to the right (light) of the peak.
2) After the partial thresholding, four groups of voxels become visible: the dark phase (B1), the middle phase (B2), the bright phase (B3) and the remaining non-selected voxels (R), so it is possible to start the middle phase PVE filtering. The PVE consists of voxels that belongs to the group B2. A spherical selection around the voxels of B2 group can be used to distinguish the voxel of the middle phase and the PVE. The standard deviation of the voxels surrounding each voxel belonging to B2 group is used to separate the real middle-phase voxels from PVE voxels. The standard deviation of voxels representing PVE is higher than the voxel representing the middle phase, causing a small increase in the highest standard deviation values in the histogram. The voxels with high standard deviation (PVE voxels) are then excluded by using a threshold filtering of standard deviation. The threshold is taken at the inflexion point on the histogram. After PVE filtering, the voxels considered as PVE are excluded from B2 and put into R.
3) The simultaneous phase growing method consists of growing all three phases into the group of the remaining voxels (R). At this stage, R represents the interface between the different phases and so they are not assigned to any phase. The growth stops when all the voxels belonging to R have at least two distinct phases as neighbours. When the phases are growing simultaneously, the first phase that "reaches" a new voxel incorporates it, but the phases do not "see" where they

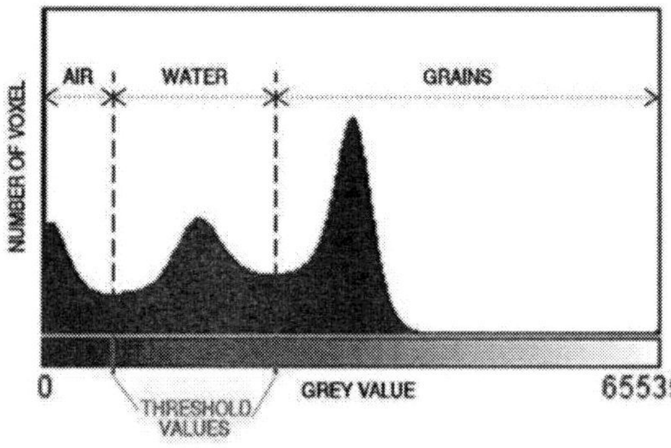

Figure 1. Histogram peaks for air, water and grains (Khaddour et al., 2013).

grow. Hence, there is a potential for overgrowing in some areas of the image. These phenomena may be limited by choosing adequate partial threshold at the beginning of the procedure. Moreover, to avoid the growth of each phase in extreme gray values, a "tolerance threshold" is chosen for air, water and grain (T_{air}, $T_{1\text{-water}}$, $T_{2\text{-water}}$ and T_{grain}). This means that the grey values where the phase is assumed as absent are excluded. These exclusions are determined by setting the gray value from which the Gaussian curve of each phase can be neglected. In this step, the PVE effect is also taken into account. After PVE filtering, some voxels (belonging to air or grain phases) of middle phase are not considered inside the partial threshold.

4) The final step of the procedure (interface filling) is to determine the phase to which the voxels of the interface are belonging. For each voxel at the interface, another sphere selects the surrounding voxels. The most present phase inside the sphere determines the phase segmentation. If two phases are in equal amount in the sphere, the considered voxels remain undecided. This procedure is repeated, and, usually after four times, the majority of the voxels have been filled. The remaining undecided voxels, instead, are defined with a simple threshold method and the phases are attributed as a function of the voxels grey level.

3. Material and methods

The sand used for the test was derived from a pyroclastic material of Southern Italy, which was formerly studied by [16], and classified as "class A ashy soil". The entire grain size distribution of the class A consists in 1.1 to 6.4% Gravel, 40.6 to 51.3% Sand, 40.9 to 53.61% Silt, 1.1 to 6.4% Clay. The grain shape is ranging from angular to sub-angular. The grain contains internal voids, so the soil specific gravity is quite small and equal to 2.55. The inter-particle porosity of undisturbed pyroclastic soil varies from 0.63 to 0.74. These high values of porosity are related to the air-fall deposition, which conferred an "open" structure to the aggregates of soil particles [17].

For the specimen preparation, the pyroclastic soil was sieved and only the fraction with grain size included between 425 μm and 600 μm was kept. A mixture of sand and water was carried out at 20% of gravimetric water content, in order to obtain the loosest specimen. The mixture was put into a bag at least for two hours to ensure the correct hydration of the grains. The cylindrical specimen was prepared directly into the X-ray cabin. When the arrangement was built, the mixture was poured into the cell with a small spoon. During the preparation of the specimen, it was better not apply any type of load, in order to obtain a loose specimen, with the grains configuration more similar to undisturbed pyroclastic specimen. Then, the excess of soil was removed and the cell was covered. The cell was placed on the rotating table and connected to the suction controlled tank.

The tests were performed through the X-ray Tomograph of the 3SR Laboratoire at the Université Joseph Fourier (Grenoble, France). The test apparatus was composed of: i) the cell; ii) the suction controller tank; iii) the rotating table; iv) the X-ray source; v) the X-ray detector (Fig.2).

The need of high-resolution images imposed a relatively small size of both the specimen and cell. Indeed, the cell had a cylindrical shape (10×10 mm, diameter and height) and was built of acrylic fabric, which facilitates the penetration of X-rays. The apparatus was formerly designed and used to investigate the water retention behaviour of unsaturated materials (such as glass beads and Hostun sand) with X-ray tomography technique [18, 13]. The acquisition of radiographs and the control are carried out through the software RX-act developed by RX solutions (Annency, France).

Tab.1 summarizes the X-ray scanning conditions. It is worth underlining that the voxel size was chosen near to technical upper limit of the X-ray source and detector, in order to acquire the images at high resolution. This was easily obtained approaching the cell and the X-ray source, but it implied that the other parameters of the test (like time of scanning, rotation rate and time of equalization) were selected in order to avoid disturbing the specimen.

During the test, the specimen was only subjected to its self-weight. The capillary pressure was controlled through the tank filled of deaired and demineralized water. The suction values were applied through the "negative water column" technique.

The first 3D image was acquired just after the specimen preparation to know the initial conditions and the internal structure of the specimen.

Then the suction was lowered from 0.5 to 0.1 kPa in 4 steps (0.5 kPa, 0.3 kPa, 0.2 kPa and 0.1 kPa).The suction between two next steps was kept constant at least for 8 hours. This time lag was necessary for the specimen to reach the equalization and to remain stable during the acquisition. Once the equilibrium reached, the 3D X-ray images were acquired. The time necessary to acquire each 3D X-ray image was about 2 hours. For this test, the rotation of the specimen around its symmetry axis was slow in order to avoid any modification of specimen structure. The 3D images were reconstructed and then the segmentation of the images was carried out obtaining images composed of three gray values (trinarized images). Macroscopic porosity was calculated using the

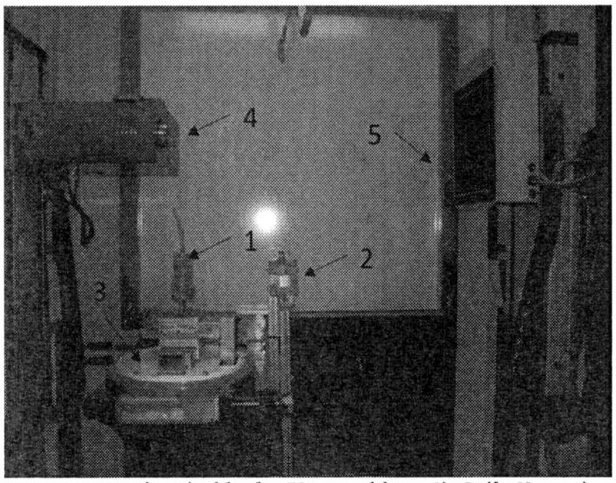

Figure 2. Device, inside the X-ray cabinet. 1) Cell; 2) suction controlled tank; 3) rotating table; 4) X-ray source; 5) detector.

DOI: 10.1051/e3sconf/20160906002

Table 1. Table of X-ray scanning conditions.

# Projection	1120
Image Averaged	5/10
Filter	Anti-ring shift
Voxel size	7.56 μm
Source Voltage	100 kV
Source Current	50 μA
Time of scanning	2 h

trinarized images. The 8-bits resulted trinarized images was of three gray values (64 for air, 128 for water, 192 for grain). The histograms of these images gave three values of the number of pixels, which were labelled for air P_a, for water P_w and for grains P_g. Through the number of pixels representing each phase, the porosity (n) was calculated through the Eq. 1.The numbers of pixels representing each phases was also used to estimate the degree of saturation (S_r) (Eq. 2) and the water content (Eq. 3).

$$n = \frac{(P_g + P_w + P_a) - P_g}{P_g + P_w + P_a} \tag{1}$$

$$S_r = \frac{P_w}{P_w + P_a} \tag{2}$$

$$\theta = S_r \cdot n \tag{3}$$

4. Results and discussion

The capillary collapse phenomenon was investigated through a test on coarse pyroclastic sand. The specimen preparation procedure allowed reproducing the typical features of collapsible soils: an open structure, a high void ratio (or high porosity), low interparticle bond strength. Moreover, the angular shape of the grains allowed preserving the open, metastable structure. The initial configuration of the specimen was characterized by several aggregates of grains. The porosity was very high, because of the macro-voids that surrounded the aggregates. The Water Retention curve, in particular the main wetting curve, was evaluated using the same cell of figure 2 and the specimens were created through the above procedure and disassembled at each suction value investigated. The initial porosity was not equal at each steps of the main wetting curve and varied from 0.59 to 0.63, this is a consequence of the specimen preparation procedure. The two curves (Fig. 3) did not overlap completely due to the different porosity of the specimens. The 3D X-ray images of the specimen at 0.5 kPa and 0.3 kPa of suction show that the increment of water content did not influence the structure of the aggregates and the presence of macro-voids (Fig. 4). The specimen at 0.2 kPa of suction presented several transformations, the aggregates were not more separated by air, the water content increased and the macro-voids had different shape. The body forces produced only a little reduction of the height, but it was not uniform. The X-ray images at the suction of 0.1 kPa show the collapse of the specimen.

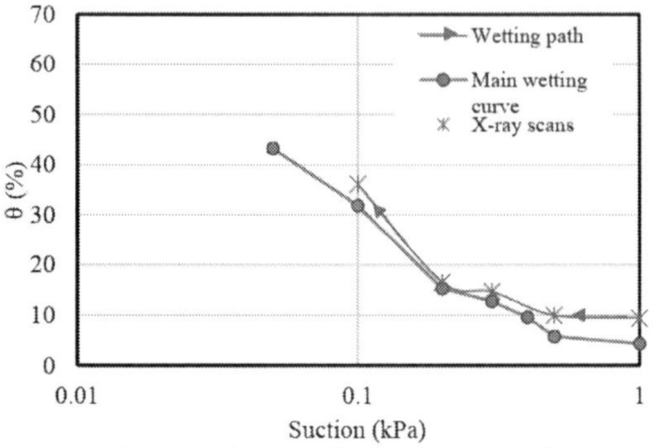

Figure 3. Wetting path during the test compared with the main wetting curve of the sand investigated.

In the correspondent image, the aggregates are not observable, but few air bubbles are still present even if the collapse happened. In figure 4, the sequence of vertical slices at different section shows the evolution of the microstructure and the disappearance of the aggregates during the wetting. The reconstructed X-ray images of the test, consisting in 1500 horizontal slices, were trinarized using a Matlab code, which implemented the procedure illustrated in the Section 2.

The trinarized images presented some small black dots inside the grains. These dots represented some noise of the image. The procedure, used to clear the images, consisted of the despeackle of noise for each phase. Particularly, the grain and the water were separated in order to obtain images with only two phases: i) water and air; ii) grain and water. Later on, the two types of images were binarized and the despeackle tool was applied. Then, the two images were unified. The unified images - clean of noise - were used to the quantitative analysis of the water content, degree of saturation and porosity. The

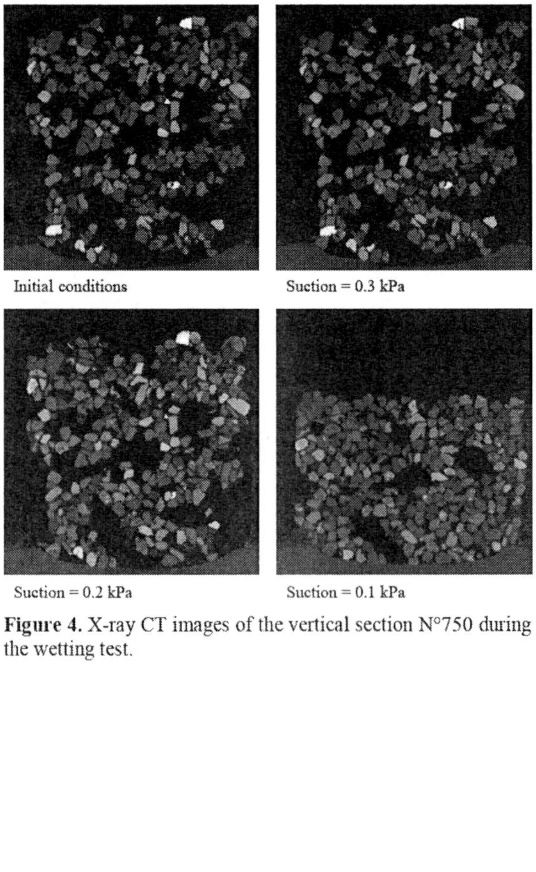

Figure 4. X-ray CT images of the vertical section N°750 during the wetting test.

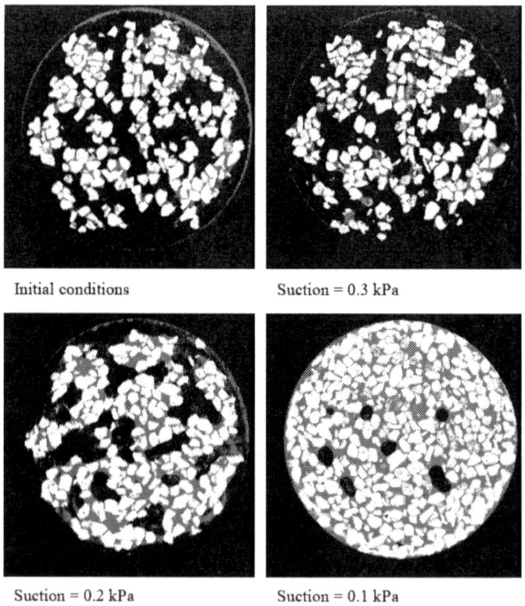

Figure 5. Trinarized images for Horizontal slice N°750, at the middle of the specimen.

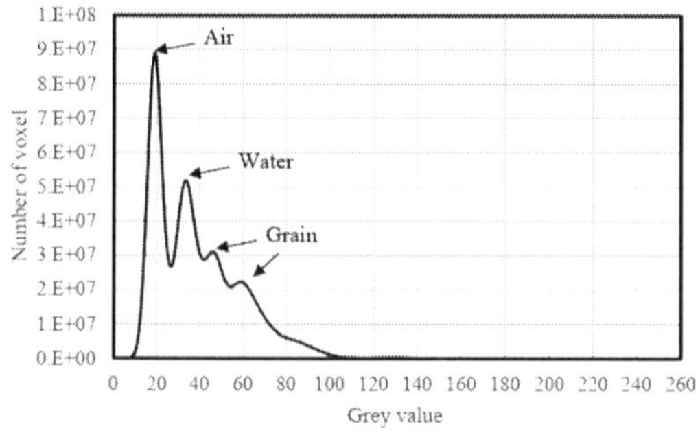

Figure 6. Histogram corresponding to the scan carried out at the collapse. The three phases and the correspondent peaks are clearly observable.

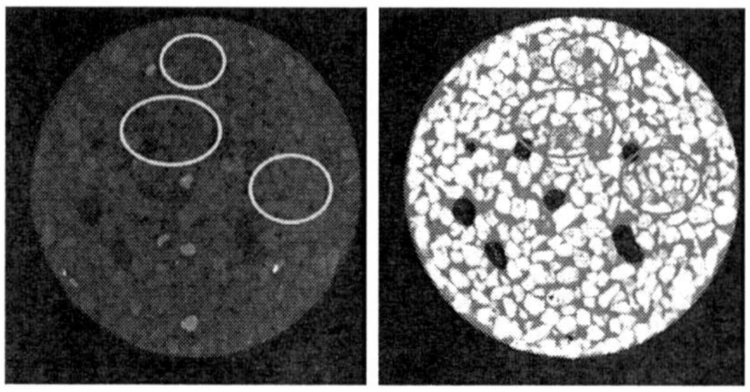

Figure 7. Comparison between the original image (left) and the trinarized image (right) of the scan carried out at 0.1 kPa of suction (scan corresponds to the collapse of the specimen).

pyroclastic sand, because of its volcanic origin, is made of grains not homogeneous, which have different "adsorption capacity" (density), so that the histogram is quite different from the typical histogram of a three phases material (Fig.1). The histogram concerning the scans at suction 0.1 kPa is plotted in figure 6 to show the

Table 2. Porosity (n), water content (θ) and saturation degree (S_r) evaluated from the X-ray images of the horizontal slice N°750 of figure 4, compared to standard measurements.

#	*Suction* (kPa)	n_{st} (%)	n (%)	θ (-)	S_r (%)
1	I.C	72	67	0.08	10
2	0.5	72	66	0.10	15
3	0.3	71	65	0.12	19
4	0.2	69	59	0.24	42
5	0.1	54	40	0.35*	88*

*Measurements through standard procedure: θ=0.39, S_r=0.71, and n_{st} column.

main differences towards a common three phases specimen. The histogram of figure 6 shows one peak for air, one for water and two peaks for the solid phase due to the different density of the grains.

The grains with a grey level close to that of the water correspond to little pumices with a high internal porosity and low specific weight. The grains with high grey value instead have a low value of internal porosity.

The presence of grains with gray level close to water phase influenced the results of the trinarization and so the estimate of the main physical properties (i.e. water content, saturation degree, porosity). For example, figure 7 shows the comparison between the original image and the result of the trinarization. The two images correspond to the horizontal section in the middle of the specimen at collapse. Yellow and red circles (Fig.7) correspond to the grains, which are entirely visible in the original image, while are partially or totally computed as water in the trinarized image. The soil porosity evaluated through the trinarization of the images of figure 5 (n_{rx}) – for the horizontal cross-section at the middle height specimen – was compared to the soil porosity calculated through the standard measurement of the specimen height during the tests (n_{st}), and they are both reported in Table 2.

The differences are attributable to: i) the presence of "mixels", which may affect the evaluation of the solid phase amount; ii) to the different procedure to evaluate the height of the specimen; iii) and, more importantly, to the inhomogeneties induced in the specimen by the wetting process.

5. Conclusions

The capillary collapse of coarse volcanic sand was investigated using the X-ray Computed Tomography. This non-destructive full-field method allowed following the internal changes of the specimen during a wetting test. The images recorded were used to obtain the degree of attenuation of the X-rays, which is related to the material composition and energy of the source. These images, after the reconstruction and segmentation procedures, furnished qualitative and quantitative information about internal structure, porosity, water content, saturation degree, but also about intergranular forces, stresses and displacements.

The qualitative description of the changes in the internal structure of the specimen and the variation of main physical properties (porosity, water content and saturation degree) during the wetting test were discussed. Particularly, the specimen showed a double porosity (void among the grains and void among the aggregates), which collapsed only during the saturation of the biggest macro-pores when capillary bridges between grain clusters merged. One peculiarity of the soil behaviour was the presence of air bubbles at collapse. This was confirmed by the saturation degree, which did not reach unity during the collapse. The test underlined that the capillary bridges were enough strong so that the aggregates preserved their structure until suction reaches 0.2 kPa, corresponding to a 30% increment of the saturation degree, from 10 to 42%. In conclusion, quantitative measurements were confidently achieved for water content and porosity, and in addition the experimental evidences outline that for the coarse pyroclastic sand the collapse occurs at a very low suction, while it is not mandatory to reach the complete saturation.

Acknowledgements

This research activity was developed within the framework of 2014-2015 Erasmus+ European Project and 2015-2016 Galileo Project Campus France "Soil mechanical behaviour from grain to specimen scale laboratory testing: towards new sustainable mitigation works against flow-like landslides and similar phenomena related to climate change" Grant n. G15-110.

References

1. Lawton, E. C., Fragaszy, R. J., Hardcastle, J. H., Stress ratio effects on collapse of Compacted Clayey sand. Journal of Geotechnical Engineering, ASCE, 117(5), 714–730 (1991).
2. Jennings, J.E.B., Burland, J.B., Limitations to the use of effective stresses in partly saturated soils. Géotechnique, 12(2), 125–144 (1962).
3. Kato, S., Kawai, K., Deformation characteristics of a compacted clay in collapse under isotropic and triaxial stress state. Soils Foundation, 40(5), 75–90 (2000).
4. Sun, D., Sheng, D., Xu,Y., Collapse behaviour of unsaturated compacted soil with different initial densities. Canadian Geotechnical Journal, 44(6), 673–686 (2007).
5. Cascini, L., Cuomo, S., Guida, D., Typical source areas of May 1998 flow-like mass movements in the Campania region, Southern Italy. Engineering Geology 96(3), 107-125 (2008).
6. Cascini, L., Cuomo, S., Pastor, M., Sorbino, G., Modelling of rainfall-induced shallow landslides of the flow-type. ASCE's Journal of Geotechnical and Geoenvironmental Engineering, 1, 85-98 (2010).
7. Cascini, L., Cuomo, S., Pastor, M., Sacco, C., Modelling the post-failure stage of rainfall-induced landslides of the flow-type, Canadian Geotechnical Journal , 50(9), 924–934 (2013).
8. Viggiani, G., Hall, S. A., Full-field measurements in experimental geomechanics: historical perspective, current trends and recent results. ALERT Doctoral School 2012 Advanced experimental techniques in geomechanics (2012). Available at http://www.alertgeomaterials.eu/internal/2012/school/2012_ALERT_school.pdf
9. Riedel, I., Characterization and micro-scale analysis of Hostun sand water retention behaviour. MSc, Université de Grenoble, 216 pages (2011).
10. Baruchel, J., Buffière, J-Y., Maire, E., Merle, P., Peix, G., X-ray Tomography in Material Science. Hermes Science Publications, Paris, (2000).
11. Arthur, J.R.F., New techniques to measure new parameters. Proceedings of Roscoe Memorial Symposium o Stress–Strain Behaviour of Soils, Cambridge, G.T. Foulis & Co., 340–346 (1971).
12. Hashemi, M. A., Khaddour, G., Francois, B., Massart, T. J., Salager, S., A tomographic imagery segmentation methodology for three-phase geomaterials based on simultaneous region growing. Acta Geotechnica, 9, 831–846 (2014).
13. Khaddour, G., Hashemi, A., Salager, S., Desrues, J., Massart, J., Francois, B., Phase segmentation in 3D X-ray images of unsaturated granular materials. Proceedings of XVI French-Polish Colloquium of Soil and Rock Mechanics, (2013).
14. Rosenfeld, A., Kak, A. C., Digital Picture Processing, Academic Press, Inc. Orlando, FL, USA (1982).
15. Adams, R., Bischof, L., Seeded region growing. Pattern Analysis and Machine Intelligence, IEEE Transactions 16(6), 641-647 (1994).
16. Bilotta, E., Cascini, L., Foresta, V., and Sorbino, G.. Geotechnical characterization of pyroclastic soils involved in huge flowslides. Geotechnical and Geological Engineering, 23, 365-402 (2005).
17. Lancellotta, R., Di Prisco, C., Costanzo, D., Foti, S., Sorbino, G., Buscarnera, G., Cosentini, R.M., Foresta, V. Caratterizzazione e modellazione geotecnica. In: Criteri di zonazione della suscettibilità e della pericolosità da frane innescate da eventi estremi (piogge e sisma). Leonardo Cascini. Composervice srl, Padova, 266-319 (in Italian). ISBN 9788890687334 (in Italian) (2012).
18. Riedel, I., Andò, E., Salager, S., Bésuelle, P and Viggiani, G., Water retention behaviour explored by X-ray CT analysis, Unsaturated Soils: Research and Applications, 1, Proceedings of the 2nd European Conference on Unsaturated Soils, 20-22 June 2012, Napoli, Italy, Springer, 81-88 (2012).

 DOI: 10.1051/e3sconf/20160906003

Microstructure analysis of laboratory and in-situ compacted silts

Giacomo Russo [1,a], Andrea Rezza [1], Claudio Mancuso[2], Vincenzo Oliviero[2], Francesca D'Onza[3], Domenico Gallipoli[4] and Simon Wheeler[5]

[1]*Department of Civil and Mechanical Engineering, University of Cassino and Southern Lazio, Italy*
[2]*Department of Department of Civil, Architectural and Environmental Engineering, University of Naples Federico II, Italy*
[3]*Italian National Agency for the New Technologies, Energy and Sustainable Economic Development, ENEA, Italy*
[4]*Laboratoire SIAME, Université de Pau et des Pays de l'Adour, France*
[5]*School of Engineering, University of Glasgow, UK*

Abstract. The paper presents and discusses some results of an experimental research aimed at analysing the influence of compaction variables (w and energy) and method on the resulting microstructure of a compacted silty soil. In particular, the experimental data here discussed allow to compare the microstructure induced by different dynamic compaction techniques, comparing that characterising specimens obtained by two laboratory methods (Proctor standard and Harvard) and that of samples compacted in-situ during the construction of an embankment built for river regimentation purposes. Both undisturbed and disturbed samples have been retrieved from the embankment, the latter one with the purpose of collecting the soil subsequently used for laboratory compaction. Microstructural analyses (SEM, MIP) performed on laboratory and in-situ compacted samples evidenced a substantial similarity of the texture induced by the various compaction techniques, highlighting that laboratory compaction is suitable to provide soil samples representative of earth in-situ compacted soil.

1 Introduction

At the phenomenological scale of laboratory tests the experimental evidence highlights that the microstructure of compacted soils depends on water content and dry density adopted for the specimen preparation. In fact, it is well known that soil compacted wet of optimum, tends to have a homogenous texture, whereas those compacted on the dry side tends to exhibit a bimodal pore size distribution, referred to as 'double structure'[1, 2, 3, 4, 5], with two or more distinct classes of pore dimensions characterising the soil texture. Typically, there are two distinct classes of pores, the larger one characterising the inter-aggregate pore space and the smaller ones typical of the aggregates, thus named intra-aggregate pores [6, 7]. The distribution of the pore classes is likely to influence the idro-mechanical behaviour of the compacted soils.

In the paper some results on microstructural analyses on in-situ and laboratory compacted samples are reported. Some of the tested samples have been retrieved from an experimental embankment, 'identical' to the real embankments used for river regimentation purposes, built in order to evaluate the evolution of suction caused by soil-atmosphere and simulated flooding in real earth structure [8] . The embankment was constructed using the same compaction techniques adopted for service embankments using a fine-grained soil of medium plasticity taken from the floodplain. Other samples (from the same soil) have been compacted in the laboratory using the Proctor standard and the Harvard compaction techniques. All the above-mentioned group of samples have been subjected to an extensive testing program in order to characterise their hydro-mechanical behaviour and to laboratory tests to analyse their microstructure. These latter tests form the focus of the present paper and are based on mercury intrusion porosimetry (MIP) performed in order to obtain a quantitative description of the microstructure of soils and the evolution of the pore size distribution (PSD) after mechanical and hydraulic paths. MIP tests allowed the comparison of pore size distributions induced by different laboratory compaction procedures and initial water contents. Scanning electron microscopy (SEM) observations have been also performed to further investigate the compaction induced fabric of the mentioned classes of samples.

2 Material and experimental procedures

2.1 Alluvial silt

The soil is an alluvial sandy silt with relatively low clay fraction, as showed by the grain size distribution reported in Figure 1. The main physical properties of the soil are reported in Table 1. Overall the soil is classified as an inorganic silt of medium/high plasticity.

[a] Corresponding author: giarusso@unicas.it

Table 1. Physical properties of alluvial silt.

Specific gravity	Gs	2.75
Plastic limit	w_P	20
Liquid limit	w_L	37
Plasticity Index	PI	17

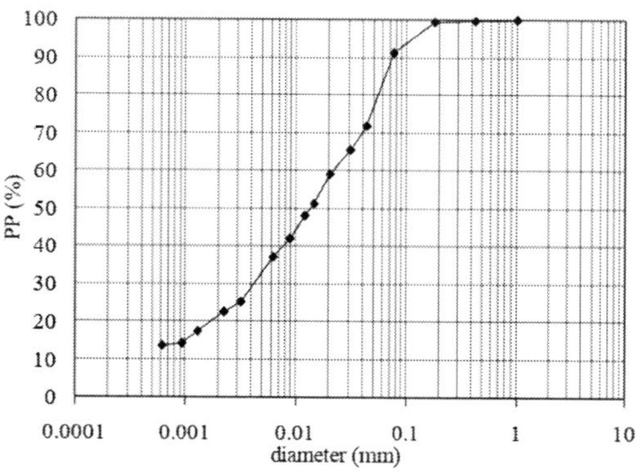

Figure 1. Grain size distribution of alluvial silt.

2.2 Compaction

The Standard Proctor compaction curve obtained for this soil by the ASTM D698-00a standard is reported in Figure 2, showing an optimum water content (w_{opt}) of 20.7% and a (maximum) dry density (γ_{dmax}) of 1.65 g/cm^3.

The analysis of the in situ undisturbed samples retrieved during the construction (here not reported) confirmed that soil construction of the embankment was performed in wet conditions [8], with water contents higher than the reference Proctor optimum. It is worth noting, however, that the data shown refer to samples retrieved from the embankment some years after its construction, thus the actual water content values are likely to be also influenced by wetting-drying cycles that the embankment itself underwent under service conditions. In order to reproduce initial conditions similar to the in-situ ones, it was decided to adopt the Harvard procedure [9] as it allows to obtain samples smaller of the Proctor ones with comparable physical properties.

As well known, the Harvard compaction apparatus consists in a small volume mould (62.4 cm^3) hold by an extension collar, a base and clamps. The tamper consists of a brass rod carrying a spring, with a wooden handle on one end. Springs of different stiffness are available to perform compaction. The soil is compacted in five layers under the load transferred by pushing down a selected number of times the handle, with the possibility of selecting the spring stiffness. The Harvard compaction procedure has been calibrated with the standard Proctor procedure following the ASTM D4609-01. A number of 35 pressures of the spring of stiffness 20 lb resulted in a dry density of the soil within the interval of ± 0.016 g/cm^3 of the Proctor standard maximum dry density at a w_{opt} = 20.7%, as requested by the ASTM procedure.

In Figure 2 the Harvard compaction curve is also reported and compared with the Proctor one. The initial state of the in-situ sample B is also shown together with some points (crosses) obtained by the Harvard compaction procedure.

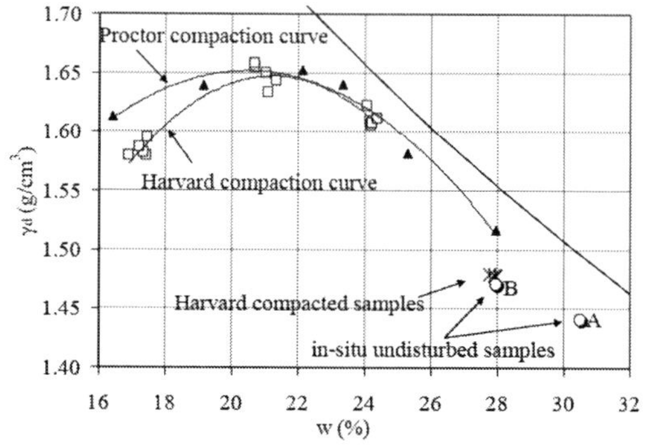

Figure 2. Standard Proctor compaction curve and undisturbed in-situ compacted samples.

2.3 Microstructure analyses

MIP tests were performed by a double chamber Micromeritics Autopore III apparatus. In the filling apparatus (dilatometer) the samples were outgassed under vacuum and then filled by mercury allowing the increase of absolute pressure up to the ambient one. Using the same unit the intrusion pressure was than raised up to approximately 200 kPa by means of compressed air. The detected entrance pore diameters ranges between 134 μm and 7.3 μm (approximately 0.01 MPa - 0.2 MPa for a mercury contact angle of 139°). After depressurisation to the ambient pressure, the samples were transferred to the high-pressure unit, where the mercury pressure was increased up to 205 MPa following a previously set intrusion program. At any intrusion step a time sufficient to observe a quasi-static penetration of mercury [10] was allowed. Corrections to the pore-size distribution due to the compressibility of the intrusion system were applied performing a blank test. The inner central region of each compacted sample was carefully cut in small pieces of approximately 1 – 2 g to obtain the MIP specimens. Sample dehydration requested before performing MIP analyses was obtained according to the freeze-drying method [11].

SEM analyses were performed on dehydrated samples in order to highlight their fabric.

3 Results

The cumulative intruded volume has been represented as a function of the entrance size of pores in terms of intruded void ratio e_{MIP}, defined as the ratio between the mercury intruded volume and the volume of solids. At the end of the intrusion stage, e_{MIP} can be directly compared to the reference void ratio of the sample e_0 determined in a traditional way by measuring the volume and water content of the sample used for the MIP tests. The initial properties of the MIP samples considered in this paper are reported in Table 2.

Table 2. Physical properties of MIP samples.

Sample	w (%)	e_0	e_{MIP}
Proctor w_opt	20.00	0.674	0.378
Harvard w_opt	20.10	0.670	0.557
in situ_A	31.36	0.909	0.745
in situ_B	25.73	0.870	0.686
Harvard in situ	26.53	0.859	0.606

In Figure 3 a comparison between MIP tests on standard Proctor and Harvard compacted samples at optimum water content is reported, evidencing the similar texture of the specimen despite the different compaction techniques. In particular, Figure 3a shows a difference in the cumulated intrusion void ratios after the intrusion stage, with a higher intruded void ratio e_{MIP} for the Harvard sample, closer to the initial void ratio e0. Both the pore size distributions show (Figure 3b) a mono-modal curve, with the most frequent value of pore entrance size between 0.3 ÷ 0.4 μm.

In Figure 4 some SEM images of the two samples are reported, confirming the similarity of their microstructural organization, independently on the compaction method. The occurrence of slightly larger pores can be observed for the Harvard sample (Figure 4b). As a result, the similar features induced by different compaction techniques at optimum water content confirmed the suitability of the Harvard technique for preparing MIP samples comparable with the Proctor standard one.

Figure 5 reports the results of the MIP tests performed on the in-situ compacted samples A and B (see Figure 2) and the Harvard one prepared at the same initial water content of B. Comparing the cumulative intruded void ratio of the in-situ samples (Figure 5a), it can be observed that for the sample A the intruded void ratio is higher than the sample B, as expected from its higher initial water content and lower dry density. The cumulative intruded volume of the Harvard sample compacted at in-situ water content is closer to the B in situ samples, as expected for the initial state of the samples. The intruded void ratio at the end of the intrusion stage e_{MIP} is lower than the initial void ratio e_0 of the samples, highlighting a similar degree of not intruded porosity of the three samples.

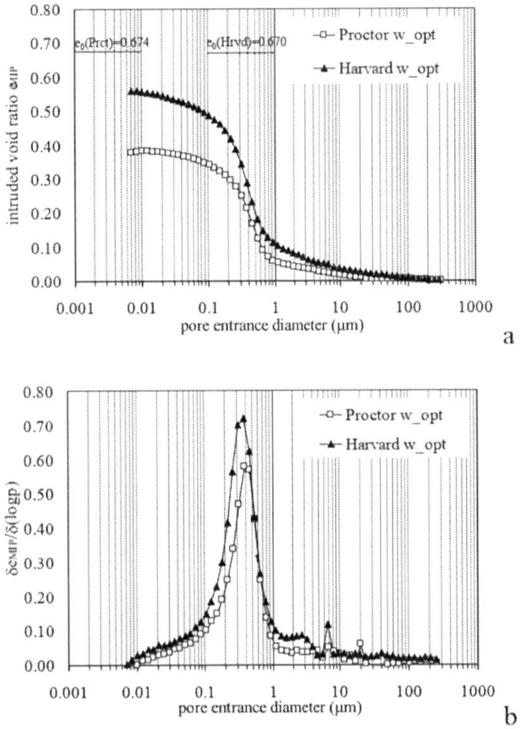

Figure 3. MIP results on Standard Proctor and Harvard compacted samples at optimum water content. a) intruded void ratio, b) pore size density function.

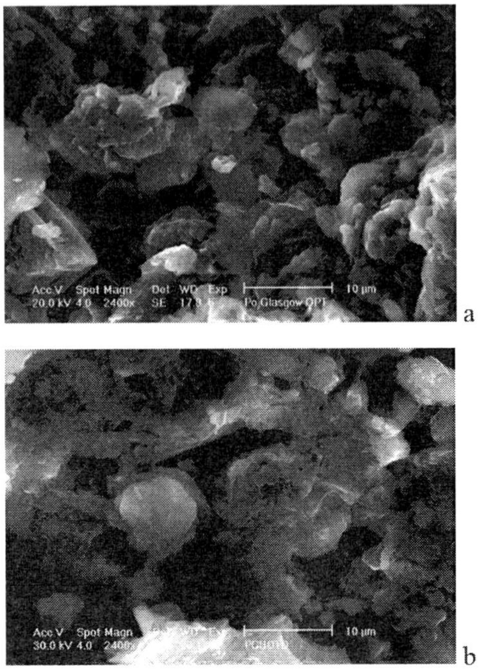

Figure 4. SEM analyses on a) Standard Proctor and b) Harvard compacted samples.

DOI: 10.1051/e3sconf/20160906003

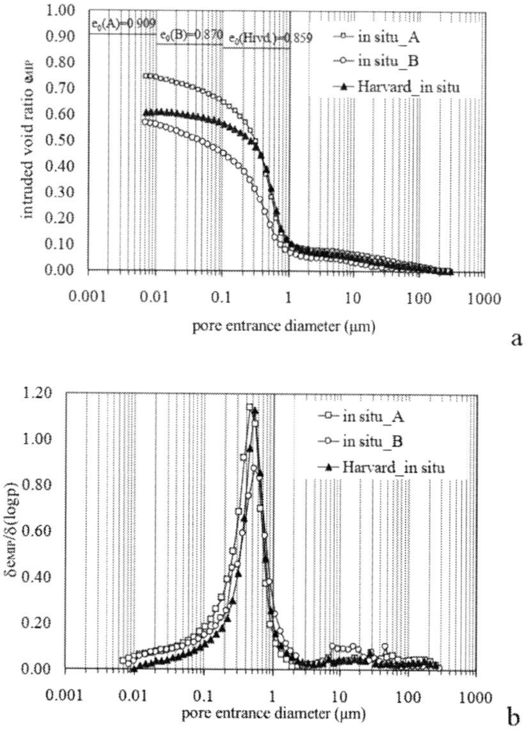

Figure 5. MIP results on in-situ samples A and B and Harvard compacted sample at in-situ water content. a) intruded void ratio, b) pore size density function.

Figure 6. SEM analyses a) on in-situ sample B and b) Harvard compacted sample at in-situ water content.

The frequency of pore entrance diameters is very similar for both the in-situ and laboratory compacted samples. The mono-modal curves of the pore size density functions showed in Figure 5b highlight a close correspondence, with the most frequent entrance pore diameter in the range 0.4–0.6 μm. A slighter difference can be found in the frequency of smaller pores (pore entrance diameter < 0.4 μm), which is systematically higher for the in situ undisturbed samples A and B with respect to the Harvard sample.

Finally, Figure 6 shows the SEM images of the in-situ B sample and Harvard one. The aggregated fabric of the two samples are quite similar apart for the detection of pores of relatively large entrance size for the sample B in the range between 1 μm and 10 μm (consistent with the slightly higher frequency evidenced by the MIP results in the same range), and an apparent less frequency of small pores inside the aggregates characterizing the fabric of the Harvard sample.

4 Conclusions

The paper reports some results of a joined research program between several institutions (see the heading of the paper) on the hydro-mechanical behaviour of a compacted alluvial silt. The results of MIP tests and SEM images from undisturbed samples taken from an experimental embankment and specimens obtained in the laboratory by different compaction techniques are presented and discussed to analyse possible effects of the compaction variables (w, energy) and method on the microstructure of the resulting soils. To this end, the microstructure of Proctor standard, Harvard and in-situ undisturbed samples has been analysed and compared. The comparison of Proctor standard and Harvard specimens at the optimum water content and density highlighted very similar soil fabrics, suggesting that the compaction technique does not play such an important role on the final texture taken on by compacted soils. The comparison between the microstructure of in situ undisturbed samples and Harvard ones compacted on the wet side at very similar water content and dry density also highlighted significant common features of the microstructure. In both the above mentioned cases, the results seem to confirm the validity of using laboratory compacted sample as reference for the analysis of the hydro-mechanical behaviour of soils used as construction materials.

Acknowledgments

The contribution of the European Commission via the Career Integration Grant SAFES (Service Assessment and Failure of Earth Structures), contract n. PCIG09-GA-2011-293727, is gratefully acknowledged.

References

1. Barden L., Sides G.R., J. Soil Found. An. 96:1171–1200, (1970).
2. Sridharan A, Altaschaeffl A.G., Diamond S. J.S.Mech.&Found.Div. ASCE, 97, 771–787, (1971)
3. Collins K, McGown A. Géotech. 24:223–254, (1974)
4. Alonso E.E., Gens A., Hight D.W. Proc. 9th ECSMFE, 5.1–5.60, (1987)
5. Cuisinier, O. & Laloui, L. Int. J. Num. &An. Meth. in Geomech., 28, 483-499 (2004)
6. Delage, P., Audiguier, M., Cui, Y.J, and Howat, D.. Can. Geotech. J. 33: 150-158, (1996)
7. Romero, E., A. Gens & A. Lloret. Eng. Geol. 54: 117-127 (1999)
8. Calabresi G., Colleselli F., Danese D., Giani G., Mancuso C., Montrasio L., Nocilla A., Pagano L., Reali E., Sciotti A. Can. Geotech. J. 50: 947–960 (2013)
9. Wilson, S., D. Eng.News Rec., (1950)
10. Romero, E., Simms, P. Geotech.l and Geol. Eng. 26:705-727, (2008).
11. Delage, P., Pellerin, F.M. Cl. Min., 19: 151-160, (1984).

Effect of pore-size distribution on the collapse behaviour of anthropogenic sandy soil deposits

Wiebke Baille[1,a], Alireza Jebeli[1] and Tom Schanz[1]

[1]*Ruhr-University Bochum, Chair of Foundation Engineering, Soil and Rock Mechanics, 44780 Bochum, Germany*

Abstract. In the former open-pit mines of the Lusatian region in Germany, several liquefaction events have occurred during the recent years in the anthropogenic deposits made of very loose sandy soils. These events are related to the rising ground water table after the stop of controlled ground water lowering. The very loose state is due to the formation of sand aggregates (pseudo-grains) during the deposition process. The pseudo-grains enclose larger voids of dimension greater than the single sand grain. Wetting induced collapse of the pseudo-grains is presumed to be one of the possible mechanisms triggering liquefaction. In the present study, the effect of larger voids on the wetting induced deformation behaviour of sandy soils is experimentally investigated by laboratory box tests. The deformation field in the sample during wetting was measured using Digital Image Correlation (DIC) technique. The results show that the observed deformations are affected by the pore size distribution, thus the amount of voids between the pseudo-grains (macro-void ratio) and the voids inside the pseudo-grains (matrix void ratio). The global void ratio of a sandy soil is not sufficient as single state parameter, but the pore size distribution has to be taken into account, experimentally as well as in modelling.

1 Introduction

In the former open-pit mines of the Lusatian region in Germany, several liquefaction events have occurred during the recent years in the anthropogenic deposits made of very loose sandy soils. These events have occurred spontaneously, i.e. they were not caused by an earthquake or other dynamic loading source. The deposits have an average thickness of about 50 m. Due to the technological process of deposition the sandy soils are of varying composition and density, leading to unknown direction and magnitude of the principal stresses inside the deposit. Mostly, the deposits are in a very loose state, in extreme cases showing a void ratio greater than e_{max}.

In general, the liquefaction events are related to the rising ground water table in the deposits due to the stop of controlled ground water lowering for about 25 years. Wetting induced collapse of larger voids inside the deposit is presumed to be one of the possible mechanisms triggering liquefaction.

The influence of soil structure on the soil behaviour is well known to be relevant for properties such as shear strength, compressibility, hydraulic conductivity, and soil-water characteristics of a soil. Soil structure changes with stress state, transfer of water and air, temperature, long-term gravimetric actions, and weathering. Natural and man-made geomaterials frequently exhibit two scales of porosity. A substantial amount of research work has been undertaken regarding the coupled hydro-mechanical behaviour of compacted clays. These soils typically are made-up of aggregates, leading to at least two different levels of pores: macro-pores between the aggregates and micro-pores inside the aggregates [1-7].

Wheeler (1988) [8] has proposed a conceptual model for describing the behaviour of marine sediments containing large discontinuous gas bubbles. Biogenic (degradation of organic matter), thermogenic or volcanogenic processes are the origin of gas in deep sediments. The soil matrix (silty, clayey sediment) is saturated, and discrete large gas bubbles having a pore radius much larger than the pores inside the soil matrix exist as long as the gas pressure is greater than the water pressure from the surrounding saturated soil matrix.

However, effects of porosity at different levels are usually not considered for coarse grained soils like sandy soils.

The following hypothesises with regard to the formation of larger gas filled voids exist: (1) formation of sand aggregates (pseudo-grains) during the technological deposition process and (2) biogenic processes inside the deposit [9]. In the former, the pseudo-grains enclose larger voids of dimension greater than the single sand grain forming a double-porosity soil system. The biogenic processes are the microbial transformation of organic matter (mainly coal) into CO_2 and the formation of CO_2 due to buffering of existing acid rich regions by carbonatic constituents. However, size and shape of the gas filled voids in-situ are not known.

[a] Corresponding author: wiebke.baille@rub.de

Based on the above considerations, an experimental program was designed to investigate the possible effect of larger gas filled voids of various amount and size on the wetting induced collapse of sandy soils. The results of the experimental study are presented in this paper. The larger gas filled voids are defined to be voids with a diameter much larger than the diameter of the grains. Thus, the large voids can be regarded as cavities inside the soil matrix. In the following, the large voids will be called bubbles. The total volume of voids inside the soil is composed of the volume of the bubble(s) and the volume of the "standard" void space inside the sand matrix. In order to investigate the possible effect of bubbles on the collapse behaviour, the amount and size of the bubbles were varied while keeping the total volume of voids constant, leading to a variation in pore size distribution inside the soil.

In the present study, the effect of amount and diameter of large voids on the wetting induced collapse of sandy soils is experimentally investigated by laboratory box tests. Five different configurations of size and amount of large voids were tested. The wetting process was observed using Digital Image Correlation (DIC) technique, which allowed for the measurement of the deformation field.

2 Material

The material used was sand named Lohsa II, sampled from a former open pit mine in Lusatian region in Germany. The grain size characteristics are given in Table 1. The maximum and minimum void ratio was measured according to the German standard DIN 18126 and was e_{max} = 1.109 and e_{min} = 0.435, respectively. Initial water content was 5%.

Table 1. Measured material properties of Lohsa II sand.

d_{10} (mm)	d_{30} (mm)	d_{50} (mm)	C_u (-)	C_c (-)	ρ_s (kg/m³)
0,13	0,21	0,35	2,8	1,01	2,63

3 Methods used

3.1. Experimental programme

Table 2 presents an overview about the experimental programme. Five tests were performed with different bubble configurations. The test name indicates the amount of bubbles and the diameter of the single bubble. The global void ratio e_{global} corresponds to maximum void ratio e_{max} and was the same for all tests. The global void ratio is composed of the volume of voids inside the sand matrix (e_{micro}) and the volume of the large bubble(s) (e_{macro}).

The deformation field in the box test during wetting was measured using Digital Image Correlation (DIC) technique. The total settlement at the sample surface was measured by displacement transducers. The box test represents a plane strain condition; therefore, the bubble void ratios (e_{macro}) were calculated based on the area of the large bubble.

Table 2. Experimental program.

Test name	Test configu-ration	e_{global} (-)	e_{bubble} = e_{macro} (-)	e_{matrix} = e_{micro} (-)	ratio e_{macro} / e_{global}
0B	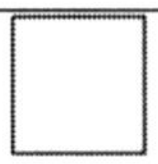 0 Bubble	1.109	0	1.109	0
1B/35	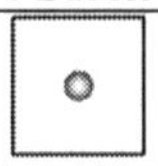 1 Bubble/ Ø35 mm	1.109	0,033	1,076	0,03
1B/50	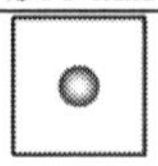 1 Bubble/ Ø50 mm	1.109	0,067	1,042	0,06
2B/35	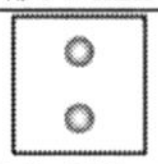 2 Bubbles/ Ø35 mm	1.109	0,067	1,042	0,06
4B/25	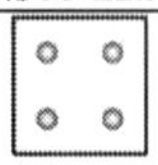 4 Bubbles/ Ø25 mm	1.109	0,067	1,042	0,06

The final settlement of the five tests was compared. Test 0B is the reference test of the sand matrix without large bubble. In this test, e_{global} is equal to e_{micro} and e_{macro} is zero. Tests 1B/35 and 1B/50 (series 1) will show the effect of increasing increasing e_{macro} while the amount of bubble and e_{global} remained constant, and e_{micro} decreases. Tests 1B/35 and test 2B/35 show the effect of increasing e_{macro} by increasing the amount of bubble (series 2). Tests 1B/50, 2B/35 and 4B/25 were performed to investigate the effect of pore size distribution of the bubbles on the wetting induced settlements, while e_{global}, e_{macro} and e_{micro} remained constant (series 3). This was realized by an increasing number of large bubbles with a corresponding decreasing bubble diameter.

3.2. Experimental set-up

Figure 1 shows a schematic of the box used for the wetting tests (Fig. 1a) and a photograph of the whole experimental set-up including DIC camera and data acquisition system (Fig. 1b). The sample dimensions were 27 cm width, 24 cm in height, and 10 cm thickness. The sample is placed on a filter paper which itself is separated by a geotextile from the stiff bottom steel plate with holes. Below the steel plate is a space which allows for an equal water level over the width of the box after start of the inflow from the water vessel (Fig. 1a).

The sample consists of 6 layers of 4 cm height; the border between each layer is marked by a thin layer of coloured sand grains. The layers are named L1 (layer 1, top of bottom layer) to L6 (layer 6, top surface), see Fig.

1a. The results of the DIC measurement are presented for selected points. These points are in the vertical centre axis on top of each layer as well as at the top and the bottom of the large bubble(s). They are marked by a red dot in Fig. 1a.

A water vessel (overflow vessel) providing a constant upper pressure head is connected by a tube to the bottom of the box. Target water level was at top of level 5 corresponding to 20 cm of initial sample height. At the end of the test, full saturation was reached in the whole specimen.

Two displacement transducers (DT) were placed on the surface of the specimen (top of layer 6). One of them was located in the centre of the top surface (DTc) and one between DTc and at border of the box (DTs). The latter is marked by a green dot in Fig. 1a. The measurements of displacement transducer DTc were used to verify the DIC measurements at top of layer 6 (L6).

Two DIC cameras (Fig. 1b) are placed in front of the box to record the deformation field over sample height and width during the wetting process. The analysis of the measured data is performed using software DaVis 7.2 of the company LaVision GmbH.

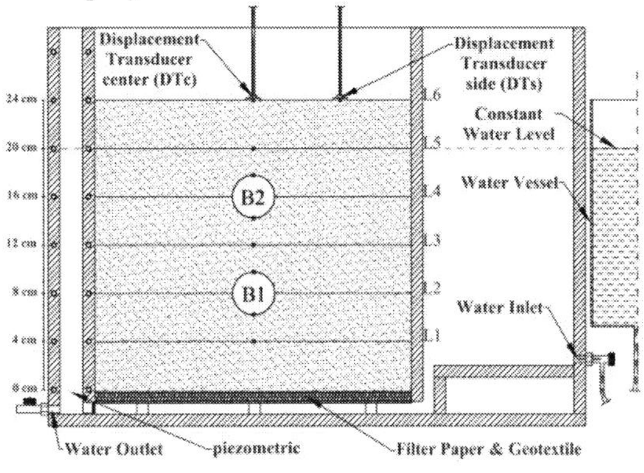

Figure 1(a). Schematic of the box and soil sample (example of test 2B/35)

Figure 1(b). Photograph of experimental set-up (example of test 1B/35)

3.3. Test procedure

The testing program as shown in Table 2 was performed. All tests were performed with 4% initial water content and the respective void ratios (e_{global}, e_{micro}, e_{macro}) as shown in Table 2.

At first, a geotextile of thickness 4 mm covered by a filter paper were placed at the base of the box. Then, the mass of sand needed for each layer of 4 cm height for the desired matrix void ratio e_{micro} was placed manually. The sand was equally distributed and slightly compacted so as to reach a homogenous sand matrix throughout the sample. At the top of each layer, black sand grains are poured to mark the boundary between two subsequent layers in order to better visualize the deformations of the respective layer.

At the desired location, prepared hollow ice cylinders were placed horizontally to serve as shuttering to create the large voids. Further, sand was placed above the large voids stabilized by the ice cylinders until the target sample height of 24 cm. The sample preparation was finalised before the cylinder was melt.

After end of sample preparation and after the ice cylinder was melt, the water vessel is connected to the box. The data acquisition system for the DIC and the displacement transducers was started. The wetting test started when the inlet valve was opened.

The water firstly distributes in the empty space below the base plate at the bottom of the specimen, and then the water level rises equally over the whole sample width. The piezometric water level can be read from the tube at the left side of the sample box (see Fig. 1a). The water level in the vessel is kept constant while the water level in the sample did increase. Thus, during the test, the hydraulic gradient was not constant. For each test, it took approximately 30 min until the target water level of 20 cm was reached and deformation process has reached equilibrium.

4 Results

In this section, the observations during the tests according to the testing program shown in Table 2 are presented. In the first part, the qualitative observations by DIC technique are described. In the second part, the measured deformations over time and the final deformations of the different test configurations are quantitatively compared.

Figures 2 to 4 present qualitative observations for test 2B/35 (i.e., 2 bubbles of diameter 35 mm embedded in the homogenous sand matrix). Figure 2 shows the initial state of the sample just before start of the wetting process

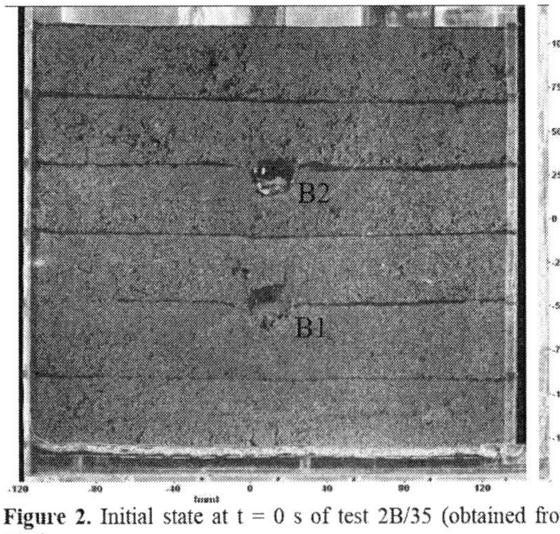

Figure 2. Initial state at t = 0 s of test 2B/35 (obtained from DIC).

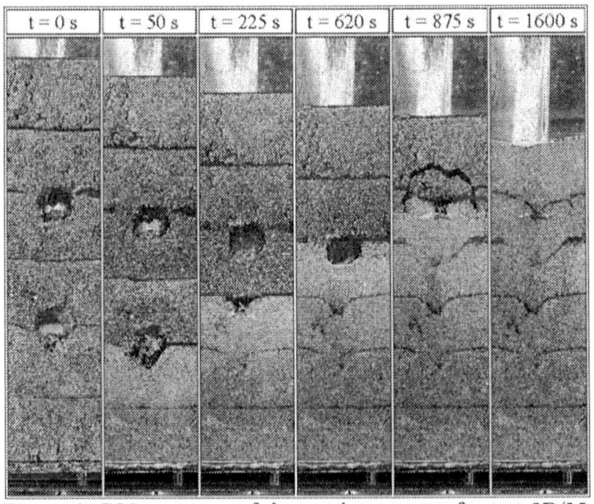

Figure 3. DIC sequence of the wetting process for test 2B/35 (blue line marks the top of capillary fringe).

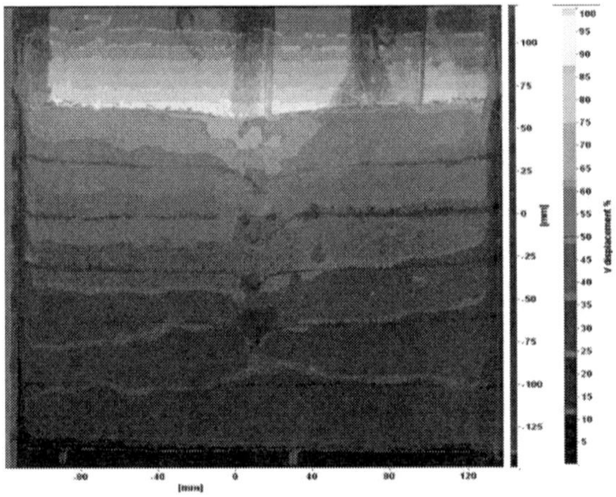

Figure 4. Vertical displacement of test 2B/35 at t= 1600 s measured by DIC (yellow indicates maximum displacement, red indicates minimum displacement)

(at t = 0). The centre of bubble 1 (lower bubble) is located at about 8 cm height from the sample bottom, the second upper bubble is located at about 16 cm height. The net distance between the circumferences of the bubbles is 4.5 cm. The melting of the ice cylinder caused only very small deformations until an arching was built up in the sand matrix beside the bubble.

Figure 3 qualitatively shows the wetting process in a sequence of six pictures for selected time steps of test 2B/35. Only the centre part of about 10 cm width including the large bubbles was shown for each time step. The blue lines highlight the location of the top of the capillary fringe.

Until t = 50 s (picture 2 of the sequence), the rise of the capillary fringe caused uniform settlements. The bubble remained intact while the capillary fringe increased at both sides of the bubble. The water did enter the bubble only when the piezometric water level was about 1 to 2 cm above the lowest point of the bubble. This is in agreement to Wheeler (1988) [8] who stated that a large gas bubble is flooded, when the pressure difference of gas pressure inside the bubble u_g and the water pressure u_w reaches a critical minimum value for the radius of curvature R_c of the menisci between the sand grains at the bubble boundary. R_c is of the order of the particle size d and is not related to the bubble radius.

Failure occurs at the top of the bubble B1, showing a mechanism similar to ground failure.

At t = 225 s (picture 3 of the sequence), the void space of bubble B1 was nearly fully closed by the settlement of the above sand layers. At this stage, the capillary fringe has reached the top of the 3rd layer. The upper three layers do not show incremental settlements, and the upper bubble (B2) is still intact.

At t = 620 s (picture 4 of the sequence), the capillary fringe has nearly reached the top of bubble B2. B2 is slightly compressed, but has not yet collapsed. At the former location of Bubble B1, the deformed line of the black coloured sand grains indicates the collapse of sand into Bubble 1 which was completely filled by sand.

At t = 875 s (picture 5 of the sequence), bubble B2 was filled by sand from the overlaying matrix. Picture 6 at t = 1600 s of the sequence represents the final state of the test, where the target piezometric water level was reached.

Figure 4 shows the final state of test 2B/35 observed by DIC including contour lines of the deformation field. It is to be noted that the upper contour lines above the soil sample should be neglected. Maximum displacement occurred at the top of the sample in the vertical centre axis above B2. The deformations in the near region around the centre axis were strongly affected by the presence of the bubbles.

Figures 5a to 5e illustrate the deformations measured

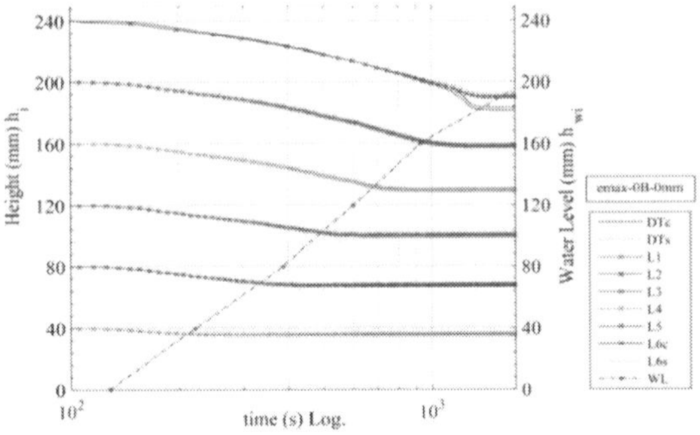

Figure 5(a). Test 0B: measured height of the measurement points versus elapsed time

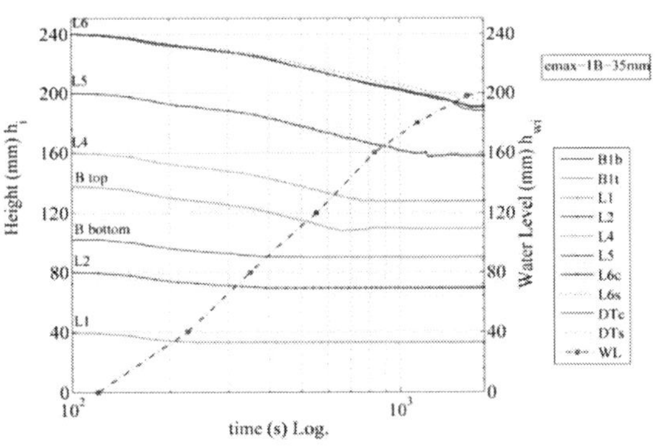

Figure 5(b). Test 1B/35: measured height of the measurement points versus elapsed time

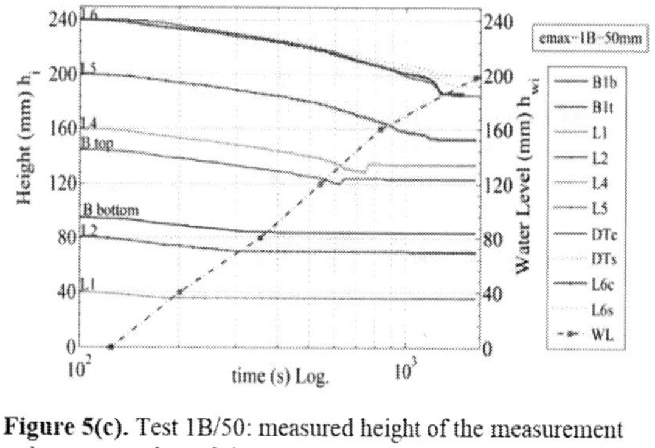

Figure 5(c). Test 1B/50: measured height of the measurement points versus elapsed time

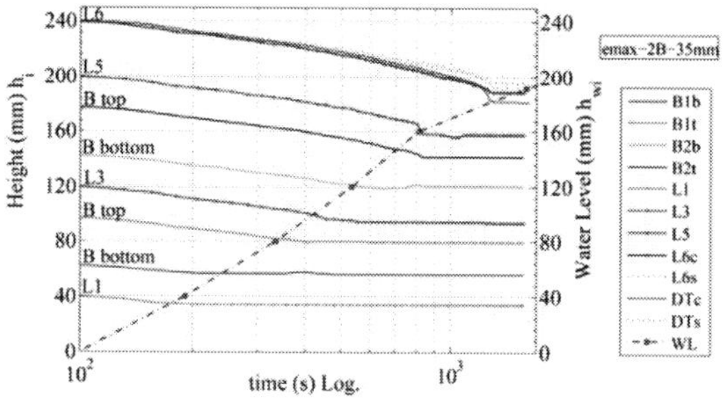

Figure 5(d). Test 2B/35: measured height of the measurement points versus elapsed time

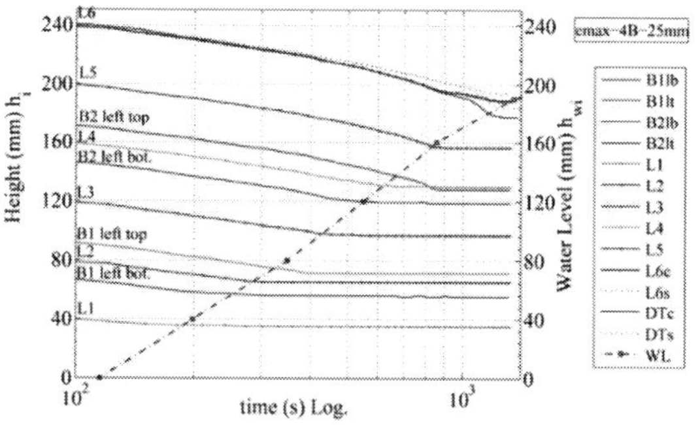

Figure 5(e). Test 4B/25: measured height of the measurement points versus elapsed time

for each test during the test duration. The left vertical axis refers to the actual height of the various measurement points, whereas the right axis refers to the piezometric water level, designated as WL in the legend of Figs. 5(a) to 5(e). DTc and DTs refer to the displacement transducers at the sample surface (see Fig. 1a). All other lines refer to measurements by the DIC system; they are designated according to their location at the layers (L1 to L6) and at top and bottom of the bubbles.

Figure 5(a) represents the behaviour of the homogenous reference sample without any large bubble. As expected, it shows smooth and continuous settlements at each measurement point, as soon as the water table together with the capillary fringe has started to rise. The cutting point of the water level (WL) with the respective measurement location (L1 to L6) indicates the point when the settlement at the respective measurement location has stopped. This means, once the water level has reached the actual location of the measurement point, no further settlement has occurred in this measurement location.

In the other 4 tests with different bubble configurations (1B/35, 1B/50, 2B/35 and 4B/25), it was observed that the settlement smoothly increased until the water level has reached the actual location of the respective measurement location. A steeper increase in settlement or sudden settlement has occurred when the water level has reached the top of the bubble(s) (Fig. 5b, 5c, 5d, 5e). In tests 2B/35 and 4B/25, this was more significant when the water level had reached the upper bubble(s).

Figure 6 compares the final maximum settlement measured by the displacement transducers for each test configuration. In all test configurations studied herein the initial global void ratio was constant (e_{global} = 1.109). However, the measured final settlement due to wetting induced collapse was found to be different for each test. In the following, the results are described for the three different series (see section 3.1).

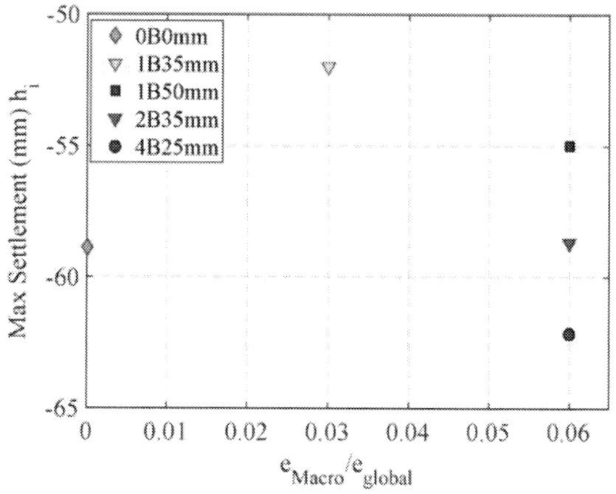

Figure 6. Maximum final settlement versus ratio e_{macro}/e_{global} for all tests

Series 1 (0B → 1B/35 → 1B/50):

The maximum final settlement of the homogeneous reference sample (0B) was about 59 mm. The final settlements of both heterogeneous tests containing a bubble (1B/35, 1B/50) were found to be smaller. This indicates that the decrease in matrix void ratio in both tests 1B/35 (e_{micro} = 1.076) and 1B/50 (e_{micro} = 1.042) as compared to the homogeneous sample 0B (e_{micro} = 1.109) had a dominating effect over the increase in macro void ratio. However, comparing tests 1B/35 and 1B/50 with each other, the further increase in macro void ratio from e_{macro} = 0.033 to e_{macro} = 0.067 by increasing the diameter of the bubble obviously dominated over the respective decrease in matrix void ratio from e_{micro} = 1.076 leading to a higher final settelement.

Series 2 (0B → 1B/35 → 2B/35):

In this series, comparing test 1B/35 to test 0B, the same explanation as in series 1 holds true (compensating effect of smaller matrix void ratio over effect of large bubble). Comparing tests 1B/35 and 1B/50 with each other (both configurations have identical e_{global}, e_{macro} and

e_{micro}), indicates that the increasing amount of bubbles lead to a higher settlement.

Series 3 (0B → 1B/50 → 2B/35 → 4B/25):

Again, in this series, global void ratio is the same for the homogeneous reference test and the three heterogeneous tests. Comparing the heterogeneous tests with each other (e_{global}, e_{macro} and e_{micro} are constant) indicates that the increase in amount of bubbles, thus the change in pore size distribution lead to an increase in the final measured wetting induced settlement.

Overall, the test results indicate that consideration of global void ratio of sand may not be the relevant state parameter for describing the soil behaviour.

5 Summary

In the present study, the effect of large voids (gas bubbles) and their pore size distribution on the wetting induced collapse of sand was experimentally investigated by laboratory box tests. Five different configurations of size and amount of large pores were tested. The wetting process was observed using Digital Image Correlation (DIC) technique and displacement transducers at the surface.

Overall, the test results indicate that for a sandy soil system with constant global void ratio, the incorporation of large gas bubbles of different size and amount affected the final measured settlement values to different extent.

In general, the higher the amount of bubbles, the greater were the settlements observed. For a smaller number of gas bubbles (up to 2 in this study), the decrease in matrix void ratio compensated the effect of the large gas bubble(s) leading to smaller or equal settlements as compared to the homogeneous reference sample. In general the study indicates that the assumption of a homogeneous soil system with a global void ratio as a state parameter may not be sufficient to fully describe volumetric soil behaviour.

However, this experimental study is a first attempt to investigate the effect of larger voids on the wetting induced collapse behaviour of anthropogenic sandy soil deposits. For better representing in-situ conditions, tests at larger scale and at representative stress level need to be performed. The test instrumentation should be enlarged to also measure water content, suction and pore water pressure.

References

1. S. Diamond, Pore size distributions in clays, Clays and Clay Minerals **18**, 7-23 (1970)
2. S. Ahmed, C. Lovell, S. Diamond, Pore sizes and strength of compacted clays, ASCE J. Geotech. Eng. Division **100**, 407-425 (1974)
3. A. Gens, E. Alonso, A framework for the bahaviour of unsaturated expansive clays, Can. Geotech. J. **29**, 1013-1032 (1992)
4. P. Delage, M. Audiguier, Y. Cui, M. Howat, The microstructure of a compacted silt, Can. Geotech. J. **33**, 150-158 (1996)
5. E. Romero, A. Gens, A. Lloret, Water permeability, water retention and microstructure of unsaturated compacted boom clay, Engineering Geology **54**, 1-2, 117-127 (1999)
6. P. Delage, D. Marcial, Y. Cui, X. Ruiz, Ageing effects in a compacted bentonite: a microstructure approach, Géotechnique **56**, 5, 291-304 (2006)
7. R. Thom, R. Sivakumar, V. Sivakumar, E. Murray, P. Mackinnon, Pore size distribution of unsaturated compacted kaolin: the initial states and final states following saturation, Géotechnique **57**, 5, 469-474 (2007)
8. S.J. Wheeler, A conceptual model for soils containing large gas bubbles, Géotechnique **38**, 3, 389-397 (1988)
9. N. Hoth, M. Ussath, A. Simon, Fluiddynamisch-hydrogeochemische Arbeitshypothesen zu rezenten Bodenverflüssigungen im Lausitzer Revier. Schriftenreihe des Instituts für Bergbau und Spezieltiefbau, TU Bergakademie Freiberg, Hrg. W. Kudla, Heft 5, Beiträge zum 2. Kolloquium Bodenverflüssigung bei Kippen des Lausitzer Braunkohlebergbaus, (2014)

Sample size choices for XRCT scanning of highly unsaturated soil mixtures

Jonathan C. Smith and Charles E. Augarde[a]

School of Engineering and Computing Sciences, Durham University, South Road, Durham, UK

Abstract. Highly unsaturated soil mixtures (clay, sand and gravel) are used as building materials in many parts of the world, and there is increasing interest in understanding their mechanical and hydraulic behaviour. In the laboratory, x-ray computed tomography (XRCT) is becoming more widely used to investigate the microstructures of soils, however a crucial issue for such investigations is the choice of sample size, especially concerning the scanning of soil mixtures where there will be a range of particle and void sizes. In this paper we present a discussion (centred around a new set of XRCT scans) on sample sizing for scanning of samples comprising soil mixtures, where a balance has to be made between realistic representation of the soil components and the desire for high resolution scanning, We also comment on the appropriateness of differing sample sizes in comparison to sample sizes used for other geotechnical testing. Void size distributions for the samples are presented and from these some hypotheses are made as to the roles of inter- and intra-aggregate voids in the mechanical behaviour of highly unsaturated soils.

1 Introduction

Soil-based construction materials (SBCMs) are receiving increasing interest due to their potential green credentials, in comparison to conventional construction materials bound with cement. *Rammed earth* and *adobe* are two popular forms of SBCM found in both heritage structures and modern buildings [1]. It is only recently however that these materials have been recognised as manufactured, highly unsaturated soils, and a number of groups around the world are now applying unsaturated soil mechanics principles to attempt an improved understanding of the mechanical behaviour of these materials [2].

SBCMs, and rammed earth in particular, are mixtures of soil fractions from clay to gravel and therefore present issues when considering sampling and testing. Investigations of the mechanical behaviour of rammed earth in uniaxial compression show that reducing sample size, and consequently having to remove the largest fractions from the mixture prior to manufacture, considerably alters the measured strength and stiffness properties [3]. X-ray computed tomography (XRCT) scanning is becoming a popular tool for geotechnical investigations, however there is a large variation in the size of samples scanned with a growing trend towards small, non-standard samples with diameters less than 12mm (to obtain higher resolutions).

In this paper we investigate the internal structure of a selected SBCM using three cylindrical sample sizes scanned using XRCT before and after loading. Comments are made on the macrostructures of the samples based on the XRCT results with the aim of establishing the roles of the different constituents within the mixtures.

2 Internal structure of soil based construction materials

While there has been considerable research into the internal structure of simple unsaturated soils, using MIP for instance, there is very little work completed for soils containing a mixture of particle sizes, from clay to gravel, as found in SBCMs.

Beckett et al. [4] used XRCT to observe the changes to the macrostructure of rammed earth following compressive loading and observed how compaction during manufacture affects the macrostructure within individual layers of rammed earth. Due to limitations on sample size, only the inter-aggregate soil matrix, formed of clay and sand particles, was investigated. The results show that the interconnectivity and size of voids significantly increases following loading and that the effect of cracking is greater within poorly compacted material. This study also shows that single layer specimens have a highly heterogeneous macrostructure whilst double layer specimens are largely homogeneous, and the macrostructure of a compacted layer is not affected by the compaction of additional material above it. This rare example provides a valuable insight into the macrostructure of a SBCM, however the analysis was only performed in 2D, preventing void volumes being determined, and the lack of gravel particles within the mix is likely to have significantly altered the void structure.

[a] Corresponding author: Charles.augarde@dur.ac.uk

DOI: 10.1051/e3sconf/20160906005

Hall et al. [5] also used XRCT to characterise the void geometry of cement-stabilized soils for earthen construction. However the addition of cement to the soil mixture will have fundamentally altered the material properties, since the cement will act as a chemical binder between the particles, and it is therefore difficult to apply results to the study of unstabilized SBCMs. This has prompted the experimental study described below.

3 Experimental programme

3.1 Sample size

Three cylindrical sample sizes, 12mm diameter x 24mm high, 38mm diameter x 76mm high and 100mm diameter x 200mm high, were used in this investigation (referred to hereafter as ϕ12, ϕ38 and ϕ100 respectively). The ϕ12 sample was chosen as it was the largest sample size for which a high resolution scan with a voxel size of less than 1μm could be easily obtained using the XRM410 scanner whilst the ϕ38 sample was close to the maximum sample size which could be easily placed and have its entire width scanned in a single XRCT image. The ϕ100 sample was used as it was close to the maximum diameter soil sample which could be scanned with the XRM410. From a material testing perspective the ϕ38 and ϕ100 samples are both standard geotechnical triaxial testing sizes, ensuring samples could be easily manufactured, and unconfined compression tests performed without the need for new testing equipment or techniques.

3.2 Materials

A single SBCM mix, 30*:60:10, was used throughout the investigation and chosen as it lay in the centre of the limits suggested in [6] and the SBCM mixes used in Smith's wider study [3]. The 30*:60:10 SBCM mix was manufactured by combining appropriate amounts of speswhite clay, sharp sand and pea gravel. Both the sharp sand and the pea gravel were oven dried at 105°C and then passed through 2mm and 10mm sieves respectively. The BS1377 Vibrating Hammer Test [7] was used to determine the optimum water content and target dry density for the SBCM mixture [8]. An OWC of 8.93% and dry density of 2060 kg/m^3 were obtained.

Five samples were manufactured for each of the ϕ12, ϕ38 and ϕ100 sample sizes. Prior to sample manufacture the soil mixture was sieved to 2mm for the ϕ12 samples and to 5mm for the ϕ38 samples, to prevent the larger particles interfering with the compaction and loading behaviour. The soil for the ϕ100 samples was not sieved in order to closely match an on-site soil mix, and had a maximum particle size of 10mm. The appropriate amount of water was added to each mixture to reach OWC and the mixture was left to equilibrate overnight. Since SBCMs are often compacted in layers within formwork, all three samples were also constructed from layers of compacted material.

The correct mass of wet soil mixture for an individual layer was added to the sample mould to ensure each sample layer would be at the target optimum dry density and was then statically compacted to achieve the required layer depth. Once the appropriate number of layers had been compacted the sample was pushed from the mould by hand, ensuring no disturbance was caused on the sides of the sample, and left to dry in a temperature monitored room at 22°C ± 2°C until a stable mass was obtained (at between 0.8% and 1.2% water content), at which point no further water was lost to the atmospheric conditions. Three filter paper test samples, each using two cylinders (50mm diameter and 20mm high), were also prepared alongside each of three sample sizes and compacted to the same target density. These were left to dry under the same conditions as the ϕ12, ϕ38 and ϕ100 samples before being placed in a sealed jar to equilibrate for ten days. The filter paper test results provided total suction values for the three different mixtures.

Three samples from each batch of five manufactured samples, those closest to the batch mean dry mass, were selected for scanning and mechanical testing. Following testing it was found that all samples were within 1.8% of the target dry density. To differentiate between different samples within the triplet, each sample was assigned a suffix A, B or C. Once the samples reached their stable dry mass, top and bottom plastic caps were placed on the samples and the sample was sealed within a latex sheath using rubber O-rings. The caps and sheath prevented any further change in water content during scanning, compression testing and re-scanning, and provided a flat surface to which the compressive load could be applied and also enabled easy secure mounting of the samples inside the XRM410.

3.3 Experimentation

All samples were scanned using the XRM410 before and after unconfined compression, constant water content tests. Displacement rates of 0.03mm/min, 0.1mm/min and 0.25mm/min for the ϕ12, ϕ38 and ϕ100 samples respectively were chosen in accordance with [9] ensuring the strain rate did not exceed 2% per minute and failure occurred within 15 minutes. Each sample was tested to failure, observed by evidence of a load peak, after which compression was stopped and the load released.

Both the pre-loading and post-loading scans involved three specific scans: a large field of view scan, to capture as much of the sample as possible; and two small field of view scans at the maximum resolution achievable for the sample size, located either side of a compaction plane to capture any differences in material structure at the top and bottom of the layers.

The ϕ12 sample's large field of view scan captured the entire sample and the ϕ38 sample's large field of view scan captured the entire sample diameter but was limited by the square x-ray detector within the XRM410 to only half the sample height, centred on the middle of the sample. The large size of the ϕ100 sample prevented the entire diameter of the sample being seen and the scan settings were chosen to enable the scanning of as much of the sample as possible, approximately 80mm centred on the interface between the bottom and middle compaction layers. The key XRCT scan parameters are given in detail in [3] as are the detailed methods used to post-process the data.

4 Results and discussion

4.1 Mechanical behaviour

Table 1 shows the compressive strengths obtained from the unconfined compression tests for all nine samples (three of each diameter). The ϕ12C sample was found to be a clear outlier as when included the ϕ12 samples' failure stresses had a standard deviation of 0.29MPa, which was reduced to 0.05MPa when the ϕ12C sample was disregarded.

The aspect ratio of all three cylindrical sample sizes was 2:1 ensuring that geometrical differences in shape did not affect the results obtained. However one significant difference in the manufacture of the ϕ100 samples was the use of three compaction layers rather than the two layers used to manufacture the ϕ12 and ϕ38 samples. This additional compaction layer was included to reduce the depth of each layer and thus the heterogeneity of the individual layers, however the three compaction layers will have introduced an additional layer interface and plane of weakness into the sample that may have resulted in the lower compressive strengths obtained.

The total suction values obtained from the filter paper tests for the ϕ12, ϕ38 and ϕ100 samples were 8.51±0.37MPa, 8.76±0.07MPa and 8.04±0.25MPa respectively. These are of the same magnitude as those measured in [2] for similar SBCMs and the difference can be attributed to the different clay material and mix composition used in this investigation.

The only remaining difference between the three samples is the sieving of the largest particles from some of the mixes. The compressive strength of the samples increases following the removal of the gravel fraction. The effect of the removal of gravel fraction on the internal structure of the SBCMs and how this may alter the compressive strength of the samples is therefore examined in detail below.

Table 1 : Strengths of samples scanned

	UCS (MPa)		
Sample	A	B	C
ϕ12	2.25	2.15	2.80
ϕ38	1.84	1.68	1.82
ϕ100	0.45	0.58	0.49

At the low water contents found in SBCMs, water will only be present in the smallest voids between the clay and sand particles, and as adsorbed layers. The clay and sand fractions in all three sample sizes are unaltered, in terms of relative proportion (1 part clay to 2 parts sand) and mineralogy, which may explain the similarity in total suction values for all three sample sizes. As the water is only present in the smallest voids within the sample when the SBCM has dried it will be those voids which most control the capillary and adsorptive suction terms and therefore the suction values measured will be largely unchanged by the removal of the gravel fraction. The decrease in compressive strength with the addition of gravel particles, despite similar measured total suction values, may therefore suggest that the gravel reduces the strength of the SBCM by introducing initial points of weakness, i.e. cracks and large voids, into the compacted clay-sand mixture.

Void size distributions were successfully obtained for all large field of view (LFoV) XRCT scans performed, and the small field of view scans for the ϕ12 and ϕ38 samples. It was not possible to obtain void size distribution plots from the small field of view scans for the ϕ100 samples as the automated analysis employed was unable to identify individual voids within the sample. This was despite the source operating at the highest power settings, the maximum exposure time being used to obtained images and the pixel size being doubled in order to ensure sufficient x-rays were detected for an image to be captured. It suggests that the ϕ100 samples were too large for the small field of view scans to be performed. The void size distributions for the ϕ12A sample are shown in Figure 1 and for the ϕ38A sample in Figures 2. The ϕ100 large field of view void size distributions for all three samples are shown in Figure 3. All the figures show the void size distribution before and after compressive loading and, where necessary, the small field of view results for the bottom of compaction layer and top of compaction layer are differentiated using `BoL' and `ToL' respectively.

The void size distribution plots used here differ considerably in form to those commonly presented when void sizes are explored using MIP. A typical void size distribution from MIP data plots the volume of intruded mercury against the void entrance diameter and it is then it is assumed that a smaller void entrance diameter represents a smaller void volume. In contrast, the VSDs presented here plot on the y-axis the cumulative void volume of all voids smaller than and equal to a given void volume, shown on the x-axis, as a percentage of the total sample volume. These plots can therefore be interpreted in a similar way to particle size distribution plots which plot on the y-axis the cumulative mass of all particles which are smaller than a given particle diameter, shown on the x-axis, as a percentage of the total sample mass. The plotting and discussion of void volumes, rather than void entrance diameters, is possible since the post-processing of the XRCT scan data identifies the volume of each void within the sample directly. For MIP data either one or two peaks (for a monomodal void structure and a bimodal pore structure respectively) will be evident which coincide with the most abundant void entrance diameters within the sample and it is assumed that these peaks indicate that a particular void volume within the sample abounds. In the case of the XRCT-obtained VSDs the **gradient** of the curve at a particular void volume can be used as an indication of the number of voids of that volume within the sample, i.e. a steep region within the curve shows there are a high number of voids of that volume and can be interpreted in a similar way as a peak in MIP data. A flat region within an XRCT-obtained void size distribution curve suggests there are no voids of the given volume within the sample.

4.2 Observations from the small field of view scan results

For all five samples in which small field of view scans were obtained it is clear that before unconfined compression, the regions scanned at the bottom of the compaction layer contained a smaller void fraction than the top of the compaction layer region. This builds upon the work described in [4] which shows that the material structure can vary through the depth of a layer, although the results presented in [4] suggest that only single-layer specimens display a highly heterogeneous material structure and double-layer compacted specimens achieve a largely homogenous material. The results from the investigation described here however clearly show that double-layer samples can also exhibit material heterogeneity and, since the ϕ12 and ϕ38 samples used in this investigation followed the same preparation technique as used by [4], differences in manufacturing procedure cannot account for the material heterogeneity of the samples investigated here.

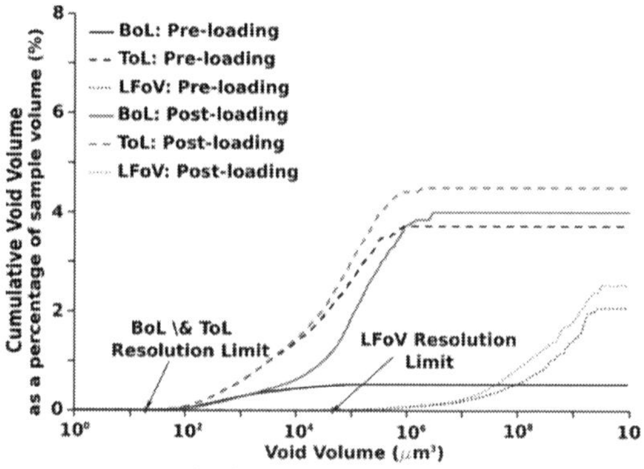

Figure 1: VSD for ϕ12A sample

It is however possible that differences in XRCT scanning and post-processing could have resulted in the difference in void structure between the top and bottom of compacted layers being observed in this investigation which were not captured in the work presented in [4]. The XRCT scans described by [4] were performed at a pixel size of 11μm and a minimum pore area of 3x3 pixels which limited the minimum void area identified within the scans to 33μm^2. Only one in every ten slices was analysed in the vertical plane limiting the overall volumetric resolution to approximately 330μm^3. In the XRCT scans performed here however the highest resolution of the small field of view scans is 20μm^3 and it is therefore suggested that the differences in the image-analysis technique employed, and the improved final resolution, had a considerable part to play in the observation of heterogeneity in the material structure of the samples used in this investigation.

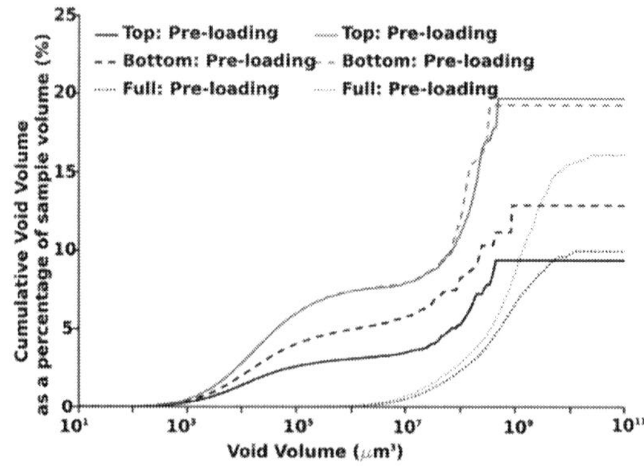

Figure 2: VSD for ϕ38A sample

The ϕ12 and ϕ38 sample small field of view scans, performed at the smallest voxel sizes possible for the sample size, provide an insight into the smallest observable voids within the SBCM samples using XRCT scanning. The steep gradient of the curve below 10^6μm^3 in the void size distribution curves of the ϕ12 samples, and the flat line visible for voids above this void volume, indicates that the voids within the ϕ12 samples are predominately less than 10^6μm^3 in volume. The void size distribution curves for the ϕ38 sample however show two clear regions where voids exist (the steep parts of the curve below 10^6μm^3 and between 10^7μm^3 and 10^9μm^3), separated by a plateau between 10^6μm^3 and 10^7μm^3 and followed by a flat line visible for voids above 10^9μm^3. This suggests therefore that the ϕ38 samples contain voids smaller than 10^6μm^3 and a second group of voids ranging in volume between 10^7μm^3 and 10^9μm^3.

Because all the scans were analysed using the same image processing techniques the lack of voids in the small field of view ϕ12 scans is not due to image processing, since voids with volumes larger than 10^6μm^3 can be seen in other void size distributions, and can be taken as a true representation of the material structure. The lower limit of the set of voids smaller than 10^6μm^3 however cannot be determined, despite the apparent flattening of the void size distribution curves at

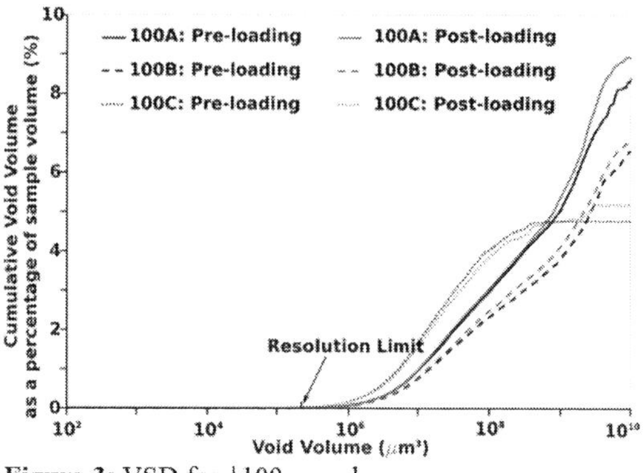

Figure 3: VSD for ϕ100 samples

$2x10^2\mu m^3$, as the resolution of the small field of view scans is limited to $20\mu m^3$. The visible plateau between $10^6\mu m^3$ and $10^7\mu m^3$, for all twelve small field of view ϕ38 scans, are not all perfectly flat or between identical void volumes which suggests the observed plateau is not a result of the image processing but from the SBCM structure and minimal voids between these volumes exist within the ϕ38 scans.

The ϕ12 samples can be considered as the soil matrix material found in SBCMs, as they contained only clay and sand fractions, and the small field of view scans were centred on regions within a layer. The voids smaller than $10^6\mu m^3$, identified in both the ϕ12 and ϕ38 samples, are therefore likely to be the intra-aggregate voids found within the assemblages of clay and sand forming the binder matrix within the broader SBCM structure.

The addition of the 2-5mm gravel particles has clearly introduced a second range of void sizes, between $10^7\mu m^3$ and $10^9\mu m^3$, in the internal structure of the ϕ38 samples. It is suggested that these voids are inter-aggregate voids which have formed between the larger aggregates now present within the mixture. Double structure models have been well-documented in unsaturated soils however predominantly focus on clay-sand mixtures in which intra-aggregate voids have been identified between clay particles, and are therefore far smaller than the resolution of the XRCT scans performed in this investigation. As such the two sets of void volumes identified in this chapter in the ϕ38 samples, below $10^6\mu m^3$ and between $10^7\mu m^3$ and $10^9\mu m^3$, are all still part of the SBCM macrostructure and a series of intra-elemental voids will also exist, undetectable using current XRCT techniques, between small groups of clay particles, i.e. the SBCM microstructure.

The lack of small field of view scans for the ϕ100 samples means it is not possible to observe this double void size macrostructure, or the changes that occur within it, however it is suggested that a similar void size distribution to the ϕ38 samples would be observed, whereby small voids, less than $10^6\mu m^3$, exist within the intra-aggregate matrix and the inter-aggregate voids are then observed between $10^7\mu m^3$ and $10^9\mu m^3$ forming around the gravel particles.

4.3 Observations from the large field of view scan results

The large field of view scans, performed at a considerably larger voxel size, are unable to capture any voids smaller than $8x10^4\mu m^3$ within the sample and do not capture many voids smaller than $10^5\mu m^3$. This can both explain why the majority of the small field of view scans show more voids within the sample and why the double void macrostructure is not observed in the ϕ38 and ϕ100 large field of view scans. This however does not prevent the large field of view results providing valuable insights.

The increasing void size distribution curve above $10^6\mu m^3$, for the ϕ12 samples' large field of view scans, indicates that between 2% and 3% of the total sample volume are voids with volumes greater than $10^6\mu m^3$. This is in contrast to the small field of view scans which show no voids larger than $10^6\mu m^3$ in volume. The ϕ12 samples were manufactured using two layers of SBCM causing the transition between the two compaction layers to occur at 50% sample depth. It is therefore suggested that larger voids identified within the large field of view scans are located within the sample at the interface between the two layers but not within the clay-sand matrix formed inside an individual layer which is scanned in the small field of view scans. Interestingly there is a much smaller increase in the large voids detected within the ϕ12 large field of view scans than in the smaller voids within the clay/sand binder matrix and this suggests that, at least for the ϕ12 samples in which only binder material is present, failure occurs within the binder and not at a layer interface.

The ϕ38 and ϕ100 large field of view scans show the largest change in void volume following compressive loading in the largest voids and this suggests that the change in internal structure during loading predominately occurs in the large voids and pre-existing flaws within the SBCM. The ϕ100 samples show a smaller increase in voids following the unconfined compression tests than the ϕ38 samples. It is however possible that the lack of evolution in the ϕ100 samples' macrostructure is due to the scanning of only one region within the ϕ100 samples. A limitation of only scanning one region around the bottom compaction layer interface means that significant macrostructure changes could have occurred around the upper compaction layer interface which would not have been captured during the XRCT scans and, if no major structural changes occurred at the bottom compaction layer interface, only a minimal increase in voids would be observed. In the ϕ38 samples however, since only one compaction layer interface was present within the sample which was at the centre of the region scanned using XRCT, any macrostructural changes which occurred during compression testing of the ϕ38 samples should have been captured.

In addition, both the ϕ38 and ϕ100 samples show a variation in the increase in void volume across the set of three samples. The reasons for this are not clear however some variation may have been introduced during the transferring of the samples from the unconfined compression testing apparatus to the XRCT scanner which could have affected the sample structure. The inability to scan the samples during the unconfined compression testing is therefore clearly a shortcoming of this study and being able to apply load to the samples within the XRCT scanner may have helped reduce the variability in the post-failure loading conditions.

It is therefore difficult to draw firm comparisons between the changes in the largest voids within the ϕ38 and ϕ100 samples, as variability may have been introduced during the transfer of samples to the scanner and not all the macrostructural changes will have been captured within the ϕ100 samples. However all six samples, the three ϕ38 samples and three ϕ100 samples, do show the same trend that the change in void volume increases with void size and this therefore suggests that these large voids are most altered during compressive loading of SBCMs.

4.4 Links between material structure and compressive strength

The void size distribution results obtained show that a double void size macrostructure exists within the SBCM samples because of the inter-aggregate voids linked to the presence of large gravel particles and intra-aggregate voids which exist within the clay-sand binder matrix. Current unsaturated soils theory also suggests that a range of smaller voids exists between clay particles, which are too small to be detected XRCT techniques, and these inter-elemental voids form a third range of voids within the internal structure of SBCMs.

Based on these findings Figure 4 illustrates a proposed internal structure of SBCM samples, using both the results from the small field of view ϕ38 samples to describe the intra-aggregate and inter-aggregate voids and the assumed existence of inter-elemental voids between the clay particles which remained undetected in this investigation. This double void size macrostructure coupled with the third range of smaller inter-elemental voids is key to explaining the compressive strength results obtained and can also help form a basic understanding of the fundamental roles of each constituent material within the SBCM mix.

The similar total suction values obtained for all three sample sizes and the similarity in the void size distributions in the voids smaller than $10^6 \mu m^3$ for the ϕ12 and ϕ38 samples suggest that it is within the clay-sand matrix that suction is developed within the SBCM. Since suction is a key source of strength it is clear that changing the clay type used affects the suction developed and hence affects the compressive strength. Studying the intra-aggregate voids below $10^6 \mu m^3$ in different clays will undoubtedly provide an interesting insight into the role of clay within SBCMs. However the results from this investigation also show that, despite similar total suctions being developed within the samples and similar void sizes existing below $10^6 \mu m^3$, the clay is not the only factor influencing compressive strength.

The reduction in compressive strength as larger particles are introduced to the SBCM mixture and the largest change in void sizes being observed in the inter-aggregate voids > $10^7 \mu m^3$ suggest that the gravel fraction within the SBCM has a large influence on the compressive strength. The ϕ12 samples show that failure of the clay-sand binder matrix occurs within voids smaller than $10^6 \mu m^3$ at approximately 1.55MPa and it is possible to assume that the clay-sand matrix within the larger samples would also fail in a similar way at a similar compressive strength. The introduction of the gravel particles to the SBCM, and the resultant presence of inter-aggregate voids within the structure, however results in a reduced strength and it is suggested that this is because the larger voids introduce points of weakness from which cracks can propagate.

The triple-structure of voids suggested here has not been identified before in SBCMs. Unfortunately the trend towards XRCT investigations focussing on idealised, narrowly graded materials and the relative infancy of investigating the internal structure of SBCMs means there exists no dataset to which these results can be compared. It is hoped therefore that these findings will provoke an interesting discussion.

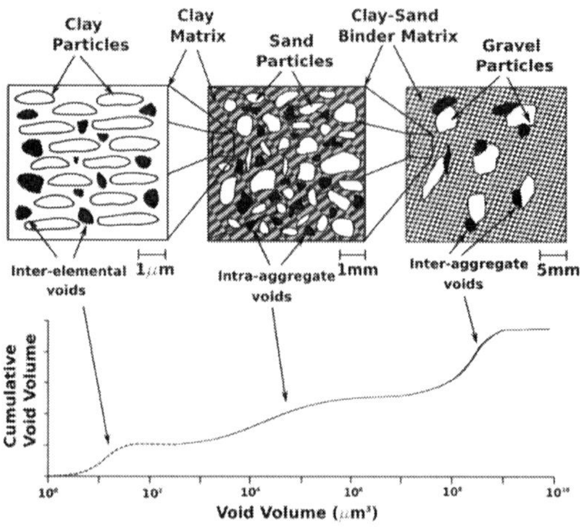

Figure 4: Speculative structure of SBCMs

5 Conclusions

This paper concerns testing of manufactured unsaturated soils, both for strength and more significantly for internal structure. Some speculations on the internal structure of these complex soil mixtures are presented. The authors acknowledge the assistance provided by the Durham XRCT Facility, which was funded in part by the UK EPSRC (EP/K036084/1 & EP/K024698/1).

References

1. P.A. Jaquin, C.E. Augarde, *Earth Building: History, science & conservation* (IHS BRE Press: Bracknell, 2012)
2. P.A. Jaquin, C.E. Augarde, D.G. Toll, D. Gallipoli Geotechnique, **59**(5):487–490 (2009)
3. J.C. Smith, Examining Soil Based Construction Materials through X-Ray Computed Tomography, PhD Thesis, Durham University, 2015
4. C.T.S. Beckett, M.R. Hall, C.E. Augarde, Acta Geotechnica, 8(4):423-438 (2013)
5. M. Hall, S.Mooney, P.Sturrock, P. Matelloni, S. Rigby, Acta Geotechnica, 8(1):67-79 (2013)
6. H. Houben, H. Guillaud, Earth construction: a comprehensive guide (ITDG Publishing, London, 1989)
7. British-Standards, EN ISO 14688, 2002.
8. J.C. Smith, C.E. Augarde, ECS Tech. Rpt, Engineering and Computing Sciences, Durham University, ECS-TR 2013/04, 2013.
9. British-Standards,BS1377-4:1990

A microstructure insight into the water retention properties of the Callovo-Oxfordian claystone

Hamza Menaceur[1,2], Pierre Delage[1,a], Anh Minh Tang[1] and Jean Talandier[3]

[1] *Ecole des Ponts ParisTech, Navier/CERMES, Marne la Vallée, France*
[2] *Now in LEMTA, Université de Lorraine, Nancy, France*
[3] *Andra, Châtenay-Malabry, France*

Abstract. The effect of changes in suction on the microstructure of the Callovo-Oxfordian claystone along wetting and drying paths was investigated by using mercury intrusion porosimetry. Based on a simplified brick model, the changes of the average thickness of the platelets that constitute the clay matrix (50%) are interpreted as the consequence of the successive adsorption of ordered layers of water molecules along the smectite surfaces, as demonstrated on pure smectites for a long time. This shows the predominant role played by the smectite phase in the overall response of the Callovo-Oxfordian claystone when submitted to suction changes under no stress.

1 Introduction

The Callovo-Oxfordian (COx) claystone is considered as a possible host rock for deep geological disposal of radioactive waste in France. To better investigate its hydromechanical properties, many investigations are carried in the Underground Research Laboratory (URL) mnaged by Andra (the French Agency for radioactive waste disposal) close to the village of Bure (Eastern France). The URL has been excavated at a depth of 490 m in a 150 m thick layer COx claystone. Many investigations are also carried out in surface research laboratories.

Desaturation of specimens of the COx claystone occur in-situ in the gallery walls due to ventilation. Also, the specimens cored appear to be desaturated due to the combined effects of stress release, and evaporation occurring during the successive phases of transport, storage and also preparation and trimming of the specimens for testing.

Claystones are known to be very sensitive to changes in water content, with significant increase in mechanical properties (stiffness and shear strength) with decreased water content and degree of saturation (e.g. Pham et al., 1). The water retention properties of the COx claystone have been investigated in detail in Wan et al. (2) who also evidenced some mechanisms governing volume changes and swelling.

In this paper, the investigation of the water retention properties of the COx specimen is completed by that of the changes in microstructure examined by using mercury intrusion porosimetry tests on freeze-dried specimens. To do so, MIP tests were carried out on specimens previously brought at various suctions, starting from the initial state. Particular attention is paid to the role played by the smectite fraction within the clay matrix that is typical from the COx claystone.

2 The Callovo-Oxfordian claystone

The COx claystone is a sedimentary rock deposited 155 millions years ago on top of a layer of Dogger limestone that was afterwards covered by an Oxfordian limestone layer. The COx claystone is composed of a clay matrix containing some detritic grains of quartz and calcite. At the depth of the Bure URL (490 m), the average mineralogical composition of COx claystone is as follows (Gaucher et al., 3): 45-50% clay fraction (mainly interstratified minerals of illite–smectites with a fraction between 50 and 70% of smectites, 20% carbonate, 22% quartz and 9% other minerals (feldspars, pyrite, dolomite and siderite). The total porosity varies in the COx layer between 14% in carbonated levels and 19.5% in the more argillaceous levels (Yven et al., 4).

The specimens were drilled from a 80 mm diameter and 300 mm long core (EST44584) parallel to bedding that was extracted at the level the URL (490 m). The water content was determined after drying in the oven at 105 °C for 48 hours. The porosity and degree of saturation were calculated from carefully measuring the sample volume by using hydrostatic weighing. The initial total suction was determined by using a dew point tensiometer (WP4, Decagon).

The specimens EST44584 appeared to be not fully saturated at initial state with a degree of saturation around 77.6 % corresponding to a suction of 34 MPa for a porosity of 17.0 %. The dry density and grain density values are 2.16 and 2.7 Mg/m^3, respectively. Partial

[a] Corresponding author: delage@cermes.enpc.fr

saturation resulted from the combined actions of coring, transport, storage and specimen preparation.

3 Experimental methods

3.1 Water retention curve

The determination of the water retention properties of the COx claystone was carried out by controlling suction through the vapour phase. This technique consists in placing the specimens (disks of 38 mm in diameter and 10 mm in height) into desiccators containing saturated saline solutions (Table 1).

Table 1. Saturated solutions used

Solution	Relative humidity (%)	Suction (MPa)
KOH	9	331
$MgCl_2$	33	150
KNO_3	93.7	9
Pure water	100	0

There is no confinement on the sample that is free to swell or shrink. Specimens were left in the desiccators until reaching mass stabilization (checked by periodic precision weighing). Once equilibrated, a suction measurement was made by using the WP4 dew point potentiometer to confirm the achievement of the desired value of suction. Afterwards, the specimens were immediately waxed using slush wax (at lowest possible temperature before solidification, see Wan et al. (2) for more detail). Careful weighing was carried out prior and after waxing, giving a good determination of the wax weight and volume. The specimen volume was obtained by subtracting the wax volume from that of the waxed specimen. The wax technique was also used to obtain the specimen volume at initial state (34 MPa of suction, degree of saturation of 77.6%). Finally, the specimens were cut into small pieces to measure their water content by oven drying. The void ratio and degree of saturation were determined by measuring the volume by means of hydrostatic weighing.

3.2. Microstructure investigation

Mercury intrusion porosimetry tests were conducted on various specimens at different suctions along the drying or wetting paths followed. To preserve the microstructure, dehydration was made by freeze-drying small pieces of claystone (1–3 g in weight) that were previously quickly frozen by immersion in slush nitrogen (– 210°C). Fast immersion in slush nitrogen is preferable because no boiling occurs around the specimen, resulting in faster freezing of water with no volume change (see Delage et al., 5). MIP tests were carried out in a Micromeritics-AutoPore IV 9500 porosimeter from a low initial pressure of 3.4 kPa up to 227.5 MPa, corresponding to entrance pore diameters of 363.6 μm and 5.5 nm respectively.

The intruded mercury porosity (n_{Hg}) was defined as the ratio of mercury intrusion volume V_{Hg} to specimen total volume V:

$$n_{Hg} = \frac{V_{Hg}}{V} \quad (1)$$

The pore entrance diameter (D) was determined from the intrusion pressure P_{Hg} by assuming a cylindrical pore shape according to the Laplace-Young equation:

$$D = \frac{4\sigma\cos\theta}{P_{Hg}} \quad (2)$$

where σ is the mercury-solid interfacial tension and θ is the mercury-solid contact angle (σ =0.484 N/m and θ =141.3° according to Diamond, 6).

3.3 Experimental program

A first series of samples were submitted to drying-wetting path from the initial water content (6.12% with a 34 MPa suction and S_r = 77.6%). Two specimens were simultaneously dried under suctions of 150 and 331 MPa along the drying path, to check repeatability. Specimens A1 and B1 were used to define the specimen characteristics and to investigate their microstructure at 150 and 331 MPa respectively. Once dried at 331 MPa, specimens A2 and B2 were afterwards wetted at 9 MPa and zero suctions, so as to determine the main wetting path. Along the wetting path, both specimens were periodically taken out from the desiccator for weighing so as to determine their water content. Their suction was also measured by means of the WP4 dew-point tensiometer.

A second series of tests corresponds to wetting-drying paths. Starting from the initial water content (6.12 %), three other specimens (C, D1 and D2) were firstly wetted under decreasing suctions (9 MPa for specimen C and zero suction for specimens D1, D2). Specimen D2 was then used to determine the main drying path up to suctions of 150 MPa and 331 MPa.

4 Water retention properties

The water retention curve expressed in terms of changes in water content versus suction is presented in Figure 1 (zero suction arbitrarily plotted at 0.01 MPa). Starting from initial state (w = 6.12%, suction 34 MPa), two points were obtained from two distinct specimens with a direct single step drying at controlled suctions of 150 and 331 MPa, respectively. Similarly, two other points were obtained with a direct single step wetting at suctions of 9 and 0 MPa. Data along the main wetting and drying paths were also obtained by suction measurements and water content determination along the wetting path from 331 to 0 MPa for specimen B2 and along the drying path from zero suction to 331 MPa for specimen D2. Good correspondence is observed along the main wetting path between the data of specimens B2 and A2 that was

DOI: 10.1051/e3sconf/20160906006

wetted between 331 and 9 MPa, providing some confidence in the quality of the data.

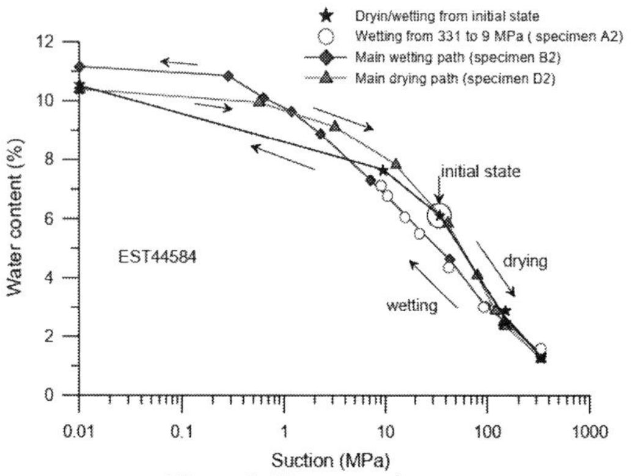

Figure 1. Water retention curve

Examination of the main drying and wetting paths shows that the water retention properties of the COx claystone are characterized, commonly, by a hysteresis effect. The data also show that the initial state resulting from desaturation from the in-situ saturated state is located, as expected, on the main drying path.

The wetting path starting from the initial state shows that the point at 9 MPa of suction (specimen C) is located below the main drying path, which is typical of scanning curves (there is unfortunately no other point between 9 MPa and zero suction along this path). The point at zero suction obtained along the main wetting path starting from the dry state at 331 MPa provides a water content (11.2%) higher than that obtained after wetting the specimen from the initial water content (10.4%). This is due to the longer period of time waited for (90 days for B2 compared to 40 days for D1 and D2.

The changes in degree of saturation with suction and water content are presented in Figure 2 and Figure 3 respectively.

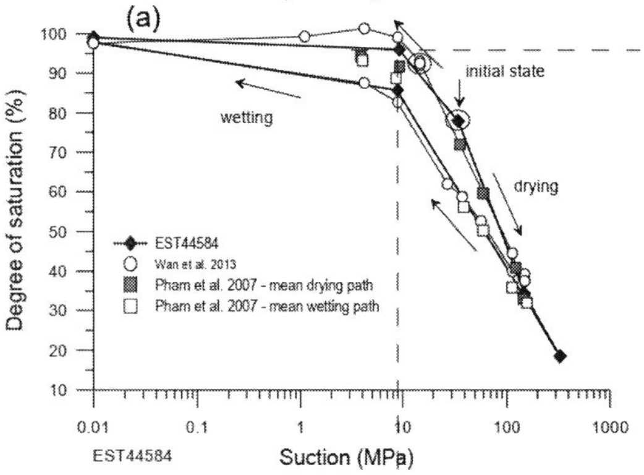

Figure 2. Change in degree of saturation with respect to suction

The data taken from similar tests by Pham et al. (1) and by Wan et al. (2) are also plotted in Figure 2, with good agreement between the data.

The specimen following the wetting path from the initial state comes close to saturation at 9 MPa (S_r = 96%, w = 7.65 %). At suction smaller than 9 MPa, the significant increase observed in water content occurs in saturated condition. Along the drying path from initial state, a linear relationship is observed between S_r and w with an average decrease of 13% in S_r for a decrease of 1% in w. The main wetting curve obtained once the specimen dried at the highest suction (331 MPa) is located below the curves of drying-wetting from initial state, confirming the hysteresis observed by Wan et al. (2).

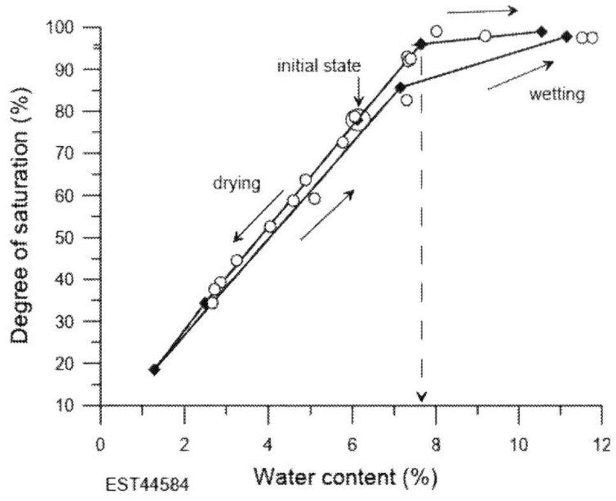

Figure 3. Change in degree of saturation with respect to water content

The volume changes with respect to suction (**Figure 4**) show that drying from initial state to 331 MPa results in a shrinkage of 2.0% while wetting to zero suction induces a swelling strain of 6.1%.

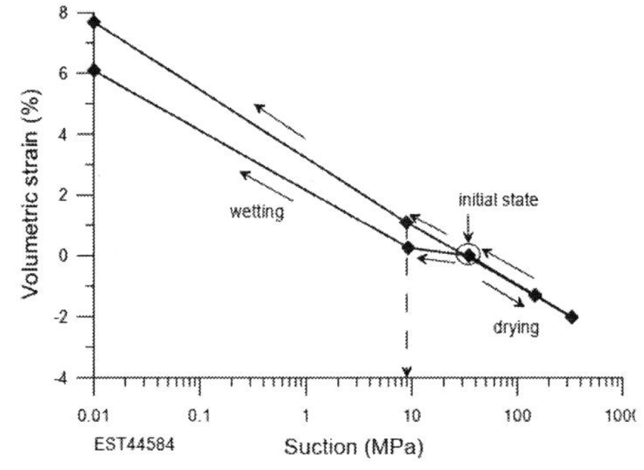

Figure 4. Volume changes with respect to water content

Figure 4 also shows that reducing suction along the wetting path from initial state occurs with little swelling (0.3%) from 34 to 9 MPa followed by significant swelling below 9 MPa (6.1% at zero suction) that occurs within a saturated specimen, as commented above. The specimen previously dried up to 331 MPa before wetting to zero suction (main wetting path) is not brought back to the same point as that of the specimen submitted to a wetting path from initial state, with some irreversible swelling strain (1.6%) and higher volume at zero suction. However, the main wetting path is close to the drying curve from initial state at higher suction (> 40 MPa) showing some reversibility in the volume change behaviour under stress free conditions at highest suctions.

DOI: 10.1051/e3sconf/20160906006

5 Pore size distributions

The results of the MIP test obtained at initial state (S_r = 77.6%, porosity n = 17.0%) are presented in **Figure 5**a in cumulative plot and **Figure 5**b in a density function plot. The total porosity n of the specimen is also plotted in the cumulative curve, showing that the mercury intruded porosity n_{Hg} is smaller than the total one, with an infra-porosity n -n_{Hg} of 4%. Pores smaller than the lower limit of 5.5 nm (corresponding to the maximum mercury pressure of 227.5 MPa) cannot be intruded.

The density function exhibits a typical monomodal curve with a well defined pore population identified by an inflection point at 32 nm defining the average entrance pore diameter within the clay matrix.

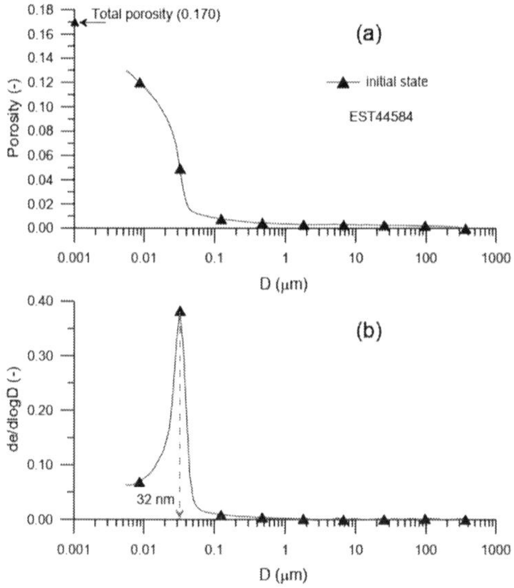

Figure 5. Pore size distribution, initial state

Figure 6 presents an enlargement between 0.001 and 0.1 μm of the PSD curves of specimens A1 and B2 dried at 150 and 331 MPa respectively from the initial state, together with the PSD data of specimens at initial state and oven-dried at 105°C during 48h. The total porosities are also given. The infra-porosities not intruded by mercury (D < 5.5 nm) are again lower than the total ones. They are smaller than that at initial state (n - n_{Hg} = 3.0% for specimens dried up to 150 and 331 MPa, compared to 4% at initial state). Conversely, the mercury intruded porosity of the oven-dried specimen (12%) appears to be quite close to the total one (13%) giving an infra-porosity n - n_{Hg} = 1% significantly smaller than at suctions of 150 and 331 MPa (3%).

The degree of saturation of each specimen is also reported in **Figure 6**, based on the hypothesis that water is contained in smaller pores due to the nature of both physico-chemical clay water interactions and capillary effects. The arrows indicate the maximum diameter of the pores saturated by water. One can observe that the main pore population is almost saturated (up to a diameter of 33 nm) at initial state (suction 34 MPa, S_r = 77.6%, mean radius 32 nm) whereas this population becomes full of air at 150 MPa suction and S_r = 34.4% with reduction of the mean radius at 28 nm and with the largest saturated pore having a diameter of 13 nm.

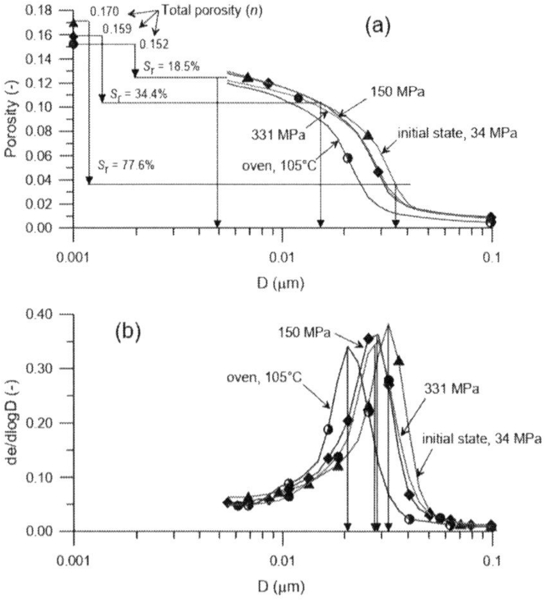

Figure 6. Pore size distributions, dried states

At 331 MPa, (S_r = 18.5%), the largest saturated pore has a diameter of around 4.9 nm with however no significant change in the mean radius value that slightly decreases to 27 nm. In other words, the desaturation of pores between 13 and 4.9 nm does not significantly affect the microstructure at the level of both the mean pore population and of the infra-porosity that remains constant (3%). Conversely, the effect of oven-drying is much stronger than drying at 150 and 331 MPa suctions, with a reduction of the mean pore diameter to 21 nm and, more importantly, a significant reduction of the infra-porosity to 1%. There is obviously a gap between the highest applied suction (331 MPa) and the total removal of water from the microstructure by oven-drying. Note that the 21 nm mean pore diameter obtained on the oven-dried specimen is comparable to the values obtained by Sammartino et al. (7) and Boulin et al. (8) on oven-dried specimens (respectively dried at 105 and 70°C).

In **Figure 7**, the pore size distribution curves of the two specimens wetted from initial state at low suctions (9 MPa and zero) and higher degree of saturation (S_r =96% and 99%, s = 9 and 0 MPa, respectively) are presented together with the data at initial state (S_r = 77,6%, s = 34 MPa). Also plotted in the Figure are the total porosities of the specimens.

The PSD cumulative and density function curves at 9 MPa suction (S_r = 96%) are quite similar to that at initial state with a only slight increase in total porosity at 9 MPa (0.2%), showing that the infra-porosity is also similar in both cases. This is compatible with the negligible volume change observed in the curve of **Figure 4**. The specimen saturation from S_r = 78% to 96% hence occurs by complete filling of the mean pore population with very

DOI: 10.1051/e3sconf/20160906006

little microstructure changes both in the porosity intruded and not intruded by mercury.

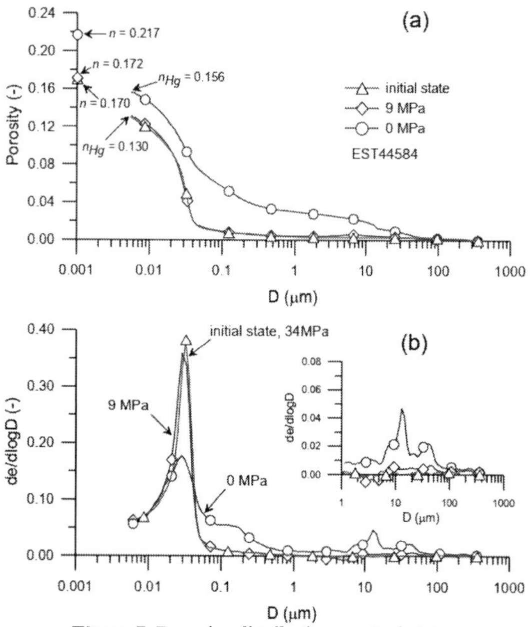

Figure 7. Pore size distributions, wetted states

The reduction in suction from 9 MPa to 0 MPa occurs at quasi-saturated state (see Figure 3) with significant 6.1% swelling giving a final water content of 10.5% and degree of saturation of 99%. Significant changes are also observed at the microstructure level, with a bimodal PSD curve at 0 MPa. Changes affect three levels of porosity:

• i) The infra porosity n - n_{Hg} that increases from 4% at 9 MPa to 6.1% at zero suction with an intruded porosity of 15.6% and a total one of 21.7%;

• ii) The large porosity with appearance of a new pore population corresponding to a proportion of 10.5% of the total porosity in the range of diameters between 7 and 100 μm with a mean diameter around 12 μm;

• iii) An enlargement of the previous pore population that moves from a narrow range between 12 and 50 nm to a wider range between 12 and 500 nm with a new plateau between 60 and 200 nm.

Note however that no change is observed in the density function curves between 5 and 20 nm.

6 Discussion

Microstructure changes with respect to suction changes are interpreted based the conceptual model of the COx claystone microstructure proposed by Yven et al. (2007) and presented in **Figure 8**. The model schematically shows how individual calcite or quartz detritic grains are embedded into a clay matrix with porosity that represents 45 – 50% of total constituents at the depth of 490m considered here.

The interpretation of the well defined single pore population defined by an average value of 32 nm observed in the PSD curve of **Figure 5** can be made by assimilating the clay matrix to an assembly of bricks made up of platelets of comparable thickness, as seen in **Figure 8**. The mean pore radius hence provides an estimate of the average platelet thickness.

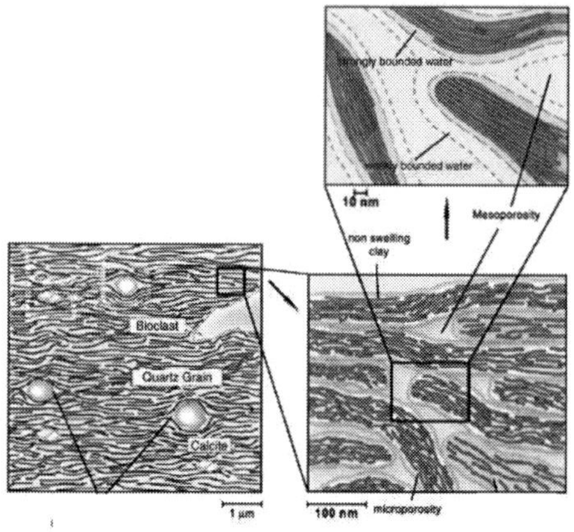

Figure 8. Microstructure model (after Yven et al., 4)

The change in PSD curves is now interpreted based on the mechanisms of hydration of smectites, which, based on the results of X-Ray diffraction investigations, is known for some time to occur by the ordered and successive placement of layers of water molecules along the clay layers, depending of the suction (e.g. Mooney et al., 9). Typically, the first layer is observed at high suctions of various hundreds of MPa, whereas the second one is observed under various tens of MPa and the third one below around 9 MPa (see also Saiyouri et al., 10). X-Ray diffraction investigations also showed that, starting from a inter-basal spacing of 9.6 Å for illite and smectite, inter-basal spacings of 12,5, 15,5 and 18,5 Å corresponded to the hydrated states of smectite with 1, 2 and 3 layers of water molecules, respectively. This adsorption occurs at the inter-layer level, within the clay platelets represented in the scheme of **Figure 8**. It hence affects the average thickness of the platelets that we detect here by measuring the average entrance pore radius within the clay matrix.

Observation of the PSD curves of **Figure 6** indicates that drying is characterized by a reduction of the mean diameter of the single pore population from 32 nm (intact) to 28 (150 MPa), 27 (331 MPa) and 21 nm (oven-dried). Based on the brick model, this corresponds to a reduction in thickness of the bricks, i.e. of the platelets. This reduction is only due to changes in the interlayer spacing of smectite, given that illite minerals are not sensitive to changes in water content. The reduction of the average diameter from 32 nm (initial state) to 28-27 nm when passing from 34MPa to 150-331 MPa can then be linked to the transition from two (inter-basal spacing of 15.6 Å) to one adsorbed layer of water molecules (inter-basal spacing of 12.6 Å).

Oven drying results in having no more water layer adsorbed. At dry state (interlayer spacing of 9.6 Å for smectites and illites) and with an average platelet thickness of 21 nm as obtained from the PSD curve of **Figure 6**, an average number of 22 layers by platelets can be roughly estimated. Considering the proportion of 50-

70% smectite minerals provided by Yven et al. (4) at the 490 m depth in the COx layer, one can conclude that the introduction of one water layer along the smectite layers would result in the placement of between 11 and 15 layers of 3 Å thick layers of water molecules, resulting in an increase in the platelet thickness between 3,3 and of 4,6 nm, from 21 to 24,3 – 25,6 nm. This is not far from the 28 nm value measured by MIP under suctions of 150 and 331 MPa. Adding another water layer would then provide a thickness between 27,6 and 30.2 nm, reasonably comparable to the 32 nm measured by MIP under 34 and 9 MPa suctions.

This indicates that the analysis based on the step hydration process evidenced in pure and compacted smectites is reasonably compatible with the analysis of the MIP data with respect to the changes in platelets thickness in the COx clay matrix with suction. Better fitting is actually obtained with the higher proportion of smectite of 70% in the clay fraction. The mechanism of hydration of smectites hence appears to be of significant interest to interpret the water retention properties and related microstructure changes of the COx claystone.

7 Conclusion

Better understanding of the changes in microstructure that occur along the wetting and drying paths of the water retention curve of the COx claystone have been gained through a microstructure study carried out by using MIP.

Some further information on the water retention properties was obtained from monitoring volume changes, showing significant swelling at saturated state at suction lower than 9 MPa through the hydration and development of saturated cracks.

The concepts developed to describe the hydration mechanisms of smectites and their dependency with respect to suction changes appear to be applicable to the hydration of the mixed-layer illite-smectite minerals that play a predominant role with respect to changes in water content and the swelling-shrinkage behaviour of the COx claystone. Releasing the suction to 9 MPa results in the saturation of the claystone but does not significantly affects the COx microstructure with no swelling and no change in the pore size distribution curve, in accordance with the stability of the two layers of interlayer water molecules in this suction range. Passing from 9 MPa to zero suction allows the placement of a third, or even a fourth layer of water molecules that results in significant changes in the inter-platelets porosity and imultaneous development of saturated cracks.

Drying the specimen at 150 and 331 MPa results in a reduction from 32 to 28-27 nm of the mean diameter of the pore size distribution curve with the placement of one layer of water molecules, which explain the small differences in microstructure observed at these two suctions. Oven-drying at 105°C results in a further decrease to 20 nm in the mean diameter of the inter-platelet porosity, a state compatible with the status of smectite minerals at dry state, with no water layer adsorbed and an interlayer space of 9.6 Å.

Finally, the mechanisms of water adsorption along smectites allows to better define the status of water in claystones and shales containing some smectites, with a distinction between the adsorbed water (located within the platelets and made up of two layers strongly bonded along the smectite faces with suctions larger than 9 MPa) and the free water (located in the inter-platelets porosity, defining the pore pressure and submitted to water transfers). The MIP curve at initial state provides an idea of the proportions of the two types of water, with around 68% of free inter-platelets water and 32 % of adsorbed intra-platelets water.

Acknowledgements

The authors are indebted to Andra, the French agency of radioactive waste disposal, who financially supported this research through the PhD thesis of the first author and provided the specimens of Callovo-Oxfordian claystone.

References

[1] Pham QT, Vales F, Malinsky L, Nguyen Minh D, Gharbi H (2007). Effects of desaturation-resaturation on mudstone. Phys. Chem. Earth 32: 646–655.
[2] Wan M, Delage P, Tang AM, Talandier J (2013). Water retention properties of the Callovo-Oxfordian claystone. Int. J. Rock Mech. Min. Sc. 64: 96–104
[3] Gaucher G, Robelin C, et al. (2004). ANDRA underground research laboratory: interpretation of the mineralogical and geochemical data acquired in the Callovian-Oxfordian formation by investigative drilling. Phys. Chem. Earth 29: 55-77
[4] Yven B, Sammartino S, et al. (2007). Mineralogy texture and porosity of Callovo-Oxfordian claystones of the Meuse/Haute-Marne region (eastern Paris Basin), Mém Soc Géol France, 178: 73-90
[5] Delage P, Marcial D, Cui YJ, Ruiz X (2006). Ageing effects in a compacted bentonite: a microstructure approach. Géotechnique 56 (5): 291–304
[6] Diamond, S. (1970). Pore size distribution in clays. Clays Clay Min. 18, 7–23
[7] Sammartino S, Bouchet A, Prêt D et al. (2003). Spatial distribution of porosity and minerals in clay rocks from the Callovo–Oxfordian formation (Meuse/Haute-Marne, Eastern France)—implications on ionics species diffusion and rock sorption capability. App. Clay Sc. 23 (1–4): 157–166
[8] Boulin PF, Angulo-Jaramillo R, Daian JF et al. (2008). Pore gas connectivity analysis in Callovo–Oxfordian argillite. App. Clay Sc. 42: (1–2) 276–283
[9] Mooney RW, Keenan AC, Wood LA (1952). Adsorption of water vapor by montmorillonite. II. Effect of exchangeable ions and lattice swelling as measured from X-ray diffraction. J. Am. Chem. Soc. 74: 1371–1374
[10] Saiyouri N, Tessier D, Hicher PY (2004). Experimental study of swelling in unsaturated compacted clays. Clay Min. 39:469–479

Study of the influence of the saline solution NaCl on the potential collapse of soil

Khelifa Abbeche [1,a] Ouassila Bahloul[1] and Azeddine Bahloul [2]

[1]*Department of Civil Engineering, University of Batna, Algeria*
[2]*CGI Technical Services, 1612 Wedding Way Redding, California, 96003 USA*

Abstract. Collapsible soils are unsaturated soils which present a potential for large strains and a complete change to the whole particle structure after wetting with or without loading. These soils are characterized with loose structures composed of silt to fine-sand-size particles.The objective of this experimental study is to illustrate that the resistance of collapsible soil can be improved. This study demonstrates that it is possible to minimize the collapsible potential C_p to an acceptable level after chemical treatment with salt (sodium chloride NaCl) at different concentrations (0.5, 1.0, 1.5 and 2.0 mole/liter) and at different compaction energies. The method used in this study is based on oedometric tests with variable normal stresses.

1 Introduction

Collapsible soils are unsaturated soils which may undergo inter-granular rearrangement with abrupt and significant volume reduction after wetting. Many research studies are developed to identify and prevent failure of collapsible soils as well as the measures to insure sufficient stability of structures to be built on such soils, [1, 2, 3, 4, 5]. The solutions proposed by various researchers to mitigate collapsible soils depend on the depth of soil layer in one hand, and the bearing capacity required by the structure foundation, on the other hand. Deep foundations, replacement or on-site thermo or mechanical treatment were part of the proposed solutions. Recently, chemical stabilization methods by salts were proposed [6-7]. However, it should be noted that there is few studies on the effect of soluble salts particularly sodium chloride (NaCl) on the mechanical behavior and microstructural characteristics of collapsible soils. It is proposed in this experimental work to study the influence of NaCl on the reduction of the potential for collapse C_p of a soil sample reconstituted from sand and kaolin. This study involves different salt concentrations, water contents, and energies of compaction based on œdometer test with different vertical stress.

2 Materials, Equipment and Test

The reconstituted soil sample is composed of 80% of sand and 20% of kaolin. The application of various criteria of collapse, as indicated by [8-9], show that this material is collapsible.

[a] Corresponding author: abbechekhelifa@hotmail.com
bahouassila@gmail.com

2.1 *Characteristics of Materials*

The sand used in the sample reconstitution comes from the Liwa River located in the region of Biskra, Algeria screened on the 2mm sieve; and its physical characteristics are given in the Table 1. The kaolin soil used in this study is extracted from Debbagh area of Guelma, Algeria. The physical characteristics are given in Table 1. Observation by scanning electron microscopy (SEM) (Figures 1) reveals that the experimental material consists of quartz and kaolin.The sieve analysis of the reconstituted sample is shown in Figure 2. The geotechnical characteristics of that sample are given in Table 1.

Table 1. Geotechnical characteristics of Materials

Materials	**Characteristics**
Sand	Sand equivalent SE = 87 % Coefficient of uniformity Cu = 3.91 Coefficient of curvature Cc = 0.95 Grain size distribution (0.08-2 mm) with 3.03% of particles < 80 µm.
Kaolin	D‹80µm Liquid limit LL = 67(%) Plastic limit PL = 39(%) Specific Gravity GS = 2.4
Reconstituted soil (80% sand + 20% kaolin)	Coefficient of uniformity Cu = 5.13 Coefficient of curvature Cc = 1.07 Liquid limit LL = 28 (%) Plastic limit PL = 16 (%) Specific gravity GS = 2.65 Maximum dry density (g/cm^3) γ_d = 1.93 Optimal water content w_{opt} = 10 (%)

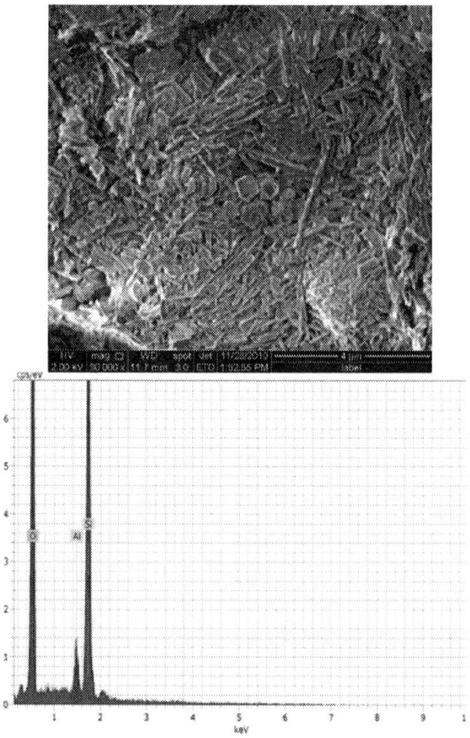

Spectrum: Acquisition 708

El	AN	Series	unn. C [wt.%]	norm. C [wt.%]	Atom. C [at.%]	Error [%]
O	8	K-series	72.59	67.73	78.60	8.9
Al	13	K-series	2.80	2.61	1.80	0.2
Si	14	K-series	31.78	29.66	19.60	1.3
		Total:	107.16	100.00	100.00	

Figure 1. Electron Photomicrograph of reconstituted sample (sand & kaolin).

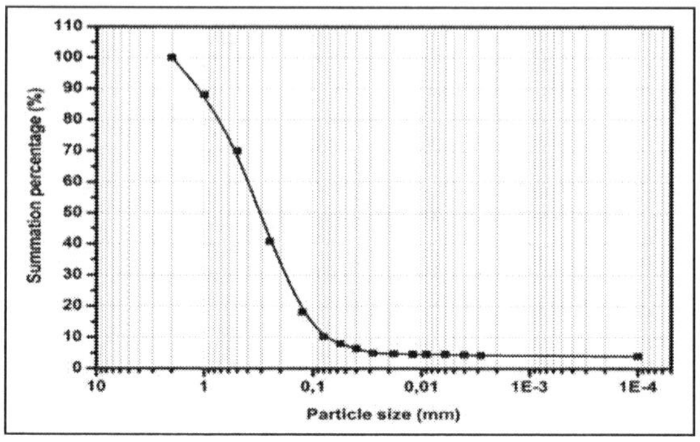

Figure 2. Sieve analysis of the reconstituted sample.

2.2 *Equipment*

The experimental equipment and tests are the same as those reported by [5], only the saline solution is different. The equipment used is shown schematically in Figure 3. It is composed of a disc having a diameter slightly smaller than that of the oedometer ring, fixed at a guidance rod and a hammer in the shape of a disc. The 152g hammer falls along the rod with a drop of 15 cm and contacts the fixed disc resulting in the compaction of sample in the oedometer ring, with the following compaction energy:

$$E_c = n.m.g.h/v \qquad (1)$$

where Ec is the compaction energy, n is the number of drops, m is the mass of hammer disc, h is the drop height, g is the acceleration of gravity and v is the volume of material before compaction.

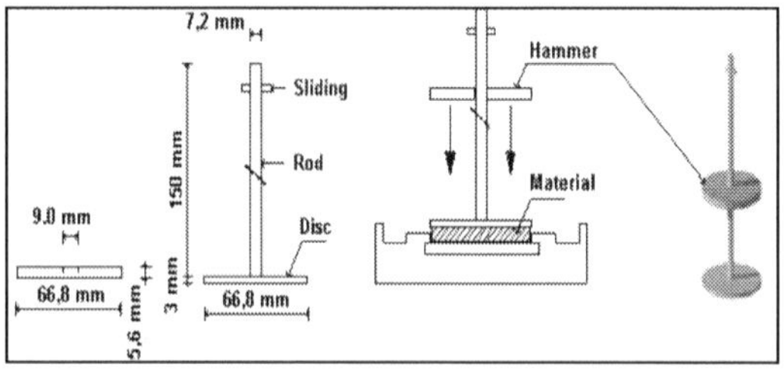

Figure 3. Experimental equipment (compaction device).

2.3 *Test and Procedure*

The object of this study is to reconstitute the soil in an oedometer mold with specific water content and dry density. The material is compacted in one layer because of the low thickness of the ring. Reconstituted soil exhibits good mechanical characteristics when it is compacted with a low initial water content (w_0=2, 4 and 6%). However, if higher moisture is introduced, even without additional load, the structure collapses and significant deformations occur. [10] proposed a procedure to estimate the collapse potential of a soil. That potential is evaluated by loading in an oedometer device two undisturbed soil samples at natural moisture content. One sample is then flooded for saturation. These tests provide the void ratios (e_1 and e_2) before and after flooding at a specific loading. This behavior can be observed in Figure 4 representing a typical curve of a double consolidation method for such soil. The collapse potential, C_p is calculated as the vertical strain similar to an abrupt settlement and thus can be expressed as

$$C_p\ (\%) = \Delta H/H_0\ 100 = \Delta e_c/(1+e_0)\ 100 \qquad (2)$$

Where ΔH is change in height of the sample upon flooding, H_0 is original height of the sample, $\Delta e_c = (e_1 - e_2)$ is the change in void ratio of the sample up on flooding, e_1 is the void ratio at dry state, e_2 is the void ratio upon of flooding and e_0 is the initial void ratio before loading.

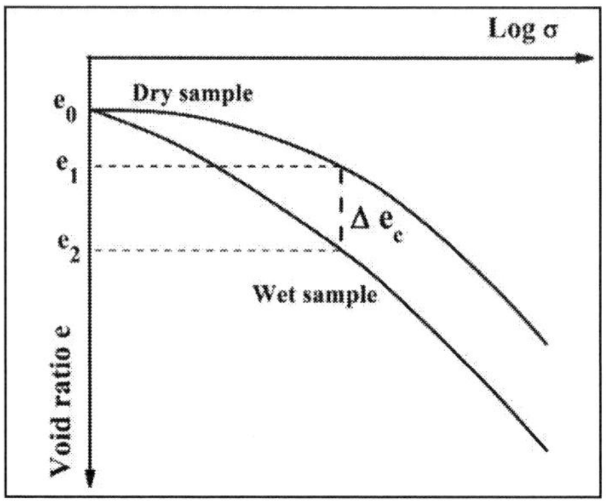

Figure 4. Typical oedometer curve for a collapsing soil [10]

Also, [10] correlated the collapse potential Cp to the severity of foundation disorder (Table 2).

Table 2. Foundation disorder level based on collapse potential C_p [10]

C_p (%)	Severity of disorders
0-1	Minor disorders
1-5	Moderate disorders
5-10	Disorders
10-20	Severe disorders
› 20	Very severe disorders

In order to obtain samples with different degrees of compaction, those samples are moistened at different water contents, then they are subject to different compaction energies (20, 40 and 60 drops) [8]. The samples thus prepared were tested using an oedometer by saturating them with solutions at different salt concentrations. Generally, work treating the effect of salts on the microstructural characteristics and hydraulic behaviour of clay materials use concentrations up to 1 mol/L [5, 11, 12]. The salt solution used for the soil treatment is sodium chloride (NaCl) with various concentrations (0.5, 1.0, 1.5 and 2 mol/L).

3 Results and Analysis

3.1. *Oedometer Results: Untreated Soils*

From the results obtained in this study, summarized in Figures 5 to 13, it is clear that the collapse of the sample without treatment correlates with collapsible soil which can cause disorders to foundation of structures [10]. Indeed, for various vertical stresses σ as well as for various compaction energies E_c, the collapse potential C_p varies from 0.98 to 12.40% for initial water content w_0=2%, from 0.82 to 10.64% for initial water content w_0=4%; and from 0.66 to 8.79% for an initial water content w_0=6%. These results correspond to minor to severe disorders (Table 2). It should be noted that the potential of collapse C_p decreases when initial water content or initial compaction energy increase. That is attributed to the reduction of the suction forces when the initial water content increases and denser structure when the energy increases. The wetting of the reconstituted sample induces a total suppression of the suction forces; which soften the material and reducing the shear resistance leading to an abrupt collapse. The collapse potential C_p is at its maximum for a vertical stress in the range of 400 kPa. This result reflects that there is a total re-arrangement of the soil skeleton under that loading, which results in the reduction of the voids. This behavior is the same regardless of the vertical stress. These results agree with those of [4, 5, 13 14]. Thus, a reconstituted soil sample has a behavior similar to those undisturbed natural collapsible soils.

3.2 *Oedometer Results: Soil Treated by Salt (NaCl)*

Results shown on figures 5 to 13 reveal a substantial reduction of the collapse potential C_p for soil samples treated with as sodium chloride (NaCl) at various ionic concentrations. Also, we note that the collapse potential C_p is at its maximum for a vertical stress in the range of 400 kPa. As stated before, this result reflects that there is a total re-arrangement of the soil skeleton under that loading, which results in the reduction of the voids. It should be noted that for a collapse potential at a vertical stress of 400 kPa, and for a low salt concentration (0.5 mol/L), the reduction ratios of C_p vary from 15% to 35%. However, for high concentrations (2 mol/L), the reduction ratios of C_p vary from 65% to 70%. The collapse reduction ratio is related to the applied vertical stress, knowing that for low concentrations (0.5 mol/L), this collapse reduction ratio varies from 15% to 25% for a vertical stress of 200 kPa, and it varies between 10% to 40% for a vertical stress of 800 kPa. For high salt concentrations (2 mol/L), the collapse reduction ratio varies between 60% to 70% for a vertical stress of 200 kPa. On the other hand for a vertical stress of 800 kPa, the values of the collapse reduction ratio vary between 30% to 65%.Thus, the behavior of a treated soil by saline solution NaCl is on one hand, affected by the initial water content and the compaction energy and on the other hand by the ionic concentration. Figures 5 to 13 show clearly that the collapse potentials in the presence of the solution NaCl at a concentration of 1.5 mo/L are close to those of 2 mol/L. As a result, we opt for an effective treatment with the saline solution at an optimal concentration of 1.5 mol/L.

DOI: 10.1051/e3sconf/20160907001

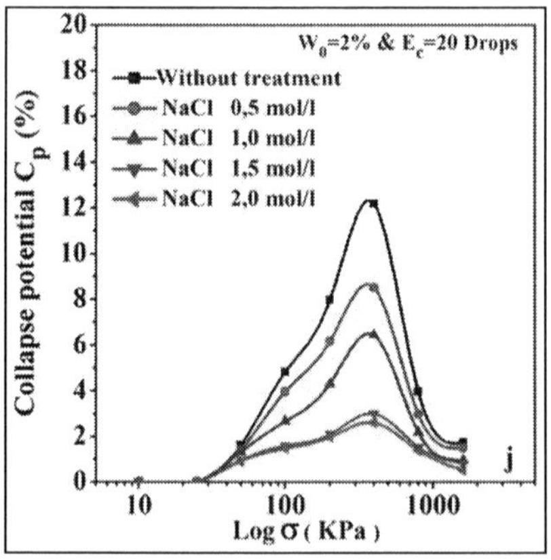

Figure 5. Variation of the collapse potential C_p with respect to vertical stress for treatment by salt NaCl (W_0=2% & E_c=20 drops).

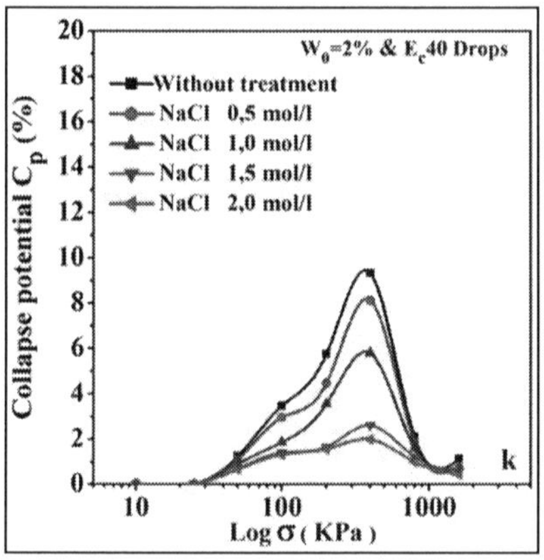

Figure 6. Variation of the collapse potential C_p with respect to vertical stress for treatment by salt NaCl (W_0=2% & E_c=40 drops).

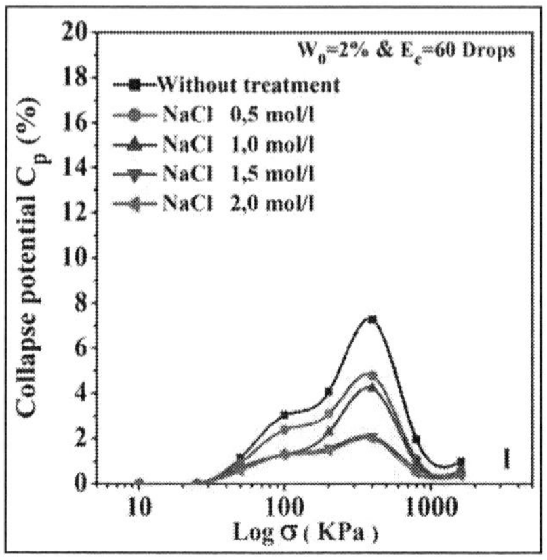

Figure 7. Variation of the collapse potential C_p with respect to vertical stress for treatment by salt NaCl (W_0=2% & E_c=60 drops).

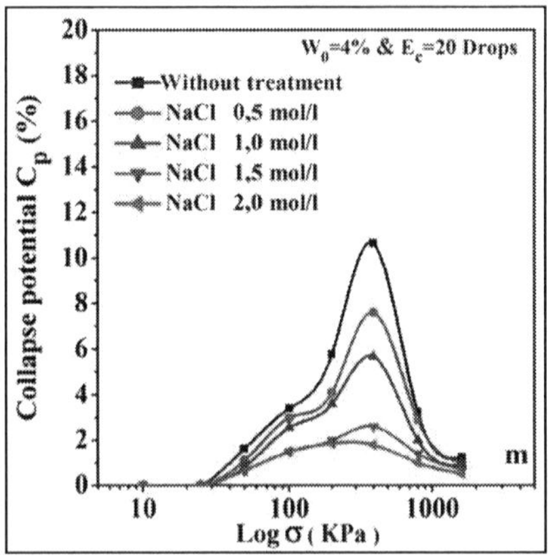

Figure 8. Variation of the collapse potential C_p with respect to vertical stress for treatment by salt NaCl (W_0=4% & E_c=20 drops).

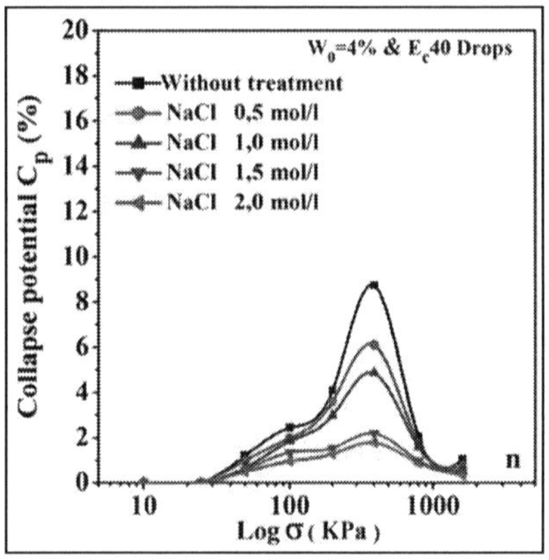

Figure 9. Variation of the collapse potential C_p with respect to vertical stress for treatment by salt NaCl (W_0=4% & E_c=40 drops).

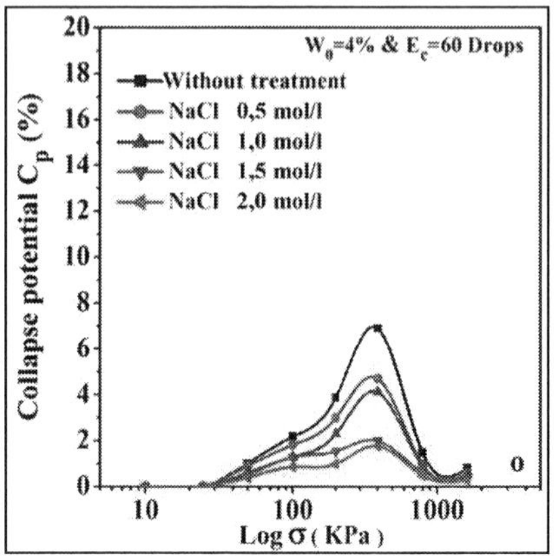

Figure 10. Variation of the collapse potential C_p with respect to vertical stress for treatment by salt NaCl (W_0=4% & E_c=60 drops).

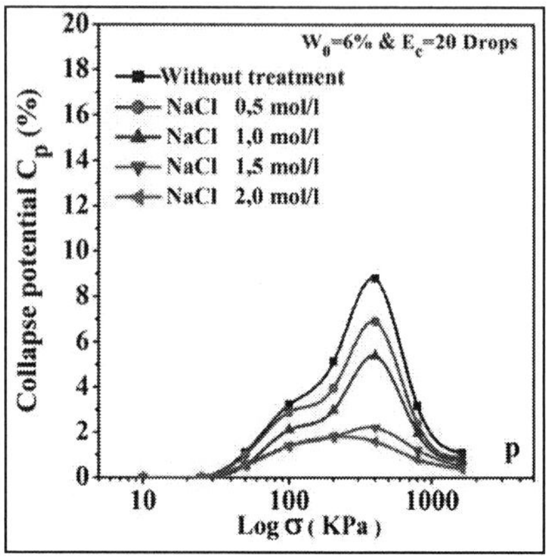

Figure 11. Variation of the collapse potential C_p with respect to vertical stress for treatment by salt NaCl (W_0=6% & E_c=20 drops).

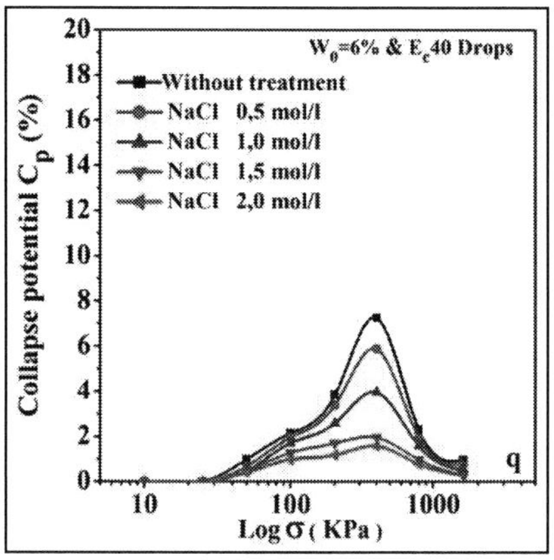

Figure 12. Variation of the collapse potential C_p with respect to vertical stress for treatment by salt NaCl (W_0=6% & E_c=40 drops).

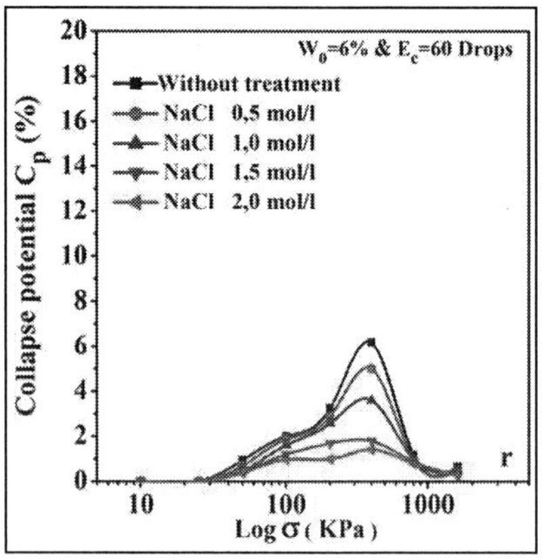

Figure 13. Variation of the collapse potential C_p with respect to vertical stress for treatment by salt NaCl (W_0=6% & E_c=60 drops).

4 Conclusions

The main observations of this study are:
The collapse potential is reduced with increase of initial water content and increase of energy of compaction;
The treatment by salt (NaCl) reduced the collapse potential and subsequently the damage associated with it to structures;
The increase in salt concentration beyond certain value (1.5 mol/L) has no significant benefit.
Thus, the physical and mechanical behavior of the reconstituted collapsible soil is related, to one hand, to the energy of compaction (Ec), the initial moisture content (w_0), vertical stress (σ).

Acknowledgements

We would like to thank the members of CGI Technical Services for their contributions to this study.

References

1. K. Rollins, M. Wayne, G. Rogers., J.G.E, **120**, (1994)
2. Y. J. Cui, P. Delage, F. Schlosser, M. Wonarowcz, XIIème C.E.M.S.G.A., **1**:337-342. (1999)
3. B.Le Runigo, V. Ferber, O. Cuisinier, D. Deneele, J. Cui., JNGG 08 Nantes, (2008)
4. T. Ayadat, A. Hanna., G.T.J, **30** (2007)
5. K. Abbeche, O. Bahloul, T. Ayadat, A. Bahloul., *Experimental and Applied Modeling of Unsaturated Soils, Proceedings of Geoshanghai, June 3-5, Shanghai, Chine.doi:10 1061/41103(396)10,* (2010)
6. C. Shao Chi, Ou. Chang, W. Ming Kuang, A.C.S, **44**, (2009)
7. S. Kaufhold, R. Dohrman., A.C.S, (2009)
8. K. Abbeche, M. Laouar, F. Messaoud., S.G.M, **XXXII** 2:.3-21, (2010)
9. AJ. Lutenegger, RT. Saber., G.T.J, **11**, 3:173-178, (1988)
10. JE. Jennings, K. Knight., *Proceedings. 4th Inter. Conf. on Soil Mechanics and Foundation Engineering, 1: 316-319,* (1975)
11. A. Halitim., Thesis, University Rennes : 383, (1985)
12. D. Tessier, J. Berrier, S.S, (1979)
13. EC. Lawton, R J. Fragaszi, H. James., J.G, ASCE, **155**,9: 1252-1267, (1989)
14. K. Abbeche, F. Hammoud, T. Ayadat., *Experimental Unsaturated Soil Mechanics, Springer Proceedings in Physics, 112: 3-9.doi:10 1007/3 540 6 9873-6-1,* (2007)

Vapor intrusion in soils with multimodal pore-size distribution

Alfaro Soto Miguel [1a] and Hung Kiang Chang [2]

[1, 2] *UNESP, Universidade Estadual Paulista, Departamento de Geologia, Rio Claro-SP, Brazil*

Abstract. The Johnson and Ettinger [1] model and its extensions are at this time the most widely used algorithms for estimating subsurface vapor intrusion into buildings (API [2]). The functions which describe capillary pressure curves are utilized in quantitative analyses, although these are applicable for porous media with a unimodal or lognormal pore-size distribution. However, unaltered soils may have a heterogeneous pore distribution and consequently a multimodal pore-size distribution [3], which may be the result of specific granulometry or the formation of secondary porosity related to genetic processes. The present paper was designed to present the application of the Vapor Intrusion Model (SVI_Model) to unsaturated soils with multimodal pore-size distribution. Simulations with data from the literature show that the use of a multimodal model in soils with such pore distribution characteristics could provide more reliable results for indoor air concentration, rather than conventional models.

1 Introduction

The contamination of subsurface soils and underground waters can contain chemicals vapors and migrate from the subsurface to nearby basements, buildings, and other enclosed spaces and move by diffusion through the soil and any covering material (concrete, floors, geo-membranes, etc.), or by advection along preferential paths (cracks, fractures, ducts, etc.).

According to USEPA [4], the model of Johnson and Ettinger [1] is designed to provide a solution for either a state-state to vapor transport (finite or non-dinimishing source) and as quasy-steady-state (finite or dinimishing source). The input of the model include chemical properties of the contaminant, properties of the saturated and unsaturated soil, and the structural characteristics or properties of a building.

In relation to the properties of soils, the model of Johnson and Ettinger [1] considers, among other factors, the functions which describe the capillary pressure curves in quantitative analyses. However, these equations were designed for porous media with a unimodal or lognormal pore-size distribution. Unaltered soils, however, may have heterogeneous systems of pores, and hence, multimodal pore-size distributions [3]. According to Mallants [5], may be the result of aggregation and/or biological processes, but have also been observed in morainic and solifluction soils

This means that predictions the transport of contaminant vapors emanating from subsurface soils into indoor spaces located above the source of contamination on present-day may be different in situations in which soil porosity involves a system of multimodal pores.

[a] Corresponding author: alfaro@rc.unesp.br

The present paper was thus designed to furnish suggestions for the determination of concentrations arising from vapor intrusion into closed environments, including transport by diffusion through concrete from contaminated soils with a multimodal pore distribution.

2 Experimental and analytic methods

2.1 Contaminant vapors from subsurface soils into indoor spaces (Johnson and Ettinger [1])

According to Johnson and Ettinger [1] and USEPA [4], the concentration of contaminating vapour in closed environments can be calculated using the following expression:

$$C_{building} = \alpha . C_{source} \tag{1}$$

Where $C_{buiding}$ (μg.m^{-3}) represents the indoor air concentration of a specific chemical under steady-state condition, C_{source} (μg.m^{-3}) concentration at the source and α represent the steady state attenuation coefficient.

$$\alpha = \frac{\left[(a)x\exp(b)\right]}{\left[\exp(b)+(a)+(c)\left[\exp(b)'-1\right]\right]} \tag{2}$$

$$a = \frac{D_T^{eff} A_B}{Q_{bulding} L_T}$$

$$b = \frac{Q_{soil} L_{crack}}{D^{crack} A_{crack}}$$

$$c = \frac{D_T^{eff} A_B}{Q_{soil} L_T}$$

Where A_B represents the area of the enclosed space below grade (m^2), L_T (m) souce buiding separation, L_{crack} (m) enclosed space foundation or slab thickness, A_{crack} (m^2) area of total cracks, $Q_{bulding}$ (m^3.s^{-1}) building ventilation rate, Q_{soil} (m^3.s^{-1}) the volumetric flow rate of soil gas entering the building and D_{crack} (m^2.s^{-1}) the effective diffusion coefficient through the cracks. On the other hand, D_T^{eff} (m^2.s^{-1}) represents the total overall effective diffusion coefficient. According to USEPA [4] the D_i^{eff} of the layer of soil in contact with the floor can be assumed to be equivalent to the coefficient of effective diffusion through the cracks (D_{crack}). The coefficient of effective diffusion in the unsaturated zone can be expressed by:

$$D_i^{eff} = D_a\left(\theta_{a,i}^{3.33}/n_i^2\right)+\left(D_w/H'_{TS}\right)\left(\theta_{w,i}^{3.33}/n_i^2\right) \quad (3)$$

$$\theta_{a,i} = n_i - \theta_{w,i} \quad (4)$$

$$\theta_{w,i} = \theta_r + \frac{\theta_s - \theta_r}{\left[1+(\alpha h)^N\right]^M} \quad (5)$$

Where D_a (m^2.s^{-1}) and D_w (m^2.s^{-1}) represent diffusivity in air and water, respectively; n_i (m3/m3) represents the total porosity in the layer i, H'_{TS} (-) Henry's law constant at the system temperature, $\theta_{a,i}$ (m^3/m^3) and $\theta_{w,i}$ (m^3/m^3) the porosity of the soil filled with air and water respectively in the layer i, θ_s (m^3/m^3) and θ_r (m^3/m^3), respectively, the saturated and residual water content, α (m^{-1}), N and M van Genuchten's [6] parameters, and h (m) pressure head.

On the other hand, the buiding ventilation rate ($Q_{bulding}$) is obtained from the following:

$$Q_{buiding} = \left(L_B.W_B.H_B.ER\right)/3600s/h \quad (6)$$

Where L_B (m), W_B (m) and H_B (m) represent the length, width and height of of the building and ER (h^{-1}) the air exchange rate.

The volumetric flow rate of soil gas entering the buiding Q_{soil} (m^3.s^{-1}) is given by:

$$Q_{soil} = \frac{2\pi.\Delta P.k_v.X_{crack}}{\mu.\ln\left(2Z_{crack}/r_{crack}\right)} \quad (7)$$

Where ΔP (Pa) represents the pressure between de soil surface and the enclosed space, and k_v (m^2) soil vapor permeability, X_{crack} (m) floor wall seam perimeter, μ the viscosity of the air (1.78x10^{-5} kg.m^{-1}.s^{-1}), Z_{crack} (m) crack depth below grade, and r_{crack} (m) equivalent crack radius. Moreover, the soil vapor permeability is given by:

$$k_v = k_i.k_{rg} \quad (8)$$

$$k_i = \frac{K_s\mu_w}{\rho_w g} \quad (9)$$

$$k_{rg} = (1-S_{te})^{1/2}(1-S_{te}^{1/M})^{2M} \quad (10)$$

Where k_i (m^2) represents the soil intrinsic permeabilty and k_{rg} (-) the relative air permeability ($0 \le k_{rg} \le 1$). Moreover, K_s (m.s^{-1}) is the soil saturated hydraulic conductivity, μ_w (g.m^{-1}.s^{-1}) the dynamic viscosity of water, ρ_w (kg.m^{-3}) density of water and g (m.s^{-2}) the acceleration due to gravity, and S_{te} (-) the effective total fluid saturation, represented by:

$$S_{te} = (\theta_w - \theta_r)/(n-\theta_r) \quad (11)$$

Based on the model of Johnson and Ettinger [1] and USEPA [4], Putzman [7] incorporates diffusion through concrete in the model formulation, allowing for the possibility to calculate the indoor air concentration by advective and diffusive transport through cracks and also considering the volatile contaminants diffusion through concrete. Thus α is represented by:

$$\alpha = \frac{\left[(a)x\exp(d)\right]}{\left[\exp(d)+(a)+(c)\left[\exp(d)-1\right]\right]} \quad (12)$$

$$d = \frac{Q_{soil}L_{crack}}{(D^{crack}A_{crack})+(D_{conc}^{eff}A_{conc})}$$

Where D_{conc}^{eff} (m^2.s^{-1}) represents the effective diffusion coefficient through concrete and other isolation layer types and A_{conc} (m^2) the area of enclosed space below grade without considering cracks area.

2.2 Contaminant vapor from subsurface soils with multimodal pore-size distribution into indoor spaces

As mentioned above, the proposals of Johnson and Ettinger [1] and USEPA [4] are based partially on soil parameters, including the concentration of vapor contaminants into indoor spaces. This can be seen from Eq.3 and Eq.7, where D_i^{eff} and Q_{soil}, respectively depend indirectly on the capillary pressure curve of the soil (Eq.5). However, some porous media have systems of pores that cannot be represented by sigmoidal curves of capillary pressure because they have secondary or multiple systems of pores.

For this paper, an equation similar to that of van Genuchten [6] was used (Eq. 5), one which is capable of describing the capillary pressure of a heterogeneous system of pores. This model was developed by Durner [3]:

$$S_{te} = \sum_{i=1}^{k} w_i\left(1+(\alpha_i h)^{N_i}\right)^{-M_i} \quad (13)$$

Where k represents the number of subsystems making up the total distribution of pores and w_i the weighting factors of the subcurves, varying from 0 to 1. Similar to the unimodal model, the parameters of the subcurves α_i [L^{-1}], N_i [-] and M_i [-] correspond to the parameters of adjustment. In this way, the coefficient of effective

 DOI: 10.1051/e3sconf/20160907002

diffusion in the unsaturated zone can consider the distribution of multimodal pores; it is expressed as:

$$D_i^{eff} = D_a\left(\theta_{ma,i}^{3.33}/n_i^2\right)+\left(D_w/H'_{TS}\right)\left(\theta_{mw,i}^{3.33}/n_i^2\right) \quad (14)$$

$$\theta_{ma,i}=n_i-\theta_{mw,i} \quad (15)$$

$$\theta_{mw,i}=\theta_r+(\theta_s-\theta_r)\sum_{i=1}^{k} w_i\left(1+(\alpha_i h)^{N_i}\right)^{-M_i} \quad (16)$$

Where $\theta_{ma,i}$ (m^3/m^3) and $\theta_{mw,i}$ (m^3/m^3) represent the soil air-filled and soil water-filled, respectively, in the layer i of a soil with a multimodal distribution of pores. Moreover, volumetric flow rate of soil gas in such a soil with a multimodal distribution of pores can be represented by the following:

$$Q_{soil}=\frac{2\pi.\Delta P.k_{vm}.X_{crack}}{\mu.\ln\left(2Z_{crack}/r_{crack}\right)} \quad (17)$$

Where k_{vm} (m^2) represents the soil vapor permeability of the soil with multimodal pore-size distribution:

$$k_{vm}=k_i.k_{rgm} \quad (18)$$

$$k_{rgm}(S_{te})=\left(\sum_{i=1}^{k} w_i(1-S_{te})\right)^{0.5}\frac{\sum_{i=1}^{k} w_i^2\alpha_i^2(1-S_{te}^{1/M_i})^{2M_i}}{\left(\sum_{i=1}^{k} w_i\alpha_i\right)^2} \quad (19)$$

The multimodal relative air permeability (k_{rgm}) has been determined here on the basis of the expression of the unsaturated hydraulic conductivity function of Mualem-van Genuchten, considering that the effective total saturations is equal to 100%:

$$S_{teg}=1-S_{te} \quad (20)$$

Where S_{teg} is the effective saturation of air in unsaturated soil.

3 Results and discussion

3.1. Parameters of soils and capillary pressure curves

Soil A is a sandy loam from the experimental field in Bekkevort, in Belgium [4], and soil B is a silty clay from the city of Londrina, in the state of Paraná in Brazil [8]; these examples were chosen because of the multimodal nature of the distribution of pores.

Figure 1 shows the results for the capillary pressure curve adjusted for non linear regression (Levenberg-Marquardt) using the model defined by Eq.5 and 16 for unimodal and multimodal (bimodal and trimodal) pore-size distribution, respectively.

Figure 1 shows the peculiar shape of the curve from the experimental data and the inadequacy of the adjustment using the unimodal model (Eq.5). On the other hand, Table 1 shows that the highest coefficients R^2 result from the trimodal adjustment for soil A, followed by the bimodal one for soil B.

3.2 Indoor air concentration of the contaminant ($C_{building}$)

A hypothetical situation of a homogeneous layer of soil with a source of contamination at a depth of 2m was considered. The indoor air concentration of the contaminants ($C_{buiding}$) was obtained by considering the coefficient of attenuation of Eq.12, i.e., considering advection via cracks and diffusion through concrete. It should be mentioned that this coefficient was obtained by the traditional unimodal method, but considering the multimodal nature of the pore-size distribution (Eq.14 to 20), as well as for different values of pressure head (0.05, 0.1, 0.5, 1 and 10 m), i.e., with decreasing values of volumetric water content from an initial level close to saturation.

The contaminant is a LNAPL (Ligth Non-aqueous Phase Liquid) composed of benzene, xylene and toluene, with concentrations of 1500, 7596 and 9300 $\mu g.m^{-3}$, respectively. Certain chemical properties such as the enthalpy of vaporization at the normal boiling point, critical temperature, normal boiling point, Henry's law constant at the reference temperature, Henry's law constant at the system temperature, diffusivity in air and water, as well as the effective diffusion coefficient through concrete were taken from USEPA [4].

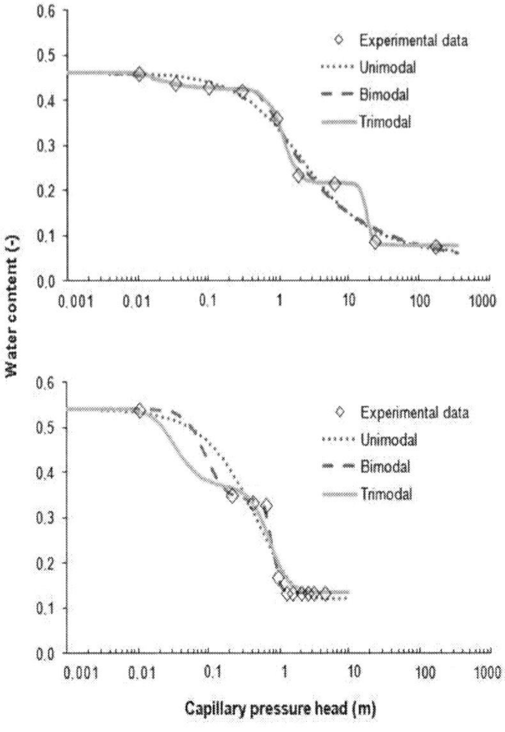

Figure 1. Experimental data (volumetric humidity x capillary pressure) and adjustment curves considering uni, bi and trimodal pore-size distributions for soils A and B.

The adjustment parameters of the capillary pressure curves in Table 1 were used for the characteristics of the porous medium. Moreover, the table includes the values of saturated hydraulic conductivity.

The dimensions of and information about the buildings were based on a hypothetical situation: length, width, and height of 10, 10 and 3m, respectively; an air exchange rate of $0,5h^{-1}$; a soil-building pressure differential of 5 Pa; a depth below grade to bottom of enclosed floor space of 0,20m and a ratio of area of cracks to total area of 0,01%.

Figure 2 compares the results of concentration obtained by the traditional methods for the chemical considered (Benzene, Xylene and Toluene) obtained for soils A and B.

Table 1. Adjustment parameters estimated for capillary pressure curves

Parameter	Soils A	Soils B
	Unimodal	
θ_r	0.0580	0.1214
α_1 (m^{-1})	0.6900	0.1900
N_1	0.9842	0.9950
M_1	0.717	10.076
R^2	0.971	0.9350
	Bimodal	
θ_r	0.0370	0.1347
w_1	0.9197	0.4989
w_2	0.0803	0.5011
α_1 (m^{-1})	8.1000	1.2800
α_2 (m^{-1})	68.400	11.330
N_1	5.7555	9.8956
N_2	2.8515	2.9527
M_1	0.0736	1.0000
M_2	0.5640	0.9723
R^2	0.9780	0.9990
	Trimodal	
θ_r	0,0780	0,1355
w_1	0.3608	0,5050
w_2	0.5432	0.0646
w_3	0.0959	0.4304
α_1 (m^{-1})	0.5400	1.4300
α_2 (m^{-1})	0.8900	1.5200
α_3 (m^{-1})	82.010	37.140
N_1	9.9996	3,3733
N_2	4.0876	3.1467
N_3	9.9994	3.0302
M_1	1.0000	1.0000
M_2	1.0000	0.9203
M_3	0.1075	0.6263
R^2	1.0000	0,9680
K_s ($m.h^{-1}$)	0.0420	0.0012

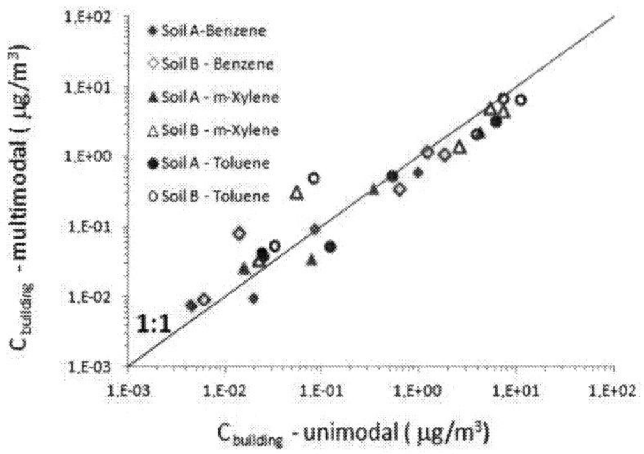

Figure 2. Comparison of $C_{bulding}$ determined for soils A and B and the different chemicals for unimodal and multimodal pore-size distribution.

This figures makes it clear that for either kind of soil (sandy loam or silty clay) and any chemical considered, the results for their concentration ($C_{buiding}$) calculated on the basis of the traditional unimodal method of calculation, generally are relatively higher (when values greater than 0.2 $\mu g/m^3$) than those which consider the multimodal pore-size distribution in the soil.

Figure 3 compares the results of the concentration found for the two soils (the sandy loam of A and the silty clay of B) determined by the unimodal and multimodal methods for the different chemicals. These results lead to the inference that the highest final concentrations $C_{buiding}$ come from the fine-textured soil B, whatever the chemical considered and method of calculation utilized. This may be due to the elevated total porosity of the soil B (0.54) in relation to that of soil A (0.46), as well as the fact that for the same capillary pressure head, the level of volumetric water content of soil A was greater than that of soil B, which reduced the flow of the contaminant in the form of vapour.

Figures 4 and 5 compares the average values of $C_{buiding}$ obtained for the different capillary pressures (0.05, 0.1, 0.5, 1 and 10m) when calculated using the two methods (unimodal and multimodal) for each chemical for the two soils (A and B; Figuras 4 and 5, respectively). As can be seen, whatever the type of soil, the traditional unimodal model overestimates the values for soils with a multimodal pore-size distribution.

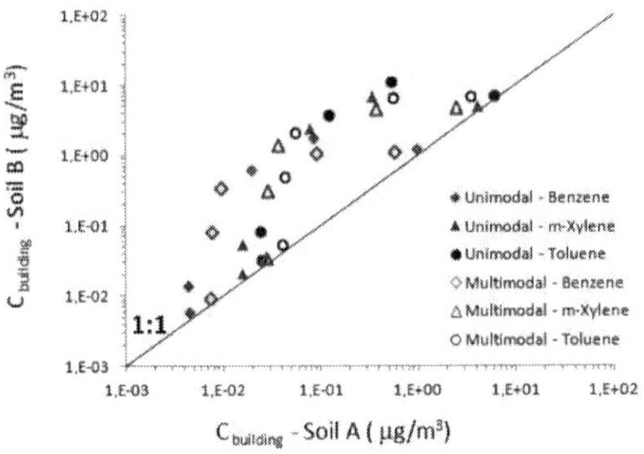

Figure 3. Comparison of $C_{buiding}$ for different chemicals from soils A and B when calculated by the traditional method and that for soils with multimodal pore-size distribution.

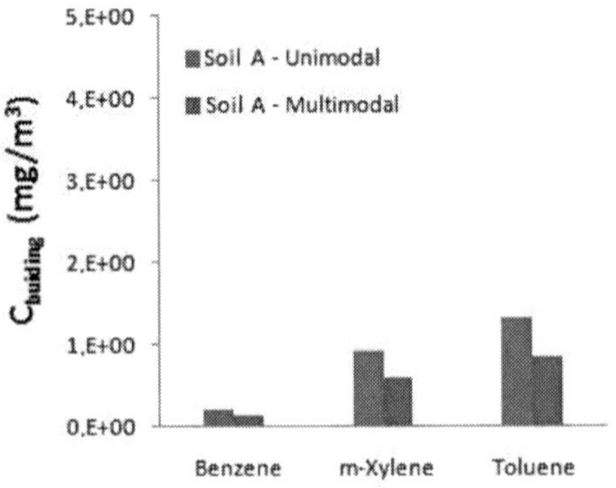

Figure 4. Comparison of $C_{bulding}$ for the different chemicals for the soil A when calculated using the traditional method and that for a multimodal pore distribution

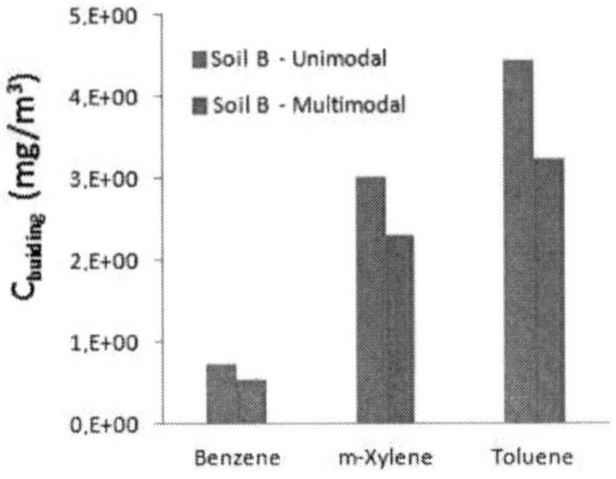

Figure 5. Comparison of $C_{bulding}$ for the different chemicals for the soil B when calculated using the traditional method and that for a multimodal pore distribution

4 Conclusions

The results obtained here make it possible to infer that the use of traditional models for the determination of the indoor air concentration of the contaminants can lead to errors of overestimation in the neighborhood of 200 to 600% as a function of the type of soil when these have a multimodal distribution of pores, such as those found in many Brazilian states.

References

1. Johnson, P.C. and Ettinger, R.A. Heuristic model for predicting the intrusion rate of contaminant vapors into buildings, Environ. Sci. Technol., **25**, 1445–1452. (1991)
2. API - American Petroleum Institute. Light Non-Aqueous Phase Liquid Distribution and Recovery Model (LDRM), API Publication 4760, (2007).
3. Durner W. Hydraulic conductivity estimation for soils with heterogeneous pore structure. Water Resources Research **30**, 211-223.(1994)
4. USEPA, User's guide for evaluating subsurface vapor intrusion into buildings. In: Enviromental Quality Management I., ed. Washington, DC. (2004)
5. Mallants, D.; Tseng, P. H.; Toride, N.; Timmerman, A.; Feyen J. Evaluation of multimodal hydraulic functions in characterizing a heterogeneous field soil. Journal of hydrology, **95**, 172-199. (1997)
6. van Genuchten M. Th. A closed-form equation for predicting the hydraulic conductivity of unsaturated soils. Soil. Sci. Soc. Am. J, **44**: 892-898. (1980)
7. Putzmann T,, de Biase C,, Loechel S,, Bittens M., Weiss H. Soil Vapour Intrusion – A new model for the determination of indoor air concentrations. II Congresso Internacional de Meio Ambiente Subterrâneo. 04 a 06/2011, São Paulo – SP, Brazil.
8. Miguel, M. G.; Teixeira, R. S.; Padilha, A. C. C. Curvas características de sucção de solo laterítico da região de Londrina-PR. Revista de Ciência e Tecnologia, **12** (24), 63-74.(2006).

 DOI: 10.1051/e3sconf/20160907003

Seasonal thermal energy storage in shallow geothermal systems: thermal equilibrium stage

Hossein Nowamooz[a], Saeid Nikoosokhan and Cyrille Chazallon[1]

[1]*INSA de Strasbourg, Civil engineering department, 67084 Strasbourg Cedex, France*

Abstract. This paper is dedicated to the study of seasonal heat storage in shallow geothermal installations in unsaturated soils for which hydrothermal properties such as degree of saturation and thermal conductivity vary with time throughout the profile. In the model, a semi-analytical model which estimates time-spatial thermal conductivity is coupled with a 2D cylindrical heat transfer modeling using finite difference method. The variation of temperature was obtained after 3 heating and cooling cycles for the different types of loads with maximum thermal load of q_{max} = 15 $W.m^{-1}$ with variable angular frequency (8 months of heating and 4 months of cooling).and constant angular frequency (6 months of heating and 6 months of cooling) to estimate the necessary number of cycles to reach the thermal equilibrium stage. The results show that we approach a thermal equilibrium stage where the same variation of temperature can be observed in soils after several heating and cooling cycles. Based on these simulations, the necessary number of cycles can be related to the total applied energy on the system and the minimum number of cycles is for a system with the total applied energy of $1.9q_{max}$.

1 INTRODUCTION

Analysis on heat transfer in subsurface is important to size the BHE that optimum performance is achieved with minimum costs. To estimate thermal response of BHE accurately, knowledge of thermal properties of the ground is needed. The subsurface soil thermal properties such as thermal conductivity, and heat capacity play a vital role in heat transfer of BHE [1-2].

The entire process of heat extraction/deposition is a transient one, due to the weather-dependent ground surface boundary conditions and heating/cooling load. The soil water content in close vicinity to the BHE can be influenced by numerous factors, such as: soil structure, temperature gradient, water gradient and irrigation [3]. In particular, the temperature gradient in the soil surrounding the BHE plays an important role in the combined heat and water flow. Structural and textural properties of the same soil sample can vary considerably with seasonal climatic conditions. Therefore, thorough knowledge of the intricate nature of soils and transport phenomena related to coupled heat and water flow in the ground is essential to both the design and the operation of ground heat pump systems [4]. Due to the very complex character of the ground, the actual design of the BHE should be based on a detailed mathematical model of simultaneous heat and water flow in soils. The parameters influencing the soil thermal properties are essential for testing the BHE length and carrying out a parametric analysis, which is very useful for estimating the long-term effects of BHE operation on the surrounding ground. The results of laboratory testing of soil samples (soil texture, soil dry density, water content) should also be available prior to the evaluation of soil thermal properties.

A few papers dealing with the influence of soil types and moisture conditions on ground heat pump performance have been published in the last decade [4, 5]. Hailey et al. [6] analyzed the behavior of the thermal conductivity of soils adjacent to the BHE. It was found that soil water content was a dominant factor responsible for seasonal thermal conductivity variations. Drown and Braven [7] monitored for several seasons the effect of soil conditions and thermal conductivity on heat transfer in ground heat storage. Increase of the soil thermal conductivity by 0.17 $W.m^{-1}.K^{-1}$ led to reduction of heat pump operating time by 1.3%.

Moreover, there exists a lack of research on the investigation of the thermal response of BHE with regard to the time periodic thermal loads and its effects on the thermal equilibrium stage.

In this study, the simulation of seasonal heat storage in the Shallow Borehole Heat Exchangers (SBHE) installed in multilayered soil subsurface in Alsace region is performed. We perform a series of 2D numerical simulations to study the effect of time periodic thermal loads with variable angular frequency (8 months of heating and 4 months of cooling) and constant angular frequency (6 months of heating and 6 months of cooling) on the temperature variations around the SBHE in Alsace region. These simulations considered the variable thermal diffusivities of the different soil layers to estimate the

[a] Hossein.Nowamooz@insa-strasbourg.fr

necessary number of cycles to reach the thermal equilibrium stage.

2 FINITE DIFFERENCE METHOD

Borehole Heat Exchanger (BHE) is considered as a geothermal system storage at shallow levels (less than 100 m). Figure 1 shows a schematic geothermal system installed in a non-homogenous soil while the groundwater level varies with time and depth.

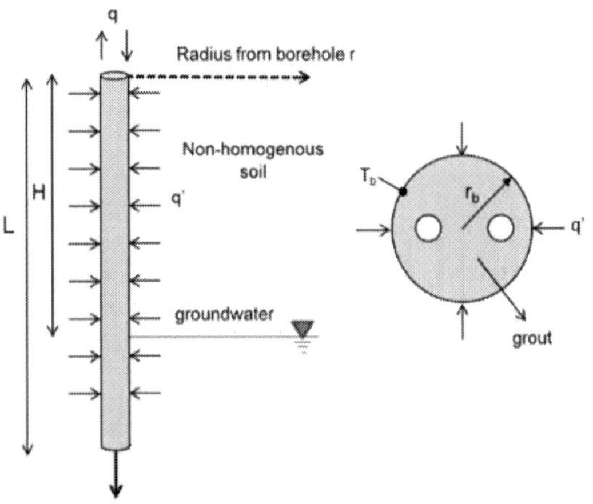

Figure 1. Schematic diagram of a shallow geothermal system installed in a non-homogenous soil with variable groundwater level.

Hydrothermal properties of the soil change with depth and time thus a 2D axisymmetric model is performed with finite difference method. The borehole is subjected to a heat flux q. We assume that such heat flux q is equal to the heat flux q' per unit length at the borehole radius.

The heat transfer problem is governed by the following partial differential equation in cylindrical coordinates:

$$\frac{1}{r}\frac{\partial}{\partial r}\left(\alpha r\frac{\partial T}{\partial r}\right)+\frac{\partial}{\partial z}\left(\alpha\frac{\partial T}{\partial z}\right)=\frac{\partial T}{\partial t} \quad (1)$$

where α is the thermal diffusivity, T is the temperature, t is the time and z is the depth.

The following equations are defined where i is the counter of r; j is the counter of z; and k is the counter of t.

$$\frac{\partial^2 T}{\partial r^2}=\frac{T_{i+1,j,k}-2T_{i,j,k}+T_{i-1,j,k}}{dr^2} \quad (2)$$

$$\frac{1}{r}\frac{\partial T}{\partial r}=\frac{T_{i+1,j,k}-T_{i-1,j,k}}{2r_i dr} \quad (3)$$

$$\frac{\partial^2 T}{\partial z^2}=\frac{T_{i,j+1,k}-2T_{i,j,k}+T_{i,j-1,k}}{dz^2} \quad (4)$$

$$\frac{\partial T}{\partial t}=\frac{T_{i,j,k+1}-T_{i,j,k}}{dt} \quad (5)$$

$$\frac{\partial \alpha(z,t)}{\partial z}\frac{\partial T}{\partial z}=\frac{\alpha_{j+1,k}-\alpha_{j-1,k}}{2dz}\frac{T_{i,j+1,k}-T_{i,j-1,k}}{2dz} \quad (6)$$

By replacing above equations 2 to 6 in Equation 1, the temperature T at time k + 1 computed in each iteration until the difference of temperature between two iteration becomes less than 10^{-4}. Normally for the convergence we should have $\alpha dt/(dr^2 + dz^2) < 0.5$.

For the boundary conditions, we consider that at t = 0, T = T_{g0}. No heat flux was also considered at large depth (z=L):

$$\left.\frac{\partial T}{\partial z}\right|_{z=L}=0 \quad (7)$$

Heat flux q' is considered to be calculated from the following equation:

$$q'(r=r_b,t)=-2\pi r_b\lambda(z,t)\frac{\partial T}{\partial r}=q' \quad (8)$$

3 ESTIMATION OF SOIL THERMAL PROPERTIES

Nowamooz et al. [2] proposed a set of relationships for the fitting functions of thermal conductivity as well as thermal diffusivity adapted to different soils in the literature [8-10].

a) Thermal conductivity

We proposed the following equation for soil thermal conductivity with regard to the degree of saturation:

$\lambda = (\lambda_{sat} - \lambda_{dry})K_\lambda + \lambda_{dry}$ (9a)

$\lambda_{sat} = ax_s + b\gamma_d$ (9b)

$\lambda_{dry} = cx_s + d\gamma_d$ (9c)

$K_\lambda=\kappa Sr/(1 + (\kappa-1)Sr)$ (9d)

$\kappa = ex_s + f$ (9e)

where Sr is the degree of saturation (-), x_s is the sand content (-), γ_d is the intial dry density (kN/m^{-3}), K_λ is the normalized thermal conductivity, λ_{dry} and λ_{sat} are the thermal conductivity of dry and saturated soils (Wm-1K-1), a and b are fitted parameters for saturated thermal conductivity equal to 0.53 Wm^{-1}K^{-1} and 0.1 Wm^{2}K^{-1}.(kN)$^{-1}$, c and d are fitted parameters for unsaturated thermal conductivity equal to 0.087 Wm^{-1}K^{-1} and 0.019 Wm^{2}K^{-1}.(kN)$^{-1}$, κ is dependent to the soil texture related to constant parameters e and f equal to 4.4 and 0.4.

b) Thermal diffusivity

We proposed the following equation for soil thermal diffusivity with regard to the degree of saturation by considering the saturated thermal diffusivity (α_{sat}) and the dry thermal diffusivity (α_{dry}):

$\alpha=(\alpha_{sat}-\alpha_{dry})(\beta Sr+(1-\beta)Sr^2)+\alpha_{dry}$ (10)

where β is a constant parameter taken equal to 4.

Results from Abu-Hamdeh [8] and Ochsner et al. [10] show that there is no effect of initial dry density on the variation of thermal diffusivity. Therefore, we propose a linear relationship for saturated and dry thermal diffusivities (α_{sat} and α_{dry}) with respect to sand content x_s:

$\alpha_{sat}=ex_s+f$ (11a)

$\alpha_{dry}=gx_s+h$ (11b)

where e, f, g and h are constant parameters of values 4.5 × 10^{-7} m^{2}s^{-1}, 3.5 × 10^{-7}m^{2}s^{-1}, 2 × 10−8 m^{2}s^{-1}and 1.8 × 10^{-7} m^{2}s^{-1}, respectively.

4 SIMULATION OF SHALLOW BOREHOLE HEAT EXCHANGER (SBHE) IN HANGENBIETEN (ALSACE REGION)

In this section, we aim to study the thermal response of the multilayered subsurface soil around the SBHE installed in Hangenbieten of Alsace region. Using the geological and geotechnical characteristics of the multilayered subsurface soil, we first estimate thermal properties of the soil with our proposed theoretical approaches. Then, we perform a series of simulations to investigate the effect of variable time periodic thermal loads on the temperature variations around the SBHE.

4.1. Geological, geotechnical, and thermal description of the multilayered subsurface

The transfer of heat between the SBHE and the adjoining soil is governed primarily by heat conduction. Therefore, soil type is an important parameter which has effects on the heat transfer. Table 1 summarizes the geological and geotechnical description of the studied field in Hangenbieten such as the soil type, the sand content and the initial dry density [11]. W_{sat} is the saturated water content calculated by the initial dry density value (γ_d). The λ_{sat} and α_{sat} of the different layers are calculated based on the equations 9 to 11 by considering their sand content x_s and initial dry density γ_d.

Soil at groundwater level in the field (z = 8m) is considered to be in saturated state (Sr = 100%). The variation of water content was reported in our previous publication (Nowamooz et al. 2015) during 3 years.

Soil type	Depth (m)	x_s (%)	γ_d (kN.m^{-3})	w_{sat} (%)	λ_{sat} (W.K.m^{-1})	α_{sat} (m^2.s^{-1})
Vegetarian soil	0-1	65	13.5	36.3	1.69	6.78×10^{-7}
Wind-blown silt	1-4	50	14	33.6	1.66	6.38×10^{-7}
Loess	4-8	40	14.5	31.2	1.66	6.12×10^{-7}
Loam	8-9	30	14.5	31.2	1.61	5.9×10^{-7}
Loess	9-11	40	15	28.8	1.71	6.09×10^{-7}
Silt loam	11-12	30	15	28.8	1.66	5.88×10^{-7}
Clay loam	12-13	40	15	28.8	1.71	6.09×10^{-7}
Loam	13-16	25	15,5	26.7	1.68	5.76×10^{-7}
Sandy clay loam	16-19	30	16	24.7	1.76	5.84×10^{-7}
Sandy Loam	19-20	80	16	24.7	2.02	6.84×10^{-7}

Table 1. Profile of the multilayered subsurface in Hangenbieten (Alsace region), geological data are adapted from BRGM geological map.

4.2 Thermal boundary conditions

In the numerical model, the following boundary conditions were applied:

- At the surface of soil, temperature variations changes with time (T_s). We have also the temperature variation during the period of 3 years at surface (z = 0).
- At large depth, there is no heat flux and soil temperature is considered as the same as the average temperature value of the ground (T_{g0}) at initial time t = 0.
- At the interface between two layers m and m+1, the continuity of temperature as well as the continuity of the heat flux was also considered.
- At the borehole radius, heat flux q_0 varies with the variable thermal conductivity (see Eq. 8).

4.3 Thermal equilibrium stage

For the simulation, we apply 3 successive heating and cooling cycles of thermal loads q(t) on the borehole installed in Hangenbieten with the height (L) of 20 m, the radius (r_b) of 5 cm, and water table level (H) of 8 m.

Thermal loads q(t) initially used for this site correspond to the same existing period of heating and cooling in Alsace region, i.e., 8 months of heating and 4 months of cooling. Three types of time periodic heat loads, each with variable angular frequency in a period of 1 year, are studied as presented in Figure 2: the first type represents 8 months of heating and 4 months of cooling with $q_{min} = -q_{max}$; the second type represents 8 months of heating and 4 months of cooling with $q_{min} = -q_{max}/3$; the third one corresponds to no cooling load during summer ($q_{min} = 0$). Table 2 presents the loads q_1, q_2, and q_3 with $q_{max} = 15$ W.m^{-1}.

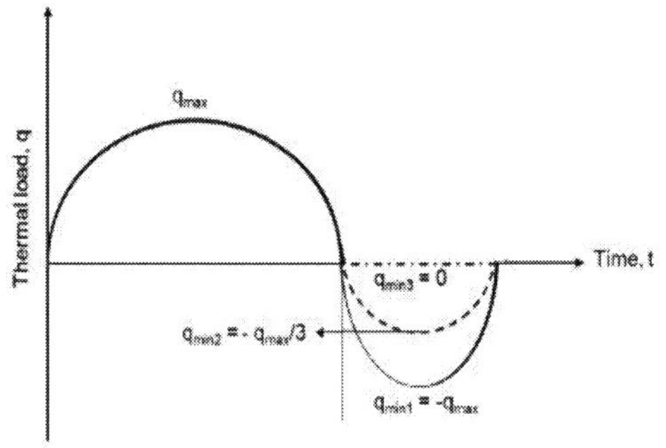

Figure 2. Schematic representation of time-periodic thermal loads with variable angular frequency: i) $q_{min} = -q_{max}$, ii) $q_{min} = -q_{max}/3$, iii) $q_{min} = 0$.

In order to better estimate the number of cycles necessary for the thermal equilibrium stage in our calculations, we take also into account the effect of load angular frequency. For this reason, three additional types of time periodic heat loads with $q_{max} = 15$ during 6 months of heating and 6 months of cooling (constant angular frequency) are studied: the first type represents a fully compensating load q_7 with $q_{min} = -q_{max}$; the second type represents a partially compensating load q_8 with q_{min}

$= -q_{max}/3$; the third kind corresponds to no cooling load q_9 during summer with $q_{min} = 0$ (see Figure 3 and Table 2).

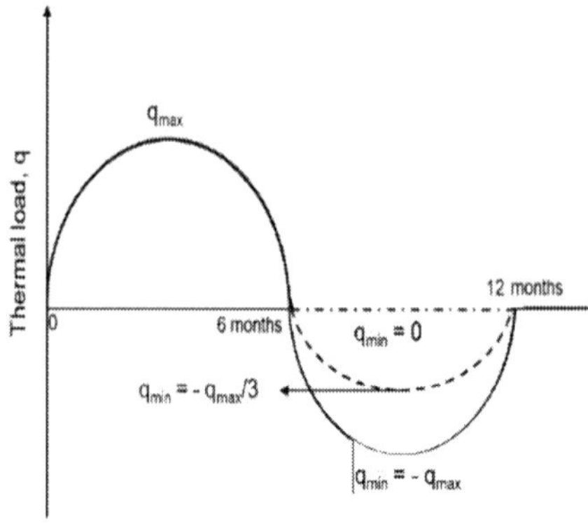

Figure 3. Schematic representation of time-periodic thermal loads with constant angular frequency: i) $q_{min} = -q_{max}$, ii) $q_{min} = -q_{max}/3$, iii) $q_{min} = 0$.

Table 2. Different thermal loads with constant and variable angular frequency.

Thermal load q (W/m)	Angular frequency ω (rad/t)	q_{max} (W/m)	q_{min} (W/m)
q_1	Variable frequency $\omega_1 = \frac{3\pi/2}{12\ months}$	15	-15
q_2		15	-5
q_3	$\omega_2 = \frac{3\pi}{12\ months}$	15	0
q_7	Constant frequency	15	-15
q_8	$\omega_1 = \omega_2 = \frac{2\pi}{12\ months}$	15	-5
q_9		15	0

Figure 4 shows the variation of temperature for the thermal loads (q_1, q_2, q_3, q_7, q_8, and q_9) with radius at the depth of 2.5 m and with depth at the radius of 0.34 m (close to the borehole) after 1 year, after 2 years, and after 3 years.

We can observe that a thermal equilibrium stage can be reached where the soil presents the same temperatures after several heating and cooling cycles. Moreover, thermal equilibrium stage can be reached more rapidly for the thermal load q_1 where $q_{min} = -q_{max}$ compared with the other two types of thermal load q_2 and q_3. For both thermal loads q_2 and q_3, a decrease of temperature can be observed after 3 years. Further calculations show that 2 or 3 additional cycles are sufficient for the thermal loads q_2 and q_3 to reach the thermal equilibrium stage.

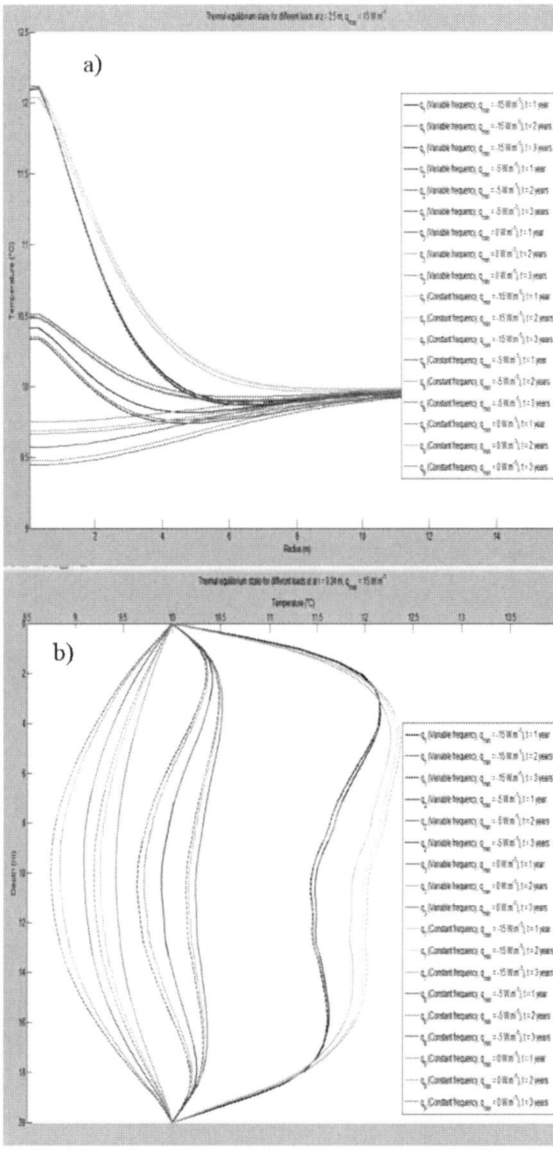

Figure 4. Comparison of temperature variations between variable and constant angular frequency thermal loads after 3 heating and cooling cycles: a) with radius at the depth of 2.5 m and b) with depth at the radius of 0.34 for the six thermal loads presented in Table 2.

We can also observe in Figure 4 that the thermal load q_8 requires less number of cycles to reach thermal equilibrium stage compared with other thermal loads q_7 and q_9. Moreover, an increase of temperature after 3 years is observed for q_7 while it decreases for q_9.

The following observations can be made on the difference of temperature variations between first and third year:

- In the case of q_1 and q_7 with $q_{min} = -q_{max}$, the difference for variable angular frequency (q_1) is almost the same between first and third year while it decreases for the constant angular frequency (q_7).

- In the case of q_2 and q_8 with $q_{min} = -q_{max}/3$, the difference for constant angular frequency (q_8) is almost the same while it decreases for the variable angular frequency (q_2).
- In the case of q_3 and q_9 with $q_{min} = 0$, the difference for both constant and variable frequencies decreases during this period.

In order to interpret our observations, we define the total applied energy E of the thermal load with the unity of W.month/m, calculated by the following equation:

$$E = \int_0^{12\ \text{months}} q(t)dt \tag{12}$$

Figure 5 shows the variation of the maximum temperature difference between first and third year, ΔT, with respect to the total applied energy E. We can observe that the maximum difference of temperature between first and third year becomes zero for a system with the total applied energy $E = 1.9q_{max}$.

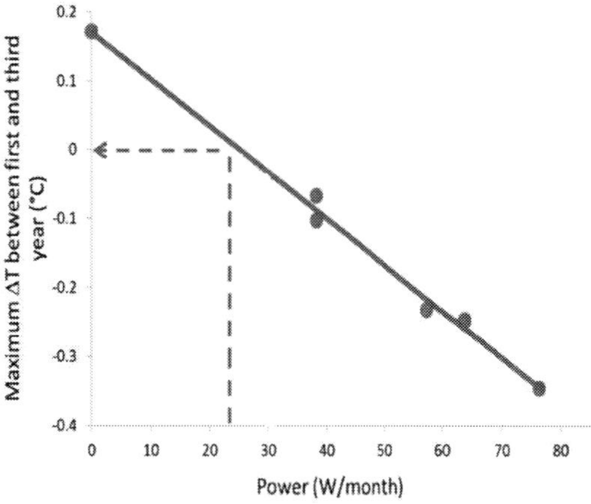

Figure 5. Variation of the maximum temperature difference between first and third year (ΔT) with respect to the total applied energy E (linear fit).

CONCLUSION

This study deals with a finite difference modeling of heat transfer in unsaturated multilayered soil while hydrothermal properties of soil differ in each layer.

A profile of thermal diffusivity within time is assigned to each node in finite difference formulation. The numerical model requires easily accessible input parameters such as water content, dry density, and sand content of the soil profile. It is coupled with the reliable theoretical approaches which estimate thermal properties of soil such as thermal conductivity and thermal diffusivity.

The variation of temperature was obtained after 3 years for the three types of loads $q_{min} = -q_{max}$, $q_{min} = -q_{max}/3$, and $q_{min} = 0$ with maximum thermal load of $q_{max} = 15\ W.m^{-1}$ with variable angular frequency (8 months of heating and 4 months of cooling).and constant angular frequency (6 months of heating and 6 months of cooling). The results show that we approach a thermal equilibrium stage where the same variation of temperature can be observed in soils after several heating and cooling cycles.

Globally, the necessary cycles to reach the thermal equilibrium stage is dependent on several factors such as the angular frequency of thermal load, the type of the thermal load, the intensity of the thermal load, and the climate conditions of the region. Based on our investigations, we propose that the total applied energy on the system can be a good alternative to estimate the required number of cycles to reach the thermal equilibrium stage.

In this study, only heat transfer was considered. In our future studies, the coupled heat and mass transfer constitutive relationships will be taken into account.

ACKNOWLEDGEMENT

This work was funded by the European Commission Initiative INTERREG IV A, Upper Rhine program (project B20, TEM3).

REFERENCES

1. Nikoosokhan S. , Nowamooz H. , Chazallon C., 2015. Temperature variations in unfrozen soils with variable hydrothermal properties, European Journal of Soil Science, 66, 378–388.
2. Nowamooz H., Nikoo-Sokhan S., Lin J., Chazallon C., 2015. Finite difference modeling of heat distribution in multilayer soils with time-spatial hydrothermal properties, International Journal of Renewable Energy, 76, 7-15.
3. Krarti, M., Claridge, D.E. & Kreider, J.F., 1995. Analytical model to predict nonhomogeneous soil temperature variation, Transacations of ASME. 117.
4. Leong, W., Tarnawski, R. & Aittomaki, A., 1998. Effect of soil type and moisture content on ground heat pump performance, International Journal of Refrigeration. 21, 595-606.
5. Song, Y. Yao, Y. & Na, W. 2006. Impacts of soil and pipe thermal conductivity on performance of horizontal pipe in a ground-source heat pump, Proceedings of the Sixth International Conference for Enhanced Building Operations, Shenzhen, China.
6. Hailey, S.M., Kast, T.P. & Drown, D.C., 1990. Thermal conductivity and soil conditions: heat transfer effects on ground source heat pumps, Proceedings of the 16th Annual Conference of Solar Energy Society of Canada, Halifax, Nova Scotia.
7. D. Drown, K & Den Braven, 1992. Effect of soil conditions and thermal conductivity on heat transfer in ground source heat pumps, Proceedings of the ASME JSES KSES International Solar Energy Conference, Maui, Hawaii.
8. Abu-Hamdeh N.H, 2003. Thermal Properties of Soils as affected by Density and Water Content, Biosystems Engineering. 86, 97-102.
9. Lu, S., Ren, T., Gong, Y. & Horton, R., 2007. An improved model for predicting soil thermal conductivity from water content at room

temperature, Soil Science Society of America Journal. 71, 8-14.
10. Ochsner, T.E., and Horton, R., & Ren, T. 2001. A New Perspective on Soil Thermal Properties, Soil Science Society of America Journal. 65, 1641-1647.
11. BRGM, 2000. report for geological map, Editions BRGM.

DOI: 10.1051/e3sconf/20160907004

Effect of temperature on volume change behaviour of statically compacted kaolin clay

Ogechi Ileme[1,a], Gabriela Medero[1] and Peter Woodward[1]

[1]*School of Energy, Geoscience, Infrastructure and Society, Heriot-Watt University, United Kingdom.*

Abstract. Several soils are subjected to high temperature due to the environment where they are located or activities around them. For instance, upper layer of soils in tropical regions, soils around geothermal structures, clay barriers around nuclear waste repository systems. Numerous studies have pointed out that high temperature affects the hydro-mechanical properties of soils. Notwithstanding already existing studies, the influence of temperature on soils is still a challenge, as most of these studies are soil specific and cannot be inferred as the behaviour of all soils. This paper presents an experimental study on the influence of temperature on the volume change behaviour of statically compacted kaolin clay. Compacted samples were tested at varying temperatures using a suction controlled oedometer cell. The influence of temperature on the magnitude of volumetric strain occurring during mechanical and thermal loading was investigated. The study showed that an increase in temperature increased the magnitude of volumetric strain of the soil on loading. Additionally, the results presented in the light of LC curve showed that an increase in temperature resulted in the contraction and a change in the position of the LC curve.

1 Introduction

The number of geo-mechanical problems involving thermal effects on unsaturated soils are on the increase due to technological advancement. Compacted and natural unsaturated soils are increasingly being used in applications where they are subjected to high temperature, which together with imposed load or/and wetting condition affect their behaviour and mechanical properties. For instance, the use of compacted clay blocks as liners for waste disposal and as barriers in nuclear waste repository [1].

Significant amount of work has been done on the influence of temperature on soils. For saturated soils, the major temperature effect reported by different authors [2 - 9] are; heating induces expansion under high OCR at low net stress and contraction under low OCR. Thermal expansion observed on heating at low vertical net stress is attributed to expansion of soil constituents; water and solid whereas the thermal contraction on heating at high vertical net stress is ascribed to weakening of contacts between aggregates [10].

For unsaturated soils, some of the findings are contradictory. As regards unsaturated expansive soils, their thermal behaviours appear to be somewhat dependent on soil type, suction value and initial density. For a thermal volume change obtained from non-isothermal path at constant suction and vertical stress, [11, 12] reported an increase in volumetric strain (expansion) of high density Boom clay with temperature increase coupled with strain accumulation. However, the magnitude of volumetric strain observed on heating decreased with increase in suction. On the contrary, [1] and [13] observed expansion of MX80 and GMZ01 bentonites respectively on heating at high suction values, and then contraction at low suction values. [13] Also reported that the effect of temperature on Boom clay is dependent on OCR value. Thermal contraction increases with decreasing OCR and the temperature at which the transition between contraction and expansion occurs decreases with OCR. Cooling was associated with an irreversible volumetric strain which reduced on subsequent cycles [11]. [13] Noted that cooling process is considered as an elastic process because it involves contraction of the constituent elements, whereas thermal contraction is considered as a plastic process. Another explanation of the phenomena surrounding the thermal behaviour of soils is given by [11]. He explained that thermally induced irreversible behaviour can be linked to structural rearrangement of particles at the macro-structural level as a consequence of thermal expansion at the microstructure.

In the case of loading paths at constant suction and temperature, [1, 11, 14, and 15] reported a decrease in yield stress and higher compressibility with temperature increase and as a result, a reduction in the size of the elastic domain. However, [12] observed an increase in yield stress with temperature. But it is worth noting that the samples used by [12] were first subjected to cycles of wetting and drying before being subjected to isothermal loading and unloading. It is then likely that the observed behaviour by [12] was influenced by the initial stress

[a] Corresponding author: oci1@hw.ac.uk

path. [1, 12] observed that the compressibility parameters are insensitive to temperature. However, [11, 14] observed from their studies that the post-yield /plastic compressibility parameters are temperature sensitive to temperature, but the pre-yield/elastic compressibility parameters are insensitive to temperature. [11] observed an increase in the compressibility parameter with temperature, whereas [14] observed a slight decrease with temperature. It is worth noting that the findings of [11] and [14] were on different soils with different initial conditions and this may have played a role in the difference. In fact, studies by [15, 16] showed that the influence of temperature on compressibility appears to be dependent on some factors such as over-consolidation ratio, plasticity, expansibility and compaction water content or suction.

For unsaturated, non-expansive soils, some of the reported behaviours are similar to those observed on expansive soils, with the direction of volume change usually contraction. However, [17] reported expansion of silty clay when heated at low suction and net applied stress. Under non-isothermal path, [17] and [18] observed an increase in contraction of sandy silt and silt respectively, with increasing temperature, although [17] observed this at high initial suction and net stress. Most findings on the effect of temperature on compressibility parameters and yield pressure are unanimous for non-expansive soils. The compressibility parameters were reported to be unaffected by temperature [18, 17]. The yield pressure reduces with increasing temperature, thereby shrinking the yield surface; a condition known as thermal softening [17, 18, 19, 20, 21, 22, and 23]. Also, the increase in temperature induces decrease in resilient modulus [21].

From literature, temperature effect on mechanical behaviours appears to be dependent on soil type and other factors. In this study, the influence of temperature on the volume change behaviour of compacted kaolin clay was investigated. Kaolin clay is a benchmark soil used for research, as a result, understanding its thermo-hydro-mechanical behaviour is essential. Most of the studies on the thermo-hydro-mechanical behaviours of kaolin clay were carried out on reconstituted samples with moisture contents on the wet side of optimum [22, 23]. Evidence has shown that a sample compacted at a water content wet of optimum is significantly different from that compacted to a dry of optimum water content in terms of their structure and mechanical behaviour [24]. Consequently, temperature effect on the behaviour of kaolin clay samples prepared at the dry side of optimum is still needed to have a full understanding of the thermal behaviour of the soil. The effect of temperature on volume change and compressibility was studied using a suction and temperature controlled oedometer cell. Two testing paths were followed;

a) Loading at constant suction and temperature.
b) Heating and cooling at constant suction and vertical net stress.

2 Material, sample preparation and method

2.1 Material

The soil used for the study is industrially manufactured kaolin clay. The soil has a liquid limit, a plastic limit and plasticity index of 59.1%, 33%, 26% respectively and a specific gravity of 2.65. The hygroscopic water content in equilibrium with the laboratory atmosphere is 0.23%. The kaolin clay constitutes 58% silt fraction and 42% clay fraction. X-ray diffraction indicated a mineralogical composition of 82% Kaolinite, 15% Quartz and 3% Illite and Mica [23]. Results from standard proctor compaction test indicated a maximum dry density of 1.41Mg/m3 and an optimum water content of 27%.

2.2 Sample preparation

Statically compacted samples; B1 and B2 were used for the study as shown in Table 1. B2 could be described as aggregated samples made of saturated macropeds [25] with large void spaces in-between the macropeds. Preliminary double oedometer tests on B2 showed high magnitude of collapse on wetting and loading. The intention in B2 was to investigate if temperature increase will have the same effect as observed on wetting or loading. Both samples were prepared at different degrees of saturation, but the same void ratio. It is worth mentioning that both B1 and B2 were considered as different samples and that results from both samples were not compared.

The samples were prepared by mixing the required mass of oven dried soil thoroughly with the required quantity of de-aired water to achieve the specified water content. Thereafter, the material for B1 was passed through a sieve of 1.18mm aperture while that of B2 was passed through a sieve of 2.8mm aperture. This was still considered adequate relative to the height of the oedometer ring (20mm). The materials for B1 and B2 were then transferred into plastic bags, sealed and left for 24hrs for moisture equilibration. Afterwards, the samples were statically compacted into oedometer rings (50mm diameter, 20mm high) using an oedometer frame. Static compaction was adopted to ensure uniform density in the samples [26, 27]. Different compaction pressures were used to compact the samples to achieve the desired void ratio. After compaction, the samples were carefully transferred into the suction controlled oedometer cell for testing.

2.3 Method

2.3.1 Suction controlled oedometer cell

The suction controlled oedometer cell used for testing was designed by [23]. General scheme of the oedometer cell is shown in Figure 1. The cell made of stainless steel has two main parts; cell body with an internal chamber for the sample and the cell cover. The cell body has an external and internal diameter of 101mm and 51mm respectively, with a height of 70mm. Inside the internal chamber is the ring that retains the sample. Suction is imposed on samples using axis translation technique. A porous stone of about 5mm thick with a diameter slightly less than the diameter of the ring sits on top of the sample. Pore air pressure is applied to the sample through this porous stone. At the base of the cell is a high air entry ceramic disc (with air entry value of 1500kPa), which acts as an interface between the sample and a water reservoir for imposing pore water pressure. Two recesses fitted with valves are located at the base of the cell; one for water inlet into the reservoir and the other for water outlet and flushing the cell.

The cell cover is fitted with a 1mm flat diaphragm attached without gap to the loading cap; the loading cap transmits load evenly on the sample by means of air pressure acting on the diaphragm. The diaphragm acts as an interface between the diaphragm chamber and the internal chamber of the cell and can withstand temperature up to 150°C [23]. A hollow stainless piston of 13mm external diameter passes through the centre of the loading cap, diaphragm as well as cell cover and is soldered to the loading cap at the lower end. The upper end of the piston is fitted with a valve through which connection is made to the air pressure system (for pore air pressure). The flat edge of the valve then provides a base for the anvil of LVDT used for monitoring the displacement in the sample. Two upper recesses are located on the cell cover, one of which is used for connection to the air pressure system, for applying diaphragm pressure. Temperature was imposed on the samples using a temperature chamber capable of maintaining temperature ranging between 0°C and 50°C. Prior to testing, the cell was calibrated to determine the actual pressure transmitted to the sample for each value of diaphragm pressure under a given value of matric suction. This was done using a calibrated proven ring with a load factor of 1.21N. The time required for the sample to reach thermal equilibrium was assessed by burying a calibrated thermocouple at the centre of a compacted sample in the cell and monitoring the time required for the sample to attain the same temperature as the temperature chamber. Details and results of the calibrations can be found in [23].

Table 1. Initial conditions

Sample	Sr_i (%)	e_i	Initial suction (kPa)	Compaction pressure (kPa)
B1	30.0	1.49	286	203
B2	58.6	1.47	109	113

2.3.2 Testing procedure

Non-isothermal path at constant suction and stress

The aim was to observe the influence of thermal cycles on the volume change of compacted samples under a constant matric suction and vertical net stress. Temperature was varied between 25°C–50°C. A vertical net stress of 55kPa was maintained in all the tests and two values of suction 30kPa and 100kPa were considered. The test was carried out on sample B1 and B2. In total four tests were carried out, namely $NI1_1$, $NI1_2$, $NI2_1$, and $NI2_2$. Both $NI1_1$ and $NI1_2$ have the same initial conditions (as in B1) and were tested at different suction values 30kPa and 100kPa respectively indicated by the subscript. Similarly, $NI2_1$, and $NI2_2$ have the same initial conditions (as in B2) and were tested at suction of 30kPa and 100kPa respectively. Prior to testing compacted sample was placed on top of the saturated ceramic disc and the cell assembled quickly to prevent the sample from taking up water from the stone. Suction was reduced to 30kPa or 100kPa from an initial value of 109kPa and 286kPa for B2 and B1 respectively. The volume change in the sample was monitored and recorded. This was maintained for 14 days for suction equalisation. The 14 days equalisation time was adopted from the study of [23]. He monitored water volume movement in compacted kaolin clay on reducing suction to 0kPa. He observed that the rate fell to less than 0.015/day after 14 days. After suction equalisation, vertical net stress was increased to 55kPa and maintained for 24 hours. Then thermal loading was imposed following three temperature steps; 25°C, 35°C and 50°C. For each step, the temperature was increased in several sub steps at a rate of 2°C per hour. Resulting in 5 hours from 25°C to 35°C and 7.5 hours from 35°C to 50°C. [19] Explained that adopting small temperature steps will reduce the build-up of excess pore water pressure that may result in an over-consolidated behaviour of the sample. The stress path is shown in Figure 2 as Non-isothermal path. Continuity of pore water pressure was always ensured during testing by flushing the water lines periodically.

Loading path at constant suction and temperature

The test was carried out on only samples B1. Two constant temperatures (25°C and 50°C) and suction values (30kPa and 210kPa) were considered during testing. The tests were named according to the suction and temperature values. L11 and L12 for suction 30kPa (first numeric value) and temperature 25°C and 50°C respectively (second numeric value). L21 and L22 for suction 100kPa and 25°C and 50°C respectively.

Following compaction, the sample was placed on the saturated stone and the cell assembled quickly. Then the sample was brought to a suction value of 30kPa or 210kPa and then subjected to the desired temperature (25°C or 50°C). A period of 14 days was allowed altogether for thermal and suction equalisation. Displacement of the sample due to suction and temperature changes were measured by the LVDT. Vertical net stresses were applied incrementally from 5kPa to 400kPa with five loading steps while maintaining the suction and temperature constant. The stress path is shown in Figure 2 as loading path. It was assumed that the number of loading steps used was enough to prevent significant pore pressure re-distribution. Increase in vertical net stress has been reported to cause re-distribution of pore water pressures [17], implying that adequate time should be allowed for re-equalization of suction. There was initial concern about allowing 14days for suction re-equalisation for each step of vertical net stress due to time limit. Since the parameter monitored during testing was volume change, deformation of the sample on application of a vertical net stress of 100kPa at 50°C was monitored for 24 hours. It was observed that the deformation ΔH fell to less than 0.004mm per hour after about 17 hours. Consequently, 24 hours of rest was adopted in all the tests, after each step of vertical net stress.

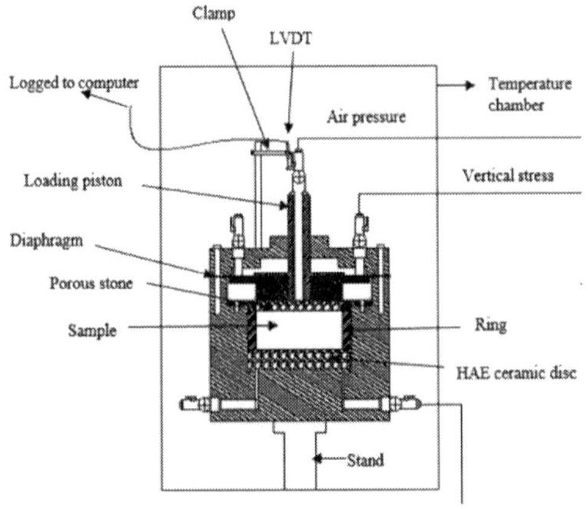

Figure 1. General scheme of suction controlled oedometer

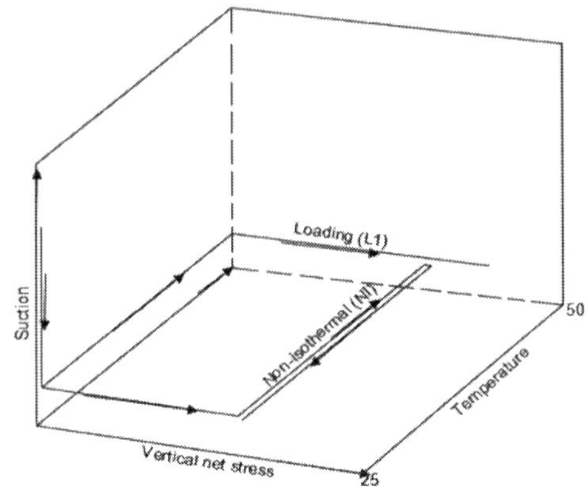

Figure 2. Stress path used during testing

3 RESULTS AND DISCUSSIONS

3.1 Non-isothermal tests

Non-isothermal tests were carried out to ascertain the volumetric behaviour in response to temperature cycles. Figure 3 shows the volumetric strain from $NI1_1$ on a complete temperature cycle. From the Figure, it is evident that heating resulted in contraction of the sample, though small volumetric strain was observed (about 7% increase by heating from 25°C to 50°C). However, on cooling, the strain was irrecoverable. Unfortunately, the test at constant suction of 100kPa was inconclusive due to equipment break down during testing. Hence, the result could not be compared with that at 30kPa.

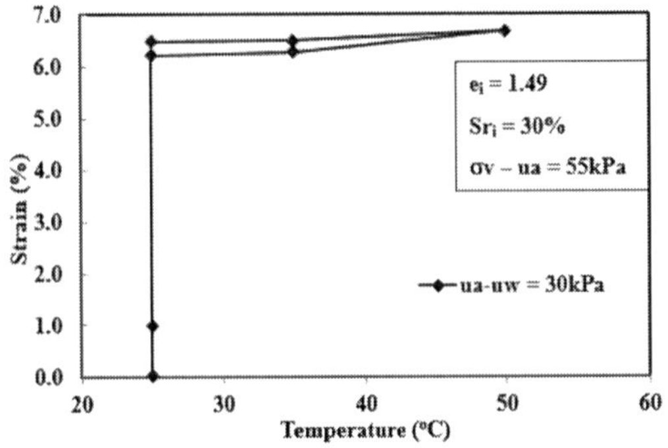

Figure 3. Soil deformation response on heating and cooling ($NI1_1$)

Figures 4a & 4b show volumetric strain from $NI2_1$ and $NI2_2$ on heating and cooling respectively. Both samples experienced substantial and irrecoverable volumetric strain (contraction) on heating and cooling. However, it could be clearly seen that $NI2_1$ tested at a constant suction of 30kPa experienced higher deformation than $NI2_2$ tested at a suction of 100kPa. This could be attributed to a reduction in strength associated with a decrease in suction [36]. Similar behaviour was also reported by [22]; the sample at a constant suction of 0kPa exhibited higher deformation on heating and cooling than the sample at a suction of 100kPa.

The magnitude of deformation on application of vertical net stress and suction could be seen to be high relative to the deformation on a complete temperature cycle. The highest magnitude of deformation occurred after an increase in vertical net stress (path 2-3) in Figure 4a & 4b.

3.2 Loading at constant suction and temperature

The loading tests L11, L12, L21 and L22 were performed at constant suction (30kPa and 210kPa) and temperature (25°C and 50°C). Figure 5 shows the variation of void ratio with vertical net stress for L11 and L12 at a constant suction of 30kPa. From the figure, it is evident that the soil compressibility increased with temperature increase. For instance, at a vertical net stress of 108kPa, the void ratio for the sample at 25°C is 1.146, while the corresponding void ratio at 50°C is 1.0637. Both NCLs are not parallel and not converging at any point. Since almost exact samples have been used in both tests, it was expected that both lines will converge at some point, probably with further increase in vertical net stress. As loading was done at a constant suction of 30kPa, it is assumed that most of the previously air filled voids were became water filled when the matric suction was relaxed from 286kPa to 30kPa. It is already established that viscosity decreases with temperature. At 50°C, water held in the voids became less viscous (note that water is not held in-between the silica-alumina layers of kaolin due to strong hydrogen bond). It is possible that the decrease in viscosity aided the consolidation of the samples by quicker expulsion of water from the voids and as a result, a reduction in the volume of voids. This trend in behaviour is not very different from those observed by [22, 23] on saturated kaolin samples. The result of this study when compared with results from [23] at the same temperature and void ratio showed a difference in the compressibility parameters and in the position of the yield stress. These differences may be due to difference in fabric induced by compaction water content and compaction pressure. Reconstituted samples were used by [23] with compaction wet of optimum.

Figure 6 shows the stress strain relationship of L11 and L12. The figure shows clearly that compressive strain increased with an increase in temperature from 25°C to 50°C. Additional strain induced in the soil by increase in temperature was maximum at an applied stress of 108kPa. Similar findings (increase in compressive strain with temperature increase) were reported by [11, 22]. The yield stresses were obtained using Cassagrand method. Yield stresses of 40kPa and 34kPa were obtained from L11 and L12 respectively, a 15% reduction due to increase in temperature from 25°C to 50°C. With the zone before yield stress known as elastic domain and zone after yield stress known as plastic domain, increase in temperature resulted in the reduction of the elastic domain and earlier onset of plastic behaviour. Similar behaviour also known as thermal softening was reported by [20, 22 and 23]. An increase in the post compressibility parameter Cc was observed with temperature. This contradicts the findings of [1, 9] which states that the slope of the normal compression line is unaffected by temperature, but is in line with the findings of [11] which states that the post-yield compressibility parameters (stiffness parameter and slope of the normal compression line) increase with temperature.

Figure 7 shows compression curves for L21 and L22 (at constant suction of 210kPa). No clear difference can be observed between the two curves along the elastic domain. However, along the plastic domain, samples tested at 50°C exhibited a slight decrease in void ratio than the sample at 25°C at corresponding values of vertical net stresses. This indicates thermal softening at the plastic domain. Similar behaviour was also observed by [22, 23] on kaolin clay at a constant suction of 300kPa, but the positions of the yield stresses were significantly different.

At a constant suction of 210kPa, the yield stress was not affected by temperature as shown in Figure 7. Additionally, the post compressibility parameters were unaffected by temperature increase, in agreement with the findings of [1, 9, 12, 23].

Finally, the effect of temperature on the soil deformation response was presented using LC curve as shown in Figure 8. From the Figure, it could be seen that an increase in temperature resulted in the contraction of the yield curve (only at a suction of 30kPa) and a change in the position i.e. a shift to the left. This is similar to the effect of a decrease in compaction stress reported by [28].

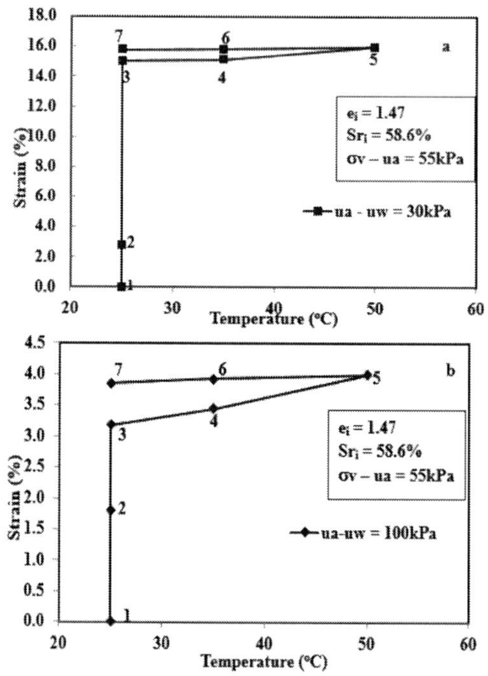

Figure 4. Soil deformation response on heating and cooling a) NI21 b) $NI2_2$

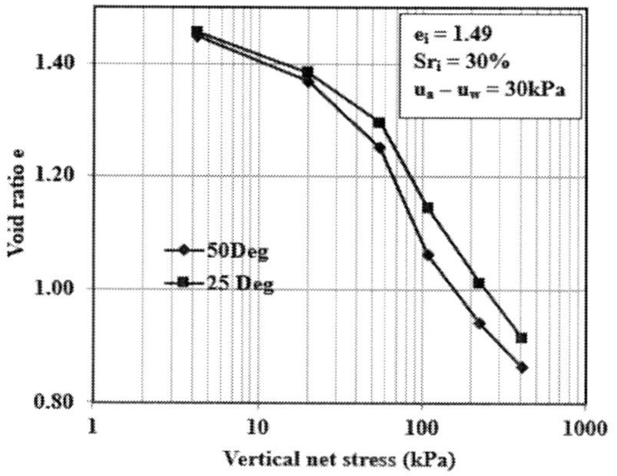

Figure 5. Compression curves for L11 and L12

Conclusions

Non-isothermal tests and loading tests were performed to investigate the influence of temperature on the volumetric behaviour of statically compacted kaolin clay. Results from non-isothermal test showed that heating induced contraction which was irreversible on cooling. The results also indicated that the volumetric strains associated with change in suction and vertical net stress are usually higher compared to strain induced by increase in temperature. Regarding the loading paths, the results showed that compressibility increases with temperature increase, resulting in thermal softening. Yield stress decreased with temperature increase (though only for sample tested at a constant suction of 30kPa) resulting in a shift of the LC curve to the left. The influence of temperature on compacted kaolin at dry of optimum and reconstituted kaolin are different. The magnitude of volume change and the change in position of the yield surface associated with heating are different for both soils.

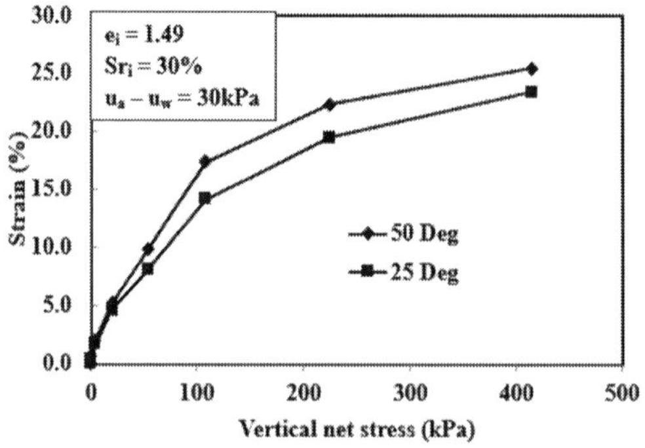

Figure 6. Stress strain relationship of L11 and L12

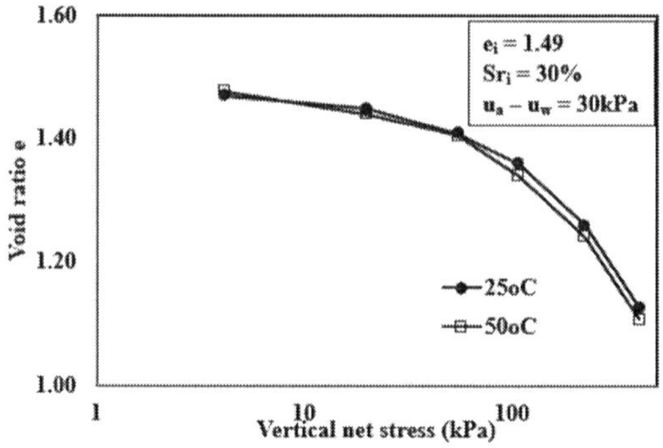

Figure 7. Compression curve for L21 and L22

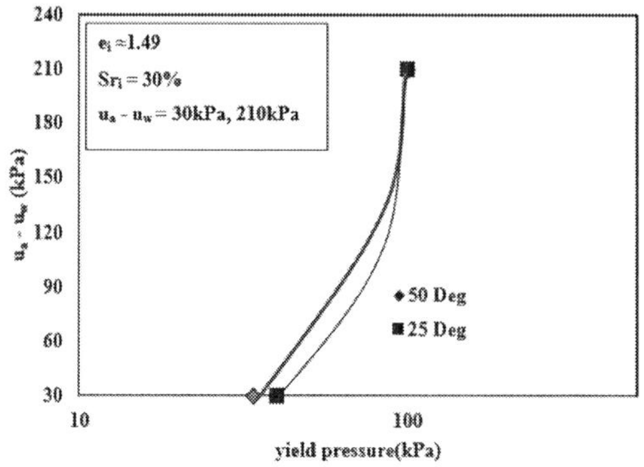

Figure 8. Temperature effect represented on LC curve

References

1. A. Tang, Y. Cui, B. Nathalie, Geotechnique 58(1), 45–54 (2008)
2. R.G. Campanella, J. R. Mitchell, J. of Soil Mech. Fnd. Div., ASCE. 94 (3), 709-734 (1968)
3. G. Baldi, T. Hueckel, R. Pellegrini, Canadian Geotech. J. 25, 807-825 (1988)
4. T. Hueckel, G. Baldi, J. Geotech. Engng 116(12), 1778–1796 (1990)
5. C. Del Olmo, V. Fioravante, F. Gera, T. Hueckel, J. C. Mayor, R. Pellegrini, Engng. Geol. 41(1–4), 87–101 (1996)
6. P. Delage, N. Sultan, Y. J. Cui, Canadian Geotech. J. 37, 343-354 (2000)
7. L. Laloui, Rev. Fr.Ge´nie Civil 5(6), 809–843 (2001).
8. N. Sultan, P. Delage, Y. J. Cui, Engng. Geol. 64 (2–3), 135–145 (2002)
9. C. Cekerevac, L. Laloui, Int. J. for Numerical and Analytical Methods in Geomechanics. 28(3), 209–228 (2004)
10. Y. J. Cui, N. Sultan, P. Delage, Can. Geotech. J. 37(3), 607–620 (2000)
11. E. Romero, A. Gens, A. Lloret, (2003), Geotechnique. 53, 65–81 (2003)
12. E. Romero, M. V. Villar, A. Lloret, Engineering Geol. 81, 255–268 (2005)
13. Y. J. Cui, M. J. Tang, J. of Rock Mech. & Geol. Engng. 5, 169-178 (2013).
14. M. W. Ye, Y. W. Zhang, Y. G. Chen, Y. J. Cui, Applied Clay Sci. 83-84, 210-216 (2013)
15. A. Lloret, M. V. Villar, Physics and Chemistry of the Earth. 32, 701–715 (2007)
16. M. V. Villar, A. Lloret, Applied Clay Sci. 26(1), 337-350 (2004)
17. A. Uchaipichat, N. Khalili, Geotechnique. 59, 339–353 (2009).
18. S. Salager, B. Francois, M. S. El Youssoufi, L. Laloui, C. Saix, Soils and Fnd. 48(4) 453-466 (2008)
19. B. François, S. Salager, M. S. El Youssoufi, D. Ubals Picanyol, L. Laloui, C. Saix, Proceedings, computer application in geotechnical engineering. 1-10 (2007)
20. C. W. W. Ng, C. Zhou, Geotechnique 64(9) 709–720 (2014)
21. A. Di Donna, L. Laloui, Engng. Geol., 190, 65-76 (2015)
22. J. P. W. Folly, PhD Thesis, Cardiff University (2001)
23. A. Haghighi, PhD thesis, Heriot-Watt University (2011)
24. A. Gens, E. E. Alonso, J. Suriol, A. Lloret, Proc. Ist Int. Conf. on Unsat. Soils, Paris. 83-88 (1995)
25. E. C. Lawton, R. J. Fragaszy, M. D. Hetherington, Journal of Geotech. Engng, 118, 1376-1394 (1992).
26. B. V. Venkatarama, Reddy, K. S. Jagadish, Geotechnique. 43(2), 337-341 (1993)
27. V. Sivakumar, PhD thesis, University of Sheffield (1993).
28. V. Sivakumar, S. J. Wheeler, Geotechnique, 50(4), 359-368 (2000)

E3S Web of Conferences 9, 07005 (2016)
DOI: 10.1051/e3sconf/20160907005
E-UNSAT 2016

Exploring ice content on partially saturated frozen soils using dielectric permittivity and bulk electrical conductivity measurements

Yanxin Mao[1], Enrique Romero[1,a] and Antonio Gens[1]
[1] *Division of Geotechnical Engineering and Geosciences, Department of Civil and Environmental Engineering, Universitat Politècnica de Catalunya, 08034, Barcelona, Spain*

Abstract. The present paper explores the use of the measurement of bulk electric conductivity and relative dielectric permittivity during freezing and thawing to estimate the ice content of two different partially saturated soils. The soils (Castelldefels fine sand and Barcelona clayey silt) were prepared at different degrees of saturation and dry densities at room temperature. Different target temperatures were used, which were reached at varying freezing rates and controlled with a thermal bath. Models were used to fit electrical conductivity and relative dielectric permittivity results under full unfrozen liquid and maximum ice conditions. The extension of these models allows estimating ice content on partially saturated soils at different porosities.

1 Introduction

When dealing with the modelling of thermo-hydro-mechanical coupled processes induced by freezing of saturated and unsaturated soils, an important issue is the definition of the freezing retention model, which links the saturation degree of the liquid phase (unfrozen water) to the temperature of the soil. The Clausius-Clapeyron-Poynting equation, which governs the condition for equilibrium between the coexisting liquid water and ice phases, together with a modified form of the van Genuchten equation to take into account the liquid-ice surface tension is usually adopted to represent this freezing retention model. Thus, determining experimentally the amount of ice content or unfrozen water in the frozen soils is necessary and helpful for verifying the correctness of the retention model and also for defining the model parameters.

For the direct measurement of ice content in frozen soil, several methods have been used, including gas dilatometry, dielectric spectroscopy and the heat pulse probe method (see for instance, [1]). Instead of measuring ice content directly, some authors combined two methods together to infer ice content and unfrozen water content in frozen soil [1, 2]. However, it is still difficult to measure the accurate ice content of frozen soil both in direct and indirect way.

The present paper explores ice content on two different partially saturated soils by controlling temperature at different liquid saturations during freezing and thawing process. The soils (a sand and a clayey silt) were prepared at different degrees of saturation and dry densities at room temperature. The technique used to measure the unfrozen liquid saturation was based on frequency domain reflectometry (FDR) measurements, as well as on electrical and thermal conductivity at different target freezing temperatures. These temperatures were reached at varying freezing rates and controlled by a thermal bath. Different models were used to fit electrical conductivity and relative dielectric permittivity results of the partially saturated soils under full unfrozen liquid and maximum ice conditions. These models were then extended for different degrees of unfrozen liquid and ice content, as well as varying porosities. This way, the combination of two methods (electrical conductivity and relative dielectric permittivity results) on a partially saturated soil of known porosity will allow inferring its ice and unfrozen water contents.

2 Description of the models

A simple model was developed by Archie (1942) relating the electrical conductivity of soils to its porosity and saline liquid saturation. For different liquid and ice saturations, the model can be written as (see for instance, [3]):

$$\sigma_{t_l} = a_l^{-1}\sigma_l n^{m_l} S_l^{p_l} \quad \text{and} \quad \sigma_{t_i} = a_i^{-1}\sigma_i n^{m_i} S_i^{p_i} \tag{1}$$

where σ_t denotes the electrical conductivity of the soil, σ represents the electrical conductivity of saline water, S the degree of saturation, n the porosity, m an exponent of the material that can be related to tortuosity or to pore shape and structure, p a saturation exponent and a a tortuosity factor. Subscripts 'l' and 'i' indicate the different phases of

[a] Corresponding author: enrique.romero-morales@upc.edu

 DOI: 10.1051/e3sconf/20160907005

the saline water, which are unfrozen liquid water and ice phase respectively.

In unsaturated frozen soils, four phases exist, i.e., soil solids, air, unfrozen water and ice. Because of the large resistance of soil solids and air compared to unfrozen water and ice, they are regarded as insulators. Thus the electrical conductivity can be expressed following an additive way as follows (in accordance to a parallel configuration in which the total current is the sum of the currents through the individual components):

$$\sigma_t = \sigma_{t_l} + \sigma_{t_i} = a_l^{-1}\sigma_l n^{m_l} S_l^{p_l} + a_i^{-1}\sigma_i n^{m_i} S_i^{p_i} \tag{2}$$

For the sake of simplicity, a similar law is adopted for (relative) dielectric permittivity:

$$\varepsilon_t = a_l^{-1}\varepsilon_l n^{m_l} S_l^{p_l} + a_i^{-1}\varepsilon_i n^{m_i} S_i^{p_i} \tag{3}$$

Chen et al. [4] also presented a simplified soil resistivity model to describe the relationships between electrical resistivity, degree of saturation and porosity, which can be expressed for liquid water and ice phases as:

$$\rho_{t_l} = \alpha_l \frac{\rho_l}{S_l n} + c_l \quad \text{and} \quad \rho_{t_i} = \alpha_i \frac{\rho_i}{S_i\, n} + c_i \tag{4}$$

where ρ_t is the electrical resistivity of the soil, α a structure factor, S the degree of saturation, n the porosity and c a constant. Subscripts 'l' and 'i' indicate the unfrozen liquid or ice phases of the water.

Electrical conductivity σ is defined as the inverse of resistivity. So for a partially saturated soil where unfrozen liquid and ice coexist, the soil electrical conductivity can be expressed as:

$$\sigma_t = \frac{S_l\, n\, \sigma_l}{\alpha_l + c_l S_l\, n\, \sigma_l} + \frac{S_i\, n\, \sigma_i}{\alpha_i + c_i S_i\, n\, \sigma_i} \tag{5}$$

For (relative) dielectric permittivity, and for the sake of simplicity, a similar expression to Eq. (5) will be also adopted:

$$\varepsilon_t = \frac{S_l\, n\, \varepsilon_l}{\alpha_l + c_l S_l\, n\, \varepsilon_l} + \frac{S_i\, n\, \varepsilon_i}{\alpha_i + c_i S_i\, n\, \varepsilon_i} \tag{6}$$

The unknown model parameters in Eqs. (1) and (4) under full unfrozen liquid $S_i = 0$ and maximum ice $S_l \approx 0$ conditions will be obtained by fitting the experimental curves using a nonlinear least square method. At known porosity, the above equations can be used to estimate ice content in partially saturated frozen soils with $S_l < 1$ and $S_i < 1$, once electrical conductivity and relative dielectric permittivity are known.

3 Materials and experimental methods

3.1 Test materials

Fine sand of Castelldefels (Barcelona) and Barcelona (BCN) clayey silt were studied [5]. The particle size of fine sand varied between 75 and 500 μm. The BCN clayey silt was modified to remove particle sizes larger than 2 mm. The solid densities of sand and clayey silt are ρ_{ss}=2.65 Mg/m^3 and ρ_{sc}=2.67 Mg/m^3, respectively. A 5% NaCl solution (mass basis) was used to perform the freezing-thawing test, and to allow better discriminating between ice and unfrozen liquid when using bulk electrical conductivity.

3.2 Experimental methods

A commercial Frequency Domain Reflectometry (FDR) sensor 5TE was used for measuring bulk electrical conductivity (EC) and (relative) dielectric permittivity [6] according to the scheme depicted in Figure 1. Good contact between the small electrodes and the soils is required for EC measurements. The dielectric constant is measured by using capacitance / frequency domain technology. In order to minimise the effects of salinity, 70 MHz frequency was used to make the 5TE sensor accurate. Two additional thermocouples were used for measuring soil temperature. The FDR sensor and thermocouples were pushed directly into the soil samples, which were contained in a cylindrical plastic container (height 109 mm, diameter 110 mm). The container was immersed into antifreeze liquid inside a thermostatically controlled freezer to conduct the freezing-thawing cycle. The temperature ranged from -20°C to 20°C. The soils with different degrees of saturation and porosities were tested during a freezing-thawing cycle to obtain the values of EC and dielectric permittivity at full liquid state $S_i = 0$ in freezing and thawing points (-3.2°C due to pore water salinity effects) and at maximum ice content $S_l \approx 0$ (-20°C).

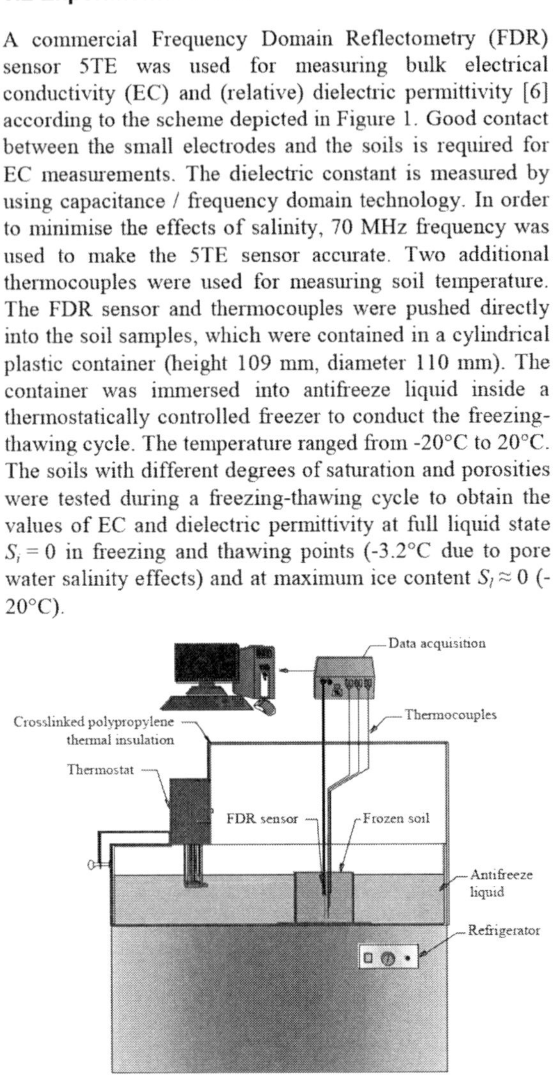

Figure 1. Scheme of the experimental setup.

4 Experimental results and analyses

4.1 Experimental results

Figure 2 shows the time evolution of bulk EC and (relative) dielectric permittivity of a sand sample prepared at $n = 0.49$ and $S_l = 32.9\%$) during a freezing-thawing cycle. It can be observed that the values of bulk EC and dielectric permittivity reduce with decreasing temperature until they reach a minimum, which indicates that the maximum ice condition had been achieved. Similar features were observed during thawing stage (correspondingly, bulk EC and dielectric permittivity increased during this stage). Short plateaus of both bulk EC and dielectric permittivity were observed at the freezing and thawing points.

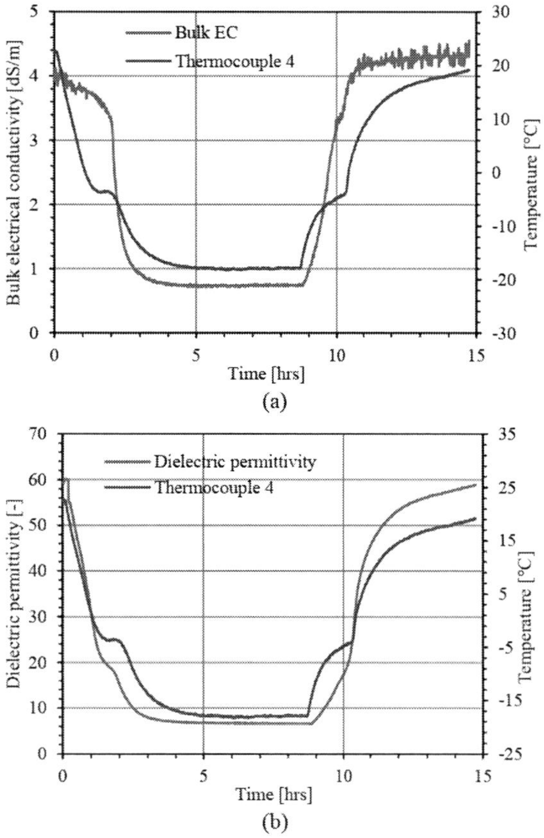

Figure 2. Time evolution of (a) bulk EC; (b) (relative) dielectric permittivity during a freezing-thawing cycle of a sandy sample ($n = 0.49$, $S_l = 32.9\%$).

The variation of bulk EC and (relative) dielectric permittivity with temperature for the two soils are presented in Figure 3. In the relatively dry sand ($n = 0.46$, $S_l = 15.5\%$), the same variation of both bulk EC and permittivity was observed during freezing and thawing. However, in the clayey silt prepared at $n = 0.42$ and $S_l = 47.9\%$, there are some differences in the variation of both bulk EC and dielectric permittivity values between freezing and thawing paths, particularly after the thawing point. Probably this indicates that the soil microstructure (more active due to the wetter state of the clayey silt) has changed during the freezing-thawing cycle.

In total, 15 sand and 12 clayey silt samples have been tested at different values of porosity and degree of saturation, as listed in Table 1. It can be observed that that the porosity of the sand was around $n \approx 0.46$ with major changes in the degree of saturation. This was not the case of the clayey silt, in which the amount of water added affected the fabrication dry density (the porosity changed from 0.32 to 0.52).

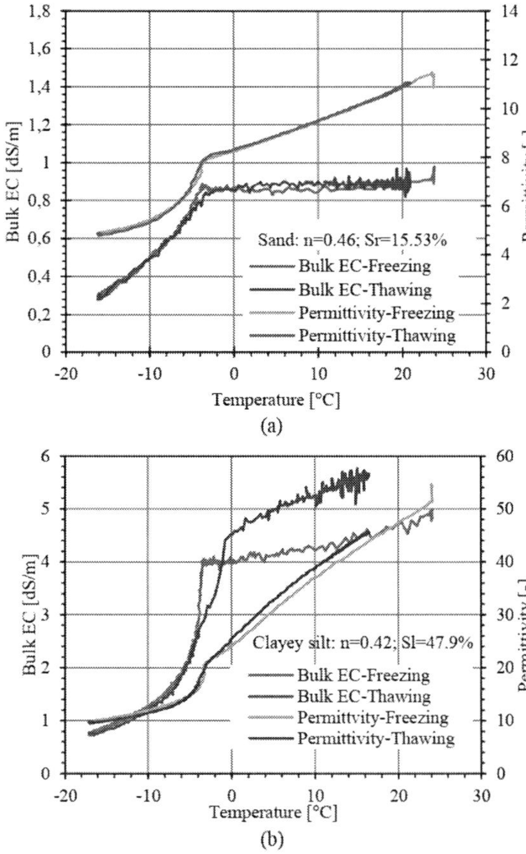

Figure 3. Variation of bulk EC and (relative) dielectric permittivity with temperature during a freezing-thawing cycle: (a) sand; (b) clayey silt.

4.2 Analysis of bulk electrical conductivity results

The bulk EC values of both partially saturated sand and clayey silt samples at maximum ice content $S_l \approx 0$ (-20°C) and at full liquid water state $S_i = 0$ in freezing and thawing point (-3.2°C) are presented in Figure 4. The 'liquid F' and 'liquid T' in the graphs denote the values measured at full unfrozen liquid state during freezing and thawing paths.

Table 1. Initial conditions of soil samples.

Sample	Dry density ρ_d / Mg/m^3	Porosity n / -	Degree of saturation S_l / %
Castelldefels sand			
1	1.43	0.460	15.5
2	1.37	0.483	23.2
3	1.38	0.479	23.9
4	1.36	0.487	32.9
5	1.41	0.468	34.2
6	1.47	0.445	34.9
7	1.40	0.472	36.9
8	1.46	0.449	41.7
9	1.44	0.457	41.7
10	1.40	0.472	43.0
11	1.45	0.453	48.9
12	1.44	0.457	51.7
13	1.47	0.445	57.6
14	1.47	0.445	64.6
15	1.48	0.442	76.0
BCN clayey silt			
16	1.29	0.517	25.7
17	1.41	0.472	35.7
18	1.53	0.427	45.8
19	1.61	0.397	47.1
20	1.53	0.427	47.9
21	1.56	0.416	50.7
22	1.63	0.390	52.0
23	1.70	0.363	56.3
24	1.82	0.318	70.3
25	1.81	0.322	70.3
26	1.79	0.330	79.2
27	1.83	0.315	89.3

Archie's law and Chen's resistivity model have been used to fit the experimental results and thereby to obtain the model parameters. The measured values of bulk EC of pure 5% NaCl solution at maximum ice content (-20°C) and at full unfrozen liquid state (freezing point) were $\sigma_i = 11.7$ dS/m and $\sigma_l = 40.0$ dS/m, respectively. The remaining model parameters can be obtained by fitting the experimental results.

The following two equations can then be used for estimating ice content of partially saturated fine sand from bulk EC readings:

$$\sigma_t = n^{1.6}\left(2.98\sigma_l S_l^{2.0} + 1.74\sigma_i S_i^{1.7}\right) \quad (7)$$

$$\sigma_t = \frac{S_l\, n\, \sigma_l}{1.8 - 0.08 S_l\, n\, \sigma_l} + \frac{S_i\, n\, \sigma_i}{2.4 - 0.32 S_i\, n\, \sigma_i} \quad (8)$$

The equivalent expressions for estimating ice content from bulk EC readings for the clayey silt are:

$$\sigma_t = 1.17\sigma_l n^{1.1} S_l^{2.0} + 0.48\sigma_i n^{1.0} S_i^{1.7} \quad (9)$$

$$\sigma_t = \frac{S_l\, n\, \sigma_l}{3.0 - 0.14 S_l\, n\, \sigma_l} + \frac{S_i\, n\, \sigma_i}{5.0 - 0.67 S_i\, n\, \sigma_i} \quad (10)$$

In Figure 4b the curves have been plotted at constant porosity ($n = 0.46$), whereas the different points display important variations in porosity (this is the reason of the deviations observed at high initial degrees of saturation that were prepared at the denser states).

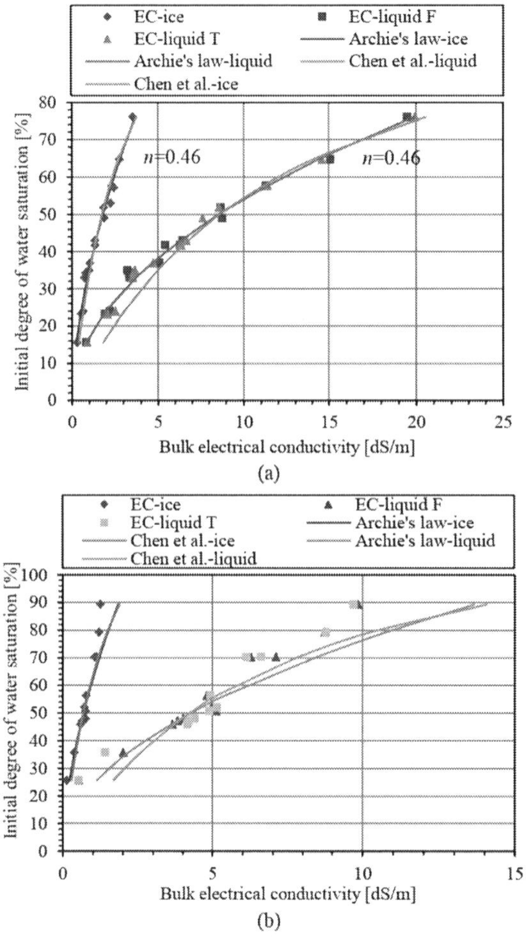

Figure 4. Bulk EC results for different initial degrees of liquid saturation for full unfrozen liquid and maximum ice content: (a) sand; (b) clayey silt.

4.3 Analysis of relative dielectric permittivity results

The (relative) dielectric permittivity values of both sand and clayey silt samples at maximum ice content $S_l \approx 0$ (-20°C) and at full unfrozen liquid state $S_i = 0$ in freezing and thawing point (-3.2°C) are presented in Figure 5. The 'liquid F' and 'liquid T' in the graphs denote the values measured at full unfrozen liquid state during freezing and thawing paths, respectively. The measured values of (relative) dielectric permittivity of pure 5% NaCl solution at maximum ice state (-20°C) and at full unfrozen liquid state (-3.2°C) were ε_i = 66.6 and ε_l = 81.0, respectively. The value at maximum ice state is very high and can be due to the salinity effect on the FDR measurement. Nevertheless, it has been accepted for fitting purpose (further tests are currently been undertaken to better define the relative dielectric permittivity value at maximum ice state of this reference 5% NaCl solution).

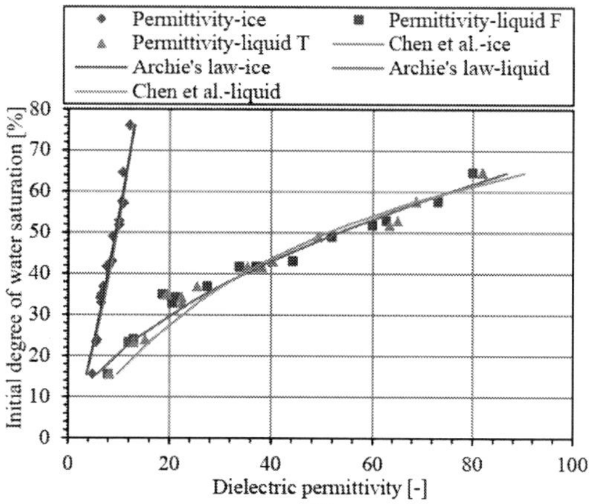

(a)

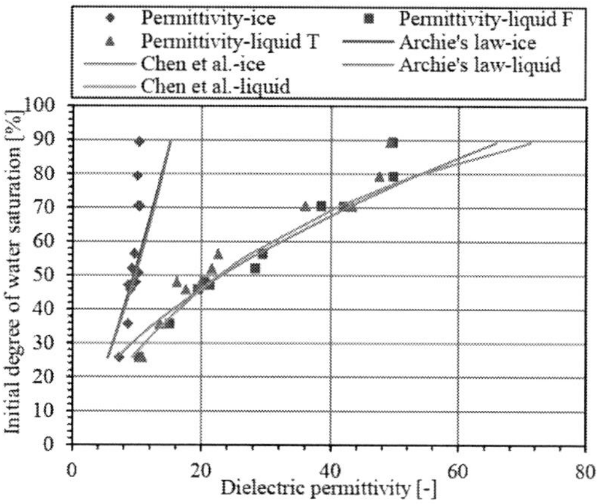

(b)

Figure 5. Relative dielectric permittivity results for different initial degrees of liquid saturation for full unfrozen liquid and maximum ice content: (a) sand; (b) clayey silt.

The expressions for estimating ice content in partially saturated sand samples from (relative) dielectric permittivity readings are:

$$\varepsilon_t = 8.50\varepsilon_l n^{1.6} S_l^{1.9} + 0.67\varepsilon_i n^{1.3} S_i^{0.8} \quad (11)$$

$$\varepsilon_t = \frac{S_l\, n\, \varepsilon_l}{0.70 - 0.018 S_l\, n\, \varepsilon_l} + \frac{S_i\, n\, \varepsilon_i}{1.15 + 0.028 S_i\, n\, \varepsilon_i} \quad (12)$$

In the case of the clayey silt, the corresponding expressions are:

$$\varepsilon_t = 1.89\varepsilon_l n^{0.7} S_l^{1.8} + 0.68\varepsilon_i n^{1.1} S_i^{0.8} \quad (13)$$

$$\varepsilon_t = \frac{S_l\, n\, \varepsilon_l}{1.10 - 0.024 S_l\, n\, \varepsilon_l} + \frac{S_i\, n\, \varepsilon_i}{1.10 + 0.020 S_i\, n\, \varepsilon_i} \quad (14)$$

Again, in Figure 5b the curves have been plotted at constant porosity (n = 0.46), whereas the different points display important variations in porosity (this is the reason of the deviations observed at high initial degrees of saturation that were prepared at the denser states).

The combination of the two measurements (bulk electrical conductivity and relative dielectric permittivity results) together with the above equations will allow inferring the ice and unfrozen water contents of a partially saturated soil of known porosity.

5 Conclusions

The modified Archie's law and Chen's resistivity model were used to fit the experimental results of both bulk electrical conductivity and relative dielectric permittivity results on two partially saturated soils. The model parameters were determined under full unfrozen liquid $S_i = 0$ and maximum ice content $S_l \approx 0$ conditions. The models were extended to cover intermediate states with $S_l < 1$ and $S_i < 1$. The fitted equations have been proposed for estimating the ice content from bulk electrical conductivity and dielectric permittivity measurements at different porosities and partially saturated states for fine sand and Barcelona clayey silt. It is planned to explore in the future some complementary methods for estimating ice content, involving the determination of specific heat capacity, thermal conductivity and permeability of frozen soils.

Acknowledgements

The first author acknowledges the financial support from China Scholarship Council (CSC).

References

1. X. Zhou, J. Zhou, W. Kinzelbach, F. Stauffer, Simultaneous measurement of unfrozen water content and ice content in frozen soil using gamma ray attenuation and TDR, Water Resour. Res. **50**, 9630–9655 (2014) doi:10.1002/2014WR015640
2. Q. Cheng, Y. Sun, Y. Qin, X. Xue, X. Cai, W. Sheng, Y. Zhao, In situ measuring soil ice content with a combined use of dielectric tube sensor and neutron moisture meter in a common access tube, Agric. Forest Meteorol. **171**, 249–255 (2013)

DOI: 10.1051/e3sconf/20160907005

3. G.E. Archie, The electrical resistivity log as an aid in determining some reservoir characteristics, Transactions of the American Institute of Mining and Metallurgical Engineers **146**, 54-62 (1942)
4. L. Chen, Z. Yin, P. Zhang, Relationship of resistivity with water content and fissures of unsaturated expansive soils, Journal of China University of Mining and Technology, **17** (4), 537-540 (2007) doi:10.1016/S1006-1266(07)60141-2
5. O. E. Cárdenas, R. C. Weber, E. Romero, A. Lloret, J. Suriol, Studying collapse behaviour of sandy silt under generalised stress conditions, Proc. 6th Int. Symposium on Deformation Characteristics of Geomaterials, IOS Press Amsterdam, 462-469 (2015) doi:10.3233/978-1-61499-601-9-462
6. Decagon Devices, Inc, Em50/Em50R/Em50G Em50 Series Data Collection System: Operator's Manual, (2015), http://manuals.decagon.com / Manuals/13453_Em50_Web.pdf.

Modelling the effects of pore-water chemistry on the behaviour of unsaturated clays

Xiaoqin Lei [1,2,3,b], Henry Wong [2,a], Antonin Fabbri [2], Ali Limam [3] and Y. M. Cheng [4]

[1] *Key Lab Mt Hazards & Earth Surface Proc, Inst Mt Hazards & Environm, CAS, Chengdu, China*
[2] *Université de Lyon/ENTPE/LGCB-LTDS (UMR CNRS 5513), Vaulx en Velin, France*
[3] *Université de Lyon/INSA Lyon/LGCIE, Villeurbanne, France*
[4] *Department of Civil and Environmental Engineering, Hong Kong Polytechnic University*

Abstract. Due to their various applications in geo-environmental engineering, such as in landfill and nuclear waste disposals, the coupled chemo-hydro-mechanical analysis of expansive soils has gained more and more attention recently. These expansive soils are usually unsaturated under field conditions; therefore the capillary effects need to be taken into account appropriately. For this purpose, based on a rigorous thermodynamic framework (Lei et al., 2014), the authors have extended the chemo-mechanical model of Loret el al. (2002) for saturated homoionic expansive soils to the unsaturated case (Lei, 2015). In this paper, this chemo-mechanical unsaturated model is adopted to simulate the chemo-elastic-plastic consolidation process of an unsaturated expansive soil layer. Logical tendencies of changes in the chemical, mechanical and hydraulic field quantities are obtained.

1 Introduction

Due to their very uncommon characteristics such as extremely low hydraulic permeability and high swelling capacity, expansive clays are widely used as sealing materials in geotechnical and geo-environmental engineering. However these clays are sensitive to pore water chemical compositions and matric suction changes. Coupled chemo-hydro-mechanical analysis are therefore required to evaluate the mechanical stability and sealing ability of these barriers during the whole life cycle of the engineering works concerned.

To address this complex problem, theoretical developments, based on the rigorous thermodynamic framework [1], were undertaken [2-4] to extend the chemomechanical model [5-6] for saturated expansive clays to account for partial saturation and suction effects. There have been other attempts following different lines of reasoning to model chemomechanical couplings [6-8], but due to limited space are not cited nor analysed here.

2 Soil structure of unsaturated smectite

The properties of unsaturated expansive soils are strongly related to their special soil structures. A typical representative structure for these soils can be described at two structural levels as shown by Figure 1. At the macroscopic level, the soil is considered as a complex of clay aggregates, separated by inter-aggregate pores filled by either gas or free water; Water in inter-aggregate pores is free to move (free water) and is dominated by capillary forces. Inside each aggregate, the microstructural system consists of several clay particles. Each particle contains several layers. Water can reside in the inter-layers or attach to the surface of these clay particles. Being strongly bonded to the clay particles, this water will not migrate nor evaporate except when subjected to very large forces (absorbed water). Accordingly, they will be treated as parts of solid phase. However, absorbed water and free water can exchange mass with each other to reach a thermodynamic equilibrium state. Here this phase change process is treated as chemical reaction.

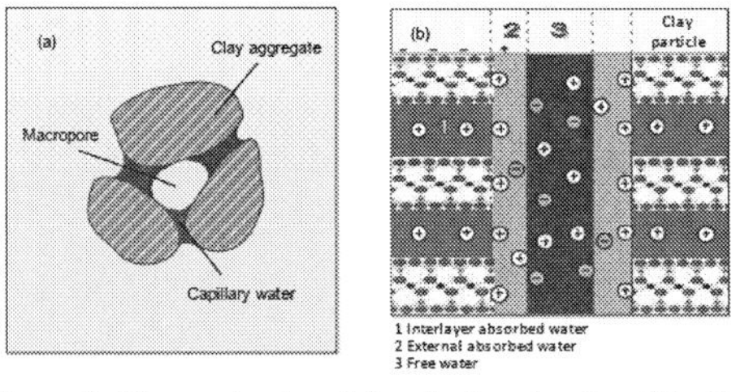

Figure 1. Macro-structure (a) and microstructure (b) of an unsaturated smectite

3 Theoretical framework

3.1 Mixture theory and volume fractions

Mathematically, the unsaturated expansive clay is treated as a three-phase multi-species porous medium. As shown

[a] Corresponding author: KwaiKwan.WONG@entpe.fr
[b] formerly PhD student co-supervised by ENTPE-INSA at the time of this work

in Figure 2, this tri-phasic porous medium, which occupies a volume $d\Omega_t$ in the current configuration, is taken as a representative elementary volume in this work. The solid phase $\mathcal{S}$, liquid phase $\mathcal{L}$, and gaseous phase $\mathcal{G}$ with their respective species can be summarised by the following sets:

$$\mathcal{S}=\{c',w',s_i'\};\mathcal{L}=\{w,s_i\};\mathcal{G}=\{g\};\mathcal{F}=\mathcal{L}\bigcup\mathcal{G} \tag{1}$$

where " c′ " is the solid clay species, " w " is the water species, " s_i " is the i -th salt species, and " g " is the air species. The superscript " ′ " indicates species that are attached to the solid phase, whereas no superscript is used to refer to species within the fluid phases. For example, " w " refers to the free water in the liquid phase, and " w′ " refers to the absorbed water attached to the solid phase. " $\mathcal{F}$ " is used to denote the set of species that compose the fluid phases.

To describe the preferential exchange of species between solid phase $\mathcal{S}$ and liquid phase $\mathcal{L}$ that arises from the physical-chemical reactions, a fictitious semi-permeable membrane is assumed to surround the cluster of clay platelets, which is a priori impermeable to clay species c′ . However, here the exchange between liquid water and water vapour is neglected by assuming that dry air is the only species in the gaseous phase. To facilitate the frequent reference of the exchangeable species, the following sets of exchangeable species are defined:

$$\mathcal{S}^{\leftrightarrow}=\{w',s_i'\};\mathcal{C}=\{w,s_i\} \tag{2}$$

Based on mixture theory, this three-phase porous medium can be treated as the superposition of three continua, one solid and two fluids. These continua are assumed to simultaneously occupy the entire physical domain but with reduced densities to maintain the overall mass balance. The volume fractions (such as porosity n) and saturations (such as the liquid saturation $S_{\mathcal{L}}$ and the gas saturation $S_{\mathcal{G}}$) are introduced to interchange these two densities, as shown in Figure 2. By the same procedure, the multi-species porous medium can also be treated as the superposition of multiple species continua.

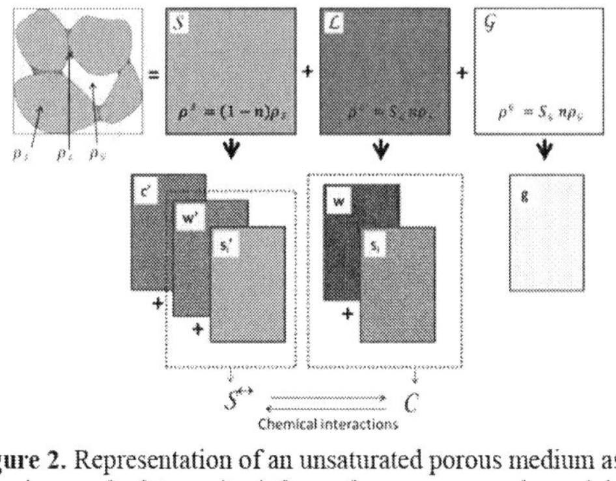

Figure 2. Representation of an unsaturated porous medium as a superimposed mixture: (top) three-phase macroscopic model of the REV; (bottom) multiple-species macroscopic model of the REV.

3.2 Irreversible thermodynamics

Irreversible thermodynamics can provide an efficient description of the coupled processes. Based on energy and entropy balance laws, the Clausius-Duhem inequality associated with the above open thermodynamic tri-phasic clay system consisting of a solid-liquid-gas mixture has been derived in Lei et al. (2014), and is presented by Equation (3). This Clausius-Duhem inequality states that the dissipation Φ of a thermodynamically viable process must be non-negative. For our multi-physical problems, we assume that this dissipation can be decomposed into dissipations associated with mechanical work Φ_m , phase change Φ_{pc} , mass transport Φ_{tr} , and thermal transport Φ_{th} . Each contribution will be required to be positive.

$$\Phi=\Phi_m+\Phi_{pc}+\Phi_{tr}+\Phi_{th}\geq 0 \tag{3}$$

With

$$\Phi_m=\boldsymbol{\sigma}:\frac{d^s}{dt}\boldsymbol{\varepsilon}+\sum_{k\in\mathcal{F}}\mu_k\frac{d^s}{dt}m_k+\sum_{k'\in\mathcal{S}^{\leftrightarrow}}\mu_{k'}\frac{d^s}{dt}m_{k'} \tag{4}$$
$$-\eta\frac{d^s}{dt}T-\frac{d^s}{dt}\Psi\geq 0$$

$$\Phi_{pc}=-\sum_{\substack{k\in\mathcal{C}\\ k'\in\mathcal{S}^{\leftrightarrow}}}\left(\mu_k-\mu_{k'}\right)\hat{m}_k\geq 0 \tag{5}$$

$$\Phi_{tr}=-\sum_{k\in\mathcal{F}}\boldsymbol{I}^k\cdot\left[\nabla\mu_k+\eta_k\nabla T\right]\geq 0 \tag{6}$$

$$\Phi_{th}=-\frac{q\cdot\nabla T}{T}\geq 0 \tag{7}$$

where, $\boldsymbol{\sigma}$ and $\boldsymbol{\varepsilon}$ are the Cauchy stress tensor and the linearized strain tensor, respectively; m_j and $\hat{m}_j$ are the mass content and the mass growth rate of species j per unit initial volume $d\Omega_0$; μ_j is the chemical potential per unit of mass of species j ; η is the total entropy per unit of overall volume of porous medium, T is the absolute temperature, q is the heat flux vector, and Ψ is the total Lagrangian free energy.

4 Model development

In what follows, a set of constitutive relationships under isothermal conditions ($\Phi_{th}=0$) will be formulated. These formulations include a chemo-elastic-plastic model for unsaturated soil skeleton, an unsaturated flow model, and a model of phase change kinematics.

4.1. Chemo-elastic-plastic model

4.1.1 Energy distribution

The dissipation Φ_m is the mechanical dissipation associated with the fluid phases and the solid skeleton, from which, the constitutive models for solid deformation can be derived. To arrive at the formulation of the solid skeleton, we decompose the total free energy Ψ and the total entropy η into two parts, one associated with the apparent solid skeleton and the other with the bulk fluids. By subtraction, the corresponding parts concerning the apparent solid skeleton can be obtained.

$$\Psi_s = \Psi - \sum_{\alpha=\mathcal{L},\mathcal{G}} \Psi_\alpha ; \eta_s = \eta - \sum_{\alpha=\mathcal{L},\mathcal{G}} \eta_\alpha \tag{8}$$

From the energy point of view, the apparent solid skeleton can be further viewed as comprising the solid skeleton and the interfaces. Each part has its own associated free energy (Coussy, 2004). To proceed, the solid skeleton is assumed to be decoupled from its attached interfaces by assuming the solid skeleton (sol) and the interfaces (int) have their own associated dissipations and free energies:

$$\Phi_m = \Phi_{sol} + \Phi_{int} ; \Psi = \Psi_{sol} + \Psi_{int} \tag{9}$$

In what follows, only the constitutive model for solid skeleton deformation will be presented. In addition, free energy Ψ may include the elastic free energy (reversible, indicated by superscript "e") Ψ^e and locked energy Ψ^p (stored during previous irreversible process, for its release must be accompanied by dissipative process, indicated by superscript "p"). Consequently, the free energy can be further decomposed into:

$$\Psi_{sol} = \Psi^e_{sol} + \Psi^p_{sol} \tag{10}$$

Based on the Clausius-Duhem inequality (4), according to the above analysis, the elastic work input stored as elastic free energies for the solid skeleton can be rewritten as (Lei , 2015):

$$d\Psi^e_{sol} = \bar{\boldsymbol{\sigma}} : d\boldsymbol{\varepsilon}^e + \bar{\mu}_w dm^e_w \tag{11}$$

In which, the generalized effective stresses $\bar{\boldsymbol{\sigma}}$ and the effective chemical potential of absorbed water $\bar{\mu}_w$ is defined as (Lei et al., 2014):

$$\bar{\boldsymbol{\sigma}} = \boldsymbol{\sigma} + p^* \boldsymbol{I};\ \bar{\mu}_w = \mu_w - \frac{p^*}{\rho_w} \tag{12}$$

With the equivalent pore pressure p^* defined by $p^* = S_{\mathcal{L}} p_{\mathcal{L}} + (1 - S_{\mathcal{L}}) p_{\mathcal{G}}$.

The following Clausius-Duhem inequality should be satisfied at the same time (Lei , 2015):

$$\Phi_{sol} = \bar{\boldsymbol{\sigma}} : d\boldsymbol{\varepsilon}^p + \bar{\mu}_w dm^p_w + \beta_{sol} d\xi_{sol} \geq 0 \tag{13}$$

Where, β_{sol} is the hardening force energetically conjugated to its corresponding hardening variable, and is defined by $\beta_{sol} = -\left(\frac{\partial \Psi^p_{sol}}{\partial \xi_{sol}}\right)_{S_{\mathcal{L}}}$.

4.1.2 Chemo-hyperelastic model

In tri-axial stress state, the strain work expression (11) can be altered accordingly:

$$d\Psi^e_{sol} = \bar{p} d\varepsilon^e_v + q d\varepsilon^e_s + \bar{\mu}_w dm^e_w \tag{14}$$

In which, the effective stresses $\bar{\boldsymbol{\sigma}}$ and strain $\boldsymbol{\varepsilon}$ are replaced with the volumetric strain ε^e_v , equivalent strain ε^e_s , effective stress $\bar{p}$, Mises equivalent stress q . For the construction of the elastic-plastic model, it is more convenient to treat the stresses as independent variables rather than the strains. To this end, the differential of free energy Ψ^e_{sol} can be transformed using a partial Legendre transform as:

$$d\bar{\Psi}^e_{sol}\left(\bar{p}, q, m^e_w\right) = \varepsilon^e_v d\bar{p} + \varepsilon^e_s dq - \bar{\mu}_w dm^e_w \tag{15}$$

which relates the generalised stress variables $\left(\bar{p}, q, \bar{\mu}_w\right)$ to their work-conjugate counterparts, the generalised strain variables $\left(\varepsilon^e_v, \varepsilon^e_s, m^e_w\right)$.

Now let $\bar{p}_0$ denote a small reference value with the pore fluid as distilled water under purely mechanical loading during the elastic deformation, and the slope of an over consolidation line (OCL) is κ^w . After chemical loading, the slope of this OCL becomes κ , and it intersects the original OCL at the point corresponding to the effective stress $\bar{p}_\kappa$. Following Loret et al. (2002), the expression of the skeleton free energy $\bar{\Psi}^e_{sol}$ is assumed as (Lei et al., 2014):

$$\bar{\Psi}^e_{sol} = \bar{p}\kappa^w \ln\frac{\bar{p}_\kappa}{\bar{p}_0} + \kappa\bar{p}\left(\ln\frac{\bar{p}}{\bar{p}_\kappa} - 1\right) + \frac{q^2}{6G} \\ -\mu^0_w m^e_w - \frac{RT}{d\Omega_0}\left(N_w \ln N_w - N_S \ln N_S\right) \tag{16}$$

In the above, R is the universal gas constant, μ^0_w is the reference chemical potential, $\mathcal{M}_w$ is the molar mass of absorbed water, and $x_w = N_w / N_S$ is the molar fraction of the absorbed water species with N_w and N_S denoting the molar quantities of absorbed water species and all solid phase species, respectively, within the original volume $d\Omega_0$. Now, an incremental chemo-hyperelastic constitutive model can be expressed as:

$$\begin{bmatrix} d\varepsilon^e_v \\ d\varepsilon^e_s \\ -d\bar{\mu}_w \end{bmatrix} = \begin{bmatrix} \frac{\partial^2 \bar{\Psi}^e_{sol}}{\partial \bar{p}^2} & \frac{\partial^2 \bar{\Psi}^e_{sol}}{\partial \bar{p} \partial q} & \frac{\partial^2 \bar{\Psi}^e_{sol}}{\partial \bar{p} \partial m^e_w} \\ \frac{\partial^2 \bar{\Psi}^e_{sol}}{\partial q \partial \bar{p}} & \frac{\partial^2 \bar{\Psi}^e_{sol}}{\partial q^2} & \frac{\partial^2 \bar{\Psi}^e_{sol}}{\partial q \partial m^e_w} \\ \frac{\partial^2 \bar{\Psi}^e_{sol}}{\partial m^e_w \partial \bar{p}} & \frac{\partial^2 \bar{\Psi}^e_{sol}}{\partial m^e_w \partial q} & \frac{\partial^2 \bar{\Psi}^e_{sol}}{\partial {m^e_w}^2} \end{bmatrix} \begin{bmatrix} d\bar{p} \\ dq \\ dm^e_w \end{bmatrix} \tag{17}$$

Based on the fact that elastic coefficient κ is dependent on the mass contents of the absorbed water m^e_w , and by assuming that Poisson's ratio is independent of m^e_w , after selected arrangement, another form taking $\boldsymbol{\Sigma} = \left[\bar{p}, q, -\bar{\mu}_w\right]^T$ as generalised stress-like variables and $E^e=\left[\varepsilon_v^e,\varepsilon_s^e,m_w^e\right]^T$ as generalised elastic strain-like variables can be obtained as

$$\dot{\Sigma}=D^e\dot{E}^e \tag{18}$$

With

$$D^e=\begin{bmatrix} D_{uu} & 0 & D_{um} \\ 0 & 3G & 0 \\ D_{mu} & 0 & D_{mm} \end{bmatrix}$$

In which,

$$D_{uu}=\frac{\bar{p}}{\kappa};D_{um}=-\frac{\bar{p}}{\kappa}\ln\frac{\bar{p}}{\bar{p}_\kappa}\frac{d\kappa}{dm_w^e}=D_{mu};$$

$$D_{mm}=\frac{\bar{p}}{\kappa}\left(\ln\frac{\bar{p}}{\bar{p}_\kappa}\frac{d\kappa}{dm_w^e}\right)^2+\bar{p}\left(1-\ln\frac{\bar{p}}{\bar{p}_\kappa}\right)\frac{d^2\kappa}{dm_w^{e\,2}}+\frac{RT}{\mathcal{M}_w}\frac{1}{x_w}\frac{dx_w}{dm_w^e}$$

4.1.3 Elasto-plastic model

Using the Cam-Clay notation, the solid skeleton dissipation (13) can be rewritten as:

$$\Phi_{sol}=\bar{p}d\varepsilon_v^p+qd\varepsilon_s^p+\bar{\mu}_w dm_w^p+\beta_{sol}d\xi_{sol}\geq 0 \tag{19}$$

Based on the above equation, the principle of maximum plastic dissipation implies the following normality flow rule

$$d\varepsilon_v^p=d\lambda\frac{\partial g}{\partial\bar{p}};d\varepsilon_s^p=d\lambda\frac{\partial g}{\partial q};dm_w^p=d\lambda\frac{\partial g}{\partial\bar{\mu}_w} \tag{20}$$

where $d\lambda\geq 0$ is the plastic multiplier, and g is the plastic potential. For simplicity, the plastic potential g is set equal to the yield function f introduced below. In addition, equation (19) also leads to the normality flow rule for the hardening variable

$$d\xi_{sol}=d\lambda\frac{\partial g}{\partial\beta_{sol}} \tag{21}$$

Consistent with the classic plasticity theory, it is assumed that a yield surface exists for the unsaturated expansive soil representing the boundary of the elastic region:

$$f=f\left(\bar{p},q,\bar{\mu}_w,\beta_{sol}\right)=0 \tag{22}$$

where the choice of the independent arguments are motivated by the dissipation inequality (19). A general form using the generalised variables can be written as

$$\dot{\Sigma}=\left(D^e-D^p\right)\dot{E} \tag{23}$$

During plastic loading, the stress point is always on the yield surface ($f=0$) and the consistency condition ($df=0$) should be satisfied at the same time. Following the classic procedure (Coussy, 2004), the plastic stiffness D^p in the above equation can be solved as

$$D^p=\frac{D^e\left[\frac{\partial g}{\partial\Sigma}\right]^t\left[\frac{\partial f}{\partial\Sigma}\right]D^e}{A+{}^t\left[\frac{\partial g}{\partial\Sigma}\right]D^e\left[\frac{\partial f}{\partial\Sigma}\right]} \tag{24}$$

where A is the hardening modulus, and is obtained by (Coussy, 2004)

$$A=-\frac{\partial f}{\partial\beta_{sol}}\frac{\partial\beta_{sol}}{\partial\xi_{sol}}\frac{\partial\xi_{sol}}{\partial\lambda}=\frac{\partial f}{\partial\beta_{sol}}\frac{\partial^2\Psi_{sol}^p}{\partial\xi_{sol}^{\ 2}}\frac{\partial g}{\partial\beta_{sol}} \tag{25}$$

The modified Cam-Clay model is adopted as the base model, with the yield surface:

$$f=\frac{q^2}{M^2\bar{p}}+\bar{p}-\bar{p}_c=0 \tag{26}$$

where M is the slope of the critical state line, and $\bar{p}_c$ is the preconsolidation stress which classically depends on the volumetric plastic strain ε_v^p. To account for the influence of suction and pore water chemical concentration, M is assumed to be dependent on pore water salt concentration as addressed by Loret et al. (2002). Concerning $\bar{p}_c$, besides the dependency on volumetric plastic strain ε_v^p, it is also strongly dependent on matric suction s and pore water salt concentration c_s:

$$\bar{p}_c=\bar{p}_{co}\exp\left(\frac{\varepsilon_v^p}{\lambda-\kappa}\right)\left(\frac{\bar{p}_\lambda}{\bar{p}_{co}}\right)^{\left(1-\frac{\lambda^w-\kappa^w}{\lambda-\kappa}\right)}\left(\frac{\bar{p}_\lambda}{\bar{p}_k}\right)^{\left(\frac{\kappa-\kappa^w}{\lambda-\kappa}\right)} \tag{27}$$

In the above hardening law, $\bar{p}_{co}$, $\bar{p}_k$, $\bar{p}_\lambda$, κ^w and λ^w are material constants, whereas κ and λ evolve with suction s and pore water salinity c_s, and control the evolution of the preconsolidation pressure.

4.2 Diffusion and mass transfer equations

To complete the theoretical model so that it can be applied to heterogeneous porous media, we also need to derive the trasnport equations of different species in the fluid phases. To do this, we need to invoke the non-negativity of the dissipations mentionned in the inequations (5) to (7). This would lead to the classic formulations of the generalised Darcy's law and Fick's law governing pore-water flow and salt diffusion, as well as to a phenomenological kinetic law governing the phase transformation between free and absorbed water. Due to space limit, these developments would not be shown here; they can be found in [2-3].

5 Experimental results and simulations

A few numerical simulations have been performed to verify the consistency, pertinence and accuracy of the theoretical model developed [2]. Concerning the case of hetergeneous porous media, [3] analysed the progressive intrusion of salt into a soil layer and its consequences ; its results were also compared to that of a previous publication. In this paper, we choose to present some simulations of the experimental data reported in [9-10]. While results are available for both saturated and unsaturated samples, attention here will be uniquely focused on unsaturated ones. On the other hand, the parameters used in the simulations, as summarized in Tables 1 ad 2, were calibrated using both data on saturated and unsaturated samples [3, 4].

	$m_{c'}$ [kg/m^3]	$m_{s'}$ [kg/m^3]	$m_{w'}$ [kg/m^3]	$x_{w'}$	κ^c	λ^c
0 M	240	0	336	0.967	0.013	0.089
6.53 M	240	0	65.573	0.853	0.007	0.080

Table 1. Parameters for interpolation of chemo-mechanical properties at saturated state

$\bar{p}_{co}$ [kPa]	$\bar{p}_{\kappa}$ [kPa]	$\bar{p}_{\lambda}$ [kPa]	s_e [kPa]	k_3	λ_3	β
50	5000	1300	10	1	5	0.084

Table 2. Parameters for chemo-mechanical model

5.1 Test series 1 and results of simulation

Two series of oedometric tests on unsaturated samples were reported in [10]. In the first, samples were prepared by mixing Boom Clay powder with $NaNo_3$ solutions at 3 different concentrations: 0 M, 4 M and 5.4 M, and were brought to an initial suction of 500 kPa. They were then subject to a compression-decompression cycle at constant suction up to a maximum axial stress of 1000 kPa. Results are plotted in Figure 3a.

Similarly to the case of saturated samples, a salinity increase reduces slightly the virgin compressibility λ, while leaving more or less unaffected the compressibility at overconsolidated states κ. The numerical simulations presented in figure 3b also appear reasonably close to experimental results.

5.2 Test series 2 and results of simulation

In the second series of oedometric tests, 2 samples were prepared with distilled water and brought to an initial suction of 500 kPa and an initial stress of 40 kPa. From this initial state and keeping constant suction and zero salinity, sample 1 was simply compressed to 1000 kPa then decompressed to around 50 kPa. By contrast, sample 2 was subject to a mixed chemo-mechanical loading path: (i) compressed to 100 kPa ; (ii) exposed to 6.54 M $NaNo_3$ solution ; (iii) further compressed to 1000 kPa ; (iv) unloaded to 100 kPa ; (v) exposed to distilled water. Results are plotted in Fig 4a.

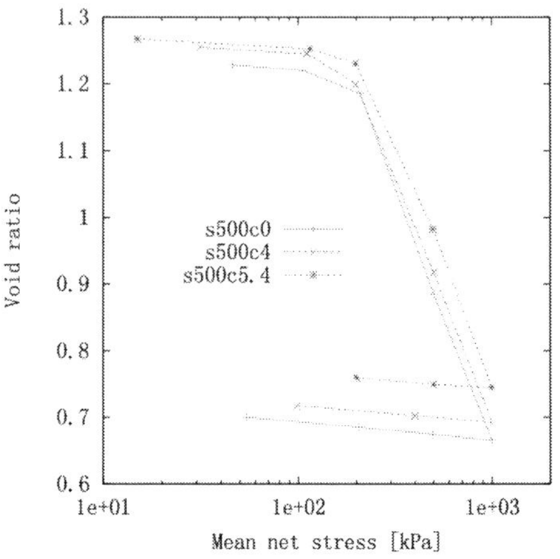

(a) Experimental results

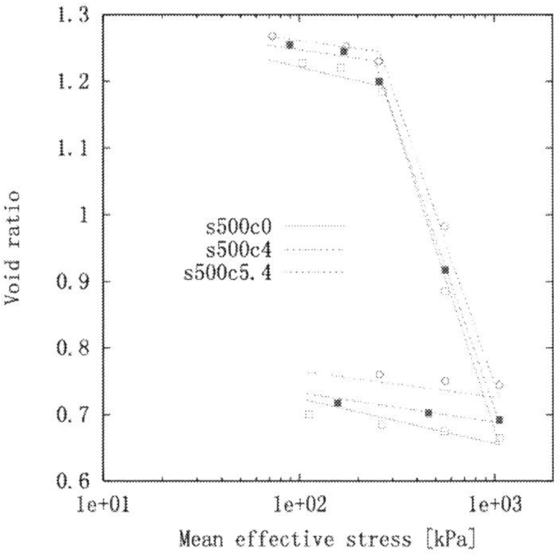

(b) Simulations

Figure 3. Oedometric tests on unsaturated Boom Clay samples (a) experimental data (b) simulations

The tendencies observed are similar to the case of saturated samples [3-4]. A normally consolidated or lowly overconsolidated clay is much more sensitive to salinity changes than a highly over-consolidated clay. This explains why exposure to 6.54 M solution during first compression (close to normally consolidated state) induces a large volume contraction while exposure to distilled water near end of unloading (overconsolidated due to previous compression) has insignificant effects. Chemically-induced plastic collapse in the first case appears to be a logical explanation. Notice also in this regard that the first exposure to 6.54 M solution does induce consistently an increase in the preconsolidation pressure.

Overall, the model seems to reproduce correctly the experimental observations, despite the complexity of the interplay of different mechanisms [4], although some local discrepancies do manifest themselves. In particular, the model appears to overestimate the volume expansion induced by exposure to distilled water at the end of unloading.

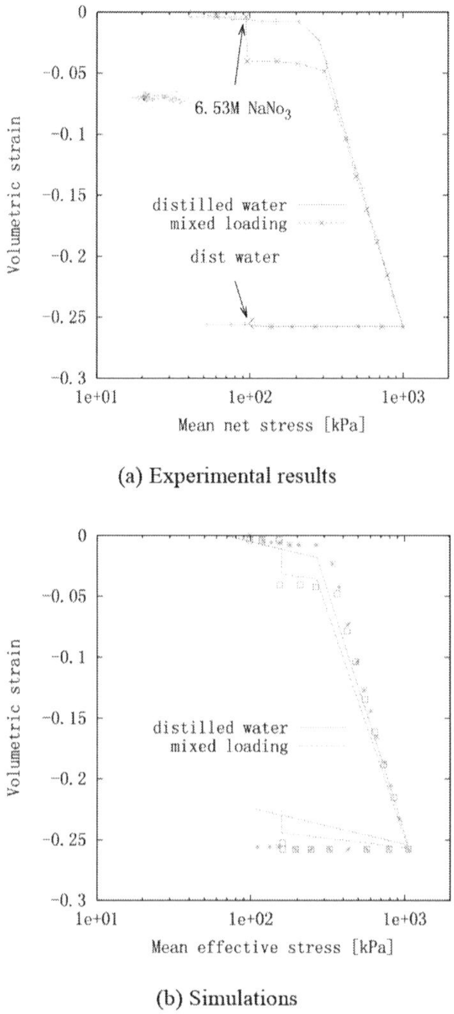

Figure 4. Mixed chemical-mechanical loading on unsaturated samples: exposure to 6.53 M solution during loading and to distilled water during unloading, both at vertical stress 100 kPa.

At the present state, the extremely limited amount of experimental data on samples subject simultaneous to suction and chemical loadings does not allow an in-depth analysis of the model which would probably yield ideas on its possible refinements. Further experimentation appears necessary.

6 Conclusion

Based on the modified mixture theory and the theory of non-equilibrium thermodynamics, a theoretical chemo-mechanical model was developed. In this model, the governing equations for chemo-elastic-plastic deformations of unsaturated skeleton, fluid species diffusion and solid-liquid mass transfer have been systematically and rigorously derived. The model was then used to simulate the response of unsaturated soil samples observed during oedometric tests when subject to different chemo-mechanical loading paths. The comparisons globally do not invalidate the model construction while thery also pointed out the necessity for further studies.

References

1. O. Coussy, John Wiley & Sons, Ltd. , (2004)
2. X. Lei, H. Wong, A. Fabbri, A. Limam, Y.M. Cheng, Comput. Geotech. **62** (2014).
3. X. Lei, PhD thesis, INSA-Lyon/ENTPE (2015).
4. X. Lei, H. Wong, A. Fabbri, A. Limam, Y.M. Cheng, I.J.Solids Str., doi:10.1016/j.ijsolstr.2016.01.008 (2016).
5. Loret, B., Hueckel, T., Gajo, A., Int. J. Solids Struct. **39** (2002).
6. Gajo, A., & Loret, B. Comput. Methods Appl. Mech. Eng. **192** (31-32) (2003).
7. L. Guimarães, A. Gens, M. Sanchez, S. Olivella. Geotechique **63**, 2 (2013).
8. Z. Liu, N. Boukpeti, X. Li, F. Collin, IJNAG **29** (2005).
9. N. Mokni, E. Romero, S. Olivella, Geotechnique **64**, (2014).
10. N. Mokni, PhD thesis, Universitat Politècnica de Catalunya (2014).

Probabilistic assessment of precipitation-triggered landslides: the role of vegetation

Estefanía Muñoz[1,2,a], Andrés Ochoa[1] and Manoel Cordão-Neto[2]

[1]*Departamento de Geociencias y Medio Ambiente, Universidad Nacional de Colombia, Colombia*
[2]*Departamento de Engenharia Civil e Ambiental, Universidade de Brasília, Brazil.*

Abstract. Landslides triggered by rainfall are one of the most common causes of disaster in tropical regions, characterized by having deep weathering soil profiles, steep slopes and high-intensity storms. The increasing number of landslides during wet season evidences the close relationship between hydro-climatic conditions as a triggering factor for the occurrence of landslides. In addition, the type of vegetation covering the slope affects the soil shear strength by the roots reinforcement, plants weight and changes in soil moisture due to transpiration and interception. This paper proposes a probabilistic methodology to study the slope stability on the long-term, considering different hydro-climatic conditions and the effect of vegetation cover in the soil moisture. The ecohydrological model developed by Rodríguez-Iturbe et al. [1] was used to determine the boundary conditions of the problem. To generate the pore pressure field, the flow equation was solved using the Finite Element Method and Finite Differences Method. Finally, the Limit Equilibrium Method was used to find the Factor of Safety. The stability of a hypothetical slope under certain hydro-climatic conditions and two types of vegetation was evaluated. The analysis showed that it is more likely that a grass-covered slope slides than a tree-covered slope, and that the average FS of the slope during wet and dry season is very similar, but the FS dispersion is higher when the probability and intensity of rainfall events increases.

1 INTRODUCTION

Landslides are closely linked to water content on surface soil layer, which depends mostly on the amount of water precipitated. However, there are many other variables (with nonlinear relationships) that make it impossible to derive a general relation between rainfall and the soil response. There are many approaches in the literature to relate cumulative rainfall and landslides, diverging in the quantity of days to be considered (e.g. [2–9]). The surface soil layers behave as dynamic systems that interact with the atmosphere and the deeper layers, making it necessary to consider the climate-soil-vegetation system, to understand the landslide problem.

The climate-soil-vegetation system is influenced by the different relations of dependency and feedback of physical processes on each subsystem. The soil water content is the result of the interaction of processes presented on atmosphere, soil and vegetation [10]. The interactions of the subsystems critically depend on the scales at which the phenomenon is studied, ecophysiological features of vegetation, pedology and climate regime [11].

Climate is an important factor, not only on diffusive processes of evaporation, but on the availability of water for vegetation [11]. Texture, mineralogical composition and arrangement of particles determine water content that the soil is allowed to store [12]. Moreover, vegetation controls energy and water flows, dividing precipitation on interception, runoff, evaporation, transpiration and stored water in the soil, and the energy on sensible and latent heat. After evaporation, transpiration of plants is the most important phenomenon on energy transfer on surface processes [13].

A probabilistic methodology was proposed in this paper to evaluate the slope stability on the long-term, considering the hydro-climatic conditions and the effect of vegetation cover in soil moisture. A modification of the ecohydrological model proposed by Rodríguez-Iturbe et al. [1] and Laio et al.[14] was used to determine boundary conditions of the problem. In order to obtain pore pressure field, the flow equation was solved by the Finite Element Method (FEM) and the Finite Difference Method (FDM). And the Limit Equilibrium Method (FEM) was used to find the Factor of Safety (FS). In addition, the stability of the hypothetical slope, proposed by Otálvaro & Cordão-Neto [15], covered by two types of vegetation during wet and dry periods was evaluated.

[a] Corresponding author: emunozh@unal.edu.co

DOI: 10.1051/e3sconf/20160908001

2 CLIMATE-SOIL-VEGETATION MODEL

Figure 1 represents the dynamics of soil water content from the mass balance on plant rooting depth, described by

$$nZ_r \frac{dS_r(t)}{dt} = \Phi[S_r(t),t] - \chi[S_r(t)] \quad (1)$$

where, n is porosity, Z_r rooting depth, S_r soil degree of saturation, Φ infiltration rate and χ water losses.

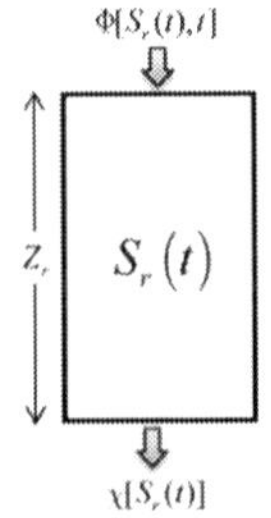

Figure 1. Conceptual model of the soil water content.

On the model initially proposed by Rodríguez-Iturbe et al. [1] and modified by Laio et al.[14], the processes describing the interaction climate-soil-vegetation evolve over time, under the influence of a variable climate.

Infiltration is the stochastic component of the balance and is described by

$$\Phi[S_r(t),t] = R(t) - I(t) - \varphi[S_r(t),t] \quad (2)$$

where, R is rainfall rate, I is the amount of rainfall lost through canopy interception and φ is the rate of surface runoff generation. Soil water losses are given by

$$\chi[S_r(t)] = ET[S_r(t)] + L[S_r(t)] \quad (3)$$

where, ET y L are evapotranspiration and percolation rates respectively, and they depend on water content stored in the soil.

2.1 Rainfall and interception

Daily precipitation was modeled as a Rectangular Pulses Poisson Model (RPPM) with arrival rate λ [16].

The distribution of times Γ between precipitation events is exponential with mean λ^{-1}, i.e.

$$f_\Gamma(\Gamma) = \lambda e^{-\lambda\Gamma} \text{ for } \Gamma \geq 0 \quad (4)$$

The depth of rainfall events was assumed to be an independent random variable h_R, described by an exponential Probability Density Function (pdf)

$$f_{H_R}(h_R) = \frac{1}{\alpha} e^{-\frac{1}{\alpha}h_R},\ h_R \geq 0 \quad (5)$$

where, α is the mean depth of rainfall events, estimated as the daily mean of days when precipitation occurs.

The mechanisms of interception by vegetation are quite complicated to model, since they depend, among others, on density and architecture of vegetation, besides rainfall intensity and duration. To simplify the model, a threshold Δ was defined, below which water reaches the ground. When the interception is evaluated, rainfall process is transformed into a new Poisson process, where the frequency of rainfall events λ' is

$$\lambda' = \lambda \int_\Delta^\infty f_{H_R}(h_R)dh_R = \lambda e^{-\frac{\Delta}{\alpha}} \quad (6)$$

With the amount of water intercepted by plants leaves, it is possible to calculate the effective precipitation R_e, defined as the amount of water that reaches the soil surface, that is to say, total precipitation minus interception.

2.2 Infiltration and runoff

It was assumed that surface runoff starts when soil hydraulic gradient equals zero, and that infiltration has a rate related to soil hydraulic conductivity k_w.

The values of k_w where obtained as a function of soil degree of saturation for each time and saturated hydraulic conductivity coefficient k_s, following the expression based on the potential law proposed by Campell [17]

$$k_w(S_r) = k_s S_r^{\,d} \quad (7)$$

where, d=2b+3, and b is a soil empirical exponent.

2.3 Evapotranspiration

Plants transpiration and soil evaporation are governed by different mechanisms, however, were modeled together, as shown on Figure 2.

Evapotranspiration ET was modeled as a function of soil water content. When soil saturation is higher than the field capacity S_{fc}, the evapotranspiration rate is maximum and equal to E_{max}. When the degree of saturation is lower than the degree of saturation of the temporary or incipient wilting point S^*, transpiration of plants is lowered by the stoma closing to avoid inner water loss. Transpiration rate continues decreasing until the moisture reaches permanent wilting point S_w. At this point, suction is so high that damages plant tissues, resulting on an irreversible wilting point, making it impossible for the plant to recover its turgidness [18]. For water content lower than S_w, there is only evaporation with a rate equal to E_w. Finally, for moistures below hygroscopic S_h, water cannot be naturally removed.

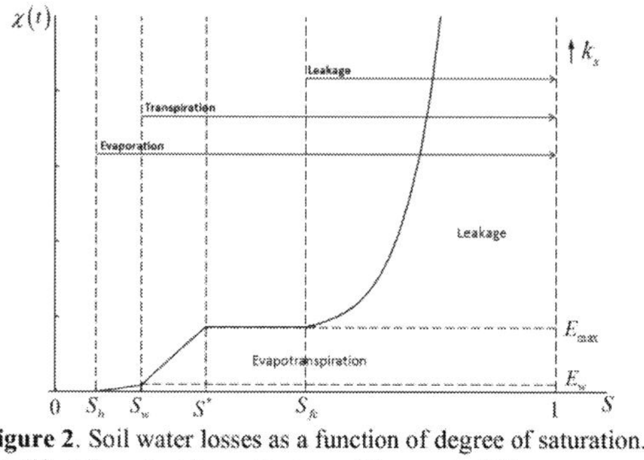

Figure 2. Soil water losses as a function of degree of saturation. Modified from Rodriguez-Iturbe and Porporato [12].

It was assumed that losses by leakage or deep infiltration L start when water content reaches a value higher than S_{fc}. The maximum leakage rate equals k_s, and lowers rapidly when soil starts to dry, following the relation [14]

$$L(S_r) = \frac{k_s}{e^{\beta(1-S_{fc})}-1}\left[e^{\beta(S_r-S_{fc})}-1\right] \text{ for } S_{fc} < S_r \le 1 \qquad (8)$$

where, β is a coefficient used to fit the equation (8) permeability function described by the equation (7). This parameter depends on type of soil and can be related to b (β=2b+4).

During the periods between rainfall events, the decay of soil degree of saturation was described from de initial values of S_r as proposed by Rodríguez-Iturbe et al. [12]. These values depend on the previous history of the complete process.

3 WATER FLOW IN UNSATURATED SOIL

For a unit length element of soil, mass conservation is mathematically expressed by

$$\frac{\partial(\rho_w n S_r)}{\partial t} + \frac{\partial(\rho_w v_i)}{\partial x_i} = 0 \qquad (9)$$

where, ρ_w is soil specific weight, v_i is velocity of water, t is time and x_i is the direction of the coordinate system.

Assuming: (i) the water incompressible, (ii) volumetric deformation and suction as state variables, and (iii) Darcy´s law is valid, it is obtained that equation that represents soil water flow is

$$-k_{wij}\frac{\partial^2 h}{\partial x_j^2} + \frac{\partial \theta_w}{\partial t} = 0 \qquad (10)$$

where, k_{wij} is the permeability matrix that is function of degree of saturation and soil voids ratio and $\partial h/\partial x$ is the hydraulic gradient.

In order to solve the flow equation spatially, the Finite Element Method (FEM) was used, and for the temporary solution the Finite Difference Method (FDM) was used as well.

As the solution of the flow problem is expressed on terms of pore pressure or hydraulic head, it is necessary to define constitutive functions that relate permeability and moisture with this variable.

To obtain the term $\beta_w = \partial\theta_w/\partial h$, that reflects facility of water inlet or outlet on an soil element due to variations on suction, it was used the function for the Soil Water Retention Curve (SWRC) proposed by Van Genuchten [19]

$$\Theta = \left[\frac{1}{1+(\alpha_{V-G} h)^{n_{V-G}}}\right]^{m_{V-G}}, \quad \left(m_{V-G} = 1 - \frac{1}{n_{V-G}}\right) \qquad (11)$$

where, h is the hydraulic head, α_{V-G}, m_{V-G}, and n_{V-G}, are fitting parameters and Θ is the normalized water content.

The constitutive relation for permeability is based on the same equation used on the model to obtain atmosphere-soil flows, defined by

$$k_w = \begin{cases} k_s & 0 \le (u_a - u_w) \le \alpha \\ k_s S_r^c & \alpha < (u_a - u_w) < u_{max} \\ k_{min} & u_{max} \le (u_a - u_w) < \infty \end{cases} \qquad (12)$$

If the pore pressure is higher than a limit value u_{max}, permeability reaches a value k_{min} that depends on type of soil.

4 CASE STUDY

The geometry of studied slope is the same as the one of the hypothetical case presented by Otálvaro & Cordão-Neto[15]. The water table was obtained through a stationary analysis, by imposing pore pressures equal to zero at the slope foot and a constant flow of $7{\cdot}10^{-7}$ m/s coming from left side. The location of water table and slope geometry are presented on Figure 3.

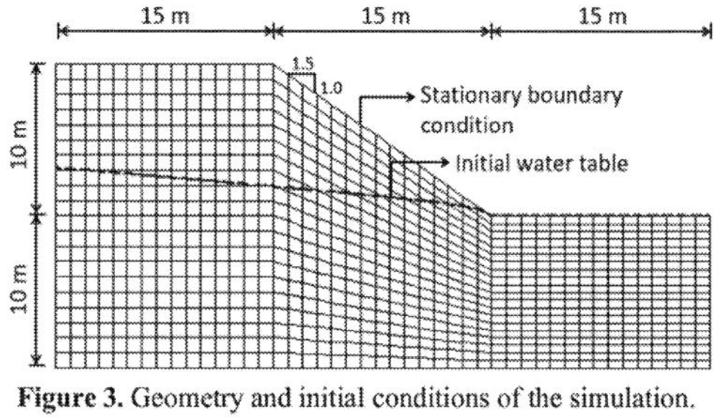

Figure 3. Geometry and initial conditions of the simulation.

Table 1 shows the parameters values used to model the soil, which is a sandy loam, the parameters of soil water retention curve and permeability curve, and the parameters of ecohydrological model.

Table 1. Mechanical soil parameters.

Parameter	Value
Specific weight [kN/m^3]	18.1
Cohesion [kPa]	12
Friction angle [°]	30
Saturated hydraulic conductivity [m/s]	2.24E-05
Air-entry value [kPa]	8
ϕ^{b*} [°]	25
$m_{V\text{-}G}$	0.06
$n_{V\text{-}G}$	1.5
b	4.6
Porosity	0.463
E_w [mm/day]	0.1
S_h	0.08
S_{fc}	0.35

*angle indicating the increase rate of shear strength relative to increase matric suction.

To evaluate the influence of vegetation type on the slope stability, the values of parameters of two types of typical vegetation cover considered on simulations by Ridolfi et al. [20] were used, and are shown in Table 2.

Table 2. Parameters of the vegetation type.

Parameter	Grass	Trees
Z_r mm]	300	700
Δ [mm]	1	2
E_{max} [mm/day]	4.7	4.3
S_w	0.10	0.11
S^*	0.24	0.22

Furthermore, simulations where made for two climate seasons, one wet and another dry, characterized by the parameters presented in Table 3. Montecarlo simulations were made for a 50-year period, with the purpose of analyzing the statistical behavior of soil moisture and slope Factor of Safety.

Table 3. Parameters of the precipitation model.

Season	λ [day^{-1}]	α [mm/day]
Wet	0.72	1.15
Dry	0.59	0.95

Figure 4 shows duration curves of Factor of Safety for two covers and two pluviometric regimes. During dry season (red lines) the slope is unstable (FS<1) 3.1% of the time for grass cover, and 0.4% for trees. During wet season (blue lines) the slope remains on stable conditions for longer time for both covers, 6.6% for grass and 1.3% for trees. The arboreal cover preserves the slope on stable conditions for a longer period compared to grass, regardless the rainfall season. The results show that stability conditions (FS>1) are presented even for longer period for trees cover during wet season than for grass during dry season.

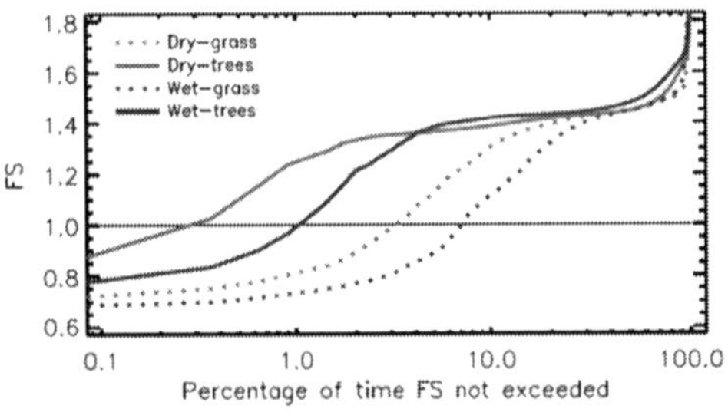

Figure 4. Percentage of non-exceedance time of FS for trees and grass during wet and dry season.

This analysis is reinforced with information of Table 4, which shows the measures of central tendency and variance of FS for all cases. The measures of central tendency change very little among climate seasons, but the dispersion changes, possibly because of the occurrence of more frequent and intense storms.

Table 4. Measures of central tendency and variance of FS for the study cases.

Case	Mean	Mode	Median	Variance
Dry-grass	1.43	1.44	1.45	0.03
Dry-trees	1.47	1.50	1.45	0.01
Wet-grass	1.41	1.44	1.45	0.06
Wet-trees	1.50	1.50	1.48	0.02

The maximum evapotranspiration (E_{max}) of grass is higher than trees, indicating that if the water content is higher than the incipient wilting point (S^*), grass evapotranspires at a higher rate than trees. However, in the model, the threshold of interception and rooting depth of trees are higher than grass, resulting on more water being retained on trees leaves, and that infiltrated water volume distributes on a larger soil volume, increasing moisture in lower proportion if porosity, initial moisture and the storm are considered the same.

On the long-term, water content tends to be higher for grass cover than for trees cover, making FS to have lower values (higher tendency to instability), due to its direct relationship with moisture.

5 CONCLUSIONS

The stability of a hypothetical slope was analyzed under two vegetation covers and two pluviometric scenarios using a mechanistic model. The results showed small differences on the measures of central tendency of slope factor of safety, but there are important differences on the dispersion. The percentage of time that the slope is unstable (FS<1) is much larger for grass cover than for tree cover. Although this result is consistent for both rainfall scenarios, the percentage of time in which FS is less than 1 are higher during wet season than during dry season.

From our point of view, trees are able to modulate the dynamics of infiltrated water into a larger volume of soil, because their roots reach greater depths. Also, they have leaves with large surfaces that can intercept more rainfall. These two features, on the long-term, make the soil water content to remain lower for the slope covered by trees than for the slope covered by grass; regardless of that, grasses can evapotranspirate to a maximum higher rate.

Considering only the effect provided by vegetation in soil moisture, we conclude that trees make the slope more stable. Outside the scope of this work, there are other matters such as the effect of weight and the apparent soil cohesion provided by roots of each vegetation cover.

ACKNOWLEDGEMENTS

The authors thank COLCIENCIAS (Colombia), CNPq (Brazil) and CAPES (Brazil) for their financial support.

REFERENCES

1. I. Rodríguez-Iturbe, A. Porporato, L. Ridolfi, V. Isham, and D. R. Coxi, Proc. R. Soc. A Math. Phys. Eng. Sci. **455**, 3789 (1999).
2. A. B. Fourie, Proc. ICE - Geotech. Eng. **119**, 211 (1996).
3. G. B. Crosta, Environ. Geol. **35**, 131 (1998).
4. U. Ratnayake and S. Herath, J. Mt. Sci. **2**, 218 (2005).
5. E. L. Harp, M. E. Reid, J. P. McKenna, and J. A. Michael, Eng. Geol. **104**, 295 (2009).
6. C. Lepore, S. a. Kamal, P. Shanahan, and R. L. Bras, Environ. Earth Sci. **66**, 1667 (2011).
7. F. Wei, Y. Guolin, L. Xiaohong, and L. Liming, in *2011 Int. Conf. Consum. Electron. Commun. Networks* (Ieee, 2011), pp. 3201–3205.
8. R. Giannecchini, Y. Galanti, and G. D'Amato Avanzi, Nat. Hazards Earth Syst. Sci. **12**, 829 (2012).
9. J. Von Ruette, P. Lehmann, and D. Or, Water Resour. Res. n/a (2013).
10. O. A. Rueda, Modelamiento de La Humedad Del Suelo En La Región Cafetera de Colombia Bajo Diferentes Coberturas Vegetales, Master, Universidad Nacional de Colombia, 2008.
11. I. Rodríguez-Iturbe, A. Porporato, F. Laio, and L. Ridolfi, Adv. Water Resour. **24**, 695 (2001).
12. I. Rodríguez-Iturbe and A. Porporato, *Ecohydrology of Water-Controlled Ecosystems* (Cambridge University Press, USA, 2004).
13. P. S. Eagleson, Proc. Am. Philos. Soc. **144**, 33 (2000).
14. F. Laio, A. Porporato, L. Ridolfi, and I. Rodríguez-Iturbe, Adv. Water Resour. **24**, 707 (2001).
15. I. Otálvaro and M. Cordão-Neto, in *Adv. Unsaturated Soils*, edited by B. Caicedo, C. Murillo, L. Hoyos, J. Colmenares, and I. Berdugo (Taylor & Francis, 2013), pp. 539–543.
16. P. S. Eagleson, Water Resour. Manag. **8**, 878 (1972).
17. G. S. Campbell, Soil Sci. **117**, (1974).
18. C. Pimentel, *A Relação Da Planta Com a Água* (Editora Universidade Federal Rural do Rio de Janeiro, São Paulo, Brazil, 2001).
19. M. Van Genuchten, Soil Sci. **44**, 892 (1980).
20. L. Ridolfi, P. D'Odorico, A. Porporato, and I. Rodríguez-Iturbe, J. Hydrol. **272**, 264 (2003).

DOI: 10.1051/e3sconf/20160908002

Seismic response of earth dams considering dynamic properties of unsaturated zone

M. Ariyan[1], G. Habibagahi[1a], E. Nikooee[1,2]

[1]*Department of Civil and Environmental Engineering, Shiraz University, Iran*
[2]*Environmental Hydrogeology Group, Utrecht University, the Netherlands*

Abstract. It is conventionally assumed in the analysis and design of earth dams that the soil located above the phreatic line, i.e. the uppermost seepage flow line, is completely dry. However, there is often an unsaturated flow of water through an unsaturated zone above this borderline and variation in moisture content in this zone results in variation of matric suction throughout this region. Variation of matric suction, in turn, results in variation of effective stresses in this zone. In this research, the seismic response of earth dams in terms of the displacement and acceleration at the crown of the dam as well as the stress distribution in the dam body is investigated. Taking into account the effect of unsaturated zone, a comparison is made to investigate the effect of conventional simplification in ignoring the dynamic characteristics of the unsaturated zone above the phreatic line and the more complicated analysis which includes the unsaturated zone. A function for the soil-water retention curve (SWRC) was assigned to the soil in the unsaturated zone to determine the variation of matric suction in this zone and analyses were made using finite difference software (FLAC). Results are then compared to the conventional method for homogeneous dams. In these analyzes the soil shear modulus was assumed to vary with the mean effective stress both for saturated and unsaturated zones. Among various results, it was notable that the history of crest x-displacement, and acceleration show higher values in models accounting for the unsaturated region. It was attributed to the considerably lower values of damping ratio in the crest region in the unsaturated models.

1 Introduction

For designing earth dams, analyzing the dynamic behavior of the dam against earthquakes is a major step. Common practice for analyzing the earth dams is to use the available numerical codes and commercial softwares. But these softwares mostly do not consider the unsaturated sate of soils.

Severe failures for dams subjected to earthquakes have been reported due to the various mechanisms such as liquefaction (e.g. the lower San Fernando Dam during 1971 San Fernando earthquake and damage of Mochikoshi Dam in 1978 Izu-Ohshim-Kinkar earthquake), slope failure and cracks leading to slope instability (i.e. Austrian Dam during 1989 Loma Prieta earthquake) and so on [1]. Although the multiphase physics of unsaturated region in dams can be very important for such failure mechanisms, there are very few studies on the seismic behavior of dams taking into account the unsaturated region above the phreatic line.

The existence of more than one fluid phase in unsaturated soils results in the capillarity phenomenon. The behavior of unsaturated soils is, therefore, a function of capillary action and the state of matric suction. Zienkiewicz et al. (1990) were first to study the dynamic behavior of unsaturated soils [2]. Their pioneering work has been followed by a couple of other studies in which different aspects of the unsaturated soils' response to the seismic excitations have been investigated [3-7]. Different constitutive equations and assumptions in these studies have been considered. For instance, Muraleetharan and Wei and Ravichandran and Muraleetharan developed the governing equations for the seismic behavior of unsaturated soils and used a two stress state approach [3,4]. Whereas, Uzuoka and Borja (2012), Shahbodagh–Khan et al. (2015), and Ghorbani et al. (2016) have used the effective stress approach and single stress framework [5-7]. It has been stated that single stress approach can eliminate the need for multiple constitutive parameters for nonlinear response of unsaturated soils. On the other hand, in the effective stress approach, the definition of the effective stress parameter has faced a debate in the recent years for which multiple formulations have been introduced. Thus, in dynamic analysis of the response of unsaturated soils, various studies have employed different assumptions for the effective stress parameter. For instance, Uzuoka and Borja (2012), and Ghorbani et al. (2016) have used degree of saturation as the effective stress parameter whereas Shahbodagh–Khan et al. (2015) have considered it to be a function of matric suction which takes the hydraulic hysteresis into account.

One importance question when applying such constitutive equations to large scale engineering problems is that whether considering the unsaturated state of the soil would affect the results of the analysis or not and how influential such considerations would be. One may ask how important is to take the unsaturated region into account. The importance of considering the difference in saturated and unsaturated soil behavior for designing earth dams has been stated before in the previous studies by Alonso and Pinyol (2008) and Alonso and Cardoso (2010) [8,9]. But the dynamic response of the earth dams considering the unsaturated state of the soil has not yet been comprehensively looked into. In particular the dynamic response of unsaturated earth dams with varying design parameters has not yet been examined in details.

Despite the slow but considerable and important progress in theoretical and numerical formulations for dynamics of unsaturated soils, the application of such models in engineering practice has faced less attention. There are

[a] Corresponding author: habibg@shirazu.ac.ir

only a few studies focusing on engineering projects such as slope and embankment. Mori et al. (2011) have modeled the dynamic behavior of a fill slope and have compared it to its monitored behavior and Matsumaru and Uzuoka (2015) have simulated the behavior of the shaking table tests on embankments [10-11]. The focus of these studies has, however, been mainly on the effects of different assumptions for pore air pressure field on the simulation results.

In order to apply unsaturated models for analysis of earth dams, more studies are required and different characteristics of dams such as different material and geometric parameters have to be considered.

To investigate the influence of considering unsaturated sate in dynamic analysis of earth dams, in this study the results of a numerical study have been presented and discussed. Dams with different heights have been considered and the differences between the dams' response to seismic excitation with and without assumption of unsaturated region above the phreatic line have been examined. In the subsequent sections, the details of the model, the theoretical background of the model and the results have been presented.

2 Model

2.1 Theoretical background

Numerical analysis were carried out using the finite difference program FLAC (Fast Lagrangian Analysis of Continua) which employs a continuum finite difference discretization based on the Langrangian approach.

In the analysis and design of earth dams, it is conventionally assumed that the soil located above the phreatic line, i.e. the uppermost seepage flow line, is completely dry. However, there is often an unsaturated flow of water through an unsaturated zone above this borderline and variation in moisture content in this zone results in the variation of matric suction throughout this region. Variation of matric suction, in turn, results in variation of effective stresses in this zone.

In order to consider dynamic properties of unsaturated zones in earth dams, the required constitutive equations were implemented in the numerical finite difference code FLAC2D using its built-in FISH language. The introduced functions were then assigned to the dam materials and seismic behavior of the dam was analysed in the plane strain condition.

Two different scenarios were used for modeling. At first similar to most of conventional studies, it was assumed that the uppermost seepage flow line, is completely dry (namely SAT-DRY condition) and in the second modeling scenario, unsaturated zone above this borderline was taken into account (SAT-UNSAT). The dynamic analysis was done in two steps. At first equivalent linear method was employed to specify shear modulus and damping ratio in each zone, then non-linear analysis was utilized to study the seismic behavior of the considered earth dams.

The Equivalent Linear method (E.L.) starts a dynamic analysis with the specified soil stiffness. In the first stage, FLAC processes the entire earthquake record and identifies the peak shear strains at each Gauss numerical integration point in each element. The shear modulus is then modified according to a specified G reduction function and the process is repeated. This iterative procedure continues until the required G modifications are within a specified tolerance.

Soils subjected to dynamic stresses tend to 'soften' in response to cyclic shear strain. In the Equivalent Linear soil model, this softening is described as a ratio relative to maximum shear modulus. This is called a G-reduction function. The cyclic shear strain is obtained from the finite element analysis. The computed shear strain together with the specified maximum shear modulus are used to compute new shear modulus values in each iteration.

Ishibashi and Zhang (1993) developed an expression for estimating the G-reduction function [12]. Where, the two main variables are PI (plasticity index) and confining pressure. Equation 1 describes the equation for G_{max} which is used in G-reduction procedure, the involved parameter in this equation are described in equations 2 to 5.

$$G_{\max} = 21.7 \times K_{2,\max} \times P_a \times \left(\frac{\sigma'_m}{P_a}\right)^{0.5} \tag{1}$$

$$\frac{G}{G_{\max}} = K(\gamma, PI) \times (\sigma'_m)^{m(\gamma,PI)-m_0} \tag{2}$$

$$K(\gamma, PI) = 0.5 \times \left\{1 + \tanh\left[\left(\frac{0.000102 + n(PI)}{\gamma}\right)^{0.492}\right]\right\} \tag{3}$$

$$m(\gamma, PI) - m_0 = 0.272\left\{1 + \tanh\left[\left(\frac{0.000556}{\gamma}\right)^{0.4}\right]\right\} \times \exp\left(-0.045 PI^{1.3}\right) \tag{4}$$

$$\begin{aligned} n(PI) &= 0.00 && for \ \ PI = 0 \\ n(PI) &= 3.37 && for \ \ 0 < PI < 15 \\ n(PI) &= 7.00 \times 10^{-7} PI^{1.976} && for \ \ 15 < PI < 70 \\ n(PI) &= 2.70 \times 10^{-5} PI^{1.115} && for \ \ PI > 70 \end{aligned} \tag{5}$$

Where G is shear modulus, G_{max} denotes max shear modulus, $K_{2,max}$ is a material coefficient which depends on porosity of soil (in our study, it was set as 60), P_a is atmospheric pressure (101 kPa), σ'_m is mean effective stress, γ specifies the shear strain (0.65 γ_{max}).

The damping ratio, ξ, is a function of the cyclic shear strain, the same as the G-reduction function. Previous studies of Ishibashi and Zhang (1993) have led to the development of an expression [12], which can be used to estimate the Damping Ratio function. The variables in the expression are the plasticity index (PI), the G modulus reduction ratio G/G_{max} and indirectly the confining pressure. The expression is as follows:

$$\xi = 0.333 \times \frac{1+\exp(-0.045PI^{1.3})}{2} \times \left[0.086\left(\frac{G}{G_{max}}\right)^2 - 1.547\left(\frac{G}{G_{max}}\right) + 1\right] \quad (6)$$

In order to account for the presence of unsaturated parts in dam, the following relationship for the effective stress has been used:

$$\sigma'_{yy} = \sigma_{yy} - u_a + \chi(u_a - u_w) \quad (7)$$

There are various formulations for the effective stress parameter. As discussed by Nikooee et al. (2013) and Pereira et al. (2010), degree of saturation is not a good candidate for the effective stress parameter, especially in fine grained soils [13, 14]. Recently Lu et al. (2010) suggested the use of effective degree of saturation for the effective stress parameter to account for the fact that the immobile water does not contribute to the mechanical behavior of unsaturated soils [15]. Furthermore, Oh and Lu (2014) have shown that the use of effective degree of saturation results in a reasonable approximation of suction stress under different hydraulic and mechanical states [16]. Therefore, in this study the effective stress parameter, χ, was assumed to be effective degree of saturation.

$$\chi = \frac{S_w - S_r^w}{1 - S_r^w} \quad (8)$$

Where σ_{yy} is vertical stress. The term $(u_a - u_w)$ denotes soil matric suction. u_a and u_w are pore air pressure and pore water pressure, respectively. S_w is wetting degree of saturation and S_r^w is the residual saturation.

For the Gmax function, the required value of the mean stress was computed based on the equations (7) to (10).

$$\sigma'_{xx} = K_0\sigma'_{yy}$$
$$\sigma'_{zz} = K_0\sigma'_{yy} \quad (9)$$
$$\sigma'_m = \frac{\sigma'_{yy} + \sigma'_{xx} + \sigma'_{zz}}{3}$$

$$K_0 = \left(\frac{\nu}{1-\nu}\right) \quad (10)$$

Where earth pressure coefficient, K_0, is defined based on Poisson's ratio, ν.

Shear strength expressed as a function of effective normal stress, and the shear strength parameters, namely, the cohesion intercept (c') and the angle of shearing resistance (ϕ').

$$\tau = c' + \sigma'_{yy}\tan\phi' \quad (11)$$

2.2 Initial conditions, and characteristics of the modelled dams

Initial condition:

In the numerical analyses, the initial condition of stress and pore pressure were considered to be a steady state condition at which earthquakes are more possible. The steady state condition was reached after a layer by layer construction analysis followed by an impounding analysis.

In fact, to have numerical stability, the element size (spatial element size), ΔL , has been suggested to be smaller than one-tenth to one-eighth of the wavelength associated with the highest frequency component of the input wave [17]. In first step of dynamic analysis, the initial value of Rayleigh damping was considered to be 3%. Then by means of fish language and programming in FLAC in each step, the damping ratio was updated based on the shear modulus ratio and plasticity index (Equation 6). Furthermore in each step, the effective stress value for the unsaturated case was calculated and used to find the value of maximum shear modulus, G_{max}. Then shear strain and G_{max} were used to update the value of the shear modulus. For this purpose, programming was performed using fish language and the required subroutine were implemented in the FLAC.

Dam characteristics:

The heights of the simulated dams were considered as 35m, 50m and 70m, respectively. Their upstream and downstream slopes were invariably 1:3.5, and foundation was considered to be 12m deep. Table 1 presents the dam soil properties. The hydraulic properties (Van Genuchten parameters) were obtained from SoilVision database, They are presented in Table 2. Where a, nvg, ad mvg are material coefficients (constant parameters), ws and wr are

DOI: 10.1051/e3sconf/20160908002

saturated and residual volumetric water content, respectively and ψ denotes matric suction.

$$w(\psi) = w_r + (w_s - w_r)\left[\frac{1}{\left[1+(a\psi)^{ng}\right]^{mg}}\right] \quad (12)$$

As mentioned earlier, to investigate the seismic response of earth dams considering dynamic properties of unsaturated zone, dam modes were analyzed in two different cases. First the dams were modeled without considering dynamic properties of unsaturated zone, therefore above water surface, soil elements were dry. In the second case the dam models were investigated while the more complicated analysis which includes the unsaturated zone was implemented. This could help to understand the effect of conventional simplification in ignoring the dynamic characteristics of the unsaturated zone above the phreatic line. For this purpose, the above mentioned function for the soil-water retention curve (SWRC) was assigned to the soil elements in the unsaturated zone to determine the variation of matric suction in this zone. The modified G-reduction for unsaturated case which considers the values of skeleton stress in unsaturated cases was also implemented.

Table 1. Earth dam body parameters

Material parameters	Type	Unit	Value
Dry density	ρd	kg/m^3	1420
Elastic modulus	E	MPa	35
Poisson's ratio	υ	-	0.3
Cohesion	C	kPa	0
Friction angle	φ	Deg	30
Permeability	K	m/s	5.8×10^{-7}
Porosity	N	-	0.5
Plastic Index	PI	-	10

The accelerogram recorded during the 1978 Tabas earthquake is used in all examples. Its duration is 30 sec and its maximum acceleration is about 0.8g. To minimize errors, this accelerogram has been adjusted according to the base line corrections. In addition this record was normalized to 0.3g and only the downstream-upstream component has been considered. Fig. 1 shows the corrected accelerogram. It has been applied uniformly along the base of the foundation.

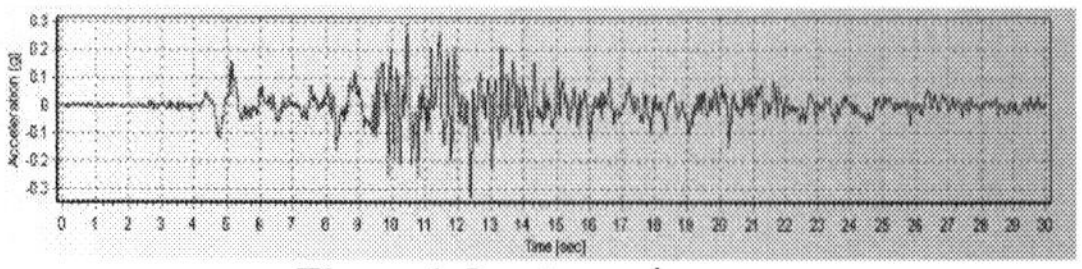

Figure 1. Input accelerogram

3 Results

Figure 2 presents the variation of the shear stress in two aforementioned scenarios and for dams of different heights. As it can be seen the shear stress in unsaturated case is higher due to the increase in the density of soil elements above water table which results in an increase in total stress. The increase in the total stress as well as the presence of the negative pore water pressure lead to the increase of the effective stress and accordingly an increase in shear stress values.

Water table (phreatic line) in dams with 35m, 50m and 70m is located in the elevation of 24m, 36m and 54m, respectively. Fig 3. depicts the shear modulus profiles of the earth dams with height. Under the ground water table in both cases, shear modulus decrease with increasing elevation but above water table (phreatic line), shear modulus increases in saturated-unsaturated scenario. The reason is that shear modulus has a direct relation with the effective stress. Fig 4. Presents the damping ratio profiles. As it can be seen damping ratio has a reverse concordance with soil stiffness represented by G/Gmax ratio. Above phreatic line effective stress increases in unsaturated condition due to suction effect leading to an increase in G/Gmax, and hence, reduction in damping ration of the soil. For the saturated dry condition, there is no suction effect and the decrease in effective stress with elevation due to surchage reduction controls the dam behavior (see also equation 6 for further insight).

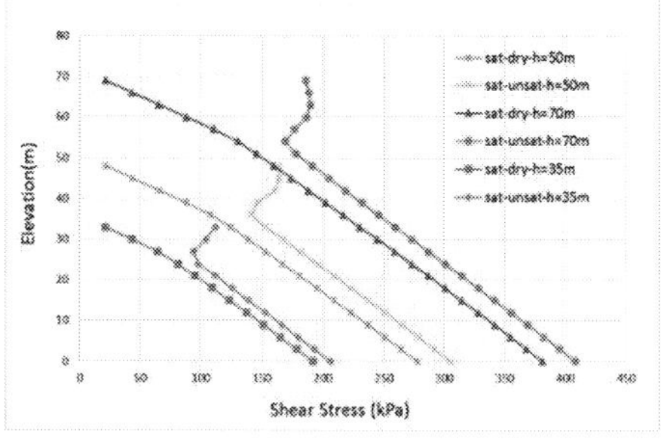

Figure 2. The variation of shear stress for 3 different heights for two modeling scenarios (dry-sat and unsat-sat models)

Table 2. Van Genuchten parameters

Van Genuchten parameters	value
a_{vg}	0.02
n_{vg}	0.54
m_{vg}	0.25
w_r	11.40%

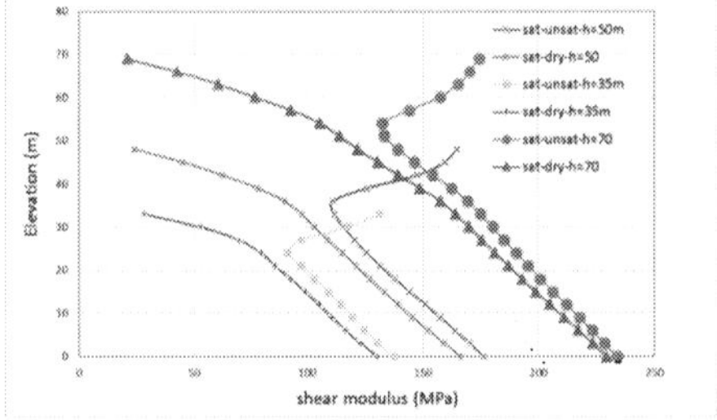

Figure 3. Shear modulus versus height in two modeling scenarios

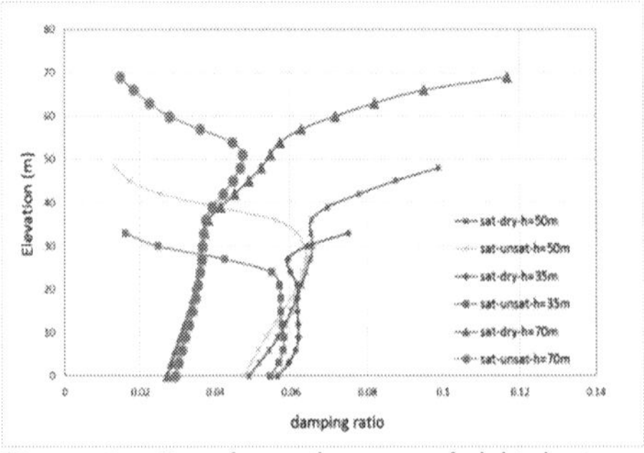

Figure 4. Damping ratio versus height in two modeling scenarios

For the reason that the soil elements in the crest region of the dam (above the phreatic lime) are unsaturated and have less damping ratio than dry case, x-displacement of unsaturated model are higher than the dry case (Fig 5. and Fig 6). Fig 7. and Fig 8. present the history of x-acceleration in dams with different elevations where greater amount especially in the peak x-acceleration in unsaturated models is observed. In addition, y-displacement of crest is less than the dry case (dry-sat scenario) because of the increase in shear modulus of the unsaturated zone (Fig 9. and Fig 10.)

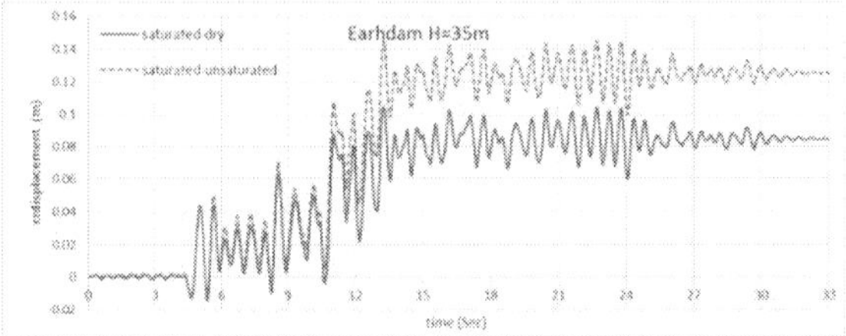

Figure 5. x-displacement history for the dam with 35 m height in two modeling scenarios

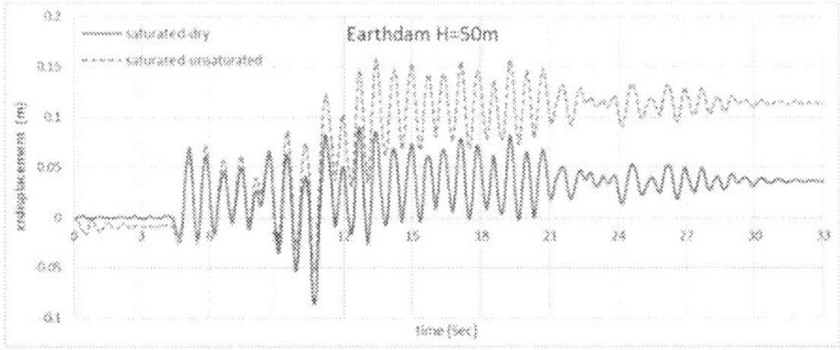

Figure 6. x-displacement history for the dam with 50 m height in two modeling scenarios

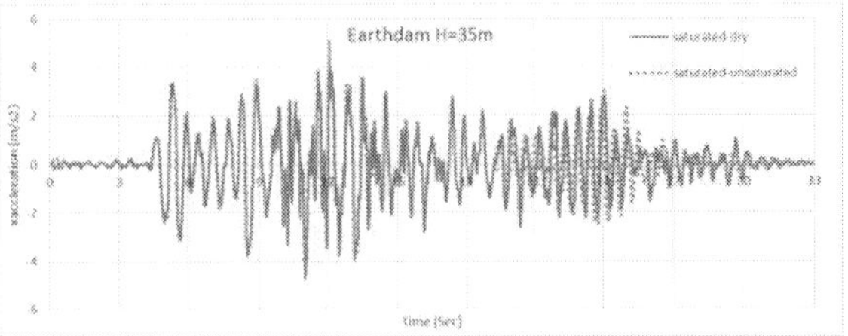

Figure 7. Crest acceleration history for the dam with 35 m height in two modeling scenarios

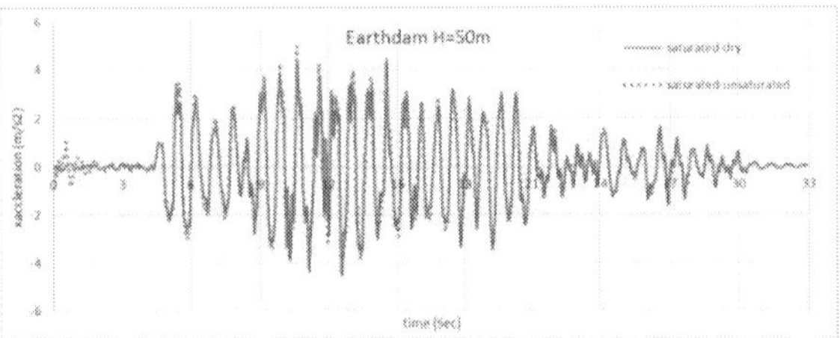

Figure 8. Crest acceleration history for the dam with 50 m height in two modeling scenarios

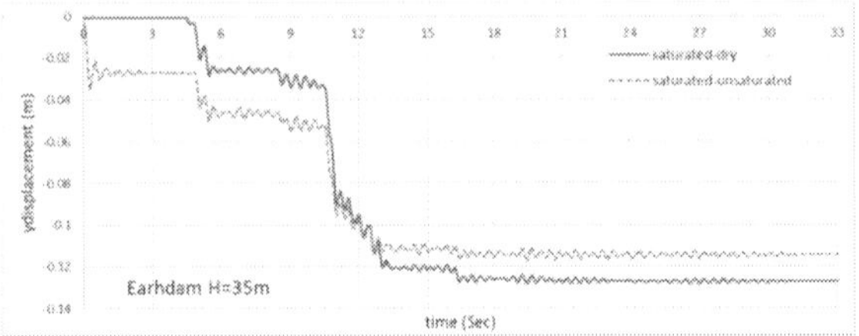

Figure 9. Crest y-displacement history for the dam with 35 m height in two modeling scenarios

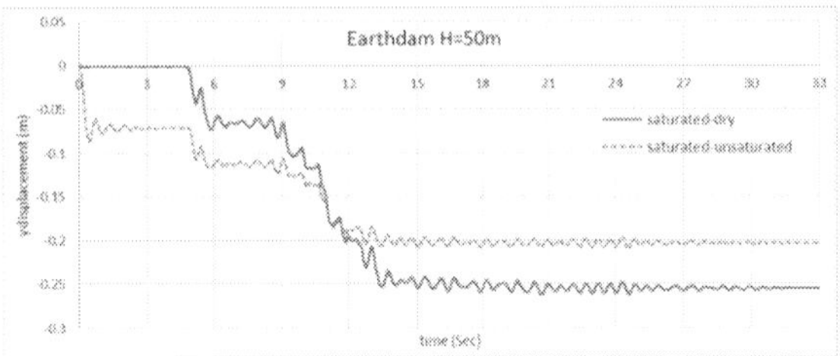

Figure 10. Crest y-displacement history for the dam with 50 m height in two modeling scenarios

4 Concluding remarks and future perspective

In this study a numerical model for seismic behaviour of unsaturated soils was presented. The modelling revealed that

- The shear stress values are increased in unsaturated-saturated scenario as compared to dry-saturated models due to the presence of negative pore pressure and the increase in the effective stress.
- The influence of suction and pore pressure on the soil shear modulus is noticeable. The shear modulus which is directly proportional to the effective stress and has a reverse relationship to the shear strain increases above water table in the unsat-sat models

contrary to the conventional modelling which neglects the unsaturated zone above phreatic line.

- The damping ratio for unsat-sat models above phreatic line is reduced by height.
- The history of crest x-displacement, and acceleration show higher values in unsat-sat models following the considerably lower values of damping ratio in the crest region.

Following two recommendations can be considered for future studies:

- In order to be able to have a good insight on seismic behavior of dams considering unsaturated region therein, it is recommended to explore various design variables and parameters. The effect of foundation stiffness, cross sectional shape and different dam body material parameters can be considered for the future studies.
- Each study carries specific assumptions which can be released or investigated in future researches. The effect of different formulations for soil water retention curve as well as the effective stress parameter can be looked into in the future studies.

References

1. K. I. Andrianopoulos, A. G. Papadimitriou, G. D. Bouckovalas, D. K. Karamitros, Insight into the seismic response of earth dams with an emphasis on seismic coefficient estimation. *Computers and Geotechnics*, **55**, 195-210 (2014).
2. O. C. Zienkiewicz, Y. M. Xie, B. A. Schrefler, A. Ledesma, N. Bicanic,. Static and dynamic behaviour of soils: a rational approach to quantitative solutions. II. Semi-saturated problems. In Proceedings of the Royal Society of London A: Mathematical, Physical and Engineering Sciences. The Royal Society. **429**, 311-321 (1990)
3. K. K. Muraleetharan, C. Wei, Dynamic behaviour of unsaturated porous media: Governing equations using the theory of mixtures with interfaces (TMI). International journal for numerical and analytical methods in geomechanics, ***23*(13)**, 1579-1608 (1999).
4. N. Ravichandran, K. K. Muraleetharan, Dynamics of unsaturated soils using various finite element formulations. International journal for numerical and analytical methods in geomechanics, ***33*(5)**, 611-631 (2009).
5. R. Uzuoka, R. I. Borja, Dynamics of unsaturated poroelastic solids at finite strain. International Journal for Numerical and Analytical Methods in Geomechanics, **36(13)**, 1535-1573 (2012).
6. B. Shahbodagh-Khan, N. Khalili, G. A. Esgandani, A numerical model for nonlinear large deformation dynamic analysis of unsaturated porous media including hydraulic hysteresis. Computers and Geotechnics, **69**, 411-423 (2015).
7. J. Ghorbani, M. Nazem, J. P Carter, Numerical modelling of multiphase flow in unsaturated deforming porous media. Computers and Geotechnics, **71**, 195-206 (2016).
8. E.E. Alonso, N.M. Pinyol, Unsaturated soil mechanics in earth and rockfill dam engineering. In, Proceedings of the 1st Europe Conference E-UNSAT, Durham, United Kindom. London: Taylor and Francis, 3–32 (2008).
9. E.E. Alonso, R. Cardoso, Behavior of materials for earth and rockfill dams. Perspective from unsaturated soil mechanics, Front. Archit. Civ. Eng. China, **4(1)**, 1–39 (2010).
10. T. Mori, R. Uzuoka, T. Chiba, K. Kamiya, M. Kazama, Numerical prediction of seepage and seismic behavior of unsaturated fill slope. Soils and foundations, **51(6)**, 1075-1090 (2011).
11. T. Matsumaru and R. Uzuoka, Two-phase and three-phase coupled analysis of embankment affected by seepage water and earthquake, *Computer Methods and Recent Advances in Geomechanics*, 1379-1384 (2015).
12. I. Ishibashi, X. Zhang, Unified Dynamic Shear Moduli and Damping Ratios of Sand and Clay, Soils and Foundations, **33**, 182-191 (1993).
13. E. Nikooee, G. Habibagahi, S. M. Hassanizadeh, A. Ghahramani, Effective stress in unsaturated soils: A thermodynamic approach based on the interfacial energy and hydromechanical coupling. Transport in porous media, **96(2)**, 369-396 (2013).
14. J. M. Pereira, O. Coussy, E. E. Alonso, J. Vaunat, S. Olivella, Is the degree of saturation a good candidate for Bishop's X parameter? *Proceedings of the 5th International Conference on Unsaturated Soils 2. DOI: 10.1201/b10526-141* (2010).
15. N. Lu, J. W. Godt, D. T. Wu, A closed-form equation for effective stress in unsaturated soil. Water Resources Research, **46(5)**, (2010).
16. S. Oh, N. Lu, Uniqueness of the suction stress characteristic curve under different confining stress conditions. Vadose Zone Journal, **13(5)**, (2014).
17. R. L. Kuhlemeyer, J. Lysmer, Finite Element Method Accuracy for Wave Propagation Problems, J. Soil Mech. & Foundations, Div. ASCE, **99**(SM5), 421-427 (1973).

The impact of soil suction variation on earthquake intensity indices

Mahnoosh Biglari[1, a], Farshad Fouladi[2] and Iman Ashayeri[1]

[1] *Assistant Professor, Razi University, Kermanshah, Iran*
[2] *Graduate student, Razi University, Kermanshah, Iran*

Abstract. Soil properties can completely change the ground motion characteristics as they travel from the bedrock to the surface because, soil as a low-pass filter, may amplify or deamplify seismic motions in some frequencies on the wave travelling path. Recent studies about the advanced unsaturated soil mechanics clearly shows that dynamic properties of soils, including small-strain shear modulus (G_{max}), shear modulus reduction (G/G_{max}), and damping ratio (D) curves are affected by changes in the soil suction level. The current study present nonlinear time-dependent analysis of three different unsaturated soils available in the literature with different ranges of nonlinear behaviour that earlier have been studied on unsaturated dynamic models. Since, the earthquake intensity parameters can be used to describe the damage potential of an earthquake, the focus of this paper is to evaluate the impact of the suction variation on the engineering ground motion parameters, including peak values of strong motion, V_{max}/A_{max}, root-mean-square acceleration, Arias intensity, characteristic intensity, cumulative absolute velocity, acceleration spectrum intensity, effective design acceleration, A95 parameter and predominant period separately under the near-field and the far-field seismicity categories.

1 Introduction

Suction affects the behaviour of soil by modifying the skeleton stress through changes in the average fluid pressure acting in the soil pores and providing an additional bonding force at the particle contacts [1], which are a major role in determination of dynamic properties of soil, such as small-strain shear modulus, normalized modulus reduction and damping ratio (G/G_0–γ and D-γ) curves. One finds that [2] presents the investigation basis of the linear ground response analysis by considering the mean net stress and suction on the initial shear stiffness (G_0) and damping ratio (D_0) of soils with the different plasticity indices. They have discussed the amplification function in terms of the amplification ratio of the surface motion and concluded that the natural frequency of a soil deposit significantly increases with the suction increase and the maximum amplification ratio is substantially reduced. More recently, [3-5] investigated the linear, equivalent linear and nonlinear seismic ground response analysis of several unsaturated soil deposits with considering the suction effects on G/G_0-γ and D-γ curves that demonstrated increasing of suction level led to increasing natural frequency. Also, it was found that the responses of the structures on the unsaturated soil profiles are more severe than the saturated ones, particularly at the higher frequencies. The difference in the dynamic material properties by changing of the suction, consequently distinct the characteristics of output motions in comparison with the input motion (Fig 1). Therefore, representation of the seismic site response of soil deposits with considering the suction would be indispensable where the water table level rises up and falls down extremely. In particular, because of the recent drought, changing of the suction profoundly affects previous calculations of the soil response especially for important structures.

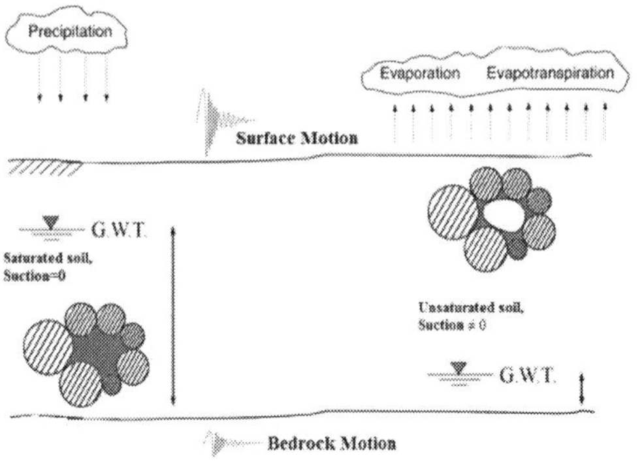

Figure 1. A visualization of differences of the output motion in the saturated and the unsaturated soils.

In this study, three different low plasticity fine grained soils that earlier have been studied on the unsaturated dynamic models are used to produce some various profiles with a variety of suction for analysis of their time domain seismic responses. The paper is studying the

[a] Mahnoosh Biglari: m.biglari@razi.ac.ir

influence of the suction variation on the seismic ground response analysis and investigates separately the effect of suction in the near field and the far field input earthquake events on strong ground motion parameters with using a computer program of DEEPSOIL.V6 that performs analyses in the frequency (linear and equivalent linear) and in the time (linear and nonlinear) domains.

2 Material properties

Three different soils, namely Zenoz kaolin, Completely Decomposed Tuff (CDT) and Po silt chosen for building of soil profiles. Zenoz kaolin as reported by [6] is a commercial Iranian kaolin from a mine in northwest Iran. It is classified as CL according to the Unified Soil Classification System (USCS). Clay and silt fraction of Zenoz kaolin (Zk) is about 18% and 60%, respectively. CDT as reported by [7] extracted from a deep excavation site in Fanling, Hong Kong can be described as clayey silt (ML) according to the USCS. The material is yellowish-brown, slightly plastic, with a grain size distribution of 24, 74 and 4 for sand, silt and clay, respectively. Po silt (PO) as reported by [8] is a clayey, slightly sandy silt and was extracted from the banks of the Po River (Italy) with a clay fraction of about 27%, and a silt fraction of about 40%. Some of the available information for the physical properties of the mentioned soil samples (i.e. ZK, Po and CDT) summarized in Table1.

Table 1. Index properties of soils used in the definition of dynamic properties of profiles.

Parameters	Zenoz kaolin	CDT	Po silt
Specific gravity	2.65	2.73	2.74
Liquid limit(%)	29	43	51
Plastic limit(%)	17	29	33
Plasticity index (%)	12	14	18
Classification (USCS)	CL	ML	ML/MH

The study were considered eleven soil profiles (four soil profiles of ZK in the suction levels of 0, 50, 150 and 300 kPa, four soil profiles of CDT in the suction levels of 0, 50, 100 and 200 kPa and three soil profiles of PO in the suction levels of 100, 200 and 400 kPa) with constant suction in depth of the profile. All of the soil profiles have 24 meter depth overlaying the stiff elastic half-space bedrock (of course it should be noted that the bedrock damping ratio has no effect in time domain analyses) with 24 same-thickness sub layers in order to satisfy the propagation condition of highest frequency in the soil layers ($f_{max} = V_s / 4H$) especially in the stiffer soils.

A plot of the shear modulus (MPa) versus the unit weight (kN/m^3) of all of the layers and all of the profiles are displayed (Fig 2). As shown in this figure for soil profiles of CDT and ZK, variation of the shear modulus and the unit weight with the suction has a greater sensitivity rather than PO. Furthermore, there is a huge difference between shear modulus variation and unit weight ranges for PO & CDT with ZK.

3 Modelling procedure

3.1 Small-strain shear modulus (Gmax)

To obtain the shear modulus at very small strain, empirical formulation of [1] (Eq. 1) which is based on the framework suggested by [9] is used.

$$G_0 = Ap_{atm}^{1-n} (p'')^n (OCR_{p''})^m f(e) h(s_r) \quad (1)$$

Where $h(s_r)$ is a non-dimensional function as Eq. (2), f(e) is a void ratio function ($f(e)=\{(2.973-e)^2/(1+e)\}$), $OCR_{P''}$ is the overconsolidation ratio and in the normally-consolidated conditions is equal to 1, P_{atm} is the atmospheric pressure, P″ is the average skeleton stress ($p'' = (p-u_a)+S_r.s$), s is suction equal to (u_a-u_w), S_r is the degree of saturation, u_a is the air pressure, u_w is the water pressure, p is the average total stress, ξ is the bonding variable defined as Eq.(3) where f(s) is a function that depends on the size of the particles and the value of the water surface tension. The value of f(s) was considered equal to 1 for the range of suctions in this study.

$$h(s_r) = 1 - a'\left[1 - \exp\left(b' \frac{\xi}{f(s)}\right)\right] \quad (2)$$

$$\xi = f(s)(1 - S_r) \quad (3)$$

The other model parameter values for CDT and PO were determined according to the previous laboratory research of [10]. Therefore, predicting of the shear modulus at very small strain according to [1] model in variation of suction and mean net stress for PO and CDT such as ZK, which was previously available became possible.

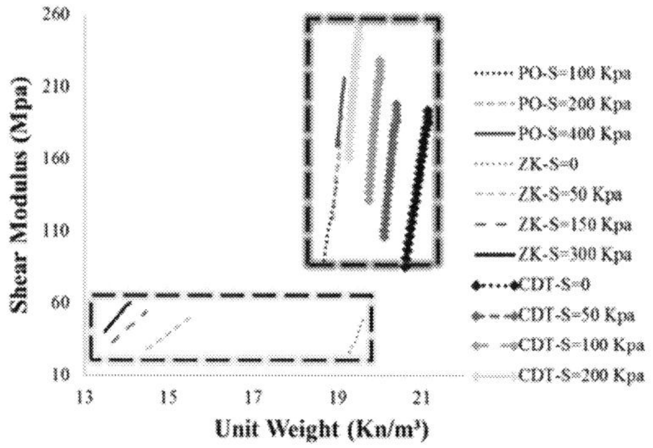

Figure 2. Variation of the unit weight with the suction.

3.2 Normalized shear modulus and damping ratio curves

[11] modified the empirical equation proposed by [12] to take into account the influence of the suction level as well as the stress state and the index properties for an unsaturated soils. Equations 4-7 represent the empirical equations for $G/G_0-\gamma$ and $D-\gamma$ curves.

$$\frac{G}{G_0} = A(\gamma, \xi, PI)\left(\frac{p''}{p_{atm}}\right)^{(n(\gamma, PI) - n_0)} \quad (4)$$

$$A(\gamma,\xi,PI)=0.5\left[1+\tanh\left\{\ln\left[\left(\frac{0.00005+0.0167\xi^{12.16}+f(PI)}{\gamma}\right)^{\left[0.26+3.61\xi^{11.6}\right]}\right]\right\}\right] \quad (5)$$

$$f(PI)=\begin{cases} 0 & \text{for } PI=0 \ (\text{sandy soils}) \\ 3.37\times10^{-6}\,PI^{1.404} & \text{for } 0<PI<15 \ (\text{low plastic soil}) \\ 7.0\times10^{-7}\,PI^{1.976} & \text{for } 15<PI<70 \ (\text{medium plastic soils}) \\ 2.7\times10^{-5}\,PI^{1.115} & \text{for } 70<PI \ (\text{high plastic soils}) \end{cases} \quad (6)$$

$$n(\gamma,PI)-n_0=0.272\left[1-\tanh\left\{\ln\left[\frac{0.000556}{\gamma}\right]^{0.4}\right\}\right]e^{(-0.0145PI^{1.3})} \quad (7)$$

$$D=f\left(\frac{G}{G_0}\right)A(PI)=\left[0.358\left\{-0.11\left(\frac{G}{G_0}\right)^2-0.587\left(\frac{G}{G_0}\right)+1\right\}\right]\left[\frac{1+e^{-0.0145PI^{1.3}}}{2}\right] \quad (8)$$

Where A (γ, ξ, PI) is the stiffness index ratio defined as Eq. 5, n (γ, PI) as defined in Eq. 7, is a stiffness coefficient accounts for the effect of p″ on the stiffness, n_0 is a stiffness coefficient accounts for the effect of p″ on the stiffness in the small strain range and PI is the soil plasticity index.

3.3 fitting process

In this paper, MRDF model was used for determining the best fit curves of shear modulus and damping ratio. This model proposed a formulation that modifies the loading-unloading criteria that result from using the Masing rules [13]. The new formulation provides better agreement with the damping curves for larger shear strains [14].

4 Input excitations

A series of 28 ground motion time history records with considering of near field and far field effects are selected from the pacific earthquake engineering research center database (PEER) as listed in Table 2. The largest horizontal component of each record was used as the input excitation. The ground motions are normalized to a target peak acceleration of 0.1g and the baseline correction technique applied for the all of the acceleration time histories. Figure 3 displays the scatter plot of the M (moment magnitude) and R (rupture distance) of the records.

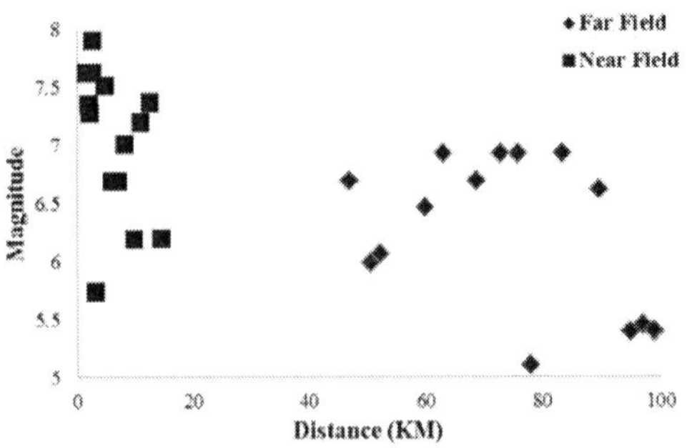

Figure 3. Details (magnitude and distance) of the input motions.

Table 2. Details of the input motions.

Earthquake Name	Year	Station Name	Magnitude	Rrup (km)
Cape Mendocino	1992	Petrolia	7.01	8.18
Chi-Chi, Taiwan-03	1999	TCU076	6.2	14.66
Coyote Lake	1979	Gilroy Array	5.74	3.11
Denali, Alaska	2002	TAPS Pump Station	7.9	2.74
El Mayor-Cucapah, Mexico	2010	CERRO PRIETO GEOTHERMAL	7.2	10.92
Kocaeli, Turkey	1999	Yarimca	7.51	4.83
Manjil, Iran	1990	Abbar	7.37	12.55
Morgan Hill	1984	Gilroy Array	6.19	9.87
Northridge-01	1994	LA Dam	6.69	5.92
Northridge-01	1994	Pacoima Dam (downstr)	6.69	7.01
Northridge-01	1994	Pacoima Dam (upper left)	6.69	7.01
Tabas, Iran	1978	Tabas	7.35	2.05
Chi-Chi, Taiwan	1999	TCU076	7.62	2.74
Chi-Chi, Taiwan	1999	TCU102	7.62	1.49
14383980	2008	Green Oak Ranch	5.39	98.95
14383980	2008	Antelope	5.39	94.97
Big Bear-01	1992	Rancho Cucamonga - Deer Can	6.46	59.87
40204628	2007	Bolinas	5.45	97
Kozani, Greece-02	1995	Grevena	5.1	78.11
Loma Prieta	1989	Piedmont Jr High School Grounds	6.93	73
Loma Prieta	1989	Point Bonita	6.93	83.45
Loma Prieta	1989	SF - Pacific Heights	6.93	76.05
Loma Prieta	1989	So. San Francisco, Sierra Pt.	6.93	63.15
N. Palm Springs	1986	Anza - Tule Canyon	6.06	52.06
Northridge-01	1994	Anacapa Island	6.69	68.93
Northridge-01	1994	Antelope Buttes	6.69	46.91
San Fernando	1971	Cedar Springs, Allen Ranch	6.61	89.72
Whittier Narrows-01	1987	Vasquez Rocks Park	5.99	50.39

5 Results

This study reviews numbers of the earthquake intensity indices in dynamics analysis of the unsaturated sediment, according to the input motions with one-dimensional nonlinear approach. To eliminate the effect of variability of the earthquake record and also consider the pure soil deposit effect, the amounts of the indices in any suction level at the surface of the profiles are normalized to the initial value (base of the profile). Then to evaluate the correlation between the existing earthquake intensity indices and the seismic responses of the unsaturated profiles with varying of the suction the mean values of each intensity parameters is calculated separately for the near-field and the far-field input excitation seismicity categories. In order to highlight the influence of the suction on the nonlinear seismic response of soil layers, soil profiles were considered with the different material properties and nonlinearities.

A wide variety of earthquake intensity indices have been presented. Some of them only describe the amplitude of motion and the other ones just interpret frequency content or duration. Also, another type of the indices exists that is influenced by involving of two or three of these important ground motion characteristics. Hereunder, some of the earthquake intensity indices in terms of the amplitude, the frequency content and the combined parameters separately have been defined. And at the end of the brief description of the parameters, the impact of the suction on the index has been evaluated.

5.1 Amplitude parameters

Peak ground acceleration (PGA) is one of the important parameter of ground motion and is defined as the absolute maximum amplitude of acceleration time history. Although, PGA is a poor indicator of the measure of damage potential of ground motion; due to its simplicity and compatibility with many seismic codes, PGA is widely used in the engineering applications.
Peak ground velocity (PGV) is the peak value of the velocity time history, which is popularly applied to characterize the damage potential cause of its sensitivity to the intermediate frequencies of a motion. In the Japanese building code, PGV is used as an index in the seismic analysis.
Effective Design Acceleration (EDA) corresponds to the peak acceleration value found after low pass filtering the input time history with a cut-off frequency of 9 Hz [16]. Since, pulses of high acceleration at the high frequencies almost induce little response, the notion of the effective design acceleration, has been proposed [15].
Mean ratio values of amplitude parameters (PGA, PGV and EDA) for ZK profiles in any distances from the earthquake source have a direct relation with the suction increasing. This will attract the attentions to the unsaturated soft soil in intermediate to high frequencies. For stiffer soil profiles, namely PO and CDT subjected to far field and near field earthquakes PGV reduced with suction increasing. While, sensitivity of PGA and EDA to suction in these stiff soil profiles in dealing with near or far field events have a completely reverse trend (Figs 4-9).

5.2 Frequency content parameters

V_{max}/A_{max} is an intensity index that involves the peak ground acceleration (PGA) and the peak ground velocity (PGV). Since for a harmonic motion, this ratio is equal to $T/2\pi$. For an earthquake motions, V_{max}/A_{max} can be interpreted as the period of an equivalent harmonic motion, thus providing an indication of which periods of the ground motion are most significant. Because the peak velocity and the peak acceleration are usually associated with motions of different frequency the ratio is related to the frequency content of the motion [15].
Predominant Period (T_p) is the period corresponds to the peak Fourier amplitude. This parameter represents the frequency content of the motion.
The two above-mentioned indices only describe the frequency content of the motion. As shown in figs 4-9 for all of the soil profiles at any suction levels, seismicity categories and the non-linearity domain, the frequency content absolutely increases with the suction increasing even in the soft soil profiles.

5.3 Combined parameters

Acceleration Spectrum Intensity (ASI) is defined as the integral of the pseudo acceleration spectral (S_a) over the period range of 0.1 to 0.5 Sec [17].

$$ASI=\int_{0.1}^{0.5} Sa(T,5\%)dT \tag{9}$$

As seen in the Figures 4-9, the mean ratio of the present index absolutely shows a same general tendency versus the suction variation. The index reflected amplitude and the frequency content of motion in short periods and insistently states potentially dangerous to suction increasing.
Cumulative absolute velocity (CAV) is defined as the integral of the absolute value of ground acceleration over the seismic time-history record [15].

$$CAV=\int_0^t |a(t)|dt \tag{10}$$

The trend of this index to the suction variation (Figs 4-9) is independent of distance from the earthquake source and only is the function of the dynamic properties of the soil profiles. The soil profiles with more non-linearity (i.e. ZK) in comparison with the stiffer profiles (i.e. PO & CDT) show a thoroughly opposite trend in the suction variation.
Arias Intensity (I_a) is defined by [18]:

$$I_a=\frac{\pi}{2g}\int_0^{\infty} a(t)^2 dt \tag{11}$$

Where I_a is the Arias Intensity in units of length per time, a(t) is the acceleration-time history and g is gravitational acceleration. Arias Intensity is an earthquake intensity index that correlates with the potential destructiveness of an earthquake.
Root-mean-square acceleration (a_{RMS}) is a measure of the average rate of energy imparted by the ground motion and is defined as:

$$a_{RMS}=\sqrt{\frac{1}{t}\int_0^t [a(t)]^2 dt} \tag{12}$$

Where a(t) is the acceleration time history and t is the total duration of the ground motion. This parameter is often useful for the engineering purposes because it incorporates the effect of duration and it is not strongly influenced by large, high-frequency accelerations, which typically occur only over a very short period. However, a_{RMS} does not provide any information about the frequency content because it is the sum of the input energy at all frequencies. Obviously, a_{RMS} depends on the method used to define strong-motion duration.

Characteristic intensity (I_c) proposed [19] and given by

$$I_c = a_{RMS}^{1.5} . t_s^{0.5} \quad (13)$$

Where t_s is the significant duration of the ground motion. Significant duration, t_s, is defined by [20] as the interval between the times at which5% and 95% of the Arias intensity is attained. The proposed intensity index has proven to correlate well with the damage potential and can be considered a reasonable representation of the destructive ground motions.

A95 parameter is another earthquake intensity index that was proposed by [21] and is the maximum value of the acceleration that corresponds to 95% of the Arias Intensity value. However, this index being less popular among the researchers, A95 such as the Arias Intensity summarizes the ground motion parameters included in amplitude, frequency content and duration in a single parameter. Four above-mentioned parameters (i.e. I_a, a_{RMS}, I_c and A95) have a close relationship together. So, it is expected that the parameters have a same tendency to the suction variation. As shown in the Figures 4-9 variation of the mean ratios of these parameters verify aforesaid statement. The tendency of these indices to the suction is sensitive to the both soil non-linearity and the distance from event source. Interpretation of the following plotted data about the indices revealed that, in the soft soil profiles (i.e. ZK) with any distance from the source, increasing in suction leads to increasing in damage potential. Also, for more dense profiles (i.e. PO & CDT), subjected to the far Field earthquakes this relationship is established.

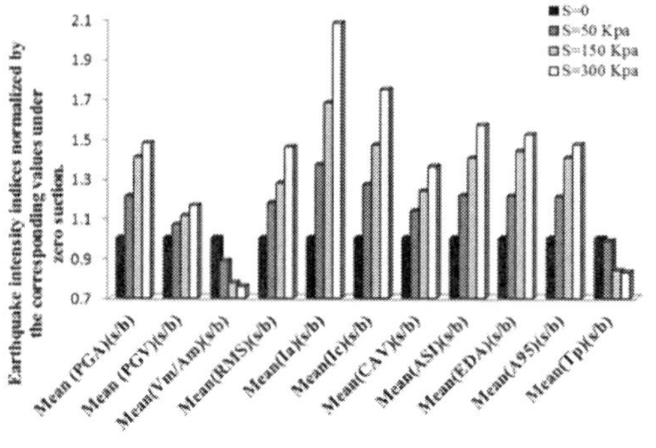

Figure 4. Normalized mean intensity indices ratios of surface (s) of ZK profiles to bedrock (b) under far field events.

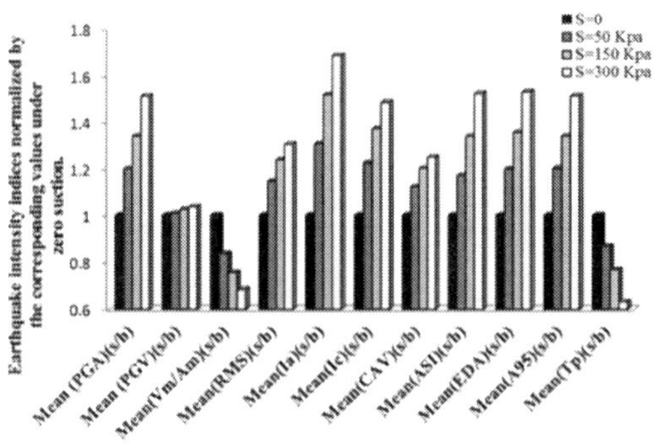

Figure 5. Normalized mean intensity indices ratios of surface (s) of ZK profiles to bedrock (b) under near field events.

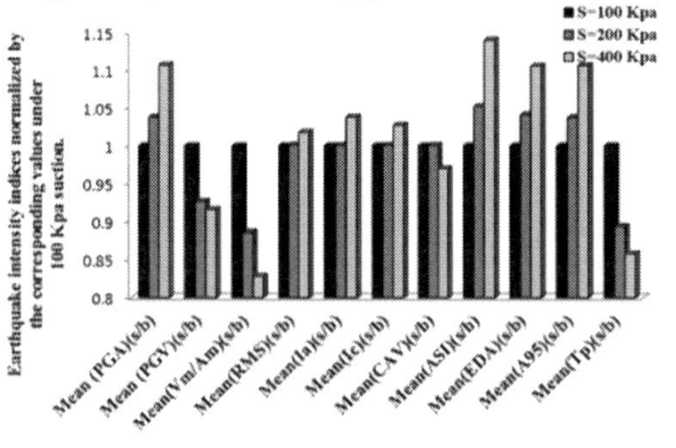

Figure 6. Normalized mean intensity indices ratios of surface (s) of PO profiles to bedrock (b) under far field events.

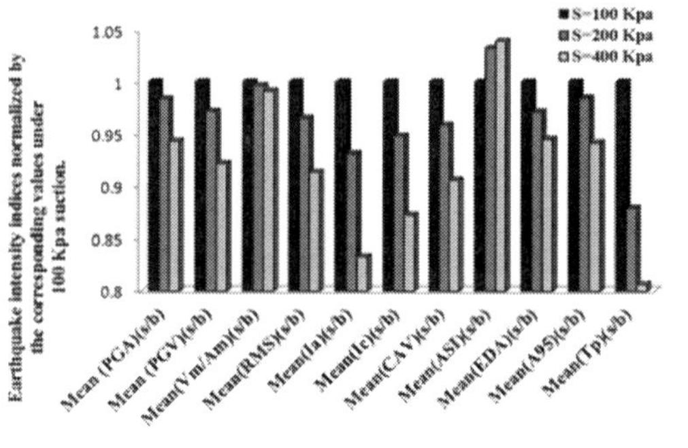

Figure 7. Normalized mean intensity indices ratios of surface (s) of PO profiles to bedrock (b) under near field events.

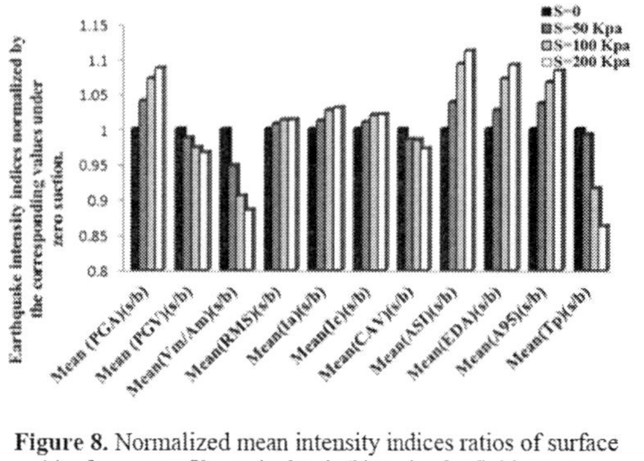

Figure 8. Normalized mean intensity indices ratios of surface (s) of CDT profiles to bedrock (b) under far field events.

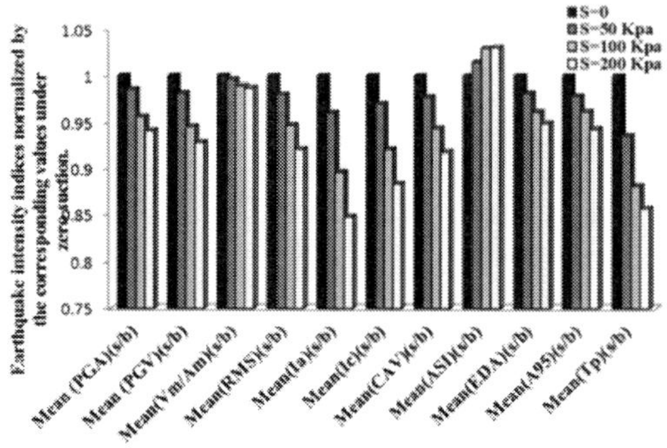

Figure 9. Normalized mean intensity indices ratios of surface (s) of CDT profiles to bedrock (b) under near field events.

6 Conclusions

In the present paper, the nonlinear site response analyses were performed for three unsaturated soil samples with varying of the suction and the different mechanical and physical properties. To avoid the interference of the different distance from the epicenter to the analyzed results, 28 near field and far field ground motions are selected. In order to emphasize on the influence of the suction variation on the characteristics of strong ground motion, a series of earthquake intensity indices, which capture the effects of amplitude, frequency content and duration of a ground motion record was investigated. The results are plotted separately as the mean ratio of each earthquake intensity index on the top to the bottom of the soil profiles at each suction level for the near field and the far field input motions. Finally, to clearly show differences at each suction level, the mean ratios are normalized to the corresponding values at the minimum suction. Response of the different soil samples (i.e. ZK, Po and CDT) shows that the dynamic response of soils in addition of the suction, is extremely affected by the important parameters in determining of the soil nonlinearity such as, plastic index, void ratio and mean net stress. Hence, some conclusions are listed below from the analysis results of the suction variation versus the earthquake intensity indices.

• In the softer profiles (i.e. ZK), the higher suction levels definitely increase damage potential of the ground motion. Also, the tendency of the suction variation to the earthquake intensity indices has no correlation with the distance from the epicentre.
• Two indices that (V_{max} /A_{max} and Predominant Period) only describe the frequency content of the motion indicates that for all of the soil profiles at any suction levels, seismicity categories and the non-linearity domain, the frequency content absolutely increases with the suction increasing even in the soft soil profiles.
• The other indices that indicate damage potential of earthquake, for CDT and PO revealed the opposite trend subjected to near field and far field. So that, potential danger of the suction increasing in the far field earthquakes, in spite of the near field ground motion will be increasing. It should be noted that for these dense soils, PGV decreased with the suction increasing in any distance from epicenter. Since, the velocity is to characterize ground motion amplitude at intermediate frequency for structures that are sensitive to loading in intermediate frequency range, suction increasing reduced potential for damage.

References

[1] M. Biglari, C. Mancuso, A. d'Onofrio, M.K. Jafari, A. Shafiee, Comp. and Geotech., **38**(5), 709-720, (2011).
[2] F. d'Onza, A. d'Onofrio, C. Mancuso, *1st European Conf on Unsat. Soils*, 531-536, (2008).
[3] M. Biglari, I. Ashayeri, Inter. Jour. of Civil Eng., **11**(2), 150-155, (2013).
[4] M. Biglari, I. Ashayeri, F. Fouladi*6th Inter. Conf. on Unsat. Soils*, 389-396, (2014).
[5] F. Fouladi, M. Biglari, I. Ashayeri, *7th Inter. Conf. on Seismology and Earthquake Eng.*, (2015).
[6] M. Biglari, A. d'Onofrio, C. Mancuso, M.K. Jafari, Can Geotech J, **49**(3), 311-322, (2012).
[7] C.W.W. Ng, S.Y. Yung, Géotech., **58**(1), 23-35, (2008).
[8] R. Vassallo, C. Mancuso, F. Vinale, Can. Geotech. J, **44**(4), 447-462, (2007).
[9] D. Gallipoli, A. Gens, R. Sharma & J. Vaunat, Géotech., **53** (1), 123–135, (2003).
[10] K.S. Wong, D. Mašín, C.W.W. Ng, Comp. and Geotech., 56, 28-39, (2014).
[11] M. Biglari, I. Ashayeri, *5th Asia-Pacific Conf on Unsat. Soils*, 591-596, (2012).
[12] I. Ishibashi, X. Zhang, Soil and Found., **33**(1), 182-191, (1993).
[13] G. Masing, *2nd Int Congress on App. Mech.*, 332-335, (1926).
[14] C. Phillips, Y.M.A. Hashash, Soil Dyn. and Earth. Eng., **29**(7), 1143-1158, (2009).
[15] S.L. Kramer, *Geotechnical earthquake engineering*, Prentice Hall, Upper Saddle River, New Jersey, (1996).
[16] JR. Benjamin. EPRI Report NP-5930, Electric Power Research Institute, Palo Alto, California, (1988).
[17] J. L. Von Thun, L. H. Rochim, G. A. Scott, and J.A. Wilson, ASCE, New York, 463–481, (1988).
[18] A. Arias, Massachusetts Inst. of Tech., Cambridge. Univ. of Chile, Santiago de Chile, 438–483, (1970).
[19] Y.J. Park, A.H.S. Ang, Y.K. Wen. Jour. of Struct. Eng., ASCE **111**(4), 740–757, (1985).
[20] M. D. Trifunac and A. G. Brady., Bull. of the Seismo. Society of America, **65**(3), 581-626, (1975).
[21] S.K. Sarma, K.S. Yang. Earth. Eng. and Struct. Dyn., **15**(1), 119–132, (1987).

DOI: 10.1051/e3sconf/20160908004

Soil-atmosphere interaction in unsaturated cut slopes

Aikaterini Tsiampousi[1], Lidija Zdravkovic[1] and David M. Potts[1]

[1]*Imperial College London, Department of Civil and Environmental Engineering, SW7 2AZ, London, UK*

Abstract. Interaction between atmosphere and soil has only recently attracted significant interest. Soil-atmosphere interaction takes place under dynamic climatic conditions, which vary throughout the year and are expected to suffer considerable alterations due to climate change. However, Geotechnical Analysis has traditionally been limited to simplistic approaches, where winter and summer pore water pressure profiles are prescribed. Geotechnical Structures, such as cut slopes, are known to be prone to large irreversible displacements under the combined effect of water uptake by a complex vegetation root system and precipitation. If such processes take place in an unsaturated material the complexity of the problem renders the use of numerical analysis essential. In this paper soil-atmosphere interaction in cut slopes is studied using advanced, fully coupled partially saturated finite element analyses. The effect of rainfall and evapotranspiration is modelled through sophisticated boundary conditions, applying actual meteorological data on a monthly basis. Stages of low and high water demand vegetation are considered for a period of several years, before simulating the effect of vegetation removal. The analysis results are presented with regard to the serviceability and stability of the cut slope.

1 Introduction

It has long been shown that pore water pressure variation affects the stability and serviceability of infrastructure slopes (cut and embankment slopes). Pore water pressures may vary due to consolidation or due to seasonal variations in the water balance under the combined effect of evapotranspiration and rainfall infiltration. It is possible that the two mechanisms act simultaneously in a number of cases. The complexity of the problem is such that use of numerical tools is essential.

Additional complications arise when the soil is unsaturated, as the behaviour of unsaturated soils is fundamentally different to the behaviour of fully saturated soils. The implications of partial saturation may be significant and Geotechnical structures in unsaturated soils may exhibit the reverse behaviour of what is expected in saturated soils. For example, Tsiampousi et al. [1] have shown that the factor of safety against failure of slopes excavated in unsaturated soils does not generally decrease with time, as in saturated soils (see fig. 1). Due to their unpredictability, careful numerical analyses are of essence when unsaturated soils are involved.

The numerical analysis of a slope excavated in an unsaturated soil is presented in this paper. Pore water pressures change in the analysis due to the combined effect of consolidation and soil-atmosphere interaction. Different stages of vegetation type and density are considered and the associated effects on the stability and serviceability of the slope are studied. The finite element (FE) code ICFEP [2], which adopts a modified Newton-Raphson solution technique with an error controlled sub-stepping stress point-algorithm, was used. First, boundary conditions relevant to soil-atmosphere interaction are briefly explained. The boundary value problem to be solved is then presented in detail and the results of the analysis are discussed. The analysis results suggest that dense, high water demand vegetation enhances the stability of the slope in expense of serviceability. Vegetation clearance, however, may induce slope instability.

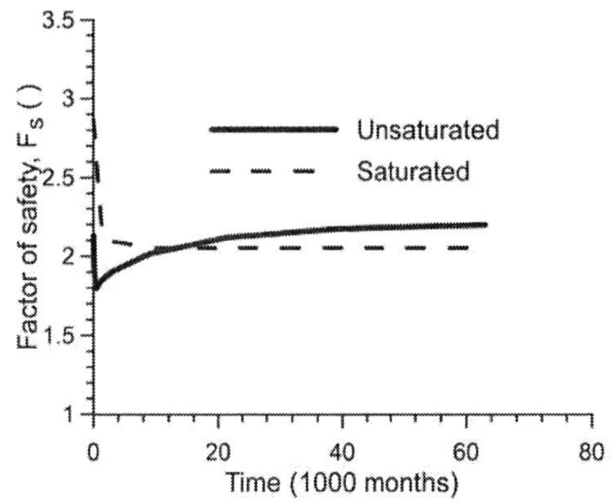

Figure 1. Variation of the factor of safety against failure of a slope cut in saturated and unsaturated soil (after[1]).

[a] Corresponding author: aikaterini.tsiampousi@imperial.ac.uk

 DOI: 10.1051/e3sconf/20160908004

2 Boundary Conditions

2.1 Vegetation

Extraction of water due to vegetation is modelled in ICFEP with the root water uptake model (RWUM), described by Nyambayo & Potts [3]. The potential evapotranspiration rate, T_p, needs to be defined for each increment of the FE analysis. From this, a sink term, S_{max}, is calculated and incorporated in the continuity equation of fluid flow [2]. S_{max} is assumed to vary linearly with depth, r, as shown in Figure 2 (a), according to the following relationship:

$$S_{max} = \frac{2 \cdot T_p}{r_{max}} \cdot \left(1 - \frac{r}{r_{max}}\right) \quad (1)$$

where r_{max} is the root depth, which is an input parameter and below which root water uptake is zero. The above equation relates to how much water could potentially be extracted if the supply of moisture in the ground were constant. To calculate the actual evapotranspiration, which is a fraction of the potential one, the sink term is multiplied by a suction-dependent parameter α [4]. The variation of α with suction is illustrated in Figure 2 (a). Suctions S1 (anaerobiosis point), S2, S3 and S4 (wilting point) are input parameters.

2.2 Precipitation

The precipitation boundary condition is a dual boundary condition which enables the simulation of rainfall infiltration and runoff; it may operate as an infiltration condition, by specifying a constant inflow rate, q_n, or as a constant prescribed pore water pressure condition, p_{fb} [5]. Values of q_n and p_{fb} have to be prescribed for every increment of the analysis that the boundary condition is to be active. At the beginning of such increments the pore water pressure at boundary nodes is compared to p_{fb}: if found to be more tensile, an infiltration boundary condition is imposed, with a flow rate equal to q_n; if found to be more compressive, a constant pore water pressure equal to p_{fb} is applied. As conditions may change during an analysis increment and a switch from one boundary condition to the other may be required, an automatic incrementation algorithm (described in Smith, 2003) is available in ICFEP.

2.3 Factor of safety calculation

To assess the stability of the cut slope, the factor of safety (FoS) against failure was calculated with the strength reduction technique explained in Potts & Zdravkovic [6]. Furthermore, the variation of the FoS with time was obtained following a similar approach to the one used by Tsiampousi et al. [1]: secondary analyses were initiated at appropriate increments of the main fully coupled analysis but consolidation was switched off and pore water pressures were not allowed to change (drained analyses). In this way, the FoS was obtained for a particular time instance. To obtain the variation of FoS with time the same process was repeated at various, appropriately selected time instances.

3 Problem definition

3.1. Geometry

A 10 m deep excavation in an unsaturated silty clay was considered (fig. 3a). The geometry of the cut slope was discretised using isoparametric, quadrilateral 8-noded solid elements, resulting in the FE mesh shown in Figure 3 (b). The problem was considered in plane strain in a fully coupled consolidation analysis. Two displacement degrees of freedom were assigned at each node of each element of the FE mesh. Pore water pressure degrees of freedom were only assigned to corner nodes.

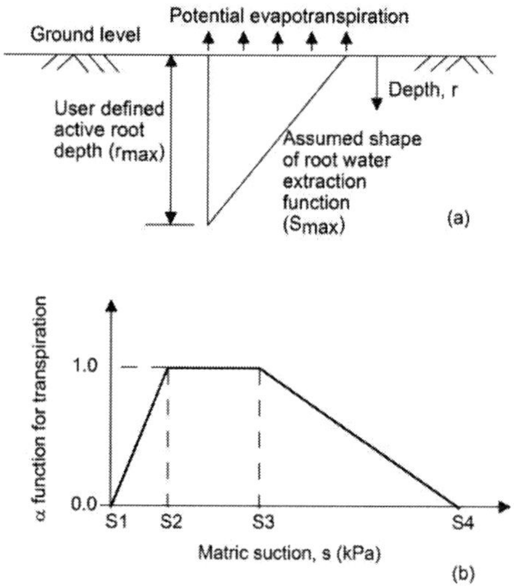

Figure 2. (a) Water extraction function when α=1; (b) Variation of the α function with suction (after Nyambayo & Potts, 2010).

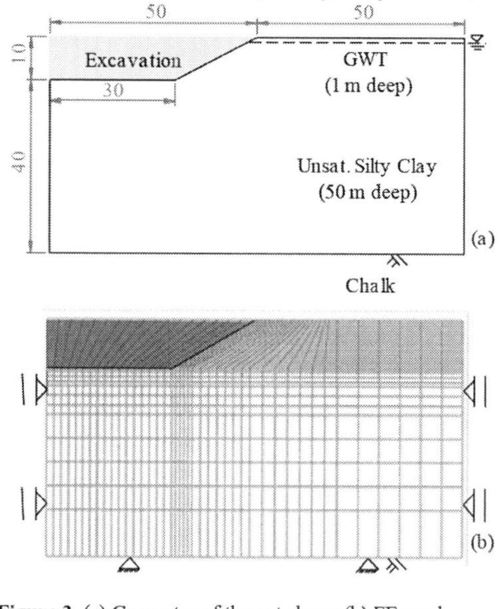

Figure 3. (a) Geometry of the cut slope; (b) FE mesh.

3.2 Soil properties and model parameters

The Barcelona Basic model [7], with modifications by Georgiadis et al. [8] and Tsiampousi et al. (2013-HV), was used to simulate the mechanical behaviour of the soil (model parameters in Table 1).The same properties were employed in a similar study by Tsiampousi et al. [9] and correspond to a silty soil tested by Estabragh & Javadi [10]. For the hydraulic behaviour, a van-Genuchten (1980) type soil-water retention curve (SWRC) was used, in combination with a permeability model in which the logarithm of permeability varies linearly with suction, from an initial saturated value, k_{sat}, corresponding to suction s_A, to a final value k_{min}, corresponding to suction s_B (Table 2). Finally, the actual transpiration properties adopted are summarised in Table 3.

Table 1. Constitutive model parameters.

Parameter	Value	Parameter	Value
α_g, α_f	0.7	$\lambda(0)$	0.086
μ_g, μ_f	0.9999	κ	0.005
M_g, M_f	0.9838	v_1	2.12
α_{HV}	0.6	p^c	1.0
n	0.5	r	0.06
β_{HV}	0.25	β	0.001
m	0.5	λ_s	0.3
s_{air} (kPa)	0.0	κ_s	0.08
s_0 (kPa)	1000	p_{atm} (kPa)	100
K_{min} (kPa)	300	k	S_r from the SWRC
G/p_0	15	OCR	5.5

Table 2. Hydraulic model parameters.

SWRC Parameter	Value	Permeab. Parameter	Value
s_{air} (kPa)	0.0	k_{sat} (m/s)	10^{-8}
α	0.015	k_{min} (m/s)	10^{-10}
m	0.95	s_A (kPa)	0.0
n	0.70	s_B (kPa)	1000
S_{r0}	0.10		

Table 3. Actual transpiration properties.

Parameter	Value	Parameter	Value
s_1 (kPa)	0.0	s_3 (kPa)	50.0
s_2 (kPa)	5.0	s_4 (kPa)	1500.0

3.3 Initial stresses

At the commencement of the analysis, soil stresses were initialised employing a unit weight of 19.1 kN/m^3 above and below the ground water table (GWT). The coefficient of earth pressure at rest K_0 was 2.1 at the surface, reducing to 0.6 at 15m below ground level. The GWT was 1 m deep (see also fig. 3a) and the pore water pressure profile was hydrostatic, with suctions developing above the phreatic surface.

3.4 Climate data

The necessary precipitation and potential evapotranspiration data were obtained from a meteorological station in Greenwich, London (Figure 4). They refer to average long-term monthly data obtained from 1971 to 2000.

As discussed in the following section, the analysis is divided into five phases, each considering a different stage of vegetation, varying from low to high water demand, and finally simulating clearance of the vegetation on the slope. For this reason, the vegetation boundary condition was altered appropriately in each phase, with regard to both the root depth and the potential evapotranspiration rates. Similarly, the precipitation boundary condition varied between phases (most notably during the excavation phase). In total, the precipitation and vegetation boundary conditions (BC) were employed in four different manners (precipitation and vegetation BC1 to BC4 in Table 4). For simplicity, the climate data in Figure 4 formed the basis for all the rates applied in the five phases, as summarised in Table 4, e.g. the potential evapotranspiration rates T_p in vegetation BC2 were reduced to 50% of those in Figure 4 in order to simulate low water demand plants.

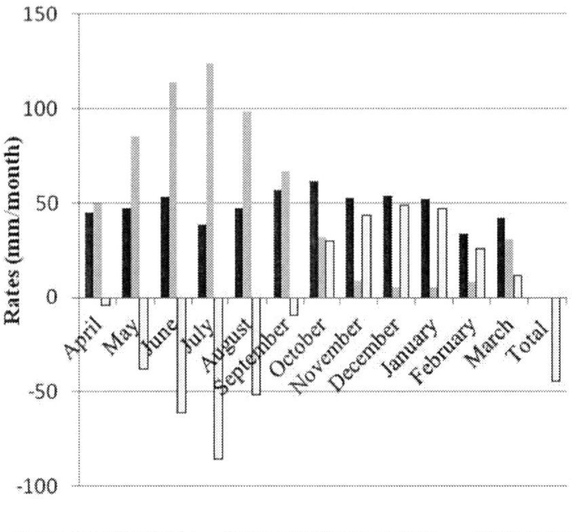

Figure 4. Precipitation and potential evapotranspiration data.

 DOI: 10.1051/e3sconf/20160908004

3.5 Vegetation stages

3.5.1 Initialisation

The first phase of the analysis refers to the time prior to the excavation, assuming that the ground surface was covered in dense, high water demand vegetation. Its purpose is to establish a pore water pressure and stress regime which would have been affected by the precipitation and evapotranspiration, prior to the excavation. Figure 5 (a) illustrates the boundary conditions applied on all four boundaries of the FE mesh during this phase. The area to be excavated subsequently is highlighted in grey. The climate boundary conditions were applied on the top horizontal boundary of the mesh. Information about the vegetation and precipitation boundary conditions can be found in Table 4. It should be highlighted that the precipitation and evapotranspiration rates correspond to those in Figure 4. They were applied monthly and the typical year, starting in April (as in Figure 4), was repeated 9 times to reproduce 9 years. This time was found to be adequate for an annual pore water pressure variation, repeating itself year after year, to be established.

3.5.2 Excavation

The excavation was performed at the beginning of the 10th year of the analysis, in a period of time short enough for essentially undrained conditions to be maintained. The hydraulic boundary conditions were altered to those illustrated in Figure 5 (b). The changes in the precipitation boundary condition are of particular interest, notably the fact that q_n was switched to zero (see also Table 4). Additionally, the vegetation boundary condition was inactive during this phase. The applied boundary conditions ensure that any further changes of the pore water pressures during the excavation were solely due to unloading and not to soil-atmosphere interaction.

Table 4. Vegetation and precipitation boundary conditions (BC).

	Precipitation BC			
	BC1	**BC2**	**BC3**	**BC4**
q_n	As in Fig. 4	0.0	0.0	50% of rates in Fig. 4
p_{fb}	10 (suction)	0.0	10 (suction)	10 (suction)
	Vegetation BC			
	BC1	**BC2**	**BC3**	**BC4**
T_p	As in Fig. 4	50% of rates in Fig. 4	As in Fig. 4	10% of rates in Fig. 4
Root depth (m)	2.0	Increasing from 0.1 to 0.5 m in 3 years	Increasing from 0.5 to 2 m in 5 years	0.1

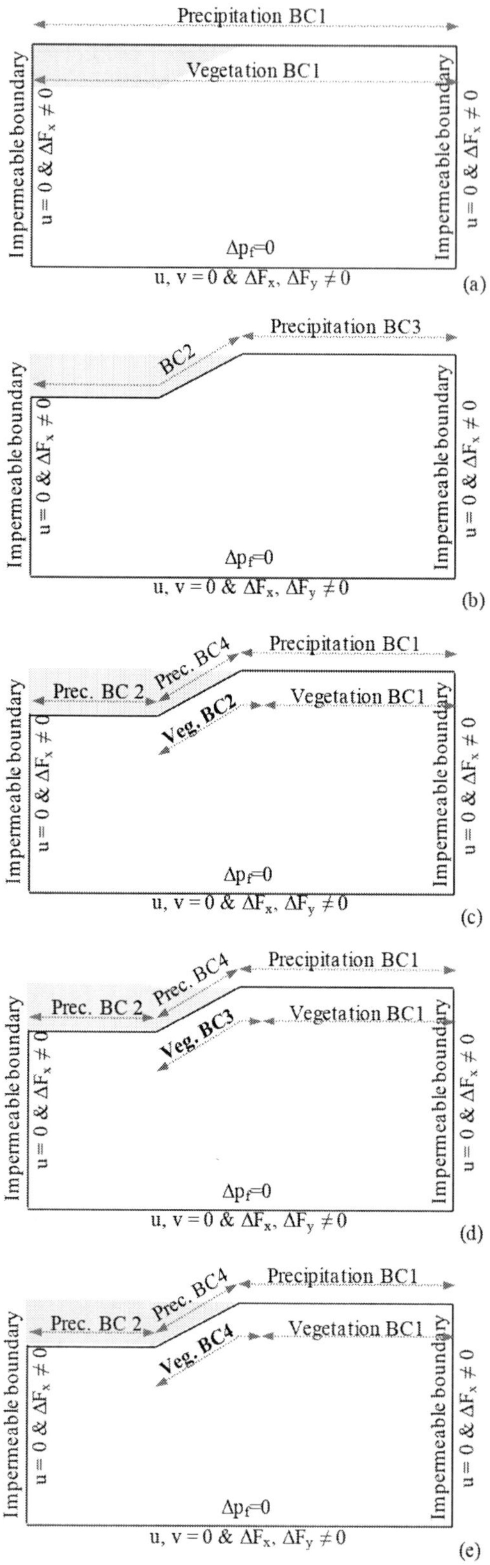

Figure 5. Boundary conditions during: (a) initialisation; (b) excavation; (c) low water demand vegetation; (d) high water demand vegetation; (e) vegetation clearance.

3.5.3 Low water demand vegetation

The subsequent phase refers to years 10 to 19 of the analysis, i.e. it extends for 10 years following the excavation. The vegetation and precipitation boundary conditions were reactivated, as illustrated in Figure 5 (c). Dense vegetation was preserved at a distance of 10m from the slope crest (vegetation BC1 in Table 4). The remaining area and up to a vertical distance of 1.5m from the bottom of the excavation was assumed to be covered in low water demand vegetation (e.g. shrubs and bushes). This corresponds to vegetation BC2 in Table 4.

The precipitation boundary condition on the horizontal ground beyond the crest of the slope was restored to its form prior to the excavation (i.e. precipitation BC1). On the slope only 50% of the precipitation rates in Figure 4 were applied (precipitation BC4), assuming that a drainage system was in place, capable of capturing and removing 50% of rainfall. Finally, at the bottom of the excavation the precipitation rate was set to zero. No further changes were made to the precipitation boundary condition for the remainder of the analysis.

3.5.4 High water demand vegetation

At the beginning of the 20th year of the analysis, growth of high water demand trees on the slope is simulated. The potential evapotranspiration rates applied on the slope are equal to those in Figure 4, with the root depth increasing gradually from 0.5 to 2m over a period of 5 years (vegetation BC3 in Table 4). This and the remaining boundary conditions are schematically explained in Figure 5 (d). Overall, this phase of high water demand vegetation on the slope lasted for a total of 10 years (i.e. years 20 to 29 of the analysis).

3.5.5 Vegetation clearance

Finally, at the end of March of Year 29 the vegetation on the slope was cleared and replaced by low and sparse vegetation (vegetation BC4 in Table 4).The total duration of the final stage of the analysis was once more 10 years (i.e. years 30 to 39 of the analysis).

4 Analysis results

4.1 Stability

The FoS was calculated bi-annually, at the end of March and August, for the 30 years following the excavation. Its variation with time, measured in months from the end of excavation, is shown in Figure 6, where the symbols correspond to results obtained from FoS analyses.

With reference to Figure 6 and for the low water demand vegetation stage, the values of FoS for August are systematically higher than the values for March. However, an overall decrease in the FoS is observed, which can be attributed to the combined effect of slope-atmosphere interaction and the dissipation of tensile excess pwp's developed during undrained excavation. By the end of this phase of the analysis, the values of FoS have more or less stabilised and oscillate about the value of 1.4. This is possibly due to the fact that average monthly rainfall and potential evapotranspiration rates were applied year after year. Although evapotranspiration data do not vary greatly from year to year, this is not so for rainfall data, which exhibit a large annual variation. It is reasonable to expect that a particularly dry year would have enhanced the stability of the slope, whereas a particularly wet year could have brought it to failure. This cannot be captured when average rainfall data are employed.

The FoS visibly increased during the next phase, where the slope was covered in high water demand vegetation. The enhanced stability can be justified by the larger suctions that have developed in comparison to the previous phase, as expected, and which contribute to a strength increase. Additionally, the values of FoS for August are systematically higher than the values for March, however, the overall increasing trend observed signifies that the beneficial effect of suction build-up during the summer months is not entirely lost during the subsequent winter months.

When the vegetation was cleared from the slope, a steady and fast decrease in the FoS was obtained. Rainfall rates are now high enough to quickly eliminate the suctions that have developed in the previous phase. It is interesting to note that the FoS no longer increases temporarily during summer and within five years the stability of the slope becomes marginal.

The formation of a slip surface about 4.5 m deep is indicated by the horizontal displacement profiles along a vertical line at mid-slope, shown in Figure 7. These displacements are sub-accumulated from the end of the previous phase and therefore represent the displacements developed due to vegetation clearance.

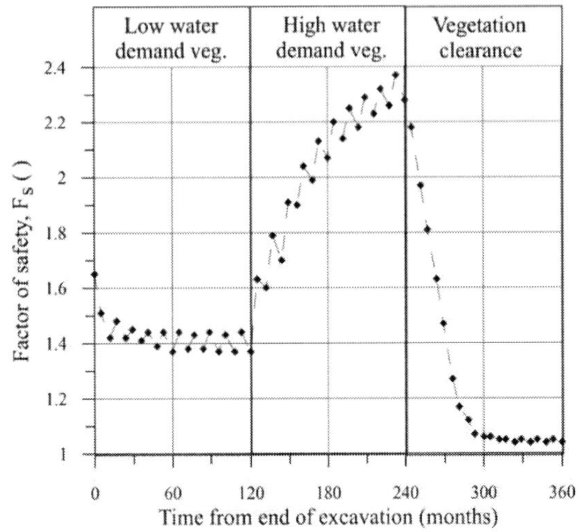

Figure 6. Factor of safety variation with time.

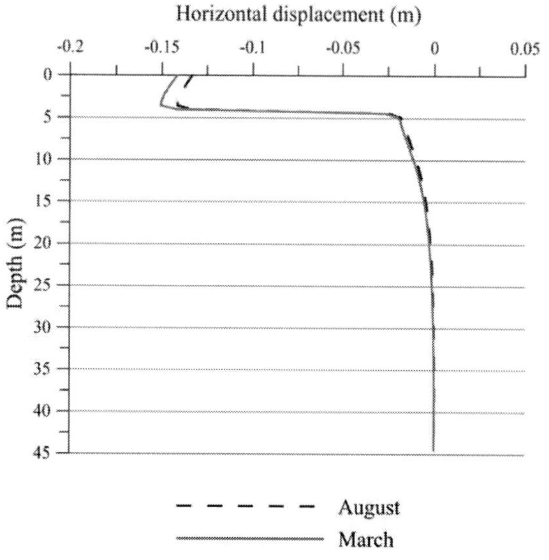

Figure 7. Horizontal displacements with depth at mid-slope for Year 34 (vegetation clearance phase).

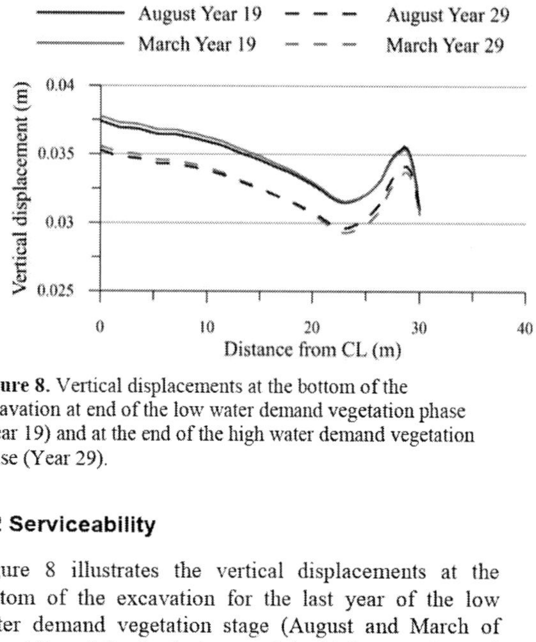

Figure 8. Vertical displacements at the bottom of the excavation at end of the low water demand vegetation phase (Year 19) and at the end of the high water demand vegetation phase (Year 29).

4.2 Serviceability

Figure 8 illustrates the vertical displacements at the bottom of the excavation for the last year of the low water demand vegetation stage (August and March of Year 19) and for the last year of the high water demand vegetation stage (August and March of Year 29). The displacements are sub-accumulated from the end of the excavation. Heave has taken place during the low water demand vegetation phase, as indicated by the curves corresponding to Year 19. Additionally, it can be observed that during the high water demand vegetation phase (Year 29) the ground settles in comparison to the previous phase (Year 19).

There is little difference between the August and March displacement profiles for both Year 19 and 29. Furthermore, although the displacements are not uniform across the bottom of the excavation the maximum difference is only about 5 mm. Serviceability problems have been reported in highly vegetated slopes excavated in fully saturated soils. The discrepancy may be due to the increased stiffness unsaturated soils generally exhibit.

5 Conclusions

This paper presents the results of a numerical study on the effect of soil-atmosphere interaction on the behaviour of a slope cut in an unsaturated soil. Under the combined effect of vegetation, precipitation and excess pore water pressure dissipation, the factor of safety exhibited an overall decrease when the slope was covered with low water demand vegetation. A seasonal fluctuation in the factor of safety was nonetheless visible. A significant increase in the overall factor of safety was obtained when high water demand vegetation was established on the slope. This type of vegetation is known to cause serviceability problems in fully saturated soils. However, differential settlements in the present analysis remained small, possibly due to the increased stiffness unsaturated soils generally exhibit. Finally, vegetation clearance was shown to cause stability issues, with the factor of safety reducing fast and a deep seated failure mechanism developing within five years.

References

1. A. Tsiampousi, L. Zdravkovic, D. M. Potts, Comp. Geotech. **48**, 2, 167-178 (2013)
2. D. M. Potts & L. Zdravkovic, *Finite element analysis in geotechnical engineering: theory* (London, Thomas Telford, 1999)
3. V. P. Nyambayo & D. M. Potts, Comp. Geotech. **37**, 175-186 (2010)
4. R. A. Feddes, P. J. Kowalk & H. Zaradny, *Simulation of field water use and crop yield* (New York, John Wiley & Sons, 1978)
5. P. G. Smith, *Numerical analysis of infiltration into partially saturated slopes* (PhD thesis, Imperial College London, 2003)
6. D. M. Potts & L. Zdravkovic, Géotech. **62**, 1053-1065 (2012)
7. E. E. Alonso, A. Gens, A. Josa, Géotech. **40**, 3, 405-430 (1990)
8. K. Georgiadis, D. M. Potts & L. Zdravkovic, Int. J. Geomech. **5**, 3, 244-255 (2005)
9. A. Tsiampousi, L. Zdravkovic, D. M. Potts, Comp. Geotech. **48**, 2, 156-166 (2013)
10. A. R. Estabragh & A. A. Javadi, Can. Geot. J. **45**, 408-420 (2008)

Some experience in numerical modelling of unsaturated slope instabilities

Josif Josifovski[1] and Stanislav Lenart[2]

[1] *Faculty of Civil Engineering, Ss. Cyril and Methodius University, Skopje, Republic of Macedonia*
[2] *Slovenian National Building and Civil Engineering Institute, Ljubljana, Slovenia*

Abstract. In the past couple years, the region of South-East Europe is subjected to gust rainfall events activating many landslides which cause significant material and human losses. To revaluate the existing risk maps and set new standards some old case histories are revaluated. This paper presents two case histories of landslide instabilities subjected to excess climatic perturbations, gust rainfall, namely the «Stanjevci» cut-slope near the railway line in North-East of Slovenia; and the «Ramina» a natural landslide in urban area near the city of Veles in Central Macedonia. They are briefly described later to be analysed using coupled thermo-mechanical calculations. They are subjected to specific short and gusting rainfall considered as possible trigger. Hence, van Genuchten's hydraulic model is used in combination with elastoplastic material models. The results are summarized with critical comments regarding the mathematical formulation used to describe atmospheric-soil interaction and the influence of different aspects on the accuracy is discussed briefly.

1 Introduction

Many new landsliding occurrences have been recorded in the past couple of years, in South-East Europe as a consequence of climate change and storm rainfall. If this trend continues, the landslides could cause significant devastation to infrastructure and human life. This requires that the existing risk maps to be revaluated thoroughly, based on advanced calculations that include the effects of saturation or partial saturation of the soil on slope stability [3].

The behaviour of the slopes is controlled by their hydro-mechanical conditions and by soil-atmosphere interaction. The pore water pressure changes within a slope in some cases significantly reducing the soil strength. Climatic factors such as precipitation, evapotranspiration and runoff may also have a substantial impact on slope stability.

In terms of numerical modelling, the stability analysis of partially saturated slopes is a complex problem with many variables. The more advanced approaches require input data obtained from the Soil Water Retention Curve (SWRC) and the unsaturated hydraulic conductivity function. Moreover, the time dependent hydro-mechanical behaviour of the soil should be described in a coupled format, taking both deformation and groundwater flow into account, in which mixed equations of displacement and pore pressure, called, have to be solved simultaneously in a coupled hydro-mechanical approach [9].

Advanced numerical model with definitions of climate-soil system is used to determine the rainfall influence on the slope failure. Conclusive discussion of the simulated hydro–mechanical numerical effects with suggestions for possible model improvements is presented at the end.

2 Site descriptions

The analysis considers two different slope instabilities, namely, the «Stanjevci» cut-slope near the railway line in North-East of Slovenia; and the «Ramina» a natural landslide in urban area near the city of Veles in Central Macedonia. In the following text they are briefly described through their geo-hydro-mechanical characteristics.

2.1 The «Ramina» case history

Macedonia is situated on the southern part of Balkan Peninsula as mountainous country 79 % are hills and mountains with mean altitude of 832 m with large areas of erodible soils and rocks, steep slopes. Most hillslopes are composed of crystalline rocks (gneiss, mica-schists, other schists), sandstones, lacustrine and river deposits – all of them highly erodible. Mean topographic slope of the country is 13.5° with 39.5 % of the area steeper than 15° [5].

The influence of Continental and Mediterranean climate in Macedonia unequally spreads the precipitations through throughout the year with intensity between 400-900 mm/year, thus long-time arid periods from (summer-autumn) and short wet seasons exist. According to the precipitation data for the period 2007-2010 there is evident increase in the precipitation (Table 1).

[a] Corresponding author: jjosifovski@gf.ukim.edu.mk

The «Ramina» is natural landslide located in highly urbanized hilly area of city Veles, Macedonia, on the left bank of the river Vardar. In the past it had been reactivated several times, last time in 2002 with major deformations leading to severe damage of the infrastructure and buildings.

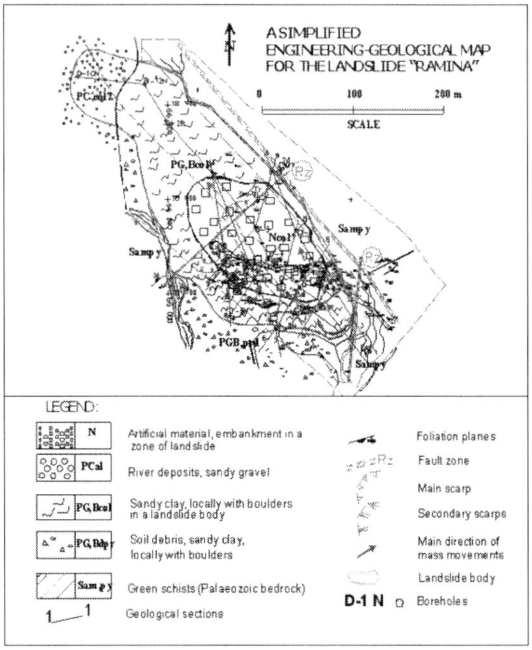

Figure 1. Simplified Engineering Geological map of «Ramina» landslide.

A season of gust rainfall in period from December to March averages with monthly precipitation of around 140 mm/month or as maximal daily precipitation of 70 mm. The elevation of the slope is around 1000 m above sea level sited on the transition of relatively steep mountain slopes to a river valley. Actually, it is attacked by a surface and ground water from wider catchment area, moreover there is no vegetation to reduce the effects of heavy rainfall events. The average slope angle is around 15° and maximal 22°. The landslide is about 500 m long with average width of about 100 m and height of 95 m (Fig. 1). The estimated maximum depth of the shear zone is 24 m. The sliding system is comprised of two parts (Fig.2), the «upper» with a length of 350 m, width 110 m, height 60 m and «lower» with length of 200 m, width 90 m, height 35 m and depth up to 18 m. The total area of the landslide is estimated to be 37,600 m^2 with around 475,200 m^3 of sliding mass, ranking it as the biggest in the Balkans, and possibly in South East Europe [4].

Table 1. Precipitations in the analysed case histories.

Site	Stanjevci, SLO		Ramina, MK	
Year	Precip. (mm/y)	Max precip. (mm/month)	Precip. (mm/y)	Max precip. (mm/month)
2000	719.1	44.4	-	-
2001	595.2	30	-	-
2002	847.5	79.2	-	-
2003	554.1	33	-	-
2004	1032.4	38	-	-
2005	851.2	79.2	-	-
2006	828.5	75	-	-
2007	895.4	79.2	498.7	140.0
2008	929.7	52	438.5	78.5
2009	-	-	642.6	104.3
2010	-	-	900.0	108.2
Avg	**805.9**	**56.7**	**626.9**	**107.7**

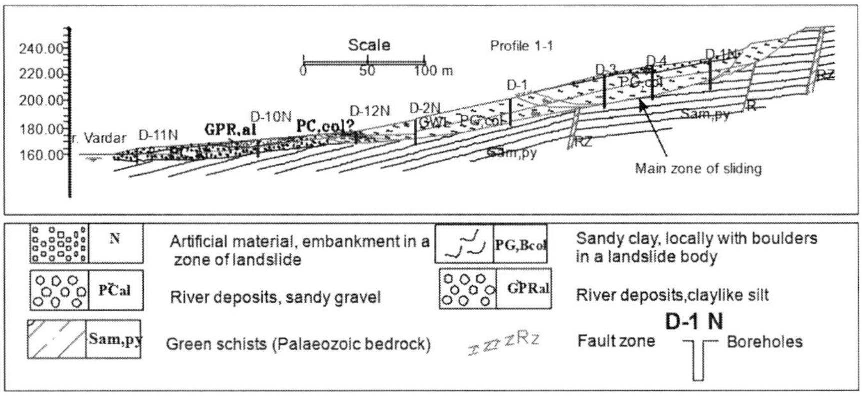

Figure 2. Typical geological profile of «Ramina» landslide.

The geotechnical investigations indicate a sliding zone on the contact between weathered bedrock (Palaeozoic amphibolitic schists, highly foliated and faulted) and low plasticity claylike mass. The hydrogeological conditions indicate that the zones with increased water contents are related to the sliding zone. Also, a slight sub-artesian effect is found in almost all boreholes [5]. The GWL has been detected at 8 m of depth while in the lower part it was around 2 m below the terrain.

2.2 The «Stajcevci» case history

Slovenia is a southern Central European country located at the junction of three important geotectonic units: the Alps at the north, the Dinaric Alps at the south and the Pannonian Basin at the east. The territory of Slovenia is mostly mountainous with a mainly continental climate. Additionally, coastal part of the country has sub-

Mediterranean climate and the north-western part has an Alpine climate. Diversity in every aspects of nature is a trade mark of country. Slovenia is located in temperate latitudes. The climate is also influenced by the variety of relief, and the influence of the Alps and the Adriatic Sea. In the Northeast, the continental climate type with greatest difference between winter and summer temperatures prevails. In the coastal region, there is sub-Mediterranean climate. The effect of the sea on the temperature rates is visible also up the Soča valley, while a severe Alpine climate is present in the high mountain regions. There is a strong interaction between these three climatic systems across most of the country [13]. Precipitation varies across the country as well. Average rainfall depth exceeds 3,500 mm/year in some Western regions and dropping down to 800 mm/year in northeastern part of the country. In the winter, also snow is quite frequent.

A seemingly almost routine railway cut slope, beside railway tunnel «Stajcevci» on the railway line between Murska Sobota and Hodoš in northeastern part of Slovenia, was constructed in the first half of early year 2000. Tunnel and cut slope passed through Pliocenne soil layers: sand with gravel, silt and clay in subhorizontal orientations [6]. Although the cut slope was made in gentle slope not steep, stability problems arose already during the construction works, particularly due to saturated silty-sand layers covered by clay and low cohesion of saturated contact between clay and silty-sand. Stanjevci tunnel and the cut slope in its vicinity are located in the northern, central part of the Goričko hills in north-eastern Slovenia, that are dominated by sediments of the middle Pliocene, both marine and brackish, as well as fluvial and lacustrine origin. Generally, sediments of Goričko hills belong to tectonic zone of the Pannonian basin. Unstable cut slope above railway line near Stanjevci tunnel is located in the middle Pliocene sediments PL2, which passes in the middle Miocene (tortonium) layers that lie discordantly on the Paleozoic metamorphic rocks. The estimated thickness of Tertiary sediments in the area of the landslide is approximately 2000 m. General dip of the layers of the Tertiary sediments and contact with Paleozoic rocks is approximately 10° towards the SSE.

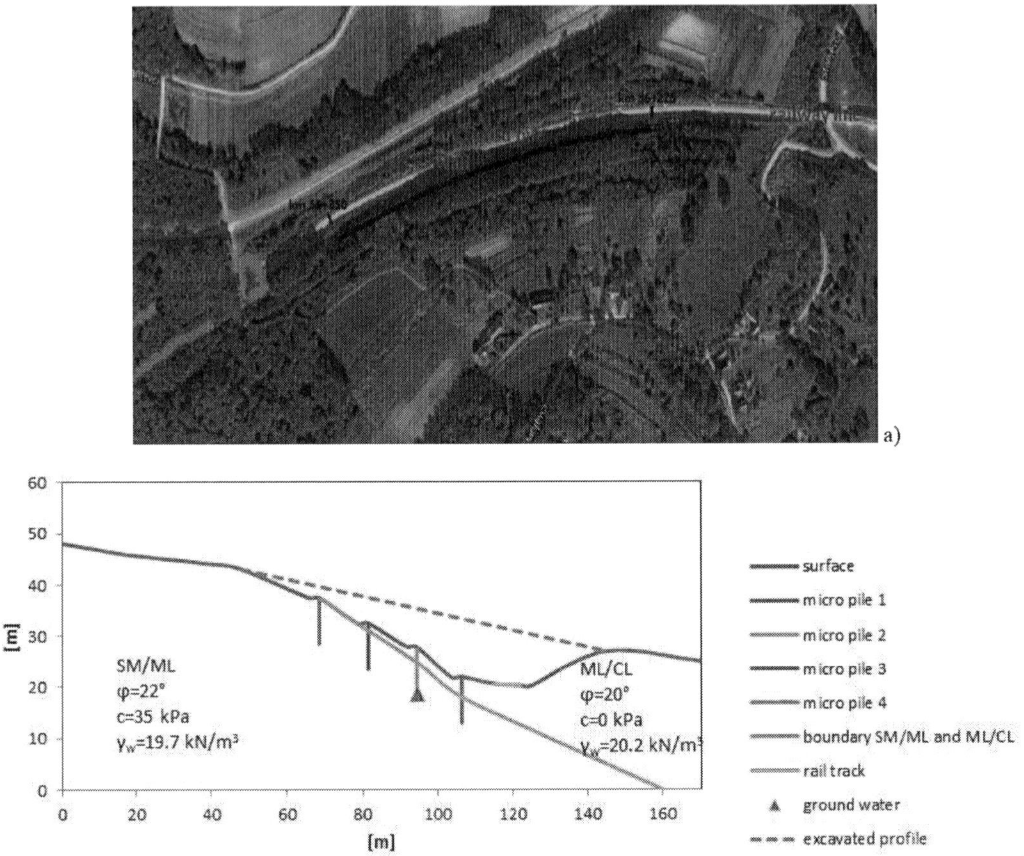

b)

Figure 3. (a) The «Stajcevci» cut-slope: (a) Plan view on cut-slope near the tunnel and (b) Simplified geological profile of the excavated slope.

The landslide is located on the hillside, where a cut slope was made with slope direction towards north. The excavation pit is in the deepest part more than 30 m deep. At the time of the pit excavation an extensive geotechnical investigation was performed to support stabilization measures. In the vicinity of the landslide there are no recorded tectonic structures. The precipitation data [12] in the period 2000-2008 was taken with maximal values in the analysis (Table 1).

3 Numerical modelling

To analyse mechanical behaviour of saturated or partially saturated soils by means of numerical method e.g. finite element method, in proper manner, it is necessary to take into account both deformation and groundwater flow. For time dependent behaviour, this leads to mixed equations of displacement and pore pressure, called coupled hydro-mechanical approach, which have to be solved simultaneously [2].

The task was to perform a coupled hydro-mechanical analysis of slope stability on the presented case histories. The calculation was performed using the finite element software Plaxis [1] which enables fully coupled hydro-mechanical analysis. This type of analysis is able properly to evaluate the effects of rainfall and water infiltration on the slope stability. The shear strength of the unsaturated soil has been examined based on effective stress concept considering suction. The finite element method with shear strength reduction technique was employed to evaluate the stability using the factor of safety (FoS). For the soil material an elastic-perfectly plastic small strain Mohr-Coulomb (MC) model was employed with special attention put on definition of mechanical and hydraulic parameters critical for realistic simulation. According to the available data, a conservative choice has been taken. For simplicity a homogeneous profile is assumed with soil parameters given in Table 2.

Table 2. Soil material parameters for MC model.

Case (material)	**Ramina** (Sandy-Clay)	**Stajcevci** (Clay)
Unit weight γ (kN/m^3)	20.14	20.2
Eff. friction angle ϕ' (^{0})	23	22
Eff. Cohesion c' (kPa)	12	15
Eff. Poisson's ratio ν' (/)	0.32	0.30
Elastic modulus E' (kPa)	10000	7000

For definition of the hydraulic models at the Slovenian National Building and Civil Engineering Institute tensiometer laboratory test were performed to determined the suction in the material samples for different water content. Through the measured data a nonlinear SWRC fitting (fig.4) after UNSODA [8] database was made.

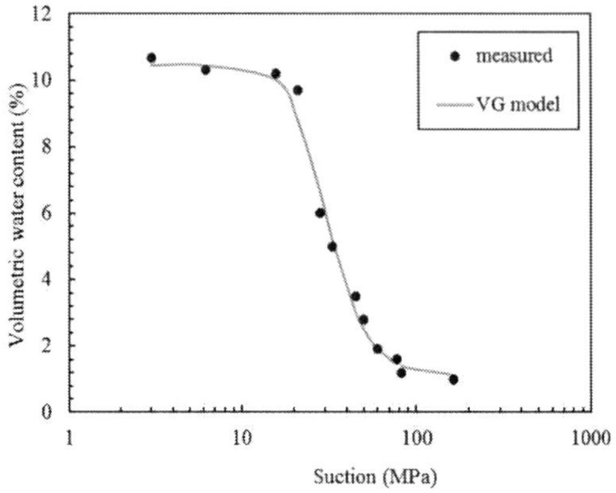

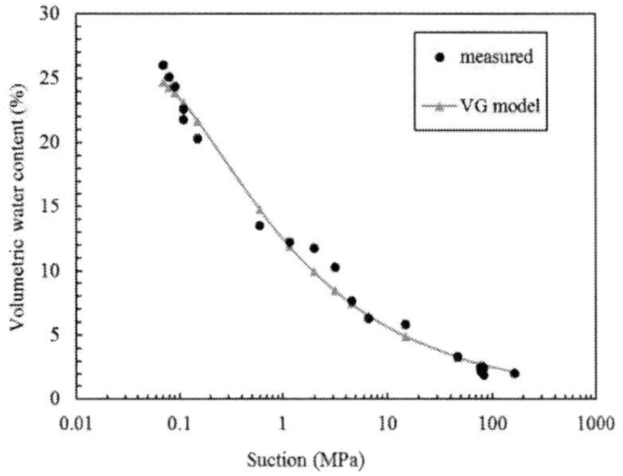

Figure 4. SWRC of a) Sandy-Clay from The «Ramina» and b) Clay from «Stajcevci» cut-slope

The hydraulic model is defined by of Van Genuchten model parameters [12] given in Table 3.

Table 3. Hydraulic data of Van Genuchten model.

Case	Soil	k_{sat} (m/s)	θ_s (-)	θr (-)	α (1/kPa)	n (-)
Ramina	Sandy-Clay	1E-6	10.45	1.1	0.355	4.17
Stajcevci	Clay	1E-8	27.8	0.001	9.5	1.35

In the «Ramina» case a constant rainfall infiltration is assumed to be 10 mm/h on the free surface plus additional 1 m^3/h on left inflow boundary. Similarly, in the «Stajncevci» case the precipitation of 15 mm/h. Both simulated scenarios run for time period of 12 h. The interaction between atmosphere-vegetation was not considered nor the temperature effects. Mesh updating has been performed, as it is often the case when due to deformation the finite elements are moving from saturated to unsaturated zone and vice versa.

4 Results

The analysis produced wide range of results clearly depicting the influence of the heavy rainfall on the hydro-mechanical material parameters. Generally, the position of phreatic level and the distribution of pore water pressure are governed climate conditions (groundwater flow boundary conditions) with suction above phreatic level where with time in the case of downward flux (i.e. precipitation) suction decreases (and degree of saturation increases) and the water level rises. Below are presented the results of the coupled flow-deformation analysis, the development of suction and displacement on the «Ramina» landslide are presented in Figure 4.

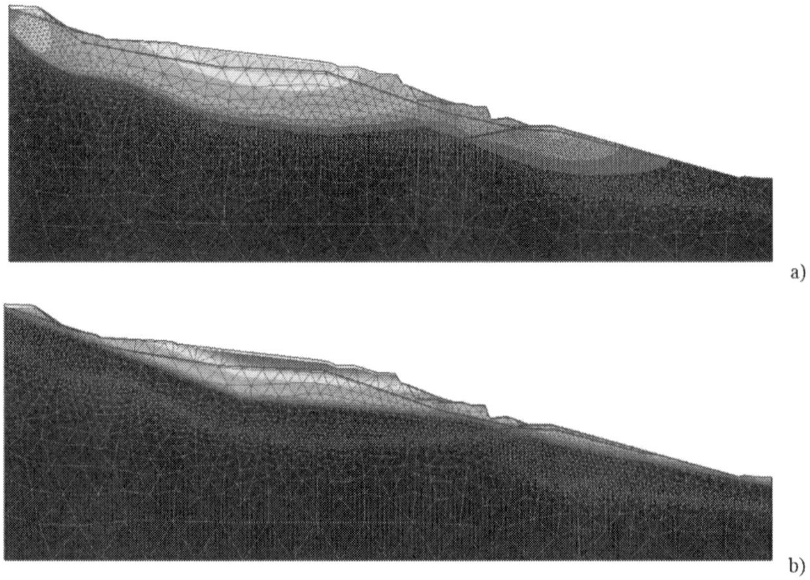

Figure 5. The «Ramina» results: (a) Total displacements and (b) Suction profile.

The mass sliding had occurred just before 12 h of rainfall triggering displacement of 0.8 m at the top of the landslide. The sliding mechanism had developed first in the upper part triggered by in PWP build-up establishing different GWL in the upper and lower part. The critical shear strength had been reached in the upper part of the landslide with PWP rise of about 50% at the bedrock interface but also with maximum suction of 169.7 kN/m^2 at top of the layer. Realistic simulation of the sliding mass behaviour (Figure 5) has been obtained where after the registered sliding a mass stabilization occurred defined by FoS = 1,3.

In the same manner the results of «Stajcevci» cut-slope are presented in Figure 5.

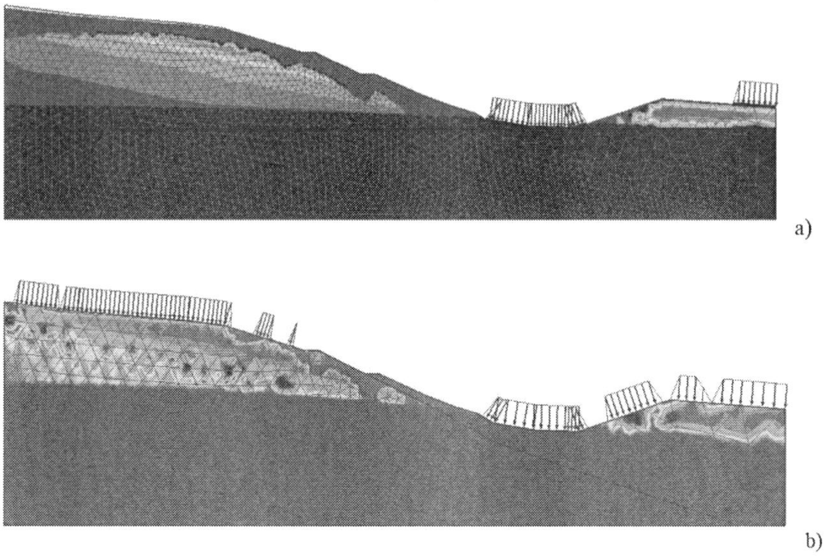

Figure 6. The «Stajcevci» results: (a) Suction and (b) Saturation profile.

The simulation of hydraulic conditions leads to a saturation profile with a raise in GWL producing 20% increase of PWP and reducing the suction to 12.81 kN/m^2 (Figure 6). This is followed by a global displacement with maximal displacement of 0.6 m being realized before the 12 h of precipitation. The ultimate limit state is and global cut-slope destabilization is ensured by FoS= 1,22.

5 Conclusion

The consequences of climate change on landslide rate, size and frequency are difficult to predict because they depend on a range of variables. The precipitation (mostly in rainfall) is among the most important factors especially severe storm would cause flash floods with intense erosion and landsliding. In addition to the above direct influences, the indirect impact of climate change on landslide will be exerted through changes of vegetation cover, quality and density, runoff, overland flow and oversaturation would increase raising the rate of erosion and frequency of landslides. Rainfall is often the main factor for landslinding. The modelling has shown that suction responses to rainfall vary on a number of different timescales [10]. On shorter timescales (ranging from minutes to hours and days) heavy rainfall which was of primary interest in this study produced very interesting results. The rainfall infiltration effects in the literature [11] are usually critical for shallow landslides, but here it proven that also the deep sited landslides with high GWL could be destabilized.

In the analysis it has been evaluated the influence in the sense of infiltration and duration of rainfall on the both slopes. It was proven that relatively short (12 h) but gust rainfall (10 - 15 mm/h) could have a significant influence on the overall (global) stability. According to the statistics this type of events are to be expected frequently. Other modelling aspects of climate-vegetation-soil system such as: influence of cracks and fissures, root water uptake were not considered, thus there are still but a few improvements in this model.

Acknowledgements

This paper is result of collaboration between Faculty of Civil Engineering in Skopje and Slovenian National Building and Civil Engineering Institute from Ljubljana in the framework of COST action TU1202 – a European network which addresses the impacts of climate change on engineered slopes for infrastructure activities. The research and knowledge transfer was enabled through short term scientific missions adequately supported.

References

1. Plaxis bv. Plaxis 2D AE. *Scientific Manuel* (2012).
2. V. Galavi, Internal report: *Groundwater flow, fully coupled flow deformation and undrained analysis in Plaxis 2d and 3D*. (2010)
3. D. Hughes, Md Rajibul Karim, K. Briggs, S. Glendinning, D. Toll, T. Dijkstra, W. Powrie and N. Dixon. XVI European Conference on Soil Mechanics and Geotechnical Engineering, At Edinburgh *A Comparison of numerical modelling techniques to predict the effect of climate on infrastructure slopes*, (2015).
4. J. Josifovski, S. Gjorgjevski, B. Susinov. 1st Regional Symposium on Landslides in the Adriatic-Balkan Region with 3rd Workshop of the Japanese-Croatian Project on 'Risk Identification and Land-Use Planning for Disaster Mitigation of Landslides and Floods in Croatia. *Ramina Landslide from a Natural Hazard to Remediation*. Zagreb (2013)
5. M. Jovanovski, I. Peshevski, B. Markoski, S. Petrusheva, B. Susinov. 1st Regional Symposium on Landslides in the Adriatic-Balkan Region with 3rd Workshop of the Japanese-Croatian Project on 'Risk Identification and Land-Use Planning for Disaster Mitigation of Landslides and Floods in Croatia. *Landslide Inventory Map of the Republic of Macedonia, statistic sand description of main historical and slide events*. Zagreb (2013).
6. S. Lenart. Slovenian National Building and Civil Engineering Institute Slovenia, Section for Earthquake Engineering. *Deformation characteristics of lacustrine carbonate silt in the Julian Alps*. Volume 26, Issues 2–4. 131–142 (2006).
7. S. Lenart, K. Fifer Bizjak. Proceedings of the 3rd International Conference on Landslides, Slope Stability and the Safety of Infra-Structures, Singapore: Ci-Premier Conference Organisation, *Numerical analysis of a seismic response of the Stože landslide*, 325-331 (2002).
8. K. Seki, Hydrol. Earth Syst. Sci. Discuss., 4*: SWRC fit - a nonlinear fitting program with a water retention curve for soils having unimodal and bimodal pore structure*. 407-437 (2007).
9. SAFELAND – *Living with landslide risk in Europe: Assessment, effects of global change, and risk management strategies*.
10. F. Casini, V. Serri, and S. M. Springman. Canadian Geotechnical Journal, 50(1): *Hydromechanical behaviour of a silty sand from a steep slope triggered by artificial rainfall: from unsaturated to saturated conditions*, 28-40 (2015).
11. S.M. Springman, C. Jommi and P. Teysseire. Géotechnique, 53(1). *Instabilities on moraine slopes induced by loss suction: a case history*, 3–10 (2003).
12. Th. M. van Genuchten. Soil Science Society of American Journal, 44: *A closed-form equation for predicting the hydraulic conductivity of unsaturated soils*. 892–898 (1980).
13. ARSO, *National Meteorological Service of Slovenia - Archive*. www.meteo.si (2015).

Modelling desiccation cracking in a homogenous soil clay layer: comparison between different hypotheses on constitutive behaviour

Cristina Jommi[1,a], Niccolò Valimberti[2], Roderick N.Tollenaar[1], Gabriele Della Vecchia[2] and Leon A. van Paassen[1]

[1]*Delft University of Technology, Department of Geoscience and Engineering, Stevinweg 1, 2628 CN Delft, The Netherlands*
[2]*Politecnico di Milano, Department of Civil and Environmental Engineering, piazza Leonardo da Vinci 32, 20133 Milano, Italy*

Abstract. Desiccation cracks are usually thought to start from the surface of an evaporating soil layer, and the available simplified models for crack initiation and propagation are based on this hypothesis. On the contrary, experimental results on a Dutch river clay showed that cracks in an evaporating soil layer may start and propagate below the surface, confirming earlier findings by other researchers. A simple one-dimensional model was set up to analyse the consequences of different hypotheses about the material behaviour on the crack onset in a homogenous soil layer undergoing surface drying. The results of the model show that dependence of the material behaviour on the rate of water content change is a necessary requirement for cracks to initiate below the surface. The conclusion suggests that, to properly understand cracking in an evaporating soil layer, an intrinsic time scale for the mechanical response must be accounted for, among all the other factors which were previously highlighted by other researchers. The key factor to predict crack onset below the surface is the dependence of the drying branch of the water retention curve of the compressible soil on the rate of drying, which would be justified by a rate dependent fabric evolution.

1 Previous findings and motivation

Cracks are a common occurrence in drying soil. They not only have effects on the integrity of geotechnical structures, but they also affect the hydraulic properties of the soil. The main driving mechanisms governing the occurrence of fractures in soil are the matric suction generated in the drying material and the internal mechanical constraints.

It is generally believed that fractures in soil start at the surface and then they propagate downwards. This is certainly the case in many situations. As described by Weinberger [1], under laboratory conditions the initiation of fractures commonly happens at the surface of the desiccating layer. However, it has been observed that it is not always the case. Evidence obtained from field observations [2] and experiments carried out on Dutch river clay in the laboratories of the Delft University of Technology and the Politecnico di Milano, indicate otherwise. During the tests, it was observed that the fractures can initiate and propagate under the surface.

Several simplified models have been created to simulate the generation of cracks in soil [3-6], but all of them are based on the hypothesis that the fracture initially starts at the top of the desiccating layer. The assumption is reasonable given that the dryer section in the soil profile is at the top, where the largest amount of suction is potentially developed. These simplified models tackle the drying process as a weakly coupled, nearly steady-state process. They use a unique relationship between the water content and the relevant variables playing a part in the mechanics of the system, and they cannot simulate cracks beginning below the surface.

It has been established experimentally that under the same environmental and initial conditions, thin clay layers dry faster than thick ones. They also exhibit less cracks, larger intact areas and wider fractures [7-12]. The phenomenon has been attributed to the difference between desiccation rates. This premise has been confirmed by studies performed by Tang et al [10-11] and Costa et al [12]. Corte and Higashi [7] also observed that the water contents at the onset of cracking decreased as the drying speed increased.

As experimental evidence shows, there is a clear rate dependency of the stresses involved during soil drying and cracking. Tamrakar et al [13] mentioned that there is also an increase in tensile strength with higher tensile pulling rates, although the trends and the magnitude of the change in the results do not seem to support such conclusion to a large extent. Actually, most of the results appear to indicate that there is no significant variation in the tensile strength with different pulling rates. This suggests that the key factor to predict crack onset below the surface must be looked for among the other hydro-mechanical properties governing the response of the compressible soil upon drying. This contribution presents an exercise aimed at analysing the consequences of different hypotheses about the material behaviour on the theoretical prediciton of crack initiation in a homogenous soil layer.

[a] Corresponding author: c.jommi@tudelft.nl

2 Material and summary of the results

An extensive investigation is being carried out at the Delft University of Technology to study the reuse of dredged clay sediments for civil applications. Part of this investigation focuses on the thorough understanding of how cracking occurs, since it is known to influence drainage, evaporation and ripening. The experimental study was performed on homogenous soil samples prepared in the laboratory, using a commercially available clay composed (by mass) of 50.2% quartz, 21% vermiculite, 16.2% muscovite, 6.8% anorthite and 5.8% calcite approximately. The soil contains 3.1% sand, 54.9% silt and 42% clay. The liquid limit of the clay is 57%, the plasticity index is 33%, and the specific gravity is 2.74. Reconstituted clay samples were prepared by mixing the commercial clay with water. The tested samples were prepared at different initial water contents.

At the Politecnico di Milano, thin vertical Perspex boards of different heights were filled with the clay in order to analyse the drying and crack formation along the vertical axis. Due to the stiffness contrast between the Perspex board and the clay, the first cracks always occurred at the side boundaries of the samples. After the first lateral cracks formed, the sample was almost free to shrink, with the base of the board offering little restraint to the shrinking process. Depending on the sample height and the evaporation rate, cracks were observed starting in the middle of the sample, even before any surface cracks appeared (Fig. 1).

At the Delft University of Technology, desiccation experiments were carried out by placing the clay at initial water contents between two and three times the liquid limit on wood, plastic and metal containers. The areas of the boxes ranged from 0.09 to 1 m^2, and the initial soil thicknesses varied between 20 and 100 mm. Detailed descriptions of the tests can be found in [14].

The clay behaviour was studied by a number of tests, which gave the following general conclusions:

(i) the tensile strength depends strongly on the water ratio, but it is almost insensitive to the rate of tensile strain;

(ii) the tensile stiffness decreases with increasing water ratio and tensile strain rate;

(iii) the air entry value upon drying depends on the initial water content, hence confirming the important role played by the soil fabric, but it is always in an order of magnitude between few to tens of kPa.

Figure 1. Evaporating clay slurry showing crack development below the surface in the clay body.

3 A simple 1-D model of crack initiation

In this section, reference is made to a homogenous isotropic deformable clay layer. Until the first crack appears, the layer can be described by a one-dimensional system with all relevant quantities varying only with depth, z. If the soil layer is initially saturated, previous research and results from the ongoing experimental investigation suggest that the first crack appears during the saturated state. Accepting this hypothesis, the water ratio $e_w = V_w/V_s$, where V_w is the volume of water and V_s is the reference volume of solids, is sufficient to describe the distribution of void ratio and water content throughout the soil profile before cracking occurs.

The water ratio profile changes in time and depth due to the evaporative flux at the top surface of the layer, and it was chosen as the driving variable of the hydromechanical model. The mechanical behaviour of the soil was described by the simplest allowable laws suggested by the experimental evidence. This was carried out to highlight the role played by different hypotheses in the response of the soil layer to the evaporation process. The hydraulic fluxes are not explicitly accounted for in the formulation, as the current water ratio distribution is sufficient for the suction profile to be calculated from the drying branch of the soil water retention curve (SWRC). In the formulation, compression stresses and strains are assumed positive. The change rate in the variables is designated by a superscript dot. The vertical and horizontal components of stress, σ, and strain, ε, are designated by subscripts z and h, respectively.

3.1 Relevant assumptions

The soil is assumed to be homogenous, isotropic and saturated until the first crack occurs. If the variation of water content with depth and time is known, the current water ratio and water ratio rate can be calculated at any depth, serving as the input of the model.

A simple incremental hypo-elastic model is assumed to describe the material stiffness:

$$\begin{bmatrix} \dot{p}' \\ \dot{q} \end{bmatrix} = \begin{bmatrix} K & 0 \\ 0 & 3G \end{bmatrix} \begin{bmatrix} \dot{\varepsilon}_{vol} \\ \dot{\varepsilon}_d \end{bmatrix} \quad (1)$$

where K is the (tangent) bulk modulus and G is the (tangent) shear modulus. The stress and strain variables are defined as:

$$\begin{aligned} p' &= \frac{1}{3}\left(\sigma'_z + 2\sigma'_h\right) \\ q &= \left(\sigma'_z - \sigma'_h\right) \\ \varepsilon_{vol} &= \left(\varepsilon_z + 2\varepsilon_h\right) \\ \varepsilon_d &= \frac{2}{3}\left(\varepsilon_z - \varepsilon_h\right) \end{aligned} \quad (2)$$

The volumetric (shrinkage) strain rate is given by:

$$\dot{\varepsilon}_{vol} = -\frac{\dot{e}_w}{1+e_w} \tag{3}$$

and it is assumed to be entirely due to a change in effective stress acting on the soil element:

$$\dot{\varepsilon}_{vol} = \frac{1}{K}\dot{p}' = \frac{1}{K}(\dot{p}-\dot{u}_w) = \frac{1}{K}(\dot{p}+\dot{s}) \tag{4}$$

where u_w is the pore water pressure and s is the suction. Under saturated conditions, $s = -u_w$. It follows that the tangent bulk modulus can be obtained as the derivative of the SWRC, starting from the state at which positive suctions are measured:

$$K = -(1+e_w)\frac{\partial s}{\partial e_w} \tag{5}$$

For simplicity, the shear modulus is obtained in the form:

$$G = \frac{3(1-2\nu)}{2(1+\nu)}K \tag{6}$$

by assuming a constant value of the Poisson's ratio, $\nu = 0.3$.

If the soil element is free to deform, no shear strains and consequently no shear stresses are developed. Under a 1-D constraint, when only vertical deformation is allowed, a shear strain must accompany the volumetric strain, so that the following lateral constraint is complied with:

$$\dot{\varepsilon}_h = 0 \tag{7}$$

By means of simple substitution, the lateral effective stress can be obtained from the previous equations in the form:

$$\dot{\sigma}'_h = \dot{s} - \frac{2}{3}G\,\dot{\varepsilon}_{vol} = \dot{s} + \frac{2}{3}G\frac{\dot{e}_w}{1+e_w} \tag{8}$$

It is worth noting that the change in total stress due to the water mass change during evaporation is disregarded in the formulation. The choice is justified by the fact that the latter term is two orders of magnitude smaller than the corresponding changes in suction. The total horizontal stress increment is given by:

$$\dot{\sigma}_h = \dot{\sigma}'_h - \dot{s} = \frac{2}{3}G\frac{\dot{e}_w}{1+e_w} \tag{9}$$

During evaporation, the horizontal stress decreases as tensile stresses are developed. In the absence of weak spots, the idealised crack initiation condition can be defined as the moment when the total tensile stress attains the same value as the tensile strength:

$$|\sigma_h| = |\sigma^T| \tag{10}$$

The choice for an assessment in terms of total stresses is justified by the available experimental data on tensile strength, which is expressed in terms of the total stress needed to trigger tensile failure.

3.2 Constitutive assumptions

The response of the soil to a given change in water ratio in the simplified 1-D model depends on: (i) tensile strength; (ii) tensile stiffness; and (iii) soil water retention properties, which link the water ratio to the suction.

The response of the soil layer is analysed first at an elementary level (Model I), assuming constant values for the soil stiffness and strength. This approach has been used in previous models that have focussed exclusively on the crack onset [3,6]. In the second analysis (Model II), the stiffness and the strength are assumed to depend on the water ratio, which is in accordance with the evidence of current and previous experimental results [15]. Eventually, a dependence of the stiffness and water retention curve on the drying rate was considered, as well (Model III). It is worth reminding that no attempt was made to formalise the constitutive laws in a comprehensive thermo-mechanical framework. Instead, non-linear hypo-elastic laws were used, interpolating the experimental information.

The tensile strength was given a power law of the type:

$$\sigma^T(e_w) = c\, e_w^{-d} \tag{11}$$

The water retention curve was described by a standard van Genuchten's equation [16] with two independent parameters, assuming $m = 1 - 1/n$. The bulk modulus $K(e_w)$ was calculated with Eq.(5) from the retention curve.

4 Model predictions and discussion

To investigate the consequences of the different assumptions on the clay layer response, a benchmark problem which replicates a typical clay board test in the laboratory was analysed.

4.1 The benchmark problem

A homogenous soil layer, 0.1 m thick, was assigned the initial state which characterised one of the laboratory tests. The water ratio profile was constant at the beginning of the tests, with a value of $e_{w0} = 2.55$. A final water content profile was assigned, based on laboratory measurements made after two weeks of drying. To simplify the analysis, a constant drying rate at each depth was assumed between the initial and final states. This allowed to define a set of water content values between the initial and final water content profiles, resulting in the water ratio evolution history in Fig. 2. The water ratio rate varied from 0.03 to 0.01 d^{-1} with depth. Table 1 summarises the values of the parameters used in the simulations to describe the tensile strength (Eq.11).

 DOI: 10.1051/e3sconf/20160908006

Table 1. Parameters used to describe tensile strength (Eq.11).

Variable	**MODEL I Constant**	**MODEL II Depending on e_w**	**MODEL III Depending on $e_w, \dot{e}_w$**
$\lvert\sigma^T\rvert$ (kPa)	0.3 kPa	c = 6.5 kPa d = 3.65	same as model II

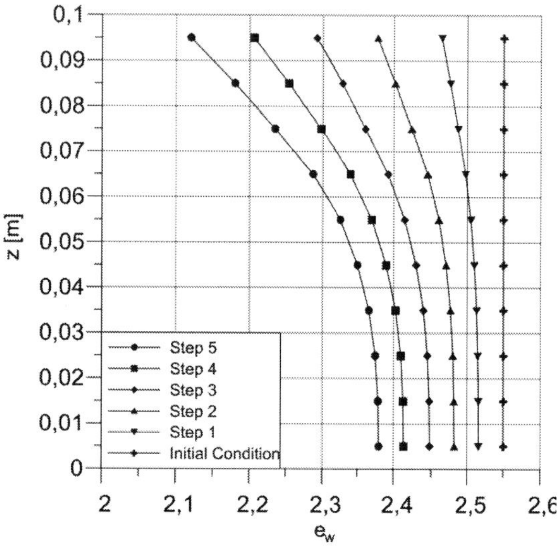

Figure 2. Water ratio profiles given as input to the 1-D model.

4.2 Results

In Fig. 3, the evolution of the suction profile is reported for Model II, showing an increase of the suction with time and attaining the maximum values at the top of the soil layer. As a consequence, the horizontal tensile stress also increases, reaching the highest value at the top of the layer (Fig. 4). Results for the simpler Model I, not shown, are qualitatively and quantitatively similar.

As depicted in Fig. 5, the tensile strength in Model II also increases during drying. In the figure, the constant average value assumed in Model I is represented by the dashed line. In spite of the tensile strength increase, the maximum tensile stress approaches the tensile strength and eventually reaches it (Fig. 6). For both Models I and II, the tensile strength is attained at the top of the layer before than at any other depth. This result is not due to the particular values assumed for the parameters, but it is a consequence of the shape of the curves, all having a similar dependence on the water ratio. If the tensile strength were to increase faster, no tensile crack would be predicted. On the contrary, if the condition for crack onset is met (Eq. 7), it would initially occur at the surface of the layer.

4.3 Dependence on the rate of water ratio

The results shown in Fig. 6 indicated that the dependence of the relevant quantities on the water ratio, as calibrated on the experimental results, was not enough to justify the occurrence of first cracks below the surface of the clay layer. Therefore, the crack initiation below the surface must be the result of some previously unaccounted factors.

The experimental data collected by Tollenaar [14] and previous research suggest that strength and stiffness might depend on the rate of change in water ratio, besides the water ratio. Nonetheless, while the rate of drying is relevant for the stiffness, the tensile strength is almost insensitive to it.

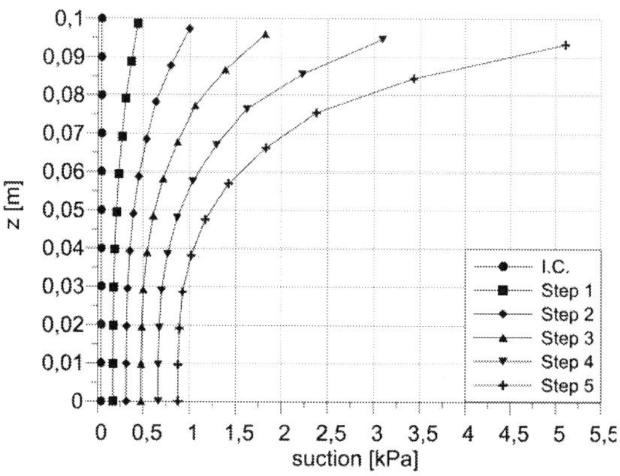

Figure 3. Suction profiles predicted by Model II.

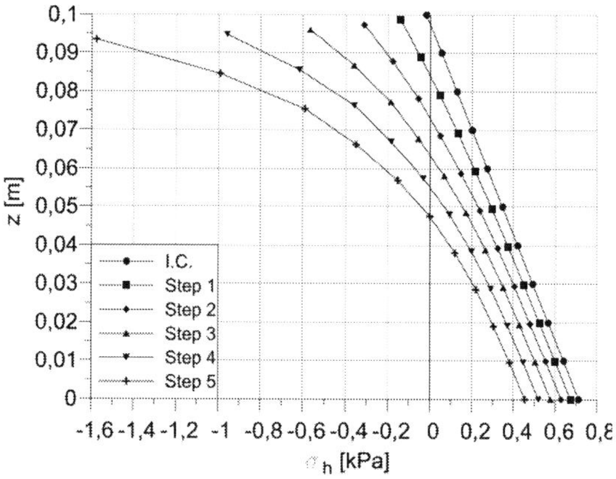

Figure 4. Horizontal total stress profile predicted by Model II.

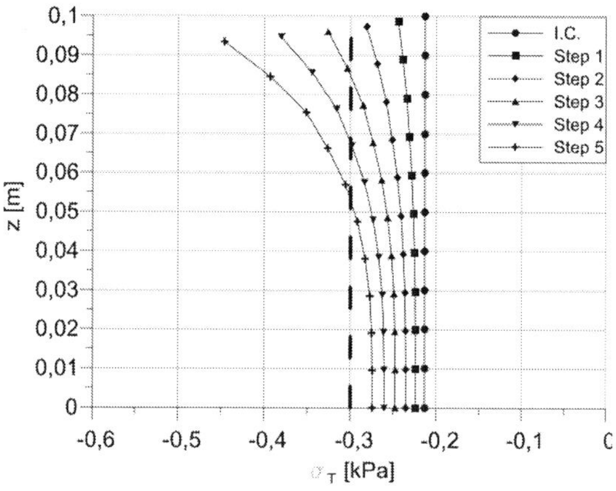

Figure 5. Tensile strength profile predicted by Model II.

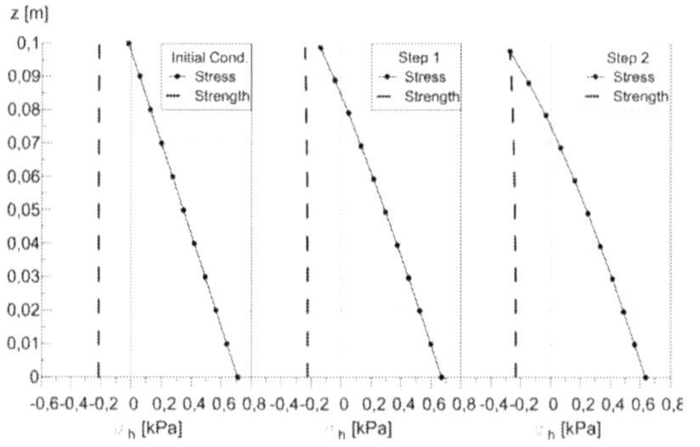

Figure 6. Comparison between the evolution of the horizontal stress and the tensile strength predicted by Model II.

Sensitivity analyses showed that the suction is the dominant factor in the shape of the horizontal stress with depth. To allow for the shape of the horizontal stress to change, the water retention curve has to be made dependent on the water ratio rate. Although this is not a common assumption, some previous work has already highlighted that this dependence may exist, either from a theoretical standpoint, or from experimental observations.

Experimental observations by Tollenaar [14] on the dependence of the tensile stiffness on the drying rate suggested that an intrinsic time scale characterises the response of the soil fabric to water content changes. A decrease in tensile stiffness at increasing drying rate can only be justified by the development of lower suction values. These lower suction values for a given water content suggest that fast drying tends to maintain an open fabric with larger pores. If drying is slow enough for the fabric to rearrange, a more massive fabric with smaller pores can develop as a consequence.

Based on this hypothesis, the water retention curve was made dependent on the rate of water ratio, by assuming a shape parameter n of the form:

$$n = \bar{n} + \bar{\eta}\,\dot{e}_w \tag{12}$$

The correspondent curves are represented in Fig. 7, where the one independent from the drying rate (Model I and Model II) is sketched with a dashed line. The parameters used for the water retention curve are summarised in Table 2. In the range of drying rates analysed, n is always higher than 1. The corresponding tangent bulk moduli are shown in Fig. 8.

Table 2. Parameters for the SWRC for Models I, II and III

Parameter	MODEL I Constant	MODEL II Depending on e_w	MODEL III Depending on $e_w, \dot{e}_w$
α (kPa^{-1})	1.0	1.0	1.0
n	1.04	1.04	$\bar{n}$ =0.96 $\bar{\eta}$ = 5.04 (d)

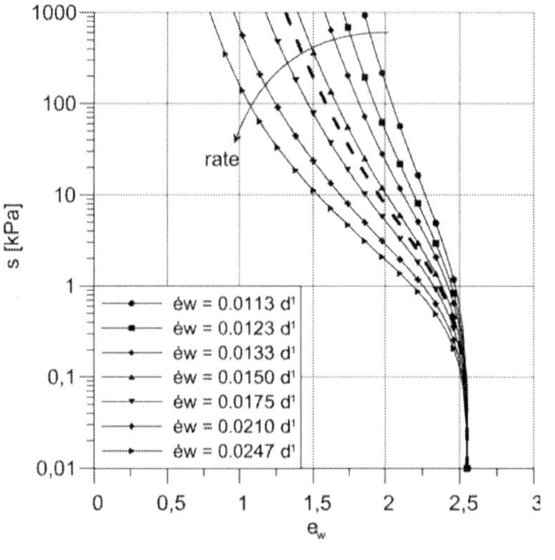

Figure 7. Suction vs water ratio and water ratio rate.

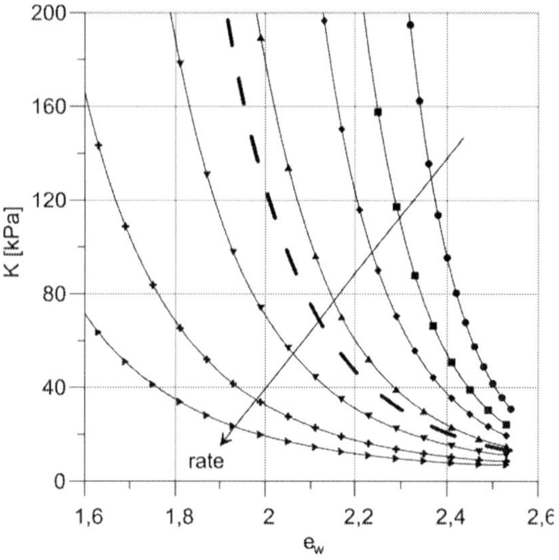

Figure 8. Bulk modulus vs water ratio and water ratio rate.

The drying rate dependence of the suction developed for a given water ratio yields a suction increase rate which is different for each depth. The implications are evident from the results of the simulation completed with Model III (Fig. 9).

The different rates at the different depths change the shape of the suction profile in time. As a result, the horizontal stress profile varies with depth, and the horizontal stresses at the top of the layer can increase less rapidly than at depth. As observed in Fig. 9, the onset of cracking is predicted for a water ratio profile occurring at the end of step 5. The critical depth is not on the surface anymore. On the contrary, the horizontal stress can reach the values of tensile strength at different depths below the surface, or even near to the bottom of the layer.

The results of this numerical simulation provide a demonstration of the possible theoretical development of cracks below the surface which were actually observed in the laboratory on the physical models.

DOI: 10.1051/e3sconf/20160908006

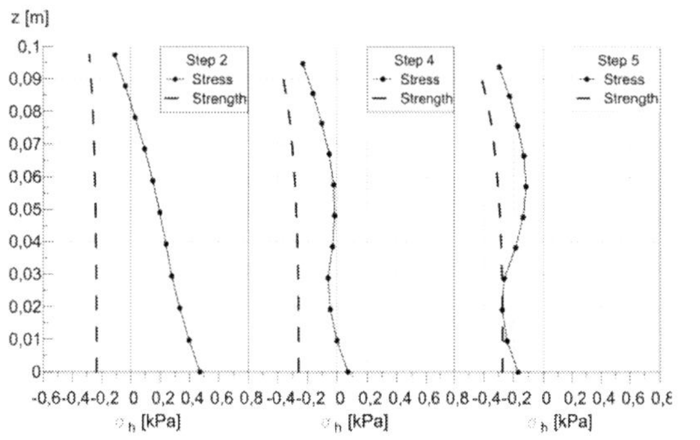

Figure 9. Comparison between the evolution of the horizontal stress and the tensile strength predicted by Model III.

5 Conclusions

The simple 1-D model for crack development was conceived to explore the consequences of various material behaviour assumptions on the response of a deformable clay layer to atmospheric desiccation. The work was motivated by the dominant role played by crack development in the mechanical and hydraulic behaviour of clay soils in engineering. Experimental observations in the laboratory have shown that cracks can begin and propagate below the surface of a desiccating soil layer. The observation was also made on clay boards specially designed to limit the influence of boundary constraints on the results of the test.

The simple models available in the literature were unable to explain the observation. This preliminary systematic study was aimed at investigating the necessary requirements for cracks to start at depth in shrinking clay, assuming the absence of weak spots and with a limited influence of external constraints.

Based on the model developed, a dependence of the water retention curve on the rate of drying seems to be the key feature necessary to reproduce the experimental evidence. This is a factor not considered in previous models for cracking. The results suggest that an intrinsic time scale for the mechanical response must be included with all the other factors being investigated. In a compressible soil, the dependence of the drying branch of the water retention curve on the rate of drying could be justified by a rate dependent fabric evolution.

Acknowledgments

The authors at the Delft University of Technology acknowledge the financial support of the Technologiestichting STW (NL) through the project Lift-up-Lowlands (project 11344). N. Valimberti acknowledges the support of the Politecnico di Milano for his study exchange period at the Delft University of Technology. Dr. Caruso is acknowledged for his help in designing the clay board experiments at the Politecnico di Milano.

References

1. R. Weinberger, Geol. Soc. Am. Bull. **113**(1), 20-31 (2001)
2. R. Weinberger, J. Struct. Geol. **21**(4), 379-386 (1999)
3. P. H. Morris, J. Graham, D. J. Williams, Can. Geotech. J. **29**(2), 263-277 (1992)
4. A. N. Abu-Hejleh, D. Znidarcic, J. Geotech. Eng. **121**(6), 493-502 (1995)
5. J. M. Konrad, R. Ayad, Can. Geotech. J. **34**(4), 477-488 (1997)
6. H. Péron, L. B. Hu, L. Laloui, T. Hueckel, *Third Asian Conference on Unsaturated Soils*, 391-396 (2007)
7. A. Corte, A. Higashi, *Technical report, U.S. Army snow and ice and permafrost research establishment* (1960)
8. J. K. Kodikara, S. L. Barbour, D. G. Fredlund, *Asian Conference on Unsaturated Soils, UNSAT ASIA 2000*, **90**, 693-698 (2000)
9. H. Nahalawi, J. K. Kodikara, Geol. Geotech. Eng. **24**(6), 1641-1664 (2006)
10. C. S. Tang, B. Shi, C. Liu, L. Zhao, B. Wang, Eng. Geol. **101**(3), 204-217 (2008)
11. C. S. Tang, Y. J. Cui, A. M. Tang, B. Shi, Eng. Geol. **114**(3), 261-266 (2010)
12. S. Costa, J. Kodikara, B. Shannon, Geotechnique **63**(1), 18-29 (2013)
13. S. B. Tamrakar, T. Mitachi, Y. Toyosawa, *Experimental Unsaturated Soil Mechanics* 207-218 (2007)
14. R. N. Tollenaar, *PhD dissertation*, in preparation (2016)
15. H. Trabelsi, M. Jamei, H. Zenzri, S. Olivella, Int. J. Numer. Anal. Meth. Geomech. **36**(11), 1410-1433 (2012)
16. M. Van Genuchten, Soil Sci. Soc. Am. J. **44**(5), 892-898 (1980)

Shakedown modeling of unsaturated expansive soils subjected to wetting and drying cycles

Hossein Nowamooz[1,a], Kai Li[1] and Cyrille Chazallon[1]

[1]*INSA de Strasbourg, Civil engineering department, 67084, Strasbourg Cedex, France*

Abstract. It is important to model the behavior of unsaturated expansive soils subjected to wetting and drying cycles because they alter significantly their hydro-mechanical behavior and therefore cause a huge differential settlement on shallow foundations of the structure. A simplified model based on the shakedown theory (Zarka method) has been developed in this study for unsaturated expansive soils subjected to wetting and drying cycles. This method determines directly the stabilized limit state and consequently saves the calculation time. The parameters of the proposed shakedown-based model are calibrated by the suction-controlled oedometer tests obtained for an expansive soil compacted at loose and dense initial states, and then validated for the same soil compacted at intermediate initial state by comparing the model predictions with the experimental results. Finally, the finite element equations for the proposed shakedown model are developed and these equations are implemented in the finite element code CAST3M to carry out the full-scale calculations. A 2D geometry made up of the expansive soil compacted at the intermediate state is subjected to successive extremely dry and wet seasons for the different applied vertical loads. The results show the swelling plastic deformations for the lower vertical stresses and the shrinkage deformations for the higher vertical stresses.

1 Introduction

Unsaturated expansive soils contain clay minerals such as smectite with the high capacity of water absorption. As the clay minerals absorb water, they expand; conversely, as they lose water, they shrink. The successive wetting and drying cycles produce the differential settlements and crack the structures built on this material [1-6]. To investigate the hydro-mechanical behavior of unsaturated expansive soils, it is necessary to determine the coupled behavior of expansive soils for the different applied loads and environmental solicitations.

Nowamooz and Masrouri [7], have performed the tests on loosely and densely compacted bentonite/silt mixtures for several wetting and drying cycles. In their work, the authors pointed out that the loose samples present a shrinkage strain accumulation while a swelling strain accumulation can be observed for dense samples before reaching a unique equilibrium state at the end of the wetting and drying cycles, regardless of the initial dry density of these expansive soil samples. In other words, the loose and dense samples will achieve a unique final state after a number of wetting and drying cycles.

Many methods for volume change prediction of unsaturated expansive soils have been proposed [8-12]. Barcelona Expansive Model (BExM) developed by Alonso et al. [12], can be mentioned as the most accepted theoretical reference. This model is able to simulate the basic behavior of unsaturated expansive soil, including the strain fatigue phenomenon during wetting-drying cycles and the prediction of final equilibrium state at the end of the wetting-drying cycles. However, this model presents a large number of parameters such as the coupling functions for micro- and macrostructural strains. The calibration of these parameters needs several experimental tests which lead to a time-consuming procedure to characterize their hydro-mechanical behaviors.

The shakedown theory has been simplified by Zarka [13,14] for kinematic hardening materials such as metals. Several authors [15-19] then applied this method to study the cyclic mechanical behavior of soils. They introduced a series of transformed internal variables to characterize the mechanical system and then constructed a local geometry in transformed internal parameter plane to estimate the stabilized limit state and its associated plastic components. This direct determination of the steady solutions in shakedown analysis is able to replace classic step-by-step method and needs less model parameters.

In this context, this paper presents a shakedown-based model for the simulation of the volume change problem in expansive soils subjected to several wetting and drying cycles. The constitutive model based on shakedown theory (Zarka method) is firstly presented. The required parameters of this model are next calibrated and validated by the experimental results obtained for an expansive soil compacted at loose, intermediate and dense initial states. After the implementation of the shakedown theory in the

[a] Hossein.Nowamooz@insa-strasbourg.fr

finite element code, the plastic strain field and the inelastic displacement field are finally calculated for a 2D geometry consisted of the intermediately compacted expansive soil and subjected to successive wetting and drying cycles.

2 Analytical modeling of expansive soils based on shakedown concept

The purpose of this section is to present a simplified shakedown-based model to simulate the hydro-mechanical behaviour of unsaturated expansive soils during the successive wetting and drying cycles.

It is generally accepted that a unique plane (net mean stress-suction) is sufficient to describe the hydro-mechanical behaviour of unsaturated soils. Figure 1 shows in this plane a rectangular yield surface representing the elastic domain. The equations of the different boundaries can be given by:

$$s = s_I \quad (1)$$

$$s = s_D \quad (2)$$

$$p = p_0 \quad (3)$$

where, s_I is the Suction Increase limit, s_D is the Suction Decrease limit and p_0 is the Preconsolidation Stress.

In this work, we take into account the width of the elastic domain is small and is not changing with suction cycles.

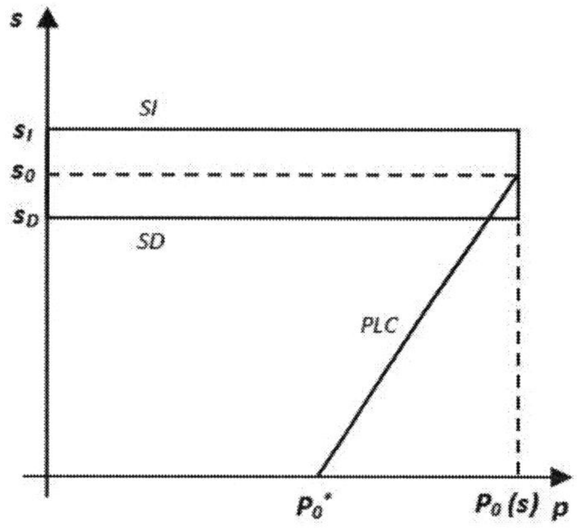

Figure 1. Rectangular shape for the yield surfaces in (suction - preconsolidation stress) plane.

The suction variation within the rectangular will result in the elastic volumetric strain:

$$d\varepsilon_{vs}^{e} = \frac{\kappa_s}{v} \cdot \frac{ds_0}{s_0} \quad (4)$$

and if the boundaries ($s=s_I$ and $s=s_D$) are activated, the following total and plastic volumetric strain will be generated,

$$d\varepsilon_{vs} = \frac{\lambda_s}{v} \cdot \frac{ds_0}{s_0} \quad (5)$$

$$d\varepsilon_{vs}^{p} = \frac{\lambda_s - \kappa_s}{v} \cdot \frac{ds_0}{s_0} \quad (6)$$

in which, κ_s and λ_s are elastic stiffness index and elastoplastic stiffness index for suction variation, respectively.

2.1 Plastic Shakedown during suction cycles

When the loading amplitude becomes very large, a stabilized limit state can be reached at the end of the suction cycles. The transformed internal parameter (y_α) as well as the preconsolidation stress (p) is presented in the same plane in Figure 2. In the (transformed internal parameter - preconsolidation stress) plane, the convex which characterizes the behaviour of the soil sample translates between the minimum suction (s_{min}) position and the maximum suction (s_{max}) position during the wetting and drying cycles. If the extreme positions of the convex have no common part on the transformed internal parameter axis, plastic shakedown will occur. Therefore, the variation of the volumetric plastic deformation ($\Delta\varepsilon_{vs}^{p}$) during suction cycles can be computed by:

$$\Delta\varepsilon_{vs}^{p} = \frac{1}{h} \cdot \Delta y_\alpha \quad (7)$$

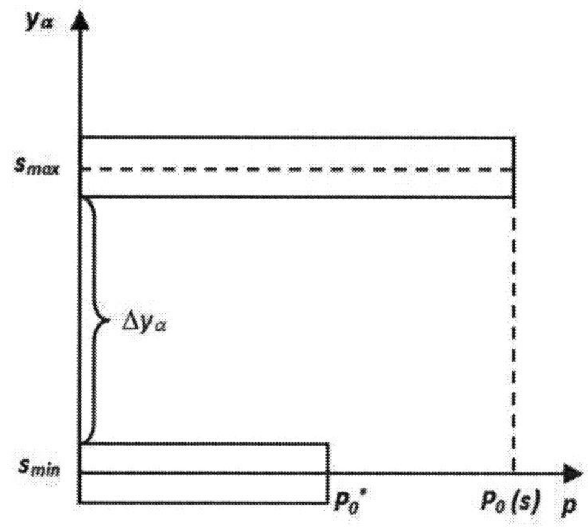

Figure 2. y_α- p (transformed internal parameter - preconsolidation stress) plane for plastic shakedown.

2.2 Elastic behaviour at the equilibrium state

After several wetting and drying cycles, the equilibrium state will be obtained at the end of the suction cycles in the (strain-suction) plane where no plastic strain accumulation can be observed. Consequently, a linear variation of the elastic strain with the suction can be supposed at the equilibrium state. It can be written as:

$$\Delta\varepsilon_{vs}^{e} = \frac{\kappa_s}{v} \cdot \Delta s \quad (8)$$

where, κ_s is the elastic stiffness index for suction variation.

Here, the resilient modulus E_r is defined and the above equation becomes:

$$\Delta\varepsilon_{vs}^{e} = \frac{1}{E_r} \cdot \Delta s \quad (9)$$

with

$$\frac{1}{E_r} = \frac{\kappa_s}{v} \cdot \frac{1}{s + p_{at}} \tag{10}$$

where, E_r can be calibrated at the elastic equilibrium state.

3 Calibration of the parameters for the shakedown-based model

Nowamooz and Masrouri [1,7,,20] and Nowamooz et al. [6] studied the hydro-mechanical behaviour of an artificially prepared mixture of 40% silt and 60% bentonite after several drying and wetting cycles. The samples were compacted with the same initial water content (w) of 15%. Three different initial dry densities (γ_d) were 1.27, 1.48 and 1.55 Mg·m^{-3}, corresponding to loose (L), intermediate (I) and dense (D) soils, respectively.

Required parameters of the proposed shakedown model are determined by the oedometer tests with suction cycles between 0 and 8 MPa under three constant vertical stresses: 15, 30 and 60 kPa.

Figure 3 and 4 show the evolution law of the inverse of the resilient modulus (1/Er) as well as the inverse of the hardening modulus (1/h) with the net mean stress (p) for loose and dense samples, respectively.

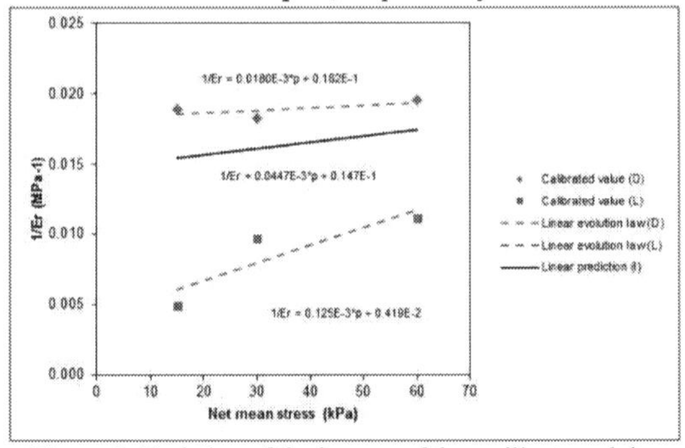

Figure 3. Variation of the inverse of the resilient modulus (1/Er) with net mean stress for the loose and dense samples as well as the prediction for the intermediate samples.

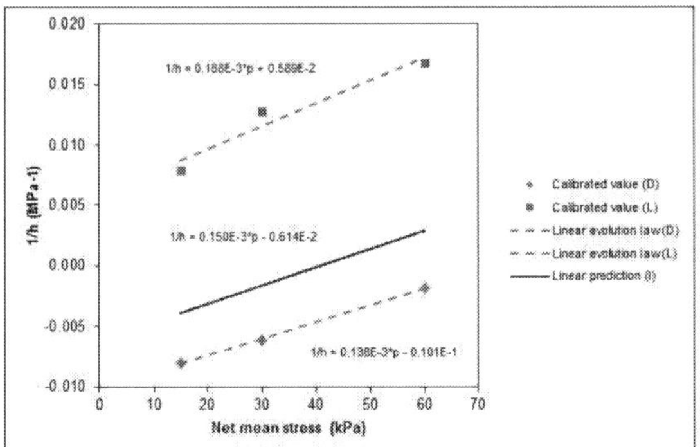

Figure 4. Variation of the inverse of the hardening modulus (1/h) with net mean stress for the loose and dense samples as well as the prediction for the intermediate samples.

The validation of the model is carried out with the test results obtained for the samples compacted at the intermediate initial state. The linear variation of the inverse of the elasticity parameter (1/Er) as well as the hardening modulus (1/h) with the net mean stresses is interpolated in Figures 3 and 4 for the these samples. For intermediate samples, the threshold value of elastic limit for suction variation (s_α) is also considered negligible.

Based on these predicted model parameters, the model validation is conducted for the intermediated samples. Figure 5 represents the comparison between the test results and the model predictions at the different net mean stresses. For these samples, the initial state is closer to the reversible line which need less suction cycles to obtain the equilibrium state. The relative tolerance varies between 5% and 8% confirming the capacity of the proposed model to estimate the accumulated plastic strains.

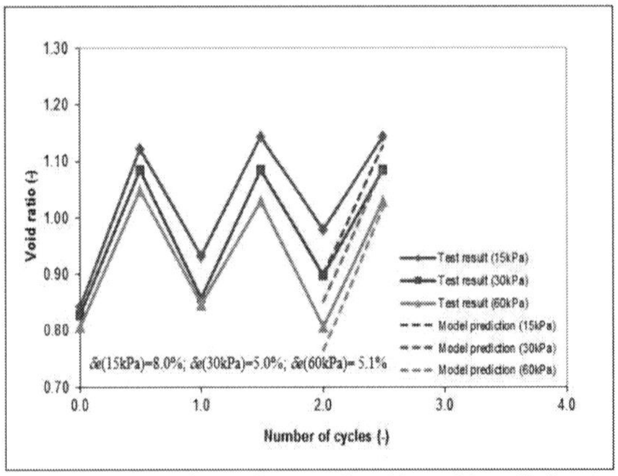

Figure 5. Comparisons of test results with model predictions for intermediate sample at the different net mean stresses.

4 Numerical simulation of the in-situ behavior of an expansive soil

In this section, the plastic strain field and inelastic displacements of an expansive soil will be simulated after several wetting and drying cycles due to the variation of the climatic conditions, till the stabilized limit state. We use the finite element code CAST3M [21] in which the proposed shakedown-based model has been implemented.

Figure 6 shows the studied geometry 600 cm×200 cm, made up of the intermediately compacted expansive soil (presented in section 3). For the simulations, the geometry is discretized into 4-node quadrilateral elements and 4800 elements have been used to analyze 2-D plane strain problem.

Before shakedown calculation, the in-situ geostatic analysis of the expansive soils is performed where the gravity and lateral stresses are applied to the model (Figure 6). This geostatic stresses are determined by the initial dry density of soil (γ_d) and the lateral stress coefficient K_0 where these values are summarized in Table 1. Young modulus E and Poisson ratio ν necessary for the elastic mechanical analysis are also given in Table 1.

DOI: 10.1051/e3sconf/20160908007

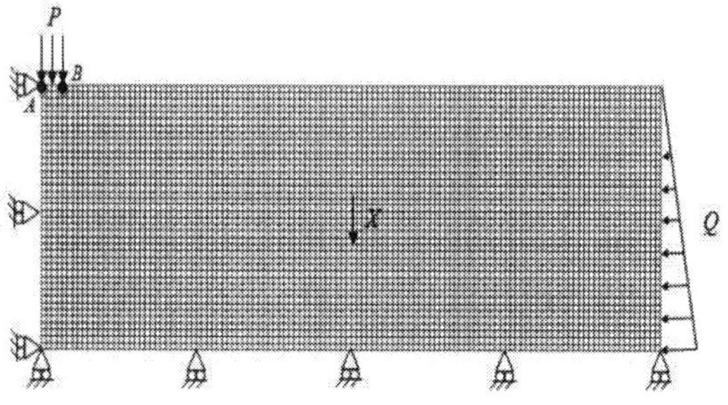

Figure 6. Finite element meshing and loading description for 2D geometry simulation.

Table 1. Mechanical parameters of the intermediate samples

Parameters	E(MPa)	ν(-)	γ_d	K_0(-)
Intermediate soil	18	0.2	1.48	0.5

Different vertical loads (P) are applied on the top of the geometry between points A and B in Figure 6 (the length of AB is 30 cm): 15, 30, 60 kPa and the limit stress of 67.85 kPa. The limit stress is 67.85 kPa produces the maximum net mean stress of 60 kPa in the finite element model. Since we obtained the evolution law of shakedown model between 15 and 60 kPa for all the studied materials, the net mean stress values beyond 60 kPa are not taken into account in the numerical modeling.

We considered a linear variation of suction with depth as described in Figure 7 in the dry season, while the soil is fully saturated (suction zero) in the wet season. Then, the finite element calculation of the limit state is performed for the wetting and drying cycles between these two extreme conditions under different vertical loads.

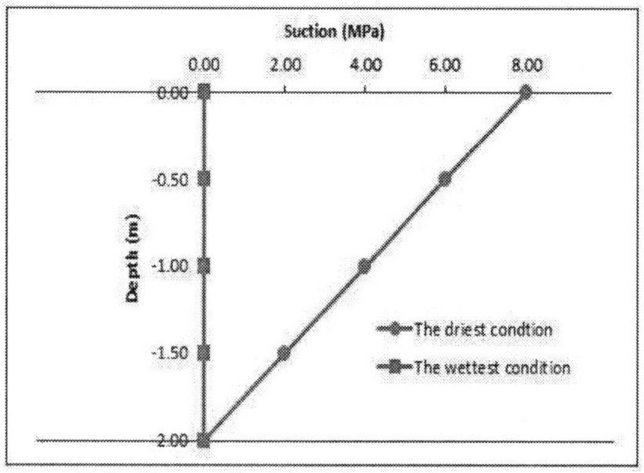

Figure 7. Suction profile at the driest and the wettest conditions for 2D geometry simulation.

Figure 8 shows the plastic strains on the soil surface (y=0 m) after several wetting and drying cycles for the different vertical stresses. It can be observed that the vertical stress mainly influences the plastic strain field between points A and B. The swelling plastic strains are produced at the vertical stresses of 15 and 30 kPa, while the accumulated shrinkage deformations are occurred for the higher vertical stresses. Outside AB, the swelling plastic strain reaches its maximum because of null external loads.

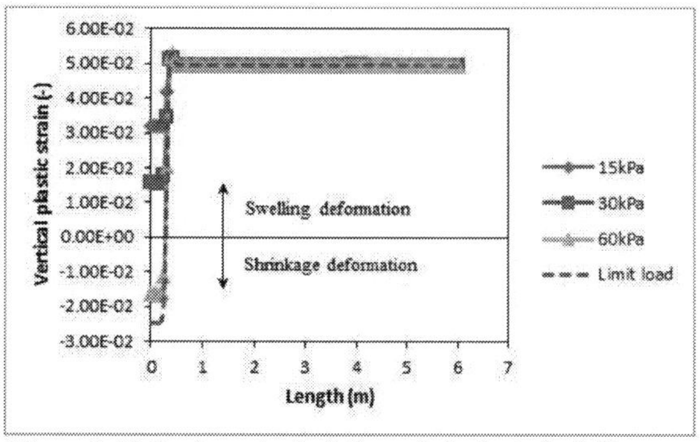

Figure 8. Finite element calculations of plastic strain field;. on the surface (y=0 m) of 2D geometry.

5 Conclusion

Several models are developed for the complex hydro-mechanical behavior of the expansive soils submitted to the successive wetting and drying cycles, but these models are based on the traditional incremental method in elasto-plasticity leading to a very large calculation time. To solve this problem, the simplified model based on Zarka shakedown method is developed for expansive soils.

The proposed shakedown model requires three parameters: the linear evolution law of elasticity parameters (1/Er) determined from the equilibrium state at the end of the wetting and drying cycles; the linear evolution law of hardening modulus (1/h) calibrated from the accumulated plastic strains during the wetting and drying cycles; and the threshold value of elastic limit for suction variation (s_α), taken zero for the sake of simplification.

With the framework of shakedown theory, finite element calculations of a 2D (plane strain) model, made up of intermediate soils was carried out. The soil is subjected to suction cycles between extremely dry and wet conditions under different vertical loads, till the stabilized limit state is achieved. The results show that the swelling plastic strains on the surface of the model (y = 0 m), are produced at the lower vertical stresses while the accumulated shrinkage deformations are occurred for the higher vertical stresses.

Our future development will deal with the activation of the PLC yield surface with the suction cycles to couple our proposed shakedown-based model with the mechanical behavior of the expansive soils.

References

[1] Nowamooz H, F,Masrouri. Hydromechanical behavior of an expansive bentonite/silt mixture in cyclic

suction-controlled drying and wetting tests. Engineering Geology 2008; 101: 154-164.
[2] Alonso EE, Romero E, Hoffmann C, Escudero EG. Expansive bentonite/sand mixture in cyclic controlled suction drying and wetting. Engineering Geology 2005; 81: 213-226.
[3] Dif AE, Blumel WF. Expansive soils under cyclic drying and wetting. Geotechnical Testing Journal 1991; 14: 96-102.
[4] Day RW. Swell-shrink behavior of compacted clay. Journal of Geotechnical Engineering 1994; 120: 618-623.
[5] Al-homoud A, Basma A, Malkawi AH, Bashabsheh MA. Cyclic swelling behavior of clays. Journal of Geotechnical Engineering 1995; 121: 562-565.
[6] Nowamooz H, Jahangir E, Masrouri F. Volume change behavior of a swelling soil compacted at different initial states. Engineering Geology 2013; 153: 25-34.
[7] Nowamooz H, Masrouri F. Influence of suction cycles on the soil fabric of compacted swelling soil. Compted Rendus Geoscience 2010; 342: 901-910.
[8] Alonso EE, Gens A, Josa A, A constitutive model for partially saturated soils. Geotechnique 1990; 40: 405-430.
[9] Cui YJ, Yahia-aissa M, Delage P. A model for the volume change behavior of heavily compacted swelling clays. Engineering Geology 2002; 64: 233-250.
[10] Wheeler SJ, Sharma RS, Buisson MSR. Coupling of hydraulic hysteresis and stress-strain behavior in unsaturated soils. Geotechnique 2003; 53: 41-54.
[11] Sun WJ, Sun DA. Coupled modeling of hydro-mechanical behavior of unsaturated compacted expansive soils. International Journal for Numerical and Analytical Methods in Geomechanics 2011; 36: 1002-1022.
[12] Alonso EE., Vaunat J, Gens A. Modeling the mechanical behavior of expansive clays. Engineering Geology 1999; 54: 173-183.
[13] Zarka J, Casier J. Elastic plastic response of structure to cyclic loading: practical rules. In: Nemat-Nasser S, editor. Mechanics Today, vol. 6. Pergamon Press: Oxford; 1979, p. 93-198.
[14] Zarka J, Frelat J, Inglebert G., Kasmai-Navidi P. A new approach in inelastic analysis of structures. M.Nijhoff, Dordrecht, 1990.
[15] Sharp R, Booker J. Shakedown of pavements under moving surface load. Journal of Transportation Engineering 1984; 110: 1-14.
[16] Chazallon C, Koval G, Hornych P, Allou F, Mouhoubi S. Modeling of rutting of two flexible pavements with the shakedown theory and the finite element method. Computers and Geotechnics 2009; 36: 798-809.
[17] Chazallon C, Allou F, Hornych P, Mouhoubi S. Finite element modeling of the long term behavior of a full scale flexible pavement with the shakedown theory. International Journal for Numerical and Analytical Methods in Geomechanics 2009; 33: 45-70.
[18] Habiballah T, Chazallon C. An elastoplastic model based on the shakedown concept for flexible pavements unbound granular materials. International Journal for Numerical and Analytical Methods in Geomechanics 2005; 29: 577-596.
[19] Allou F, Chazallon C, Hornych P. A numerical model for flexible pavements rut depth evolution with time. International Journal for Numerical and Analytical Methods in Geomechanics 2007; 33: 1-22.
[20] Nowamooz H, Masrouri F. Mechanical behavior of expansive soils after several drying and wetting cycles. Geomechanics and geoengineering 2010; 5(4): 213-221.
[21] CAST3M. See website http://www-cast3m.cea.fr/. (CAST3M is a research FEM environment; its development is sponsored by the French Atomic Energy Commission).

DOI: 10.1051/e3sconf/20160908008

Investigation into the isotropic compression of wet granular soils using discrete element method

Vinh-Du Than[1,2,a], Jean-Noël Roux[1], Anh-Minh Tang[1] and Jean-Michel Pereira[1]

[1] *Université Paris-Est, Laboratoire Navier (UMR 8205), ENPC, CNRS, IFSTTAR, 77455 Marne-la-Vallée Cedex 2, France*
[2] *The University of Danang, College of Technology, Department of Civil Engineering, Danang City, Vietnam*

Abstract. In this paper, the authors aim to present a 3D model using the Discrete Element Method (DEM) to study the quasistatic response of very loose assemblies of frictional spherical grains to an isotropic compression in the presence of a small amount of an interstitial liquid, which gives rise to capillary menisci and attractive forces. A reduced pressure P^* is defined, comparing the confining pressure to the tensile strength of contacts. The effects of initial assembling process and various micromechanical parameters on the plastic compression curves are first studied. The influences of rolling resistance at contacts and the size polydispersity along those compression curves are also mentioned.

1 Introduction

Bonded granular materials are present in many natural and industrial processes, in which this bonding may have different physical origins, and proceed for instance from capillary effects, from solid precipitations, or from artificial solid bridges at the intergranular contacts. In geotechnical engineering, for the case of granular soils, the presence of water menisci between soil particles plays a key role in the overall behavior. Hence, the cohesive materials exhibit some specific features that do not appear in non-cohesive materials, such as the ability to form stable structures at very low density, and they are very sensitive to stress intensity as well as the external force. So far, there are a lot of studies in the macro-micromechanical behavior of bonded granular materials, such as: cohesive powders [1,2], wet beads [3], loessic soils [4,5], wet sand [6,7], etc. The contact networks are hardly accessible to experiments, and intergranular forces are also inaccessible to measurements. Generally, the behavior of materials under the growing of external load (oedometric or isotropic tests) is characterized by the compression curve which is the irreversible compaction under increasing pressure.

The first numerical simulation method of the "discrete element" type was introduced 35 years ago by [8]. Since then, the DEM has become a valuable and efficient tool to investigate the microscopic mechanisms and to classify mechanical properties of granular systems. Yet, cohesive granular materials, especially in loose/very loose states, have less frequently been investigated by numerical simulation than cohesionless ones. Based on the 2D model for cohesive powders of Gilabert *et al.* [1,2], a 3D model [9] is developed to investigate the mechanical behavior of a model bonded granular soils (glass beads with capillary bonds in the pendular state). Results are presented for simulations of isotropic compression cycles, both for the macroscopic behavior (plastic compression curve) and the evolution of the arrangement of microstructure upon growing of pressure.

2 Model material

2.1 Force model

The elastic and frictional forces are jointly implemented in contacts as in [10]. The simplified Hertz-Mindlin-Deresiewicz force model is used for calculating the contacts between grains, relating the elastic normal force F^H_N to the normal deflection h, as described in [11,9].

The attractive capillary force is only present if grains have been in contact and the meniscus has not broken since. A meniscus of volume V_m remains until separation distance reaches its rupture value, $D_r \approx V_m^{1/3}$ [12]. The Maugis approximation [13] is chosen for capillary forces:

$$F^{cap} = -F_0\left[1-\frac{1}{\sqrt{1+\dfrac{4V_m}{\pi a D^2}}}\right]. \quad (1)$$

D denotes here the distance between the surfaces of the particles joined by the liquid meniscus. $F_0 = \pi a\gamma\cos\theta$ is the maximum tensile force, involving surface tension γ of the water-air interface (7.27x10^{-2} N/m at 20°C), and θ is the wetting angle, with $\theta = 0$ for a perfectly wetting liquid. Most calculations here are carried out with $V_m/(a^3) = 10^{-3}$, and thus the force range extends to $D_r = a/10$

[a] Corresponding author: vinh-du.than@cermes.enpc.fr

(corresponding to a very small capillary force). This ratio of meniscus volume V_m to a^3 is one of the dimensionless control parameters in the present study, determining the number of liquid bonds and the range of the corresponding capillary attraction. This work is limited to the pendular state of isolated menisci [6].

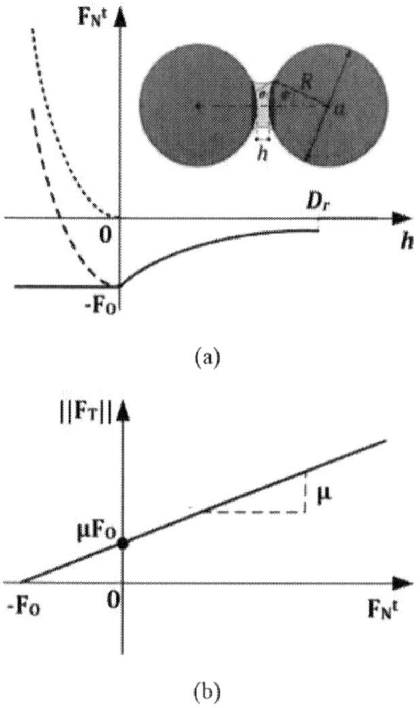

Figure 1. (a) Static normal force versus deflection h or distance $D = -h$. (b) Coulomb cone limiting the value of tangential force.

The total static normal force combines the elastic normal force and the capillary force, $F^t_N = F^H_N + F^{cap}$, as shown in Figure 1-a. A viscous force is added in contacts, opposing the normal relative velocity of grains as in [14], in order to damp the vibrations about equilibrium states – with no notable influence on quasistatic rheology.

Two particles moving away from each other after a collision will only separate if their receding relative velocity is large enough to overcome the capillary force, i.e. larger than a threshold proportional to V^* [1], with

$$V^* = \sqrt{\frac{F_0 D_r}{m}} \qquad (2)$$

The influence of *rolling resistance* (RR) at contacts, as investigated in 2D in [2], is also studied. For simplicity, we only implement this feature with linear contact elasticity. The existence of RR is related to the particle surface roughness, such that contact regions are larger than the ones deduced from contact elasticity. Four additional parameters are necessary: (i) a *rolling spring constan* K_R which expresses the proportion between relative rotation and rolling moment (in the tangential plane), as long as the rolling friction threshold is not reached; (ii) a *pivoting spring constant* K_P (chosen equal to K_R), relating similarly the pivoting moment to the pivoting (i.e., along the normal direction) relative rotation angle; (iii) a *rolling friction coefficient* μ_R with the dimension of a length, setting the maximum norm of the rolling moment $\|\Gamma_R\|$ to $\mu_R F^H_N$, proportional to the elastic part of the normal force; and (iv) a *pivoting friction coefficient* μ_P (chosen equal to μ_R), requesting similarly the absolute value of the pivoting moment to stay below $\mu_P F^H_N$. In most simulations with RR, the values of $\mu_R/a = 0.02$, $K_R/(K_N a^2) = 10^{-4}$ while $K_T/K_N = 1$ and $K_N/a = 4$ GPa.

2.2 Stress control and equilibrium

Numerical samples comprise N=4000 monodisperse spherical beads of diameter a and mass m, with no gravity, within a cubical cell with periodic boundary conditions. The cell edges are parallel to the coordinate axes $(x_\alpha)_{\alpha=1,3}$. Their lengths $(L_\alpha)_{\alpha=1,3}$ vary simultaneously with the grain positions until a mechanical equilibrium state is achieved for all particles with externally imposed values $(\Sigma_\alpha)_{1\leq\alpha\leq3}$ of principal stresses:

$$\sigma_{\alpha\alpha} = \frac{1}{\Omega}\left[\sum_i m_i v_i^\alpha v_i^\alpha + \sum_{i<j} F_{ij}^{(\alpha)} r_{ij}^{(\alpha)}\right]. \qquad (3)$$

Here, $\Omega = L_1 L_2 L_3$ is the sample volume, $r^{(\alpha)}_{ij}$ is coordinate α of vector $\mathbf{r}_{ij}$ joining the centers of neighbor beads i and j and $F^{(\alpha)}_{ij}$ the corresponding force coordinate. Velocities $\mathbf{v}_i$ of grain centers comprise, in addition to a periodic field, an affine term corresponding to the global strain rate. Equations of motion for dimensions L_α are written in addition to the ordinary equations for the dynamics of a collection of solid objects, and they drive the system towards an equilibrium state in which conditions $\sigma_{\alpha\alpha} = \Sigma_\alpha$ are satisfied [14].

A typical intergranular force value $F_1 = \max(F_0, Pa^2)$ sets the tolerance levels for individual grain equilibrium, where P is the applied pressure. A configuration is deemed equilibrated when the following conditions are simultaneously satisfied: (i) the net force (the total force) on each spherical grain is less than $10^{-4}F_1$; (ii) the total moment on each sphere is lower than $10^{-4}F_1 a$; (iii) the difference between imposed and measured stresses is less than $10^{-4}F_1 a$; and (iv) the kinetic energy per grain is less than $10^{-7}F_1 a$.

3 Simulation procedures

3.1 Specimen preparation

A hard sphere event-driven method is first used to prepare disordered, low density configurations of 4000 grains. All particles are then launched with Gaussian-distributed random velocities with quadratic mean V_0. They collide and stick to one another within a cell of constant size, forming larger and larger aggregates. Finally, all grains are connected to one another by cohesive contacts and reach an equilibrium position. All specimens are prepared in the pendular state, just to create liquid bridges at the contact points of grains. Cohesive forces act through the liquid bridges.

This initial structure depends on one dimensionless parameter, characterizing the agitation intensity in the assembling stage. It is defined as the ratio of V_0 to the minimum receding velocity V^* introduced in Eq. (2), this ratio is also called initial coordination number and it will be used priorly. As in [1], larger initial agitation levels are observed to produce better connected structures. The initial parameters are listed in Table 1. These values Φ_0=0.30, V_0/V^*=0.2041 and V_m/a^3=10^{-3} will be set for the standard values.

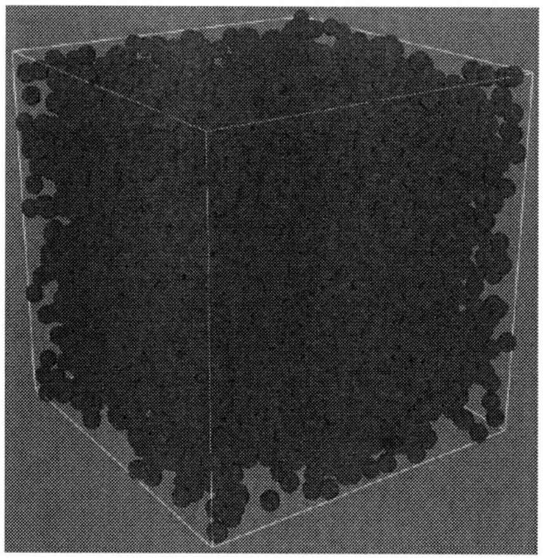

Figure 2. A cubical specimen before compaction process.

Table 1. Initial parameters with initial solid fraction Φ_0, initial coordination numbers of contacts z^0_c, and of distant interactions z^0_d, agitation level V_0/V^*, meniscus volume V_m/a^3.

Φ_0	V_0/V^*	z^0_c	z^0_d	V_m/a^3
0.25	0.2041	4.0850	0	1.00x10^{-3}
0.30	0.2041	4.0780	0	1.0x10^{-3} ÷ 7.8x10^{-6} (factor of 1/2)
	0.4082	4.1110	0	
	1.2247	4.1530	0	
	4.0825	4.3715	0.0120	
	12.2474	4.5965	0.0820	
	40.8248	4.7480	0.2510	
0.32	0.2041	4.0730	0	1.00x10^{-3}
0.35	0.2041	4.0780	0	1.00x10^{-3}
0.40	0.2041	4.0790	0	1.00x10^{-3}
0.45	0.2041	4.0785	0	1.00x10^{-3}

In this table, the level of initial agitation in the end of the assembling stage, expressed by ratio V_0/V^*, determines the initial solid fraction.

3.2 Compaction process

The final packing structures are compacted under growing external isotropic pressure. The effect of the external pressure should be compared to the adhesive strength F_0 in contacts. This defines a dimensionless reduced pressure P^*, $P^* = Pa^2/F_0$, in which F_0 is already given in connection with Eq. (1), $F_0 = \pi a \gamma$ (assuming a null contact angle), so $P^* = Pa/(\pi a \gamma)$. In the present study, $P^* = 1$ means P = $\pi\gamma/a \approx 2$ kPa, with beads of diameter 118μm in water. For $P^* << 1$, initial loose structures are stabilized by adhesion and the relevant force scale is F_0. For large reduced pressures $P^* >> 1$, externally imposed forces dominate the adhesive strength, the relevant force scale is Pa^2 [9].

The main objective of the present paper is the study of the effect of a gradual compression, starting from cohesion-dominated loose states at small P^*, and ending in confinement-dominated denser states at large P^*. A stepwise pressure-controlled loading path is applied. In each compression step, pressure P^* is multiplied by a constant factor $10^{1/4}$, and one waits until the new equilibrium configuration is reached, with the criteria stated in Sec. 2.2. The compression program is pursued until P_{max}^* (maximum value 10^4), well beyond complete plastic collapse is obtained. Then, the effect of P^* back to 10^{-3} from its highest value is also simulated.

4 Simulation results

4.1 Effect of initial solid fraction

All the tests are run with low packing solid fraction, from Φ_0=0.25 to Φ_0=0.45. In this study, the results are priorly shown in the conventional compression – decompression curves in soil mechanics e-log(P^*), in which e = 1/Φ - 1. Figure 3-a illustrates the compression and decompression curves for different values of initial solid fraction Φ_0 with V_0/V^*=0.2041 and V_m/a^3=10^{-3}.

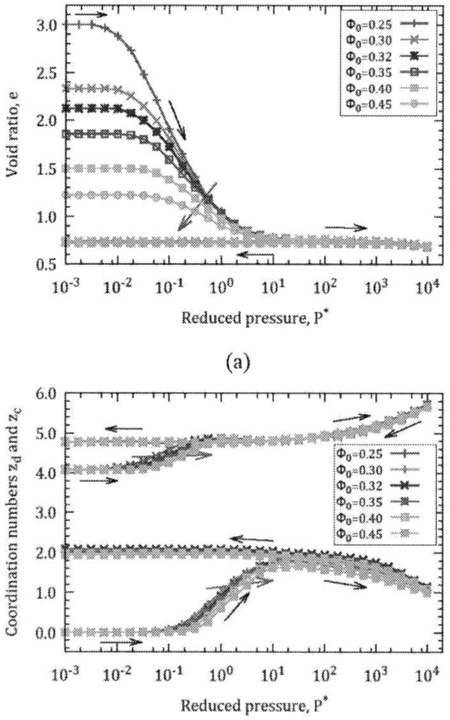

(a)

(b)

Figure 3. (a) Void ratio e and (b) coordination numbers z_c and z_d versus P^* in compression cycle for different values of Φ_0.

Three regimes can be classified in the compression curve, depending on the range of the pressure P^* [9]. A first regime, thereafter called regime **I**, is observed in the range of low reduced pressures P^*. In this regime, the initial structure can sustain the increasing pressure without rearrangement and void ratio e remains nearly constant. The regime **II** starts when void ratio starts to decrease sharply. Finally, void ratio e decreases slightly in regime **III** and approaches the minimum void ratio e_{min} at the end of the loading process. Upon decompression, e increases slightly, remaining very close to e_{min}. It can be observed from the decompression that the compaction cycle is an irreversible process. For different initial solid fraction Φ_0, the denser systems are able to support larger pressures before rearranging, whence larger regime **I** plateaus. However, in regime **II**, these compression curves start to converge and very nearly coincide in regime **III**.

The initial states of structures are strongly influenced by external pressures while the microstructures seem to have no effect on the microstructural arrangement. An important microstructural characteristic, the coordination number z, defined as the average number of interactions per grain, is the sum of the contact coordination number z_c and the coordination number of distant interactions, through menisci joining non-contacting grains, z_d.

Figure 3-b plots z_c, z_d versus P^* in the compression cycle. Initially, one has $z_d = 0$, as the same minimal initial coordination number in the aggregation process, are low, and do not allow contact opening. z_c and z_d remain unchanged in regime **I**. Both coordination numbers start to increase as the structure collapses and reorganises in regime **II**. They exhibit little change in regime III, showing that the increase of the number of contacts is mainly due to the closing of narrow gaps between pairs of grains joined by a meniscus, as the structure is further compressed -- a moderate effect, partly reversed upon unloading. Further unloading back to small pressures takes place with nearly constant coordination numbers of both types, reflecting the stability of the dense structure formed at high P^*.

4.2 Effect of initial coordination number

Ratio V_0/V^*, characterizing the intensity of initial agitation and its ability to break adhesive contacts, strongly influences the initial assembling process and the resulting coordination number.

Figure 4-a shows that this parameter mainly affects the beginning of the compression curve. Six different values of V_0/V^* are used, the difference between minimum and maximum values is two-hundredfold. Under low P^*, the wider regime **I** plateaus correspond to the higher initial coordination numbers. If initial velocity V_0 is of the order of V^*, the initial packing structure is strongly perturbed. In other words, the larger the initial coordination numbers, the stronger the initial structures.

Moreover, the relatively strong increases in initial coordination number of z_c and z_d (under low pressure P^*) prove that more contacts are created when large initial coordination numbers are considered (see Figure 4-b). Conversely, with low intensity of initial coordination numbers ($V_0 < V^*$), grains stick gently to one another and clusters of aggregated grains are not disturbed. In other words, lower initial coordination numbers induce more tenuous aggregates.

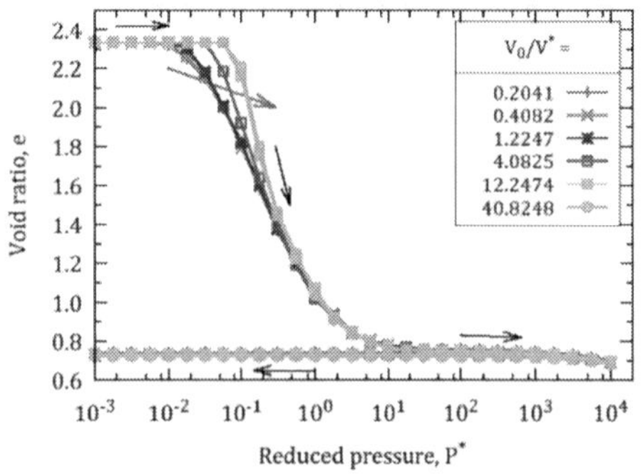

(a)

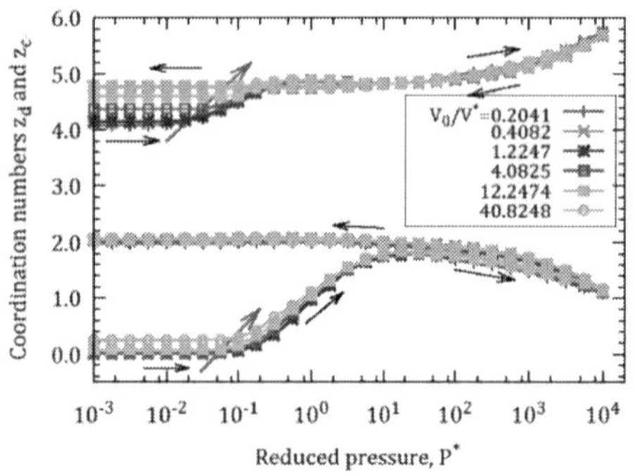

(b)

Figure 4. (a) Compression and decompression curves and (b) coordination numbers z_c and z_d for standard values of Φ_0 & V_m/a^3, different values of V_0/V^*.

4.3 Effect of meniscus volume

The investigations of the influence of meniscus volume V_m/a^3 are now reported, which is kept low enough for the material (with a saturation hardly exceeding 1%) is maintained in the pendular state.

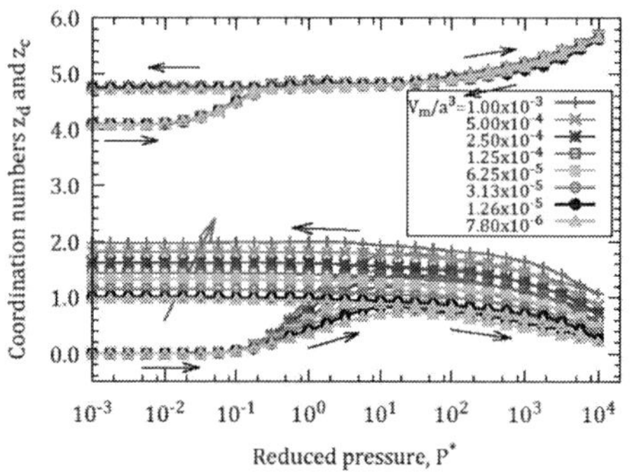

Figure 5. Coordination numbers z_c and z_d for standard values of Φ_0 & V_0/V^*, different values of V_m/a^3.

DOI: 10.1051/e3sconf/20160908008

Despite the importance of capillary bridges in the stabilization of loose structures, changing the meniscus volume (or liquid content) seems to have no effect on the macroscopic behavior of the granular specimens under isotropic compression (no apparent change in the compression curve, e versus $\log(P^*)$). This is corroborated by the fact that very little change in the coordination number of contacts z_c is observed. Only the coordination number of distant interactions z_d is notably influenced by such a change, especially in regimes **II** and **III**, and in the decompression process, as shown in Figure 5. As menisci break at larger interparticle distance with larger meniscus volume, more liquid bonds are present.

4.4 Effect of rolling resistance

Figure 6 illustrates the influence of initial coordination numbers V_0/V^* and rolling/pivoting friction coefficients $\mu_R = \mu_P = 0.02$ on the initial assembling process and coordination number for standard values of Φ_0 and V_m/a^3. For the smallest initial coordination number, the appearance of RR at contacts between grains creates weaker system under low P^*, with smaller coordination.

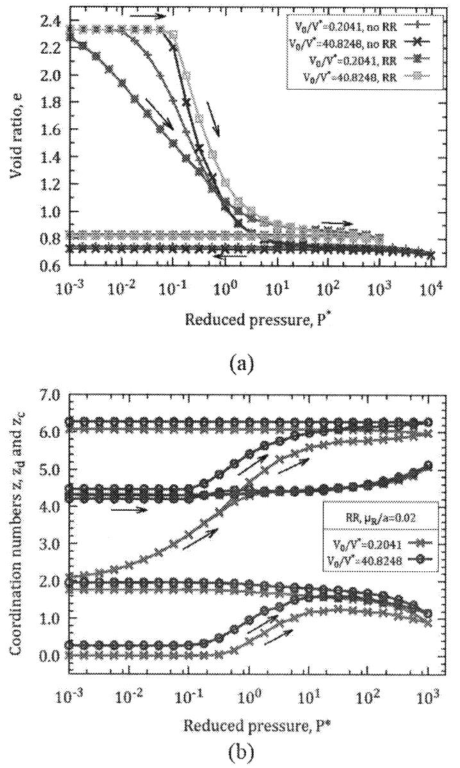

Figure 6. (a) Void ratio e and (b) z_c and z_d versus P^* for standard values of Φ_0 & V_m/a^3 in systems with and without RR.

Unlike the initial loose structures formed without RR, which can sustain a small non-vanishing pressure in regime **I**, the compression curve of the tenuously connected (with coordination number approaching 2, i.e., virtually no loop) initially formed with RR, and a low coordination number, responds by some plastic collapse at the first pressure increment, as shown in Figure 6-a. Its regime I, if it exists, is confined too much smaller pressures. It reappears on the pressure scale of Figure 6-a in systems assembled with larger initial coordination numbers, which are better coordinated (Figure 6-b), and collapse more abruptly in plastic compression. Interestingly, the systems assembled with the lower coordination numbers first collapse quickly by plastic compression (i.e., the structure is very fragile with RR), as z_c is gradually increased meanwhile z_d remains equal to zero (no contact opening creating a distant interaction). More complete analyses of the corresponding structural changes are being carried out, and their results will be published in forthcoming paper [15].

4.5 Effect of the size polydispersity

In granular materials, polydispersity means "several sources of dispersion", which is conventionally not only summarized by particle size distribution but also by particle shape and density. In this study, the effect of particle size distribution is only mentioned in order to compare with the monodisperse case. The specimen was also prepared in the same procedure and initial parameters, for 4000 frictional spherical grains with a uniform distribution of diameters ($d_{max} = 2d_{min}$) by volume fractions. In the present study, a cubical specimen is created with $d_{max} = 2$, $d_{min} = 1$, as shown in Figure 7.

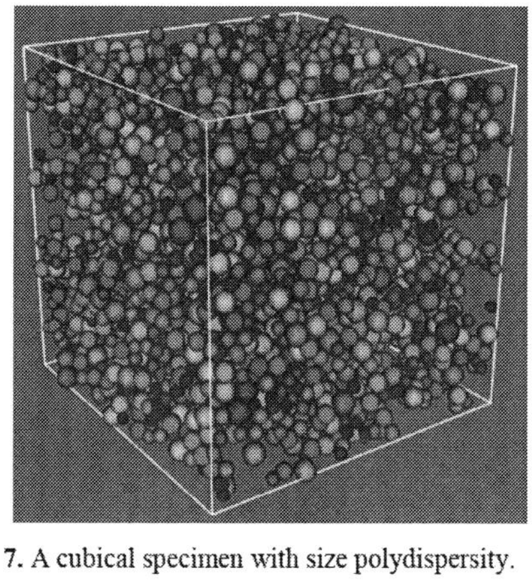

Figure 7. A cubical specimen with size polydispersity.

The model of Soulié *et al.* [16] is applied to calculate the capillary force. The new reduced pressure is redefined for polydisperse case [15], as follows:

$$P^*_{po} = \frac{P}{\pi\gamma}\frac{d_{max}d_{min}}{d_{max}-d_{min}}\ln\left(\frac{d_{max}}{d_{min}}\right) \tag{4}$$

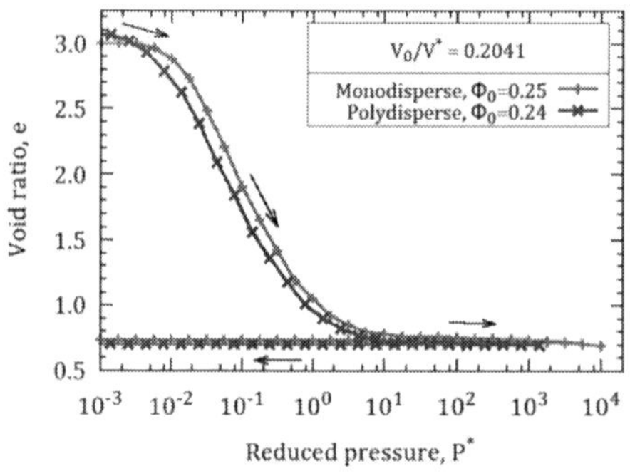

Figure 8. A comparison between monodisperse and polydisperse cases.

Figure 8 illustrates a comparison of compression and decompression curves between the monodisperse and polydisperse grains. This result confirms that, with the same initial parameters, the effect of the particle size distribution is negligible for the monodisperse grain, at least in the macroscopic behavior. More detailed analyses in the microstructural changes will be mentioned in forthcoming paper [15].

5 Conclusions and Perspectives

These results show that the simple model of wet granular materials exhibit the familiar plastic compression properties of cohesive soils or powders, with a well-defined limit of gently assembled (minimally coordinated) structures, and low initial densities, such that the behavior does not depend on dynamical model ingredients. A small level of rolling and pivoting friction is enough to produce strong differences in the plastic behavior, with the collapse under growing confining pressure becoming notably more gradual. Moreover, the change of particle size seems to be no effect on the macroscopic behavior.

The research program developed here can be extended in several directions. For example, the quasistatic behavior of a 3D model accounting for the effect of rolling resistance under oedometer and triaxial tests, in pendular state, is a straightforward extension of the present work. A comparison with an experimental investigation will be mentioned for macro-microscopic behavior. The role played by solid bridges instead of capillary bonds could also be investigated.

Acknowledgments

The first author of the paper is supported by the Scholarship (the Project No 911) of Ministry of Education and Training, the Vietnamese Government to do this work.

References

[1] F. Gilabert, J.-N. Roux, and a. Castellanos, Phys. Rev. E **75**, 011303 (2007).
[2] F. Gilabert, J.-N. Roux, and a. Castellanos, Phys. Rev. E **78**, 031305 (2008).
[3] Z. Fournier, D. Geromichalos, S. Herminghaus, M. M. Kohonen, F. Mugele, M. Scheel, M. Schulz, B. Schulz, R. Seemann, and A. Skudelny, J. Phys. Condens. Matter **17**, 477 (2005).
[4] J. Munõz-Castelblanco, P. Delage, J.-M. Pereira, and Y. J. Cui, Géotechnique Lett. **1**, 17 (2011).
[5] M. Jiang, H. Hu, and F. Liu, Can. Geotech. J. **1157**, 1147 (2012).
[6] N. Mitarai and F. Nori, Adv. Phys. **00**, 1 (2006).
[7] J.-F. Bruchon, J.-M. Pereira, M. Vandamme, N. Lenoir, P. Delage, and M. Bornert, Granul. Matter **15**, 783 (2013).
[8] P. A. Cundall and O. D. L. Strack, Géotechnique **29**, 47 (1979).
[9] V.-D. Than, S. Khamseh, A.-M. Tang, J.-M. Pereira, F. Chevoir, and J.-N. Roux, J. Eng. Mech. 1 (2015) (accepted).
[10] I. Agnolin and J.-N. Roux, Phys. Rev. E **76**, 061302 (2007).
[11] S. Khamseh, J.-N. Roux, and F. Chevoir, Phys. Rev. E **90**, 022201 (2015).
[12] S. Herminghaus, Adv. Phys. **54**, 221 (2005).
[13] D. Maugis, *Contact, Adhesion, and Rupture of Elastic Solids* (2000).
[14] I. Agnolin and J.-N. Roux, Phys. Rev. E **76**, 061303 (2007).
[15] V.-D. Than, J.-N. Roux, A.-M. Tang, and J.-M. Pereira, (2015) (to be published).
[16] F. Soulié, F. Cherblanc, M. S. El Youssoufi, and C. Saix, Int. J. Numer. Anal. Methods Geomech. **30**, 213 (2006).

Numerical modelling of desiccation cracking of clayey soil

Thi Dong Vo[1], Ahmad Pouya[2], Sahar Hemmati[1] and Anh Minh Tang[2]

[1]*IFSTTAR, Marne-la-Vallée, France*
[2]*Ecole des Ponts ParisTech, Marne-la-Vallée, France*

Abstract. The formation and propagation of desiccation cracks in soil is an extremely complex phenomenon because of the coupling between hydraulic and mechanical behaviour of soil, which are constituted here by the presence of capillary forces and discontinuities. The formation of a cracks network strongly influences the mechanical and hydraulic properties of soil. The main objective of this research is to study the evolution of suction and strain fields, the initiation and propagation of cracks under the effect of drying, using the finite element method. A simulation of a soil sample with four cohesive joints shows the results similar to experimental data. In addition, a simulation of multi-joints shows that cracks does not open in all potentials positions and it gives similar spacing.

1 Introduction

The desiccation cracking is a common phenomenon in soils and rocks. The essential mechanisms of initiation and propagation of desiccation cracking are not yet well understood, and consequently, the method to predict, control or avoid cracking is inadequately developed. In soils, especially in clayey soils, the water content decreases due to evaporation. This reduction of water content is accompanied with increase of suction and effective stress which causes shrinkage. If the shrinkage is restrained (by frictional boundary conditions, concentration of stress or gradient of humidity...), it will generate a tensile stress. When this tensile stress reaches the tensile strength of the material, a crack network can be formed. The desiccation cracks increase infiltration of fluid and influence the transport of pollutions [1][2] . They affect also the performance of coverage of waste disposal and decrease the strength of earth embankments [3][4].

The formation and evolution of desiccation cracks have been studied by many authors experimentally and numerically. Some numerical simulations have been carried out by discrete element method and others by finite difference method. The objective of this research is to study the evolution of suction and strain fields during desiccation, as well as the propagation of desiccation cracks with taking into account the interaction between them. The simulation is carried out by using the finite element (FEM) 2D code Porofis [5].This code allows modelling hydro-mechanical coupling and fluid mass exchanges between fracture/matrix, fracture/fracture, and fracture/boundary. The hydro-mechanical coupling in this code is realized by an iterative process passing from the hydraulic to the mechanical problem and vice versa (weak numerical coupling). In addition, the cohesive fracture law [6] allows simulating crack propagation under drying process. Two examples of simulation with four cohesive joints and multi-joints will be presented in the following in order to illustrate first the soil surface settlement and then initiation and propagation of desiccation cracks.

2 Model descriptions

A heterogeneous porous body Ω containing a set of joints is considered. Both the porous matrix and joints are subjected to fluid flow and deformation (Figure 1)

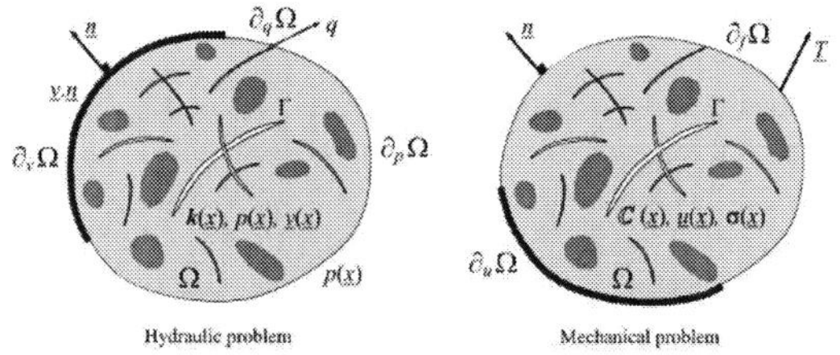

Figure 1. Hydraulic and mechanical problems

2.1. Hydraulic diffusion around the crack

The flow in the soil around the crack is governed by the Darcy's law:

$$\underline{v} = -k\nabla p \qquad (1)$$

with $k = k_m / \mu$ where k_m (m^2) is the intrinsic permeability, p (MPa) the pore pressure and μ (Pa.s) the fluid dynamic viscosity. In absence of source term,

[a] Corresponding author: thi-dong.vo@cermes.enpc.fr

for an unsaturated medium the balance of fluid mass becomes:

$$\frac{d(\rho_f \phi S)}{dt} - div(\rho_f k \nabla p) = 0 \qquad (2)$$

Here, the fluid mass m_f is calculated in unit volume and $m_f = \rho_f \phi S$ where ϕ is the porosity, ρ_f is the fluid density and S is the degree of saturation of medium. To make possible a standard formulation for simulation by FEM, the variation of m_f is written as a function of the nodal pressure p as follows:

$$\frac{dm_f}{m_f} = \frac{d\rho_f}{\rho_f} + \frac{d\phi}{\phi} + \frac{dS}{S} \qquad (3)$$

The variation of fluid density ρ_f is related to the pressure by the following relation with K_f is the fluid compressibility.

$$\frac{d\rho_f}{\rho_f} = \frac{dp}{K_f} \qquad (4)$$

The variation of porosity ϕ is related to the pressure by using the poromechanical model:

$$d\phi = b d\varepsilon_v + \frac{dp}{N} \qquad (5)$$

with b the Biot's coefficient and N the Biot modulus. Theses parameters are linked with the bulk modulus of skeleton solid grain, K_s, by the following relationship:

$$b = 1 - K / K_s \qquad 1/N = (b - \phi_0)/K_s \qquad (6)$$

In reality, the increase of suction is accompanied with increase of bulk modulus and the tensile strength of soil. But in the present work, a constant Young modulus is assumed.

During drying, the degree of saturation decreases. This value is related to suction by the water retention curve :

$$\frac{S - S_r}{1 - S_r} = \frac{1}{(1 + (\alpha p)^n)^m} \qquad (7)$$

In this equation α, n are the coefficients, $m = 1 - 1/n$ and S_r is the residual degree of saturation of soil. In this work, S_r assumed equal to 2%. From Eq.(7), we can deduct:

$$\frac{dS}{S} = \frac{dS}{dp}\frac{dp}{S} = (1 - S_r)\left\{-m.\left[1 + (\alpha p)^n\right]^{-m-1}.\left(\alpha^n n p^{n-1}\right)\right\}\frac{dp}{S} \qquad (8)$$

During drying, the change of permeability is considered as a function of its permeability in the saturated state and the degree of saturation. The following expression is adopted for this study:

$$k = k_s \left(\frac{S - S_r}{1 - S_r}\right)^3 \qquad (9)$$

Replacing Eq. (3),(4), (5), (8) in Eq. (2), we found the final equation for hydraulic problem:

$$C_M \frac{dp}{dt} + Sb\frac{d\varepsilon_v}{dt} - div(k\nabla p) = 0 \qquad (10)$$

with $C_M = S\left(\frac{\phi}{K_f} + \frac{1}{N} + \frac{\phi \, dS/dp}{S}\right)$. In this work, the effect of volumetric strain of solid skeleton to the hydraulic diffusion, i.e. the term $Sb\frac{d\varepsilon_v}{dt}$ is neglected. The Eq.(10) after this simplification becomes:

$$C_M \frac{dp}{dt} = div(k\nabla p) \qquad (11)$$

2.2 Fluid flow in the crack

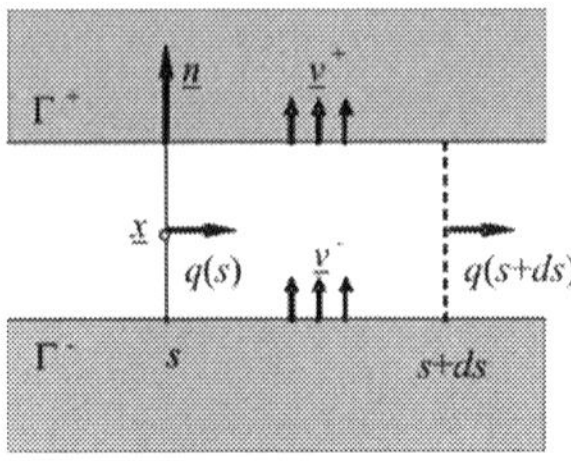

Figure 2. Masse exchange between crack and matrix [7]

We assumed that the cross conductivity of crack is infinite which imply that the pressure is the same on two sides of crack and maybe presented by only value p that is variable along the crack. The flow rate q in the crack is governed by the cubic law:

$$q(s) = -c(s)\partial_s p \qquad (12)$$

where c_f is the conductivity tensor in the crack plane and s presents curvilinear abscise along the crack. In the most cases c_f is supposed isotropic in the tangent plane and represents by a scalar c. This hydraulic conductivity can be related to crack opening e by the Poiseuille's law $c = e^3 / 12\mu$. Flow in crack taking into account the fluid exchange between crack and matrix, crack and crack, crack and boundary is usually a difficulty of modelling of cracking. In most of the researches, this problem isn't considered. The balance of fluid mass in the crack taking into account the mass exchange with surrounding matrix is written as:

$$\partial_s(\rho_f q) + \|\rho_f \underline{v}\| . \underline{n} + e\frac{d\rho_f}{dt} = 0 \qquad (13)$$

where $\|\underline{v}\|$ presents the fluid velocity discontinuity or the velocity jump between the two sides of the fracture, and $\underline{n}$ is the unit normal to the fracture surface (Figure 2). We neglected the change of thickness, i.e. term $\partial e/\partial t$, in solving the Eq.(13), but this opening cracks e is accounted in updating the hydraulic conductivity of cracks (Poiseuille's law). Also the final equation is obtained by replacing Eq. (12) in Eq.(13) and in using the compressibility of fluid:

$$C_F \frac{dp}{dt} = \partial_s(c\partial_s p) - \|\underline{v}\| . \underline{n} \qquad (14)$$

with $C_F = e / K_f$

2.3. Mechanical behaviour

In this work, the soil is assumed to have a poroelastic linear behaviour reads:

$$\varepsilon_{ij} = \frac{1+\upsilon}{E}\sigma_{ij} - \frac{\upsilon}{E}\sigma_{kk}\delta_{ij} + \frac{b}{3K}p\delta_{ij} \qquad (15)$$

with $\varepsilon_v = \varepsilon_{kk}$ the volumetric strain and $K = E/3(1-2\upsilon)$ the drained bulk modulus.

In this model the joints are simulated as elements of zero thickness with a very small hydraulic conductivity and an important stiffness in the first. For the mechanical behaviour of joint elements, the cohesive fracture law [6] is applied. A damage variable D is added in the model to present the process of damage through decrease of crack stiffness and evolution of load surface. Under the effect of evaporation, the tensile stress increases with suction and as soon as this tensile stress reaches the tensile strength, the crack initiates. In the code, the crack opening is calculated by the displacement discontinuity on the two sides of crack and represented by relative displacement vector $\underline{u}$. It is related to the stress by a crack stiffness tensor k and the damage variable D by the relationship: $\sigma = (1-D)ku$. The criterion of damage includes the evolution of mechanical strength as well as process of damage:

$$F(\sigma, D) = \tau^2 - \sigma_n^2 \tan^2\varphi + 2g(D)\sigma_c\sigma_n - g^2(D)C^2 \qquad (16)$$

where $\tau = \sigma - \sigma_n n$. In this equation, C represents the cohesion of intact (undamaged) joint, φ the friction angle and σ_c an auxiliary constant parameter related to C, φ and the tensile strength σ_R of intact joint by the following relation:

$$\sigma_c = \frac{C^2 + \sigma_R^2 \tan^2\varphi}{2\sigma_R} \qquad (17)$$

$$g(D) = (1-D)(1-\beta\ln(1-D)) \qquad (18)$$

The function *g(D)* plays a key role in this constitutive model. It takes the value 1 for the initial intact state and zero for the ultimate failure or completely damaged state.

In this model, the coefficient β presents the ductility of the material. The ductility of material increases with β. The damage evolution is calculated through the evolution of damage variable D, by using the equation of associated consistence $dF = 0$. The damage evolution law must ensure the positivity of energy dissipation in the damage process. The completed damage evolution is then following

$$\begin{cases} dD = 0 & if\ F < 0 \\ dD = 0 & if\ F = 0\ and\ (\partial F/\partial\sigma)\sigma < 0 \\ dD = -(\partial F/\partial D)^{-1}(\partial F/\partial\sigma)\sigma & elsewhere \end{cases}$$

2.4. Hydro-mechanical coupling

Eq.(11) and Eq.(14) are the base of hydraulic simulation in the code Porofis. The suction which is a result in the hydraulic problem is introduced after in the mechanical model to find the stress and strain fields. The opening of cracks calculated in the mechanical problem plays the role to update the cracks hydraulic conductivity like Poiseuille's law. The hydro-mechanical coupling in the code is carried out as showing in the Figure 3:

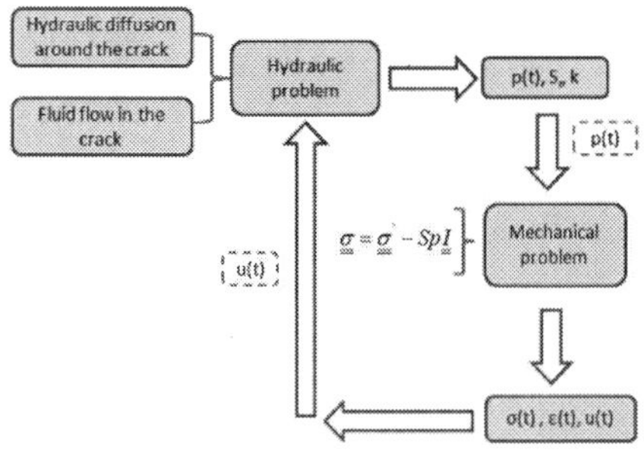

Figure 3 : Hydro-mechanical coupling in the code Porofis

3 Numerical simulations

3.1 Description of test [8]

Sanchez et al. [8] carried out a desiccation test in laboratory. A plate circular silt of 100 mm of diameter and 13 mm of thickness in slurry state is prepared and let in chamber condition with temperature 19.5° (±0.3 °C) and relative humidity 37 %(±4%). By using the 2D profile laser technique, the characteristics soil in drying are observed (volume change, water loss...). Especially, the evolution crack network's morphology (thickness, depth) is observed in the vertical plane. The main results as settlement, degree of saturation and crack development are used to discus in this study of numerical simulation.

3.2 Parameters determination

The water retention curve is fitted with experimental data [9] on the same type of soil, but in the compacted state (see Figure 4). We kept the air-entry value and used the Van-Genuchten model to determine the curve. The distance of two curves presents the difference between slurry and a compacted soil. The coefficients in Van-Genuchten model are found: $\alpha = 9.81$ (1/MPa) and $n = 1.60$

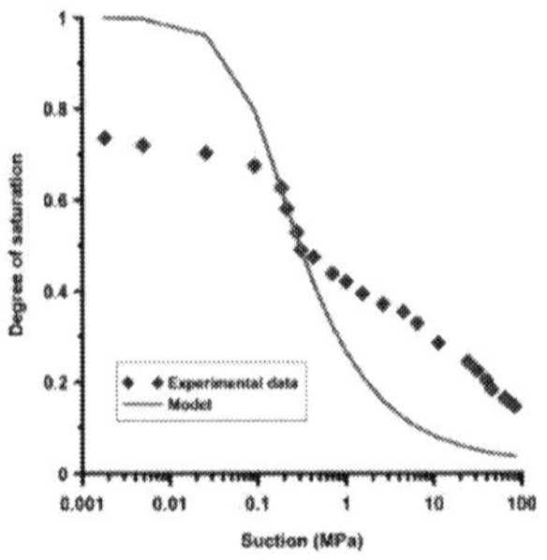

Figure 4 : Water retention curve

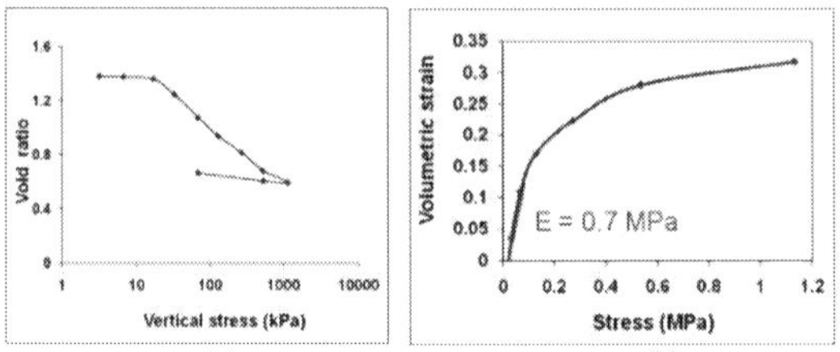

(a) (b)

Figure 5 : (a) Odometer curve of compacted soil [10] and (b) Stress-volumetric strain curve

The soil is assumed linear elastic and isotropic with $\upsilon = 0.3$, $\phi = 0.3$, $K_s = 30(GPa)$, $k = 5\times10^{-11}$. *(m/s)*.The elastic modulus is found from odometer curve [10] (Figure 5a). This odometer curve is translated to volumetric strain/stress curve (Figure 5b). The modulus **E = 0.7 MPa** is so deducted equal to the slope on the stress-volumetric strain curve.

The fluid characteristics are: $\rho_f = 1000(kg/m^3)$, $\mu = 10^{-3}(Pa/s)$, $K_f = 2.2(GPa)$

The parameters of joints are: $K_{tt} = 1$ MPa/mm; $K_{nn} = 700$ MPa/mm; $K_{tn} = K_{nt} = 0$ MPa/mm, $C = 0.06$ MPa, $\varphi = 25°$; $\beta = 1.5$; $e_0 = 0.0002$ mm.

3.3 Numerical model

The first simulation is a rectangle of $100mm \times 13mm$ (similar to experimental test) including four cohesive joints of *10 mm* of length (from Y = 3 mm to Y = 13 mm) in $X = 5,15,50,80$ mm (the same positions observed in the experimental test) under plane strain condition (Figure 6). This simulation allows observing two phases of desiccation and evolution of cracks opening.

For hydraulic boundary conditions, evaporation rate with condition of temperature 20° and relative humidity of 40% is applied on the top surface sample. This rate evaporation will decrease with time, as a function to suction on the top surface like the relation proposed by Wilson, 1990 [11]. For mechanical boundary conditions, the horizontal and vertical displacements are blocked on the left, right and bottom edges to present a perfect cohesion between the soil sample and the mold.

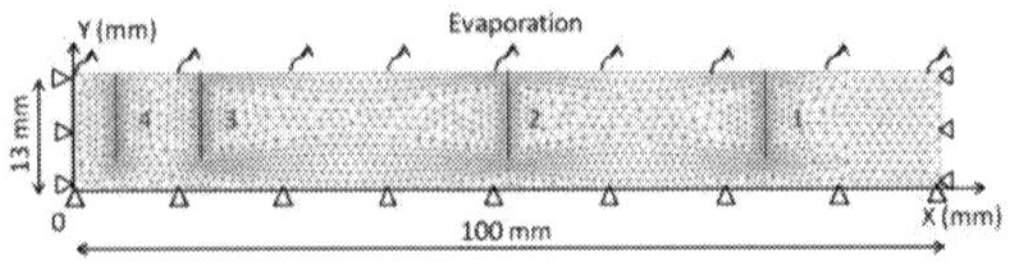

Figure 6 : Geometry and boundary conditions of simulation

3.4 Results

The Figure 7 presents profile of suction including the deformation evolution of the sample with time. We distinguish clearly two phases in the drying process. In the first phase (t = 0 to t = 4 h), the soil settlement increases, but all the joints remain close. We can imagine that in this time, there is only soil consolidation without any crack initiation. Then, at t = 4.2h, the first crack opened and during drying, the other cracks opened gradually.

The Figure 8 presents the suction profile in time for the section X = 5 mm (position of joint 4). Because the thickness of sample is very small (13mm), we find also that from t = 0 to t = 4h (before onset of cracks), the suction gradient is almost homogenous and continuous from top to bottom. After the opening of this crack, this profile suction is devised in two part at Y = 3 mm, the limit between joint and soil matrix. Because as soon as the crack is opened, its hydraulic conductivity increases according to the Poiseuille's law. From this moment, the suction is the same and increases more quickly in the crack (10 mm of length from Y = 3 mm to Y = 13 mm) than in the soil matrix (from Y = 0 mm to Y = 3 mm).

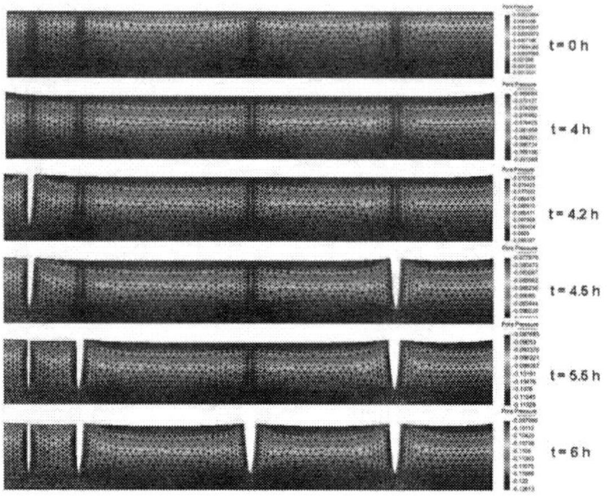

Figure 7 : Evolution of deformation sample in time

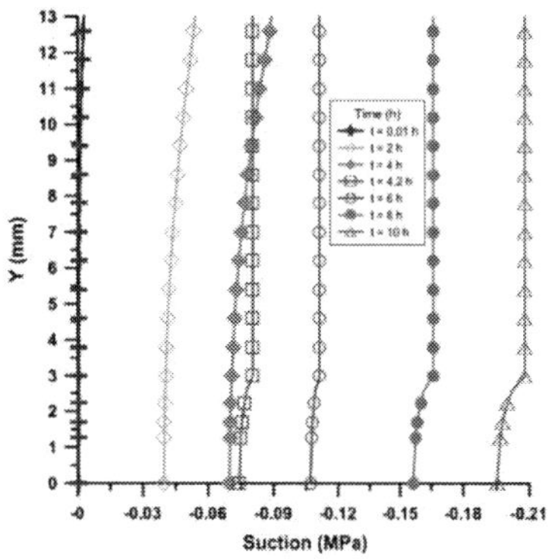

Figure 8 : Suction profile in the section X = 5 mm

Evolution of settlement at the middle surface of sample (X = 50 mm, Y = 13 mm) is considered. The Figure 9 presents the comparison between this evolution in time for numerical model and experimental data [8]. It shows that there isn't the difference between the ones. Additionally, the deformable form before cracking (t=4 h in Figure 7) is similar to the experimental observation.

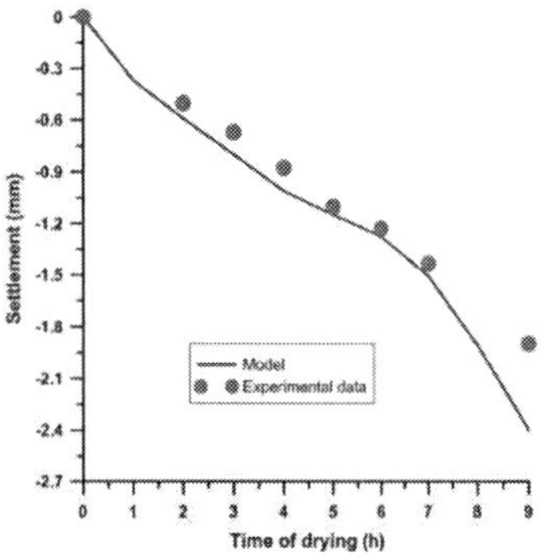

Figure 9 : Settlement in time for numerical simulation and experimental data [8]

In the FEM code, the opening of a crack is calculated as the displacement discontinuity of two sides of the crack. The Figure 10 presents the increase of opening for 4 cracks. Until t = 4 h, the opening of all cracks is zero, corresponding the consolidation phase without cracking and elastic behavior for joint elements. When the tensile strength was reached, the damage phase began and we obtained the displacement jump on opening cracks. The interaction between cracks 3 and 4 is displayed in the Figure 10. We observe that the crack 4 is onset before crack 3. But when the crack 3 is opened (t = 4.5h), the opening of crack 4 starts to be reduced, but the crack 1 and crack 2, far from the crack 4, and the zone between them were not influenced.

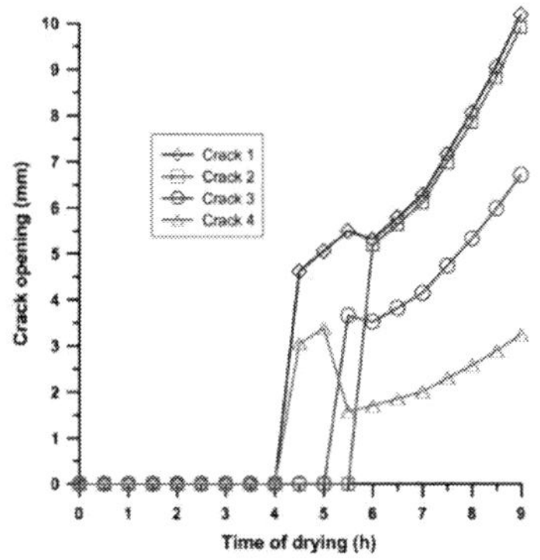

Figure 10 : Development of opening cracks in time

3.5 Simulation with multi-joint elements

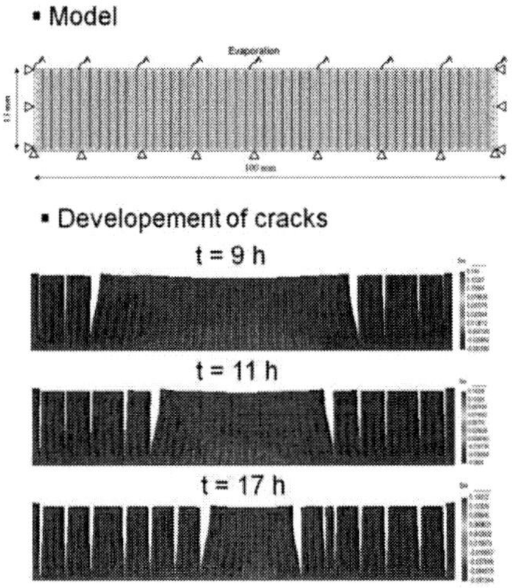

Figure 11 : Simulation with 50 joints

For further discussion, a preliminary result using 50 cohesive joints is showed in the Figure 11. In this simulation, 50 cohesive joints (one joint every 2 mm) are set up to present the potential cracks (Figure 11). The result of this simulation showed that the crack doesn't onset on all positions or joints simultaneously. In this simulation, the cracks were generated with an average spacing of 6 mm. In addition, this simulation showed clearly the interaction between cracks. The opening of a new crack induces the closure of other nearby cracks.

This simulation shows also that the cohesive fracture law applied could allow simulating the desiccation cracking and estimating an average spacing between cracks that will be studied in the future work.

6 Conclusions

This work presents an approach for desiccation cracking simulation talking into account the hydro-mechanical coupling: flow and deformation in presence of cracks by using the FEM code Porofis. The simulation including the cohesive joints shows that this approach allows simulating two phases of desiccation process: the consolidation one without cracking and the cracks development one. In addition, the simulation allows observing the evolution of soil parameters during drying as: suction, degree of saturation, settlement and cracks opening. The comparison of the numerical simulation with the experimental data [8] allowed evaluating the model.

This work showed that the cohesive fracture model is a good tool to simulate the evolution of desiccation cracks. Study of spacing and depth of cracks during drying will be investigated in future work in order to better understand the mechanisms of desiccation cracking.

Acknowledgements

The first author would like to thank to the COST Action TU1202 for fruitful exchange during the Workshop "Vadose Zone Workshop" from 5th July 2015 to 8th July 2015 that have been useful for this study.

References

[1] E. Drumm, D. Boles, and G. Wilson, "Desiccation cracks result in preferential flow," *Environ. Geotech.*, pp. 22–26, 1997.

[2] M. Flury, H. Flühler, W. A. Jury, and J. Leuenberger, "Susceptibility of soils to preferential flow of water: A field study," *Water Resour. Res.*, vol. 30, no. 7, pp. 1945–1954, 1994.

[3] W. H. Albright, C. H. Benson, G. W. Gee, T. Abichou, S. W. Tyler, and S. a. Rock, "Field performance of a compacted clay langfill final cover at a humid site," *J. Geotech. Geoenvironmental Eng.*, vol. 132, pp. 1393–1403, 2006.

[4] P. D. Jackson, K. J. Northmore, P. I. Meldrum, D. a. Gunn, J. R. Hallam, J. Wambura, B. Wangusi, and G. Ogutu, "Non-invasive moisture monitoring within an earth embankment - A precursor to failure," *NDT E Int.*, vol. 35, pp. 107–115, 2002.

[5] A. Pouya, "A finite element method for medeling coupled flow and deformation in porous fractured media," *Int. J. Numer. Anal. Methods Geomech.*, vol. 32, 2015.

[6] A. Pouya and P. Bemani, "A damage-plasticity model for cohesive fractures," *Int. J. Rock Mech. Min. Sci.*, vol. 73, pp. 194–202, 2015.

[7] A. Pouya and S. Ghabezloo, "Flow Around a Crack in a Porous Matrix and Related Problems," *Transp. Porous Media*, vol. 84, pp. 511–532, 2010.

[8] M. Sanchez, A. Atique, S. Kim, E. Romero, and M. Zielinski, "Exploring desiccation cracks in soils using a 2D profile laser device," *Acta Geotech.*, vol. 8, no. 6, pp. 583–596, Oct. 2013.

[9] G. El Mountassir, M. Sánchez, E. Romero, and R. a a Soemitro, "Behaviour of compacted silt used to construct flood embankment," *Proc. ICE - Geotech. Eng.*, vol. 164, no. 3, pp. 195–210, Jun. 2011.

[10] G. El Mountassir, M. Sánchez, and E. Romero, "An experimental study on the compaction and collapsible behaviour of a flood defence embankment fill," *Eng. Geol.*, vol. 179, pp. 132–145, Sep. 2014.

[11] G. W. Wilson, "Soil Evaporative Fluxes for Geotechnical Engineering Problems," 1990.

Propagation of cohesive crack in unsaturated porous media by use of XFEM method

Alireza Mokhtari[1] , Ehsan Motevalli Haghighi[1] ,Behrouz Gatmiri[1,a]

[1]*School of Civil Engineering,University of Tehran,Tehran,Iran*

Abstract. Various damage models have been presented to model the cracks. These models don't simulate the discontinuity in displacement field; just see its effect on an upper level on stress field through damage parameter. Damage parameter is used to modify other fields like pressure. This paper tries to model crack itself through fracture mechanics. The opening of the crack is modeled and is connected to stress field by use of cohesive crack model that relates plasticity of crack tip to its opening. When crack opens, there exists discharge of water and gas through crack that are calculated by use of multiscale methods and incorporated in weak form of governing equations.Due to complexity of proposed model, it can't be solved analytically and extended finite element method (XFEM) is used to solve it. Xfem method which is based on partition of unity property of FEM adds some terms to conventional interpolation functions to model discontinuity of fields adjacent to the crack. This method simulates crack in ordinary mesh of FEM and doesn't need remeshing around the crack when it propagates.

1. Introduction

In the present study the propagation of crack in unsaturated porous media is investigated. To describe the behavior of unsaturated porous media Gatmiri's equations [1-3] are used. The differential forms of these equations are written based on the two independent variables of net stress and suction.

Cohesive crack model [4-6] is used for mechanical interaction of porous media and crack. In this model, the resisting force against opening of the crack is determined based on an opening-force relation. Magnitude of this force varies from tensile strength of material for no opening to zero for critical opening (w_c). Critical opening is the opening that cohesive force is zero after that.

It is assumed that water pressure has weak discontinuity in vicinity of the crack. Thus pressure gradient would be discontinues in neighborhood of the crack. This discontinuity leads to water flow through the opening. Deduction of this flow through the opening needs rewriting of mass equilibrium of water around the crack. Micro and macro coupling is obtained by adding the derived term for flow through the crack to the integral form of mass balance equation of water.

XFEM [7-9] method is used to descretize the resulted equations in space. This method which is based on partition of unity property of finite element shape functions adds some enrichment to ordinary shape functions and doesn't need any remeshing after crack propagation and allows modeling the discontinuity in a regular FEM mesh. Finite difference method is used for temporal descretization.The resulted equations are nonlinear and Newton-Raphson method is used to solve them.

2. The Strong Form of the Governing Equations for Unsaturated Porous Media

2.1. Water phase Equations

2.1.1. Water Phase Transfer Equation

Considering the generalized Darcy equation for movement of Water phase in porous media, the equation of water flow is defined as:

[a] Corresponding author: behrouz.gatmiri@andra.fr

$$U = \frac{q_w}{\rho_w} = K_w(-\nabla s + \nabla Z) \quad (1)$$

Where U is the liquid velocity and q_w is the liquid flow vector. ρ_w and K_w are defined as water density and water permeability. s is the suction (P_g-P_w) and Z is the gravity potential. K_w is total water permeability that is written in the following form:

$$K_w = k_{int}(n) k_r(S_w) \frac{1}{\upsilon} \quad (2)$$

$k_{int}(n)$ is the intrinsic permeability, $k_r(S_w)$ is the relative permeability and υ is viscosity of the fluid. The intrinsic permeability reflects properties of solid matrix. Intrinsic permeability is isotropic and depends on porosity of the material:

$$k_{int\,jenab} = k_{int}(n)\delta \quad (3)$$

$$k_{int}(n) = k_{w_0} 10^{\alpha_w e}$$

In above equation δ is identity tensor of order two, e the void ratio, α_w a model parameter and k_{w_0} is reference permeability. $k_r(S_w)$ is the relative permeability that shows fluid properties. It may be computed based on a state surface [10, 11] or more simply a retention curve[12]. The following formula is used for relative permeability:

$$k_r(S_w) = \left(\frac{S_w - S_{w,r}}{1 - S_{w,r}}\right)^d, 3 < d < 4 \quad (4)$$

υ is the viscosity of fluid which is considered to be equal to $9.94 \times 10^{-7} N.S.m^{-2}$.

2.1.2. Water Mass Conservation Equation:

Considering a representative volume element (RVE) of porous media, time derivative of included mass of water becomes equal to divergence of mass flow of water. Mathematical form of this equilibrium may be expressed as:

$$\frac{\partial(nS_w\rho_w)}{\partial t} = -div(\rho_w U) \quad (5)$$

Where n, S_w, ρ_w and u are defined as the porosity, saturation degree, water density and water velocity respectively.

The left side of mass balance equation can be expanded as:

$$\frac{\partial(nS_w\rho_w)}{\partial t} = S_w\rho_w\frac{\partial n}{\partial t} + n\rho_w\frac{\partial S_w}{\partial t} + nS_w\frac{\partial \rho_w}{\partial t} \quad (6)$$

Solid particles are incompressible so the derivative of porosity is equal to the volumetric strain that is defined as sum of normal strains:

$$\dot{n} = \dot{\varepsilon}_v = \delta_{ij}\dot{\varepsilon}_{ij} \quad (7)$$

The density of water is assumed to be function of water pressure, so its time derivative becomes dependable on pressure time derivative as:

$$\frac{\partial \rho_w}{\partial t} = \frac{\partial \rho_w}{\partial P_w} \cdot \frac{\partial P_w}{\partial t} = \beta_P . \dot{P}_W \quad (8)$$

For calculation of saturation degree, the Van Genuchten [12] model that relates saturation and suction is considered here:

$$S_w = \left[(1 - S_{w,r})\left(1 + (\alpha|s|)^n\right)^{-(1-\frac{1}{n})} + S_{w,r}\right] \quad (9)$$

S_{wr} is the residual saturation degree that saturation degree can't reduce to a value less than it. s is the suction and α and n are the constants of the model.

The derivative form of the water saturation degree is written as:

$$\dot{S}_w = g_4\dot{s} = g_4(\dot{P}_g - \dot{P}_w) \quad (10)$$

$$g_4 = \frac{\partial S_w}{\partial s} =$$

$$sgn(s)(1-n)\alpha^n(1-S_{w,r})|s|^{n-1}(1+(\alpha|s|)^n)^{-\left(2-\frac{1}{n}\right)}$$

By substituting (6), (7), (8) and (10) for (5) water mass conservation equation on the basis of principle variables is concluded:

$$S_w\rho_w(\dot{\varepsilon}_v) + (nS_w\beta_p - n\rho_w g_4)\dot{P}_w - n\rho_w g_4\dot{P}_g = \quad (11)$$

$$div(K_w\rho_w.(\nabla(P_w - P_g) + \nabla Z))$$

2.2. Solid Skeleton Behavior:

Considering balance of momentum equation for unsaturated media as:

$$(\sigma_{ij} - \delta_{ij}P_g)_{,j} + P_{g,i} + b_i = 0 \quad (12)$$

$\sigma_{ij} - \delta_{ij}P_g$ is the net stress and b is the volumetric force. Constitutive law of solid skeleton in unsaturated porous medium can be presented on the basis of net stress and suction that are two independent variables:

 DOI: 10.1051/e3sconf/20160908010

$$\sigma_{ij} - \delta_{ij} P_g = D\varepsilon - F(P_g - P_w) \tag{13}$$

D is the deformation module. Linear elastic behavior is considered for bulk material. $s = P_g - P_w$ is the suction and $F = S_w.m = S_w.[1 \quad 1 \quad 0]$.

3. Hydromechanical Coupling

The force which is exerted on the body of the crack is the combination of two factors: the first is the plasticity of cracked solid medium and the second is suction of unsaturated media. This force might be demonstrated as:

$$(\sigma_{ij} - \delta_{ij} P_g).\mathrm{n}_{\Gamma_d} = \mathrm{t}_d - \mathrm{S}_w (P_g - P_w).\mathrm{n}_{\Gamma_d} \ \text{ on } \Gamma_d \tag{14}$$

In above equation $(\sigma_{ij} - \delta_{ij} P_g)$ is the net stress, t_d is the cohesive force which is calculated based on the opening-force relation shown on **Figure2**, $\mathrm{S}_w(P_g - P_w)$ is the suction force that is assumed to be similar on opposite faces of the crack, n_{Γ_d} is normal to the crack and Γ_d shows the body of the crack.

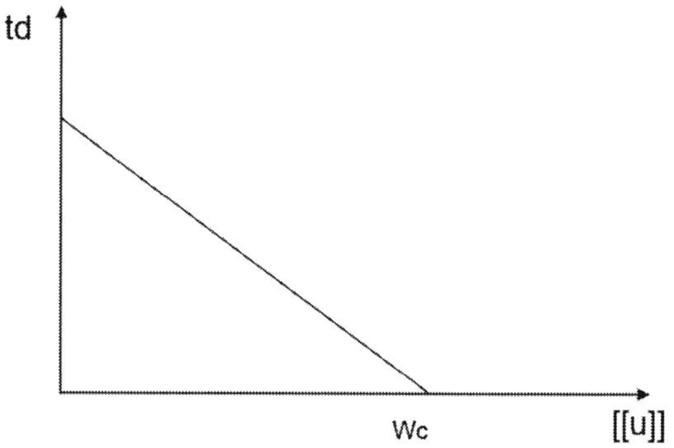

Figure1. Linear cohesive force versus opening

Discharge of water through the crack is calculated similar to [13, 14].It is supposed that the length of the crack is much more greater than its height and the changes of S_w, $\rho_w, \dot{P}_w, \dot{P}_g$ and n_d is negligible on the height of the crack. With these assumptions, flow of water through the crack is calculated as:

$$\tilde{q}_{wd} = -2hS_w\rho_w < \frac{\partial V_z}{\partial x} > -S_w\rho_w [|v_z|] - 2hn_d(S_w\beta_p - \rho_w g_4)\dot{P}_w$$
$$-2hn_d\rho_w g_4 \dot{P}_g - 2h\frac{\partial}{\partial x}\left[\rho_w (K_{wd}.\nabla(\mathrm{P}_w - \mathrm{P}_g) + K_{wd}.\nabla Z)\right] \tag{15}$$

Where $<*>$ and $[|*|]$ represent the average and difference of solid velocity at the two opposite sides of the crack, respectively. $2h$ shows opening of the crack and subscript d indicates that the parameter is specific to the crack. Mechanical and hydraulic coupling terms are connected through n_d that is porosity of damaged porous media and is defined as:

$$n_d = \frac{w}{w_c} \tag{16}$$

In above equation w is opening of the crack. To calculate the coupling terms, two Gauss points are placed on every segment of crack path which is bounded in an element (**Figure2**).

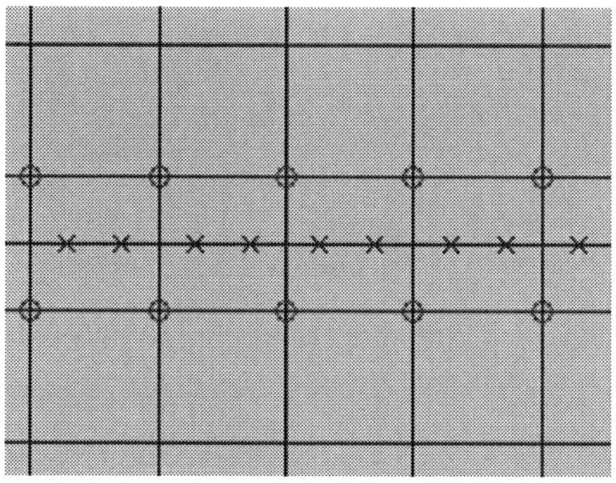

Figure2. Finite element mesh. Blue circles show enriched nodes and crosses show Gauss points on the crack path.

4. Integration and Descritization of Governing Equations

To Obtain the integral form of governing equations, u^* (allowable virtual displacement) and P_w^* (allowable virtual liquid pressure) are multiplied in equations (11) and (12) respectively. The gas pressure is assumed to be equal to the atmospheric pressure. After using the Green theorem and some manipulation, final form of integrated equations are resulted:

(17)
(a)

$$\int_\Omega D\varepsilon.\varepsilon^* d\Omega - \int_\Omega m.S_w \mathrm{P}_w .\varepsilon^* d\Omega$$
$$-\int_\Omega m.(1-S_w)\mathrm{P}_g .\varepsilon^* d\Omega - \int_{\Gamma d} [[u]].(t_d - t_{suction}.n_d)\,\mathrm{d}\Gamma$$
$$= \int_{S2x} \sigma.n u^* d\Omega + \int_\Omega b u^* d\Omega$$

(b)

$$\int_{\Omega} S_w \rho_w (\dot{\varepsilon}_{ii}).\mathrm{P}_w^* \, d\Omega + \int_{\Omega} (nS_w \beta_p - n\rho_w g_4)\dot{P}_w . \mathrm{P}_w^* \, d\Omega$$
$$+ \int_{\Omega} n\rho_w g_4 \dot{P}_g . \mathrm{P}_w^* \, d\Omega = \int_{\Gamma} \mathrm{P}_w^* \, q_w \, d\Gamma$$
$$- \int_{\Omega} (\rho_w K_w . \nabla(\mathrm{P}_g - \mathrm{P}_w) + \rho_w K_w . \nabla Z) . \nabla \mathrm{P}_w^* \, d\Omega$$

In preceding equations, $\int_{\Gamma d} [[u]].(t_d - t_{suction}.n_d)\,\mathrm{d}\Gamma$ shows the cohesive and suction forces exerted on the opened crack body ($[[u]]$ means opening of the crack). Rewriting the first term of the right hand side of equation (17) (b) as:

$$\int_{\Gamma} \mathrm{P}_w^* \, \tilde{q}_w d\Gamma = \int_{\Gamma} \mathrm{P}_w^* \, q_w d\Gamma + \int_{\Gamma d} \mathrm{P}_w^* \, \tilde{q}_{wd} d\Gamma \qquad (18)$$

The The flow through the crack is involved in integral form of governing equations either.

Spatial discretization of governing equations is done through XFEM. XFEM method presumes the interpolation function of unknown variables consists of standard and enriched parts, so the unknown variables can be given as:

(19)

$$u(x_i, t) = \sum_{i \in \Omega} N_u(x_i) u(t)$$
$$+ \sum_{i \in \Omega_{enr}} N_u(x_i)(H_{\Gamma d}(x_i) - H_{\Gamma d}(x_I))\tilde{u}(t)$$
$$P_w(x_i, t) = \sum_{i \in \Omega} N(x_i) P_w(t) +$$
$$\sum_{i \in \Omega_{enr}} N(x_i)(D_{\Gamma d}(x_i) - D_{\Gamma d}(x_I))\tilde{P}_w(t)$$

N is the standard shape function. $H_{\Gamma d}(x_i) - H_{\Gamma d}(x_I)$ and $D_{\Gamma d}(x_i) - D_{\Gamma d}(x_I)$ are defined as transferred heaviside function and transferred distance function, respectively. $u(t)$ and $P_w(t)$ are the normal degrees of freedom and $\tilde{u}(t)$ and $\tilde{P}_w(t)$ are the enriched ones. The shape functions are substituted for equilibrium and mass conservation equation. Considering implicit finite difference method, temporal discretization can be done.

$$(\dot{.}) = \frac{(.)^{t+\Delta t} - (.)^t}{\Delta t} \qquad (20)$$

After applying the spatial and temporal discretization, the matrix form of governing equations becomes:

(21)

$$\begin{bmatrix} R_{uu} & R_{u\tilde{u}} & R_{uw} & R_{u\tilde{w}} \\ R_{\tilde{u}u} & R_{\tilde{u}\tilde{u}} & R_{\tilde{u}w} & R_{\tilde{u}\tilde{w}} \\ R_{wu} & R_{w\tilde{u}} & R_{ww} + Q_{ww}\Delta t & R_{w\tilde{w}} + Q_{w\tilde{w}}\Delta t \\ R_{\tilde{w}u} & R_{\tilde{w}\tilde{u}} & R_{\tilde{w}w} + Q_{\tilde{w}w}\Delta t & R_{\tilde{w}\tilde{w}} + Q_{\tilde{w}\tilde{w}}\Delta t \end{bmatrix} \begin{bmatrix} u(t_{n+1}) \\ \tilde{u}(t_{n+1}) \\ P_w(t_{n+1}) \\ \tilde{P}_w(t_{n+1}) \end{bmatrix}$$
$$- \begin{bmatrix} 0 \\ F_{\tilde{u}}^{int}(t_{n+1}) \\ F_w^{int}(t_{n+1}) \\ F_{\tilde{w}}^{int}(t_{n+1}) \end{bmatrix} = \begin{bmatrix} F_u^{ext}(t_{n+1}) \\ F_{\tilde{u}}^{ext}(t_{n+1}) \\ F_w^{ext}(t_{n+1}).\Delta t \\ F_{\tilde{w}}^{ext}(t_{n+1}).\Delta t \end{bmatrix}$$
$$+ \begin{bmatrix} 0 \\ 0 \\ R_{wu} u(t_n) + R_{w\tilde{u}} \tilde{u}(t_n) + R_{ww} P_w(t_n) + R_{w\tilde{w}} \tilde{P}_w(t_n) \\ R_{\tilde{w}u} u(t_n) + R_{\tilde{w}\tilde{u}} \tilde{u}(t_n) + R_{\tilde{w}w} P_w(t_n) + R_{\tilde{w}\tilde{w}} \tilde{P}_w(t_n) \end{bmatrix}$$

Since the existence of the terms representing internal force and discharge of water through crack, the system of equations become nonlinear and it becomes necessary to use a repetitive procedure like Newton-Raphson to solve the equations.

5. Example

In this section, the ability of model to simulate crack propagation in unsaturated porous media is investigated. A square plate with dimension of 250×250 mm and initial notch length of 50 mm along its centerline is considered. The upper and lower edges of plate are stretched with fixed vertical velocity $\dot{u} = 2.35\ \frac{\mathrm{mm}}{\mathrm{s}}$. Figure 3. shows the geometry of crack and the properties of media are mentioned in Table1. The edges of plate are impermeable to water and the plate is considered saturated, initially.

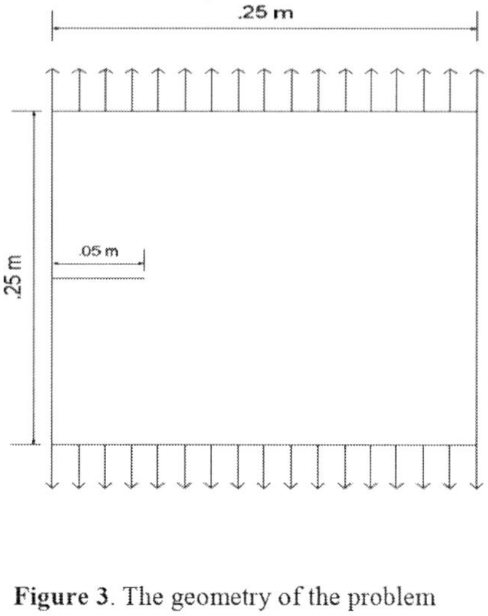

Figure 3. The geometry of the problem

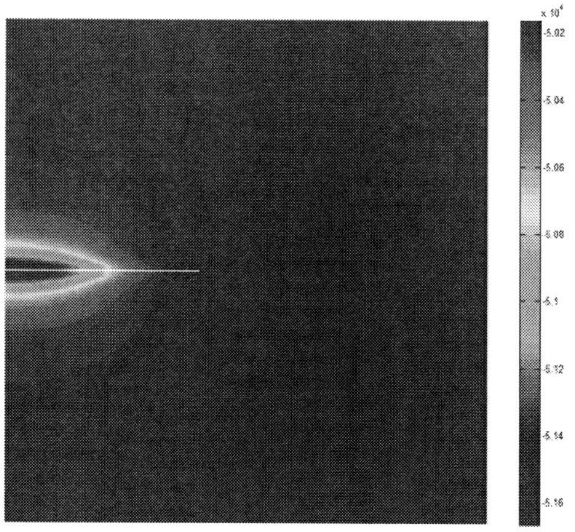

Figure 4. Suction distribution in plate

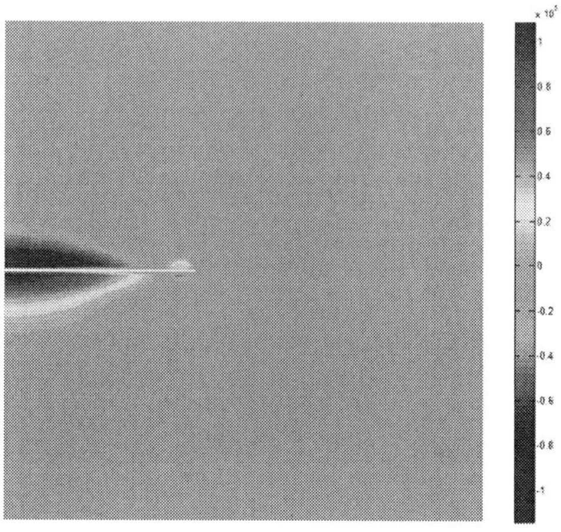

Figure 5.Gradient of suction distribution in plate

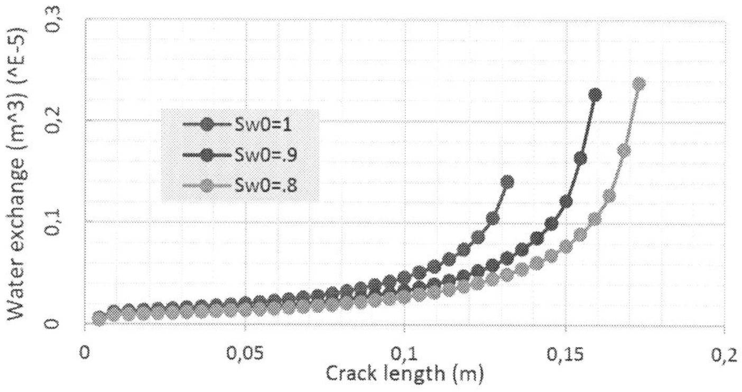

Figure 6. Comparison of water exchange in various initial saturation degree at a specified time.

E	25.85E9
Passion ratio	0.18
porosity	0.2
K_{w0}	3E-15 m.s^{-1}
α	3
S_{w0}	0.9
S_{wr}	0
α_{vg}	3E-5
n_{vg}	1.1

Table1. The properties of material

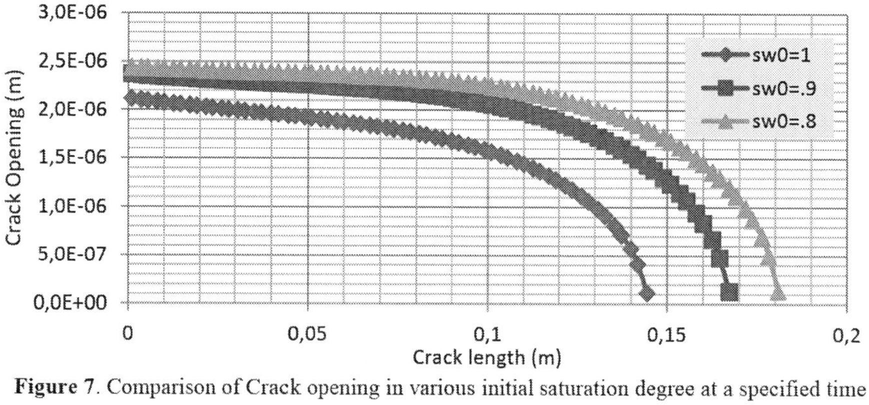

Figure 7. Comparison of Crack opening in various initial saturation degree at a specified time

The suction distribution is shown in Figure 4. Suction increases around the crack body and because of the gradient of suction which is shown in Figure 5 water is absorbed to the crack. For having better assessment of the model performance, a comprehensive analyze is done. The influence of initial suction degree on water discharge at a specified time is shown in Figure 6 . Water exchange increases with decrease in initial saturation degree because when initial saturation degree decreases, suction increases and crack has more inclination to absorb water.

Two mechanisms are at work in discharge through crack: The first is the porosity change and the second is plane Poiseuille flow which relates the cube of opening to discharge of water through crack. Here just the first mechanism is taken into account, therefore discharge of water increases in proximity of the crack. The influence of initial saturation degree on crack opening is shown in Figure 7. As it is shown in Figure 7, the opening of crack increases when the initial saturation degree decreases. Due to higher suction in lower initial saturation degree, the bearing capacity of the plate increases and it leads to higher stresses in the plate. These higher stresses cause the crack to propagate faster. Consequently at a specific time, length of crack and its opening is more for lower initial saturation degree.

6. Conclusion

It was shown the XFEM method has the potential to model crack in unsaturated porous media in a regular FEM mesh. A formulation is outlined to model unsaturated porous media and the terms related to crack were added to weak form of this formulation. A problem was solved and it was shown how the suction concentrates around the crack and not only helps to close the opening but also absorbs the water to the crack.

References

1. B. Gatmiri, in Proc of the 3rd Iranian Int. Conf. on Geotech Eng and Soil Mech. (2002)

2. B. Gatmiri, A. Hoor, Phys. and Chem. of the Earth, Parts A/B/C. **32**(8): p. 947-956, (2007)

3. B. Gatmiri, P. Delage, *Int. j. numer. anal.* meth. geomech., **21**(3): p. 199-225, (1997)

4. D.S. Dugdale,. J. Mech. and Phys. Solids, p. 100-108.(1960)

5.Barenblatt, G.I., The mathematical theory of equilibrium of cracks. Advances in applied mechanics, 1962. 7(1): p. 55–129.

6.HILLERBORG, A., M. MODEER, and P.G. PETERSON, Analysis of crack formation and crack growth by fracture mechanics and finite elements. Cement and Concrete Research, 1976. 6: p. 773–786.

7.Belytschko, T. and T. Black, Elastic crack growth in finite elements with minimal remeshing. International Journal for Numerical Methods in Engineering, 1999. 45(5): p. 601-620.

8.Moës, N., J. Dolbow, and T. Belytschko, A finite element method for crack growth without remeshing. International Journal for Numerical Methods in Engineering, 1999. 46: p. 131–150.

9.Babuška, I., G. Caloz, and J.E. Osborn, Special finite element methods for a class of second order elliptic problems with rough coefficients. SIAM Journal on Numerical Analysis, 1994. 31(4): p. 945-981.

10.Gatmiri, B. and P. Delage. Nouvelle formulation de la surface d'état en indice des vides pour un modele non linéaire élastique des sols non saturés. in Proceedings of the 1st International Conference on Unsaturated Soils. 1995.

11.Gatmiri, B., P. Delage, and M. Cerrolaza, UDAM: A powerful finite element software for the analysis of unsaturated porous media. Advances in Engineering Software, 1998. 29(1): p. 29-43.

12.Van Genuchten, M.T., A closed-form equation for predicting the hydraulic conductivity of unsaturated soils. Soil science society of America journal, 1980. 44(5): p. 892-898.

13.Réthoré, J., R. Borst, and M.A. Abellan, A two scale approach for fluid flow in fractured porous media. International Journal for Numerical Methods in Engineering, 2007. 71(7): p. 780-800.

14.Réthoré, J., R. de Borst, and M.A. Abellan, A two-scale model for fluid flow in an unsaturated porous medium with cohesive cracks. Computational Mechanics, 2008. 42(2): p. 227-238.

DOI: 10.1051/e3sconf/20160908011

Simple Numerical Model to Simulate Penetration Testing in Unsaturated Soils

Pegah Jarast S.[1] and Majid Ghayoomi, PhD[a, 1]

[1] *Department of Civil and Environmental Engineering, University of New Hampshire, Durham, NH, USA*

Abstract. Cone penetration test in unsaturated sand is modelled numerically using Finite Element Method. Simple elastic-perfectly plastic Mohr-Coulomb constitutive model is modified with an apparent cohesion to incorporate the effect of suction on cone resistance. The Arbitrary Lagrangian-Eulerian (ALE) remeshing algorithm is also implemented to avoid mesh distortion problem due to the large deformation in the soil around the cone tip. The simulated models indicate that the cone resistance was increased consistently under higher suction or lower degree of saturation. Sensitivity analysis investigating the effect of input soil parameters on the cone tip resistance shows that unsaturated soil condition can be adequately modelled by incorporating the apparent cohesion concept. However, updating the soil stiffness by including a suction-dependent effective stress formula in Mohr-Coulomb material model does not influence the cone resistance significantly.

1 Introduction

Unsaturated soils are available in surficial parts of the earth and above the water table. Dealing with unsaturated soils is inevitable in many geotechnical projects including shallow foundations, fills, embankment, dams, pavements, slopes, and landfills. The behaviour of unsaturated soils is complex and influenced by many factors including stress state, soil type, soil fabrics, mineralogy, density, and most importantly, inter-particle suction stresses within the pore space [1]. For instance, suction stress alters strength and stiffness of geomaterial by increasing the effective stress acting on soil [2, 3]. Thus, a suction-dependent soil characterization approach should be implemented in order to interpret the results of laboratory and field investigations in unsaturated soils.

Cone Penetration Testing (CPT) is one of the popular field investigation methods to evaluate soil properties. The measured cone resistance (q_c) and sleeve friction (f_s) can be linked to the physical, mechanical, and hydraulic properties of soils using empirical and analytical relations [4-6]. CPT results (q_c, f_s) are in fact controlled by soil shear strength, stiffness, compressibility, and permeability [7].

Suction stresses can be significant in unsaturated soil and influence CPT results directly or indirectly by changing the effective stress, and in turn, affecting the cone resistance. Therefore, using correlations that have been developed for dry or saturated soil conditions without considering the effect of degree of saturation may lead to misrepresentation of soil properties and requires further scrutiny.

Experimental studies by Hryciw & Dowding [8] and Lehane et al. [9] showed that cone resistance (q_c) in an unsaturated soil condition can be increased up to two times compared to that of dry and saturated soil conditions under certain circumstances. Also, results of CPTs conducted by Vanapoli & Fathi [10] in a specially designed tank showed that cone resistance and bearing capacity of sand can be influenced by suction. Russell and Khalili [11] also showed that the cone penetration resistance is comparable to the required pressure for expansion of a spherical cavity in an infinite sand, and it could be increased due to presence of suction in an unsaturated condition. Further, CPT results on saturated and unsaturated sands in a suction-controlled calibration chamber conducted by Pournagiazar et al. [12] showed that the effect of suction is more significant for low chamber confining stresses and low relative density values. This emphasizes the importance of suction in cone penetration into the shallow ground where soils are mostly unsaturated.

Numerical modelling of cone penetration process is complex due to large deformations in the model and sever mesh distortion. Several attempts have been made by different researchers to overcome the challenges involved in solution schemes and mesh generation [13-16]. Recently, Nazem et al., Sheng et al, and Yi et al. [17-19] used ALE method to avoid mesh distortion problem in soil. Thus far, this technique combined with adaptive remeshing showed the most stable responses.

Implementing these numerical modelling techniques to simulate cone penetration in unsaturated soil requires further examination as previous works mainly focused on dry or saturated soil conditions. In this paper, a simple

[a] Corresponding author: majid.ghayoomi@unh.edu

numerical model for simulating CPT in unsaturated soil using Abaqus/Explicit is presented. Results of a set of CPT in unsaturated sand inside a model calibration chamber are discussed. In addition, sensitivity analyses were performed to investigate the effect of various soil parameters on the penetration response.

2 Numerical modelling

2.1 Model Specification

Penetration of a 500 mm long cone with a 16 mm diameter into a sandy soil inside a calibration chamber was modelled using axisymmetric Finite Element (FE) in Abaqus/Explicit. Model geometry, boundary conditions, and mesh configuration are shown in Figure 1. Soil domain is 160 mm-wide and 400 mm-high. The side boundary was located at radial distance of 20 times the cone radius to reduce the boundary effects [20]. The bottom boundary was completely fixed while a constant normal stress boundary, equal to the confining pressure in a calibration chamber, was considered for the top and sides of the model. Two confining pressures of 50 and 100 kPa were considered in this study. In order to model the penetration process, a downward vertical velocity of 250 mm/s was imposed to the cone's top nodes, which were all coupled together. Penetration was considered to begin from the soil surface.

4-node bilinear, axisymmetric quadrilateral reduced integration elements (CAX4R) were used for both soil domain and cone and they were both considered as deformable bodies. To prevent mesh distortion problem in soil elements, the remeshing method, ALE, was used which keeps the mesh size fine enough during penetration and allows independent movement of mesh from material while maintaining the topology (the element and connectivity) of the mesh.

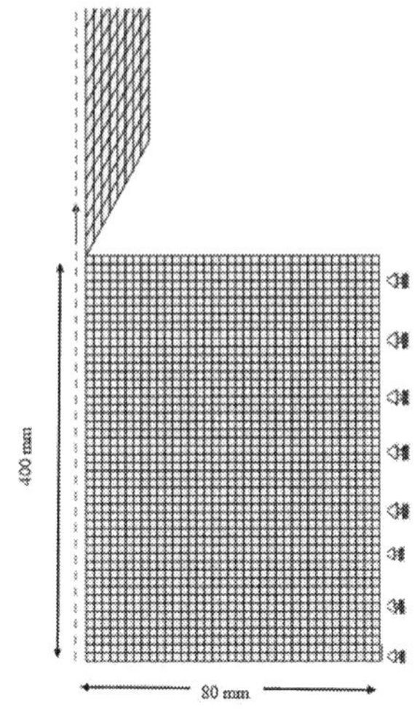

Figure 1. Geometrical schematic of Abaqus models

2.2 Material model

Simple elastic-perfectly plastic Mohr-Coulomb (MC) model was chosen as the soil constitutive model. The MC model parameters were estimated based on the properties of Sydney sand [21] in Tables 1 and 2.

Table 1 Properties of Sydney sand [12]

Parameter	value
USCS clasification	SP
G_s (gr/cm^3)	2.65
Maximum void ratio, e_{max}	0.92
Minimum void ratio, e_{min}	0.6
Minimum dry density, ρ_{min}	1.38
Maximum dry density, ρ_{max}	1.66
Critical state friction angle, $\emptyset'_{cs}$	36.6
Poisson's ratio, υ	0.3

Table 2 Mohr-Coulomb model parameters

Parameter	Value	Value
	P=50kPa	P=100kPa
Density	1593 kg/m^3	1593 kg/m^3
Elastic modulus, E	10 MPa	20 Mpa
Poisson's ratio, ν	0.3	0.3
Dilation angle, ψ	10°	10°
Friction angle, φ	41.6°	40.4°
Cohesion, C	0.0000001 Pa	0.0000001 Pa

The very small value of cohesion for this sand was to avoid numerical instability in the system. Equation 1 proposed by Bolton [22] was used to update soil friction angle for different confining pressures.

$$\emptyset' - \emptyset'_{cs} = 3\left\{Dr\left[3.7 - ln\left({}^{p'}/_{p_a}\right)\right] - 0.9\right\} \qquad Eq. 1$$

where $\emptyset'_{cs}$ is critical state friction angle, Dr is the soil relative density, p' is the mean effective stress before shearing and p_a is the reference atmospheric pressure.

2.3 Interaction model

The cone-soil interaction was considered using a surface-to-surface contact model with pure master-slave algorithm in tangential direction. Here, the cone was chosen as the master surface while the soil domain was considered to be slave node region. To avoid slave node penetration master surface and make the interaction model work properly, slave node region (the soil elements) should have coarser mesh compared to that of master surface (cone element) [23].

The penalty method with no limit for transmitted shear stress was also used as friction formulation. The friction angle at the cone-soil interface was assumed equal to δ=25° which resulted in an interface friction coefficient of μ=tan δ=0.43, based on suggested value of soil-pile friction angle for medium sand (API, [24]). "Finite Sliding Algorithm" was also used for tracking

 DOI: 10.1051/e3sconf/20160908011

contact nodes in which the nodes and elements paired between master and slave surfaces are tracked and kept linked until the calculation stops.

2.4 Remeshing method

The mesh distortion due to the large deformation around the cone could be avoided by implementing very fine mesh in the soil zone near the cone and using an adaptive remeshing technique. ALE remeshing technique is an efficient method to control the mesh distortion problem when large none-recoverable deformation occurs. It combines the features of pure Lagrangian analysis and pure Eulerian analysis. The ALE method allows independent movement of mesh form material while it doesn't alter the topology (element and connectivity) of the mesh. In this study, ALE method with 'improve aspect ratio' remeshing algorithm was used for soil elements. The remeshing frequency was set to be at every time step to improve the quality of results unless it involves high computational cost.

3 Unsaturated Soil Modelling

In order to capture the deformation behaviour of soils in unsaturated soil condition, a simple modification was applied to Mohr-Coulomb model. In this method, suction stress was incorporated in Bishop's effective stress equation (i.e. Eq. 2).

$$\sigma' = \sigma - u_a + \chi(u_a - u_w) \qquad Eq. 2$$

where 'χ' is the effective stress parameter, which is a function of degree of saturation and varies between zero and unity, '$(u_a - u_w)$', also known as 'ψ', is the matric suction (i.e. the difference between pore air and pore water pressure), and '$(\sigma - u_a)$' is the net normal stress. $\chi(u_a - u_w)$, is called suction stress and is the contribution of matric suction to effective stress [25]. Then, this modified effective stress formula was used to estimate the shear strength, as in Eq. 3.

$$\tau_f = c' + [(\sigma - u_a) + \chi(u_a - u_w)] \tan \varphi' \qquad Eq. 3$$

where 'c'' is effective soil cohesion and φ' is the effective angle of internal friction.

The extra shear strength gained from the addition of suction stress called "apparent cohesion" (i.e. $c'' = c' + \chi(u_a - u_w) \tan \varphi'$) was regarded as an input cohesion in the material constitutive model. The effective stress parameters, χ, was estimated using the empirical equation proposed by Russell & Khalili [26], shown in Eq. 4.

$$\chi = \begin{cases} 1 & for \ (^s/_{s_e}) \leq 1 \\ (^s/_{s_e})^{-0.55} & for \ 1 \leq (^s/_{s_e}) \leq 25 \\ 25^{0.45} (^s/_{s_e})^{-1} & for \ (^s/_{s_e}) \geq 25 \end{cases} \qquad Eq. 4$$

where 's_e' is the suction value separating saturated from unsaturated states, i.e. the air entry value. This expression was found based on laboratory strength and volume change data for Sydney quartz sand.

Considering S_e equal to 7 kPa for Sydney sand, the effective stress parameter, χ, and the apparent cohesion c'', and friction angle values for each suction level under different confining pressures can be calculated using Eqs. 1, 3, and 4, which are shown in table 3.

Table 3 Apparent cohesion and friction angle for different confining pressures and suction values

Confining pressure (kPa)	50			100		
Suction value (kPa)	25	100	200	25	100	200
Apparent cohesion (kPa)	11	20.6	26.5	10.6	19.7	25.4
Friction angle (degree)	41.6	41.6	41.6	40.4	40.4	40.4

4 Results

In the following sections, the results of numerical model CPTs for different saturation conditions and confining pressures are presented.

4.1 Cone resistance

Cone resistance, q_c, profiles for CPT on saturated and unsaturated sands are shown in Figures 2(a) and 2(b) for confining pressures of 50 and 100 kPa, respectively. All models had an initial relative density of 61% and suction values were 25, 100, and 200 kPa.

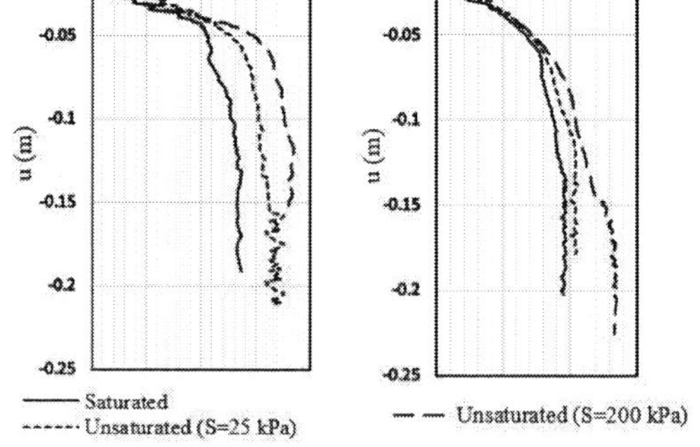

Figure 2. CPT results for saturtaed and unsaturated sand (D_r=61%), (a) P=50 kPa, (b) 100 kPa

The numerically estimated cone resistance in depth was increased gradually until it gets to steadily constant value, where the simulation was stopped. This is consistent with observations from calibration chamber testing. The results indicated that suction has significantly influenced and increased q_c comparing with the saturated soil condition. However, the effect of suction is more noticeable in lower confining pressures, which is evident where suction contribution to the effective stress and overall resistance is more pronounced. For example, for suction equal to 25 kPa, the q_c was increased by 20 % and 30 % for confining pressures of 50 kPa and 100 kPa, respectively, while for suction equal to 200 kPa, it only increased by 9% and 26 %, respectively.

4.2 Stress distribution

The distribution contours of stress components after 50, and 150 mm of penetration are shown in Figure 3. It indicates that the vertical and horizontal stresses around the cone tip were increased greatly due to penetration and were higher near the upper end of the cone than near the tip. This shifting of stresses to the upper end of the tip has likely been resulted from the frictional interface between the cone and the sand and the cone corner geometry, which led to stress built-up [14].

In order to evaluate stress distribution in different degrees of saturation, generated horizontal and vertical stresses versus distance from the cone tip during the penetration through the depth of 0.15 m for different confining pressures and saturation conditions were obtained and presented in Figure 4 and 5 respectively. For both confining pressures, there is no change in horizontal stress at and beyond a distance approximately 6 times the cone radius. This distance is the same for CPTs on unsaturated sand with different suction levels and under the two confining pressures. Presence of suction in unsaturated sand increased the induced horizontal stress inside the zone of influence while the stress values converged outside this zone. However, for vertical stress, the stress influenced zone was about 2 times the cone radius, which is less than the one for horizontal stress.

4.3 Sensitivity Analysis

The suction stress in unsaturated soil affects the effective stress in the soil. This may change other soil properties such as friction angle and soil stiffness. However, we only considered the suction effects in an apparent cohesion, which is a strength parameter in MC constitutive model. In order to investigate the significance of carrying the other parameters, numerical CPTs were performed on unsaturated sands while the elastic modulus and the friction angle were altered and the cone resistance results were compared with the case with a confining pressure of 50 kPa and suction value of 200 kPa.

The suction-dependent effective stress was calculated using Equation 4. Then, the friction angle was modified using Equation 1 by considering the updated effective stress for unsaturated sand. The elastic modulus was also modified considering a typical square root of effective stress variation profile, as follows

$$\frac{E_1}{E_2} = \sqrt{\frac{\sigma'_{v,1}}{\sigma'_{v,2}}} \qquad Eq.5$$

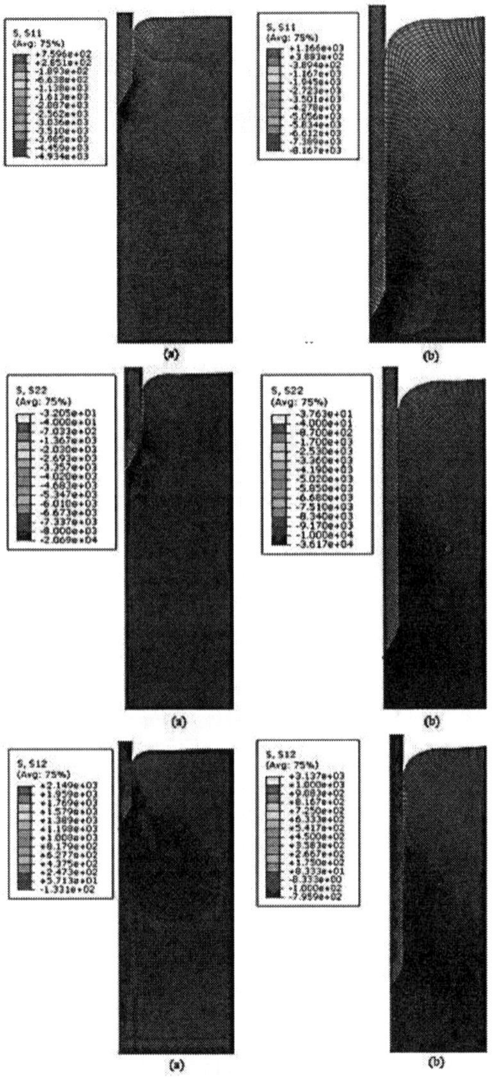

Figure 3. Stress components after distribution (a) 50 mm and (b) 150 mm cone penetration. (s11 : Horizontal Stress, S22 : Vertical Stress, S12 : Shear Stress)

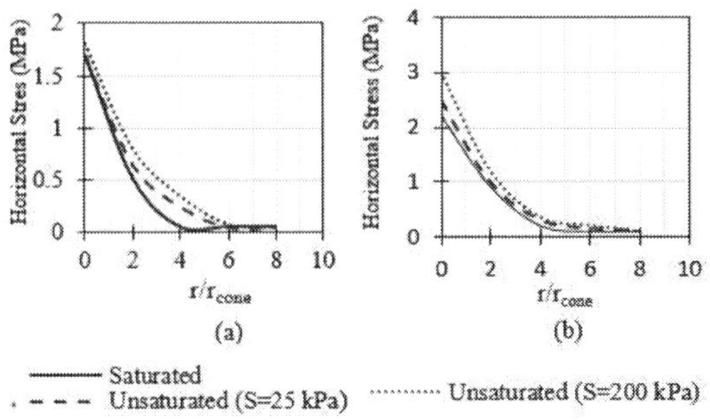

Figure 4. Comparison of horizontal stress versus distance ratio (distance from cone tip/ cone radius) for different unsaturated condition (a) P=50 kPa (b) P=100 kPa

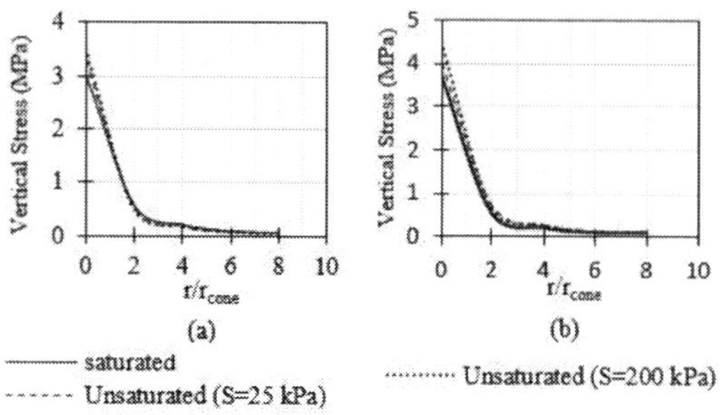

Figure 5. Comparison of vertical stress versus distance ratio (distance from cone tip/ cone radius) for different unsaturated condition (a) P=50 kPa (b) P=100 kPa

The modified material parameters are summarized in Table 4. The cone resistance profiles resulted from these alternate parameters were obtained and compared with the one from the reference case, shown in Figure 6. Both plots demonstrate a slight change in cone resistance that could be ignored. The small change in the friction angle is due to the very low sensitivity of the friction angle to the induced amount of increase in the effective stress, otherwise friction angle, which is a strength parameter, play a major role in the cone penetration response modeled using MC constitutive behavior. On the other hand, the negligible effect of elastic modulus on the cone resistance is attributed to the fact that the soil during the penetration process becomes plastic. Thus, for a perfectly plastic model like MC the effect of elastic modulus on the cone resistance turns out to be insignificant.

Table 4. Modified material model values

Suction stress (χs), (s=200 kPa)	29.8 kPa
Effective confining pressure (p+ χs)	79.8 kPa
Modified Friction angle (φ) (Eq. 1)	40.8 Degree
Modified Elastic modulus (E) (Eq. 6)	7.9 MPa

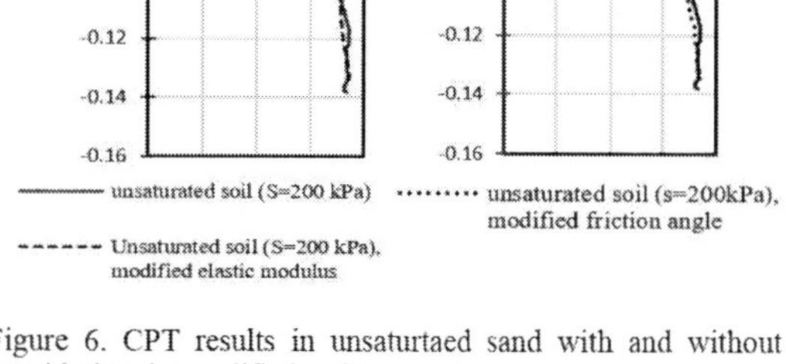

Figure 6. CPT results in unsaturtaed sand with and without considering the modified soil parameters

5 Conclusions

A numerical approach was successfully implemented to investigate the cone penetration process in unsaturated sand inside a soil calibration chamber. The application of ALE and adaptive remeshing in Abaqus/Explicit avoided the mesh distortion in this large deformation problem. Apparent cohesion concept was employed in suction-dependent strength characterization based on Mohr Coulomb constative behaviour.

Suction in unsaturated soil influenced the cone resistance significantly while the apparent cohesion efficiently was able to capture this effect. However, modifying the soil elastic modulus and friction angle based on the effective stress equation in unsaturated soil did not affect the cone resistance profiles partly due to the choice of the constitutive model and partly due to the low sensitivity of the parameters to the range of the varied effective stress.

Although the simple frictional model implemented in this study was relatively successful, it can be improved to better simulate the penetration. These include but not limited to using a stress-dependent interaction model, considering the effect of soil compressibility during the penetration, or using more sophisticated elastoplastic constitutive models.

References

1. Loret B, Khalili N. An effective stress elastic-plastic model for unsaturated porous media. *Mech. Mater* 2002; **34(2)**:97-116.
2. Lu N, Likos WJ. Suction stress characteristic curve for unsaturated soil, *J. Geotech. Geoenviron. Eng.*2006; **132(2)**: 131-142.
3. Ghayoomi M, McCartney JS. Measurement of the Small-Strain Shear Moduli of Partially-Saturated Sands During Infiltration in a Geotechnical Centrifuge. *Special Issue on Advances in Experimental Characterization of Unsaturated Soils Volume 1, ASTM Geotechnical Testing Journal* 2011; **34(5)**:503-513.
4. Baldi G, Bellotti R, Ghionna V, Jamiolkowski M, Pasqualini E. Design parameters for sand from CPT. *Proceedings of the 2nd European symposium on penetration testing*, Amsterdam, the Netherlands 1982; **2**:425–438.
5. Robertson PK, Campanella RG. Interpretation of cone penetration tests, Part I: Sand. *Can. Geotech. J.* 1983a; **20(4)**:718–733.
6. Been K, Crooks JHA, Becker DE, Jefferies MG. The cone penetration test in sands: part I, state parameter interpretation. *Géotechnique* 1986; **36(2)**:239–249, http://dx.doi.org/10.1680/geot.1986.36.2.239.
7. Robertson PK. Interpretation of cone penetration tests-a unified approach. *Can Geotech J* 2009; **(46)**: 1337-55.
8. Hryciw RD, Dowding CH. Cone penetration of partially saturated sands. *Geotech. Test. J.* 1987; **10(3)**:135–141.

9. Lehane B, Ismail MA, Fahey M. Seasonal dependence of in situ test parameters in sand above the water table. *Geotechnique* 2004; **54 (3)**:215-218.
10. Vanapalli SK, Fathi MOM. Bearing capacity and settlement of footings in unsaturated sands. *Int. J. of GEOMATE*.2013; **5(1)**:595-604.
11. Russell AR, Khalili N. On the problem of cavity expansion in unsaturated soils. *Comput. Mech.* 2006b; **37(4)**:311-330.
12. Pournaghiazar M, Russell AR, Khalili N. The cone penetration test in unsaturated sands, *Geotechnique* 2013; **63(14)**:1209-1220
13. Van den berg P. Analysis of soil penetration. Ph.D. thesis 1994, Technische Universiteit Delf
14. Susila E, Hryciw RD. Large displacement FEM modeling of the cone penetration test (CPT) in normally consolidated sand. *International Journal for Numerical and Analytical Methods in Geomechanics* 2003; **27(7)**:585-602.
15. Huang W, Sheng D, Sloan SW, Yu HS. Finite element analysis of cone penetration in cohesionless soil. *Computers and Geotechnics* 2004; **31(7)**:517-528.
16. Ahmadi MM, Byrne PM, Campanella RG. Cone tip resistance in sand: modeling, verification and applications. *Canadian Geotechnical Journal* 2005; **42**:977-993.
17. Nazem M, Sheng D, Carter JP. Stress integration and mesh refinement for large deformation in geomechanics, *International Journal for Numerical Methods in Engineering* 2006; **65**:1002-1027
18. Sheng D, Nazem M, Carter JP. Some computational aspects for solving deep penetration problems in geomechanics. *Comput Mech* 2009; **44**:549-561.
19. Yi JT, Goh SH, Lee FH, Randolph MF. (2012). A numerical study of cone penetration in fine-grained soils allowing for consolidation effects. *Géotechnique* 2012; 62(8):707–719, http://dx.doi.org/10.1680/geot.8.P.155.
20. Bolton MD, Gui MW, Garnier J, Corte JF, Bagge G, Laue J, Renzi R. Centrifuge cone penetration tests in sand. *Géotechnique* 1999; **49(4)**:543-552
21. Russell AR, Khalili N. A bonding surface plasticity model for sands exhibiting particle crushing. *Can. Geotech. J.* 2004; **41(6)**:1179-1192.
22. Bolton MD.. The strength and dilatancy of sands. *Géotechnique* 1986; **36(1)**: 65–78.
23. Abaqus-User's Manual- Version 6.13. Dassualt Systems Simulia Crop; 2013.
24. American Petroleum Institute. Recommended practice for planning, designing and constructing fixed offshore platforms-working Stress Design. *20th edition API RP 2A-WSD, American petroleum Institute*, Washington D.C.1993; 59-61.
25. Lu N, Likos WJ, Suction stress characteristic curve for unsaturated soil, *J. Geotech. Geoenviron. Eng.*2006; **132(2)**:131-142.
26. Russell AR, Khalili N. A unified bounding surface plasticity model for unsaturated soils. *Int. J. Numer. Analyt. Methods Geomech* 2006a; **30(3)**:181–212.

DOI: 10.1051/e3sconf/20160908012

Inverse analysis of cyclic constitutive models for unsaturated soil under consideration of oscillating functions

Alalade Muyiwa E.[1,a], Kafle Binod[2,b], Wuttke Frank[2,c], and Lahmer Tom[3,d]

[1] *Department of Civil Engineering, Bauhaus-Universität Weimar, Germany*
[2] *Geomechanics and Geotechnics, Kiel University, Germany*
[3] *Institute of Structural Mechanics–Stochastics and Optimization, Bauhaus-Universität Weimar, Germany*

Abstract. In order to assess the probability of foundation failure resulting from cyclic action on structures and to minimize the prediction error, various existing constitutive models considering cyclic loaded dry soils were extended to unsaturated soil conditions by the authors, thus requiring further calibration during application on existing slightly variable soil condition as well as the soil heterogeneities. The efficiency and effectiveness of these models is majorly influenced by the cyclic constitutive parameters and the soil suction. Little or no details exist in literature about the model based identification and the calibration of the constitutive parameters under cyclic loaded soils. This could be attributed to the difficulties and complexities of the inverse modeling of such complex phenomena. A wide variety of optimization strategies for the solution of the sum of least-squares problems as usually done in the field of model calibration exists, however the inverse analysis of the unsaturated soil response under oscillatory load functions has not been solved up to now. This paper gives insight into the model calibration challenges and also puts forward advanced optimization methods for the inverse modeling of cyclic loaded foundation response on unsaturated soils.

1 Introduction

The increased awareness in the impact of cyclic actions from natural and man-made sources on a structure during its lifespan necessitates understanding the complex soil structure interaction during these cyclic loads. The stresses induced by these cyclic loads lead to failure patterns such as flow-type failure (liquefaction), cyclic mobility or plastic strain accumulation. The failure type experienced depends on the initial stress state and degree of stress reversal as observed by [1]. In order to design structures such that the effects of these stresses are minimized, various constitutive models based on different formulations such as –plasticity, elasto-plasticity, hypoplasticity and so on, to describe the soil structure interaction during complex cyclic loading on dry, saturated and unsaturated soils have been developed. The quality of these model responses is a function of the model-based parameter identification from experimental data. Various strategies based on different concpets exist in finding out the values of these parameters, however the complexities involved in modeling the soil response to cyclic loading limits the application of some of these strategies specifically with regards to the quality and computation effort required for the inverse analysis.

The last three decades heralded the development and extension of many constitutive models some of which includes the elasto-plastic strain hardening law model by [2] for monotonic loads which served as a foundation for many other models such as the model by [3] for cyclic loads and model by [4, 5] for monotonic and cyclic loaded footings on unsaturated soil which takes into account the effect of matric suction on the bearing capacity of the soil as observed by [6]. The models in [3] and [5] for cyclic loading is capable of simulating the foundation response under different loading conditions and soil states. However, the quality of the model's response is strongly influenced by certain constitutive parameters, thus in order to apply the models for the actualization of economical and efficient structural design, we have to bridge the gap between model simulation and experimental response. This is achieved by the calibration and optimization of the models' sensitive 'cyclic constitutive parameters' with experimental results with the aim of identifying parameters which are unique and robust with respect to change in experimental data. Thus, the implementation of an efficient, stable and robust algorithm is required for the solution of the complex ill-posed inverse problem for unsaturated soils under oscillatory load functions with minimum error and computational cost.

The importance of these constitutive models cannot be over-emphasized, especially when considering the ease with which it can be incorporated into finite element software for structural analysis, structural design and struc-

[a] e-mail: muyiwa.alalade@uni-weimar.de.
[b] e-mail: kafle@gpi.uni-kiel.de
[c] e-mail: fw@gpi.uni-kiel.de
[d] e-mail: tom.lahmer@uni-weimar.de

tural health monitoring as well, therefore in order to yield reliable model responses an in-depth knowledge on the cyclic parameter behavior is required in addition to selecting an efficient optimization strategy.

2 Theory

2.1 Mathematical Model

The elasto-plastic strain hardening model proposed by [3] for response to cyclic loading on dry soil is an extension of the model proposed by [2] which was developed for monotonic loading. The plastic strain as a result of monotonic load is expressed in terms of generalized force and displacement as shown in Equation 1, where Λ according to [?] is the plastic multiplier and it depends on the positive change in loading, the loading history, ρ_c and the macro-element through the hardening modulus, H.

$$dq = \Lambda \frac{\delta g}{\delta Q} \tag{1}$$

The generalized force $\vec{Q}$ consists of the vertical force V, horizontal force H and moment M, normalized with the maximum centric force the soil can resist V_m. The generalized displacement $\vec{q}$ consists of the vertical displacement v, the horizontal displacement u, and the rotations θ normalized by V_m.

$$\vec{Q} = \begin{Bmatrix} \xi \\ h \\ m \end{Bmatrix} = \frac{1}{V_m} \begin{Bmatrix} V \\ \frac{H}{\mu} \\ \frac{M}{\psi B} \end{Bmatrix} \tag{2}$$

$$\vec{q} = \begin{Bmatrix} \eta \\ \epsilon \\ \zeta \end{Bmatrix} = V_m \begin{Bmatrix} v \\ \mu u \\ \psi B \theta \end{Bmatrix} \tag{3}$$

In order to simulate cyclic loading, the model in [3] permits the formation of plastic strains inside the bounding surface. The magnitude of plastic strains induced is a function of the distance between the stress point and the image point I_p on the bounding surface, the memory parameters (ρ_c and ρ_k) and the cyclic constitutive parameters expressed in Φ in Equation 4. The parameters $\Lambda(Q_I)$ and $\frac{\delta g}{\delta Q}(Q_I)$ are calculated on the image point and not on the current stress point as would be the case for monotonic loading

$$dq^{pl} = \Lambda(Q_I)\ \Phi(\delta, \rho_k)\ \frac{\delta g}{\delta Q}(Q_I) \tag{4}$$

The term Φ which is a function of δ and ρ_k is a diagonal matrix (Equation 5) which acts as a weighting function during cyclic loading. Larger values of δ leads to small diagonal terms and also smaller plastic displacements [3]

$$\Phi = \begin{bmatrix} \exp^{-\left(\varsigma_\eta \sqrt{\frac{\delta}{\frac{\xi}{\rho_c}}} + \kappa_\eta \rho_k\right)} & 0 & 0 \\ 0 & \exp^{-\left(\varsigma_\epsilon \sqrt{\frac{\delta}{\frac{\xi}{\rho_c}}} + \kappa_\epsilon \rho_k\right)} & 0 \\ 0 & 0 & \exp^{-\left(\varsigma_\zeta \sqrt{\frac{\delta}{\frac{\xi}{\rho_c}}} + \kappa_\zeta \rho_k\right)} \end{bmatrix} \tag{5}$$

Plastic strain caused by cyclic loading is sensitive to the constitutive parameters ς_i and κ_i (where $i = \eta,\ \epsilon,\ \zeta$) in the Φ matrix. These parameters regulate the rate at which shakedown occurs in the model and also the the coupling between the horizontal loading and the corresponding vertical displacement at constant vertical force.
The extension of these model to incorporate the multiphase soil behaviour was proposed by [5]. It was observed that not only the intrinsic soil properties like maximum centric vertical force V_m and initial slope of load deformation curve (R_0) are sensitive to matric suction but also the cyclic constitutive parameters. The proposed model expressed in Equation 6 is similar to Equation 4, except that the parameters are a function of 'ψ'

$$dq^{pl}(\psi) = \Lambda_I(\psi)\ \Phi(\delta, \rho_k, \psi)\ \left(\frac{\delta g(\psi)}{\delta Q(\psi)}\right)_I \tag{6}$$

2.2 Calibration and optimization

The quality of any prognosis depends on the quality of the identified cyclic constitutive parameters. Besides accuracy, efficiency is also an important issue as dynamic non-linear models tend to be complex and time consuming, particularly during calibration. Therefore it was necessary to study the cyclic constitutive parameter dependency using contour lines before selecting an optimization strategy. A strong dependency between some parameters was observed and is expressed in Figure 1. The presence of many valleys (indicating minima) can be traced to the structure of the Φ matrix which has addition operations in each of the diagonal terms as seen in Equation 5. Thus different pairs of cyclic constitutive parameters in each diagonal term in the Φ matrix can yield similar results with regards to model response.

Solving these obvious non-convex inverse problem by minimizing the conventional sum of squared error between the calculated vertical cyclic displacement, η [3] from the forward model and the experimental response in Equation 7 yields non-unique parameters which are a function of the initial value used to initiate the optimization process. The Nelder-Mead method ([10]), which is heuristic based, deterministic and known to be robust with respect to noisy cost function is applied in identifying these cyclic constitutive parameters.

$$C_f(p) = \sum_i^n (\eta^{mod}_{(t_i,p)} - \eta^{exp}_{(t_i)})^2 \tag{7}$$

where:

 DOI: 10.1051/e3sconf/20160908012

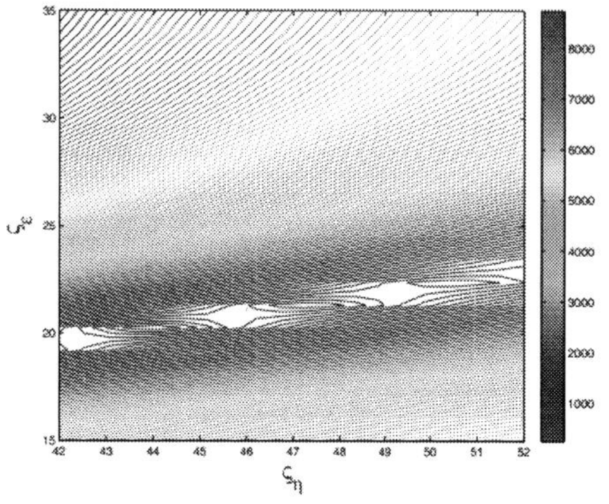

Figure 1. Dependencies between cyclic constitutive parameter ς_η and ς_ϵ [9]

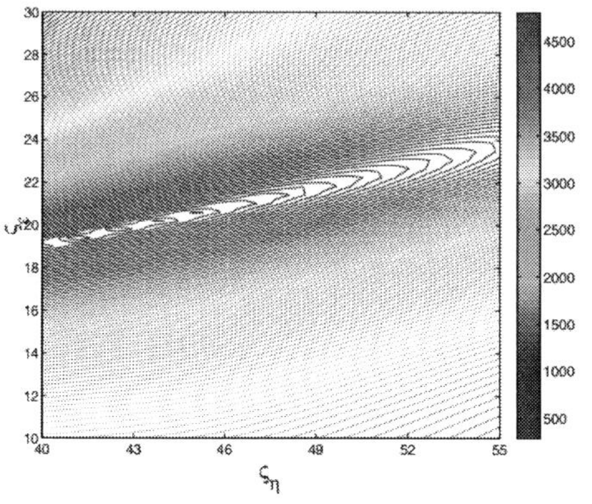

Figure 2. Dependency between ς_η and ς_ϵ after regularization

- C_f is the cost function to be minimized
- η^{mod} is the vertical component of the generalized cyclic plastic displacement vector calculated by the model
- η^{exp} is the vertical component of the generalized cyclic plastic displacement vector obtained from the experiment
- p is a vector of $\Lambda_I(\psi)$, $\Phi(\delta, \rho_k, \psi)$ and $\left(\frac{\delta g(\psi)}{\delta Q(\psi)}\right)_I$
- t_i is the time step at each point
- n is the total number of data points

In order to increase the probability of reaching the global minima and getting best quality results with the Nelder-Mead simplex algorithm, it was necessary to vary the starting point of the optimization until the best fit or lowest value of C_f is attained.

However, an attempt to obtain a more reliable lowest function value and unique parameters from the optimization, convexification of the problem (as seen from the multiple minimas in Figure 1) can be achieved by introducing a penalty term to the cost function (Equation 7)in line with Tikhonov regularization. This regularization/penalty process is influenced also by a priori knowledge and can be observed in Equation 8.

$$C'_f(p) = \sum_i^n (\eta^{\,mod}_{\,(t_i,p)} - \eta^{\,exp}_{\,(t_i)})^2 + \Gamma \sum_{k=1}^4 (\Psi^0_{(k)} - \Psi^{mod}_{(k)})^2 \quad (8)$$

where:

- C'_f is the cost function to be minimized
- η^{mod} is the vertical component of the generalized cyclic plastic displacement vector calculated by the model
- η^{exp} is the vertical component of the generalized cyclic plastic displacement vector obtained from the experiment
- p is a vector of $\Lambda_I(\psi)$, $\Phi(\delta, \rho_k, \psi)$ and $\left(\frac{\delta g(\psi)}{\delta Q(\psi)}\right)_I$
- t_i is the time step at each point
- n is the total number of data points
- Γ is a scaling factor for the penalty term
- $\Psi^{\,0}$ is a vector of a priori cyclic constitutive parameters
- $\Psi^{\,mod}$ is a vector of calculated cyclic constitutive parameters
- k is the total number of cyclic constitutive parameters

The effect of this regularization can be observed in Figure 2. The value of Γ should be carefully chosen because it greatly influences the properties of the regularized solution. A large value of Γ results in over-smoothing and also large residuals, whereas very small Γ values (under-smooothing) may give a good fit but the solutions would be dominated by data errors. An optimal value for Γ is obtained from a plot of the residual norm and the solution norm (L-curve) according to [11].

3 Discussion of results

Model parameter identification was carried out by calibrating the model to experimental data obtained from [5]. Before the model calibration, it was necessary to filter out measurement errors from the experimental data. The model parameter identification was carried out using the Nelder-Mead algorithm on saturated and unsaturated soils, with outcomes shown in Table 1 and Figure 3 to Figure 4, where a comparison between the experimental result is made with the optimized model response.

It is worth noting that as a result of the strong dependency between the cyclic constitutive parameters and the application of a deterministic local optimization strategy, the values used to initiate the optimization plays a major role in the computational time required and the quality of the output result obtained. Although values from literature (values published in [12]) were first used to initiate the optimization process, the quality of fit was not satisfactory, thus other randomly generated values were used as initial values until a best fit was attained. Furthermore, the outcome of the inverse modeling revealed a lower sensitivity to the parameters ς_ϵ and κ_ϵ. This is as a results of

Table 1. Parameters obtained after model optimization with and without regularization

State	ψ $[kPa]$	V_m $[kN]$	R_0 $[kN/mm]$	ς_η	κ_η	ς_ϵ	κ_ϵ	F.val	Iter.
Sat. ($\Gamma = 0$)	0	1.12	0.252	54.3857	0.3435	−59.3968	74.9093	50.77	482
Sat.($\Gamma \neq 0$)	0	1.12	0.252	52.2103	0.1825	27	55	47.11	328
Unsat.($\Gamma = 0$)	1.7	6.4	0.822	7.1696	3.3187	12.9429	109.6090	1.143	121

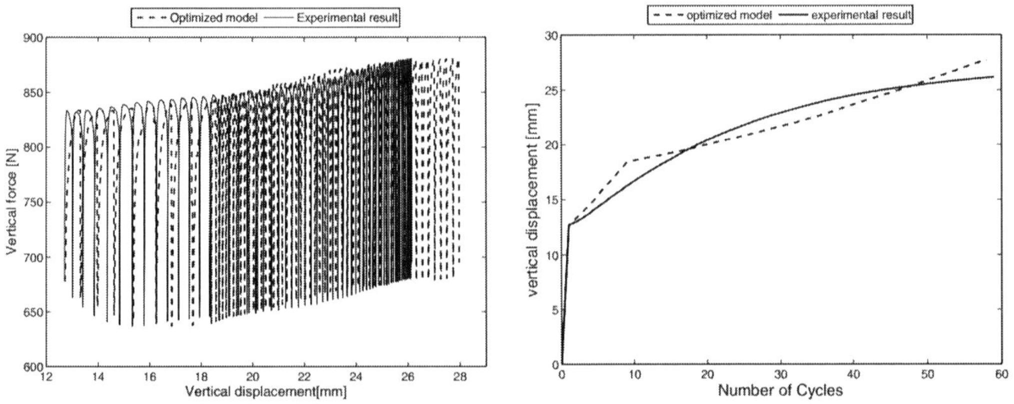

(a) Cyclic force-displacement plot
(b) Accumulated plastic displacement plot

Figure 3. Comparison between the optimized model response and experimental result for saturated soil (0[kPa])

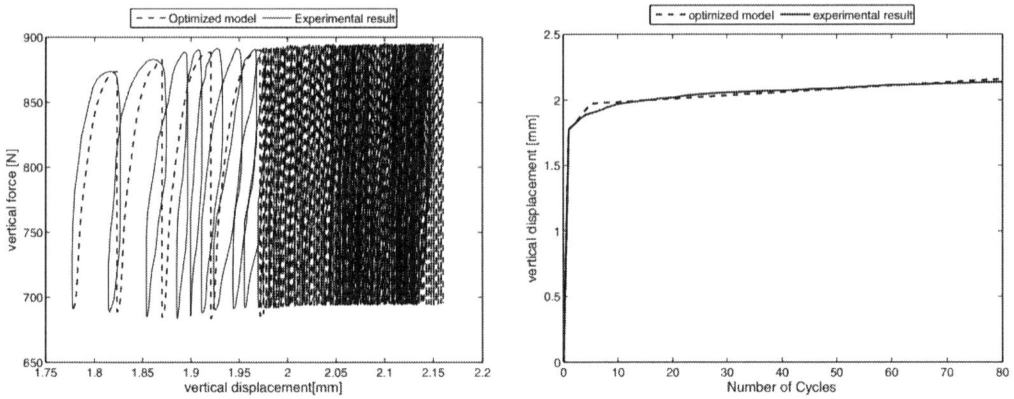

(a) Cyclic force-displacement plot
(b) Accumulated plastic displacement plot

Figure 4. Comparison between the optimized model response and experimental result for unsaturated soil (1.7[kPa])

Table 2. Initial value and optimized constitutive parameters

	ς_η	κ_η	ς_ϵ	κ_ϵ	F. val	Iter.
Init val. 1	1	10	10	10	385.15	0
Opt. val. 1	7.16	3.31	88.80	13.37	1.143	385
Init val. 2	5	10	10	10	6.51	0
Opt. val. 2	7.16	3.31	14.47	23.97	1.143	179
Init. val. 3	10	10	10	10	9.2926	0
Opt. val. 3	7.16	3.31	12.94	10.96	1.143	103
Init val. 4	15	10	10	10	16.26	0
Opt. val. 4	7.16	3.31	17.51	17.32	1.143	114

the experimental data being dominated by vertical oscillatory loads as obtained in Table 2 where although different values were used to initiate the optimization algorithm, the cyclic constitutive parameters ς_η and κ_η related to η in the Φ matrix (Equation 5) remained constant while parameters related to ϵ and the computational cost varied.

Moreover, in some particular cases the solution to the inverse problem yields a good quality response, however the identified parameters are unrealistic as observed for the soil sample with $\psi = 0$ kPa and $\Gamma = 0$ in Table 1. An attempt to yield realistic parameters was made by implementing a modified Nelder-Mead algorithm to constrain the resulting optimized parameters. Although realistic values of parameters were obtained, a decrease in the quality of the model response was observed. An efficient and robust strategy to tackle this issue was to incorporate the regularization term Γ into the objective function as obtained in Equation 8 and explained in subsection 2.2. The process of convexification not only tackles the issue of unrealistic parameters but it also reduces the computational costs and yields unique and stable parameters irrespective of the values used to initiate the optimization.

4 Conclusion

As observed in Figure 3 and Figure 4 and also corroborated in literature, the unsaturated soil yields lower plastic displacements when compared to other soil states, in addition to this and with respect identifying the model parameters it can be concluded that

- Measurement error (noise) affects quality of fit during calibration and thus have to be filtered out before parameter identification
- The initial guess is a key factor in determining the outcome (quality of parameters obtained) and efficiency of the the inverse modeling computation.
- The model exhibited a higher sensitivity towards parameters are related to the vertical component of the model. This is as a result of the experimental data being dominated by vertical oscillatory loads.
- Though the Nelder–Mead algorithm may converge to a local minima, the process of using random initial values and comparing the quality of fit shows the application of this algorithm to find the global minima with a high probability.
- Better quality fit was observed for unsaturated soil condition with minimal computational effort. This can be attributed to the fact that shakedown occurs earlier and as such the effect of the slight reduction in the accumulated plastic displacement during unloading by soil rebound which the model does not account for is minimized.
- With regularization, calibration yields more reliable results with minimal computational effort for these complex dynamic soil models as seen in Table 1.

References

[1] J. Yang, H. Sze, Journal of Geotechnical and Geoenvironmental Engineering **137**, 1275 (2011)
[2] R. Nova, L. Montrasio, Géotechnique **41**, 243 (1991)
[3] C. Di Prisco, C. Zambelli, Revue francaise de genie civil **7**, 881 (2003)
[4] F. Wuttke, B. Kafle, Y. Lins, T. Schanz, International Journal of Geomechanics **13**, 557 (2013)
[5] B. Kafle, F. Wuttke, Advances in unsaturated soils, Taylor and Francis pp. 521–526 (2013)
[6] S. Vanapalli, D. Fredlund, D. Pufahl, A. Clifton, Canadian Geotechnical Journal **33**, 379 (1996)
[7] S.K. Vanapalli, F.M.O. Mohamed, in *Experimental Unsaturated Soil Mechanics*, edited by T. Schanz (Springer Berlin Heidelberg, 2007), Number 112 in Springer Proceedings in Physics, pp. 483–493, ISBN 978-3-540-69872-2, 978-3-540-69873-9, `http://link.springer.com/chapter/10.1007/3-540-69873-648`
[8] T. Schanz, Y. Lins, S. Vanapalli, *Bearing capacity of a strip footing on an unsaturated sand* (2011), Vol. 2, pp. 1195–1200, ISBN 9780415604307
[9] M.E. Alalade, Master's thesis, Bauhaus Universität Weimar (2014)
[10] S. Singer, J. Nelder, Scholarpedia **4**, 2928 (2009)
[11] P.C. Hansen, *The L-curve and its use in the numerical treatment of inverse problems* (IMM, Department of Mathematical Modelling, Technical Universityof Denmark, 1999)
[12] C. Di Prisco, R. Nova, A. Sibilia, NUMOG **8**, 637 (2002)

DOI: 10.1051/e3sconf/20160908013

Effect of hydraulic hysteresis on dynamic response of unsaturated soils

B. Shahbodagh[a], N. Khalili, G. Alipour Esgandani

School of Civil and Environmental Engineering, The University of New South Wales, Sydney, NSW 2052, Australia

Abstract. Starting from the conservation laws, field equations governing the dynamic behaviour of unsaturated soils are presented. The coupling between solid and fluid phases is enforced according to the effective stress principle taking suction dependency and volume change of the effective stress parameters into account. The hydraulic hysteresis is accounted for through the effective stress parameter and the soil water characteristic curve. The spatial discretization of the governing equations is achieved using finite element method, whereas the time integration is conducted using the Newmark scheme. The numerical results are then presented, and the effect of hydraulic hysteresis on dynamic response of unsaturated soils is particularly emphasized.

1 Introduction

Despite the widespread distribution of unsaturated soils throughout the world, the development of constitutive and computational models for the dynamic behaviour of unsaturated soils has significantly lagged behind similar developments for dry and saturated soils. This has mainly been due to the inherent complexities associated with the dynamic behaviour of unsaturated soils, i.e. simultaneous flow of air and water through the porous materials, the presence of inertia forces in different phases, highly nonlinear deformation behaviour of the soil matrix, and complex interaction of fluid flow and deformation fields.

In recent years, several contributions have been made to dynamics of unsaturated soils. In the studies, two different approaches have generally been followed to enforce the coupling between solid and fluid phases. In the first approach, a single effective stress variable is used to describe the mechanical behaviour of unsaturated soils, e.g. see [1-7], while in the second approach two stress state variables, i.e. net stress and suction, are adopted to cast the constitutive relationships of the soil, e.g. see [8,9]. A major difficulty with the second approach is that it requires determination of two sets of material parameters, one for each of the stress state variables, which in fact may not be independent, and can lead to intractable stress strain relationships. In the effective stress based models developed for coupled dynamic analysis of unsaturated soils, the degree of saturation has mostly been used as the effective stress parameter. From experimental evidence, this parameter is strongly dependent on the soil structure and cannot be uniquely determined by the degree of saturation [10]. In addition, the effect of hydraulic hysteresis was ignored in the studies, which can significantly alter the response of unsaturated soils under dynamic loading conditions that can invariably involve complex cycles of strain-induced wetting and drying.

In this paper, a coupled flow-deformation model is presented for the dynamic/seismic analysis of unsaturated soils. The theoretical approach adopted is based on the model proposed by Khalili et al. [11], addressing the suction dependency and volume change of the effective stress parameter. The effect of hydraulic hysteresis on the effective stress parameter and soil water characteristic curve is accounted for using the model proposed by Khalili and Zargarbashi [12]. The spatial discretization of the governing equations is achieved using finite element method, whereas the time integration is conducted using the Newmark technique. The numerical results are then presented, demonstrating the performance of the proposed approach, and the effect of hydraulic hysteresis on static and dynamic response of unsaturated soils is particularly emphasized.

2 Basic Concepts

2.1. Effective Stress

The concept of effective stress plays a key role in quantitative assessment of response in saturated and unsaturated soils. In its most common form, the effective stress can be expressed as [13]

$$\sigma'_{ij} = \sigma_{net\ ij} - \chi s \delta_{ij} \tag{1}$$

where $\sigma_{net\ ij} = \sigma_{ij} + p_G \delta_{ij}$ is the net stress, $s = p_G - p_W$ is the matric suction, χ is the effective stress parameter, σ_{ij} is the total stress, p_W and p_G are the pore water and pore gas pressures, respectively, and δ_{ij} is the Kronecker

[a] Corresponding author: b.shahbodagh@unsw.edu.au

DOI: 10.1051/e3sconf/20160908013

delta. The incremental form of the effective stress equation can be expressed as

$$\dot{\sigma}'_{ij} = \dot{\sigma}_{net\ ij} - \psi \dot{s} \delta_{ij} \tag{2}$$

where $\psi = \partial(\chi s)/\partial s$ is the incremental effective stress parameter, and superimposed dot denotes the time derivative with respect to the solid phase. Following the approach proposed by Khalili and Khabbaz [14] and Khalili et al. [15], the effective stress parameter is defined as

$$\chi = \begin{cases} 1 & \text{for } s \le s_e \\ (s/s_e)^{-\Omega} & \text{for } s > s_e \end{cases} \tag{3}$$

where Ω is a material parameter with the best fit value of 0.55, and s_e is the suction value marking the transition between saturated and unsaturated states. For wetting process, s_e is equal to the air expulsion value, s_{ex}, whereas for drying process, s_e is equal to the air entry value, s_{ae}. s_e is a priori a function of the specific volume or volume change of the solid skeleton. This leads to a shift to the right of the effective stress parameter curve and soil water characteristic curve with increasing density, see Figure 1.

Khalili and Zargarbashi [12] experimentally studied the effect of hydraulic hysteresis on the effective stress parameter, and proposed the following relation for suction reversals:

$$\chi = \begin{cases} \left(\frac{s_{rd}}{s_{ae}}\right)^{-\Omega} \left(\frac{s}{s_{rd}}\right)^{\zeta} & \text{for drying path reversal } \left(\frac{s_{ex}}{s_{ae}}\right)^{\frac{\Omega}{\Omega+\zeta}} s_{rd} \le s \le s_{rd} \\ \left(\frac{s_{rw}}{s_{ex}}\right)^{-\Omega} \left(\frac{s}{s_{rw}}\right)^{\zeta} & \text{for wetting path reversal } s_{rw} \le s \le \left(\frac{s_{ae}}{s_{ex}}\right)^{\frac{\Omega}{\Omega+\zeta}} s_{rw} \end{cases} \tag{4}$$

where ζ is the slope of the transition line between the main wetting and main drying paths in a $\ln\chi \sim \ln s$ plane, and s_{rw} and s_{rd} are the points of suction reversal on the main wetting and main drying paths, respectively.

2.2. Soil Water Characteristic Curve

Another important concept in the mechanics of unsaturated soils is the soil water characteristic curve (SWCC), which is used to determine the volumetric deformation of the water phase with respect to change in matric suction. In this formulation, the SWCC model proposed by Brooks and Corey [16], extended to include hydraulic hysteresis effect [11], is adopted as

$$S_{eff} = \begin{cases} 1 & \text{for } s \le s_e \\ (s_e/s)^{\lambda_p} & \text{for } s > s_e \end{cases} \tag{5}$$

$$S_{eff} = \begin{cases} \left(\frac{s_{ae}}{s_{rd}}\right)^{\lambda_p} \left(\frac{s_{rd}}{s}\right)^{\xi} & \text{for drying path reversal } \left(\frac{s_{ex}}{s_{ae}}\right)^{\frac{\lambda_p}{\lambda_p-\xi}} s_{rd} \le s \le s_{rd} \\ \left(\frac{s_{ex}}{s_{rw}}\right)^{\lambda_p} \left(\frac{s_{rw}}{s}\right)^{\xi} & \text{for wetting path reversal } s_{rw} \le s \le \left(\frac{s_{ae}}{s_{ex}}\right)^{\frac{\lambda_p}{\lambda_p-\xi}} s_{rw} \end{cases} \tag{6}$$

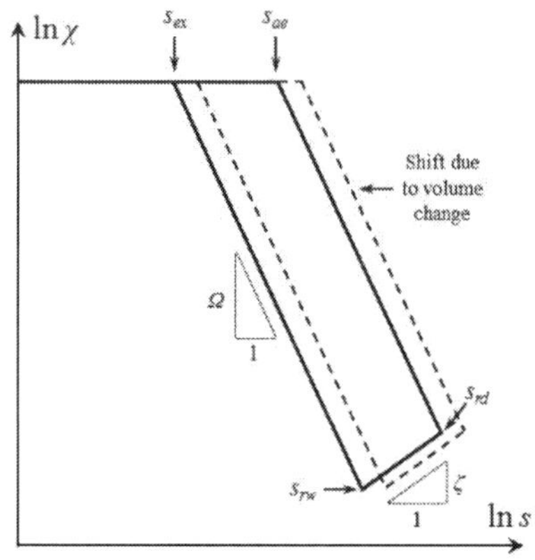

Figure 1. Evolution of the effective stress parameter χ with hydraulic hysteresis

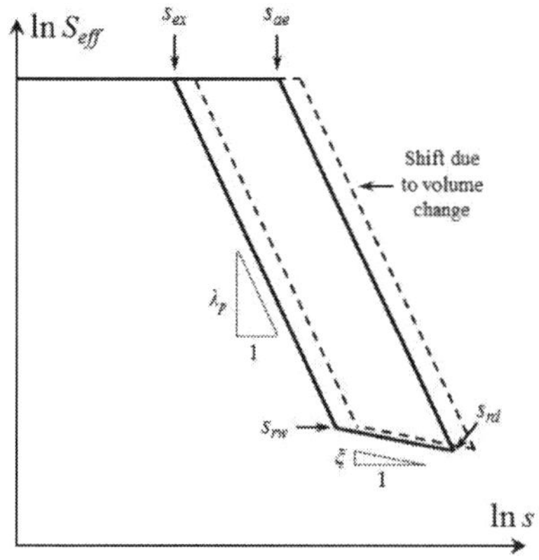

Figure 2. Soil water characteristic curve including hydraulic hysteresis

where λ_p is the pore size distribution index, $S_{eff} = (S_r - S_{res})/(1 - S_{res})$ is the effective degree of saturation, and S_{res} is the residual degree of saturation, see Figure 2.

3 Governing Equations

3.1. Conservation of Mass

Unsaturated soils consist of three phases, i.e. solid (S), water (W), and gas (G), which within the context of theory of mixtures are assumed to be continuously distributed throughout representative elementary volume. By ignoring mass exchanges among the phases, the conservation of mass for each phase can be expressed by

$$\frac{d_\alpha}{dt}(n^\alpha \rho_\alpha) + n^\alpha \rho_\alpha v_{\alpha\,i,i} = 0 \qquad (\alpha = S, W, G) \tag{7}$$

where $v_{\alpha i}$ is the velocity vector of phase α, and $d_\alpha()/dt = \partial()/\partial t + ()_{,i} v_{\alpha i}$ is the material time derivative with respect to phase α. The material time derivative of volume fraction of fluid phases n^β $(\beta = W, G)$ with respect to the solid phase is obtained as

$$\frac{d_S n^\beta}{dt} = \frac{1}{V}\left(\frac{d_S V_\beta}{dt} - n^\beta \frac{d_S V}{dt}\right) = \frac{1}{V}\frac{d_S V_\beta}{dt} - n^\beta v_{S\,i,i} \quad (8)$$

in which V_α is the volume of constituent α, and V is the total volume. Assuming barotropic fluid, and substituting (8) into (7), we obtain

$$-\frac{1}{\rho_\beta}(\rho_\beta \dot{\omega}_{\beta i})_{,i} = n^\beta c_\beta \frac{d_S p_\beta}{dt} + \frac{1}{V}\frac{d_S V_\beta}{dt} \quad (9)$$

in which $\dot{\omega}_{\beta i} = n^\beta (v_{\beta i} - v_{S i})$ is the relative velocity of fluid phases with respect to the solid phase, and c_β is the compressibility coefficient for phase β. Invoking existence of elastic and plastic potentials, it can be shown that [11]

$$\frac{1}{V}\frac{d_S V_W}{dt} = \psi \dot{\varepsilon}_v + n \frac{\partial S_r}{\partial s}\dot{s} \quad (10)$$

$$\frac{1}{V}\frac{d_S V_G}{dt} = (1-\psi)\dot{\varepsilon}_v - n \frac{\partial S_r}{\partial s}\dot{s} \quad (11)$$

Substituting (10)-(11) into (9), the mass balance equations for the water and gas phases, respectively, become

$$-\frac{1}{\rho_W}(\rho_W \dot{\omega}_{W i})_{,i} = n^W c_W \frac{d_S p_W}{dt} + \psi \dot{\varepsilon}_v + n \frac{\partial S_r}{\partial s}\dot{s} \quad (12)$$

$$-\frac{1}{\rho_G}(\rho_G \dot{\omega}_{G i})_{,i} = n^G c_G \frac{d_S p_G}{dt} + (1-\psi)\dot{\varepsilon}_v - n \frac{\partial S_r}{\partial s}\dot{s} \quad (13)$$

3.2. Conservation of Momentum

The conservation of linear momentum of phase α gives

$$\sigma^\alpha_{ji,j} + \rho_\alpha n^\alpha b_i + \sum_{\gamma \neq \alpha} h_i^{\alpha\gamma} = \rho_\alpha n^\alpha a_{\alpha i} \quad (14)$$

where σ^α_{ij} is the partial Cauchy stress, $\gamma = S, W, G$, b_i is the body force per unit mass, $h_i^{\alpha\gamma} (= -h_i^{\gamma\alpha})$ is the interaction force per unit volume exerted by phase γ on phase α, and $a_{\alpha i}$ is the acceleration vector. Ignoring the relative accelerations between the phases and summing the conservation of momentum equations of the three phases yields

$$\sigma_{ji,j} + \rho b_i = \rho a_{Si} \quad (15)$$

where $\rho = \sum_\alpha \rho_\alpha n^\alpha$ is the density of the mixture, and ρ_α is the intrinsic density of phase α.

From the incorporation of the momentum balance equations into the mass balance equations, the continuity equations for the fluid phases become

$$\frac{1}{\rho_W}\left(\frac{k_W}{g}\left(p_{W,i} - \rho_W b_i + \rho_W a_{Si}\right)\right)_{,i} = n^W c_W \frac{d_S p_W}{dt} + \psi \dot{\varepsilon}_v + n \frac{\partial S_r}{\partial s}\dot{s} \quad (16)$$

$$\frac{1}{\rho_G}\left(\frac{k_G}{g}\left(p_{G,i} - \rho_G b_i + \rho_G a_{Si}\right)\right)_{,i} = n^G c_G \frac{d_S p_G}{dt} + (1-\psi)\dot{\varepsilon}_v - n \frac{\partial S_r}{\partial s}\dot{s} \quad (17)$$

3.3. Constitutive Equations

Constitutive relationships for volumetric deformations of fluid phases can be expressed as [11]

$$\frac{\dot{V}_W}{V} = \psi \dot{\varepsilon}_v + a_W \dot{s} \quad (18)$$

$$\frac{\dot{V}_G}{V} = (1-\psi)\dot{\varepsilon}_v + a_G \dot{s} \quad (19)$$

which relates the changes in pore fluid volumes to rates of suction and volumetric strain $\dot{\varepsilon}_v$. Due to highly nonlinear behaviour of soils, the stress-strain relationship of the soil skeleton is written in the incremental format as

$$\dot{\sigma}'_{ij} = C_{ijkl}\dot{\varepsilon}_{kl} \quad (20)$$

where C_{ijkl} is the constitutive tensor, and $\dot{\varepsilon}_{kl}$ is the soil skeleton strain tensor. The constitutive model adopted to express C_{ijkl} must be able to explain the deformation characteristics of the soil skeleton under cyclic loading conditions [11, 17-18].

4 Spatial and Temporal Discretization

For a quantitative solution, the weak forms of the equations of motion and flow continuity equations are obtained by applying the Galerkin method and spatially discretized using the finite element procedure. The primary unknown variables in the whole domain can be expressed in terms of the nodal values as

$$\{u\} = [N]\{u_N\} \quad , \quad p_\beta = \{N_h\}^T \{p_{\beta N}\} \quad (21)$$

where $[N]$ and $\{N_h\}$ represent the displacement shape functions for the solid phase, and the shape functions for the pore fluid pressures, respectively. The spatially discretized form of the governing partial differential equations can be written as

$$[M]\{\ddot{u}_N\} + [K]\{\Delta u_N\} - \chi[K_v]\{p_{WN}\} - (1-\chi)[K_v]\{p_{GN}\} = \{F\} - \{R\}_t \quad (22)$$

$$\rho_W \tilde{k}^W [K_v]^T \{\ddot{u}_N\} - \psi [K_v]^T \{\dot{u}_N\} - \tilde{k}^W [K_h]\{p_{WN}\} \\ - c_{11}[K_n]\{\dot{p}_{WN}\} + c_{12}[K_n]\{\dot{p}_{GN}\} = -\tilde{k}^W \{q^W\} \quad (23)$$

$$\rho_G \tilde{k}^G [K_v]^T \{\ddot{u}_N\} - (1-\psi)[K_v]^T \{\dot{u}_N\} - \tilde{k}^G [K_h]\{p_{GN}\} \\ - c_{22}[K_n]\{\dot{p}_{GN}\} + c_{12}[K_n]\{\dot{p}_{WN}\} = -\tilde{k}^G \{q^G\} \quad (24)$$

where $c_{11} = n^W c_W - n\partial S_r / \partial s$, $c_{22} = n^G c_G - n\partial S_r / \partial s$, $c_{12} = -n\partial S_r / \partial s$, $\tilde{k}^\beta = k^\beta / \rho_\beta g$, $[M]$ is the mass matrix, $[K]$ is the element stiffness matrix, $[K_v]$ is the coupling matrix, $[K_h]$ is the flow matrix, $[K_n]$ is the pore mass matrix, $\{q^W\}$ and $\{q^G\}$ are vectors of nodal fluxes of the water and gas flows, respectively, and $\{F\}$ and $\{R\}_t$ are the vectors of nodal forces. The definition of the finite element matrices can be found in [19]. The Newmark scheme is used for the time integration of the discretized governing equations, with the lowest allowable order for each primary variable. The unconditional stability of the Newmark scheme is achieved when $2\beta \geq \gamma \geq 0.5$ in which β and γ are the Newmark's parameters.

5 Numerical Examples

For the numerical analysis, a boundary value problem consisting of an unsaturated porous medium of 10 m height and 10 m width with an initial suction of 20 kPa is considered, see Figure 3. The mesh is composed of 400 mixed quadrilateral elements of dimensions 0.5 m×0.5 m. The upper boundary is drained and partially subjected to a load of width B=3.0 m with a uniform intensity f(τ), while the remaining boundaries are impervious. As the displacement boundary, the side boundaries are supported horizontally, and the bottom surface is vertically constrained. Table 1 shows the material parameters adopted for the analyses. The parameters for the fully saturated state are the same as those used by Simon et al. [20]. For the case with hydraulic hysteresis effect, the air entry and air expulsion values of s_{ae}=10 kPa, s_{ex}=5 kPa, along with ξ=0.04, and ζ=0.15 are used in the simulations, whereas for the case without hysteresis effect $s_{ae} = s_{ex}$ =10 kPa are adopted. The initial suction is assumed to be placed on the main drying path. In the case with hydraulic hysteresis, wetting and drying occurs along the scanning and main drying paths, respectively, whereas without hysteresis the wetting-drying cycles occur along the main drying path. The analyses are performed under a harmonic loading condition with angular frequency ω=62.83 rad/sec. A comparison of numerical results of the proposed hysteretic approach with those of the non-hysteretic approach is demonstrated in Figure 4 where $\hat{u}$ $\left(= uV_p \gamma_W s_{Ini} / (k_{Ws} \sigma_{\max} s_{ae})\right)$ is the normalized displacement and $\tau\left(= t\gamma_W / (\rho_{sat} k_{Ws})\right)$ is the normalized time. By taking the effect of hydraulic hysteresis into account, a deviation of the center line of

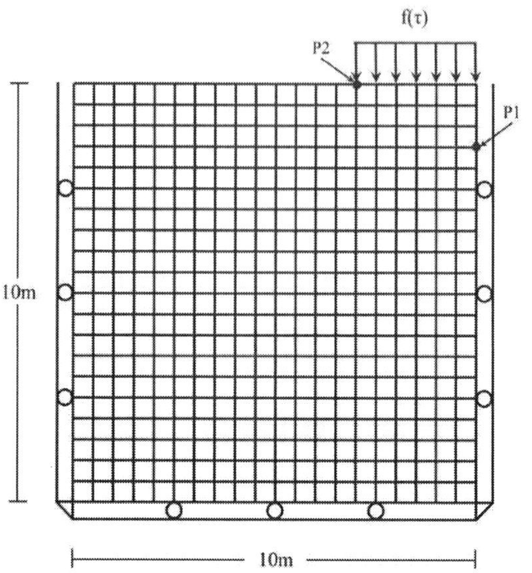

Figure 3. Finite element mesh and boundary conditions

the response from $\hat{u} = 0$ line is observed. Also, considering the hydraulic hysteresis in the analysis induces significantly larger variations in suction level during dynamic loadings.

6 Conclusions

A rigorous framework based on the theory of mixtures is presented for describing the dynamic behaviour of unsaturated soils. The coupling between solid and fluid phases is established through the effective stress principle taking suction dependency and volume change of the effective stress parameter into account. The hydraulic hysteresis is accounted for through the effective stress parameter and the soil water characteristic curve. The change in the degree of saturation due to a change in the volume of the pore at constant suction is also taken into account. The numerical results demonstrate that the hydraulic hysteresis markedly alter the response of unsaturated soils under dynamic loading condition.

Table 1. Material parameters considered for the analysis.

Initial void ratio	e_0	0.5
Lame's constant *(kPa)*	λ	833.3
Lame's constant *(kPa)*	μ	1250.0
Permeability of liquid at S_r=1 *(m/s)*	k_{Ws}	0.01425
Permeability of gas at S_r=0 *(m/s)*	k_{Gs}	0.05
Density of saturated mixture *(t/m³)*	ρ_{sat}	1.8
Pore size distribution index	λ_p	0.15
Residual degree of saturation	S_{res}	0.2

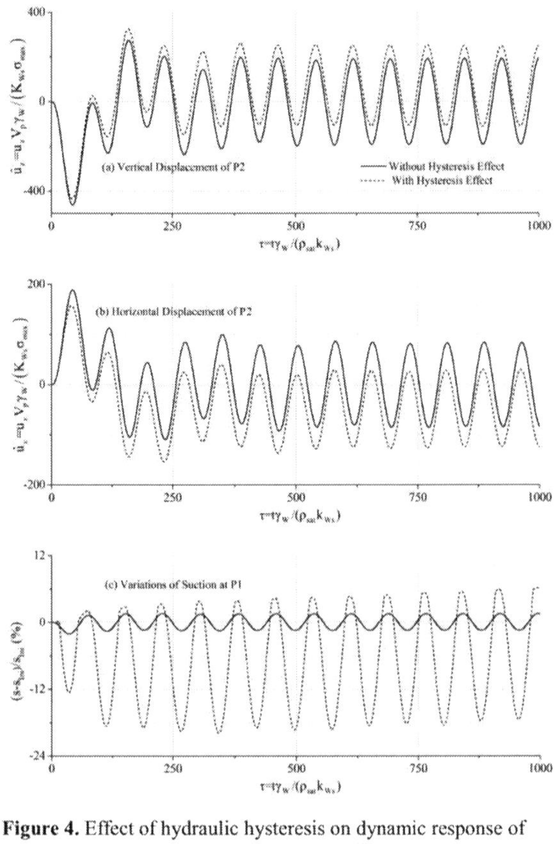

Figure 4. Effect of hydraulic hysteresis on dynamic response of the unsaturated porous medium

References

[1] Zienkiewicz OC, Xie YM, Schrefler BA, Ledesma A, Biĉaniĉ N. Static and dynamic behaviour of soils: a rational approach to quantitative solutions. II. Semi-saturated problems. *Proceedings of the Royal Society of London. Series A, Mathematical and Physical Sciences* 1990; **429**: 311-321.

[2] Meroi EA, Schrefler BA, Zienkiewicz OC. Large strain static and dynamic semisaturated soil behaviour. *International Journal for Numerical and Analytical Methods in Geomechanics* 1995; **19**: 81-106.

[3] Schrefler BA, Scotta R. A fully coupled dynamic model for two-phase fluid flow in deformable porous media. *Computer Methods in Applied Mechanics and Engineering* 2001; **190**: 3223–3246.

[4] Shahbodagh Khan B. Large deformation dynamic analysis method for partially saturated elasto-viscoplastic soils, *PhD dissertation*, Kyoto University, 2011.

[5] Uzuoka R, Borja RI. Dynamics of unsaturated poroelastic solids at finite strain. *International Journal for Numerical and Analytical Methods in Geomechanics* 2012; **36**: 1535-1573.

[6] B Shahbodagh, G Esgandani, N Khalili. Large deformation dynamic analysis of unsaturated soils. *The 6th International Conference on Unsaturated Soils, UNSAT 2014, Sydney*, 2014; 755 - 760.

[7] B Shahbodagh, N Khalili. A u-p formulation for fully coupled dynamic analysis of flow and deformation in unsaturated soils. *The 2nd Australasian Conference on Computational Mechanics, Brisbane*, 2015.

[8] Muraleetharan KK, Wei C. Dynamic behaviour of unsaturated porous media: governing equations using the Theory of Mixtures with Interfaces (TMI). *International Journal for Numerical and Analytical Methods in Geomechanics* 1999; **23**: 1579-1608.

[9] Ravichandran N, Muraleetharan KK. Dynamics of unsaturated soils using various finite element formulations. *International Journal for Numerical and Analytical Methods in Geomechanics* 2009; **33**: 611–631.

[10] Zerhouni MI. Application des réseaux de Petri continus à l'analyse dynamique des systèmes de production. *Thèse de Doctorat de l'INP Grenoble*, France; 1991.

[11] Khalili N, Habte MA, Zargarbashi S, A fully coupled flow deformation model for cyclic analysis of unsaturated soils including hydraulic and mechanical hystereses, *Computers and Geotechnics* 2008; **35**(6): 872-889.

[12] Khalili N, Zargarbashi S. Influence of hydraulic hysteresis on effective stress in unsaturated soils. *Geotechnique* 2010; **60**(9): 729-739.

[13] Bishop, A.W. 1959. The principle of effective stress. *Teknisk Ukeblad*;**106**(39): 859–863.

[14] Khalili N, and Khabbaz MH. A unique relationship for shear strength determination of unsaturated soils, *Geotechnique* 1998; **48**(5): 681-688.

[15] Khalili N., Geiser F, Blight G. Effective Stress in Unsaturated Soils: Review with New Evidence. *International Journal of Geomechanics* 2004; **4**(2): 115–126.

[16] Brooks RH, Corey AT. *Hydraulic properties of porous media: Hydrology Papers*. Colorado State University, Colorado, 1964.

[17] Oka F, Kimoto S. Computational modeling of multiphase geomaterials. CRC press, Taylor & Francis group, Boca Raton, London and New York, 2012.

[18] Kimoto S, Shahbodagh Khan B, Mirjalili M, Oka F. A cyclic elasto-viscoplastic constitutive model for clay considering the nonlinear kinematic hardening rules and the structural degradation, *International Journal of Geomechanics* 2013; **15**(5), A4014005.

[19] Shahbodagh Khan, B., Mirjalili, M., Kimoto, S. and Oka, F. 2013. Dynamic analysis of strain localization in water-saturated clay using a cyclic elasto-viscoplastic model, *International Journal for Numerical and Analytical Methods in Geomechanics* 2014; **38**(8), 771-793.

[20] Simon BR, Zienkiewicz OC, Paul DK. An analytical solution for the transient response of saturated porous elastic solids. *International Journal for Numerical and Analytical Methods in Geomechanics* 1984; **8**: 381-398.

DOI: 10.1051/e3sconf/20160908014

An XFEM Model for Hydraulic Fracturing in Partially Saturated Rocks

Saeed Salimzadeh[1,a] and Nasser Khalili[2]

[1]*Department of Earth Science and Engineering, Imperial College, London, United Kingdom*
[2]*School of Civil and Environmental Engineering, University of New South Wales, Sydney, Australia*

Abstract. Hydraulic fracturing is a complex multi-physics phenomenon. Numerous analytical and numerical models of hydraulic fracturing processes have been proposed. Analytical solutions commonly are able to model the growth of a single hydraulic fracture into an initially intact, homogeneous rock mass. Numerical models are able to analyse complex problems such as multiple hydraulic fractures and fracturing in heterogeneous media. However, majority of available models are restricted to single-phase flow through fracture and permeable porous rock. This is not compatible with actual field conditions where the injected fluid does not have similar properties as the host fluid. In this study we present a fully coupled hydro-poroelastic model which incorporates two fluids i.e. fracturing fluid and host fluid. Flow through fracture is defined based on lubrication assumption, while flow through matrix is defined as Darcy flow. The fracture discontinuity in the mechanical model is captured using eXtended Finite Element Method (XFEM) while the fracture propagation criterion is defined through cohesive fracture model. The discontinuous matrix fluid velocity across fracture is modelled using leak-off loading which couples fracture flow and matrix flow. The proposed model has been discretised using standard Galerkin method, implemented in Matlab and verified against several published solutions. Multiple hydraulic fracturing simulations are performed to show the model robustness and to illustrate how problem parameters such as injection rate and rock permeability affect the hydraulic fracturing variables i.e. injection pressure, fracture aperture and fracture length. The results show the impact of partial saturation on leak-off and the fact that single-phase models may underestimate the leak-off.

1 Introduction

Flow through fractures in deforming porous media has been the subject of great interest in many engineering disciplines. Typical examples are: fractured oil reservoirs [1], hydraulic fracturing for enhanced hydrocarbon production, tight gas reservoirs [2], weakly consolidated offshore sediments [3], soft coal bed methane extraction [4], geothermal energy [5,6], isolation of hazardous waste [7], measurement of in situ stresses [8], fault reactivation [9], and remediation of soil and ground water aquifers [10] to name a few.

Hydraulic fracturing is a complex multi-physics problem. A fluid is injected into the host rock; the fracture initiates and propagates due to the induced hydraulic loading. The simulation of the process requires several components: (i) flow of the fracturing fluid within the fracture; (ii) the mechanical deformation of the porous medium due to the applied hydraulic load; (iii) the fracture propagation; (iv) the leak-off flow to the surrounding porous media (host rock); and (v) the fluid flow through the host medium. The first three components are widely addressed in the literature, while last two are considered only in few studies albeit with some restrictions.

Based on the energy-dissipation mechanism and the ability of the rock matrix to dissipate fracturing fluid, four combined asymptotic regimes are defined: storage-viscosity, storage-toughness, leak-off-viscosity and leak-off-toughness [11]. Asymptotic solutions provide fundamental understanding of the hydraulic fracturing process, and provide benchmarking cornerstones for numerical models. However, analytical solutions are restricted to simplified fracture geometries in homogenous media and are typically constrained to a set of fixed boundary conditions. Standard geometries are 2D plane-strain fractures (PKN fracture [12, 13]; KGD fracture [14-16], and radial (penny-shaped) fractures [17-19].

In majority of the available models, the flow through rock matrix and the fluid exchange between fracture and rock matrix, is either ignored by assuming impermeable rock formations, e.g. [20], or simplified using one-dimensional analytical leak-off model, e.g. [21]. The impermeable matrix assumption is proven to be an unrealistic assumption by the substantial evidence of conductive rock matrix [1, 4]. One-dimensional leak-off model, on the other hand, is based on Carter's one-dimensional leak-off model [22], which is represented as a sink term in the mass balance equation of fracture flow. This approach has several shortcomings including one-dimensional assumption of flow, time-dependency of flow instead of pressure-dependency and more importantly ignoring the poroelastic effects. Flow

[a] Corresponding author: s.salimzadeh@imperial.ac.uk

dissipation through rock matrix is multi-dimensional unless the permeability in the direction normal to the fracture plane is significantly higher than in other directions [23]. Also, the fluid exchange between fracture and matrix depends on the pressure gradient, while in Carter's leak-off model the fluid leak-off rate at a given location within the fracture depends only on the fracturing fluid pressure history at this position and does not depend on the pressure in the adjacent regions. As time elapses, the leak-off rate at each position decreases proportionally to $1/\sqrt{t}$, where t is time; therefore, a scenario of fracture arrest is not possible [24]. Finally, the seepage of the fracturing fluid into the rock formation increases fluid pressure in the matrix causing dilation of rock matrix. Dilated matrix applies stresses back on the fracture, known as 'back-stress' in hydraulic fracturing context, tending to close the fracture [19, 25, 26]. These shortcomings have also affected the available semi-analytical solutions for leak-off-dominated regimes that use simplified one-dimensional leak-off model in deriving solution equations. These solutions fail to predict hydraulic fracturing parameters in leak-off-dominated regimes accurately as shown in [23], and in [27] for single-phase flow, and in [24] for two-phase flow in two-dimensional case.

In this paper, a fully coupled extended finite element model for hydraulic fracturing in porous media is proposed. Two fluids: fracturing fluid and host fluid are considered. The independent fluid flow for the porous host medium makes the present model capable of simulating multidimensional leak-off fluid flow through high/low permeable, homogeneous/inhomogeneous host medium with transient pressure boundary condition at the fracture-rock boundary. The fracture discontinuity in the deformation model is handled using XFEM technique, while cohesive crack model is used as fracture propagation criteria. The model is verified against several benchmark solutions from the literature. It is shown that leak-off is higher when partial-saturation of rock matrix is considered.

2 Governing equations

The model proposed comprises five phases and three models: rock skeleton (deformation model), wetting and non-wetting fluids in fracture (fracture flow model), and wetting and non-wetting fluids in matrix (matrix flow model). The differential equation describing poroelastic deformation of rock matrix can be written as [24]

$$\mathrm{div}(1/2\,\mathbf{D}(\nabla d\mathbf{u}+d\mathbf{u}\nabla)-\beta_w dp_{\mathrm{pw}}\mathbf{I}-\beta_{nw} dp_{\mathrm{pnw}}\mathbf{I})+d\mathbf{F}=0 \quad (1)$$

where $\mathbf{u}$ donates the displacement vector of the porous medium, β is the incremental effective stress parameter, p is the fluid pressure, $\mathbf{I}$ is the second-order identity tensor, and F is the body force per unit volume.

Following the cohesive fracture mechanics, a traction-separation law governs the nonlinear behaviour of the fracturing medium in the cohesive zone. In addition to cohesive traction, in fluid saturated fracture discontinuities, the hydraulic loading is applied to the fracture planes. The hydraulic loadings can be either positive for pressurised fluid within the fracture or negative due to suction in the lag region. Therefore, the natural boundary conditions for loading at the fracture discontinuity is expressed as

$$d\boldsymbol{\sigma}.\mathbf{n}_c=d\mathbf{t}-dp_f\,\mathbf{n}_c \quad (2)$$

where $\boldsymbol{\sigma}$ is the total stress, $\mathbf{t}$ is the cohesive tractions, p_f is the average fluid pressure within the fracture, and $\mathbf{n}_c$ is the outward unit vector normal to the discontinuity (Figure 1). Notice that in the above equation Biot's coefficient is not applied to fracture loading, and that the fluid pressure in the fracture is treated identical to an external pressure applied to fracture planes. Such a pressure does not require scaling according to Biot's coefficient.

The flow through planar fracture is commonly modelled using lubrication theory [28] for an incompressible Newtonian fluid obeying cubic law. Lubrication equation is derivable from general Navier-Stoke's equation for the flow between two parallel plates [29]. The fracturing fluid can have properties substantially different from those of the host fluid. Furthermore, depending on the in-situ stresses, the fracturing fluid front and the fracture front may not coincide, giving rise to the so-called fluid lag phenomenon [1, 30, 31]. In the lag region, the host fluid enters the fracture under the developed suction. Therefore, a second fluid is required to complete the fracture flow model.

Independent fracture flow model is considered for fracture discontinuity in this research. Fracture pressures are calculated directly, which may be linked to multiple pore pressures within the porous medium. The objective is to obtain a more realistic representation of fracture flow compared to current models presented in the literature in which gradient of fracture flow is calculated entirely based on the enriched component of the fluid pressures [32, 33, 34]. The extended lubrication equation considering compressibility of fluid constituent and fluid transfer between fracture and matrix (leak-off) can be written as [24]

$$\partial/\partial l\,[S_{f\xi}\,k_{fr\xi}\,w^3/(12\mu_\xi)\,\partial p_{f\xi}/\partial l] = S_{f\xi}\,wc_\xi\,\partial p_{f\xi}/\partial t + S_{f\xi}\partial w/\partial t + w\partial S_{f\xi}/\partial t + k_{p\xi}/\mu_\xi\,(p_{f\xi}-p_{p\xi})/y) \quad (3)$$

where w is the fracture aperture, μ is the fluid dynamic viscosity, $S_{f\xi}$ is the degree of saturation of the fluid ξ, k_{fr} is the relative permeability, c_ξ is the fluid compressibility, $p_{p\xi}$ is the fluid pressure in the adjacent matrix, , y is the distance between fracture and adjacent matrix, and l is the dimension along the fracture. The change in saturation is linked to the change in capillary pressure (suction) through available capillary pressure-suction curves. In this study the relationships proposed by van Genuchten-Mualem (VGM) model are utilised for capillary pressure-saturation and for relative permeability-saturation interpolations. Details are given in [24]. Note that fracture pressure loading on the fracture walls (Eq. 2) and fracture aperture in Eq. 3 provide symmetric coupling between mechanical deformation and fracture flow models.

A mathematical description of fluid flow through host porous medium can be derived by combining linear momentum balance equation (Darcy's law) with the mass conservation equation of the fluid. Such partial differential equation can be written as

$$\mathrm{div}[k_{pw}/\mu_w\,(\nabla p_{pw} + \rho_w g)] = \beta_w\,\partial(\mathrm{div}\,\mathbf{u})/\partial t + \lambda_{pw,pw}\,\partial p_{pw}/\partial t + \lambda_{pw,pnw}\partial p_{pnw}/\partial t + k_{pw}/\mu_w\,(p_{pw}-p_{fw})/y \quad (4)$$

$$\mathrm{div}[k_{pnw}/\mu_{nw}\,(\nabla p_{pnw} + \rho_{nw} g)] = \beta_{nw}\,\partial(\mathrm{div}\,\mathbf{u})/\partial t + \lambda_{pnw,pnw}\,\partial p_{pnw}/\partial t + \lambda_{pnw,pw}\partial p_{pw}/\partial t + k_{pnw}/\mu_{nw}\,(p_{pnw}-p_{fnw})/y \quad (5)$$

$$\lambda_{pw,pw} = nS_{pw}c_w + (\beta_w - nS_{pw})c_s - n\partial S_{pw}/\partial s_p \quad (6)$$

$$\lambda_{pnw,pnw} = nS_{pnw}c_{nw} + (\beta_{nw} - nS_{pnw})c_s - n\partial S_{pw}/\partial s_p \quad (7)$$

$$\lambda_{pw,pnw} = \lambda_{pnw,pw} = -n\partial S_{pw}/\partial s_p \quad (8)$$

where, n is the porosity and s_p is the capillary pressure(suction) in the porous medium. The pressure field is assumed continuous across the fracture discontinuity, with the discontinuity in the Darcy flow velocity normal to the fracture taken into account through leak-off loading. Note that leak-off terms in Eqs. 3 and 4 provide symmetric coupling between two flow models.

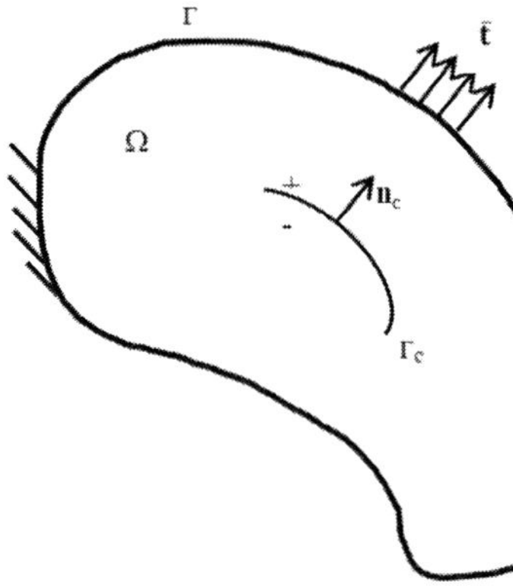

Figure 1. Schematic representation of the problem with the body Ω, boundary Γ, fracture boundary Γ_c.

3 Numerical approximation

Spatial discretisation has been performed using the standard Galerkin method with displacements and fluids pressures defined as primary variables. Extended finite element method (XFEM) is used to model the fracture discontinuity within displacement field at element level. Arbitrary discontinuities can be handled using XFEM without the need for remeshing. In XFEM, the displacement field consists of two parts, one continuous and the other discontinuous. The continuous part is the standard displacement field corresponding to the situation without any cracks. The discontinuous displacement field (also known as enriched part) is based on local partitions of unity and allows the element to include a discontinuity.

Heaviside step function, also known as 'jump' function, is typically used to enrich the nodes in fully cracked elements if their support was cut by the crack into two disjoint pieces [35]. To obtain numerical solution of the governing equations, the rate form of the discretised equations is integrated over the time domain. The time integration is performed using a generalized trapezoidal method.

The set of discretised equations are highly nonlinear. Nonlinearities arising from fluid pressure dependency of constitutive coefficients are handled using Picard method such that the coefficient matrices appearing in the stiffness matrix as nonlinear functions of unknown variables are updated in every iteration. The four-node quadrilateral elements together with two-point Gaussian quadrature rule are used for numerical approximation and numerical integration. Numerical integration of the elements bisected by the fracture discontinuity is done using common method of element partitioning in which numerical integration is done separately within each sub-domain. The model has been implemented in a well-known code (Matlab) and verified against several analytical solutions and published results [27, 36].

4 Hydraulic fracturing in partially saturated medium

In these examples, a viscous fracturing fluid is injected into a fracture with initial length of 1 m. The host medium is assumed to be saturated with gas as non-wetting fluid at atmospheric pressure. The model dimensions and mesh configuration is shown in Figure 2. The model parameters used are: Young's modulus E=50 GPa, Poisson ratio υ=0.25, tensile strength f_t=1000 kPa, and fracture energy G_f=0.1 KN/m. The fracture is assumed to have plane-strain behaviour (KGD fracture) with fracture height of 100 m.

The fracturing fluid with dynamic viscosity $\mu_w=1\times10^{-4}$ kPa.s and compressibility coefficient $c_w=1\times10^{-6}$ kPa^{-1} is injected into the fracture at the rate Q_w=1 m^3/min. The dynamic viscosity of gas is assumed as $\mu_{nw}=1\times10^{-8}$ kPa.s.

The simulated examples in this section include: i) the case with impermeable rock matrix (k_p=0) to validate current model against available semi-analytical solutions, ii) the case with low matrix permeability ($k_p=1\times10^{-13}$ m^2) and iii) the case with high matrix permeability ($k_p=1\times10^{-12}$ m^2). The medium porosity in all examples is n=0.01. Geertsma and de Klerk [14] obtained an approximate analytical solution for the case without leak-off and with negligible toughness (KGD model). Later, Spence and Sharp [15] extended the solution to include fracture toughness.

The results from present study for the net well pressure, the fracture aperture and the fracture half-length are shown in Figures 3 to 5, respectively. Included in these figures are the analytical solutions by Geertsma and de Klerk [14] and Spence and Sharp [15]. In Figure 5 the solution by Carbonell et al. [37] for the limiting case of storage-viscosity regime and the solution by Adachi and

Detournay [16] for leak-off dominated regime are also included. The leak-off dominated solution is calculated for the most permeable case with k_p=1×10^{-12} m^2 which represents the most suitable case for this solution.

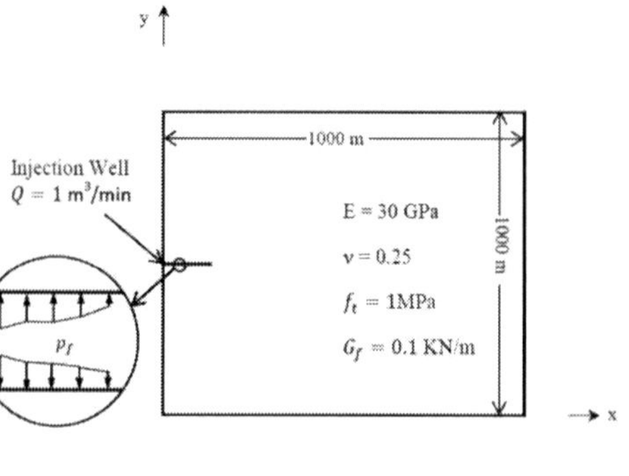

Figure 2. Model used for hydraulic fracturing simulations.

For the impermeable case very good agreement is found between present model results and KGD analytical solutions for net well pressure and fracture aperture at the wellbore. KGD solution is for viscosity dominated regime (zero toughness), and the parameters used in the present simulation falls in to the viscosity-dominated regime. This is verified through computing the dimensionless viscosity and toughness: M=308 and K=0.239 [38]. Viscosity-dominated regime is in accordance with actual hydraulic fracturing practice, where the energy dissipation associated with the flow of viscous fracturing fluid often dominates over the energy dissipation associated with the rock damage [19]. Spence and Sharp [15] calculated higher net well pressure due to including finite toughness. The fracture half-length (distance from fracture origin to the fracture tip) is larger than the analytical solution due to the pressure lag ahead of the fracturing fluid front as shown in Figure 6. The size of the lag region can be affected by the in situ stresses [11, 30]. The in situ stresses are considered zero in these simulations to allow the formation of lag zone. Negative pressure (suction) is induced in the lag region. If the position of the fracturing fluid front is considered instead of the fracture tip, as shown in Figure 5, a perfect match is observed between present model and KGD results. Again, Spence and Sharp [15] calculated slightly lower fracture length, while Carbonell et al. [37] calculated higher fracture length for the limiting case of viscosity-storage regime.

When the leak-off is allowed by considering permeable host medium, a significant portion of the injected fracturing fluid is dissipated into the surrounding media. This causes smaller fracture aperture and lower fracture propagation. The net well pressure on the other hand remains higher and maintains a steady leak-off flow towards rock matrix which becomes equal to the injected flow rate as time elapses. The higher the medium permeability, the sooner the equilibrium. For the case of higher medium permeability, the equilibrium is reached within five minutes, consequently, the fracture aperture and half-length are stabilised and no fracture propagation is observed. This implies that all the injected fracturing fluid are dissipated into the host rock. The model results for the high permeable case are in good agreement with the analytical solution by Adachi and Detournay [16]. The solution is for leak-off-dominated-regime based on the 1D Carter's leak-off model. However, in late time response the two models diverge as the current model approaches an equilibrium in which there is no further fracture growth and the entire injected fracturing fluid is dissipated into the host medium, while the analytical solution predicts fracture growth proportional to square root of time. Note that the Carter's leak-off model is for single-phase flow.

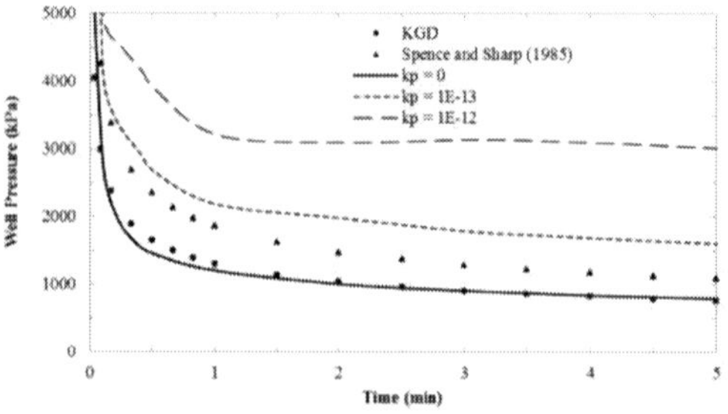

Figure 3. Well pressure versus elapsed time.

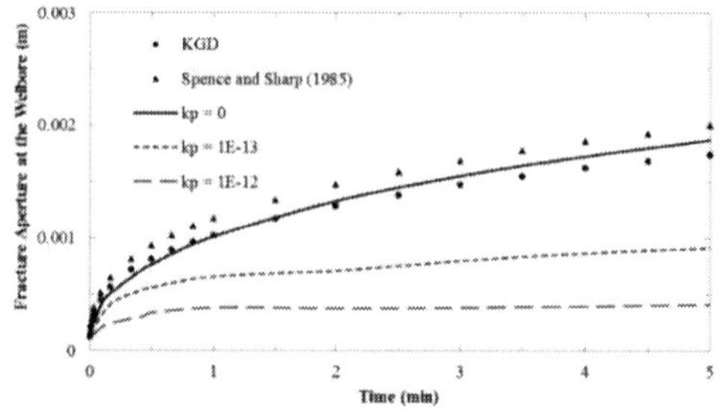

Figure 4. Fracture aperture at the well versus elapsed time.

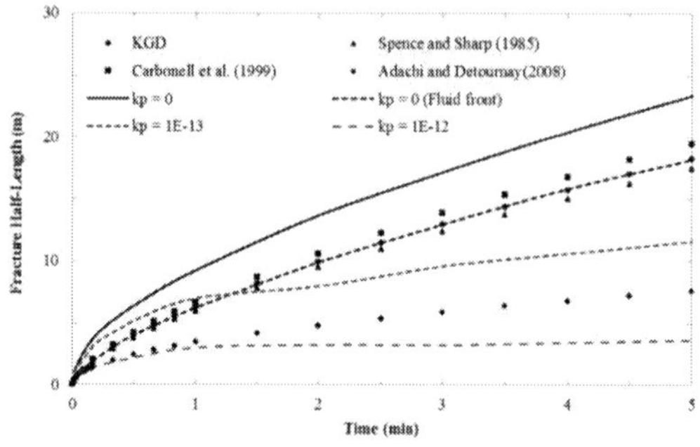

Figure 5. Fracture half-length versus elapsed time.

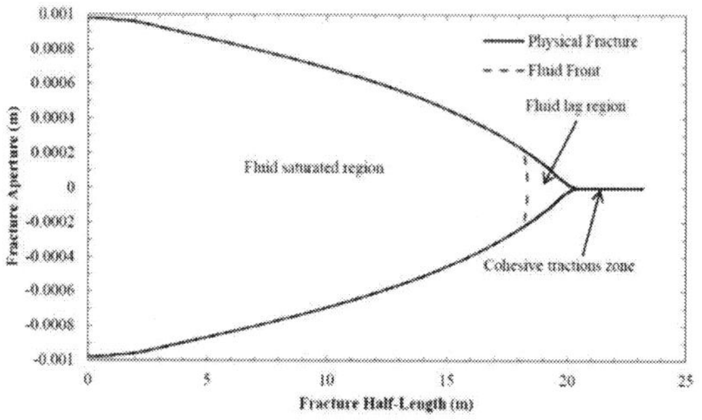

Figure 6. Fracture opening, fluid front, fluid lag, and cohesive zone.

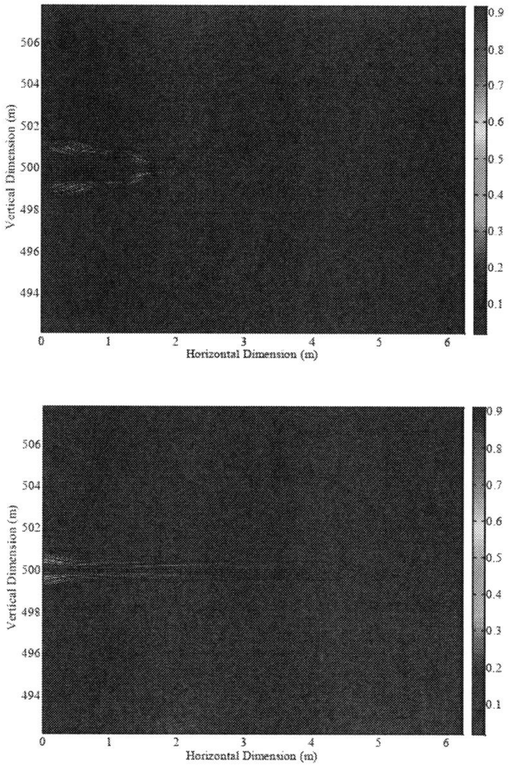

Figure 7. The invaded zone showed by contours of degree of saturation of hydraulic fracturing fluid within host medium, top: $k_p=1\times10^{-12}$ m^2, bottom: $k_p=1\times10^{-13}$ m^2.

The shape of the invaded zone, i.e. the zone within host medium saturated by fracturing fluid, also depends on the permeability of the host medium. Figure 7 shows the contours of the degree of saturation of fracturing fluid within host medium for the two examples with permeable rock matrices. For the case with higher permeability, lower pressure gradient is required to push fracturing fluid into the porous medium so the invaded zone shape is short and dispersed, whereas for the case with the lower permeability it is elongated and thin. In the latter case, higher pressure-gradient is required to push fracturing fluid into the porous medium, so the higher fracture pressure is developed which causes longer fracture propagation and the preferred path for the fracture flow is along the fracture.

5 Conclusions

A fully coupled XFEM model for hydraulic fracturing in partially saturated medium is presented. Five phases are identified: solid skeleton, wetting and non-wetting fluids in fracture, and wetting and non-wetting fluids in rock matrix. Simulation results show that fluid dissipation is higher than what is predicted by asymptotic solutions. One-dimensional analytical leak-off model cannot accurately capture multidimensional leak-off. Furthermore, leak-off is higher when the matrix partial saturation is accounted for.

References

1. J. Adachi, , Siebrits, E., Peirce, A., J. Desroches, Int. J. of Rock Mech. & Mining Sciences, **44**, 739–757 (2007).
2. N.R. Warpinski, SPE 18259 (1990).
3. M.G. Mack, N.R.Warpinski, *Reservoir stimulation. 3rd ed.* (2000).
4. M.J. Economides, K.G. Nolte, U. Ahmed, *Reservoir Stimulation* (1989).
5. S. Sasaki, Tectonophysics, 289, 171-188 (1998).
6. S. Berumen, D. Tiab, F. Rodriguez, J Pet Sci Engg, **25**, 49-58 (2000).
7. B.W. Hainey, R.G. Keck, M.B. Smith, K.W. Lynch, J.W. Barth, SPE PRODUCTION & FACILITIES, **14**, 88-93 (1999).
8. A.M. Raaen, E. Skomedal, H. Kjorholt, P. Markestad, D. Okland, Int. J. Rock Mech and Mining Sciences, **38**, pages: 529-541 (2001).
9. M. Board, T. Rorke, G. Williams, N. Gay, *Proceedings of the 33rd U.S. symposium on rock mechanics*, 111–20 (1992).
10. L.C. Murdoch, W.W. Slack, J Geotechnical and Geoenvironmental Engineering, **128**, 479-487 (2002).
11. D. Garagash, E. Detournay, J. Adachi, *J. Fluid Mech.*, **669**, 260-297 (2011).
12. T.K. Perkins, L.R. Kern, *J Pet Tech* **13**(9):937–949 (1961).
13. R.P. Nordren, SPE J;**12**(8):306–14 (1972).
14. J. Geertsma, F. de Klerk, *J Pet. Tech.,* **21**, 1571–1581 (1969).
15. D.A. Spence, P. Sharp, *Proc. R. Soc. London A.*, **400**:289-313 (1985).
16. J. Adachi, E. Detournay, *Engng Frac. Mech.* **75**, 4666–4694 (2008).
17. A. Savitski, E. Detournay, *IJSS* **39** (26), 6311–6337 (2002).
18. A. P. Bunger, E. Detournay, D. I. Garagash, *Intl J. Fract.* **134**, 175–190 (2005).
19. Y. Kovalyshen, *Fluid-driven fracture in poroelastic medium, Ph.D. thesis*; University of Minnesota (2010).
20. A. Dahi-Taleghani, *Analysis of hydraulic fracture propagation in fractured reservoirs: an improved model for the interaction between induced and natural fractures, PhD dissertation*, The University of Texas at Austin, Austin, Texas (2009).

21. L. Zhou, Z. Hou, Y. Gou, M. Li, J Petrol Sci Eng **120**:119–129 (2014).
22. E. Carter, Drill Prod Pract, 261–70 (1957).
23. B. Carrier, S. Granet, Eng. Frac. Mech. **79**, 312-328 (2012).
24. S. Salimzadeh, N. Khalili. Computers and Geotechnics. 69, 82-92 (2015).
25. J. Hagoort, B.D. Weatherill, A. Settari, Soc. Pet. Eng. J., 293–301 (1980).
26. S. Mathias, M. van Reeuwijk, Transp Porous Med, **80**:499–518 (2009).
27. S. Salimzadeh, N. Khalili, Int. J. of Geomechanics, (2015).
28. G.K. Batchelor, *An introduction to fluid dynamics,* (1967).
29. R.W. Zimmerman, G.S. Bodvarsson, J Transp Porous Media, **23**:1–30 (1996).
30. D.I. Garagash, E. Detournay, ASME J Appl Mech, **67**:183–92 (2000).
31. E. Gordeliy A. Peirce, Comput. Methods Appl. Mech. Engrg., **253**, 305-322 (2013).
32. R. de Borst, J. Rethore, M. Abellan, Arch. Appl. Mech., 75: 595-606 (2006).
33. J. Rethore, R. de Borst, M.A. Abellan, Computational Mechanics, **42**, 227–238 (2008).
34. A.R. Khoei, M. Vahab, E. Haghighat, S. Moallemi, Inj. J. Fract., **188**:79-108 (2014).
35. N. Moës, J. Dolbow, T. Belytschko, International Journal for Numerical Methods in Engineering, **46**, 131-150 (1999).
36. S. Salimzadeh, N. Khalili, J of Geo-engineering Sciences, **2**, 67-82 (2014).
37. R. Carbonell, J. Desroches, E. Detournay, Intl J. Solids Struct. **36** (31–32), 4869–4888 (1999).
38. E. Detournay, Int J Geomech, **4**:1–11 (2004).

Experimental and numerical approaches of the hydro-mechanical behaviour of a quasi-saturated compacted clayey soil

Zhong-Sen Li[1 a], Dino Mahmutovic[2,3], Jean-Marie Fleureau[1], Luc Boutonnier[2], Jacques Monnet[4]

[1] Centrale-Supélec, Châtenay-Malabry, France
[2] EGIS-Géotechnique, 3, rue Docteur Schweitzer, 38180, Seyssins, France
[3] UNIV. GRENOBLE ALPES, 3SR, F-39800 Grenoble, France
[4] GAIATECH, 22 rue Antoine Chollier, 38170, Seyssinet, France

Abstract: The present research is funded by the French National Project « TerreDurable », which is dedicated to the study of soils in quasi-saturated conditions (close to saturation) for the analysis of stability and settlement of earth structures such as embankment, dams. A global presentation of the drying-wetting test shows the volume change, air entry and soil-water characteristics of the soil at slurry and oven-dried conditions. Unsaturated undrained triaxial test was carried out in order to investigate the variation of pore-water pressure from quasi-saturated domain to saturation. The experimental results of the triaxial test are then modeled using a two-dimensional explicit finite difference program (Flac 2D). A constitutive law developed in the TerreDurable project allows better understanding the behaviour of quasi-saturated soils using the water retention curve of quasi-saturated domain proposed by Boutonnier (2007, 2010). A simple effective stress model is used (Cam Clay) by taking into account both the suction and the compressibility of equivalent fluid (water + air). The results from numerical calculation and experimental measurements are compared.

1 Introduction

1.1 Problem

In practice, due to the complexity of microstructure, heterogeneity and soil suction, it is usually difficult for the engineers to measure and control effectively the parameters that affect the engineering behaviour of compacted clayey soils. Consequently, experimental studies of these soils in laboratory are useful to a better understanding of their hydro-mechanical behaviour and then to guide the practice.

1.2 Related work

Compacted soils are unsaturated soils, for which the Proctor optimum water content and maximum density correspond to a degree of saturation comprised between 70 and 90%. During their life cycle, under the combined effects of suction and stresses, they may be subjected either to wetting up to saturation or to drying. This domain, between 70 and 90% of degree of saturation, has been the subject of numerous studies.

Since 1950s, much attention was paid, both experimentally and numerically, to the study of the variation of volume and pore-water pressure in the framework of dam constructions (Hilf, 1948; Skempton, 1954; Bishop, 1954). As regard to the laboratory experiments of compacted soils, the difficulty was to measure or control the pore water pressure, which is negative and cannot be measured directly by a traditional pressure transducer when the suction is below -80 kPa. In 1956, Hilf firstly proposed the axis-translation method to increase the pore-water pressure (u_w) from negative to positive by applying an external air pressure (u_a). This method made it possible to measure and control the soil suction (s), which is equal to the difference between the pore-air and pore-water pressures ($s = u_w - u_a$). Application of axis-translation method has made much progress in the unsaturated soil mechanics, especially for the study of pore-water variation (Bishop et Donald, 1961; Gibbs and Coffey, 1969; Fredlund et al., 1976; Taibi, 1994).

With the development of theory and testing device, Soil Water Characteristic Curve (SWCC, also called Soil Water Retention Curve, SWRC) was proposed to describe the relationship between the hydro-mechanical parameters and suction. A complete description of SWCC in four diagrams (w-e, w-Sr, s-w, s-Sr) was firstly presented by Biarez et al. (1987), and then developed into five diagrams (w-e, w-Sr, s-e, s-Sr, s-w) by Fleureau et al. (1993, 2002). In the meantime, many researchers began to correlate the main parameters of SWCC with the intrinsic parameter of soils such as liquid limit (Fleureau et al., 2002), and special attention

[a]Corresponding author: zhongsen.li@centralesupelec.fr

was paid recently to the study of SWCC under mechanical loading (Lu and Likos, 2006).

More recently, works have highlighted the specific behaviour of soils close to saturation (Boutonnier 2007), with a high compressibility of the interstitial fluid due to the presence of occluded air bubbles.

1.3 Attributions

In this paper, we attempt to present 1) some experimental results such as the SWCC, unsaturated undrained triaxial test of a compacted clayey soil whose degree of saturation is close to unit; 2) a numerical approach based on the Cambridge-model to model the experimental results; 3) a comparison of the experimental results and the numerical model.

2 Material and methods

2.1 Material

The material of this study was taken from a site on the Tours - Bordeaux High-speed Railway Line in the south-west of France at a depth of 1.5 - 7.8 meters. The material is made mainly of clayey soil but also with some sharp gravel with diameters of several centimetres, which were eliminated manually before the laboratory tests. Its initial water content was equal to 39%.

The main geotechnical properties of the material are presented in Table 1. The material can be classified as A3 in the French classification (AFNOR, 1992) and A-7-5 in the American Association of State Highway and Transport Officials (AASHTO) classification (ASTM, 2004).

Table 1. Main properties of the material

Material	Châtaignier Clay	Gs [kN/m^3]	28.3
w_L [%]	70.5	γ_{dmax} [kN/m^3]	13.3
w_P [%]	37	u_{copt} [kPa]	1100
I_p [%]	33.5	< 80 µm [%]	88
w_{opt} [%]	37.5	< 2 µm [%]	70

2.2 Experimental Methods

2.2.1 Preparation of test specimen

The material was dried under laboratory condition (20 ± 1 °C) for several weeks, and then sieved at 2 mm. The passing particles were dried in an oven at 105 °C for about 48 hours, and mixed afterwards with the required quantity of distilled water. Then, the mixed soil was kept in a sealed plastic bag for more than 48 hours. Experiments with a similar soil showed that this time was sufficient to reach equilibrium. The moist soil was compacted, according to the French standard (AFNOR, 2014), in a standard Proctor mold. The compaction was carried out in 3 layers with 25 blows per layer, which corresponds to energy of 0.6 MJ/m^3.

Concerning the saturated oedometer test, the well mixed slurry was carefully poured into a 60-mm-diameter oedometer cell, where the circulation of water from the bottom to the top was possible. Loading was applied by step from 1 to 80 kg, with each load lasted more than 2 days. The slurry with water content of 1.5 times of liquid limit was prepared for the drying-wetting test. As regard to the triaxial test, the prepared compacted sample at water content of 47.5% (w_{SPO} + 10%) was cut into 35 mm in diameter and about 30 mm in height.

2.2.2 Unsaturated undrained triaxial test

The experimental device for the unsaturated triaxial test by axis-translation method is presented in Figure 1, which contains mainly 1) a loading system; 2) three Pressure/Volume controllers, which control the confining pressure, pore-water pressure and the external pore-air pressure of the sample; 3) several transducers, two of them measure the pore-air and pore-water pressures, the other transducers which are not shown in the figure to measure the axis displacement, axis force, etc.; 4) a traixial cell, a small standard cell for the specimen of 35 mm in diameter was modified by equipping the base with a high air-entry value (HAEV) ceramic disk, whose air-entry pressure was 1.5 MPa.

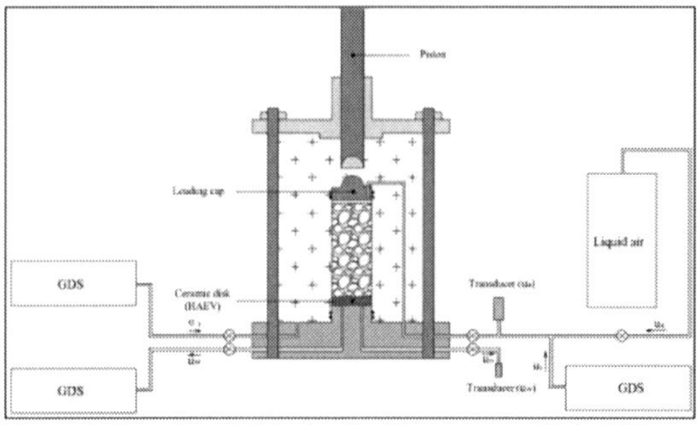

Figure 1. Experimental device for the unsaturated triaxial test by axis-translation method

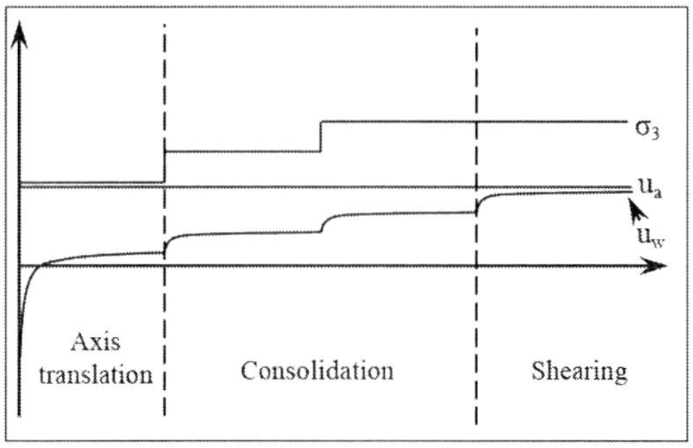

Figure 2. Schematic diagram of the axis translation method

To carry out the unsaturated triaxial test, a non-standard procedure is used and described as follows:

- Saturation of ceramic disk;

- Installation of the specimen and acquisition system;
- Axis translation and consolidation: During the axis translation process, an external air pressure (u_a) was applied by liquid air and pressure/volume controller from the circuit at the top, followed by an increase of confining water pressure (σ_3 = u_a + 20) simultaneously until the pore-water pressure became positive. After that, the air pressure was constant, while the confining pressure was increased by steps (Figure 2);
- Shearing: the confining pressure and the air pressure was hold and the sample was sheared at a speed of 0.5 %/h.

2.2.3 Drying-wetting test

During drying-wetting test, suction was controlled by the sintered tensiometric plates, osmotic technique and salt desiccator for low, medium and high suctions, respectively. More details of these methods can be found in the work of Fleaureau et al. (2001).

2.3 Numerical Methods

2.3.1 Model used

The model reported by Boutonnier (2007) is adopted in this work. More details are available in the works of Mahmutovic et al. (2014), Boutonnier et al. (2015) and Andrianatrehina et al. (2015). In this model, as shown in Figure 3, the soil can be divided into 4 domains of saturation (D1 to D4 - unsaturated to saturation).

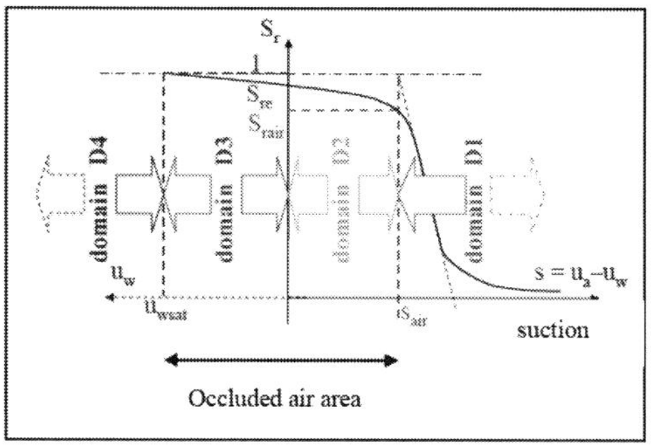

Figure 3. Division of retention curve in 4 domains of saturation (Boutonnier 2007)

In the quasi-saturated areas, the model is based on the compressibility coefficient of the interstitial fluid c_f (Equation (1) for domain D2 and Equation (2) for domain D3). A particular attention is paid to these areas as they are the domains concerned in most of earthworks like motorway, railway, embankments and earth dams.

$$c_f = \frac{1}{S_r} \cdot \frac{1}{s_{air}} \left(S_{re} - S_{rair} \right) + c_w \quad (1)$$

$$c_f = \frac{1 - S_r + h.S_r}{u_w + r_{bm} + P_a - u_{wg}} + c_w \quad (2)$$

Quasi-saturated domains can be modeled using the Cam Clay constitutive relational law with an assigned compressibility coefficient (c_f) for air/water phases. The c_f value depends on the degree of saturation, which itself depends upon the pore pressure.

2.3.2 Geometry and boundaries condition

The Flac 2D software was chosen for the numerical simulation due to its wide use both in academic and professional practice. This software includes many pre-existing constitutive laws and enables new theoretical models to be defined.

An axisymmetric calculation is used because of the axial symmetry of the triaxial test. The geometry of the numerical model is drawn with the dimensions of 70 mm in height and 35 mm in diameter which corresponds to the size of the triaxial sample (D = 35, H = 30 in this study). No displacement is allowed in x-y directions at the base of the model. A stress condition is applied on the top and at the lateral boundary of the model.

2.3.3 Simulation parameters

Cam Clay parameters

Figure 4 presents the results of two oedometer tests for the slurry under saturated drained condition. The main parameters are listed in Table 2. For the reason of heterogeneity of the samples, only the average value of the two tests was taken for the later simulation.

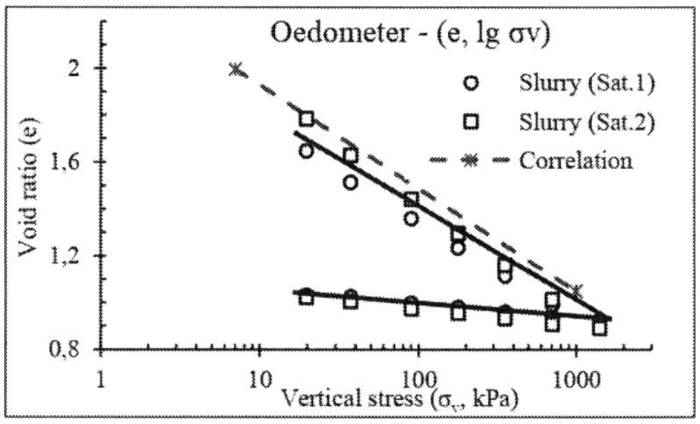

Figure 4. Results of oedometer test for the slurry sample under saturated drained condition

Values of s_{air} change with void ratio (Vanapolli 1998). The chosen value corresponds to the value of suction at the Optimum Proctor: 1200 kPa (Boutonnier 2007).

Table 2. Mechanical parameters of the material

Test	λ	κ	φ	M	s_{air} desat [kPa]	s_{air} resat [kPa]
Oedo.	0.191	0.026	-	-	-	-
Triaxial	-	-	22.3°	0.87	-	-
Drying-wetting	-	-	-	-	5000	1700

Quasi-saturated parameters

The proposed model contains 5 parameters: the degree of saturation for a null suction S_{re}, the degree of saturation at air entry S_{rair}, the air entry suction s_{air}, the mean radius of air bubbles r_{bm}, and the Henry's constant h (Table 3).

The value of air entry suction coincides with the value of suction at the optimum Proctor (Boutonnier 2007), and S_{rair} the value of the degree of saturation at the optimum Proctor. h is neglected because of it low influence in the calculus. S_{re} is deduced thanks to the Equation (3) and the understanding of the degree of saturation at 48% of water content (Sample tested in the triaxial test).

$$S_r = S_{re} - u_w \frac{(S_{re} - S_{rair})}{u_{wair}} \qquad (3)$$

Table 3. Parameters of the proposed model

Test	S_{re}	S_{rair}	s_{air} [kPa]	r_{bm} [m]	h
Oedo	0.92	0.90	1200	2 e^{-6}	0

2.3.4 Aim of the simulation

The main goal of the numerical simulation is to reproduce the response of the soil sample in terms of pore pressure. The standard triaxial test is modeled, and not the axis-translation triaxial test because of the leakage into the air circuit when the pore-water pressure exceeds the applied air-pressure.

3 Results and comparison

3.1 Experimental results

Figure 5 presents the drying-wetting path for the slurry at 1.5 w_L and oven-dried slurry, the main parameters are listed in Table 2. For the SWCC, the drying path and wetting path are reversible in the (w, e) plan, whereas an obvious hysteresis can be observed in the other four plans. This general observation is coincident with the previous works by Taibi (1994), Fleureau et al. (1993, 2002) for the soils under slurry condition.

Figure 6 and Figure 7 present the results of the unsaturated triaxial test under undrained condition.

The saturation step (Figure 6 - A) shows that the ceramic disk with a diameter of 33 mm and depth of 7 mm can arrive at saturation in six hours for the applied water pressure of 1500 kPa.

Figure 6 - B shows that the initial suction of the sample (w = 49.25%) is about 20 kPa, where the net confining pressure (σ_3 - u_a) is 20 kPa. The pore-water pressure (u_w) in the sample stabilised in about 2 days after the application of the pore-air pressure (u_a) for the compacted sample with a diameter of 35 mm and depth of 30 mm. The measured suction decreased slightly from 25 to 100 kPa by increasing simultaneously the pore-air pressure (u_a) from 50 to 200 kPa and confining pressure (σ_3) from 70 to 220 kPa. This means that, by increasing simultaneously the pore-air (u_a) and confining pressure (σ_3).

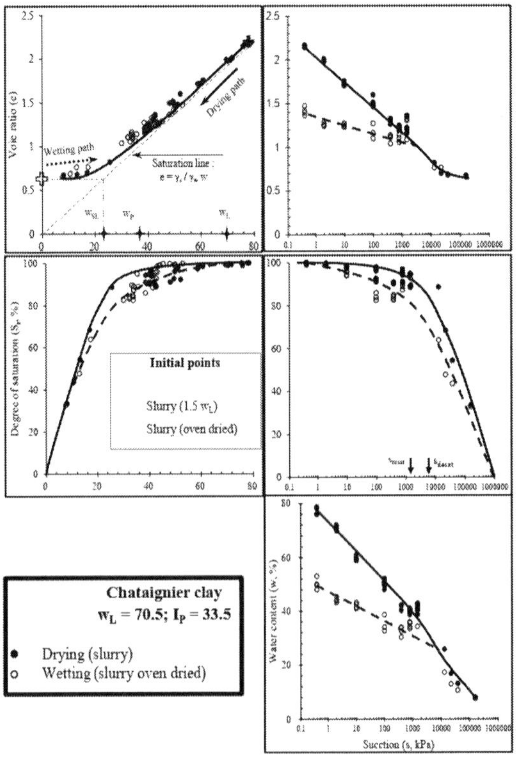

Figure 5. Drying-wetting path of the Chataignier clay

Figure 6 - C presents the results of the consolidation step for the specimen N.1, where axis translation method is applied. Confining pressure was increased from 250 to 400 kPa, whereas the pore-air pressure was constant (200 kPa). The pore-water pressure in the sample increased and exceeded the applied pore-air pressure, where the water began to drain into the pore-air circuit, which means the test was no longer under undrained condition.

Figure 6 - D presents the consolidation results of the specimen N.2, who has the same initial water content and size. But instead of applying a pore-air pressure, the specimen was submitted a relative high confining pressure (300 kPa) so that the pore-water pressure can increase from negative to positive pressure. Then, the confining pressure was continued to increase by 100 kPa each step from 300 to 1000 kPa. As it is noted in the figure, the temperature had an influence on the measurement of the pore-water pressure, and it is necessary to take into account the effect of the temperature for some loading steps in order to calculate the B parameter.

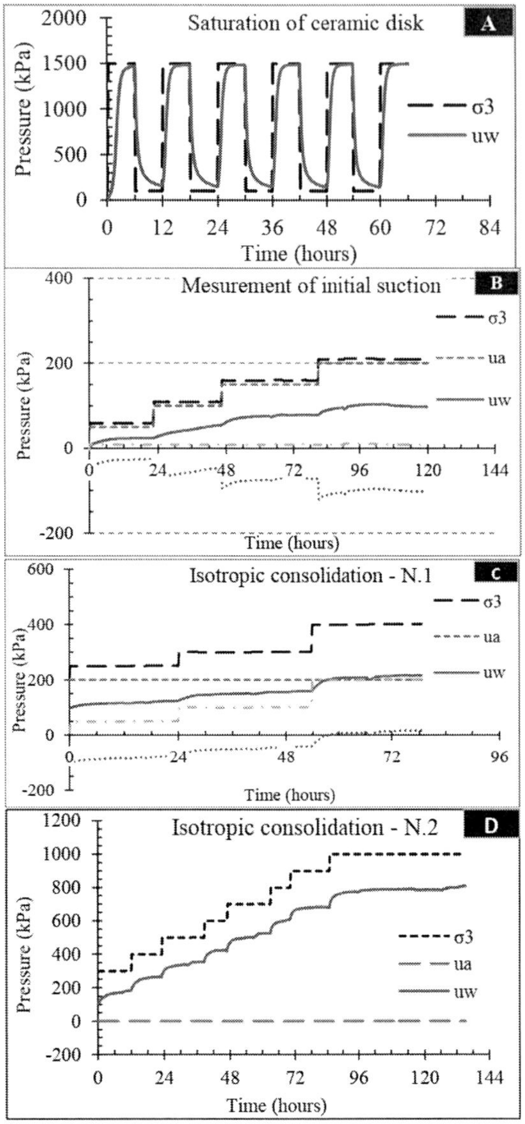

Figure 6. Results of undrained unsaturated triaxial test by axis-translation method (figures A, B, C and D).

By integrating the previous results (Figure 6), the relationship between the variation of suction, B and the net confining pressure is presented in the Table 4 and Figure 7. For the net confining pressure smaller than 200 kPa, it seems that the suction increases with the net confining pressure, but more and more slowly. This observation is consistent with that of Taibi (1994) for the compacted Vieuxpré Clay whose suction is lower than 80 kPa. The B is about 0.75 when the net confining pressure was increased from 300 to 900 kPa, and then B goes up to 0.91 for the net confining pressure at 1000 kPa which means that the specimen was close to saturation under the loading condition.

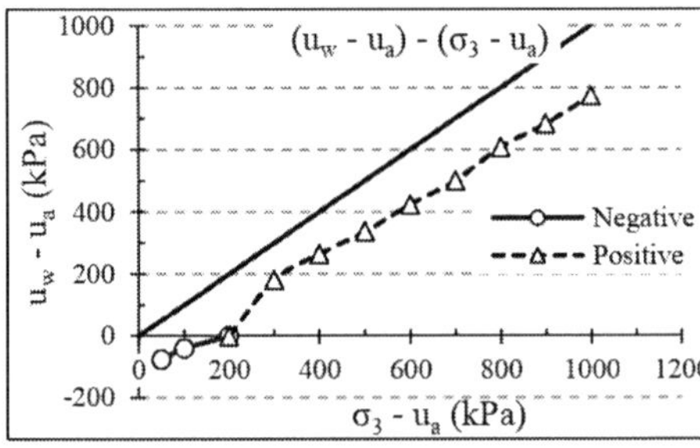

Figure 7. Relationship between net confining pressure (σ_3 - u_a) and suction ($u_w - u_a$)

Table 4. Variation of (u_w - u_a) and B in terms of net confining pressure (σ_3 - u_a) - comparison between experimental results and numerical simulations

(σ_3 - u_a) (kPa)	B	(u_w - u_a) Experimental (kPa)	(u_w - u_a) Simulation (kPa)	Remarque
50	x	-77	-	
100	0.54	-41	-	(u_w - u_a) is negative
200	0.4	-1	-	
300	–	181	200	
400	0.83	264	270	
500	0.72	336	342	
600	0.71	423	427	(u_w - u_a) is positive
700	0.77	500	516	
800	0.81	607	607	
900	0.75	682	706	
1000	0.91	773	800	

3.2 Numerical results and comparison

Results of numerical simulation are plotted in Figure 8 (triaxial test without axis translation). Values of (u_w-u_a) pressure are also reported in the table 4.

Three numerical curves are plotted 1) the curve with the parameters of the table 3; 2) the curve with the parameters of the table 3 less 3%; 3) the curve with the parameters of the table 3 more 3%. The third curve is the closest of the experimental results. Measurement uncertainties of the degree of saturation explain the necessity of this margin of error.

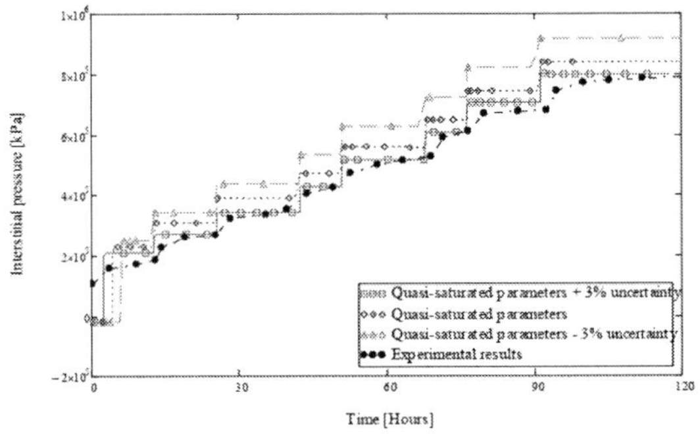

Figure 8. Evolution of pore pressure with net confining pressure in triaxial test – comparison between experimental results and numerical simulation

Conclusion

Evolution of the pore pressure during net confining stress (σ_3 - u_a) in a triaxial cell is presented. The tested sample is in the quasi-saturated domain and close to saturation. The triaxial test by axis-translation method is complemented by a standard triaxial test for high values of (σ_3 - u_a) with a null value of u_a. Indeed, in quasi-saturation, with the axis-translation method, water is drained into the air circuit when the pore-water pressure exceeds the applied air-pressure. Results of these experimental study permit to obtain the evolution of the Skempton Coefficient B in terms of net confining pressure and observe the progressive saturation of the sample under mechanical loading.

A numerical simulation was carried out with the model developed in the ANR project TerreDurable for the standard triaxial test. This model use a simple effective stress based on the Cambridge model with the coefficient of compressibility c_f of the interstitial fluid (water + occluded air) which depends on the degree of saturation.

A good agreement is observed between theoretical and experimental results on the triaxial test without axis-translation, which validates the proposed model. The small discrepancies may result from the influence of temperature during the test.

References

Andriantrehina, R., Mahmutovic, D., Boutonnier, L., Taibi, S., Fleureau, J. M., Fry, J. J., Monnet, J. (2015). Observation of craking evolution during free dessication - experimental and numerical approaches. Symposium international SEC 2015. Marne-la-vallée.

AFNOR (French Association of Standardization). (1992). "Earthworks - Classification of materials for use in the construction of embankments and capping layers of road infrastructures." NF P11-300, France.

ASTM (American Society for Testing and Materials). (2004). "Standard Practice for Classification of Soils and Soil-Aggregate Mixtures for Highway Construction Purposes." D2487-00, West Conshohocken, PA.

Biarez, J., Fleureau, J. M., Zerhouni, M. I. & Soepandji, B. S. (1987). Variations de volume des sols argileux lors de cycles drainage-humidification. Revue française de Géotechnique. No. 41, Page 63-71.

Bishop, A. W. (1954). The use of pore-pressure coefficients in practice. Géotechnique, Volume 4, Issue 1, Page 148-152.

Bishop, A. W. & Donald, I. B. (1961). The experimental study of partly saturated soil in the triaxial apparatus; Proceedings of the 5th International Conference of Soil Mechanics, Paris, Volume 1, Page 13-21.

Boutonnier, L. (2007). Comportement hydromécanique des sols fins proches de la saturation cas des ouvrages en terre: coefficient B, déformations instantanées et différées, retrait / gonflement. Thèse de doctorat, Institut National Polytechnique de Grenoble, France.

Boutonnier, L., Mahmutovic, D., Fry, J. J. (2015). Génération de pression interstitielle dans les fondations argileuses des barrages en remblai. Colloque CBR 2015.

Fleureau, J. M., Kheirbek-Saoud, S., Soemitro, R., & Taibi, S. (1993). Behavior of clayey soils on drying-wetting paths. Canadian geotechnical journal, 30(2), 287-296.

Fleureau, J. M., & Kheirbek-Saoud, S. (2001). Retrait et Gonflement des sols soumis à un cycle de drainage humidification.

Fleureau, J. M., Verbrugge, J. C., Huergo, P. J., Correia, A. G., & Kheirbek-Saoud, S. (2002). Aspects of the behaviour of compacted clayey soils on drying and wetting paths. Canadian geotechnical journal, 39(6), 1341-1357.

Fredlund, D. G & Morgenstern, N. R. (1976). Constitutive relations for volume change in unsaturated soils. Canadian Geotechnical Journal, Volume 13, Issue 3, Page 261-276.

Gibbs, H. J., Coffey, C. T. (1969). Techniques for pore pressure measurements and shear testing of soil ; Proceedings of the 7th International Conference of Soil Mechanics and Foundation Engineering, Mexico, Volume 1, Page 151-157

Hilf, J. W. ; 1948 ; Estimating Construction Pore Pressures in Rolled Earth Dams ; Proceedings of the 2nd International Conference of Soil Mechanics and Foundation Engineering, Rotterdam, Volume 3, Page 234-240.

Lu, N. & Liko,s W. (2006). Suction stress characteristic curve for unsaturated soil ; Journal of Geotechnical and Geoenvironmental Engineering, Volume 132, Issure 2, Page 131-142.

Mahmutovic, D., Monnet, J., Loret, B., Ple, O., Pelizzari, B., Boutonnier, L., Lai, B. (2014). A model for weakly unsaturated soils taking into account the three phases (air, water and soil). Unsat 2014. Sydney.

Skempton, A. W. (1954). The pore-pressure coefficients A and B. Géotechnique, Volume 4, Issue 1, Page 143-147.

Taibi, S. (1994). Comportement mécanique et hydraulique des sols soumis à une pression interstitielle négative - Etude expérimentale et modélisation. Thèse de doctorat, Ecole Centrale Paris, France.

Taibi, S., Fleureau, J. M., Abou-Bekr, N., Zerhouni, M. I., Benchouk, A., Lachgueur, K., & Souli, H. (2011). Some aspects of the behaviour of compacted soils along wetting paths. Géotechnique, 61(5), 431-437.

 DOI: 10.1051/e3sconf/20160908016

Elastoplastic model for unsaturated, quasi-saturated and fully saturated fine soils

Ba Tien Lai[1], Antonin Fabbri[1,a] , Henry Wong[1] Tuan Anh Bui[1] and Denis Branque[1]

[1] Université de Lyon/ENTPE/LGCB-LTDS (UMR CNRS 5513) 3 rue Maurice Audin, 69120 Vaulx en Velin, *France*

Abstract. In unsaturated soils, the gaseous phase is commonly assumed to be continuous. This assumption is no more valid at high saturation ratio. In that case, air bubbles and pockets can be trapped in the porous network by the liquid phase and the gas phase becomes discontinuous. This trapped air reduces the apparent compressibility of the pore fluid and affect the mechanical behavior of the soil. Although it is trapped in the pores, its dissolution can take place. Dissolved air can migrate through the pore space, either by following the flow of the fluid or by diffusion. In this context, this paper present a hydro mechanical model that separately considers the kinematics and the mechanical behavior of each fluid species (eg liquid water, dissolved air, gaseous air) and the solid matrix. This new model was implemented in a C++ code. Some numerical simulations are performed to demonstrate the ability of this model to reproduce a continuous transition of unsaturated to saturated states.

1 Introduction

An unsaturated soil is conceptually a tri-phasic medium, composed by a solid skeleton (S) and a porous network partially saturated by a liquid phase (L) while its remaining space is filled by a gas phase (G). At low to moderate water contents, both gaseous and liquid phases are continuous. The interaction between these two phases and the solid grain leads to a well-known suction effect which has the effect of increasing the shear strength and the rigidity of soils. In the recent years, numerous researches ([1], [2], [3], [4], [5], [6]) have focused on the modeling of unsaturated soil behavior. They have led to enhance classical elasto-plastic models (like CamClay) by bringing the influences of non-saturation. However few researches have studied the transition zone between the unsaturated and the fully saturated domains.

Experiments results on natural fine soils in temperate region or in compacted soils at the Optimum Proctor show that, , when the degree of saturation reaches values of 75-85%, air bubbles and pockets are trapped within the continuous water phase. This domain is called "quasi-saturated" by authors such as [8], [9],[10].

In the literature, some approaches to describe the coexistence of saturation domain such as [11], [12], [13] exist. However, the above models cannot account for the coupled hydromechanical behaviors and the continuous transition between different regimes of saturation.

A rigorous theoretical framework that is well-adapted to the formulation of coupled multi-physics problems in porous media would be necessary to simulate the existence of different regimes of saturation, as well as the evolution relative to space and time of the corresponding physical domains. To that end, an original model is proposed in this paper. It is part of an ongoing research project named Terredurable, which aims to provide a clearer understanding of coupled hydromechanical responses of civil engineering structures at close-to-saturation states.

2 The new description of saturation domain

In this approach, we account for three different domains of saturation: the unsaturated domain where both liquid and gas phases are continuous (US); the quasi-saturated domain where the gas phase in the form of bubbles is discontinuous (QS); and the fully saturated domain where air is only present in dissolved form (FS).

In order to arrive at a consistent description of a continuous transition across different saturation domains, the porous space occupied by the gaseous phase will conceptually be divided into two sets that can freely exchange within each other. The first, named "displaceable air" is the part that will be invaded by water during the wetting process before the loss of connectivity

[a] Corresponding author: antonin.fabbri@entpe.fr

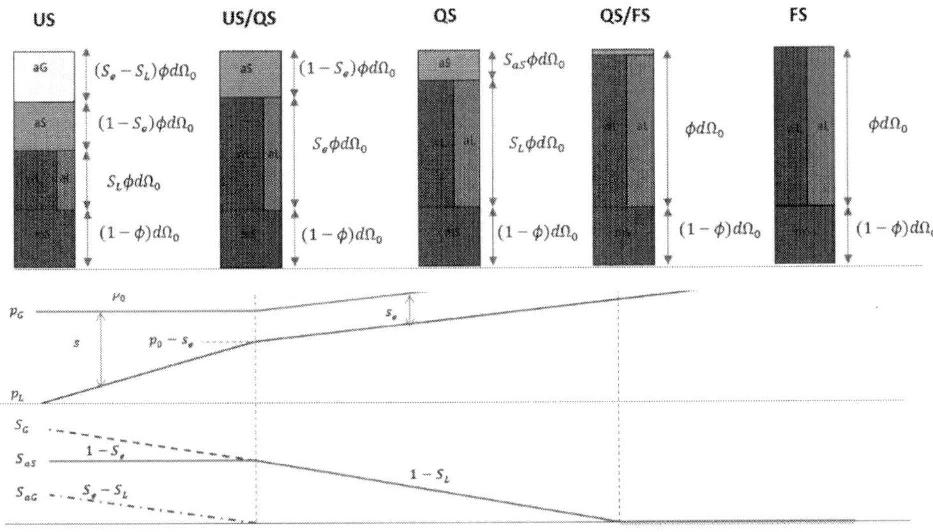

Figure 1. Transition between different domains of saturation during wetting.

of the gaseous phase; this volume is noted "aG". The remaining part, called "non-displaceable air", corresponds to locations where the air is trapped when the saturation attains the threshold S_e The air within this volume is noted "aS". As mentioned previously, the capital "S" is used to emphasize the fact that the air within this volume has the same kinematics as the solid matrix. A simplified description of the saturation domains and the continuous transition assumed here is summarized in Figure 1.

In the US-domain, the gaseous phase is continuous and connected to outside; the difference between the gas pressure (equal to atmospheric pressure here) and the liquid pressure, defined as matric suction, is higher than the air-entry suction. The two sets of gaseous air, "aS" and "aG", are simultaneously present, physically connected to each other and share a common pressure equal to the atmospheric pressure:

$$p_0 - p_L \geq s_e; p_G = p_{aG} = p_0 \tag{1}$$

This implicitly implies that the flow of gas towards the external boundary surface is instantaneous. Note that, by hypothesis, the degree of saturation corresponding to the air set "aS" remains constant in this phase: $S_{aS} = 1 - S_e$. Suction and degree of saturation are linked to each other via the celebrated equation of [14] for saturation ratios below S_e : $S_L = S_e(s/s_e)^{-\alpha}$.

In the QS-domain, suction is below the air-entry value and the gaseous phase is discontinuous; air is only present in the form of bubbles.

Being non-connected to the outside, pressure in the air bubbles can now exceed the atmospheric pressure. Note that the gas pressure is continuous at the transition US/QS, equal to p_0. In summary:

$$p_0 - p_L < s_e; S_{aG} = 0; S_{aS} > 0; p_G = p_L + s_e > p_0 \tag{2}$$

The QS-domain is reached when all the trapped air is dissolved, the material becomes fully saturated. The mathematical conditions is $S_{aS} = 0$.

3 Mathematical formulations

3.1 Rheological constitutive equations

In the model, the classic Camclay variables are used, namely the mean effective stress $p' = tr(\boldsymbol{\sigma}')/3$, and the equivalent deviatoric stress $q = \sqrt{(3/2)\boldsymbol{s}:\boldsymbol{s}}$ where $\boldsymbol{s} = \boldsymbol{\sigma}' - \mathbf{1} \cdot p'/3$ is the deviatoric stress tensor. $\varepsilon_p = tr(\boldsymbol{\varepsilon})$ is the volumetric strain while $\varepsilon_q = ((2/3)\boldsymbol{e}:\boldsymbol{e})^{1/2}$ is the equivalent deviatoric strain, with $\boldsymbol{e} = \boldsymbol{\varepsilon} - \mathbf{1} \cdot \varepsilon_p / 3$.

The effective stress concept is used:

$$\boldsymbol{\sigma} = \boldsymbol{\sigma}' - \mathbf{1} \cdot \pi \tag{3}$$

where $\boldsymbol{\sigma}$ is the total stress tensor, $\mathbf{1}$ the second order identity tensor and π the equivalent pore pressure. Several definitions for π have been proposed, for example [15], [16]. The following form, proposed by Bishop, is used in this study:

$$\pi = -p_G + S_L(p_G - p_L) \tag{4}$$

We also assume that under isotropic virgin compression, the volumetric strain will vary linearly with the logarithm of the mean effective stress with a slope equal to:

$$\lambda(s) = \begin{cases} \lambda_0 [(1-r)] e^{-\beta s} + r & : \ s > s_e \\ \lambda_0 & : \ s \leq s_e \end{cases} \tag{5}$$

where β, r are model parameters

 DOI: 10.1051/e3sconf/20160908016

We therefore adopt a simple isotropic logarithmic elastic behaviour like Camclay. The strain rates are entirely defined by the stress rates:

$$\begin{pmatrix} d\varepsilon_p^e \\ d\varepsilon_q^e \end{pmatrix} = \begin{pmatrix} dp'/K \\ dq/(3G) \end{pmatrix} \quad ; \quad K = \frac{1+e_0}{\kappa} p' \, ; G = \frac{3(1-\nu)}{2(1+\nu)} K \tag{6}$$

where where K and G are the state-dependent bulk and shear moduli, ν the Poisson ratio and κ a material constant.

The loading surface adopted, is described by:

$$f(p',q,p_c) = q^2 - M^2 p'(p_c - p') \tag{7}$$

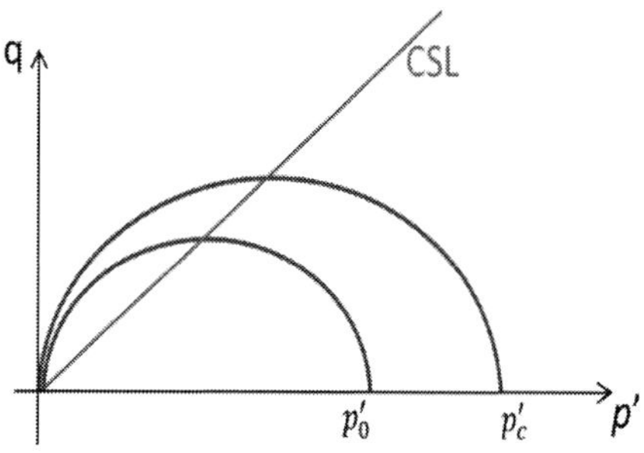

Figure 2. Loading surface in the plan (p',q).

The dependence of p_c on suction can be described by:

$$p_c = \begin{cases} p_{ref}\left(\dfrac{p_0}{p_{ref}}\right)^{\frac{\lambda_0-\kappa}{\lambda(s)-\kappa}} & : \; s > s_e \\ p_0 & : \; s \le s_e \end{cases} \tag{8}$$

p_0 is the hardening variable in the case of full saturation, similar to Camclay, verifying:

$$dp_0 = \frac{1+e_0}{\lambda_0 - \kappa} p_0 d\varepsilon_p^p \tag{9}$$

It is postulated that plastic strain rates write:

$$\begin{pmatrix} d\varepsilon_p^p \\ d\varepsilon_q^p \end{pmatrix} = d\lambda \begin{pmatrix} \dfrac{\partial f}{\partial p'} \\ \dfrac{\partial f}{\partial q} \end{pmatrix} \tag{10}$$

where $d\lambda = \dfrac{1}{H}\left(\dfrac{\partial f}{\partial p'}dp' + \dfrac{\partial f}{\partial q}dq + \dfrac{\partial f}{\partial s}ds\right)$

Where $d\lambda$ plastic multiplier and H plastic modulus.

$$H = \frac{\nu p'^3 \left(M^4 - \eta^4\right)}{\lambda(s) - \kappa} \tag{11}$$

3.2 Masse balances equations

There are three independent mass balance equations, on liquid water, gaseous air and dissolved air. The mass fraction of dissolved air being much lower than that of liquid water, the mass balance of liquid water writes:

$$\frac{\partial S_L}{\partial t} + \frac{\phi S_L}{K_L}\frac{\partial p_L}{\partial t} + S_L \frac{\partial \phi}{\partial t} = \nabla.\left(D_L \nabla p_L - D_L \rho_L^0 \mathbf{g}\right) \tag{12}$$

where ρ_L^0 the liquid density, D_L the permeability.

The mass balance of dissolved air is due to three main contributions: (a) the advective transport of dissolved air within the liquid phase; (b) the diffusion of the dissolved air within the liquid phase; and (c) exchange with gaseous air via the dissolution/bubbling process. After some detailed developments, we get:

$$\phi S_L \frac{\partial m}{\partial t} + m\left(\frac{\phi S_L}{K_L}\frac{\partial p_L}{\partial t} + \phi \frac{\partial S_L}{\partial t} + S_L \frac{\partial \phi}{\partial t}\right) = \nabla.\left[m D_L \left(\nabla p_L - \rho_L \mathbf{g}\right) + D_e \nabla m\right] + \frac{\hat{m}_{G\to aL}}{\rho_L} \tag{13}$$

where D_e is the effective diffusion coefficient of dissolved air, $\hat{m}_{G\to aL}$ the mass rate of dissolution/bubbling of air within the liquid phase. $m = m_{aL} / m_{wL}$ the mass of dissolved air per unit of liquid water mass. When the air in the gaseous state still exists within the REV, S_L<1 we have $m = m^{eq}$ where m^{eq} is the mass of dissolved air per unit of liquid water mass at equilibrium with the entrapped air pockets. m^{eq} is calculated by the Henry's law:

$$p_{aS} = K_H \frac{M_L}{M_G} m^{eq} \tag{14}$$

To derive the equation of mass balance for gaseous air, we assume that the in-pore air phase follows the law of perfect gas:

$$m_{aG} = \frac{\phi S_{aG} p_{aG} M_G}{RT} \, ; \; m_{aS} = \frac{\phi S_{aS} p_{aS} M_G}{RT} \tag{15}$$

With R the universal gas constant and T the absolute temperature. Neglecting evaporation and condensation phenomena, the general form of the mass conservation of gaseous air is:

$$\frac{\partial (m_{aG} + m_{aS})}{\partial t} = \nabla \cdot \boldsymbol{\omega}_G - \hat{m}_{G\to aL} \tag{16}$$

where $\nabla \cdot \boldsymbol{\omega}_G$ stands for the mass in-flux due to the Darcean flow of gas through the porous network. To advance further, we need to inject the expressions of various quantities into equations (12), (13) and (16) and simplify. After a lengthy manipulation, we obtain the following three equations:

$$A_1\frac{\partial p_L}{\partial t}+A_2\frac{\partial e}{\partial t}-n\frac{\partial S_{aS}}{\partial t}=\nabla\cdot\left(D_L\nabla p_L\right) \tag{17}$$

$$\phi S_L\frac{\partial m}{\partial t}+A_3\frac{\partial p_L}{\partial t}+A_4\frac{\partial e}{\partial t}+\phi\left(\beta_1\frac{\rho_G}{\rho_L}-m\right)\frac{\partial S_{aS}}{\partial t}$$
$$=\nabla\cdot\left[mD_L\nabla p_L+D_e\nabla m\right]$$
$$\phi\left(m_{eq}-\beta_1\frac{\rho_G}{\rho_L}\right)\frac{\partial S_{aS}}{\partial t}$$
$$=\beta_2\left[A_5\frac{\partial p_L}{\partial t}+A_6\frac{\partial e}{\partial t}-\nabla\cdot\left[mD_L\nabla p_L+D_e\nabla m\right]\right]$$

With

$$A_1=\phi\left[\frac{S_L}{K_L}-\frac{dS_{aG}}{dp_L}\right];\ A_2=\phi\left[-\frac{dS_{aG}}{de}+\frac{S_L}{e}\right]; \tag{18}$$
$$A_3=\phi\left(m\frac{S_L}{K_L}-m\frac{dS_{aG}}{dp_L}+\beta_1\frac{\rho_G}{\rho_L}\frac{S_{aS}}{p_G}\right);$$
$$A_4=\phi\left[m\frac{S_L}{e}-m\frac{dS_{aG}}{de}+\beta_1\frac{\rho_G}{\rho_L}S_{aS}\left(\frac{1}{e}\right)\right];$$
$$A_5=\phi\left[\left(m_{eq}S_L+\frac{\rho_G}{\rho_L}S_{aS}\right)\frac{\beta_1}{p_G}+m_{eq}\frac{S_L}{K_L}-m_{eq}\frac{dS_{aG}}{dp_L}\right];$$
$$A_6=\phi\left[\left(m_{eq}S_L+\frac{\rho_G}{\rho_L}S_{aS}\right)\left(\frac{1}{e}\right)-m_{eq}\frac{dS_{aG}}{de}\right]$$

4 Numerical results

To illustrate the consistency and the applicability of this model, we present in this section a few numerical examples. We consider a 1D configuration of a homogeneous soil sample of 5 cm thick in an oedometric cell. Initially, the sample is in thermodynamic and mechanical equilibrium and is in an unsaturated state (US). The sample is then subjected to a monotonic increase of liquid pressure at the "bottom" boundary surface (x=0), from -250 [kPa] to 650 [kPa] in 150[s]. At the top boundary (x=5cm), there is no flux. The total stress is kept constant equal to its initial value σ_0 =1MPa. Radial displacement and radial flow are null. This simulation corresponds to a progressive wetting of a sample. The parameters used for the simulation are summarized in Appendix. Finally, we assume that the mass concentration of dissolved air in the incoming liquid water is, at any time, equal to the concentration of dissolved air within the pores of the sample which are directly in contact with the external surfaces. In summary, we have:

$$\sigma_x\left(x=0,t\right)=\sigma_0\ ;\ u_x\left(x=L,t\right)=0;$$
$$\nabla m\cdot\boldsymbol{e}_x\left(x=0,t\right)=\nabla m\cdot\boldsymbol{e}_x\left(x=L,t\right)=0;$$
$$p_L\left(x=0,t\right)=p_{L0}+\delta p_L\frac{t}{t_{final}}$$

For t=35 [s] the whole sample is unsaturated (US). The liquid saturation profile is directly linked to the liquid pressure (which is synonymous with suction, given the constant air pressure) through the retention curve. In this domain, the entrapped air saturation ratio is constant. (c.f. Figures 3 & 4).

For t=85[s] a transition between the unsaturated and the quasi-saturated domain is observed at, x=0.0105 [m]. This transition can easily be observed in Figure 3, where S_{aS} remains equal to 1-S_e in the unsaturated domain, and progressively decreases from the US-QS.

Finally, at t=150[s] all three domains of saturation coexists: the unsaturated state in $0\le x\le 0.002$ the quasi-saturated state in $0.002[\mathrm{m}]<x<0.0255[m]$ and the fully saturated state for $x\ge 0.0255\ [\mathrm{m}]$.

The variation of void ratio with liquid pressure and effective stress at x=0 [m] are presented in the Figure 5. Three different behaviours, delimited by the points A, B C and D in Figure 5, can be observed. From A to B, the material at x=0 is in the unsaturated domain (the value of liquid pressure is less than p_0-s_e), the behaviour of soils is elastic; from B to C, it is in the quasi-saturated domain, the behaviour of soils is elastic; From C to D, a variation of the slope of the relation σ'-e is observed, the behaviour of soil becomes elastoplastic.

Furthermore, in the unsaturated domain, the void ratio variations with the increase of the liquid pressure remain limited. On the opposite, in the quasi-saturated state, an increase in a fluid pressure causes a sharp variation of the void ratio. This numerical result is consistent with experimental results of quasi-saturated soils of [17].

5 Conclusions

This paper presents a detailed study of the hydromechanical behaviours of compacted fine soils close to saturation. These soils are characterized by the presence of entrapped air. An original model capable of simulating the various couplings between physical and hydromechanical phenomena intervening in such type of soils is presented in this paper. The model developed is able to simulate a consistent description of the simultaneous presence of different domains of saturation, as well as their evolution relative to space and time.The numerical example allows checking the consistency of the model.

Acknowledgement

This work has been supported by the French National Research Agency through the projet TerreDurable (ANR-11-VILD-0004).

DOI: 10.1051/e3sconf/20160908016

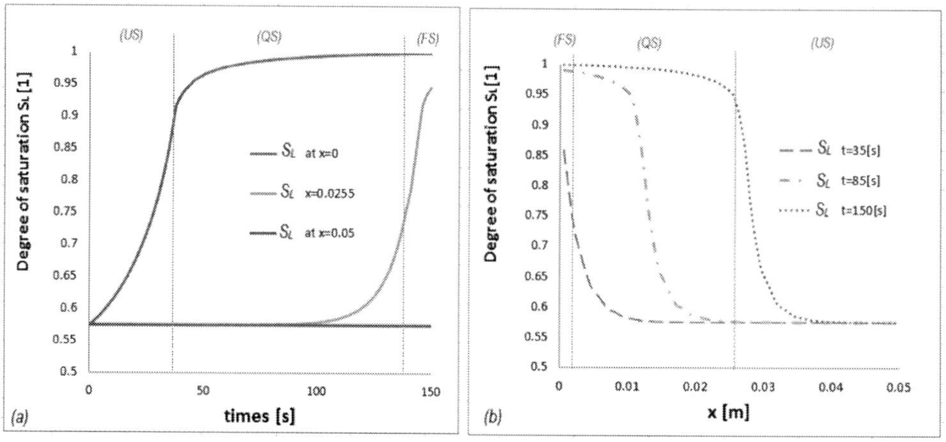

Figure 3. Variation of degree of saturation with (a) times (b) space

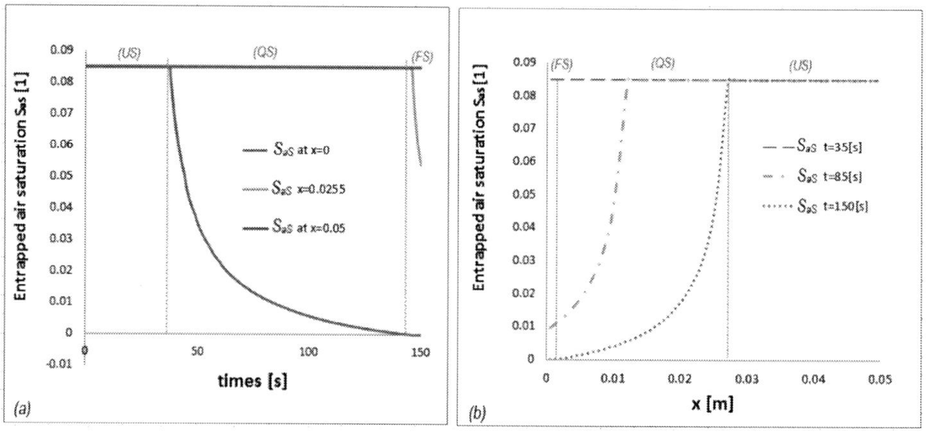

Figure 4. Variation of saturation of entrapped air with (a) times (b) space

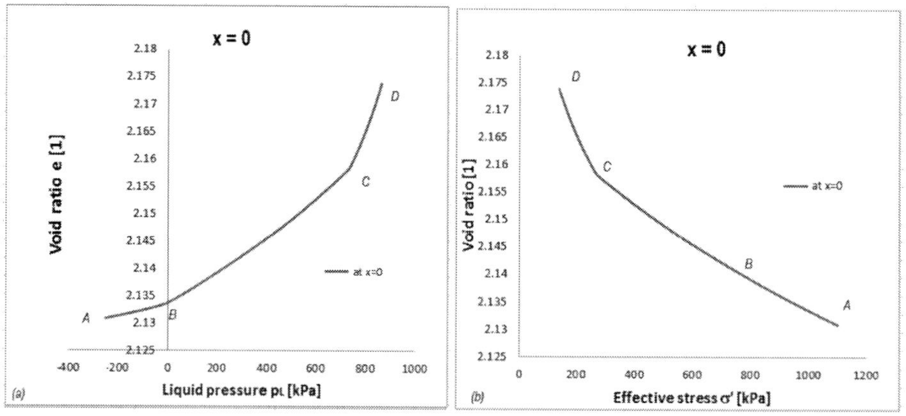

Figure 5. Variation of void ratio at x=0 with (a) liquid pressure (b) effective stress

Appendix 1

Parameters	Description	Value [unit]
K_H	Henry' constant for air	7326 [MPa]
R	ideal gas constant	8.314 [J/(mol.K)]
M_L	Molar mass of water	18 [g/mol]
M_G	Molar mass of air	29 [g/mol
p_0	Atmospheric air pressure	100 [kPa]
T	Absolute temperature	293 [K]
K_L	Bulk modulus of water	1970 [MPa]
η	Dynamic viscosity of water	10-3 [kg/(m.s)]
D_L	Permeability	8.10-9 [m/s]
e_0	Initial void ratio	1.97 [1]
S_e	Degree of saturation at air-entry	0.915 [1]
s_{e0}	Initial air-entry suction	110 [kPa]
α	Coefficient of retention curve	0.4 [1]
κ	Elastic stifness parameters	0.05[1]
λ_0	Stiffness parameter in virgin state of saturated soils	0.6[1]
β	Model parameter	12.5[kPa^{-1}]
r	Model parameter	0.75[1]
p_{L0}	Initial liquid pressure	-250 [kPa]
σ_0	Initial total stresse	1000 [kPa]
S_{L0}	Initial degree of saturation	0.566 [1]
m_0	Initial mass fraction of dissolved air	22.10^{-6}[1]

References

[1] E. E. Alonso, A. Gens, A. Josa, Géotechnique **40**, 405-430 (1990).
[2] A. Modaressi, N. Abou-Berk, Proceedings of Numerical Methods in Geotechnical Engineering. (1994)
[3] H.Q. Pham, D.G. Fredlund, S.L. Barbour, Géotechnique, ***53***(2), 293-298 (2003)
[4] H.Q. Vu, D.G. Fredlund, Canadian Geotechnical Journal, **41**(4), 713-737 (2004)
[5] E.E. Alonso, Revue Française de Géotechnique, **111**, 33-43 (2005)
[6] J-M. Pereira (2005). LGM, DGCB, ENTPE (2005)
[8] D. A. Stonestrom and J. Rubin J, Water resources research, **25**(9), 1947-1958 (1989)
[9] B. Faybishenko, Water Resources Research, **31**(10), 2421-2435 (1995)
[10] B.T. Lai, A. Fabbri, H. Wong, D. Branque, Computers and Geotechnics **69**, 627-640 (2015)
[11]A. Denoth, Journal of Glaciology ***28***(99) 357-364 (1980)
[12]L. Boutonnier, Thèse à l'Institut National Polytechnique de Grenoble (2007)
[13]S.K. Vanapalli, D.G. Fredlund, & D.E.Pufahl, Geotechnique, **49**(2), 143-159 (1999)
[14]R. Brooks,A. Corey, *Hydraulic properties of Porous Media*. Colorado State University, (1964)
[15]O. Coussy, *Poromechanics*. John Wiley & Sons.(2004),
[16] A. W. Bishop, Teknisk Ukeblad, **106**(39), 859-863 (1963)
[17]J. M. Fleureau, rapport du LMSS.(2003).

The determination of model dimension for an embankment to study soil atmosphere interaction with Finite Element Method

Ni An[1, a], Sahar Hemmati[2], Yu-jun Cui[1] and Anh-minh Tang[1]

[1] *Laboratoire Navier/CERMES, Ecole des Ponts ParisTech, 77455 Marne la Vallée, France*

[2] *IFSTTAR - French institute of science and technology for transport, development and networks, 77447 Marne la Vallée, France*

Abstract: In order to study the interaction between soil and atmosphere, Héricourt experimental embankment was instrumented with different equipments for collecting the meteorological data, runoff, soil temperature and volumetric water content, etc. The corresponding water flux and heat flux boundary conditions of the Hydro-thermal coupled model of unsaturated soil can be calculated based on the energy and mass balance. However, in Finite Element Method, a suitable model dimension of experimental embankment needs to be set in advance, which will diminish the effect of bottom and lateral boundaries. Therefore, in this paper, four different dimension models are calculated respectively and compared in two cases with different boundary conditions. For Case 1, top water flux boundary condition is set as infiltration and evaporation happening on the soil surface alternately. In Case 2, top heat flux boundary condition is assumed based on daily solar radiation value. Comparing the results of four model dimensions, temperature and volumetric water content distributions in four dimensions lead nearly consistent results. Therefore, with consideration of computational time and accuracy, dimension 2 or 3 is suggested for Hydro-thermal study of Héricourt embankment case.

1 Introduction

For numerical calculation, it is important to set suitable dimension to describe the real situations of problem. In literature, a typical dimension value of embankment model for mechanical numerical calculation has been suggested by Azizi [1]. Rujikiatkamjorn et al. [2] proposed a 3D model based on Finite Difference Method in Hexagonal Wire Mesh by FLAC 3D, which was applied to simulate the vertical settlement, excess pore pressures and lateral displacements of an embankment on soft Bangkok clay. With an experienced dimension, Yue [3] simulated the thermal regime of embankment and berm in permafrost based on 2D embankment model with finite element method. Kalliainen et al. [4] studied the effect of embankment dimensions on the mechanical behavior of railway track, showing that the embankment dimensions had a slight effect on the accumulated permanent deformations of the embankment on a very stiff subgrade. However, on more flexible subgrade, the rate of accumulated permanent deformation is remarkable and the effect of embankment dimension on accumulated permanent deformation of the embankment is also clearly visible. In the study of mechanical behavior of

[a] Corresponding author: ann@cermes.enpc.fr

embankment, it is subgrade stiffness that has a remarkable effect on its deformation behaviour, not the embankment dimension.

With the experiences in literature, in most cases, model dimension was set for hydraulic or mechanical study, which asks for more attention on the variation of soil water content, strain and stress over time. However, for Hydro-thermal coupled model of unsaturated soil, no related suggestion of model dimension was proposed for numerical calculation. In this paper, in order to get an ideal model dimension for field Héricourt embankment, four different dimensions are compared. By adding water flux and heat flux boundary conditions respectively, the soil volumetric water content and temperature variations are calculated by Finite Element Method. Based on the comparison results, with considering computational accuracy and time consumed, the suitable dimension value is suggested to be applied in further study about the interaction between atmosphere and unsaturated soil.

2 Material and method

2.1 Soil parameters

Studied Héricourt embankment was constructed by lime/cement treated silt. Relevant experimental work was conducted in order to get its parameters.

The thermal conductivity of treated silt is set to be linear to volumetric water content as proposed by De Vries [5]. Fig. 1 presents the variation of thermal conductivity as function of volumetric water content. Based on the test results, the fitting curve is expressed as,

$$\lambda = 2.1818\theta + 0.808 \quad (1)$$

where λ is thermal conductivity (W/mK), θ is volumetric water content.

With the testing data shown in Fig. 2, the fitting water retention curve based on Van Genuchten model could be drawn. In the same model, accompanied with the saturated hydraulic conductivity measured, its unsaturated hydraulic conductivity curve is drawn in Fig. 3. Accordingly, soil water retention curve and hydraulic conductivity are written as equation (2) and (3) respectively,

$$S_e = \frac{\theta - \theta_r}{\theta_s - \theta_r} = \left[\frac{1}{1 + (\partial\varphi)^n}\right]^m \quad (2)$$

$$K = K_s S_e^{0.5}\left[1 - \left(1 - S_e^{1/m_1}\right)^{m_1}\right]^2 \quad (3)$$

where saturated volumetric water content $\theta_s = 0.4$, residual volumetric water content $\theta_r = 0.004$, parameter $\alpha = 0.003$ kPa^{-1}, $m = 0.18$, $n = 1.8$, $m_1 = 0.5$, saturated hydraulic conductivity $K_s = 10^{-9}$ m/s.

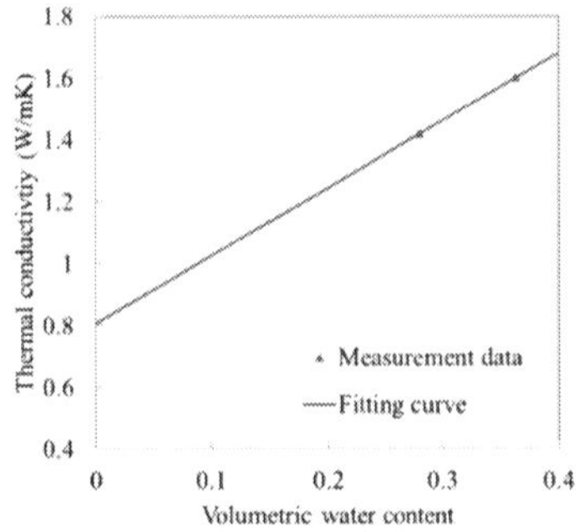

Figure 1. Thermal conductivity of treated silt

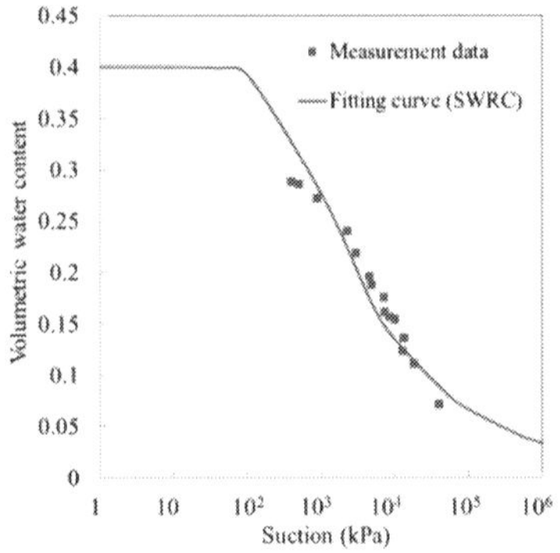

Figure 2. Soil retention curve of treated silt

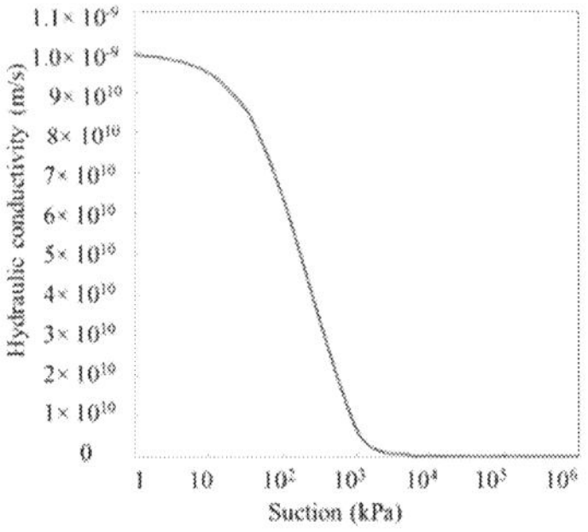

Figure 3. Hydraulic conductivity curve of treated silt

2.2 Model dimension

Normally, because of symmetry, further simplification is made by modeling only half of the whole embankment. To avoid the spurious influence from boundary conditions, four different dimensions of the finite element mesh are tested. Their details are listed in Table 1, Fig. 4 and Fig. 5. Dimension 4 is set based on the typical dimension of embankment model for mechanical numerical calculation suggested by Azizi [1]. Three other dimensions are set in order to be compared for studying the effect of boundary conditions. Dimension 1 is the original embankment size with part of basement. Dimension 2 and 3 are extended in different sizes respectively.

In addition, section A and B as shown in Fig. 4 are chosen to record the influence of different dimensions on numerical results. Six points with different depths from soil surface in section A and section B are selected (P1-0.00 m, P2-0.25 m, P3-0.50 m, P4-0.75 m, P5-1.00 m, P6-1.50 m). The groundwater table is assumed at the depth of 5 m from the ground as shown in Fig. 4. The bottom boundary condition is set as with constant 10°C temperature. No water and heat transfer happen to the left and right boundaries.

Two Cases with different top boundary conditions are conducted for dimension analysis. Case 1 is calculated with just hydraulic boundary conditions (heat flux boundary is zero) and Case 2 is conducted with only thermal boundary conditions (hydraulic flux boundary is zero). The details are explained in the next chapter. In terms of the initial condition, the uniform initial temperature of treated silt is assumed as T_0 = 10 °C. Besides, the soil above groundwater table is in unsaturated state whose suction is expressed as function U_0 = -y (Fig. 4), where y is soil depth from top surface of embankment.

Table 1. The details of four dimensions

Item	L_e	H_e
Dimension 1	0	5 m
Dimension 2	0.5L	5 m
Dimension 3	1.5L	3H
Dimension 4	> 3L	≈5H

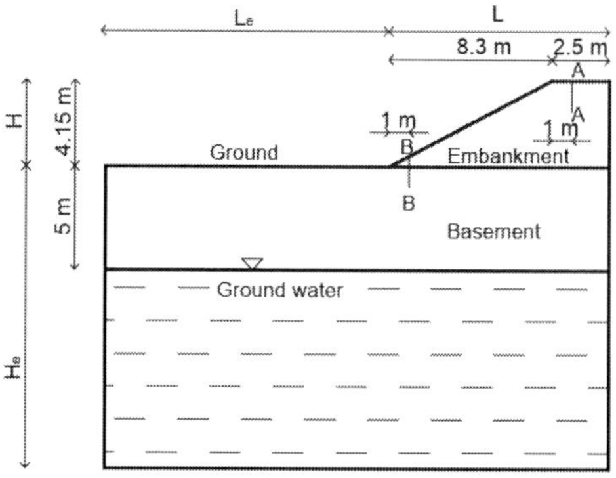

Figure 4. Embankment sketch and two section positions

3 Results and analysis

3.1 Case 1

The top boundaries of this embankment consist of top, slope surface of embankment and ground surface. For Case 1, it is supposed that evaporation (10^{-7} m/s) and infiltration (5×10^{-8} m/s) happen alternately during six days. Water flux top boundary condition as shown in Fig. 5 and zero heat flux top boundary condition are applied at the same time.

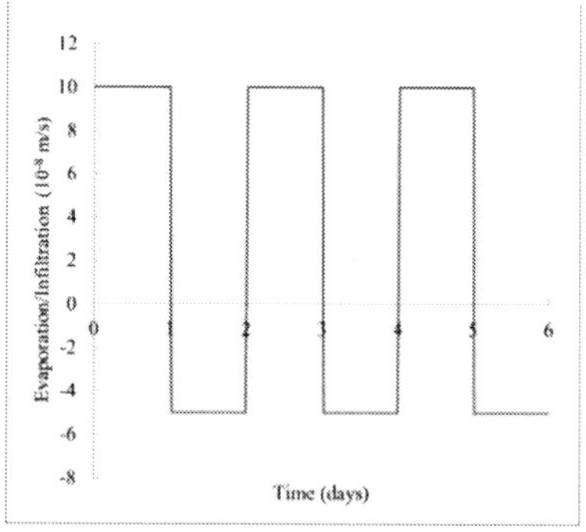

Figure 5. Water flux top boundary condition

Based on the Hydro-thermal coupled model of treated silt, only alternate infiltration and evaporation in six days are added as top boundary condition. With the volumetric water content variations of six points in section A and B recorded continuously. It can be seen that the values in six points are nearly consistent in four different dimensions (Fig. 6 et Fig. 7, 1-P1 represents point 1 in dimension 1, 2-P2 represents point 2 in dimension 2, etc.). However, the

DOI: 10.1051/e3sconf/20160908017

temperature of soil keeps as the initial value when no heat flux exchange happens.

The results of section A in Fig. 6 show that volumetric water content of each point changes regularly as the water flux changes on the top boundary. When evaporation happens, volumetric water content of surface point P1 decreases to about 0.323 at the first day, then goes back to nearly saturated situation at the second day, continuously drops to 0.312, 0.301 in the third and fifth day with rebounding to 0.387, 0.381 in the fourth and last day respectively. For point P2, it shows the similar going up and down circulation as the surface point P1. While for point P3, P4, P5 and P6, they show the decrease tendency continuously. Point P3, P4, P5 and P6 are so far from top surface that no apparent rebound phenomenon exists. It can be inferred that water flux boundary effects will not be reflected in soil layers whose depth is larger than 1m in this embankment. Furthermore, compared with point P1, it is observed that obvious time lag to reach the peak volumetric water content exists for point P2 because of its deeper position to the surface than point P1. Generally, these six lines show that the effect of top surface water flux boundary condition on soil volumetric water content becomes weaker in deeper positions.

Compared with other three dimensions, dimension 1 doesn't consider the effect of left ground to embankment slope toe. In Fig. 7, for section B that is located 1 m away to the slope toe of embankment, the results of dimension 1 show a little difference with other three dimensions. The minimum values for surface point P1 reach 0.329, 0.319, 0.309 respectively in the first, third, fifth day, which all are a little larger than those values in section A. It is caused by the gravity effect of liquid flow. Thus higher volumetric water content exists on the slope surface than the top surface of embankment.

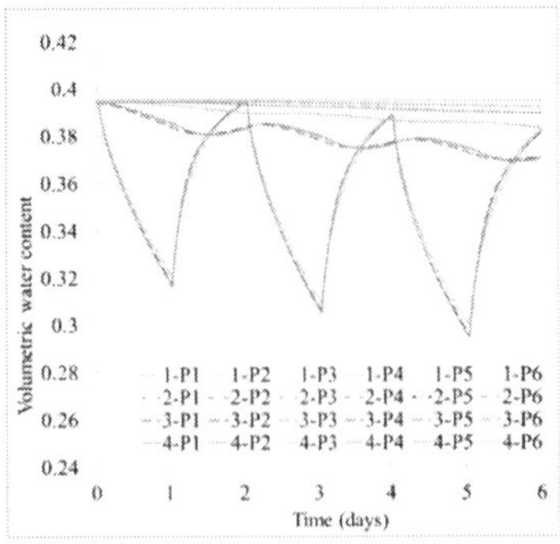

Figure 6. Volumetric water content change of six points in section A in four dimensions

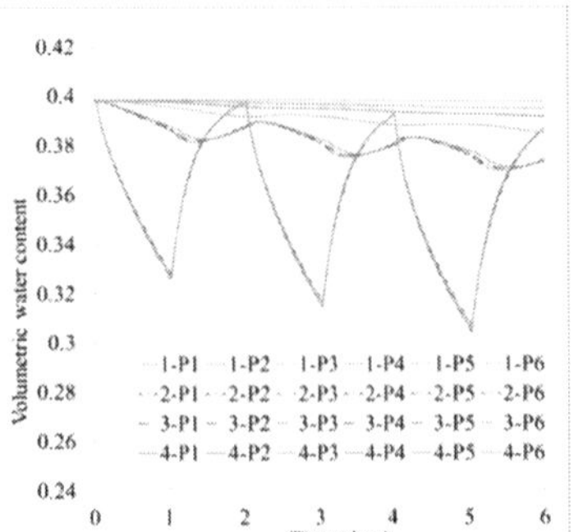

Figure 7. Volumetric water content change of six points in section B in four dimensions

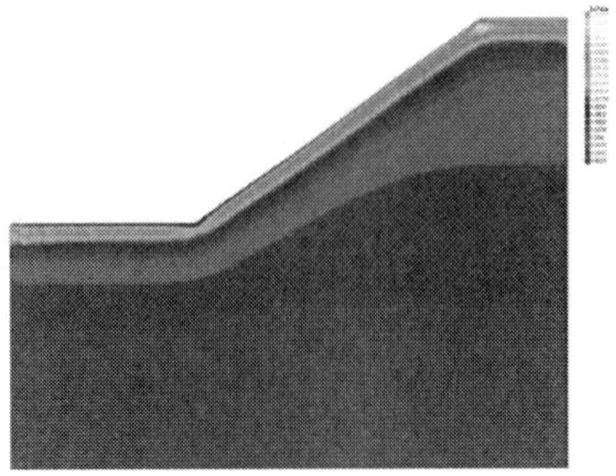

Figure 8. Volumetric water content distribution of dimension 2

As shown in Fig. 8, the contour line of volumetric water content of dimension 2 is parallel to the surface line of embankment. The influencing region by the top boundary condition of water flux is in the first meter of depth for this embankment. In fact, the distribution of volumetric water content in the region near surface of dimension 1, 3 and 4 are similar to dimension 2. What needs to be mentioned is that, because of the gravity effect, the region of left ground near slope toe cannot be neglected in numerical calculation. It will lead a more reasonable result with considering slope toe in our model.

3.2 Case 2

In the second case, based on the real solar radiation, heat boundary condition is assumed as the top boundary condition, positive 45 W/m^2s for 6 hours in day time and negative 15 W/m^2s for 18 hours during night time (Fig. 9a et Fig. 9b).

The six points of two sections are nearly consistent in four different dimensions (Fig. 10 et Fig. 11). As shown in Fig. 10, in the first nine hours of nighttime when soil heat is negative as 15 W/m^2s, surface point P1 decreases quickly to 8.69 °C and goes up to 12.65 °C in the daytime when receiving positive soil heat 45 W/m^2s. During six days, it keeps nearly the same minimum and maximum temperature in each day. The similar situation exists in point P2, which is in temperature variation range 9.80~10.20 °C. Obvious time delay of 2.4 to 4.8 hours could be observed for points P2 compared with point P1 because of deeper position of point P2. Point P3, P4, P5 and P6 keep nearly constant temperature as the initial value, which means that they are located out of the influencing region of surface heat flux. This influencing depth is closely related to the thermal conductivity of treated soil.

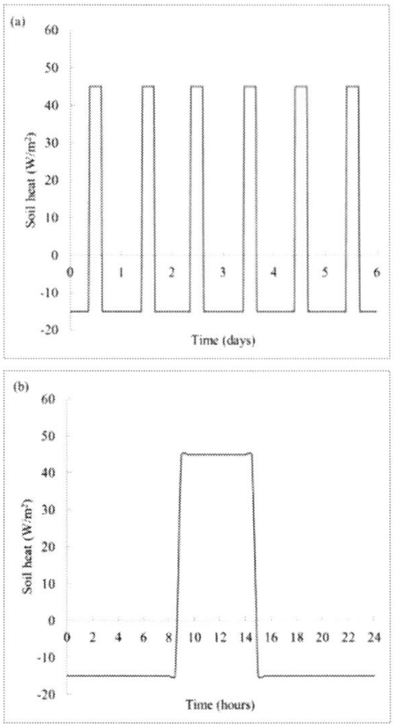

Figure 9.

(a) Heat flux top boundary condition in six days

(b) Details of heat flux variations in 24 hours

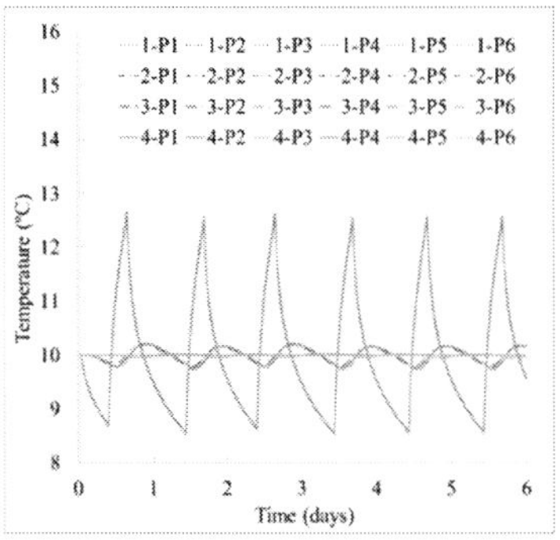

Figure 10. Temperature change of six points in section A in four dimensions

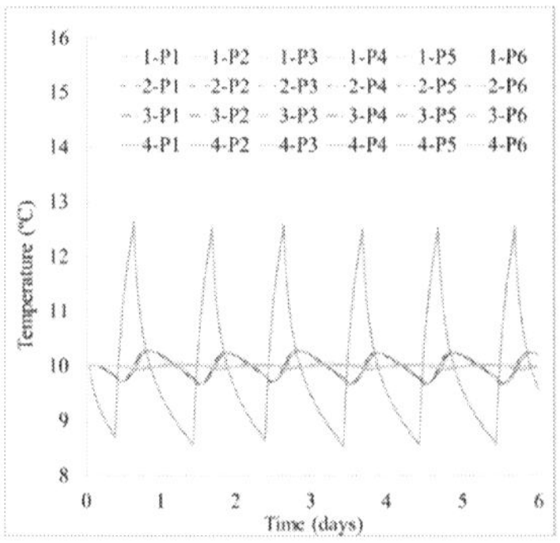

Figure 11. Temperature change of six points in section B in four dimensions

Figure 12. Temperature distribution of dimension 2

In Fig. 11, for section B, it shows nearly all the same results in values and variation patterns as section A. The other three dimensions have the same temperature distribution as dimension 2 shown in Fig. 12. The contour line of temperature distribution is parallel to the surface line. The influencing region of soil surface heat flux is in about 0.5 m of depth. In addition, the consideration of left ground effect on slope toe doesn't change the general temperature distribution of embankment.

In terms of Case 2, what is different with Case 1 is that the temperature gradient changes volumetric water content distribution. However, this effect could be neglected because of the small differences of values they caused. The volumetric water content is influenced by gravity initially and varies slightly, then keep almost stable in the six days when water flux boundary condition is set as zero.

In this study, it is preferred to focus on Hydro-thermal behavior of unsaturated soil. Dimension 4 could be regarded as reliable because that its size is based on mechanical analysis, which is highly sensitive to dimension selection. However, apart from left ground effect on slope toe the volumetric water content results of section B in dimension 1, the results of other three dimensions show that different dimensions will not lead large differences on volumetric water content and temperature distribution in this field embankment. Therefore, the effect of model dimension is not significant for Hydro-thermal coupled unsaturated soil model.

The soil surface temperature and volumetric water content studied are controlled by the interaction between atmosphere and unsaturated soil. The contour lines of volumetric water content and temperature distribution are always parallel to the surface line. Consequently, more attention needs to be paid for the top boundary conditions.

4 Conclusions

In order to study the interaction between atmosphere and embankment, the Hydro-thermal coupled unsaturated soil model is used to calculate the reaction of soil under instantaneous boundary conditions. With abundant meteorological data, the corresponding water flux and heat flux boundary condition could be obtained conveniently relying on the energy and mass balance. With Finite Element Method, a suitable dimension of this embankment needs to be set in advance, which takes the possible influencing region from atmosphere into consideration in the numerical calculation.

In sum, four different dimension models are calculated respectively in two Cases. The first one is applied with water flux boundary condition, considering boundary heat flux as zero. Another one is applied with heat flux boundary condition based on the daily solar radiation values with no water flux exchange. Compared three dimensions with the results of the reliable model dimension (dimension 4), the temperature and volumetric water content values and variation patterns in these dimensions are nearly consistent. The contour lines of volumetric water content and temperature distribution are parallel to the surface line. It can be concluded that the effect of model dimension to field embankment is not significant for Hydro-thermal coupled unsaturated soil research. In dimension 1, the effect of left ground on embankment slope toe is not fully considered, which is not advised based on the analysis in this study. Therefore,

DOI: 10.1051/e3sconf/20160908017

with consideration of less demanding of computational time and memory, dimension 2 or 3 are suggested to be adopted in the further study.

References

1. F. Azizi, *Applied Analyses in Geotechnics* (E & FN Spon Press, 2000)
2. C. Rujikiatkamjorn, B.Indraratna, D. T.Bergado, 3D numerical modelling of hexagonal wire mesh reinforced embankment on soft Bangkok silt, Geo Congress, 2263-2272 (2012)
3. Z.R. Yue, Thermal regime simulation of embankment and berm in permafrost, Geomechanics and Geoengineering, **3** (2), 97-105 (2008)
4. A. Kalliainen, P. Kolisoja, A. Nurmikolu, Modeling of the effect of embankment dimensions on the mechanical behavior of railway track, Joint Rail Conference **1**, 389-398 (2010)
5. D. A. DeVries, Thermal Properties of Soils, *Physics of Plant Environment* (North-Holland Publishing Co., Amsterdam, 210-235, 1963)

 DOI: 10.1051/e3sconf/20160908018

Analysis of passive earth thrust in an unsaturated sandy soil using discontinuity layout optimization

Bestun Shwan[a]

Department of Civil & Structural Engineering, University of Sheffield, Mappin Street, Sheffield S1 3JD, UK.

Abstract. This paper addresses a numerical study into passive earth pressure in an unsaturated sandy soil. The computational limit analysis method, discontinuity layout optimization (DLO), is extended to take into consideration the effect of saturation and suction on strength. The extended analysis was utilised to model a retaining wall case study using a sandy soil as a simulated backfill material and the results were compared with Rankine equations which were modified to take into account the capillary rise effect. The numerical results demonstrated that an increase of the total passive thrust up to 47%, for a frictionless wall, at $\phi' = 30°$ due to the effect of partial saturation in the sandy soil.

1 Introduction

Many historical geotechnical structures such as railway embankments and cuttings were designed before the modern science of soil mechanics was developed. In many cases these structures only stand up due to the strength imparted to them through partial saturation and the effects of surface tension (water suction) acting in the soil pores which holds the soil particles together.

With improved understanding of unsaturated soil mechanics in such conditions, it may be possible to utilise the additional strength due to partial saturation in conventional design e.g. with engineered controls on the saturation, under a risk based framework, in temporary works, or in assessing cumulative cyclic loading effects through the seasons.

Significant efforts during the last two decades have been focused on the field of unsaturated soil mechanics and this has led to the formulation of several constitutive models (for example, [1-4]).

In contrast, the application of unsaturated soil mechanics theory to geotechnical design is much more limited (e.g. [5] and [6]).

Such studies are limited by the nature of what can be dealt with using hand calculations. Recent developments in computational limit analysis (CLA) (e.g. [7]) have extended the scope of such analytical methods, so that they can deal with any geometry and loading configuration, and have been applied in many areas to ultimate limit state (ULS) design.

There is a significant scope to extend CLA to include unsaturated soil behaviour, providing a tool that can find the collapse load for a wide range of problems such as retaining walls or foundation stability without the simplifications inherent in hand calculations or the complexity of the elasto-plastic finite element (FE) method.

The aim of this paper is therefore to extend the application of the CLA method, discontinuity layout optimization (DLO), to model the effects of partial saturation on the passive earth pressure exerted by an unsaturated sandy soil.

2 Theory

2.1 Discontinuity layout optimization (DLO)

The DLO procedure is a computational limit analysis method that directly identifies the collapse load for stability problems. The procedure allows either the determination of load or strength for any stability problem. The concept of DLO, which is based on the upper bound theorem of plasticity proposed by [7], is built on determining the critical failure mechanism that results in the least amount of energy dissipation.

The basic principle of the DLO is based on recognizing a critical layout of lines of discontinuity to create failure mechanism. These lines are the slip-boundaries between the rigid blocks due to the applied loads. A wide range of different failure mechanisms can be produced by utilising high numbers of nodes, then discretizing these nodes by connecting them each to another as shown in Fig. 1.

[a] bjnshwan@gmail.com

2.2 Energy dissipation and work equation in DLO

Following [7] and [8] in the presence of water, the rate of internal energy dissipation and work done against body forces for the problem of a Mohr-Coulomb material with self-weight γ, cohesion c and angle of friction ϕ collapsing as a set of sliding blocks where each discontinuity (or interface) i between adjacent sliding blocks has relative shear and normal displacement jumps of s_i and n_i can be written as:

$$E = \sum_{i=1}^{m}(c_i\, l_i s_i + U_i n_i + W_i s_i \sin\theta_i + W_i n_i \cos\theta_i) \quad (1)$$

where m is the number of interfaces and U_i, and W_i are respectively the pore water force on, and weight of the strip of soil above interface i and l_i, θ_i are the length of interface i and the angle of interface i to the horizontal. For a limit analysis approach, $n_i = |s_i|\tan\phi'$.

where W is weight of the soil above a discontinuity (see Fig.2).

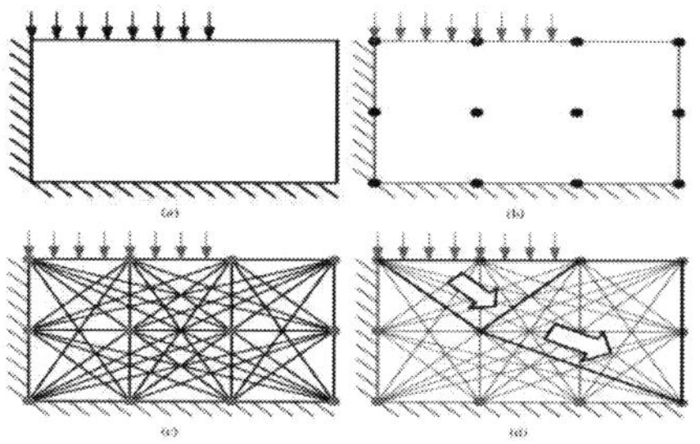

Figure 1. DLO stages procedure, after [8].

2.3 Incorporation of suctions into DLO

Stanier and Tarantino [6] proposed equations for shear strength, for compacted aggregated soils in partially saturated conditions, and for suction as follows:

$$\tau = (\sigma + s\ S_r)\tan\phi' \quad (2)$$

$$s = \gamma_w (H_w - z) \quad (3)$$

where τ is the shear strength (kPa), σ is the total normal stress (kPa), s is the suction (kPa), S_r is the degree of saturation %, γ_w is the unit weight of the water (kN/m^3), H_w is water table depth (positive downward) (m), z is vertical coordinate (positive downward) (m). Equation (3) assumes full water continuity within the soil.

Shwan and Smith [10] proposed equations for S_r as follows:

$$S_r = 1.0 \qquad s \le s_o \quad (4)$$

$$S_r = e^{-a(s - s_o)} \qquad s > s_o \quad (5)$$

where a is a fitting parameter (kPa^{-1}), s_o is the air entry value of the soil (kPa) and it is related to the height of the capillary rise (full saturation) H_c as follows:

$$H_c = \frac{s_o}{\gamma_w} \quad (6)$$

In order to compute the value of U in Eq. (1) for use in the DLO formulation in a partially saturated soil, the following integration is required:

$$U = \int_0^L s S_r .dl \quad (7)$$

where L is length of discontinuity as shown for example in Fig. 2.

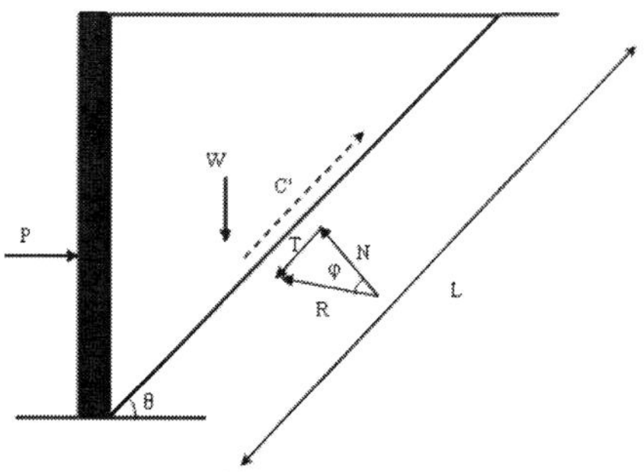

Figure 2. Geometry and features of the total passive earth pressure problem.

Figure 2 shows a simple Coulomb wedge analysis. By equating the work done and the internal energy dissipated, the total passive thrust (P_p) in Fig. 2 can be determined as follows:

$$P_p\ \delta - W\,\delta\tan(\theta + \phi') = c'\,L\cos(\phi')\frac{\delta}{\cos(\theta + \phi')} \quad (8)$$

where δ is the block displacement, W is the weight above the discontinuity (kN) and c' is the cohesion (kPa).

3 Total passive earth pressure analysis

3.1 Case study

A non-dimensional analysis of the total passive earth pressure for a wall with a levelled backfill is modelled. The boundaries of the backfill are 1m height (height of the wall) and 6.2 m length. Three types of walls are modelled as frictionless, frictional wall with $\delta = (2/3)\ \phi'$ and fully frictional wall $\delta = \phi'$ and they are named throughout the context of this paper as series FL, 0.67FW and FW, respectively. For series 0.67FW and FW, the depth of the soil is extended 0.8 m below the base of the wall to prevent restriction of the failure mechanism by the bottom boundary (see Figs. 5b and c).

In this analysis, the soil is assumed to be fully saturated below the capillary rise height (determined using Eq. 6), while the average unit weight, between dry and saturated unit weight, is utilised above the capillary rise height. Water table height (Y_w) is varied from 1 m (fully saturated) to -3 m below the base of the wall. This represents a degree of saturation range from 100% to ≈ 0% and suction from 0 to 39.2 kPa for the sandy soil used according to its soil water characteristic curve SWCC (see Fig. 3).

The soil properties are shown in Table 1 for the sandy soil. Dry and saturated unit weights are assumed as 1.5 and 1.9 of the unit weight of water. This is to keep the effect of the unit weight constant while studying the effect of the degree of saturation. A range of internal friction angle values from 30° to 45° is utilised for this parametric study.

Parameter a and s_o in Table 1 are obtained using best fit (using Eqs. 4 and 5) for the actual SWCC obtained from the work of [11] as shown in Fig. 3.

Table 1. Soil properties and unsaturated parameters of the sandy soil.

	Material	Sandy Soil
Soil properties	c' (kPa)	0
	γ_w (kN/m^3)	9.81
	γ_{sat} (kN/m^3)	18.64
	γ_{dry} (kN/m^3)	14.72
	$\gamma_{average}$ (kN/m^3)	16.68
Unsaturated parameters	a (kPa^{-1})	0.1
	s_o (kPa)	5
	H_c (m)	0.509

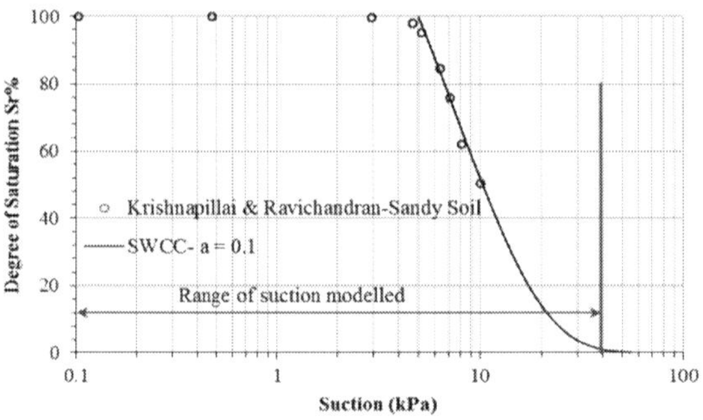

Figure 3. SWCC for sandy soil (after [11]) with the best fit using Eqs. 4 and 5 and a range of the suctions modelled.

3.2 Results

Figures 4a, b and c give an example of the design chart in which the x-axis represents normalised total passive thrust ($P_P/H^2\gamma_w$) and the y-axis represents normalised water table height ($Y_w/(s_o\ a\ H)$) for a range of internal friction angles for FL, 0.67FW and FW series. The results are compared with the Rankine method which takes into account capillary rise height.

The total passive earth thrust equations for a frictionless wall, with a levelled backfill, using the Rankine method for the fully saturated and a case when water table is below the soil surface and capillary rise occurs to the surface are given as follows:

$$P_p = \frac{1}{2}k_p\gamma' H^2 + \frac{1}{2}\gamma_w Y_w^2 \quad (9)$$

$$P_p = (k_p - 1)\times\gamma_w H_c^2 + \frac{1}{2}H_c[k_p\gamma_{sat}H_c - (k_p-1)\gamma_w H_c] + k_p\gamma_{sat}H_cY_w + \frac{1}{2}k_p\gamma' H_w^2 + \frac{1}{2}\gamma_w Y_w^2 \quad (10)$$

where k_p is passive earth pressure coefficient based on Fig. 4d for all the series, γ' is buoyant unit weight (kN/m^3), H is height of the wall (m), γ_w is unit weight of water (kN/m^3), Yw is water table height (m) (positive upwards), H_c is distance from the water table to the capillary rise line and γ_{sat} is saturated unit weight (kN/m^3). The derivation of Eq. 10 is based on full continuity of water between the soil and the wall. A capillary rise height H_c = 0.509 m is used shown in Table 1 based on Eq. 6.

It can be seen from Figs. 4a, b and c that an increase of about 46.8%, 54.2% and 54.2% can be obtained in ($P_P/H^2\gamma_w$) for the case of ϕ' = 30° at Y_w = - 0.6m (corresponding to $Y_w/(s_o\ a\ H)$ = -1.2 in Figs. 4a, b and c) for the series FL, 0.67FW and FW when compared to their counterpart total passive thrust result using the Rankine method. The height Y_w = - 0.6 m corresponds to a maximum normalized Pp value, to a hydrostatic suction of 15.696 kPa (1.6×9.81) and degree of saturation of about 34% (see Fig. 3).

At depth Y_w =-3 m ($Y_w/(s_o\ a\ H)$ = -6 in the Figs. 4a, b and c), the ($P_P/H^2\gamma_w$) for the unsaturated case at ϕ' = 30° is closer to the result of the Rankine method. No further drop of the water table beyond 3 m below the base of the wall is carried out as the overall trend of the curves is clear.

Sharper curves can be seen for the higher ϕ' values at the same Y_w. For example, for series FL (Fig. 4a) at $Y_w/(s_o\ a\ H)$ = -1.2 for ϕ' = 45°, higher total passive thrust can be seen when compared to the case of ϕ' = 40° at the same Y_w. This increase is inherently due to the effect of ϕ' in the term (sS_r) tan ϕ' (see Eq. 2).

Figures 5a, b and c show the failure mechanisms obtained by the modified DLO method for the FL, 0.67FW and FW series at Y_w = - 0.6 m and ϕ' = 45°. The FW series reveals a wider and deeper failure mechanism compared to the other two cases due to the effect of the friction of the wall.

4 Conclusions

A parametric study on the total passive earth thrust analysis using a simulated sandy soil backfill material was investigated. The following conclusions were demonstrated:

1. A theoretical extension to the Discontinuity Layout Optimization (DLO) procedure to allow the modelling of partially saturated soils has been described, and includes the combined effects of suction and saturation.
2. The influence of partial saturation on the passive earth problem was investigated by utilizing a range of suction profiles and ϕ' values. A non-linear relationship for the total passive earth pressure with the water table depth was determined with total passive pressure initially increasing with increased suction and then followed by a reduction as the saturation of the soil fell.

DOI: 10.1051/e3sconf/20160908018

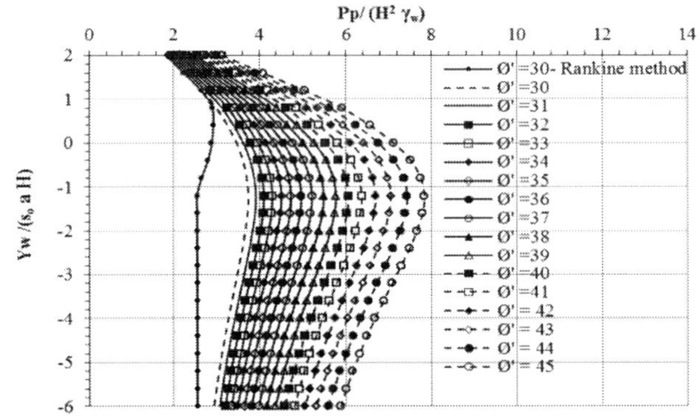

(a)

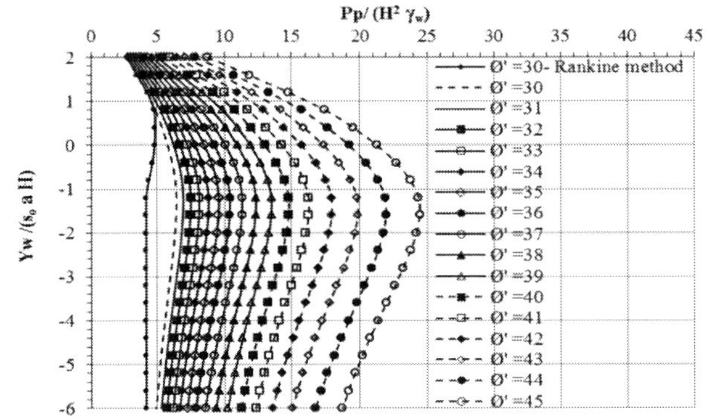

(b)

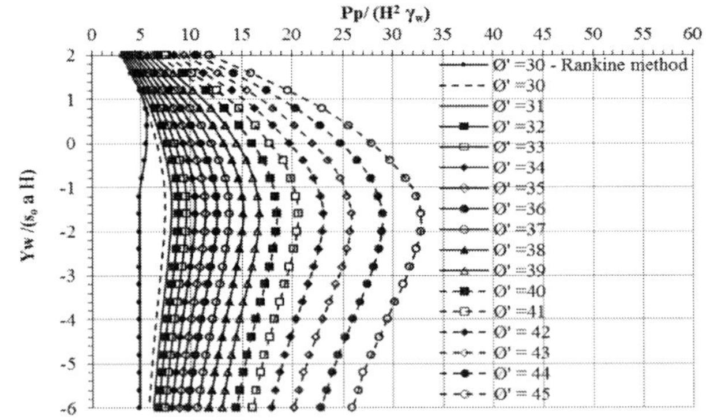

(c)

 DOI: 10.1051/e3sconf/20160908018

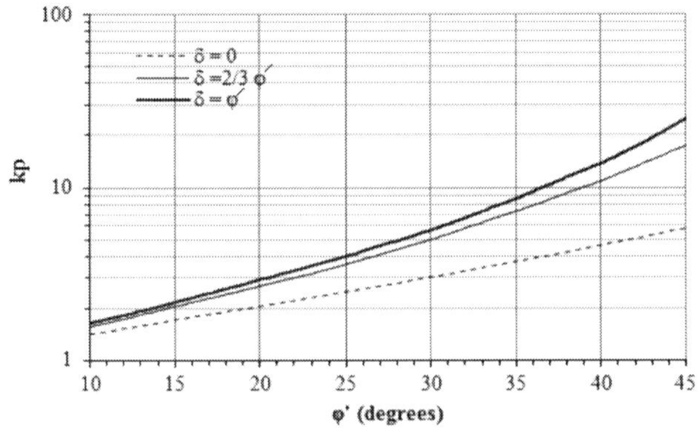

(d)

Figure 4. Normalised ($P_P/H^2 \gamma_w$) versus normalised $Y_w/(s_o\ a\ H)$ at a range of the internal friction angle values for (a) FL (b) 0.67FW (c) FW (d) Passive earth pressure coefficient for various design values of ϕ' for all three series FL, 0.67FW and FW based on [12].

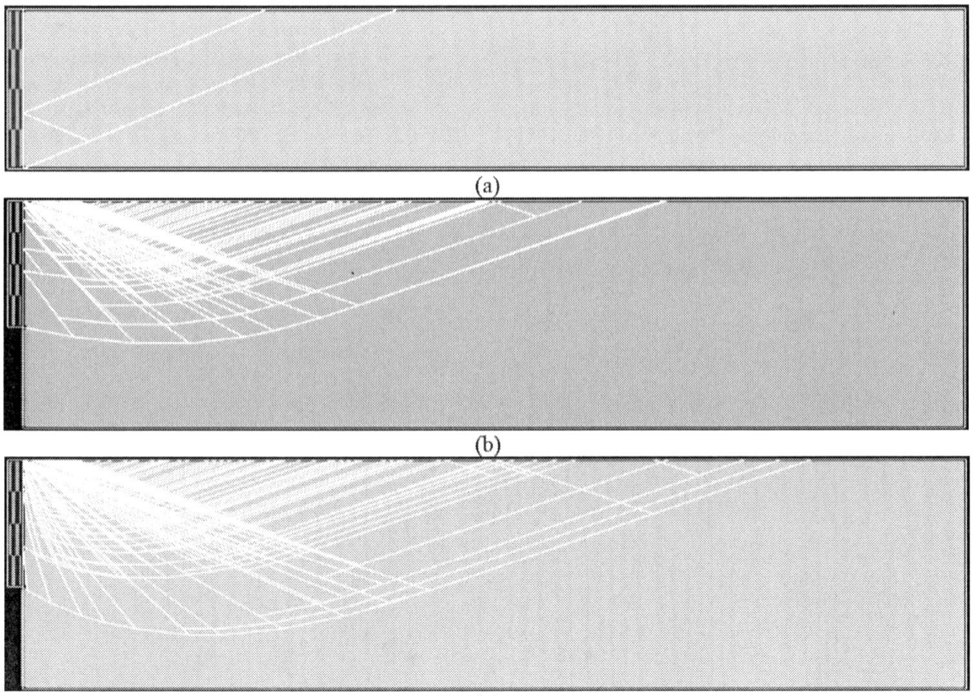

(a)

(b)

(c)

Figure 5. Failure mechanism obtained by the DLO method at Y_w = - 0.6 m, ϕ' = 45° for (a) FL (b) 0.67FW (c) FW.

Acknowledgments

The author wishes to acknowledge the financial support which was provided by the Ministry of Higher Education-Kurdistan (Iraq) for this work.

References

1. E.E. Alonso, A. Gens, A. Josa, *A constitutive model for partially saturated soils*. Ge'ot. J. **3,** pp 405-430 (1990).
2. S.J. Wheeler, *An alternative framework for unsaturated soil behaviour*. Ge'ot. J. **41**(2), pp 257–261, (1991).
3. S.J. Wheeler, V. Sivakumar, *An ealso-plastic critical state framework for unsaturated soil*. Ge'ot. J. **45**(1), 35–53 (1995).
4. A. Tarantino, *A possible critical state framework for unsaturated compacted clay*. Ge'ot. J. **57,** pp 385–389, (2007).
5. L.H. Zhao, Q. Luo, L. Li, F. Yang, X.L. Yang, *The upper bound calculation of passive earth pressure based on shear strength theory of unsaturated soil*. Proc. of Slope Stability, Retaining walls and Foundatios, ASCE, pp 151-157, (2009).
6. S. Stanier, A. Tarantino, *Active earth force in 'cohesionless' unsaturated soils using bound theorems of plasticity*, In: Alonso, E.E., Gens, A. (Eds.), Proc. of 5th Int. Conf. on Unsat. Soils, CRC Press, Barcelona, Spain, pp 1081–1086, (2010).
7. C. C. Smith, M. Gilbert, *Application of discontinuity layout optimization to plane plasticity problems*. Proc. R. Soc., A 463, pp 2461–2484, (2007a).
8. C. C. Smith, M. Gilbert, *New upper bound solutions for layered soil bearing capacity problems using discontinuity layout optimazation*. In: Proc. 10th Australia New Zealand Conf. Geo. Mechs., Brisbane, (2007b).
9. M. Gilbert, C. Smith, I. Haslam, T. Pritchard, *Application of discontinuity layout optimization to geotechnical limit analysis problems*. In: Proc. 7th Eur. Conf. Num. Meths. Geo. Eng., Trondheim, Norway, pp 169-174, (2010).
10. B. J. Shwan, C.C. Smith, *Application of limit analysis in unsaturated soils: numerical and experimental study of bearing capacity*. In Unsat. Soils: Res. and Appl. - Proc. 6th Int. Con. Unsat. Soils, Sydney, Australia, pp 1757–1762, (2014).
11. S. H. Krishnapillai, N. Ravichandran, *New soil-water characteristic curve and Its performance in the finite-element simulation of unsaturated soils*. Int. J. Geo., **12,** pp 209-219, (2012).
12. B. Standard, Eurocode 7: *Geotechnical design,* (2004).

Modelling large deformation problems in unsaturated soils

Alba Yerro [1], Eduardo Alonso[2] and Núria Pinyol[2,3]

[1] *Department of Engineering, University of Cambridge, Cambridge CB2 1PZ, UK*
[2] *Department of Geotechnical Engineering and Geo-Sciences (ETCG), UPC, 08034 Barcelona, Spain*
[3] *Centre Internacional de Metodes Numerics en Enginyeria (CIMNE), UPC, 08034 Barcelona, Spain*

Abstract. Applications of unsaturated soil mechanics often involve large deformations and displacements. This is the case of collapse behaviour of low density soils or the unrestrained swelling of expansive clays. Rain induced instability of unsaturated slopes is a further example of large displacements that cause important damages around the world every year. Since standard lagrangian Finite Element methods are not well suited to model large deformations, particle-based methods are under development. This is the case of the Material Point Method (MPM), which offers an interesting alternative. Recently, the MPM has been extended to model unsaturated soil problems, where the soil is understood as a unique medium integrated by three distinct phases (solid, liquid and gas). In this paper, the MPM computational cycle for unsaturated soils is described. In addition, a validation of the 3-phase MPM approach is presented by means of the modelling of a one-dimensional infiltration problem. Finally, the applicability of MPM to solve slope instabilities is presented. The simulation of the whole instability process of an embankment subjected to rain infiltration is analysed in detail.

1 Introduction

Many applications associated with unsaturated soils involve large deformations in history-dependent constitutive models. This is the case of landslides events and slope instabilities which are often triggered by heavy rainfall. They cause important damage around the world every year. Swelling problems in expansive clays or collapse behaviour of low density materials are other examples of problems involving large deformations.

Since standard lagrangian Finite Element methods are not well suited to model large deformations due to limitations on mesh tangling, particle-based methods are under development. This is the case of the Material Point Method (MPM) [1], which offers an interesting alternative. MPM discretizes the media into a set of lagrangian material points which move attached to the material and carry all the soil properties. The governing equations are solved incrementally at the nodes of a computational grid that remains fixed throughout the calculation. This dual description of the media prevents mesh distortion problems hence re-meshing techniques are not required.

Recently, Yerro et al. [2] extended MPM to model problems in unsaturated soils by means a coupled formulation that considers 3 phases (solid, liquid and gas) within each material point. In this work, the validation of such formulation in an infiltration problem is presented. Finally, the whole instability process (from failure initiation to final stabilisation) of an embankment slope subjected to rain infiltration will be analysed.

[a] Corresponding author: eduardo.alonso@upc.edu

2 Basis of MPM

Material Point Methods (MPM) are rapidly evolving in the geotechnical field. This is due to their capability to analyse large deformation problems in a continuum framework.

Inspired by particle methods, MPM [1] discretizes the media by means of a set of material points. Each point moves attached to the portion of continuum which carries all the material information such as strain, stress, and mass. The main governing equations are solved at the nodes of a computational mesh that covers the whole domain and it is maintained fixed throughout the calculation. Domain integration is very similar to the operational rules used in FEM but integration points are now the material points instead of the familiar Gauss points in FEM. Linear shape functions are typically used in order to transfer information between both spatial discretizations (nodes and material points).

In this manner, most of the expertise learned from FEM can be extended to the MPM with the additional improvement that large deformations can be simulated without the requirement of re-meshing algorithms.

MPM was initially developed for fluid mechanics [3]. However, the necessity of solving hydro-mechanical problems, such as undrained loading, consolidation or changes in water table, has prompted several authors to extend the method to solve multiphase problems. Two different approaches can be distinguished to solve the interaction between solid skeleton and fluids within a porous media or even with free water. The first one is a "single-layer" strategy which is based on representing the porous medium as a unique continuum by means of one set of material points that moves attached to the solid skeleton. It has been adopted by several authors to solve geotechnical problems under saturated conditions (two-phases) [4-6]. It has also been extended to unsaturated problems in [2] (three-phases). The information of pore fluids is carried as an internal variable at each material point. This approach is appropriate to simulate seepage problems but it is not suitable to model free liquid water. A second strategy was developed for modelling fully saturated soils [7,8]. It is a "multi-layer" approach in which each constituent -grains and water in saturated soils- is described separately by means of different sets of material points. This MPM formulation has the capability of modelling both water within the pores and free water as a unique continuum which allows the simulation of fluid-structure interactions. However, the number of material points is much higher and the approach requires an additional computational cost.

3 MPM formulation for unsaturated soils

This work is based on the MPM approach presented by Yerro et al. [2]. Here the soil, understood as a unique continuum, is a mixture of three distinct phases: solid (s), liquid (l) and gas (g) (Figure 1). In order to facilitate computations, only one component is considered within each phase (for instance, no air in liquid and no water in gas).

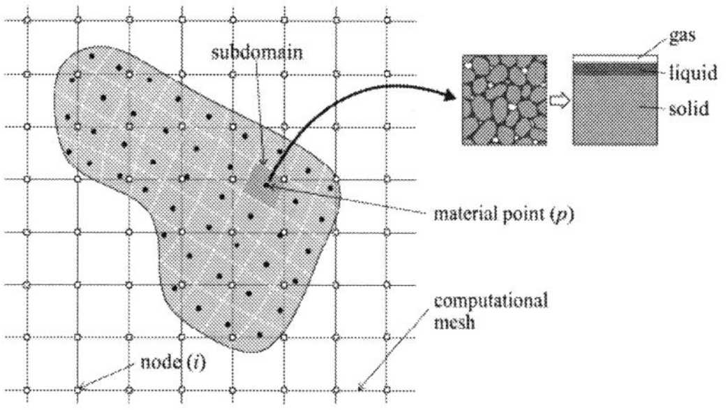

Figure 1. Scheme of the MPM discretisation for unsaturated soils.

The main governing equations are the momentum balances of the gas, the liquid and the mixture. These are integrated into the domain and discretised at the nodes of the computational mesh. At the beginning of each time step, information carried by the material points is mapped to the mesh in order to calculate nodal mass, nodal velocities, internal and external forces and dragging terms. Because it is a fully dynamic formulation, it leads to a system of equations in which liquid, gas and solid nodal accelerations are the principal unknowns of the problem.

An additional term can be included in the system in order to reduce dynamic effects and numerical instabilities. This is defined as a damping force proportional to the corresponding unbalanced force (α: proportional factor) and opposite to the phase velocity [2]. In dynamic problems, where the accelerations have an important role in the course of the calculation, this factor should be very small (α = 0-0.05).

Once the nodal accelerations are calculated, an Euler-Cromer scheme is used to update velocities, displacements and strains at the material points by means of shape functions.

Afterwards, in order to ensure mass conservation of each phase, the mass balance equations for the solid, liquid and gas are imposed in the material points. Considering the solid grans incompressible, the mass balance of solid becomes the material derivative of the porosity n:

$$\frac{Dn}{Dt} = (1-n)\nabla \cdot \mathbf{v}_s \tag{1}$$

where t is time and $\mathbf{v}_s$ is solid velocity of a material point.

Taking into account (1) and assuming that liquid and gas pressures (P_l and P_g) are the state variables, liquid and gas mass balances can be written as follows,

$$n\frac{\partial(\rho_l S_l)}{\partial P_l}dP_l + n\frac{\partial(\rho_l S_l)}{\partial P_g}dP_g = \nabla \cdot \left[nS_l\rho_l(\mathbf{v}_l - \mathbf{v}_s)\right] - nS_l\rho_l\nabla \cdot \mathbf{v}_s \tag{2}$$

$$n\frac{\partial(\rho_g(1-S_l))}{\partial P_l}dP_l + n\frac{\partial(\rho_g(1-S_l))}{\partial P_g}dP_g = \nabla \cdot \left[n(1-S_l)\rho_g(\mathbf{v}_g - \mathbf{v}_s)\right] - n(1-S_l)\rho_g\nabla \cdot \mathbf{v}_s \tag{3}$$

$\mathbf{v}_l$ and $\mathbf{v}_g$ being liquid and solid velocities at a material point, S_l the degree of saturation, and ρ_l and ρ_g liquid and gas densities.

Equations (2) and (3) provide the relationships to find liquid and gas pressures rates (dP_l and dP_g).

This MPM approach is formulated in terms of two stress variables: net stress $\bar{\boldsymbol{\sigma}}$ and isotropic suction (s). Writing such variables in the following convenient manner (4) and (5), in which $\boldsymbol{\sigma}$ is the total stress tensor and $\mathbf{I}$ is the identity matrix, the general form of a suitable stress-strain relationship can be written incrementally as follows in equation (6):

$$\bar{\boldsymbol{\sigma}} = \boldsymbol{\sigma} - \max\left\{P_g, P_l\right\}\mathbf{I} \tag{4}$$

$$s = \max\left\{P_g - P_l, 0\right\} \tag{5}$$

$$\Delta\bar{\boldsymbol{\sigma}} = \mathbf{D} \cdot \Delta\boldsymbol{\varepsilon} + \mathbf{h}\Delta s \tag{6}$$

where $\Delta\boldsymbol{\varepsilon}$ is the strain increment vector. $\mathbf{D}$ and $\mathbf{h}$ are, respectively, the tangent matrix and a constitutive vector.

Then, the stress and also other soil properties are updated at the material points; for instance, the degree of saturation, intrinsic permeability and porosity. Finally, the material points carry all the updated information and time is updated.

4 Infiltration problem

The equation that represents the movement of water in unsaturated soils is the Richard's equation [9], which is derived from the water mass balance and the Darcy's equation (quasi-static liquid momentum balance). However, because the nonlinearities of soil hydraulic parameters (for instance, permeability depends on degree of saturation, and degree of saturation depends on fluid

 DOI: 10.1051/e3sconf/20160908019

pressures), it is not easy to obtain an analytical solution to describe the unsaturated flow.

With the aim of validating the hydro-mechanical MPM formulation outlined in the previous section, an analytical solution for the one-dimensional infiltration problem is required. To do that, and according to Alonso and Lloret [10], Richard's equation has been simplified following these assumptions:

- vertical liquid flow
- deformability of the solid skeleton is neglected
- incompressible solid grains
- constant permeability
- constant liquid density
- constant total stress field
- linear water retention curve

Linearizing the retention curve according to (7), a_s being a constant, the mathematical expression which describes the vertical water flow within an unsaturated soil can be written as equation (8):

$$S_l = 1 - a_s\left(P_g - P_l\right) \quad (7)$$

$$\frac{\partial P_l}{\partial t} = C_i \frac{\partial^2 P_l}{\partial z^2} \quad (8)$$

The previous expression is essentially the diffusion equation, where z is the infiltration direction and C_i corresponds to

$$C_i = \frac{k_l}{n\gamma_l a_s} \quad (9)$$

k_l and γ_l being the intrinsic permeability and specific weight of liquid, respectively.

A dimensionless time T can be defined depending on C_i as follows, where h is the infiltration length:

$$T = \frac{C_i t}{h^2} \quad (10)$$

At this stage, an MPM model has been developed in order to compare numerical and analytical solutions. A soil column 1 m high is considered. The material is linear elastic and all properties are summarised in Table 1. Whereas contours are permeable for gas, liquid can only flow vertically. Initially, the sample is unsaturated with a constant suction value (s_0=0.5 MPa). Because the bottom is impervious for liquid, the infiltration length is 1 m. Gravity is neglected and the solid skeleton cannot deform. The wetting process starts when suddenly s=0 kPa (saturation condition) is imposed at the upper boundary. This condition is maintained throughout the calculation.

Table 1. Material properties.

Material property	Value
Porosity [-]	0.3
Solid density [kg/m^3]	2700
Liquid density [kg/m^3]	1000
Young's modulus [MPa]	10
Liquid bulk modulus [MPa]	80
Liquid viscosity [kg/sm]	0.001
Intrinsic permeability liquid [m^2]	$5 \cdot 10^{-11}$
a_s [kPa^{-1}]	0.001

Applying such boundary conditions, the analytical solution that comes out from equation (8) is equivalent to the one-dimensional consolidation problem in saturated media, the well-known Terzaghi's solution [11].

The analytical solution for such problem is shown in Figure 2, in which the suction profile is presented along the depth z of the sample for different times during the infiltration process. It is clear that suction decreases as the sample gets wet.

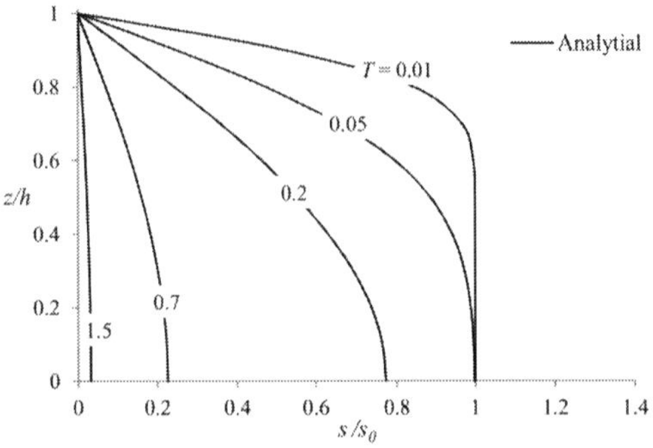

Figure 2. Analytical solution. Suction evolution along depth z.

Different simulations have been carried out in order to analyse the effect of number of material points placed within each element. In Figure 3, the numerical results considering 1 and 4 material points are presented. In both cases, the numerical solution approaches the analytical solution. However, even if the case with 4 points fits perfectly well the analytical solution (Fig.3a), some numerical oscillations are observed when only 1 material point is considered (Fig.3b). Note that in this cases no artificial damping is considered (α = 0.0).

On the other hand, two more calculations are presented in Figure 4, considering 4 points in each element and different damping factors: α = 0.05 (Fig.4a) and α = 0.75 (Fig.4b). See also Figure 5, in which the suction evolution of a material point located at z = 0.49 m is presented for different damping factors (α = 0, α = 0.05, α = 0.75) .

From Figures 4 and 5, it is clear that the inclusion of artificial damping (α) has a direct effect on the infiltration rate. Large values of α lead to slowing the infiltration process. For small values, α = 0 and α = 0.05, numerical results fits the analytical expression (see Fig.3b and Fig.4a).

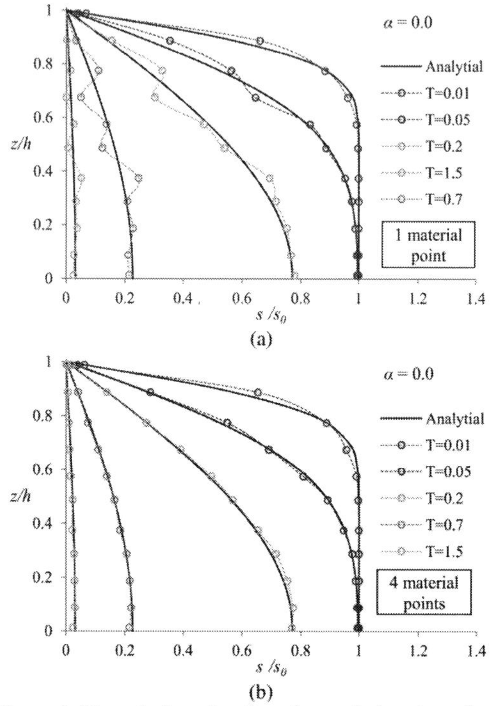

Figure 3. Numerical results of suction evolution along depth z. Damping factor $\alpha = 0$. (a) 1 material point in each element; (b) 4 material points in each element.

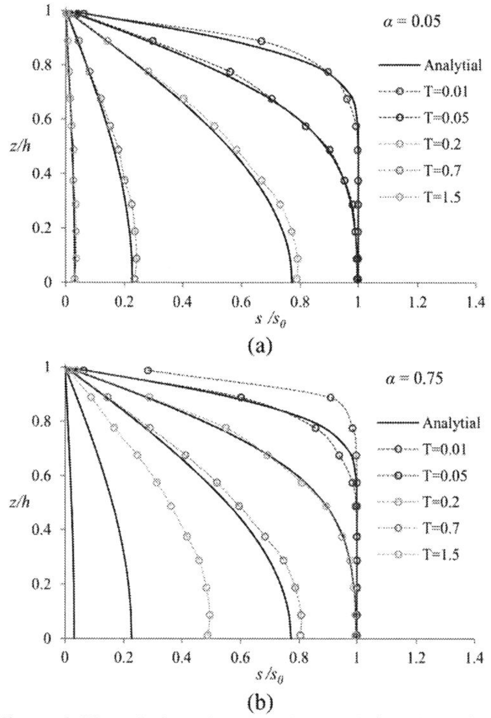

Figure 4. Numerical results of suction evolution along depth z. 4 material points are considered in each element. Comparison between different damping factors: (a) $\alpha = 0.05$; (b) $\alpha = 0.75$.

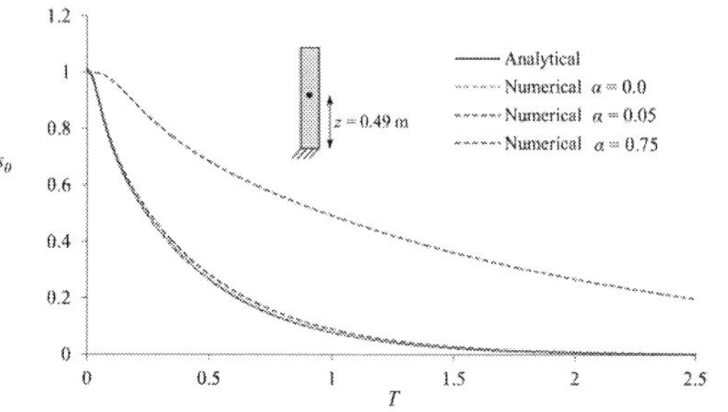

Figure 5. Suction evolution in a material point located at z = 0.49 m. Comparison between analytical solution and numerical results obtained different damping factors and 4 material points per element.

Note that over-damping (α = 0.75) leads to a gross error in the calculated spatial and temporal evolution of suction. A small damping (α = 0.05) is able to stabilize the MPM solution and it reproduces accurately the analytical solution

Finally, two additional calculations have been carried out in order to emphasise the importance of taking adequate assumptions in the mass balance calculations when unsaturated fluxes are modelled. Consider, in particular, the hypothesis of neglecting the spatial variations of water and air masses. Assuming that gradients of liquid and gas masses can be neglected equations (2) and (3) can be rewritten as follows:

$$n\frac{\partial(\rho_l S_l)}{\partial P_l}dP_l + n\frac{\partial(\rho_l S_l)}{\partial P_g}dP_g = -(nS_l\rho_l)\nabla\cdot\mathbf{v}_l - (1-n)S_l\rho_l\nabla\cdot\mathbf{v}_s \quad (11)$$

$$n\frac{\partial(\rho_g(1-S_l))}{\partial P_l}dP_l + n\frac{\partial(\rho_g(1-S_l))}{\partial P_g}dP_g = n(1-S_l)\rho_g\nabla\cdot\mathbf{v}_g - (1-n)(1-S_l)\rho_g\nabla\cdot\mathbf{v}_s \quad (12)$$

The results presented in Figure 6 have been calculated assuming that equations (11) and (12) are correct. In particular Figure 6a shows the results maintaining the same properties considered in Figure 3b. It is clear that for this analysis, numerical solution moves away from the analytical one, especially for T=0.2 and T=0.7, which means that spatial variations of fluid masses are relevant and cannot be neglected.

One of the constitutive equations that play a significant role in the spatial distribution of the fluids masses is the water retention curve. In this example, it is a linear function that depends on a_s. According to equation (7), a small value of a_s implies that small changes in suction correspond to small change of S_l. All parameters used in calculating Figure 6a coincide with values listed in Table 1 (a_s = 0.001 kPa^{-1} in particular), whereas in Figure 6b a_s is a smaller value (a_s=0.0001 kPa^{-1}). Comparing these two plots, it can be concluded that the larger the a_s, the larger the committed error. Therefore, advective fluxes due to spatial variations of fluids masses cannot be neglected in materials such as sand, although such terms will be less relevant in clay.

Another parameter that influences the spatial distribution of fluid mass is the porosity. If materials with very different porosities are in contact, the spatial gradients of fluid mass along the contact cannot be neglected.

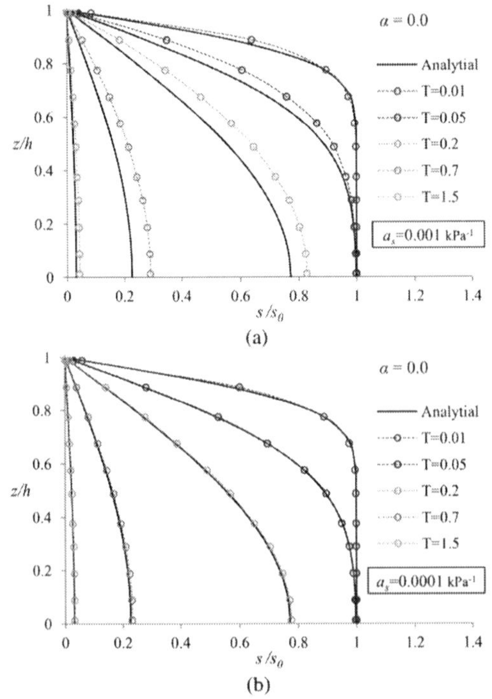

(a)

(b)

Figure 6. Numerical results when advective fluxes due to spatial variations of fluids masses are neglected. Comparison between different a_s values: (a) a_s=0.001 kPa^{-1}; (b) a_s=0.0001 kPa^{-1}.

5 Rainfall effects on an embankment slope

The slope stability problem presented here is inspired by a real case in which several road embankments of medium height (6–8 m) became unstable and shallow failures developed after a heavy rainfall, immediately after the end of construction. The estimated run-out of the slides was 2–4 m. The embankments were built in summertime and the soil, a low to medium plasticity sandy clay, was compacted dry of optimum.

The numerical model presented here is a 7 m high slope with an angle of 32.5°.

The elastoplastic constitutive model presented in [2] is considered in this work to model the effect of suction in the stability of the embankment. The shear strength is written according to the Mohr-Coulomb criterion and strength parameters (c and φ) are be written as follows,

$$c = c' + \Delta c_{\max}\left(1 - e^{-B(s/P_{atm})}\right) \quad (13)$$

$$\varphi = \varphi' + A(s/P_{atm}) \quad (14)$$

First terms (c' and φ') are the values for saturated conditions, which in this analysis are considered c'=1kPa and φ'=20°. Second terms in equations (13) and (14) introduce suction effects and provide an additional strength. It has been accepted that cohesion increases from c' to a maximum value $c'+\Delta c_{max}$. P_{atm} is the atmospheric pressure and B is a constant parameter that controls the rate of apparent cohesion. Although changes in friction angle are typically less relevant, it is considered that it has linear dependence with suction depending on A. In the model Δc_{max}=15 kPa, B=0.7 and A=0.1.

The water retention curve considered in modelling is based on field measurements [12]. Other material properties are summarized in Table 2.

Table 2. Material properties.

Material property	Value
Porosity [-]	0.35
Solid density [kg/m^3]	2700
Liquid density [kg/m^3]	1000
Gas density [kg/m^3]	1
Liquid bulk modulus [MPa]	100
Gas bulk modulus [kPa]	10
Liquid viscosity [kg/sm]	10^{-3}
Intrinsic permeability liquid [m^2]	10^{-10}
Poisson ratio	0.33
Young's modulus [MPa]	10

The computational mesh is formed by tetrahedrons. The initial distribution of material points is presented in Figure 7a. A damping factor α = 0.05 is adopted.

The lower boundary is fixed and horizontal displacements along vertical contours are prevented. Lateral and bottom contours are impervious for the liquid phase. A constant zero gas pressure in excess of atmospheric pressure is prescribed in all the boundaries (P_g=0 kPa).

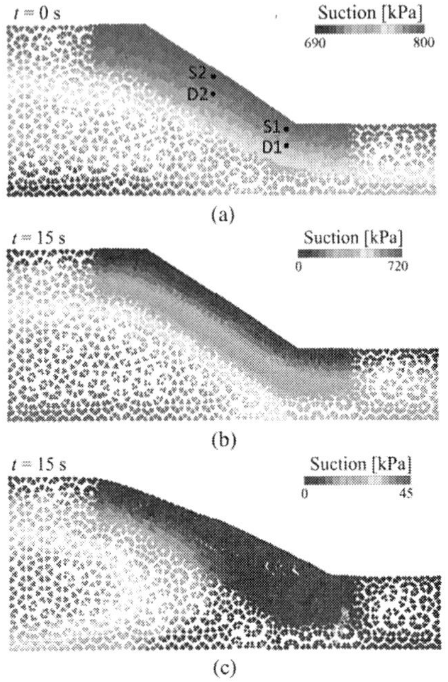

(a)

(b)

(c)

Figure 7. Suction evolution at three different times: (a) t=0 s, (b) t=15 s, (c) t=200 s. The initial location of 4 control material points is indicated (S1, S2, D1, D2).

The initial stresses and pore pressures of the slope are in equilibrium with the gravity force and the prescribed suction (s_0=800 kPa) distributed along the slope surface, which is in contact with the atmosphere (see initial suction distribution in Fig.7a)

The heavy rainfall is modelled by applying a reduction of suction on the ground surface from 800 to 0 kPa during 10 seconds. Afterwards, the saturated boundary condition is maintained constant during the entire simulation.

The embankment response is presented in Figures 7, 8 and 9 in terms of suction, deviatoric strain and total displacements respectively. Figure 7 shows the evolution of suction as a result of the imposed wetting at 3 different times. In Figure 8 the contours of deviatoric plastic strain at times t = 15s and t = 200s are plotted. High shear strains begin to develop at the slope toe soon after the beginning of wetting due to a strength softening. A shear band defining a potential shallow failure surface at an average depth of 1.5 m is already defined at t = 15s (Fig.8a), although the slope remains stable. A few seconds later, the slope becomes unstable, large deformations are involved and a failure mechanism is well defined.

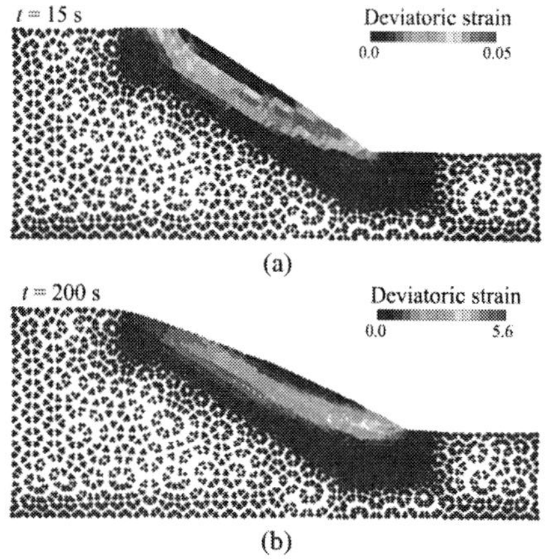

Figure 8. Deviatoric strain profiles at different times: (a) failure initiation t=15 s; (b) t=200 s.

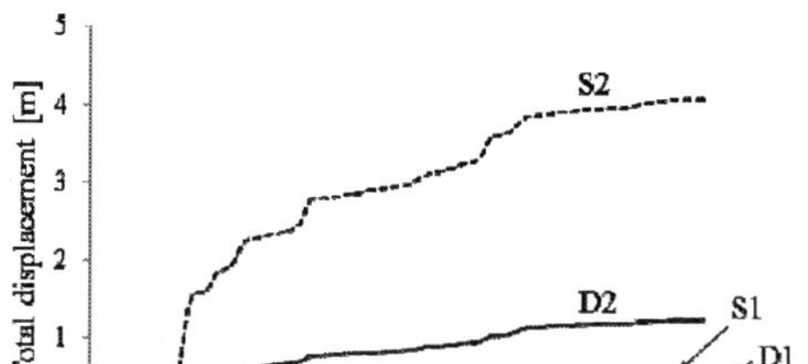

Figure 9. Total displacements of material points S1, S2, D1, D2.

In Figure 9 the total displacements of 4 material points (indicated in Fig.7a) are presented. Material points located in the central part of the slope (S2, D2) slide down and pass over the material points located in the slope toe (S1) which experiences small displacements. The lowest point, D2, remains motionless because it is located below the shear band.

The final run-out, defined as the distance between the initial toe of the slope and the toe of the final geometry, is 2.5 m.

6 Conclusions

A three-phase formulation (solid, liquid and gas) of the MPM is described in the paper. Material points are assumed to carry all the necessary information for the three phases

The MPM formulation for unsaturated soils presented in [2] has been validated by modelling an infiltration problem. A complete analysis has been carried out and the effect of some numerical aspects (number of material points, artificial damping term) and modelling assumptions was discussed.

In the second part of the paper an embankment slope instability induced by heavy rains has been simulated. An elastoplastic suction dependent Mohr-Coulomb model formulated in terms of two stress fields (net stress and suction) has been used in order to model strength variations due to suction changes. Suction decrease in the slope results in a marked strength softening. The result is a complex motion, which simulates observations in rainfall-induced instabilities.

Other large deformation problems, such as wetting induced collapse or swelling may be analysed by the same approach but they will require the consideration of a different constitutive model. However, the general 3-phase MPM algorithm will remain unchanged.

References

1. D. Sulsky, Z. Chen, H.L. Schreyer. Comput. Methods Appl. Mech. Engrg., **118**, 1–2, 179–196 (1994)
2. A. Yerro, E. Alonso, N. Pinyol. Géotechnique, **65**, 3, 201–217 (2015)
3. D. Sulsky, S.J. Zhou, H.L. Schreyer. Comput. Phys. Comm., **87**, 1–2, 236–252 (1995)
4. H.W. Zhang, K.P. Wang, Z. Chen. Comput. Methods Appl. Mech. Engrg., **198**, 17-20, 1456–1472 (2009)
5. F. Zabala, E.E. Alonso. Géotechnique **61**, 9, 795–808 (2011)
6. I. Jassim, D. Stolle, P. Vermeer. Int. J. Numer. Anal. Meth. Geomech. **37**, 15, 2502–2522 (2013)
7. K. Abe, K. Soga, S. Bandara. J. Geotech. Geoenviron. Eng. **140** ,3, 1–16 (2014)
8. S. Bandara, K. Soga. Comput. Geotech. **63**, 1, 199–214 (2015)
9. L.A. Richards. Physics, **1**, 5, 318–333 (1931)
10. E. Alonso, A. Lloret. *Fourth International Conference on Applications of Statistics and Probability in Soil and Structural Engineering*, P. Editrice, ed., Firenze (Italy) (1983)
11. K. Terzaghi. *Theoretical soil mechanics*. New York: Wiley (1943)
12. E.E.Alonso, N.M. Pinyol, A.M. Puzrin *Geomechanics of Failures. Advanced Topics*, Springer (2010)

Suction-controlled dynamic simple shear apparatus for measurement of dynamic properties of unsaturated soils

Khoa Le[1] and Majid Ghayoomi[1,a]

[1] *University of New Hampshire, Department of Civil and Environmental Engineering, Durham, NH, United States of America*

Abstract. State variables such as confining pressures and the degree of saturation can have a profound effect on the dynamic properties of a soil. These dynamic properties are essential when performing seismic response analysis of geotechnical systems in the phreatic zone. In order to accurately obtain these properties, laboratory testing using a Dynamic Simple Shear system is often implemented. This paper concentrates on the advancements and modifications of a custom-built dynamic simple shear system at the University of New Hampshire to accommodate soils with unsaturated conditions by employing axis translation technique. Tests could be performed under drained (constant suction) or undrained (constant water content) conditions. The procedure for preparing an unsaturated soil sample and testing is discussed, followed by the methods for interpreting the data and the challenges involved. Preliminary data confirms the ability of the system to control and track suction during the cyclic simple shear test. Suction in unsaturated soil increased the shear modulus and decreased the damping ratio comparing with those in dry and saturated conditions.

1 Introduction

The surficial part of the earth is mostly partially saturated except where the water table is at shallow depths. The dynamic response of unsaturated/partially saturated soils is complex and differs from that of dry or saturated soils. Thus, characterizing the dynamic material properties of soils with various degrees of saturation is crucial in seismic analysis of geotechnical systems.

The Dynamic Simple Shear (DSS) apparatus at the University of New Hampshire (UNH) has been recently modified and upgraded for unsaturated soil testing and more accurate data acquisition. A summary of these developments, testing procedure, and experimental results verifying the performance of the system is presented and discussed in this paper.

2 Background

2.1 Dynamic Properties of Soils

The stress-strain dynamic response of soils is non-linear where the induced strain controls the soil stiffness modulus [1-3]. At extremely small strains soils tend to behave elastic while at medium to larger strains, however, soils lose a significant portion of their stiffness. The shear modulus, G, is the ratio between the shear stress and strain imposed to a particular soil element. The small-strain shear modulus, G_{max}, is basically the initial slope of the shear stress-strain curve. Both G and G_{max} have been investigated extensively resulted in several empirical equations. Further, the damping ratio represents the dissipated energy in the system. However, it increases in higher shear strains with a minimum at the small strain range. Accurate estimation of shear modulus and damping for varying induced shear strain levels during an earthquake is critical in better understanding and simulation of seismic response of geotechnical systems.

The reduction of shear modulus by increasing the shear strain is typically presented using normalized shear modulus reduction functions. This modulus reduction function has been improved over the past decade starting from a simple hyperbolic form. The most recent version of this equation, developed based on compilation of numerous dynamic laboratory test results, has been presented by Oztoprak and Bolton [4], as shown in Equation 1.

$$\frac{G}{G_{max}} = \frac{1}{\left[1+\left(\frac{\gamma-\gamma_e}{\gamma_r}\right)^a\right]} \tag{1}$$

where γ_r represents a reference shear strain when G/G_{max} is equal to 0.5. ; γ_e is the elastic threshold strain that signifies when the soil starts to experience stiffness degradation; and a is a curvature parameter [4].

In addition, damping ratio functions were also defined where damping increases proportional to the modulus reduction. Measuring the shear modulus and damping in medium to large strain requires accurate testing and measurement approaches that may differ between the

[a] Corresponding author: majid.ghayoomi@unh.edu

testing apparatus such as cyclic triaxial, cyclic torsional shear, and dynamic simple shear tests.

2.2 Dynamics of Partially Saturated Soils

Both shear modulus and damping mainly depend on the effective stress in the soil. Thus, changes in effective stress due to fluctuation of degree of saturation could lead to different stiffness and damping values. Specifically, increasing the suction as a result of soil desaturation increases the effective stress; according to Bishop's effective stress formula presented in Equation 2.

$$\sigma' = (\sigma - u_a) + \chi(u_a - u_w) \tag{2}$$

This equation is compliant with Terzaghi's classical effective stress equation. The parameter ($\sigma - u_a$) represents the net normal stress and ($u_a - u_w$) represents the matric suction applied to the soil element. The term χ is often called Bishops effective stress parameter; it is a function of the soil elements degree of saturation and typically ranges from 0 (completely dry) to 1 (fully saturated) [5, 6].

In recent years, empirical equations have been developed to correlate the effective stress parameter to different hydraulic properties of soils such as van Genuchten Soil-Water-Retention-Curve parameters [6, 7]. For example, Lu et al. [8] proposed Equation 3 based on suction stress concept in unsaturated soils.

$$\chi = \left\{\frac{1}{1+[\alpha(u_a-u_w)]^n}\right\}^{1-\frac{1}{n}} \tag{3}$$

where α and n are SWRC fitting parameters.

Recently, owing to the advancement of unsaturated soil mechanics, effect of suction on dynamic properties of soils has received more attention. This includes estimating the small strain shear modulus of unsaturated soils using resonant column or bender element tests [e.g. 9-12] and determining the strain-dependent modulus using cyclic triaxial systems [e.g. 13,14]. These studies consistently reported a higher modulus in unsaturated soils. However, less of these work have been on strain-dependent dynamic properties, which require further investigation.

2.3 Dynamic Simple Shear Apparatuses

The development of the Dynamic Simple Shear (DSS) apparatus was introduced by the Swedish Geotechnical Institute in the mid-20th century [15]. The objective of the project was to create a device in which uniform shear strains could be imparted onto soil samples that were subjected to shear loads. Prior to this time, most samples were tested using the standard direct shear box. The advantage of DSS apparatus over the direct shear test has been shown to impart uniform shear strains along the whole sample specimen instead of a single forced plane of shear that is applied in the direct shear test [16, 17].

In recent years, variations of these systems have been developed to produce different field conditions and loading patterns to the soil samples. Developers at UCLA have been able to effectively produce multidirectional horizontal loading using complex PID controls [18]. Additionally, the development of a double specimen DSS apparatus has also been developed to eliminate errors due to potential mechanical inaccuracies [3].

3 Description of the UNH DSS

3.1 Current System

The Dynamic Simple Shear system at the University of New Hampshire was developed in 1992 [19]. The system was initially built to study the small strain behaviour of Holliston 00 sand in relation to thixotropic effects. The system was then further modified in 1998 to effectively produce even smaller strains that could be applied to reconstructed Gulf of Mexico clay samples [20].

The system is comprised of multiple parts including (1) the framework, (2) vertical and horizontal actuators/movement, (3) a control system hardware/software and data acquisition (DAQ) system, and (4) a series of valves, piping, and a flow pump to provide saturation and various suction pressures. The framework of the system is comprised of a steel frame based on top of a steel table. Two sets of Thomson ball slides provide guides for the bottom platen of a soil sample system to be loaded in a horizontal direction, and a top platen to be loaded in the vertical direction. The soil sample cell is based off of the SGI-DSS configuration in which Teflon coated aluminium rings and membrane confine the soil. The soil cell is 10.16 cm (4 inches) in diameter and approximately 2.54 cm (1 inch) in height. The system schematic is shown in its entirety in Figure 1.

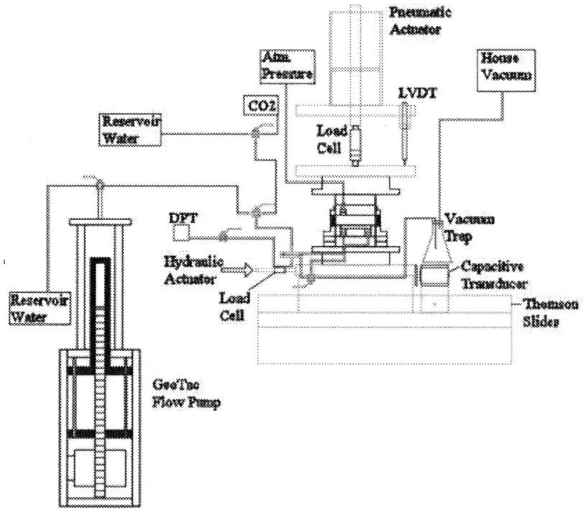

Figure 1. Schematic of the UNH DSS

One of the implications of using the DSS is that the lateral/horizontal confining pressures cannot be readily determined. This becomes critical when determining the mean effective stress that should be applied in the small strain modulus equation. The assumption of using the at-rest (K_0) pressure for the sides is often used to estimate the horizontal effective stress. In theory, this assumption would be correct if the soil were tested in a static condition. However, since the soil sample is subjected to a horizontal cyclic load, the horizontal confining pressures would be in the active or passive condition depending if it is loaded or unloaded, respectively. Additionally, the confining rings cannot provide a coupling vertical shear force along the sides of the soil sample, thus limiting the capability to impart uniform shear forces along the sides of the sample [17].

3.2 Modifications for Unsaturated soil Testing

Figure 2 shows the general setup for most of the modified soil cells for laboratory testing of partially saturated soils. The High Air Entry Value (HAEV) ceramic disk that is often used allows for water to pass through the disk, while prohibiting the flow of air through (past a certain threshold value). The HAEV disk utilized in the UNH-DSS system is rated at ½ bar (50 kPa). A schematic and picture of the modified cell in UNH DSS system is shown in Figure 3.

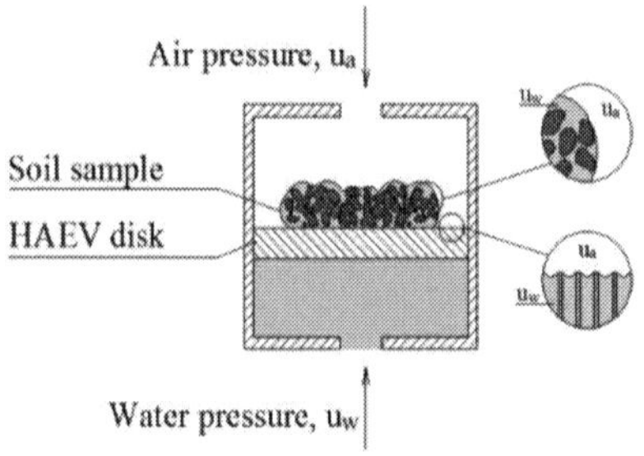

Figure 2. Soil sample in a hypothetical soil cell [6]

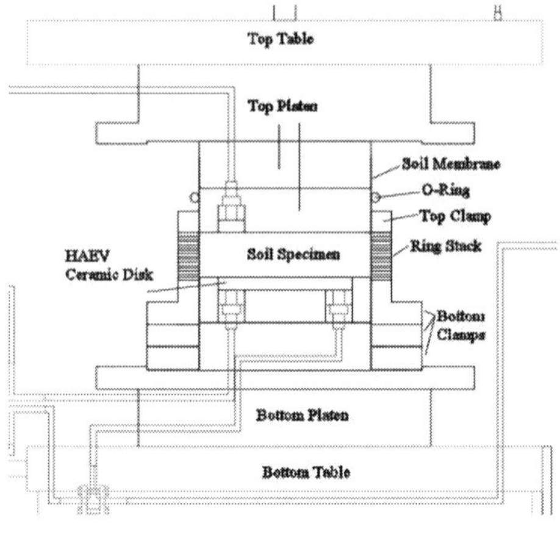

Figure 3. Schematic of modified cell in UNH DSS system.

The axis translation technique was implemented to control the matric suction in a soil sample. In order to use this technique a differential pressure transducer was installed to read the pressure difference between the air at the top of the sample (could be taken as the atmospheric pressure) and the water pressure in the soil sample. The reference axis used for this project was established at the mid height of the sample. As the water level was lowered in the specimen through the use of a Geo-Tac flow pump, the matric suction in the sample was increased. It is critical to ensure that when the matric suction is measured that the sample is at a steady state condition and that the flow of water is nearly non-existent.

4 Testing program

4.1. Testing material

F-75 Ottawa sand, a fine, uniform poorly graded (SP), silica sand was used in this study. The grain size distribution curve was created by performing sieve analysis and is shown in Figure 4. A summary of the geotechnical properties of the tested sand is shown in Table 1 [21, 22].

Prior to running dynamic tests SWRC were determined using the axis translation set up in the system both on wetting and drying paths, as shown in Figure 5. The results are consistent with previously reported SWRC for the same sand [23], thus verifying the success of the system in suction control.

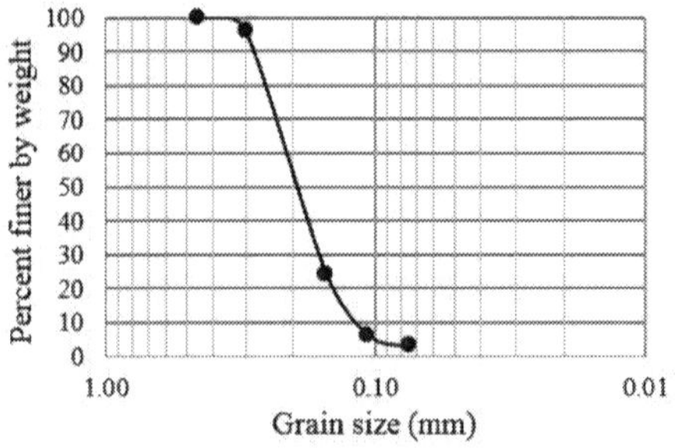

Figure 4. Grain size distribution of the tested material

Table 1. Properties of F75 – Ottawa Sand

Property	Value
Specific Gravity	2.65
e_{min}, e_{max}	0.805, 0.486
ρ_{min}, ρ_{max} (kg/m^3)	1468, 1781
Coefficient of Uniformity	1.83
Coefficient of Curvature	1.09
Specimen Relative Density	45%
Friction Angle (°)	40
van Genuchten's fitting parameters:	
Alpha	0.25
N	8
Residual Water Content (%)	7.15
Saturated Water Content (%)	38.88

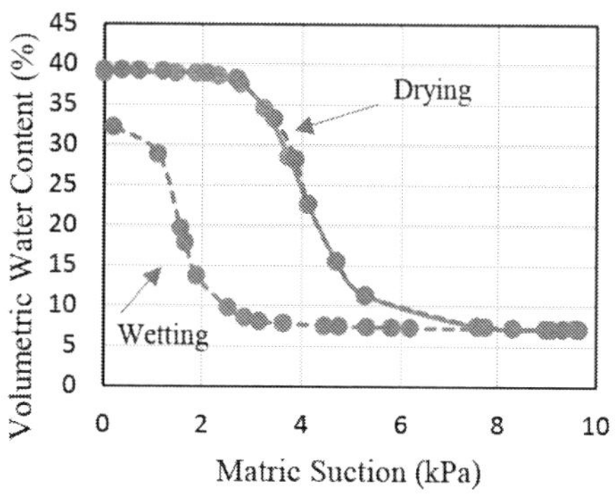

Figure 5. Soil Water Characteristic Curve for F-75 Sand

4.2 Sample Preparation

The soil sample was constructed on top of a bottom platen. Various rings were attached to the platen and were used to clamp a soil membrane around the base of the soil sample. Vacuum grease and O-rings were attached to the membrane to create a seal. A stack of Teflon rings were then inserted onto the sample and made flush with the base. A vacuum mold was installed onto the platen and the membrane was stretched over the top of the mold. A picture of the assembled bottom system before sample preparation is shown in Figure 6.

Figure 6. Bottom platen of cell with assembled vacuum mold

A piece of filter paper was then inserted inside the membrane. Ottawa sand was then dry pluviated into the mold to the desired density. It is then levelled off and another piece of filter paper was inserted on top of the sand. A regular porous disk was placed on top of the filter paper and guiding rods were then installed around the edges of the bottom platen. The top platen was then slowly lowered onto the guide rods and on top of the porous disk. An annular clamp was then installed to secure the membrane to the top platen. The vacuum mold was then removed. The soil sample was inserted into the DSS system and clamped into place using T-Clamps that were embedded into the top and bottom platens of the soil chamber. The guide rods were then removed at the end. A picture of fully assembled DSS system ready for testing is shown in Figure 7.

Figure 7. Assembled soil sample chamber in UNH-DSS

4.3 Testing procedure

After preparing the specimen, the sample was vertically confined using a pneumatic actuator and regulator. The vertical load and displacement were manually measured and recorded. The sample was then fully saturated by connecting the sample to a reservoir tank and letting the water flow slowly through the base of the sample and up through the top of the soil sample. After the soil was saturated target matric suction was applied and controlled using the flow pump, which extracts water from the sample. The differential pressure transducer provides an indication of where the water/suction level is compared to the reference axis that was previously mentioned.

After the target matric suction level is reached and considered to be at a steady state condition, the horizontal cyclic motion was applied to the sample using the hydraulic actuator. Tests were performed under drained (constant suction) condition where the pump maintains the suction on the specimen.

In order to provide the cyclic motion, the hydraulic actuator requires a few preliminary steps to function properly. Since the capacitive transducer provides feedback for the PID loop, the initial position of the sensor was placed approximately .025 inches from a target sensor that is mounted on the bottom table. The hydraulic pump was turned on and the control program was activated. The motion parameters (i.e. number of cycles, amplitude, and frequency) for the cyclic motion were inputted into the program. The program was then executed and the displacements and forced were measured and recorded using the displacement transducers and load cells.

4.4 Test program

Although the majority of the system controls have been established and calibrated, the UNH DSS is still in the process of fine tuning the apparatus to provide reasonable results. Three tests were conducted consecutively on a prepared F-75 Ottawa sand and were tested at a strain level of 0.032%. A completely dry, fully saturated, and four partially saturated samples were prepared with matric suctions of 4, 6, 8, and 10 kPa.

A vertical confining pressure of 50 kPa (7.25 psi) was exerted onto the sample. This would correspond with a soil element that would approximately be 3.2 meters

DOI: 10.1051/e3sconf/20160909001

(10.5 feet) deep (at a relative density of 45%). Each sample was also subjected to 5 cycles of horizontal cyclic loading at a frequency of 1 Hz.

5 Results and Data Analysis

The data that was recorded by the Data Acquisition System (DAQ) was in terms of electrical voltages. Proper calibration techniques were utilized to ensure that the correct stress and strain values were obtained when converting the electrical signals to actual measurements. The data were then corrected to account for the initial stresses and strains that were in the system prior to cyclic horizontal loading. The normal and shear stresses and strains were estimated from the measured displacements and loads throughout the cyclic loading.

As a result, shear stress–shear strain hysteresis loops, similar to the one shown in Figure 8, were obtained from the analysed data.

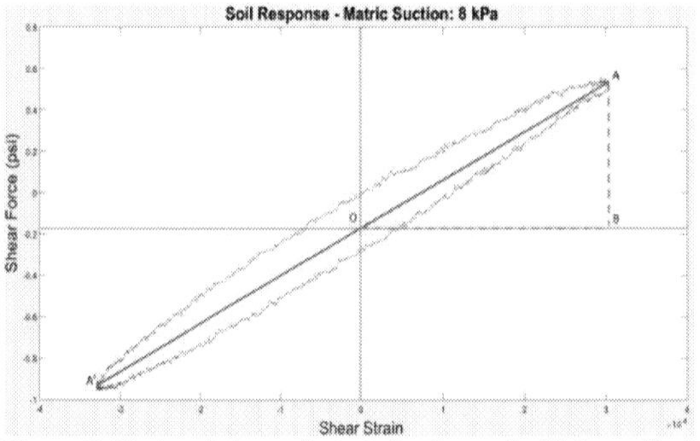

Figure 8. Hysteresis loop of a tested soil sample (8kPa suction)

In this figure, the secant shear modulus was obtained through taking the slope of line A – A'. Points A and A' represent the top and bottom peaks of the cycle. The damping ratio was calculated through the following equation.

$$\zeta = \frac{1}{4\pi} * \frac{Areaof\ Triangle\ OAB}{Area\ of\ Hysteresis\ Loop} \tag{4}$$

Using the methods proposed above the secant shear modulus and damping were calculated for cycles 2-4 in each test and then averaged. The changes of shear modulus with suction and degree of saturation are shown in Figures 9 and 10, respectively. In order to avoid confusion between fully saturated and dry data (in representing soils not subjected to suction), the results of the dry tests are not presented in Figures 9. The shear modulus increases under higher suction (lower degrees of saturation) up to the point of residual water content (at about 8-10 kPa suction). However, the modulus is lower in dry sand comparing with partially saturated ones. This is consistent with previously observed pattern in shear modulus [11].

Further, the damping ratio for tests on different suction and degrees of saturation over cycles 2-4 were estimated and averaged based on Equation 4, and shown in Figures 11 and 12, respectively. Expectedly, stiffer soil in unsaturated condition comparing with the dry and saturated soils resulted in lower damping.

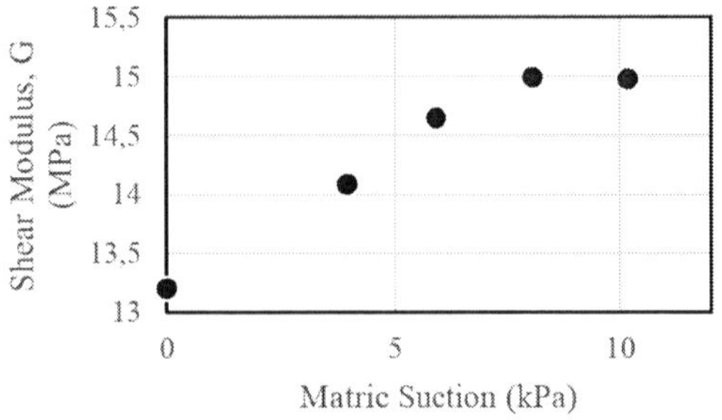

Figure 9. Effects of Matric Suction on Shear Modulus

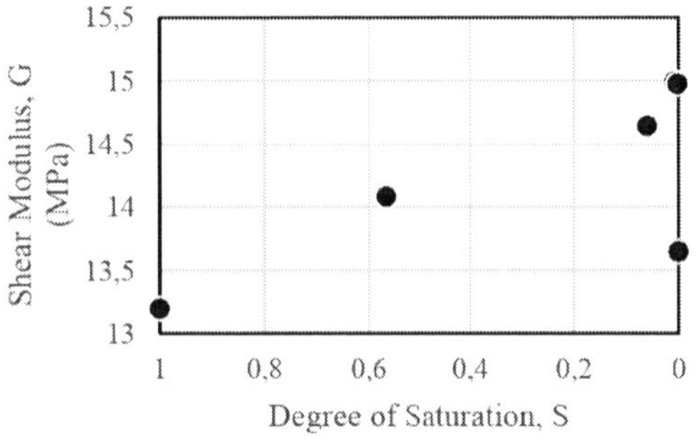

Figure 10. Effects of Degree of Saturation on Shear Modulus

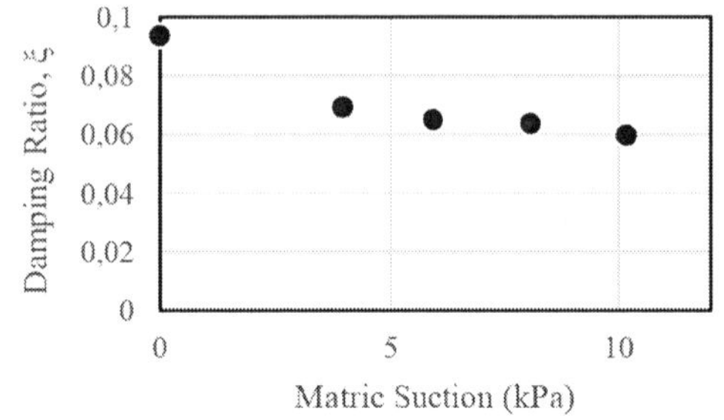

Figure 11. Effects of Matric Suction on Damping Properties

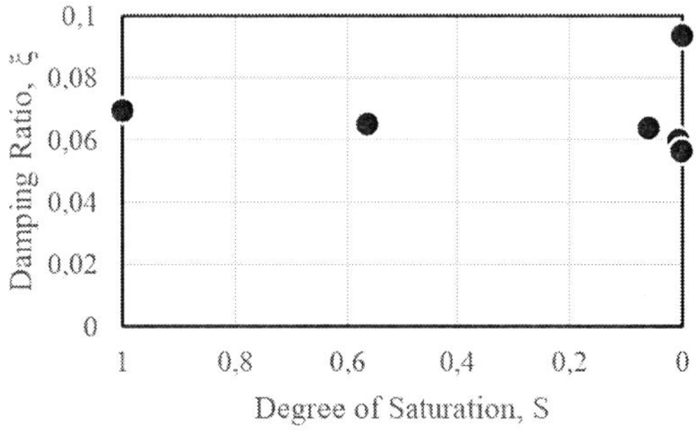

Figure 12. Effects of the Degree of Saturation on Damping Properties

The vertical strain after each test was monitored to track the changes of relative density. The total changes of relative density before the first and after the last test was less than 5%, which is in the acceptable range considering the accuracy of the sample preparation relative density. Although these tests were performed under constant suction condition, some changes in pore pressure during cyclic loading was expected. However, due to the relatively small shear strain and the resulting minor volume change very little suction change was measured in DPT.

6 Conclusions

The UNH-DSS system was upgraded with new DAQ and modified for unsaturated soil testing using axis translation technique. The data that was obtained through the use of this system has been shown to provide consistent results. A set of constant suction cyclic simple shear test was performed on a sandy soil. The suction in unsaturated soil increased the shear modulus comparing with that of dry and saturated soils. In addition, stiffer unsaturated soil resulted in lower damping ratio. Despite imposing drained condition, changes in suction would be expected in higher shear strain cyclic loads, but it was minimal in low strain cyclic tests presented in this paper.

References

1. H.B. Seed, I.M. Idriss, Berkeley: Earthquake Engineering Research Center. *Soil Moduli and Damping Factors for Dynamic Response Analyses.* **70** (1970)
2. B.O Hardin, V.P. Drnevich, Journal of Soil Mechanics and Foundations Division, Proceedings of ASCE. *Shear modulus and damping in soils: design equations and curves.* **98** 667 - 692 (1972)
3. M. Doroudian, M. Vucetic. Geotechnical Testing Journal. *A Direct Simple Shear Device for Measuring Small-Strain Behavior.* **18** 69 – 85 (1995)
4. S. Oztoprak, M.D. Bolton, Géotechnique. *Stiffness of sands through a laboratory test database.* **63** 54-70 (2013)
5. A.W. Bishop. Teknisk Ukeblad I Samarbeide Med Teknikk. *The principle of effective stress.* **106** 859 – 863 (1959)
6. N. Lu, W.J. Likos, *Unsaturated Soil Mechanics (2004)*
7. M.A. van Genuchten, Soil Science Society of America Journal. *Closed Form Equation for Predicting the Hydraulic Conductivity of Unsaturated Soils.* **58** 647- 652 (1980)
8. N. Lu, J. Godt, D. Wu, Water Research Journal. *A closed-form equation for effective stress in unsaturated soil.* **46** 1-14 (2010)
9. Marinho, E.A.M., Chandler, R.J., and Crilly, M.S. 1995. "Stiffness Measurements on an Unsaturated High Plasticity Clay using Bender Elements." Proc. of the First International Conference on Unsaturated Soils, UNSAT '95, Paris, France, 6– 8 September 1995. A.A. Balkema, Rotterdam, the Netherlands. Vol. 2. 535–539.
10. Mancuso, C., Vassallo, R., and d'Onofrio, A. (2002). "Small Strain Behavior of a Silty Sand in Controlled-Suction Resonant Column-Torsional Shear Tests." Canadian Geotechnical Journal. CGJ. Vol. 39, No. 1, 22-31.
11. Ghayoomi, M. and McCartney, J.S. 2011. "Measurement of Small-Strain Shear Moduli of Partially Saturated Sand during Infiltration in a Geotechnical Centrifuge." ASTM Geotechnical Testing Journal, 34(5), 1-12.
12. Hoyos, L.R., Suescun-Florez, E.A., and Puppala, A.J. 2015. "Stiffness of Intermediate Unsaturated Soil From Simultaneous Suction-Controlled Resonant Column and bender Element Testing." Engineering Geology, 188, 10-28.
13. Biglari, M., Jafari, M.K., Shafiee, A., Mancuso, C., and d'Onofrio, A. 2011. "Shear Modulus and Damping Ratio of Unsaturated Kaolin Measured by New Suction-Controlled Cyclic Triaxial Device", ASTM Geotechnical Testing Journal, 34 (5), 1-12.
14. Ghayoomi, M., Suprunenko, G., and Mirshekari, M. (Under Review) Cyclic Triaxial Test to Measure Strain-Dependent Shear Modulus of Unsaturated Sand, ASCE International Journal of Geomechanics.
15. W. Kjellman. Géotechnique. *Testing the Shear Strength of Clay in Sweden.* **2** 225-232 (1951)
16. K. Terzhaghi, R.B. Peck, *Soil Mechanics in Engineering Practice.* (1948)
17. K.H. Roscoe. Proceedings of the Third International Conference on Soil Mechanics and Foundation Engineering. *An Apparatus for the Application of Simple Shear to Soil Samples.* 186 – 191 (1953)
18. P.M. Duku, J.P. Stewart, D.H Whang, R. Venugopal, L.D. Suits, T.C. Sheahan. Geotechnical Testing Journal. *Digitally Controlled Simple Shear Apparatus for Dynamic Soil Testing.* **30** (2007)
19. H.J. Miller, *Development of Instrumentation to study effects of aging on the small strain behavior of sands.* Dissertation – UNH. (1994)
20. A. Dunstan, *Laboratory Simulation of Earthquake Loading on Clay.* M.Thesis – UNH. (1998)
21. J. Hargy, *Measuring Residual Strength of Liquefied Soil with the Ring Shear Device.* M.Thesis – *UNH.* (2011)
22. M. Ghayoomi, J.S McCartney. Special Issue on Advances in Experimental Characterization of Unsaturated Soils, Volume 1. ASTM Geotechnical Testing Journal. Measurement of the Small-Strain Shear Moduli of Partially-Saturated Sands during Infiltration in a Geotechnical Centrifuge. 34 503-513 (2011)
23. Suprunenko, G. and Ghayoomi, M. (2015) Suction-Controlled Cyclic Triaxial Apparatus for Measurement of Dynamic Properties of Unsaturated Soils, the XV Pan-American Conference on Soil Mechanics and Geotechnical Engineering, Buenos Aires, Argentina.

 DOI: 10.1051/e3sconf/20160909002

Optimisation of in-situ dry density estimation

Mathilde Morvan[1,a] and Pierre Breul[1]

[1]*Institut Pascal, Université Blaise Pascal, Clermont-Ferrand, France*

Abstract. Nowadays, field experiments are mostly used to determine the resistance and settlements of a soil before building. The needed devices were heavy so they cannot be used in every situation. It is the reason why Gourves et al (1998) developed a light dynamic penetrometer called Panda. For this penetrometer, a standardized hammer has to be blown on the head of the piston. For each blow, it measures the driving energy as well as the driving depth of the cone into the soil. The obtained penetrogram gives us the cone resistance variation with depth. For homogeneous soils, three parameters can determined: the critical depth z_c, the initial cone resistance q_{d0} and the cone resistance in depth q_{d1}. In parallel to the improvement of this apparatus, some researches were lead to obtain a relationship between the dry density of soil and the cone resistance in depth q_{d1}. Knowing dry density of soil can allow to evaluate compaction efficiency for example. To achieve this point, a database of soils was initiated. Each of these soils was tested and classified using laboratory tests, among others, grain size distribution, proctor results, Atterberg limits. Penetrometer tests were also performed for three to five densities and for three to five water contents. Using this database, Chaigneau managed to obtain a logarithmic relation linking q_{d1} and dry density. But this relation varies with the water content. This article presents our recent researches on a mean to obtain a unified relation using water content, saturation degree or suction. To achieve this point, at first we studied the CNR silt responses with saturation degree and water content. Its water retention curve was realised using filter paper method so we can obtain suction. Then we verified the conclusion of this study to seven soils of the database to validate our hypotheses.

1 Introduction

Panda penetrometer is a light dynamic penetrometer that allows to obtain soil cone resistance for first soil meters.

Previous researches showed that a relationship between dry density of tested materials and cone resistance exists. For Panda [1, 2, 3], this unique relationship was enlightened for a given soil and a given water content. Five water contents were studied depending on proctor optimum: very dry, dry, medium, wet, very wet. The aim of this paper is to propose a new relationship valid for any water content in order to reduce uncertainty of the in-situ dry density evaluation. Moreover climatic changes push researcher to study the bearing capacity variation due to drying-wetting cycles. This new relationship will allow to have a new evaluation of bearing capacity after a change of water content. In the future, this evaluation could give us a mean to evaluate suction in-situ using other parameters that can be easily obtained in laboratory (for example: grain size distribution, proctor results, Atterberg limits).

2 Panda and data base

Panda is a light dynamic penetrometer using variable energy. Due to its small size (cone of 4 cm^2 and rods diameter of 15,4 mm) and its lightness, it gives the opportunity to examine first meters of most soft soils even when it is hard to get to.

The dynamic beating is realized manually by the mean of a standardized hammer. The beating energy and the cone displacement are recorded for each blow. Thanks to this simultaneous measurement, the device provides a penetrogram (cone resistance variation versus depth) with a cone resistance measurement every each 5 mm of cone displacement on average (Fig.1).

From the recorded penetrogram, several parameters are deduced for homogenous soil [1]:

- Critical depth z_c, depth from which cone resistance stays constant
- Initial cone resistance q_{d0}
- Average cone resistance below z_c , q_{d1}.

Like all penetrometers, Panda device has been developed in order to obtain the in situ mechanical soil properties during geotechnical soil investigations. But considering that the dynamic resistance of a soil is strongly connected to the state of density of the tested

[a] Corresponding author: mathilde.morvan@univ-bpclermont.fr

materials it can also be used for the control of compaction [4], as for example in the control of tailing deposits in Chile [5].

In order to obtain an indirect evaluation of the dry density of the studied soil, relations linking the cone resistance in depth (q_{d1}) measured with the Panda to the material dry density were established. For that purpose, a base of "model soils" (containing 1 to 3 materials for each class of soils defined in the French classification (GTR)) was built [6]. The objective of this base of "model soils" is to be able to characterize any soil studied in situ by referring to "model soils" of similar properties. For each material, Panda tests have been carried out in a calibration chamber (80.6 cm height and diameter of 37.5 cm), for five levels of density, varying from the bulk density to about 110% of the standard Proctor density and for three different water content corresponding to a wet, medium, and dry state.

Chaigneau [4] showed that for one material class at given water content, the relation linking the cone resistance to the dry density was:

$$\gamma_d = \alpha \ln(q_{d1}) + \beta \quad (1)$$

This relation being valid for one water content, it would be interesting to be able to generalize it for all ranges of water content of a soil in order to estimate the in situ dry density of a material only from the knowledge of the geotechnical soil classification and from a penetrometric test.

We thus undertook to study the influence of the water content on this relation and on the soil parameters.

3 Study of CNR silt

In our database, we can find silt named CNR silt and classified in A1-2 for the French GTR classification. Its characteristics are:

- D max=2.6mm
- Passing at 80microns: 77.76%
- Grain volumetric weight: 26.62 kN/m^3.
- Liquidity limit: 25.8%
- Plasticity limit: 22.5%
- Normal proctor results: wopt=14.2% and optimum dry volumetric weight 17.98kN/m3

This database also gives us results of dynamic penetrometer Panda tests for each soil at different hygrometry and density. For each test, average dry density, cone resistance in depth and average water content are measured. Each test is repeated three times, presented results are the average of these three tests. Figure 1 presents an example of the obtained penetrogram for wet CNR silt. Thanks to this penetrogram we obtained z_c=0.26m, q_{d1}=4.05MPa, q_{d0}=2.33MPa. In this study, only q_{d1}, the cone resistance in depth, will be used.

The objective of the presented study was to characterize the influence of pore water on cone resistance.

In order to verify water content effects on cone resistance, we present results in (q_{d1}, γ_d) plane without taking water content variations into account (Fig 2). We can notice that the obtained relationship is not accurate enough and the dependency to water content is to be studied.

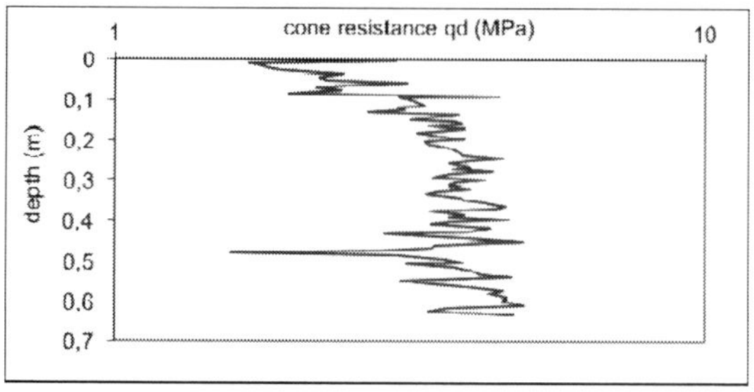

Figure 1: example of CNR silt penetrogram

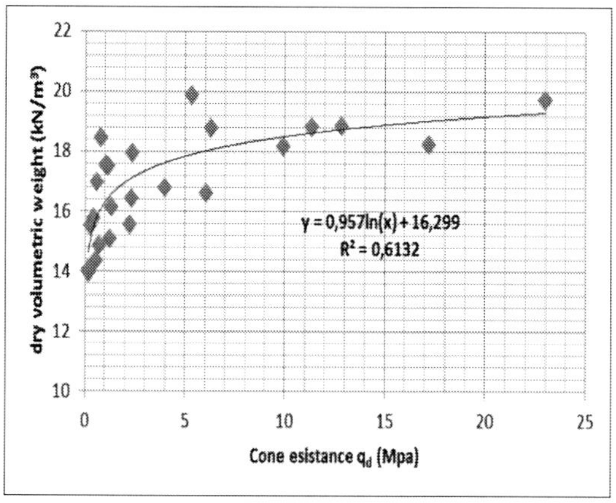

Figure 2: Calibration points (q_{d1} vs γ_d) plotted together without water content distinction

Calibration curves for each water content tested are presented in fig 3. With these calibration curves, knowing water content of a soil, we can approximate a $\gamma_d = f(q_{d1})$ law. No generalized law has been found yet. This is what we will study in this article.

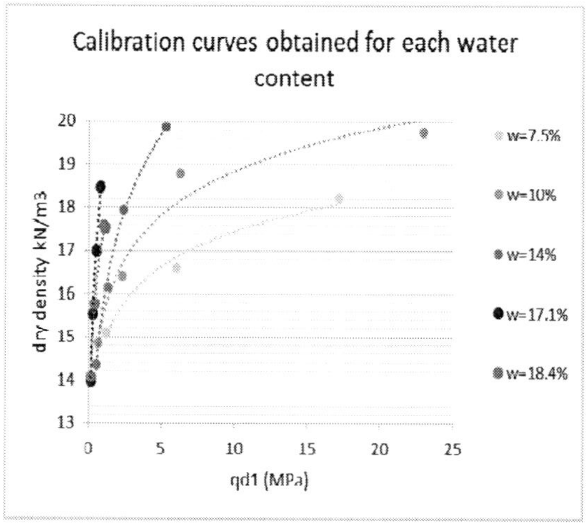

Figure 3: Calibration curves (q_{d1} vs γ_d) for different water contents.

These curves lead to the following relationship for each water contents:

$$\gamma_d = \alpha(w)\ln(q_{d1}) + \beta(w) \quad (2)$$

Table 1: variation of α and β parameters with water content

w (%)	α(kN/m³)	β (kN/m³)
7,5	1,1714	14,748
10,0	1,4646	15,468
13,1	1,8637	14,113
14,0	2,4699	15,721
17,1	2,7297	18,896
18,4	1,9625	17,413

Table 1 summarizes obtained parameters. Fig 4 shows values of α and β parameters versus water content. It is difficult to conclude on the dependency of these parameters on water content. Any correlation cannot be found. So we decided to study another variable.

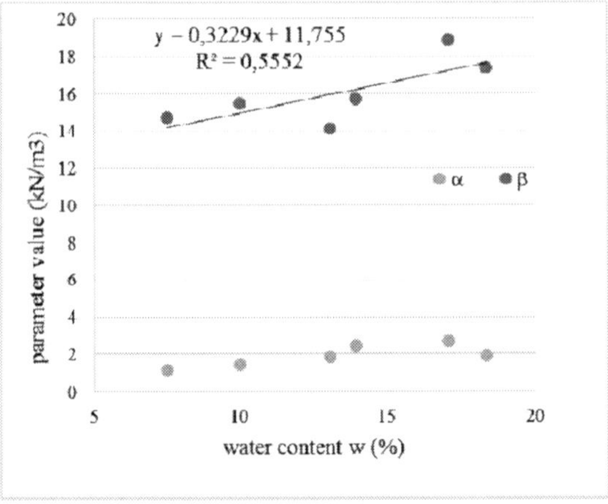

Figure 4: variation of α and β parameters with water content

To describe the mechanical behaviour of unsaturated soils, water content is not usually taken to represent pore water but instead it is saturation degree Sr. Sr can be obtained using water content and dry density:

$$S_r = \frac{w}{\frac{\gamma_w}{\gamma_d} - \frac{\gamma_w}{\gamma_s}} \quad (3)$$

So we chose to study the $\gamma_d = f(q_{d1})$ relationship with saturation degree (Fig. 5). This new relationship can be written in the following form:

$$\gamma_d = A(S_r)\ln(q_{d1}) + B(S_r) \quad (4)$$

Results of A and B for different saturation degrees are presented in table 2 and plotted in Fig 6.

Variation of A with S_r is not obvious but we observed a logarithmic type relationship between B and S_r.

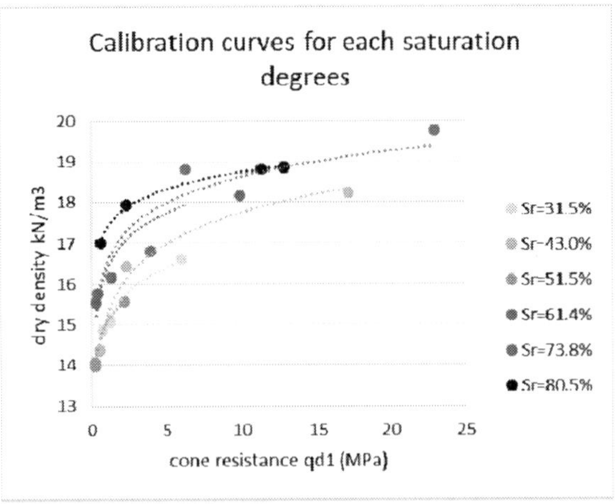

Figure 5: Calibration curves (q_{d1} vs γ_d) for different saturation degrees

Table 2: variation of A and B with Sr

A(kN/m³)	B (kN/m³)	Sr (%)
0,83	15,07	31,57
1,11	15,19	42,98
0,61	15,07	51,46
0,87	16,35	61,43
0,92	16,52	73,86
0,59	17,37	80,47

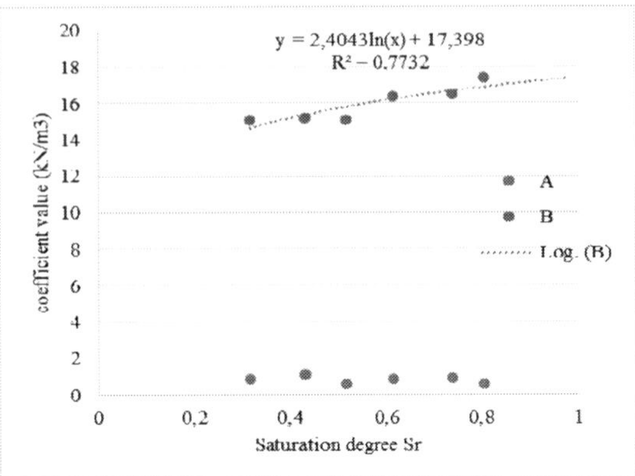

Figure 6: variation of A and B with Sr

Fig 6 shows a logarithmic relationship between B and saturation degree S_r. We considered that A remains constant. So we proposed a new relationship:

$$\gamma_d = A\ln(q_{d1}) + B' + C'(S_r) \quad (5)$$

Before going further in applying this new formula to the database soils, we wanted to validate our results using other researchers' studies.

 DOI: 10.1051/e3sconf/20160909002

4 First step to validate the obtained relationship

Among previous studies on cone resistance, Pournaghiazar, M., Russell, A. R. & Khalili, N. [7, 8] formulated a relationship for CPT. This relation is given in eq.6.

$$q_c = Ap'^B e^{CD_r} \quad (6)$$

With p' confining effective stress and D_r, relative density ($D_r = \frac{e_{max} - e}{e_{max} - e_{min}}$).

We can rewrite this equation to get closed to the previously proposed relationship:

$$\gamma_d = A''\ln(q_{d1}) + B'' + C''(p') \quad (7)$$

If we consider the simplified form of effective stress given by Bishop:

$$\sigma' = \sigma - u_a + sS_r \quad \text{so} \quad p' = p - u_a + sS_r \quad (8)$$

We made the assumption that calibration curves were realized with no confining pressure and that air pore pressure is zero. So eq5 can be simplified and gives:

$$\gamma_d = A''\ln(q_{d1}) + B'' + C''(sS_r) \quad (9)$$

To verify this equation, we measured the retention curve of CNR silt. This curve gives suction for each saturation degree. To obtain this curve we used the filter paper technique [9, 10, 11]. Normalised Whatman 42 filter papers were used. For each point, water content and void ratio of the soil were measured as well as the water content of the filter paper. This method is based on the suction equilibrium between the sample and the filter paper with a well-known retention curve. Suction of the filter paper can be obtained using these equations:

For w<45.3%

$$\log(s) = 5.327 - 0.0779w \quad (10)$$

For w>45.3%

$$\log(s) = 2.412 - 0.0135w \quad (11)$$

To verify the obtained results, a simulation was performed using a model developed by Aubertin et al [12]. This simulation, mainly based on grain size distribution, is compared to our experimental results in Fig. 7)

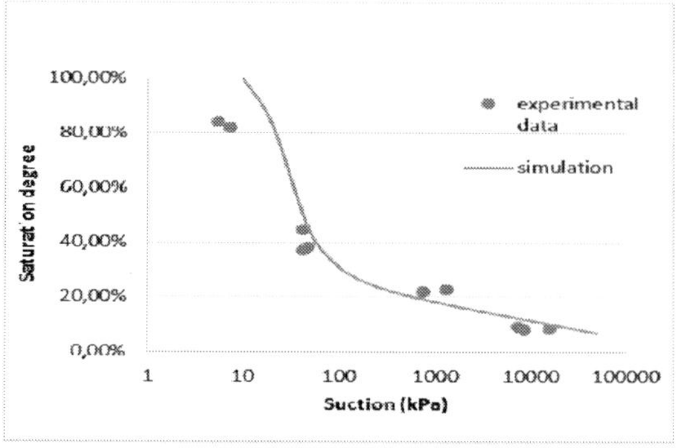

Figure 7: water retention curve of CNR silt

Samples where realised at a constant void ratio of 0.80. Fig. 8 shows the experimental points obtained and the simulation results using a Brooks and Corey law (12).

$$S_r = \left(\frac{s_e}{s}\right)^k \quad (12)$$

Where s_e, air entry suction and k are material parameters linked to particle size distribution and particle chemistry.

For CNR silt, we obtained se=5kPa and k=0.25.

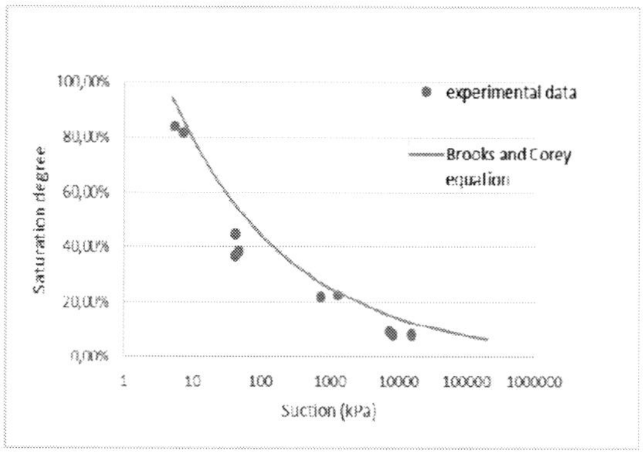

Figure 8: water retention curve of CNR silt

Fig 9 and Table 3 show the dependency of B parameter toward sS_r.

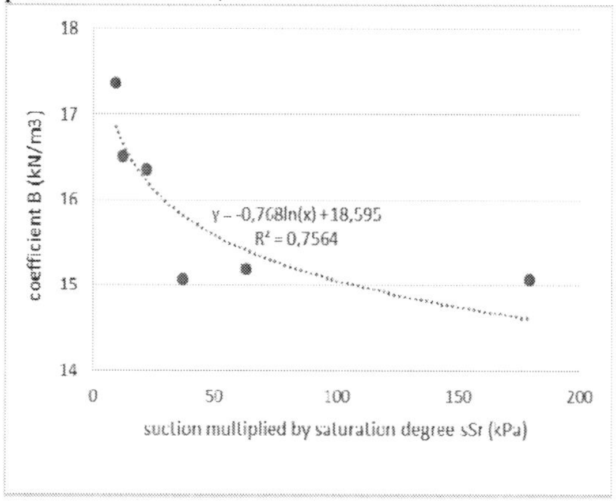

Figure 9: variation of and B with sSr

Table 3: variation of A and B with suction and sSr

A	B	Succion (kPa)	sSr (kPa)
0,83	15,07	567,73	179,23
1,11	15,19	147,12	63,23
0,61	15,07	72,14	37,13
0,87	16,35	35,76	21,97
0,92	16,52	16,97	12,54
0,59	17,37	12,01	9,67

The obtained curve global equation is:

$$\gamma_d = 0.82\ln(q_{d1}) + 18.595 - 0.768(sS_r) \quad (13)$$

Using Brooks and Corey equation (eq.12), we get:

$$\gamma_d = A''\ln(q_{d1}) + B'' + C''\ln(s_e S_r^{\frac{-1}{k}}) + C''\ln(S_r) \quad (14)$$

That can be expanded:

$$\gamma_d = A''\ln(q_{d1}) + B'' + C''\ln(s_e) - \frac{C''}{k}\ln(S_r) + C''\ln(S_r) \quad (15)$$

If we simplify the previous equation:

$$\gamma_d = A\ln(q_d) + B' + C'\ln(S_r) \quad (16)$$

With A'=A''; B'=B''+C''*ln*(s$_e$) ; $C' = C''\left(1 - \frac{1}{k}\right)$

Equation (16) has the same form as equation (5). The proposed equation is in agreement with bibliography. In order to verify this equation on other kind of soils, we performed a global study on seven soils of our database. For CNR, using obtained B'' and C'' with retention curve parameters, we get B'=17.48 and C'=2.4. These values are close from those obtained from the best fit regression.

5 Results obtained on seven soils

We chose in our previously presented database different kinds of soils: sands, silts and clays. Among them, we chose three sands: Sayat sand, Sermentison sand and Hauterverne sand. For silts, we selected two: CNR silt and DGA silt. Lachamps and Allier clays were used in this study. Their characteristics are summarized in Table 4 and Table 5.

Table 4: Sands characteristics

Soil	Sayat sand	Hauteverne sand	Sermentison sand
D_{max} (mm)	2.5	8	6
Passing at 2mm (%)	90.5	>70	62.5
Passing at 80microns (%)	4.9	37.29	12.90
w_l (%)	22.4	-	26.2
w_p(%)	17.9	-	-
Classification	B2-1	A1-3	B5-2
w_{opt}(%)	11	13.9	11.1
γ_{opt}(kN/m^3)	19.30	18.57	19.76

Table 5: Silts and clays characteristics

Soil	CNR silt	DGA silt	Lachamps clay	Allier clay
D_{max} (mm)	0.26	0.37	0.08	0.71
Passing at 2mm (%)	99.8	98.5	99.3	99.5
Passing at 80microns (%)	77.76	91.3	96.73	67.9
w_l (%)	25.8	27.9	42.7	36.1
w_p(%)	22.5	22.9	27.6	23.2
Class	A1-2	A1-1	A2-1	A2-2
w_{opt}(%)	14.2	13.3	15.8	14.9
γ_{opt} (kN/m^3)	17.98	18.35	18.08	17.60

This study is based on more than 250 tests performed with the dynamic penetrometer PANDA in our calibration chamber.

Results of the three parameters A', B', C' calculated for each soil are given in Table 6.

These parameters are close for both CNR and DGA silts as well as for both Lachamps and Allier clay. Classification of these two silts is A1 and for these two clays A2. It seems that coefficients are close for soils of the same characteristics. These three selected sands do not have the same classification and we can notice that the observed parameters varies a lot for parameter C'.

Table 6: Obtained parameters for 7 studied soils

	A'	B'	C'	class
Sayat sand	0,65	17,52	1,41	B2-1
Hauteverne sand	0,67	18,76	4,26	A1-3
Sermentison sand	1,04	18,85	3,56	B5-2
CNR silt	0,82	17,40	2,40	A1-2
DGA silt	1,22	16,78	2,32	A1-1
Lachamps clay	1,06	16,59	5,33	A2-1
Allier clay	1,29	15,56	4,25	A2-2

It is very uneasy to compare the previously used method using approximated calibration curves and our three parameters law.

Table 7 presents the precision of dry density calculated with the previously developed law (eq. 2) and our three parameters law (eq. 5).

We can notice that the new relationship improves the average difference by a factor two and the maximum difference by about three. This first improvement is significant. Moreover, the three parameters needed could be kept constant for the same type of soil. This hypothesis has to be verified for sand but it seems verified for A1 silts and A2 clays.

Table 7: Improvement due to new relationship

	Previous method	Three parameters law
RMS between measured γ_d and calculated γ_d	1.03 kN/m^3	0.48 kN/m^3
Maximum difference between measured γ_d and calculated γ_d (kN/m^3-%)	3,00 kN/m^3-24.01%	1.23 kN/m^3-8.69%
Average difference between measured γ_d and calculated γ_d (kN/m3-%)	0.81 kN/m^3-4.95%	0.37 kN/m^3-2.31%

6 Conclusion

To conclude with, we proposed a new $\gamma_d = f(q_{d1})$ relationship. This law used with the variable energy panda penetrometer to estimate dry density is closed to the one used on CPT [8]. This law was verified on different kind of soils. Parameters seem to be constant for each kind of soils regarding the French GTR classification. So these parameters depend on grain size distribution, Atterberg limits and proctor results. With this new relationship it is now possible to evaluate accurately dry density of soils. And in our climate changes context, it can also be used to predict cone resistance variation with hydrometric changes in soils. So it can lead to more sustainable road designs.

References

1. Gourvès R., Richard B., Le pénétromètre dynamique léger PANDA, Proceeding of ECSMFE, Copenhagen, Denmark, pp 83 – 88, (1995).
2. R. Gourvès, F. Oudjehane et S. Zhou, The in-situ characterization of mechanical properties of granular media with the help of penetrometers, proceeding of Powders and Grains, Durham, USA, pp. 57-60 (1997)
3. Breul, P.; Benz, M.; Gourvès, R. and Saussine, G. Penetration test modelling in a coarse granular medium. International conference on micromechanics of granular media, Powder and Grain, AIP Conf. Proc., Golden, Colorado, USA vol 1145, pp. 173-176 (2009).
4. Chaigneau, L., Gourvès, R., Bacconnet, C., Penetration test coupled with a geotechnical classification for compaction control, GeoEng2000 An international conférence on geotechnical & geological Engineering, Melbourne, Australia, 19-24 November (2000)
5. Villavicencio G. Méthodologie pour évaluer la stabilité des barrages de résidus miniers, Thèse de l'Université Blaise Pascal, Clermont-Ferrand (2009).
6. Chaigneau, L. Caractérisation des milieux granulaires de surface à l'aide d'un pénétromètre. Thèse de l'Université Blaise Pascal, Clermont-Ferrand (2001).
7. Pournaghiazar, M., Russell, A. R. R. R. & Khalili, N. 2011. Development of a new calibration chamber for conducting cone penetration tests in unsaturated soils. Canadian Geotechnical Journal, 48, 314-321.
8. Pournaghiazar, M., Russell, A. R. & Khalili, N. 2013. The cone penetration test in unsaturated sands. Géotechnique, 63, 1209-1220.
9. International, A. 2003. Standart Test Method for Measurement of Soil Potential (Suction) Using Filter Paper. D5298-03 ©ASTM International.
10. Muñoz-Castelblanco, J. A., Pereira, J.-M., Delage, P. & Cui, Y.-J. Year. Suction measurements on a natural unsaturated soil: A reappraisal of the filter paper method. In: Unsaturated Soils - Proc. Fifth Int. Conf. on Unsaturated Soils, 2010-09-06 2010 Barcelona, Spain. CRC Press, 707-712.
11. Bicalho, K. V., Gomes Correia, A., Reis Ferreira, S., Fleureau, J.-M. & Marinho, F. Year. Filter paper method of soil suction measurement. In: 13th Panamerican Conference on Soil Mechanics and Geotechnical Engineering, 2007-07-16 2007 Isla de Margarita, Venezuela.
12. M. Aubertin, M. Mbonimpa, B. Buissière and R.P. Chapuis, A model to predict the water retention curve from basic geotechnical properties, Canadian Geotechnical Journal, 40, 1104-1122 (2003).

Effect of particle size on the measurement of the apparent contact angle in sand of varying wettability under air-dried conditions

Yunesh Saulick[1, a], Sergio Lourenço[1] and Béatrice Baudet[1]

[1]*The University of Hong Kong (HKU), Pokfulam, Hong Kong, Hong Kong SAR, China*

Abstract. Changes in the wettability of soil are known to affect several processes such as infiltration and the shear strength of soil. In this study, the wettability of a medium to fine sand was chemically modified by using different concentrations of dimethyldichlorosilane (DMDCS). The sessile drop method (SDM) was used for the assessment of wettability of hydrophobised Leighton Buzzard Sand (LBS). The results demonstrate that beyond a concentration of 2 g per kg of LBS, the finer fraction had its apparent contact angle (ACA) increased up to 115° while the maximum ACA attained by the coarser fractions was 100°. At such high concentration of DMDCS, the effect of trapped air, which is known to increase the ACA, was found to be either small or insignificant. The standard deviations of the ACAs agreed well with past studies. The most important factors contributing to the water-repellent behaviour of chemically synthesised sand were attributed to the characteristics of the particles; these include surface area and particle shape.

1 Introduction

The wettability of natural soils is known to alter significantly with time due to numerous reasons such as organic matter and soil texture [1, 2, 3]. For instance, a decrease in particle size due to crushing will cause a significant increase in the surface area of the particles, which will in turn require more organic matter for coating. A decrease in the wettability of soils causes increased surface runoff and enhance preferential flow in soils [3-4]. Several techniques to assess the wettability of soils have been established throughout the literature; they include direct methods of measurement of the apparent contact angle (ACA) such as the sessile drop method (SDM) and indirect methods such as the capillary rise methods.

In the field of geotechnical engineering, especially where unsaturated soils are involved, the wetting characteristic of soils is important because they have a significant influence on suction. Lourenço *et al.* [5] demonstrated that for a mixture of sand and clay, its ability to retain water decreased as the ACA approached 90°. Shear strength of non-wettable soils as evaluated by direct shear tests performed by Bardet *et al.* [6] using soils coated by wax demonstrated that there was no change in the angle of shear resistance compared to wettable samples. However, Lee *et al.* [7] showed that a decrease in the peak friction angle (tests carried out with samples having same initial void ratios) could be observed when their samples were made less wettable by means of a silane coating.

The influence of wettability on the behaviour of unsaturated soils is relevant in studies of porous and granular materials down to the grain-size level. Here, we investigate the effect of the grain-size on the wettability of a medium to fine commercial sand made repellent to different extents by means of an organosilicon compound.

2 Materials and methods

2.1 Soil material

The soil used in the study was Leighton Buzzard Sand (LBS), which originates from quarries in southern UK. Due to its rich content in silica (mainly in the form of quartz comprising nearly 98% of the sand), LBS exhibits excellent properties such as chemical inertness. The LBS sample was sieved by means of a mechanical shaker to obtain four sieve fractions namely 63-212, 212-300, 300-425 and 425-600 μm. The samples were initially air-dried, the water content of which were between 1 and 2%. Table 1 shows the particle shape characteristics extracted from a dynamic shape analyser (QICPIC; Sympatec GmbH, Clausthal-Zellerfeld, Germany). For each of the aforementioned sieve sizes, a minimum of 10,000 images was analysed. From the shape parameters extracted, a general increase in sphericity, aspect ratio and convexity was observed as the size of the particles increased. The sphericity as defined by the QICPIC is the ratio of the equivalent perfect circle to the actual perimeter for a given area. The value of the aspect ratio is

[a] Corresponding author: yunesh@connect.hku.hk

given by the ratio of of the minimum to maximum Feret diameters while convexity is calculated as a ratio of the actual to the convex area. The convex area refers to the circumscribed borderline to the 2-D outline of the particle.

Table 1. Particle shape characteristics of the different sieve fractions obtained from a dynamic shape analyser.

Particle attributes	63-212 µm	212-300 µm	300-425 µm	425-600 µm
Sphericity	0.8806	0.8884	0.8926	0.8942
Aspect Ratio	0.7294	0.7454	0.7611	0.7599
Convexity	0.9200	0.9381	0.9487	0.9578

2.2 Hydrophobization reaction

To a 40g sample of LBS comprising of an equal proportion of the aforementioned sieve fractions, was added an incremental amount of Dimethyldichlorosilane (molecular weight of 129.06 & density 1.06 gcm^{-3} from Acros Organics, New Jersey, USA) drop by drop from a single channel pipette (Eppendorf Research® plus) with constant and gentle stirring. The end of the coating reaction between the sand and the chemical solution was indicated by the formation of hydrogen chloride fumes. The next steps were to leave the samples in the fume hood for half an hour and then keeping them in Ziploc bags for 24 hours prior to any measurements.

2.3 Apparent contact angle (ACA) measurement

The ACAs were determined by a goniometer (Drop Shape Analyser 25 from KRÜSS GmbH) where the motion of a 10-µl drop of de-ionised water was recorded by a 1394 AVT CCD camera upon contact on a 76 x 26 x 1 mm slide to which the soil sample was pasted via a a double-sided tape following the method proposed by Bachmann *et al.* [8]. Similarly to Shang *et al.* [9] in their study of porous materials, the initial ACA was considered the most representative measure of wettability. The initial ACA was taken from the first stable frame from the movie (approximately 50 ms after contact with the substrate) as the contact angle hysteresis due to volume loss is least likely to occur at this stage. The magnitude of the ACAs was obtained by developing a semi-automated method combining the low-bond axisymmetric drop shape analysis (LBADSA) method developed by Stalder *et al.* [10] and a manual adjustment of the baseline. All ACAs measurements were performed at a temperature of 23 ± 1 °C and relative humidity of 55 ± 5 %.

3 Results and discussions

The relationship between the different size fractions and the ACAs is presented in Figure 1. The standard deviations of the ACAs were at most ± 6° and in line with the studies of Bachmann *et al.* [11]. It can be seen that at relatively small concentrations of DMDCS, the change in ACAs within the different size fractions is small although the lowest ACAs are usually associated with the finer materials. As the amount of DMDCS increases, the difference between the sizes becomes more apparent as indicated by the steeper slopes. A possible justification for such behaviour is that the smaller size fractions require a larger amount of DMDCS to be coated due to their larger surface areas [12]. The overlapping lines at the two highest concentrations, which were investigated (Figure 1), are probably due to the difficulty in extracting the values of the ACAs. This arises when adjusting the baseline and the optics settings on the image output from the Drop Shape Analyser 25 especially at high ACAs.

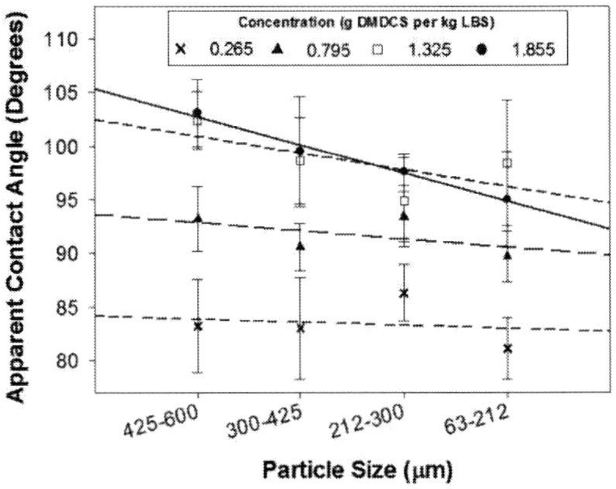

Figure 1. Apparent contact angles as a function of particle size in samples with different concentration of DMDCS.

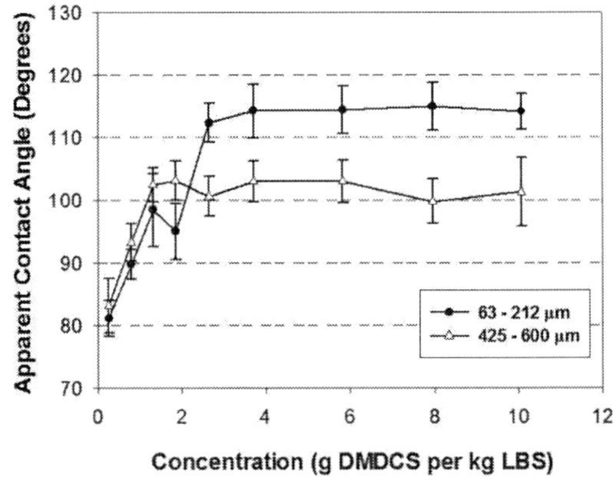

Figure 2. Maximum apparent contact angles attained by the coarsest and finest fractions.

To determine the percentage of particles coated, the maximum ACA by the different fractions was first identified. The results are shown in Table 2 and Figure 2 illustrates the plot for the 63-212 µm and 425-600 µm sieve fractions. From Figure 2, it can be seen that for a concentration of DMDCS of up to 2 g per kg of LBS, it is unclear which of the two fractions show the larger ACA due to the overlapping error bars. However, as the concentration increases beyond this, the maximum ACA attained by the 300-425 and 425-600 μm sieve fractions is 105.5 ± 5.38° while a maximum value of 116.6 ± 2.86° was reached for the 63-212 and 212-300 μm sieve fractions. In addition, the difference in the maximum ACAs between the two sieve-fractions in Figure 2 demonstrates that the assumption of a larger fraction of air being trapped in the coarser particles (which usually enhances ACAs) compared to the finer ones is not justified for the relative higher ACA in fine sand.

Table 2. ACA with increasing concentration of DMDCS.

Concentration (g DMDCS / kg of Soil)	ACA (Degrees)			
	63-212 μm	212-300 μm	300-425 μm	425-600 μm
0.265	81.1 ± 2.86	86.3 ± 2.62	83.0 ± 4.78	83.2 ± 4.34
0.795	89.7 ± 2.37	93.4 ± 2.34	90.6 ± 2.21	93.2 ± 3.03
1.325	98.4 ± 5.86	94.9 ± 4.35	98.6 ± 4.02	102.4 ± 2.71
1.855	95.0 ± 4.51	97.6 ± 1.33	99.5 ± 5.13	103.1 ± 3.15
2.650	112.3 ± 3.12	100.4 ± 1.74	102.0 ± 4.14	100.6 ± 3.21
3.710	114.2 ± 4.26	111.7 ± 4.18	92.4 ± 3.73	103.0 ± 3.24
5.830	114.4 ± 3.77	110.8 ± 4.98	103.7 ± 4.00	103.0 ± 3.45
7.950	114.9 ± 3.80	116.6 ± 2.86	103.4 ± 4.07	99.8 ± 3.54
10.070	114.1 ± 2.84	113.9 ± 2.28	105.5 ± 5.38	101.3 ± 5.44

The fully methylated surfaces of the coarse and fine sand samples were taken as 100° and 115° respectively. Using these values and the Cassie equation (Eq. 1), the percentage of sand coated within each of the four sieve fractions for concentration up to 1.855 g kg^{-1} has been calculated and illustrated in Figure 3. The larger shape parameters associated with the coarser sieve fractions showed a higher tendency to coat by the DMDCS at and below this value.

$$\cos ACA = f_1 \cos \theta_1 + f_2 \cos \theta_2 \quad (1)$$

In Eq. 1, f_1 and f_2 represent the fraction of soil having intrinsic contact angle θ_1 and θ_2 respectively. For the un-methylated soil fraction, θ_1 was taken as 0° and θ_2 denotes the fully methylated soil surface.

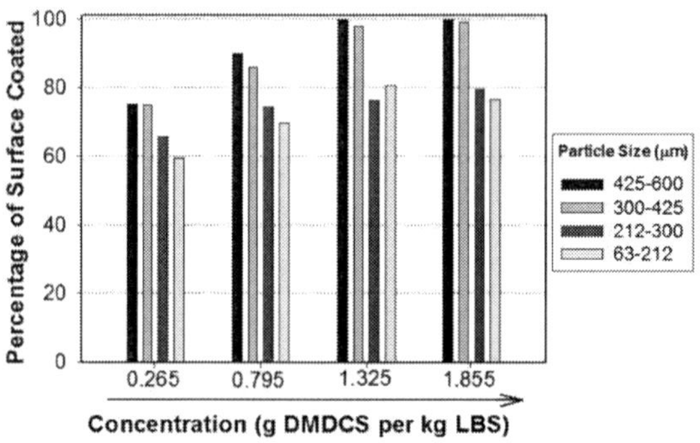

Figure 3. Percentage of surface coated within each sieve fraction.

4 Conclusions

This study shows that particle size is a significant parameter to be considered if the wetting properties of soils are to be exploited in the field of geotechnical engineering. It was found that the chemically synthesised water repellent soils examined were more water repellent with the finer fractions for a concentration above 2 g DMDCS per kg of LBS. Below this value, it is not distinguishable which of the two sieve sizes had a larger ACA. Furthermore, the influence of particle size on the wettability of chemically synthesised sand seem to suggest that there are threshold values regarding the particle size and DMDCS concentration which dictate how and to which extent the ACA will vary.

The maximum ACA values reached by the coarser and finer fractions were around 100° and 115° respectively. At high concentration of DMDCS, it was concluded that the effect of trapped air within soil particles is small as compared to the particle characteristics such as their surface area and shape.

Acknowledgement

This research work has been possible through the funding support provided by The University of Hong Kong. Help and comments by Mr. Shuang Zheng as well as laboratory support from Mr. N.C. Poon are much appreciated.

References

1. R.D. Bond, Trans. 9th Int. Congr. Soil Sci. **1,** 339-347 (1968)
2. L.F. DeBano, L.D. Mann, D.A. Hamilton, SOIL SCI. SOC. AMER. PROC. **34,** (1970)
3. S. Doerr, R. Shakesby, R. Walsh, Earth Sci. Rev. **51,** 33-65 (2000)
4. C. Ritsema, L. Dekker, Aust. J. Soil Res. **34,** 475-487 (1996)
5. S.D.N. Lourenço, N.C. Jones, S.H. Doerr, R. Bryant, Vadose Zone J. **14,** (2015)
6. J.P. Bardet, J. Jesmani, N. Jabbari, EJGE. **16,** 451-461 (2011)

 DOI: 10.1051/e3sconf/20160909003

7. C. Lee, H. Yang, T.S. Yun, Y. Choi, S. Yang, Vadose Zone J. **14**, (2015)
8. J. Bachmann, A. Ellies, K.H. Hartge, J HYDROL. **231-232,** 66-75 (2000)
9. J. Shang, M. Flury, M. Harsh, J.B. Harsh, R.L. Zollars, J COLLOID INTERF SCI. **328,** 299-307 (2008)
10. A.F. Stalder, T. Melchior, M. Muller, D. Sage, T. Blu, M. Unser, COLLOID SURFACE A. **364,** 72-81 (2010)
11. J. Bachmann, S.K. Woche, M.-O. Goebel, WATER RESOUR RES. **39,** 1353-1366 (2003)
12. I.E. Dubois, S. Holgersson, S. Allard, M.E. Malmström, PROC MONOGR ENG WATE. 717-720 (2010)

Characterization of the secondary swelling of compacted Maryland clay

Shengyang Yuan[1], Xianfeng Liu[1,a], Scott Sloan[1], Olivier Buzzi[1]

[1]*Priority Research Centre for Geotechnical and Materials Modelling, School of Engineering, The University of Newcastle, NSW, Australia.*

Abstract. This paper presents the result of a multi-scale experimental investigation on the phenomenon of secondary swelling. Specimens of compacted Maryland clay were prepared under different initial conditions (void ratio and water content), loaded at different levels of vertical stress and inundated until secondary swelling was observed. One test was also performed with suction control and incremental swelling. These tests were complemented by micro structural analyses via mercury intrusion porosimetry. The coefficient of secondary swelling, defined as the gradient of the evolution of void ratio with logarithm of time was calculated and correlated to the macro scale strain. Several trends were observed and it was showed that, under low confinement, the secondary swelling mostly takes place in the macro pores.

1 Introduction

Due to financial and environmental constraints, a number of roads projects in Australia have seen problematic expansive soils incorporated in road embankments. For example, an embankment of the Hunter expressway, located in the hunter valley, west of Newcastle, New South Wales, is made of about nine meters of expansive material over which about two and half meters of a capping layer have been placed. This layer is meant to provide a surcharge against swelling and prevent moisture ingress [1-2]. However, fluctuations of the water table level might lead to moisture movement in the embankment and swelling or collapse of the expansive material. In the core of the embankment, i.e. away from the areas sensitive to seasonal influence, the wetting process is likely to be slow and monotonic, similar than under a slab or a foundation [3]. With a tolerance to swelling, imposed by the road and maritime services, close to zero, it becomes relevant to consider both primary and secondary swelling in the analysis/prediction of the long term response of the embankment. To date, researchers attention has largely been focused on the primary swelling and there is little data on the secondary swelling [4].

This study investigates the swelling behaviour of compacted specimens of an Australian expansive soil with an emphasis on the secondary swelling phase. A large number of one-dimensional swelling tests under different vertical stress and initial conditions were performed. The response during incremental swelling, with a controlled reduction in suction, was also investigated. The macro scale observations were completed by a number of micro structural analyses.

2 Materials

A natural expansive clayey soil, coming from the field site of the University of Newcastle in Maryland, NSW, Australia, was used for this experimental investigation (see details in [5]). The soil is highly reactive with about 10% in mass of interlayered illite-smectite clay. The plasticity index of the material is about 45%. More details regarding the characterization of the clay can be found in [5]. Crumbs of natural soil were dried at 105 °C and crushed to particles smaller than 1.18 mm. Three batches of soil were then sprayed with water in order to reach three different target water contents (13.2 %, 17.8% and 21.4 %). These values of water content correspond to suctions of 6.38 MPa, 2.57 MPa and 0.82 MPa, respectively. After an equilibration period of two weeks, the specimens were compacted at 0.5 mm/min in an oedometer ring (diameter of 45 mm, height of 19 mm) using a load frame, in order to reach specific void ratios.

3 Experimental facilities

This study involves 3 series of tests. The first series investigated the effect of vertical stress, ranging from 6 kPa to 400 kPa on the secondary swelling phase. For these tests, the void ratio was constant (e_o = 0.62) but three initial suctions were used. The second series of tests looked at the effect of the initial void ratio. Here, only one stress value was applied (100 kPa) and one value of initial suction was used (6.38 MPa). All tests were conducted using standard oedometers by flooding the specimens with distilled water. The final test to be conducted involves swelling with suction control by the

[a] Corresponding author: Xianfeng.Liu@newcastle.edu.au

DOI: 10.1051/e3sconf/20160909004

osmotic method and a multistage reduction of suction. This test was conducted under a constant vertical stress of 6 kPa with an initial void ratio of 0.62 and an initial suction of about 0.86 MPa. This last test was performed using the specific device depicted in Figure 1 with the suction generated by the osmotic system duly calibrated as in [6]. The apparatus was placed in a controlled temperature environment (20°C±0.1°C) and the suction generated was estimated using the equation provided in [7]:

$$s = 11\,c^2 \tag{1}$$

where s is the suction in MPa and c is the PEG concentration in grams of PEG per gram of water. A thin layer of paraffin oil was placed on top of the PEG solution to prevent evaporation.

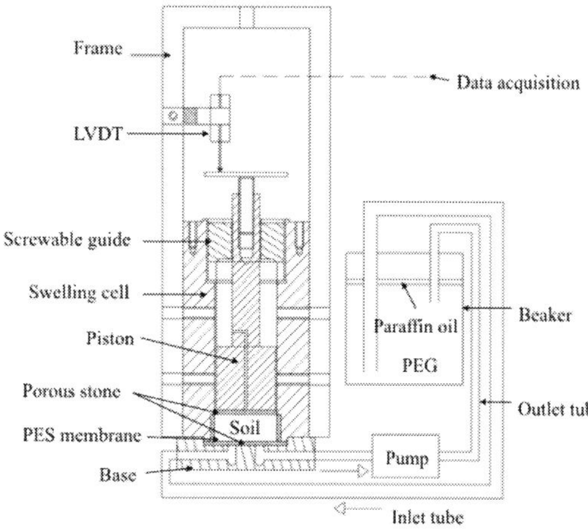

Figure 1 Sketch of the swelling cell with osmotic control of suction.

The microstructural investigations were performed using mercury intrusion Porosimetry (MIP), which is based on the Laplace-Young's law, an equation that relates the capillary pressure P_c to the capillary radius r:

$$P_c = \frac{2\gamma\cos\theta}{r} \tag{2}$$

where θ is the contact angle between the liquid mercury and the solid surface (140°), and γ is the mercury surface tension ($\gamma = 0.482 N/m$). The MIP apparatus is a Micromeritics AutoPore IV 9500 with a maximal pressure of 228 MPa (yielding an equivalent entrance pore diameter of about 6.5 nm, as per equation 2).

Some Scanning Electron Microscope (SEM) images were obtained using a Zeiss Sigma VP FESEM (Bruker light element SSD EDS detector). For this part, specimens were broken rather than cut in order to avoid a "smearing effect". Also, for better imaging contrast, the undisturbed surfaces of specimens were coated with gold under vacuum. It is essential to remove all moisture from the soil specimens for meaningful SEM and MIP analysis. This was achieved by a freeze-drying process where the "freeze" phase consists of immersing small (0.5 cm^3) soil pieces in liquid nitrogen while the "drying phase" consists of sublimating the amorphous ice formed in the freeze phase using a freeze-dryer (Christ ® ALPHA 1-2). This latter process lasts at least 48 hours, after which no moisture uptake should be permitted. Consequently, all specimens were sealed in airtight plastic containers bottles and stored in a desiccator containing dessicant gel.

4 Results and discussion

4.1 Inundation tests

Figure 2 shows the typical evolution of swelling strain with time for compacted specimen swelling under different loads. The phases of primary and secondary swelling are clearly visible and, consistent with other studies of the literature, the higher the vertical stress, the lower the swelling strain. Note that collapse was observed for vertical stresses of 300 and 400 kPa for initial suctions of 2.57 MPa and 0.82 MPa. This is later reported as a negative swelling strain.

The response illustrated in Figure 2 can be used to compute the coefficient of secondary swelling $C_{\alpha s}$ defined as:

$$C_{\alpha s} = \frac{\Delta\varepsilon}{\Delta Log_{10}(t)} \tag{3}$$

where $\Delta\varepsilon$ is the change in swelling or compression strain between the end of primary (EOP) swelling and any hydration time t larger than the time required to reach EOP.

Figure 3 shows the evolution of $C_{\alpha s}$ with the magnitude of vertical stress for three different initial suctions (Figure 3a). Consistent with the general response of Figure 2, the higher the stress, the lower $C_{\alpha s}$. For a given stress level, $C_{\alpha s}$ is also found to increase as the initial suction increases. The effect of the void ratio is shown in Figure 3b. $C_{\alpha s}$ progressively drops from 1.4 to about 0 as the void ratio increases from 0.47 to 0.82. These tests suggest that the behaviour in the secondary swelling is fully consistent with that of the primary swelling. Based on these results, an attempt was made to correlate the coefficient of secondary swelling to the swelling strain at the end of primary swelling (EOP). Fitting the data (see Figure 4) returned a R^2 coefficient of 0.88:

$$C_\alpha = 0.135 e^{\left(\frac{\varepsilon_{EOP}}{3.9}\right)} - 0.08 \tag{4}$$

Note that the magnitude of final strain reflects the initial specimen conditions (void ratio and suction) and the level of confinement (vertical stress). Consequently, these parameters do not need to appear explicitly in the equation. The correlation can then be used to estimate the magnitude of swelling occurring in the secondary phase

DOI: 10.1051/e3sconf/20160909004

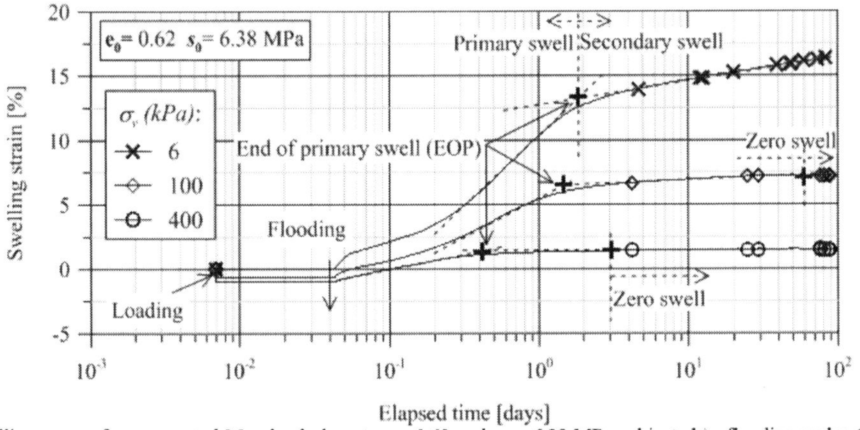

Figure 2. Swelling curves for compacted Maryland clay at e_0 = 0.62 and s_0 = 6.38 MPa subjected to flooding under 6, 100 and 400 kPa.

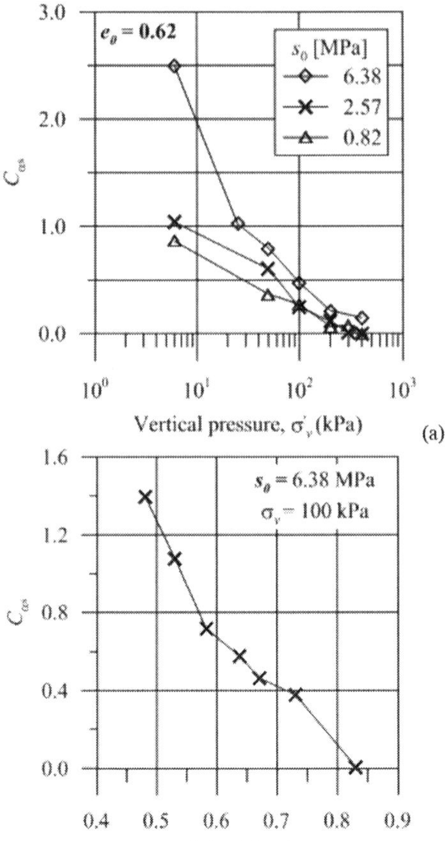

Figure 3: Evolution of $C_{\alpha s}$ with applied vertical pressure (a) and with the initial void ratio (b).

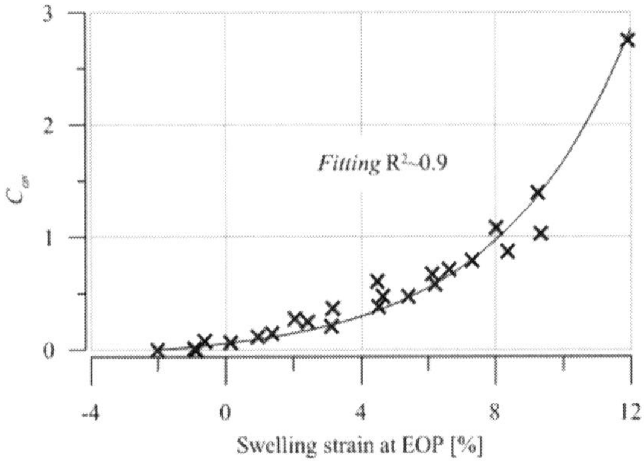

Figure 4 Evolution of $C_{\alpha s}$ with swelling strain at the end of primary swelling (EOP) for compacted ML samples at different initial states.

4.2 Incremental swelling

The incremental swelling test, performed under 6 kPa, also shows occurrence of secondary swelling (see Figure 5). Note that the first stage of the test (for a suction of 860 kPa) is an equilibration phase not resulting in swelling. The results presented in Figure 5 are consistent with existing data of the literature (e.g. [8]).

The coefficient of secondary swelling was estimated at the various stages of hydration and plotted as a function of the suction at the beginning of each stage and swelling strain occurring for each stage. Results are showed in Figure 6. The coefficient of secondary swelling appears to be increasing linearly with the imposed suction. Again, this is consistent with the response during primary swelling (see Figure 5). However, the $C_{\alpha s}$ values associated with incremental swelling do not align with those coming from the inundation tests. This is evidenced in Figure 6b where Equation 2 is represented by a dashed line. This might be due to different swelling mechanisms occurring at different stages of the overall swelling process but of the changes in hydration procedure. At this stage, this is not fully explained yet.

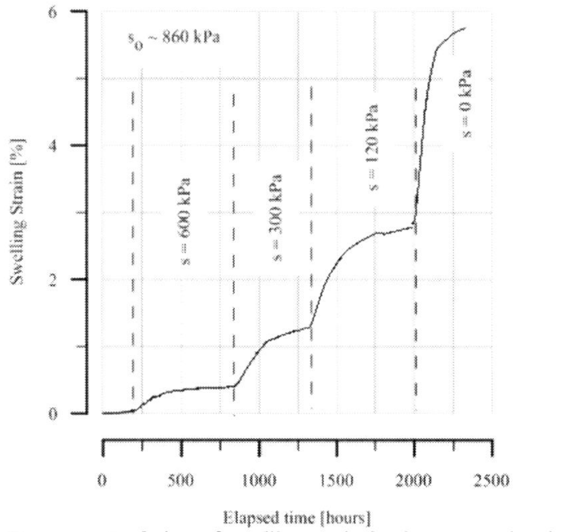

Figure 5. Evolution of swelling strain in time as suction is incrementally reduced from 860 kPa to 0 kPa. Initial void ratio e_o = 0.62.

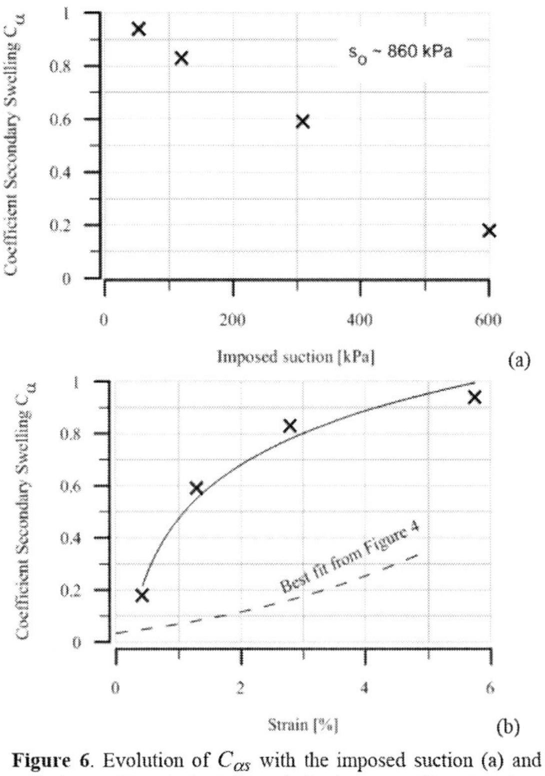

Figure 6. Evolution of $C_{\alpha s}$ with the imposed suction (a) and with the swelling strain at the end of primary swelling at various stages of the incremental swelling test (b). Initial suction of about 860 kPa and initial void ratio of 0.62.

4.3 Evolution of pore size distribution

The initial pore size distributions (PSD) pertaining to three different initial states are shown in Figure 7. The curves show a clear influence of the initial suction on the compaction process and the resulting bimodal microstructure. Increasing the initial suction tends to reduce the size of the predominant pores, both at macro and micro levels. However, the effect of initial suction on the frequency of predominant pores is opposite at macro and micro scale.

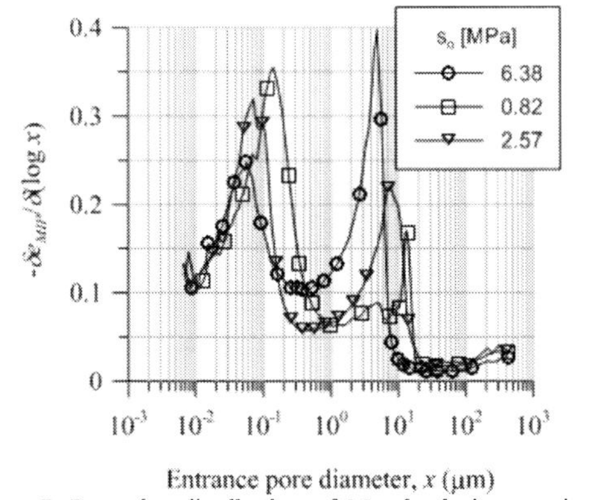

Figure 7. Pore size distribution of Maryland clay specimens compacted to a void ratio of 0.62 with an initial suction of 0.82, 2.57 and 6.38 MPa.

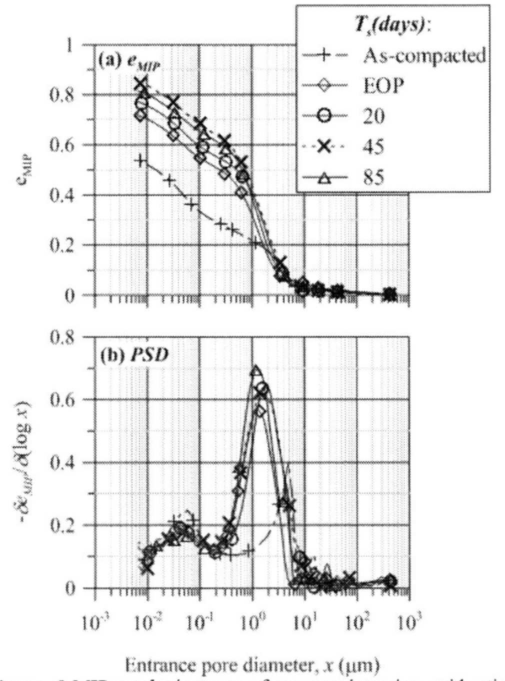

Figure. 8 MIP results in terms of mercury intrusion void ratio, e_{MIP} (a) and pore size distribution (b) for compacted samples prepared at e_o=0.62, s_o=6.38 MPa and subjected to a vertical stress σ_v of 6 kPa.

One swelling test was conducted with several parallel specimens in order to conduct MIP analysis at various stage of the swelling process. This test pertains to the following initial conditions: e_o=0.62, s_o=6.38 MPa and σ_v=6 kPa. The results show a clear decrease in predominant pore size as a result of the swelling process, which has already been observed in the literatures. Interestingly, this peak does not seem to shift during secondary swelling. Also, most of the microstructural changes seem to be occurring in the macro pores after EOP.

Finally, some SEM images taken prior and after 2 days (EOP) and 85 days swelling tests show a strong modification of the specimen structure. While the initial structure is clearly aggregated (see Figure 9a), the structural aggregation tends to be significantly reduced at end of primary swelling (Figure 9b) until that the final structure appears to have been smeared. The aggregates have swollen and filled the inter-aggregates pores yielding a more uniform structure (Figure 9c). A similar final structure was observed from a vertical stress of 100 kPa. Note that the final soil structure, as shown in Figure 9, is the result of the final swelling process, i.e. primary and secondary phases.

Figure 9. SEM images taken on compacted specimens (e_0 = 0.62 and s_0 = 6.38 MPa) right after compaction (a), at end of primary swelling (b) and after 85 days hydration under 6 kPa (c).

4.4 Discussion on multiscale coupling behaviour

The mineralogy analysis shows that ML clay consists of both swelling and non-swelling minerals [5]. When the compacted ML samples were subjected to a decrease in imposed suction, the swelling minerals, for instance illite-smectite interlayered clay, within the clay aggregates expanded, leading to the expansion of the whole aggregate and in turn a decrease in the magnitude of structural aggregation (Figure 9). Consequently, the dominant pore size of intra-aggregate pores increases while that of inter-aggregate pores decreases due to the invasion of aggregates (Figure 8). As a result of the swelling occurring at the microscale, the swelling strain progressively developed (Figure 2 and 5). More swelling strain seems to take place at low suction range (Figure 5). The changes associated with swelling at both micro- and macro-scale appear to have occurred mainly during the primary swelling stage (prior to EOP). Moreover, the subsequent hydration during the secondary swelling stage leads to a microstructural homogenization (Fig.9c) and only a small increase in peak density of PSD without affecting much the dominant pore size.

As shown by the previous studies [9-13], the occurrence of swelling of expansive minerals consists of the development of crystalline swelling and osmotic swelling, which is associated with the hydration of inter-layer and interparticle pore spaces, respectively. In case of Na dominated bentonites, the crystalline swelling may widen the basal spacing, accommodating up to 4 layers of water molecules, while the osmotic swelling is mainly related to the formation of electrical diffuse double-layers [10-12]. These swelling mechanisms may cause a significant breakdown of the clay particles made up of several hundreds of stacked unit layers when the adsorbed interlayer water is more than two [11-12]. The complete separation of two swelling phenomena appears to be difficult since they can coexist as shown by Saiyouri et al. [11-13]. The predominance of the one over the other highly depends on the hydration conditions. For instance, in case of the incremental swelling (Figure 5), the crystalline swelling tends to fully develop at high suction range, which is followed by the occurrence of the osmotic swelling at low suction range. In order words, the water adsorption seems to take place in a progressive manner initially at interlayer space and then at interparticle space. However, in case of inundating the sample, the ingress of water may be initiated by the osmotic swelling at the boundaries of aggregates, likely resulting in a quick increase in swelling strain during the primary swelling phase. This process is followed by the crystalline swelling through water diffusion from inter-aggregate to intra-aggregate pore space, which plays an important role in the secondary swelling phase. The different swelling mechanisms involving in the two types of swelling tests may explain the difference in the swelling strain rate during the secondary swelling phase. For instance, the incremental swelling likely induces more important breakdown the clay particles and in turn an increase in the exposed surface sites of water adsorption. This may further enhance the swelling magnitude occurring during the secondary swelling phase.

5 Concluding remarks

This study investigates the response of compacted specimens of an expansive soil during the secondary swelling phase. A series of inundation swelling tests under constant vertical stress were performed on specimens prepared at different initial suctions and void ratios. These tests were complemented by one incremental swelling test with a suction control device. The behaviour during the secondary swelling was found to be consistent with the response during primary swelling. A very strong correlation was found between the coefficient of secondary swelling and the strain at the end of primary swelling, both for swelling until saturation and partial swelling. However, the trends pertaining to incremental swelling and inundation tests do not agree, which is not fully explained at this stage. The micro structural analyses show a strong effect of the initial suction on the pore size distribution of the compacted specimens. Also, it seems that, under 6 kPa of normal stress, most microstructural changes occur in the macro pores during the secondary swelling phase.

References

1. O. Buzzi O, S. Fityus, R. Kingsland, S. Aryal ANZ 2012 Conf. proceedings, Melb. pp 1075–1080 (2012)
2. S. Aryal, R. Kingsland, D. Och et al. ANZ 2012 Conf. proceedings, Melb. pp 1304–1309 (2012)
3. J. Nelson and D. Miller. Expansive Soils: Problems and Practice in Foundation and Pavement Engineering. John Wiley and Sons, New York (1997)
4. G. Mesri, C.R. Ullrich, Y.K. Choi Géotechnique **28**:281–307 (1978).
5. X.F. Liu, O. Buzzi, S.Y. Yuan, J. Mendes and S. Fityus. Can. Geo. Journal, **53**: 1–17 (2016).
6. S.Y. Yuan, X.F. Liu and O. Buzzi. Paper submitted to the 3rd European Conference on Unsaturated Soils, Paris (2016).
7. P. Delage, M. Howat and Y.J. Cui. Eng. Geology, **50** (1-2): 31-48 (1998).
8. V. Escario and J. Saez. In: Proceedings of the Third International Conference on Expansive Soils. Vol. 1. Haifa, Israel, pp. 195–200 (1973).
9. R. Pusch and R. Yong. Microstructure of smectite clays and engineering performance, Taylor and Francis, New York (2006).
10. L. Liu. Prediction of swelling pressures of different types of bentonite in dilute solutions. Colloids Surf. A. Physicochem. Eng. Asp., 434, 303–318 (2013).
11. N. Saiyouri, P. Y. Hicher and D. Tessier. Microstructural approach and transfer water modelling in highly compacted unsaturated swelling clays. Mech. Cohesive-frictional Mater., 5(1), 41–60 (2000).
12. N. Saiyouri, P. Y. Hicher and D. Tessier. Experimental study of swelling in unsaturated compacted clays. Clay Miner., 39(4), 469–479 (2004).
13. P. Delage, D. Marcial, Y.J. Cui and X. Ruiz. Ageing effects in a compacted bentonite: a microstructure approach. Géotechnique, 56:291–304 (2006).

 DOI: 10.1051/e3sconf/20160909005

Stress-Strain response of Malaysian granitic residual soil grade V, according to Rotational Multiple Yield Surface Framework

Pooya Saffari [1,a], Mohd Jamaludin Md. Noor [2] and Yasmin Ashaari [2]

[1] Ph.D. Candidate, Department of Civil Engineering, Universiti Teknologi Mara, Shah Alam, Malaysia
[2] Associate Professor, Department of Civil Engineering, Universiti Teknologi Mara, Shah Alam, Malaysia

Abstract. In soil mechanics characterizing the soil volume change behavior is based on the effective stress concept. Conventional soil settlement models which have been developed by this principle, are empirical in nature and being verified against experimental or field observations. However, they are not still able to explain and formulate some unusual soil volume change behaviors such as wetting collapse by this concept. The consequence of shear strength in the interaction of effective stress should be integrated into volume change frameworks in such cases in order to puzzle out these weird behaviors. The principle has been incorporated in an anisotropic soil settlement model know as Rotational Multiple Yield Surface Framework (RMYSF). The framework has been developed from the standpoint of the soil stress-strain response within anisotropic stress condition. The anisotropic soil volume change behavior is described from the interaction between applied stress (presented by the Mohr-Coulomb circle) and shear strength (in terms of curved-surface mobilized shear strength envelope). The applicability of the model has been examined for limestone gravel from the United Kingdom and granitic residual soil grade VI from Kuala Kuba Baru, Malaysia before, however still needs to be tested for another kind of soils. In this paper, a series of triaxial tests were conducted on Malaysian granitic residual soil grade V, in saturated and unsaturated condition. The stress-strain curves and volume change behavior of the soil were obtained from laboratory test results. Besides that, the stress-strain response of the soil has been predicted using RMYSF framework. The predicted stress-strain curves were compared with the test result and the accuracy of the framework was examined. The obtained and predicted curves had a fairly good match.

1 Introduction

The stress-strain behavior in a partially saturated soil is highly dependent on the relevant stress variables. Being mostly in unsaturated state, the surface soils are under the influence of evaporation and rainfall. Basically, the engineering properties of unsaturated soil vary from the saturated state. In this respect, various problems related to slope failure and inundation settlement (wetting collapse) can be simply solved using the approaches that take into account of unsaturated condition.

The conventional volume change models [1-4] were piled on Terzaghi effective stress concept [2]. These models are developed in the fact that the increment of effective stress is linked to settlement. When total stress increases or pore water pressure is expelled within a soil mass, the effective stress increases, thereby settlement occurs. However, the settlement would terminate at some point because of rearrangement of soil particles to a denser position and increase of the soil stiffness. This rise in soil stiffness is a sign of the increment of the soil strength. This indicates the importance of shear strength in controlling the settlement behavior. Nevertheless, incorporating the shear strength in a settlement model is not an easy effort and most of such frameworks are the empirical type. In additional, the existence of some unusual soil volume change behaviors such as wetting collapse which are in contrast with effective stress concept makes the situation more complicated.

Some researchers [5-7] reported that wetting collapse happens during the wetting process of soil mass under steady surcharge. Before inundation, soil mass initially is partially saturated and pore water pressure is negative. Soil particles are pushed together due to a surface tension (suction). After infiltration, according to the Terzaghi effective stress concept, pore water pressure changes from being negative to positive, thereby a decrease in effective stress is reached. As a result, wetting collapse happens as a reduction in effective stress. The phenomenon cannot be only be characterized relying on conventional settlement models. Therefore, besides effective stress, a second factor can affect the soil volume change of unsaturated soils. This factor is the reduction of shear strength [8]. The volume change behavior of unsaturated soil can be even more complicated while massive settlement near saturation [9] or extensive wetting collapse at low net stress compared with higher net stress [10] exists.

[a] Corresponding author: saffari.pooya@gmail.com

 DOI: 10.1051/e3sconf/20160909005

The main objective of the current study is to examine the accuracy of a new semi-empirical volume change model know as Rotational Multiple Yield Surface Framework (RMYSF) against the volume change behavior of Malaysian granitic residual soil grade V. The applicability of the model has been examined for limestone gravel from the United Kingdom [11] and granitic residual soil grade VI from Kuala Kuba Baru, Malaysia [12] before, however still needs to be tested for another kind of soils. A series of triaxial tests were conducted on Malaysian granitic residual soil grade V, in saturated and unsaturated condition. The stress-strain curves and volume change behavior of the soil were obtained from laboratory test results. Besides that, the stress-strain response of the soil has been predicted using RMYSF framework. Finally, the experimental result and the predicted result were compared and the accuracy of the framework was examined.

2 Rotational Multiple Yield Surface Framework (RMYSF)

Md. Noor and Anderson [11] proposed an anisotropic soil volume change model known as Rotational Multiple Yield Surface Framework (RMYSF). It has been established on the extended concept of interaction between mobilized shear strength and effective stress. The concept defines settlement as the influence of a reduction in shear strength (wetting collapse) or an increase in effective stress (wetting collapse).

The framework was introduced relying on elastic-plastic soil stress-strain response. The authors claimed that this semi-empirical model is able to predict settlement either for wetting collapse or loading collapse, through a set of laboratory and theoretical concepts. While soil is in the saturated condition, stress state is defined as effective stress, $(\sigma - u_w)$, and in unsaturated condition is taken as net stress, $(\sigma - u_a)$. Stress state is presented by Mohr circles and considered as driving variables. On the other hand, as presented in Figure (1), the curved-surface mobilized shear strength envelope of [13], shows the state of mobilized shear strength and behaves as resisting variable in the soil mass.

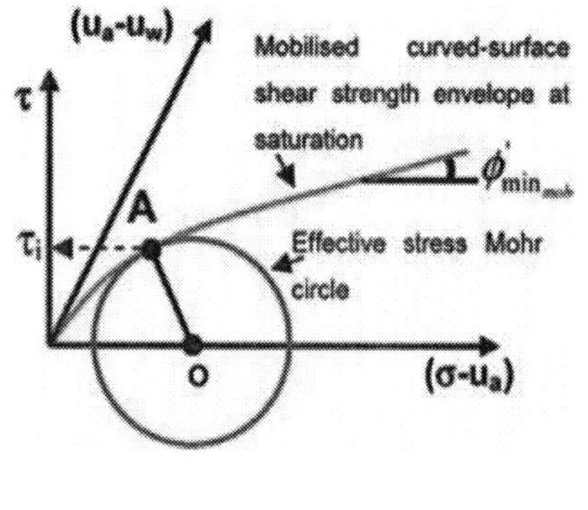

Figure 1. Mobilized shear strength envelope (the Mohr circle representing the state of applied stress and the curved-surface envelope representing the state of shear strength) [12].

Based on the model, the settlement occurs while the state of stress (net or effective) extends beyond the mobilized shear strength envelope. In this situation, the driving variable is greater than resisting variable. While a compression process is in place, soil elements shift to a denser state. At this stage, the mobilized shear strength withstands against this transformation by rotating anticlockwise about the origin towards the mobilized shear strength envelope at failure and providing the frictional resistance. At this point, an addition in shear strength of the soil reaches. The process will move forward till the resisting and driving variables come to an equilibrium state. When this state is reached, the Mohr circle is located under mobilized shear strength. Therefore, the driving variables are smaller than resisting variables and soil is stable. Adversely, if the Mohr circle is located over the mobilized shear strength, it replicates that the soil mass is not stable and it is subjected to the settlement. However, wetting collapse takes a place under constant stress. In such case during inundation, the shear strength of the soil reduces due to a reduction in suction. Decrease in suction causes a non-linear decrease in apparent shear strength (c′) [13] while the size of the Mohr circle remains constant (a variation on suction does not any effect on the net stresses). At this point, if the Mohr circle relocates beyond the mobilized shear strength envelope, settlement occurs. Mobilized shear strength envelope rotates into a position with a bigger mobilized friction angle.

Later on, it was reported that there is a unique relationship between the minimum mobilized friction angle, $Ø'_{min\ mob}$, and axial strain, ε_a [14] . Applying the deviator stress rotates the mobilized shear strength into a new position regardless of the value of effective or net stress in accordance with axial strain either for saturation or unsaturation condition. RMYSF model is based on this unique relationship. The authors claimed that this relationship is able to predict the stress-strain response of a specific soil at every desired net stress or effective stress. Equation 1, was introduced by [14] in order to obtain the best fit curve for the minimum mobilized friction angle, $Ø'_{min\ mob}$ and axial strain, ε_a.

$$\varepsilon_a = \frac{ln\left[\frac{Ø'_{min\,f} - Ø'_{min\,mob}}{Ø'_{min\,f}}\right]}{\Omega} \quad (1)$$

Ω is a fitting parameter and known as soil coefficient of anisotropic compression. The value of Ω depends on the soil structure.

3 Stress-strain response and Mobilized shear strength envelopes for Malaysian granitic residual soil grade (V)

3.1 Triaxial test results

The soil tested in this study was granitic residual grade V from Seremban, Selangor, Malaysia. The specimen consists of 51.25% sand, 45.23% gravel, 1.18% silt and

2.25% clay and classified as well graded SAND. Consolidated drained triaxial tests in fully saturated and partially saturation condition were conducted on specimens with 50 mm diameter and 100 mm height. In terms of partially saturated condition a modified double wall triaxial apparatus with suction control was used. The suction of 70 kPa (residual suction obtained from SWCC test) was applied to the specimen using axis translation technique [14]. The triaxial test has been conducted at effective stresses of 50 kPa, 100 kPa, 200 kPa and 300 kPa. From the shearing stage, the stress-strain curves for the saturated and unsaturated condition were plotted as shown in Figures 2 and 3 respectively. Shear strength parameters of the tested soil were obtained following CSESSSM model of [13] and are given in Table 1.

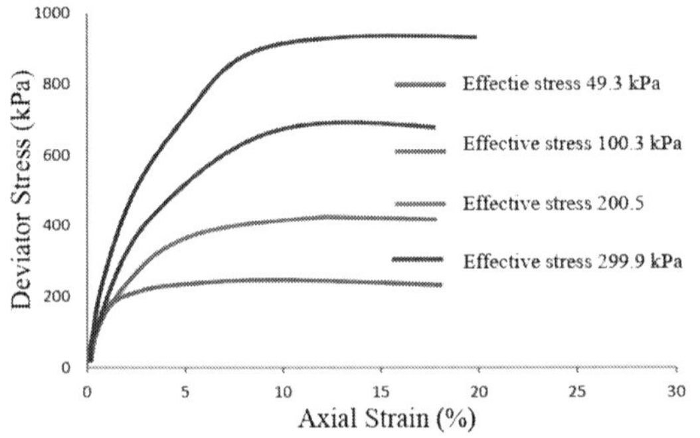

Figure 2. Stress-strain curves in fully saturated condition from the triaxial test.

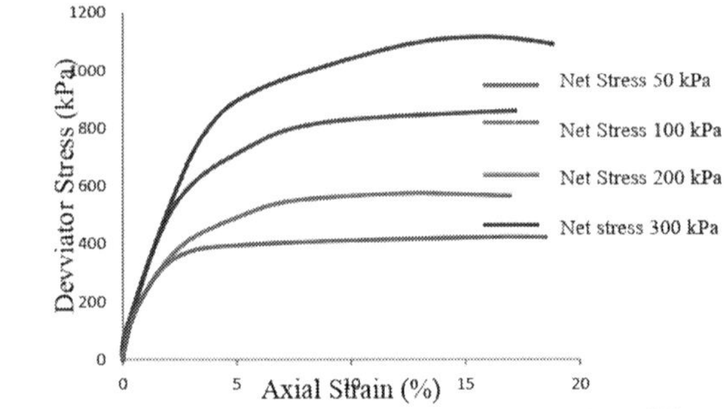

Figure 3. Stress-strain curves in unsaturated saturated condition (70 kPa suction) from the triaxial test.

Table 1. Shear strength parameters using CSESSSM model

Shear strength parameters	Value
$\phi'_{min\,f}$	34°
$(u_a-u_w)_r$	70 kPa
$(u_a-u_w)_u$	250 kPa
$(\sigma-u_a)_t$	122 kPa
τ_t	105 kPa
C_s^{max}	31 kPa

The value of axial strain in the stress-strain curves (Figures 2 and 3) indicates the specimen axial deformation due to the related deviator stress. For example, for 10% strain at 200 kPa effective stress in the saturation condition, the value of deviator stress is 550 kPa. On the other hand, if the specimen loaded to 550 kPa deviator stress at 200 kPa effective stress in the fully saturation condition, it would settle 10% of its initial height.

Mobilized shear strength envelopes in various axial strains in fully saturated condition are presented in Figures 4-7. Mohr circles were plotted under the effective stress of 49.3, 100.3, 200.5 and 299.9 kPa and their related deviator stress in various axial strains. From the Mohr circles, the mobilized shear strength envelopes and their mobilized friction angles in various axial strains were obtained according to the curved surface envelope of [13].

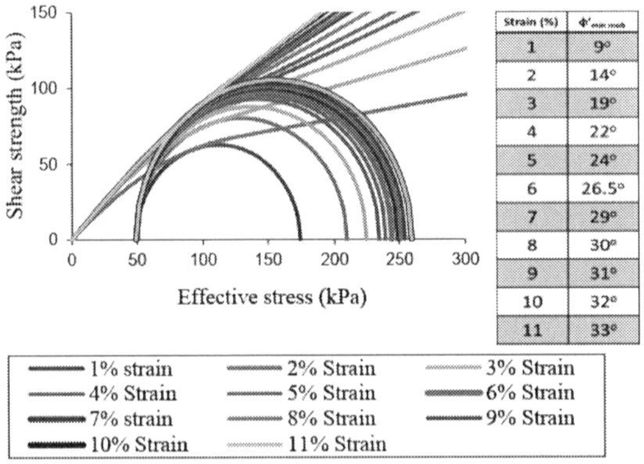

Figure 4. Mobilized shear strength envelopes in various axial strains in fully saturated condition and 49.3 kPa effective stress.

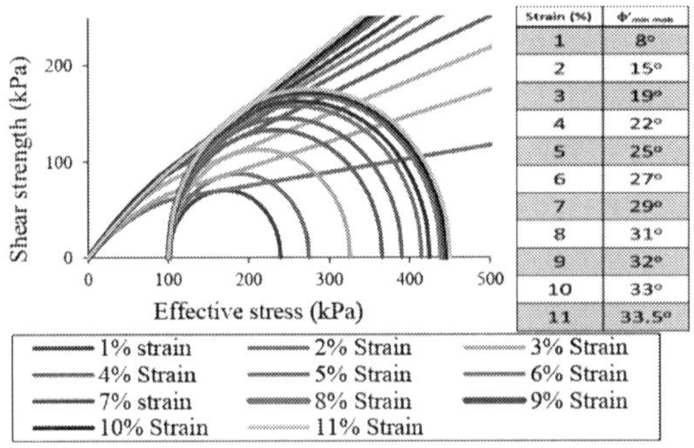

Figure 5. Mobilized shear strength envelopes in various axial strains in fully saturated condition and 100.3 kPa effective stress.

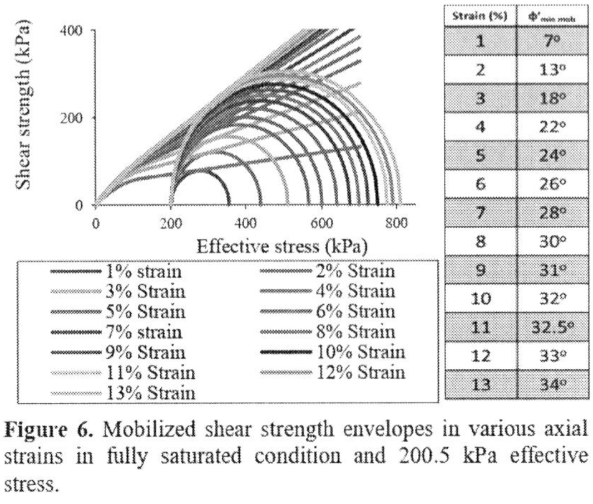

Figure 6. Mobilized shear strength envelopes in various axial strains in fully saturated condition and 200.5 kPa effective stress.

 DOI: 10.1051/e3sconf/20160909005

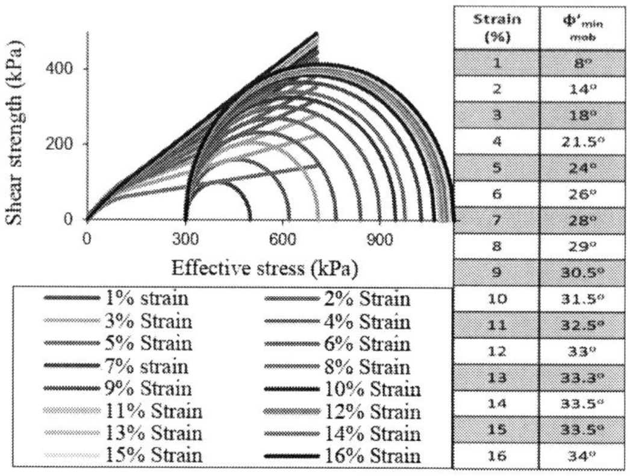

Figure 7. Mobilized shear strength envelopes in various axial strains in fully saturated condition and 299.9 kPa effective stress.

As illustrated in Figure 8, it can be observed that there is a relationship between each specific axial strain and $Ø'_{min\ mob}$. In fact, the position of mobilized shear strength presents a specific amount of axial strain irrespective of effective stress. This is the identical unique relationship which has reported by [15]. This unique relationship for Malaysian granitic residual soil, at different effective stresses, is shown in Figure 8. In additional the unique relationship is plotted using the equation 1 with the value of 0.277 for the Ω. It can be observed that there is a good agreement between obtained $Ø'_{min\ mob}$ from the test result and the equation.

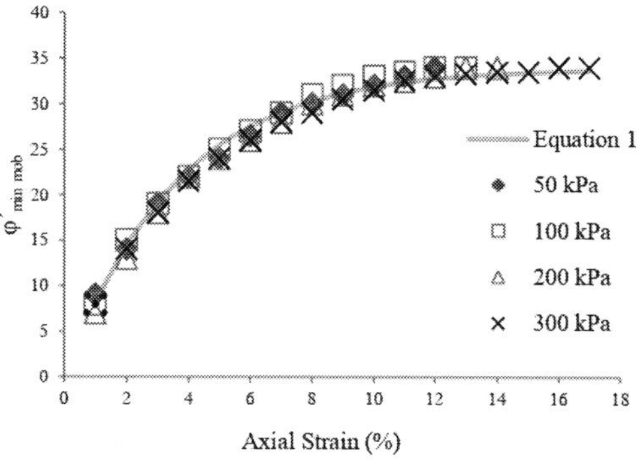

Figure 8. The unique relationship for Malaysian granitic residual soil grade V at different effective stresses using experimental test results and Equation 1.

3.2 Prediction of soil stress-strain response for granitic residual soil in unsaturated condition (70 kPa suction), using RMYSF model

As claimed by RMYSF, the unique relationship presented in Figure 8, is an intrinsic property of the soil. It is valid for the whole range of net stress and effective stress applied during the test. In other words, it is possible to predict the stress-strain curve of the soil at any desired net stress or effective stress up to the peak point of the curve. This prediction is carried out first by drawing Mohr circle, starting at desire $\sigma'3$. The Mohr circle has to touch the particular mobilized shear strength envelope and ends up at $\sigma'1$. It should be noted that in unsaturated condition, the value of apparent shear strength, c', related to the desire suction in accordance with CSESSSM model should be applied to the mobilized shear strength envelope. The diameter of the Mohr circle determines the respective deviator stress, q. By knowing the predicted deviator stress and axial strain presented by the mobilized shear strength envelope, the stress-strain can be predicted. The predicted deviator stresses related to various axial strains in the net stress of 50, 100, 200 and 300 kPa for Malaysian granitic residual soil grade V at 70 kPa suction are given at Table2. Figure 9 presents a comparison between predicted deviator stresses which are presented by dotted points and the actual stress-strain curves obtained from experimental results. It indicates that the predicted data is in an acceptable agreement with experimental results.

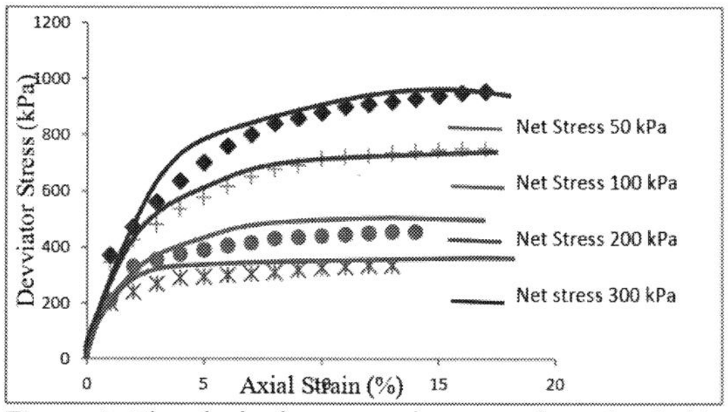

Figure 9. The obtained stress-strain curves from the double wall triaxial test against the predicted ones using RMYSF model.

Table 2. Predicted deviator streses for granitic residual soil grade V in unsaturated condition (70 kPa suction).

Axial Strain (%)	50 kPa	100 kPa	200 kPa	300 kPa
1	200	210	345	370
2	240	330	425	470
3	270	355	480	560
4	290	375	535	635
5	295	390	575	700
6	300	405	615	760
7	305	415	650	800
8	310	430	675	840
9	320	435	690	860
10	325	440	715	880
11	330	445	720	900
12	335	450	725	910
13	335	455	735	920
14		455	740	930
15			745	940
16			750	950
17			750	955

DOI: 10.1051/e3sconf/20160909005

4 Conclusion

The following conclusion can be observed from this study:

- From the experimental test results it can be observed that there is a unique relationship between minimum mobilized friction angle, $Ø'_{min\ mob}$ and axial strain, ε_a, regardless of effective stress for Malaysian granitic residual soil grade V. This relationship approves the concept of RMYSF model.
- The result also shows, using Equation 1 can provide a perfect fit for the unique relationship.
- RMYSF model has the ability to predict the stress-strain response of Malaysian granitic residual soil grade V in unsaturated condition with a good accuracy.

References

[1] N. Janbu, L. Bjerrum, and B. Kjaernsli, "Veiledning ved løsning av fundamenteringsoppgaver," 1956.

[2] K. Terzaghi, *Theoretical soil mechanics* vol. 18: Wiley New York, 1943.

[3] E. de Beer and A. MARTEN, "Method of computation of an upper limit for the influence of hetero-geneity of sand layers on the settlement of bridges'. 4th ICSMFE," ed: London, 1951.

[4] J. H. Schmertmann, P. R. Brown, and J. P. Hartman, "Improved strain influence factor diagrams," *Journal of the Geotechnical Engineering Division,* vol. 104, pp. 1131-1135, 1978.

[5] J. Jennings and J. Burland, "Limitations to the use of effective stresses in partly saturated soils," *Géotechnique,* vol. 12, pp. 125-144, 1962.

[6] R. Tadepalli, H. Rahardjo, and D. G. Fredlund, "Measurements of matric suction and volume changes during inundation of collapsible soil," *Geotechnical Testing Journal,* vol. 15, pp. 115-122, 1992.

[7] R. Blanchfield and W. Anderson, "Wetting collapse in opencast coalmine backfill," *Proceedings of the ICE-Geotechnical Engineering,* vol. 143, pp. 139-149, 2000.

[8] S. Wheeler and V. Sivakumar, "An elasto-plastic critical state framework for unsaturated soil," *Géotechnique,* vol. 45, pp. 35-53, 1995.

[9] E. E. Alonso, A. Gens, and A. Josa, "A constitutive model for partially saturated soils," *Géotechnique,* vol. 40, pp. 405-430, 1990.

[10] M. Md.Noor, R. Jidin, and M. Hafez, "Effective Stress and Complex Soil Settlement Behavior," *EJGE,* vol. 13, 2008.

[11] M. J. Md.Noor and W. F. Anderson, "A Qualitative Framework for Loading and Wetting Collapses in Saturated and Unsaturated Soils," in *16th South East Asian Geotechnical Conference*, Kuala Lumpur, Malaysia, 2007.

[12] M. J. Md.Noor and I. B. Mohamed Jais, "Rotational multiple yield surface framework and the prediction of stress-strain response for saturated and unsaturated soils," in *5th International conference on unsaturated soils* Barcelona, Spain, 2010.

[13] M. J. Md.Noor and W. Anderson, "A comprehensive shear strength model for saturated and unsaturated soils," in *Unsaturated Soils 2006*, 2006, pp. 1992-2003.

[14] P. Saffari, M. J. Md.Noor, B. A. Hadi, and S. Motamedi, "Laboratory Test Methods for Shear Strength Behavior of Unsaturated Soils Under Suction Control, Using Triaxial Apparatus," in *InCIEC 2014*, ed: Springer, 2015, pp. 567-576.

[15] I. B. Mohamed Jais and M. J. Md.Noor, "Establishing Unique Relationship between Minimum Mobilised Friction Angle and Axial Strain for Anisotropic Soil Settlement Model," presented at the Asia Pacific Conference on Unsaturated Soil, Newcastle, Australia, 2009.

DOI: 10.1051/e3sconf/20160909006

Interpreting measurements of small strain elastic shear modulus under unsaturated conditions

Ahmed Hasan[1] and Simon Wheeler[1, a]

[1] University of Glasgow, School of Engineering, Rankine Building, G12 8LT, Glasgow, UK

Abstract. Bender element testing of unsaturated isotropically compacted speswhite kaolin samples was used to investigate the variation of small strain elastic shear modulus G under unsaturated conditions. Testing was performed in a suction-controlled triaxial cell and involved combinations of isotropic loading and unloading stages and wetting and drying stages. Analysis of the experimental results indicated that the variation of G could be represented by a simple expression involving only mean Bishop's stress p^* and specific volume v, with the only significant mis-matches between measured and predicted values of G occuring at the end of final unloading. No significant improvement of fit was achieved by incorporating additional dependency on degree of saturation S_r or a bonding parameter ζ. The proposed expression for G reverts to a well-established form for saturated soils as S_r tends to 1.

1 Introduction

Bender element transducers [1] can be used to measure shear wave velocity V_s and hence elastic shear modulus G at very small strains (less than 0.001%), where G is related to V_s and the bulk density ρ of the soil by:

$$G = \rho V_s^2 \tag{1}$$

Under saturated conditions and isotropic stress states, many authors, e.g. [2, 3, 4], have found that the value of small strain shear modulus G for a particular soil is controlled by the mean effective stress p' and void ratio e. A commonly proposed form of expression for G under these saturated conditions and isotropic stress states is:

$$G = C f(e)(p'/p_a)^n \tag{2}$$

where C and n are soil parameters, p_a is atmospheric pressure and $f(e)$ is a function of void ratio [2, 4, 5].

Based on experimental evidence from tests on saturated sands, silts and clays at very small strains, many authors, e.g. [2, 6, 7], suggested a value of 0.5 for the exponent n in Eq.(2). Various expressions have been proposed for the function $f(e)$ in Eq.(2), but the most common, as proposed by Shibuya et al. [6], is:

$$f(e) = (1+e)^{-m} = v^{-m} \tag{3}$$

where v is specific volume and m is a soil parameter. Oztoprak & Bolton [4] suggested m=3.

For unsaturated soils, under isotropic stress states, based on experimental results from resonant column and bender element tests, Mancuso et al. [8] and Ng & Yung [9] proposed expressions for small strain elastic shear modulus G in terms of conventional unsaturated state variables of mean net stress $\bar{p}$ (where $\bar{p} = p - u_a$), matric suction s (where $s = u_a - u_w$) and void ratio e. It was, however, unclear whether such expressions, which include no direct dependency on degree of saturation S_r, could properly capture variation of G during wetting-drying cycles (where hysteresis in the water retention behaviour implies that S_r is not uniquely related to suction s).

Authors such as [10,11,12] interpreted measurements of small strain shear modulus G in terms of alternative unsaturated state variables, such as mean Bishop's stress p^*, defined by Wheeler et al. [13] as:

$$p^* = p - S_r u_w - (1 - S_r)u_a = \bar{p} + S_r s \tag{4}$$

For example, Chao [12] proposed that, under unsaturated conditions and isotropic stress states, G could be related to specific volume v, mean Bishop's stress p^* and a bonding parameter ζ introduced by Gallipoli et al. [14] (which depends upon both S_r and s) by the following expression:

$$G = C_1 v^{-3} [(p^*/p_a)^{0.5} + C_2 \zeta^{0.5}] \tag{5}$$

where C_1 and C_2 are soil parameters.

It is unclear whether Eq.(5) is unnecessarily complicated, in including dependence on the bonding parameter ζ. This bonding parameter represents the role of meniscus water bridges on mechanical behaviour [14], but Wheeler et al. [13] in the development of their large strain elasto-plastic constitutive model argued that these meniscus water bridges, while having a vital role in the yield behaviour of an unsaturated soil, would be relatively unimportant for elastic behaviour.

[a] Corresponding author: Simon.Wheeler@glasgow.ac.uk

The aim in this paper was therefore to re-examine bender element measurements of G from tests on three unsaturated compacted kaolin samples previously reported by Hasan & Wheeler [15], to see if these could be successfully interpreted in terms of only $p*$ and v. The tests included wetting-drying cycles as well as isotropic loading and unloading paths. A previous attempt [15] to interpret the same experimental measurements in terms of conventional unsaturated state variables $\bar{p}$, s, v and S_r had been rather unsatisfactory, because it had indicated dependency on all four variables and apparently contradictory results in terms of the influence of S_r (in addition to the influence of s).

2 Test programme

2.1 Sample preparation

Three unsaturated samples of speswhite kaolin clay (LL = 68%, PL = 36%) were prepared by compaction at a water content of 25% (4% dry of the optimum from the standard Proctor compaction test). An isotropic form of static compaction was used [15]. The uncompacted soil mix was sealed within a large cylindrical rubber membrane and then subjected to an isotropic compaction stress of 390 kPa in a triaxial cell. A smaller, 50 mm diameter, 100 mm height, triaxial test sample was then cored from the larger body of soil. Samples were subsequently tested in a suction-controlled triaxial cell fitted with bender elements.

2.2 Bender element testing

Two pairs of bender elements were used to measure shear wave velocity V_s. Both pairs transmitted horizontally across the sample, at the sample mid-height. One pair was aligned to give shear waves of vertical polarization and the other was aligned to give shear waves of horizontal polarization, providing measurements of shear wave velocities V_{shv} and V_{shh} and hence elastic shear moduli, G_{shv} and G_{shh} , respectively (where the second and third subscripts represent the transmission direction and polarization direction respectively).

Transmitter bender elements were excited by a single sinusoidal pulse, using a frequency of 20 kHz, corresponding to wavelengths of approximately 10mm. This frequency conformed to the recommendation of Leong et al. [16] that wavelengths should be less than 30% of the transmission path length and preliminary tests [17] indicated that it provided reliable and consistent values of shear wave velocity.

Values of wave velocity were calculated from the tip-to-tip distance between transmitter and receiver bender elements and the measured travel time. Travel time was determined from transmitted and received signals, using a peak-to-first-peak method in the time domain. Preliminary tests [17] investigated four possible methods for determining travel time (using both time and frequency domains) and concluded that the selected method was the most reliable.

2.3 Stress paths

The stress paths followed in Tests A, B and C are plotted in terms of $\bar{p}$ and s in Fig. 1. In all three tests the as-compacted suction was approximately 650 kPa (point X in Fig. 1). After mounting in the triaxial cell, a mean net stress of 10 kPa was applied (point Y in Fig. 1) and then samples were wetted to a suction of 300 kPa (points A_1, B_1 and C_1 in Fig. 1).

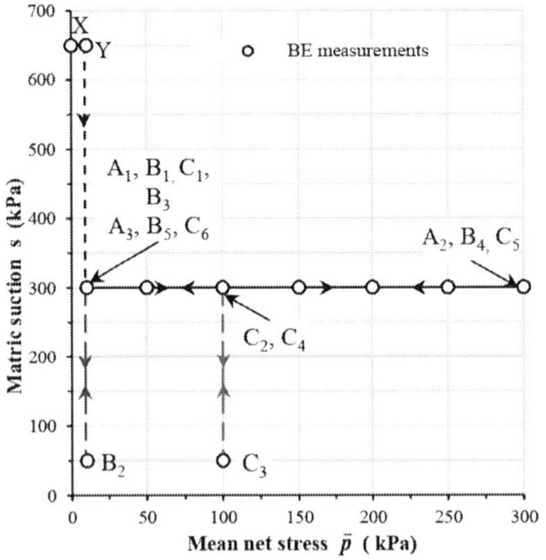

Figure 1. Stress paths for Tests A, B and C in s: $\bar{p}$ plane [15].

Test A involved isotropic loading (A_1– A_2) at constant suction (s=300kPa) to $\bar{p}$ =300kPa, followed by isotropic unloading (A_2–A_3) to $\bar{p}$ =10kPa. Test B involved a wetting-drying cycle (B_1–B_2–B_3 in Fig. 1) at $\bar{p}$ =10kPa to a minimum s of 50 kPa, followed by isotropic loading-unloading (B_3–B_4–B_5) at s=300kPa. Finally, Test C involved isotropic loading (C_1–C_2) at s=300kPa to $\bar{p}$ =100kPa, followed by a wetting-drying cycle (C_2–C_3–C_4) at $\bar{p}$ =100kPa to a minimum s of 50 kPa and then isotropic loading and unloading (C_4–C_5–C_6) at s=300kPa.

Loading and unloading stages were performed by increasing or decreasing $\bar{p}$ at a fixed rate of 2 kPa/hour, whereas wetting and drying stages were performed by applying a rapid change of s to the sample boundary and then waiting until inflow or outflow of water reached an appropriately low rate. Bender element (BE) tests were performed at intervals throughout each test (see Fig. 1).

The purpose of the wetting-drying cycles in Tests B and C was to ensure that during subsequent isotropic loading and unloading the values of S_r were different for samples A, B and C (due to hysteresis in the retention behaviour), even though the suction value was identical at 300 kPa in all three cases. For sample C the wetting was expected to produce collapse compression, whereas collapse compression was not expected in the wetting performed in Test B, so that the subsequent loading stages of Tests B and C were expected to be at different values of v.

DOI: 10.1051/e3sconf/20160909006

3 Test results

3.1. Variation of specific volume v

Fig. 2 shows the variations of v for all three tests. The two data points at the end of each isotropic loading stage represent the start and end of a 24 hour rest period. There were small differences in the measured initial values of v for the three samples (after initial wetting to s = 300kPa). Investigation over a large number of tests suggested that this was mainly attributable to small errors in measurements of initial sample volume rather than to variability between samples. Results were therefore adjusted so that initial values of v were identical for all three samples (see A_1, B_1 and C_1 in Fig. 2).

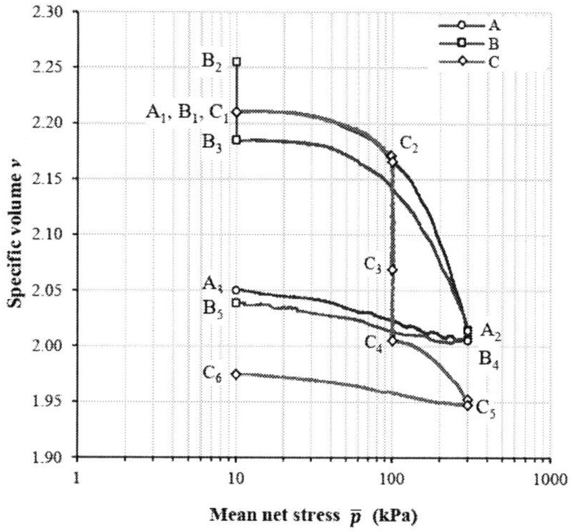

Figure 2. Variation of specific volume [15].

The variation of v during Test A (see Fig. 2) showed, as expected, significant irreversible compression during the isotropic loading-unloading cycle (A_1–A_2–A_3), with the suggestion of yielding during loading.

During Test B, swelling was observed during the wetting stage (B_1–B_2) (see Fig. 2), with no indication of any wetting-induced collapse compression. This was followed by shrinkage during the drying stage (B_2–B_3), with a net reduction of v over the wetting-drying cycle (B_1–B_2–B_3). During the isotropic loading stage (B_3–B_4) the compression curve gradually converged with the curve from Test A, and during subsequent unloading (B_4–B_5) the variation of v was very similar to Test A.

For Test C, Fig. 2 shows that the variation of v during initial loading (C_1–C_2) was, as expected, almost identical to that from Test A. The wetting stage (C_2–C_3) produced significant reduction of v (collapse compression), and this was followed by shrinkage in the drying stage (C_3–C_4). The final isotropic loading (C_4–C_5) and unloading (C_5–C_6) stages produced irreversible compression, with the suggestion of yielding during loading. By the end of the loading stage (C_4–C_5) the compression curve had not fully converged with the curves from Tests A and B, and the final value of v after unloading to C_6 was still significantly lower than in Tests A and B.

3.2 Variation of degree of saturation S_r

Fig. 3 shows the variations of degree of saturation S_r for all three tests. The initial values of S_r shown in Fig. 3 for the three samples (at A_1, B_1 and C_1) were all very similar.

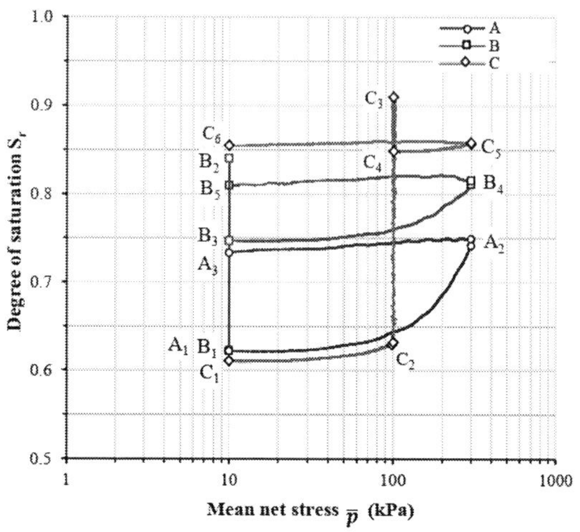

Figure 3. Variation of degree of saturation [15].

The variation of S_r during Test A (see Fig. 3) showed significant irreversible increase of S_r during the loading-unloading cycle (A_1–A_2–A_3), with the suggestion of yielding during the loading stage.

In Test B there was significant net increase of S_r during the wetting-drying cycle (B_1–B_2–B_3), as a consequence of hysteresis in the water retention behaviour, so that the value of S_r at the start of the subsequent loading stage was higher than in Test A (compare B_3 and A_1 in Fig. 3). By the end of the loading stages (B_4 and A_2 in Fig. 3) the difference between the values of S_r in Tests B and A had reduced, and this difference in S_r then remained almost unchanged during the final unloading stages (A_2–A_3 and B_4–B_5).

In Test C the variation of S_r (see Fig. 3) during the first loading stage (C_1–C_2) was very similar to that in Test A, as expected. A large increase of S_r occurred during the wetting stage (C_2–C_3), with a smaller reduction of S_r during the drying stage (C_3–C_4), as a consequence of hysteresis in the water retention behaviour. Little further change of S_r occurred during the final loading stage (C_4–C_5) and unloading stage (C_5–C_6), and the value of S_r remained higher than in Tests A and B.

3.3 Variation of elastic shear modulus G

The variations of small strain elastic shear moduli G_{hv} and G_{hh} for all three samples A, B and C are plotted in Fig. 4. These values of G were determined from the corresponding values of V_s, using Eq.(1), where the value of sample bulk density ρ varied throughout each test, as a consequence of changes of v and S_r. Inspection of Fig. 4 shows that the values of G_{hv} and G_{hh} were always very similar, confirming that the isotropic compaction procedure and subsequent isotropic stress history resulted in soil samples with isotropic properties.

DOI: 10.1051/e3sconf/20160909006

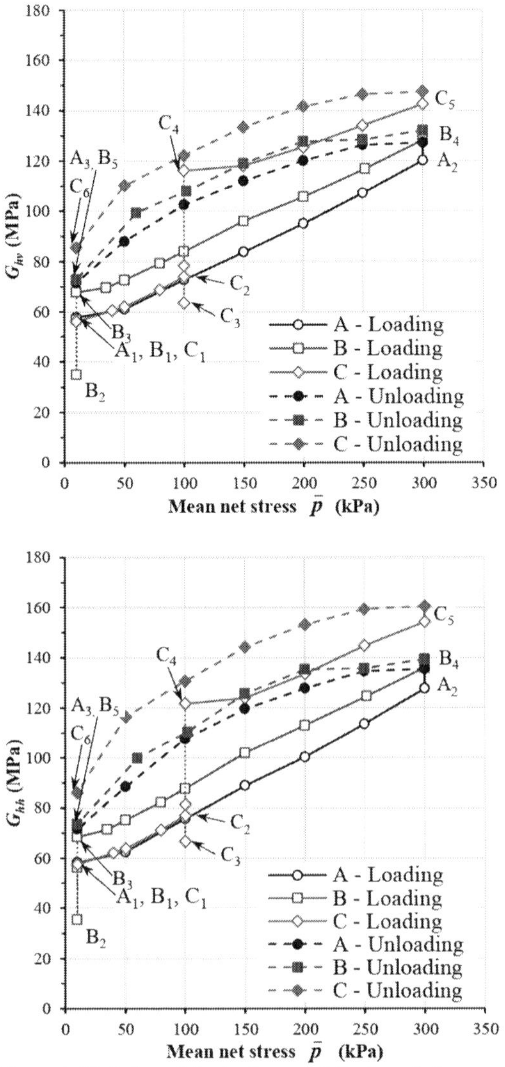

Figure 4. Variation of elastic shear moduli G_{hv} and G_{hh}.

Results from Test A in Fig. 4 show that the elastic shear modulus G increased during the loading stage (A_1-A_2). During the subsequent unloading stage (A_2-A_3), values of G were higher than at the same values of $\bar{p}$ in the loading stage, showing the combined influence of reduced values of v (Fig. 2) and increased values of S_r (Fig. 3) during unloading.

For Test B, inspection of Fig. 4 shows that a significant decrease of G occurred during the wetting stage (B_1-B_2), with a larger increase of G during the subsequent drying stage (B_2-B_3). There was therefore a net increase of G over the wetting-drying cycle, attributable to the combined influence of a net decrease of v (Fig. 2) and net increase of S_r (Fig. 3).

The values of G during the subsequent loading stage (B_3-B_4) of Test B were higher than during the loading stage of Test A, attributable to the combined influence of lower values of v and higher values of S_r in Test B (see Figs. 2 and 3). Interestingly, the difference in G values between Tests B and A remained approximately constant throughout the loading stages (see Fig. 4), even though the values of v converged (see Fig. 2) and the difference in values of S_r reduced (Fig. 3). During the 24 hour rest period at the end of the loading stages, the difference in values of G between Tests B and A reduced, and the values of G in the two tests then remained very similar during the final unloading stages, despite the higher values of S_r in Test B (see Fig. 3).

Values of G during the unloading stage (B_4-B_5) of Test B were generally higher than at the same values of $\bar{p}$ during the loading stage, as in Test A. However, by the end of the loading-unloading cycle in Test B (B_3-B_4-B_5) there had been little net change of G over the cycle (compare B_5 and B_3 in Fig. 4), despite the net changes of v and S_r over the cycle.

For Test C, Fig. 4 shows that the variation of G during the initial loading stage (C_1–C_2) was, as expected, very similar to that observed in Test A (up to this point the two samples had followed the same stress path, see Fig.1). During the wetting stage (C_2-C_3) of Test C there was a decrease of G, followed by a significantly larger increase of G during the subsequent drying stage (C_3-C_4). The net increase of G over the wetting-drying cycle of Test C (C_2-C_3-C_4) was much greater than during the wetting-drying cycle of Test B (B_1-B_2-B_3), because the net decrease of v over the cycle was much greater in Test C, due to the occurrence of collapse compression during wetting (see Fig. 2).

During the second loading stage (C_4-C_5) of Test C the values of G were higher than in Tests A and B, due to the combined influence of lower values of v (Fig. 2) and higher values of S_r (Fig. 3) in Test C. During the unloading stage (C_5-C_6) of Test C the values of G were higher than at the same values of $\bar{p}$ in the loading stage (consistent with observations from Tests A and B).

4 Interpretation

The aim in this paper was to investigate whether the experimentally observed variations of small strain shear modulus G could be interpreted solely in terms of mean Bishop's stress p^* (defined in Eq.(4)) and specific volume v. Fig. 5 shows the calculated variation of mean Bishop's stress p^* for one of the tests (Test C), as an example. The value of p^* increases during wetting stages (e.g. Y-C_1 and C_2-C_3) and decreases during drying stages (e.g. C_3-C_4). Over a wetting-drying cycle (e.g. C_2-C_3-C_4) there is a net increase of p^*, due to the increase of S_r arising from hysteresis in the water retention behaviour.

4.1 Proposed expression for G

Based on experience from saturated soils (see Eqs. (2) and (3)), a possible expression relating the variation of G to p^* and v for isotropic soils under unsaturated conditions and isotropic stress states is:

$$G = C\,v^{-m}\,(p^*/p_a)^{0.5} \qquad (6)$$

where C and m are soil parameters.

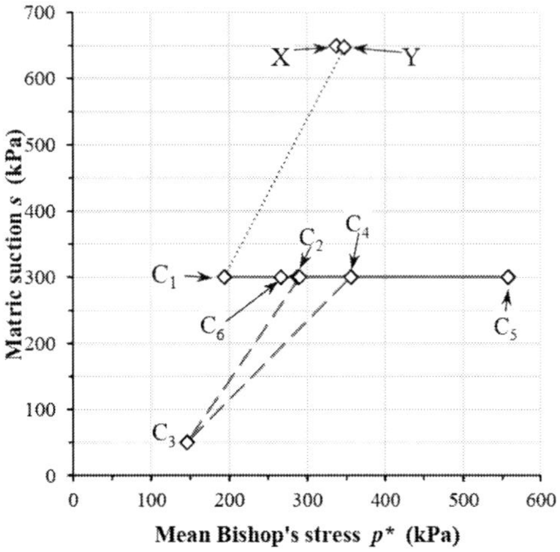

Figure 5. Stress path for Test C in s: p^* plane.

Fig. 6 examines the validity of Eq.(6) against the experimental data from Tests A, B and C, by plotting $G/(p^*/p_a)^{0.5}$ against v on a log-log plot. Each experimental value of G in Fig. 6 was taken as the average of G_{hv} and G_{hh} from Fig. 4. Inspection of Fig. 6 shows that the test data from all three tests fit well to a single straight line in this plot. The 3 data points plotting furthest below the best-fit straight line in Fig. 6 correspond to the three final points at the end of unloading (A_3, B_5 and C_6 in Fig. 4) and the single data point plotting significantly above the best-fit line in Fig. 6 corresponds to the end of the wetting stage in Test B (B_2 in Fig. 4). The gradient and intercept of the best-fit straight line in Fig. 6 correspond to m=3.60 and C=721 MPa respectively in Eq.(6).

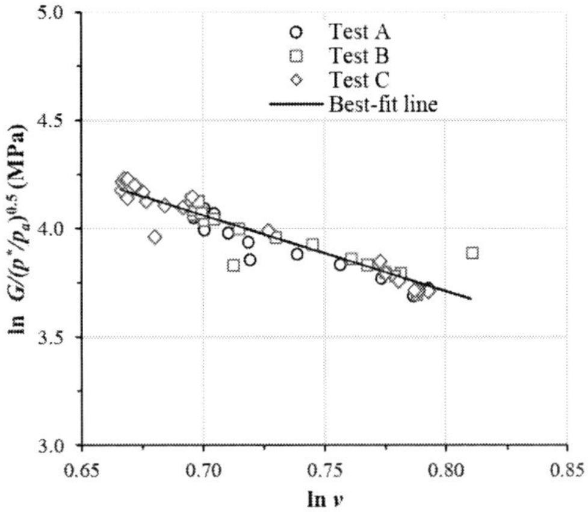

Figure 6. Variation of $\ln G/(p^*/p_a)^{0.5}$ *against ln v.*

4.2 Predicted variations of G

Fig. 7 shows the predicted variations of G from Eq.(6), with m=3.60 and C=721 MPa, plotted against p^* for Tests A, B and C, together with the corresponding experimental values of G.

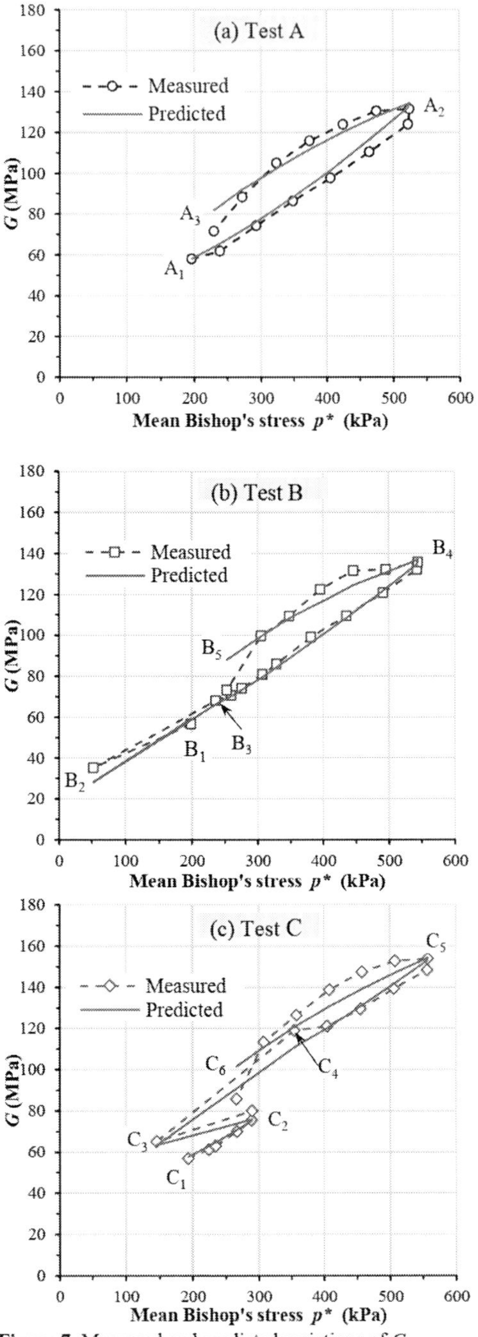

Figure 7. Measured and predicted variations of G.

Fig. 7a shows a comparison between measured and predicted values of G for Test A. It is clear from Fig.7a that there was excellent match between measured and predicted values of G during the loading stage (A_1-A_2). Eq.(6) also correctly predicted that values of G were higher during the unloading stage (A_2-A_3) than at the same values of p^* during the loading stage (because of the lower values of v during unloading). The match between predicted and experimental values of G was, however, less good during unloading than during loading and, in particular, the final value of G at the end of unloading was over-predicted by Eq.(6).

For Test B (Fig. 7b), the predicted variation of G during the wetting-drying cycle (B_1-B_2-B_3) matches well the observed variation, with Eq.(6) predicting a significant reduction of G during wetting (B_1-B_2), because of the reduction of p^* and the increase of v (wetting-induced swelling). Eq.(6) also correctly predicts a net increase of G over the wetting-drying cycle (compare B_3 and B_1 in Fig. 7b), because of a net increase of p^* over the cycle (see Eq.(4) and the values of S_r in Fig. 3) and a net decrease of v over the cycle (see Fig. 2). The predicted value of G at the end of wetting (B_2) did, however, under-predict the corresponding measured value of G.

Inspection of Fig. 7b shows that values of G were well-predicted during the loading stage (B_3-B_4) of Test B, but were less accurately predicted during the subsequent unloading stage (B_4-B_5). In particular, the final value of G at the end of unloading (B_5) was again over-predicted by Eq.(6).

For Test C (Fig. 7c), the predicted variation of G during initial loading (C_1-C_2), wetting (C_2-C_3), drying (C_3-C_4) and subsequent loading (C_4-C_5) matched well the observed behaviour. Eq.(6) correctly predicted significantly greater net increase of G over the wetting-drying cycle of Test C (C_2-C_3-C_4) than over the wetting-drying cycle of Test B (B_1-B_2-B_3 in Fig. 7b), because the net decrease of v over the cycle was much greater in Test C, due to the occurrence of collapse compression during wetting. The variation of G during the final unloading stage of Test C was not predicted as accurately as during the rest of the test (see Fig. 7c) and the final value of G at the end of unloading (C_6) was again over-predicted.

Eq.(6), which involves dependency on only p^* and v, provides reasonable predictions of the variations of G throughout Tests A, B and C. Further investigation showed that incorporating the bonding parameter ζ as an additional variable, through use of Eq.(5), as proposed by Chao [12], but with a best-fit value for the exponent of v, rather than enforcing a value of -3, led to no significant improvement in fit over Eq.(6). The bonding parameter ζ depends mainly on S_r, and further examination of the data from Tests A, B and C indicated that it was unlikely that the mis-matches between predicted and observed variations of G during the final unloading stages in Fig. 7 could be significantly reduced by including dependency on S_r as well as p^* and v in any proposed expression for G.

5 Conclusions

An initial investigation, based on bender element data from three tests on unsaturated compacted kaolin samples, which involved isotropic loading and unloading together with wetting and drying stages, indicated that Eq.(6), which includes dependency on only mean Bishop's stress p^* and specific volume v, provides a good match to the observed variations of small strain shear modulus G. No significant improvement of fit was achieved by incorporating additional dependence on degree of saturation S_r or bonding parameter ζ. Eq.(6) reverts to a well-established form for saturated soils at S_r=1 (when $p^* = p'$). Further validation against a wider range of experimental data is now required.

References

1. R. Dyvik and C. Madshus, *Proceedings ASCE Annual Convention Advances in the art of testing soils under cyclic conditions, Michigan*, 186-197 (1985)
2. B. O. Hardin, W. L. Black, *Journal of Soil Mechanics & Foundation Divisions*, ASCE, 94, 353-369, (1968)
3. G. Viggiani & J. Atkinson, *Géotechnique*, 45, 2, 149-154 (1995)
4. S. Oztoprak, M. Bolton, *Géotechnique*, **63**, 1, 54-70, (2013)
5. D. C. Lo Presti, O. Pallara, D. Costanzo, M. Impavido, *Pre-failure Deformation of Geomaterials*, 1, 11-16 (1995)
6. S. Shibuya, S.C. Hwang, T. Mitachi, *Géotechnique*, 47, 3, 593-601, (1997)
7. G. T. Houlsby & C. P. Wroth, *Géotechnique*, **31**, 3, 138-143 (1991)
8. C.Mancuso, R. Vassallo, A. d'Onofrio, *Canadian Geotechnical Journal*, **39**,1, 22-31 (2002)
9. C. W. W. Ng, S. Yung, *Géotechnique*, **58**, 1, 23-35. (2008)
10. M. Biglari, C. Mancuso, A. d'Onofrio, M. K. Jafari, A. Shafiee, *Computers and Geotechnics*, **38**, 5, 709-720 (2011)
11. K. S. Wong, D. Mašín, C. W. W. Ng, *Computers and Geotechnics*, **56**, 28-39 (2014)
12. Z. Chao, *PhD thesis*, Hong Hong University of Science and Technology (2014)
13. S. J. Wheeler, R. S. Sharma, M. S. R. Buisson, *Géotechnique*, **53**, 1, 41-54 (2003)
14. D. Gallipoli, A. Gens, R. Sharma, J. Vaunat, *Géotechnique*, **53**, 1, 123-135 (2003)
15. A. Hasan, S.J. Wheeler, *6th Asia-Pacific Conf. Unsaturated Soils*, Guilin, China, 191-196 (2015)
16. E.C. Leong, S.H. Yeo, H. Rahardjo, *Geotechnical Testing Journal, ASTM*, **28**, 488-498 (2005)
17. A. Hasan, S.J. Wheeler, *16th ECSMGE*, Edinburgh, UK, **6**, 3171-3176 (2015)

Low-cost alternatives to estimate unsaturated soil shear strength parameters

Michael Bardanis[1,a] and Sofia Grifiza[1]

[1]*Edafos Engineering Consultants S.A., Laboratory, 2 Therapion st., 10444 Athens, Greece*

Abstract. The paper presents experimental results form unconfined compression and splitting tensile tests used in combination in order to estimate the cohesion intercept of unsaturated soils. Samples of five different soils in the form of slurry were subjected to various suction values using the axis translation technique in a pressure plate extractor and then these samples were removed and subjected to unconfined compression and splitting tensile tests. Mohr' circles for the two loading conditions were drawn and the tangents plotted in order to obtain the cohesion intercept for each soil and each suction. Cohesion intercept was then plotted against suction for each soil and the angle of shear strength increase due to suction increase was estimated and compared to the value of the angle of shearing resistance of the fully saturated soil. The method allowed insight into the evolution of unsaturated soil shear strength of slurried soils, and seems to be an interesting low-cost alternative compared to established controlled suction methods to estimate unsaturated shear strength, yet it is based on two assumptions: first that the failure envelope is linear in the range between zero vertical stress and the stress corresponding to unconfined compression loading, and second that the suction applied to the samples does not change significantly from the moment a sample is removed from the pressure plate extractor until it is subjected to loading. The former assumption seems a fair one given the range of stress involved; the latter was investigated by conducting an unconfined compression tests with suction measurement during loaded. This test indicated that the latter assumption is a fair one too.

1 Introduction

The shear strength of unsaturated soils is of importance in many engineering problems. Test methods allowing its measurement that have been established for saturated soils, have also been extended for unsaturated soils. Yet this usually ends in methods that are quite expensive, time-consuming and presently quite limited in very few research centres and even fewer industry partners. The paper presents a low-cost alternative to estimate unsaturated soil shear strength parameters. The method proposed combines the Mohr's circles obtained from unconfined compression and splitting tensile (brazilian) tests on samples of soils subjected to a suction value in a pressure extractor (or a controlled relative humidity chamber if one wishes to extend the range of applied suction values). The method makes two assumptions: first that suction does not change significantly during the loading tests, and second that the generalised Mohr-Coulomb failure criterion for unsaturated soils [1] is fairly linear in the range of stress involved in unconfined compression and splitting tensile tests. The latter assumption is used to extrapolate the tangent to the two Mohr's circles backwards to the zero vertical stress axis in order to obtain an estimate of the cohesion intercept corresponding to the suction value applied to the samples prior to loading. The paper discusses both assumptions by presenting tests on five different soils tested in slurry form, in one unconfined compression test with suction measurement during loading too in order to exhibit the validity of the first assumption as well.

2 The soils tested

The soils tested were the Corinth and Chalkoutsi Marls, the Parnitha weathered siltstone, the Maroussi Clay and the Skiros weathered Phyllite. Index properties, soil-water characteristic curves and results from unconfined and indirect tensile strength tests for the first four soils mentioned have already been reported [2]. In this paper, the same data for Skiros weathered Phyllite are also presented, along with the saturated shear strength measured for all five soils in slurry form by means of the direct shear test. Finally, the combination of unconfined compression and indirect tensile strength test results is presented and the results are compared to the fully saturated shear strength parameters of the soils.

The index properties of the soils are summarised in Table 1. The Corinth Marls are a very well-studied formation [3] through which the Corinth Canal was excavated in late 19th century. Various experimental investigation programmes have been undertaken in order to investigate the properties of this material including the SWCC of the undisturbed marl and its recomposed

[a] Corresponding author: mbardanis@edafos.gr, lab@edafos.gr

counterpart [4], where recomposed is the material consolidated from a slurry condition to a desired value of void ratio (preferably that of the undisturbed soil) and then unloaded to zero stress. The Chalkoutsi Marl is a formation found 35km north of Athens. High, steep slopes are formed in the formation by sea erosion of their toe with occasional landslides occurring along the 5km coast that the formation outcrops in the highest slopes. The SWCC of the undisturbed material and its recomposed counterpart have already been reported [5]. Parnitha weathered siltstone is found in mount Parnitha 30km to the north of Athens close to the ground surface. Samples of Maroussi Clay came from boreholes drilled as part of a site investigation in the suburb of Maroussi in Athens close to the 2004 Olympic Games complex. Finally, samples of Skiros weathered phyllite came from a low road-cut in the north part of the island of Skiros, Greece.

Table 1. Index properties of the soils tested.

Soil	Clay (%)	Silt (%)	w_L (%)	I_P	φ′	USCS
Corinth Marl (CM)	6.5	84.7	30.5	5.5	27.5	ML
Chalkoutsi Marl (ChM)	20.5	64.2	51	30	25.0	CH
Parnitha weathered siltstone (PWS)	27.5	47.1	33	17	24.7	CL
Maroussi clay (MC)	40.1	41.9	47	28	19.2	CL
Skiros weathered phyllite (SWP)	1.3	71.6	N.P.		35.3	ML

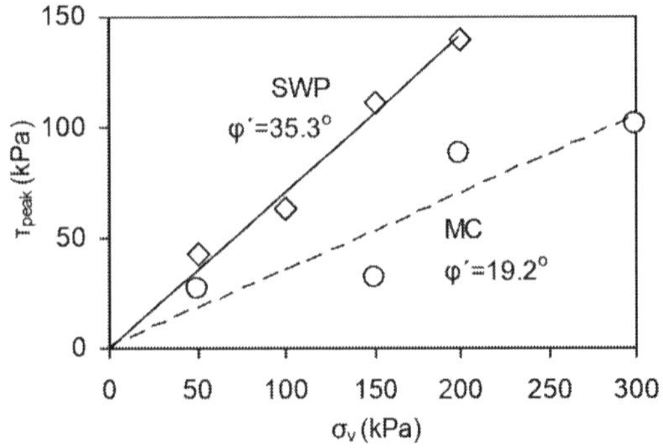

Figure 1. Shear strength versus vertical stress for Skiros weathered Phyllite (SWP, highest φ′) and Maroussi Clay (MC, the soil with the lowest φ′, and the largest variation in results).

3 Experimental method

The axis translation technique as applied in a Soilmoisture Equipment Corp. pressure extractor with 1500 kPa air-entry value porous ceramic disks was used for matric suction control. Tests were performed on samples prepared in the form of slurry. All slurries were prepared at an initial water content of 1.5 × w_L using deaired, deionised water, left for hydration for two days in a humidity chamber with occasional stirring in order to avoid sedimentation of coarser particles in the slurry and then placed for at least half an hour under vacuum for removal of air. Samples were placed in lubricated plastic tubes taped on the porous stone of the pressure extractor. Given that samples were slurries in their initial condition, volume decrease during their drying was large. Careful lubrication of the inner surfaces of the tubes was critical so as to avoid cracking of the samples during shrinkage as any inhibition of the diameter decrease by adhesion to the inner surface of the tubes resulted in cracking.

Regarding shear strength, direct shear tests were performed on all soils tested, with the samples in the form of slurries. Angles of shearing resistance measured are summarised in Table 1. The experimental points from the tests on samples yielding the highest and lowest angle of shearing resistance are shown in Fig. 1. The larger the plasticity of the soils (the smaller therefore the angle of shearing resistance), the larger was the scatter of experimental points as slurries were becoming progressively more and more difficult to handle, despite the w_o/w_L ratio being the same for all soils. This often resulted in quantity of the material flowing out of the gap between the top cap and the cell (Fig. 2), creating problems summing up to the larger scatter observed for higher plasticity soils. The measurement of the angle of shearing resistance of the fully saturated soils was considered essential in order to interpret the unconfined compression and splitting tensile tests on the samples subjected to various values of suction. As shown in Fig. 3, Mohr's circles from these two types of tests were plotted for each suction and the tangent to the circles was drawn in order to obtain the cohesion intercept corresponding to the applied suction on the assumption that the shear strength

Figure 2. Examples of slurries flowing out of the gap between the top cap and the cell (top Chalkoutsi marl, bottom Maroussi clay). Tests with problems so severe were repeated.

envelope is linear. Mohr's circles for splitting tensile strength tests were plotted using established calculations for minimum and maximum stress during this type of loading condition [6]. The same tangent may be used to

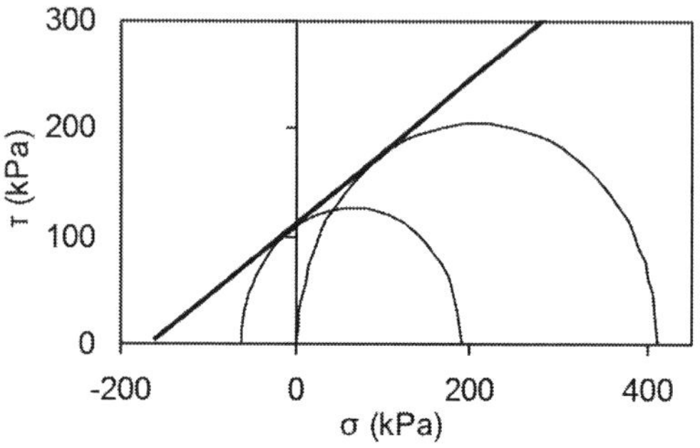

Figure 3. An example of the combination of Mohr's circles for splitting tensile test (circle on the left) and unconfined compression test (circle on the right) with the tangent to the two circles.

obtain the angle of shearing resistance; still as the two loading conditions (unconfined compression and splitting tensile loading) correspond to very low stress, these values were expected to be higher and were not used (it was only checked for each soil that they are higher than those in Table 1).

4 Suction evolution investigation

In order to investigate the assumption that suction does not change significantly from the time a sample is removed from the pressure plate extractor until the actual loading test is performed, one unconfined compression test was performed with an MPS-2 porous block frequency domain reflectometry (FDR) sensor (Decagon Devices, Inc.) installed in the sample (previously calibrated in the laboratory). A common tensiometer was initially considered as an option but it was finally rejected in favour of the MPS-2 as it has a limited range for measuring suction (0-90 kPa) although admittedly is the instrument with the fastest response during suction measurement. Yet, as newer evidence [7] suggests that porous blocks have considerably fast response, the MPS-2 sensor was selected for its greater range of suction measurement that allowed suctions to be applied that definitely lie beyond the air entry pressure of even the least plastic and coarser soil of those tested. This was the Skiros weathered Phyllite. A plastic tube was cut in order to install the MPS-2 sensor prior to placing the slurry of the soil in the tube (Fig. 4a & c). The sensor was stabilised in position using plasteline and the inner surface of the tube was lubricated in order to minimise friction while the whole tube with the sensor was sealed on a metal plate again using plasteline. The whole array was placed on a scale for continuous monitoring of moisture loss due to evaporation and then the weathered phyllite slurry was poured into the tube and drying commenced (Fig. 4b). Once the sample dried enough to detach from the inner wall of the tube, the plastic tube was removed (Fig. 4c). Finally, when suction reached 520 kPa, the sample was lifted from the scale and taken to the loading frame for the loading under unconfined conditions to begin (Fig. 4d). Suction was monitored during the loading stage too. Axial stress vs axial strain is plotted in Fig. 5a and suction vs axial strain is plotted in Fig. 5b. It may be seen that the suction measured at the end of drying of 520 kPa increased to only 524 kPa at the beginning of the loading stage (approximately 20 min difference) and reached a maximum of 556 kPa during the unconfined compression test. Even if one accepts that this is true increase of suction in the soil, it corresponds *to only 7% increase in a non-plastic sandy silt*. However it seems that this increase is not wholly due to soil drying but to cracking, allowing some air to come in contact with the sensor. As seen in Fig. 5a, a first yield occurred at 1.2% axial strain which corresponds to the initiation of

Figure 4. a) Empty plastic tube, cut to receive the MPS-2 sensor (green plasteline used to replace the part of the tube removed), b) same piece of tube filled with slurry of weathered phyllite, c) sample with the sensor installed after considerable drying had occurred that allowed removal of the plastic tube, and d) sample with sensor on the frame during loading.

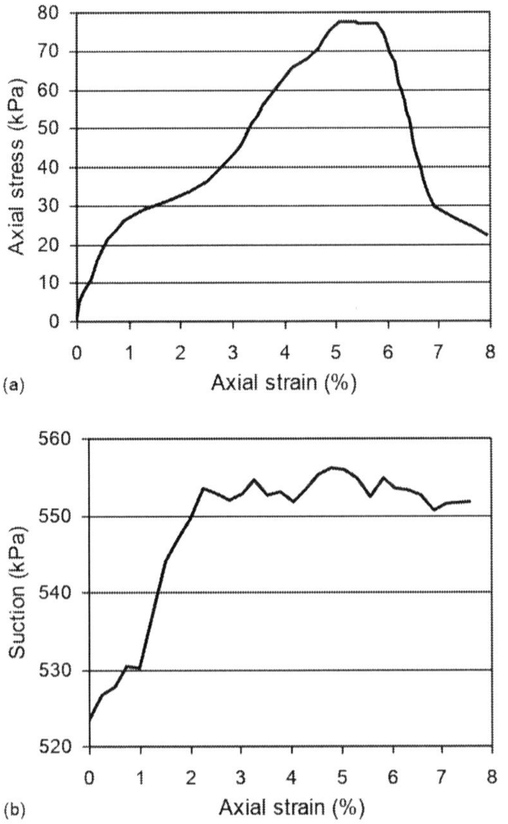

Figure 5. a) Axial stress vs axial strain, and b) suction versus axial strain during unconfined compression test on Skiros weathered Phyllite with and MPS-2 sensor installed.

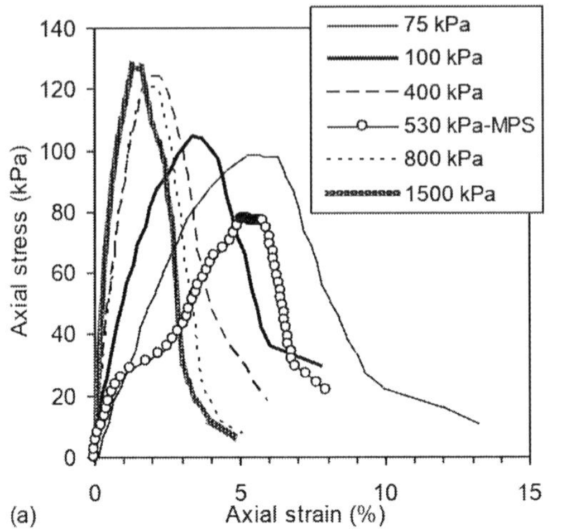

Figure 6. Axial stress vs axial strain during unconfined compression tests on Skiros weathered Phyllite for all suctions applied and for the sample with the MPS-2 sensor installed.

the crack seen at the top of the sample in Fig. 4d. This axial strain value corresponds to the beginning of a rapid suction increase in Fig. 5b. This seems to suggest that when the crack appeared some air reached the sensor leading to the suction increase observed. Therefore, the

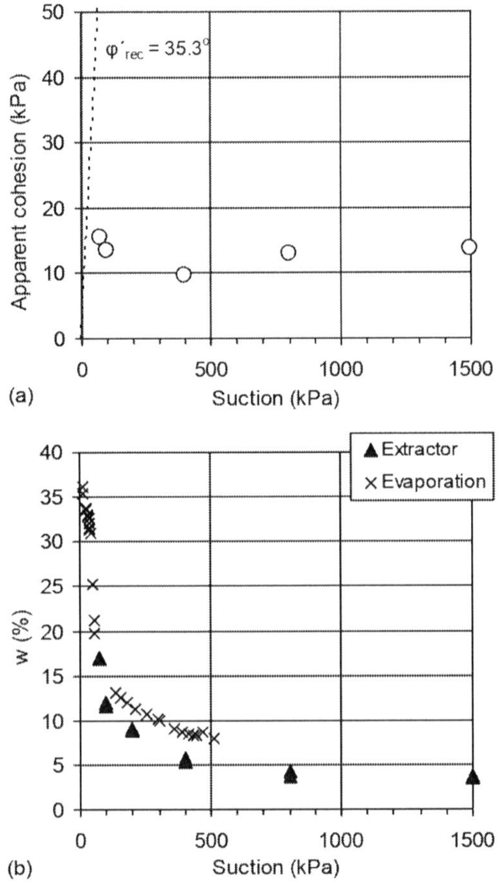

Figure 7. a) Apparent cohesion vs suction with the line corresponding to φ΄ from direct shear on slurry, and b) water content vs suction, for Skiros weathered Phyllite.

previously mentioned 7% increase could be overestimated and the actual increase of suction is probably half that. The first yield observed in the stress-strain curve is associated with the crack appearance (due to the presence of the sensor itself as the difference in the shape of the stress-strain curve of this test in comparison to all other tests without a sensor shown in Fig. 6 indicates). In conclusion, this test exhibits that the assumption that suction changes insignificantly from the end of drying until the unconfined compression test is performed is a valid one, as in the coarsest of the soils tested, the difference in suction was only 3-4%. For the other soils that were more fine-grained and plastic (let alone splitting tensile tests which are quicker) this difference is anticipated to be even smaller.

5 Unconfined compression and splitting tensile tests and their combination

For all samples tested at each value of suction, Mohr's circles were plotted as shown in Fig. 3 (for samples of Maroussi clay under 1000 kPa of suction) and the cohesion intercept was determined. This was then plotted vs suction for each soil along with the line corresponding to

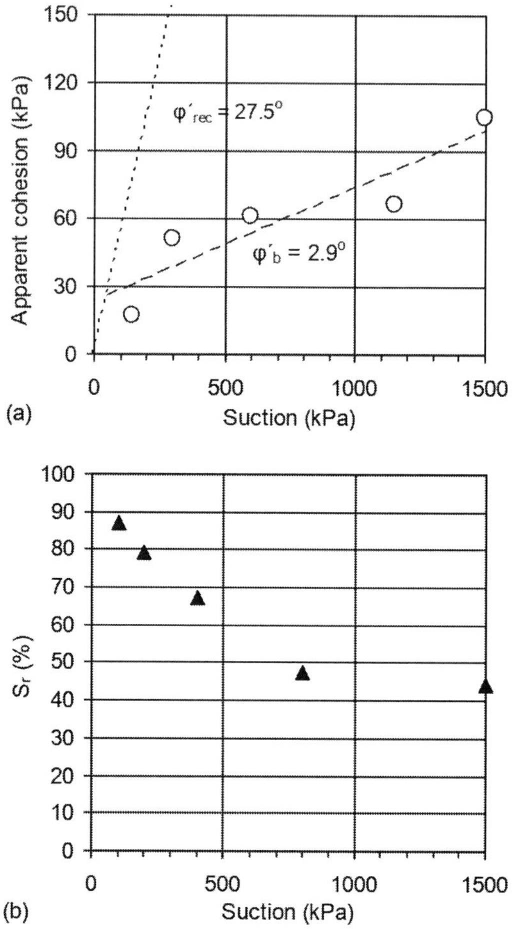

Figure 8. a) Apparent cohesion vs suction with the line corresponding to φ′ from direct shear on slurry, and b) degree of saturation vs suction, for Corinth marl.

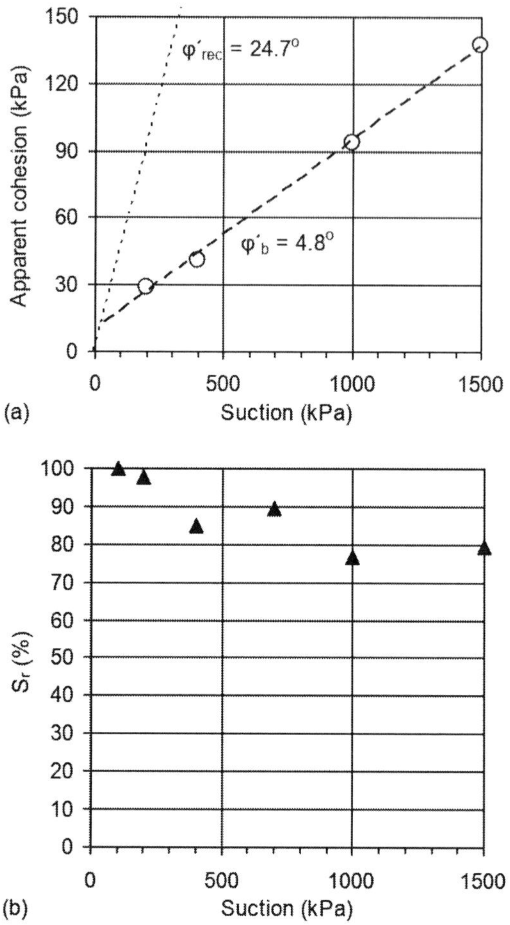

Figure 9. a) Apparent cohesion vs suction with the line corresponding to φ′ from direct shear on slurry, and b) degree of saturation vs suction, for Parnitha weathered siltstone.

the angle of shearing resistance of the same soil reconstituted and fully saturated (φ'_{rec}). These plots are presented in Figures 7 to 11 for Skiros weathered Phyllite, Corinth marl, Parnitha weathered siltstone, Chalkoutsi marl and Maroussi clay respectively. Figures denoted (a) present cohesion intercept vs suction, and figures denoted (b) present the part of the soil-water characteristic curve in the same range for the slurry of the soil with suction in linear scale. An exception was made for Skiros weathered Phyllite (Fig. 7) where given that the continuously monitored water content during drying from the sample with the MPS-2 sensor was available too, water content vs suction is plotted in Fig. 7b, rather than degree of saturation vs suction as in all other plots. These plots were used in accordance with the Mohr-Coulomb failure criterion for unsaturated soils [1] presented in Eq. 1, where c is the cohesion intercept of the fully saturated soil, φ is the angle of shearing resistance and φ_b is an angle, the tangent of which expresses the increase of strength τ with suction s.

$$\tau = c + \sigma \cdot tan\varphi + s \cdot tan\varphi_b \quad (1)$$

As seen in Fig. 7, Skiros weathered Phyllite desaturates at very low suction. Cohesion intercept due to suction reaches a value corresponding to the shear strength that would be obtained from the angle of shearing resistance of the fully saturated soil very close to the value corresponding to desaturation and stabilises (it may be suggested that it even decreases). In Fig. 8, Corinth marl is desaturated at 100 kPa of suction and cohesion intercept values vs suction yield a straight line corresponding to φ_b 2.9°. Given that these samples were all slurries placed in the pressure extractor in plastic tubes attached to the porous plate in a manner similar to that described for the plastic tube in Fig. 4, the lowest values of suction applied had to be high enough for the samples to gain enough strength to be handled, the lowest suction applied was not sufficient for some soils to obtain accurately the suction correspondding to desaturation. One case was the Corinth marl in Fig. 8b. Still the line drawn for cohesion intercepts in Fig. 8a, intersects the line corresponding to the angle of shearing resistance of the fully saturated Corinth marl at 50 kPa, a value quite reasonable for air-entry pressure given the subsequent points in Fig. 8b. Similarly, for Parnitha weathered siltstone (Fig. 9) an air-

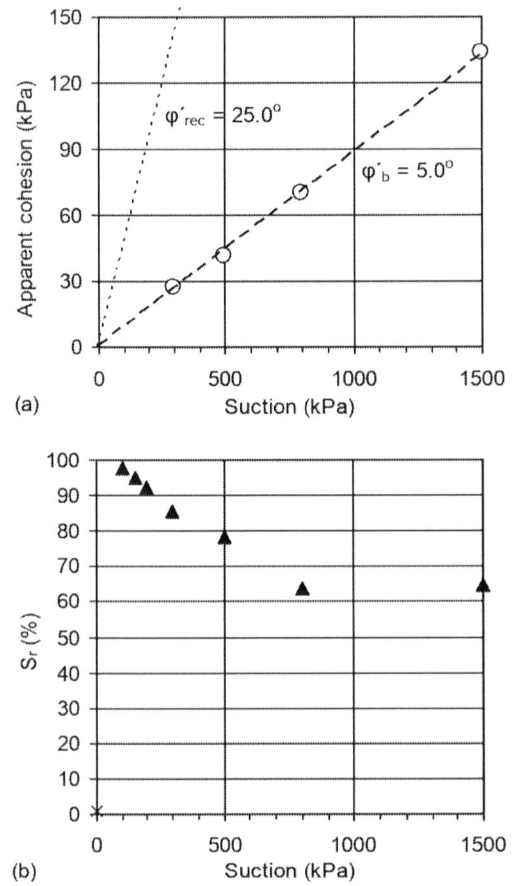

Figure 10. a) Apparent cohesion vs suction with the line corresponding to φ′ from direct shear on slurry, and b) degree of saturation vs suction, for Chalkoutsi marl.

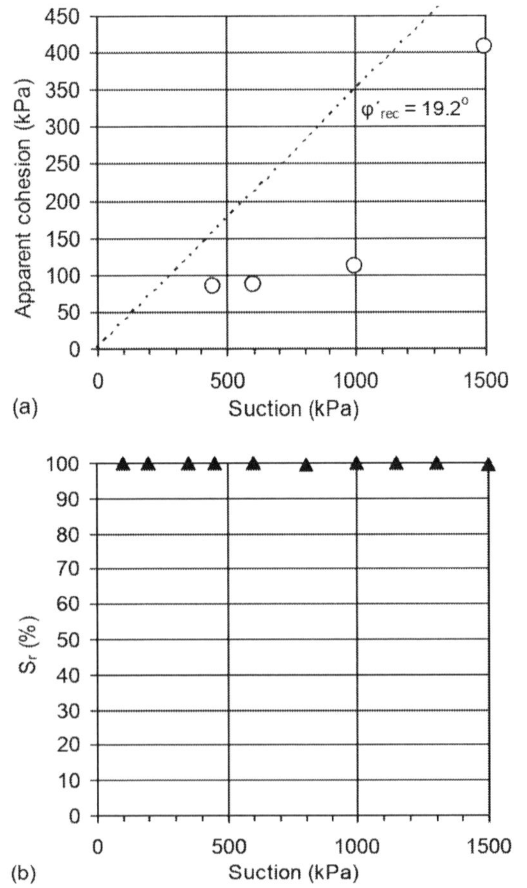

Figure 11. a) Apparent cohesion vs suction with the line corresponding to φ′ from direct shear on slurry, and b) degree of saturation vs suction, for Maroussi clay.

entry pressure of 100 kPa was obtained at which cohesion intercept change with suction is described by a value of φ_b of 4.8°. For Chalkoutsi Marl as well (Fig. 10) an air-entry pressure of less than 100 kPa was obtained and a value of φ_b of 5.0°. Yet this line seems to pass through zero indicating a very small value of the air entry pressure for this material in the form of slurry. Finally, for Maroussi clay (Fig. 11) air-entry pressure is higher than 1500 kPa and as a result no straightforward interpretation of the evolution of the cohesion intercept with suction may be obtained.

6 Conclusions

Unconfined compression and splitting tensile tests on samples of slurries subjected to various values of suction were combined in order to obtain the cohesion intercept of these soils for various values of suction. This method yielded values of φ_b in all cases except for one soil that did not desaturate in the range of suctions applied. A special test with an MPS-2 sensor installed in a sample of the coarsest and non-plastic soil during unconfined compression loading exhibited that the negligible suction change assumption behind this method is a valid one.

References

1. D. G. Fredlund, N. R. Morgenstern, R. A. Widger, 1978, Can. Geot. J., Vol. 15, pp. 313-321
2. M. E. Bardanis, S. Grifiza, Unsaturated Soils: Research and Applications, Mancuso et al. (eds), Proc. 2nd Eur. Conf. on Unsaturated Soils, Naples, Italy, 20-22 June, 2012, Vol.1, Springer, pp. 249-256
3. M.J. Kavvadas, A.G. Anagnostopoulos, V.N. Georgiannou, M.E. Bardanis, 'Characterisation and Engineering Properties of Natural Soils', Singapore, 2002, Tan et al (eds.), 1435-1459.
4. M. E. Bardanis, M. J. Kavvadas, Unsaturated Soils: Advances in Geo-Engineering, Toll et al. (eds), Proc. 1st Eur. Conf. on Unsaturated Soils, Durham, UK, 2-4 July, 2008, pp. 263-269.
5. M. E. Bardanis, S. Grifiza, Proceedings of the 15th Eur. Conf. on Soil Mechanics and Geotech. Engng, A. Anagnostopoulos et al. (eds.), IOS Press, 2011, Vol. 1, pp. 609-614.
6. G. Hondros, Aust. J. Applied Science, Vol. 10, 1959, 243.

The cone penetration test in unsaturated silty sands

Hongwei Yang[1,a] and Adrian Russell[2]
[1]*Department of Civil Engineering, The University of Hong Kong, Pokfulam Road, Hong Kong*
[2]*Centre for Infrastructure Engineering and Safety, School of Civil and Environmental Engineering, The University of New South Wales, Sydney, NSW, Australia, 2052*

Abstract. Very little is known about how to interpret the cone penetration test (CPT) when performed in unsaturated soils. The few published studies on the CPT in unsaturated soils have focused on either clean sands or a silt. In this study new results of laboratory-controlled CPTs in an unsaturated silty sand are presented. The silty sand exhibits hydraulic hysteresis and suction hardening. Suction is observed to have a pronounced affect on measured cone penetration resistance. For an isotropic net confining stress of 60 kPa it is observed that higher suctions give rise to cone penetration resistances that are 50% larger than those for lower suctions. A semi-theoretical correlation is presented that links measured cone penetration resistances to initial relative density and mean effective stress. For this silty sand it is shown that failing to account for suction may result in significant overestimations and unsafe predictions of soil properties from measured cone penetration resistances.

1 Introduction

The cone penetration test (CPT) is widely used for in-situ characterization of saturated or dry soils for which interpretation methods are well established (e.g. [1]). The CPT is also performed in unsaturated soils yet very little is known about how to interpret the recovered results.

There is evidence that the cone penetration resistance can be significantly influenced by the presence of suction, as observed in [2-3]. However, engineers are left to interpret the CPT conducted in unsaturated soils using correlations developed for saturated and dry soils. This will inevitably lead to unknown errors in estimations of soil properties ([2,4]). Limited research has been done on the CPT conducted in unsaturated soils and considerable further research is needed to identify the effect of unsaturation on the interpretation of CPT results, both theoretically and experimentally. This is the motivation for the work presented in the paper.

2 The test soil: Lyell silty sand

2.1 Index properties

The soil used in this study is a decomposed granite from the catchment area of Lyell dam, NSW, Australia. It is classified as a silty sand (SM) according to the Unified Soil Classification System. Index properties are listed in the Table 1, and a particle size distribution curve is shown in Figure 1.

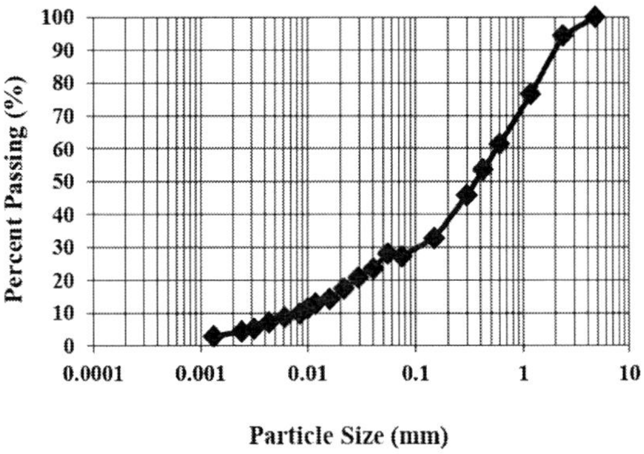

Figure 1. Particle size distribution curve

Table 1. Index properties of Lyell silty sand

Property	Value
Liquid Limit, %	15.2
Plastic Limit, %	N/A
Plasticity Index, %	N/A
Specific Gravity	2.55
Gravel content, %	0
Sand content, %	73
Fines content, %	27
Clay Size Fraction, %	4.4
Unified Soil Classification	SM

2.2 Soil-water characteristic curve

The relationship between void ratio (e), degree of saturation (S_r) and suction (s) is defined through a soil-

---

[a] Corresponding author: yanghhw@hku.hk

water characteristic curve (SWCC). S_r on a main drying or wetting curve is ([5]):

$$S_r = \begin{cases} 1 & \text{for} \quad s \le s_e \\ \left(\dfrac{s}{s_e}\right)^{\alpha} & \text{for} \quad s \ge s_e \end{cases} \tag{1}$$

where s_e is the s value separating saturated from unsaturated states. For a state on the main drying curve, $s_e = s_{ae}$, where s_{ae} is the air entry value. For a state on the main wetting curve, $s_e = s_{ex}$, where s_{ex} is the air expulsion value. α is a negative constant. α depends on the pore size distribution. On scanning curves S_r is:

$$S_r = \begin{cases} \left(\dfrac{s_{ae}}{s_{rd}}\right)^{-\alpha}\left(\dfrac{s_{rd}}{s}\right)^{-\beta} & \text{for drying path reversal} \\ \left(\dfrac{s_{ex}}{s_{rw}}\right)^{-\alpha}\left(\dfrac{s_{rw}}{s}\right)^{-\beta} & \text{for wetting path reversal} \end{cases} \tag{2}$$

where β is the slope and s_{rd}, s_{rw} are the points of suction reversal on the main drying and wetting curves. s_{ae} depends on e through $s_{ae} = C_1 e^{-\gamma}$, in which C_1 is a positive constant with units of stress and γ is a constant. A fixed ratio between s_{ae} and s_{ex} is assumed, $s_{ae} = C_2 s_{ex}$, in which C_2 is a constant. For fractal soils $\gamma = D_s$, where D_s is the fractal dimension of the particle size distribution ([6]). As Lyell silty sand is a fractal soil, having fractal particle and pore size distributions, it is found that $s_{ae} = 1.5e^{-D_s}$ kPa, $D_s = 2.61$ and $s_{ae} = 30s_{ex}$. The fractal dimension of the pore size distribution (D_p) is 2.48, meaning $\alpha = D_p - 3 = -0.52$. Scanning curves have slopes $\beta = -0.2$.

2.3 Effective stress and strengths

The effective stress approach is adopted here. The effective stress is ([7-8]):

$$p' = p_{net} + \chi s \tag{3}$$

where $p_{net} = p - u_a$ is the net stress, $s = u_a - u_w$ (u_a and u_w are pore air and water pressures), χ is the effective stress parameter (having a value of 1 for saturated soils and 0 for dry soils) and $\psi = d(\chi s)/ds$. When the state is on a main wetting or drying curve χ is:

$$\chi = \begin{cases} 1 & \text{for} \quad \dfrac{s}{s_e} \le 1 \\ \left(\dfrac{s}{s_e}\right)^{-\Omega} & \text{for} \quad \dfrac{s}{s_e} \ge 1 \end{cases} \tag{4}$$

where Ω is a parameter with a best fit value of 0.55. When the hydraulic state is located on a scanning curve χ is defined as ([9]):

$$\chi = \begin{cases} \left(\dfrac{s_{rd}}{s_{ae}}\right)^{-\Omega}\left(\dfrac{s}{s_{rd}}\right)^{-\zeta} & \text{for drying path reversal} \\ \left(\dfrac{s_{rw}}{s_{ex}}\right)^{-\Omega}\left(\dfrac{s}{s_{rw}}\right)^{-\zeta} & \text{for wetting path reversal} \end{cases} \tag{5}$$

A triaxial critical state friction angle (ϕ'_{cs}) of 35.7° was determined for both saturated and unsaturated conditions. The peak friction angle (ϕ'_p) observed in the triaxial compression tests (performed with constant cell pressures) on unsaturated samples with constant suctions may be estimated using the expression:

$$\phi'_p - \phi'_{cs} = 3\{D_r(9.5 - \ln p') - 1\} \tag{6}$$

where D_r is the relative density and p' is the mean effective stress at the start of shearing with units of kPa.

3 Cone penetration tests

The calibration chamber detailed by [10] was used to conduct the laboratory controlled tests. Samples (460mm diameter and 760 mm height) were prepared using static compaction of soil cured at certain moisture contents ([11]). All the tests were conducted with a penetration rate of 10 mm/s at boundary conditions of constant vertical and horizontal stresses.

Test conditions and results are summarized in Table 2 ([12]). The average values are for depths between 0.3 m to 0.5 m where q_c profiles are approximately constant, as shown in Figure 2.

Table 2 Tests conducted and relevant conditions

Test number	Compression Type	Applied vertical net stress (kPa)	Average moisture content	Average s (kPa)	Average q_c (MPa)
UL60SAC	Isotropic	60	4.65%	24	2.17
UL60S100	Isotropic	60	5.34%	37	3.26
UL60S300	Isotropic	60	3.93%	72	3.36
UL120SAC	Isotropic	120	4.50%	31	4.50
UL120S100	Isotropic	120	5.65%	37	5.24
UD120SAC	Isotropic	120	6.40%	13	7.80
UD120S100	Isotropic	120	4.83%	55	10.50
UL240SAC	Isotropic	240	4.40%	38	8.98
UL240S100	Isotropic	240	5.41%	47	8.73
K60S100-1	σ_h/σ_v= 1.67	60	5.30%	42	5.45
K60S100-N	σ_h/σ_v= 1.67	60	5.68%	34	4.97

Suction equilibrium was not reached in samples subjected to axis translation ([13]), the SWCC with hydraulic hysteresis and the moisture contents of the samples measured after a CPT was conducted were used to infer suctions. In other samples, the suction values were measured by three vibrating wire piezometers.

Comparing the results for various net confining stress magnitudes reveals that the contribution of suction to

 DOI: 10.1051/e3sconf/20160909008

cone penetration resistance becomes more significant as the confining stress decreases ([14]).

Saturated CPTs in samples at void ratios comparable to those used in unsaturated CPTs were not conducted. This would have required an impractical amount of time to saturate a sample due to its low hydraulic conductivity (being around 3×10^{-7} m/s at saturated states) ([13]). Also as unsaturated Lyell silty sand collapses in volume when it becomes saturated this would render the application of vertical stress in the calibration chamber inaccurate due to contact loss between the chamber top plate and the sample. However, a theoretical analysis provides some compensation for this lack of data for the saturated condition, which will be elaborated on in the following sections.

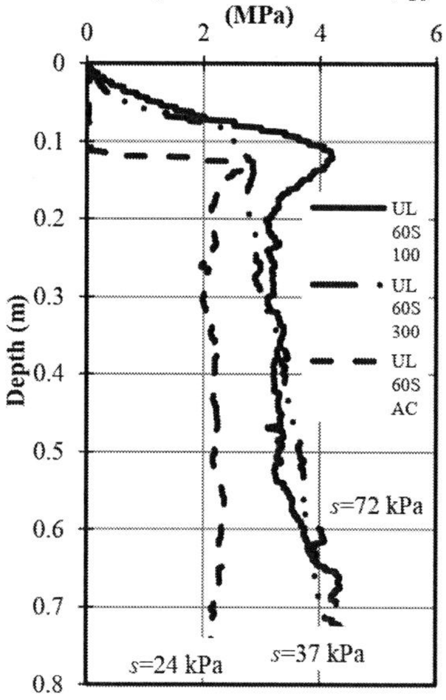

Figure 2. Cone penetration resistance (q_c) versus depth for samples at isotropic confining pressures of 60 kPa with different suction values.

4 Interpretation of the CPT results

4.1 Cavity expansion solutions

The cavity expansion problem is an analog to the CPT. The pressure at the wall of an expanding spherical cavity (σ_L) in a soil is related to the q_c measured in a CPT. The findings of [14] for cavity expansions in unsaturated Lyell silty sand are adopted.

[14] studied the effects of three different drainage conditions (constant suction, constant moisture content and constant χs) on σ_L. For each drainage condition they observed that a significant change of void ratio occurs around an expanding cavity in the unsaturated soil, and analogously occurs around the tip of a penetrating cone. They found that, for a constant moisture content condition and full hydro-mechanical coupling, the changes to χ and s mostly counteract each other and constant χs may be assumed to give a reasonable approximation. For constant moisture content an increase in e results in a decrease in s but an increase in χ and vice versa. The assumption of χs being a constant will be used to simplify interpretation of the CPT results.

[14] also studied the influence of where the initial hydraulic state was located on the SWCC on σ_L. They found when results were presented in the $p'_0 \sim \sigma_L$ plane, where p'_0 is the initial mean effective stress in the soil, unique relationships were observed irrespective of the initial hydraulic state locations. This observation will also be used to simplify interpretation of the CPT results.

The initial p'_0 and e_0 values used in generating the cavity expansion results are shown in Figure 3. Also shown are the limiting isotropic compression lines (LICLs) determined using a range of laboratory compression testing ([14]).

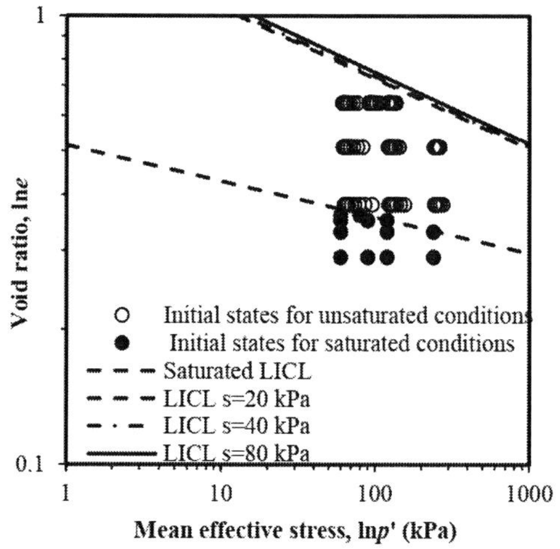

Figure 3. Initial states for the cavity expansion analysis and associated LICLs.

Results for expanding spherical cavities in saturated Lyell silty sand under drained conditions are shown in Figure 4.

Fitted to the results in Figure 4 is a correlation, similar to that of [15], relating σ_L to D_r and p'_0, with errors less than 5%. The correlation is:

$$\sigma_L = 0.26 p_0'^{0.65} \exp(6.18 D_r) \tag{7}$$

Notice that $\sigma_L \approx \sigma'_L$ since $\sigma_L >> u_w$.

Results for expanding spherical cavities under constant χs conditions can be fitted with a modified correlation, with errors less than 15%, as shown in Figure 5. The modified correlation is:

$$\sigma_L = 18 p_0'^{0.65} \exp(2.6 D_r) \tag{8}$$

Significant differences between Equation (6) and (7) are due to the presence of suction hardening. On the cavity wall, where the soil is at a critical state, significantly elevated pressures act for an unsaturated condition due to the suction induced shifts of the LICLs and CSLs in the ln*e*~ln*p*′ plane ([4, 14]).

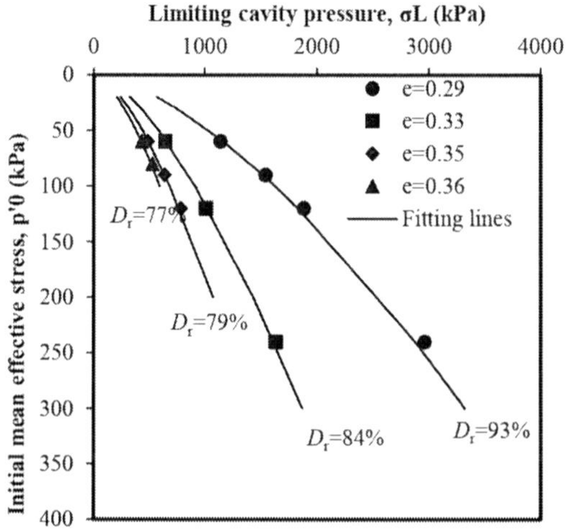

Figure 4. The limiting cavity pressure (σ_L) from saturated drained spherical cavity expansion analysis fitted with Equation (6).

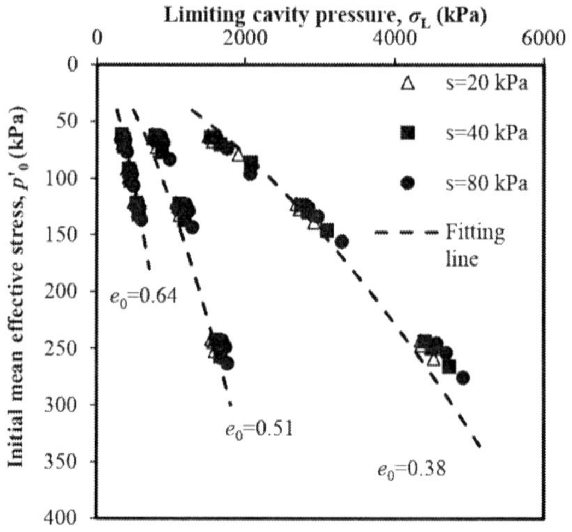

Figure 5. The limiting cavity pressure σ_L fitted with Equation 7.

Shown in Figure 6 is σ_L divided by $p'^{0.65}_0$ versus D_r for spherical cavity expansions in saturated and unsaturated Lyell silty sand. This Figure illustrates the effect of suction hardening. As can be seen, for unsaturated conditions, the results for different suction values are very similar. However, there is a large difference between results for unsaturated and saturated conditions. This is due to suction hardening and the suction dependent shifts of the LICLs and CSLs, features incorporated in the constitutive model used to generate the cavity expansion results.

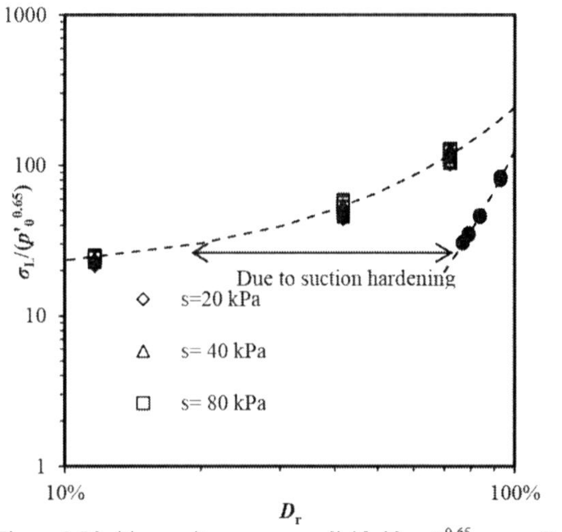

Figure 6. Limiting cavity pressure σ_L divided by $p'^{0.65}_0$ versus D_r for spherical cavity expansions conceptual illustration of suction hardening effects.

4.2 Comparison of cavity expansion results with CPT results

Cavity expansion analysis results were generated for soil states used in the calibration chamber tests. It is acknowledged that the cone penetration resistance measured in the calibration chamber can be slightly different from that measured in the field. Several methods have been proposed to account for the finite chamber size effect. Some are empirical (e.g., [16]) and some are numerical or analytical (e.g., [17-18]). The vertical stress which goes into the computation of the initial mean net stress and then to compute the cavity wall pressures is corrected through ([19]):

$$\sigma_{vc} = \sigma_v - \frac{q_c}{(R_D)^2} \tag{9}$$

where σ_v is the applied vertical pressure at the chamber base and R_D is the ratio of diameters for the chamber and cone, being 460/16 = 28.8 here.

Following [16], a linear proportionality between q_c and σ_L can be observed, as shown in Figure 7, and for unsaturated Lyell silty sand and void ratios ranging from 0.51 to 0.65 is $q_c = 9\ \sigma_L$. It follows that:

$$q_c = 162(p_0 + \chi s)^{0.65} \exp[2.6 D_r] \tag{10}$$

As shown in Figure 8 and Figure 9 the error associated with Equation (10) to obtain q_c is less than 30%. For unsaturated Lyell silty sand a rearranged Equation (10) gives:

$$D_r = \frac{1}{2.6} \ln\left[\frac{q_c}{162(p_0 + \chi s)^{0.65}}\right] \tag{11}$$

This enables determination of D_r from q_c, p_0 and χs.

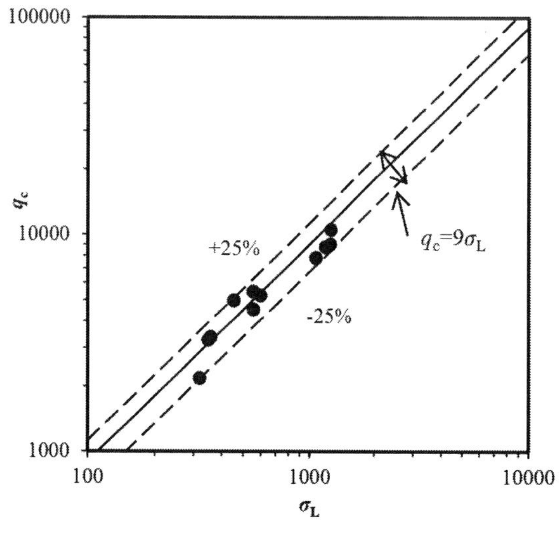

Figure 7. A possible correlation between q_c measured from chamber and σ_L from spherical cavity expansion.

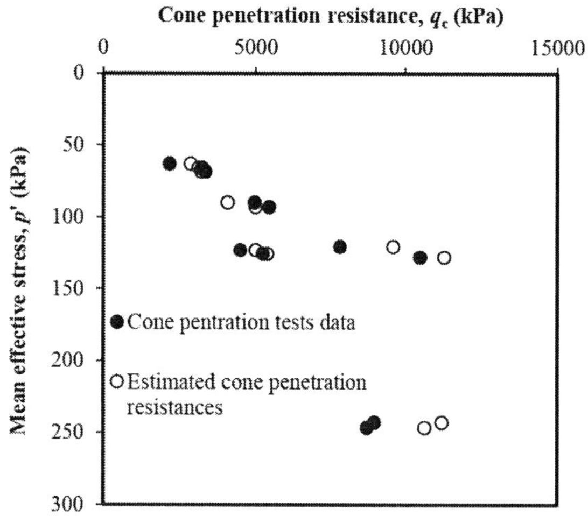

Figure 8. Comparison of cavity expansion data and estimations using Equation (10).

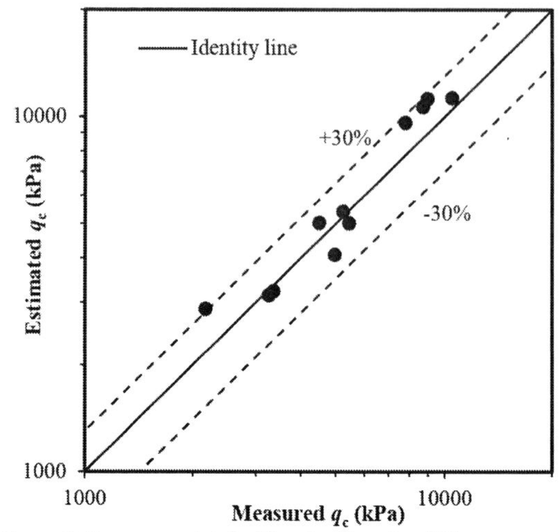

Figure 9. Errors associated with using Equation (10).

4.3 Example of application

A change in suction will alter the strength of the soil. Ignoring the effect of suction may lead to an overestimation of D_r and the corresponding ϕ'_p from Equation (6). A descriptive example is now used to highlight this. Suppose q_c = 5530 kPa was measured at a location in a profile of Lyell silty sand where p_0 = 25 kPa and χs = 25 kPa. D_r = 0.38 is obtained from Equation (11) which corresponds to ϕ'_p = 39.1° (Equation (6)). If suction influence is ignored and p'_0 = 25 kPa is incorrectly assumed then D_r = 0.55 is obtained corresponding to ϕ'_p = 43.1° (an overestimation of 4.0°).

The correlation between q_c, D_r and p'_0 presented here for unsaturated Lyell silty sand is similar to those presented by [15] and [2] for saturated and unsaturated clean sands, e.g. Sydney sand. As Sydney sand does not exhibit suction hardening the same correlation applies for saturated and unsaturated conditions. However, for soils which contain a significant amount of fines, like Lyell silty sand, a correlation which works for unsaturated conditions will not work for saturated conditions. This is due to the very different soil responses around the cone tip for saturated and unsaturated conditions and, in some cases, to the presence of suction hardening.

5 Conclusions

The results of laboratory-controlled CPTs were presented for unsaturated Lyell silty sand, a soil which exhibits suction hardening. The CPTs were conducted to study the effects of suction on cone penetration resistance for a range of confining stresses and densities.

The data was interpreted taking guidance from results of a recent cavity expansion analysis in Lyell silty sand. The cavity expansion analysis indicated a constant χs condition can be assumed to simplify the data interpretation without loss of significant accuracy. The cavity expansion analysis also indicated that the location of the initial hydraulic state on the SWCC has negligible influence on the relationship linking p'_0 to q_c.

For an isotropic net confining stress of 60 kPa a suction of 72 kPa was associated with a q_c 55 % larger than the value for a suction of 24 kPa. For samples subjected to an isotropic net confining stress of 120 kPa a suction increase from 13 kPa to 55 kPa caused an increase in q_c by about 35%. There was a diminishing affect of suction on q_c as net confining stress increased.

A correlation was introduced, inspired by cavity expansion results and the observation of a strong linear proportionality between q_c and the wall pressures of expanding spherical cavities. The correlation links q_c to p_0 and χs (through p'_0) and D_r. It resembles other correlations proposed for saturated and poorly graded sands. The resemblance is due to cone penetration in unsaturated Lyell silty sand being accompanied by significant soil deformation around the cone tip, as also occurs around penetrations in saturated poorly graded sands. However, a rather different correlation would be needed for saturated Lyell silty sand in which penetration would occur under undrained conditions.

An example was given to demonstrate the effect of suction. It was found that failing to account for the effect of suction may result in significant overestimation of relative density and peak friction angle.

The results and conclusions presented in this paper are only relevant to unsaturated Lyell silty sand. Many further studies of the CPT in a wider range of unsaturated soils are needed before CPT interpretation methods can be generalised.

References

1. T. Lunne, P.K. Robertson, J.M. Powell, *Cone penetration testing in geotechnical practice*. Blackie Academic & Professional, London (1997)
2. M. Pournaghiazar, A.R. Russell, N. Khalili, Geotechnique **63**(14): 1209-1220 (2013)
3. B. Lehane, M.A. Ismail, M. Fahey. Geotechnique **54**(3): 215-218 (2004)
4. A.R. Russell, N. Khalili. Comput. Mech **37**(4): 311-330 (2006)
5. A.R. Russell, O.Buzzi, Geotechnique **62**(3): 269-274 (2012)
6. A.R. Russell, Geotechnique **64**(5): 379-390 (2014)
7. A.W. Bishop, The principle of effective stress. Teknisk Ukeblad. 106: 859-863 (1959)
8. N. Khalili, M.A. Habte, S. Zargarbashi, Comput. Geotech. **35**(6): 872-889 (2008)
9. N. Khalili, S. Zargarbashi, Geotechnique, **60**(9): 729-734 (2010)
10. M. Pournaghiazar, A.R. Russell, N. Khalili, Can. Geotech. J. **48**(2):314-321 (2011)
11. H. Yang, A.R. Russell, In *Advances in Unsat. soils* pp: 191-197 (2013)
12. H. Yang, A.R. Russell. Can. Geotech. J. 53(3): 431-444 (2016).
13. H. Yang, A. Khoshghalb, A.R. Russell, Geotech. Lett. **4**: 1-10 (2014)
14. H. Yang, A.R. Russell. Int. J. Numer. Anal. Met. **39**(18): 1975-2016 (2015)
15. G. Baldi, R. Bellotti, V. Ghionna, M. Jamiolkowski, E. Pasqualini, In *Proceedings of the Fourth International Geotechnical Seminar* pp. 143–156 (1986)
16. R. Cudmani, V.A. Osinov, Can. Geotech. J. **38**(3): 622-638 (2001)
17. R. Salgado, J. Mitchell, M. Jamiolkowski, J. Geotech. Geoenviron. **124**(9): 878-888 (1998)
18. M. Pournaghiazar, A.R. Russell, N. Khalili, Int. J. Numer. Anal. Met. **37**(4): 331-352 (2013)
19. L.D. Wesley, Geotechnique **52** (4): 289-293 (2002)

Thermal Volume Change of Unsaturated Silt under Different Stress States and Suction Magnitudes

John S. McCartney[1,a]

[1] *University of California San Diego, Department of Structural Engineering, 9500 Gilman Dr, La Jolla, CA 92093-0085, USA*

Abstract. This paper presents an evaluation of the thermal volume change of compacted specimens of the same type of silt under a wide range of stress states, initial void ratios, and suction magnitudes. Stress states include both isotropic and anisotropic conditions with varying principal stress ratios, as well as normally consolidated and overconsolidated conditions. Initial void ratios range from 0.60 to 0.86, spanning very dense to loose conditions. Suctions evaluated range from saturated conditions, to low suctions in the funicular range, to suctions corresponding to residual saturation conditions. For the same soil, wide variations in thermal volume change are observed. Thermal contraction is observed for normally consolidated conditions regardless of the initial degree of saturation. Different mechanisms of thermal volume change can be used to explain the results, ranging from thermally-induced pore water pressure dissipation, to thermal collapse, to thermally-accelerated creep.

1 Introduction

Several studies have recently been performed to characterize the drained thermal volume change response of compacted specimens of Bonny silt under different initial conditions and stress states using different experimental approaches (Coccia and McCartney 2012; Vega and McCartney 2015; Coccia and McCartney 2015, 2016b; Alsherif and McCartney 2015, 2016; Shanina and McCartney 2016), including both saturated and unsaturated conditions. The results from these different studies have yielded thermal volume changes that are within the same order of magnitude of each other, but differences in the trends are observed in some cases. This paper provides a critical review of the reasons leading to different responses in the tests from the studies mentioned above. This includes an assessment of the role of the initial void ratio, the role of compaction conditions (wet or dry of optimum), the role of stress history (i.e., the overconsolidation ratio, OCR), the role of anisotropic stress state (the principal stress ratio $K=\sigma_3/\sigma_1$), and the initial degree of saturation.

2 Background

The thermo-mechanical behavior has been characterized for a wide range of saturated soils (Campanella and Mitchell 1968; Demars and Charles 1982; Hueckel and Baldi 1990; Towhata et al. 1993; Burghignoli et al. 2000; Sultan et al. 2002; Cekerevac and Laloui 2004; Abuel-Naga et al. 2007; Coccia and McCartney 2012; Vega and McCartney 2015) and unsaturated soils (Romero 1999; Saix et al. 2000; Romero et al. 2005; Tavalli et al. 2007; Salager et al. 2008; Tang et al. 2008; Uchaipichat and Khalili 2009; Coccia and McCartney 2015, 2016b, Alsherif and McCartney 2015, 2016, Shanina and McCartney 2016). The results from some these studies have been compared by McCartney (2012) and Laloui et al. (2015), who found that the soil mineralogy and the OCR were the primary variables that affected the magnitude and sign of the thermal volume changes. Contraction is expected to occur for normally-consolidated soils and expansion is expected for heavily overconsolidated soils, with a transitional behavior as the OCR increases.

Vega and McCartney (2015) confirmed trends in thermal volume change with stress history observed in the literature, but found that cyclic heating and cooling may lead to continued thermal volume changes in the direction observed during the initial heating cycle (i.e., either expansion or contraction). In addition to stress history, Coccia and McCartney (2012) found that anisotropic stress states may lead to different magnitudes of thermal axial strains in saturated soils under different principal stress ratios. Further, if the major principal stress is close to normally consolidated conditions, contraction may occur in this direction. However, as the principal stress ratio decreases, it is possible that expansion can occur in the minor principal stress direction. In addition to the stress state, the path of heating and stress changes may affect the magnitude and direction of thermal volume changes. Similar observations have been observed for unsaturated silt by Shanina and McCartney (2016). Alsherif and McCartney (2015) found that heating at low suctions leads to contraction while heating at high suctions leads to

[a] Corresponding author: mccartney@ucsd.edu

expansion for similar stress states and temperature changes. This behavior was explained using different hardening and softening mechanisms by Alsherif and McCartney (2016). The thermal volume change of unsaturated soils considering changes in degree of saturation during heating was recently considered by Coccia and McCartney (2016a, 2016b). Inconsistent trends in thermal contraction were observed for normally consolidated specimens with decreasing degree of saturation, similar to the lack of trend with this variable observed by Uchaipichat and Khalili (2009). Coccia and McCartney (2016a) proposed that thermal volume change may instead be due to thermally-accelerated secondary creep under the previously applied mechanical stress.

3 Material

The relevant geotechnical properties of Bonny silt are summarized in Table 1. The liquid and plastic limits of the soil are 25 and 21, respectively, measured in accordance with ASTM D4318, and the fines content is 83.9%. Accordingly, the silt is classified as ML (inorganic low plasticity silt) according to the Unified Soil Classification System (USCS, ASTM D2487).

Table 1. Bonny silt characteristics.

Parameter	Value
D_{10}	< 0.0013 mm
D_{30}	0.022 mm
D_{50}	0.039 mm
% Fines smaller than 75 μm	83.9%
% Sand size	16.1%
Liquid Limit, LL	25
Plastic Limit, PL	21
Plasticity Index, PI	4

All of the Bonny silt specimens evaluated in this study were prepared using static compaction to different dry unit weights at different compaction water contents. The compaction curves corresponding to the standard and modified Proctor compaction energies are shown in Figure 1 for reference. The specimen sizes were different in each of the studies assessed, as summarized in Table 2. The specimen size is not expected to play an important role as the silt is relatively fine-grained, but there may be some differences in the transient response.

Table 2. Specimen sizes evaluated.

Study	Height (mm)	Width or Diameter (mm)
Coccia and McCartney (2015, 2016b)	25.4	63.5
Alsherif and McCartney (2015) S-T	71.1	35.6
Alsherif and McCartney (2015) T-S	71.1	35.6
Vega and McCartney (2015)	25.4	83.0
Shanina and McCartney (2016)	177.8	177.8

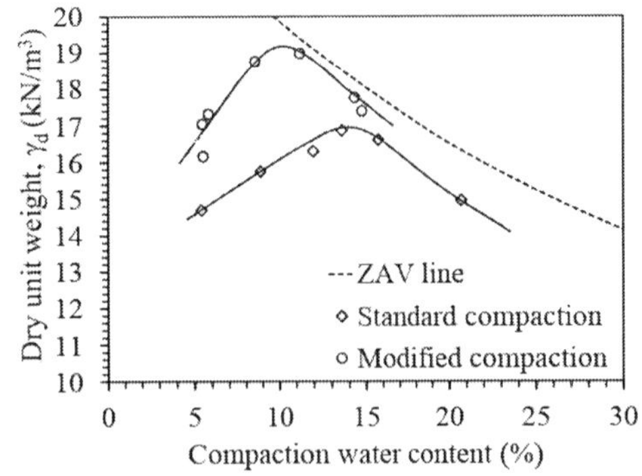

Figure 1. Compaction curves for Bonny silt

The compaction conditions and initial conditions at the start of heating from the different studies is presented in Table 3. A wide range in densities and initial water content values has been assessed, ranging from wet to dry of optimum.

Table 3. Initial compaction conditions.

Study	w_c (%)	$\gamma_{d,c}$ (kN/m^3)	$e_{initial}$
Coccia and McCartney (2015, 2016b)	13.9	14.2	0.80
Alsherif and McCartney (2015) S-T	10.5	15.7	0.68
Alsherif and McCartney (2015) T-S	10.5	15.7	0.68
Vega and McCartney (2015)	15.0	16.6	0.60
Shanina and McCartney (2016)	15.8, 18.2	16.2	0.64

The different compaction conditions were found to correspond to different hydraulic properties for the same soil. Specifically, the parameters of the soil-water retention curve (SWRC) model of van Genuchten (1980) (α_{vG}, n_{vG}) reported by each of the studies are summarized in Table 4. All of the studies observed or assumed that Bonny silt has a residual degree of saturation $S_{r,res}$ of 0.06. It is likely that differences in the hydraulic conductivity and thermal conductivity are also present, but these were not measured in all of the studies.

Table 4. Hydraulic properties for the specimens with different compaction conditions.

Study	α_{vG} (kPa^{-1})	N_{vG}
Coccia and McCartney (2015, 2016b)	0.20	1.30
Alsherif and McCartney (2015) S-T	0.33	1.61
Alsherif and McCartney (2015) T-S	0.33	1.61
Vega and McCartney (2015)	NR	NR
Shanina and McCartney (2016)	0.11	1.40

 DOI: 10.1051/e3sconf/20160909009

4 Experimental Approaches

Each of the different studies presented in the tables above followed different testing approaches, which are discussed in more detail in the respective papers. Coccia and McCartney (2015, 2016b) used an isotropic thermal cell with internal deformation measurements along with a flow pump used to control the suction in drained heating tests and track changes in the degree of saturation. Alsherif and McCartney (2015, 2016) used a modified triaxial cell with suction control using the vapor flow technique to study the thermal volume change and shear strength of soils under high temperatures and suction magnitudes. Vega and McCartney (2015) used a thermal oedometer test and evaluated the cyclic heating and cooling response of saturated silt specimens. Shanina and McCartney (2016) used a thermal true-triaxial cell with suction control on one face of a cubical specimen and principal stress control on the other five faces to study the role of stress-induced anisotropy.

The initial stress state evaluated in each of the studies was different, as summarized in Table 5. It is clear that the void ratios at the start of heating are lower than those induced by compaction due to the application of the target stress state or changes in the initial suction.

Table 5. Initial stress state prior to heating.

Study	OCR range	Principal stress ratio	$e_{heating}$	$\psi_{heating}$ (kPa)
Coccia and McCartney (2015, 2016b)	1	1	0.754 to 0.796	30
Alsherif and McCartney (2015) S-T	12 to 22	1	0.40 to 0.57	168, 291
Alsherif and McCartney (2015) T-S	1 to 3	1	0.55 to 0.62	40
Vega and McCartney (2015)	1 to 31	1	0.51 to 0.6	0
Shanina and McCartney (2016)	1	1.0, 0.7, 0.5	0.576	20, 10

Note: Principal stress ratio = (Minor principal stress/Major principal stress), $\psi_{heating}$ and $e_{heating}$ denote the suction and void ratio at the start of heating

5 Results

5.1 Coccia and McCartney (2015, 2016b)

The thermal volume changes for a saturated, compacted silt specimen reported by Coccia and McCartney (2015) are shown in Figure 2, along with the results from a similar isotropic true-triaxial test presented by Coccia and McCartney (2012) for comparison. Both the heating and cooling cycles are shown in the figure, and a similar plastic contraction behavior during heating and elastic contraction during cooling is observed in both tests. This confirms that thermal volume change experiments on specimens with very different shapes and sizes are similar.

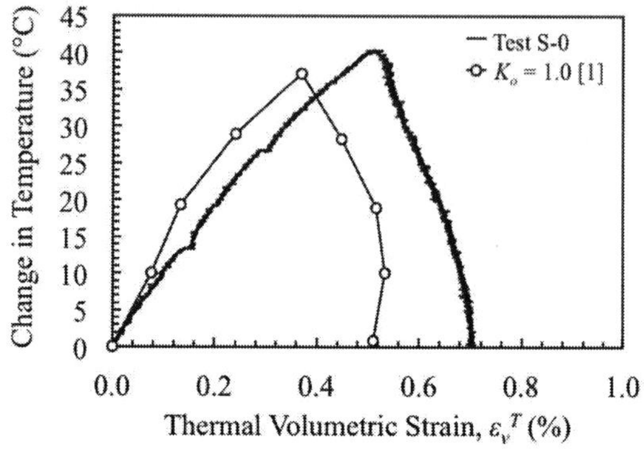

Figure 2. Thermal volume changes for isotropic tests on saturated, compacted silt specimens performed by Coccia and McCartney (2012) [1] and Coccia and McCartney (2015)

The results from tests on specimens reported by Coccia and McCartney (2016b) having different initial degrees of saturation are shown in Figure 3. The thermal volumetric strains for a change in temperature of 40 °C ranges from 0.3 to 0.75%. Although no clear trend in the thermal volume change with the initial degree of saturation is observed for these tests, this may be because the degree of saturation is not the only variable controlling the magnitude of thermal volume change. Coccia and McCartney (2016a, 2016b) proposed that the thermal volume change may not be due to the dissipation of thermal excess pore water pressures, which would naturally be associated with the quantity of water in the specimen. Instead, they proposed that the thermal volume change is due to thermally-accelerated secondary compression under the previously applied change in mean stress.

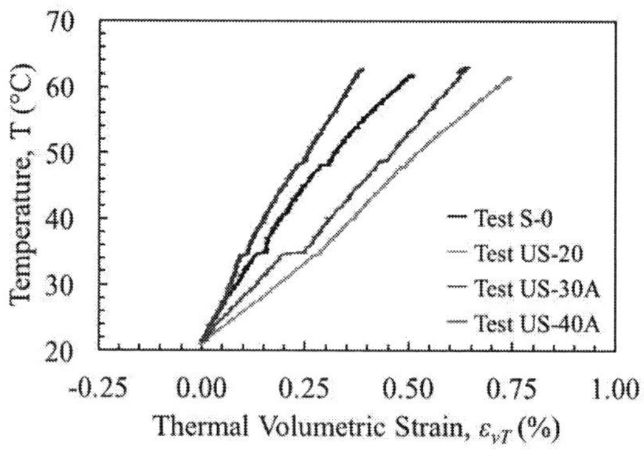

Figure 3. Thermal volume changes for specimens with different initial degrees of saturation (Coccia and McCartney 2016b)

5.2 Alsherif and McCartney (2015)

The thermal volume change of specimens that were heated under relatively low suctions of 40 kPa and different confining stresses are shown in Figure 4. Contraction is observed in all of the tests, with a magnitude similar to those observed by Coccia and McCartney (2016b) despite the lower initial degree of saturation and lower initial void ratio. The applied confining stress does not have a major effect on the thermal contraction. The results for specimens that were heated under high suction magnitudes of 291 MPa and different confining stresses are shown in Figure 5. In this case, expansion is observed in all of the tests. This indicates that the application of high suction magnitudes induces a hardening effect that has a similar effect to a high OCR.

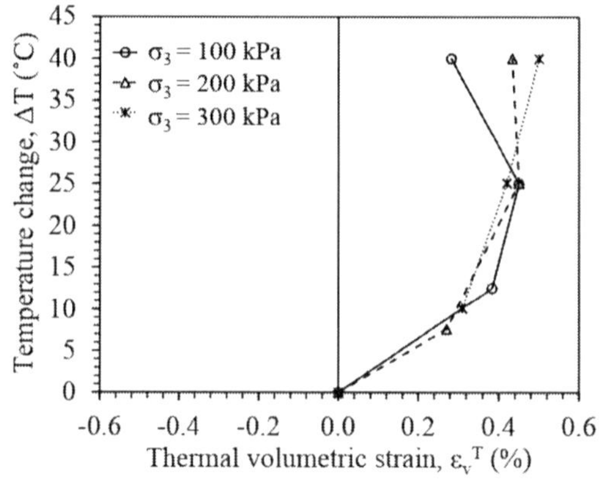

Figure 4. Thermal volume changes of specimens heated under a suction magnitude of 0.4 MPa (Alsherif and McCartney 2015)

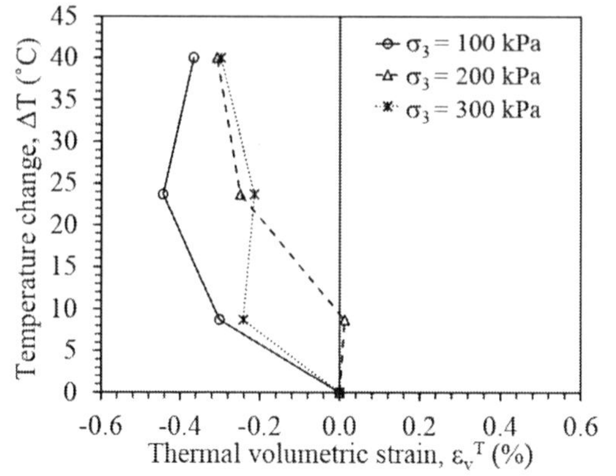

Figure 5. Thermal volume changes of specimens heated under suction magnitudes of 291 MPa (Alsherif and McCartney 2015)

5.3 Vega and McCartney (2015)

The thermal volume changes measured by Vega and McCartney (2015) are shown in Figure 6. For normally consolidated specimens (OCR = 1), contraction was consistently observed during heating. As the OCR increased, a transitional behavior to expansion was observed. As heating and cooling cycles were applied, a slight increase in expansion or contraction was observed in each of the tests. This is consistent with observations of Campanella and Mitchell (1968) and Burghignoli et al. (2000). The cyclic increase is greater for specimens that are closer to normally consolidated conditions. This is a feature that cannot be captured using thermo-elasto-plastic models, but may be considered using a thermal creep model.

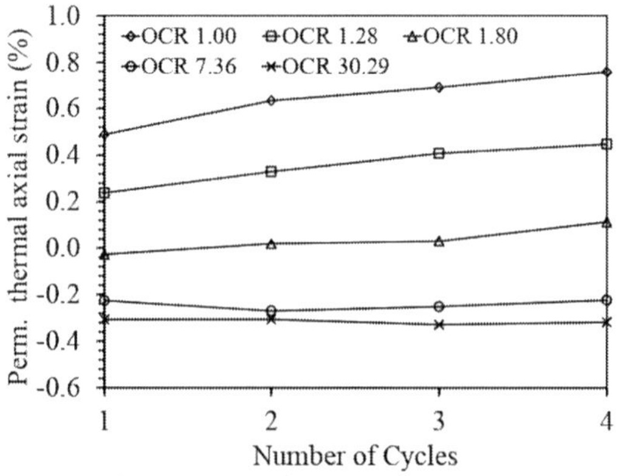

Figure 6. Thermal volume changes for specimens with different initial degrees of saturation (Vega and McCartney 2015)

The rate of thermal axial strain with change in temperature for the results from Vega and McCartney (2015) are compared with those from Alsherif and McCartney (2015) in Figure 7. In order to make the comparison in this figure, the mean effective stress and mean effective preconsolidation stress for the high suction tests of Alsherif and McCartney (2015) were back-analysed by Alsherif and McCartney (2016) using a thermo-elasto-plastic model. When plotted in terms of the effective stress OCR, similar behavior is noted for the saturated and unsaturated specimens, regardless of the fact that the specimens tested by Alsherif and McCartney (2015) were under very high suction magnitudes and those tested by Vega and McCartney (2015) are saturated. This indicates that an effective stress analysis of thermal volume change tests is critical to correctly interpret trends in the data.

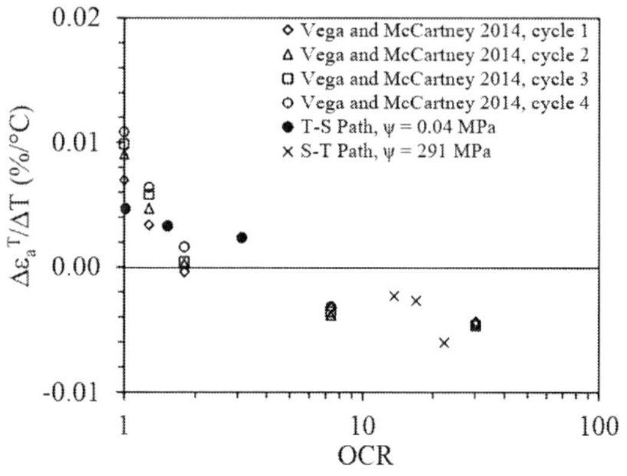

Figure 7. Thermal volume changes for specimens with different initial degrees of saturation (Alsherif and McCartney 2016)

5.4 Shanina and McCartney (2016)

The thermal axial strains measured by Shanina and McCartney (2016) for different initial degrees of saturation and principal stress ratios are shown in Figure 8. The specimens were compressed isotropically to the same mean effective stress corresponding to normally consolidated conditions, after which the specimens were loaded further in the major principal stress direction while the minor principal stress was maintained constant to reach different principal stress ratios.

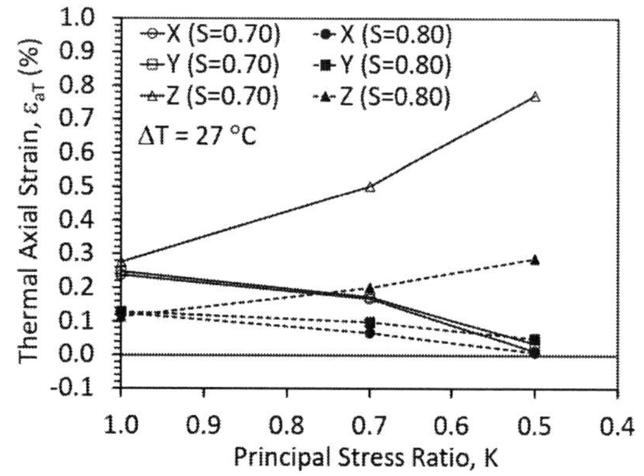

Figure 8. Thermal axial strain changes for specimens with different initial degrees of saturation and principal stress ratios (Shanina and McCartney 2016)

As the principal stress ratio decreases, the thermal axial strain in the major principal stress direction for the same change in temperature increases. The opposite is true for the thermal axial strains in the minor principal stress direction. This behavior may have interesting effects on thermally active structures in the field, as the geostatic stresses are typically anisotropic. This implies that expansion may occur in the horizontal direction (minor principal stress direction) while contraction may occur in the vertical direction (major principal stress direction).

The thermal volumetric strains for the same tests described in Figure 8 are shown in Figure 9. Different from the results of Coccia and McCartney (2016b), a clear increasing trend is observed in the magnitude of thermal volume change with decreasing degree of saturation. However, a difference between the tests of Shanina and McCartney (2016) and Coccia and McCartney (2016b) is that Shanina and McCartney (2016) performed tests on specimens compacted to different initial degrees of saturation, while Coccia and McCartney started with the same compaction conditions for all tests and used the axis translation technique to change the degree of saturation. Accordingly, the trends in thermal volume change with different initial degrees of saturation observed in Figure 9 may be due to the initial soil structure induced by compaction. At the same time, all of the specimens evaluated in this figure are all compacted wet of optimum, so this inconsistency may still be due to thermally-accelerated creep.

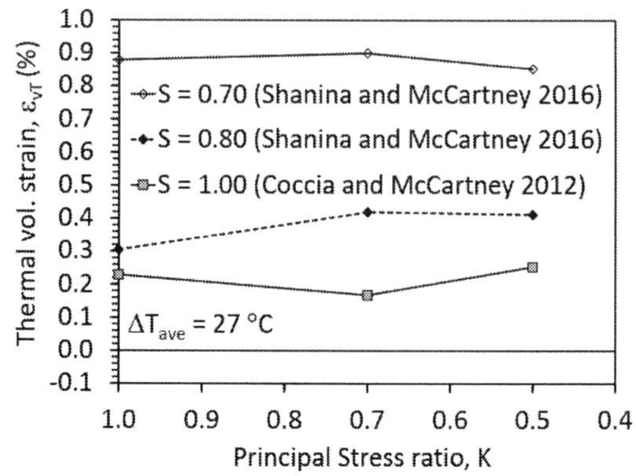

Figure 9. Thermal volumetric strain changes for specimens with different initial degrees of saturation and principal stress ratios

An interesting observation from the results in Figure 9 is that the thermal volumetric strains are not sensitive to the principal stress ratio. This indicates that the opposite trends in the thermal axial strains in the major and minor principal stress directions offset so that the overall volumetric response does not change with initial stress anisotropy. The data from Coccia and McCartney (2012) for saturated specimens of the same silt are also shown in Figure 9 for comparison, and similar conclusions can be drawn. The results in The implication of the insensitivity of the thermal volumetric strain to anisotropic stress states is that the results obtained from isotropic thermal volume change experiments can be combined with an empirical model for the role of the principal stress ratio to account for the impact of stress-induced anisotropy (Shanina and McCartney 2016).

6 Discussion and Conclusions

The comparison of results from the same soil but different initial conditions presented in this paper confirms that the thermal volume change of compacted, unsaturated soils is complex and can be affected by many different parameters and testing paths. The initial void ratio and compaction conditions do not strongly affect the thermal volume change results. Although some of the results appear to follow trends indicating that they may be evaluated using an elasto-plastic type model with suction hardening and thermal softening, other results are not consistent. Specifically, the lack of a trend in thermal volume change with the initial degree of saturation and the continued thermal volume change upon multiple heating and cooling cycles indicates that other analyses may be needed to predict the thermal volume change of unsaturated soils, including thermally-accelerated creep models. Another common theme in the comparison of results is that thermal volume change tests must be interpreted in terms of effective stress. Interpretation of results in terms of effective stress was observed to permit unification of the thermal volume change behavior of the soils in both saturated and unsaturated conditions.

Acknowledgements

This author would like to acknowledge financial support of this work from NSF grants CMMI-0928159 and CMMI-1054190.

References

1. Alsherif, N.A. and McCartney, J.S. (2015). "Nonisothermal behavior of compacted silt at low degrees of saturation." Géotechnique. 65(9), 703-716. DOI: 10.1680/geot./14-P-049.
2. Alsherif, N. and McCartney, J.S. (2016). "Yielding of silt at high temperature and suction magnitudes." Geotechnical and Geological Engineering. 1-14. 10.1007/s10706-015-9961-x.
3. Coccia, C.J.R. and McCartney, J.S. (2012). "A thermo-hydro-mechanical true triaxial cell for evaluation of the impact of anisotropy on thermally-induced volume changes in soils. " ASTM Geotechnical Testing Journal. 35(2), 227-237.
4. Coccia, C.J.R. and McCartney, J.S. (2015). "High-pressure thermal triaxial cell for evaluation of the impact of temperature on soil volume change mechanisms." ASTM Geotechnical Testing Journal. In press.
5. Coccia, C.J.R. and McCartney, J.S. (2016a). "Thermal volume change of poorly draining soils I: Critical assessment of volume change mechanisms." Computers and Geotechnics. In review.
6. Coccia, C.J.R. and McCartney, J.S. (2016b). "Thermal volume change of poorly draining soils II: Constitutive modelling." Computers and Geotechnics. In review.
7. Shanina, M. and McCartney, J.S. (2016). "Impact of stress-induced anisotropy on the thermal volume change of unsaturated silt." Soils and Foundations. In review.
8. Vega, A. and McCartney, J.S. (2015). "Cyclic heating effects on thermal volume change of silt." Environmental Geotechnics. 2(5), 257-268.
9. Abuel-Naga, H.M., Bergado, D.T., Bouazza, A. and Ramana, G.V. (2007). "Volume change behaviour of saturated clays under drained heating conditions: Experimental results and constitutive modeling. " Canadian Geotechnical Journal. 44, 942-956.
10. Baldi, G., Hueckel, T., and Pelegrini, R. (1988). "Thermal volume changes of the mineral-water system in low-porosity clay soils." Canadian Geotechnical Journal. 25, 807-825.
11. Burghignoli A., Desideri A., and Miliziano S. (2000). "A laboratory study on the thermomechanical behaviour of clayey soils." Canadian Geotechnical Journal. 37, 764-780.
12. Campanella, R.G., and Mitchell, J.K. (1968). "Influence of temperature variations on soil behavior." Journal of the Soil Mechanics and Foundation Engineering Division. 94(SM3), 9-22.
13. Cekerevac, C., and Laloui, L. (2004). "Experimental study of thermal effects on the mechanical behavior of a clay." International Journal for Numerical Analytical Methods Geomechanics. 28, 209–228.
14. Cui, Y.J., Le, T.T., Tang, A.M., Delage, P., and Li, X.L. (2009). "Investigating the time-dependent behaviour of boom clay under thermomechanical loading." Géotechnique. 59(4), 319-329.
15. Cui, Y. J., Sultan, N., and Delage, P. (2000). "A thermomechanical model for clays." Canadian Geotechnical Journal, Vol. 37, No. 3, pp. 607–620.
16. Demars, K.R., and Charles, R.D. (1982). "Soil volume changes induced by temperature cycling." Canadian Geotechnical Journal. 19, 188–194.
17. Hueckel, T., and Baldi, M. (1990). "Thermoplasticity of saturated clays: Experimental constitutive study." Journal of Geotechnical Engineering. 116(12), 1778–1796.
18. Romero, E. (1999). Characterization and Thermo-Hydro-Mechanical Behavior of Unsaturated Boom Clay: An Experimental Study. Ph.D Thesis, Technical University of Catalonia, Department of Geotechnical Engineering and Geosciences, Spain.
19. Romero, E., Villar, M.V., and Lloret, A. (2005). "Thermo-hydro-mechanical behaviour of two heavily overconsolidated clays." Engineering Geology. 81, 255-268.
20. Saix, C., Devillers, P., and El Youssoufi, M.S., (2000). "Éléments de couplage thermomécanique dans la consolidation de sols non saturés." Canadian Geotechnical Journal. 37, 308- 317.
21. Salager, S., François, B., El Youssoufi, M.S., Laloui, L., and Saix, C. (2008). "Experimental investigations of temperature and suction effects on compressibility and pre-consolidation pressure of a sandy silt." Soils and Foundations. 48(4), 453-466.
22. Sultan, N., Delage, P. and Cui, Y.J. (2002), "Temperature effects on the volume change behavior of Boom clay." Engineering Geology. 64, 135-145.
23. Tang, A.M., Cui, Y.J., and Barnel, N. (2008). "Thermo-mechanical behaviour of a compacted swelling clay." Géotechnique. 58(1), 45-54.
24. Tavallali, A., Tang, A.M., and Cui, Y.J. (2007). "Thermo-hydro-mechanical behaviour of compacted bentonite." Experimental Unsaturated Soil Mechanics. Vol. 112, 259-265.
25. Towhata, I., Kuntiwattanakul, P., Seko, I., and Ohishi, K. (1993). "Volume change of clays induced by heating as observed in consolidation tests." Soils and Foundations. 33(4), 170–183.
26. Uchaipichat, A., and Khalili, N. (2009). "Experimental investigation of thermo-hydro-mechanical behavior of an unsaturated silt." Géotechnique. 59(4), 339–353.
27. Burghignoli A., Desideri A. and Miliziano S. (2000). "A laboratory study on the thermomechanical behaviour of clayey soils." Canadian Geotechnical Journal. 37: 764-780.

Stress path during pressuremeter test and link between shear modulus and Menard pressuremeter modulus in unsaturated fine soils

Vincent Savatier[1], Maxime Savatier[2]

[1]EGIS Geotechnique 33-43 Avenue G. Pompidou BP 13115-31131 Balma Cedex France
[2] 1 rue des remparts 31450 Montgiscard

Abstract: This paper presents the stress path before and during a pressumeter test using a Modified Cam-Clay model. This model have been used for unsaturated clay with continuous water phase (D2 domain according to Boutonnier 2007) and for silty to clayey soils with discontinuous water phase (D1 domain – ibid.). After reminding the stress distribution around a pressuremeter cell and the meaning of the Menard pressuremeter modulus which is linked to the soil shear modulus, we explain the stress state during pressuremeter modulus measurement. Then, we show that this stress path diagram can be used to explain classical observations made using Menard pressuremeter tests in continental area and typically:

- Measurement of high ratio "Menard pressuremeter modulus / pressuremeter limit pressure" in clay subjected to drought.
- Limit pressure and Ménard pressuremeter modulus variation with respect to drought in silty clay situated in water-table fluctuation area.

Finally, we will try to show the consequences of these stress paths for performing and interpreting pressuremeter test in unsaturated and collapsible soils.

1. Stress state around pressiometer cell in Cambriges axes

In the Cambridge axe p is the average soil stress $p=\frac{\sigma_1+\sigma_2+\sigma_3}{3}$ and $q=\sigma_1-\sigma_3$ the stress deviator.
The stress state around a pressuremeter test look like the stress state around an expending cylindrical cavity. Therefore in elastic state expansion the stress path is a pure shear (isovolumetric). In figure 1, z is vertical axe; Ko the "at rest earth pressure coefficient", h is test depth ; rf the cavity radius; r_o the cavity radius at the pressuremeter cell introduction.; p_{c0} soil pressure at the borehole limit.

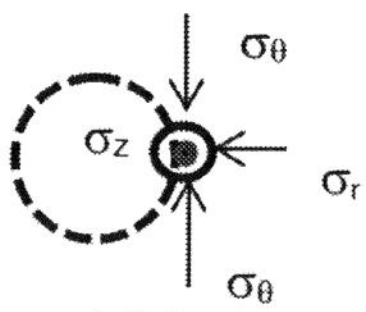

Figure 1. Stress definition around pressuremeter cell

In elastic state the stresses equations around a cylindrical hole with radius r_0 are:

$$\sigma_r=\sigma_0-(\sigma_0-p_{co})\frac{r_f^2}{r^2} \quad (1)$$

$$\sigma_\theta=\sigma_0+(\sigma_0-p_{co})\frac{r_f^2}{r^2} \quad (1')$$

$$\sigma_z=\sigma_v;\ avec\ \sigma_0=K_O.\ \sigma_v \quad (1'')$$

- Before drilling : $\sigma_z=\gamma.h$; $\sigma_\theta=\sigma_r=K_o.\gamma h$; therefore:

$$pi=\gamma h(1+2K_0)/3\ ;\ qi=\gamma h.(1-K_0) \quad (2)$$

- During drilling if remaining in elastic state (this is the most frequent situation for unsaturated soil) the stress equations at the hole limit are :

$$\sigma_z=\gamma.h$$
$$\sigma_r=p_{c0};$$
$$p=(1/3)(\sigma_z+\sigma_r+\sigma_\theta)=\gamma h(1+2K_0)/3$$

Therefore :

$$p=(1/3)(\gamma.h+p_{c0}+\sigma_\theta)=\gamma h(1+2K_0)/3 \quad (2')$$

Thus $\sigma_\theta=\gamma h(2Ko)-p_{c0}$; $q=max(\sigma_z-\sigma_r;\ \sigma_r-\sigma_\theta)$.

For an empty hole that gives:

$$q=max\ (\gamma.h\ ;\ 2(\gamma h(K_0))\ ;p=\gamma h(1+2K_0)/3 \quad (3)$$

- During test while staying in elastic state stress equation are the same that equations (2) replacing p_{c0} with pc pressure at the hole limit:

$$q=max\ (\gamma.h-p_c;\ 2(\gamma h(K_0)-p_c). \quad (4)$$

In the following we will consider the most frequent case where $\gamma.h-p_c<2(\gamma h(K_0)-p_c$ (always observed when $K0>0{,}5$). Thus we have:

$$q=2(\gamma h(K_0)-p_c)\ ;\ p=\gamma h(1+2K_0)/3 \quad (5)$$

The deformation of the radius r follows in elastic state follows (G is the shear modulus):

$$du=dp_c.\ \frac{ro+u}{G}$$

This could be written; (Vo is initial cell volume at the contact with hole limit and V is cell volume during the test): $dV=2\ dp_c\ \frac{V_0+V}{G}$

that gives: $$G=2(V0+V)\ \frac{dp}{dV}$$

Or with Menard pressuremeter Modulus:

$$E=2\ (1+v).(V0+V)\ \frac{dp}{dV} \quad (6)$$

For over-consolidated soil for each average stress p noted σ'_m; the oedometric modulus can be defined with the following formula:

[a]Vincent.savatier@egis.fr

$$Eoed = \frac{2{,}3(1+e_i).\sigma'_1}{Cs} = \frac{4{,}6(1+e_i)\sigma'_m}{3.Cs} \qquad (6')$$

Cs is the LOG_{10} swelling index; σ'_m the average stress which value during an oedometric test is $\sigma'_m=2/3\ \sigma'_1$.
σ'_1 is the principal major stress.

Using the relationship between oedometric and shear modulus, the following relation between G and σ_m is obtained:

$$G = \frac{(1-2\nu)}{(1-\nu)} . \frac{4{,}6(1+e_i).\sigma'_m}{3.Cs} \qquad (7)$$

Thus shear modulus is proportional to average stress.

- <u>During test when plastic state is reached</u>

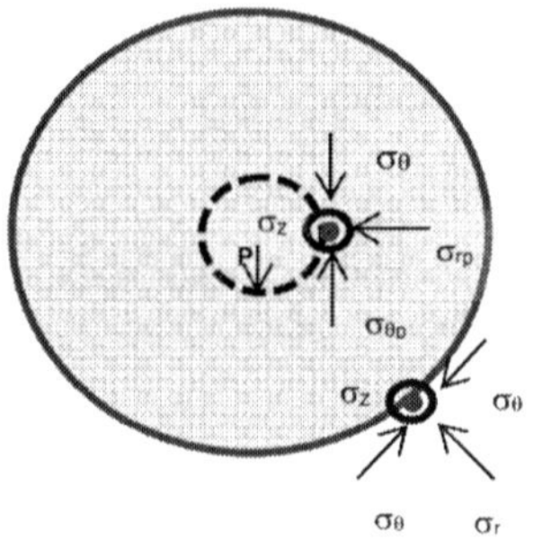

Figure 2. Plastic stress definition around pressuremeter cell

It appears around the cell a soil crown in which limits the values of p and q, can be found with:

- The equation of the limit line for normally consolidated soil with hydric state that allows consolidation:

$$q=(\sigma'_{rp} - \sigma'_{\theta p})=M\sigma'_m = M.p' \qquad (8)$$

- The equation of the yeld flow surface for over-consolidated clay that do not allows consolidation:

$$q'^2-M^2p'(\sigma'_{mpo}-p)=0 \qquad (9)$$

with $M=\dfrac{6\sin\phi'}{3-\sin\phi'}$. Therefore:

$$(\sigma'_{rp} - \sigma'_{\theta p})^2-M^2\sigma'_m(\sigma'_{mpo}-\sigma'_m)=0 \qquad (10)$$

- σ'_{mpo} is the isotropic pre-consolidation stress
- ϕ' soil effective angle of internal friction

According to Wood (1990) ref [1], when consolidation yield flow surface (equation (9)) is reached; the effective stress follows the equation:

$$\frac{p'i}{p'} = \left\{ \frac{M^2 + \left(\dfrac{q}{p'}\right)^2}{M^2 + \left(\dfrac{qi}{p'i}\right)^2} \right\}^{\Lambda} \qquad (11)$$

with M as above ; $\Lambda = \dfrac{\lambda-\kappa}{\lambda} = \dfrac{Cc-Cs}{Cs}$;
Cc is the LOG_{10} compression index; λ the compression index; κ the swelling index; pi and qi are the initial values of p et q before putting the cell in the borehole.

We will call:

$$STi=\left[M^2+\left(\frac{qi}{p'i}\right)^2\right]pi^{\frac{1}{\Lambda}}=\left[M^2+\left(\frac{3(1-Ko)}{1+2Ko}\right)^2\right]pi^{\frac{1}{\Lambda}}$$

By deriving equation (11) it can be shown that:

$$dp'=\left\{\frac{2.q.dq}{STi.\left(\dfrac{\lambda-2\kappa}{\lambda-\kappa}\right)\left(\dfrac{1}{p'}\right)^{\frac{3\kappa}{\lambda-\kappa}}-2M^2p'}\right\} \qquad (12)$$

At the outside limit of this crown, the soil follows equation [1].

According to D. Rangeard ref [2] for a clayey saturated soil case (with permeability below 10^{-6} m/s) the pressuremeter test is fast enough to have iso-volumetric plastic deformations . Thus the following stress paths are obtained. They are given considering two cases:

- If $\sigma'_{mpo} > p=\gamma h(1+2K_0)/3$ deformation will be with effective stress increase and negative interstitial water pressure (U<0)
- *If* $\sigma'_{mpo} < p=\gamma h(1+2K_0)/3$ deformation will be with effective stress decrease and positive interstitial water pressure (U>0)

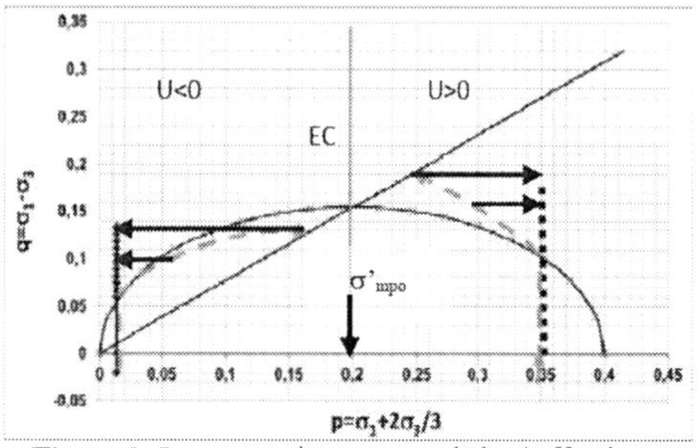

Figure 3. Stress way in a saturated clay(effective stresses paths are in grey with dotted line and total stresses paths are in black).

2. Pressuremeter behavior of unsaturated soils

[a]Vincent.savatier@egis.fr

2.1. Saturations variation

We will consider two of the unsaturated soil states defined by Boutonnier [2007] ref [3] :

- D1 : for this state the gas phase is continuous (suction is over air-entry suction s_{air} ($s>s_{air}$) and saturation degree is over air entry saturation S_{lair} (usually between 0,7 to 0,8).Additionally suction has no effect on volumetric forces and his variation effect is limited to inter-grain forces.

- D2 : for this state, the gas phase is discontinuous and subdivided in bubbles in contact with several soil grains. Suction is under s_{air} and saturation is Sr with S_{lair}<Sr< Sle. Sle is the shrinkage limit.

The water content is $W=\frac{Pw}{Ps}$; with Pw water weight; Ps solid part of soil weight. Considering a soil change from a state *i* to a state *f*, we have the following properties:

$$W_i=\frac{Pw_i}{Ps}=\frac{Vw_i}{Vv_i}.\frac{Vv_i}{Vs}.\frac{\gamma w}{\gamma s}=S_{li}*e_i\frac{\gamma w}{\gamma s}$$

$$W_f=S_{lf}*e_f\frac{\gamma w}{\gamma s}$$

Vw_i is initial water volume, Vs solid grain volume; e_i et e_f are respectively the initial and final void ratio

During a pressuremeter test, deformation will be with constant water content for both D1 and D2 states: In D1 soil because of the water discontinuity; in D2 soil because of the relative high velocity of the test compared with water pressure dissipation. According that: $Wi=Wf$ that gives: $S_{li}*e_i\frac{\gamma w}{\gamma s}=S_{lf}*e_f\frac{\gamma w}{\gamma s}$

With Sl_i et Sl_f initial and final liquid saturation.

Therefore : $Sl_i*e_i=Sl_f*e_f$

Volumetric deformation between start and end of deformation is :

$$\varepsilon_V=\frac{e_f-e_i}{1+e_i}=\frac{\frac{S_{l_i}}{S_{lf}}ei-ei}{1+ei}=\frac{e_i}{1+e_i}\left(\frac{S_{l_i}}{S_{lf}}-1\right)$$

thus :

$$Sl_f=\frac{S_{l_i}}{1+\left(\frac{1+ei}{ei}\right)\varepsilon_V} \quad (13)$$

We will consider a deformation that leads from a stress p'_i to a stress p'_f. Beyond this point, we will note them respectively σ'_{mi} et σ'_{mf} to avoid confusion with pressuremeter cavity pressure. During this deformation:

- For over-consolidated soil :

$$\varepsilon=-\kappa.\ln\left(\frac{p'_f}{p'_i}\right)=-\frac{Cs}{(1+e_i)}Log\left(\frac{\sigma'_{mf}}{\sigma'_{m_i}}\right) \quad (14)$$

From (13) and (14), the evolution of saturation during a pressuremeter deformation is:

$$S_{lf}=\frac{S_{l_i}.ei}{ei+\left(-Cs.Log\left(\frac{\sigma'_{mf}}{\sigma'_{mi}}\right)\right)} \quad (15)$$

- For normally consolidated soil:

$$\varepsilon=-\lambda.\ln\left(\frac{p'_f}{p'_i}\right)=-\frac{Cc}{(1+e_i)}Log\left(\frac{\sigma'_{mf}}{\sigma'_{m_i}}\right) \quad (16)$$

From (13) and (16) the evolution of saturation during a pressuremeter deformation is in this case:

$$S_{lf}=\frac{S_{l_i}.ei}{ei+\left(-Cc.Log\left(\frac{\sigma'_{mf}}{\sigma'_{mi}}\right)\right)} \quad (17)$$

2.2. Relationship between total stress and effective stress for D1 or D2 soil states

According to its definition, the variations of the effective stress follow the relationship:

$$d\sigma'_m=d\sigma_m-(1-S_l)\,dP_g-S_l\,dPl \quad (18)$$

Sl :liquid saturation ; *dPg* : gas pressure variation ; *dPl* : liquid pressure variation

2.2.1. D1 soil case

In D1 case the total stress variation will be applied directly on the hydric bridges between the soil grains and gas pressure will not change. Otherwise, liquid pressure variation *dPl* in hydric bridges, will be proportional to the effective stress variation. The proportionality ratio Ksg will depend of grain surface state and will be above or equal to one. Thus we have the following relationship :

$$d\sigma'_m=K_{sg}\,dPl \quad (19)$$

Equation (18) gives then $d\sigma'_m=-d\sigma_m-K_{sg}S_l\,d\sigma'_m$ and then:

$$d\sigma_m=(1+K_{sg}S_l)\,d\sigma'_m \quad (20)$$

2.2.2. D2 soil case

In D2 case, the gas phase is discontinuous and divided in bubbles in contact with several soil grains. Since gas has a higher deformability than liquid, it will absorb most of the deformation and liquid pressure will stay already constant.

Gas pressure Pg will follows Mariotte et Henry laws. This gives:

$$Pg.\,Vg+H(t)\,dPg=Pgo\,Vo \quad (21)$$

[a]Vincent.savatier@egis.fr

- *Pgo* and *Pg*: gas pressure respectively before and after deformation
- *Vg* et *Vgo* : gas volume before and after deformation
- *H(t)* Henry constant that gives dissolved gas quantity variation according to gas pressure variation and time.

$$H(t=0)=0\ ;\ H(t=\infty)=0{,}02(T/To) \qquad (22)$$

T=temperature in K ;$T0$=293° K

As pressuremeter test is fast, dissolved gas concentration has no time to change and (22) can be simplified as :

$$Pg.Vg+Pgo.Vgo \qquad (23)$$

According to the saturation degree definition (23) :

$$Pg.(1\text{-}Sl)=Pgo.(1\text{-}Slo)$$

And then $Pg= Pgo\dfrac{1-Sl_o}{1-Sl_i}$ that gives by derivation:

$$dPg= -Pgo(1\text{-}SLo).\ \frac{dSl}{(1-Sl)^2} \qquad (24)$$

As $dPl=0$ equation (18) gives: $d\sigma'_m= d\sigma_m -(1\text{-}S_l)\ dP_g\text{-}S_l$ and then:

$$dPl= d\sigma_{m+}\ Pgo(1\text{-}SLo).\ (1\text{-}S_l)\ \frac{dSl}{(1-Sl)^2} \qquad (25)$$

$$d\sigma_m= d\sigma'_m\text{-}\ Pgo(1\text{-}SLo).\ \frac{dSl}{(1-Sl)}. \qquad (26)$$

2.3. Plastic stress variation around pressiometer

The following figure provides the stress distribution in a thin crown of soil around a pressiometric cell that has reached the plastic state.

The increase of P2 the pressure in this cell drives the soil from a state n to a state n+1.

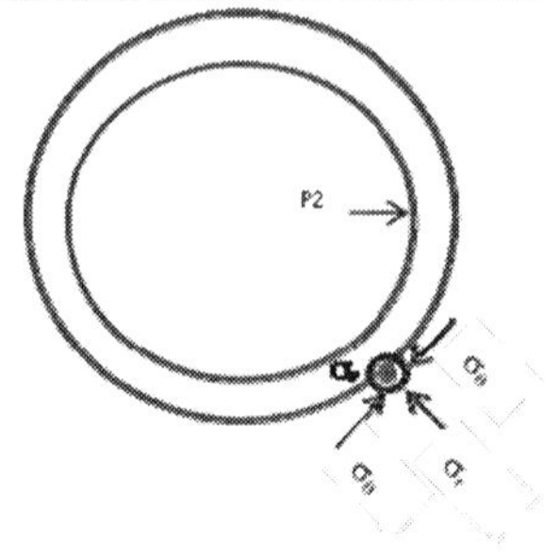

Figure 4. Stress distribution around a pressiométrique cell

We will assume that the thickness of the crown is thin enough to have constant shear strength in all its area between P2 to P2+dP2. In these conditions:

$$\sigma_r\text{-}\sigma_\theta=2(P2\text{-}Po)=2\ Su \qquad (27)$$

Su : undrained shear strength
Po : soil pressure before the test.

The notations, for the transformation from n to n+1 state, are the following:

state n	gives	**state n+1**
$P2$	gives	$P2+dP2$
$2\tau_n=\sigma_r\text{-}\sigma_\theta=2(P2\text{-}Po)$	gives	$2\tau_{n+1}= 2(P2\text{-}Po)+2dP2$
$\sigma'm_n$	gives	$\sigma'm_{n+1}= \sigma'm+ d\sigma'm$
σm_n	gives	$\sigma m_{n+1}= \sigma m+ d\sigma m$

- The application of equation (8) between n and n+1 in a normally consolidated soil (likes collapsible loess) gives:

$$\sigma'm_{n+1}= \sigma'm_n+\ 2MdP2 \qquad (28)$$

- Considering that $q==\sigma_r\text{-}\sigma_\theta=2(P2\text{-}Po)$; $dq=2dP2$, *and* $p'=\sigma'm_n$,the application of the equation (12) for an over consolidated soil gives*:*

$$d\sigma'm=\left\{\frac{2.q.dq}{STi.\left(\dfrac{\lambda-2\kappa}{\lambda-\kappa}\right)\left(\dfrac{1}{p'}\right)^{\frac{3\kappa}{\lambda-\kappa}}-2M^2p'}\right\} \qquad (29)$$

Note that all the elementary crowns that are in plastic state will be on the plastic deformation curve and follows from n to n+1states, equation (28) for loessic soil and equation (29) for overconsolidated clay.

Thus, the particular crown situated at the cell contact will follows equation (28) to (29) for each soil type.

3. Stress way of pressuremeter test in unsaturated soil

3.1. Unsaturated clay

The soil will be in D2 state and follows modified cam clay model. Thus on elastic state the crown of soil around the pressuremeter cell will follows an iso-volumetric shear. The effective stress $\sigma'm$ will be above total stress, with a constant difference to *so* the initial suction. $\sigma'm=p'$ follows the equation:

$$q'=q;\quad p'=p+S_0=p+(u_a\text{-}u_w) \qquad (30)$$

with: u_a air pressure ; u_w water pressure .
Saturation degree S_0 will stay constant.
Shear modulus will be proportional to $\sigma'_m=p'$ according to equation (7).
We remind that the pressuremeter modulus will follow:

$$Em=(1+\upsilon)\left[\frac{(1-2\upsilon)}{(1-\upsilon)}\frac{4{,}6(1+e_i)(\sigma_m+so)}{3Cs}\right] \qquad (31)$$

---

[a]Vincent.savatier@egis.fr

Therefore an increase of suction; will increase pressuremeter modulus.

In the plastic state, for each increase of stress path in the crown of soil around the pressuremeter :

- Saturation degree will follows equation (17)
- Effective variation $d\sigma'm=\sigma'm_{n+1}-\sigma'm_n$ will follows equation (29) that will allow effective stress calculation step by step.
- σm_n calculation from $\sigma'm_n$ will be done by equation (26)

The following figures give the stress path for a less overconsolidated clay with high suction (fig 5); and for a very overconsolidated clay with a low suction (fig 6).

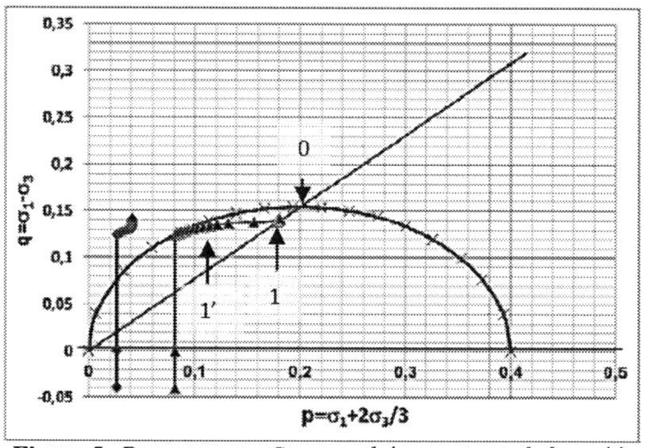

Figure 5. Pressuremeter Stress path in unsaturated clay with low suction compared with σ'p

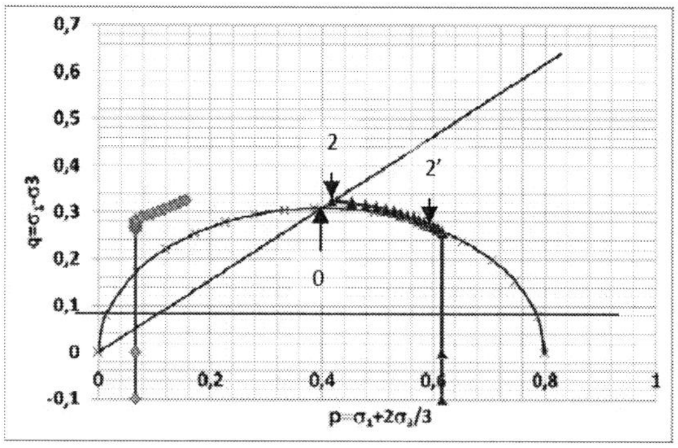

Figure 6. Pressuremeter stress path in unsaturated clay with high suction compared with σ'p

Depending on initial saturation *So*, initial suction s_o and soil deformability, the crown at cell contact will reach the critical point (point 1 or 2 on figures) or will stop before (point 1'or 2' on figures).

It results from the figures 4 and 5 that $q=\sigma'_r-\sigma'_\theta$ and therefore the limit pressure value will be similar to the limit pressure in a saturated soil with the same $\sigma'p$. Consequently, an increase of suction due to dryness will increase the *Em/pl**ratio proportionally to $\sigma m+so$ and when wet weather decrease suction *Em/pl** ratio will also decrease in the same way. This is an explanation of the experimental observation of high rapport *Em/pl** in clayey sol between 0 to 3m depth during dry periods.

The existence of different path of stress depending on initial saturation suction and total stress shows that interpretation of pressuremeter test using finite elements calculation based on saturated Cam clay model may fail because of these parameters influence.

3.2. Unsaturated silts collapsible soil

Soil state will be D1 and he follows a pure friction law or a normally consolidated law with consolidation during deformation.

On the elastic crown, as in clayey soil, shear will be iso-volumetric with an effective stress linked with total stress by the same formula: $q'=q$; $p'=p+so$. Note that s_o is here a function of saturation degree So and will stay constant during elastic step in this case too. Therefore, pressuremeter modulus will increase with suction increase in the same way than in clayey soil .

On the plastic crown, for each loading path:

- Saturation degree will follows equation (17)
- Effective variation $d\sigma'_m=\sigma'm_{n+1}-\sigma'm_n$ will follows equation (28). This will allows the calculation of effective stress step by step.
- σm_n calculation from $\sigma'm_n$ will be done by using equation (20) as long as saturation will be below a limit saturation S_{lim} , which is linked to a limit suction. Over this value, soil will pass in D2 state and follows equation (26).

The following figures gives the stress path for a collapsible soil with initial saturation S=50% and initial average stress of 66 kPa with an initial suction of 150 kPa (fig 7) and with a higher initial suction of 350 kPa (fig 8).

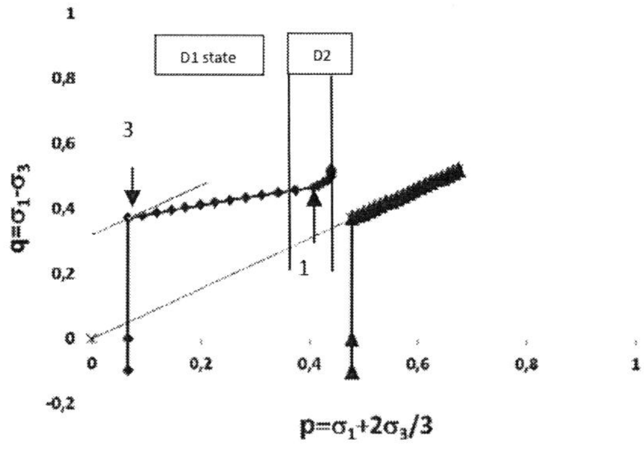

Figure 7. Pressuremeter stress path in unsaturated collapsible loess with low suction

[a]Vincent.savatier@egis.fr

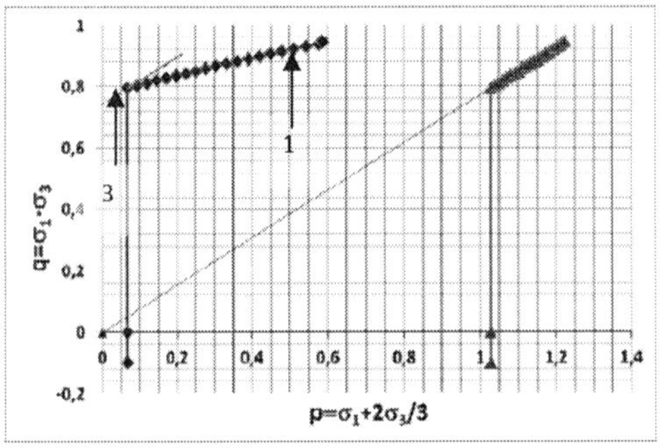

Figure 8. Pressuremeter stress path in unsaturated collapsible loess with high suction

It appears that pressuremeter modulus Em will increase proportionally to the amount of average initial total stress with suction $\sigma_{mo}+s_o$ while $q=\sigma'_r-\sigma'_\theta$ will increase proportionally to $M(\sigma_{mo}+s_o)$. Thus *Em/pf** ratio (*pf** corresponding to point 3 on the figures) will be constant. For limit pressure (point 1 on the figures 7 and 8) the stress path will depend on initial saturation degree and suction.

For low saturations, soil will stay in D1 state and *Em/pl** ratio will stay constant with suction variation. The *EM/pl** ratio will depend, in this case, on compressibility coefficient Cc and on initial saturation So.

For higher saturation, *Em/pl** variation will depend on initial suction. Over a limit value of initial suction, soil will also stay in D1 state with a constant *EM/pl** ratio. Below this limit value, a part of the plastic crown will pass in D2 or D3 state. In that case, soil consolidation will not be possible during the test and *pl** increase with suction will not continue to be linear. *Em/pl** ratio will therefore increase with suction.

This helps to explain the pressuremeter modulus and limit pressure variations in silt in the vadose area. In the case of water table level increase, soil suction will decrease or disappear, leading to a high decrease of pressuremeter modulus and limit pressure (divided by two for the study case of the figures 7 and 8). On the contrary, a water level decrease and a growing of evapotranspiration will increase suction and therefore limit pressure and modulus with a *Em/pl** ratio constant or slightly growing.

Note also that as seen above, for the clayey soil, the stress paths depend on initial suction and saturation. Therefore, estimation for soil of effective friction ratio or of shear strength using finite elements models based on pure friction soil model may also fail if suction and saturation incidence is not low.

4. Conclusion

The estimation of saturation and suction variations in unsaturated soil around a pressuremeter cell allows drawing qualitatively the stress path during a pressuremeter test in clayey and collapsible loessic soils. It shows that stresses path are depending on initial saturation degree and suction. This must lead to be very careful when using finite elements calculation with saturated model or pure friction model in order to estimate cam clay soil parameter or modulus, friction angle and dilatancy estimation. It seems necessary to check that the soil is perfectly saturated or perfectly dry to avoid calculation disturbance due to saturation and suction variation.

Onshore, in the very frequent case of unsaturated soil, these methods will need to take into account the effect of initial saturation and suction.

This could be done by using the presented method or similar. For this purpose, it will be necessary to sample soil for each pressuremeter test in order to measure in situ suction and saturation. This may be done by using a core cutter sampler in order to do the test cavity excavation and the soil sampling at the same time.

References

1. Y. Bentaiebi ; thèse Dep. Génie Civil Géol. et Mine Ecole polytechnique de Montréal (2011)
2. D. Rengeard ; thèse école centrale de Nantes ; *www.theses.fr/2002NANT2092;* 37- 57 (2002)
3. L. Boutonnier ; Comport. hydromécanique des sols fins proche de la saturation thèse de docteur de l'INP de Grenoble 95-100 (2007)
4. V.Savatier ; P. Garnier ; JNGG 2014 Sols effondrables interprétation théorique et amélioration de l'essai double œdomètre Roumain 108 (2014)

[a]Vincent.savatier@egis.fr

The influence of saturation on the cracking process in compacted desiccating clays

Maria Noack[1,a], Paul Winkler[1] and Karl Josef Witt[1]

[1]*Bauhaus-Universität Weimar, Faculty of Civil Engineering, Chair of Geotechnical Engineering, 99423 Weimar, Germany*

Abstract. Tensile failure in unsaturated cohesive soils during desiccation is important for the design of geotechnical applications such as capping systems of landfills and sealing material of dikes. This study presents the results of different initial parameters of compacted clay samples such as gravimetric water content, dry density and degree of saturation. These parameters are varied systematically for each test to find the correlation between those parameters and the tensile failure. The tensile failure for all tests occurred by a comparable constant change of the degree of saturation. The soil specific saturation ratio $S_{r,s}$ is defined as the quotient of saturation changes ΔS_r to initial saturation $S_{r,0}$. This parameter related to the total suction shows an equal course for all results. All in all, the course of the soil specific saturation ratio $S_{r,s}$ is independent of all initial parameters. These results provide a physical and hydraulic-mechanical description for modelling the desiccation process. To demonstrate the initiation and progress of tensile failure, the experimental results are modelled with a Discrete Element Method (DEM) approach.

1 Introduction

In desiccation processes of cohesive soils due to evaporation or transfer of heat the loss of the fluid phase leads to an increase of suction in the soil which is accompanied by a noticeable change of volume. The soil shrinks. A number of researchers studied in decades this shrinkage behaviour with different focuses.

Methods of desiccation tests of clays can be divided in free (without development of cracks) and restrained (exceeding of tensile soil strength leads to cracks) shrinkage test. Most tensile strength experiments are induced by external loading and not by the internal desiccation stresses [1] or fail to incorporate stage-wise drying of soils [2], which is essential for investigations related to cracking characteristics in clays. Commonly studies investigate full saturated conditions, few researchers performed experiments on initially unsaturated and compacted clays.

This paper investigates the influence of different initial compaction parameters on shrinkage process of clays, in particular the significance of the degree of saturation. The tests were systematically conducted with samples compacted on the dry and wet side of the optimum of the standard proctor curve (σ_c = 600 kPa). To study the hydraulic-mechanically coupled correlations suction tests, free shrinkage tests, shrinkage tensile tests and restrained shrinkage tests were performed. The purpose of the test program is to provide a variety of parameters which can be put in correlation to crack initial state. In addition the cracking behaviour is modelled with Discrete Element Method (DEM).

[a] Corresponding author: maria.noack@uni-weimar.de

2 Soil properties and preparation

In the test program a clay called "Plessa" was used, which is classified as a high plasticity clay with a liquid limit of w_l = *54.5%*, a plastic limit of w_p = *21.0%* and a saturated shrinkage limit at w_s = *16.7%*. The specific gravity is ρ_s = *2.691 g/cm³*. The amount of clay size particles (< *2 μm*) is *42.8%* and *57.2%* of the particles are between *2 μm* and *2mm*. The main minerals are kaolin with *18.0%* and illite with *9.0%*.

The preparation of the soil affects strongly the compaction properties of the material. To ensure reproducible and comparable test results, the clay was homogenized in accordance to the advice of Birle et al. [3]. Additionally the clay with the required water content was pushed through a sieve of *4mm* mesh size for a consistent initial sample structure.

To choose the initial conditions of dry density, water content and saturation for the experimental program three moisture-density compaction curves associated with compaction vertical stresses σ_c = 210, 400 and 600 kpa have been carried out. The results are shown in Figure 1, where contours of equal degrees of saturation are indicated.

Test point D1 is located at the dry side of the Proctor Optimum, while the test points A2, B1, B2 and B3 represent the properties at the wet side. The test points D1 (w = *17.0%*, ρ_d = *1.512 g/cm³*), B2 (w = *21.0%*, ρ_d = *1.512 g/cm³*) and A2 (w = *25.0%*, ρ_d = *1.512 g/cm³*) have different initial water contents but the same dry density. Same initial water content with different dry density have the points B1 (w = *21.0%*, ρ_d = *1.615 g/cm³*), B2 (w =

21.0%, ρ_d = *1.512 g/cm³*) and B3 (*w* = *21.0%*, ρ_d = *1.345 g/cm³*). Moreover the points B1 (S_r = *84.8%*) and A2 (S_r = *86.3%*) have a comparable initial saturation.

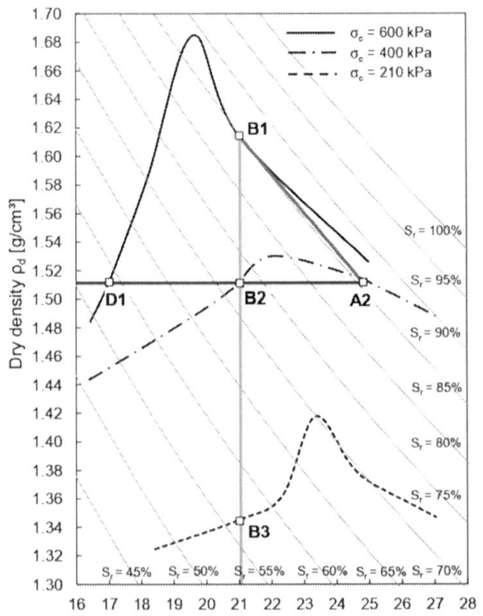

Figure 1. Static compaction curves for the vertical stresses σ_c = 210, 400 and 600 kPa

3 Setup and experimental procedure

3.1 Free Shrinkage Tests

For each of the five different compacted mounted state three samples were prepared with a diameter of *d* = *7.15 cm* and a height of *h* = *2.00 cm*. The samples dried without previous saturation in constant environmental conditions. During the desiccation process the changes of volume and mass were regularly measured. At the end of the experiments the samples were dried in oven and the dry densities were determined by the wax displacement method. From these experiments the changes of the volume, the dry density, the void ratio and the saturation in desiccation process are derived.

3.2 Suction Measurements

The relation between the total soil suction and the changes in water content, while the desiccation, were experimentally determined by the use of a chilled-mirror dew-point hygrometer (WP4-T by Decagon). To determine the total suction, this equipment uses the thermodynamic relation between the total suction, the relative humidity and the temperature according to the Kelvin's equation.

From large prepared free shrinkage samples three discs with a diameter of *d* =*3.88 cm* and a thickness of *h* = *0.50 cm* has been cut. The suction measurements were performed on these discs. After each measurement the samples were weighted and air-dried to the next water content. After a homogenisation of 24 h in an airtight chamber the measurements were continued. At the end the water content was determined by oven drying. As a result besides the suction-water content-relation the pore size distribution was analyzed. The volumetric water content and the degree of saturation were determined with the dry densities of the free shrinkage tests.

3.3 Shrinkage Tensile Tests (STT)

To investigate the effect of desiccation process on the initiation and propagation of tensile cracks for each initial condition a minimum of 3 shrinkage tensile tests were performed. An illustration of the experimental setup is shown in Figure 2. The soil samples were compacted in a mould shaped like the number eight but with similar dimensioned loops. The dimension of the samples were *l* = *7.21 cm* in length and *h* = *1.00 cm* in height. Each end of the soil sample was covered with steel claws. While at one claw the tensile stresses during the test were simultaneously recorded by a load cell, the other claw was fixed. During shrinking process the change in water content was measured continuously. At the moment of crack initiation the dry density was determined by the wax displacement method. The purpose of these experiments was the determination of the water content, dry density, void ratio and saturation at failure state. Furthermore the duration of time until failure was determined.

It should be noted that the tensile stress measured in this setup is strongly influenced by other physical parameters like friction or transverse forces. So the measurement of the tensile stress is qualitative in character and useful to find the peak of the stress and a definition of failure.

Further investigations with an improved and advanced Shrinkage Tensile Test apparatus will be conducted.

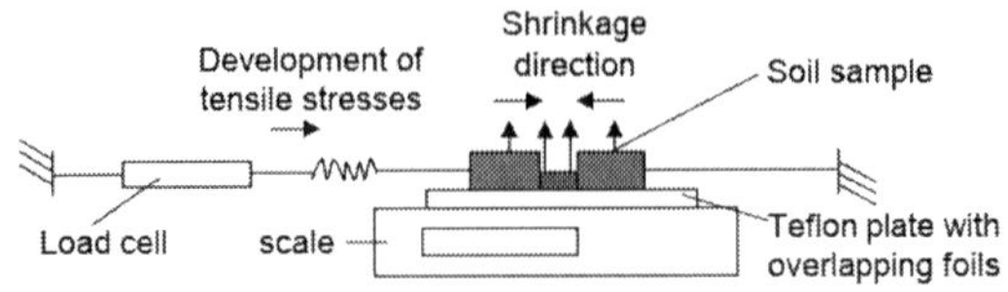

Figure 2. Shrinkage Tensile Test (STT).

3.4 Restrained Ring Tests (RRT)

An image of the setup is presented in Figure 3. At the beginning of the experiment an area of 20 cm x 35 cm of soil was compacted to the required initial parameters with a purpose-built compaction plate. The sample height was set to *h* = *1.00 cm*. After the compaction two samples with an outer diameter of d_o = *15.02 cm* and an inner diameter of d_i = *7.62 cm* were cut out with steel rings. The remaining soil was used to determine the initial parameters of water content, dry density, void ratio and saturation. Both samples each in two steel rings were put on a Teflon foil to reduce friction at the bottom. One setup was placed on an electronic balance with an accuracy of *0.01 g* and a recording interval of 5 values per minute. Each five minutes a high resolution digital camera (Canon PowerShot SX50 HS) recorded the changes of area in desiccation process. To prevent disturbing influence of shadows a LED ring light was direct screwed with the lens. The second sample was placed nearby and was equipped with a height sensor at the surface to determine the settlements continuously. Around the setup there are steady environmental conditions, recorded on a climate data logger.

The shrinkage displacements were calculated by geoPIV (White et al. 2003 [4]), a Matlab based particle image velocimetry (PIV) analysis. This code has been used effectively in several geotechnical engineering researches to analyze soil deformations (e.g. [1],[5]). A contrasting stochastic pattern was sprinkled randomly on the sample. From a mesh of patches of 32 by 32 pixels with a path size of less than *2.4 mm* by *2.4 mm* (1 pixel ~ *0.075 mm*) the displacements and strains were determined by the software.

To analyze the changes in area the images were convert to binary pictures, based on threshold by Matlab image processing. Geometric parameters were evaluated by the detected pixel information, which was provided by the software code.

As results of these test series the displacement, strain, volumetric change, change in dry density, water content, void ratio, saturation over the whole shrinkage process and in particular at failure can be documented.

First findings relating to the topic are presented in this paper.

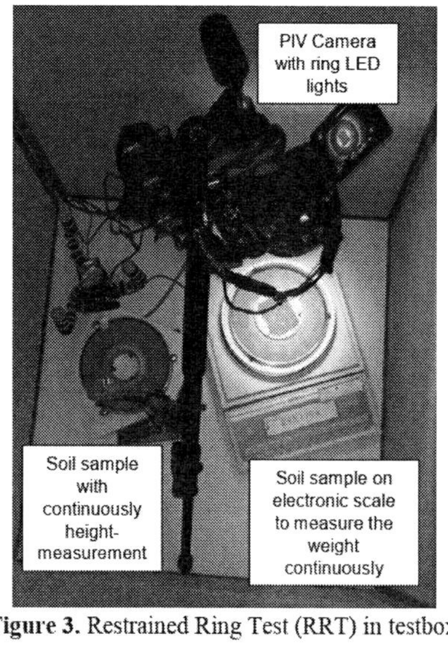

Figure 3. Restrained Ring Test (RRT) in testbox.

4 Experimental results

The total soil suction properties could be investigated in a range of *0.5-300 MPa*, which is caused in the limited determination range of the dew-point potentiometer, where capillary forces < *0.5 MPa* cannot be analysed in pores between aggregates (interaggregat pores). In Figure 4b the saturation properties over the suction data are presented. It shows the influence of different initial water contents and dry densities as well as the initial saturation. Samples with the same initial saturation (A2 and B1) show a very similar drying path independent of the different initial water content and dry density, respectively the different initial pore structures. An increase in the initial dry density at the same water content (from B3 to B2 to B1) implies on the structural properties a decrease in the bulk void ratio through the reduction of the amount of the greater pore diameters (interaggregat pores), while the amount of intraaggregat pores (pores in aggregates with smaller diameters) is continuous similar [6]. The changes in hydraulic properties are obviously in the increase of the saturation degree (from B3 to B2 to B1). These initial changes of the structural and hydraulic properties do not affect the initial suction measurements. All three samples start with equal suctions. The well known influence of increasing initial water content on same dry density (from D1 to B2 to A2) to the increase of the total suction is becoming apparent in Figure 4b.

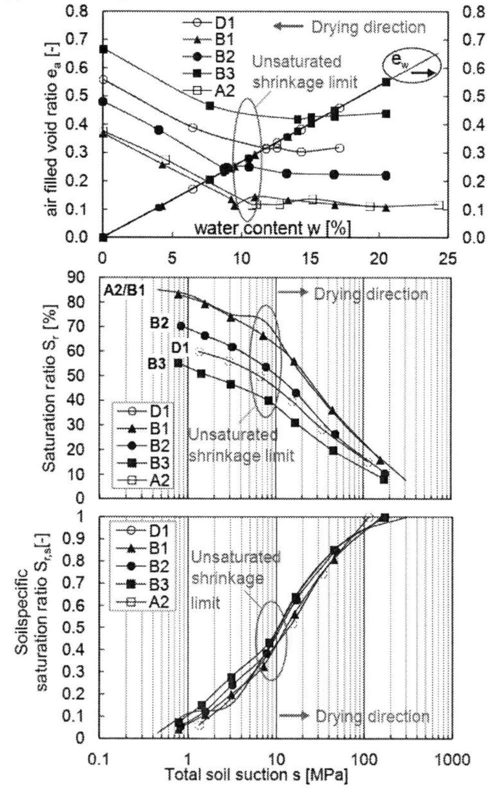

Figure 4 a, b, c. Results of free shrinkage and suction tests.

It can be seen that during the drying process the amount of changes in saturation (ΔS_r) in all samples are similar until the shrinkage limit is reached (see Figure 4b). This can be explained by the development of the air filled (left vertical axis) and water filled (right vertical axis) void ratio in Figure 4a. The water filled void ratio is the same line for all samples. Until the shrinkage limit is reached the changes of the void ratio are in accordance with the amount of changes of the water filled pores. As the relative humidity of the environmental air is for all samples similar, they dewater in the same amount,

DOI: 10.1051/e3sconf/20160909011

independent of different structural properties. The amount of the air filled void ratio remains approximately constant. With occurrence of the shrinkage limit the air filled void ratio increase in the same value the water filled void ratio decrease. This behaviour is comparable identically for all samples despite several initial structural and hydraulic properties. This suggests a soil specific property over the complete desiccation process, which is graphically highlighted in the soil specific saturation ratio $S_{r,s}$ in Figure 4c:

$$S_{r,s} = \Delta S_r / S_{r,0} = (S_{r,0} - S_r) / S_{r,0} \quad (1)$$

Where S_r is the current saturation ratio and $S_{r,0}$ is the initial saturation ratio.

In the following, these properties of the compacted samples are related to the crack behaviour investigated by shrinkage tensile and restrained ring test. The determining of the parameters of these experiments has already been described (see 3.3 and 3.4). Just the development of the total soil suction has been adopted by the free shrinkage tests. In Figure 5 sample of the results of PIV analysis and the corresponding binary image in restrained ring test is illustrated.

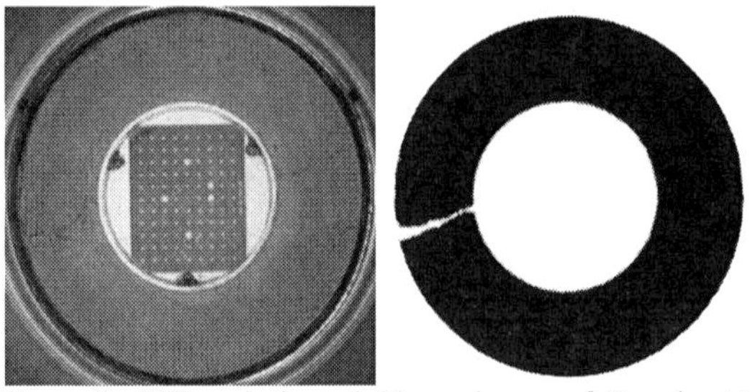

Figure 5. PIV analysis and binary image of Sample A2 complete cracked.

Table 1 highlights the most important findings subjected to the drying process and the comparison of the STT and RRT A2 results. Interesting remarks can be made about the changes of the hydraulic parameters. In particular all samples cracked approximately if the water content changed around *2-3%*. The results of STT and RRT are similar. So the occurrence of the crack appears comparable in both types of experiments.

Further conclusions can be achieved due to the addition of the results which are presented in Figure 6, where the combination of free shrinkage tests and STT results are illustrated. It is supposed that the failure state is a function of the changes in water content along the free shrinkage curve, which can be expressed in the changes of saturation. In four out of five (B1, B2, B3, A2) cases in STT series the changes of saturation from initial to failure state are between *3.69* and *4.51 %*.

The experimental determination of soil specific saturation ratio $S_{r,s}$ for the samples B1, B2, B3 and A2 is in good accordance to the measurements of free shrinkage - soil suction behaviour (Figure 4c), expressed in results of the last line in Table 1.

The reasons for divergences are the comparison of three different test series with possible fluctuations of the external impacts (differences in: human resource, compaction instruments, environmental conditions, soil properties). It should be noted that the factor time is not mentioned. Compaction conditions and crack initiation have an influence on the rate of desiccation.

Table 1. Results from initial to failure state: STT and RRT (A2)

	STT					RRT
	D1	**B1**	**B2**	**B3**	**A2**	**A2**
Δw [%]	2.84	2.17	2.36	2.10	2.89	2.74
ΔS_r [%]	5.51	4.52	3.69	3.78	3.72	3.51
$S_{r,s}$ [-]	0.093	0.052	0051	0.067	0.043	0.043
$s_{failure}$ [MPa][a]	2.861	1.107	1.072	1.065	0.731	0.731
$s_{failure}$ [MPa][b]	1.850	1.000	1.000	0.950	0.500	0.500

[a] determined by suction tests
[b] determined by diagram in Figure 4c

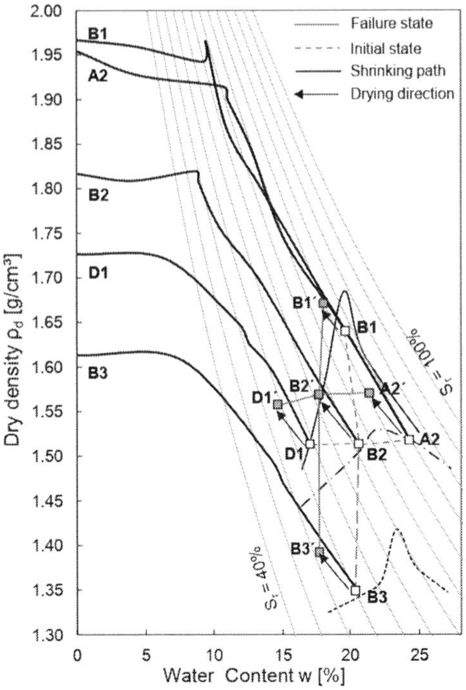

Figure 6. Results of free shrinkage test and STT.

DOI: 10.1051/e3sconf/20160909011

5 Numerical simulation

The shrinking behaviour of the soil and the resulting desiccation cracks are modelled in a numerical simulation for a better understanding and the development of a theoretical model. Granular material at all and cracking in particular is not well represented in methods of continuum mechanics [7]. Therefore particle simulation with grid free methods, especially Distinct- or Discrete-Element-Method (DEM) is used.

The DEM is a numerical method, computing the motion and effect of a large number of particles [8]. It gives the opportunity to determine the discrete parameters of a particle assembly. Although DEM is related to molecular dynamics, the method is generally distinguished including rotational degrees of freedom as well as stateful contact. The application of the DEM in geomechanics was proposed by Cundall and Strack [9]. Today the DEM is well established [7] and can provide insight into the soils structure [10]. The method is computationally intensive. Several DEM codes using parallel processing to be able to simulate models with higher number of particles and longer duration. With advances in computing power and numerical algorithms for nearest neighbour sorting, it has become possible to increase the number of particles with simulations of longer duration.

Different programs (e.g. PFC3D, EDEM, LAMMPS, LIGGGHTS, Yade, WhoDEM) are available for DEM simulations. The authors use LIGGGHTS, because it is effective for high particle numbers and there is a version free of charge. LIGGGHTS is based on the molecular dynamics simulator LAMMPS and extended to support granular media [11]. The model for the simulation is described in a scripting language, which allows easy modifications of the model. The program is open source, the code can be modified and extended for individual users' needs. It is possible to import complex geometry from computer aided design (CAD) into the simulation. For further investigations a coupling with the CFD solver OpenFOAM is available [12].

The biggest disadvantage of the DEM is its computational effort for high particle numbers. Since clay consists of very small particles it is neither a granular material nor is the number of clay particles in a range which could be handled by current simulation hardware in a reasonable time. Microscopic observations show that the clay particles clump together in aggregates of various scales like a fractal structure [13]. The swelling and shrinking of clay can be explained as a behaviour of the aggregates [14]. The small clay particles gathered to this much bigger aggregates makes it possible to modell the shrinking and swelling of clay with the DEM in a numerical simulation [15], [16]. To obtain reliable results in a reasonable time, the particle size of the aggregates is varied over a range of 1 to 4 mm. This decreases the necessary particle number from 190,000 to 3,000 and results in a 60 times faster computation for testing purpose.

The Simulation consists of three steps, 1. Particle Insertion, 2. Compaction of the Sample, 3. Desiccation. In the first step a flat ring shaped bin with the measure-

Table 2. Parameters for Numerical Simulation.

Parameter	Value	Unit
Contact Model	Hertz	-
Boundaries	Periodic	-
Simulation Domain	0.1 × 0.1 × 0.1	m
Gravity	9.81	m/s²
Time Step Length	10E-7	s
Particle Size	0.001 – 0.004	m
Particle Density	2500	kg/m³
Skin depth	0.02	m
Volume Fraction Insert	0.2	-
Young's Modulus	50	GPa
Poisson's Ratio	0.35	-
Coef. of Restitution	0.3	-
Coef. of Friction	0.1	-
Coef. of Rolling Friction	0.5	-
Cohesion Energy Density	1E8	J/m³

Figure 7. With the RSA inserted particles.

Figure 8. Compacted sample.

Figure 9. Sample with desiccation crack

ments of the sample volume is created and the simulation parameters are set up with the values given in Table 2. The clay aggregates are inserted (Figure 7) as spherical particles with the Random-Sequential-Addition-Algorithm (RSA) [17]. In this algorithm, particles are sequentially placed in the bin by random. If they overlap with existing particles they will be discarded, else they were kept. The resulted packing has a low density. In step two the particles settle by a gravitational field. For a higher packing density and an even surface of the sample a compaction plate was lowered on top of the particles (Figure 8). The test sample is now prepared like the one in the laboratory test. The porosity is about 36%, what is typically observed for random close packings of monodisperse spheres [18]. It is less than the porosity of the clay samples which is caused by the more chaotic order of the clay particles (edge-to-face flocculated fabric) [19] compared to the assembly of the isotropic shaped spheres. In the third step the desiccation of the clay is represented by the shrinkage of the clay aggregates. The loss of the volume of the water due to desiccation correlates with the decrease of the particle diameter. Whilst the particles are shrinking they stick together by the force of cohesion. The cohesion is modelled with the Simplified-Johnson-Kendall-Roberts model (SJKR) using an additional normal force contribution. Two particles in contact will have an additional normal force maintaining the contact which writes:

$$F_n = kA \quad (2)$$

where A is the particle contact area and k is the cohesion energy density (Table 2). Therefore during the desiccation process the particles move closer to each other. Where this particle rearrangement is restricted due to geometric boundaries or by the particles itself, restrained shrinking will occur. The force of cohesion will cause tensile stresses. Since the fabric of the particles is not regular like in a crystal, but stochastic heterogeneous, the distance, and therefore the forces, between the particles variegate. So locally particles lose contact and if the assembly is not able to cope with the rising tensile stresses in some area an incipient crack will start and propagate whilst desiccation, a crack appears (Figure 9).

6 Conclusions

A study of desiccation cracking in compacted clay based on experimental and numerical determination is presented. It is found that the failure state for all tests occurred by a comparable change of the saturation ratio ΔS_r. The soil specific saturation ratio $S_{r,s}$ presents the ΔS_r in proportion to the initial adjusted saturation ratio $S_{r,0}$. Independent of the investigated initial properties $S_{r,s}$ follows the same path in correlation to the total suction. The DEM simulation offers the suitable possibility for the grid free numerical modelling of desiccation induced cracks. Further extensive experimental and numerical research in this field will be performed.

References

1. B. Shannon, J. Kodikara, P. Rajeev, Geotechnical Testing Journal 38, 1, 98-112, (2015)
2. K. V. Uday, P. N. V. Jayanthi, D. N. Singh, Drying Technology, 32: 869-876, (2014)
3. E. Birle, D. Heyer, N. Vogt, Acta Geotechnica, 3: 191-200, (2008)
4. D. J. White, W. A. Take, M. Bolton, Geotechnique, 53, 7: 619-631, (2003)
5. N. I. Thusyanthan, W. A. Take, S. P. G. Madabhushi, M. D. Bolton, Geotechnique, 57, 6: 581-594, (2007)
6. E. Birle, Geohydraulische Eigenschaften verdichteter Tone unter besonderer Berücksichtigung des ungesättigten Zustandes. Doctoral Thesis, München, (2011)
7. C. O'Sullivan, Particulate discrete element modelling, Hoboken, NJ: Taylor & Francis, (2011)
8. P. A. Cundall, A computer model for simulating progressive large scale movements in blocky rock systems. Proceedings of the Symposium of the International Society of Rock Mechanics, Nancy. (1971)
9. P. A. Cundall, O. D. L. Strack O.D.L., Géotechnique, 29, 47 –65, (1979)
10. P. Winkler, M. R. Sadaghiani, H. Jentsch, K. J. Witt, Granular packing generation using DEM–Modified Force-Biased-Algorithm. Proc. 7th Int. Conf. on Scour and Erosion, Perth, Australia, 2-4 (p. 345). CRC Press, (2014)
11. C. Kloß, C. Goniva C., LIGGGHTS an Open Source Discrete Element Simulations of Granular Materials based on LAMMPS, Proc. TMS Annual Meeting, San Diego., (2011)
12. C. Goniva, C. Kloß, A. Hager, S. Pirker S., An Open Source CFD-DEM Perspective, Proc. of OpenFOAM Workshop, Series OpenFOAM Workshop, (2010)
13. V. Ferber, J. C. Auriol, Y. Cui, Can. Geotechnical Journal, 45, 252-65, (2008)
14. E. E. Alonso, J. Vaunat, A. Gens, Eng Geol, 54, 173-83, (1999)
15. J. Sima, M. Jiang, C. Zhou, Computers and Geotechnics, 56, 168–180, (2014)
16. A. Amarasiri, J. Kodikara, S. Costa, Int. J. Numerical and Analytical Methods in Geomechanics, DOI: 10.1002/nag.894., (2010)
17. R. Jullian, A. Pavlovitch, P. Meakin P., J. Phys. A: Math. Gen., 25, 4103–4113, (1992)
18. J. D. Bernal, Nature 183, 141 – 147, (1959)
19. J. K. Mitchell; K. Soga, Fundamentals of soil behaviour, Hoboken, NJ : Wiley, (2005)

 DOI: 10.1051/e3sconf/20160909012

Characterization of compacted silty sand via relative humidity-controlled triaxial testing

Ujwalkumar D. Patil[1,a], Laureano R. Hoyos[2], and Anand J. Puppala[3]

[1]*Faculty Research Associate, University of Texas at Arlington, Arlington, Texas 76019, USA.*
[2]*Professor, University of Texas at Arlington, Arlington, Texas 76019, USA.*
[3]*Professor, University of Texas at Arlington, Arlington, Texas 76019, USA.*

Abstract. Rather limited experimental evidence is available of triaxial shear-induced response of unsaturated soils at suction states far beyond residual suction. In this work, a fully automated, relative-humidity (Auto-RH) control unit is adapted to a newly implemented double-walled triaxial cell to test compacted silty sand specimens under considerably high total suction states via the vapor-pressure technique. The work is intended to gain critical insight into some of the most essential hydro-mechanical features of densely compacted intermediate geomaterials, such as post-peak softening and strain-induced dilatancy, under suction-controlled monotonic shearing. In general, peak strength is followed by large strain-induced softening until critical state is apparently reached. Strain-softening is observed to become considerably more pronounced with increasing total suction. The slope of critical state lines, however, remains virtually constant, regardless of induced total suction, in agreement with critical state-based constitutive frameworks previously postulated for unsaturated soils.

Key words: Unsaturated soils, vapor-pressure technique, double-wall triaxial cell, critical state line.

1 Introduction

Recently, researchers have shown growing interest in assessment of soil shear strength and stiffness behaviors in the high suction range ([8], [17], [2], and [9]). However, use of the axis-translation technique to impose high soil suction is limited (to about 1500 kPa) by the air-entry value of the ceramic disk. Changes in moisture content in unsaturated soil specimens at suction greater than approximately 3,000 kPa can be controlled using vapor phase equilibrium [9]. This method has been used to measure triaxial shear strength and volume change behavior of a non-plastic silty soil at high suction range, under controlled relative humidity environment, by regulating the inflow rate of air ([11] and [10]).

Although, the oedometer and direct shear tests are comparatively simple to be conduct, the triaxial testing facilitates soil testing along a wide variety of stress paths, thereby simulating different field conditions encountered in the geotechnical practice [2]. A substantial portion of the data available on unsaturated soil behavior has been obtained using the axis-translation technique (i.e. low-to-medium suction range). However, research using osmotic or vapor pressure based techniques in the higher suction range (i.e. above residual suction range), is very limited. This study is one of the few attempts made on high suction-based soil strength tests as it pertains to intermediate geomaterials.

An attempt has been made at Geomechanics laboratory of University of Texas at Arlington to develop a new apparatus by accommodating the RH triaxial equipment, in conjunction with the triaxial device, to directly control the relative humidity inside the specimen. Shear induced strength/volume change behavior in high total suction range is studied in current research. Same equipment was previously used to accommodate axis-translation technique to test soil specimen up to matric suction of 1500 kPa [13]. Present modifications have enhanced the capabilities of this triaxial equipment to test soil specimen up to total suction as high as 300 MPa. Efficacy of this equipment to reproduce similar test results at low, medium and high suction range has been performed and documented by [14].

2 Test material

The test soil comprised of 55% sand, 37% silt and 8% clay and is classified as silty sand (SM) according to the Unified Soil Classification system (USCS). Properties of compacted test soil specimen prepared by static compaction method are summarized in Table 1.

[a] Corresponding author: ujwalkumar.patil@mavs.uta.edu

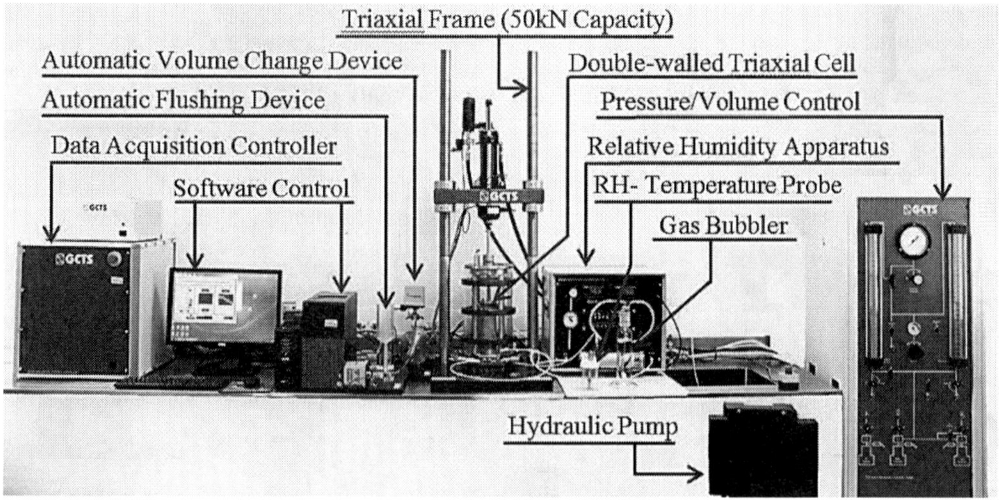

Figure 1. Panoramic view of entire Auto-RH/Triaxial test setup.

Table 1. As-compacted soil specimen properties.

Dry unit weight, ρ_d	1.8 g/cm^3
Moisture content, w	14.2 %
Air entry value (AEV)	10 kPa
Initial voids ratio ($e_{initial}$)	0.46-0.49
Diameter	2.8 in
Height	5.6 in

3 Specimen preparation(s)

Each specimen was statically compacted in nine equal layers. Each layer was compacted at a constant rate of 1 mm/min to a total vertical stress of 1600 kPa to produce overall homogenous specimen [14]. The convention used to designate the specimen is CD_{x-y} where "CD" denotes the consolidated drained test; "x" represents the net confining pressure (σ_3-u_a), while "y" represents the imposed constant matric suction (u_a-u_w).

4 Auto-RH triaxial setup

The Auto-RH system [6] and [7], designed is capable of automatic control of relative humidity (RH) between ~ 1% RH and ~ 99% RH. This corresponds to a total suction range of 600 MPa to 1.4 MPa. The thermodynamic relationship between relative humidity of pore water vapor and total suction Ψ_t (kPa) is given by Kelvin's equation (1):

$$\psi_t = -\frac{RT}{v_{wo}\,\omega_v}\ln\left(\frac{u_v}{u_{vo}}\right) = -\frac{RT}{v_{wo}\,\omega_v}\ln(\mathrm{RH}) \qquad (1)$$

where u_v is partial pressure of water (e.g., soil pore-water) vapor (kPa); u_{vo} = saturation pressure of pure water vapor (kPa); R = universal gas constant (8.31432 J mol^{-1} K^{-1}); T = absolute temperature (K); v_{wo} = specific volume of water (reciprocal of density, m^3/kg); and, ω_v = molecular mass of water vapor (18.016 kg/kmol).

Main features of the fully integrated system, also referred to as the Auto-RH/Triaxial system in this work, include: (1) Auto-RH control unit; (2) Gas bubbler, desiccant, and temperature probe; (3) Double-walled triaxial cell; (4) Automated volume-change device; and (5) Automated flushing device. Figure 1 shows a panoramic view of entire RH-Triaxial setup.

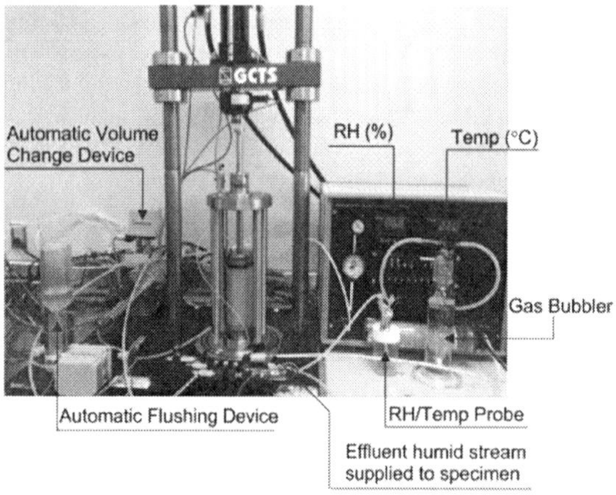

Figure 2. Triaxial setup with adapted Auto-RH control unit.

The double-walled chamber includes an inner cell subjected to same internal and external pressures, thus avoiding differential pressures and hence minimizing cell expansion and/or water leakage. The pressure and flow of water into the inner cell is controlled by a pressure and volume controller. The total change in volume experienced by the soil is equivalent to the amount of water flowing from (or into) the inner cell and into (or from) the outer cell. The total soil volume change was monitored by a volume-change device with a rolling diaphragm to minimize the sliding friction that normally occurs in conventional volume-change devices [14].

Humidity in the pore-air phase of the test soil can be ramped along paths of increasing or decreasing relative humidity, ranging from ~2% RH to ~95% RH, which corresponds to total suction states of ~500,000 kPa (500 MPa) to ~10,000 kPa (10 MPa); and is typically accomplished in step increments of ~10% RH. The "forced-flow" nature of the mixed-flow system significantly reduces the required pore-fluid equalization time [7].

5 Test procedure

5.1 Suction equalization

Statically compacted specimens prepared for triaxial testing were first air-dried under laboratory environment (24°C) until the monitored water content came reasonably close to that corresponding to the desired total suction state, a process that took between 4-5 days. The specimen was then transferred to the calibration chamber connected to RH-apparatus for preconditioning of the pore-fluids prior to triaxial testing. The RH was incrementally stepped up/down to a desired value by proportioning the "wet" to "dry" gas flows, under constant feedback from the RH/Temp probe. It took between 8-10 days for the desiccated specimens to attain equilibrium (pore-fluid equalization under constant soil mass), irrespective of the magnitude of the total suction to be induced (20-300 MPa).

The preconditioned specimen was then immediately mounted onto the bottom pedestal of the cell and O-ring-sealed with a latex membrane. Target total suction states in triaxial specimens, were automatically attained by supplying vapor-saturated air from the Auto-RH unit into the specimen through bottom pedestal via ¼ in nylon tubing. The mounted specimen was finally allowed to equilibrate under target RH value for at least 15 additional days [14].

5.2 Isotropic consolidation

Upon reaching equilibrium, the triaxial double-walled cell was assembled and filled with water under a water pressure of 10 kPa. The RH/Temp and the related total suction was continuously monitored from the suction equalization stage till shearing. This monitoring provided an indirect knowledge of total suction inside specimen. Volume changes experienced by the soil throughout testing stages were recorded via an automatic volume-change device, which allowed for accurate dimensions of the specimens to be accounted for prior to initiating each stage.

The next stage consisted of isotropic consolidation under controlled suction. This was done by increasing the cell pressure at the rate of 5 kPa/hr., while keeping the circulation of relative humidity from bottom to top of the specimen. It should be noted that the bottom of the specimen was connected to the RH equipment; hence, no water back pressure was applied. The top of specimen was connected to the chamber with a RH probe that had a vent open to the atmosphere for the effluent (Fig. 2). Thus, the air pressure in the specimen was at atmospheric pressure (or reference zero). Depending upon the final consolidation pressure, $(\sigma_3 - u_a)$ = 100, 200, or 300 kPa, the application of desired isotropic consolidation pressure took 18, 38, and 58 hrs, respectively. Each specimen was kept for at least 24 hours after the consolidation pressure was applied to ensure complete dissipation of pore air pressure (no change in volume of specimen).

5.3 Shearing under constant suction

Once the isotropic consolidation process was complete, the total suction inside the specimen and the net confining pressure, $(\sigma_3 - u_a)$ = 100, 200, or 300 kPa, was kept constant and the specimen was sheared at a constant shearing rate of 0.0009%/min. Independent studies were conducted to determine appropriate shearing rate of 0.0009%/min [12].

6 Shear-induced soil response

The newly developed Auto-RH equipment was used to conduct a series of consolidated-drained CTC tests on nine identically prepared specimens of compacted SM soil under either saturated (s = 0) or constant total suction states of 20 MPa or 300 MPa. Specimens were sheared under initial net confining pressures of 100, 200, or 300 kPa. Figure 3a and 3b show the stress-strain and volume-change response of compacted SM soil from suction-controlled CTC tests conducted at initial net confinement, $(\sigma_3 - u_a)$ = 100 kPa. Similar tests were conducted at initial net confinement, $(\sigma_3 - u_a)$ = 200 kPa and 300 kPa with total suction, s = 0, 20, and 300 MPa.

Fig 3a clearly indicates an increase in soil stiffness (tangent modulus), peak strength, and soil brittleness with increasing total suction. In general, peak strength is followed by large strain-induced softening, until critical state is apparently reached. Thereby strain-hardening type response, observed in saturated specimens, progressed towards strain-softening type response with the introduction of higher soil suction. This strain-softening, in turn, is observed to become considerably more pronounced with increasing total suction: The specimens failed at lower strains under highest total suction of 300 MPa, then featuring a sudden drop in deviator stress, until eventually reaching critical state with a relatively small change in strain.

Figure 3b clearly manifests the change in shear-induced volumetric response from initial compressive to dilational type when the soil saturation state changed from saturated to unsaturated state with the introduction of total suction of magnitude 20 and 300 MPa.

All the specimens showed initial compression, followed by a stress-induced dilatancy-type response that increased with an increase in the suction induced. Such types of stress-strain and volumetric response are typical of dense or overconsolidated soils. All the specimens showed brittle type failure, without any bulging (Fig. 4b). Similar observations were made from testing SM soil at net confining pressure, $(\sigma_3 - u_a)$ = 200 kPa [14]. Figure 4a shows the homogenous compacted specimen with no visible cracks prior to testing while Figure 4b shows failed specimen at the end of test.

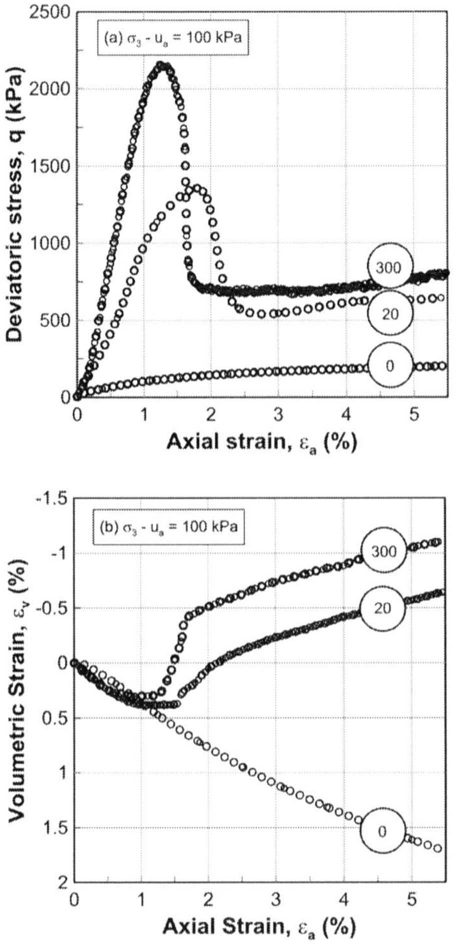

Figure 3. Response of compacted silty sand at net confining pressure, $(\sigma_3 - u_a) = 100$ kPa, and total suctions of 0, 20, and 300 MPa: (a) stress-strain response, (b) volume change.

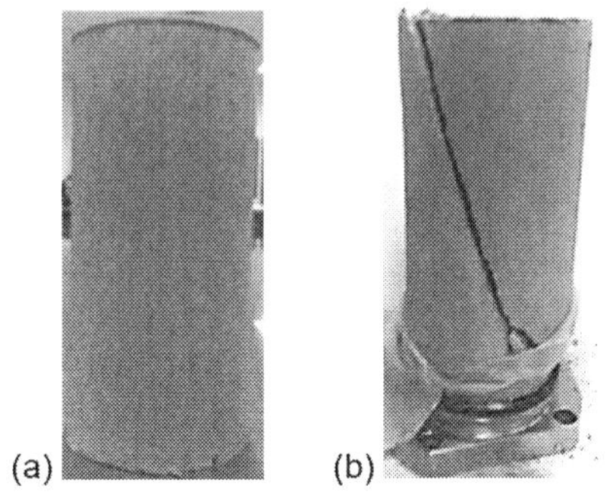

Figure 4. Typical features of as-compacted and failed silty sand specimens: (a) Desiccated specimen before testing, (b) Brittle type failure after CDRH300-300MPa triaxial test.

On the other hand, a considerable increase in net confining pressure, from 100 to 300 kPa, appears to have inhibited the relatively large amount of shear-induced dilation that is expected from SM soil when tested under higher total suction (300 MPa).

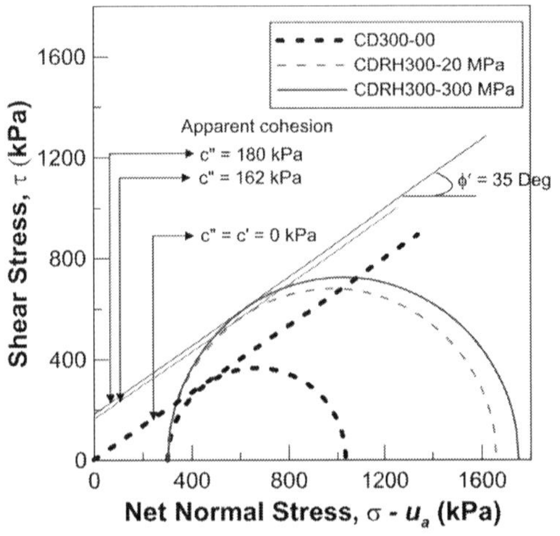

Figure 5. Mohr circles at critical state condition under net confining pressure, $(\sigma_3 - u_a) = 300$ kPa, and total suctions of 0, 20, and 300 MPa.

This can be attributed to large particle crushing under such high net confinements, which yields a particle gradation that renders a soil less dilational in nature. These observations substantiate some of the key findings reported from previous works on similar types of intermediate geomaterials [3] and [18].

Figure 5 shows Mohr's stress circles drawn at critical state failure. The angle of internal friction $\varphi' = 35$ deg. remains unchanged with increase in total suction up to 300 MPa. In fact, the failure envelope obtained at s = 300 MPa appeared to be very similar to that obtained by [5]. Clearly, there is no effect of change in soil suction on angle of internal friction. Figure 6 manifests an upward shift in critical state lines (CSL) with increase in soil suction. It also illustrates that slope of CSL is independent of increase in suction inside the specimen.

7 Effect of suction on critical state lines

Figures 6 shows the best-fit critical state lines (CSLs) obtained from triaxial testing at soil suction s = 0, 20 and 300 MPa via vapor-pressure technique. The critical state line is shifted upward with an increase in total suction, thereby, causing an increase in suction stress. The slope of all critical state lines, however, remains virtually constant, in close agreement with the constitutive, critical state based framework postulated by [1] and similar to [15 and 16].

8 Concluding remarks

A newly developed Auto-RH/Triaxial system was used to conduct Consolidated-drained CTC tests on specimens of compacted SM soil under total suctions of 0-300 MPa. In general, shear strength at peak and critical state increased with increasing total suction. Peak strength is followed by large strain-induced softening, until critical state is apparently reached. Strain-softening is also observed to be considerably more pronounced with increasing suction.

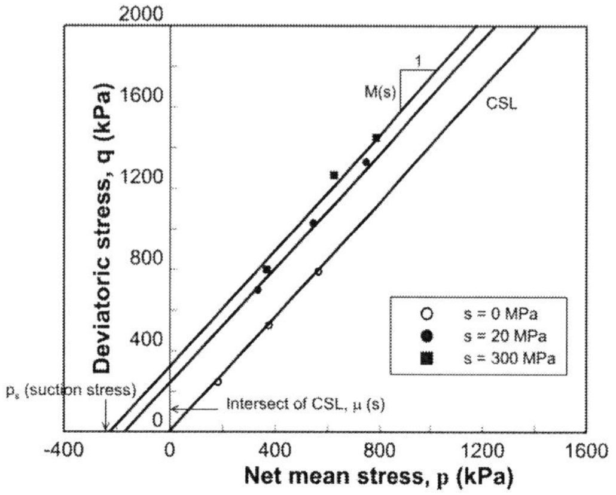

Figure 6. Critical state lines (CSLs) under net confining pressures, (σ_3 - u_a) = 100, 200, and 300 kPa, and total suctions of 0, 20, and 300 MPa.

Test results clearly showed a marked change in shear-induced volumetric response from initially compressive to purely dilational type when total suction was increased from 0 (saturated) to 20 or 300 MPa. An increase in net confinement from 100 kPa to 300 kPa tends to inhibit the shear-induced dilation of SM soil under constant total suction. The slope of all critical state lines (CSLs) obtained via vapor-pressure technique remains virtually constant, in close agreement with the constitutive, critical state based framework originally postulated by [1].

Experimental results such as the one obtained by testing densely compacted geomaterials at low, medium and high suction will help in the calibration, verification and fine-tuning of constitutive models. Modeling such shear-induced continuous stress-strain response along entire suction range (0 to 300 MPa) is currently being studied at The University of Texas at Arlington.

Acknowledgements

The experimental work described in this paper is part of an ongoing research project funded by National Science Foundation (NSF-MRI award ID # 1039956). This support is gratefully acknowledged. Any findings, conclusions, or recommendations expressed in this material are those of the authors and do not necessarily reflect the views of the National Science Foundation.

References

1. Alonso, E.E., Gens, A., and Josa, A. A constitutive model for partially saturated soils. Géotechnique, **40**(3), 405-530, (1990).
2. Blatz, J. A., Graham, J., and Chandler, N. A. "Influence of suction on the strength and stiffness of compacted sand-bentonite." Canadian Geotechnical Journal, **39**, 1005-1015, (2002).
3. Estabragh, A. R., Javadi, A. A., and Boot, J. C. "Effect of compaction pressure on consolidation behavior of unsaturated silty soil." Canadian Geotechnical Journal, **41**(3), 540-550, (2004).
4. Hoyos, L. R., Diego D. Perez-Ruiz and Puppala, A.J. Modeling unsaturated soil response under suction-controlled true triaxial stress paths. International Journal of Geomechanics, ASCE. **12** (3), 292-308, (2012).
5. Jotisankasa, A., Coop, M., and Ridley, A. "The mechanical behaviour of an unsaturated compacted silty clay." Géotechnique, **59**(5), 415-428, (2009).
6. Likos, W. J. "User Manual automated relative humidity control system," (2012).
7. Likos, W. J. and Lu, N. "Automated humidity system for measuring total suction characteristics of clay." Geotechnical Testing Journal, ASTM, (2), 1-12, (2003).
8. Mahalinga-Iyer, U. and Williams, D. J. "Unsaturated strength behavior of compacted lateritic soils." Géotechnique, **45**(2), 317-320, (1985).
9. Nishimura, T., Toyota, H., Vanapalli, S. K. and Won, O. T. "Determination of the shear strength behavior of an unsaturated soil in the high suction range using vapor pressure technique." *Proc. of the Ist European Conference on Unsaturated Soils. Unsaturated Soils: Advances in Geo-Engineering*, Toll et al. (eds) CRC Press: Taylor and Francis Group, 441-447, (2008).
10. Nishimura, T. and Vanapalli, S. K. "Volume change and shear strength behavior of an unsaturated soil with high soil suction." *16th International Conference on Soil Mechanics and Geotechnical Engineering*, 563-566, (2005).
11. Nishimura, T. and Fredlund, D. G. "A new triaxial apparatus for high total suction using relative humidity control." *12th Asian Regional Conference on Soil Mechanics and Geotechnical Engineering*, Leung et al. (eds), World Scientific Publishing, (2003).
12. Patil, U. D., Hoyos, L. R., and Puppala, A. J. "Suitable shearing rate for triaxial testing of intermediate soils under vapor controlled medium to high suction range." *IFCEE 2015*, ASCE, 2141-2150, (2015).
13. Patil, U. D., Puppala, A. J., and Hoyos, L. R. "Assessment of suitable loading rate for suction-controlled triaxial testing on compacted silty sand via axis-translation technique." *Geo-Congress 2014*, GSP 234, ASCE, 1307-1316, (2014).
14. Patil, U. D. "Response of unsaturated silty sand over a wider range of suction states using a novel double-walled triaxial testing system." Ph.D. dissertation, Univ. of Texas at Arlington, Arlington, TX, (2014).
15. Toll, D.G., and Ong, B.H. "Critical state parameters for an unsaturated residual sandy clay." Géotechnique, **53**(1), 93-103, (2003).
16. Toll, D.G."A framework for unsaturated soil behaviour." Géotechnique, **40**(1), 31-44, (1990).
17. Vanapalli, S. K., and Fredlund, D. G. Comparison of different procedures to predict unsaturated soil shear strength." *Advances in Unsaturated Geotechnics*, 195-209, (2000).
18. Zhang, X., and Li, L. "Limitations in the constitutive modeling of unsaturated soils and solutions." Int. J. Geomech., **11**(3), 174-185, (2011).

Reconciling soil-water retention properties and shear strength parameters of compacted clayey soils via suction-controlled ring shear testing

Laureano R. Hoyos[1,a], Jairo E. Yepes[2], and Anand J. Puppala[3]

[1] *Professor, University of Texas at Arlington, Arlington, Texas 76019, USA*
[2] *Teaching Assistant, University of Texas at Arlington, Arlington, Texas 76019, USA*
[3] *Professor, University of Texas at Arlington, Arlington, Texas 76019, USA*

Abstract. A preliminary experimental program has been undertaken to assess suction-controlled peak and residual shear strength properties of a statically compacted, moderate plasticity clayey soil. The experimental program was accomplished in a fully servo/suction-controlled ring shear apparatus, suitable for testing unsaturated soils under large deformations via the axis-translation technique. The test results highlight the important role played by matric suction on residual shear strength behaviour of compacted clayey soils. For the range of net normal stresses and suction states investigated, the increase in peak shear strength with increasing suction was found to be significantly nonlinear for the clayey soil. A distinct correspondence was also observed between the nonlinear nature of peak shear strength envelope, with respect to increasing matric suction, and the soil-water retention curve. The residual failure envelope, however, remains reasonably linear for the range of suction values induced in the present work via axis-translation.

1 Background and importance

A vast majority of geotechnical infrastructure made of compacted soil, or resting on unsaturated ground, involves a wide range of deformations. Calculation of foundation settlement, for instance, requires a good estimation of soil stiffness at relatively small strains. Analysis of earth slopes, embankments, and soil bearing capacity, on the other hand, requires good estimations of shear strength from peak to residual. Assessment of residual shear strength parameters at relatively low net normal stresses and suction states, as reflected by the range of experimental variables considered in this work, is of critical importance in slope stability analyses involving potentially shallow failures triggered by rainfall at relatively high degrees of saturation. To date, however, there is limited experimental evidence of unsaturated soil behaviour at large deformations as the soil is subjected to controlled suction states. Research efforts have been deterred in the past by the lack of suitable testing tools and techniques. It is in this context that a suction-controlled ring shear (RS) apparatus plays a fundamental role in a thorough characterization of this type of materials.

This paper introduces a servo/suction-controlled RS apparatus that is suitable for testing unsaturated soils under large deformations via axis-translation technique. The design of its core system is essentially based upon the original Bromhead ring shear apparatus (Bromhead 1979). A detailed description of its full development, including its main components and thorough performance verification against the original Bromhead device, is presented by Hoyos et al. (2011, 2014). The step-by-step assembling process of the RS system can be summarized as follows:

1. The actuators and DA/PC system are switched on to allow the instruments to come into equilibrium and minimize the influence of temperature offsets.
2. A small piece of wet filter paper is placed over the top of 5-bar ceramics, prior to soil compaction, to ensure phase continuity between pore-water in the soil and water in the saturated 5-bar ceramics: Figure 1(a).
3. A 15-mm (0.59-in) thick specimen is statically compacted directly into a bottom annular platen: Figure 1(b). The specimen is then transferred to the RS frame and the platen tightly fixed onto the bottom base plate: Figure 1(c).
4. The vertical load shaft is brought up via a servo controller and the upper annular platen affixed to the top of the piston shaft: Figure 1(c). A vertical sitting load of 25 N is applied in order to bring the upper platen in full contact with the soil.
5. All drainage and flushing lines are filled with de-aired water and flushed several times to avoid any trapped air in the whole system.
6. The main RS cell is set into place and the top cover plate affixed to the cell wall: Figure 1(d). A pore-air pressure u_a line, from a model PCP-15U panel, is connected to the top RS cover plate via a quick connector.

[a] Corresponding author: lhoyos@uta.edu

DOI: 10.1051/e3sconf/20160909013

7. Readings of the load-torque transducers are reset while one LVDT and angular deformation sensors are re-zeroed prior to RS testing.
8. The specimen is then subjected either to a suction-controlled single-stage or multi-stage RS test using the $s = u_a$ testing approach (i.e., $u_w = 0$).
9. When the test is finished, all pressures are gradually reduced back to atmospheric pressure and the failure surface examined via microscopic digital imaging. Key items of the isometric view shown in Figure 2 are described by Hoyos et al. (2011).

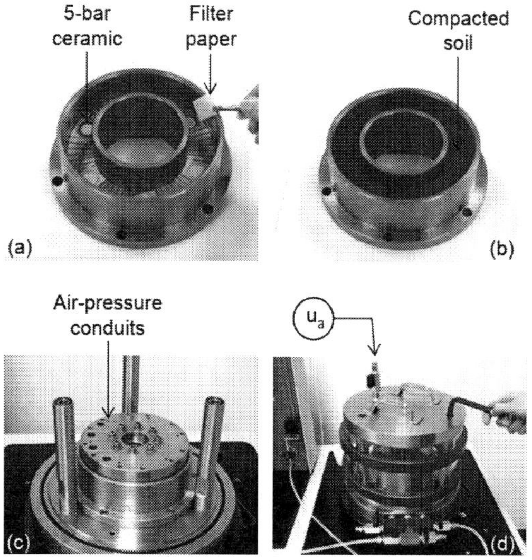

Figure 1. Assembly of suction-controlled RS device: (a) lower platen, (b) test specimen, (c) top platen, (d) main cell.

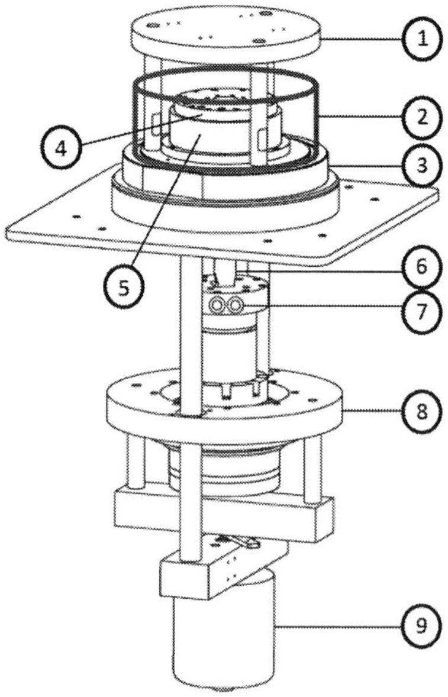

Figure 2. Isometric view of RS core system. (Adapted from Hoyos et al. 2011.)

2 Test soil and procedures

The soil used in this investigation classifies as silty clay (CL) according to the USCS, with 18% sand, 50% silt, and 32% clay; maximum dry unit weight, $\gamma_{d\text{-max}} = 1.77$ g/cm^3; optimum moisture content, OMC = 17%; liquid limit, LL = 37%; plasticity index, PI = 20%; and specific gravity, $G_s = 2.72$. During compaction, the upper platen is used to compress one single lift of a loose soil-water mix to target dry unit weight of 95% of $\gamma_{d\text{-max}}$ under a constant displacement rate of 1.25 mm/min, which corresponds to initial degree of saturation of 88% and matric suction of 100 kPa. Further details, including induced compaction energy, are presented by Hoyos et al. (2011, 2014).

In the present research work, all suction-controlled RS tests were conducted at an equivalent horizontal shear displacement rate of 0.025 mm/min (equivalent to a rotational speed of 0.023°/min), which is slightly lower than that recommended for suction-controlled RS testing on clayey soils (Vaunat et al. 2006, 2007; Infante Sedano et al. 2007). All RS tests were performed by following a multi-stage testing scheme in which all residual strength assessments were made at three or more values of net normal stresses, $(\sigma_n - u_a)$ = 25, 50, 75, 100, and 200 kPa; under constant matric suction, $s = (u_a - u_w)$ = 25, 50, 75, 100, 200, or 300 kPa. Shearing was stopped when it was readily apparent that a residual state had been reached.

3 Soil-water retention properties

Figure 3 shows the drying loop of the corresponding soil-water retention curves (SWRC) for three different types of intermediate soils considered by the authors for current and future investigations, which have been assessed via both pressure plate and filter paper techniques. The figure also shows the best-fit model curves as per Fredlund and Xing (1994). These SWRCs were obtained to assist the authors in defining the appropriate range of suction states (i.e., beyond the air-entry value of each test soil) to be induced on compacted specimens of CL, SC-SM and SM soils, via the axis-translation technique.

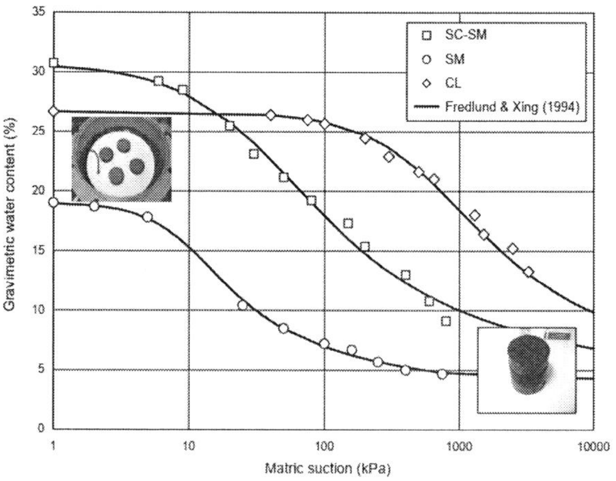

Figure 3. Soil-water retention curves for CL, SC-SM and SM soils via pressure plate and filter paper techniques.

[a] Corresponding author: lhoyos@uta.edu

4 Shear strength parameters

Figure 4 shows the shear stress vs. equivalent horizontal shear displacement response of compacted CL soil, during first shearing stage, at matric suctions, s = 75, 100, 200, and 300 kPa; and net normal stress, $(\sigma_n - u_a)$ = 25 kPa. Specimens exhibit a steady increase in shear stress up to a peak value, followed by a gradual decrease until an apparent residual state is attained. The peak is clearly more pronounced at higher suctions. It can also be readily noticed that suction has a significant effect on residual strength, with a considerable increase for s = 300 kPa.

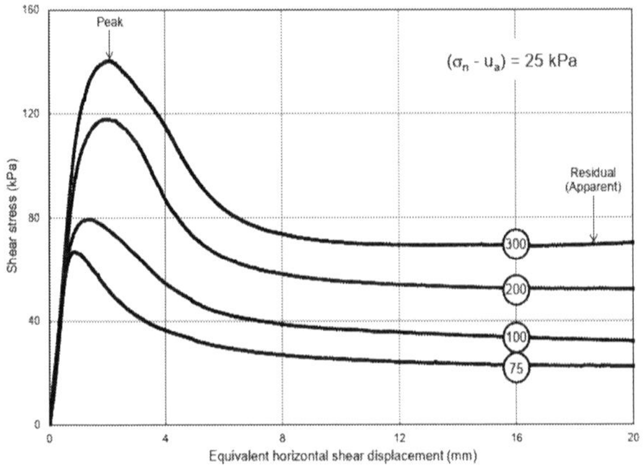

Figure 4. Peak and residual response of CL soil from suction-controlled ring shear testing.

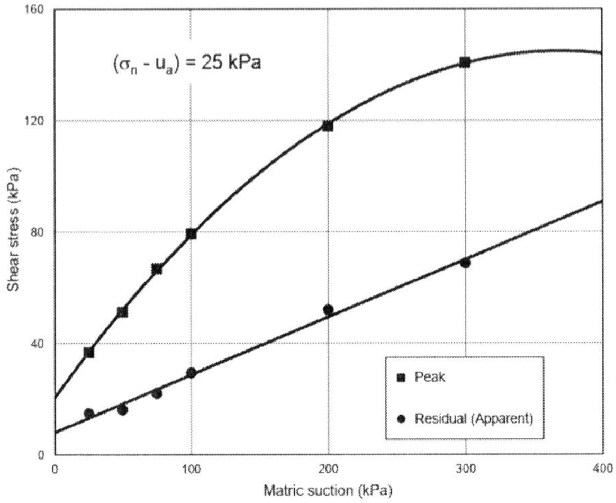

Figure 5. Effect of matric suction on peak and residual failure envelopes from CL soil.

Figure 5 above shows the effect of matric suction on both the peak and the residual failure envelopes obtained from suction-controlled RS testing of compacted CL soil. Results appear to substantiate the direct correspondence that has been observed between the nonlinear nature of peak failure envelopes (threshold value of matric suction beyond which they become nonlinear) and the air-entry value of the test soil as obtained from the SWRC, which is about 100 kPa for CL soil (Fig. 3). The residual failure envelope, however, remains reasonably linear for the range of suction values induced via axis-translation in the present work.

Several authors, including Vanapalli et al. (1996) and Lu and Likos (2004), consider that there is a direct correspondence between the nonlinear nature of the peak shear strength envelope, with respect to increasing matric suction, and the SWRC. Within the regime of relatively low suction, and prior to the air-entry pressure, the soil pores remain essentially saturated, the shear strength envelope is reasonably linear, and the beta angle ϕ^b is effectively equal to friction angle ϕ'. As the soil becomes unsaturated, the reduction in the volume of pore-water within this regime effectively reduces the contribution of matric suction toward increasing the shear strength.

This particular feature of hydro-mechanical behaviour of unsaturated soils might provide the most rational basis for understanding the patterns observed in Figures 4 and 5 above.

References

1. Bromhead, E.N. (1979). "A simple ring shear apparatus." Ground Engineering, 12(5), 40-44.
2. Fredlund, D.G., and Xing, A. (1994). "Equations for the soil-water characteristic curve." Canadian Geotechnical Journal, 31, 521-532.
3. Hoyos, L.R., Velosa, C.L., and Puppala, A.J. (2014). "Residual shear strength of unsaturated soils via suction-controlled ring shear testing." Engineering Geology, Elsevier, 172(2014), 1-11.
4. Hoyos, L.R., Velosa, C.L., and Puppala, A.J. (2011). "A servo/suction-controlled ring shear apparatus for unsaturated soils: Development, performance, and preliminary results." Geotechnical Testing Journal, ASTM, 34(5), 413-423.
5. Infante Sedano, J.A., Vanapalli, S.K., and Garga, V.K. (2007). "Modified ring shear apparatus for unsaturated soil testing." Geotechnical Testing Journal, ASTM, 30(1), 1-9.
6. Lu, N., and Likos, W.J. (2004). *Unsaturated soil mechanics*, John Wiley & Sons, Hoboken, N.J.
7. Vanapalli, S.K., Fredlund, D.G., Pufahl, D.E., and Clifton, A.W. (1996). "Model for the prediction of shear strength with respect to soil suction." Canadian Geotechnical Journal, 33(3), 379-392.
8. Vaunat, J., Merchán, V., Romero, E., and Pineda, J. (2007). "Residual strength of clays at high suctions." Proceedings of the Second International Conference on Mechanics of Unsaturated Soils, Weimar, Germany, vol. 2, 151-162.
9. Vaunat, J., Amador, C., Romero, E., and Djeren-Maigre, I. (2006). "Residual strength of a low plasticity clay at high suctions." Proceedings of Fourth International Conference on Unsaturated Soils, Carefree, Arizona, vol. 1, 1279-1289.

[a] Corresponding author: lhoyos@uta.edu

DOI: 10.1051/e3sconf/20160910001

Building the UPPA high capacity tensiometer

Joao Mendes[1,a], Domenico Gallipoli[1], Fabian Boeck[2], Georg von Unold[2] and Alessandro Tarantino[3]

[1]*Laboratoire SIAME, Equipe Géomatériaux et Structures du Génie Civil, Université de Pau et des Pays de l'Adour, 64600 Anglet, France.*
[2]*UMS GmbH, Gmunder Str. 37 D-81379 München, Germany.*
[3]*Department of Civil and Environmental Engineering, University of Strathclyde, Glasgow, UK.*

Abstract. High capacity tensiometers (HCTs) are sensors capable of directly measuring tensile pore water pressure (suction) in soils. HCTs are typically composed of a casing that encapsulates a high air entry value ceramic filter, a water reservoir and a pressure sensing element. Since the creation of the first HCT by Ridley and Burland in 1993 at Imperial College London, HCTs have been almost exclusively built and used in academic research. The limited use in industrial applications can be explained by a lack of unsaturated soil mechanics knowledge among engineering practitioners but also by the technical difficulties associated to the direct measurement of tensile water pressures beyond the cavitation limit of -100kPa. In this paper, we present the recent design and manufacture of a new HCT at the Université de Pau et des Pays de l'Adour (UPPA) in France. Different prototypes were tried by changing the main components of the device including the type of ceramic filter, pressure transducer and geometry of the external casing. In particular, two ceramic filters of distinct porosity, three pressure transducers with distinct materials/geometries and four casing designs were tested.

1 Introduction

High capacity tensiometers (HCT) can directly measure tensile pore water pressure (suction) in soils, which is an important variable for the hydro-mechanical characterization of these materials. The development of HCTs has so far been restricted to university research and has resulted in the development of prototypes capable of measuring relatively high values of suction up to 2000 kPa. The first HCT was developed in the 90's by Ridley and Burland [1] at Imperial College London. This HCT, whose design is shown in Figure 1, was able to measure values of suction greater than 1500 kPa. Since then, other HCTs with similar design have emerged, such as those described in [2-9] among others.

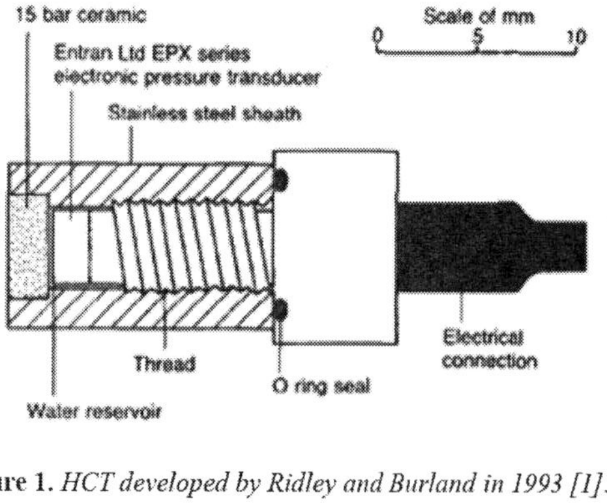

Figure 1. *HCT developed by Ridley and Burland in 1993 [1].*

In general, HCTs are composed of a high air entry value (HAEV) ceramic filter, a water reservoir and a pressure transducer, all housed inside a casing made of stainless steel. Each of the above components can have a direct impact on the performance of HCTs.

The ceramic filter has the strongest influence on the measuring range [10] because the highest measurable value of suction strongly depends on the air entry value of the ceramic, which in turn increases with decreasing size of the largest ceramic pore.

As for the pressure transducer, ideally this should be symmetrical with respect to the sensing membrane so that the same calibration equation can be adopted regardless of the direction of deflection of the sensing membrane. Symmetry is desirable because calibration of HCTs is usually performed in the positive (compressive) pressure range and then extrapolated to the negative (tensile) pressure range [5]. Furthermore, the measuring range of the pressure transducer should be at least equal to the air entry value of the ceramic filter.

Mendes and Buzzi [9] showed that, contrary to previous suggestions [1], the volume of the water reservoir does not need to be very small in order to measure high values of suction. Nevertheless, HCTs should always be thoroughly saturated by de-aired water and a particularly careful saturation procedure should be adopted when starting from an initially dry HCT, as further discussed later.

In general, the external casing is the component that has the least influence on the performance of HCTs, provided that this is sufficiently stiff to protect the sensor

[a] Corresponding author: joaopaulo.duarte@univ-pau.fr

from external loads. In some prototypes, the casing is in direct contact with the water reservoir, whose dimensions therefore depend on the design of the casing (see Figure 1).

This article presents five different HCT prototypes developed at the Université de Pau et des Pays de l'Adour (UPPA) in France within the framework of a collaborative industry-academia project on Monitoring systems to Assess Geotechnical Infrastructure subjected to Climatic hazards (MAGIC). The project, which involves three universities and four small-medium enterprises across Europe, is funded by the European Commission through a "Marie Curie" Industry Academia Partnerships and Pathways (IAPP) network (http://www.magic-iapp.com/).

Various combinations of ceramic filters, pressure transducers, casing designs and water reservoir sizes were experimented in this work. In particular, the different prototypes make use of: a) two different ceramic filters with distinct air entry values of 700 kPa and 1500 kPa, respectively, b) three different pressure transducers, namely a flush diaphragm ceramic transducer, a cavity diaphragm ceramic transducer and a flush diaphragm stainless steel transducer and c) four different stainless steel casing dimensioned to accommodate the transducer and the ceramic as well as to create the water reservoir in four out of five prototypes. The maximum attainable suction of each prototype was measured by means of evaporation tests during which the face of the ceramic filter was exposed to the atmosphere so that water could evaporate and suction could increase. As expected, the prototypes incorporating ceramic filters with an air entry value of 1500 kPa were able to measure suctions greater than 1500 kPa while those incorporating a ceramic filter with an air entry value of 700 kPa attained suctions of around 700 kPa.

2 Materials

The porosity characteristics of the two ceramic filters (referred as ceramics A and G in the following) were studied with reference to the measuring range of the respective HCTs. Ceramic A is made of typical pottery clay, mainly kaolinite, while ceramic G contains predominantly alumina. The microstructure of the two ceramics was studied by mercury intrusion porosimetry (MIP) and nitrogen absorption (NA).

Table 1 summarizes the porosimetry data of the two ceramics. Ceramic G has significantly lower porosity and higher density than ceramic A. This is also consistent with the fact that ceramic G is much harder to cut or break than ceramic A [10].

The results of the MIP and NA tests are presented in Figure 2, which also shows for comparison the pore sizes corresponding to the air entry values of 500 kPa and 1500 kPa (estimated according to the Young-Laplace equation). The pore sizes of ceramic G cover the range from 100 nm to 300 nm and are mono-disperse around a value of 230 nm, while the pore sizes of ceramic A cover a smaller range from 1 nm to 190 nm and are poly-disperse without a dominant value.

Table 1. Porosimetry data of ceramics A and G.

Ceramic type	Largest pore diameter nm	Bulk density[#] g/cm³	Apparent (skeletal) density g/cm³	Porosity %
A	165	1.31	1.88	30.1
G	250	2.68	3.11	13.7

- at 1.45kPa of absolute pressure.

Table 2. Pressure transducer characteristics.

Designation	Ceramic flush diaphragm	Ceramic cavity diaphragm	Stainless steel flush diaphragm
Short name	CFD	CCD	SSFD
Sensor type	Piezoresistive		Strain gauge
Pressure reference	Relative / Sealed relative	Relative	Sealed relative
Accuracy (%FS)	0.5		1
Pressure range (kPa)	2000		

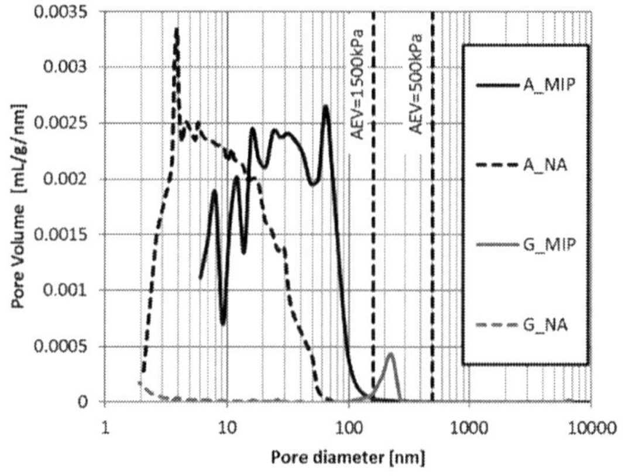

Figure 2. *Pore size distributions of ceramics A and G by mercury intrusion porosimetry (MIP) and nitrogen adsorption (NA)*

Three different types of transducers, all capable of measuring relative pressures with a range of 2MPa, were used in this work: a) a ceramic flush diaphragm transducer (CFD), b) a ceramic cavity diaphragm transducer (CCD) and c) a threaded stainless steel flush diaphragm transducer (SSFD). The characteristics of these three pressure transducers are summarized in Table 2 while their incorporation in the different designs is shown in Figure 3.

All prototypes casings were machined with a lathe from SS316L low carbon stainless steel to prevent corrosion in those cases where the water reservoir was in direct contact with the casing.

DOI: 10.1051/e3sconf/20160910001

3 HCT prototypes

Figure 3 shows the schematics of the five prototype designs pursued in this work (referred in the following as prototypes 1-5). The first prototypes to be built were number 1 and 2. These prototypes are very similar in conception, i.e. they employ identical CFD pressure transducers and stainless steel casings. However, prototype 1 was fitted with ceramic G while prototype 2 was fitted with ceramic A. The transducers and the ceramics were sealed in place by gluing around their perimeter inside the respective slots. The casings were designed to accommodate a small water reservoir of about 40mm^3 between the pressure transducer and the ceramic by making the transducer slot slightly deeper than the transducer itself.

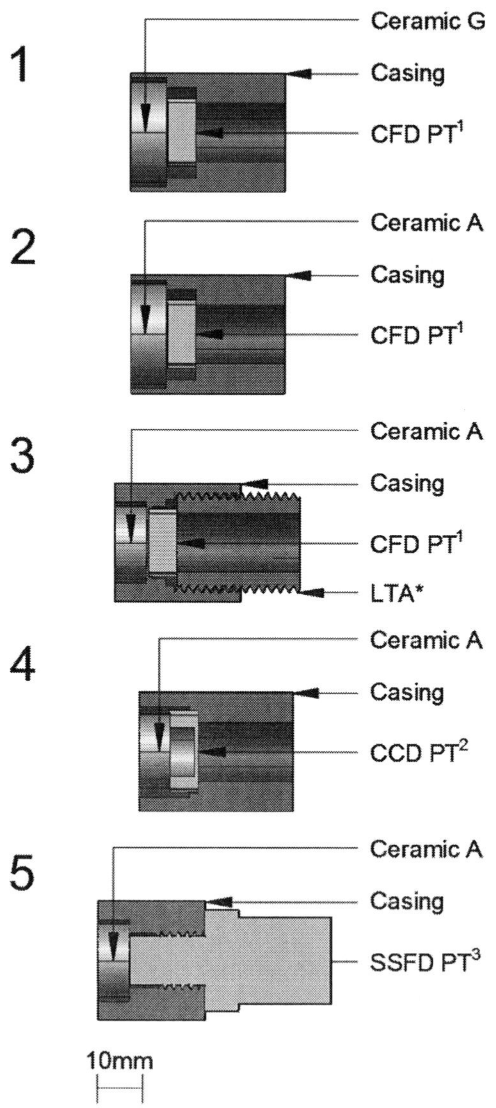

[1] – ceramic flush diaphragm pressure transducer
[2] – ceramic cavity diaphragm pressure transducer
[3] – stainless steel flush diaphragm pressure transducer
* – locking thread adaptor

Figure 3. *Schematics of HCT prototypes*

Prototype 3 was designed as an improvement over prototypes 1 and 2. In this case, a CFD pressure transducer and ceramic A were used. The water reservoir was integrated in the design of the casing by machining a thin protruding lip that separates the ceramic filter from the transducer, thus resulting in a water reservoir of about 40mm^3. To prevent dislodgement of the transducer when subjected to large positive pressures during saturation, a threaded adaptor (LTA in Figure 3) was screwed on the back of the casing to tightly secure the transducer in place by pushing it against the protruding lip.

In prototype 4, ceramic A was directly glued on a ceramic cavity diaphragm transducer (CCD). This transducer incorporates a small cavity inside its body. It is therefore this cavity that forms the water reservoir rather than the external casing as in other prototypes. This also means that, in this design, the water reservoir has a relatively large volume of about 430 mm^3. Unlike previous prototypes, here the external casing has the only purpose of protecting the transducer and attached ceramic filter and is not in direct contact with the water reservoir.

In prototype 5, ceramic A was used in association with a threaded stainless steel flush diaphragm pressure transducer (SSFD). This prototype resembles closely the designs developed by [1] and [11]. Similarly to prototypes 1-3, the water reservoir has a size of about 40mm^3 and is in direct contact with the casing. The size of the reservoir can be controlled during construction by adjusting the position of the transducer with its thread.

4 Saturation and calibration of the HCTs

4.1 Saturation

Good saturation is crucial for the performance of HCTs. As suggested by Marinho and Chandler [12], initial saturation of HCTs is particularly important and should include application of high vacuum followed by water flooding and pressurization. Application of high vacuum is necessary to minimise the amount of trapped air inside both the ceramic and the reservoir, while subsequent water pressurization helps dissolving any remaining air pocket.

In this work, the initial saturation of all prototypes followed the above recommended procedure. The dry HCTs were first exposed to a high vacuum for about 1 hour to evacuate any trapped air. Afterwards, they were flooded by de-aired water and pressurized at 2990kPa overnight to dissolve any remaining air pocket. For the HCTs with a small water reservoir (prototypes 1, 2, 3 and 5), this procedure was more than adequate to achieve full saturation. However, for prototype 4, it was necessary to increase the duration of the pressurization stage to 96 hrs because of the larger reservoir size (this duration will also depend on the level of vacuum that is applied prior to water pressurization).

4.2 Calibration of HCTs

Following previous research [5], the calibration of all HCTs was performed in the positive (compressive)

 DOI: 10.1051/e3sconf/20160910001

pressure range and then extrapolated to the negative (tensile) pressure range.

A positive pressure cycle of 2990kPa→50kPa→2990 kPa was imposed to all five prototypes and the resulting regression lines of pressure against voltage are shown in Figure 4. The calibrations are linear in the positive pressure range with typical hysteresis of 0.04% for prototypes 1, 3, 4 and 5 and 0.02% for prototype 2. This suggests that, if the HCTs are properly saturated, the effect of hysteresis can be almost neglected. It is also interesting to note that, regardless of the particular prototype design (large/small reservoir, type of ceramic filter or pressure transducer), the magnitude of hysteresis was very similar.

5 Results

After saturation and calibration, all prototypes were subjected to evaporation tests to determine the measuring range. In evaporation tests, a fully saturated HCT is exposed to the atmosphere so that the face of the ceramic filter (i.e. the measuring face) is allowed to dry. As the ceramic filter dries, water is pulled from the reservoir and the HCT reads increasingly larger values of tensile water pressures over time until cavitation. When cavitation occurs, air pockets form into the water reservoir resulting in a sudden reduction of the tensile pressure to around -100kPa. The value of the largest tensile pressure measured immediately before cavitation is taken as the limit of the measuring range.

5.1 Ceramic A versus ceramic G

Prototypes 1 and 2 incorporate identical pressure transducers, casings and water reservoir sizes but different ceramic filters. Prototype 1 used ceramic G while prototype 2 used ceramic A. Typical evaporation tests on both prototypes are shown in Figure 5, where it can be observed that prototype 2 has a higher measuring range (about -2000kPa) than prototype 1 (about -700kPa). As previously mentioned, this difference is due to the distinct ceramic filters and, in particular, to the distinct sizes of their largest ceramic pores. As shown in Figure 2 and Table 2, the size of the largest pore is smaller in ceramic A than in ceramic G. Ceramic A can therefore sustain higher water tensions and, hence, achieves a larger measuring range compared to ceramic G. For this reason, subsequent prototypes 3, 4 and 5 were all built with ceramic A.

5.2 Casing design for CFD pressure transducer

The construction of further HTCs based on prototype 2 was rather unsuccessful, with five out of eight HCTs not working correctly. The HCTs failed during re-pressurization at 2990 kPa after the first evaporation test because water leaked along the sides of the CFD transducer into the back. Probably, the relatively high water tension measured during evaporation pulled the transducer towards the reservoir, thus damaging the glue seal around the perimeter of the transducer. To prevent this, prototype 3 was developed by designing a casing that incorporated a thin lip protruding inside to separate the ceramic filter from the transducer, thus forming a gap for the water reservoir. The transducer is locked against this protruding lip by means of a threaded adaptor screwed on the back of the casing. The protruding lip prevents movement of the transducer towards the reservoir when tensile pressures are present while the threaded adaptor impedes movement away from the reservoir during positive pressurization. This means that the construction of prototype 3 HCTs is slightly more complicate than prototypes 1 and 2. A typical evaporation test performed on prototype 3 is shown in Figure 6, which suggests a measuring range of -1800 kPa.

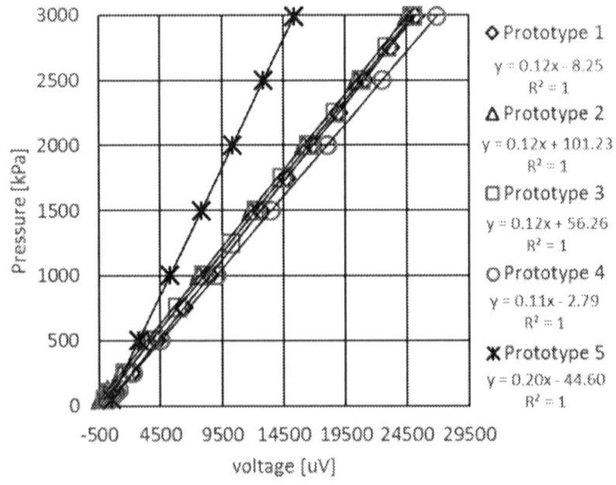

Figure 4. Calibration curves of the five prototypes.

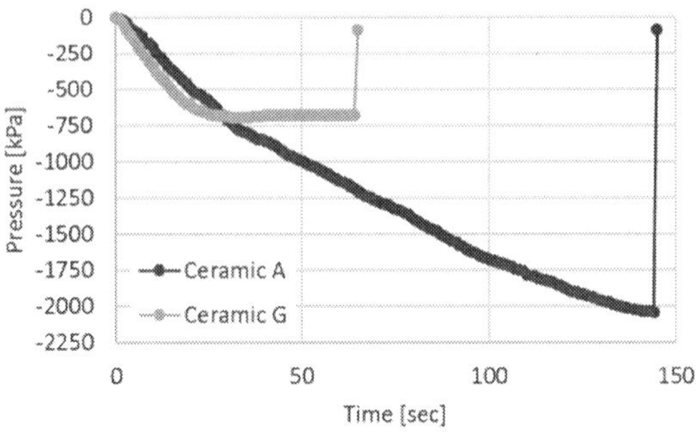

Figure 5. *Evaporation tests of prototype 1 (ceramic G) and prototype 2 (ceramic A).*

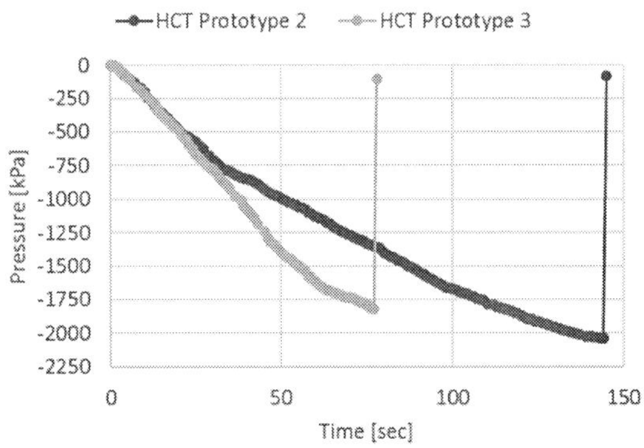

Figure 6. *Evaporation tests of prototype 2 and prototype 3.*

Prototype 3 seems to react quicker than prototype 2, though it has a slightly smaller measuring range. This

 DOI: 10.1051/e3sconf/20160910001

difference is not believed to be a cause of the design, but rather of the inherent variability of the ceramic filters used in each prototype.

5.3 Large reservoir versus small reservoir

Unlike prototypes 1, 2 and 3, the water reservoir of prototype 4 is an integral part of the CCD transducer and is therefore relatively large (430mm³). This also means that construction of prototype 4 is much simpler as the ceramic filter is directly glued to the pressure transducer, eliminating all problems related to potential water leaks when using CFD transducers.

Figure 7 shows the results of a typical evaporation test performed with prototype 4. Inspection of Figure 7 suggests a measuring range of around -1800kPa, which is comparable to that of prototypes 2 and 3. This is consistent with the findings of Mendes and Buzzi [9], who suggested that the reservoir size does not have a marked influence on measuring range. However, the response during evaporation was slower for prototype 4 than for other prototypes. In prototype 4, cavitation occurred after 275 seconds, compared to 145 seconds and 80 seconds for prototypes 2 and 3, respectively.

5.4 Ceramic pressure transducer versus stainless steel pressure transducer

Most commercially available pressure transducers are made of either metal or ceramic. In order to study the influence of the material of the transducer on the response of HCTs, prototype 5 was built by using a stainless steel flush diaphragm transducer (SSFD). Prototype 5 was designed to have a water reservoir of similar dimensions as in prototypes 2 and 3 (i.e. about 40mm³). However, prototypes 2 and 3 incorporated a ceramic flush diaphragm transducer (CFD) instead of a stainless steel one.

Figure 8 shows some typical evaporation tests performed with prototypes 2, 3 and 5. Figure 8 suggests that all prototypes have similar measuring ranges varying between -1800kPa and -2100kPa. The response time was found to be similar for all three prototypes up to a suction of -800kPa. After that, the response of prototype 2 became gradually slower as suction increased while prototypes 3 and 5 continued to dry at an approximately constant rate. In general, it appears that there are no significant differences in terms of both measuring range and response time between stainless steel and ceramic pressure transducers.

6 Conclusions

Five different high capacity tensiometer prototypes were designed and manufactured at the Université de Pau et des Pays de l'Adour in France within the framework of an EU-funded collaboration between industry and academia.

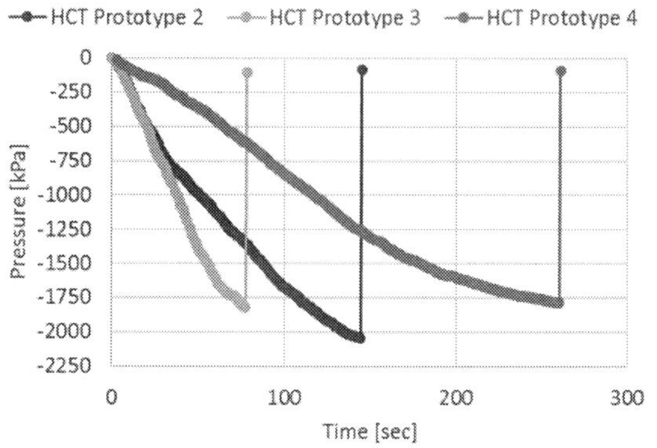

Figure 7. *Evaporation tests with small water reservoir (prototypes 2 and 3) and large water reservoir (prototype 4).*

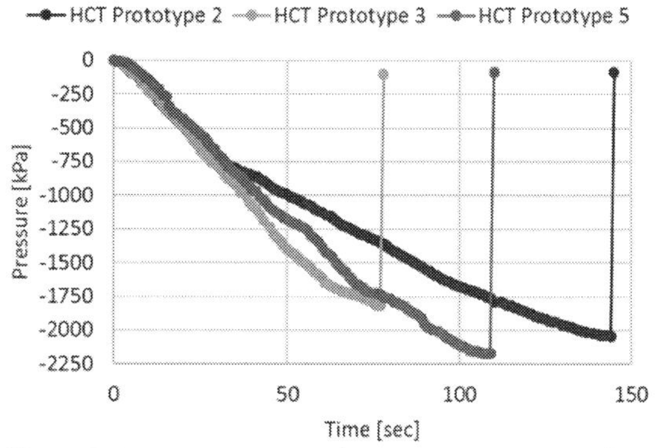

Figure 8. *Evaporation tests with ceramic pressure transducer (prototypes 2 and 3) and stainless steel pressure transducer (prototype 5).*

The prototypes differ in the choice of: a) ceramic filters (two ceramic filters were used with air entry values of 700 kPa and 1500 kPa, respectively), b) water reservoir size (two reservoir sizes equal to 40 mm³ and 430 mm³ were investigated), c) geometry of the pressure transducers (two transducer geometries were explored consisting of a flush sensing diaphragm and a cavity sensing diaphragm) and d) material of the transducer (two different transducer materials were considered, i.e. ceramic and stainless steel). This allowed a detailed study of the influence of the above factors on the performance of high capacity tensiometers and, in particular, on the measuring range and response time during evaporation tests.

Results show that the type of ceramic filter has the strongest influence on measuring range (Figure 5) and, in particular, the measuring range increases as the air entry value of the ceramic increases. Therefore, the tensiometer prototypes incorporating a ceramic filter with an air entry value of 700kPa measured pore water tensions up to about -700 kPa while the prototypes incorporating a ceramic filter with an air entry value of 1500kPa exhibited a much larger measuring range in the region of 1800-2100kPa. Mercury intrusion porosimetry and nitrogen adsorption tests also showed that the air entry value of a ceramic filter tends to increase as the size of the largest pore within the ceramic decreases.

The volume of the water reservoir has a noticeable effect on response time, with a larger reservoir tending to produce a slower response during evaporation tests. In addition, a larger reservoir requires significantly longer pressurization times during initial saturation of the tensiometer. Finally, the geometry and material of the pressure transducer have a relatively small influence on performance in terms of both measuring range and response time.

In terms of ease of manufacture, not all prototypes showed the same success rate during repeated construction. In particular, the first two prototypes making use of a ceramic flush diaphragm transducer tended to leak when re-pressurized after cavitation. This was probably due to the pull exerted by water tension on the transducer. This pull is transferred from the transducer to the adjacent glue joint, which has the double purpose of fixing the transducer to the casing and sealing the gap around its perimeter. The repeated pulling action during subsequent cavitations produces shearing of the surrounding glue layer and eventually damages it resulting in the creation of small pathways that allow water to leak from the front to the back of the transducer. This problem was overcome in a third prototype by blocking the movement of the pressure transducer with a locking system that held the transducer in place without relying on the strength of the glue. The last two prototypes did not suffer from similar malfunction because, in one case, the ceramic filter was directly attached to a cavity diaphragm transducer and, in the other case, the transducer incorporated a thread and could therefore be securely locked inside the casing.

The preliminary study has shown that, independently of design, all prototypes were able to reach values below the nominal air entry value of the porous ceramic during evaporation tests. However, further studies, i.e. soil suction measurements, are still required to properly evaluate and access the best design for a high capacity tensiometer.

Acknowledgements

The financial contribution of the European Commission to this research through the Marie Curie Industry-Academia Partnership and Pathways Network MAGIC (Monitoring systems to Assess Geotechnical Infrastructure subjected to Climatic hazards) – PIAPP-GA-2012-324426 – is gratefully acknowledged.

References

1. A.M. Ridley and J.B. Burland, Géotechnique **43**(2): 321-324 (1993).
2. A.M. Ridley and J.B. Burland, *Advances in Site Investigation Practice*. 510-520 (1995).
3. Y. Guan and D.G. Fredlund, Can. Geotech. J., **34**(4): 604-614 (1997).
4. I. Meilani, H. Rahardjo, E.C. Leong and D.G. Fredlund, Can. Geotech. J. **39**: 1427-1432 (2002).
5. A. Tarantino and L. Mongiovi, Proc. 3rd Int. Conf. on Unsaturated soils, Recife, Brasil **1**: 319-324 (2002).
6. W.A. Take and M.D. Bolton, Géotechnique **54**(3): 159-172 (2003).
7. S.D.N. Lourenço, D. Gallipoli, D.G. Toll, C. Augarde, F. Evans and G.M. Medero, Géotechnique **58**(8): 659-668 (2008).
8. Y.J. Cui, A. Tang, A. Mantho, and E. De Laure, Geotech. Test. J. **31**(1): 95-100 (2008).
9. J. Mendes and O. Buzzi, Can. Geotech. J. **50**(5): 550-556 (2013).
10. J. Mendes, D. Gallipoli, F. Plantier and D. Grégoire, *Rencontres Universitaires de Génie Civil*, Bayonne, France (2015) <hal-01167597>.
11. J. Mendes and O. Buzzi, *Proc. UNSAT 2014, Unsaturated Soils: Research & Applications*, Sydney, Australia: 1611-1616 (2014).
12. F.A.M. Marinho and R.J. Chandler, Géotechnique **44**(3): 551-556 (1994).

Correspondence between electrical resistivity and total suction in compacted kaolin considering the presence of salt

Ana Sofia Dias[1], Vikas Gingine[1] and Rafaela Cardoso[1a]

[1] *CERis/ICIST, Instituto Superior Técnico, University of Lisbon, Av Rovisco Pais, 1 1049-001 Lisbon, Portugal*

Abstract. Prospection techniques based on measuring the electrical resistivity of geomaterials are being used mainly for geophysical characterization, for evaluating soil contamination and the extension of contaminated areas, in addition to the compaction control in geotechnical works. This technique is based on the fact that contrasts in electrical resistivity along the profiles analysed help identifying zones of transition, because this property is affected by the presence of water and dissolved salts in the electrolyte, as well as by soil structure. However, *in situ* data interpretation has some difficulties because electrical resistivity is affected by changes in the unsaturated state of the soil, as well as by the presence of salt compounds (natural or from contamination). The correspondence between electrical resistivity and total suction in compacted kaolin considering the presence of salt was investigated in the study presented. The samples tested were compacted for the same voids ratio and water content, however one type was prepared with distilled water and the other with a NaCl solution (0.5 M). This is a relatively small concentration of salt, to which there were no significant changes in the liquid and plasticity limits, neither on zeta potential if pH is equal to 7, however the quantity of ions supplied was enough to affect the electrical resistivity of the compacted material. The water retention curve of the two kinds of samples was determined using a water dewpoint device and through vapour equilibrium. The electrical resistivity was measured in samples to which suction was known, allowing to find a relationship between total suction and electrical conductivity, and therefore between water content and electrical resistivity. The results found are commented considering the electrical transport through the liquid phase and its continuity in the porous media, which depends on the conductivity of the electrolyte and on the degree of saturation.

1 Introduction

Prospection techniques based on measuring the electrical resistivity of geomaterials are being used mainly for geophysical characterization, for evaluating soil contamination and the extension of contaminated areas and for compaction control [1-7].

This technique is based on the fact that contrasts in electrical resistivity along the profiles analysed help identifying zones of transition because this property is affected by the presence of water and dissolved salts in the pore water, as well as by soil structure. However, *in situ* data interpretation has some difficulties because electrical resistivity is affected by changes in the unsaturated state of the soil and by the presence of salt compounds (natural or from contamination). The correspondence between electrical resistivity and total suction in compacted kaolin considering the presence of salt was investigated in the study presented in order to understand how these values would be affected by the presence of salt. Such study has several engineering applications as relations like this may help in better interpretation of *in situ* tests based on measuring electrical resistivity, as well as in laboratory tests where resistivity sensors may be used.

The samples tested were compacted for the same voids ratio and water content, however one type was prepared with distilled water and the other with a NaCl solution (0.5 M). This salt was chosen because it is well known that its presence affects electrical resistivity and also because it is not harmful and for this reason no special requirements are necessary to ensure safety during the experimental tests. The results were interpreted taking into account soil structure and its changes in the presence of salt, as well as how salt concentration in the electrolyte changes during wetting and drying cycles.

2 Material studied

A commercial white kaolin was used, having 68% of silt-size particles (size between 0.075 mm and 0.002 mm) and 31% of clay-size particles (smaller than 0.002 mm). Solid particles density is 2.61.

The Atterberg limits are presented in Table 1 considering kaolin mixed with distilled water or with the NaCl solution of 0.5 M. It can be seen that salt presence

[a] Corresponding author: rafaela@civil.ist.utl.pt

reduces the plasticity of the kaolin because classification changes from MH to CL (Unified Soil Classification adopted, USC). This indicates chemical interaction between clay minerals and the ions provided by the salt solution, which reduce dispersive forces between the clay minerals.

Table 1. Atterberg limits for the two fluids studied.

Fluid	Liquid limit (%)	Plastic limit (%)	Plasticity index (%)	USC
Distilled water	51	30	21	MH
0.5 M NaCl (pH= 8.2)	48	25	23	CL

Dispersive forces are maximum when distilled water is used and are expected to reduce with increasing concentrations of NaCl. This effect can be identified also by comparing the zeta potential of the clay studied when dispersed in distilled water or in a salt solution. Indeed, as shown in Figure 1, for pH= 7 the zeta potential in water is equal to that in 0.5 M NaCl solution, however for this pH the value in water is higher than the values measured for smaller concentrations of NaCl. This indicates the reduction of the thickness of the layer of water adsorbed by the charged surfaces (thickness of the double layer) and a larger tendency for particles to aggregate in the presence of salt, reducing plasticity. This effect prevails when pH is acid (pH< 7), as presented in Figure 1, while for basic solutions the particles will tend to aggregate better in water rather than when salt is present in the fluid [9,10].

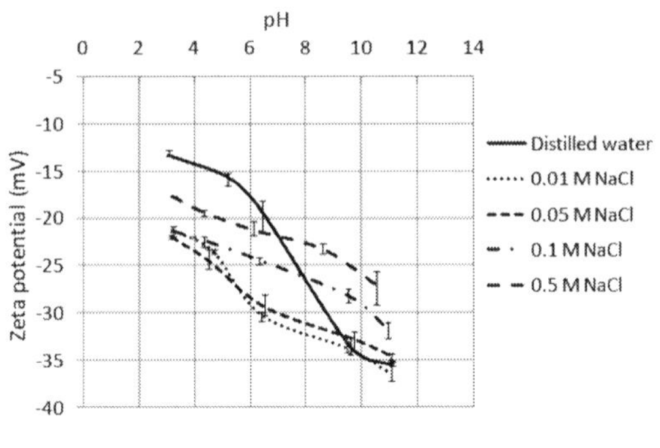

Figure 1. Zeta potential curves and the effect of different concentrations of NaCl [8].

This effect may also be observed when analysing pore size distribution measured in Mercury intrusion porosimetry tests (MIP) presented in Figure 2. The larger peaks identified correspond to the size of micropores (diameters between 40 to 50 nm) and reflect different dispersive forces for the two fluids. In this case dispersion is higher for the NaCl because the pores are slightly larger than those found for the samples prepared with water. Even if pH is 7 for both solutions, for the solution prepared with salt probably its concentration increased during drying for MIP tests and for this reason dispersive forces increased following the tendency identified in Figure 1 for the zeta potential increasing with salt concentration.

The samples used for the MIP plots were compacted with the corresponding fluid in order to have voids ratio of 0.9 and water content of 25%. As shown in the photograph of Figure 3, the compacted material is a double porous material, where the micropores (detected by MIP) are those of the clay aggregates and the macropores are very large (in the order of millimetres, not detected by MIP). These are the voids between the aggregates [11].

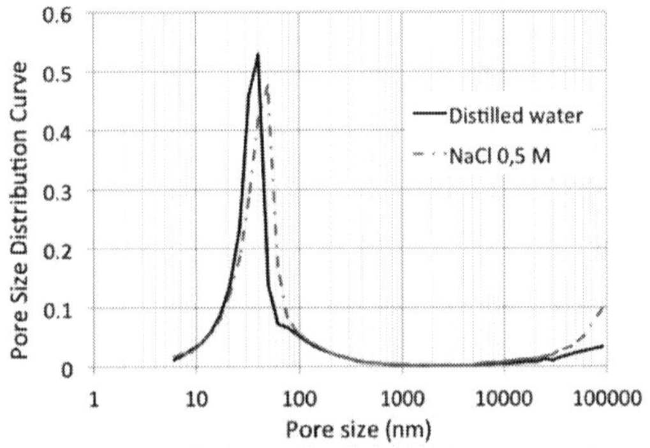

Figure 2. MIP of the samples prepared with distilled water and NaCl 0.5 M solution.

Figure 3. Compacted sample (prepared with water).

The compaction conditions adopted intend to create similar soil structures independently from the fluid used. From the MIP tests it can be seen similar microstructures, and because voids ratio is the same it can be assumed similar macrostructures as well.

3 Experimental setup

The water retention curves were measured for the samples prepared with the distinct fluids using a Dewpoint Potential Meter (WP4C) equipment. The same sample was used for the entire curve, compacted as shown in Figure 3 and subjected to one drying-wetting cycle.

For the electrical resistivity tests, the samples were compacted in a mould (Figure 4), in which four steel electrodes were inserted. These samples were subjected to vapour equilibrium technique (Figure 5) so that each sample have a different water content following a wetting path or a drying path. The samples that followed the wetting path were dried in contact with the air before the use of the vapour equilibrium (s=87 MPa). The samples of the drying branch are saturated before vapour equilibrium. Volume changes due to this preliminary treatment were not considered in the analysis because previous studies [12] indicated that the material has low

DOI: 10.1051/e3sconf/20160910002

swelling potential. Partially saturated solutions of NaCl were used to apply different suctions (1, 5, 15 and 39 MPa), prepared as suggested by [13].

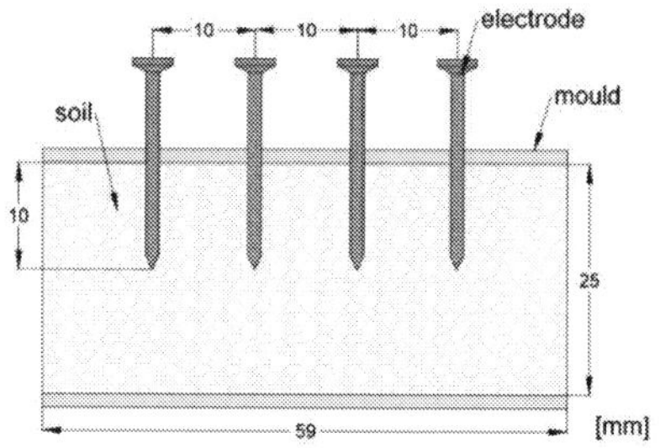

Figure 4. Setup for the electrical resistivity test.

Figure 5. Samples in vapour equilibrium.

Electrical resistivity, suction and water content were measured after reaching vapour equilibrium. Electrical resistivity measurements were done according with Wenner method configuration [14], by applying a DC current in the exterior electrodes and measuring potential in the inner ones. The value of electrical resistivity was computed through Equation 1 (geometric factor from Wenner method [15]), where L1 and L2 are the spacing and depth of the electrodes, respectively, and R is the electrical resistance. This parameter corresponds to the slope of a linear trendline determined for several values of applied current and measured potential.

$$\rho = \frac{4\pi L_2 R}{1 + \frac{2L_2}{\sqrt{{L_2}^2 + 4{L_1}^2}} - \frac{L_2}{\sqrt{{L_2}^2 + {L_1}^2}}} \quad (1)$$

4 Results and discussion

4.1 Water retention curve

The water retention curves are presented in Figure 6.

The curves were adjusted using van Genuchten expression (Equation 2, [16]), where e is voids ratio, G_s is the density of the solid particles, s is total suction, w is water content and P and λ are calibration constants (Table 2). In the absence of data it was assumed constant volume during wetting and drying.

$$S_r = \frac{G_s}{e} w = \left[1 + \left(\frac{s}{P}\right)^{\frac{1}{1-\lambda}}\right]^{-\lambda} \quad (2)$$

The residual state of saturation [17] can be determined. This value marks the discontinuity of the liquid phase. Above this value it can be assumed that the macrostructural water has no longer influence on suction and the consideration of electrons flow through the liquid phase is no longer realistic. The values found are presented in Table 3.

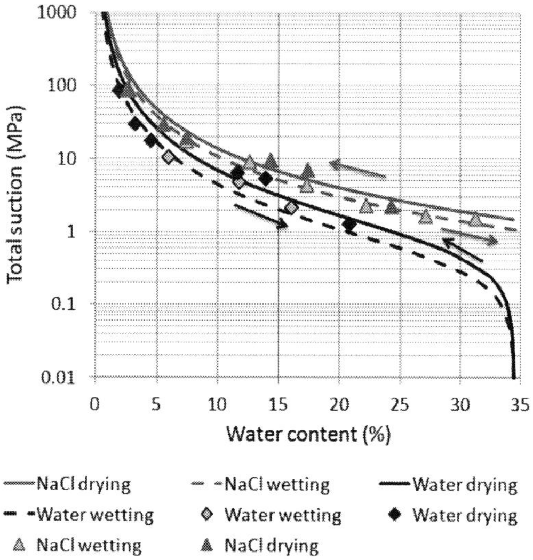

Figure 6. Water retention curves.

Table 2. Calibration parameters of Equation 2.

Solution	Distilled water		NaCl 0.5 M	
Branch	**Wetting**	**Drying**	**Wetting**	**Drying**
***P* (MPa)**	0.45	0.70	0.16	0.19
λ	0.350	0.350	0.353	0.360

Table 3. Residual state of saturation.

Solution	Distilled water		NaCl 0.5 M	
Branch	**Wetting**	**Drying**	**Wetting**	**Drying**
***w* (%)**	2.8	3.2	4.5	4.5
S_r (%)	8.12	9.28	13.1	13.1

It can be observed in Figure 6 that the curves measured using water are below those measured using the NaCl solution mainly due to the contribution of the osmotic component. This is because viscosity and surface tension of the solution prepared with NaCl is not that different from the curve found for distilled water.

For the sample prepared with NaCl it was not possible to measure null suction at saturation. This lowest suction was approximately 1.2 MPa. This value is similar to the contribution of the osmotic component, which minimum value occurs to the minimum concentration of NaCl allowed by the maximum amount of water in the pores, assuming no volume change. At preparation, the solution prepared with 0.5 M applies s=2.2 MPa (25°C) but the soil was not saturated. Keeping the same mass of NaCl, at saturation its concentration reduces to approximately

0.35 M, corresponding to s=1.53 MPa. This last value is closer to the value measured and error can be attributed to some swelling deformation during saturation.

This rule can also be applied for the highest suctions. In this case the maximum suction measured should be 39MPa, the value applied by a full saturated NaCl solution. This suction is verified at the residual state of saturation, confirming that osmotic suction depends on the continuity of the liquid phase, and therefore on the amount of free water. This can be considered to be macrostructural water. Above this suction it can be assumed that water in the microstructure is more important, and therefore osmotic effects due to the presence of dissolved salts are no longer relevant.

To conclude, to calibrate the WRC of the samples with NaCl using Equation 2 an artificial water content at saturation had to be adopted. This value was adjusted to achieve a good fitting [12].

4.2 Electrical resistivity

Table 4 presents the values measured for the electrical resistivity (ρ) for each water content. The electrical resistivity of the samples prepared with NaCl is lower than the values found for the samples prepared with water in the entire suction range, because of the presence of ions in the solution. Both electrical conductivity of distilled water and 0.5M NaCl were measured (6.5 μS/cm for water and 45.7 mS/cm for the salt solution), however they cannot be directly related with the values of electrical resistivity measured at full saturation because the latter are affected by soil structure (mainly porosity and tortuosity) [1] and by the salts present in the soil.

The values from Table 4 allowed the calibration of Equation 3 [4], where ρ_{sat} is the electrical resistivity measured at saturation and B is a calibration parameter. Both values are presented in Table 5. The values found for parameter B are within the ranges of values expected for clays [4]. The electrical conductivity of the clay minerals was not considered in this equation assuming the validity of the equation only for the cases where liquid phase is continuous. This is a simplification.

$$\frac{\rho}{\rho_{sat}} = S_r^{-B} \tag{3}$$

The curves found for each B are presented in Figure 7. Hysteresis exists, as it is observed in the water retention curve. Electrical resistivity and degree of saturation are related because the electrical current flows preferentially through the liquid phase. It is also observed that, for high degrees of saturation, the electrical resistivity tends to a common value which is dependent on the type of pore fluid.

A considerable decrease in the electrical resistivity is observed when the specimens are dried below the residual state of saturation (Table 3). This is consistent with the assumption that electrical current flows through the liquid phase, because it is expected to become more difficult as soon as this path loses continuity. Such result was observed in other clays [18]. Nevertheless, electrical current in clays can also flow through their electrically charged surfaces, and probably this is the reason why the electrical resistivity could be measured above the residual state of saturation and the values were not very high.

Table 5. Values of B and ρ_{sat}.

Fluid	B		ρ_{sat} (kΩm)
	Wetting	Drying	
Distilled water	1.70	2.76	0.079
NaCl 0.5 M	1.69	3.27	0.007

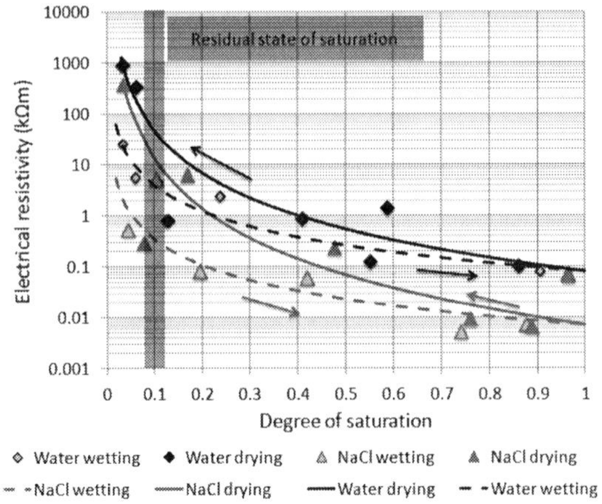

Figure 7. Electrical resistivity relation with degree of saturation and residual state of saturation.

Table 4. Electrical resistivity and degree of saturation of the samples.

Fluid	Branch	Readings	Applied suction (MPa)					
			0	1	5	15	39	87
Distilled water	Wetting	S_r (%)	90.6	46.6	23.9	10.7	6.2	3.4
		ρ (kΩm)	0.079	15.106	2.357	4.621	5.492	25.544
	Drying	S_r (%)	86.1	58.9	41.0	12.8	6.2	3.4
		ρ (kΩm)	0.098	1.410	0.877	0.772	328.098	879.398
NaCl solution	Wetting	S_r (%)	87.6	74.2	42.0	19.5	10.5	4.5
		ρ (kΩm)	0.007	0.005	0.060	0.081	0.531	5.150
	Drying	S_r (%)	88.8	76.1	47.6	17.1	8.0	3.6
		ρ (kΩm)	0.064	0.069	0.129	0.227	0.464	4.193

4.3 Electrical resistivity versus suction

A direct relationship between electrical resistivity and suction was found by joining Equations 2 and 3 to obtain Equation 4.

$$\rho = \rho_{sat}\left[1 + \left(\frac{s}{P}\right)^{\frac{1}{1-\lambda}}\right]^{\lambda B} \quad (4)$$

The curves relating electrical resistivity and suction can be defined using the calibration parameters previously presented in Tables 2 and 5. They are presented in Figure 8. This figure includes the experimental points to show the adjustment achieved. It is not very good for the samples prepared with distilled water and measurements done in the drying branch.

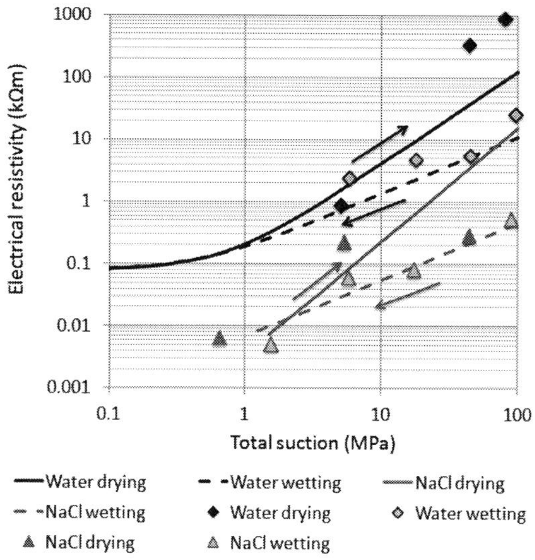

Figure 8. Electrical resistivity versus total suction using Eq. 4.

However Equation 4 is not practical, and for this reason Equation 5 can be proposed as a simpler alternative to Equation 4. In Equation 5, a and b are constants calibrated using the data from Table 4 and ρ and s were already defined. The values of the constants found are in Table 6.

$$\rho = as^{b} \quad (5)$$

Table 6. Values of a and b.

Fluid	a		b	
	Wetting	Drying	Wetting	Drying
Distilled water	0.5197	0.0146	0.7627	2.5643
NaCl 0.5 M	0.0046	0.0164	1.0749	0.9021

It can be seen in Figure 9 that this alternative expression provides a good adjustment for the experimental points as well. The comparison between the quality of curve fitting achieved using the definition of the WRC or the simpler alternative provided by Equation 5 cannot be done such straightforward because it is strongly affected by the equation used for the definition of WRC.

Simpler equations such as Equation 5 can be used to calibrate electrical resistive sensors for soil suction measurement, which may be considered as an alternative to WP4 or other existing equipment.

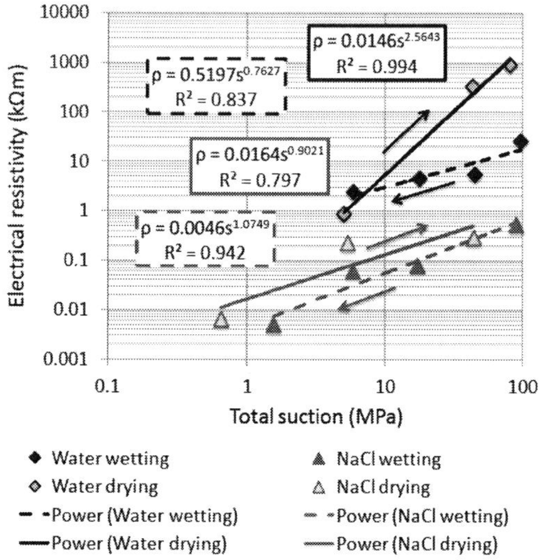

Figure 9. Electrical resistivity versus total suction using Eq. 5.

5 Conclusions

The experimental results show that the presence of NaCl affects the WRC and electrical resistivity.

Concerning the effect of the presence of salt in the WRC, suction value measured at saturation suggest that it corresponds to osmotic suction. It corresponds to the value applied by a solution having the minimum concentration of salt in the pore fluid when the pores are completely saturated. On the other hand, suction measured at residual state of saturation is closer to the value applied by the saturated NaCl solution.

Concerning the effect of the presence of salt in electrical resistivity, it reduces this parameter because of the presence of ions dissolved in the water. Even if electrical current may flow through the electrically charged surfaces, the results show that the liquid phase provides an important path. It can be seen an increase of resistivity when the degree of saturation is lower than the residual degree of saturation, and also hysteresis similar to that observed in the WRC.

Good relationships between total suction and electrical resistivity were found. The quality of the fitting when using the expression for the WRC may be improved by adopting a better definition of this curve. The equation proposed by van Genuchten was adopted in this paper because it is very simple, however it may not be the best for this material.

The relations found between suction and electrical resistivity can be used for soil sensors development and for the interpretation of geophysical prospection tests when the evaluation of soil strength is being investigated. However they require the inclusion of the electrical conductivity of the clay minerals, which is under progress.

Acknowledgments

The authors thank INESC-MN and Prof. Susana Freitas for lending the equipment used for the electrical resistivity measurements. Acknowledgement is also due to Portuguese Foundation for Science and Technology, FCT, for the funding (EXPL/ECM-GEO/0109/2013).

References

1. G.E. Archie. Petroleum Transactions of AIME, **146** (1), pp. 54-62 (1942).
2. W. Kelly, R. Frohlich. Ground Water, **23** (2), pp. 182-189 (1985).
3. Sharma, P. *Environmental and engineering geophysics*. Cambridge University Press, pp. 207-261 (1997).
4. Z. Abu-Hassanein, C. Benson, L. Boltz, Journal of Geotechnical Engineering, **122**(5), pp. 397-406 (1996).
5. M. Khalil, F. Santos. Developments in Hydraulic Conductivity Research, pp. 49-70. (2011).
6. J. Santamarina, K. Klein, M. Fam. (2001). *Soils and Waves*. John Wiley & Sons, pp. 336-341.
7. D.A. Gunn, J.E. Chambers, S. Uhlemann, P.B. Wilkinson, P.I. Meldrum, T.A. Dijkstra, F. Haslam, M. Kirkham, J. Wragg, S. Holyoake, P.N. Hughes, R. Hen-Jones, S. Glendinning. Construction and Building Materials, **92**, pp. 82-94 (2014).
8. V. Gingine, R. Cardoso, Soil Structure Influence on Electrokinetic Dewatering Process. *Electrokinetics Across Disciplines and Continents*, pp. 19-42, (2015)
9. G.H. Bolt. Géotechnique, **6**(2), pp. 86–93 (1956).
10. Y. H. Wang, D. Xu. J. Geotech. and Geoenv. Eng., Vol. **133**(7), (2007).double layer e swelling potential
11. E. Romero, A. Gens, A. Lloret. Engineering Geology, 54, pp. 117-127 (1999).
12. A. S. Dias, *Electrical resistivity of compacted kaolin and its relation with suction*, MSc Thesis, Instituto Superior Técnico, University of Lisbon (2015)
13. E. Romero, 4° Simpósio Brasileiro de Solos Não Saturados, pp. 535-542 (2001).
14. BS 1377-3. Methods of test for Soils for civil engineering purposes. Chemical and electro-chemical tests (1990).
15. R. Herman. Am.J.Phys., **69** (9). pp 943-952 (2011).
16. M. van Genuchten. Soil Sci. Soc. Am. J., Vol. 44, No. 5, pp. 892-898 (1980).
17. S. K. Vanapalli, D. G. Fredlund, D. E. Pufahl, Géotechique, **49** (2), pp. 143-159 (1999).
18. M. Fukue, T. Minato, H. Horibe, and N. Taya, Eng. Geol., **54**, pp. 43-53 (1999).

E3S Web of Conferences 9, 10003 (2016)
DOI: 10.1051/e3sconf/20160910003
E-UNSAT 2016

Study of the variation of B with Sr

Mathilde Morvan[1a], Mathilde Vernay[1] and Pierre Breul[1]

[1] *Institut Pascal, Polytech Clermont-Ferrand, Université Blaise Pascal, Clermont-Ferrand, France*

Abstract. Skempton coefficient B is commonly used to verify the saturation of a sample before triaxial testing. This coefficient is obtained during undrained isotropic consolidation and is defined as the ratio between the increment of pore pressure u measured and the imposed increment of isotropic stress. This coefficient varies between 0 for dry soils and 1 for saturated soils. Many studies on liquefaction of unsaturated soils were published using Skempton coefficient B to represent saturation degree Sr of soil. On the first hand, this variation of B coefficient with saturation degree is mostly due to the compressibility of air in the pores. On the second hand, we also know that the presence of air as a fluid phase gives birth to suction after equilibrium is reached inside the sample. The higher the suction, the stiffer the soil skeleton. These two phenomena are opposite. Their effects in laboratory testing depend on the experimental apparatus. For example, if we consider an unsaturated triaxial device, we will have to take suction into account. On the contrary if we plan to break the menisci just before measuring B, suction equilibration will not occur. Experimental tests were performed to show the difference between these two cases and to study the equilibrium phase. Based on these observations, this article presents new relationships that permit to calculate saturation degree with a given Skempton coefficient with different hypotheses and with different experimental devices. These results are confronted to the commonly used relation given by Lade and the difference between all these calculations is studied.

1 Introduction

Nowadays, Skempton coefficients A and B [1] are commonly used in earth dam study. They were proposed by Skempton (1954) in order to anticipate pore pressure changes due to mechanical stresses:

$$\Delta u = B[\Delta\sigma_3 + A(\Delta\sigma_1 - \Delta\sigma_3)] \qquad (1)$$

Practical use of these parameters was exposed by Bishop [2].

Coefficient B is linked to saturation degree so it can also be used to verify saturation of a sample during triaxial testing. Coefficient B can be defined as the ratio between increment of pore water pressure and increment of isotropic pressure when a sample is subjected to an increment of isotropic pressure. So we have:

$$B = \frac{du_w}{dp} \qquad (2)$$

For a saturation degree of 100%, we get B=1 if we consider water and skeleton grain as incompressible.

Some authors try to give a unique relation between B and Sr. The usual one was determined by Skempton [1]. This equation is widely used during undrained triaxial tests and liquefaction tests for example [5, 6, 7, 8].

$$B = \frac{du_w}{dp} = \frac{1}{1 + \dfrac{nC_v}{C_c}} \qquad (3)$$

Where C_c is the compressibility of skeleton, C_v fluid compressibility and n sample porosity.

During their study of membrane penetration, Lade and Hernandez [3] used the equation giving B in function of Sr using the results of Hilf [4] and Skempton [1].

Schuurman [9] modified this equation taking equilibrium of air bubbles into account. To perform these calculations, he assumed that the number of bubbles is constant during the test. This hypothesis was not verified [10].

More recently, some researchers give new equations to obtain this relationship. Hasan and Fredlund [11] developed a method to obtain B_w and B_a depending on the studied fluid (water or air) for unsaturated state. They considered two compressibility parameters for skeleton and air [12]. These parameters and the obtained relationship do not offer an easy calculation.

Boutonnier [13] developed a methodology to obtain B for different saturation degree ranges.

In this article we propose to study this relationship between B and Sr and parameters related firstly using a classical approach, then an approach considering an equilibrium before the application of the isotropic pressure increment and finally to consider the elastoplastic behavior of the soil skeleton.

[a] Corresponding author: mathilde.morvan@univ-bpclermont.fr

2 Usual simulation

In this first part, we used Skempton and Lade [1, 3] hypotheses so we can obtain a simple relationship between B and Sr. We made the assumption that effective stress of saturated soils is still valid and that stress variation is small so we have an elastic behavior of soil. We also consider that no phase change occurs.
To measure B coefficient, triaxial cell is used. So in this simulation and the following ones, we considered a revolution symmetry and used following parameters:

$$p = \frac{1}{3}(\sigma_1 + 2\sigma_3) \; ; \; q = \sigma_1 - \sigma_3 \; ;$$

$$\varepsilon_p = \varepsilon_1 + 2\varepsilon_3 \; ; \; \varepsilon_q = \frac{2}{3}(\varepsilon_1 - \varepsilon 3) \qquad (4)$$

Where subscript 1 is linked to vertical direction and subscript 3 is linked to radial direction.
Mechanical behaviour can be described by:

$$d\varepsilon_p = -\frac{dV_a}{V_0} = \frac{\kappa}{p'(1+e_0)}dp' \qquad (5)$$

With $p' = p - u_w, u_a = u_w$ and κ represents the slope of elastic loading lines in a (*e*-log p) plane.
We also considered that water (subscript w) and grains (subscript s) are uncompressible: $dV_w = dV_s = 0$
Air pore pressure is calculated using Boyle's law:

$$P_a V_a = P_{a0} V_{a0} = (P_{a0} + dP_a)(V_{a0} + dV_a) \qquad (6)$$

With P_a absolute air pore pressure and V_a air volume.
So we get

$$(P_{a0} dV_a) + (V_{a0} dP_a) = 0 \qquad (7)$$

Considering that $dV_a dP_a$ is negligible,
Knowing that:

$$V_a = \frac{e_0}{1+e_0} V_0 (1 - S_r) \qquad (8)$$

We can obtain:

$$B = \frac{du_w}{dp} = \frac{du_a}{dp} \qquad (9)$$

$$B = \frac{\kappa}{\kappa + \frac{p_0' V_{a0}(1+e_0)}{V_0 P_{a0}}} = \frac{1}{1 + \frac{p_0' e_0 (1-S_r)}{\kappa P_{a0}}} \qquad (10)$$

Parameters used for this first simulation are summarized in Table1.

Table 1. Parameters used in usual simulation.

Parameter	Value
κ	0.006
p_0	100 kPa
$u_{a0} = u_{w0}$	30 kPa
e_0	0.8
p_{a0}	130 kPa

These hypotheses give the relationship shown in Fig.1. We can see that the relationship between B and Sr is monotonic and that for B>0.2, Sr>95%. So even if B is low, saturation degree is still high. So B is a good mean to evaluate near saturation value of saturation degree.

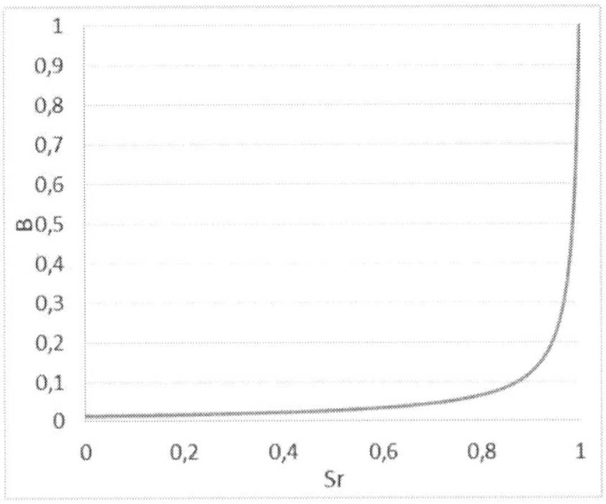

Figure 1. Evaluation of B using simple method

Two parameters seems to influence this relationship: isotropic elastic modulus and initial void ratio. So we propose to study their influence.
Fig.2 presents variation of B, Sr relationship with kappa parameter. As we could forecast, this parameter has a significant effect on the curves. The more rigid is the soil, the lower B is for a same saturation degree. For example, Skempton coefficient at a saturation degree of 90%, for κ=0.06 we get B=0.58 and for κ=0.0006, B=0.01. This huge difference show that the determination of this parameter is essential to determine B.

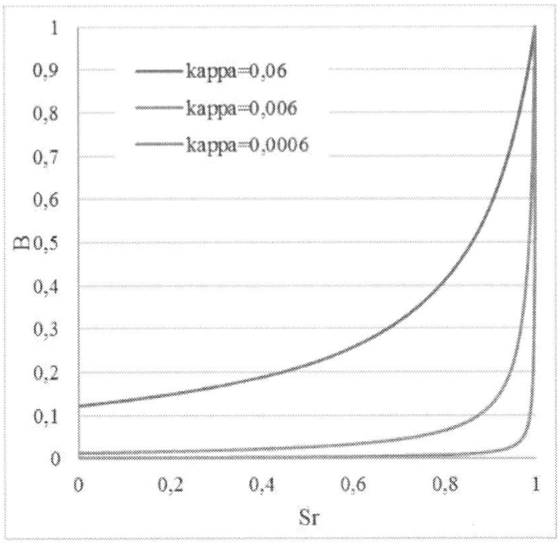

Figure 2. Effect of kappa on B simulations

If we consider the relationship variations with e_0 (Fig.3) we can see that for a void ratio from 0.6 to 1.2 observed difference is low. Skempton coefficients for a saturation degree of 90% for e_0=0.6, B=0.15 and e_0=1.2, B=0.08.

 DOI: 10.1051/e3sconf/20160910003

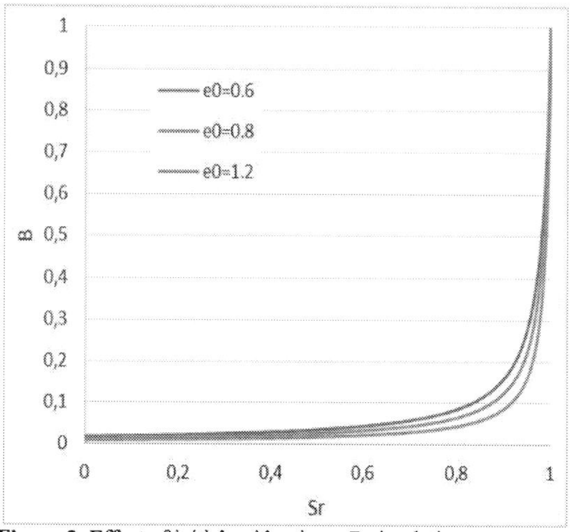

Figure 3. Effect of initial void ratio on B simulations

3 Simulation with initial suction equilibrium

When soil is unsaturated, voids are filled with two fluids. At equilibrium it causes a suction responsible for changes of mechanical behavior of soil skeleton. Suction is given as the difference between air pore pressure and water pore pressure:

$$s = u_a - u_w \tag{11}$$

If we measure B after waiting the equilibrium, calculation have to take the effects of suction into account. So we have an increase of sample rigidity with saturation degree decrease. In this simulation, parameters will depend on mechanical behavior of soil, but also on hydromechanical behaviour. Two main parameters can be taken into account, these two parameters govern water retention curve and vary with kind of soil.

Table 2. Parameters used in elastic simulation

Parameter	value
κ	0.006
p_0	100 kPa
e_0	0.8
p_{a0}	130 kPa
s_e	20
α	0.8

An incremental calculation for each saturation degree is needed to obtain Skempton coefficient for this simulation.

For each saturation degree we can calculate suction thanks to Brooks and Corey law [14]. Parameters are s_e and α. s_e is air entry suction and depends on grain size and α depends on grain size distribution uniformity. These two parameters vary with the type of soil. These simulation were performed using a bijective relationship between suction s and saturation degree Sr.

This obtained suction leads us to initial water pore pressure. We verified experimentally that when equilibrium is reached, water pore pressure is negative and air pore pressure remains equal to atmospheric pressure.

So we get:

$$u_w = u_a - s \tag{12}$$

Initial effective stress is calculated using enlarged effective stress concept with a simplified law [15].

$$\sigma' = \sigma - u_a + sS_r = \sigma - u_a + \chi \tag{13}$$

For mechanical behavior, we choose a classical law with an isotropic modulus depending on effective mean stress as in previous simulation.

The main difference between this new simulation and the previous one is the effect of suction that causes effective stress variation whereas stress increment remains the same. We verified that *dp* has no influence on the simulation results.

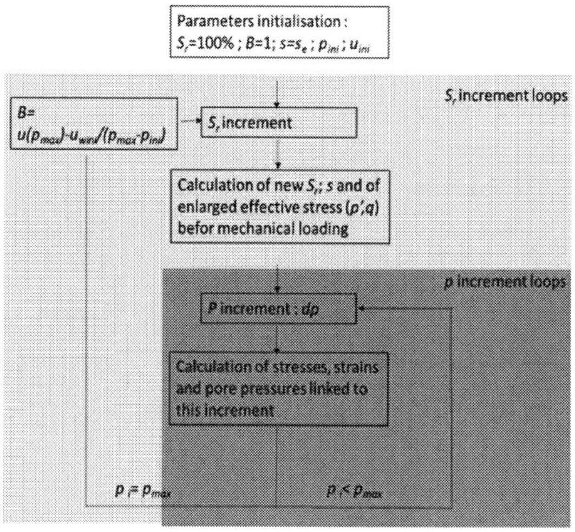

Figure 4. Diagram of simulation algorithms

Then to study the influence of parameters of the water retention curve, we made some simulations for different α and air entry suction s_e. For a fine grain size distribution, s_e will be high and on the contrary for coarse grain size distribution it can be lower than 10 kPa.

Fig. 5 presents the results of simulations for four values of s_e: 5 kPa, 20 kPa, 100 kPa, 1000 kPa. We can see here that s_e has also a significant effect on *B-S_r* relationship. The higher s_e, the lower *B* is. Previous relationship curve is close to the curve we can obtain for a sand which behaviour is little influenced by suction. So for a granular material, previous relationship giving B variation with Sr is accurate enough. For fine soils, this relationship can lead to a significant underestimation of saturation degree for a given B.

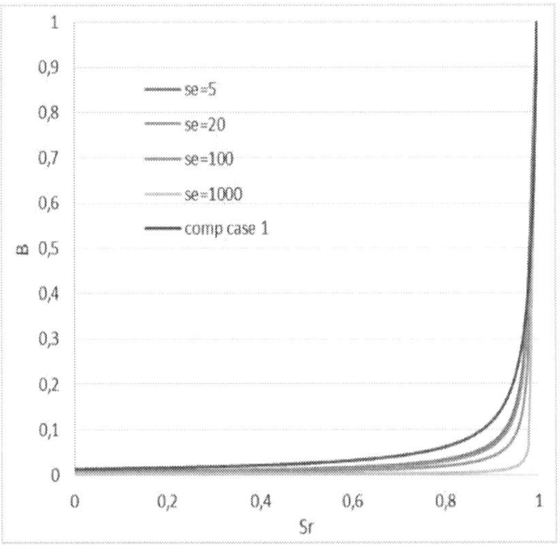

Figure 5. Dependency of B on air entry suction for second simulation

4 Elastoplastic simulation

We wanted to be more accurate in this study of B versus Sr relationship. So we propose a third simulation taking elasto-plasticity into account with enlarged effective stress. Initial state of sample will give birth to different consolidation state. A normally consolidated sample will quickly develop plastic strains and an overconsolidated sample would develop mainly elastic trains. Triaxial apparatus allows us to rebuild samples but also to work on undisturbed samples, so each case can occur.

We choose BDNS model [16] for this calculation. In this model, enlarged effective stress concept is used as well as the bounding surface theory that permits a smooth transition between elastic and plastic domains. It gives us a matrix $\boldsymbol{A}$ that linked the strain vector to the stress vector where suction is added. So it gives:

$$\begin{pmatrix} d\varepsilon_p \\ d\varepsilon_q \end{pmatrix} = \begin{pmatrix} A_{11} & A_{12} & A_{13} \\ A_{21} & A_{22} & A_{23} \end{pmatrix} \begin{pmatrix} dp' \\ dq \\ ds \end{pmatrix} \quad (14)$$

As for the previous simulation, we need two incremental loops: the first one is based on an increment of saturation degree and the second one, for each saturation degree increment, an elastoplastic calculation is performed for a given isotropic pressure increment.

For this calculation, after isotropic pressure increment, we consider that isotropic pressure increment is imposed too fast to reach suction equilibrium in this loop. So we consider that χ remains constant. It could be interesting to study the influence of rate of load application (equilibrium can be reached or not).

To obtain the result of dp isotropic pressure on pore pressure increment, we have to separate this matrix law. To achieve this point, some hypotheses are necessary. For this calculation, we make the assumption that suction remains constant so it gives $du_a= du_w$. Saturation degree increment is inserted in strain increment vector. Variation of air pore pressure is obtained using Boyle's law. It adds another dimension to the matrix with du_a as stress variable and dV_a as strain variable.

$$\begin{pmatrix} d\varepsilon_p \\ d\varepsilon_q \\ dS_r \\ dV_a \end{pmatrix} = \begin{pmatrix} D_{11} & D_{12} & D_{13} & D_{14} \\ D_{21} & D_{22} & D_{23} & D_{24} \\ D_{31} & D_{32} & D_{33} & D_{34} \\ D_{41} & D_{42} & D_{43} & D_{44} \end{pmatrix} \begin{pmatrix} dp \\ dq \\ ds \\ du_a \end{pmatrix} \quad (15)$$

So we obtained a matrix $\boldsymbol{D}$ with 16 unknown variables that links $(d\varepsilon_p, d\varepsilon_q, dS_r, dV_a)$ to (dp', dq, ds, du_a).

To study the influence of dp, first we have to separate mechanic and hydric variables. Using the enlarged effective stress definition, we get:

$$dp' = dp - du_a + d\chi = dp - du_a + s dS_r + S_r ds \quad (16)$$

Relationship between ds and dS_r gives us:

$$dS_r = -S_r \frac{\alpha}{s} ds = D_{33} ds \quad (17)$$

So we have:

$$dp' = dp - du_a + S_r(1-\alpha)ds \quad (18)$$

Using Boyle's law incremental form:

$$dV_a = -\frac{V_a}{P_a} du_a = D_{44} ds \quad (19)$$

Increment of volumetric strain is given by:

$$d\varepsilon_p = D_{11}dp + D_{12}dq + D_{13}ds + D_{14}du_a \quad (20)$$

With $D_{11} = A_{11}$; $D_{12} = A_{12}$, $D_{13} = A_{13} + A_{11}(1-\alpha)S_r$; $D_{14} = -A_{11}$

We made the assumption that $dq = d\chi = 0$

So volumetric strain increment can be simplified:

$$d\varepsilon_p = D_{11}dp + D_{14}du_a \quad (21)$$

We also have:

$$du_a = -\frac{P_a}{V_a} dV_a = \frac{P_a}{V_a} V_0 d\varepsilon_p \quad (22)$$

So we get:

$$d\varepsilon_p = \frac{A_{11}}{1 + A_{11}\frac{P_a}{V_a}V_0} dp \quad (23)$$

With $du_a = du_w$:

$$du_a = du_w = \frac{P_a}{V_a} V_0 \frac{A_{11}}{1 + A_{11}\frac{P_a}{V_a}V_0} dp \quad (24)$$

For this simulation, elastic parameters are the same as in previous simulations. Every parameters are given in table 3. Parameters Γ, ρ and M defined the shape of the bounding surface. λ_0 is the plastic isotropic modulus. A_{ini} gives the initial position of the surface.

Simulations for a normally consolidated sample and for an overconsolidated sample are given Fig. 6. For these two simulations the difference is not located at high saturation degrees but also at low saturation degrees. For a saturation degree of 90%, we obtained B of 0.4 for a normally consolidated sample and B equal to 0.04 for an overconsolidated sample.

Table 3. Parameters used in elasto-plastic simulation

Parameter	value
κ	0.006
p_0	100 kPa
e_0	0.8
p_{a0}	130 kPa
s_e	20
α	0.8
λ_0	0.2
Γ	1.85
ρ	2
M	1.6
A_{ini}	0 or 10kPa

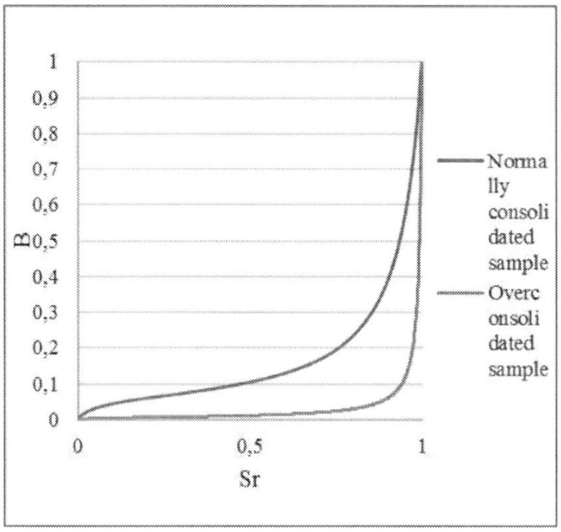

Figure 6. Dependency of B toward consolidation state

5 Conclusion

To conclude, this article proposed three simulations to obtain Skempton coefficient B knowing saturation degree Sr. We enlighten that the initial relationship depends principally on two parameters. The two other simulations show that two other parameter could be taken into account: water retention curve and consolidation state of studied soil. These two parameters need to be further studied and these results need to be compared with experimental data.

References

1. A. W. Skempton, The Pore-Pressure Coefficients A and B. Géotechnique, **4**, 143-147 (1954).
2. A. W. Bishop, The Use of Pore-Pressure Coefficients in Practice. Géotechnique, **4**, 148-152 (1954).
3. P. V. Lade, & S. B. Hernandez, Membrane penetration effects in undrained tests. Journal of the geotechnical engineering division, **103**, 109-125 (1977).
4. J.W. Hilf, Estimating construction pore pressure in rolled earth dams. Proceedings of 2nd International Conference on Soil Mechanics and Foundation Engineering. Rotterdam, vol. **3**, p.234-240 (1948)..
5. K. Mitsuji, Numerical simulations for development of liquefaction. Countermeasures by use of partially saturated sand. The 14th World Conference on Earthquakes Engineering. Beijing, China (2008).
6. Y. Yoshimi, K. Tanaka, K. Tokimatsu, Liquefaction resistance of a partially saturated sand. Soils and Foundations, **29**, 157-162 (1989).
7. N. Della, Laboratory testing of the Monotonic behavior of partially saturated sandy soil. Earth Sciences Research Journal, **14**, 181-186. (2010)
8. M. E. Raghunandan, & A. Juneja, A study on the liquefaction resistance and dynamic properties of de-saturated sand. Electronic Journal of Geotechnical Engineering, 16, 109-123 (2011).
9. I. E. Schuurman. The Compressibility of an Air/Water Mixture and a Theoretical Relation Between the Air and Water Pressures. Géotechnique, **16**, 269-281 (1996).
10. A.Bendi-Ouis, N. Abou-Bekr, S. Taibi,, J. M. Fleureau,Prédiction du coefficient de pression interstitielle B: Etat de l'art. Colloque International Sols Non Saturés et Environnement UNSAT09. Tlemcen, Algérie (2009).
11. J. U. Hasan, D. G. Fredlund, Pore pressure parameters for unsaturated soils. Canadian Geotechnical Journal, **17**, 395-404 (1980).
12. D. G. Fredlund, N. R. Morgenstern, Constitutive relations for volume change in unsaturated soils. Canadian Geotechnical Journal, **13**, 261-276 (1976).
13. L. Boutonnier, Comportement hydromécanique des sols fins proches de la saturation. Cas des ouvrages en terre : coefficient B, déformations instantanées et différées, retrait/gonflement. PhD Thesis, INPG, Grenoble (2007).
14. R. Brooks, A. Corey, Hydraulic Properties of Porous Media, Colorado State University (1964).
15. A. W. Bishop, G. E. Blight, Some aspects of effective stress in saturated and partly saturated soils. Géotechnique, **13**, 177-197 (1963).
16. M. Morvan, H. Wong, D. Branque, An unsaturated soil model with minimal number of parameters based on bounding surface plasticity. international journal for numerical and analytical methods in geomechanics, **34**, 1512-1537 (2009).

Validation of a new sensor to measure the concentration of polyethylene glycol for unsaturated soil testing

Buzzi[1,a] O., Routley[2] B., Yuan[1] S.

[1]*Priority Research Centre for Geotechnical and Materials Modelling, School of Engineering, The University of Newcastle, NSW, Australia.*
[2] *School of Electrical Engineering and Computer Science, University of Newcastle, Callaghan, Australia*

Abstract. Accurate measurement of PEG concentration is primordial when using the osmotic method to control soil suction. Because PEG is an inert molecule, common chemistry methods are not adequate to measure its concentration and researchers resort to the measurement of Brix Index. Such measurement can be manual, which is not very accurate and time consuming but cheap. On the other hand, more advanced sensors do exist but they tend to be quite expensive. This paper presents a new sensor for the indirect measurement of PEG concentration based on a tuneable diode laser and fibre optics. The laser signal is shone through the PEG and the reduction in signal voltage is correlated to the PEG concentration. The system has been assessed for temperature stability and accuracy. Such sensor offers the possibility of multipoint measurement, which significantly reduces the capital cost of the equipment.

1 Introduction

Accurate and reliable control of suction is of paramount importance to produce high quality data in experimental unsaturated soil mechanics. The osmotic technique is one of the methods available to control soil suction [1-3]. The technique relies on a semi permeable membrane separating a solution of dissolved polyethylene glycol (PEG) and the soil specimen to be tested. The difference of PEG concentration across the membrane generates suction within the soil specimen. The osmotic technique is relatively simple to use and works for suctions up to 9 or 10 MPa, which is well beyond the capability of the axis translation technique (i.e. 1.5 MPa), another method allowing suction control. One technical challenge with the osmotic method resides in the measurement of the PEG concentration. Indeed, PEG is an inert molecule and, consequently, common chemistry methods cannot be used. A simple alternative is to measure the refractive index, also called Brix index, of the PEG solution. Indeed, the Brix index is directly correlated to the concentration and is independent of the PEG molecular weight [4]. Brix Index measurement can be done with hand held refractometers, which is not very costly but not very accurate, or using an automatic in line process refractometer, which is quite an expensive piece of equipment. This study aims at developing and validating a new sensor for PEG concentration. The sensor is based on the measurement of absorbance of the PEG solution using a tunable diode laser and a photodetector. The present paper presents the proof of concept and an estimation of errors and temperature sensitivity.

2 Experimental setup

The experimental setup is presented in Figure 1. Note that, in this validation phase, the measurements were not made in situ. Instead, some PEG solutions (details given in Table 1) were placed in micro cuvettes and a near infrared laser was shone through the cuvette and the PEG solution. A controller (Model ITC4001 from Thorlabs) was used to supply the laser with a constant power of 42 mA. The laser beam was transported via a fibre optic to the cuvette holder where the cuvettes containing the PEG were placed. A photodetector (Thorlabs model PDA 50B-EC) was then used to record the voltage of the output signal, required to compute the absorbance A:

$$A=log_{10}(I_o/I) \quad (1)$$

where I_o and I are the input and output voltage, respectively. Note that this absorbance actually characterises the system (cuvette + PEG) since the laser beam shines through 8.5 mm of glass and 4 of PEG solution. Note also that A has no unit. The whole setup, but the computer, was placed in a laboratory fridge for a control of temperature at ± 0.1 degree Celsius.

[a] Corresponding author: Olivier.Buzzi@newcastle.edu.au

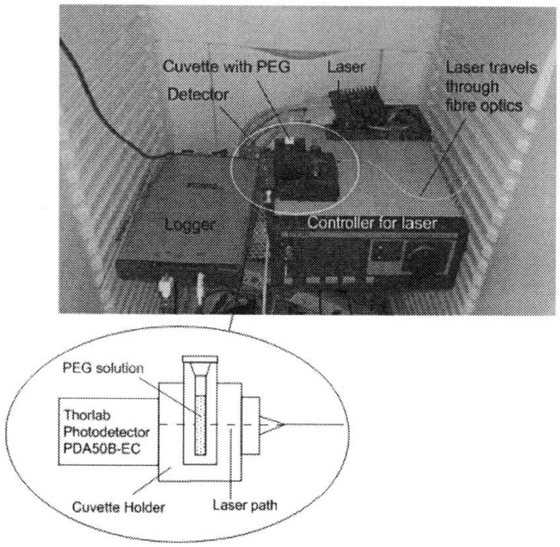

Figure 1. Sketch of the experimental setup used to validate the new sensor.

3 Experimental program

The absorbance of a number of PEG solutions was measured using the setup described in Figure 1. First, the existence of a correlation between absorbance and PEG concentration was investigated using three PEG types (PEG6000, 20000 and 35000 noted PEG6k, 20k and 35k, respectively) and up to 6 concentrations (test series #1). The range of concentration corresponds to suction values ranging from about 30 kPa to about 3 MPa (using $s=11\ c^2$, [2]). These tests were conducted at 20 degrees. Then, a series of tests were conducted to assess the repeatability of measurement, accuracy and possible drift (test series #2). A third series (#3) focused on the sensitivity to temperature. For these three series of tests, the solutions were prepared by dissolving a determined mass of PEG in about 20 mL of deionised water. Masses were recorded using a high precision scale (to the 1/10,000th of a gram), hence minimizing the error on the concentration.

Table 1. Experimental program. All concentrations C are in grams of PEG per grams of water.

	T [°C]	PEG6k	PEG20k	PEG35k
Series 1: Correlation Concentration - Absorbance	20	C = 0.05, 0.1, 0.2, 0.3, 0.4, 0.5	C = 0.05, 0.1, 0.2, 0.3, 0.4, 0.5	C = 0.05, 0.1, 0.2, 0.3, 0.4
Series 2: Repeatability, drift, detection	20		C = 0.05, 0.5	
Series 3: Effect of temperature	20, 23, 26, 29		C = 0.05, 0.5	
Series 4: Effect of chemical species	20		C = 0.05, 0.5	

A fourth series investigates the possibility of interference in the measurement should there be other chemical species in solution. Tadza [5] showed that some of the cations and anions naturally present in soils can diffuse in the PEG solution. Considering the wide range of chemical species and concentrations in natural soils, it has been decided to dissolve PEG in synthetic seawater (composition given in Table 2) and check whether the absorbance measurement is affected by the presence of the chemical species.

Table 2. Chemical composition of artificial seawater used for this study. Concentrations in g/L

$NaCl$	24.53	$NaHCO_3$	0.201
$MgCl_2$	5.2	KBr	0.101
Na_2SO_4	4.09	H_3BO_3	0.027
$CaCl_2$	1.16	S_rCl_2	0.025
KCl	0.695	NaF	0.003

4 Results

4.1 Test series #1

Figure 2a depicts the evolution of absorbance in time for one concentration of PEG 20k. It shows that the equilibrium is not instantaneous; it takes some minutes to reach equilibrium. Similar curves were observed for all tests and data were recorded after equilibrium was reached. Figure 2b shows the absorbance measurements pertaining to the three PEG types and whole range of concentration. It can clearly be seen that the higher the PEG concentration, the lower the absorbance and the correlation can reasonably be considered linear. It is also remarkable that, for a given concentration, the absorbance value is independent from the PEG molecular mass.

Optical absorption is theoretically described using Beer Lambert's law which states that the absorbance is proportional to the concentration of attenuating species:

$$A=a \times b \times c \qquad (2)$$

where A is the absorbance, a is the attenuation coefficient, b is the path length and c is the concentration of attenuating species.

Here, the laser wavelength was chosen so that the laser beam be absorbed by the water rather than the PEG. In other words, water is the attenuating species and the experimental results are fully consistent with Beer Lambert's law. Indeed, as the PEG concentration increases, the water concentration (in grams of water per grams of solution) decreases and the absorbance decreases.

The results presented in Figure 2b validate the idea of inferring PEG concentration from absorbance measurement.

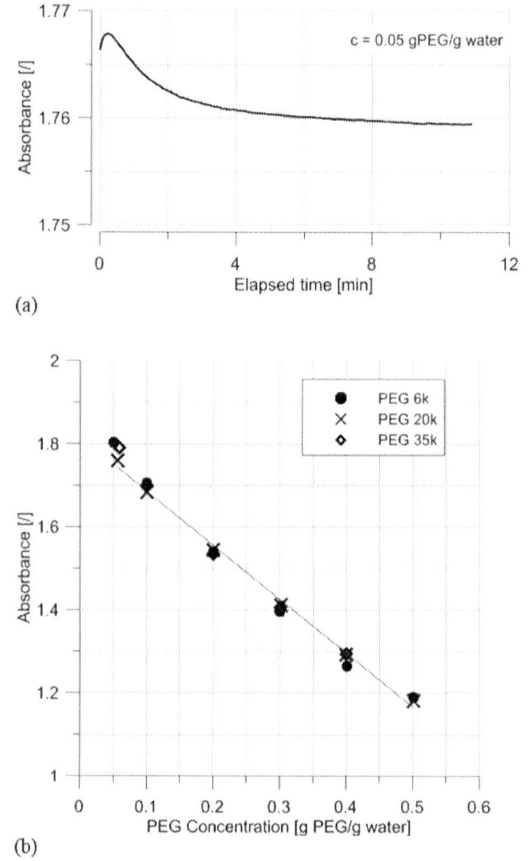

Figure 2. (a): evolution of absorbance with time for a PEG20k at 0.05 g PEG/g water. (b): Results of absorbance values (at equilibrium) for different concentrations of PEG6K, 20k and PEG35k.

4.2 Test series #2

This section deals with the estimation of errors on the absorbance measurement and on the estimation of PEG concentration. Several sources of uncertainties have been estimated, as summarised below:

-At any point in time, the resolution is 0.001 [no unit], yielding an absolute error of $\Delta A_{resolution} = \pm 0.0005$

-A repeatability exercise showed that the largest scattering (max value – min value = 0.009 [no unit]) occurs at the highest concentration (0.5 g PEG/g H_2O), yielding an absolute error of $\Delta A_{repeat} = \mp 0.0045$ [no unit].

-A drift exercise (continuous measurement of the same solution for 2 days) showed an amplitude of fluctuation (max value – min value) of about 0.007 [no unit], yielding an absolute error of $\Delta A_{drift} = \mp 0.0035$

These uncertainties are independent and an equivalent uncertainty, referred to as the equivalent resolution of the sensor, can be estimated as:

$$\Delta A = \sqrt{\Delta A^2_{resolution} + \Delta A^2_{repeat} + \Delta A^2_{drift}}$$
$$\Delta A = \pm 0.0057 \tag{3}$$

Note that the first drift exercise was conducted with the PEG solution in the laboratory fridge but the laser was kept outside the fridge. Significant variations were observed, which can be attributed to room temperature fluctuation (See Figure 3). Following this, the whole setup was placed in the temperature-controlled environment. This proved to reduce the fluctuation to a magnitude of 0.007.

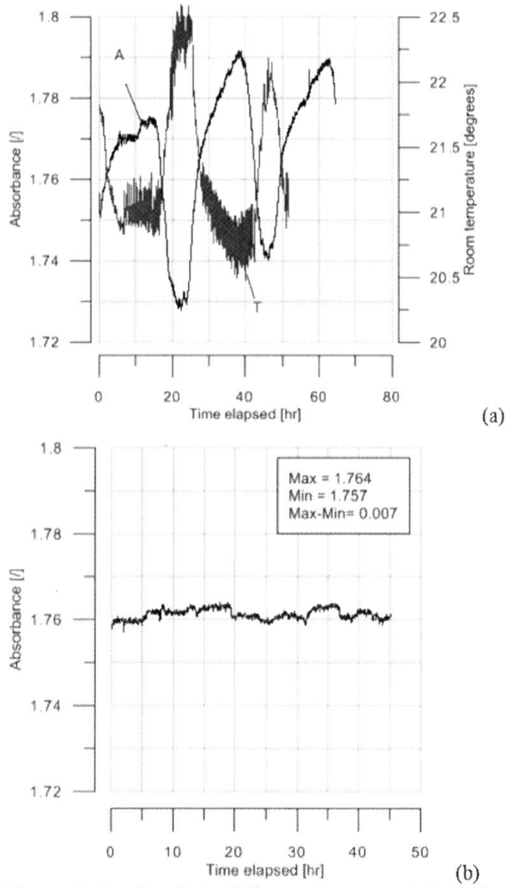

Figure 3. Results of the drift experiment: evolution of room temperature and absorbance with time with the laser and ITC controller outside (a) the temperature controlled environment. (b): evolution of absorbance with all equipment placed inside the temperature controlled environment.

Estimating the error made on the actual PEG concentration requires an assessment of the goodness of fit of the calibration curve, i.e. the best fit of the calibration data. The data showed in Figure 2 can be fitted with the following equation:

$$C = -0.726A + 1.33 \tag{4}$$

DOI: 10.1051/e3sconf/20160910004

where A is the absorbance of the cuvettes + PEG solution, and C is the PEG concentration. Such a fit returns a R^2 of 0.988 and a residual mean square of 2.98×10^{-4}.

The standard error being the square root of the residual mean square, it yields an absolute error of ±0.017 gr PEG/gr water. At the highest concentration of PEG (0.9 gr PEG/ gr water, see the Brix Index - concentration data by the authors in Figure 4), this represents a relative error of ±2%.

In comparison, the E Line 90 hand held refractometer by Bellingham + Stanley Ltd has a resolution of 0.2 unit of Brix Index (absolute error of ±0.1 unit of Brix Index) but the absolute error associated with the goodness of fit of the "Concentration-Brix Index" data is about 0.11 gr PEG/gr Water (from a residual mean square of 0.0117, Figure 4). At the highest concentration of PEG (0.9 gr PEG/ gr water), this represents a relative error of ±12%.

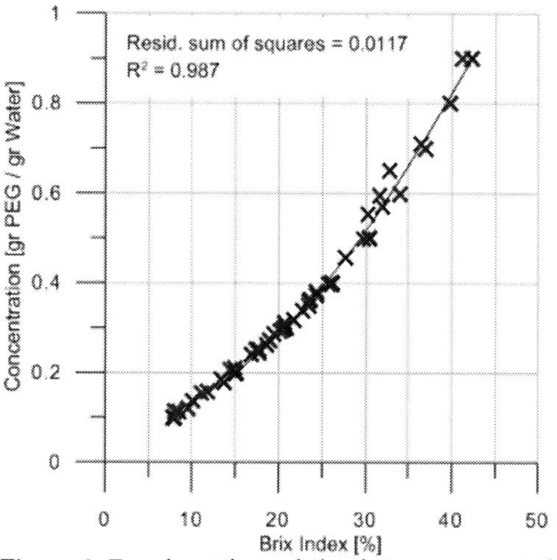

Figure 4. Experimental correlation between concentration of PEG solution and Brix Index measured with a hand held refractometer.

The standard error, as estimated from the residual mean square is an average value and does not necessarily reflect the actual accuracy of measurement, especially at the low end of concentrations.

In order to better assess accuracy at low concentrations, a detection test was performed, where a small amount of water was added to a PEG solution in order to reduce its concentration from 0.0562 to 0.0558 gr PEG/gr water. This is a difference of concentration of 0.0004 gr PEG/gr water. The absorbance changed from 1.759 to 1.803, which represents in increase of 0.044, well beyond the equivalent resolution of the sensor of 0.0057 (see section 4.2). This result suggests that the change in concentration was detected by the sensor.

In conclusion, this new sensor provides an indirect measurement of PEG concentration via absorbance measurement with a much better accuracy than hand held refractometer.

4.3 Test series #3

The drift experiment (Figure 3) demonstrated that temperature fluctuations do influence the absorbance measurement of a given PEG solution. Hence, additional tests were conducted at about 23, 26 and 29 degrees to assess this effect. Effectively, the results confirm the temperature sensitivity. Figure 5 shows that the absorbance progressively increases when the temperature rises. The magnitude of the increase is in the order to 8% for 9 degrees of heating (from 20 to 29 degrees). Such variations are consistent with data that can be found in the literature [6].

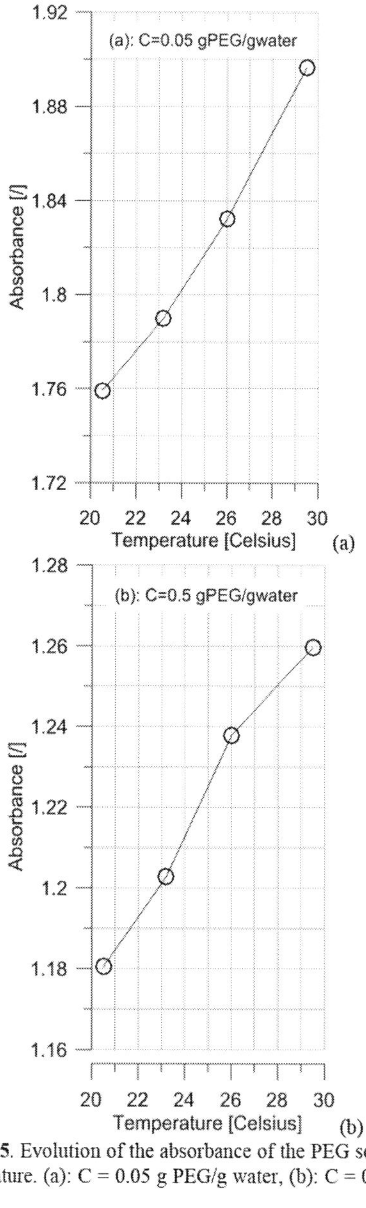

Figure 5. Evolution of the absorbance of the PEG solution with temperature. (a): C = 0.05 g PEG/g water, (b): C = 0.5 g PEG/g water.

4.4 Test series #4

Soil specimens often have chemical species dissolved in their pore water. The semi permeable membranes typically used for the osmotic method do not typically prevent these chemical species (cations and anions) from diffusing into the PEG solution because their molecular weight cut off is too high. Consequently, it is possible to find some cations and anions in the PEG solution post testing [5]. Test series #4 was conducted to assess whether these cations and anions could interfere with the measurement of PEG concentration with the new sensor. The absorbance measurements made in seawater are plotted in Figure 6 together with the line representing the best fit of the data showed in Figure 2. Note that the concentration values were corrected to be expressed in g PEG/g of water. With the seawater composition given in Table 2, 0.5 g PEG/g seawater is equivalent to 0.518 g PEG/g water. Figure 6 shows that the presence of cations and anions (at a relatively high concentration compared to typical pore water) does not affect the measurement of PEG concentration in a significant manner. The data points do not fall exactly on the line but this is attributed to natural scattering as visible in Figure 2.

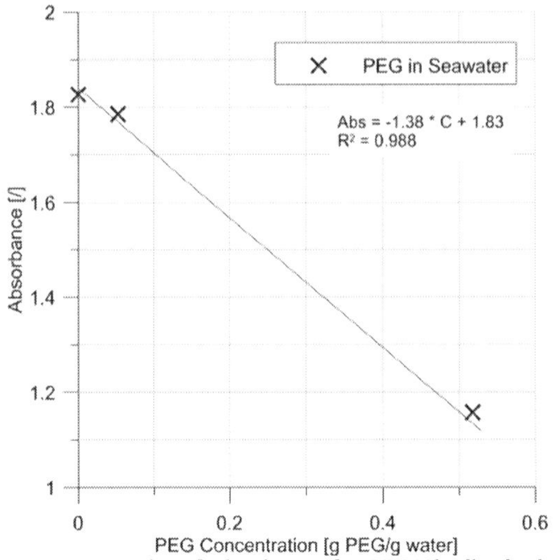

Figure 6. results of absorbance for PEG20k dissolved in seawater. The line represents the best fit of all data presented in Figure 1. Equation of the line is: Abs = -1.38C+1.83

5 Conclusions

This paper presents the validation and characteristics of a new sensor for the indirect measurement of PEG concentration. The sensor uses a laser, the intensity of which is absorbed by the water in the PEG solution. The input and output voltages are used to compute the absorbance, a quantity shown to be strongly correlated to the PEG concentration. The equivalent resolution of the new sensor is ±0.0057 while the error on the estimation of concentration is about ±2% at the high end of PEG concentration. A temperature dependence was observed and quantified. Finally, the tests conducted suggest that the chemical species that could diffuse from the soil to the PEG solution are unlikely to interfere in a significant manner with the measurement of PEG concentration.

References

1. G. Kassiff and A. Ben Shalom Géotechnique, **21**, 245-255 (1971)
2. P. Delage, M. Howat and Y.J. Cui. Eng. Geology, **50** (1-2) : 31-48 (1998).
3. R. Monroy, A. Ridley, K. Dineen, L. Zdravkovic (2007). ASTM Geotech. Test. J. 30, 1–7.
4. J. Lagerwerff, G. Ogata and H. Eagle, Science (80). 133, 1486–1487 (1961).
5. M. Tadza PhD Cardiff University (2011).
6. J.R. Collins. Phys Rev. **26**, 771-779 (1925)

Time response of two types of porous blocks for suction measurement

Michael Bardanis[1,a]

[1] *Edafos Engineering Consultants S.A., Laboratory, 2 Therapion st., 10444 Athens, Greece*

Abstract. Various types of porous blocks have been used for the indirect measurement of soil suction by various researchers, the usual experience being the importance of appropriate calibration, relatively low accuracy, and slow response. Two types of commercially available porous blocks were used. They were placed in samples undergoing either drying or inundation. Porous blocks were placed always with a tensiometer so that at the range of suctions measured by the tensiometer a comparison could be made between the response of the tensiometer and the porous block. The tests performed involved both slow drying in atmospheric conditions and fast inundation due to wetting. The experience gained from the tests performed is that in the three cases examined and contrary to common experience reported in the literature, the response of the porous blocks used was comparable to that of conventional tensiometers if not in fact faster.

1 Introduction

Earlier work by other researchers has indicated that porous block sensors can be used for the indirect measurement of suction in a wide range, yet often with relatively low accuracy and more importantly with quite slow response to suction changes actually occurring in the soil [1, 2, 3]. Three different tests are presented in the paper indicating that the two porous blocks that were used are quite accurate (in the range of suction values that a common tensiometer was used simultaneously) when calibrated sensors are used and have quite a fast response, certainly faster than that reported in the literature. Reference is first made to the porous blocks sensors used; then the tests performed are presented; finally the results are discussed in an attempt to identify common characteristics that allowed the performance of the sensors observed.

2 The porous blocks used and the tests performed

Two types of commercially available, off-the-shelf, porous blocks were used; the 5201 gypsum block by Soilmoisture Equipment Corp. and the MPS-2 frequency domain reflectometry (FDR) sensor by Decagon Devices, Inc. These sensors were installed in samples in the laboratory and in the field in combination with the Soilmoisture Equipment Corp. 2100F laboratory tensiometer and the 2725ARL jet-fill tensiometer respectively. Both tensiometers can measure suction in the range 0-90 kPa provided they have been carefully prepared with ceramic porous tips saturated and the tubes carefully deaired prior to their use.

Three tests have been performed and are reported in this paper. The first test involved the installation of gypsum blocks and a tensiometer in broken coarse sand undergoing inundation. The second test involved the installation of gypsum blocks, a tensiometer and other sensors in alluvial coarse sand undergoing inundation. The third test involved the installation of gypsum blocks, an MPS-2 sensor, a tensiometer and various other sensors in silty sand with gravel in the field, where both actual field variations, response after inundation and subsequent drying were monitored.

2.1 Gypsum blocks in a small-size sample of broken coarse sand undergoing inundation

The first test was performed in coarse sand. This was broken material of predominantly quartzitic sand with grains of limestone as well, passing through ASTM sieve No 4 (4.75mm) with only a small fraction of fines (0.5-1.5%) and usually poorly graded (Cu = 4-5, Cc = 1-2). The purpose of the test was to familiarise personnel with the use of the equipment and gain experience on the performance of the sensors used, in this case two gypsum blocks and the 2100F laboratory tensiometer.

The sample was hand-compacted using a hand operated compactor with standard Proctor test energy to an initial dry unit weight of 17 kN/m^3 at an initial water content of 3%. It was prepared in a 4 inch Proctor compaction mould with soil extending into the mould collar up to just a few millimetres below the end of the collar. Using steel tubes of the appropriate diameter, holes were opened from the sample surface for the gypsum blocks and the tensiometer to be installed approximately at the middle of the sample. Holes were

[a] Corresponding author: mbardanis@edafos.gr, lab@edafos.gr

backfilled using the material from the opening of the holes. Once the sensors were installed and the holes backfilled, the sample was covered with molten paraffin wax (Fig. 1a) and left for paraffin wax setting and insulation of the top surface of the sample despite cables and plastic tubing coming out (Fig. 1b). The sample was then left for sufficient time for sensor readings equilibration.

Once the readings of all sensors had come to equilibrium, the sample was inundated, bottom to top. As soon as the water surface covered the sample, small bubbles started coming out of a small hole at the periphery of the collar of the mould indicating the entrapment of air as inundation was taking place only from the bottom of the sample (Fig. 1c). The layer of the paraffin wax was therefore torn and removed (Fig. 1d) revealing collapse of the soil that had taken place under the paraffin wax layer. Removal took place within approximately two minutes from the beginning of the inundation. Suction readings from the tensiometer and the gypsum blocks are plotted in Fig. 2 with time. In Fig. 2a gypsum block readings have been turned into suction by use of the manufacturer's universal calibration curve for the gypsum blocks and in Fig. 2b by use of a sensor-specific calibration curve developed in the laboratory [4]. It can be seen that the manufacturer's universal calibration curve does not necessarily give values close to actual suction in the soil (assuming that the tensiometer in this particular range of suction values measures accurately). A sensor-specific calibration curve however gives values of suction fairly close (at least on average) to those measured by the tensiometer. Regarding time response, irrespective of the calibration curve used, equilibration with the new value after inundation corresponding to zero was comparable (if not faster) to that of the tensiometer. Gypsum blocks equilibrated within 15-16 minutes whereas the tensiometer equilibrated within 25 minutes.

It must be pointed out that the range of suction at the beginning of the inundation process in the particular test corresponded to a range of saturation close to full saturation for the gypsum block. Also, both blocks had been carefully saturated prior to their installation in the sample. The two gypsum blocks were placed in deionised water for 48 hrs prior to their installation in the sample. As it may be seen in Fig. 3, gypsum residue was observed at the bottom of the glass jar holding the gypsum blocks which at the time had become softer enough to require very careful handling during their installation in the sample. Finally, the whole test took place in a laboratory environment with the corresponding care, time, and personnel competence, that cannot necessarily be expected in field applications related to everyday agricultural or geotechnical practice. On the other hand the very fast response observed came as a surprise, given the experience reported in the literature, and served as motivation for further study of the time response of this particular type of sensors and other porous blocks currently available.

a) b) c) d)

Figure 1. a) Proctor test mould with the paraffin wax immediately after it was poured over the coarse sand, and b) after it set, c) after inundation of the whole mould (bubbles from a tiny hole can be seen), and d) after the wax layer was cut and partially removed (settlement of the sand pointed by arrow).

 DOI: 10.1051/e3sconf/20160910005

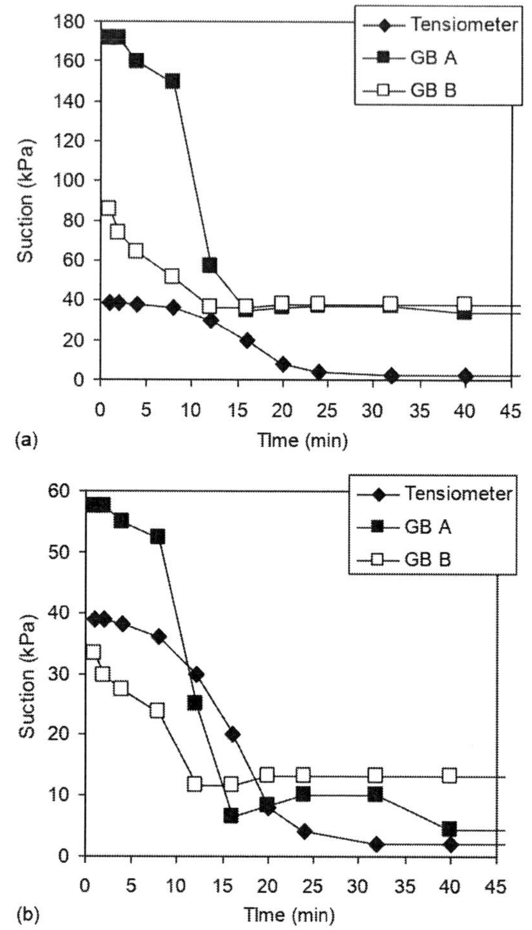

Figure 2. Suction measured by tensiometer and gypsum blocks GB A and GB B using a) manufacturer's universal calibration, curve and b) a sensor-specific calibration curve obtained in the laboratory.

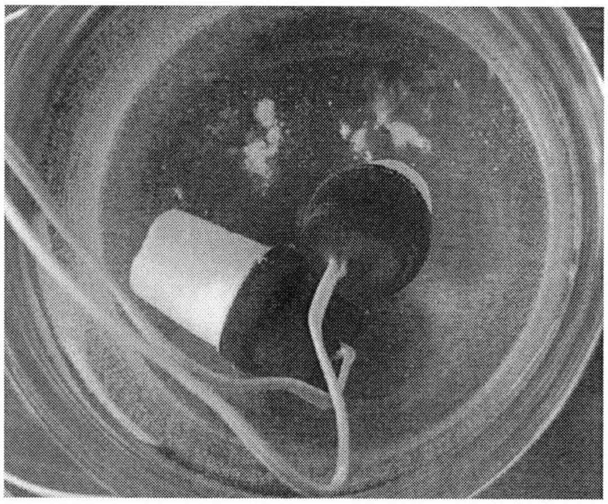

Figure 3. Gypsum blocks used in the first test at the end of their saturation prior to their installation in the sample. Gypsum residue may be clearly seen at the bottom of the glass jar holding the gypsum blocks in deionised water.

2.2 Gypsum blocks in a large sample of alluvial coarse sand undergoing inundation

The second test performed involved the installation of gypsum blocks, a tensiometer and other sensors in alluvial coarse sand undergoing inundation. The sample was of much larger size than in the previous test as it was prepared inside the cell of a large diameter direct shear device (45cm diameter by 30cm height). This device has already been presented as well as results from tests with it [5, 6, 7]. In the previous tests, the material used was broken sand from a quarry. In this test the readily available at the time fraction of the Ilarion dam rockfill passing through ASTM sieve No 4 was used. This material is an alluvium from Aliakmon river in northern Greece with grains rounded to well-rounded with varying lithological origins. In order to evaluate the performance of the sensors, vertical stress in the device was kept very low so as not to affect the sensors at this stage of the research (only 50 kPa). For similar reasons, compaction was limited only to that achieved by hand-compacting the material and not using an electric compactor used in the particular device in order to achieve high densities [5, 6].

A sample of the material was prepared in the cell of the apparatus with initial dry unit weight of 15 kN/m^3, and an initial water content of 5%. In the middle of the sample were installed: a laboratory tensiometer, two gypsum blocks and a Delta-T Devices volumetric water content sensor (ML-2 known as 'Theta Probe'). The gypsum blocks were again saturated prior to their use and then immersed in a slurry prepared by the non-plastic fines of the sample they would be installed in (Fig. 4a & 4b) in order to achieve better contact with the surrounding material (in a manner similar to the treatment of the porous tips of vibrating wire piezometers before installation). Once covered by the slurry, the gypsum blocks were carefully installed in the holes opened (Fig. 4b) and all holes were backfilled after sensors' installation before the rest of the sample was prepared in the cell (Fig. 4c).

After installation of the sensors and preparation of the sample, time was left for all sensors to come in equilibrium. Suction was small, measured 22-23 kPa by the tensiometer and 19-20 kPa by the gypsum blocks, and volumetric water content was 7.0 to 7.5%. These values exhibited negligible change due to the application of 50 kPa of vertical stress. After application of the vertical stress for 24hrs, inundation of the sample took place by filling the carriage of the direct shear device with water and the settlement due to collapse of the sample and the values of properties in the various sensors were monitored. In Figure 5a the vertical strain calculated from the settlement measured by displacement transducers placed on the top plate above the sample and the height of the sample before commencing inundation is plotted with time since commencing inundation of the sample. In Figure 5b, suction measured from both types of suction measurement sensors and volumetric water content are plotted against time during the same timeframe. The evolution of vertical strain due to inundation, suction and volumetric water content are in absolute agreement between them. In fact the time interval that rapid changes

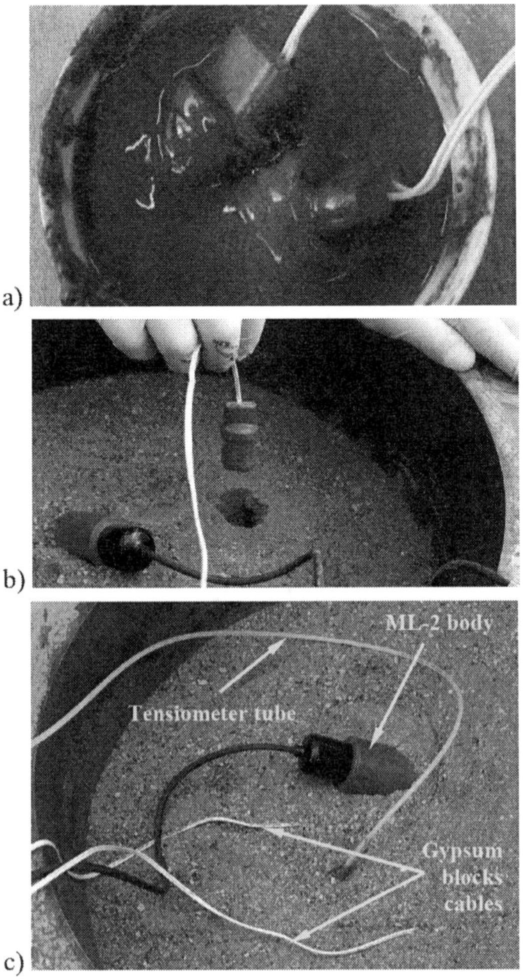

Figure 4. a) Gypsum blocks placed in a slurry of the fines of the sample after their saturation, b) slurry-covered gypsum block installed in the sample, and c) surface of sample at the level of sensors' installation.

of suction and volumetric water content occur is more narrow than the time interval that strain increases from 0 to 1.65%. This is attributed to the fact that the displacement transducers record strain from the beginning of the inundation when parts of the sample relatively far away from the location of the sensors start collapsing due to wetting, while the sensors record the changes essentially only when the wetting front reaches their location. In fact, the time the changes start to be recorded by the sensors is the time that the level of water in the carriage outside the cell has approximately reached the middle of the sample height where the sensors are located. Therefore, the macroscopic observation of collapse strain occurring due to inundation in the particular sample is in full agreement with the loss of approximately 20 kPa of suction and an increase of the volumetric water content from 7.5% to 35.5%. In this test too therefore the response of the gypsum blocks was very fast, again as in the previous test, at least as fast if not faster than that of the laboratory tensiometer.

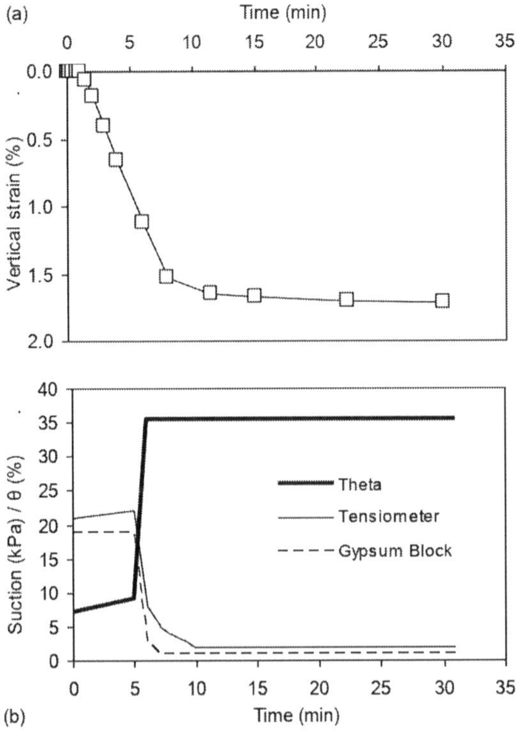

Figure 5. Evolution with time after start of inundation during the second test of a) collapse vertical strain, and b) suction from gypsum block and tensiometer and volumetric water content.

2.3 Porous blocks in silty sand with gravel undergoing inundation and drying in the field

The third test involved the installation of gypsum blocks, an MPS-2 sensor, a tensiometer and various other sensors in silty sand with gravel in the field, where both actual field variations, response after inundation and subsequent drying were monitored. A field array for measuring suction and volumetric water content was installed in shallow depth in silty sand with gravel at a semi-urban area close to Athens, Greece, Artemida region with climate typical of central Greece and in general of areas around the Mediterranean sea. Suction and volumetric water content were monitored for approximately one month in August 2013. The sensors installed included one gypsum block, one MPS-2 sensor, a jet-fill tensiometer, and two volumetric water content measurement sensors; the Delta-T Devices ML-2 'Theta Probe' and the Decagon Devices GS-3 (Fig. 6a & 6b). These constituted the main array. A smaller array only with one MPS-2 and one GS-3 was installed very close for reference of the conditions of the dry soil in the field during the wetting test in the main array. All sensors were installed at a depth of 25 to 30cm from ground surface.

The material the sensors were installed in is a silty sand with gravel and only traces of clay. Classification tests were performed on three samples taken during excavation of the main and the reference array location in order to remove the sensors from the ground. Gravel was mostly fine gravel with traces of coarse gravel amounting to 22% on average. Sand was on average 46% practically evenly shared between fractions of fine, medium and coarse sand. The fines were on average 32% and the clay content was found only 2.5-3.0% using the hydrometer test. Specific gravity was 2.68 and organic content 2.5-3.0%. The material is non-plastic. Classification according to USCS is SM. At the time of sensors removal that samples were taken, gravimetric water content of all samples was found 4% on average, with values ranging between 1.5 and 6.0%. Using the sand-cone method the dry unit weight was found to be 14 kN/m^3.

Prevailing meteorological conditions of the specific site have already been reported in detail [8]. These generally led to measured values of suction during monitoring period 2.5 MPa which never dropped below 1.4 MPa (Fig. 7a). Daily fluctuation of suction was in the order of 300 kPa, not matched by similar fluctuation in volumetric water content given the nature of the soil the sensors were installed in. A complete wetting test at the surface (Fig. 6c) revealed a time period of 11 days until prevailing suction and volumetric water content prior to wetting were obtained again (Fig. 7a).

What is of particular interest is again the response of the porous blocks used for suction measurement. Inundation took approximately 30 minutes. For the first twenty, none of the suction sensors recorded any change (as well as the volumetric water content sensors) indicating that the wetting front from the surface had not yet reached the depth of the sensors. Then, within a matter of 10 minutes, all sensors recorded zero values of suction and volumetric water content sensors practically the maximum values they ever recorded. Once again therefore both the gypsum blocks and the MPS-2 sensor responded as fast as the tensiometer during inundation, in this case a field one.

In this test however there was not only inundation but subsequent drying as well. As seen in Fig. 7b where the suction measurements from all sensors are plotted with time in the range of suction corresponding to the range measured by the tensiometer, the MPS-2 sensor responds similarly to the tensiometer with the measurements matching closely those of the tensiometer without the fluctuations observed in the tensiometer, while the gypsum block shows a similarly fast response but with the values of suction obtained from the manufacturer's calibration curve lying at a constant distance from the suction values obtained from the tensiometer. Once again therefore and in this case, two types of porous block sensors responded as fast as the tensiometer and in this case not only during inundation but also during drying.

3 Discussion of results

If one attempts to identify common features in the response and general performance of the porous block sensors used in the tests presented these may be:

- Both types of sensors responded as fast as the tensiometer (either laboratory or field one) both during inundation in three different cases presented and during drying in one case presented. In fact it may be supported on the basis of the two first examples that

a) b) c)

Figure 6. a) Open hole for the installation of the first set of sensors in the Artemida field array (top: MPS-2, middle: GS-3, bottom: Gypsum Block), b) dry surface of the field array before wetting, and c) after wetting (water seeped outside the plastic ring used to contain the free surface of water).

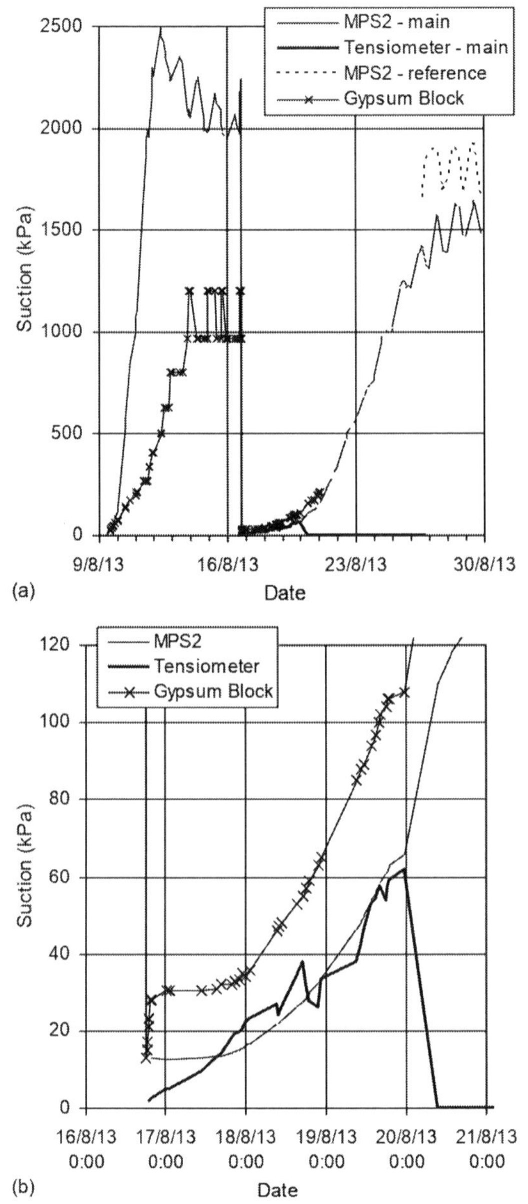

Figure 7. Suction measurements from all sensors: a) full scale, and b) reduced scale.

the gypsum blocks responded a little faster than the laboratory tensiometer.

- The MPS-2 sensor obtained values of suction very close to those of the field tensiometer in the third test. The gypsum blocks on the other hand did not obtain values close to the tensiometer (laboratory or field one) when using the manufacturer's calibration curve. Using a sensor specific calibration curve obtained for the particular sensors in the first test allowed much closer values of suction to be obtained on average from the two gypsum blocks installed.
- It must be pointed out that observed performance of the sensors was obtained using slurry-covered, carefully saturated sensors prior to their installation. Also all tests presented involved relatively coarse grained soils. Given however that expected changes in these soils are faster than those in clayey materials, it is anticipated that the response of the sensors used should be adequate –if not better– for more fine-grained materials too.
- Finally it is also pointed out that all the tests presented, although long in duration, did not reach the timeframes mentioned in the literature as insufficient for continuous use of gypsum blocks and required replacement of the sensors with new ones (approximately 2-3 years [2]).

4 Conclusions

Two types of commercially available, off-the-shelf, porous blocks were used in three different tests (two in the laboratory and one in the field). All sensors used were carefully fully saturated prior to their installation. All the tests also involved relatively coarse-grained soils (coarse sand and even more coarse-grained than that). Both types of sensors responded at least as fast as the tensiometers installed with them for comparison; the MPS-2 yielding values of suction fairly close to those of the tensiometer; the gypsum block not so when using the manufacturer's universal calibration curve, although was much better when using a sensor-specific calibration curve.

References

1. G. D. Aitchison, B. G. Richards, Moisture in Soils beneath Covered Areas, Butterworths, Australia, 1965, pp. 198-204.
2. A. J. Pullin, Comparative investigation of various soil water content measurement techniques for small scale agriculture, Seminar, School of Bioresources Engineering and Environmental Hydrology, University of KwaZulu-Natal, Pietermaritzburg, South Africa, 2004.
3. A. M. Ridley, 1993, PhD Thesis, Imperial College of Science, Technology and Medicine, University of London, 281 p.
4. M. E. Bardanis, 2016, PhD Thesis, in progress.
5. M. E. Bardanis, S. Cavounidis, G. Dounias, A new direct shear testing device for coarse grained soils and soils with large-size agglomerates. Measurement of shear strength and volume changes, Proc. 2nd Hell. Conf. on Dams and Reservoirs, Athens, Greece, 7-8 November 2013 (in greek).
6. M. E. Bardanis, Collapse of coarse grained materials due to inundation under one-dimensional compression conditions, Proc. 6th Int. Conf. on Unsaturated Soils, Sydney, Australia, 2-4 July 2014.
7. M. E. Bardanis, S. Grifiza, P. Kokkali, Shear strength of very coarse-grained materials and measurement of their coefficient of earth pressure at rest, Proc. 7th Hell. Conf. on Geotech. Engineering, Athens, Greece, 5-7 November 2014 (in greek).
8. M. E. Bardanis, Proc. Int. Symp. Shrink-swell processes in soils - Climate and constructions, 18-19 June 2015, Marne-la-Vallee, France, pp. 463-472.

E3S Web of Conferences 9, 10006 (2016) DOI: 10.1051/e3sconf/20160910006
E-UNSAT 2016

Effects of testing techniques on the SWRC of a partially saturated soil

Xiwei Zhang[1,2], Maria Mavroulidou[1,a] and Michael J. Gunn[1]

[1]*London South Bank University, SE1 0AA, UK*
[2]*Northeastern University, Shenyang, Liaoning, 110819, P.R. China*

Abstract. The paper studies the laboratory measurement of the Soil Water Retention Curve (SWRC) of statically compacted London Clay focusing on the effects of different specimen preparation methods and testing techniques to acquire a better understanding of how measurements can be affected by these factors. Testing methods include filter paper, two modified pressure plate systems, and a ceramic pressure membrane extractor. A discussion on the repeatability of the measurements is also made.

1 Introduction

In view of its major importance in the context of unsaturated soil mechanics, the soil water retention behaviour of natural soils in various forms (mostly after compaction or in reconstituted form) as expressed by their soil water retention curve (SWRC) has been extensively investigated. The form of the curve was shown to depend on a number of factors including soil structure, stress history and state, initial water content and void ratio of compacted soils, chemistry of the pore-fluid, volume change of the soil through swelling, shrinkage, loading or unloading, and wetting or drying paths.

In this study we focus the investigation on the possible effects of a number of factors related to experimental procedures and equipment, namely a) saturation procedures of compacted specimens prior to drying SWRC testing; b) comparison between filter paper and pressure plate measurements and (c) comparisons between different pressure plate type equipment. The latter comparisons also discuss repeatability of measurements; the discussion is then extended to filter paper measurements; as these rely on manual procedures, they were found to produce a large scatter of the data with conflicting reports on the accuracy of the technique. For instance as reported in Ridley [1] the BRE mentioned a value of +/- 25% of the mean, whereas Ridley [1] found an accuracy of +/-10%. This paper investigates this aspect further.

2 Materials and experimental methods

2.1 London Clay

The London Clay Formation is a well-developed marine geological formation found in the London Basin and Hampshire Basin, UK reaching an average thickness of 130 m. Up to five sedimentary cycles were identified within the London Clay (Divisions A –from the bottom- to E –to the top-), each linked to an initial marine transgression followed by a gradual shallowing of the sea. In its natural state, London Clay is a stiff overconsolidated clay, as in most parts of the London Basin, substantial erosion has taken place in the late Tertiary and Pleistocene times, removing the upper parts of the London Clay, and any other overlying Tertiary. London Clay is a material extensively encountered in construction in the London area and the South Eastern England (a very densely populated area with intensive industrial activity) including pavement construction, airports (e.g. Heathrow Terminal 5), underground railway (an example of recent engineering works being the Crossrail project), embankment and building foundation construction.

The soil used in this study was London Clay taken from an excavation at Westminster Bridge in the city of London and depths corresponding to B2 stratigraphic unit. Although at varying percentages according to location, a typical composition of London Clay with respect to its main three constituent minerals would be illite (70%), kaolinite (20%) and montmorillonite (10%) although exact composition changes according to location [2]. Due to the presence of illite and also, in particular, montmorillonite, London Clay is a shrinking clay with annual surface movements due to moisture variations typically of 50mm or more [3]. To identify the mineralogy of the particular London Clay used in this study XRD analysis was performed. The results are shown in Figures 1 and 2; the first figure discusses the mineral components as a whole, whereas the second focuses in particular on the clay composition. The summary of the XRD results are provided in Table 1.

The tests in this study were performed on the portion passing through a BS 425 μm sieve of air dried material

[a] Corresponding author: mavroum@lsbu.ac.uk

pulverised and ground in a soil grinder. The particle size distribution testing of the selected fraction shown in Figure 3 was performed in duplicate. The plasticity, specific gravity and standard Proctor compaction characteristics of this soil fraction as determined in this research are shown in Table 2.

Table 1. Mineralogical composition of the tested soil.

Mineral	%
Smectite	14.2
Illite	27.3
Kaolinite	8.19
Chlorite	4.91
Quartz	33.4
Feldspar	5.3
Plagioclase	4.1
Gypsum	1.2
Hematite	1.4

Table 2. Basic properties of the London Clay soil.

w (%)	w_L (%)	w_P (%)	G_s	w_{opt} (%)	ρ_{dmax} (g/cm^3)
31%	63.5	25.8	2.75	25.5	1.43

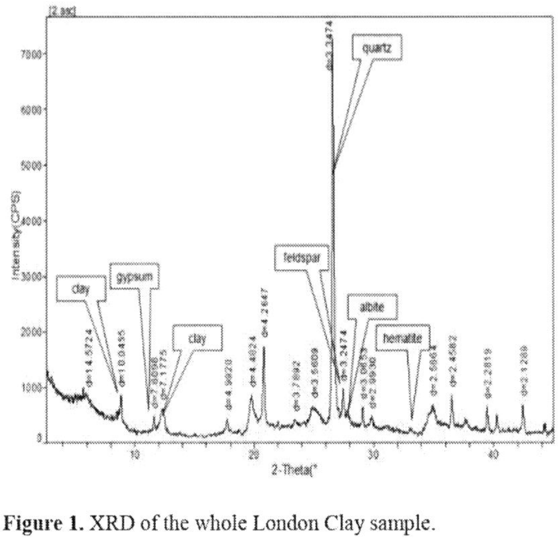

Figure 1. XRD of the whole London Clay sample.

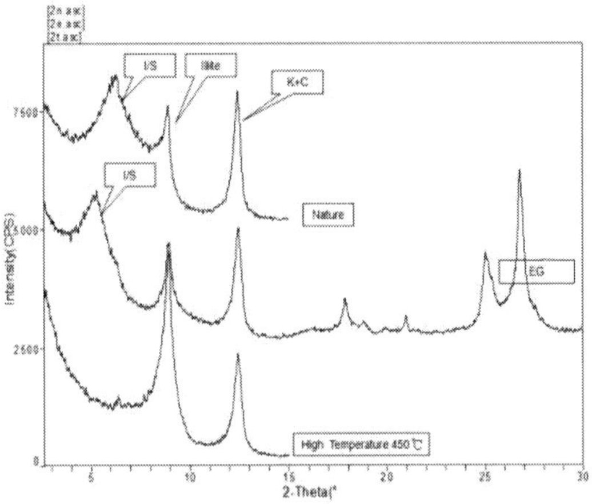

Figure 2. XRD analysis of the clay fraction.

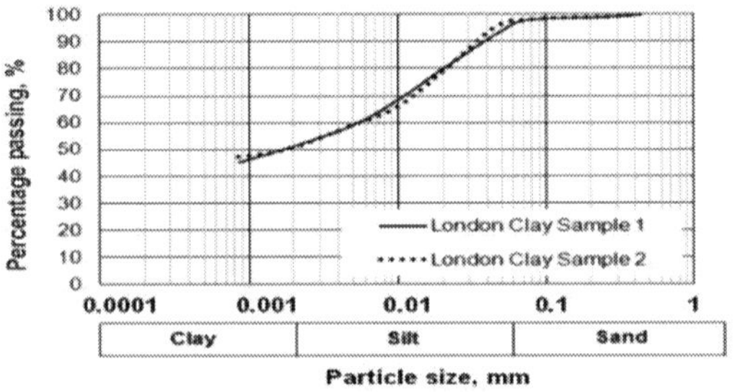

Figure 3. Particle size distribution of the tested soil.

2.2 Specimen preparation

The clay powder was thoroughly mixed with water at the required water content and left to hydrate in sealed bags for 72 h. A number of statically compacted specimens were prepared in two layers at the standard Proctor maximum density of the clay. For this, a standard loading frame at a loading rate of 1mm/min was used. Subsequently most specimens were subject to saturation whereas a few others were tested as compacted. We used the following different saturation conditions, whose effect is assessed in this paper:

-Some specimens were left in the rings to swell freely while absorbing water through underlying porous stones standing in water.

-Some specimens were kept within standard oedometer rings (75mm diameter and 20mm height approximately) and sandwiched between filter papers and porous stones in a confining frame immersed in water, especially designed to suppress any swelling during saturation, thus ensuring consistent void ratios at the beginning of each test (Fig 4). Note that it was still difficult to obtain totally identical void ratios because of some slight expansion after extruding the specimens from the cutting rings.

-To improve saturation of confined specimens some specimens (held in the confining frame) were placed in a sealed bigger cell and connected to a suction pump that vacuumed the sample for 15 min and water was filled slowly from the bottom of the cell. This saturation process lasted for 7 days and enabled degrees of

DOI: 10.1051/e3sconf/20160910006

saturation of as high as 95-99% to be reached.

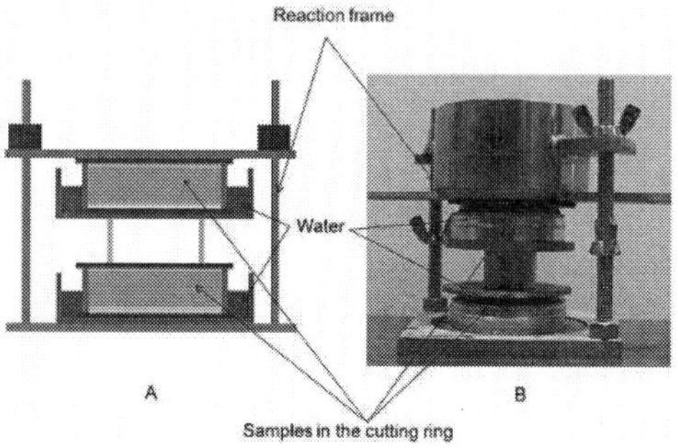

Figure 4. Saturation frame.

2.3 Testing techniques

The filter paper procedure used in the present research was based on ASTM D-5298 [4] with the following two main differences based on procedures developed at Imperial College, London, and presented in a number of doctoral theses (e.g. [5,6]): (i) one single soil specimen of oedometer size was used, sandwiched between two Perspex disks; three filter papers were used on each side of the specimen between the soil and the Perspex disks (ASTM D-5298 uses two equal pieces soil between which three filter papers are sandwiched); (ii) the calibration formulae according to Chandler and Gutierrez [7] and Dineen [4] were used for initially dry and wet filter paper respectively, as shown in Table 3. To obtain the drying curves the soil specimen was left to air-dry slowly to the desired water content for each measurement point, until its residual water content was reached. One single soil specimen (and not multiple 'identical' specimens at different initial water contents) was used to determine each SWRC. For the wetting curves, three wet filter papers held by tweezers were placed on either side of the specimen. The papers were soaked in distilled water until a mass of 2 g of water in total was absorbed by the papers. This amount of water on each side of the specimen was found to be appropriate to alleviate measurement errors due to evaporation or excessively small changes in suction.

Table 3. Calibration relationships for the filter paper.

	Filter paper water content, w_f, %	**Matric suction, *s*, kPa**
Initially dry paper	$w_f \leq 47$	$s = 10^{(4.842-0.0622w_f)}$
	$w_f > 47$	$s = 10^{(6.05-2.48\log_{10} w_f)}$
Initially wet paper	$w_f \leq 15.47$	$s = 10^{(4.842-0.0622w_f)}$
	$15.47 < w_f \leq 57.2$	$s = 10^{(4.573-0.0449w_f)}$
	$w_f > 57.2$	$s = 10^{(2.904-0.0158w_f)}$

The pressure plate apparatus system designed at London South Bank University (LSBU) for this study, consisted of a rigid wall permeameter cell with a modified pedestal, incorporating a High Air Entry Value (HAEV) disk for the application of axis translation suction control technique. To prevent air leakage the HAEV disk was integrated into the base pedestal plate of the apparatus. The system was completed with a burette and a flushing device (Fig. 5).

The ceramic disc was saturated before testing. After assembling the cell the target suction was achieved by changing the air pressure accordingly using an air pressure regulator. The initial reading of the burette, was then recorded and the water outflow in the burette monitored until equilibrium was achieved (typically, 3-7 days). After suction equilibrium for each suction level the specimen was taken out of the cell for mass and volume measurements (using Vernier callipers). Following these, the specimen was placed back into the cell for the next suction level. By repeating this procedure with a stepwise increase or decrease in suction the drying or wetting branch of the SWRC were obtained accordingly.

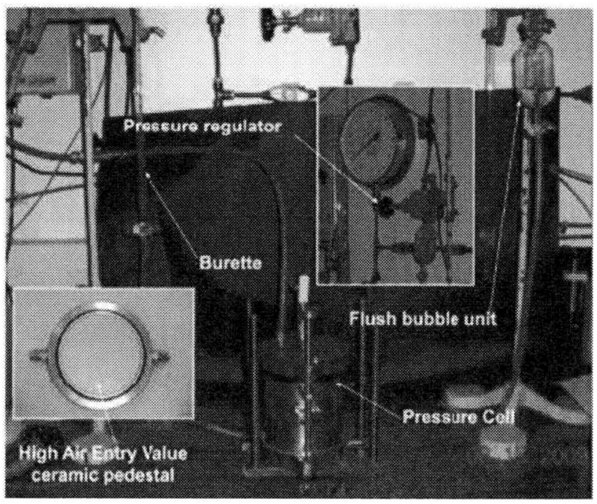

(a)

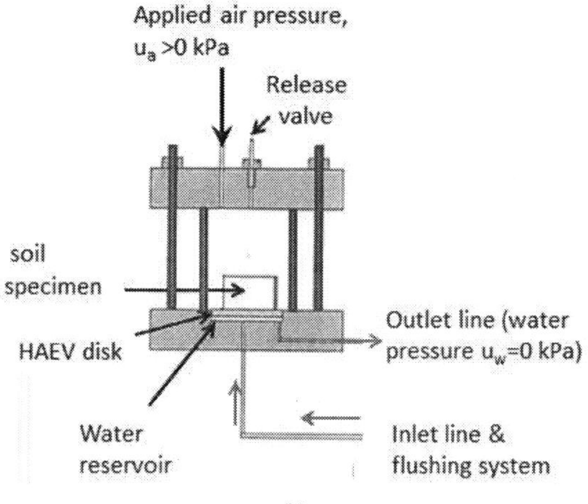

(b)

Figure 5. Pressure plate: (a) apparatus designed at LSBU; (b) schematics with applied boundary conditions.

Two additional devices available at Beijing Jiaotong University were also used for the sake of comparisons, namely (a) a ceramic pressure membrane extractor (the most usual method of determining the SWRC in the field of soils science and hydrology); the apparatus has a membrane to the base and side of the HAEV disk, especially designed for water drainage. A wire mesh is placed between the membrane and the disk to keep an open space for water to flow. The membrane is connected to an outflow tube through a special fitting thus enabling the collected water to escape from the system (b) SWC-150 Fredlund SWCC Device, a pressure plate type apparatus described in Padilla et al. [8].

3 Experimental results

For the sake of brevity in the following presentation of the results the notation *w-s*, S_r-*s*, and *e-s* will refer respectively to the relationships between (a) gravimetric water content and matric suction, (b) degree of saturation and matric suction and (c) void ratio and matric suction. The first set of comparisons refers to the different saturation conditions used (Fig 6 (a)-(c)). These were confined saturation (accompanied by suction/back pressure through a GDS Instruments controller/pump) vs. unconfined saturation. For the sake of comparison results from a third specimen subjected to drying from as compacted state (i.e. without saturation) are shown. The results of the two methods of saturation are very different for all curves plotted. The *w-s* curve of the specimen saturated confined and subjected to back-pressure plots higher than that of the unconfined saturation; this is due to the fact that air was removed by vacuuming allowing better saturation of the former specimen and higher water contents (it is commonly seen that upon wetting the curves of specimens may not achieve degrees of saturation of 100% due to entrapped air; this was overcome by the application of suction). The *e-s* curves of the confined saturation specimen plot clearly lower than those of the specimen saturated under unconfined conditions. Whereas this could have been attributed to the fact that drying of the confined specimen started at lower void ratios as no expansion was allowed during wetting, it can be observed that all void ratios of this specimen plot lower that those of the specimen subject to drying from as-compacted state, (note that the drying curve of the unconfined specimen converges with that of the as-compacted specimen). A different explanation for the lower void ratios of the confined specimen justifying the discrepancy could perhaps be the effect of vacuuming the specimen prior to applying a back-pressure which could have induced some shrinkage to the swelling-shrinking London Clay soil. As such saturation conditions are not consistent with in situ conditions and result in differences in the SWRC curves, it could be recommended that they should be avoided, even if they can lead to higher degrees of saturation.

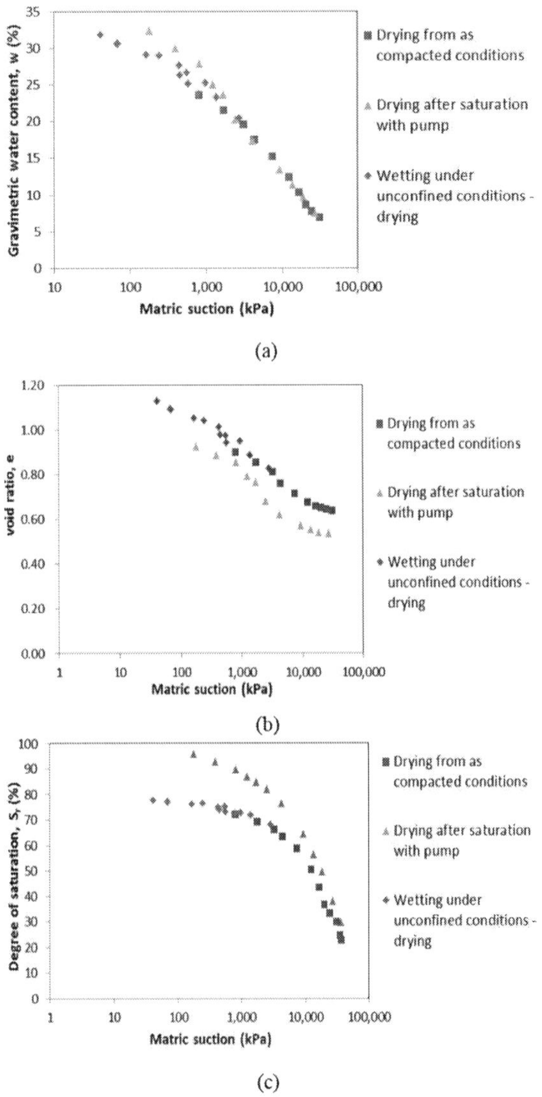

Figure 6. Comparison between different saturation conditions: (a) w-s; (b) e-s; (c)S_r-s.

Figure 7 shows a comparison between filter paper and pressure plate results (LSBU pressure plate apparatus). It can be seen that although the curves in general tend to converge at lower suctions, they diverge in general after air entry/maximum curvature points (see Fig 7(a) and (c)). Whereas for the *w-s* wetting curves this difference around the suction reversal could have been attributed to some scanning effect (as the path followed was a drying-wetting path) this cannot be said for the drying curves too. On the other hand it is notable that the *e-s* drying curves coincide in the common suction region but the wetting curves do not. Namely, the void ratio changes less along the pressure plate wetting curve than is for the filter paper specimen. This trend tends to extend far beyond the reversal points. It is possible that there is some effect of the boundary conditions /air pressure on the lower void ratio change during wetting. The above differences are reflected in the S_r-s curves of the two specimens (Fig 7(c)).

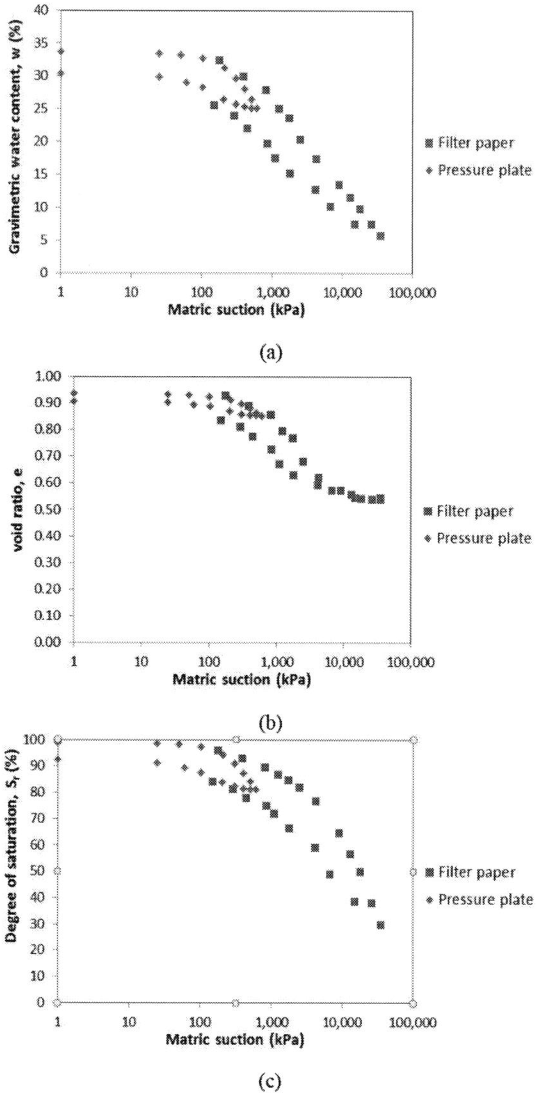

Figure 7. Comparison between pressure plate and filter paper results: (a) w-s; (b) e-s; (c)S_r-s.

The final investigations concerned the repeatability of the measurements (Fig 8 (a)-(c) and 9(a)-(b)). First, we present results of four filter paper specimens compacted identically so that at the end of compaction they had compaction void ratios differing by a maximum of 0.008 and compaction water contents differing by a maximum of 0.4% (Fig 8(a)-(b)). The repeatability of the *w-s* measurements is found to be excellent, with maximum differences of +/- 1.1% from the mean. On the other hand, greater differences in the *e-s* curves can be seen, with the maximum differences from the mean being +/- 4.5%. In addition to human error during calliper reading this can also be due to the uncontrolled expansion of the specimens when extracted out of the rings (note that these specimens were saturated under constant volume conditions, which initially helped consistence in the void ratios across specimens). The maximum differences in the S_r-s values are approximately +/- 2.8%, which are very small compared to previously reported values.

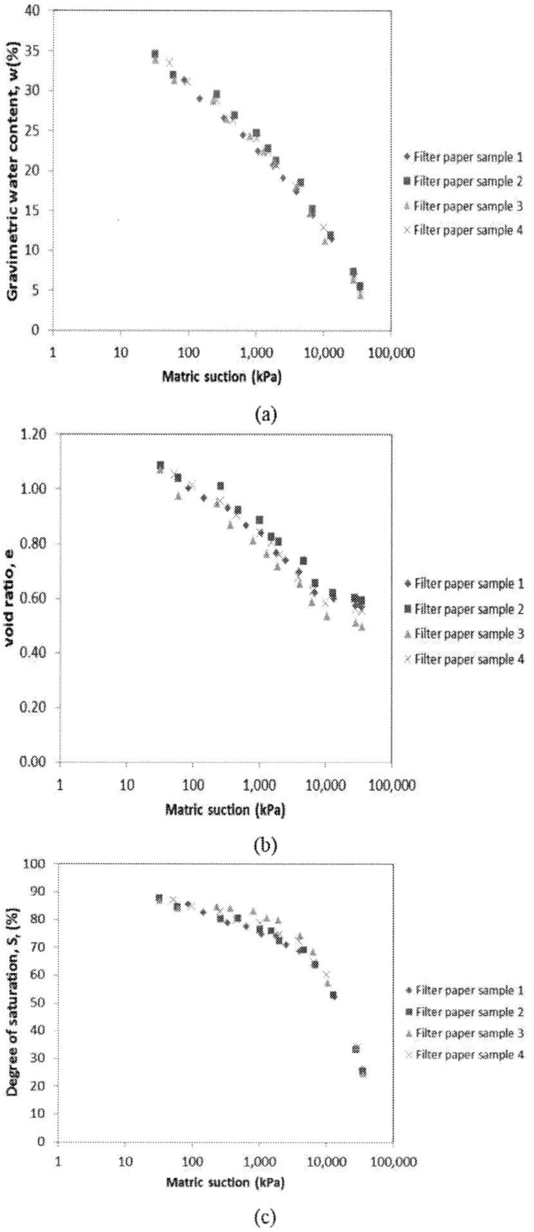

Figure 8. Repeatability of filter paper results: (a) w-s; (b) e-s; (c)S_r-s.

Repeatability assessment was also performed using different pressure plate equipment of similar but not identical design (Fig 9(a)). Comparing LSBU pressure plate with SWC 150 Fredlund SWRC system it can be seen that the available results in terms of *w-s* are almost identical of maximum gravimetric water content differences of approximately 0.9% (Fig 9(a)); interestingly, the results of a pressure membrane extractor and pressure plate shown in Figure 9(b) are also very similar with maximum differences in the values of gravimetric water content of less than 1%.

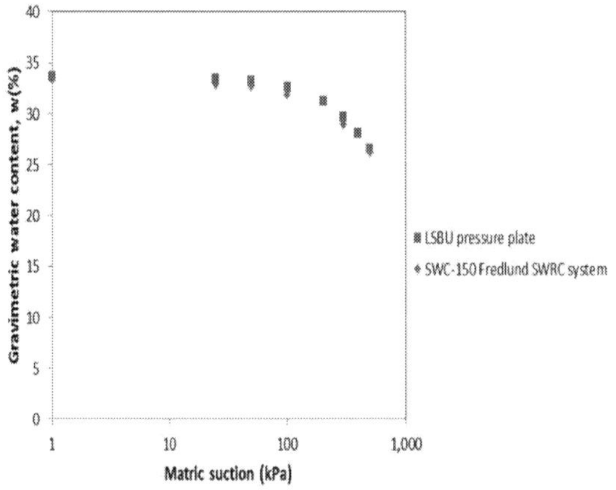

(a)

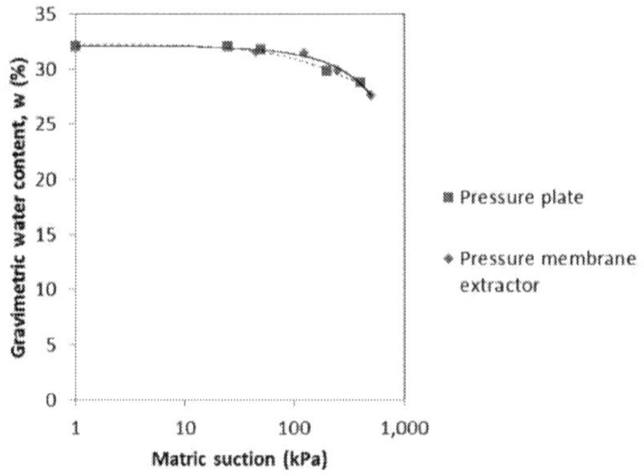

(b)

Figure 9. Repeatability of measurements using different axis translation systems: (a)LSBU pressure plate vs SWC 150 system; (b) pressure membrane extractor vs pressure plate.

4 Further developments

In order to directly measure suction during SWRC testing, a high suction capacity mini suction probe nominally calibrated to measure pressures of +/- 1500 kPa originally developed by GDS Instruments is currently being tested at London South Bank University. The probe has an overall diameter of 6.4 mm, an inner pressure chamber of 10 mm^3, a maximum nominal voltage of 125 mV and an accuracy of 0.5% in terms of Full Scale Output (FSO). Current testing is proving the probe during London Clay and silt testing in axis translation controlled systems.

5 Conclusions

The paper studied the laboratory measurement of the Soil Water Retention Curve (SWRC) of statically compacted London Clay focusing on the effects of different specimen preparation methods and testing techniques to acquire a better understanding of how measurements can be affected by these factors. It was found that the repeatability of the measurements performed by the same experienced operator and following rigorous and consistent procedures was excellent. This was also the case when similar but not identical devices were used. On the other hand different saturation methods and filter paper testing versus pressure plate type measurements resulted in differences further demonstrating the need to adopt procedures close to in situ conditions, consistent with discussions in the unsaturated soil literature. To this effect, the use of a novel high capacity tensiometer is currently being tested as an alternative suction measurement technique for future tests.

Acknowledgements

This study constituted part of a larger research programme, funded by the UK Engineering and Physical Sciences Research Council (EPSRC) through grant EP/E037305/1. All tests described in this study were conducted by the first author during his doctoral studies at London South Bank University under the supervision of the co-authors. Two sets of results were conducted at Beijing University of Technology. The authors would like to thank Prof. Liu Jiankun, Beijing Jiaotong University and Prof. Yao Aijun, for their permission to conduct the latter tests in their laboratory and for their advice and support during the related experimental work.

References

1. A. M. Ridley, *The measurement of soil moisture suction*. PhD thesis (Imperial College, University of London, 1993).
2. P.G.H. Boswell, The Contribution of Clay Mineralogy to the Study of the Diagenesis of Sediments. Clay Minerals **1** (8): 246-251 (1951)
3. R. Whitlow, *Basic Soil Mechanics*, 4th Ed. (Prentice, Harlow, 2001)
4. American Society for Testing and Materials (ASTM), *Standard Test Method for Measurement of Soil Potential (Suction) Using Filter Paper, ASTM D 5298: 1994* (ASTM, West Conshohocken, PA 19428, 1994)
5. K. Dineen, *The influence of soil suction on compressibility and swelling*. PhD Thesis (Imperial College, University of London, 1997)
6. M. Melgarejo-Corredor, *Laboratory and numerical investigations of soil retention curves*. PhD thesis (Imperial College, University of London, 2004)
7. R. J. Chandler, C. L. Gutierrez, The filter paper method of suction measurement, Géotechnique, **36**(2): 265-268 (1986)
8. J.M. Padilla, Y.Y. Perera, W.N. Houston, D.G. Fredlund, A new soil-water characteristic curve device In A. Tarantino, E. Romero & Y.J. Cui (eds), *Advanced Experimental Unsaturated Soil Mechanics - Proceedings of an International Symposium, Trento, Italy, 27-29 June 2005*:15-22. (Taylor & Francis Group, London, 2005).

Investigating soil water retention characteristics at high suctions using relative humidity control

Vasileios Mantikos[1,a], Steven Ackerley[1], Andrew Kirkham[1], Aikaterini Tsiampousi[1], David M.G. Taborda[1] and Jamie Standing[1]

[1]*Department of Civil and Environmental Engineering, Imperial College, London, United Kingdom*

Abstract. A technique for controlling relative humidity (RH) is presented, which involves supplying a sealed chamber with a continuous flow of air at a computer-regulated RH. The desired value of RH is achieved by mixing dry and wet air at appropriate volumes and is measured for servo-control at three locations in the chamber with capacitive RH sensors and checked with a sensitive VAISALA sensor. The setup is capable of controlling RH steadily and continuously with a deviation of less than 0.2% RH. The technique was adopted to determine wetting soil-water retention curves (SWRC) of statically compacted London Clay, under both free-swelling and constant volume conditions. The RH within the chamber was increased in a step-wise fashion, with each step maintained until vapour equilibrium between the chamber atmosphere and the soil samples was established. Independent filter paper measurements further validate the method, while the obtained retention curves complement those available in the literature for lower ranges of suction.

1 Introduction

The soil-water retention curve (SWRC) relates suction to water content or degree of saturation and has long been shown to be highly non-linear and hysteretic. Traditionally, measurement of the SWRC in the laboratory in the laboratory involves exposing soil samples to a certain suction and recording their mass through a sequence of suction levels. The results are commonly presented by plotting the calculated degree of saturation or volumetric water content against suction in a semi-logarithmic plot.

The retention capacity of soils is specific volume dependent [1, 2]. Ideally volume should be either controlled or measured while wetting/drying, especially for expansive clays, where the volume changes may be significant. Alternatively, the SWRC can be presented in terms of gravimetric water content, although it does not provide any information on volumetric change effects.

To obtain the SWRC, knowledge of the current value of suction is necessary. Suction can be measured in a direct or indirect manner. Measurement techniques suitable to determine suction include dew-point hygrometer, transistor/thermistor psychrometer, thermo-couple psychrometer, filter paper method, tensiometers, electrical/thermal conductivity sensors, resistance/capacitive sensors [3, 4]. Many methods for measuring total suction actually acquire a value for relative humidity which is then converted to suction using Kelvin's thermodynamic law.

The basic concepts behind suction control can be classified in three general types. A common method for controlling suction is the axis-translation technique, where the air and water pressures in the system are increased by the same amount keeping matrix suction constant, until water pressure is positive and can be measured conventionally [5, 6]. Another way to control suction is the osmotic technique, where the sample is in contact with a salt solution (e.g. PEG, polyethylene glycol) through a semi-permeable membrane, achieving a value of matrix suction of up to 12 MPa [7, 8]. Finally, using Kelvin's law in reverse, it is possible to control the total suction in a soil specimen by controlling the relative humidity of its environment [9]. This is usually achieved by means of thermodynamic equilibrium between the air in a closed chamber and an unsaturated acid or a saturated salt solution. Another method for RH control described by Lu and Likos [4] is the divided-flow method which relies on the proportionate mixing of saturated and dry air. Variations of the latter method have also been adopted in sorption balances and gravimetric analysers in Material sciences and the food industry.

This paper focuses on the development of a suction control equipment in the geotechnical laboratory of Imperial College London. Suction values in the range of 10 MPa to 300 MPa are achieved inside sealed chambers through servo-control of relative humidity (RH). A particular application of the technique relates to the determination of SWRCs of expansive clays over high suction ranges, e.g. compacted bentonite used as buffer in nuclear waste disposal. The divided-flow method and the

[a] Corresponding author: vasileios.mantikos12@imperial.ac.uk

capacitive sensors were selected to control and measure RH respectively. A brief description of the development of the equipment is presented in the current paper, followed by the results of its application for two common soils. A comparison with the filter paper method is also carried out.

2 Development of equipment

2.1 Description

Pictures of the components and a schematic depiction of the arrangement are presented in Figures 1 and 2 respectively. The equipment consists of a manostat, two water excluders, a water filled saturation chamber, a perforated steel tube (bubbler), plastic tubes, two valves, a sealed chamber, a perforated tray, a fan, a rod with three attached temperature and RH capacitive sensors, a constant volume cell, an auxiliary brass cell for compaction, a micro-computer control system, a data logger and a computer.

2.2 Air-flow system

The divided-flow principle is employed in order to apply suction in samples in the range of 10 MPa to 300 MPa. For reasons of convenience the circulating gas is compressed air from the main laboratory supply at a RH of approximately 5%. The air supply is reduced to a safe pressure with a manostat. The air is routed through a water exclusion chamber, where any condensed moisture is collected at its base. The flow is then divided into two streams. One is injected through a perforated metal tube through a water bath to be saturated. The "wet" stream is routed through a second water excluder, before both "dry" and "wet" streams enter the sealed chambers at flows individually controlled by two valves, at a computer-controlled wet/dry ratio. Temperature and relative humidity in the chambers are recorded using capacitive probes connected to data loggers, giving feedback to the computer program controlling the valves. The chamber is sealed, but for a small diameter vent to prevent potential pressure build up.

Having an open system instead of recycling the air and circulating it back into the system improved control and reduced the response time of the system. Having a continuous flow of controlled-humidity air provided better control rather than injecting dry or wet air only for corrections.

The total air-flow into the first excluder is maintained constant using a manual manostat. The manostat reduces the pressure of the supplied air to 15-20 kPa. Higher air pressure introduced in the system might cause failure of the seals at the connections of various components, with the valves being the most sensitive parts. If, on the other hand, the pressure is too low, humidity flow might reverse, making the system sensitive to changes in the laboratory environment. Maintenance and cleaning of all tubes and components should be performed regularly to avoid blocking and clogging of the flow path.

To avoid condensation, water excluders were introduced in the system, the length of tubes was reduced, sagging of the tubing was avoided and metal tube connectors were replaced with plastic ones.

Figure 1. Two chambers in operation.

2.3 Constant volume cell

A perforated steel cell was developed to keep the volume constant during wetting, while allowing moisture to travel through the sample. The cell concept is similar to those used in CIEMAT [10]. The cell was modified at Imperial College so that the sample is placed in a ring and confined at the top and bottom with perforated discs fitted with porous stones (Figure 3).

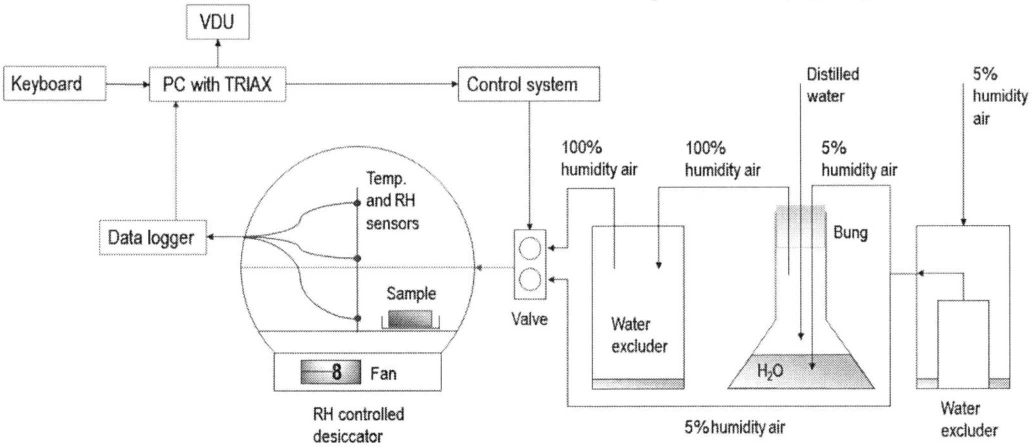

Figure 2. Arrangement of the IC RH Control system [11].

Confining rings of different heights can be employed according to the needs of the experiment. The initial ring was manufactured at a 15mm height. Six screws hold the discs and ring together, but also raise the base of the cell above the perforated tray thus allowing circulation of air. The steel used for the construction of the cell was the corrosion resistant AMINOX-AS-1®, to counter the corrosive properties that some soils may exhibit, e.g. bentonite. The choice of 85mm diameter cell allows simultaneous fitting of four samples in the chamber. Considering that the available scale has a resolution of 0.01g and a capacity of 3 kg, the cell provides a sample large enough to obtain a resolution of 0.015% in water content measurements assuming a minimum 1.3 kg/cm^3 dry density, while the total mass of soil and cell does not exceed 2.5kg. Considering the use of the cell for expansive soils, it was found that 10 mm thick walls would deform less than 0.05 mm under 15MPa of swelling pressure. This leads to the cell design shown in Figure 3.

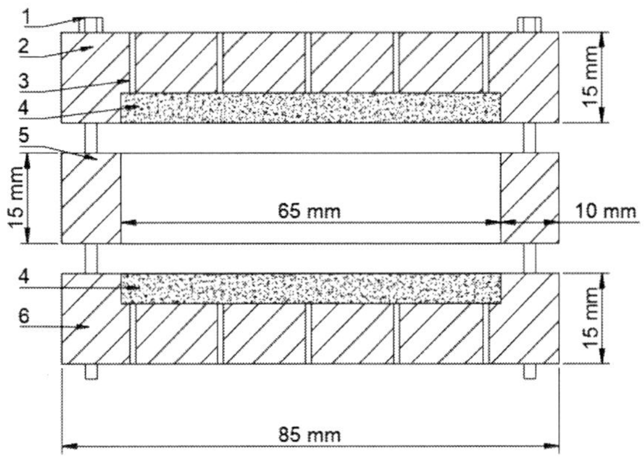

Figure 3. Dimensions of the IC constant volume cell. 1. Long screws, 2. Top perforated disc, 3. Perforations, 4. Coarse porous stones, 5. Confining ring, 6. Base perforated disc.

3 Validation tests

3.1 Control operation

A series of increasing/decreasing humidity cycles has been performed to check the response of the system. The parameters examined are the standard deviation of RH once at equilibrium, the system's stability over changes in laboratory humidity and temperature, and stability against loss of services (e.g. power or air supply and computer communication errors).

The humidity in the chamber is compared to the humidity in the laboratory in Figure 4. It is observed that variations in the environment of the laboratory do not affect the RH conditions in the chamber. The spikes in the chamber measurements represent occasions of opening the lid of the chamber. During a 5-day period with no computer control after a computer communication system crash, the RH in the chamber remained unchanged, suggesting that the valves continued regulating RH successfully at the previously set target value.

3.2 Capacitive sensors

RH and temperature are recorded using common capacitive sensors of 1% RH accuracy and 0.001% RH resolution. The sensors are calibrated against a pre-calibrated high-accuracy VAISALA sensor. To reduce uncertainty three sensors were used simultaneously in the chambers. The fluctuation of a single sensor was limited to 0.2% RH. The deviation of RH change among the sensors was less than 0.1% RH, showing a similar trend for each sensor with 0.1% deviation (Figure 5).

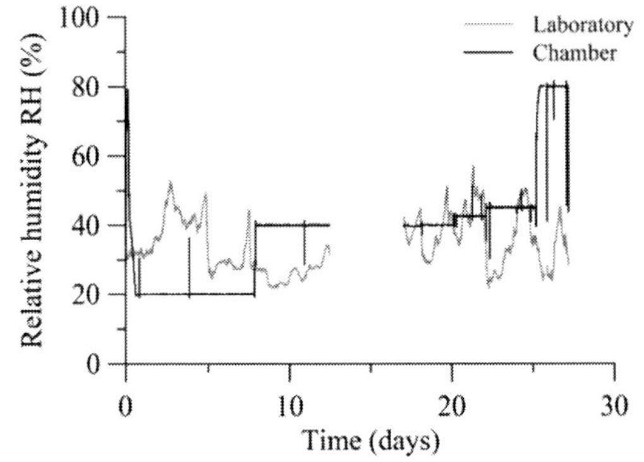

Figure 4. Comparison between RH of laboratory and environmental chamber.

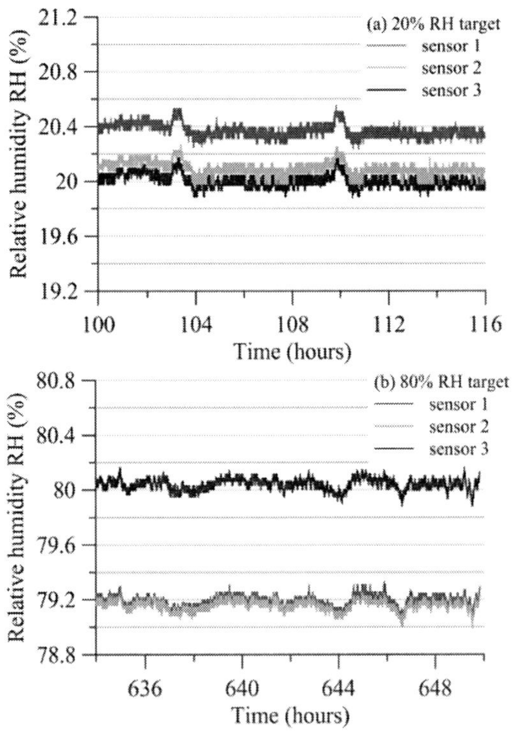

Figure 5. RH readings inside chamber at a target value of (a) 20% RH and (b) 80% RH.

DOI: 10.1051/e3sconf/20160910007

In Figure 5 the RH readings of the three sensors in a chamber are presented, with sensor 1 being the highest. Sensor 3 was positioned at the level of the perforated tray, at the same level that the soil samples would be, and therefore it was chosen as the control sensor for feedback to the computer program. While sensor 3 kept reading a value around the target RH (black line), the other two sensors drifted either towards a higher (Figure 5a) or a lower (Figure 5b) value of RH, for low and high target RH respectively.

3.3 Constant volume cell

A soil sample was compacted and placed in the constant volume cell to investigate its effectiveness in terms of hydration. The soil used was oven-dried, pulverised London Clay, compacted at 0.4% water content under a 1.8 MPa compaction load.

The cell was placed in the RH chamber to equilibrate at 80% RH. The limited free surface through which the cell can exchange water with the humid environment of the chamber, the sample being oven-dried and the large change in suction levels are factors that were expected to delay the hydration process. After 32 days the sample was still absorbing moisture, indicating that equilibrium had not yet been achieved.

An estimation of the required time for equilibrium was thought to be 40 days. This value was derived by fitting a straight line in the semi-logarithmic plot of rate of change in mass with time shown in Figure 6. It should be noted that the above time value represents the maximum expected time for equilibrium to be reached for the examined soil, as the suction step applied was the maximum allowed by the equipment.

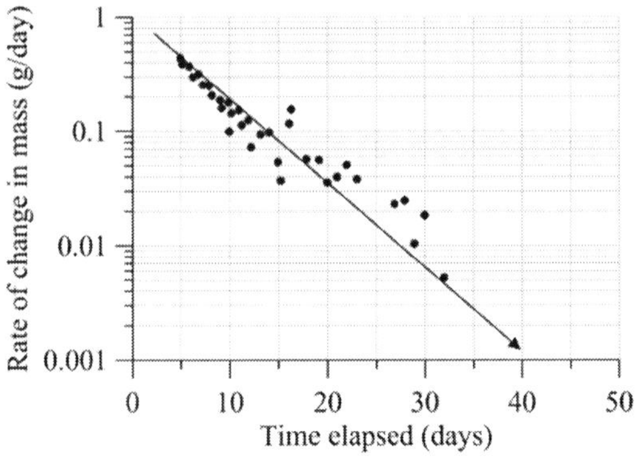

Figure 6. Change of mass of soil sample inside cell for a change in RH from 0% to 80%.

4 Experimental results

4.1 Black cotton soil

A soil sample of black cotton clay, previously studied by Al Haj [12], was tested in the RH chamber. The soil sample was loosely placed in the chamber on a tray at an initial water content of 58.4%. The initial RH target value was set at 20%, gradually increasing up to 80%. The range of applied RH corresponds to an initial suction of 200MPa (i.e. 20% RH), followed by wetting steps up to a suction of 30 MPa (i.e. 80% RH).

The mass of the sample was frequently measured using a 0.01g resolution scale. This involved removing samples from the chamber, sealing the trays with a top cap, taking the mass measurement quickly and carefully and placing the sample back into the chamber. Equilibrium was assumed to have been achieved when the change in mass was less than 0.01g for 12 hours. The derived gravimetric water content is plotted against total suction in Figure 7. The SWRC derived by Al Haj [12] using the filter paper method, with the Chandler & Guiterez [13] calibration for Whatman No.42 filter papers for suction values up to 30MPa, is also included for comparison. The wetting path measured with the new system is in good agreement with Al Haj's wetting curve.

4.2 London Clay

Subsequently, a series of tests was performed using London Clay, similar to the one tested by Monroy [14]. The initial material was air-dried, ground into powder and sieved before being stored in a large container in the storage room where RH was estimated to be about 63%. The Atterberg limits were measured using the cone penetrometer and the standard rolling method, as described in BS 1377-2:1990 [15]. The particular London Clay was found to be of very high plasticity (PL=0.28, PI=0.52). According to particle density tests using the glass jar method, the specific gravity, G_s, has a value of 2.75 [11].

Eight 65mm-diameter samples were compacted at 1.8 MPa, at an initial water content of 7.3%. Four of the samples were subjected to wetting, with RH ranging from 60% to 80% (70 MPa to 30 MPa of suction) while the other four were oven dried prior to being subjected to a wetting path from 10% to 70% RH (310 MPa to 50 MPa of suction). Two additional samples of pulverised London Clay were loosely placed on a tray and were subjected to a range of 10% to 85% RH (310 MPa to 20 MPa of suction), with one of them being initially oven-dried. Because of their uneven shape, the volume of these samples, and therefore their void ratio values, could not be calculated.

The SWRCs are plotted in Figure 8 in terms of gravimetric water content. As expected, the curves for the oven-dried samples (0% w/c) plot at a lower water content than the air-dried samples (6.1% & 7.3% w/c). Also, compacted samples plot above the pulverised loosely placed samples, possibly due to different void ratios obtained during preparation.

Also plotted for comparison are SWRCs for a reconstituted London Clay by Dineen [7], consolidated under 200kPa pressure and subjected to a cycle of drying (grey curve in Figure 8) and wetting (black curve in Figure 8). Additionally, Mavroulidou et al. [16] obtained two drying curves for an untreated pulverised London Clay. The samples were compacted at a dry density of 1.43 g/cm^3 at 27% (dry of optimum) and at 32% (wet of optimum) water contents. Both Dineen and Mavroulidou et al. used the filter paper method with the Chandler & Guiterez [13] calibration for Whatman No.42 filter papers.

The curves derived by Dineen for the reconstituted London Clay plot below the ones for the compacted samples obtained by Mavroulidou et al., perhaps due to differences in preparation. The two drying paths from Mavroulidou et al. plot close to the ones obtained in the present study, despite the differences in the compaction method and the origin of the soil.

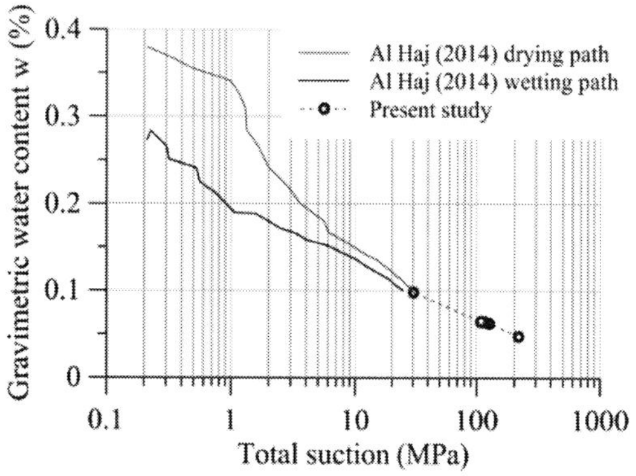

Figure 7. Soil Water Retention Curves for black cotton clay.

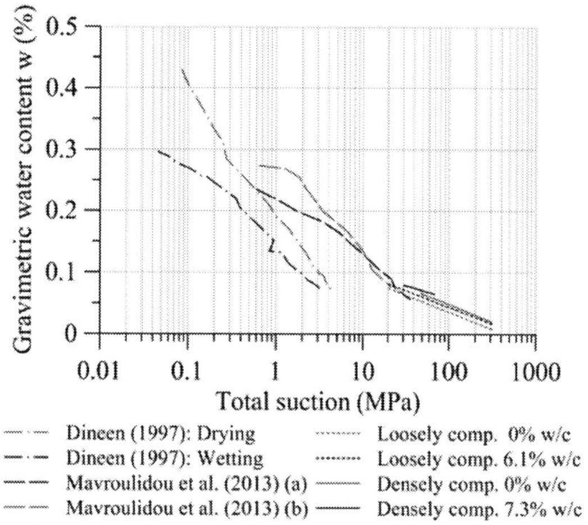

Figure 8. Soil Water Retention Curves for London Clay.

4.3 Comparison with filter paper measurements

This series allowed calibration of the RH – w/c curves for the particular filter papers used (Whatman No.42) at suctions higher than 30 MPa (~80% RH), which was then used to compare the two techniques. Therefore, two series of tests involving filter papers were conducted. In the first series, the subject of the tests was the filter papers themselves, while in the second series filter papers were used for indirect measurement of suction within soil samples.

More specifically, in series A, the filter papers were directly placed in a RH chamber and were exposed to different levels of RH. Three filter papers were placed in chamber A (fp_A2, fp_A3 & fp_A4) and subjected to consecutive RH values from 10% to 85% (310 MPa to 20 MPa of suction), while an additional filter paper was placed simultaneously in chamber B (fp_B2) and was subjected in a single step to a change in RH from 10% to 85% to examine any equilibration time effects. These are compared to the following calibration curves from the literature in Figure 9:

Chandler and Gutierrez [13] for up to 30 MPa,

$$w = (4.842 - \log \psi) / 0.0622 \quad (1)$$

Fawcett and Collis-George [17]:

$$w = (5.327 - \log \psi) / 0.0779 \quad (2)$$

and Likos and Lu [18]:

$$w = (5.5 - \log \psi) / 0.13 \quad (3)$$

Of the three calibration curves, the one proposed by Likos and Lu gives a better fit to the experimental points of series A, predicting drier filter papers by approximately 1% water content. An additional curve was custom fitted to the results of series A (Figure 9), as an upper boundary to the obtained experimental data:

$$w = (5.7 - \log \psi) / 0.13 \quad (4)$$

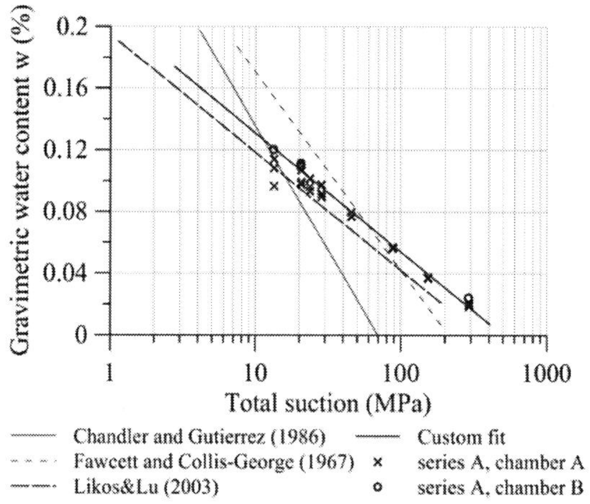

Figure 9. Filter paper calibration data and comparison with calibration curves found in the literature.

In series B the filter paper technique was used to measure the suction of soil samples (London Clay, 2MPa, 7.3% w/c) treated in the RH chamber at different levels of RH, namely 70% RH, 75% RH, 77.5% RH and 80% RH. One sample was prepared at each RH level. The suction applied to the samples is plotted against their gravimetric water content (symbol **x** in Figure 10).

The procedure recommended by Ridley et al. [3] and described in Kokkinou [19] was followed after the samples were removed from the chamber to determine the water content of the filter papers. The predictions of the aforementioned calibration curves (Eq. 1-4) for each sample are included in Figure 10.

The calibration of Likos and Lu (Eq. 3) compares relatively well to the suction estimated from the thermodynamic law for the equilibrium RH at the time of removal of the soil samples from the chamber. Eq. 4, on the other hand, overestimates significantly the suction measured in the sample, in this range (50MPa to 30MPa).

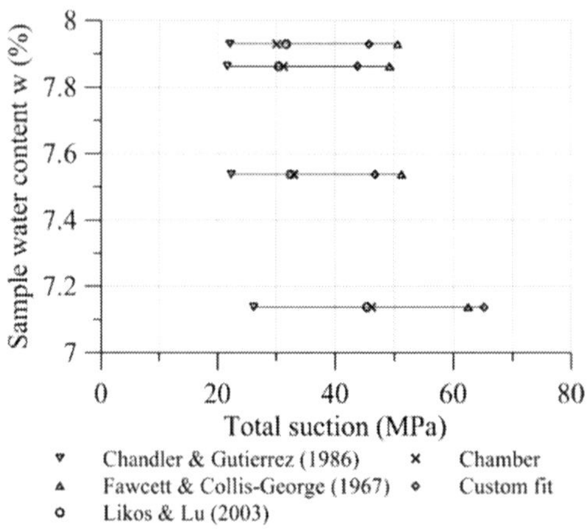

Figure 10. Comparison of suction prediction for series B using different filter paper calibration curves.

5 Summary

A method of controlling RH in chambers is presented and has been shown to provide an accurate control (0.2% RH standard deviation) of relative humidity in an environmental chamber for a range of 10% RH to 85% RH. The control is fully automatic, but the recording of mass is still manual. The RH conditions can be successfully held constant for long periods without any effect from the laboratory humidity.

The SWRCs derived for two soils using this method show good agreement with data presented in the literature for similar soils. An upper boundary calibration for Whatman No.52 filter paper for suction values larger than 30 MPa is suggested, but the Likos & Lu calibration curve fits better to the measurements of sample suction.

Acknowledgements

The first author is supported by a Skempton scholarship and the project is partly funded by Radioactive Waste Management ltd (RWM). The method has been developed thanks to the engineering skills of Duncan Parker.

References

1. A. Tsiampousi, L. Zdravkovic, and D.M. Potts, *A three-dimensional hysteretic soil-water retention curve*. Geotechnique, (2013). **63**(2): p. 155.
2. D. Gallipoli, S.J. Wheeler, and M. Karstunen, *Modelling the variation of degree of saturation in a deformable unsaturated soil*. Geotechnique, (2003). **53**(1): p. 105.
3. A.M. Ridley, K. Dineen, J.B. Burland, and P.R. Vaughan *Soil matrix suction: some examples of its measurement and application in geotechnical engineering*. Géotechnique, (2003). **53**, 241-253.
4. N. Lu and W.J. Likos, *Unsaturated Soil Mechanics*. (2004), Hoboken, New Jersey: John Wiley & Sons.
5. J.W. Hilf, *An investigation of Pore-Water pressure in compacted cohesive soils*. (1956), Thesis, University of Colorado.
6. P. Delage, M.D. Howat, and Y.J. Cui, *The relationship between suction and swelling properties in a heavily compacted unsaturated clay*. Eng.Geol., (1998). **50**(1-2): p. 31.
7. K. Dineen, *The influence of soil suction on compressibility and swelling*. (1997), PH.D Thesis, Imperial College.
8. R. Monroy, A. Ridley, K. Dineen, and L. Zdravkovic, *The suitability of the osmotic technique for the long-term testing of partly saturated soils*. Geotechnical Testing Journal, (2007). **30**(3): p. 220.
9. P. Delage, E. Romero, and A. Tarantino, *Recent developments in the techniques of controlling and measuring suction in unsaturated soils*. Unsaturated Soils: Advances in Geo-Engineering, (2008): p. 33-52.
10. M.V. Villar. Reseacrcher. (Personal communication, September 2013)
11. A. Kirkham, *Novel experimental technique for the measurement of hydraulic properties of geomaterials*. (2015), MEng Dissertation, Imperial College London.
12. K. Al Haj, *Mechanical response of two plastic clay soils from Sudan*. (2014), Ph.D Thesis, Imperial College London.
13. R.J. Chandler and C.I. Gutierrez, *Filter-paper method of suction measurement*. Geotechnique, (1986). **36**(2): p. 265-268.
14. R. Monroy, *The influence of load and suction changes on the volumetric behaviour of compacted London clay*. (2006), Ph.D Thesis, Faculty of Engineering Imperial College London.
15. British Standards Institute, *BS 1377-2:1990 Methods of test for soils for civil engineering purposes. Classification tests*. (1990), British Standards Institute: London.
16. M. Mavroulidou, Z. Xiwei, M.J. Gunn, and Z. Cabarkapa, *Water Retention and Compressibility of a Lime-Treated, High Plasticity Clay*. Geotech.Geol.Eng., (2013). **31**(4): p. 1171-85.
17. R. Fawcett and N. Collis-George, *A filter-paper method for determining the moisture characteristics of soil*. Australian Journal of Experimental Agriculture, (1967). **7**(25): p. 162-167.
18. W.J. Likos and N. Lu, *Automated humidity system for measuring total suction characteristics of clay*. Geotechnical Testing Journal, (2003). **26**(2): p. 179-190.
19. A. Kokkinou, *Investigation of soil water retention curves of compacted clays*. (2014), MSc Dissertation, Imperial College London.

E3S Web of Conferences 9, 10008 (2016) DOI: 10.1051/e3sconf/20160910008
E-UNSAT 2016

Application of fitting parameters in best fit equation

Zhai Qian[1], Harianto Rahardjo[1,a]

[1]*School of Civil and Environmental Engineering, Nanyang Technological University, 50 Nanyang Avenue, Singapore*

Abstract. Soil-water characteristic curve (SWCC) contains key information for the application of unsaturated soil mechanics principles to engineering practice. SWCC variables such as air-entry value, residual suction and residual saturation are commonly referred to for estimation of other unsaturated soil properties. Though there are clear definitions for these SWCC variables, it is difficult to measure them directly in laboratory and these SWCC variables are commonly estimated from SWCC. Hydraulic conductivity of unsaturated soil is the major property that controls water flow in soil. The laboratory measurement of hydraulic conductivity is time consuming and costly. Therefore, hydraulic conductivity is also commonly estimated from SWCC. Normally, SWCC is represented by a best fit equation which is typically governed by few fitting parameters. These fitting parameters can be obtained from a best fit procedure and are proven to be dependent on each other. Fredlund and Xing's (1994) equation is one of the popular equations commonly used by researchers and there are only three fitting parameters such as "a", "n", "m" in this equation. Correlation equations between SWCC variables, hydraulic conductivity and the fitting parameters such as "a", "n" and "m" are presented in this paper. In addition, the variability in the determined SWCC variables and hydraulic conductivity due to the variation in these fitting parameters is also discussed in this paper.

1 Introduction

SWCC is normally determined from several experimental data and a best fit equation is adopted to provide a continuous model. The best fit equation is normally governed by a few fitting parameters and these fitting parameters are typically determined using a curve fitting technique. As the fitting parameters are determined from a regression procedure, these fitting parameters are mathematical solutions rather than physical soil properties.

SWCC variables such as air-entry value, residual suction and residual saturation are commonly referred to for the description of unsaturated soil properties (e.g., shear strength such as Vanapalli et al.'s 1996 equation, constitutive model such as Sheng et al. 2008). These SWCC variables are conventionally determined using the graphical method with subjective interpretation. In addition, permeability function or hydraulic conductivity of unsaturated soil is the essential information required for seepage analysis. As the experimental measurement of hydraulic conductivity of unsaturated soil is tedious and costly, the unsaturated hydraulic conductivity is commonly estimated from SWCC. The mathematical equations to correlate the SWCC variables, hydraulic conductivity and the fitting parameters in Fredlund and Xing's (1994) equation are derived and illustrated in this paper. The variability in the determined SWCC variables and hydraulic conductivity due to variation in these fitting parameters is also discussed in this paper.

2 Literature review

Various researchers developed different equations to describe soil-water characteristic curve (SWCC) (Brooks and Corey, 1964; Gardner, 1958; Farrel and Larson, 1972; van Genuchten, 1980; Williams et al., 1983; Fredlund and Xing, 1994; Kosugi, 1994 and Satyanaga et al., 2013). Studies by Leong and Rahardjo (1997a) and Zapata (1999) concluded that Fredlund and Xing's (1994) equation performed best among different equations to best fit SWCC. Leong and Rahardjo (1997a) suggested that Fredlund and Xing (1994) equation can be used for a wide range of soil over the entire range of matric suction. Zhai and Rahardjo (2012a, 2013a) derived equations for determination of SWCC variables from fitting parameters of Fredlund and Xing's (1994) equation. Zhai and Rahardjo (2015) derived equations for estimation of hydraulic conductivity from SWCC using capillary model. In addition, fitting parameters are commonly obtained with the accompanying residual error after the regression procedure. Fitting parameters are then used while the residual error is always discarded. In fact, the residual error is helpful for the estimation of variance of fitting parameters and confidence limits of the best fitted SWCC. Zhai and Rahardjo (2013b) proposed equations for the estimation of variance of these fitting parameters from the residual error using the first-order error method. Confidence limits of SWCC were also proposed by Zhai and Rahardjo (2013b) using variance of fitting parameters. Consequently, the variability in SWCC variables and hydraulic conductivity can also be

[a] Corresponding author: chrahardjo@ntu.edu.sg

DOI: 10.1051/e3sconf/20160910008

estimated from variance of the fitting parameters and is discussed in this paper.

By introducing the concept of pore-size distribution function, Fredlund and Xing (1994) proposed the best fit equation as follows:

$$\theta = C(\psi)\frac{\theta_s}{\left\{\ln\left[e+\left(\frac{\psi}{a}\right)^n\right]\right\}^m} = \left[1-\frac{\ln\left(1+\frac{\psi}{C_r}\right)}{\ln\left(1+\frac{10^6}{C_r}\right)}\right]\frac{\theta_s}{\left\{\ln\left[e+\left(\frac{\psi}{a}\right)^n\right]\right\}^m} \quad \text{---- (1)}$$

where,

a, n, m: fitting parameters;

C_r: parameter related to residual suction that can be selected as C_r=1500 kPa for most cases (Fredlund and Xing 1994, Zhai and Rahardjo 2012a, 2012b).

θs : saturated volumetric water content.

Zhai and Rahardjo (2012a) presented the graphical method for the estimation of SWCC variables from SWCC as illustrated in Fig. 1.

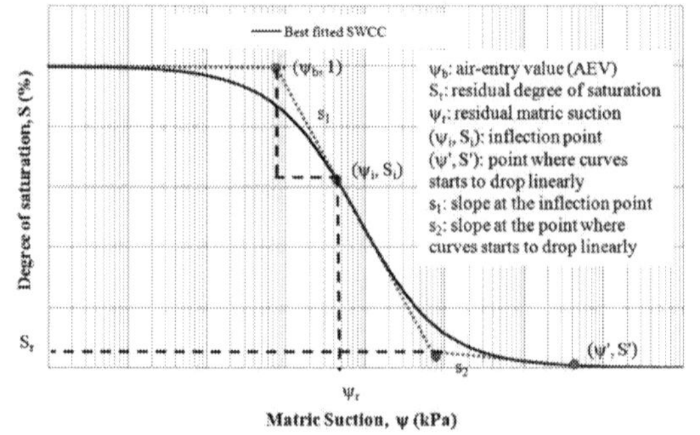

Figure. 1 Definitions of SWCC variables after Zhai and Rahardjo (2012a)

Zhai and Rahardjo (2015) presented that SWCC in the form of degree of saturation could be considered as analogous to the pore-size distribution function or pore-suction distribution function. The capillary model was illustrated by dividing SWCC into different intervals (Δψ) to represent capillary tubes with different radii as shown in Fig. 2. Zhai and Rahardjo (2015) derived an equation for the estimation of hydraulic conductivity from SWCC using the theory of random connections and Poiseuille's law

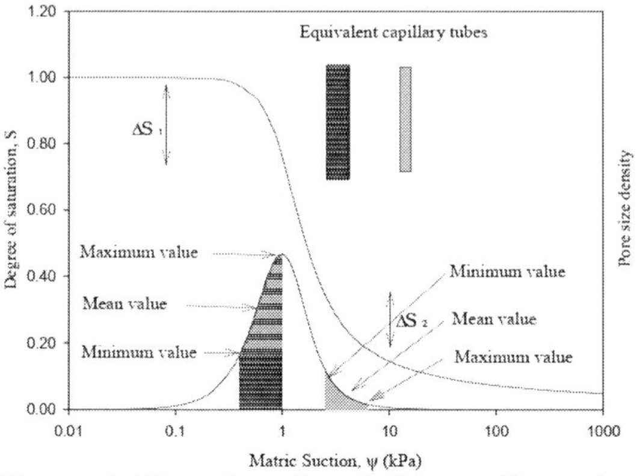

Figure. 2 Illustration of equivalent capillary tubes associated with the pore-suction distribution function.

3 Theory

The equations for computation of SWCC variables, unsaturated hydraulic conductivity from the SWCC fitting parameters are derived in this section.

3.1 Determination of the inflection point.

SWCC is commonly plotted in a log-arithmetic scale rather than an arithmetic scale. The slope of the plot in the log-arithmetic scale is different from the slope of the plot in the arithmetic scale. The relationship between the slopes of the plots in two different scales can be obtained from Equation (2).

$$s = s'\ln(10)\psi \quad \text{----- (2)}$$

where,

s = slope plotted in log-arithmetic scale,

s'= slope plotted in arithmetic scale

ψ = matric suction.

The inflection point is a point on a curve at which the curve changes from being concave (concave downward) to convex (concave upward), or vice versa. If x is an inflection point for the function f(x) then the second derivative, f''(x), should be equal to zero if the inflection point exists. Therefore, the inflection point on the SWCC can be obtained by solving the following differential equation as illustrated in Equation (3).

$$\frac{d^2S}{d(\log(\psi))^2} = \ln(10)\psi\left[\frac{d^2S}{d\psi^2}\ln(10)\psi + \frac{dS}{d\psi}\ln(10)\right] = 0 \quad \text{----- (3)}$$

3.2 Determination of air-entry value, residual suction and residual saturation.

The slope of SWCC using Fredlund and Xing's (1994) equation can be obtained from Equation (4) as follows:

$$s = \left\{\frac{1}{\ln\left(1+\frac{10^6}{C_r}\right)\left(1+\frac{\psi}{C_r}\right)C_r\left\{\ln\left[e+\left(\frac{\psi}{a}\right)^n\right]\right\}^m} + \frac{mn\left(\frac{\psi}{a}\right)^{n-1}\left[1-\frac{\ln\left(1+\frac{\psi}{C_r}\right)}{\ln\left(1+\frac{10^6}{C_r}\right)}\right]}{a\left[e+\left(\frac{\psi}{a}\right)^n\right]\left\{\ln\left[e+\left(\frac{\psi}{a}\right)^n\right]\right\}^{m+1}}\right\}\ln(10)\psi \quad \text{----- (4)}$$

Zhai and Rahardjo (2012a) presented that air-entry value (AEV), residual suction (ψ_r) and residual saturation (S_r) could be obtained from Equations (5) to (7) as follows:

$$AEV = \psi_i 0.1^{\frac{1-S_i}{s_1}} \quad \text{----- (5)}$$

DOI: 10.1051/e3sconf/20160910008

$$\psi_r = 10^{\frac{S_i - S' + s_1 \log(\psi_i) - s_2 \log(\psi')}{s_1 - s_2}} \qquad \text{----- (6)}$$

$$S_r = S_i - s_1 \log(\psi_r / \psi_i) \qquad \text{----- (7)}$$

where:
AEV = air-entry value
ψ_i = suction corresponding to the inflect ion point
S_i = degree of saturation corresponding to the inflection point
s_1 = slope at the inflection point.

Estimation of hydraulic conductivity
Zhai and Rahardjo (2015) derived the equation for the prediction of permeability function using capillary model and Poiseuille's law. By substituting fitting parameters "a", "n" and "m" into equation (8) as proposed by Zhai and Rahardjo (2015), hydraulic conductivity can be computed.

$$k(\psi_{m+i}) = k(\psi_m)\frac{n_{m+i}^2}{n_m^2} \left\{ \begin{array}{l} (S(\psi_{m+i}) - S(\psi_{m+i+1}))^2 r_{m+i}^2 + \\ \sum_{j=m+i+1}^{N} \left[(S(\psi_{m+i}) - S(\psi_j))^2 - (S(\psi_{m+i}) - S(\psi_{j-1}))^2 \right] r_j^2 \end{array} \right\} \left\{ \begin{array}{l} (S(\psi_m) - S(\psi_{m+1}))^2 r_m^2 + \\ \sum_{i=m+1}^{N} \left[(S(\psi_m) - S(\psi_i))^2 - (S(\psi_m) - S(\psi_{i-1}))^2 \right] r_i^2 \end{array} \right\}^{-1} \qquad \text{------- (8)}$$

where,
$k(\psi_{m+i})$, $k(\psi_m)$ = hydraulic conductivity corresponding to suction ψ_{m+i} and ψ_m, respectively;
ψ_{m+i}, ψ_m = suction in the soil, ψ_{m+i}, $>\psi_m$.
n_{m+i}, n_m = porosity corresponding suction of ψ_{m+i} and ψ_m, respectively;
r_{m+i}, r_m = equivalent pore radius of pores corresponding to suction of ψ_{m+i} and ψ_m using Kelvin's capillary law, respectively;
$S(\psi_{m+i})$, $S(\psi_m)$ = degree of saturation in the soil corresponding to suction of ψ_{m+i} and ψ_m, respectively;
Estimation of variability in determined SWCC variables and hydraulic conductivity
By adopting the first-order error method, Zhai and Rahardjo (2013b) proposed equations (9) and (10) to estimate the confidence limits of SWCC as follows:

when $0 < \psi < a_{max}$,

$$\theta_{upper} = C(\psi)\frac{\theta_s}{\left\{ \ln\left[e + \left(\frac{\psi}{a_{max}} \right)^{n_{max}} \right] \right\}^{m_{min}}}$$

when $\psi > a_{max}$,

$$\theta_{upper} = C(\psi)\frac{\theta_s}{\left\{ \ln\left[e + \left(\frac{\psi}{a_{max}} \right)^{n_{min}} \right] \right\}^{m_{min}}} \qquad \text{------- (9)}$$

when $0 < \psi < a_{min}$,

$$\theta_{lower} = C(\psi)\frac{\theta_s}{\left\{ \ln\left[e + \left(\frac{\psi}{a_{min}} \right)^{n_{min}} \right] \right\}^{m_{max}}}$$

when $\psi > a_{max}$,

$$\theta_{lower} = C(\psi)\frac{\theta_s}{\left\{ \ln\left[e + \left(\frac{\psi}{a_{min}} \right)^{n_{max}} \right] \right\}^{m_{min}}} \qquad \text{------- (10)}$$

where,

$$a_{max} = a + t_{\alpha/2}\sqrt{Var(a)} \quad a_{min} = a - t_{\alpha/2}\sqrt{Var(a)}$$
$$n_{max} = n + t_{\alpha/2}\sqrt{Var(n)} \quad n_{min} = n - t_{\alpha/2}\sqrt{Var(n)}$$
$$m_{max} = m + t_{\alpha/2}\sqrt{Var(m)}$$
$$m_{min} = m - t_{\alpha/2}\sqrt{Var(m)}$$

a, n and m are fitting parameters.

The variability in SWCC variables and hydraulic conductivity can be estimated from the confidence limits of SWCC (using the proposed equations with a_{max}, a_{min}, n_{max}, n_{min}, m_{max}, m_{min} instead of a, n and m).

4. Applications and Discussions

To evaluate the correlations among SWCC variables, unsaturated hydraulic conductivity and the fitting parameters of SWCC, soils with available experimental data of both SWCC and permeability function are selected in this paper. Four types of soil such as Volcanic Sand, Glass Beads, Fine sand and Touch Silt Loam from Brooks and Corey (1964) are selected for this study. The experimental data of SWCC and relative hydraulic conductivity for these four types soil are illustrated in Figures. 3 and 4.

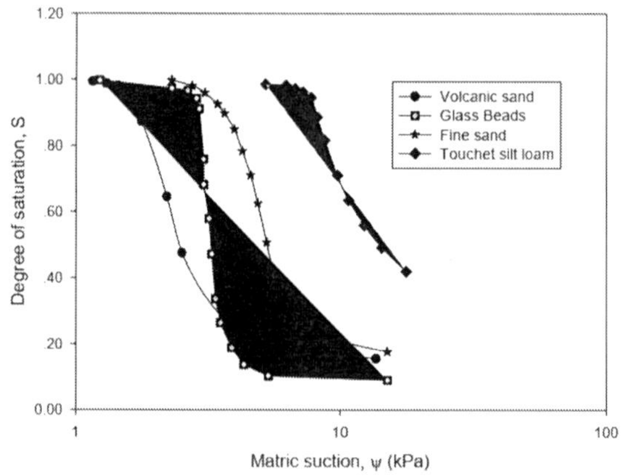

Figure. 3 SWCC of four types of soil from Brooks and Corey (1964)

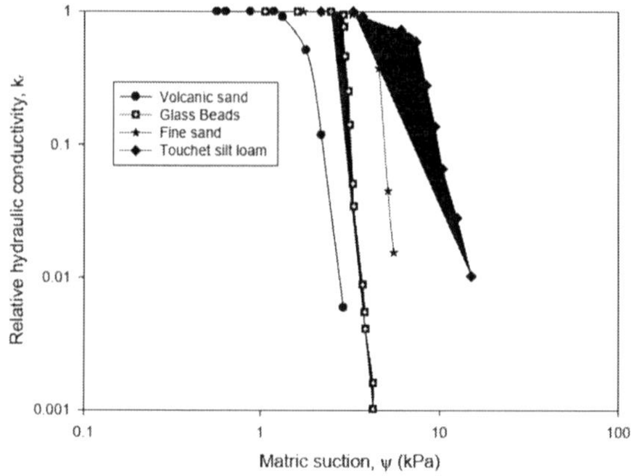

Figure. 4 SWCC of four types of soil from Brooks and Corey (1964)

The fitting parameters for the four types of soil, using the input value C_r=1500 kPa, obtained from the best fit procedure are summarized in Table 1. The SWCC variables computed from the fitting parameters using the equations proposed in this paper are illustrated in Table 2. The comparison of the estimated relative hydraulic conductivity and the experimental data for the four types of soil are illustrated in Figure. 5.

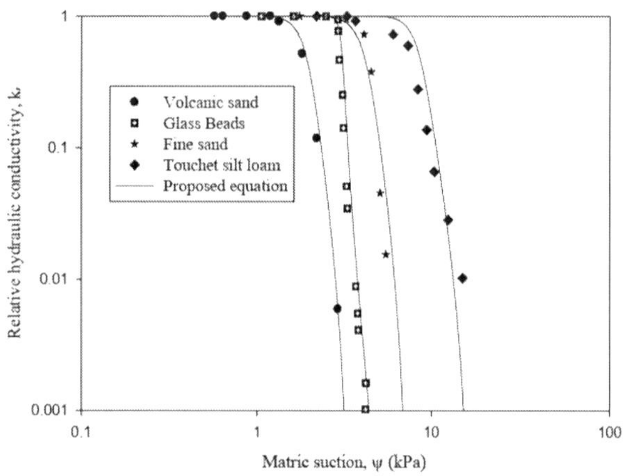

Figure. 5 Comparison of estimated hydraulic conductivity and experimental data for four types of soil

Table 1: Fitting parameters for these four types of soil

	a(kPa)	n	m	R^2(%)	SSE
Volcanic sand	1.87	9.14	0.68	99.96	1.6e-3
Glass beads	3.03	28.73	0.82	99.65	1.15e-2
Fine sand	4.23	8.25	0.85	99.82	5.60e-3
Touch silt loam	8.68	6.49	0.75	99.97	1.26e-3

Table 2: Computed SWCC variables for four types of soil

	AEV (kPa)	Residual suction (kPa)	Residual saturation
Volcanic sand	1.64	3.34	0.12
Glass beads	2.00	4.37	0.02
Fine sand	3.51	8.11	0.07
Touch silt loam	7.23	17.93	0.12

Table 2 indicates that SWCC variables can be calculated from the SWCC fitting parameters using Zhai and Rahardjo's (2012) equations. Figure. 5 indicates that prediction results from Zhai and Rahardjo's (2015) equation agree very well with the experimental results. In other words, the permeability function can also be calculated directly from the fitting parameters.
The confidence limits of SWCC for the four types of soil as computed using Zhai and Rahardjo's (2013b) equations are illustrated in Figure. 6. On the other hand, the permeability function for these four types of soil as computed from the best fitted SWCC and the confidence

DOI: 10.1051/e3sconf/20160910008

of limits of the SWCC using Zhai and Rahardjo's (2015) equation are illustrated in Figure. 7.

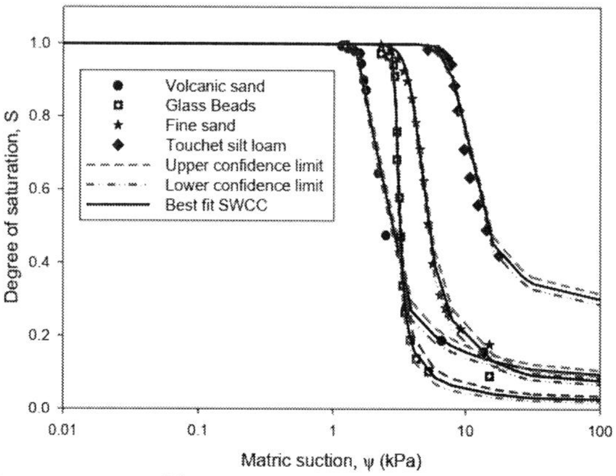

Figure. 6 Confidence limits of SWCC for four types of soil using Zhai and Rahardjo's (2013b) equations

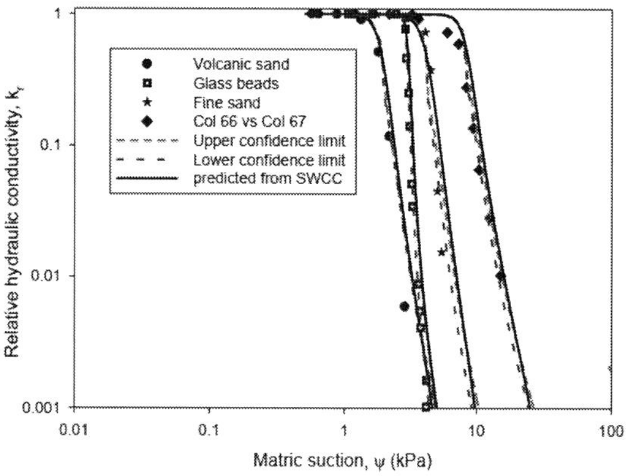

Figure. 7 Confidence limits of permeability function for four types of soil using Zhai and Rahardjo's (2015) equation

Figure. 6 indicates that uncertainty in SWCC can also be estimated from the fitting parameters and the residual error using Zhai and Rahardjo's (2013b) equations. Figure. 7 indicates that variation in permeability function can also be estimated from the confidence limits of SWCC.

5 Conclusions

The SWCC variables such as air-entry value, residual suction and residual saturation can be computed from the SWCC fitting parameters using Zhai and Rahardjo's (2012) equations. The permeability function can be predicted from the SWCC fitting parameters using Zhai and Rahardjo's (2015) equation. On the other hand, the uncertainty in the SWCC can also be estimated from the fitting parameters and the residual error using Zhai and Rahardjo's (2013b) equations. Subsequently, the variation in permeability function can also be estimated from the confidence limits of SWCC. In other words, the equations proposed by Zhai and Rahardjo (2012a, 2012b, 2013, and 2015) provide a convenient method to estimate the SWCC variables, permeability function and their variations using an electronic spreadsheet.

References

1. Brooks, R.H., and Corey, A.T. "Hydraulic properties of porous media."Hydrology Paper, No. 3 Colorado State Univ., Fort Collins Colo. (1964)
2. Farrel. D.A. and Larson W.E. "Modeling the pore structure of porous medium. " Water Resour. Res. 3 699-706 (1972)
3. Fredlund, D.G. and Xing, A. "Equations for the soil-water characteristic curve." Canadian Geotechnical Journal, 31(3): 521-532. (1994)
4. Gardner, W.R "Mathematics of isothermal water conduction in unsaturated soils."Highway Research Board Special Rep. No. 40, International Symposium on physic-Chemical Phenomen in Soils, Washigton, D.C. 78-87. (1958)
5. Kosugi K "Three-parameter lognormal distribution model for soil water retention". Water Resources Research Vol 30 No. 4 Pages 891-901. (1994)
6. Leong, E.C. and Rahardjo H. "Review of soil-water characteristic curve equations." J. Geotech. Geoeniron. Eng 123(12), 1106-1117. (1997)
7. Satyanaga, A., H. Rahardjo, E.C. Leong and J.Y. Wang. "Water Characteristic Curve of Soil with Bimodal Grain-size Distribution".Computer and Geotechnics, January, Vol. 48, pp. 51-61. (2013)
8. Sheng, D., Fredlund, D.G., and Gens, A. " A new modeling approach for unsaturated soils using independent stress state variable" ". Canadian Geotechnical Journal, Vol. 45, pp. 511-534. (2008)
9. van Genuchten, M.T. "A close form equation predicting the hydraulic conductivity of unsaturated soil." Soil Sci. Soc Am. J 44, 892-898. (1980)
10. Vanapalli, S. K., Fredlund, D. G., Pufahl, D. E. and Clifton, A. W. "Model for the prediction of shear strength with respect to soil suction", Canadian Geotechnical Journal, Vol. 33, pp. 379-392. (1996)
11. William. J., Prebble.R.E., Williams W.T. and Hignett, C.T. "The influence of texture, structure and clay mineralogy on the soil moisture characteristics." Australian J. of Soil Res. 21, 15-32 (1983)
12. Zapata, C. E. "Uncertainty in Soil-Water Characteristic Curve and Impacts on Unsaturated Shear Strength Predictions", Ph.D. Dissertation, Arizona State University, Tempe, United States. (1999)
13. Zhai Q, Rahardjo H. "Determination of soil–water characteristic curve variables." Comput Geotech 2012;42:37–43. (2012a)
14. Zhai Q, Rahardjo H. " Reply to the discussion by Bellia et al. on "Determination of soil–water characteristic curve variables"by Zhai Q, Rahardjo H. Comput Geotech 2012;42:37–43 " Comput Geotech Volume 45, Pages 151–152 (2012b)
15. Zhai Q, Rahardjo H. " Soil-Water Characteristic Curve Variables." Conference paper, Proceeding of

DOI: 10.1051/e3sconf/20160910008

the International Symposium on Unsaturated Soil Mechanics and Deep Geological Waste Disposal, Shanghai, China, 07-10 July (2013a)
16. Zhai Q, Rahardjo H. "Quantification of uncertainty in soil-water characteristic curve associated with fitting parameters" Engineering Geology 163 144-152 (2013b)
17. Zhai Q., and Rahardjo H. "Estimation of permeability function from Soil-Water Characteristic Curve". Engineering Geology 199 :148-156 (2015)

 DOI: 10.1051/e3sconf/20160910009

Estimation of hydraulic conductivity function in unsaturated pyroclastic soils

Marianna Pirone[1a], Raffaele Papa[1], Marco Valerio Nicotera[1] and Gianfranco Urciuoli[1]

[1] *University of Naples 'Federico II', Department of Civil, Architectural and Environmental Engineering, Via Claudio 21, 80125 Naples*

Abstract. A full understanding of slope failure conditions in unsaturated pyroclastic slope needs a fair analysis of groundwater flow and, therefore, a proper hydraulic soil characterization in terms of either soil water retention curves (SWRC) and hydraulic conductivity function (HCF). A practical method to detect the HCF by processing in situ data is provided, thus, the identification of the HCFs operating on site by applying the Soil Water Balance (SWB) to the pyroclastic soil cover is carried out. In this regard data monitoring (matrix suction and volumetric water content) collected at experimental field of Monteforte Irpino in Southern Italy over four years have been used. Moreover a comparison between the saturated hydraulic conductivity from lab test and from in situ measurements is carried out and the results confirm that the effective hydraulic conductivity operating at the site scale is higher than the hydraulic conductivity obtained in the laboratory for the whole range of suction measured at site.

1 Introduction

Flowslides in granular soils undoubtedly constitute a major threat to human life, man made structures and the environment in general. In unsaturated soils, rainfall is the most usual triggering cause, due to rainwater infiltrating into the superficial soil, which causes a decrease in matric suction and consequently in shear strength. In this regard, Early Warning Systems (EWSs) are widely used as measures for rapid landslide risk mitigation and can be set up by using physically-based models able to reproduce the hydro-mechanical slope behaviour through numerical analyses [1], [2]. However the weakness in forecasting rainfall-induced landslides is often due to uncertainties about hydraulic soil characterisation at the site scale [3]. Here a method to detect a hydraulic conductivity function by processing in situ data is provided, in particular, the identification of the HCFs operating on site by applying the Soil Water Balance (SWB) to a pyroclastic soil cover is carried out. These features have been investigated by processing data from the test site at Monteforte Irpino [4] where meteorological data, matric suction and volumetric soil water content measurements were collected for about four years. Moreover the saturated hydraulic conductivity and the hydraulic conductivity function obtained in the laboratory [5] are compared with the unsaturated hydraulic conductivity values derived from in situ measurements to show that laboratory testing results are not always representative of the effective hydraulic conductivity operating at the site scale. Lastly some measurements of saturated hydraulic conductivity gained at site via double ring infiltrometer tests [6] are reported.

[a] Corresponding author: marianna.pirone@unina.it

2 Test site

The test site at Monteforte Irpino (40°54'13.11" N, 14°40'24.21"E), about 40 km East of Naples, was selected as being representative of other pyroclastic slopes in Campania subjected to rapid landslides (e.g. Pizzo D'Alvano, Monti di Avella and Monte Partenio). In the test site area, the slope is quite regular and has an average angle of 25–30°, with local values reaching 35–40°. An area of about 230 m^2 was chosen on the slope and twenty instrumented vertical sections were set up along three longitudinal alignments, sections A–A', B–B', C–C'. The stratigraphic profile consists of an unsaturated pyroclastic soil cover a few metres thick (3–5.5 m), deposited by a series of eruptions of Mt Somma–Vesuvius on top of the limestone bedrock. The test site was monitored from 2006 to 2012. The monitoring equipment consisted of: (i) 94 traditional vacuum tensiometers, i.e.: jet-fill tensiometers (SoilMoisture Equipment Corp.) and SMS (Soil Measurement system) tensiometer tubes (SDEC France); (ii) 40 TDR (Time Domain Reflectometry) probes 15 cm long; (iii) 6 Casagrande piezometers; (iv) a weather station. The tensiometers were installed at different depths along all the vertical sections, TDR and Casagrande piezometers along the verticals only at the central section, B–B'. However, the test site itself has been extensively described elsewhere; readers can refer to [7] and [4] for further detailed information about the instrumented area.
The simplified mean soil profile obtained from experimental investigation through some trenches and boreholes, mean physical properties (specific gravity G_s,

DOI: 10.1051/e3sconf/20160910009

porosity Φ, dry soil unit weight γ_d) and instrumentation installed along the central longitudinal section, B–B', of the test site are reported in Figure 1.
To apply the water balance calculation to soil cover, the data collected from the verticals located at the central section, B–B', will be used and reference will henceforth be made to a simplified mean soil profile of three layers of ash soils separated by two pumiceous layers (soils 3 and 5): the superficial ash layer comprises soils 1 and 2; the intermediate ash layer consists of soil 4; the deep ash layer contains soils 6 and 8 (Figure 1).

3 Hydraulic conductivity functions

3.1. Determination from laboratory tests

Nicotera et al. (2010), [5], presented saturated hydraulic conductivity determined in the laboratory by constant head hydraulic conductivity tests and hydraulic conductivity functions determined by inverse modelling of a sequence of testing phases (i.e. a constant head permeation test, a forced evaporation test, and finally a drying test in a pressure plate apparatus to be conducted on a single undisturbed soil sample) for all the ash soils recovered at the test site. The saturated hydraulic conductivities measured for each soil and the mean values of the parameters of the Mualem–van Genuchten [8] equation modeling the soil water retention curves along main drying paths obtained in the laboratory for each soil are summarized in Table 1 (volumetric water content at saturation, residual volumetric water content, α, n, *l* Mualem–van Genuchten parameters, k^{sat}). In Figure 2 the saturated hydraulic conductivity and the hydraulic conductivity function on the plane k (conductivity)-θ (volumetric water content) are reported for ash soils, 1-2, 4, 6 and 8. The saturated hydraulic conductivity measurement decrease from the ground surface to the bottom of soil cover (starting from soils 1&2 to 8). Indeed at fixed interval of volumetric water content, i.e. 0.3 to 0.5, the largest range of the hydraulic conductivities is observed in the surficial soils [9].
According to the literature [10]-[11], measurements of hydraulic conductivity are approximately log-normally distributed. Here the lognormal distribution is applied to the lab measurements of saturated hydraulic conductivity for ash soils (see Figure 4); the number of determinations (N), the standard deviation, σ, of the logarithm of measurements, and the modal and median values resulting from the distribution are reported in Table 2.
Moreover the hydraulic conductivity functions corresponding to median value of saturated hydraulic conductivity from lab tests are reported for ash soils in Figure 3a-d (grey lines).

3.2 Determination from in situ measurements

The procedure to estimate the hydraulic conductivity function exploits the in situ measurements of volumetric water content and matric suction collected during autumn and uses the assessment of soil water balance (SWB). By assuming the monitored section of slope as infinite [3], the SWB can be applied to the one-dimensional simplified soil profile comprising all the soil layers along the direction normal to the slope surface (Figure 1) and written as (modified from [12]):

	Mean Thickness	Gs	Φ	$\gamma_d \left(\frac{kN}{m^3}\right)$
1-2: topsoil & ashy soil	0.80	2.65	0.69	7.90
3: pumiceous deposits of Avellino eruption	0.70	2.55	0.80	4.80
4: palaeosoil	0.80	2.57	0.71	7.11
5: pumiceous deposits of Ottaviano eruption	0.60	2.55	0.82	4.63
6: palaeosoil	0.60	2.57	0.72	7.13
8: highly weathered fine-grained ashy soil	0.30	2.49	0.58	10.64

fl: fractured limestone

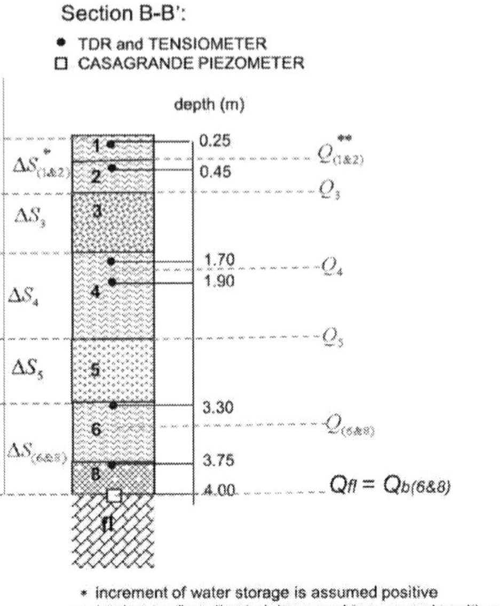

Figure 1. Simplified soil profile with mean soil physical properties and the layout of the instrumentation installed along each monitored vertical section along section B–B'. The symbols used to indicate soil water volume storage and groundwater flow in each layer are reported: ΔS water storage variation, Q_i flow through layer i (modified from [3]).

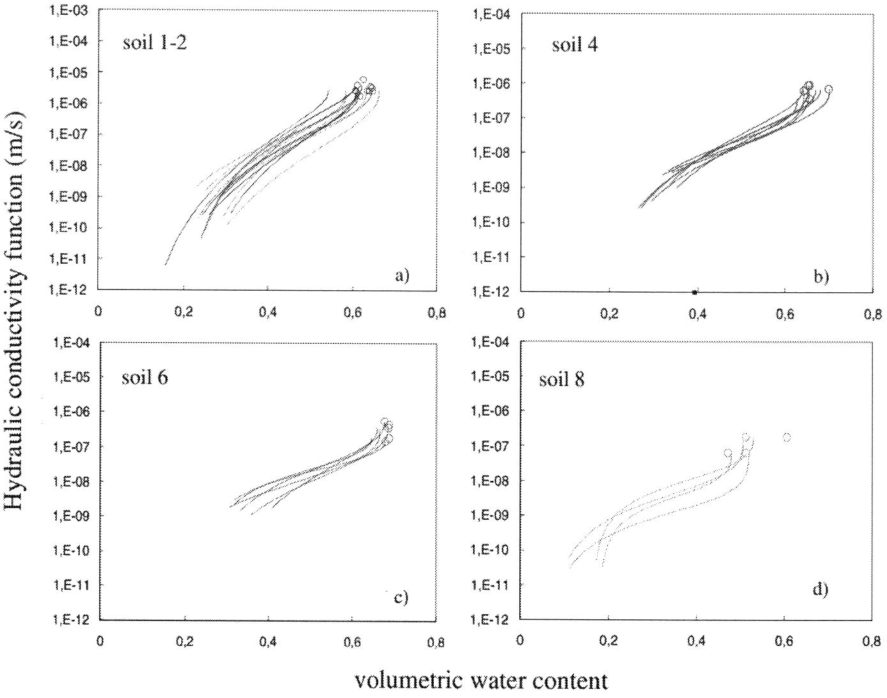

Figure 2. Hydraulic conductivity function (continous line) and hydraulic saturated permeability (symbol) obtained in the laboratory on samples from soil 1-2 (a); soil 4(b); soil 6(c); soil 8(d) [5].

$$I - ET_C = \sum_{i=1}^{n_L} \Delta S_i + Q_{fl} \quad (1)$$

$$I - ET_C = Q_{1\&2} \quad (2)$$

where I is the infiltration, ET_C the crop evapotraspiration [13], ΔS_i, the variation in soil water storage in the i-th layer (the superficial, intermediate and deep ash layers being denoted by, respectively, 1 & 2, 4 and6 & 8), Q_{fl} is the water exchange between the entire soil cover and the fractured limestone bedrock, n_L is the number of layers within the soil cover, $Q_{1\&2}$ is the amount of water flowing normal to the slope surface across the top boundary of the superficial ash layer. The physical quantities involved in the SWB are also summarized in Figure 1.

Table 1 Parameters of Mualem van Genucthen model.

	θs	θr	α (kPa^{1})	n	l	k_{sat} (m/s)
1	0.565	0.135	8.08	1.716	-1.052	2.13×10^{-6}
2	0.617	0.143	8.72	1.602	-1.054	3.04×10^{-6}
4	0.659	0.164	9.30	1.495	-2.850	6.85×10^{-7}
6	0.669	0.198	13.10	1.645	-2.698	3.08×10^{-7}
8	0.508	0.120	12.20	1.390	-5.707	1.08×10^{-7}

According to [3] it is reasonable to assume that Q_{fl} is exactly zero during October (when the soil groundwater flow into deeper layers was parallel to the slope) and approximately zero during November and December (when the normal flux $Q_{6\&8}$ was just sufficient to provide a moderate variation of the water volume storage in the deeper layers). As a consequence of these observations, in these three months, the amount, $Q_{1\&2j}$, of water flowing normal to the slope surface across the top boundary of the superficial ash layer along the jth vertical section in the time interval from t to t + Δt can be calculated as the volume of water needed to supply the water storage in the whole soil cover during the time interval Δt. Thus eqs. (1)-(2) applied at each single instrumented vertical j-th become:

$$Q_{1\&2j} = \sum_{i=1}^{n_L} \Delta S_{ij} \quad (3)$$

where

$$\Delta S_{ij} = \left[S_{ij}(t + \Delta t) - S_{ij}(t) \right] \quad (4)$$

For each layer i and vertical profile j, $S_{ij}(t)$ is calculated as the difference between the volumetric soil water contents $\theta_{ij}(t)$ and $\theta_{ij}(0)$ measured respectively at time t and at the beginning of monitoring, multiplied by the thickness of the layer d_{ij} (i.e. assuming that the soil water content is uniform along the layer, [4]):

$$S_{ij} = \left[\theta_{ij}(t) - \theta_{ij}(0) \right] \cdot d_{ij} \quad (5)$$

Therefore, the fluxes crossing the top of the k-th soil layer at the jth vertical section can be written trivially as:

$$Q_{kj} = \sum_{i=k}^{n_L} \left[S_{ij}(t + \Delta t) - S_{ij}(t) \right] \quad (6)$$

Thus the mean value of the current hydraulic conductivity for the k-th soil layer at the jth vertical section during the time interval from t to t + Δt can be estimated from measurements of volumetric water content as:

$$k_{kj} \approx -\frac{1}{i_{kj}^n} \cdot \frac{Q_{kj}}{\Delta t} \quad (7)$$

However, the ground water flux infiltrating into the limestone may not always be zero in November and December and the distribution of volumetric water content along the first 0.25 m is not always uniform as is assumed instead for the calculation of water volume storage [3]. Thus, the total amount of water flowing across the k-th layers may exceed Q_{kj} obtained from Eq. (6), and hence Eq. (7) provides only a lower estimate for the actual value of hydraulic conductivity. However in [3], it is shown the lower estimate of hydraulic conductivity function matches the upper one proving the robustness of the method proposed here.

Mean values of hydraulic conductivity were calculated for each vertical profile and for each sampling day between October and December. These calculations were performed for: the superficial ash layer comprising soils 1 and 2, $k_{1\&2,j}$; the intermediate ash layer consisting of soil 4, $k_{4,j}$; and the deep ash layer comprising soils 6 and 8, $k_{6\&8,j}$. The calculated values of k_{kj} are plotted in Fig. 3a–d (black-filled symbols) against the corresponding mean volumetric water content.

For all ash soils apart from soil 6, the values of unsaturated hydraulic conductivity estimated from in situ measurements lie above the hydraulic conductivity curve

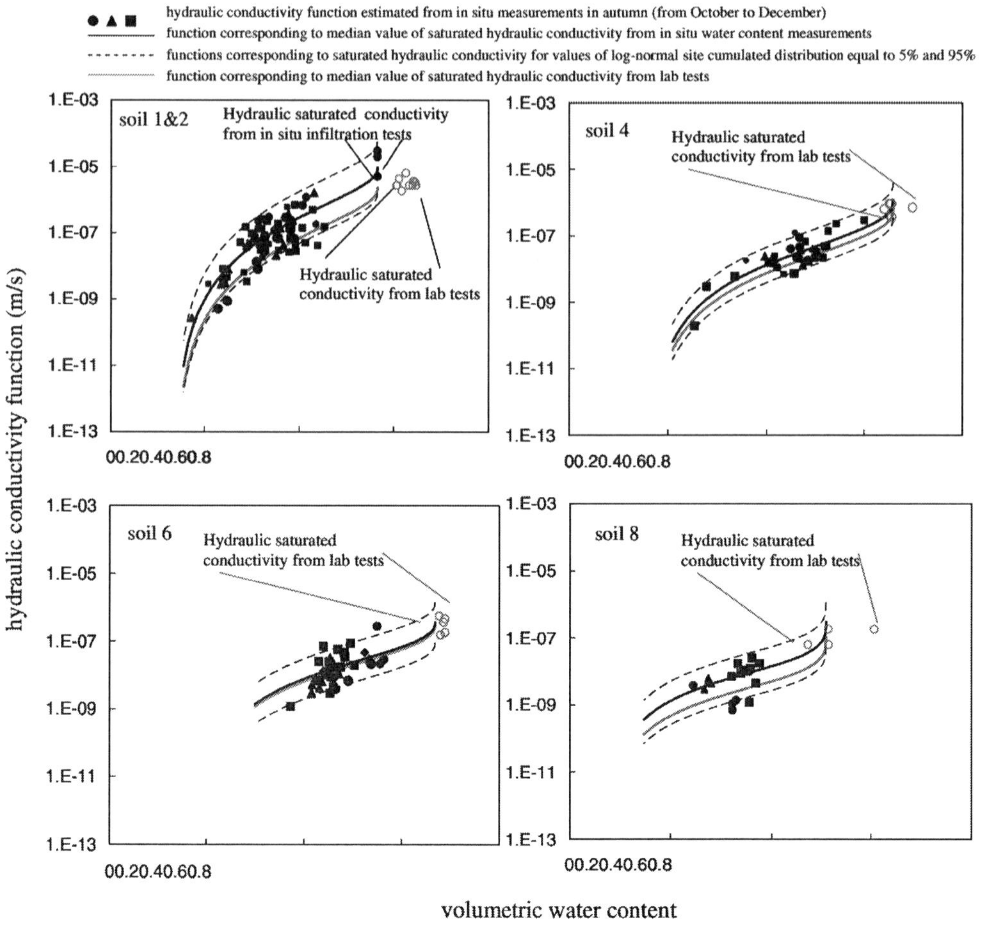

Figure 3. Hydraulic conductivity of the ash soils: saturated hydraulic conductivities from lab tests (grey circles) and in situ infiltration tests (black circles); unsaturated hydraulic conductivities determined from in situ measurements of soil water content via Eq. (7)(black symbols), hydraulic conductivity functions estimated on the basis of laboratory tests (grey line) and in situ measurements (black line).

determined in the laboratory, likely due to a poor representativeness of the soil laboratory sample [14]-[15]-[16], to all the uncertainties related to the estimation of in situ permeability and to the presence of some cracks on the soil surface where vegetation is lacking. By observing Fig. 3a–d (black-filled symbols), the variability of k_{kj} derived from field data, evaluated at constant volumetric water content, ranges between 0.5 and 1.5 orders of magnitude. The largest variability results in the surficial layers where ground–atmosphere interaction occurs and local topographic irregularities can affect the hydraulic soil behaviour. Nevertheless, the data from in-situ measurements appear to follow the same shape as the conductivity functions determined in the laboratory, albeit simply shifted towards higher hydraulic conductivity values, especially in the superficial soils for which more data are available. In order to elaborate a log-normal distribution also from in situ measurements, a unique value k^{sat}_{kj} of the saturated hydraulic conductivity of the k_{th} soil layer at the j_{th} vertical is estimated for each k_{kj} data point collected in situ and reported in Fig. 3a–d; k^{sat}_{kj} was calculated by imposing that the curve from the Mualem–van Genuchten model with the shape parameters fixed at the mean values from the lab and the saturated hydraulic conductivity at the chosen value of k^{sat}_{kj} could best fit the data points in hand. In Table 2 the parameters of the log-normal distributions of k^{sat}_{kj} thus obtained are compared to those of the laboratory measurements. In this regard it is worth noting that the median value of the log-normal distribution of hydraulic conductivity from in situ measurements always exceeds that of hydraulic conductivity measured in the laboratory: their ratio is 4 for soils 1 and 2, 2 for soil 4, 1.2 for soil 6, and 3 for soil 8 (see also Figure 4). Therefore, in Fig. 3a–d the continuous black line represents a *reasonable hydraulic conductivity function operative at the site*, obtained through the Mualem–van Genuchten model with a saturated hydraulic conductivity equal to the median value of log-normal distribution of values estimated from in situ measurements and the shape parameters fixed at the mean values obtained from the laboratory tests.

Table 2 Statistical parameters of log-normal distribution of saturated hydraulic conductivities.

Soil	method	N (#)	σ (m/s)	$\bar{\mu}$ (m/s)	mode (m/s)	median (m/s)
1&2	Lab	10	0.94	2.9E-06	9.0E-07	2.3E-06
	Site	73	1.04	1.6E-05	2.7E-06	9.6E-06
4	Lab	5	0.39	7.2E-07	6.0E-07	6.8E-07
	Sito	36	0.75	1.6E-06	6.2E-07	1.2E-06
6	Lab	5	0.56	3.4E-07	2.2E-07	3.0E-07
	Site	36	0.83	5.2E-07	1.9E-07	3.6E-07
8	Lab	4	0.60	1.2E-07	7.0E-08	1.1E-07
	Site	21	0.99	4.20E-07	1.1E-07	3.00E-07

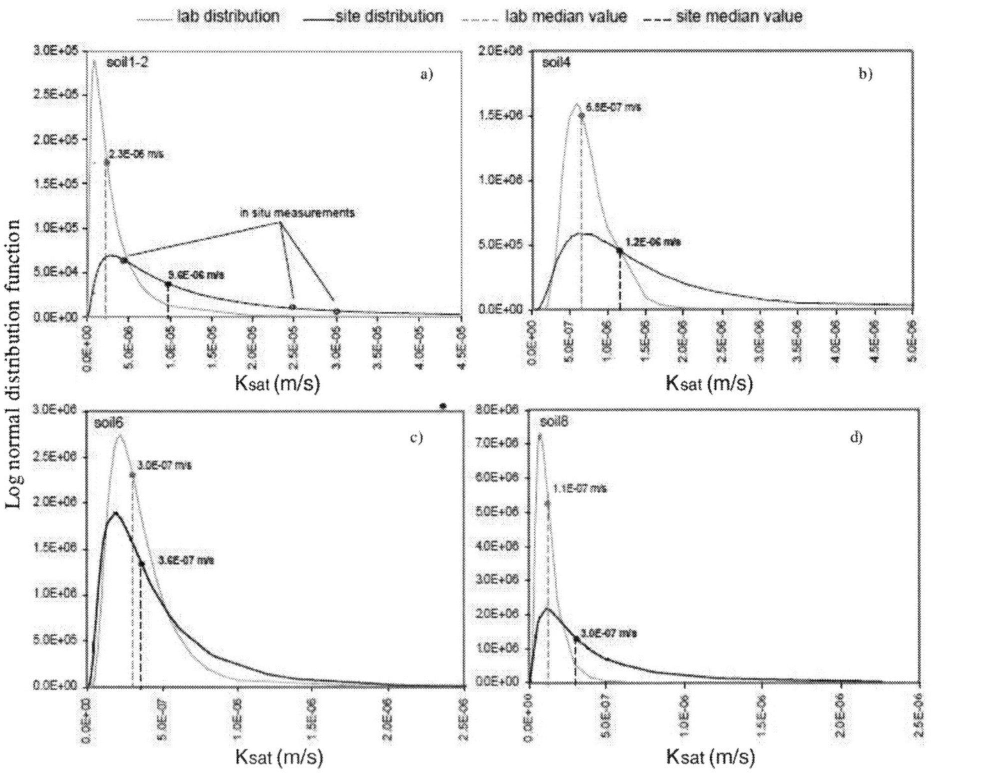

Figure 4. Log-normal distribution functions of saturated hydraulic permeability measured in the laboratory and estimated in situ for soil 1-2 (a); soil 4 (b); soil 6 (c); soil8 (d).

Table 3 Values of saturated conductivity estimated via different methods for each ash soil (modified from [3])

Soil	method	description	K_{sat} (10^{-7}m/s)
1&2	in lab measurements	Range between the minimum and maximum measurements	1.82-63.7
	estimation from in situ measurements	Range between the values corresponding to the cumulative log-normal distribution of 5% and 95%	15.80- 580
	in situ infiltration tests	Range between the minimum and maximum measurements	50.00-300
4	in lab measurements	Range between the minimum and maximum measurements	3.64-9.08
	estimation from in situ measurements	Range between the values corresponding to the cumulative log-normal distribution of 5% and 95%	3.40-41.00
6	in lab measurements	Range between the minimum and maximum measurements	1.52-5.47
	estimation from in situ measurements	Range between the values corresponding to the cumulative log-normal distribution of 5% and 95%	0.92-14.00
8	in lab measurements	Range between the minimum and maximum measurements	0.64-1.82
	estimation from in situ measurements	Range between the values corresponding to the cumulative log-normal distribution of 5% and 95%	0.58-15.00

3.3 In situ measurements of saturated hydraulic conductivity

Some direct measurements of saturated hydraulic conductivity obtained in situ via double ring infiltrometer tests [6] confirm, at least for the shallower soil layer, that the saturated hydraulic conductivity is higher in situ than in the laboratory (see Fig. 3a, black circles). However, the range of the saturated conductivities from infiltration tests contains the median value derived from elaboration of the monitoring data (see Table 2). In Table 3 the range of saturated conductivity determined for each soil and the type of tests performed are summarised. Hence it seems acceptable to perform several in situ measurements of saturated conductivity and some laboratory evaporation tests to obtain the parameters of the Mualem–van Genuchten hydraulic conductivity function operating at the site scale.

4 Conclusion

A practical method to set an appropriate hydraulic conductivity function is suggested. It is enough to perform several in situ measurements of saturated conductivity and some evaporation tests in accordance with the procedure proposed by [5] to obtain the parameters to be used in the Mualem–van Genuchten model. As regards the test site, the saturated conductivities estimated from in situ measurements prove higher than those determined in the laboratory, especially for the superficial ash layer, the amplification ratio being 4. Thus the non-representativeness of the laboratory samples was proved.

References

1. L. Pagano, L. Picarelli, G. Rianna, G. Urciuoli, Landslides **7**, 3, 273–289 (2010)
2. M. Pirone, R. Papa, M.V. Nicotera, G. Urciuoli, Engineering Geology for Society and Territory **2**, Landslide Processes, 1069-1073 (2015)
3. M. Pirone, R. Papa, M.V. Nicotera, G. Urciuoli, Journal of Hydrology **528**, 63–83 (2015)
4. M. Pirone, R. Papa, M.V. Nicotera, G. Urciuoli, Landslides **12**, 2, 259–276 (2015)
5. M.V. Nicotera, R. Papa, G.Urciuoli, Geotech. Test. J. **33**, 4, 263–285 (2010)
6. ASTM D5126-90, Standard Guide for Comparison of Field Methods for Determining Hydraulic Conductivity in the Vadose Zone. doi: 10.1520/D5126-90R04 (2004)
7. R. Papa, M. Pirone, M.V. Nicotera, G. Urciuoli, Geotechnique **63,** 5, 420–426 (2013)
8. M.Th. van Genucthen, Soil Sci. Soc. Am. J. **44**, 892-898 (1980)
9. M. Pirone, R. Papa, M.V. Nicotera, G. Urciuoli, Volcanic Rocks and Soils – Rotonda et al. (eds) © 2016 Taylor & Francis Group, London, ISBN 978-1-138-02886-9, 273-280 (2016)
10. R.J. Hoeksema, P.K. Kitanidis, Water Resour. Res. **21**, 4, 563–572 (1985)
11. E.A. Sudicky, Water Resourc. Res. **22**, 13, 2069–2083 (1986)
12. G.E. Blight, Géotechnique **47**, 4, 715–767 (1997)
13. R.G. Allen, L.S. Pereira, D. Raes, M. Smith, Crop evapotranspiration: Guidelines for computing crop requirements. Irrig. Drain. **56**, FAO, Rome, Italy (1998)
14. R. Greco, L. Comegna, E. Damiano, A. Guida, L. Olivares, L. Picarelli, Earth Syst. Sc. **17**, 10, 4001–4013 (2013)
15. A. Basile, G. Ciollaro, A. Coppola, Water Resour. Res. **39**, 12, 1–12 (2003)
16. M. Pirone, R. Papa, M.V. Nicotera, G. Urciuoli, Procedia Earth Planet. Sci. **9**, 163–170 (2014)

[a] Corresponding author: marianna.pirone@unina.it

Response of different osmotic membranes and calibration of a coupled model capturing the temperature effect for the osmotic method

Shengyang Yuan[1], Xianfeng Liu[1,a], Olivier Buzzi[1]

[1]*Priority Research Centre for Geotechnical and Materials Modelling, School of Engineering, The University of Newcastle, NSW, Australia.*

Abstract: Reliable suction control is of paramount importance for experimental unsaturated soil mechanics. The osmotic method, based on polyethylene glycol (PEG) and semi-permeable membranes, is one of the possible methods to control suction. Although this method is quite simple to implement, reliable results can only be obtained with adequate calibration curves linking suction and PEG concentration, and a good understanding of the factors influencing the method. Several studies have investigated different aspects of this method, including the osmotic potential generated by given PEG concentrations, but the effect of temperature is not yet fully captured. This paper presents the results of a comprehensive study of the osmotic method looking at the long-term response of three different types of membranes, the effect of temperature, PEG molecular weight (MW) and the combination of different MW cut off (MWCO) on the calibration curves. Finally the data were used to calibrate a model capturing the temperature effect, in a coupled manner, in order to provide a robust calibration.

1 Introduction

The osmotic technique is one of the techniques available to control soil suction in experimental unsaturated soil mechanics [1-4]. The technique relies on an osmotic solution and semi-permeable membranes preventing the chemical species of the osmotic solution from diffusing into the soil tested. The suction generated within the soil specimen depends on the concentration of polyethylene glycol (PEG) and, in order to obtain reliable results, it is crucial to have a reliable and quality correlation between suction generated, PEG concentration, temperature and any other parameters that might influence the magnitude of suction. This correlation is referred to as a calibration curve.

Many authors have investigated the relationship between suction and PEG concentration but the effect of temperature has not been investigated comprehensively and is not accounted for explicitly in the calibration curves published in the literature. In the 1970s, Michel and Kaufmann showed that suction generated by PEG solutions decreases as temperature increases. This was recently confirmed by Tang and co-workers al. [5]. Such experimental observations are not in agreement with thermodynamics based models such as Van't hoff equation, which tend to predict an increase of suction as temperature increases. The data presented in this paper are part of a study aiming at ascertaining the quality of calibration of the osmotic technique and providing comprehensive dataset to capture and predict the effect of temperature on the magnitude of suction generated by the osmotic method. Three membrane types were tested, and their long-term response was studied. After deciding upon the most adequate membrane types, other tests were performed to establish the relationship between PEG concentration, PG molecular weight and temperature.

2 Materials and Methods

Four different types of PEG were used for this study (PEG 6000, 12000, 20000 and 35000). All were supplied by Sigma Aldrich, Germany. The PEG solutions were prepared by mixing the appropriate mass of PEG in 60 grams of deionized water. A magnetic stirrer was used to ensure good mixing. In terms of membranes, three types were used:

- Polythersulfon membranes (noted PES, supplied by Sartorius Stedim Biotech GmbH, Germany). Molecular weight cut off (MWCO) of 1000, 5000 and 10000 were used.
- Cellulose Acetate membrane (noted CA, supplied by Sigma Aldrich, Germany). MWCO of 8000.
- A ceramic membrane (noted CM, supplied by Sterlitech) with a single MWCO of 5000.

The set up used to measure the suction generated by the osmotic system in shown in Figure 1a. It involves a high capacity tensiometer [7] directly in contact with the semi permeable membrane. The whole set up was placed in a controlled temperature environment (± 0.1 °C). Unlike the PES and CA membranes that look like a supple plastic film, the ceramic membrane is made of a hard ceramic substratum with an active layer on one side

[a] Corresponding author: Xianfeng.Liu@newcastle.edu.au

(Figure 1c). The pore size distribution of the ceramic membrane shows a dominant pore size of about 5 microns (Figure 1b), which translates into an air entry in the order of 300 kPa. This implies that the active layer should be placed in contact with the soil specimen (or the tensiometer) and the ceramic part has to be immersed in the PEG solution. Otherwise, there is a risk of desaturation of the ceramic during testing. Moisture transfer would then occur via the vapour phase, which known to be a slow process.

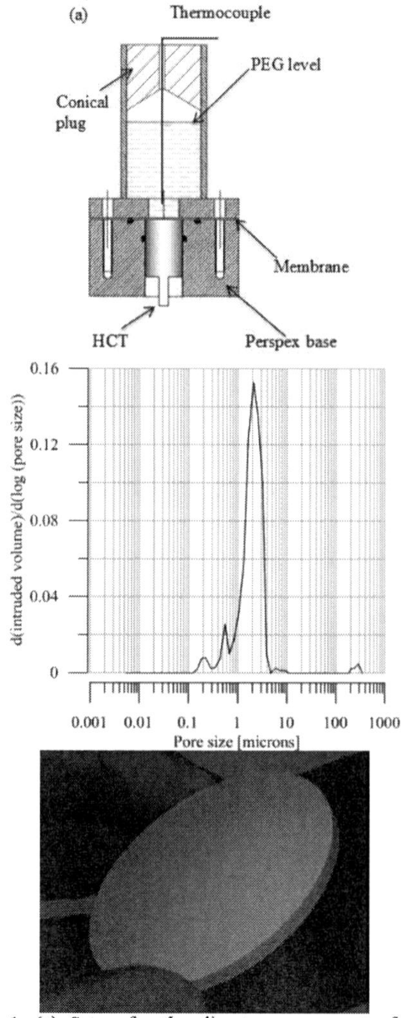

Figure 1. (a) Setup for the direct measurement of suction generated by the osmotic system. (b): Pore size distribution of the ceramic membrane obtained by mercury porosimetry. (c): View of the ceramic membrane and its active layer.

3 Testing Program

The test undertaken for this study are summarised below:

- Test series #1 focuses on the general response of the CA, PES and CM membranes. Only one PEG concentration per membrane, one temperature (20 degrees), one PEG molecular weight (MW) and one molecular weight cut off (MWCO) were used.
- Test series #2 investigates the suction generated by a combination of PES membranes, PEG molecular weight and different temperatures. For this part, four PEG MW (6000, 12000, 20000 and 35000), four PEG concentrations (about 0.15, 0.20, 0.25 and 0.35 g PEG/g water) and five temperatures (20, 25, 30, 35 and 40 degrees) were selected.

4 Results

4.1 Equilibration time and durability of membranes

Figure 2 shows the response of the three membrane types for testing times in excess of 700 hours. Three different behaviours are clearly visible. Equilibration is reached very quickly for the CA membrane (about 4 to 5 hours), after which the suction measured by the tensiometer drops progressively and continuously until a stabilization at about 500 hours. Such response was observed for all tests with the CA membrane. This reflects an issue of stability of the membrane, a phenomenon pointed out by Tarantino and Mongiovi [3] and Tripathy and co-workers [8], among others.

The PES membrane requires more than 50 hours for the equilibration to be reached. Afterwards, the suction is sustainably maintained for about 24 hours. A slight increase in suction can be observed in Figure 2 from 70 hours onwards, which is the result of a slight evaporation of the PEG solution, yielding an increase in concentration and, hence, in suction. This was proved from a change in Brix index, from 20.3% before testing, to 21.9% at the end. This equates to a change of PEG concentration from 0.3 to 0.33 g PEG/g water. Regardless of the slight evaporation, the PES membrane remained functional for about 800 hours before the test was terminated.

As far as the authors are aware, there is no data in the literature on ceramic membranes. The preliminary results showed here suggested that the equilibration time is much longer than that of PES and CA membranes. Indeed, it takes more than 100 hours to reach equilibrium. Testing with this membrane showed no sign of degradation and this type of membranes might be more suited when long term testing is required (e.g. control of suction for soil testing with multiple cycles).

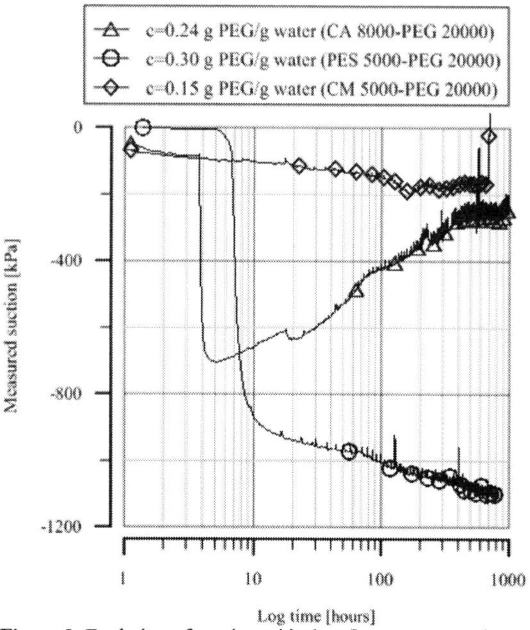

Figure 2. Evolution of suction with time for CA, PES and CM membranes.

At this stage, testing of the ceramic membrane is in progress and the rest of the paper will focus on new data pertaining to PES membranes which exhibit a relatively short equilibration time and good stability.

4.2 Relationship between concentration, suction and temperature

Figure 3 displays the data obtained at 20 degrees, for a range of PEG MW and two different membranes molecular weight cut off.

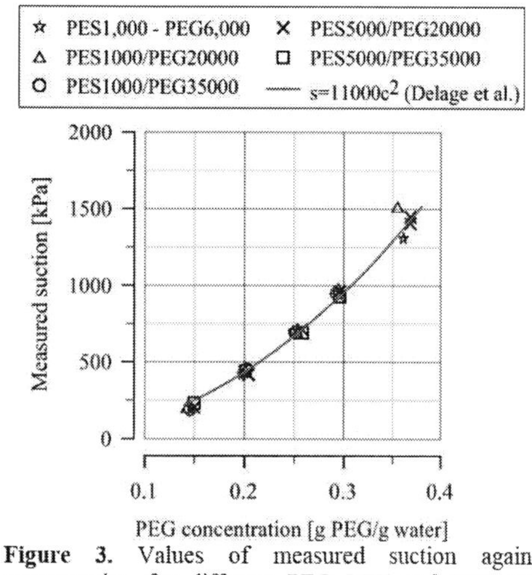

Figure 3. Values of measured suction against PEG concentration for different PEG MW and PES membranes MWCO.

All measured suctions are found to fall very close to the equation proposed by Delage and co-workers [4] that reads:

$$s = 11000\, c^2 \quad (1)$$

with c in g PEG/g water and s in kPa.

These data confirm that, regardless of the molecular weight and at 20 degrees, equation (1) applies to a number of combinations of PES membranes and PEG molecular weights, as long as no diffusion of PEG molecule occurs across the membrane. Such a result that was already observed by other researchers shows the robustness of equation (1) as a calibration equation.

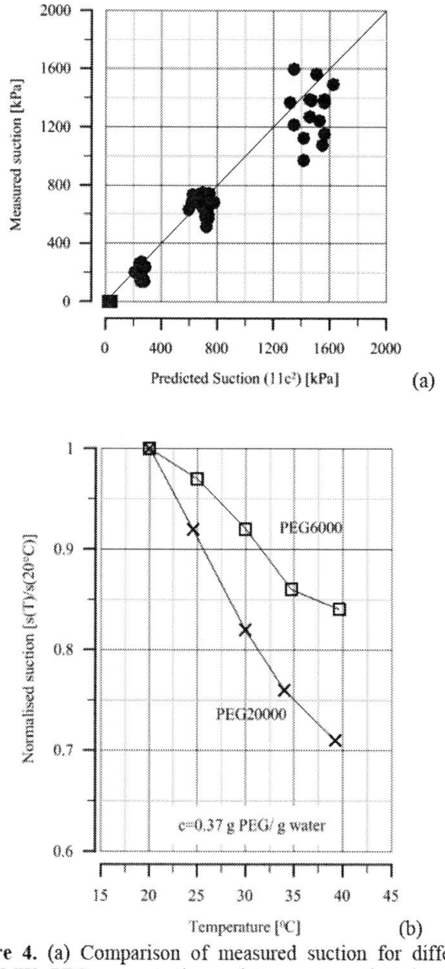

Figure 4. (a) Comparison of measured suction for different PEG MW, PEG concentration and temperature plotted against the predicted value, using Equation 1. (b): Reduction of suction (expressed in a normalised manner) with increasing temperature for PEG6000 and PEG20000. The normalised suction is the suction at temperature T over the suction at 20 degrees, which is about 1500 kPa.

Increasing the temperature produces an interesting outcome. Figure 4 plots the measured suction (at different temperatures) against the predicted suction (using Equation (1)). This time, a large scattering of data, more

DOI: 10.1051/e3sconf/20160910010

pronounced for high values of suction, is visible. In particular, the measured suction is systematically lower than its predicted counterpart. This trend is confirmed by Figure 4b where the normalised suction drops as temperature increases. Interestingly, the magnitude of suction generated by the osmotic system is now dependent on the PEG molecular weight.

These experimental results are consistent with existing data of the literature [5, 6] but do not agree with thermodynamics based models, such as the virial model [9] or Van't Hoff equation, used to estimate osmotic pressure. Indeed, such models generally predict an increase of osmotic pressure as temperature rises:

$$\pi / c \cong RT\left(A_1 + A_2 c + A_3 c^2 + \cdots\right) \quad (2)$$

The present results suggest that a complete calibration of the osmotic method should account for the temperature, PEG molecular weight as well as PEG concentration.

4.3 A coupled model to predict the suction generated by the osmotic system

Michel and Kaufmann [6] proposed an empirical equation that captures the observed experimental results with terms coupling concentration and temperature:

$$s = A_1 c + A_2 c^2 + A_3 cT + A_4 c^2 T \quad (3)$$

where s is the suction [kPa], c is the PEG concentration [g PEG/g water], T is the temperature [^{0}C], A_1, A_2, A_3, A_4 are fitting coefficients.

For the PEG20000, the empirical model is calibrated as:

$$s = -1500 \cdot c + 21000 \cdot c^2 + 48 \cdot cT - 365 \cdot c^2 T \quad (4)$$

Figure 5 shows the calibrated model (lines) against the experimental data points. The coupled equation adequately captures the experimental trends and can be considered an efficient tool to correct suction values in unsaturated soil testing should temperature fluctuate. Note that using a different PEG molecular weight would require re-calibrating the model.

5 Conclusions

This experimental study investigated the response of three types of semi-permeable membranes. Both CM and PES membranes seem durable but the equilibration time of the CM membrane is much longer than for the PES membrane. At this stage, testing on the ceramic membrane is to continue. The series of tests on the PES membranes conducted at 20 degrees Celsius show that the results are largely in agreement with the equation proposed by Delage and co-workers [4] and are independent of the PEG molecular weight and membrane molecular weight cut off (as long as no diffusion of PEG molecules occur across the membrane). However, it was found that the PEG molecular weight does influence the suction generated when the temperature is increased. A consistent reduction in suction was observed with a clear dependence on the PEG molecular weight. The experimental results were used to calibrate a coupled empirical model that can capture the evolution of suction generated as a function of PEG concentration and temperature hence providing a robust solution to adjust values of suction should the temperature fluctuate during testing.

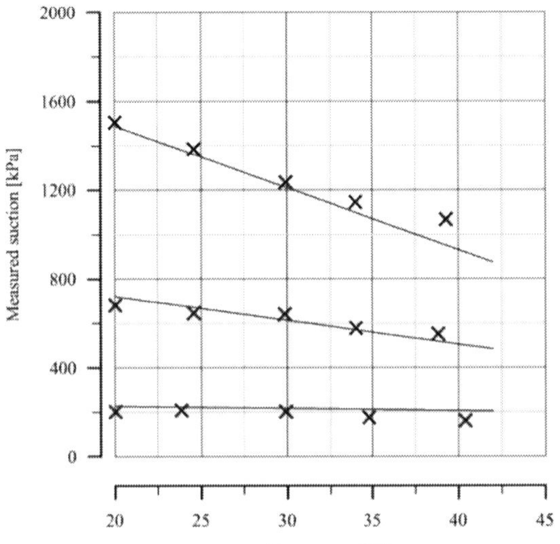

Figure 5. Calibrated model (Equation 4, for PEG20000) represented by the lines and experimental values of measured suction for different temperatures and PEG concentrations. Highest suction values correspond to 0.35 g PEG/g water, the middle ones to 0.25 g PEG/g water and the lowest ones to 0.15 g PEG/g water.

References

1. G. Kassiff and A. Ben Shalom Géotechnique, **21**, 245-255 (1971)
2. R. Monroy, A. Ridley, K. Dineen, K. and L. Zdravkovic ASTM Geotech. Test. J. **30**, 1-7 (2007)
3. A. Tarantino, L. Mongiovi, In: Toll, D.G., Leong, E.-C. (Eds.), Unsaturated Soils for Asia. Balkema, pp. 303–308 (2000)
4. P. Delage, M. Howat and Y.J. Cui. Eng. Geology, **50** (1-2) : 31-48 (1998).
5. A.M. Tang, Y.J. Cui, L.X. Qian, P. Delage and W.M. Ye. Can. Geotech. J. **47**, 359–365, 2010.
6. B.E. Michel and M.R. Kaufmann, Plant Physiol. **51**, 914–6. (1973)
7. J. Mendes and O. Buzzi. Can. G. J. **50**, 550-556 (2013)
8. S. Tripathy, M.Y.M. Tadza and H.R. Thomas Géotechnique Lett. **1**, 47–51. (2011)
9. P.J. Flory. Principles of polymer chemistry. Cornell University Press (1953).

 DOI: 10.1051/e3sconf/20160910011

Erodibility of soil above the groundwater level: some test results

Anna Shidlovskaya[1,a], Jean-Louis Briaud[2], Mabel Chedid[2], Keshavarz M.[2]

[1]*National Mineral Resources University (University of Mines), St. Petersburg, Russia (shidanna2013@gmail.com)*
[2]*Texas A&M University, College Station, TX, USA*

Abstract. The paper presents a study on erodibility of soil above the groundwater level where the water is in tension. Such soils particularly clays are very sensitive to moisture and temperature changes and can be eroded significantly by water flow. The erosion of clay and sand samples from the US National Geotechnical Experimentation Site at Texas A&M University is studied. Two sets of experiments are done with the clay and the sand. The first set was performed on sample collected in November 2014 and the second set on samples from June 2014. The depth of the samples varied from 0.6 to 3.6 m where water content and density changes. A series of erosion tests was performed in the Erosion Function apparatus (EFA) with the intact clay and then with the sand reconstructed to the field density and field water content. The erosion tests are performed at different flow velocities varying from 0.5 m/s to 5.5 m/s. The erodibility is quantified by the relationship between the erosion rate and the water velocity called the erosion function. Some relationships between the critical velocity and common soil properties are discussed. The collapse of the clay structure when inundated (soaking) is studied.

1 Introduction

Erosion of soil is a phenomenon which affects many engineering infrastructure elements. Some examples include bridge failures by scour during storms, levee failures by overtopping during hurricanes or floods, internal erosion of dams, suffusion in a soil in the foundation, meander migration in rivers, river banks erosion and slope failure associated with that, cliff and shore erosion, and surface erosion of highway embankments. This paper addresses the erosion of soils at shallow depth above the water table where the water is in tension. Erosion laboratory tests are presented as well as geotechnical properties with the goal of gaining some insight on the relationship between erodibility and soil properties.

2 Definition of erodibility

Erosion is the removal of soil particles by flowing water causes soil structure collapse. The water applies the load on the soil which provides a level of resistance to the erosion process. The water also affects soil not only mechanically but also chemically. The chemical processes between water and soil are leaching and dissolution of soil that depends on the type of structure and bonding as well as salt composition of soil. Erodibility is the potential for erosion and can be quantified by the relationship between the erosion rate $\dot{z}$ (mm/hr) and the water velocity v (m/s); therefore erodibility is not a single number like the friction angle of a soil for example but a curve like a stress strain curve. In fact the $\dot{z}$ vs. v curve is called the erosion function and plays the role of constitutive law for erosion processes. One problem is that the velocity varies in the water body and is theoretically zero at the interface between the water and the soil where the erosion takes place. So the water velocity, while a simple and convenient parameter, is not an ideal parameter for the erosion function. Instead many researchers have been using the shear stress τ imparted by the water at the interface. This shear stress is extremely small but so is the soil resistance in many instances. This is why erosion can be devastating in many cases. In this paper, the erodibility of a soil will be characterized by the erosion function expressed as the erosion rate $\dot{z}$ vs. water velocity v. This erosion function is often described by using the following model:

$$\dot{z} = \alpha_v (v - v_c)^{\beta_v} \quad (1)$$

Where $\dot{z}$ is the erosion rate in mm/hr, α_v is soil constant corresponding to the erosion rate when $v-v_c$ is equal to 1 m/s, v is the mean depth flow velocity, v_c the critical velocity, β_v the slope of the velocity based erosion function in log log space.

[a] shidanna2013@gmail.com

3 Erosion function apparatus or EFA

The Erosion Function Apparatus test or EFA test was developed in the early 1990s to measure the erodibility of soils and soft rocks. The principle is to go to the site where erosion is being investigated, collect samples within the depth of concern, bring them back to the laboratory and test them in the EFA. The 75 mm outside diameter sampling tube is placed through the bottom of the conduit where water flows at a constant velocity. The soil or rock is pushed by a piston out of the sampling tube only as fast as it is eroded by the water flowing over it. The test result consists of the erosion rate $\dot{z}$ versus shear stress τ curve and erosion rate $\dot{z}$ versus mean flow velocity v curve. This paper presents only the results of the erosion in terms of the erosion rate versus water flow velocity.

For each flow velocity v, the erosion rate $\dot{z}$ (mm/hr) is simply obtained by dividing the length h of sample eroded by the time t required to do so.

$$\dot{z} = \frac{h}{t} \tag{2}$$

The velocity v is obtained by measuring the flow Q and dividing by the flow area A.

4 Erodibility parameters

In order to characterize the erodibility of a soil with a single number, it becomes necessary to isolate one of the elements of the erosion function or to associate certain ranges of erosion functions with a classification number.

Note that the erosion function is often plotted on a log log space to capture the extreme values with precision; this technique also has the advantage that the erosion function is often close to a straight line on this set of axes. Elements of the erosion function which are commonly used are the critical velocity v_c, the critical shear stress τ_c, and the slope of the curve, β_v and β_τ. The critical velocity is the mean depth velocity of the water flow below which no erosion takes place. Practically, "no erosion" in the erosion soil test is defined as an arbitrarily low finite value of 0.1 mm/hr. The initial slope of the erosion curve defines how sensitive the soil erosion is to an increase in water velocity. It should be noted that even if water velocity tends to zero or water does not flow soil could soak in steady water that could be considered as a type of erosion.

5 Erosion factors

The erodibility of soils varies significantly from one soil to the next; therefore erodibility depends on the soil properties such as structure, lithology, grain size and chemical composition, type of bonding, physical and mechanical properties, and presence of fissures that has an effect on the full-scale field behavior of hard clay. It also depends on the properties of the water flowing over the soil. In this paper the full-scale behaviour of soil and the properties of the water are not considered as only samples are tested.

The soil above and below the groundwater level could erode in different ways depending on whether it is saturated or partially saturated with both air and water in the pores. The wet-dry cycles due to the weather and associated moisture migration in the soil profile transported by a thermal gradient during the year can also have an impact on the soil erodibility. For example, the formation of shrinkage cracks and then water flowing through the cracks can erode a soil significantly. In this case, erodibility will be affected by both the water velocity and the density of the shrinkage cracks pattern. This is particularly important at shallower depths with problems such as overtopping of levees during hurricanes or floods, river banks erosion, surface erosion of highway embankments and so on.

6 Soil profile and properties

The erodibility of soils above the groundwater level and clay in particular is very sensitive to moisture changes and water flow. For this research the soil was obtained from the National Geotechnical Experimentation Sites at Texas A&M University. A total of 7 borings were drilled at the Clay site and at the Sand site. Three borings were conducted in November 2014 at both sites and four were conducted in June 2015 at both sites as well. The soil was sampled every 0.6 m in the interval of 0.6-3.6 m depth with 76 mm pushed thin wall steel tubes.

The ground elevation at the Clay Site is about 77 meters above MSL and the Sand site is about 79 m above MSL. The clay is terrace deposits of Pleistocene age. More precisely it is a calcium-carbonate-cemented light-brown and reddish-brown clay containing nodules of calcareous and organic matter. The calcareous nodules and the calcium-carbonate-cemented clay mass show a strong reaction to acid. The amount of carbonates nodules in the clay decreases with depth from 25 % of the sample volume at 0.6-1.8 m to 10-15 % at 1.8-2.4 m and 5 % at 2.4-3.6 m. The clay appears multi-fissures at the depth from 0.6 to 3.6 m. The major clay minerals for the clay are primarily smectite group (very likely montmorillonite) with some illite. The unit weight of solids varies from 24.7 kN/m^3 at the 0.6-1.8 m to 25.9 kN/m^3 at the 1.8-3.6 m.

The sand at the Sand Site is from fluvial and overbank deposit of Pleistocene age. This sand is mottled light red silty sand slightly cemented by capillary water in tension. For testing, the sand samples from depths 0.6-3.6 m was remolded and reconstructed while matching the field density and the water content.

Figure 1 presents grain size distribution curves for the clay. The clay composed primary of clay-size and colloidal particles. The amount of clay particles varies from 62% to 82% through the clay profile. The sand consists of silt, sand and some gravel particles. The coefficient of uniformity varies from 3.1 to 4. Figure 2 shows grain size distribution curves for the sand taken from the Clay Site in November 2015. The plasticity index of the clay varies from 22 to 36. Table 1 shows the water content and total unit weight of the clay samples in November 2014 and June 2015. An expected fluctuation of water content and total unit weight between the winter (November) and the summer (June) is observed within the depth of 0.6-3.6 m (Fig. 3).

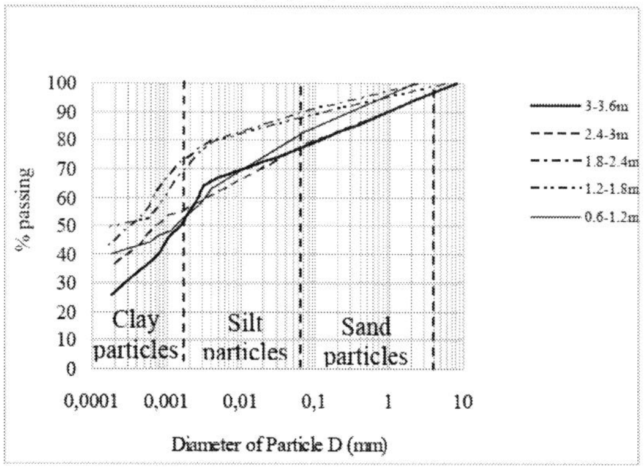

Figure 1. Grain size distribution of clay.

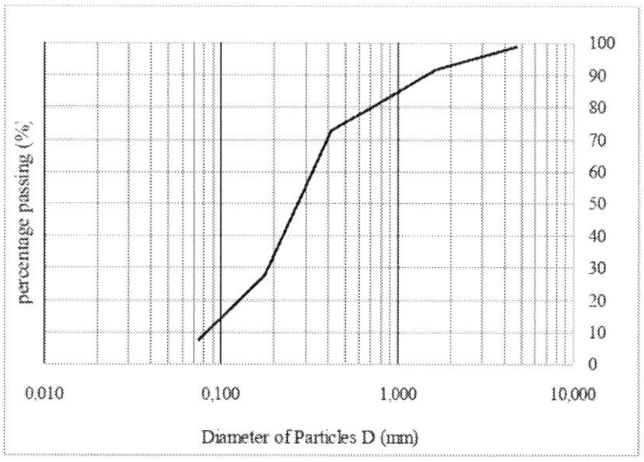

Figure 2. Grain size distribution of sand.

Table 1. Classification and properties of clay (National Geotechnical Experimentation Clay Site).

Soil classification	Depth, m	Water content, %	Total unit weight, kN/m^3
CH/fat clay	0.6-1.2	NA* / 20.5**	NA* / 19.7**
CH/fat clay	1.2-1.8	17.6* / 21.8**	20.9* / 21.4**
CH/fat clay	1.8-2.4	29.9* / 22.8**	19.3* / 20.8**
CH/fat clay	2.4-3.0	NA* / 22.5**	19.7* / 20.1**
CH/fat clay	3.0-3.6	27.0* / 22.7**	20.0* / NA**

* - water content and total unit weight of the samples obtained in November 2014; ** - water content and total unit weight of the samples obtained in June 2015.

Three zones in the clay profile from the top to the bottom can be identified based on the June results. The first zone is a zone of low water content (0-1.2 m); the second zone is a zone of increasing water content compared to the first zone (1.2-1.8 m); the third zone is a zone of higher and constant water content (1.8-3.6 m). Such variation may be due to the likely fluctuation of the groundwater level over the years. The GWL at the Clay Site is currently expected at a depth below 6 meters at the Sites

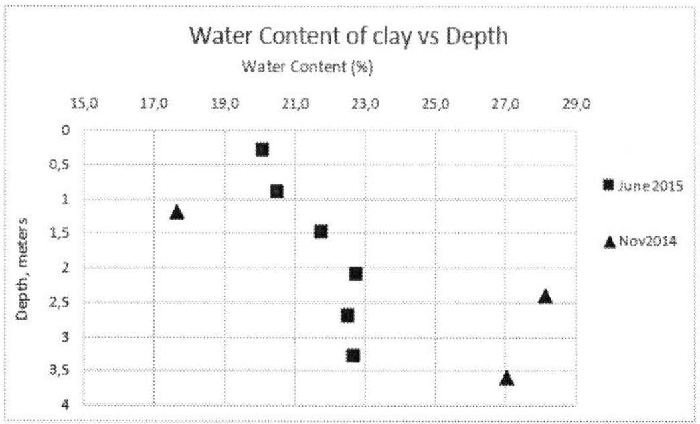

Figure 3. Water content of clay vs. depth

Groundwater is transported by thermal gradient which has an impact on the water content at shallow depth down to 3 m. At a depth greater than 3 m at this location the soil temperature remains relatively constant throughout a year. However it is expected that the soil temperature could change below 3 m with an increase in a depth of shrink fissures. Measurements made in Houston indicate that the temperature variation between the ground surface and a depth of 3 m can reach 8 degree Celsius in the summer and 6 Celsius degrees in the winter. Within that zone, the thermal gradient can vary from 2 degree Celsius per meter in November to 2.6 degree Celsius per meter in June.

7 Water soaking durability test results

Water soaking of clayey soils is a phenomenon which affects the clay structure and causes its complete destruction in some cases. The reaction to soaking of clays varies from one soil to the next and depends mostly on the type of bonds between particles. These bonds include the increase in effective stress due to the tension in the water at the contact between particles. They are easily lost when the clay is immersed in water. Other bonds also play a role in this process. The clay tested is cemented with water-colloidal and crystallization bonding between particles. The first type of bonding is due to electromagnetic forces between water molecule and soil particles particularly colloids with a size of less than 0.0001 mm. Because of this type of bonding, the strength of the dry clay decreases when getting wet. The second type of bonding between particles depends on the composition of the cementing material between particles. As was mentioned above, the clay is formed by calcium carbonates and likely gypsum bonding due to salt crystallization during drying periods. This type of bond is very likely destroyed by water especially if water contains aggressive CO_2 and cannot be recovered even if the clay dries again.

During the water soaking durability test, the clay samples were first air dried and then completely immersed in water; they all collapsed within 5 to 7 minutes after immersion. The effect of "melting sugar" and air bubbles formation was observed during the soaking of the clay.

8 Erosion results in the EFA

A series of erosion tests were performed in the Erosion Function apparatus (EFA) on the clay and sand samples collected in November 2014 and June 2015. The clay samples were undisturbed samples but the sand samples were reconstituted to match the estimated field density and measured water content. The erosion tests were performed at different flow velocities varying from 0.5 m/s to 5.5 m/s. These tests were not repeatable because of the limit number of soil samples. The erodibility of the soil was quantified by the relationship between the erosion rate and the water velocity called the erosion function. Figs. 4 and 5 show the erosion functions of the clay and of the sand respectively. As seen in Fig. 4, all clay samples behaved mostly as medium erodibility soils with a general increase in erosion rate with an increase in velocity as expected. The erratic behavior of the sample from 0.6-1.2 m clay is due to a highly heterogeneous sample with different layers being eroded at different velocities.

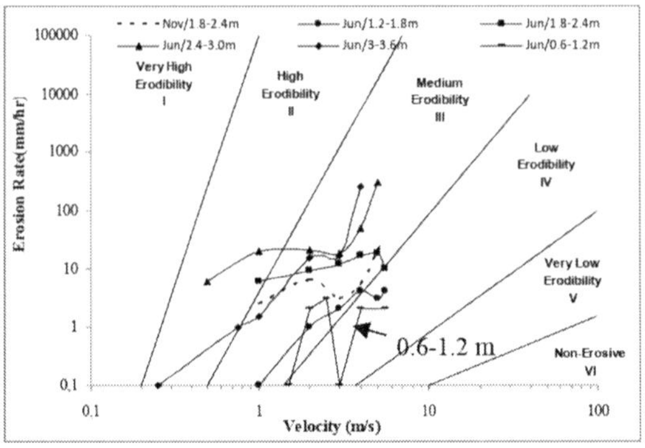

Figure 4. The erosion function of clay.

The relationship between the erosion rate and the water velocity of the sand is more consistent than the clay. The erosion rate increases with an increase in water velocity (Fig. 5). It was also observed that for the clay, the collapse of the clay structure when soaking in water correlated well with the erosion behavior of the clay. The clay samples were put in stationery water and left there for 24 hours. The clay structure for some samples depending on the soil water content and density collapsed even in 5-10 minutes after running the test. It was indicated that the more sensitive to water soaking the clay was, the more erodible it was in the EFA.

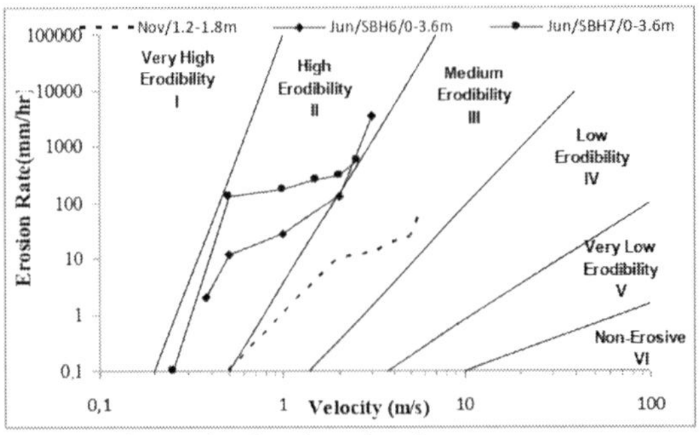

Figure 5. The erosion function of sand.

9 Erosion and properties for Clay Site

The relationship between the critical velocity and the soil properties is the only relationship investigated in this paper due to complexity of the relationship between erodibility and soil properties as well as the limited extent of the test results. The critical velocity is defined as the velocity corresponding to an arbitrary low erosion rate of 0.1 mm/hr.

A comparative analysis of the clay results at a depth of 1.8-2.4 m in November 2014 and June 2015 leads to the following observations (Fig. 6 and 7). The critical velocity for the clay with low water content (22.8%) and higher dry density (16.9 kN/m^3) in June is 3.3 times less than the critical velocity of the same clay in November with higher water content (29.9%) and lower dry density (14.8 %). It is clear that this clay with lower water content and higher dry density is more sensitive to water flow and erodes more. The erosion function (Fig. 4) also indicates that the erosion rate of the clay at a depth of 1.8-2.4 m is higher in June than in November. The possible reason is that when the clay with the low water content and the higher water tension in the pores comes into contact with the water at the beginning of the erosion process, the water rushes into the pores. The air is compressed because it cannot escape and the particles are loosened and detached by this process. By comparison the clay with the higher water content has already come to water tension equilibrium and is less impacted by that phenomenon.

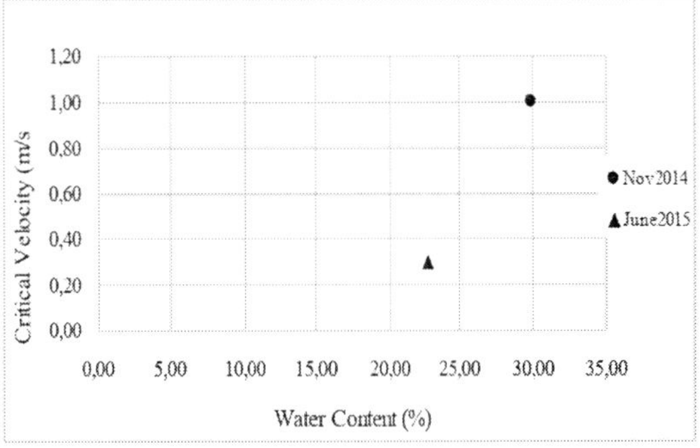

Figure 6. Critical velocity of 1.8-2.4 m depth clay vs. water content.

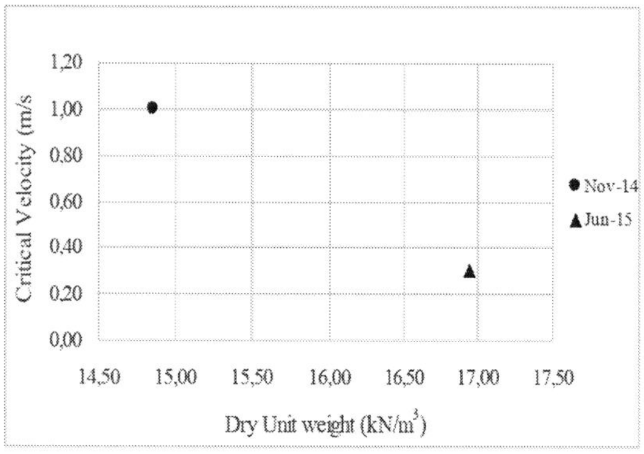

Figure 7. Critical velocity vs. dry unit weight of 1.8-2.4 m depth clay.

Other relationships between the critical velocity and clay properties were investigated without much success. The content of clay and silt particles did not show an identifiable relationship (Figs. 8 and 9).

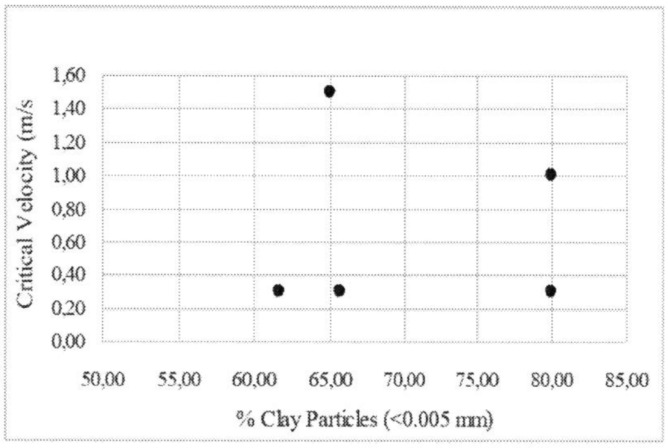

Figure 8. Critical velocity vs. percent of clay particles of clay.

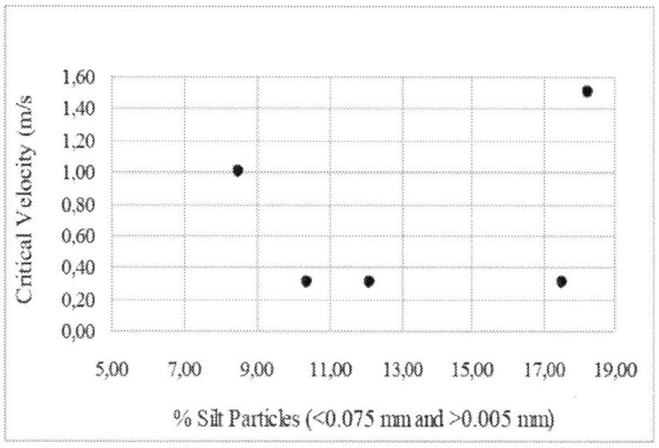

Figure 9. Critical velocity vs. percent of silt particles of clay.

It would seem reasonable to expect that the plasticity of clay would impact its erodibility with more erosion resistance for higher plasticity clays. This is mildly observed in Fig. 10 where there is a possible trend upward in the data between critical velocity and plasticity. This confirms that the mineral composition of clay is very important to understand its resistance to erosion.

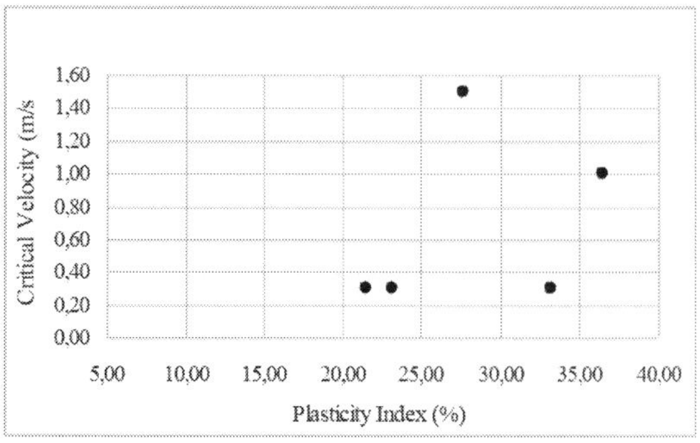

Figure 10. Critical velocity vs. plasticity index of clay.

The comparison on Fig. 11 between the critical velocity and the undrained shear strength does not show a strong correlation.

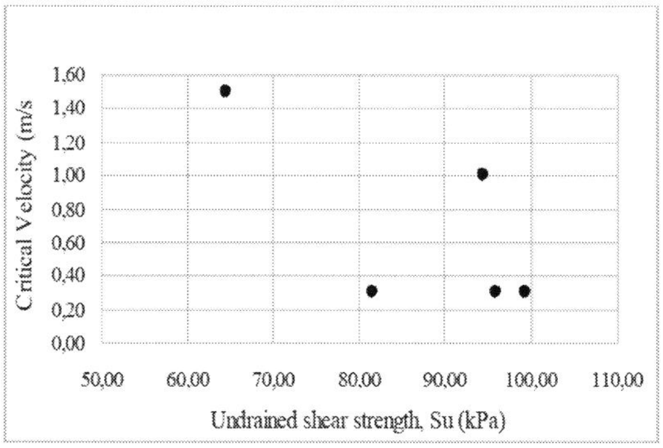

Figure 11. Critical velocity vs. undrained shear strength of clay.

10 Erosion and properties for Sand Site

The results of the tests performed with the sand samples confirm that the sand-water interaction is very different from the clay-water interaction. Figs. 12 and 13 show that the critical velocity of the sand decreases with an increase in water content and a decrease in dry density. This is contrary to the findings for the clay. The reason is that a denser particle arrangement helps resist the water flow better than a loose arrangement. In this sense compaction will help increase the critical velocity.

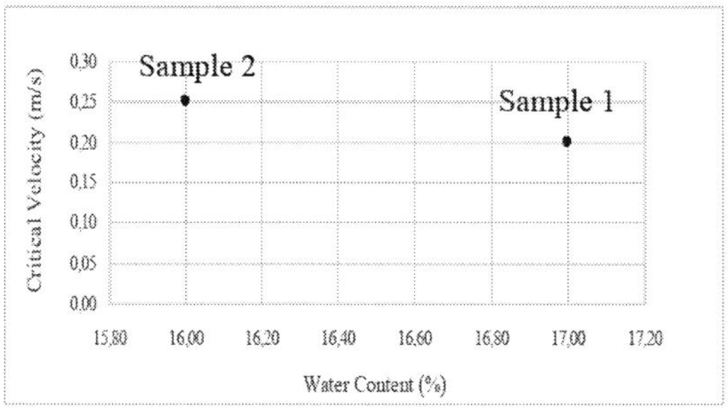

Figure 12. Critical velocity vs. water content of sand.

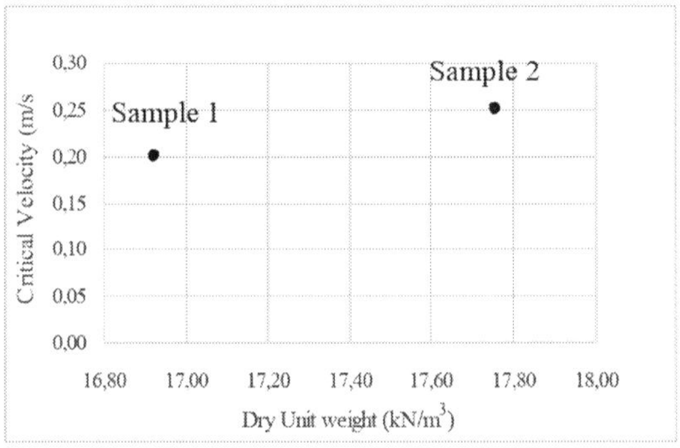

Figure 13. Critical velocity vs. dry unit weight of sand.

Fig. 14 shows that the critical velocity decreases as the sand becomes more uniform. Indeed the more uniform the sand is the higher the void ratio and the harder it is to reach a high level of compaction. Fig. 15 confirms that the critical velocity increases with an increase in the mean grain size D50 as is well known. Indeed sand behavior is controlled by the weight of the particles unlike clays. The heavier the particle is the more difficult it is to move it by water flow.

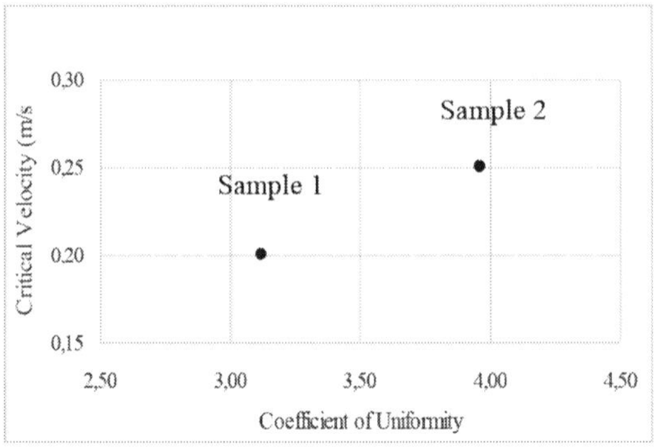

Figure 14. Critical velocity vs. coefficient of uniformity of sand.

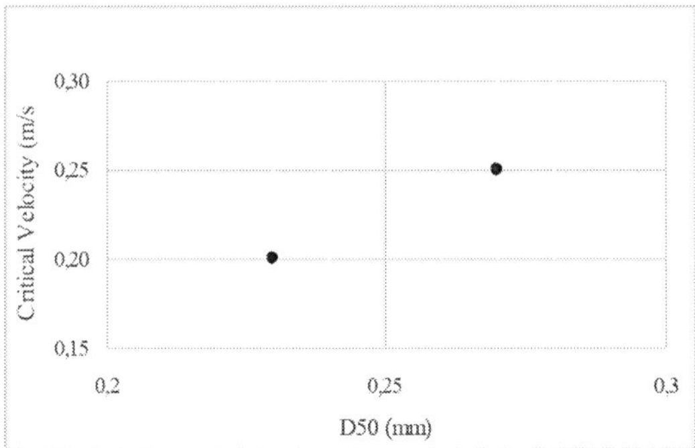

Figure 15. Critical velocity vs. D50 of sand.

11 Conclusion

This paper presents the results of a first set of experiments on the erosion characteristics of the soils obtained at shallow depths at the National Geotechnical Experimentation Site at Texas A&M University. Clay and sand samples were obtained above the groundwater level where water is in tension and transported by thermal gradient during a year. The clay at the Clay Site and the sand at the Sand Site were tested for erodibility in November 2014 and June 2015. The samples had a range of field water content and field density. The erosion test was the Erosion Function Apparatus (EFA) test. The clay samples were tested undisturbed with field water content while the sand samples were reconstituted to their natural water content and density.

The erosion function for the clay samples and the sand samples were obtained. They gave the critical velocity as well as the complete erosion function. The possible relationships between the critical velocity and the soil properties were studied. It was found that the erodibility of clays is more complex and has more influencing factors than the erodibility of sand. Sometimes opposite trends are found between the two soil types. Much more data needs to be collected before stronger conclusions can be reached. This is the purpose of a new NCHRP project in which the authors are involved.

Acknowledgements

The authors wish to thank Fugro and especially Scott Marr for their help with drilling and sampling. The authors also wish to thank colleagues and graduate students at Texas A&M University for their contributions to the laboratory and field work.

References

1. Briaud J.-L., *Geotechnical Engineering: unsaturated and saturated soils*, John Wiley and Sons, 1000 pp. (2013)
2. Briaud, J.-L. The National Geotechnical Experimentation Sites at Texas A&M University: Clay and Sand; Data collected until 2002; Report NGES –TAMU-001, 311 p., Civil Engineering, Texas A&M University, College Station, Texas, USA (2003)
3. Cao Y., Wang J. , Briaud J.L., Chen H.C., Li Y., Nurtjahyo P. "EFA Tests and the influence of Various Factors on the Erodibility of Cohesive Soils", *Proceedings of the First International Conference on Scour of Foundations*, Texas A&M University, Dpt. of Civil Engineering, College Station, TX, USA (2002)
4. Croad, R.N. Physics of Erosion of Cohesive Soils, PhD Thesis, Department of Civil Engineering, University of Auckland, Auckland, New Zealand (1981)
5. Dashko, R.E., Shidlovskaya, A.V. Physical and Chemical Genesis of Swell and Osmotic Shrinkage of Clay Soils in Construction's Base by Results of Experimental Research". Notes of the University of Mines. St. Petersburg. Russia. V. **200**, pages 4-14. (2012)
6. Dashko, R.E., Shidlovskaya, A.V. "Engineering Geological and Geotechnical Approach to Clay Soils

DOI: 10.1051/e3sconf/20160910011

of Different Genesis". *Proceedings of the Eighth Russian Conference on the Engineering Survey for Construction.* Moscow, Russia, organized by PNIIIS, 13-14 December 2012, published by Geomarketing, Pages 12-14. (2012)

7. Reddy, G.B. "An experimental investigation of subsurface ground temperature in Texas: a complete study", International Journal on Ambient Energy. Volume **21**, Issue 4. (2000)

Measurement of soil suction using moist filter paper

Eng-Choon Leong[a], Richard Kizza and Harianto Rahardjo

Nanyang Technological University, School of Civil and Environmental Engineering, Blk N1, 50 Nanyang Avenue, Singapore 639798

Abstract. Measurement of soil suction is important as soil suction is one of the two stress-state variables that control both the hydro and mechanical behaviour of unsaturated soils. One of the soil suction measurement techniques is the filter paper method. Even though the filter paper technique has been standardized, the experimental procedure can still be improved in order to reduce the inconsistencies that are often reported with the filter paper technique. This paper explores usage of both initially dry and initially moist Whatman No. 42 filter paper for matric and total suction measurements. Residual soils from the Bukit Timah Granite in Singapore were compacted at various water contents from dry to wet of optimum using standard Proctor energies and their suctions were measured using both contact and non-contact filter papers. Results of the suction measurements showed hysteresis between the initially dry and the initially moist filter papers for both the contact and non-contact methods. Suction measurement by the contact filter paper method is more consistent than suction measurement by the non-contact filter paper method. Condensation causes the non-contact filter papers to have higher water contents. The results showed that the non-contact initially dry filter paper follows the total suction curve but the non-contact initially moist filter paper may follow the matric suction curve instead.

1 Introduction

In order to understand fully the behavior of unsaturated soils, measurement of soil suction is essential. Suction influences both the hydro and mechanical behavior of unsaturated soils. Despite its importance, measurement is difficult largely due to the reliability of the existing techniques. One such measurement technique uses the filter paper. The filter paper is economical, easily accessible, easy to use, relatively accurate if a strict protocol is followed and measures a wide range of soil suction. However, despite being standardized in ASTM D5298-10 [1], the experimental procedures are still controversial and research is being performed to better understand the technique. Substantial contributions to the filter paper suction measurement technique have been made by Al-Khafaf and Hanks [2], Houston et al. [3], Bulut et al. [4], Likos and Lu [5], Leong et al. [6], Marinho and Oliveira [7], Bicalho et al. [8], Bulut and Leong [9], Haghighi et al. [10] and Marinho and da Silva Gomes [11].

Houston et al. [3], Bulut et al. [4] and Leong et al. [6] suggested different calibration curves for total and matric suction measurements but Marinho and Oliveira [7] suggested that the calibration curve is unique for both total and matric suction measurements. The different calibration curves for the Whatman No. 42 filter curve suggested by Leong et al. [6] are shown in Figure 1. Marinho and Olivera [7] attributed the differences to equilibrium times as shown in Figure 2. Based on their findings, Marinho and Olivera [7] recommended different equilibrium times for the non-contact filter paper based on the total suction range reproduced in Table 1. The recommended equilibrium time for the filter paper method in ASTM D5298-10 [1] is 15 days and this was an increase from the previously recommended equilibrium time of 7 days.

Some investigators have pointed out the existence of hysteresis between the drying and wetting filter papers [6, 12, 13]. However, very little experimental data exist to show the effects of hysteresis of the filter paper on the suction measurement. According to Leong et al. [6], the hysteresis between a wet and dry filter paper ranges between 1 and 5% of the filter paper water contents. This amount of hysteresis is small but may be considerable when used to measure suction in the low range.

[a] Corresponding author: cecleong@ntu.edu.sg

 DOI: 10.1051/e3sconf/20160910012

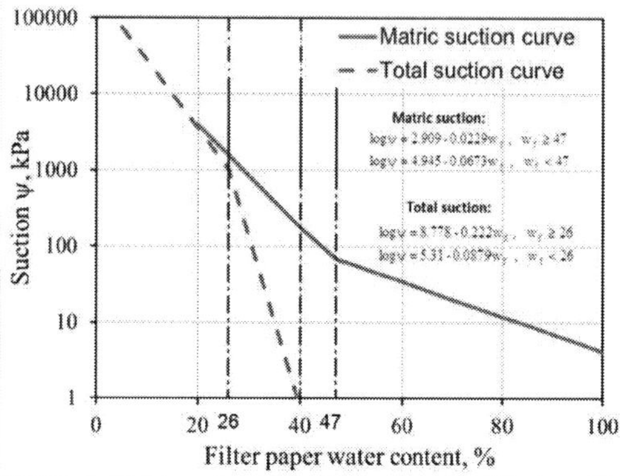

Figure 1. Matric and total suction curves for Whatman No. 42 filter paper.

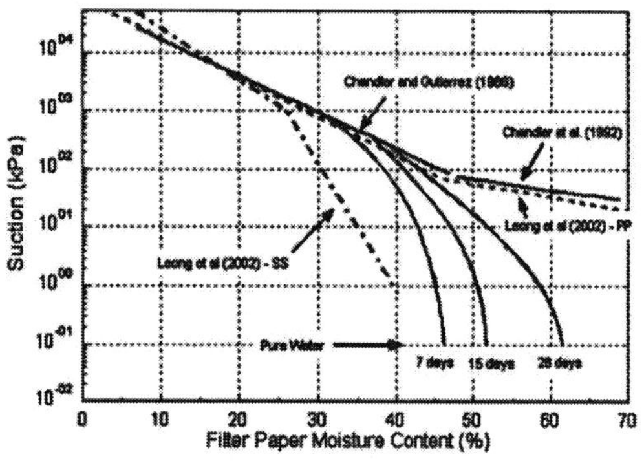

Figure 2. Effect of equilibrium time and distance from the water source (from [7]).

To address the above, this paper investigates the use of initially dry and initially moist Whatman No. 42 filter paper for suction measurement in the range from 0 to 1000 kPa corresponding to filter paper water content of more than 20% using dynamically compacted soils as the suction source. The use of initially moist filter papers will reduce the equilibrium time if it is indeed affecting the differences in calibration equations for the Whatman No. 42 filter paper.

Table 1. Suggested equilibrium time for total suction measurement as function of the suction level for Whatman No. 42 filter paper (from [7]).

Total Suction Range (kPa)	Equilibrium Time Suggested
0–100	Not determined, but certainly more than 30 days
100–250	30 days
250–1000	15 days
1000–30000	7 days

2 Materials and methods

2.1 Residual soils

Suction measurements were performed on compacted residual soil samples. The residual soils are from the Bukit Timah Granite of Singapore [14]. The residual soil was air dried and sieved through No. 4 sieve (4.75 mm). Only the soils passing through the No. 4 sieve were used in the study. The basic soil properties are liquid limit, w_L = 58, plastic limit ,w_P = 31, specific gravity, G_s = 2.7. The grain size distribution revealed that the soil consists of 27% sand and 73% fines. According to the Unified Soil Classification System, the soil is classified as MH (inorganic silt). The standard Proctor compaction curve of the residual soil is determined in accordance with ASTM D698-12 [15] and shown in Figure 3. For the compaction, distilled water was used. The maximum dry unit weight is 1.68Mg/m3 and the optimum water content is 16%.

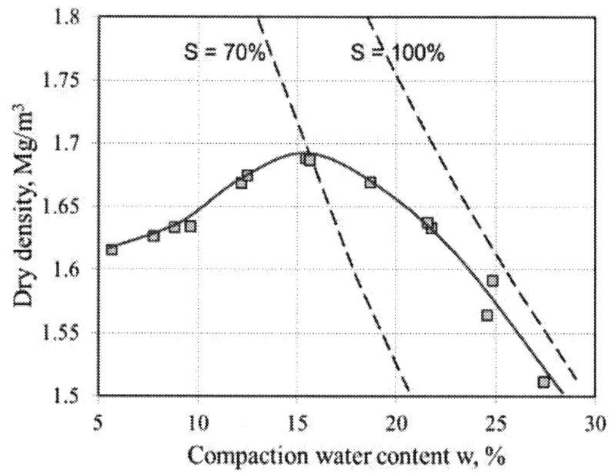

Figure 3. Standard Proctor compaction curve.

2.2 Filter paper methods

In this study, both the contact and non-contact filter papers were used to measure the matric and total suctions, respectively. Both initially dry and initially moist filter papers were used. The initially dry filter papers were used as obtained directly from their boxes while the initially moist filter papers were initially soaked in distilled water for about 15 minutes and then left to dry to a moist state for two hours in a room whose relative humidity was between 60 and 70% and room temperature was between 24 and 26°C. In all cases, the filter papers were handled using a clean pair of tweezers with gloved hands to ensure no contamination of filter papers.

A number of compacted soil samples were made at various water contents from dry to wet of optimum to provide a range of suctions for measurements. Approximately cylindrical soil specimens were carefully trimmed from the compacted samples such that they occupy at least 75% of the volume of the air-tight containers with internal diameter of 85 mm and height of 70 mm. Next, the cylindrical specimens were sliced at the middle into two approximately equal pieces. The contact filter paper was sandwiched between the two soil slices and the joint was sealed off using electrical tape to ensure no moisture loss. The soil slices were then further wrapped in three layers of cling wrap on its sides and at its bottom before placing them into the air-tight container. A PVC ring (diameter 40mm and thickness 5mm) was placed on top of the soil specimen onto which the non-contact filter paper was placed. The test specimen was then sealed in the air-tight container. All the containers containing the test specimens were transferred into a styrofoam box and stored for up to 21 days. At the end of 21 days, the contact and no-contact filter papers were retrieved and their water contents were determined following the procedures suggested by Bulut et al. [4]. An equilibrium period of 21 days was chosen as it exceeds the equilibrium period of 15 days recommended ASTM D5298-10 [1]. In addition, the maximum suction of the compacted soil samples was expected to be less than 1000 kPa and hence 21 days also exceeded the equilibrium period of 15 days recommended by Marinho and Oliveira [7].

3 Results and discussion

3.1. Contact filter paper

Figure 4 shows the results of the contact filter papers. Figure 4(a) shows both initially dry and initially moist filter paper water contents plotted against compaction water contents. Figure 4(b) shows the matric suction obtained using the equation for the matric suction calibration curve shown in Figure 1. The initially dry and initially moist filter papers display a distinct hysteresis in their water contents. A difference of about 5% water content is observed between the initially dry and initially moist filter papers for all the soils tested, in agreement with the observations of Leong et al. [6]. It is expected that the compacted soil specimens at the same water content will have the same matric suction. Thus, the matric suction calibration curve for the contact filter paper method should be different for the initially dry and the initially moist filter papers. The filter paper water content increases with increase in the compaction water content. Increase in compaction water content represents a decrease in matric suction [16, 17].

3.2 Non-contact filter paper

Figure 5 shows the results of the non-contact filter papers. Figure 5(a) shows both initially dry and initially moist filter paper water contents plotted against compaction water contents. Figure 5(a) shows that the hysteresis is less distinct when compared with that for the contact filter paper method but it is still present with the initially moist filter papers registering higher water contents compared to the initially dry filter papers. In addition, Figure 5(a) shows that the filter papers with water contents above 40% showed wider scatter for both initially dry and initially moist filter paper. Comparing Figures 4(a) and 5(a) suggests that the water contents are similar with compaction water content and thus the matric suction curve in Figure 1 should also be used to infer the total suction from the non-contact filter papers. Figure 5(b) shows the total suction obtained using the

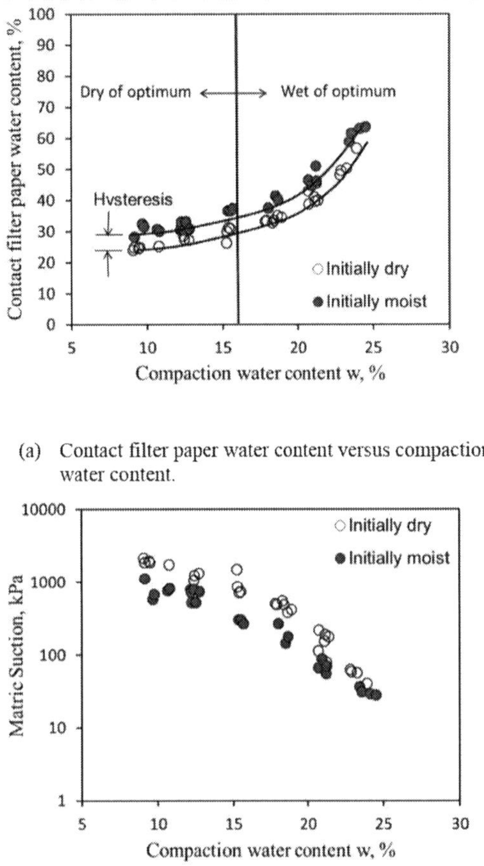

(a) Contact filter paper water content versus compaction water content.

(b) Matric suction versus compaction water content

Figure 4. Measurement of suction using contact filter papers.

equation for the matric suction curve shown in Figure 1. From Figure 1, it is expected that a non-contact initially dry filter paper follows the total suction curve and will not attain water contents higher than 40%The above seems to support the suggestion by Marinho and Oliveira [7] that the filter paper calibration curve should be unique and there should not be a total suction curve for filter paper. However, it will be shown later why Figure 5b is misleading.

3.3 Initially dry and initially moist filter papers compared

The contact and non-contact initially dry filter paper results are compared in Figure 6. The contact and non-contact filter paper water contents of the initially dry filter paper seem to agree with each other following a similar trend with the compaction water content. However, above 40% filter paper water content the non-contact filter paper shows a greater scatter. This observation seems logical as the compacted residual soil specimens were compacted with distilled water and if salt was absent from the soil, osmotic suction is absent and matric suction should be the same as total suction. Hence, the water contents of the contact and non-contact filter paper water contents should be similar.

The contact and non-contact initially moist filter papers are compared in Figure 7. Similar to Figure 6, the contact and non-contact filter paper water contents of the initially moist filter paper seem to agree with each other following a similar trend with the compaction water content. However, a greater scatter in the non-contact initially moist filter paper is observed especially when the filter paper water content is greater than 40%.

Based on Figures 6 and 7, it appears that the calibration equation for the filter paper should be unique and independent of the filter paper method, that is, contact or non-contact. To validate this observation, the filter paper calibration curves shown in Figure 1 need to be recovered from the non-contact filter paper measurements.

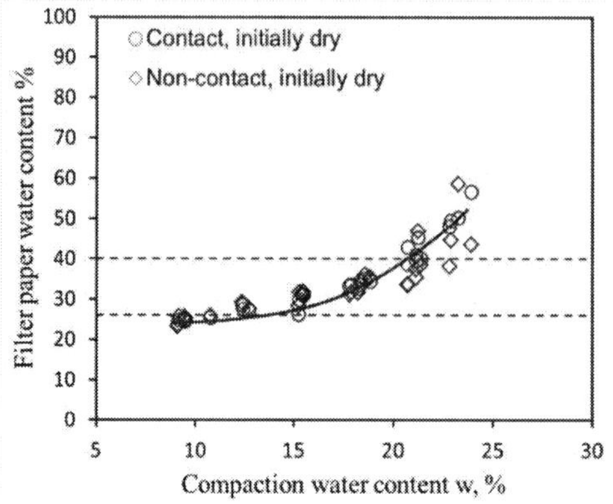

Figure 6. Contact and non-contact initially dry filter paper water content versus compaction water content.

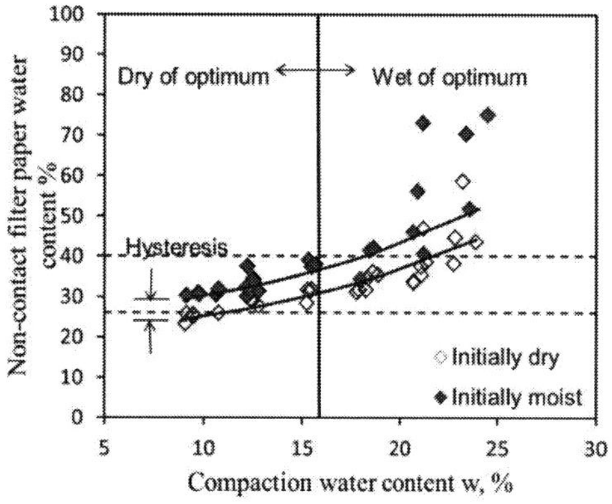

(a) Non-contact filter paper water content versus compaction water content.

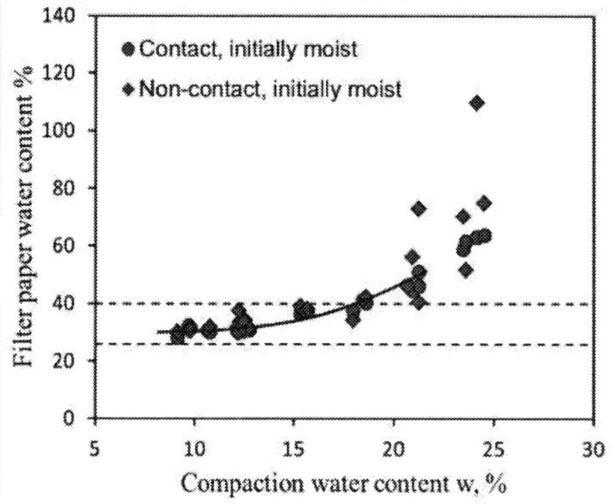

Figure 7. Contact and non-contact initially moist filter paper water contents versus compaction water content.

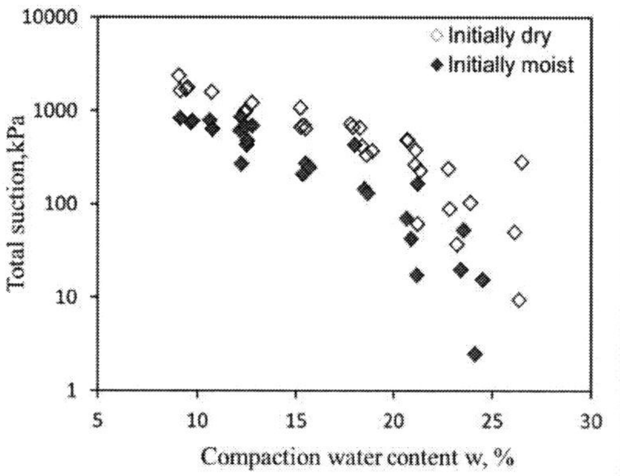

(b) Total suction versus compaction water content

Figure 5. Measurement of suction using non-contact filter papers.

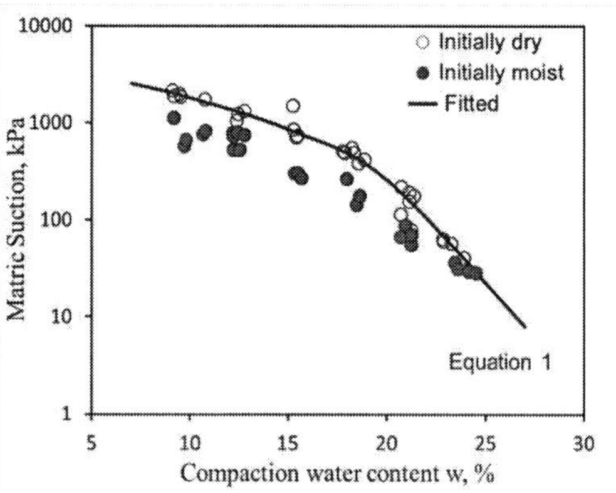

Figure 8. Relationship between soil suction and compaction water content of 7-27% of the soil.

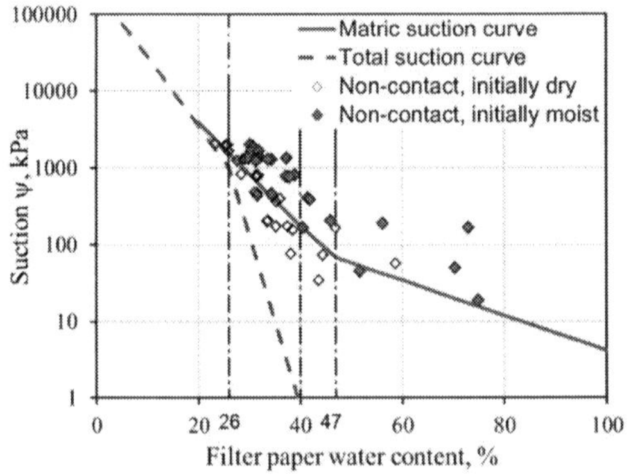

Figure 9. Suctions from non-contact initially dry and initially moist filter papers.

Figure 4a shows that the suction of the compacted soil can be reliably measured using the contact filter paper method. As the filter paper calibration curves were obtained from initially dry filter papers, the suctions shown by the initially dry filter papers in Figure 4b should give the matric and total suctions of the compacted soils. The relationship between suction and compaction water content of 7 – 27% for the soil is:

$$\psi = 7w^2 - 365w + 4760 \qquad (1)$$

where ψ is soil suction and w is compaction water content in %. Equation 1 is shown in Figure 8. Once the matric suction of the compacted soil for any compaction water content is known, it can be treated as a suction source for the non-contact filter paper method. Figure 9 shows the plot of the soil suction versus the non-contact filter paper water contents.

The non-contact initially dry filter paper water contents are expected to straddle the total suction curve but the data are mostly between the total suction and the matric suction curves. For the non-contact initially moist filter paper water contents, there could be two possible outcomes: The water content of the non-contact initially moist filter paper is expected to approach either the matric suction curve or the total suction curve. However, almost all of the non-contact initially moist filter paper water contents are above the matric suction curve.

The above observations can only be explained by the person conducting the experiments. At total suctions less than 1000 kPa, the relative humidity at 25°C is above 99% and condensation was frequently observed as small beads of water on the inside of the lid of the airtight container above the non-contact filter paper. Therefore, there was a high possibility of the water from the condensation falling onto the non-contact filter paper giving a higher than expected filter paper water content as observed in Figure 8. Thus, the water contents of the initially dry filter paper were higher than the total suction curve and the water contents of the initially moist filter paper were higher than that of the initially dry filter paper. While it is shown that the total suction calibration curve should be used for the non-contact initially dry filter paper, it has not been established irrefutably if a different calibration curve is applicable for the non-contact initially moist filter paper. Currently, experiments are ongoing to provide the answer.

4 Conclusion

All filter paper calibration curves were developed using initially dry filter papers. In this paper, both initially dry and initially moist filter papers were used to measure both matric and total suctions. The measurement results showed that there was hysteresis between the initially dry and the initially moist filter papers with the initially moist filter papers giving correspondingly higher water content for the same compacted soil sample. For the non-contact filter paper method, both the initially dry and initially moist filter papers showed a higher than expected water content due to condensation. The calibration curve that should be used for the non-contact initially dry filter paper should be the total suction curve. However, it is not clear if the calibration curve for the non-contact initially moist filter paper should be the matric suction curve. Tests are ongoing to provide the answer in the near future.

Acknowledgement

The second author acknowledges the scholarship from the Singapore International Graduate Award (SINGA) administered by the Agency for Science, Technology & Research (A*STAR) and the Nanyang Technological University (NTU).

References

1. ASTM D5298-10, ASTM International, www.astm.org (2010).
2. S. Al-Khafaf, R. Hanks, Soil Sci., **117**(4), 194-199, (1974).
3. S.L. Houston, W.N., Houston, A.M. Wagner, 1994, Geotech. Test. J., **17**(2), 185–194 (1994).
4. R. Bulut, R.L. Lytton, W.K. Wray, W. K. (2001) *Expansive clay soils and vegetative influence on shallow foundations*, 243-261 (2001).
5. W.J. Likos, N. Lu, N. (2002) TRR: Journal of the Transport. Res. Board **1786**(1), 120-128 (2002).
6. E.C. Leong, L. He, H. Rahardjo, ASTM Geotech. Testing J., **25**(3), 322-333 (2002).
7. F.A.M. Marinho, O.M. Oliveira, O. M. ASTM Geotech. Testing J., **29**(3), 250-258 (2006).
8. K.V. Bicalho, A.G. Correia, S.R. Ferreira, J.-M. Fleureau, F.A.M. Marinho, *Proc. XIII Pan-Am. Conf. Soil Mech. Geotech Eng.*, 215-219 (2007).
9. R. Bulut, E.C. Leong, Geotech Geol. Eng., **26**(6), 633-644 (2008).
10. A. Haghighi, G.M. Medero, F.A.M. Marinho, B. Mercier, P.K. Woodward, ASTM Geotech. Testing J., **35**(1), 83-90 (2012).
11. F.A. Marinho, J.E. da Silva Gomes, J. E. ASTM Geotech. Testing J., **35**(1), 172-181 (2012).

DOI: 10.1051/e3sconf/20160910012

12. G.E. Swarbrick, *Proc. 1st Int. Conf. Unsaturated Soils*, **2**, 701-708 (1995).
13. B. Harrison, G. Blight, *Proc. of the 2nd Int. Conf. Unsaturated Soils*, **1**, 362-367 (1995).
14. E.C. Leong, H. Rahardjo, S.K. Tang, *International workshop on characterisation and engineering properties of natural soils, Singapore*, **2**, 1279-1304 (2002).
15. ASTM D698-12, ASTM International, www.astm.org (2012).
16. E.C. Leong, S. Tripathy, H. Rahardjo, ASTM Geotech. Testing J., **27**(3), 322 – 331 (2004).
17. E.C. Leong, S. Widiastuti, C.C. Lee, H. Rahardjo, Geotechnique, **57**(6), 547-556 (2007).

Soil water retention behaviour for an instrumented embankment

D.G. Toll[1,a], P.N. Hughes[1], J.D. Asquith[1]

[1]*School of Engineering and Computing Sciences, Durham University, Durham DH1 3LE, UK*

Abstract. The research project iSMART (Infrastructure Slopes: Sustainable Management And Resilience Assessment) has been established as a collaboration between six UK academic partners and 11 asset owners and industrial partners to investigate the impacts of weather and climate on infrastructure slopes. One of the sites being monitored within the iSmart project is an instrumented embankment to investigate the response to changing climatic conditions. The BIONICS embankment was built at Nafferton farm in North East England. The fill material was a glacial till (Durham Lower Boulder Clay), a common fill material in North East England and hence representative of earthwork construction. The fill material can be classified as a sandy clay of intermediate plasticity. The soil water retention curves (SWRC) for the BIONICS soil have been measured in the laboratory using novel high suction tensiometer based equipment. These are compared to field observations of suction and water content obtained in the field. The first cycle of drying and wetting compares reasonably well between the laboratory and the field. However, subsequent drying in the field shows a significantly different path to that observed in the laboratory.

1 Introduction

The water retention behaviour of soils is an essential aspect of understanding the unsaturated behaviour of soil materials [1]. It represents the relationship between suction and water content (which can be gravimetric or volumetric) or may be represented in terms of degree of saturation. This paper reports results obtained using a relatively new method of measurement for soil water retention curves, using high capacity tensiometers.

Results are presented for a sandy clay fill material obtained as part of an investigation of the impacts of climate on infrastructure slopes. The project iSmart (Infrastructure Slopes: Sustainable Management And Resilience Assessment: www.ismartproject.org) has been established as a collaboration between six UK academic partners (Universities of Newcastle, Durham, Loughborough, Queen's Belfast, Southampton and British Geological Survey) and 11 asset owners and industrial partners to investigate the impacts of weather and climate on infrastructure slopes.

The soil used in this study was a glacial till (Durham Lower Boulder Clay) obtained from fill material used in the construction of an experimental embankment (called the BIONICS embankment) constructed at Nafferton Farm in North East England [2]. The fill material is a glacial till (Durham Lower Boulder Clay), a common fill material in North East England and hence representative of earthwork construction in North East England. Field measurements are being obtained from a number of sites (embankments and cuttings) across the UK. Laboratory measurements are needed to interpret and understand the field observations and provide input parameters for numerical modelling.

Field measurements have been carried out at the BIONICS embankment that allow determination of the field soil water retention curves. The laboratory measurements obtained using the tensiometer technique are compared to field observations of suction and water content obtained from the BIONICS embankment.

2 Material tested

A sandy clay soil used in the construction of the BIONICS embankment at Nafferton Farm in North East England was prepared by sieving through a 2.8mm sieve to remove the larger particles to reduce the variation in properties. The sieved material comprised 30% sand, 35% silt and 35% clay, i.e., a sandy clay soil. The Liquid Limit was 43.3% and the Plastic Limit was 23.7%, resulting in a Plasticity Index of 19.6 [3]. The particle density was 2.66 Mg/m^3.

Specimens for testing were prepared by compaction into a 100mm diameter mould using the equivalent compactive effort of the standard Proctor test (BS light compaction, [4]). Samples were compacted wet of optimum at a water content of 24%. This resulted in specimens close to saturation (degree of saturation, Sr>95%). From the larger compacted samples, smaller specimens having a diameter of 75mm and a thickness of 20mm were trimmed from the 100mm diameter compacted samples for testing in the tensiometer equipment.

[a] Corresponding author: d.g.toll@durham.ac.uk

3 The Durham SWRC Equipment

Measurements of the soil water retention curves (SWRC) have been carried out using the Durham SWRC equipment. The experimental apparatus allows continuous measurements of water content, suction and volume change [5, **Erreur ! Source du renvoi introuvable.**]. The apparatus is made up of a PVC frame (Figures 1 & 2) placed on an electronic balance to determine the change in sample weight and hence water content (as used by Lourenço et al. [7, 8]).

Figure 1. The Durham SWRC equipment, showing the frame sitting on the electronic balance and connected to the data-logging system.

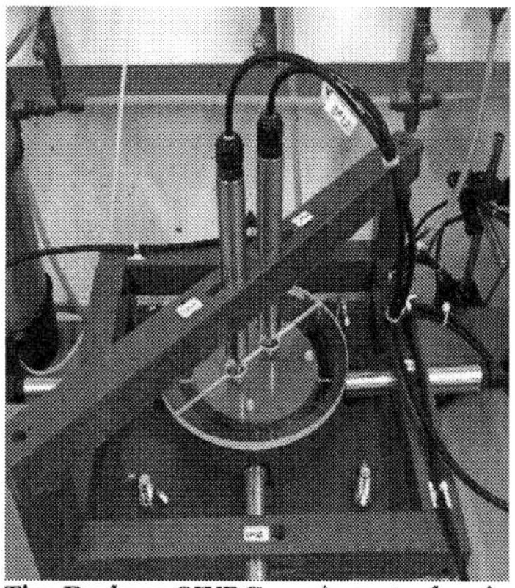

Figure 2. The Durham SWRC equipment, showing the PVC frame holding six LVDTs to measure volume change.

For volume change measurements, four displacement transducers were installed through the four outside beams of the frame to measure radial displacement of the specimen and two more displacement transducers were fitted through the upper beam to measure axial displacement (change in height) (Figure 2). Volume change of the specimen could then be calculated from the radial and axial deformations.

A high suction tensiometer developed at Durham University [9, 9] was used to measure suction. These devices have been used for direct measurement of suction as large as 2000kPa. The tensiometer was fitted through a hole in the support plate, with a tight fitting rubber O-ring to secure it in place. All transducers were connected to a real-time data acquisition system [11].

Using the tensiometer technique, suction can be measured in samples either dried continuously while exposed to the atmosphere (continuous procedure) or by drying in stages (stage procedure). In the stage procedure, the specimen is sealed and allowed to equalise internally after each period of drying. Both approaches are quicker than traditional methods for obtaining SWRCs (e.g. pressure plate).

4 Field measurements of water retention

Glendinning et al. [12] describe field measurements of suction and water content measured in the BIONICS embankment. Suctions were measured using MPS-1 dielectric water potential sensors produced by Decagon Ltd. These devices can measure soil suction in the range of 10 to 600 kPa and are accurate to 20% of total soil suction measured.

Volumetric water content was measured using ML2 Theta probes produced by Delta-T Devices. The sensors for suction and water content were positioned at the base, mid-point and top of the slope at depths of 0.5 and 1m and measurements were taken hourly.

The field measurements for suction and water content at a depth of 0.5m for March-June 2009 (a drying period), August-November 2009 (a wetting period) and April-June 2010 (a drying period) were interpreted as gravimetric water content vs. suction curves [12]. These data will be compared with the laboratory measurements of SWRC obtained from tensiometer devices in the laboratory.

5 Comparison of laboratory and field SWRCs

A series of laboratory measured soil water retention curves obtained using the tensiometer technique is plotted in Figure 3 [6, 13]. The majority of results were obtained by continuous drying. Two tests shown were determined by stage drying. Good agreement is shown between all sets of results over the range 1-1000 kPa.

One set of tests followed the SWRC over two complete cycles of drying and wetting [13]. Figure 3 shows that for each full cycle (drying and wetting) the hysteretic behaviour is quite different. For the first cycle, the difference between the primary drying curve and the subsequent wetting curve is much larger than for the second cycle. The last three paths (wetting-drying-wetting) lie very close to each other.

The field data obtained by Glendinning et al. [12] is plotted in Figure 3 for comparison with the laboratory data. It can be seen that the drying period (March-June 2009) follows a path roughly parallel to the laboratory measurements. It should be noted that the embankment was constructed in 2007, so this period in 2009 would not represent a primary drying curve. The embankment had already been through some drying and wetting history before the instruments were installed. The data is likely to represent a drying scanning curve. While there are differences between the field and laboratory, the general pattern of behaviour seems consistent.

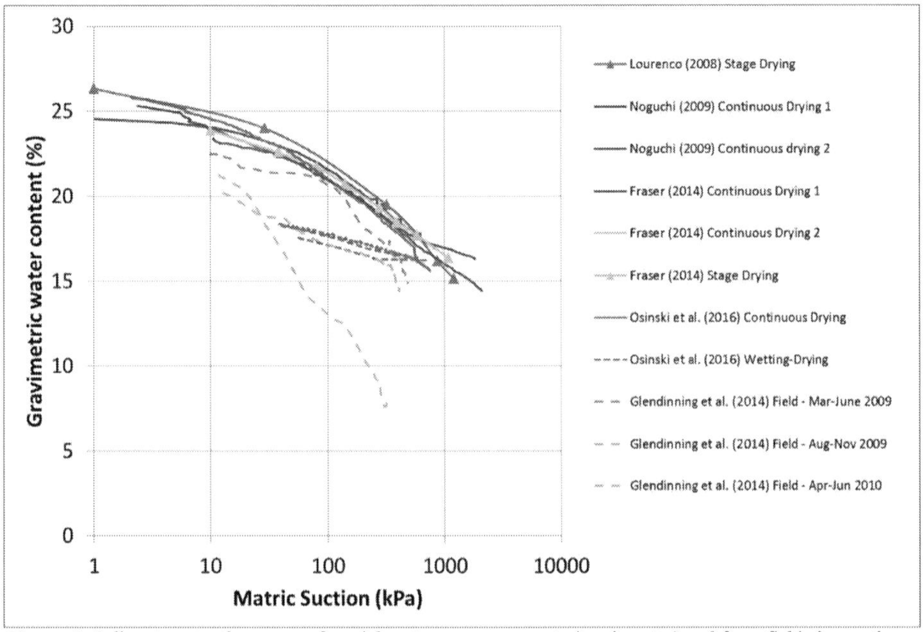

Figure 3. Soil water retention curves from laboratory measurements (tensiometer) and from field observations.

For the wetting period (August-November 2009) the field SWRC seems to follow well the laboratory determined wetting curves. The field wetting continues to a higher water content (~20%) and a lower suction (~10 kPa) than was achieved in the laboratory tests.

However, the second drying period (April-June 2010) shows the field data diverging very significantly from the laboratory measurements. The laboratory data shows consistent wetting-drying-wetting paths, showing very little hysteresis between the first and second wetting paths. The field data shows a much steeper path i.e. a much greater water content change for the suction increase to 300 kPa.

Further investigations are needed to explain the divergence between the field and laboratory data for this second field drying (April-June 2010). The data will need to be studied in terms of volumetric water content, to see if there are differences in volume change response between the field and laboratory data. As noted by Toll et al. [6], differences in volumetric response can explain differences between types of test to measure water retention behaviour.

Acknowledgements

The authors wish to acknowledge funding from the Engineering and Physical Sciences Research Council for the BIONICS project (GR/S87430/01) and the iSMART project (EP/K027050/1).

References

1. Fredlund, D.G. Use of the soil-water characteristic curve in the implementation of unsaturated soil mechanics. *Proc. Third Int. Conf. on Unsaturated Soils, UNSAT 2002, Recife, Brazil.* (eds. J.F.T. Juca, T.M.P. de Campos, and F.A.M. Marinho. London: Taylor and Francis, Vol. 3, pp. 887–902. 2002.
2. Hughes, P.N., Glendinning, S., Mendes, J., Parkin, G., Toll, D.G., Gallipoli, D., Miller, P. Full-scale testing to assess climate effects on embankments. *Special Issue of Engineering Sustainability: Proc. Institution of Civil Engineers*, 162, 2, pp. 67-79. 2009.
3. Mendes, J. Assessment of the impact of climate change on an instrumented embankment: an unsaturated soil mechanics approach. *PhD Thesis, Durham University*. Available online: http://etheses.dur.ac.uk/612/. 2011.
4. BS1377. Methods of test for Soils of Civil Engineering Purposes, British Standards Institute, Milton Keynes. 1990.
5. Liu, G., Toll, D.G., Kong, L. & Asquith, J.D. Matric suction and volume characteristics of compacted clay soil under drying and wetting cycles, *submitted to ASTM Geotechnical Testing Journal*. In press.
6. Toll, D.G., Asquith, J.D., Fraser, A., Hassan, A.A., Liu, G., Lourenço, S.D.N., Mendes, J., Noguchi, T., Osinski, P. & Stirling, R.A. Tensiometer techniques for determining soil water retention curves. *Proc. 6th Asia-Pacific Conference on Unsaturated Soil. Guilin, China.* London: Taylor & Francis (2015)
7. Lourenço, S.D.N., Gallipoli, D., Toll, D.G., Evans, F.D. & Medero, G.M. Determination of the Soil Water Retention Curve with Tensiometers, *in Experimental Unsaturated Soil Mechanics*, (ed. T. Schanz), Springer, pp. 95-102. 2007.
8. Lourenço, S.D.N., Gallipoli, D., Toll, D.G., Augarde, C.E. & Evans, F. 2011. A new procedure for the determination of the Soil Water Retention Curves by continuous drying using high suction tensiometers. *Canadian Geotechnical Journal*, 48, 2, pp. 327-335.
9. Lourenço, S.D.N., Gallipoli, D., Toll, D.G. and Evans, F.D. Development of a Commercial

Tensiometer for Triaxial Testing of Unsaturated Soils, *Proc. 4th International Conference on Unsaturated Soils, Phoenix, USA, Geotechnical Special Publication No. 147*, Reston: ASCE, Vol.2, pp. 1875-1886. 2006.

10. Toll, D.G., Lourenço, S.D.N. and Mendes, J. Advances in suction measurements using high suction tensiometers, *Engineering Geology*, 165, pp 29–37. 2013.
11. Toll, D.G. 1999. A data acquisition and control system for geotechnical testing, *in Computing Developments in Civil and Structural Engineering* (eds. B. Kumar and B.H.V. Topping), Edinburgh: Civil-Comp Press, pp 237-242.
12. Glendinning, S., Hughes, P.N., Helm, P., Chambers, J., Mendes, J., Gunn, D., Wilkinson, P. & Uhlemann, S. Construction, management and maintenance of embankments used for road and rail infrastructure: implications of weather induced pore water pressures. *Acta Geotechnica* 9(5): 799. 2014.
13. Osinski, P., Toll, D.G. & Koda, E. Comparison of Soil Water Retention Curves for sandy clay obtained using different laboratory testing methods. *Proc. 3rd European Conference on Unsaturated Soils*, Paris. 2016.

Analytical reliability analysis of soil-water characteristic curve

A.Johari[1,a], A. Hooshmand Nejad[1], A.A. Javadi[2]

[1]*Department of Civil and Environmental Engineering, Shiraz University of Technology, Shiraz, Iran*
[2]*Computational Geomechanics Group, College of Engineering, Mathematics and Physical Sciences, University of Exeter, Exeter, EX4 4QF, UK*

Abstract. The Soil Water Characteristic Curve (SWCC), also known as the soil water-retention curve, is an important part of any constitutive relationship for unsaturated soils. Deterministic assessment of SWCC has received considerable attention in the past few years. However the uncertainties of the parameters which affect SWCC dictate that the problem is of a probabilistic nature rather than being deterministic. In this research, a Gene Expression Programming (GEP)-based SWCC model is employed to assess the reliability of SWCC. For this purpose, the Jointly Distributed Random Variables (JDRV) method is used as an analytical method for reliability analysis. All input parameters of the model which are initial void ratio, initial water content, silt and clay contents are set to be stochastic and modelled using truncated normal probability density functions. The results are compared with those of the Monte Carlo (MC) simulation. It is shown that the initial water content is the most effective parameter in SWCC.

1. Introduction

The success of any unsaturated analysis like unsaturated slope stability greatly depends on the input data such as SWCC that defines the relationship between the amount of water in a soil (i.e. gravimetric water content w, volumetric water content θ or degree of saturation S and matric suction ψ [1]).

Many empirical equations have been proposed to curve fit the SWCC [2–7]. Another approach is to estimate the SWCC from the grain-size distribution curve [8,9]. The other method is artificial intelligence methods to predict the SWCC such as artificial Neural Network (ANN) [10], Genetic-Based Neural Network (GBNN)[11], Genetic programming (GP)[11], evolutionary polynomial regression (EPR)[12], and Gene Expression Programming (GEP) [13]. These methods use some basic soil properties to predict the SWCC, instead of some time consuming tests such as pressure plate test.

Reliability analysis is an area of growing importance in geotechnical engineering. The success of numerical modelling of geotechnical systems greatly depends on the uncertainty modelling, among which uncertainty in soil properties is a crucial input of the analysis.

Most of the previous statistical studies on the SWCC-fitting parameters focus on the mean, standard deviation, and coefficient of variation [14]. Some researchers observed the correlations between curve fitting parameters for hydraulic properties of unsaturated soils[15]. The grain size distributions is used to estimate SWCC curves and the SWCC parameters are determined graphically from the curves [16]. This work results in a mean and standard deviation for each SWCC parameter. It was attempted to construct a joint probability distribution function for the van Genuchten, (1980) parameters [17]. The key step is to relate the empirical joint nonnormal distribution of the van Genuchten, (1980) parameters to a standard joint normal distribution. When the experimental data on hand is restricted to only marginal distributions and correlations, this translation approach is popular [18]. The suitability of a lognormal random vector in predicting the parameters of van Genuchten, (1980) model was demonstrated [19]. As well as this, The Bayesian analysis is applied to derive the PDF of the model

[a] Ali Johari : johari@sutech.ac.ir

parameters in various forms of van Genuchten equation using the observed data of sand, sandy loam and silty loam [20].

Most of previous reliability analysis of SWCC focuses on SWCC-fitting models. However, in this research the reliability analysis of an artificial intelligence model of SWCC with acceptable accuracy is performed. The reliability analysis is regarding to uncertainties of independent input parameters of model which are based on grain size distribution curve and inherent soil properties. This analysis is carried out at different suctions using an analytical method and results were verified by Monte Carlo simulation (MCs).

2. Corresponding SWCC model used for reliability analysis

In this research, one of the artificial intelligence models is selected for reliability analysis. The model which is namely GEP model [13] has acceptable accuracy to predict the SWCC. Table 1 shows the correlation coefficient and error values of this model to predict the SWCC based on SoilVision (2002) databank which contains more than 180 pressure plate tests results performed on clay, silty clay, sandy loam, and loam soil.

Table 1. Performance of GEP model (Johari and Hooshmand , 2015)

	Training Data			Testing Data		
Model	ARE (%)	MSSE	R^2	ARE (%)	MSSE	R^2
GEP	25.59	0.0014	0.94	28.73	0.0014	0.94

GEP model use the soil properties as inputs to predict SWCC and does not use the curve-fitting parameters. The Johari and Hooshmand (2015) GEP model for predicting soil-water characteristic curve is as the following:

$$\omega=\frac{-1}{S+2Cl-2.202\frac{S^4}{Cl^2}-7.285}-\omega_0\left(S+0.062(Si+e)^2-1\right) \quad (1)$$

Where:

e Initial void ratio;

ω_0 Initial water content

S log [suction (kPa)/p_a] where pa= Atmospheric pressure (taken as 100 kPa)

Cl Clay content (%)

Si Silt content (%)

ω Predicted gravimetric water content

The comprehensive explanation of this method is available in Johari and Hooshmand (2015), however, Fig. 1. Shows a prediction of SWCC by GEP model for a typical soil sample of databank with the following properties: Void ratio: 0.71; Initial water content: 26.89%; Clay content 28.71%; Silt content: 63.57%.

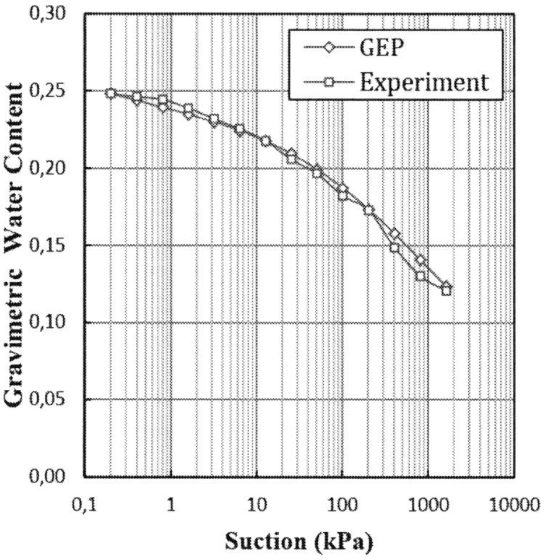

Figure 1. A typical SWCC prediction of GEP model

3. Reliability analysis of SWCC by JDRV method

Jointly Distributed Random Variables (JDRV) method is an analytical probabilistic method. In this method probability density function of input variables are expressed mathematically and jointed together by statistical relations [21–23]. Due to the nature of analytical methods, utilizing this method for some geotechnical problems with complex governing relationships may require complicated analysis and a single expression in this method may become rather complex. Some recent researches have been made to apply this method to geotechnical problems [24–26].

The GEP model of SWCC has five input parameters namely void ratio, initial water content, logarithm of suction normalized with respect to atmospheric air pressure, clay content and silt content. To account for the uncertainties in soil-water characteristic curve, all four input parameters have been defined as stochastic variables at every specific suctions of SWCC. Therefore, selected parameters are void ratio (e), initial water content (ω_0), clay content (Cl), and silt content (Si). All of these parameters are modeled using a truncated normal probability distribution function (pdf).

By considering the stochastic variables within the range of their mean plus or minus four times standard deviation, 99.994% of the area beneath the normal density curve is covered.

For reliability assessment of soil-water characteristic curve using JDRV method, Eq. (1) is rewritten into terms of K_1 to K_{11} as shown in Eq. (2). The terms K_1 to K_{11}, are introduced in Eq. (3). The probability density function of each term is derived separately. Using the mathematical functions for K_1 to K_{11} and $fK_1(k_1)$ to $fK_{11}(k_{11})$ a computer program was developed (coded in MATLAB) to determine the probability density function curve of SWCC.

$$\frac{-1}{k_7+2k_4}-k_3\left(S+0.062k_1^2-1\right)= \quad (2)$$

$$\frac{-1}{k_7+k_8}-(k_3\times k_5)=k_{10}-k_6=\omega$$

Where:

$$\begin{cases} k_1 = Si + e \\ k_2 = 0.062\ k_1^2 \\ k_3 = \omega_0 \\ k_4 = Cl \\ k_5 = k_2 + S - 1 \\ k_6 = k_3 \times k_5 \\ k_7 = S - 7.285 - \dfrac{2.202\ S^4}{k_4^2} \\ k_8 = 2\ k_4 \\ k_9 = k_7 + k_8 \\ k_{10} = \dfrac{-1}{k_9} \\ k_{11} = k_{10} - k_6 \end{cases} \quad (3)$$

In addition, for comparison, determination of the SWCC using the Monte Carlo simulation was also coded in the same computer program. In order to verify the results of the presented method with those of the Monte Carlo simulation, the final probability density function for gravimetric water content are showed for the same suction (3.0 kPa). Fig. 2 shows the obtained probability density function of gravimetric water. Table 2 presents the properties of obtained gravimetric water content distribution resulted from JDRV and MCs which are perfectly compatible.

Table. 2. Properties of gravimetric water content distribution from JDRV and MCs

Method	Mean	Mode	Variance	Skewness
JDRV	0.5224	0.5353	0.0337	1.0676
MCs	0.5115	0.5349	0.0334	1.0740

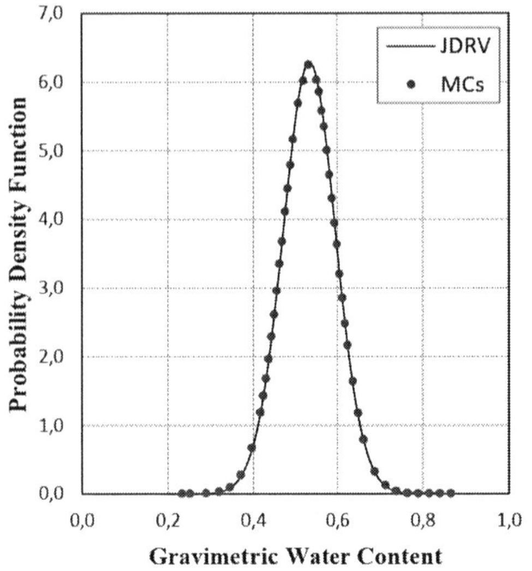

Figure 2. Probability density function of gravimetric water content at 3.0 kPa suction

3.1. Reliability analysis of SWCC at various suctions

Probability density functions of gravimetric water content in various suctions (from 0.2 kPa to 52428.8 kPa) are shown in Fig. 3.Ten arbitrary suctions from 0.2 kPa are selected where each one four times as much increased relative to previous one. The 2D plot of probability density functions of gravimetric water content are shown in Fig. 4. It is clear that increasing suction cause decrease in gravimetric water content, therefore, the pdf shifted leftward. It can be seen that the sample with lower suction has a greater dispersion (more uncertainties) of gravimetric water content.

Fig.5 shows the distribution of gravimetric water content at various suctions that initiate from 0.2 kPa and doubled until 104857.6 kPa. Normal distribution of SWCC, in fact is pdf of gravimetric water content at different suctions. Two boundaries are shown in this figure which are the upper bound and lower bound of gravimetric water content at the specific suction. The values of lower bound is the limit of minimum gravimetric water content at each suction and similarly, upper bound shows the limit for maximum of gravimetric water content. It can be seen that the distance between the upper and lower bound (D) is decreased as the suction increase. Furthermore, the mean curve consists of gravimetric water contents values that have cumulative density equal to %50 at considered suctions, which in fact correspond with the peak of pdf diagram. The values of upper bound, lower bound, mean and 'D' are given in Table 3.

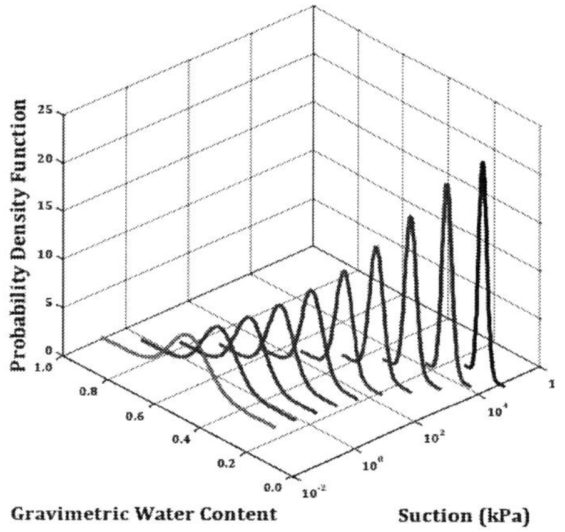

Figure 3. Probability density function of gravimetric water content in various suctions

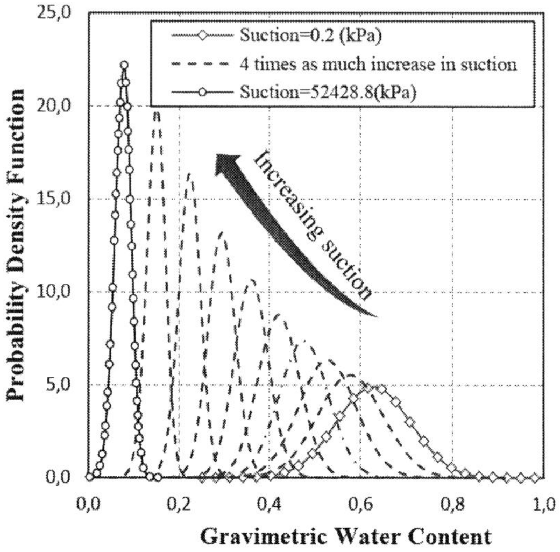

Figure 4. Probability density function of gravimetric water content in various suctions

4. Conclusion

In this research, a Gene Expression Programming (GEP)-based SWCC model is employed to assess the reliability of SWCC. For this purpose, the Jointly Distributed Random Variables (JDRV) method is used as an analytical method for reliability analysis. Uncertainties in SWCC can be evaluated in terms of the uncertainties in geotechnical input parameters. To account for the uncertainties in SWCC, all input parameters of GEP model which are initial void ratio, initial water content, silt and clay content have been defined as stochastic variables at every specific suctions of SWCC. The results showed that the determined probability density function of SWCC by JDRV method as analytical approach is very close to prediction of the Monte Carlo simulation. Furthermore, the upper bound, lower bound and the mean value of gravimetric water content at various suctions are determined. It can be seen that the determined probability distribution of gravimetric water content has greater dispersion (more uncertainty) in lower suctions.

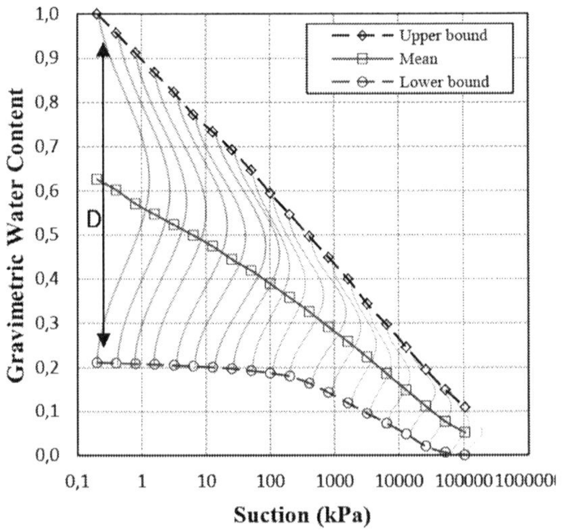

Figure 5. Boundaries of predicted initial drying curve

Table 3. Values of boundaries, mean and distance boundaries gravimetric water content for initial drying

Suction (kPa)	Upper bound	Mean	Lower bound	D
0.2	1.000	0.614	0.234	0.766
0.4	0.986	0.589	0.231	0.755
0.8	0.938	0.565	0.228	0.710
1.6	0.891	0.541	0.225	0.666
3.2	0.843	0.516	0.221	0.622
6.4	0.7946	0.492	0.218	0.5766
12.8	0.746	0.466	0.213	0.533
25.6	0.681	0.439	0.208	0.473
51.2	0.645	0.412	0.201	0.444
102.4	0.592	0.383	0.193	0.399
204.8	0.538	0.354	0.181	0.357
409.6	0.4873	0.321	0.159	0.3283
819.2	0.432	0.289	0.134	0.298
1638.4	0.382	0.254	0.111	0.271
3276.8	0.330	0.218	0.09	0.240
6553.6	0.279	0.184	0.07	0.209
13107.2	0.232	0.148	0.045	0.187
26214.4	0.190	0.111	0.020	0.170
52428.8	0.150	0.076	0.001	0.150
104857.6	0.117	0.042	0.000	0.117

References

1. D. G. Fredlund and H. Rahardjo, *Soil mechanics for unsaturated soils*. John Wiley & Sons, (1993).
2. R. H. Brooks and A. T. Corey, *Hydraulic Properties of Porous Media*. Colorado State University, (1964).
3. D. G. Fredlund and A. Xing, "Equations for the soil-water characteristic curve," *Can. Geotech. J.*, vol. 31, no. 4, pp. 521–532, (1994).
4. W. R. Gardner, "Some steady-state solutions of the unsaturated moisture flow equation with application to evaporation from a water table," *Soil Sci.*, vol. 85, no. 4, pp. 228–232, (1958).
5. C. R. McKee and A. C. Bumb, "The importance of unsaturated flow parameters in designing a monitoring system for hazardous wastes and environmental emergencies," in *Proc. Hazardous Materials Control Research Institute Natl. Conf., Houston, TX*, pp. 50–58, (1984).
6. M. T. van Genuchten, "A Closed-form Equation for Predicting the Hydraulic Conductivity of Unsaturated Soils1," *Soil Sci. Soc. Am. J.*, vol. 44, no. 5, p. 892, (1980).
7. J. Williams, R. E. Prebble, W. T. Williams, and C. T. Hignett, "The influence of texture, structure and clay mineralogy on the soil moisture characteristic," *Soil Res.*, vol. 21, no. 1, pp. 15–32, (1983).
8. L. M. Arya and J. F. Paris, "A physicoempirical model to predict the soil moisture characteristic from particle-size distribution and bulk density data," *Soil Sci. Soc. Am. J.*, vol. 45, no. 6, pp. 1023–1030, (1981).

[9] M. D. Fredlund, D. G. Fredlund, and G. W. Wilson, "Prediction of the soil-water characteristic curve from grain-size distribution and volume-mass properties," in *Proc., 3rd Brazilian Symp. on Unsaturated Soils*,(1997), vol. 1, pp. 13–23.

10. A. Johari and A. A. Javadi, "Prediction of a soil-water characteristic curve using neural network," *5thInternationalConference Unsaturated Soils Barrcelona Spain*, vol. 1, pp. 461–466, (2011).
11. A. Johari, G. Habibagahi, and A. Ghahramani, "Prediction of soil-water characteristic curve using a genetic based neural network," *Sci. Iran.*, vol. 13, no. 3, pp. 284–294, (2006).
12. A. Ahangar-Asr, A. Johari, and A. A. Javadi, "An evolutionary approach to modelling the soil–water characteristic curve in unsaturated

soils," *Comput. Geosci.*, vol. 43, pp. 25–33, Jun. (2012).

13. A. Johari and A. Hooshmandnejad, "Prediction of soil-water characteristic curve using gene expression programming," *Iran. J. Sci. Technol. Trans. Civ. Eng.*, vol. 39, no. C1, (2015).
14. W. S. Sillers and D. G. Fredlund, "Statistical assessment of soil-water characteristic curve models for geotechnical engineering," *Can. Geotech. J.*, vol. 38, no. 6, pp. 1297–1313, (Dec. 2001).
15. R. G. Hills, D. B. Hudson, and P. J. Wierenga, "Spatial variability at the Las Cruces trench site," *Indirect Methods Estim. Hydraul. Prop. Unsaturated Soils Univ Calif. Riverside*, pp. 529–538, (1992).
16. M. Fredlund, G. F. N. Gitirana, Jr., and H. Pham, "A methodology for applying probability theory to unsaturated hydraulic proper- ties for seepage analysis," *Proc GeoCongress 2008 Chall. Sustain. Geoenvironment Geotech. Spec. Publi- Cation 178 ASCE Rest. Va*, (2008).
17. R. F. Carsel and R. S. Parrish, "Developing joint probability distributions of soil water retention characteristics," *Water Resour. Res.*, vol. 24, no. 5, pp. 755–769, (1988).
18. K.-K. Phoon, "Modeling and simulation of stochastic data," in *GeoCongress 2006@ sGeotechnical Engineering in the Information Technology Age*, pp. 1–17, (2006).
19. K.-K. Phoon, A. Santoso, and S.-T. Quek, "Probabilistic analysis of soil-water characteristic curves," *J. Geotech. Geoenvironmental Eng.*, vol. 136, no. 3, pp. 445–455, (2010).
20. C. F. Chiu, W. M. Yan, and K.-V. Yuen, "Reliability analysis of soil–water characteristics curve and its application to slope stability analysis," *Eng. Geol.*, vol. 135–136, pp. 83–91, (May 2012).
21. P. G. Hoel, S. C. Port, and C. J. Stone, *Introduction to Statistical Theory*. Houghton-Mifflin, (1971).
22. K. M. Ramachandran and C. P. Tsokos, *Mathematical Statistics with Applications*. Academic Press, (2009).
23. H. Tijms, *Understanding Probability: Chance Rules in Everyday Life*. Cambridge University Press, (2007).
24. A. Johari and A. A. Javadi, "Reliability assessment of infinite slope stability using the jointly distributed random variables method," *Sci. Iran.*, vol. 19, no. 3, pp. 423–429,(2012).
25. A. Johari and A. R. Khodaparast, "Modelling of probability liquefaction based on standard penetration tests using the jointly distributed random variables method," *Eng. Geol.*, vol. 158, pp. 1–14, (May 2013).
26. A. Johari, A. A. Javadi, M. H. Makiabadi, and A. R. Khodaparast, "Reliability assessment of liquefaction potential using the jointly distributed random variables method," *Soil Dyn. Earthq. Eng.*, vol. 38, pp. 81–87, (2012).

Influence of structure in the soil-water characteristic curves of two residual soils of granite

Marcelo Heidemann[1,a], Luiz A. Bressani[2], Wai Y.Y. Gehling[2], Juan A. A. Flores[3] and Mattheus S. Porto[2]

[1]*Federal University of Santa Catarina, Mobility Engineering Center, Joinville - SC, Brazil*
[2]*Federal University of Rio Grande do Sul, Porto Alegre - RS, Brazil*
[3]*Federal University of Santa Catarina, Institute of Geosciences, Florianopolis - SC, Brazil*

Abstract. This paper discusses the influence of structure in the soil-water characteristic curves (SWCC) of two residual soils of granite formed in a subtropical environment. One soil has a saprolitic nature (named GrSp) and the other shows lateritic behavior (named GrLt). Both materials occur on a slope which presents a extensive history of landslides in the municipality of São José, Southern Brazil. SWCC of undisturbed and remolded specimens were determined using the filter-paper technique. Undisturbed specimens were collected and remolded specimens were produced by static compression of disintegrated soil to the same void ratio and moisture content of undisturbed specimens. SWCC of saprolitic soil showed curves which were modelled as unimodal but the lateritic soil required the use of bimodal curves because of the structure developed in micro and macro levels. Suction levels measured in GrLt soil were higher than in saprolitic soil. Remolding process changed the levels of suction achieved, which is due to the structure and more specifically to the size of pores in the remolded soils. Hysteresis were verified in both materials, but were far more pronounced in the lateritic soil. Analysis of mercury intrusion porosimetry were carried out in GrSp soil. They showed that remolding generates a different pore distribution from the existing in the undisturbed soil.

1 Introduction

Geomechanical properties of residual soils may be affected by the action of macro and microstructure features developed by weathering processes [1]. Soil structure refers to combined effect of particle, groups of particles and pores arrangement, composition and forces between particles [2]. Similar definition for soil structure was proposed by [3, 4].

While structure origins may be complex, the effects themselves may be described in a simple and general way [5].

The term structure is used by [5,6] to refer to those soil features which are peculiar to the soil in its undisturbed condition, as inter-particles bonding or cementation, and are eliminated when the soil is remolded.

Thus, investigation of macrostructure effects on soil behavior may be carried out through evaluation of possible differences exhibited in tests carried out using undisturbed and disturbed specimens. Undisturbed (or natural) specimens must be obtained through high quality sampling. Disturbed specimens should be obtained by remolding, compaction or reconstitution, in order to remove structural elements of soil.

In [7, 8] is mentioned that water retention depend of many features, such as amount, size and distribution of pores, soil texture and mineralogy. Thus, the structure developed with the formation of residual soils should influence on suction levels that should be mobilized in such materials, when they are unsaturated.

Although there are many papers reporting the determination of soil-water retention curves (SWCC) for residual soils, most of them carried out test in compacted specimens [8-13] or undisturbed specimens only [14-19, 7].

There are few studies discussing the effects of structure loss on SWCC. In this sense [20] should be mentioned. The authors present SWCC of volcanic soils from Honk Kong both in undisturbed and recompacted specimens.

This paper presents the results of SWCC for two tropical residual soils of granite from Southern Brazil, obtained though filter-paper tests in undisturbed and remolded specimens. One of these soils is a saprolite and the other exhibits some lateritic features.

2 Characterization of studied soils

Both soils are residual of the same granite rock and occur in São José (SC) municipality (Figure 1). These soils are found in a slope 120 m height, which has suffered successive instabilities. The coordinates of the sampling area are 732.370 E and 6.948.100 S with an average altitude of about 53 m above sea level.

[a] Corresponding author: marcelo.heidemann@gmail.com

DOI: 10.1051/e3sconf/20160911002

The elevations of this region are formed by Neoproterozoic granites that forms Dom Feliciano Belt [21,22], which extend from Santa Catarina state, Brazil, to Uruguay.

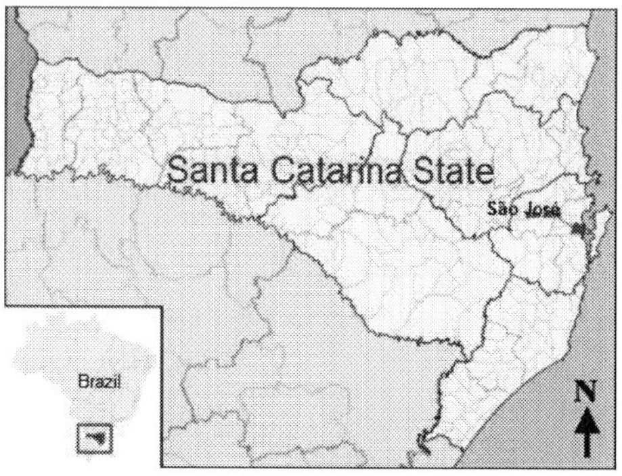

Figure 1. Localization of São José Municipality

The saprolitic residual soil is named GrSp. This soil occurs in the whole area of the slope but mostly as a layer subjacent to the soil here named GrLt. The GrLt soil shows some features which are typical of lateritic soils, but it is not a laterite. The manner in which these soils occur is illustrated in Figure 2 while Figure 3 shows each soil in detail.

Figure 2. A profile view of both studied soils.

Figure 3. Close view of GrSp soil (left) and GrLt (right).

Physical index and classification of these soils are presented in Table 1 and grain size distributions are shown in Figure 4.

Table 1. Physical properties and classification indexes of studied soils.

	GrSp	GrLt
G	2,625	2,699
w_{nat} (%)	29,0	30,0
γ_{nat} (kN/m³)	16,84	15,65
γ_d (kN/m³)	13,04	12,05
γ_{sat} (kN/m³)	17,88	17,39
e	0,97	1,20
Sr (%)	78,5	67,0
wL	42	60
PI	18	24
USCS	ML	MH

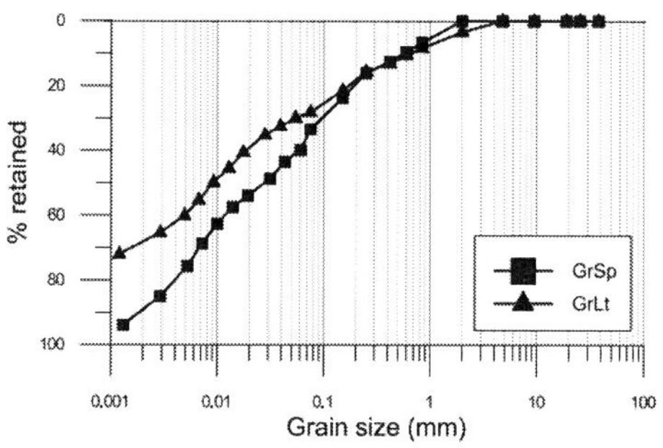

Figure 4. Grain size distribution of studied soils.

The GrLt soil presents a larger amount of fine particle than GrSp although the clay particles naturally occur as concretions or clusters resulting in a rough texture similar to a sandy soil. However, the clayey nature of the soil is reflected in its grain size distribution curve and in its high PI.

From the mineralogical point of view these materials show different compositions (Table 2), despite the same origin rock. This highlights the different weathering level.

Table 2. Mineralogical composition of studied soils (semi quantitative data from X-ray diffraction analysis)

	GrSp	GrLt
Quartz	13 %	30 %
Plagioclase	68 %	24 %
Biotite	-	19 %
Kaolinite	13 %	27 %
Chlorite	8 %	

 DOI: 10.1051/e3sconf/20160911002

3 Experimental program

3.1. Soil-water characteristic curves (SWCC)

Determination of soil-water characteristic curves (SWCC) was carried out through the filter-paper technic based in recommendations made in [23] as well as suggestions of [24] based in the experience with tropical soils from Southern Brazil.

The SWCC of GrSp and GrLt were obtained using undisturbed and remolded specimens which 50 mm in diameter and 20 mm height. These remolded specimens were produced in order to achieve void ratio, density and moisture content similar to undisturbed specimens.

Whatman n° 43 filter paper were used. According to [25] it is able to reach moistures between 5% and 175% (suctions between 29,000 kPa and 3 kPa). The correlation between suction (Ψ) and paper moisture (w_f) is given by Equation 1 (when w_f <45,3) and Equation 2 (when $w_f \geq$ 45,3) as proposed in [23].

$$log\psi = 5{,}327 - 0{,}0779 \cdot w_f \quad (1)$$

$$log\psi = 2{,}413 - 0{,}0135 \cdot w_f \quad (2)$$

Although [23] mentions that the equations are only valid for measurement of the total suction [26] demonstrated that there is only one calibration curve for the filter-paper method regardless the type of suction that is being measured.

Differently of proposed in [23] the tests were carried out using one filter-paper positioned in contact on just one soil specimen side and not between two specimens. This procedure has the advantage of using a smaller quantity of soil for testing and the ease in controlling the moisture of a single specimen. This proposition was presented by [24].

An interface paper was also used between the soil and the paper used to measure suction. This interface paper was not exchanged during the test. The paper was kept in contact with soil for seven days protected from the light and temperature variations.

The first suction level measurement was made in natural moisture condition. After that it was performed a wetting cycle followed by a drying cycle and a new wetting cycle. Changes of soil moisture were made in steps of 5% in saturation degree.

3.2 Pore-size distribution

Mercury intrusion porosimetry technic (MIP) has been employed by many researchers in order to investigate soil structural features as [27-31] and others. Distribution, size and connectivity of pores may be used to define mechanical and hydraulic properties of soils [32, 33].

Two specimens of GrSp soil were submitted to MIP analysis. One of them was extracted from an undisturbed block. The second one was obtained from a specimen remolded from disaggregated soil similar to those used for the determination of the SWCC (similar physical indices as in undisturbed condition).

The specimens were air-dried and after that oven-dried at 40° C. Some authors [28-30, 15] report that in clays oven drying is not appropriated because causes shrinkage and pore diminution, mainly when samples are prepared as slurry. In this case, according to such authors, sample freezing and subsequent sublimation for water extraction is a more appropriate drying procedure.

In this research the option for air-drying is justified because when the determination of SWCC is carried out, the soil is air-dryed. Therefore, suctions under specific degree of saturation are associated to soil volumetric changes. Besides, because the specimens are not prepared as slurry, there was little significant shrinkage during drying.

The equipment used to MIP analysis was an AutoPore IV – Micromeritics following procedures described in ISO 15901/2005. Measurements of intruded volume were done through stepwise mode. The physical parameters of mercury used for the interpretation of analysis were density of 13.54, superficial tension of 0,485 N/m and contact angle of 130°

4 Results and Discussions

Figure 5 shows SWCC obtained for GrSp soil in undisturbed and remolded conditions and experimental points.

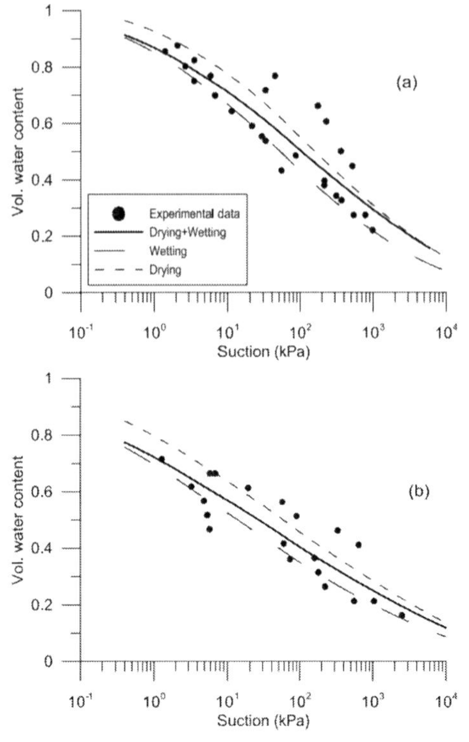

Figure 5. SWCC of GrSp soil (a – undisturbed, b – remolded).

For undisturbed and remolded specimens there are differences between the curves obtained from drying and

 DOI: 10.1051/e3sconf/20160911002

wetting data points indicating hysteresis. Drying curves are situated above wetting curves. For the undisturbed soils, suction values less than 1 kPa occur when volumetric water content is higher than 90% (w=31% and Sr=78%). For the remolded soil such suction level stands to volumetric water content of 70%.

When the behavior of undisturbed and remolded are comparedit seems that degradation of soil structure by remolding reduces suction levels that are mobilized. Thus, under same moisture content undisturbed soil develops higher suction levels than the remolded specimen.

SWCC to express the relationship between suction and saturation for GrSp soil showed in Figure 5 follow the model proposed by [34], according Equation 3.

$$\theta = \theta_s \left[\frac{1}{ln\left[e + \left(\frac{\psi}{a}\right)^n\right]} \right]^m \cdot \left[1 - \frac{ln\left(1 + \frac{\psi}{\psi_r}\right)}{ln\left(1 + \frac{10^6}{\psi_r}\right)} \right] \quad (3)$$

Where : θ – volumetric water content θ_s – volumetric water content for saturation; ψ – suction; ψ_r – suction under residual moisture; *a, m, n* – fitting parameters.

The parameters used to adjust the curves for GrSp soil are shown in Table 3.

Table 3. Fitting parameters for GrSp.

	Undisturbed			Remolded		
	W/D	Dry.	Wet.	W/D	Dry.	Wet.
a	72,51	74,72	71,00	13,68	13,68	13,68
n	0,351	0,416	0,375	0,295	0,321	0,287
m	0,976	0,854	1,171	0,886	0,784	1,078
θ_{sat}	1,056			1,011		

The SWCC for GrLt soil in undisturbed and remolded specimens are shown in Figure 6 including experimental data points.

In this Figure it is possible to verify a strong hysteresis while in remolded specimen this phenomenon is less marked. In undisturbed condition for volumetric water content levels higher than 75% (degree of saturation about 65%), small suctions are measured, in general below 10 kPa.

Comparing the results of SWCC for undisturbed and remolded sample is possible verifies that fitted curves refers to the development of higher suction in remolded specimen than in undisturbed one. These results are opposite to the obtained in GrSp soil.

The behavior of GrLt soil is not well reproduced by Fredlund and Xing (1994) model and it necessary the use of a bimodal model. In this sense [35-36] pointed out that unimodal curves could not be used to reproduce the behavior of many residual or sedimentary soils originated in tropical or subtropical environments.

According to [24] this occurs because these soils often exhibit well-defined micro and macro structures being composed for clay particles which are aggregated forming clay clusters of silty and sandy size and may have some behavior of those.

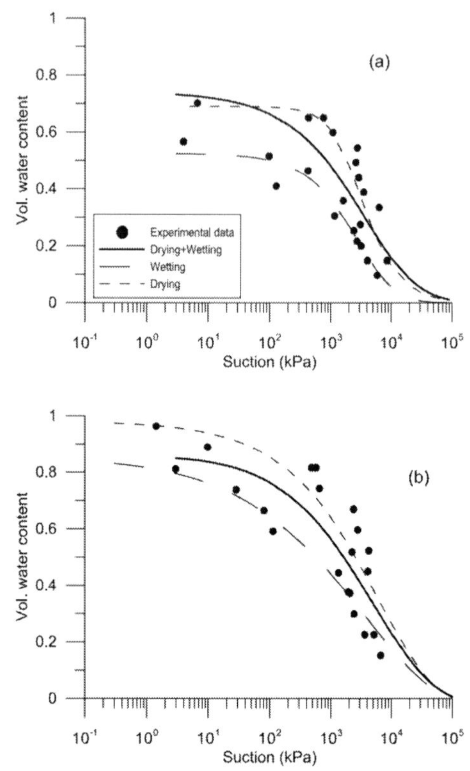

Figure 6. SWCC of GrLt soil (a – undisturbed, b – remolded).

The shapes of SWCC in these materials suggest a bimodal pores distribution: macro pores between clay clusters and micro pores inside the clusters. So, a model proposed by [37] was used to modeling the curves for GrLt soil (Equation 4)

$$\theta = \theta_s \left[\left[\frac{s_1}{ln\left[e + \left(\frac{\psi}{a_1}\right)^{n_1}\right]^{m_1}} \right] - \left[\frac{1 - s_1}{ln\left[e + \left(\frac{\psi}{a_2}\right)^{n_2}\right]^{m_2}} \right] \right] \cdot \left[1 - \frac{ln\left(1 + \frac{\psi}{\psi_r}\right)}{ln\left(1 + \frac{10^6}{\psi_r}\right)} \right] \quad (4)$$

Where: s – saturation degree that divide the curve in low and high suction levels; θ – volumetric water content; θs – volumetric water content in saturation; ψ – suction; ψr – suction under residual moisture; a_i, m_i, n_i – fitting parameters.

The parameters used to adjust the curves for GrLt soil are shown in Table 4.

Although a bimodal model is useful it was not possible to fit a curve over a wide suction range so that it could had a "classic" bimodal curve. This is because: (i) under moisture contents larger than 27% the suction drops to levels below 1 kPa were the filter paper method is not valid; (ii) suction levels higher than 8 MPa were not achieved in the tests, requiring another techniques for that.

Comparing Figures 5 and 6, one verifies a more pronounced hysteresis in soil GrLt than in GrSp both in undisturbed and remolded condition. There is no clear evidences about the cause, but is possible correlates a less pronounced hysteresis between wetting and drying curves with a smaller amount of fines and consequently lower level of weathering.

Figure 7 shows the SWCC for both soils plotted in the same space. These curves were fitted using both drying and wetting experimental data.

Table 4. Fitting parameters for GrLt.

	Undisturbed			Remolded		
	W/D	Dry.	Wet.	W/D	Dry.	Wet.
a_1	24000	2283	10000	8000	15000	10000
n_1	0,600	1,800	0,907	0,5	0,5	0,364
m_1	3,00	0,550	3,00	1,2	1,45	1,416
a_2	45000	45000	45000	44689	44633	40000
n_2	0,08	0,08	0,08	0,104	0	0,05
m_2	25	25	25	0,927	0,747	0,744
S_1	0,5	0,46	0,35	0,770	0,793	0,762
ψ_{res}	12000	12000	12000	12000	12000	12000
θ_{sat}		1,483			1,465	

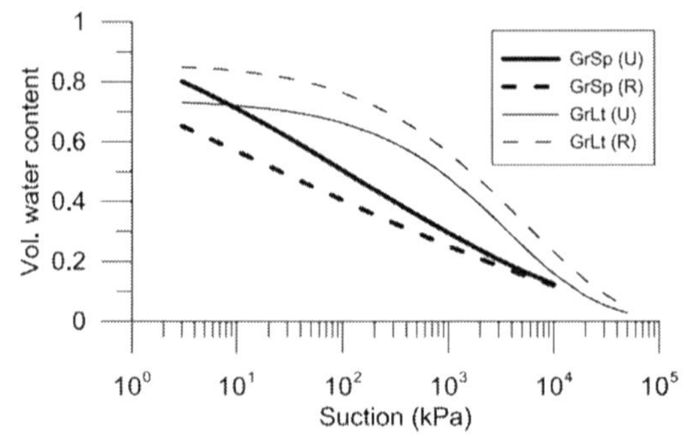

Figure 7. Fitted SWCC for GrSp and GrLt.

It is noted that the mobilized suction levels are proportional to the fines content and degree of weathering for undisturbed and remolded conditions. However, for a volumetric water content higher than 70% GrSp soil was able to maintain suctions slightly higher than GrLt, if compared results from undisturbed specimens.

Figure 8 shows the pore size distribution for the GrSp soil both in undisturbed and remolded conditions.

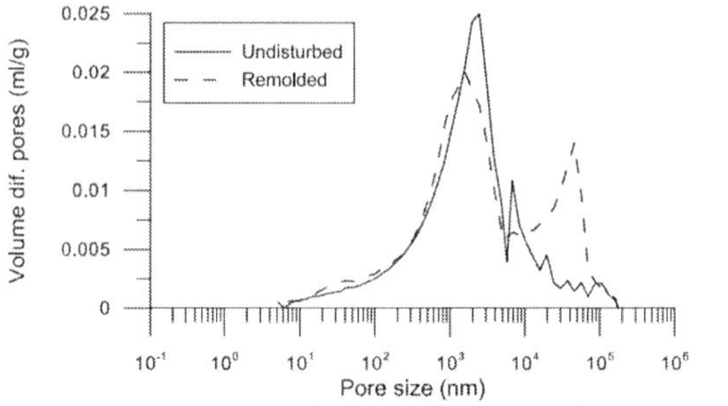

Figure 8. Pore size distribution in GrSp specimens.

Figure 8 shows a clear difference between pore size distributions in both samples. The number of pores in the range of 10^3 and 10^4 nm are reduced when the soil is remolded but a large amount of pores in 10^4 and 10^5 nm range are developed. It means that if the soil is remolded a larger number of bigger pores are produced than it is found in the undisturbed specimen. Thus, larger pores desaturate at lower suction levels and consequently SWCC of remolded sample indicate lower suctions than in undisturbed sample for the same volumetric water content.

5 Conclusions

The SWCC obtained for saprolitic soil (GrSp) is unimodal but for the lateritic (GrLt) a bimodal curve fitting was required. This seems to be due the structure developed in a macro and micro levels along the lateritization process. Suctions measured in lateritic soil are, in general, higher than in the GrSp.

The remolding process affects the suction curves obtained. This was caused by changes in soil structure specifically related to pores and clayey clusters size developed after remolding. The GrLt soil is more sensible to remolding and in this condition higher suction levels are achieved than when undisturbed. In the GrSp soil the opposite was observed.

Mercury intrusion porosimetry (MIP) analyses in GrSp soil showed that remolded specimensat the same voids ratio, density and moisture as the undisturbed, showed a distinct pore configuration. In remolded condition, soil has larger pores and therefore the suctions are lower than those measured in undisturbed specimen. This behavioral change is caused by structural changes since the relationship between pore size and suction is very strong as shown in [17, 14, 38].

References

1. Irfan, T.Y. Fabric Variability and index testing of a granitic saprolite. In Int. Conf. on Geomechanics in Tropical Soils, v.1. (1988)
2. Mitchell, J.K. Soga, K. Fundamentals of Soil Behavior, 3rd ed. John Wiley and Sons, New York (2005).
3. Fookes, P.G. Tropical residual soils. The Geological Society, London. (1997)
4. Collins. K.; McGowen, A. The form and function of microfabric features in a variety of natural soils. Géotechnique, E **24**. (1974)
5. Leroueil, S.; Vaughan, P.R. (1990) The general and congruent effects of structure in natural soils and weak rocks. Géotechnique, E **40**, 3 (1990)
6. Wesley, L.D. Influence of structure and composition on residual soils. J. Geot. Eng. E **116**, 4 (1990)
7. Miguel M.G., Vilar, O.M. Study of the water retention properties of a tropical soil. Can. Geotech. J. **E** 46. (2009)
8. Vanapalli, S.K., Fredlund, D.G., Pufahl, D.E. The influence of soil structure and stress history on the soil-water characteristics of a compacted till. Géotechnique, **E** 49, 2 (1999)

9. Oh, S., Lu, N., Kim, Y.K., Lee, S.J., Lee, S.R. Relationship between the Soil-Water Characteristic Curve and the Suction Stress Characteristic Curve: Experimental Evidence from Residual Soils. J. Geotech. Geoenv. Eng. **E** 138, 1. (2012)
10. Lee, I.M., Sung, S.G., Cho, G.C. Effect of stress state on the unsaturated shear strength of a wheathered granite. Can. Geotech. J. **E** 42 (2005)
11. Hossain, M.A., Yin, J.H. Shear strength and dilative characteristics of an unsaturated compacted completeuly decomposed granite soil. Can. Geotech. J. **E** 47. (2010).
12. Rahardjo, H., Heng, O.B, Choon, L.E. Shear strength of a compacted residual soil from consolidated drained and constant water content triaxial tests. Can. Geotech. J. **E** 41. (2004)
13. Kim, C.K, Kim, T.H. Behavior of unsaturated weathered residual granite soil with initial water contents. Engineering Geology, **E** 113 (2010)
14. Aung, K.K., Rahardjo, H., Leong, E.C., TOLL, D.G. Relationship between porosimetry measurement and soil-water characteristic curve for an unsaturated residual soil. Geotech. Geol. Eng. **E** 9. (2001)
15. Samingan, A.S., Leong, E.C., Rahardjo, H. A flexible wall permeameter for measurements of water and air coefficients of permeability of residual soils. Can. Geotech. J. **E** 40, 3 (2003)
16. Gan J.K, Fredlund, D.G. Shear strength characteristics of two saprolitic soils. Can. Geotech J. **E** 33. (1996)
17. Miguel, M.G., Bonder, B.H. Soil–Water Characteristic Curves Obtained for a Colluvial and Lateritic Soil Profile Considering the Macro and Micro Porosity. Geotech. Geol. Eng. **E** 30, 6. (2012)
18. Agus, S.S., Leong, E.C., Rahardjo, H. Soil-water characteristic curves of Singapore Residual soisl. Geotech. Geolog. Eng. **E** 19. (2001)
19. Futai, M.M., Almeida, M.S.S. An experimental investigation of the mechanical behaviour of an unsaturated gneiss residual soil. Géotechnique **E** 55, 3. (2005)
20. Ng, C.W.W., Pang, Y.W. Experimental investigations of the soil-water characteristics of a volcanic soil. Can. Geotech. J. **E** 37. (2000)
21. Bitencourt, M.F. et al. Estratigrafia do Batólito Florianópolis, Cinturão Dom Feliciano, na Região de Garopaba-Paulo Lopes, SC. Revista Pesquisas em Geociências **E** 35, 1 (2008). In Portuguese
22. Philipp, R.P. Machado R. Suítes graníticas do Batólito Pelotas no Rio Grande do Sul: petrografia, tectônica e aspectos petrogenéticos. Revista Brasileira de Geociências, **E** 31, 3 (2001). In Portuguese.
23. ASTM, Standard test method for measurement of soil potential using filter paper: D5298. (2010)
24. Feuerharmel, C. Shear strength and hydraulic conductivity of unsaturated colluvium soils from Serra Geral Formation. Phd thesis. Federal University of Rio Grande do Sul. Porto Alegre. (2007). In portuguese.
25. Marinho F.A.M. Soil suction measurement in soils and porous materials. Short course on unsaturated soils, Geodenver 2000. ASCE. Denver. (2000)
26. Marinho, F.A.M., Oliveira, O.M. The filter paper method revisited. Geotech. Testing J. **E** 29, 3. (2006)
27. Diamond, S. Pore size distributions in clays. Clays and Clay Minerals. **E** 18. (1970)
28. Sasanian, S. Newson, T.A. Use of mercury intrusion porosimetry for microstructural investigation of reconstituted clays at high water contents. Eng. Geology. **E** 158. (2013)
29. Arel, E., Onalp, A. Diagnosis of the Transition from Rock to Soil in a Granodiorite. J. Geotech. Geoenv. Eng. **E** 130 (2004)
30. Griffiths, F.J., Joshi, R.C., Change in pore size distribution due to consolidation of clays. Geotechnique. **E** 39, 1 (1989)
31. Lapierre, C., Leroueil, S., Locat, J. Mercury intrusion and permeability of Louisville clay. Canad. Geot. J. **E** 27. (1990)
32. Tugrul A. Change in pore size distritbution due to weathering of basalts and its engineering significance. Proc. Int. Sym on Engineering Geology and Environment. **E** 1. (1997)
33. Romero, E., Gens, A., Lloret, A. Water permeability, water retention and microstructure of unsaturated compacted Boom clay. Eng. Geology. **E** 54. (1999)
34. Fredlund, D.G., Xing A. Equations for the soil-water characteristic curve. Can. Geot. J. **E** 31. (1994)
35. Feuerharmel, C., Gehling, W.Y.Y., Bica, A.V.D. The use of filter paper and suction-plate methods for determining the soil-water characteristic curve of undisturbed colluvium soils. Geotech. Testing J. **E** 29, (2006)
36. Carvalho, J. C., Leroueil, S. Normalizing models for soil retention curves. Proc. 32nd Pavements Meeting, Brasilia. (2000). In Portuguese.
37. Fredlund, M.D. The role of unsaturated soil property functions in the practice of unsaturated soil mechanics. PhD thesis, University of Saskatchewan, Saskatoon. (1999)
38. Sillers W.S., Fredlund D.G., Zakerazadeh, N. Mathematical attributes of some soil-water characteristic curve models. Geot. Geol. Eng., **E** 19. (2001).

Acknowledgements

The authors would like to thank the Municipality Civil Defense, CNPq for its research grant (476360/2012-9), PPGEC/UFRGS and CAPES for scholarship of the first author.

E3S Web of Conferences 9, 11003 (2016)
DOI: 10.1051/e3sconf/20160911003
E-UNSAT 2016

Effect of in situ and laboratory compaction on the retention behaviour of a clayey soil

Francesca D'Onza[1,a], Vincenzo Oliviero[2], Domenico Gallipoli[3], Claudio Mancuso[2] and Simon J. Wheeler[4]

[1]*ENEA, Infrastructures and Services Directorate, 80055 Piazzale E. Fermi, Portici, Italy*
[2]*UNINA, Department of Civil, Architectural and Environmental Engineering, 80137 Via Claudio 21, Napoli, Italy*
[3]*Université de Pau et des Pays de l'Adour, Laboratoire SIAME, France*
[4]*University of Glasgow, School of Engineering, G12 8LT Oakfield Avenue, Glasgow, UK*

Abstract. A clayey soil has been tested in the laboratory in order to investigate the influence of the compaction procedure on the soil retention behaviour. In common engineering practice, data available for modelling are those of the soil compacted in the laboratory and soil behaviour during the earth structures lifecycle is predicted on that basis. This practice, however, seems to overlook the fact that construction procedures in the field might differ significantly from the compaction techniques used in the laboratory and this may induce considerable differences in material texture and therefore in the soil behaviour. The investigation shown in the present work aims to provide further insight into this aspect and to help endorsing or refuting the validity of such practice.

1 Introduction

The engineering properties of compacted soils can be designed by controlling particle grading, compaction method, water content and plasticity [1, 2]. This is very important for earth structures, which are almost invariably built by using locally sourced soils to minimize transportation costs and carbon footprint. If local soils show poor engineering characteristics, appropriate techniques must be used before and during emplacement of the fill to enhance mechanical performance through compaction. Although the behaviour of laboratory compacted soils is relatively well understood, there are still considerable uncertainties about the influence of field compaction techniques on soil properties. Further research in this respect is therefore required to maximize efficiency and cost-effectiveness of engineering design.

Earth embankments are usually built by compacting soils at the optimum water content as determined in the laboratory by means of a Standard Proctor test. Current geotechnical practice assumes that the energy applied to the soil during field compaction of earth embankments is comparable to that applied in the laboratory during Standard Proctor compaction. This means that the in-situ properties of earth fills are also similar to those of laboratory samples compacted at the same water content and according to the Standard Proctor method. Based on this assumption, laboratory compacted samples can be used to characterize the mechanical and water retention properties of earth fills.

This practice, however, overlooks the fact that field construction might differ significantly from laboratory compaction and this can induce considerable differences in material texture and, hence, engineering properties between the two materials.

The present study aims to investigate this issue by testing the same soil compacted both in situ (according to usual field construction procedures) and in the laboratory (according to the Standard Proctor method). In particular, the paper focuses on the study of water retention, which is an essential aspect of soil behaviour for the prediction of the deformation, strength and permeability of engineering fills.

2 Experimental apparatus and testing program

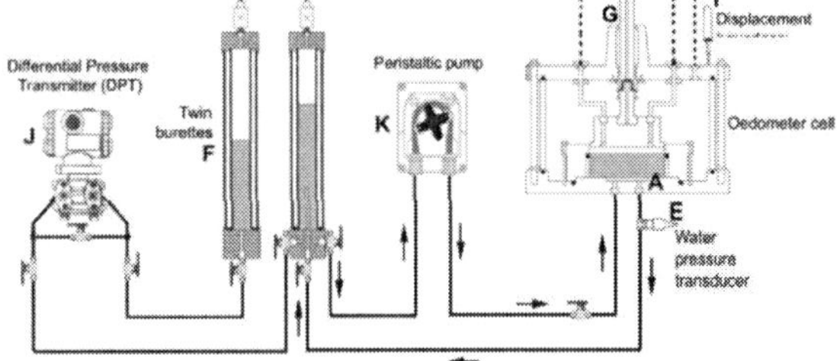

Figure 1. Wissa oedometer modified for testing unsaturated soils.

[a] Corresponding author: francesca.donza@enea.it

The experimental apparatus used in this work is a Wissa oedometer that can accommodate specimens with a diameter of 79.8 mm and a height of 25 mm (Fig. 1).

As well as oedometric CRS tests, the apparatus allows running either constant net stress or constant volume suction controlled tests. A personal computer connected to a load cell and to a digital ram regulates the vertical stress level using a feed-back procedure. Only constant net stress tests are presented in this paper.

Suction is controlled by the axis translation technique. The pore water pressure is applied at the base of the specimen, while the pore air pressure is applied at the top. The high air entry value porous stone, which is placed between the base of specimen and the water drainage circuit, has an air entry value of 1500kPa. Nevertheless, because of limitations to the maximum pressure achievable by the pore air supply, the largest applicable suction is only 850 kPa. The axial strains, and therefore the volumetric strains, are monitored by an external LVDT.

Water content changes are monitored by measuring the volume of water flowing in or out of the specimen through a measuring burette. The level inside the measuring burette is recorded by means of a differential pressure transducer connected to on the other side to a second burette with a constant water level used as a reference.

The saturation of the pore water line is ensured by periodically flushing the drainage circuit through a peristaltic pump (K in figure 1) that pushes water through a spiral groove cut inside the base pedestal. This forces any potential air bubble into the measurement burette where it is eliminated by buoyancy through the free water surface. The arrows in Figure 1 show the flushing path.

A thin layer of silicon oil is placed upon the water surface in both measurement and reference burettes to keep evaporation losses within acceptable limits. In addition, any potential evaporation should be the same in both burettes due to the symmetry of the system.

Evaporative losses through the air drainage line cannot be predicted as they depend on several factors including sample water content, air humidity and pressure. These evaporative losses are therefore measured at the end of each suction equalization stage when the sample is assumed to be in equilibrium. At this point, all loss of water is attributed to air evaporation and is fitted by a linear relationship with time, which is then used for correcting water content measurements.

3 Experimental procedure

The soil tested in this work is a clayey silt from a flood defence embankment of the Po River (Italy). Basic soil properties such as Atterberg limits, specific gravity, and grain size distribution were determined on several samples.

The soil has a specific gravity equal to 2,740, a uniformity coefficient, Uc, ≈ 15, a clay fraction of about 25%, and a silty fraction of about 60%. On average, the material has a liquid limit (w_L) of 48.9%, a plastic limit (w_P) of 27.1% and, therefore, a plasticity index (I_P) of 21.8%. In the Casagrande's chart, the experimental points lie below the "A" line around w_L = 50%, thus the material is classified as inorganic silt of medium/high compressibility.

A twin set of samples was tested in this work. The first set of samples were compacted in the laboratory at the optimum value of water content according to Standard Proctor method. The second set of samples were instead retrieved, with the least possible disturbance, from the field embankment. The embankment had been built by compacting soil at the same optimum water content of the Standard Proctor through the use of conventional heavy machinery. It is important to mention, however, that the undisturbed samples were retrieved few years after construction of the embankment and are therefore likely to have experienced at least some wetting/drying cycles between construction and retrieval from the field.

The samples compacted in the laboratory were produced by using soil collected from the embankment at the same depth of the undisturbed samples. This soil was dried in open air and broken down by means of a mechanical grinder. The soil was then compacted at the optimum water content using the Standard Proctor procedure [3].

The initial values of water content (w), specific water volume (v_w=1+e_w where e_w is water ratio that is equal to S_r*e), specific volume (v), degree of saturation (S_r) and dry density (ρ_d) are summarized in Table 1 for all samples tested in this work. The initial letter of the sample name (S or L) indicates in-situ or laboratory compacted samples while the numbers (50, 100, 200 or 500) at the end of the sample name indicates the constant net vertical stress (in kPa) at which the water retention test was performed.

Table 1. Initial values of tests.

Test code	w (%)	v_w	v	S_r (%)	ρ_d (g/cm^3)
S-E1tvn50	31.27	1.86	1.92	93.41	1.43
S-E3tvn100	30.51	1.84	1.91	92.43	1.44
S-E4tvn200	27.99	1.77	1.88	88.09	1.47
L-E7tvn50	20.34	1.56	1.63	88.36	1.69
L-E8tvn200	20.32	1.56	1.65	85.87	1.67
L-E9tvn500	19.75	1.54	1.63	86.94	1.69

Fig. 2 shows two Standard Proctor curves determined by two different research laboratories (UNINA and UNICAS) together with the points of the samples tested in this work. Fig. 3 shows some grading curves determined on different samples of soil tested.

Prior to imposing any water retention path, each specimen was loaded to a different vertical net stress (σ_v - u_a) to investigate the influence of the applied stress, and hence of the deformation, on the retention behaviour. Each test consisted of a drying-wetting cycle with step changes of suction at 5, 10, 20, 50, 100, 200, 300, 400,

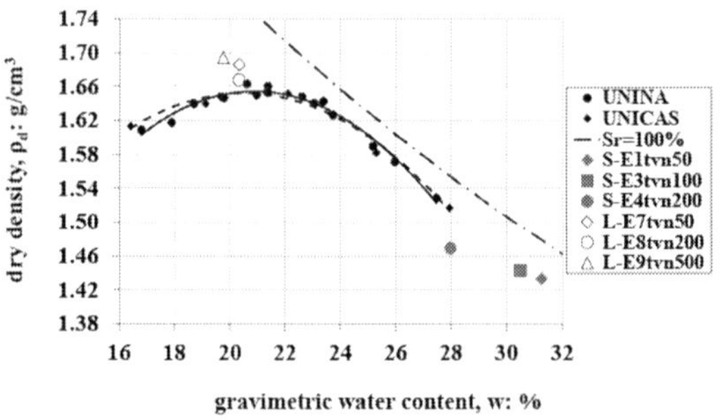

Figure 2. Soil proctor curves and points representative of the samples tested.

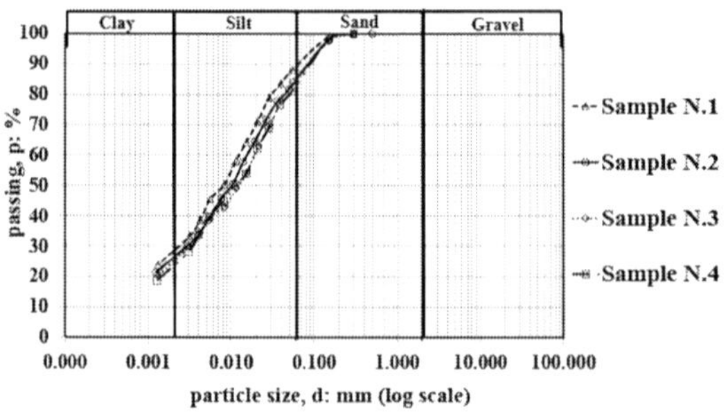

Figure 3. Soil grading curves determined on different samples of tested soil.

600 and 850 kPa (pore air pressure was changed while pore water pressure was maintained constant and equal to 50 kPa). An initial equalization stage at a suction of 5 kPa (10 kPa only for test S-E1tvn50) was followed by a step drying path until a maximum value of suction that was 400 kPa for tests S-E1tvn50, S-E3tvn100, S-E4tvn200, 500 kPa for test L-E7tvn50 and 850 kPa for tests L-E8tvn200, L-E9tvn500. All tests ended with a wetting path from the maximum suction level down to a suction of 5kPa (100 kPa only for test S-E1tvn50).

4 Experimental results

Figure 4 shows the variation of degree of saturation versus suction for the tests on the in-situ compacted samples.

The slopes of the three drying curves are relatively similar, though a slight dependency on the vertical net stress can be appreciated. This is consistent with the fact that samples subjected to higher stresses and thus higher reduction of void ratio exhibit a smaller variation of degree of saturation for the same suction increment due to smaller pores dimensions.

A significant hydraulic hysteresis is observed upon suction reversal with all wetting paths characterized by similar slopes.

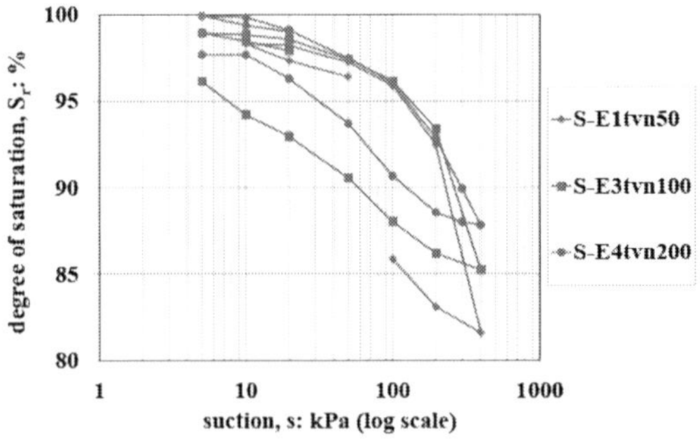

Figure 4. Tests on samples compacted in situ in the S_r : s plane.

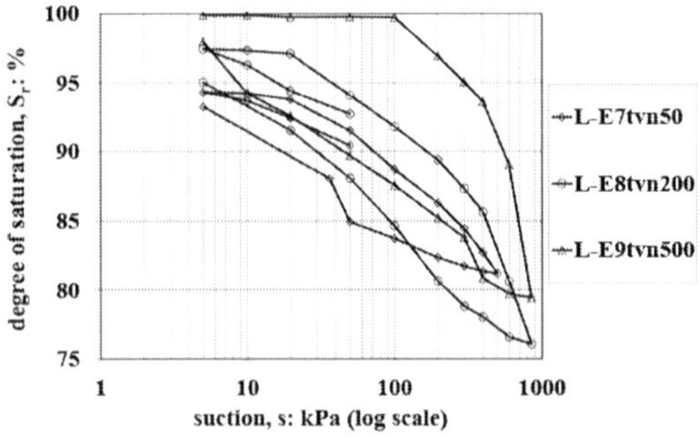

Figure 5. Tests on samples compacted in the laboratory in the S_r : s plane.

Figure 5 presents the retention curves of the three samples compacted in the laboratory. These tests present larger differences between them compared to the three tests on the samples compacted in-situ. This is probably because the samples compacted in-situ have experienced already several wetting/drying cycles in the field, which have hardened the soil and made it less sensitive to the applied vertical net stress during testing.

An additional reason of such differences could be that the in-situ compacted samples start drying from relatively high values of degree of saturation (above 97.5%) and are, presumably, on a main drying curve since the very beginning of the drying path. Instead, the laboratory compacted samples start drying from slightly lower

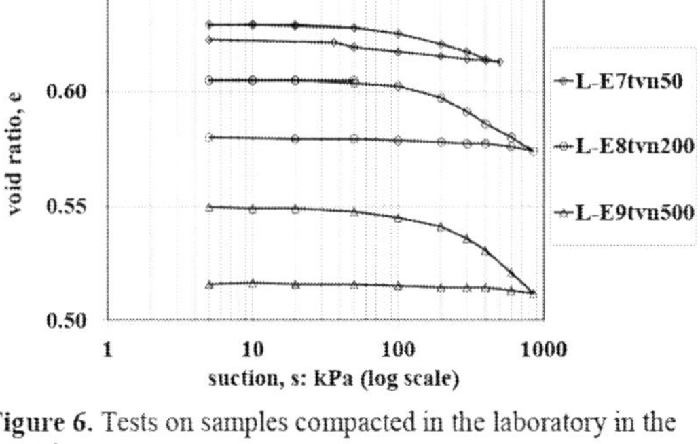

Figure 6. Tests on samples compacted in the laboratory in the e : s plane.

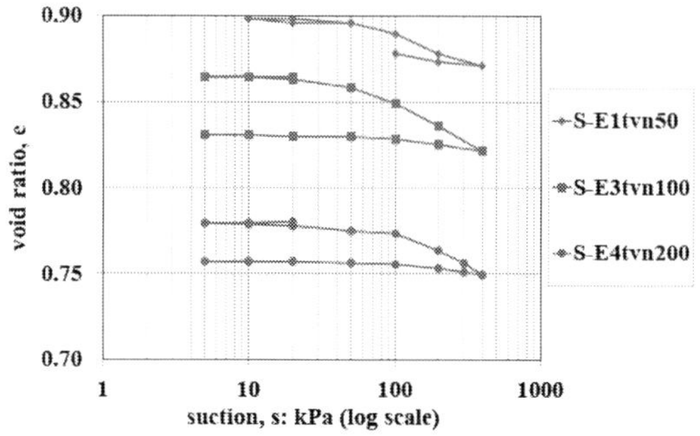

Figure 7. Tests on samples compacted in situ in the e : s plane.

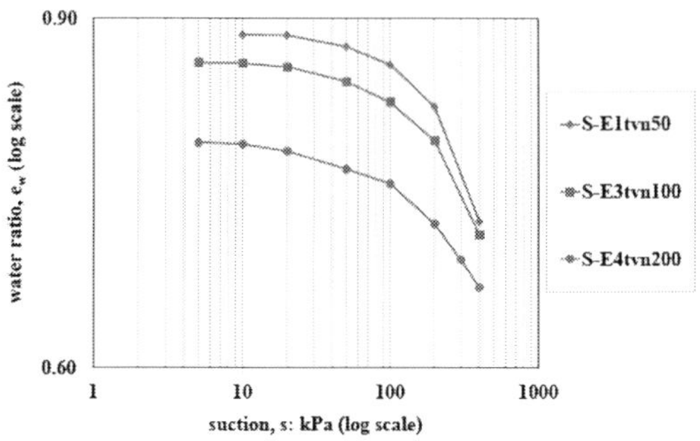

Figure 9. Tests on samples compacted in situ in the e_w : s plane.

values of degree of saturation, which means that they are probably on a scanning curve for the first part of the drying path.

Figures 6 and 7 show the variation of void ratio versus suction for the samples compacted in the laboratory and in situ, respectively. A slight dependency of elastic deformation during wetting on the vertical net stress is observed. For both in-situ and laboratory compacted samples, the magnitude of elastic deformation during suction variation increases at decreasing values of vertical net stress. This behaviour seems not to depend on the compaction technique as similar sensitivity of void ratio to suction variation have been observed in the two sets of tests.

Elasto-plastic strains occur during drying paths but their magnitude is not very sensitive to the applied vertical net stress. Different responses are however observed for the samples compacted in situ and in the laboratory with the former samples experiencing larger strains under similar variations of suction.

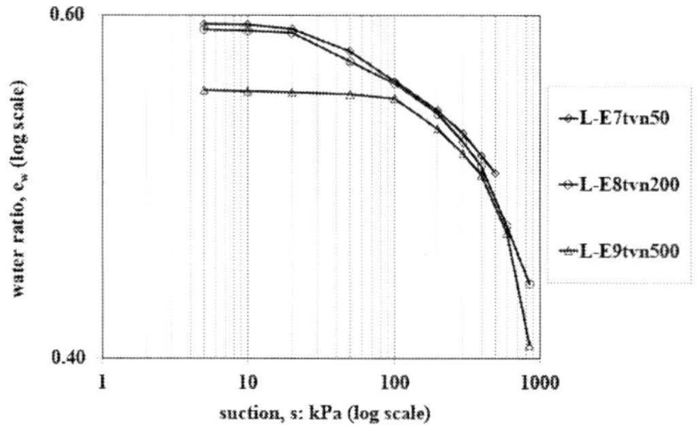

Figure 8. Tests on samples compacted in the laboratory in the e_w : s plane.

One interesting aspect emerges when the different drying curves are recast in the loge$_w$-logs plane. As suction grows large, the water ratio tends to a linear asymptote that is independent of void ratio and reduces to a unique relationship between water ratio and suction. This can be observed in Figures 8 and 9 where all curves seem to tend to a unique asymptote. The uniqueness of this relationship between water ratio and suction at high suction (and low saturation) has already been experimentally observed in compacted silts and clays with different degrees of activity [4-6].

The above observations are also in agreement with the retention model proposed by Gallipoli (2012) [7] and Gallipoli et al. (2015) [8], who suggested that any mechanically-induced change of void ratio will only affect the largest pores while leaving the smaller ones almost unchanged. At high suction (and low saturation), water withdraws in the smallest pores so that any mechanically-induced change of void ratio is unable to cause a variation of water ratio. This also means that, as suction grows large, the drying curves at different levels of void ratio will tend towards a unique relationship between water ratio and suction, which is dictated by the retention properties of the smallest pores.

5 Interpretation of experimental results

5.1 Model used

Experimental results have been interpreted with the water retention model by Gallipoli 2012 [7].

As well known, this water retention model is capable of predicting the hysteretic response of soils during both wetting-drying cycles at constant void ratio and compression-swelling cycles at constant suction. The model formulation is capable of capturing important soil features such as, for example, the influence of hydraulic hysteresis and deformation on the variation of degree of saturation and the dependency of water retention behaviour during compression on previous wetting-drying history. Moreover, during main wetting or main drying at high suction (i.e. at low saturation), the model correctly predicts a "virgin" retention line that uniquely relates water ratio and suction regardless of the current value of void ratio.

In the model two main surfaces, i.e. the main wetting and main drying surfaces, enclose the domain of attainable soil states in the degree of saturation – suction – void ratio space. Inside this domain, the variation of degree of saturation is governed by a reversible scanning law, which describes the transition from one main surface to the other.

From the physical point of view:

- a soil state is assumed to lie on the main drying surface if there is at least one infinitesimal change of suction

DOI: 10.1051/e3sconf/20160911003

and/or void ratio which causes irreversible drainage of initially flooded pores;
- a soil state is assumed to lie on the main wetting surface if there is at least one infinitesimal change of suction and/or void ratio which causes irreversible flooding of initially drained pores.

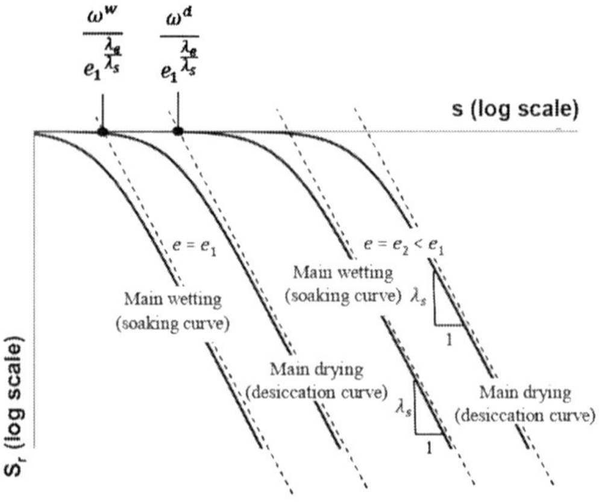

Figure 10. Main hysteretic loops in S_r : s plane corresponding to different e_1 and e_2 (with $e_1 > e_2$)

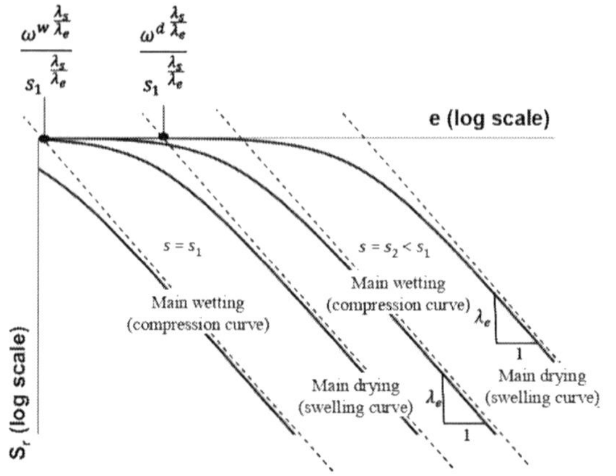

Figure 11. Main hysteretic loops in S_r : e plane corresponding to different s_1 and s_2 (with $s_1 > s_2$)

Some terminology is introduced to refer to the cross-sections of the main drying and wetting surface with planes at constant e (log S_r – log s plane, figure 10) and constant s (log S_r – log e plane, figure 11). In particular, the intersection of main (wetting or drying) surface with generic log S_r – log s plane at constant e generates:
- a Soaking curve (in case of main wetting surface);
- a Desiccation curve (in case of main drying surface).

The intersection of a main (wetting or drying) surface with generic log S_r – log e plane at constant s generates:
- a Compression curve (in case of main wetting surface);
- a Swelling curve (in case of drying surface).

The two main wetting and the main drying surfaces are defined by a total of 8 parameters (4 for each surface):
- main drying: { λ_{sd} λ_{ed} ω_d m_d }
- main wetting{ λ_{sw} λ_{ew} ω_w m_w }

The number of parameters is reduced by assuming that:
- identical asymptotic slopes for main drying and wetting surfaces in log Sr – log s plane:

$$\lambda_{sd} = \lambda_{sw} = \lambda_s \text{ or } m_d * n_d = m_w * n_w \quad (1)$$

- identical asymptotic slopes for main drying and wetting surfaces in log Sr – log e plane:

$$\lambda_{ed} = \lambda_{ew} = \lambda_e \text{ or } \psi_d * m_d * n_d = \psi_w * m_w * n_w \quad (2)$$

- identical shapes for main drying and wetting surfaces when departing from asymptotic planes:

$$m_d = m_w = m \quad (3)$$

The above assumptions imply, as a consequence :

$$n_d = n_w = n \quad (4)$$

$$\psi_d = \psi_w = \psi \quad (5)$$

A family of surfaces scanning the space between the two main drying and wetting surfaces is also defined. If a soil path on the main drying surface is reversed, the soil state departs from the main drying surface, moves along the adjacent scanning surface, spanning across the space between the two main surfaces, until it joins the opposite main wetting surface. Similarly, if a soil path on the main wetting surface is reversed, the soil state departs from the main wetting surface, moves along the adjacent scanning surface, spanning across the space between the two main surfaces, until it joins the opposite main drying surface. Soil behaviour is assumed to be reversible along a scanning surface.

5.2 Model calibration

Experimental data shown in section 4 have been interpreted with the model by Gallipoli 2012 [7] described in the previous section. Only data relative to the drying paths have been modelled as calibration of main surfaces is more crucial, with respect to the scanning surfaces, for an appropriate prediction of soil behaviour in earth structures for engineering purposes.

The first hypothesis made is that the experimental data relative to the drying paths lie on the main drying. This is a reasonable assumption because, for each tested sample, after the first wetting stage at 5kPa of suction, degree of saturation is close to 100%, and in any case lower than 94.5%.

The equation of main drying proposed in the model is:

$$S_r = \left(1 + \left(\frac{e^{\psi} * s}{\omega}\right)^n\right)^{-m} \quad (6)$$

As proposed by the model, as suction grows large the water ratio tends to a log-planar asymptote whose expression in the model is given by:

$$\log e_w = m * n * \log\omega - (m * n * \psi - 1) * \\ * \log e - m * n * \log s \quad (7)$$

Assuming the uniqueness of the relationship between water ratio and suction at high suction/low saturation levels, as observed in the literature (see section 4), the product m * n * ψ is equal to 1. In this hypothesis, in fact, the log-planar asymptote becomes independent of void ratio and reduces to a unique relationship between water ratio and suction.

Indeed, a calibration of the main wetting surface performed by Gallipoli et al. (2003b) [9] for a different material (i.e. the compacted Speswhite kaolin tested by Sivakumar (1993) [10]) yielded a similar value of 1.1 for the parameters product m * n * ψ. It is also interesting to note that, by following an alternative modelling approach, Sheng (2010) [11] advocated a choice of parameter values that satisfies the condition m *n * ψ = 1 to ensure consistency with the intrinsic phase relationship in the model by Gallipoli et al. (2003b) [9]. Moreover, inspection of figures 8 and 9 indicates that it is reasonable to assume the product m * n * ψ equal to 1 for the tested soil.

Under the above assumption, the number of independent parameters defining the main drying surface reduces from four to three, namely (ω, m, n). Indeed ψ is taken equal to 1/(m*n) and it is possible to rewrite equation (1) as:

$$Sr = \left(1 + \left(\frac{e^{\frac{1}{m*n}*s}}{\omega}\right)^{n}\right)^{-m} \quad (8)$$

Values of ω, m and n have been determined by fitting experimental points of drying paths with equation 8.

5.3 Results interpretation

Two separate parameters determination have been done for the laboratory and in-situ compacted samples.

For each set of samples the model parameters defining the main drying curve have been determined on the basis of experimental points of drying paths relative to two of the three tests performed.

The parameters determination done for the laboratory compacted samples refers to the tests performed under a vertical net stress of 200 and 500 kPa. The above choice has been made on the basis of the higher degree of saturation at the starting of the drying and therefore the more confidence for that data to be on a main surface.

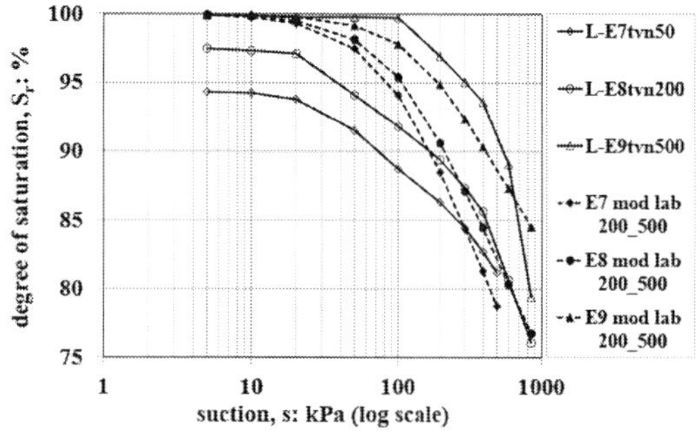

Figure 12. Tests on samples compacted in the laboratory in the S_r : s plane and model predictions with calibration done using tests L-E8tvn200 and L-E9tvn500.

In figure 12 experimental data relative to soil compacted in the laboratory together with model predictions are shown. The model seems not to predict particularly well the soil behaviour, especially for the curve not used in the fitting process. In reality, part of the misprediction could be due to the fact that degree of saturation at the starting of the drying is not particularly high and therefore the experimental data points could be not on a main surface in the first part of the test.

The model parameters and the starting conditions obtained from calibration have been then used to simulate the behaviour of samples compacted in situ (Fig. 13). This is because, in common engineering practice, data available for modelling are those of the soil compacted in the laboratory and soil behaviour during the earth structures lifecycle is predicted on this basis.

The model seem to predict reasonably well the position of the main drying surface and the sensitivity to the applied stress, thus to the current void ratio, but not the slope.

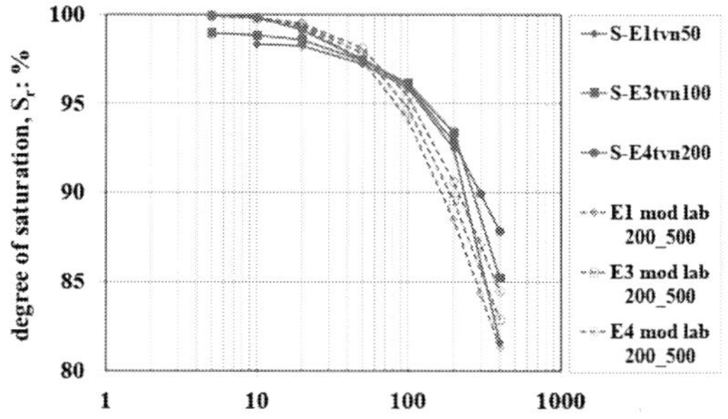

Figure 13. Tests on samples compacted in situ in the S_r : s plane and model predictions with calibration done using tests L-E8tvn200 and L-E9tvn500.

In order to investigate whether the weakness is in the model or in the difference in the compaction procedures, the model has been also calibrated with data relative to the tests on samples compacted in situ.

The experimental points chosen for the fitting procedure are those of tests performed under a vertical net stress of 50 and 200 kPa. In figure 14 experimental data relative to soil compacted in situ together with model predictions are shown.

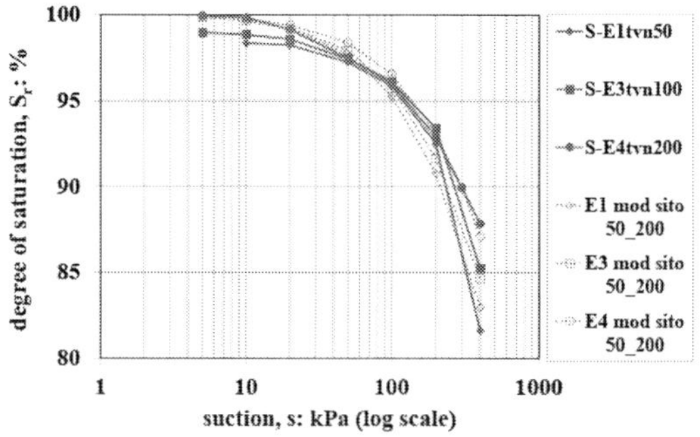

Figure 14. Tests on samples compacted in situ in the S_r : s plane and model predictions with calibration done using tests S-E1tvn50 and S-E4tvn200.

The model seems to predict reasonably well the experimental behaviour in terms of position, slope and sensitivity to applied stress, even for the curve not used in the fitting process.

The good model prediction can be attributable to the consistency of data of different tests and to the very high degree of saturation at the starting of each drying path.

6 Concluding remarks

The present work has investigated the influence of compaction procedure on the water retention behavior of a clayey silt. With this aim, a twin set of samples has been tested. Samples compacted in the laboratory at optimum Proctor standard and samples compacted in situ by conventional heavy machinery.

Samples compacted in situ show an higher degree of saturation and a lower dry density. This could be due to the stress paths experienced from the time of the embankment construction to the time of retrieval and/or to an under compaction and/or to a compaction at a too high water content. If the first is the case, the soil compacted in situ and in the laboratory should still have the same texture as the stress paths modify the status but not the fabric of the soil and therefore its behaviour.

Tests on samples compacted in the laboratory present larger differences between them compared to the tests on the samples compacted in-situ. This is probably because the samples compacted in-situ have experienced already several wetting/drying cycles in the field, or because the in-situ compacted samples start drying from relatively high values of degree of saturation in comparison to laboratory compacted ones and are therefore on a main drying curve since the very beginning of the drying path. Elasto-plastic strains occur during drying paths but their magnitude is not very sensitive to the applied vertical net stress. Different responses are however observed for the samples compacted in situ and in the laboratory with the former samples experiencing larger strains under similar variations of suction.

Experimental data relative to the drying paths of laboratory compacted samples have been interpreted with the model by Gallipoli 2012. The model parameters and the starting conditions obtained from calibration have been then used to simulate the behaviour of samples compacted in situ as it is done in common engineering practice where only data on laboratory compacted samples are available for design of earth structures. The model seems to predict reasonably the position of the main drying surface and its sensitivity to the applied stress, thus to the current void ratio, but not its slope. In order to investigate whether the weakness is in the model or in the dissimilar soil behaviour when compacted with different procedures, the model has been calibrated with data relative to the tests on samples compacted in situ. The model seems to predict reasonably the soil behaviour in terms of position, slope and sensitivity to applied stress.

In conclusion a dissimilar behaviour has been observed on the two sets of samples. Caution should be therefore used in common engineering practice when assuming that properties of earth fills are similar to those of laboratory compacted samples under the same prescriptions. This is for two reasons: the first is that control procedures of in situ compaction is not always accurate and could produce compaction at a different water content and different energy with respect to the prescriptions. The second reason is that different compaction procedures could induce different soil fabrics. Both the above reasons could cause the dissimilar behaviour observed in the present work.

Acknowledgments

The contribution of the European Commission via the Career Integration Grant SAFES (Service Assessment and Failure of Earth Structures), contract n. PCIG09-GA-2011-293727, is gratefully acknowledged.

References

1. S.J. Wheeler, V. Sivakumar. *Influence of compaction procedure on the mechanical behaviour of an unsaturated compacted clay. Part 1: wetting and isotropic compression.* (Géotechnique **50**, 4, 2000).
2. S.J. Wheeler, V. Sivakumar. *Influence of compaction procedure on the mechanical behaviour of an unsaturated compacted clay. Part 2: shearing and constitutive modelling.* (Géotechnique **50**, 4, 2000).
3. ASTM 2005. *D0698-00 AE01 test method for laboratory compaction characteristics of soil using standard effort (12 400 ft·lbf/ft^3 (600 kN·m/m^3)).* Vol. 04.08. In ASTM Book of Standards. ASTM, Philadelphia, Penn.
4. E. Romero, A. Gens, A. Lloret. *Water permeability, water retention and microstructure of unsaturated Boom clay.* (Eng. Geol. **54**, 117-127, 1999).
5. E. Romero, J. Vaunat. *Retention curves of deformable clays.* (Proc. Int. Workshop on Unsaturated Soils, Trento, Italy, 91-106, 2000).
6. E. Romero, G. Della Vecchia, C. Jommi. *An insight into the water retention properties of compacted clayey soils.* (Géotechnique **61**, 4, 2011).
7. D. Gallipoli. *A hysteretic soil-water retention model accounting for cyclic variations of suction and void ratio.* (Géotechnique **62**, 7, 2012). http://dx.doi.org/10.1680/geot.11.P.007.
8. D. Gallipoli, A.W. Bruno, F. D'Onza, C. Mancuso. *A bounding surface hysteretic water retention model for deformable soils.* (Géotechnique **65**, 10, 2015). http://dx.doi.org/10.1680/geot.14.P.118.
9. D. Gallipoli, S.J. Wheeler, M. Karstunen. *Modelling the variation of degree of saturation in a deformable unsaturated soil.* (Géotechnique **53**, 1, 2003b). http://dx.doi.org/10.1680/geot.2003.53.1.105.
10. V. Sivakumar. *A critical state framework for unsaturated soils.* (Ph.D. Thesis, U. Sheffield, U.K.,1993).
11. D. Sheng. *Constitutive modelling of unsaturated soils: Discussion of fundamental principles.* (Proc. 5th Int. Conference on Unsaturated Soils, Barcelona, Spain, 1, 2010).

Experimental determination of the hysteretic behaviour of soil-water retention curve of silty sands

Ludwig Angerer[1,a], Emanuel Birle[1]

[1] *Zentrum Geotechnik, Department of Civil Engineering, Technical University Munich, Baumbachstrasse 7, 81245 Munich, Germany*

Abstract. Shear behaviour of unsaturated silty sands is closely linked to the soil-water retention curve (SWRC). The SWRC depends beside the pore structure mainly on the hydraulic path. It hence shows hysteretic behaviour for wetting and drying cycles. This paper presents the results of the experimental determination of the boundary drying and wetting curve for two silty sands. The measurements were carried out using tensiometers at low suctions and the dew point method at high suctions. The influence of water content during compaction was studied by measuring the boundary drying and wetting curves for one silty sand compacted at two different water contents. The experimental results were compared to the predictions of 3 empirical models. The study showed that all models compare well to the experimental results for water contents above the residual water content. A modification was suggested to improve the performance at low water contents.

1 Introduction

The soil-water retention curve (SWRC) represents a useful function to predict constitutive behaviour of unsaturated soils. So it is e.g. well accepted to relate the SWRC to the increase of shear strength of unsaturated soils compared to the saturated state [1, 2, 3, 4, 5]. Beside the pore structure, the SWRC is mainly dependent on the hydraulic path and shows a hysteretic behaviour. That means that at the same water content the suction during wetting processes is less than during drying process.

It is common practice to determine the boundary drying curve experimentally while usually less effort is made to measure the boundary wetting curve [6]. The hysteretic behaviour is then often predicted with empirical or physically based models [7, 8, 9, 10, 11].

The present investigations are carried out as a part of the examination of slope stability of railway embankments out of silty sands. Since suction and hence shear strength can decrease significantly during wetting processes special emphasis is put on the determination of the hysteretic behaviour of the SWRC of silty sands. For this purpose the boundary drying curves as well as the boundary wetting curves of two statically compacted silty sands are measured in the laboratory. Furthermore, to study the influence of the water content during compaction, the hysteresis of one silty sand is experimentally determined for two compaction water contents. The measurements are carried out using tensiometers at low suctions and the dew point method at high suctions, as described in [6, 12].

Based on the results of the experimental studies, three empirical models [7, 10, 11] are tested with respect to a suitable prediction of the boundary wetting curve. Presupposing the boundary drying curve is known.

[a] Ludwig Angerer: l.angerer@tum.de

2 Experimental investigations

2.1 Soil properties

The paper examines the SWRC of two silty sands obtained from a railway embankment in Germany. Disturbed and undisturbed soil samples were collected and tested for particle size distribution, specific gravity, in-situ void ratio, in-situ water content and classified by the United Soil Classification System (USCS). Prior to testing the sands were passed through a 4 mm sieve. Both soil samples were silty sands (SM) with fine content of 28 % and 35.6 %, respectively. The particle size distributions of both silty sands are shown in Figure 1. The USCS texture, index properties and in-situ conditions are summarized in Table 1.

Table 1. Index properties of the soils and in-situ conditions

Name	USCS	Particle size distribution by weight			Gs	e	w
		Clay [%]	Silt [%]	Sand [%]	[-]	[-]	[%]
Sand 1	SM	5	23	70	2.66	0.56	14.25
Sand 2	SM	15.6	20	64	2.67	0.62	4.75

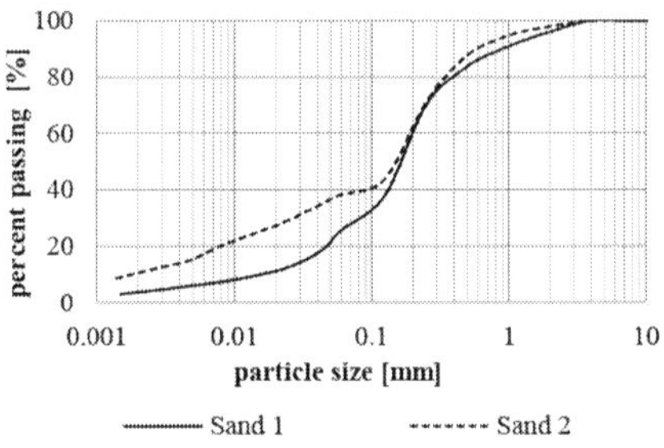

Figure 1. Particle size distribution curves of the silty sands

2.2 Testing methods and sample preparation

Suction was obtained by direct measurements with tensiometers, manufactured by UMS, at low suctions (< 100 kPa) and by indirect measurements with the dew point method at high suctions (> 1000 kPa). The dew point method was conducted by using the Decagon devices WP4 and WP4C. According to [6] osmotic suction can be neglected for the present soils. For this reason it is assumed to measure only matric suction with both methods.

The boundary drying and wetting curves were obtained from statically compacted specimens. Static compaction was carried out to approximately obtain the in-situ soil fabric at the in-situ void ratio at the in-situ water content. Since Sand 1 seemed to be in a relatively wet state an additional set of drying and wetting curve were measured in which the water content during compaction was decreased from $w_{comp.} = 14.25$ % to $w_{comp.} = 5$ %.

2.2.1 Sample preparation for tensiometers

The soil samples for the measurements with tensiometers were statically compacted in two layers directly into 8 suction cells with a diameter of 10 cm and a height of 3.5 cm as shown in Figure 2. Compaction was carried out by a hydraulic press with one piston compacting the soil on the top. The interfaces between the layers were roughened carefully.

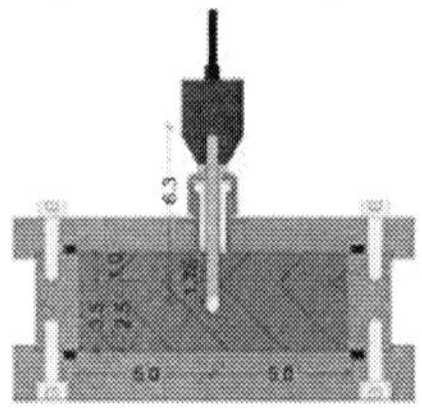

Figure 2. Suction cell with tensiometer

In order to determine the drying curves, the samples were initially placed into a water basin for at least 3 days. For that, the top caps of the cells were removed and the bottoms were replaced with filter plates. Saturation was assumed to be complete when no more change in weight was observed. After reaching the maximum degree of saturation (quasi-saturation) the samples were closed for one day to achieve homogeneity and placed in a climate chamber with a controlled air temperature of 20°C and a relative humidity of 60 %. Afterwards first measurements were carried out. Subsequently the specimens were air dried stepwise to the desired water content and measurements were performed. Prior each measurement a homogenisation time of 2 - 5 days was provided.

The wetting curves were obtained by initially desaturating the samples under a heating lamp for several days until no more changes in weight was observed. Subsequently water was carefully added until the desired water content was reached and measurements were carried out. The time of homogenisation was about 2 - 7 days.

2.2.2 Sample preparation for dew point method

The soil samples for the measurements with the chilled mirror dew-point hygrometers were statically compacted in one layer of 0.6 cm height into sample cups with a diameter of 4 cm and a height of 1.2 cm. The sample cups are provided by Decagon for WP4 measurements.

To obtain the drying curves the samples were initially wetted by carefully adding water with a pipette. Subsequently the specimens were dried under a heating lamp stepwise to the desired water content and measurements were carried out. For each step a homogenisation time of 3 - 7 days was provided.

To determine the wetting curves the samples were initially desaturated by a heating lamp for several days until no more changes in weight was observed. Subsequently water was added until the desired water content was reached. After a homogenisation time of 3 - 7 days measurements were carried out.

2.3 Results

The measured data were used to estimate the soil suction over the entire range of 0 kPa to $1 \cdot 10^6$ kPa by using the Fredlund and Xing equation [13]. The best-fit parameters were obtained by a nonlinear regression analysis. The results for all sands and water contents during compaction are shown in Figures 3 to 5.

Table 2. Air entry values ψ_{AE} and residual conditions ψ_r

SWRC	ψ_{AE} [kPa]	w_{max} [%]	ψ_r [kPa]	w_r [%]
Sand 1 ($w_{comp.} = 14.25$ %) - d	1.0	19.0	70	4.2
Sand 1 ($w_{comp.} = 14.25$ %) - w	0.8	19.0	45	3.6
Sand 1 ($w_{comp.} = 5$ %) - d	1.9	17.7	22	7.0
Sand 1 ($w_{comp.} = 5$ %) - w	1.1	17.7	19	5.4
Sand 2 - d	1.9	22.0	25	7.3
Sand 2 - w	0.9	21.9	20	5.3

All SWRCs show a similar behaviour with a flat transition zone. The air entry values are between approximately 0.8 kPa and 1.9 kPa and residual conditions, as described by [4], are reached at suctions higher than approximately 19 kPa to 70 kPa (Table 2). The water content at quasi-saturated conditions w_{max} remains constant for each drying and wetting cycle, whereas the residual water content of the wetting branch is lower than the one of the drying branch.

In a semilogarithmic scale the shift between the boundary drying curve and the boundary wetting curve seems to be approximately parallel over the entire range of water content. At the inflection point of the boundary wetting curve the matric suction decreases about 5 kPa to 20 kPa compared to the boundary drying curve.

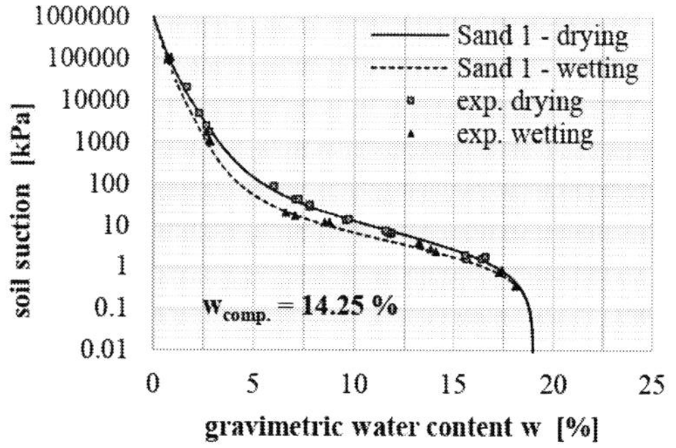

Figure 3. SWRCs of Sand 1 ($w_{comp.}$ = 14.25 %)

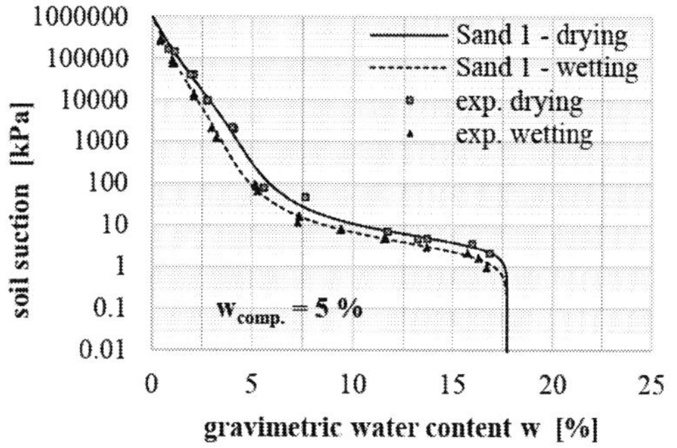

Figure 4. SWRCs of Sand 1 ($w_{comp.}$ = 5 %)

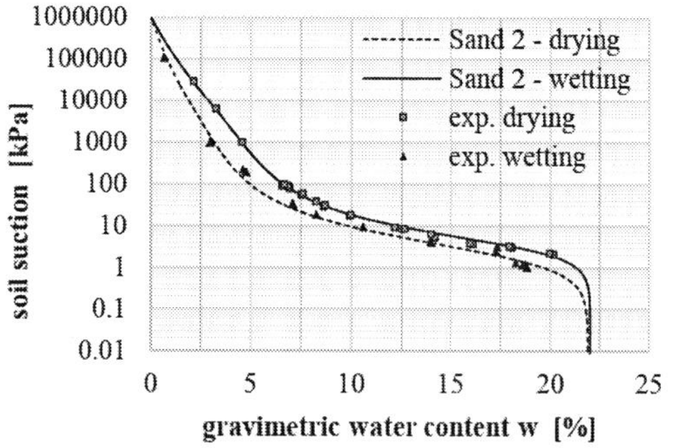

Figure 5. SWRCs of Sand 2

As can be seen by comparing the measured SWRCs of Sand 1 static compaction at different water contents influences the soil fabric of the tested samples. Figure 6 reveals that the compaction water content affects the water content at quasi-saturated conditions. That can be explained with the increase of inter-aggregate pores with decreasing compaction water content [14, 15]. The water in the comparatively large inter-aggregate pores drains as soon as the suction cells are lifted from the water basin and hence the measurable maximum water content at quasi-saturated conditions decreases. The air entry value and the transition zone remain mainly unaffected from the compaction water content. Slight differences in this range are supposed to be caused by small inhomogeneities within the samples. However, the residual zone seems to be unexpectedly influenced by the compaction water content. The measured suctions are higher for samples compacted at a water content of 5 % than for samples compacted at 14.25 %, what is in contrast to the results obtained by [14]. The reasons for this cannot be clearly identified but it is assumed that due to the preferential building of aggregates at lower water contents and the comparatively small measures of the sample cups of the WP4 devices the fine content differs between the tested samples. Therefore, the samples compacted at water contents of 5 % could comprise a slightly higher fine content than the samples compacted at 14.25 % which leads to higher suctions. Despite the dependency on water content during compaction the drying and wetting curves of Sand 1 show a similar hysteretic behaviour and for this both sets of curves are taken into account for testing the three empirical models for the prediction of hysteretic behaviour.

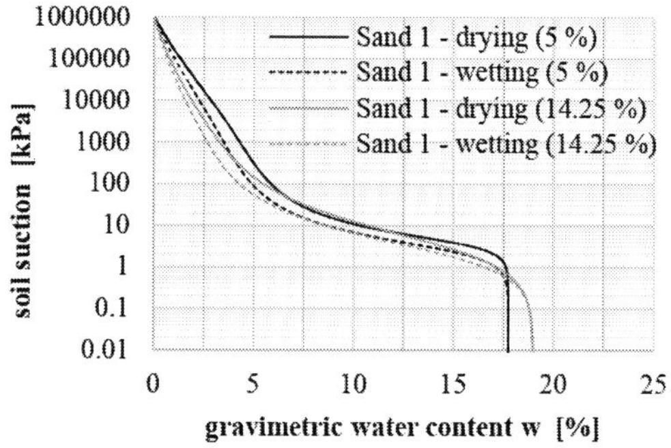

Figure 6. Comparison of the Fredlund and Xing best-fit curves of measured data of Sand 1 ($w_{comp.}$)

3 Prediction of hysteretic behaviour

3.1 Pham et al. (2003) model

The Pham et al. (2003) model [10] uses the following equation to describe the SWRC:

$$w(\psi) = \frac{w_u b + c\psi^d}{b + \psi^d} \tag{1}$$

Based on the best-fit parameters d_d, b_d, w_u and c of the drying branch, they suggested approximate locations of two suctions on the wetting branch, which should be measured to predict the entire boundary wetting curve:

$$\psi_{1w} = \left(\frac{b_d}{10}\right)^{\frac{1}{d_d}} \tag{2}$$

$$\psi_{2w} = \psi_{1w} - 2\left\{\left[\frac{b_d(w_u - w_{1w})}{w_{1w} - c}\right]^{\frac{1}{d_d}}\right\} - b_d^{\frac{1}{d_d}} \tag{3}$$

From those two points the parameters for the boundary drying curve d_w and b_w can be calculated:

$$d_w = \frac{\log\left[\frac{(w_{1w} - c)(w_u - w_{2w})}{(w_u - w_{1w})(w_{2w} - c)}\right]}{\log\left(\frac{\psi_{2w}}{\psi_{1w}}\right)} \quad (4)$$

$$b_w = \frac{\psi_{1w}^{d_w}(w_{1w} - c)}{w_u - w_{1w}} \quad (5)$$

The parameters c and w_u are the same for both the drying curves and wetting curves.

3.2 Pham et al. (2005) model

The Pham et al. (2005) model [7] presented a method to predict the above mentioned parameter of equation (1) of the boundary drying curve without any further measurements. Furthermore they added the Fredlund and Xing correction factor [13] to the equation (1) in order to achieve a better prediction of the behaviour of the SWRC in the residual zone:

$$w(\psi) = \frac{w_u b + c\psi^d}{b + \psi^d} \cdot \left[1 - \frac{\ln\left(1 + \frac{\psi}{\psi_r}\right)}{\ln\left(1 + \frac{10^6}{\psi_r}\right)}\right] \quad (6)$$

$$d_w = \frac{d_d}{R_{SL}} \quad (7)$$

$$b_w = \frac{b_d}{(10^{D_{SL}})^{\frac{1}{R_{SL}}}} \quad (8)$$

In equation (7) R_{SL} represents the ratio of the slopes of the boundary wetting and boundary drying curves in semilogarithmic coordinate system. In equation (8) D_{SL} is the distance between both boundary curves. Based on a study of numerous soils the parameters for sands can be set to $R_{SL} = 1.0 - 2.0$ and $D_{SL} = 0.2 - 0.35$ [7].

[7] suggested to estimate the residual suction with:

$$\psi_r = (2.7 \cdot b)^{\frac{1}{d}} \quad (9)$$

The parameters c and w_u remain, similar to the Pham et al. (2003) model, constant for the drying and wetting branch.

3.3 Fredlund et al. (2011) model

As the boundary drying curve and the boundary wetting curve seem to be congruent to each other, Fredlund et al. (2011) [11] suggested to adjust the best-fit parameter a_d of the Fredlund and Xing equation, which controls the shift of the SWRC in direction of the suction in order to predict the boundary wetting curve:

$$a_w = a_d \cdot 10^{-\frac{x}{100}} \quad (10)$$

The exponent x in equation (10) represents the shift of the inflection point between the boundary drying and wetting curve expressed in percent of a log-cycle. For sands [11] suggest values for x varying between 15 % and 35 %. The best-fit parameters m_d and n_d of the boundary drying curve remain constant for the wetting curve.

3.4 Results and discussion

The results of the three models are shown in Figure 7 together with the Fredlund and Xing best-fit curves of the measured data.

The Pham et al. (2003) model compares very well to the Fredlund and Xing curves for water contents higher than the residual water content, which is controlled by the parameter c. The asymptotic behaviour of the model cannot reproduce the measured data at water contents bellow c. This could be overcome by introducing the Fredlund and Xing correction factor. A disadvantage is the necessity of experimental determination of two points on the boundary drying curve to obtain the model parameters.

For the Pham et al. (2005) model the parameter $R_{SL} = 1.0$ was chosen as the measured boundary drying curves turned out to be approximately parallel to the boundary wetting curves. The parameter D_{SL} was set to 0.2, as suggested by [7] for sands. Those parameters yielded the best results. The model fits the measured data for water contents above the residual water content very well but tends to slightly overestimate the wetting curve. However, due to the fact that the parameter c remains constant for both, the drying and the wetting curve, the model cannot predict hysteretic behaviour at water contents bellow the residual water content properly. This can be overcome by modifying the parameter c as it is shown in chapter 4.

The results of the Fredlund et al. (2011) model are shown in Figure 7 for x = 15 %, 25 % and 35 in order to cover a possible area in which the boundary wetting curve can occur. The measured data lie for high water contents between the predicted boundary curves (x = 15 % and 35 %) but are overestimated for water contents approximately bellow the residual water content. The model is not able to predict the boundary wetting curve with a single value of x but gives a good estimation for expectable suctions.

4 Modified Pham et al. (2005) model

In order to achieve a suitable estimation of the boundary wetting curve for low water contents with the Pham et al. (2005) model, the parameter c_w of the wetting curve has to differ from the best-fit parameter c_d of the drying curve. As shown in Table 2 the residual water content w_r of the wetting curve tends to decrease compared to the drying branch. The SWRC is shifted in this area towards lower water contents. The difference of the residual water contents seem to be of the same order as the shift of suction in a semilogarithmic coordinate system. This can be considered by modifying the parameter c_w for the wetting curve of the Pham et al. (2005) model:

$$c_w = c_d \cdot (1 - D_{SL}) \quad (11)$$

 DOI: 10.1051/e3sconf/20160911004

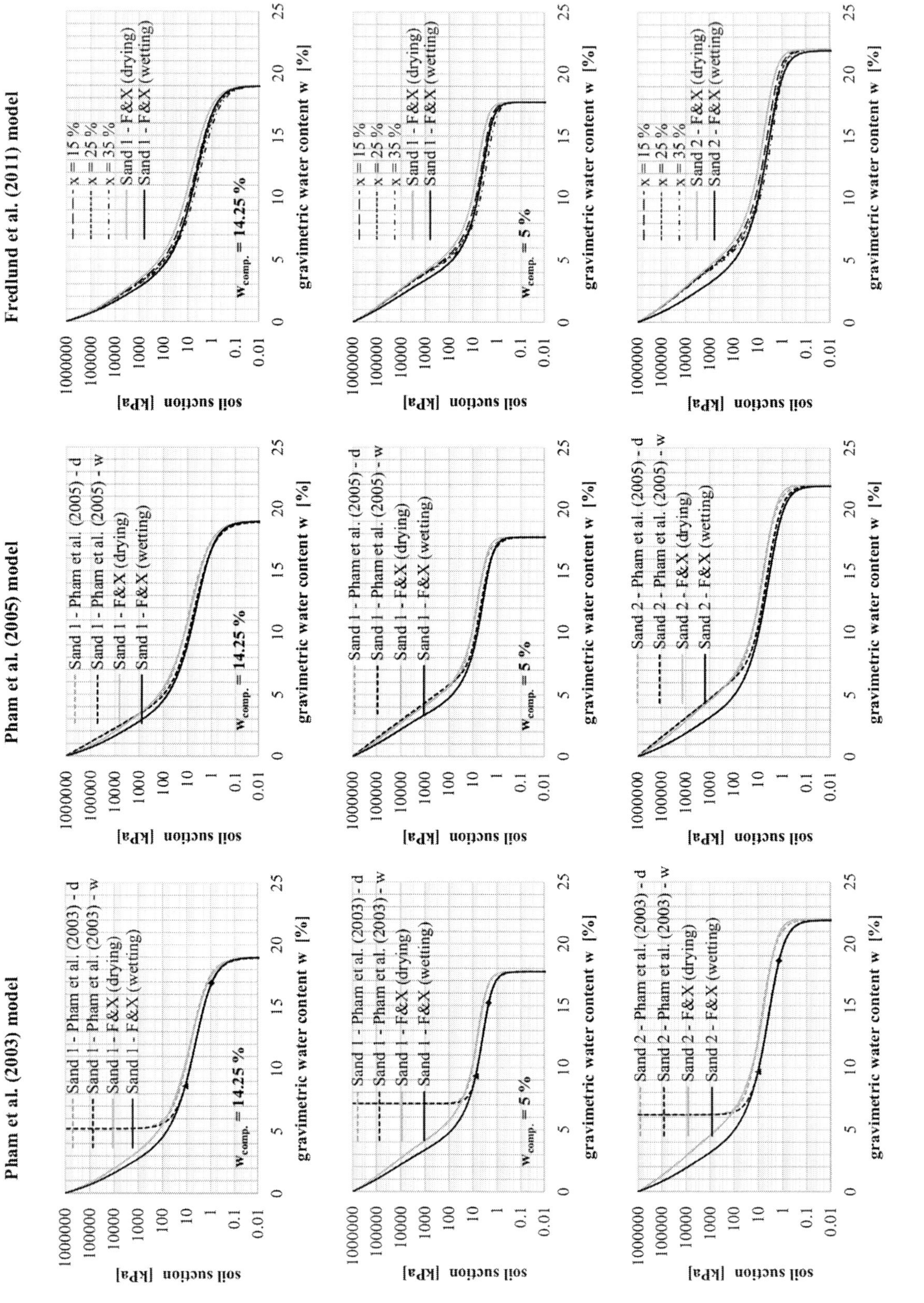

Figure 7. Results of the models of Pham et al. (2003), Pham et al. (2005) and Fredlund et al. (2011)

The results of the modified model are shown in Figures 8, 9 and 10. They indicate that hysteretic behaviour of the tested silty sands at low water contents can be satisfactorily predicted by adjusting the parameter c_w as suggested in equation (11).

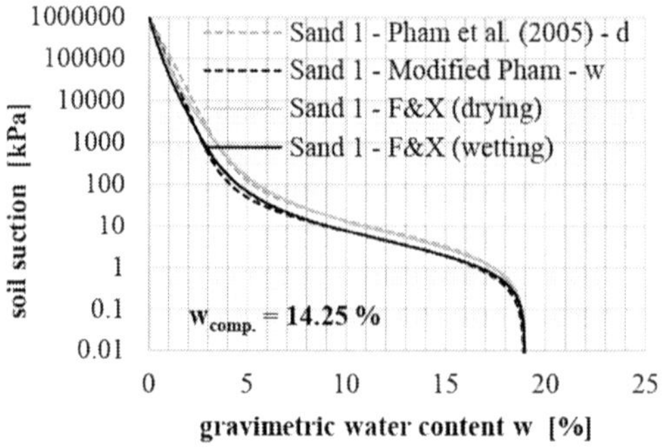

Figure 8. Modified Pham et al. (2005) model for Sand 1 ($w_{comp.}$ = 14.25 %)

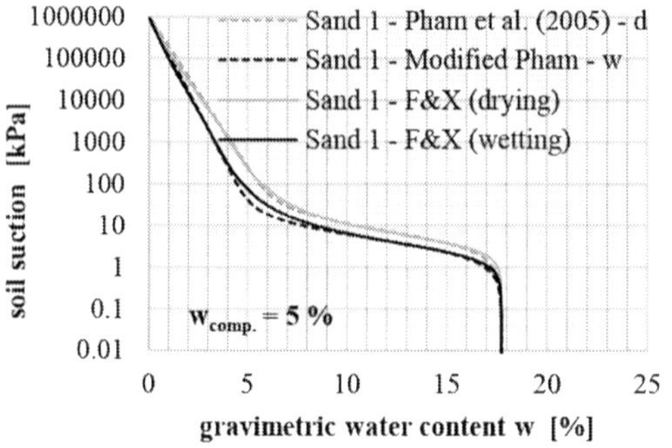

Figure 9. Modified Pham et al. (2005) model for Sand 1 ($w_{comp.}$ = 5 %)

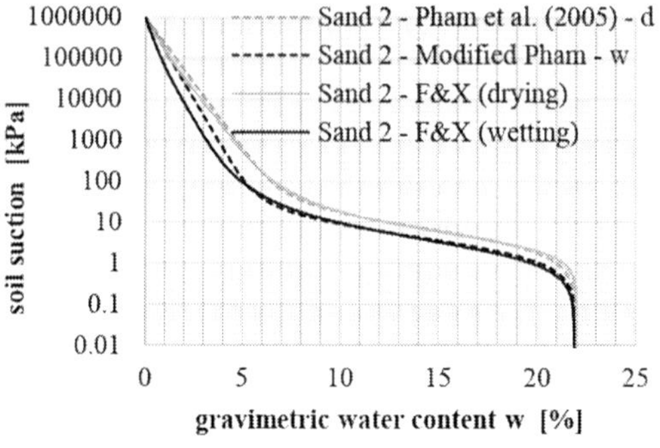

Figure 10. Modified Pham et al. (2005) model for Sand 2

The proposed modification can only be regarded as a first approach for the estimation of the wetting branch based on the drying curve of silty sands and should be checked by further experiments.

5 Conclusions

In this paper the hysteretic behaviour of two silty sands was experimentally determined. The measured SWRCs showed a hysteretic behaviour over the entire range of suction. The water content during compaction affects the behaviour of the SWRC of the tested silty sands with a fine content of 28 %. Compaction at decreasing water content leads to an increase of inter-aggregate pores which decreases the maximum water content measured at quasi-saturated conditions. The transition zone remains mainly unaffected. In the residual zone an increase in suction was observed with decreasing water content during compaction which cannot be satisfactorily explained but is assumed to be linked to differences of fine content between the samples tested with the WP4 devices.

The boundary drying curves were compared to the prediction of three empirical models which need the boundary drying curve as input parameter. All three models yielded reasonable results for water contents higher than the residual one. The Pham et al. (2003) model turned out to give the best results for high water contents but needs additional experimental data as input parameters. A suggestion was made to modify the Pham et al. (2005) model in order to achieve a better prediction of the wetting branch at low water contents. This modification has yet to be validated.

References

1. I.B., Donald, Proc. Of the sec. Australian-New Zealand Conf. on Soil mechanics and foundation engineering (1956)
2. N. Khalili, M. H. Khabbaz, Géotechnique, **48, No. 5**, pp. 681-687, (1998).
3. S. L. Houston, J. Geotech. Geoenviron. Eng., **134**, pp. 1619-1632, (2008)
4. S.K. Vanapalli, W.S. Sillers, M.D. Fredlund, 51st Geotech. Conf., (1998)
5. D. Sheng, A. Zhou, D.G. Fredlund, Geotech. Geol. Eng., **29**, pp. 145 – 159, (2011)
6. D.G. Fredlund, H. Rahardjo, M.D. Fredlund, *Unsaturated Soil Mechanics in Engineering Practice*, (2012)
7. H.Q. Pham, D. G. Fredlund, S. L. Barbour, Can. Geotech. J., **42**, pp. 1548–1568, (2005)
8. L. M. Arya, J. F. Paris., Soil Sci. Soc. Am. J., **45**, pp. 1023-1030, (1981)
9. D. M. Pedroso, D. J. Williams, Comput. Geotech., **37, No. 3,** pp. 374-380, (2010)
10. H. Q. Pham, D. G. Fredlund, S. L. Barbour, Géotechnique, **53, No. 2**, pp. 293–298, (2003)
11. D. G. Fredlund, D. Sheng, J. Zhao, Can. Geotech. J., **48**, pp. 186–198, (2011)
12. P. Delage, 3rd Intern. Conf. on Unsat. Soils, pp. 973-996, (2002)
13. D.G. Fredlund, A. Xing, Can. Geotech. J., **31**, pp. 521-532, (1994)
14. E. Birle, Sec. Europ. Conf. on Unsat. Soils, pp. 145-152, (2012)
15. P. Delage, M. Audiguier, Y.J. Cui, M.D. Howat, Can. Geotech. J., **33**, pp. 150-159 , (1996)

Modelling of the water retention characteristic of deformable soils

Yu Wang[1,a], Daichao Sheng[2], Keith Ross[1], David G. Toll[3]

[1] *School of Computing, Science & Engineering, University of Salford, Manchester M5 4WT, UK*
[2] *Centre for Geotechnical and Materials Modeling, University of Newcastle, NSW 2308, Australia*
[3] *School of Engineering, University of Durham, UK*

Abstract. A recently proposed water retention model has been further developed for the application on unsaturated deformable soils. The physical mechanisms underpinning the water retention characteristic of soils was at first described in terms of traditional theories of capillarity and interfacial physical chemistry at pore level. Then upscaling to macroscopic level of material scale in terms of average volume theorem produces an analytical formula for the water retention characteristic. The methodology produces an explicit form of the water retention curve as a function of three state parameters: the suction, the degree-of-water-saturation and the void-ratio. At last, the model has been tested using experimental measurements.

1 Introduction

In geotechnical engineering applications, it has been widely accepted that the water retention characteristic (WRC) of deformable soils should explicitly take into account of the coupling effects of the hydraulic and mechanical behaviours [1], such as the volume change or the void ratio change in soils as a function of the water content or the variation of suction, or both of these effects. As a result, the water retention characteristic of deformable soils presents a curve in the 3D space of suction/water-content/void-ratio [2] or in the equivalent space of suction/the-degree-of-water-saturation/specific-volume [3]. In addition, for any particular deformable soil, its water retention characteristic also depends on it initial state, particular its initial porosity or void ratio. This fact means that the water retention characteristic of a deformable soil presents a surface which is called the soil-water retention surface (SWRS) [2], and any specific water retention curve of the soil should in the surface. Many researchers have contributed to the description of the SWRS of deformable soils. Gallipoli et al. [3] suggested to take one of the parameters in the original van Genuchten [4] formula to be a function of the specific volume. Stange & Horn [5] assumed that the parameters in the original van Genuchten formula depend on the void ratio. Tarantino [6] proposed a power function for the suction in terms of the water ratio and combined this with the original van Genuchten formula. Wheeler et al. [7] used a bilinear function between the degree of saturation and the logarithmic modified suction. Mbonimpa et al. [8] proposed a modified Kovacs (MK) model which incorporates the void ratio into the relationship between the degree of water saturation and the suction. Marsin [9] used a suction rate as a function of the deformation rate. Sheng & Zhou [10] suggested a general differential form of the relationship between the degree of water saturation and suction and voids ratio. In addition to these empirical approaches, a fractal method has also been adopted [11,12].

In hydrology and soil physics, much progress has been made adopting the approach based on fundamental physical description. Assouline et al. [13] proposed a conceptual model to relate the pore volume to the particle volumes in terms of a power function, then adopt the capillary equation to define the water retention curve. Taking account of the effect of the water film on empty pore surfaces, Tuller et al. [14] proposed a unit cell model for water retention characteristic modeling [15]. Chertkov [16-18] proposed a physically based model to define the suction of soil matrix as a product of two physical factors which stand for the adsorption-capillary effect and the influence of variation of soil-porosity, respectively. Generally, these models describe WRC in the plane of water saturation and suction (S_w-s).

Wang et al. [19] proposed a physical-chemical WRC model based on the classical theory of capillarity, interfacial physical chemistry and the average volume theorem. This paper is to investigate the application of the model on deformable unsaturated soils.

2 A Water Retention Model

From the point view of interfacial physical chemistry, a fluid in a porous medium can be regarded as being adsorbed by the pore surface [20], where the fluid is called as the adsorbate and the pore surface is called the substrate which exerts adsorptive forces on the adsorbate due to atom and molecular interaction at the interfacial

[a] Corresponding author: y.wang@salford.ac.uk

region [21]. The adsorptive forces will modify the chemical potential of the molecules of the adsorbate at the interfacial region in the reference to the state when the bulk adsorbate fluid is under a free and open atmospheric condition [22]. In general, the chemical potential change depends on the distance of the molecule of the adsorbate above the surface of the substrate [14]. In unsaturated soils, the water and vapour phases have different individual pressures due to their different chemical potential changes [19]. Under the equilibrium state of a mechanical balance, the pressure difference of the two fluid phases in the pore system follows the Laplace equation [23], i.e.:

$$p_c = p_w - p_v = -\frac{2\sigma\cos\theta}{r} \quad (1)$$

where p_c is called the capillary pressure or matric potential at pore scale, and its absolute value is called as suction; p_w and p_v are the microscopic intrinsic pore pressures of the water and vapour phases, respectively; σ is the surface tension on the meniscuses, r indicates the pore radius at the position of meniscuses, θ is the contact angle between the meniscus and the pore surface.

The chemical potential change of the adsorbates on a substrate surface can be evaluated using the Kelvin equation [20, 14], as the result, the intrinsic pressure of the adsorbates on substrates can be expressed as [19]:

$$p_f = \frac{\bar{u}_f}{V_f} + p_0 \exp\left(\frac{\bar{u}_f}{RT}\right) \quad (2)$$

where p_f is the average intrinsic pressure of the fluid on the surface; $\bar{u}_f$ is the average intrinsic molar potential change of the fluid of a thickness h, which is defined as $\bar{u}_f = \frac{1}{h}\int_0^h u_f(z)dz$, where $u_f(z)$ is the fluid molar molecule potential change at position z above the substrate surface; V_f is the molar volume of the liquid phase. R is the gas constant; T is the temperature; p_0 is a normal pressure.

Eq. (2) consists of two terms: the first term takes account of the intrinsic chemical potential change, while the second term is it physical effect, which is called the gauge pressure to be measurable directly. Practically, the physical state of the fluid on substrates can be expressed using the gauge pressure, i.e.:

$$p_f = p_0 \exp\left(\frac{\bar{u}_f}{RT}\right) \quad (3)$$

In terms of the mean-field model [24], Tuller et al. [14] suggested that the chemical potential change $\bar{u}_f$ in Eq. (3) can be estimated in terms of the fluid film thickness (t_f) on the surface of substrates using the following equation:

$$\bar{u}_f = \frac{A^m}{t_f^3} + \frac{A^e}{t_f^2} + \frac{A^s}{t_f} \quad (4)$$

where the constant A^m relates to disjoining force due to van der Waals molecular interaction force, A^e relates to the interfacial electrostatic effect and A^s depends on the structure of substrate surface. In addition to the molecular interaction, in terms of the statistical molecular mechanics for the liquid-gas surface [25], the total chemical potential change of the wetting fluid film on substrate surfaces should consist of two components of the 'intrinsic' and the 'external' natures. The intrinsic component may be defined in terms of Eq. (4), while the external component depends on the other physical effects rather than molecular interfacial interactions. As an improvement, the Eq. (4) may be revised into the following form:

$$\bar{u}_f = \frac{A^m}{t_f^3} + \frac{A^e}{t_f^2} + \frac{A^s}{t_f} + A^0 \quad (5)$$

where A^0 is a new constant introduced for the external effect such as the curvature of pore wall surface. Mathematically, Eq. (5) may be characterised into the following form:

$$\bar{u}_f = \frac{A_1}{t_f^b} + A_0 \quad (6)$$

where A_1, A_0 and b are three new constants.

For unsaturated soils, representing the pores in soils using a bundle of capillary cylinders (the BCC modelling), a complicated pore system can be equivalent to a capillary tube of a characterised 'hydraulic radius', which is defined as ratio of total pore volume to total pore surface area. Based on the characterised capillary tube, the bulk water phase in the water filled part and the coexisting vapour phases in the empty part will have their respective chemical potential changes and physical pressures, which can be described using the Eq. (6) and (3). For example, for the water phase, its chemical potential change can be expressed as:

$$\begin{aligned}\Delta\bar{u}_w &= \frac{1}{A_S}\int_{A_w} \bar{u}_w dA \\ &= \frac{1}{A_S}\int_{A_w} \left(\frac{A_{1,w}}{r_h^{b_w}} + A_{0,w}\right) dA \\ &= \left(\frac{A_{1,w}}{r_h^{b_w}} + A_{0,w}\right)\frac{A_w}{A_S} + \bar{u}_0 \\ &= \bar{u}_w S_w + \bar{u}_0\end{aligned} \quad (7)$$

where r_h is the radius of the capillary tube or the hydraulic radius of the porous medium. $A_{1,w}$, $A_{0,w}$ and b_w are three constants depending on the nature of the material and the hydraulic radius of the porous medium. A_S is the total surface area of the tube or the pores of the porous media represented. A_w is the area of the part of the surface occupied by the bulk water phase. As a result, $\bar{u}_w = \frac{A_{1,w}}{r_h^b} + A_{0,w}$ is the local average of the chemical potential change along the radius of the capillary tube. S_w is the degree of water saturation or the fraction of the volume of the capillary tube occupied by the bulk water

DOI: 10.1051/e3sconf/20160911005

phase. $\overline{u}_0$ is the initial chemical potential change when bulk water starts to fill the tube due to capillary condensation, which depends on the initial water molecular film on the totally empty pore surface [26]. Substituting the Eq. (7) into (3) for $\overline{u}_f$ produces the following Eq. (8) which describes the pressure of the bulk water in the tube in terms of the water saturation.

$$\begin{aligned} p_w &= p_0 \exp\left(\frac{\Delta\overline{u}_w}{RT}\right) = p_0 \exp\left(\frac{\overline{u}_w S_w + \overline{u}_0}{RT}\right) \\ &= p_0 \exp\left(\frac{\overline{u}_0}{RT}\right)\exp\left(\alpha S_w\right) \end{aligned} \tag{8}$$

where $\alpha = \frac{\overline{u}_w}{RT}$. A similar approach may be applied on the coexisting vapour phase in unsaturated soils, and the vapour pressure can be estimated in terms of the degree of water saturation as:

$$p_v = p_0 \exp\left(\frac{\overline{u}_0}{RT}\right)\exp\left(\beta(1 - S_w)\right) \tag{9}$$

where $\beta = \frac{\overline{u}_v}{RT}$ and $\overline{u}_v = \frac{A_{1,v}}{r_h^{b_v}} + A_{0,v}$ is a local average of the chemical potential change of the vapour phase along the radius of the capillary tube. $A_{1,v}$, $A_{0,v}$ and b_v are taken as three constants.

Following the volume average theorem [27, 28], the Darcy-scale capillary pressure can be expressed as:

$$\begin{aligned} &\langle p_c \rangle = \langle p_w \rangle - \langle p_v \rangle = \\ &\frac{1}{V_{REV}}\int_{V_w} p_w dV - \frac{1}{V_{REV}}\int_{V_v} p_v dV \end{aligned} \tag{10}$$

where $\langle p_c \rangle$ indicates the macroscopic volume average of the capillary pressure. Substituting Eqs. (8) and (9) into (10) for p_w and p_v, respectively, generates the following definition of the average capillary pressure in terms of the degree of water saturation [29]:

$$\langle p_c \rangle = \lambda\left[\frac{1}{\beta} - \frac{1}{\alpha} + \frac{1}{\alpha}\exp(\alpha S_w) - \frac{1}{\beta}\exp(\beta(1 - S_w))\right] \tag{11}$$

where, $\lambda = \varepsilon p_0 \exp(\frac{\overline{u}_0}{RT})$, $\varepsilon = V_{pore}/V_{bulk}$ is the porosity.

3 Modelling Water Retention Characteristic of Deformable Soils

Using the Eq. (11) to describe the SWRS in the 3D space, we need to know the intrinsic relationship between the three parameters (λ, α and β) and the change of pore structure (porosity or void ratio). According to Eqs. (7)-(9), α and β depend on the characterised hydraulic radius, $r_h = \frac{V_{pore}}{A_S}$. Assuming the total pore surface area A_S and the total volume of solid phase V_{solid} do not change, we can have $r_h = \frac{V_{solid}}{A_S} \times \frac{V_{pore}}{V_{solid}} = C \cdot e$, where C is a constant and $e = V_{pore}/V_{solid}$ is the void ratio. Taking the intrinsic relationships that porosity $\varepsilon = \frac{e}{1+e}$ and water saturation $S_w = \frac{w}{e}$, where $w = V_{water}/V_{solid}$ is water ratio, Eq. (11) can be rewritten into the following form:

$$\langle p_c \rangle = \frac{\lambda e}{1+e}\left[\begin{aligned} &\frac{1}{\frac{B_2}{e^{B_3}} + B_1} - \frac{1}{\frac{A_2}{e^{A_3}} + A_1} + \frac{1}{\frac{A_2}{e^{A_3}} + A_1}\exp\left(\left(\frac{A_2}{e^{A_3}} + A_1\right)\frac{w}{e}\right) - \\ &\frac{1}{\frac{B_2}{e^{B_3}} + B_1}\exp\left(\left(\frac{B_2}{e^{B_3}} + B_1\right)(1 - \frac{w}{e})\right) \end{aligned}\right] \tag{12}$$

where the parameters λ, A_is and B_is are redefined by merging with other constants. A_is relate to the bulk water phase in pore spaces, while B_is relate to the coexisting vapour phase in pore spaces. the term RT has been taken into A_is and B_is. In the next section, the Eq. (12) is used to model the SWRS of three unsaturated soils reported in references by fitting all of the measured residual shrinkage SWRCs of each soil in the space of *s-w-e*.

4 Modelling Examples and Discussion

In this paper, the experimental data are taken from previously reported experiments on the deformable soils under draining hydraulic loadings only. Three series of experimental data of soil under draining shrinkage are chosen from the published references. They are two sets of undisturbed samples from a field site and one set of samples reconstituted in laboratory to different porosities. The modelling of each set of samples is conducted in the space of matric suction, water ratio and void ratio (*s-w-e*) for all of the measured shrinking curves of different initial void ratios.

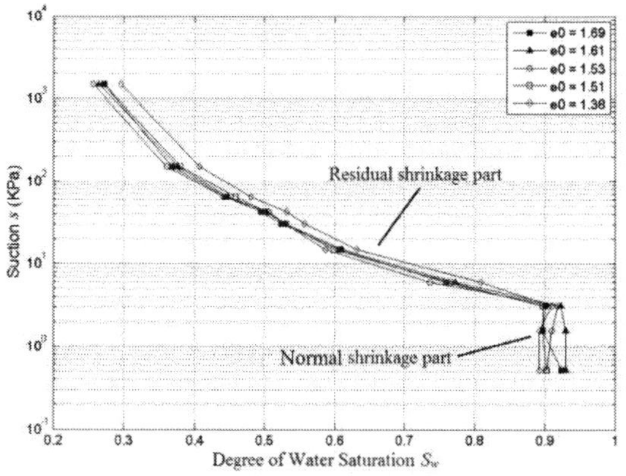

Figure 1. The plot, in S_w-s plane, of the SWRCs of the Stagnic Chernozem soil samples at A_p horizons (0-30cm) [5]

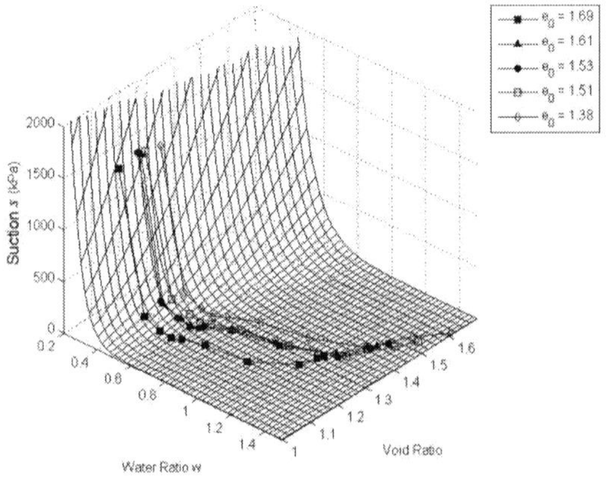

Figure 2. The modelling result for the SWRS of A_p

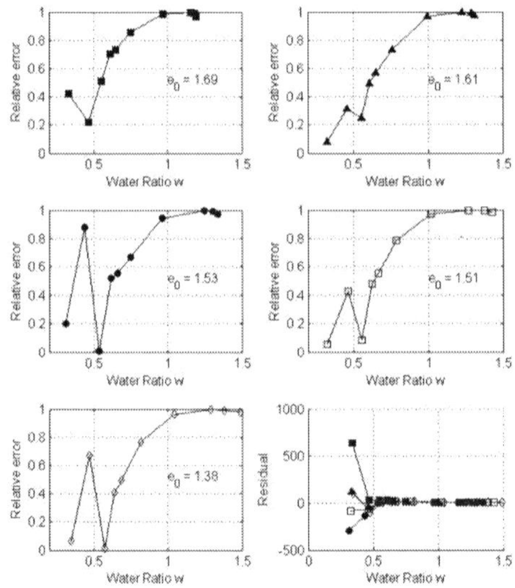

Figure 3. The modelling relative errors and residuals for the individual SWRCs in Fig. 2

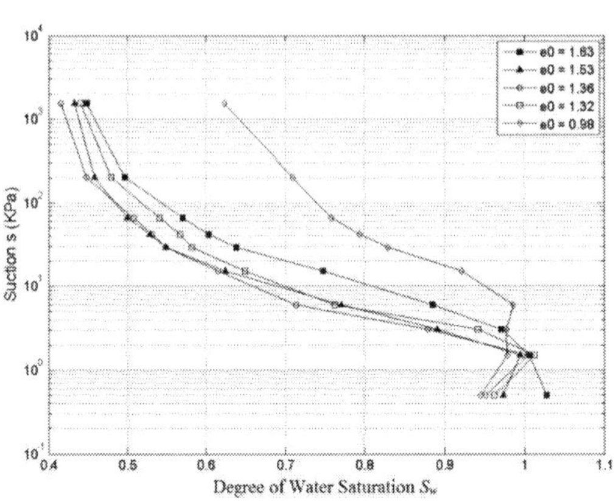

Figure 4. The plot, in S_w-s plane, of the SWRCs of the Stagnic Chernozem soil samples at A_h horizons (30-50cm) [5]

Stange and Horn [5] reported an experiment (Fig. 1) on a deformable soil which was taken from a site called Stagnic Chernozem in Germany. The soil samples of different initial void ratios, e_0, were from a shallow range of horizons A_p (0-30 cm). They were exposed to a series of desaturation to different pressure head/suction values without any external mechanical loading except for the self-gravimetric effect. The SWRCs in Fig. 1 have displayed two distinct parts which represent two types of deformation, i.e. the normal shrinkage part, where saturation has little change, when suction $s < 3.1$ kPa or the degree of water saturation $S_w \cong 0.9$ and the residual shrinkage part elsewhere ($S_w < 0.9$). It can be seen that, in the normal shrinkage part, the degree of water saturation S_w changes little while the void ratio e of the soil samples decreases. This normal shrinkage part is known as the consolidation phase of soils. It also shows that the air entry value of the suction also changes little in the range of the initial void ratio e_0 - from 1.38 to 1.96. Because Eq. (12) derives from the original form of Eq. (11) for the relation of the matric suction and the degree of water saturation, it does not work for the normal shrinkage part. So it is only used to model the residual shrinkage part in this paper.

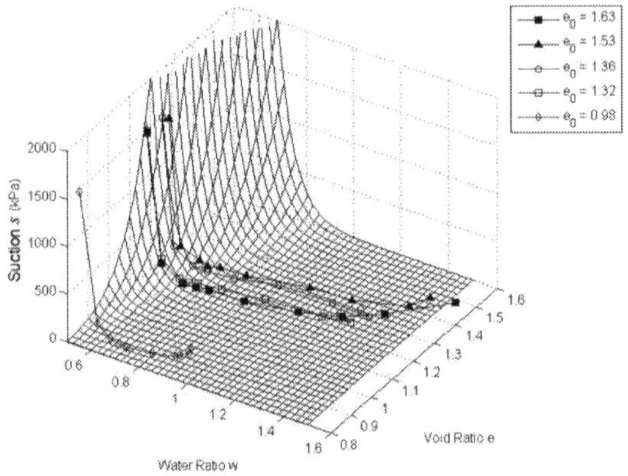

Figure 5. The modelling result for the SWRS of A_h

Fig. 2 replots these curves in the 3D space of *s*-*w*-*e* while shows the predicted SWRS result using the Eq. (12) to fit all of these curves. The corresponding fitting data are listed in Table 1. Fig. 3 shows the relative errors ($\frac{|F_i - f_i|}{f_i}$) and the residuals ($F_i - f_i$) of the predicted suction to the measurements, where F_i is the modelling result at the measured point i, and f_i is the measurement at that point. Both Figs. 2 and 3 have shown that Eq. (12) gives a reasonably good modelling result. The predicted SWRS fits all of the measured SWRCs, in the whole range of the variation of s, w and e, reasonably well in terms of the shape and the calculated errors. In terms of the residual in Fig. 3, a non-ideal point presents at the highest suction or the lowest water ratio of the curve of e_0 = 1.69. The worst accuracy at the point may relate to two reasons. Firstly, the accurate point of the highest suction corresponding to the lowest water content is very hard to be precisely measured. Secondly, because the Eq. (12)

takes an assumption of a non-changed pore surface area, in the case if the soil skeleton collapses due to relatively slide or relocation of particles, the collapse of pore structure could cause new blockages or inaccessible pores that will change the total pore surface areas taken into account by Eq. (12). When samples are under high suction and of low water content, the collapse of pore structure is highly possible.

Fig. 4 shows the measured SWRCs of another set of samples which were taken from a relatively deep range of horizons A_h (30-50 cm) of the Stagnic Chernozem soil [5]. Fig. 5 shows the predicted SWRS result using the Eq. (12) to fit these curves in the 3D space of *s-w-e*. The corresponding fitting data are listed in Table 1. The calculated relative errors and residuals of the modelling suction results against the measurements (Fig. 6) have also showed that the Eq. (12) produces a reasonably good modelling for most of these curves. It can be seen that the curve of the soil sample e_0 = 0.98 presents an inconsistence with the others. Noticed that its initial porosity is much smaller than that of the others, it is expected to possess a distinctively different pore structure of significantly different pore surface area.

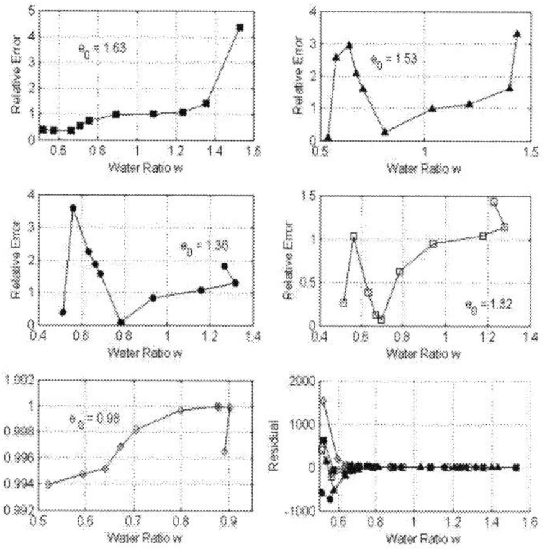

Figure 6. The modelling relative errors and residuals for the individual SWRCs in Fig. 5

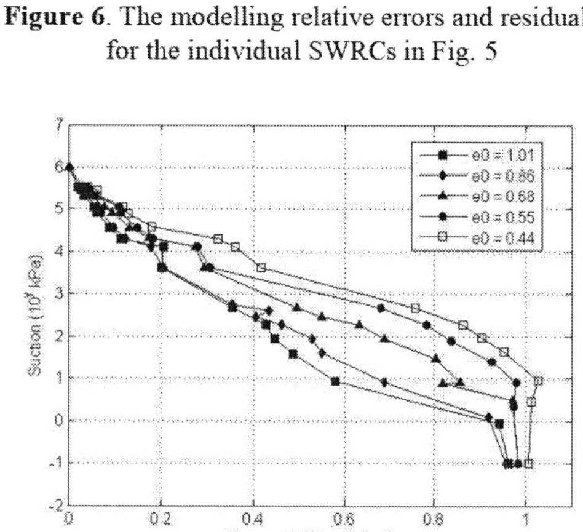

Figure 7. The plot, in S_w-s plane, of the SWRCs of a clayey-silty-sand soil samples under a desaturation process [2]

Salager et al. [2] reported an experiment (Fig. 7) and modelling of the SWRS for a reconstituted soil, which consists of clay (10%), silt (18%) and sand (72%). Fig. 8 shows the SWRS modelling results using the Eq. (12) to fit all of the SWRC measurements in the space of *s-w-e*. The relative errors and residuals of the modelling result against to the experimental data have showed that a reasonably good predicting result has been obtained again. For each curve, the relative error (Fig. 9) at the low water content (high suction) is lower than that at high water content (low suction), but the residual at low water content is higher than that at high water content. This is because of that the magnitude of suction is much greater at low water contents.

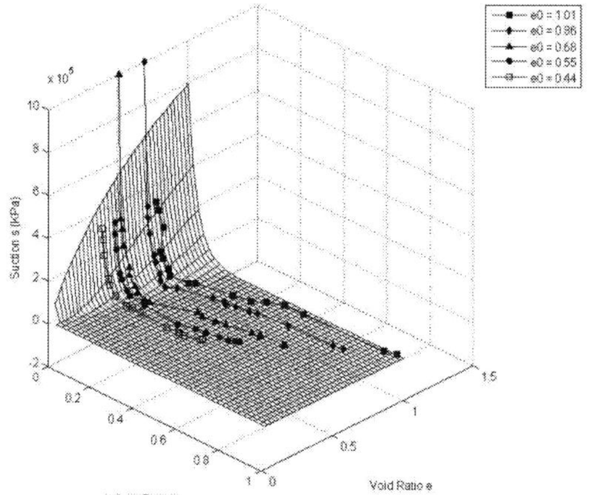

Figure 8. The modelling result for the SWRS of a clayey-silty-sand soil [2]

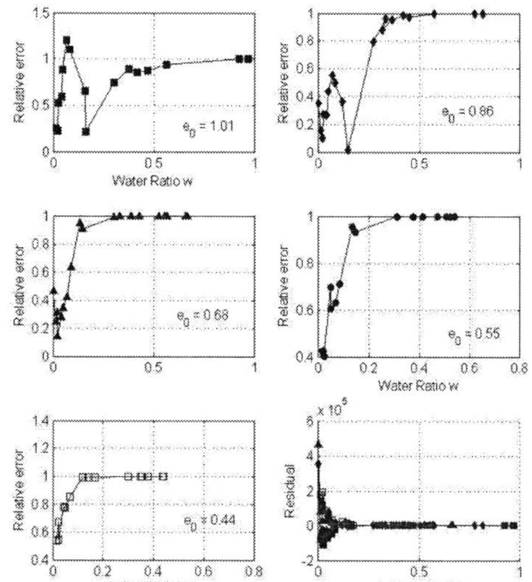

Figure 9. The modelling relative errors and residuals for the individual SWRCs in Fig. 8

From all the modelling examples we can see that, in terms of the relative error, the model has a better performance in high suction region than in low suction region. This is explained due to the modelling of the vapour phase, where we neglect the effect of the wetting water film on the wall of empty pores, which increases with the increase of the degree of water saturation or vice versa, and take the parameter, β, as constant for the vapour phase.

In the current work, the water film has been effectively treated with the bulked condensed capillary water together. If the film thickness is very small such as at high suctions, the effect of the water film will not be significant. Iwamatsu and Horii [24] have demonstrated that the capillary condensation radius is more than one order bigger than the wetting film thickness). However, at low suctions, particularly when the pore radius is bigger than that of significant capillary condensation effect, such the effective treatment of the water film brings in the deterioration of accuracy.

Table 1. The fitting parameters data using Eq. (12)

Samples	A_h	A_p	clayey-silty-sand soil
λ	-0.03927	-0.09463	-8.658e-4
A_1	6.581	225	1.392e-2
A_2	-25.39	-1794	-1611
A_3	9.107	0.4636	233.4
B_1	25.2	17.97	24.53
B_2	-1.49e-3	-8.304e-3	-2.855e-4
B_3	6.271e-5	1.532e-3	1.186e-2

5 Conclusions

This paper reports a model of the water retention characteristic of unsaturated porous media and its application on the residual shrinkage of deformable soils. The model is based on an equation of the state of the fluid in porous media established on the classical capillary theory. It explicitly take account of the effect of deformation and can be used to fit the measured SWRCs of different initial void ratios of a soil and constructs a general SWRS in the space of *s-w-e*. The tests on three soils have shown that the model works well in terms of the shape of the modelled SWRS and the calculated relative errors and residuals for modelling accuracy. The explanation on its performance is in agreement with the fundamental concepts and the assumptions of the model. By now, the model has been proved to work particularly well for high suction region when the capillary condensation plays a significant role. A further work to improve its performance at low suction region needs the consideration of the effect of the wetting water film on the empty pore surface. On another hand, it needs to be pointed out as well that modelling the water retention characteristic of deformable soils in terms of the soil-water retention surface is not enough. Further investigation in future needs to be able to explicitly describe each of the specific water retention curves of different initial void ratios in the space of *s-w-e*.

References

1. D. Sheng, *Computers and Geotechnics*, **38**, (2011)
2. S. Salager, M.S. El Youssoufi, C. Saix, *Can. Geotech. J.* **47**, (2010)
3. D. Gallipoli, S.J. Wheeler, M. Karstunen, *Geotechnique* **53**, (2003)
4. M.T. van Genuchten, *Soil Science Society of America Journal* **44**, (1980)
5. F. Stange, R. Horn, *Vadose Zone Journal*, 4, (2005)
6. A.Tarantino, Geotechnique, **59**, (2009)
7. S.J. Wheeler, S.R. Sharma, R.S.R. Buisson, 2003, *Geotechnique* **53**, (2003)
8. M. Mbonimpa, M. Aubertin, A. Maqsoud, B. Bussiere, *Journal of Geotechnical and Geoenvironmental Engineering* **132**, (2006)
9. D. Masin, *International Journal for Numerical and Analytical Methods in Geomechanics* **34**, (2010)
10. D. Sheng, A.N. Zhou, *Canadian Geotechnical Journal* **48**, (2011)
11. A.R. Russell, *European Journal of Soil Science* **61**, (2010)
12. N.R.A. Bird, E. Perrier, M. Rieu, *European Journal of Soil Science* **51**, (2000)
13. S. Assouline, D. Tessier, A. Bruand, *Water Resources Research* **34**, (1998)
14. M. Tuller, D. Or, L.M. Dudley, *Water Resources Research* **35**, (1999)
15. D. Or, M. Tuller, *Water Resources Research* **35**, (1999)
16. V.Y. Chertkov, *Journal of Hydrology* **286**, (2004)
17. V.Y. Chertkov, 2010, *The Open Hydrology Journal* **4**, (2010)
18. V.Y. Chertkov, Advances in Water Resources **44**, (2012)
19. Y. Wang, S.M. Grove, M.G. Anderson, *Advances in Water Resources* **31**, (2008)
20. N.V. Churaev, *Liquid and Vapour Flow in Porous Bodies: Surface phenomena*, (Gordon and Beach Science Publishers, Amsterdam, 2000)
21. A.W. Adamson, *Physical chemistry of surfaces* (John Wiley & Sons, 1982)
22. F. Rouquerol, J. Rouquerol, K. Sing, *Adsorption by Powers & Porous Solids - Principles, Methodology and Applications* (Academic Press, London, 1999)
23. F.A.L. Dullien, *Porous media fluid transport and pore structure*, (Academic Press Inc., USA, 1991)
24. M. Iwamatsu, K. Horii, *Journal of Colloid Interface Science* **182**, 1996
25. J.S. Rowlinson, B. Widom, *Molecular Theory of Capillarity*, (Oxford University Press, NY, 1984)
26. Y. Wang, *Journal of Porous Media* **13**, (2010)
27. J.M. Nordbotten, M.A. Celia, H.K. Dahle, S.M. Hassanizadeh, *Water Res. Res.* **43**, (2007)
28. C.L. Tucker III, R.B. Dessenberger, *Flow and rheology in polymer composites manufacturing* (Elsevier Science, 1994)
29. Y. Wang, X.Y. Wang, M. Scholz, D.K. Ross, *Cons and Built Mats* **35**, (2012)

DOI: 10.1051/e3sconf/20160911006

Estimation of water retention behaviour of MX-80 bentonite partially saturated with saline solution

Auli Kuusela-Lahtinen[1], Gowthaman Sinnathamby[1], João Mendez[2], Wojciech T. Sołowski[1,a], Domenico Gallipoli[2], Xavier Pintado[3] and Leena K. Korkiala-Tanttu[1]

[1]*Department of Civil and Environmental Engineering, School of Engineering, Aalto University, P.O.Box 12100, FI-00076, Espoo, Finland*
[2]*Université de Pau et des Pays de l'Adour, UFR Sciences Côte Basque - Bâtiment ISABTP, Laboratoire SIAME, Allée du Parc Montaury,64600, Anglet, France*
[3]*B+Tech Oy, Laulukuja 4, FI-00420 Helsinki, Finland*

Abstract. The water retention behaviour of partially saturated MX-80 bentonite with pure water is relatively well investigated. However, in practical cases, the water contains a number of chemical compounds which affects the bentonite behaviour. In particular, bentonite is used in buffer applications of geological repositories for spent nuclear fuel, where the concentration of total dissolved solids in groundwater is expected to increase significantly with time. This paper investigates water retention behaviour of MX-80 type bentonite partially saturated with deionised water as well as 1M NaCl solution. In the experiments the specific volume of bentonite has been kept approximately constant.

1 Introduction

A wide range of bentonite and other swelling clay materials have been studied extensively, partially because of their potential for creating nearly impermeable self-healing barriers. Those barriers are useful for isolating toxic waste material, such as nuclear waste. Due to the very long service life of such barriers, the understanding of the barrier material must be very thorough. Likewise, the accuracy of the barrier material characterisation must be exceptional.

The present study describes the water retention behaviour of MX-80 bentonite partially saturated with: (1) deionised water and (2) 1 molar solution of NaCl. The paper adds to significant existing body of knowledge on MX-80 bentonite water retention behaviour e.g. [1-8].

Establishing the water retention curve is essential for determination of its behaviour upon wetting and drying. Water retention must be known not only for deionised water, but also for wide range of saline solutions. This study uses mainly filter paper technique to estimate the matric and total suctions of partially saturated bentonite. Filter paper data has been compared to the psychrometer measurements for selected bentonite samples.

2 Materials and Methods

2.1 Material

MX-80 bentonite has been formed as a result of hydrothermal alteration of volcanic ash during the

Figure 1. MX-80 bentonite from Wyoming, USA: a) raw bentonite powder, b) hydrated with 1M NaCl solution (target saturation after compaction 40%), and c) hydrated with deionized water (target saturation after compaction 40%).

Cretaceous period. No characterisation of the material has been performed in this study. Based on [1], MX80 bentonite is composed mainly of montmorillonite (65–82%), quartz (4–12%), feldspars (5–8%), and smaller quantities of cristobalite, calcite and pyrite. The liquid limit and plastic limits of the MX-80 bentonite are 350%-570% and 70%, respectively. The density of the bentonite grains in the study was taken as 2.82 Mg/m^3 [1].

The MX-80 bentonite utilised in the study was supplied in a powder form with the original gravimetric

[a] Corresponding author: wojciech.solowski@aalto.fi

DOI: 10.1051/e3sconf/20160911006

water content being around 11-12%. Deionised water and 1 M NaCl water solution were used for hydrating the samples to the target saturation levels.

2.2 Methods

2.2.1 Sample Preparation

In order to prepare the cylindrical samples (50mm diameter, 20mm height), the oven dried (105 °C) MX-80 bentonite powder was mixed with water (pure or saline, as required). The amount of water has been chosen such that the desired target degree of saturations and specific volume should be reached after pressing the samples to the correct height.

Upon mixing, the saline water yielded a fairly uniform mixture of MX-80 with small lumps, while the deionized water produced larger lumps (Fig. 1). The mixture was then placed in an air-tight plastic bag and sealed. The lumps were crushed manually inside the sealed plastic bag to ensure a uniform distribution of moisture within the mixture. Once the mix was uniform, it was left in the sealed bag for two weeks to equilibrate at room temperature.

The pressing procedure has been the same for all the samples. Following the equalisation period of at least two weeks, carefully weighted amount of bentonite powder at water content correct for the targeted degree of saturation has been put into the steel mould. The mould was coated with a very thin layer of PTFE lubricant. The top and the bottom of the sample were isolated from the piston and background by thin plastic sheets (so almost no moisture could escape the bentonite during pressing).

Then the samples have been uniaxially pressed at a loading rate of around 1mm /min till reaching the desired height of 20 mm (Fig. **2** a,b). Due to noticeable relaxation of the samples after the initial pressing (around 0.6 mm), they were pressed multiple-times. The target dry density of all the samples was 1.6 g/cm3. See Figure 3 for typical stress-displacement curves.

Samples have been measured multiple times in order to estimate their volume. Knowing the sample volume, water content and wet weight, dry densities were computed (Fig. 4). However, despite using high accuracy press and significant care, some variation of dry density (as indicated in Fig. 4) was unfortunately impossible to avoid. However, the achieved uniformity of dry density is similar to that of previous studies [1]. The dry-density variation seems to be correlated with the sample degree of saturation, as well as with the salinity of pore fluid. To achieve more uniform dry density distribution of samples, instead of following the same preparation procedure for all the samples, the procedure perhaps should be modified for drier and wetter samples. The drier samples should be pressed to lower height and the wetter samples to higher height.

In a conventional filter paper test, a single soil specimen is cut into two halves and the 'contact filter paper' for matric suction measurements is placed in between. However, since the bentonite samples were almost impossible to cut into two pieces, as an alternative, both halves were pressed separately (the target height of each was 20 mm) to produce quite identical samples. Pressed samples were then sealed in air-tight plastic bags and left for moisture equalisation for at least one night before commencing the filter paper test.

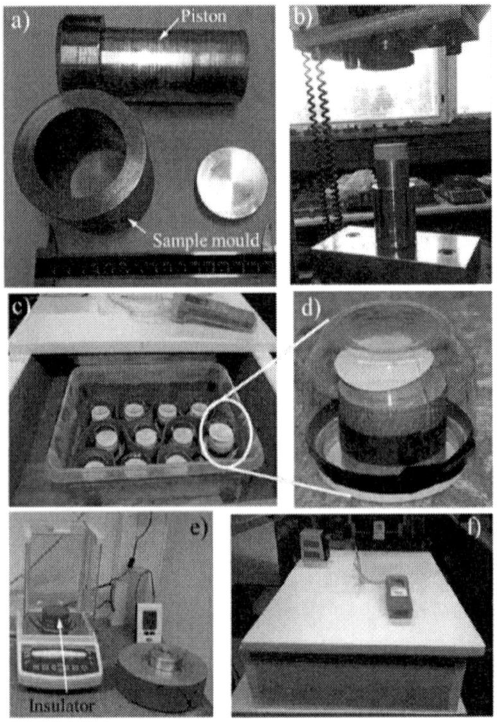

Figure 2. a) tools used for sample pressing, b) compressed specimen being extracted from the sample mould, c) samples kept in the constant-temperature water bath, d) filter paper test assembly, e) insulation placed between the balance and moisture tin, and f) well insulated water bath with temperature control.

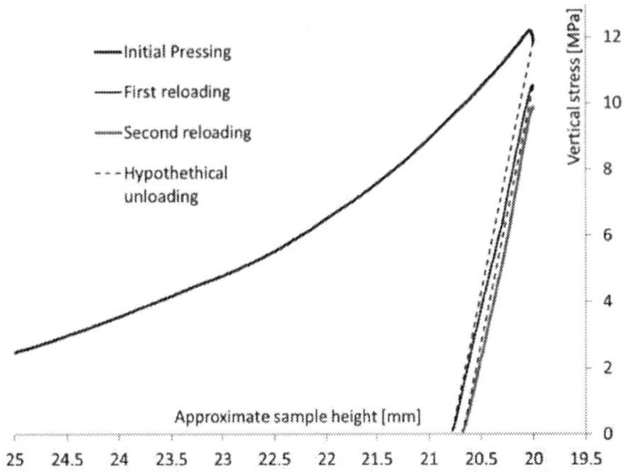

Figure 3. Vertical stress and displacement during sample pressings. No unloading data is available – the unloading paths are hypothetical. The vertical stress value is based on the press force readout. The target sample height was 20 mm. Pressing resulted in sample of Sr=0.59 and dry density 1.56 g/cm^3

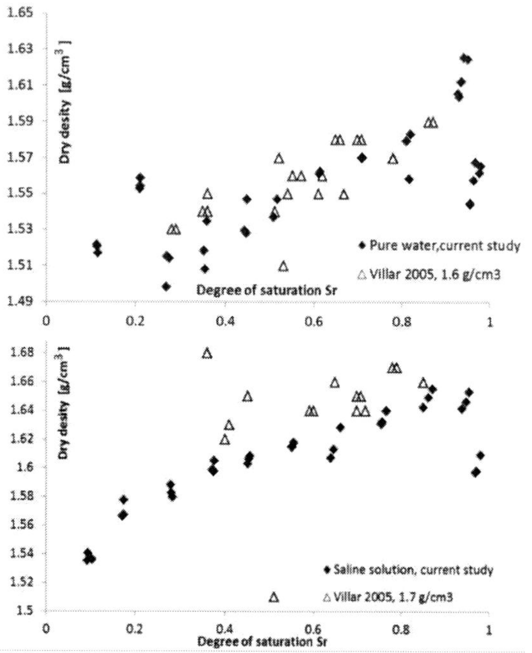

Figure 4. Variation of dry density in the current study and in [1]. For comparison, data for samples with target dry density of 1.6 and 1.7 g/cm^3 are shown [1].

2.2.2 Filter Paper Technique

Soil suction measurement with filter paper technique is well established method of estimation of suction. The major advantage of this technique is the ability to measure both total and matric suction quite reliably without expensive equipment.

The filter paper technique measures suction indirectly by relating the water absorbed by the filter paper with suction by means of a calibration curve for the particular filter paper used. The total suction is measured with a filter paper not being in contact with the sample; hence the paper is wetted via the water vapour. The contact filter paper measures matric suction.

The Whatman® Grade 42 type filter paper was used in the present study. The filter paper was cut to fit the diameter of the bentonite samples. The experimental procedure was very similar to these in [9,10].

In the filter paper test, a contact filter paper, sandwiched between the upper and lower halves of the soil sample, measures the matric suction. The contact filter paper is enclosed by two high-permeable protective filter papers to avoid any contamination. The non-contact filter paper, placed on top of the soil specimen measures the total suction (Fig. 5).

Once the contact-filter paper was in position, an electrical tape was used to tape both halves of the samples together to enforce firm contact between the filter papers and the samples. The filter paper assembly was then immediately enclosed in an air-tight transparent plastic jar and the lid was taped with the electrical tape to prevent any possible exchange of humidity with the environment. The jars in the present study were oriented upside-down (Fig. 4d) (lid on the floor). This orientation facilitates the emplacement and removal of the bentonite sample and the non-contact filter paper without much disturbance in shorter period of time compared to the orientation proposed in [9].

The jar was then stored inside an air-tight plastic container (Fig. 2c). The plastic container was subsequently placed in a constant temperature water-bath (+25°C ±0.2°C) for equalisation over a period of two weeks (Fig. 2f). The equilibration period was longer than often recommended 1 week (e.g. [9]) to make sure the equilibrium is fully reached.

Once the equalisation time is over, the assembly was removed from the water bath and the gravimetric water content of the filter papers were determined with a high precision scale of 0.0001g accuracy (Fig. 2e). Insertion of a piece of thermal insulation material between the metal plate of the balance and the aluminium tin containing the wet filter paper helped to minimise the waiting time for a stable reading. Corresponding suction values were obtained from the calibration curves (see section 2.2.3).

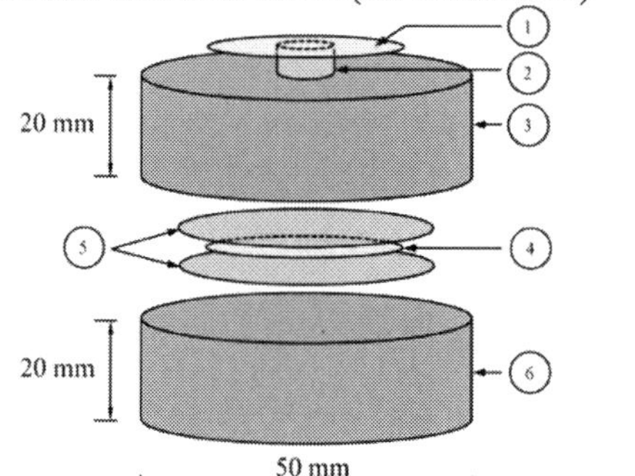

1. Non-contact filter paper for total suction measurement
2. Ring support for non-contact filter paper
3. Upper-half of the bentonite sample
4. Contact filter paper for matric suction measurement
5. Two protective filter papers sandwiching contact filter paper
6. Lower-half of the bentonite sample

Figure 5. Schematic of the filter paper total and matric suction measurement techniques.

2.2.3 Calibration curves

To determine the matric suction of the MX-80 specimens, calibration curves for Whatman 42 filter paper by Habmlin [11] (eq. 1), Leong et al. [12] (eqs. 2 & 3) alongside the standard ASTM [10] (eqs 4 & 5) as well as using calibration using van Genuchten [13] curve (eq. 6, [14]) were tested in the present study:

$$\log \Psi = 8.022 - 3.683 \log w_f \quad (1)$$

$$\log \Psi = 2.909 - 0.0229\, w_f \,,\ w_f \geq 47 \quad (2)$$

$$\log \Psi = 4.945 - 0.0673\, w_f \,,\ w_f < 47 \quad (3)$$

$$\log \Psi = 5.327 - 0.0779 w_f \,,\ w_f < 45.3 \quad (4)$$

$$\log \Psi = 2.412 - 0.0135 w_f \,,\ w_f > 45.3 \quad (5)$$

$$\Psi = 0.051\, [(248/w_f)^{9.615} - 1]^{0.473} \quad (6)$$

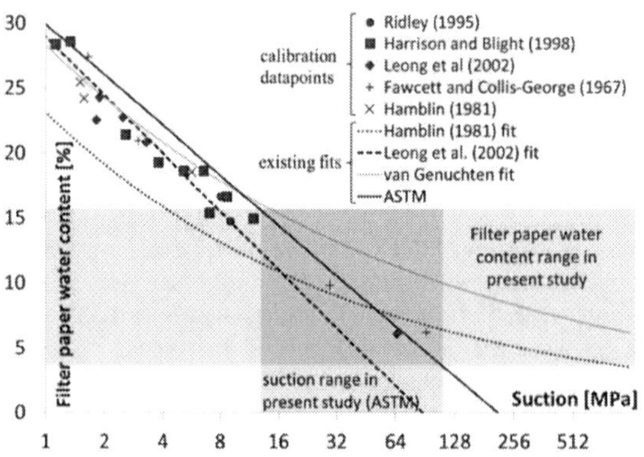

Figure 6. Calibration curves for filter paper and available calibration points matched with approximate range of suctions and filter paper water content in current study. Data courtesy Ms Acikel, after [15-16]. Calibration data points taken from [12,17-19] and fits from [10-12, 14].

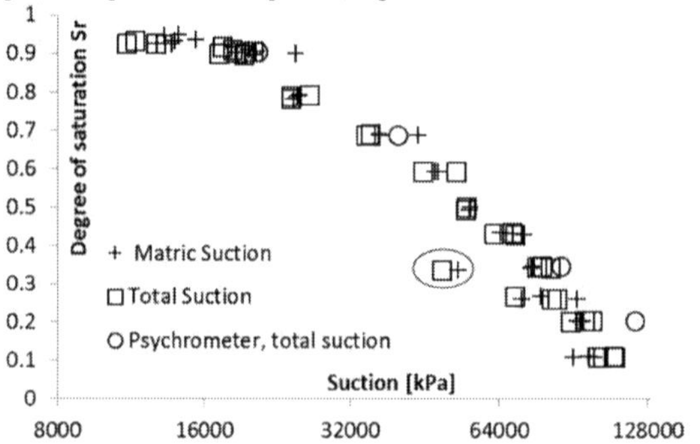

Figure 7. Obtained total and matric suctions for MX-80 bentonite partially saturated with deionised water. Points in the ellipse not taken into account when calculating errors.

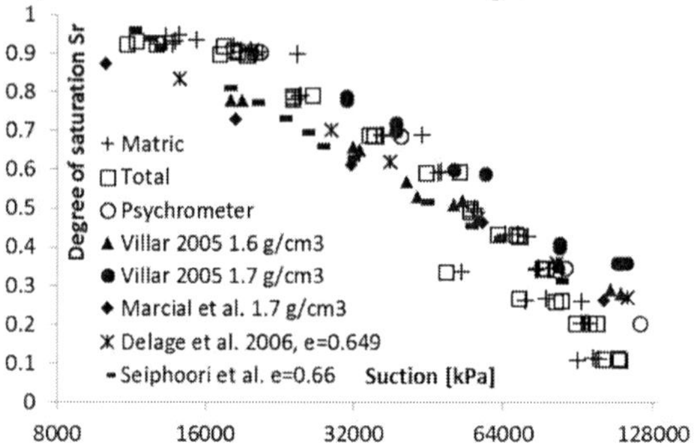

Figure 8. Obtained water retention curve vs literature data on the same material [1, 3, 5-6].

where Ψ is soil suction (kPa) and wf is filter paper water content (%). Unfortunately, there is relative lack of calibration data in high suction ranges typical for bentonite (corresponding to suctions above 10 MPa), as shown in Figure 6. In the calculations, ultimately the ASTM calibration has been used.

Filter paper, when initially wetted by a saline solution and subsequently dried, will change its mass due to the salt residual from absorbed water. To take that effect into account, we note that in each dm3 of 1M saline solution there is 58.44g of NaCl and that the density of such

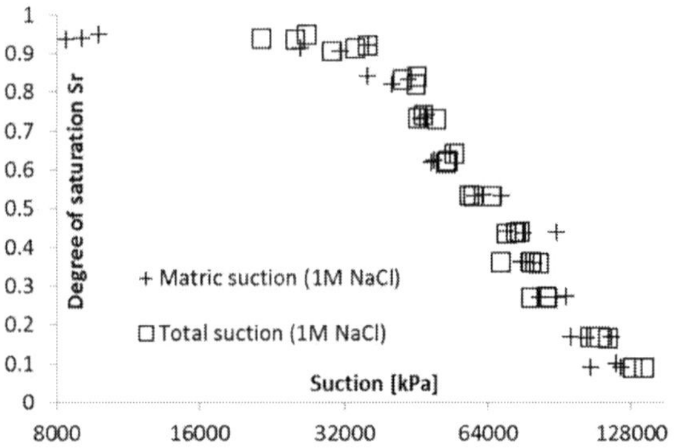

Figure 9. Obtained water retention curve for MX-80 bentonite partially saturated with 1 molar solution of NaCl.

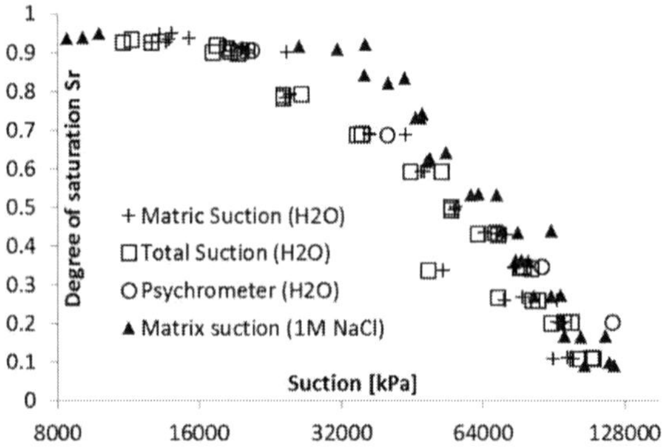

Figure 10. Differences between suctions for samples partially saturated with deionised (H2O) and salty water (NaCl)

solution at 25 °C is approximately 1042.54 g/dm3. Hence, for 1M solution of NaCl, the salt mass remaining in the filter paper after evaporation of water is equal to the evaporated water mass multiplied by [58.44 / (1042.54-58.44)]. The filter paper water content was than calculated as the quotient of the evaporated water mass and the clean dry filter paper mass, which is, in turn, the difference of the weighted dry filter paper mass and the mass of salt remaining on the filter paper:

$$wf = water\ mass/(dry\ filter\ paper\ mass - salt\ mass) \quad (7)$$

3 Results and discussions

In the experimental program 36 samples partially saturated with pure water and 33 samples partially saturated with 1 molar solution of NaCl were tested. The degree of saturation of samples varied from approximately 10% to above 95%.

3.1. Measured suction

The results for bentonite saturated with pure water are shown in Figure 7. Figure 8 shows that those suctions are within the range of the typical published values, e.g. [1, 3, 5-6] and are further confirmed by psychrometer readouts (Table 1). The water retention curve for bentonite partially saturated with 1 molar NaCl (Fig. 9) is

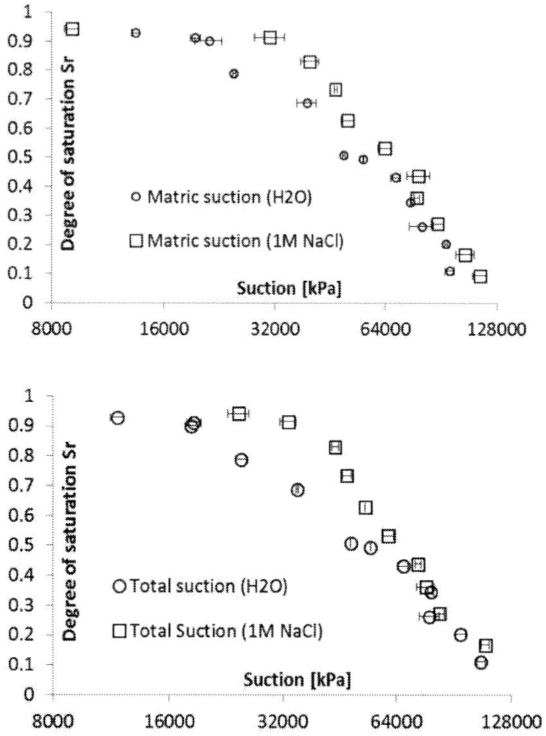

Figure 11. Standard errors (computed based on 3 samples for each average value) for matric and total suctions.

Table 1. Comparison of filter paper and psychrometer results.

Sample	Sr [%]	Suction [MPa]		Difference [%]
		Filter Paper	Psychrometer	
51_w_20_3	20.3	98.7	120.3	18
11_w_40_5	34.8	78.0	85.1	8.4
35_w_80_5	68.8	35.2	39.9	11.9
45_w_100_3	90.5	20.2	20.7	2.3

somewhat different to that of deionised water. Matric suctions for pure and salty water are shown in Figure 10. It is quite clear that most significant difference in the water retention behaviour for pure and salty water is at higher saturations. The likely reason is that saturating the bentonite with the salty solution leads to creation of different miscrostructure and pore sizes in the material. That, in turn, leads to quite different water retention characteristic. However, some part of the differences may be attributed to the variation in dry densities of the samples saturated with pure water and salty solution.

3.2. Accuracy of the results

As a first check of the filter paper technique accuracy, the obtained suctions have been compared against values from psychrometer tests. Suctions obtained with the filter paper technique seem to match the psychrometer results reasonably well (see Table 1), with the biggest difference at very high suction where filter paper measurements are known to be less reliable.

Furthermore, it is expected that, on average, the difference between total and matric suction in case of samples saturated with deionised water should be low (as the osmotic suction component should be zero assuming no salt is present in the natural bentonite; very small amounts of salt may be present though [20]). That is indeed confirmed – the average difference between the matric and total suction measurements (for all the samples partially saturated with water) is just 0.083 MPa. However, the standard error of that mean value is significant and equal to 1 MPa. In case the error in for each value is computed in % of the total suction, the standard error of the mean is approximately 2%. Similarly, for salty solution the average difference between matric and total suction is 3.95 MPa. That value matches the estimated value of osmotic component of suction (4.6 MPa [14]) somewhat poorer than in the case of deionised water. The standard error of the difference between total and matric suction is also higher (approximately 2.05 MPa). Hence the obtained mean value of 3.95 MPa is within reasonable statistical error. Finally, omitting the correction for salt residue in the dry filter paper decreases the average value of the osmotic suction to 3.67 MPa (standard error remains the same). All the measurements were made for three samples with the same target degree of saturation. Those samples have some unavoidable variation in dry density (see Fig. 4). That variation, unfortunately, affect the suction values in the samples. Nonetheless, having sets of three samples for each degree of saturation, it is possible to compute the average value and standard deviation of each set (Fig. 11), even though such error estimates based only on three data-points should be treated with caution.

Averaging the average standard error for each three samples set gives approximately 1.5 MPa (matric suction, water), 1.4 MPa (total suction, water), or approximately 3% and 2,7% of the suction value, respectively. Those values for matric (2.9 MPa, 4.6%) and total suction (2.7 MPa and 3.7%) in samples partially saturated with 1M NaCl solution are again higher. It is unclear why, despite following exactly same procedure for samples saturated with pure water and salty solution, the inaccuracies are larger for salty solution.

The standard error of the degree of saturation value is very small for most of the samples. The averaged values for all the samples are 0.0023 (0.47%) and 0.0039 (0.73%) for samples partially saturated with pure water and saline solution, respectively. The errors are so small that they could not be shown in Figure 11.

3.3. Discussion of the results

In general, the obtained water retention characteristic of bentonite fit reasonably well into the known body of data. However, there are some interesting features that attract attention.

In particular, there is somewhat significant variance between the matric and total suction values close to saturation seen in samples partially saturated with each liquid (in case of salty solution, beyond the expected value of osmotic suction) which are difficult to explain (Figs **7** & **9**).

Furthermore, there are clear differences in matric suctions in the region of high saturations between samples saturated with pure and saline liquid (Fig. 10). Those differences may be due to differences in microstructure caused by saline solution, as well as some variation of dry density. Should they be confirmed in further studies, they may be one reason of the altered bentonite behaviour when saturated with salty solutions.

4 Conclusions & future work

The paper provides further data on water retention behaviour of MX-80 bentonite. The aim of the study was to obtain high quality consistent set of water retention data for bentonite. Here only the results from the first part of the testing program is shown – the investigation of the effects of various saline solutions will be continued in summer 2016. Presented data will also be further confirmed by psychrometer measurements of total suction. Subsequent microstructural investigations are also planned.

The planned testing program will offer further insights on how salinity affect the water retention behaviour of bentonite, as well as perhaps provide some further link between changes in microstructure, salinity of wetting liquid and water retention behaviour. These results, in turn, will allow for more accurate prediction of long term behaviour of bentonite barriers. Providing such predictions is the final goal of this research program.

Acknowledgements

The authors would like to gratefully acknowledge that the presented research has been funded by KYT2018 Finnish Research Programme on Nuclear Waste Management via THEBES project.

References

1. M.V. Villar. MX-80 *Bentonite. Thermo-Hydro-Mechanical Characterisation Performed at CIEMAT in the Context of the Prototype Project.* Informes Técnicos Ciemat, 1053. (2005).
2. M.V. Villar, Lloret A. Influence of temperature on the hydro-mechanical behaviour of a compacted bentonite. Applied Clay Science **26,** pp. 337-350 (2004).
3. P. Delage, Marcial, D., Cui, Y. J. & Ruiz, X. Ageing effects in a compacted bentonite: a microstructure approach. Geotechnique **56** (5), pp. 291–304, (2006).
4. M.V. Villar. Water retention of two natural compacted bentonites. Clay Clay Miner. **55** (3), pp. 311–322, (2007).
5. D. Marcial, P. Delage, Y.J. Cui. Hydromechanical couplings in confined MX80 bentonite during hydration. *Unsaturated Soils: Advances in Geo-Engineering – Toll et al. (eds)*, pp. 249-255, (2008).
6. A. Seiphoori, A. Ferrari, L. Laloui. Water retention behaviour and microstructural evolution of MX-80 bentonite during wetting and drying cycles. Géotechnique **64**(9), pp. 721–734, (2014).
7. S. Saba, J.D. Barnichon, Y.J. Cui, A.M. Tang, P. Delage. Microstructure and anisotropic swelling behaviour of compacted bentonite/sand mixture. J. Rock Mech. Geotech. Eng. **6**(2), pp. 126–132. (2014)
8. Gatabin C., Talandier J., Collin F., Charlier R., Dieudonné A-C. Competing effects of volume change and water uptake on the water retention behaviour of a compacted MX-80 bentonite/sand mixture. Applied Clay Science, **121-122**, pp. 57-62 (2016).
9. R. Bulut, R. Lytton, W. Wray. Suction measurements by filter paper method. American Society of Civil Engineers Geotechnical Special Publication No.115 pp 243-261 (2001).
10. ASTM D 5298: Standard test method for measurement of soil potential (suction) using filter paper. West Conshohocken, PA, USA: ASTM International, (2010).
11. A.P. Hamblin. Filter paper method for routine measurements of field water potencial. Journal of Hydrology, 53, pp. 355-360 (1981).
12. E.C. Leong, L. He, H. Rahardjo. Factors Affecting the Filter Paper Method for Total and Matric Suction Measurements. Geotechnical Testing Journal, 25, 3 pp 321-332 (2002).
13. M.T. van Genuchten. A closed-form equation for predicting the hydraulic conductivity of unsaturated soils. Soil Sci. Soc. Am. J., pp. 892-898 (1980).
14. D.G. Fredlund, H. Rahardjo, M.D. Fredlund. *Unsaturated Soil Mechanics in Engineering Practice.* Wiley, Hoboken, (2012).
15. A.S. Acikel, R.M. Singh, A. Bouazza, W.P. Gates, R.K. Rowe. Applicability and accuracy of the initially dry and initially wet contact filter paper tests for matric suction measurement of geosynthetic clay liners. Géotechnique **65**(9), pp. 780–787, (2015).
16. J.A. Munoz-Castelblanco, J.M. Pereira, P. Delage, Y.J. Cui. The water retention properties of a natural unsaturated loess from northern France. Géotechnique **62**(2), pp. 95–106, (2012).
17. R. Fawcett, N. Collis-George. A filter-paper method for determining the moisture characteristics of soil. Aust. J. Expl. Agric. 7(25), pp. 162–167, (1967).
18. A.M. Ridley. Discussion on 'Laboratory filter paper suction measurements' by S. L. Houston, W. N. Houston and A. M. Wagner. Geotech. Testing J. **18**(3), 391–396, (1995).
19. B.A. Harrison, G.E. Blight. The effect of filter paper and psychrometer calibration techniques on soil suction measurements. *Proceedings of the 2nd international conference on unsaturated soils*, Beijing, China, pp. 362–367, (1998).
20. S. Tripathy, M.Y. Tadza, H.R. Thomas. Soil-water characteristic curves of clays. Can. Geotech. J. **51**, 869–883 (2014).

A theoretical retention model for unsaturated uniform and graded soils

Jacques Monnet[1,a], Dino Mahmutovic[2, 3] Luc Boutonnier[3] Saïd Taïbi[4] and Ngoc Lan Hoang[5]

[1]*Gaiatech, 22 rue Antoine Chollier, 38170, Seyssinet, France*
[2]*L3S-R, Université Joseph Fourier, Grenoble, BP53, 38041, Grenoble, France*
[3]*EGIS Géotechnique, 3, rue Docteur Schweitzer, 38180, Seyssins, France*
[4]*LOMC, CNRS UMR 6294, Université du Havre, BP 540, 75058 Le Havre, France*
[5]*GCD,Ecole Nationale des Travaux Publics de l'Etat.3 rue Maurice Audin,69518, Vaulx en Velin France*

Abstract. Modelling the relation between the degree of saturation and the suction (ie retention curve) is an important challenge for geotechnical engineering. It has a huge influence on the behavior of large soil constructions as levees, embankments, road earthworks. We present here a theoretical model of retention curve which considers physical relations of unsaturated soils. With this approach, there is no need to assume particular shapes of the retention curves, which are a consequence of the physical assumptions. The present study is focused on a theoretical model based on elastic spherical particle arrangement. As a first step a uniform model is presented with a single diameter of soil particle. A second step extends the use of the model to graded soils. The model uses only 5 physical parameters. It is compared with the experimental retention curve of two different samples of glass uniform particles and two different graded soils, a graded glass sample and the Livet-Gavet loam. It shows its ability to model the experimental curves and a better agreement than the former theory of Brooks and Corey (1966). This current publication is funded by the French National Project « Terredurable» (ANR 2011),.

1. Introduction

Development of numerical calculations and increasing power of microcomputers allow determining the behavior of large soil construction as levees, embankments, road earthworks. All these constructions are compacted at the Proctor optimum with a degree of saturation between 80 to 98% and an unsaturated soil. For a precise estimation of the final state of density of the soil, it is needed to model the compaction of the soil along a wetting path. Furthermore when the earth construction is achieve, it supports drying and wetting events linked to the rainy and sunny different days. These events need to develop a model able to simulate drying and wetting paths

There are several ways for modelling retention curve. The first method is to use experimental relations. This approach had many successes with the model (1) of Brooks and Corey (1966) which needs the determination of air entry suction s_{air} and the experimental parameter δ. Later Van Genuchen (1980) used a new experimental relation (2) able to simulate larger retention curve with 3 experimental parameters (α, n, m). Gallipoly et al. (2003) follow the same research scheme and proposed a complete experimental relation (3) with 4 experimental parameters (Φ, Ψ, n, δ). The increase of precision is paid by an increase number of parameters.

The second approach is to simulate the retention curves by an incremental elasto-plastic model (Arairo., 2013). This approach needs the knowledge of the experimental curve of saturation and imbibition to predict the behavior of other retention curve. It uses the parameters of Van Genuchen (1980) and two limit curves

The third approach is to consider physical modelling. With this approach, there is no need to assume particular shapes of the retention curves, which is a consequence of the physical assumptions. It is the approach used in this paper. The present study is focused on a theoretical model based on elastic spherical particle arrangement. This approach follows a new research way so that the physical phenomenon will be explained and the number of experimental parameters will be limited.

$$S_r = \left(\frac{s}{s_{air}}\right)^{-\delta} \quad (1)$$

$$S_r = [1 + (\alpha . s)^n]^{-m} \quad (2)$$

$$S_r = \left(\frac{1}{[\phi.(v-1)^{\psi}.s]^n}\right)^{-\delta} \quad (3)$$

2. The expression of the theoretical retention curve for the unsaturated soil-study

2.1. Hypothesis

Many authors have mentioned the existence of four areas of saturation, each with a distinct behaviour. This

[a] Corresponding author: monnet.jacques@wanadoo.fr

assumption is resumed in the design of our model, mostly based on the work of Boutonnier (2007).

- *Domain D1* : $s \geq s_{air}$ and $S_r \leq S_{rair}$

The gaseous phase is continuous in the soil. This state gives a suction s higher than the air entry suction and a lower degree of saturation than the degree of saturation of air entry.

- *Domain D2* : $s \leq s_{air}$ and $S_{rair} \leq S_r \leq S_{re}$ and $u_w \leq 0$

In this domain, free air disappeared. The air is occluded in the soil. The air is in contact with the soil particles. The suction has the effect of increasing the strength of interparticle contacts. The water pore pressure is negative

- *Domain D3* : $S_{re} < S_r < 1$ and $u_w > 0$

Air is occluded into the soil sample and is considered as to be independent air bubbles from the skeleton. Capillary strength has no effect on the contact forces between soil particles. We consider here D3 is corresponding with the case of positive pore pressures with a degree of saturation lower than 1.

- *Domain D4* : $S_r = 1$

There is no air in gaseous state in the soil. The soil is saturated. The boundary between D3 and D4 can also be expressed through the pore pressure with a degree of saturation equal to 1.

2.2. Deformation of the spherical particles and volume of meniscus: Domain D1

To complete the original model of non-saturation (Monnet and Boutonnier, 2012) written for domain D2 to D4, it was decided to model the D1 domain.

Under the action of the water at the interparticle contact, there is suction on the cross section of the meniscus (Fig.1) and a surface tension as assumed previously (Taibi, 1994). The model of soil is composed of uniform spherical particles of radius R.

According to Laplace's law, suction is the product of the surface tension T_c by the sum of two principal curvatures of radius r of the meniscus and the radius b of the torus corresponding to the wetted area (4). Under the action of a normal force F_N on the contact plane (5), the two balls deforms elastically. The distance between their centers decreases of the quantity α and the contact is no more punctual, but is a circular plane area of radius (a Fig.1). We assume that the balls remain spherical except for the flat portion of the contact. The values of a and α are given by equations (6) to (8) of the elastic Hertz contact between the two spheres. The value of the normal force F_N to the contact plane is given by equation (9) which takes into account le value of the suction and the value of the surface tension T_c

$$s = T_c.(1/r - 1/b) \quad (4)$$

$$\alpha = 2.R.(1-m) \quad (5)$$

$$a = R.\sqrt{1-m^2} = \left[\left(3.\left(1-\vartheta_p^2\right).R.f_N\right)/\left(2.E_p\right)\right]^{\frac{1}{3}} \quad (6)$$

$$\alpha = a^2/R \quad (7)$$

$$m = 1 - \alpha/(2.R) \quad (8)$$

$$F_N = s.\pi.b^2 + 2.\pi.b.T_c \quad (9)$$

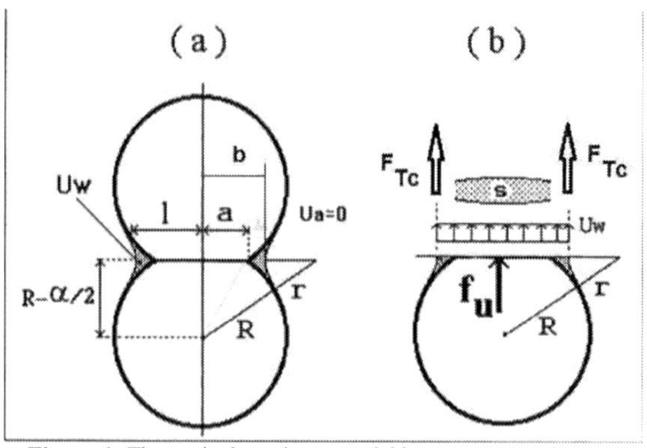

Figure 1. Theoretical meniscus model between two spheres (S.Taïbi, 1998)

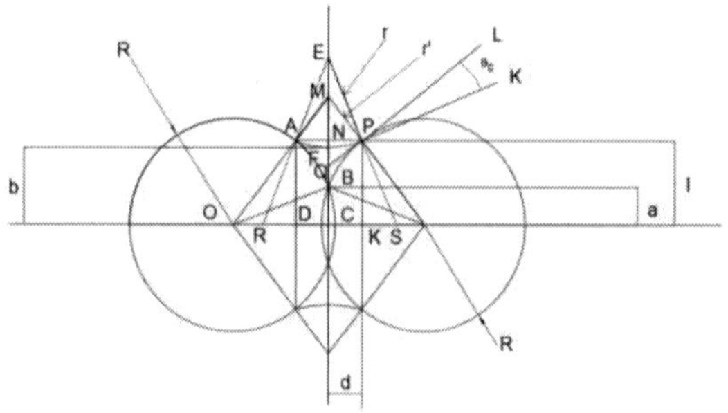

Figure 2. Geometrical model for the meniscus calculation of 2 spherical particles with a wetting angle θ_c

2.3. Analytical calculation of meniscus volumes with a wetting angle meniscus: Domain D1

The contact between the particles and the water has a wetting angle of θ_c which varies in case of drainage or humidification (Gras, 2011).

The calculation of the volume of the meniscus is made with a rotational assumption around the horizontal axis through the centre of the two particles (Fig. 4) and a relative horizontal symmetry with the EC plane contacting the particles. The volume of the meniscus is the difference between the volume of CDAE frustum, the volume of the soil flattened sphere CDAB, and the volume of air of the torus of AFE area (10). The main variables of the calculation are the angles θ_1 (11) to θ_5 (12), l the radius of the contact circle (13), b radius of the meniscus torus (14), r' the radius of the circle which connect the two contact circles (15), R1 the radius of the frustum (18).

$$V_{men} = 2.\left(V_t - V_p - V_{air}\right) \quad (10)$$

$$\cos(\theta_1) = \frac{R - \alpha/2}{R + r'} = \frac{R - \alpha/2}{R + r.\left(\cos(\theta_c) - tg(\theta_1).\sin(\theta_c)\right)} \quad (11)$$

$$\theta_5 = \widehat{AEF} = \frac{\pi}{2} - \theta_1 - \theta_c \quad (12)$$

$$l = R_2 = R.\sin\theta_1 \quad (13)$$

$$b = R.\sin(\theta_1) + r.\sin(\theta_1 + \theta_c) - r \quad (14)$$

$$r' = r.cos(\theta_1 + \theta_c)/cos(\theta_1) \tag{15}$$

The main variables of the problem can be defined, as the volume of the frustum (16), the volume of the air (17) and the volume of the particles (18).

$$V_t = \frac{1}{3}.\pi.(R_1^2.H_1 - R_2^2.H_2) \tag{16}$$

$$V_{air} = \pi.r^2.CE.\theta_5 - \frac{2}{3}\pi.r^3.sin\theta_5 \tag{17}$$

$$V_p = \frac{1}{3}.\pi.\left\{ \begin{array}{c} R^3.[(1-cos\theta_1)^2.(2+cos\theta_1)] \\ -\left[\left(\frac{a}{2}\right)^2.\left(3.R-\frac{a}{2}\right)\right] \end{array} \right\} \tag{18}$$

The meniscus volume was simulated by a Solidworks® model which allows measuring the numerical volume of the meniscus, and comparing with the analytical value of the meniscus (9). The comparison shows that the analytical relations allow finding the meniscus volume of the Solidworks® model.

Figure 3. Tetrahedral arrangement, e = 1.315 ; 4 contacts per particle ; 8 menisci

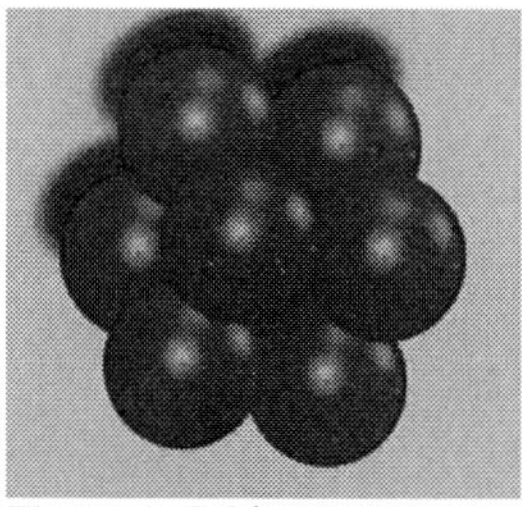

Figure 4. Cubic arrangement ; e = 0.910 ; 6 contacts per particle ; 3 menisci

Figure 5. Octahedral arrangement ; e = 0.470 ; 8 contacts per particle ; 8 menisci

Figure 6. Dodecahedral arrangement ; e = 0.343 ; 12 contacts per grain ; 17.6 menisci

2.4. Different arrangements of soil particles

The different arrangements of uniform soil particles give a variation of void ratio between 1.315 to 0.343 (Taibi, 1994). The SoildWorks® program was used to a precise modelling of the four possible arrangements, tetrahedral, cubic, octahedral, dodecahedral (Fig.3 to 6). It shows that the void ratio is independent of the radius of the particles, but depends only of the arrangement between particles. The radius of particles has only an influence on the size of the REV (Representative Volume Element). The variable which describes the different phenomenon along the drying path is the ratio between the air bubble and the particles .

The Solidworks model allows determining:

- The number and the angle of contact for each particle
- The number of total menisci per arrangement

For the drying path at nucleation, the model finds:

- The radius of the bubble which appears at nucleation
- The number of bubbles able to appear inside VER
- The degree of saturation at nucleation
- The suction at nucleation

For the drying path at percolation, the model finds:

- The radius of the bubble which is able to percolate
- The suction at percolation

The theoretical calculation of the relation between suction and degree of saturation can be performed when the menisci are independent (Fig.7). When the meniscuses merge, it is assumed that the suction remains unchanged at wetting. At the drying path nucleation appears first and percolation of air follows.

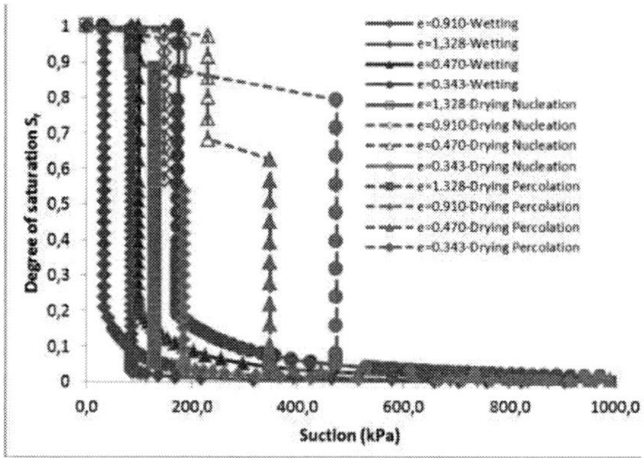

Figure 7 : Theoretical retention curves calculated for different arrangement, deformable spheres of 4µm diameter,

2.5. Calculation of the theoretical retention curve for a uniform soil: Domain D1

For a uniform soil, with only one diameter of soil particles, there are different possible arrangements (tetrahedral, cubic, octahedral, dodecahedral). The relative position of the particles is known as well as the REV (eq.21 to 24) and the void ratio and the number and orientation of the contacts in the arrangement.

$$V_w = S_r.e.REV/(1+e) \tag{19}$$

$$V_{men} = S_r.e.REV/[(1+e).Nb_{menisci}] \tag{20}$$

$$V_{men} = S_r.e.REV/[(1+e).Nb_{menisci}] \tag{21}$$

$$REV_{cubic} = 8.R^3 \tag{22}$$

$$REV_{Octahedral} = 64.R^3/(3.\sqrt{3}) \tag{23}$$

$$REV_{Dodecahedral} = 40.R^3.\sqrt{2}/3 \tag{24}$$

On a reverse way, the knowledge of the void ratio allows identifying the type of arrangement of soil particles. The knowledge of the water content or the saturation allows calculating the total volume of water in the REV (19). The knowledge of the number of contacts in the REV and the number of complete menisci allows calculating the volume a single meniscus (20). The volume of the meniscus depends of the radius of the meniscus, the radius of the grain and the suction. This allows calculating the suction associated with a void ratio and a water content (Fig.7).

2.6. Compactness and homogeneity of granular mixture for a graded soil

The theoretical model is organized from the larger to the smaller of n different Di diameters for the soil squeleton particles, such as (25). The symbol C is the compactness which is equal to the solid volume of the soil sample (26). The relative compactness to the total sample of the class i is noted Ci (27). As a consequence, the total compactness C of the soil sample is given by (28). In the theory, Ri corresponds to the volume refusal on the sieve of diameter Di , based on the total volume of the sample grains (29). Assuming a single density for all aggregates, so this term is the refusal by sieve reduced to the total mass of the sample

$$D_1 \geq \cdots D_i \ldots \geq D_n \tag{25}$$

$$C = {V_s}/{V_t} = 1 - n \tag{26}$$

$$C_i = {V_{si}}/{V_t} = 1 - n_i \tag{27}$$

$$C = \sum_{i=1}^{n} C_i \tag{28}$$

$$R_i = {C_i}/{\sum_{j=1}^{n} C_j} = {M_{di}}/{\sum_{j=1}^{n} M_{dj}} \tag{29}$$

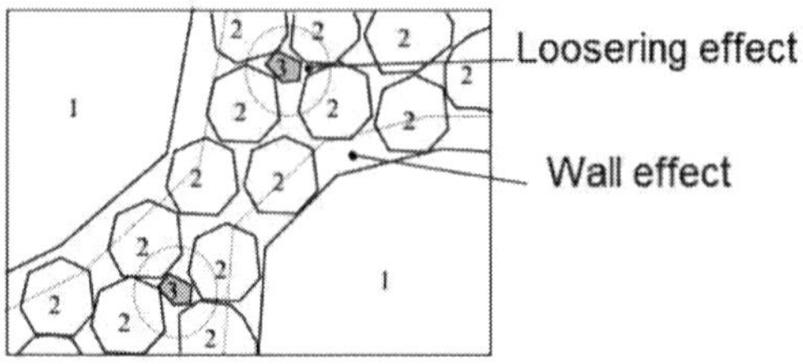

Figure 8. The perturbations exerted on the middle class over the large grains and the small grains (De Larrard, 1999)

2.7. Mixture scattered – general case for a graded soil

With a mixture of 3 parts a loosening effect (Fig. 8) and wall effect may appear. If the class 2 prevails the grains are subjected to a loosening effect exerted by the smaller grains from class 3 and a wall effect exerted by class 1. In

This can be generalized for n different classes (30) by finding the virtual C compactness of the mixture, considering that the class i is dominant in equation (31). The relative density corresponding to βi coefficient is given by (32). The experimental βi coefficient is found by interpolation of the experimental curve (Fig.9): This is the ratio between the density of the mixture to the particle density versus the particle diameter Di.

$$C = inf_{1 \leq i \leq n} C_i \tag{30}$$

$$C_i = \frac{\beta_i}{1 - \sum_{j=1}^{i-1}\left[1 - \beta_i + b_{ij} \cdot \beta_i \left(1 - {1}/{\beta_j}\right)\right] \cdot R_j - \sum_{j=i+1}^{n}\left(1 - a_{ij} \cdot {\beta_i}/{\beta_j}\right) \cdot R_j} \tag{31}$$

$$\beta_i = \frac{\rho_{di}}{\rho_s} = -0.03147 . [Log(D_i)]^2 - 0.13539 . Log(D_i) + 0{,}4060 \tag{32}$$

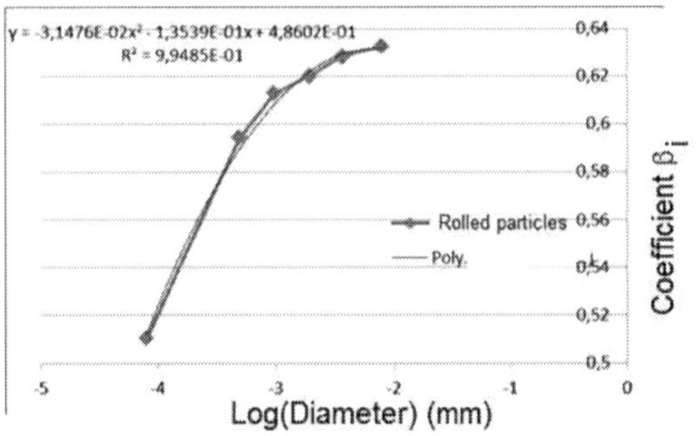

Figure 9. βi compactness coefficient of rolled grains (De Larrard, 1999)

2.8. Determination of the compactness of each size class; graded soil

In a mixture of n different diameter organized according to equation (30) from the largest to smallest, the theory of De Larrard (1999) allows finding the diameter D_i dominant, with its compactness, which allows calculating the porosity of class i of grains by equation (27) and the corresponding void ratio e_i.

At a first step of calculation, the total compactness C_m of the mixture and the porosity n_m are known. Now consider the total size curve (Fig.10). The dominant diameter separates the curve into two portions:

- At the right of the dominant diameter, large elements floating into small elements
- At the left of dominant diameter, small items fall into the gaps left by the dominant grains

Each portion has a density of its own and depends only on the prevailing diameter. Then we can consider two separate samples for which two new dominant diameters appear with their own characteristics of compactness, porosity and void ratio. We can determine from part to part all theoretical porosity of all diameters of the particle size curve.

Let us consider the total mixture of n different particles. For a single rank of aggregate i, the volume of the void V_{vi} into the mixture is given by the equation (37). Under these conditions, the void ratio $e_{théo}$ of the theoretical mixture is known. This void ratio is different from the measured void ratio e_m of the mixture, since the former refers to a method of standardized compaction where the grains grow to different relative densities. We assume the proportionality between the void ratio of the different classes. According to equation (39) e_{mi} is the theoretical void ratio of class i; e_{essai}.is the known value of mixture measured during the test; $e_{théo}$ is the void ratio of the mixture compacted at the standard energy (Fig.8), and e_i is the void ratio of each diameter into the compacted sample. This void ratio e_{mi} allows knowing the arrangement of the particle class.

$$V_{vi} = e_i . R_i = \frac{n_i}{1-n_i} . R_i \tag{33}$$

$$e_{théo} = e_m = \frac{V_v}{V_s} = \frac{\sum_{i=1}^{n} \frac{n_i}{1-n_i} . R_i}{\sum_{i=1}^{n} R_i} \tag{34}$$

$$e_{mi} = \frac{e_{essai}}{e_{théo}} . e_i \tag{35}$$

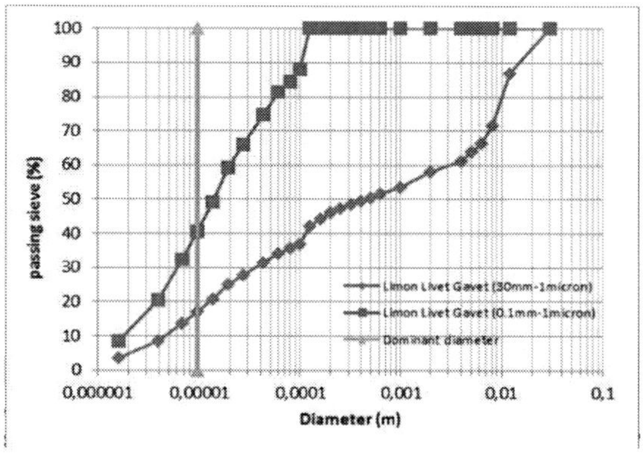

Figure 10. The grade curve of Livet-Gavet loam separated into two portions, large elements on the right and left small particles

Table 1 : The 5 physical parameters of calculation (w) for Wetting path (d) for Drying path on Glass specimen

γ_s kN/m^3	E_p GPa	ν_p	θ_c Deg.	R_p µm	T_c 10^{-5} kN/m^3
29	65	0.25	20d 25w	80	7.28
29	65	0.25	20d 25w	300	7.28
29	65	0.25	20d 25w	4 to 140	7.28

2.9. Determination of the grain diameter corresponding to the air inlet and the retention curve

We assume that small pores are saturated ($S_{ri} = 1$), and that large pores are dry ($S_{ri} = 0$). Under these conditions, the water content can be decomposed as a function of particle size (40), with dry pores ($S_r = 0$) for j <i (large diameter) and saturated pores ($S_r = 1$) for j> = i (small diameters). The rank of the pores involved in the air inlet is i,. In relation (40), the only unknown is the rank i. We can define the diameter of the grains involved in the air inlet, and builds a theoretical retention curve for mixtures. The knowledge of e_{mi} and the water content w_{NSi} allows finding the suction of class i (Fig.7).

$$w_{NS} = \frac{\rho_w}{\rho_s} . \left[w . \frac{\rho_s}{\rho_w} - \sum_{j=i}^{n} e_{mi} . R_i \right] = w - w_{atr} \tag{36}$$

Table 2 : The 5 physical parameters of calculation (w) for Wetting path (d) for Livet Gavet Loam from paste of soil

γ_s kN/m^3	E_p GPa	ν_p	θ_c Deg.	R_p	T_c 10^{-5} kN/m^3
26	107	0.2	0d 5w	1.6 µm to 3cm	7.28

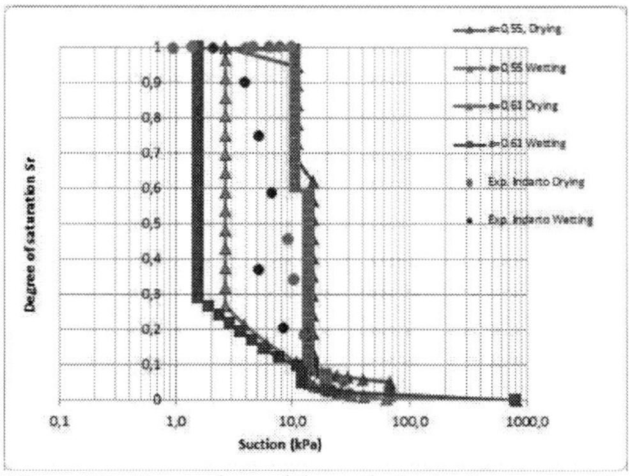

Figure 11 : Comparison of theoretical retention curve with the experimental results Indarto (1991); void ratio 0.55; 80µm diam. glass; wetting angle, 0° at drying 5° at wetting

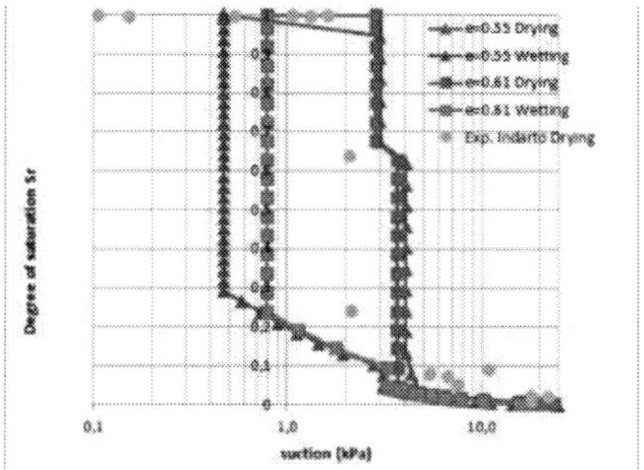

Figure 12 : Comparison of theoretical retention curve with the experimental results Indarto (1991); void ratio 0.55; 300µm-diam. glass; wetting angle, 0° at drying 5° at wetting

3. Results

3.1. Comparison with an experimental uniform sample of glass

The theoretical retention curves shows a large variation of degree of saturation for single value of suction. For the wetting path, there is coalescence of menisci at a precise value of the suction and an unstable situation between the state of separated menisci and complete saturation of void between particles. For the drying path the air bubbles are allowed to percolate between particles at a precise value of suction. when the air percolates, it reaches larger void between particles which allowed sudden decrease of the water content and sudden decrease of the degree of saturation. The vertical parts of the retention curves are linked to the fact that the grade curve is known only by measurements points and not by a continuous curve.

The simulation of the experience on two glass particle samples (Indarto, 1991) is made along drainage and wetting paths. The samples had respectively a uniform diameter of 80μm and 300μm, a unit weight of 18,7kN/m² (e = 0.550; dodecahedral). The unit weight of the glass particle is assumed to be 29kN/m3. The comparison is made for the relation between the suction and the degree of saturation (Fig.11-12). The parameters of the theoretical model are shown (Table 1). It appears that the new model allows finding a correct estimation of the experimental curve on the drying path (cor.0.852-0.695) or on the wetting path (cor.0.843). The difference between the new theory and the experience at 50% of saturation is only 4.2-0.8kPa on drying path and 2.7kPa on wetting path

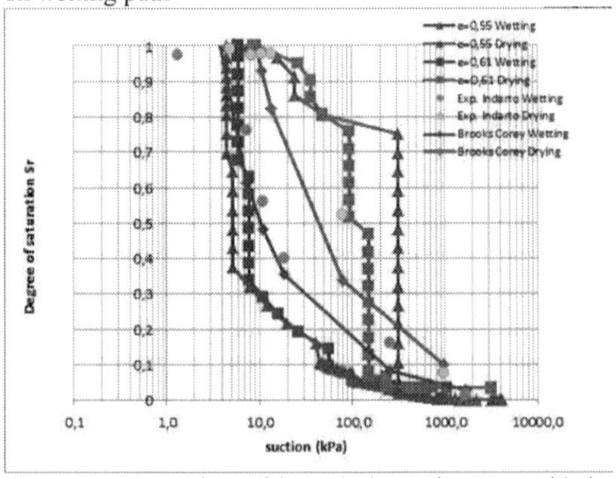

Figure 13 : Comparison of theoretical retention curve with the experimental results Indarto (1991); void ratio 0.55; 4μm-140μm diam. glass; wetting angle, 20° at drying 25° at wetting

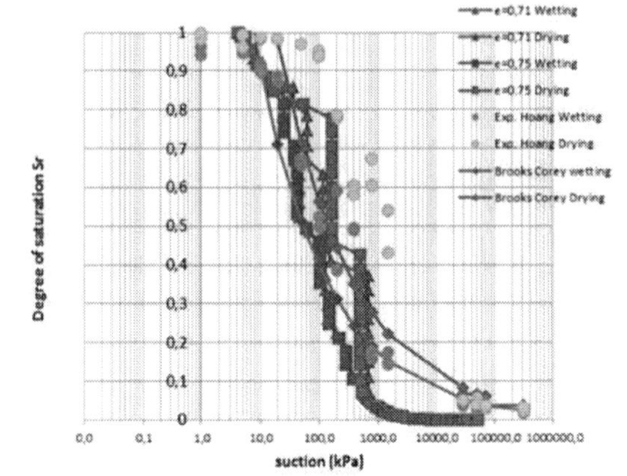

Figure 14 : Comparison of the theoretical curve of retention and the experimental results; void ratio 0.71; Livet Gavet Loam; wetting angle, 0° at drying, 5° at wetting

3.2. Comparison with an graded sample of glass and the Livet-Gavet Loam

The simulation of the experience on a grade glass particle sample (Indarto, 1991) and on the Livet-Gavet loam is made along drainage and wetting paths. The sample are respectively graded between 4-140μm and 1.6μm-50mm of respectively unit weight 18,7kN/m² (e = 0.55) and 15,2kN/m² (e = 0.71)..The comparison is made for the relation between the suction and the degree of saturation (Fig.13-14). The parameters of the theoretical model are shown (Table 1-2). It appears that the new model allows finding a correct estimation of the experiment curve on drying path (cor.0.954-0.875) and on the wetting path (cor.0.885-0.946). The difference between the new theory and the experience at 50% of saturation varies between 20 to 600kPa on drying path and between 5.6 to 50kPa on wetting path

4. Conclusion

A new theory is presented for the retention curve of the soil along drying and wetting path. It explain the difference between these two paths by the saturation of the soil particles with water menisci, which are independent at low saturation, then coalescent at high saturation on the saturation path and by the apparition of air nucleation followed by percolation on the drying path

This new model uses only 5 physical parameters (mineralogy, void ratio, particle radius and granulometry, wetting angle, surface tension of water) and needs no additional experimental parameters always difficult to determine. It allows finding the retention curve along drying and wetting path.. It has been successfully used to determine the retention curve of uniform or graded glass sample and the natural Livet-Gavet loam.

References

1. Arairo W., Influence des cycles hydriques de la dessiccation et de l'humidification sur le comportement hydromécanique des géomatériaux non saturés, Thèse INSA Lyon (2013)
2. Boutonnier L., Comportement hydromécanique des sols fins proches de la saturation. Cas des ouvrages en terre, Thèse INPG Grenoble, (2007)
3. Brooks R.T., Corey A.T, Properties of porous media affecting fluid flows, J. Irrig. Drain. Div., ASCE, 92 (IR2), p.6188-. (1966)
4. De Larrard F. Structure granulaire et formulation des bétons, Etudes et rech. des LCPC, n° OA 34 (1999)
5. Gallipoli D.; Wheeler S.; Karstunen (2003) Modelling the variation of degree of saturation in a deformable unsaturated soil., Geotechnique, Vol. 53, No. 1, 2003, p. 105-112
6. Gras J.P. Approche micromécanique de la capillarité dans les milieux granulaires, Thèse Univ. Montpellier (2011)
7. Indarto, Comportement mécanique et compactage des matériaux de barrage, Thèse Ecole Centrale Paris (1991).
8. Monnet J. , Boutonnier L., Calibration of an unsaturated Air-Water-Soil model, Archives of Civil and Mech. Eng., Vol. 12, Issue 4, 5, p.493-49, (2012)
9. Taibi S., Comportement mécanique et hydraulique des sols soumis à une pression interstitielle négative, Thèse Ecole Centrale Paris (1994)
10. Van Genuchen, A closed form equation for predicting the hydraulic conductivity of unsaturated soils, *Soil Scienc.Soc. Am. J.*, Vol.44, p.892-898 (1980)

E3S Web of Conferences 9, 11008 (2016)
E-UNSAT 2016

DOI: 10.1051/e3sconf/20160911008

Comparison of Soil Water Retention Curves for sandy clay, obtained using different laboratory testing methods

Piotr Osinski[1a], David Toll[2], Eugeniusz Koda[1]

[1]*Department of Geotechnical Engineering, Faculty of Civil and Environmental Engineering, Warsaw University of Life Sciences, Nowoursynowska Street 159, 02-776 Warsaw, Poland.*
[2]*School of Engineering and Computing Sciences, Durham University, South Road, DH1 3LE Durham, UK*

Abstract. The design of engineering structures is still a challenging task when considering unsaturated soil mechanics behaviour. The present paper focuses on analysing Soil Water Retention Curves (SWRC) for clayey soil samples collected from a purpose built experimental embankment. SWRC determination is crucial when analysing geotechnical parameters of the soil. During the measurements, soil samples were subjected to wetting and drying cycles using a stage procedure (where the process is carried out in stages to allow equalisation) and continuous processes. The methods used for obtaining the suction curves were high suction tensiometer, attached at the bottom of the sample, and pressure plates with suctions applied up to 800 kPa. The study compares laboratory methods on measuring suction for sandy clay samples. It proves that the wetting and drying history have an impact on the hydraulic behaviour of tested material. In order to give a wider picture of soil behaviour it also presents the outputs as a function of matric suction and volumetric and gravimetric water content, and degree of saturation.

1 Introduction

Unsaturated soil mechanics play a significant role when designing earth structures like railway and road embankments, dams, dykes or shallow foundations. They are commonly constructed in the vadose zone (Figure 1). Their mechanical properties are subjected to the hydraulic changes which are caused by the different climate events. Soil Water Retention Curves (SWRC) determination is crucial when analysing geotechnical parameters of the soil, especially when analysing rainfall triggered slope failures [1-3]. The present paper focuses on analysing SWRC for compacted clay soil samples, prepared at different moisture contents and subjected to wetting and drying cycles. The reason for applying hydraulic cycles was to reflect different climate scenarios [2]. As the soil suction has an impact on mechanical strength of the fill material for earth structures, it is crucial to determine and understand the hydraulic behaviour of the soil.

SWRC generally reflects soil behaviour for adsorption and desorption processes. It is usually presented as a relationship between water content (gravimetric or volumetric), or degree of saturation against soil suction which can be expressed in kPa. There are three main stages that can be distinguished when obtaining the SWRC. At the beginning of the drying process the soil may be close to full saturation. As drying proceeds, causing an increase in suction, air starts to enter the pores of the soil defining the air entry value of the soil. Beyond the air entry value, the stage of desorption process is the transition zone, where most of the desaturation takes place. This is followed by residual suction, what means that the only water present in the soil is that which is closely bound to soil particles [2, 4]. All these stages can be observed for all type of soils, however shapes of SWRC and suctions values could vary significantly. This is due to the fact that soil suction is also influenced by other factors such as void ratio, soil structure, stress history and stress state [5-7].

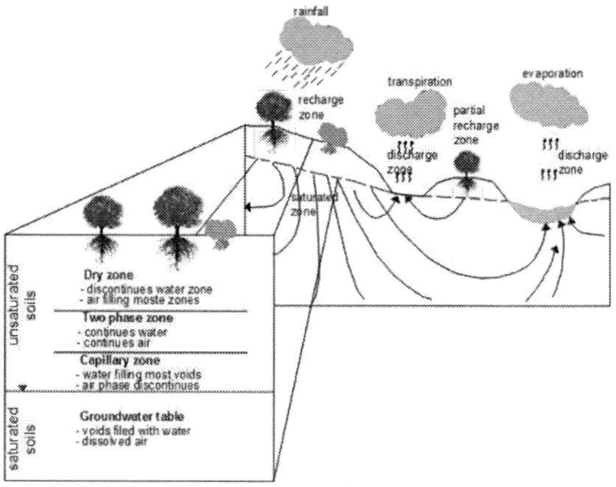

Figure 1. Unsaturated soil zone location.

Another crucial feature of a SWRC is its hysteretic character. That means that suction curves for wetting and drying are not reversible. That is why analysing hydraulic

[a] Corresponding author: piotr_osinski@interia.pl

behaviour and its influence on the geotechnical parameters is so challenging [8-11].

The present paper focuses on determining and evaluating SWRC for sandy clay samples using different testing methods and approaches. The methods used for obtaining the suction curves were the tensiometer, attached at the bottom of the sample, and the axis translation technique. For the first method a high capacity suction probe was used, for the latter one, the pressure plate method. There were also different approaches applied. Compacted samples were subjected to stage drying and wetting cycles, but also the suctions were directly measured for continuous repeated wetting and drying paths that provides the best option in terms of evaluating the hysteresis of different hydraulic paths.

2 Materials and methods

2.1 Soil sample preparation

The soil material used for the study was classified as sandy clay. The grain size distribution of the material was obtained by using sieving method. The material was collected from a purpose built experimental embankment located near Newcastle, UK [12]. First the physical parameters of the soil were tested in the laboratory and then specimens were prepared. All the physical tests were conducted by using methods specified in British Standards [13]. The results for physical parameters testing are provided in Table 1.

Table 1. Physical parameters of the tested material.

Soil type	Light compaction Proctor test		Atterberg limits			ρ_s [g/cm^3]
	W_{opt} [%]	ρ_{dmax} [g/cm^3]	LL [%]	PL [%]	PI [%]	
Sandy clay	15	1.73	41.5	20.1	21.4	2.60

The samples used for the SWRC determination were prepared "as compacted" at 25 % of water content. First the soil was oven dried and sieved to achieve a density that was as homogenous as possible. Then portions of soil were placed into the plastic bags and water was added. The prepared soil was left for 24 h to allow even moisture distribution through the soil.

The samples were reconstituted in a 100 mm diameter by 200 mm high split mould, and compacted using light compaction Proctor tests (Figure 2). The core of the soil was extruded and trimmed to the required dimensions. After trimming each sample, specimens for water content measurements were also collected to ensure the water content was uniform throughout the soil sample profile. There were two sizes of samples prepared for the purpose of the study. The first one, used for the stage drying and wetting was 50 mm diameter by 20 mm high and the other was 75 mm diameter by 20 mm high. The latter size was used for samples subjected to continuous hydraulic path determination and for the pressure plate method. Trimming to the desire sizes was achieved by pressing stainless still rings into the soil sample core.

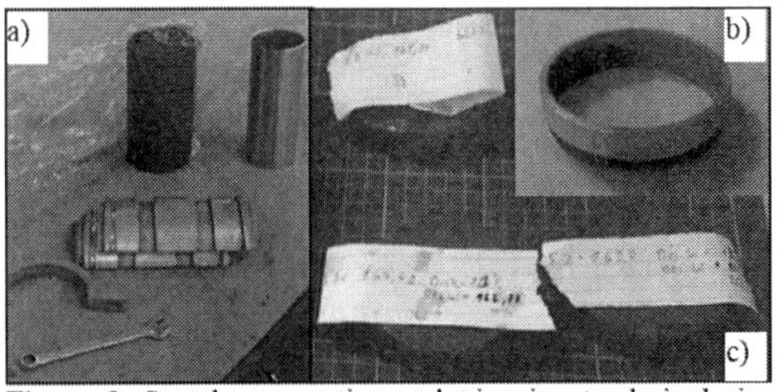

Figure 2. Sample preparation and trimming to desired size: a) sample core extruded after Proctor test; b) stainless still ring; c) soil specimens, size of 20/75 mm.

2.2 Pressure plate method for SWRC determination

The pressure plate method uses the axis translation technique. It allows controlling and maintaining the desired value of suction rather than measuring it. The system consists of a sealed chamber where a High Air Entry Value Porous Disc (HAEVPD) is placed. The soil suction is created by imposing an air pressure on samples placed on the HAEVPD, and by changing the water pressure of tested samples. The matric suction for the axis translation technique is expressed as (u_a-u_w) which is a difference between the imposed air pressure (u_a) and the water pressure (u_w) of the soil sample, both controlled during the test. The amount of suction that could be generated by using the pressure plate system depends on the Air Entry Value of the ceramic disc used for the test. Usually this value ranges from 500 to 1500 kPa.

The system that was used for the purpose of this study was modified (Figure 3). Instead of using a single porous ceramic disc, there were three HAEVPD that allowed independent measurement for each soil sample. Each disc of 1500 kPa air entry value was connected to a different water volume measurement.

Figure 3. Pressure plate system, with three independent 1500 kPa HAEV ceramic disks connected to water volume controllers.

The air pressure in the chamber was applied in stages, and the water pressures where adjusted accordingly, to achieve the desired value of suction. Then the system was left to equalise; the next stage was commenced when no volume change of the water coming out of each specimen was recorded. The stages of imposing air pressure were repeated, the volume changes of water were measured thus the SWRC could be plotted. Measurements of sample volume were obtained using a Vernier calliper and the mass change was taken by weighing manually after equalisation at each suction level (assuring that no soil mass was left on the porous plate). The whole system was connected to the data acquisition system which was supported by TRIAX software developed at Durham University [14].

2.3 High suction probe for SWRC determination

2.3.1 Description of high suction probe used in the study

The tensiometer used for the study was the one designed in Durham University by Lourenço et al. [15]. Since the first development of a high suction probe by Ridley and Burland in 1993 [16] there have been a number of improvements and modifications applied. The main characteristic that distinguishes pore pressure transducers from each other is the air entry value of the ceramic disc and the form of construction. Figure 4 shows the instrument used for the study consisting of (from the top): 1500 kPa porous stone, small water reservoir, spacer, ceramic transducer and electrical connection. The entire components are enclosed in a stainless steel casing (Figure 4). Another successful design was also reported by Tarantino and Mongiovì [17], however this used a strain gauged diaphragm instead of a ceramic transducer.

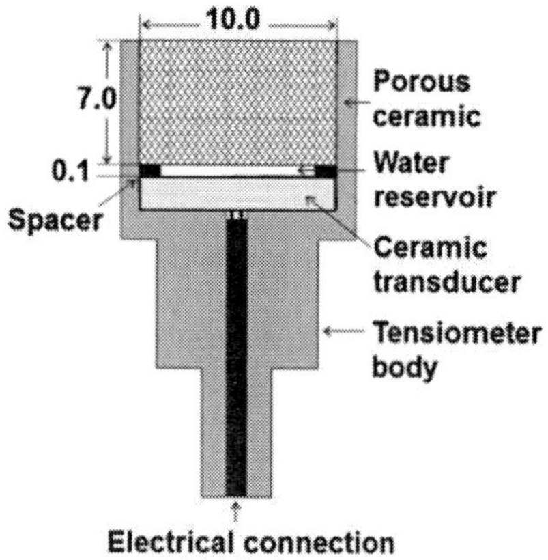

Figure 4. Design of tensiometer used in the study [18].

The great advantage of the high suction probe is its almost immediate response when measuring the suction. Prevention of cavitation is achieved through using a small water reservoir. The maximum negative pore water pressure measured with the use of the tensiometer developed in Durham University was 2.1 MPa [4, 15].

2.3.2 Stage drying and wetting approach

The first approach used for determination of SWRC using the high suction probe was a stage drying and wetting procedure. The samples were prepared as compacted, by using light Proctor tests at the initial water content of 25%. The samples were trimmed to the desired dimensions of 20 mm height and 50 mm diameter. The sample was then placed on a pedestal with the tensiometer attached at the bottom of the supporting plate and sealed with an O-ring [4]. To ensure permanent contact with the device the sample was gently pressed down. Then the sample was covered by a sealed container and left to reach equilibrium of suction reading. After recording the suction value the sample was removed from the pedestal, and the measurements such as weight and dimension changes were taken manually. In the next stage, depending on the hydraulic path determination, the sample was dried (open to atmosphere) or wetted (by using a syringe to add water) to reach the new desired water content. Then the sample was wrapped in cling film, closed in an air tight container and left for 24h to reach water distribution equilibrium, then again the suction was measured. The procedure was repeated for a number of water contents from 25% to 16 %. The suction measurements were taken in stages.

2.3.3 Continuous drying and wetting approach

In contrast to the stage approach, for the continuous method all the measurements such as mass, volume and suction changes are taken continuously. All the readings are digitally recorded by the data acquisition system supplied with TRIAX software [14]. The major component of the system is a PCV framework with high suction probe fixed at the bottom of the supporting plate, and six local small strain transducers (LVDT) attached to each side of the frame. Two of them are vertically placed at the top of the specimen and 4 at the side walls in the horizontal direction (Figure 5). The entire system rests on the digital balance.

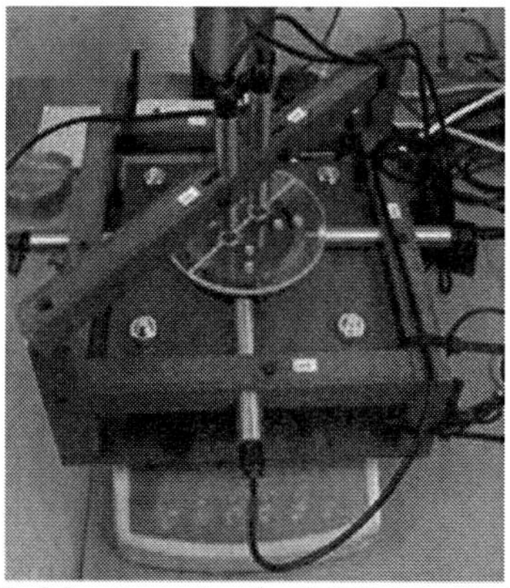

Figure 5. The laboratory set up for SWRC determination applying continuous hydraulic path changes (with tensiometer attached at the bottom).

The system is also equipped with a portable, split cover to avoid too rapid evaporation or drying of a sample during the measurement. The sample preparation procedure is the same as the one proposed for stage approach, however, the diameter was larger (20/75 mm). When the hydraulic path followed is drying the sample is simply left on the pedestal and enclosed within the split

DOI: 10.1051/e3sconf/20160911008

cover. When the procedure is reversed for the wetting path, water drops are applied by using a syringe through holes drilled in the Perspex cover, as shown on Figure 5.

3 Results and discussion

3.1 Comparison of SWRC testing methods

The results presented in the paper focus on determination and evaluation of SWRC obtained by the pressure plate system (applying axis translation technique) and direct suction measurement by using high suction probe. The results are plotted as a function of gravimetric water content (commonly used by engineers) and suction (Figure 6), but also the SWRCs are presented in terms of changes in volumetric water content (Figure 7) and degree of saturation (Figure 8). This is to give a wider picture of the hydraulic behaviour of the tested sandy clay material.

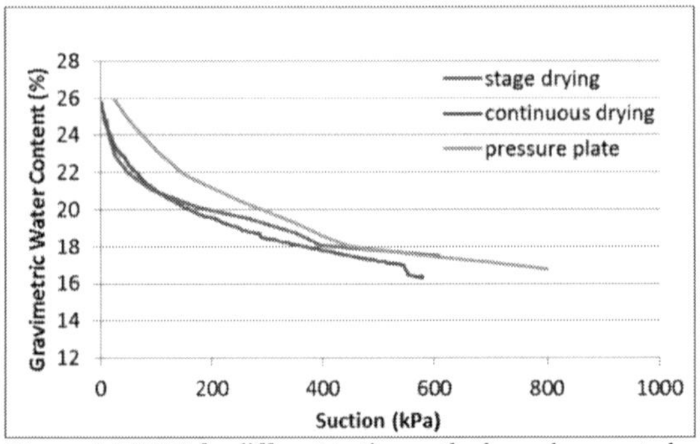

Figure 6. SWRC for different testing methods used, presented as gravimetric water content against suction.

From the graph (Figure 6) it can be seen that the suction measurements for the stage and continuous drying match each other, especially at the initial stage. The line inclination for the continuous approach is more consistent, which could be the effect of simultaneous reading of mass losses with suction increase. The readings could also be influenced by the differences in sample dimensions. The second graph (Figure 7) presents SWRC as a function of volumetric water content.

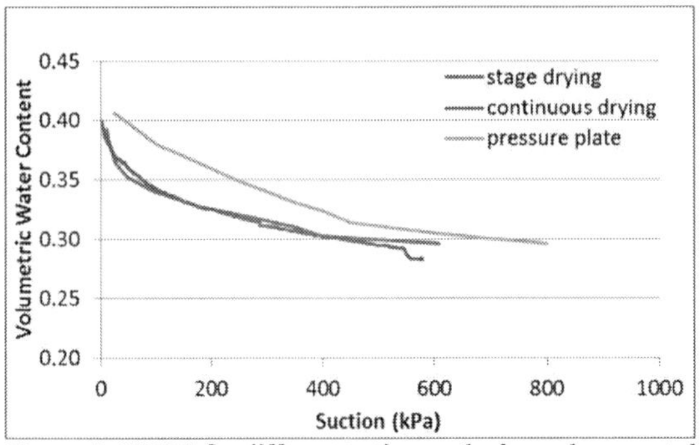

Figure 7. SWRC for different testing methods used, presented as volumetric water content against suction.

When calculating the volumetric water content, the changes in density of the sample is taken into account, From this graph it can be noticed that the stage and continuous approach shows even better agreement, confirming the correctness of the results obtained. The third plot (Figure 8) presents the relationship between suction and degree of saturation (Sr).

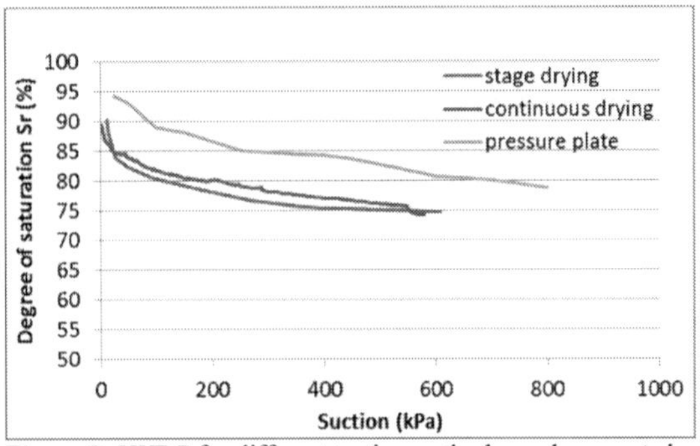

Figure 8. SWRC for different testing methods used, presented as degree of saturation against suction.

As for the previous figures there is again a quite good alignment for suction probe measurements (stage and continuous drying). However, it can be seen that the pressure plate test specimen started from a higher initial degree of saturation (94% rather than 90%).

There is one main conclusion deriving from those results. It is clear that the pressure plate results plot above the lines recorded for tensiometer readings. An explanation put forward by Toll et al. [18] was that cavitation was prevented in the pressure plate tests (using axis translation) whereas cavitation could take place during natural drying. However, Toll et al. [19] has shown that this discrepancy is due to different volumetric paths followed in each test.

3.2 Continuous hydraulic paths for silty clay

The next stage of the research was to investigate how the soil will respond when subjected to repeated continuous wetting and drying cycles (Figure 9).

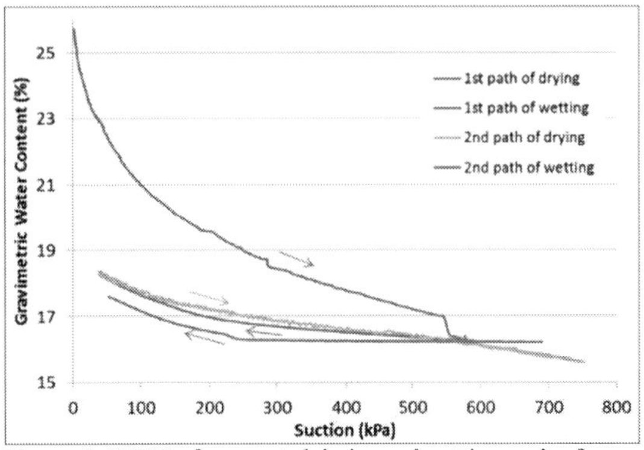

Figure 9. SWRCs for repeated drying and wetting cycles for sandy clay, using tensiometer measurements (continuous approach).

For that purpose the sample was prepared and subjected to drying as described earlier. When a certain value of suction was achieved then a continuous wetting path was commenced. The specimen was subjected to four hydraulic paths as presented on Figure 9. The graph shows that for each full cycle (drying and wetting) the hysteretic behaviour is quite different. For the first cycle the difference between the primary drying curve and the scanning (wetting) curve is much larger than for the second one. It shows that the three last paths lie very close to each other. Another practical observation was that the continuous approach was much less time consuming than the stage approach and the pressure plate method. A single path could be accomplished in 2 days whereas for pressure plate method it could be weeks.

4 Conclusions and remarks

The present study describes techniques allowing precise and reliable suction measurements. Two methods were compared and discussed. The pressure plate method which employs axis translation technique can be successfully used, however it is seen that this shows differences to the tensiometer method.

It was confirmed that the continuous tensiometer approach is a very efficient and effective method of measuring the water retention behaviour. It shows good agreement with the stage approach which could take 10 times longer to finish the complete test.

Another important conclusion deriving from the study is that the wetting and drying history have a significant impact on the hydraulic behaviour of tested material. It is a common knowledge that the wetting and drying paths for soil are not reversible. However, it is shown that the major differences occur in the first cycle and smaller differences are seen in subsequent scanning curve paths.

Acknowledgments

The authors would like to acknowledge EU COST Action TU1202 "Impact of climate change on engineered slopes for infrastructure" and iSMART project "Infrastructure slopes sustainable management and resilience assessment" (EPSRC EP/K027050/1), for the financial support of the study. The work described was carried out in the Unsaturated Soil laboratory at Durham University.

References

1. P.N. Hughes, S. Glendinning, J. Mendes, G. Parkin, D.G. Toll, D. Gallipoli, P. Miller. 2009. Full-scale testing to assess climate effects on embankments. *Special Issue of Engineering Sustainability: Proc. Ins. of C. Eng*, **162, 2**, 67-79 (2009)
2. *S. Au. Rain-Induced Slope Instability in Hong Kong, Eng. Geo, 51, 1, 1-36 (1998)*
3. P. Osinski, R.J. Rickson, M.J. Hann, E. Koda. Assessment of slope stability influenced by vegetation cover and additional loads applied. *Ann. Warsaw Univ. of Life Sci. – SGGW, Land Reclam.* **46, 2**, 81-91 (2014)
4. S.D.N. Lourenço. Suction Measurements and Water Retention in Unsaturated Soils, *PhD Thesis, Durham University* (2008)
5. D.G. Fredlund, H. Rahardjo. Soil mechanics for unsaturated soils. *New York: Wiley* (1993)
6. J. Mendes. Assessment of the impact of climate change on an instrumented embankment: an unsaturated soil mechanics approach. *PhD Thesis, Durham University* (2011)
7. D.G. Fredlund. Use of the soil-water characteristic curve in the implementation of unsaturated soil mechanics. *Proc. Third Int. Conf. on Unsaturated Soils, UNSAT 2002, Recife, Brazil.* **3** (2002)
8. D.G. Toll, M.S. Rahim, M. Karthikeyan, I. Tsaparas. Soil atmosphere interactions for analysing slopes in tropical soils, *Proc. 14th Int. Conf. IACMAG, Kyoto Japan* (eds. F. Oka, A. Murakami, R. Uzuoka), *London: Taylor & Francis*, 1333-1338 (2014)
9. S.J. Wheeler, R.J. Sharma, M.S.R Buisson. Coupling of hydraulic hysteresis and stress–strain behaviour in unsaturated soils. *Geotechnique* **53, 1**, 41-54 (2003)
10. E. Koda, P. Osiński. Assessment of bio-mechanical reinforcement materials influencing slope stability, based on numerical analyses. *Proc. of the 18th Int. Conf. on Soil Mech. and Geotech. Eng. Press des Pontes, Paris.* **3**, 2525-2528 (2013)
11. Z. Skutnik, M. Biliniak. Suction controlled triaxial testing of unsaturated medium sand and sandy clay. *Ann. Warsaw Univ. of Life Sci. – SGGW, Land Reclam.* **46, 4**, 301-307 (2014).
12. P.N. Hughes, S. Glendinning, D.G. Toll. Report on the design and construction of the BIONICS embankment. University of Newcastle upon Tyne – Durham University (2005)
13. British Standard Institute, BS 1377: Methods of test for Soils of civil engineering purposes. Part 1-4, BSI, Milton Keynes (1990)
14. D.G. Toll. A data acquisition and control system for geotechnical testing, in Computing Developments in Civil and Structural Engineering (eds. B. Kumar and B.H.V. Topping), *Edinburgh: Civil-Comp Press*, 237-242 (1999)
15. S.D.N. Lourenço, D. Gallipoli, D.G. Toll, C.E. Augarde, F.A. Evans. A new procedure for the determination of the Soil Water Retention Curves by continuous drying using high suction tensiometers. *Can. Geo. J.*, **48, 2**, 327-335 (2011)
16. A.M. Ridley, J.B. Burland. A new instrument for the measurement of soil moisture suction. *Geotechnique* **43, 2**, 321–324 (1993)
17. A. Tarantino, L. Mongiovì. Calibration of tensiometer for direct measurement of matric suction *Geotechnique*, **53, 1**, 137–141 (2003)
18. D.G. Toll, S.D.N. Lourenço, J. Mendes. Advances in suction measurements using high suction tensiometers, *Eng. Geo*, **165**, 29–37 (2013)
19. D.G. Toll, J.D. Asquith, A. Fraser, A.A. Hassan, G. Liu, S.D.N. Lourenço, J. Mendes, T. Noguchi, P. Osinski, R.A. Stirling, Tensiometer techniques for determining soil water retention curves. *Proc. 6th Asia-Pacific Conference on Unsaturated Soil.* Guilin, China. *London: Taylor & Francis* (2015).

Classification, soil-water characteristic curve and swelling/collapse behaviour of the Nicosia marl, Cyprus

Dimitrios Loukidis[1,a], Michael Bardanis[2] and Georgia Lazarou[1]

[1]*University of Cyprus,Civil & Environmental Engineering Department, 75 Kallipoleos st., P.O. Box 20537, 1678 Nicosia, Cyprus*
[2]*Edafos Engineering Consultants S.A., Laboratory, 2 Therapion st., 10444 Athens, Greece*

Abstract. The city of Nicosia in the island of Cyprus is situated on a sedimentary basin, the surface layers of which are dominated by an expansive clayey soil called Nicosia marl. The high potential for volume change of this soil due to the seasonal changes of water content is the leading cause of foundation related damage to the buildings and infrastructure of the city. This paper presents preliminary results of a study focusing on the investigation of the Nicosia marl mechanical characteristics pertaining to moisture changes. Samples of Nicosia marl were collected from a borehole and a trial pit in the University of Cyprus campus. The pit which was subsequently used for installing an in-situ array of sensors monitoring suction and volumetric water content at depths inside the active zone. The Soil-Water Characteristic Curve (SWCC) was measured in the laboratory for the entire range of degree of saturation. The potential of swelling/collapse at various values of vertical stress was studied through a series of oedometer tests.

1 Introduction

Expansive soils are responsible for damage to buildings and other civil engineering works, costing tens of billion dollars annually worldwide. Expansive clays cover wide areas of the island of Cyprus. They can be found throughout the Mesaoria basin between the Troodos and Kerynia mountain ranges and along the southern coastal region. Cyprus' climate is characterized by hot, dry summers and rainy winters. Buildings founded on the expansive soils of the island develop various types and degrees of damage due to expansion upon wetting during the winter months or due to water leakage from utility lines, as well as due to shrinkage in the summer months or due to transpiration caused by vegetation roots. In less severe cases, this damage may be limited to cracks on walls and other nuisance of architectural nature. In more extreme cases, soil expansion may even lead to failure of structural members.

This paper presents a preliminary experimental study of the physical and mechanical characteristics of the Nicosia marl, which is outcropping in most of the part of the central Mesaoria basin. The study focuses mainly on the properties pertaining to the volume expansion/ shrinkage associated to moisture changes, such as the soil-water characteristic (SWCC) curve and the swelling/collapse vs. vertical stress curve.

2 Geological setting

The Nicosia marl is a calcareous fine-grained soil belonging to the Nicosia geological formation, which has a thickness of several hundreds of meters. Nicosia marl is a marine sediment deposited during the Pliocene, when the Mesaoria basin was covered by a shallow sea. The Nicosia marl geological unit can be divided in two parts: the older grey marl and the more recent, overlying brown (khaki) marl [1]. The grey marl is the least expansive of the two. This study deals solely with the brown marl. The Nicosia marl is expansive because, although substantially calcareous, its montmorillonite content is significant and in certain parts of Nicosia reaches 25%.

The location of the study is the new University of Cyprus (UCY) campus in the southeastern outskirts of Nicosia. The area is in the middle of the island of Cyprus, at a minimum distance of 20km from the sea in any direction. Soil samples were collected from one borehole (drilled in May 2015) and one trial pit (excavated in July 2015) in a narrow zone where the Nicosia marl outcrops (excluding any thin top soil cover). Fig. 1 shows the surface geology of the UCY campus and the locations of the borehole (BH1) and pit (P1), which are 70m apart and at an elevation of 137m. The northern part of the campus is a terrace, where the ground elevation steps to 140m and the marl is covered by gravel cemented in a clay matrix (fanglomerate). In the southern part, the marl is covered by recent alluvial deposits (clayey sandy silt).

Fig. 2 shows the soil stratigraphy in the investigated area. The brown marl in the upper part of the profile is more sandy and stiffer than the marl found deeper than 5.5m. This is due to the higher percentage of calcium carbonate ($CaCO_3$) rather than overconsolidation or dessication. In fact, the $CaCO_3$ content determined using Bernard calcimeter [2], is 46% for the samples taken

[a] Corresponding author: loukidis@ucy.ac.cy

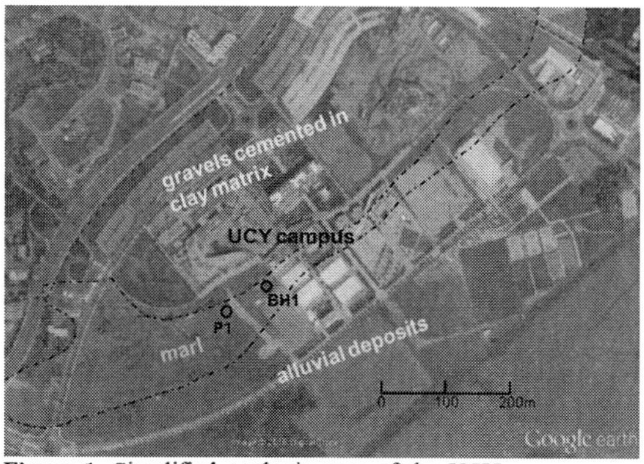

Figure 1. Simplified geologic map of the UCY campus area showing locations of sampling points (P1 & BH1).

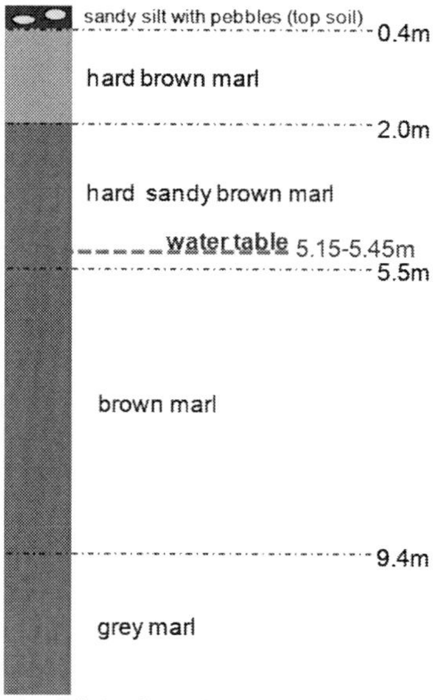

Figure 2. Simplified soil stratigraphy based on pit and borehole data.

from the trial pit (depth range 0.85 to 1.5m) and 43% at 4.5m depth (borehole sample), while it is only 35% at 6m depth (borehole sample). The underlying grey marl was encountered from 9.4m depth to the bottom of the borehole at 36.5m depth. Measurements from an open-type piezometer installed in BH1 show that the water table depth varies seasonally from 5.15m to 5.45m, which roughly corresponds to an elevation consistent with a nearby creek further south (300m away).

3 The recent experimental programme

3.1 Trial pit sampling and sensor installation

One trial pit approximately 1.5m deep and 2m wide on either side was excavated in July 2015. The purpose of the trial pit was to obtain a clear view of the freshly excavated Nicosia marl, collect samples from the sides and the bottom of the pit, and install volumetric water content measurement sensors and porous blocks for suction measurement, for permanent monitoring of the seasonal fluctuations. The ground surface at the location of the pit is generally flat, sloping about 2° to the south. Up to approximately 40cm of top soil consisting of sandy silt with pebbles was found overlying the upper stiff brown Nicosia marl at the location of P1 (Fig. 3).

Figure 3. Western side of the trial pit excavated during July 2015, after installation of the sensors and before backfilling.

The samples taken were trimmed samples from the sides and bottom of the trial pit, and an intermediate bench created at 0.85cm depth. These samples were protected by wrapping in cling film and plastic bags. In addition, undisturbed samples were taken using steel tube samplers, the ends of which were covered with molten paraffin wax. The sensors installed were the GS3 frequency domain reflectometry (FDR) volumetric water content sensor and the MPS-6 porous block FDR sensor measuring matric suction, both by Decagon Devices, Inc. A pair of sensors of each type were installed at a depth of 0.75m and another one at 1.5m. Fig. 3 shows the western side of the trial pit after installation of the sensors, before backfilling and compacting the pit. Care was taken in using for backfill as much as possible of the excavated material, targeting in having final density in the backfill similar to that of the undisturbed marl. The installed sensors, which are connected to an automatic Em50 data logger (Fig. 4) and measure continuously, are expected to shed light in the future into the seasonal variation of suction and volumetric water content in the Nicosia marl inside the active zone.

3.2 Index properties and classification

The properties from soil classification tests are summarised in Table 1. Nicosia marl at shallow depths (trial pit) was found in the form of a light brown very stiff material. USCS classification varied between CL and CH, i.e. clay of either low or high plasticity. On average, the liquid limit was 47.6%, the plasticity index (I_p) 23.2

Figure 4. Trial pit backfilled and fenced to protect the sensor data acquisition box.

Table 1. Index properties and classification of Nicosia marl from samples taken during the 2015 geotechnical investigation.

Sample	clay (%)	silt (%)	w_L (%)	I_P	G_s	USCS
1 (pit)	20.3	62.4	53.9	28.8	2.69	CH
2 (pit)	20.5	62.7	52.3	25.4	2.68	CH
3 (pit)	24.5	62.5	44.9	19.4	2.66	CL
4 (pit)	16.3	61.6	39.2	19.1	2.72	CL
BH1 6m	30	60	65.6	35.5	2.68	CH

and the specific gravity G_s 2.685. The water content ranged from 10% to 14% and the dry unit weight from 14.7 to 16.0kN/m^3. Particle size distribution tests showed that, on average, the clay-size fraction was 20.4%, silt-size fraction 62.3%, and fine sand 17.3% (only traces of medium and coarse sand were measured). The sand content of the borehole samples of brown marl below 5.5m depth is even smaller. Organic content was present in all trial pit samples, on average found to be 2.4%. The latter is not considered to be a generally observed property of the Nicosia marl formation, but it certainly exists in samples coming from shallow depths as in the case of the trial pit. One linear shrinkage test was performed on Sample 1 yielding a value of 12.6%.

Based on the data shown in Table 1, the potential expansiveness of the shallow hard brown marl samples from the trial pit ranges from medium to high according to the South African chart [3, 4]. On the other hand, the softer brown marl found at 6m depth is categorized as very highly expansive based on the same chart. The profile of degree of saturation S_r in the summer of 2015 can be seen in Fig. 5, plotted using data from both the trial pit and the borehole. The marl was found to be fully saturated at 3m depth, 1.45m above the water table. In the depth range of the trial pit, the marl has a degree of

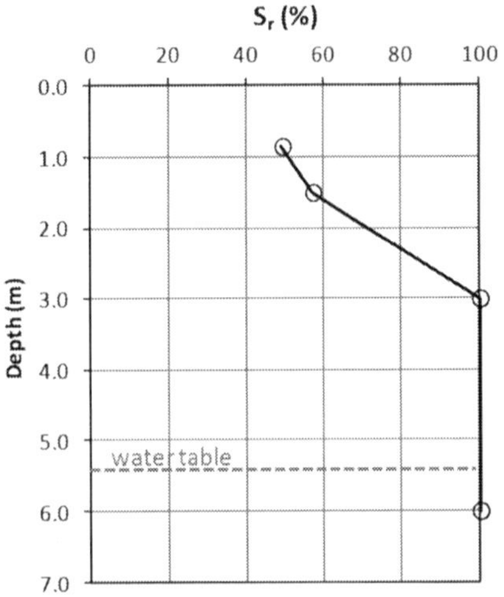

Figure 5. Profile of degree of saturation in July 2015.

saturation around 50%. Hence, the active zone in the region of the study appears to be 2 to 3m.

3.3 Hydraulic conductivity

The hydraulic conductivity of two Nicosia marl samples, one from the bench of the trial pit (0.85m depth) and one from the borehole at 6m depth was determined in a triaxial cell (flexible wall permeameter) [5]. The specimens (3.5cm diameter, 7cm height) were first back pressure saturated (a back pressure of 800kPa was needed to reach B-values larger than 0.95) and then consolidated to a mean effective stress p'=120kPa for the pit sample and 90kPa for the borehole sample, before applying various hydraulic gradients ranging from 15 to 31. The coefficient of hydraulic conductivity *k* of the shallow hard marl (pit) was 2.6×10^{-7} cm/s, while for the softer marl at depth turned to be smaller and equal to 9.8×10^{-8} cm/s, despite having an initial void ratio of 0.79 (γ_d=14.7kN/m^3), much higher than that of the hard marl (e=0.67). This is due to the fact that the shallow hard marl is less plastic and less clayey and has experienced desiccation cycles in its geological history.

3.4 One-dimensional compression

An one-dimensional compression test was performed on a trial pit sample in a conventional 15kN capacity pneumatic oedometer, but with a modified 30mm diameter cell. The height of the sample was 15mm in order to keep a diameter to height ratio of 2. The cell has a fixed ring of smooth inner surface slightly lubricated prior to the test. The one-dimensional compression curve for the particular sample (w_L=53.9%/I_p=28.8) is shown in Fig. 6 along with the Casagrande graphical method for yield stress determination. The yield stress according to this method was found to be 3.3 MPa, indeed a very high value even if one takes into account the cementation of the particular marl and its initial partial saturation at the

 DOI: 10.1051/e3sconf/20160911009

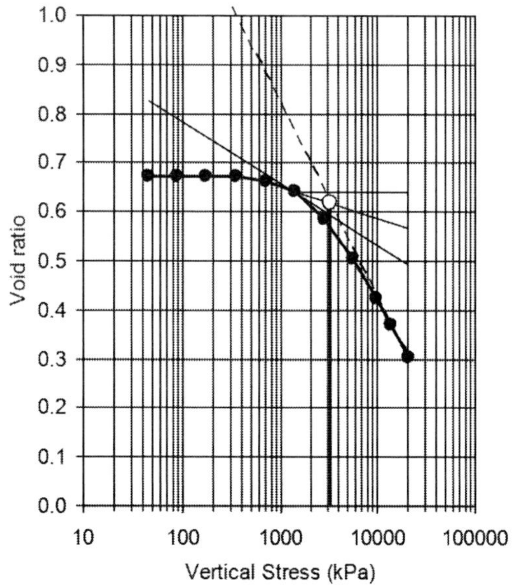

Figure 6. One-dimensional compression curve of hard Nicosia marl (pit sample) up to 21.2 MPa of vertical stress with the Casagrande graphical method for yield stress estimation.

beginning of the test. A compression index value of 0.39 was obtained from the tangent of the post-yield linear part of the one-dimensional curve in the semi-logarithmic plot between 9.9 and 21.2MPa.

3.5 Swelling/collapse under one-dimensional conditions

A series of swelling/collapse tests were performed using conventional 15kN capacity pneumatic oedometers. Specimens were trimmed from the larger samples obtained from the trial pit, placed on the oedometers and subjected to various values of constant vertical stress. Once the observed deformations had ceased the samples were inundated and the deformation due to inundation were measured. The strains calculated from these tests are plotted against vertical stress in Fig. 7. The vertical stress corresponding to zero deformation upon inundation was 85kPa, a relatively small value that in the particular occasion could not have caused any alarm in foundation engineering practice, although it is associated with significant swelling strain for lower vertical stress, such as those applied by mat foundations, and similarly significant collapse strain for higher vertical stress.

Swelling/collapse tests were also performed on specimens of compacted brown marl obtained from 6m depth, which is more plastic and expansive than the marl encountered in the trial pit. The specimens were compacted at various values of initial dry unit weight γ_d and degree of saturation S_r. Fig. 8 shows the result of this series of tests, grouped in ranges of γ_d and S_r. As observed in previous studies [6,7], the tendency of expansion increases with increasing initial γ_d and decreasing initial S_r. Moreover, the expansiveness seems to be much more sensitive to the initial dry unit weight than the degree of saturation.

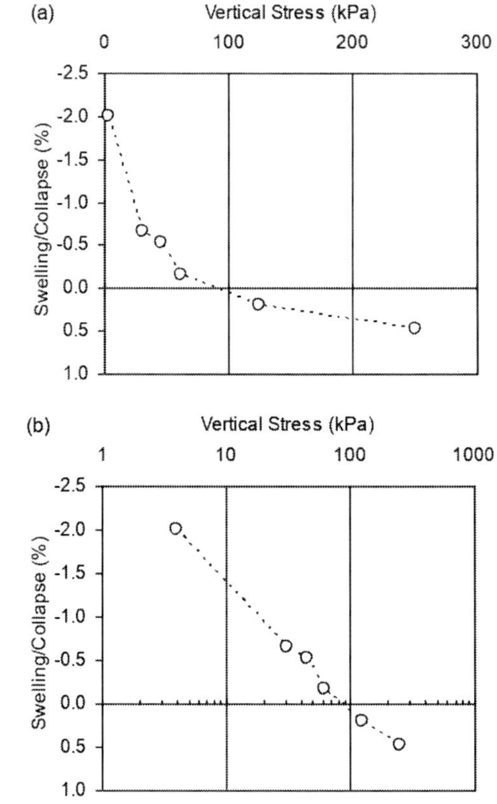

Figure 7. Strain due to inundation versus vertical stress for hard Nicosia marl (pit samples) as a function of applied constant vertical stress in a) linear scale, and b) logarithmic scale.

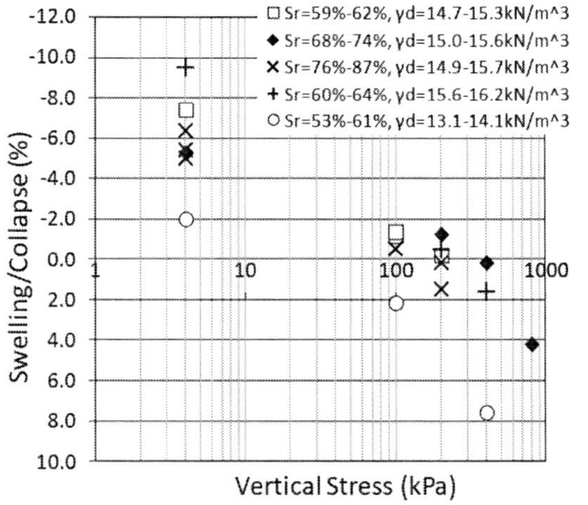

Figure 8. Strain due to inundation versus vertical stress for compacted brown Nicosia Marl sampled from 6m depth as a function of vertical stress.

For the initial states tested in this series, the vertical stress corresponding to zero deformation upon inundation ranges from 12kPa to 200kPa and the "free" swelling strain (swelling at vertical stress of 4kPa) from 2% to 9.5%. As in Fig. 7b, the swelling/collapse versus vertical stress loci could be approximated by straight lines, which appear to have roughly the same inclination, provided that the horizontal axis is in logarithmic scale.

3.6 Soil-water characteristic curve (SWCC)

Suction was measured on the sides of the trial pit and in the samples taken and brought to the laboratory. The use of Soilmoisture Equipment Corp. Quickdraw field tensiometer and 2100F laboratory tensiometer revealed suction values higher than 90kPa; in fact much higher given the rate of desaturation of the porous ceramic tips of both instruments. The Decagon Devices, Inc. chilled mirror hygrometer was therefore used and total suctions between 5.4 and 8.7MPa were measured (Fig. 9). This was a remarkable finding corresponding to very high values of suction (even if the anticipated osmotic component were subtracted from the values measured with the chilled mirror hygrometer which measures total suction). The trial pit was opened in July 2015, a very dry and hot season in the area. The block samples taken were protected by cling film and plastic bags and tube samples were sealed with molten paraffin wax. Although there was concern at the beginning that the samples covered only with cling film and plastic bags may dry until they are brought in the laboratory, the highest suction was actually measured in an undisturbed sample from tube sampler and covered with paraffin wax, indicating thus that the values measured were little affected by drying.

After initial suction measurements, several samples were left to dry in order to measure the rest of the drying portion of the SWCC, and others were gradually wetted in order to measure (at least part of) the wetting portion of the soil-water characteristic curve from the particular initial conditions. Drying was applied by leaving samples in the atmospheric conditions of the laboratory, then sealing them for equilibrium of moisture conditions throughout the samples, and then unsealing them for suction measurement using the chilled mirror hygrometer and total volume measurement using immersion in molten paraffin wax. For comparison, some samples were dried by placing them in controlled relative humidity chambers using saturated salt solutions [8,9] of $BaCl_2$, KCl, NaCl, $Mg(NO_3)_2$ and $MgCl_2$. No difference was observed between the two methods; the salt solutions method allows though for targeted and more controlled values of applied total suction.

The soil-water characteristic curves obtained are shown in Fig. 9 and the shrinkage curves in Fig. 10. The combination of suction, water content, degree of saturation and void ratio position the material, in its initial state, close to the residual void ratio branch of the shrinkage curve (Fig. 10a) and the soil-water characteristic curve expressed in terms of void ratio versus suction (Fig. 9c). The material can certainly dry more as exhibited by the drying portions from the natural state of the soil-water characteristic curve, as shown in Fig. 9a,b. Still it is highly unlikely that the material could really shrink more, as exhibited by the drying portions from the natural state of the soil-water characteristic curve as shown in Fig. 9c and 10a. Consequently, the drying portion from the natural state of the shrinkage

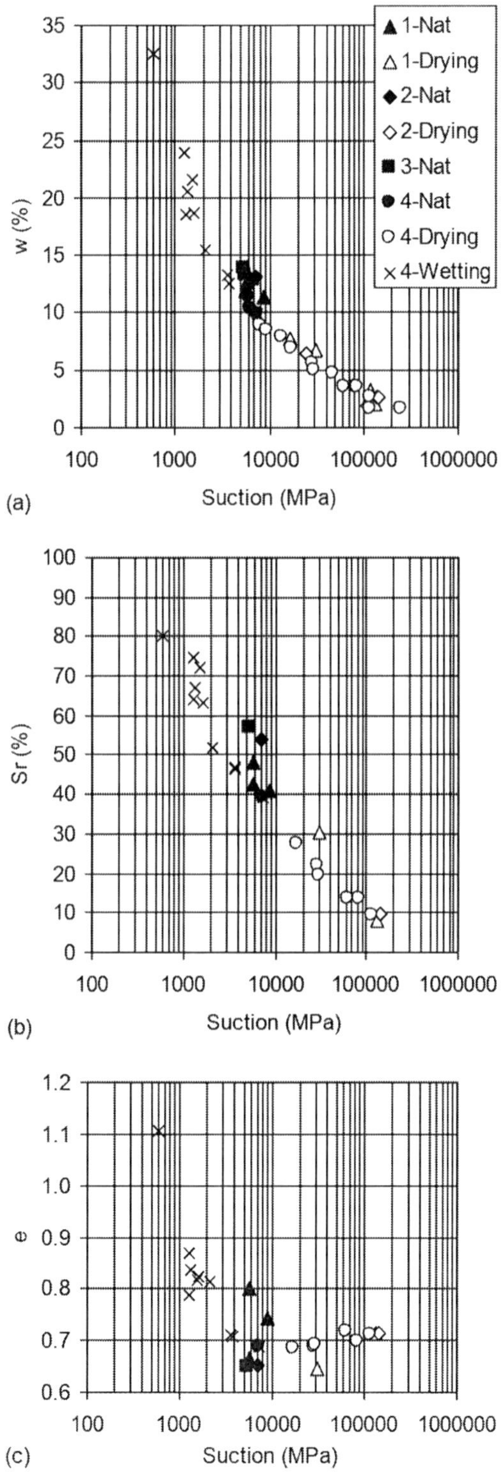

Figure 9. Soil-water characteristic curve of hard Nicosia marl in terms of a) water content, b) degree of saturation, and c) void ratio against suction. Filled points correspond to the natural state, empty points to drying from the natural state, and crosses to wetting from the natural state.

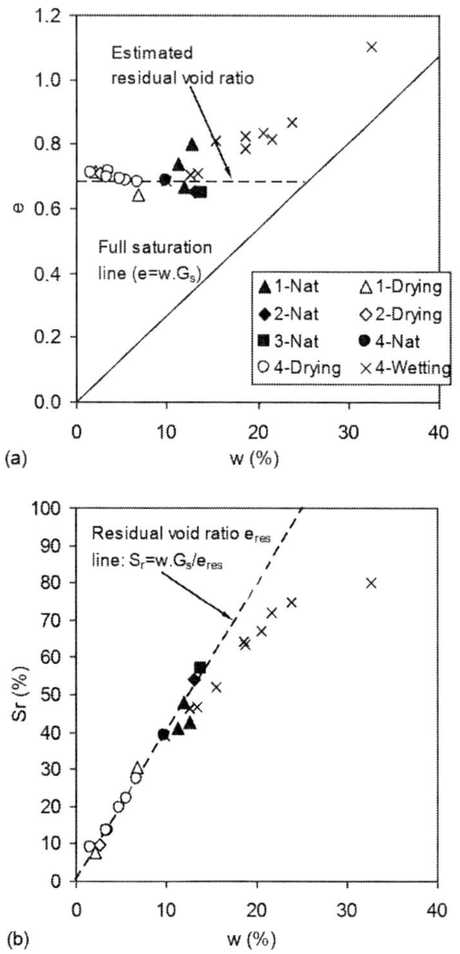

Figure 10. Shrinkage curve of Nicosia Marl in terms of a) void ratio versus water content, and b) degree of saturation versus water content. Filled points correspond to the natural state, empty points to drying from the natural state, and crosses to wetting from the natural state.

curve expressed in terms of degree of saturation versus water content in Fig. 10b lies on the residual void ratio line ($S_r = w{\cdot}G_s/e_{res}$), corresponding to an average residual void ratio, e_{res}, of 0.68 and the average G_s value of 2.685. Given the very high values of suction in the material in its natural state at the time of sampling, the first inflection point of the soil-water characteristic curve was not observed and the second inflection point seems to lie close to the natural state itself (filled points in Fig. 9).

Regarding wetting from the natural state, this was achieved by gradual addition of water to the samples, trimmed from the larger block samples from the trial pit. These samples were again sealed for equilibrium and then unsealed for suction measurement using the chilled mirror hygrometer and total volume measurement using immersion in molten paraffin wax. The material yielded large increases of void ratio (in fact without even complete extinction of suction). It is of interest that even at the highest value of water content during rewetting (32.5%) the soil did not obtain full saturation with the experimental point lying at considerable distance from the full saturation line in Fig. 10a and the S_r=100% line in Fig. 10b. The measured value of total suction using the chilled mirror hygrometer at this moisture condition was 600kPa. Even if we assume very high values of the osmotic suction component, such a high value of total suction cannot correspond to zero matric suction.

4 Conclusions

The findings of a preliminary study on the physical and mechanical characteristics of the expansive Nicosia marl were presented. The soil water characteristic curve and shrinkage curves were determined. A series of oedometer tests were performed in order to explore the potential for swelling/collapse upon inundation as a function of the applied vertical stress.

Total suction values inside the active zone during summer are very high (several MPa in magnitude). The natural state of the Nicosia marl at shallow depths was found to be very close to its residual void ratio. The Nicosia marl has a relatively large potential for expansion or collapse despite a relatively low swelling pressure that seldom exceeds 200kPa. Finally, the potential of expansion for a given soil is more sensitive to the values of initial dry unit weight than to the initial degree of saturation.

Acknowledgements

The authors would like to thank the personnel of the Technical Services of the University of Cyprus for their support in field work and making available previous geotechnical investigation data. EDAFOS S.A. laboratory technicians Mr Ch. Vagenas and Mr M. Tsoukaladakis assisted with testing.

References

1. G. Constantinou, I. Panagides, K. Xenophontos, S. Afrodisis, P. Michaelides, S. Kramvis, *The geology of Cyprus* (Bulletin No. 10, Cyprus Geological Survey, 2002)
2. J. Hulseman, J. of Sedimentary Petrology, **36**, 622-625 (1966)
3. D.H. Van der Merwe, *Proc. 6th Reg. Conf. Africa on Soil Mech. and Found. Engin.*, Vol. 2, 166-167 (1975)
4. A.A. Williams, G.W. Donaldson, *Expansive Soils*, 834-844 (1980)
5. K.H. Head, R.J. Epps, *Manual of Soil Laboratory Testing, Volume III: Effective Stress Tests* (Whittles Publishing, 2014)
6. H. Komine (2004). Engin. Geol., **71**, 265-279 (2004)
7. O. Buzzi, S. Fityus, A. Giacomini. Géotechnique, **61**, 271-277 (2011)
8. D.G. Fredlund. *Proc. 3rd Int. Conf. on Unsat. Soils*, 20-23 (2002)
9. A.M. Tang, Y.J. Cui. Canadian Geot. J., **42**, 287-296 (2005)

AUTHOR INDEX

AUTHOR INDEX

AUTHOR INDEX

AUTHOR INDEX

AUTHOR INDEX

EDP Sciences – Web of Conferences
17, Avenue du Hoggar
Parc d'Activité de Courtabouf
BP 112
F-91944 Les Ulis Cedex A
France

ISBN 978-1-5108-3018-9

9 781510 830189

CURRAN ASSOCIATES INC.
proceedings
.com

3rd European Conference on Unsaturated Soils (E-UNSAT 2016)

E3S Web of Conferences Volume 9 (2016)

Paris, France
12 - 14 September 2016

Part 2 of 2

edp sciences

3rd European Conference on Unsaturated Soils (E-UNSAT 2016)

E3S Web of Conferences Volume 9 (2016)

Paris, France
12 - 14 September 2016

Part 2 of 2

Editors:

P. Delage
Y.-J. Cui
S. Ghabezloo
J.-M. Pereira
A.-M. Tang

ISBN: 978-1-5108-3018-9

Printed from e-media with permission by:

Curran Associates, Inc.
57 Morehouse Lane
Red Hook, NY 12571

Some format issues inherent in the e-media version may also appear in this print version.

Printed by Curran Associates, Inc. (2016)

For additional information, please contact EDP Sciences – Web of Conferences
at the address below.

EDP Sciences – Web of Conferences
17, Avenue du Hoggar
Parc d'Activité de Courtabœuf
BP 112
F-91944 Les Ulis Cedex A
France

Phone: +33 (0) 1 69 18 75 75
Fax: +33 (0) 1 69 28 84 91

contact-edps@webofconferences.org

Additional copies of this publication are available from:

Curran Associates, Inc.
57 Morehouse Lane
Red Hook, NY 12571 USA
Phone: 845-758-0400
Fax: 845-758-2633
Email: curran@proceedings.com
Web: www.proceedings.com

TABLE OF CONTENTS

PART 1

PART 2

DOI: 10.1051/e3sconf/20160911010

A void ratio dependent water retention curve model including hydraulic hysteresis

Amin Y. Pasha [1], Arman Khoshghalb [1] and Nasser Khalili [1,a]

[1] *School of Civil and Environmental Engineering, UNSW Australia, Sydney, NSW, 2052, Australia*

Abstract. Past experimental evidence has shown that *Water Retention Curve* (WRC) evolves with mechanical stress and structural changes in soil matrix. Models currently available in the literature for capturing the volume change dependency of WRC are mainly empirical in nature requiring an extensive experimental programme for parameter identification which renders them unsuitable for practical applications. In this paper, an analytical model for the evaluation of the void ratio dependency of WRC in deformable porous media is presented. The approach proposed enables quantification of the dependency of WRC on void ratio solely based on the form of WRC at the reference void ratio and requires no additional parameters. The effect of hydraulic hysteresis on the evolution process is also incorporated in the model, an aspect rarely addressed in the literature. Expressions are presented for the evolution of main and scanning curves due to loading and change in the hydraulic path from scanning to main wetting/drying and vice versa as well as the WRC parameters such as air entry value, air expulsion value, pore size distribution index and slope of the scanning curve. The model is validated using experimental data on compacted and reconstituted soils subjected to various hydro-mechanical paths. Good agreement is obtained between model predictions and experimental data in all the cases considered.

1 Introduction

A crucial step in many constitutive modellings of unsaturated soils is the determination of the *water retention curve* (WRC) or alternatively called *soil water characteristic curve* (SWCC). Conventionally, WRC has been shown in terms of degree of saturation versus suction. Experimental evidence have shown that WRC is affected not only by matric suction but also by void ratio; that is, the hydraulic and mechanical states of the soil both have significant effects on the degree of saturation. Hence, at equilibrium conditions, the interplay of degree of saturation-suction-void ratio of a soil constitutes a state surface [1-3]. The state surface for the main drying path is different from that for the main wetting path, and these two unique surfaces serve as the upper and lower boundaries of all the possible states attainable for a soil.

Numerous relationships have been presented in the literature for the mathematical representation of WRC (e.g. [4-6]). However, a vast majority has been for non-deformable soils ignoring the impact of volume change on WRC [7].

The main contributions to the evolution of WRC with volume change have been due to [1, 8-13]. The available models are empirical in nature and mainly rely on fitting parameters. Excessive experimental time and cost related to the quantification of the embedded parameters in the existing models is a major issue in the process of the application of such models in the geotechnical engineering practice. More importantly, the influence of volumetric deformations on the evolution of the scanning path of WRC, i.e. the hysteresis effect, has been rarely investigated. This is despite the fact that essentially in the majority of cases, either in laboratory investigations on unsaturated soils or in real-life, the loading/unloading of soils are being done on the scanning path of WRC.

In this paper, we present the outline of an analytical model for capturing the evolution of the whole WRC hysteretic loop. The model has been developed based on the theoretical work of [14] that requires a few input parameters with clear physical meanings and more importantly the hysteresis effect in water retention behavior is incorporated. Particular attention is given to the scanning behaviour of the soil and the dependency of pore size distribution to void ratio. The approach proposed is general in nature and allows prediction of the shift in WRC with volume change entirely based on the form of WRC at a reference void ratio with no additional parameters required. The applicability of the model to experimental data is demonstrated.

2 Outline of the model

The model is developed based on the fundamental contribution of [14], which relates the water volumetric strain ($\dot{\varepsilon}_{vw} = \frac{dV_w}{V}$ where V_w and V are water volume and

[a] Corresponding author: n.khalili@unsw.edu.au

total volume of soil, respectively) to the total volumetric strain ($\dot{\varepsilon}_{vw} = \frac{dV}{V} = \frac{dV_v}{V}$ where V_v is the volume of voids) in constant suction loading conditions through the effective stress parameter in incremental from, ψ , as

$$\dot{\varepsilon}_{vw} = \psi \dot{\varepsilon}_v \quad (1)$$

Using the definition of the degree of saturation combined with (1) gives (assuming solids are incompressible)

$$\dot{S}_r = (\psi - S_r)\frac{\dot{e}}{e} \quad (2)$$

where e is void ratio. ψ can be calculated in (1) and (2) from

$$\psi = \frac{\partial(\chi s)}{\partial s} = \left(1 - \frac{\partial ln\chi}{\partial lns}\right)\chi \quad (3)$$

where χ is the effective stress parameter in total form which describes the contribution of suction to the effective stress of the soil solid skeleton. Note that in (3), $\frac{\partial ln\chi}{\partial lns}$ is taken as positive when χ decreases with an increase in suction. Earlier definitions of the effective stress parameter assumed direct correlation with the degree of saturation, $\chi = S_r$ [15, 16]. However, it was shown in many recent works (e.g. [17-20]) that only when the work of air-water interface is ignored, the effective stress parameter may be taken as the degree of saturation. Based on back-calculation of unsaturated shear test data for many soil types, Khalili and Khabbaz [21] proposed an equation for the effective stress parameter. In this work, The hysteretic effective stress parameter, χ, is quantified based on the contributions of [21-23].

For main wetting and drying paths

$$\chi = \begin{cases} 1, & s < s_e \\ \left(\frac{s_e}{s}\right)^{\Omega}, & s \geq s_e \end{cases} \quad (4\text{-}1)$$

For drying path reversals

$$\chi = \left(\frac{s_{ae}}{s_{rd}}\right)^{\Omega}\left(\frac{s_{rd}}{s}\right)^{\zeta}, \quad \left(\frac{s_{ex}}{s_{ae}}\right)^{\frac{\Omega}{\Omega-\zeta}} s_{rd} \leq s \leq s_{rd} \quad (4\text{-}2)$$

For wetting path reversal

$$\chi = \left(\frac{s_{ex}}{s_{rw}}\right)^{\Omega}\left(\frac{s_{rw}}{s}\right)^{\zeta}, \quad s_{rw} \leq s \leq \left(\frac{s_{ae}}{s_{ex}}\right)^{\frac{\Omega}{\Omega-\zeta}} s_{rw} \quad (4\text{-}3)$$

where the air entry, s_{ae}, and air expulsion, s_{ex}, values denote the points of transition from saturation to unsaturation and vice versa along the main drying and main wetting paths, respectively. Ω is the material parameter, with the best fit value of 0.55, ζ is the slope of the transition path between the main drying and wetting paths in a lnχ~ln$s$ plane. s_{rd} and s_{rw} are the points of suction reversal on the main drying and main wetting paths, respectively. A schematic representation of (4) is shown in Figure 1.

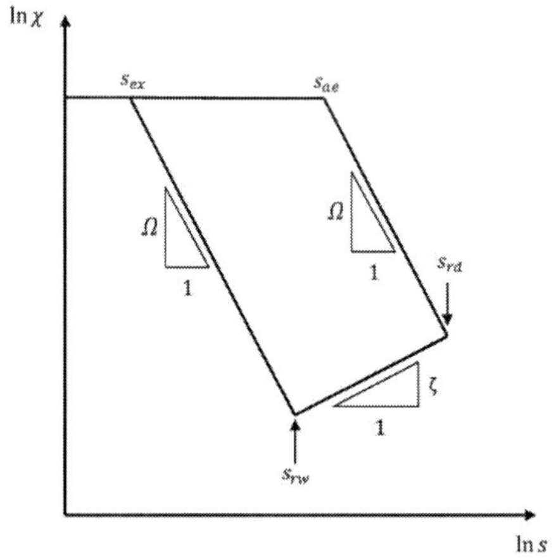

Figure 1. Effective stress parameter in terms of suction including hydraulic hysteresis

The step by step procedure in the model for quantifying the evolved form of the hysteretic WRC for a change in void ratio starting from a reference state is described below (all updated parameters are denoted by asterisk). All WRC parameters used in the model and the evolution with void ratio in $lnS_r - lns - e$ space is shown in Figure 2. The details of the derivations can be found in [24].

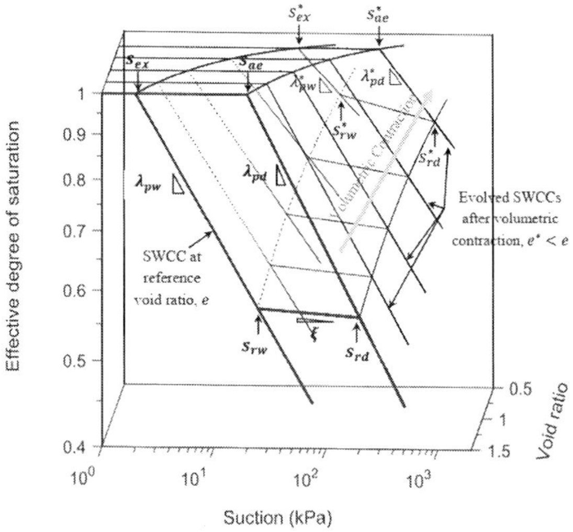

Figure 2. Evolution of water retention curve with void ratio (parameters at the evolved states are denoted with asterisk)

STEP 1. Evaluate the input parameters of the model including the WRC parameters at the reference void ratio, s_{ae}, s_{ex}, λ_{pd}, λ_{pw}, and ξ (see Figure 2), and the

DOI: 10.1051/e3sconf/20160911010

effective stress parameters, Ω and ζ (see Figure 1). Since compression is related to wetting and expansion is related to drying, for simulations including volumetric contraction, only use s_{ex0} and λ_{pw0}, and for simulations including volumetric extension, only use s_{ae0} and λ_{pd0}. The parameters associated with the WRC at the reference void ratio can be obtained by fitting the Brooks and Corey model [4] model extended by [14] to include hydraulic hysteresis effects as follows to the WRC at the reference void ratio.

For main wetting and drying paths

$$S_{eff} = \begin{cases} 1, & s < s_e \\ \left(\frac{s_e}{s}\right)^{\lambda_p}, & s \geq s_e \end{cases} \tag{5-1}$$

For drying path reversals

$$S_{eff} = \left(\frac{s_{ae}}{s_{rd}}\right)^{\lambda_p} \left(\frac{s_{rd}}{s}\right)^{\xi}, \quad \left(\frac{s_{ex}}{s_{ae}}\right)^{\frac{\lambda_p}{\lambda_p-\xi}} s_{rd} \leq s \leq s_{rd} \tag{5-2}$$

For wetting path reversal

$$S_{eff} = \left(\frac{s_{ex}}{s_{rw}}\right)^{\lambda_p} \left(\frac{s_{rw}}{s}\right)^{\xi}, \quad s_{rw} \leq s \leq \left(\frac{s_{ae}}{s_{ex}}\right)^{\frac{\lambda_p}{\lambda_p-\xi}} s_{rw} \tag{5-3}$$

where S_{eff} = effective degree of saturation, is the normalised form of the total degree of saturation, S_r, with respect to residual degree of saturation, S_{res}, $(S_{eff} = \frac{S_r - S_{res}}{1 - S_{res}})$.

STEP 2. Update the air entry value, s_{ae}, or air expulsion value, s_{ex}, for an infinitesimal change in void ratio, $\dot{e}$, using

$$s_e^* = s_e \left(1 - \frac{\Omega}{(1-S_{res})\lambda_p} \frac{\dot{e}}{e}\right) \tag{6}$$

where s_e is either air entry or air expulsion value.

STEP 3. Update the pore size distribution index pertaining to main drying, λ_{pd}, and/or main wetting, λ_{pw}, from

$$\lambda_p^* = \lambda_p [1 - \frac{3\left((1-\Omega)\left(2^{(1-\Omega/\lambda_p)}-1\right)-Sres\right)}{2(1-S_{res})} \frac{\dot{e}}{e}] \tag{7}$$

STEP 4. For the impact of void ratio on the scanning path, update the slope, ξ, using

$$\xi^* = \xi[1 + \frac{(1-\zeta)(\zeta-\xi)}{\xi} S_{eff,r}^{\frac{\Omega}{\lambda_p}-1} \frac{\dot{e}}{e}]$$
$$= \xi[1 + \frac{(1-\zeta)(\zeta-\xi)}{\xi} \left(\frac{s_e}{s_r}\right)^{\Omega-\lambda_p} \frac{\dot{e}}{e}] \tag{8}$$

where s_r and $S_{eff,r}$ are the suction and effective degree of saturation at the point of suction reversal. If the soil is subjected to volumetric contraction, use s_{rw} and the respective effective degree of saturation, otherwise use s_{rd}.

STEP 5. Use (2) - (4) to calculate the updated degree of saturation at each step.

STEP 6. If the point of interest is located on the scanning path, use the following equation to update the suction at the point of suction reversal and use this to check whether the transition from scanning to main path of WRC due to the change in void ratio has occurred

$$s_r^* = \left(\frac{(s_e^*)^{\lambda_p^*}}{s^{\xi^*} S_{eff}^*}\right)^{\frac{1}{(\lambda_p^*-\xi^*)}} \tag{9}$$

where S_{eff}^* is the updated effective degree of saturation at the point of interest on the scanning path. Comparing the suction at the point of interest, s, to s_r^* from (9) at each step determines whether the soil hydraulic state has remained on the scanning path or it has been pushed to the main path due to the change in void ratio.

STEP 7. Repeat the procedure for incremental void ratio steps until reaching the target void ratio.

3 Model predictions

3.1 Evolution of main path of WRC with void ratio

***Municipal Boom clay* [25]**

Drying WRC test results on municipal Boom clay was presented by [25]. Four tests at different void ratios 0.59, 0.69, 0.8, and 0.93 were performed as replotted in Figure 3. For the purpose of prediction, the following parameters are evaluated from the reference WRC at the void ratio of 0.59: s_{e0} = 600 kPa; λ_{p0} = 0.23; S_{res} = 0; $\Omega = 0.55$; e_0 = 0.59. The water retention behavior at the other three void ratios is predicted using these parameters and plotted in Figure 3 along with the experimental data points. The figure shows that the model has yielded good prediction of the evolved main drying path of WRC with void ratio. It is also inferred from these results that the air entry value and the pore size distribution index of the soil are well-captured through the equations of the presented framework.

 DOI: 10.1051/e3sconf/20160911010

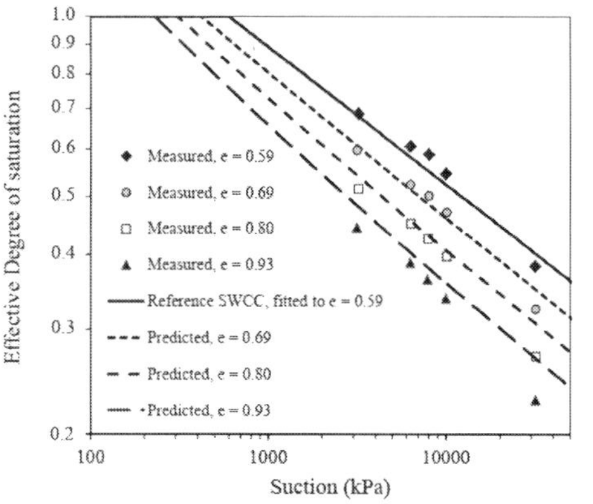

Figure 3. Measured and predicted evolution of the main drying path of WRC with void ratio for municipal Boom clay – data after [25]

3.2 Variation of degree of saturation at constant-suction loading/unloading tests

***Highly expansive sand-bentonite mixture* [26]**

Results from constant suction compression tests on four specimens of a mixture of bentonite with Toyoura sand reported by [26] and [27] were used here for further model validation. The specimens were prepared by static compaction with nearly similar initial conditions at void ratios between 0.55 and 0.56, and the water contents in the range of 9.2% to 10%. The reported initial suction of the samples was around 2500 kPa. The specimens were first wetted to different target suction values of 300 kPa, 600 kPa, 1200 kPa, and 1500 kPa, for Tests 1 to 4, respectively, under a constant vertical stress of 10 kPa. The process resulted in swelling of all four samples. After wetting, the suction was held constant and the specimens were subjected to vertical loading.

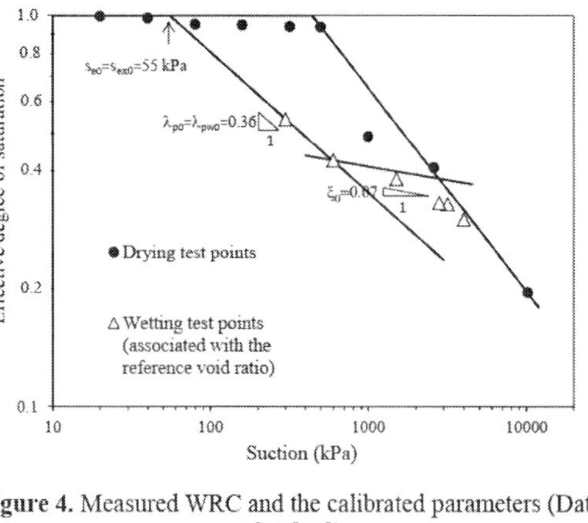

Figure 4. Measured WRC and the calibrated parameters (Data after [26])

To predict the variation of degree of saturation with void ratio during the loading, the measured reference WRC, replotted in Figure 4, were utilised with the following model parameters: s_{e0}=55.0 kPa; λ_{p0} = 0.36; ξ_0 = 0.07; S_{res} = 0.1; e_0 = 0.587. Note that, due to the lack of sufficient data points reported for the main wetting path of WRC (see Figure 4), the main drying data points of the WRC test are also plotted in Figure 4. This enabled us to compare the line fitted to the main wetting path with that fitted to the main drying path and therefore avoid possible errors in evaluation of the input parameter λ_{p0} as the pore size distribution index of the main wetting path (the values of λ_{pd} and λ_{pw} are generally close for most of the soils). The effective stress parameters used in the model are Ω = **0.55** and ζ= -0.1. The value of ζ= -0.1 was back-calculated from Test 4 and the same value was used in predicting other tests. To ascertain the initial hydraulic state of the samples; i.e. at the start of the loading test, the main wetting curve of the reference WRC in Figure 4 was adjusted for the volumetric swell/shrinkage experienced by the samples during the preparation stage as shown in Figure 5. By comparing the location of the points related to the start of the loading tests to the corresponding adjusted main wetting WRCs in Figure 5, all the samples were found to be located on the scanning wetting path.

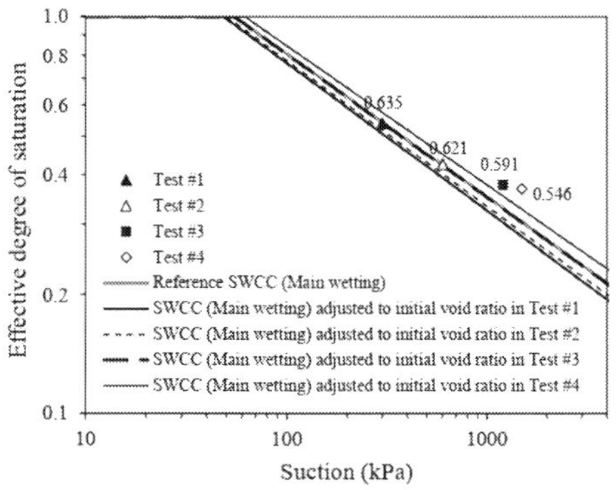

Figure 5. WRC main wetting curves at the reference void ratio and adjusted to the void ratio at the start of the constant suction loading tests (shown as numbers). Points are related to the start of the loading tests shown to compare the samples state with respect to the corresponding main wetting paths at the start of loading

Presented in Figure 6 are the model predictions together with the corresponding experimental results. Arrows on the plots depict the predicted points of transition from scanning wetting to main wetting path for the tests. The initial hydraulic state and the hydraulic path(s) followed in each loading test are also shown on the figure. As can be seen, the model has accurately predicted the variation of degree of saturation with void ratio for all four tests performed at different suctions and void ratios using a single set of reference model parameters. The results show that an increase in the rate of change in degree of saturation during a test typically relates to transition of the hydraulic path from scanning to main wetting. For tests conducted at constant suctions of 300 kPa and 600 kPa this transition occurs close to the start of the test with a clear increase in the rate of increase in the degree of saturation with decrease in the void ratio, whereas it is not detected in tests performed at suctions of

1200 kPa and 1500 kPa. These are reasonable as the initial hydraulic state of the tests performed at smaller suctions was relatively close to the main wetting path compared to the tests performed at higher suction levels. The ability of the model to capture such a subtle transition from scanning to wetting clearly shows the robustness of the approach proposed. The results also indicate that accurate assessment of the soil's initial hydraulic state (i.e. main or scanning) based on sample preparation prior to loading is important in subsequent prediction of the behavior as this directly affects the rate of change in degree of saturation with void ratio. This is particularly seen in this example for test Nos. 1 and 2.

4 Conclusions

The outline of an analytical model for taking into account the dependency of the WRC to void ratio was presented which includes both branches of main wetting and main drying as well as transition from wetting to drying and vice versa. Unlike similar models presented in the literature being mainly empirical, in the proposed model the void ratio dependency of the WRC is captured adopting energy approach; hence no additional soil parameter is needed in order to characterize the dependency of WRC on volume change. Good agreement was observed between the results of simulations and the experimental data showing that the proposed model is able to effectively capture the volume change effect on the WRC and can be incorporated into numerical simulations of unsaturated soils.

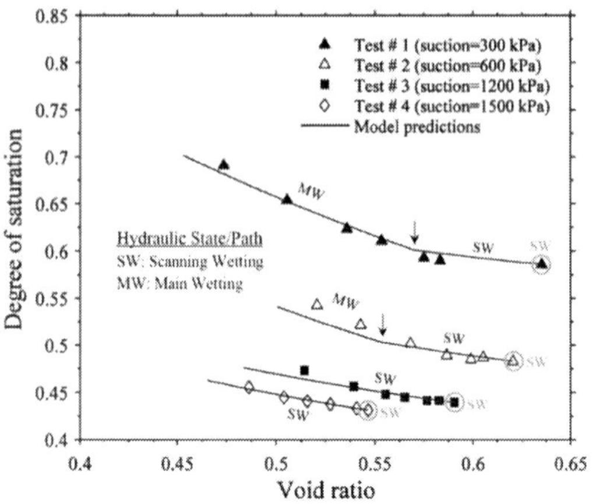

Figure 6. Measured and predicted variation of degree of saturation with void ratio at constant suction oedometric compression paths on highly expansive sand-bentonite mixture – data after [26]. Arrows show the predicted points of transition from scanning to main wetting path

References

1. Gallipoli, D., Wheeler, S. J., & Karstunen, M. 2003. Modelling the variation of degree of saturation in a deformable unsaturated soil. *Géotechnique* 53(1): 105-112.
2. Tarantino, A., and Tombolato, S. 2005. Coupling of hydraulic and mechanical behaviour in unsaturated compacted clay. Geotechnique, 55(4): 307-317.
3. Tarantino, A. 2009. A water retention model for deformable soils. Geotechnique, **59**(9): 751-762.
4. Brooks, R.H., and Corey, A.T. 1964. Hydraulic properties of porous media. Hydrology Papers, Colorado State University(March).
5. van Genuchten, M.T. 1980. A closed-form equation for predicting the hydraulic conductivity of unsaturated soils. Soil Science Society of America Journal, **44**(5): 892-898.
6. Fredlund, D.G., and Xing, A. 1994. Equations for the soil-water characteristic curve. Canadian Geotechnical Journal, **31**(4): 521-532.
7. Pasha, A.Y., Khoshghalb, A., and Khalili, N. 2016. Pitfalls in interpretation of gravimetric water content based soil water characteristic curve for deformable porous media. International Journal of Geomechanics, ASCE - In print
8. Wheeler, S.J., Sharma, R.S., and Buisson, M.S.R. 2003. Coupling of hydraulic hysteresis and stress-strain behaviour in unsaturated soils. Geotechnique, **53**(1): 41-54.
9. Gallipoli, D. 2012. A hysteretic soil-water retention model accounting for cyclic variations of suction and void ratio. Geotechnique, **62**(7): 605-616.
10. Nuth, M., and Laloui, L. 2008. Advances in modelling hysteretic water retention curve in deformable soils. Computers and Geotechnics, **35**(6): 835-844.
11. Mašín, D. 2010. Predicting the dependency of a degree of saturation on void ratio and suction using effective stress principle for unsaturated soils. International Journal for Numerical and Analytical Methods in Geomechanics, **34**(1): 73-90.
12. Romero, E., Della Vecchia, G., and Jommi, C. 2011. An insight into the water retention properties of compacted clayey soils. Geotechnique, **61**(4): 313-328.
13. Zhou, A.N., Sheng, D., and Carter, J.P. 2012. Modelling the effect of initial density on soil-water characteristic curves. Geotechnique, **62**(8): 669-680.
14. Khalili, N., Habte, M.A., and Zargarbashi, S. 2008. A fully coupled flow deformation model for cyclic analysis of unsaturated soils including hydraulic and mechanical hystereses. Computers and Geotechnics, **35**(6): 872-889.
15. Bishop A.W. 1959. The principle of effective stress. Teknisk Ukeblad; **106**(39):859–863.
16. Bishop, A.W., and Blight, G. 1963. Some aspects of effective stress in saturated and partly saturated soils. Geotechnique, **13**(3): 177-197.
17. Hassanizadeh, S.M., and Gray, W.G. 1990. Mechanics and thermodynamics of multiphase flow in porous media including interphase boundaries. Advances in Water Resources, **13**(4): 169-186.

18. Houlsby, G.T. 1997. The work input to an unsaturated granular material. Geotechnique, **47**(1): 193-196.
19. Borja, R.I. 2006. On the mechanical energy and effective stress in saturated and unsaturated porous continua. International Journal of Solids and Structures, **43**(6): 1764-1786.
20. Likos, W.J. 2014. Effective stress in unsaturated soil: Accounting for surface tension and interfacial area. Vadose Zone Journal, **13**(5).
21. Khalili, N., and Khabbaz, M.H. 1998. A unique relationship for χ for the determination of the shear strength of unsaturated soils. Geotechnique, **48**(5): 681-687.
22. Russell, A.R., and Khalili, N. 2006. A unified bounding surface plasticity model for unsaturated soils. International Journal for Numerical and Analytical Methods in Geomechanics, **30**(3): 181-212.
23. Khalili, N., and Zargarbashi, S. 2010. Influence of hydraulic hysteresis on effective stress in unsaturated soils. Geotechnique, **60**(9): 729-734.
24. Pasha, A.Y., Khoshghalb, A., and Khalili, N. 2015. Analytical model for the evolution of soil water characteristic curve with void ratio including hydraulic hysteresis - Journal of Engineering Mechanics - under review.
25. Romero, E. 1999. Characterisation and thermo-hydro-mechanical behavior of unsaturated Boom clay: An experimental study. PhD thesis, Universitat Polite´cnica de Catalunya, Barcelona, Spain.
26. Sun, D.A., and Sun, W.J. Elastoplastic modelling of hydraulic and mechanical behaviour of expansive soils. *In* Unsaturated Soils - Proceedings of the 5th International Conference on Unsaturated Soils2011, pp. 973-978.
27. Sun, D.A., Sun, W.J., and Yan, W. Hydraulic and mechanical behaviour of sand-bentonite mixture. *In* Proceeding of International Symposium on Unsaturated Soil Mechanics and Deep Geological Nuclear Waste Disposal, Shanghai. 24–25 Aug. 2009, pp. 90-97.

Water repellent soils: the case for unsaturated soil mechanics

Christopher Beckett[1,a], Andy Fourie[1] and David Toll[2]

[1]*The University of Western Australia, Perth, WA.*
[2]*Durham University, Durham , UK.*

Abstract. Water repellent (or "hydrophobic" or "non-wetting") soils have been studied by soil scientists for well over a century. These soils are typified by poor water infiltration, which leads to increased soil erosion and poor crop growth. However, the importance of water repellence on determining soil properties is now becoming recognised by geotechnical engineers. Water repellent soils may, for example, offer novel solutions for the design of cover systems overlying municipal or mine waste storage facilities. However, investigations into factors affecting their mechanical properties have only recently been initiated. This purpose of this paper is to introduce geotechnical engineers to the concept of water repellent soils and to discuss how their properties can be evaluated under an unsaturated soils framework. Scenarios in which water repellent properties might be relevant in geotechnical applications are presented and methods to quantify these properties in the laboratory and in the field examined.

1 Introduction

Soil water repellency (also known as "hydrophobicity" or "soil non-wetting") has been a concern of soil scientists for well over a century. Water repellency is now recognised to be a worldwide phenomenon affecting soils under both natural and agricultural vegetation [1]. For example, in Western Australia, it is estimated that as many as 7 *million* hectares of agricultural land are affected by or are under risk of water repellency. Expressed another way, lost production on this land due to water repellence alone is of the order of AUD250-300 million per annum [2].

Soil water repellency develops due to a variety of causes. The most common is the deposition of organic matter derived from decaying plants, however secretion of plant oils, growth of humic substances and treatment with wastewater ("grey water") are also strong contributors [1,3]. Excessive drying, for example as occurs during wildfire events or during the summer periods in arid regions also give rise to short-term water repellency [4]. Water repellency results in poor infiltration and increased runoff and erosion during rainfall or irrigation, which accelerates pesticide and nutrient transport to groundwater and reduces the efficiency of irrigation strategies [5].

Water repellency has generally been dismissed by geotechnical engineers due to the common assumption that soils are always wettable [6]. However, the effects of water repellency are now becoming recognised as being important in governing soil infiltration and stability [7,8]. The use of vegetated solutions in geotechnical design (e.g. slope stability, waste site rehabilitation) is also becoming more popular [9-11]. The effects of water repellency on such systems must therefore be examined.

This paper discusses the effects of water repellency on the properties of unsaturated soils and presents a case for water repellency to be considered by geotechnical engineers. Effects of water repellent behaviour on soil retention properties on the micro and macro scales are examined. Examples of consequences of water repellency on geotechnical design are then presented and methods to quantify water repellency in the laboratory and in the field briefly discussed.

2 Effect of water repellency on soil water retention

2.1 Microscopic level

A water-repellent surface is defined as one whose contact angle (θ) is $\geq 90°$ [12]. The contact angle is fundamental to unsaturated soil mechanics and can be linked to suction through the familiar capillary tube analogy of the Young-Laplace equation (Equation 1), which gives the matric suction (ψ_m) arising due to a curved air-water interface

$$\psi_m = \frac{2\gamma}{r}, r = \frac{R}{\cos\theta} \qquad [1]$$

where γ is the air-water surface tension, r is the radius of the concave meniscus and R is the radius of the capillary tube (a pore size analogue).

[a] Corresponding author: christopher.beckett@uwa.edu.au

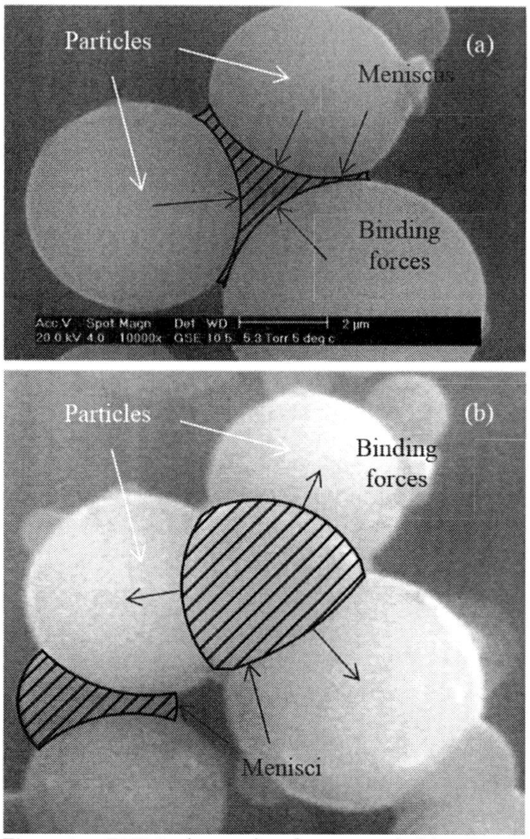

Figure 1. ESEM micrographs of water menisci between: a) hydrophilic particles; b) the same particles in a water repellent state. Magnification as per (a). Modified from [13].

An ESEM micrograph of concave menisci between hydrophilic silica spheres (a simulated soil) is shown in Figure 1a. By Equation 1, the concave shape gives rise to positive suction (i.e. negative pressure), acting to bind the particles together. However, for water repellent soils, $\theta > 90°$ gives rise to *convex* menisci, as shown in Figure 1b. In a highly simplified sense (as the meniscus geometry is not spherical), by Equation 1 suction in these menisci would be expected to be *negative*, i.e. positive water pressures would be generated despite saturation being <1. Geotechnical engineers are already familiar with the effect of changes in contact angle affecting suction due to suction hysteresis; advancing or retreating contact angles during wetting or drying (respectively) give rise to significantly different suction values for a given degree of saturation [14]. It is reasonable, then, that this effect might be heightened in water repellent soils, so that suction may enter a negative domain (e.g. as suggested by [15] and [16] for different liquid bridge or meniscus configurations). The concept of negative 'suctions', however, requires further investigation to be validated.

2.2 Macroscopic level

Retention properties of hydrophilic and water-repellent soils were investigated by [17]. A granular soil was artificially hydrophobised by treating with dichlorodimethylsilane (a process called "silanisation"). Samples were prepared at an initially-wet state close to saturation and suction was applied via a hanging column.

Results found by [17] are summarised in Figure 2, where S_r is the degree of saturation. A reduction in air entry value (AEV) was found for the water repellent material, and retention curves for the water repellent and hydrophilic soils differed significantly in the AEV<ψ<residual suction range. Notably, hydrophilic and water-repellent retention curves were similar in the low (<AEV) and high (greater than residual) suction ranges. Similar results were found by [16] and [18]; lower gravimetric water contents were found for hydrophobised granular soils than for the same soils in a hydrophilic condition for a given suction.

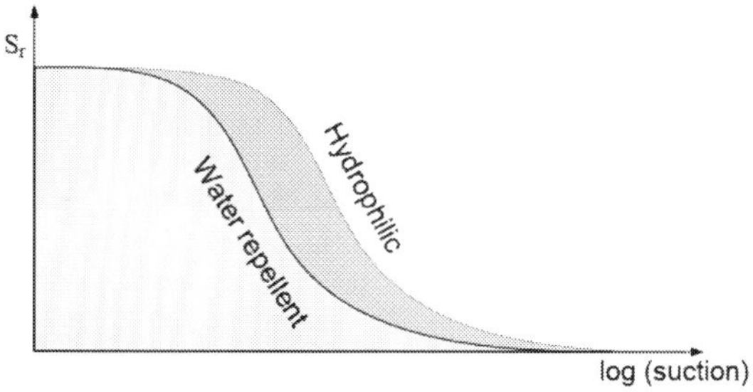

Figure 2. Simplified drying soil water retention curves for hydrophilic and water repellent soils tested in [16].

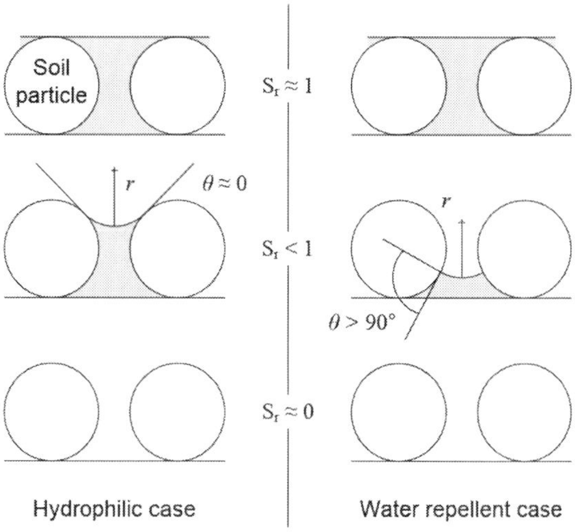

Figure 3. Desaturation of hydrophilic (left) and water repellent (right) granular soil. Shaded region represents pore water.

Possible desaturation mechanisms for a hydrophilic and water repellent granular soil are shown schematically in Figure 3. For the hydrophilic and water repellent soil, similar pore water volumes are likely to be present at high and low degrees of saturation. However, higher contact angles in the water repellent soil result in lower pore water volumes in the intermediate suction range for a given suction value (i.e. given value of r in Equation 1). The same result can be obtained by changing θ in

DOI: 10.1051/e3sconf/20160911011

Equation 1; higher values of θ (restricted to $\theta \leq 90°$, for which Equation 1 is defined) give smaller R values for given values of r, demonstrating smaller stored volume. It should be noted, however, that this is not a reduction in actual pore size, merely the storage capacity of given pore sizes [19].

Notably, tested soils in [16] to [18] displayed *positive* suctions. Soils in [17] were not tested in wetting, which is more likely to produce higher contact angles. Samples in [16] and [18] were tested in both drying and wetting. However, hydrophobising reagents were selected to mimic natural water repellent substances, whose contact angles were either <90 degrees or reduced from $\theta = 90°$ in the presence of water [4]. The behaviour of naturally water repellent soils is discussed later in this paper. It is therefore not possible to conclusively say from these results whether negative 'suctions' may develop in macroscopic cases. However, it is evident that water repellency has a significant effect on retention properties.

3 Example effects of water repellency on soil behaviour

3.1 Shear strength

It is well known that the mechanical properties of unsaturated soils are intimately linked to changes in soil suction. Changes in soil water retention on the advent of water repellency will therefore also result in a change in mechanical behaviour.

[7] examined the shear strengths of glass beads (an idealised granular soil) under hydrophilic and water repellent conditions. The hydrophobisation process used was similar to that used in [16] and produced a permanently water repellent material. Contact angles for the two materials were 10 and 100° respectively.

Results from [7] for direct shear tests on hydrophilic and water repellent material at 0 and 5% saturation are given in Table 1. Although tested on a relatively narrow normal stress range, water repellency resulted in a reduction in ϕ' and c at both tested S_r values. Unfortunately, retention properties for the materials were not determined, nor were suction measurements made during testing. It is therefore possible that both suction and S_r changed during the test due to sample volume changes [20]. However, a decrease in strength between the hydrophilic and water repellent samples at a nominally-constant value of S_r is consistent with results summarised in Figure 2; for a given value of S_r, a lower suction, and so a lower strength, would be expected for the water repellent material.

[21] performed a similar testing program to [7] using a natural, uniform sand. Retention properties for the hydrophilic sand were determined and direct shear tests performed on hydrophilic and hydrophobised (again, using silanisation) samples under suction-monitored conditions. Results are presented in Table 2. As for [7], [21] found a reduction in ϕ' between the hydrophilic and water repellent material. That little change was found between air-dry and moist samples was likely due to a combination of the low suction values and high normal stresses present during the test. Notably, little difference was found between the dilative behaviour of hydrophilic samples across the tested normal stress range (all samples dilated), however water repellent sample behaviour differed significantly; samples dilated at low S_r ($S_r <$ 0.4) but contracted at higher S_r. Such changes could have been due to slight differences in initial sample void ratio.

Table 1. Summary of shear strength results reported in [7]. Note that c is suction dependent (i.e. values have not been related to zero suction).

S_r (%)	Condition	ϕ' (degrees)	c (kPa)	Normal stress range (kPa)
0	Hydrophilic	29.3	2.83	1-20
	Water repellent	10.6	0.73	1-20
5	Hydrophilic	40.4	2.28	1-10
	Water repellent	28.9	0.73	1-10

Table 2. Shear strength results reported in [21]. [a] hydrophilic suction values.

S_r (%)	Suction range (kPa)	Condition	ϕ' (degrees)	c (kPa)	Normal stress range (kPa)
Air dry	n/a	Hydrophilic	36.1	0	20-80
Air dry	n/a	Water repellent	27.9	0	20-80
0.36-0.64	2-3	Hydrophilic	36.1	0	50
0.27-0.62	2-3[a]	Water repellent	27.9	0	50

Calculation of stress parameter ($\tan\phi^b$ or χ) values for results in tables 1 and 2 would not be appropriate due to the sparsity of data [22,23]. However, it is noteworthy that both [7] and [21] found reductions in ϕ' in *dry* materials. This suggests that the silanisation process resulted in a reduction in interparticle friction, likely due to particle coatings reducing surface roughness.

3.2 Infiltration and slope stability

Reduction in soil suction and shear strength due to water repellency increases the risk of slope instability. Although examples discussed so far have involved the use of artificially-prepared water repellent soils, excessive drying, for example during summer periods in arid regions, can also lead to the development of water repellent properties on a wide scale. An extreme case is in the event of wildfires, where soil surface temperatures can reach upwards of 300°C. In addition to drying, wildfires also induce water repellency through infiltration of vaporised organic substances (e.g. plant oils and waxes) and the introduction of ash to the upper soil layers [1]. Regions that are affected by wildfires are also generally threatened by post-summer rainy seasons. There is therefore a significant risk of slope instability due to changes in infiltration behaviour [24]. Effects of changes in water repellency on slope stability may also be of concern for deposition processes of treated material, for example mine tailings.

Slope stability and debris flow mechanisms in slopes of varying degrees of water repellency were investigated by [25]. Slope models were prepared by mixing hydrophilic and water repellent sand (d_{50}=0.13 mm) in known ratios from 0 to 100%, in 10% increments, creating soils of heterogeneous wettability. Water repellent sand was prepared via silanisation, as per [16,18]. Results are shown schematically in Figure 4.

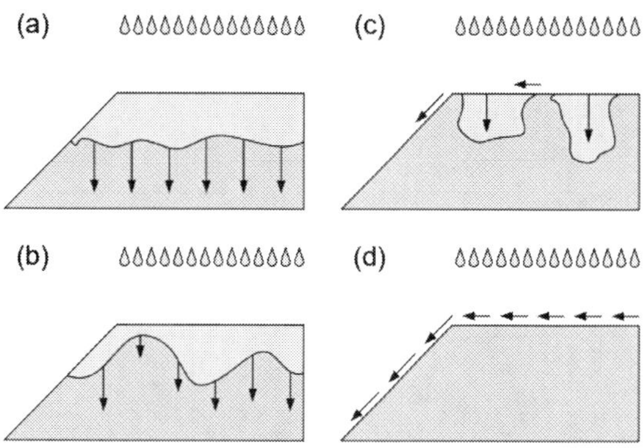

Figure 4. Rainfall infiltration in [25] as affected by water repellency: a) hydrophilic soil (parallel wetting front); b) marginally water repellent (fingered flow); c) severely water repellent (fingered and surface flow); d) fully water repellent (surface flow).

For hydrophilic soil (Figure 4a, water repellent soil content <10%), infiltration produced a wetting front parallel to the infiltration surface. Slope failure was due to a series of retrogressive slumps due to a build-up of pore water pressure at the slope toe. Introduction of water repellent material (Figures 4b and c) produced a "fingered" flow regime. Fingered flow is a well-documented phenomenon in water repellent soils and is characterised by an unstable wetting front, with infiltration following paths of least resistance (here, paths of less water repellent material) [26]. Transition to a fingered flow regime occurred for soils with 10 to 60% water repellent soil content by mass. Slope failure under these conditions was due to a combination of slumping and debris flow via surface erosion. For models with 70% water repellent material and above, no infiltration was recorded and slope failure was due to erosion only (Figure 4d).

It is important to note that, although tests in [25] used a permanently water repellent soil, water repellency in natural materials varies over time as a function of water content and water exposure (i.e., prolonged wet periods may break down water repellent surface coatings) [27]. An initially-water repellent soil may therefore become more hydrophilic as water contents increase, for example with changing seasons [4]. This is an important consideration for the geotechnical engineer, as different stability conditions must be considered depending on the time of year. A further point of consideration is that all results discussed here are for granular soils. Although the majority of soils affected by water repellency are granular (due to their smaller specific surface areas, e.g. [2]), how the presence of fine particles affects material properties is a necessary question for further research. Indeed, it is common practice in agriculture to add clay to a water repellent soil to increase its water storage capacity [1].

4 Methods to quantify soil water repellency

Water repellency has been shown to significantly affect unsaturated soil behaviour. Therefore, it is useful to be able to quantify the degree of water repellency for a given material. The following methods may be useful for geotechnical laboratory or field applications. It is neither the intention here to provide the full procedures for each method, nor to provide an exhaustive list; rather, these methods are listed to demonstrate a range of techniques that can easily be applied in laboratory or field applications.

4.1 Sessile Drop Method (SDM)

The SDM is suitable for all likely apparent contact angles (θ^*, between 0 and 180°) and is performed on the material fine fraction (particles <250µm). Contact angles are "apparent" as the test is performed on multiple soil grains, rather than individually, and so values are affected by surface roughness, entrapped air and material water content (due to changes in the soil surface's affinity for water) [29].

Soil particles are tamped onto a glass slide, covered on one side with double-sided adhesive tape. A deionised water droplet of roughly 1µL is then placed on the surface and images of the droplet and surface captured over a required time interval (typical times are 1, 2 and 5 seconds). θ^* is the instantaneous angle between the droplet and the soil surface at the droplet extremities at that specific time, as shown in Figure 5 [13,28].

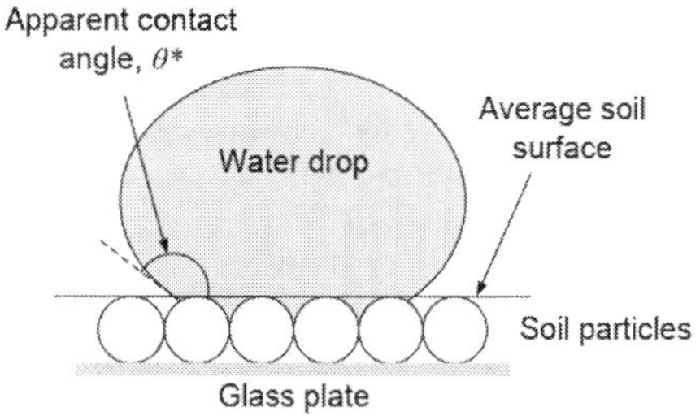

Figure 5. Apparent contact angle measurement using the sessile drop method.

4.2 Capillary Rise Method (CRM)

The CRM is suitable for soils with likely apparent contact angles $\theta^* < 90$ degrees. Modifications to the CRM are available for $\theta^* > 90°$ [28].

A capillary tube of known height and diameter (typical values are 60 mm and 9 mm respectively) and with a sintered glass base is filled with soil under the required conditions to the desired density. A water reservoir is then slowly lifted until it is in contact with the capillary tube and any mass change, w, due to capillary uptake recorded. θ^* is then calculated via the Washburn equation, substituting mass for capillary head:

$$\cos\theta^* = \frac{\eta w^2}{\rho^2 \gamma c t} \quad [2]$$

where η is dynamic viscosity, ρ is the density of water, t is time (of capillary rise) and c is a soil-specific factor [28].

4.3 Water Drop Penetration Time Test (WDPT)

The WPDT is an index test that provides a rough assessment of water repellency for bulk samples [25,28]. A sample of roughly 100g (particle size <2 mm) is prepared at the desired water content and density. Water droplets >1µL are then placed onto a smoothed sample surface at random locations and the time for the drops to infiltrate the sample recorded. Little guidance is provided as to sample dimensions. However, [17] used cylindrical samples of height 10 mm and 54 mm diameter. It is suggested here that droplets are not placed closer than $^r/_2$ from the sample edge, where r is the sample radius.

Relationships between penetration time and soil wettability are given in Table 3, based on results in [30,31]. Equivalent θ^* for sand, loam and silt loam are also given, as found by [17] using the SDM for artificially hydrophobised (again via silanisation) material tested at different water contents. θ^* are not given as exact equivalents, merely as a method to relate WPDT results to likely soil properties.

Table 3. WDPT soil wettability versus penetration time [31]. Equivalent θ^* from [17]. [a] very broad range as θ^* increases rapidly in sand with slight decreases in water content.

Penetration time (s)	Water repellency	Equivalent θ^* (degrees)		
		Sand	Loam	Silt loam
<5	Hydrophilic	0	0-20	0-20
6-60	Slight	0-80[a]	20-100	20-60
60-600	Moderate	80-110	100-120	60-120
600-3600	Severe			
>3600	Extreme	>110	>120	>130

4.4 Equivalent Cross Section (ECS)

The ECS is a method for testing water repellency in the field, based on the phenomenon of fingered flow. ECS quantifies the degree of preferential flow in a soil profile according to the area occupied by preferential pathways through a layer of soil of uniform depth. ECS is defined as the minimum fractional area responsible for 90% of the total flow through the profile. Example flow conditions for a hydrophilic and water repellent soil are shown in Figure 6.

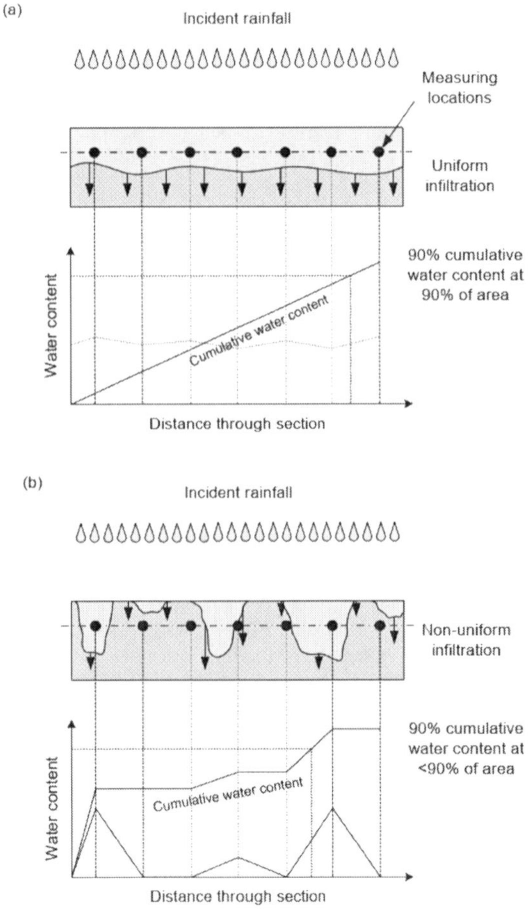

Figure 6. Flow through an instrumented soil profile: a) hydrophilic soil, uniform water content across the section during infiltration; b) water repellent soil, fingered flow and non-uniform water content across the section.

ECS assumes that water content is monitored at uniformly-spaced positions, so that each measurement is representative of flow through the same area. Flow in a given time at a point x during a rainfall event is defined as the change in water content between the minimum and peak value over the duration of the flow, W_x. Fractional contribution to total flow per point is then given by

$$f_{x,z} = \frac{W_x}{\sum_{x=1}^{n} W_x} \quad [3]$$

where n is the number of measurements (i.e. the number of sensors). $f_{x,z}$ values are then sorted from highest to lowest to determine points most responsible for flow. Cumulative flow is then given by $\sum_1^m f_{x,z}$, $1 \leq m \leq n$ For uniform flow, 90% of the monitored area will be responsible for 90% of the flow (Figure 6a). As water repellency increases, fractional area responsible for total flow reduces (Figure 6b). ECS can therefore be used to determine changes in water repellency over a given period, which may be important when assessing likely material behaviour or stability, as previously discussed.

5 Conclusions

The study of water repellency has been the domain of soil scientists for over a century. However, implications of water repellence on soil behaviour are now becoming recognised by geotechnical engineers. The purpose of this paper was to show how water repellent soil properties can be interpreted using an unsaturated soil mechanics framework.

This paper discussed how water repellence affects the properties of unsaturated soils on a micro and a macro scale. Results from literature investigations were used to demonstrate that differences between the two material conditions are significant. Implications of these behavioural changes were then examined for the cases of soil shear strength (in dry and unsaturated cases) and slope stability, as affected by changes in infiltration patterns. Finally, several methods by which water repellence can be quantified were presented for use in the laboratory and in the field.

Acknowledgements

The first author is supported by ARC Linkage Grant LP140100375, AMIRA P1087 Integrated Tailings Management Project, funded through AMIRA International by Anglo American, Freeport-McMoRan, Gold Fields, Total E&P Canada, Newmont, Shell Canada Energy, BASF, Nalco and Outotec, and MRIWA M430 Integrated Tailings Management. The first author would like to thank Margaret Roper and Phil Ward at the WA CSIRO and Daniel Murphy and Ken Flower at UWA for their comments on this work.

References

1. DeBano, L.F. Journal of Hydrology **231-232**, 4-32 (2000).
2. GRDC. https://www.grdc.com.au/Resources/Publications/2014/07/Combatting-non-wetting-soils. (2014)
3. Travis, M.J., Weisbrod, N., Gross, A. Science of the Total Environment, **394**, 68-74 (2008).
4. Rye, C.F., Smettem, K.R.J. Vadose Zone Journal, **13**, 4, 1-12 (2015).
5. Blackwell, P.S. Journal of Hydrology, **231-232**, 384-395 (2000).
6. Lourenço, S.D.N., Woche, S.K., Bachmann, J., Saulick, Y. Géotechnique Letters, 5, 173-177 (2015).
7. Byun, Y.-H., Tran, M.K., Yun, T.S., Lee, J.-S. Geotechnical Testing Journal, **35**, 1, 1-8.
8. Bardet, J.-P., Jesmani, M., Jabbari, N. Géotechnique, 64, 5, 341-350 (2014).
9. Rahardjo, H., Harnas, F.R., Indrawan, I.G.B. *et al.* Urban Forestry and Urban Greening, **13**, 2, 355-364 (2014)
10. Eab, K.H., Likilersuang, S., Takahashi, A. Soils and Foundations, **55**, 5, 1270-1281 (2015).
11. DeJong, J., Tibbett, M., Fourie, A.B. Environmental Earth Sciences, **73**, 1067-1082 (2015)
12. Atkins, P., De Paula, J. *Physical Chemistry*, Oxford University Press (2006)
13. Lourenço, S.D.N., Gallipoli, D., Augarde, C.E., Toll, D.G., Fisher, P.C., Congreve, A. Géotechnique, **62**, 3, 193-199 (2012).
14. Emerson, W.W., Bond, R. D. Australian Journal of Soil Research, **1**, 1, 9-16 (1964).
15. Likos, W. and Lu, N. Journal of Engineering Mechanics, **130**, 646-655 (2004).
16. Czachor, H., Doerr, S.H., Lichner, L. Journal of Hydrology, **380**, 104-111 (2010).
17. Liu, H., Ju, Z., Bachmann, J., Horton, R., Ren, T., Soil Science Society of America Journal, **76**, 342-349 (2012).
18. Lourenço, S.D.N., Jones, N., Morley, C., Doerr, S.H., Bryant, R. Vadose Zone Journal. DOI: 10.2136/vzj2014.07.0088.
19. Beckett, C.T.S., Augarde, C.E. Canadian Geotechnical Journal, **50**, 4, 435-450 (2013).
20. Tarantino, A., Tombolato, S. Géotechnique, 55, 4, 307-317 (2005).
21. Kim, B.S., Takeshita, Y., Kato, S., Park, S.W. *Unsaturated Soils: Research and Applications*, **1**, 159-165 (2014).
22. Fredlund, D.G., Morgenstern, N.R., Widger, R.A., Canadian Geotechnical Journal, **15**, 3, 313-321 (1978).
23. Khalili, N., Khabbaz, M.H., Géotechnique, **48**, 6, 681-687 (1998).
24. Suradi, M., Fourie, A.B., Beckett, C.T.S., Buzzi, O. *Unsaturated Soils: Research and Applications*, **2**, 1459-1464 (2014).
25. Lourenço, S.D.N., Wang, G.-H., Kamai, T. Engineering Geology, **196**, 47-58 (2015).
26. Bauters, T.W.J., DiCarlo, D.A., Steenhuis, T.S., Parlange, J.-Y. Soil Science Society of America Journal, **62**, 1185-1190.
27. Wijewardana, Y.N.S., Kawamoto, K., Komatsu, T., Hamamoto, S., Subedi, S., Moldrup, P. *Unsaturated Soils: Research and Applications*, **1**, 255-260 (2014).
28. Bachmann, J., Woche, S.K., Goebel, M.-O., Kirkham, M.B., Horton, R. Water Resources Research, **39**, 12, SBH11-1-14 (2003).
29. Cassie, A.B.D., Baxter, S. Transactions of the Faraday Society, **40**, 546-551 (1944).
30. Dekker, L.W., Ritsema, C.J. Water Resources Research, **30**, 9, 2507-2517 (1994).
31. Doerr, S.H., Shakesny, R. A., Dekker, L.W., Ritsema, C.J. European Journal of Soil Science, **57**, 741-754 (2006).

Water retention behaviour of compacted bentonites: experimental observations and constitutive model

Anne-Catherine Dieudonne[1,2,*], Claude Gatabin[3], Jean Talandier[4], Frédéric Collin[1] and Robert Charlier[1,*]

[1]*University of Liege, Quartier Polytech 1, Allée de la Découverte 9, 4000 Liege, Belgium*
[2]*Ecole Polytechnique Fédérale de Lausanne, Station 18, 1015 Lausanne, Switzerland*
[3]*Commissariat à l'énergie atomique et aux énergies nouvelles (CEA), Centre de Saclay, Gif-sur-Yvette, 91191 cedex, France*
[4]*Agence nationale pour la gestion des déchets radioactifs (ANDRA), rue Jean-Monnet 1/7, 92298 Châtenay-Malabry cedex, France*

Abstract. Bentonite-based materials are studied as potential barriers for the geological disposal of radioactive waste. In this context, the hydro-mechanical behaviour of the engineered barrier is first characterized by free swelling conditions followed by constant volume conditions. This paper presents an experimental study conducted in order to characterize the water retention behaviour of a compacted MX-80 bentonite/sand mixture. Then, based on observations of the material double structure and the water retention mechanisms in compacted bentonites, a new water retention model is proposed. The model considers adsorbed water in the microstructure and capillary water in the aggregate-porosity. The model is calibrated and validated against the experimental data. It is used for better understanding competing effects between volume change and water uptake observed during hydration under free swelling conditions.

1 Introduction

The water retention capacity of compacted bentonites is one of the fundamental properties required for predicting the behaviour of unsaturated engineered barriers. In the context of deep geological repositories for nuclear waste, bentonite-based materials are generally manufactured and emplaced at their hygroscopic water content. Under repository conditions, the engineered barrier experiences hydration from the surrounding geological formation. Because of the existence of unavoidable technological gaps, the periphery of the engineered barrier swells under free conditions first. During this stage, technological gaps are progressively filled by bentonite and the effective density of the buffer decreases. When contact between the geological formation and the engineered barrier is reached, the global volume constraints imposed to the bentonite buffer are close to constant volume conditions and a swelling pressure develops on the gallery wall. The objective of the engineered barrier is to form a tight contact with the surrounding formation and to create a zone of low permeability that is able to limit water flow around the excavated galleries, thereby delaying the release of radionuclides to the biosphere.

Because of the evolution of the volume constraints during the hydration process of the bentonite buffer, a good characterization of the material under both free and restricted swelling is required. To date, few experimental studies have investigated both volumetric and water retention behaviours of compacted bentonite with the aim of obtaining suction – degree of saturation relationships [1-4].

In this paper, the water retention properties of a compacted bentonite/sand mixture are first investigated under both constant volume and free swelling conditions. Then, based on observations of the material double structure and the water retention mechanisms in compacted bentonites, a new retention model is proposed. The model considers adsorbed water in the microstructure and capillary water in the aggregate-porosity. The model is validated against the experimental data. It is used for better understanding competing effects between volume change and water uptake observed during hydration under free swelling conditions.

2 Material and methods

2.1 Material and sample preparation

The material used in the investigation is a mixture of Wyoming MX-80 bentonite and quartz sand with a respective proportion of 7/3 in dry mass. The MX-80 bentonite contains 92% of montmorillonite and other minerals including quartz and feldspars [5]. The solid density of MX-80 bentonite is 2.78 Mg/m³ and that of the quartz sand is 2.65 Mg/m³.

The initial water content of the bentonite/sand mixture was 7.08%. Compacted samples were prepared by uniaxial static compaction (57 mm in diameter, 12 mm in height) to a target dry density of 2.0 Mg/m³. This dry density was selected in order to reach a swelling pressure of about 7 MPa in the PGZ2 in situ experiments performed by ANDRA in its underground research

* Corresponding authors: anne-catherine.dieudonne@epfl.ch, Robert.charlier@ulg.ac.be

laboratory, accounting for 12% of technological gaps. The vertical net stress required to reach the target dry density was about 80 MPa. For this compaction pressure, the optimum water content is about 8%, so that the samples used in the investigation are compacted on the dry side of the optimum (although not far). The initial characteristics of the samples are presented in Table 1.

Table 1. Initial characteristics of the samples used for the determination of the water retention properties.

Compaction pressure	~ 80 MPa
Dry density	2.0 Mg/m³
Void ratio	0.37
Water content	7.08%
Degree of saturation	52.25%

2.2 Experimental methods

Water retention curves of the compacted samples were determined under both constant volume and free swelling conditions by using the vapour equilibrium technique. The experiments were performed in the Laboratoire d'étude du comportement des bétons et argiles (LECBA) of the CEA in Saclay, France. After compaction, the samples were carefully removed from the compaction mould and either transferred to a constant volume cell (having the same diameter as the compaction cell) for the measurement of the water retention curve under isochoric conditions, or placed in a glass Petri dish for the determination of the sorption curve under free swelling conditions. Constant volume cells were specially designed for the measurement of the water retention curve under isochoric conditions (Figure 1). Each cell includes a titanium containment ring and two stainless steel porous discs put on both faces of the compacted sample and allowing vapour exchanges. Two perforated flanges hold the entire assembly.

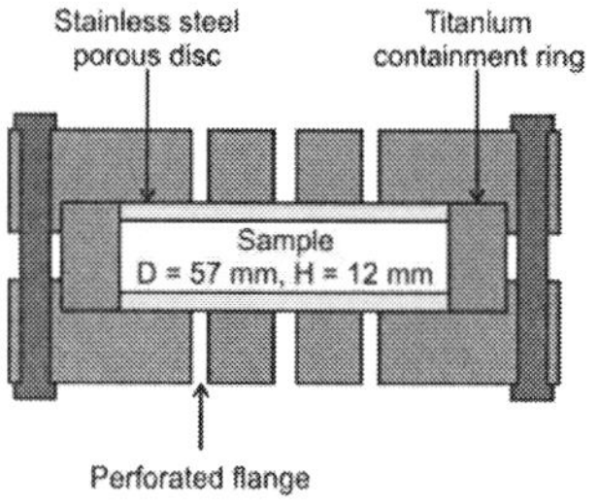

Figure 1. Constant volume cell for water retention curve determination.

All samples were placed in desiccators in which the relative humidity was controlled by using different saturated saline solutions. For the purpose of the experiment, the desiccators were placed within an environmentally controlled room with a temperature of 20°C (+/- 0.2°C). At the end of the test, the mass of each sample was measured and the gravimetric water content was determined by oven drying at 105°C. In addition, the heights and diameters of the samples wetted under unconfined conditions were measured with a precision calliper. Details on the experimental protocol can be found in [4].

3 Experimental results

Figure 2 presents the water retention curves obtained under constant volume and free swelling conditions, expressed in terms of water content. At high suctions (above around 10 MPa), the water content appears to be relatively independent of the volume constraints. In the high suction range, water is indeed essentially stored by adsorption at the surface of the clay particles. Therefore water uptake is mainly controlled by the physicochemical properties of the clay minerals, in particular smectite, and is independent of the current void ratio. On the contrary, in the lower suction range, bentonite swelling affects the water storage capacity of the material [6-12]. The water content reached under free swelling conditions at suction of 3 MPa is much higher than it is under prevented swelling (25.3% compared to 14.8%).

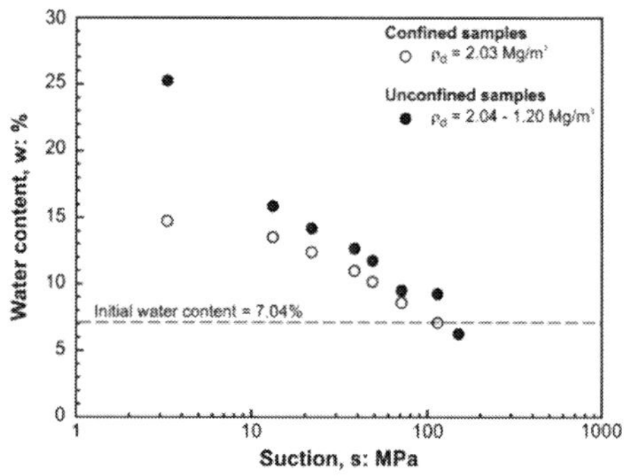

Figure 2. Water retention curves expressed in terms of water content versus suction.

The experimental data presented in Figure 2 can be expressed in terms of degree of saturation. As observed in Figure 3, the imposed volume constraint significantly impacts the water retention behaviour of compacted bentonite-based materials. For samples wetted under isochoric conditions, hydration led to an increase of the degree of saturation. On the contrary, the decrease of suction did not significantly impact the degree of saturation of the samples wetted under free conditions.

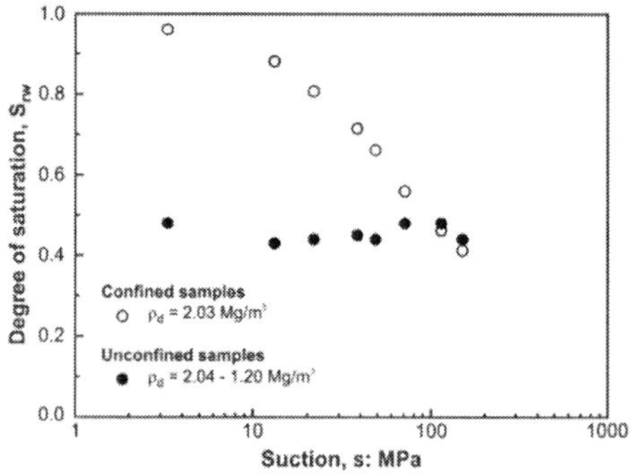

Figure 3. Water retention curves expressed in terms of degree of saturation versus suction.

This observation is a direct consequence of the important hydro-mechanical coupling in compacted bentonites. Indeed, hydration under unconfined conditions is associated with both an increase in the water content and an increase of the sample volume.

4 Water retention model

Water retention models have traditionally been formulated as a unique relationship between suction and the degree of saturation or water content [13,14]. If this approach has been successfully applied for granular soils and low activity clay materials, it cannot describe accurately the water retention behaviour of highly expansive materials such as bentonites. Indeed, in the case of compacted bentonites, the material swells significantly upon wetting, resulting in important changes in dry density. Consequently, the dependency of the water retention curve on the dry density of the material is a major issue and explains why classical water retention models have failed in describing the water retention properties of expansive materials. In the last years, a large number of constitutive models have been proposed to tackle the effect of dry density on the water retention behaviour [see for instance 9,15-18]. However, the particularity of expansive clays such as bentonite is their very important volume changes along wetting or drying paths.

Compacted bentonite-based materials are characterized by an aggregated structure which can be observed by mercury intrusion porosimetry or environmental scanning electron microscopy, among others. Aggregates are clusters or clay particles and retain water essentially by adsorption (in the micropores, which include inter-particle pores and interlayer spaces). They are separated by inter-aggregate pores (also called macropores) in which the water retention mechanism may be assimilated to capillary (Figure 4).

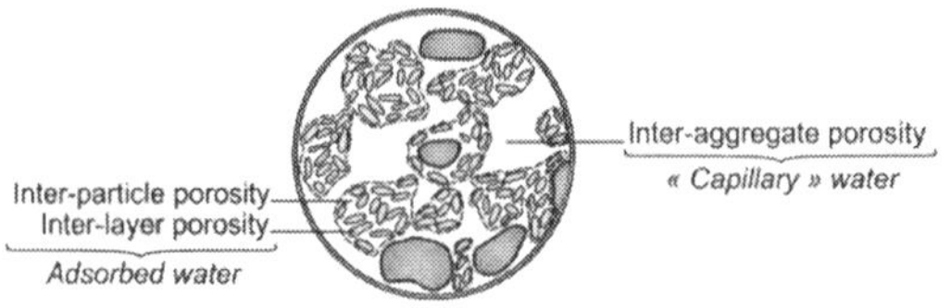

Figure 4. Conceptual representation of the structure of compacted bentonite-based materials and the different water retention mechanisms.

Accordingly, the total water ratio e_w (defined as the volume of water over the volume of solids) contains two contributions from the microstructure and macrostructure

$$e_w = e_{wm} + e_{wM} \tag{1}$$

where e_{wm} represents the water adsorbed in the microstructure and e_{wM} the water stored in the inter-aggregate pores or macrostructure.

Water retention models for both the microstructure and the macrostructure should be defined. Moreover, the structure created upon compaction is strongly influenced by hydromechanical loads [6,7]. The evolution of the microstructure during wetting or drying should therefore be considered.

4.1 Microstructural water retention model

Water in the microstructure is mainly stored by adsorption. In this paper, a generalized expression of Dubinin's isotherm is used [19]. The microstructural water ratio e_{wm} is given by

$$e_{wm} = e_m \exp\left[-\left(C_{ads} s\right)^{n_{ads}}\right] \tag{2}$$

where e_m is the microstructural void ratio (volume of micropores over volume of solids), and C_{ads} and n_{ads} are material parameters.

4.2 Macrostructural water retention model

The van Genuchten water retention model [13] is extended to represent water retention in the macropores. It is expressed as

$$e_{wM} = \left(e - e_m\right)\left[1 + \left(\frac{s}{\alpha}\right)^n\right]^{-m} \tag{3}$$

where $(e\text{-}e_m)$ represents the macrostructural voids ratio (volume of macropores over volume of solids), α is a parameter related to the air-entry pressure, and m and n are material parameters. In order to represent the influence of bentonite structure on the air-entry value, the parameter α is assumed to depend on the macrostructural void ratio. The following law is adopted

$$\alpha = \frac{A}{e - e_m} \tag{4}$$

where A is a parameter of the model.

4.3 Microstructure evolution model

The microstructure of compacted clayey soils is known to evolve along hydromechanical stress path. In particular, good correlations have been found between the microstructural void ratio and the water content. In this work, the evolution law of the microporous volume with water content has been introduced through the equation proposed by [20]

$$e_m = e_{m0} + \beta_0 e_w + \beta_1 e_w^2 \tag{5}$$

where β_0 and β_1 are two parameters quantifying aggregate swelling and e_{m0} is the intra-aggregate void ratio for the dry material. Note that such a relationship, relating the microstructural void ratio to the total water content (not the microstructural water content), is justified by the fact that, in compacted bentonites, the capillary water represents only a small fraction of the total water stored. The parameters of the model for the compacted mixture were calibrated based on experimental pore size distributions presented in [21,22]: $e_{m0} = 0.29$, $\beta_0 = 0.1$ and $\beta_1 = 0.18$.

5 Validation of the water retention model

Figure 5 presents the calibration of the water retention model on the experimental water retention curve determined under confined conditions. The parameters used for the model are: C_{ads} = 0.0053 MPa^{-1}, n_{ads} = 0.79, A = 0.2, n = 3 and m = 0.15.

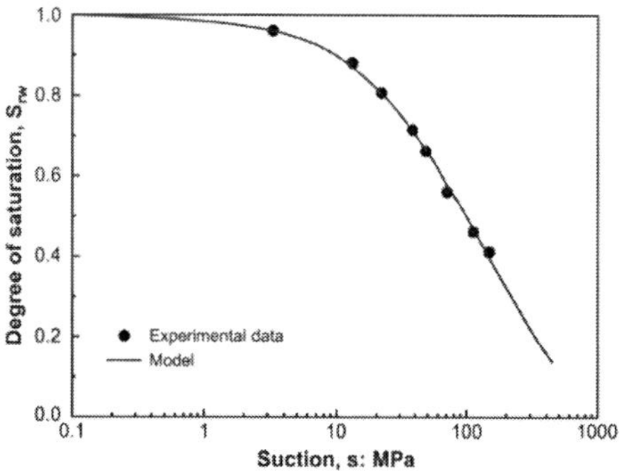

Figure 5. Comparison between experimental data and model predictions. Wetting path under confined conditions.

The model is validated against experimental data on wetting paths under unconfined conditions. Figure 6 presents the evolution of the degree of saturation of the compacted mixture upon wetting under free swelling conditions. In this case, the values of void ratio determined experimentally were used to compute the degree of saturation. The model remarkably succeeds in tracking the evolution of the degree of saturation over the whole range of investigated suctions. For the sake of completeness, the water retention curves predicted for the different current void ratios are also represented. As observed in Figure 6, an important decrease in the air entry pressure is associated with the important swelling of the material.

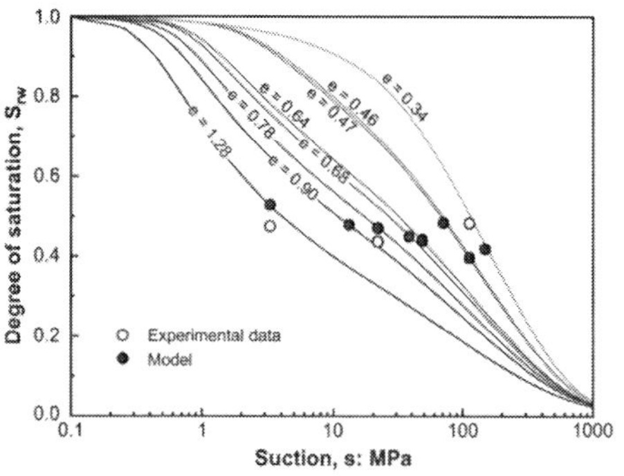

Figure 6. Comparison between experimental data and model predictions. Wetting path under unconfined conditions. Experimental data and model predictions are superposed at suctions of 150 MPa, 71 MPa, 48 MPa, 38 MPa and 13 MPa.

In order to better understand the mechanisms behind the competing effects, the evolution of the total, microstructural and macrostructural void ratios upon wetting is analysed. Under confined conditions, the total void ratio is constant while wetting leads to an increase of the microstructural void ratio (following the evolution predicted by Equation (5)), hence a decrease of the macroporous volume (Figure 7).

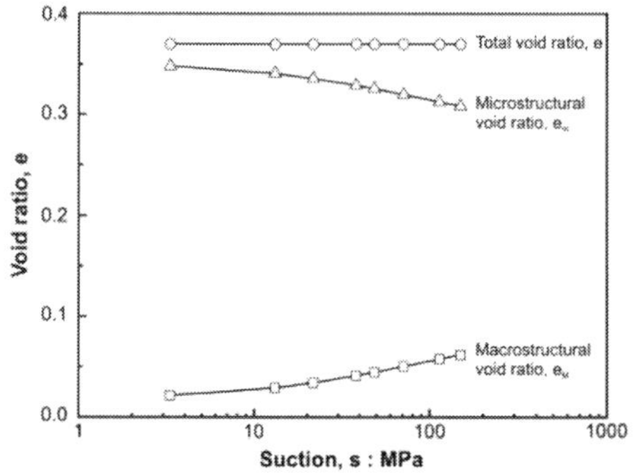

Figure 7. Model predictions for the evolution upon wetting of the total, microstructural and macrostructural void ratios. Wetting under confined conditions.

On the other hand, when the sample is wetted under free swelling conditions, the overall swelling of the sample is more important than the development of the microstructure (Figure 8). The increase in the total void ratio is indeed more significant than the micropores swelling. An increase in the inter-aggregate volume is therefore predicted. Unfortunately, no experimental information on the evolution of pore size distributions upon wetting under unconfined conditions is available in the literature.

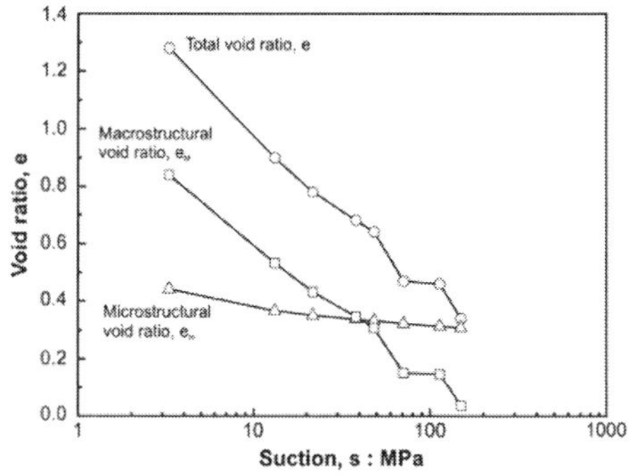

Figure 8. Model predictions for the evolution upon wetting of the total, microstructural and macrostructural void ratios. Wetting under unconfined conditions.

6 Conclusions

Water retention curves under constant volume and free swelling conditions were determined. For high suction values, the wetting branch in terms of water ratio appeared to be void ratio independent. On the contrary, the boundary conditions significantly influenced the water uptake capacity of the bentonite/sand mixture in the lower suction domain. The quantity of water stored under free swelling conditions is much higher than under constant volume.

In terms of degree of saturation, the samples at high initial density wetted under free volume conditions do no exhibit significant changes of degree of saturation. This is explained by the important competing effects of bentonite swelling and water uptake along wetting paths, as a consequence of strong hydro-mechanical coupling in compacted bentonites. Indeed the high-density samples exhibit important swelling, i.e. mechanical drying when suction decreases.

In order to complement the interpretation and model the experimental results, a water retention model was developed. This model explicitly takes into account the aggregated and evolving double structure of compacted bentonite-based materials. The model is calibrated and validated against the experimental data. The model succeeds in reproducing the stability of the degree of saturation upon wetting under unconfined conditions. The model provides a better understanding of the influence of the complex hydromechanical processes on the water retention curve. The influence of the microstructure evolution is highlighted.

While classical water retention models relating suction to the degree of saturation are generally used in numerical modelling (their parameters being determined from constant volume water retention curves), they are not able to represent the observed behaviour. In this case, the predicted degree of saturation of a bentonite plug under in situ conditions can be considerably overestimated if technological voids are important and not taken into account in modelling. A good representation of the water retention properties of compacted bentonites is thus a fundamental issue.

References

1. P. Delage, M. Howat, Y.J. Cui. The relationship between suction and swelling properties in a heavily compacted unsaturated clay. Engineering Geology 50, 31–48 (1998)
2. A. Dueck. Laboratory results from hydro-mechanical tests on a water unsaturated bentonite. Engineering Geology 97 (1–2), 15–24 (2008)
3. M. Ajdari, G. Habibagahi, F. Masrouri. The role of suction and degree of saturation on the hydro-mechanical response of a dual porosity silt-bentonite mixture. Applied Clay Science 83–84, 83–90 (2013)
4. C. Gatabin, J. Talandier, F. Collin, R. Charlier, A.C. Dieudonné. Competing effects of volume change and water uptake on the water retention behaviour of a compacted MX-80 bentonite/sand mixture. Applied Clay Science 121–122, 57–62 (2016)
5. A.M. Tang, Y.J. Cui, T.T. Le. A study on the thermal conductivity of compacted bentonites. Applied Clay Science 41(3–4), 181–189 (2008)
6. E. Romero, A. Gens, A. Lloret. Water permeability water retention and microstructure of unsaturated compacted Boom Clay. Engineering Geology 54, 117–127 (1999)
7. C. Loiseau, Y.J. Cui, P. Delage. The gradient effect on the water flow through a compacted swelling soil. In: J.F.T. Juca, T.M.P. de Campos, F.A.M. Marinho (Eds), Unsaturated Soils, Proceedings of the 3rd International Conference on Unsaturated Soils (UNSAT2002), Recife, Brazil, 395–400 (2002)
8. M.V. Villar. Water retention of two natural compacted bentonites. Clays and Clay Minerals 55(3), 311–322 (2007)
9. E. Romero, G. Della Vecchia, C. Jommi. An insight into the water retention properties of compacted clayey soils. Géotechnique 61(4), 313–328 (2011)
10. S.S. Agus, Y.F. Arifin, S. Tripathy, T. Schanz. Swelling pressure – suction relationship of heavily compacted bentonite–sand mixtures. Acta Geotechnica 8(2), 155–165 (2013)
11. A. Seiphoori, A. Ferrari, L. Laloui. Water retention behaviour and microstructural evolution of MX-80 bentonite during wetting and drying cycles. Géotechnique 64(9), 717–734 (2014)
12. G. Della Vecchia, A.C. Dieudonne, C. Jommi, R. Charlier. Accounting for evolving pore size distribution in water retention models for compacted clays. International Journal of Numerical and Analytical Methods in Geomechanics 39(7), 702–723 (2015)
13. M.T. Van Genuchten. A closed-form equation for predicting the hydraulic conductivity of unsaturated soils. Soil Science Society of America Journal 44(5), 892–898 (1980)
14. R.H. Brooks, A.T. Corey. Hydraulic properties of porous media. Hydrological papers (Colorado State University) 3 (1964)
15. D.A. Sun, D. Sheng, L. Xiang, S.W. Sloan. Elastoplastic prediction of hydro-mechanical behaviour of unsaturated soils under undrained conditions. Computers and Geotechnics 35(6), 845–852 (2008)
16. D. Masin. Predicting the dependency of a degree of saturation on void ratio and sucion using effective stress principle for unsaturated soils. International Journal for Numerical and Analytical Methods in Geomechanics 34, 73–90 (2010)
17. S. Salager, M. Nuth, A. Ferrari, L. Laloui. Investigation into water retention behaviour of deformable soils. Canadian Geotechnical Journal 50, 200–208 (2013)
18. D. Gallipoli, A.W. Bruno, F. D'Onza, C. Mancuso. A bounding surface hysteretic water retention model for deformable soils. Géotechnique 65(10), 793–804 (2015)
19. M.M. Dubinin, L.V. Radushkevich. Equation of the characteristic curve of activated charcoal. Proceedings of the academy of Sciences, Physical Chemistry Section, USSR 55, 331–333 (1947)
20. A.C. Dieudonne, S. Levasseur, R. Charlier, G. Della Vecchia, C. Jommi. A water retention model for compacted clayey soils. In: S Pietruszcak, G.N. Pande (Eds.) Computational Geomechanics COMGEO III, Proceeding of the 3rd International Symposium on Computational Geomechanics, 23–31 (2013)
21. Q. Wang, A.M. Tang, Y.J. Cui, P. Delage, J.D. Barnichon, W.M. Ye. The effects of technological voids on the hydromechanical behaviour of

DOI: 10.1051/e3sconf/20160911012

compacted bentonite-sand mixture. Soils and Foundations 53(2), 232–245 (2013)
22. S. Saba, P. Delage, N. Lenoir, Y.J. Cui, A.M. Tang, J.D. Barnichon. Further insight into the microstructure of compacted bentonite-sand mixture. Engineering Geology 168, 141–148 (2014)

Aggregate size effect on the water retention properties of a lime-treated compacted silt during curing

Yejiao Wang[1], Yu-Jun Cui[1,a], Anh Minh Tang[1], and Nadia Benahmed[2]

[1] *Ecole des Ponts ParisTech, U.R. Navier/CERMES, 6 – 8 av. Blaise Pascal, Cité Descartes,Champs – sur – Marne, 77455 Marne – la – Vallée cedex 2, France*
[2] *IRSTEA, Groupe de Recherche "Ouvrages hydrauliques", 3275 route Cézanne, CS 40061, 13182 Aix En Provence Cedex 5, France*

Abstract. The main drying paths of the water retention curves of lime-treated soils were measured. Aggregate size effect on the water retention property was emphasised. Four soil powders with different maximum aggregate sizes (D_{max} = 5, 2, 1 and 0.4 mm) were prepared and mixed with 2% quicklime (by weight of dry soil). Samples were prepared by static compaction at dry side of optimum water content (w = 17%) with a dry density of 1.65 Mg/m^3. Suction measurement was performed by a dew-point hygrometer at different curing periods (t = 7, 28 and 90 days). The results obtained show that: i) aggregate size effect is insignificant on the water retention curve of untreated soil in the studied suction range (ca 500 kPa ~ 55 MPa); ii) lime treatment gradually improves the soil water retention capacity with the increasing of curing time; iii) aggregate size effect on water retention property of lime-treated soil becomes significant in the suction range from ca 860 kPa to 9 MPa on the long curing term: treated soils prepared with smaller aggregate size (S0.4 and S1) have a higher water retention capacity rather than soils prepared with larger aggregate size (S5).

1 Introduction

Lime treatment could effectively improve both the workability and mechanical behaviours of problematic soils [1-6]. When quicklime (CaO) is added into humid soil and mixed together, hydration of quicklime takes place immediately with heat release. Then the calcium ions (Ca^{2+}) from ionization reactions of the calcium hydroxide ($Ca(OH)_2$) in the pore water would react with clay subsequently. There are two main different reactions could take place in the soil-lime-water system: cation exchange in short curing time and pozzolanic process on the long term. Cation exchange induces the flocculation and agglomeration of soil particles, which modify the soil workability. Cementitious compounds generated in the pozzolanic process, significantly contribute to the strength gain by bonding the soil particles together. This process is time-dependent.

Recently, more attentions have been paid to the efficiency and durability of lime treatment. Loss of efficiency and poor durability of treated soils in the field conditions were reported. Generally, field samples have higher hydraulic conductivity, lower strength and stiffness [5, 7, 8]. Besides of several environmental factors, such as heat transfer and wetting/drying cycles, another essential factor can be the aggregate size [5]. Indeed, nature soils in the laboratory are usually first air-dried, and then grounded into few millimetres. After that the prepared soil powders are mixed with lime and water. Nevertheless, it is important that the soil aggregate size in the field condition could reach several centimetres before lime treatment. Tang et al. [5] also pointed out that the lime-treated soil prepared with larger aggregate had a lower stiffness and higher sensitive to the wetting/drying cycles simulating the climate changes.

Water retention property is an essential parameter to describe the hydro-mechanical behaviour of untreated soils. Numerous studies showed that the water retention capacity of lime-treated soil was increased with curing time [2, 9, 10]. However, still few works focusing on the aggregate size effect on the water retention capacity of treated soils have been reported. This constitutes the main objective of this study.

2 Materials and methods

Studied soil was a plastic silt, with 27 % clay-size fraction. Its main minerals are quartz (55%), kaolinite (12%), feldspars (11%), illite (10%), goethite (6.5%), montmorillonite (4%), chlorite (1%) and rutile (0.5%) [11]. More details about the geotechnical properties of this silt are indicated by Wang et al. [12]. Nature soil was first air-dried, ground and passed through four target sieves in order to get four kinds of powders with different maximum aggregate sizes, namely S5, S2, S1 and S0.4 (as shown in Figure 1). In detail, D_{max} is 5 mm for S5, 2 mm for S2, 1mm for S1, and 0.4 mm for S0.4. Initially, dry soils were mixed with 2 % of quicklime thoroughly, and then the mixture were humidified with distilled water

[a] Corresponding author: yu-jun.cui@enpc.fr

to reach the target water content (w = 17 %, dry of optimum) and compacted statically to reach the target density (ρ_d = 1.65 Mg/m^3). Immediately after compaction, samples were carefully covered by plastic films, sealed in a plastic box and cured for a certain period (t = 7, 28 or 90 days). At each curing time, one sample was cut into more than 8 small pieces. They were air-dried for different durations, in order to obtain different suction values. Afterwards, they were covered again for over-night water homogenisation. Suction measurement was performed at the second day by using the WP4 dew-point hygrometer. After suction measurement, water content was determined by 105°C oven-drying.

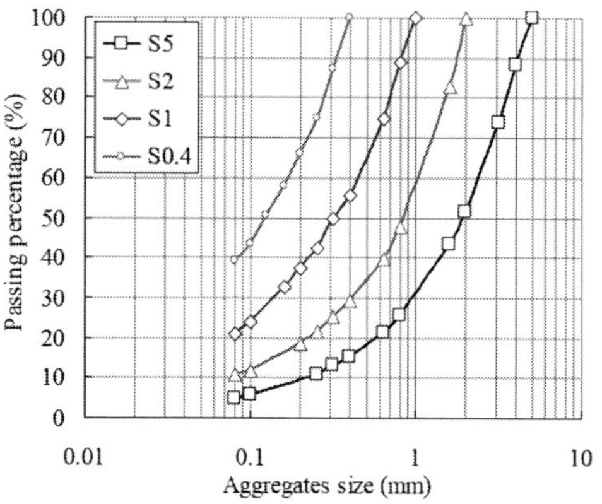

Figure 1. Aggregate size distributions of the four soil powders (S0.4, S1, S2 and S5) by the dry sieving method.

3 Experimental results

As shown in Figure 2, water retention curves of four different untreated soils are plotted together. Different soils with different aggregate sizes present a similar linear water retention curve in terms of water content in the semi-logarithmic scale. It indicates that aggregate size effect is insignificant for untreated soil during this measured suction range (from ca 500 kPa to 60MPa).

By contrast, water retention curves of treated soils at different curing durations are plotted in Figure 3a, b and c, respectively. At the curing time of t = 7 days, aggregate size effect is still not so obvious to be distinguished. However, the water retention curves of treated soils after 7 days' curing are located just slightly above that of untreated soils, in a suction range of ca 1300 kPa to 5500 kPa. This indicates that the water retention capacity of lime-treated soil after 7 days' curing is slightly increased.

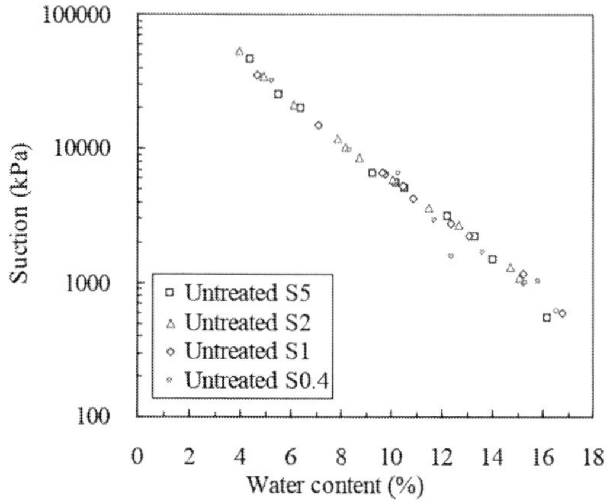

Figure 2. Water retention curves of untreated soils with different aggregate sizes.

With the increase of curing time, water retention abilities of treated soils are continually improved. Simultaneously, the water retention curves gradually form an S-shaped character. At the curing time of t = 28 days, suction difference in the range from ca 960 kPa to 7200 kPa is getting larger between the treated soils and untreated soils. Moreover, this suction difference deepens at a curing time of t = 90 days. Obviously, aggregate size effect is significant on the long curing term: samples prepared with smaller aggregates (S0.4 and S1) have a relatively higher water retention capacity rather than that with larger aggregates (S5), especially in the suction range from ca 860 kPa to 9 MPa. Nevertheless, in the higher suction range (more than 13 MPa), no evident aggregate size effect is observed, neither a visible difference between the water retention curve of treated soils and that of untreated soils.

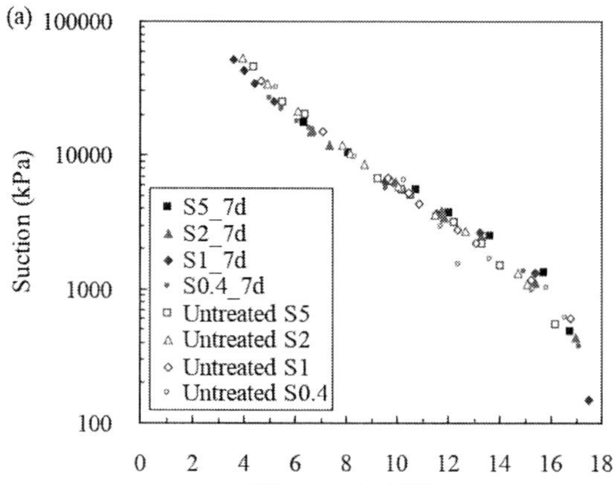

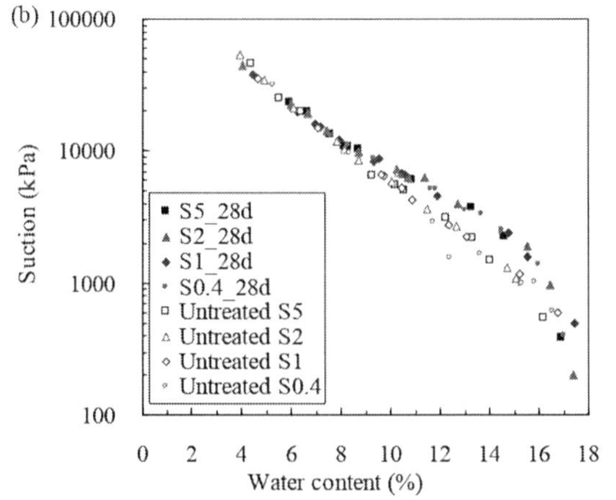

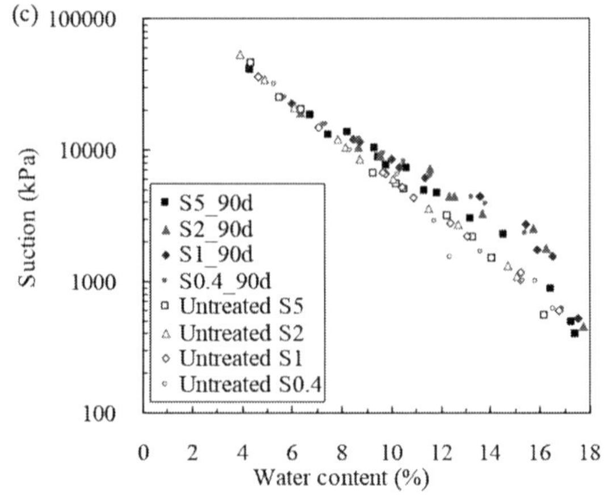

Figure 3. Water retention curves of 2% lime-treated soils during different curing times in comparison with that of untreated soils.

4 Discussions

The main drying paths of water retention curves measured on untreated soils with different sizes of aggregates show a similar behaviour. Romero et al. [13] explained that low water content range is related to intra and inter-cluster water, and the corresponding linear relationship between the logarithm suction and water content is controlled by the soil intra-aggregate microstructure. Wang et al. [12] indicated that the aggregate size mainly affect the macro-pores (inter-aggregate pores) rather than the micro-pores (intra-aggregate pores). In fact, different aggregate sizes form different contacts among soil aggregates and different size of inter-aggregate pores, which govern the lower suction range and air entry value. When at lower water content levels, main de-saturation takes place in the intra-aggregate pores (referring to the remove of inter-cluster and intra-cluster water), which is controlled by the clay structure.

The increase of water retention capacity of lime-treated soils observed during curing can be attributed to the production of cementitious compounds during curing. These cementitious compounds gradually filling the inter-aggregate pores over time, in the way of coating on the surface of soil aggregates [14-15]. Thus, the size and connectivity of the inter-aggregate pores are modified; some entrances of small intra-aggregate can also be blocked, changing some non-constricted pores to constricted pores and increasing the possibility of ink-bottle pores. These pores modified by the cementitious compounds can improve the water retention capacity of soil by retaining water in the wide inner pores upon drying. And the smaller the narrow openings of the link bottle pores, the higher the suction values needed to de-saturate the soil [12, 16-18].

Pozzolanic reactions are largely time-dependent. Very slight increase of water retention capacity is observed on the lime-treated soils at the curing time of t = 7 days, owing to the small quantity of cementitious compounds generated during that short curing time. With the increase of curing time, water retention capacity is continuously increased. At the curing time of t = 90 days, the water retention curves of treated soils become an S-shaped character with a higher suction values in the range from ca 860 kPa to 13 MPa. In the long curing period, larger quantity and better distribution of cementitious compounds are allowed to occur, resulting in an evident increase in the water retention capacity of soil.

Aggregate size effect is observed significant on the treated soils after a long time of curing (t = 90 days). Whereas, this effect is not evident in the short curing duration (t = 7 and 28 days). This is also in agreement with the time-dependent pozzolanic reactions. In short period, only a small quantity of cementitious compounds is generated, increasing the water retention capacity very slightly. Under this circumstance, it is difficult to distinguish the different effects induced by different size of aggregates.

In the soil prepared with small aggregates, there are more contact areas between soil aggregates and lime particles, compared with that with large aggregates. Under this circumstance, more cementitious compounds are expected to be generated or a better distribution of cementitious compounds is formed in a long curing period, which is beneficial to the modification of microstructure as discussed above, resulting in a larger improvement on the water retention capacity.

In the high suction range (more than 13 MPa), the water retention properties of untreated soils and that of treated soils even in the long curing period are similar. Note that the cementitious compounds mainly coating on the surface of soil aggregates. Even though some entrance of intra-aggregate pores can be blocked and constrict pores can be developed, most inter- and intra-cluster water may not be modified. When de-saturation takes place in this suction range, most inter-cluster water and some intra-cluster water begin to remove, leads to a similar behaviour between untreated soils and treated soils.

5 Conclusion

Water retention curve measurement on both lime-treated soil and untreated soil compacted dry of optimum were performed. Aggregate size effect on the water retention property of treated soil was emphasized in this study. Based on the obtained results, the following conclusions can be drawn:

• Aggregate size effect is insignificant on the water retention curve of untreated soil in the studied suction range, which can be mainly controlled by the intra-aggregate pores.

• Lime addition improves the soil water retention capacity with the increasing of curing time, due to the generation of cementitious compounds which modify the soil microstructure.

• In a long curing period, aggregate size effect on water retention property of lime-treated soil becomes significant in the suction range from ca 860 kPa to ca 9 MPa: treated soils prepared with smaller aggregate size (S0.4 and S1) have a relatively higher water retention capacity rather than that prepared with larger aggregate size (S5). This attributes to more quantity and better distribution of cementitious compounds which are expected in the soil of small aggregates, because of the large contact areas between aggregates and lime.

Acknowledgements

The authors would like to thank the China Scholarship Council (CSC) and Ecole des Ponts ParisTech for their financial support. They would also gratefully acknowledge the helpful support provided by the European Commission via the Marie Curie IRSES project GREAT - Geotechnical and geological Responses to climate change: Exchanging Approaches and Technologies on a world-wide scale (FP7-PEOPLE-2013-IRSES- 612665).

References

1. D.I. Boardman, C.D.F. Rogers, S. Glendinning, Géotechnique, **51**, 533–543 (2001)
2. G. Russo, *Advanced Experimental Unsaturated Soil Mechanics* (Taylor & Francis, London, 391–396. 2005)
3. N.C. Consoli, L.S. Lopes, K.S. Heineck, J. Mater. Civ. Eng., **21**, 210–216 (2009)
4. M. Al-Mukhtar, A. Lasledj, J.F. Alcover, Appl. Clay. Sci., **50**, 191–198 (2010)
5. A.M. Tang, Y.J. Cui, M.N. Vu, Géotechnique, **61**, 421–429 (2011)
6. T.D. Tran, Y.J. Cui, A.M. Tang, M. Audiguier, R. Cojean, J. Rock Mech. Geotech. Eng. **6**, 399–404 (2014)
7. I. Bozbey, E. Guler, Waste Manage., **26**, 1277–1286, (2006)
8. A. Kavak, A. Akyarh, Environ. Geol., **51**, 987–997 (2007)
9. D.V. Tedesco, *PhD Thesis at the University of Cassino*, (Cassino, Italy, 2006)
10. S. A. A. Khattab, L.Kh.I. Al-Taie, *Unsaturated Soils*, ASCE, 1671–1682 (2006)
11. D. Deneele, K. Lemaire, *Terrassement Durables – Ouvrages En Sols Traités, 2008-2012* (2012)
12. Y. Wang, Y. J. Cui, A. M. Tang, C. S. Tang, N. Benahmed, Geotech. Lett., **5**, 269-274 (2015)
13. E. Romero, A. Gens, A. Lloret, Eng. Geol., **54**, 117–127 (1999)
14. K. Lemaire, D. Deneele, S. Bonnet, M. Legret, Eng. Geol., **166**, 255–261 (2013)
15. B. Shi, Z. Liu, Y. Cai, X. Zhang, J. Mater. Civ. Eng., **19**, 99–104 (2007)
16. D.V. Tedesco, G. Russo, *Unsaturated Soils: Advances in Geo-Engineering*, Taylor & Francis, London, 277–282 (2008)
17. M. Cecconi, G. Russo, *Unsaturated Soils: Advances in Geo-Engineering*, Taylor & Francis, London, 271-276 (2008)
18. G. Russo, S. Vecchio, G. Mascolo, *Experimental Unsaturated Soil Mechanics*, Springer Berlin Heidelberg, 49–56 (2007)

Study of the hydric behaviour of double-layered columns of unsaturated soils - Analysis of water and suction profiles during infiltration tests

Ibrahim Alimi Ichola[1,a] and Laouni Gaidi[2]

[1]*LGCIE, INSA-LYON, 69621 Villeurbanne, France*
[2]*Faculté des Sciences et Technologie, Université de Mascara, Algérie*

Abstract. Systems of sealing of containers or retention pounds are often associated to drainage layers for leachate recovery. The drainage layer can be installed either above the sealing layer or below it. In this paper, double layered columns of alluvia and clayey sand or of alluvia and clay are made up to run infiltration tests. A follow-up of the water content is implemented during the infiltration in multi-layer columns by using TDR probes. The values and changes in infiltration rate are different if the drainage layer is above or below the sealing layer. It is found that the drainage layer saturation is delayed when the sealing layer is above but the equilibrium water content remains the same for the both configuration. The electrical measurements in the drainage layers prove that the increase rate of the solute accumulation is the same for the both configuration and the same equilibrium solute concentration is reached. The measured coefficient of permeability of the double-layered becomes five times higher when the drainage layer is below the sealing layer. The position of the drainage layer influences the infiltration rate, the water and solute migration. The change in coefficient of permeability must be taken into account for the assessment of the flux of pollutant across a barrier system.

1 Introduction

The systems of sealing of containers or retention pounds are often associated with a drainage layer for leachate recovery. The drainage layer can be installed either above or under the sealing layer. The purpose of the tests carried out in this study is to observe the influence that the presence of the drainage layer has on the evolution of the hydrodynamic parameters of the sealing layer during the saturation process.

Columns made up either of a layer of alluvium and of clayey sand or of a layer of alluvium and of clay are prepared for the infiltration tests. The unsaturated soils are compacted in individual rings to ensure the homogeneity of the density of each layer of the double-layered or the multi-layer one. Each layer is made up of one or more rings according to the final height of the column. A follow-up of the water content is implemented during infiltration by using TDR probes. The knowledge of the moisture characteristic curve of the soils and the use of the van Genuchten model, allows the layout of the profiles of suction. The results of the tests allow the comparison of the curves of infiltration of full-layer and multi-layered of unsaturated soil, to describe the profile forms in the double-layered ones and to compare the coefficients of permeability. By using TDR measurement of electrical conductivity, pollutant migration and retention is described.

2 Experimental device and materials

To study the hydric behaviour of multi-layer of unsaturated soils, soils of different geotechnical nature are associated in two or three layers. Tests of identification and compaction are carried out on each type of soil. These tests are supplemented by permeability tests on saturated soil to compare their resistance to the flow.

During the infiltration, the water content and impedance of soil in each layer are measured using TDR method. The infiltration head pressure is applied by using Mariotte burette.

2.1 Geotechnical characteristic of tested soils

The identification test results e.g. particle size distribution and Atterberg limits, indicate that the used soils are clayey sand (SC), clay and gravel (Gp). Standard permeability tests were performed on compacted soil at different dry unit weight γ_d.

[a] Corresponding author: ibrahimalimiichola8@msn.com

Table 1. Soil geotechnical characteristics.

Soils	w (%) natural	LL	IP	w_{opt} (%)	$(\gamma_d/\gamma_w)_{opt}$
Clayey sand	12	30	19	12.5	1.85
Clay	15.1	40	21	17.5	1.77
Alluvia	6	-	-	6.5	2.06

Table 1 presents the average values of the geotechnical parameters of the tested soils. Suction measurements on the three soils allow the determination of the moisture characteristic curves. The Van Genuchten model (Fig. 1), is used to describe the measured points. Table 2 presents the values of the parameters defining the model by fitting the experimental measured points.

$$\theta(\psi) = \theta_r + \frac{\theta_s - \theta_r}{\left(1 + (\alpha\psi)^n\right)^m} \tag{1}$$

θ volumetric water content, ψ suction potential
θ_r residual water content, θ_s equilibrium water content

The model formulas are used to compute the suction potential profiles from water content profiles.

Table 2. Parameters of van Genuchten model.

Soils	θr (m³/m³)	θs (m³/m³)	α(cm⁻¹)	m	n
Clayey sand	0.024	0.403	0.01	0.28	1.389
Clay	0.125	0.42	0.003	0.25	1.333
Alluvia	0.024	0.29	0.0012	1.97	0.864

2.2 Methods and test devices

The infiltration tests are carried out on columns made up of several rings. The soils are compacted in each ring with the chosen dry density and water content. Each ring is equipped with a TDR probe of three rods of 81mm of length. The rings used have a diameter of 106mm and of 50mm height. The diagram of the assembly in column is presented in Figure 1. Because of the duration of test, only three rings are used for the column made up of clay and alluvia. In the assembly of the column into multi-layer, the layer having the lowest permeability represents the barrier layer. When the bed of clay or GCL (Geotextile Clay Liner) represents the barrier, a layer of alluvia is installed on both sides of the sealing layer.

Table 3. Initial conditions of soils in layered column.

Soils	w_i %	γ_d/γ_w	height (cm)	head pressure(cm)
Clayey sand	6.5	2.04	35	10
Clay	9.3	1.6	20	10
Alluvia	20	1.66	20	10

When one studies the hydric behaviour of double-layered soil, the influence of the position of the layer of drainage onto the sealing layer is examined. The initial conditions of the tests are presented in the table 3.

TDR waveforms are produced by cable tester Tektronix 1502C (OR) and collected on computer using our own signal acquisition program. Further analysis of the collected TDR waveforms allows the determination of the dielectric constant of the soil. Using the relationship proposed by Topp et al.[7], the volumetric water profiles are determined for each multi-layer column. Applying the formula of van Genuchten of each soil, the profiles of the suction potential are computed.

The coefficient of permeability of the multi-layer is determined using the outflow rate.

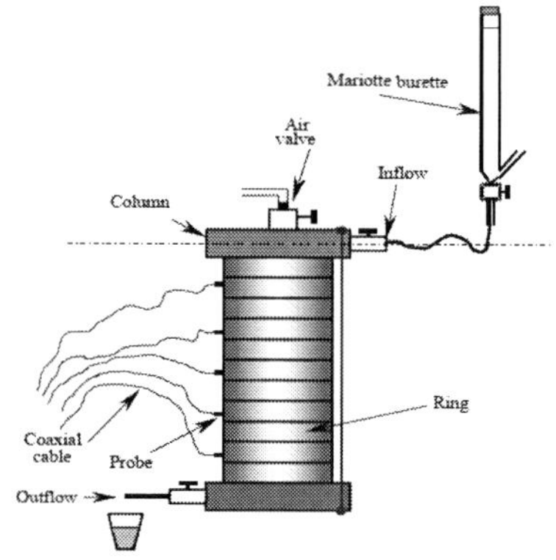

Figure 1. Schematic of soil column used for infiltration tests

3 Experimental results

To highlight the hydric behaviour of the unsaturated multi-layer soil, the measurement results are used to compute the water content profiles, the electrical conductivity profiles and that of the water potential during moisture movement. The infiltration rate is used to observe the influence of the drainage layer above or below the sealing layer. When the percolated outflow occurs, the coefficient of permeability of the multi-layer soil is computed. The electrical conductivity profiles computed from soil impedance, allow the description of the pollutant movement.

3.1. Description of the infiltration rate in multi-layer

To investigate the difference in infiltration rate in layered soils, three columns are built up with alluvia on clayey sand, alluvia below the clayey sand and clayey sand only. The three columns have the same height. A 10cm head pressure is used to launch the infiltration.

Three types of rate are used to describe the water inflow in the three columns. The rate v_i is determined at early time of the infiltration. It is called initial rate of infiltration. The second rate used v_0 is defined when the steady flow is reached and the third rate v_s is computed when the water outflows the column.
The values of these different rates are presented in the table 4.

Table 4. Different rates characterising the infiltration in layered columns.

	Column 1 Clayey sand	Column 2 Alluvia on clayey sand	Column 3 Alluvia under clayey sand
Rate v_0 (m/s)	3.36 10^{-6}	5.63 10^{-6}	1.33 10^{-5}
Rate v_i (m/s)	3.67 10^{-5}	5.55 10^{-5}	4.26 10^{-5}
Rate v_s (m/s)	5.10 10^{-6}	9.6 10^{-6}	1.63 10^{-5}

It is observed that the steady infiltration is reached more quickly in multi-layer alluvia-clayey sand; in 2hours when 4hours are needed for clayey sand column. It is shown in Table 4 that the steady rate is higher during the infiltration test on the multi-layer columns. It is also found that this rate increases when the drainage layer is located below the sealing layer. However, the initial infiltration rates in the three columns are slightly different. The outflow rates are also higher for the multi-layer columns. This rate is larger when the drainage layer is below the sealing layer.

3.2 Influence of the multi-layer on the soil permeability

Figure 2 presents the trends of the hydraulic conductivity during the process of saturation of the three columns. It is found that the hydraulic conductivity function decreases during the columns saturation and the values remain higher when the drainage layer is under the sealing layer. The position the different curves of the hydraulic conductivity function proves that the sealing trends are better when there is no drainage layer and it is relevant to install the drainage layer over the sealing layer.

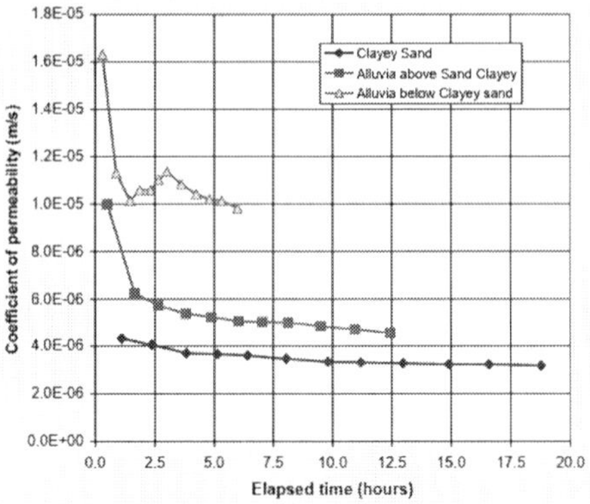

Figure 2. Comparison of permeability coefficient of the three columns

Determined by the standard permeability test, the range of the coefficient of permeability of alluvia layer is 2.10^{-5} m/s. That of clayey sand layer is 3.3 10^{-6} m/s.
The equivalent permeability k_{eq} of the two layers alluvia – clayey sand can be computed by the following formula:

$$k_{eq} = \frac{\sum_i L_i}{\sum_i \frac{L_i}{k_i}} \quad (2)$$

L_i thickness of each layer
k_i permeability coefficient of each layer

Table 5. Comparison of calculated equivalent permeability of multi-layer and measured coefficient.

Equivalent permeability	Alluvia on clayey sand	Alluvia under clayey sand
	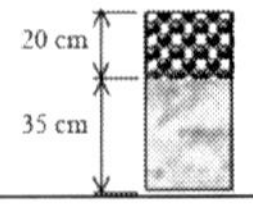	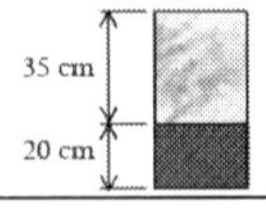
Measured (m/s)	5.0 10^{-6}	1.0 10^{-5}
Calculated (m/s)	4.75 10^{-6}	4.75 10^{-6}

In table 5 are presented the values of measured and calculated equivalent permeability of double layers soils of different permeability. The formula (2) does not take into account the position of the drainage layer, whereas it is found that the measured value is different according its position. So it is noticed that the drainage layer changes the permeability of the sealing layer and its position also. In this sealing system, the hydraulic conductivity increases. To improve the sealing action, it is better to install the drainage layer above the sealing layer. To complete the hydraulic behaviour of the sealing system, the profiles of water and suction must be examined.

3.3 Description of water content profiles during multi-layer saturation

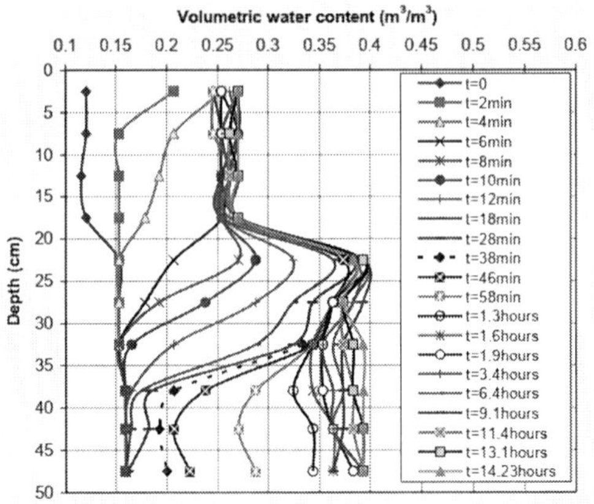

Figure 3. Water content profiles: Alluvia above clayey sand

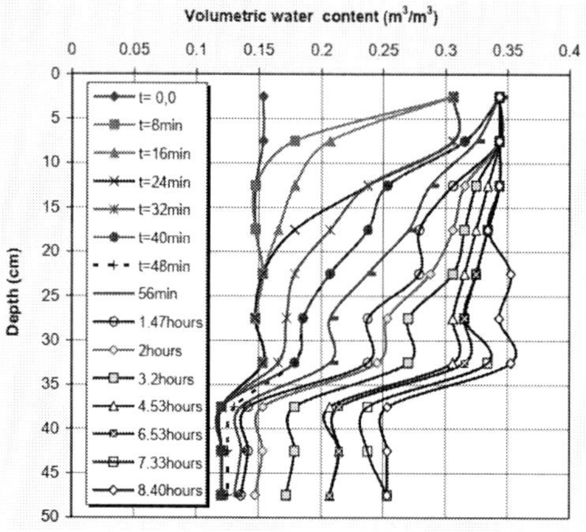

Figure 4. Water content profiles: Alluvia below clayey sand

Figures 3 and 4 present the profiles monitored during the saturation of a double layered soil. The profile corresponding to water outflow is indicated by dot curve. When the alluvia layer is above the sealing layer, the saturation is reached in 10 minutes and the water outflow begins after 38 minutes of infiltration. The value of the water content of alluvia layer remains small ($0.125m^3/m^3$), during the water outflow when the drainage layer is below the sealing layer. The sealing layer delays the saturation of the drainage layer which is not reached before the water outflows. That is why the outflow rate increases and increases the permeability of the sealing system. It is noticed that sealing layer saturation is also delayed when it is above the drainage layer (Fig. 4). The value of the water content reached in the sealing layer at the saturation state is less when the sealing layer is above the drainage layer ($0.35m^3/m^3$ instead of $0.40m^3/m^3$). So above the sealing layer, the drainage layer favours its saturation. In the single column of clayey sand, the equilibrium water content reached is $0.36m^3/m^3$, value which is related to gravity effect.

3.4 Description of suction profiles during multi-layer saturation

To describe water potential evolution in the double layered soil, during saturation process, the function (3) is used [6]. Suction potential profiles computed from water content profiles are presented in figures 5 and 6.

$$Se = \frac{\theta - \theta_r}{\theta_s - \theta_r} = \left[\frac{1}{1+(\alpha\,|h|)^n} \right]^m \quad \text{if } h < 0 \tag{3}$$

$$Se = 1 \qquad \text{if } h \geq 0$$

where, θ_r: residual volumetric water content

θ_s: saturated volumetric water content (h = 0)

α, n and m: parameters of the model

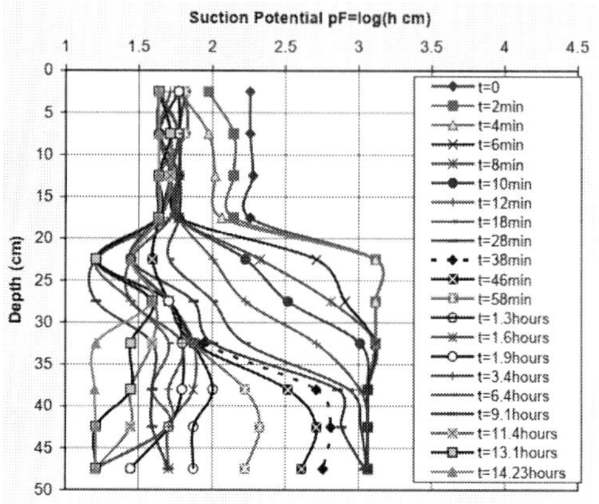

Figure 5. Suction profiles: Alluvia above clayey sand

When the alluvia layer is above the clayey sand layer, water potential value at saturation is reached in 8 minutes. So only a convective flow in the drainage layer provides water in the sealing layer. The drainage layer behaves like a reservoir providing water in the sealing layer. It is noticed that the water potential remains high in bottom of the double layered column (pF2.75, 56kPa), when the outflow begins (Fig. 5). This high retention pressure allows a better saturation of the sealing layer reducing the outflow volume.

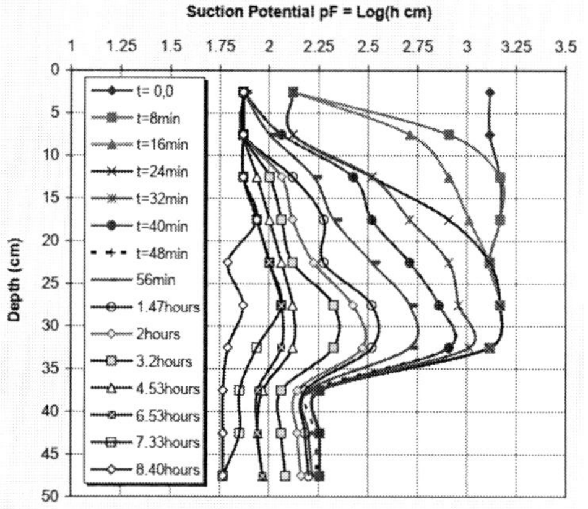

Figure 6. Suction profiles: Alluvia below clayey sand

The retention pressure at the bottom of the double layered soil column remains also high (pF2.24, 17kPa) when water outflows from the drainage layer below the sealing layer. The value of the retention pressure in the drainage layer remains two hours with little change whereas it decreases quickly in the sealing layer. The gravity flow prevents the accumulation of water in the drainage layer. This explains the high outflow rate when the drainage layer is below the sealing layer.

It is noticed that the value of the equilibrium water content in the drainage layer is nearly the same whatever its position. It is found nearly the same value of the equilibrium suction in the double layered column when the drainage layer is below the sealing layer. This means that the drainage layer controls the water flow inside the column by its position.

3.5 Description of the double layered soil as pollutant barrier

To describe the flow of polluted water and the retention of pollutant in double layered soil, lead slag leachate is percolated through the soil column. The used lead slag leachate is an acid leachate (pH6.3) with an electrical conductivity of 53mS/cm. Meddahi et al. [4] have proved that the TDR measurements can be used to obtain the breakthrough curves of non-reactive solute. So, supposing a linear relation between the soil electrical conductivity and the concentration of the soil pore water, it is shown that relation between the solute mass and the soil electrical conductivity is linear [3]. So the transfer of the solute in the soil during infiltration tests can be followed by monitoring the electrical conductivity of the soil.

Electrical conductivity measurement by the TDR method is described by Nader et al. [5]. Using the geometric constant of the TDR probes and the measured soil impedance profiles, electrical conductivity profiles and the breakthrough curves of the solute can be computed [1].

3.5.1 Description of solute retention in the double layered soil by electrical conductivity

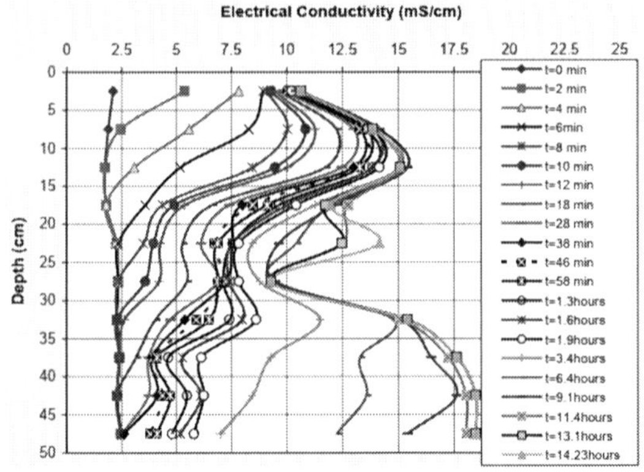

Figure 7. Electrical conductivity profiles: Alluvia above clayey sand

The changing profiles of electrical conductivity provide an input on the retention of the solute in the soil. When the alluvia layer is above the sealing layer (Fig. 7), the last value of the electrical conductivity remains at 10mS/cm and 15mS/cm respectively on the top and bottom of the drainage layer. The values decrease at the entrance of the clayey sand layer proving that the leachate is depleted by flowing in depth. While the equilibrium water content is practically achieved by 6 minutes, the electrical conductivity decreases in the thickness of the drainage layer before the equilibrium value is reached. The shape of the profiles in the drainage layer shows that the solute diffuses in depth during the first 8 minutes of infiltration. This solute transfer mode is also observed in the sealing layer to the release of the leachate. Then the transfer of the solute in the clayey sand is a convective transfer and the solute accumulation at the bottom of the column after 1.9 hours of infiltration, accelerates.

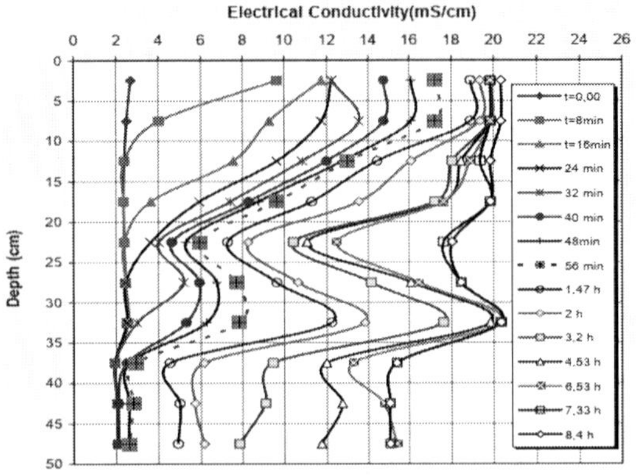

Figure 8. Electrical conductivity profiles: Alluvia below clayey sand

When the alluvia layer is below the sealing layer, the solute transfer begins in the clayey sand layer by diffusion and after 24 minutes of infiltration, convective-diffusive transfer is observed. Solute accumulation is noticed at the bottom of the sealing layer after the leachate begins to flow out. The electrical conductivity reaches 20mS/cm when it does not exceed an equilibrium value of 18.5mS/cm when the drainage layer is above. The solute transfer in the drainage layer begins with the leachate outflow in convective mode and reaches equilibrium value of 15mS/cm.

The position of the drainage layer changes both the transfer mode of the solute in the sealing layer, the retention capacity of the two layers and the time required for the output of the leachate from the double layered column.

Gaidi and Alimi-Ichola [2] have computed a relationship between leachate solute concentration and leachate electrical conductivity. The equation (4) gives the scaled value of the solute concentration (FS/FS_0) when the scaled value of the electrical conductivity of the leachate (σ/σ_0) is measured. FS_0 is the initial solute concentration in the leachate and σ_0 its electrical conductivity. Measuring the electrical conductivity of the leachate during the percolation of the barrier system, the change in leachate concentration is computed using the relationship (4).

$$\frac{FS}{FS_0} = 5.3747\left(\frac{\sigma}{\sigma_0}\right)^4 - 8.8419\left(\frac{\sigma}{\sigma_0}\right)^3 + 4.1908\left(\frac{\sigma}{\sigma_0}\right)^2 + 0.4095\left(\frac{\sigma}{\sigma_0}\right) - 0.0159 \quad (4)$$

Using relationship (4), it is noticed that the maximum solute concentration retained in the drainage layer is 27% of solute contained in the leachate, whatever its position. The proportion of the solute retained in the clayey sand layer reaches 37% when the drainage layer is below and 34% when it is above.

3.5.2 Description of solute breakthrough curves

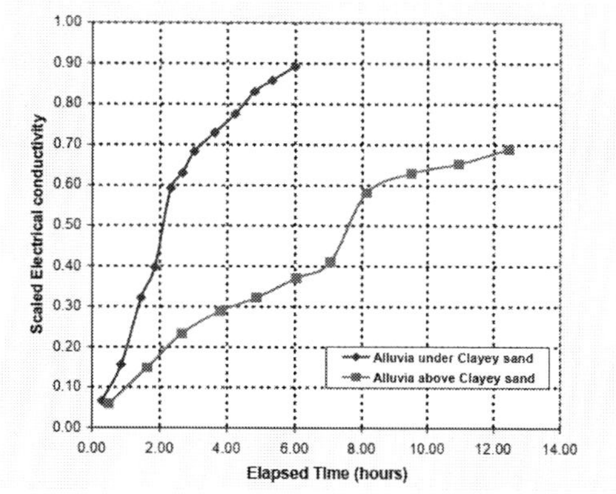

Figure 9. Breakthrough curves during leachate outflow described by the electrical conductivity

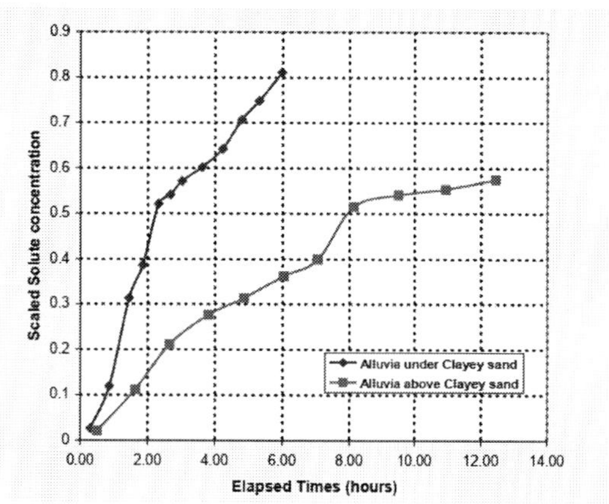

Figure 10. Breakthrough curves during leachate outflow (Solute)

Figures 9 and 10 present the breakthrough curves described by the electrical conductivity and computed by relation (4). The proportion of solute in output leachate is greater when the drainage layer is below the sealing layer. After recovering a volume of leachate equal to 1.8 volumes of the pores of the soil in the column, 70% of solute outflows when the drainage layer is above. Then 80% of solute concentration is reached when the drainage layer is below after 1.2 volumes of the pores is recovered. The sealing system is a better pollutant barrier when the drainage layer is above. Its performance is provided by the delay it imposes on the outflow of the pollutant.

4 Conclusion

The fight against flooding or groundwater pollution involves the setting up of basins fitted either for infiltration or to prevent pollution. The infiltration system or the sealing system, made up of natural materials, is unsaturated soil layer. The study shows that the hydric behaviour of samples soil layer composed of layers of different permeability is not only described by the equivalent permeability coefficient. The liquid flow through the multi-layer sample depends on the arrangement of the layers of different permeability within it. A layer of high permeability above a layer of low permeability slows down the flow. This appears in infiltration ponds, zones of permeability, described as restricted zones which promote clogging of soil surface. But to ensure the performance of a sealing layer, it is better to install the drainage layer on the low permeability layer.

The delay of the water propagation in a multi-layer is explained by the rapid decrease of the potential value of suction during saturation, when the drainage layer is below.

The transfer of solute from the leachate to the soil is also influenced by the superposition layers of different permeability coefficient. By continuous infiltration of leachate with constant solute concentration, through the multi-layer, it is found that the retention capacity depends on the position of the layer having the higher permeability coefficient. Solute retention is better if this layer is placed below the barrier layer. It is noted that the outflow of the solute is faster when the drainage layer is below and the multi-layer quickly reaches its retention capacity. The change in the coefficient of permeability must be taken into account for the assessment of the flux of pollutant across a barrier system.

References

1. I. Alimi-Ichola, Use of Time Domain Reflectrometry probes for monitoring unsaturated soil pollution. Environmental Geotechnics. Proc. 3rd Int. Cong. Lisboa, Portugal, Sept. 1998, 107-112 (1998)
2. L. Gaidi, I. Alimi Ichola, Utilisation des sondes TDR pour estimer la masse de soluté retenue par une couche d'argile lors d'une infiltration (Bull. Eng. Geol. Env 62:47-56) (2003)
3. R.G. Kachanoski, E. Pringle, A.L. Ward, Field Measurement of solute Travel Times using Time Domain Reflectrometry, Soil Sci. J., vol. 56, 47-52 (1992)
4. M.E. Meddahi, D. Mallants, J. Feyen, M. Badji, H. Vereecken, Paramètres essentiels d'écoulement des solutés dans un milieu poreux saturé : Méthode de mesure et modèle d'interprétation, Science du sol, 31, (4), 233-250 (1993)
5. A. Nader, S. Dasberg, I. Lapid, Time Domain Reflectrometry Measurement of water content and electrical conductivity of layered soil column, Soil Sci. Am. J. 55 938-943 (1991)
6. M.Th. Van Genuchten, A closed-form equation for predicting the hydraulic conductivity of unsaturated soils, Soil Science of America Journal, vol. 44, pp. 892-898 (1980)
7. G. C. TOPP, J. L. DAVIS, and A. P. ANNAN, Electromagnetic Determination of Soil Water Content: Measurement in Coaxial Transmission Lines. Water Resour. Res. 16:574-582 (1980)

Numerical modelling of infiltration profiles in the silt Tlemcen (Algeria)

Abdeldjalil Zadjaoui[1,a]

[1]*Departement of Civil Engineering; RisAM Research Laboratory; University AbouBekr Belkaid Tlemcen Algeria*

Abstract. It is obvious that for geotechnical knowledge of soil properties namely, suction-water content and water content hydraulic-conductivity relationship is essential to solve the problems associated with flow in unsaturated soils. The best way to evaluate these properties is to make direct measurements. However, the hydraulic-conductivity water content and suction-conductivity relationship are complex and it is often very difficult to measure for practical reasons, economic and spatial variability. Indeed, the used equipment is expensive and the tests are slow. Of more over the spatial variability of this property that makes the steps number necessary to characterize a site is relatively large. Furthermore, when this relationship is determined, usually it has only limited information and it is always desirable to extrapolate the experimental curve. Facing these difficulties, this given research targets computer tool to assess this relationship from other measurable parameters. In this paper, we used the software "flow line temperature" to achieve numerical and graphical simulations of the water flow in a soil initially unsaturated non-deformable modeling the evolution of the water content and suction in function of depth. The use of this software allowed us to make comparison with the experimental results [18] as well as a qualitative assessment of the drought penetrating in an unsaturated soil.

1. Introduction

The knowledge of soil properties, suction-water content and water content-hydraulic conductivity relationship is essential to solve the problems associated with flow in unsaturated soils. The best way to evaluate these properties is to make direct measurements. That during the hydraulic conductivity water content and suction conductivity relationships are complex and it is often very difficult to measure for practical reasons, economic and spatial variability. Indeed, the equipment used is expensive and tests are long, particularly because most of the measuring systems do not cover a large range of suction. Furthermore, the spatial variability of hydraulic properties makes the number of steps necessary to characterize a site relatively large. Also, once this relationship is determined, only limited information is available and it is always necessary to extrapolate the experimental curve. To face these difficulties, research is oriented towards to the use of digital tools in engineering practice. It is clear that the few necessary parameters in this case are easily measurable.

After a historical overview, concerning particularly the existing tools for identification, a theoretical overview of the infiltration problem is given. Subsequently, a description of the used experimental model and of the data needed for the simulations is exposed. An analysis using the software named "Flow line temperature" was performed with numerical simulations of the water flow in a column of a soil initially unsaturated and supposed to be non-deformable. The changes in water content and suction with respect to depth ad time were computed to allow comparisons with the experimental results.
The findings in relation to the practice of geotechnical engineering are finally outlined.

[a] Corresponding author: a.zadjaoui@gmail.com

2. Background

The description of water flow in unsaturated soils, described by the generalization of Darcy's law (1856) in unsaturated media, was mathematically summarized in the Richards' equation in 1931. The theory and basic assumptions are described in detail in the works of soil physics [11].

However, the problem of the non-uniqueness of the solution (humidification path and different drainage path) and the dependence of the conductivity with respect to the water content make it highly nonlinear. No analytical solution has been found to solve the Richard's equation.

Soil physicists have conducted several researches and proposed different resolution patterns [6, 1, 5, 8] under different initial and boundary conditions. All works provide numerical resolutions of the partial differential equations that governs the spatial and temporal changes of water content or pressure during a infiltration in a soil column. The major difference between these algorithms is their accuracy with respect to the mass balance error generated by the numerical approximations adopted.

Indeed, the use of these models often requires some information not always available in current engineering practice. This is why some software packages have

 DOI: 10.1051/e3sconf/20160911015

recently been developed that allows easier use of flow model [20, 16, 15]. For example, [15] only discusses, in the package HYDRUS, various analytical expressions [17], while [20] includes in the new package SWIM different analytical expressions. Simunek and al. [15] introduced advanced procedures for two-dimensional flow and solute transport. The hypothesis of deformable medium was not mentioned in these references.

In this study one uses a new package that allows the one-dimensional simulation of a vertical flow in the laboratory or in situ [3, 4]. The choice made partly relies on the easiness of use and on the existence of a dynamic display feature of moisture redistribution in the soil. Note that its use is free for academic [12].

3. Theory and simulations description

3.1. Flow model

According to [12], the model used to simulate the flow of water is established at macroscopic scale (a laboratory infiltration column) for a continuous and isotropic material. The flow is assumed to be vertical. The effects of air are neglected (assuming constant air pressure equal to atmospheric pressure) and the effects of chemical gradients (solute transport, chemical reaction with the ground) and temperature gradients are not considered. The simulation is then restricted to the vertical water transfer of water, in a homogeneous medium made up of several homogeneous layers. Modelling is based on four elements. To run, it requires:

The soil characteristics (general characteristics and water properties);

The initial water profile describing the amount of water contained in the soil (θ_i)

The algorithm solving the flow equation and the evolution over time of the boundary conditions applied to the soil (surface and deep boundary conditions).

If all data are available, each simulation will provide some information as follows:

- The simulated values of the water profile obtained by space discretization and by solving the flow equations discretized with respect to time
- The variables useful for understanding or validation of simulations launched.

3.2. Distribution of water in the soil during the infiltration

The distribution of the water in the ground at a given time is described by the water profile that depends on the physical properties of the soil and on the weather conditions. The infiltration of water at the surface causes a perturbation of the fluid profile that is characterized by 4 zones [11]. In the case of a homogeneous profile under constant infiltration, these zones are arranged as follows:

a. The surface saturation zone extends generally at a depth of 1cm; in the case of clay soil, this area may be subjected to a swelling phenomenon, which leads to very high water contents;

b. The transmission area, almost saturated, is characterized by a very rapid decrease in water content to values of 70 to 80% saturation; this area becomes longer during infiltration. The saturation is apparently uniform in this zone which "z_s" is called depth of the wetting front.

c. The humidification area in which the water content changes rapidly with depth and time. It ends with the humidification brow; the distance between the edges at the surface is called the depth of the humidifying edge denoted "z_f".

d. The humidifying front is a region of a very high moisture gradient that represents a visible limit moisture penetration especially when the soil is initially dry.

Musy and Soutter [11] observe a transition zone between the vadose zone and the transmission zone. There is a zone of a few centimetres of depth through which the water content was reduced from saturation to a value characteristic of the transmission area. Subsequent research concerned the existence of both a zone of saturation and a distinct transition zone of the transmission area. An air trapping term explanation has been given, but it is possible that the saturating zone and the transition from a simple structure of soil disturbance which is unstable and subject to collapse in the vicinity of the surface by the action of the rain [7].

During infiltration, water penetrates the soil and gradually moves downwards by establishing successive water content profiles. The water distribution in the soil depends on the size of the grains. Periodic examination of the moisture profile during infiltration shows that the transmission area lengthens continuously, and that the humidification zone and the front move down. The front becomes less clear in depth. The suction gradient, initially high in a dry soil, decreases during infiltration to unit values.

4. Application and comparison with experimental results

4.1. Application of the model

To model infiltration in unsaturated soils, we used data obtained experimentally [18] on the Tlemcen silt with an initial water content of 13.8% and a permeability at saturation of 1.2 x 10^{-7} m/s for a hydraulic head of 120 cm. The initial profile content was determined by using the filter paper technique. The model of Van Genuchten for the water retention curve and the Brooks-Corey model for hydraulic conductivity function were used.

4.2. Comparison of water content profiles and moistening profiles

The experimental work was performed at the Central Laboratory in Paris. The equipment used, the experimental protocol, the sample preparation technique and the acquisition system are detailed in [21]. Figure 1 shows the set of curves of evolution of θ = f (t) for

DOI: 10.1051/e3sconf/20160911015

different depths. In this case, we note that there is a shift in the experimental curves with respect to those obtained by simulation. This discrepancy (eg z = 3 cm; t = 7min) can be related to the fact that the initial conditions of laboratory water content obtained were difficult because the sample should have been left for more than 7 days to have a homogeneous water content profile. This finding is also observed at other depths.

Another important point is that the water supply is provided through a valve which provides once opened a large amount of water to the first ring placed just in contact. However, this difference is not very large compared to the time set for the specimen to reach saturation.

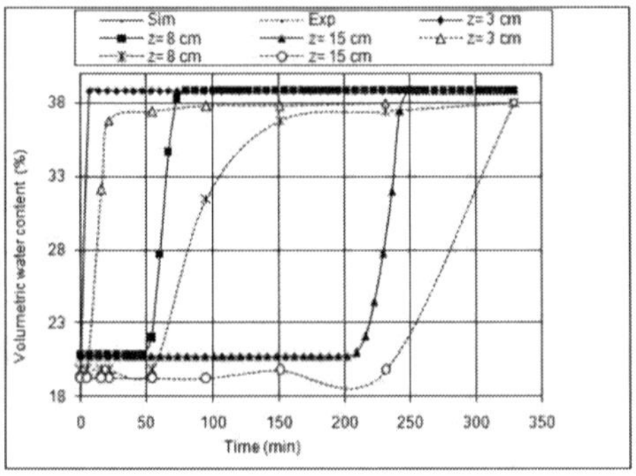

Figure 1. Comparison of levels of volumetric water.

The comparison of water profiles is shown in Figure 2. Figure 2a shows the profiles obtained experimentally while Figure 2b represents the simulated profiles using the software "Flow line temperature".

It is clear that the time taken for the measurement of θ experimentally are different from those obtained by simulation. Only recent instants (t simulation) are those which give a difference in the water status in relation to the previous state. Nevertheless, it is thought likely that the choice of time Δt has a considerable influence on the result.

It should be noted that there is a difference between t_{fexp}=330 min and t_{fsim}=265min although $\theta_{fexp}=\theta_{fsim}$=38%.

This difference may be related to the diffusion phenomenon slowed in the laboratory during the vertical infiltration due to gravity effects.

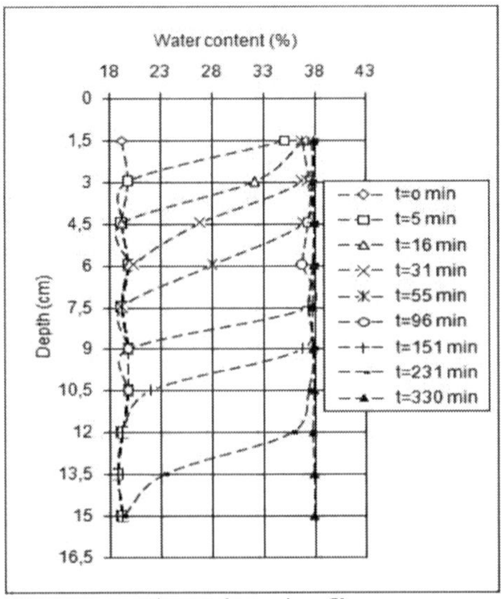

a) Experimental profile

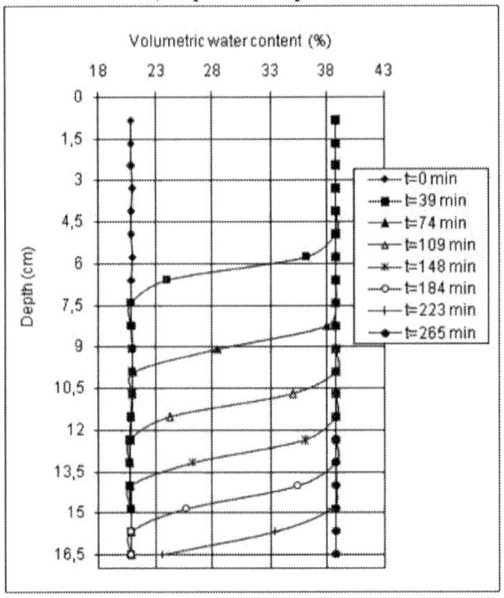

b) Simulation profile

Figure 2. Comparison between the contents profiles volumetric water experimentally obtained and those obtained by simulation.

4.3. Comparison of cumulative infiltration

Generally, infiltration increases monotonically until a limit corresponding to maximum infiltration. Figure 3 shows this increase in both the test and the simulation. One can see that the difference between simulation and experimental is increasing with larger periods of time.

One also observes that the saturation time is different for the two cases (a difference of 24.5%). This is consistent with observations made by [9].

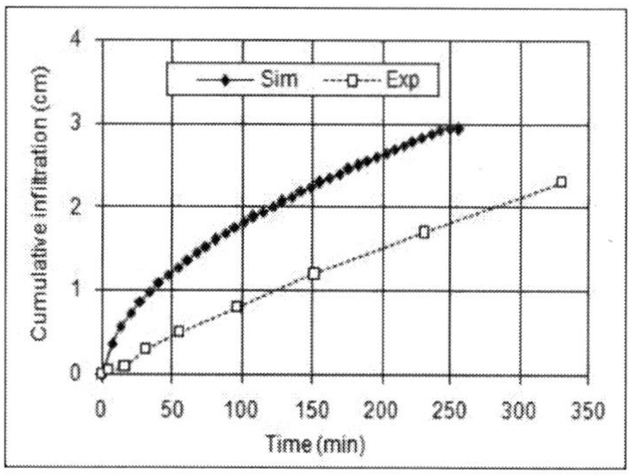

Figure 3: Comparison between the cumulative infiltration experimentally obtained and that obtained by simulation.

4.4. Comparison of the rate of infiltration

The description of the physical phenomenon of infiltration shows that the rate of infiltration is directly proportional to the humidity state of the soil. In the case of dry soil the rate of infiltration is relatively high.

Figure 4 is a graphic representation of the evolution of the infiltration rate with time. There is a decrease until the outflow corresponds to the complete saturation of the specimen.

The results obtained show a slight difference, that is larger for small values of time.

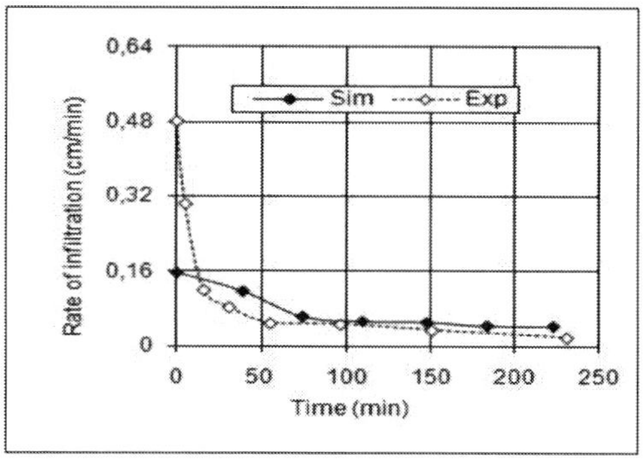

Figure 4. Comparing the rate of infiltration experimentally obtained with that obtained by simulation.

4.5. Comparing the depth of the humidification front

Figure 5 shows the evolution of the wetting front during infiltration. The observed trend is comparable to that of the cumulative infiltration. The progress of the front is directly related to the amount of water present in the sample. The difference between the experimental and the numerical range from 20% at t = 31 min and 28% at t = 223 min. The difference is smaller compared to that of the infiltration.

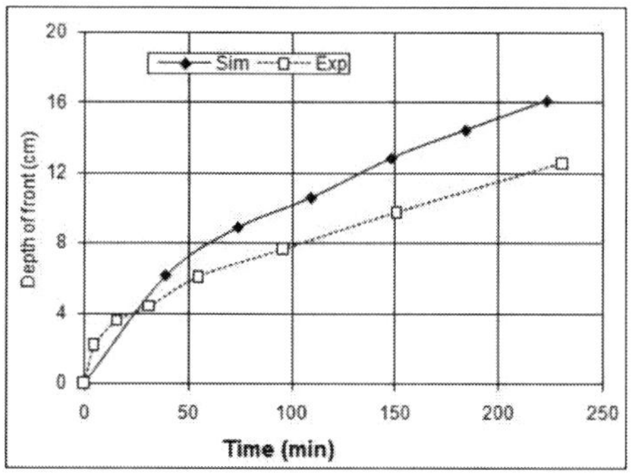

Figure 5. Comparing the depth of the humidity front obtained experimentally with that obtained by simulation.

4.6. Comparison of the rate of advance of the front

The evolution of the speed of the front with time is shown in Figure 6. This figure shows a difference between the experimental results and numerical which is especially important at the beginning. This difference may be related to the representativeness of the value of the volumetric water content θ and suction ψ determined by the filter paper measuring depth (top part, bottom or middle of the inter-ring) [2].

But we note that the value of the progression speed during saturation is almost the same for both models. This explains the validity of Darcy's law and the exponential decay with respect to the root of time justifying the main assumption in the numerical solution procedure.

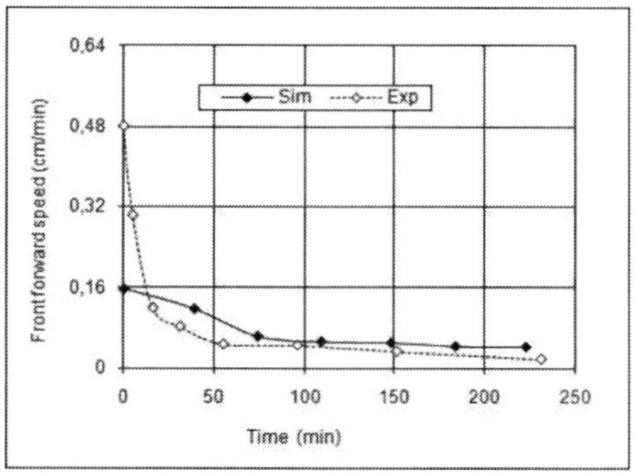

Figure 6. Comparison between the forward speed of the front and those obtained experimentally by simulation.

An indication of the variation as a function of drainage time given by simulation is shown in Figure 7. At a height h_0 of 120 cm a change is obtained at t = 208 min.

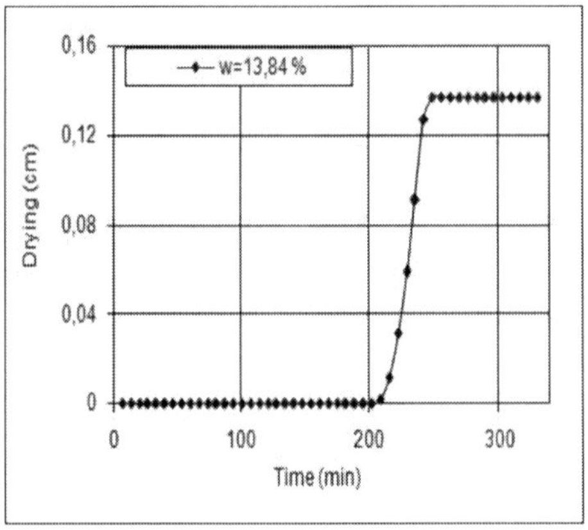

Figure 7. Cumulative drainage obtained by simulation for h_0=120cm.

5. Conclusion

The flow line temperature software is an interactive software that simulates the vertical flow of water in unsaturated soils. It is a convenient tool to predict the evolution of hydric soil profile with a wide range of possible boundary conditions, through a system of conventional numerical solution of the Richards equation. Flow must be characterized by water retention curves and hydraulic conductivity curves, the form may be selected from various analytical expressions of the literature. When parameters are not known, an inverse method is provided to estimate parameters from experimental data.

Comparing the numerical results with the experimental ones showed a slight difference that can be related to several causes including the mode of compaction used (static compaction), some errors in the measurements (error balance 0,001g) and also the influence of the filter papers measurements carried out during the infiltration.

Predicting the progression of the humidification front can be used together with the water content method for the prediction of swelling and shrink proposed by [19] and with the method of water profiles for the prediction of permeability of fine grained soils [18]. The ongoing work on models show whether this approach is good or not in the practice of engineering and geotechnical engineering.

References

1. Celia, M.A., Bouloutas, E.T., Zarba, R.L. A general mass-conservative numerical solution for the unsaturated flow equation. *Wat. Res. Res.* **26** (7), pp. 1483– 1496. (1990)
2. Didier G., Norotte V., Bentoumi O., Alimi-Ichola I. et Cazaux D. Protocole opératoire de mesure de la perméabilité en laboratoire- Cas des sols fins non saturés. Inst. Nat. Sci. App. Lyon. (1996)
3. Garnier, P. Détermination des caractéristiques hydrodynamiques de sols déformables par la méthode inverse. Ph.D. Dissertation, Nancy. (1996)
4. Garnier, P., Perrier, E., Angulo-Jaramillo, R., Baveye, P. Numerical model of 3-dimensional anisotropic deformation and 1-dimensional water flow in swelling soils. *So. Sci.* **162** (6), 410–420. (1997).
5. Gottardi, G., Venutelli, M. Richards: Computer program for the numerical simulation of one dimensional infiltration into unsaturated soil. Com. Geo. **19** (9), 1239–1266. (1993)
6. Haverkamp, R., Vauclin, M., Touma, J., Wierenga, P.J., Vachaud, G. A comparison of numerical simulation models for one dimensional infiltration. *Soil Sci. Soc. Am. J.* **41** (2), 285–294. (1977)
7. Hillel D. L'eau et le sol: Principes et processus physiques. Collection PEDASUP 5. (1988)
8. Ju, S.H., Kung, K.J.S., Mass types, element orders and solution schemes for the Richards equation. *Com. Geos.* **23** (2), 175–187. (1997)
9. Lafhaj Z., Samara M., Mania J., Sam R. et Hajnajib B. Mesures expérimentales de la perméabilité in situ des limons en zone urbaine au moyen de l'infiltromètre à double anneau et du perméamètre de Guelph. Coll. *GeocityNet*, Lille. (2005)
10. Leclerc, C., Le logiciel ECOUL, Mémoire de DESS Informatique. Université Henri Poincaré, Nancy. (1996).
11. Musy A. et Soutter M. Physique du sol. Presses Polytechniques et Universitaires Romande, *Collection Génie de l'environnement*, **6**. (1991)
12. Perrier E., Garnier P. et Leclerc C. ECOUL: An interactive computer tool to study hydraulic behavior of swelling and rigid soils. Com. Geos. **28** 1107–1118. (2002)
13. Perrier, E., Garnier, P., Leclerc, C. Profile-scale simulation of water flow: A software package to visualize and estimate soil hydraulic properties effects, poster and demonstration, In: van Genuchten, M.Th., Leij, F.J., Wu, L. (Eds.), *Proc. Int. Work. on Characterization and Measurement of the Hydraulic Properties of Unsaturated Porous Media*. University of California, pp. 349–354. (1997)
14. Richards L.A. Capillary conduction of liquids through porous medium. *Physic* **1**, pp.318-333. (1931)
15. Simunek, J., Van Genuchten, M.Th., Sejna, M., Using the HYDRUS-1D and HYDRUS-2D codes for estimating unsaturated soil hydraulic and solute. In: Van Genuchten, M.Th., Leij, F.J., Wu, L. (Eds.), *Proceedings of the International Workshop on Characterization and Measurement of the Hydraulic Properties of Unsaturated Porous Media*. University of California, pp. 1523–1536. (1999)
16. Van Dam, J.C., Hugen J., Wesseling J.G., Feddes R.A., Kabat P., van Walsum, P.E.V., Groenendijk, P., van Diepen, C.A., Theory of Swap version 2.0. Simulation of water flow, solute transport and plant growth in the soil–water–atmosphere–plant environment. Report 71, *Water Resources, Wageningen University, Technical Document 45*, Alterra Green World Research, Wageningen. (1997)
17. Van Genuchten, M.T. A closed form equation for predicting the hydraulic conductivity of unsaturated

soils. *Soil Science Society American Journal* **44**, 892–898. (1980)
18. Zadjaoui A., Bekkouche A. et Magnan J.P. La méthode des profils hydriques pour évaluer la perméabilité des sols fins ». *Symposium International : Sécheresse et constructions. SEC'08*, Marne-la-vallée, pp 287-294. (2008).
19. Briaud J.L. and Zhang X. Water content method for movement prediction of swell shrink soils. *Symposium International FondSup*, Vol. **2**. Magnan (éd.). (2004).
20. Verburg, K., Ross, P.J., Bristow, K.L. SWIMv2.1. User manual. *Division Report, CSIRO Division of Soil, CSIRO Information Services*, East Melbourne, Australia, 130. (1997).
21. Zadjaoui, A. Étude du transfert hydrique dans les sols non saturés : Échange sol – atmosphère. Ph.D. Dissertation, Tlemcen (Algeria). (2009).

Network modelling of fluid retention behaviour in unsaturated soils

Ignatios Athanasiadis, Simon J. Wheeler and Peter Grassl

School of Engineering, University of Glasgow, Glasgow, UK

Abstract. The paper describes discrete modelling of the retention behaviour of unsaturated porous materials. A network approach is used within a statistical volume element (SVE), suitable for subsequent use in hydro-mechanical analysis and incorporation within multi-scale numerical modelling. The soil pore structure is modelled by a network of cylindrical pipes connecting spheres, with the spheres representing soil voids and the pipes representing inter-connecting throats. The locations of pipes and spheres are determined by a Voronoi tessellation of the domain. Original aspects of the modelling include a form of periodic boundary condition implementation applied for the first time to this type of network, a new pore volume scaling technique to provide more realistic modelling and a new procedure for initiating drying or wetting paths in a network model employing periodic boundary conditions. Model simulations, employing two linear cumulative probability distributions to represent the distributions of sphere and pipe radii, are presented for the retention behaviour reported from a mercury porosimetry test on a sandstone.

1 Introduction

Realistic modelling of fluid retention behaviour of porous materials is important in many fields. Selection of parameter values within conventional analytical expressions for retention curves, such as [1], is normally based on back-fitting experimental retention curves. This does not provide insight to improve the understanding of the influence of microstructure on the fluid storage properties of the material. Discrete numerical modelling of fluid storage and transport in porous materials at the microstructural scale can improve this understanding.

Since the pioneering work of [2], network models have been widely used for fluid retention and hydraulic conductivity modelling of porous materials, particularly in fields other than geotechnical engineering. Authors such as [3] and [4] investigated the influence of mean coordination number, sphere and pipe size distributions and spatial correlation on retention curves and relative permeability curves, including the influence of phenomena such as trapping and snap off.

[2]-[4] aimed to improve qualitative understanding of the influence of microstructure on macroscopic fluid retention and transport properties of porous materials. Subsequently [5] proposed a calibration technique to quantitatively match experimental mercury intrusion and extrusion curves. This approach managed to reproduce retention curves for values of degree of saturation between 0.25 and 0.6. This calibration approach was then extended by [6], where the main objective was to deduce the material microstructure from experimental fluid retention curves and then evaluate the conductivity curves. Later this model was used in [7] for investigation of CO_2 storage capability of sandstones due to trapping.

Previous studies that attempted to quantitatively reproduce retention properties of porous materials by network modelling used computationally demanding algorithms and also resulted in relatively poor matching towards the tails of retention curves. Also, most models involved the generation of 3-D networks based on cubic arrays that induce a structural regularity that could produce problems for potential usage in hydro-mechanical analyses. Also periodic boundary conditions (PBCs) were applied on only 4 lateral faces of a statistical volume element (SVE), so that fluid could be injected from one of the remaining faces.

In the present work, a novel computationally efficient irregular network model for achieving quantitatively realistic fluid retention modelling is described. Original features of the model include a new way of setting up analyses with full PBCs and a novel pore volume scaling technique, which avoids the need for excessively large networks to represent materials with a wide range of void sizes. The approach is suitable for future extension to hydro-mechanical and multi-scale numerical modelling.

2 Network model

The discrete numerical model for the analysis of fluid retention is based on a three-dimensional network model of one-dimensional pipes that connect spheres, with the spheres representing soil voids and the pipes representing inter-connecting throats.

The network construction is based on a dual Delaunay-Voronoi tessellation. The domain is initially saturated with points placed randomly but with a restriction on the minimum separation distance d_{min}. The

Delaunay tessellation (consisting of irregular tetrahedra with their vertices at the points) is then applied and consequently its dual Voronoi tessellation is generated (consisting of irregular polyhedra, with each face of a polyhedron perpendicular to a Delaunay edge and intersecting this Delaunay edge at its mid-point). The spheres are then placed on the polyhedra vertices and the pipes along the polyhedra edges. This is a suitable approach for subsequent hydro-mechanical analyses, where the polyhedra are idealisations of the porous material grains surrounded by voids and throats, as discussed in [8]. Grain interactions can be modelled by placing mechanical elements along the Delaunay edges. A schematic presentation of four Voronoi polyhedra and their corresponding Delaunay edges is presented in figure 1, with each polyhedral grain scaled down in the figure in order to provide a clearer visual impression (in the actual network the grains are in contact).

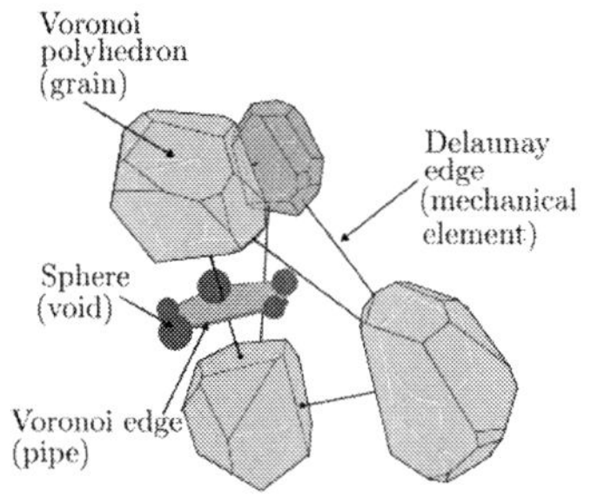

Figure 1. Schematic presentation of four Voronoi polyhedra and the corresponding Delaunay tetrahedron. The spheres and pipes corresponding to a Voronoi polyhedron face are also presented.

The work presented in this paper was restricted to analysis of fluid retention behaviour in a non-deforming material i.e. it did not include mechanical elements.

Spheres and pipes each have a distribution of radii, each with a specified probability distribution function. Sphere radii are allocated first, on a random basis, with no spatial correlation. Pipe radii are then allocated to locations, with the smallest radii pipes allocated first, commencing with locations connecting to the smallest spheres.

2.1 Periodic boundary conditions

The numerical analysis of fluid retention presented in this paper was conducted on a network representing a cubical statistical volume element (SVE). Periodic boundary conditions (PBCs) were applied on all 6 faces of the SVE, so that the SVE was considered to repeat indefinitely in all directions. Traditional techniques for PBCs, of Lagrange multipliers or four driving nodes, are not well suited for irregular Voronoi and Delaunay networks. Instead a different PBC technique was used. This technique was presented in [9] for mechanical lattice elements in a 2-D domain, and in the current work it was extended to 3-D transport networks. This PBC approach requires special treatment of the network construction, because the elements are allowed to cross the SVE boundaries. An example of a 3-D periodic transport network is presented in figure 2. It can be observed that the pipes, presented in grey, cross the SVE boundaries indicated by the black edges of the cell.

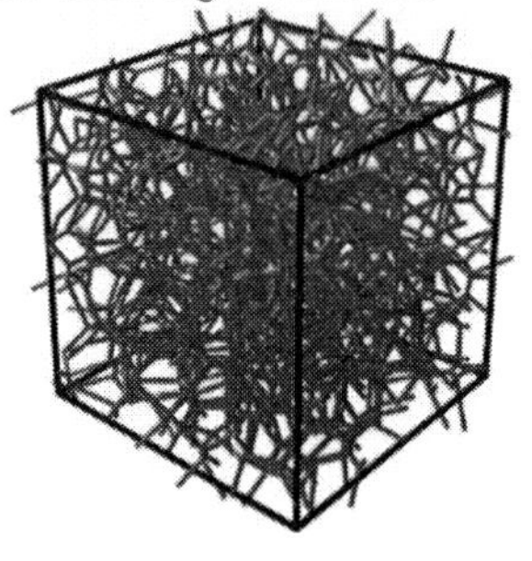

Figure 2. Example of a 3-D fully periodic network where the grey and the black edges represent the pipes and the SVE edges, respectively.

2.2 Pore volume scaling

For a material with a wide range of void sizes, a SVE would require a very large number of smaller voids in order to also include a statistically representative number of larger voids. With a standard network, this would require a SVE with an impractically large number of spheres. In addition, a standard network could not capture the fact that the average centre-to-centre spacing of smaller voids should be much less than the corresponding spacing for larger voids, otherwise adjacent small voids would be unreasonably far apart (suggesting an exceptionally low local porosity) and two adjacent large voids would be impossibly close together (suggesting overlap of the two voids). In order to avoid these problems, a new approach was introduced, where each sphere within the network represents a scaled number of voids, with this scaling number increasing as the sphere size reduces.

In the pore volume scaling technique, each sphere of radius r_s is taken to represent not a single void but a non-integer number N of voids, each of radius r_s, where N is given by

$$N = \left(\frac{r_{sm}}{r_s}\right)^3 \quad (1)$$

where r_{sm} is the mean radius of the spheres. Eq.1 means that each sphere of radius smaller than the mean represents more than a single void, whereas each sphere of radius greater than the mean represents less than one void. This scaled number of voids means that each sphere represents a volume V_v of voids given by

$$V_v = N\frac{4}{3}\pi r_s^3 = \frac{4}{3}\pi r_{sm}^3 \quad (2)$$

Eq.2 shows that each sphere, whatever its radius, represents the same volume of voids. The volume of pipes is ignored in calculating the porosity or degree of saturation of the SVE and therefore the porosity n of a cubical SVE of side length l_{SVE} containing a total of m spheres is simply given by

$$n = \frac{4\pi m r_{sm}^3}{3 l_{SVE}^3} \quad (3)$$

The advantage of the pore volume scaling approach is that the network model is able to represent a SVE containing a large number of small voids without excessive computational effort. In addition, local values of porosity internally within the SVE are realistic, rather than being unreasonably low in regions around smaller spheres and impossibly high in regions around the largest spheres. A disadvantage of the pore volume scaling approach is, however, that modelling of connectivity between voids, which is always imperfectly represented in a network model, may be represented less realistically than without scaling. This can be illustrated by considering the case of a sphere that is significantly smaller than the mean sphere radius r_{sm}.. This sphere represents a substantial number N of voids, rather than a single void. The scaling technique means that local connectivity between these N voids is overstated, as they are all assumed to be at the same location in the network and are therefore perfectly connected to each other. In contrast, external connectivity between these N voids and other voids is understated, as only 4 pipes connect the single sphere representing all these N voids to other spheres. Reverse arguments apply to voids that are larger than the mean radius.

3 Modelling retention behaviour

Capillary suction P_c , applied uniformly throughout the SVE, was the driving variable for drying and wetting of the SVE during modelling of retention behaviour. The applied value of P_c was monotonically increased during a drying process and monotonically decreased during a wetting process.

3.1 Representing drying and wetting processes

The rules that determine, for a given value of P_c, which spheres are filled with the wetting fluid and which are filled with the drying fluid are based on the Young-Laplace equation

$$P_c = 2\,\gamma \cos\theta \,/\, r \quad (4)$$

where γ is the surface tension and θ is the contact angle formed between the fluid-fluid interface and the solid (measured on the wetting fluid side). r is the radius of a pipe during a drying process and the radius of a sphere during a wetting process, to represent the so-called "ink bottle" effect. A drying or wetting process is modelled by considering the intruding fluid moving from a void already filled with that fluid to a connected void not previously filled with that fluid i.e. direct connection to a void already filled with the intruding fluid is imposed as a requirement for a void to fill with the intruding fluid. The drying fluid is the intruding fluid during a drying path, whereas the wetting fluid is the intruding fluid during a wetting path.

Figure 3 illustrates the situation during a drying process. The hatched area corresponds to the wetting fluid. The drying fluid has already intruded through pipe 1 into sphere A at a value of P_c corresponding to the application of Eq.4 with r as the radius of pipe 1. As the value of P_c is increased further during the drying process, the next consideration is when the drying fluid will intrude further from sphere A along either pipe 2 or pipe 3. Pipe 2 is of larger radius than pipe 3, and therefore the first action to occur is intrusion of the drying fluid along pipe 2 and into sphere B, at a value of P_c corresponding to the application of Eq.4 with r as the radius of pipe 2. If either pipe 4 or pipe 6 was larger than pipe 2 the intrusion would continue along this pipe into an additional sphere without need for further increase of P_c.

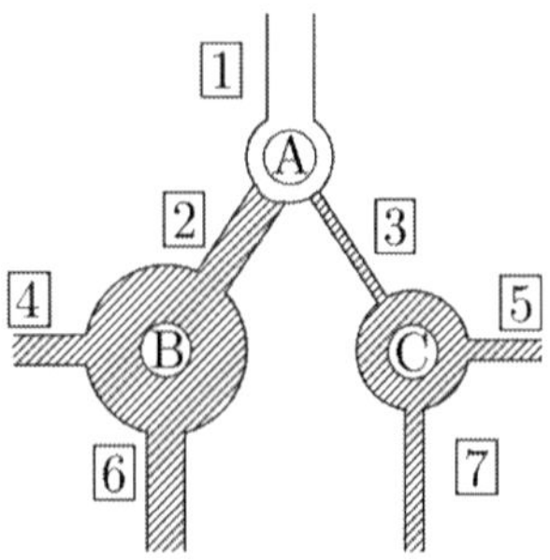

Figure 3. A simple 2-D network subjected to drying. The hatched areas represent the wetting fluid. The drying fluid is allowed to enter the network from pipe 1.

Figure 4 illustrates an equivalent situation for a wetting process, with the hatched area again representing the wetting fluid. Wetting fluid has already entered sphere A from pipe 1, at a value of P_c corresponding to the application of Eq.4 with r as the radius of sphere A, and immediately moved into pipes 2 and 3. As the value of P_c is reduced further during the wetting process, the next consideration is when the wetting fluid will intrude into sphere B or sphere C. As sphere C is smaller than sphere B, the first thing that happens is intrusion of the wetting fluid into sphere C, at a value of P_c corresponding to the application of Eq.4 with r as the radius of sphere C, and immediate further movement of the wetting fluid into pipes 5 and 7. If either pipe 5 or pipe 7 connected to an additional sphere that was smaller than sphere C, the wetting fluid would continue into this sphere without need for further decrease of P_c.

At each value of P_c during a drying or wetting process, the degree saturation S_r (expressed in terms of the wetting fluid) is calculated by considering the proportion of spheres filled with the wetting fluid (bearing in mind that each sphere, whatever its radius, represents the same volume of voids). Hence, if m is the total number of spheres in the SVE and m_w is the number of these spheres filled with the wetting fluid, the degree of saturation S_r is given by

$$S_r = \frac{m_w}{m} \quad (5)$$

DOI: 10.1051/e3sconf/20160911016

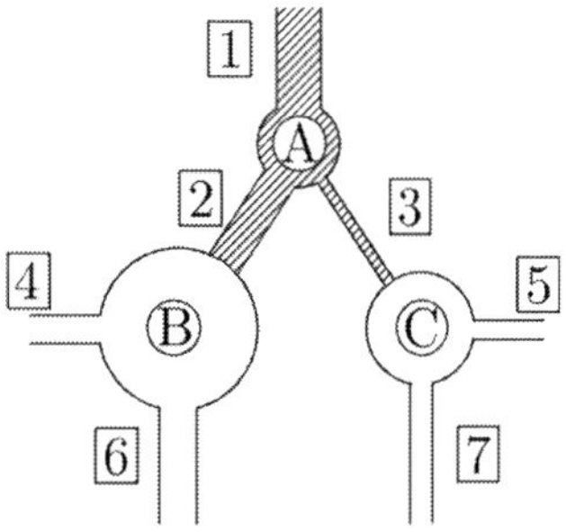

Figure 4. A simple 2-D network subjected to wetting. The hatched areas represent the wetting fluid. The wetting fluid is allowed to enter the network from pipe 1.

In analyzing drying and wetting processes, no possibility of trapping of the non-intruding fluid is considered, because there is no consideration of whether there is appropriate connectivity to provide an exit route for the non-intruding fluid (connectivity requirements are applied to the intruding fluid, but not to the non-intruding fluid). In practice, trapping of the non-intruding fluid is not an issue if the non-intruding fluid can escape by diffusing through the intruding fluid and provided that drying or wetting is performed sufficiently slowly for the diffusion process to dissipate excess pressure in the trapped non-intruding fluid.

3.2 Initiation

With the use of PBCs, it is not possible to start a drying path from a fully saturated condition ($S_r = 1$) or a wetting path from a fully dry condition ($S_r = 0$), because there is no external boundary from which to introduce the intruding fluid. Instead, a drying path must start with at least one sphere within the SVE already filled with the drying fluid and a wetting path must start with at least one sphere already filled with the wetting fluid.

Investigation showed that realistic modelling of a drying or wetting path could not be achieved by starting with only a single sphere filled with the intruding fluid, because of the extremely low connectivity (only 4 pipes) from a single sphere and the consequent possibility that extreme changes of P_c might be required for the intruding fluid to break out from this first sphere (dependent on only the radii of the 4 pipes in the case of a drying path or the radii of the 4 spheres on the other ends of these pipes in the case of a wetting path) or to break out from the local region immediately surrounding the first sphere. In order to achieve realistic modelling, drying or wetting paths should start with a small number of spheres already filled with the intruding fluid and these "seeding" spheres should not simply be selected at random. Instead, to achieve realistic modelling of a drying path, it should always be preceded by modelling of a previous wetting path finishing with a small number of spheres (the seeding spheres) still filled with the drying fluid. A similar condition applies for realistic modelling of a wetting path, which should always be preceded by modelling of a previous drying path to leave an appropriate number of seeding spheres filled with the wetting fluid.

Investigation of this initiation process allowed an appropriate value to be selected for the number of seeding spheres (and hence the appropriate maximum starting value of S_r for realistic modelling of a drying path or appropriate minimum starting value of S_r for realistic modelling of a wetting path). This was confirmed by the fact that on subsequent cycling between these maximum and minimum values of S_r each cycle concluded with exactly the same configuration of spheres filled with the drying fluid.

4 Application

The network model, including the pore volume scaling technique and the initiation procedure, both described above, was used to study the retention behaviour reported by [5] from a mercury porosimetry test on sandstone.

4.1 Procedure and input parameters

In a mercury porosimetry test, air is the wetting fluid and mercury is the drying fluid, with mercury intrusion and extrusion corresponding to drying and wetting paths respectively. The contact angle θ measured on the air (wetting fluid) side is 40° (see Table 1).

The square data points in figure 5 show the experimental results of the wetting (mercury extrusion) and drying (mercury intrusion) curves respectively in S_r-P_c space, where S_r is the degree of saturation of the air (the wetting fluid). For the wetting path, the maximum value of S_r reported in the experimental results was 0.6, because the authors in [5] stated that trapping of mercury prevented significant further increase of S_r, presumably because the wetting (mercury extrusion) was performed insufficiently slowly to allow escape of trapped mercury by evaporation and subsequent diffusion through the air.

With the pore volume scaling technique described in Section 2.2, the fact that each sphere represents the same volume of voids means that the drying curve predicted by the network model (in S_r-P_c space) depends upon only the distribution of pipe (throat) sizes and the network connectivity i.e. it does not depend on the distribution of sphere (void) sizes. Similarly, the wetting curve predicted by the network model depends upon only the distribution of sphere sizes and the network connectivity i.e. it does not depend upon the distribution of pipe sizes. Probability distribution functions for the pipe and sphere radii were therefore selected by interpretation of the experimental results of the drying and wetting curves respectively.

Figure 6 shows the experimental results re-plotted as S_r against back-calculated sphere or pipe radius r, by using Eq.4 to convert values of P_c to values of r. Of course, even for the idealised case of a network model simulation where the pore volume scaling technique was employed, these plots of S_r against back-calculated values of r would not represent the true cumulative probability distribution of sphere or pipe radii, because of the influence of network connectivity. For example, a network model wetting curve simulation presented in this way would only give the cumulative probability distribution of sphere sizes used in the network if there had been perfect connectivity, with access of the wetting fluid to all spheres throughout the wetting process, so that spheres could be filled with the wetting fluid in the exact sequence of increasing sphere size. Similarly, a network model drying curve simulation of S_r against back-calculated value of r would only correspond to the true cumulative probability distribution of pipe radii used in the network if network connectivity allowed filling of pipes in the exact sequence of decreasing pipe size and if each pipe then provided access to a sphere that was not previously filled with the drying fluid (the latter is not true because some spheres will already have been filled with drying fluid entering by another route). Despite these issues, it is informative to use the experimental results of S_r against back-calculated radius, shown in figure 6, to select the distributions of sphere and pipe radii for the network model.

Table 1. Main input parameters.

Parameters	Value	Units
Mercury contact angle θ	40	°
Mercury surface tension γ	0.48	N m^{-1}
a_p	2.0e+4	m^{-1}
b_p	2.7e-5	
a_s	2.9e+3	m^{-1}
b_s	3.9e-6	
d_{min}	0.756e-3	m
l_{SVE}	5.4e-3	m

Inspection of figure 6 shows that the experimental plots of S_r against back-calculated value of r (from Eq.4) for both drying and wetting paths can each be approximated by a straight line, at least up to a certain value of S_r. For the drying curve this is true up to approximately $S_r = 0.8$ and for the wetting curve it is true up to the final experimental point ($S_r = 0.6$). Two best-fit straight lines were fitted to these parts of the two experimental sets of data, as shown in figure 6, to give an equation for the cumulative probability distribution of pipe or sphere sizes for input to the network model:

$$f = ar - b \qquad (6)$$

Values of the coefficients a and b for pipes and spheres are given in Table 1. The linear cumulative probability distribution of Eq.6 corresponds to a uniform probability distribution between a minimum radius of b/a and a maximum radius of $(1 + b)/a$, and a mean radius of $(1 + 2b)/(2a)$. With the values of a and b presented in Table 1, the pipe radii varied between 1.35 e-9 m and 5.00 e-5 m, the sphere radii varied between 1.35 e-9 m and 3.44 e-4 m and the mean sphere radius r_{sm} was 1.72 e-4 m.

Having defined the mean sphere radius r_{sm}, the values of l_{SVE} (the SVE side length) and d_{min} (the minimum distance between the randomly positioned points forming the Delaunay tessellation) were selected to give an appropriate number m of spheres within the SVE and to match the reported porosity of the sandstone of 0.21. The values of l_{SVE} and d_{min} shown in Table 1 produced a SVE containing 1560 spheres.

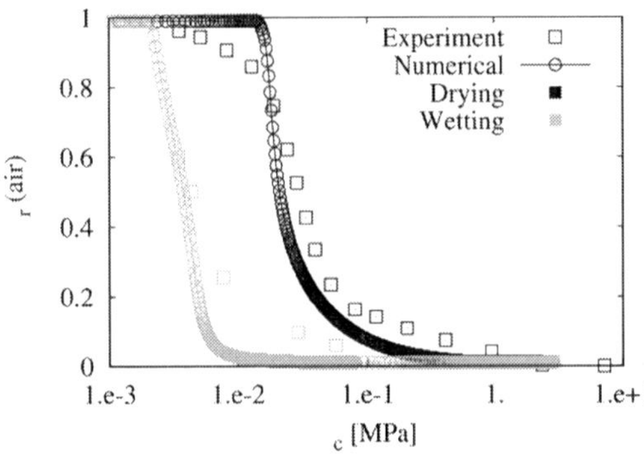

Figure 5. Comparison of the experimental results ([5]) to the mean numerical results in S_r-P_c space.

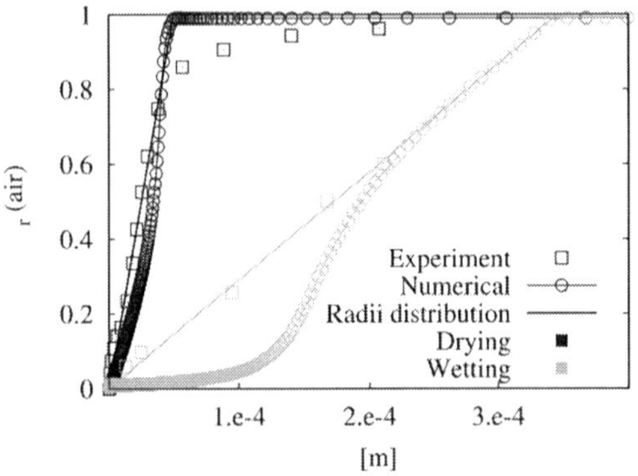

Figure 6. Comparison of the experimental results ([5]) and the cumulative radii distribution curves to the mean numerical results in the S_r-r space.

A series of 100 analyses was performed. The same network geometry was used for all 100 analyses, but different sphere and pipe radii generation and allocation were undertaken for each analysis. The number of the seeding spheres used for the initiation procedure described in Section 3.2 was 1% of the total number of spheres, so that drying and wetting curves were simulated over a range of S_r from 0.99 to 0.01. The same increments of suction were applied for all 100 analyses.

4.2 Results

The circular data points in figure 5 show the results of the network model simulations in S_r-P_c space. Values of S_r represent the average from the 100 analyses. Comparison with the experimental results shows that the network model simulation of the drying curve overestimates values of S_r at low values of P_c and underestimates values of S_r at high values of P_c. The network model simulation of the wetting curve underestimates values of S_r at all but the lowest experimental value of P_c.

The circular data points in figure 6 show the results of the network model simulations re-plotted in the space of S_r against back-calculated radius r (where Eq. 4 was used to convert values of P_c to corresponding values of r). Figure 6 provides insights into some of the mismatches between model simulations and experimental results in figure 5.

In figure 6, comparison of the network model simulations of S_r against back-calculated r with the cumulative probability distributions of spheres and pipes used as input for the network model (shown by the straight lines in figure 6) highlights the influence of network connectivity on wetting and drying curves. For example, the network model simulation of the wetting curve predicts lower values of S_r than those corresponding to the cumulative probability distribution of sphere sizes, because network connectivity means that some smaller spheres cannot be intruded with the wetting fluid at their expected value of P_c, because access of the wetting fluid to these spheres is shielded by some larger spheres. As the degree of saturation increases, the network model simulation converges with the cumulative probability distribution of spheres, because this shielding effect reduces as more spheres are filled with the wetting fluid.

The model simulations in figure 5 could be tuned to better match the experimental results by adjusting the values of some of the input parameters in Table 1. This would mean, however, that the method had reverted to no more than a curve fitting exercise. Proper assessment of the achievements and limitations of network model simulations of retention behaviour can be achieved only by using inputs for the sphere and pipe probability distribution functions derived from independent imaging of the material microstructure, rather than from back-analysis of retention behaviour from mercury porosimetry tests.

One significant point to emerge from figure 5 is that the match achieved between model simulations and experimental results was better than in previous network model simulations [10] which did not use the pore volume scaling technique, even though these earlier simulations included substantial tuning of input parameter values to try to achieve the best possible match. This provides encouraging evidence on the usefulness of the pore volume scaling technique.

5 Conclusions

A network model was used to study retention behaviour in unsaturated materials. Original aspects of the modelling included a form of periodic boundary conditions applied for the first time to a 3-D transport network, a new pore volume scaling technique and a new procedure for initiating drying or wetting paths in a network model employing periodic boundary conditions. The model is suitable for future extension to hydro-mechanical and multi-scale analysis. The model was applied to the simulation of a mercury porosimetry test on sandstone, with linear cumulative probability distribution functions selected for both sphere and pipe sizes, based on back-analysis of the experimental test results. Comparison of model simulations with both experimental results and the cumulative probability distributions used for sphere and pipe sizes provided useful insights into the roles of network connectivity on drying and wetting retention curves predicted by network models.

Acknowledgments

The authors acknowledge funding received from the UK Engineering and Physical Sciences Research Council (EPSRC) under Grant EP/I036427/1, together with funding from Radioactive Waste Management Limited (RWM) (http://www.nda.gov.uk/rwm), a wholly-owned subsidiary of the Nuclear Decommissioning Authority. RWM is committed to the open publication of such work in peer reviewed literature, and welcomes e-feedback to rwmdfeedback@nda.gov.uk. The simulations were performed with the object-oriented finite element package OOFEM (www.oofem.org) extended by the present authors.

References

1. M.T. Van Genuchten, Soil. Sci. Soc. Am. J., **44**, pp. 892-898 (1980)
2. I. Fatt, Pet. Trans. AIME, **207**, pp. 103-130 (1956)
3. C.D. Tsakiroglou, A.C. Payatakes, J. Colloid Sci. J., **137**, pp. 315-339 (1990)
4. G. R. Jerauld, S.J. Salter, Transp. Porous Media, **5**, pp. 103-151 (1990)
5. P.G. Matthews, C.J. Ridgway, M.C. Spearing, J. Colloid Interface Sci., **171**, pp. 8-27 (1995)
6. R. Campos, I. Barrios, J. Lillo, Energy Rep., **1**, pp. 71-79 (2015)
7. A. Johnson, I.M. Roy, G.P. Matthews, D. Patel, Eur. J. Soil. Sci., **54**, pp. 477- 489 (2003)
8. G. Manson, J. Colloid Sci. J., **35**, pp. 279- 287 (1971)
9. P. Grassl, M. Jirásek, Int. J. Solids Struct., 47, pp. 957-968 (2010)
10. I. Athanasiadis, S. Wheeler, P. Grassl, *International Symposium on Geomechanics from Micro to Macro* (2014)

On the mechanisms and kinetics of evaporation of a compacted silt

Régis Mpawenayo[1,a], Frédéric Debaste[2], Madjid Douzane[1], Pierre Gerard[1]

[1] *Université libre de Bruxelles (ULB), BATir Department, Laboratory of GeoMechanics, Brussels, Belgium*
[2] *Université libre de Bruxelles (ULB), TIPs Department, Brussels, Belgium*

Abstract. In order to define a convective evaporation model able to reproduce the evaporation rates according to the atmospheric conditions, drying tests have been performed on cylindrical samples of a compacted silt in a closed chamber where the relative humidity of the air is controlled by means of saturated saline solutions. Calibrating such models using experimental results from small sample sizes can indeed be an interesting alternative to simplify the experimental procedures and to reduce the test duration in comparison with the standard in-situ tests (e.g. lysimeter). However the influence of the size of the samples on the kinetics of evaporation has to be investigated. Samples with different diameters and heights are therefore tested. The results show that the observed mechanisms of drying are similar to those at the scale of the geostructure if the dimensions of the sample are higher than a critical height and a critical diameter. We show then that the experimental critical height is consistent with the concept of critical length proposed by Lehman et al. (2008), who expresses analytically when the hydraulic connection between the sample surface and the drying front is upset. Finally the kinetics of evaporation recorded under different relative humidities highlight that the mass transfer coefficient defined in the convective evaporation model does not depend on the relative humidity.

1 Introduction

Soil-atmosphere interactions relate to the moisture transfers occurring between a geomaterial and its surroundings. These transfers of moisture between the soil and atmosphere include infiltration, evaporation, evapotranspiration and runoff. The understanding and quantification of these transfers is a crucial issue in many civil engineering applications: slope stability problems, soil cover systems for waste materials or constructions built on expansive clay. In this paper, we focus on the mechanisms of evaporation in soils.

It is now admitted that the experimental evaporation rates measured can be successfully reproduced numerically by means of vapour flux at the soil surface equal to the product of a driving force (the difference between the vapour density at the interface and of the surroundings) and a mass transfer coefficient characterizing the resistance of a boundary layer where the exchanges with the surroundings take place [1, 2]. Some correlations between the evaporation rates and the climatic conditions, as the wind speed, the air relative humidity and temperature, have been determined [3, 4, 5, 6]. Many of these correlations are calibrated using tests carried out in situ or in an environmental chamber [7, 8, 3]. The main drawback of field-scale experiments is that atmospheric conditions are naturally changing, which makes difficult separating the influence of ambient conditions and soil properties on the evaporation rate measurement. To avoid this drawback, it is proposed to perform drying tests in laboratory where atmospheric conditions are controlled. Environmental chamber provides a relevant solution [7], but the large volume of soil (1 m^3) tested increases the time needed for the soil preparation and installation, as well as the test duration. It is why it is proposed to perform laboratory drying tests on small samples with different sizes. The question of the influence that could have the sample sizes (the evaporation surface and the height of the sample) on the kinetics of evaporation is not well documented. This raises the question of knowing the minimal dimensions of the samples needed to obtain relevant evaporation rates useable for large-scale applications.

This paper proposes to carry out an experimental campaign focusing on the effect on the influence of the sample dimensions on the kinetics of evaporation and especially on the mass transfer coefficients. A drying chamber where the relative humidity is controlled by means of saturated saline solutions is developed and allows performing evaporation tests without any forced convection. This paper proposes also new insights into the evaporation process at soil surface and the understanding of the boundary layer mechanisms.

[a] rmpawena@ulb.ac.be

2 Material and methods

2.1 Material and sample preparation

The soil examined in this research is a sandy silt from Marche-les-Dames (Belgium) (Unified Soil Classification System (USCS)). Its index properties are: liquid limit (w_L) = 32.5 %; plasticity index (IP) = 15%. The grain-size distribution is: 13 % of clay, 64 % of silt and 26 % of sand. The permeability of the saturated silt compacted first at the Optimum Proctor compaction conditions (initial water content w_0 = 14.8%; initial dry density ρ_{d0} = 18.40 kN/m³ - see [9] for more details) and then resaturated is equal to 10^{-10} m/s. The determination of the retention curve of the silt is based on the use of 2 techniques: the pressure plate apparatus [10] and the vapour transfer technique [11]. Experimental data are presented in Figure 1, where a van Genuchten model [12] has been fitted on the data.

$$S_r = \left[1 + \left(\frac{s}{P_r}\right)^n\right]^{-(1-1/n)} \quad (1)$$

with P_r = 1000 kPa and n = 1.4.

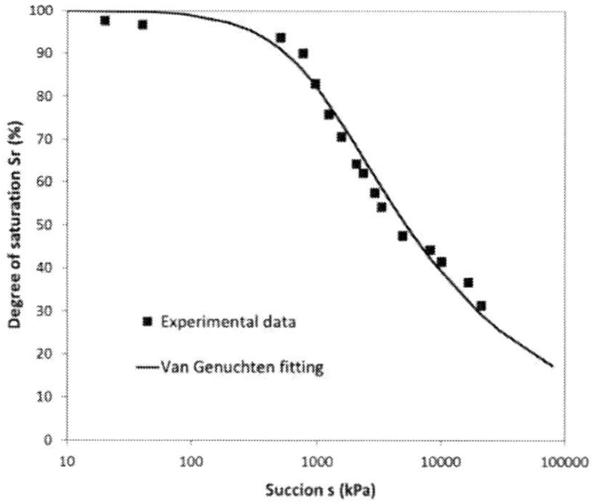

Figure 1. Retention curve of Marche-les-Dames silt: experimental data and fitting of van Genuchten relationship.

Soil samples with different dimensions (diameter D=15; 35; 50 mm – height H=15; 25; 45; 70 mm) are prepared at the optimal compaction conditions. For samples with a final height of 15 mm specimen are statically compacted in oedometer cell (D=50 mm; H=15 mm) and then saturated in oedometer cell. When saturated conditions are met the samples are cut with hollow cylinder cutter of different diameters. For samples with a final height higher than 15 mm specimen are statically compacted in triaxial mold (D=35 mm; H=70 mm). Samples are then saturated in a triaxial cell with a confining pressure of 60 kPa and a water pressure of 50 kPa applied on the top of the sample. The saturation is reached when the flowrate of water injected in the sample through lower porous stone is equal to the flowrate of water measured through the upper porous stone.

In order to create drying conditions as close as possible to 1D flow all the core of saturated samples are covered with a plastic film and aluminium foil and finally wrapped with a paraffin layer to prevent evaporation from the base and lateral surfaces of the samples and to minimize the lateral water flow (Figure 2).

Figure 2. Imposition of 1D vapour transfers within the sample.

2.2 Experimental methods

The principle of the determination of the mass transfer coefficients is based upon measurements of mass loss of a sample during drying test. So, a drying chamber without any forced convection has been developed. A cylindrical sample (initially saturated) is introduced inside a hermetic chamber in which the relative humidity (RH) is maintained. A saturated saline solution has been used to control the relative humidity in the chamber thanks to vapour transfer technique [11]: $MgCl_2$ (RH=33%). The samples are placed on a stand that hangs above the saline solution and that is connected to a weighing machine (Figure 3). Some tests are performed on very small samples, which implies the use of a high-precision weighing machine (accuracy 0.0001 g). The mass loss is recorded every 5 min.

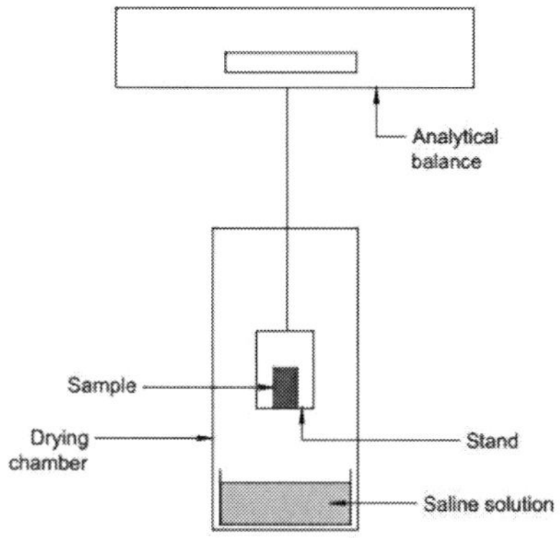

Figure 3. Schematic representation of the set-up for drying without any forced convection.

2.3 Data treatment

Experimental data of evolution of sample mass $M(t)$ as a function of time t are treated to present the Krischer curve, i.e. the drying rate q (kg.m^{-2}.s^{-1}) as a function of the water content w. The drying rate q at a given time is calculated as follows:

$$q = \frac{M(t) - M(t - \Delta t)}{A \Delta t} \quad (2)$$

where Δt the time interval between two measurements, A is the area of the external surface of the sample that contributes to the evaporation.

The drying rate calculated based on equation (2) is filtrated using Lanczos method [13] to reduce noise generated by the subtraction between two mass measurements that only differ from an order of magnitude slightly larger than the weighing measurement error. The obtained curve highlights the two classical distinct drying stages [14, 15]. The stage I, for water content between the initial water content w_0 and a lower limit value called the critical water content w_c, shows a constant drying rate q_{max}. During this stage, the soil surface is at or near saturation and the evaporation rate is limited by the diffusion in the external boundary layer. The second period, for water content lower than the critical value, shows a falling drying rate that corresponds to a limiting and growing internal mass transfer resistance. Stage II of drying starts when the internal conductive properties of the soil (permeability, diffusivity) do no longer allow a sufficient water flow to the soil surface to maintain the maximal evaporation rate. Therefore, the transition between the two periods, around the critical water content, expresses the progressive development of an internal mass transfer resistance in series to the previously limiting external mass transfer resistance [13, 16].

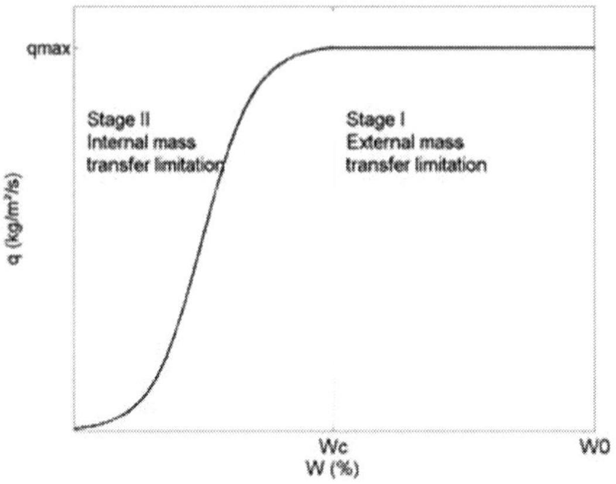

Figure 4. Typical Krischer curve highlighting Stage I (w>wc) and stage II (w<w_c).

From the Krischer curves obtained, the maximal drying rates and the critical water contents w_{crit} are deduced. For stage I, the constant mass transfer coefficient α can be evaluated based on the value of q_{max} and assuming a saturated surface at the sample top [17].

$$q_{max} = \alpha(\rho_{v,0,sample} - \rho_{v,air}) \quad (3)$$

So the mass transfer coefficient is given by:

$$\alpha = \frac{q_{max}}{\left(\rho_{v,0,sample} - \rho_{v,air}\right)} \quad (4)$$

with $\rho_{v,0,sample}$ the saturated vapour density at soil surface which depends only on the temperature of the interface.

Various empiric or thermodynamics expressions allow to evaluate this dependence. In this work, the correlation from [18] is used:

$$\frac{1}{\rho_{\upsilon,0}} = 194.4 \exp(-0.06374T + 0.1634\,10^{-2}T^2) \quad (5)$$

with T the temperature (°C).

As evaporation is an endothermic process, the local interfacial temperature is not necessarily equal to surrounding temperature. When the evaporation rate remains constant (the maximum evaporation rate q_{max}), the temperature at the sample surface is considered equal to the wet bulb temperature. The wet bulb temperature T_h is calculated with the relationship [19]:

$$T_h = \frac{\left[1810.9\left[p_{v,0}(T_h) - p_v(T_{air})\right]\right] - T_{air}\left[p_{atm} - p_{v,0}(T_h)\right]}{2p_{v,0}(T_h) - p_v(T_{air}) - p_{atm}} \quad (6)$$

with T_{air} the air temperature (°C), p_{atm} the atmospheric pressure, $p_{v,0}$ the saturated vapour pressure and p_v the vapour pressure.

3 Results and discussions

7 cylindrical samples of compacted silt with different diameters D and heights H have been dried in the drying chamber without any forced convection under air relative humidity RH=33%. The air temperature in the room where the tests are performed is constant and is equal to 24°C. From the Krischer curves the maximal drying rates and the critical water contents w_c are deduced. From equation (4) the mass transfer coefficients are calculated. Those different values characterizing the drying kinetics are reported in Table 1.

 DOI: 10.1051/e3sconf/20160912001

Table 1. Experimental conditions and results of tests without forced convection.

N°	Diameter D (mm)	Height H (mm)	Air relative humidity RH (%)	Maximal drying rate q_{max} (kg.s^{-1}.m^{-2})	Mass transfer coeffiient α (m/s)	Critical water content w_c (-)
1	50	15	33	2.4 10-5	$5.32\ 10^{-3}$	0.08
2	35	15	33	3.0 10-5	$6.06\ 10^{-3}$	0.09
3	15	15	33	5.50 10-5	$1.11\ 10^{-2}$	0.10
4	15	15	33	5.0010-5	$1.01\ 10^{-2}$	0.10
5	35	70	33	3.0 10-5	$6.82\ 10^{-3}$	0.14
6	35	45	33	2.8 10-5	$6.37\ 10^{-3}$	0.11
7	35	25	33	3.0 10-5	$6.82\ 10^{-3}$	0.095

3.1 The influence of the sample surface on the kinetics of evaporation

Soil samples with the same height H=15 mm, prepared under the same conditions but with different evaporation surfaces, were dried in the same drying chamber in the same relative humidity RH=33% (Table 1, N°1 to 4). The results show that the drying rate is higher for the samples with diameter D=15 mm than for samples with diameter D=35 mm. But the evaporation flows for sample with D=35 mm and sample with D= 50 mm are similar (Figure 5). The mass transfer coefficients also decrease from diameter D=15 mm to D=35 mm, but then reach a plateau (Figure 6). The mass transfer coefficients remain indeed constant when diameter increases to 50 mm. Physically, this can be explained by the development of the boundary layer around the sample surface, and especially along the edges. At the edge of the sample, evaporation can take place in both horizontal and vertical directions, while only vertical evaporation is allowed at the center of the top surface. The thickness of the boundary layer (and so the value of the local mass transfer coefficient) is therefore not constant all along the top surface, but the influence of this edge effect becomes negligible when the diameter increases (Figure 7). This is why the mass transfer coefficients determined experimentally on samples with diameters of 35 and 50 mm are equivalent, which is not the case with smaller diameters.

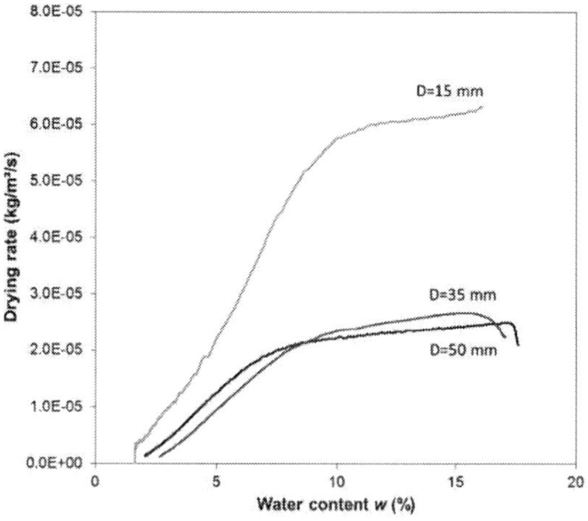

Figure 5. Influence of the sample surface on the drying rate (H=15 mm – RH=33%)

The same trend is observed on the values of the critical water contents (Table 1), where the values for small diameters (15 mm) are always higher than for the other dimensions. As the evaporation is high on the edges of the sample, it is there that the soil properties will first limit the evaporation rate. Once the soil properties begin to limit the evaporation rate at the sample edges, the second evaporation phase begins, which allows defining the critical water content. This decrease of the

evaporation rate is not uniform on entire evaporation surface because in the middle of the evaporation surface, there is still enough water to maintain maximum evaporation rate. The effect of the edge on evaporation rate is negligible for sample with large surface of evaporation.

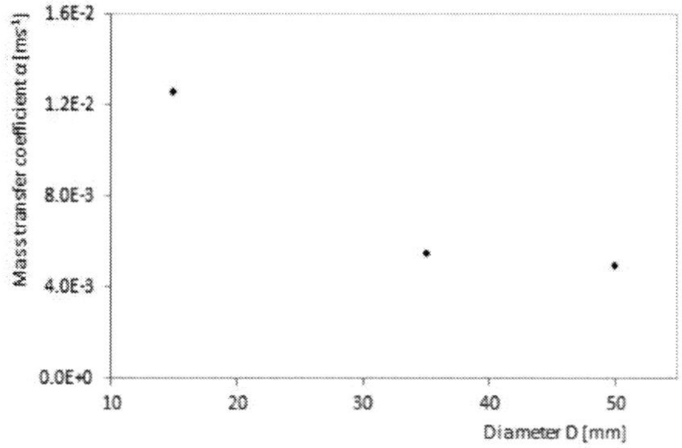

Figure 6. Influence of the sample surface the mass transfer coefficient

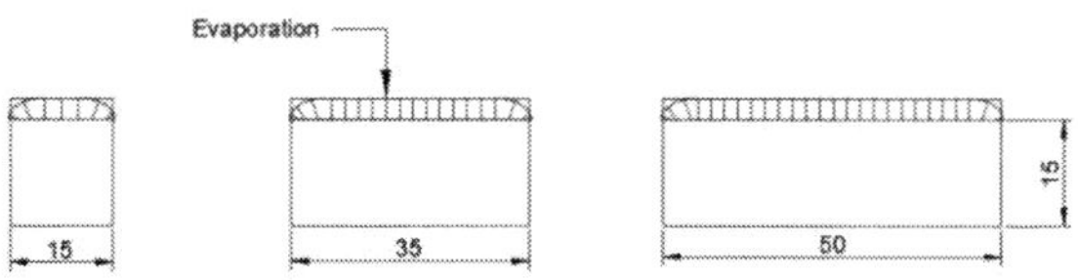

Figure 7. Development of the boundary layer at the surface – Edge effect

3.2 The influence of the sample height on the kinetics of evaporation

The influence of the sample height is analyzed by the comparison of the drying rate of 4 samples with the same diameter (D=35 mm) and submitted to the same drying conditions (RH=33%), but with different heights (H=15, 25, 45, 70 mm – Tests 2 , 5 to 7 in Table 1). Figure 8 and 9 present the drying rates of the different tests according to the time and the water content respectively. First the results show that the maximum drying rate q_{max} is not dependent on the sample height. On the other hand the critical water content w_c is influenced by the sample height, as confirmed by the data from Table 1. Finally the temporal analysis suggests that the curves drying rate vs. time are the same for heights of 45 and 70 mm.

Those results are consistent with the concept of critical length affecting drying of porous media proposed by [20]. They considered that the transition between Stage I and Stage II of drying is explained by a modification of the fluid transfers mechanisms within the sample, from a liquid-flow based first stage to a second stage controlled by vapor diffusion through the sample and the receding of a drying front. The critical length determines the maximal extent along which water can flow to ensure a hydraulic connection between the sample surface and this internal drying front. They proposed a relation to estimate this critical length based on the equilibrium between upwards capillary forces and downwards viscous and gravity forces to whom are submitted the interstitial water.

$$\Delta h_{cap} = \Delta h_{visc} + \Delta h_{grav} \tag{7}$$

with Δh_{cap}, Δh_{visc}, Δh_{grav} the capillary, viscous and gravity head differences respectively.

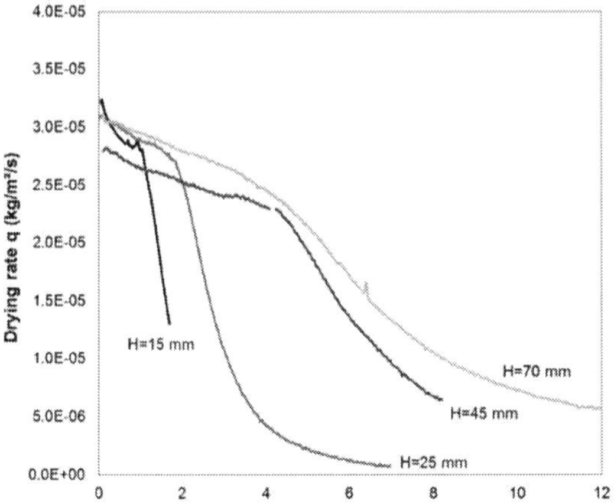

Figure 8. Drying rate according to the time for different sample heights (D=35 mm; RH=33%).

The capillary head difference is affected by the soil microstructure. [20] proposed to consider the slope of the tangent to the retention curve at the inflection point to estimate the capillary pressure difference. Using a van Genuchten retention model (equation1), the capillary head difference can be obtained from (more details in [20]) by equation 8.

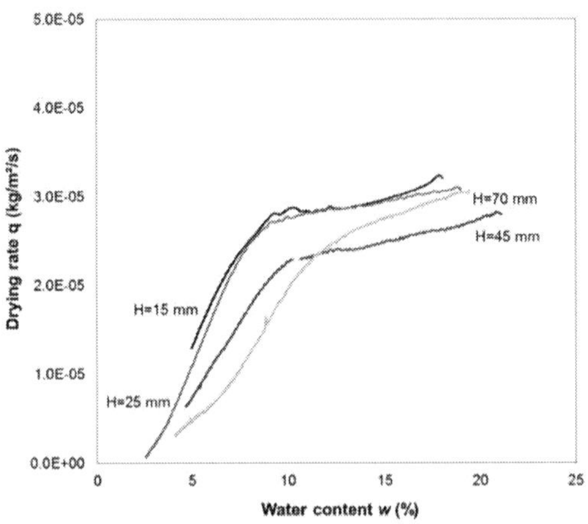

Figure 9. Drying rate according to the water content for different sample heights (D=35 mm; RH=33%).

$$\Delta h = \frac{P_r}{n-1}\left(\frac{2n-1}{n}\right)^{(2n-1)/n}\left(\frac{n-1}{n}\right)^{(1-n)/n} \tag{8}$$

with P_r and n the parameters of the van Genuchten retention model (equation 1).

The viscous forces are estimated from the Darcy law for unsaturated flow. They depend on the maximal drying rate, which is itself influenced by the atmospheric conditions.

$$\Delta h_{visc} = \frac{L_{crit} q_{\max}}{\overline{\overline{K(S_r)}}\rho_w} \tag{9}$$

with L_{crit} the critical length, q_{max} the maximum drying rate, ρ_w the water density, $\overline{\overline{K(S_r)}}$ the mean hydraulic permeability (which depends on the degree of saturation S_r) between the surface and the drying front assuming a hydrostatic water content distribution above the drying front.

The gravity forces depend only on the distance between the drying front and the surface, which corresponds to the critical length.

$$\Delta h_{grav} = L_{crit} \tag{10}$$

The characteristic length can be finally expressed as:

$$L_{crit} = \frac{\Delta h_{cap}}{1 + \dfrac{q_{\max}}{\rho_w \overline{\overline{K(S_r)}}}} \tag{11}$$

For a diameter of 35 mm, the maximal drying rate is about 3 10^{-5} $kg.s^{-1}.m^{-2}$ whatever the sample height (Table 1). From the equation 1 and the permeability of the soil under saturated conditions, the mean hydraulic permeability can be obtained from the Mualem relationship [21], which expresses the permeability as a function of the degree of saturation S_r:

$$K(S_r) = K_{sat}\sqrt{S_r}\left[1-(1-S_r^{1/(1-1/n)})^{1-1/n}\right]^2 \qquad (12)$$

with K_{sat} the permeability under saturated conditions, n a parameter equal to the equivalent van Genuchten parameter (equation 1).

If we apply the equation 11 to the drying experiments presented in this paper, we find a characteristic length of the hydraulic connection L_{crit} about 40 mm. Owing to the assumptions used in the Lehman's methodology (capillary forces estimated from the retention curve, water content distribution between the drying front and the surface, etc.), only an order of magnitude of the characteristic length can be obtained with this relationship. However this value of 40 mm is consistent with the experimental observations in Figure 8 and 9:

i. When the height of the samples is larger than the characteristic length, the Stage II of drying is explained by the end of the hydraulic connection within the sample. The critical water content is therefore dependent on the amount of water available within the sample when the hydraulic connection is broken. Because the characteristic length is not dependent on the sample height, the critical water content increases with the height as observed in Figure 9 (H= 45 mm; H= 70 mm).
On the other hand the same duration of stage I is expected when the sample height is larger than the characteristic length, because the maximal drying rate is not influenced by the height. This is confirmed experimentally in Figure 8 (H= 45 mm; H= 70 mm).
ii. When the height of the samples is smaller than the characteristic length, capillarity forces are strong enough and gravity and viscous forces have no direct impact on the critical water content. Rather, the hydraulic connection is lost when a first region of the bottom of the sample is dried: the breakthrough [22]. For cylindrical samples with no lateral evaporation like in our case, this breakthrough should appear at a water content that varies only slowly with the sample height [22]. Thus the critical water content should be independent on the sample height. Even if the conclusions on drying kinetics drawn from samples with a height smaller than the critical length are consistent (Figure 7 for example), values of critical water contents from those tests can thus probably not directly transferred to large-scale evaporation kinetics problems.

4 Conclusions

Drying tests have been performed to study the influence of sample dimensions on the kinetics of evaporation. For cylindrical soil sample, we showed that the drying surface does not have any influence on the maximum drying rate and the mass transfer coefficient for diameters greater than 35 mm. The results highlight also that the slenderness of the sample does not have any influence on the drying rate and the mass transfer coefficient, but affects the transition from the stage I to the stage II (critical water content). These observations allow us proposing a set of guidelines on the dimensions of the soil samples to test in order to reproduce drying mechanisms similar to those observed on geostructures at large field-scale:
- When the analysis of the drying kinetics is related to the maximum evaporation flux and mass transfer coefficient, the optimal sample must have a diameter greater than or equal to 35 mm, regardless of the sample height.

- If the study is also related to the time and the critical water content, the samples must have, in addition to a diameter greater than 35 mm, a sample height higher than the critical length is needed. This critical length depends on the soil type and the drying conditions.

References

1. D.M. Gray, Can. Nat. Com. for the Intern. Hydr. Dec., (1970)
2. G.W. Wilson, D.G. Fredlund, and S.L. Barbour, Can. Geotech. J. 31(2), 151-161 (1997)
3. H. L. Penman, Math., Math. Phys. Eng. Sci. 193, 120-145 (1948)
4. T. Yamanaka, A. Takeda, F. Sugita, Water Resour. Res. 33(9), 2117-2128 (1997)
5. D.E. Blight, Géotechnique, 47(4), 715-767.
6. J. Teng, N. Yasufuku, Q. Liu, S. Liu, Nat. Hazards, 73, 1405-1418 (2014)
7. A-N. Ta, Phd thesis (2009)
8. G.E. Harbeck Jr., M.A. Kohler, G.E. Koberg, U.S. Dept. of the Interior, Geological Survey, (1958)
9. P. Gerard, M. Mahdad, A. Robert McCormack, B.Francois, Constr. Build. Mater. 95, 437-447 (2015)
10. L.A. Richards, Soil Sci. 51, 377-385 (1941)
11. P. Delage, M.D. Howat, Y.J. Cui, Eng. Geol. 50, 31-48 (1998)
12. M.T. Van Genuchten, Soil Sci. Soc. Am. J. (1980)
13. N. Prime, Z. Housni, L. Fraikin, A. Léonard, R. Charlier, S. Levasseur, Transport Porous Med. 106(1), 47-72 (2015)
14. I.C. Kemp, B.C. Fyhr, S. Laurent, M.A. Roques, C.E. Groenewold, E. Tsotsas, A.A. Sereno, C.B. Bonazzi, J.J. Bimbenet, M. Kind, Dry. Technol. 19(1), 15-34 (2001)
15. G.W. Wilson, D.G. Fredlund, and S.L. Barbour, Can. Geotech. J. (1994)
16. A. Léonard, Phd Thesis (2002)
17. P. Gerard, A. Léonard, J.P. Masekanya, R. Charlier, F. Collin, Int. J. Numer. Anal. Met. (2010)
18. Ewen & Thomas, Géotechnique. 3, 455-470 (1989)
19. Nadeau & Puiggali, Technique et Documentation - Lavoisier: Paris (1995)
20. P. Lehmann, S. Assouline, D. Or, Phys. Rev. E. 77, 056309 (2008)
21. Y. Mualem, Water Resour. Res. 12(3), 513-522 (1976)
22. M. Prat, Int. J. Multiphas. Flow. 21(5), 875-892. (1995)
23. J.B. Laurindo, M. Prat, Chem. Eng. Sci. 51(23), 5171-5185 (1996)

Implications of the atmosphere-soil interaction for the design of earth retaining structures

Juan Carlos Ruge[1a], Julio E. Colmenares[1], Renato P. da Cunha[2] & David Mašín[3]

[1] *Universidad Nacional de Colombia, Bogotá D.C.*
[2] *University of Brasilia, Brasilia*
[3] *Charles University, Prague.*

Abstract. The performance of most geotechnical structures is highly governed by environmental factors, particularly in tropical regions where there are very pronounced dry and wet seasons. Design of earth retaining structures generally tend to be too conservative due to the uncertainty generated by the incorporation of environmental variables. Those variables control the soil unsaturated response and in addition to the known insufficiency of the basic models used in traditional designs they are responsible for conservative designs. Rainfall is the main aspect that affects the soil properties of a particular site. It modifies the soil suction potential, according to the degree of saturation caused by the soil-atmosphere interaction. Currently, state-of-the-art numerical tools allow to simulate the influence of those variables in the behaviour of earth retaining structures. This paper analyses the possible implications of the use of numerical simulations for the design, which include, in the mathematical formulation, the suction as a main parameter. The hypoplastic model for unsaturated response was used. Numerical simulations performed with the use of traditional and modern constitutive models obtained encouraging results that reveal the importance of include suction in design processes.

[a] jcruge@unal.edu.co

1 Introduction

Various kind of geotechnical structures have their behaviour dependant on environmental variables such as their own configuration and interaction between elements and soil, especially those that are exposed to rain (i.e. slopes, excavations, embankments, fillings, etc.)

These conditioning factors make the structures have a critical condition during the most of its useful life, since in short-term load solicitations and excess of water pore-pressure control the response, and in long-term the environmental aspects can affect the soil behaviour. Likewise, the designers have considered these variables in their computations, but there is an uncertainty involved due to the nature of the environmental factors. This uncertainty is reflected in the overestimation of geotechnical parameters [1].

The objective of this research is to analyse the possible implications in design, considering the unsaturated response of the soil in excavations. Through numerical tools that include the rainfall effect in the computational model, as well as the influence on the suction and on the effective stress, was observed the evolution of hypoplastic unsaturated parameters [2] that can be implied in the design of these kind of structures. Also, the useful life of the project is considered on the results, since some retaining structures may be conceived as temporary or permanent prior to design stage.

2 Methodology

2.1 Site Analysis

The site chosen for this study was located at the SHN (Setor Hoteleiro Norte) of Distrito Federal (Brasilia), where it is possible to identify two geotechnical levels, according to soil investigations. The first layer of around 11 meters is composed by collapsible porous red clay with silt and gravels contents of soft to medium consistency, underlined for clayey silt with lens of sand medium to hard. The retaining structure in the analysis section consisted in a tangent pile curtain of 18 meters of length, 60 centimeters of diameter and embedding depth of 5 meters. Also the contention is reinforced by means of soil nailing.

The relative area of the retaining structure has a characteristic profile of the clayey red lateritic soil layer, so-called Brasilia porous clay. This material presents low strength to penetration (N_{SPT} 1-6 blows), low strength tip (*CPT* varying of 0.6 to 2.3 MPa), low bearing capacity, low level of saturation and high permeability (10^{-3} to 10^{-4} m/s), (details in [3]).

Due to the high porosity and kind of cementation, it presents a structure highly unstable when is subjected to increased wetting and/or alteration of stress state, presenting almost always a significant reduction of volume, denominated collapse [4].

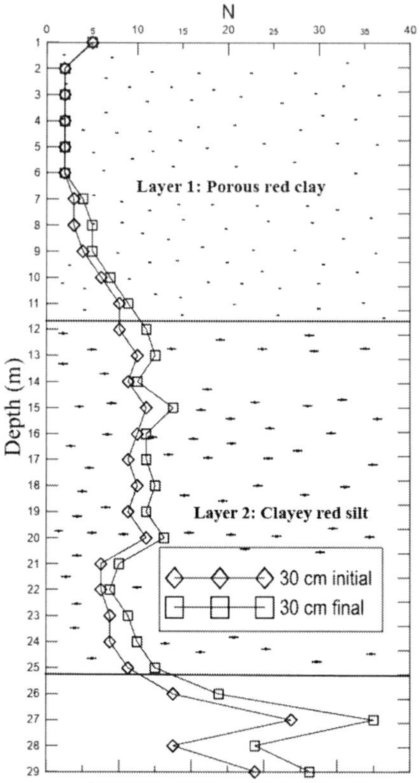

Figure 1. SPT results at the site.

2.2 Characterization of the material

It is obvious that an analysis involving atmosphere-soil interaction requires a more demanding characterization of the soil in terms of mechanical and hydraulic behaviour. For this reason is necessary to carry out, a series of basic and special tests in order to obtain parameters important to reproduce the unsaturated response of the soil in an adequate way. Some of those tests will be discussed below.

2.2.1 Basic index characterization

In order to obtain the basic characterization, index tests were performed on disturbed samples, as it is shown in Table 1.

Table 1. Basic characterization of the material

Sample	1B	2A
Depth (m)	6,0	9,0
w (%)	26,8	19,6
γ_s [kN/m³]	28,01	28,02
γ_d [kN/m³]	11,39	12,88
γ [kN/m³]	14,70	15,51
Gs	2,81	2,82
LL (%)	48	50
PL (%)	29	30
IP (%)	19	20
IC	1,13	1,55

[a] jcruge@unal.edu.co

2.2.1 Microstructure characterization

In this regard, it is also important to define the microstructure predominance in the soil, since much of the suction potential is dictated by the presence of macropores or micropores in the system. Through a Mercury Intrusion Porosimetry test, the Pore Size Distribution (PSD) was obtained; it shows the large amount of macropores indicated by the peak in Figure 2.

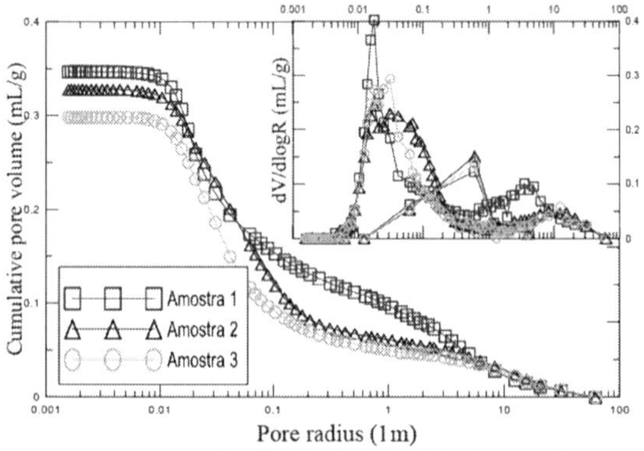

Figure 2. Pore size distribution (PSD) in the sample.

The typical particle size distribution was obtained too, showing the bimodal behaviour characteristic of this kind of lateritic soil. The real distribution depends of the use of deflocculant in the sample, as the actual granulometry results are shown in Figure 3.

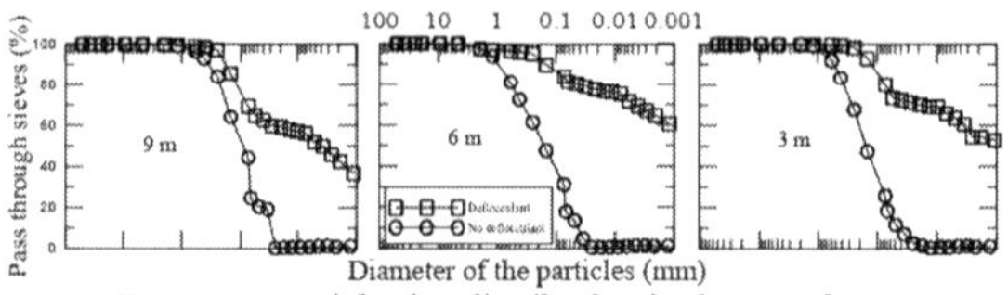

Figure 3. Particle size distribution in the samples

2.2.2 Geotechnical characterization (strength tests)

Regarding to the strength tests, the objective of the research is related with finding the initial unsaturated response in unaltered samples, and then trying to correlate with the mechanical behaviour of a retaining structure using a numerical tool. The results of unsaturated triaxial tests for different levels of suction are shown in Figure 4. Results demonstrated the suction phenomenon, at least in laboratory terms.

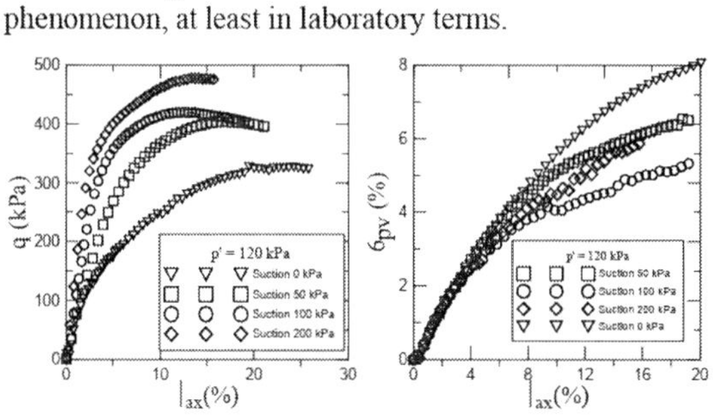

Figure 4. Unsaturated triaxial test on sample (9 m)

2.3 Reference constitutive model

This previous stage of the numerical modelling is based in the analysis of the reference constitutive model. The model selected was the hypoplasticity for the mechanical response of unsaturated soils. The constitutive equations of this law were developed based on a critical state concept and the effective stress principle, considering the effect of increasing stiffness, governed by the variation of the suction on the mechanical response and the collapse phenomenon by wetting [5].

2.3.1 Constant suction model

The current reference model is supported in the concept proposed by [5]. The phenomenon of suction occurs inside of the soil skeleton influencing the effective tension and the normal forces on the interparticle contacts increasing the stiffness of the soil, since the particles are closer to each other. In other terms, this behavior increases the state boundary stress (*SBS*) analogously, like are bonded the soil particles in cemented saturated geomaterials [6].

Therefore, the *SBS* size for unsaturated soils is controlled by the *NCL* (Normal Compression Line) based on the Butterfield compression law [7], for the critical state line as it is showed in equation 1.

$$\ln(1+e) = N(s) - \lambda^*(s)\ln\frac{p}{p_r} \quad (1)$$

In equation 1 e is the void ratio, denoted like state variable in the constitutive law. The expressions $N_{(s)}$ and $\lambda^*_{(s)}$ reproduce the position of the inclination of the virgin line compression in the space $\ln(p/p_r)$ *vs* $ln(1+e)$ for a suction s. The model parameters N and λ^* define the values for saturated conditions.

Equation 2 defines the expression for equivalent tension of Hvorslev on the NCL for a given suction [8].

$$p_e = p_r \exp\left[\frac{N(s) - \ln(1+e)}{\lambda^*(s)}\right] \quad (2)$$

[9] Adjusted the values of the barotropy f_s and picnotropy f_d (eq. 3), since revealed that the inclusion of the virgin compressibility in the intercept $N_{(s)}$, affects the first values proposed by Gudehus (1996).

$$f_d = \left(\frac{2p}{p_e}\right)^\alpha \quad (3)$$

p_e is estimated by using the equation 2, and the barotropy factor is:

$$f_s = -\frac{tr\mathrm{T}}{p_e}(3 + a^2 - 2^\alpha\sqrt{3})^{-1} \quad (4)$$

The scalar factor α is in function of the parameters λ^* and κ^*. Information about this variable can be consulted in [5], as well as for factor a.

2.3.2 Model for any state of overconsolidation

These considerations are used to complement the constitutive law for any state and load solicitation:

- The suction controls the stability of contacts between particles, increasing the suction under constant effective tension.
- The more open is the soil structure, the lower is the number of interparticle contacts; hence, each contact must transmit larger shear forces [5].

$$\overset{\circ}{\mathrm{T}} = f_s(L:\mathrm{D} + f_d \mathrm{N}\|\mathrm{D}\|) + f_u \mathrm{H} \quad (5)$$

With

$$\mathrm{H} = \frac{\mathrm{T}}{p_e}\frac{\partial p_e}{\partial s}\left\langle \dot{s} \right\rangle \quad (6)$$

Where s is introduced to study the first consideration, and f_u the new picnotropy factor that controls the collapse in the soil, reproduce the second. When the factor is 1 the stress state is in the SBS (in this case the structure is as open as possible and the collapse being controlled by **H** only).

Instead f_u tends to zero for $OCR = p_e/p$ (no wetting induced interparticle slippage occurs in highly overconsolidated soil). The following expression for the factor f_u satisfying these requirements is proposed:

$$f_u = \left(\frac{p}{p^{SBS}}\right)^m \quad (7)$$

Where p^{SBS} is the effective mean stress at the SBS corresponding to the current normalized stress **T**/*tr***T** and the current void ratio e. Likewise, m is a model parameter governing the influence of overconsolidation on the wetting-induced collapse.

2.3.3 Parameters of the model

Besides the parameters of the hypoplastic model for saturated soils (ϕ_c, N, λ^*, κ^*, r). The extension of the proposed model needs to consider the influence of the suction on N and λ, respectively denoted as $N_{(s)}$ and $\lambda_{(s)}$.

m controls the collapse of the structure along the wetting paths and s_e is the value of suction at the air entry and/or air expulsion.

Similarly, n and l are scalar quantities included inside of the formulation of $N_{(s)}$ and $\lambda_{(s)}$ controlling the unsaturated NCL [5].

$$N(s) = N + n\ln\left(\frac{s}{s_e}\right) \quad (8)$$

$$\lambda^*(s) = \lambda^* + l\ln\left(\frac{s}{s_e}\right) \quad (9)$$

2.3.4 Calibration of the parameters

It is necessary to adjust the parameters obtained at the lab through the constitutive law selected for the numerical phase. The calibration of the saturated parameters (λ^*, κ^* and N) are shown in Figure 5 [2]. The friction angle of the critical state is obtained (ϕ_c) of results of the strength peak in triaxial tests.

The unsaturated parameters m, l and n of the model can be computed from the saturated and unsaturated normal compression line, using isotropic compression tests and the equations 8 and 9.

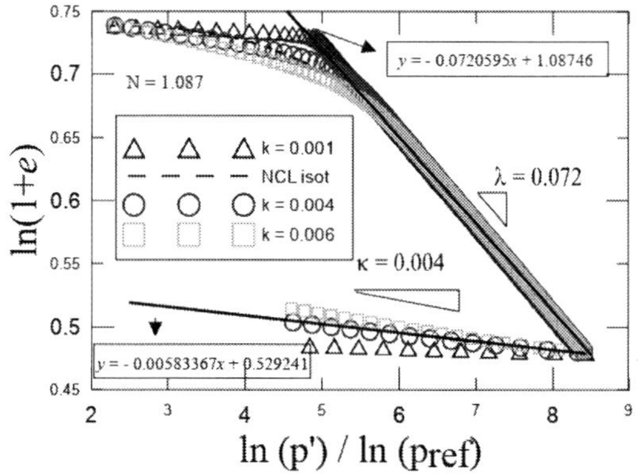

Figure 5. Calibration of the parameters λ^*, κ^* and N [2]

2.4 Numerical predictions

One of the main challenges of the designer is about trying to predict the behaviour of the geotechnical structures by means of numerical tools. The essence of the matter consists in anticipating, at least in a qualitative way, the mechanical response of the soil or geotechnical elements before construction. However, it is also valid to have continuous complements for certain kind of structures as deep excavations, which require periodic control during the excavation sequences.

Table 2 shows the classification of predictions according the moment of occurrence.

Table 2. Predictions in geotechnical structures [10]

Prediction	Moment in the event	Results
A	Before	----------
B	During	Unknown
B1	After	Known
C	After	Known
C1	After	Known

In this case, the numerical predictions involve not only specific factors about the behaviour of the soil, but also external variables as environmental factors (i.e. rainfall), which govern the global behaviour of the excavation under analysis.

To get an adequate approach including different factors, it is necessary to use a modern constitutive model that reproduce variables like suction, cementation, stress history, among others, into the numerical implementation.

2.4.1 Computational model

Fitting an appropriate computational model is a key job in the numerical modelling step. It is important to consider substantial aspects and to discard some that are not essential in the simulation [11].

Figure **6** shows the computational model developed for the geotechnical problem under analysis. The numerical modelling was performed taking into account the atmosphere-soil interaction, inserting real rainfall at the site. It is important to mention that the software also considers the soil water characteristic curve (SWCC) of the material, whereby it is possible to know the suction changes during the excavation affected by the rainfall too.

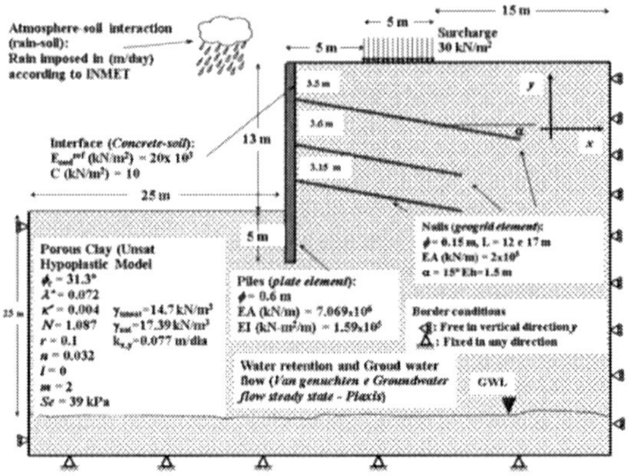

Figure 6. Computational model of the geotechnical problem

The groundwater level is located at 25 m deep from the last level of excavation. The position of the GWL indicates the high susceptibility of the upper layer to develop suction in a dry season, in the site under study.

3 Results

The results of the predictions were compared with the real behaviour of the retaining structure, monitored by means of instrumentation to control the deformations of the structure. Figure 7 shows the deformations occurred in the retaining during the period of instrumentation (223 days), as well as the numerical predictions (rain and dry).

These predictions were developed just to consider the atmosphere-soil interaction in the geotechnical problem (rain prediction). In the case of the dry prediction, a global retaining structure was simulated to observe the capacity of the constitutive model to take in account the suction. Rain prediction is a realistic scenario, while the dry prediction is conservative (high suction).

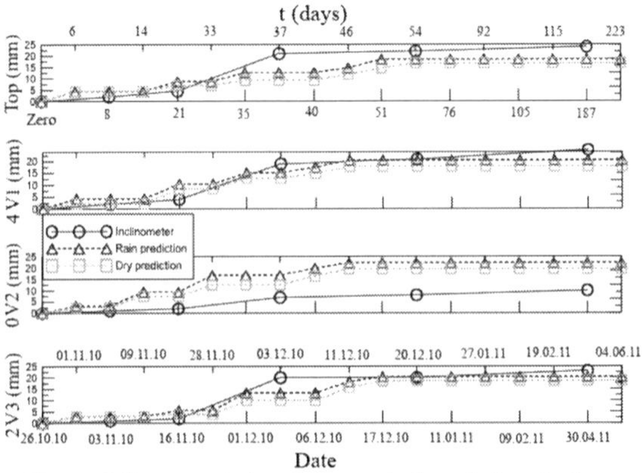

Figure 7. Instrumentation and numerical predictions in the geotechnical problem selected.

Figure 8 shows the deformations in the retaining structure measured by inclinometer (Inc) and modelled using a hypoplastic model (HP). This figure reveals the time-dependent deformation in the retaining system, it is important to analyse the behaviour in relation with the timeline to understand the response in dry and raining season.

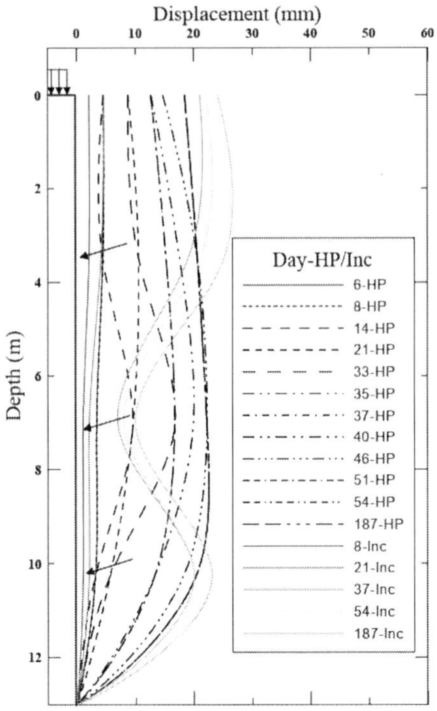

Figure 8. Instrumentation and numerical predictions in the geotechnical problem selected.

DOI: 10.1051/e3sconf/20160912002

Due to the fact that the main scope of this research was related with the implications in design of the unsaturated response of a retaining structure, it is important to analyse the problem using a traditional Mohr-Coulomb model. In Figure 9 shows a numerical simulation by means of the option "ϕ-c reduction" in the software. This option permits to reduce the geotechnical parameters until the failure of the material. The safety factor obtained for this option was 2.

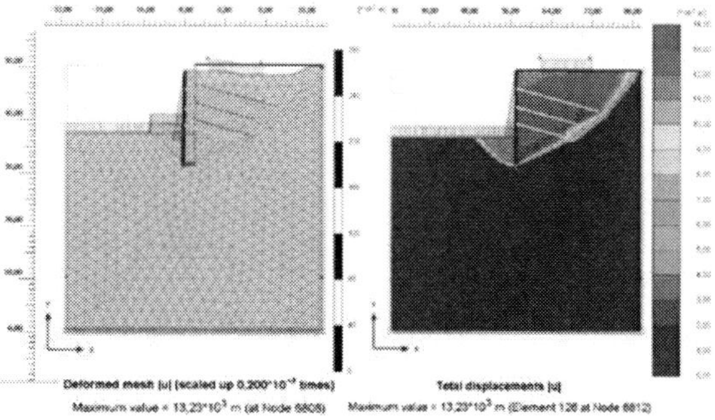

Figure 9. Prediction using Mohr-Coulomb mode.

3.1 Analysis of the results

It has always been evident in the geotechnical field that the designs are very conservatives due to the use of traditional constitutive models (i.e. no suction), which within their implementation overestimated parameters. The numerical simulation using "*ϕ-c reduction*", allows to reach such conclusions.

The modern constitutive models for unsaturated soils, which include suction into the mathematical implementation can be an essential tool in this kind of research. Some retaining structures are of temporary nature, due to the fact that their lateral loads are transferred to basement beams of the vertical structure at the end of the excavation.

In the dry season, the suction developed in the soil is higher. When the soil is close to be dry, the suction potential is very high, hence the effective stress and shear strength is also increased.

Therefore, building these structures during the dry period allows the design of retaining structures to be less conservative. Of course using more adequate geotechnical parameters. It is possible, as long as the constitutive model has within its implementation geotechnical parameters that simulate the unsaturated response of the soil.

4. Conclusions

The numerical predictions using modern constitutive models are an essential tool in the designing stage of any geotechnical project. However, consulting firms currently do not use this models since specialized tests, that are not time and cost efficient are required. Suction is a parameter that must be included in geotechnical designs, since it can offer a possibility to work with more realistic geotechnical scenarios.

The fact of building the retaining structure during the dry season must coincide with a susceptibility of the soil at the site, to generate high potential suctions by the atmosphere-soil interaction. In regions where there is a marked contrast between dry and raining seasons, it is easier to guarantee the time for the construction of retaining structures.

Some excavations are contained by using retaining structures as piles, diaphragm wall, sheetpiles, etc., reinforced with anchors, soil nailing, and shotcrete. Most numerical predictions show that in some cases it is not necessary to use additional engineering techniques to support the retaining structure.

References

1. J.C. Ruge. Analise do comportamento de cortina de estacas executada em solo poroso metaestável mediante o uso de um modelo constitutivo hipoplástico considerando a resposta não saturada. PhD Thesis. University of Brasilia (2014).
2. Ruge J. C., Cunha R. P. a Masin D. (2013). "Results of unsaturated tests on metastable soils". Advances in unsaturated soils. Proceedings of the First Pan-American Conference on Unsaturated Soils, Cartagena, Colombia, pp. 469-474.
3. R. Cunha, N. Jardim, and J. Pereira. In situ characterization of a tropical clay via dilatometer tests. Geo-Congress 99 on Behavior Characteristics of Residual Soils, ASCE Geotechnical Special Publication 92, Charlotte. 113-122 (1999).
4. N. Mota. Ensaios avançados de campo na argila porosa não saturada de Brasilia: interpretação a aplicação em projetos de fundação. Ph.D. Thesis, University of Brasilia (2003).
5. D. Mašín, and N. Khalili. A hypoplastic model for mechanical response of unsaturated soils. International Journal for Numerical and Analytical Methods in Geomechanics, 1(32), 1903-1926 (2008).
6. D. Mašín, and I. Herle. State boundary surface of a hypoplastic model for clays. Computers and Geotechnics, 6 (32), 400-410. (2005)
7. R. Butterfield. A natural compression law for soils. Géotechnique, 29(4), 469–480. (1979).
8. M. Hvorslev. Physical components of the shear strength of saturated clays. In Shear Strength of Cohesive Soils, proc. ASCE Research Conf. Boulder. (1960).
9. D. Mašín. A hypoplastic constitutive model for clays with meta-stable structure. Canadian Geotechnical Journal, 44(3), 363–375. (2007).
10. T. Lambe. Predictions in soil engineering. Géotechnique, 23(2), 151–202. (1973).
11. D. Kolymbas. Advanced mathematical and computational geomechanics. Springer, 1st ed (2003).

Effect of wetting-drying cycles on soil desiccation cracking behaviour

Chao-Sheng Tang[1], Yu-Jun Cui[2], Bin Shi[1], Anh-Minh Tang[2], Ni An[2]

[1]*School of Earth Sciences and Engineering, Nanjing University, 163Xianlin Avenue 210023, China*
[2]*Ecole des Ponts ParisTech, UR Navier/CERMES, 6 et 8, avenue Blaise Pascal, Cité Descartes, Champs-sur-Marne, 77455 Marne-la-Vallée CEDEX 2, France*

Abstract. Better understanding the desiccation cracking process is essential in analysing drought effects on soil hydraulic and mechanical properties through consideration of the atmosphere-ground interaction. Laboratory tests were conducted to investigate the consequence of wetting-drying cycles on the initiation and propagation characteristics of desiccation cracks on soil surface. Initially saturated slurry specimens were prepared and subjected to five subsequent wetting-drying cycles. Image processing technique was employed to quantitatively analyze the morphology characteristics of crack patterns formed during each drying path. The results show that the desiccation cracking behaviour of soil is significantly affected by the wetting-drying cycles. Before the third wetting-drying cycle is reached, the surface crack ratio and the average crack width increases while the average clod area decreases with increasing the number of wetting-drying cycles. The number of intersections and crack segments per unit area reaches the peak values after the second wetting-drying cycle. After the third wetting-drying cycle is reached, the effect of increasing wetting-drying cycles on crack patterns is insignificant. Moreover, it is observed that the applied wetting-drying cycles are accompanied by a continual reconstruction of soil structure. The initial homogenous slurry structure is completely replaced with aggregated structure after the third cycles, and a significant increase in the inter-aggregate porosity can be observed.

1 INTRODUCTION

The formation of desiccation cracks on soil surface due to loss of water is a common phenomenon in nature. This behavior significantly affects the performance of soil in various geotechnical, geological and environmental applications. Generally, the presence of cracks in soil would increase the compressible and reduce the overall mechanical strength [1]. The hydraulic properties of soil are directly influenced by crack networks in soil [2]. Many researchers have found that the hydraulic conductivity of cracked soils is typically several orders of magnitude greater than that of uncracked soils [3]. Therefore, this issue is a major concern in designing and constructing low permeable structures as clay buffers and barriers for nuclear waste isolation, liners and covers for landfill, etc. So far, a number of laboratory experiments have been conducted to investigate the initiation and propagation of desiccation crack in soils [4, 5, 6-9]. In addition, some modeling and theory studies have also been conducted on this issue [10, 11]. Nonetheless, the prediction of cracks initiation and the associated crack network propagation behaviour still faces several challenges.

Desiccation cracking is mainly due to water loss by evaporation which results in the generation of soil suction. Cracking is likely to occur if the surface tensile stress induced by increasing suction exceeded the bonding strength of soil grains [12]. However, the intrinsic mechanism of desiccation cracking is very complex. It is related to the physical, chemical, mechanical and hydraulic properties of soil, including clay proportion, mineral composition, compaction condition, initial water content, etc. [1, 3]. Especially, climatic situations and surrounding geo-environments are known as important factors affecting soil desiccation cracking behaviour. Tang et al. (2007) and Tang et al. (2008) confirmed that the influences of environmental temperature and layer thickness on desiccation cracking behavior of clayey soil [6, 13]. Rodríguez et al. (2007) analyzed the desiccation phenomena of a mining waste at different relative humidity conditions, and found that the time at onset of crack initiation decreases with decreasing relative humidity [14]. Moreover, the soil in field experiences diurnal or seasonal rainy and sunny climate changes, and undergoes periodical wetting-drying cycles. Several studies have been performed to investigate the effect of wetting and drying on soil physical properties [15]. However, there is a lack of data covering the desiccation cracking behavior of soil subjected to wetting-drying cycles.

The aim of this work is to determine how wetting-drying cycles affect the desiccation cracking behavior of soil. Initial homogenous slurry specimens were prepared

[a] Corresponding author: tangchaosheng@nju.edu.cn

in laboratory and subjected to five wetting-drying cycles. The initiation and propagation of cracks on specimen surface were monitored during the cycles. The morphological characteristics of surface crack pattern were discribed and quantified based on image analysing method. Some interesting phenomena observed in the tests were discussed.

2 Material and methods

The used soil was taken from the Est of Paris. The physical properties are presented in Table 1. The soil was air-dried, crushed and sieved at 2 mm. Saturated slurry specimens were prepared by mixing soil powder with distilled water at a water content of 170 %. The desired quantity of slurry was then poured into glass cups (117 mm in diameter). To remove entrapped air bubbles in the slurry, these cups were placed on a vibration device for few minutes. Finally, the cups containing prepared slurries were sealed with plastic membrane for 3 days. The final thickness of the settled slurry was about 10 mm.

The prepared specimen was exposed in constant room temperature (25 ± 1 °C, 50 ± 5 % of relative humidity, *RH*) to be dried, until the residual water content was reached. Note that this is the first wetting-drying cycle. The subsequent wetting cycle was started with adding distilled water immediately into the glass cups. In this process, sufficient water was provided to ensure full saturation of the soil. Three days were allowed for the equilibration of soil and water.The glass cups were also sealed with plastic membrane to prevent evaporation of moisture. Then, the specimens were exposed in room temperature to be dried again. By repeating the above procedures, a total of five wetting-drying cycles were applied to the specimens.

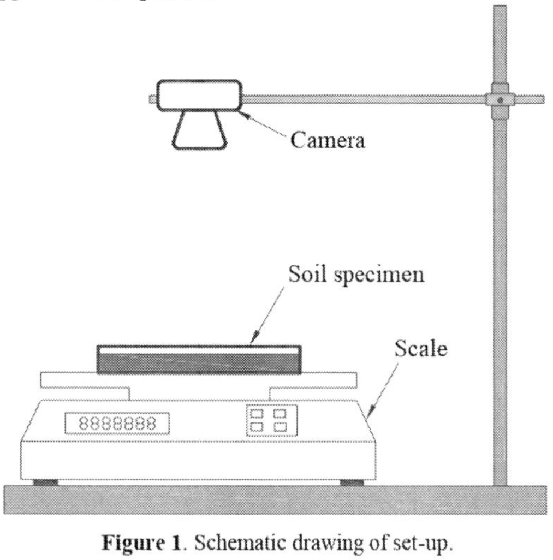

Figure 1. Schematic drawing of set-up.

During wetting and drying, a digital camera was mounted above the specimens to record the change of surface crack pattern, as the schematic drawing of the set-up shown in Fig. 1. By applying computer image processing technique, the surface crack ratio (R_{sc}) which is the ratio of the surface area of cracks to the total surface area of specimen, was defined to quantify soil cracking extent during drying. In addition, some other parameters (i.e. the number of intersections, crack segments per unit area and the average width and clod area) that can be used to describe the geometric characteristics of crack pattern were also determined. The details and procedures of digital crack image processing were introduced by Tang et al. (2008) [6]. During the tests, the water evaporation and water content changes of specimens were monitored using a scale (Fig. 1).

Table 1. Physical properties of soil

Soil properties	Values
Specific gravity	2.79
Liquid limit	77 %
Plastic limit	40 %
Plasticity index	37
USUC classification	CH
Clay (<2μm)	79 %
Clay composition	Illite and smectite
Specific surface area	340 m/g

3 Results and Discussion

3.1 The first wetting-drying cycle

Fig. 2 depicts the determined cracking curve (R_{sc} versus w) during the first drying cycle. It can be observed that the R_{sc} increases quickly with decreasing w in the early stage of cracking. However, with further drying, the increment of R_{sc} decreases quickly once w is lower than a specific value. From the cracking curve, the following two parameters can be identified: the cracking water content (w_c) which corresponds to the initiation of desiccation cracking, and the final critical water content (w_{fc}) which corresponds to the transition point where R_{sc} trends to stable value. It can be seen that cracking occurred at very high water content of about 38.1%, that is closed to the plastic limit (40%) of the used soil (Table 1). The final stabilized R_{sc} is about 15.8 %, means that there is 15.8 % surface area was covered by cracks. As indicated by Tang et al. (2010a), the formed crack on soil surface is directly attributed to the shrinkage of pore volume [7]. When the water content reaches soil shrinkage limit, all particles are in contact, further water loss will not result in any more volume reduction. Accordingly, the cracks propagation also trends to end and reaches stabilization. They therefore concluded that the physical signification of w_{fc} determined from the cracking curve is identical with the shrinkage limit determined from shrinkage curve. For the tested soil, the shrinkage limit can be evaluated as about 11.3%, based on the determined w_{fc} shown in Fig. 2.

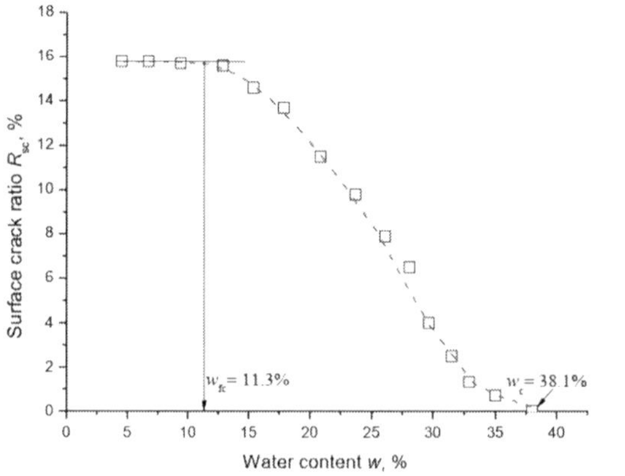

Figure 2. The cracking curve of specimen during the first drying cycles.

The initiation of desiccation crack on soil surface can be considered as tensile fracture of soil material. The w_c is therefore an very important critical parameters related to the mechanism of this phenomenon. However, even for a given soil, the w_c is not unique while conditioned by a lot of factors. Corte and Higashi (1960) and Nahlawi and Kodikara (2006) found that w_c generally increased as soil layer thickness increased [5, 12]. Rodríguez et al. (2007) confirmed that the w_c significantly depends on relative humidity [14]. Tang et al. (2010a) indicated that the w_c increases with increasing temperature [7]. According to the unsaturated soil mechanics theory, a given soil may initiate cracks at various critical suction levels due to the various w_c. Because of the uncertainty of critical fracture conditions, it is currently difficult to construct an appropriate failure criterion to predict soil cracking behaviour. This is also why the precise mechanism of soil desiccation cracking is still not perfectly understood.

After the first drying cycle, a typical crack pattern is presented in Fig. 3. It can be seen that the specimen surface is split to several separate clods by the crack network. The shapes of clods and crack segments are relatively regular and smooth. Most of the clods are quadrangles and pentagons. A close observation of the crack pattern shows that most of the crack segments are perpendicular to each other.

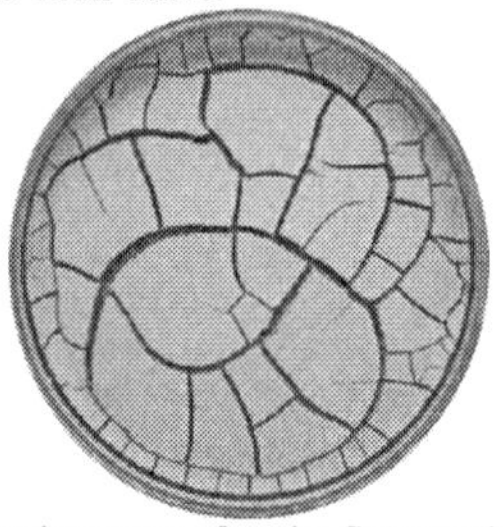

Figure 3. The crack pattern after the first wetting-drying cycle.

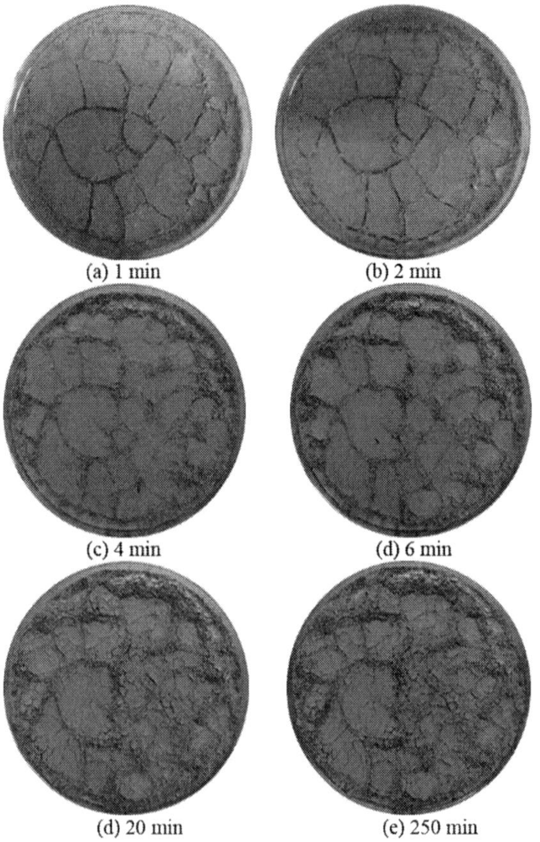

Figure 4. The evolution of crack pattern during the second wetting path with elapsed time.

3.2 The second wetting-drying cycle

A series of 6 time-images are shown in Fig. 4 to describe the evolution of crack pattern during the second wetting path. After the water was put into the glass cup, immediate slaking of the clods was observed (Fig. 4 (a) and (b)). The clods broke down into several small aggregates. Some new microcracks appeared on specimen surface. This phenomenon can be attributed to the following factors:

i) The added water weakened particle bonding strength and softened soil structure.

ii) Internally generated forces such as differential swelling pressur resulted in the successive disintegration of the clods.

iii) Entrapment and pressurization of air bubbles inside the clods intensified the damage of soil structure. When water was put into the glass cup, it penetrated inside the clods driven by capillary forces. This rather brutal hydration process can lead to entrapment and pressurization of air bubbles (Fig. 5). Because capillary first develops in smallest pore spaces, the pore air is extruded and gradually gathers in macropores. On one hand, with successive gathering of pore air, the air bubble volume increases and leads to soil destruction. On the other hand, as hydration carries on, the pressure P_a inside

 DOI: 10.1051/e3sconf/20160912003

the air bubbles increases due to the meniscus imbibition (Fig. 5). Once P_a exceeds the bonding strength of surrounding particles, it results in miniexplosions of soil structure, and thus leads to general damage. Some times, pressure equilibrium will be reached between the air bubbles and soil structure, these air bubbles will be more likely to be trapped and not escape to outside. This phenomenon was confirmed in present investigation, as the picture shown in Fig. 6. It is evident that the entrapped air bubble intensified the disaggregation of surrounding soil.

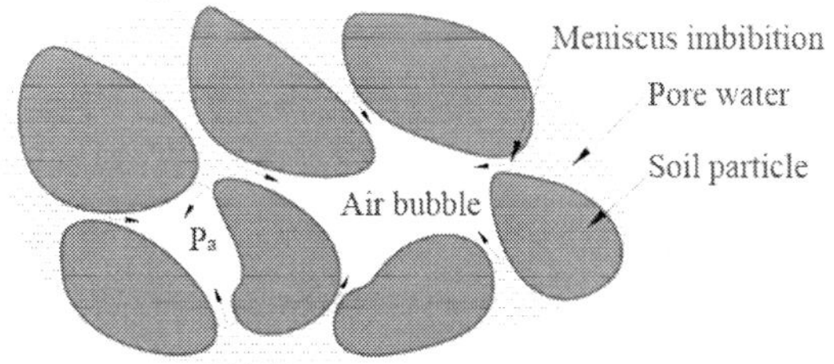

Figure 5. Meniscus imbibition during wetting.

Figure 6. The entrapped air bubble inside the soil during wetting cycle

From Fig. 6 (a), it is interesting to find that the slaking or destruction of clods firstly occurred on the clods edges, especially on the corner positions of clods. This is because the hydration firstly occurred in these positions. With increasing wetting time, the original cracks became narrower and narrower due to clods collapse and swelling of soil particles, and were fully closed in 4 min (from Fig. 6 (a) to Fig. 6 (c)). But the new appeared microcracks became more and more with the elapsed wetting time (from Fig. 6 (d) to Fig. 6 (e)). It was found that the aggregates and the microscracks were always maintained during the whole wetting period, even after about 3 days later they still exist.

Sequences of wetting-drying cycles can lead to the rearrangement of soil particles and the modification of pore system. The reversibility of the soil structure during wetting-drying cycles is known to influence the hydraulic and mechanic behavior, including the volume shrinkage and swelling properties. Yong and Warkentin (1975) revealed that shrinkage during the first drying cycle caused irreversible fabric changes, particle bonds might be broken due to desiccation cracking [16]. These broken bonds may attract water and become weakest locations where are the preferential zones of cracking. Yesiller et al. (2000) and Tang et al. (2008) have observed that cracks appeared at the same locations as the first drying cycle in the second and third cycles [6, 17]. However, this observation was not evident in present research after comparing Fig. 7 with Fig. 3. After the second drying path, the shapes of clods were more irregular and the cracks were much coarser (Fig. 7). Most interestingly, the clods were degraded significantly. The wetting-drying cycles caused dramatic aggregate formation from initially non-aggregated soil. The clod surface is also not smooth any more.

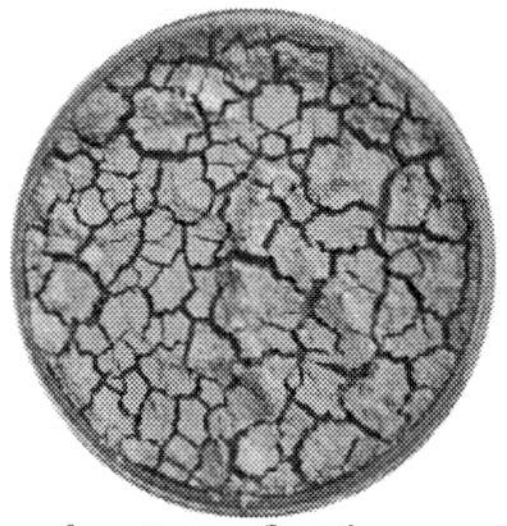

Figure 7. The crack pattern after the second wetting-drying cycle.

3.3 Subsequent wetting-drying cycles

Typical image of specimen during the third wetting path is presented in Fig. 8. The observations are different from that during the second wetting path (Fig. 6). After water were put into glass cup, clods degraded rapidly. Only about 1 min later, the original cracks were completely disappeared, and no new microcracks were observed. This is because the stability of clod during the third wetting path was mainly controlled by the bonding strength between the formed aggregates in the second wetting-drying cycle. As the image shown in Fig. 8, the specimen is composed of large amount of individual aggregates. It is proved that the degradations of clods in the third wetting path are the results of breakdown of the bonds between aggregates. During the fourth and fifth wetting path, no significant difference was observed with the third wetting path, and no image is presented.

Figure 8. The typical crack pattern during the third wetting path (after about 1 min wetting).

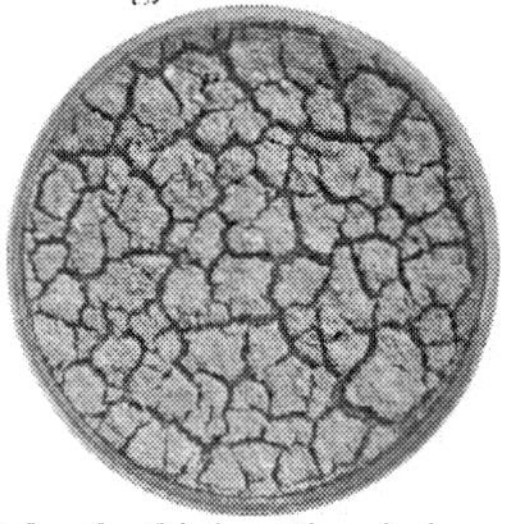

(a) After the third wetting-drying cycle

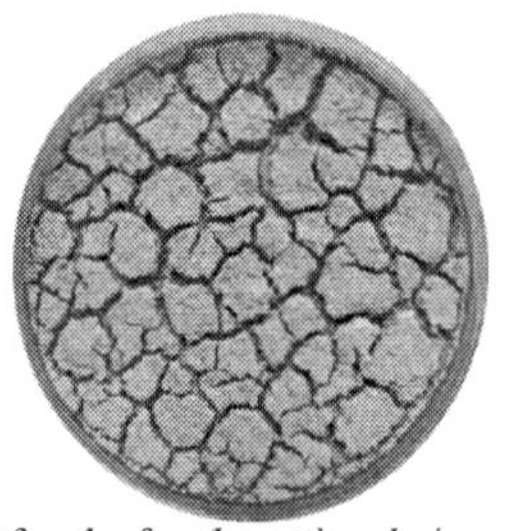

(b) After the fourth wetting-drying cycle

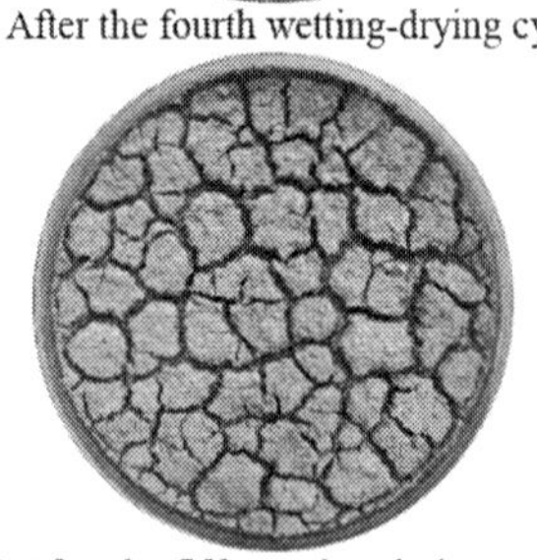

(c) After the fifth wetting-drying cycle

Figure 9. Crack patterns after the third, fourth and fifth wetting-drying cycle.

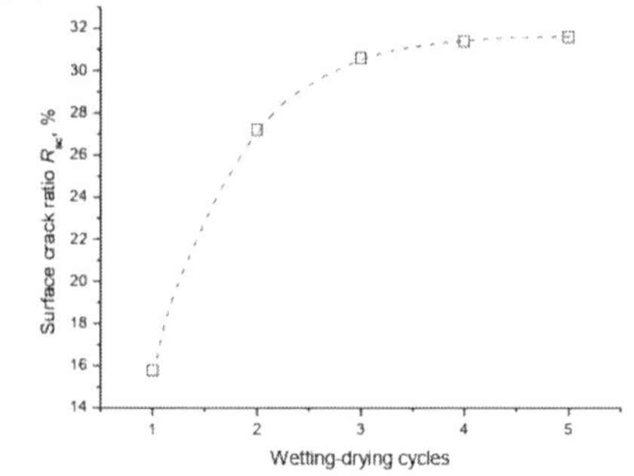

(a) Surface crack ratio

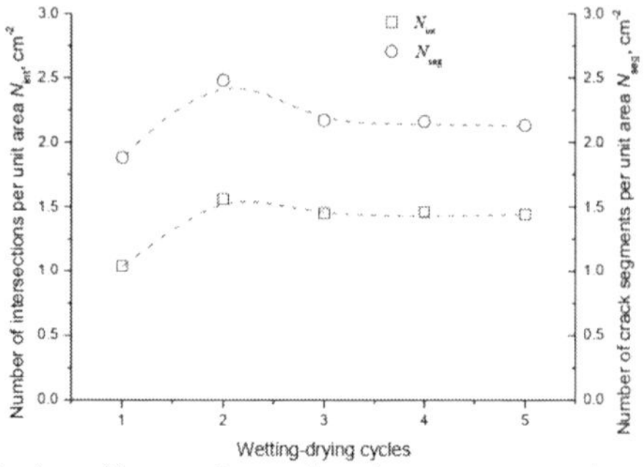

(b) Number of intersections and crack segments per unit area

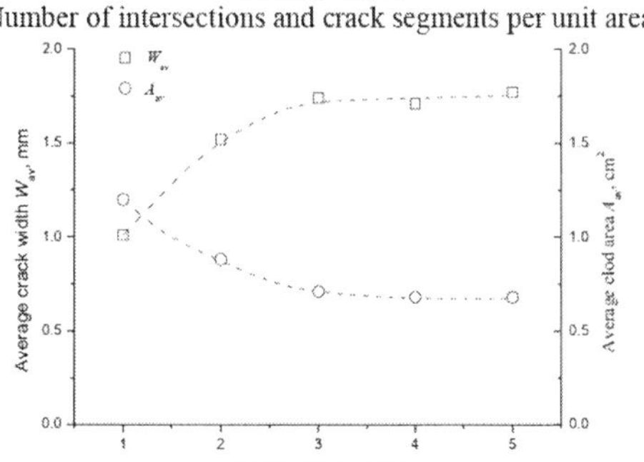

(c) Average crack area and clod area

Figure 10. Variation crack quantitative parameters with the number of wetting-drying cycles.

The final crack patterns for the last three wetting-drying cycles are shown in Fig. 9. They show similar morphology characteristics, but they are different from the crack pattern obtained from the first wetting-drying cycle (Fig. 3). Fig. 10 present the variation of some crack quantitative parameters with the number of wetting-drying cycles, including surface crack ratio R_{sc}, number of intersections per unit area N_{int}, number of crack segments per unit area N_{seg}, average crack width W_{av} and clod area A_{av}. Before the third wetting-drying cycle was reached, it can be seen that an increase in number of wetting-drying cycles was accompanied by increase in R_{sc} and W_{av}. For N_{int} and N_{seg}, they reached the peak values after the second cycle. However, A_{av} generally decreases with increasing number of wetting-drying cycles. After the third wetting-drying cycle, these crack quantitative parameters almost keep constant, and the effect of increasing cycles is insignificant. Based on the quantification results of the crack patterns, it can be concluded that the soil cracking behaviour reached equilibrium state after the soil underwent three wetting-drying cycles. The soil properties will not be significantly influenced by the subsequent cycles.

3.4 Effect of wetting-drying cycles on structure

By comparing the final crack patterns shown in Figs. 3, 7 and 9, it can be found that the wetting-drying cycles can significantly change soil structure and give rise to intensification of material heterogeneity. In the first wetting-drying cycle, the specimen was subjected to drying from relative homogenous slurry state. The final clods are dense and the surfaces are smooth (Fig. 3). However, after the second and third cycles, the apparent homogenous slurry structure was completely replaced with aggregated structure (Figs. 7 and 9). A strong increase in the inter-aggregate porosity can be observed after sequences of wetting-drying cycles. These aggregates and the corresponding macro-pores were mostly generated during the second wetting path, and they were maintained during the subsequent drying paths.

The laboratory test results obtained in the present work can be probable used to highlight the natural regeneration of soil. Actually, in the agricultural soil mechanics point of view, mechanical impacts of agricultural machineries and grazing animals can induce excessive soil compaction; that alters the structure of cultivated soil and thus changes a number of key soil properties for crop production and the environment, and results in degradation of soil physical qualities [18]. To regenerate the deteriorated soil has emerged as a major issue. Recently, McHugh et al. (2009) assessed changes in soil properties over a 22 month period on a site which had been used for crop and pasture production for 100 years under conventional tillage practices [19]. It was found that the cropping zone bulk density decreases of 0.23% per month in the zone isolated from wheel traffic and soil tillage. However, due to the consideration of the environmental factors and energy consumption, soil in compacted zones is expected or even encouraged to regenerate naturally (through natural climatic conditions) other than through the traditional deep tillage. After Drewry (2006), the mechanisms of natural regeneration of soil are: soil shrinkage and swelling, cracking or micro-cracking from rapid wetting to reduce soil tensile strength, the wetting and drying process, freezing and thawing, frost penetration, plant root penetration, earthworm burrowing [20]. The results in the present work can be then useful when studying the soil regeneration by wetting/drying process and by cracking. In general, under unconfined conditions, wetting-drying cycles can give rise to a decrease in soil bulk density and an increase in crack extend and also aggregation, as the obtained results presented in the above sections. The soil permeability, macro-pore space and continuity may therefore be enhanced, so as to improve soil aeration and rapid drainage of water and solutes through soil.

4 Conclusions

Desiccation cracking behavior of soil subjected to five wetting-drying cycles was investigated. The effect of wetting-drying cycles on the morphology characteristics of crack pattern were quantitatively analyzed and discussed.

During the first wetting-drying cycle, it was found that cracking water content w_c at the onset of surface cracking is close to the plastic limit of the tested soil. The extent of surface cracking was quantified with surface crack ratio R_{sc}, which increased with decreasing water content and reached stabilization around shrinkage limit.

The subsequent wetting-drying cycles especially the second cycle leaded to significant rearrangement of soil particles and modification of pore system. In most of the cases, these structure changes were not reversible. During the second wetting path, rapid clods slaking phenomenon was observed due to softening of particle bonds, developed differential swelling pressure and entrapment of pore air. The original cracks were closed or healed immediately and large amount of new microcracks were appeared on soil surface, which divided the clods into several aggregates. A typical aggregated structure was constructed after the second wetting path and significantly enhanced soil heterogeneity.

Wetting and drying of soil with an increasing number of cycles have been shown to reach an equilibrium state, where the soil system nearly behaves in a reversible manner. For the tested soil, before the third wetting-drying cycle was reached, the surface crack ratio and the average crack width increased while the average clod area decreased with increasing the number of wetting-drying cycles. The number of intersections and crack segments per unit area reached the peak values after the second wetting-drying cycle. After the third wetting-drying cycle was reached, the effect of increasing wetting-drying cycles on crack patterns is insignificant. It indicated that the third cycle seems to be the transition point where the structure changes resulted from the subsequent cycles were generally reversible.

Acknowledgements

This work was gratefully supported by the National Natural Science Foundation of China for Excellent Young Scholars (Grant No. 41322019), National Natural Science Foundation of China (Grant No. 41572246), Key Project of National Natural Science Foundation of China (Grant No. 41230636), and the Fundamental Research Funds for the Central Universities. The authors also wish to acknowledge the support of the European Commission via the Marie Curie IRSES project GREAT - Geotechnical and geological Responses to climate change: Exchanging Approaches and Technologies on a world-wide scale (FP7-PEOPLE-2013-IRSES- 612665).

References

1. P.H. Morris, J. Graham, D.J. Williams, Can. Geotech. J. **29**, 15 (1992)
2. V.Y.Chertkov, Soil Sci. Soc. Am. J. **64**, 4 (2000)
3. B. A.Albrecht, C.H. Benson, J. Geotech. Geoenviron. **27**, 9 (2001)
4. C.J. Miller, H. Mi, N. Yesiller, J. Am. Water Resour. As. **34**, 10 (1998)
5. H. Nahlawi, J. Kodikara. Geotech. Geol. Eng. **24**, 24 (2006)
6. C.S. Tang, B. Shi, C. Liu, et al., Eng. Geo. **101**, 14(2008)
7. C.S. Tang et al., Eng. Geo. **114**, 6 (2010a)
8. C.S. Tang et al., J. Mater. Civil Eng. **23**, 6 (2010b)
9. C.S. Tang et al., Appl. Clay Sci. **52**, 9 (2011)
10. A.N. Abu-Hejleh, D. Znidarčić, J. Geotech. Eng. **121**, 10 (1995)
11. H. Péron, J.Y. Delenne, L. Laloui, M.S. El Youssoufi, Comput. Geotech. **36**, 9 (2009)
12. A. Corte, A. Higashi. Wilmette, Illinois. (1960)
13. C.S. Tang, B. Shi, C. Liu, et al., Chinese J. Geotech. Eng. **27**, 6 (2007)
14. R. Rodríguez, M. Sánchez, A. Ldesma, A. Lioret, Can. Geotech. J. **44**, 15 (2007)
15. S.M. Rao, B.V.V. Reddy, M. Muttharam. Eng. Geo. **60**, 11 (2001)
16. R.N Yong, B.P. Warkentin, Elsevier, Amsterdam, (1975)
17. N. Yesiller, C.J. Miller, G. Inci, K.Yaldo, Eng. Geo. **57**, 17 (2000)
18. K.Y.Chan, A.Oates, A.D. Swan, et al, Soil Till. Res., **89**, 9 (2006)
19. A.D. McHugh, J.N Tullberg, D.M. Freebairn, Soil Till. Res. **104**, 9 (2009)
20. J. J. Drewry, Agr. Ecosys. Environ. **114**,11 (2006)

Evaluation of the effect of land vegetation cover on water and energy balance of an unsaturated pyroclastic cover

Alfredo Reder[1,2a], Guido Rianna[2] and Luca Pagano[1]

[1]*DICEA (Dipartimento di Ingegneria Civile, Edile e Ambientale), University "Federico II", via Claudio 21, 80123, Naples (Italy)*
[2]*REMHI(Regional Models and geo Hydrological Impacts), CMCC, via Maiorise, Capua (Italy)*

Abstract. Land cover features can strongly affect the actual soil wetting state; for example, the presence of a vegetated cover can induce relevant variations in the main components of soil-atmosphere water balance: interception and evapotranspiration. For these reasons, its proper characterization represents a crucial point in typical geotechnical issues as, for example, the assessment of slope stability conditions. To this aim, in the work, in two different hydrological year albeit characterized by comparable weather patterns, the variations induced by varying land cover have been investigated for water and heat balance of a soil volume constituted by pyroclastic soil. These soils, mantling a large part of slopes in Campania Region, are particularly interesting since they have been often affected, in last years, by landslide phenomena causing huge damages and victims. Observations are retrieved by a physical model constituted by a soil layer exposed to weather forcing and instrumented to record weather variables, component of soil surface heat balance and soil state properties within the layer. The two antipodal considered land cover conditions are bare soil and full natural vegetation (mixing of Lolium perenne, Festuca rubra and Poa pratensis); on Campania slopes, the first condition can be retrieved on slopes in man-modified areas (road cuts, quarries) while the second ones are plant species commonly found on Mediterranean slopes. The analysis focused on Summer season where in clearer way, different behaviour can be detected stressing the variations induced by the vegetation cover not only on the potential evaporative demand but also on the actual values.

1 Introduction

As well known, frequency, magnitude and consequences of rainfall induced landslides are highly related to the land use (and then the land cover) of the affected areas [1]. Land use, in fact, can directly influence the dynamics inducing an event by increasing or decreasing slope stability.

Concerning vegetated surfaces, Sidle & Ochiai [1] recognize four mechanisms as the main hydrological effects for shallow landslides recognizing:

1) rainfall and snow are intercepted by canopies and they can directly evaporate from them inducing subsequent reduction of infiltration rates;
2) root systems extract water from soil via transpiration, leading to lower average soil moisture levels;
3) roots, stems and organic litter increase ground surface roughness and hydraulic conductivity in the shallower soil layers;
4) reduction of soil moisture may cause desiccation cracks, resulting in higher infiltration capacity and short-circuiting of infiltrating water to a deeper failure plane.

The Authors state that the first two mechanisms result beneficial for the slope stability whereas the second two ones can be assumed marginally adverse.

Actually, a vegetation cover can be due to presence of natural vegetation, agricultural activities, pastureland, coppice or high forest: then beyond the general benefits/disadvantages above reported, the management and maintenance activities carried out in such contexts play a key role: proper clearcutting cycles for forested areas, conversion from natural vegetation to agricultural lands or grazed lands (more frequent in developing countries) or abandonment of agricultural activities no longer profitable (frequent in developed countries) can lead to substantial variations of associated maintenance activities. In this perspective, quantifying how the presence of a vegetated surface can influence the soil water balance and the water distribution within the soil (and then slope stability) in a specific geomorphological context represents a crucial issue. It is important also to properly assess the consequences of different options in land use planning. In this regard, for example, in recent years, the adoption of "green infrastructures" as mitigation tool for reducing the occurrence of landslide events or as adaptation countermeasure to cope with the effect of potential climate change have aroused a growing interest [2].

In this work, the effect of a vegetated surface overlying a pyroclastic soil layer has been investigated; in the specific, a comparison between the behaviour of a soil

[a] Corresponding author: alfredo.reder@unina.it

layer exposed to the atmosphere respectively in bare and vegetated surface conditions has been carried out.
It has been performed for the same season (JJA) in two different years characterized by comparable weather forcing; the attention focused mainly on variations induced by vegetation in evaporative losses.
The investigated layer is of particular interest since coming from slopes of Campania Region (Southern Italy), frequently affected in recent years, by landslide events able to induce huge economic damages and fatalities [3].
In effect, a large part of slopes in the region is covered by ashes and pumices produced by several eruptions of the two volcanic centres in the Region, Campi Flegrei and Somma Vesuvius [4]. The formed covers lie in partially saturated conditions continuously modified by the water exchanges between atmosphere and soil surface. These water exchanges affect the degree of saturation and then suction within the soil, entailing in reduction or increasing in the slope stability [5-8]. Suction drops following heavy/prolonged rainfalls can induce reductions of soil resistance until attaining failure conditions.
In last years, researchers have focused their activities mainly on quantifying the variations of suction due to precipitation events through laboratory and field studies [9-10]. Only recently the role of soil wetting state at the beginning of potential triggering event have been recognized as relevant for these events. Back analysis of the events occurred on Lattari Mts. (Gragnano and Pozzano, 1997; Nocera Inferiore, 2005) have pointed out the role of antecedent rainfall as predisposing conditions over a time periods also several months long [5,7,11-13]. On these intervals, however, not only wetting events but also evaporative losses (low but persisting outward flows) have a significant influence in determining the predisposing conditions.
Regarding Campanian pyroclastic covers, some studies have been conducted focusing on the effects of soil evaporation under bare soil conditions [14-18]; while the effects of plant transpiration have been scarcely investigated [19-21].
The experimentation of the present paper has been conducted over the hydrological period 2010-2014 under bare soil conditions, while since September 2015 the soil surface is covered with a full natural vegetation commonly found on Mediterranean slopes.
The paper initially describes the physical model, its monitoring instrumentation and the two different cover systems (§2). Subsequently it displays the different evaporative patterns and interprets their trends (§3.1). After the work shows some difference in terms of radiation fluxes between the two different cover systems trying to quantify the albedo values (§3.2). Finally, it highlights the effects of the different evaporative patterns in terms of water content (§3.3).

2 Physical model and investigated soil

The monitoring activity has been performed on a layer contained in a wooden tank (surface area = 1.15 x 1.15 m; thickness = 0.75 m) (Figure 1). The layer is a non-plastic silty sand (Figure 1a) taken from the stratigraphic series of Monteforte Irpino and widely characterized in its physical properties, mechanical and field hydrological behaviour [8,10,22]. Moreover, it displays consistent similarities with ashes mantling slopes of Nocera Inferiore area affected by failure events in last years [5].
The stratum has been deposed through dry pluviation technique [23] ensuring high porosity (about 70%) close to that in situ.

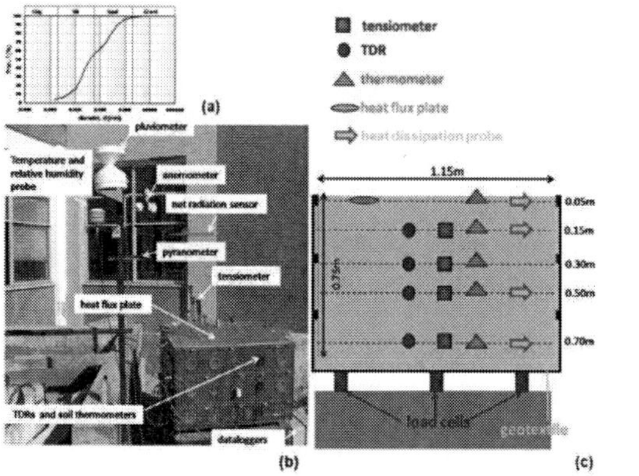

Figure 1. (a) Grain-size distribution of the soil layer; (b) picture of the physical model and of its devices; (c) monitoring devices installed within the soil layer (modified from [14] and [17]).

The monitoring system (Figure 1c) is conceived to quantify water and energy fluxes and the effects of such ones within the layer. Changes in water storage are monitored through three shear load cells; suction, water content and temperature within the soil are measured at 5 depths respectively using tensiometers (0-70kPa)/heat dissipation probes (15-500kPa) [16-24], TDR probes and thermistors. Finally, a well-equipped weather station (Figure 1b) provides observations related to main meteorological variables (precipitation, air temperature, wind speed, humidity, air pressure) and main energetic components (net/solar radiation, soil heat flux).
The soil surface has been monitored in both bare (Figure 2a) and vegetated conditions (Figure 2b).

Figure 2. (a) Bare soil surface; (b) vegetated soil surface.

In the first phase, albeit not matching all the usual conditions met on site, the bare soil surface allowed investigating in simplified condition the effect of the soil evaporation.
After four years, a vegetation cover, constituted by a mixing of Lolium perenne, Festuca rubra and Poa pratensis has been planted.
The comparison between the monitoring data obtained under bare soil and vegetated conditions permits to

DOI: 10.1051/e3sconf/20160912004

illustrate and roughly discriminate the effects of plant transpiration on soil surface water balance and on state conditions within the layer.

3 Evaluation of evaporative processes for different land cover conditions

3.1 Water balance

During the five years of hydrological monitoring, the layer was subjected to natural sequences of precipitation and evaporation. This allowed the characterization of the soil hydrological response under different weather conditions and for different levels of suction and water content.

Hydrological behaviours observed under bare and vegetated conditions may be obviously be compared only if evaporative processes have been driven in the two cases by similar weather forcing and same initial soil states in terms of water content values.

Among the several sequences recorded during the four years of observation in bare conditions, the period ranging from 6th June to 31th August of the first hydrological year was selected. It resulted, in fact, the most similar in weather forcing and initial soil state to the corresponding period (falling in the fifth hydrological year) with vegetated cover.

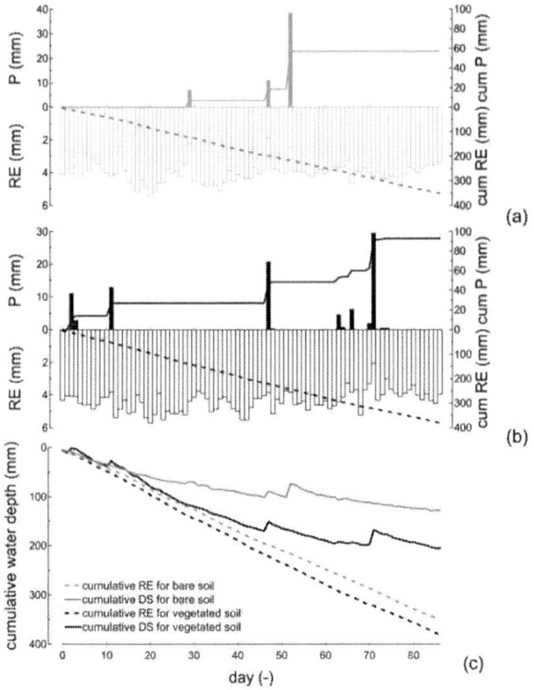

Figure 3. Daily and cumulative precipitation (P) and daily and cumulative reference evaporation (RE) for bare (a) and vegetated soil cover conditions (b); trends of cumulative reference evaporation and of cumulative water storage variations (DS – positive values correspond to water loses) for bare and vegetated soil cover conditions (c).

Firstly, the weather observations retrieved during this time window for bare soil and vegetated cover have been compared in order to assess the weather likeness of the two investigated periods. Specifically, daily and cumulative precipitation (P) and daily and cumulative reference evaporation (RE) [25] are displayed for the bare soil in Figure 3a and for the vegetated cover in Figure 3b (the time instant t = 0 day corresponds to 6th June and the time window lasts 86 days).

Reference Evaporation (RE) is used instead of Potential Evaporation (PE), according to Allen et al. [25]:

$$RE = \frac{0.408\Delta(R_n - G) + \gamma \frac{900}{T+273} u_2 (e_s - e_a)}{\Delta + \gamma(1 + 0.34u_2)} \tag{1}$$

where:

- RE reference evaporation
- R_n net radiation
- G soil heat flux
- Δ slope vapour pressure against temperature curve
- γ psychrometric constant
- T mean daily air temperature at 2 m height
- u_2 wind speed at 2 m height
- e_s saturation vapour pressure
- e_a actual vapour pressure

The reference evaporation takes into account, on the one hand, the role of available energy at the soil surface provided in terms of net radiation and soil heat flux and, on the other hand, the drying power associated to air movement depending on mean air temperature, relative humidity and wind speed.

During the investigated time windows, the evaporative demand estimated during the fifth year (380.5 mm) is higher than that estimated during the first year (350.3 mm). On the same way, the total precipitation recorded during the fifth year (93 mm) is higher than that collected during the first year (57.2 mm).

At t = 0 the layer water storage is for the bare soil equal to 326 mm, corresponding to an average volumetric water content θ = 0.434 whereas for the vegetated cover it is equal to 330 mm, corresponding to an average volumetric water content θ = 0.440.

Figure 3c displays for both cases the trends of cumulative reference evaporation and the trends of cumulative water storage variations (DS).

Worth noting is that at the end of the reference period the spread between the two cases is in reference evaporation (30.2 mm) below the half that in water storage (77 mm). The layer in fact reduces its weight of 128 mm under bare soil conditions and of 205 mm under vegetated soil conditions.

In bare soil conditions cumulative reference evaporation departs from cumulative water storage at the beginning. The two trends are observed parallel only over few days following rainy events.

In vegetated soil conditions, however, cumulative water storage follows the cumulative reference evaporation for about 30 days. Subsequently, the two trends diverge.

Even for this case the two curves return parallel over few days following precipitation events.

To better quantify the effect of plant transpiration, the trends of cumulative water storages have been cleaned of variations due to precipitation. This correction corresponds to eliminate the effects of precipitation on evaporation, consisting in the attainment of the potential value over few days.

Figure 4 compares the two cleaned curves.

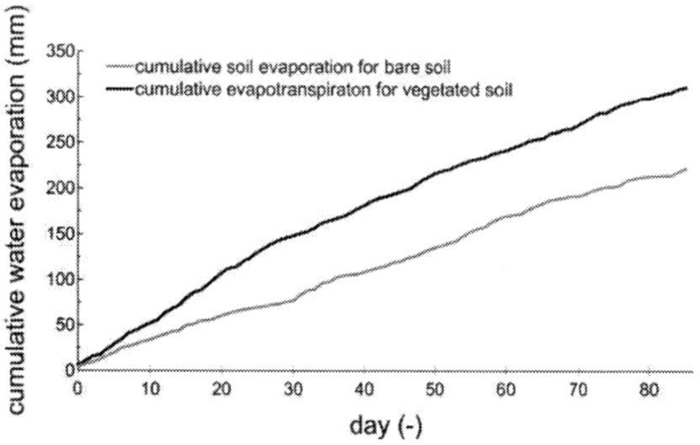

Figure 4. Trends of cumulative evaporation for bare and vegetated cover

Evaporation cumulating in bare conditions (222 mm) is much less than that cumulating in vegetated conditions (306 mm). The effect of plant transpiration may be obtained by computing the difference (306-222) and subtracting either the difference between reference evaporation values (30mm). It can be stated that the presence of vegetation induces a reduction in total water storage equal to 54 mm (on average, 1mm/day).

3.2 Energy balance

The difference in reference evaporations (30 mm) cumulating in the two cases at the end of the reference period may be itself related to differences in surface characteristics and consequent differences in available energies.

To this aim, Figure 5 shows the trends of cumulative energetic fluxes acting just above the soil surface during the investigated time window for the two cases. These fluxes are related to each other through the radiation balance that states the energy available at the soil surface (or net radiation) R_n in the following way:

$$R_n = (1-a)\, R_s - R_{nl} \qquad (2)$$

where:

R_n net radiation
R_s solar radiation
R_{nl} net longwave radiation
a albedo

Through the weather station placed near to the layer, solar and net radiation can be directly measured; on the other hand, the net longwave radiation can be assessed by adopting the empirical relationship proposed by Allen et al [25]. It requires, as inputs, air temperature, relative humidity and cloudiness; these variables are provided by the weather station. In this way, the only unknown of (2) is the albedo. The albedo is a dimensionless coefficient that measures the reflectivity of an incident light depending on roughness, colour and inclination of the soil surface. It represents an important indicator for investigating the evaporative processes since it could influence the energy available at the soil surface and then could affect the evaporative demand.

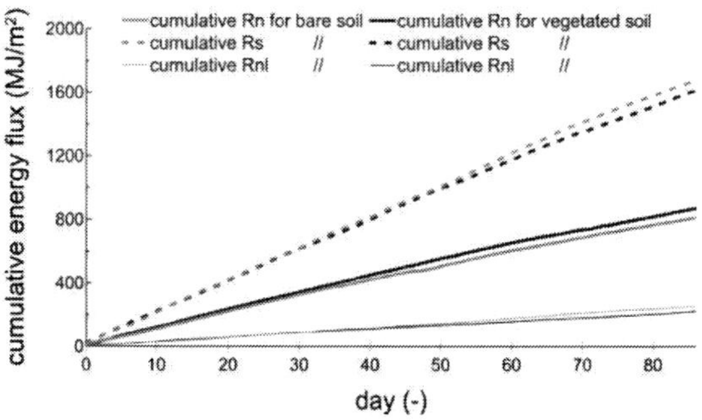

Figure 5. Trends of cumulative net radiation (Rn), solar radiation (Rs) and net longwave radiation (Rnl) quantified under bare and vegetated condition.

The energy provided by the sun during the entire monitoring time span under bare soil condition is slightly greater than that measured under vegetated condition (1675 MJ/m^2 against 1611 MJ/m^2). Even the net long wave radiation quantified under bare soil condition is slightly higher than that obtained under vegetated condition (253 MJ/m^2 against 221 MJ/m^2) probably due to higher soil surface temperature detectable for bare soil. Nonetheless, the vegetated soil associates with a greater amount of energy available for the evaporation processes, quantified by net radiation measurement (R_n):In fact, the net radiation for bare soil is 811 MJ/m^2 while for vegetated condition it is 871 MJ/m^2.

Net, solar and net long wave radiation allows the quantification of the albedo for the two cover conditions. For bare soil the mean value of albedo retrieved is about 0.36 while for vegetated cover conditions it is about 0.32. Such values are consistent with those retrievable in literature [26]. Indeed, for its roughness and colour, the vegetated surface is characterized by lower albedo then the bare soil. Under the same solar and long wave radiation, the vegetated soil has therefore a more available energy than the bare soil for the evaporation.

3.3 Water content variations within the layer

The major evaporative flows occurring under vegetated conditions due to the plant transpiration turns, as expected, into more pronounced water content reductions throughout the layer.

In this regard, Figure 6 shows the isochrones of water content recorded under bare soil (Figure 6a) and vegetated soil conditions (Figure 6b) measured at several times.

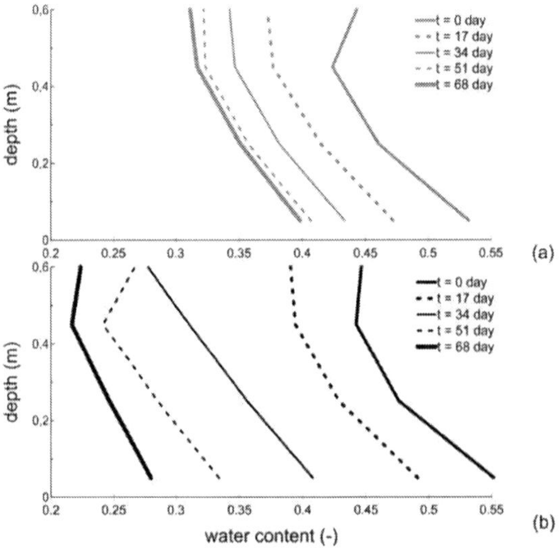

Figure 6. Isochrones of water content recorded under bare soil (a) and vegetated cover (b) conditions.

The data are related to four depth (15, 30, 50 and 70 cm) within the layer.

In general, the water content isochrones show a rigid translation with a reduction of water content at all depths. Such behaviour is typically exhibited by the layer during the evaporative processes as shown in [14-16].

During the bare soil experiment, moving from water content ranging between 0.45 and 0.55 at t = 0 day, the layer reduces at all depths its water content as a consequence of the evaporative process. At t = 68 day, the water content is about 0.3 at depth of 15 cm and about 0.4 at depth of 70 cm.

From an analogue initial distribution of water content, the isochrones monitored under vegetated cover condition exhibit after only 17 days a trend similar to that recorded after 68 days for bare soil case. Under vegetated cover condition, at t = 68 day, the evaporative fluxes induce a strong desaturation of the soil layer; at this stage, the water content ranges between 0.22 and 0.27.

This sharp difference observed in the two experimentations is mainly attributable to the effect of plant transpiration. A rough assessment between final profiles show how in the deeper 60 cm the differences worth about 65 mm (0.11) ; as difference with total value (89 mm), the spread in the shallower 15 cm can be estimated as about 24 mm (θ=0.16).

These results show how the presence of roots permits pulling out water from the cover in more effective way also from greater depths.

4 Conclusion

The works displays the main findings related to the monitoring of a pyroclastic soil volume under two different soil surface conditions during two summer seasons characterized by comparable weather forcing. Focusing on evaporative losses, the summer season allows evaluating in more evident way the different behaviours. Since the results, the added value of vegetated surface clearly arises; however, albeit they are related to summer season, it is reasonable that also in other seasons the vegetation can induce higher evaporative fluxes even if under lower values. In this way, the presence of vegetation on the slopes could contribute to reducing average water content values within the soil and improving slope stability conditions producing "less predisposing conditions" when an heavy rainfall event occur. It represents an issue of great interest for the soils adopted in monitoring, coming from Campanian slopes affected in last years by several landslide events. Furthermore, it should not be omitted that higher evaporative losses represent only one of the beneficial effects of the vegetation on the slope stability acting also with interception in reducing water inputs and providing through roots a further mechanical support for the shallower soil layers.

References

1. R. C. Sidle, H. Ochiai, *Landslides: Processes, Prediction, and Land Use* (American Geophysical Union, 2006)
2. EEA (European Environment Agency), *A Exploring nature-based solutions: The role of green infrastructure in mitigating the impacts of weather- and climate change-related natural hazards* (EEA Technical report No 12/2015, 2015)
3. A. Vallario, *Il dissesto idrogeologico* (CUEN, 2001)
4. L. Picarelli, A. Santo, G. Di Crescenzo, L. Olivares, X Int. Symp. Land. **2**, 1951-1958 (2008)
5. L. Pagano, L. Picarelli, G. Rianna, G. Urciuoli, *Landslides* **7(3)**, 273–289 (2010)
6. E. Damiano, L. Olivares, L. Picarelli, Eng. Geol. **137–138**, 1–12 (2012)
7. L. Cascini, G. Sorbino, S. Cuomo, S. Ferlisi, Landslides, **11(5)**, 779-792 (2014)
8. M. Pirone, R. Papa, M. V. Nicotera, G. Urciuoli G, Landslides **12(2)**, 259–276 (2015)
9. E. Damiano, L. Olivares, Nat. Haz. **52(2)**, 329-350 (2010)
10. M. V. Nicotera, R. Papa, G. Urciuoli, Geotech. Testing J. **33 (4)**, 263–285 (2010)
11. F. Rossi, G. B. Chirico, *Definizione delle soglie pluviometriche d'allarme* (G.N.D.C.I.-C.N.R., Italy, 1998)
12. P. De Vita, V. Piscopo, Nat. Haz. Earth Sys. Sc. **2**, 27-35 (2002)
13. G. Di Crescenzo, A. Santo, Geom. **66**, 225-276 (2005)
14. G. Rianna, L. Pagano, G. Urciuoli, Eng. Geol. **174**, 22-35 (2014)
15. G. Rianna, L. Pagano, G. Urciuoli, J. Hydrol. **510**, 480–492 (2014)
16. L. Pagano, A. Reder, G. Rianna, It. Geot. J. **1**, 62-76 (2014)
17. L. Pagano, A. Reder, G. Rianna, Proc. Earth Plan. Sc. **9**, 14-22 (2014)
18. A. Reder, G. Rianna, L. Pagano, Ad. Env. Geol. Sc. Eng. 273-280 (2014)

19. M. Pirone, E. Damiano, L. Picarelli, L. Olivares, G. Urciuoli, It. Geot. J. **3**, 29-50 (2012)
20. R. Greco, L. Comegna, E. Damiano, A. Guida, L. Olivares, L. Picarelli, Hydrol. Earth Syst. Sci. **17**, 4001–4013 (2013)
21. L. Comegna, E. Damiano, R. Greco, A. Guida, L. Olivares, L. Picarelli, Proc. Env. Sc. **19**, 922-931 (2013)
22. R. Papa, A. Evangelista, M. V. Nicotera, G. Urciuoli, Proc. E-UNSAT, 917-923 (2008)
23. M. C. Zingariello, *Progettazione, messa a punto e primi risultati di un modello fisico di pendio* (Phd thesis, Naples, 2006)
24. A. Reder, G. Rianna, L. Pagano, Proc. Earth Plan. Sc. **9**, 171-179 (2014)
25. R. G. Allen, L. S. Pereira, D. Raes, M. Smith, *FAO Irrigation and Drainage. Paper No. 56: Crop Evapotranspiration (Guidelines for Computing Crop Water Requirements)* (1998)
26. T. R. Oke, *Boundary Layer Climates* (Routledge, 2nd edition, 1987)

Description of a field test involving cracking in a drying soil

Josbel Cordero[1], Agustín Cuadrado[1], Pere Prat and Alberto Ledesma[1a]

[1]*Department of Civil and Environmental Engineering, UPC-BarcelonaTECH, Barcelona, Spain*

Abstract. The analysis of cracking in desiccating soils is a research topic that can be addressed by using concepts of Unsaturated Soil Mechanics. In this context, the use of physical models constitutes a promising tool to understand the mechanisms involved in this problem. In fact, previous works by the authors included the use of a laboratory environmental chamber controlling temperature and relative humidity. This paper, however, describes a field experiment consisting of a large container (3 m by 3 m and 0.5 m height) with a soil mass undergoing desiccation in an open environment near Barcelona. The container is continuously weighed to monitor the water loss evolution (or water uptake in case of rain). Basic soil variables are monitored as well: suction, water content, temperature and heat flux at different points inside the soil mass. Environmental variables, including temperature, relative humidity and wind speed close to the soil surface are also recorded. The test started early in January 2015 and the paper presents the preliminary results corresponding to the first few months. Due to the weather regime, the soil has undergone desiccation and some single rainy events. Crack patterns change dramatically when applying suction cycles to the soil.

1 Introduction

Degradation of soils due to cracking may become a serious problem in soil structures, both from the mechanical and from the hydraulic point of view. In this paper, cracking produced by desiccation is considered. Within this context, cracks appear on the soil surface and their evolution is closely related to the soil-atmosphere interaction. Some authors [1-3] have already suggested that environmental variables as wind velocity, air relative humidity or solar radiation have a strong influence on the evaporation and infiltration of water through the soil surface. Therefore, although laboratory experiments are always useful, the development of field tests has a valuable outcome in this case.

Clayey soils are particularly prone to volume changes when water content varies. During dry periods, evaporation from the soil surface takes place resulting in volume changes and eventually cracking. Further rainy events may close some cracks and create new ones having a great impact on infiltration. The analysis of drying-wetting cycles on the cracking pattern has been carried out in the laboratory in previous works and seems to be quite complex [4, 5]. Those laboratory works include environmental chambers, where Relative Humidity (RH) and/or temperature cycles can be imposed. In the experiments, a tray of initially wet soil is continuously weighed and monitored while air at a particular RH and temperature is injected into the chamber. The evolution of the tray weight along time is an indicator of the water loss/gain in the soil sample.

This paper presents an experiment that reproduces most of the features of the tests carried out in environmental chambers, but with the application of natural boundary conditions. A large tray has been installed in the field following the experience of the authors in the laboratory. The aim was to compare measurements from laboratory and field tests in order to detect if any environmental variable usually not considered in the laboratory had a significant influence on the evolution of any soil variable. Cracking was obviously a phenomenon to analyse in detail as well.

The paper describes the design and the initial installation of the field experiment. Preliminary measurements are presented, including a description of the evolution of the main variables involved as a response to environmental actions. However, a detailed interpretation of the measurements is out of the scope of this paper.

Field conditions include some variables that are usually not considered when using environmental chambers. In particular, effects of wind speed and solar radiation appear to be significant in the field and, therefore, differences between soil desiccation produced in a closed chamber or in the field could be expected. The final purpose of this field test is to identify those differences and to state their role in the context of soil desiccation and cracking. Duration of at least one year is planned for the test, with the aim of including different weather conditions from rainy to sunny periods. The test location is close to Barcelona, so a typical Mediterranean climate is expected during that period.

[a] Corresponding author: alberto.ledesma@upc.edu

 DOI: 10.1051/e3sconf/20160912005

2 Experimental model

A large field experiment has been designed based on the experience from environmental chambers developed in the laboratory. The soil is confined in a large tray, in order to monitor continuously its weight, as it was performed in the laboratory. Changes in weight are directly related to changes in water content within the soil mass, at least at a global scale.

The main differences of the test with respect to the laboratory experiments are basically the size of the soil sample involved and the natural environment considered (open atmosphere). The size of the sample involved plays an important role in soil cracking as shown in previous works [6], but also applying actual atmospheric conditions becomes very important when investigating real scenarios.

The experimental setup was completed with several sensors to measure soil variables and external variables. As the test represents a soil-atmosphere interaction experiment, sensors were installed both inside the soil mass and outside, to monitor atmospheric variables. A particular effort was devoted to the measurement of variables close to the soil-air interphase, where most of the hydraulic and mechanical changes are expected. This is quite often a difficult task, as the sensors themselves may constitute a preferential path for water and air flow, and a starting point for soil cracking. In order to avoid that interaction, remote sensing of some variables has been attempted as well.

2.1 Equipment

2.1.1 Overview of the field test

A high-density polyethylene (PEHD) container of 3 m by 3 m and 0.5 m of height was installed at open atmosphere in the Agrópolis open research laboratory field of UPC. The location is near Barcelona airport in an area surrounded by farms and crops. A general view of the test is shown in Figure 1, where the external sensors are highlighted.

The container may include up to 18000 Kg of soil, involving a volume of 4.5 m^3. On the upper part of the container, outdoor sensors monitoring atmospheric variables were installed. A weather station nearby (Viladecans weather station located 1.5 km away) was used as a reference. In addition to that, a reflectometer antenna developed by the RSLab from UPC (Department of Signal Theory and Communications) was also installed with the aim of measuring in a remote manner the surface water content of the soil as part of a companion research project on remote sensing leaded by that group.

The installation required the design of a steel structure to support the filled container and a shallow foundation to hold the whole system. A software controlled digital camera was used to get pictures of the soil surface every hour. All sensors installed in the soil mass or externally to the container were automatically monitored by means of a data logger Campbell CR1000. In the following sections, a detailed description of the internal and external sensors installed in the test is presented.

Figure 1. Overview of the field test. 1. Load cells; 2. Data recording system; 3. Anemometer; 4. VP-3 (Relative humidity, air temperature and vapour pressure sensors); 5. Digital camera; 6. IR-120 (infra-red remote temperature sensor); 7. RSLab reflectometer.

2.1.2 External sensors

The external sensors are not in direct contact with the soil mass. The weight of the container is monitored in order to obtain the evolution of the global gravimetric water content. This is useful when a balance of water is carried out. Internal distribution of water content can be obtained by means of internal sensors and both types of sensors are required to check the consistency of the water balance. Number #1 in Figure 1 indicates load cells used to that purpose: a total of 4 cells were used.

Environmental variables obtained from the nearby weather station include: precipitation, wind velocity (value and direction), global solar radiation, air temperature and air relative humidity. Those variables are supplied by Meteocat (Catalan Meteorological Service). In addition to that information, on top of the container, just a few centimetres above soil surface, there are two sensors VP-3 Decagon, measuring temperature, relative humidity and water vapour pressure at two different levels. There is a Davis Cup Anemometer measuring wind velocity as well (value and direction). Numbers #3 and #4 in Figure 1 indicate where they were installed.

VP-3 Decagon is an integrated sensor to measure air relative humidity (by means of a capacitance sensor) and air temperature (by using a thermistor) at the same point. A microprocessor computes the vapour pressure from temperature and relative humidity values. A small teflon sheet protects the sensor from liquid water and dust without limiting water diffusion. An additional protection against solar radiation was employed and finally, two sensors at different heights were installed.

Davis Cup anemometer records wind speed by means of bowls of wind and a magnetic switch, while the direction is measured by a weather vane and a potentiometer.

Remote measurement of temperature at the soil surface is obtained by using an IR-120 Infra-red remote temperature sensor, which detects the infra-red radiation emitted by the surface. The sensor contains a number of thermocouples connected in series which detect thermal radiation. Some of them are exposed to the source of radiation and the rest are used for reference. The output voltage is proportional to the balance of thermal energy between the exposed surface and the sensor itself. A calibrated internal thermistor measures the reference temperature and finally the processing of that information provides the surface temperature of the soil. Number #6 in Figure 1 shows the location of that sensor.

The reflectometer developed by the RSLab group attempts to measure volumetric water content of the upper part of the soil (about a few centimetres). Measurements are global but may constitute a good procedure to obtain water contents on the soil-air interphase without disturbing the soil mass.

Finally, a reflex digital camera provides with pictures of the soil surface at any convenient time and in particular, every hour. During night, an automatic lighting system is activated while taking the picture. The camera is inside a box to protect from rain and is totally software controlled.

All variables are recorded in a continuous manner (see number #2 in Figure 1).

2.1.3 Internal sensors

Internal sensors are located inside the soil mass. They were installed in the container prior to soil placement. The soil is initially poured in a liquid state, so the sensors must be fixed to a vertical bar and located at different depths, in particular, 15 cm, 25 cm and 40 cm.

This type of sensors measure volumetric water content (VWC), electric conductivity (EC), temperature, suction (Ψ) and heat flow in the soil (G). To measure VWC, EC and temperature, 3 sensors type 5TE Decagon were used. They use Frequency Domain Reflectometry (FDR) to measure VWC [7], and a thermistor to measure temperature. EC is obtained by using two small stainless steel electrodes.

Suction up to 100 MPa, according to the available range of MP-6 Decagon sensors, is measured at 8 different locations. They have a porous ceramic stone of known water retention curve that becomes in equilibrium with the surrounding soil. Using the FDR technique, indirect values of suction are obtained.

Two sensors HFP01SC are used to measure G, a useful variable to perform the energy balance. This sensor is a heat flux plate including a thermopile measuring the temperature gradient across the plate. The output voltage is proportional to the heat flux.

2.2 Material properties

The soil used in the experiment is a natural soil from the area close to the emplacement of the experiment. That zone corresponds to the Llobregat river delta (south of Barcelona) and the surficial soil has a mixture of different grain sizes, including clay particles, which may be prone to cracking when desiccated. The material was brought to the Soil Mechanics laboratory to perform standard classification tests. Preliminary experiments suggested that the material was convenient for this type of research and not very different from the Barcelona silty clay used in previous experiments by the group at UPC laboratory involving an environmental chamber [5].

The natural soil has a substantial amount of sand and silt sizes, although its geotechnical classification is a low plasticity clay (CL), likely due to almost 10% of clayey components. However, the sand and silt content is quite important. The most relevant parameters regarding soil classification are shown in Table 1.

Table 1. Basic properties of the soil tested.

Parameter	Experimental value
γ_s (unit weight of solid particles)	27.1 kN/m^3
w_L (liquid limit)	28.9%
w_P (plastic limit)	16.5%
Sand content	48.3%
Silt content	42.1%
Clay content	9.6%

DOI: 10.1051/e3sconf/20160912005

2.3 Specimen preparation

The material to be used in the experiment was selected from the indicated natural soil, sieving particles smaller than 2 mm. To facilitate the placement of the soil, enough water was added to get a liquid state. Approximately, initial water content ($w_{initial}$) was about 1.5 times the soil liquid limit.

The preparation of the material required a toilsome work. First, the natural soil was excavated using a tractor, then 40 mm, 20 mm, and 2 mm large sieves were used to obtain 4 m^3 of material. Particles passing the 2 mm sieve were finally used. After that, the sieved soil was mixed with water at about 45% of $w_{initial}$ into a truck mixer, and the resulting slurry was spilled into the experimental container filling about 4.5 m^3 of total volume. A liquid consistency was very important in order to get a homogeneous mixture and to avoid any damage to the sensors already installed in the container.

Before filling the container, a pervious geotextile net was placed on its bottom, to establish a limit with a well-defined boundary condition regarding equilibration of water pressures. Figure 2a shows a particular moment when installing that geotextile. Also, a metallic wire net was installed between the bottom of the container and the geotextile, as a reference plane for the reflectometer antenna, for calibration purposes. Figure 2b shows a picture of the container just after filling with the soil, the day when the test started.

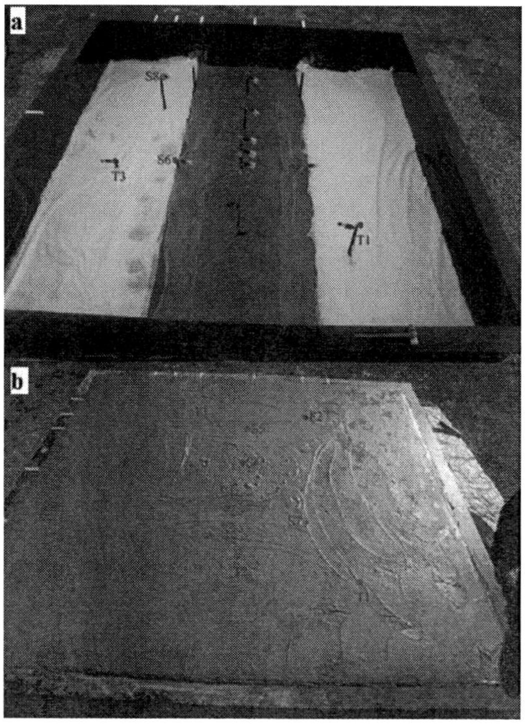

Figure 2. Distribution of the internal sensors. (S1-S8) Suction sensors; (T1-T3) VWC, temperature and EC; (F1-F2) Heat flux. **(a)** Container, geotextile & indoor sensors. **(b)** Initiation of the test.

2.4 Test program

All physical variables from the sensors are recorded and stored digitally. The system was programmed to work for at least one year. However, periodic checks were planned (every 2 - 3weeks approximately) and they were used to download the measurements to a laptop for further analysis. Also some small samples of the soil close to the surface were taken in each visit, in order to measure the water content in the laboratory to compare with in situ measurements.

It should be pointed out that some of the sensors may become out of range during such a long time period of monitoring, but still it was considered important not to change the geometry by replacing the sensors, as any defect or any odd object within the soil mass may lead to a point of crack initiation.

The construction of the setup was carried out during fall 2014, and the installation of the sensors and preparation of the soil were performed in early winter 2015. It is planned to dismantle the test after one year of recorded data. Then there are plans for further experiments changing some of the boundary conditions involved.

3 Preliminary results

Preliminary results presented here refer to the first 160 days of test, starting in winter 2015. The main trend suggests a global desiccation of the soil mass, as shown in Figure 3, where the weight of the soil mass and the precipitation are depicted as function of time.

Drying dominates the environmental action in an area with a Mediterranean climate and as a consequence of that, cracks start to form at day 22. Note that water was always lost, unless some precipitation occurred. Soil gravimetric water content when cracks initiated was about 35%. The loss of weight is almost linear at the beginning, but then slows down when cracks develop. Small rainfall events do not change significantly the global tendency of weight loss.

It has been observed, that the effects of meteorological changes (i.e., daily changes) are more sensitive in the sensors closer to the surface.

Cracking patterns change with time and may be affected by rainfall events. Figure 4 shows two pictures taken at days 30 and 150. It can be seen that the morphology of cracks is different, regarding thickness, length, area and shape of cells. Rainfall events may close some small cracks, but they may generate new ones, as stated in [5]. This is partly due to the loss of cohesion of the surficial soil layer when saturated.

Cracks after 150 days of drying are quite deep and are difficult to see in nature, at least in the area where the soil was taken. The main difference of the experiment with respect to the real scenario is the container, avoiding any water input from the ground. In actual conditions, groundwater table keeps a deep boundary condition different from the one applied in the field test.

 DOI: 10.1051/e3sconf/20160912005

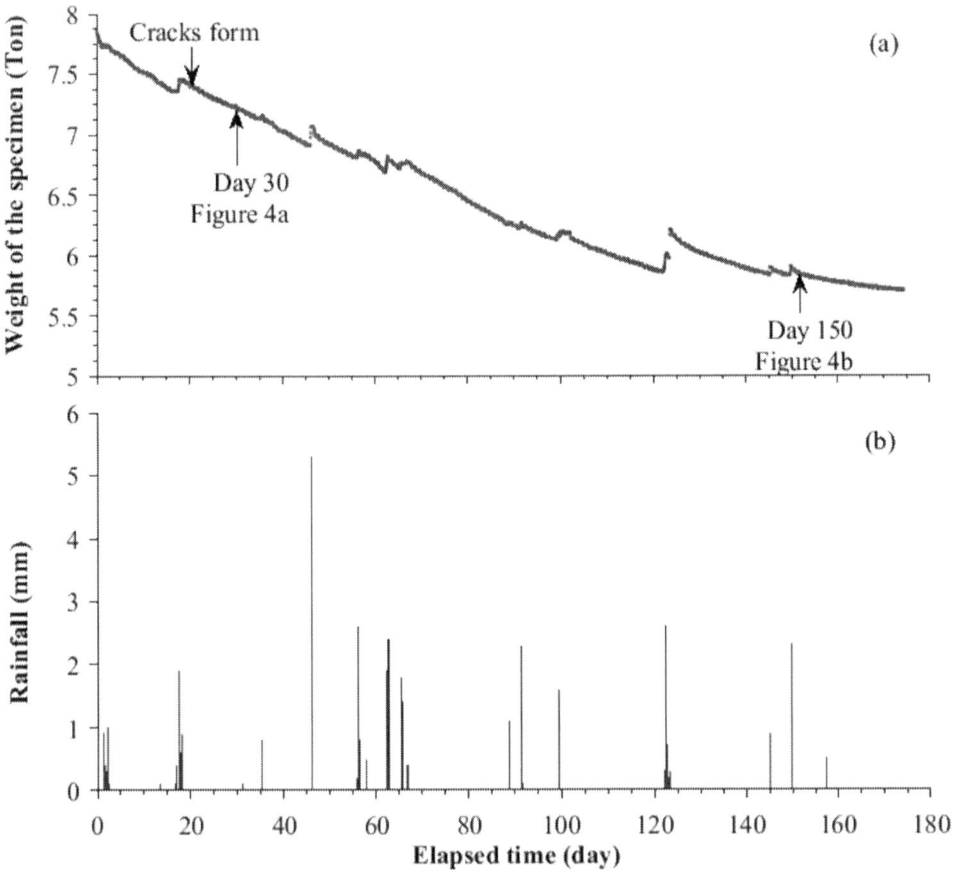

Figure 3. Natural wetting plotted against time. **(a)** Weight of the specimen evolution. **(b)** Rainfall by weather station.

Figure 4. Crack pattern obtained during the time elapsed. **(a)** Crack pattern at day 30 and **(b)** Crack pattern at day 150.

Some of the internal sensors reached their measurement range, either due to the intensity of the evaporation rate at the beginning of summer or because they became exposed to open atmosphere due to the cracks. However, the records obtained during their measurement range will be useful in future analyses.

Cracking in desiccating soils has been described at three different levels: macro-scale continuum, meso-scale of grain and pore clusters, and micro-scale of individual pore structure or grains with liquid bridges [8]. All scales have been observed or suggested in this field test thanks to the diversity of environmental actions applied to the material and it seems that Unsaturated Soil Mechanics can be a useful tool to explain the behaviour of these processes.

[a] Corresponding author: alberto.ledesma@upc.edu

4 Conclusions

This paper describes the design and installation of a field test devoted to the analysis of cracking due to desiccating soil. The test constitutes a natural evolution of previous laboratory experiments on an environmental chamber, where soil trays were desiccated under controlled temperature and relative humidity. In those experiments, the water content evolution was recorded by weighing the soil tray and that information became an important data to analyse the whole test. The same strategy was adopted for the field test, where a large container filled with silty clay was exposed to open atmosphere conditions. The time evolution of the container weight and the main physical variables involved are automatically recorded. That includes meteorological data as well as internal variables in the soil mass, like suction, water content, temperature and heat flux. A particular effort was devoted to measure variables close to the soil surface, as large gradients of those physical variables are expected there.

Preliminary results corresponding to the first 160 days of test have been presented. The data suggest a global desiccation of the soil mass and subsequent cracking, which is consistent with the Mediterranean climate in Barcelona. Small rainfall events have not changed that global tendency. The experiment will last at least for one year and it is expected that the collected data, after interpretation, will contribute to improve the understanding of the processes involved in soil desiccation at a natural scale.

References

1. G. Blight, "Interactions between the atmosphere and the earth," *Géotechnique*, vol. **47**, no. 4, pp. 715–767, (1997).
2. G. Blight, "Solar heating of the soil and evaporation from a soil surface," *Géotechnique*, vol. **59**, no. 4, pp. 355–363, Jan. (2009).
3. Y. J. J. Cui, Y. F. F. Lu, P. Delage, and M. Riffard, "Field simulation of in situ water content and temperature changes due to ground atmospheric interactions," *Géotechnique*, vol. **55**, no. 7, pp. 557–567, (2005).
4. C.-S. Tang, Y.-J. Cui, B. Shi, A.-M. Tang, and C. Liu, "Desiccation and cracking behaviour of clay layer from slurry state under wetting–drying cycles," *Geoderma*, vol. **166**, no. 1, pp. 111–118, Oct. (2011).
5. J. Cordero, A. Cuadrado, A. Ledesma, and P. C. Prat, "Patterns of cracking in soils due to drying and wetting cycles," in *Unsaturated Soils: Research and Applications - Proceedings of the 6th International Conference on Unsaturated Soils, UNSAT 2014*, pp. 381–387, (2014).
6. Lakshmikantha, M.R., Prat, P.C., Ledesma, A. (2012). "Experimental evidence of size effect in soil cracking". *Canadian Geotechnical Journal,* vol. 49, p. 264-284.
7. G. C. Topp, J. L. Davis, and A. P. Annan, "Electromagnetic determination of soil water content: Measurements in coaxial transmission lines," *Water Resour. Res.*, vol. **16**, no. 3, p. 574, (1980).
8. T. Hueckel, B. Mielniczuk, M. El Youssoufi, L. Hu, and L. Laloui, "A three-scale cracking criterion for drying soils," *Acta Geophys.*, vol. **62**, no. 5, pp. 1049–1059, Jan. (2014).

Desirable leaf traits for hydrological reinforcement of soil

D. Boldrin[1,2], A. K. Leung[1] and A. G. Bengough[1,2]

[1]*School of Science and Engineering, University of Dundee, Dundee, UK*
[2]*The James Hutton Institution, Invergowrie, Dundee, UK*

Abstract. Vegetation has an important influence on slope hydrology and hence slope stability via plant transpiration. Little is known about the relative merit of evergreen versus deciduous shrubs in maintaining suctions through the year. This study aims to quantify the soil-plant-water relations of two shrub species and to identify relevant plant traits that correlate with hydro-mechanical properties of vegetated soil. *Corylus avellana* L. (Hazel) and *Ilex aquifolium* L. (Holly) were chosen as contrasting deciduous and evergreen broadleaf species. For each species, three replicates were planted in separated pots of sandy loam soil. Each pot was irrigated until the soil was saturated and then was left to transpire for 20 days. Soil suction, leaf conductance to water vapour (g_L) and soil penetration resistance were recorded. After testing, some key plant traits were determined. It was found that Hazel dried soil faster than Holly. The mean suction induced by Hazel (82.9±1.5 kPa) was 2.7 times greater than that induced by Holly (30.6±8.2 kPa), as Hazel has significantly higher g_L and specific leaf area. Both suction and soil penetration resistance were strongly correlated with the total leaf area, but not the total leaf biomass.

1 Introduction

1.1 Hydrological reinforcement

Vegetation influences natural and engineered slopes due to root growth and root-water uptake driven by plant transpiration [1, 2, 3]. The associated change in the matric suction of soil will affect the stability of these slopes. Pollen-Bankhead and Simon [2] showed that changes in matric suction due to tree transpiration provided the greatest potential benefit to the streambank stability during the growing season. The study also showed that while mechanical root reinforcement increased the streambank factor of safety (*Fs*) by 25%, the effect of transpiration on matric suction translated in a much more significant increase in *Fs* by 52% during growing season. Pollen-Bankhead and Simon [2] introduced a term, hydrological reinforcement, to describe the increase in soil strength due to plant-induced suction in analogy with the well-recognised mechanical root reinforcement.

A more recent field study conducted by Rahardjo et al. [3] showed that shrub and grass covers preserved significant matric suction in an instrumented slope during rainfall, maintaining the slope stability. While the control fallow slope had 25.9% drop in *Fs* after 24 h of rainfall, the slopes covered with the shrub and the grass had a decrease of only 5.9 and 6.2% in *Fs*, respectively.

1.2 Plant water relations

It is apparent that plants can help stabilise slopes hydrologically, but there is limited knowledge of hydro-mechanical properties of vegetated soil, such as soil strength enhancement due to the increase in transpiration-induced suction, and their correlation with plant functional traits. There is thus a need to better integrate the knowledge from environmental plant physiology with unsaturated soil mechanics.

Water transport from the soil, through the plant, to the atmosphere, takes place in a soil-plant-air continuum (SPAC) that is interconnected by a continuous film of liquid water maintained in tension by a water potential gradient [4]. Leaves and, in particular, stomata largely control plant water relations because of the steep gradient of water potential between leaf and air [5]. Stomatal conductance depends on various plant and environmental factors, including species, plant water status and relative humidity [6]. Leaf conductance to water vapor (g_L) limits transpiration rate and reflects the regulatory control exerted by the stomata as well as mesophyll and cuticle conductance. gL can be expressed as the ratio of the rate of transpiration to the driving force for evaporation, which is the pressure gradient in water vapor from the intercellular spaces in the leaf to the atmosphere. Maximum g_L varies among plant functional groups with minimum and maximum values recorded in succulents and in plant of wet habitats, respectively [5]. Changes in stomatal opening and g_L can strongly affect soil water depletion by plants and hence the magnitude of suction developed in

[a] Corresponding author:aleung@dundee.ac.uk

wide vegetated areas. Decrease in transpiration due to the closure of stomata slows the rate of soil water depletion. For example, Hussain et al. [7] show that the decrease in evapotranspiration (ET) by 9% in Maize's field induced a soil moisture increase by 5% – 10%.

Deciduous and evergreen phonologies in cold-temperate climate have mainly evolved as response to freezing-induced cavitation and consequent interruption of SPAC [8]. Evergreens, transpiring when frost occurs, have small vessels that are less likely to cavitate, but have limited vessel conductance and hence low transpiration rates. On the contrary, the stress avoidance strategy adopted by deciduous species generally compensates for a shorter growing season by large-diameter vessels with high conductance and hence transpiration. [9].

1.3 Aim and objectives

In order to identify suitable species for improved slope stabilization, it is necessary to evaluate and select relevant, well-defined and measurable plant functional traits at the individual level and use them as screening criteria for fair comparison across multiple species [10].

Although soil hydrology has been identified to be one of the major factors that govern slope failure [11], the hydrological effect of vegetation is generally ignored in most of the plant selection processes. The current process mainly considers mechanical root reinforcement based on a limited number of biomechanical traits such as root tensile strength and root architecture [12]. One of the major difficulties in incorporating the hydrological effects in such selection process is attributed to a lack of data that reports any correlation between plant functional traits and hydrological reinforcement of the soil.

This study thus aims to quantify the soil-plant-water relations of contrasting deciduous and evergreen broadleaf shrubs native to Europe. Plant functional traits that can correlate with hydro-mechanical properties of vegetated soil such as soil strength are identified.

2 Materials and Methods

2.1 Plant materials

In this study, two species, *Corylus avellana* L. (Hazel) and *Ilex aquifolium* L. (Holly), were chosen as contrasting deciduous and evergreen broadleaf species that would grow into shrubs or small trees. These species are both common in Europe. Both species were supplied by British Hardwood Tree Nursery, Gainsborough, UK and they presented similar dimensions, but different age, approximately 1 and 2 years old, respectively.

2.2 Soil and planted pots

Pots with a volume of 4L were packed with soil collected from Bullion field, The James Hutton Institute, Dundee, UK. The soil was a sandy loam comprised of sand 71%, silt 19%, clay 10%, with liquid and plastic limits of 32% and 23%, respectively [13]. The soil (sieved < 10 mm; water content 0.14-0.16 g/g) was dynamically compacted in four layers to obtain an initial dry density of 1200 kg m^{-3}. The surface of each layer was abraded to achieve a better contact between each successive layer. After compacting the third layer in each pot, a bare root plant was planted in each pot and finally the fourth layer of soil was added. A total of 20 pots for each species were prepared. After transplanting, the soil surface was covered with wecult® discs to decrease evaporation. All plants were maintained in a glasshouse and abundantly irrigated two times a week for two months to favour plant establishment. Plants were considered established when stable canopy development was observed.

Soil water retention curves (SWRCs) of three cores of sandy loam (55 mm diameter; 40 mm height) were measured at the same dry density. Each of them was subjected to suctions ranging from 1 to 1500 kPa using a tension table (1 – 50 kPa) and a pressure plate (50 – 1500 kPa; ELE International, Hemel Hempstead, UK).

2.3 Monitoring of plant-induced matric suction

After plant establishment, three replicates of each species were randomly selected from the 20 pots. A miniature tensiometer (SWT-5, ΔT devices, Cambridge, UK) was installed in each pot at 0.1 m below the soil surface for monitoring any matric suction induced by each replicate under glasshouse conditions (average daily temperature 15 – 18°C and daily relative humidity between 50% – 80%). Initially, each pot was irrigated until the soil was saturated (as indicated by 0 kPa by the tensiometer) and it was then left to transpire for 20 days from May 29th to June 17th 2015. In order to measure any water loss due to transpiration, each pot was weighed every two days on a balance with an accuracy of 0.1 g. Immediately after each measurement of water loss, matric suction was recorded using the tensiometer.

2.4 Soil penetration resistance

To quantify the hydrological reinforcement to the soil due to transpiration-induced suction, soil penetration resistance tests [14] were carried out in each pot using a portable penetrometer (Basic Force Gauge, Mecmesin, UK; cone diameter: 1.94 mm; cone angle: 30°). Maximum soil resistance was determined by penetrating the cone to 30 mm depth from the soil surface. The cone diameter and penetration depth were chosen according to the dimension of the pot [15]. The measurements were taken at three different points for each replicate after 15 days of transpiration. The major advantage of such testing method is that the hydrological reinforcement due to transpiration-induced suction can be isolated from mechanical reinforcement of roots. The use of penetrometer also offers a relatively quick and less destructive way to determine soil strength using simple pots. Due to the simplicity of the testing method, multiple penetration tests can be carried out using the same pot, hence reducing the variability of test results. Soil penetration resistance has been used as a parameter to indicate the mechanical or/and hydrological effect of plants on slope stability by Osman and Barakbah [16].

2.5 Leaf conductance to water vapor

To quantify plant transpiration, leaf conductance to water vapor (g_L) was measured on at least one mature leaf on 19 plants per species using a portable porometer (AP4, ΔT devices, Cambridge, UK). This device is a dynamic diffusion porometer, in which part of the leaf is enclosed at the base of a cup containing a humidity sensor. Dry air is then flushed through the cup until a pre-selected lower level of relative humidity is achieved. The flushing then stops and the transit time required for a small, fixed increase in relative humidity is measured. The time taken for the humidity to increase over the fixed interval is related to g_L via a calibration curve. Before measurement, the porometer was calibrated using a perforated plate with known diffusive conductance to water vapor. Measurements were performed on a sunny day (June 5th) on plants in the absence of water stress, which is ideal to prevent stomatal closure and hence negatively affect g_L measurements. Note that the soil moisture condition for the measurements of g_L in the 19 pots was equivalent to that in the three suction testing pots after four days of transpiration when small suction was developed.

2.6 Plant traits

A number of functional leaf traits, including specific leaf area (SLA), total leaf biomass and total leaf area, were determined to correlate with the hydrological responses of the vegetated soils. SLA is defined as the one-sided area of a fresh leaf divided by its oven-dry mass, expressed in $m^2\ kg^{-1}$ [17]. Ten fully expanded leaves per species were collected at early morning (June 16th) from five plants. Leaves were then scanned and the area was measured using an open source, image analysis software, ImageJ (NIH, USA). Each leaf sample was placed in an oven at 60 °C for 72 h to obtain a constant weight measured using an electronic balance with an accuracy of 0.0001 g. SLA was hence calculated by dividing the leaf area by the corresponding leaf dry weight ($m^2\ kg^{-1}$).

After the suction monitoring of each pot for 20 days, total leaf biomass of the three replicates was estimated by multiplying the number of leaves by the mean leaf dry biomass, while the total leaf area was calculated by multiplying the number of leaves with the mean leaf area. These testing methods avoid destructive measurements.

2.7 Statistical analysis

Statistical analysis was performed using GenStat17th Edition (VSN International) and SigmaPlot12.5 (Systat Software Inc). The significance of differences between experimental treatments was assessed by Student's t-test. The significance of correlations established in this study was tested using regression analysis. Results were considered statistically significant at p-value ≤ 0.05.

3 Results

The SWRC of the sandy loam is shown in Fig. 1 (van Genuchten-Mualen regression; m=1-1/n; r^2=0.99).

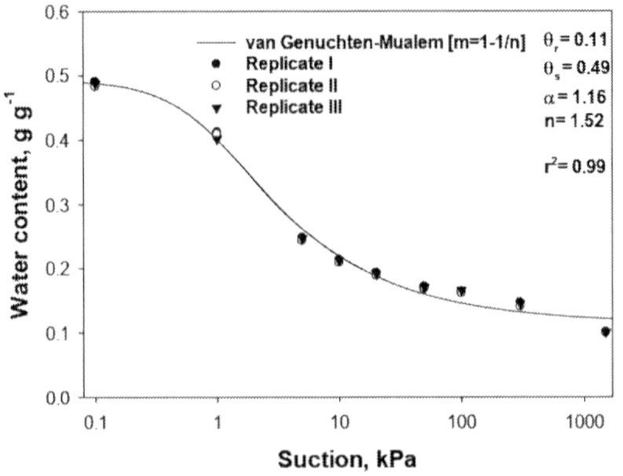

Figure 1. Soil water retention curve of Bullion field soil sieved to 10 mm and compacted to 1200 kg m^{-3}. Van Genuchten model parameters are reported in the graph, as well as r^2.

Fig. 2 shows the water loss from planted pots during time from May 28th (soil saturation) to June 17th. The difference in water loss between Holly and Hazel pots translated in a large difference in matric suction (Fig. 3). After 20 days of transpiration, the mean suction induced by Hazel (82.9 ± 1.5 kPa) was 2.7 times greater than that induced by Holly (30.6 ± 8.2 kPa). The difference of induced suction between Holly and Hazel is found to be statistically significant (p-value = 0.003).

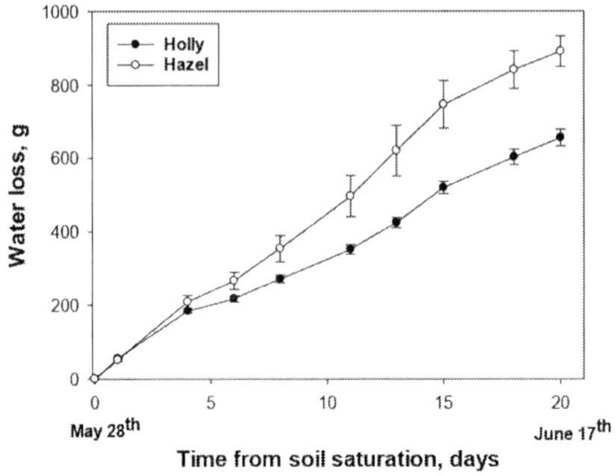

Figure 2. Measured water loss in Holly and Hazel replicate pots during the experimental period, May 28th – June17th. Error bars represent the standard error of mean.

Soil penetration resistance between Holly and Hazel pots is significantly different (p-value = 0.035) after 15 days of transpiration, with the minimum value of 1.57 MPa in Holly pot and maximum value of 9.66 MPa in Hazel pot. Note that these resistances mainly serve as indicators for the relative comparison of the hydrological reinforcement between the two species. A strong correlation (r^2 = 0.93; p-value = 0.001) between suction

DOI: 10.1051/e3sconf/20160912006

and soil penetration resistance is sought, highlighting the significance of hydrological reinforcement (Fig. 4).

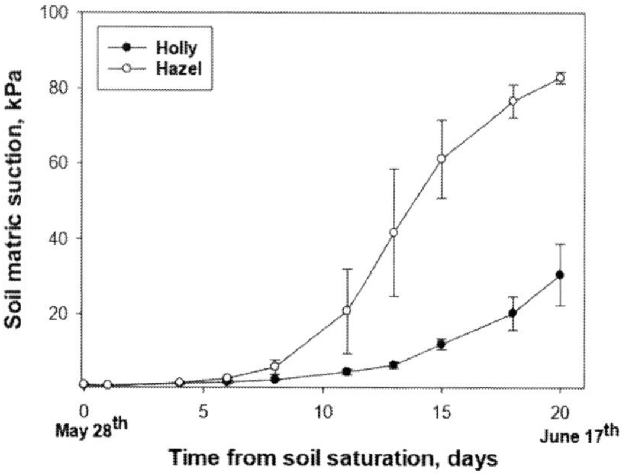

Figure 3. Measured matric suction in Holly and Hazel replicate pots during the experimental period, May 28th – June 17th. Error bars represent the standard error of mean.

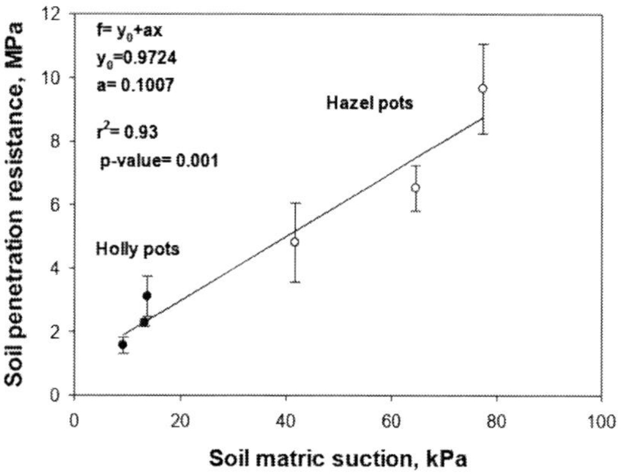

Figure 4. Correlation between matric suction and soil penetration resistance in Hazel and Holly pots, after 15 days of transpiration. Linear regression equation, coefficients y_0 and a are reported in the graph, as well as r^2 and p-value. Error bars represent the standard error of mean.

The highly significant difference of SLA between Holly and Hazel (Fig. 5) highlighted their difference in leaf biomass investment and allocation. This is because the mean leaf area of Hazel was 2.2 times higher than that of Holly, whereas the mean leaf mass of Hazel was 1.5 times lower (Table 1).

Table 1. Leaf area and dry mass of Hazel and Holly leaves. Means are reported ± standard error of mean (n=10).

Species	Leaf area, mm^2	Leaf mass, g
Hazel	2787.33 ± 229.02	0.15 ± 0.02
Holly	1260.65 ± 109.39	0.23 ± 0.02

Strong correlations were found between the total leaf area and both matric suction (r^2 = 0.76; p-value = 0.01; Fig. 6A) and soil penetration resistance (r^2 = 0.89; p-value = 0.003; Fig. 6B). The total leaf area can explain the 76% of matric suction and 89% of penetration resistance variability in soil. Although there is no correlation between matric suction and the total leaf biomass when both species are considered (Fig. 7), there is a positive trend between total leaf biomass and matric suction if species are individually considered.

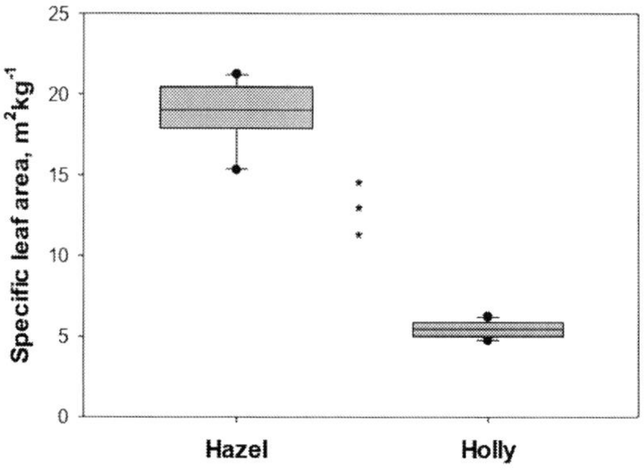

Figure 5. Box plots of Holly's and Hazel's SLA. *** represents a highly significant difference (p-value ≤ 0.001; Student's t-test; n=10) between the SLA of Holly and Hazel.

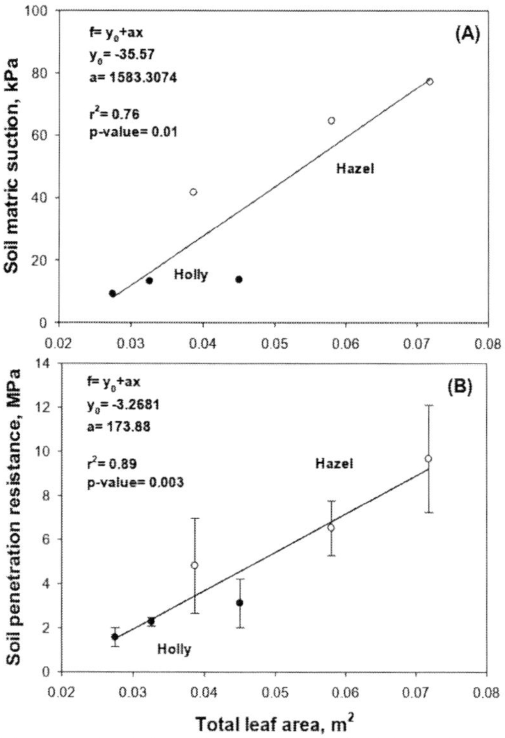

Figure 6. Correlation of total leaf area with (A) matric suction and (B) soil penetration resistance in Hazel and Holly pots, after 15 days of transpiration. Linear regression equation, coefficients y_0 and a are reported in the graph, as well as r^2 and p-value. Error bars represent standard error of mean.

DOI: 10.1051/e3sconf/20160912006

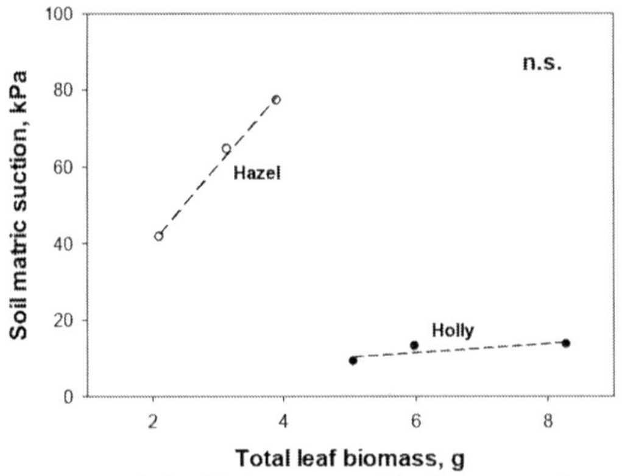

Figure 7. Relationship between total leaf biomass and matric suction in Hazel and Holly pots, after 15 days of transpiration. Dash lines represent positive trends. n.s.: not statistically significant correlation

The leaf conductance to water vapor (g_L), recorded in nineteen Holly and Hazel replicates under the same environmental condition, highlights a highly significant difference between the two species (Fig. 8). Hazel has a mean g_L (84.3 ± 7.5 $mmolm^{-2}s^{-1}$) about 3.2 times higher than Holly (26.1 ± 3.8 $mmolm^{-2}s^{-1}$).

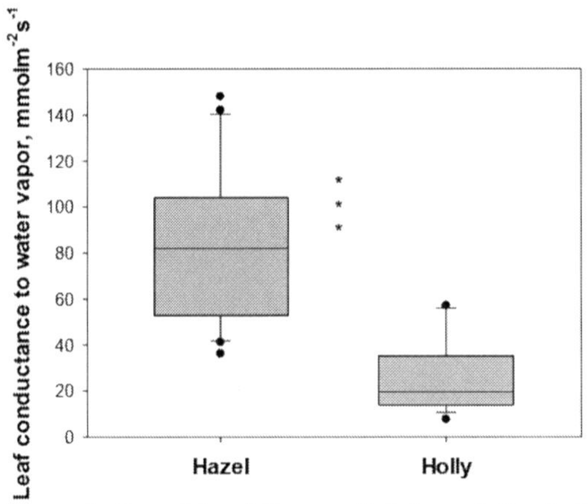

Figure 8. Box plots of the g_L of Holly and Hazel. *** represents a highly significant difference (p-value ≤ 0.001; Student's t-test; n=19) between the g_L of Holly and Hazel.

4 Discussion

The monitoring of water loss and matric suction (Figs 2 and 3) highlighted that Hazel influenced the soil hydrology more than Holly. The greater hydrological effects of Hazel translated in a three times higher penetration resistance in Hazel pots (Fig. 4).

The regression study shows that total leaf area is an important leaf trait that has strong correlations with plant water uptake capability (i.e. in terms of matric suction; Fig. 6A) and mechanical strength of vegetated soil (i.e. in terms of soil penetration resistance; Fig. 6B). Another important finding is that plant water uptake is not driven by the leaf biomass (Fig. 7), but more by its allocation and investment as leaf area. SLA has been demonstrated to be positively correlated with net photosynthesis [18], potential growth rate [19] and transpiration rate [20]. Differences in SLA between species are the result of different leaf design, which represents the plant investment in leaf tissue [21]. A small SLA, and hence large leaf dry matter per area, translates into more resistant leaves with high leaf life span and hence a slow-return of initial energy investment in the leaf. On the other hand, large SLA is related to a fast-return of energy investment. In a temperate climate, generally deciduous species, such as Hazel, are characterized by high values of SLA in contrast to evergreens, being the maintenance of canopy leaves during unfavourable seasons a higher investment of nutrients and dry mass than the cost of new leaves [22]. Therefore, species like Hazel maximises the growth and hence the water uptake during the summer growing season, while the evergreen Holly has a slow-return of energy investment and hence low water use and growth rate. Bochet and García-Fayos [23] identified SLA as an important plant trait associated with the establishment success on road-side slopes in semi-arid environments, reflecting plant competitiveness. Our study highlights the possibility of using SLA as one of the quick and easily-measured screening traits to identify suitable species for hydrological reinforcement of soil.

The higher water uptake of Hazel was also explained by its higher g_L, as compared with Holly (Fig. 8). As g_L is directly proportional to transpiration, this parameter is thus an important driver of plant transpiration to induce suction. Hungate et al. [24] demonstrated that a reduction of g_L in scrub-oak woodland can directly reduce the ET of the ecosystem and hence result in relatively wet soil. On natural and engineered slopes, the decrease of ET can cause a decrease in matric suction, and consequently slope stability [25]. Differences in transpiration of Hazel and Holly can be explained by their different phenology: deciduous and evergreen, respectively. While deciduous trees generally compensate for a shorter growing season by producing leaves that have higher photosynthetic rates per unit mass and hence transpiration, evergreen trees are known to have lower hydraulic conductance and an architectural feature that promotes water saving and enhances tolerance during dry and frost periods [9].

As g_L is significantly different between Hazel and Holly, this parameter can be a useful indicator of hydrological reinforcement. However, it must be pointed out that g_L varies with several environmental factors such as water availability, irradiance and relative humidity. We thus suggest the use of g_L in plant screening only if plants are under the same environmental conditions and in well water status to avoid stomatal closure.

Although both Hazal and Holly are suggested for their use on road landscapes by the Scottish Government (http://www.gov.scot/Publications), Hazel was shown to be a more efficient species, as far as slope stability is of concern, as this species provides greater hydrological reinforcement due to its high transpiration rate. The high water use of Hazel was also reported in the natural environment by Herbst et al. [26], who found that Hazel determined most of the understory forest ET (75%; 50 mm) in a South England forest. Bischetti et al. [27] also showed that Hazel roots had the highest tensile strength among different forest species in Italy. Therefore, it seems that Hazel can be a suitable species in terms of hydro-mechanical reinforcement of road-side slopes.

5 Conclusions

Our study uses leaf traits to correlate and explain the hydrological reinforcement of vegetated soils due to transpiration-induced suction. Some conclusions may be drawn from this study:

- There are significant differences of hydrological reinforcement to the soil between Hazel and Holly. Hazel induced a mean suction (82.9 ± 1.5 kPa) that was 2.7 times greater than Holly (30.6 ± 8.2 kPa), translating to a higher soil penetration resistance.
- Strong correlation between total leaf area and both suction (r^2 = 0.76) and soil penetration resistance (r^2 = 0.89) existed, regardless of the species.
- Specific Leaf Area (SLA) may be an indicator of plant water relations as significant differences in SLA were found between Hazal and Holly (p-value ≤ 0.001).
- No correlation was found between total leaf biomass and the matric suction induced by transpiration, when both species are considered.
- Under controlled conditions, g_L measurements might be a useful parameter for plant screening for hydrological reinforcement associated with plant-induced suction, as the g_L of Hazel and Holly were significantly different (p-value ≤ 0.001)

The initial results suggest that there is substantial scope to choose species to manipulate hydro-mechanical properties of soil for engineering and ecological functions. Future work is needed to increase the number of species and traits, as well as to quantify seasonal variation in evergreen versus deciduous species in relation to their effects on soil hydrology and resistance.

References

1 A. Stokes, G. B. Douglas, T. Fourcaud, F. Giadrossich, C. Gillies, and e. al., Plant Soil, **377**, 1, 1 (2014).
2 N. Pollen-Bankhead and A. Simon, Geomorphology, **116**, 3-4, 353 (2010).
3 H. Rahardjo, A. Satyanaga, E. C. Leong, V. A. Santoso, and Y. S. Ng, Soils and Foundations, **54**, 3, 417 (2014).
4 E. Steudle, Annu Rev Plant Phys, **52**, 847 (2001).
5 H. Jones, *Plants and Microclimate A Quantitative Approach to Environmental Plant Physiology*, Cambridge 3ed. (2013).
6 K. Aasamaa and A. Sober, Environ Exp Bot **71**, 1, 72 (2011).
7 M. Z. Hussain, A. Vanloocke, M. H. Siebers, U. M. Ruiz-Vera, R. J. C. Markelz, A. D. B. Leakey, D. R. Ort, and C. J. Bernacchi, Glob Change Biol **19**, 5, 1572 (2013).
8 H. Lambers, F. S. Chapin, and T. L. Pons, *Plant Physiological Ecology*, Springer 2 ed. (2008).
9 D. Baldocchi, S. Ma, S. Rambal, L. Misson, J.-M. Ourcival, J.-M. Limousin, J. Pereira, and D. Papale, Ecol appl **20**, 6, 1583 (2010).
10 A. Stokes, C. Atger, A. G. Bengough, T. Fourcaud, and R. C. Sidle, Plant Soil, **324** (1-2), 1 (2009).
11 D. G. Toll, S. D. N. Lourenco, J. Mendes, D. Gallipoli, F. D. Evans, and e. al., Q J Eng Geo Hydrog, **44**, 1, 23 (2011).
12 S. De Baets, J. Poesen, B. Reubens, K. Wemans, J. De Baerdemaeker, and B. Muys, Plant Soil, **305**, 1-2, 207 (2008);
13 K. W. Loades, A. G. Bengough, M. F. Bransby, and P. D. Hallett, Plant Soil, **370**, 1-2, 407 (2013).
14 K. Weaich, A. Cass, and K. L. Bristow, Soil Till Res, **25**, 2-3, 149 (1992).
15 R. K. Misra and F. D. Li, Soil Till Res, **38**, 1-2, 59 (1996).
16 N. Osman and S. S. Barakbah, Ecol Eng, **28**, 1, 90 (2006).
17 N. Pérez-Harguindeguy, S. Díaz, E. Garnier, S. Lavorel, H. Poorter, and e. al., Aust J Bot, **61**, 3, 167 (2013).
18 P. B. Reich, M. B. Walters, and D. S. Ellsworth, PNAS, = **94**, 25, 13730 (1997).
19 J. P. Grime, K. Thompson, R. Hunt, J. G. Hodgson, J. H. C. Cornelissen, I. H. Rorison, and e. al., Oikos, **79**, 2, 259 (1997).
20 P. B. Reich, D. S. Ellsworth, M. B. Walters, J. M. Vose, C. Gresham, J. C. Volin, and W. D. Bowman, Ecology, **80**, 6, 1955 (1999).
21 I. J. Wright, P. B. Reich, M. Westoby, D. D. Ackerly, Z. Baruch, F. Bongers, and e. al., Nature, **428**, 6985, 821 (2004).
22 K. Bai, C. He, X. Wan, and D. Jiang, AoB PLANTS, **7**, 1, (2015).
23 E. Bochet and P. García-Fayos, Ecol Eng, **83**, 444 (2015).
24 B. A. Hungate, M. Reichstein, P. Dijkstra, D. Johnson, G. Hymus, J. D. Tenhunen, C. R. Hinkle, and B. G. Drake, Glob Change Biol, **8**, 3, 289 (2002).
25 J. A. Smethurst, K. M. Briggs, W. Powrie, A. Ridley, and D. J. E. Butcher, Geotechnique, **65**, 11, 869 (2015).
26 M. Herbst, P. T. W. Rosier, M. D. Morecroft, and D. J. Gowlng, Tree Physiol, **28**, 6, 959 (2008).
27 G. B. Bischetti, E. A. Chiaradia, T. Simonato, B. Speziali, B. Vitali, P. Vullo, and A. Zocco, Plant Soil, **278**, 1-2, 11 (2005).
http://www.gov.scot/Publications accessed 14/09/2015

Influence of atmospheric conditions on the strength of unstabilized earthen constructions

Pierre Gerard[1,a], Mohamed Mahdad[1], Alexandre Robert[1], Bertrand François[1]

[1]*Université libre de Bruxelles (ULB), BATir Department, Laboratory of GeoMechanics (LGM), Brussels, Belgium*

Abstract. Uniaxial compression tests and indirect tensile tests are performed on compacted clayey silt samples upon varying suctions in order to assess the influence of changes in the relative humidity conditions on the strength of unstabilized rammed earthen building materials. The results show that suction plays an important role on the strength of the material. Also the ability of the Belgian clayey silt to develop sufficient mechanical strength to be used as an unstabilized earthen construction material is demonstrated whatever the relative humidity conditions, excepted the fully water saturated state. The experimental data are interpreted in the context of unsaturated soil mechanics using the generalized effective stress concept. This constitutive framework allows defining a unified failure criterion predicting the strength of the earthen building material as a function of the environmental hygroscopic conditions.

1 Introduction

Earthen construction is an ancient technique that is experiencing a renascence today thanks to the hygrothermal regulation and the energy performance of this material and its potential for recycling. The material is locally available and the energy use for its manufacturing is very limited [1]. Soil is one of the most predominant materials on earth and can be considered universal to some extents that avoid dependency on importation.

In order to provide appropriate mechanical properties, the earthen materials must be installed in a proper way in order to optimize the density and the water content. Among different kinds of earthen constructions, rammed earth is the technique that consists in forming the wall by compacting moist soil between temporary forms. For "unstabilized" rammed earth, the system does not require any additional binder elements. A part of the cohesion is brought by the clayey materials in combination with the compaction process that provides the required density. In addition to the natural binding effect of the clayey material, capillary cohesion contributes also to the total strength of the materials [2,3]. However, after the construction, external rammed earth walls can be subject to changes in humidity and incident wetting from rainfall. Those modifications of environmental conditions induce continuous changes of the water retention conditions of the wall that affect the durability of the constructions.

Only a few studies focus directly on the impact of the change in relative humidity on the internal suction, and so on the strength of the wall. Unconfined compression tests have been performed on earthen materials samples air-dried to different target water contents [4], but the samples water content varied between 5.5% and 10.2%, while the water content of an unstabilised rammed-earth construction subject to atmospheric conditions is generally lower (1-2 %) [5]. The unconfined strength of different soil samples (sand, clay) have been determined with a greater range of water content (from a wet state after manufacturing w = 11 % to a dry state in atmospheric conditions w = 1-2%) [6]. This study confirmed that suction plays an important role on the strength of the material, but highlighted also that a slight increase in moisture content of dry rammed-earth walls (water content not exceeding 4%) due to rainfall or change of relative humidity in the atmosphere is not followed by a sudden drop in the wall strength.

This literature review reveals therefore first the scarcity of experimental procedure replicating the prevailing climatic conditions in different regions of the world, and quantifying the evolution of the unstabilized earthen materials strength with the atmospheric conditions. On the other hand no unified failure criterion has been formulated to characterize the effect of the suction and capillary cohesion on the strength of these materials.

The objective of the study is to evaluate the ability of a representative Belgian clayey silt to develop sufficient mechanical strength under variable relative humidity conditions to be used as an unstabilized rammed earthen building materials. To do so, the evolution of the strength as a function of the atmospheric relative humidity is characterized through uniaxial compression tests and indirect tensile tests. Then, a constitutive framework is proposed, able to predict the failure criterion of unstabilized rammed earthen materials, including the

[a] Corresponding author: piergera@ulb.ac.be

DOI: 10.1051/e3sconf/20160913001

effect of capillary and intrinsic cohesion through the generalized effective stress approach for unsaturated soils.

2 Materials

The soil examined in this research is a clayey silt from the region of Marches-Les-Dames (Belgium). Its index properties are: liquid limit (w_L) = 32,5 %; plasticity index (IP) = 15%. The clayey fraction represents 13 %, the silty one about 64 % and the sandy one about 26 %. A series of recommendations on the grain-size distribution of soils particularly well adapted for earthen constructions has been published in [7]. Even though most of the guidelines for a suitable rammed earth particle size distribution recommend an inert aggregate fraction of gravel, the Marche-Les-Dames silt used in this study does not contain any gravel. Nevertheless the particle size distribution of the Marche-les-Dames silt approaches the one proposed by [8] and will show his relevance for earthen constructions. Those good properties are due to the spread grain size distribution of this natural clayey silt that induces a good interlocking of grains after compaction. The normal Proctor optimum water content w is 15 % and the normal Proctor optimum dry density ρ_d is 18,40 kN/m^3.

Even if the normal Proctor compaction test constitutes a widely used standard for the geotechnical earthworks, this method is generally not considered in the context of earthen materials. Indeed, the targeted properties of an earthen wall largely differ from the properties expected for geotechnical works. It is why another compaction process has been used in this research. The soil was dynamically compacted by sequentially ramming the soil in layers with a 2.5kg Proctor hammer directly inside two kinds of mould: 36 mm in diameter and 72 mm in height for the uniaxial compression test (compacted in 3 layers) and 36 mm in diameter and 22 mm in height for indirect tensile test (compacted in 1 layer). The compaction of each layer was achieved until the handle of the hammer "rings" when dropped onto the compacted soil, which is considered as the indication of full compaction having being attained [7].

To determine the optimum initial water content samples were prepared at 8 different water contents (dried soil, 2%, 4%, 6%, 8%, 10%, 12%, 14%). The samples at a theoretical 0% and 2% of water contents were unusable as the samples became laminated and quickly dismantled at the limits of their compacted layers. The samples were then tested upon uniaxial compression loadings in order to evaluate their uniaxial compression strength (UCS) as a function of their water contents at compaction. For each compression, 3 to 4 samples were tested under unconfined compression for statistical consistency. It is to mention that the spreading of the strength never exceeds 10% of the mean value, which demonstrates a good reproducibility of the results. The obtained mean UCS is reported in Figure 1 together with the obtained dry density. For the sake of comparison, the mean UCS obtained on samples compacted at much lower dry density, 1732 kg/m^3, at different water contents are reported in Figure 2. This dry density is close to the optimum Proctor dry density of the soil.

First of all, the results show the relevance of the compaction method adapted for earthen construction. The process for earthen construction (Figure 1) provides much denser samples than the normal Proctor method with the consequence that the water content of rammed earth is lower that the normal Proctoroptimum water content. On the other hand the comparison between Figures 1 and 2 clearly demonstrates the drastic effect of soil density on its strength. This 'as-compacted UCS' reaches barely 1.4 MPa upon a dry density of 1732 kg/m^3 at 6% of water content. On the contrary, when the soil is compacted until bouncing of the hammer, the samples develop an 'as-compacted UCS' that may reach 3.8 MPa for a dry density of 2020 kg/m^3 when compacted at 8% of water content.

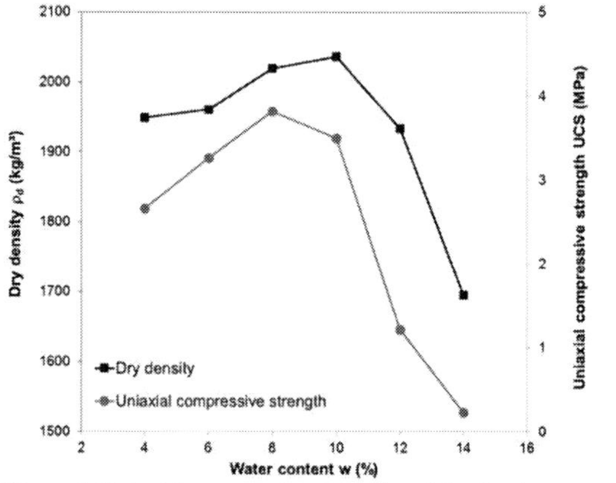

Figure 1. Uniaxial compressive strength and dry density vs. initial water content obtained on samples compacted with the process for earthen construction.

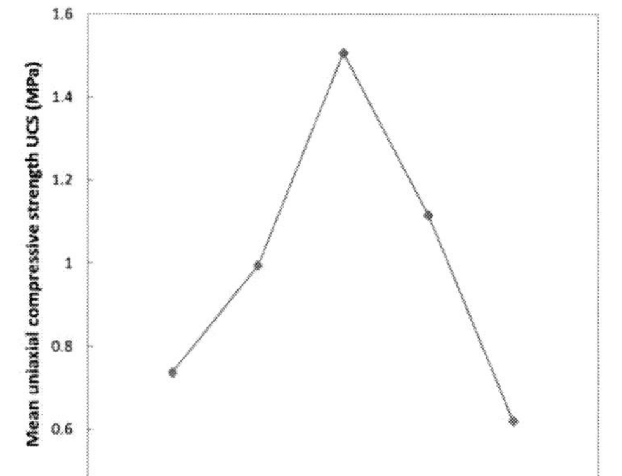

Figure 2. Mean uniaxial compressive strength according to the initial water content obtained on samples compacted at a dry density of 1732 kg/m^3.

In the next sections, the optimized conditions of compaction (w=8% and ρ_d = 2000 kg/m^3) have been selected to determine the effect of the hygroscopic variations on the strength. A target dry density slightly lower than the maximum density reached in Figure 1 was applied in order to take into account some possible losses of efficiency of the compaction process in in-situ conditions.

3 Methods

3.1 Control of the suction

To influence the relative humidity of air surrounding the sample, saturated saline solutions were used at a constant temperature of 20°C [9]: K_2SO_4 (*RH*=97%), KNO_3 (*RH*=92%), KCl (*RH*=85%), NaCl (*RH*=75%). Also, *RH*= 40% was reached by placing the samples under ambient conditions in a room of the laboratory in which temperature and humidity remain constant. According to the relative humidity of the air, water exchanges occurring by vapour transfers between the sample and its surrounding induce a given suction at equilibrium within the sample (Kelvin's law):

$$s = \frac{\rho_w RT}{M_w} lnRH \quad (1)$$

Also, a last hydraulic condition was obtained by sample saturation. A confining pressure of 200 kPa was applied in the confining cell while a water pressure of 190 kPa was applied on the lower porous stone. The saturation is reached when the flowrate of water injected in the sample through lower porous stone is equal to the flowrate of water measured through the upper porous stone. After dismantling the system, the degree of saturation is also checked by measuring and weighting the sample.

3.2. Uniaxial compression test

The scope of the uniaxial compression test is to determine the uniaxial compressive strength of soil samples with unrestricted horizontal deformation. The displacement of the piston was at 0.0667 mm/min and the stress and strain were monitored every second. The uniaxial compressive strength σ_{UCS} is calculated as:

$$\sigma_{UCS} = \frac{F}{A_c} \quad (2)$$

where A_c is the equivalent section estimated under the assumption of constant volume of the sample during loading, and *F* is the force applied by the piston.

It is worth to note that the size of the samples (*D*=36 mm; *H*=72 mm) is consistent with the recommendations on the geometry of the samples [10]. A slender ratio (*H*/*D*) of 2 is indeed necessary for the determination of the uniaxial compressive strength of earthen materials. On the other hand Ciancio & Gibbings have studied the influence of the sample size on its strength [11]. They tested several samples manufactured with different diameters and a constant slender ratio of 2. While an inverse relationship between strength and specimen size is generally observed on concrete samples [12], they showed that the strength of earthen materials is not influenced by the size.

3.3. Indirect tensile test

In order to determine the uniaxial tensile strength of earthen materials, indirect tensile tests have been conducted. The test is valid for the materials exhibiting relative brittle failure which is the case for highly compacted earthen materials [13]. Under a compressive force F along its length L, the minor principal stress is the horizontal stress, in tension, which corresponds to the tensile strength σ_t:

$$\sigma_3 = \sigma_h = -\frac{2F}{\pi dL} = \sigma_t \quad (3)$$

with *d* the diameter of the sample.

4 Experimental results

4.1. Water retention curve

The amount of water stored in the specimen affects the capillary cohesion and has, in turn, a direct influence on the strength. So we focus first on the water retention curve.

Just after compaction, the initial water content is the imposed water content at compaction (w_0=8%) while the suction *s* has been evaluated at 2.4 MPa by the filter paper method. This suction corresponds to the 'as-compacted' state of the material before any effect of the relative humidity of the atmosphere.

Then the soil suction was imposed through the control of relative humidity. Soil samples prepared at the two different dimensions and compacted according to the methodology described above were placed in desiccators with saturated salt solutions (or at ambient laboratory conditions for RH=40%) and were weighted every day until reaching a constant weight. Figures 3 reports the water retention curve under the assumption of a constant volume of the sample upon drying.

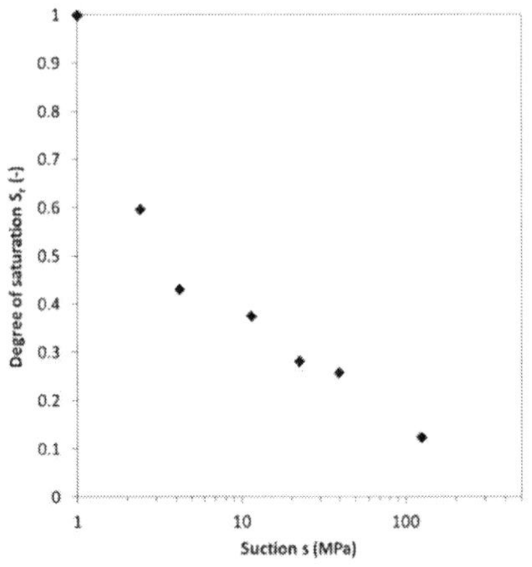

Figure 3. Retention curve of Marche-les-Dames silt

DOI: 10.1051/e3sconf/20160913001

4.2. Uniaxial compressive strength

The results (available in details in [14]) exhibit the following trends:

- The UCS increases with suction: dryer is the soil and higher is the strength.
- However, in the same time, the brittleness also increases with suction: the material is less and less ductile when the amount of water in the soil decreases.
- Upon saturated conditions, the strength is almost zero while it is very ductile.

Figure 4 summarizes the obtained UCS as a function of the applied suction. Suction is reported on a logarithmic scale. Saturated conditions (for which suction is almost zero) were arbitrary linked to suction of 1 MPa, for the convenience of the representation. Excepted for the results at 4.17 MPa of suction, a logarithmic regression curve provides a good estimation of the obtained UCS. The points out of the trend at 4.17 MPa (*RH*=97 %) can probably be explained because of the precision of the relative humidity technique at low suctions. This is a well-known drawback of this technique which is not adequate to apply low suctions. A small error in the imposed relative humidity induces a large approximation in the imposed suction (e.g. 1% of relative uncertainty on *RH* leads to 1.38 MPa of absolute uncertainty for the suction) [9].

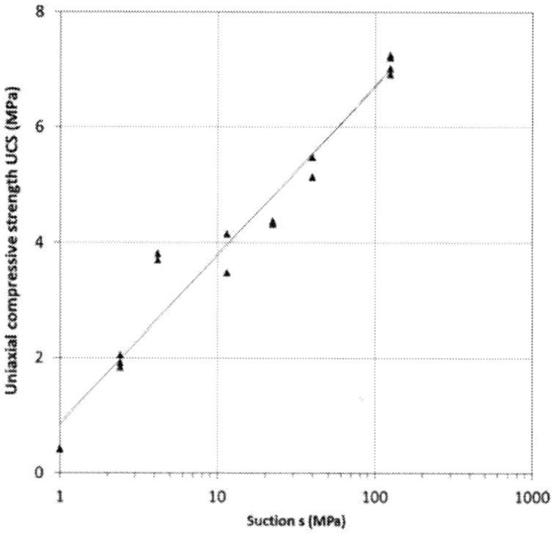

Figure 4. Uniaxial compressive strength according to the imposed suction.

4.3. Tensile strength

The ultimate tensile strength is reported in Figure 5 according to the imposed suction. Again, logarithmic regression curve provide a good estimation of the evolution of the strength as a function of suction, except for the points at 4.17 MPa (*RH*=97 %), for the same reasons than exposed before.

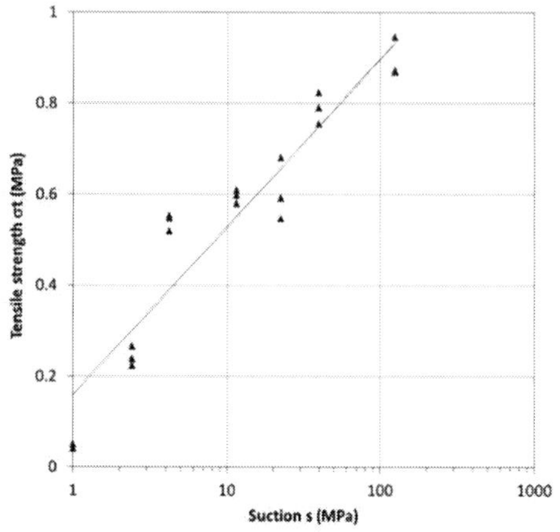

Figure 5. Tensile strength according to the imposed suction.

5 Constitutive framework

The objective of this section is to find an expression of generalized effective stress that permits to obtain a unique failure criterion that is unaffected by suction and degree of saturation, when it is expressed in this new stress reference. In other words, the dependency of the water retention conditions on the strength is directly included in the stress definition and is not explicitly taken into account in the expression of the failure criterion. To do so, it is proposed to start from the general expression of generalized effective stress σ'_{ij} as originally proposed by Bishop [15]:

$$\sigma'_{ij} = \sigma_{ij} + \chi s \delta_{ij} \tag{4}$$

where χ, called the effective stress parameter, varies with the degree of saturation [15, 16], from zero for dry soil to unity for fully saturated conditions. σ_{ij} is the total stress tensor and δ_{ij} is the Kronecker symbol. This Bishop's effective stress implies that the mechanical response is directly linked to the water retention behavior through the parameter χ. Consequently, generalized effective stress state experienced by specimen during both uniaxial compression tests and indirect tensile tests can be reported in the Mohr plane with Mohr circles which are shifted to the right by an amount equal to χs. The objective is to find a suitable expression of χ as a function of the degree of saturation S_r that permits a shift of the obtained Mohr circles at failure that fulfills a unique failure criterion.

The first step towards the definition of a unified failure criterion for the studied rammed earth material consists in the determination of the intrinsic strength parameters, friction angle φ and cohesion c. To do so, 4 consolidated and undrained (CU) triaxial tests have been performed on saturated samples at different confining pressures σ_3 (50, 100, 200 and 300 kPa), with measurement of the pore pressures evolution during shearing. Considering the maximum ratio σ'_3/σ'_1 as the ultimate stress state, this leads to a friction angle φ=36.5° and a cohesion c = 6.2 kPa (more details in [14]).

DOI: 10.1051/e3sconf/20160913001

Secondly, when the intrinsic criterion is defined, the purpose being to obtain an unique failure criterion, the objective is to find the value of χ, expressed as a function of S_r, that allows to translate horizontally the corresponding Mohr circle at the ultimate stress state expressed in total stress on the failure criterion. The horizontal distance between the Mohr circle expressed in total stress and the failure criterion is the value χs. Knowing that σ'_3 is expressed as $\sigma'_3 = \chi s$ for the uniaxial compression test and $\sigma'_3 = \sigma_h + \chi s$ for the indirect tensile tests, the value χs can be obtained geometrically from (Figure 6):

$$\sigma'_3 = \frac{R - c\cos\varphi}{\sin\varphi} - R \qquad (5)$$

with $R = (\sigma_1 - \sigma_3)/2$

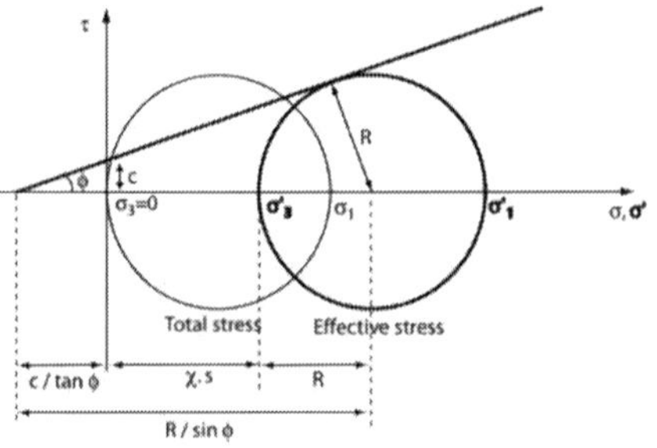

Figure 6. Mohr circle at failure for uniaxial compression test expressed in total stress and effective stress.

This method can be applied to every single test result to obtain one value of χ per test. Because for each test, the value of the degree of saturation of the sample is measured, it is possible to obtain a set of points in the χ-S_r plane, as reported in Figure 7. It is then decided to find a relation in the form:

$$\chi = (S_r)^{\alpha} \Rightarrow \log\chi = \alpha \log S_r \qquad (6)$$

where α is a material parameter which is the slope of the regression line in the plane $\log\chi$-$\log S_r$ reported in Figure 8.

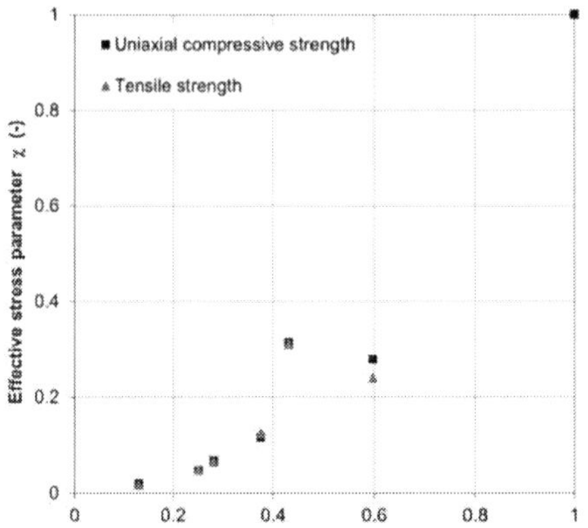

Figure 7. Degree of saturation vs. effective stress parameter χ from both uniaxial compressive strength and tensile strength.

In such a way, we find α = 2.08. The obtained relationship between χ and S_r is compared, in Figure 9, with other relationships proposed for several soils and summarized by Zerhouni [17]. Our relationship is relatively similar to the ones obtained by Zerhouni [17] on the Sterrebeek silt and by Jennings and Burland [18] on a silty clay. This is totally consistent because those materials have rheological properties very similar to the ones of the soil tested in this study. Moreover, the obtained trend follows the conceptual model of Alonso et al. [19] with a limited increase of the effective stress parameter for low degree of saturation while the increase is more significant for higher degree of saturation. The compaction process at relatively low water content induces a double structure to the material with two characteristic sizes of pores (large pores between aggregates and small pores inside aggregates). Consequently, upon low degree of saturation, the water is essentially stored inside the aggregates and this "intra-aggregate" water does not contribute to the macroscopic stress. Upon higher degree of saturation, water floods the "inter-aggregate" voids which contributes more to the increase of the internal stress.

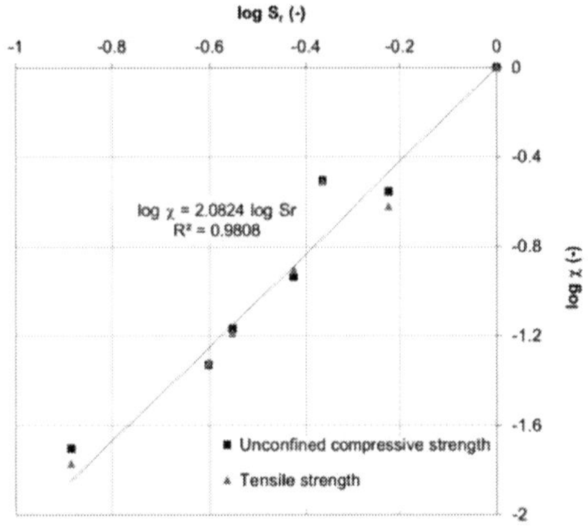

Figure 8. Degree of saturation vs. effective stress parameter χ from both uniaxial compressive strength and tensile strength in logarithmic plane.

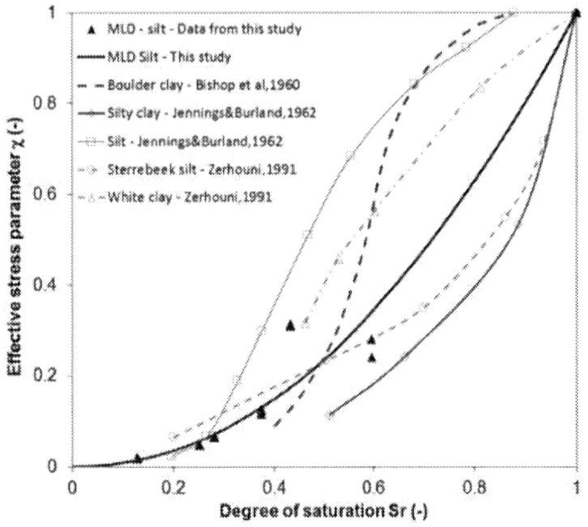

Figure 9. Effective stress parameter according to the degree of saturation for different soils.

Finally, the set of Mohr circles at failure obtained from uniaxial compression tests and indirect tensile tests are reported in Figure 10 and 11 using the generalized effective stress reference defined previously:

$$\sigma'_{ij} = \sigma_{ij} + S_r^{2.08} s\delta_{ij} \qquad (7)$$

As expected, the obtained Mohr circles at failure fulfill a unique failure criterion based on the cohesion and the friction angle deduced from triaxial tests under saturated conditions. It is worth mentioning that both uniaxial compression and indirect tensile tests upon saturated conditions do not fit well the failure criterion. It is explained by the probable undrained conditions of the saturated samples under such loadings that makes the interpretation in effective stress inappropriate.

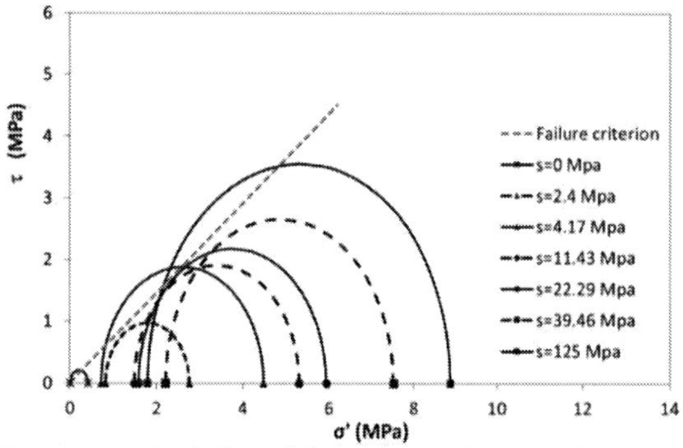

Figure 10. Mohr circles at failure expressed in terms of generalized effective stress for uniaxial compression tests at different initial suctions.

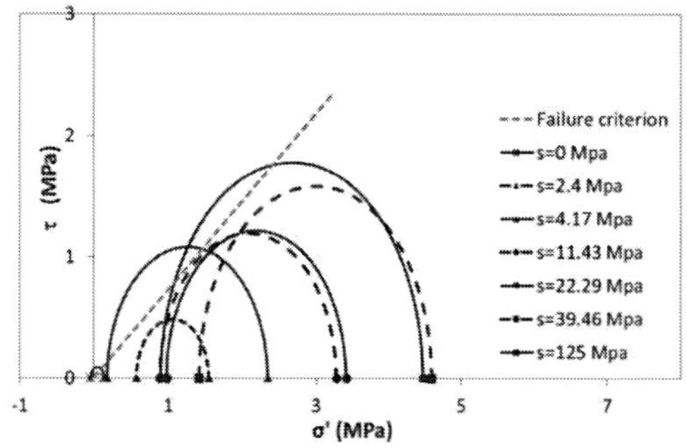

Figure 11. Mohr circles at failure expressed in terms of generalized effective stress for indirect tensile tests at different initial suctions.

5 Conclusions

The tensile and compressive strengths of unstabilized rammed earthen building materials are strongly affected, not only by the density of the material, but also by the relative humidity conditions. In this study, the hygroscopic behavior is taken into account through the water retention curve that relates the degree of saturation to the suction. This curve has a significant impact on the internal stress state of the material which, in turn, affects its mechanical behavior. We demonstrated that a framework based on the concepts of unsaturated soil mechanics using a generalized effective stress is particularly well-suited to provide a unified failure criterion including the effect of the suction inside the stress state. Consequently, suction provides additional strength to the material by mobilizing internal friction between grains. Moreover, we demonstrate the uniqueness of the obtained failure criterion in the sense that it gathers the strengths obtained from uniaxial compression tests and indirect tensile test into a single criterion. A practical outcome of this unified failure criterion is the estimation of the rammed earth

References

1. J.C. Morel, A. Mesbah , M. Oggero, P. Walker, Build Environ **36** (2001)
2. D.G. Fredlund, H. Rahardjo, *Soil Mechanics for Unsaturated Soils* (Wiley, 1993)
3. C. Beckett, C. Augarde, *Unsaturated soils: Research and Applications* (C. Mancuso, C. Jommi, F. D'Onza (Eds.), Springer, 2012)
4. P.A. Jaquin, C.E. Augarde, D. Gallipoli, D.G. Toll, Geotechnique **55**, 5 (2009)
5. Q.B. Bui, J.C. Morel, S. Hans, N. Meunier, Mater Struct **42**, 8 (2009)
6. Q.B. Bui, J.C. Morel, S. Hans, P. Walker, Constr Build Mater **54** (2014)
7. P. Walker, R. Keable, J. Martin, V. Maniatidis, *Rammed Earth: Design and Construction Guidelines* (BRE Bookshop, Watford, 2005)
8. P.J. Alley, New Zeal Eng **3**, 6 (1948)
9. P. Delage, M.D. Howat, Y.J. Cui, Eng Geol **50** (1998)
10. J.E. Aubert, A. Fabbri, J.C. Morel, P. Maillard, Constr Build Mater **47** (2013)
11. D. Ciancio, J. Gibbings, Constr Build Mater **28** (2012)
12. J.H. Bungey, S.G. Millard, *Testing of concrete in structures* (Taylor & Francis, 1996)
13. D. Li, L.N.Y. Wong, Rock Mech Rock Eng **46**, 2 (2013)
14. P. Gerard, M. Mahdad, A. Robert McCormack, B. François, Constr Build Mat **95** (2015)
15. A.W. Bishop, Ukeblad **39** (1959)
16. L. Laloui, M. Nuth, Comput Geotech **36**, 1-2 (2009)
17. M.I. Zerhouni, *Rôle de la pression interstitielle négative dans le comportement des sols - application au calcul des routes* (PhD Thesis. Ecole Centrale Paris, 1991).
18. J.E.B. Jennings, J.B. Burland, Geotechnique **12** (1962)
19. E.E. Alonso, N.M. Pinyol, A. Gens, Geotechnique **63**, 6 (2013)

Improved climatic chamber for desiccation simulation

Catalina Lozada[1,a], Bernardo Caicedo[1] and Luc Thorel[2]

[1]*Civil and Environmental Department, Universidad de Los Andes, Cr 1 #18A-12 Bogotá, Colombia*
[2]*LUNAM Univ. IFSTTAR, GERS, Geomaterials and Modeling in Geotechnics Lab.44340 Bouguenais, France*

Abstract. The climatic chamber at the Universidad de Los Andes was improved for modeling desiccation in soil layers. This chamber allows the measurement of different environmental variables. In this research, evaporation tests were conducted in water imposing boundary conditions for drying, and then these tests were performed in a soil layer. The soil was prepared from a slurry state and was drying controlling the temperature, the infrared radiation, the wind velocity, and the relative humidity. In the first part of this paper, a description of the climatic chamber, operation ranges and theoretical work principles of the climatic chamber are presented. Then, the second part shows the results for desiccation in water and soil. The desiccation tests performed with the climatic chamber allow simulating all environmental conditions accurately during drying coupling the effect of all environmental variables. As a result, the evaporation rate increases with infrared radiation in soil and water. The rate at the beginning of the desiccation tests in clays is the same as in water. However, this evaporation rate decreases as the soil becomes desiccated.

1 Introduction

Cracks appear in clay deposits when water evaporates from the surface during dry seasons. The saturation degree decreases with time and soil shrinks producing cracks. Many authors have investigated the formation of these cracks; principally, these studies deal with experimental tests of cracking process controlling environmental conditions [1-7]. However, little research has been made in laboratory tests controlling all environmental variables at the same time.

The first version of a climatic chamber was developed at Los Andes University in 2010 [8] for physical modelling applications in centrifuge. This apparatus has shown possibilities of controlling some atmospheric variables but also had limitations on power. This second chamber is an improved version of the first one ; it can simulate adequately soil desiccation and was used for modelling the evolution of cracks in a thin layer of clay controlling all environmental variables.

The climatic chamber has different subsystems allowing the control of temperature, relative humidity, solar radiation, and wind velocity independently.

This paper presents three main items: (i) the characteristics of the climatic chamber of the Universidad de Los Andes; (ii) preliminary results of potential evaporation of water; (iii) preliminary tests of drying in clayey soil.

2 The climatic chamber for modelling soil desiccation

A new climatic chamber was designed to perform desiccation tests in clayey soil. This chamber has different subsystems allowing the air to follow a particular thermodynamic path through a close circuit (Fig. 1). The subsystems, included in the climatic chamber, are described in next items and depicted in Figure 2.

1) Two cold plate coolers condensate water to obtain a specific relative humidity in the chamber (Fig. 2a). A condensate water reservoir is located following the plates.
2) After the air is cooled, it passes through two air heaters to impose the temperature in the chamber (Fig. 2a).
3) To propel the air from the heaters to the chamber and to simulate wind velocity above the soil samples, a set of six fans were installed (Fig. 2b).
4) After the air passes above the soil sample, it is propelled to the cold plate coolers through a set of four fans (Fig. 2b).
5) Afterwards, the circuit starts again.

2.1 Instruments of measurement

As is shown in Figure 2b, the adiabatic box, where is located the soil sample, includes a robotic XY module. This module has a set of sensors to measure the environmental variables [8]. Also, temperature and relative humidity are measured before and after air cross through the air heaters (points a and b in Figure 1). The

[a] Corresponding author: c.lozada29@uniandes.edu.co

 DOI: 10.1051/e3sconf/20160913002

list of sensors included in the climatic chamber is shown in Table 1.

Table 1. Specifications of the sensors of the climatic chamber.

Quant	Type sensor	Range / Precision	Position in the chamber
1	IR temperature	-40-600°C / ±2°C	XY module
1	Atm. pressure	15-115kPa / ±1.7kPa	XY module
2	Wind velocity	0-10m/s / ±0.2m/s	XY module
3	Relative humidity	0-100% / ±3.5%	(1)XY module (2)after and before heaters
3	Temperature RTD	-200-800°C / ±0.1°C	(1)XY module (2)after and before heaters
1	Laser displacemen t	0-50mm / ±10 μm	XY module

2.2 Sub-systems of the climatic chamber

2.2.1 Solar radiation

Soil and air are additionally heated with two Philips infrared lamps simulating solar radiation. To compare the irradiance spectra of the sun and the lamps, Figure 3 shows the normalized irradiance spectra for the black body temperature of the sun and the lamps. This temperature is 5523K of the sun and 2450K of the lamps.

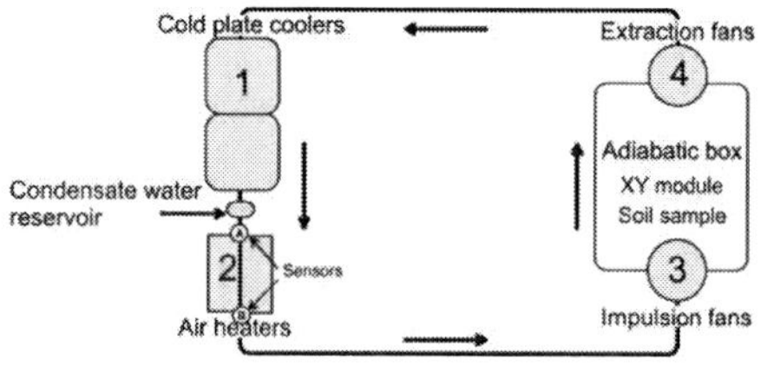

Figure 1: Air circuit in the climatic chamber

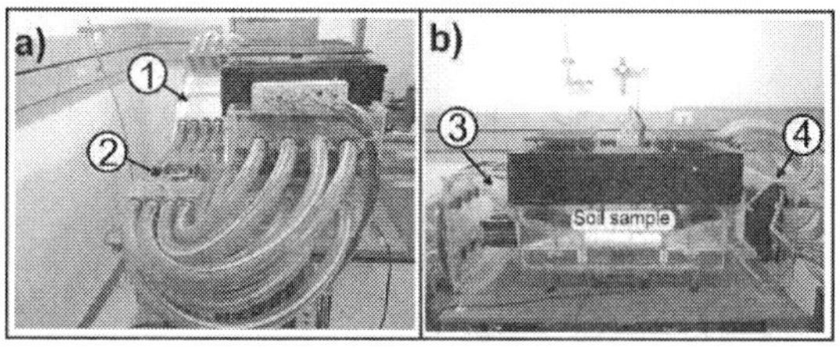

Figure 2: Air conduction components of the climatic chamber. a) Lateral view b) Frontal view.

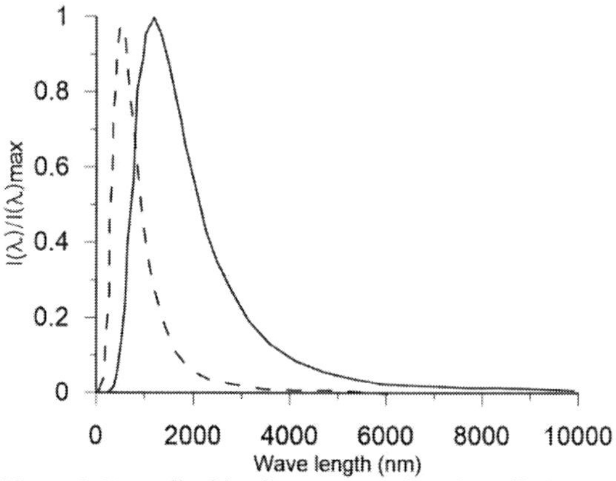

Figure 3: Normalized irradiance spectra for solar radiation and infrared lamps calculated with the Planck's law [8].

The direct infrared radiation I_{IR} at the soil surface is calculated with a geometrical relationship for heat exchange by radiation [8,9].

$$I_{IR} = 0.5^2 \sum_{i=1}^{2} \frac{I_{E(0.5)}(\alpha R)}{r_R^2} \quad (1)$$

where $I_E(0.5)$ is the emission angle of the lamp at a distance of 0.5m, αR is the emission angle, and r_R is the distance between the lamp and the surface of the soil. Figure 4 presents the direct infrared radiation of the two lamps installed in the climatic chamber calculated with the previous equation. The lamps are separated 40cm, and the distance between the lamps and the surface is 30cm.

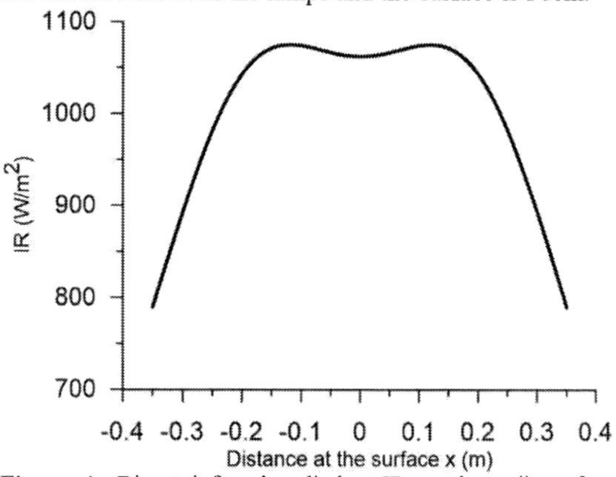

Figure 4: Direct infrared radiation IR at the soil surface calculated for the two infrared lamps

2.2.2 Wind velocity

Six fans were installed at the entrance of the adiabatic box to conduct the air above the soil and to simulate the wind velocity. This velocity is measured by the sensor placed in the XY module. The distance between the soil and the sensor is 1.5cm. The wind velocity was measured at the points A, B, and C (Fig. 5).

These measurements were made varying the power supply of the fans (Fig. 6). During the tests, the extraction fans have a constant power supply of 100% (Fig. 2b).

DOI: 10.1051/e3sconf/20160913002

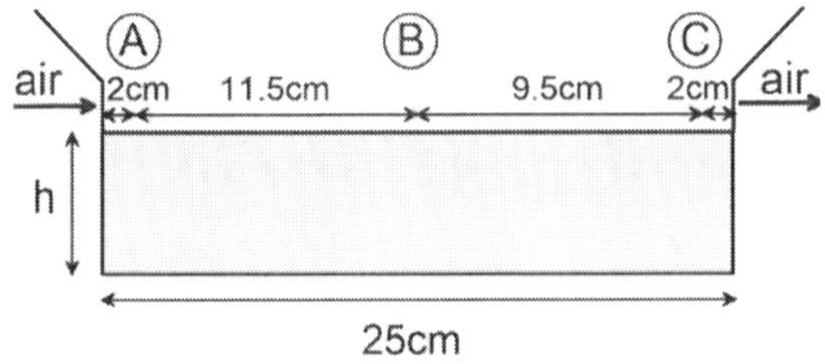

Figure 5: Location of measurements of wind velocity above the soil

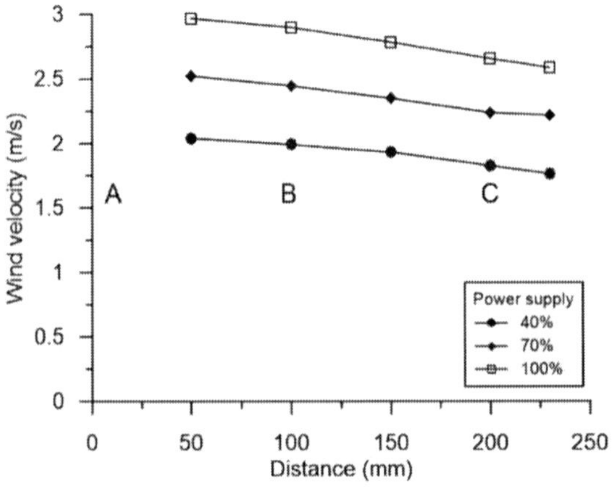

Figure 6: Wind velocity in function of the distance of the soil sample

2.2.3 Temperature and relative humidity subsystem

To obtain specific conditions of relative humidity and temperature in the chamber, the temperature of the two cold plate coolers and the temperature of the two heaters is imposed. The temperature of the cold plate coolers allows the water to condensate reaching 100% of saturation. Therefore, the temperature applied by the cold plate coolers is compared with the theoretical dew point temperature.

The theoretical dew point temperature is calculated by the integration of the Clausius Clayperon equation considering a volume of moist air cooled isobarically [10].

$$T_d \approx T_c + \frac{RT_c^2}{\Delta H_v M_w} \ln(RH) \quad (2)$$

where T_c is the temperature in the chamber, R is the ideal gas constant, ΔH_v is the latent heat for water evaporation, M_w is the molecular weight of water, and RH is the relative humidity in the chamber.

Figure 7 depicts dew point temperatures in function of relative humidity for a temperature of the heaters of 22°C and the applied dew point temperature i.e the temperature of the cold plate coolers.

The theoretical values are higher than the applied temperature, in the left part of the curve, because the loss of heat in the chamber. Contrary, the theoretical values are lower at the right of the curve because of the limitations of relative humidity in the laboratory.

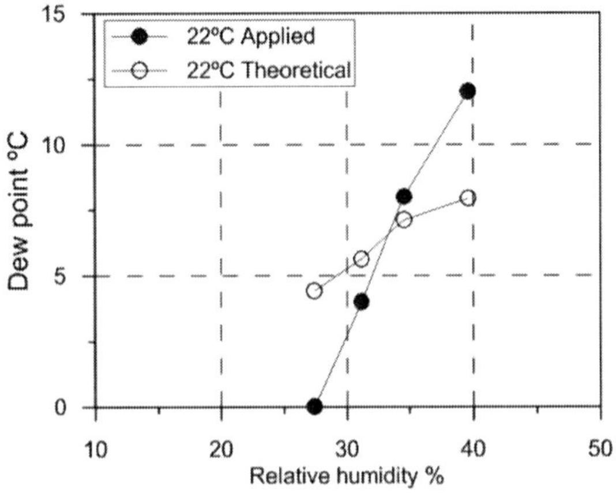

Figure 7: Dew point temperature imposed and theoretical.

Fifteen tests were made varying the temperature of the heaters and the temperature of the cold plate coolers to calibrate the range of conditions of relative humidity and temperature in the climatic chamber. Relative humidity and air temperature were measured in the adiabatic box (Fig. 8a and Fig. 8b). The operation ranges of the chamber for a temperature of the cold plates between 0 and 12 °C and a temperature of the heaters between 22°C to 39°C are: relative humidity [18% - 40%] and temperature [23°C - 38°C].

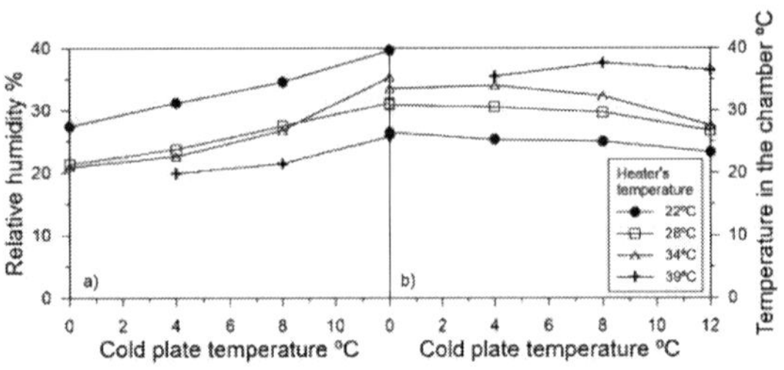

Figure 8: Environmental conditions in the adiabatic box controlling the temperature of the heaters and the cold plate coolers.

3 Performance of the apparatus

3.1. Theoretical working principle

Fundamental principles of physics regarding evaporation and condensation of water, and heating by radiation and convection have been considered in the design of the climatic chamber [8], Variables of moist air such as: the saturation water vapour pressure, *ew*; the partial vapour pressure, *e*; the relative humidity, *Uw*; the dew point temperature *Td*; and the mixing ratio *r* (relation between the mass of water vapour and the mass of dry air); are used to describe the state of vapour in the artificial atmosphere created in the environmental chamber. The Mollier diagram summarizes these different humidity functions in one chart: i.e. the mixing ratio as a function of air temperature.

The thermodynamic cycle applied by the cold plate coolers and heaters used to control relative humidity and temperature is illustrated in Figure 9. This cycle is as follows:

a) Point C represents the thermo-physical properties of air in its initial state C in Fig. 9.

b) The temperature of the cold plate coolers is reduced resulting in a decrease of the air temperature and an increase in relative humidity. If cooling continues, the mixing ratio will decrease, and the air will reach the saturation curve. At this point water vapour condenses. The process follows the C-D-E path in Fig. 9.

c) The water condensate is channelled into a container allowing the calculation of the volume.

d) Air heats again when flowing below the heaters, air temperature increase and relative humidity decrease along path E-A in Fig. 8.

e) Finally, path A-B-C in Fig. 8 indicates that when air circulates over the soil its temperature and relative humidity increase or decrease, depending on the properties of the soil at the surface. After this, the cycle starts again.

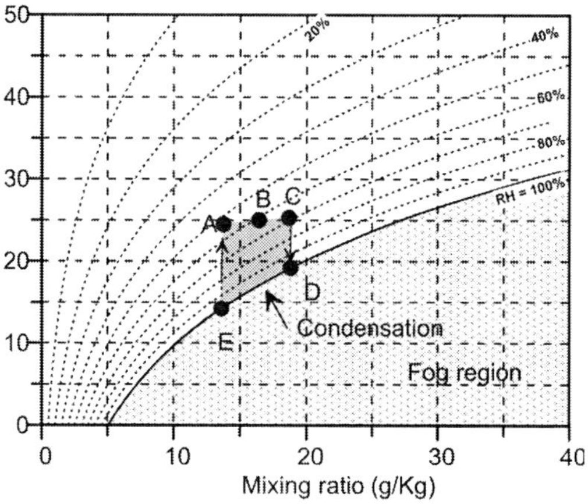

Figure 9: Thermodynamic cycle applied by the cold plate coolers and heaters.

4 Water evaporation

To assess the potential evaporation, tests in water were made controlling the conditions of wind velocity, temperature, and relative humidity.

4.1. Test procedure

Tests T2 and T3 were conducted to determine the effect of the infrared radiation on the evaporation rate at the same wind velocity and the same initial relative humidity (see Table 2). The test procedure is divided into three parts:

(i) environmental conditions were imposed in the climatic chamber without water during 30 minutes.

(ii) 500g of water were added to a square glass container and get into the chamber.

(iii) Evaporation tests started, and measurements of relative humidity, temperature, atmospheric pressure, and wind velocity were registered each 15 minutes at the points A, B, C (Fig. 5). Additionally, evaporated water was weighted with an electronic balance.

Table 2 Environmental conditions for evaporation tests in water.

Test	Wind velocity (m/s)	T heaters (°C)	IR (W/m^2)	T Peltier (°C)
T2	2.75	18.5	371.7	0
T3	2.75	18.5	0	0

4.2. Experimental results

Figure 10 shows the thermodynamic cycle for the test T2 and Figure 11 shows the thermodynamic cycle for the test T3.

Point A, B, and C are located above the water surface, and point E corresponds to the temperature of the cold plate coolers. Additionally, a point D represents a midpoint in the path of the air when passes through the cold plate coolers.

The increase of infrared radiation leads the rise of air temperature and a decrease in relative humidity. For this reason, the thermodynamic cycle expanded as radiation increases (Fig. 11). In this figure, point B presents lower values of temperature. This expansion occurs as Infrared radiation emitted by the lamps is maximum in this position as shown in Figure 4.

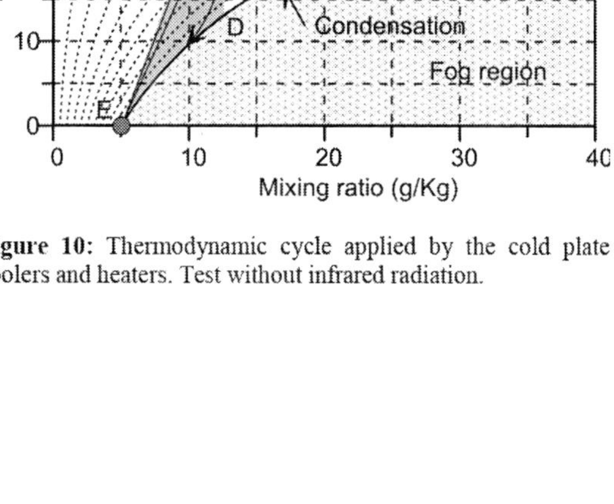

Figure 10: Thermodynamic cycle applied by the cold plate coolers and heaters. Test without infrared radiation.

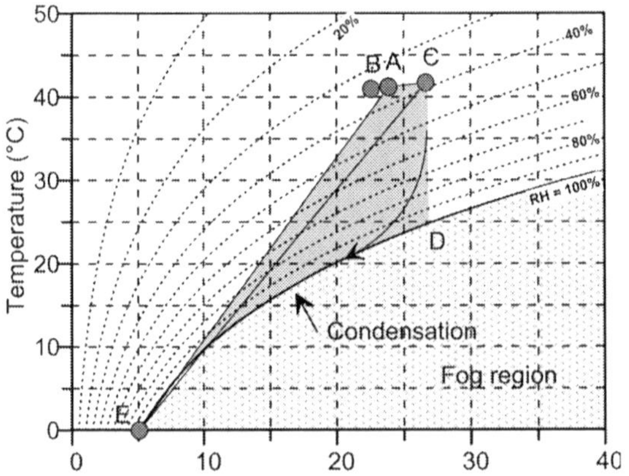

Figure 11: Thermodynamic cycle applied by the cold plate coolers and heaters. Test with infrared radiation.

The evaporation rate of water depends on environmental variables as a whole. As relative humidity decreases and temperature increase with infrared radiation, the capacity to contain water of the simulated atmosphere increase. For that reason evaporation rate is higher for the test T2 than for the test T3 as shown in Figure 12.

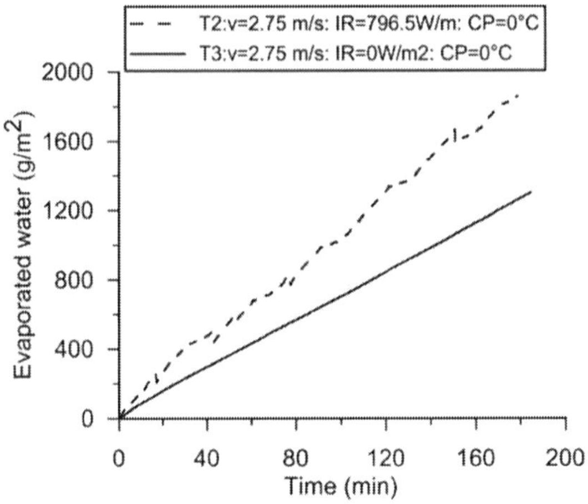

Figure 12: Effect of infrared radiation on evaporated water.

5 Preliminary tests in clayey soil

Same tests, T2 and T3, were conducted on clay. The experimental procedure is the same as the tests in water. Additionally, each hour a photo was taken to observe the soil behaviour during evaporation of water.

The soil used in this research is Speswhite kaolin clay and is prepared with a water content of w=1.5 LL. The liquidity limit of the Speswhite clay is LL=55%, so the initial water content is w_o=82.5%.

The evaporated water in the tests with clay is shown in Figure 13. At the beginning of the tests, the evaporation rate shows the same value as in the tests conducted with water. However, as the soil is desiccated the evaporation rate decreases until the soil is completely dried.

Figure 14 and Figure 15 show the soil sample at the end of the drying test. It seems that soil contraction is higher in the tests exposed to less severe environmental conditions without cracks evidence.

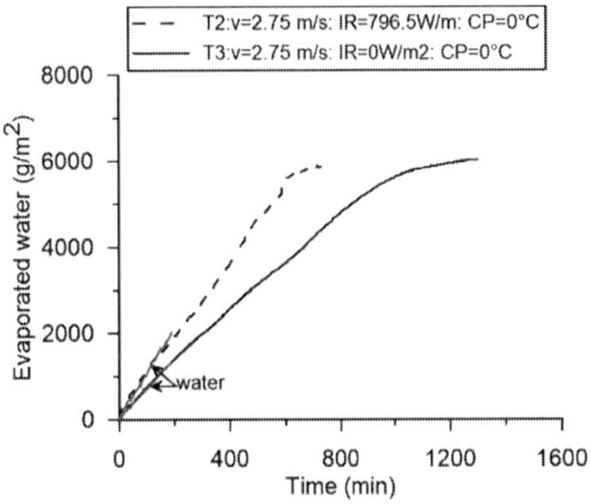

Figure 13: Effect of infrared radiation on evaporated water in clay.

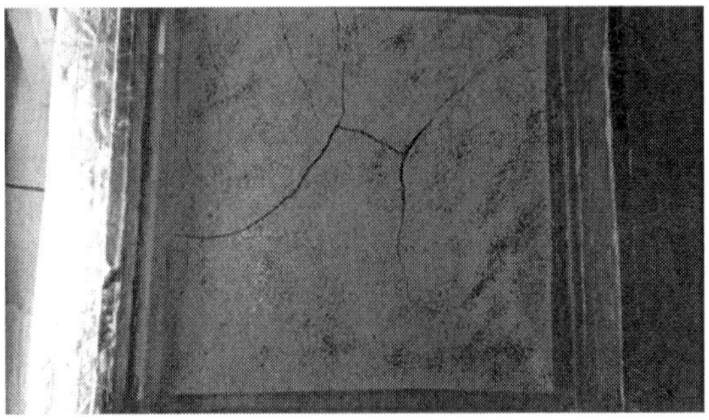

Figure 14: Soil sample at the end of desiccation. Test 2.

Figure 15: Soil sample at the end of desiccation. Test 3.

6 Conclusions

The climatic chamber at the Universidad de Los Andes was redesigned in order to perform desiccation tests in soil layers. This chamber allows the user to impose specific environmental conditions through a close circuit of air. Preliminary results of water evaporation lead to state the next conclusions:

1. The main effect of infrared radiation on evaporation is the decrease of relative humidity in the air and the increase of air temperature. These conditions, and the increase of water temperature allow the increase of the evaporation rate.

2. The evaporation rate in soil is the same as water at the beginning of the test. As desiccation occurred, evaporation rate decreases.

This device will be used in the future to conduct parametric study on the evaporation phenomenon in coupled environmental condition (humidity, temperature, wind, infrared radiation).

References

1. A. Corte, & A. Higashi, No. CRREL-RR-66 Final Rpt. 5. (1964).
2. C. J. Miller, H. Mi, & N. Yesiller, J. Am. Water Resour. Assoc. 677-686 (1998).
3. P. C. Prat, A. Ledesma, & M. R. Lakshmikantha, *In Fracture of Nano and Engineering Materials and Structures*. 1373-1374. (2006).
4. H. Nahlawi, & J. K. Kodikara, Geotech. Geol. Eng., **24**, 6. 1641-1664. (2006).
5. M. R. Lakshmikantha, P. C. Prat, & A. Ledesma, Geotech. Test. J. **32**(6), 505-51 (2009).
6. C. S. Tang, Y. J. Cui, A. M. Tang, & B. Shi, Eng. Geol.**114**, 3, 261-266. (2010).
7. C. S. Tang, B. Shi, C. Liu, W. B. Suo, & , L. Gao, Appl Clay Sci, **52**, 1, 69-77. (2011).
8. J. Tristancho, B. Caicedo, L. Thorel, & N. Obregon, Geotech. Test. J. **35**, 1, 159-171. (2012)
9. F. Kreith, R. Manglik, & M. Bohn, *Principles of heat transfer*. (2010)
10. J. H. Seinfeld, & S. N. Pandis, *Atmospheric chemistry and physics: from air pollution to climate change*. (2012).

 DOI: 10.1051/e3sconf/20160913003

Ageing effect on swell, shrinkage and flexural strength of sand and waste marble powder stabilized expansive soil

Serife Oncu[1], Huriye Bilsel[2]

[1]*Eastern Mediterranean University, Famagusta, N. Cyprus*
[2]*Cyprus International University, Nicosia, N. Cyprus*

Abstract. Expansive soils in semi-arid climates must be mitigated to minimize potential structural damage to the overlying structures due to swell-shrink behavior caused by climatic changes. The expansive soil in this study was amended by sand which reduced the swell-shrink potential significantly. As a secondary additive, a waste by product of construction industry was selected, marble powder. Recycling this material would minimize its accumulation and covering a large space in landfill areas. In this study, waste material was assessed as a possible partial replacement for sand and due to its high calcite content, curing effect on its cementitious characteristic was also evaluated. It was observed that for mitigating the swell-shrink potential, 10% marble powder by dry mass was the optimum amount which was observed to gain improved characteristics with curing. The flexural strength, however, slightly reduced with marble powder addition, the soil mixture displayed a brittle behavior. Therefore the utilization of this material is recommended to be restricted to soils exposed to lower flexural loads, such as light traffic. The correlation between shrinkage strains and flexural strength parameters suggested herein, could be a potential empirical approach to predict the flexural strength based on shrinkage behavior.

1 Introduction

The swell and shrinkage properties of expansive soils due to climatic effects cause serious damages to the light structures. Many researchers studied the factors which influence the swell-shrink behavior of compacted expansive soils. The influence of compaction dry density and moisture content on the swelling of desiccated clay was investigated by Rao et al. (2000) [1] and concluded that it is actually the initial void ratio which affects the swell potential of the desiccated soils, and not the compaction water content. Conversely, initial moisture content is an important factor for both swelling and shrinkage. According to Chen and Ma (1987) [2] initial dry density is a controlling factor on swelling but not on shrinkage. Day (1994) [3] studied the effect of cyclic wetting-drying on volume change of the compacted silty clay and observed that the swell potential increased which was attributed to the change in clay structure from dispersed to flocculated state with the repeated wetting-drying cycles. Detrimental effects of swelling-shrinking soils on structures are well documented in the literature [4]. To alleviate these effects, various materials such as lime, Portland cement and bitumen are used in the stabilization of swelling soils of which most commonly used are lime and cement [5].

Recently utilization of industrial waste materials such as fly ash, blast furnace slag, cement kiln dust, limestone dust and marble dust as additives became very popular for both recycling and soil improvement purposes. Accumulation of these wastes presents considerable environmental and economical problems, therefore their utilization as stabilizing agent impedes their hazardous effects on the environment and reduces construction costs as well as accumulation of the waste in areas where landfill capacities are restricted. Gandhi (2013) [6] compared marble dust and rice husk ash in the stabilization of expansive soils and showed that marble dust was more effective than rice husk ash. Agrawal and Gupta (2011) [7] studied with varying proportions of marble dust (0, 5, 10, 15, 20, 25 and 30%) for the expansive soil treatment. They investigated the effect of marble dust on the physical properties of the stabilized soil and reported that liquid limit, plasticity index and shrinkage index decreased after using marble dust. On the other hand, plastic limit and shrinkage limit increased. Abdulla and Majeed (2014) [8] added marble powder with different percentages (10, 20, 30%) by weight of soil. Liquid limit, plastic limit and plasticity index values reduced with the marble dust inclusion. Effect of marble dust addition on compaction characteristics were evaluated by [9, 10]. Çimen et al. (2011) [9] stated that marble content affect the maximum dry density but optimum water content did not change. Sabat and Nanda (2011) [10] stated that maximum dry density showed a decreasing trend and optimum moisture content showed vice versa with the addition of marble dust. Many investigations showed that swelling of soils reduced with

[a] Corresponding author: serife.oncu@emu.edu.tr

the utilization of marble dust [7-12]. Cömert et al. (2010) [13] investigated of using fly ash, marble dust and waste sand as additives for two types of soils. They prepared specimens in different percentages and found that 40% waste materials can be used as a road sub-base. On the other hand, according to unconfined compression test results optimum ratios were found 20% for fly ash, 5-10% for marble dust and 5% for waste sand. Fırat et al. (2012) [14] recommended fly ash, marble dust and waste sand are effective stabilizers in road subbases. They indicate that these additives improved swelling ratio, CBR and water conductivity. Gupta and Sarma (2014) [15] used sand, fly ash and marble dust with different proportions in order to improvement of black cotton soil. They concluded that this mixture can be used as a sub-grade material for construction of flexible pavements in rural roads with low traffic volume.

This study is focused on the effect of curing time on shrinkage and flexural strength of compacted silt-sand with marble powder additive as a replacement for sand. Flexural strength is an indirect tensile test yielding a measure of toughness and ductility of materials, which are required characteristics of compacted soils mainly as road base. This study also shows a relationship between shrinkage properties and flexural strength.

2 Materials and methods

Expansive soil used in this study is classified as silt with high plasticity (MH) according to the Unified Soil Classification System (ASTM D2487), consisting of 42% clay, 48% silt and 10% sand. Sand used in stabilization of expansive soil is poorly graded sand (SP) taken from the eastern coast of Cyprus, with effective diameter D_{10}, mean diameter D_{50}, coefficient of uniformity C_u and coefficient of curvature C_c values of 0.16, 0.19, 1.25 and 1.013 respectively. Waste marble powder (MP) utilized as a secondary additive, was collected from a marble plant which was freshly accumulated after a cutting process. It is constituted of 6% clay size, 32% silt size and 62% sand size particles. The chemical composition of the expansive silt and waste marble powder obtained using X-ray flourescence spectrometer (XRF) are depicted in Table 1.

Table 1. Chemical composition of natural silt and waste marble powder.

Element	Expansive soil	Waste marble powder
SiO_2 (%)	36.5	0.178
CO_2 (%)	17.9	42.6
CaO (%)	16.2	55.7
Al_2O_3 (%)	11.8	0.0708
Fe_2O_3 (%)	6.87	0.0641
MgO (%)	6.26	1.22
Other elements	4.47	0.1671

One-dimensional swell test (ASTM D4546-08) was applied to the compacted soil samples which were at 75 mm diameter and 15 mm height. Upon completion of swelling, the specimens were drained and stored in a temperature controlled room for volumetric shrinkage test. Diameter and height of the soil samples were measured at different time intervals on shrinkage path using a digital Vernier caliper, until volume change ceased.

For flexural strength test (ASTM C348-14), soil specimens (beams) were prepared using a special mold which was 40 mm width, 40 mm depth and 160 mm height. These beams were compacted by using static compaction according to maximum dry density and optimum water content values. All samples which were used in swell, consolidation, shrinkage and flexural test were cured for 0, 7 and 28-days. Soil specimens were packed and stored in a desiccator.

3 Experimental Results

3.1 Physical properties

Soil specimens were categorized in five different groups. Group 1 is expansive silt which is denoted by N. Group 2 is 50% silt and 50% sand (NS), Group 3 is 47.5% silt, 47.5% sand and 5% marble powder (NS5%MP), Group 4 is 45% silt, 45% sand and 10% marble powder (NS10%MP), Group 5 is 40% silt, 40% sand and 20% marble powder (NS20%MP).

Table 2. Physical properties of soil groups.

	N	NS	NS5% MP	NS10% MP	NS20 %MP
LL (%)	65	32	30	29	26
PL (%)	36	20	17	19	15
PI (%)	29	12	13	10	11
LS (%)	18	9	10	10	9
w_{opt} (%)	26	17.5	15.5	15	15.3
ρ_{drymax} (kg/m^3)	1380	1690	1740	1790	1781

Physical properties including Atterberg limits, linear shrinkage, optimum water content and maximum dry density values from standard Proctor compaction test of all soil groups are presented in Table 2. It can be observed that there is a significant reduction in consistency limits with sand addition which remains almost the same with marble inclusion. Sand and marble powder inclusion increase the maximum dry density while considerably reduce the optimum water content.

3.2 Volume Change

3.2.1 One-dimensional swell

One-dimensional swell test (ASTM D4546-08) was conducted on all of the soil groups at curing times of 0, 7 and 28 days. Primary swell potential of silt was as high as 4.95%. Swell potential decreased to 0.67% after addition of sand. Studying the swell percentages presented in Figure 1, it can be observed that addition of 5% and 10%

DOI: 10.1051/e3sconf/20160913003

MP to silt-sand mixture initially increased the swell potential which followed by a considerable reduction with time. 10% and 20% MP additions have been very effective, keeping the swell potential below the critical limit of 1.5% at all curing times.

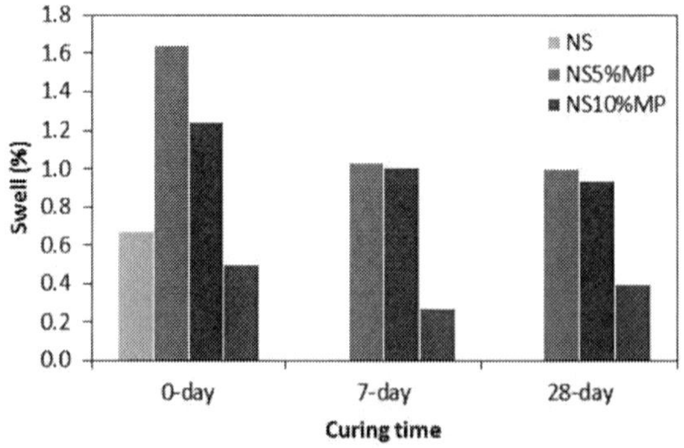

Figure 1. Primary swell percentage change with curing time.

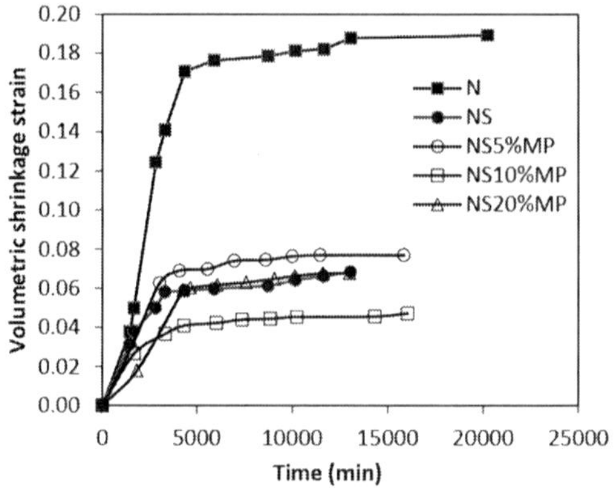

Figure 2. Volumetric shrinkage strain versus time.

3.2.2 Volumetric shrinkage

Shrinkage test results were studied as volumetric shrinkage strain ($\Delta V/V_0$) versus time as plotted in Figure 2. Maximum volumetric shrinkage strain of N was 0.19 which reduced significantly to 0.068 when sand was added. This value reduces furthermore to 0.045 with 10% MP inclusion after 28 days curing, whereas 5% and 20% MP yielded higher volumetric strains. As it can be seen from test results the lowest and most convenient value was obtained after addition of 10% marble powder. Shrinkage test results were also modelled by SoilVision software [16] using a hyperbolic model given in Equation (1).

$$e(w) = a_{sh}\left[\frac{w^{c_{sh}}}{b_{sh}^{\ c_{sh}}}+1\right]^{\left(\frac{1}{c_{sh}}\right)} \tag{1}$$

where, a_{sh}: fitting parameter (minimum void ratio), b_{sh}: fitting parameter (shrinkage limit), c_{sh}: fitting parameter (curvature of the hyperbola), w: gravimetric water content.

Table 3 shows the fitting parameters for all soil groups. The results reveal that 10% MP gives the best result in shrinkage, as the shrinkage limit (b parameter) and the final void ratio (a parameter) remain the highest after 28 days curing period.

3.2.3 Prediction of ultimate swelling and shrinkage

The shape of swell (%) and volumetric shrinkage strain versus time graphs resemble the shape of a rectangular hyperbola. Kondner (1963) [17] suggested that the non-linear stress-strain curves of soils represented by a rectangular hyperbolic equation, could be linearized by plotting the results in a modified form. Therefore, swell-time and volumetric shrinkage strain-time relationships could also be linearized [18-20]. Hence using this concept the ultimate swell and volumetric shrinkage strain could be predicted using the hyperbolic model given in Equation (2). Komine and Oggata (1994) [21] proposed to obtain the maximum values by finding the limiting value at infinite time as given in Equation (3).

$$S(t) = \frac{dh(t)}{h_o} = \frac{t}{(a+bt)} \tag{2}$$

$$S_{max} = \lim_{t\to\infty}\left(\frac{1}{\frac{a}{t}+b}\right) = \frac{1}{b} \tag{3}$$

where t is the time from the start of water inundation, S(t) is the vertical swell at time t, and a and b are constants obtained from straight line fits giving the highest R^2 value.

Theoretically, it takes infinite time to reach the ultimate swell value, which cannot be practically measured in the laboratory. Therefore, prediction from the initial stages of swell is advantageous instead of continuing till equilibrium. Addition of sand and MP have yielded both predicted swell and shrinkage percentages slightly bigger than the measured values as

Table 3. Hyperbolic fitting parameters of the shrinkage curves.

Material	0-day			7-day			28-day		
	a	b	c	a	b	c	a	b	c
NS	0.401	0.143	9.469	-	-	-	-	-	-
NS5%MP	0.410	0.164	8.522	0.480	0.190	46.81	0.472	0.174	7.38
NS10%MP	0.439	0.173	14.66	0.476	0.179	29.92	0.503	0.181	25.5
NS20%MP	0.456	0.191	6.031	0.422	0.146	16.52	0.425	0.150	6.74

can be observed in Figure 3 (a) and (b).

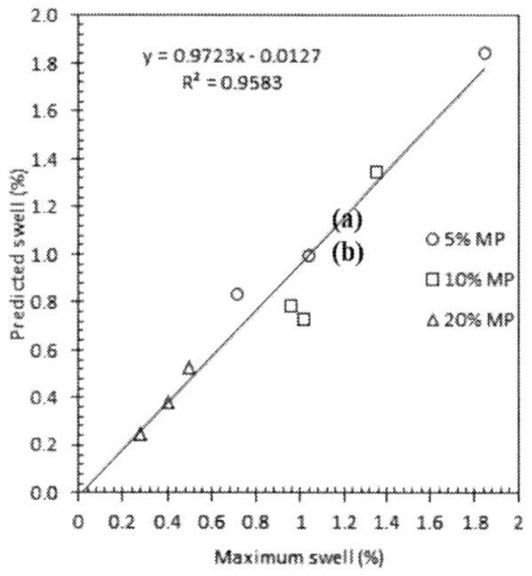

(a)

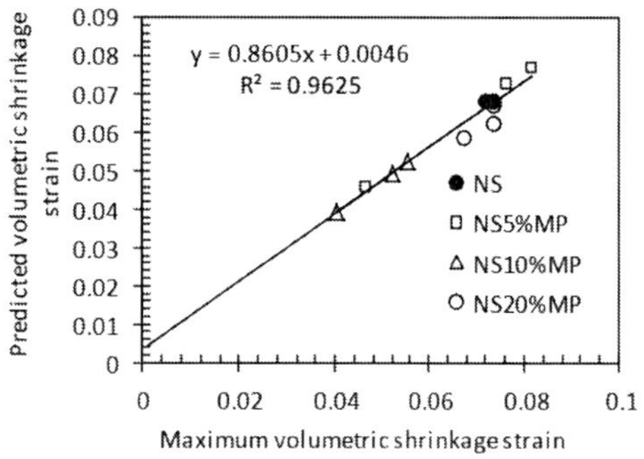

(b)

Figure 3. Predicted swell (a) and volumetric shrinkage strain (b) versus measured maximum swell and shrinkage strains.

3.3 Flexural Strength

Flexural strength tests were performed for all soil groups at different curing times, and load-deflection curves were obtained. The peak flexural load (P_f) of each mixture at different curing times are obtained and displayed in Table 4. The flexural performance of compacted stabilized soil mixture displays either deflection-softening or deflection-hardening behavior. From the flexural load versus deflection curves, the point at which the linearity of the load-deflection curve ends (P_1), the initial peak, also known as the limit of proportionality (LOP) as described in ASTM C1609-10 is determined and compared with the peak flexural load (P_f). In this study both values are almost equal, therefore, the flexural behavior is deflection-softening, conversely if P_1/P_f is greater than 1, deflection-hardening occurs. This can further be justified by determining the equivalent flexural strength ratio given in Equation (4).

$$R_{T,150}^{D} = \frac{T_{150}^{D}}{P_1 \frac{L}{150}} \quad (4)$$

where $R_{T,150}^{D}$ is the flexural strength ratio, T_{150}^{D} is toughness which is the area under the load-deflection curve from 0 to L/150 (0.6 mm) deflection.

Flexural strength values are obtained by using Equation (5).

$$f = 1.5\frac{PL}{bd^2} \quad (5)$$

where, P is the flexural load, L is the span length, b is the width and d is the depth of the beam.

Toughness is the ability of a material to absorb energy during fracture [22, 23]. The equivalent flexural strength ratio represents the efficacy of energy absorption of the material from the beginning to the deflection of L/150. If this value is less than 100%, deflection-softening, and if greater than 100% deflection-hardening behavior occurs. The latter indicates a high toughness material. Table 4 summarizes the flexural strength performance data from which it can be deduced that the toughness value after 28-day curing period is the highest for NS10%MP specimen. This is also justified by the flexural strength ratio, $R_{T,150}^{D}$, which yields the highest value with 10%MP usage after the curing period of 28 days. However all samples display a strain-softening behaviour, hence utilization of waste marble powder made soil specimens more brittle.

With an attempt to correlate volumetric and axial ($\Delta H/H_0$) shrinkage strains and the flexural strength parameters, the relationships in Figure 4 were produced. There appears to be a logarithmic relationship between the shrinkage strains and the flexural strength and toughness values. Figure 4 depicts that higher the flexural strength and toughness lower will be the desiccation deformations in a semi-arid climate where unsaturated soils prevail. It can also be noted from these findings that the volumetric shrinkage strains are almost twice the axial shrinkage strains.

 DOI: 10.1051/e3sconf/20160913003

Table 4. Flexural performance data.

Flexural properties	NS	NS5%MP			NS10%MP			NS20%MP		
		0-d	7-d	28-d	0-d	7-d	28-d	0-d	7-d	28-d
	78.91	50.86	69.66	92.7	80.18	98.98	80.18	46.4	65.13	69.66
P_1 (N)	35	22	30	40	34	42	34	20	28	30
Flexural strength (kPa)	78.91	50.86	69.66	92.7	80.18	98.98	80.18	46.36	65.13	69.66
Toughness (Nmm)	11	6.3	10.1	7.3	13.2	12	9.5	7.5	6.7	6.8
$R^D_{T,150}$(%)	52	48	56	30	65	48	47	63	40	38

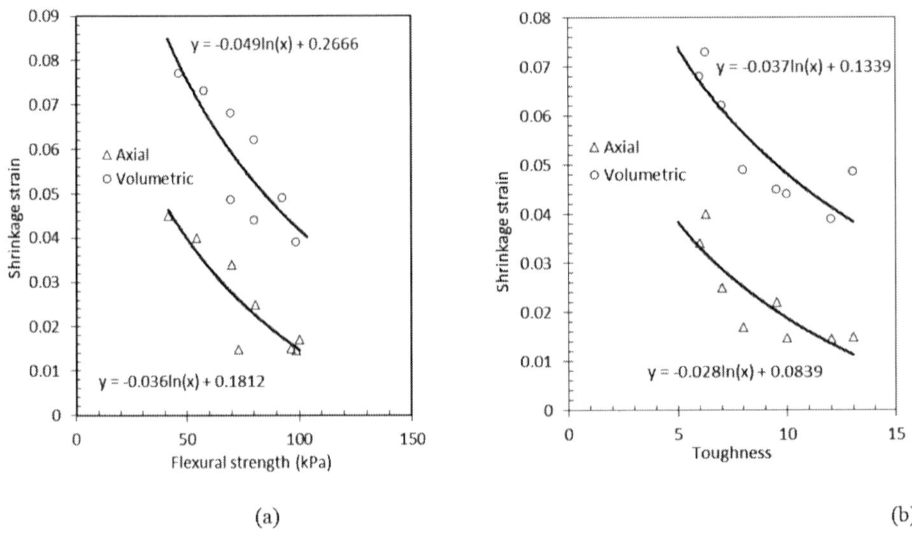

Figure 4. Relationship between volumetric and axial shrinkage strains and (a) flexural strength and (b) toughness.

4 Conclusions

This study summarizes the initial findings of a research which aims to utilize a waste product for recycling purpose, as well as mitigating the expansive soils which cause structural distresses in a semi-arid climate, mainly to light structures. The marble powder produced as a waste product in construction industry has been used to partially replace the sea sand used to mitigate an expansive silt. The research was aimed to study the efficiency of this material in soil stabilization. The experimental findings demonstrated that swell and shrinkage properties improved by at least 10% MP usage and that the effect of curing has been rather significant. However, regarding the flexural strength properties, it was observed that the MP addition increased the brittleness of the soil mixtures, hence could be recommended for soils to bear light loads. The predictive determinations of maximum swell and shrinkage values were also tested and proved to be very effective. Furthermore, a correlation could be formed between volumetric shrinkage and flexural strength parameters which clearly needs further work to ascertain its applicability to different soils.

References

1. K. S. S. Rao, S. M. Rao, S. Gangadhara, Swelling Behavior of a Desiccated Clay. Geotechnical Testing Journal 23(2): 193-198 (2000)
2. F. H. Chen, G. S. Ma, Swelling and shrinkage behavior of expansive clays. *Proceedings 6th International Conference on Expansive Soils*, Vol. **1**, New Delhi, India, pp. 127-129 (1987)
3. R. W. Day, Swell-Shrink Behavior of Compacted Clay, Journal of Geotechnical Engineering, ASCE, 120(**3**), pp. 618-623 (1994)
4. F. H. Chen, *Foundations on Expansive Soils*, second ed. Elsevier, New York (1988)
5. P. T. Sherwood, *State of the art review- Soil Stabilization with Cement and Lime*, HMSO publication, London (1993)
6. K.S. Gandhi, Stabilization of expansive soil of Surat region using rice husk ash and marble

dust. International Journal of Current Engineering and Technology. Vol. **3**, No. 4, pp. 1516-1521 (2013)

7. V. Agrawal, M. Gupta, Expansive soil stabilization using marble dust. International Journal of Earth Sciences and Engineering, Vol. **4**, No. 6, pp. 59-62 (2011)
8. R. S. Abdulla, N.N. Majeed, Some physical properties treatment of expansive soil using marble waste powder. International Journal of Engineering Research & Technology. Vol. **3** (1), pp. 591-600 (2014)
9. Ö. Çimen, S. N. Keskin, S. Seven, E. Erişkin, D. Güllü, The effect of waste marble pieces on swelling pressure at compacted clay. *Fourth Geotechnical Symposium*, (in Turkish), Çukurova University, December 1-2, pp 206-211, Adana, Turkey (2011)
10. A. K. Sabat, R. P. Nanda, Effect of marble dust on strength and durability of rice husk ash stabilised expansive soil. International Journal of Civil and Structural Enginering Vol. **1**, No. 4, pp. 939-948 (2011)
11. O. Başer, E. Çokça, Stabilization of Expansive Soils Using Waste Marble Dust. *13th National Conference on Soil Mechanics and Foundation Engineering*, September 30- October 1, Vol. **1**, (in Turkish), pp. 143-152, Istanbul, Turkey (2010)
12. I. Zorluer, M. Usta, Stabilization of Soils by Waste Marble Dust. *Fourth National Marble Symposium*, December 2003, pp. 305-311 (in Turkish), Afyon, Turkey (2003)
13. T. Cömert, S. Fırat, G. Yımaz, M. Sümer, Reuse of fly ash, marble dust and disposal sand for road subbase fill. *13th National Conference on Soil Mechanics and Foundation Engineering*, September 30- October 1, Vol. **1**, pp. 153-162 (in Turkish), Istanbul, Turkey (2010)
14. S. Fırat, G. Yılmaz, A. T. Cömert, M. Sümer, Utilization of marble dust, fly ash and waste sand (silt-quartz) in road subbase filling materials. KSCE Journal of Civil Engineering. Vol. **16**, No. 7, pp. 1143-1151 (2012)
15. C. Gupta, R. K. Sharma, Influence of marble dust, fly ash and beas sand on sub grade characteristics of expansive soil. Journal of Mechanical and Civil Engineering, pp. 13-18 (2014)
16. SoilVision Systems Ltd., SoilVision Theory Guide, Version 3.34, 2nd edition, Saskatoon, Saskatchewan, Canada (2001)
17. R. L. Kondner, Hyperbolic Stress-Strain Response: Cohesive Soils. Journal of Soil Mechanics and Foundations Division, ASCE, Vol. **89**, No. SM1. Proc. Paper 3429, pp. 115-143 (1963)
18. A. S. Muntohar, Swelling and Compressibility Characteristics of Soil- Bentonite Mixtures, Dimensi Teknik Sipil Vol. **5**, No. 2, pp. 93 – 98 (2003)
19. H. Nagaraj, M. Munnas, A. Sridharan, Swelling behavior of expansive soils. International Journal of Geotechnical Engineering, Vol. **4**, Issue 1, pp. 99-110 (2010)
20. M.A. Dafalla, M.A. Al-Shamrani, Swell and aging effect on a bentonite sand mixture, Unsaturated Soils: Theory and Practice, Kasetsart University, Thailand, ISBN 978-616-7522-77-7 1 (2011)
21. H. Komine, N. Ogata, Experimental study on swelling characteristics of compacted bentonite. Canadian Geotechnical Journal, **31** (4), pp. 478-490 (1994)
22. S. Onyejekwe, G.S. Ghataora, Effect of fiber inclusions on flexural strength of soils treated with nontraditional additives. Journal of Materials in Civil Engineering, ASCE, 04014039: pp. 1-9 (2014)
23. P. Jamsawang, P. Voottipruex, S. Horpibulsuk, Flexural strength characteristics of compacted cement-polypropylene fiber sand. Journal of Materials in Civil Engineering, ASCE, 04014243: pp. 1-9 (2014)

DOI: 10.1051/e3sconf/20160913004

Feasibility study of a new unsaturated three-layer landfill cover system

Jason Lim Coo[1,a], Pui San So[1], Bruce Chen[1], Zhou Chao[1] and Charles Wang Wai Ng[1]

[1]Department of Civil and Environmental Engineering, Hong Kong University of Science and Technology, Kowloon, Hong Kong

Abstract. As an improvement of the two-layer cover with capillary barrier effect (CCBE) (i.e. fine-grained soil overlying a coarse-grained soil), a new three-layer landfill cover system is proposed and investigated for humid climate. This new system is to add a fine-grained soil (i.e., clay) underneath a two-layer CCBE (i.e., a silt overlying a gravelly sand layer). The feasibility of this proposed cover system was investigated by conducting a one-dimensional water infiltration test. In addition, transient seepage simulations were carried out to back-analyse the test results and investigate the importance of hydraulic properties of the CCBE on the proposed cover. Based on the infiltration experiment and numerical back-analysis, it is found that no percolation was observed after 48 hours of ponding, which is equivalent to a rainfall return period of greater than 1000 years. However, the upper two-layer CCBE is only effective for a rainfall return period of about 35 years. This implies that the proposed bottom clay layer is needed for humid climate. Numerical parametric simulations reveal that increasing the saturated permeability of the upper fine-grained soil by two orders of magnitude ($1.4x10^{-6}$ m/s to $2.1x10^{-4}$ m/s), the wetting front is still within the clay layer after 12 hours of constant water ponding (>1000 year rainfall) and no percolation occurred.

1 Introduction

Covers that are commonly placed over landfills have been identified as an important component in preventing rainfall infiltration to the buried solid waste after their closure [1-2]. Most modern landfill cover systems utilize geotextile composites and geomembranes to meet the permeability criteria of 10^{-9} m/s [3]. It is also common to rely on naturally occurring low permeability materials such as clays. However, landfill covers composed of naturally occurring materials are prone to desiccation induced cracking, which can compromise their integrity and performance [4-5].

Recently, more attention has been given to covers with capillary barrier effects (CCBEs) as an alternative landfill cover system because of their ease of construction, long service life and relatively low cost [6-7]. Even around 1000 years ago, CCBE has already been applied in burial mounds to avoid water infiltration into the remains [8]. The principle of CCBE utilizes the capillary barrier effect between the two soil layers which consist of a layer of fine-grained soil over a layer of coarse-grained soil. The amount of water infiltrating the system can be minimized due to the coarse-grained soil being less permeable than the fine-grained soil under relatively dry conditions. Although more attention has been paid to CCBEs as an alternative cover system in semi-arid and arid regions, the performance of CCBE under humid climates has so far been unsatisfactory [9-12].

A new three-layer cover system is proposed and explored to improve a capillary barrier for humid climatic conditions such as Hong Kong, Philippines, Columbia and Brazil where annual rainfall of 2,000 mm is not uncommon. This new system consists of three layers in which a coarse-grained soil layer is placed in between two fine-grained soil layers. The innovation of this three-layer cover system is based on the theory of unsaturated soil mechanics [13] leading to possibly eliminate the use of traditional flexible impermeable membrane for minimizing rainfall infiltration. The feasibility and effectiveness of this three-layer cover system are investigated by conducting a one-dimensional water infiltration test. The experiment is back-analysed and comparisons between experimental data and numerical results are discussed. Moreover, a numerical parametric study is carried out to investigate the importance of saturated permeability of the upper fine-grained soil on the effectiveness of the proposed three-layer system.

2 Theoretical considerations of the new unsaturated three-layer landfill cover

Schematic diagrams of a two-layer CCBE and the proposed landfill cover are shown in Fig. 1. As shown in Fig. 1a, CCBE consists of a two-layer soil system that utilizes the distinct differences in the unsaturated hydraulic properties between a fine-grained soil (i.e. silt) and a coarse-grained soil (i.e. gravelly sand). Comparatively, the newly proposed landfill cover system is a three-layer cover system, which consists of a soil layer with a relatively low permeability (i.e., compacted clay layer) that is added underneath a CCBE, as shown in

[a] Corresponding author: jlcoo@connect.ust.hk

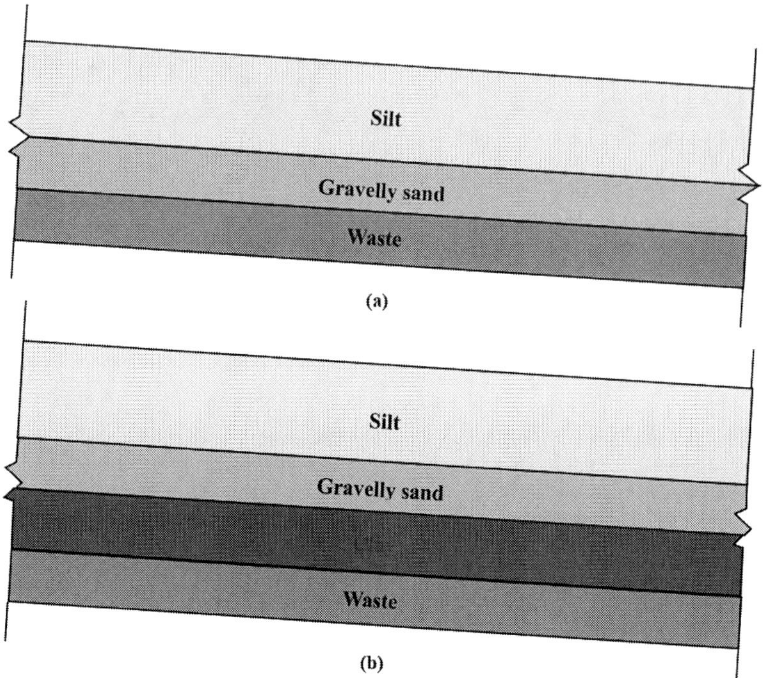

Figure 1. Conceptual diagram of landfill covers : (a) Conventional capillary barrier cover ; (b) Newly proposed landfill cover

Fig. 1b. In this proposed system, water infiltration is minimised by utilizing the capillary barrier effects provided by the two upper soil layers and the low water permeability of the bottom clay layer when it is nearly saturated. On the other hand, the bottom clay layer is protected by the upper two soil layers from desiccation during dry seasons because the upper two soil layers have low water permeability at high suctions (i.e., low relative humidity). The prevention of desiccation crack occurrence on the clay layer can assure the good performance of this alternative cover system for minimising water infiltration. Additionally, the high saturation of the clay layer also minimizes landfill gas emission [14].

2.1 Principle of reducing water infiltration

Fig. 2 shows a typical relationship between water permeability and matric suction (i.e. permeability function) of each soil layer.

a. At semi-arid or arid climates, soil suction in the three-layer landfill cover is larger than point S_1 (shown in the figure), indicating that the cover system is relatively dry. Water permeability of silt is larger than that of gravelly sand and implies that the upper two-layer CCBE is functional. Hence, infiltrated water is then stored in the silt layer and finally removed by evaporation and/or lateral draiange.
b. At humid climates, soil suction in the landfill soil cover is less than point S_1 (shown in the figure) under heavy or prolonged rainfalls, the cover system is nearly saturated. Water permeability of gravelly sand layer is the highest while that of clay layer is the lowest. Water breakthrough of the upper CCBE occurs when the suction at the interface between the first two layer drop below the water entry value of the gravelly sand layer. Hence, rainfall is able to infiltrate into the gravelly sand layer. However, infiltrated water is prevented by the clay layer due to its low water permeability (see Fig. 2). Infiltrated water may subsequently be drained through the gravelly sand layer and collected into a nearby storage location where water can either be evaporated, discharged or recycled back to the landfill cover.

The combination of the underlying compacted clay layer and the overlying CCBE makes the proposed landfill cover applicable to all weather conditions.

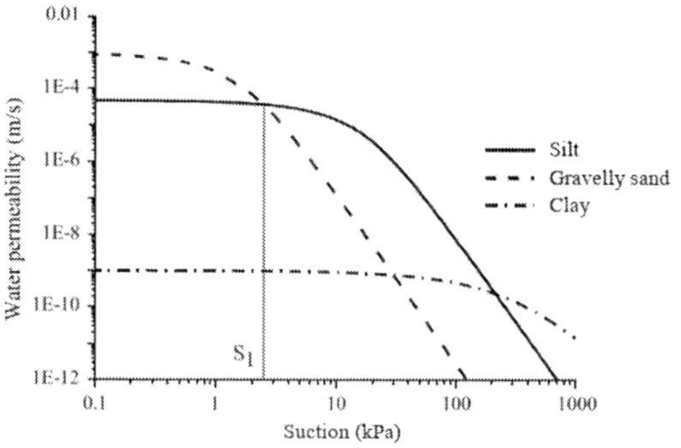

Figure 2. Typical permeability functions of silt, gravelly sand and clay (After Ng et al. [15])

3 Experimental apparatus and instrumentation

3.1 1-D soil column

The performance of the proposed three-layer landfill cover system was assessed using a 1300 mm height and 140 mm diameter soil column made of transparent acrylic. The photo of the 1D soil column for water infiltration test is shown in Fig. 3. It is designed to control and measure both the top and bottom boundary flow conditions. At the top boundary upon ponding, constant head infiltration can be achieved by using an electronic weighing scale and a Mariotte's bottle [16]. For the bottom boundary, a valve is connected at the base of the soil column to measure periodically any percolation that occurs during a test.

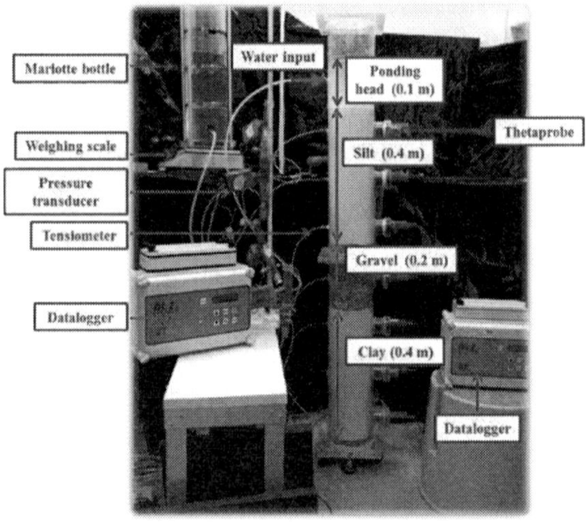

Figure 3. Photo of 1D soil column water infiltration experiment

3.2 Instrumentation

The instrumentation consisted of small tip tensiometers and heat dissipation matric water potential sensor to measure soil suctions. Theta-probe soil moisture probes were also used to measure volumetric water content (VWC). Measurements were made at depths of 75 mm, 225 mm, 375 mm, 525 mm, 675 mm, 825 mm and 975 mm above the base of soil column. Prior to installation, all transducers were properly calibrated.

4 Test materials and specimen preparation

4.1 Test materials

In this study, silt, gravelly sand and clay were used to form the three-layer landfill cover system. The basic properties of these three soils are shown in Table 1. Compaction curves for the silt and clay were determined in accordance with test method ASTM D698 while ASTM D4253 was used for the gravelly sand. The saturated permeability (k_s) of the silt and clay was measured by flexible wall permeameter as described in test method ASTM D5084 while constant-head method as described in test method ASTM D2434 was used for gravelly sand.

Table 1. Basic properties of the soils used in this study.

	Silt	Gravelly sand	Clay
Unified soil classification system (USCS)	ML	SP	CH
Specific gravity, G_s	2.61	2.62	2.52
Liquid limit, LL	22	-	59
Plastic limit, PL	16	-	32
Plasticity Index, PI	6	-	27
Maximum dry density, ρ_d (kg/m^3)	1771	1494	1264
Optimum moisture content (%)	14	-	36
Saturated permeability, k_s (m/s)	1.4x10^{-6}	9.7x10^{-3}	5.7x10^{-9}

Water retention curve (WRC) is the relationship between volumetric water content and matric suction of soil. WRC is an important parameter which dictates the water storage of unsaturated soil. In this study, both the wetting and drying WRCs of the soils were obtained by using modified pressure plate apparatus [17]. Subsequently, the wetting WRC data is fitted using the van Genuchten [18] WRC equation. The van Genuchten WRC equation is as follows:

$$\theta_w = \theta_r + (\theta_s - \theta_r)/[(1+\alpha\psi)^n]^m \quad (1)$$

where θ_w is the volumetric water content at a matric suction (ψ), θ_s is the saturated water content, θ_r is the residual water content, and α, n, and m are fitting parameters. The parameter m was constrained as $1\text{-}1/n$ following the standard convention. The WRC data and the fitted wetting WRC are shown in Fig. 4. Best fitting parameters for the wetting WRC based on the van Genuchten equation are presented in Table 2. The water-entry value of silt and gravelly sand from the WRCs is found to be about 15 kPa and 0.6 kPa, respectively. The ratio of water-entry value between these two soils is larger than 10 which creates the formation of a capillary barrier [12].

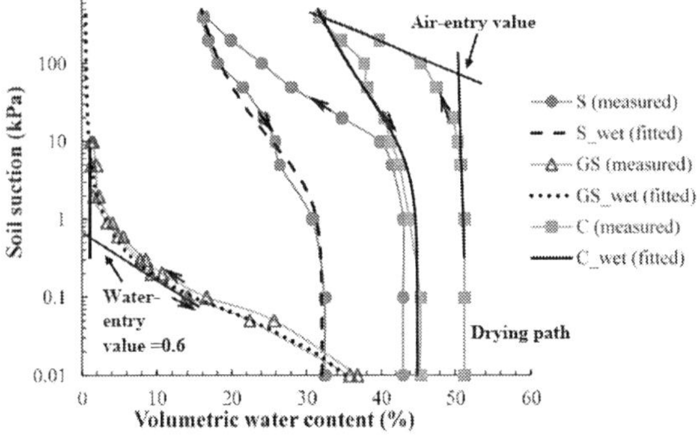

Figure 4. Measured and fitted WRCs of silt (S), gravelly sand (GS) and clay (C)

Table 2. Fitting parameters of wetting WRC using the van Genuchten [18] equation.

Soil type	Fitting parameters				
	θ_s	θ_r	α (kPa^{-1})	n	$m = 1\text{-}1/n$
Silt	0.33	0.03	0.5	1.17	0.15
Gravelly sand	0.37	0.01	25	1.9	0.47
Clay	0.45	0.09	0.11	1.13	0.12

4.2 Specimen preparation

The three soil layers; namely, a clay layer, a gravelly sand layer and a silt layer, were compacted successively from bottom to the top of the soil column. The thickness of each soil was 0.4 m, 0.2 m and 0.4 m for clay, gravelly sand and silt, respectively. The soils were initially mixed with water to reach the optimum moisture content as given in Table 1. The soils were then compacted at their targeted degree of compaction (DOC) or relative density (RD) which is 95 DOC, 90 RD and 90 DOC for clay, gravelly sand and silt, respectively (see Table 1). A thin film of vacuum grease was also applied to the walls of the column before compaction to minimize any occurrence of preferential flow path during infiltration.

5 Experimental procedures

After sample preparation, the soil column was then subjected to water infiltration. The infiltration stage involved imposing a 0.1 m constant head ponding controlled by a constant-head water supply system. The soil surface was covered during infiltration to minimize evaporation. The test involved measurements of the volume of infiltrated water and water outflow rate as well as profiles of pore water pressure and volumetric water content. Since the soils were initially mixed with water to achieve the respective moisture content, initial suctions were expected to be different between each soil layer in accordance to its WRC (see Fig. 4). The average initial suctions for silt, gravelly sand and clay were found to be 55 kPa, 20 kPa and 90 kPa, respectively. It should be noted that the water infiltration is one dimensional and has no lateral drainage, thus representing a worst case scenario for a landfill cover system under water infiltration.

Most engineering design guidelines are based on rainfall return period [19]. Rainfall return periods at various ponding duration are back-calculated by using the relationship between rainfall depth and duration according to the "Hong Kong Stormwater Drainage Manual" [20]. Further details regarding the conversion of water ponding to equivalent rainfall return period are given in Ng et al. [21].

6 Experimental results

6.1 Distribution of hydraulic head profiles

The measured hydraulic head distributions along the soil depth for the water infiltration test are shown in Fig. 5.

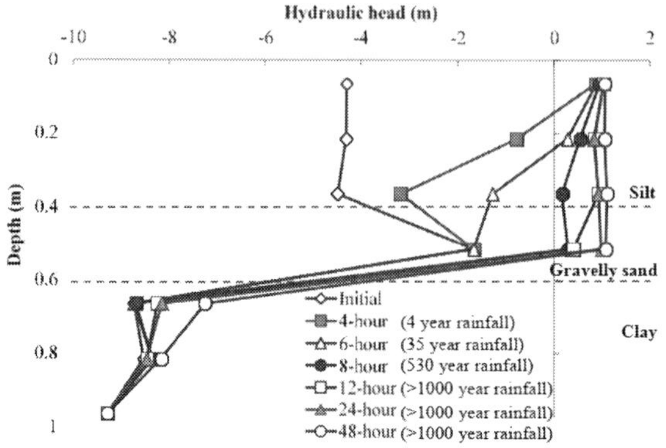

Figure 5. Measured hydraulic head distributions in water infiltration test

The bottom of the soil column, 1 meter below the ground surface, was selected as datum for the hydraulic head calculation. At the initial condition, hydraulic head at the gravelly sand layer is highest followed by the silt layer and then the clay layer. After 4 to 6 hours of ponding (4 and 35 year rainfall), the hydraulic head profile in the silt layer shifted to the right. However, there was no observable hydraulic head change in the gravelly sand. This implies that the gravelly sand layer served as a capillary barrier and prevented the downward flow of water due to its relatively low permeability at this duration. After 8 hours of ponding (530 year rainfall), hydraulic head at the gravelly sand layer shifted to the right while the hydraulic head at the silt layer also shifted further towards the right nearing hydrostatic condition. This shows that water infiltrated into the gravelly sand layer due water breakthrough across the fine-coarse soil interface of the CCBE [6, 22]. The upper two-layer CCBE was no longer functional after this ponding duration. This phenomenon has been demonstrated by researchers for two-layer CCBEs under one-dimensional conditions [22-24]. After 24 hours of ponding (>1000 year rainfall), a hydrostatic condition seems to develop above the clay layer. This is because downward water infiltration is blocked by the inherently low permeability of the clay layer. Therefore, a perched water table occurs and also saturates the soil above the clay layer. Which is consistent with the measured volumetric water content. Further details regarding the measured volumetric water content profile is given in Ng et al. [21]. It is also observed that even with application of 0.1 m constant head ponding for 48 hours (>1000 year rainfall), no percolation was measured.

7 Numerical back-analysis of the infiltration test

The laboratory experimental data was simulated numerically using CODE_BRIGHT (Coupled Deformation Brine, Gas and Heat Transport) [25]. The model was one-dimensional with the same geometry as the actual experimental conditions. Similar boundary conditions as those applied in the experimental column test were imposed. The numerical simulation results of the infiltration tests were verified by comparing them with the laboratory experimental data. The wetting WRC (see Fig. 4 and Table 2) and unsaturated permeability functions (see Fig. 2) of the soils were used in the simulation.

The unsaturated permeability functions were predicted from the measured wetting WRC by using in conjunction the van Genucthen-Mualem equation [18, 26]. The equation is as follows:

$$k_r(\psi) = k_w(\psi)/k_s = \{1-(\alpha\psi)^{n-1}[1+(\alpha\psi)^n]^{-m}\}^2/[1+(\alpha\psi)^n]^{m/2} \quad (2)$$

where $k_w(\psi)$ is the unsaturated coefficient of permeability at a matric suction (ψ), k_s is the saturated permeability coefficient and α, n, and m are fitting parameters from the van Genucthen [18] equation best fit to the WRC.

7.1 Comparison between experiment and computed results

Fig. 6 shows comparisons between the pore water pressure profiles, which were obtained from the finite element analysis and the measured results during the experiment. The numerical simulation results show a good agreement with those observed in the laboratory experiments. Upon application of 0.1 m constant ponding head for the first 6 hours, infiltration occurred mainly in the silt layer but not in the gravelly sand layer as reflected by the fact that measured and simulated pore-water pressure remained unchanged in the gravelly sand layer. After 8 hours of ponding, the simulated pore-water pressure at the gravelly sand layer increased, which is consistent with the movement of pore-water pressure measured from the experimental test. This infers that water has already infiltrated into the gravelly sand layer and the upper two-layer capillary barrier is no longer effective. Due to continuous application of ponding it can be observed that after 24 hours a hydrostatic condition seems to develop above the clay layer. After 48 hours of ponding, both simulation and measurement demonstrates that only the upper portion of the clay layer shows an increase of pore-water pressure. This indicates that due to the addition of clay layer underneath a two-layer CCBE, infiltrated water require a rather long duration to reach the deeper portion of the clay layer. Thus percolation can be prevented even for rainfalls greater than 1000 year return period.

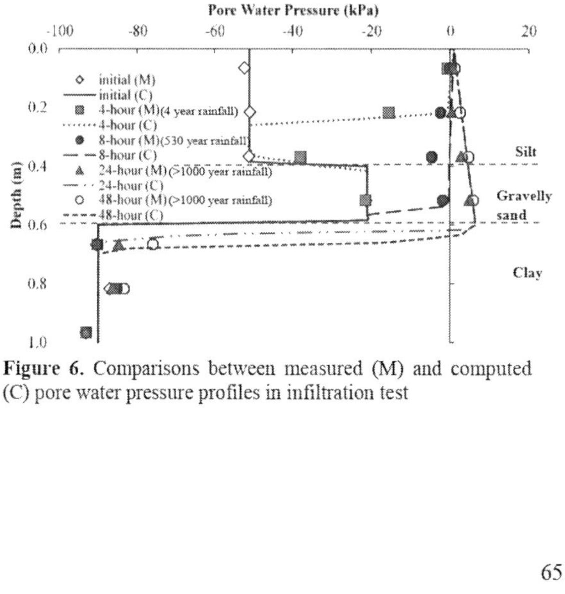

Figure 6. Comparisons between measured (M) and computed (C) pore water pressure profiles in infiltration test

8 Parametric study

According to Albrecht [27], who carried out laboratory permeability test of landfill covers, it is reported that saturated permeability of surface soils in landfills increases after it undergoes several wetting and drying cycles due to desiccation cracks. Hence, an additional parametric study is carried out to investigate the importance of permeability of the upper fine-grained soil on the effectiveness of the proposed three-layer system. The intention of this study was to take into account the influence of desiccation cracks on the upper fine-grained soil. It should be noted that this is just a simple sensitivity study and two additional higher values of k_s typically utilized in the upper fine-grained soil were chosen from the literature. The saturated permeability (k_s) of the fine-grained soil were $1.4x10^{-6}$ m/s (this study), $2.0x10^{-5}$ m/s [23] and $2.1x10^{-4}$ m/s [28]. The shape of the permeability function has been kept the same within this study.

8.1 Influence of upper fine-grained soil saturated permeability

Fig. 7 shows a series of computed pore water pressure profiles of constant head ponding at 12 hours with different upper fine-grained soil k_s ($1.4x10^{-6}$ m/s, $2.0x10^{-5}$ m/s and $2.1x10^{-4}$ m/s).12 hours (>1000 year rainfall) represent the time that the upper two layer CCBE is no longer effective and for the wetting front to reach clay layer with respect to the reference upper fine-grained soil k_s used in the experiment. For all k_s values, a positive pore water pressure develops at the upper two-layer CCBE. The k_s of the upper fine-grained soil has some influence on the pore water pressure distribution. The larger the value of k_s, the downward flow of water is easier, and hence water breakthrough of the CCBE occurs earlier [10]. After this event, water starts to infiltrate into the gravelly sand layer and eventually into the clay layer. Thus, a deeper wetting front in the clay layer is observed as the value of k_s becomes larger. Additionally, due to earlier water breakthrough of the CCBE, a hydrostatic condition seems to develop for the two additional values of k_s ($2.0x10^{-5}$ m/s and $2.1x10^{-4}$ m/s) with respect to the reference k_s ($1.4x10^{-6}$ m/s). Nonetheless, the wetting front

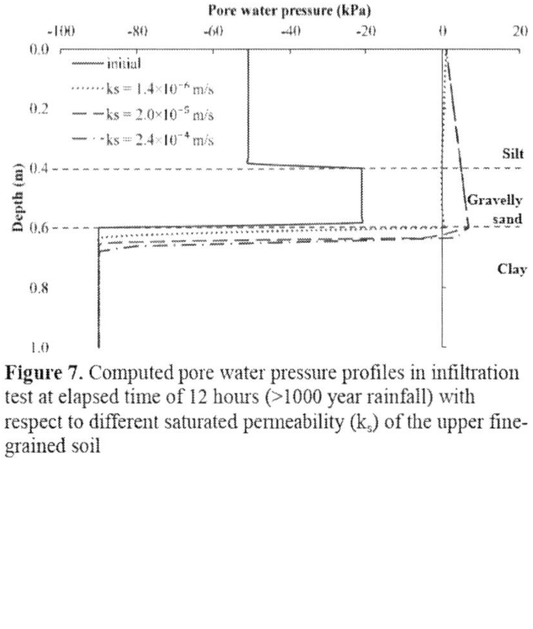

Figure 7. Computed pore water pressure profiles in infiltration test at elapsed time of 12 hours (>1000 year rainfall) with respect to different saturated permeability (k_s) of the upper fine-grained soil

for the largest value of k_s ($2.1x10^{-4}$ m/s) is still within the clay layer and is approximately 700 mm below the ground surface. This suggests that the clay layer had yet to be fully saturated and thus no percolation happened. This further indicates that the three-layer landfill cover system performs satisfactorily despite the increase of k_s for the upper fine-grained soil.

9 Summary and conclusions

An alternative three-layer landfill cover system was proposed and explored for regions with humid climates where annual rainfall exceeds 2000 mm. The fundamental modification of this alternative system is the addition of a clay layer underneath a conventional two-layer capillary barrier system. This bottom clay layer is intended to function as an impeding layer to minimise percolation into the underlying waste when water breakthrough occurs into the upper two-layer capillary barrier during heavy and prolonged rainfalls. The results of the numerical analyses and infiltration tests using one-dimensional soil column showed that no percolation was observed after 48 hours of constant ponding, which is equivalent to a rainfall return period of greater than 1000 years. However, the upper two-layer capillary barrier is only effective for a rainfall return period of about 35 years. This means that the proposed bottom clay layer is necessary for humid climates. Numerical parametric simulations reveal that increasing the saturated permeability of the upper fine-grained soil by two orders of magnitude ($1.4x10^{-6}$ m/s to $2.1x10^{-4}$ m/s), the wetting front is still within the clay layer after 12 hours of constant water ponding and no percolation occurred. Both experimental investigation and numerical simulations showed that the proposed three-layer cover system is a promising alternative landfill cover system for humid climates.

Acknowledgment

The authors would like to acknowledge the research grant (2012CB719805) from the National Basic Research Program (973 Program) provided by the Ministry of Science and Technology of the People's Republic of China and research grant HKUST6/CRF/12R provided by the Research Grants Council (RGC) of the Hong Kong Special Administrative Region.

References

1. U.S. EPA, *Design and Construction of RCRA/CERCLA Final Covers* (Seminar publication, EPA/625/4-91/025, 1991)
2. C.H. Benson, M.A. Barlaz, D.T. Lane, J.M. Rawe, Waste Manage. **27(1)**, 13-29 (2007)
3. U.S. EPA, *Solid Waste Disposal Facility Criteria* (Technical manual, EPA530-R-93-017, Washington DC, 1993)
4. S. Melchior, Land Contam. Reclam. **5(3)**, 209-216 (1997)
5. W.H. Albright, C.H. Benson, G.W. Gee, T. Abichou, E.V. McDonald, S.W. Tyler, S.A. Rock, Geotech. Geoenviron. Eng. **132(11)**, 1393-1403 (2006)
6. B. Ross, Water Resour Res. **26(1)**, 2625-2629 (1990)
7. S.F. Dwyer, *Water Balance Measurements and Computer Simulations of Landfill Covers* (PhD Thesis, The University of New Mexico, 2003)
8. K. Watanabe, *Proc., Joint International Waste Management Conference, Vol 1*, 567-571 (1989)
9. C.E. Morris, J.C. Stormont, J. Geotech. Geoenviron. Eng. **123(12)**, 1057-1065 (1999)
10. M.V. Khire, C.H. Benson, P.J. Bosscher, J. Geotech. Geoenviron. Eng. **126(8)**, 695-708 (2000)
11. W.H. Albright, C.H. Benson, G.W. Gee, J. Environ. Qual. **33(6)**, 2317-2332 (2004)
12. H. Rahardjo, D. Tami, E.C. Leong, *Proc., 2nd International Conference on Problematic Soils*, 39-54 (2006)
13. C.W.W. Ng, B. Menzies, *Advanced Unsaturated Soil Mechanics and Engineering* (Taylor and Francis, London and NY, 2007)
14. C.W.W. Ng, Z.K. Chen, J.L. Coo, R. Chen, C. Zhou, Waste Manage. **44**, 155-163 (2015)
15. C.W.W. Ng, J. Liu, R. Chen, J.L. Coo, Environ. Earth Sci. **74(5)**, 4419-4429 (2015).
16. E.L. McCarthy, Science **80(100)**
17. C.W.W. Ng, Y.W. Pang, J. Geotech. Geoenviron. Eng. 126(2), 157-166 (2000)
18. M.T. van Genucthen, Soil Sci. Soc. Am. J. **44(5)**, 892-898 (1980)
19. GEO, *Geotechnical Manual for Slopes* (Geotechnical Engineering Office, Hong Kong, 2011)
20. DSD, *Stormwater Drainage Manual* (Drainage Services Department, Hong Kong Government, 2013)
21. C.W.W. Ng, J.L. Coo, Z.K. Chen, R. Chen, J. Environ. Eng. (Accepted)
22. H. Yang, H. Rahardjo, E.C. Leong, J. Hydrol. Eng. **41(5)**, 908-920 (2006)
23. J.S. McCartney, J.G. Zornberg, Can. Geotech. J. **47(11)**, 1201-1213 (2010)
24. J.C. Stormont, C.E. Anderson, J. Geotech. Geoenviron. Eng. **125(8)**, 641-648 (1999)
25. S. Olivella, J. Carrera, A. Gens, E.E. Alonso, Transp. Porous Media **15(3)**, 271-293 (1994)
26. Y. Mualem, Water Resour. Res. **12(3)**, 513-522 (1976)
27. B. Albrecht, *Effect of Desiccation on Compacted Clays* (MS thesis, University of Wisconsin-Madison, 1996)
28. S.W. Webb, Water Resour. Res. **22(8)**, 1855-1859 (1997)

Experimental study of gas breakthrough and emission in an unsaturated clay landfill cover

Z. K. Chen[1,a], J. L. Coo[1] and C. W. W. Ng[1]

[1] *Department of Civil and Environmental Engineering, Hong Kong University of Science and Technology, HKSAR*

Abstract. Decomposition of municipal solid waste (MSW) generates a large amount of landfill gases (LFG) which can pose a high risk to the health of nearby residents and significantly promote global warming when they are released to the atmosphere. A commonly used method to control LFG emission is to place an engineering cover with a gas collection system over MSW. LFG pressure increases in landfill covers when the efficiency of gas collection systems is low, resulting in large gas emissions from cover surface. Compacted clay has been commonly used in landfill cover systems as a hydraulic barrier. However, the effectiveness of compacted clay as a gas barrier, especially resisting gas breakthrough and reducing emission rate under controlled conditions is not fully understood. In this paper, gas breakthrough pressure and emission rate were determined through a one-dimensional (1D) column over a wide range of landfill gas pressures under various degrees of saturation, thicknesses and degrees of compaction. Experiment results show that under a gas pressure of 10 kPa, a minimum of 0.4 m thick clay layer is able to prevent gas breakthrough at degree of saturation of about 60% (in humid regions). The reduction of gas emission rate by increasing degree of compaction is most prominent in the range of intermediate degree of saturation.

1 Introduction

Waste decomposition from municipal solid waste (MSW) generates a large amount of landfill gases (LFGs), namely CO_2 and CH_4. The LFG may emit from the landfill and hence intensify global warming. Although gas collection systems have been proposed to install in the landfill, LFGs cannot be collected completely [1]. To minimize emission of LFG to the atmosphere, a common method to isolate MSWs from the environment is to place an engineered cover over landfills. However, LFG will break through the cover when certain gas pressure is reached [2, 3]. Compacted clay, due to its relatively low saturated water permeability, has been widely used for landfill final cover systems [4, 5]. In reality, however, landfill covers are unsaturated in service [6]. Infiltrated rainfall water can only increase their degrees of saturation but it is extremely not likely to saturate them. This means that it would be non-conservative to consider saturated compacted clay for preventing gas since the unsaturated environments favour gas flows than the extreme saturated conditions. The feasibility of unsaturated compacted clays as gas barriers in a landfill cover is thus worthwhile to be explored over a wide range of landfill gas pressures under various degrees of saturation and degrees of compaction.

For the gas breakthrough pressure of clay at ultimate limit state (high gas pressure), water-saturated swelling clay has been widely tested [7, 8], due to wide applications of swelling clay in nuclear waste disposal which generally locates hundreds of meters below the ground surface. The measured gas breakthrough pressure of saturated swelling clay is relatively high (in the order of MPa). The reported gas pressure may not be applicable to compacted clay in the landfill cover due to several reasons. Firstly, compacted clay in the landfill cover is usually unsaturated. Effects of degree of saturation on gas breakthrough pressure in the context of landfill cover are not understood. Secondly, although LFG pressure may reach 50 kPa at some extreme condition [9], field monitoring results suggest that LFG pressure is typically less than 10 kPa [1]. Compared with gas pressure in nuclear waste disposal, gas pressure in municipal waste landfill is much lower. The huge difference in gas pressure may affect test results.

For the gas emission rate at serviceability limit state (low gas pressure), Moon et al. [10] carried out a series of gas permeability tests on oven-dried clayey soil using a permeameter. Based on measured gas permeability, they concluded that compacted clay may not be sufficient to prevent landfill gas emission. It should be pointed out that there is no standard for gas permeability for designing landfill cover but gas emission rate is considered as an indicator for designing the landfill cover in the available guidelines. For example, Australian guideline [11] allows for a maximum methane emission rate of 63 $ml/m^2/min$. The gas emission rate should be strongly affected by not only gas permeability of soil, but

[a] Corresponding author: zchenam@connect.ust.hk

 DOI: 10.1051/e3sconf/20160913005

also the thickness, degree of compaction and degree of saturation of soil layer. Moon et al. [10] only tested dry triaxial size soil specimens, which are much smaller than the typical thickness of compacted clay (or fine-grained soil) layer in a landfill (i.e., 0.3 to 1.5 m) [12]. To fully understand efficiency of unsaturated compacted clay on limiting gas emission, it is necessary to study effects of clay thickness, degree of compaction and degree of saturation.

In this study, two series of gas breakthrough tests at pressure ranged from 0 to 50 kPa and two series of gas emission rate tests at pressure range of 0 to 20 kPa were carried out on unsaturated compacted Kaolin clay using one-dimensional soil column. Three clay degrees of saturation (40%, 60% and 80%), two degrees of compaction (90% and 100%) and two clay thicknesses (0.4 m and 0.6 m) were considered.

2 Experimental program and procedures

Two types of tests (gas breakthrough test and gas emission rate test) were carried out in this study by using 1-D soil column. The first type of tests was conducted to determine gas breakthrough pressure in unsaturated clay in high gas pressure ranged from 0 to 50 kPa. In each test, gas pressure was increased in a stepwise manner at constant time intervals until gas breakthrough occurs. Breakthrough pressure was defined to be the pressure at which the gas flow rate has a dramatic increase. For this test type, two series of tests (BSr and BTh) were conducted to identify gas breakthrough pressures considering three degrees of saturation (S_w = 60%, 80% and 100%) and two clay thicknesses (H = 0.4 and 0.6 m).

The second type of tests was conducted to study the gas emission rate through the unsaturated clay in low gas pressure range of 0 to 20 kPa. In each test, constant gas pressure was applied at certain intervals until gas flow rate reaches a stable value. There are also two series of gas emission rate tests (CSr and CTh) in this study. The objectives of CSr and CTh are to investigate influence of degree of saturation, degree of compaction and thickness of unsaturated clay layer on the gas emission rate, respectively. More details of the test program are given in Table 1.

Table 1. Testing program.

Test series	Degree of saturation at compaction S_w (%)	Thickness H (m)	Degree of compaction DOC (%)	Gas pressure P_g (kPa)
BSr	60, 80, 100	0.4	100	0~50
BTh	80	0.6	100	0~50
CSr	40, 60, 80	0.4	90, 100	1, 5, 10, 20
CTh	40, 60, 80	0.6	90, 100	1, 5, 10, 20

2.1 Testing apparatus

A 1-D soil column testing system including a transparent acrylic cylinder and a loading system was developed and used in this study. The photograph of this system is shown in Figure 1. The height, inner diameter and outer diameter of the cylinder are 700, 140 and 152 mm, respectively. Soil sample can be directly compacted inside the column and detailed descriptions of sample preparation are given later.

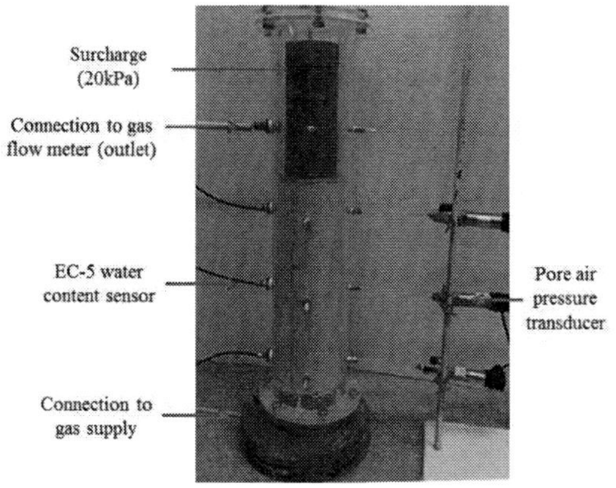

Figure 1. Soil column device for gas breakthrough and emission rate tests.

The bottom and top boundaries of soil sample are controlled to explore gas flow in unsaturated compacted soil. At the bottom of the column, inflow gas was controlled by a high-precision gas pressure regulator. The pressure of inflow gas was also double checked using an air pressure transducer with an accuracy of 0.2 kPa. Furthermore, to avoid drying effects of gas on soil sample, inflow gas passed through a gas saturator before being applied to soil sample. To ensure uniformity of gas pressure at the bottom boundary, a gravel layer with a height of 20 mm was placed. On the other hand, the top of soil column was connected to the atmosphere, with the outflow rate of gas monitored using a flow meter. In addition, a constant load correspond to a pressure of 20 kPa is applied at the top of soil sample to simulate the weight of the upper soil layers overlying the compacted clay layer in landfill final covers.

The gas type used for the experiment was nitrogen (N_2) (μ = 1.76×10^{-5} Pa·s, ρ = 1.15 kg/m^3 at 20°C) due to its relatively inert property and low solubility in water. Another important reason for selecting this gas is to avoid the risk of fire and explosions. Although the main compositions of landfill gas in an actual landfill are CO_2 and CH_4, it is supposed not to give significant different results by using N_2 for the tests [13].

As shown in the figure, the soil column testing system was equipped with various instruments, including pore air pressure transducers, EC-5 water content sensors and gas flow meter. Pore air pressure transducers were installed at 75, 225 and 375 mm above the bottom of soil column to measure pore air pressure distribution in unsaturated compacted soil sample. The measuring range and accuracy of the transducer are 0 to 100 kPa and ±0.2%

respectively. EC-5 water content sensors were installed at the same elevation with pore air pressure transducers to measure soil volumetric water content. Furthermore, depending on gas emission rate in each test, two types of gas flow meter with different measuring ranges (0 to 100 ml/min and 0 to 30 L/min) and accuracies (0.5 ml/min and 5 ml/min) were considered and used in this study. When high gas emission is expected, the one with wider measuring range is chosen. Otherwise, the one with narrower measuring range but higher accuracy is used. Prior to each test, all transducers were carefully calibrated.

2.2 Testing material

The test soil in this study is Kaolin clay. Some basic properties including Atterberg limits, specific gravity, compaction curve, water saturated permeability of the soil were determined following ASTM standard test methods [14, 15, 16, 17]. The test results are summarized in Table 2 [2]. Figure 2 shows water retention curve (WRC) of the compacted clay during the drying process from fully saturated state. It was measured using a modified pressure plate apparatus [18]. From the relationship between volumetric water content and suction, air-entry value of the compacted clay is estimated to be about 70 kPa. Furthermore, the measured WRC was fitted using van Genuchten's equation [19]:

$$\theta_w = \theta_r + (\theta_s - \theta_r)[1+(s/a)^n]^{-m} \quad (1)$$

where θ_w is the volumetric water content of unsaturated soil; θ_s is the saturated volumetric water content; θ_r is the residual volumetric water content; s is soil suction (kPa); a is a parameter related to the air-entry value of soil (kPa); n and m are fitting parameters related to the shape of WRC. According to van Genuchten [19], the parameter m was simplified to be $(1-1/n)$ in the fitting process to reduce the number of model parameters. In this study, two methods were used to prepare soil samples, depending on their degree of saturation. For the tests at degrees of saturation of 40%, 60% and 80%, oven-dried soil was initially mixed with de-aired water to reach the targeted water content. The mixed soil was then placed in a temperature and humidity controlled room for 48 h for moisture homogeneity. After that, each sample (140 mm in diameter and 400/600 mm in height) was dynamically compacted in 20 sub-layers.

To achieve desired degree of compaction, 15 hammer drops was required for each sub-layer. It should be pointed out that a thin layer of vacuum grease was placed on the inner surface of soil column to minimize preferential flow paths along the gap between soil and the sidewall of soil column.

For the tests at degree of saturation of 100%, soil samples were firstly prepared at the optimum water content which is equivalent to degree of saturation of 80%, following the methods described above. Then, each sample was subjected to bottom-up saturation to achieve degree of saturation of about 100%. All the soil samples were allowed to equalize after preparation.

Table 2. Basic properties of the soil.

Property	Value
Unified soil classification system	CH
Specific gravity	2.52
Liquid limit (%)	59
Plastic limit (%)	32
Plasticity Index (%)	27
Maximum dry density (kg/m^3)	1264
Optimum moisture content (%)	36.18
Saturated water coefficient of permeability (m/s)	5.2x10^{-9}
Water content at saturation (%)	42.06

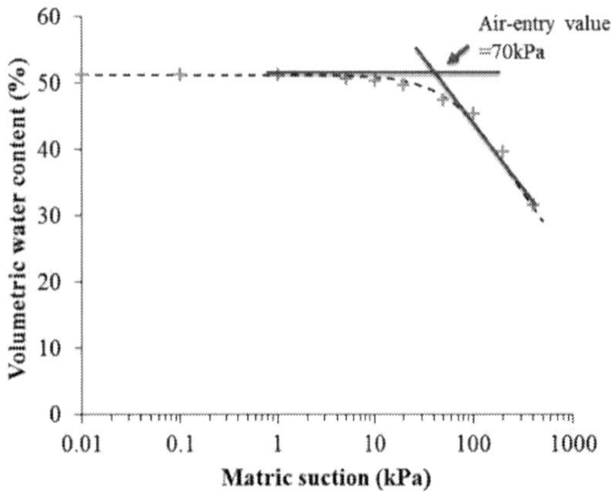

Figure 2. Measured and fitted WRC.

2.3 Testing procedures

For the gas breakthrough tests (series BSr and BTh), after soil preparation, a constant gas pressure of 5 kPa was applied at the base of soil sample for 3 days. If no gas breakthrough occurs, gas pressure was increases from 5 to 10 kPa and then maintained at constant value for another 3 days. Then, gas pressures of 15, 20, 30, 40 and 50 kPa were subsequently applied to soil sample until gas breakthrough occurs.

For the gas emission rate tests (series CSr and CTh), after soil preparation, constant gas pressures were applied at the bottom of soil until a stable gas flow is achieved. When the stage at gas pressure of 1 kPa was finished, the gas pressure was increased to 5 kPa and then kept constant until a stable gas flow is achieved. Then, 10 and 20 kPa gas pressures were applied by following a similar procedure.

During the tests, pore air pressure distributions were measured using pore air pressure transducers along the depth of soil column. Water content and density of soil samples were checked before each test to ensure the desired degree of saturation was reached. Gas emission rate was monitored at the top of soil sample using a flow meter. Furthermore, all tests were conducted in a

DOI: 10.1051/e3sconf/20160913005

temperature-controlled room (20±1°C) where density and viscosity of gas were considered to be constant.

3 Results and discussion

3.1 Gas breakthrough pressure at different degrees of saturation

Figure 3 shows the relationship between applied gas pressure and gas emission rate at three different clay degrees of saturation (S_w = 60%, 80% and 100%). The figure clearly reveals two types of soil response. At degree of saturation of 60% and 80%, the gas emission rate is almost zero in the lower range of gas pressure. When the gas pressure reaches a threshold value, the gas emission rate suddenly increases. According to the definition of gas breakthrough pressure mentioned, it can be seen from the figure that gas breakthrough pressure is estimated to be 22 and 38 kPa at degree of saturation of 60% and 80%, respectively. On the other hand, for the test at degree of saturation of 100%, gas emission rate is almost zero in the whole range of gas pressure (0-50 kPa). This suggests that gas breakthrough pressure is higher than 50 kPa at this testing condition. Comparisons between the three curves illustrate that the gas breakthrough pressure increases with increasing degree of saturation. The influence of degree of saturation is most likely because more pores are occupied by soil water in the compacted sample at higher degree of saturation. A higher pressure is therefore required to form an interconnected air-filled channel for gas flow.

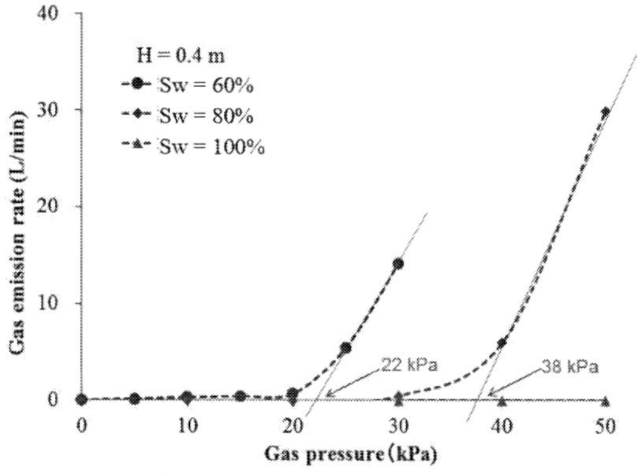

Figure 3. Gas breakthrough pressure at different degrees of saturation.

Based on field measurements, McBean et al. [1] reported that landfill gas pressure in the field is generally less than 10 kPa. Under typical landfill gas pressures, compacted clay layer with thickness of 0.4 m or above is able to prevent gas breakthrough when clay degree of saturation is larger than 60%, as illustrated in Figure 6. This finding suggests that in humid regions where soil degree of saturation is generally larger than 60% [6], a 0.4 m-thick clay layer is sufficient to satisfy design requirements at ultimate limit state.

3.2 Influence of clay thickness on gas breakthrough pressure

Figure 4 shows results of gas breakthrough tests at two clay thicknesses (H = 0.4 and 0.6 m). Gas breakthrough pressure of about 38 kPa was estimated for 0.4 m-thick clay. When the clay thickness increases to 0.6 m, the gas emission rate still remains almost zero even for gas pressure up to 50 kPa. This suggests that the gas breakthrough pressure can be higher than 50 kPa for a 0.6 m-thick clay layer. The observed increase of gas breakthrough pressure with increasing clay thickness is mainly because as clay thickness increases, the possibility of forming interconnected air-filled flow paths along the whole sample depth decreases. Similar observations have been reported by Hildenbrand et al. [8]. Based on the tests in a limited range of sample thickness (less than 0.03 m), they also found that gas breakthrough pressure increases as soil thickness increases under same gas pressure. These results suggest that the capacity of compacted clay to prevent gas breakthrough can be significantly improved by increasing clay thickness. According to the results in Figure 3 and Figure 4, in arid regions where soil degree of saturation may drop below 60%, a 0.4 m-thick clay layer may fail to prevent the occurrence of gas breakthrough. One possible solution to this problem is to increase the thickness of clay layer in the landfill cover.

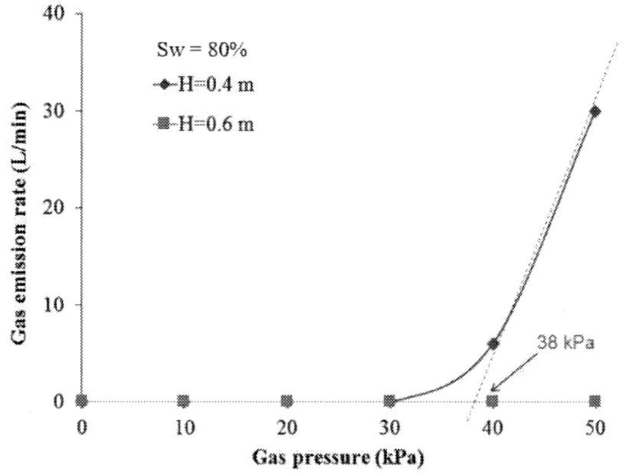

Figure 4. Effect of clay thickness on gas breakthrough pressure.

3.3 Gas emission rates at different degrees of compaction and degrees of saturation

Figure 5 shows the relationship between gas emission rate and gas pressure at two degrees of compaction (90% and 100%) and three degrees of saturation (40%, 60% and 80%). For 40% and 60% degrees of saturation, as expected, gas emission rate increases with an increase of gas pressure for both degrees of compaction. This is because the gas pressure increases induce a continuous increase of gas emission rate with increasing pressure gradient. For degree of saturation of 80%, gas emission rate was almost zero even for gas pressure up to 20 kPa for both degrees of compaction. This is consistent with the result of gas breakthrough tests. As illustrated in Figure 3, the gas breakthrough pressure at degree of

DOI: 10.1051/e3sconf/20160913005

saturation of 80% is up to 38 kPa and gas emission rate is almost negligible at gas pressures less than 20 kPa. Furthermore, the effect of degree of compaction on gas emission rate is also illustrated in the figure. At low degree of saturation (i.e., 40%), gas emission rate for 90% degree of compaction is similar to that of 100% degree of compaction within the whole applied gas pressure range (0~20 kPa). When the degree of saturation increases up to 60%, the effect of degree of compaction on gas emission rate is distinct. For a given gas pressure (i.e., 10 kPa), gas emission rate can be reduced by 65% when the degree of compaction increases from 90% to 100% under a soil degree of saturation of 60%. This may be due to the aggregation effect of compacted clay. During compaction, soil aggregates are much easier to compress at a relatively high initial degree of saturation than a low initial degree of saturation. When the degree of saturation further increases up to 80%, gas emission rate obtained from both degrees of compaction becomes almost zero. This is because more clay aggregates at this high degree of saturation are compressed compared to the intermediate degree of saturation (i.e., 60%) and low degree of saturation (i.e., 40%). Comparisons between the six curves (corresponding to degree of saturation of 40%, 60% and 80% and two degrees of compaction of 90% and 100%) clearly reveal that at the same gas pressure, gas emission rate is significantly lower at higher degree of saturation. This is because a higher degree of saturation induces smaller gas permeability and thus lower gas emission rate. The effect of degree of compaction on reducing gas emission rate dominates only in relatively high degree of saturation due to the clay aggregation effect.

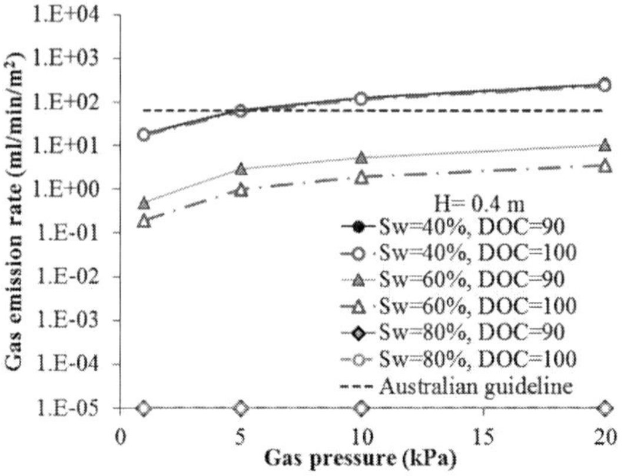

Figure 5. Gas emission rate at different degrees of compaction and degrees of saturation (H = 0.4 m).

In the figure, methane emission limit proposed by the Australian government guideline [11] is also included for comparison. According to this guideline, methane emission rate in a landfill should be less than 63 ml/m^2/min. To meet the requirement of the gas emission rate limit set by the Australian guideline, a clay layer with thickness of 0.4 m at degree of saturation of 60% and 80% is sufficient in the range of typical landfill gas pressure (less than 10 kPa) observed in field [1]. However, at degree of saturation of 40%, a 0.4 m-thick clay layer is sufficient only if the gas pressure is less than 5 kPa. A possible solution to this problem is to increase the thickness of clay layer. More discussions of effects of clay thickness are given later. It should be pointed out that although this guideline is proposed for methane emission, it should be applicable to some other gases (such as N_2). This is because the physical properties including viscosity of methane and nitrogen are similar.

3.4 Influence of clay thickness on gas emission rate

Figure 6 shows the gas emission rate corresponding to the gas pressure applied. Three degrees of saturation (40%, 60% and 80%) and two degrees of compaction (90% and 100%) were considered. The clay thickness used herein is 0.6 m. As expected, gas emission rate increases with increasing gas pressure and decreasing degree of saturation. Similar phenomena was observed as compared with Figure 5. Clay degree of compaction has prominent effect on reducing gas emission rate only in the intermediate degree of saturation (i.e., 60%). For example, at a given gas pressure (i.e., 10 kPa), 75% increment for gas emission rate was observed when the degree of compaction increase from 90% to 100% under a soil degree of saturation of 60%. As compared with the maximum gas emission rate allowed by Australian guideline [11], when clay thickness increase up to 0.6 m, the gas emission limit can be met under a gas pressure of less than 10 kPa. Comparison between this figure and Figure 5 clearly demonstrated that gas emission rate decreases significantly with an increase of clay thickness.

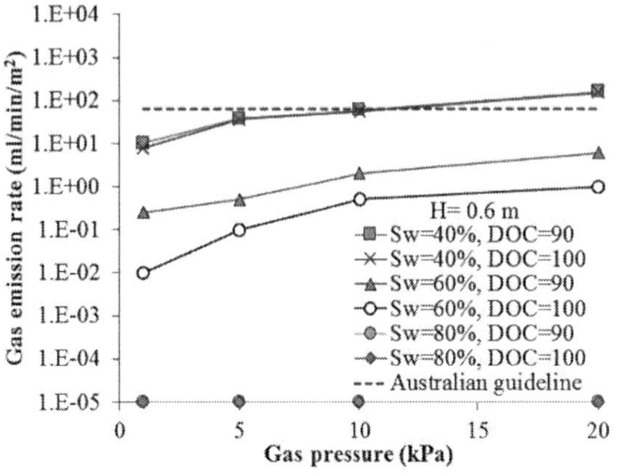

Figure 6. Effect of clay thickness on gas emission rate (H = 0.6 m).

To meet the requirement of the gas emission rate limit set by the Australian guideline, landfill designer should locate the clay layer at the bottom of landfill cover to eliminate moisture loss. Moreover, a thicker clay layer should be adopted in relatively arid area in order to minimize gas emission rate.

4 Conclusions

Gas breakthrough pressures and gas emission rates of unsaturated compacted clay were measured at different degrees of saturation and soil thicknesses. Gas breakthrough pressure of unsaturated compacted clay increases as the degree of saturation and thickness of clay increase. Under a gas pressure of 10 kPa (the upper bound limit of typical landfill gas pressure), a minimum of 0.4 m thick clay layer is able to prevent gas breakthrough at degree of saturation of about 60% or higher (in humid regions). A thicker clay layer is required if clay degree of saturation is lower than 60%. The reduction of gas emission rate by an increase in the degree of compaction is remarkable only in the range of intermediate degree of saturation. For low degree of saturation (i.e., 40%), degree of compaction has almost no influence on gas emission in the gas pressure range from 0 to 20 kPa.

Acknowledgements

The authors would like to acknowledge the research grant (2012CB719805) from the National Basic Research Program (973 Program) provided by the Ministry of Science and Technology of the People's Republic of China and research grant (HKUST6/CRF/12R) provided by the Research Grants Council (RGC) of the Hong Kong Special Administrative Region.

References

1. E. A. McBean, F. A. Rovers, G. J. Farquhar, *Solid Waste Landfill Engineering and Design* (Prentice Hall PTR, Englewood Cliffs, 1995)
2. C. W. W. Ng, Z. K. Chen, J. L. Coo, R. Chen & C. Zhou, Waste Manag. **44,** 155-163 (2015)
3. C. W. W. Ng, J. Xu, and R. Chen, U.S. Patent No. 9,101,968 B2 (2015)
4. C. H. Benson, M. A. Barlaz, D. T. Lane, J. M. Rawe, Waste Manag. **27**, 1, 13-29 (2007)
5. K. D. Barnswell, D. F. Dwyer, Waste Manag. **32**, 12, 2336-2341 (2012)
6. G. Zhan, J. Keller, M. Milczarek, J. Giraudo, Mine Water Environ. **33**, 195-205 (2014)
7. J. F. Harrington, S. T. Horseman, Swedish Nuclear Fuel and Waste Management Company, Stockholm (Sweden) (2003)
8. A. Hildenbrand, S. Schlomer, B. M. Krooss, Geofluids **2,** 1, 3-23 (2002)
9. H. Y. Wei, Ph.D. dissertation, Zhejiang University, Hangzhou, China (in Chinese) (2007)
10. S. Moon, K. Nam, J. Y. Kim, S. K. Hwan, M. Chung, Waste manag. **28,** 10, 1909-1914 (2008)
11. Carbon Farming Initiative, Guidelines for calculating regulatory baselines for legacy waste landfill methane projects, Australian Government (2013)
12. EPA, U.S. Research Triangle Park, NC. EPA-453R/96-004 (1999)
13. A. Bouazza, T. Vangpaisal, Geotext. Geomembranes **21,** 2, 85-101 (2003)
14. ASTM, D4318, West Conshohocken, PA (2010)
15. ASTM, D2487, West Conshohocken, PA (2011)
16. ASTM, D698, West Conshohocken, PA (2012)
17. ASTM, D5084, West Conshohocken, PA (2010)
18. C. W. W. Ng, Y. W. Pang, J. Geotech. Geoenviron. Eng. ASCE **126,** 2, 157-166 (2000)
19. M.T. van Genuchten. Soil Sci. Soc. Am. J. **44**, 5, 892-898 (1980)

Testing surfactants as additives for clay improvement: compaction and suction effects

Nathan Jones[1], Sérgio D.N. Lourenço[2,a] and Alison Paul[3]

[1]*Amey, United Kingdom*
[2]*Department of Civil Engineering, The University of Hong Kong, Hong Kong SAR, China*
[3]*School of Chemistry, Cardiff University, United Kingdom*

Abstract. This paper presents an exploratory study on surfactants as additives to improve soil properties. It is hypothesized that surfactant molecules populate the air-water interfaces reducing surface tension and suction thus allowing a control of the mechanical response of the soil. Suction measurements by means of a high suction tensiometer, compaction tests and Atterberg limits were conducted in mixtures of sand and kaolin, with and without a surfactant solution. The results revealed a prominent effect on suction, but to a lesser extent on the Atterberg limits and compaction behavior (the maximum dry density). This targeted effect of the surfactants suggests its molecules populate, not only the air-water interfaces decreasing surface tension, but may be adsorbing to the clay particles and forming micelles in the pore water as well. Therefore the interplay between the three may influence the soil behavior.

1 Introduction

1.1 Surfactants

Surfactants molecules consist of one polar, hydrophilic head group (water-loving) and one non-polar, hydrophobic tail group (water-hating). Surfactants have an affinity for a surface, forming into an orientated monolayer at the interface and reducing the interfacial tension. When the surface has no more space they begin to self-associate to form aggregates inside the bulk of the liquid (micelles). Micelles typically contain between 20-100 surfactant molecules. The concentration of surfactants, at which micelles begin to form, is defined as the critical micelle concentration (CMC). The CMC depends on surfactant type/chemistry, temperature (Rosen, 1989) among other factors.

The chemical affinity of a surfactant with a clay depends on the basic structure of the surfactant. Surfactants are classified into three groups according to the nature of the hydrophilic portion of the molecule: (1) anionic surfactants (negatively charged head group); (2) cationic surfactants (positively charged head group); (3) nonionic surfactants (non-charged) (Rosen, 1989). Sanchez-Martin et al. (2008) found that polar surfactant adsorption towards clays would stabilize, whereas non-ionic surfactant adsorption (including Triton X-100) would continue to increase above the CMC. Sanchez-Martin et al. (2008) also found cationic surfactants to be the most adsorbent across a variety of clay soil types. Wang et al. (1999) studied cationic surfactants and their adsorption behaviour with kaolinite and found that once the surfactant CMC had been reached adsorption levels would level off, suggesting that micelle formation can interrupt further adsorption. Instead, the highest adsorption levels occurred at concentrations just below the CMC. The chain length of surfactant tail groups and the chain structure within these groups may also influence the adsorption potential of surfactants. Clay surface charge distribution is sensitive to changes in chain lengths. Single chained cationic surfactants, for example, have shown a significantly greater adsorption potential towards kaolinite in comparison with double chained cationics at similar concentrations (Wang et al, 1999).

Surfactants control surface tension. Surface tension is the measure of the cohesive energy present at an interface. It is quantified as the force per unit length acting on an imaginary line drawn in the interface. For liquid water, at room temperature, the surface tension is 72 mNm^{-1}. Surface tension is important in unsaturated soils because matric suction depends, not only, on the radii of small water menisci at the interparticle contacts but also on the surface tension, where a reduction in surface tension equates to a reduction in suction. Since suction is a key variable controlling unsaturated soil behavior, being able to regulate suction via surface tension offers a series of new opportunities to use surfactants as additives for ground improvement (reduced volumetric changes, for instance).

[a] Corresponding author: lourenco@hku.hk

DOI: 10.1051/e3sconf/20160913006

1.2 Surfactants and soil behaviour

Among the relevant literature on the effect of surfactants on soil behaviour, it has been shown that surfactants can influence fluid density and viscosity, thus speeding up or slowing down the apparent hydraulic conductivities of soils (Lee et al, 2002). Depending on the properties and concentration of a surfactant used and clay interlayer electrostatics, Akbulut et al. (2012) revealed that surfactants can cause both flocculation and dispersion of clay particles. For example, cationic surfactants CTAC and QEFA were found to produce a dramatic change in clay plasticity. Within the effect of surfactants on soil compaction, Berney et al. (2003) assessed the influence of a wetting agent on compaction, and found that the soil with the surfactant solution achieved a higher maximum dry density at the optimum water content than the sample only with water. The Authors justified it based on a reduction of suction. However, no evidence was provided. More recently, Gluck (2010) concluded that the observed changes in the maximum dry density on various compacted granular samples with and without a commercial detergent were within possible experimental error.

In order to test the potential benefits of surfactants for ground improvement, such as an increase in maximum dry density during compaction, their effects on various soils and properties were tested in the laboratory. The remit in this study is to investigate the effect Merpol A, a non-ionic surfactant, on some soil parameters (consistency, compaction and suction) on different compositions of kaolin and sandy soils at concentrations of 0.05% and 1.0%. The composition of kaolin to sand ranged from 100:0, to 50:50 and 25:75, respectively.

2 Materials and Methods

2.1 Surfactant and soil

Merpol A is described as a nonionic, polyethylene glycol phosphate ester, non-foaming wetting agent (Sigma-Aldrich, 2010). A surface tension of 26mN/m^{-1} at 0.1% aqueous volume, and critical micelle concentration of 0.005% is provided in the product specification.

All tests were conducted with samples of Speswhite kaolin (100% kaolin) and mixtures of sand and kaolin at ratios of 50% kaolin : 50% sand and 25% kaolin : 75% sand, at surfactant concentrations of 0.05% and 1%, both above the critical micelle concentration. Prior to all tests, both the sand and clay were oven-dried.

2.2 Consistency and compaction tests

To determine the consistency behavior, the liquid limit (LL) was obtained following the Cone Penetrometer Method (BS1377:Part 2:1990:4.3) with LL corresponding to a penetration of 20 mm while the plastic limit (PL) also followed BS1377 with PL obtained for a water content of a thread of soil that crumbles at 3 mm.

To determine the compaction behavior, testing followed BS1377, with dynamic compaction in three layers at increasing water contents. Selected compacted samples were kept in the compaction mould with the bottom and upper surfaces sealed with paraffin wax for suction measurement.

2.3 Measurement of matric suction

Suction was measured with a high suction tensiometer (Ridley & Burland, 1993), manufactured at Cardiff University, with a 1500 kPa air-entry value ceramic stone (Figure 1). The tensiometer used in this research measured up to 1000 kPa, its suction at cavitation.

The tensiometer was pressurized to the range 2-3 MPa, and the measurement range tested by leaving the stone to dry to the atmosphere. The tensiometer was considered saturated, if suction at cavitation reached 1000 kPa. If not, the instrument was placed back in the pressure vessel and repressurized. No calibration in the negative range was conducted. The calibration from the positive range was extended to the negative range.

Suction was measured at constant water content conditions by smearing the stone against the soil and waiting for equilibrium. For drier samples, the stone was covered on thin film of water to force the contact with the soil. For accuracy, measurements were conducted with two separate tensiometers. Prior to the measurements, the tensiometer was removed from the vessel and left in free water for 2 to 3 days to allow the metal body to recover from the pressure vessel confinement. Further details and justifications for the procedures are provided in Lourenço et al (2009).

Figure 1. The high suction tensiometer used (diameter = 14mm; length = 22mm) (Image: Dr Xiangwei Wang, Rolls-Royce Aviation & Aerospace).

3 Results and Discussion

3.1 Consistency behaviour

Figure 2a and 2b show the effect of surfactant concentration on the plastic and liquid limit, respectively. The increasing concentration of surfactant, from 0% to 1%, increased the plastic limit for the three soil mixtures. For the kaolin sample, the plastic limit increased from 35.1% to 42.6% while for the 25% kaolin sample, the increase was from 14.5% to 18.2%. For the liquid limit, the results were not conclusive, with LL remaining unchanged for the kaolin sample, increasing for the 50% kaolin sample and increasing and decreasing for the 25% kaolin sample.

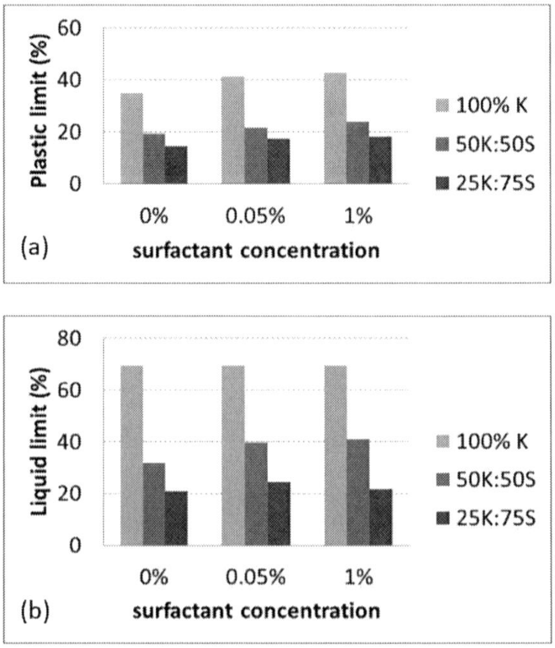

Figure 2. Effect of the surfactant concentration in the consistency behavior of three soil mixtures (K=kaolin, S=sand). (a) Plastic limit; (b) Liquid limit.

3.2 Compaction behaviour

Figures 3a, 3b and 3c show the compaction curves at different surfactant concentrations (0%, 0.05%, 1%) for 10% kaolin, 50% kaolin and 25% kaolin, respectively. Figure 3a, for kaolin, shows tests at two further concentrations at 0.1% and 4.5%. The compaction curves revealed no particular trend, with the curves overlapping for the 50% and 25% kaolin samples. Differences in the maximum dry density and optimum water content between the three figures are due to the material, rather than surfactant concentration. The only exception is for the kaolin samples, where the maximum dry density dropped by 0.05 Mg/m^3 for the samples prepared at higher surfactant concentrations (1% and 4.5%). This finding contradicts the results by Berney et al. (2003) where the maximum dry density was increased with the surfactants.

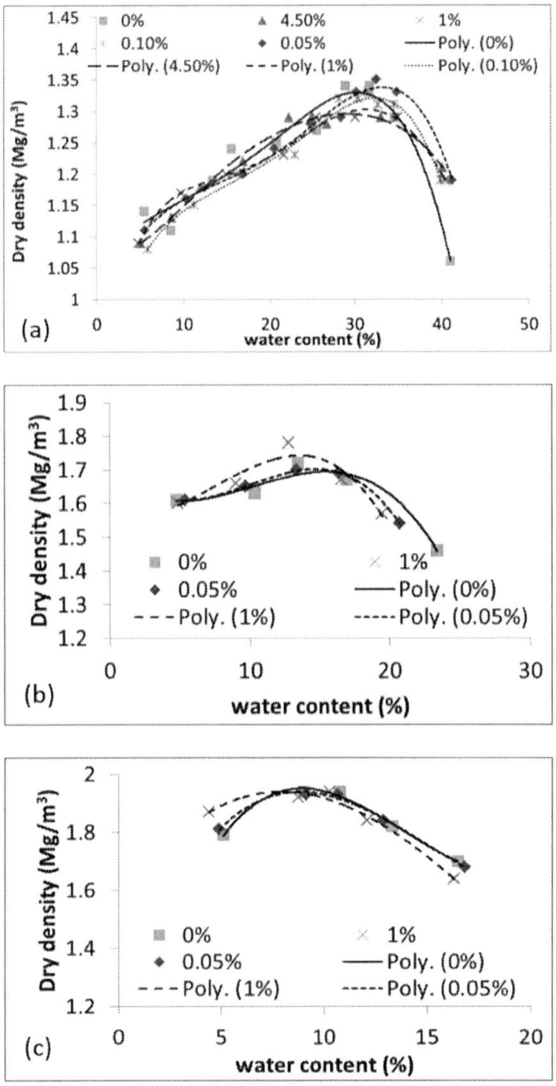

Figure 3. Compaction curves at different surfactant concentrations for (a) 100% kaolin, (b) 50% kaolin : 50% sand, (c) 25% kaolin : 75% sand.

3.3 Suction behaviour

Figure 4a shows suction against the surfactant concentration for a 50:50 sample at 10% and 15% water content. At 10% water content, there was a reduction of 235kPa by adding 0.05% surfactant and 800kPa when adding 1% surfactant solution. At 15% water content, the reduction in suction was smaller because of the higher water content. Figure 4b shows suction against the surfactant concentration for a 75:25 sample at 5% and 10% water content. The results are similar to Figure 2, except that given the higher sand content, the water content was smaller (5%) to achieve high suctions (1000kPa for the sample without surfactant). These results confirm that the surfactants reduce suction, likely linked to a reduction in the surface tension.

DOI: 10.1051/e3sconf/20160913006

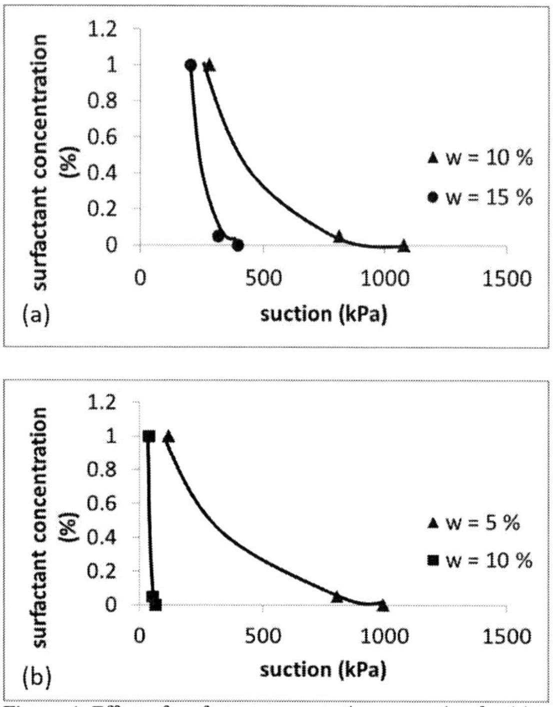

Figure 4. Effect of surfactant concentration on suction for (a) 50% kaolin : 50% sand, and (b) 25% kaolin : 75% sand.

3.4 Summary

This preliminary study revealed no common trends between the three sets of results (consistency, compaction and suction), suggesting that the surfactants have specific effects in certain soil properties. The presence of surfactants had an effect in the plastic limits and suction behaviour, but was inconclusive in the compaction behaviour and liquid limit. Increasing surfactant concentrations decreased suction and increased the plastic limits. A reduction in the maximum dry density did occur but only at high surfactant concentrations.

The compaction results revealed that the explanation proposed by Berney et al. (2003) whereby the maximum dry density is increased due to a reduction in suction does not hold for all soils and conditions. The maximum dry density changed little even when suction was reduced by 80% at 1% surfactant concentration.

This study also revealed that the CMC, an important property of surfactants, does not act as a threshold in the soil behavior. If it did, soil properties would not change for concentrations higher than the CMC. But, as shown for all datasets, increasing concentrations past the CMC had an effect on suction, plastic limit and compaction. This suggests that surfactant molecules populate, not only the air-water interfaces decreasing surface tension, but may be adsorbing to the clay particles and forming micelles in the liquid phase. Therefore the interplay between the three, may affect soil behavior.

4 Conclusions

The paper presents an experimental study on the behavior of unsaturated soils with surfactant solutions. Suction measurements by means of high suction tensiometers, compaction tests and the Atterberg limits were conducted in mixtures of sand and kaolin, with and without surfactant solutions. Compaction showed a decrease in the maximum dry density for the 100% kaolin sample at higher surfactant concentrations. The change in the maximum dry density data between the different concentrations was negligible or inconclusive. Plastic limits grew significantly with increasing surfactant concentration. Suction consistently reduced with increasing surfactant concentrations, even past the CMC, which suggests that surface tension cannot be the only mechanism responsible for decreasing suction. Other factors, such as the adsorption of surfactant molecules to the clay particles and micelle formation in the pore water may also affect soil behaviour.

Acknowledgements

Funding to develop the high-suction tensiometer was provided via a UK Royal Society Research Grant RG090044. Manufacturing was conducted by Dr. Xiangwei Wang (Manufacturing Engineering Centre, Cardiff University, UK, *now* Rolls-Royce Aviation & Aerospace, Bristol, UK).

References

1. Z. Akbulut, N. Kurt, S. Arasan. Earth Sciences Resource Journal **16** (2):13-19 (2012)
2. E. Berney, J. Peters, K. Newman, D. Smith. Transportation Research Record: Journal of the Transportation Research Board, 1819:57-62 (2003)
3. O. Gluck. MSc. Dissertation, Cardiff University, pp. 117 (2010)
4. D.-H. Lee, R.D. Cody, D.-J. Kim, S. Choi. Environment International **27**, 681-688 (2002)
5. S.D.N. Lourenço, D. Gallipoli, D. Toll, F. Evans. ASTM Geotechnical Testing Journal **32**, 6, 1-7 (2009)
6. A. Ridley, J. Burland. Géotechnique **43** (2), 321-324 (1993)
7. M.J. Rosen. Surfactants and interfacial phenomena, 2nd edition, John Wiley & Sons Inc. (1989)
8. M.J. Sánchez-Martín, M.C. Dorado, C. del Hoyo, M.S. Rodríguez-Cruz. Journal of Hazardous Materials **150**, 115-123 (2008)
9. J. Wang, B. Han, M. Dai, H. Yan, Z. Li, R.K. Thomas. Journal of Colloid and Interface Science **213**, 596-601 (1999)

Study of the microstructure of a collapsible soil flooded with NaCl saline

Ouassila Bahloul [1,a] Khelifa Abbeche [1] and Azeddine Bahloul [2]

[1]*Department of Civil Engineering, University of Batna, Algeria*
[2]*CGI Technical Services, 1612 Wedding Way Redding, California, 96003 USA*

Abstract. Collapsible soils are unsaturated soils which may undergo inter-granular rearrangement with abrupt and significant volume reduction after wetting. The large settlement resulting from the soil collapse pushed for deep studies of these types of soils. The experimental study consists to illustrate the influence of the salt concentration, compaction energy and moisture content on the change of microstructural characteristics of the reconstituted soil and reduction of the collapse potential.

1 Introduction

The collapse of soils is at the origin of several engineering problems, such as excessive settlement of large civil construction and public works projects works in many regions of the world. The collapse can occur in the case of a natural soil (aeolian deposits, alluvial deposits or residual soils) or compacted soils (embankments compacted on the dry side of the Proctor optimum curve). Generally, collapsible soils are defined as metastable soils that contain voids within their structures composed often grains ranging from silt to fine sand [1].The present study aims to illustrate that it is possible to decrease the potential of collapse of a soil reconstituted in the laboratory to an acceptable level. Also, to study the effect of a salt solution (sodium chloride NaCl) at various concentrations as well as the effect of energy of compaction on the potential of collapse of the same soil, while varying the vertical stress. We present a set of evaluations with scanning electron microscopy (SEM) of samples of untreated reconstituted and on samples treated by various salts, to correlate the macroscopic evaluations to the results obtained at the microscopic level obtained by oedometer tests, and to discuss the effects of the addition of salts on the microstructure and the physical properties of the reconstituted.

2 Materials, Equipment and Test

The reconstituted soil sample is composed of 80% of sand and 20% of kaolin. The application of various criteria of collapse, as indicated by [2-3], show that this material is collapsible.

[a] Corresponding author: *bahouassila@gmail.com*
abbechekhelifa@hotmai.com

2.1 Characteristics of Materials

The sand used in the sample reconstitution comes from the Liwa River located in the region of Biskra, Algeria screened on the 2mm sieve; and its geotechnical characteristics are given in the Table 1. The kaolin soil used in this study is extracted from Debbagh area of Guelma, Algeria.

Table 1. Geotechnical characteristics of Materials

Materials	Characteristics
Sand	Sand equivalent SE = 87 % Coefficient of uniformity Cu = 3.91 Coefficient of curvature Cc = 0.95 Grain size distribution (0.08-2 mm) with 3.03% of particles < 80 μm.
Kaolin	D‹80μm Liquid limit LL = 67(%) Plastic limit PL = 39(%) Specific Gravity GS = 2.4
Reconstituted soil (80% sand + 20% kaolin)	Coefficient of uniformity Cu = 5.13 Coefficient of curvature Cc = 1.07 Liquid limit LL = 28 (%) Plastic limit PL = 16 (%) Specific gravity GS = 2.65 Maximum dry density (g/cm^3) γ_d = 1.93 Optimal water content w_{opt} = 10 (%)

In order to obtain samples with different degrees of compaction, those samples are moistened at different water contents, then they are subject to different compaction energies (20, 40 and 60 drops) [2], The samples thus prepared were tested using an oedometer by saturating them with solutions at different salt concentrations. Generally, work treating the effect of salts on the microstructural characteristics and hydraulic behaviour of clay materials use concentrations up to 1 mol/L [4, 5, 6]. The salt solution used for the soil treatment is sodium chloride (NaCl) with various concentrations (0.5, 1.0, 1.5 and 2 mol/L).

2.2 Characteristics of Consistency of the reconstituted Soil

The literature revealed that a soil is likely to collapse if at least one of the following criteria is verified: activity of soil (A_c<1), liquid index (I_l<0), plasticity index (PI≤20), consistence index (I_c>1) and manageability index (I_w≤1) [3-7], The results presented in Table 2 show that the soil of our study is likely to collapse, and that four of the five criteria mentioned above are checked, which confirms the results of [8],

Table 2. Characteristics of consistency of the reconstituted soil

w_0*	Ac	PI	I_l	I_c	I_w
2			-1.17	2.17	0.16
4	2.60	12	-0.67	2.00	0.33
6			-0.83	1.83	1.50

•Initial water content

2.3 Physico-Chemical Characteristics of the Reconstituted Soil

It was found that the Cation Exchange Capacity (CEC) of the sample is low according to Table 3 with a saturated adsorption complex. The CEC is essentially determined by the nature and content of the clay fraction, which in our case consists of kaolin mixed with smectites impurities. The pH indicates that the material is alkaline and the electrical conductivity (EC) is low.

Table 3. Characteristics of consistency of the reconstituted soil

Adsorption complex			Solution soil		
			pH	7.36	
			EC	1.2	ms/cm
Ca^{++}	0.21	meq/100g	CO_3H^-	3.88	meq/ l
Mg^{++}	0.10	meq/100g	SO_4^{--}	2.5	meq/ l
K^+	0.13	meq/100g	Cl-	5.52	meq/ l
Na^+	2.07	meq/100g	Ca^{++}	4.50	meq/ l
CEC	2.51	meq/100g	Mg^{++}	3.70	meq/ l
			Na^+	3.00	meq/ l
			K^+	0.80	meq/ l

3 Oedometer results

The results obtained illustrate the collapse potential C_p varies from 0.98 to 12.40% for initial water content w_0=2%, from 0.82 to 10.64% for initial water content w_0=4%; and from 0.66 to 8.79% for an initial water content w_0=6%. After treatment, we constate a substantial reduction of potential of collapse C_p for al soils treated with sodium chloride (NaCl) with various concentrations. The maximum potential of collapse C_p is obtained of vertical stress 400 kPa. Figure 1 show that the potential C_p decreases almost linearly when the water content or the compaction energy increase, for vertical stress of 400 kPa. It should be noted that the potential of collapse C_p decreases when initial water content or initial compaction energy increase. That is attributed to the reduction of the suction forces when the initial water content increases and denser structure when the energy increases.

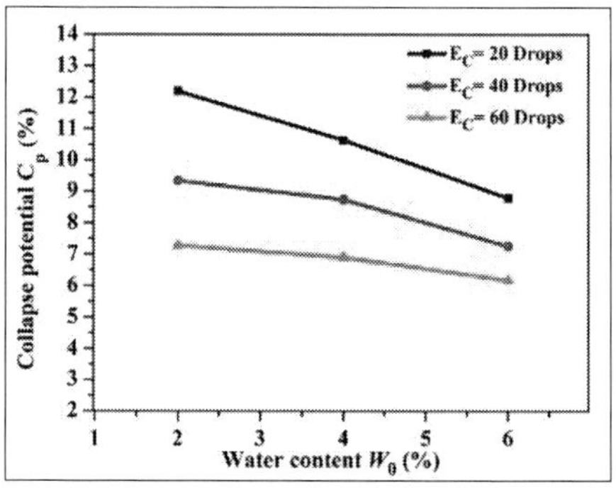

Figure 1. Variation of the collapse potential Cp with respect to water content for untreated soil at a vertical stress of 400 KPa

4 Observations under the Scanning Electron Microscope (S.E.M.)

4.1 Quartz Structure: Case of the Untreated Soil

In the case of the untreated soil under compacting force 20 drops, we notice a loose contact between kaolinite crystallites and the quartz crystals (Figure 2). Kaolinite crystallites do not present any preferential orientation compared to the quartzose skeleton (Figure 3). This type of microstructure was reported by various authors [9, 10, 11].

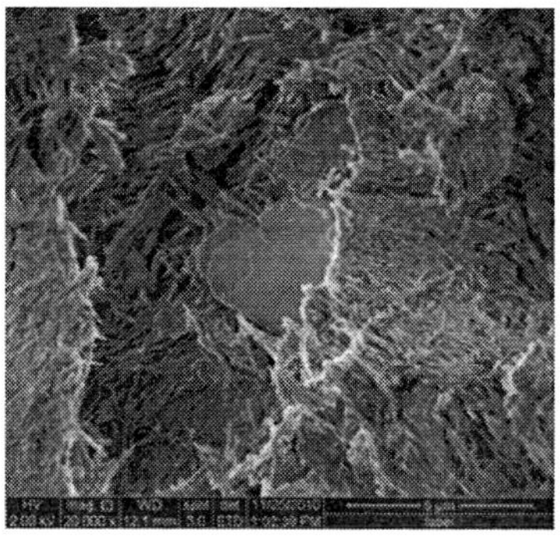

Figure 2. Electron photomicrograph of reconstituted untreated soil sample (20 drops).

Figure 3 shows that the material compacted with 40 drops is a little more packed than the sample subjected to a compacting force of 20 drops. Quartz crystals are coated with crystallites of kaolinite with more coverage than the sample that has undergone slight compaction energy (20 drops).

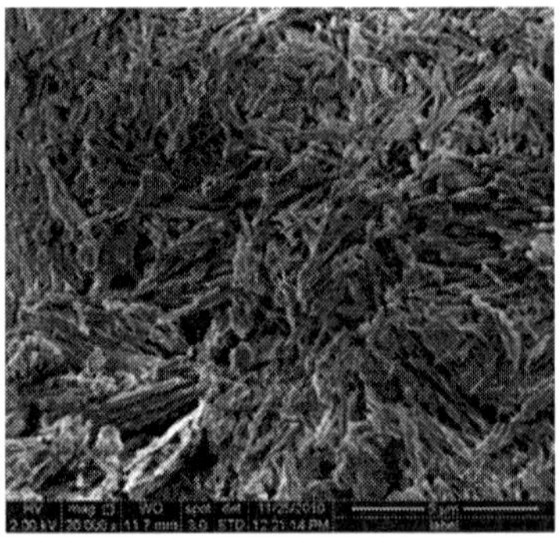

Figure 3. Electron photomicrograph of reconstituted untreated soil sample (40 drops).

Figure 4 shows that the material compacted under the effect of a force of 60 drops has a denser coating than the previous samples compacted under 20 and 40 drops and has a low porosity. The crystallites of kaolinite without define boundaries cover the quartz crystals.

Figure 4. Electron photomicrograph of reconstituted untreated soil sample (60 drops).

4.2 Relation Salt-Fine particles: Case of NaCl

The treated sample presents cubic NaCl crystals (halite) relatively crystallized, coated with clay in the form of sticks or rods in mixture with a smectite and quartz crystals. A figure 5 and 6 indicates that a space exists between the clay fraction and the NaCl crystals with certain fracture porosity. X-rays analysis reveals a higher concentration of silicon (Si) versus aluminum (Al) due to the presence quartz and kaolin, which reveals the presence of three-layer clay beside the Kaolin. Halite is highlighted in the X-ray analysis by peaks of sodium (Na) and chlorine (Cl) (Figure 7).

The mechanical behavior of the reconstituted sample in the presence of soluble salts is attributed to the effects of the salts on the interstitial solution of the soil and the modification of the exchange complex ionic structure of the clay.

First of all, because of the increase in the ionic force of the solution of the soil, the diffuse double layer of clays is compressed, which results in a flocculation of the material and its stabilization compared to reference material (untreated soil). As for the effects of the adsorbed cations, their nature plays an important part in the physical and mechanical behavior of the cohesive material. The alkaline cation monovalent Na^+ compresses the diffuse double layer particularly when the solution of the soil presents a weak ionic force; thus supports with high water content the swelling and the dispersion of the clay particles [12].That has as a consequence in obstructing the pores of the material and its densification.

An important characteristic of the electrostatic forces is their instability in connection with the water content. Indeed, when the material is very wet, the distance between the particles increases and the electrostatic force decreases. At the extreme, the electrostatic forces disappear; thus, a repulsion of the elementary particles occurs. Therefore, the sodic materials (rich in adsorbed sodium) with low hydric potential retain water

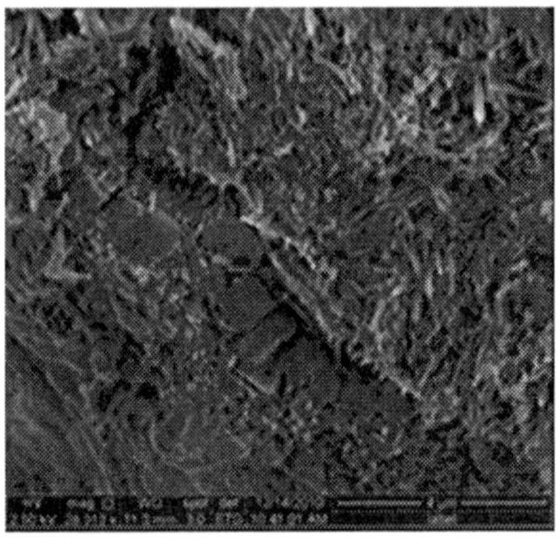

Figure 5. Electron photomicrograph of treated soil by NaCl Smectite and quartz crystals.

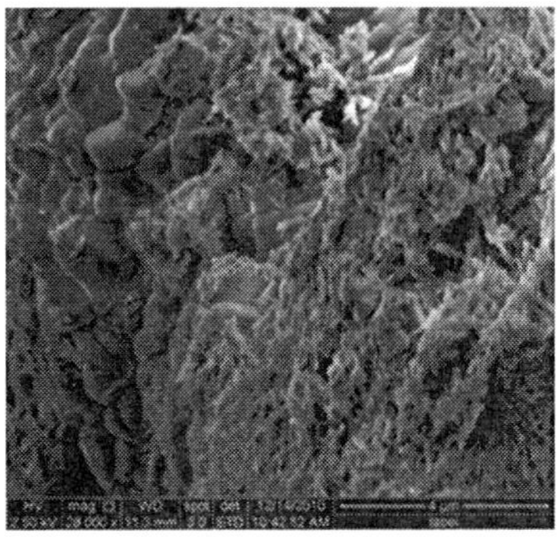

Figure 6. Electron photomicrograph of treated soil by NaCl quartz crystals,fracture porosity.

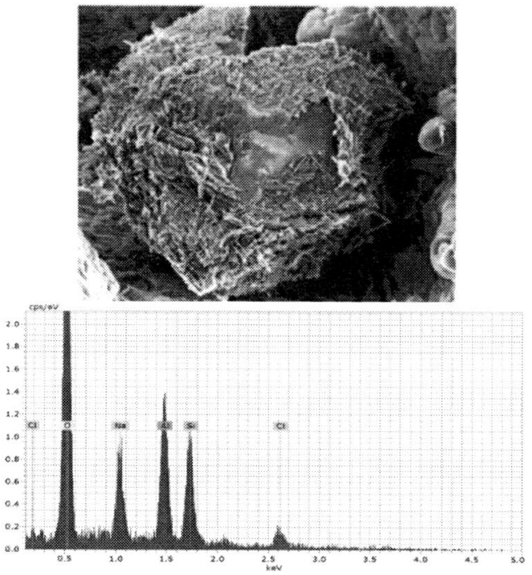

Spectrum: Acquisition 777

El	AN	Series	unn. C [wt.%]	norm. C [wt.%]	Atom. C [at.%]	Error [%]
O	8	K-series	74.97	66.52	77.67	10.8
Na	11	K-series	4.74	4.21	3.42	0.3
Al	13	K-series	9.86	8.75	6.06	0.6
Si	14	K-series	16.62	14.75	9.81	0.9
Cl	17	K-series	6.52	5.78	3.05	0.5
		Total:	112.71	100.00	100.00	

Figure 7. X-ray output spectrum of treated soil by NaCl.

5 Conclusions

The collapse of soils is a very complex phenomenon involving a large number of intrinsic and environmental parameters. It is mainly due to the rearrangement of the soil particles after flooding.
The main observations of this study are:

- The collapse potential is reduced with increase of initial water content and increase of energy of compaction;
- The treatment by salt (NaCl) reduced the collapse potential and subsequently the damage associated with it to structures;
- The study of the microstructure by the SEM showed that the reconstituted soil is characterized by a porous structure based on loose skeleton of grains with 80 % of sand (quartz) and a clayey fraction (20 % of kaolin with a small proportion of smectite) is unevenly distributed
- The crystalo-chemistry of its constituents; and on the other hand to the physico-chemical properties in relation with the composition of pore solution and the ion structure of the exchange complex.
- Effect of the cation Na^+ has a dispersing effect on the colloids but being also translated by a dense and compact material. The cation favors the formation of a stable material.

Acknowledgements

We would like to thank the members of CGI Technical Services for their contributions to this study.

References

1. J.H. Dudley., *JSMF.* Div, ASCE 96:925-947. (1970)
2. K. Abbeche, M. Laouar, F. Messaoud., S.G.M, **XXXII** 2:.3-21, (2010)
3. AJ. Lutenegger, RT. Saber., G.T.J, **11**, 3:173-178, (1988)
4. D. Tessier, J. Berrier, S.S, (1979)
5. A. Halitim., Thesis, University Rennes : 383, (1985)
6. K. Abbeche, O. Bahloul, T. Ayadat, A. Bahloul., *Experimental and Applied Modeling of Unsaturated Soils, Proceedings of Geoshanghai, June 3-5, Shanghai, Chine.doi:10 1061/41103(396)10,* (2010)
7. T. Ayadat, A. Hanna., G.T.J, **30** (2007)
8. K. Abbeche, M. Laouar, F. Messaoud., S.G.M, **XXXII** 2:.3-21, (2010)
9. K. Rollins, M. Wayne, G. Rogers., J.G.E, **120**, (1994)
10. O. B, Grabowska., B.I.A.E.G, Krefeld. (1975)
11. EC. Lawton, R J. Fragaszi, H. James., J.G, ASCE, **155**,9: 1252-1267, (1989)
12. A. Halitim, M. Robert, D. Tessier, R. Prost., Agronomie, 4 :451-459. (1984)

Influence of saturation degree and role of suction in unsaturated soils behaviour: application to liquefaction

Mathilde Vernay[1], Mathilde Morvan[1] and Pierre Breul[1]

[1]*Institut Pascal, Polytech Clermont-Ferrand, Université Blaise Pascal, Clermont-Ferrand, France*

Abstract. The effect of the pore fluid compressibility on liquefaction has been studied by various authors. But few papers have been published about the role of suction in cyclic behavior of unsaturated soils. Most of these works use Skempton coefficient B as a reference in terms of saturation degree to analyze their results. The use of B in experimental conditions is convenient, but is not accurate when studying liquefaction behavior, since effects of suction are neglected. In this paper, the influence of saturation degree on mechanical behavior of a soil under dynamic loads is studied. Cyclic undrained triaxial tests were performed on sand samples, under various levels of saturation. Soil-water characteristic curve was used, in order to study influence of suction. The first results confirm that when the degree of saturation decreases, the resistance increases. Initial positive suction tends to stiffen the soil. It also appears that the presence of air delays the occurrence of liquefaction, but doesn't prevent it. Indeed, liquefaction is observed, whether the soil is saturated or not.

1 Introduction

Since the 1950', unsaturated soils' behavior study has witnessed a growing interest within scientific community. Volumetric behavior and shear strength have been studied and gave birth to adapted models for unsaturated soils behavior, but we observed a lack of data on liquefaction.
Liquefaction has already caused tremendous damages, particularly in seismic areas: Niigata, Japan [1] (1964), Anchorage, Alaska [2] (1964), or more recently in the Tokyo Bay area, Japan [3] (2011). During earthquake, or any vibratory disturbance, local collapses of the soil structure occur, meanwhile pore water pressure within the soil increases, and equals in some cases the total pressure. It leads to a zero effective stress state, and loss of the shear strength. It is commonly admitted that unsaturated soils are riskless towards liquefaction, and the study of unsaturated soils behavior under cyclic loading suffers from a lack of data.
Meanwhile, climate changes affects soils' hydric conditions; they undergo more pronounced wetting-drying cycles. As a result, we are likely to be more and more confronted to unsaturated soils, but their hydro-mechanical behavior under cyclic loading is not entirely understood. This lack of knowledge leads to an unappropriated risks management policy, and eventually to an understatement of liquefaction risks.

Various studies have already shown interesting results regarding the liquefaction instability of unsaturated soils ([4], [5] and [6]). It is now admitted that the saturation degree influences liquefaction resistance: the lower the saturation degree, the higher the resistance to liquefaction. But the majority of these works use the Skempton coefficient B to quantify the level of saturation of the sample. The use of B in experimental work is convenient, but is not satisfactory when studying influence of saturation on cyclic behaviour. It does not provide a simple relation between physical and mechanical parameter of the soil.

B depends on the pore air compressibility, while we know that the presence of air within the soil creates suction, which tends to stiffen the material. Pore compressibility and suction have opposite effects; it is then not accurate to analyse potential of liquefaction of unsaturated soil considering only volume compressibility of the soil skeleton. The use of B as saturation level reference would be relevant if air compressibility was the only difference between saturated and unsaturated soil. But many authors ([7], [8], [9]) have used the fact that another state variable is needed to differentiate unsaturated soils behavior from saturated one: suction. The use of B only as an indicator of saturation in results' analysis is not enough considering suction, whose effects on hydro-mechanical behaviour are proved.
Some authors, like Yoshimi et al. [10], or Arab et al. [11] used a theoretical relation between B and Sr, introduced by Lade et al. in 1977 [12]. But it appeared that this theoretical relation was not entirely accurate. Moreover, if these studies were innovative because they provided a

[a] Corresponding author: Mathilde.vernay@univ-bpclermont.fr

relation between a value of B and resistance to liquefaction, the only parameter considered is pore fluid compressibility.
Another approach was developed by Okamura et al. [6]: they quantified a saturation degree measuring the volume of air within the specimen, estimated from Boyle's law with the measured volume of the water introduced in the sample. They establish a unique relationship between liquefaction resistance and the potential volumetric strain, but again neglecting other effects of partial saturation.
As far as we know, only few experimental campaigns aiming to study cyclic behaviour of unsaturated soils have already been performed with suction-controled conditions. Unno et al. [13, 14] conducted triaxial cyclic tests, under various suction conditions, using axis translation technique. They observed that the air entry value of the soil characteristic curve was a major factor of influence towards liquefaction instability of the material. The air entry value appears to be a critical value. Such studies, exploring the effects of suction on cyclic behaviour of unsaturated soils, are few and further investigations need to be carried out.

In this context, and through this paper, we want to:

- confirm that initial saturation degree does have an influence to liquefaction resistance,
- show that unsaturated soils can liquefy.

Contrary to most of the studies, we will quantify saturation degree in terms of Sr, and not only in terms of B. This will provide us to relate observed effects, and values of initial suction, which is rarely done. This study is a preliminary work, first part of an overall project. The final goal will be to give a framework regarding liquefaction risks evaluation: link physical and mechanical parameter, set critical limits in terms of Sr, in order to improve liquefaction risks management.

2 Influence of initial saturation degree

It is noteworthy to specify that this work is preliminary; indeed, it aims to highlight the influence of initial saturation degree and initial suction on the cyclic behaviour of sands, using predictive model of the B-Sr relation, and the soil-water characteristic curve. Further tests are about to begin in our laboratory, realized under controlled initial suction, and suction monitoring conditions. We will be able to:

- confirm the accuracy of our numerical predictions;
- realize triaxial cycling tests on unsaturated specimens quantifying saturation degree in terms of Sr.

2.1 Material and testing system

In this study, a fine clean sand was used as the testing material to study the influence of saturation degree on cyclic behaviour of unsaturated soil. The material is known as Fontainebleau sand, and is commonly used in experimental works, especially when liquefaction instability is studied ([15], [16]).

Cyclic triaxial tests are performed, using a Bishop and Wesley triaxial cell (Bishop et al., [17]). The samples are reconstituted in laboratory, by the wet tamping technique (initial moisture of 3%). This process was chosen in order to satisfy low density criteria ($e > 0.85$). The saturation process consists in 30 minutes of CO_2 circulation through the sample, followed by deaired water circulation. The sample is consolidated under a pressure $\sigma_c' = 100$ kPa.

The level of saturation is experimentally quantified by the Skempton coefficient B. It is then associated to a value of Sr, like it will be introduced in the next paragraph.

The cyclic loading is then applied, with a 0.017Hz frequency, and amplitude of 50 kPa.

2.2 Initial experimental conditions

Thanks to a numerical model developed in our laboratory, we are able to accurately relate values of B, measured before triaxial testing, and saturation degree. Three triaxial cyclic tests have been realized, and the initial parameters of the sample are presented in Table 1:

Table 1. Initial experimental parameters.

Test	B	Sr
triaxcyc100CUfontainSr1	0.98	1
triaxcyc100CUfontainSr98	0.53	0.98
triaxcyc100CUfontainSr95	0.22	0.95

2.3 Results of undrained cyclic shear test

2.3.1 Pore water pressure ratio

Pore-water pressure ratios stands for the pore water pressure, normalized by the total stress σ_c. When pore-water pressure ratio equals 1, liquefaction occurs. Figure 1 shows the pore-water pressure ratio evolution versus number of cycles. It shows that independently from the saturation degree, the pore-water pressure ratio increases under undrained cyclic loading. Nevertheless, differences are remarkable between each sample tested: initial increasing rate of pore-water pressure ratio depends on the initial saturation degree. The higher the saturation degree, the higher the increase of pore-water pressure within the soil sample. Indeed, for 20 loading cycles, the sample initially totally saturated has reached a 0.91 pore-water pressure ratio value; the sample initially 98%-saturated has reached a 0.74 pore-water pressure ratio value, while the sample initially 95%-saturated has only reached a 0,65 pore-water pressure ratio value.

 DOI: 10.1051/e3sconf/20160914002

Figure 1 also shows that initial saturation degree affects the number of cycles needed to trigger liquefaction: for Sr = 1, liquefaction occurs after 40 cycles. For Sr = 0,98, liquefaction occurs after 51 cycles and for Sr = 0,95, liquefaction occurs after 54 cycles. It depicts the fact that lower initial saturation degree delays the occurrence of liquefaction. Moreover, the most significant difference in terms of applied cycles before liquefaction is between saturated sample and unsaturated samples. It seems that only small quantity of air is needed to observe significant differences. But pore-water pressure ratio eventually equals 1 for the three tested samples, meaning that unsaturated soils can liquefy.

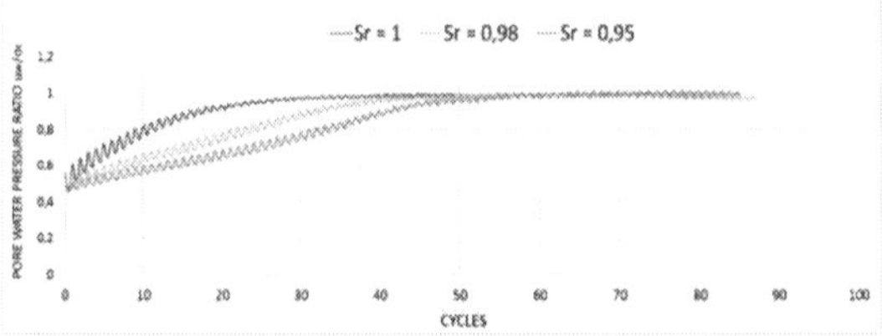

Figure 1. Pore-water pressure ratio evolution with number of loading cycles applied, for three different initial saturation degree

2.3.2 *Axial strain*

Figure 2 depicts axial strain evolution with number of loading cycles applied, for each tested sample.

It shows that presence of air within the soil affects strain development: if we consider a fixed strain level of 5%, totally saturated sample reaches this rate after 36 cycles. For Sr = 0,98, this 5%-rate is reached after 47 cycles, and for Sr = 0,95 after 53 cycles. That is, the smaller the saturation degree, the faster the development of significant strain rates. But, eventually, for the same number of applied cycles (80 cycles), the three samples have reached the same final level of strain (19%).

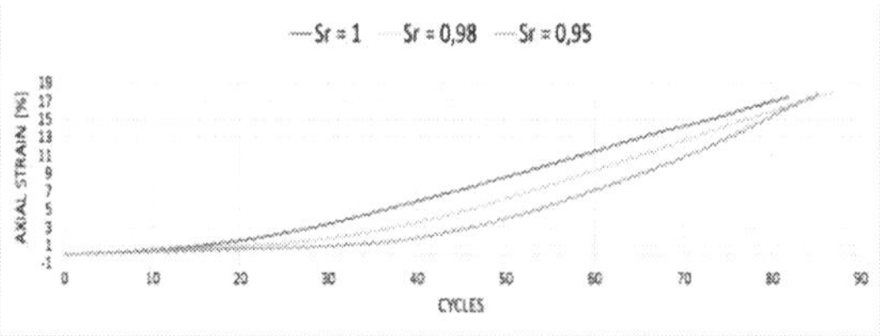

Figure 2. Axial strain evolution with number of loading cycles applied, for three different initial saturation degree

Table 2 summarize the previous graphic results:

Table 2. Summary table for the results obtained on three samples with different initial saturation degree.

Sr	Nb of cycles before liquefaction	Nb of cycles before reaching 5% axial strain
1	40	36
0,98	51	47
0,95	54	53

2.3.3 *Stress-strain relationship and effective stress paths*

Figure 3 (a), (b) and (c) and Figure 4 (a), (b) and (c) depicts the stress-strain relationship for the three tested samples, corresponding to three different initial saturation degree. Figure 5 (a), (b) and (c) depicts the effective stress paths for the three tested samples.

Firstly, Figure 3 (a), (b) and (c) show that deviator peak decreases with increasing axial strain during the cyclic loading, regardless the initial saturation degree. We can also highlight the fact that stiffness decreases with the application of the undrained cyclic loading.

Despite the stress-feedback control, deviator stress is not maintain at a constant value during the cyclic loading. This is due to the fact that under cyclic loading, sand loses its resistance, and deforms plastically. The loading device, controlled under pressure-volume conditions, is no longer able to achieve stress and velocity feedback control. But this material limitation underlines the fact that for lower saturation degree, the initial deviator stress reaches the 50kPa-setpoint value, while saturated one only reaches 40 kPa. It shows that the lower saturation degree, the higher cyclic shear resistance.

Comparing Figure 4 (a), (b) and (c) shows that initial stiffness is higher for lower initial saturation degree. Indeed, stress-strain curve's initial slope for unsaturated samples is greater than totally saturated one. Moreover, the lower the saturation degree, the tighter the stress-strain cycles, which shows that decreasing saturation degree decreases residual (plastic) strain between each cycle. But if stiffness decreases in each loading cycle for every sample, it appears that for important strain rates (from about 10%), unsaturated samples are more affected; indeed, the strain increment between each cycles becomes more important for unsaturated samples than for saturated one, for significant axial strain rates (Figure 3 (a), (b) and (c)). These results show that the presence of air, implicitly meaning positive suction, increases initial soil stiffness. But they also underlined that major plastic strain appeared for unsaturated samples, meaning that suction state is being modified during the loading.

DOI: 10.1051/e3sconf/20160914002

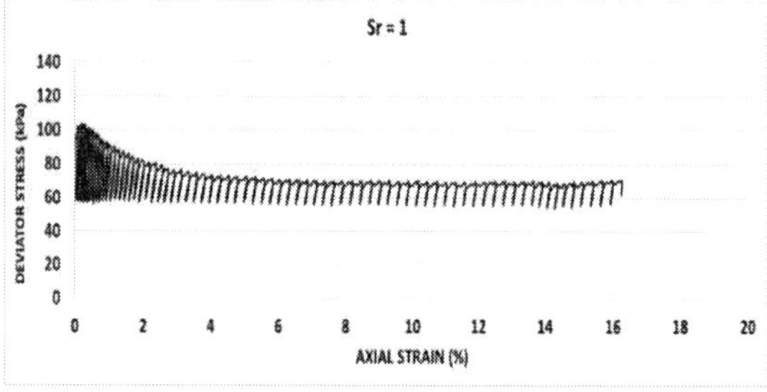

(a)

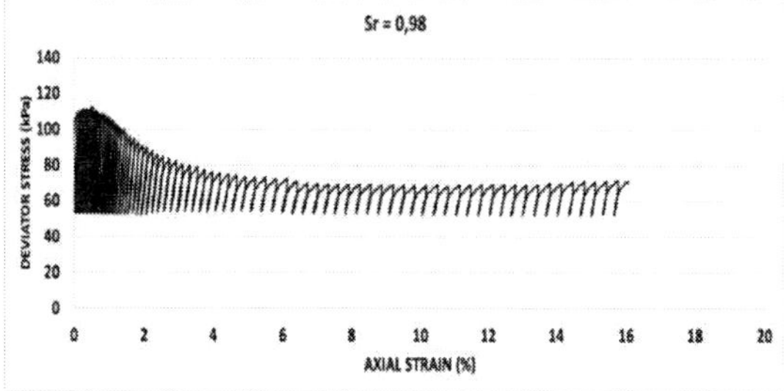

(b)

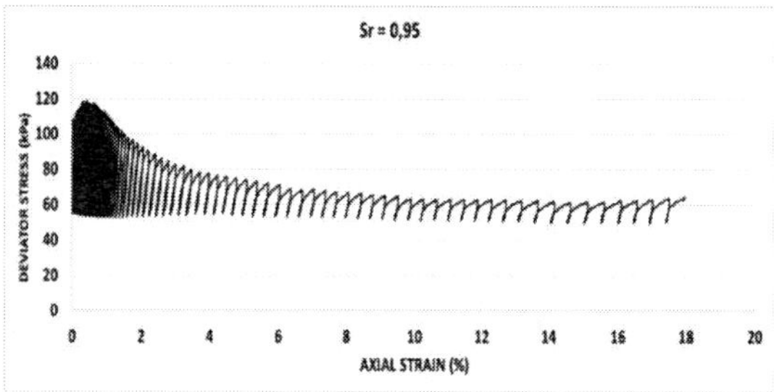

(c)

Figure 3. Stress-strain relationship for undrained cyclic triaxial tests

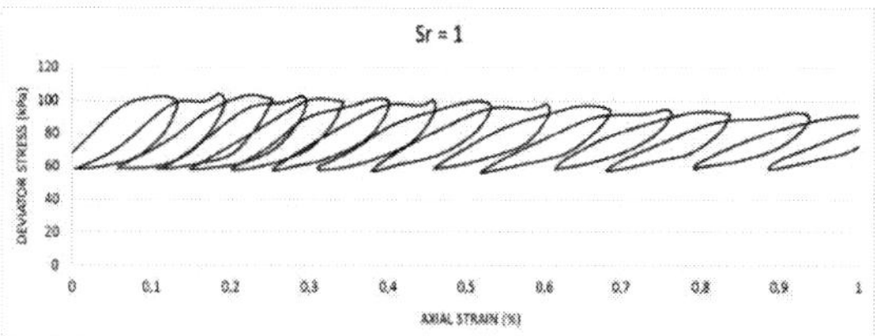

(a)

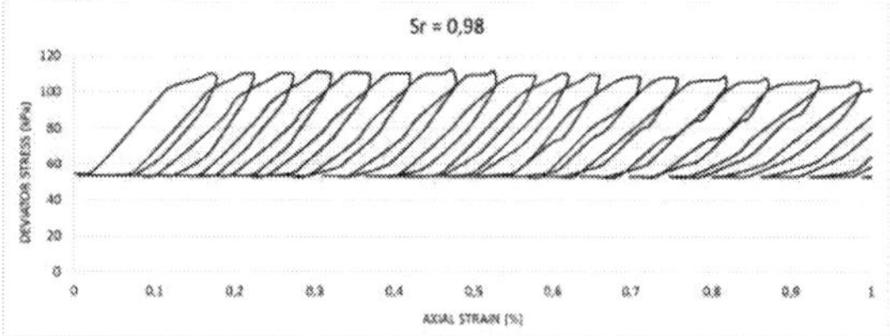

(b)

Figure 4. Focus on small strain behaviour

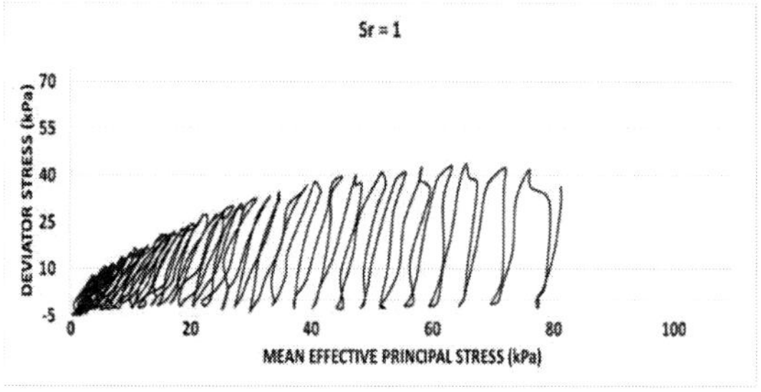

(a)

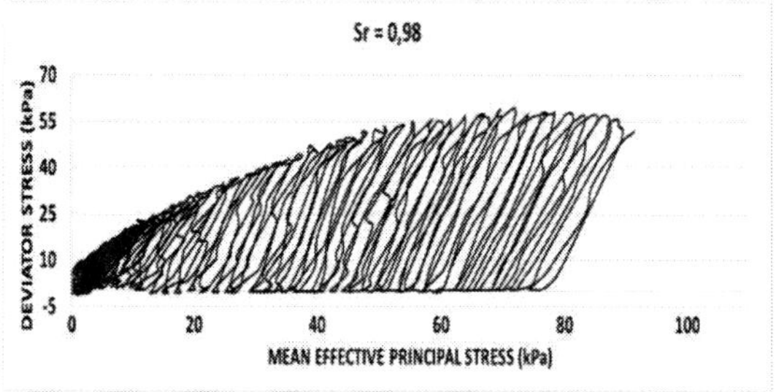

(b)

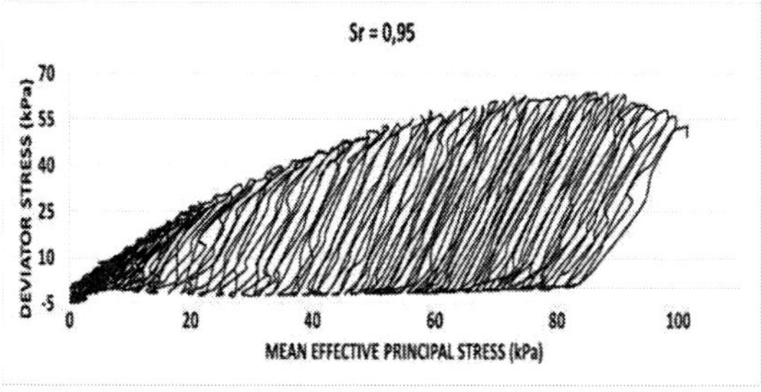

(c)

Figure 5. Effective stress paths for undrained cyclic triaxial tests

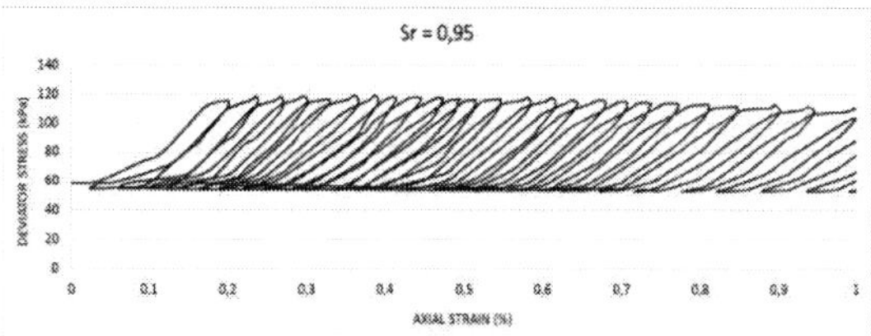

(c)

Comparing Figure 5 (a), (b) and (c) shows that unsaturated samples maintain a higher initial shear resistance level: totally saturated sample undergoes quick decreasing of main effective stress, with vertical effective stress path. The loss of mean effective stress between two cycles is less important for unsaturated samples.

However, for the three tested samples, the effective stress path reaches zero, which means that liquefaction occurred, disregarding initial saturation degree.

This paragraph showed influence of initial saturation degree on cyclic shear behaviour of triaxial sand samples. The lower initial saturation degree, the later the occurring of liquefaction, and development of important strain within the soil sample. It appeared that initial saturation degree equally has an impact on initial soil stiffness: unsaturated samples showed higher initial stiffness. But it was degraded under cyclic loading, and residual strain between each cycle becomes more important for unsaturated samples. Finally, it was pointed out that if the loss of effective stress between each cycles was less important for unsaturated sample than for saturated one, the effective stress path reached zero for all the tested samples. This may be caused by volumetric strain under cyclic loading, inducing increasing saturation degree, and leading to a state close to total saturation of the sample. This may be studied by measuring local volumetric strain during the loading.

It confirms that complex mechanisms are engaged in cyclic behaviour of unsaturated soils, and particularly suction. The influence of this state variable will be discussed in the next paragraph.

3 Influence of initial suction

Through previous results, it appeared that initial saturation degree affects the cyclic soils' behaviour. Various authors studied the effect of decreasing saturation degree on liquefaction behaviour, but level of saturation was quantified by the Skempton coefficient B. In this study, we suggest that using Sr would be more appropriate, in order to include suction values to our analysis.

3.1 Soil-water characteristic curve

The relation between Sr and suction is obtained by the soil-water characteristic curve (Figure 6). The experimental curve is obtained by the filter paper method (ASTM Standards [18]; Bicalho et al., [19]). The model curve arises from a numerical model developed by Aubertin et al. [20].

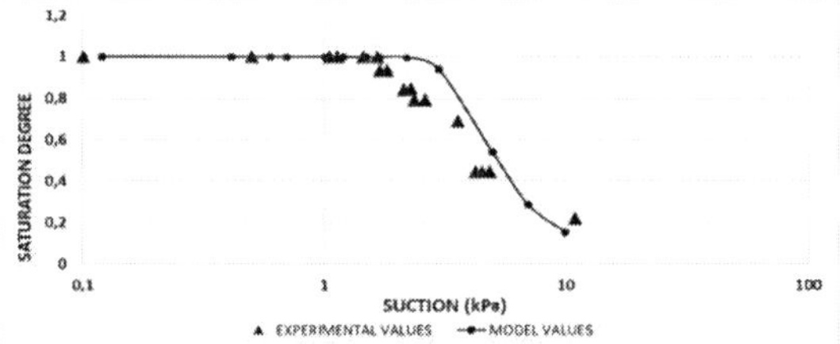

Figure 6. Soil-water characteristic curve – Experimental and model

Filter paper method is a simple and convenient method to determine the soil-water characteristic curve. But it is limited in terms of accuracy, arising from the filter paper boundaries. In this regard, it is noteworthy that experimental values are consistent with predictive model values. In parallel, simulation was realized using a Brooks and Corey law (Equation 1) [21]:

$$S_r = \left(\frac{s_e}{s}\right)^{\alpha} \qquad (1)$$

The simulation gave the following parameters: s_e = 2kPa, and α = 0.35.

Table 3 summarizes saturation degrees and corresponding suctions.

Table 3. Corresponding saturation degrees and suction obtained through soil-water characteristic curve (filtrer paper method)

Sr	Suction (kPa)
1	0 - 2
0.98	2
0.95	2,2

3.2 Discussion

The suction increases with decreasing saturation degree. Meanwhile, we know that suction increases soil stiffness. Relating those values with experimental results presented in previous paragraph, we confirm that initial soil stiffness increases with positive suction value. Moreover, initial suctions of each samples are extremely close from each other. It leads us to the conclusions that even small changes in suction state can induce significant modification in soil's behaviour.

The suction air entry value (SAEV) of the soil, obtained from the model is approximately 2 kPa, meaning that the tested samples have initial suction that bracket the air entry value of the soil. It is pertinent, since Unno et al. [14] have showed that SAEV stands for a critical value regarding liquefaction instability.

In his paper, Unno [13] showed that samples with initial suction lower than SAEV of the soil liquefied; while samples with initial suction greater than SAEV did not. Our results do not allow conclusions to be drawn regarding Unno's results. Further tests need to be done, under higher initial suction.

4 Conclusions and perspectives

Influence of saturation degree has been studied. Thanks to a model developed in our laboratory, we were able to quantify level of saturation in terms of Sr, and not only in terms of B like it is commonly done. We showed that:

- The higher the saturation degree, the faster liquefaction instability appear;
- The presence of air within the soil increases initial stiffness;
- Increasing of soil stiffness is due to suction, but the mechanism, especially evolution of this parameter under cyclic loading needs to be investigated;
- But liquefaction was observed for all tested samples, regardless the initial saturation degree.
- Regarding suction values of the tested samples, it was highlighted that very small changes in initial suction state lead to significant change in behaviour toward cyclic loading.

Through this work, we aimed to highlight the fact that liquefaction of unsaturated soil exist, contrary to what is commonly admitted. Lack of knowledge about the subject has leaded to a misunderstanding of mechanisms, and an unappropriated risks management in practice. This work stands for a solid basement in the overall project of liquefaction risks management, aiming to give a practical and convenient framework, and pertinent criteria to anticipate better liquefaction risks of all type of soils. Further tests will be realized, under lower saturation degree.

Acknowledgment

Authors want to thank the Association Universitaire de Génie Civil for their financial support, permitting us to attend the conference.

References

1. N.Y. Chang, Liquefaction susceptibility of fine-grained soils preliminary study report, Colorado University at Denver Dept of Civil and Urban Engineering (1987).
2. W.R. Hansen, Effect of the earthquake of March 27, 1964 at Anchorage, Alaska, U.S. Geological Survey Professional Paper (1965)
3. S. Yasuda, K. Harada, K. Ishikawa and Y. Kanemaru, Characteristics of liquefaction in Tokyo Bay area by the 2011 Great East Japan Earthquake, Soils and Foundations, **52**, 793-810 (2012).
4. M.A. Sherif, I. Ishibashi and C. Tsuchiya, Saturation effect on initial soil liquefaction, Journal of the Geotechnical Engineering Division, **103**, 914-917 (1977).
5. R.C. Chaney, Saturation effects on the cyclic strength of sands, Volume 1 of earthquake engineering and soil dynamics. Proceedings of the ASCE Geotechnical Engineering Division Specialty Conference, California (1978).
6. M. Okamura and Y. Soga, Effect of pore fluid compressibility on liquefaction resistance of partially saturated sand, Soils and Foundations, **46**, 695-700 (2006).
7. Y.J. Cui and P. Delage, Mécanique des sols non saturés, Techniques de l'ingénieur, 2000.
8. Y.J. Cui, Etude du comportement d'un limon compacté non saturé et de sa modélisation dans un cadre élasto-plastique, Thèse de Doctorat, ENPC (1993).
9. D.G. Fredlund ans H. Rahardjo, Soil mechanics for unsaturated soils (1993).
10. Y. Yoshimi, K. Tanaka and K. Tokimatsu, Liquefaction resistance of partially saturated sand, Soils and Foundations, **29**, 157-162 (1989).
11. A. Arab, I. Sharour and L. Lancelot, A laboratory study of liquefaction of partially saturated sand, Journal of Iberian Geology, **37**, 29-36 (2011).
12. P.V. Lade and S.B. Hernandez, Membrane penetration effects in undrained tests, Journal of the Geotechnical Engineering Division, **103**, 109-125 (1977).
13. T. Unno, M. Kazama, N. Sento and R. Uzuoka, Cyclic shear behavior of unsaturated volcanic sandy soil under various suction conditions, The 4th International Conference on Unsaturated Soils, Arizone, U.S.A. (2006).
14. T. Unno, M. Kazama, R. Uzuoka and N. Sento, Liquefaction of unsaturated sand considering the pore air pressure and volume compressibility of the soil particle skeleton, Soils and Foundations, **48**, 87-99 (2008).
15. J.C. Dupla, Application de la solicitation d'expansion de cavité cylindrique à l'évaluation des caractéristiques de liquéfaction d'un sable, Thèse de Doctorat, ENPC (1995).
16. N. Benahmed, Comportement mécanique d'un sable sus cisaillement monotone et cyclique : application aux phénomènes de liquéfaction et mobilité cyclique, Thèse de Doctorat, ENPC (2001).
17. A. W. Bishop and L. D. Wesley, A hydraulic triaxial apparatus for controlled stress path testing, Geotechnique, **25**, 657-670 (1975).
18. ASTM 1992. Standard test method for measurement of soil potential (suction) using filter paper. Annual Book of ASTM Standards, vol.15.09.
19. K.V. Bicalho, A. Gomes Correia, S. Ferreira, J.M. Fleureau and F.A.M. Marinho, Filter paper method of soil suction measurement, Panamerican Conference on Soil Mechanics and Geotechnical Engineering (2007).
20. M. Aubertin, M. Mbonimpa, B. Buissière and R.P. Chapuis, A model to predict the water retention curve from basic geotechnical properties, Canadian Geotechnical Journal, **40**, 1104-1122 (2003).
21. R.H. Brooks and A.T. Corey, Hydraulic properties of porous media and their relation to drainage design, Transactions of the ASAE (1964)

Effects of loading frequency on cyclic parameters of unsaturated Zenoz kaolin

Mohsen Mojezi[1,a], Mahnoosh Biglari[2], and Mohammad Kazem Jafari[1]

[1]*International Institute of Earthquake Engineering and Seismology (IIEES), Tehran, Iran*
[2]*Razi University, Kermanshah, Iran*

Abstract. According to the considerable effects of the loading frequency on the cyclic behaviour of the soils and lack of the sufficient and precise data in this field for the unsaturated soils, more attention needs to be paid to the experimental investigations in various loading paths and frequencies. Meanwhile the unsaturated cyclic tests require advanced equipment and need more precision and time. In this paper, the results of two groups of tests which were performed on an unsaturated fine grained soil in the same initial condition are presented and compared during specific stress paths (pre-cycles of loading and during cycles). The tests are performed using the same suction-controlled cyclic triaxial apparatus in the same stress path before loading cycles, but two conditions of the cyclic loading frequencies (slow cyclic tests with the constant suction condition during the cycles and rapid cyclic tests with the unsteady suction condition during the cyclic path). The results show that increase in the loading frequency leads to the considerable increase in the shear modulus of the samples in the same strain levels.

1 Introduction

The loading frequency (the cyclic loading rate) could be subjected to the machine foundation cases with variable loading rate. The fast cyclic loading rate can be unequalised in the suction distribution. Hence, understanding the effect of the loading rate is useful for the precise interpretation of the behaviour.

Several researches have been conducted for investigating the dynamic parameters of the unsaturated soils, especially small-strain shear stiffness (e. g. [1] to [8]).

In spite of the various studies on the cyclic and dynamic behaviours of the unsaturated materials in the small strain levels, the experimental data in the medium to large strain levels of the unsaturated materials are limited. For the first time, [9] studied the shear modulus and damping ratio of the unsaturated Zenoz kaolin materials using the new suction controlled triaxial apparatus at the medium to the large strains in a constant suction during the slow loading rate cycles. Since, the effects of cyclic loading frequency on the cyclic parameters are quite unexplored, recently, [10] studied the cyclic parameters by the same apparatus on the same material at the higher cyclic rate.

In this paper the results of two groups of the tests are compared. These tests have been performed on the unsaturated Zenoz kaolin under the identical stress paths including (i) equalization (ii) isotropic consolidation (iii) slow or fast rate cyclic loading.

2 Testing Apparatus and materials

In this research the suction-controlled cyclic triaxial apparatus developed by [9] in International Institute of Earthquake Engineering and Seismology (IIEES) was used (Figure1) (More detailed data about the apparatus could be found in [9]). Suction is applied and controlled by Axis Translation Technique via HAVEdiscs from top and bottom of the specimens in suction controlled triaxial cyclic device.

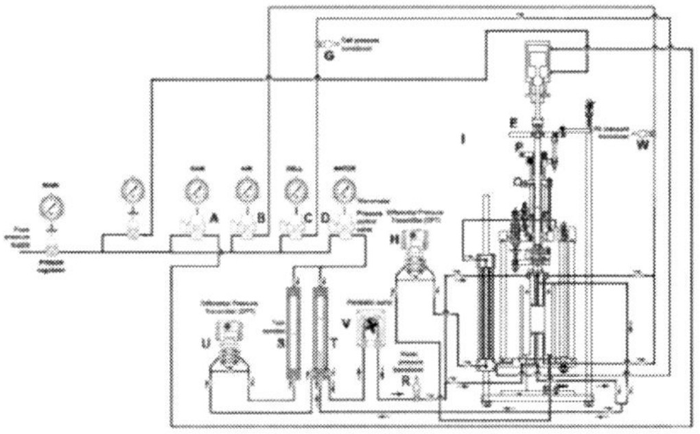

Figure 1. Suction controlled cyclic triaxial device [9]

In this research Zenoz kaolin which is a commercial fine-grained soil was studied in the unsaturated condition. The physical specifications of this soil are presented in Table 1.

[a] Corresponding author: m.mojezi@iiees.ac.ir

Table 1. Physical specification of the tested soil

Classification (USCS)	CL
Clay Fraction (%)	18
Silt Fraction (%)	60
Plasticity Index: PI (%)	12
Liquid Limit: LL (%)	29
Plastic Limit: PL (%)	17
Specific Gravity: G_S	2.65
Optimum Water Content: ω_{opt} (%)	15.4
Maximum Dry Unit Weight: $\gamma_{d(max)}$ (g/cm^3)	1.74
Initial Void Ratio: e_0	1.19
Initial Degree of Saturation: Sr_0 (%)	27

To achieve the normally consolidated condition, the samples were compacted with 11.9% of water content (3.5% less than the optimum water content) and dry unit weight of 12 kN/m3 (69% of the maximum dry unit weight, obtained by the standard proctor test).The specimens were compacted using under compaction method proposed by [11], in seven layers with diameter of 38 mm and final height of 76 mm.

3 Stress Paths

In the both series of tests, the samples were equalized in the suction levels of 0, 150 and 300 kPa, and then were isotropically consolidated up to the mean net stress of 100 kPa. Finally, the specimens were subjected to two-way deviatoric cyclic loading of 18, 42 and 81 kPa in each suction level. In the slow group of the tests (i.e. [9]), in order to keep the suction condition constant during the loading cycles, the cycles were performed with very low rate (8 kPa/hr: i.e. 3.09E-5 Hz for the deviatoric stress amplitude of 18 kPa, 1.32E-5 for the deviatoric stress amplitude of 42 kPa, and 6.86E-6 for the deviatoric stress amplitude of 81 kPa). In the higher cyclic loading rate group of the tests ([10]) the cycles were done with the frequency of 0.003 Hz while the water content was kept constant (consequently, the suction is not constant during the cyclic loading).
The stress paths of the both groups of the performed tests are schematically plotted in Figure 2. In this paper the tests are named as a term like "S-number-D-number", in which the number after "S" refers to the suction level and the number after "D" refers to the deviatoric stress amplitude.

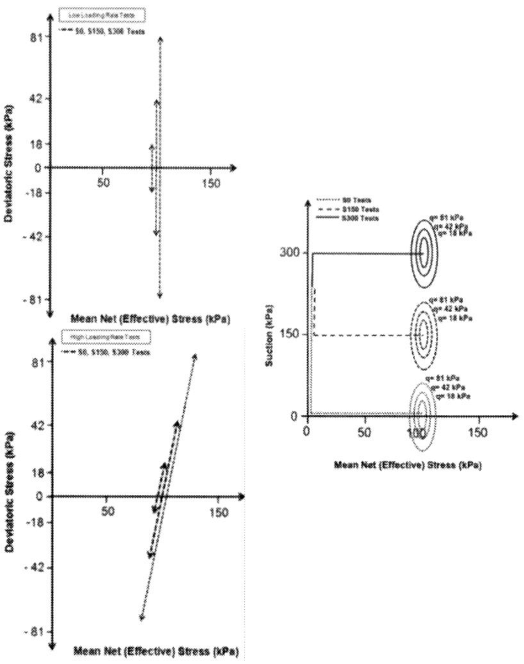

Figure 2. Stress paths of the low and high cyclic rate tests

In the slow rate tests, the loading speed was kept in a way that the suction equalization condition of the samples was not changed during all of the stress paths.
In the high rate tests, however, the suction equalization condition is not maintained during the cyclic loading(due to higher loading frequency), hence a separate sample with the same pre-cyclic condition was tested for each amplitude of the deviatoric stresses (i.e.18, 42 and 81 kPa) and each suction levels (i.e. 0, 150 and 300 kPa).

4 Results

Following the changes of two parameters, specific volume (v= 1+e) and specific water volume (v_w= 1+S_r.e), are presented versus time and mean net stress (p-u_a).
Figures 3 and 4 show the specific water volume changes (dv_w) versus time for the tests for the suction of 150 kPa and 300 kPa in the slow and high rate groups during the initial suction equalization process, respectively. According to these results, in both groups, the samples which were subjected to the suction level of 150 kPa absorbed water whereas the samples with the suction level of 300 kPa lost water. It means that the compacted specimen has initial suction level higher than 150 kPa and lower than 300 kPa. Initial suction of these ideal samples were measured by [6] at 240 kPa.

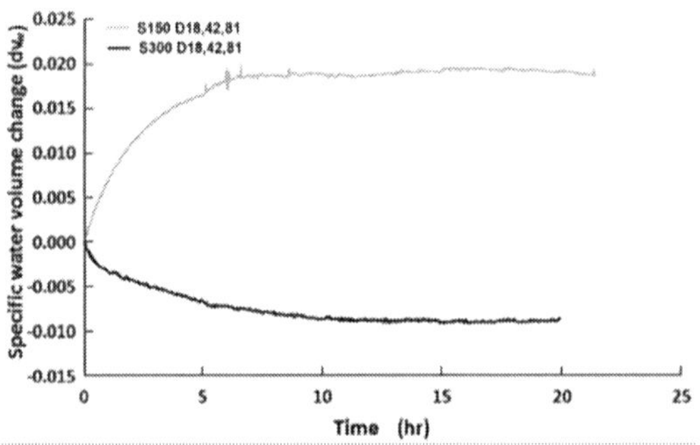

Figure 3. Water volume changes (dv_w) during initial suction equalization process, in the slow rate tests [9]

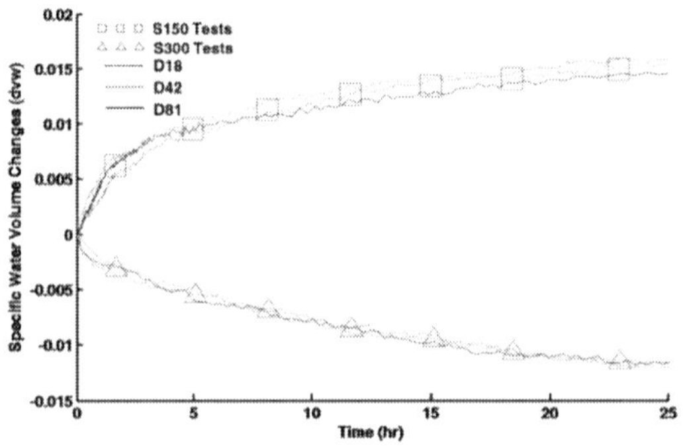

Figure 4. Water volume changes (dv_w) during initial suction equalization process, in the high rate tests [10]

In Figures 5 and 6 the specific volume changes (dv) of the samples are plotted versus the time in the suction levels of 150 kPa and 300 kPa for the slow and the high rate tests, respectively. As seen in these figures all of the samples lost the volume due to compaction for the suction level of 300 kPa and collapse for the suction level of 150 kPa during the equalization process.

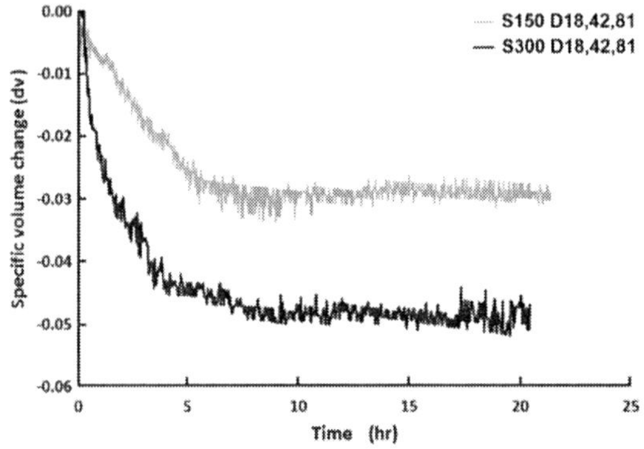

Figure 5. Specific volume changes (dv) versus time in the slow rate tests [9]

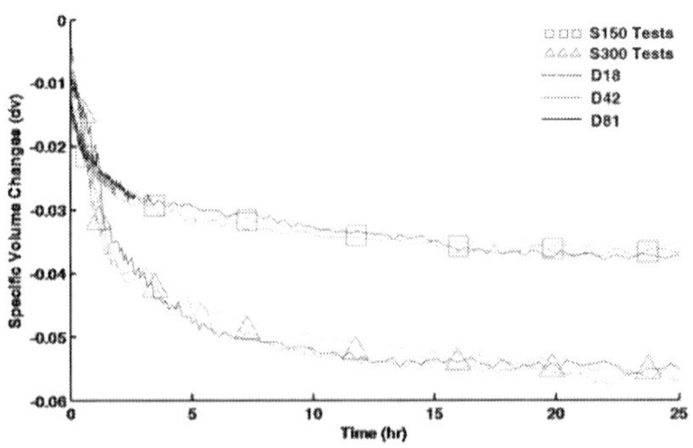

Figure 6. Specific volume changes (dv) versus time in the fast rate tests [10]

The specific volume variations of the specimens (v=1+e) versus the mean net stress ($p-u_a$) are presented in Figures 7 and 8 for the slow and the high rate tests, respectively.

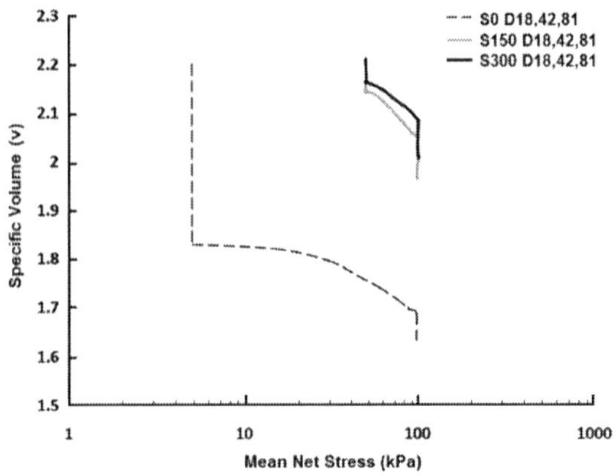

Figure 7. Specific volume (v) versus Mean net stress, during the slow tests [9]

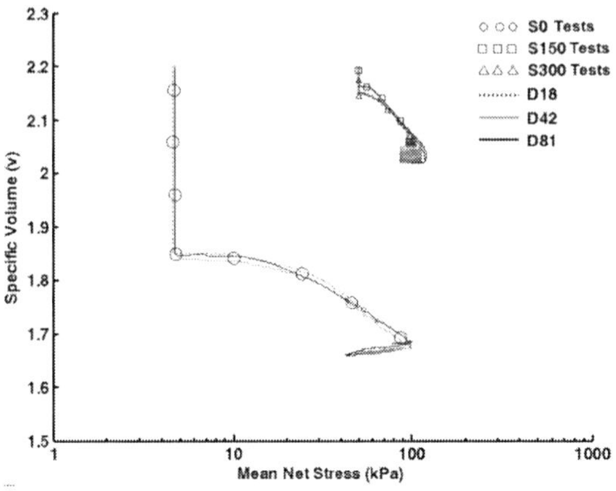

Figure 8. Specific volume (v) versus Mean net stress, during the fast tests

In both test groups, after the mean net stresses of 100 kPa, the cyclic deviatoric stress loading was applied up to the specific amplitudes of 18, 42, and 81 kPa. The cyclic loadings were applied with the rate of 8 kPa/hr in the

slow rate group and the frequency of 0.003 Hz in the high rate tests. In the slow rate group due to very time consuming (e. g. almost 40 hours for the deviatoric amplitude of 81 kPa/hr), 5 cycles were performed but in the high rate tests the samples were subjected to 40 cycles in each amplitude of the deviatoric stress for each suction level. Although in some of the high rate tests due to the failure which were occurred for the samples during the cyclic loading (especially for the higher amplitudes), which the tests were stopped in the less cycle numbers.
The compatibility of the volumetric results during the equalization and the isotropic compaction process show that the specimens in the both groups are identical. At deviatoric cyclic loading phase, due to increase in the pore water pressure in the high rate test group, similar to the saturated liquefaction evaluation tests the mean net stress decrease slightly in the zero suction levels.
The variations of the specific water volume (v_w) of the samples versus the mean net stress ($p-u_a$) are shown in Figure 9 for the slow rate tests and in Figure 10 for the higher rate tests. In the zero suction level tests (saturated tests), v_w was increased during equalization process due to the larger amount of the water which were entered the samples and then decreased due to the isotropic consolidation process. In the tests with suction level of 150 kPa, the samples were absorbed a little amount of water and vw was slightly increased up to achieve the constant suction of 150 kPa and was approximately remained constant. For 300 kPa suction the trend is reverse which is happened due to the increase in the initial suction level.

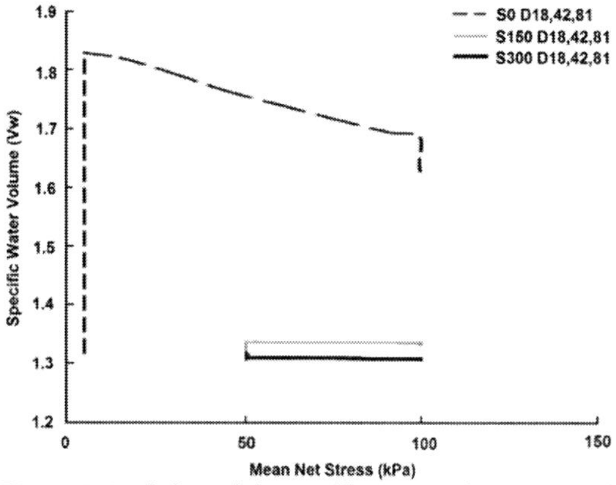

Figure 9. Variation of the specific water volume versus the mean net stress in the slow rate tests [9]

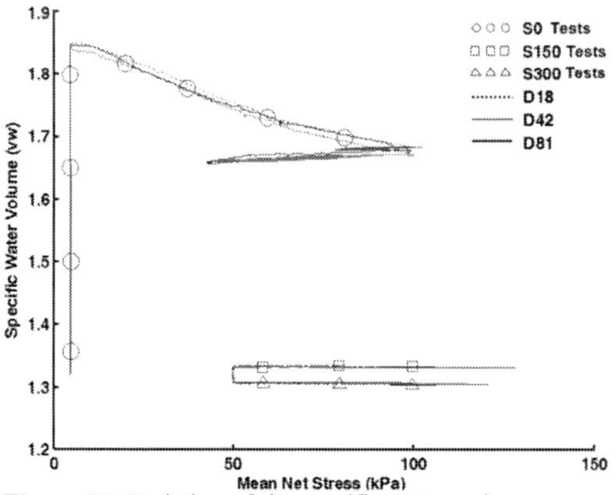

Figure 10. Variation of the specific water volume versus the mean net stress in the fast rate tests

The variations of the shear modulus (G) versus the shear strain are presented in Figures 11 and 12 for the slow rate and the high rate tests, respectively. The variations of the damping ratio (D) for the slow and the high cyclic loadings are shown in Figures 13 and 14, respectively. The values related to the slow rate tests are presented for 1, 2, 3, 4, 5 cycle numbers and the results of the high rate tests are presented for 1, 3, 5, 10, 15, 20, 25, 30, 35, 40 cycle numbers in the various suction levels and the various deviatoric stress amplitudes. The values of the first cycle are connected to each other by the dotted line for the saturated tests, the grey line for the suction level of 150 kPa, and the black line for the suction level of 300 kPa.

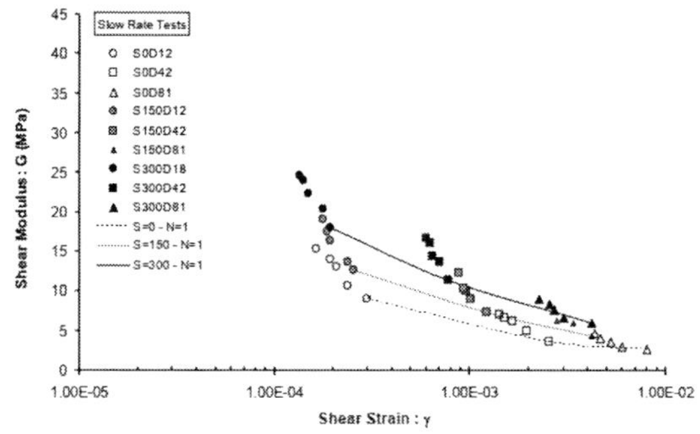

Figure 11. Variations of the shear modulus (G) versus the shear strain in the slow rate tests [9]

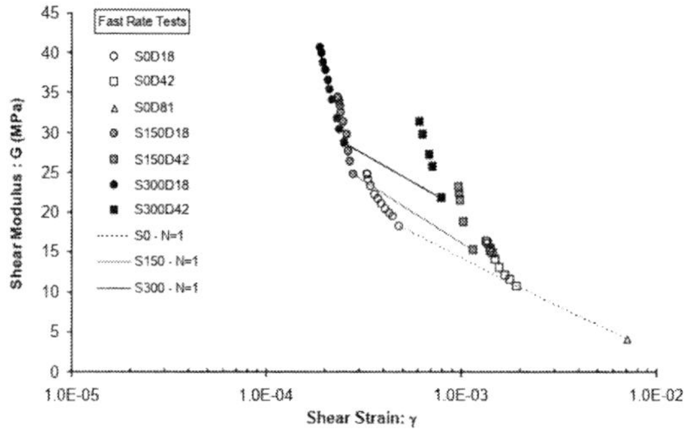

Figure 12. Variations of the shear modulus (G) versus the shear strain in the fast rate tests [10]

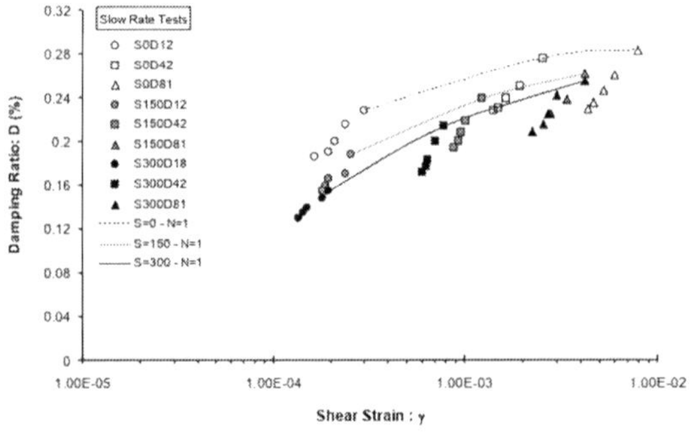

Figure 13. Variations of the damping ratio (D) versus the shear strain in the slow rate tests [9]

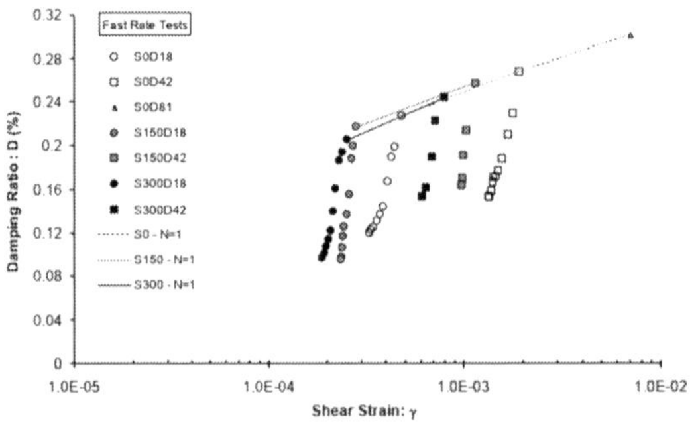

Figure 14. Variations of the damping ratio (D) versus the shear strain in the fast rate tests [10]

According to the Figures 11 and 12, in the same strain level and cycle's number, the increase in the suction level has been led to the increase in the shear modulus. On the contrary, Figures 13 and 14 show decrease in the damping ratio due to the increase in the suction level in same strain level and cycle's number.

Also, the increase in the number of cycles in all of the suction levels induces the sample stiffness and is followed by the shear modulus increase and the damping ratio decrease. This issue is similar for both groups of the tests but the trend is sharper in the higher rate tests.

On the other hand, the comparison between the presented results in Figures 11 and 12 is shown that in the same strain levels, the shear modulus of the high rate tests significantly are more than the slow ones. However, the increase in the rate of the cyclic loading could has led to this increase, it would be happened by uncontrolled variation in the suction due to the high rate of the cyclic loading.

Meanwhile, the results of the damping ratio in the both groups of the tests do not show a considerable different.

5 Conclusions

The effects of the loading rate on the cyclic parameters of the unsaturated Zenoz kaolin are studied.

The results showed the increase in the suction level has been led to increase in the shear modulus and decrease in the damping ratio due to the virtual cohesion which has been induced by the suction. Furthermore, increase in the cycle's number has been resulted in the shear modulus increase and the damping ratio decrease for all of the suction levels and loading rates.

In summary, the cyclic loading rate variation were significantly changed the cyclic parameters of the fine grained unsaturated soils. Hence, in the same suction level and equal strain, the shear modulus of the higher rate tests were more than the slow ones. Meanwhile, there were not observed considerable differences in damping ratio. Hence, the loading frequency effects need more consideration in modelling of the cyclic behaviour of unsaturated.

References

1. EAM. Marinho, R.J. Chandler, M.S. Crilly. The 1st Intern. Conf. on Unsat. Soils, Paris; p. 535–539 (1995).
2. C. Mancuso, R. Vassallo, A. d'Onofrio. Can Geotech J; 39(1):22–31 (2002).
3. R. Vassallo, C. Mancuso, F. Vinale. Can Geotech J; 44(4):447–462 (2007).
4. C.W.W. Ng, S.Y. Yung. Géotechnique; 58(1):23–35 (2008).
5. M. Biglari, C. Mancuso, A. d'Onofrio, M.K. Jafari, A. Shafiee. Comput. and Geotech.; 38(5): 709-720 (2011).
6. M. Biglari, A. d'Onofrio, C. Mancuso, M.K. Jafari, A. Shafiee, I. Ashayeri. Can. Geotech. J.; 49(3): 311-322 (2012).
7. A. Heitor, B. Indraratna, C. Rujikiatkamjorn. Can. Geotech. J.; 50(2):179-188 (2013).
8. K.S. Wong, D. Mašín, C.W.W. Ng. Comput. and Geotech.; 56:28-39 (2014).
9. M. Biglari, M.K. Jafari, A. Shafiee, C. Mancuso, A. d'Onofrio. Geotech. Test. J; 34(5):525-536 (2011).
10. M. Mojezi, M.K. Jafari, M. Biglari, Int. J. Civil Eng.; 13(3) (Transaction B: Geotech. Eng.)(2015)
11. R.S. Ladd. Geotech. Test. J.;; 1(1):16–23 (1978).

Effect of very high compaction pressures on the physical and mechanical properties of earthen materials

Agostino Walter Bruno[1], Domenico Gallipoli[1], Céline Perlot[1] and Joao Mendes[1]

[1]*Université de Pau et des Pays de l 'Adour, Laboratoire SIAME, 64600 Anglet, France*

Abstract. The paper compares the effect of different compaction procedures on the porosity and mechanical characteristics of earthen materials. In a first series of tests, a very high static pressure, up to 100 MPa, is applied to the soil for a sufficient period of time to allow consolidation (hyper-compaction). This method produces materials with very high densities up to 2300 kg/m^3 and good mechanical properties, which are suitable for the construction of masonry structures. In a second series of tests, the soil is compacted according to the standard Proctor method, which is the reference compaction procedure for the design of geotechnical fills such as dams and embankments. In the Proctor method, the soil is dynamically compressed by means of a much lower effort compared to hyper-compaction. All specimens are equalized under identical hygro-thermal conditions and subjected to unconfined compression tests to measure stiffness and strength. It is shown that hyper-compacted specimens exhibit physical and mechanical properties that are comparable with those of traditional building materials for masonry structures. Finally, mercury intrusion porosimetry tests are performed to study the effect of the compaction method on the pore structure of the material.

1 Introduction

Earthen materials are widely used in civil engineering. Dams and embankments are among the most common earthen structures but dwellings made of raw earth are also spread worldwide. In all cases, the compaction process represents a crucial stage during material manufacturing. Compaction also affects the hydraulic behaviour by changing the porosity and fabric of the earthen material.

Current compaction methods produce mechanical properties that are weaker than those of conventional materials for masonry construction (e.g. fired bricks, stabilized earth or concrete blocks). These methods must therefore be significantly enhanced if unstabilized earthen materials are to be widely used in the construction of large buildings.

A considerable number of studies have analysed the influence of compaction effort on the mechanical properties of earthen materials [1-5]. These studies agree that a higher compaction effort increases the dry density and, consequently, the stiffness and strength of the material. These studies also show that, for a given compaction effort, there is an optimum value of water content for which the highest density and the best mechanical properties are achieved.

Past studies have focused on the effect of compaction on the porosity and fabric of fine-grained soils, such as silty clays [6] or swelling clays [7], which are most relevant to geotechnical applications.

Relatively few studies however exist about the compaction of coarser soils used in the construction of raw earth dwellings.

In this context, the present work investigates the effect of different compaction methods on both the mechanical and microstructural properties of a relatively coarse soil suitable for raw earth construction. Two compaction methods are compared, namely the standard Proctor method, which represents a norm in geotechnical construction, and an innovative static compaction method applying very high levels of pressure (hyper-compaction). As expected, the stiffness and strength of the material vary widely depending on the chosen compaction procedure, which confirms that material properties can be tailored to different applications by selecting an appropriate compaction procedure. For example, hyper-compaction generates mechanical properties that are comparable with those of conventional building materials. The effect of compaction procedure on pore fabric is also analysed by means of mercury intrusion porosimetry tests. Results from these tests confirm that, when compaction effort increases, the porosity reduces and the mechanical properties of the material improve.

Finally, it is observed that a higher compaction effort reduces the variability of physical and mechanical properties between specimens, thus facilitating the quality control of material properties.

[a] Corresponding author: agostinowalter.bruno@univ-pau.fr

DOI: 10.1051/e3sconf/20160914004

2 Material and methods

2.1 Material

The soil used in the present work has been provided by a brickwork factory from the region of Toulouse in France. Table 1 shows some of the relevant properties of the soil. The grain size distribution has been determined by both wet sieving and sedimentation in compliance with the norms XP P94-041 [8] and NF P 94-057 [9]. The plasticity properties of the fine fraction (i.e. fraction smaller than 400 μm) have been measured in agreement with the norm NF P94-051 [10]. In particular, the liquid limit, plastic limit and plasticity index have been determined as the average of four independent tests. The specific gravity of solid particles has been obtained by means of the pycnometer test according to the norm NF P 94-054 [11].

Bruno et al. [12] show that the grain size distribution and the plasticity properties of the material used in this work satisfy the requirements for raw earth construction [13-16].

Clay activity, defined as the ratio between the plasticity index and the clay fraction (i.e. the fraction smaller than 2 μm), is equal to 0.79. This classifies the clay fraction as normally active [17], which is consistent with mineralogy information from the soil provider that indicates a predominantly illitic material with a small quantity of montmorillonite. Illite is a three-layers clay with good bonding characteristics and a limited swelling potential upon wetting, which makes it particularly suited to raw earth construction [18].

2.2 Compaction procedure

Hyper-compacted samples were statically compressed at pressure levels of 25 MPa, 50 MPa and 100 MPa. The lowest pressure level of 25 MPa is comparable to that of the most powerful presses currently available on the market for the manufacture of compressed earth blocks. The other two pressure levels (chosen according to a geometrical progression with ratio of two) are, to the authors' knowledge, the highest ever applied during production of medium-scale compressed earth samples.

Prior to compaction, 500 grams of dry soil were mixed with the desired amount of water by using an electrical planetary mixer for at least 15 minutes. This time is sufficient to ensure a good distribution of moisture throughout the soil [5]. The moist soil was subsequently placed inside two plastic bags to prevent evaporation and left to equalize for at least one day so that moisture could redistribute across the soil. After this, the soil was placed inside a cylindrical mould with a diameter of 50 mm and compressed to the required pressure by using a load-controlled Zwick press with a capacity of 250 kN. Pressure was applied to the soil by two cylindrical aluminium pistons acting at the top and bottom of the specimen (Fig. 1). This double-piston compression action reduces the friction between the mould and the sample, thus increasing the uniformity of stresses inside the soil. A finely perforated aluminium disk and filter papers were placed between the top and bottom surfaces of the specimens and the respective pistons to facilitate drainage of pore air/water from the soil during compaction. Also, eight longitudinal fine grooves were cut along the lateral surfaces of the two pistons to create a preferential path for drainage at the interface between the outer surface of the pistons and the inner surface of the mould. More details about this compaction technique are available in Bruno et al. [12].

After compaction, specimens were cut down to a height of 100 mm by trimming the excess soil from the top and bottom extremities. The water content of the trimmed soil was measured by drying at 105°C until attainment of constant weight according to norm NF P 94-050 [19]. The water content of the specimen was then taken as the average of the top and bottom values. The water contents measured from the top and bottom trimmings were very similar suggesting that distribution moisture is uniform across the entire specimen.

Proctor compacted samples were obtained in compliance with the norm NF P 94-093 [20]. A fixed mass of 2250 gr of dry soil was mixed with the desired amount of water by means of a planetary mixer for at least 15 minutes and then stored for at least one day in two plastic bags. The moist soil was subsequently compacted in a standard Proctor mould in three layers.

Table 1. Material properties.

Grain size distribution		
Gravel	> 2 mm	0.4 %
Sand	0.063 – 2 mm	40.4 %
Silt	0.002 – 0.063 mm	42.9 %
Clay	< 0.002 mm	16.3 %
Plasticity properties		
Liquid limit, w_L		33.0 %
Plastic limit, w_P		20.1 %
Plasticity index, I_P		12.9 %
Activity A		0.79
Specific gravity of soil solids		
G_s		2.66

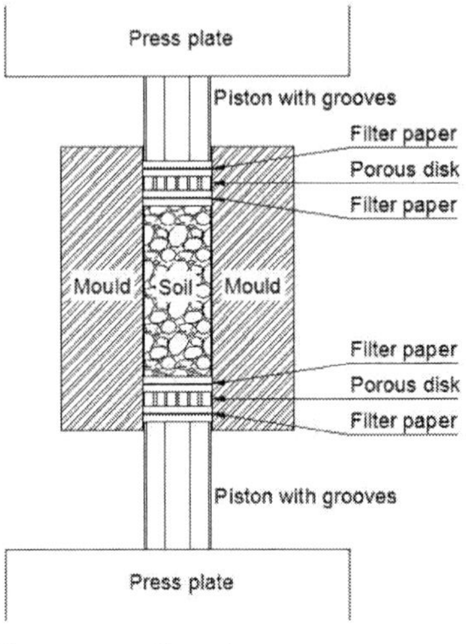

Figure 1. Hyper-compaction set-up

Each layer was compacted by 25 blows of a 2.490 kg hammer falling from a fixed height of 305 mm. After compaction, a cylindrical specimen of 50 mm diameter was cored from the larger Proctor sample. The cored specimen was then cut down to a height of 100 mm by trimming the excess soil from the top and bottom extremities. The remaining soil from the larger Proctor sample was used to determine the water content. Three samples of about 50 gr each were taken at three different heights of the initial Proctor sample and dried at 105°C until weight became constant [19]. The water content was then determined as the average of these three measurements.

For each specimen, three measurements of diameter were taken at different heights and three measurements of height were taken at different angles. The volume of the sample was then calculated from the average values of diameter and height. The mass of the sample was finally measured by using a scale with a resolution of 0.01 g. Based on the measured values of mass, water content, volume and specific gravity, it was possible to calculate bulk density, dry density, porosity and degree of saturation.

3 Results

3.1 Compaction curves

For each compaction level, the experimental values of dry density against the corresponding water contents are plotted in Fig. 2 together with the respective interpolating curves. For each compaction curve, the highest dry density corresponds to the optimum value of water content. The optimum water content becomes progressively smaller, while the corresponding density becomes progressively larger, as the compressive energy increases from Proctor compaction to static compaction at 25 MPa, 50 MPa and 100 MPa. This means that, as the effort increases, the compaction curve shift towards the theoretical point in which all porosity is erased and the dry density becomes equal to the density of the soil particles. The optimum dry density increases less than linearly with compaction pressure, i.e. the increase in dry density from 25 MPa to 50 MPa is greater than the increase in dry density from 50MPa to 100MPa. It would therefore be necessary to apply an unfeasibly high pressure to attain the theoretical "no porosity" point.

Prior to mechanical testing, the compacted specimens were equalized inside a climatic chamber at a temperature of 25°C and relative humidity of 62%. Equalization took about 15 days and was considered complete when the specimen mass changed less than 0.1% over at least one week. This equalisation stage prior to mechanical testing was considered essential to avoid any influence of different hygro-thermal conditions on the measured mechanical properties [21; 22].
During equalization, all samples experienced desaturation and shrinkage as water content reduced to about 3.5% and dry density increased, especially for the wetter samples. The specimens compacted at the highest pressure of 100 MPa showed very similar values of dry density at the end of equalization (Fig. 3). This suggests that application of a high compaction pressure reduces the dependency of the material properties measured after equalization on the compaction water content, thus resulting in better quality control of the final product. On the contrary, samples manufactured at smaller pressures show, after equalization, variable values of dry densities depending on the water content at the time of compaction.

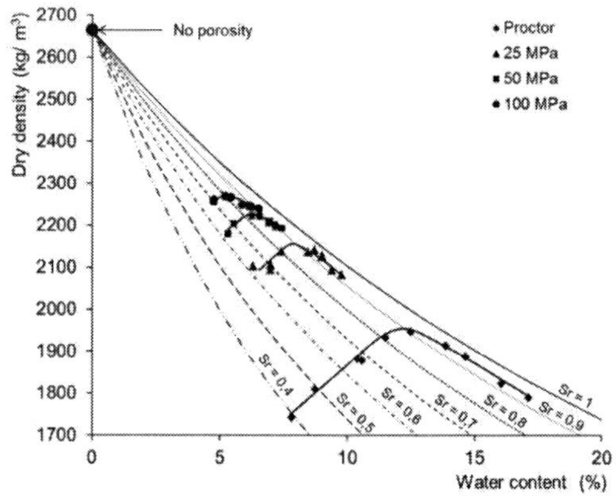

Figure 2. Comparison between standard Proctor compaction curve and hyper-compaction curves at 25 MPa, 50 MPa and 100 MPa.

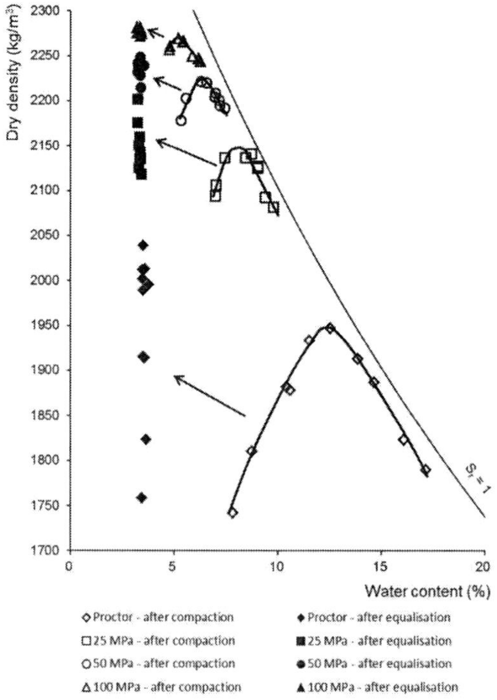

Figure 3. Change in dry density and water content during equalization.

3.2 Measurement of stiffness

Young modulus was measured by performing five unconfined loading-unloading cycles, at a loading rate of 0.005 MPa/s, between one ninth and one third of the estimated compressive strength (strength was estimated as the average value from two tests on randomly chosen samples for each compaction level). Axial displacements were measured between two points at a distance of 50 mm by means of two transducers placed on diametrically opposite sides of the specimen.

Due to the hysteretic response of the material during the loading-unloading cycles [5], the Young modulus was determined as the average slope of the unloading branches of the five cycles. This was based on the assumption that material behaviour is elasto-plastic during loading but essentially elastic during unloading.

Fig. 4 shows the values of Young modulus plotted against dry density for all specimens. Note that the values of dry density in Fig.4 are those after equalisation (and hence before testing), which are shown in Fig. 3. Inspection of Fig.4 confirms the inadequacy of the Proctor compaction method to manufacture earthen materials with sufficient stiffness to be used in masonry construction. Conversely, the statically compacted specimens showed much higher values of Young modulus with a variation of one order of magnitude between the specimens compacted at 100 MPa and those compacted according to Proctor.

Interestingly, Young modulus grows more than linearly with increasing dry density. Therefore, any small increase of dry density beyond the current maximum value of 2280 kg/m^3 would produce a significant augmentation of Young modulus. Of course, the dry density of the compacted specimens cannot be higher than the density of the soil particles (corresponding to the "no porosity" point in Fig. 2). In addition, the attainment of this theoretical limit is problematic from a practical point of view as previously discussed.

3.3 Measurement of compressive strength

The same specimens tested to measure Young modulus were subsequently loaded to failure to measure the unconfined compressive strength.

The accuracy of strength measurements can be affected by sample slenderness and friction with press plates [23, 24]. In this study, the slenderness ratio was equal to two (i.e. cylindrical specimens with height of 100 mm and diameter of 50 mm were used), which is considered sufficient to avoid measurements errors. In addition, Teflon spray was employed to reduce friction between the press plates and the specimen top and bottom surfaces. To avoid any dynamic effect, all tests were performed with a displacement rate of 0.001 mm/s, which was the slowest rate that could be applied by the press.

Fig. 5 shows the variation of the measured peak strength with dry density after equalization. Similar to the measurement of Young modulus, compressive strength increases more than linearly with increasing dry density. Thus, a small increase of dry density beyond the maximum value measured in this study could lead to a significant augmentation of compressive strength. Bruno et al. [12] showed that the compressive strength of specimens compacted at a pressure of 100 MPa is already comparable with that of conventional masonry materials such as stabilised earth blocks and fired earth bricks.

3.4 Mercury intrusion porosimetry tests

After the measurement of compressive strength, small fragments of about 4 grams were taken from the failed specimens to perform mercury intrusion porosimetry (MIP) tests. These small specimens were re-equalised for a week at a temperature of 25°C and a relative humidity of 62% to avoid any potential change of material fabric caused by the environmental conditions inside the laboratory during testing. The small specimens were subsequently freeze-dried [25, 26] to remove all pore water without altering material fabric.

Freeze-drying consisted in dipping the specimens in liquid nitrogen (-196°C) until boiling finished and then exposing them to vacuum at -50°C of temperature for at least two days in order to sublimate the frozen pore water. After this, specimens were subjected to MIP tests and the results were interpreted according to Washburn equation :

$$\Delta P = 4\,\gamma\, cos\theta\,/d_{pore} \quad (1)$$

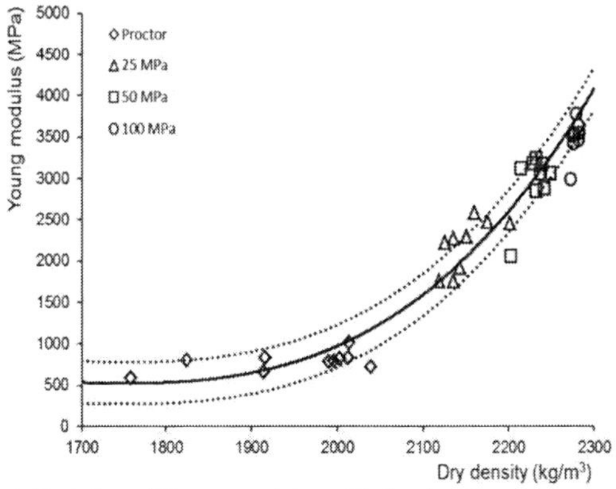

Figure 4. Variation of Young modulus with dry density

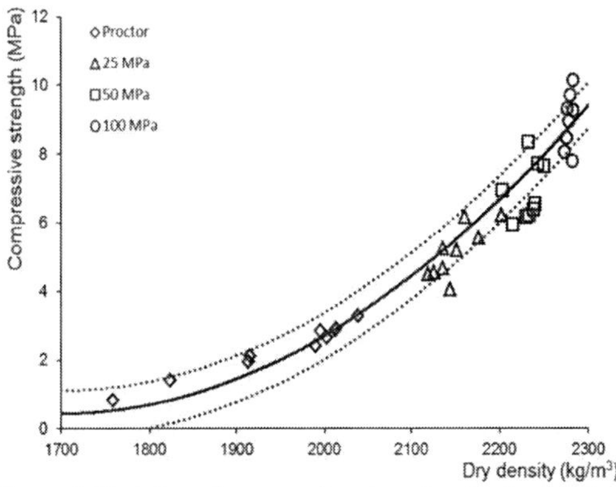

Figure 5. Variation of compressive strength with dry density

DOI: 10.1051/e3sconf/20160914004

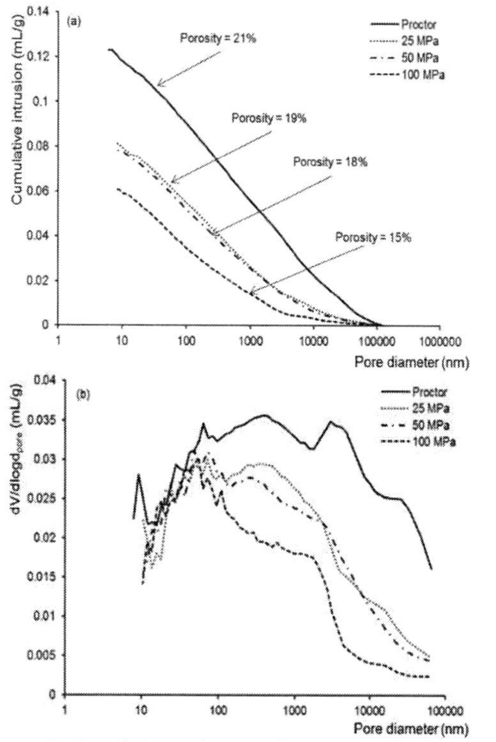

Figure 6. Cumulative volume and pore size distribution : comparison Proctor standard and high-pressure compaction.

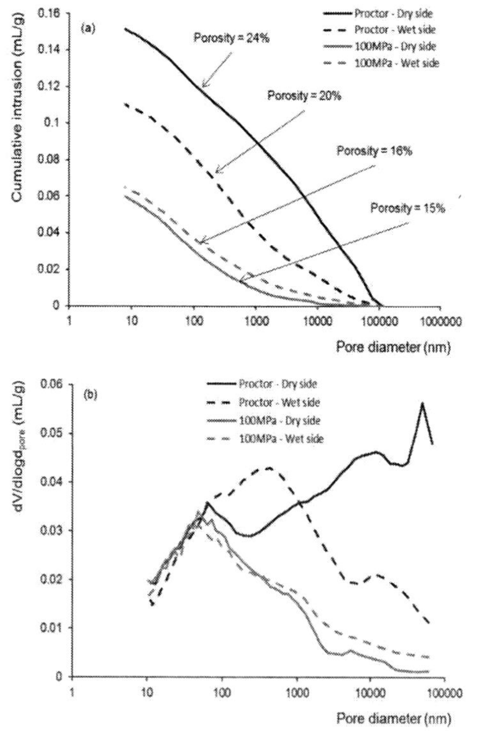

Figure 7. Cumulative volume and pore size distribution : comparison Proctor and 100 MPa.

By assuming cylindrical pores, Equation 1 allows the calculation of the pore diameter d_{pore} as a function of the pressure difference across the mercury interface ΔP, the mercury surface tension γ and the mercury contact angle θ. A surface tension of 485 dyn/cm and a contact angle of 147° are assumed as suggested by Diamond [27] for illitic soils.

Figs. 6a and 6b show the cumulative intruded volume and pore size distribution of specimens compacted, according to the different procedures, at their respective optimum water contents. Inspection of Fig. 6a indicates that the increase of compaction effort induces a large reduction in porosity, which is consistent with the observed differences in terms of mechanical properties. The increase of compaction effort affects the biggest pores (i.e. pores with diameter larger than 100 nm) while the influence on smallest pores (i.e. pores with diameter smaller than 100 nm) is very limited.

Figs. 7a and 7b compare the cumulative intruded volume and pore size distribution of two specimens compacted according to the Proctor method and two specimens statically compacted at a pressure of 100 MPa. For each compaction method, the two specimens were chosen at the same dry density but on the dry and wet sides of the optimum water content, respectively. The two Proctor specimens exhibit different pore fabrics depending on the compaction water content. In particular, the specimen compacted wet of optimum shows a lower porosity than the one compacted dry of optimum. This is due to the significant shrinkage during equalisation. Conversely, the two specimens compacted at 100 MPa show an almost identical pore fabric. This is consistent with the very similar dry densities and mechanical properties exhibited after equalization by all specimens compacted at 100 MPa regardless of their compaction water content. This confirms that application of a high compaction pressure can standardize the material properties and reduce the variability associated to the compaction water content.

4 Conclusions

The paper has studied the influence of the compaction procedure on the mechanical and microstructural properties of unsaturated earthen materials. The standard Proctor compaction method has been compared with a high pressure "hyper-compaction" method developed by the authors for the manufacture of masonry bricks.

Results have showed that stiffness and strength increase more than linearly with increasing dry density. Hyper-compaction is capable of producing specimens with very high levels of dry density, never attained before for unstabilized earth materials. Mechanical tests confirmed that the hyper-compacted specimens exhibit values of strength and stiffness comparable to traditional building materials such as stabilized earth or fired bricks. Conversely, the lower compaction effort of the Proctor method is insufficient to generate mechanical properties suitable for masonry construction.

MIP tests have confirmed that, for a given compressive effort, the water content at the time of compaction affects the porosity of the material. However, this effect become progressively less significant as the compaction effort is increased from the Proctor standard to static compaction at 25 MPa, 50 MPa and 100 MPa. This means that the application of a very high pressure might lead to the manufacture of the same material regardless of the compaction water content.

Further analysis of the durability properties of the present soil are also planned in the continuation of this study.

Acknowledgements

The authors gratefully acknowledge the financial support from the Aquitaine Region (France) and the Agglomération Côte Basque-Adour (France).

References

1. M. Olivier and A. Mesbah. Bulletin de liaison du Laboratoire des Ponts et Chaussé, Paris, France, **146**, 37-43 (1986).
2. B. V. Venkatarama Reddy and K. S. Jagadish. Géotechnique, **45**, 363-367 (1995)
3. M.F. Attom. Appl. Clay Sci., **12** (1-2), 61-71 (1997).
4. A. Mesbah, J. C. Morel and A. Olivier. Mater. Struct. **32**, 687-694 (1999).
5. C.H. Kouakou, J. C. Morel. Appl. Clay Sci., **44**, 27-34 (2009).
6. I. Garcia-Bengochea, A. G. Altschaeffl and C. W. Lovell. J. Geotech. Eng.-ASCE, **105** (7). 839–856 (1979).
7. H. Nowamooz and F. Masrouri. Eng. Geol., **114**, 444-455 (2010).
8. AFNOR. XP P 94-041 (1995)
9. AFNOR. NF P 94-057 (1992).
10. AFNOR. NF P 94-051 (1993).
11. AFNOR. NF P 94-054 (1991)
12. A. W. Bruno, D. Gallipoli, C. Perlot , J. Mendes and N. Salmon. ICBBM, Clermont-Ferrand, France (2015).
13. M.C. Jiménez Delgado and I. Cañas Guerrero. Constr. Build. Mater., **21**, 237-251(2007).
14. AFNOR. XP P13-901 (2001).
15. CRATerre-EAG, CDI, Brussels, Belgium (1998).
16. MOPT. Bases Para el Diseño y Construcción con Tapial. Madrid, Spain (1992).
17. A.W. Skempton. Proceedings 3rd International Conference of Soil Mechanics and Foundation Engineering, 1, 57-60 [1953].
18. K. Dierks and C. Ziegert. In Steingass, P. : Moderner Lehmbau. Tagungsband, Fraunhofer IRB (2002).
19. AFNOR. NF P 94-050 (1995).
20. AFNOR. NF P 94-093 (1999).
21. C. T. S. Beckett and C. E. Augarde. In Proceedings of 2nd European Conference on Unsaturated Soils; 287-292. Naples, Italy, Springer, ISBN: 978-3-642-31342-4 (2012).
22. Q.-B. Bui, J. C. Morel, S. Hans and P. Walker. Constr. Build. Mater., **54**, 163-169 (2014).
23. J. C. Morel, A. Pkla and P. Walker. Constr. Build. Mater., **21**, 303-309 (2007).
24. D. Ciancio and J. Gibbings. Constr. Build. Mater., **28**, 294–304 [2012].
25. E. Romero, A. Gens and A. Lloret. Eng. Geol. **54**, 117-127 (1999).
26. O. Cuisinier and L. Laloui. Int. J. Numer. Anal. Met., **28**, 483-499 (2004).
27. S. Diamond. Clay Clay Miner **18**, 7-23 (1970)

E3S Web of Conferences 9, 14005 (2016)
DOI: 10.1051/e3sconf/20160914005
E-UNSAT 2016

Lime stabilization of expansive soil from Sergipe - Brazil

Rafaella Leite[1,a], Rodrigo Cardoso[1], Carlos Cardoso[1], Erinaldo Cavalcante[2] and Osvaldo de Freitas[3]

[1]*Universidade Federal de Sergipe, Department of Civil Engineering, Av. Marechal Rondon, 49100-000, São Cristóvão, Brazil*
[2]*Universidade Federal de Campina Grande, Department of Civil Engineering, R. Aprígio Veloso, 58429-900, Campina Grande, Brazil*
[3]*Universidade Federal do Rio Grande do Norte, Department of Civil Engineering, Lagoa Nova, 59078-970, Natal, Brazil*

Abstract. Expansive soils are characterized by volumetric changes caused by variations in moisture. They can cause several damages to civil constructions, especially to lightweight structures, including cracks and fissures. Chemical stabilization through addition of lime is one of the most effective techniques used to treat this type of soil. Due to cationic exchanges, lime can significantly reduce swell potential. This research studied a disturbed sample of expansive soil collected in Nossa Senhora do Socorro – Sergipe, Brazil, through the following laboratory tests: sieve and hydrometer tests, Atterberg Limits, compaction, free swell and swell pressure. All direct and indirect methods mentioned in this paper indicated that the natural soil presented high to very high degree of expansion, which reached approximately 20% of free swell and nearly 200 kPa of swell pressure. In order to evaluate the effect of lime, the same tests were conducted in soil-lime mixtures, using lime contents of 3%, 6% and 9%. The results confirmed the efficiency of lime stabilization. It was noted that, as lime content increased, there was reduction of clay fraction and increment of silt fraction; plasticity index decreased to nearly its half; compaction curve was displaced; and free swell and swell pressure reduced significantly.

1 Introduction

Expansive soils are recognized by great changes in volume (swell and shrinkage), under same state of stress, upon variations in water content. They are widespread throughout world, especially in semi-arid areas, where climate has significant impact.

They can cause many structural and aesthetic problems to civil constructions, such as roads, airports, underground utilities, lightweight residential buildings and other facilities. In terms of economical loss, damages caused by expansive soils exceed the combined average annual damages from floods, hurricanes, earthquakes and tornadoes [1].

In order to enable construction of lightweight structures on top of expansive soils, special attention must be given to geotechnical designs. Foundations whose superstructures may be isolated from effects of expansive soils (e.g. pier and grade beam support), or foundations rigid enough to resist differential movement without causing damages to their superstructure (e.g. posttensioned slab-on-grade), should be selected.

Other alternative is soil stabilization. The treatment may be chemical, by adding products such as lime, cement and fly ashes, or mechanical, through addition of fibres.

Lime stabilization has proved to be one of the most efficient techniques used to mitigate swell potential. This paper intends to evaluate the effects on geotechnical properties of an expansive soil collected in Northeast of Brazil, after mixed with selected percentages of lime.

2 Literature Review

Clay particles are characterized by negative charges on their plates, caused by isomorphic substitutions. In order to balance these charges, cations are attracted to their surface and are surrounded by polar molecules of water, which form a diffuse double layer, leading to expansion.

Montmorillonite is the most unstable clay mineral, since it presents higher amount of isomorphic substitutions and intercrystalline expansion.

Beside type and content of clays, many intrinsic and external factors may affect swell potential of expansive soil, such as dry unit weight, plasticity, initial water content, climate and surcharge loading.

There are several indirect methods to predict swell potential. They usually classify soils as having low, medium, high or very high swell potential. Some of them are charts based on soil activity [2, 3]. Other methods include ranges of clay content and plasticity index [1].

2.1 Lime Stabilization

Stabilizing expansive soil by adding lime is an ancient art and an age old practice, which has been followed all over the world [4].

[a] Corresponding author: rafaellapradoleite@gmail.com

The short-term phase of lime stabilization is characterized by cationic exchanges: the addition of lime to clayey soils provides an abundance of calcium and magnesium ions, which tend to displace other common monovalent cations, such as sodium and potassium present on clay mineral plates [5]. This leads to a reduction of diffuse double layer water thickness which surrounds soil particles.

After cationic exchanges occur, several differences on geotechnical properties are expected. Increment of optimum water content and decrease in maximum dry unit weight of soil by adding lime was observed in several researches [5-8]. Reductions in clayey fraction and increase in percentage of silt due to aggregation and flocculation have been incontestably verified [7-9]. Decrease of plasticity index was also proven [6-8]. However, behaviours of plastic limit and liquid limit themselves may vary depending on type of soil. Both free swell and swell pressure have typically been reduced [5-9]. Nonetheless, low contents of lime may not be able to induce great reactivity, leading to reverse effects [7].

In addition, previous studies have shown that lime content and curing time of soil-lime mixtures have important effect on swell potential of soil [5].

3 Materials and Methods

3.1. Materials

The sample of soil used in this research is known as "massapê", recognized by its typical expansive properties. It was extracted in Nossa Senhora do Socorro, in the state of Sergipe (Northeast of Brazil). Since the sample was collected during the raining period, it did not show resistance to excavation.

Hydrated lime CH-I was used due to its high availability in Sergipe (this type of lime is classified according to chemical and physical properties proposed by Brazilian standard NBR 7175).

Lime contents of 3%, 6% and 9% were selected.

3.2 Methods

All tests were performed seven days after the mixture, allowing lime to react and cation exchanges to occur. It is important to highlight that, due to the need of removing the time factor from the variables, homogeneity of samples was a little compromised.

All tests were performed at least twice for each lime content, in order to enhance their reliability.

3.2.1 Soil-lime mixture

The amount of lime mixed with natural soil was based on the selected lime contents, which were related to the dry mass of pure soil. The mixtures were performed manually, and they were kept in plastic bags during a seven-day curing period.

3.2.2 Sieve and hydrometer tests

Sieve and hydrometer tests were performed in order to obtain grain size distributions of natural soil and soil-lime mixtures. The procedures of this test followed the recommendations indicated by Brazilian Standard NBR 7181.

3.2.3 Atterberg Limits

Liquid limit, plastic limit and shrinkage limit tests were performed following the procedures indicated by Brazilian Standards NBR 6459, NBR 7180 and DNER-ME 087/94, respectively.

3.2.4 Compaction

Compaction tests were performed following the recommendations in Brazilian Standard NBR 7182. Standard Proctor Energy was applied and the small mould was selected. These tests had the goal of providing the parameters of compaction for free swell and swell pressure remoulded samples.

3.2.5 Free Swell

These tests followed the procedures indicated by ASTM D 4546-03. Method A was selected for their performance. At first, the specimen was moulded with water content and dry unit weight similar to the parameters obtained from compaction curves. Later, the sample was placed in a consolidometer apparatus, under a seating pressure of at least 1 kPa. Within 5 minutes, the extensometer device was adjusted and the specimen was inundated from top and bottom and allowed to swell vertically.

Deformation readings were made typically at 0.1, 0.2, 0.5, 1.0, 2.0, 4.0, 8.0, 15.0 and 30.0 minutes and 1, 2, 4, 6, 24, 48 and 72 hours, until invariability was achieved, after primary and secondary swell. Then, the percent of free swell was related to the specimen's initial height.

Preparation of specimens is shown in Figure 1.

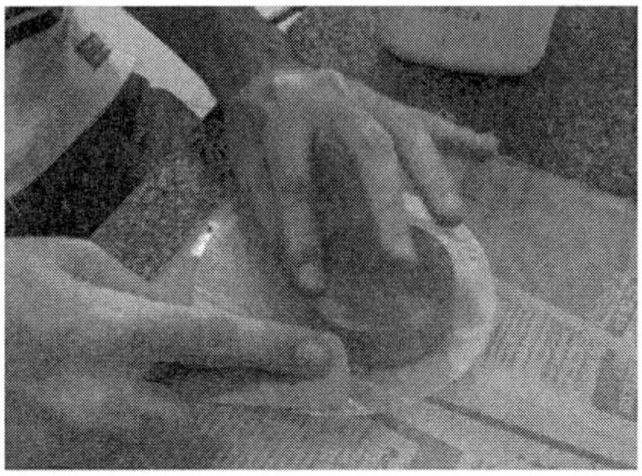

Figure 1. Moulding of specimens for Free Swell and Swell Pressure tests.

DOI: 10.1051/e3sconf/20160914005

3.2.6 Swell Pressure

Swell pressure tests had similar procedures to free swell tests. After primary swell was achieved, vertical pressure was applied in increments on the top of the specimen, until initial void ratio was reached, i.e., no swell was registered by the extensometer.

Placement of specimen in a consolidometer apparatus and load increments are shown in Figure 2.

Figure 2. Swell Pressure Test.

4 Results and Discussion

4.1 Natural Soil Characterization

Sieves and hydrometer tests performed on natural soil revealed predominance of fine-grained particles (approximately 94% was finer than sieve #200) and clay content around 38%. Grain size distribution is shown in Table 1, considering MIT classification.

Liquid limit and plastic limit values of natural soil were 57 and 23, respectively, which results in a plasticity index of 34.

Indirect methods [1-3] predicted natural soil to present high to very high degree of expansion.

Table 1. Grain Size Distribution of Natural Soil.

Classification	Grain Size	%
Gravel	2 mm – 6 in	1.8
Sand	60 μm – 2 mm	13.2
Silt	2 μm – 60 μm	47.2
Clay	< 2 μm	37.8

4.2 Hydrometer and sieves tests

Grain size distribution of natural soil and soil-lime mixtures are shown in Figure 3. The results show significant decrease in clay fraction and increment of silt fraction when lime was added, in accordance with previous studies. Sand fraction increased slightly.

This behaviour can be justified by the aggregation and flocculation of soil particles, due to cationic exchanges induced by lime treatment.

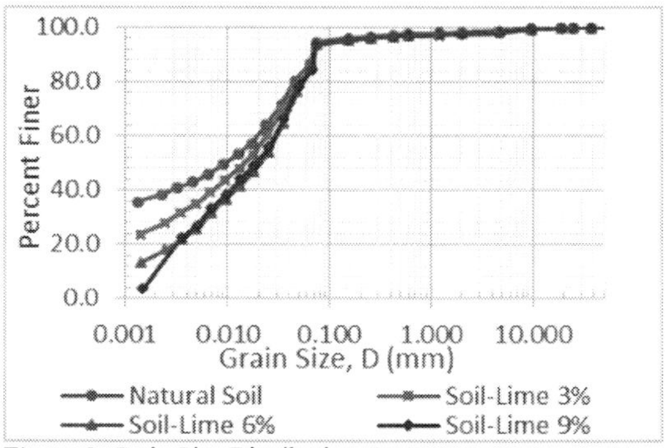

Figure 3. Grain Size Distribution Curves.

4.3 Atterberg Limits

The results of Atterberg Limits from natural soil and soil-lime mixtures are given in Figure 4.

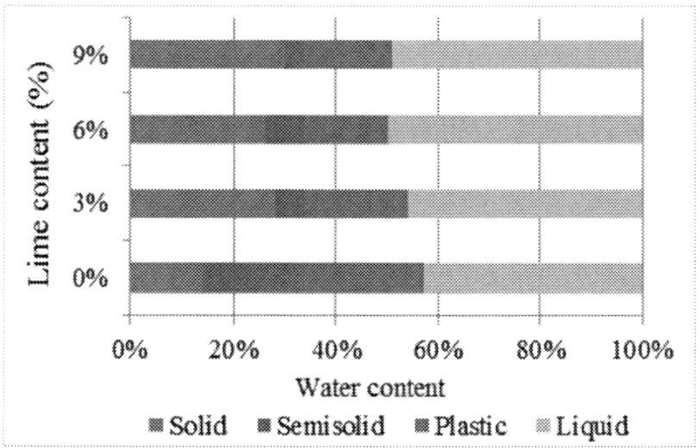

Figure 4. Atterberg Limits Results.

As expected, plasticity index decreased significantly (to nearly its half) when lime was added. There was a slight decrease in liquid limit, whereas plastic limits increased substantially when 3% of lime was mixed. However, higher contents of lime had almost no additional effect on plastic limit and plasticity index.

The reduction of soil plasticity also leads to an improvement of soil workability.

Besides, an increment of shrinkage limit can be noticed, leading to a smaller range of water content in which cracks and fissures are formed.

4.4 Soil Classification

According to Unified Soil Classification System, this expansive soil can be classified as a clay of high plasticity (CH), since its liquid limit is higher than 50 and it is situated above the line A from Casagrande's Plasticity Chart. Due to reduction of Plasticity Index, the addition of lime turned the pure soil into a highly compressible silt, for all contents used. This difference can be visualized in Figure 5.

DOI: 10.1051/e3sconf/20160914005

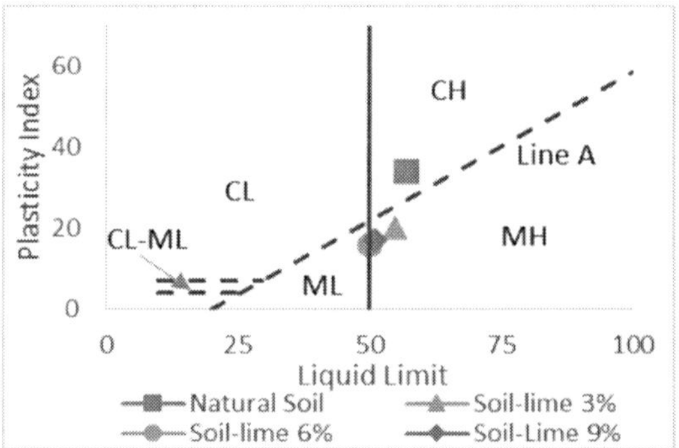

Figure 5. Classification of natural soil and soil-lime mixtures.

4.5 Compaction

The results of compaction tests performed on natural soil and soil-lime mixtures can be visualized in Figure 6.

In accordance with literature, the results show that there was reduction in maximum dry unit weight and increment of optimum water content, proportionally to the lime percentage added. Optimum water content increased from 17.9% up to 21.2%, and maximum dry unit weight varied in the range of 16.87 kN/m^3 to 15.90 kN/m^3.

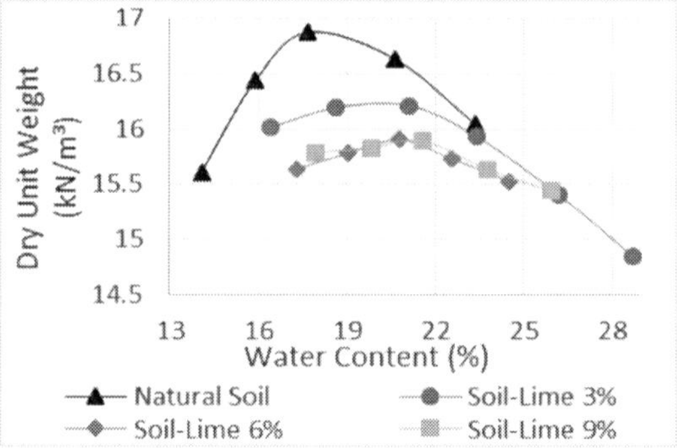

Figure 6. Compaction Results.

This behaviour can be explained by the flocculation induced by adding lime. Initial void ratio increases, and the soil fabric is strong enough to resist the same Proctor energy, reducing the maximum dry unit weight of the mixture. In addition, higher volume of voids requires larger amount of water to be fulfilled, raising the optimum water content.

It is also possible to observe that compaction curves tend to reduce their concavity as lime is added, which leads to lower sensitivity from changes in moisture.

4.6 Free Swell

The final results of all free swell tests performed on natural soil and soil-lime mixtures at seven-day cure time are shown in Figure 7. As mentioned in literature, swell potential reduces significantly when lime is mixed with expansive soil. It can be noted that by adding 3% of lime, the percentage of swell decreased from 20% to approximately 1.85%. Lime contents of 6% and 9% presented free swell close to 1.0% and 0.5%, respectively, which are below the brazilian imposed limits for pavement subgrades and base.

The tests performed on natural soil took six days to be completed. On the other hand, the final results from soil-lime mixtures were achieved after only 2 days, which probably indicates that addition of lime also affects permeability of soil.

In Figure 7, we can also observe that the results from both samples of each lime content are quite similar, diverging up to 10% approximately.

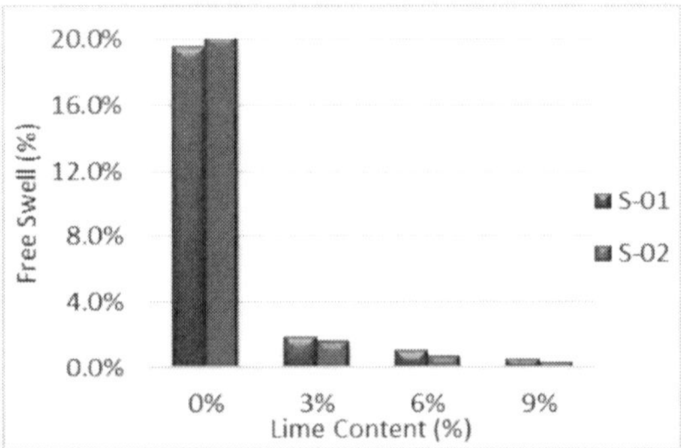

Figure 7. Free Swell Results.

4.7 Swell Pressure

The final results of all swell pressure tests performed on natural soil and soil-lime mixtures at seven-day cure time are given in Figure 8.

As expected, there was significant reduction in swell pressure due to lime treatment. The addition of only 3% of this chemical stabilizer reduced swell pressure from nearly 200 kPa to an average value of 37. Lime contents of 6% and 9% presented mean swell pressures of 32 kPa and 23 kPa, respectively.

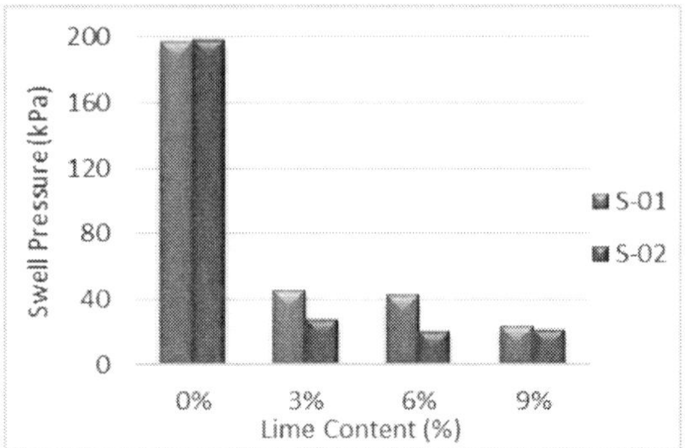

Figure 8. Swell Pressure Results.

The most accepted reason to explain this behaviour is the decrease in maximum dry unit weight by adding lime, which causes weaker repulsion forces between the particles of soil [7].

The results from both tests of soil-lime mixtures of 3% and 6% presented small discrepancy. It could be caused by inaccuracies from the test itself; by small differences in compaction parameters when moulding the samples; or by the loss of homogeneity induced by the need to assure the same curing time for all samples. However, these differences are not significant enough to modify the conclusions obtained from these tests.

5 Conclusions

Due to cationic exchanges, which lead to agglomeration and flocculation of soil, lime stabilization causes significant effects in many properties of expansive soils, such as:

1. Reduction of clay fraction and increase of silt and sand fractions;
2. Plasticity index of soil reduced significantly by adding lime, due to a slight decrease in liquid limit and to an expressive increment of plastic limit.
3. Compaction curves were displaced. By mixing lime and natural soil, optimum water content increased and maximum dry unit weight decreased.
4. Addition of lime caused drastic reduction on swell potential. Free swell reduced from 20% to 1.85%, 1.0% and 0.5%. Meanwhile, swell pressure decreased from 200 kPa to 37 kPa, 32 kPa and 23 kPa, when 3%, 6% and 9% of lime were blended, respectively.

References

1. F. H. Chen, *Foundation on Expansive Soils* (Elsevier Scientific Publishing Co., New York, 1975)
2. H. Seed, R. J. Woodward Jr., R. Lundgren, ASCE, 88, 53-88 (1962)
3. D. H. Van der Merwe, Trans. SA Ins. Civ. Eng, **6**, 103-107 (1964)
4. S. Bhuvaneshwari, R. G. Robinson, S. R. Gandhi, IGJ, **44,** 278-293 (2014)
5. A. R. Estabragh, H. Rafatjo, A. A. Javadi, Geosynth Int, **21,** 233-243 (2014)
6. J. J. J. M. Garcia, *Masters Thesis* - Universidade Federal do Rio de Janeiro (1973)
7. M. A. M. Silva Júnior, RBGEA, **2**, 2 (2012)
8. A. L. I. K. Abass, JEAD, **17,** 1 (2013)
9. V. Barbosa, *Masters Thesis* - Universidade Federal de Pernambuco (2013)

Unconfined strength of an unsaturated residual soil struck by replicated lightning

Thiago de S. Carnavale[1a], Mariana F.B. Motta[1], Tácio. M.P. de Campos[1], Haimon D.L. Alves[2], Antônio R.M.B. de Oliveira[1] and Hélvio J.A. Martins[3].

[1]*PUC-Rio, Civil Engineering Department, 22451-900 Rio de Janeiro, Brazil*
[2]*UERJ, Physics' Department, 20550-900 Rio de Janeiro, Brazil*
[3]*CEPEL, LabDig, 21941-911 Rio de Janeiro, Brazil*

Abstract It is well known that different triggering factors are related to landslides occurrence. However, in many cases, it is not possible to identify main factors that may contribute to start a landslide. Following that, lightning phenomena is herein considered as a possible factor that may promote changes in the structure, and eventually, in the strength of soils. The current study aims to analyse the influence of laboratory simulated lightning in the structure of undisturbed granite-gneiss residual soil samples. The main focus is to compare the peak strength of unsaturated samples that were not struck by replicated lightning with the peak strength of soil samples struck by replicated lightning. The methods used are: Soil sampling and physical characterization; unconfined compression strength tests on unsaturated undisturbed samples; submission of soil samples to replicated lightning; unconfined compression strength tests on samples struck by replicated lightning and micro tomography of samples submitted to lightning. As results, it is seem that lightning may cause a hole with irregular geometry inside the soil. Analysing the tests of the samples struck by laboratory simulated lightning, a peak strength reduction with the charge incidence was observed. Comparing the variation of soil matric suction on the peak strength of the soil that was not struck by replicated lightning with that of the soil struck by the higher charge of the replicated lightning, it is observed that the samples struck by high-voltage presents lower values of peak strength.

1 Introduction

In January of 2011, the mountainous region of the state of Rio de Janeiro, Brazil, suffered what was considered as one of the worst natural disasters of the Brazilian history, counting hundreds of landslides that caused more than 900 deaths (see Figure 1).

In this event, failure mechanisms were not fully understandable, opening a wide discussion on the theme and suggesting the existence of triggering factors not considered before.

Interviews with local-people indicated a large occurrence of lightning and of ground shacking at the flashing moments. Based on that, investigations were started at PUC-Rio trying to understand the lightning-soil interaction.

Lightning is a natural phenomenon that involves a huge transfer of energy. Rakov and Uman [1] highlights that an overall flash has 200-300 ms time duration and its temperature is of circa 30.000 °C [2] of magnitude. According to these authors, lightning can propagate through the air with velocity of 220.000 km/h.

The objective of the present work is to analyse if laboratory replicated lightning may cause strength reduction in an unsaturated soil higher than a strength reduction promoted by a decrease of the matric suction of the soil due, for instances, an increase in its degree of saturation.

Figure 1. Rio de Janeiro mountain region. The landscape after disaster

[a] Thiago de S. Carnavale: thicarnavale@gmail.com

2 Material

The material chosen for testing is from a site named Condomínio do Lago – Nova Friburgo – RJ (coordinates 742961.31 m E; 7539347.23 m S), which was destroyed in the 2011 event.

It comprises a granite-gneiss residual soil, with e = 0.715, ρ_d = 1,49 g/cm³ and Gs = 2.66. Average grain size of the material is shown in Table 1. The soil is classified as SM - silty sand, according to the Unified Soil Classification System.

Table 1. Distribution of particle sizes

Gravel (%)	Sand (%)	Silt (%)	Clay (%)
1.4	60.5	31.9	6.2

The filter-paper [3], Haines' funnel [4] and potentiometer WP4C (Decagon Devices Inc.) techniques were employed to define the soil water retention curve of the undisturbed soil (Figure 2). It can be seen in this figure that a quite reasonable unimodal relationship was obtained using these three different techniques.

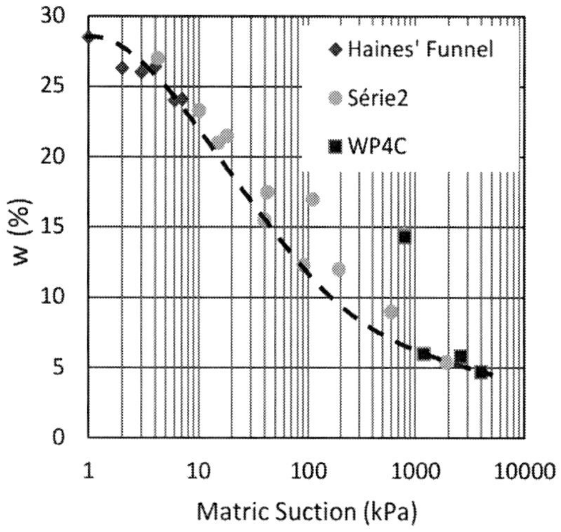

Figure 2. Water retention curve (WRC)

3 Methods

Cylindrical samples, with 38 mm of diameter and 76 mm in height were used in the experiments, comprising:

a) pulse generator submission, where the granite-gneiss soil samples were struck by different charges of replicated lightning;
b) test on a computerized micro tomographer, to analyse structural characteristics of the samples [5];
c) unconfined compression strength tests performed under different soil matric suction.

3.1 Pulse generator

To analyse possible lightning influences, a cooperation between the Pontifical Catholic University of Rio de Janeiro - PUC-Rio and the Electrical Energy Research Center - CEPEL-ELETROBRAS was developed. Replicated lightning were created in the CEPEL laboratory, enabling to submit the soil samples to high-voltage electrical pulses.

Figure 3 shows the overall pulse generator schema containing the layout of the set composed of seven stages and the place where soil samples were located along the tests.

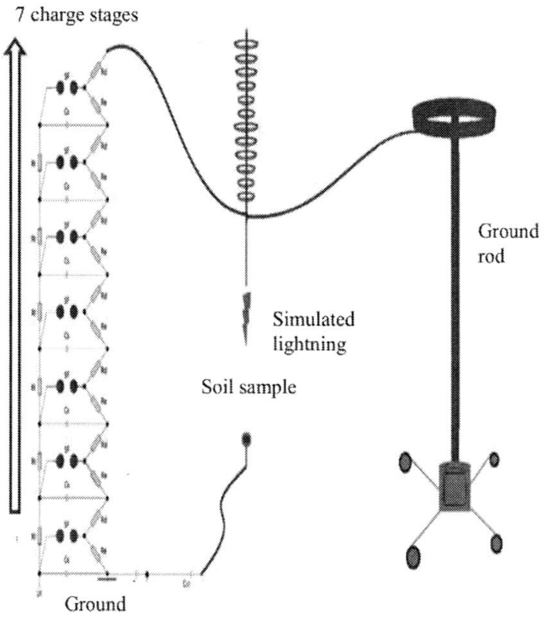

Figure 3. High-voltage pulse generator scheme

The first step of the experimental programme was to submit the undisturbed soil to simulated lightning. Following the pulse generator Series L from HIGHVOLT [6], the samples were struck by 300 kV, 450 kV and 600 kV electric charges.

It is important to note here that the discharges induced by the electrical impulses in the laboratory are considered to replicate lightning, as they are in agreement with the time to reach the peak (1,2μs) and tail (50 μs) of the impulses [7].

Figure 4 shows the response emitted by the lightning equipment simulator, where channel 1 presents the kV (voltage) response and channel 2 presents A (amperage) response along time. The data sheet also reveals the time (in μs) and the negative characteristic of the charge, which represents a negative lightning.

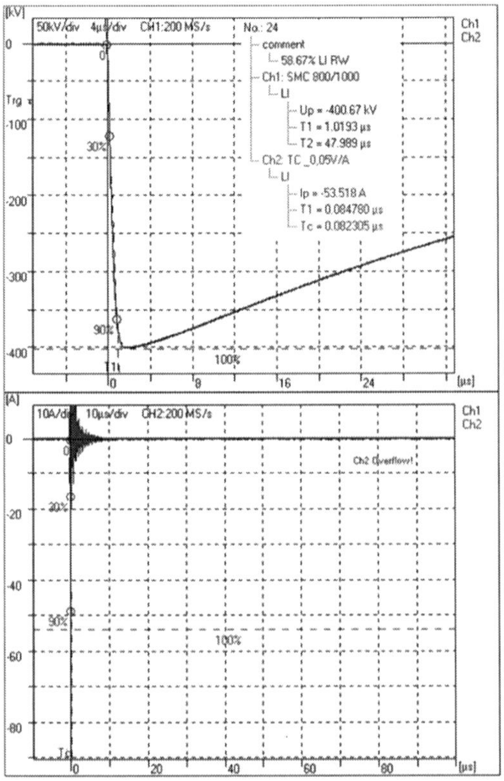

Figure 4. Simulated lightning developed in the lab

If the electrical signal is higher than the air resistance, it is formed a heated luminous channel that crosses the soil sample, discharging the high-voltage in the ground rod. At this time, it is possible to see the replicated lightning (Figure 5).

Figure 5. Simulated lightning developed in the lab

It is interesting to notice here that saturated samples were initially tested. However, they presented catastrophic failure under the incidence of replicated lightning (Figure 6). So it was chosen to work just with unsaturated samples, under their natural water contents.

Figure 6. Catastrophic failure of saturated samples

3.2 Micro tomography

The effects of atmospheric impulses in the soil samples are until now undiscovered. Thus to better understand what happens with the soil when it is struck by a huge electric energy, this study used computerized micro tomography, with 22.12 resolution, to obtain 3-D internal images (Figure 7)

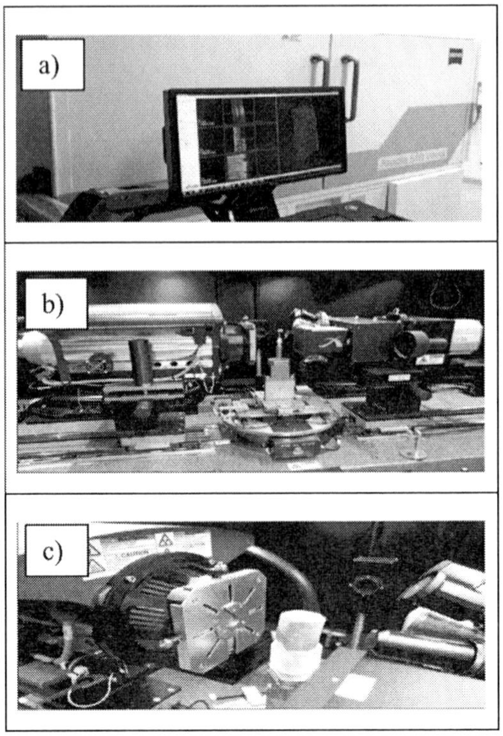

Figure 7. Micro tomography equipment and internal configuration a) Computer graphic analysis, b) X-ray emission mechanism and c) Soil cylindrical sample on the adapted support

3.3 Unconfined compression strength tests

Unconfined compression strength tests (Figure 8) were carried out under a strain rate of 0.4 mm/min.

Undisturbed samples non-struck by replicated lightning were wetted and dried before the unconfined strength tests. Table 2 shows the initial moisture content of such samples. The samples submitted to replicate lightning were at their natural water content before being struck by 300 kV, 450 kV and 600 kV. Such water contents are also included in Table 2.

Figure 8. Unconfined compression strength tests

Table 2. Initial moisture content of the samples struck and non-struck by lightning

Sample			Moisture w (%)
Struck by Lightning	1	300 kV	15,83
	2	300 kV	16,11
	3	450 kV	15,8
	4	450 kV	15,77
	5	600 kV	16,26
	6	600 kV	16,13
Non-struck	7		22,1
	8		19,89
	9		17,68
	10		12,3
	11		10
	12		6,4
	13		6,33

4 Results and discussion

On using the micro tomography technique it was observed that the electric discharge affected only the samples struck by 450 kV and 600 kV. The simulated lightning caused an irregular tortuous hole with some radial cracks (Figure 9) in the central area of the sample (which was the selected principal path of the lighting experiment|).

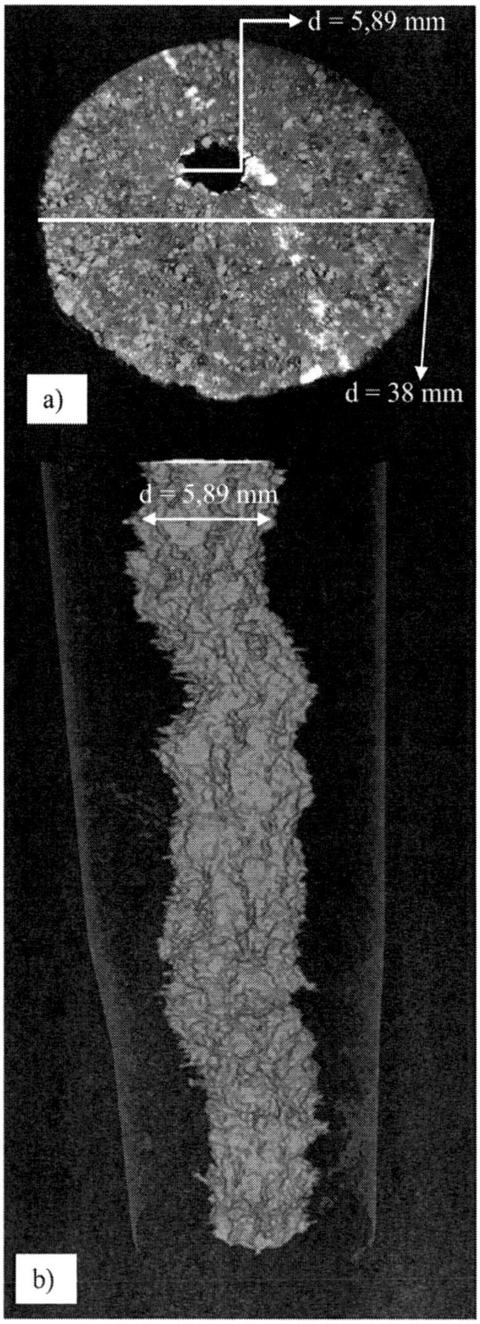

Figure 9. Micro tomography. (a) Top and (b) central hole images

For the 600 kV sample, the hole size distribution was analysed and it was possible to observe a higher size in the sample top. Analysing the figure 10 it is possible to observe a 6 mm maximum variation of the hole size diameter. Another observation is related to the spiral behaviour shown in Figure 9, which indicates that the simulated lightning crossed the soil sample following a zig-zag path.

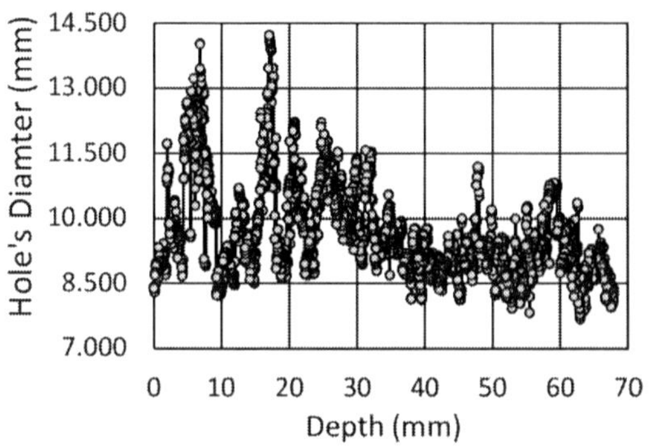

Figure 10. Variation of the hole size diameter

Figure 11 shows the results of the unconfined strength tests performed on the samples non-struck by replicated lightning under variable initial matric suction. Such suction (MS in Figure 11) was estimated based on the measured initial moisture content of the samples and on the water retention relationship shown in Figure 2. As expected, higher peak strengths were obtained in the tests performed in the samples with lower degree of saturation (or moisture content).

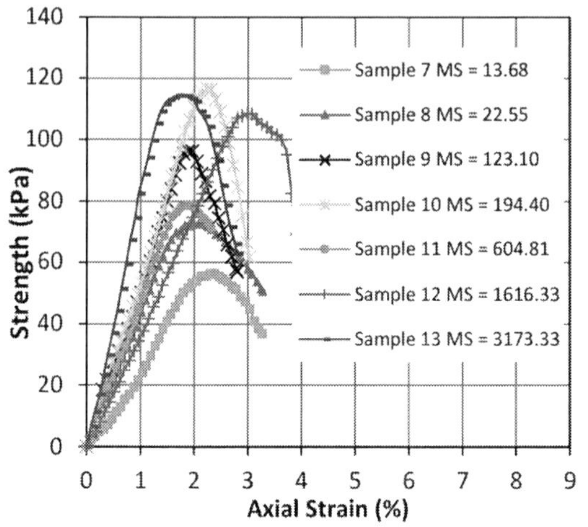

Figure 11. Unconfined strength tests for the samples non-struck by replicated lightning

Analysing the unconfined strength tests results for the samples struck by replicated lightning (Figure 12), it is notable that, for the samples in which holes were formed (samples 3 to 6), the peak strengths were much lower than the peak strengths of the samples submitted to 300 kV replicated lightening (samples 1 and 2), which did not show a continuous hole in their central portion.

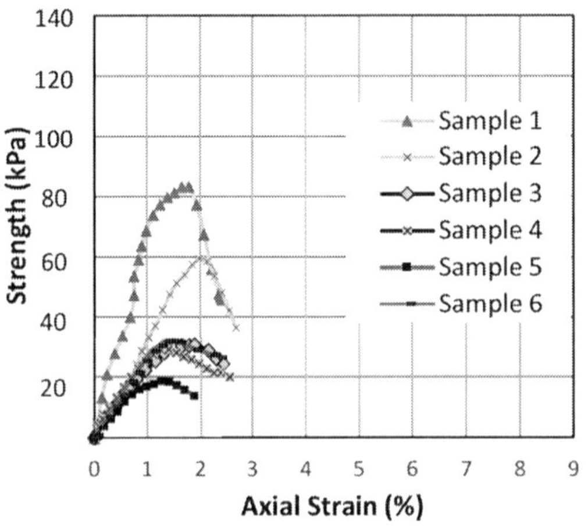

Figure 12. Unconfined strength tests for the samples struck by replicated lightning

Table 3 shows the unconfined strength of all tested samples and the matric suction related to each one of the samples before lightning struck and / or strength testing, estimated as mentioned before. Based on these data and on the data shown in Table 2, Figures 13 and 14 were produced.

Table 3. Unconfined compression strength results

Sample			Strength (kPa)	Matric Suction (kPa)
Struck by Lightning	1	300 kV	85	41,21
	2	300 kV	60	35,49
	3	450 kV	32,8	42,33
	4	450 kV	29,9	41,61
	5	600 kV	36	34,57
	6	600 kV	19,5	36,32
Non-struck	7		56,3	13,68
	8		73,1	22,25
	9		96,8	123,10
	10		118	194,40
	11		79,5	604,81
	12		109	1616,33
	13		116	3173,33

Taking as a basis the range of initial soil suction of the tested samples, Figures 13 and 14 indicates that the matric suction (plotted in log scale) decreases in an approximately linear way with an increase in moisture content (plotted in natural scale) for the non-struck samples. In that case, shear strength (plotted in natural scale in Figure 13 and in log scale in Figure 14) decreases with moisture content, following a law not clearly defined. Figure 13 also indicates that the unconfined strength tends to decrease with an increase of the lightning charge (in natural scale) for the samples with initial moisture content of circa 16%, that corresponds to an initial suction of circa 40 kPa.

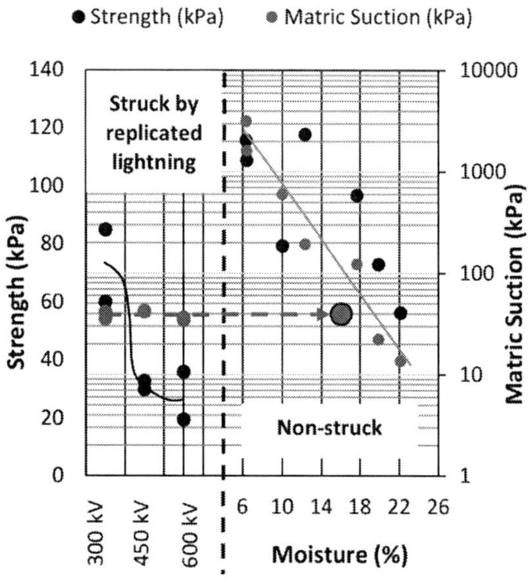

Figure 13. Peak strength and matric suction data versus replicated lightning charges (300 kV, 450 kV and 600 kV) and moisture content.

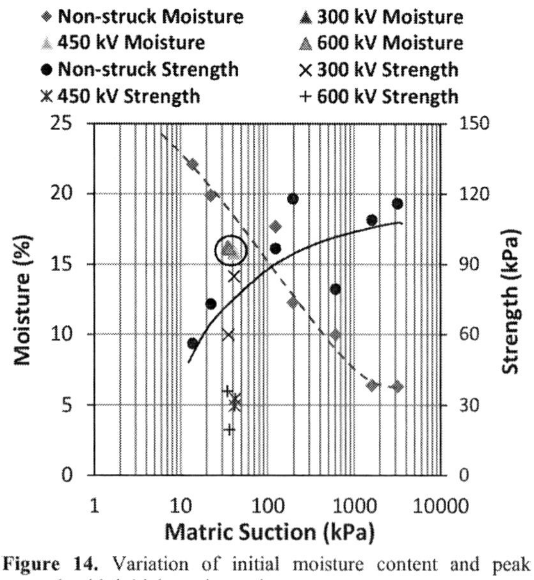

Figure 14. Variation of initial moisture content and peak strength with initial matric suction

Samples struck by a 300 kV simulated lightning showed an unconfined strength higher than the strength of non-struck samples at equivalent initial suction (see Figure 14). As in the present stage of the investigations it was not yet possible to evaluate the potential change in suction that might be promoted by the lightning inside the soil sample, such result must be seen with care.

5 Conclusion

This work presents preliminary results obtained in a large investigation under way at PUC-Rio on effects of lightning on the shear strength of unsaturated soils. These results indicate that lightning tend to promote a decrease of the shear strength of the soil; the larger the lighting energy, the lower the strength.

Also, it was found that depending on the lightning energy, a sinuously hole, with a nearly constant radius, occurs in the soil sample.

Further investigations are required in order to understand the hole of suction in the soil response after being struck by lightning.

Acknowledgements

This work has been developed within a PRONEX project supported by the Brazilian (CNPq) and the Rio de Janeiro (FAPERJ) research financing agencies.

References

1. V.A. Rakov, M.A. Uman, *Lightning: Physics and Effects*. 2003. 3.ed. New York: Cambridge, v.**II**. (2005)
2. M.M. Newman, J.R. Stahmann, J.D. Robb, E. A. Lewis, S. G. Martin; S. V. Zinn, *Triggered lightning strokes at very close range*. J. of Geoph. R. **72**, DOI: 10.1029/JZ072i018p04761 (1967).
3. F. Marinho, O. Oliveira, *The Filter Paper Method Revisited*, Geotech. Testing J. **29**, No. 3, pp.1-9, http://dx.doi.org/10.1520/GTJ14125 (2006)
4. F.A.M. Cassaro et al., *Modified Haines' funnel: soil water retention curves of soil samples near saturation*. Rev. Bras. Ciênc. Solo., v. **32**:6 – pp.2555-2562. http://dx.doi.org/10.1590/S0100-06832008000600032 (2008).
5. S. Paciornik, T. Bernthaler, V.R. Santos, M.J. Monteiro, M.H.P. Mauricio, A. Bracarense, E. Pessoa, *Characterization of Pores and Cracks in Underwater Welds by μCT and Digital Optical Microscopy*. In: 1st International Conference on 3D Materials Science, 2012, Seven Springs. Hoboken: John Wiley & Sons, Inc., pp. 177-182. (2012).
6. HIGHVOLT, *Impulse Voltage Generators* – Data Sheet no 3.11/5 (2005).
7. Associação Brasileira de Normas Técnicas "*ABNT NBR 5418: Instalações elétricas em atmosferas explosivas*" (In Portuguese). Rio de Janeiro, Brazil (2005).

Experimental study of the hydro-mechanical behaviour of unsaturated argillaceous rocks

Jairo Martin Espitia[1,2,a], Bernardo Caicedo[1] and Luis Vallejo[3]

[1]*Department of Civil Engineering, Universidad de Los Andes, Bogotá, Colombia*
[2]*Mining engineering department, Universidad Pedagógica y Tecnológica de Colombia, Sogamoso, Colombia*
[3]*Department of Civil & Environmental Engineering, University of Pittsburgh, Pittsburgh, United States.*

Abstract. The aim of this paper is to study the influence of total suction on some physical/mechanical properties related to argillaceous rocks from the Colombian Andes. A wetting path was applied to the rocks using the vapor transfer technique with continuous recording of mass and deformations. Uniaxial compression tests were conducted on specimens at different levels of suction. In addition, microstructural observations are carried out using a Mercury Intrusion Porosimeter Apparatus. Results from the laboratory tests indicate that anisotropic deformations took place during the wetting path. Also, total suction contributed to a considerable and non-linear reduction in the compressional wave velocity, peak compressive stress, and elastic modulus of the tested samples. Microstructural analysis indicated the influence of the size and number of the pores in the samples on their high levels of suction.

1 Introduction

A substantial portion of rocks forming part of the earth's crust are argillaceous nature and are usually present in many mining and infrastructure projects[1]. For example, in Colombia a significant part of the population, infrastructure and mining projects are located in the eastern Andes mountain range. In this area, some argillaceous rocks outcrop causing significant challenges for the stability of rock excavations. Indeed, the mineralogical composition of this material makes it sensitive to environmental conditions, such as variation in water content [2]. Furthermore, its mechanical behaviour is the result of the combination of sedimentation, gravitational compaction (consolidation), uplift/unloading (overconsolidation) and cementation/bonding [3].

The hydro-mechanical behaviour of argillaceous rocks has been studied recently using the vapor transfer technique [4-6]. Some features of the behavior of this material in unsaturated conditions can be highlighted as follows: (I) during the application of hydric cycles (without mechanical loading) it is observed hysteresis in the water retention curve, anisotropic deformations and permanent volumetric strains in the material; (II) elastic properties and strength are controlled by the hydric state and the hydric history of the material; (III) the material is degraded as a consequence of hydric cycles.

These studies have focused on evaluating the hydro-mechanical behavior of rocks from geologic formations of Europe. Nonetheless, only a few experimental studies have been performed on South American argillaceous rocks.

Thus, the aim of this paper is to study the hydro-mechanical behaviour of an argillaceous rock from the eastern Andes mountain range in Colombia in an attempt to establish relations between total suction and some mechanical properties of the material.

2 Materials and methods

2.1 Material

The material studied is a lower Hauterivian Barremian argillaceous rock, which is an indurated argillite that has caused stability problems in surface and underground rock excavations in some areas of the eastern Andes mountain range. Rock samples were taken from a limestone quarry of the Holcim Company located in Nobsa (Colombia). The average mineralogical composition of this rock is: 53.2% kaolinite and illite, 29.0% calcite, 11.0% quartz, 4.0% opaque minerals 4.0% and 1.0% Muscovite [7]. Table 1 summarizes some properties of the rock.

[a] Corresponding author: jm.espitia132@uniandes.edu.co

Table 1. Index properties of the argillaceous rock [7].

Property	Value
Bulk density	2570 kg/m^3
Specific gravity	2.76
Water content	1.50%
Liquid limit	25
Plastic limit	15

Irregular block rock samples were taken from a slope at depths greater than 0.5m. A detailed description of handling and specimen preparation is described in [8]. Specimens, of 20 mm in diameter and 40 mm in height perpendicular to bedding plane, were bored with air pressure in order to minimize alteration caused by liquid water. The dimensions of the specimens were chosen in order to minimize the time needed to achieve moisture equilibrium within a controlled humidity atmosphere. Samples tested were obtained from three rock blocks (rock block 1-B1, rock block 10-B10 and rock block 17-B17).

The water retention curve (WRC) under unconfined conditions was determined using a chilled-mirror dewpoint psychrometer [9]. The drying path was obtained by setting the rock specimen in a silica gel environment while vapor transfer technique using distilled water was applied for obtaining the wetting path. Figure 1 shows the WRC along a drying-wetting-drying cycle of the undisturbed rock. The WRC exhibits hysteresis during the cycle.

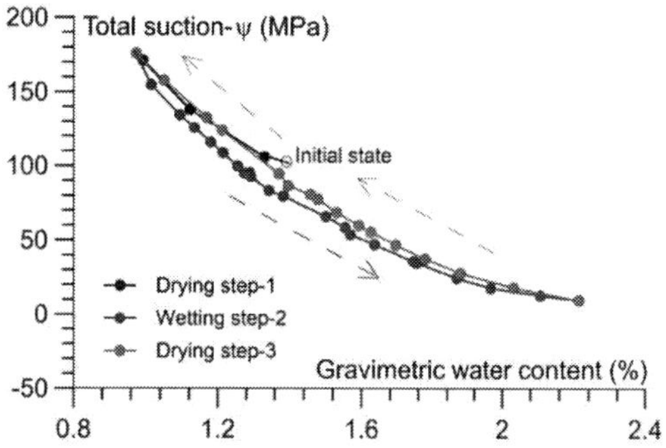

Figure 1. Water retention curve.

Figure 2 illustrates an SEM micrograph of the undisturbed material (perpendicular to the bedding). This figure shows the laminar structure of the material.

Figure 2. SEM micrograph of the undisturbed material

2.2 Suction control

Suction was controlled using vapor transfer technique. This technique is implemented by controlling the relative humidity in a sealed system [10]. The experimental program consisted of the application of a wetting path to the rock specimens. For this, the cylindrical specimens were set in a hermetic acrylic box in which saline solutions imposed relative humidity (RH). We select five saline solutions to cover a range of relative humidity between 33% and 95%. The values of relative humidity and their corresponding suction are given in Table 2.

During the wetting path, mass, relative humidity, and temperature were recorded. We also measured axial and radial deformations using strain gauges glued to the specimens. At the end of each hydric state, compressional wave velocity was measured on one specimen. During the measuring period, the temperature stayed approximately constant at 22.9±1.0°C.

Table 2. Saline solutions used.

Saline solution	Relative humidity (%)	Suction (MPa)
$CaCl_2$. $6H_2O$	33	151
K_2CO_3. $2H_2O$	47	102
$NaBr.2H_2O$	57	78
NaCl	72	46
K_2SO4	95	7

The wetting path was imposed on a group of 15 specimens from B1, B10 and B17 rock blocks. The moisture equilibrium of the rock specimens in the controlled humidity atmosphere was assumed when weight and strains were stabilized. In general, after eight days the stabilization was reached. Once specimens reached equilibrium, mechanical tests were performed on three specimens of each rock block (Figure 3).

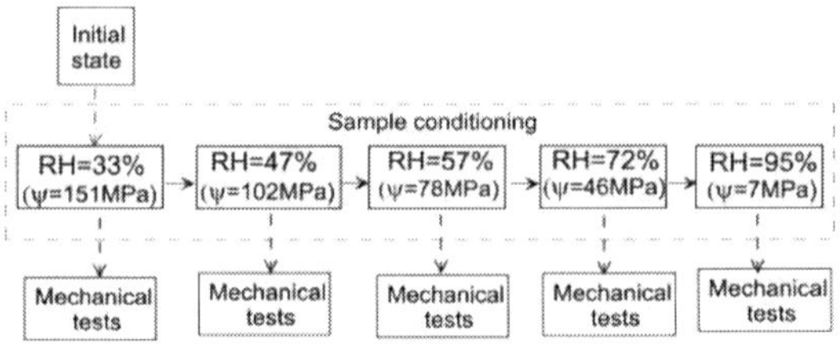

Figure 3. Specimen conditioning experiment.

2.3 Mechanical tests

Uniaxial compression tests were conducted on conditioned specimens in a controlled displacement load frame. The axial displacement rate was 0.1 mm/min. Strain gauges were fixed to measure the local strains in the axial and radial axis directions. The rock moisture was assumed to remain constant during the test due to their short duration (20 min).

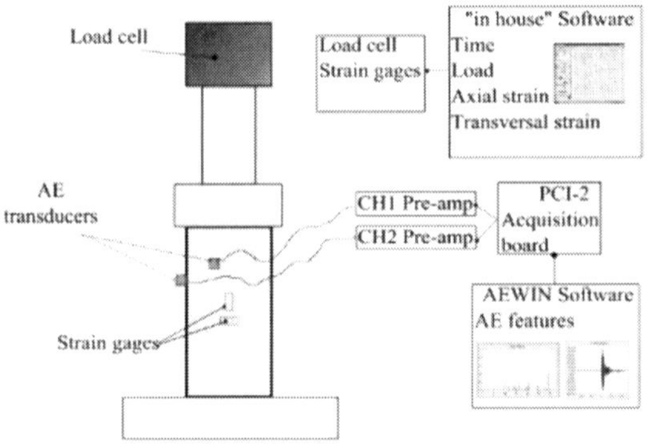

Figure 4. Experimental setup.

3 Experimental results

The purpose of the experimental tests was to obtain a better understanding of the hydromechanical behaviour of some argillaceous rocks from the Colombian Andes mountains. In order to achieve this goal, a wetting path was imposed on the rock specimens using the vapor equilibrium technique.

3.1 Evolution of parameters during the wetting path without mechanical loading

In this section, we present the evolution of some parameters during the wetting path imposed without mechanical loading. Figure 5 displays the evolution of relative humidity, mass variation, axial and radial strains.

The dimensions of the sealed box provide a ratio between the volume air in the system and the free surface area of the saline solution lower than 25 cm^3/cm^2, according to the suggestion in ASTM E 104-02 [11]. According to Figure 5a, the condition of relative humidity for each hydric step was achieved shortly after three days and remained approximately constant after that.

In general, specimens attained stabilization in mass and deformations after eight days (see Figure 5b. and 5c.). In similar studies [4, 5], for specimens of 33 mm in diameter and 76 mm in height the equilibrium took longer time (approximately 18 days). Therefore, the dimension of the specimens adopted for this study (20 mm in diameter and 40 mm in height) allowed a significant reduction of the equilibrium time. Figure 5 (b) indicates that the main mass variation occurred in the last two hydric states (RH 72% and RH 95%). Thus, it is reasonable to suppose that the principal changes will occur at these hydric stages.

Regarding deformations due to swelling, Figure 5(c) indicates anisotropy strains in the material. Also, the rock experience the majority of deformations at the same hydric states indicated for mass variation. Figure 6 represents the axial and radial strains at equilibrium for each suction value. An analysis of Fig. 6 shows a nonlinear behaviour and a ratio that ranges from 2.5 to 3 between axial and radial strains.

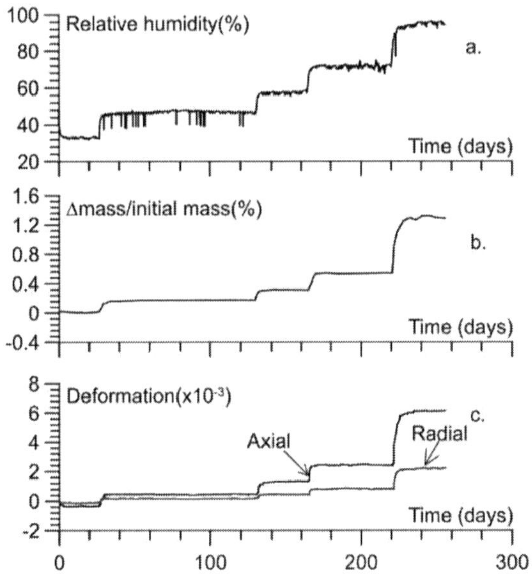

Figure 5. Parameters evolution versus time. (a) Relative humidity. (b) Mass variation. (c) Strains.

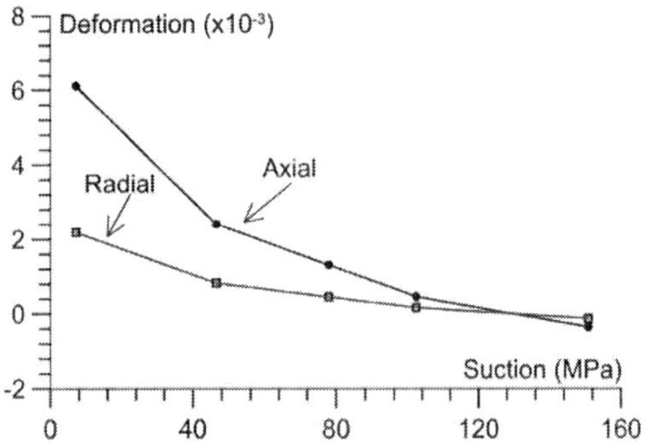

Figure 6. Strains evolution versus suction.

We measure the compressional wave velocity (Vp) at the end of each hydric state on the same specimen. The variation of compressional wave velocity versus suction is given in Figure 7. We observe from Figure 7 a reduction in Vp of 25% from high suction (151 MPa) to low suction (7 MPa) and a non-linear relationship. However, this reduction occurs mainly at the first hydric step (from 151 MPa to 102 MPa).

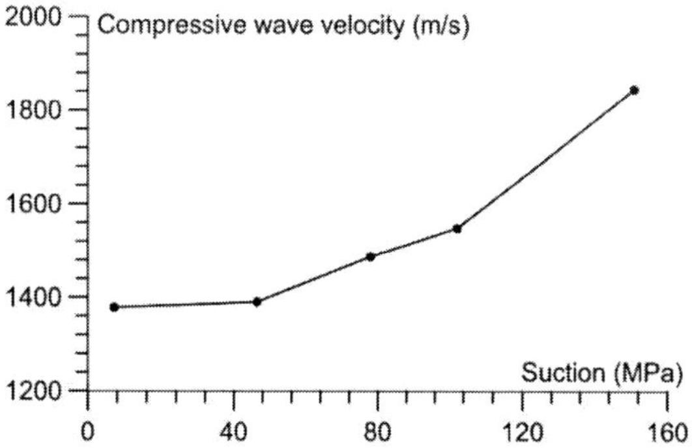

Figure 7. Variation of compressional wave velocity versus suction.

3.2 Uniaxial compression tests

In this section, we report data from the uniaxial compression tests carried out on 15 specimens (3 for each hydric state and from B1, B10, and B17 rock blocks). We execute the test with the mechanical load perpendicular to the bedding. Figure 8 shows the stress-strain curves of the wetting path of specimens from rock block B1.

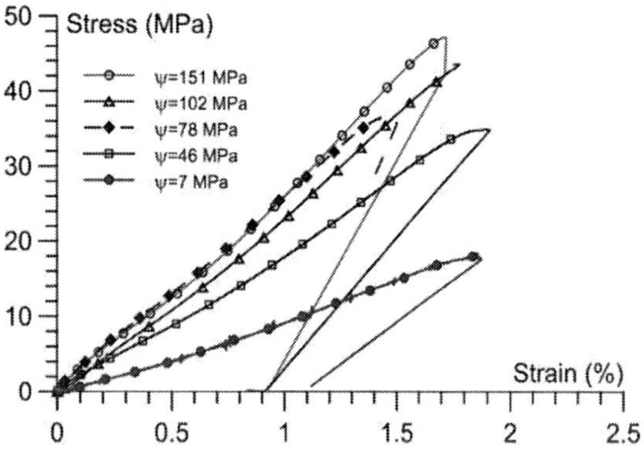

Figure 8. Stress-strain curves for rock block 1 (B1).

Figure 9 contains the peak stress versus suction for all of the tested specimens. A significant reduction in the peak stress can be identified. Comparing the peak stress for the higher (151 MPa) and lower suction (7 MPa), reduction of 61%, 58%, and 67% are obtained for B1, B10 and B17 respectively. It is important to notice that this reduction occurs with a small variation in the water content of the material, from 1% for the driest condition to 2.3 % for the wettest condition (see Figure 1). Therefore, in the context of rock excavations in this material, the considerations of this features are essential to conceive safer designs. Besides, more than 70% of the uniaxial compressive strength reduction is observed for the last hydric state as expected. An additional observation is a non-linear relation between peak stress and suction.

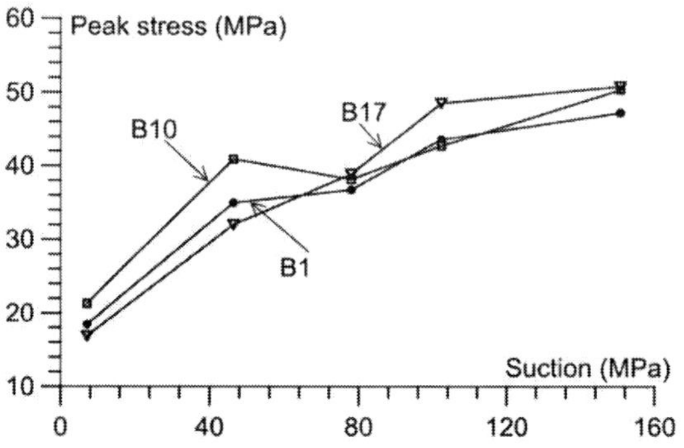

Figure 9. Variation of peak stresses versus suction.

We apply the moving point regression technique to improve the analysis of strain gage data. This procedure utilizes a "sliding window" approach to moving through an x, y data set, fitting a straight line over a user-defined interval. The slope at each point is calculated over the interval and recorded, the process being repeated at successive points [12]. We calculate the slope for each point considering an interval of twenty values, ten before and ten after the point of calculation. Figure 10 presents for rock block B1 the average axial stiffness versus stress for different suction levels. In general, a decrement in the mean axial stiffness with the reduction of suction can be noticed.

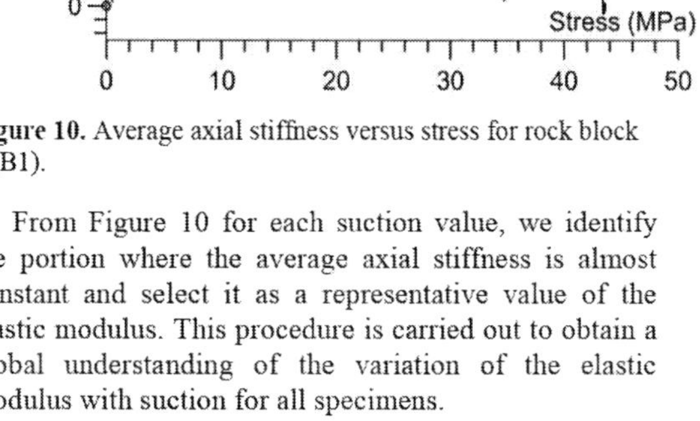

Figure 10. Average axial stiffness versus stress for rock block 1 (B1).

From Figure 10 for each suction value, we identify the portion where the average axial stiffness is almost constant and select it as a representative value of the elastic modulus. This procedure is carried out to obtain a global understanding of the variation of the elastic modulus with suction for all specimens.

Accordingly, Figure 11 summarizes the values of elastic modulus versus suction. Some data dispersion could be noticed from this Figure for suction values of

77 MPa and 102 MPa. Despite this, a general non-linear decrease in the elastic modulus can be observed in Figure 11. For the extreme conditions of suction, elastic modulus decreases 58%, 49% and 35% for B1, B10 and B17 rock blocks. Further, more than 78% of the elastic modulus reduction occurs for the last hydric state as expected.

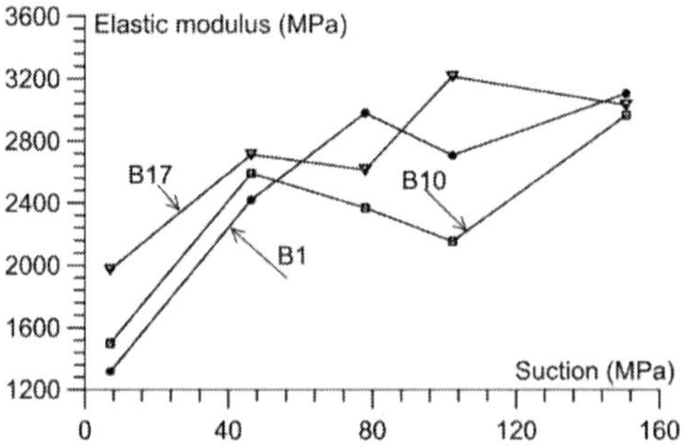

Figure 11. Variation of elastic modulus versus suction.

Data from uniaxial compression tests confirm the influence of suction on the mechanical behaviour of argillaceous rocks. The general trend is a reduction in both peak stress and Elastic Modulus along the wetting path.

3.3 Microstructural observations

At the end of each hydric state, a freeze-dried specimen was used to perform a microstructural analysis. The aim was to evaluate variations in the pore size distribution of the material. To do this, we use an 'AutoPore IV 9500-Micrometrics Instrument Corp.' porosimeter.

Figure 12 plots pore size distribution (PSD) curves for specimens at different hydric states. The dominant pore for the undisturbed material is about 40 nm and exhibits some macropores. PSD curve for the higher suction (ψ=151 MPa) indicates a dominant pore of approximately 34 nm while a considerable increase in the dominant pore of the PSD curve for the lower suction (7 MPa) is observed (180 nm). These features are in agreement with the swelling that was detected in the material. In general, for this wetting path applied are not identifying substantial changes in the macroporosity of the material. Therefore, the main observations in its behavior may be attributed to variations in its microporosity and mesoporosity.

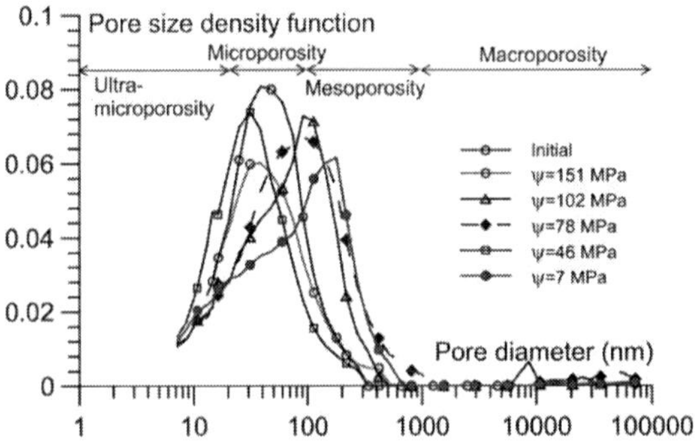

Figure 12. Variation of pore size distribution versus suction.

4 Conclusions

The mechanical properties of Argillaceous rocks from the Colombian Andes mountains are influenced by their moisture content which is the result of its high clay content (53%).

As reported for other argillaceous rocks, the water retention curve represented a hysteresis loop. Dimensions of the specimens tested (20 mm in diameter and 40 mm in height) helped to reduce the time to obtain moisture equilibrium for each hydric state. Axial and radial strains due to the wetting path imposed without mechanical load were found to be clearly anisotropic with a ratio between 2.7 and 3.

Physical and mechanical properties of the rock are dependent on the hydric state and exhibit a non-linear relation with suction. For the compressional wave velocity, a reduction of 25% is obtained for the extreme conditions. On the other hand, peak stress displays a decrease of at least 58%, while the reduction in elastic modulus is of at least 35%. It is important to mention that these changes result from small variations in the water content.

Regarding microstructural observations from MIP analysis, modifications in the pore size distribution were recorded at the micro and meso scale. Therefore, the macroscopic behavior of the samples tested seems to be influenced by the number and size of the pores present in the samples.

 DOI: 10.1051/e3sconf/20160914007

References

1. J. A. Pineda, E. E. Alonso, and E. Romero, Géotechnique **64**, 64-82 (2014)
2. A. G. Corkum and C. D. Martin, Int. J. Rock Mech. Min. **44**, 196-209 (2007)
3. A. Gens, *15th European Conference on Soil Mechanics and Geotechnical Engineering*, (2013)
4. Q. T. Pham, F. Vales, L. Malinsky, D. Nguyen Minh, and H. Gharbi, Phys. Chem. Earth. **32**, 646-655 (2007)
5. F. Valès, D. Nguyen Minh, H. Gharbi, and A. Rejeb, Appl. Clay. Sci., **26**, 197-207 (2004)
6. D. S. Yang, M. Bornert, S. Chanchole, H. Gharbi, P. Valli, and B. Gatmiri, Int. J. Rock Mech. Min. **53**, 45-55 (2012)
7. J. Espitia and B. Caicedo, *6th International Symposium on Deformation Characteristics of Geomaterials*, 342-349 (2015)
8. J. Espitia and B. Caicedo, *UNSAT 2014*, 1597-1603 (2014)
9. I. Decagon-Devices, *WP4 Dewpoint PotentiaMeter for models WP4 and WP4-T Operator's Manual Version 5*, (2007)
10. P. Delage, E. Romero, and A. Tarantino, *1st European Conference on Unsaturated Soils*, (2008)
11. ASTM, *E104 − 02*, (2012)
12. E. Eberhardt, D. Stead, B. Stimpson, and R. S. Read, Can. Geotech J. **35**, 222-233 (1998)

 DOI: 10.1051/e3sconf/20160914008

Rheology of wet granular materials in shear flow: experiments and discrete simulations

Michel Badetti[1], Abdoulaye Fall[1], Jean-Noël Roux[1a]

[1]Laboratoire Navier, 2 Allée Kepler, 77420, Champs-sur-Marne, France

Abstract. The behaviour of wet granular media in shear flow is characterized by the dependence of apparent friction μ^* and solid fraction Φ_S on the reduced pressure P^* and the inertia number I. Reduced pressure, $P^* = \sigma_{22}a^2/F_0$, compares the applied normal stress σ_{22} on grains of diameter a to the tensile strength of contact F_0 (proportional to the surface tension Γ of the liquid and the beads diameter). A specifically modified rotational rheometer is used to characterize the response of model wet granular material (beads with diameter of submillimetric range) to applied shear rate $\dot{\gamma}$ under controlled normal stress σ_{22}. Discrete Element Method (DEM) simulations in 3D are carried out in parallel and numerical results are compared with experimental ones. As P^* is reduced, an increase of the apparent friction coefficient $\mu^* = \sigma_{12}/\sigma_{22}$, measured in the critical state and in slow flows with inertial effects is observed. While the agreement between experiments and simulations is good for dry materials as well as for wet materials in the quasistatic limit ($I \to 0$), some differences appear at finite I, for which some possible origins are discussed.

1 Introduction

In their steady state, shear flows of granular materials under controlled normal stress σ_{22} are well described by the solid fraction Φ_S and the apparent friction μ^* of the grains defined as:

$$\mu^* = \frac{\langle \sigma_{12} \rangle}{\sigma_{22}} \quad (1.a)$$

$$\Phi_S = \frac{\Omega_S}{\Omega} \quad (1.b)$$

where $<\sigma_{12}>$ denotes the time average of the shear stress, σ_{22} the normal confinement stress, Ω the sample volume and Ω_S the volume of the grains.

Flows of dry granular materials are commonly characterized [1,2] by the inertial number I, defined as:

$$I = \dot{\gamma} a \sqrt{\frac{\rho}{\sigma_{22}}} \quad (2)$$

where $\dot{\gamma}$ is the shear rate, a is the grain diameter and ρ its mass density (e.g. 1050 kg.m^{-3} for polystyrene). In the frequent case when elastic contact deflections are negligible, I fully characterizes the state of the material in steady homogeneous shear flow. Constitutive laws express Φ_S and μ^* as functions of I alone. Fig. 1 shows how μ^* and Φ_S vary throughout the flow regimes [1], in the quasi-static limit ($I < 10^{-3}$) μ^* and Φ_S are constant. In the inertial regime ($10^{-3} < I < 10^{-1}$) μ^* increases and Φ_S decreases for increasing values of I.

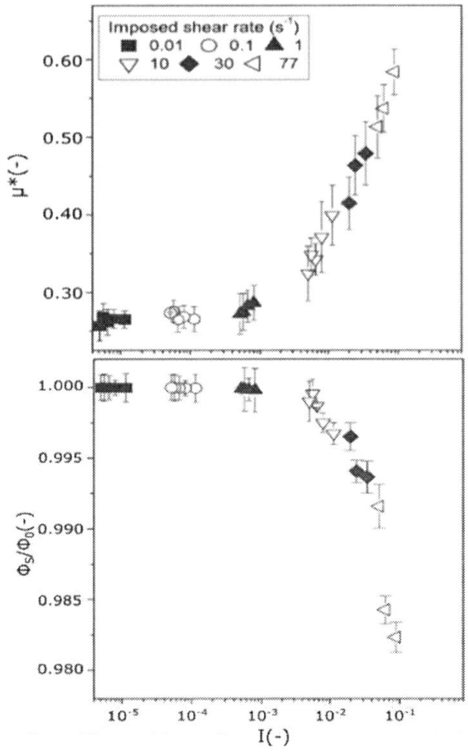

Figure 1: flow curve from [1], μ^* and Φ_S/Φ_0 versus I (Φ_0 being the solid fraction in the quasistatic limit). Notice the plateau for small values of I corresponding to the quasistatic limit.

[a] Corresponding author: jean-noel.roux@ifsttar.fr

In a cohesive material, intergranular contacts exhibit a tensile strength F_0 and the description of granular flow requires an additional dimensionless parameter, the reduced pressure P^*, which compares the cohesive effect with the normal stress :

$$P^* = \frac{\sigma_{22}a^2}{F_0} \tag{3}$$

Previous DEM studies [3] proposed constitutive laws in the form of Φ_S and μ^* as functions of both I and P^*. Despite some investigations of the quasistatic limit [4] there are very few experimental studies of wet granular flow.

The aim of the present communication is to compare DEM simulations of wet grain assemblies to experimental rheometry using a normal stress controlled shear cell, both in the quasistatic limit and in dense inertial flows.

In section 2 the numerical model is presented, a particular focus is made on the cohesive force expression. Section 3 presents the experiments with descriptions of the model wet granular sample, the shear cell and the measurement procedure. In section 4, experimental and numerical results are compared. Further developments are discussed in section 5, stressing the potential interest of X-ray microtomographic investigations of the microstructure of sheared wet granular media.

2 Numerical method

2.1 Principle

Simple plane shear of a 3D box filled with N spheres (4000 or 8800) of diameter a is simulated. The box is periodic in all three directions $\mathbf{x_1}$, $\mathbf{x_2}$, $\mathbf{x_3}$. Fixed values of lengths $L_1 = L_3$ are imposed and σ_{22} is kept constant using a retroaction on length L_2. The shear originates from Lees-Edwards boundary condition, imposing a small displacement $\Delta/2$ and $-\Delta/2$ respectively to the top and bottom cell as illustrated on figure 2. For homogeneous plane shear, shear deformation γ is:

$$\gamma = \frac{\Delta}{L_2} \tag{4}$$

Contact and capillary forces are modelled as in [2,3].

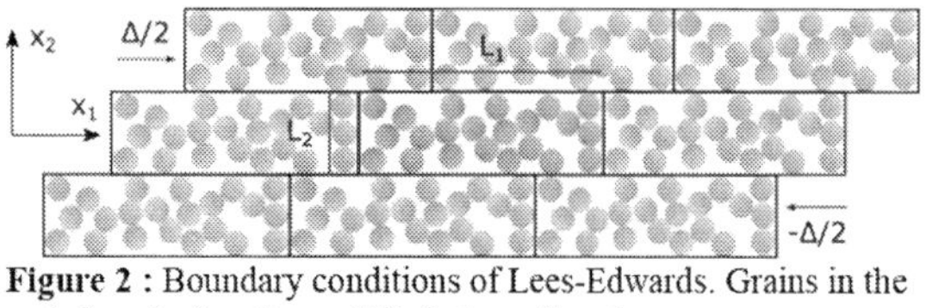

Figure 2 : Boundary conditions of Lees-Edwards. Grains in the main box (colored) are shifted along direction $\mathbf{x_1}$.

2.2 Intergranular forces

Contact forces are implemented following a simplified Hertz-Mindlin-Deresiewicz force model as in [3,5]. The friction in contact satisfies Coulomb's law with a friction coefficient $\mu = 0.09$. This coefficient is a parameter that has been adjusted to fit the dry behaviour of our material in the quasistatic limit (details are given in section 4).

Working within the limits of the pendular state where the saturation degree S_w lies roughly between 1% and 10%, only simple liquid bridges between beads are present (see figure 3). Assuming a perfectly wetting fluid, the cohesive force F_{cap} in a bridge is approximated with the Maugis formula [3,6]:

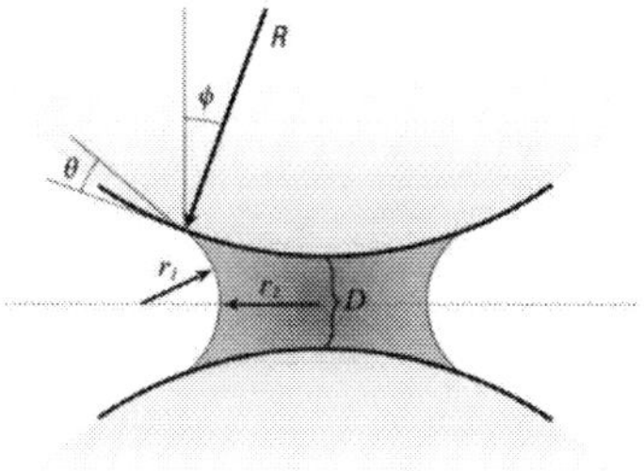

Figure 3: Simple cohesive bridge.

$$F_{cap}(d) = \pi\Gamma a\left[1 - \frac{1}{\sqrt{1 + \frac{4V}{\pi a d^2}}}\right] \tag{5}$$

where d denotes the distance between two spheres, a the beads diameter, Γ the surface tension and V is the meniscus volume. Liquid bridges are assumed to form only once beads are in contact, however, the bridge breaks once a certain rupture distance D_{rupt} is reached. A good approximation of this rupture distance is given by [3,6]:

$$D_{rupt} = V^{1/3} \tag{6}$$

For negative values of d (interpenetration) F_0 is assumed to be a constant equal to $F_{cap}(0)$. Most parameters discussed above can be chosen according to the experimental material (see section 3.1). Setting $V/a^3 = 10^{-3}$ corresponds to S_w about 1%.

2.3 Numerical measurement

Simulation runs are performed until a steady state is reached (μ^* and Φ_S are constant), γ is roughly a few units. In homogeneous plane shear the velocity v is linear with coordinate x_2, a plot of these parameters is proposed on figure 4, the linearity of v with x_2 ensures a good measurement.

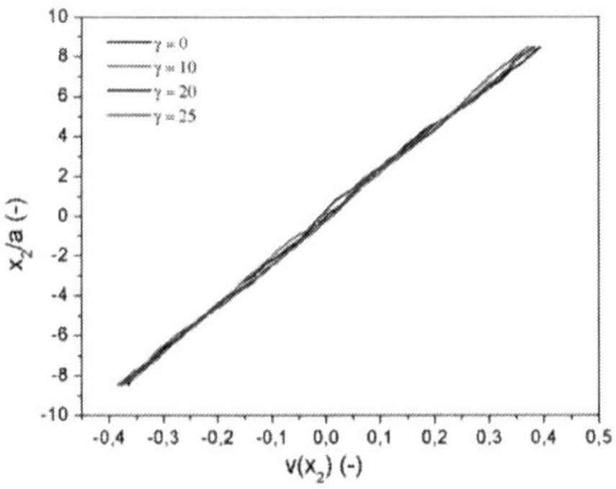

Figure 4: velocity profiles v versus x_2, for different values of γ obtained with $I = 10^{-1}$, $P^*=1$, N=4000 and $V=10^{-3}$.

Values of Φ_S and σ_{ij} are determined as in [3]. Averages and error bars are calculated using the method of block transform estimates [8] in steady state only. Figure 5 presents time series of normal and shear stresses and solid fraction in the steady state.

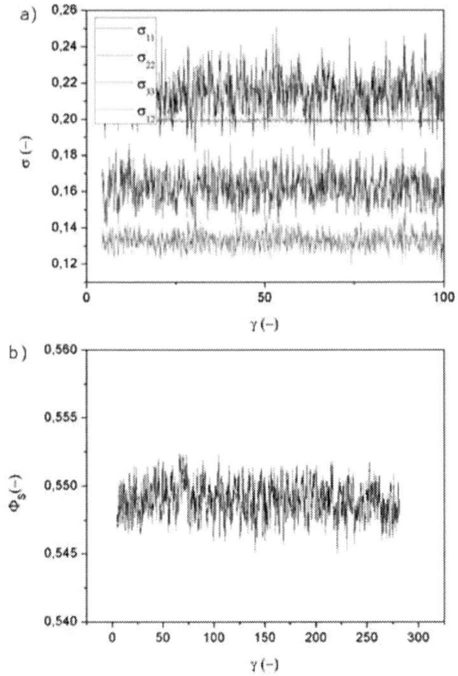

Figure 5: a) Normal stresses and shear stress, b) solid fraction versus γ. Time series obtained with $I=10^{-1}$, $P^*=1$, N=4000 and $V=10^{-3}a^3$.

3 Macroscopic experiments

3.1. Materials and samples

The model material used is an assembly of dry polystyrene beads of diameter a = 500µm mixed together with a non-volatile, wetting, Newtonian silicon oil with a surface tension Γ = 47mN.m^{-2} and viscosity η_L = 500mPa.s. In practice, silicon oils almost perfectly wets polystyrene. The mass ratio between the oil and beads m_L/m_S is controlled by weighing the ingredients before they are mixed. Knowing this ratio enables the control of the liquid saturation S_w as long as the sample volume and mass are known.

3.2. Rheological measurements

Experimental devices must allow the control of the normal stress σ_{22} and the shear rate $\dot{\gamma}$ and enable the measure of the shear stress σ_{12}. To this extent, an annular shear cell is used (see figure 6), it is assumed to produce a simple plane shear in the $\mathbf{x_1}$, $\mathbf{x_2}$ plane, similar to a plane-plane geometry truncated in the center. Defining the inner R_i and outer R_e radii, the sample height H, the rotational velocity ω, the normal force F_n and the torque T, the averages of σ_{12}, σ_{22} and $\dot{\gamma}$ can be expressed as:

$$\sigma_{12} = \frac{3\,T}{2\,R_e^3\left[1-\left(\frac{R_i}{R_e}\right)^3\right]} \quad (7.a)$$

$$\sigma_{22} = \frac{F_n}{\pi R_e^2\left[1-\left(\frac{R_i}{R_e}\right)^2\right]} \quad (7.b)$$

$$\dot{\gamma} = \frac{3R_e}{4H}\frac{\omega}{\left[1-\left(\frac{R_i}{R_e}\right)^3\right]} \quad (7.c)$$

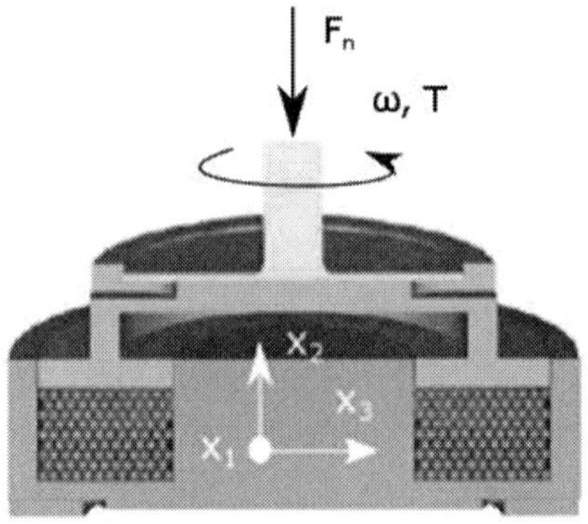

Figure 6: Experimental shear cell. The rotor (yellow) is bound to the top (blue), this ensemble can exert a pressure and a shear on the beads (red) which are confined in the base (grey). Perfect slipping is assumed on the sides (smooth polyoxymethylene), perfect sticking is assumed on top and bottom (square based polystyrene pyramids of 500µm in height and side).

In this study, 2 annular cells are used; they differ only by inner and outer radii:

- A1 – R_i = 2.1cm, R_e = 4.5cm
- A2 – R_i = 3.8cm, R_e = 6.2cm

Edge effects, such as local ordering around the top and bottom (see figure 6) and imperfect slip or loosening along the sides, are assumed negligible. Note that with imperfect slip, a dependence of μ^* on H should appear. Results on dry materials [1] do not exhibit such dependence. The friction at the wall μ_w = 0.05 [1] is also

 DOI: 10.1051/e3sconf/20160914008

small compared with the values of μ^* found experimentally (see section 4).

The order of magnitude of the normal stress in the experiment is 1kPa. Considering the Young modulus of polystyrene beads, the stiffness modulus is well above 10^4. With such order of magnitude of stiffness, it is reasonable to assume that beads behave as perfectly stiff.

Measurements are performed in the steady state, each sample is tested 6 times. For one material there are at least 2 samples used. Averages are deduced from all these measures, error bars are taken as the standard deviation from the averages calculated that way. Raw data from a measurement on a dry material are presented on figure 7. As well as simulations, the measures are performed on a γ interval of several units at least.

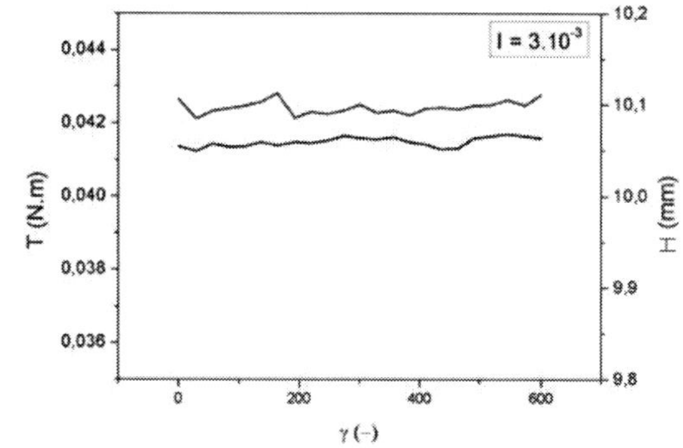

Figure 7: Torque (left axis, black curve) and cell height (right axis, blue curve) versus shear strain. Shear cell A1 is used with a pressure of 900Pa.

4 Results

Figure 8 shows both the numerical and the experimental results obtained for flow curves of dry granular materials, i.e. μ^* and Φ_S versus I. Note that the value of the intergranular friction coefficient μ used in the numerical model is chosen such that $\mu^* = 0.25$ in the quasistatic limit, which is close to the values obtained for dry grains in experiments. This choice is comforted by the very close experimental and numerical values of Φ_S in this flow regime.

For dry materials, the agreement between the 2 cells (A1 and A2) and the simulations is very good. Differences are at most 4% for μ* and 0.7% for Φ_S between A1 and A2, 7% and 0.7% between A1 and simulations, 10% and 0.3% between A2 and simulations. The good agreement between cells A1 and A2 indicates that assuming a constant shear stress is reasonable. The good agreement between experiments and simulations indicates that the experimental device is adequate to measure our material properties.

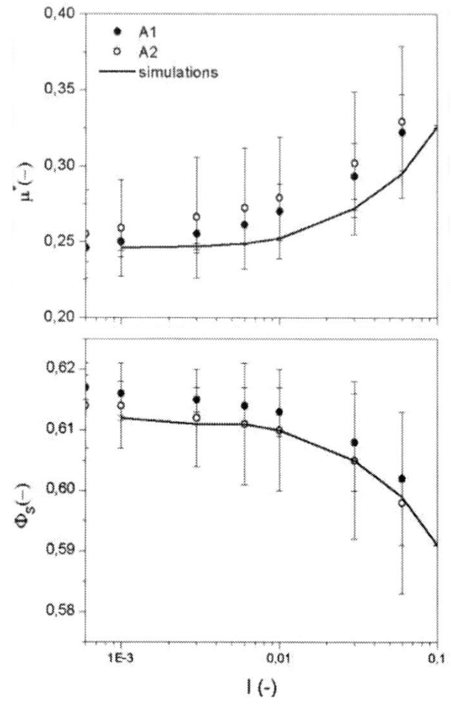

Figure 8: Flow curves of dry grains. Open and closed circles are experiments, solid line are numerical results.

Using now the A1 shear cell, friction coefficient and solid fraction, depending on P^* and I are shown and compared to numerical results on figure 9. The agreement in the quasistatic limit seems good, with $I = 10^{-3}$:

- For $P^* = 2$ - $\mu^* = 0.340$ and $\Phi_S = 0.595$ in experiments versus $\mu^* = 0.364$ and $\Phi_S = 0.597$ in simulations.
- For $P^* = 3$ - $\mu^* = 0.299$ and $\Phi_S = 0.606$ in experiments versus $\mu^* = 0.294$ and $\Phi_S = 0.605$ in simulations.

As remarked in [3], in the quasistatic limit the increase of μ^* for decreasing P^* is well predicted with an effective pressure argument, i.e. assuming the wet grains assembly behaves as a dry material with an additional pressure equal to the capillary pressure.

The agreement between simulations and experiments remains good when comparing the solid fractions for dense flows, the maximal relative difference between experiments and simulations amounts to 0.5% for $P^* = 2$ and 0.8% for $P^* = 3$. There is, however, a notable discrepancy when looking at the values of μ^* in the inertial regime. For higher values of I (above 3.10^{-3}), the relative difference rises up to 30% for P^*=3 and up to 17% for P^*=2. In fact, it is particularly clear that μ^* increases much faster with I in the experiments than in the numerical simulations.

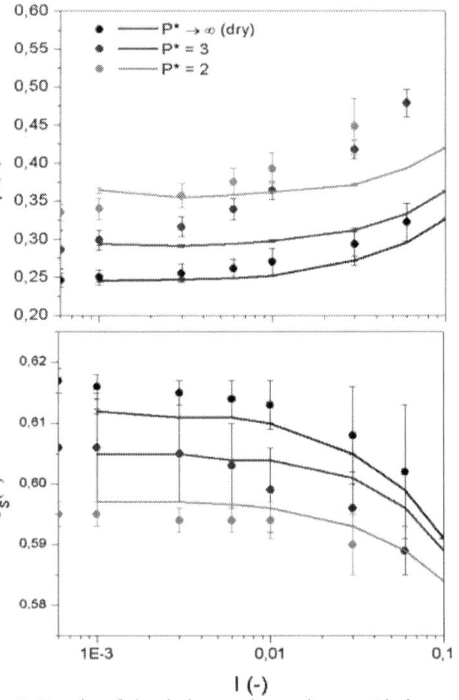

Figure 9: Results of simulations and experiments. Circles are experimental values, plain lines represent numerical values.

Although we do not have an explanation yet, possible clues are provided by the following 4 remarks.

• The fluid viscosity may impact the behaviour of the packing. With increasing viscosity, liquid bridges should take either a stronger force or longer time to break [9] giving rise to a more frictional packing.
• Heterogeneization (drainage, local ordering) may occur during experimental measurements. It is unclear how such occurrence would modify the macroscopic results obtained experimentally.
• Shear localization may occur during experiments [2]. In the case of simple shear, shear bands in the $\mathbf{x}_1$, $\mathbf{x}_3$ plane can be expected. Such localization would lead to an experimentally underestimated shear rate $\dot{\gamma}$, i.e. an underestimated I, which is consistent with the faster increase of μ^* with I found in experiments.
• Some menisci may exhibit a complex morphology despite working with small values of liquid saturation (S_w about 5%) [10]. Characterizing the cohesive effect in such case is much less trivial as the Laplacian pressures can differ greatly from one morphology to another.

5 Further developments

New experimental samples, using different oil viscosity, as well as a thorough study of an eventual gap (defined in section 3.2) effect should give enough clues to assess whether the fluid viscosity or localization have an effect on the experimental measures.

X-ray microtomography experiments are also designed to help answer the questions raised in section 4. They are performed on the samples described in section 3.1. 3D images are taken at different deformation steps. The images are composed of successive slices of materials in the plane perpendicular to the rotation axis (see Fig. 10).

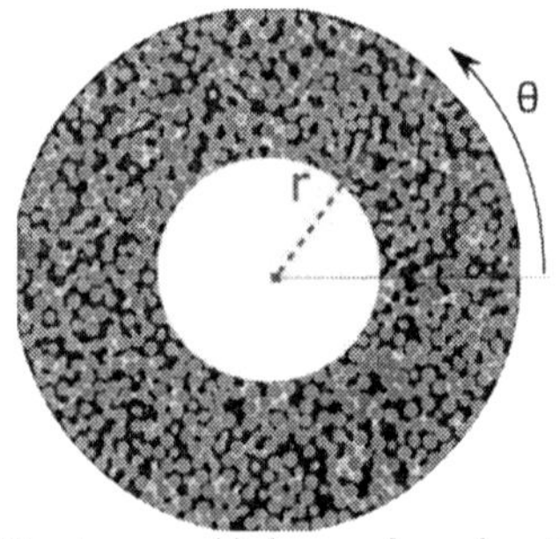

Figure 10: Microtomographic image of sample cut orthogonally to the axis of rotation.

Figure 11 is an example of 2D plot of the liquid saturation and solid fraction in the azimuthal plane (r, z), obtained by averaging over θ from 0 to 2π (see Fig. 10). These pictures show evidence of local ordering and drainage (note the dryer region marked "-" in the top part of the sample).

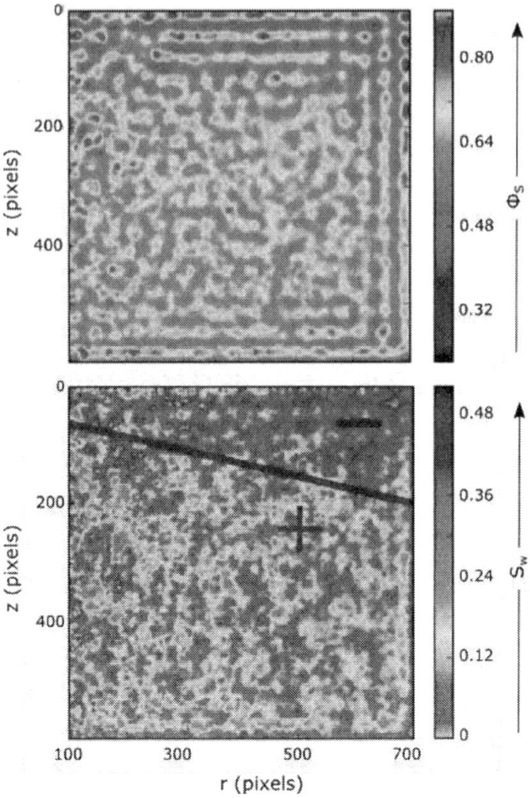

Figure 11: Map of solid fraction (top) and liquid saturation (bottom, global average is 10% here) as averaged over θ, in (r, z) plane (z oriented downwards).

References

1. A. Fall, G. Ovarlez, D. Hautemayou, C. Mézière, J.-N Roux, F. Chevoir, J. of Rheol., **59**,4,(2015)
2. B. Andreotti, Y. Forterre, O. Pouliquen, *Physics,* (EDP Science, 2011)
3. S. Khamseh, J.-N Roux, F. Chevoir, Phys. Rev. E 92, 022201, (2015)
4. V. Richefeu, M. S. El Youssoufi, F. Radjaï, Phys. Rev. 73, 051304, (2006)
5. I. Agnolin, J.-N Roux, Phys. Rev. E 76, 061302, (2007)
6. O. Pitois, P. Moucheront, X. Chateau, J. Coll. Interf. Sci., **231**, 26-31, (2000)
7. I. Agnolin, J.-N Roux, Phys. Rev. E 76, 061302, (2007)
8. H. Flyvbjerg, H. G. Petersen, J. Chem. Phys., **91**, 1, (1989)
9. G. Lefebvre, P. Jop, Phys. Rev E, **88**, 032205, (2013)
10. M. Scheel, R. Seemann, M. Brinkmann, M. Di Michiel, A. Sheppard, S. Herminghaus, Jour. Of Phys., **20**, 494236, (2008)

E3S Web of Conferences **9**, 14009 (2016) DOI: 10.1051/e3sconf/20160914009
E-UNSAT 2016

Settlements in unsaturated granular soils induced by changes in saturation and suction

Marius Milatz[1,a], Tom Törzs[1] and Jürgen Grabe[1]

[1]*Institute of Geotechnical Engineering and Construction Management, Hamburg University of Technology, Hamburg, Germany*

Abstract. In this contribution the hydro-mechanically coupled behaviour of a sand is experimentally investigated with the focus on settlements induced by changes in degree of saturation and suction. This phenomenon, referred to as collapse behaviour, is attributed to rearrangements of the grain skeleton due to changing capillary effects on wetting of the soil. For the experimental investigation of the collapse behaviour of a medium coarse sand cyclic drying-wetting tests are performed under oedometric conditions. In the test set-up a sand specimen, subjected to a constant small vertical stress, is cyclically dried and wetted by volume control of the pore water, whereas matric suction is measured using a tensiometer implemented to the specimen loading plate (topcap tensiometer). The test procedure, originally designed to investigate the hysteretic nature of the soil-water characteristic curve of the sand, allows to measure the one dimensional volume change of the specimen as evoked by the applied hydraulic paths under constant net stress. By varying the specimen void ratio the impact of density on the collapse behaviour can be assessed. The test data is important for the development of a mechanical constitutive model which can take the volume change behaviour due to suction changes into account.

1 Introduction

The hydraulic and mechanical behaviour of unsaturated soils is closely related due to capillary effects. These effects, which can either be attributed to an averaged macroscopic capillary pressure or to micromechanical menisci forces caused by surface tension of pore water, influence the stress state of the grain skeleton and therefore also the shear strength and volume change behaviour of the soil. When the soil encounters a change in degree of saturation or suction, e. g. due to infiltration, the capillary effects vary, which has an impact on the mechanical behaviour. Especially for cohesive soils a pronounced volume change is encountered due to changes of the hydraulic state with the soil swelling on wetting and shrinking or even cracking on drying. This volume change behaviour is an important research issue when structural damage to foundations or leakage through low permeability barrier materials is of concern.

An early approach to capture the volume change behaviour of unsaturated cohesive soils was presented by Matyas and Radhakrishna (1968), who described volume change by state surfaces in the net stress-suction-void ratio space. From the results a metastable structure with collapse behaviour induced by changes in matric suction and an influence of hysteresis could be deduced (Fredlund & Rahardjo 1993).

The volume change behaviour of cohesive materials has also been investigated experimentally by Sharma and Wheeler (2000) with a special focus on drying-wetting cycles. The results indicate irreversible compression during drying of the investigated specimens and an influence of hysteresis.

The volume change effects of granular soils are typically less pronounced, although settlement behaviour has been observed for loose sands due to wetting, which is referred to as *capillary collapse*. The observed volume change is attributed to microstructural effects. These are given for example, when the grain skeleton adopts a more stable state as the capillary forces change.

Bruchon et al. (2013) used X-ray Computer Tomography (CT) to analyze changes in the three-dimensional grain skeleton structure upon wetting of a sand. In their tests a small cylindrical specimen of loose sand was wetted from the bottom with a hanging water column technique and simultaneously scanned in a CT-device. The results indicated that the local collapse behaviour is related to the coalescence of capillary bridges in the grain skeleton upon wetting.

In order to describe the volume change behaviour of unsaturated soils by constitutive laws, an effective stress relation is needed. With the enhanced effective stress principle for unsaturated soils, presented by Bishop (1959) in the form of equation 1, the partial pressures of pore air, u_a, and pore water, u_w, can be incorporated. Weighted by the Bishop-parameter χ, the influence of capillary pressure or suction, $s = u_a - u_w$, on effective stress can be considered.

$$\sigma' = \sigma - u_a + \chi(u_a - u_w) \quad (1)$$

[a] Corresponding author: marius.milatz@tuhh.de

Bishop stress is widely used to describe the stress state of unsaturated soils and has also been applied to calculate shear strength with the Mohr-Coulomb limit state equation. It is known to predict the increase in shear strength well for some soils, which has been investigated by triaxial testing (Bishop, 1961; Bishop & Blight, 1963). However, an assumption for the parameter χ has to be made. This parameter is usually assumed to be equal to the degree of saturation ($\chi = S_r$) or the effective degree of saturation ($\chi = S_e$). In other cases different and more complex relationships are assumed. The role and shape of the χ-parameter is frequently discussed in literature (e. g. Pereira et al., 2011).

As it was found that Bishop stress was not always able to capture shear strength and volume change at the same time, Fredlund and Morgenstern (1977) developed the principle of two independent stress state variables. The use of either one or two stress state variables for the modelling of the constitutive behaviour of unsaturated soils is discussed by Jommi (2000). When two stress state variables, e. g. net stress, $\sigma - u_a$, and suction, $u_a - u_w$, are used, the influence of changes in net stress and suction on the constitutive behaviour can be separated. This allows to model volume change at constant net stress induced by suction changes.

Up to today the coupling mechanisms are not fully understood and no model has been found to describe both stress state and volume change behaviour properly at the same time for different soil types Jommi (2000).

In this contribution an experimental method is presented to study the coupling effects in granular soils leading to settlements as a result of changes in saturation and suction, in order to gather more data on this phenomenon. The measured results will be interpreted in terms of changes of the effective stress for unsaturated soils and may be used in further studies to develop or improve a constitutive model to take the hydro-mechanical coupling into account.

2 Experimental investigation of settlement behaviour

2.1 Investigated soil

In continuation of a research project on unsaturated granular soils as subgrade materials for sites of container handling in ports and the influence of repeated vehicle loads (Grabe and Milatz, 2014) a medium coarse sand from Container Terminal Burchardkai (CTB) at the Port of Hamburg was investigated. This sand is of sedimentary origin from the river Elbe at Hamburg and is used as backfilling material for quay wall construction. In different experiments the hydraulic and hydro-mechanical behaviour of this sand has been investigated (Milatz, 2013; Milatz and Grabe, 2014; Milatz and Grabe, 2015; Milatz, 2015). The same sand will be investigated in the studies presented in this contribution with the focus on hydro-mechanical coupling and settlements induced by changes in the hydraulic state. Selected soil parameters of CTB-sand are summarised in table 1.

Table 1. Selected soil parameters of CTB-sand.

ρ_s	e_{min}	e_{max}	d_{10}	d_{max}	U
(kg/m³)	[-]	[-]	[m]	[m]	[-]
2650	0.41	0.86	1.6e-4	2e-3	2.4

ρ_s: grain density, e_{min}: minimum void ratio
e_{max}: maximum void ratio
d_{10}: grain diameter at 10% passing
d_{max}: maximum grain diameter, U: uniformity index

The main findings on the hydraulic behaviour of CTB-sand are its low capillarity with a void ratio-dependent air entry value (AEV) of the soil-water characteristic curve of 2.5 to 3 kPa and a pronounced hysteretic behaviour with air entrapment on drying-wetting cycles (Milatz, 2015; Törzs, 2015).

2.2 Test set-up and testing procedure

In Milatz and Grabe (2014) and Milatz and Grabe (2015) a simple shear device was presented that allows to control and measure matric suction in unsaturated sand specimens on shearing. For this purpose a vacuum method for suction control was applied and a topcap tensiometer implemented to the simple shear apparatus. This method was enhanced to study the drying-wetting behaviour in cylindrical sand specimens under oedometric stress conditions for cyclic hydraulic paths in Milatz (2015) and Törzs (2015). In the executed hydraulic tests on sand specimens with a diameter of 50 mm and initial height of 20 mm a pore water volume change was prescribed with a computer-controlled enhanced precision pore water pressure controller (GDS Instruments). The specimen matric suction response due to the applied changes in saturation was measured with a topcap tensiometer. With the help of this testing procedure the hysteretic paths of the soil-water characteristic curve could be investigated. The test results show a negligible relationship between the applied flow velocity and the measured hydraulic paths (Törzs, 2015). In the hydraulic tests presented in Törzs (2015) the original flat ceramic disc of the topcap tensiometer used in the simple shear tests was modified and changed to a pointy ceramic tip that pierces into the core of the cylindrical sand specimens. This method, meant to improve hydraulic contact and to measure a mean suction value, was adopted in the studies presented in this paper. The full test setup is represented in figure 1. A comparison of the two existing topcap tensiometers is shown in figure 2.

The sand specimens are dewatered through the base, in which a microporous membrane (Pall Corp.) with pore diameter of 0.8 µm is clamped. This filter membrane method represents an alternative to ceramic discs and has already been used by Nishimura et al. (2012) to measure the SWCC. In the case of this study the filter membrane is important, because the pore water volume-controlled testing requires a high hydraulic conductivity of the specimen interface to the dewatering system. The dewatering system connected to the pore pressure controller is initially filled with deaired water. The filter membrane is also deaired and wetted by placing it in a vacuum chamber. The test set-up allows to prescribe hydraulic paths to the investigated sand specimen, in order to investigate different paths of the soil-water characteristic curve. When a sand specimen is dryed from the bottom, the pore space is filled with air which can pass through holes in the topcap. Therefore, it is assumed that the pore air pressure is equivalent to the atmospheric pressure (u_a = 0), resulting in a matric suction equivalent to the negative pore water pressure (s = $-u_w$). The negative pore water pressure is then directly measured with the topcap tensiometer.

Figure 1. Test set-up for the hydraulic tests on CTB-sand. 1: T5 Laboratory tensiometer (UMS GmbH, Munich), 2: Topcap with pointy ceramic tensiometer tip, 3: LVDT, 4: Cylindrical sand specimen in aluminium cylinder and latex membrane, 5: Membrane holder.

Figure 2. Topcaps for tensiometer connection to measure matric suction in cylindrical sand specimens. Left: Toothed topcap with a flat ceramic disc used in simple shear tests. Right: Topcap with a pointy ceramic tip used for the tests on hydro-mechanical coupling in the present study.

Typical results of a hydraulic test with cyclic hydraulic loading using the topcap tensiometer with the flat ceramic disc from Milatz (2015) are represented in figure 3. It can be seen that the test results from conventional methods to obtain the soil-water characteristic curve are well reproduced. Furthermore, the hysteretic behaviour can be investigated thoroughly.

In this study, however, the hydro-mechanical coupling is of interest. In order to capture specimen volume change induced by changes in the hydraulic state, an LVDT with high accuracy of ±0,005 mm is used (compare figure 1).

The sand specimens are prepared by pluviating dry sand into the water filled test container. For looser specimens with a high initial void ratio e_0 no compaction was applied, whereas specimens with higher initial density were achieved by compacting them on a vibrating table during specimen preparation. The topcap is then lowered down onto the specimen top using a load-controlled docking procedure. The docking is completed, when a vertical stress of 10 kPa is reached. This small stress level was found to be enough for the tensiometer tip to penetrate into the middle of the sand specimens as shown in figure 1. In a second test step the vertical stress level is linearly ramped to 25 kPa for all specimens and then kept constant in the following test steps. This small stress level is selected in order to reduce the initial settlements and changes in void ratio due to consolidation.

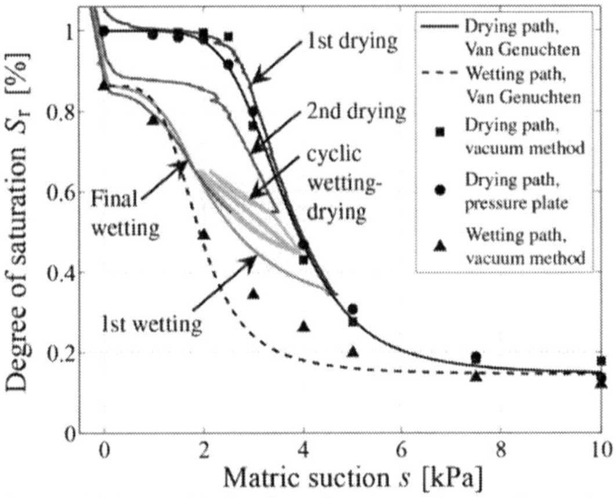

Figure 3. Measured hydraulic paths compared to test data and calculated data using the Van Genuchten (1980) model for CTB-sand with e_0 = 0.5 from Milatz (2015).

After application of vertical stress and equilibration of all settlements, three drying-wetting cycles are prescribed. For this purpose the pore water volume is removed and added linearly over time using the pore water pressure controller and a flow velocity of 0.5 to 5 mm³/s. The hydraulic test steps begin with a first drying stage, in which S_r is reduced to 0.3. It was found that this degree of saturation seems to represent a limit for accurate tensiometer measurements due to a loss of hydraulic contact with the tensiometer tip and the pore water phase. In this case a further drying of the sand specimens leads to oscillating tensiometer measurements. Furthermore, test results indicate that this failure of the topcap tensiometer appears to be more pronounced for very dense specimens.

On rewetting of the sand specimens significant air entrapment was discovered with an air degree of saturation $S_a = 1 - S_r \approx 0.18$ reached after the first wetting path (Milatz, 2015). Therefore, the sand specimens are rewetted only up to S_r = 0.82 to avoid over-saturation. However, this air degree of saturation turns out not to be a constant, because the volume of entrapped air appears to increase during cyclic drying-wetting and to converge asymptotically to a residual air degree of saturation. The target values of S_r are used to calculate the pore water volume change to be prescribed by the pressure controller. In these calculations as well as in the overall data evaluation a volume change of the specimens due to settlements is taken into account, because it leads to changes in S_r during the test. The drying-wetting cycles with $0.3 \leq Sr \leq 0.82$ are repeated twice resulting in a total of three drying-wetting cycles.

In the selected tests presented in this contribution, the void ratio was varied, in order to capture the influence of density on hydro-mechanical coupling due to cyclic changes in suction and degree of saturation. The initial void ratios e_0 and void ratios after the initial consolidation stage e_1, which represent the true initial values at the start of the hydraulic test stage, are summarized in table 2.

Table 2. Specimens of CTB-sand with different initial void ratios for tests on the hydro-mechanical coupling.

Specimen no.	e_0 [-]	e_1 [-]
1	0.69	0.66
2	0.64	0.62
3	0.56	0.54

e_0: initial void ratio, e_1: void ratio after consolidation

3 Test results

3.1 Suction response of topcap tensiometer

Before the tests the topcap tensiometer was filled with deaired water and checked for response time and accuracy by applying negative pore water pressure steps with a vacuum controller. The results indicate a quick sensor response and good accuracy at low suction, as represented in figure 4.

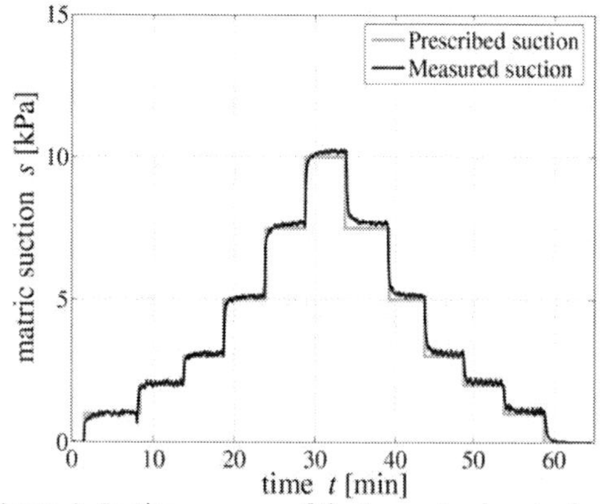

Figure 4. Suction response of the topcap tensiometer in a test procedure using a vacuum controller.

The prescribed changes in degree of saturation and a typical measured suction response versus test time for sand specimen 1 at $e_0 = 0.69$ are represented in figure 5. It can be seen that a characteristic suction response behaviour is measured. Under the assumption that the degree of saturation, calculated from the outflow and inflow data, and the measured matric suction are representative values for the whole sand specimen, the hydraulic paths of the soil-water characteristic curve can be plotted as represented in figure 6.

The soil-water characteristic curve shows the typical hysteretic behaviour. It can be seen that the entrapped air volume increases after each wetting cycle. Different studies (Milatz, 2015; Törzs, 2015) demonstrate that the soil-water characteristic curve of the investigated CTB-sand is well reproduced compared to conventional methods, such as the pressure plate technique or vacuum method. The method can be used to investigate different hydraulic paths, which can be easily prescribed with the available hydraulic functions for water volume flow of the pore water pressure controller. However, the method seems to be limited to $S_r \geq 0.3$. Unfortunately, a good suction response was not given in all tests. It was noticed that for specimen 3 with the smallest initial void ratio the tensiometer showed oscillations and often the hydraulic contact appeared to be lost.

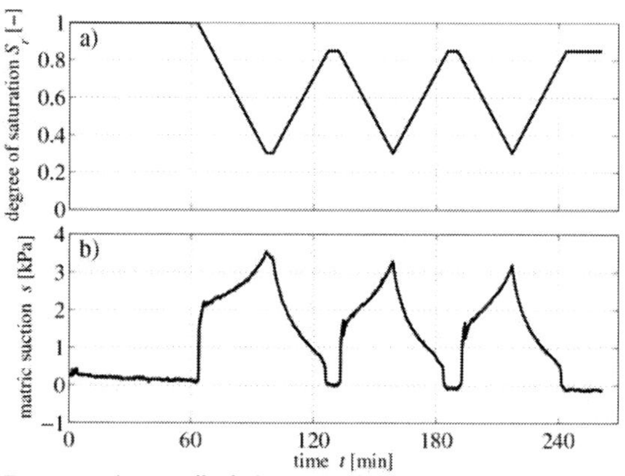

Figure 5. a) Prescribed change in degree of saturation and b) measured matric suction versus time in the hydraulic test on a sand specimen with $e_0 = 0.69$.

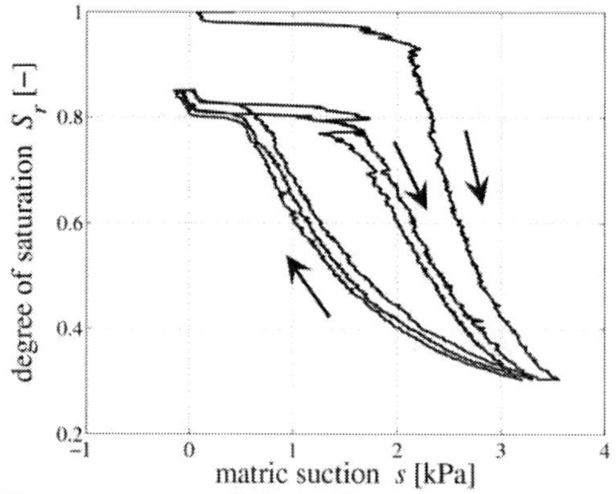

Figure 6. Measured hydraulic paths of the soil-water characteristic curve in the s-S_r plane for the sand specimen with $e_0 = 0.69$.

3.2 Settlements induced by changes in saturation and suction

The hydraulic test steps allowed to measure the volume change behaviour during cyclic drying and wetting. In all cases a typical volume change pattern was noticed as represented in figure 7 for sand specimen 1 with $e_0 = 0.69$. It can be seen that the main changes in void ratio due to settlements under constant vertical stress occur on rewetting of the specimen shortly before $S_r = 0.82$ as the maximum degree of saturation after air entrapment is reached. When plotted versus measured suction, the volume changes closely coincide with a suction drop to zero. For a second and third rewetting the volume change becomes smaller. This behaviour could be observed for all selected specimens.

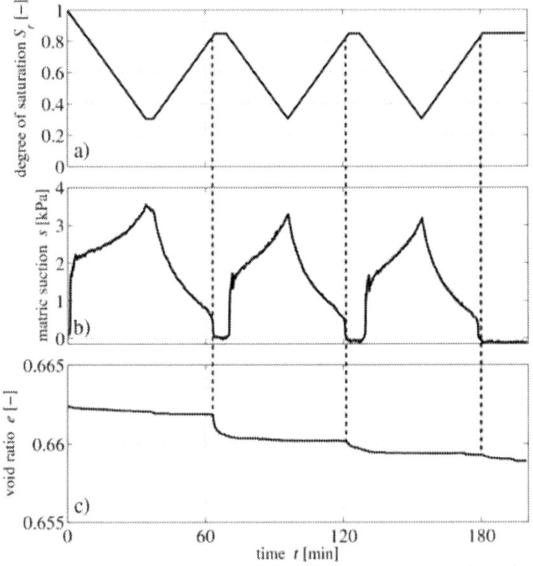

Figure 7. a) Prescribed paths of degree of saturation, b) measured matric suction and c) changes in void ratio calculated from measured settlements for a sand specimen with $e_0 = 0.69$.

The influence of changes in degree of saturation and measured matric suction on volume change of the sand are further emphasized when the axial strain of different specimens is plotted over degree of saturation or matric suction as represented in figures 8 and 9. The recorded volume change can be directly related to changes in S_r and s. For all specimens a very slight increase in compressive axial strain (negative mathematical sign) can be noticed during drying. However, the compressive axial strain mainly increases during wetting. The increase in axial strain is not equally pronounced in all cases and a clear relationship to the initial void ratio cannot be found, when the presented tests are considered. An example of this is the relatively small volume change for specimen 1 with the highest initial void ratio of $e_0 = 0.69$ which may represent an outlier.

When the change of theoretical Bishop effective stress $\Delta\sigma' = s\chi$ with $\chi = S_r$ according to equation 1 is considered in figure 10, the resulting paths in the $\Delta\sigma'$-ε-plane resemble oedometric loading and unloading paths. However, the mechanical response seems to be inversed compared to the observed behaviour when total stress is applied to the specimen. In the case of small changes in $\Delta\sigma'$ a very high stiffness is observed for loading (increase of $\Delta\sigma'$) and a small stiffness for unloading (decrease of $\Delta\sigma'$).

Although the applied stresses and measured axial strains are very small and maybe suffer from errors of measurement or specimen preparation, the observed behaviour indicates some intergranular phenomena occurring due to changes in degree of saturation and suction. The results clearly show that an increase in $\Delta\sigma'$ has a smaller effect on volume change than its decrease. Obviously, also the hydraulic hysteresis is of importance. On the micromechanical scale the results maybe an indication that the rewetting of the grain skeleton leads to restructuring of singular grains, because the menisci or water-air-interfaces change. With a further reduction of void ratio the hydro-mechanical coupling effects seem to be reduced, which can be deduced from the decreasing axial strains. The product of matric suction and degree of saturation $\Delta\sigma' = s\, S_r$, referred to as "suction stress" by Lu and Likos (2006), can be used to describe the micromechanical changes in intergranular stress on a macroscopic scale.

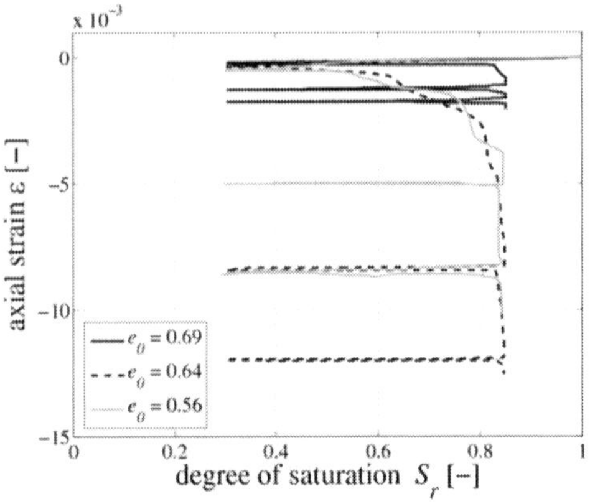

Figure 8. Axial strain versus degree of saturation for sand specimens with different initial void ratios in the hydraulic test stage with three drying-wetting cycles.

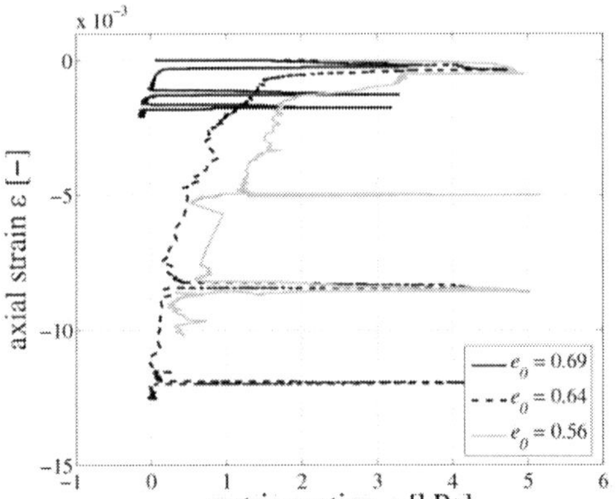

Figure 9. Axial strain versus measured matric suction for sand specimens with different initial void ratios in the hydraulic test stage with three drying-wetting cycles.

It is interesting to notice that unlike most cohesive soils, the investigated sand specimen only encounters a negative volume change due to increasing settlements. The effect of swelling on rewetting, as described and modelled e. g. by Wheeler et al. (2003) for cohesive soils, is irrelevant for the sand. Compared to a cohesive soil, where the electro-chemical bonds are additionally influenced by changes in degree of saturation, leading to a pronounced volume change behaviour, the sand encounters only weak capillary effects, which directly interact with the grain skeleton. Under gravity and constant normal stress, coarse grain skeletons will develop states of higher density which are more stable compared to looser initial grain structures, when single particles are restructured due to changes in the hydraulic state.

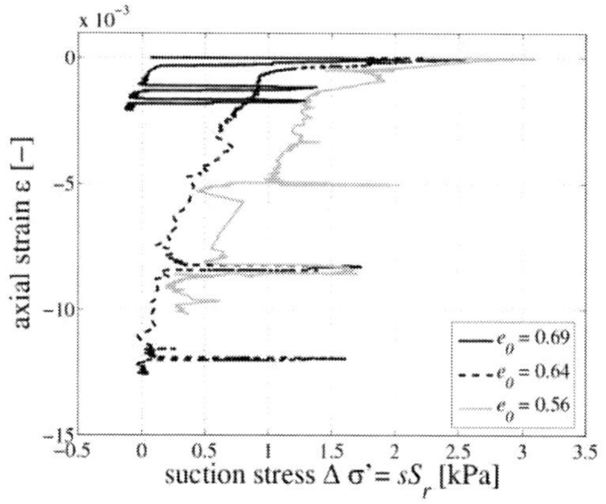

Figure 10. Axial strain versus change in effective stress $\Delta\sigma' = s\chi$ ($\chi = S_r$) for sand specimens with different initial void ratios in the hydraulic test stage with three drying-wetting cycles.

4 Summary and outlook

In this contribution selected results of different tests on the hydro-mechanical coupling of an unsaturated medium coarse sand have been presented. In the presented tests sand specimens under oedometric stress state have been investigated by prescribing pore water volume-controlled changes in degree of saturation and by applying simultaneous suction measurements with a tensiometer. It could be shown that the presented test method, originally meant to investigate the hysteretic nature of the soil-water characteristic curve of sandy soils, can be expanded to study specimen settlements induced by changes in degree of saturation and suction.

The results indicate that changes in the unsaturated stress state lead to settlements on rewetting. These can be related to changes in the contribution of suction and degree of saturation to Bishop effective stress. The settlements due to an increase in suction stress $\Delta\sigma' = sS_r$ on drying of the sand specimens are small compared to the effect of wetting paths. The results show a collapse-like behaviour on rewetting, which may be due to a restructuring of single grains as the capillary menisci change on wetting.

In the future it is planned to evaluate further hydraulic paths controlled with the presented test setup to investigate the influence of void ratio and also normal stress state which was not varied in the selected tests presented in this paper. The expected further results may help to better understand the volume change behaviour of coarse grained soils in the unsaturated state and can be used for the enhancement of existing constitutive models or for the design of new ones that can take the observed coupled effects into account.

References

Bishop, A. W. (1959). The principle of effective stress. *Teknisk Ukeblad*, 106 (39), 859–863.

Bishop, A.W. (1961). The experimental study of partly saturated soil in the triaxial apparatus. In: *Proc. of 5th International Conference on Soil Mechanics, Paris*, pp. 13–21.

Bishop, A. W. & G. E. Blight (1963). Some aspects of effective stress in saturated and partly saturated soils. *Géotechnique* 13, 177–197.

Bruchon, J.-F., J.-M. Pereira, M. Vandamme, N. Lenoir, P. Delage, & M. Bornert (2013). Full 3d investigation and characterisation of capillary collapse of a loose unsaturated sand using x-ray ct. *Granular Matter* 15, 783–800.

Fredlund, D. G. & N. R. Morgenstern (1977). Stress state variables for unsaturated soils. *ASCE J. Geotech. Eng. Div. GT5 103*, 447–466.

Fredlund, D. G. & H. Rahardjo (1993). *Soil mechanics for unsaturated soils*. John Wiley & Sons.

Grabe, J. & M. Milatz (2014). The change of matric suction due to heavy vehicle crossing. In: *Proc. of 6th International Conference on Unsaturated Soils (UNSAT 2014), 2-4 July 2014, Sydney, Australia*, Volume 2, pp. 1431–1437.

Jommi, C. (2000). Remarks on the constitutive modelling of unsaturated soils. In: A. Tarantino and C. Mancuso (Eds.), *Experimental evidence and theoretical approaches in unsaturated soils*., pp. 139–153. Rotterdam: Balkema.

Lu, N. & W. J. Likos (2006). Suction stress characteristic curve for unsaturated soil. *Journal of Geotechnical and Environmental Engineering* 132 (2), 131–142.

Matyas, E. L. & H. S. Radhakrishna (1968). Volume change characteristics of partially saturated soils. *Géotechnique* 18, 432–448.

Milatz, M. (2013). On the control of low negative water pressures in laboratory tests on unsaturated sand. In: *Proc. of 5th International Young Geotechnical Engineers' Conference (iYGEC), Paris, France*, pp. 55–58.

Milatz, M. (2015). An experimental method to study the drying-wetting behaviour of a sand. In: *Proc. of 6th Asia-Pacific Conference on Unsaturated Soils (AP–UNSAT 2015)*. (Paper accepted for publication).

Milatz, M. & J. Grabe (2014). Triaxial and simple shear tests on unsaturated sand specimens with negative pore water pressures. In: *Proc. of 6th International Conference on Unsaturated Soils (UNSAT 2014), 2-4 July 2014, Sydney, Australia*, Volume 2, pp. 1631–1637.

Milatz, M. & J. Grabe (2015). A new simple shear apparatus and testing method for unsaturated sands. *Geotechnical Testing Journal* 38, 9–22.

Nishimura, T., J. Koseki, D. G. Fredlund, & H. Rahardjo (2012). Microporous membrane technology for measurement of soil-water characteristic curve. *Geotechnical Testing Journal* 35 (1), 1–8.

Pereira, J. M., O. Coussy, E. E. Alonso, J. Vaunat, & S. Olivella (2011). Is the degree of saturation a good candidate for bishop's χ-parameter? In: E. E. Alonso and A. Gens (Eds.), *Proc. of 5th International Conference on Unsaturated Soils*, pp. 913–919.

DOI: 10.1051/e3sconf/20160914009

Sharma, R. S. & S. J. Wheeler (2000). Behaviour of an unsaturated highly expansive clay during cycles of wetting and drying. In: *Proc. of 1st Asian Conference on Unsaturated Soils (UNSAT-ASIA 2000), Singapore*, pp. 721–726.
Törzs, T. (2015). *Experimental and numerical investigations on moisture transport in unsaturated soils.* Master thesis, Institute of Geotechnical Engineering and Construction Management, Hamburg University of Technology. (Unpublished).
Wheeler, S. J., R. S. Sharma, & M. S. R. Buisson (2003). Coupling of hydraulic hysteresis and stress-strain behaviour in unsaturated soils. *Géotechnique* 53 (1), 41–54.

Relating shear strength of unsaturated soils with capillary water retention curve

Annan Zhou[1,a] and Daichao Sheng[2]

[1]*Royal Melbourne Institute of Technology, Melbourne, Vic 3001, Australia*
[2]*The University of Newcastle, Callaghan, NSW 2308, Australia*

Abstract. This paper proposes a new water retention model for unsaturated soils, which takes into account capillary condensation of adsorbed water. In the proposed water retention model, the degree of saturation of a soil is separated into that based on capillary water and that based on adsorbed water. Through the analysis of a partially saturated two-cylinder system, a new shear strength criterion for unsaturated soils is proposed, in which only the degree of saturation based on capillary water contributes to the variation of shear strength with suction. The proposed shear strength criterion is justified against thermodynamic principles. The proposed strength criterion is compared against existing criteria in the literature, which shows that it provides a much improved prediction of the experimental data, for a wide range of suction values.

1 Introduction

A number of shear strength criteria for unsaturated soils have been proposed in the literature during past three decades or so. It is often recognised that the shear strength of an unsaturated soil is closely related to the suction or the degree of saturation in the soil. Therefore, the degree of saturation that can be obtained from the water retention curve for a given suction has been widely embedded into strength criteria for unsaturated soils, such as [1]; [2]; [3] and [4]; etc.

With a given suction, water retained in soil pores due to different processes contributes differently to the shear strength of the soil. Pore water in an unsaturated soil can be divided into two categories: capillary water and adsorbed water. Only the capillary water contributes to the shear strength. Capillary water exists among soil particles and the pressure of capillary water affects the contact stress (σ') among soil particles [3], [5]. Compared with capillary water, the contribution from adsorbed water to the shear strength of the soil is very limited [6-9]. In principle, this is because adsorbed water wraps the surface of each soil particle and the thus pressure of adsorbed water almost does not affect the contact stress (σ') among soil particles.

On the other hand, water retention equations for unsaturated soils, such as [10] and [11], are commonly based on simple regression of experimental data, and overlook the physical mechanisms behind. These water retention equations usually assume that all pore water exists as capillary water by simplifying soil pores to capillary cylinders. No specific surface area is explicitly considered in these equations and hence the existence of adsorbed water is ignored. When the degree of saturation is embedded in the shear strength criteria [3, 12], it leads to significant overestimation of shear strength of unsaturated soils, especially for fine-grained soils and high suction values [13]. Using the effective degree of saturation to replace the degree of saturation in shear strength criteria [2, 8, 9] can somewhat improve the prediction, but still overestimates the shear strength for a wide range of suction values [13]. The use of the effective degree of saturation somewhat acknowledges the existence of adsorbed water, as the residual water content, that exists mostly in form adsorbed water, comes into play. However, in the definition of the effective degree of saturation, the volume of residual water or microscopic water (i.e., the water attached to the particles or trapped in the micro-pores) is usually assumed to be either independent of suction [2, 9, 14-16] or increasing monotonically along with the decrease of suction [8, 17-19]. The assumption of constant adsorbed water conflicts with surface physics principles [5]. Besides, adsorbed water cannot always increase along with the decrease of suction because of capillary condensation [20]. Other strength criteria use an additional fitting parameter to improve their predictions [1, 4] According to a comparative study by [13], shear strength criteria with additional fitting parameters usually fit observed data better, but the challenge is to determine these additional parameters.

The objective of this study is to derive the shear strength criterion for unsaturated soils by considering that pore water consists of capillary water and adsorbed water. The physical mechanisms of water retention in porous media are reviewed first. A new water retention

[a] Corresponding author: annan.zhou@rmit.edu.au

model is proposed, which takes into account (i) the difference between capillary water and adsorbed water and (ii) the possibility of capillary condensation. A simple shear strength criterion that only involves capillary water is derived based on the analysis of a partially saturated two-cylinder system and is justified against thermodynamic principles. The proposed model is compared with existing shear strength criteria, against experimental data in the literature.

2 Theoretical backgrounds

Based on the qualitative discussion of the distribution of water on the surface of two soil particles [21], [22] developed a quantitative unitary approach to capillarity and adsorption in porous media. The liquid-vapor interface is considered as a surface of either constant partial specific Gibbs free energy (Ψ) or constant chemical potential (μ) which consists of an adsorptive component and a capillary component. The capillary component (C) of the potential depends on the liquid-vapor interface mean curvature (κ), and the adsorptive component (A) depends on the normal distance of the interface from the solid (h). [20] modified the unitary approach of [22] by establishing a linkage between modern interface science concepts and equations for adsorption developed by [5] and [23]. A considerable simplification was also introduced by [20], by considering the absorptive and capillary contributions separately, which is termed the shifted Young-Laplace equation. The most important advantage of the shifted Young-Laplace approach is that this approach is able to separate adsorbed water from capillary water at a given chemical potential, which makes the derivation of the capillary water retention equation possible.

The total potential of water in soil is equivalent to the chemical potential of soil water at a chosen temperature and atmospheric pressure [24]. Given that water is incompressible, the negative total potential is equal to the total suction ($\psi_t = u_a - u_w$, u_a being the pore air pressure and u_w the pore water pressure) divided by the water density (ρ). Total suction consists of matric suction and osmotic suction, where the matric suction represents the interaction of soil water with both the mineral soil skeleton and the gas phase, and the osmotic suction depends on the salt concentration in the bulk water [3]. For non-saline soils, it is reasonable to assume that the matric suction is much greater than the osmotic suction, and the total suction is equal to matric suction approximately.

The increase of adsorbed water film due to suction decrease can be explained by the absorptive term in shifted Young-Laplace equation, which states $h^3 \propto 1/\psi$. Theoretically, the adsorptive water film can increase to 7×10^{-8}m when the suction is equal to 0.01kPa [25], which is only valid for the simplest condition that no mutual attraction between two adjacent surfaces exists (see Figure 1Erreur ! Source du renvoi introuvable.a). The mutual attraction between two adjacent surfaces affects the stability of adsorbed liquid films. At a certain film thickness (h_c), two water films can form a liquid bridge, followed by a spontaneous filling up of the gap between two adjacent surfaces (see Figure 1c). This phenomenon is known as the capillary condensation or the 'snap-off', which has been observed experimentally with the surface force apparatus by [26]. For a system consisting of two planar surfaces parallel with each other (see Figure 1Erreur ! Source du renvoi introuvable.b), [20] concluded that the critical film thickness (h_c) at the point of capillary condensation is approximately equal to one third of the distance between two surfaces (H). Therefore, the critical film thickness may be much smaller than the theoretical thickness due to the capillary condensation. When the thickness of water films is less than the critical film thickness, the liquid can stably be retained on the surface by adsorption. As shown in Figure 1b, along with decreasing suction (namely an imbibition process), the adsorbed water film thickness increases first and the quantity of adsorbed water increases correspondingly. Once the thickness of adsorbed water film reaches the critical film thickness, capillary condensation occurs (see Figure 1c) and the liquid retained in soil pores is governed by the capillarity only [20].

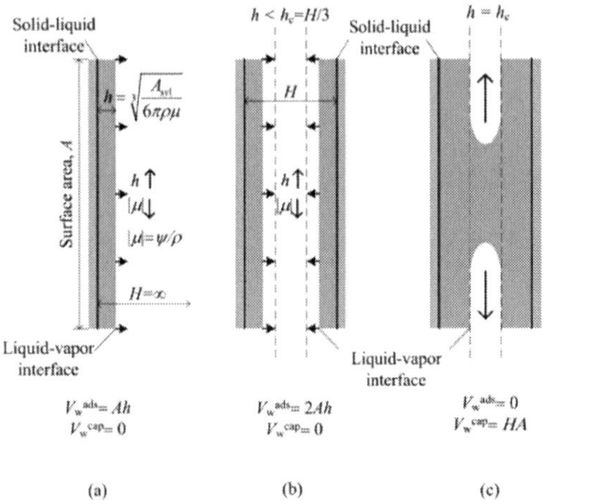

Figure 1 Increase of adsorbed liquid film and capillary condensation.

3 Water retention model for capillary and adsorbed waters

Pore space geometry is traditionally simplified as a 'bundle of cylindrical capillaries' (BCC), such as in [27]. However, the BCC model only takes into account the cylindrical capillarity and ignore the role of specific surface area and liquid films adsorbed on the surface. Recently, [20] suggested a new pore geometry model to facilitate the consideration of both adsorption and capillary processes. The new pore geometry model consists of a series of capillary pores (either angular or circular) to represent capillary-dominated processes and a series of slit-shaped spaces to represent adsorption-dominated phenomena. The adsorption process is governed by the adsorptive term of the shifted Young-Laplace equation and capillary process is governed by the capillary term of the shifted Young-Laplace equation. The theoretical water retention curve considering both capillarity and adsorption was firstly proposed by [28]. Compared with other water retention models in the literature, the most important feature of the model by Or and Tuller is its ability to separate the individual contributions of absorptive and capillary processes from a given chemical potential and take into account capillary condensation as well. However, the equation of the theoretical water retention curve derived by [28] is very complicated and difficult for practical applications.

In addition to theoretical water retention models, some practical models were proposed to describe the capillary/adsorbed water retention behaviours based on the theoretical understandings. For example, [18] proposed a water retention model that retains the form of the widely-used empirical water retention equations, such as the Brooks-Corey model and the van Genuchten model, in the wet range and the form of an adsorption equation developed by [17] in the dry range. Similarly, [19] applied the adsorption equation developed by [17] to a water retention model developed by [29].

$$\theta = (\theta_s - \theta_a)C(\psi) + \theta_a, \quad C(\psi) = \frac{1}{2}\mathrm{erfc}\left(\frac{\ln(\psi/\psi_m)}{\sqrt{2}\zeta}\right), \tag{1}$$

where θ is the volumetric water content, θ_s the volumetric water content at fully saturated state, θ_a the volumetric water content ascribed to adsorption, erfc() the complementary error function, ψ_m the suction that corresponds to the median pore radius, and ζ the standard deviation of the log-transformed pore radius. According to [17], θ_a can be written as

$$\theta_a = \theta_0 A(\psi), \quad A(\psi) = \left(1 - \frac{\ln\psi}{\ln\psi_d}\right), \tag{2}$$

where θ_0 is the volumetric water content due to adsorption at a suction of 1kPa and ψ_d suction at oven dryness. Experimental results have shown that oven dryness generally corresponds to a finite suction of 10^6 kPa. Equation (2) was widely adopted in the literature [18, 19, 30]. However, equation (2) does not take into account capillary condensation. The neglect of capillary condensation results that the volumetric water content due to adsorption keeps increasing along with the decrease of suction, which conflicts with the experimental observation described in the previous section.

Sharing the theoretical concept delivered by [28] and practical method by [19] , a simple equation of water retention curve is proposed here to consider capillarity and adsorption separately, which can be written as follows:

$$S_r = S_r^{cap} + S_r^{ads}, \tag{3}$$

where S_r is the degree of saturation, S_r^{cap} the capillary component of degree of saturation and S_r^{ads} the adsorptive component of degree of saturation. Following [19], the two-parameter equation proposed by [29] to quantify the capillary component (i.e., capillary water retention curve, CWRC):

$$S_r^{cap} = \left(1 - S_r^{ads}\right)C(\psi), \tag{4}$$

The adsorptive component S_r^{ads} (i.e., adsorbed water retention curve, AWRC) can be described by the following equation:

$$S_r^{ads} = \frac{\theta_a}{\theta_s}(1 - P_{cc}), \tag{5}$$

where θ_a is the volumetric water content due to adsorption ignoring capillary condensation, which can be described by equation (2) and P_{cc} stands for the possibility of capillary condensation ($0 \le P_{cc} \le 1$). Obviously, $P_{cc} = 1$ if soil is fully saturated ($S_r = S_r^{cap} = 1$) and $P_{cc} = 0$ if capillary water is equal to zero ($S_r^{cap} = 0$). The simplest equation meets the above requirement of P_{cc} can be written as: $P_{cc} = S_r^{cap}$. Therefore, the adsorbed water retention curve can be specified as

$$S_r^{ads} = \alpha A(\psi)(1 - S_r^{cap}), \tag{6}$$

where α is a parameter ($=\theta_0/\theta_s$) that is related to the theoretical degree of saturation due to adsorption (without considering capillary condensation) when the suction is equal to 1kPa. Combining equations (3), (4) and (6) yields close-form equations for WRC, CWRC and AWRC:

$$\begin{cases} \text{WRC:} & S_r = S_r^{cap} + S_r^{ads} = \dfrac{C(\psi) + \alpha A(\psi) - 2\alpha C(\psi)A(\psi)}{1 - \alpha C(\psi)A(\psi)} \\ \text{CWRC:} & S_r^{cap} = \dfrac{C(\psi) - \alpha C(\psi)A(\psi)}{1 - \alpha C(\psi)A(\psi)} \\ \text{AWRC:} & S_r^{ads} = \dfrac{\alpha A(\psi) - \alpha C(\psi)A(\psi)}{1 - \alpha C(\psi)A(\psi)} \end{cases} \tag{7}$$

where $C(\psi)$ and $A(\psi)$ can be found from equations (1) and (2), respectively.

Erreur ! Source du renvoi introuvable. shows a typical WRC, the corresponding CWRC and AWRC predicted by the proposed equations in the space of degree of saturation and suction. A similar plot can be found in [28]. As shown in Figure 1, the CWRC is a typical s-shape curve which indicates more capillary pores are dried along with increasing suction. The size of a drying pore is inversely proportional to the suction (or the negative chemical potential), which is quantitatively governed by the capillary term of the shifted Young-Laplace equation. The AWRC shows a peak value at an intermediate suction, which implies the degree of saturation due to adsorption decreases after an initial increase, with decreasing suction. This hill-shape curve can be explained by surface adsorption and capillary condensation jointly. At the beginning, the capillary water quantity is very limited at the dry end and thus the possibility of capillary condensation is very low. Therefore, along with decreasing suction, the adsorbed water film thickness increases freely and the degree of saturation based on adsorbed water also increases accordingly (see Figure 1b). However, the increase of the adsorbed water film thickness leads to the increase of the possibility of capillary condensation, which will result in the decrease of adsorbed water quantity (see Figure 1c). When soil is fully saturated, adsorbed water is completely removed because of capillary condensation.

4 Shear strength criterion for unsaturated soils

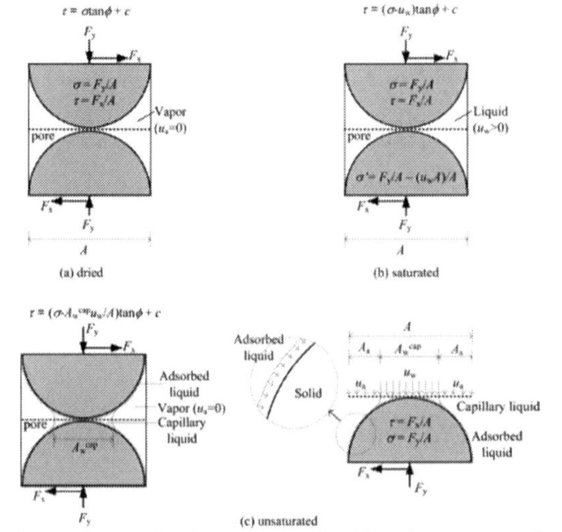

Figure 2 Contribution of pore liquid and vapor on the shear strength.

Soil is a frictional material and its shear strength can be attributed to the friction (ϕ) and the inherent cohesion (c) between soil particles. For example, for a system consisting of two dry cylinders in the atmosphere (see Figure 2a), the shear strength (τ) of this system is determined by the apparent contact stress (σ), for a given material (i.e., both ϕ and c are given). If the system is fully saturated with a liquid of a bulk liquid pressure u_w (Figure 2b), the true contact stress (σ') is equal to the apparent contact stress (σ) subtracts the liquid pressure (u_w) and the shear strength (τ) is determined by the effective contact stress (σ'). When the system is partially saturated, liquid-vapor interface is formed. The pore liquid retained by suction consists of capillary liquid that exists between two particles and adsorbed liquid that wraps the surface of each particle, according to the shifted Young-Laplace equation. If this unsaturated system is virtually separated at the potential slip plane, only the pressure of the capillary liquid can be exposed and affects the contact stress (σ') between two particles. The pressure of adsorbed water, which is caused by the van der Waal's forces between liquid molecules and solid molecules, is always treated as an internal stress and does not affect the contact stress (σ') between two particles. Therefore, only the pressure of capillary water affects the shear strength of this partially saturated system. In general, if neither u_a nor u_w is equal to zero, the shear strength can be written as

$$\tau = \left(\sigma - u_a + (u_a - u_w) \frac{A_w^{cap}}{A} \right) \tan\phi + c \tag{8}$$

where A_w^{cap}/A is the normalised area of capillary water. If the adsorption is ignored and only capillary liquid is taken into account, the above equation can be rewritten through replacing A_w^{cap} by A_w:

$$\tau = \left(\sigma - u_a + (u_a - u_w) \frac{A_w}{A} \right) \tan\phi + c \tag{9}$$

where A_w/A is the normalised area of water.

In the past several decades, adsorbed water was broadly ignored in unsaturated soil mechanics, equation (9) were widely adopted, such as in [2]; [3]; [31] and [12]. The normalised area of water (A_w/A) is assumed to be in direct proportion to the normalised volumetric water content (i.e., degree of saturation, S_r) in the soil by applying Greens theorem [2, 32]. Therefore, equation (9) can be written as

$$\tau = (\sigma - u_a + (u_a - u_w) S_r) \tan\phi + c \tag{10}$$

Compared with equation (9), equation (10) is more practical and can be directly applied to predict the shear strength of unsaturated soils. Equation (10) highlights the effect of hydraulic behaviour (water retention) on mechanical behaviour (shear strength) of unsaturated soils. In other words, the shear strength of an unsaturated soil is related directly to its water retention curve ($S_r - \psi$). Equation (10) is widely adopted in the literature. However, as shown in a competitive study on eight shear strength criteria [13], equation (10) usually much overestimates the shear strength of unsaturated soils, especially for clayey soils at a high suction level.

To achieve a better prediction to shear strength behaviour of unsaturated soils, a revision to equation (10) has been suggested through replacing the degree of saturation (S_r) by the effective degree of saturation (S_e) [2, 9, 14, 16, 33]. Pore water is divided into two categories: free water and residual water. Residual water is considered to be attached to the particles or trapped in the micropores. The intrinsic velocity of residual water is also equal to that of the soil solids, whereas the remaining water fraction, referred to as free water, is able to flow through the pores with distinct intrinsic velocity [7]. The volume of residual water is assumed to be a constant and cannot be removed by increasing suction. Only free water contributes to shear strength of unsaturated soils. By employing the effective degree of saturation, equation (10) can be revised as

$$\tau = (\sigma - u_a + (u_a - u_w) S_e) \tan\phi + c, \quad S_e = \frac{S_r - S_r^{res}}{1 - S_r^{res}} \tag{11}$$

where S_e is the effective degree of saturation and S_r^{res} is the residual degree of saturation. Equation (11) implies that the shear strength of an unsaturated soil is related directly to its effective water retention curve ($S_e - \psi$), EWRC, other than the WRC. Equation (11) was widely used in unsaturated soil mechanics recently, such as in [33]; [14]; [9], [16] and [15]. Compared with equation (10), equation (11) gives a better prediction to the observed shear strength of unsaturated soils [13]. However, for a clayey soil with a high suction, equation (11) still overestimates shear strength. In addition, dividing the pore water into a changeable free water and a constant residual water suffers from several shortcomings. Firstly, all the liquid retained in soil pores, including the water trapped in the micropores, can be fully removed if zero water content refers to oven dryness (namely, 10^6 kPa suction). Secondly, the quantity of water attached to soil particles (i.e. adsorbed water) can also change with the variation of suction and the occurrence of capillary condensation.

Equations (10) and (11) can be further revised by including an additional strength fitting parameter κ, such as in [1], [2] and [4] :

$$\tau = \left(\sigma - u_a + (u_a - u_w) S_r^{k}\right)\tan\phi + c \quad \text{or} \quad = \left(\sigma - u_a + (u_a - u_w) S_e^{k}\right)\tan\phi + c , \qquad (12)$$

Compared with equations (10) and (11), equation (12) can provide a better prediction [13]. However, the determination of the additional parameter (k) in equation (12) requires more experimental data, which limits its application to practice. Another limitation is that this fitting parameter k is completely based on the regression of experimental results and has no clear physical meaning.

A fundamental but effective way to achieve a better prediction to the shear strength data is to acknowledge the difference between capillary water and adsorbed water. In analogy to the relationship between equations (9) and (10), a new shear strength criterion is proposed here based on equation (8), by assuming that the normalised area of capillary water (A_w^{cap}/A) is in direct proportion to the normalised volumetric capillary water content (i.e., degree of saturation due to capillary component, S_r^{cap}):

$$\tau = \left(\sigma - u_a + (u_a - u_w) S_r^{cap}\right)\tan\phi + c , \qquad (13)$$

Equation (13) implies that the shear strength of an unsaturated soil is related directly to its CWRC ($S_r^{cap} - \psi$) other than either WRC or EWRC. The capillary water retention curve can be determined by equation (7) with given water retention test results.

For triaxial conditions, the deviator stress (q) can be calculated via $q = M(p-u_a-p_s)$, with $M=6\sin\phi/(3-\sin\phi)$, p being the total mean stress, and p_s the apparent tensile strength that is equal to $-S_r(u_a-u_w)$, $-S_e(u_a-u_w)$ or $-S_r^{cap}(u_a-u_w)$, corresponding to equations (10), (11) or (13), respectively.

5 Validation and comparison

In this section, different shear strength criteria including equations (10), (11) and (13) are compared with each other against experimental data. The common ground of the above three equations is that: (1) all of them involve water retention equations (i.e., WRC / EWRC / CWRC) and (2) all of them do not include any additional fitting parameter that has no clear physical meaning. The value of S_r involved in equation (10) and S_r^{cap} in equation (13) are calculated by the 3-parameter (ψ_m, ζ and α) water retention model developed in this paper (i.e., equation (7)). The value of S_e in equation (11) is calculated by 3-parameter (a, n and S_r^{res}) VG model developed by [11].

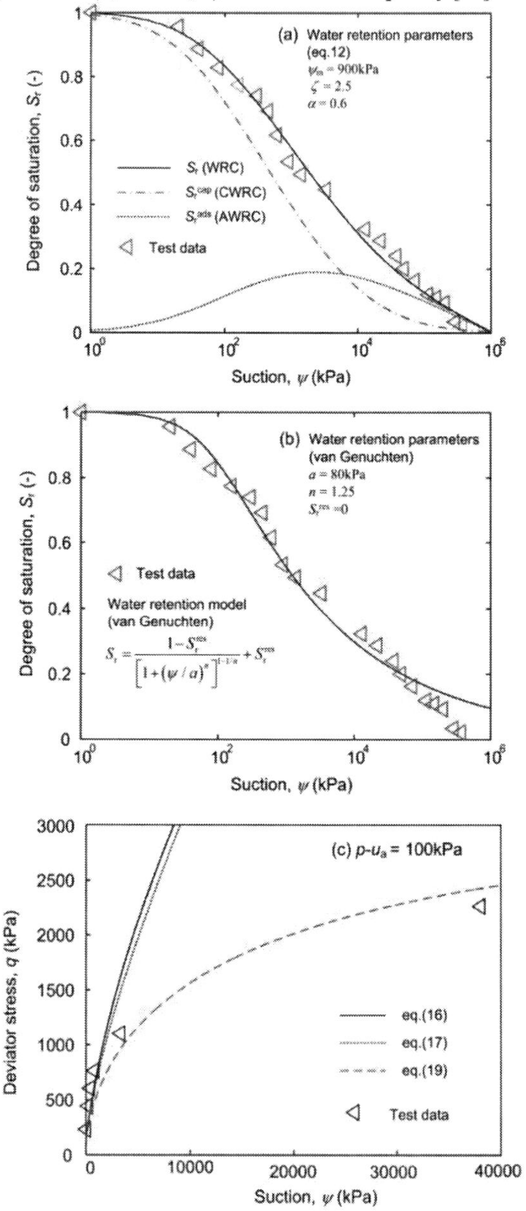

Figure 3 Measured water retention behaviour and shear strength for Nanyang expansive soil (data after Zhang et al., 2015) and model predictions.

Zhang et al. conducted a series of triaxial tests on Nanyang expansive soil under various suctions (0, 200, 400, 800, 3290, 38000kPa) [34]. Nanyang expansive soil is a weakly expansive soil with a liquid limit of 38.8% and a plasticity index of 21.6. X-ray diffraction analysis indicated that the predominant minerals of the soil were 62% quartz, 12% albite, 10% microcline, 9.9% illite and 5.8% montmorillonite, with a small percentage of kaolinite (0.3%). The critical state parameter was measured by saturated triaxial tests (M = 1.05). The water retention behaviour of Nanyang expansive soil was measured by drying tests for almost the entire saturation range (S_r = 1~0.02). For the low suction range (0~1000kPa), the axis translation technique (ATT) was employed for measuring the water retention behaviour, whereas the vapor equilibrium technique (VET) was adopted to impose suction beyond 1000kPa. The suction of 3290 kPa was achieved by VET using K_2SO_4 and the suction of 38000kPa by VET using NaCl. The air entry value of this weakly expansive soil is about 15kPa. As shown in Figure 3a and 3b, the proposed model (equation(7)) and the VG model are used to simulate the observed water retention behaviour, respectively. For Nanyang expansive soil, the proposed water retention curve gives a better simulation to the experimental results than the VG model. The shear strength data of Nanyang expansive soil with different suctions (0~38000kPa), at a net confining pressure of 100kPa, are replotted in Figure 3c as well as the predictions from the three models. The comparison (see Figure3c) shows that the proposed model predicts the measured shear strength reasonably well, especially for high suctions (such as 3290kPa and 38000kPa).

6 Conclusion

Based on the theoretical analysis provided by Tuller et al. [20], this paper proposed a new water retention model which is able to separate the capillary water from the adsorbed water and consider capillary condensation. A new shear strength criterion for unsaturated soils, which involves the capillary part of the proposed water retention model, was then derived and validated.

References

1. Fredlund, D.G., et al. Canadian Geotechnical Journal, 1996. 33(3): p. 440-448.
2. Vanapalli, S.K., et al. Canadian Geotechnical Journal, 1996. 33: p. 379-392.
3. Oberg, A. and G. Sallfors. Geotechnical Testing Journal, ASTM, 1997. 20: p. 40-48.
4. Toll, D.G. and B.H. Ong. Geotechnique, 2003. 53(1): p. 93-103.
5. Derjaguin, B.V., N.V. Churaev, and V.M. Muller, *Surface Forces*. 1987, New York: Plenum Publishing Corporation.
6. Baker, R. and S. Frydman. Engineering Geology, 2009. 106(1-2): p. 26-39.
7. Fuentes, W. and T.H. Triantafyllidis. Geotechnique, 2013. 63(16): p. 1451-1455.
8. Konrad, J.-M. and M. Lebeau. Canadian Geotechnical Journal, 2015(In press): p. DoI: 10.1139/cgj-2014-0300.
9. Lu, N., J.W. Godt, and D.T. Wu. Water Resources Research, 2010. 46(5): p. W05515.
10. Brooks, R. and A. Corey, *Hydraulic properties of porous media*, 1964, Colorado State University: Fort Collins, Colorado.
11. van Genuchten, M.T. Soil Science Society of America Journal, 1980. 44: p. 892- 898.
12. Lu, N. and W. Likos. Journal of Geotechnical and Geoenvironmental Engineering, 2006. 132(2): p. 131-142.
13. Sheng, D., A.N. Zhou, and D.G. Fredlund. Geotechnical and Geological Engineering, 2011. 29(2): p. 145-159.
14. Alonso, E.E., et al. Geotechnique 2010. 60(12): p. 913-925.
15. Zhou, A.N. and D. Sheng. Computers and Geotechnics, 2015. 63: p. 46-66.
16. Zhou, A.N., et al. Computers and Geotechnics, 2012. 43: p. 178-187.
17. Campbell, G.S. and S. Shiozawa, *Prediction of hydraulic properties of soils using particle-size distribution and bulk density data.* , in *Indirect methods for estimating the hydraulic properties of unsaturated soils*, M.T. van Genuchten, F.J. Leij, and L.J. Lund, Editors. 1992: Univ. of California, Riverside. p. 317-328.
18. Fayer, M.J. and C.S. Simmons. Water Resources Research, 1995. 31(5): p. 1233-1238.
19. Khlosi, M., et al. Vadose Zone Journal, 2008. 7(1): p. 87-96.
20. Tuller, M., D. Or, and L.M. Dudley. Water Resources Research, 1999. 35(7): p. 1949-1964.
21. Edlefsen, N.E. and A.B.C. Anderson. Hilgardia, 1943. 15: p. 31-298.
22. Philip, J.R. The Journal of Chemical Physics, 1977. 66(11): p. 5069-5075.
23. Iwamatsu, M. and K. Horii. Journal of Colloid and Interface Science, 1996. 182(2): p. 400-406.
24. Iwata, S., T. Tabuchi, and B.P. Warkentin, *Soil-water interaction: mechanisms and applications (2nd edition)*. 1995, New York: Dekker.
25. Tuller, M. and D. Or. Water Resources Research, 2001. 37(5): p. 1257-1276.
26. Christenson, H.K. Physical Review Letters, 1994. 73(13): p. 1821-1824.
27. Mualem, Y. Water Resources Research, 1976. 12(3): p. 513-522.
28. Or, D. and M. Tuller. Water Resour. Res., 1999. 35(12): p. 3591-3605.
29. Kosugi, K. Water Resources Research, 1996. 32(9): p. 2697-2703.
30. Lebeau, M. and J.-M. Konrad. Water Resources Research, 2010. 46(12): p. W12554.
31. Xu, Y.F. Journal of Geotechnical and Geoenvironmental Engineering, ASCE, 2004. 130(3): p. 264-273.
32. Fung, Y.C., *A frist course in continuum mechanics*. 1977, Englewood Cliffs, New Jersey.
33. Tarantino, A. and S. Tombolato. Geotechnique, 2005. 55(4): p. 307-317.
34. Zhang, J., et al. Canadian Geotechnical Journal, 2015(in press): p. doi:10.1139/cgj-2014-036

E3S Web of Conferences 9, 14011 (2016) DOI: 10.1051/e3sconf/20160914011
E-UNSAT 2016

Evaluation of swell behaviour of expansive clays from specific moisture capacity

Alejandro Pino[1,a], Aravind Pedarla[1], Anand Puppala[1], and Laureano R Hoyos[1]

[1] *The University of Texas at Arlington, Department of Civil Engineering, Arlington, Texas 76019 USA*

Abstract: Current swell characterization techniques used to interpret the mechanical volume change occurring during the swelling process are not successful due to lack of inclusion of influential properties. Accurate prediction of swelling behaviour allows us to design more efficiently and with better reliability. This research aims at developing a more comprehensive framework to predict swell potential. Laboratory studies are conducted on five natural expansive soils with different degree of expansiveness. Initial studies include determination of basic soil characterization, swell strains and swell pressures at their compacted state along with their inherent mineralogy. Later, replicate samples were studied for soil water characteristic curves using standard pressure cell apparatus and filter paper techniques. The path traversed by the specimen during swelling process is representative of the soil water characteristic curve of the same specimen. Hence, studies are pursued to understand the relationship between degree of expansiveness and the specific moisture capacity relative to that particular range of suction head. Test results showed that the degree of expansion represented by swelling strain or swelling pressure is proportional to the specific moisture capacity determined during the swelling process.

1 Introduction and background

Inherent chemical nature of expansive soils leads to the complex swell/shrink behaviour thereby leading to premature failures in civil infrastructure. Robust design of structures on these soils requires accurate assessment of the swelling behaviour [1]. In the last few decades, the swelling potential of soils has been linked to parameters like void ratio, moisture content, liquid limit and activity through empirical methods. However, studies have been performed and proved the incompleteness of the soil behaviour characterization when correlations based on such parameters are applied [2].

Swelling behaviour of expansive clays is highly influenced by their moisture/suction state, mineralogy, pore distribution and in-situ stress state. Several semi-empirical models based on matric suction have been proposed including volume change constitutive relationships or statistical correlations [3, 4]. Many researchers including Berniner et al [7] and Delage et al. [8] studied the relationship between matric suction and swelling behavior of expansive clays. Pedarla et al. [9] conducted suction studies on two expansive clays with different mineralogy and found that mineralogy plays an important role in the swell-shrink and suction characteristics of the clay.

In the current work, the identification of swelling behaviour of compacted clay specimens is relied on the initial matric suction, the air/water-entry suction and the slopes of logarithm of soil suction-void ratio relationship. These factors are considered to play an important role in the volume change potential. The slope of the SWCC or specific moisture capacity has been used to determine the swell behaviour for natural compacted clays.

2 Experimental study

For this study, five natural expansive soils were collected from different parts of the United States and named after the locations from where they were sampled. Basic soil characterization on these test samples is conducted and the details of the results are presented in Table 1. Also, Standard Proctor tests were conducted to establish the compaction relationships of the soils.

Samples were compacted at maximum dry density (MDD) and 95% of maximum dry density using a constant strain compactor method prescribed in the AASHTO T-307 standard intending to minimize the soil layering effects induced by the impact compaction method. Samples compacted at 95% MDD were prepared at the same optimum moisture content condition as at MDD. From Table 1 it can be observed that Colorado soil exhibited highest plasticity indices value whereas Plano soil exhibited least indices.

One dimensional swell strain and swell pressure tests were performed in a conventional consolidometer setup following Method C (loading-after-wetting test) normalized in the ASTM D4546 – 08 standard. Appliance correction factors were applied to each of the

[a] Corresponding author: alejandro.pinobravo@mavs.uta.edu

DOI: 10.1051/e3sconf/20160914011

swell pressure tests for the determination of corrected swell pressure [5].

Table 1. Laboratory tests results.

Soil	LL	PI	MDD (kN/m^3)	OMC (%)	USCS Classification
Burleson	55	37	1633	19	CH
Colorado	63	42	1649	19	CH
Grapevine 2	46	26	1693	19	CL
Mansfield	67	38	1489	26	CH
Plano	24	12	1462	27	CL

A typical swell strain plot is depicted in Figure 1. The plot shows the vertical swell strain readings versus elapsed time period of a compacted Burleson soil specimen at MDD and 95% MDD. Also, a typical load-back swell pressure plot of the compacted Burleson soil specimen is shown in Figure 2. The sample is allowed to free swell from an initial void ratio 'e'. After the completion of swell process, the sample is loaded systematically to allow the sample to reach its initial void ratio 'e'. The figure depicts void ratio versus overburden pressure at 95% MDD. As expected, the specimens of the soils compacted at 95% MDD condition exhibited lower values of swell strains and pressures than those exhibited by the specimens compacted at MDD.

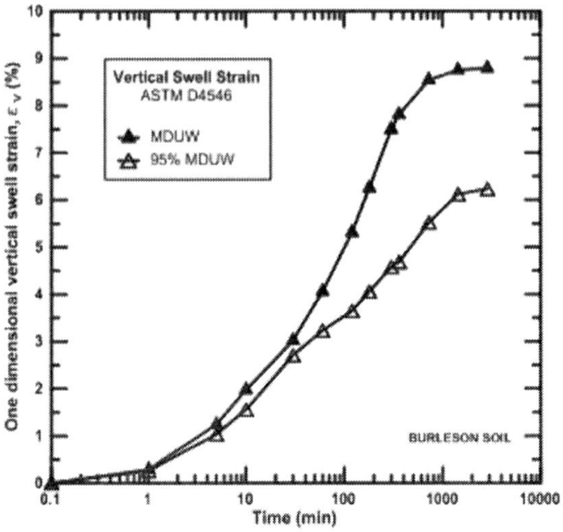

Figure 1. One dimensional swell strains for Burleson soil.

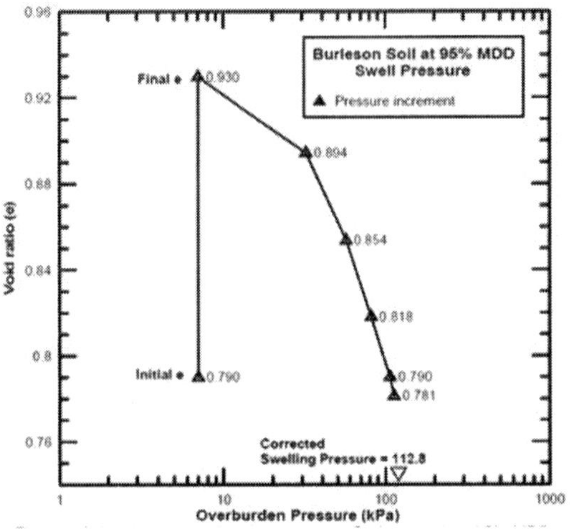

Figure 2. Load-back swell pressure test on Burleson soil at 95% MDD.

Table 2 presents a summary of the results obtained from similar testing procedures conducted on the remaining 4 sampled soils along with their plasticity index property.

Table 2. Laboratory tests results.

Ranking		1-D Swell strain (%)		Swell Pressure (kPa)	
Soil	PI	MDD	95% MDD	MDD	95% MDD
Colorado	42	12.0	8.2	194.0	137.7
Mansfield	38	11.7	9.5	164.2	138.9
Plano	12	9.1	7.7	158.0	108.0
Burleson	37	8.8	5.8	183.4	112.8
Grapevine 2	26	6.2	4.3	88.3	82.7

Drying and wetting soil water characteristic curves (SWCCs) for the 5 test soils are determined using filter paper technique and dew point potentiometer method [6]. SWCCs are determined for samples compacted at 95% MDD condition. Therefore, the model proposed herein will be developed based solely on the tests results acquired for the 95% MDD condition.

SWCC for Burleson soil is presented in Figure 3. Similar studies were conducted for the rest of the test soils. Van Genuchten (1980) fitting model is utilized to fit the experimental data. Associated curve fitting parameters 'a', 'n' and 'm' for all the test soils are presented in Table 3.

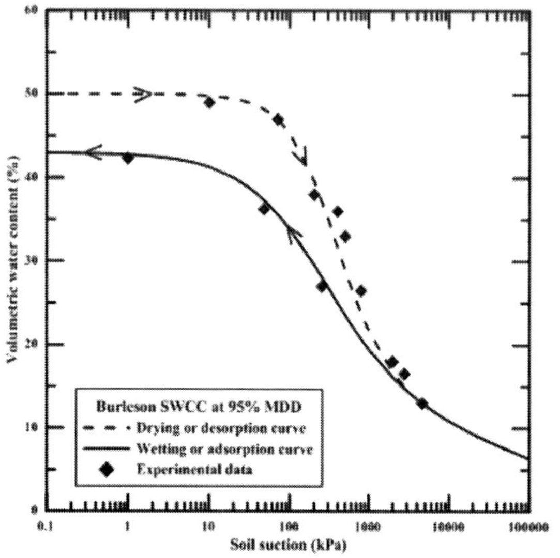

Figure 3. Soil water characteristic curves of Burleson soil at 95% MDD.

Table 3. Curve fitting parameters for SWCCs based on Van Genuchten (1980) model.

Soil	Drying SWCC			Wetting SWCC		
	a	n	m	a	n	m
Burleson	0.005	1.55	0.35	0.033	1.15	0.13
Colorado	0.004	1.60	0.37	0.004	1.60	0.38
Grapevine 2	0.007	1.46	0.32	0.068	1.12	0.10
Mansfield	0.002	1.58	0.367	0.027	1.11	0.10
Plano	0.013	1.27	0.300	0.025	1.11	0.10

3 Modified specific moisture capacity (α) model

The present model relies on the correlation between the specific moisture capacity, α (i.e. slope of the SWCC) and the one dimensional swell strain and pressure obtained from the loading-after-wetting test (Method C ASTM D4546 – 08) performed only on 95%MDD compacted specimens of expansive clay.

Physico-mechanical attributes of the soil swelling behaviour are characterized by the cumulative variation of void ratio, matric suction and moisture content undergone during the swell process. Also, hydro mechanical behaviour of swelling soil during the heaving process may be detailed through the hysteretic paths of the drying and wetting SWCCs.

Figure 4 depicts the assumed idealized plot linking the void ratio to the matric suction which allows for the determination of the herein called modified specific moisture content, α. The idealized and actual paths exhibited by a soil specimen during hydration are shown in the Figure 4. In all the tested soils, the moisture content and matric suction relationships from the SWCCs are correlated to the corresponding void ratios and moisture contents, thus, void ratio – log matric suction or void ratio – matric suction in pF plots are acquired. During the swelling process, soil specimen undergoes a decrement in its matric suction and a corresponding void ratio increment. In the herein proposed model, this soil matric suction - void ratio variation is assumed to follow a linear trend (i.e. modified specific moisture capacity).

The slope of the soil matric suction (X-axis) and void ratio (Y-axis) plots (i.e. modified α) is therefore considered as a significant parameter in representing the variation of void ratio of the soil specimen with soil matric suction during swelling process.

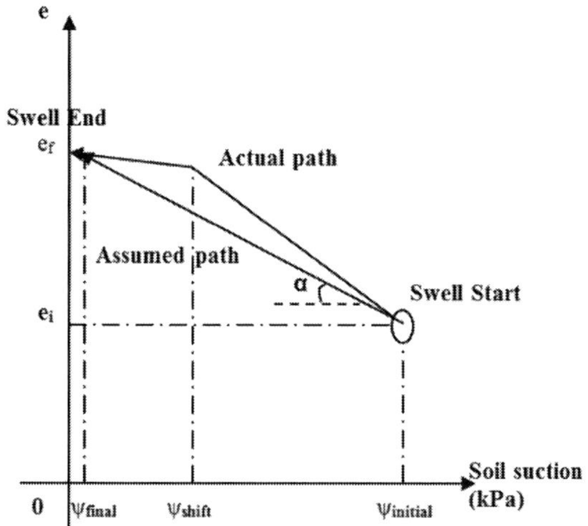

Figure 4. Schematic for the measurement of slope of void ratio and logarithmic of matric suction.

In Figure 4, the e_i is the initial void ratio (i.e. e when free swell starts), e_f is the void ratio when the soil specimen is fully saturated (i.e. e when free swell ends), $\psi_{initial}$ is the initial matric suction in the soil specimen before swell testing (i.e. existing suction in the sample after compaction at 95% MDD and optimum moisture content), ψ_{shift} is the matric suction when a shift on the wetting SWCC is noticed (i.e. small changes in saturation condition take place), ψ_{final} is the final matric suction at almost full saturation condition and α is the slope of idealized suction - void ratio path. Initially the soil specimen has high matric suction ($\psi_{initial}$) and a lower value of void ratio (e_i).

The difference between $\psi_{initial}$ and ψ_{final} values generates a negative slope value of the modified α, however it is neglected since the actual 'Mechanical-Hydro' coupling effect is based only on the magnitude of the modified α. Swelling process induces an increment in moisture content and void ratio in the soil specimen. As the degree of saturation increases, the void ratio of the soil specimen will increase consequently following a matric suction reduction. This behaviour is clearly noticed in the e-w-log ψ and e-w-ψ plots of the five expansive soils. The mentioned plots for one of the soils will be shown in the following sections.

The slope of the idealized matric suction-void ratio path (i.e. modified specific moisture capacity) is obtained from the wetting SWCC of each soil and is given by the following Equation 1.

$$\alpha = \frac{e_{final} - e_{initial}}{\log \frac{\psi_{final}}{\psi_{initial}}} \quad (1)$$

The swell behaviour exhibited by Burleson soil specimen on an e-w-logψ and e-w-ψ plots is presented in Figure 5 and Figure 6, respectively.

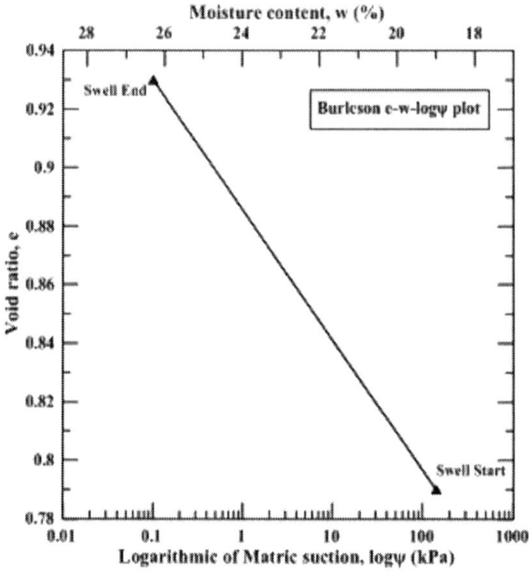

Figure 5. Behaviour of Burleson soil specimen at 95% MDD on e-w-log ψ plot.

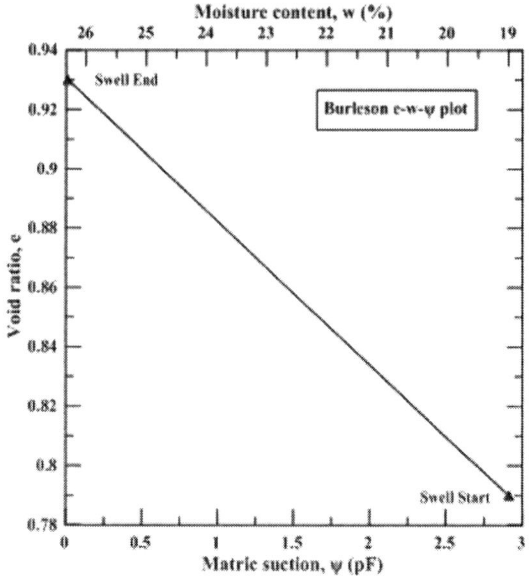

Figure 6. Behaviour of Burleson soil specimen at 95% MDD on e-w-ψ plot.

The slope of logarithm of matric suction-void ratio and matric suction in pF-void ratio plots (i.e. modified specific moisture capacity, α) is presented in Table 4.

Table 4. Modified specific moisture capacity, α, for all the expansive clays.

Soil	α from e-w-log ψ plot	α from e-w-ψ plot
Burleson	0.0444	0.0481
Colorado	0.0606	0.0593
Grapevine 2	0.0264	0.0259
Mansfield	0.0690	0.0693
Plano	0.0529	0.0528

The relationship established between the measured one dimensional swell strain and pressure and the modified specific moisture capacity, α, as obtained from the logarithm of matric suction-void ratio plot and the matric suction in pF -void ratio plot (Table 4), is presented in the following Figure 7 and Figure 8.

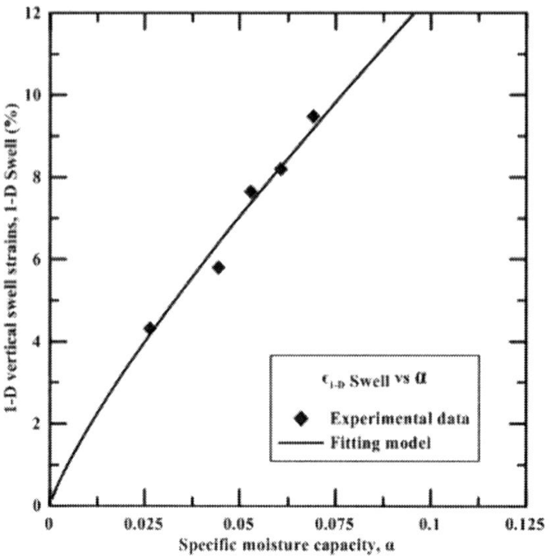

Figure 7. 1 D Swell Strain vs α at 95% MDD condition on e-w-log ψ plot.

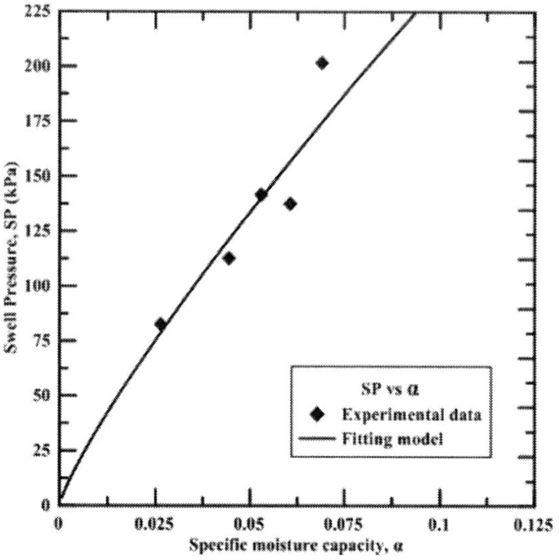

Figure 8. 1 D Swell Pressure vs α at 95% MDD condition on e-w-log ψ plot.

It is possible to notice only a slight variation on the values of the modified specific moisture capacity, α, found through the two studied approaches (Table 4), thus, the exponential fitting curves yielded close formulae as presented in Equation 2 pertaining to the one dimensional swell strains and Equation 3 for swell pressure.

$$\epsilon_{1-D\ Swell}(\%) = 82.56 \times \alpha^{0.821} \quad (2)$$

$$SP\ (kPa) = 1596.48 \times \alpha^{0.828} \quad (3)$$

The model for 1-D swell strains and pressures showed a good fit with the measured experimental data. The coefficients of determination (R^2) for these models were found to be 0.97 and 0.92 in the swell strain case and 0.89 and 0.87, in the swell pressure case. Overall, a direct relationship between the swell strains and swell pressure is observed with modified specific moisture capacity, α. Further investigations on diverse test soils are required to refine these formulations derived.

4 Summary

A parameter designated herein as the modified specific moisture capacity, α, which represents the slope of an idealized path on the soil water characteristic curve (SWCC). This parameter couples the combined effects of the hydro-mechanical related properties of an expansive soil has been used in this research for swell strain and pressure prediction assessments on specimens of clay specimens compacted at 95% MDD. The samples were tested for one dimensional swell strain and swell pressure using a conventional consolidometer setup following ASTM D-4546 method C. A model correlating one dimensional swell strains and pressures to the specific moisture capacity was derived. The model was established using two equivalent approaches for finding the specific moisture capacity of soils, corresponding to the slope of the logarithm of the matric suction-void ratio plot and the slope of the matric suction in pF-void ratio plot. The proposed model for the prediction of one dimensional swell strains and swell pressures provided reliable results with both of the equivalent approaches, though, the logarithm of the matric suction-void ratio approach yields a higher coefficient of correlation. Similar variation of the proposed model results obtained for soil specimens compacted at 95% MDD are expected and assumed to occur for specimens compacted at MDD condition.

References

1. A.J. Puppala, B. Katha, L.R. Hoyos, Geot. Tes. J. ASTM, **27**, 6 (2004)
2. Y.M. Mowafy, Trans. Res. Rec., **1032**, 23-33 (1985)
3. R.G. McKeen, Proc. 7th Int. Conf. on Exp. Soils, **1**, 1-6 (1992)
4. E. Yusuf, O. Erol, Eng. Geol., **92**,133-145 (2007)
5. D.G. Fredlund, Proc. 2nd Intern. Conf. on Exp. Clay Soils, **1**,435-456 (1969)
6. D.G. Fredlund, J. of Geot. And Geoenv. Eng., **132(3)**, 286-321 (2006)
7. F. Bernier, Volckaert, G., Alonso, E., and Villar, M. Engineering Geology, vol. 47, pp. 325-338. (1997)
8. P. Delage, Howat, M.D., Cui, Y.J. Engineering Geology, vol. 50, pp. 31–48. (1998)
9. A. Pedarla, Puppala A.J., Hoyos, L.R., Zapata, C.E. and Chittoori B. Unsaturated soils: research and applications, UNSAT 2014, Australia, V 2, pp 951-957 (2014)

Hydro-mechanical paths within unsaturated compacted soil framed through water retention surfaces

Benjamin Pelizzari[1, 2, a], Apostolia Kontogianni[1, 2], Simon Salager[1, 2] and Olivier Ple[3, 4]

[1]*Univ. Grenoble Alpes, 3SR, F-38000 Grenoble, France*
[2]*CNRS, 3SR, F-38000 Grenoble, France*
[3]*Univ. Savoie Mont Blanc, LOCIE, F-73000 Chambéry, France*
[4]*CNRS, LOCIE, F-73000 Chambéry, France*

Abstract. Compaction is a key issue of modern earthworks... From sustainable development, a need arise of using materials for compaction under given conditions that would normally be avoid due to unpredictable pathologies. The application of compaction on fine grained soils, without a change of gravimetric water content, lead to very important modifications of the void ratio and hence suction. Therefore the hydro-mechanical behaviour of fine grained soil need to be rendered around three variables: suction, void ratio, saturation degree or water content. The barring capacity of the soil is assessed through Penetrometers (In-situ manual penetrometer, CBR) in order to assess gains through compaction. The three states variables are then assessed for in situ and frame through water retention surfaces, realized from Proctor tests, in which compaction effect and path could be described.

1 Introduction

Earthwork relies in great part on compaction for the realisation of key infrastructures. As simple as compaction seems, its methods might differ greatly depending on the region it is executed. More other, compaction of fine grained soils without use of hydraulic binder is greatly disregarded, even not recommended by standards. This study as part of the French ANR project TerreDurable [1] aims to better understand the behaviour of marl under compaction with different energies, water content and compactors. The effect on the three main state variable of the soil are then assessed (density, suction, degree of saturation). In this regard, full scale on site compaction tests are studied from samples, and then are compared to laboratory compaction tests (Proctor) [2].

2 Full scale compaction tests

2.1 Studied Material

The fined grained soil studied is marl extracted from the earthworks of the A304. This Highway aims to link Belgium to Charleville-Mézières in France. The materials are defined as A2 by French standard regulation (GTR) [3]. Its undersize at 85 μm is more than 35 % and its plasticity index is 24 (Tab. 1). Their geological origin lies in the Pliensbachian and/or the Toarcian. Some materials will be tested undisturbed; those aim to simulate supports soils. Others will be reworked, either on site as they would for earthworks, either with standard compaction tests.

Table 1. Reference Values for the marl

Propriety	Values
Liquidity limit	46%
Plasticity limit	22%
Plasticity index	24%
Natural water content	20%
Grains density	2.665gcm^{-3}

2.2 Layout and goals

The parameters and their variations (Tab.2) are defined aiming to see compaction pathologies. Hence the need arise to use dry and wet materials that respectively increases the possibilities for lamination and wave effect. Tamping shoe on rammer tends to overcome those two pathologies while the flat roller facilitates them (with help from the vibration). The two other parameter variations are here to assess the effects of the use of pulvimixer (mostly not regarded if the soil is non-treated) and over-compaction energy.

The test is carried out in three layers. The two first layer are used as a compaction base for the third layer that will receive on one side a normal compaction (2 times back

[a] Corresponding author: benjamin.pelizzari@3sr-grenoble.fr

and forth), and the other side over-compaction (6 times back and forth). The water contents are attained by natural drying (scarification/sun/wind) and wetting by injection. Each set of parameters on the compaction site represented a compacted bloc of soil of 15m length, 10 m wide (4 compactor rammers width) and three layers about 30cm. Figure 1 shows test realization. After compaction, different measurements are carried out on the layer (manual dynamic penetrometer, gammadensimeter). Samples are then made for laboratory test by force driving a tube in the compacted soil and extracting it.

Table 2. Parameter variation for on-site experimentations

Parameters	Variations
Compaction Energy	Increase of the number of passes
Compactor Type	Flat roller and tamping roller
Material Humidity	Medium, dry side and wet side of normal Proctor
Material grinding	mixed and raw material

Figure 1. In situ compaction test construction.

2.3 Results of the full scale test

The effect of the gridding is not visible in any of cases studied. It seems that the clods were not hard enough to remain uncrushed. This is corroborated by penetrometer measurements executed in the soil depot before compaction. As well, the compactor type is indecisive. Though Wave effect appears on one on the wet bloc compacted when using flat roller. On the opposite, tamping roller does not trigger it. The most interesting results come from over-compaction. A loss in mechanical resistance is observed on the wet side with the increase in number of passes while the dry material gained in strength. The medium humidity soil showed transition behaviour with gain and losses depending of the compaction set.

2.4 Core samples analysis

2.4.1 Methodology

The core samples are opened and then three slices of ~1cm are cut at the top part and at a 20cm depth. The goal is to try to compare the evolution of density, saturation and suction at two depth of the compaction layer, densification being stronger at the top of the layer. The slices are wrapped with filter papers for suction measurements. After the filter weighing, density and saturation degree are obtained through hydrostatic weighing of the slices in a non-wetting liquid. The water content is finally obtained by stove drying.

2.4.2 Results

Before analysing the results, on site experimental details must be discussed. The material was extracted at water content close to the optimum proctor, though, to obtain wet material, water injection was needed. A good homogeneity of the material after such a process is hard to attain in such huge proportions. Looking at the measurements of the samples, this heterogeneity shows, and the results have to be studied accordingly.

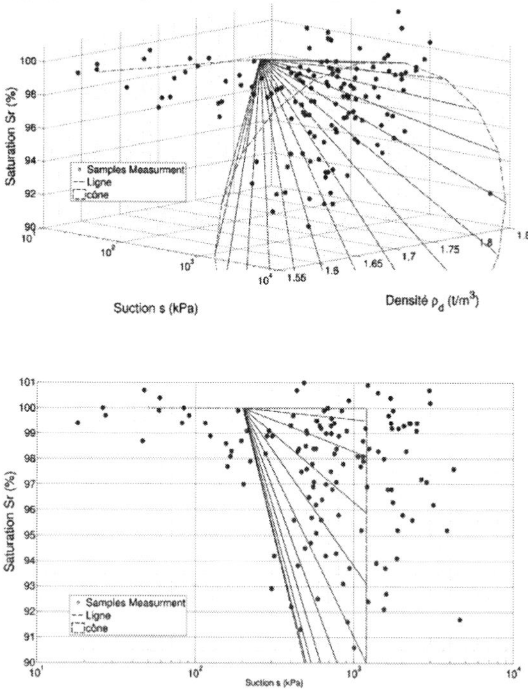

Figure 2. Core sample point spread, 3D and 2D projection.

The compaction test points seem to be spread widely but some sense can make from the process even though the heterogeneity. The global results are presented in figure 2. First, the graph above should not be treated as a retention diagram but as a compaction results. Hence no drying or wetting path appears. The points for low suction and density present higher degrees of saturation. A minimal air entry value of the material then appears for suction around 200kPa. After that point, there is a spread of the data in a circumferential surface. All those points belong to volume in which could be drawn series of water retention surfaces each depending of its own density. The aim is then to try to frame those data using laboratory testing, and generating water retentions surfaces for the materials. As a reminder, it should be noted that the samples are all very close to saturation (Sr > 90%).

Table 3 present the evolution of the suction and the density regarding compaction test depending of the preparation water content.

Table 3. Density and suction variation with compaction

Initial position	Normal compaction		Over-compaction	
	Suction	Density	Suction	Density
Wet side	1263	1.710	832	1.650
Optimum	1231	1.740	658	1.770
Dry side	728	1.725	180	1.730

An element that arises from this table is that, whatever the water content, density is increased by compaction except on the wet side. On the opposite, suction values change drastically. The dryer the material, the less effect over-compaction will have on suction. On wet material the losses of suction are important. Increase in compaction energy does not change the suction after compaction on the dry side, but does affect it on the wet side of the proctor [4]. These sudden changes in suction and density may explain the mechanical response of the soils as it depends mostly on those two values. A loss of suction will lower the strength of the material while higher density will increase it.

3 Laboratory compaction tests

3.1 Experimental layout

The goal of the lab tests is to frame the results obtained with in situ experimentations regarding the three state variables. Soils taken from the compaction test is stove dried, then sieved under a 5mm size, then wetted to different water content and finally compacted in an automatic proctor machine. Before compaction, one week was left to the material to reach homogeneity. Two compaction energies are used: normal and modified, in order to represent the two in situ degrees of compaction. The moulds are lubricated before test. After compaction, left other of the uncompacted soil is placed in an air tight recipient with a filter paper to measure the matric suction before compaction. The Proctors are then submitted to a CBR test. Finally, small samples are extracted from Proctors and are submitted to a drying wetting cycle. Suction is measured thanks to filter papers and a water retention curve for the 100%-70% saturation degree is established. Density and saturation degree are obtained by hydrostatic weighing.

3.2 Mechanical response

The mechanical response, observed through CBR tests, showed the same tendencies as the penetrometer on the full scale compaction test. The CBR showed better results for modified proctor on the dry side whereas the resistance was lost on the wet side of the normal Proctor.

Same results were obtained by Sawangsuriya et al. (2000) the elastic part of the mechanical response using bender elements [5]. The hypothesis of the direct dependence of the mechanical response to suction and density is confirmed, whatever the energy or the compaction method, if density and suction are the same for two different samples, the mechanical response will be the same [6]. The increase of energy plays an important role on suction depending on the water content regarding the normal optimum. It is to be noted that those variations of suction appear only when fine grained soils are in sufficient proportions [7].

3.3 Evolution of suction during compaction

One of the final elements of this series of test was to compare the suction before and after compaction. Figure 3 presents those values. The variations in suction are nearly negligible when on the dry side. The wetter the material, the more it will lose suction by the effect of compaction. Obviously, it confirms that increased compaction energy might lower even more the suction than normal compaction. Hence bearing capacity is lost, and might lead to disorders.

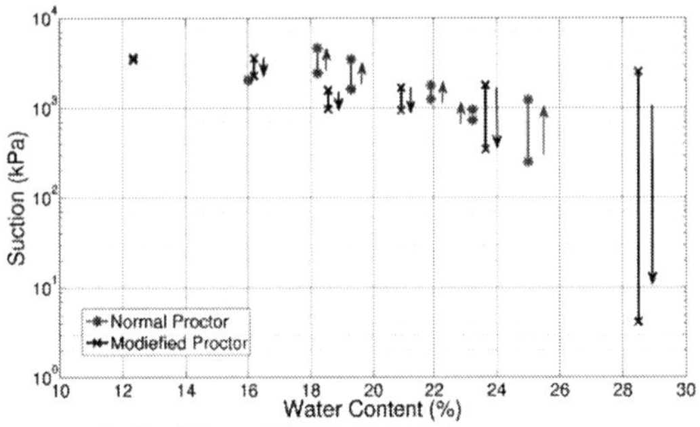

Figure 3. Paths followed during compaction

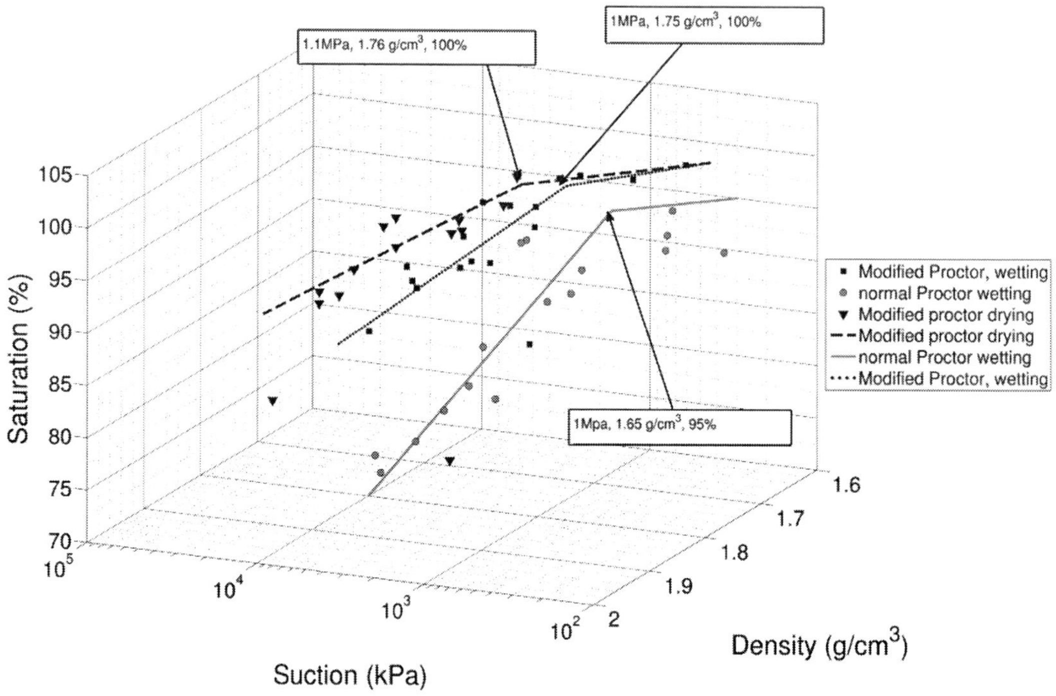

Figure 4. Water retention curves for normal and modified Proctor.

3.4 Retention curves

After extraction of soil samples from the different Proctors, water retention curves where realized and are shown in figure 4. The graph presents three curves. Two represents the drying and wetting curves of the modified Proctor. The third one is the wetting curve of the normal Proctor. The curves are defined and fitted as bilinear. Since in situ compaction gives saturation degrees above 90%, the retention curves are only processed to values around 70%. It is possible to establish a retention surface between the pair of wetting and drying curves [8].

Each compaction energy gives a totally different water retention curve, mostly due to the variation of the density of compaction. It seems that the water retention curves relatively well align. Whatever the water content before compaction, only one wetting and one drying curve appears per water content and energy pair when several would have been expected (one per initial density). The estimated air entry point fairly evolves with the compaction energy, giving a value of 1.1MPa for the modified proctor, and a relatively lower value under 1MPa for the normal proctor. This type of result is corroborated by other studies [9, 10] find in the literature. The main difference between the two curves comes from their respective densities. The density of the material will define its skeleton structure and shape (pore size) and will directly influence the suction and saturation at air entry. The water retention at lower suction should not be affected by compaction [11].

4 Crossing analysis

4.1 Proctor graph with core samples data

Showing all points issued from in situ test samples and the proctor points obtained in laboratory with the stove drying, an anomaly arises in figure 5. It seems that all the points of the compaction test present a density greater than the modified proctor, with a greater saturation degree. This result is unusual for earthworks, though appears sometimes on site with fine grained soils, mostly with clays and marl. Looking at the drying method, it seems that the stove dries the material too much at 105°c, removing part of the tightly adsorbed water and tends to offset the proctor curve on the dry side. It should be noted that the French standard for the Proctor recommend 50°C. For the same gridding of the material, same optimum density is obtained but with a different optimum water content. The modification of the grinding before compaction tends to modify the optimum density. Proctor tests prepared with natural drying shows a better harmony with the in situ points of compaction.

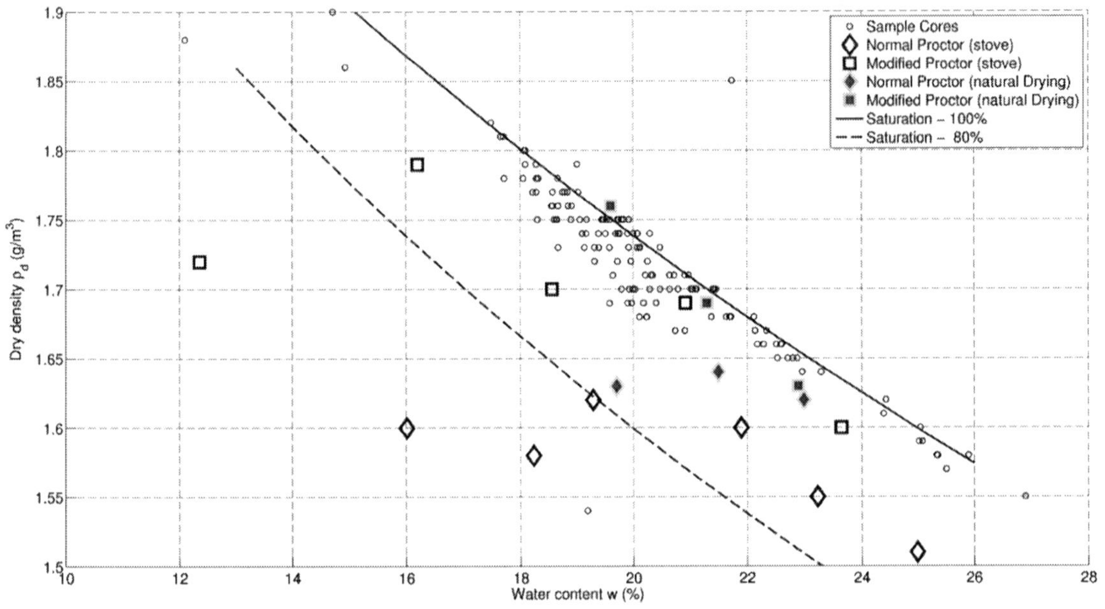

Figure 5. Proctor domain with sample values and Proctors

4.2 Core samples point saturation framing

Even if the proctor is offset, the comparison between the points seems to give a good parallel. Taking the retention lines drawn in figure 4, and placing them on the samples points in figure 6, a good framing of the samples is obtained. The compaction points are nearly all contained in between the retention curves. With several energies, a batch of curves could be drawn and a retention surface could be extracted. The air entry of the modified proctor aligns well with drier points of the compaction test.

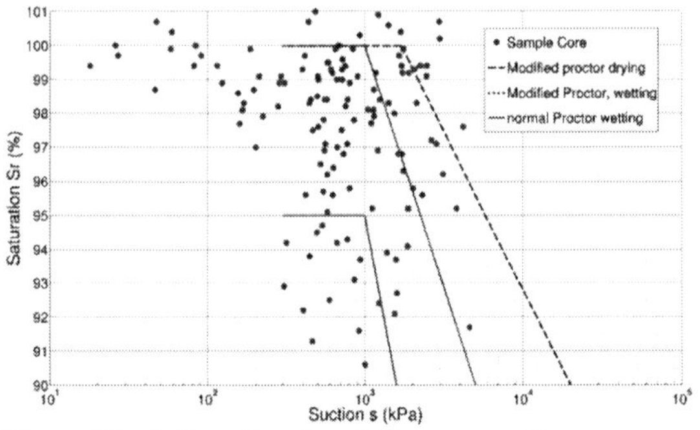

Figure 6 Evolution of suction during compaction

Nearly all the points of the in situ test are under the drying curve of the modified Proctor. Hence this retention curve gives the maximum air entry obtainable while compaction occurs on our material at water content between 14% and 26% of water content. This is confirmed as the points have reach 100% of saturation and any more energy applied would not be able to increase the density. More other, the lower limit given by the wetting curve of the normal proctor gives a fair lower limit; even if it seems that a lower energy would be able to provide a lower curve.

5 Conclusions

The in situ compaction test showed impact of the compaction energy used during earthworks. Important losses of mechanical strength depending of the water content are visible. These results were then confirmed by laboratory experimentations.

Globally over compaction, for fine grained soils, reduces suction on the wet side of the proctor while rest mostly unchanged for dryer material. At the same time, density is roughly maintained or increased whatever the water content. These effects lead to a competition between loss of resistance by suction lowering and a gain of resistance by densification of the material that will define the mechanical response of the soil.

Moreover, it is possible to frame the in situ compaction results in a water retention volume, defined from water retention curve from proctor at different energies. The results allow assessing the evolution of the air entry in the material depending of its compaction degree.

The test carried out showed also that the representativity of laboratory tests can be questioned on work site. The material dried using a stove showed a modification of its proctor curve manifested by an offset of the optimum water content on the dry side by a non-negligible 3 points but with same optimum density, as the grinding was the same for both.

Acknowledgements

The authors would like to thank Vinci Construction Terrassement for financing our laboratory in the context of the French ANR project TerreDurable

References

1. L. Boutonnier, J. Monnet, J.-M. Fleureau, J.-J. Fry, B. Loret, J.-P. Magnan, O. Plé, K. Wong, TerreDurable *Terrassements économiques écologiques et durables*, Agence Nationale de la Recherche – Programme Bâtiments et Villes durables (2012) [ANR-11-VILD-0004]
2. Camp S., Plé O. et Gourc J.P. *Proposed protocol for characterizing a clay layer subjected to bending* Geotechnical Testing Journal : **32, 3** (2009).
3. LCPC et SETRA. *Guide des Terrassements Routiers, Réalisation des remblais et des couches de forme, Guide technique* (GTR). T. 1, 2. Ministère de l'équipement du logement et des transports français, (2000).
4. E. Romero, A. Gens, A. Lloret. *Water permeability, water retention and microstructure of unsaturated compacted boom clay*. Engineering Geology **54** p.117-127 (1999).
5. S. Tripathy, E. C. Leong, H. Rahardjo. *Suction of compacted residual soils,* Unsaturated Soils : Experimental Studies, Springer, p.111-122 (2005).
6. V. Sivakumar, J. Wheeler. *Influence of compaction procedure on the mechanical behaviour of an unsaturated compacted clay Part 1: Wetting and isotropic compression*. Géotechnique **50**, 4 p.359-368 (2000)
7. A. Sawangsuriya, T. B. Edil, P. J. *Bosscher. Modulus-Suction-Moisture Relationship for compacted soils in postcompaction state*. Journal of Geotechnical and geoenvironmental engineering **135**, 1, p.1390-1403 (2009).
8 S. Salager, M. S. El Youssoufi, C. Saix. *Definition and experimental determination of a soil-water retention surface*, Can. Geotech. J. **47** p.609-622 (2010).
9. S. K. Vanaplly, D. G. Fredlund, D. E. Pufahl *Influence of soil structure and stress history on the soil-water characteristics of a compacted till.* Géotechnique **49**, 2 p143-159 (1997).
10. F A. M. Marinho , M. M. Stuermer ,*The Influence of the compaction energy on the SWCC of a residual soil*, Advances In Unsaturated Geotechnics p.125-141 (2012).
11. S. Frydman, R. Baker. *Theoretical soil-water characteristic curves based on adsorption, cavitation, and a double porosity model*. International journal of Geomechanics 96 p.250-257 (2009).

Hydromechanical behavior of a quasi-saturated compacted soils on drying-wetting paths-experimental and numerical approaches

Soanarivo Rinah Andriantrehina[1,a], Dino Mahmutovic[2,3], Saïd Taibi[1], Luc Boutonnier[2] and Jean-Marie Fleureau[4]

[1]*LOMC, CNRS UMR 6294, Université du Havre, 53, rue Prony, BP 540, 76058, Le Havre, France*
[2]*EGIS-Géotechnique, 3, rue Docteur Schweitzer, 38180, Seyssins, France*
[3]*UNIV. GRENOBLE ALPES, 3SR, F-39800 Grenoble, France*
[4]*Laboratoire MSSMat, CNRS UMR 8579, CentraleSupélec, Grande voie des vignes, 92295 Châtenay-Malabry, France.*

Abstract. This paper presents an experimental and numerical investigation funded by the French National Project "Terredurable", which is devoted to the study of soils in quasi-saturated state. The experimental study is focused on the behavior of compacted soils on drying-wetting paths and the macroscopic effect of the drying path on shrinkage and cracking. Furthermore, a protocol for image analysis of crack in drying tests was developed. Two approaches are used for the measurement of surface strains and identification of the ultimate stress before the formation of the first crack, using VIC-2D software, and for the monitoring of crack evolution, using ImageJ software. The aim of the numerical approach is to reproduce the drying experiments with a finite difference code (FLAC 3D), in order to understand the stress conditions that can explain crack initiation, without modeling the crack formation itself.

1 Introduction

Compacted soils used in geotechnical engineering are the result of the densification of soil by application of mechanical energy. They are often in an unsaturated state and subjected to suction. In the laboratory, the characterization of compacted soil is generally carried out by standardized Proctor tests [1, 2, 3]. In the literature, many studies are devoted to the behavior of compacted soils on drying wetting paths [4, 5] and the induced cracking desiccation [6, 7].

This paper presents an experimental and numerical investigation carried out on a compacted marl on drying-wetting paths. The experimental study is focused on the behavior of compacted soils on drying-wetting paths and the macroscopic effect of drying on shrinkage and cracking. The aim of the numerical approach is to reproduce the drying experiments with a finite difference code (FLAC 3D). The originality of this paper is to treat the problem of desiccation cracking of compacted soils related to the suction which is responsible for the appearance of the first cracks

2 Material

The study was performed on a high-plasticity marl whose properties are summarized in Table 1. The standard Proctor optimum, water content and maximum density are consistent with the values given by the correlations of [5] (Fleureau et al. 2002), and [8] (Biarez and Favre 1975), based on the liquid limit of the material, as shown in Table 1. According to the USCS classification [9], the marl is a highly plastic clay (CH) with a percentage of $CaCO_3$ ranging from 30 to 40%. In the French road classification GTR 92 [10], the marl is classified as A4, and in the American Association of State Highway and Transport Officials (AASHTO) classification [11], it is classified as A7.

Table 1. Properties of the marl.

Grain size distribution <80 μm (%) <2 μm (%)	 94 36
Cu	6.25
Cc	0.93
w_L (%)	80
w_{SPO} (%) Measured Fleureau et al. 2002 Biarez and Favre 1975	 24.5 31 32
γ_d / γ_w at SPO Measured Fleureau et al. 2002 Biarez and Favre 1975	 1.50 1.37 1.35
$CaCO_3$ content (%)	30-40

[a] Corresponding author: andrians@univ-lehavre.fr

3 Methods

3.1 Drying-wetting paths

Several methods were used to control suction in the samples [5]. For suction lower than 20 kPa, the samples were placed on tensiometric plates, and a negative pressure was applied to water, the air pressure being atmospheric. For suctions ranging from 50 to 8000 kPa, the osmotic technique was used. In this technique, the sample is in contact with a solution of organic macromolecules of polyethylene glycol PEG (molar weights of 20 000 and 6 000 for negative pressures ranging from 50 to 1 500 kPa, and 3 MPa to 8 MPa, respectively) through a semi-permeable membrane which only allows water molecules to pass. For higher suctions, saturated salt solutions were used, the transfer of water then occurring in vapor phase. Samples were placed in desiccators in which the relative humidity of the atmosphere was controlled by several salt solutions.

Once the equilibrium was reached, the final properties of the samples were measured. The samples were weighed, and then immersed in non-wetting oil (commercial kerdane). Their external volume was derived from the difference between their weight after wetting and their apparent weight while immersed in kerdane. Finally, their dry weight was measured after evaporation of both water and oil in an oven at 105°C for 24 h.

3.2 Free drying tests with image analysis

The drying tests with image analysis were carried out on compacted soil samples with a water content equal to 27.5% (w_{SPO} + 3%). The sample (2 cm thick) was compacted in a CBR mold, then put on a precision balance. Drying tests were performed under controlled relative humidity (30% ± 5%) and temperature (22°C ± 1°C) conditions. The experimental device (Figure 1) consists of a sample support placed on a precision balance and a camera taking photos at regular time intervals during the free drying test [12, 13]. The water content measured via the precision balance is an average value in the sample over the total dry mass.

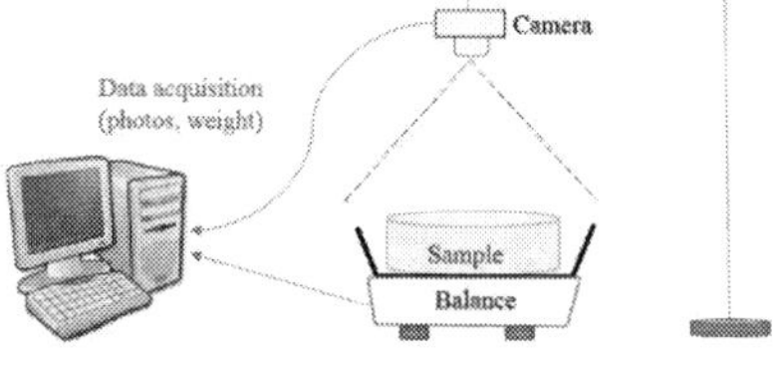

Figure 1. Schematic of experimental setup.

Two approaches were used for the measurements [13, 14]:

- Before failure (cracking), measurement of surface strains and identification of the ultimate stress before the formation of the first crack, using VIC-2D software for digital image correlation. At regular time intervals during the free drying tests, photos of the soil surface are taken and analyzed to provide two-dimensional strain maps of the complete planar specimen.

- Study of crack evolution, geometrical measurements and percentage of cracks as a function of water content. Crack evolution is analyzed using ImageJ software. The technique consists in identifying the cracks on a grey scale image using a threshold. The crack ratio, called CR, is defined as the ratio between the total area of cracks and the area of the soil sample:

$$CR = \frac{\text{Area of cracks}}{\text{Area of the sample}}$$

Suction is derived from water content on drying path, at the times corresponding to crack formation.

3.3 Numerical approach of the free drying test

The aim of the numerical approach is to reproduce the drying experiments with the finite difference code FLAC 3D (Fast Lagrangian Analysis of Continua in 3D). The objective is to understand the stress conditions that can explain crack initiation, without modeling the crack itself.

Calculation assumptions

Geometry

The geometry of the numerical model is that of the experimental sample: it is a circular disk, 2 cm thick and 15 cm in diameter. However, square mesh was preferred to minimize its influence on the results. Therefore, the shape is approached through a fine mesh using 30000 elements (Figure 2).

Figure 2. Overview of the sample modeled with square mesh (30000 elements) - Initial state corresponds to the sample.

Initial state

The model reproduces the initial conditions of the experimental test (compacted sample in Proctor test, drying test). The preconsolidation stress of 200 kPa is deduced from an oedometer test carried out on a sample identical to that of the drying test. The initial suction value of 60 kPa is also derived from the experimental measurements.

Boundary conditions

Two boundary conditions are defined:

- No movement of the sample base: the translation along the z axis is blocked, those in the other directions are permitted. The support of the sample for the experimental test validates this approach.
- A flow boundary condition is applied at the ends of the model in contact with air. In fact, water in contact with the atmospheric pressure progressively evaporates and creates a flow transfer. We tried to reproduce this phenomenon by imposing a transfer of water equivalent to that measured in the experimental test. Table 2 summarizes the water flow recorded during the drying test for different time periods. Numerical modeling aims to reproduce this phenomenon.

Table 2. Calculation of water flow transfers imposed on the numerical model boundary conditions.

Phase	Periods	Durations (s)	Water loss (g)	Lateral area (m^2)	Disk area (m^2)	Total area (m^2)	Flow (l/s)	Flow (l/s.m^2)
1	t0-t1	2700	1,54E+00	9,43E-03	1,77E-02	2,71E-02	7,12E-07	2,63E-05
2	t1-t2	900	3,86E-01	9,43E-03	1,77E-02	2,71E-02	5,36E-07	1,98E-05
3	t2-t3	20700	1,07E+01	9,43E-03	1,77E-02	2,71E-02	6,44E-07	2,38E-05
4	t4-t5	78300	3,77E+01	9,43E-03	1,77E-02	2,71E-02	6,02E-07	2,22E-05

Mechanical and hydraulic parameters

Mechanical constitutive law and associated parameters

The constitutive law plays an important role in predicting stresses in this simulation. However, it has no effect on volumetric strains. Indeed, drying is studied for suctions lower than the air entry pressure (desaturation of the sample). In this domain volumetric strains correspond to normal shrinkage, i.e. volume decrease caused by evaporation of water. Beyond that domain, the model is no longer valid.

Two different constitutive laws were tested: an elastoplastic law (modified Cam-Clay) and a perfect plastic law (Mohr Coulomb). The first one correctly reproduces the overall behavior but it is unstable in areas where strain softening develops. Under these conditions, Mohr Coulomb helps to treat the areas of heterogeneities where cracks can potentially develop. The mechanical parameters of these two laws were derived from the experimental tests (triaxial and oedometer tests).

Table 3. Mechanical parameters of the used constitutive laws

Cam-Clay parameters		Mohr Coulomb parameters	
λ	0,3	K	1517 kPa
κ	0,04	G	910 kPa
P'_0	120 kPa	Phi	13°
M	0,7	c	40 kPa
vp1	2,8		

Flow law, hydraulic parameters

The used numerical model is a coupled model that considers the flows. Darcy law was used, with a measured permeability of 10^{-11} m/s. The sample is considered as quasi-saturated. Indeed, air is present in the sample in the form of occluded air bubbles in water, forming a homogeneous and compressible pore fluid. The behavior of this fluid is considered in the modeling through a saturation equal to 1 and a compressibility factor taken equal to 6.10^{-5} kPa^{-1} (based on feedback for this soil type [15, 16, 17]).

4 Results

4.1 Drying-wetting paths

Figures 3 and 4 show the drying-wetting paths. Graphs A and C illustrate the void ratio and the degree of saturation as a function of water content. Graphs B, D, and E show void ratio, degree of saturation, and water content as a function of suction.

Slurry samples

Graph A shows the shrinkage curve. On drying path, the soil follows the saturation line, with the equation:

$$e = w.(\gamma_s/\gamma_w) \quad (1)$$

down to the shrinkage limit water content, after which the void ratio becomes constant. The shrinkage limit w_{SL} = 20% is positioned at the intersection of the saturation line with the asymptote of the curve for w = 0.

Graph B shows the compressibility of the soil (void ratio as a function of suction). This graph present a domain characterized by a significant decrease in void ratio, followed by a stabilization of the void ratio. The shrinkage suction s_{SL} ($s_{SL} = 3*10^4$ kPa) is the suction of the point at the beginning of this stabilization, related to the shrinkage limit water content of the soil. NC and OC correlation curves [8], derived from the w_L of the material, are shown in this graph to compare with the position of the wetting and drying paths of the material. The drying path follows the Cc correlation line and the wetting path follows the Cs correlation line. Graph C shows that the material was initially saturated. The degree of saturation decreases slowly up to the air entry point, then decreases sharply with the water content. In graph D, the soil continues to be quasi-saturated up to the desaturation suction s_d =1.5 MPa. This domain corresponds to an irreversible behavior for the drying and wetting paths of the material. Then, the degree of saturation decreases quickly. This rapid decrease in the degree of saturation results in the fall of the water content (Graph E). Graph E [log s; w] corresponds to the usual retention curves graph, in which the axes are reversed to highlight the correspondence between the changes in void ratio, degree of saturation, water content and suction. For suctions lower than the desaturation suction, variations in water content and void ratio are related, while for higher suctions, the water content varies with the degree of saturation. In the domain of high suctions where the water content and the void ratio vary little, the hysteresis between drying and wetting paths is negligible. The domain where the increase in water content and degree of saturation is more important, and where the variation of the void ratio is not significant, is an intermediate phase of resaturation. In this domain, the hysteresis between the drying and wetting paths appears in the graphs D and E. This hysteresis is mainly the result of "ink-bottle" effect. In the domain of low suctions, the material is quasi-saturated even if the water content and void ratio continue to grow. The maximum degree of saturation is less than 1. Thus, the behavior is not reversible in this domain.

DOI: 10.1051/e3sconf/20160914013

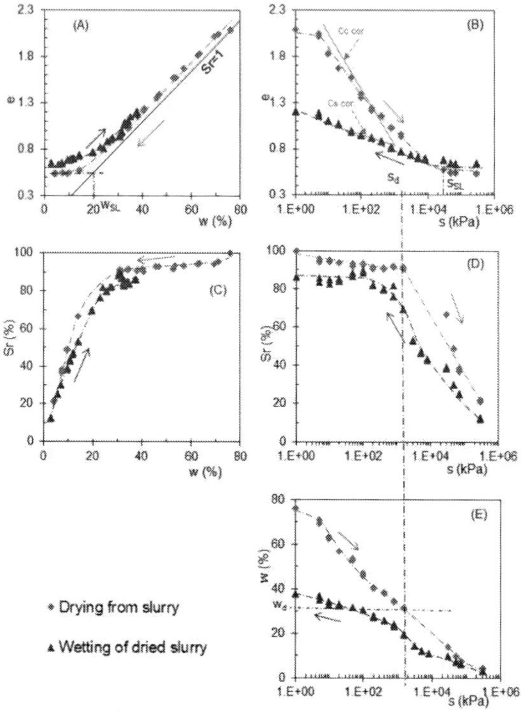

Figure 3. Drying-wetting paths of samples of slurry.

Compacted samples

At the SPO, the material water content is 24.5% and the void ratio is 0.77. So, graphs B and E give a value of the suction s_{SPO} of approximately 800 kPa.

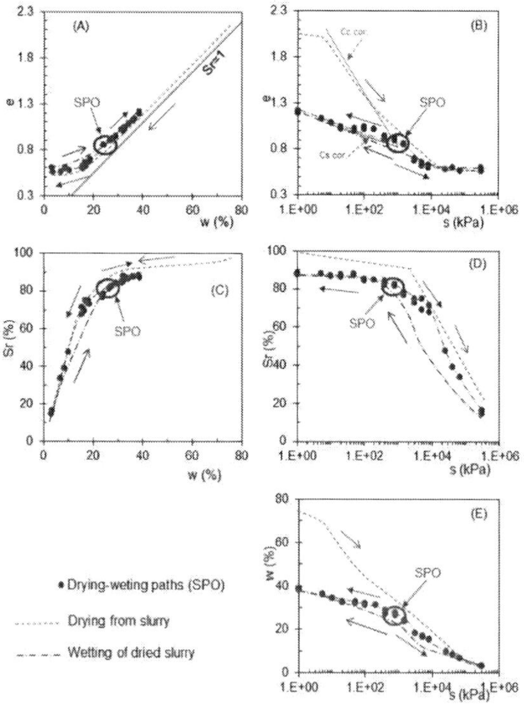

Figure 4. Drying-wetting paths of compacted samples.

As the s_{SPO} value is smaller than the value of the shrinkage limit suction and the void ratio e_{SPO} is larger than the shrinkage limit void ratio of the samples of slurry, the drying path rejoins the Cc line, down to the shrinkage limit void ratio of the paste, then follows the same path. Wetting path follows the Cs correlation line. The drying and wetting paths are very close to the wetting path of dry slurry samples.

4.2 Free drying tests with image analysis

The free drying test on the compacted marl (Figure 5a) was followed up to stability of the water content (residual water content). The formation of cracks was observed, with a more or less random distribution. The photo below (Figure 5a) shows the difference between the initial state (left) and final state (right) of the material.

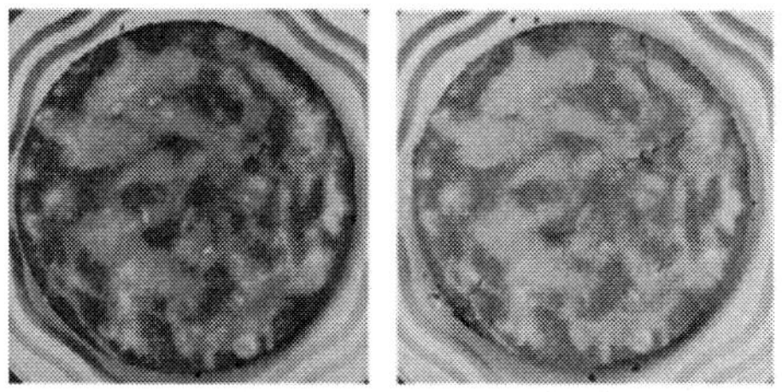

Figure 5a). Free drying of the compacted marl.

During the first 40 hours, the rate of evaporation of water is constant (0.037 g/min) and the water content decreases down to 12 %, called the critical water content. After 100 hours, evaporation stops and the residual water content is equal to 5.7%, as shown in Figure 5b. The first crack appears after 45 min at an average water content of 27.1%.

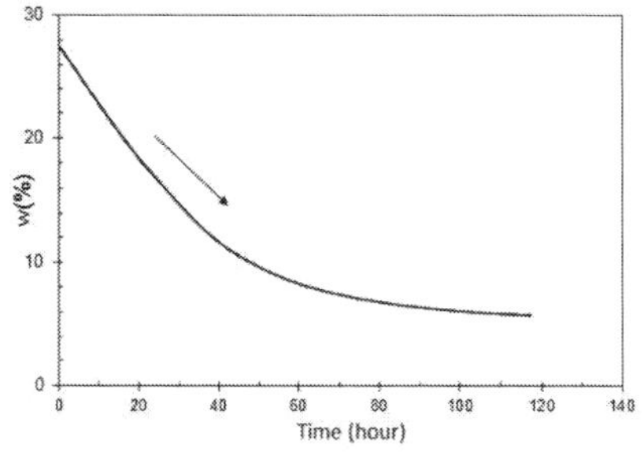

Figure 5b). Variation of water content function of time.

Analysis of the strain field on the sample surface deduced from image correlation (Figure 6) shows a first localization of large strains at certain locations, which then results in the formation of cracks in the first hours of desiccation. This network of cracks spreads and stabilizes when the water content is close to the material shrinkage limit water content (w_{SL} = 16%). This value is confirmed by the drying-wetting test shown above, interpreted in terms of volumetric strains (Figure 6). According to this test, the volume strain increases linearly (contractancy) when the water content decreases, up to an ε_v value of about 15%, before reaching a constant value.

 DOI: 10.1051/e3sconf/20160914013

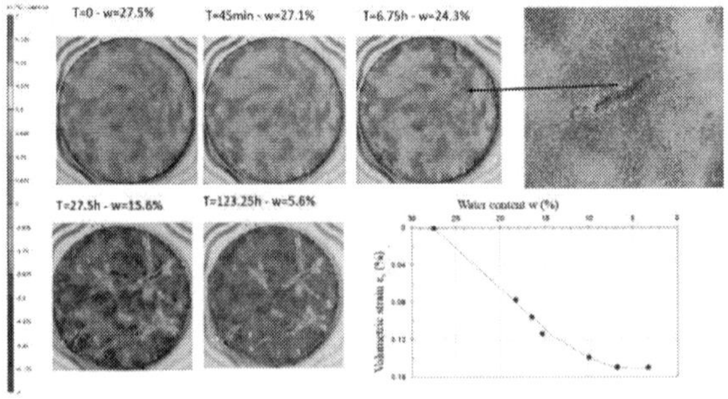

Figure 6. Evolution of the strain field versus time - evolution of the volumetric strain as a function of the water content.

Figure 7 presents the variation of the crack ratio derived from Image J. Starting from a water content of about 27.5%, the sample begins to dry without cracking. The first crack occurs at an average water content of about 27.1%. As the average water content decreases, there is a sudden increase in crack ratio when water content is about 21.7%, and then a trend toward stabilization is observed during the rest of drying.

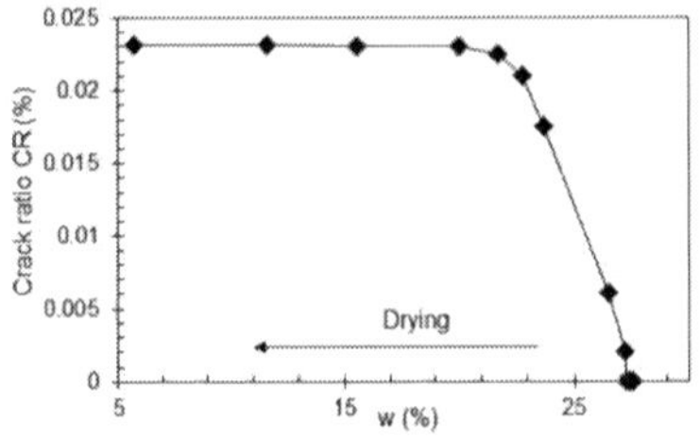

Figure 7. Crack ratio.

Figure 8 shows the drying-wetting paths of compacted soil presented above, where the initial state of the sample in the free drying test is indicated, as well as the formation of the first crack and the point where the cracks stabilize. The first crack appears at a suction around 400 kPa and cracking stabilizes (ie when the cracking no longer changes, while drying continues) around 4 MPa.

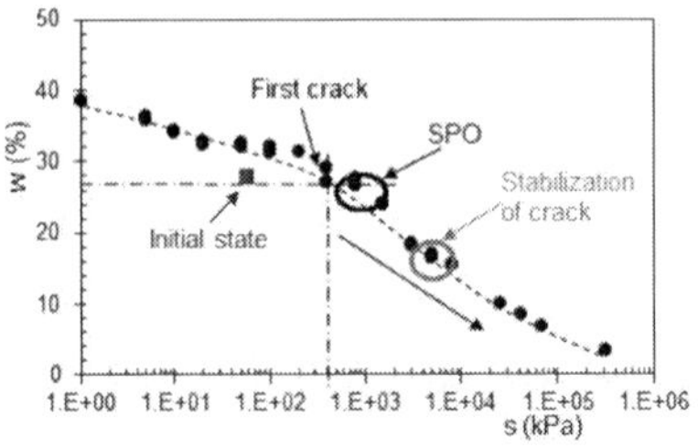

Figure 8. Suction values deduced from drying path.

4.3 Numerical approach of the free drying test

Several phenomena were followed by: (1) the volumetric deformation of the entire sample due to evaporation of water at the borders of the model - (2) changes in pore pressure within the sample - (3) areas for the development of cracks.

Volumetric strains

The change in total volume was followed during numerical simulation (Figure 9). The soil sample progressively shrinks in time due to water transfer; the decrease in volume corresponds to the volume of water evaporated. The simulations were stopped at a water content close to that of air entry (16%). It was also considered that normal shrinkage continues to the end of the total deformation of the sample, without consideration of the residual deformation. This assumption is acceptable because cracks appear and develop mainly in the normal shrinkage area [18, 19, 20].

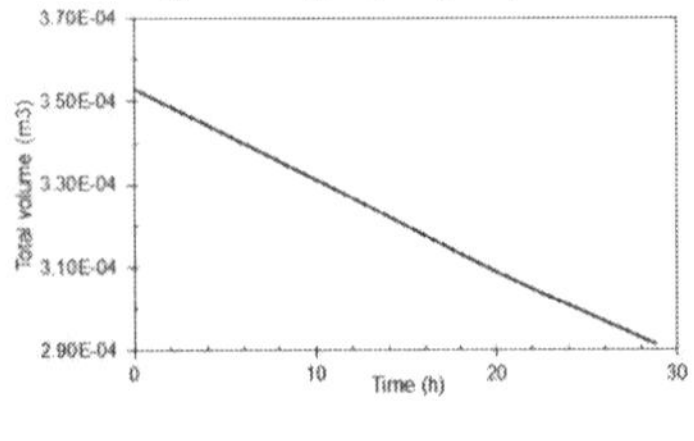

Figure 9. Evolution of the total volume by numerical approach.

Suction prediction

The flow imposed at the ends of the model avoids taking water into account. The degree of saturation of the sample always remains constant and equal to 1. The effect of that is a decrease in volume and an increase in suction of the sample. FLAC allows to visualize the evolution of suction in the sample over time (Figure 10). The final suction was not measured experimentally in the sample. However, the values obtained, though qualitative, are consistent with the soil retention curve and the phenomenon of contractancy.

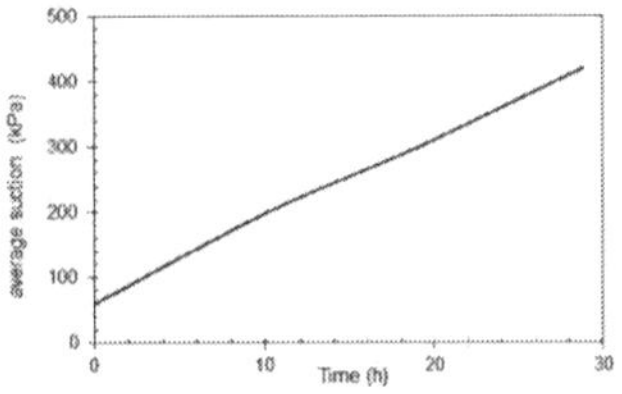

Figure 10. Evolution of the suction by numerical approach.

Prediction of areas that can be potentially cracked

The occurrence of cracks is conditioned by the presence of heterogeneities in the sample (carbonated elements, mainly shells, dispersed in the skeleton.

The aim of modeling is to highlight the specific and local behavior of the soil around areas of weakness or high resistance. So, three stiff areas were introduced in the numerical model to simulate the cracks observed in the experimental test results (Figure 11). These areas are intended to model the presence of discontinuities so as to see if the obtained stress state is the more conducive to surface cracking. Drying is free and the sample does not undergo the effect of any load. Its total average stress should be zero. Now it can be seen that areas of high tensions in total stresses appear in the right area of the stiff areas. These areas may lead to the formation of cracks and their development. This observation can also be made with respect to shear forces, in which one can see the influence of the stiff zones.

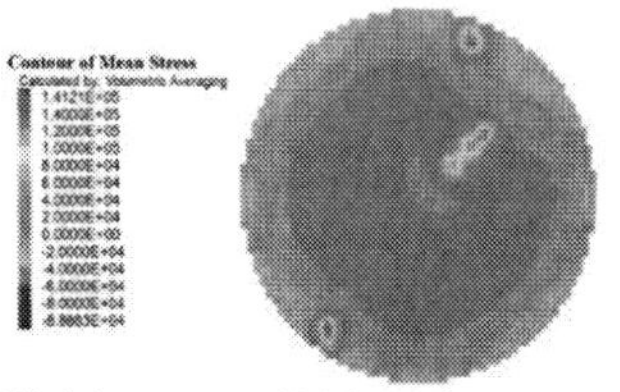

Figure 11. Highlighting areas of high tensions.

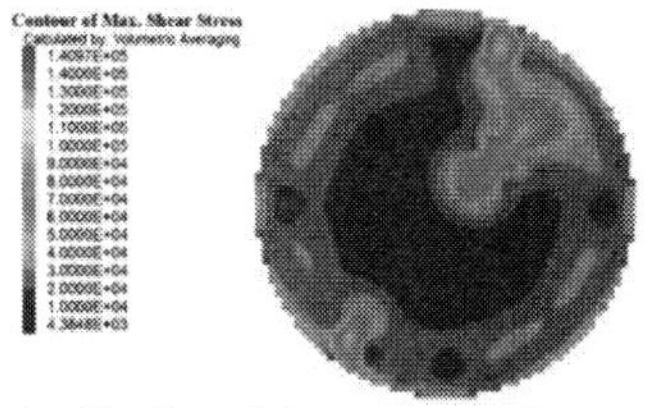

Figure 12. Visualization of shear stress to the sample surface - Highlighting high shear zones around the hard points.

5 Conclusion

Experimental and numerical approaches are presented in this paper concerning the behavior and cracking of a remolded compacted high plasticity marl on drying-wetting paths. The results show the interest of a global representation of the state of materials in drying-wetting paths, allowing to follow the evolution of soil saturation and link shrinkage characteristics and desaturation to drying properties. Free drying tests highlighted the initial conditions of crack formation (water content, suction ...) which occurs in the quasi-saturated domain of the material. The numerical approach allows to model the phenomena and proves that cracks are initiated close to saturation and for suctions lower than the desaturation suction.

Acknowledgment

The authors acknowledge the French ANR project TerreDurable (ANR 2011 VILD 004 01 convention, programme Bâtiments et villes durables).

References

1. AFNOR 1999 - NF P 94-093. Détermination des références de compactage d'un matériau, Essai Proctor normal - Essai Proctor modifié.
2. ASTM D 698-12. Standard Test Methods for Laboratory Compaction Characteristics of Soil Using Standard Effort (600 kN-m/m^3).
3. ASTM D 1557-12. Standard Test Methods for Laboratory Compaction Characteristics of Soil Using Modified Effort (2,700 kN-m/m^3).
4. S. Kheirbek-Saoud, "*Comportement mécanique de la couche de fondation d'une voie ferrée*", Thèse de doctorat, Ecole centrale Paris (1994).
5. J-M. Fleureau, J. C. Verbrugge, P.G. Huergo, A. Gomes Correia, S. Kheirbek-Saoud, "Aspects of the behaviour of compacted clayey soils on drying and wetting paths", Can. Geotech. J., **39**, pp. 1341-1357, (2002).
6. C.J. Miller, H. Mi, N. Yeiller, "Experimental analysis of desiccation crack - Propagation in clay liners", JAWRA **34** (3), pp. 677-686.
7. R. Auvray, S. Rosin-Paumier, A. Abdallah, F. Masrouri, "Etude de la fissuration due à la dessiccation des sols gonflants", *21éme Congrès Français de Mécanique*, Bordeaux, (2013).
8. J. Biarez and J. L. Favre J L. Parameters filing and statistical analysis of dada in soil mechanics. *Proceedings of the 2nd Int. Conf. on Application of Statistics and Probabilities in Soil Mechanics*, Aachen, Vol. **2**, pp. 249-264 (1975).
9. ASTM D 2487- 06. Standard Practice for Classification of Soils for Engineering Purposes (Unified Soil Classification System).
10. AFNOR 1992 - NF P11-300. Exécution des terrassements - Classification des matériaux utilisables dans la construction des remblais et des couches de forme d'infrastructures routières.
11. ASTM D 3282-09. Standard Practice for Classification of Soils and Soil-Aggregate Mixtures for Highway Construction Purposes.
12. X. Wei. Etude micro-macro de la fissuration des argiles soumises à la dessiccation, Thèse de doctorat, Ecole Centrale Paris (2014).
13. J. Eid, S. Taibi, J-M. Fleureau, M. Hattab, Drying, cracks and shrinkage evolution of a natural silt intended for a new earth building material. Impact of reinforcement, J. Con. Build. Mat. **86**, 120-132 (2015).
14. R. Andrianatrehina, D. Mahmutovic, L. Boutonnier, S. Taibi, S. J-M. Fleureau, J-J. Fry, J. Monnet. Observation of craking evolution during free dessication – experimental and numerical approaches. *Symposium international SEC* (2015). Marne-la-vallée.
15. L. Boutonnier. Comportement hydromécanique des sols fins proches de la saturation. Cas des ouvrages en terre : coefficient B, déformations instantanées et différées, retrait / gonflement. Thèse INPG Grenoble (2007).
16. L. Boutonnier. *Mechanics of unsaturated geomaterials*, Chap.14: Coefficient B, Consolidation and swelling in Fine Soils near saturation in Engineering Practice. Ed. L. Laloui, J. Wiley. (2010).
17. L. Boutonnier, G. Mahmutovic, J-J. Fry. Génération de pression interstitielle dans les fondations argileuses des barrages en remblai. *Colloque CFBR* (2015).
18. H. Péron, L. Laloui, T. Hueckel, L. Hu. Experimental study of desiccation of soil. *Proceedings of the Fourth International Conference on Unsaturated Soils*, (Vol **1**, pp. 1073-1084). Reston, VA: American Society of Civil Engineers. (2006).
19. H. Péron, T. Herchel, L. Laloui, L. Hu. Fundamentals of desiccation cracking of fine-grained soils: experimental characterization and mechanisms identification. Can. Geotech. J. **46**, 1177-1201. (2009).
20. H. Nahlawi, J.K. Kodikara. Laboratory experiments on desiccation cracking of thin soil layers. Geotech. Geol. Eng. **24**, 1641-1664. (2006).

Hydro-mechanical behaviour of sandy silt under generalised stress conditions

Enrique Romero[1a], Octavio E. Cárdenas[2], Antonio Lloret[1] and Rodrigo C. Weber[1]

[1]*Department of Civil and Environmental Engineering, Universitat Politècnica de Catalunya, Spain*
[2]*Facultad de Ingeniería Civil, Universidad Autónoma de Coahuila, Torreón, México*

Abstract. This paper presents results on the deformation response of an artificially prepared sand-silt mixture in a hollow cylinder apparatus. The wetting stage was performed under constant mean net stress (p'=200 kPa) and deviatoric stress (q=200 kPa) but at different intermediate principal stresses (controlled through the principal stress parameter $b=(\sigma_2-\sigma_3)/(\sigma_1-\sigma_3)$ and with values b=0, 0.5 and 0.8). Shear strength tests were first performed at constant mean net stress, different Lode angles and water contents (as-compacted and saturated) to ensure that the aforementioned stress state could be applied at the as-compacted water content. Consistent shear strength results were obtained when compared to triaxial compression and extension results at different water contents, which allowed defining the variation of the critical state line with Lode angle and suction. The soaking results indicated that collapse under constant mean and deviatoric stresses was larger when the intermediate stress coincided with the minor one, i.e. under conventional axi-symetric triaxial compression state (b=0). This is a consequence of the dominant shear strains that occurred during saturation when the stress point reached the critical state line at b=0.5 and 0.8.

1 Introduction

There are relatively few experimental results on the influence of the intermediate principal stress on the deformation response of unsaturated soils. Hoyos [1] and [2] used a true triaxial apparatus, whereas [3, 4] a hollow cylinder device, to perform tests under constant suction to study the influence of this constitutive stress variable on soil behaviour under generalised stress conditions.

Nevertheless and to the author's knowledge, the effect of intermediate stress and the rotation of principal stresses while suction is varying -particularly on the deformation response on soaking- have not been experimentally studied in depth. Suction changes may occur on compacted earthwork constructions and slopes, in which these effects (principal stress direction and intermediate principal stress) have been widely recognised (see for instance, [4-6]).

The authors of the present paper consider that this limitation is related to experimental difficulties, manly associated with the homogenisation of the hydraulic field (this is particularly important when working with a large sample as in the hollow cylinder apparatus and when liquid contact/pressure is applied at the lower boundary) and to the large deformations undergone on soaking. Deformation non-uniformities occur on soaking in the specimen as a result of its curvature and end-restraint effects, and this is important when selecting the type of transducers and the way to measure the displacement field and thus precisely determine the volume change response.

To get further experimental insight into the behaviour of unsaturated soils under generalised stress conditions, an experimental program has been launched to explore the deformation response of a lightly compacted and artificially prepared soil mixture during soaking at constant mean net stress (p'=200 kPa) and deviatoric stress (q=200 kPa) and at different intermediate principal stresses. A mixture of 30% sand and 70% clayey silt by dry mass has been statically compacted at low dry density and low water content to induce high collapsibility on suction reduction. Shear strength tests in the hollow cylinder were also performed at constant mean net stress, different Lode angles and water contents (as-compacted and saturated) to define the position of the critical state line with Lode angle and suction, as well as to ensure that the aforementioned stress state could be applied under the as-compacted water content.

2 Artificially prepared soil used in the investigation

The soil used in the research was a mixture of two materials: fine sand of Castelldefels (Barcelona) and Barcelona (BCN) clayey silt [7, 8]. The particle size of the sand varied between 150 μm and 1.18 mm, with a density of solids of 2.65 Mg/m^3. The clayey silt presented

[a] Corresponding author: enrique.romero-morales@upc.edu

a liquid limit w_L=36%, a plastic index PI=19% and a density of solids of 2.67 Mg/m^3.

The mixing ratio was studied to find a suitable combination of shear strength, water permeability –given the large size of the sample–, measurable matric suction with tensiometer, and high collapsibility on soaking. Finally, a mixture of 30% sand and 70% clayey silt by dry mass was selected. The mixture with a water content of 4.0% (suction around 10 MPa) was statically compacted to a dry density of 1.55 Mg/m^3 (void ratio e=0.716, degree of saturation S_r=0.15). The vertical stress applied to compact the sample was below 200 kPa.

Since the sand grains are immersed in the dominant clayey/silt matrix, it is important to know the real void ratio of the clayey silt. This real void ratio is e_c=1.02 (dry density 1.32 Mg/m^3), which is large enough to induce an important collapse on wetting, to present a relatively high water permeability k=5.85×10^{-7} m/s and to display suction values that are measurable by high-range tensiometer readings (at least for water contents above 8% [8]).

A preliminary laboratory program was carried out with oedometer, direct shear cell and triaxial apparatus to explore the stress level associated with the maximum collapse zone (around a vertical stress of 200 kPa), pre and post-yield compressibility, and the shear strength properties. The water retention properties of the mixture on wetting were determined with psychrometer and high-range tensiometer readings. The water retention curves reported by [8] were used to estimate matric suction changes during the soaking process.

3 Hollow cylinder apparatus

The compacted sample was installed in the hollow cylinder apparatus (HCA) and featured 50 mm in external radius r_o, 30 mm in inner radius r_i, and 200 mm in height H_o (Figure 1). The capacity of the axial load cell and the torque cell were 10 kN and 100 Nm, respectively. The principal stress parameter $b=(\sigma_2-\sigma_3)/(\sigma_1-\sigma_3)$ [9] was automatically controlled during the tests (σ_i are principal stresses, being σ_1, σ_2 and σ_3 the major, intermediate and minor ones, respectively). Both the base and top caps presented conventional porous discs. Three pressure/volume controllers were used for the application and control of the inner and outer cell pressures (P_0 and P_i in Figure 1), as well as for the measurement of the soil water volume changes. Figure 1 shows a scheme of the sample geometry, axial force W, torque T and inner/outer pressures applied, as well as the stress state of the soil with the definition of the deviatoric stress q, mean stress p and Lode angle θ invariants. The procedure to obtain the principal stresses from the applied force and pressures has been described in [10].

The axial strain ε_z was evaluated directly from the vertical displacements controlled by the stepper motor and the initial height of the sample. The determination of volumetric ε_p, radial ε_r, and circumferential ε_θ strains needed the knowledge of the magnitude of the changes in the internal and external radii w_i and w_o, respectively, indicated in the figure (see for instance, [10]). Due to the lack of local instrumentation and the relatively high volumetric deformation (volumetric strain on soaking was around 10%), it was necessary to obtain these deformations from alternative recorded values. The changes in internal and external radii were evaluated from inner and outer chamber volume changes, which were measured by automatic pressure/volume controllers, together with height changes that were registered with stepper motor counts [8]. Outer chamber volume changes were corrected to take into account the deformability of the cell with pressure changes. The measurement of the final water content under saturated conditions, as well as the initial and final dry densities of the sample, allowed better assessing void ratio changes and volumetric deformation determination.

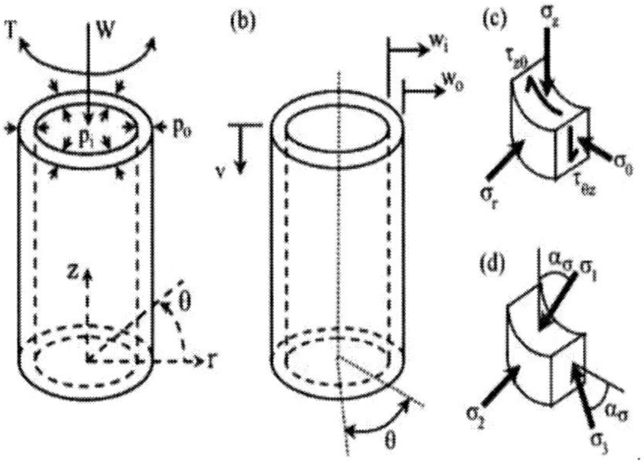

$$q = \frac{1}{\sqrt{2}}\sqrt{(\sigma_1-\sigma_2)^2+(\sigma_2-\sigma_3)^2+(\sigma_1-\sigma_3)^2}$$

$$p = \frac{(\sigma_1+\sigma_2+\sigma_3)}{3}$$

$$\theta = tan^{-1}\frac{2b-1}{\sqrt{3}}$$

$$\sigma_z = \frac{W}{\pi(r_o^2-r_i^2)} + \frac{P_o r_o^2 - P_i r_i^2}{r_o^2-r_i^2}$$

$$\sigma_r = \frac{P_o r_o + P_i r_i}{r_o + r_i}$$

$$\sigma_\theta = \frac{P_o r_o - P_i r_i}{r_o - r_i}$$

$$\tau_{\theta z} = \frac{3T}{2\pi(r_o^3 - r_i^3)}$$

Figure 1. Sample geometry and stress state in the HCA.

4 Shear strength results. Influence of intermediate principal stress

Drained shear strength tests in the HCA were performed at constant mean net stress (p''=200 kPa), different Lode angles (θ=-30°, 0°, 19.1° and 30°, corresponding to b=0, 0.5, 0.8 and 1, respectively), as well as different water contents (full saturation and as-compacted state). Drained shear strength tests were also complemented in the triaxial cell (compression tests TXC at θ=-30° and extension tests TXE at θ=30°).

Figure 2 shows the evolution of the normalised shear strength results with Lode angle. The normalised shear strength at ultimate failure conditions (equivalent to the slope of the critical state line *M*) is expressed as $q_f/(p''+c'/\tan\phi')$, in which c' is the cohesion (that is affected by suction) and ϕ' the drained friction angle (that is considered not affected by suction). As observed, consistent results were obtained when compared to TXC and TXE results at different water contents, which allowed defining the variation of the slope of the critical state line $M(\theta)$ with Lode angle and the variation of $c'(s)$ with suction. The variation of $M(\theta)$ with Lode angle has been fitted to the following expression [11],in which M_e=6sinϕ'/(3+sinϕ') and M_c=6sinϕ'/(3-sinϕ')

$$M(\theta)=\frac{2\mu}{(1+\mu)-(1-\mu)\cos 3(\theta+30°)}M_c \qquad (1)$$

$$\mu=\frac{M_e}{M_c}=0.724\ ;\ M_c=1.14$$

Fitting test results allowed obtaining ϕ'=28.7° (the same for saturated and partially saturated states), a drained cohesion $c'(0)$=0 under saturated condition and $c'(s)$=36 kPa at s=10 MPa. Direct shear tests on as-compacted and saturated states also reported similar results ϕ'=29.7° (not affected by suction), $c'(0)$=0 under saturated condition and $c'(s)$=31 kPa at as-compacted state.

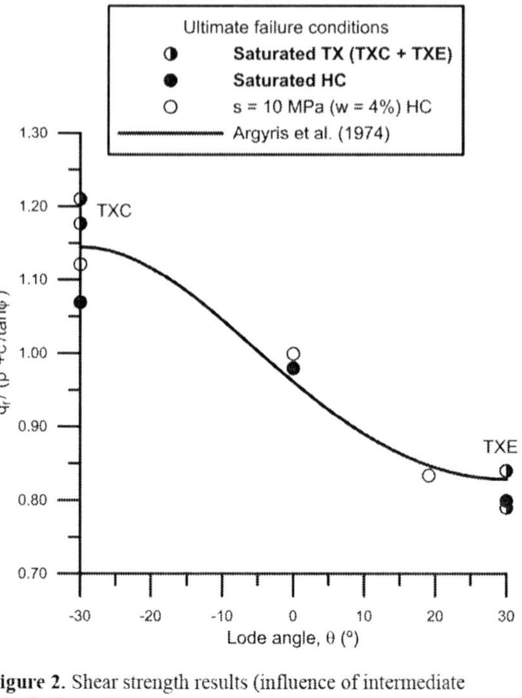

Figure 2. Shear strength results (influence of intermediate principal stress).

5 Collapse at constant mean and deviatoric stresses. Influence of intermediate principal stress

To determinate the effect of the intermediate principal stress σ_2 on collapse, three tests were performed by changing the value of parameter *b*, which was set to 0, 0.5 and 0.8 (θ=-30°, 0° and 19.1°, respectively). The tests were carried out by maintaining constant the mean net stress p''=200 kPa and deviatoric stress q=200 kPa during the saturation stage. The test started with an isotropic compression stage by applying approximately the same inner and outer chamber pressures up to a maximum mean net stress p''=200 kPa at constant water content (w=4.0%) and atmospheric pressure of air in soil. A rate of 2.5 kPa/hour was selected for the application of the external P_o and internal P_i pressures during isotropic compression. After this initial stage, a small increase of deviatoric stress q=30 kPa was applied before setting parameter *b*.

Afterwards, the value of *q* was increased at a rate of 15 kPa/hour up to approximately 200 kPa by maintaining $p''\approx$200 kPa and approximately constant *b* parameter. The shearing stages were carried out at w=4.0%. The paths in terms of principal strains ε_1, ε_2 and ε_3 followed during the deviatoric stress increase are shown in Figure 3. It should be remarked that values of ε_2 changed from negative (expansion) for the case of b=0, to positive (compression) for b=0.8. This behaviour was equivalent to the one observed in tests performed using a true triaxial equipment [1].

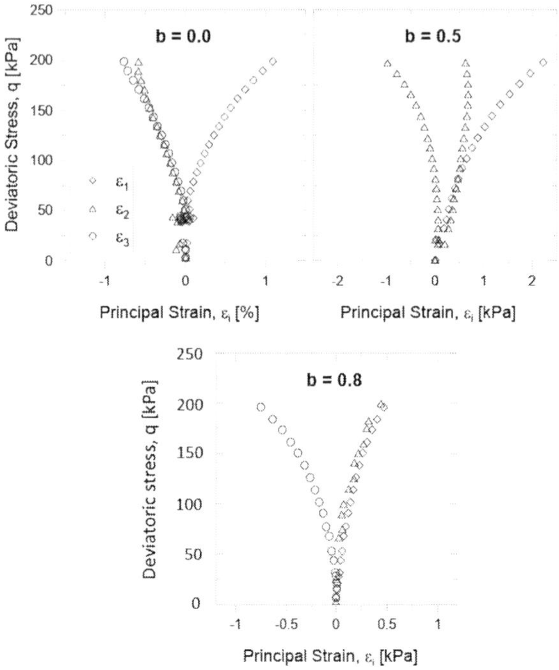

Figure 3. Evolutions of principal strains during deviatoric stress increase at constant p''=200 kPa and w=4.0% for different values of *b*.

Once the target stress state was reached, the samples were soaked under constant stress state by setting a small water pressure at the base (around 10 kPa). Water volume changes on soaking were monitored by an automatic pressure/volume controller. Table 1 presents the principal stresses and stress invariants at different b values, which were kept constant along the wetting process. Figure 4 shows the measured and simulated evolutions of the global degree of saturation during the soaking process at constant b=0. The simulated evolution has been performed using a 1D infiltration model. Based on this model and considering the sample as a 'point test', the evolution of the (average) matric suction has been estimated by the water retention curve reported in [8]. The time evolution of this (average) matric suction has been also plotted in the figure.

Table 1. Principal stresses and stress invariants applied during the three soaking tests.

b(-)	σ_1 (kPa)	σ_2 (kPa)	σ_3 (kPa)	q (kPa)	p (kPa)	θ (°)
0	340	125	125	215	197	-30
0.5	320	205	90	200	205	0
0.8	290	245	60	211	198	19.1

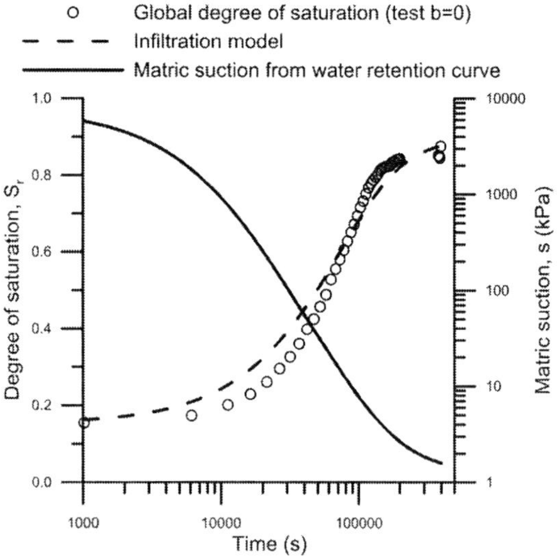

Figure 4. Time evolution of the global degree of saturation of the sample along the soaking process (test at b=0). Time evolution of average matric suction (sample considered as a 'point test').

Figure 5 shows the time evolution of the volumetric strain ε_p (associated with collapse) for different values of b during the saturation stage at constant p'≈200 kPa and q≈200 kPa. Important collapse strains were recorded along the three soaking tests. Final ε_p values around 20%, 9.6% and 8.4% were measured at the end of these soaking stages for values of b = 0, b = 0.5 and b = 0.8, respectively. These results seemed to indicate that collapse under constant mean and deviatoric stresses was larger when the intermediate stress coincided with the minor one, i.e. under conventional axi-symetric triaxial compression state.

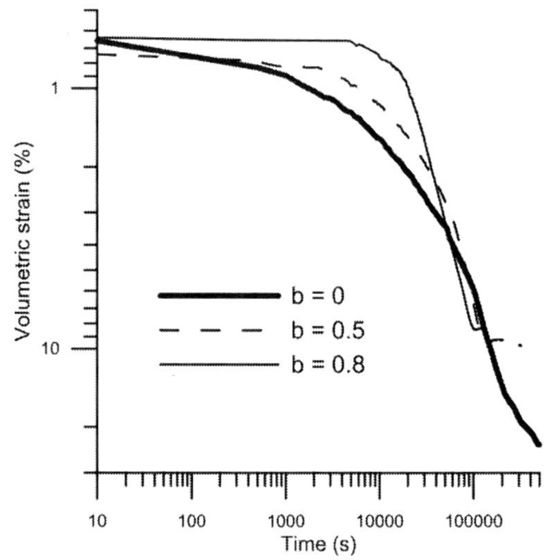

Figure 5. Time evolutions of volumetric strains (collapse) during soaking at different b.

The stress state (p'=200 kPa and q=200 kPa), which is kept constant during the soaking process, is plotted in Figure 6 for different b values. The position of the initial (as-compacted) critical state line, together with the initial cap yield surface, is also indicated in the figure (thick solid lines). This cap yield surface has been plotted according to Barcelona Basic Model (BBM) [12]. As observed, the stress state at a normally consolidated condition lies on the yield surface of the cap. The distance between the stress point and the initial (as-compacted) critical state line reduces as parameter b increases (this is a consequence of the reduction of the slope $M(\theta)$ as the Lode angle increases).

During soaking, the slope of the obliquity of the stress point $q/(p'+c'/\tan\phi')$ (dashed lines in Figure 6 at the top) increases due to the reduction in c', which is affected by suction decrease. As a first approximation, a linear variation of $c'(s)$ with suction has been considered between $c'(0)$=0 at s=0 and $c'(s)$=36 kPa at s=10 MPa. In addition, the position of the critical state lines is shifted towards a reduction in the elastic domain during suction reduction (thin solid lines with the same slope $M(\theta)$ as the initial ones). In fact, the final position of the critical state line reaches the stress point in the case of b=0.5 and 0.8 during saturation. In neither of these cases, c' does not reach a value of 0. This means that the deformational response on soaking is dominated by shear strains ε_q in

 DOI: 10.1051/e3sconf/20160914014

the case of b=0.5 and 0.8 (besides the occurrence of collapse deformations). This can be better appreciated in the direction of the plastic flow vectors depicted in the figures and plotted according to BBM [12] (vectors plotted in thick lines represent the initial conditions before the soaking takes place, whereas vectors in thin lines are associated with the final saturated condition).

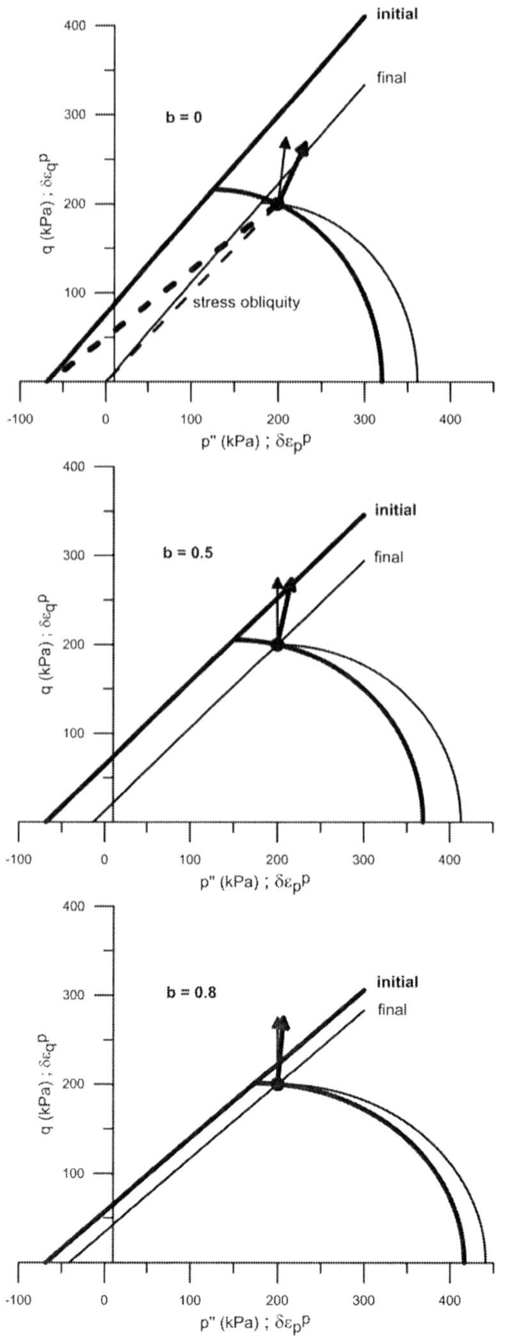

Figure 6. Stress state during soaking. Initial and final cap yield loci and critical state lines. Initial and final plastic flow vectors.

Figure 7 shows the evolution of the inclination of the plastic vectors (measured and modelled through BBM [12]) along the soaking process. As observed, the inclination of the plastic vectors stabilises at b=0, which is not the case of b=0.5 with dominant plastic shear strains. A relatively good agreement between measured and modelled responses can be observed.

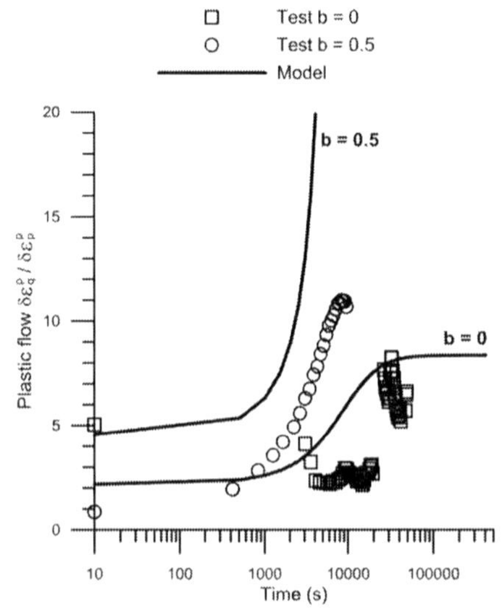

Figure 7. Time evolutions of the inclination of plastic vectors for different b values. Experimental data and model results.

The experimental normalised volumetric strains (with respect to the maximum value at b=0) are presented in Figure 8 for different Lode angles. The figure also includes model predictions. As observed, BBM predicts a reduction consistent with measured values at increasing Lode angles.

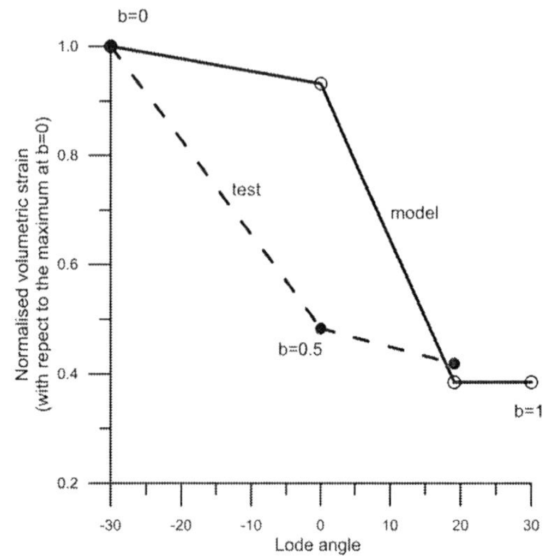

Figure 8. Evolution of normalised volumetric strain during soaking (with respect to the maximum value at b=0) for different Lode angles. Experimental data and model results.

6 Summary and concluding remarks

A mixture of 30% sand and 70% clayey silt by dry mass was statically compacted at low dry density (1.55 Mg/m^3) and low water content (4.0%) to induce high collapsibility on suction reduction. A preliminary laboratory program was carried out under oedometer conditions to explore the stress level associated with the maximum collapse zone of the mixture (around a vertical stress of 200 kPa). With this information, a laboratory test program was started aimed at studying the influence of intermediate principal stress on the deformation response of the mixture at an equivalent mean stress to that of the maximum collapse under oedometer conditions. A conventional HCA was a suitable equipment to perform these tests, provided that changes in inner and outer diameters of the specimen were correctly evaluated from volume changes of inner and outer chambers and by taking into account the initial and final dry densities of the material. The HCA was also used to explore the changes of the slope of the critical state line with Lode angle, as well as the variation of cohesion at different water contents (as-compacted and saturated).

Deformation results on soaking at constant mean and deviatoric stress (p'=200 kPa and q=200 kPa) showed to be influenced by the intermediate principal stress. Test results indicated that collapse was larger when the intermediate principal stress was equal to the minor one (b=0 and Lode angle θ=-30°), as in conventional axi-symmetric triaxial compression conditions. A systematic decrease of collapse strain was detected as the intermediate principal stress tended to the major principal one (increasing b or Lode angle).

Test results were interpreted by considering the slope of the obliquity of the stress point $q/(p''+c'/\tan\phi')$, which increased due to the reduction in c' (affected by suction decrease). In addition, the position of the critical state line was shifted towards a reduction in the elastic domain during suction reduction. At b=0.5 and 0.8, the critical state line reached the stress point inducing a deformational response governed by dominant shear strains (besides collapse deformation) during the saturation episode.

Acknowledgements

The fourth author acknowledges the financial support of Capes Scholarship Proc. Bex13299/13-1 (Brazil).

References

1. L.R. Hoyos, *Georgia Institute of Technology PhD Thesis*, (1988)
2. H. Matsuoka, D.A. Sun, A. Kogane, N. Fukuzawa, W. Ichihara, Can. Geotech. J. **39**, 608-619, (2002)
3. H. Toyota, N. Sakai, T. Nishimura, Soils and Foundations **41**, 1 13-24 (2001)
4. H. Toyota, K. Nakamura, W. Sramoon, Soils and Foundations **44**, 5 1-13 (2004)
5. F.Z. Zerfa, B. Loret, Soil Dynamics and Earthquake Engineering **23**, 435-454 (2003)
6. L. Zdravkovic, D.M. Potts, D.W. Hight, Géotechnique **52**, 6 447-457 (2002)
7. E. Romero, Engineering Geology **165**, 3-19 (2013)
8. O. E. Cárdenas, R. C. Weber, E. Romero, A. Lloret, J. Suriol, *Proc. 6th Int. Symposium on Deformation Characteristics of Geomaterials*, IOS Press Amsterdam, 462-469 (2015)
9. A.W. Bishop, *Géotechnique* **16**, 2 91-128 (1966)
10. D.W. Hight, A. Gens, M.J. Symes, Géotechnique **33**, 4 355-383 (1983)
11. J.H. Argyris, G. Faust, J. Szimmat, E.P. Warnke, K.J William, Nuclear Engrg. and Design **8**, 42-75 (1974)
12. E.E. Alonso, A. Gens, A. Josa, Géotechnique **40**, 3 405-430 (1990)

 DOI: 10.1051/e3sconf/20160914015

An experimental investigation of the independent effect of suction and degree of saturation on very small-strain stiffness of unsaturated sand

Arianna Pagano[1,a], Alessandro Tarantino[1], Meghdad Bagheri[2], Mohammad Rezania[2], and Philippe Sentenac[1]

[1]*EDP University of Strathclyde, Department of Civil and Environmental Engineering, Glasgow, UK*
[2]*EDP Nottingham Centre for Geomechanics, University of Nottingham, Nottingham, UK*

Abstract. The paper presents an experimental investigation of very small strain stiffness of unsaturated sand. A triaxial test apparatus was equipped with bender elements and compression discs in order to assess the stiffness at very small strains by measuring the velocity of propagation of shear and compression waves through an unsaturated sample. The negative water column method was adopted to apply suction at the base of the sample. The experiments were designed to investigate the independent effect of suction and degree of saturation on the wave propagation velocities. This was achieved by testing the sand sample on both the drying and wetting path.

1 Introduction

Stiffness is an essential characteristic of soil response and a critical parameter in geotechnical design. Soil stiffness exhibits a strong non-linear behaviour with strain level. Strains occurring in the ground for structures designed to be far from failure (Serviceability Limit State design) are generally small. Therefore, a better knowledge of stiffness parameters at small and very small strain level is important for realistic predictions of ground movements that may affect the performance of the geotechnical structure itself and the adjacent geotechnical structures. Furthermore, the small-strain analysis should take into account the unsaturated condition of the soil for the case where shallow geotechnical structures interact with the atmosphere. In these cases, the use of 'conventional' saturated soil stiffness parameters might lead to erroneous predictions of ground movement.

Bender element (BE) and compression disc (CD) methods, first developed by Shirley and Hampton (1978), are simple techniques to obtain small strain stiffness parameters by measuring the velocity of propagation of shear and compression waves through a soil sample. The paper presents an experimental investigation on the effect of unsaturated condition on small-strain stiffness of well-graded sand under isotropic stress states via the propagation of mechanical waves. Bender element and compression disc tests have been carried out in a modified triaxial apparatus.

2 Experimental setup and material properties

A triaxial cell apparatus was modified by incorporating bender and compression elements into the pedestal and top cap to transmit and receive mechanical waves respectively. A function generator and an oscilloscope allowed the input signal to be sent and both input and output signals to be recorded respectively. Figure 1 shows a schematic view of the modified triaxial platens and manufactured pieces for accommodation of bender elements and compression discs.

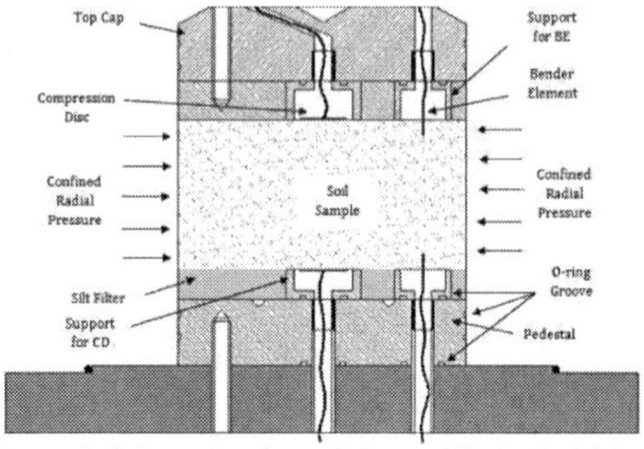

Figure 1. Schematic view of the modified triaxial base and platens (Bagheri 2011).

The tests were carried out on a cylindrical sand sample of 90 mm height and 100 mm diameter, with a well-graded particle size distribution. A well-graded particle size distribution leads to a well-graded pore size distribution and, hence, to a gradual desaturation process. A smoother water retention curve minimises differences in the degree of saturation between the top and the bottom of the sample.

[a] Corresponding author: arianna.pagano@strath.ac.uk

Table 1. Sample parameters.

Height	9.0	cm
Diameter	10.0	cm
Total volume	706.86	cm^3
Specific gravity	2.63	-
Dry density	1.77	g/cm^3
Void ratio	0.49	-
Porosity	0.33	-
Dry mass	1250.98	g

A cylindrical (10×100 mm) silt filter rather than a conventional porous stone was installed at the base to control pore-water pressure. The silt layer provides a high air-entry interface to allow transmission of the negative pore-water pressure while preventing the ingress of air into the water drainage system. Furthermore, it prevents the development of large pores at the sample/filter interface (*wall effect*) which is likely to occur in coarse grained materials. The filter was prepared using a crushing quartz stone known by the commercial name of *Silica* (mean particle size of 25.2 μm). The silt filter parameters are shown in Table 2.

Table 2. Silt filter parameters.

Height	1.0	cm
Base radius	10.0	cm
Total volume	78.54	cm^3
Specific gravity	2.7	-
Dry density	1.24	g/cm^3
Void ratio	1.16	-
Porosity	0.54	-
Dry mass	98	g

Pore-water pressure inside the sample was controlled with the aid of a water reservoir connected to the bottom drainage of the sample. The negative water column method was used to consolidate the silt filter and to determine the soil water retention curve on both drying and wetting paths.

3 Testing procedures

3.1. Silt filter preparation

A 100 mm diameter cylindrical splitting mould was used to give the required shape to the sample and the silt filter. Filter paper was first placed on the pedestal in order to prevent silt loss through the hydraulic drainage. A latex membrane was placed inside the mould, fixed to the cell pedestal with two O-rings, stretched to the top of the mould and held open by applying a vacuum. A silt slurry (w=500%) was gently poured into the mould and allowed to settle for 48 hours. The amount of silt used to prepare the slurry (98 g) was calculated in order to obtain a 10 mm thick filter. After sedimentation occurred, excess clear water was drained out of the mould by adjusting the water reservoir height until its water level was up to the top of the silt surface.

The negative water column method was used to consolidate the silt filter by applying suction up to approximately 6 kPa. Suction was applied in steps by lowering the water reservoir so that its water level was below the bottom of the silt filter. Consolidation was allowed upon each suction step until hydraulic equilibrium was reached. The consolidation process was monitored by recording the water exchange between the silt filter and the reservoir using a balance.

After applying the last suction step, the water reservoir was raised in steps until its water level was at the top of the silt filter.

3.2. Sample preparation

A sandy material was prepared by mixing different sizes of fine, medium and coarse grained sand. The required mass of sand for each diameter (sieve size) was calculated in order to obtain a well-graded grain size distribution by using a modified Fuller equation:

$$P = \frac{\sqrt{\frac{D}{D_{max}}} - \sqrt{\frac{D_{min}}{D_{max}}}}{1 - \sqrt{\frac{D_{min}}{D_{max}}}} \times 100 \quad (1)$$

where P is the passing percentage, D is the anticipated grain (sieve) size, D_{max} is the maximum grain (sieve) size and D_{min} is the minimum grain (sieve) size. D_{max} and D_{min} were equal to 0.425 mm and 0.063 mm respectively. Particle size distribution is shown in Figure 2.

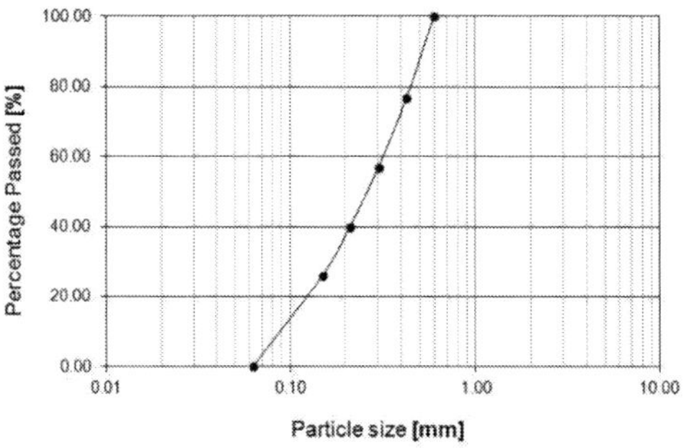

Figure 2. Well-graded particle size distribution

Once mixed, the material was oven-dried and placed into the mould with the aid of a specially manufactured pluviator in order to obtain a homogeneous, repeatable sample. A schematic view of the pluviator is shown in Figure 3. The perforated plates and sieves were purposely designed to obtain a flat and even top surface of the sample after pluviation. With the two perforated plates misaligned ("closed" position), the upper part of the pluviator was filled with oven-dried sand. Then, the top perforated plate was allowed to slide and align with the bottom perforated plate ("open" position). Sand was then pluviated into the mould in four steps until a 90 mm height sample was obtained. The triaxial cell top cap was then covered with filter paper and placed on the top of the sample.

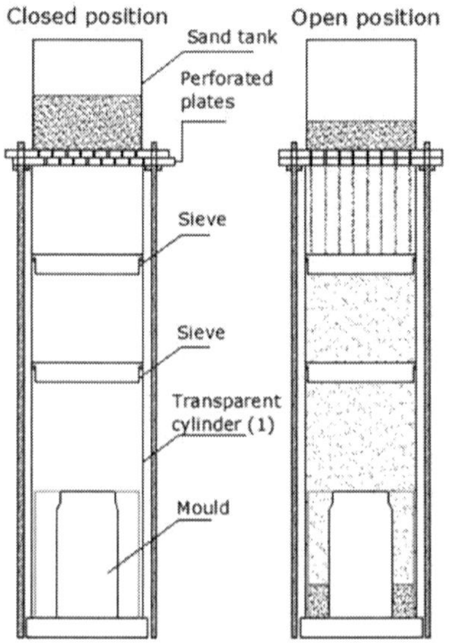

Figure 3. Schematic view of the purposely manufactured pluviator.

3.3 Water retention curve development and BE/CD measurements

After sample preparation, the relationship between suction and degree of saturation was assessed by using the negative water column method. The sample was first saturated by adjusting the water level up to the top of the sample. The amount of water needed to saturate the sample was recorded. Then, suction steps were applied by lowering the water reservoir. For each step, the change in mass of water in the reservoir due to the water exchange induced by the suction change was recorded using a balance and used to calculate the degree of saturation S_r:

$$S_r = \frac{M_w}{V_t - M_d/G_s} \times 100 \quad (2)$$

where M_w is the mass of water in the sample, V_t is the sample total volume, M_d is the dry mass of sand and G_s is the sample specific density.

The value of suction for each step was calculated as:

$$s = (H - h) \times g \quad (3)$$

where s is the applied suction, H is the distance between the top of the sample and the water level in the reservoir, h is the sample height and g is the acceleration due to gravity. The water retention behaviour was determined in a suction range between 0 and approximately 6 kPa for both drying and wetting paths, under different stress conditions. At each suction step shear and compression wave measurements were taken after hydraulic equilibrium was reached. A sinusoidal input wave having a frequency of 5 kHz was applied to the BE and CD transmitters. Output signals from BE and CD receivers were simultaneously recorded and then filtered in order to remove the background noise and enable identification of the first wave arrival. Once the wave travel time Δt was determined, the wave velocity V and the correspondent stiffness modulus K (Shear or Young modulus) can be calculated as follows:

$$V = \frac{L_{tt}}{\Delta t} \quad (4)$$

$$K = \rho \times V^2 \quad (5)$$

where L_{tt} is the travel length, usually considered as the tip-to-tip distance between the transmitter and the receiver bender element, and ρ is the dry density of the soil. It is worth noting that the travel time Δt was simply taken as the time distance between characteristic points of the input and output signals. The chosen characteristic point is the first peak (positive or negative) of the wave. The identification of the first arrival is however a crucial issue in bender and compression element tests interpretation. Although the simplified approach used in this research may lead to erroneous absolute values of stiffness moduli, it still allows identification of the relative effect of suction and degree of saturation.

4 Experimental results

Two tests were conducted on the same sample at two different stress state conditions. The first test was conducted on the unloaded sample (test I), which was left inside the mould while suction steps were applied. The second test was carried out under isotropic loading conditions by applying a cell pressure of 10 kPa (test II). In order to make the sample self-standing before applying the cell pressure, the splitting mould was removed after the application of the last suction step on the unloaded sample. The triaxial cell was then assembled on the cell base and filled with de aired water. Water pressure inside the cell was then increased by means of a bladder (air-water interface) connected to a pressure panel.

4.1. Unloaded sample

Figure 4 shows the water retention curve obtained during test I. The unloaded condition led to a poor contact between the soil and the bender/compression elements. No waves could then be recorded by the oscilloscope.

4.2 Loaded sample

Figure 5 shows the degree of saturation plotted against the applied suction in test II. The soil stiffness parameters measured at some critical points are also reported in the figure. In particular, the stiffness is compared at the suction of about 3 kPa along the drying and wetting path, i.e. same suction and different degree of saturation (points A and C). The stiffness is also compared at the degree of saturation of about 67% along the drying and wetting path, i.e. same degree of saturation and different suction (points B and C).

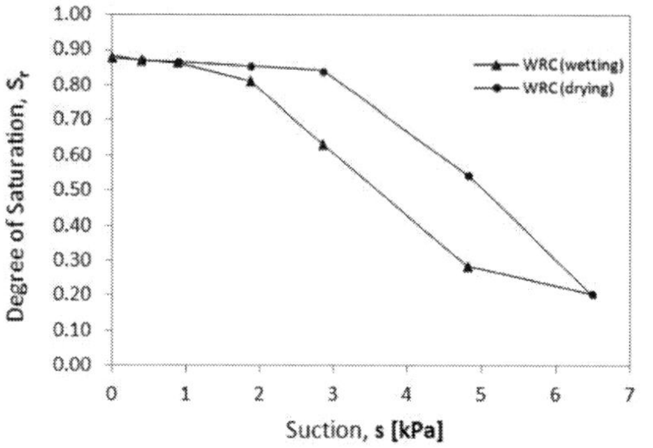

Figure 4. Variation of degree of saturation with applied suction in test I, cell pressure 0 kPa (drying and wetting paths).

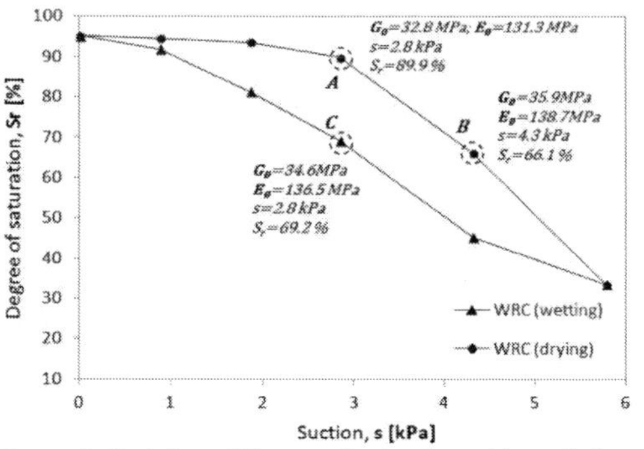

Figure 5. Variation of degree of saturation with applied suction and BE/CD measurements in test II, cell pressure 10 kPa (drying and wetting paths).

As expected, the stiffness moduli increase from A to B along the drying path. From A to B, the suction increases and the degree of saturation decreases and it is difficult to discriminate between these two effects.

The role of degree of saturation appears to be clearer when comparing stiffness at same value of suction (points A and C). The lower the degree of saturation (wetting path), the higher is the wave velocity and hence the soil stiffness. Measurements of stiffness along the entire drying and wetting paths (which are not reported in the figure) confirm this trend.

The role of suction appears to be clearer when comparing stiffness at same value of degree of saturation (points B and C). The higher the suction (drying path), the higher is the wave velocity and hence the soil stiffness. Measurements of stiffness along the entire drying and wetting paths (which are not reported in the figure) confirm this trend.

This evidence is in contrast with the common assumption that unsaturated soil stiffness is controlled by average skeleton stress σ'' (e.g. Alonso *et al.* 2010):

$$\sigma'' = \sigma + s \times S_r \tag{6}$$

If this was the case, lower values of degree of saturation at the same value of suction (wetting path), which lead to lower values of average skeleton stress, should have led to lower (and not higher) values of stiffness.

On the other hand, at the same value of degree of saturation, the increase in stiffness associated with an increase in suction appears to be marginal. This seems to suggest that the degree of saturation has a dominant role on soil stiffness.

As regards the effect of cell pressure, a series of tests carried out at a cell pressure of 50 kPa (not reported in the paper) have shown that the effect of partial saturation becomes less significant as the cell pressure (total isotropic stress) increases. This effect is somehow captured by Equation 6 as an increase in total stress tends to offset the effect of partial saturation.

5 Conclusions

An experimental program has been carried out using a modified triaxial cell to investigate unsaturated soil behaviour in the very small strain range. Tests have been conducted on well-graded sand in order to investigate the independent effect of suction and degree of saturation on soil stiffness parameters.

The experimental study led to the conclusion that the effect of suction and degree of saturation on soil stiffness cannot be interpreted in terms of average skeleton stress. Although suction leads to an increment in soil stiffness, degree of saturation appears to have a dominant role. In addition, the effect of degree of saturation is not captured by the average skeleton stress, not even at qualitative level.

Further investigation is needed to identify a conceptual model capable of predicting the independent effect of suction and degree of saturation on very small strain stiffness.

References

1. Alonso EE, Pereira J-M, Vaunat J, and Olivella S (2010). A microstructurally based effective stress for unsaturated soils. Géotechnique, 60 (12): 913–925.
2. Bagheri M (2011). Small strain stiffness measurement of unsaturated sand using bender and compression elements. Unpublished MSc thesis, Department of Civil and Environmental Engineering, University of Strathclyde, Glasgow.
3. Shirley DJ, Hampton LD (1978). Shear-wave measurement in laboratory sediments. Journal of the Acoustical Society of America, 62(2): 607-613.

Influence of effective stress on swelling pressure of expansive soils

Wiebke Baille[1,a], Linzhi Lang[1], Snehasis Tripathy[2] and Tom Schanz[1]

[1]*Ruhr-Universität Bochum, Chair of Foundation Engineering, Soil and Rock Mechanics, 44780 Bochum, Germany*
[2] *Geoenvironmental Research Centre, School of Engineering, Cardiff University, Newport Road, Cardiff, CF24 3AA, Wales, UK*

Abstract. The volume change and shear strength behaviour of soils are controlled by the effective stress. Recent advances in unsaturated soil mechanics have shown that the effective stress as applicable to unsaturated soils is equal to the difference between the externally applied stress and the suction stress. The latter can be established based on the soil-water characteristic curve (SWCC) of the soil. In the present study, the evolution of swelling pressure in compacted bentonite-sand mixtures was investigated. Comparisons were made between magnitudes of applied suction, suction stress, and swelling pressure.

1 Introduction

The volume change and shear strength behaviour of soils are controlled by the effective stress. The excess pore water pressure is equal to the difference between the externally applied stress and the effective stress in case of saturated soils. However, the impact of the negative pore water (or soil suction) on the effective stress for unsaturated soils is not straight-forward since the distribution of external stress within unsaturated soil systems occurs either via the water phase, or via the air phase, or even via both water and air phases depending upon the range of suction under consideration.

Recent advances in unsaturated soil mechanics have shown that the effective stress as applicable to unsaturated soils is equal to the difference between the externally applied stress and the suction stress [1]. The suction stress embraces the van der Waals attractive force, the double-layer repulsive force, the surface tension, and the solid-liquid interface forces due to pore water pressure.

The suction stress of unsaturated soils either can be measured in the laboratory from tensile strength tests or can be calculated based on the shear strength tests, or even can be calculated based on theoretical considerations. The suction stress characteristic (SSCC) curve of a soil (i.e., the relationship between suction stress and degree of saturation or suction or water content) can be established based on the soil-water characteristic curve (SWCC) of the soil [2].

The suction stress approach has been applied to determining the shear strength behaviour of unsaturated soils [2, 3] and the volume change behaviour of soils during the drying process [4]. At this stage, very limited studies are available on the swelling pressure development of unsaturated soils as affected by the suction stress, particularly in the context of a change in the suction stress on account of a change in suction and its influence on the effective stress of swelling soils.

The main objective of the study was to explore the impact of changes in the effective stress for swelling soils during the wetting process based on the suction stress approach.

2 Methods

Laboratory tests were carried out on highly compacted bentonite-sand mixtures [5]. Compacted specimens were prepared by mixing bentonite and sand in equal proportions by dry mass (50% bentonite and 50% sand). The Material properties of the bentonite and the sand used in the mixtures are presented in Table 1 and Table 2, respectively.

The specimens were prepared at predetermined water contents and dry densities. The compacted bentonite/sand specimens (dia. = 50 mm and a height = 20 mm) had the following compaction conditions: (i) dry density = 2.056 Mg/m³, water content = 9.1%, and suction = 22.7 MPa and (ii) dry density = 2.042 Mg/m³, water content = 9.1%, and suction = 22.7 MPa. Three out of the five specimens were further oven dried prior to testing them in a specialised oedometer. Oven drying caused an increase in the dry density and suction of the specimens.

The specimens were then step-wise wetted by decreasing suction. Both axis-translation and vapour equilibrium techniques were adopted for this purpose. The swelling pressure at each applied suction was mea sured.

Table 1. Properties of Calcigel bentonite

Specific gravity of soil solids (-)	Liquid limit (%)	Plastic limit (%)	Plasticity index (%)
2.65	130	97	33

[a] Corresponding author: wiebke.baille@rub.de

Table 2. Properties of the sand

Specific gravity of soil solids (-)	D_{10} (mm)	D_{30} (mm)	D_{60} (mm)
2.65	0.25	0.40	0.70

Suction controlled tests on compacted bentonite specimens were carried out using several devices developed at the Technical University of Catalonia, Barcelona (or UPC), Spain [6]. The device used is also known as the UPC-isochoric cell. The device with an exchangeable pedestal (either a corrosion-resistant porous disk for vapour equilibrium technique or a high-air-entry ceramic disk (air-entry value = 1,500 kPa) for axis-translation technique), can be used to control suction of specimens during the wetting process under isochoric condition. Table 3 shows details of the bentonite specimens tested (i.e., dry density, water content, suction, and the techniques used to control suction).

Table 3. Details of the 50% bentonite/50% sand specimens (50B) tested

Sp. No.	[a]Initial compaction conditions; [b]Specimen conditions before testing: Dry density (Mg/m3)	Water content (%)	Suction (MPa)	Suction control technique step-wise used (Vapour equi-librium technique (VET) or Axis-translation technique (ATT))
1	2.056 [a] 2.056 [b]	9.0 [a] 9.0 [b]	22.7 [a] 22.7 [b]	VET for suctions 10.5, 5.5, 4.25, 3.3 and 2.2 MPa and water circulation (suction = 0 MPa)
2	2.038 [a] 2.038 [b]	9.3 [a] 9.3 [b]	22.7 [a] 22.7 [b]	ATT for suctions 0.5, 0.3, 0.1 and 0.05 MPa and water circulation (suction = 0 MPa)
3	2.042 [a] 2.072 [b]	9.1 [a] 0 [b]	22.7 [a] 1000 [b]	VET for suctions 10.5, 5.5, 4.25, 3.3 and 2.2 MPa and water circulation (suction = 0 MPa)
4	2.005 [a] 2.072 [b]	9.3 [a] 0 [b]	22.7 [a] 1000 [b]	VET for suctions 39.4, 10.5, and 5.5 MPa
5	2.052 [a] 2.072 [b]	9.1 [a] 0 [b]	22.7 [a] 1000 [b]	ATT for suctions 0.7, 0.2, and 0.05 MPa

3 Results

3.1 Presentation of results

The set of equations used for the calculation of effective stress is presented in the following paragraph. Based on the degree of saturation values of the soil specimens at various applied suctions, the effective degree of saturation values were calculated from eq. (1). The suction – degree of saturation SWCCs of the soils were then transformed into the suction – effective degree of saturation SWCCs. The suction – effective degree of saturation SWCCs were best-fitted using van Genuchten (1980) [7] model (eq. 2). The air-entry fitting parameter (α) and the pore-size parameter (n) were obtained. The parameters α and n were used to calculate the suction stress as a function of the effective degree of saturation using the closed-form expression (eq. 3) [2]. Equation (3) is applicable for S_e between 0 and 1. The effective stress corresponding to any suction stress can be calculated from eq. (4).

For the bentonite/sand specimens, the applied stress during the isochoric swelling pressure tests was zero; however, a decrease in suction caused a development of total stress (i.e., swelling pressure). In this case, at any applied suction, the suction stress is subtracted from the total stress (i.e., the measured swelling pressure) to obtain the effective stress.

$$S_e = \frac{S - S_r}{100 - S_r} \quad (1)$$

$$S_e = \left\{\frac{1}{1 + [\alpha(\psi)]^n}\right\}^{\frac{1}{1-n}} \quad (2)$$

$$\sigma^s = -\frac{S_e}{\alpha}\left[S_e^{\,n/(1-n)} - 1\right]^{1/n} \quad (3)$$

$$\sigma' = (\sigma - u_a) - (\sigma^s) \quad (4)$$

3.2 Discussion of results

The swelling pressure (i.e., total stress) at each applied suction was measured. Fig. 1 presents the measured swelling pressure vs. applied suction plots for the bentonite/sand specimens (50B-specimens) of this study. For comparison, the corresponding results of pure bentonite (100B specimens) are also shown.

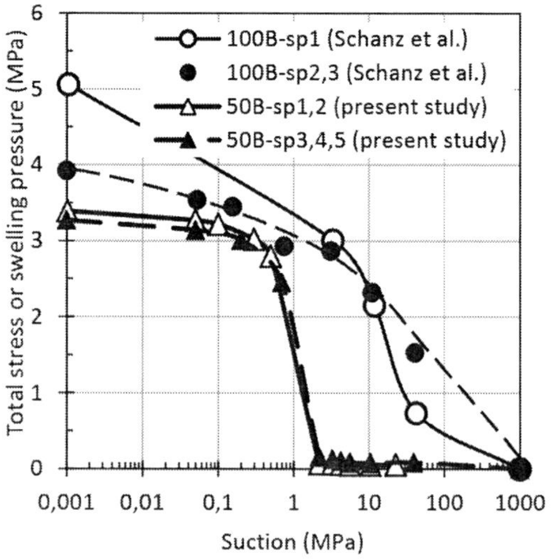

Figure 1. Suction control swelling pressure test results of bentonite/sand specimens 1 to 5 of the present study (50B) and of pure bentonite specimens (100B) (for comparison).

For the bentonite/sand mixtures (50B-sp1,2 and 50B-sp3,4,5) swelling pressure was found to slightly increase due to a decrease in suction to about 2 MPa, whereas at suctions smaller than 2 MPa, swelling pressure was found to significantly increase until reaching the maximum swelling pressure at zero suction corresponding to the saturated state. For the 100-B specimens, an increase in swelling pressure was observed over the full suction range.

Figures 2 and 3 show the suction – degree of saturation SWCCs (Figs. 2a and 3a), the suction – effective degree of saturation SWCCs (Figs. 2b and 3b), the SSCCs in terms of the effective degree of saturation (Figs. 2c and 3c), and the SSCCs in terms of suction (Figs. 2d and 3d) for the combined results of specimens 1 and 2 and specimens 3, 4 and 5 of Calcigel bentonite/sand mixture (50B/50S), respectively.

The fitting parameters obtained from the best-fit SWCCs are shown in Table 4. The results presented in Figs. 2(d) and 3(d) show that the suction stress increased (i.e., the magnitude of suction stress decreased) with a decrease in suction for the bentonite/sand specimens.

Table 4. SWCC fitting parameters for the bentonite/sand specimens.

Parameter	Specimen 1 and 2	Specimens 3, 4 and 5
Sr (-)	0.769	0.000
α (-)	5.071	0.444
n (-)	1.245	1.326

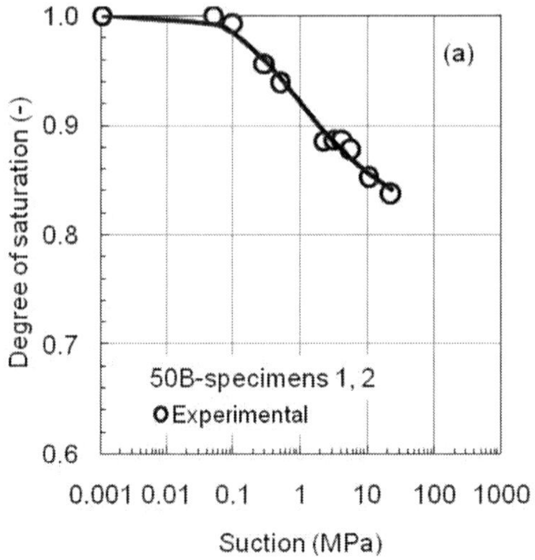

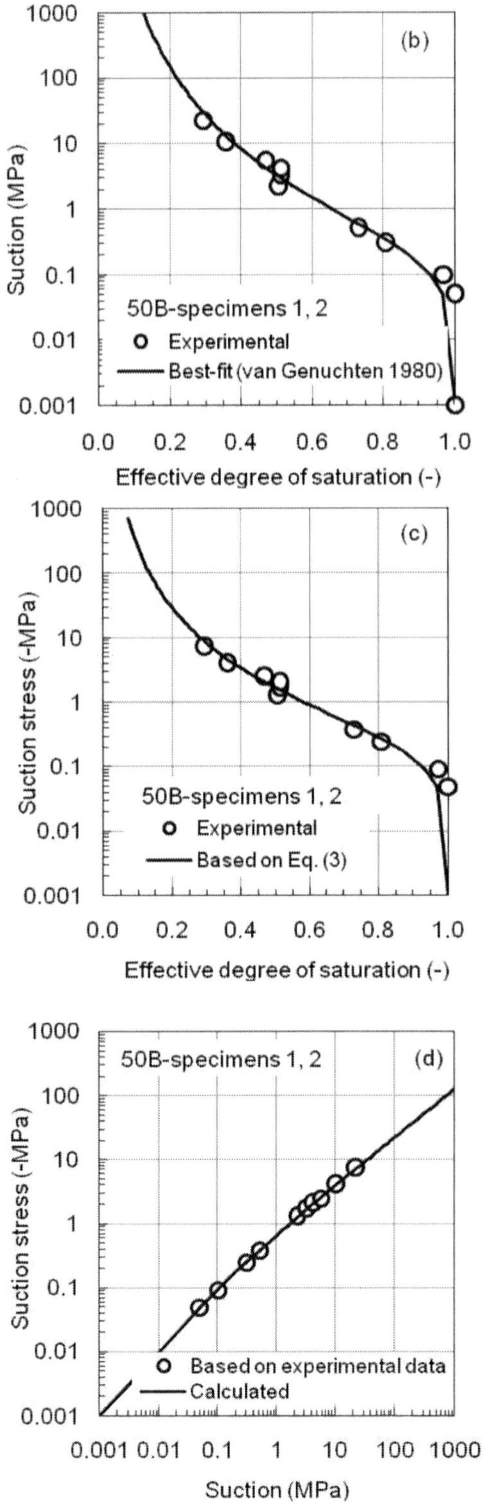

Figure 2. SWCCs and SSCCs of Calcigel bentonite/sand (50B-specimens 1, 2): (a) suction – degree of saturation SWCC, (b) Suction – effective degree of saturation SWCC, (c) SSCC in terms of effective degree of saturation, and (d) SSCC in terms of suction (log-scale)

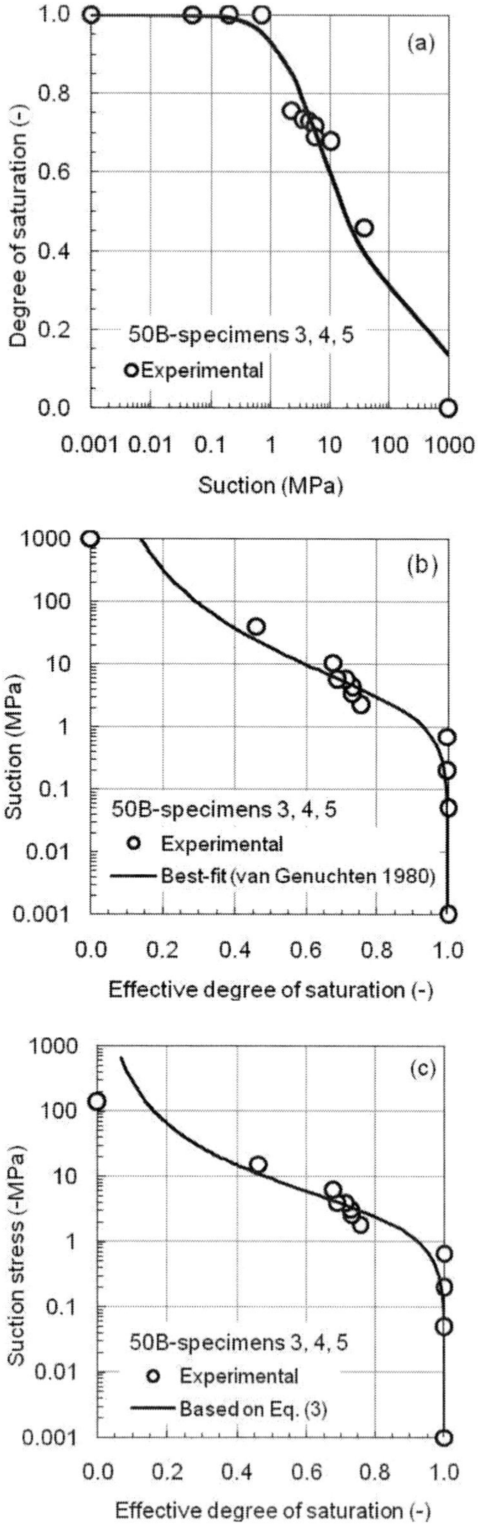

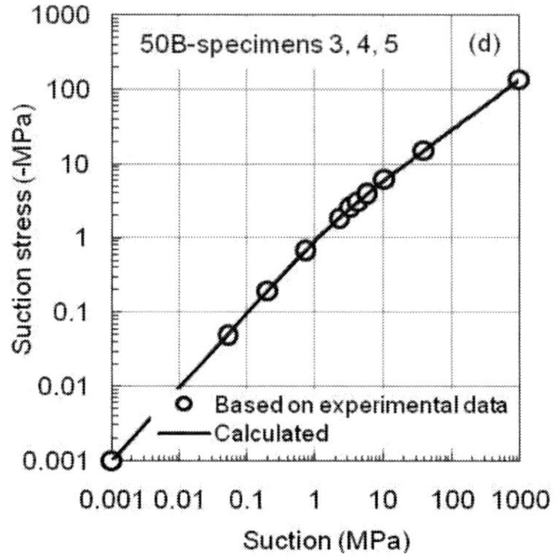

Figure 3. SWCCs and SSCCs of Calcigel bentonite/sand (50B-specimens 3, 4, 5): (a) suction – degree of saturation SWCC, (b) Suction – effective degree of saturation SWCC, (c) SSCC in terms of effective degree of saturation, and (d) SSCC in terms of suction (log-scale)

Baille et al. (2014) stated that for bentonites, a decrease in suction causing an increase of suction stress leads to decreasing the effective stress. Figures 4 and 5 show the combined plots of suction versus suction stress and effective stress versus swelling pressures (total stress) for compacted Calcigel bentonite/sand (50/50) specimens. The equivalent plot of results for pure bentonite is presented in Fig. 6. The results presented in Figs. 4 and 5 show that, at any applied suction the magnitude of suction stress is greater than the magnitude of swelling pressure. At saturation (i.e., at zero suction), the swelling pressure (i.e., the total stress) and effective stresses are equivalent since the excess pore water pressure was zero and the suction stress reduced to zero.

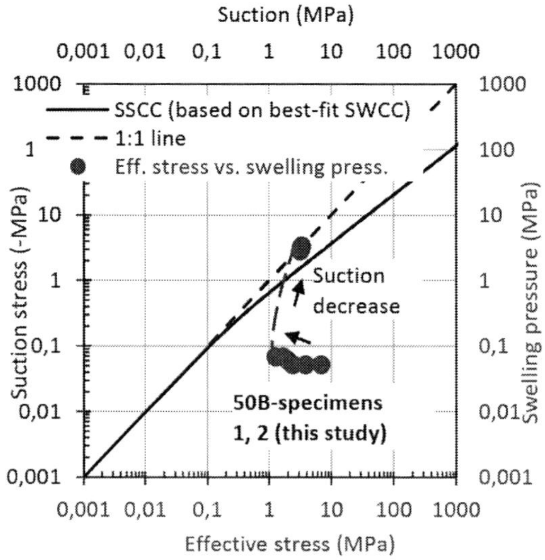

Figure 4. Influence of a change in suction on suction stress and effective stress of Calcigel bentonite/sand mixture (50B-specimens 1, 2).

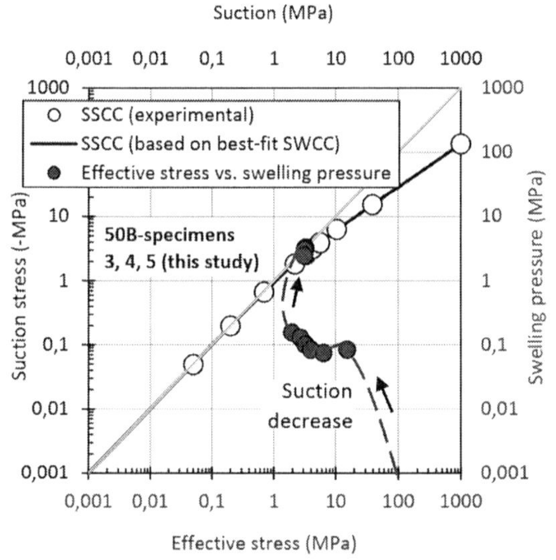

Figure 5. Influence of a change in suction on suction stress and effective stress of Calcigel bentonite/sand mixture (50B-specimens 3, 4, 5).

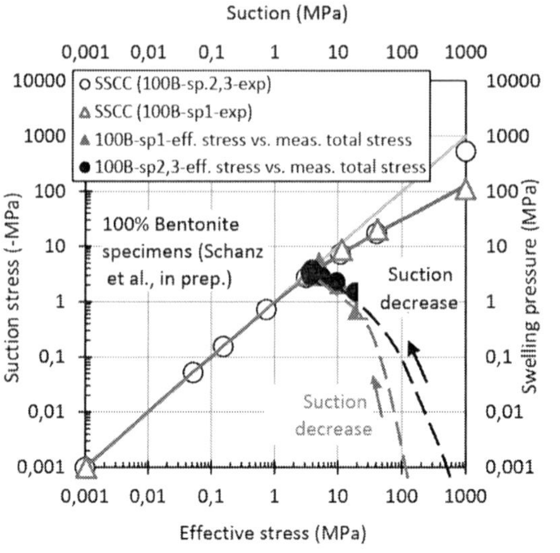

Figure 6. Influence of a change in suction on suction stress and effective stress of pure Calcigel bentonite (100B-specimen 1 and 100B-specimens 2, 3) from Schanz et al., submitted.

This observation is similar for the pure bentonite (100B) and the 50/50 bentonite/sand mixture. The effective stress of the pure bentonite decreased continuously upon wetting which was manifested on the development of swelling pressure (see Fig. 6). It is also visible in Fig. 6 that for similar scaling of the suction stress axis and the swelling pressure axis, the final swelling pressure at saturated state reached the suction stress line at the point where the latter deviates from the 1:1 line, corresponding to the air-entry value. This was also found for the specimens 3, 4, 5 shown in Fig. 5, whereas the final measured swelling pressure was found to be greater than suction stress at air-entry for specimens 1, 2 in Fig. 4. The dashed lines in Figs. 4, 5 and 6 indicate the effective stress-swelling pressure path during suction decrease. For the pure bentonite (Fig. 6), a continuous increase in swelling pressure was manifest by a decrease in effective stress. For the bentonite/sand mixtures, a different effective stress-swelling pressure path was observed. In Fig.4 an increase in swelling pressure was accompanied by a significant decrease in effective stress for a suction stress smaller than 1.2 MPa. For a suction stress greater than 1.2 MPa, both the swelling pressure and the effective stress increased, which is not consistent. In Fig. 5, after an initial increase in swelling pressure, it slightly increased at suction stresses between 15 and 1.8 MPa accompanied by a decrease in effective stress. Again, for suction stress greater than 1.8 MPa, an inconsistency was observed; both the swelling pressure and effective stress increased. Considering the global behaviour of the bentonite/sand mixtures from the initial to the final point of the swelling pressure-effective stress path (i.e., ignoring the intermediate points), a decrease in effective stress manifest by an increase in swelling pressure was observed, which is concurrent with the behaviour observed for the pure bentonite. However, the interpretation of the various stages between initial and final state need further consideration.

4 Conclusions

The investigation clearly showed that, at high suctions, the swelling pressures developed were smaller than the corresponding applied suctions. However, at smaller suctions, the swelling pressure and the magnitude of suction stress were found to be similar indicating that at zero water pressure the effective stress is equivalent to the swelling pressure for highly plastic clays.

References

1. N. Lu, W.J. Likos, J. Geotech. and Geoenviron. Eng. ASCE, **132**, 131-142, (2006).
2. N. Lu, J.W. Godt, D.T. Wu, Water Resources Research, **46**, W05515 (2010).
3. S. Oh, N. Lu, Y.K. Kim, S.J. Lee, S.R. Lee, J. Geotech. Geoenviron. Eng. ASCE **138**, 47-57 (2012).
4. W. Baille, S. Tripathy, T. Schanz, Vadose Zone Journal. **13**, doi:10.2136/vzj2013.06.0112, (2014).
5. S.S. Agus, Y.F. Arifin, S. Tripathy, T. Schanz, Acta Geotechnica, DOI 10.1007/s11440-012-0189-0, (2012)
6. E. Romero, Characterisation and thermo-hydro-mechanical behaviour of unsaturated Boom clay: an experimental study, PhD thesis, Universitat Polytècnica de Catalunya (1999).
7. M. T. van Genuchten, Soil Sci. Soc. Am. J. **44**, 892–898 (1980).
8. T. Schanz, S. Tripathy, W. Baille, Consequences of effective stress changes on the volumetric behaviour of unsaturated soil. In prep.

The influence of wetting and drying paths on the shear strength of a silty clayey compacted soil

Matheus Matuella[1,a], Washington Núñez[1], Wai Ying Yuk Gehling[1], Rodrigo Weber[2], Wiliam Fedrigo[1], Mattheus Porto[1] and Lucas Delongui[1]

[1]*Universidade Federal do Rio Grande do Sul, Escola de Engenharia, Departamento de Engenharia Civil, Laboratório de Geotecnologia, Av. Osvaldo Aranha, n° 99/302, Porto Alegre – RS, Brazil.*
[2]*Universitat Politècnica de Catalunya, Departament d'Enginyeria del Terreny, Cartogràfica i Geofísica, C. Jorni Girona, 1-3, Módulo D2, Barcelona, Spain.*

Abstract. The vast majority of geotechnical projects are designed in saturated condition, aiming safety. However, when some conditions might be achieved, designing in unsaturated conditions may end up being economically interesting, while still safe. Therefore, this study aims at evaluating the shear strength behaviour of a silty clayey soil, with optimum moisture content of 16% and dry unit weight of 16,5 kN/m³ from south Brazil. From direct shear tests, performed in saturate and unsaturated conditions, with three different starting moisture conditions – 14%, 16% and 18%, and the same unit weight, this paper shows the influence of the wetting and drying paths on this material and its contribution to evaluate shear strength. The results showed that a variation of ±2% around the optimum, in drying and wetting paths, can significantly influence the behaviour of the studied material. The wetting process caused a small variation in comparison with the optimum, while the drying process resulted in increased shear resistance for all the analyzed samples. With this study, it was possible to obtain a better understanding of the behaviour of an unsaturated soil from south Brazil, when evaluating tradeoffs in project's safety and economics.

1 Introduction

The layers that form a pavement structure are compacted at optimum moisture content (ω), therefore in unsaturated conditions. These layers' moisture content (and of the subgrade) may vary during the construction of a pavement, due to poor technical control or even deliberate not following standardization. It's known that the characteristics of the soil where the pavement is settled directly influence its lifetime; following Brazilian standardization DNIT 137/2010 [1], there is a tolerance of ±2% around optimum moisture content (ω_0). Even though the moisture content may be altered during construction and also after the structure is opened to traffic, seasonal weather conditions may also influence the saturation level, compromising the pavement's performance.

Considering the importance of saturation conditions of the soils when building these structures, the evaluation of the shear strength – tied up to the moisture content at time of compaction, and also to wetting and drying paths, induced by climate conditions – may influence significantly pavements analysis and design. Several papers highlight the importance of evaluating pavements' mechanical behaviour and the relevance of knowing the materials' properties, such as Rodrigues (1997, [2]), Zaman and Khoury (2007, [3]), Chittoori et al. (2012, [4]), Weber (2013, [5]) and Matuella et al. (2014, [6]). By this mean, the present study aims at evaluating the shear strength of a soil against variations in the moisture content after its compaction.

2 Material and methods

The experimental program consisted in two main steps: the first was to choose the soil to be analyzed, based on its applicability as a pavement material; the second was to arrange test procedures that would make possible to evaluate moisture content variations during and after compaction and its influence in the soil shear strength. It is assumed that those variations might be related to climate changes and to possible construction problems, which could be reproduced in laboratory by wetting and drying paths.

2.1 Material of study

The studied soil was collected in the Brazilian highway BR-116, near the city of Porto Alegre, located in the state of Rio Grande do Sul. Figure 1 shows the location where the studied soil has been collected from. This soil was

[a] Corresponding author: m.matuella@gmail.com

chosen because it was part of a quarry that would be used in the highway's duplication. Some characteristics of the studied soil are presented in Table 1. Figure 2 shows the compaction curve obtained for the studied soil while Figure 3 presents its grain size distribution curve. The soil specimens were compacted using standard Proctor effort and then shaped in order to carry out direct shear tests and suction tests.

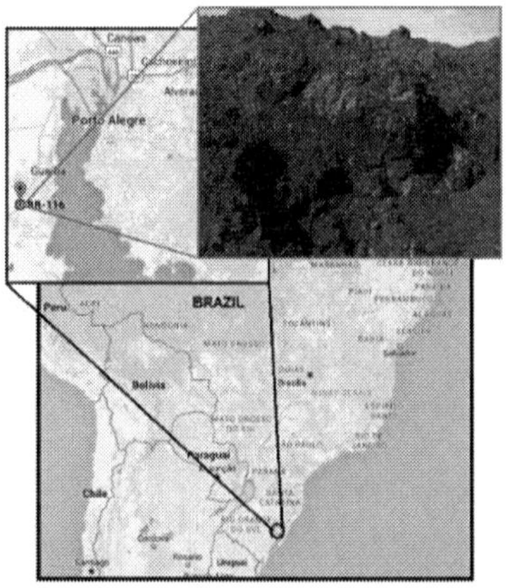

Figure 1. Location where the soil was collected from.

Table 1. Soil physical properties.

Grain size distribution	Size	San d	Silt	Clay
	With dispersant	36%	19%	45%
	Without dispersant	31%	65%	4%
Compaction test results (Standard Proctor)	ω_0	16%		
	γ_d	16,5 kN/m³		
Atterberg Limits	Liquid limit	36%		
	Plastic limit	27%		
	Plasticity index	9%		

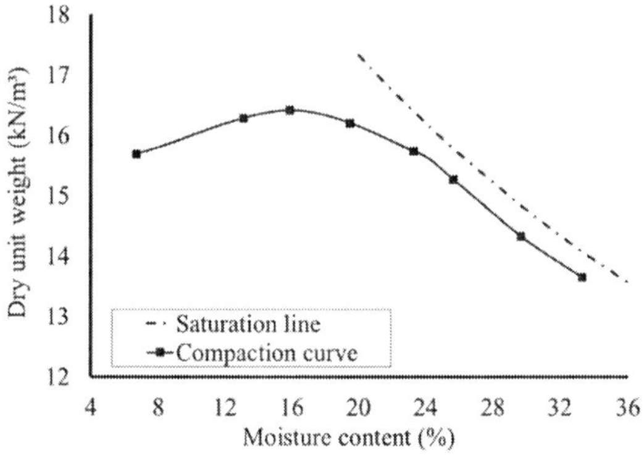

Figure 2. Compaction curve.

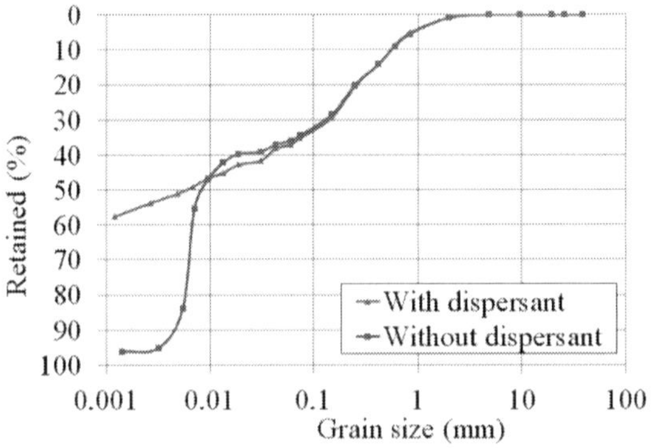

Figure 3. Grain size distribution curve.

2.2 Direct Shear Test

The direct shear test was adopted because of two main reasons: relatively fast results and simple testing procedures, following ASTM D3080 [7].

Cylindrical specimens with 6 cm in diameter and 2 cm in height were compacted with moisture contents equal to 14%, 16% and 18% (within the allowable limits established by DNIT 137/2010). Some samples were then wet, some of them dried and some of them remained at the same moisture content. The specimens in the drying path were air dried and water was added to the specimens in the wetting path. In both cases, the moisture content was controlled by mass variation; after achieving the correct content, the specimens were wrapped in plastic film and stored in order to provide homogeneity in moisture. Table 2 synthesizes the moisture content at compaction and while the tests were run.

Table 2. Variations of the friction angle as function of the moisture content at compaction and wetting/drying paths.

Path	ω at compaction	ω at testing
No path	14	saturated
	16	saturated
	18	saturated
	14	14
	16	16
	18	18
Wetting	14	16
	16	18
	18	20
Drying	14	12
	16	14
	18	16

When tested under saturated conditions, the soil specimens were compacted at desired moisture content, placed at the Wykeham Farrance equipment for direct shear test, shown in figure 4, where they were flooded, thickened up to stabilization, and then tested.

DOI: 10.1051/e3sconf/20160914017

Figure 4. Wykeham Farrance equipment for direct shear test.

During the test, it was observed that the moisture content was constant in the unsaturated condition. On the other hand, the saturated condition tests were carried out after keeping the specimens in water during a predefined period of time in order to achieve complete saturation. The applied normal stresses were 25 kPa, 50 kPa and 100 kPa and two hypotheses were assumed as rupture criteria: the shear stress stabilization or a horizontal displacement of 10% the specimen diameter.

3 Results and analysis

From the stress x strain curves obtained at the direct shear test to the set of conditions studied, the following shear stress (T) x normal stress (σ) graphics, shown in Figures 5 to 8, were plotted.

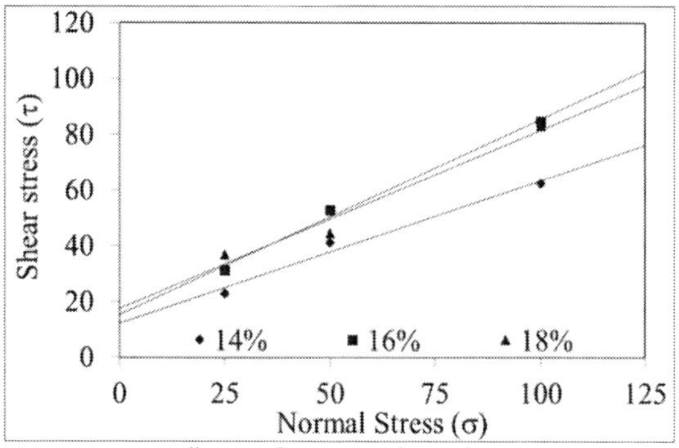

Figure 5. . T *x* σ diagram for saturated samples.

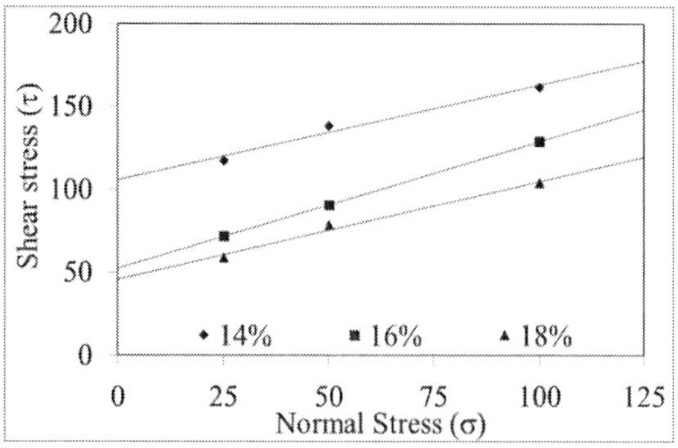

Figure 6. . T *x* σ diagram for unsaturated samples.

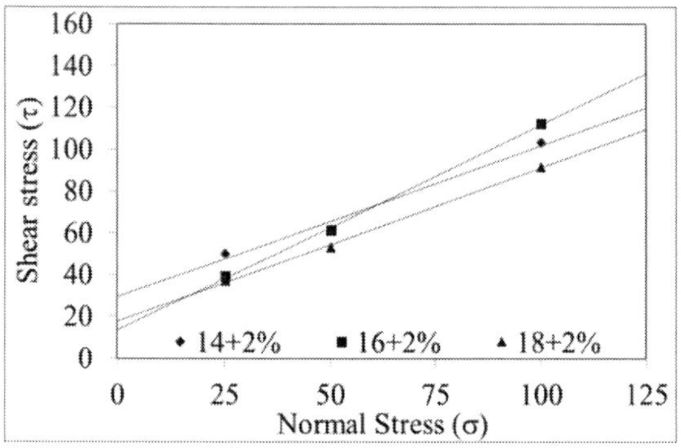

Figure 7. . T *x* σ diagram for unsaturated samples under a wetting path.

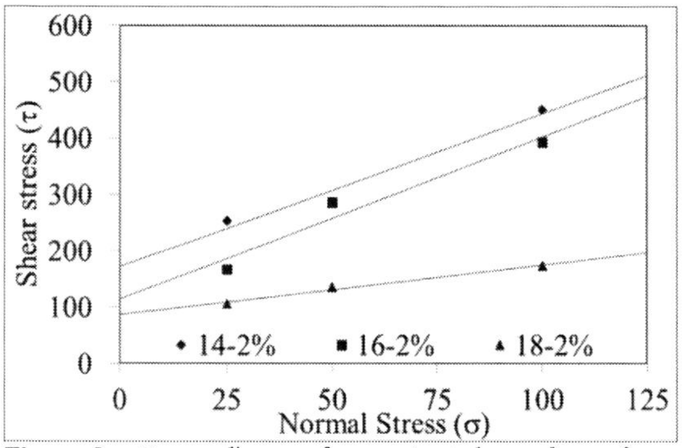

Figure 8. . T *x* σ diagram for unsaturated samples under a drying path.

The results were then summarized; table 3 shows the variation of the friction angle, while table 4 expresses the variations in the cohesion, both related to moisture content of the samples at the moment they were tested.

Table 3. Variations of the friction angle, related to the moisture content at compaction and wetting/drying paths.

ω at compact-tion	friction angle – ϕ (º)			
	Satu-rated	Unsa-turated	Wet (+2%)	Dried (-2%)
14%	27,1	29,7	35,8	69,8
16%	35,1	37,4	44,5	70,9
18%	32,6	30,5	36,3	41,4

Table 4. Variations of the cohesion as function of the moisture content at compaction and wetting/drying paths.

ω at compact-tion	Cohesion (kPa)			
	Satu-rated	Unsa-turated	Wet (+2%)	Dried (-2%)
14%	12,2	105,8	29,6	172,0
16%	15,3	52,5	13,8	113,7
18%	17,5	45,8	17,8	86,8

After obtained, these values were plotted in order to help in the analysis of the soil's mechanical behaviour. In figure 9, the variation of the cohesion is presented,

 DOI: 10.1051/e3sconf/20160914017

related to the variations of moisture contents, while figure 10 shows the friction angle as function of the moisture content.

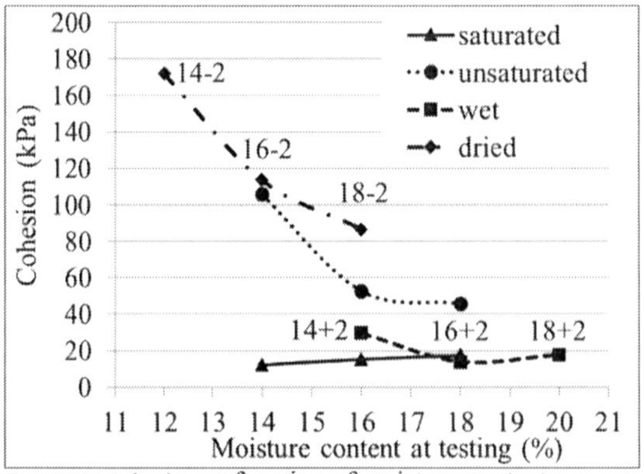

Figure 9. Cohesion as function of moisture content.

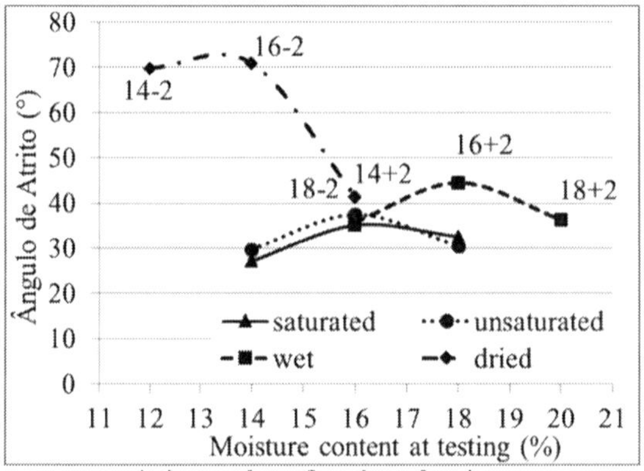

Figure 10. Friction angle as function of moisture content.

About the mechanical behaviour of this soil, as most relevant, it was observed that:

- Under saturated and under unsaturated conditions, the friction angle tends to have a maximum value, related to optimum moisture content at compaction. When under wetting paths, the samples compacted at optimum moisture content presented higher friction angle, compared to those which did not have its moisture changed. When not compacted at ω_0, values of the same order of magnitude were obtained, as the ones resulted when the samples were tested at ω_0, with no moisture variations. When analysing the samples that underwent the drying process, all the three cases showed an increase in the friction angle. In this case, the ones compacted at 14% and 16% presented exceedingly high values for this parameter. There is still lack of results to evaluate the behaviour of the drying paths, at lower moisture contents, to determine if the friction angles will stabilize, decrease or still show a peak value.
- The soil's cohesion, under saturated condition, presents a slight increment, as the moisture content at compaction increases, but this increment is not as notable as the cohesion obtained in unsaturated conditions. When compared directly, for every value of moisture content at compaction, cohesion under unsaturated conditions are roughly at least 3 times higher. As expected, the wetting path proved to be a worse condition when compared to all unsaturated samples. It was observed that the drying paths may endow the soil substantial improvements, in terms of cohesion (46 kPa to 87 kPa, for samples compacted at 18%, 52 kPa to 114 kPa, for the samples compacted at 16% and 105 kPa to 172 kPa, for those compacted at 14% of moisture content).

4 Conclusion

The experiments developed in this work allowed the evaluation of the behaviour of the shear strength of a compacted soil at different moisture contents, which were conducted through wetting and drying paths.

It was noticed that when this soil is compacted near optimum moisture content and, after compaction, it suffer loss of moisture, we can achieve considerable increases in shear strength. When the effect is the opposite and the moisture content of the material increases after compaction, there may be great resistance loss, particularly when the material is compacted below optimum moisture content.

These observations are of great relevance to the practical applications of geotechnical engineering. With proper control of soil moisture, we might save financial resources and get benefits in terms of security.

We can, for example, in long-term construction work, plan a certain excavation or cutting a slope, which should be done in a drier season when the soil is possibly less humid and hence support higher shear stresses and remain stable on steeper slopes. Another example to be taken into account is the construction of a highway: if we manage to control the moisture of the pavement support layers by installing devices which lead the water out of the pavement structure, or by using layers which allow the percolation of the water in a way that it wouldn't accumulate within the structure, we prevent these layers from loosing strength due to high humidity.

References

1. Departamento Nacional De Infraestrutura De Transporte, *DNIT 137/2010. Pavimentação – Regularização do subleito – Especificação de serviço*. Rio de Janeiro. [2010]
2. M. R. De Rodrigues, *Influência da sucção no módulo de resiliência de solos típicos de subleito de pavimentos do Rio Grande do Sul*, Dissertação de Mestrado, Programa de Pós-Graduação Engenharia Civil, Departamento de Engenharia Civil, Universidade Federal do Rio Grande do Sul, 106 p. [1997]
3. N. N Khoury, M. Zaman, *Environmental Effects on Durability of Aggregates Stabilized with Cementitious Materials*. ASCE JMCE, Vol. **19** (1), pp. **1-8**, [2007]
4. A. J. Puppala, T. Manosuthkij, S. Nazarian, L. R. Hoyos, B. Chittoori, *In Situ Matric Suction and Moisture Content Measurements in Expansive Clay*

During Seasonal Fluctuations. GTJ Vol. **35.1** [2012]: **1-9**.

5. R. C. Weber, *Avaliação das trajetórias de umedecimento e secagem na deformabilidade elástica de solos compactos*. Dissertação de Mestrado, Programa de Pós-Graduação Engenharia Civil, Departamento de Engenharia Civil, Universidade Federal do Rio Grande do Sul, 139 p. [2013]
6. M. F. Matuella, W. P. Núñez, W. Y. Y. Gehling, R. C. Weber, *Avaliação dos efeitos do teor de umidade de compactação e variações de umedecimento e secagem na resistência ao cisalhamento de um solo argilo-siltoso*. In: Congresso Brasileiro de Mecânica dos Solos e Engenharia Geotécnica, 2014, Goiânia. COBRAMSEG 2014, [2014].
7. American Society For Testing And Materials, *ASTM D3080. Standard Test Method for Direct Shear Test of Soils Under Consolidated Drained Conditions*. USA, [1998].

Characterizing the variation of small strain shear modulus for silt and sand during hydraulic hysteresis

Ali Khosravi[1,a], Mehrzad Rahimi[2], Parisa Shahbazan[2], Ali Pak[3], and Amin Gheibi[2]

[1] *Assistant Professor, School of Civil Engineering, Sharif University of Technology, Tehran, Iran*
[2] *Graduate Student, School of Civil Engineering, Sharif University of Technology, Tehran, Iran*
[3] *Professor, School of Civil Engineering, Sharif University of Technology, Tehran, Iran*

Abstract. Experimental studies have indicated that the small strain shear modulus, G_{max}, of unsaturated silt and clay has a greater amount during imbibition than during drainage, when presented as a function of matric suction. However, due to material properties and inter-particle forces, different behavior is expected in the case of sand. Although considerable research has been devoted in recent years to characterize the behaviour of G_{max} of sand during drainage, rather less attention has been paid to the effect of hydraulic hysteresis on G_{max} and its variations during imbibition. In the study presented herein, an effort has been made to compare the G_{max} behavior of specimens of silt and sand during hydraulic hysteresis. In this regard, a series of bender element tests were carried out in a modified triaxial test device with suction-saturation control to evaluate the impact of hydraulic hysteresis on G_{max} for specimens of silt and sand. Trends between the G_{max} and matric suction for unsaturated sand were found to be different from those for silty specimens. The variations in G_{max} showed an up and down trend in both drainage and imbibition paths for sandy specimens, where plotted as a function of matric suction. Results also indicated smaller magnitudes of G_{max} upon imbibition than those during drainage; a behavior which is believed to be attributed to variations in suction stress with matric suction. In silty specimens, a stiffer response was measured during imbibition which was hypothesized to be due to drainage-induced hardening experienced by the specimens that was not fully recovered during imbibition.

1 Introduction

The knowledge of small strain shear modulus, G_{max}, in unsaturated condition is of particular interest to properly understand the seismic response of geotechnical systems. A significant body of studies have been developed on the measurement of G_{max} and its relationship with matric suction and mean net stress for unsaturated silt and clay. Former studies showed the importance of concurrent consideration of degree of saturation, S_r, and matric suction, ψ, in characterizing the behavior of G_{max} for silt and clay during drainage and imbibition [1-4]. Results of previous studies indicated that variations in G_{max} during drainage had a non-linear increasing trend with matric suction with greater values upon imbibition. This behavior was postulated to be due to plastic hardening the specimens experienced during drainage as a result of changes in degree of saturation and specimen volume.

In contrast to fine grained soils in which a clear and specified trend for G_{max} with respect to degree of saturation was reported, the G_{max} of sands was observed to behave differently during drainage with no unique trend between G_{max} and degree of saturation or matric suction [5-9]. Furthermore, the variations in G_{max} during imbibition for sand has not been fully investigated.

In the work presented herein, an effort has been made to gain a better understanding of the behavior of G_{max} during drainage and imbibition for silty and sandy soils. In this regard, a series of bender element tests were carried out in a modified triaxial apparatus with suction-saturation control to measure variations in G_{max} under different hydro-mechanical loadings. The modified test setup allowed precise control over matric suction in magnitudes as low as 0.1 kPa corresponding to 10 mm of suction head. In this setup, the degree of saturation of the specimens during hydro-mechanical loading was obtained through measurements of changes in water level in a burette connected to the bottom of the specimen, while the specimen volume change was measured using a modified Digital Image Processing (DIP) technique.

2 Tested material and specimen preparation

Bender element tests were carried out on two different types of soils: Bonny silt and silica sand. Bonny silt was obtained from borrow materials of the Bonny dam near the Colorado-Kansas border and is described as ML in accordance with the Unified Soil Classification System, USCS, (ASTM D 2488-00). The silica sand is a poorly

[a] Corresponding author: khosravi@sharif.edu

graded fine sand with sub-rounded to rounded particles. Typical properties of these two soils are summarized in Table 1.

Table 1. Geotechnical properties of tested soils.

Bonny silt		Silica sand	
Activity (%)	29	Mineral	99% SiO_2
γ_{dmax} (kN/m^3)	16.3	G_s	2.65
w_{opt} (%)	13.6	e_{max}	0.9
D_{50} (mm)	0.039	e_{min}	0.62
D_{10} (mm)	0.0013	D_{50} (mm)	0.19
Clay fraction (%)	14	C_u	1.29
LL, PL, PI	25, 21, 4	D_r (%)	67
$e_{Initial}$	0.85	$e_{Initial}$	0.71

Cylindrical silty and sandy specimens with 60 mm diameter and 120 mm height were prepared using the static compaction and aerial pluviation (sand raining) techniques, respectively. To prepare the silty specimens, the de-aired water was added and uniformly mixed with a desired amount of dried soil. Thereafter, the soil was statically compacted in 6 layers in a cylindrical mold to reach an initial void radio, e, of 0.85. After compaction, the specimen was removed from the mold and immediately placed in the triaxial testing apparatus. For the sandy specimens, specific amount of oven-dried sand was pluviated from a guide tube through a series of holes into a membrane expanded over a split mold, placed on the pedestal of the testing setup. The height of the sand free-fall and its deposition intensity were then controlled to obtain specimens of $e = 0.71$.

3 Experimental setup and testing procedure

A bender element test setup with suction-saturation control was developed to measure variations in G_{max} of specimens of silt and sand along different paths of drainage and imbibition under different levels of mean net stress. A schematic layout of the testing system is presented in Figure 1. The experimental approach includes the use of bender elements for G_{max} measurements, the axis translation technique [10] to apply and control the desired matric suction, Digital Image Processing (DIP) technique for specimen's volume change measurement during each step of loading, and a high pressure hydraulic pump for high mean net stress testing for silt.

The shear wave velocity and the corresponding small strain shear modulus were measured using a pair of bender elements, designed and built at Sharif University of Technology, one as a source and the other as a receiver, installed inside the bottom platen and the top cap, respectively. A signal generator, Model TTI, TGA-1241, was used to produce a sinusoidal pulse input signal to the transmitting bender element at the bottom, and a two-channel digital oscilloscope, Pico ADC-212, high speed analogue to digital converter was used to record the signals from both the transmitting and receiving elements. An amplifier was used to amplify the voltage supplied to the transmitting bender element and enhance the response of the bender element soil system.

In this setup, the axis translation technique was used to separately control the air and water pressures, and the corresponding suction value in the specimen during the tests. Water that flows out of/into the setup in response to the applied suction and mechanical load was measured using visual observation of water levels in graduated burettes connected to the water drainage lines from the specimen and confining cell. The Digital Image Processing, DIP, technique was used to measure the volume change of the soil specimens directly. In this technique, a digital camera was placed precisely at the horizontal direction, where its shooting direction perfectly captures any possible change through the height of the specimens. To control the temperature of the testing room, all tests were performed in an isolated humidity and temperature controlled room which was constructed in the advanced soil testing lab at Sharif University of Technology for unsaturated testing.

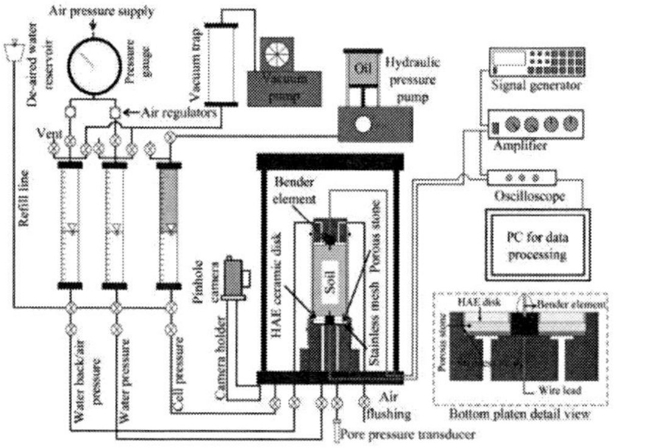

Figure 1. Schematic of the modified triaxial test device equipped with bender elements

After placing the soil specimens within the triaxial setup, the saturation process was initiated using the backpressure technique under 40 kPa net stress for silty specimens and applying vacuum to the top of the sandy specimens following ASTM D 6836-02. When the specimens were saturated, they were imposed to the desired confining pressure and the first step of the bender element test was performed to measure the value of G_{max} corresponding to the saturated condition. After the specimens were saturated, the desired suction was introduced at the boundaries of the specimens. The desired level of suction in silty specimens was imposed by decreasing the water pressure at the bottom while keeping the air pressure constant at the top. However, in the case of sandy specimens, the matric suction application was performed by elevating the magnitude of air pressure at the top and keeping the water pressure constant through the rest of the test. For the sandy specimens, simultaneous and equal changes in applied confining pressure were made to hold the mean net stress ($p_n = p-u_a$) constant for each incremental change in matric suction. Depending on the level of suction and the soil type, each equalization stage needed 2 to 15 days to be completed. After reaching hydraulic equilibrium, the bender element test was conducted to measure shear wave velocity and hence G_{max} related to each step of unsaturation process. After defining the drainage path of the SWRC and the corresponding G_{max} values, the decrements of matric suction were applied to the specimens in order to define the trends of S_r and G_{max} during imbibition.

4 Results

The SWRCs obtained from tests on silty and sandy specimens under different levels of mean net stress are shown in Figures 2 and 3, respectively. The SWRCs denote initial drainage from a saturated condition up to a desired matric suction of 110 kPa for silt and 15 kPa for sand. As shown in these figures, there is a remarkable difference between the SWRCs subjected to different levels of mean net stress in both types of soils, indicating that the mean net stress and the type of material may have a great effect on the water retention ability of soils. As presented, the mean net stress increase led to an increase in the ability of soil to retain water, resulting in the SWRCs shifting to the right. This behavior can be attributed to rearrangement of soil particles or specimen's volume change due to mean net stress increase. Similar observations were also reported by [1, 2] for silty specimens.

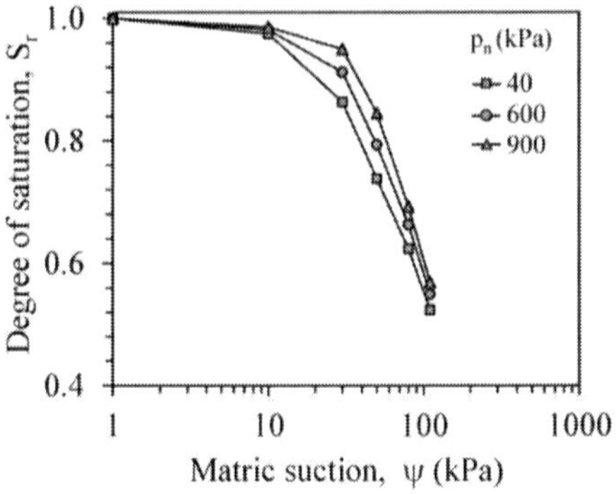

Figure 2. The SWRCs of Bonny silt specimens under different mean net stresses

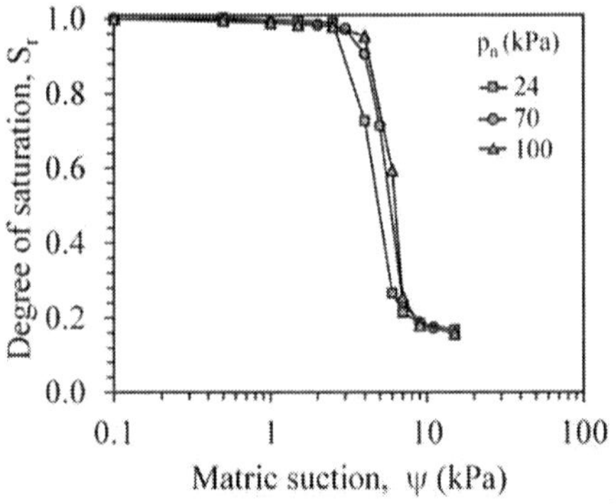

Figure 3. The SWRCs of silica sand specimens under different mean net stresses

The results of six bender element tests carried out on specimens of silt and sand during hydraulic hysteresis are presented in Figures 4 and 5, respectively. As in the case of sandy specimens a narrower range of matric suction was applied, the curves for sandy specimens were plotted in a natural scale, while those of silty specimens were presented in a logarithmic scale. As shown in these figures, the mean net stress has a pronounced effect on G_{max} values with greater values of G_{max} at higher net stresses. Moreover, the G_{max} of silts and sands were observed to behave differently during hydraulic hysteresis. Variations of G_{max} of silt along the drainage path showed an increasing trend when plotted as a function of matric suction, while an up and down trend for G_{max} was observed in the case of sandy specimens.

The results of tests on silty specimens shown in Figure 4 indicated two distinct zones for the variation of G_{max} along the drainage path of the SWRC. For the range of suction below the air entry value, negligible changes in degree of saturation of the specimens with suction increase were resulted in slight increase in G_{max} during drainage. For suctions ranging from air entry suction to a matric suction of 110 kPa, a greater increase in G_{max} was measured during drainage, although a limiting value was not expected for silty specimens because of the relatively high degree of saturation.

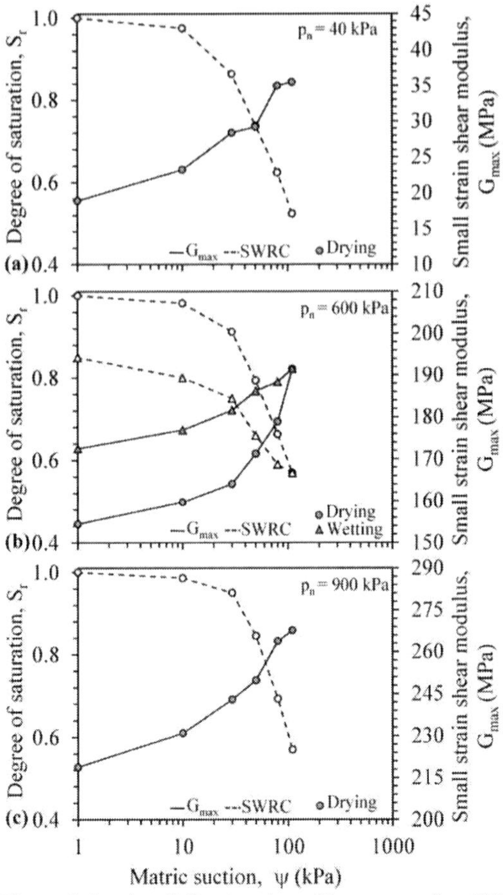

Figure 4. Results of G_{max} together with corresponding SWRC for Bonny silt specimens at various values of mean net stress: (a) P_n = 40 kPa; (b) P_n = 600 kPa; (c) P_n = 900 kPa

DOI: 10.1051/e3sconf/20160914018

In the case of sand, as presented in Figure 5, the measured G_{max}-ψ trends showed an up-and-down characteristic for suctions ranging from zero to the value corresponding to the residual condition. For the range of suction from zero to the air entry value, matric suction increase led to a relatively linear increase in G_{max}. However, for suctions beyond the air entry value, depending on the magnitude of applied mechanical load, the G_{max} trends showed a little difference in behavior. At lower mean net stresses of 24 kPa and 70 kPa, G_{max} initially increased to a peak value at a suction value of 4 kPa (a value beyond the air entry suction), after which a sudden drop in its magnitude was observed with suction increase. For the specimen under the mean net stress of 100 kPa, after reaching a maximum value at the air entry suction, the G_{max}-ψ curve followed a decreasing path toward a minimum value at residual suction. For suctions greater than the residual value, regardless of the magnitude of applied mean net stress, the rate of changes in degree of saturation was negligible and G_{max} experienced a slight decrease in magnitude with suction increase.

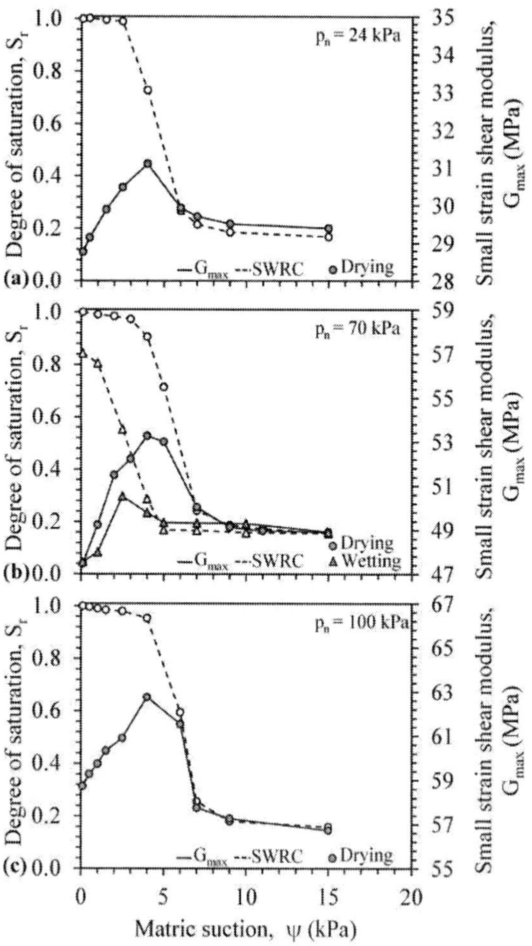

Figure 5. Results of G_{max} together with corresponding SWRC for silica sand specimens at various values of mean net stress: (a) P_n = 24 kPa; (b) P_n = 70 kPa; (c) P_n = 100 kPa

To investigate the effect of hydraulic hysteresis on the G_{max} measurements of silt and sand, the specimens under the mean net stresses of 600 kPa (for silt) and 70 kPa (for sand) were subjected to subsequent imbibition after drainage. In this regard, after an initial drainage, the specimens were imbibed thorough the imbibition paths of the SWRCs from an initial matric suction of 110 kPa (for silt) and 15 kPa (for sand) down to the zero matric suction. The results of these tests are presented in Figures 4(b) and 5(b) for silty and sandy specimens, respectively. As shown in these figures, for both tested materials, similar to the SWRC curves, there was a hysteretic behavior in G_{max} during imbibition. However, trends for G_{max} along the imbibition path were observed to be different for silty and sandy specimens.

Upon imbibition, G_{max} of the silty specimen experienced a reduction in magnitude with decreasing suction, a trend similar to that during drainage. The value of G_{max} was observed to have a slight decrease in magnitude with suction decrease when above or near the water entry. With further decrease in suction, significant amount of water was absorbed by the specimen, leading to a greater decrease in G_{max} during imbibition. The rate of changes in G_{max} during imbibition was observed to be smaller than that during drainage, resulting in the measurement of higher values of G_{max}. For the sandy specimen, similar to the drainage path, an up-and-down trend was noted for G_{max} along the imbibition path. However, for the same magnitudes of equilibrium suction, the values of G_{max} for sand during imbibition were consistently smaller than those during drainage.

5 Analysis

5.1 Characterization of the G_{max} behavior for silty specimens

The hysteretic behavior of G_{max} with matric suction during cycles of drainage and imbibition for silty soils has been postulated to be as a result of drainage-induced hardening which was not fully recovered during imbibition. Accordingly, the values of G_{max} along the imbibition path has been observed to be higher than those measured during drainage. In recent studies on G_{max} of unsaturated silt, the hardening effects associated with changes in degree of saturation or volume change were considered through the term mean preconsolidation stress, p_c', which can be described using an expression proposed by [2] as follows:

$$\frac{p_c'}{p_{c0}'} = \exp\left[\frac{\Delta e^p}{\lambda - \kappa} + b(S_{e0} - S_e)\right] \quad (1)$$

where p_{c0}' is the mean preconsolidation stress at saturation, Δe^p is a plastic change in void ratio, λ and κ are the slopes of the virgin compression and elastic rebound parts of the isotropic compression curve, b is a hardening parameter which controls the rate of change in p_c' with changes in S_e, and dS_e is variations in the effective saturation. S_e is defined as follows:

$$S_e = \frac{S_r - S_{r,res}}{1 - S_{r,res}} \tag{2}$$

where S_r is the current degree of saturation, and $S_{r,res}$ is the value of S_r in residual condition. A comparison of the trends in p_c' with the values of G_{max} measured in the bender element tests during hydraulic hysteresis is presented in Figure 6. In this figure, the values of mean preconsolidation stress were obtained from Eq. 1 using a b value of 2.15, a $S_{r,res}$ value of 0.05, a p'_{c0} of 180 kPa at saturated condition, and zero Δe^p. To obtain the b value, a series of isotropic compression tests were conducted on specimens with the same initial conditions under different levels of matric suction. For the range of suctions used in this study, the specimens were observed to experience relatively little volume change during hydraulic hysteresis.

As presented in this figure, similar to G_{max} measurements, the predicted drainage and imbibition mean preconsolidation stresses are in a continuous form nonlinear curve when plotted in terms of matric suction, with a smooth transition at the air entry suction. Results also indicated higher values of p_c' during imbibition than those during drainage, a behavior similar to that noted for G_{max}.

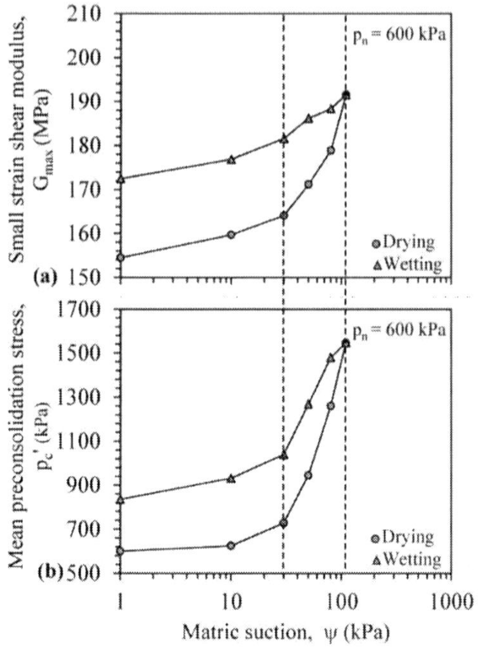

Figure 6. Hysteretic behavior during drainage and imbibition for Bonny silt specimens at mean net stress of 600 kPa: (a) G_{max}; (b) p_c' (LC yield curve)

5.2 Characterization of the G_{max} behavior for sandy specimens

While the G_{max} measurements of the silty specimens showed a strong correlation to the mean preconsolidation stress, there are still some difficulties in interpreting trends for G_{max} of unsaturated sand during hydraulic hysteresis. Based on the results of previous work, the small strain shear modulus for an unsaturated sand is greatly dependent on the effective stress [6, 11, 12]. While the well-known effective stress-dependency of G_{max}, correlations between the small strain shear modulus and effective stress are not well understood. In this study, an attempt was made to define relationships between G_{max} and effective stress using the concept of suction stress introduced by [13]. The suction stress is a stress variable that incorporates the effect of different inter-particle stresses arising from chemical cementation, van der Waals attraction, capillarity, and double layer repulsion into the effective stress definition for unsaturated soils [13-15]. The suction stress can be easily defined from shear/tensile strength tests data or using a closed-form mechanical expression proposed by [16] as follows:

$$p_s = \psi \times \left[\frac{1}{1 + (\alpha \times \psi)^n} \right]^{1-\frac{1}{n}} \tag{3}$$

where p_s is the suction stress, and α and n are van Genuchten SWRC model parameters.

The values of G_{max} obtained from the bender element tests are presented in Figure 7(a) and Figure 7(b) show the predicted trend of suction stress with matric suction (described as suction stress characteristic curves, SSCC) for the specimen of sand under the p_n of 70 kPa during hydraulic hysteresis. α and n were obtained by fitting the van Genuchten SWRC model to the experimental curve presented in Figure 4(b) (α = 0.18 and n = 8.1). A comparison of the trend in p_s with the measured values of G_{max} indicated that G_{max} of the sandy specimen followed an up and down trend during hydraulic hysteresis complying with the shape of SSCC.

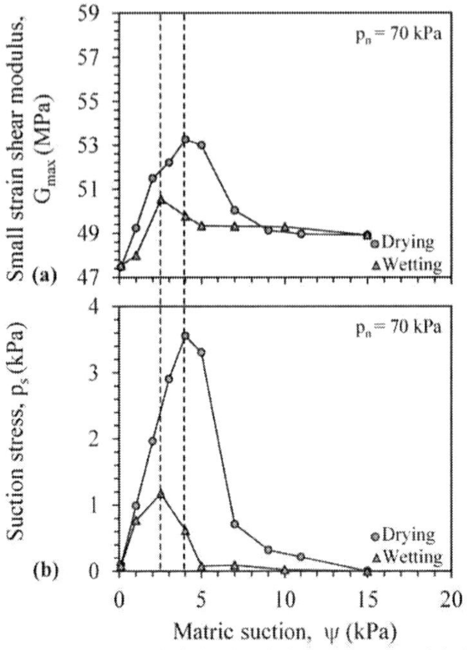

Figure 7. Hysteretic behavior during drainage and imbibition for silica sand specimens at mean net stress of 70 kPa: (a) G_{max}; (b) SSCC

Based on the results presented in Figure 7(b), the SSCC followed an initial increase with matric suction, reaching a maximum value at a suction corresponding to the maximum small strain shear modulus. When increasing suction beyond this specific value, the suction stress values dropped nonlinearly to a magnitude of zero at a suction stress of 15 kPa. During imbibition, similar to the experimental G_{max} data, a hysteresis behavior in p_s was observed, with values of suction stress being lower than those during drainage. The predicted SSCC is consistent with those reported by [14].

6 Conclusion

In the current study, variations in the small strain shear modulus during hydraulic hysteresis were investigated for specimens of silt and sand. The combination of matric suction control using the axis translation technique and shear wave velocity measurement using a pair of bender elements housed in the bottom platen and top cap of the setup allowed the measurements of the small strain shear modulus along different cycles of drainage and imbibition.

According to the results of this study, different trends for G_{max} were observed during hydraulic hysteresis in the case of unsaturated silt and sand. Based on the performed analysis, hardening of soil caused by drainage played a key role in the dynamic response of unsaturated silty soils at small strains, with G_{max} and p_c' having similar evolving trends during hydraulic hystersis. While the well-known continuous increasing trend for G_{max} of silt with matric suction, G_{max} of sand varied in an up and down manner with respect to matric suction; a behaviour complying with the shape of the suction stress characteristic curve, SSCC.

References

1. C.W.W. Ng, J. Xu, S.Y. Yung, Can. Geotech. J., **46**(9), 1062–1076 (2009)
2. A. Khosravi, J.S. McCartney, ASCE J. Geotech. and Geoenviron. Eng. **138**(11), 1326–1333 (2012)
3. K.S. Wong, D. Masın, C.W.W. Ng, J. Comput. Geotech., **56**, 28–39 (2014)
4. A. Khosravi, A. Gheibi, M. Rahimi, J.S. McCartney, S.M. Haeri, *The 15th Asian Regional Conference on Soil Mechanics, 15ARC* (Fukuoka, Japan, 2015)
5. X. Qian, D.H. Gray, R.D. Woods, Geotech. Test. J. **14**(3), 266–275 (1991)
6. A. Khosravi, M. Ghayoomi, J.S. McCartney, H.Y. Ko, *Geo-Florida 2010: Advances in Analysis, Modeling & Design,* (West Palm Beach, FL, 2010)
7. M. Ghayoomi, J.S. McCartney, Geotech. Test. J. **34**(5), 1-11 (2011)
8. M. Asslan, F. Wuttke, Unsaturated Soils: Research and Applications, Springer Link, PP. 355-362 (2012)
9. W.T. Oh, S.K. Vanapalli, Geotech. Geol. Eng. **32**(2), 259-271 (2014)
10. J.W. Hilf, Technical Memorandum No. 654, United State Department of the Interior Bureau of Reclamation, Design and Construction Division, Denver, Colorado, USA (1956)
11. A. Sawangsuriya, T.B. Edil, P.J. Bosscher, J. Geotech. Geoenviron. Eng. **132**(2), 131–142 (2009)
12. Y. Dong, N. Lu, J.S. McCartney, J. Geotech. Geoenviron. Eng. (2015)
13. N. Lu, W.J. Likos, J. Geotech. Geoenviron. Eng. **132**(2), 131-142 (2006)
14. N. Lu, T. H. Kim, S. Sture, W. J. Likos, J. Eng. Mech. **135**(12), 1410-1419 (2009)
15. S. Oh, N. Lu, Vadose Zone J. **13**(5) (2014).
16. N. Lu, J. W. Godt, D. T. Wu, Water Resour. Res. 46.5. (2010)

Study on the stress path dependency of collapse behavior of Gorgan loess implementing unsaturated oedometer devices

S. Mohsen Haeri [1,a], Majid Beigi[2], Saeed Saberi[2] and Amir Akbari Garakani[3]

[1] *Professor, Department of Civil Engineering, Sharif University of Technology, Tehran, Iran*
[2] *M.Sc, Department of Civil Engineering, Sharif University of Technology, Tehran, Iran*
[3] *Ph.D, Department of Civil Engineering, Sharif University of Technology, Tehran, Iran*

Abstract. Collapsible soils such as loess are particular types of problematic soils, whose collapse potential have caused severe damages to the buildings and structures. Loess is susceptible to large and sudden volume reduction upon wetting in constant normal stress or upon increasing normal stress in constant water content or a combination of these two stress paths. In this research both of these collapse induced paths: wetting and stress paths are investigated on undisturbed specimens of a Loess to observe the effect of various parameters on the collapse behavior of the tested soil., Seven controlled matric suction unsaturated oedometer tests have been conducted on undisturbed specimens taken from loessial soil of Gorgan, a city in Golestan province in Iran to study the collapse mechanisms of this type of soils. Two identical sets of unsaturated oedometer apparatuses designed and developed at Sharif University of Technology, with the capability of controlling matric suction have been utilized. In these tests, the matric suction, vertical net stress, volumetric water content, and vertical deformation of the samples have been controlled and recorded. Tests have been performed under two separate stress paths: constant matric suction with varying vertical net stress (loading and unloading) and constant vertical net stress with varying matric suction (wetting and drying). Obtained results show the high collapse potential of the tested soil and significant effects of stress path, matric suction, and vertical net stress on the collapse behavior of the tested soil.

1 Introduction

Collapsible soils are known as soils with open and metastable structures. Increase of water content in normal and expansive soils leads to increase in volume or expansion, whereas in soils with open and metastable structure it leads to collapse if enough stress is applied on the soil as well. Most researchers [e.g. 1-4] suggest four for collapse in a soil: 1. An open- metastable and unsaturated fabric 2. A required bonds or cementing agent and/or suction for stability in unstructured state 3. A required net total vertical stress for triggering metastable condition 4. An increase in water content, which subsequently results in collapse or sudden reduction of volume of the soil mass. Unlike the saturated soils, unsaturated soils have three phases: air, water and soil skeleton. The behaviors of unsaturated soils in general appear to be relatively complex in comparison to those of fully saturated soils. Since 1950's, significant developments have been made in understanding unsaturated soil behavior. Early attempts were aimed at treating unsaturated soil behavior in terms of a single effective stress parameter [5]. Jennings and Burland [6] recognized the limitation of this approach, and study the unsaturated behavior with total stress and suction variables which later led to the independent stress framework. Subsequently, many researchers proposed the frameworks in terms of two independent stress variables, net stress (σ-u_a) and matric suction (u_a-u_w) [7]. One of the most widely used elasto-plastic frameworks was first proposed by Alonso [8] as Barcelona model. This type of framework was unable to predict some behavior observed in the laboratory, for example, coupled hydro-mechanical soil response under drying/wetting cycles and influence of these cycles on subsequent behavior. More recent approaches [e.g. 9, 10] shifted towards a new choice of stress variables, which have implemented the corporation of both suction and degree of saturation to the soil grain stability and to the effective stress, similar to the original Bishop [5] approach. Later some other researchers proposed different models to describe the behavior of unsaturated soils [e.g. 11-13]. In brief, a variety of frameworks have been proposed for modelling of the behavior of unsaturated soils, most of which are still at the stage of development, and thus need experimental validation.

Based on observations on the experimental results presented in this research and some other studies [14-22], stress path dependency is one of the major aspects of the behavior of collapsible soils and a model or framework capable of predicting such behavior is needed. In this paper the hydro-mechanical stress path dependency of tested loess is highlighted, however, the framework for

[a] Corresponding author: smhaeri@sharif.edu

DOI: 10.1051/e3sconf/20160914019

modelling such behavior is out of the scope of this paper and will be discussed elsewhere.

In this regard, two sets of suction-controlled oedometer apparatuses implementing axis-translation method [23] have been developed at Sharif University of Technology (SUT) and used to study the behavior of undisturbed specimens of highly collapsible loessial samples taken from a semi-arid region in Iran, Gorgan [14].

2 Tested Soil Properties

The studied soil in this research is a loessial soil taken intact in waxed monolithic boxes from the city of Gorgan in Golestan province of Iran. The oedometer undisturbed specimens with 2 cm height and 5 cm, diameter have been taken from the intact samples. For more information the reader is referred to [15, 16]. The soil is mainly silt size and the soil index tests reveal that the tested soil is ML in accordance with Unified Soil Classification System (USCS). Some other index properties of the tested soil are listed in Table 1.

As presented in Table 1. The average value of the initial degree of saturation of the tested soil is about 25%. This value corresponds to an initial matric suction of 750 kPa according to the results of filter paper tests performed on this soil.

3 Unsaturated Oedometer Devices

To study the complex behavior of the collapsible soils in unsaturated soil mechanics framework, two oedometer devices with capability of controlling matric suction implementing axis translation method [23] have been designed, built and utilized in Advanced Soil Mechanics Laboratory of SUT as sketched in Figure 1.

In this research a 5cm diameter high air entry value ceramic disk (HAEV) was fixed to the base of the oedometer cell and de-aired water pressure has been applied to the bottom of the HAEV disc. Monitoring of the height of the water column in the burette interconnected to this system specifies the absorbed or desorbed water to or from the specimens during the test. The require air pressure has been applied with vent on the top cap to keep the matric suction under control. The changes in the height of the specimen have been controlled with a strain gauge fixed at the top of the cap of the specimen (Figure 1).

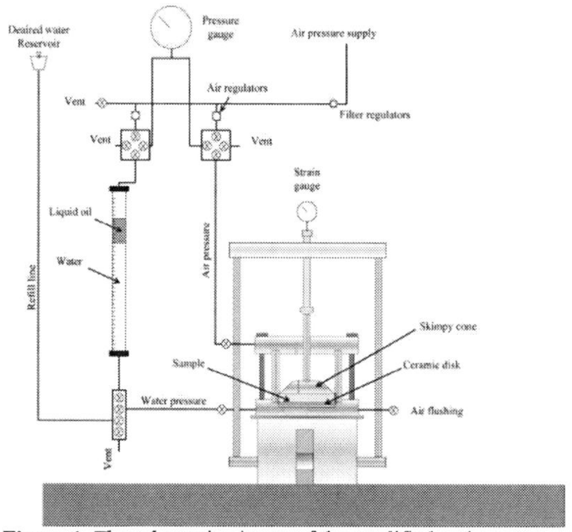

Figure 1. The schematic picture of the modified oedometers

4 Stress and Wetting Paths and Test Procedures

In order to investigate the collapsibility and stress-suction paths dependency of the experiments in addition to the mechanical behavior of the tested loess, a total of seven tests, in two groups: 1) constant matric suction under varying vertical net stress (loading and unloading) which has been named as vertical compression induced collapse tests, and 2) constant vertical net stress under varying matric suction (wetting and drying) which has been named as wetting induced collapse tests have been performed. The sitting load of 25 kPa has been applied to the samples at the beginning of each test to reach to the required initial matric suction before taking any step toward the designed stress path. The stress path of each test is shown in Figure 2. During the tests the volume change and the amount of absorption (or desorption) of water for each specimen have been monitored.

5 The Tests Results

A summary of the tests results are shown in the following sections. Both variations of specific volume and degree of saturation of the specimens in each stress path versus vertical net stress and matric suction are presented and analyzed in Figures 3 and 4. In order to achieve the suction equilibrium within the specimens, each vertical compression induced collapse test lasted about 25-45 days and each wetting induced collapse test lasted about 65-75 days.

Table 1. Index properties of the selected "undisturbed" loessial soil specimens.

	Initial void ratio	Initial dry unit weight	Initial moisture content	Initial degree of saturation	Pass no. 4 sieve	Pass no. 200 sieve	Clay Frac (<2 μm)	Atterberg limits		USCS
	e_o	γ_{do} (kN/m^3)	w_o (%)	S_{ro}	d_4 (%)	d_{200} (%)	d_c (%)	w_L (%)	w_p (%)	
Avg	0.76	15.09	7	0.25	100	96.5	26	32.7	25.5	ML
Stnd. dev.	0.006	0.005	0.119	0.005	0	1.1	3.3	0.3	0.8	

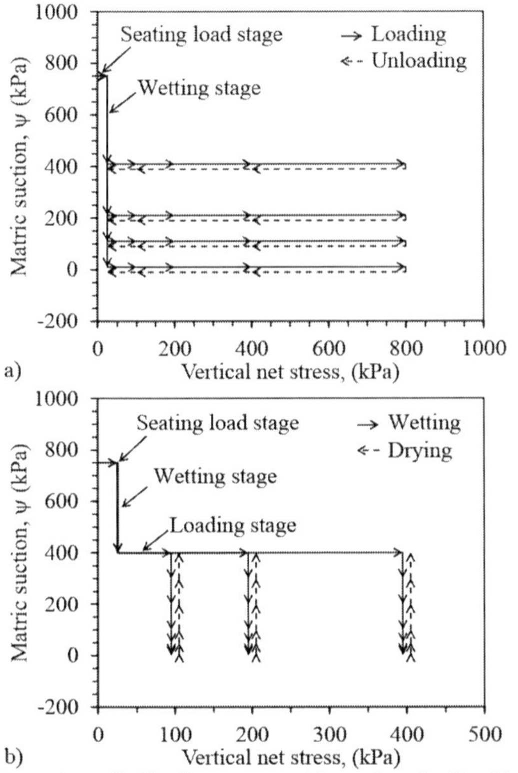

Figure 2. Applied loading stages used for testing of collapsible soils: a) vertical compression induced collapse tests at different constant matric suctions and b) wetting-induced collapse tests at different constant vertical net stresses.

5.1 Stress Variables

The variations of specific volume (v) versus vertical net stress (P_{vn}) corresponding to the vertical compression induced collapse tests are depicted in Figure 3.a. P_{vn} (vertical net stress) is obtained from the Equation 1, where P_v is the total vertical stress, u_a is the air pressure and ψ is the matric suction which is defined in Equation 2, in which u_w is the pore water pressure. The variation of specific volume with matric suction in the wetting induced collapse tests are presented in Figure 4.a as well:

$$p_{vn}=p_v-u_a \quad (1)$$

$$\psi=u_a-u_w \quad (2)$$

5.2 Vertical Compression Induced Collapse Tests

The changes in specific volume of the specimens in vertical compression induced collapse tests are plotted against vertical net stress for both loading and unloading paths in Figure 3.a. Yield stress in these figures is defined as the stress in which significant volume change or collapse starts to take place in loading paths. As presented in Figure 3.a the amounts of change in specific volume before the yield point are much smaller in tests with higher matric suctions than those for the tests with lower matric suctions. This mild slope of volume changes versus vertical net stress before yield point especially in dryer soils is indicative of strong bonds between the soil grains. These bonds will break at specified yield stress that is a function of applied matric suction. In tests with smaller matric suctions the yield stress has smaller values which represent the lower threshold stresses for the bond breakage in lower matric suctions. Other observation from the tests results is that the slope of compression curves is not unique for different matric suctions and depends on the value of the tested matric suction. This fact is true both in pre-collapse (before yield stress) and in main collapse (after yield stress) stages of the tests meaning that the soil behaves elasto-plastic in both parts of the loading path. In other words, the behavior of collapsible soil in compression is neither elastic nor unique before pre-compression pressure (yield stress).

The changes of the degree of saturation with vertical net stress are presented in Figure 3.b for vertical compression induced collapse tests. As can be observed, in loading stage, the degrees of saturation of the specimens increase with increase in vertical net stress because of the pore collapse and reduction in pore size, especially after the yield stress.

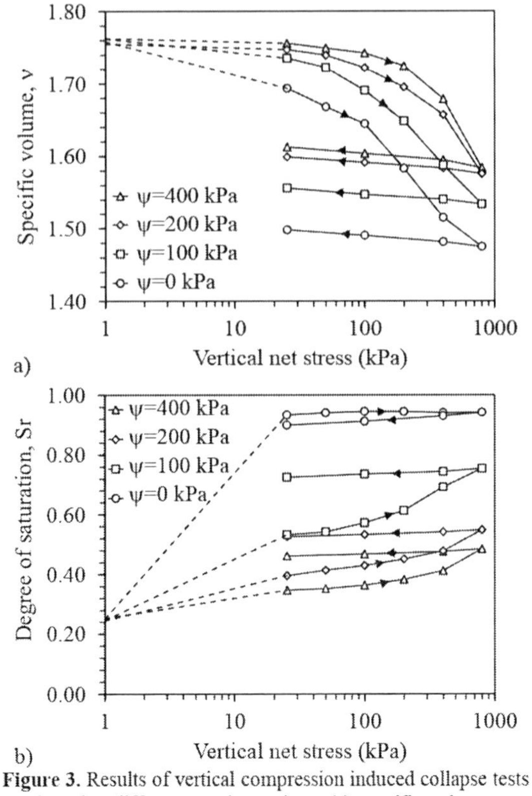

Figure 3. Results of vertical compression induced collapse tests conducted at different matric suctions: (a) specific volume as a function of applied vertical net stress, and (b) degree of saturation as a function of applied vertical net stress.

In unloading steps no significant decrease in degree of saturation happens because the volume reduction due to collapse is unrecoverable and the soil behaves elastically at this stage.

5.3 Wetting Induced Collapse Tests

In Figure 4.a. the changes in specific volume of the wetting induced collapse tests have been plotted for various applied matric suctions. The initial specific volume and matric suction for all specimens is a point which is shown in the figure with a cross symbol. The initial changes in specific volume for reaching to the start of each test at ψ=400 kPa for various vertical net stresses are shown by dashed lines in Figure 4.a as well. In the main part of the tests which starts from ψ=400 kPa the matric suction decreases (wetting path) under constant vertical net stresses to get to the saturation state and then drying path is followed as shown by arrows in this figure.

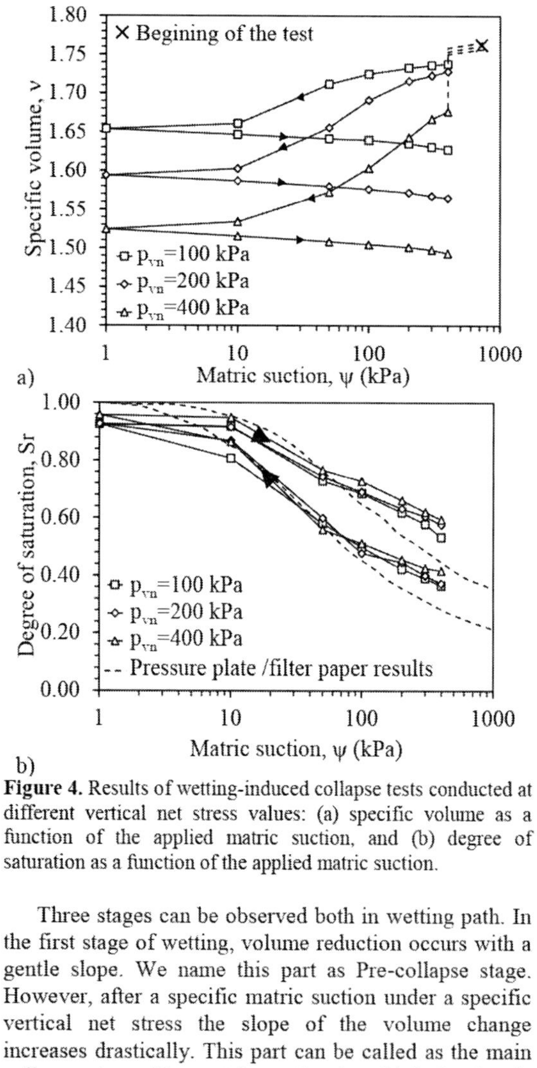

Figure 4. Results of wetting-induced collapse tests conducted at different vertical net stress values: (a) specific volume as a function of the applied matric suction, and (b) degree of saturation as a function of the applied matric suction.

Three stages can be observed both in wetting path. In the first stage of wetting, volume reduction occurs with a gentle slope. We name this part as Pre-collapse stage. However, after a specific matric suction under a specific vertical net stress the slope of the volume change increases drastically. This part can be called as the main collapse stage. The matric suction in which the drastic change in the volume or main collapse starts is called as the "yield point" which increases with increase in the applied constant vertical net stress. When the main collapse takes place under a specific net stress with decreasing suction, the third stage starts and the slope of the volume change will significantly decrease which in turn results in small volume reduction with further wetting or decrease in matric suction. The drying path which follows the wetting path will cause very small additional volume reduction which can be related to increase in effective stress due to increase in matric suction during drying process. This fact indicates that the collapse associated with wetting path or wetting induced collapse is not recoverable with drying.

The changes in the degree of saturation with matric suction or SWRC for wetting induced collapse tests on the tested loessial soil, under various vertical net stresses, are presented in Figure 4.b. As shown in this figure, the SWRCs are very similar for different values of vertical net stresses and are very close to the SWRC of tested loess using filter paper, shown in the figure by dashed lines, especially in low to medium matric suctions. However the SWRCs resulted from wetting induced collapse oedometer tests on the tested collapsible soil for matric suctions of 100 kPa and more deviate from the SWRC obtained from filter paper that is measured under zero effective stress [16].

5.4 The Hydro-mechanical Stress Path Dependency of the Collapse Behavior

So far the results shown and discussed in sections 5.2 and 5.3 clearly indicate the hydro-mechanical stress and stress path dependency of the tested collapsible soil. In the following, this fact is more elaborated. The results of the same stress states obtained from vertical compression induced collapse and wetting induced collapse tests are presented as variation of volumetric strain with matric suction in Figure 5.

As shown in this figure, the amounts of volumetric strains for wetting induced collapse tests are smaller than those for vertical compression induced collapse tests. The exception is the point associated with ψ=200 kPa in vertical compression induced collapse tests for which the reason is not known and can be an experimental error. These differences are more significant in the middle or main collapse parts of the paths. However, at high matric suction values the soil structure is more stable and has stronger bonds which make the volume changes more independent to the applied stress and wetting paths. Also at matric suction values near the saturation state, most of the bonds have been broken and the volumetric strain due to bond breakage meaning that at low matric suctions the behavior is less path dependent.

The variation in degree of saturation against matric suction (SWRC) of the specimens of both group of tests are plotted in Figure 6 as well. The inter-related behavior of the variation of degree of saturation and volumetric strain is noticeable in Figures 5 and 6. The amounts of volumetric strain of specimens in wetting induced collapse tests for a particular matric suction is smaller than that for vertical compression induced collapse tests as discussed above and shown in Figure 5. However, the degree of saturation at that particular matric suction would be larger for the specimen tested following the vertical compression induced collapse compare to that tested under wetting induced collapse condition as can be observed from Figure 6. The variation of the slopes of the curves given in Figures 5 and 6 are closely linked as well.

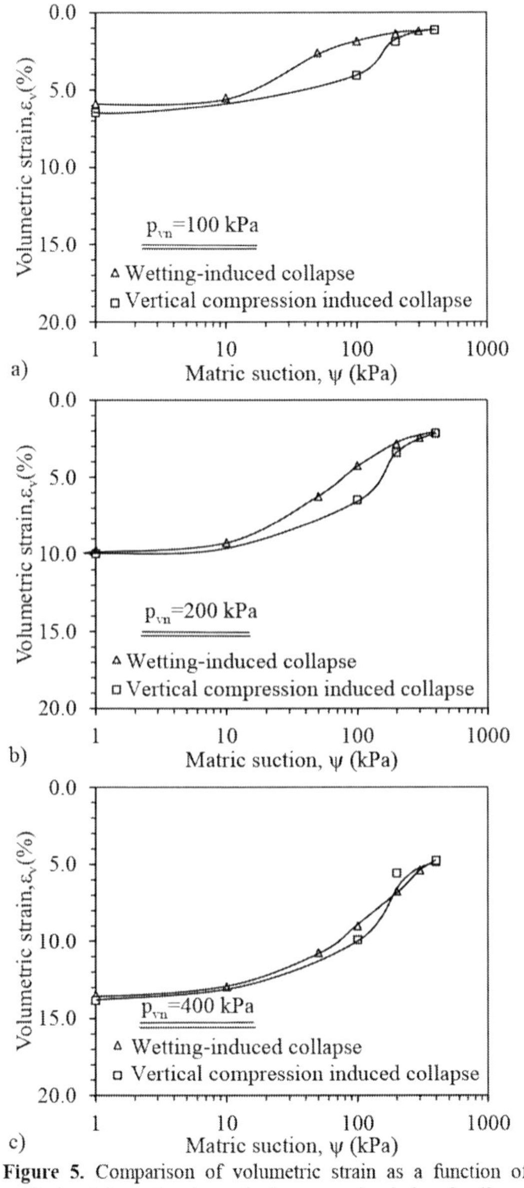

Figure 5. Comparison of volumetric strain as a function of applied matric suction in vertical compression induced collapse and wetting-induced collapse tests on tested loess at vertical net stresses of a) 100 kPa, b) 200 kPa and c) 400 kPa

5.5 Stress and Stress Path Dependency of the SWRC

The SWRCs for various compression induced collapse tests are calculated from the test results and are given in Figure 7. As can be observed from SWRCs are not unique and are completely stress dependent. Also there is no match between SWRC obtained from filter paper under Zero stress and those obtain by compression induced collapse tests at various vertical net stresses.

Stress path dependency of SWRC for the tested collapsible soil is obvious when the results shown in Figure 7 is compared with that shown in Figure 5.b which is the SWRC curves for the same soil tested under wetting induced collapse path.

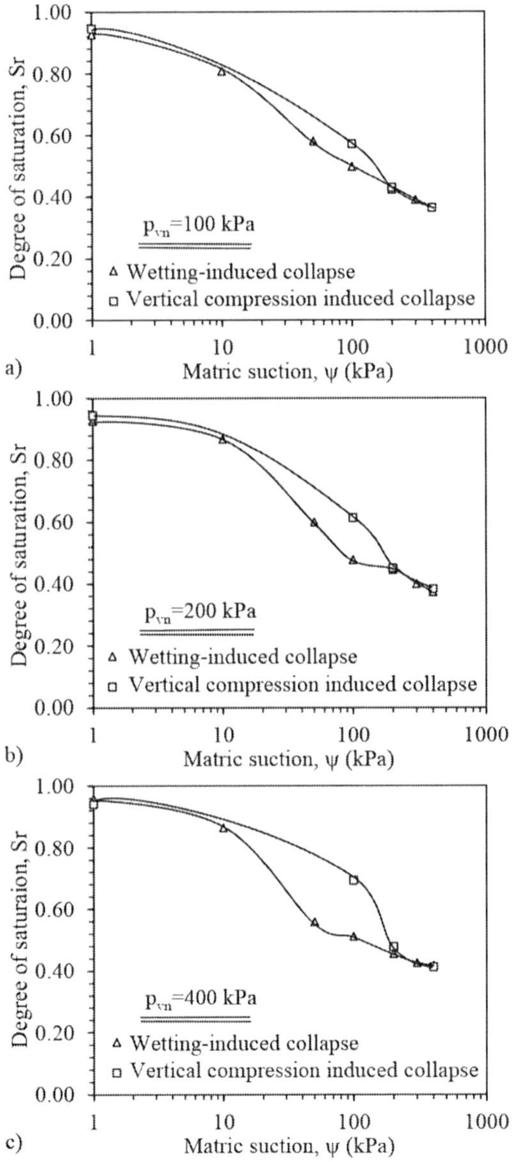

Figure 6. Comparison of degree of saturation as a function of the applied matric suction from vertical compression tests and wetting-induced collapse tests at vertical net stresses for a) 100, b) 200 and c) 400 kPa

If the result shown in Figure 5.b which are the SWRC curves for wetting induced collapse tests implementing unsaturated oedometer devices, are compared with the SWRCs obtained from wetting induced collapse tests on the same soil, using triaxial device under isotropic compression [15, 16], one can observe the differences between these two cases. The reason could be difference in stress states. During tests on specimens using oedometer device an intrinsic shear stress is induced in the specimen due to k_o condition, whereas in triaxial compression tests the soil is under an isotropic pressure with zero shear stress. This fact emphasizes the stress path dependency of the unsaturated behavior of collapsible soils.

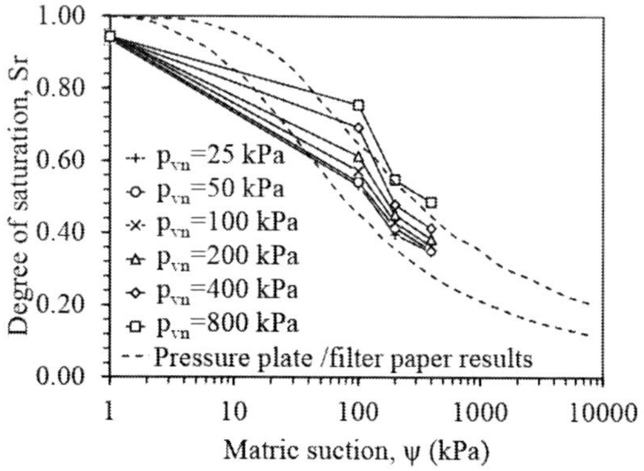

Figure 7. SWRC which has obtained from vertical compression tests results

6 Conclusions

In this study, the hydro-mechanical behavior of a collapsible loessial soil is investigated by conducting unsaturated odometer tests under applying two different stress paths namely, "vertical compression induced collapse tests" and "wetting induced collapse tests". Results of the vertical compression induced collapse tests reveal that collapsible behavior of the tested soil is elastoplastic throughout the loading stage. Also it is shown that the yield stress of the soil during vertical compression induced collapse tests, increases by increasing the matric suction. In the same line, in wetting induced collapse tests, by increasing vertical net stress the value of the yield point of the soil increases. Also it is observed that the dependency of the hydro-mechanical behavior of the tested soil to the stress path is more profound for the middle parts of the wetting induced collapse tests where main collapse of the structure of the soil happens in this part. For both types of the applied stress path, the SWRCs of the soil are highly dependent on the level of applied matric suction and the amount of vertical net stress.

References

1. L. Barden, A. O. Madedor, G. R. Sides. J Soil Mech Found Div, ASCE **95**, 33-51 (1969)
2. J. H. Dudley. J Soil Mech Found Div **97**, (1900)
3. L. Barden, A. McGown, K. Collins. Engineering Geology **7.1**, 49-60 (1973)
4. J. K. Mitchell, K. Soga. New York: Wiley **422** (1976)
5. A. W. Bishop. Teknisk Ukeblad I Samarbeide Med Teknikk, **39**, 859–863 (1959)
6. J. E. B. Jennings, J. B. Burland. Geotechnique **12**, 125-144 (1962)
7. D. G. Fredlund, H.Rahardjo. John Wiley & Sons (1993)
8. E. Alonso, A. Gens, A. Josa, Geotechnique, London, **40(3)**, 405-430 (1990)
9. S. J. Wheeler, R. S. Sharma, M. S. R. Buisson. Geotechnique **53**, 41-54 (2003)
10. D. Gallipoli, S. J. Wheeler, M. Karstunen. Geotechnique **53**, 105-112 (2003)
11. D. G. Toll. Geotechnique **40**, 31-44 (1990)
12. J. A. Blatz, J. Graham. Geotechnique **53**, 113-122 (2003)
13. N. Khalili, F. Geiser, G. E. Blight. International Journal of Geomechanics **4.2**, 115-126 (2004)
14. M. Beigi. M.Sc. thesis,, Sharif University of Technology, Tehran, Iran (2015)
15. A. A. Garakani. Ph.D. thesis, Department of Civil Engineering, Sharif University of Technology, Tehran, Iran (2013)
16. S. M. Haeri, A. A. Garakani, A. Khosravi, C. L. Meehan. ASTM, Geotechnical Testing Journal, **37(2)** (2014)
17. S. M. Haeri, A. A. Garakani. Research and Applications. Springer Berlin Heidelberg, 293-300 (2012)
18. S. M. Haeri, S. Saberi, M. Beigi, A. A. Garakani. 15th Pan-American Conference on Soil Mechanics and Geotechnical Engineering, Buenos Aires, Argentina, (2015)
19. S. M. Haeri, A. Khosravi, A. A. Garakani, S. Ghazizadeh. ASCE, International Journal of Geomechanics (under review) (2016)
20. S. M. Haeri. 15th Asian Regional Conference on Soil Mechanics and Geotechnical Engineering, Fukuoka, Japan (2015)
21. S. M. Haeri, A. A. Garakani. 15th Asian Regional Conference on Soil Mechanics and Geotechnical Engineering, Fukuoka, Japan (2015)
22. A. A. Garakani, S. M. Haeri, A. Khosravi, G. Habibagahi. Engineering Geology **195**, 28–41 (2015)
23. J.W. Hilf. Ph.D. thesis, University of Colorado Boulder, Boulder, CO, USA. (1956)

Suction controlled drying and wetting cycle effects on the volumetric behaviour of a lime-treated high plasticity clay

Marco Rosone[1,a], Camillo Airò Farulla[1], Alessio Ferrari[1], Claudio Torta[1] and Clara Celauro[1]

[1]*Department of Civil, Aerospace, Environmental and Materials Engineering (DICAM) - University of Palermo, Palermo, Italy*

Abstract. The paper presents some experimental results collected on samples recovered from an experimental embankment obtained by compacting a lime-treated clay. Samples were collected soon after the in situ compaction and they were cured in controlled environmental conditions for at least 18 months. Mercury intrusion porosimetry tests (MIP) were carried out on freeze-dried specimens to characterize the microstructure of the material. In order to assess the durability of the improved material, laboratory tests focused on the effects of cyclic variations of the degree of saturation on the water retention properties and the volumetric behaviour of the stabilized clay. Collected results show that the lime-treated clay undergoes an almost irreversible volumetric behaviour; this irreversible contraction is associated to severe drying processes, while wetting paths do not produce significant volumetric deformations.

1 Introduction

Lime stabilization of fine soils is an advanced technology among those that promote sustainable use of natural resources. The technique is aimed at improving the physical, chemical and mechanical properties of the clays that otherwise would not be suitable for these purposes. Then, it allows the re-use of clayey soils in the construction of transport infrastructures, minimizing the need for suitable materials from borrow-pits and the need to transport waste soils to these lands. Furthermore, this technique permits, when well managed, a significant improvement in both workability and the mechanical properties of the treated clay.

As proved by many studies in this field, lime treatment induces a quick modification of the chemical and physical characteristics of the treated clay due to the absorption of Ca^{2+} ions by the clay particle surfaces by cation exchange.

Afterwards, more complex chemical reactions develop with time in the lime-soil mixture: in such a high *pH* environment the dissolution of the clay silica and alumina (SiO_2, Al_2O_3) and their reaction with the lime calcium produces the development of calcium-silicate-hydrates (*CSH*) and calcium-aluminate-hydrates (*CAH*), thus creating cementitious bonds within the soil [1]. These bonds are responsible for the improvement of the mechanical properties of the treated soil in terms of an increase in shear strength and reduction of the compressibility [2].

Basic reactions in the lime treatment process are quite well understood and so is the consequent mechanical improvement, mainly in terms of bearing capacity, shear strength and compressibility of the treated clays [3-6] Nonetheless, some aspects may be of concern mainly during the construction phase due to variability of the clay characteristics. Serious attention should be paid when defining the in-field execution procedures in order to guarantee the required performances [7].

Furthermore, despite the quite wide use of the lime treatment technique, so far not many studies deal with the long term behaviour of the treated clay or the effect of repeated loading or variation in the boundary hydraulic conditions [8]. In particular, a very interesting aspect to be clarified is the one related to the durability in time of the mechanical properties gained by the treatment, in relation to the repeated variations in the degree of saturation of the material as laid in situ.

The results presented in the paper refer to a wide experimental programme carried out during the construction of a main extra-rural state road in Sicily. A high plasticity clay, available in large quantities after the excavation work, was studied in order to define the technical and economic feasibility of the treatment. A research programme was defined based on a field trial specifically made for defining the correct execution procedures with regard to each single production phase that could affect the final performance of the structure. Indeed, a complete geotechnical characterization, an evaluation of the mechanical properties as well as one of the deformation characteristics of the material as extracted from the field trial, was carried out.

With the aim to characterize the microstructure of lime-treated high plasticity clay, mercury intrusion porosimetry (*MIP*) tests were carried out on a specimen sampled soon after the in situ compaction and treated in

[a] Corresponding author: marco.rosone@unipa.it

DOI: 10.1051/e3sconf/20160914020

controlled environmental conditions (T = 20°C and $U_r \geq$ 90%) for at least 18 months.

Finally, in order to assess the stability in time of the geotechnical characteristics, the paper presents test results of repeated wetting and drying cycles carried out in a suction-controlled oedometric device and by means of the vapour equilibrium technique.

2 Material properties and experimental programme

The material was recovered from an experimental embankment built in the Favarella district, in the province of Caltanissetta, Sicily. Some geometric and constructive details of the experimental embankment are reported in [9]. Before the treatment and the compaction stage (corresponding to the Standard Proctor energy level), the clay was partly disaggregated by means of a motor grader, to obtain a material with apparent grain size distribution through d = 31.5 mm and d = 4.76 mm sieves, respectively, equal to $p_{31.5}$ = 100% and $p_{4.76}$ = 63%.

Undisturbed samples used for laboratory experiments were taken soon after the construction of the embankment, by means of metal thin-walled tubes inserted at depths corresponding to 0.30 ÷ 0.60 m and 0.90 ÷ 1.20 m from the finished surface of the embankment. Spreading controls carried out with measurement of the mass of lime spread on the surface, have shown that the quantity of lime considered as actually used in the layers is equal to 2.3 % (Airò Farulla et al., 2014). The initial consumption of lime, obtained from laboratory tests was slightly less than 2% [10, 11]. The tested samples, recovered at the time of the construction of the embankment, were wrapped in multiple layers of plastic film, and treated for a period of 18 months in an air-conditioned cabinet with a temperature T = 20°C and relative humidity $U_r \geq$ 90 %. The following properties were determined for the tested samples: grain size distribution consisting of gravel fraction f_{gravel} = 1 ÷ 6%, sandy fraction f_{sand} = 9 ÷ 17%, silty fraction f_{silt} = 39 ÷ 59%, clayey fraction f_{clay} = 24 ÷ 52%; liquid limit w_l = 51÷53%, plasticity index PI = 24 ÷ 28% and activity index I_a = 0.54 ÷ 1.00; soil specific weight γ_s = 26.3 ÷ 26.4 kN/m^3; water content w = 17.6 ÷ 21.8%; dry unit weight γ_d = 15.9 ÷ 17.4 kN/m^3; void ratio e_0 = 0.52 ÷ 0.65.

MIP tests were performed using a porosimeter (Pascal 140–240 series, Thermo Scientific Corp.) attaining a maximum intrusion pressure of 200 MPa, which corresponds to an entrance pore diameter of approximately 7 nm. Macropores were detected at the beginning of the tests in the low pressure unit operating between 0 and 400 kPa. The advancing non-wetting contact angle between mercury and the clay minerals was assumed to be 140° [12].

MIP tests were carried out on an untreated sample, compacted at optimum standard Proctor condition (γ_d =16.1 kN/m^3 and w = 20.3 %), and on a lime treated clay sample coming from the embankment.

Samples for the *MIP* tests were dehydrated by means of a freeze-drying technique consisting in quick freezing the samples by dipping them in liquid nitrogen (boiling temperature -198°C) and sublimation with vacuum pump at -60°C for 24 h.

In order to assess the water retention properties of the treated material in a wide range of suction, both in drying and in wetting paths, various techniques for suction control were used. The axis translation technique (air overpressure method) applied in an oedometric apparatus was used to control matric suction in the range of 0.01 ÷ 0.80 MPa, whereas the vapour equilibrium technique was used to impose total suction in the range of 2 ÷ 110 MPa. When using the vapour equilibrium technique, the stationary condition was considered to have been achieved when the difference in weight of the specimen between two successive measurements, carried out at a distance of one week, was lower than 0.1 %.

A fluid displacement technique was used to measure the volume of each tested specimen after equalization at the imposed suction with the vapour equilibrium technique.

3 Results analysis

The most relevant results of the MIP are reported in terms of the cumulative volume of intruded mercury and the value $\Delta(V_i/V_{tot})/\Delta logd$, which expresses the frequency of the pores, as a function of equivalent diameter d. Figure 1 shows how the untreated compacted clay presents a typical double porosity pore size distribution, characterized by a very well marked peak in the field of micropores (d = 0.5 μm) and a uniform distribution in the macropore field, that is the range of diameter between 3 and 100 μm. In particular, assuming as a boundary limit the diameter d = 1 μm, from the cumulated volume curve (Fig. 1a) it can be calculated that macropores characterize a little less than 20 % of the total intruded porosity. These results are typical for natural clay compacted in optimum conditions [13].

The intruded volume in the treated sample increases due to the lower state of compaction (γ_d =16.5 kN/m^3). The treated clay has a bimodal pore size distribution, with a peak slightly more marked for $d \geq$ 10 μm, as could be expected for an aggregated structure. The modal value of the diameter in the field of the macropores is equal to about 60 μm, while the total volume of macropores is more greatly increased (about 75 mm^3/g) than the untreated clay (about 30 mm^3/g) .

Minor variations in terms of intruded volume are found in the field of micropores even though the pore distribution is somewhat different. In the field of micropores, untreated clay shows the modal value 0.41 μm while the treated clay has a much lower modal value (d = 0.06 μm). Furthermore, treatment with lime reduces the micropores intruded volume (from 159 to 147 mm^3/g).

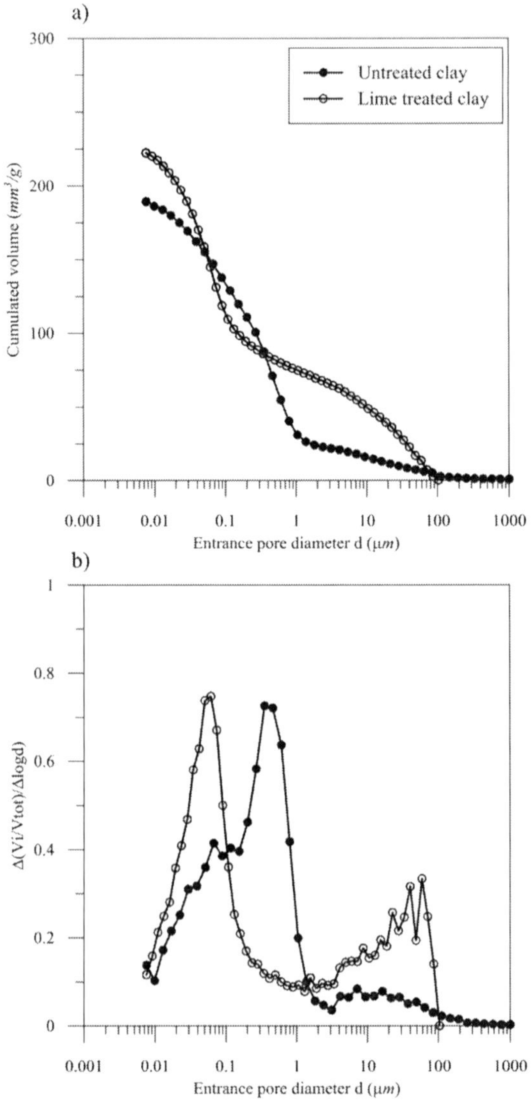

Figure 1. Cumulated intruded volume (a) and pore size frequency (b) as a function of equivalent pore diameter for untreated clay and lime treated clay.

The results of the mechanical tests carried out during the early stage of the research, highlight the already well-known beneficial effects of the treatment on the strength and deformability of the clay. For instance in Figs. 2 and 3 the oedometric tests results are reported with the purpose of highlighting the effect of saturation and the successive loading/unloading paths on the volumetric behaviour of treated and untreated clay compacted with laboratory procedure. During the initial phase of saturation, obtained by subjecting the specimen immersed in water to a total vertical pressure σ_v = 10 kPa, the untreated material underwent a remarkable swelling, corresponding to axial strain, ε_a, equal to -5.82% and a void ratio variation Δe = 0.07, while the embankment material essentially kept its volume intact (ε_a = -0.45%, Δe = 0.01) (Fig. 2).

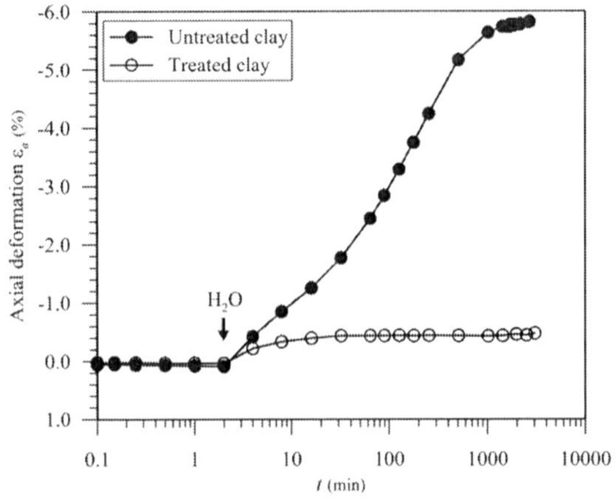

Figure 2. Swelling during saturation stage under oedometric condition (σ_v = 10 kPa).

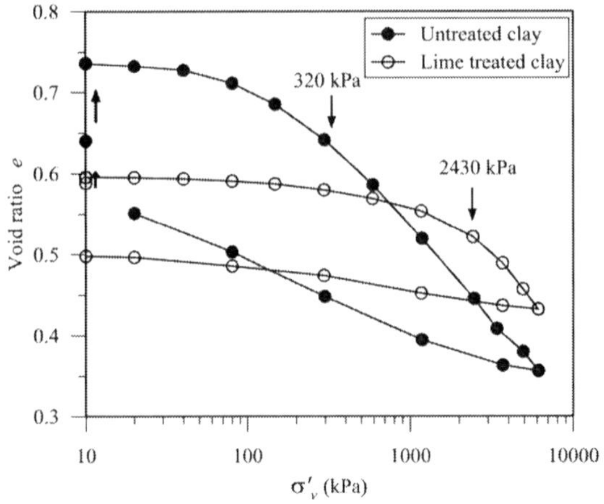

Figure 3. Oedometric curves for compacted untreated and in situ treated clay.

The results of the subsequent stages of loading and unloading expressed in terms of *e-log* σ'_v are reported in Fig. 3. The yield stress, determined by the Casagrande method, undergoes a strong increase, from $\sigma'_{v,max}$ = 320 kPa for the untreated material, to $\sigma'_{v,max}$ = 2430 kPa for the embankment material. However, it is observed that, passing the yield strength, the slope of the straight portions of the two oedometric curves is practically the same (c_c = 0.249 ÷ 0.258), even if the swelling coefficient c_s of the compacted untreated clay (c_s = 0.078) is about three times that of treated clay (c_s = 0.024).

Although the improvement of the properties of the treated clay is a well-known result, the stability of such properties in time, especially with seasonal variation in weather, at the moment appears to be unclear and, at the same time, a key topic to evaluate the durability of the work.

Fig. 4 shows the results, expressed in terms of void ratio and degree of saturation S_r, as a function of applied matric suction *s*, of the cycling matric suction tests performed at constant net vertical pressure ($\sigma_{v,net}$ = 50

 DOI: 10.1051/e3sconf/20160914020

kPa). The range of variation of suction during the test, s = 0.01 ÷ 0.8 MPa, was selected because minimum variation of water content (Δw = -0.08%) at constant void ratio was measured during first equalization at 0.8 MPa. During the first cycle, the variation of suction was performed by steps while in subsequent cycles, the maximum variation of suction was applied instantaneously.

The evolution of the void ratio as applied matric suction varied (Fig. 4a) shows that during the wetting and drying cycles the sample undergoes a very low void ratio variation (Δe = 0.002), reversible in nature. During the first wetting, as well as during the subsequent, the specimen reaches almost complete saturation while in the following drying it starts significantly to desaturate for suction higher than 0.05 MPa.

Variations in the degree of saturation take a different course varying the considered cycle. During the first cycle, the degree of saturation differences, at constant matric suction, are linked to the effect of the hydraulic hysteresis [14]. In subsequent cycles, the cyclical variations of degree of saturation is reduced a lot, although it should be noted a slight tendency to reduce the degree of saturation in the third and fourth round of drying (ΔS_r = -0.003). Then, it can be concluded that the degree of saturation S_r, as it is shown in Fig. 4b, cyclically varies in the range between 0.92 and 1.00 without significant variation in successive cycles of suction.

On the basis of the collected results, it is possible to claim that the processes inducing a reduction of suction do not intervene significantly on the hydro-mechanical behaviour of such material. Different behaviour can be observed in the case of wide cyclic variation covering a suction range greater than the level of suction operating on the specimen as a result of the compaction process and following treatment.

In order to highlight this peculiar behaviour, the results of several cycles of wetting and drying applied by varying total suction between 2 and 110 MPa (Fig. 5) and 32 and 110 MPa (Fig. 6) are reported. The evolution of the volumetric strain ε_v (Figs. 5a and 6a) due to cyclic suction variations is characterized by the accumulation of significant deformations of shrinkage. Always in the same diagrams, it can be observed that these deformations occur, mainly, in the first cycle of wetting and drying. Such behaviour tends to become reversible in nature, although a tendency to cumulate shrinkage deformations during the last cycle of the sample subjected to 32 and 110 MPa is recognized.

The evolution of the water content w of the specimens (Figs. 5b and 6b) is characterized by different behaviour. During the first equalization stage at 2 MPa in the series of cycles between 2 and 110 MPa, the specimen slightly decreases the water content (Δw = -0.01%) at constant volume and only after the first equalization stage at total suction equal to 110 MPa does it significantly reduce the water content (Δw = -15.7%). The specimen subjected to cycles between 32 and 110 MPa continuously reduces the water content after the first equalization to 32 MPa (Δw = -13.3%). These results highlight that the initial total suction can be assumed to be equal to 2 MPa.

However, the resulting reduction of water content is generally very significant during the equalization stage of the first cycle while it is more modest, but still significant in the second and the third cycles.

The irreversible variations of water content occurring almost exclusively in the first cycle of suction, are primarily due to the hydraulic hysteresis phenomenon.

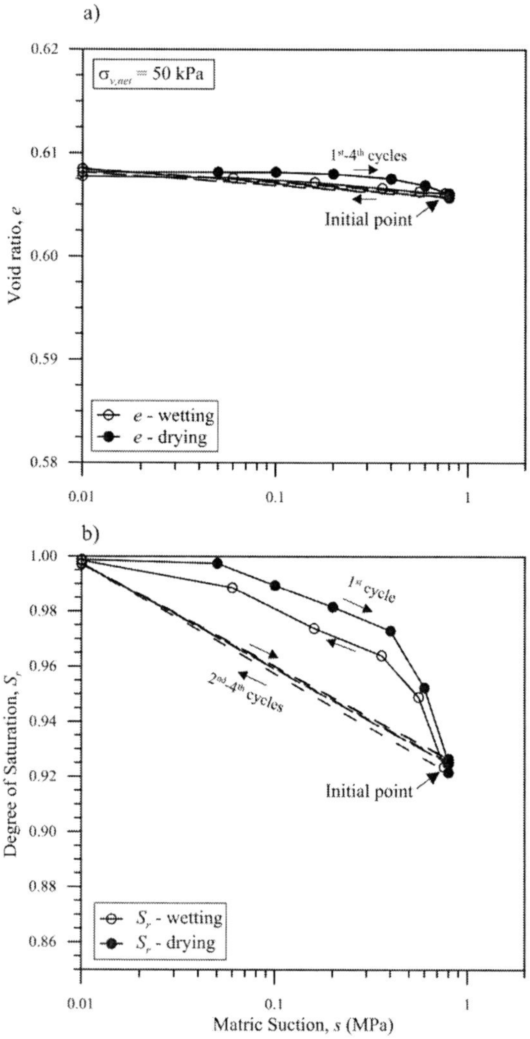

Figure 4. Void ratio, e, (a) and degree of saturation, S_r, (b) during suction cycle between 0.01 and 0.80 MPa.

Volumetric deformation variation measured in the phases of wetting and drying the individual cycles are represented in Figs. 5c and 6c. In the first equalization step at the suction of 32 MPa (Fig. 6c) the sample undergoes a pronounced deformation of volumetric shrinkage of the order of 4%. As already reported, the specimen subjected to the suction of 2 MPa (Fig. 5c) does not undergo a significant volumetric shrinkage ($\Delta\varepsilon_v$ = 0.02%).

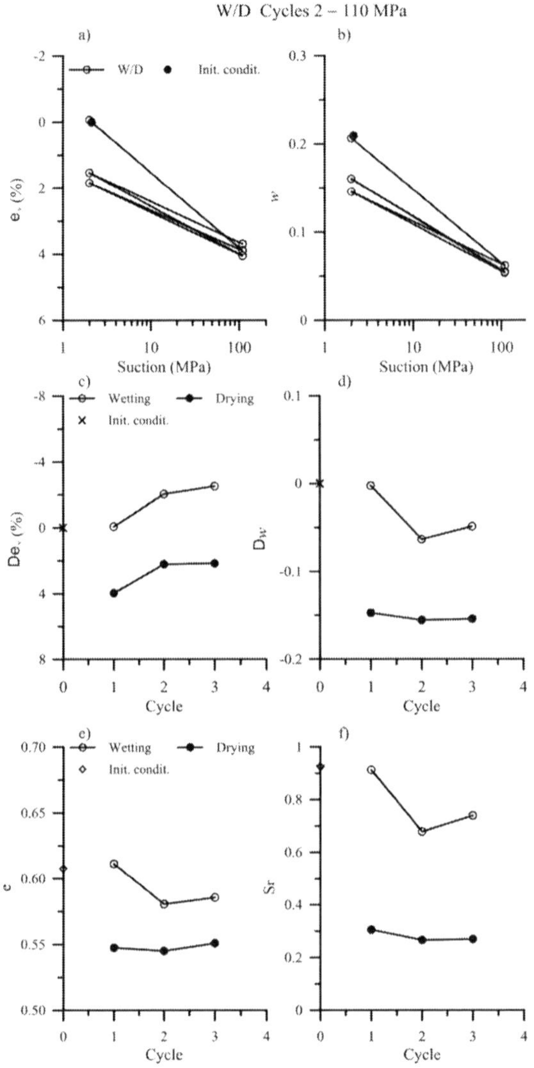

Figure 5. Evolution of Volumetric deformation, ε_v, (a), water content w (b), volumetric deformation changes $\Delta\varepsilon_v$ (c), water content variations Δw (d), void ratio e (e) and degree of saturation S_r (f) during suction cycles between 2 and 110 MPa.

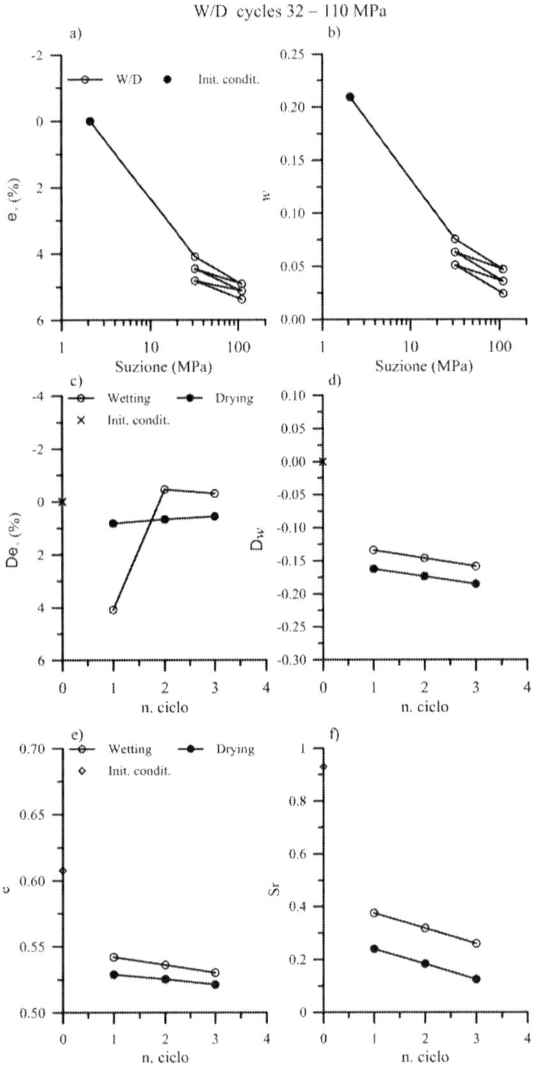

Figure 6. Evolution of Volumetric deformation, ε_v, (a), water content w (b), volumetric deformation changes $\Delta\varepsilon_v$ (c), water content variations Δw (d), void ratio e (e) and degree of saturation S_r (f) during suction cycles between 32 and 110 MPa.

In the following drying at 110 MPa a further deformation (volumetric shrinkage) occurs that obviously deeply influences behaviour in the subsequent cycles. Drying at 110 MPa produces volumetric shrinkage of about 5%, which is not totally recovered in later wetting stages. In fact, after the second cycle, volumetric deformation changes are greatly reduced and they show opposite signs, i.e. swelling deformation in wetting and shrinkage deformation in drying. In addition, absolute values of deformation are very close and are further reduced in the third cycle.

However, it seems that volumetric deformation increases with increasing amplitude of the suction cycle and the tendency of the treated material to move towards an equilibrium characterized by reversible deformations, does not appear evident at higher suction levels.

Figs. 5c and 6c represent the evolution of water content variation Δw as a function of the number of cycles assuming the following convention: the volumes of water expelled from the sample (i.e. water content reductions) are considered negative and those absorbed (water content increasing) positive.

In the first equalization step a considerable loss of water content was observed for the specimen equalized to the suction of 32 MPa (Fig. 6d). Water content variations Δw are negligible for the specimens equalized to the suction of 4 MPa (Fig. 5d). In the first cycle, the reduction in w during the drying step is higher (in absolute value) than that of the phase of wetting, with a progressive reduction of w, as effect of the hydraulic hysteresis [14]. At the same time, despite the strong volumetric shrinkage, a reduction in the degree of saturation with the continuation of the cycles occurs (Figs. 5f and 6f).

In the subsequent cycles, the volumes of water exchanged tend to decrease, although a tendency to reversible behaviour, particularly in terms of the degree of saturation, is not less clear.

The evolution of the void ratio with the cycles of suction (Figs. 5e and 6e) reflects, of course, the evolution of the volumetric strain already examined with the diagrams in Figs. 5c and 6c. After the second cycle series, the volume variations are quite moderate.

4 Conclusions

The analysis of obtained results from tests performed in the suction controlled oedometer suggests that the clay treated with lime does not suffer significantly from cyclic processes of wetting and subsequent drying which develop in a range of matric suction, between 0.8 to 0.01 MPa, lower than the initial matric suction of the specimen.

Conversely, the treated clay undergoes in a particular way, the processes of drying developing in a range of total suction higher than the initial total suction of the specimen. In fact, cyclic variations of total suction, which determine a significant drying, give rise to significant irreversible deformations of shrinkage because during the wetting stages, the material is unable to recover most of the deformations developed in the previous drying stages.

The volumetric behaviour undergoes an almost reversible pattern when the specimens are subjected to cycles between 2 and 110 MPa while they tend to accumulate shrinkage deformation even when the total suction imposed in the wetting phase is maintained at high values (32 MPa).

The observed behavior may be interpreted with reference to the various mechanisms that control the volume response of the material to the double porosity, then considering the interactions between the microstructure and the macrostructure [15]. The triggering mechanism is due to the mutual sliding of the aggregates for the reduction of the shear resistance along the areolae of contact. This reduction can be determined by the breaking of the bonds of cementation between the aggregates of clay particles, triggered by the great loss of water, which is fundamental for the formation and permanence of cementing pozzolanic products and, only after, by the disappearance of the menisci and of their stabilizing effect.

References

1. D.I. Boardman, S. Glendinning, C.D.F. Rogers. Géotechnique **51**,6:533-543 (2001)
2. P. Croce, G. Russo G., Proc. XIII European Conf. of Soil Mechanics and Geotech. Engineering (ECSMGE) (2003)
3. F.G. Bell, Eng. Geol. **42**(4):223-237 (1996)
4. J. Locat, M.A. Berube, M. Choquette. Can. Geotech. J. **27**(3), 294-304 (1990)
5. S. Horpibulsuk, R. Rachan, A.Chinkulkijniwat, Y. Raksachon, A. Suddeepong, Construction and Building Materials, **24**, 10, 2011-2021 (2010)
6. X. Zhang, M. Mavroulidou, M.J. Gunn, Eng. Geol., **193**, 320-336 (2015)
7. B. Celauro, C. Celauro, C., *Construction and Building: Design, Materials, and Techniques*, 1-36 (2015)
8. G. Stoltz, O. Cuisinier, F. Masrouri. Applied Clay Science, **61**, 44-51 (2012)
9. C. Airò Farulla, B. Celauro, C. Celauro, M. Rosone, Ingegneria Ferroviaria **69**(9), 729-752 (2014)
10. G. H. Hilt D.T. and Davidson Highway Research Board Bull., 262, 20-32 (1960)
11. J. Eades, J. R. Grim, Highway Research Record 139, 61-72 (1966)
12. E. Romero, P. Simms, P. Geotechnical and Geological Engineering 26,705-727 (2008)
13. P. Delage, M. Audigier, Y. Cui, M.D. Howat, M. D. Canad. Geotech. J. **33**, 1, 150–158 (1996)
14. C. Airò Farulla, A. Battiato, A. Ferrari. In Proc. 5th Int. Conf. on Unsaturated soils, Barcelona, 417–422 (2011)
15. E.E. Alonso, J. Vaunat, A. Gens, A. Eng. Geol., **54**(1–2), 173–183. (1999)

DOI: 10.1051/e3sconf/20160914021

Anisotropic behaviour of compacted clayey silt subjected to hydro-mechanical paths

Carlos Buenfil[1], Enrique Romero[2,a], Antonio Lloret[2] and Antonio Gens[2]

[1]Universidad Autónoma de Campeche and Instituto Tecnológico de Campeche, Mexico
[2]Universitat Politècnica de Catalunya, Spain

Abstract. Compaction induces anisotropy on soil deformational response due to preferential straining associated with the fabrication process. An experimental insight into the stress-strain response of Barcelona clayey silt is provided here. The material was statically compacted at low dry density to induce high collapsibility upon wetting (dry density 1.48 Mg/m^3, water content 12%). Two types of controlled-suction tests were carried out: a) constant suction radial paths on as-compacted samples following different stress ratios (deviator stress / mean net stress: q/p''), including isotropic and K_0 stress paths; and b) the same type of radial stress paths but performed after saturation under low confining stress. These paths were intended to define the as-compacted and saturated yield surfaces. The experimental data on as-compacted state is satisfactorily interpreted adopting an anisotropic yield surface with an inclination linked to the compaction under oedometer conditions. Development of shear strain was observed during isotropic loading of the compacted samples. This anisotropy that the soil exhibits at the end of the compaction vanishes after collapse by saturation and as plastic volumetric straining develops.

1 Introduction

The effect of the anisotropy derived from the compaction process is one aspect of the behaviour of unsaturated compacted that has received relatively less attention. Despite the significance in practical applications, this initial anisotropy and its evolution along general stress paths have not been usually taken into account in the constitutive formulations proposed to model the behaviour of compacted soils. This article presents the results of a study on the deformational behaviour of a clayey silt, which was compacted under oedometer conditions and was subjected to different stress paths and suction changes.

Compaction can induce anisotropy on soil response due to important and preferential straining associated with the fabrication process, particularly when using static compaction under oedometer conditions. This anisotropic deformation pattern induces a transversally isotropic structure, resulting in larger stiffness and strength in the axial direction.

There are a number of experimental studies on unsaturated compacted soils [1-5]. Cui and Delage [1] were probably the first authors to present results from anisotropic loading tests on a compacted silt showing yield surfaces at constant suction, which were inclined with the major axis coinciding with the K_0 (at rest earth pressure coefficient) line. This behaviour is similar to that observed in natural saturated soils. Romero and Jommi [5] showed and modelled the distortional effects that were expected to accompany volumetric strains of one-dimensionally compacted soils during wetting processes under isotropic stress conditions. The changes in the direction of plastic strain increment vectors observed experimentally by these authors upon wetting contributed to a better definition of the anisotropy evolution.

2 Characterization of the tested soil

Laboratory tests were performed on a low plasticity clayey silt from Barcelona. It has a liquid limit of w_L=28%, a plastic limit of w_P=19%, a clay-size fraction ≤ 2 μm of 19% and a silty fraction of 47%. Density of solids is 2.67 Mg/m^3.

Triaxial samples (38 mm diameter and 76 mm high) at a prescribed water content of 12.0% were prepared at a dry density of ρ_d=1.48 Mg/m^3 (degree of saturation of S_r=40%), by one-dimensional static compaction. Maximum fabrication vertical net stress was 0.27 MPa. Lateral stresses were measured by an active lateral stress system, resulting in a lateral stress coefficient at rest of K_0=0.48 [6]. Suction after compaction, s=270 kPa, was measured using a high-range tensiometer. A relatively low dry density was selected with the aim of inducing an open structure, which was susceptible to undergo important void ratio changes on loading and in turn to induce important changes of the hydro-mechanical properties of the soil.

Retention curves of the clayey silt were obtained using a controlled-suction oedometer cell. Air entry

[a] Corresponding author: enrique.romero-morales@upc.edu

 DOI: 10.1051/e3sconf/20160914021

pressure values ranged between 8 kPa and 20 kPa for dry densities of 1.48 and 1.70 Mg/m^3, respectively [6, 7].

3 Experimental program

3.1. Controlled-suction equipment

Controlled-suction tests were performed using an automatic triaxial cell [5, 6]. The deformation response was monitored with local axial (miniature LVDTs attached to the membrane) and radial (electro-optical laser system mounted on two diametrically opposite sides) transducers.

Suction was applied simultaneously via axis translation technique on both ends of the sample, maintaining a constant air pressure and modifying the water pressure. Both top and bottom platens presented a combination of two porous discs: a peripheral annular coarse one connected to the air system and an internal disc with a high air-entry value ceramic (1.5 MPa) connected to the water system.

Two stepper motors operating air pressure regulators were used to continuously control the deviator and confining stresses. The stepper motors and the measurements of 14 transducers were managed by an automatic data acquisition and control system that allowed applying net stress and suction paths, as well as performing strain controlled tests [6].

Tests on saturated specimens were conducted in a conventional automatic triaxial apparatus.

3.2 Stress paths followed

Various types of tests were carried out to provide results to study different aspects of soil behaviour minimising the testing time.

Equalisation stages to bring the as-compacted suction (A) to the different target suctions (B) are show in Figure 1. Equalisation was assumed to be achieved once water flow was inducing water content rates lower than 0.04%/day [5]. All the equalisation stages were performed under a constant mean net stress $p''=(p-u_a)=23$ kPa and a deviator stress $q=10$ kPa.

Constant suction drained compression tests following different stress ratios, $\eta=q/p''$, including isotropic ($\eta=0$) and K_0 ($\eta=0.75$) stress paths, were performed. These paths were intended to define the as-compacted yield surfaces, as well as the direction of the plastic strain vectors (paths B-C in Figure 1). The increase of mean net stress was applied at a stress rate of 1.8 kPa/hr, which was assumed adequate to avoid suction changes.

Other types of tests were performed following similar stress paths but after saturating the specimens under low confining stress (9 kPa for test SI1 and 21 kPa for the remaining tests). The stress paths followed under saturated conditions are indicated in Figure 2.

4 Results

4.1. As-compacted behaviour

Figure 3a shows the evolution of void ratio and water ratio (water volume to solid volume: $e_w=\rho_s w/\rho_w$) undergone by the soil during the different isotropic compression paths at constant suction (s=10, 100, 600 kPa and saturated state). It can be observed that the evolution of void ratio displays clear pre- and post-yield domains. Yield stresses increase at higher suctions in accordance with the elastoplastic model of [9]. The post-yield compressibility in the figure reduces at higher suctions, also in accordance with the model. For the different tests, the compressibility in the elastic domain, $\kappa(s)$, does not exhibit significant changes with suction. In contrast to the test under saturated conditions, very small changes of water ratio were observed during isotropic compression for suctions of 10, 100 and 600 kPa. The initial position of the loading-collapse yield locus (LC curve in [9]) that corresponds to the as-compacted state is shown in Figure 3b.

Figure 4 shows the evolution of the shear ε_s and volumetric ε_v strains, as well as the ratio between increments in shear strain and volumetric strain $\delta\varepsilon_s/\delta\varepsilon_v$ of the different samples during the continuous isotropic loading paths under constant suction. These isotropic compression paths were initiated on samples that showed no significant volume changes and remained within the elastic region during the initial suction equalisation stage, and therefore keep a fabric close to what they had at the final state of compaction but at different suctions and water contents.

The evolution of $\delta\varepsilon_s/\delta\varepsilon_v$ with $(p-u_a)$, developed during isotropic compression, can be used to characterise the degree of anisotropy of the compacted soil. The results plotted in Figure 4 suggest a behaviour closer to the isotropic one when suction decreases. Observed values of $\delta\varepsilon_s/\delta\varepsilon_v$ at yielding states show a consistent pattern of response with suction ($\delta\varepsilon_s/\delta\varepsilon_v$ of -0.082 for s=0 kPa, -0.114 for s=10 kPa, and -0.206 and -0.214 for s=100 kPa and s=600 kPa), suggesting a slightly greater transverse anisotropy when the soil state is closer to the initial compacted condition (s=270 kPa).

Strain ratio $\delta\varepsilon_s/\delta\varepsilon_v$ decreases when $(p-u_a)$ increases, and it converges to very small values at the end of the isotropic loading path (-0.004 and -0.076 for suctions s=0 and s=10 kPa, respectively), suggesting that the soil reaches nearly isotropic deformational response as plastic volumetric strain accumulates. A test performed s=100 kPa fails to stabilise, ending with a strain ratio of about -0.12. This pattern of behaviour can be explained using the concepts of elasto-plasticity in soils with transverse anisotropy (e.g. [5, 10]), where an isotropic stress increase induces the expansion and rotation of the yield surface towards a more isotropic orientation.

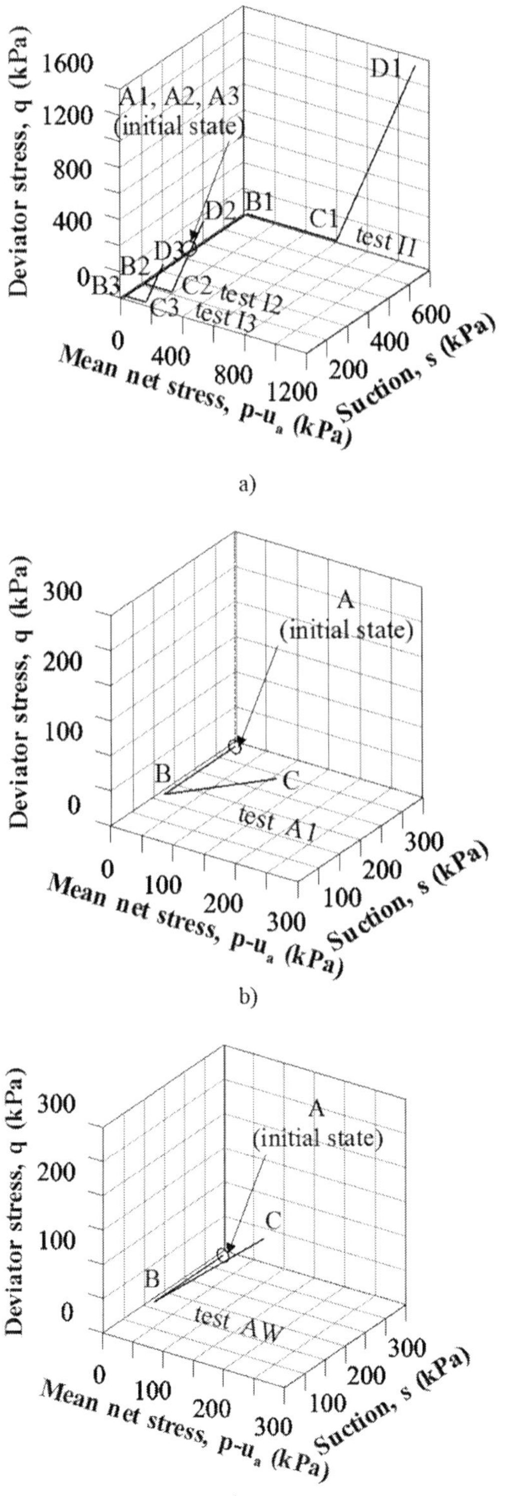

Figure 1. Stress paths in $(p\text{-}u_a)\text{-}s\text{-}q$ space: a) anisotropic η=constant loading after initial isotropic compression at constant suction (I1, s= 600 kPa; I2, s=100 kPa and I3, s=10 kPa); b) test A1, anisotropic loading (η=0.38) at s=100 kPa; c) test AW, anisotropic loading (η=0.75) at s=100 kPa.

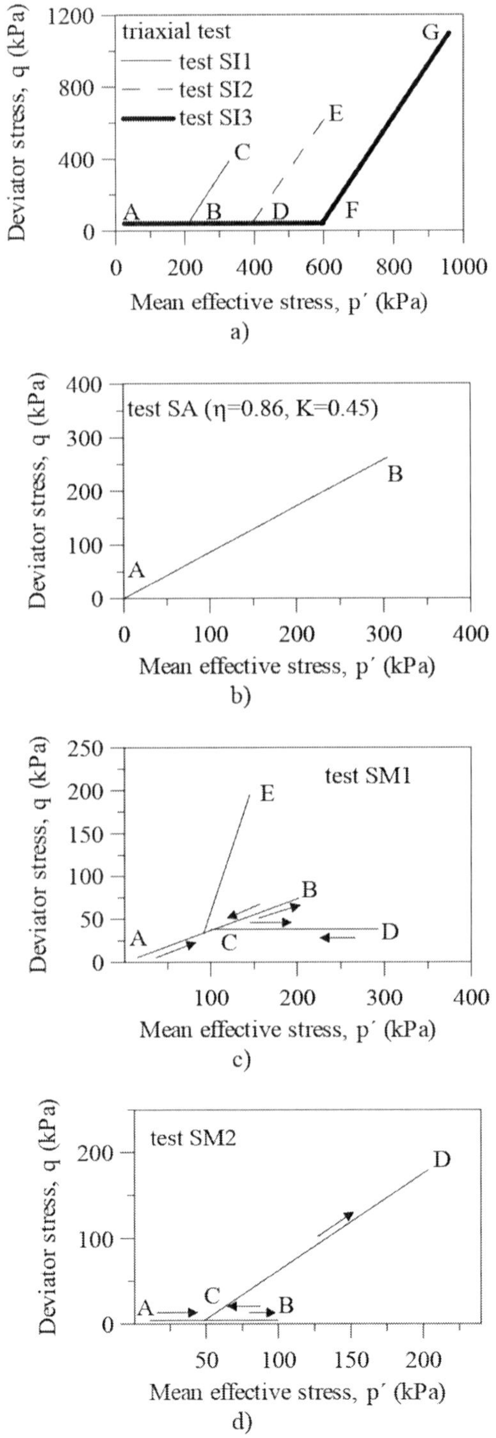

Figure 2. Stress paths in saturated tests. a) SI1, SI2 and SI3 conventional CD triaxial tests ($\Delta q/\Delta p'$=3); b) test SA, anisotropic loading (η=0.86); c) test SM1, anisotropic loading A-B (η=0.36), unloading B-C (η=0.36), isotropic loading-unloading (C-D-C) and conventional loading ($\Delta q/\Delta p'$=3); d) test SM2, isotropic loading-unloading (A-B-C) and anisotropic loading (η=0.86).

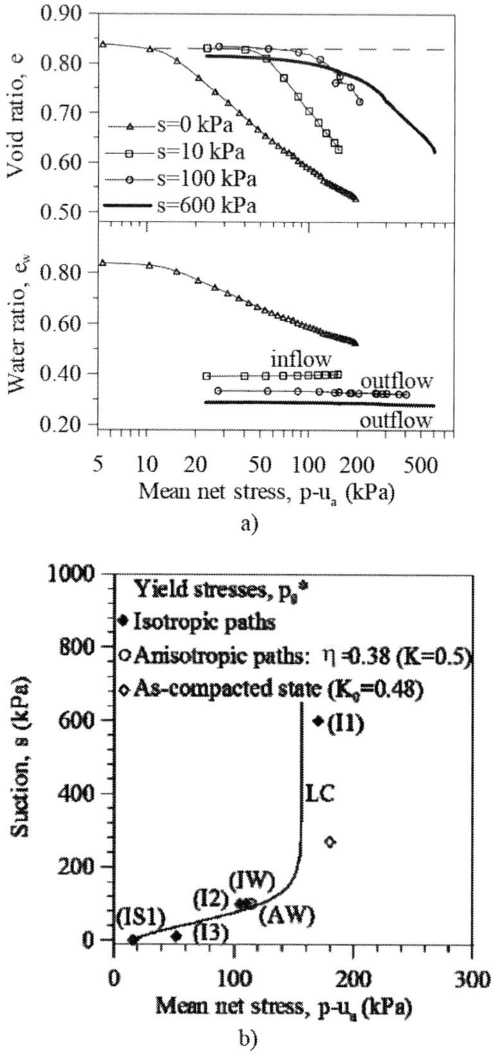

Figure 3. a) Changes in void ratio and water ratio during isotropic compression paths; b) yield stresses and loading-collapse LC yield curve for as-compacted state.

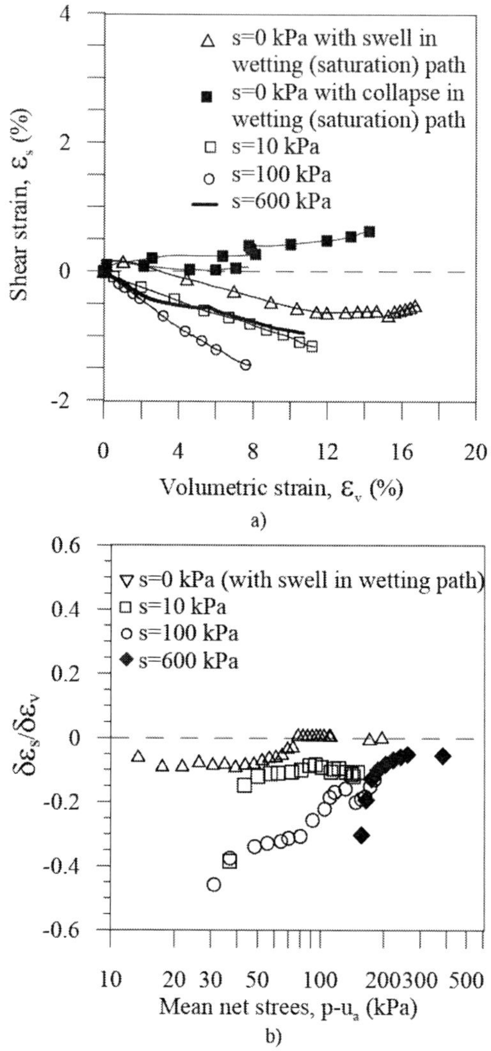

Figure 4. a) Shear and volumetric strains evolutions for tests at different suctions; b) evolution of $\delta\varepsilon_s/\delta\varepsilon_v$ during the isotropic compression paths at different suctions.

Figure 5 shows results from anisotropic (A1 and AW) and isotropic (I2) compression tests under a constant suction of 100 kPa. As expected, increasing the ratio η, increases the axial strain and thus increases the shear strains ε_s, which change from negative values at η=0 to positive values at η=0.75. In the compression test at η=0.38, the observed shear strains are close to 0 and $\delta\varepsilon_s/\delta\varepsilon_v$ stabilise at values close to zero when the yield stress is reached. In the compression path at η=0.75 (close to K_0 condition) radial strains are lower than the axial ones, with $\delta\varepsilon_s/\delta\varepsilon_v$ around 0.5 at yield stress (close to the condition of ε_r=0 and $\delta\varepsilon_s/\delta\varepsilon_v$=2/3, which are characteristic of oedometer conditions). During the isotropic compression paths the radial strains are greater than the axial ones with $\delta\varepsilon_s/\delta\varepsilon_v$= -0.21 at the yield point.

To describe the shape of the yield surface, the following expression can be used in terms of mean net stresses [12]:

$$[q-\alpha(p''+p_s)]^2 = M^2 (p''+p_s)(p_m-p'') \qquad (1)$$

where α and M indicate the slope of the yield surface and the critical state line, respectively, and p_s and p_m indicate the minimum and maximum values of the mean net stress p''. Parameters p_s and p_m increase with suction and the angle α decreases with plastic strains.

As indicated above, for s=100 kPa the soil exhibits a behaviour with near-zero values of shear strains in the path with stress ratio of η=0.38. This suggests that the path approaches the angle of inclination η=α of the yield surface [9, 10]. Figure 6 shows the location of the yield points obtained in the compression tests performed at s=100 kPa. In the same figure, an estimation of a yield surface is included. Values of 30 and 114 kPa can be estimated for parameters p_s and p_m, respectively, which are also plotted in the figure.

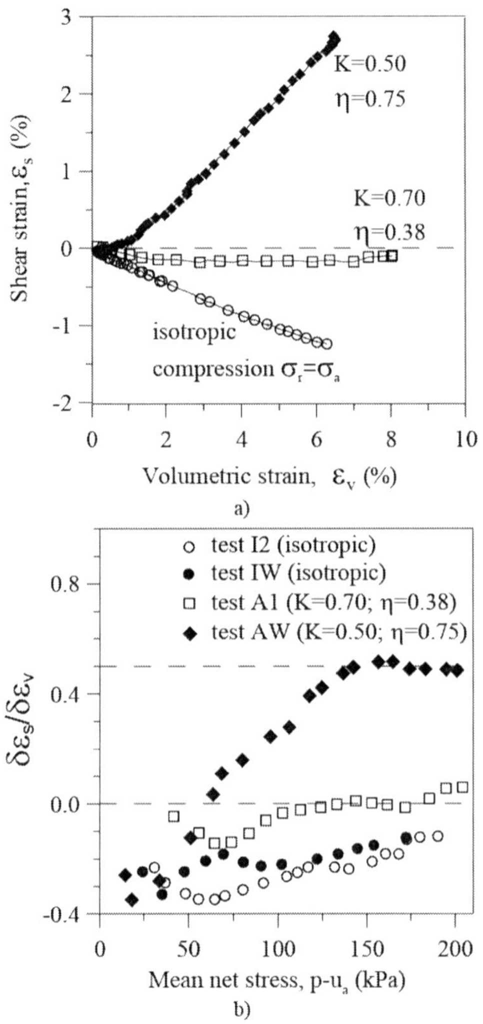

Figure 5. a) Evolution of shear and volumetric strains for tests at different stress ratios (η) and s=100 kPa. b) Evolution of $\delta\varepsilon_s/\delta\varepsilon_v$ during the compression paths at s=100 kPa and for different stress ratios (η).

4.2 Behaviour of soil after saturation

Volumetric strains were observed during the saturation process under low confining stress. In test SI at a mean net stress of 9 kPa a swelling strain of -0.5% was measured on soaking. In the remaining tests, where the saturation process was made under a higher confinement stress (21 kPa), collapse volumetric strains ranged between 1.6% and 4.9 %.

Figure 7a presents the yield points in the space p'-q for different paths under saturated conditions. These data are associated with a yield surface f_0, which corresponds to the initial state after saturation. The shape of yield surface is consistent with that expected for a soil with isotropic mechanical behaviour. In the tests in which the sample collapse during saturation, values of effective yield mean stress p'_0 ranging between 29 kPa and 31 kPa were obtained. On the other hand a lower value p'_0=11 kPa was obtained in test SI1 that underwent swelling during saturation. This suggests that collapse strains induced an increase in p'_0 associated with hardening [9].

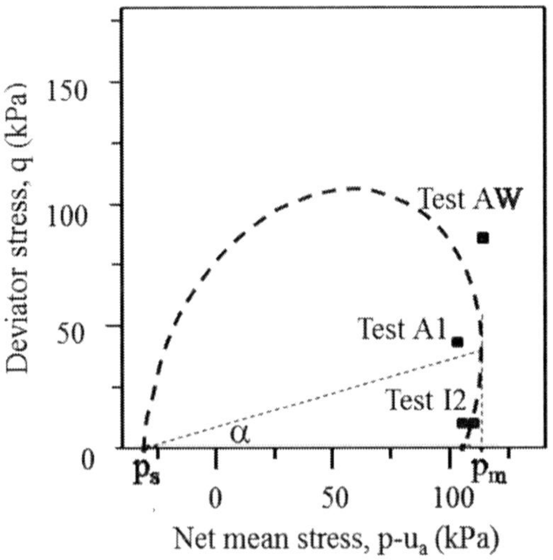

Figure 6. Yield points and estimated yield surface for different compression paths at a controlled suction of 100 kPa.

The modified Cam-Clay model has been used to plot the yield surfaces using a value of M=1.2 (drained friction angle of 30°), which corresponds to the value obtained in conventional drained triaxial tests performed on saturated samples [6]. The good agreement suggests that, after collapse, soil behaves like an isotropic material. The anisotropy that exhibits the soil at the end of the compaction (as suggested by the mechanical behaviour observed in unsaturated samples) appears to be erased after collapse by saturation.

Figure 7b provides information on the evolution of $\delta\varepsilon_s/\delta\varepsilon_v$ with the increase of mean effective stress p' from the samples that collapse during saturation. All of the p'-$\delta\varepsilon_s/\delta\varepsilon_v$ curves tend to stabilise at a certain value that depends on the value of the applied stress ratio η. The values of $\delta\varepsilon_s/\delta\varepsilon_v$ are different from those obtained in the paths at s=100 kPa (see Figure 5b).

In the isotropic compression test, the value of $\delta\varepsilon_s/\delta\varepsilon_v$ is approximately 0.1 at yielding at $p'\approx30$ kPa, but it stabilises at values close to 0 when it reaches $p'\approx50$ kPa. In the anisotropic compression test with η=0.36, the value of $\delta\varepsilon_s/\delta\varepsilon_v\approx0.59$, which was measured at yield point (p' = 26 kPa), decreases up to about 0.25 at $p'\approx40$ kPa. The estimated value of $\delta\varepsilon_s/\delta\varepsilon_v$ using the Cam-Clay model for η=0.36 is 0.55. This value is consistent with the one obtained experimentally at yield state. However, its decrease and subsequent stabilisation towards 0.25 cannot be explained by this model. The same occurs for the path at η=0.86 where a value of $\delta\varepsilon_s/\delta\varepsilon_v$=2.46 is predicted by the Cam-Clay model and smaller values around 0.8 are obtained in the tests.

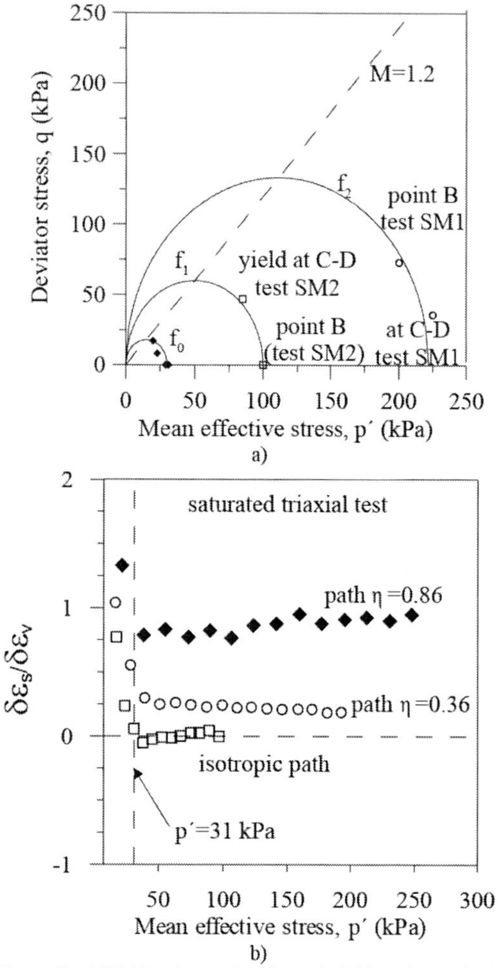

Figure 7. a) Yield points and estimated yield surfaces obtained from saturated samples. b) Evolution of $\delta\varepsilon_s/\delta\varepsilon_v$ with effective mean stress (p').

5. CONCLUSIONS

Performance of suction-controlled triaxial tests on a silty clay has shed light on the effects of the anisotropy set up by the static compaction process. It has been found that the behaviour of the compacted material is well represented by an anisotropic yield surface with an inclination linked to the oedometer conditions applied duding compaction. Significant shear strains develop during the isotropic compression of compacted specimens indicating unambiguously the presence of anisotropy.

It has also been observed that the inclination of the yield surface evolves during the application of radial stress paths and the development of plastic volumetric strains. The changes in the direction of plastic strain increment vectors observed experimentally upon wetting contribute to a better definition of the anisotropy evolution.

It is also interesting to note that the anisotropy that exhibits the soil at the end of the compaction appears to fade after collapse by saturation.

References

1. Y. Cui, P. Delage, Géotechnique **46** (2), 291-311 (1996)
2. S.J. Wheeler, V. Sivakumar, Géotechnique **50** (4), 369-376 (2000)
3. G. Della Vechia, C. Jommi, E. Romero, Int. J. Numer. Anal. Meth. Geomech. **37**, 503-535 (2013)
4. A.M. Hansan, S.J. Wheeler, *Proceedings of the Sixth International Conference on Unsaturated Soils, UNSAT 2014*. Sydney, Australia (2014)
5. E. Romero and C. Jommi. Water Resour Res; **44**,W12412, 1-16 (2008)
6. C. Buenfil, *Caracterización experimental del comportamiento hidromecánico de una arcilla compactada* (in Spanish), Tesis de Doctorado, Universidad Politécnica de Cataluña, Barcelona, España, (2007)
7. C. Buenfil, E. Romero, A. Lloret, A. Gens, *The International Conference from Experimental Evidence Towards Numerical Modeling of Unsaturated Soils. Unsaturated Soils: Experimental Studies,* Weimar, Germany, Springer, ISBN: 3-540-21121-7, 331-342 (2004)
8. C. Rampino, C. Mancuso, F. Vinale, Can. Geot. J. **36,** 1-12 (1999)
9. E.E. Alonso, A. Gens, A. Josa, Géotechnique **40** (3), 405-430 (1990)
10. S.J. Wheeler, A. Naatanen, M. Karstunen M. Lojander, Can. Geot. J. **40**, 403-418 (2003)
11. M. A. Al-Sharrad, *Evolving anisotropy in unsaturated soils: experimental investigation and constitutive modelling*. PhD thesis, University of Glasgow, UK. (2013)

Predicting the volumetric variation due to changes in suction, applied stress and swelling pressure

Roberto Galaviz-González[1,a], Jaime Horta[1], Eduardo Rojas[1], María de la Luz Pérez-Rea[1] and Jaime Robles-Sotelo[2]

[1]DIPFI, Universidad Autónoma de Querétaro, 76010, Santiago de Querétaro, Querétaro, México.
[2]División de Ingenierías, CUCEI, Universidad de Guadalajara, 44430, Guadalajara, Jalisco, México.

Abstract. The settlements produced by the load transmitted to the structure on expansive soils, and those calculated by the classical theories of soil mechanics, are different because the swelling pressure acts inversely to the applied stress. In this paper we describe a procedure to determine a volumetric variation coefficient by hydration (α_h) which considers the expansive soil behaviour. In order to do this, it is necessary to know the soil's initial water content, the swelling pressure, and the applied stress. Soil suction and swelling pressure were measured with filter paper technique and a mechanical oedometer, respectively. Unsaturated undisturbed samples of expansive soil were used. The water content was varied, starting from 0%, with increments of 5.5% to reach 38%. Furthermore, we present a set of curves that show the magnitude of the coefficient (α_h) associated to a water content, and the relationship between applied stress and swelling pressure. The results show that the variation of the coefficient under different initial water contents ranges between 0% to 22%. This is because water is strongly attracted by clay minerals, but this attraction decreases as water layers are father from the surface of clay minerals, thus decreasing its swelling potential.

1 Introduction

Lightweight buildings such as houses, can be damaged if they are built over expansive soils [1]. This is because water content variations induce volume strains on soils. This has led to overestimations and consequently to expensive foundations. This is because soil mechanics classical methodologies consider soils as if they were saturated, hence, they do not consider swelling pressure.

Some methods predict volumetric strains and optimize the design of foundations. These methods were proposed by Zeevaert [2], Nelson and Miller [3], Lytton [4], and ASTM D4546 [5]. Such methods require the swelling pressure as an input parameter. Therefore, various researchers have developed methods to predict the soil swelling pressure. Some of these methods use basic soil properties (indirect methods) and some others need a physical property which is measured in the laboratory (direct methods) [6]. However, prediction of volumetric changes rely heavily on the correct estimation of the swelling pressure.

This paper presents the determination of a coefficient (α_h) associated to an initial water content (ω_i) and the relationship between the applied stress and swelling pressure ($\sigma_{app}/\sigma_{swe}$) measured on conventional laboratory equipment, which can be introduced into soil-structure interaction methods (SSI) in order to facilitate the prediction of soil settlement.

2 Background

2.1 Expansive soils

Expansive soils contain expansive clay minerals such as montmorillonite [7] that exhibit swelling behavior. This volumetric behavior has been recognized to compromise the stability of structures [8]. Light structures built on these soils usually contain structural damages as a result of the volumetric strains of the soils [1, 9]. As these soils exhibit large volumetric strains, it is difficult to predict their behavior [8]. Swelling and expansive potential of soils mainly depend on the type and amount of clay minerals and the amount of water they absorb. When their water content increase, expansive soils swell affecting the structure stability [10].

2.2 Swelling pressure estimation

Some authors have proposed methodologies to estimate soil expansion that require the swelling pressure as an input parameter. Therefore, direct and indirect methods to estimate the swelling pressure have been proposed [6].

2.2.1 Indirect methods

[a] Corresponding author: galaviz.gonzalez.r@gmail.com

The swelling pressure is correlated with: liquid limit, plasticity index, the clay activity, water content, dry density and void ratio [6, 11].

Komornik and David (1969) proposed the following equation to correlate swelling pressure with water content, dry density and liquid limit for natural unaltered clayey soils:

$$\log P_s = -2.132 + 0.0208 w_L + 0.00665 \gamma_d - 0.0269 w_i \quad (1)$$

where: P_s swelling pressure, w_L liquid limit, γ_d initial dry density, and w_i initial water content.

Nayay and Christensen (1974) proposed a statistical relationship to predict swelling pressure as:

$$P_s = \left(2.5x10^{-1}\right)\left(PI\right)^{1.12}\left(C^2/w_i^2\right) + 25 \quad (2)$$

where: P_s swelling pressure, PI plasticity index, C clay content, and w_i initial water content.

Lytton [4] proposed equations (16) and (17) to determine soil volumetric strains $\Delta V/V$.

$$\sigma_v = \left(\frac{2}{3}\right)\sigma_i 10^{\frac{2\varepsilon_v}{\gamma_\sigma(1-f)}}\left(\frac{h_i}{h_f}\right)^{\frac{\gamma_h}{\gamma_\sigma}} - \frac{\gamma_\pi z}{2} \quad (3)$$

$$\varepsilon_v = f(\Delta V/V) \qquad f = (1/Sr) - 1 \quad (4)$$

where: σ_v is the swelling pressure; σ_i is the initial mean stress; h_i and h_f are the initial and final matric suction values respectively; γ_h, γ_σ, and γ_π are compression indexes related to matric suction, mean stress and osmotic suction, respectively. z is the thickness of the soil layer and Sr is the degree of saturation.

Erzin and Erol (2004) carried out standard constant volume swelling tests on a series of statically compacted clayey samples:

$$\log P_s = -5.020 + 0.01383 PI + 2.356 \gamma_d \quad (5)$$

Rani and Rao (2009) studied the influence of the liquid limit, plastic index, water content and dry density on the swelling pressure of four types of clayey soils and proposed the following relationship:

$$\log P_s = -4.3341 + 0.0071 PI + 0.0006 Log \gamma_d + \ldots \\ \ldots + 51.2802 m_c + 0.0037 S_i \quad (6)$$

where: P_s swelling pressure, PI plasticity index, γ_d dry density, m_c water content, and S_i initial surcharge pressure.

2.2.2 Direct methods

Direct methods usually involve laboratory procedures such as one-dimensional settlement measurement devices (oedometer) [12]. There are a number of procedures that can be reviewed in Shuai [12]. The most commonly ones are: 1) Free swell oedometer, 2) Loaded swell oedometer, and 3) constant volume oedometer.

In the free swell oedometer test, the soil specimen is brought in contact with water and allowed to swell freely with an applied load. Then the soil is gradually consolidated back to its original volume. The swelling pressure is defined as the necessary stress to consolidate the specimen back to its original volume [12].

In the loaded swell oedometer test, a number of identical specimens are subjected to different initial applied loads and allowed to swell freely. The resulting final volume strains are then plotted against the corresponding applied stresses. The stress corresponding to zero volume change is termed the swelling pressure [12].

In the constant volume test, a specimen is loaded and immersed in water. The specimen volume is maintained constant throughout the first part of the test by varying the applied load as required. This procedure is continued until there is no further tendency for swelling. The applied stress at this point is referred to as the "uncorrected" swelling pressure [12].

The free swell oedometer test usually results in an upper bound value for the swelling pressure. The loaded swell test gives the lowest value, and the constant volume method gives intermediate results [12].

After determining the swelling pressure, it is possible to estimate the volume strains with existing methodologies. However, the prediction of the volume strains depends heavily on the accuracy of the procedure used to estimate the swelling pressure. Baeza [13] made a comparison between the direct loaded swell oedometer and the indirect method proposed by Lytton, and concluded that the direct method is more feasible to use but, the indirect method provides consistent results.

2.3 Volumetric variation prediction

The total applied stress $\Delta\sigma$ to the soil-structure interface is defined as $\Delta\sigma = \sigma_{swe} - \sigma_{app}$. Here, we have three different possibilities:

Case a) When stress applied on soil mass is less than swelling pressure $\Delta\sigma_{app} < \sigma_{swe}$, then the settlements are negative. Here, the contact surface moves upwards [14].

Case b) The system is in equilibrium and the deformations at the contact surface are theoretically zero $\Delta\sigma_{app} = \sigma_{swe}$ [14].

Case c) When structural stress applied on soil mass is higher than swelling pressure $\Delta\sigma_{app} > \sigma_{swe}$, settlements will be positive but smaller in magnitude than in a non-expansive soil [14]. In this case the coefficient α is determined by the theory of "Recompression by load" from Zeevaert [2]

When a load due to the weight of the building is placed on the foundation, the soil is recompressed and its elastic, plastic and visco-plastic components of deformation can be expressed as [2, 15]:

$$\Delta\varepsilon_c = \Delta\varepsilon_{eo} + \Delta\varepsilon_{po} + \Delta\varepsilon_{vpo} \quad (7)$$

For unloading and total recompression $\Delta\sigma_{oi}$ the secant module M_{co} will be expressed as:

$$M_{co} = M_{eo}\left(1 + k_{po} + k_{vpo}\right) = \frac{\Delta\varepsilon_{eo}}{\Delta\sigma_{co}}\left(1 + \frac{\Delta\varepsilon_{po}}{\Delta\varepsilon_{eo}} + \frac{\Delta\varepsilon_{vpo}}{\Delta\varepsilon_{eo}}\right) \quad (8)$$

For the case of an increment of stress $\Delta\sigma_i$ due to unloading or recompression, the secant modulus can be expressed as:

$$M_{ci} = M_{ei}\left(1 + k_{pi} + k_{vpi}\right) \quad (9)$$

If expressions for M_{ci} and M_{co} are considered, then, taking into account that parameters k_{po} and k_{vpo} are proportional to each other, then for the level of stresses $\Delta\sigma_i$ the expression (10) can be derived to relate these parameters with the applied stresses for loading and unloading:

$$\frac{k_{pi}}{k_{po}} \cong \frac{\Delta\sigma_i}{\sigma_{co}} \qquad \frac{k_{vpi}}{k_{vpo}} \cong \frac{\Delta\sigma_i}{\sigma_{co}} \quad (10)$$

The recompression factor ρ_c is defined as:

$$\rho_c = \rho_{ei}\frac{1 + \left(k_{po} + k_{vpo}\right)\left(\Delta\sigma_i/\Delta\sigma_{co}\right)}{1 + \left(k_{po} + k_{vpo}\right)} \quad (11)$$

where: ρ_{ei} represents the swelling factor, $\Delta\sigma_{co}$ the increment of total stress and $\Delta\sigma_i$ is the partial stress increment.

The secant modulus for soil recompression is

$$M_{ci} = \rho_{ci} \cdot M_{co} \quad (12)$$

and the volumetric variation coefficient α_{ci} for a layer of thickness d_i is:

$$\alpha_{ci} = M_{ci} \cdot d_i \quad (13)$$

Nelson and Miller [3] proposed an equation for determining the expansion through the compressibility curve:

$$\rho = \frac{C_s z_i}{\left(1 + e_o\right)}\log\left(\frac{\sigma'_f}{\sigma'_{sc}}\right) \quad (14)$$

where: ρ represents the swelling deformation, z is the layer thickness, C_s is the swell index, σ'_f stands for the final effective stress, σ'_{sc} is the corrected swelling pressure measured in a test at constant volume and e_o is the initial void ratio.

Furthermore, Nelson and Miller [3] also proposed an expression to estimate the soil expansion, which involves the matric suction:

$$\rho = \frac{z_i}{\left(1 + e_o\right)}\left[c_m \Delta\log\left(u_a - u_w\right) + c_t \Delta\log\left(\sigma - u_a\right)\right] \quad (15)$$

where: c_m is a matric suction index, c_t goes for the effective stress index, σ is the net stress applied over the soil, u_a represents the air pore pressure, u_w is the water pore pressure, and is the $(u_a - u_w)$ matric soil suction.

Lytton [4] suggested that for the small increases in the volume change in constitutive surface (suction, mean stress, and volume) (h, σ, V), the volumetric strain $\Delta V/V$ is linearly related.

$$\frac{\Delta V}{V} = \gamma_h \log\left(\frac{h_f}{h_i}\right) - \gamma_\sigma \log\left(\frac{\sigma_f}{\sigma_i}\right) - \gamma_\pi \log\left(\frac{\pi_f}{\pi_i}\right) \quad (16)$$

$$\gamma_\sigma = \frac{C_c}{1 + e_o} \qquad \gamma_h = \frac{\gamma_\sigma}{\dfrac{1}{1 + \left(\dfrac{h}{\theta\left(\partial h/\partial\theta\right)}\right)}} \quad (17)$$

where: $\Delta V/V$ is the volumetric strain; h_i and h_f are the initial and final matric suction values; σ_i and σ_f are the initial and final mean stresses; π_i and π_f are the initial and final osmotic suction values; γ_h, γ_σ, and γ_π represent the compression indexes for matric suction, mean stress and osmotic suction, respectively. If we consider nil osmotic suction variations ($\pi_i = \pi_f$) the last right hand side term of (16) is zero.

Finally, Arrieta and Sánchez [16] presented a set of equations to compute volumetric strains for any stress state under the foundation slab:

$$\varepsilon_v = \varepsilon_{v\max}\left[1 - \frac{\log\left(\sigma_{eq}/\sigma_v\right)}{\log\left(\sigma_{eq}/\sigma_o\right)}\right] \quad (18)$$

$$\delta e = \varepsilon_v \cdot d_i \quad (19)$$

where: ε_{vmax} is the maximum unitary volumetric strain, σ_o represents the normal stress, σ_{eq} stands for the expansive equilibrium stress and d_i is the layer thickness.

3 Materials and methods

For this study, undisturbed soil samples were obtained from the region of Jurica in Santiago de Queretaro, Mexico, at depths between 0.60 m and 0.80 m. Subsequently, tests to determine index properties were performed to establish the plasticity properties and grain size distribution curve of the soil samples.

For the determination of the volumetric variation coefficient, laboratory tests were performed with a mechanical oedometer using samples of unsaturated unaltered expansive soil with different initial water contents. A load of equal magnitude to that found at the site was placed on the specimen until the deformation was consistent. The saturation process began immediately and continued until the swelling deformation was developed, following the procedure outlined by method B in ASTM D4546 [5]. This swelling reflects the properties of expansive soil and depends solely on the initial water content.

Once the soil was saturated, typical consolidation tests were conducted, creating a hysteresis cycle in load and total unload [5]. The swelling pressure was measured in unaltered specimens with an initial water content rising in increments of 5.5% from 0% up to 38%. These water contents represent a degree of saturation that rises in increments of 10% from 0% up to 70%, which is because no volume increase neither swelling pressure acts with a degree of saturation equal to 70% was found. The initial strain caused by swelling soil is considered an important part of the elastic strain sustained by the material.

Suction tests with undisturbed specimens were also performed using filter paper technique (using Whatman® 42 filter paper) [17], in order to obtain relationship between water content (or degree of saturation) and soil suction (see Figure 1). The test started with dry samples (0% of the degree of saturation) whose degree of saturation was increased by steps up to nearly full saturation.

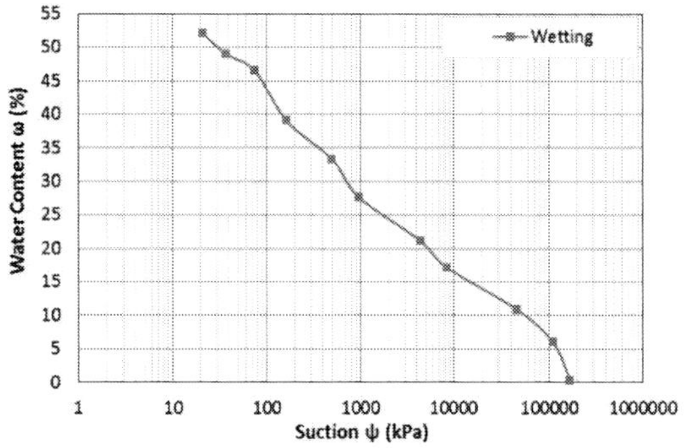

Figure 1. Soil-water retention curve (main wetting path).

3.1 Estimation of volumetric variation coefficient by hydration α_h

The volumetric variation coefficient by hydration α_h depends on two types of stresses. The first one is soil suction increments $\Delta\sigma\psi$ and refers to variations of the soil moisture generating an effective stress relief that results in a swelling. The other stress is caused by applied external stresses $\Delta\sigma_{app}$ which are transmitted to the soil mass through the foundation.

Here, we propose a procedure to determine the volumetric variation coefficient α_h associated with an initial water content ω_i and the relationship between applied stresses and swelling pressure ($\sigma_{app}/\sigma_{swe}$). Because a stress relief due to unloading is similar to a stress relief induced by soil hydration [18], the coefficient α_h is determined in the same way to that described by Zeevaert [2] for the coefficient α_{ci} in his theory "swelling by unloading".

Assuming that the relationship between the stresses induced by suction, and the corresponding deformations is linear, the maximum swelling secant modulus M_{ho} due to changes in degree of saturation can be defined as:

$$M_{ho} = \frac{\Delta\varepsilon_{ho}}{\Delta\sigma_{ho}} \quad (20)$$

and the total strains due to variations in soil suction.

$$\Delta\varepsilon_{ho} = a(\Delta\sigma_{ho})^C \quad (21)$$

combining (21) with (20) the secant modulus M_{ho} is expressed as:

$$M_{ho} = a(\Delta\sigma_{ho})^{C-1} \quad (22)$$

Similarly, partial stress relief can be expressed as:

$$M_{hi} = a(\Delta\sigma_{hi})^{C-1} \quad (23)$$

combining (22) and (23):

$$\frac{M_{hi}}{M_{ho}} = \left(\frac{\Delta\sigma_{hi}}{\Delta\sigma_{ho}}\right)^{C-1} \quad (24)$$

Thus, the swelling factor by suction variations will be:

$$\rho_h = \frac{M_{hi}}{M_{ho}} = \left(\frac{\Delta\sigma_{hi}}{\Delta\sigma_{ho}}\right)^{C-1} \quad (25)$$

Assuming that applied stresses are proportional to suction [19], it follows that:

$$\frac{\Delta\sigma_{hi}}{\Delta\sigma_{ho}} = \frac{\Delta\psi_{hi}}{\Delta\psi_{ho}} \quad (26)$$

Therefore, stresses can be replaced by the soil suction in (25), where the swelling factor is:

$$\rho_h = \left(\frac{\Delta\psi_{hi}}{\Delta\psi_{ho}}\right)^{C-1} \quad (27)$$

In (27), C is obtained from the plot in Figure 3 (swelling pressure against unitary strain), and represents its gradient:

$$C = \frac{\log(\varepsilon_f/\varepsilon_i)}{\log(\sigma_{swef}/\sigma_{swei})} \quad (28)$$

Combining (26) and (27) the secant modulus for unitary strains M_{hi} at a depth z_i corresponding to the response to a partial stress relief induced by swelling is defined as:

$$M_{hi} = \rho_{hi} \cdot M_{ho} \quad (29)$$

The volumetric variation coefficient by hydration α_{hi} for each layer thickness d_i is:

$$\alpha_{hi} = M_{hi} \cdot d_i \quad (30)$$

4 Results and discussion

In Figure 2 we show the relationship between the water content and swelling pressure obtained through an Oedometer tests. The relationship between swelling pressure σ_{swe} and volumetric strain ε can thus be inferred from Figure 3.

Using a fitted equation to the experimental data in Figure 3 parameters ε_i, ε_f, σ_{swei} and σ_{swef} can be determined.

They represent the minimum volumetric strain, maximum unitary volumetric, minimum swelling pressure and maximum swelling pressure respectively. Finally using (28) Zeevaert's index C it can be computed as:

$$\varepsilon_i = 0.00617 \qquad \varepsilon_f = 0.28901$$

$$\sigma_{swei} = 1.05kPa \qquad \sigma_{swef} = 126.00kPa$$

$$C = \frac{\log(\varepsilon_f / \varepsilon_i)}{\log(\sigma_{swef} / \sigma_{swei})} = 0.8034 \frac{1}{kPa} \tag{31}$$

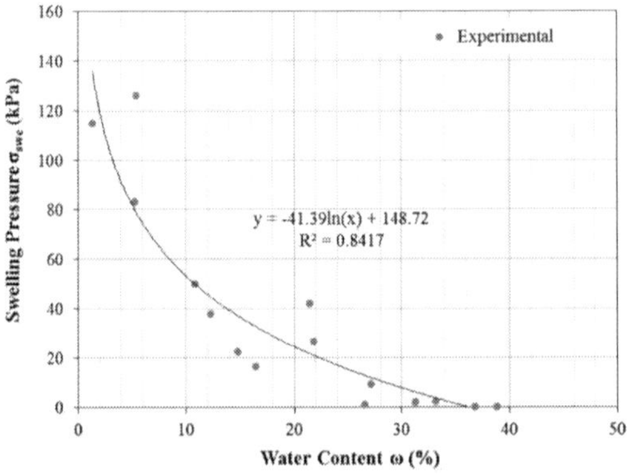

Figure 2. Relationship between water content and swelling pressure of Jurica's soil.

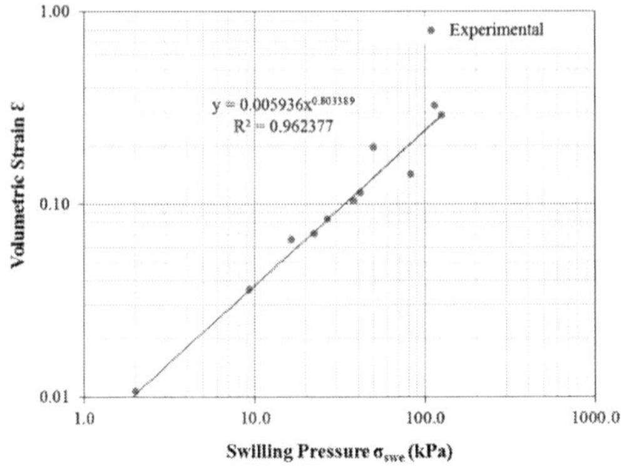

Figure 3. Relationship between swelling pressure and volumetric strain of Jurica's soil.

The maximum suction stress change induced by hydration $\Delta\psi_{ho}$, occurs when the soil is in an initial low water content and then nearly reaches full saturation. The following parameters can be obtained by inspection from Figure 3:

$$\omega_i = 0.35\% \qquad \omega_f = 52.07\%$$

$$\psi_o = 168{,}352.85kPa \qquad \psi_f = 21.08kPa$$

$$\Delta\psi_{ho} = \psi_i - \psi_f = 168{,}331.77kPa \tag{32}$$

According to (20) and (26), the secant modulus of unitary strain due to a total stress relief induced by water content variations M_{ho} is:

$$M_{ho} = \frac{\Delta\varepsilon_{ho}}{\Delta\sigma_{ho}} = \frac{\varepsilon_f - \varepsilon_i}{\psi_i - \psi_f} = 1.68x10^{-6} \frac{1}{kPa} \tag{33}$$

Then, the value of partial suction changes $\Delta\psi_{hi}$ is computed by subtracting the initial suction ψ_i from the next one ψ_{i+1}. This process will continue until the final suction ψ_f is reached. The secant modulus for unitary strains related to a partial stress relief due to increase of water content of soil M_{hi} is calculated with (29). With (27) the swelling factor ρ_{hi} is computed. The process ends with the determination of the volumetric variation coefficient α_{hi}.

Figure 4 shows the behavior of the variation volumetric coefficient α_h. Starting with a water content of 0% and subjected to 1% increments. Notice that α_h shows the largest variations when the water content goes from 0% to 22% (8.21x10^{-3} a 4.45x10^{-3} m/kPa). Whereas, when the water content varies from 22% to 55% α_h remains practically unchanged. A polynomial fitting function has strong correlation (R^2=0.9926) with experimental data; this indicates that if the water content is above 0%, α_h can be obtained with the procedure described earlier, however, lower values of the water content are difficult to find in common practice. It can be inferred that α_h can be computed accurately if the water content ranges between 0 to 100%. For that reason, in order to determine the behavior of the variation volumetric coefficient α_h at different water content. Figure 4 also shows that, if the water content is between 0% and 22%, the volumetric variation coefficient α_h takes its highest values (between 9.41x10^{-3} and 4.45x10^{-4} m/kPa). If the water content is above 22%, α_h remains practically unchanged in any change of thereof.

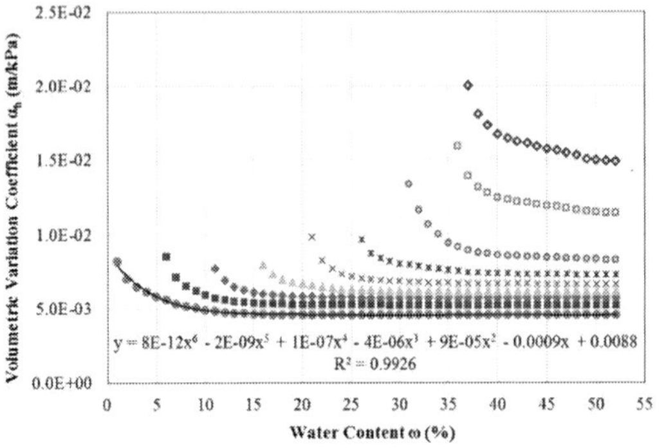

Figure 4. Behavior of volumetric variation coefficient with different initial water content.

Considering the previous ideas, the next step is to build up a data set for the ratio of applied stress and swelling pressure ($\sigma_{app}/\sigma_{swe}$) ranging from 0 to 1. The data set will be related to a curve with a specific initial water content ω, this in turn represents a volumetric variation coefficient α_h in the x-axis.

With Figure 5 it can be noticed that the critical water content ranges from 0% to 22%. Here α_h varies significantly, this is a difficult condition for the Jurica soil to meet as its water content ranges from 17% to 41% [20, 21].

Then, using Figure 5 is quite simple as the only parameters that are needed is the initial water content ω, the applied stress σ_{app}, and the swelling pressure σ_{swe} of soil. After that, the ratio $(\sigma_{app}/\sigma_{swe})$ is needed to be computed to input its value in the y-axis. Then an horizontal line is drawn to intersect the curve corresponding to the initial water content, and a vertical line is also drawn to cross the x-axis where α_h will be found.

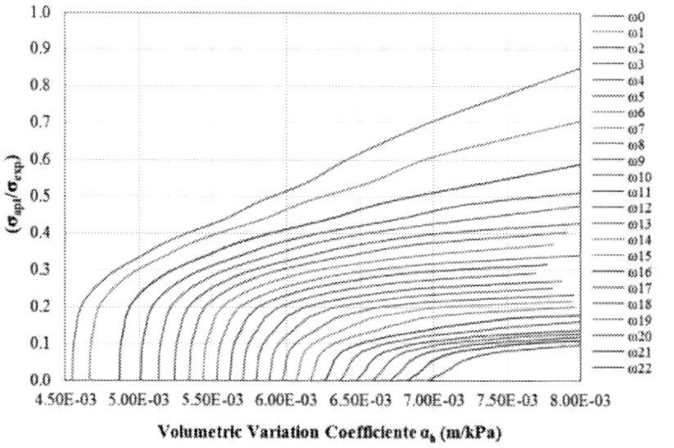

Figure 5. Volumetric variation coefficient associated to an initial water content (from 0% to 22%), the relationship between applied stress and swelling pressure (from 0 to 1).

5 Conclusions

The volumetric variation coefficient shows its greatest fluctuations between 0% and 22% of the water content, but from 22% up to 55%, the changes in its magnitude are not significant. This is because soil particles with water contents near zero experience strong attractions to water molecules. These forces tend to decrease as the water molecules move away from clay minerals surface decreases its swelling potential.

In determining the volumetric variation coefficient it is required to have good temperature humidity control and excellent testing conditions during any lab procedure. Small variations may induce great uncertainties.

The assumption that the swelling of a soil mass (induced by a stress relief by hydration) (changes in suction) has shown to be precise and provides an easy and simple tool that takes into account the soil expansive behavior in the analysis and design of foundations on these soils.

The results presented herein (such as Figure 5) cannot be generalized to the use of any expansive soil since the curves can be different for each soil due to a number of factors such as natural dispersion, different solid minerals, cracking planes, alteration due to sampling techniques, etc. Currently the authors are working with several other reported lab data coming from many different soils with similar characteristics. This, in order to establish relationships between them and get to use the results in common practice in the prediction of the coefficient.

References

1. K. Kayabali, S. Demir, Can. Geotech. J. **48**, 3, 354-364 (2011)
2. L. Zeevaert, *Interección suelo-estructura de cimentación* (D.F. Mexico: Limusa, 1980)
3. J.D. Nelson, D. J. Miller, *Expansive soils. Problems and preactice in foundation and pavement engineering* (John Wiley & Sons, Inc., 1992)
4. R.L. Lytton, Geotechnical Special Publication. **2**, 1827-1845 (1994)
5. ASTM D4546, Standard test methods for one-dimensional swell or collapse of soils (2014).
6. M.E. Zumrawi, University of Khartoum Engineering Journal. **3**, 2, 35-39 (2013).
7. H. King, [On line]. Available: http://geology.com/articles/expansive-soil.shtml. [accesed: January 28, 2015].
8. J.P. Puppala, T. Manossuthikij, C.S. Chittoori, Engineering Geology. **164**, 187-194 (2013).
9. E. Jahangir, O. Deck, F. Masrouri, Comput. Geotech. **54**, 16-32 (2013).
10. L.D. Jones, [On line]. Available: http://core.ac.uk/download/pdf/386982.pdf.2015. [accesed: March 25, 2015].
11. A.T. Sudjianto, K.B. Suryolelono, A. Rifa, I.B. Mochtar, *International Journal of Engenieering & Technology*. **11**, 6, 142-148 (2011)
12. F. Shuai, *Simulation of swelling pressure measurements on expansive soils* (University of Saskatchewan, 1996)
13. A. J. Baeza, *Diseño de zapatas continuas de cimentación sobre suelos expansivos mediante la metodología Lytton-Mitchel caso Querétaro* (Universidad Autónoma de Querétaro, 2012)
14. M.L. Pérez-Rea, O. Cabrera, *XXI Reunión Nacional de Mecánica de Suelos* (Sociedad Mexicana de Mecánica de Suelos, 2002)
15. M.L. Pérez-Rea, T.I. Ayala, *XIII Conferencia Panamericana de Mecánica de Suelos e Ingeniería Geotécnica* (2007)
16. L. Arrieta, N.E. Sánchez, *Mater. Cosntrucc.*, **53**, 271-272 (2003)
17. ASTM D5298, Standard test method for measurement of soil potential (suction) using filter paper (2010)
18. R. Galaviz-González, *Predicción de asentamientos de cimentaciones superficiales mediante análisis de interacción suelo expansivo-estructura de cimentación* (Universidad Autónoma de Querétaro, 2013)
19. M.L. Pérez-Rea, *Succión y comportamiento esfuerzo-deformación en suelos expansivos de Jurica y Tejeda* (Universidad Autónoma de Querétaro, 1993)
20. T. López-Lara, *Resistencia al esfuerzo cortante en arcillas expansivas de Jurica* (Universidad Autónoma de Querétaro, 1996)
21. A.O. Cabrera, *Determinación de la expansión volumétrica por hidratación en un suelo expansivo y su influencia en el analisis y diseño de cimentaciones* (Universidad Autónoma de Querétaro, 2001)

Creep response in shear of clayey geo-materials under saturated and unsaturated conditions

Nor Shahidah Mohd Nazer[1,2,a] and Alessandro Tarantino[1]

[1]*Department of Civil and Environmental Engineering, University of Strathclyde, Glasgow*
[2]*Faculty of Science and Technology (Geology Programme), National University of Malaysia, Bangi, Selangor, Malaysia*

Abstract. Translational slides in clays are often characterized by long-lasting intermittent movements associated with the fluctuations of pore-water pressure. Physically-based models designed to support hazard analysis of landslide movements and early warning systems require the integration of time-dependent (viscous) constitutive models for the shear displacements because landslide movements are typically controlled by the viscous behaviour of the clay geo-material. This paper presents an investigation of the viscous response of a clay geo-material under saturated and unsaturated conditions. Creep and relaxation tests have been first carried out on saturated clay samples by means of direct shear box. To gain a conceptual understanding of the viscous response of the clay in shear, mechanical analogues were considered based on combinations of springs and dashpots. Preliminary tests on unsaturated samples were finally carried out to gain a first insight into the viscous response of the clay under unsaturated conditions.

1 Introduction

Landslides induced by rainfall often lead to disastrous hazard. Heavy rainfall accompanied by slope displacement may cause damage in terms of serviceability to houses, buildings, roads, as well as other man-made structures and brings danger to life. Translational slides in clay are usually intermittent and happening in long run. They exhibits long-term deformation and proper land management is therefore important in area prone to slope failure.

Early warning systems (EWS) are generally designed to forecast landslide hazard and detect hazard and risk zones. Designing simple EWS is important so as to avoid confusion and loss of time during emergencies [1]. EWS must incorporate field monitoring data and should be underpinned by properly designed mechanical models to allow for accurate projection of future movements. Understanding mechanisms that lead to failure and investigating its interplay with landslide kinematics is therefore vital.

When considering landslide's failure mechanism, relationships between pore-water pressure and rainfall become the subject of interest due to their inevitable contribution to landslide movements [2-4]. When rain water infiltrates into the ground, it causes fluctuations of pore water pressure in the upper portion of the soil profile (unsaturated zone). Seasonal fluctuations in pore water pressure cause significant decrease in soil shear strength as a result of pore pressure build up and may trigger slope movements. The Sarno landslide in town Campania, Italy is one example of landslides which cyclically mobilized by pore-water pressure fluctuations [5].

Movements of slopes along the slip surface are controlled by viscous behaviour mainly in the form of creep. Creep is causing slow slope movements over time under constant effective stress. Studies have shown that creeping slope may moves at imperceptible rate (several mm per year) and the movements may continue for long periods without reaching failure [2].

A large number of creep models in the literature, associated with early warning system for rainfall-induced landslides, are based on empirical models rather than physically-based models [6]. Empirical models are known to be limited to a specific boundary condition and did not provide clear conceptual understanding of landslide kinematics. Empirical models therefore offer limited application and also lack in theoretical basis [7].

Lack of creep studies in relation with unsaturated soil has raised great interest among researchers to investigate the time-dependence behaviour of unsaturated geo-materials by accounting the effects of matric suction. Among them, Lai et al. [8] established a linear relationship between suction and parameter from a model built for saturated soils and developed a creep model with consideration for suction effects.

In this research, we aim to simulate viscous response in shear of clayey geo-material using a mechanical

[a] Nor Shahidah Binti Mohd Nazer: ashiqamanullah@gmail.com

analogue model built upon the combinations of spring and dashpots. In this way, creep and relaxation can be simulated using a single model, i.e. using a single set of parameters, and not treated separately as often happens when empirical models are used. Using a single model to capture both creep and relaxation is important in modelling the kinematics of a landslide since viscous response of clay geo-materials that controls landslide movements is never purely 'creep' mode [9]. Experimental tests to gain insight into viscous responses in shear were carried out under both saturated and unsaturated conditions.

2 Experimental Criteria

2.1 Sample preparation

2.1.1 Saturated samples

Samples were prepared from slurry by mixing Ball clay powder with demineralised water to achieve 100% water content (about 1.5 w_l). The slurry was placed in the 60 mm side shearbox and initially let to consolidate under the weight of the loading cap (~2.7 kPa) for 1 day. To ensure that the sample remains saturated, the shear box container was completely filled with water. Afterwards, samples were consolidated in steps to 100kPa normal stress. Tests were performed in a temperature-controlled room.

Saturated samples in the residual state were prepared by 8 reverse shearing (8mm displacement for each stroke). This was sufficient to ensure the attainment of residual state

2.1.2 Unsaturated samples

Unsaturated samples were prepared by consolidating the slurry in the same manner but without the presence of water in the shear box container. Once done, samples were removed from the shear box frame and air-dried to a target water content estimated by weighing the samples. Prior to cutting and trimming, samples were then wrapped with parafilm and sealed in polythene bag to allow moisture equalization. The samples were cut and trimmed using a 60 mm x 60 mm square cutter. The cutter was advanced slowly into the sample using a clamp. Since the samples were stiff, cutting could be achieved with ease.

Prior to moving the specimen into the shear box, silicon grease was placed between the bottom plate and the lower frame and between the lower and upper frames of the shearbox. The specimen was then transferred into the 60 mm side shearbox

The loading cap was put in place and parafilm tape was used to cover the gap between the loading cap and the upper frame) in order to prevent evaporation of the sample. The silicone grease and the parafilm tape were aimed at ensuring constant the water constant throughout the test. The specimen was then ready for the shearing test.

2.2 Tests for viscous response in shear

2.2.1 Displacement -controlled test

Displacement controlled-test was performed using conventional direct shear box. Shear rate of 0.05mm/min was used in this research. This rate was sufficiently low to ensure drained conditions during shearing. A horizontal displacement was applied until a target shear stress was reached. The horizontal displacement was then maintained constant for approximately one day to allow for the shear stress to 'relax'.

2.2.2 Stress-controlled test

Stress controlled-test was performed using a modified direct shear box. The modification involved an additional pulley system which was set in series with the load cell. The tension force applied by the pulley system to the upper frame converted into the shear force applied to the sample. The test was performed by applying the shear force in steps to reach a target shear stress. This was then maintained constant for approximately one day to allow the displacement to 'creep'.

3 Experimental results and modelling

3.1 Viscous response in saturated clay

3.1.1 Pre-peak

Stress relaxation and creep response in saturated soil are shown in Figure 1. Specimens S_50 were sheared to a target shear stress of 27 kPa either by controlling the displacement for stress relaxation test (Figure 1a) or by controlling the shear stress for creep test (Figure 1b). The target shear stress was about 50% of the ultimate shear strength. To model viscous response of the clay, analogue mechanical models built upon combinations of springs and dashpots, i.e. generalised Kelvin and Maxwell models, were considered. This is because these models allow capturing the creep and relaxation response using a single set of model parameters.

The mechanical model considered is shown in Table 1 (M5). It was selected because of its ability to reproduce stress decay (relaxation) or displacement evolution (creep) that tends to level off. In other words, this model allows reproducing, at least qualitatively, the response observed experimentally. Model parameters were determined by the least-square method to minimise errors between experimental data and simulation for both tests simultaneously.

Because the loading system is not infinitely stiff, another spring was added in series to the model M5 to take into account the compliance of the system. The stiffness of the system was inferred from a combination of tests on dummy samples and the defamation of the system measured during relaxation.

The performance of the M5 model in simulating the relaxation response (Figure 1a) and the creep response

(Figure 1b) can be considered satisfactory considering that the model only includes a single dashpot. A more accurate simulation of the time-dependent response could be achieved by adding another dashpot into the mechanical model. It is worth noticing that a single set of parameters for the M5 model appears to satisfactorily capture the final viscous response in both relaxation and creep mode. Overall, it appears that the M5 model able to reproduce both types of response, creep and relaxation respectively, using only a single set of parameters.

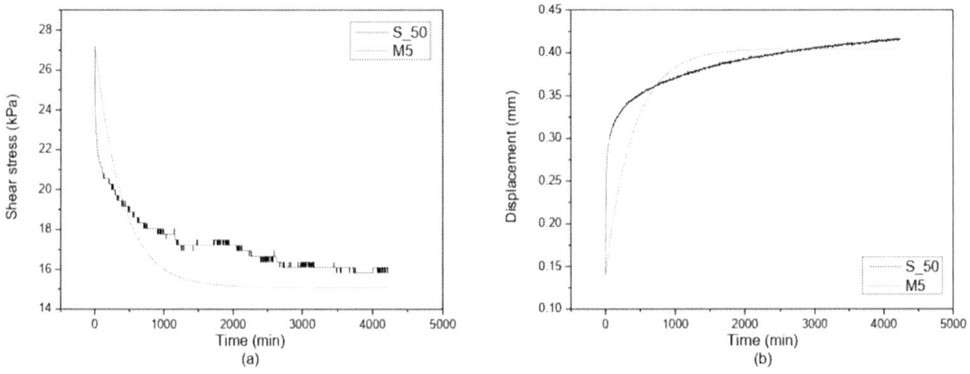

Figure 1. Viscous responses in saturated soil and simulation curve from M5 and M6 model: (a) stress relaxation (b) creep

Table 1. Creep and relaxation responses for various viscoelastic models after [10]

MODEL	NAME	CONSTITUTIVE EQUATION	CREEP COMPLIANCE, *J(t)*	RELAXATION MODULUS, *G(t)*
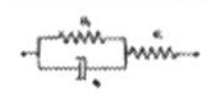	M5	$G_1G_2\gamma + G_1\eta_2\dot{\gamma} = (G_1+G_2)\sigma + \eta_2\dot{\sigma}$	$\frac{e^{-t/\tau}}{G_1}+\frac{G_1+G_2}{G_1G_2}\left(1-e^{-t/\tau_2}\right)$, or $\frac{1}{G_1}+\frac{1}{G_2}\left(1-e^{-t/\tau_2}\right)$	$G_1e^{-t/\tau'}+\frac{G_1G_2}{G_1+G_2}\left(1-e^{-t/\tau'}\right)$, or $\frac{G_1}{G_1+G_2}\left(G_2+G_1e^{-t/\tau'}\right)$ where $\tau'=\frac{\eta_2}{G_1+G_2}$

3.1.2 Residual

Figure 2 shows the viscous response in saturated soil at the residual (post-peak) state. Specimens R_60 were sheared to a target shear stress of 30 kPa, equivalent to 60% of ultimate shear strength. Stress relaxation response (Figure 2a) and creep response (Figure 2b) are presented herein. Using the same set of parameters applied in the pre-peak modelling exercises, M5 model was used to simulate the experimental data for the residual state. As shown in the figure, the simulation seems to overestimate both creep and stress relaxation observed experimentally. The simlyation does not appear to be satisfactory even if the model parameters are revaluated by fitting the experimental data (M5_R in Figure 2).

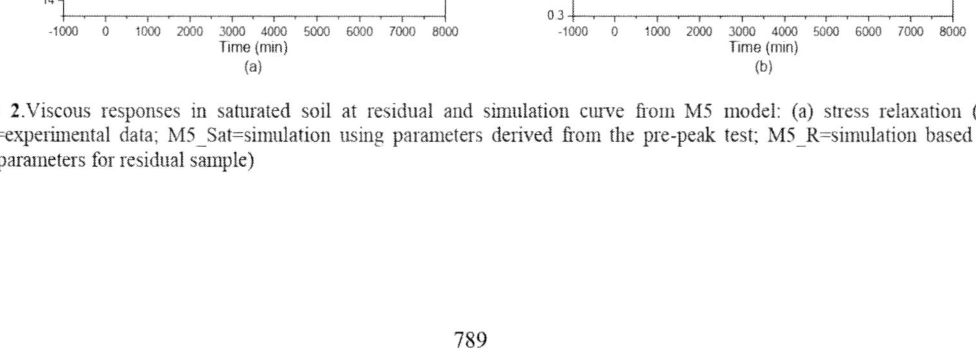

Figure 2.Viscous responses in saturated soil at residual and simulation curve from M5 model: (a) stress relaxation (b) creep (R_60=experimental data; M5_Sat=simulation using parameters derived from the pre-peak test; M5_R=simulation based on best-fitting parameters for residual sample)

DOI: 10.1051/e3sconf/20160914023

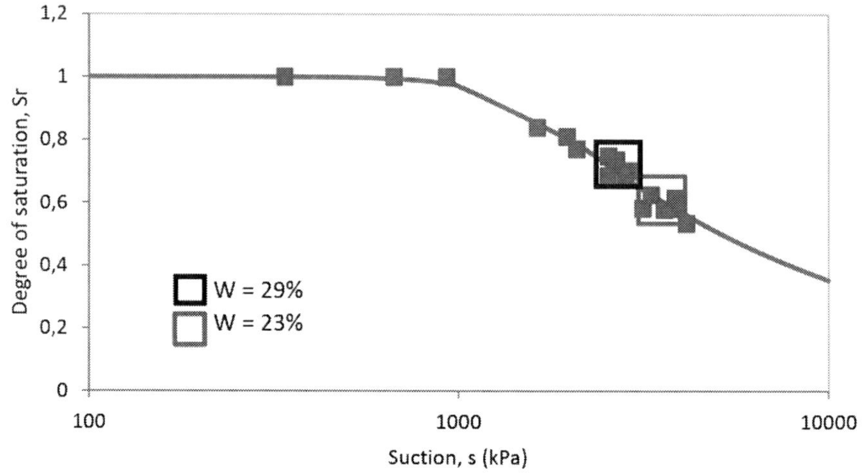

Figure 3. Water retention curve for ball clay

3.2 Viscous response in unsaturated clay

The water retention curve of the Ball clay is shown in Figure 3. The air entry value was found to be around 1000 kPa. Two tests at different water content (23% and 29%) were performed. The degree of saturation of these two specimens is shown in Figure 3, and these were equal to 0.5 and 0.75 for the specimens having water content of 23% and 29% respectively.

Differences in water content values before and after shearing were observed suggesting that some evaporation did take place throughout the shearing and viscous stages (Table 2). Although the corresponding change in suction was not very significant, it clearly appeared that the anti-evaporation system should be improved.

Stress relaxation and creep response in unsaturated soil are shown in Figure 4. Specimen US-23_30 refers to unsaturated soil sheared at water content 23% and to a target shear stress of 78 kPa, equivalent to 30% of ultimate shear strength. Similarly, US-29_30 refers to a specimen sheared at 29% water content and to a target shear stress of 76 kPa, corresponding to 30% of ultimate shear strength.

Viscous responses in unsaturated soils were showing similar curve trend as in saturated soils and, from these preliminary observations, viscous responses in unsaturated soil does not seem to vary too much with the degree of saturation.

Simulation of M5 model, based on the parameters derived from the tests on saturated samples in the pre-peak stage, seems to overestimate both creep and relaxation response (M5_Sat in Figure 4). However, a more satisfactory simulation can be obtained if the model parameters are revaluated by fitting the experimental data for the unsaturated samples (M5_US in Figure 4).

Again, it appears that the M5 model able to reproduce both types of response, creep and relaxation respectively, using only a single set of parameters (though model parameters appear to be different between saturated and unsaturated states).

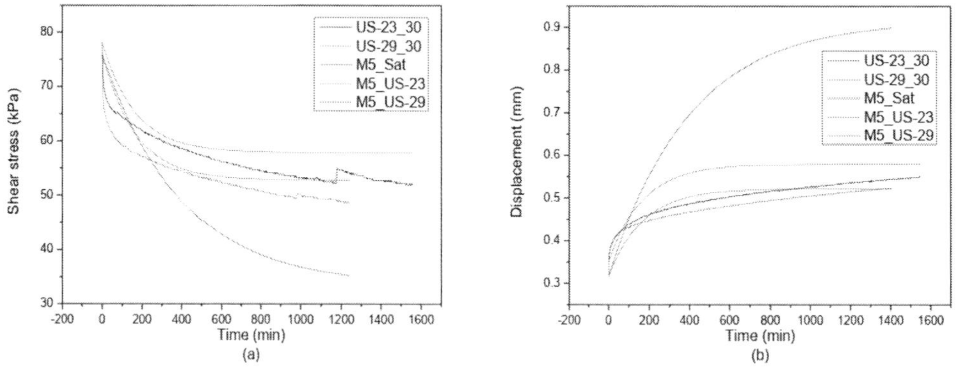

Figure 4. Viscous responses in unsaturated soil and simulation curves from M5 model: (a) stress relaxation (b) creep. (US=experimental data; M5_Sat=simulation using parameters derived from the pre-peak test; M5_US=simulation based on best-fitting parameters for unsaturated sample).

 DOI: 10.1051/e3sconf/20160914023

Table 2. Matric suction and water content for unsaturated soil

		Relaxation				Creep			
		Before		After		Before		After	
Water content	tau	Water content	suction (MPa)	Water content	suction (MPa)	Water content	suction (MPa)	Water content	suction (MPa)
0.29	30%	25.98	-2.68	24.75	-2.95	26.41	-2.55	24.37	-2.84
0.23	30%	22.07	-3.33	20.42	-3.61	21.72	-3.86	18.86	-4.13

4 Conclusions

The paper has presented experimental data on the viscous response of clay in different states (pre-peak state under saturated conditions, residual state under saturated conditions, and pre-peak state under unsaturated conditions). The viscous response has been investigated in terms of both creep and relaxation on samples starting from the same initial state. A mechanical analogue viscous model (M5) has been used to attempt modelling creep and relaxation simultaneously.

For samples tested in the pre-peak state, the model selected (M5) has proven successful in modelling creep and relaxation response with a single set of parameters. However, the model parameters appear to change from saturated to unsaturated states. In particular, the partial saturation appears to dampen the viscous response in terms of both creep and relaxation.

The residual state appears more difficult to model with a single set of parameters. The previous loading history (reverse shearing) appears to have a different effect on creep and relaxation.

References

1. E. Intrieri, G. Gigli, N. Casagli, F. Nadim, Brief communication "Landslide Early Warning System: toolbox and general concepts", Nat. Hazards Earth Syst. Sci. **13**, 85-90 (2013)
2. L. Picarelli, G. Urciuoli, C. Russo, Effect of groundwater regime on the behaviour of clayey slopes. Can. Geotech. J. **41**(3), 467-484 (2004).
3. W. H. Schulz, J. P. McKenna, J. D. Kibler, g. Biavati, Relations between hydrology and velocity of a continuously moving landslide—evidence of pore-pressure feedback regulating landslide motion?. Landslides, **6**(3), 181-190 (2009)
4. S. Bernardie, N. Desramaut, J. P. Malet, M. Gourlay, G. Grandjean, Prediction of changes in landslide rates induced by rainfall. Landslides, 1-14 (2014)
5. G. Capparelli, P. Versace, Analysis of landslide triggering conditions in the Sarno area using a physically based model. Hydrol Earth Syst Sci. **18**, 3225-3237 (2014)
6. G. Furuya, K. Sassa, H. Hiura, H. Fukuoka, Mechanism of creep movement caused by landslide activity and underground erosion in crystalline schist, Shikoku Island, southwestern Japan. Engineering geology, 53(3), 311-325 (1999).
7. W. Huang, D. Y. Liu, B. Y. Zhao, Y. B. Feng, Y. C. Xia, Study on the Rheological Properties and Constitutive Model of Shenzhen Mucky Soft Soil. J . Eng. Sci. Technol. Rev, **7**(3), 55-61. (2014)
8. X. L. Lai, S. M. Wang, W. M. Ye, Y. J. Cui, Experimental investigation on the creep behavior of an unsaturated clay. Can. Geotech. J. **51**(6): 621–628 (2014)
9. R. I. Borja, Generalized creep and stress relaxation model for clays, J. Geotech. Eng, **118**(11), 1765-1786 (1992)
10. *Continuum Mechanics for Engineers*, Second Edition. (1999)

Application of a multi-scale form of Terzaghi's effective stress principle for unsaturated expansive clays to simulate hydro-mechanical behavior during hydration

Julia Mainka[1,a], Marcio A. Murad[2], Van Duy Tran[1], Tien Dung LE[1] and Christian Moyne[1]

[1]*LEMTA - Université de Lorraine - CNRS - UMR 7563, Vandœuvre-lès-Nancy, France*
[2]*LNCC, avenida Getúlio Vargas 333, 25651-075 Petrópolis, RJ, Brasil*

Abstract. Our recently developed multi-scale form of Terzaghi's effective stress principle for unsaturated swelling clays that was rigorously derived by periodic homogenization starting from micro- and nano-mechanical analyses is applied to numerically simulate one-dimensional swelling pressure tests of compacted bentonites during hydration. The total macroscopic stress captures the coupling between disjoining forces at the nanoscopic scale of clay platelets and capillary effects at the microscopic scale of clay aggregates over the entire water content range. The numerical results allow to draw conclusions on the water transfer mechanism between inter- and intra-aggregate pores during hydration and consequently on the evolution of the external swelling pressure resulting from the competition between capillary and disjoining forces. In addition, such application highlights the abilities and the limits of the electrical double-layer theory to compute the disjoining pressure in the nano-pores. For large platelet distances, in the range of osmotic swelling, the nature of the disjoining pressure is electro-chemical and can be computed from Poisson-Boltzmann theory. Conversely, at small distances, in the crystalline swelling, a solvation component has to be added to account for the molecular nature of the solvent. As a first improvement of the nano-scale description the solvent is treated as a hard sphere fluid using Density Functional Theory.

1 Introduction

A variety of experimental [1,2] and modelling efforts [3-5] have been undertaken to analyze the hydro-mechanical behaviour of expansive clays during water infiltration by means of one-dimensional swelling pressure tests. The basic results are: (i) The swelling pressure increases during hydration and reaches values of some MPa at the fully saturated state. The final value is basically linked to dry density [2]. (ii) After an initial fast increase, the swelling pressure exhibits a temporary decrease followed by a new increase until reaching a stationary value being of the same amplitude as the first maximum [1,2,6,7]. This behaviour is associated with the collapse of the macroscopic solid structure when reaching the loading-collapse (LC) line [2].

It is now well recognized that the microstructure plays a crucial role in the evolution of the overall swelling pressure. In order to enhance the understanding of the processes underlying the experimental observations, several modelling efforts were undertaken. Among them we may highlight double-porosity models [3-5,8-10]. Recent works include the hydro-mechanical behaviour of non-expansive [10] and expansive unsaturated clays [2-5,9]. When accounting for elasto-plastic deformation, the evolution of the swelling pressure could be reproduced including its temporary drop [2], as well as the reorganization of the macrostructure [3-5,10]. All models are able to link the mechanical behaviour to capillary pressure drop in the porous space. On the other hand, to the best of our knowledge they are all based on an empirical approach to account for micro- and nanoscopic contributions to the overall swelling pressure.

This work tries to fill this gap. Our double-porosity model that was rigorously derived by micro-mechanical analyses [9] is applied to numerically simulate the swelling pressure evolution during hydration at fixed volume. Starting from a formulation of the mechanical equilibrium at the nano-scale of the electrical-double layer (EDL) and the micro-scale where capillary effects take place, the coupling between these two effects is incorporated over the entire saturation range. As will be shown, the model provides furthermore insight on the water transfer mechanism from the micro- to the nano-porous space and the linked evolution of the disjoining pressure during hydration. In addition, we show the limits of the electrical-double layer theory when computing the disjoining pressure at low hydration levels. Note that only elastic deformation is considered at the micro- and macro-scale. Ongoing work is in progress to incorporate elasto-plasticity in the multi-scale approach.

[a] Corresponding author: julia.mainka@univ-lorraine.fr

DOI: 10.1051/e3sconf/20160914024

2 Theory

2.1 Isotropic form of the multi-scale model

The detailed derivation of our multi-scale Terzaghi's effective stress principle and all model equations are presented in a former paper [9]. In the following, we present directly the final macroscopic model equations in an isotropic medium approximation recalling the principle ideas underlying their derivation.

The model is based on a given representation of the microstructure of expansive clays which is characterized by two porosity levels (referred to nano- and micro-pores tied up their size [11]) and three disparate length scales as shown on Figure 1.

- At the nano-scale the medium is composed of incompressible linear-elastic clay platelets of constant surface charge, $\sigma<0$. The platelets are separated by a nano-porous network saturated by a binary electrolyte solution (considered as continuous in classical EDL approach) with monovalent completely dissociated ions considered as point charges.
- At the intermediate micro-scale the platelets form assemblies of swollen clay aggregates (or particles) constituting the solid phase at that length scale. The particles are separated by a micro-porous network (free of EDL effects regarding their size) which is filled by a mixture of bulk water and air.

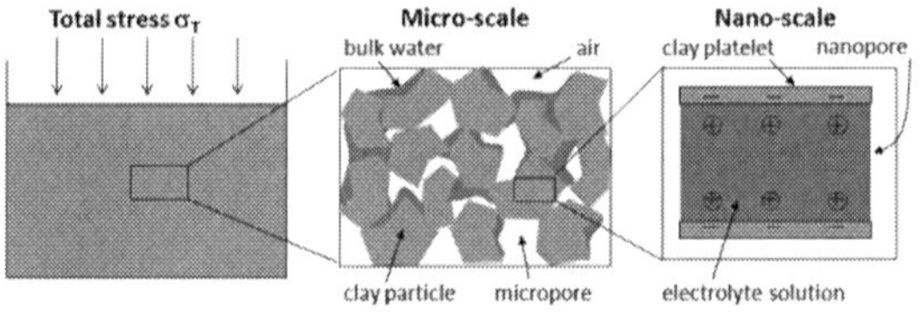

Figure 1. Multi-scale structure of swelling clays.

Starting from the formulation of the mechanical equilibrium at the local scales, the overall total stress is obtained using periodic homogenization [12]. Supposing no preferential orientation of the local clay fabric, all stresses reduce to pressures and the total stress is given by:

$$\boldsymbol{\sigma}_T = -P_{ext}\mathbf{I} = \sigma_E\mathbf{I} - P_f^{eff}\mathbf{I} - \Pi_{eff}\mathbf{I}\,, \tag{1}$$

with P_{ext} the external total pressure and $\mathbf{I}$ the identity tensor. From the above equation we may note that P_{ext} is given by the sum of three terms (i) a net contact stress σ_E, (ii) an effective fluid pressure P_f^{eff} accounting for capillary effects arising from the micropores and (iii) an effective disjoining pressure Π_{eff} strongly tied up with the action of the electrolyte solution on the clay platelets at the nano-scale. The elastic contact stress $\sigma_E = \mathrm{tr}(\boldsymbol{\sigma}_E)/3$ is linked to the volumetric macroscopic strain $\varepsilon = \mathrm{tr}(\boldsymbol{\varepsilon})$ by Hooke's law $\sigma_E = \tilde{K}\varepsilon$. The macroscopic bulk modulus $\tilde{K}$ is expressed as a function of the micro-porosity n_f, the microscopic bulk modulus of the clay particles K_s and the microscopic Poisson ratio ν_s using self-consistent medium approximation according to [12]. The equivalent fluid pressure reads as

$$P_f^{eff} = P_a - \chi^{eff} P_c\,, \tag{2}$$

with P_a the air pressure and P_c the capillary pressure defined by the negative water pressure $P_c=P_a-P_w$. P_c can be linked to the overall water saturation S_w by a Van Genuchten type law. χ^{eff} is the volumetric component the effective Bishop type tensor which is linked to clay fabric properties and saturation in the micropores according to

$$\chi^{eff} = 1-(1-\chi)\alpha\,, \tag{3}$$

where χ ($0\leq\chi\leq1$) is the microscopic Bishop parameter which can be expressed as a function of saturation, for example according to [13]. In addition, α is the macroscopic Biot coefficient defined in a classical way by $\alpha = 1-\tilde{K}/K_s$ represented in the form

$$\alpha = 1-\frac{2(1-n_f)(1-2\nu_s)}{(1+\nu_s)n_f + 2(1-2\nu_s)}\,. \tag{4}$$

The macroscopic result of the nanoscopic disjoining pressure Π_d is finally given by

$$\Pi^{eff} = (1-\alpha)\Pi_d\,. \tag{5}$$

We may observe that α plays the role of a scaling factor accounting for the transmissibility of disjoining forces between adjacent clay particles. Π_d is generally expressed as a function of ion concentration c_b and inter-platelet distance 2H using Poisson-Boltzmann (PB) equation [8]

$$\Pi_d = 2c_b RT\left(\cosh\frac{F\varphi_0}{RT} - 1\right)\,, \tag{6}$$

with R the perfect gas constant, T the absolute temperature, F the Faraday constant and $\varphi_0 = \varphi_0(c_b,H)$ the electric potential at the midplane between two adjacent platelets.

This is the classical approach for osmotic swelling where disjoining forces are driven by EDL effects. For small inter-platelet distances however, in the crystalline swelling regime, the molecular nature of solvent and ions is not more negligible and the disjoining pressure should be dominated by particle size effects [14] which could be included using *Density Functional Theory* (DFT) [15]. DFT allows to obtain particle density distribution $\rho(z)$ (with z the distance to the platelet surface for isotropic fluids) at equilibrium as a function of the excess free energy functional $F^{ex}[\rho]$ by minimizing the appropriate thermodynamic potential, here the grand canonical potential.

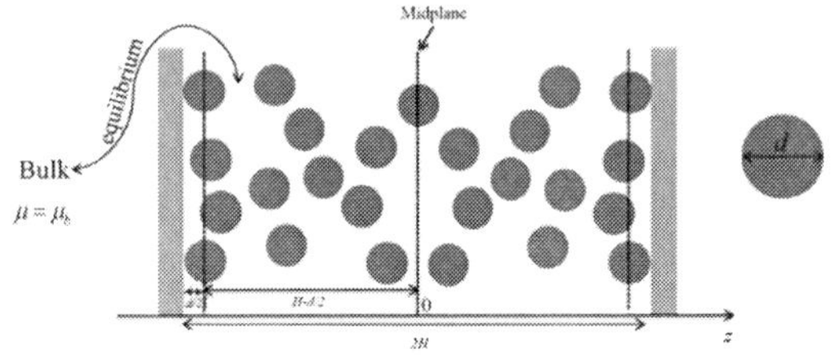

Figure 2. Improved nano-scale description considering the electrolyte solution as a hard sphere fluid confined between two planar hard walls.

As a first improvement of the nano-scale description, we consider in the present work the molecular nature of the solvent only (ion size and concentration are negligible compared to those of water). By considering the electrolyte solution as a simple hard sphere (HS) fluid of diameter d confined between two planar hard walls (Figure 2), $F^{ex}[\rho]$ is expressed by *Fundamental Measure Theory* (FMT) according to [16]. The disjoining pressure being defined as the force acting on the walls to keep them at a fixed distance 2H, it can be expressed using contact theorem which tells that, at mechanical equilibrium, Π_d can be computed from the difference between the pressure the fluid acts on the wall P and the bulk pressure P_b:

$$\begin{aligned}\Pi_d(2H) = P(2H) - P_b &= k_B T\left[\rho_{contact}(2H) - \rho_{CS}(\rho_b)\right] \\ &= k_B T\left[\rho(z = d/2, 2H) - \rho_{CS}(\rho_b)\right]\end{aligned}, \quad (7)$$

with k_B the Boltzmann constant. The density distribution of the bulk fluid ρ_{CS} is expressed in terms of the bulk density ρ_b by Carnahan-Starling equation [17].

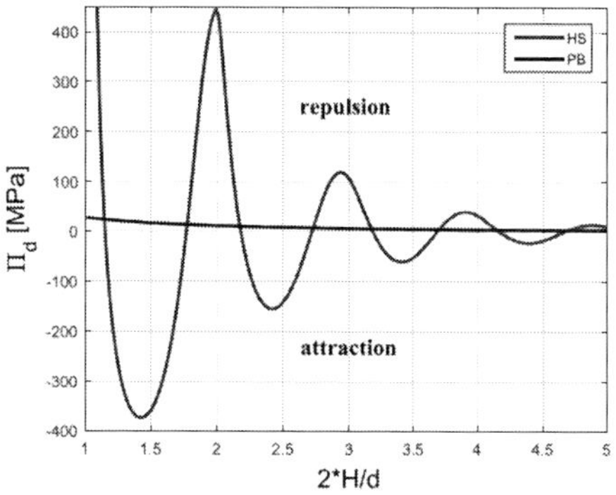

Figure 3. Disjoining pressure profiles of a PB fluid ($c_b=10^{-4}$ mol/l) and a HS fluid ($\rho_b d^3=0.73$) as a function of normalized inter platelet distance 2H/d (d_{H2O}=2.8Å).

Figure 3 compares the profiles of the disjoining pressure obtained by PB theory ($c_b=10^{-4}$ mol/l) and for a HS fluid as a function of normalized inter platelet distance 2H/d. For the HS fluid, the water molecule diameter and normalized bulk concentration are set to d_{H2O}=2.8Å and $\rho_b d^3=0.73$ (corresponding to a mass density $\rho_{H2O}=10^3$kg/m^3). The HS fluid pressure allows to account for attraction ($\Pi_d<0$) and repulsion ($\Pi_d>0$), while PB is a purely repulsive decreasing function of the platelet distance. According to the expectations [14], the oscillatory profile of the HS fluid exhibits maxima at entire multiples of d whose amplitude decreases rapidly with the platelet distance. Note that only negative slope parts represent equilibrium positions. In addition, the HS law reaches somewhat higher values of the disjoining pressure at small distances than PB theory.

The coupling of the different physical phenomena is made through the porosities at the micro- and nano-scale ϕ being defined as the respective pore volume over the total volume. Assuming a stratified local arrangement, the nano-porosity is expressed by the inter-platelet distance according to $\phi = H/(H+H_s)$, with $2H_S$ the platelet thickness. For an isotropic homogeneous distribution of the clusters, the two porosities are linked by a scalar version of the overall solid mass balance

$$(1-n_f)(1-\phi) = (1-\bar{n}_f)(1-\bar{\phi})\exp(-\varepsilon), \quad (8)$$

where the overbars refer to a stress free reference state which can be chosen arbitrarily. Similarly, using self-consistent approach, the micro-porosity is given by an integral form of the fluid mass balance

$$n_f = \bar{n}_f + \frac{n_f(1-n_f)}{\frac{2(1-2\nu_s)}{1+\nu_s} + n_f}\left(\varepsilon + \frac{(1-\chi)P_c}{K_s} - \frac{\Pi_d}{K_s}\right). \quad (9)$$

2.2 Application to an oedometric swelling test

In the following, we present the governing equations that underlie the computation of the swelling pressure P_{ext} at different stages of wetting, for the dry and fully saturated states and along with intermediate saturations. The numerical application refers to an experimental swelling pressure test on FoCa clay 50/50 weight percent bentonite powder/pellets mixture performed at the CIEMAT and presented in [1,2]. The clay sample we refer to is *MGR7* with a dry density ρ_{dry}=1.45g/cm^3 which corresponds to the ratio of the mass of the solid phase (clay cluster) at the dry state m_c^i over the total volume V_T. The cylindrical sample (100mm in diameter and 50m in height) was hydrated from the bottom of an oedometric cell. The swelling pressure applied to the cell to prevent deformation during water intake was recorded for mean water contents w varying from w_i=0.8 to w_f=0.33.

2.2.1 Initial "dry state"

Initially, we assume all water to be confined in the nanopores so that the micropores are only filled by air implying that χ^i=0. The gas/liquid interface should thus be at clay cluster surface (cf. Figure 4). This hypothesis agrees with a rough estimation of the pore diameter linked to the initial capillary pressure of about 200MPa using Laplace equation $2R = 4\gamma/P_c \approx 2\text{nm}$ (with water/air surface tension γ(298K)=72x10^{-3}N.m) which is clearly superior to the nanopores.

The initial state represents the reference state for this application, so that $\bar{n}_f = n_f^i$ and $\bar{\phi} = \phi^i$. As sample deformation is prevented during water intake no overall elastic stress appears with respect to the initial state and σ_E=0. Initially there is no pressure applied to the cell P_{ext}^i=0 and it follows from (1) in combination with (2), (3) and (5) that all capillary effects are compensated by the disjoining pressure $\Pi_d^i = P_c^i = 200MPa$.

As w_i=0.8, it is reasonable to assume crystalline swelling [18] so that Π_d^i is computed according to the HS fluid law in (7). The inter platelet distance estimated with this law, $2H^i$=3.12Å, corresponds to 1-2 water layers which is consistent with crystalline swelling. Assuming a platelet thickness $2H_s$=1nm, the corresponding nano-porosity is $\phi^i=H^i/(H^i+H_s)$=0.24.

It is important to note that using PB theory, the disjoining pressure is not able to compensate the initial capillary pressure (cf. Figure 3). Even if the approximation of the electrolyte solution by a simple HS fluid requires further improvement including ionic and dipolar interactions [19], the proposed pore treatment presents an improved nano-scale description at low hydration where the molecular nature dominates over EDL effects and where small pore diameter deviations imply important disjoining pressure variations.

For the estimation of the micro-porosity we introduce the platelet density $\rho_{ss}=m_{sc}/V_{sc}=2670 kg/m^3$ [20], with m_{sc} and V_{sc} the mass and the volume of the solid phase in a clay cluster. All water being located in the nanopores the water content is $w^i=m^i_{H20,c}/m_{sc}$ and it follows

$$\frac{\rho_{dry}}{\rho_{ss}\left(1-\phi^i\right)}=\frac{m_c^i}{V_T}\frac{V_{sc}}{m_{sc}}\frac{V_c^i}{V_{sc}}=\frac{V_c^i}{V_T}\frac{m_{sc}+m^i_{H20,c}}{m_{sc}}=\left(1-n_f^i\right)\left(1+w^i\right) \quad (10)$$

From this relation we obtain $n_f^i=0.39$ which is quite similar to the experimental value given in [2].

2.2.2 Final "saturated state"

At full saturation the capillary pressure vanishes and from (2) it follows that the fluid pressure contribution becomes negligible $P_f^{eff,f}=P_a\approx 0$. From the mechanical equilibrium (1) it can be concluded that the swelling pressure corresponds to the effective disjoining pressure

$$P^f_{ext}=\Pi^f_{eff}=\left(1-\alpha^f\right)\Pi^f_d \; , \quad (11)$$

where Π_d^f is a function of H (or ϕ) expressed by classical PB theory according to (6). This hypothesis seems reasonable as disjoining stresses may originate from osmotic swelling at full saturation.

2.2.3 Intermediate saturations

At intermediate saturations we chose a representation of the overall medium by a division into two parts (cf. Figure 4): a saturated part of height x+Δ at the bottom and a part of height L-x-Δ (L being the sample height) at the top being at the initial state. Δ is the local macroscopic deformation due to the difference in water content between both parts. This corresponds not necessarily to the real configuration, as no information is given about the advancement of the water front during infiltration. The advantage of this approximation is that it does not require defining microscopic saturation and capillary pressure which can differ from their macroscopic values.

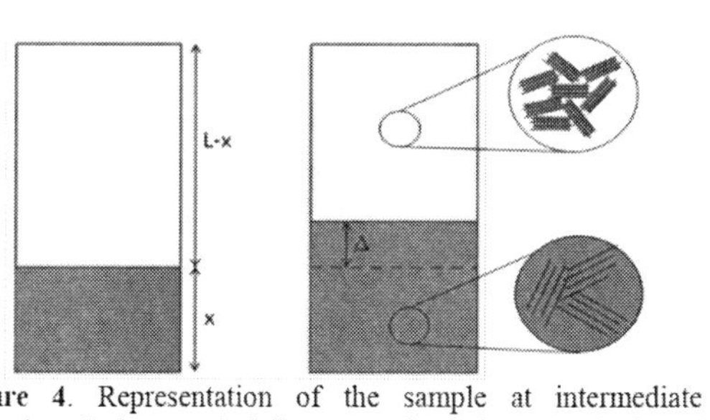

Figure 4. Representation of the sample at intermediate saturation: the bottom is fully saturated and the top at the initial state. The overall saturation is obtained by the volume ratio of the two parts. On the left: configuration without deformation; on the right: with deformation Δ due to different water contents.

For a constant sample diameter, the volumetric fractions of the two parts are given by the ratio of their respective height to the sample height, *i.e.* $\theta_b=x+\Delta/L$ and $\theta_t=1-\theta_b$. Volume fraction is linked to the overall saturation by

$$S_w=\theta_b+\left(1-\theta_b\right)S_w^i \; , \quad (12)$$

with $S_w^i=w^i/w^f=0.26$. Even if overall sample deformation is prevented, local vertical deformation occurs in both parts $\varepsilon=\{0,0,\varepsilon\}$ which are expressed as

$$\varepsilon_b=\frac{\Delta}{x} \text{ and } \varepsilon_t=-\frac{\Delta}{L-x} \; , \quad (13)$$

where the coupling with the saturation is made by (12). The corresponding oedometric elastic pressures are $\sigma_E^{b/t}=\tilde{M}^{b/t}\varepsilon^{b/t}$ with $\tilde{M}^{b/t}=\tilde{K}^{b/t}+4/3\mu^{b/t}$ the oedometric bulk moduli. Using (4) in combination with the relation $\mu=3(1-2\nu)/[2(1+\nu)]\tilde{K}$ for the Lamé parameter μ by supposing a constant macroscopic Poisson ratio ν, the oedometric bulk modulus is linked to the micro-porosity via the Biot coefficient as follows

$$\tilde{M}^{b/t}=\frac{3\left(1-\nu\right)}{1+\nu}K_s\left(1-\alpha^{b/t}\right). \quad (14)$$

Water intake being very slow the system can be assumed at mechanical equilibrium instantaneously so that the total pressure of each part corresponds to the external swelling pressure. The bottom part being completely saturated $P_f^{eff,b}=0$, whereas at the top part we have $P_f^{eff,t}\approx-\left(1-\alpha^t\right)P_c^i=-\left(1-\alpha^t\right)\Pi_d^i$. The respective vertical components of the total pressure are written as

$$-P_{ext}=\sigma_E^b-\Pi^b_{eff}=\tilde{M}^b\varepsilon^b-\left(1-\alpha^b\right)\Pi^b_d \; , \quad (15)$$

and

$$-P_{ext}=-P_f^{eff,t}+\sigma_E^t-\Pi^t_{eff}=\tilde{M}^t\varepsilon^t+\left(1-\alpha^t\right)\left(\Pi^i_d-\Pi^t_d\right). \quad (16)$$

The disjoining pressure at the sufficiently hydrated bottom part is computed by PB law (6), whereas at the much drier top part it is given by the HS law in (7).

In a similar way, the respective micro-porosities are obtained from (9) which yields

$$n_f^b=n_f^i+\frac{n_f^b\left(1-n_f^b\right)}{\frac{2\left(1-2\nu_s\right)}{1+\nu_s}+n_f^b}\left(\varepsilon^b-\frac{\Pi_d^b}{K_s}\right), \quad (17)$$

and

$$n_f^t=n_f^i+\frac{n_f^t\left(1-n_f^t\right)}{\frac{2\left(1-2\nu_s\right)}{1+\nu_s}+n_f^t}\left(\varepsilon^t+\frac{\Pi_d^i-\Pi_d^t}{K_s}\right). \quad (18)$$

The nano-porosities are expressed starting from (8) and read as

$$\begin{aligned}\left(1-n_f^b\right)\left(1-\phi^b\right)&=\left(1-n_f^i\right)\left(1-\phi^i\right)\exp\left(-\varepsilon^b\right)\\&=\left(1-n_f^i\right)\left(1-\phi^i\right)\frac{x}{x+\Delta}\end{aligned} \; , \quad (19)$$

and

$$\left(1-n_f^t\right)\left(1-\phi^t\right)=\left(1-n_f^i\right)\left(1-\phi^i\right)\exp\left(-\varepsilon^t\right)$$
$$=\left(1-n_f^i\right)\left(1-\phi^i\right)\frac{L-x}{L-x-\Delta} \quad (20)$$

To derive the above expressions we adopted an approximation of the elementary deformation dε as a function of the elementary variation of the sample height dl as follows dε=dl/l. Integration of this equation yields

$$\left[\varepsilon\right]_0^{\varepsilon^b}=\left[\ln l\right]_x^{x+\Delta}\rightarrow\varepsilon^b=\ln\frac{x+\Delta}{x}, \quad (21)$$

and

$$\left[\varepsilon\right]_0^{\varepsilon^t}=\left[\ln l\right]_{L-x}^{L-x-\Delta}\rightarrow\varepsilon^t=\ln\frac{L-x-\Delta}{L-x}. \quad (22)$$

We obtain a system of 6 unknowns {P_{ext}, x (or Δ), $n_f^{b,t}$, $\phi^{b/t}$} depending on S_w parameterized by K_s, ν and ν_s.

3 Results and Discussion

We now depict the numerical results of the simulation of the swelling pressure during hydration and the related evolution of the microscopic properties giving further insight to the local rearrangement. As the present study has a qualitative character, without any loss of generality we assume identical Poisson ratios at the macro- and micro-scale, ν_s=ν=0.25. The numerical results are analyzed in terms of K_s playing the role of a scaling parameter accounting for the transmissibility between nanoscopic disjoining and microscopic capillary effects.

3.1 Swelling pressure

Figure 5 shows the external swelling pressure for saturations between S_w^i and S_w^f=1. The numerical results for different values of K_s (1MPa, 5MPa and 100MPa) are compared to the experimental values taken from [1].

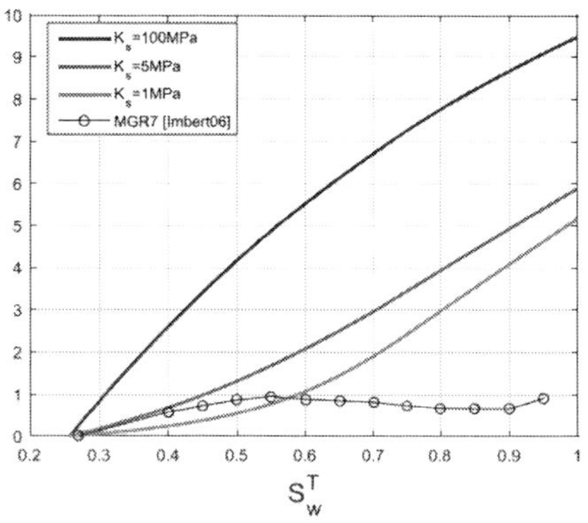

Figure 5. External swelling pressure vs. saturation parameterized by the microscopic bulk modulus and comparison with experimental data from [1].

In agreement with the experimental results, P_{ext} increases during water intake. Owing to the separation of the different contributions (contact stress, capillary pressure and disjoining forces) to the total pressure (1), the model shows that the increase in the swelling pressure is not due to an increase of the disjoining pressure.

As shown on Figure 6 for K_s=5MPa, Π_d decreases during water infiltration. The red dots correspond to the top part of the sample being at initial saturation where Π_d is computed by a HS law (7), whereas the blue dots correspond to the bottom part being at full saturation and where Π_d is given by PB theory (6). As capillary pressure vanishes at full saturation and deformation is restricted, the external pressure is completely determined by the disjoining pressure which is given by PB law as the overall medium is now composed of the bottom part. In agreement with the expectations [18], the completely saturated state corresponds to about 3 water layers in the nanopores.

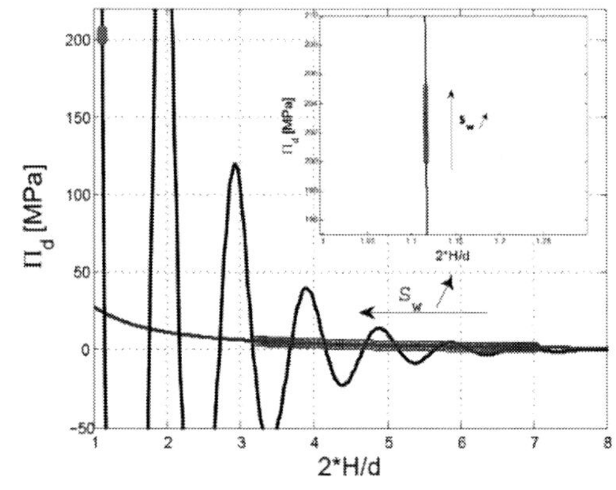

Figure 6. Evolution of the disjoining pressure with saturation: in the top dry part (red dots) computed by a HS law (7) and in the bottom part (blue dots) by PB theory (6); K_s=5MPa.

Figure 5 highlights also the limits of the present model. The best fit to the experimental data is observed for K_s=5MPa. After a good agreement at low saturations, the numerical values remain clearly superior to the experimental ones. This discrepancy is even more pronounced for higher values of K_s. Setting K_s to 1MPa allows reducing the swelling pressure at full saturation, but the evolution of P_{ext} during hydration does not enhance the fit to the experimental observations. Notwithstanding these shortcomings it can be concluded that in order to obtain swelling pressures in the MPa range, K_s has to be of the same order of magnitude which is clearly inferior to the global values of the bulk modulus given in the literature and varying between 10-50GPa [21,22]. Since K_s being solely linked to the connectivity of the solid structure inside the clay agglomerates, we conclude from this observation that the overall rigidity may be ensured by the action of the disjoining forces originating from the nanoscopic clay platelet scale.

It is noteworthy that, conversely to the numerical data, after in initial increase the experimental swelling pressure decreases to increase again. Such behaviour stems from the collapse of the macrostructure when reaching the LC line in the frame of elasto-plastic deformation [2] which is not accounted for in the present model approach. This multi-scale model allows to give insight to the contribution of local phenomena to the macroscopic mechanical behaviour. As a next step one has to incorporate macroscopic plastic deformation.

3.2 Porosities at the fully saturated state

Figure 7 shows the nano- and the microporosities at the fully saturated state as a function of the microscopic bulk modulus. In agreement with the expectations, the microporosity decreases between the initial dry (n_f^i=0.39) and the final saturated state, which is linked to capillary pressure decrease [3,6]. The volume being fixed impeding any macroscopic deformation, the nanoporosity increases during water intake. We remind that initially ϕ^i=0.24. This confirms the ability of the multi-scale model in capturing water transfer from the microscopic inter particle to the nanoscopic inter platelet pores during hydration. Such change in the porous space is more pronounced for a soft solid matrix being characterized by low values of K_s. Note finally that the increase of the inter platelet distance is at the origin of the decrease of the disjoining pressure during hydration.

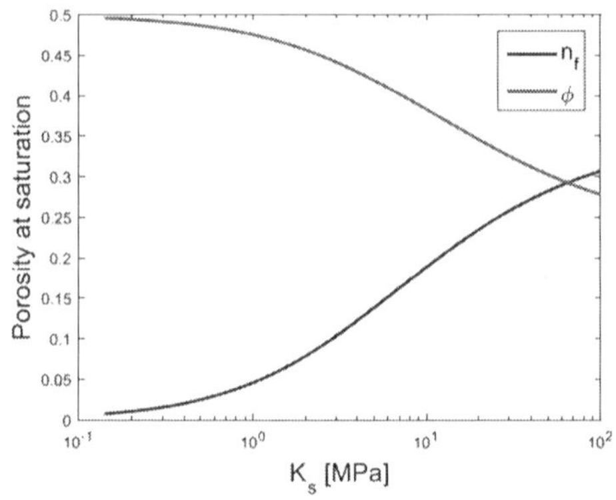

Figure 7. Nano- and microporosities at the fully saturated state as a function of the microscopic bulk modulus K_s.

4 Conclusion

Our former developed multi-scale version of Terzaghi's effective stress principle for unsaturated expansive clays accounting for electro-chemico-hydro-mechanical coupling at different length-scales is applied to numerically simulate swelling pressure tests at constant volume. The comparison between our numerical results and experimental data performed at the CIEMAT highlight the ability of our model in capturing the contribution of microscopic capillary pressure and nanoscopic disjoining forces to the overall swelling pressure. The main conclusions are:

- The model is able to capture water transfer from microscopic inter particle to nanoscopic inter platelet pores during hydration. As a consequence the increase of the swelling pressure is the result of two phenomena: Initially the high capillary pressure is completely compensated by the disjoining pressure. At full saturation where capillary pressure vanishes the swelling pressure corresponds to the effective disjoining pressure with a somewhat lower magnitude due to a higher inter platelet distance within the clay agglomerates at the nano-scale.
- In spite of classical PB theory remaining valid at high saturation, the molecular character of the electrolyte solution has to be taken into account for accurately modelling the disjoining pressure at small inter-platelet distances. By selecting a nanoscopic description of a simple HS fluid confined between two planar hard walls the disjoining pressure has been computed using classical DFT. Notwithstanding the simplicity of our approach it leads to an improvement of the nanoscopic description particularly at low hydration.
- The microscopic bulk modulus plays the role of a scaling parameter between disjoining and capillary effects. The numerical results show that in order to obtain swelling pressures in the MPa regime, K_s has to be of the same order of magnitude, which is significantly lower than the values of the overall bulk modulus typically found in the literature. The overall rigidity of the medium may thus be ensured by nanoscopic disjoining forces.

Some shortcomings still persist such as the overestimation of the swelling pressure and the loss of accuracy in reproducing the decrease of the swelling pressure during water intake due to a rearrangement of the clay clusters. Such behaviour is tied up to a collapse of the macrostructure when reaching the LC line. Further work is ongoing to include plastic deformation.

References

1. C. Imbert, M.V. Villar, Appl. Clay Sci. **32**, 197 (2006)
2. A. Gens, B. Valleján, M. Sánchez, C. Imbert, M.V. Villar, M.Van Geet, Géotechnique **61(5)**, 367 (2011)
3. A. Gens, E.E. Alonso, Can. Geotech. J. **29**, 1013 (1992)
4. E.E. Alonso, J. Vaunat, A. Gens, Eng. Geol. **54**, 173 (1999)
5. M. Sánchez, A. Gens, L. do Nascimento Guimarães, S. Olivella, Int. J. Numer. Anal. Meth. Geomech. **29**, 751 (2005)
6. I.J.A. Brackeley, Int. *Conf. Expansive Soils* 1, **169** (1973)
7. H. Komine, N. Ogata, Can. Geotech. J. **31**, 478 (1994)
8. C. Moyne, M.A. Murad, Transport Porous Med. **62**, 333 (2006)
9. J. Mainka, M.A. Murad, C. Moyne, S.A. Lima, Vadose Zone J. **13(5)** (2014)
10. E.E. Alonso, A. Gens, A. Josa, Géotechnique **40(3)**, 405 (1990)
11. A. Lloret, M.V. Villar, M. Sánchez, A. Gens, X. Pintado, E.E. Alonso, Géotechnique **53(1)**, 27 (2003)
12. J.-L. Auriault, C. Boutin, C. Geindreau, *Homogenization of Coupled Phenomena in Heterogeneous Media* (John Wiley & Sons, Hoboken, NJ, 2009)
13. J.-M. Pereira, H. Wong, P. Dubujet, P. Dangla, Int. J. Numer. Anal. Meth. Geomech. **29(11)**, 1127 (2005)
14. P.F. Low, Langmuir **3**, 18 (1987)
15. A. Olesky, J.-P. Hansen, J. Chem. Phys. **132**, 204702 (2010)
16. R. Roth, J. Phys.: Condens. Matter **22**, 063102 (2010)

DOI: 10.1051/e3sconf/20160914024

17. J.-P. Hansen, I.R. McDonald, *Theory of Simple Liquids* (Elsevier Science, 3rd edition, 2006)
18. E. Ferrage, B. Lanson, L.J. Michot, J.-L. Robert, J. Phys. Chem. C **114**, 4515 (2010)
19. R. T. Cygan, J-J. Liang, A.G. Kalinichev, J. Phys. Chem. B **108**, 1255 (2004)
20. A.-M. Tang, *Applied Unsaturated Soil Mechanics*, MUSE School, Paris, May (2006)
21. B.E. Hornby, L.M. Schwartz, J.A. Hudson, Geophysics **59(10)**, 1570 (1994)
22. C.M. Sayers, Geophys. Prospect. **61**, 416 (2013)

DOI: 10.1051/e3sconf/20160915001

Rainfall-induced landslides in Quaternary soils in Norway

Håkon Heyerdahl[1,2,a]

[1]*PhD student, University of Oslo, Norway*
[2]*Norwegian Geotechnical Institute (NGI), Oslo, Norway*

Abstract. Increased intensity of rainfall in later years seems to result in increased frequency of rainfall-induced landslides in natural slopes with glacial Quaternary deposits, which cover large parts of Norway. Many slides hit railway and road infrastructure, and sometimes dwellings. Typically, the soil matrix has a high content of intermediate soils (sand and silt). Natural soil slopes may have inclinations above the effective friction angle of the soils. Slope stability hence must rely on some cohesion, which often is "apparent", i.e. caused by negative pore-water pressure (suction). Dissipation of suction during short- or long-term rainfall hence results in reduced shear strength of the slope, and may lead to slope failure. There is a scarcity of data for Norwegian soils for thorough analysis of landslide triggering based on unsaturated geomechanics. More data may result in improved landslide warning. In this paper, results from unsaturated shear box testing of samples of a silty sand taken from a landslide site in Eastern Norway are presented and discussed.

1 Introduction

Soils encountered in Norway are predominantly of Quaternary origin. Marine clay and silt were deposited in the sea. The former seabed has later been lifted onshore by isostatic uplift when the glaciers covering Northern Europe gradually melted. Above the marine limit of today's landscape, other types of glacial sediments are encountered, primarily till and glacio-fluvial deposits, and to some extent glacio-lacustrine deposits. In addition to these two groups, post-glacial deposits of all varieties are encountered, like fluvial deposits, colluvium, scree/talus, landslide debris and sometimes eolic sediments.

2 Landslide morphology

Rainfall-induced landslides in Norway often occur in deposits where the soil matrix consists of intermediate soils (silt and sand). Above the marine limit, slides occur frequently in lateral moraine deposits on slopes of U-shaped valleys. Although the content of coarse material may be substantial in moraines, finer grains in the silt and sand fraction are abundant and generally govern the behaviour. Initial slides released by rainfall are often shallow or superficial with typical depths of 1-2 meters below the ground surface. Slides initiated on open slopes or along small creeks may quickly develop into highly mobile debris slides or debris flows due to high water content and erosion and entrainment along the moraine-covered valley sides. The landslide debris may be very mobile, and runout often affects transportation infrastructure or buildings down-slope of the release zone, resulting in e.g. closed railroads and highways.

Sandy and silty slopes are vulnerable to subsequent erosion, if surface water finds its way into the landslide scar, and especially so if the vegetation cover first has been damaged or removed. Slides occur in fluvial, glacial, glacio-fluvial and glacio-lacustrine deposits with varying composition. Below the marine limit, rainfall-induced slides occur in marine silts, and to some extent also marine clays.

A study of meteorological conditions leading to the release of a high number of slides in primarily marine sediments in Eastern Norway during the autumn year 2000 indicated that long-term rainfall rather than intense scours was the primary landslide trigger [1].

Slides in coarser sediments, e.g. glacial deposits on steep valley sides, generally are released as the result of short-term rainfall with high intensity, or as a combination of short- and long-term rainfall. The critical intensity of infiltration for triggering of debris flows in glacial deposits has been observed to depend on duration of infiltration, i.e. combined effect of rainfall and snowmelt [2]. Release of slides in fine-grained soils generally requires prolonged rainfall, rather than short-term intense scours.

The effect of the vegetation's root system on the soil stability should be considered and may be of substantial importance where the root zone goes down to some depth. The shear strength of roots is difficult to quantify, but advances are being made [3]. For grassy slopes, however, the root zone can be of limited thickness, maybe only a few centimeters or somewhat more (Figure 1), and critical failure planes may cut underneath it.

[a] Corresponding author: hhe@ngi.no

Natural slopes are exposed to rainfall infiltration depending on the climatic conditions of the region where the slope is located. Slopes will eventually find a geometry somewhat in balance with the local climatic and hydrological conditions. Alternatively, if the slope is not in balance with the prevailing climatic/hydrologic conditions, the slope will be subject to continuous processes as erosion and landslides.

Both for glacial and non-glacial sediments, natural deposits may form natural slopes that may be very steep. In areas with glacio-lacustrine silt, vertical soil slopes of tenths of meters are encountered. Such slopes may keep stable as long as the slope surface is protected towards direct infiltration from rainfall, and suction may continue to keep the slope stable without dissipating. When extreme rainfall occasionally hits such a slope, the result may be that the slope surface fails.

Suction is assumed to contribute in keeping many steep natural slopes stable, but the role of unsaturated geomechanics for a wide range of Norwegian soil types has not been investigated extensively. In the following sections, data from laboratory testing of a sand from Eastern Norway is presented.

3 Landslide site

Samples of a silty sand for laboratory testing were collected manually at a landslide site at the farm Negarden Sander in the municipality of Eidsvoll, Eastern Norway (Figure 1). The foundations of the nearest building were almost undermined, but the building was saved due to repair works of the landslide scar (Figure 1).

Unsaturated shear box tests were performed on intact specimens from sample Sand A1 (Table 1).

Table 1. Sand sample. Fractions of sand, silt and clay.

Sample	Depth	Sand	Silt	Clay	C_u
ID	m	%	%	%	-
Sand A1	0.55-0.80	71	~27	~2	3.1

Figure 1. Landslide in sand and silt, Negarden Sander, Eidsvoll in Eastern Norway in the autumn of year 2000. Seepage of ground water visible from the foot of the landslide scarp.

The samples were brought to Barcelona for testing in the geotechnical laboratory at the Catalonian Technical University (UPC).

The landslide was released in the unusually rainy autumn of year 2000 [1] in a slope consisting of 5 m layered sand and silt overlying a thick marine clay deposit. The landslide is typical of many landslides in the region, often released in silty and sandy deposits. The landslide debris was highly mobile (Figure 2).

Main soil layering consists of silty sand and sandy silt layers down to 5 m, followed by a thick deposit of marine clay to large depth. Ground water level is assumed to lie at approx. 5 m depth, i.e. at the top of the clay layer (Figure 1). The top section down to 1.5 m depth was inspected from an excavated pit. Main layers are described as follows: Silty sand above 0.8 m and below 1.3 m depth, a clay lens from 0.8 to 1.0 m and sandy silt between 1.0-1.3 m (Figure 3).

Figure 2. Highly mobile runout from landslide in silty sand at Negarden Sander, Eidsvoll in Eastern Norway, in the autumn of year 2000.

In situ tensiometer measurements down to 1.2 m at the time of soil sampling confirmed that suction up to approx. 25 kPa was present in the soil (Figure 3), in the order of 10 kPa in the sand specimens tested (depth 0.6-0.7 m), and just below 25 kPa in the silt layer.

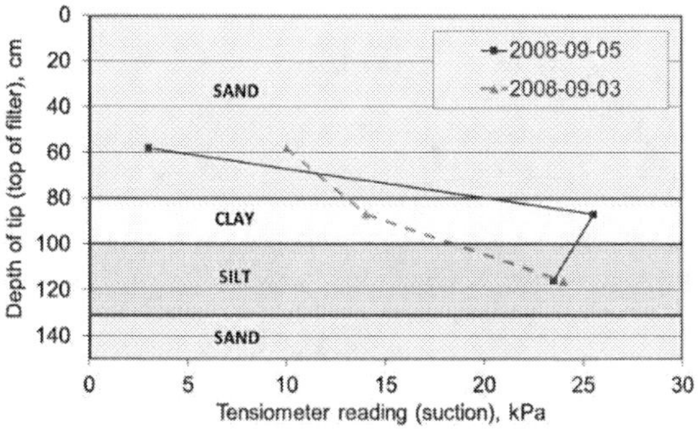

Figure 3. Measurement of suction during soil sampling.

A rainfall on the day prior to sampling had significant effect on the suction profile. Reduced suction in the sand layer indicates increased saturation due to infiltration of water in the upper part of the profile. However, suction in the clay layer increased in the same period, indicating some time lag. For the deepest tensiometer, no significant suction change was measured. The tensiometer tip was positioned in the silt layer below the clay layer. Infiltration in this layer may be prevented by the cap layer of less permeable clay, and suction remains almost unchanged after the rainfall (Figure 3).

4 Retention curve

The retention curve of the sand was measured combining several laboratory test methods: grain size distribution, high capacity tensiometer, mercury intrusion porosimetry, psychrometer and negative water column [4]. The main drying branch of the retention curve is presented in Figure 4. The air entry value is 5 kPa, and the residual water content is 3.35 % (for suction 27 kPa).

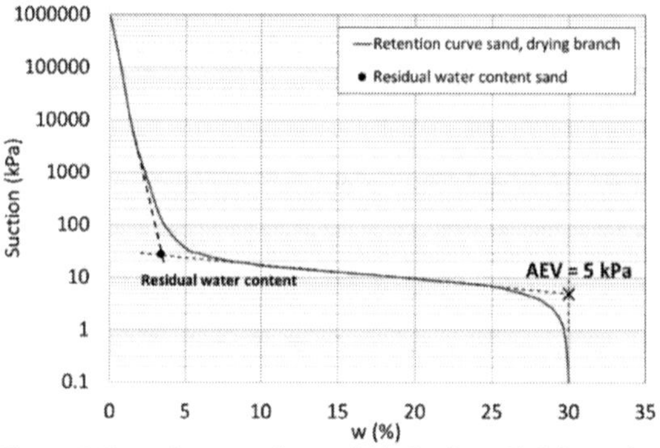

Figure 4. Retention curve for sand (drying branch) [4]. Suction versus gravimetric water content.

5 Shear tests

Results from two multi-stage unsaturated shear box tests on intact sand specimens are presented in the following. The tests were performed using an unsaturated shear box at the geotechnical laboratory at UPC, Barcelona. The tests were performed as an introductory part of a test program, primarily focussing on unsaturated shear strength of silt sediments [5], and do not give a complete picture of the unsaturated strength of the sand layer. However, the data still give some indications of interesting features for the unsaturated strength of sand.

The shear box apparatus is similar to the one described in [6], however instead of one pressure chamber, the apparatus has one upper and one lower pressure chamber. Pressure in the upper chamber is used to control vertical load. The shear box is placed inside the lower pressure chamber on top of a porous disk with air entry value 500 kPa. A water outlet exits from the underside of the disk. The water outlet may be connected to a GDS pump in order to control matric suction by axis translation, in combination with controlled air pressure in the lower test chamber. Volume of water in and out of the specimen is found by monitoring volume change with time in the GDS pump. Air diffusion through the porous disk will affect the total volume change. The rate of air diffusion through the disk was measured together with the water permeability of the disk in a separate test (Figure 5). When the added water has flowed through the porous disk, the volume change continues at a lower rate, representing air diffusion. From the rate of air diffusion, net water flow during a test can be computed from measured total volume change in the GDS pump, excluding the volume related to air diffusion.

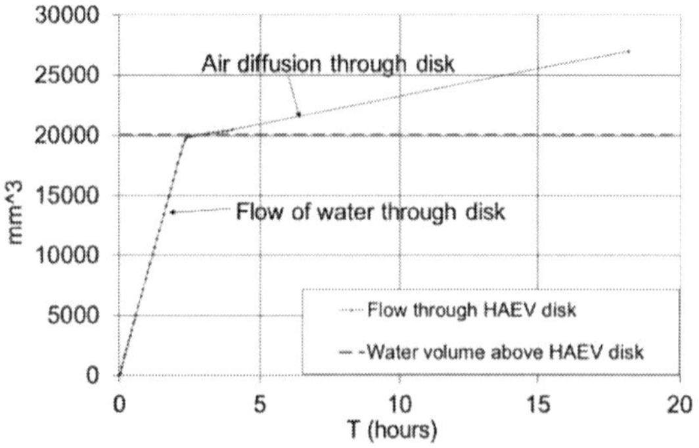

Figure 5. Measurement of water permeability and air diffusivity through porous disk of unsaturated shear box apparatus.

Test conditions and horizontal displacement rates during shear tests are given in Figure 6 and Table 2. In test Sand S2, vertical net stress was changed from step to step while suction was kept constant. Suction was controlled by the axis translation technique [7]. In test Sand S3, the specimen was saturated prior to the shear test. The saturated test was performed with water inside the test chamber. In the next test step, suction was applied by use of negative water column: A water-filled tube was connected to the water outlet from the water compartment below the high air entry value disk, which is used to control suction when applying axis translation. Suction was established by allowing the tube to suck water from the specimen from one day to the next, before shearing the specimen.

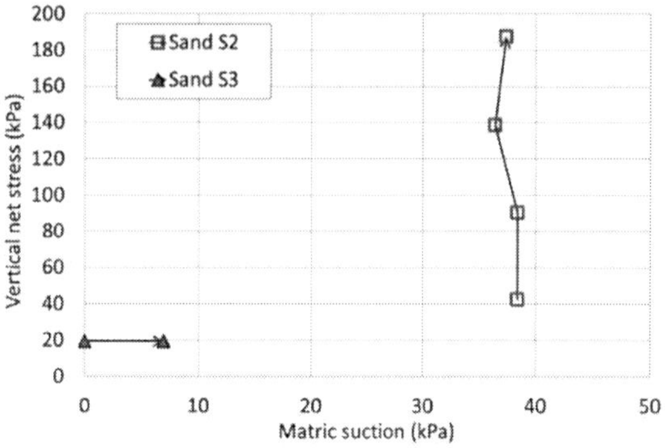

Figure 6. Load paths for unsaturated shear tests. Arrow tips indicate last load stage in each test.

Table 2. Unsaturated shear tests on sand.

Test	Depth	Initial condition	Displacement rate
ID	m	-	mm/min
Sand S2	0.62	$w_{\text{in situ}}$	0.005
Sand S3	0.66	$w_{\text{saturated}}$	0.015

The measured peak shear stress in test Sand S2 increases for increased vertical net stress (Figure 7). The vertical compression is positive throughout all test steps. Applied vertical net stress in test Sand S2 is higher than in situ stress, which makes compression expected.

Test Sand S3 is performed at vertical net stress only slightly higher than estimated in situ stress. The behaviour is still compressive during the shear phase (Figure 8). Peak shear stress increased from the first to the second load step, when suction was applied. Because the shear box had to be disassembled to remove water between the two load steps of test Sand S3, the track of total displacements was lost. For total displacements it was assumed elastic unloading/reloading cycle between the two load steps (Figure 8).

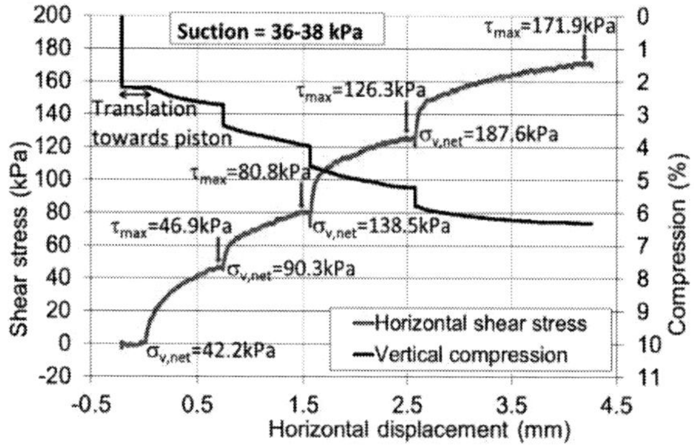

Figure 7. Shear test Sand S2, shear stress versus horizontal displacement. Four load steps increasing vertical net stress at approx. constant matric suction (36-38 kPa).

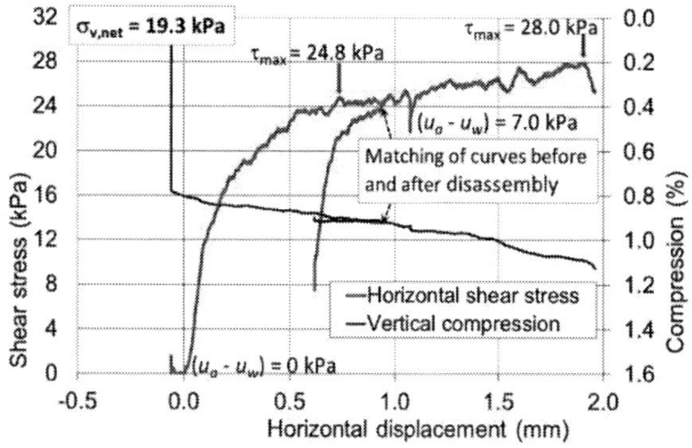

Figure 8. Shear test Sand S3, shear stress versus horizontal displacement. Two load steps, starting with saturated specimen. Approx. constant vertical net stress (19.3 kPa).

Measured peak shear stress was plotted versus vertical net stress (Figure 9) and suction (Figure 10), respectively. Shear tests were performed at different levels of applied vertical net stress (Figure 6). Peak shear stress values presented in Figure 9 and Figure 10 may therefore not be compared directly, since shear strength varies with both vertical net stress and matric suction. To isolate effect of suction on shear strength, measured peak shear stress values were translated to the zero vertical net stress plane [8] using the effective friction angle, φ' = 41°, interpreted from test Sand S2 (Figure 9). An effective cohesion *c'* of 8 kPa was interpreted from the saturated load step of test Sand S3 (Figure 11). The second load step of test Sand S3 shows a substantial increase in shear strength when a suction of 7 kPa is applied, with approx. 40 % higher peak shear stress than in the saturated load step.

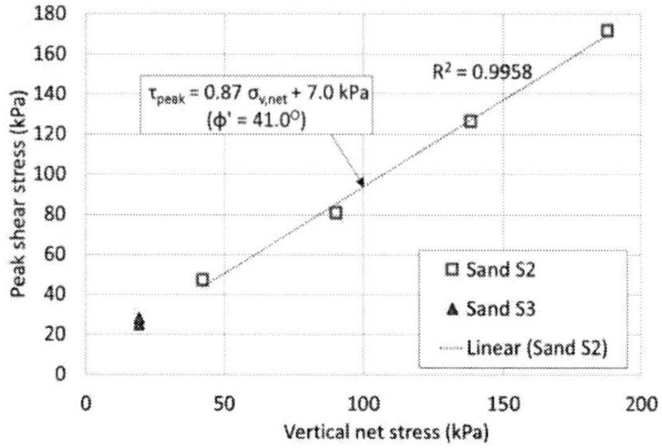

Figure 9. Peak shear stress versus vertical net stress for shear tests on sand.

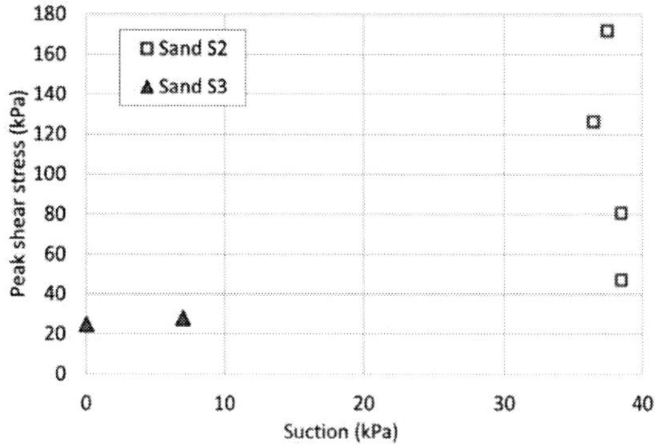

Figure 10. Peak shear stress versus matric suction for shear tests on sand.

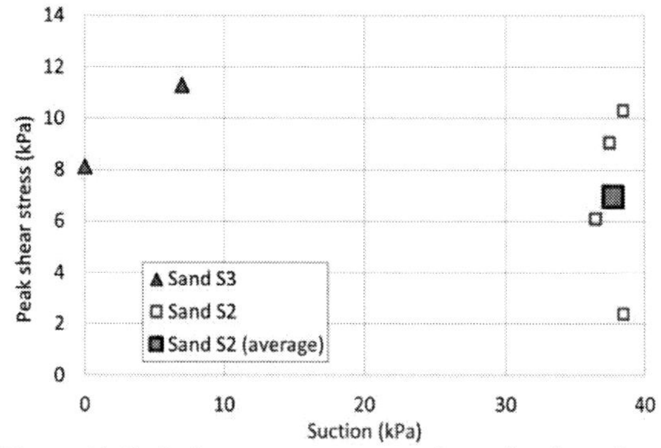

Figure 11. Peak shear stress versus matric suction from shear tests on sand, translated to zero vertical net stress. Average value of four measured values from test S2 is shown.

For test Sand S2, translation of comparatively large vertical net stress values to the zero vertical net stress plane results in some data spread, because the relationship between peak shear stress and vertical net stress is not completely linear (Figure 9). As a result, the back-calculated peak shear stress values for zero vertical net stress are not equal, although applied suction is the same (+/-1 kPa) for the four data points. The mean value of the four data points from test S2 is approx. 7 kPa (Figure 11), indicating that the effect of increased suction is small (or even negative) when compared to measured peak shear stress in test Sand S3. Similar observations of decreased unsaturated shear strength above some limiting suction in fact was observed already in early unsaturated research on sand [9].

6 Odometer test

The relatively high effective cohesion (for a sand) found from shear test Sand S3 (Figure 11) could be related to over-consolidation and mechanical interlocking of grains. A conventional odometer test with incremental stepwise loading was performed on an intact sand specimen (Table 3). In situ vertical stress is estimated to approx. 5 kPa. The results may indicate that the soil actually is overconsolidated (Figure 12). For comparison, vertical strains from consolidation steps of shear tests Sand S2 and Sand S3 (values from before shearing) are also included in Figure 12. However, care should be taken with respect to interpretation of overconsolidation from odometer tests on sand.

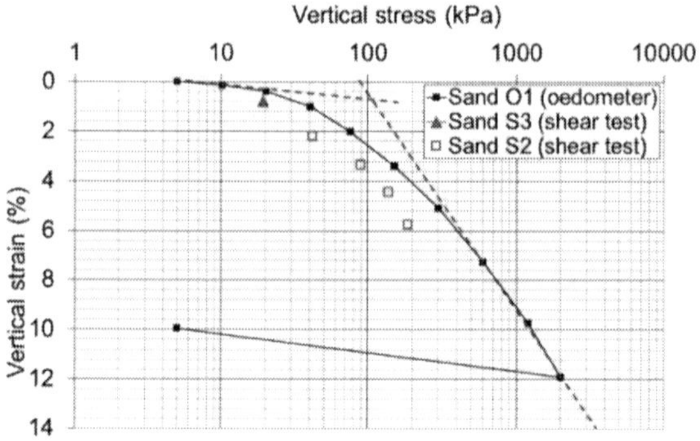

Figure 12. Saturated odometer test O1. Vertical strain versus vertical stress at end of step (logarithmic 1st axis). Also shown: Vertical strain at end of consolidation from shear tests Sand S2 and Sand S3 (vertical strain before shearing).

No notable wetting collapse (volume reduction) was observed during initial saturation of the initially unsaturated odometer specimen. Since vertical net stress was low (< 1 kPa) in the specimen during saturation, collapse during wetting would expectedly be moderate, compared with saturation performed at a larger vertical net stress [10]. On the other hand, the sand has already experienced a large number of wetting and drying cycles in situ, due to the climatic setting and small in situ depth (0.2-0.3 m), and thereby expectedly less prone to wetting collapse. Suction stress resulting from drying could also result in over-consolidation [11].

Two important differences could be found from a comparison of the observed behaviour of these intact soil specimens with the collapsing behaviour of loess sediments experiencing wetting: Firstly, loess deposits are extremely well sorted materials, which may be deposited as metastable structure relying on capillary suction in menisci to keep stable (and sometimes some cementation). The soil skeleton will collapse when the stabilizing forces from the menisci at the grain contacts disappear during wetting. Secondly, the climate under which such eolic sediments are often encountered, e.g. in north-central China, however is very dry, which will keep the sediments stable and prevent sliding [12], except for during extreme rainfall events.

7 Predicted unsaturated shear strength

The unsaturated strength τ_f was predicted from the two commonly used unsaturated stress state variables, i.e. vertical net stress $(\sigma_n - u_a)$ and matric suction $(u_a - u_w)$ [13]. The retention curve (Figure 4) and effective shear strength parameters interpreted from shear test S2 (Figure 9, Figure 11) were used for the prediction, applying "Vanapalli's 1st method" [14] (Eq. 1).

$$\tau_f = [c' + (\sigma_n - u_a)\tan\varphi'] + (u_a - u_w)[\Theta^\kappa \tan\varphi'] \qquad (1)$$

in which $(\sigma_n - u_a)$ is net normal stress on the failure plane at failure, normalized water content Θ is given by $\Theta = \theta/\theta_s$ where θ and θ_s are volumetric and saturated volumetric water contents, respectively, κ is an exponent accounting for the relation between normalized water content and the normalized area of water in the pores, a_w, where $a_w = \Theta^\kappa$. The normalized water content is a function of suction and of the retention curve (Figure 4). Good correlation between experimental results and predicted shear strength for a value of κ equal to 2.2 was found for shear tests on a compacted glacial till [14].

A comparison between measured and predicted shear strength gives good agreement (Figure 13). This is not surprising, since the effective friction angle φ' was interpreted from the same data set. Vertical net stress is relatively high and dominates the total value of measured shear strength, in particular for test Sand S2.

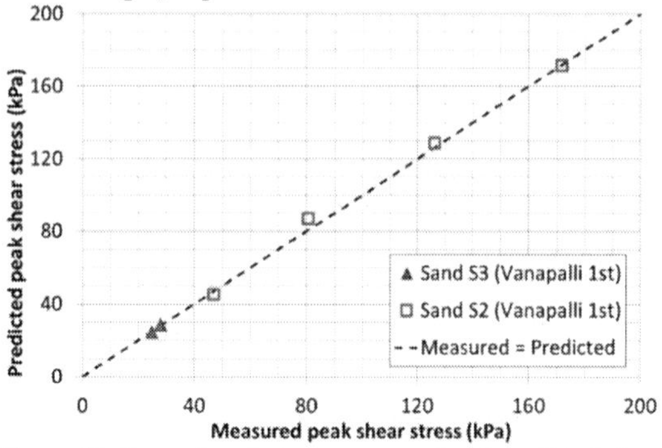

Figure 13 Comparison between measured and predicted shear strength for tests Sand S2 and Sand S3 using prediction method "Vanapalli 1st" [14].

For evaluation of suction effects on unsaturated shear strength, measured shear strength data translated to the zero vertical net stress plane (Figure 11) were compared with predictions applying Eq. (1) (Figure 14).

The predicted shear stress has a good match with measured peak shear stress (translated to zero vertical net stress) at suction equal to 7 kPa (Figure 14). Applied suction is slightly higher than the air entry value of 5 kPa (Figure 4). The predicted shear strength is non-linear below the air entry value, but no test data are available for this suction range.

The predicted non-linearity of the shear strength curve in Figure 14 is a result of the saturation rate being lower than unity below the air entry value (Figure 4). The tangential definition of the air entry value [15] characterizes the retention curve in a consistent way, but the air entry value generally does not correspond with the value where drainage of water from the soil actually starts. From an initial steep increase of shear strength as suction increases at a high saturation rate, the inclination of the predicted shear strength curve drops gradually as water is drained from the soil. The predicted peak shear strength is located close to the peak shear stress from the second load step in shear test Sand S3, for which the suction was 7 kPa. Increasing suction further, the predicted shear stress drops gradually to a value just above the effective cohesion. This means that increased suction above some limiting value gives practically no contribution to the shear strength. Test Sand S2 gives some confirmation of this effect; considering either the average value of the four load steps of test Sand S2 or each of the four data points (translated to zero vertical net stress), they all lie below predicted and measured peak shear strength for a suction value of 7 kPa (Figure 14).

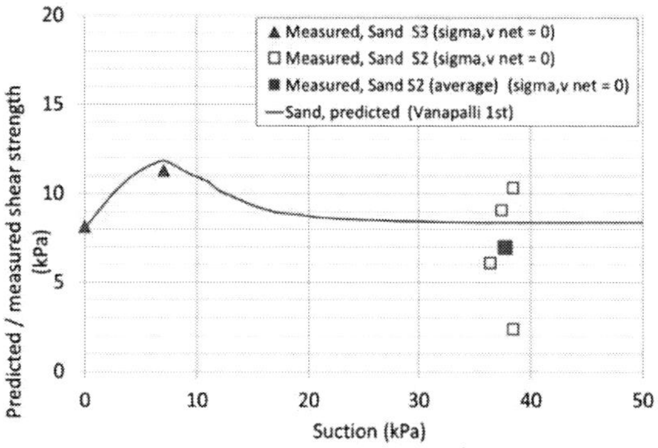

Figure 14 Predictions with the "Vanapalli 1st" method [14] and measured shear strength versus suction for shear tests on sand, translated to zero vertical net stress. Average point of four test steps at constant suction (37-38 kPa) included for test Sand S2.

The ultimate shear strength for large suction (above suction corresponding to residual water content, Figure 4) seems to approach the effective cohesion, which on the other hand is equal to the shear strength for zero suction (and zero vertical net stress). The highest shear strength along the suction axis is reached for some suction value depending directly on the shape of the retention curve; in general, the maximum shear strength coincides with the maximum of the product $(u_a-u_w)\Theta^{\kappa}$ [14]

For evaluation of slope stability, the results above indicate that the shear strength, and hence the slope stability, will be about the same for completely dry and for fully saturated conditions. In both cases, there is no effect of suction to stabilize the soil, and the shear strength contribution as a function of suction is reduced to the value of the effective cohesion. In this case, the low suction range (0-15 kPa) in fact will give the best theoretical stability. This however holds only as long as pore-water pressure remains negative (i.e. as long as there is soil suction in situ). If a soil is close to full saturation, positive pore-water pressures may quickly develop as a result of infiltration, which will then act to reduce also the shear strength contribution from the other stress state variable, i.e. vertical net stress, since σ_n-u_a for saturated soil will be equal to σ_n-u_w or Terzaghi's effective stress. The stabilizing effect of suction on slope stability will consequently be reversed with saturation. Hence, for slope stability evaluations, the dry side of the peak is still a safer place than the wet side, although the theoretical suction effect on stability may be marginal, at least for the sandy soil presented herein.

References

1. C. Jaedicke, A. Kleven, Hydr. Proc. **22** (2008)
2. F. Sandersen, S. Bakkehøi, E. Hestnes, K. Lied, Proc. 7th ISL (1996)
3. G. J. Meijer, A. G. Bengough, J. A. Knappett, K. W. Loades, B. C. Nicoll, Proc. XVI ECSMGE **4** (2015)
4. H. Heyerdahl, T. Pabst (in prep.)
5. H. Heyerdahl, in prep.
6. V. Escario, J. Saez, Geot. **36**, 3 (1986)
7. J.W. Hilf, US Bureau of Reclamation, *Tech. memo.*, **654** (1956)
8. D. G. Fredlund, H. Rahardjo, J. K. M. Gan, Proc. 6th Int. Conf. On Expansive Soils **1** (1987)
9. I. B. Donald, Proc. 2nd Australia-New Zealand Conf. Soil Mech. Found Eng. (1956)
10. A. Tarantino, Mech. of unsaturated geomaterials (2010)
11. E. E. Alonso, A. Gens, A. Josa, Geot. **40**, 3 (1990)
12. E. Derbyshire, T. A. Dijkstra, I. J. Smalley, Y. Li, Quat. Int. **24** (1994)
13. D. G. Fredlund, N. R. Morgenstern, J. of the Geot. Eng. Div., ASCE **103**, GT5 (1977)
14. S. K.Vanapalli, D. G. Fredlund, D. E. Pufahl, A. W. Clifton, Can. Geot. J. **33**, 3 (1996)
15. D. G. Fredlund, H. Rahardjo, Soil mechanics for unsaturated soils (1993)

Table 3. Summary of volumetric properties for tested sand specimens

Sample	*Specimen*	*Test*	*Depth*	γ_d	n	e	$w_{in\ situ}$	$S_{r,in\ situ}$	$\gamma_{in\ situ}$	w_{sat}	γ_{sat}
ID	ID		m	kN/m³	%	-	%	-	kN/m³	%	kN/m³
A1	Sand S1	Shear box	0.56-0.58	13.80	46.9	0.88	15.81	0.47	15.98	33.36	18.40
	Sand S2		0.61-0.63	14.25	45.2	0.82	12.81	0.41	16.08	31.10	18.68
	Sand S3		0.65-0.67	14.50	44.2	0.79	12.59	0.42	16.32	29.92	18.83
P4	Sand O1	Odometer	0.24-0.28	15.48	40.5	0.68	-	-	-	25.64	19.44

 DOI: 10.1051/e3sconf/20160915002

Effects of soil-vegetation-atmosphere interaction on the stability of a clay slope: a case study

Giuseppe Pedone[1,a], Aikaterini Tsiampousi[2], Federica Cotecchia[3] and Lidija Zdravkovic[2]

[1] *Geotechnical Consulting Group, London, UK*
(formerly Technical University of Bari, Department of Civil, Environmental, Territorial, Building and Chemical Engineering, Bari, Italy)
[2] *Imperial College London, Department of Civil & Environmental Engineering, London, UK*
[3] *Technical University of Bari, Department of Civil, Environmental, Territorial, Building and Chemical Engineering, Bari, Italy*

Abstract. Deep and slow landslide processes are frequently observed in clay slopes located along the Southern Apennines (Italy). A case study representative of these processes, named Pisciolo case study, is discussed in the paper. The geo-hydro-mechanical characteristics of the materials involved in the instability phenomena are initially discussed. Pluviometric, piezometric, inclinometric and GPS monitoring data are subsequently presented, suggesting that rainfall infiltration constitutes the main factor inducing slope movements. The connection between formation of landslide bodies and slope-atmosphere interaction has been demonstrated through a hydro-mechanical finite element analysis, whose results are finally reported in the work. This analysis has been conducted employing a constitutive model that is capable of simulating both saturated and unsaturated soil behaviour, as well as a boundary condition able to simulate the effects of the soil-vegetation-atmosphere interaction.

1 Introduction

Deep and slow landslides are recurrently observed in clay slopes in the Southern Apennines (Italy) [1, 2]. Despite the fact that the landslide movements are slow, the displacement rates show seasonal variations, with trends similar to those characterising the pore pressures. Hence, the landslide activity appears to be influenced by the hydraulic regime of the slope, which in turn depends on the soil-vegetation-atmosphere interaction.

The correlation between the landslide movements, the pore pressure changes and the slope-atmosphere interaction was assessed using numerical analyses [3, 4]. In particular, fully coupled hydro-mechanical analyses of the Pisciolo case study [5], which is representative of the above mentioned processes, were performed. The present paper briefly introduces the case study and reports some results of the numerical analysis.

2 Pisciolo case study

The Pisciolo slope has been intensively studied in the last decade, as severe damage has been caused on existing infrastructures from landslide bodies (Fig. 1) [5]. In particular, the high landslide risk associated with a 2 m diameter pipeline has prompted a wide geotechnical investigation campaign, aimed at identifying the landslide causes and designing appropriate stabilisation measures.

The geo-morphological setting of the slope (Fig. 1) has been defined from stereoscopic analyses of aerial photos and monitoring data interpretation. It is characterised by the presence of slow to extremely slow roto-translational landslides [6], whose shear bands vary from intermediate to deep (up to 60 m of depth) [7]. The deepest shear bands occur in the southern portion of the unstable area (Fig. 1) and appear to be characterised by a retrogressive activity [5].

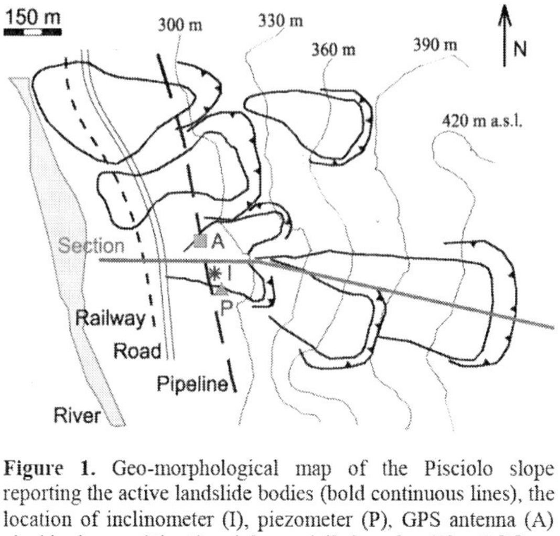

Figure 1. Geo-morphological map of the Pisciolo slope reporting the active landslide bodies (bold continuous lines), the location of inclinometer (I), piezometer (P), GPS antenna (A) cited in the text (Fig. 2) and the modelled section (Fig. 4) [5].

[a] Corresponding author: g.pedone@gcg.co.uk

Failure mainly involves the fissured clay part of a turbiditic formation, which is a transition between the Numidian Flysch and the Red Flysch. The transition soils have a clay fraction of 40-60% and are characterised by high activity (0.85 on average), low mechanical strength (c'=0 kPa and $\varphi'=20.7°$ for $p'<300$ kPa) and high saturated permeability ($k_{in\ situ}>10^{-9}$ m/s) [5, 8]. Such hydro-mechanical behaviour is primarily due to the fissured meso-structure of the clay, which impoverishes the mechanical strength properties and increases the hydraulic conductivity with respect to that of the same material when unfissured [1-4, 8-11]. Fissuring also affects the unsaturated soil behaviour [4, 5, 8], as observed when subjecting the Pisciolo clay to drying-wetting cycles, during which the suctions have been measured by means of the filter paper technique [12] and the Imperial College high capacity tensiometers [13].

The movements of the landslide bodies have been monitored by means of both inclinometers and GPS sensors (Fig. 2), that recorded the maximum velocities at the end of winter. The landslide activity appears to be connected to the hydraulic regime of the slope, because the highest displacement rates are observed concurrently with the maximum pore pressures (Fig. 2). In turn, the hydraulic regime seems to be strictly connected to the pluviometric regime of the area, as observed by analysing the 6-month cumulative rainfall (Fig. 2). The latter clearly shows the seasonal character of the pluviometric events in the South of Italy, where rainfalls are more frequent from late autumn to mid spring, while they reduce drastically during late spring and summer [5].

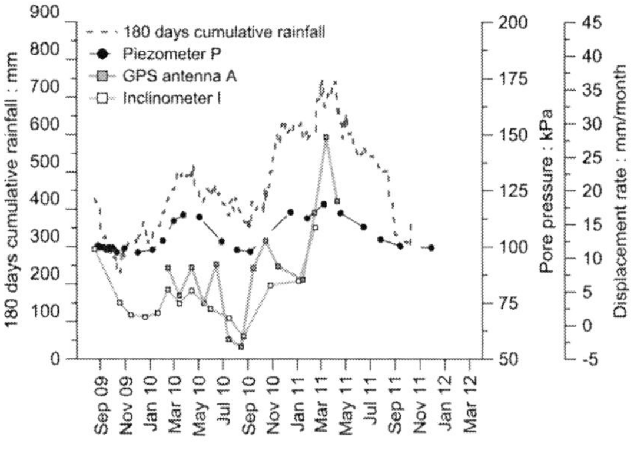

Figure 2. 180 days cumulative rainfall measured at a weather station located in Melfi (Italy), pore pressures measured at 15 m depth (piezometer P in Fig. 1), displacement rates measured at 18 m depth (inclinometer I in Fig. 1) and at the ground level (GPS antenna A in Fig. 1) [4, 5].

The monitoring data (Fig. 2) suggest that the slope-atmosphere interaction represents the main triggering factor of the instabilities. Rainfall infiltration appears to induce an increase in pore pressure that generates a reduction of the available shear strengths, which in turn produces an acceleration of the landslide movements.

3 Numerical modelling

The effect of slope-atmosphere interaction on the stability of the Pisciolo slope was investigated by means of a 2D fully coupled hydro-mechanical analysis [3, 4], conducted with the finite element code ICFEP [14]. The code has fully developed capabilities with reference to unsaturated soil behaviour [15].

The numerical simulation was conducted using the constitutive model proposed by Tsiamposi et al. [16], which was developed within the framework introduced by Alonso et al. [17]. On the wet side of critical state, the yield and plastic potential functions implemented in the model correspond to those reported by Lagioia et al. [18]. On the dry side, in order to avoid overestimation of shear strength, a Hvorlsev surface is employed, whose shape is defined by means of two parameters, named n_{HV} and α_{HV} [16]. The former is a fitting parameter, while the latter represents the inclination of the Hvorslev surface at critical state (Fig. 3).

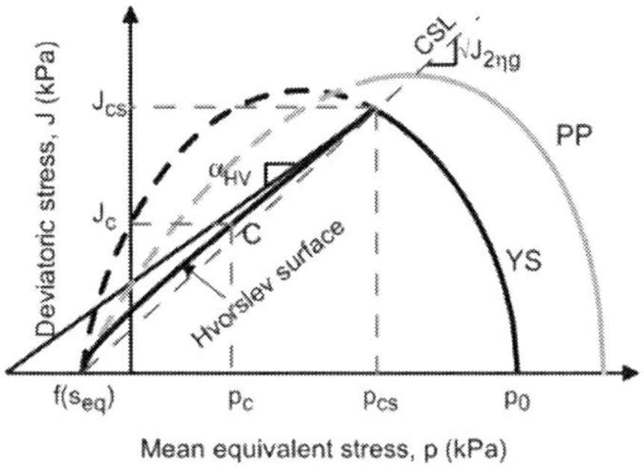

Figure 3. Yield surface (YS) and plastic potential (PP) for the constitutive model proposed by Tsiamposi et al. [16].

Additionally, a loading-collapse curve and a suction increase yield surface are employed, similar to Alonso et al. [17]. Desaturation occurs when suction exceeds the air entry value, which has been considered equal to 1000 kPa in the present work. Under unsaturated conditions, apparent cohesion changes as a function of suction and degree of saturation [16]. Finally, the elastic coefficient of compressibility due to changes in suction, κ_s, is dependent on the degree of saturation, S_r, according to the expression proposed by Tsiamposi et al. [19]:

$$\kappa_s=\chi(S_r)^{\omega} \tag{1}$$

where χ and ω are fitting parameters.

The calibration of the constitutive model, with particular reference to the saturated soil behaviour, was performed through the simulation of oedometer and triaxial tests [3-5]. The oedometer tests allowed to identify the compressibility behaviour of the Pisciolo clay (well represented by the parameters $\lambda=0.0914$ and $\kappa=0.0365$), while the triaxial tests allowed to define the inclination of the critical state line (assumed as $M=0.689$), as well as the parameters that describe the shape of the yield function (equal to $n_{HV}=0.45$ and $\alpha_{HV}=0.3$).

 DOI: 10.1051/e3sconf/20160915002

With reference to the unsaturated soil behaviour, the model was employed in a simplified form, as only unconfined drying tests were available for the calibration process [4, 5, 8]. In particular, a linear loading collapse curve has been considered and all the suction-induced deformations have been assumed elastic, albeit non-linear, as the effect of degree of saturation on compressibility was taken into account when fitting the results of the unconfined drying tests (obtaining χ=0.1372 and ω=8.0). As reported by Cotecchia et al. [5], these tests have been also used to calibrate the retention behaviour of the material, which has been modelled using the equation proposed by van Genuchten [20].

Unsaturated hydraulic conductivities of the Pisciolo clay have not been measured, so a model similar to the one proposed by van Genuchten [20] has been used to estimate the hydraulic conductivity function. On the other hand, the saturated hydraulic conductivity has been defined according to the in situ permeability measurements reported by Cotecchia et al. [5] ($k_{in\ situ}$>10^{-9} m/s at depths lower than 30 m).

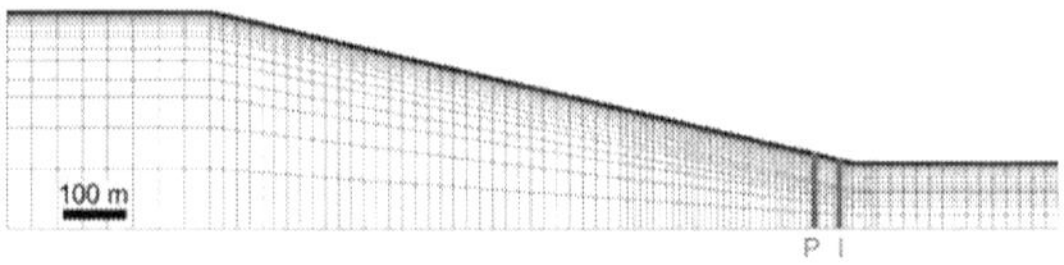

Figure 4. Finite element mesh used for the fully coupled hydro-mechanical analysis of the Pisciolo slope (modelled section in Fig. 1) [3, 4].

The numerical simulation has been conducted with reference to a longitudinal section representative of the most active portion of the slope (see Fig. 1). The finite element mesh (Fig. 4) was generated using isoparametric, quadrilateral, 8-noded elements, with two displacement degrees of freedom (DOF) on each node and pore water pressure DOF only on corner nodes.

The lateral boundaries of the finite element model have been horizontally constrained, while the bottom boundary has been fully fixed. As for the hydraulic boundary conditions, both lateral and bottom boundaries have been assumed as impervious, while net rainfalls have been applied on the top boundary of the mesh, using a dual hydraulic boundary condition [15, 21], which is capable of changing automatically from a prescribed flow to a prescribed pore water pressure boundary condition. This boundary condition makes it possible to account for run off, as rainfall infiltration stops when the pore pressures at the boundary nodes approach zero [15, 21].

Net rainfalls (Fig. 5) were calculated as the difference between gross rainfalls and evapo-transpiration fluxes, the latter estimated by means of the FAO Penman-Monteith method [22] according to temperature measurements only. Both gross rainfalls and temperatures have been measured at a weather station located in the town of Melfi (Italy), few kilometres away from the Pisciolo slope.

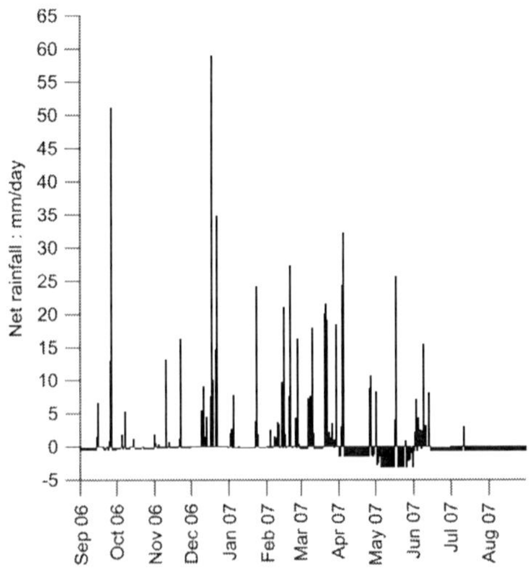

Figure 5. Net rainfalls used as boundary condition for the fully coupled hydro-mechanical analysis of the Pisciolo slope (Fig. 4) [3, 4].

The numerical model is able to reproduce seasonal pore pressure variations similar to those observed in situ (Fig. 6). These seasonal changes of the pore pressures induce the propagation of a shear band within the slope and the consequent formation of a landslide body that has geometrical features similar to those characterising the first-time failures observed within the Pisciolo slope (Fig. 7) [3,4].

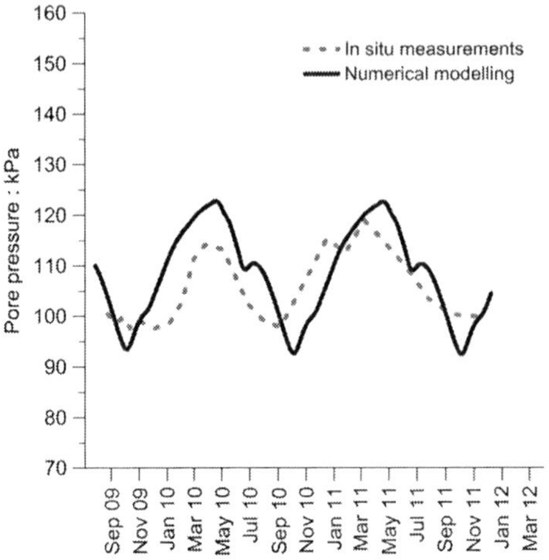

Figure 6. Comparison between numerical results (Fig. 4) and in situ measurements (Fig. 2) with reference to the pore pressures measured at the piezometer P located at 15 m depth (Fig. 1) [3].

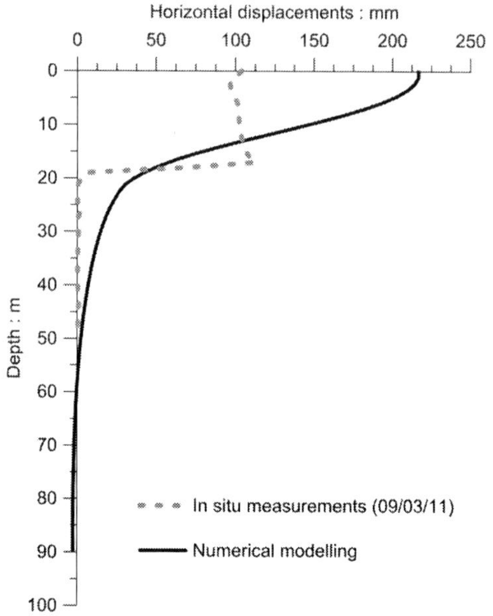

Figure 7. Comparison between numerical results (Fig. 4) and in situ measurements (Fig. 2) with reference to the inclinometer I (Fig. 1) [3].

4 Conclusions

The paper briefly introduces the Pisciolo case study [5], which is representative of the deep and slow landslide processes frequently observed in the Southern Apennines (Italy) [1-2]. A fully coupled hydro-mechanical finite element analysis confirmed that cumulated rainfall infiltration is the main factor triggering the instabilities. The numerical modelling was conducted with a constitutive model that is capable of simulating both saturated and unsaturated soil behaviour [16, 19], combined with a boundary condition able to simulate the effects of the slope-atmosphere interaction [15, 21]. In particular, the results of the numerical simulation prove that the soil-vegetation-atmosphere interaction is responsible for the formation of shear bands within the slope [3, 4].

Acknowledgements

Special thanks go to the Apulian Aqueduct Spa, that funded the research activities and provided precious information regarding the case study. The authors would also like to express their gratitude to researches and technicians of Technical University of Bari, CNR-IRPI of Bari and Imperial College London that supported the research project.

References

1. F. Cotecchia, C. Vitone, F. Santaloia, G. Pedone, O. Bottiglieri, Land. **12**, 5, 877-893 (2015)
2. F. Cotecchia, F. Santaloia, P. Lollino, C. Vitone, G. Pedone, O. Bottiglieri, Eur. J. Environ. Civ. Eng. **DOI**: 10.1080/19648189.2014.968744 (2014)
3. G. Pedone, A. Tsiampousi, F. Cotecchia, L. Zdravkovic, *First-time failures in clay slopes triggered by soil-vegetation-atmosphere interaction: numerical modelling of a case study in Southern Italy*, in prep.
4. G. Pedone, *Interpretation of slow and deep landslides triggered by slope-atmosphere interaction in slopes formed of fissured clayey turbidites* (PhD thesis, Technical University of Bari, 2014)
5. F. Cotecchia, G. Pedone, O. Bottiglieri, F. Santaloia, C. Vitone, It. Geotech. J. **1**, 34-61 (2014)
6. D. M. Cruden & D. J. Varnes, *Landslides: investigation and mitigation, special report* **247**, 36-75 (1996)
7. J. N. Hutchinson, *Proc. 5th Int. Symp. Land., Lausanne* **1**, 3-35 (1988)
8. G. Pedone, F. Cotecchia, O. Bottiglieri, J. Vaunat, *Hydraulic characterisation of a fissured clay*, in prep.
9. C. Vitone & F. Cotecchia, Géotech. **61**, 12, 1003-1018 (2011)
10. C. Vitone, F. Cotecchia, G. Viggiani, S. A. Hall, Int. J. Num. An. Meth. Geomech **37**, 1510-1534 (2013)
11. C. Vitone, F. Cotecchia, G. Viggiani, S. A. Hall, Ac. Geotech. **8**, 247-263 (2013)
12. F. A. M. Marinho & O. M. Oliveira, Geotech. Test. J. **29**, 3, 250-258 (2006)
13. A. M. Ridley & J. B. Burland, Géotech. **43**, 2, 321-324 (1993)
14. D. M. Potts & L. Zdravkovic, *Finite element analysis in geotechnical engineering: theory* (London, Thomas Telford, 1999)
15. P. G. Smith, *Numerical analysis of infiltration into partially saturated slopes* (PhD thesis, Imperial College London, 2003)
16. A. Tsiampousi, L. Zdravkovic, D. M. Potts, Comp. Geotech. **45**, 2, 156-166 (2013)
17. E. E. Alonso, A. Gens, A. Josa, Géotech. **40**, 3, 405-430 (1990)
18. R. Lagioia, A. Puzrin, D. M. Potts, Comp. Geotech. **19**, 3, 171-191 (1996)
19. A. Tsiampousi, L. Zdravkovic, D. M. Potts, Géotech. **63**, 2, 155-164 (2013)
20. M. T. van Genuchten, Soil Scie. Soc. Amer. J. **44**, 892-898 (1980)
21. P. G. Smith, D. M. Potts, T. I. Addenbrooke, *Proc. 1st Eur. Conf. Unsat. Soils*, 773-778 (2008)
22. R. G. Allen, L. S. Pereira, D. Raes, M. Smith, *FAO Irr. Dran. Pap.* **56** (1998)

Estimation of the variation of matric suction with respect to depth in a vertical unsaturated soil trench associated with rainfall infiltration

Won Taek Oh[1,a], Sai K. Vanapalli[2], Shunchao Qi[2], and Zhong Han[2]

[1]*Department of Civil Engineering, University of New Brunswick, Fredericton, NB, Canada*
[2]*Department of Civil Engineering, University of Ottawa, Ottawa, ON, Canada*

Abstract. Soil trenching is extensively used in geotechnical, mining, tunneling and geo-environmental infrastructures. Safe height and stand-up time are two key factors that are required for the rational design of soil trenches. Rainfall infiltration has a significant influence on the safe height and stand-up time of unsaturated soil trenches since it can significantly alter the shear strength of soils by influencing the matric suction. In other words, predicting the variation of matric suction of soils associated with rainfall infiltration is vital to the design of unsaturated soil trenches. In this paper, finite element analysis is carried out to reproduce the variation of matric suction profile in unsaturated soil trenches associated with rainfall infiltration using the published results of a full scale instrumented test trench at the site of BBRI at Limelette, Belgium. The analysis results showed that the variation of matric suction in unsaturated soil trenches can be reliably estimated using the information of environmental factors such as the rainfall measurements.

1 Introduction

Soil trenching is extensively used in geotechnical, mining, tunneling and geo-environmental infrastructures (Brachman and Krushelnitzky 2005, Oshati et al. 2012, Widisinghe and Sivakugan 2014). Extensive precaution should be taken in the design and excavation of soil trenches. This is because trenches of relatively shallow depths (less than 3 m) can also pose short- and long-term safety hazards due to their failure (OSHA Data Base 1985 - 1989). Provinces in Canada have published guidebooks that emphasize occupational health and safety concerns associated with trench failures (e.g. New Brunswick - Excavation Trenching; Manitoba - Guideline for Excavation Work; Ontario - Trenching Safety - Introduction to Trenching Hazards). Nonetheless, a large number of worker deaths are still reported each year associated with soil trench failures. Safe height (i.e. maximum depth of a trench that can be excavated without failure; H_{safe}) and stand-up time (T_{stup}) are two key factors that are required for the rational design of soil trenches, which can be estimated based on shear strength parameters of soils. The existing approaches to estimate H_{safe} and T_{stup} in various guidebooks may not be reasonable since they do not consider local environmental factors such as rainfall infiltration, which contributes to the significant decreases in the shear strength of soils (Tomboy et al. 2008, Van Alboom and Whenham 2003, Whenham et al. 2007, Vanapalli et al. 2009, Vanapalli and Oh 2012). In other words, estimation of the variation of matric suction associated with rainfall infiltration is vital to the rational design of unsupported unsaturated vertical trenches. In the present study, an attempt is made to simulate the variation of matric suction in unsaturated soil trenches using commercial finite element software, SEEP/W (product of GeoStudio 2012, Geo-Slope Int. Ltd.). The variation of matric suction associated with rainfall infiltration simulated from the finite element analysis (FEA) were compared with published results of a full scale instrumented test trench at the site of BBRI at Limelette, Belgium (Whenham et al. 2007). The comparison showed that the variation of matric suction in unsaturated soil trenches can be reliably estimated using the information of environmental factors such as the rainfall measurements.

2 Background

2.1 Full scale instrumented test trench

Whenham et al. (2007) investigated the influence of rainfall infiltration on the stability of an unsupported vertical trench (3 m deep, 6 m wide and 20 m long). The trench was excavated at a site that was initially in a state of unsaturated condition. Both weekly averaged rainfall measurements and averaged suction values with depth were recorded over 12 months. The variation of matric suction profile with time were monitored using tensiometers installed at different depths. The first local failures and general failures were observed in January and February 2005, respectively, due to the decrease in

[a] Corresponding author: woh@unb.ca

matric suction values associated with the precipitation activity. Figure 1 shows the variation of matric suction values at the depth of 1m, 1.5m, 2.5m, and 3.5m for initial condition (before excavation), first local failures, and general failures. The data from this field study is used in the finite element analysis to estimate the variation of matric suction with depth due to rainfall infiltration into the trench.

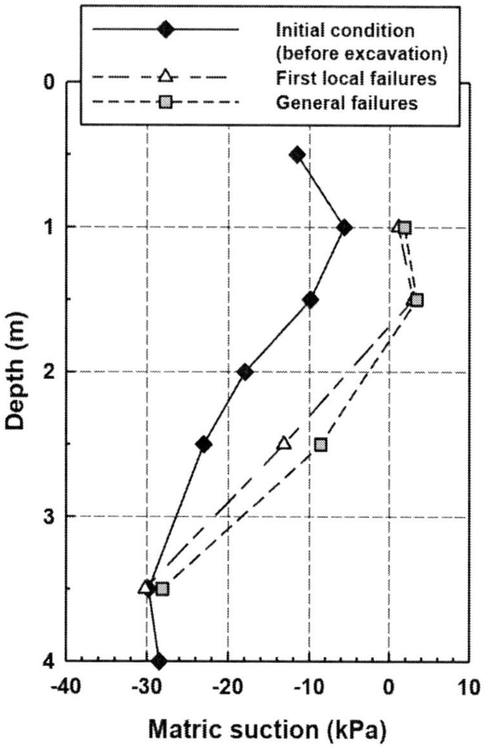

Figure 1. Variation of matric suction with respect to depth for initial condition (before excavation), first local failures and general failures

2.2 MODIFIED EFFECTIVE AND MODIFIED TOTAL STRESS APPROACH

Vanapalli and Oh (2012) analyzed stability of the trench detailed in 'Section 2' taking account of the variation of matric suction associated with rainfall infiltration extending the mechanics of unsaturated soils. Two different approaches, namely, Modified Effective Stress Approach (*MESA*) and Modified Total Stress Approach (*MTSA*), were used for the stability analysis. *MESA* and *MTSA* were originally developed to estimate the bearing capacity of unsaturated soils (Oh and Vanapalli 2013). Vanapalli and Oh (2012) suggested that stability of the test trench can be more reliably analyzed using *MTSA* rather than *MESA*.

2.2.1 Modified effective stress approach

MESA uses the effective shear strength parameters, c', ϕ' and ϕ^b (rate of increase in shear strength with matric suction) assuming both the pore-air and pore-water are in drained conditions. The active earth pressure at a depth z ($\sigma_a(z)$) in a trench excavated into an unsaturated soil can be calculated using Eq. (1) extending *MESA*.

$$\sigma_a(z) = \gamma z K_a - 2\left[c' + (u_a - u_w)\tan\phi_e^b\right]\sqrt{K_a} \quad (1)$$

where $(u_a - u_w)$ = matric suction, and $\phi_e{}^b = \phi^b$ for effective stress approach

By adopting the model to predict the nonlinear behavior of $\tan\phi^b$ (Eq. (2); Vanapalli et al. 1996), Eq. (1) can be rewritten as Eq. (3).

$$\tan\phi_e^b = S^\kappa \tan\phi' \quad (2)$$

$$\sigma_a(z) = \gamma z K_a - 2\left[c' + (u_a - u_w)(S^\kappa)\tan\phi'\right]\sqrt{K_a} \quad (3)$$

where γ = unit weight of a soil, z = depth of interest, K_a = coefficient of active earth pressure, S = degree of saturation, and κ = fitting parameter that is a function of plasticity index (Garven and Vanapalli, 2006).

2.2.2 Modified Total Stress Approach

MTSA is based on the assumption that the pore-air and pore-water are in a mode of drained and undrained condition, respectively. This approach requires results from constant water content (*CW*) tests (Thu et al. 2006). Hence, if total cohesion, c from the *CW* test (i.e. c_{CW}) is used, the active earth pressure at a depth z in an unsaturated soil trench can be estimated using Eq. (4) extending *MTSA*.

$$\begin{aligned}\sigma_a(z) &= \gamma z K_a - 2\left[c + (u_a - u_w)\tan\phi_t^b\right]\sqrt{K_a} \\ &= \gamma z K_a - 2c_{CW}\sqrt{K_a}\end{aligned} \quad (4)$$

where $\phi_t{}^b = \phi^b$ for total stress approach

As can be seen in Eq. (3) and Eq. (4), matric suction value is the key factor for analyzing the unsaturated soils behavior using *MESA* and *MTSA*. In the present study, an attempt is made to simulate the variation of matric suction in unsaturated soil trenches based on the rainfall measurements using a commercial finite element software, SEEP/W (product of GeoStudio 2012, Geo-Slope Int. Ltd.).

3 Finite element analysis

3.1 Numerical model

Figure 2 shows the numerical model in SEEP/W used for the rainfall infiltration analysis for the test trench. The Soil-Water Characteristic Curve (*SWCC*), which is the main tool to estimate the variation of mechanical properties of unsaturated soils with matric suction, was measured at the depths of 1.5m, 2.5m and 3.5m. Hence, the soil was divided into three layers; namely, 0 to 2m, 2 to 3m, and 3m to bottom. Rainfall was simulated by applying 'unit flux, q' boundary on the top surface based the measured rainfall.

In SEEP/W, initial (positive or negative) pore-water pressure can be specified by either (i) drawing the initial water table, (ii) using activation values, or (iii) using spatial function (SEEP/W manual, 2012 edition). Option (i) is useful to define hydrostatic variation of pore-water pressure with distance above and below the water table. Option (ii) can be used when a new soil region become active with a certain initial pore-water and pore-air pressure. Option (iii) allows the users to directly assign non-hydrostatic pressure heads to the soils at different depths using the feature 'Spatial Functions'. Hence, in the present study, negative pressure heads were assigned to the soils by using option (iii) based on the initial matric suction distribution profile shown in Figure 1 (Figure 3). Matric suction values between specified points are linearly interpolated in SEEP/W. The measured initial matric suctions values with depth and those simulated in SEEP/W using spatial functions are compared in Figure 4.

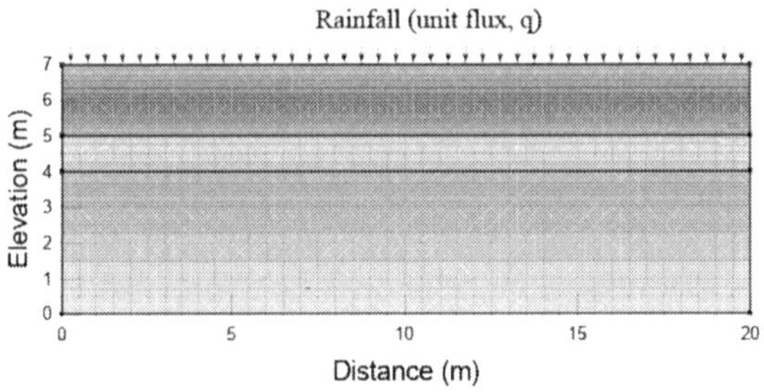

Figure 2. Numerical model of soil trench

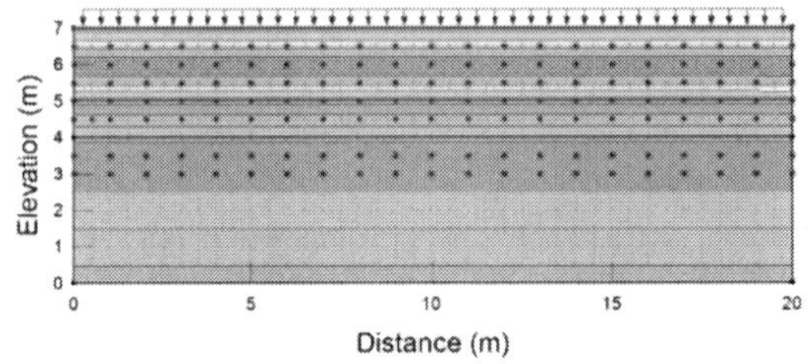

Figure 3. Assignment of non-hydrostatic pore-water pressure distribution using 'Spatial Functions' in SEEP/W

3.2 Soil-Water Characteristic Curve

Soil-Water Characteristic Curves (*SWCC*) from the three depths (1.5m, 2.5m and 3.5m) in the trench are shown in Figure 5. Minor differences were observed between the *SWCC*s.

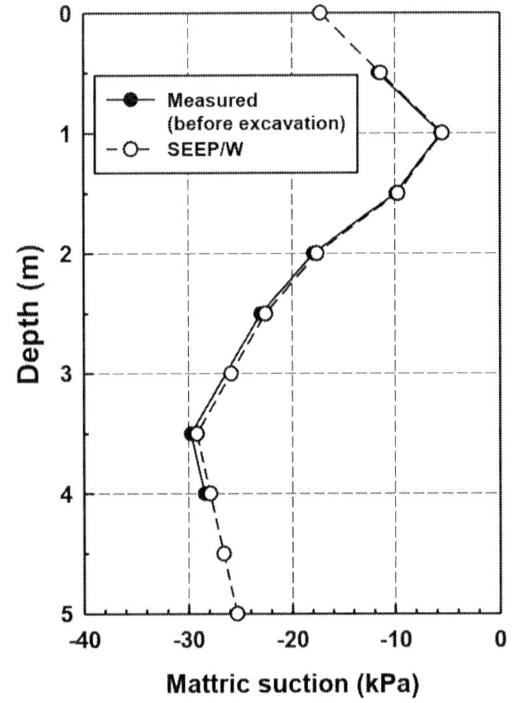

Figure 4. Measured initial matric suction values with depth and those simulated using SEEP/W (before excavation)

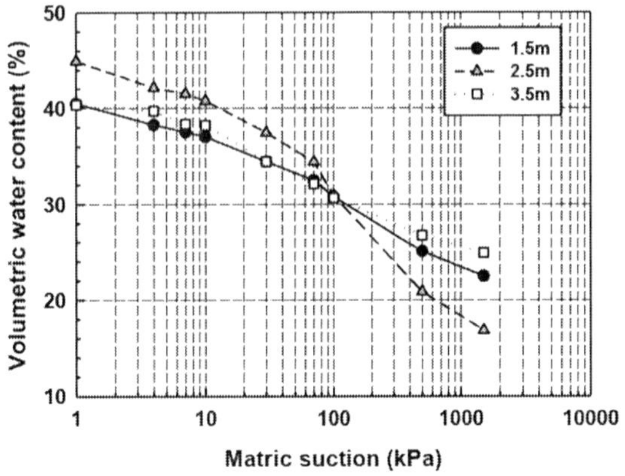

Figure 5. Soil-Water Characteristic Curves (from Whenham et al. 2007)

3.3 Hydraulic conductivity function

The variation of hydraulic conductivity with respect to matric suction (i.e. hydraulic conductivity function) was estimated using the model proposed by Fredlund et al. (1994) (Eq. (5)). Figure 6 shows the hydraulic functions generated in SEEP/W using the *SWCC*s (Figure 5) the hydraulic conductivity for saturated condition (i.e. k_{sat}). The similarities in the *SWCC*s and hydraulic functions for three different depths indicate that the soils consisting of the trench are relatively homogeneous within the investigated depth in terms of hydraulic properties.

$$k_{unsat} = k_{sat} \frac{\sum_{i=j}^{N} \frac{\Theta(e^{y}) - \Theta(\Psi)}{e^{y_i}} \Theta'(e^{y_i})}{\sum_{i=1}^{N} \frac{\Theta(e^{y}) - \Theta_{sat}}{e^{y_i}} \Theta'(e^{y_i})} \quad (5)$$

where:

k_{unsat} = the calculated conductivity for a specified water content or negative pore-water pressure,
k_{sat} = the measured saturated conductivity,
Θ_{sat} = the saturated volumetric water content,
e = the natural number 2.71828,
y = a dummy variable of integration representing the logarithm of negative pore-water pressure,
i = the interval between the range of j to N,
j = the least negative pore-water pressure to be described by the final function,
N = the maximum negative pore-water pressure to be described by the final function,
Ψ = the suction corresponding to the jth interval, and
Θ' = the first derivative of the equation.

4 Analysis results

Whenham et al. (2007) provided 'weekly averaged rain measurements' in litres per square meter (l/m^2). These rainfall measurements may not be useful when carrying out seepage analysis in the trench since the duration of each rainfall measurement was not provided. Hence, the seepage analyses were conducted by assuming that each rain measurement was collected for (i) one day intervals (i.e. daily intensity; (Figure 7)) and (ii) one hour intervals (hourly intensity; measured for one hour from midnight to 1:00 AM; Figure 8).

Figure 9 shows the comparison between measured and estimated matric suction distribution profiles based on the assumed one day intervals (Figure 7). General failures in the trench occurred approximately 8 months after the excavation was initiated. Hence, the estimated matric suction distribution profile from SEEP/W that corresponds to 8 months were used for comparison purposes. As can be seen in Figure 9, negligible change in matric suction values was observed from the seepage analysis when compared with initial matric suction distribute profile.

Figure 10 shows the comparison between measured and estimated matric suction distribution profiles based on the assumed one hour intervals (Figure 8). At the depths of 2.5m and 3.5m the measured and estimated matric suction values were approximately the same. The estimated matric suction distribution profile well captured the trend of matric suction variation with depth with small discrepancy differences between the measured and estimated matric suction values at the depths of 1m and 1.5m. The results in Figure 10 clearly indicates that seepage analysis based on detailed rainfall intensity data can provide more reliable matric suction values with time.

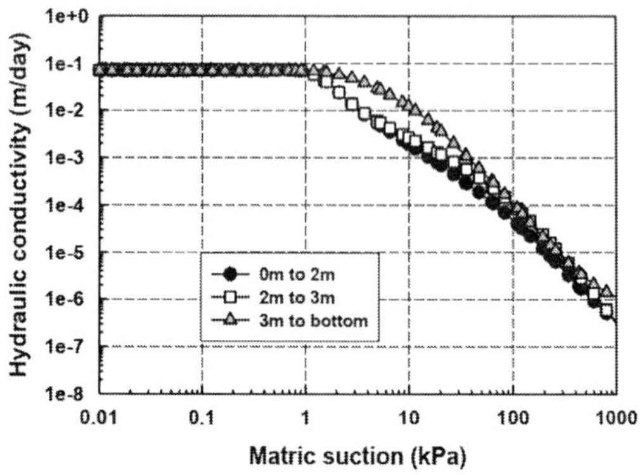

Figure 6. Hydraulic conductivity functions

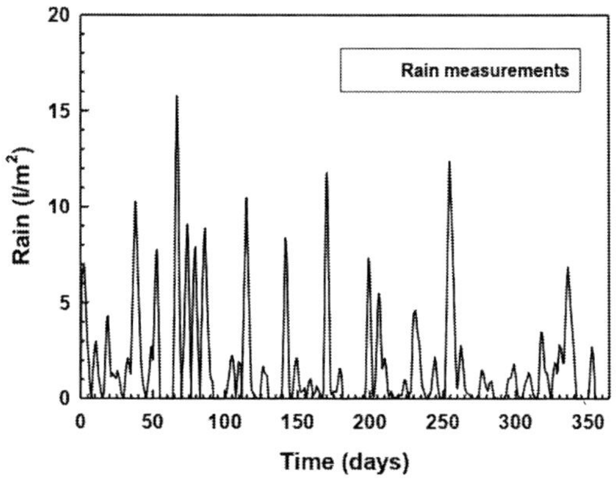

Figure 7. Assumed daily rain intensity used in the finite element analysis

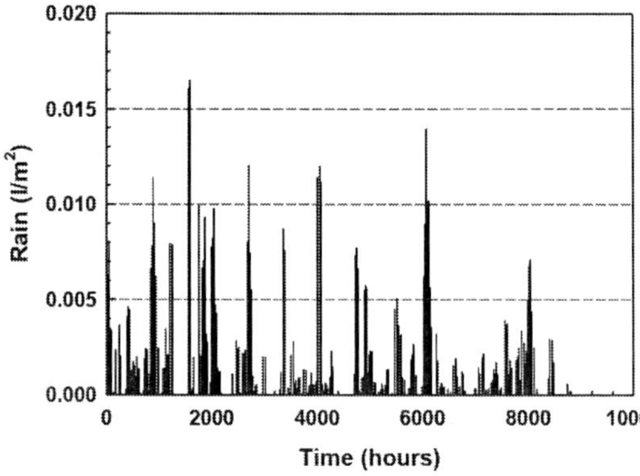

Figure 8. Assumed hourly rain intensity used in the finite element analysis

Additional seepage analysis was undertaken assuming average rain intensity of 10 mm/hr allowing ponding at the top boundary (Figure 11). For 40 days, the change (i.e. increase) in matric suction values took place within only relatively shallow depth (i.e. less than 1m from the soil surface), which is not realistic compared with the measured matric suction distribution profiles.

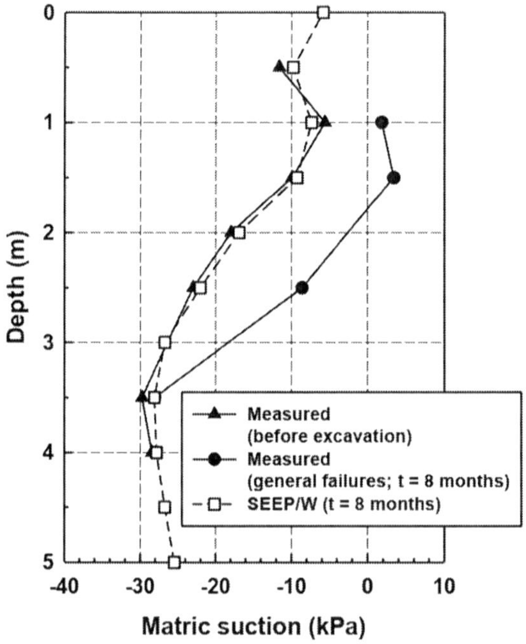

Figure 9. Comparison between measured and estimated matric suction distribution profiles based on assumed daily rain intensity

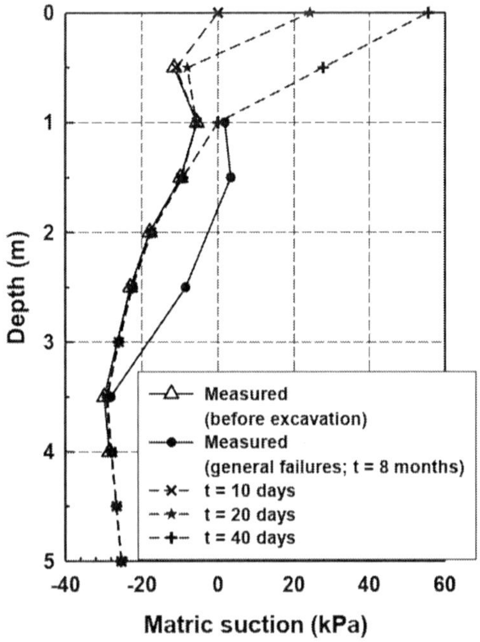

Figure 11. Comparison between measured and estimated matric suction distribution profiles assuming 10 mm/hr rain

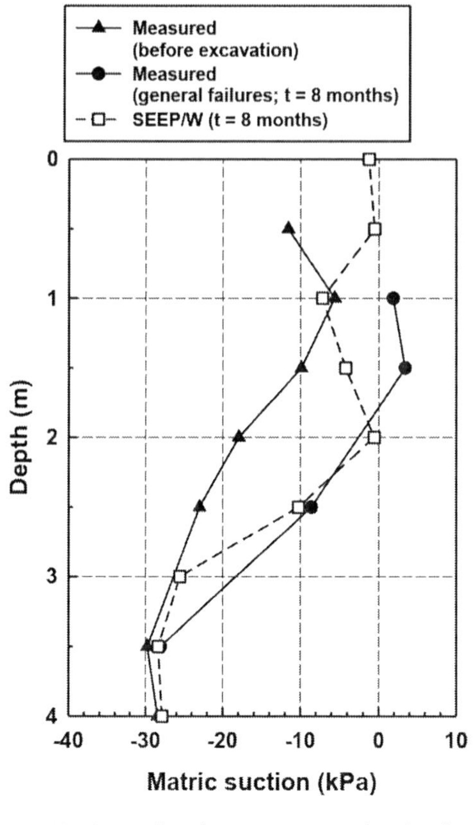

Figure 10. Comparison between measured and estimated matric suction distribution profiles (based on assumed hourly rain intensity)

5 Summary and conclusions

The instability of unsaturated soil trenches can be attributed to the reduction of matric suction associated with rainfall infiltration. Hence, estimating the variation of matric suction due to rainfall infiltration is fundamental to analyze the stability of unsaturated soil trenches. In this study, commercial finite element software, SEEP/W was used to simulate the variation of matric suction associated with rainfall infiltration for a test trench excavated in an unsaturated soil (Whenham et al. 2007). The analysis results showed that the variation of matric suction distribution profile with time in a trench can be reliably estimated by using rainfall measurements in the finite element analysis. However, matric suction distribution profile estimated assuming constant rainfall intensity (i.e. 10 mm/hr in the present study) provided unrealistic matric suction distribution profile.

References

1. Brachman, R.W.I. and Krushelnitzky, R.P. 2005. Response of a landfill drainage pipe buried in a trench. Canadian Geotechnical Journal, 42(3): 752-762.
2. Fredlund, D.G., Xing, A. and Huang, S. 1994. Predicting the permeability function for unsaturated soils using the soil-water characteristic curve. Canadian Geotechnical Journal, 31(3): 521-532.
3. Occupational Safety and Health Administration (OSHA). 1990. Analysis of construction fatalities - OSHA Data Base 1985-1989.

4. Garven, E. and Vanapalli, S.K. 2006. Evaluation of empirical procedures for predicting the shear strength of unsaturated soils. Proc. of the 4th International Conference on Unsaturated Soils, Carefree, Ariz., 2-6 April 2006, 2: 2570–2581.
5. Oh, W.T. and Vanapalli, S.K. 2013. Interpreting the bearing capacity of unsaturated fine-grained soils using modified effective and total stress approaches. International Journal of Geomechanics, 13(6): 769-778.
6. Oshati, O. S., Valsangkar, A. J., Schriver, A. B. 2012. Performance of two cast-in-place box culverts under high embankments. Canadian Geotechnical Journal, 49(12): 133-346.
7. Seepage modeling with SEEP/W (2012). Geo-Slope International Ltd.
8. Tomboy, O., De Vos, M., Charlier, R., Maertens, J., and Verbrugge, J.-C. Shear strength of unsaturated soil and its influence on slope stability. Proc. of the 1st European Conference, E-UNSAT 2008, 2-4 July 2008, Durham, United Kingdom: 459-464.
9. Tu, X.B. Kwong, A.K.L., Dai, F.C. Tham, L.G. and Min, H. 2009. Field monitoring of rainfall infiltration in a loess slope and analysis of failure mechanism of rainfall-induced landslides. Engineering Geology, 105(1-2): 134-150.
10. Van Alboom, G. and Whenham, V. 2003. Soil investigation campaign at Limelette (Belgium): Results. Proc. of the 2nd Symposium on Screw Piles, May 7, Brussels, Belgium, 1: 21-70.
11. Vanapalli, S.K., Fredlund, D.G., Pufahl, D.E. and Clifton, A.W. 1996. Model for the prediction of shear strength with respect to soil suction. Canadian Geotechnical Journal, 33(3): 379-392.
12. Vanapalli, S.K. and Oh, W.T. 2012. Stability analysis of unsupported vertical trenches in unsaturated soils. Proc. of Geo-Congress 2012, ASCE, 25-29 March 2012, Oakland, CA: 2502-2511.
13. Whenham, V., De Vos, V., Legrand, C., Charlier, R., Maertens, J. and Verbrugge, J.-C. 2007. Influence of soil suction on trench stability. Proc. of the 2nd International Conference on Mechanics of Unsaturated Soils, Weimar, Germany March 7-9, 2007: 495-501.
14. Widisinghe, S. and Sivakugan, N. 2014. Vertical stress isobars for trenches and mine stopes containing Granular Backfills. International Journal of Geomechanics, 14(2): 313-318.

Effects of rainfall characteristics on the stability of tropical residual soil slope

Harianto Rahardjo[1,a], Alfrendo Satyanaga[1], Eng Choon Leong[1]

[1]*School of Civil and Environmental Engineering, Nanyang Technological University, 50 Nanyang Avenue, Singapore*

Abstract: Global climate change has a significant impact on rainfall characteristics, sea water level and groundwater table. Changes in rainfall characteristics may affect stability of slopes and have severe impacts on sustainable urban living. Information on the intensity, frequency and duration of rainfall is often required by geotechnical engineers for performing slope stability analyses. Many seepage analyses are commonly performed using the most extreme rainfall possible which is uneconomical in designing a slope repair or slope failure preventive measure. In this study, the historical rainfall data were analyzed and investigated to understand the characteristics of rainfall in Singapore. The frequency distribution method was used to estimate future rainfall characteristics in Singapore. New intensity-duration-frequency (IDF) curves for rainfall in Singapore were developed for six different durations (10, 20, 30 min and 1, 2 and 24 h) and six frequencies (2, 5, 10, 25, 50 and 100 years). The new IDF curves were used in the seepage and slope stability analyses to determine the variation of factor of safety of residual soil slopes under different rainfall intensities in Singapore.

1 Introduction

Climate change has a significant effect on the statistical distribution of weather patterns over periods of time ranging from decades to millions of years. The study by Stern [1] showed that climate change affects many factors, such as, increase in groundwater table, ice melting, ecosystems, intense rainfall, high temperature, strong wind and floods. Among all factors, rainfall is the main factor that affects slope stability around the world. Previous studies indicated that rainfall events have become more intense in a number of countries, i.e. United States [2], Canada [3], Australia [4], South Africa [5], United Kingdom [6] and Europe [7]. Zhai et al. [8] reported increasing trends in daily extreme rainfalls in western China, and also parts of southwestern and southern China using the data from 1951 to 2000. Fujibe [9] found significant upward trends in heavy daily rainfalls in Japan using the data from 1901 to 2006. Groisman et al. [10] found an upward trend in the number of days with heavy rainfalls (greater than 99.7 percentile) in south eastern Australia. Gallant et al. [11] investigated rainfall trends including heavy rainfall events (95th and 99th percentiles) over two periods (i.e. 1910-2005 and 1951-2005) in southwest and east of Australia. The study by Rahardjo et al. [12] indicated that Singapore experienced the third highest amount of rainfall for one day (366 mm) in 75 years.

Rainfall characteristics are often used to design water-related structures. Reviewing and updating rainfall characteristics (i.e. Intensity-Duration-Frequency (IDF) curves) for future climate scenarios is necessary [13]. Studies on the rainfall IDF curves are important to prevent flooding and rainfall-induced slope failures. Some research works have been carried out to evaluate extreme rainfall data for different return periods and durations. Endreny and Imbeah [14] used two separate rainfall data sets in the country of Ghana with two different probability distribution frequency analysis methods to estimate IDF parameters. The results indicated that it is essential to combine meteorological and tropical rainfall measuring mission (TRMM) satellite data for IDF generation in Ghana. Madsen et al. [15] used a regional model for estimation of extreme rainfall characteristics in Denmark and updated the regional model with data from the augmented rain database for the period (1997–2005). Their studies indicated a typical annual increase of 10 % in rainfall intensity with durations between 30 minutes and 3 hours (for 10 years return periods) which is applicable for most urban drainage designs.

Many researchers have developed various formulae for design storm based on the construction of IDF curves. Chen [16] derived a generalized IDF formula for any location in the United States using three basic rainfall depths, that is, R110 (1 h, 10-year rainfall depth), R2410 (24 h, 10-year rainfall depth), and R1100 (1 h, 100-year rainfall depth). Pao-Shan [17] developed regional Intensity–duration–frequency (IDF) formulas based on

[a] Corresponding author: author@e-mail.org

the scaling theory for sites without rain gauges. Forty-six recording rain gauges over northern Taiwan provide the data set for analysis. Three homogeneous regions were classified by different scaling regimes and regional IDF scaling formulas were developed in each region.

Toll et al. [18], Alonso et al. [19] and Rahardjo et al. [20] observed that rainfall-induced slope failures commonly occurred around the world, especially in areas with tropical climatic conditions like Singapore. Therefore, analyses of rainwater infiltration into slopes in many geotechnical and geoenvironmental applications, such as: embankment and landfill cover system become a very important topic of study. Most of these constructions are located above a thick unsaturated zone above a deep groundwater table [21, 22]. The rate of water infiltration into slope is governed by permeability of the unsaturated zone above the groundwater table. As a result, changes in negative pore-water pressure or soil suction and shear strength of unsaturated soil are a function of water content [23].

In this study, the frequency distribution method (Gumbel method) was used to develop a new IDF curve for rainfall design in Singapore for six different durations (10, 20, 30 min and 1, 2 and 24 h) and six frequency periods (2, 5, 10, 25, 50 and 100 years). The new IDF curve was used in the seepage and slope stability analyses to determine the variation of factor of safety of residual soil slopes under different rainfall intensities in Singapore.

2 Rainfall Conditions in Singapore

Climate in Singapore is characterized by uniform temperature and pressure, high humidity and particularly, abundant rainfalls. Based on National Environmental Agency of Singapore data, the climatic condition in Singapore is divided into two main seasons, the wetter Northeast Monsoon season from December to May and the drier Southwest Monsoon season from June to November. Maximum rainfall is usually observed between December and January, whereas June is noted as the driest month [24]. The annual extreme values of precipitation for six different recorded durations, namely: 10, 20, 30 min and 1, 2, 24 h have been extracted from historical rainfall records of Singapore island for the period from 1985 to 2010. Meteorological services Singapore has a network of meteorological stations which provide real-time observations across Singapore. There are over 60 automatic weather stations which are complemented with the electronic tipping bucket rain gauges across the island. The analyses of rainfall in this paper utilized rainfall data from 15 rainfall stations island-wide as shown in Figure 1 [25].

3 Intensity-Duration-Frequency-Curve

Intensity-duration-frequency (IDF) curves are a tool to relate rainfall and its frequency. The IDF curve is commonly used to obtain "design rainfall" for different water system projects. Chow et al. [26] and Singh [27] suggested three steps in the estimation of IDF curves. The first step included a best fitting function to each group comprising the data for a specific duration. In the second step, the new rainfall intensities for each duration and a set of selected return periods (e.g. 2, 5, 10, 25, 50, 100 years, etc.) were estimated. This was performed by using the best fitting functions of the first step. In the third step, the final IDF relationship could be obtained in two different ways: either (a) for each selected return period the rainfall intensities were computed and a graphical relationship of intensity and duration for different return periods was established, or (b) the rainfall intensity was functionally related to the rainfall duration and the return period using mathematical and regression analyses to derive such equation.

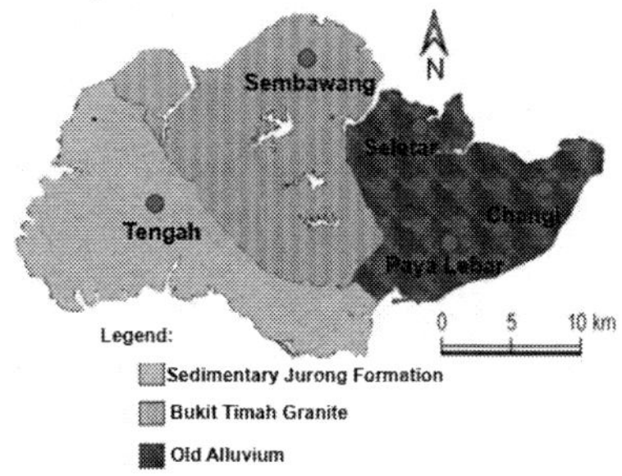

Figure 1. Location of rainfall station across Singapore island

The frequency distribution method, namely Gumbel distribution function [28] was used in this study. Gumbel method was the most widely used distribution function for analyses of IDF curve owing to its suitability for modelling maxima. The annual maximum rainfall depth (P) for different durations, t_d (5 min, 10 min, 30 min, 60 min, 120 min and 600 min) was tabulated to obtain the average (Pave) and standard deviation (S) of the maximum rainfall depth for periods between 1985 and 2010. These variables were required in computing rainfall intensity (IT) for specific frequency (return) period using frequency distribution methods. Annual historical maximum rainfall depth, P (in mm) values for the above six durations were extracted from 5 main meteorological stations in Singapore and summarized in Table 1. The selection of maximum rainfall for each duration in Table 1 was carried out based on the "continuous rainfall" method. If the duration of particular rainfall from meteorological data was less than the designated duration (i.e. 10 min, 20 min, 30 min, etc) in the paper, the rainfall was not used in this study.

Table 1. Annual maximum rainfall (P) from 5 different rainfall stations in Singapore

Year	P(mm)					
	5 min	10 min	30 min	60 min	120 min	600 min
1985	24	103	123	101	250	10
1990	58	128	129	115	99	107
1995	138	95	135	124	22	452
2000	88	66	126	117	28	11
2005	74	96	122	137	48	106
2010	26	80	147	106	34	111
P_{ave}	**68**	**95**	**130**	**117**	**80**	**133**
S	**43**	**21**	**9**	**13**	**88**	**163**

The first step in the construction of IDF curves is best fitting certain function to the maximum rainfall amounts for a number of fixed durations. Then, the relationships of the rainfall intensity for any duration and return period can be derived. Figure 2 shows the original IDF curves of Singapore for 2, 5, 10, 25, 50 and 100 years return periods. The IDF curves show similarity with the upper half portion of the sigmoidal curve. Therefore, a logistic function was adopted for best fitting the IDF curve of Singapore. The proposed equation for best fitting IDF curve of Singapore is shown in Equation 1. The parameters of equation 1 used in best fitting the original IDF curve of Singapore are summarized in Table 2.

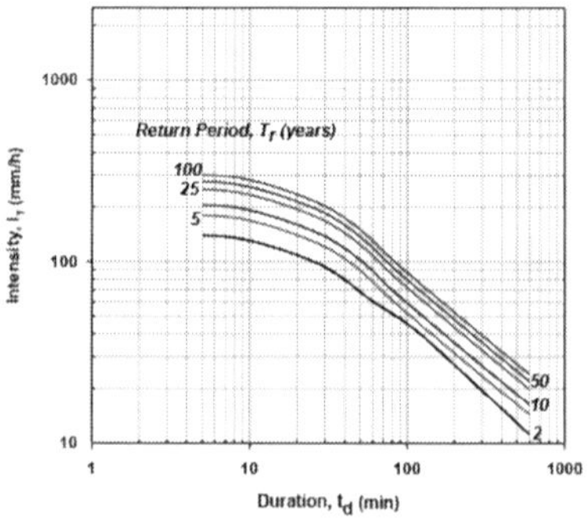

Figure 2. Original intensity-duration-frequency curves of Singapore

$$I_T = \left[\frac{I_{tr}}{\ln\left(\exp + \left(\frac{t_d}{b} \right)^c \right)} \right]^d \quad (1)$$

Note:
I_T = rainfall intensity (mm/h)
t_d = duration (min)
I_{tr} = the maximum rainfall intensity at particular return period (mm/h)
b = fitting parameter related to the bending in the IDF curve (min)
c, d = fitting parameters related to the shape of the IDF curve

Table 2. Parameters of the proposed equation in best fitting the original IDF curve of Singapore

T_r (years)	I_{tr}	b	c	d
2	139	32.4	1.48	1.74
5	180	32.4	1.48	1.74
10	205	32.4	1.48	1.74
25	249	32.4	1.48	1.74
50	275	32.4	1.48	1.74
100	301	32.4	1.48	1.74

The second step in the construction of IDF curves is the estimation of a new rainfall intensity for different durations and return periods in Singapore. In this method, the frequency of rainfall, PT (mm) for any rainfall duration t_d (hour) with specified return period Tr (years) is computed using Equation 2 [29].

$$P_T = P_{ave} + K_T S \quad (2)$$

where:
P_{ave} = the average of maximum rainfall depth which can be calculated using Equation 3.

$$P_{ave} = \frac{\sum_{i=1}^{n} P_i}{n} \quad (3)$$

where P_i is the maximum rainfall depth for particular year and for certain rainfall duration (mm) and n is the number of years, respectively.
S = the standard deviation of maximum rainfall depth which can be calculated using Equation 4.

$$S = \sqrt{\frac{1}{n} \sum_{i=1}^{n} (P_i - P_{ave})^2} \quad (4)$$

K_T is Gumbel frequency factor which can be calculated using Equation 5.

$$K_T = -\frac{\sqrt{6}}{\pi} \left\{ 0.5772 + \ln\left[\ln\left(\frac{T_r}{T_r - 1} \right) \right] \right\} \quad (5)$$

Then, the rainfall intensity, I_T (mm/h) for return period T_r can be obtained from Equation 6.

$$I_T = \frac{P_T}{t_d} \quad (6)$$

where t_d is duration in hour (1/6, 1/3, 1/2, 1, 2, 24 h).

The calculated frequency and intensities of rainfall for different durations and return periods based on the historical annual maximum rainfall from 1985 to 2010 in Singapore are indicated using dotted symbols in Figures 3 and 4. The final step in the construction of IDF curves is best fitting the calculated new rainfall intensities using Equation 1 as proposed in this study. The best fitted line for the new IDF curves of Singapore are shown in Figures 3 and 4. The parameters of Equation 1 in describing the new IDF curve are summarized in Table 3.

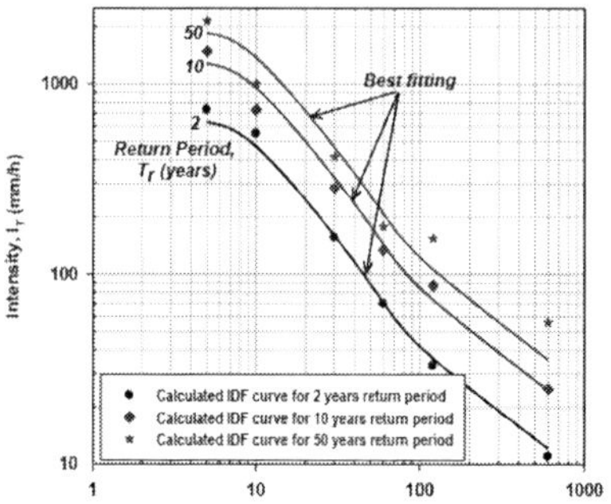

Figure 3. Modified intensity-duration-frequency curve of Singapore for 2, 10 and 50 years return periods as calculated using Gumbel method

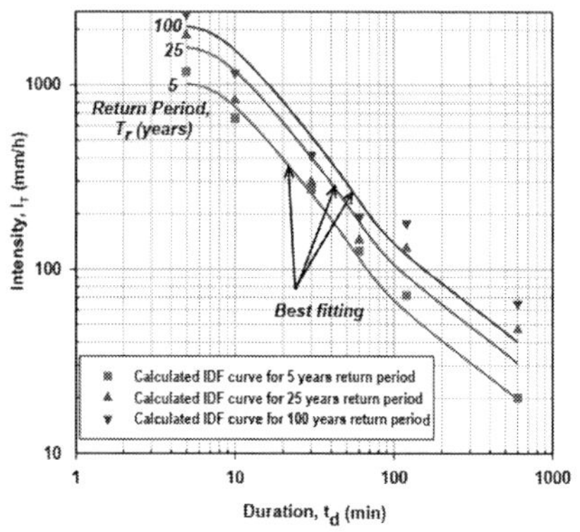

Figure 4. Modified intensity-duration-frequency curve of Singapore for 5, 25 and 100 years return periods as calculated using Gumbel method

Table 3. Parameters of Equation 1 in describing the new IDF curves of Singapore

T_r (years)	I_{tr}	b	c	d
2	631	12	1.8	2.1
5	1020	12	1.8	2.1
10	1279	12	1.8	2.1
25	1606	12	1.8	2.1
50	1848	12	1.8	2.1
100	2088	12	1.8	2.1

Parameters I_{tr} in Table 3 indicated that the rainfall intensity for short duration of rainfall (5, 10, 20 minutes) in the new IDF curves were much higher than the one in the original IDF curves of Singapore for all return periods. The same values of parameters b, c and d in Table 3 show that the new IDF curves have the same shape for different return periods which are also observed in Table 2. The lower b value in the new IDF curve (12 min) as compared to that in the original IDF curve (32.4 min) shows that the increase in the rainfall intensity occurs in the shorter duration of rainfall. In the new IDF curve, the rainfall intensity has increased significantly for durations higher than 12 minutes whereas in the original IDF curve, the rainfall intensity only increases significantly for durations higher than 32.4 minutes. The higher values of parameters c and d in Table 3 as compared to those in Table 2 indicate that the new IDF curve is steeper than the original IDF curve. As a result, the rate of increase in rainfall intensity in the new IDF curve is higher than that in original IDF curve.

4 Seepage and Slope Stability Analysis

In this paper, numerical analyses were carried out to observe the variations of factor of safety under different frequencies of rainfall (i.e. 30 minutes and 10 hours) for 50 year return periods. Rainfall frequency of 30 minutes was selected to represent short duration of rainfall ($\leq$ 60 minutes) whereas rainfall frequency of 10 hours was selected to represent long duration of rainfall (> 60 minutes). Transient seepage analyses were conducted using the finite element software Seep/W (Geoslope International Ltd., 2007a) and slope stability analyses were conducted using Slope/W (Geoslope International Ltd., 2007b). Based on the original IDF curve, rainfall intensities of 22 mm/h and 185 mm/h were observed in conjunction with rainfall durations of 10 hours and 30 minutes, respectively. Based on the new IDF curve, rainfall intensities of 36 mm/h and 465 mm/h were observed in conjunction with rainfall durations of 10 hours and 30 minutes, respectively. Therefore, four cases of seepage and stability analyses were carried out in this study (Table 4). Case no 1 included numerical analyses of residual soil slope under 22 mm/h of rainfall for 10 hours. Case no 2 included numerical analyses of residual soil slope under 36 mm/h of rainfall for 10 hours. Case no 3 included numerical analyses of residual soil slope under 185 mm/h of rainfall for 30 minutes. Case no 4 included numerical analyses of residual soil slope under 465 mm/h of rainfall for 30 minutes.

Table 4. Case study for numerical analyses in this study

Case	Rainfall intensity (mm/h)	Duration (min)
1	22	600
2	36	600
3	185	30
4	465	30

The initial condition for all cases was taken as hydrostatic pore-water pressure conditions. The finite element model and the boundary conditions for all cases were similar as shown in Figure 5. The initial position of groundwater table (6 m and 1 m depths below the crest and the toe of the slope, respectively) as well as the height (15 m) and the angle (27°) of the slope for all cases analyzed in this study were obtained from Rahardjo et al. (2010) to represent typical position of groundwater table and slope geometry of residual soil slopes in Singapore. Soil-water characteristic curve (SWCC) and permeability function used in this study were extracted from Satyanaga et al. (2014) for residual soil slopes from Bukit Timah Granite in Singapore as presented in Figures 6 and 7, respectively.

Stability analyses of the residual soil slope were carried out by incorporating the pore-water pressures from different time increments as obtained from the seepage analyses. The factors of safety at different time increments were calculated using Bishop's simplified method of slices. The saturated and unsaturated shear strengths were incorporated in Slope/W for analyzing the stability of the residual soil slope under different rainfalls. Effective cohesion, c', effective friction angle, ϕ' and rate of change in shear strength due to matric suction change, ϕ^b angle used in this study were 16.5 kPa, 26° and 13°, respectively. Figure 8 shows the variation of factor of safety of the residual soil slope under rainfall intensities of 185 mm/h and 465 mm/h for 30 minutes. Figure 9 shows the variation of factor of safety of the residual soil slope under rainfall intensities of 22 mm/h and 36 mm/h for 10 hours.

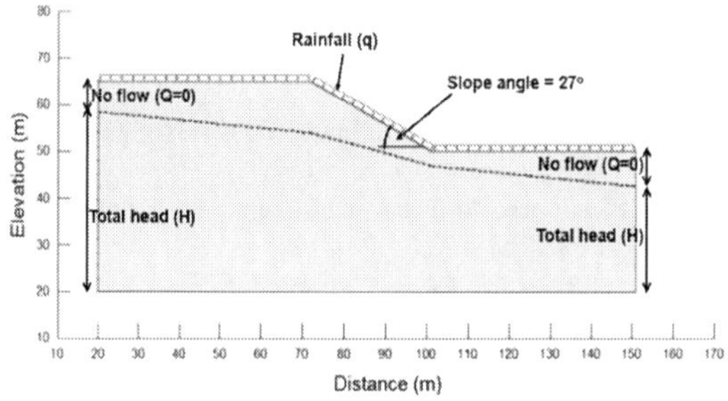

Figure 5. Slope model for seepage analyses of all cases in this study

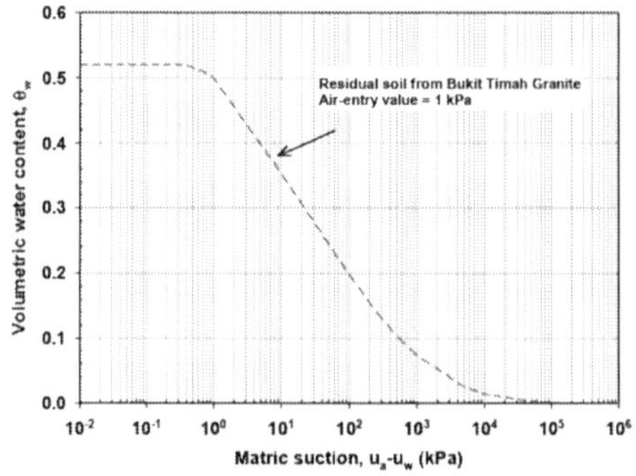

Figure 6. Soil-water characteristic curve used in this study

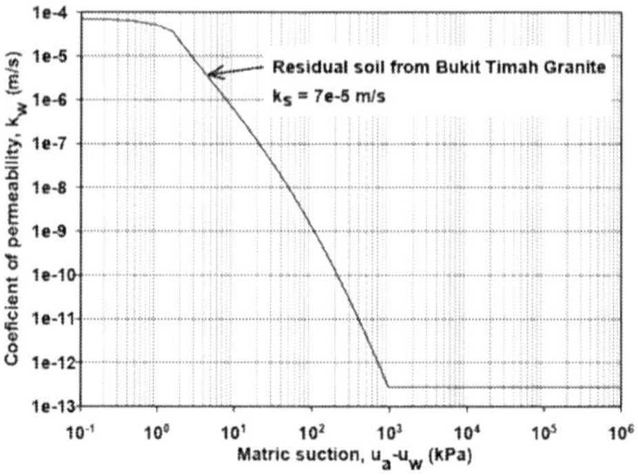

Figure 7. Permeability function used in this study

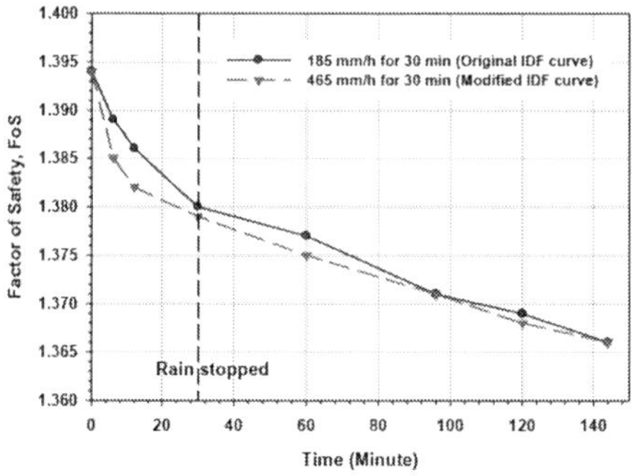

Figure 8. Variation of factor of safety for the investigated residual soil slope under 30 minutes of rainfall

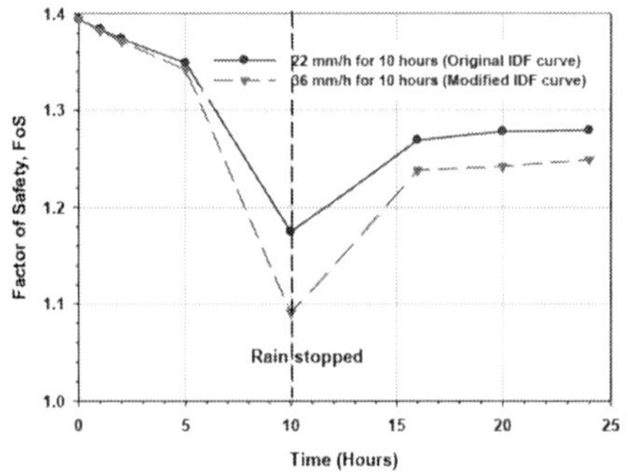

Figure 9. Variation of factor of safety for the investigated residual soil slope under 10 hours of rainfall

Comparing rainfall intensities and saturated permeability (k_s) of soil used in the analyses, rainfall intensity in case 3 is lower than k_s whereas rainfall intensity in case 4 is higher than k_s. However, Figure 8 shows that the variations in factor of safety with respect to time during and after rainfalls of 185 mm/h and 465 mm/h were similar. These could be attributed to a short duration of rainfall loading on the surface of the slope, resulting in small changes of pore-water pressures near the surface. Hence, the differences in the variations of factor of safety for residual soil slopes under rainfall intensities obtained from the original and the new IDF curves were very small. Figure 9 indicates that factor of safety of the residual soil slope decreased more significantly during rainfall of 36 mm/h as compared to that during rainfall of 22 mm/h. These happened since the rainfall frequency was high (10 hours), resulting in the higher total rainfall and higher water flow within the soil layer under 36 mm/h rainfall as compared to those associated with 22 mm/h rainfall. Figure 9 also indicates that it is important to analyze the stability of the slope using the new IDF curve since the stability analyses using the new IDF curve produced a lower minimum factor of safety of the slope.

5. Conclusions

New IDF curves were proposed in this paper based on Gumbel distribution function. A new equation was also proposed to best fit the IDF curves of Singapore. The results of the numerical analyses indicated that the rainfall intensity from the new IDF curves generated a lower minimum factor of safety as compared to the rainfall intensities from the original IDF curve of Singapore. The study presented in this paper is limited to one soil type. Further analyses must be carried out to investigate the effect of changes in IDF curve on the factor of safety for different soil type.

Acknowledgement

The authors gratefully acknowledge the assistance of Handoko, Christopher and Zhao Yan, School of Civil and

DOI: 10.1051/e3sconf/20160915004

Environmental Engineering, NTU, Singapore during the data collections.

References

1. N. Stern. Cambridge University Press., 712 pp. (2007)
2. K.E. Kunkel. Natural Hazards, **29**, pp 291–305 (2003)
3. D. Francis, H. Hengeveld. Environment Canada, (1998)
4. R. Suppiah, K.J. Hennessy. Int. J. Climatology, **18**, pp 1141–1164 (1998)
5. S.J. Mason, P.R. Waylen, G.M. Mimmack, B. Rajaratnam, J.M. Harrison. Climatic Change, **41**, pp 249–257 (1999)
6. T.J. Osborn, M. Hulme, P.D. Jones, T.A. Basnett. Int. J. Climatology, **20**, pp 347–364. (2000)
7. E.J. Forland, H. Alexandersson, A. Drebs, I. Hamssen-Bauer, H. Vedin, O.E. Tveito. Norwegian Meteorological Institute, pp 1–55 (1998)
8. P. Zhai, X. Zhang, H. Wan, X. Pan. J. Climate, **18**, pp 1096-1108 (2005)
9. F. Fujibe. Long-Term Changes in Precipitation in Japan. Journal of Disaster Research, 3(1), 51-52 (2008)
10. P.Y. Groisman, R.W. Knight, D.R. Easterling, T.R. Karl, G.C. Hegerl, V.N. Razuvaev. J. Climate, **615**, 18, pp 1326-1350 (2005)
11. A.J.E. Gallant, K.J. Hennessy, J. Risbey. Australian Meteorological Magazine, **56**, pp 223-239 (2007)
12. H. Rahardjo, A. Satyanaga, E.C. Leong. J. Eng. Geology, Special Issue on Unsaturated Soils: Theory and Applications, **165**, pp. 133-142 (2013)
13. G. Mirhosseini, P. Srivastava, L. Stefanova. Regional Environmental Change, **13**, 1, pp 25–33 (2013)
14. T.A. Endreny, N. Imbeah. J. Hydrology, **371**, 1-5, 182-191 (2009)
15. H. Madsen, K. Arnbjerg-Nielsen, P.S. Mikkelsen. Atmos. Res. **92**, 3, 343–349 (2009)
16. C.L. Chen. ASCE, J. Hydro. Eng. **109**, 1603-1621 (1983)
17. Yu, Pao-Shan, Yang, Tao-Chang, Lin, Chin-Sheng. (2004). J. Hydro. **295**, 1-4,10, 108-123.
18. D.G. Toll, H. Rahardjo, E.C. Leong. Proc., 2nd International Conf. on landslides, slope stability and the safety of infrastructure (1999)
19. E.E. Alonso, A. Gens, C.H. Delahaye. Hydrogeology J. **11**, pp. 174-192 (2003)
20. H. Rahardjo, A. Satyanaga, E.C. Leong. Proc. Australia New Zealand Conference on Geomechanics "Common Ground". Brisbane, Australia. 2: 704-709 (2007)
21. J. Gasmo, K.J. Hritzuk, H. Rahardjo, E.C. Leong. Geo. Test. J. **22**, 2, 128-137 (1999)
22. H. Rahardjo, A. Satyanaga, E.C. Leong. Proc. The International Seminar on Civil and Infrastructure Engineering 2008 for Environmental Sustainability. pp. 1-20 (2008)
23. D.G. Fredlund, H. Rahardjo, M.D. Fredlund. *Unsaturated Soil Mechanics in Engineering Practice*. 926 pp (2012)
24. H. Rahardjo, A. Satyanaga, E.C. Leong, J.-Y. Wang. Procedia Earth and Planetary Science, **9**, pp. 23-43 (2014)
25. National Environment Agency Meteorological Services Data. Singapore: National Environment Agency. (2010)
26. V.T. Chow, D.R. Maidment, L.W. Mays. *Applied Hydrology*. (1988)
27. V.P. Sigh. Elementary Hydrology. Prentice-Hall, New Jersey (1992)
28. E.J. Gumbel. Statistics of Extremes. Columbia University Press, New York (1958)
29. E.M. Wilson. Engineering Hydrology, fourth ed. Macmillan, London (1990)
30. Geoslope International Ltd., Seep/W User's Guide for Finite Element Analysis, (2007a)
31. Geoslope International Ltd., 2007b. Slope/W User's Guide for Finite Element Analysis, (2007b)
32. H. Rahardjo, A. Satyanaga, E.C. Leong, Ng, Y.S. J. Geo. Geoenviro. Eng., ASCE. **136**, 11, 1555–1564 (2010)
33. A. Satyanaga, H. Rahardjo, E-.C. Leong, Y.H. Chia, J-.Y. Wang. Proc. Int. Conf. Unsaturated Soils. (2014)

The impact of hydrological parameters on modelling slope safety factor towards shallow landslides: a case study from Oltrepò Pavese

M. Bordoni[1a], C. Meisina[1], R. Valentino[2], M.G. Persichillo[1], M. Bittelli[3] and S. Chersich[1]

[1]*University of Pavia, Department of Earth and Environmental Sciences, 27100 Via Ferrata 1 Pavia, Italy*
[2]*University of Parma, Department of Civil and Environmental Engineering and Architecture, 43100 Viale G.P. Usberti 181/A Parma, Italy*
[3]*University of Bologna, Department of Agricultural Sciences, 40127 Viale Fanin 44 Bologna, Italy*

Abstract. Hydrological monitoring of slope susceptible to shallow landslides allows for identifying the triggering conditions of shallow failures and implementing slope stability analysis at site-specific scale. In this work, a case study of a long-term hydrological monitoring in a slope susceptible to shallow landslides of Oltrepò Pavese (Northern Italy) is presented. The triggering mechanism develops in wet seasons (winter and spring) due to the uprise of a perched water table at about 1 m from ground surface, in consequence of the most intense rainfalls (about > 60 mm in 48 h). Unstable conditions (safety factor < 1.0) are correctly modeled on the basis of both water content and pore water pressure, with a better prediction considering hysteresis effects. Safety factor on the basis of water content can correctly assess the triggering conditions for unsaturated and completely saturated soils. It is possible estimating shallow landslides triggering caused by positive pore water pressures only considering this parameter.

1 Introduction

Rainfall-induced shallow landslides involve the slope materials till less than 2.0 m from ground. Due to their proximity to urbanized areas, they cause significant damages to structures and infrastructures and, sometimes, human losses [1].

In the last years, several authors identified different soil hydro-mechanical behaviors leading to shallow landslides triggering through the application of monitoring tools [2-4]. Moreover, monitoring data revealed very useful to be implemented in different kinds of physically-based methods for the assessment of slope stability at site-specific and local scales [5-6]. Monitoring devices improve the correct identification of shallow landslides triggering moments and of the areas which could be threatened by this kind of instability.

In this work, a case study of a long-term hydrological monitoring of a slope susceptible to shallow landslides is presented. The research aimed at investigating the triggering mechanism and implementing slope stability analyses. The study area is located in Oltrepò Pavese (Northern Italy). The monitoring period covers more than 3 years (42 months). The main objectives of the research are: 1) analyzing the main hydrological behaviors of the test-site slope soil based on field monitoring, in particular for identifying the soil hydrological conditions which can lead to shallow failures; 2) evaluating the effect of the soil hydrological parameters on modeling the safety factor at site-specific scale, using the data collected through the monitoring equipment.

2 Material and methods

2.1 The study site

The monitored slope is located in north-eastern Oltrepò Pavese, in correspondence of the northern termination of Pavese Apennines (Fig. 1). This slope is representative of other sites in this area prone to shallow landslides.

The topographic gradient is medium-high (26-35°). The slope is east-facing. Its elevation ranges from 210 to 170 m a.s.l., with the monitoring station located at 185 m. a.s.l. The land use is mainly constituted by grass and shrubs. The climatic regime is temperate/mesothermal with a mean yearly air temperature of 12 °C and a mean yearly rainfall amount of 684.4 mm. The bedrock is made of gravel, sand and poorly cemented conglomerates. Superficial soils, derived by bedrock weathering, are prevalently clayey-sandy silts and clayey-silty sands with different amount of pebbles and carbonate soft nodules. Soil thickness ranges between 0.5 to 1.85 m, increasing from the top to the bottom of the slopes.

The area where the study slope is located is characterized by a high density of past rainfall-induced shallow landslides (Fig. 1). The first and most significant event occurred on 27-28 April 2009, due to an extreme rainfall of 160 mm of cumulated rain in 62 h [1]. Further shallow landslides occurred in the period between March and April 2013, and between 28 February and 2 March 2014 [7]. In particular, the study slope was affected by shallow failure on 27-28 April 2009 and 28 February-2

[a] Corresponding author: massimiliano.bordoni01@universitadipavia.it

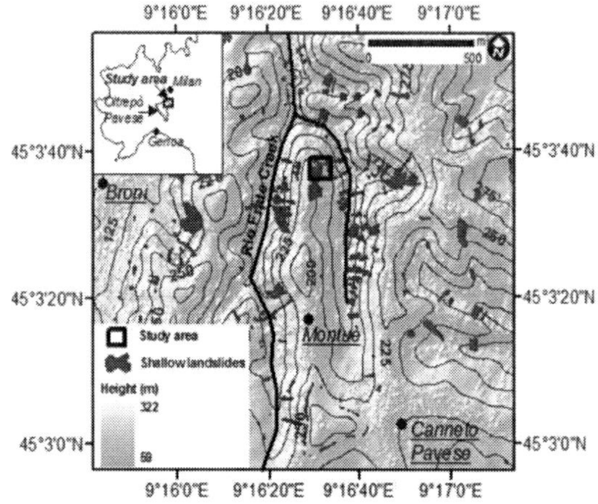

Figure 1. Location of the study area.

March 2014 events. The last event occurred 15 m far from the monitoring station (Fig. 1) during the monitoring activities. The source area had the same slope angle of the zone where the monitoring station is located (30°), and the failure surface developed at 1.0 m from the ground.

2.2 Soil properties of the test-site slope

A multidisciplinary study was carried out to characterize the superficial soils and the weathered bedrock of the study slope. A pedological description of the representative soil was performed in correspondence of the monitoring station. The soil is classified as Calcic Gleysol (Abruptic, Siltic) [8], with five main soil horizons (Tabs. 1, 2). The weathered bedrock (R) was identified at 1.3 m from the ground level.

Table 1. Grain size distribution of test-site slope soils and weathered bedrock.

	Representative depth	Gravel (> 2 mm)	Sand (2 mm-0.75 mm)	Silt (0.75 mm-2 μm)	Clay (< 2 μm)
	m	%	%	%	%
C	0.2	12.33	12.50	53.92	21.25
D	0.4	1.50	11.40	59.42	27.68
E	0.6	8.47	13.23	51.10	27.20
F	1.0	2.40	12.20	56.40	29.00
G	1.2	0.50	7.50	65.63	26.37
R	1.4	0.20	75.00	24.80	0.00

Soil horizons are calcareous (13.7-16.1%). A calcic horizon was detected at 1.2 m from the ground (G level). In correspondence of this depth, the carbonate content is of 35.3%. Soil horizons have a high silty content (51.1-65.6%), that grows slightly with depth, and a clay content higher than 21.3% (Tab. 1). Sand and gravel contents are always low in the soil levels. They have lowest values in the G horizon (0.5 and 7.5%, respectively). At 1.4 m from ground, the weathered bedrock (R) has a sandy texture, with a sand content of 75.0% (Tab. 1). According to USCS classification, soil horizons are prevalently non-plastic or slightly plastic soils (CL). Liquid limit (w_L) ranges from 38.5 to 41.9%, while plasticity index (P_I) ranges between 14.3 and 17.2% (Tab. 2).

Table 2. Main geotechnical features of test-site slope soils and weathered bedrock.

	Representative depth	w_L	P_I	γ	φ'	c'
	m	%	%	kN/m^3	°	kPa
C	0.2	39.79	17.18	17.00		
D	0.4	38.46	14.25	16.70		
E	0.6	40.32	15.65	16.70	31	0
F	1.0	39.15	15.94	18.60	33	0
G	1.2	41.85	16.54	18.25	26	29
R	1.4	-	-	18.06		

Unit weight (γ) trend is characterized by a significant increase in correspondence of the F horizon. This parameter passes from 16.7 to 18.6 kN/m^{3}at 1.0 m from ground and, then, keeps rather steady with depth (Tab. 2). Peak shear strength parameters were reconstructed for the E, F and G horizons through triaxial tests (Tab. 2). E and F horizons have friction angle (φ') between 31° and 33° and nil effective cohesion (c') (Tab. 2) The G horizon is characterized by a friction angle of 26° and an effective cohesion of 29 kPa (Tab. 2).

2.3 The monitoring equipment

A more detailed description of the monitoring station is reported elsewhere [7]. In this paper, the necessary information required for completeness is provided. The station collects data with a time resolution of 10 min. It measures: rainfall amount, air temperature, air humidity, atmospheric pressure, net solar radiation, wind speed and direction. Six Time Domain Reflectometer (TDR) probes measure the soil water content at 0.2, 0.4, 0.6, 1.0, 1.2 and 1.4 m from the ground level. A combination of three tensiometers and three Heat Dissipation (HD) sensors measure the soil pore water pressure at 0.2, 0.6 and 1.2 m. The HD sensors only allow for acquiring pore water pressures lower than -10^1 kPa [3]; thus, tensiometers are installed to measure pore water pressures above -10^1 kPa. The field data are collected by a datalogger powered by a photovoltaic panel. In this work, field data acquired between 27 March 2012 and 27 October 2015 are analyzed.

2.4 Model for slope stability analysis

Lu and Godt's model (Eq. 1; [5]) was selected for slope stability analysis at site-specific scale.

$$F_s = \frac{\tan\phi'}{\tan\beta} + \frac{2c'}{\gamma z \sin 2\beta} - \frac{\sigma^s}{\gamma z}\left[\left(\tan\beta + \cot\beta\right)\tan\phi'\right] \quad (1)$$

, where F_s = slope safety factor; c' = effective cohesion; γ = unit weight; z = depth below ground level in which a

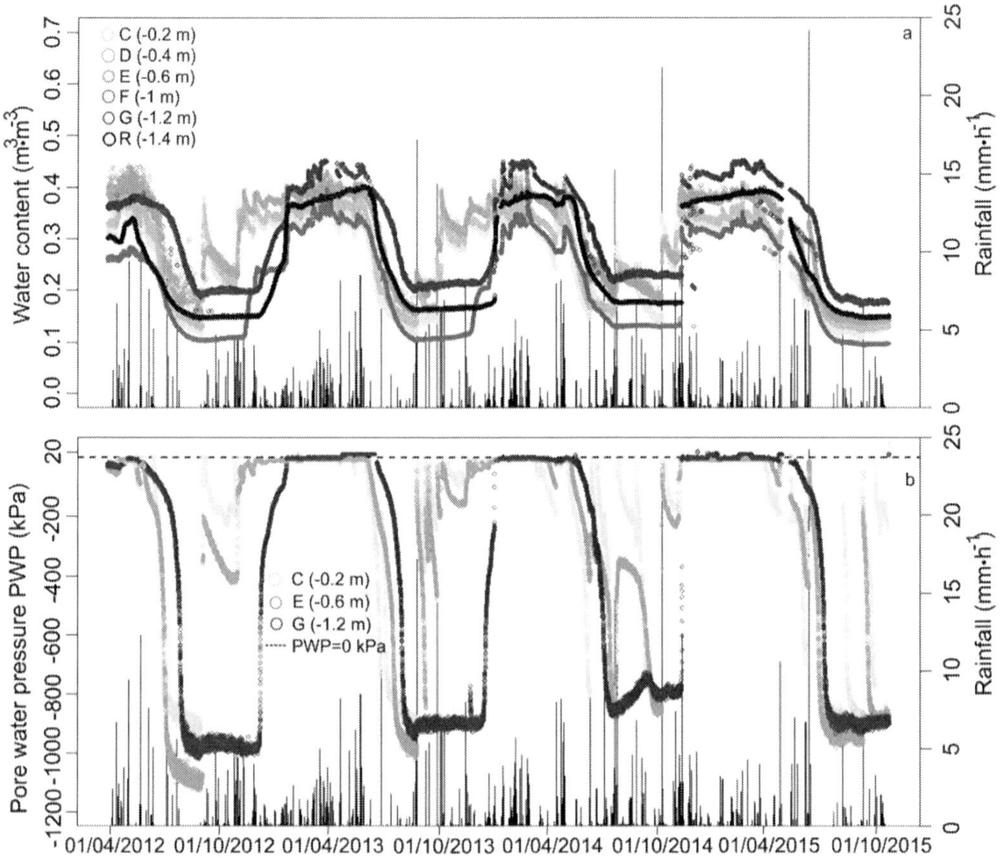

Figure 2. Water content (a) and pore water pressure (b) dynamics in relation with rainfalls for the analyzed time span.

potential sliding surface could develop; β = slope angle; φ' = the friction angle; σ^s = suction stress.

The safety factor can be calculated alternatively by considering water content or pore water pressure and hydrological hysteresis, thanks to the term of suction stress (Eqs. 2-3). Thus the effects of these parameters on safety factor calculation can be highlighted

For modelling F_s alternatively using water content or pore water pressure, suction stress was inserted in Eq. 1 through the following equation (Eqs. 2-3; [9]):

$$\sigma^s = -\frac{S_e}{\alpha}\left(S_e^{\frac{n}{1-n}} - 1\right)^{\frac{1}{n}} \tag{2}$$

$$\sigma^s = \begin{cases} -\dfrac{(u_a - u_w)}{\left(1 + (\alpha(u_a - u_w))^n\right)^{\frac{n-1}{n}}} & (u_a - u_w) > 0 \\ (u_a - u_w) & (u_a - u_w) \le 0 \end{cases} \tag{3}$$

, where α and n = fitting parameters of the Soil Water Characteristic Curve (SWCC) [10]; θ_s = saturated water content (measured through SWCC fitting); θ_r= residual water content (measured through SWCC fitting).

3 Results and discussions

3.1 Soil hydrological behaviors

Average hourly values of collected data were analyzed to reproduce water content and pore water pressure dynamics both in the soil profile and in the weathered bedrock (Fig. 2). Due to breakage of the tensiometer, pore water pressure in the range higher than -10^1 kPa was not measured at depth of 0.2 m since November 2012.

In the analyzed period, water content ranged between 0.10 and 0.45 $m^3 \cdot m^{-3}$ in the topsoil, and between 0.15 and 0.38 $m^3 \cdot m^{-3}$ in the weathered bedrock. Instead, pore water pressure ranged from positive values, till 12.7 kPa in the G horizon, to values in the order of -10^3 kPa.

Soil horizons till 0.6–0.7 m from the ground level had quick response than the deepest ones to long dry (summer months) or long wet (winter-spring months) periods (Fig. 2). On the other hand, only prolonged rainy periods, especially during winter months, can provoke an increase of the pore water pressure and of the water content in correspondence of soil horizons deeper than 0.6–0.7 m and of the weathered bedrock (Fig. 2).

During summer and early autumn rainfalls, the re-wetting of the soil horizons till 0.6–0.7 m is very rapid (Fig. 2). Moreover, in correspondence of concentrated and moderately intense summer rainstorms, as on 27 June 2013 (13.3 mm in 2 h), on 26 August 2013 (16.5 mm in 3 h), and on 14 June 2015 (39.7 mm in 6 h), pore water pressure increase due to re-wetting is not coupled with a

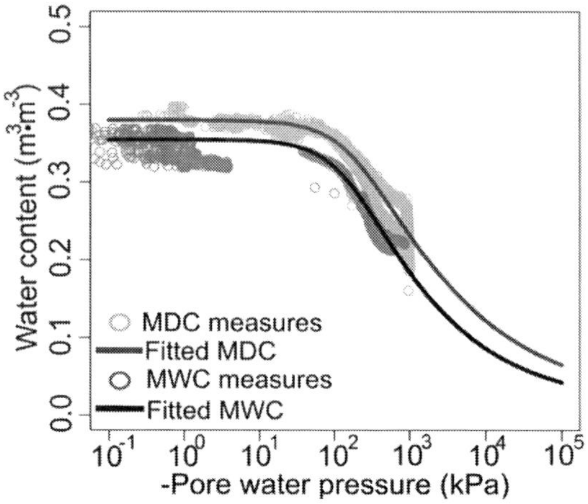

Figure 3. Example of field SWCC (G horizon).

Table 3. Van Genuchten equation fitting parameters of field fitted MDCs for the studied soil horizons.

	MDC				
	α_d	n_d	θ_{sd}	θ_{rd}	RMSE
	kPa^{-1}	-	$m^3 \cdot m^{-3}$	$m^3 \cdot m^{-3}$	$m^3 \cdot m^{-3}$
C	0.016	1.30	0.370	0.01	0.0159
E	0.007	1.35	0.425	0.01	0.0165
G	0.004	1.20	0.390	0.01	0.0092

Table 4. Van Genuchten equation fitting parameters of field fitted MWCs for the studied soil horizons.

	MWC				
	α_w	n_w	θ_{sw}	θ_{rw}	RMSE
	kPa^{-1}	-	$m^3 \cdot m^{-3}$	$m^3 \cdot m^{-3}$	$m^3 \cdot m^{-3}$
C	-	-	-	-	-
E	0.010	1.43	0.395	0.01	0.0104
G	0.006	1.26	0.370	0.01	0.0103

correspondent increase in water content (Fig. 2). This behavior could be linked to non-equilibrium processes due to the fast infiltration phase [11] in which pore water pressure or water content trend lags behind each other by the SWCC equilibrium.

The months between December and May are the most wet periods. Frequent precipitations increase the soil wetness till it approaches or reaches saturated conditions. Completely saturated conditions were reached in the G horizon in winters and springs, as testified by the values of pore water pressure around 0 kPa (Fig. 2b). Rather prolonged rainy periods with many rainfall events could cause a further increase in pore water pressure, as occurred at the end of April–first day of May 2013, when pore water pressure reached 12.7 kPa in the G horizon (Fig. 2b).

In wet periods, only the E horizon approaches saturated conditions, with pore water pressure increase till values of −3 kPa, also in agreement with the highest values of measured soil water content at the same depth (around 0.40 $m^3 \cdot m^{-3}$). During wetting periods, water content in the weathered bedrock was lower than in the overlying G horizon (Fig. 2a). According to the monitored data, it could be supposed that during wet periods (winter and springs) a perched water table forms in the G horizon. The upper soil levels could be saturated later, in correspondence of intense rainfalls (>60 mm in 48 h), as observed in correspondence of 28 February–2 March 2014. In correspondence of this event, the F horizon, which is positioned immediately above the G level, was affected by a significant increase of water content, passing from 0.32 m^3 m^{-3} to values similar of those measured in the G horizon (Fig. 2). This increase leaded to the complete saturation of the F horizon as well, till 1.0 m from ground. This condition can be considered the triggering mechanism of rainfall-induced shallow landslides in the test-site slope.

Coupling contemporaneous field measurements of pore water pressure and water content allowed for reconstructing SWCCs for C, E and G layers. A Main Drying Curve (MDC) and a Main Wetting Curve can be identified (MWC) for both the E and the G layers, with a not-closed hysteretic behavior (Fig. 3). The drying path corresponded to the data measured in summer and autumn before the soil re-wetting, while the wetting path includes the values measured during the wet months (winter and spring).

Field measured data were interpolated by using Van Genuchten equation [13] through Marquardt algorithm [13] for evaluating the fitting parameters of the curves. Curve parameters (α, n, θ_s, θ_r) changed passing from MDC to MWC, with a general increase in n and a decrease of α and θ_s for the not-closed hysteresis (Tabs. 3, 4). The reliability of the fitting procedure was good, as testified by the low values of Root Mean Square Error (RMSE) statistical index (0.01-0.02 $m^3 \cdot m^{-3}$) (Tab. 3, 4).

3.2 Slope stability analysis

Slope stability analysis was performed for the period between 28 February and 2 March 2014. Safety factor (F_s) trends were reconstructed either from water content (Eq. 2) or pore water pressure (Eq. 3), considering, in both cases, the fitting parameters of drying or wetting curves (Fig. 4). These analyses were performed at 0.6 (E horizon), 1.0 (F horizon) and 1.2 m (G horizon) from ground level. For the F horizon, only the F_s through water content could be reconstructed, for the loss of pore water pressure measurements at that depth.

The input soil geotechnical properties (γ, φ', c') were provided in Tab. 2, while, the input hydrological properties were provided in Tabs. 3 and 4, by assigning to the F horizon the same values of the E level due to the similar grain size distribution and physical properties (Tabs. 1, 2). The safety factor at a depth of 1 m decreases till unstable conditions on 1 March 2014 when calculated on the basis of the field measured water content only by using MWC parameters (Fig. 4a, b). This is consistent with the fact that shallow landslide occurred in wetting conditions of rainwater infiltration, thus the wetting parameters are more suitable than the drying ones.

At 1.2 m from the ground level, safety factor calculated from water content keeps constant for both MDC and MWC parameters due to the completely

 DOI: 10.1051/e3sconf/20160915005

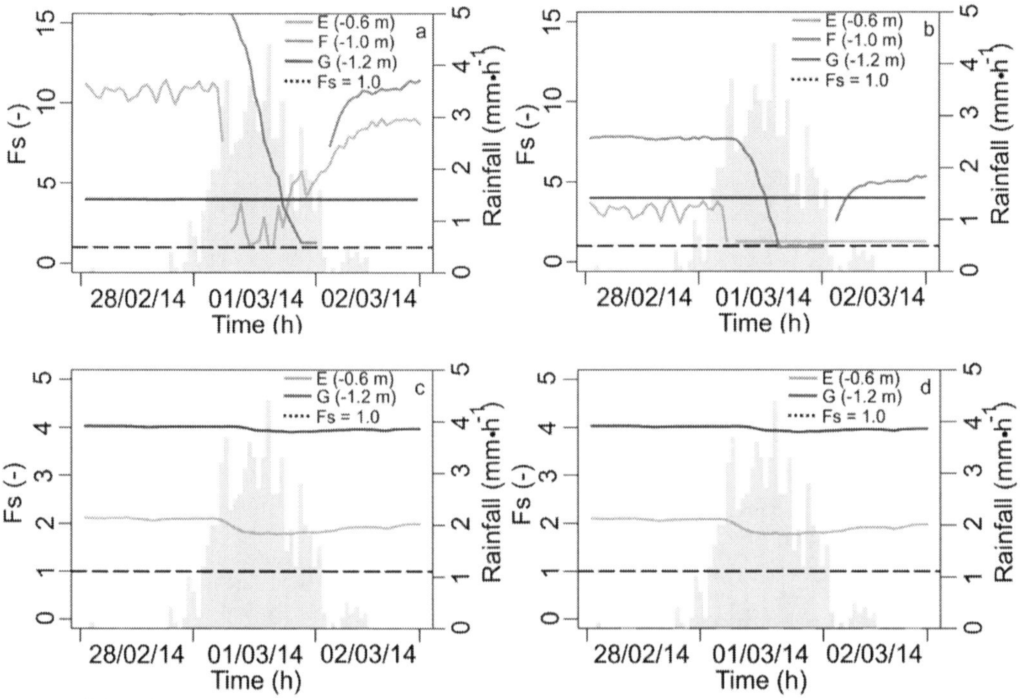

Figure 4. Modeled safety factor (F_s) trends for the period of 28 February-2 March 2014: a) from water content with MDC parameters; b) from water content with MWC parameters; c) from pore water pressure with MDC parameters; d) from pore water pressure with MWC parameters.

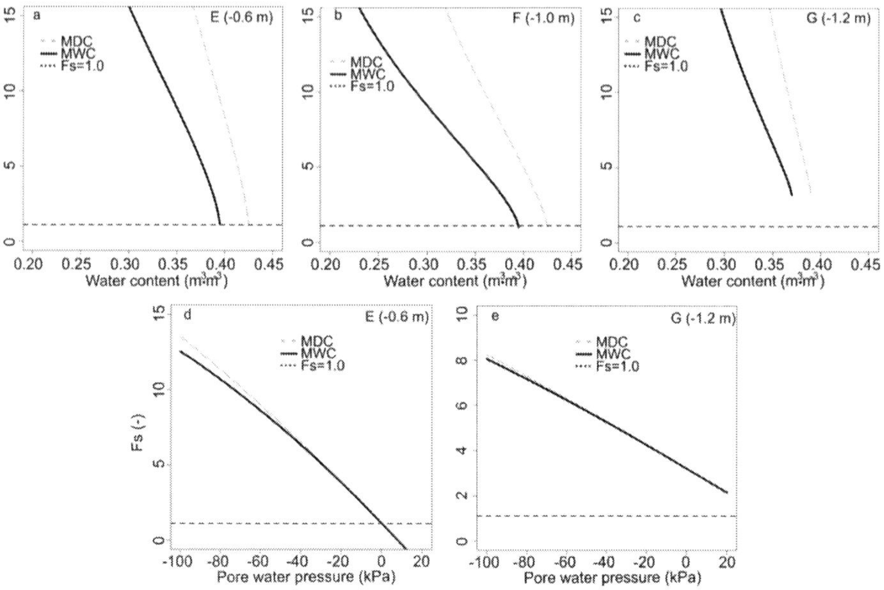

Figure 5. Lu and Godt's model safety factor (F_s) as a function of water content (a, b, c) or pore water pressure (d, e) for different soil levels of the test-site slope.

saturated condition of this horizon (Figs. 4a, b). At the same depth, the reconstructed safety factor on the basis of pore water pressure shows little variations, linked to little changes in field measured pore water pressure. This trend is similar by considering or not the hydrological hysteresis effects (Fig. 4c, d). In both cases the high soil cohesion (29 kPa) is enough to keep the safety factor values in the stable range ($F_S > 1$).

At 0.6 m from the ground level, in the E horizon, safety factor calculated on the basis of field water content suction estimated by using MWC parameters decreases till values slightly higher than the unstable condition (Fig. 4b). Instead, by considering the reconstructions made on the basis of pore water pressure by using both MDC and MWC parameters, the safety factor keeps in the stable range with values around 2 (Fig. 4c, d).

DOI: 10.1051/e3sconf/20160915005

The effect of hydrological hysteresis seems to be more marked if the safety factor is estimated on the basis of water content data. In fact, at the same depth, some relevant differences (till more than 5.0 for safety factor values) between the reconstructions obtained by using either MDC or MWC parameters are evident if water content data are used (Fig. 4a, b). On the other hand, the differences given by considering or not the hysteresis in the modeling based on pore water pressure data are very limited, and are in the order of the second decimal term for the safety factor (Fig. 4c, d). In both cases, safety factor evaluated on the basis of drying parameters seems to model a higher contribution to the stability than the reconstructions on the basis of wetting parameters.

3.3 Effects of hydrological parameters on safety factor trends

Safety factor was calculated also from the other measured values of water content and pore water pressure along the monitoring time span (Fig. 5). The differences considering or not hydrological hysteresis are more evident when safety factor is modeled on the basis of soil water content (Fig. 5a, b, c). At values of soil water content close to saturation, MDC reconstructed safety factor is at least 3.0 times higher than that in wetting conditions. Moreover, safety factor curves as a function of soil water content do not connect at the lowest value, due to the presence of a not-closed hysteresis, with θ_s of wetting curve lower than θ_s of drying path (Fig. 5a, b, c).

The effects of taking into account the MDC properties with respect to the MWC ones are less evident for pore water pressure modeling (Fig. 5d, e). The gaps decrease till a complete convergence of the curves when pore water pressure reaches nil and then positive values (Fig. 5d, e). This different behavior is due to the absence of θ_s parameter on the formulation of suction stress estimation from pore water pressure (Eq. 3).

The safety factor estimated from pore water pressure can continue to decrease in the presence of positive pressure (Fig. 5d, e). Instead, the safety factor evaluated on the basis of the water content reaches a minimum constant value for soil water content equal to θ_s. For this reason, modeling the safety factor on the basis of water content can make a correct prediction of the failure at site-specific scale only when a shallow landslide develops in unsaturated conditions or in conditions of water content equal to the saturated water content, so that pore water pressure is less or equal to 0 kPa. In triggering conditions linked to the development of positive pressures, it is more appropriate modeling safety factor from pore water pressure. In this case, the modeling based on water content can only indicate conditions which lead to shallow landslides triggering, without a correct prediction of the failure instant.

4 Conclusions

Monitoring of a slope susceptible to shallow landslides is fundamental for identifying the main soil hydrological behaviors and the conditions which lead to soil failure. In the presented case study, the main triggering mechanism develops especially during wet seasons (winter and spring) and it is linked to an increase in water content till about 1 m from ground in correspondence of the most intense rainfalls (about > 60 mm in 48 h).

Slope stability analysis using monitoring data correctly models the time instant and the depth of sliding surface formation. A better prediction occurs if hydrological hysteresis is taken into account, especially if safety factor is modeled using soil water content.

Modeling the safety factor on the basis of water content can make a correct prediction until a shallow landslide develops in conditions of water content equal to the saturated water content. In triggering conditions linked to the development of positive pore pressure, it is more appropriate modeling safety factor from pore water pressure, while the modeling based on water content can only indicate conditions which lead to the triggering, without a correct prediction of the failure instant.

These analyses at site-specific scale highlight the necessity of a correct choice of the models used for the slope stability analysis, also according to the most probable landslide triggering mechanism. For this reason, further developments will be the application of physically-based models based on water content and pore water pressure also in other monitored slope in different geological and environmental contexts, and also the application of these models at both local and regional scales.

References

1. D. Zizioli, C. Meisina, R. Valentino, L. Montrasio, Nat. Hazards Earth Syst. Sci. **13**, 559–573 (2013)
2. T.T. Lim, H. Rahardjo, M.F. Chang, D.G. Fredlund, Can. Geotech. J. **33**, 618–628 (1996)
3. M. Bittelli, R. Valentino, F. Salvatorelli, P. Rossi Pisa, Geomorphology **173–174**, 161–173 (2012)
4. E. Damiano, L. Olivares L. Picarelli, Eng. Geol. **137–138**, 1–12 (2012)
5. N. Lu, J.W. Godt, Water Resour. Res. **44**, http://dx.doi.org/10.1029/2008WR006976 (2008)
6. S.M. Springman, A. Thielen, P. Kienzler, S. Friedel, Geotechnique **14**, 1177–1193 (2013)
7. M. Bordoni, C.Meisina, R. Valentino, N. Lu, M. Bittelli, S. Chersich, Eng. Geol. **193**, 19-37 (2015)
8. IUSS Working Group WRB, *World Reference Base for Soil Resources* (2014)
9. N. Lu, J.W. Godt, D.T. Wu, Water Resour. Res. **46**,. http://dx.doi.org/10.1029/2009WR008646 (2010)
10. M.T. Van Genuchten, Soil Sci. Soc. Am. J. **44**, 892–898 (1980)
11. H.J. Vogel, U. Weller, O.J. Ippisch, J. Hydrol. **393**, 20–28 (2010)
12. D.W. Marquardt, SIAM J. Appl. Math (1963

Field instrumentation for performance assessment of Geobarrier System

Harianto Rahardjo[1,a], Zhai Qian[1], Alfrendo Satyanaga[1], Eng Choon Leong[1], Chien Looi Wang[2], Johnny Liang Heng Wong[2]

[1]*School of Civil and Environmental Engineering, Nanyang Technological University, 50 Nanyang Avenue, Singapore*
[2]*Building Research Institute, Housing and Development Board, HDB Hub, 460 Toa Payoh, Singapore*

Abstract: The saturated hydraulic conductivity of a coarse-grained material is normally higher than that of a fine-grained material. However, the unsaturated hydraulic conductivity of the coarse-grained material decreases much faster than that of the fine-grained material with increasing matric suction and consequently the coarse-grained material can be less permeable than the fine-grained material in unsaturated condition. A capillary barrier system with a fine-grained layer over a coarse-grained layer will act as a hydraulic barrier to minimize rainwater infiltration into the original soil. Geobarrier system (GBS) which can function as both a cover system (i.e. prevention of rainwater infiltration) and a retaining structure was designed and will be constructed at Orchard Boulevard in Singapore. Numerical simulations were carried out to determine the optimallocations of instruments that can be used to assess the performance of GBS during dry and wet periods. The monitoring results from tensiometers, earth pressure cells and weather station would be used to validate the results from the numerical analyses.

1 Introduction

Many factors can potentially determine slope failures, including climatic conditions, seismic activities, geological features, topography, vegetation and a combination of these factors [1]. In tropical regions, rainfall has been identified as the main cause of slope failures. Studies by Chowdhury et al. [2] indicated that the effects of rainfall have to be considered in landslide hazard assessments. Researchers have reported that most landslides occur in unsaturated residual soils during the wet season and they can potentially cause damages to infrastructure and human casualties [3, 4, 5]. Residual soils frequently exist in an unsaturated state in regions where the groundwater table is usually deep. Soil layers near the slope surface, which are initially unsaturated during dry seasons, have negative pore-water pressures, which are a major contributor to the shear strength of soils and to the stability of soil slopes. However during the wet seasons, rainwater infiltrates into the residual soil and increases the water content, thereby significantly increasing the pore-water pressures. Under these circumstances, the increase in pore-water pressure greatly reduces the shear strength of the soil. As a result, the slope may lose its original equilibrium and a slope failure will likely occur [6].

Singapore is located in the tropics where heavy rainfalls and high temperatures are conducive for rapid in-situ chemical and mechanical weathering that results in deep residual soil profiles. Because of the climatic conditions and geological features, slope instabilities are common in this region. To protect slopes against the possibility of rainfall-induced slope failures, preventive measures are necessary to ensure the safety of nearby buildings or public facilities. Capillary barrier system (CBS) is constructed on the surface of a slope in order to minimize rainfall infiltration into the slope and it can act as a slope preventive measure. A capillary barrier is an earthen cover system consisting of a fine-grained layer of soil overlying a coarse-grained layer of soil (Stormont, 1996). Previous research works have indicated the effectiveness of the capillary barrier system as a soil cover in reducing rainfall infiltration [7, 8, 9, 10, 11, 12].

The land space is very limited in Singapore. The construction of slopes with a steep inclination will have the advantage in intensifying of land usage. Previous studies by Rahardjo et al. [11, 12] indicated that CBS materials tend to slide down a steep slope surface if they are used solely as an added layer of slope cover system. Retaining structures are commonly used to retain the soil of a steep slope. Majority of materials that are used for the construction of retaining system comprise reinforced concrete or precast concrete elements, anchors embedded into soil and geosynthetic [13, 14]. In this paper, a new retaining structure that also acts as slope cover system, Geobarrier System (GBS) is presented. The GBS acts as earth retaining structure and slope cover by minimizing rainwater infiltration into the underlying soil layer. The surface materials of GBS can be vegetated to blend in with the surrounding environment.

Monitoring of slope is required to observe the current condition of the slope (stability, seepage, deformation) and to evaluate the effectiveness of slope preventive measures [15, 16, 17, 18]. Real-time monitoring will provide the early warning whether a slope is likely to fail or preventive measures are unsuccessful. It can also be

[a] Corresponding author: chrahardjo@ntu.edu.sg

used to verify the results of numerical analyses using finite element method. The collected data will also determine whether a single high value of measurement at a particular time is within a normal range for the instrumented slope. In this paper, the results from numerical analyses of slope supported with GBS are presented. The pore-water pressure variations from seepage analyses and the stress distributions from stress-strain analyses were used to determine the location of instruments that would be used to assess the performance of GBS in minimizing rainwater infiltration and retaining the lateral earth pressures of the soil from the steep slope.

2 Geobarrier System

Geobarrier system (GBS) consists of a fine-grained layer of material overlying a coarse-grained layer of material. The principle of the geobarrier system is based on the contrast in unsaturated hydraulic properties (soil-water characteristic curves and permeability functions) of these materials. Based on unsaturated soil mechanics theory, the coefficient of permeability as well as the water content of the coarse-grained soil can be much lower than that of the fine-grained soil at high suction values (Figure 1).

Under unsaturated conditions, the difference in permeability between the fine-grained layer and the coarse-grained layer limits the downward movement of water through capillary barrier effect. The mechanism of GBS can be explained using Figure 1b. At high matric suction values, the coefficient of permeability of the coarse-grained layer is much lower than that of the fine-grained layer. Under this condition, water will not flow into the coarse-grained layer. The infiltrated water is stored in the fine-grained layer by capillary forces. This infiltrated water is ultimately removed by transpiration, evaporation and lateral drainage through the slope or percolation into the drainage at the toe of the slope. When percolation (breakthrough) takes place, the capillary barrier no longer impedes water from infiltrating into the slope. Figure 2 shows the mechanism of GBS in minimizing the infiltration of rainwater into the slope.

In this study, a GBS will be constructed on a residual soil from Bukit Timah Granite in Singapore. Fine recycled concrete aggregate (RCA) will be used as the fine-grained material that will be placed inside Geobag within GBS whereas coarse RCA will be used as the coarse-grained material within GBS. In order to facilitate the planting of vegetation above GBS, the approved soil mixture (ASM) will be placed inside Geobag that will be placed as the most outer layer of GBS. Gravel layer will be placed at the bottom of GBS to drain out the water from GBS into the main surface drainage. The GBS will be constructed with 70 degrees of slope angle and 4 meters of slope height (Figure 3). The unsaturated properties which include the soil-water characteristic curves and permeability functions of ASM, fine and coarse RCA, and residual soil are presented in Figure 4.

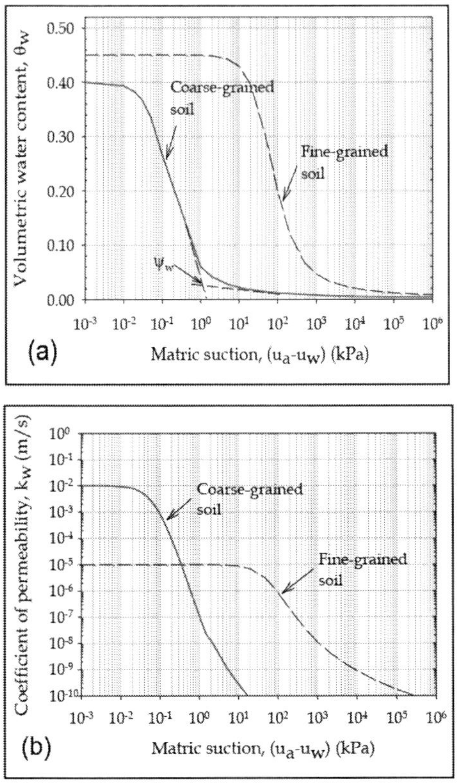

Figure 1. (a) Soil-water characteristic curves and (b) permeability functions of CBS materials

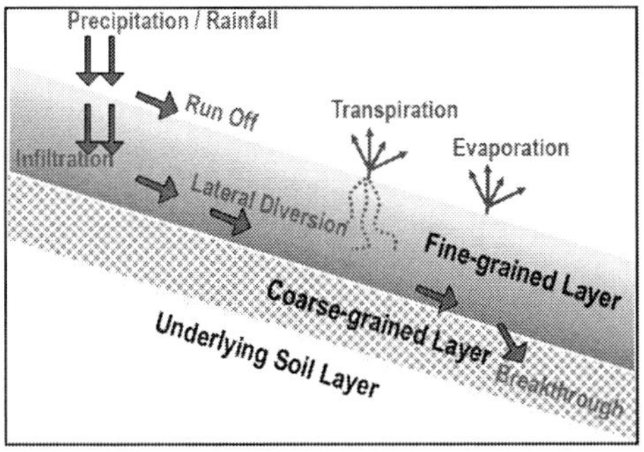

Figure 2. Capillary barrier system

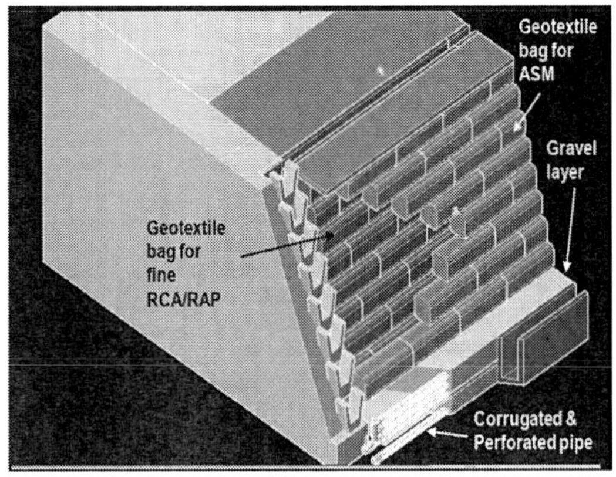

Figure 3. Schematic diagram of Gebarrier system

DOI: 10.1051/e3sconf/20160915006

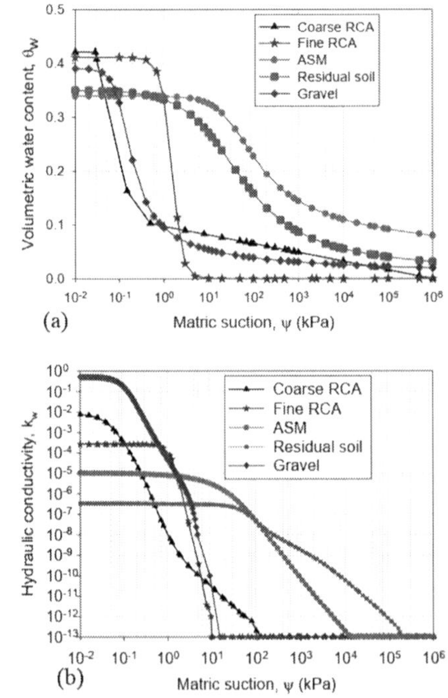

Figure 4. (a) SWCCs and (b) permeability functions of RCA, ASM and residual soil from Bukit Timah Granite

3 Numerical Analysis

Two-dimensional seepage analyses were carried out using finite element software, SEEP/W [19] and stress-strain analyses were conducted using SIGMA/W [20]. Two cases of numerical analyses were performed in this study. In the first case, the numerical model followed the geometry of the original slope consisting of a slope with an angle of 45 degrees and a height of 4 meters. In the second case, the numerical model followed the geometry of the GBS consisting of a slope with an angle of 70 degrees and a height of 4 meters. The groundwater table was observed at a depth of 1.86 meters and 4.54 meters below the ground surface on the top and bottom of the slope, respectively.

The boundary conditions applied to the finite element model are illustrated in Figure 5. The distance between the slope and the side of the slope model was set to three times the height of the slope to avoid the influence of the side boundary conditions. No flow boundaries were simulated by assigning a nodal flux, Q, equal to zero at the bottom and along the sides of the slope model above the groundwater table. A constant total head, hw, on each side was applied as the boundary condition along the sides of the slope model below the groundwater table. Rainfall was applied to the slope surface as a flux boundary, q. Ponding was not allowed to occur on the slope surface. This meant that when a flux greater than the permeability of the soil was applied to the top boundary, the seepage model would not allow pore-water pressures greater than 0 kPa to build up at the ground surface. This simulated the actual field conditions where the excess rainfall at the slope surface is removed as runoff.

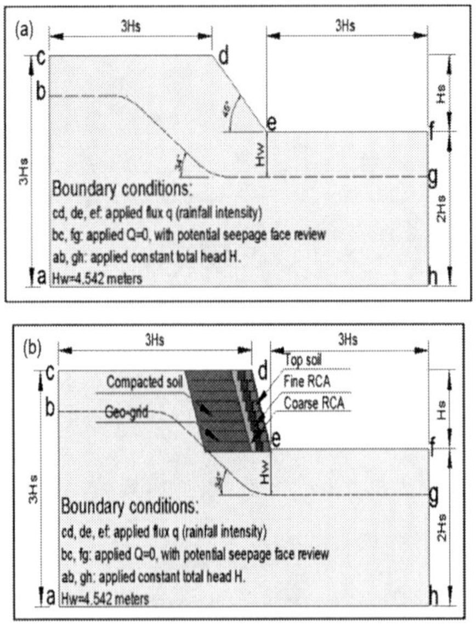

Figure 5. Geometries and boundary conditions of (a) existing slope with slope with slope angle of 45° and (b) 70° slope using GBS

The initial condition of the slope was taken based on the position of groundwater table in the field. The finite element model within the GBS and residual soil layer down to 5 m below the slope surface had a mesh size of approximately 0.5 m, smaller than elements in other parts of the slope, in order to obtain accurate results within the infiltration zone.

In the seepage analyses, a rainfall intensity of 18 mm/h was applied to the slope for 10 hours duration. The materials in the analyses were modelled as a linear elastic material. The parameters of materials and geo-grid used in the stress-strain analyses are summarized in Table 1. Several points within the numerical model were selected to illustrate the changes in pore-water pressures and stress with time as obtained from the seepage and the stress-strain analyses of cases 1 and 2 (Figures 6 and 7, respectively). Points TA and TC were located within the fine-grained layer, point TB was located within the residual soil below the GBS layer, point TD was located within the coarse-grained layer.

The Young's Modulus and Poisson ratio of soils were summarized in Table 1. Pore-water pressures computed from SEEP/W were incorporated into SIGMA/W to calculate the stress in a selected position in three directions (i.e., x, y, z directions).

 DOI: 10.1051/e3sconf/20160915006

Table 1. Parameters of materials and Geo-grid used for stress-strain analysis

Soil Description	γ (kN/m^3)	E (kPa)	μ
Approved soil mixture	18	5000	0.45
Fine RCA	20	120000	0.35
Coarse RCA	21	200000	0.35
Residual soil	18	6000	0.40
Gravel	21	200000	0.34
Precast drain	24	200000	0.34

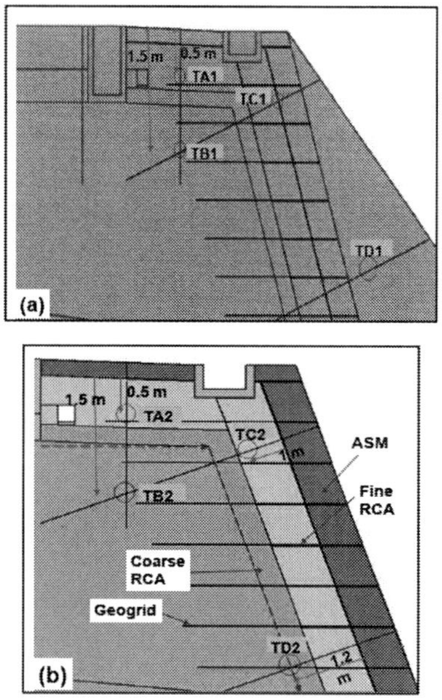

Figure 6. Location of points within GBS and residual soil for monitoring of pore-water pressure variations within (a) original slope and (b) slope retained with GBS during and after rainfall

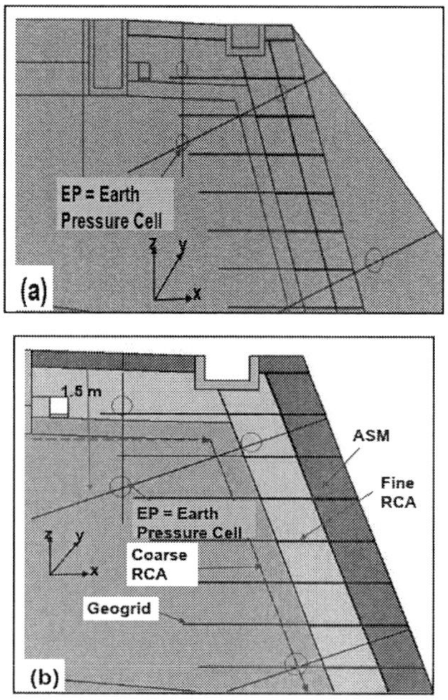

Figure 7. Location of point EP within GBS and residual soil for monitoring of lateral earth pressure within (a) original slope and (b) slope retained with GBS during and after rainfall

4. Results and Discussions

The pore water pressures variations with time at points TA, TB, TC and TD (see Figure 6) from the seepage analyses of case 1 (original slope) and case 2 (slope retained with GBS) are shown in Figures 8 and 9. Figure 8a shows that the pore-water pressure at point TA1 within the original slope increased significantly during rainfall and decreased after rain stopped. On the other hand, the pore-water pressure at point TA2 at the bottom of fine RCA within the slope retained with GBS remained constant around -65 kPa during rainfall. This indicated that GBS performed well in reducing rainwater infiltration into the underlying soil layer below GBS. Tensiometers and soil moisture sensors can be installed at point TA to assess the performance of GBS and to verify the results from the seepage analyses at the crest of the slope during and after rainfall. Figure 8b shows that the pore-water pressure at point TB1 within the original slope increased significantly during rainfall and decreased after rain stopped. On the other hand, the pore-water pressure at point TB2 within the residual soil behind GBS increased gradually from – 55 kPa to – 20 kPa during rainfall. This indicated that the surface drainage behind the GBS layer at the crest of the slope was deep enough to avoid the excessive increase in pore-water pressure behind GBS. Tensiometers and soil moisture sensors can be installed at point TB to assess the performance and to validate the results of seepage analyses of the surface drainage behind GBS at the crest of the slope.

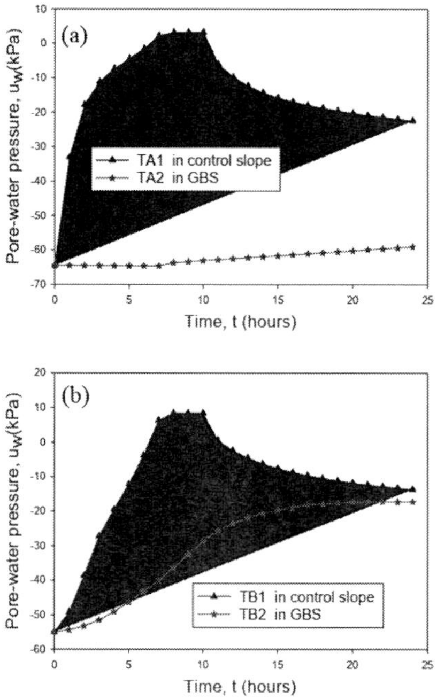

Figure 8. Variations of pore-water pressures with time at points (a) TA and (b) TB within original slope and slope retained with GBS

 DOI: 10.1051/e3sconf/20160915006

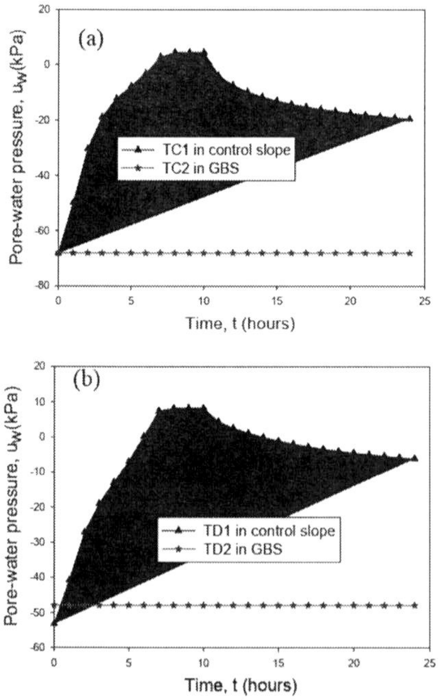

Figure 9. Variations of pore-water pressures with time at points (a) TC and (b) TD within original slope and slope retained with GBS

Figs 9a and 9b show that the pore-water pressures at point TC1 and TD1 within the original slope increased significantly during rainfall and decreased after rain stopped. On the other hand, the pore-water pressure at point TC2 and TD2 at the bottom of fine RCA and at the bottom of coarse RCA, respectively, within the slope retained with GBS remained constant around -70 kPa and -48 kPa, respectively during and after rainfall. This indicated that GBS performed well in reducing rainwater infiltration into the underlying soil layer below GBS. Tensiometers and soil moisture sensors can be installed at points TC and TD to assess the performance of GBS and to validate the results from the seepage analyses of GBS comprising fine RCA and coarse RCA layers during and after rainfall.

Total stresses at point EP in x, y and z directions within the original slope and the slope retained with GBS (Figure 7) were plotted against pore-water pressures and presented in Figure 10. Figure 10 indicates that total stresses in all directions remained constant when the pore-water pressures were negative. On the other hand, total stresses in all directions increased drastically when the pore-water pressures became positive. In the original slope, the soil was saturated by the infiltrated rainwater and total stresses in all directions increased significantly after 6 hours of rainfall. On the contrary, the soil under the GBS remained in unsaturated condition during the rainfall period and there were no changes in total stress. As the non-linear contributions from matric suction to effective stress are not yet accommodated by SIGMA/W, the total stress computed from SIGMA/W in this study did not incorporate the effect of changes in suction. However, the results from SIGMA/W can describe the differences in total stresses between soils with and without suction as presented in Figure 10. Figure 10 indicates significant changes in total stresses at point EP in Figure 7 for the soil within the original slope. As a result, the earth pressure cell can be installed at point EP in the original slope and in the slope retained by GBS to illustrate the effectiveness of GBS during rainfall.

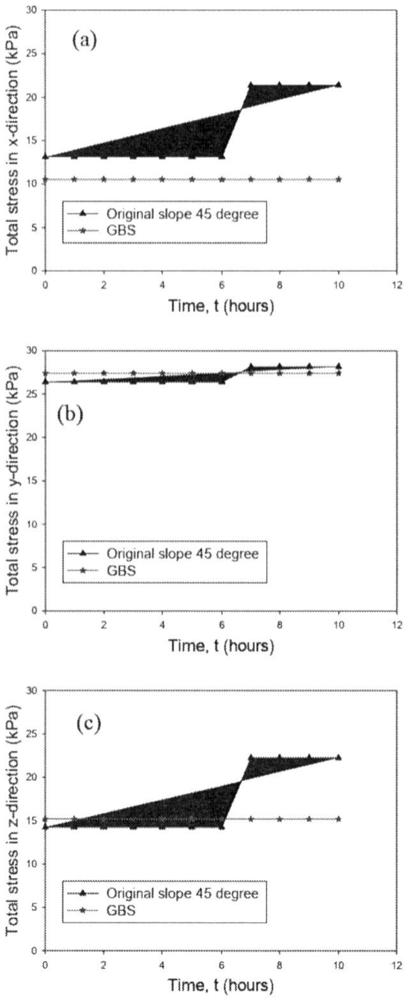

Figure 10. Variations of total stress with pore-water pressures in (a) x-direction, (b) y-direction and (c) z-direction within original slope and slope retained with GBS

5 Locations of Instruments

Numerical analysis results indicated that the proposed locations of tensiometers (TA, TB, TC and TD) as shown in Figure 6 were appropriate for monitoring the pore-water pressures changes due to rainwater infiltration. In addition, earth pressure cells should be located at the proposed location EP as shown in Figure 7 to monitor the stress changes due to rainwater infiltration. In the field, three slopes will be retained with GBS consisting of different types of recycled materials (Figure11). Slope 1 will comprise fine and coarse recycled concrete aggregate (RCA) as fine- and coarse-grained materials within GBS, respectively. Slope 2 comprises fine and coarse reclaimed asphalt pavement (RAP) as fine- and coarse-grained materials within GBS, respectively. Slope 3 comprises fine RCA and coarse RAP as fine- and coarse-grained materials within GBS, respectively. Slope 4 is an original slope that will be used as a reference slope.

Tensiometers (TA, TB, TC and TD) and earth pressure cells (EPA at x-direction, EPB at y-direction, EPC at z-direction) will be installed at the same depth in the proposed locations shown in Figures 6 and 7. In addition, soil moisture sensors (SA, SB, SC and SD) will be installed following the same arrangement as those for tensiometers (Figure 11). Two piezometers will be installed at the crest and toe of the slope to monitor the variations of groundwater table during dry and wet periods. A weather station will be installed near the toe of the slope to monitor the variations of rainfall and evaporation with time.

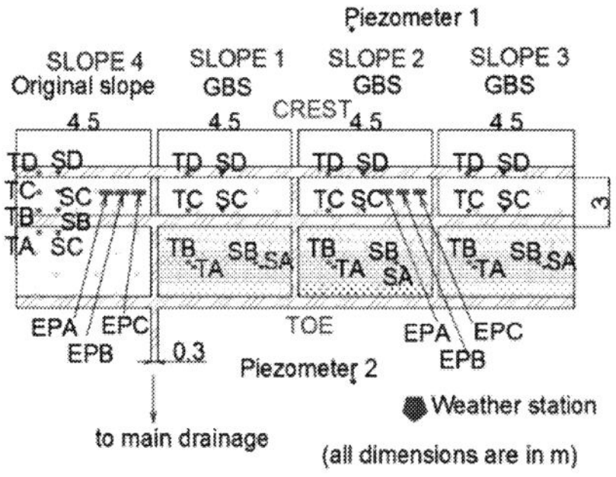

Figure 11. Proposed plan view of field instrumentation for assessment of GBS

6 Conclusions

A new slope protection system named as Geobarrier system (GBS) is proposed from this study. The GBS can be used as a retaining structure that utilizes capillary barrier mechanism for controlling rainfall infiltration into the original soil behind the retaining structure. The results of the numerical analyses indicated the optimal locations for tensiometers, soil moistures sensors and earth pressure cells within the original slope and the slope retained with GBS. The monitoring results of the instruments can be used to assess the performance of GBS as a retaining structure and slope stabilization system.

Acknowledgement

The findings published in this paper are based on the project "GeoBarrier System for Use in Underground Structure" (Grant No. SUL2013-3) that is supported by a research grant from the Ministry of National Development Research Fund on Sustainable Urban Living, Singapore

References

1. A. Basile, G. Mele, F. Terribile. Geoderma, **117**, 3-4, 331-346 (2003)
2. R. Chowdury, P. Flentje, G. Bhattacharya. *Geotechnical Slope Analysis* (2010)
3. D.G. Fredlund, H. Rahardjo. *Soil Mechanics for Unsaturated Soils*, pp. 517 (1993)
4. H. Rahardjo, A. Satyanaga, E.C. Leong, Y.S. Ng. J Geo & Geoenviro Eng, ASCE, **136**, 11, 1555–1564 (2010)
5. H. Rahardjo, A. Satyanaga, E.C. Leong. J. Eng. Geo., Special Issue on Unsaturated Soils: Theory and Applications, **165**, 133-142 (2013a)
6. D.G. Fredlund, H. Rahardjo, M.D. Fredlund. *Unsaturated Soil Mechanics in Engineering Practice*, 926 pp (2012)
7. C.E. Morris, J.C. Stormont. J. Envir. Engrg., ASCE, **123**, 1, 3-10 (1997)
8. M. Khire, C. Benson, P. Bosscher. J. Geo. Geoenvir Eng., ASCE, **126**, 8, 695-708 (2000)
9. D. Tami, H. Rahardjo, E.C. Leong, D.G. Fredlund. Geo. Test. J., ASTM, **27**, 2, 173-183 (2004a)
10. D. Tami, H. Rahardjo, E.C. Leong, D.G. Fredlund.. Can. Geo. J., **41**, 5, 814 – 830 (2004b)
11. H. Rahardjo, V.A. Santoso, E.C. Leong, Ng., Y.S. C.J. Hua. J. Geo. & Geoenvi Eng., **138**, 4, 481 – 490 (2012)
12. H. Rahardjo, V.A. Santoso, E.C. Leong, Y.S. Ng, C.P.H. Tam, A. Satyanaga. Can. Geo. J., **50**, 1-12 (2013b)
13. AASHTO. *Standard Specifications for Highway Bridges*, 7th ed., 689 (2002)
14. NCMA. *Design Manual for Segmental Retaining Walls*, 3rd ed., 282 (2010)
15. J. Gasmo, K.J. Hritzuk, H. Rahardjo, E.C. Leong. Geo. Test. J., **22**, 2, 128-137 (1999)
16. G. Gitirana Jr., D.G. Fredlund, M.D. Fredlund. Proc. 58th Can. Geo. Conf., 516-523 (2005)
17. H. Rahardjo, V.A. Santoso, E.C. Leong, Y.S. Ng, C.J. Hua. Soils and Foundations, **51**, 3, 471-482 (2011)
18. H. Rahardjo, A. Satyanaga, E.C. Leong, J.-Y. Wang Procedia Earth and Planetary Science, **9**, 23-43 (2014)
19. Geoslope International Ltd., *Seep/W User's Guide for Finite Element Analysis*. (2007a)
20. Geoslope International Ltd., *Sigma/W User's Guide for Finite Element Analysis,* (2007b)

Degree of saturation effect on the grout-soil interface shear strength of soil nailing

Qiong Wang[1], Xinyu Ye[1], Shanyong Wang[a,1], Scott William Sloan[1], Daichao Sheng[1]
[1]*ARC Centre of Excellence for Geotechnical Science and Engineering, The University of Newcastle, Callaghan, 2308 NSW, Australia*

Abstract. In the grouted soil nailing system, the bonding strength of cement grout-soil interface offers the required resistance to maintain the stability of whole structure. In practice, soil nailing applications are often placed at unsaturated conditions, such as soil slopes, shallow foundations, retaining walls and pavement structures. In these cases, the water content in the soil nail zone may increase or decrease due to rain water or dry weather, and even cannot become saturated during their design service life. In this study, the effect of water content (degree of saturation) on the shear strength of interface between cement grout and sand are experimentally investigated by means of direct shear test. Meanwhile the water retention curve was determined and interface microstructure was observed. Experimental results show that the shear strength of interface changes non-monotonously with degree of saturation when the interface was prepared, due to the non-monotonousness of the cohesiveness between soil particles. The less the cohesiveness between sand particles, the more grout was observed been penetrated into the voids, and thus the larger the interface shear stress.

1 Introduction

The fundamental mechanism of grouted soil nail is the development of tensile forces in the passive zone. The tensile forces generated are transferred to the soil through friction mobilized at the grout/soil interface, and tends to restrain the reinforcements from pull-out [1, 2]. Therefore, the pull-out shear resistance between the grout and the surrounding soils is of primary importance for design and safety assessment of soil nailing system [3].

Traditionally, the design of soil nail is assumed in the conditions of saturated soils, where the effect of matric suction or the negative pore water pressure in the unsaturated zone is neglected. However, soil nailing applications are often placed at unsaturated conditions in the practice, such as soil slopes, shallow foundations, retaining walls and pavement structures. In these cases, the water content in the soil nail zone may increase or decrease due to rain water or dry weather, and even cannot become saturated during their design service life. The soil mechanic behaviour is significantly influenced by the water content changes (degree of saturation which in turn leads to different grouting behaviour and interface performance.

In this study, the effect of soils degree of saturation (water content) when the grout was injected, on the interface shear stress of cement grout-sand was investigated by means of direct shear test. Soil water retention curve was accordingly determined as it demonstrates the unsaturated soil behaviour. While the cement-sand interface obtained at different water content was observed by using microscope, which helps to explain the differences in shear stress.

2 Materials and Methods

2.1 Materials

The soil tested is silica sand obtained from Stockton Beach, Newcastle, Australia. It consists 98.82% of quartz, 0.8 % of rock fragments, 0.21 % of Zircon, 0.11% of Ilmenite and 0.06% of Rutile [4]. With a specific gravity of 2.67 Mg/m^3, the maximum and minimum densities are 1.71 Mg/m^3 and 1.46 Mg/m^3, respectively. The grain size distribution curve determined by dry sieving is shown in Figure 1, which is characterized as by a uniformity coefficient C_u of 3.56 and a D50 close to 0.9 mm.

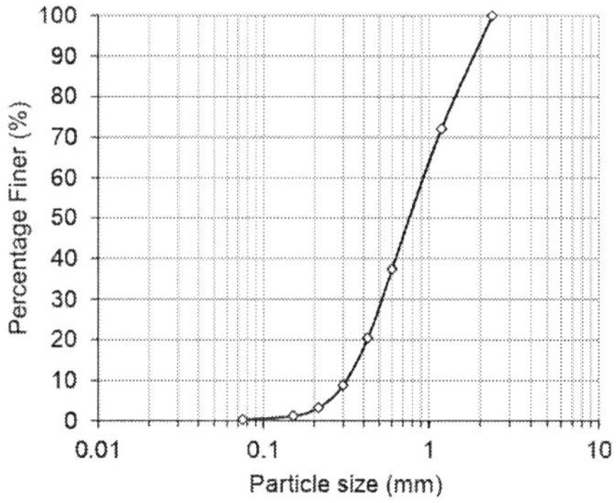

Figure 1 Particle size distribution of Stockton sand

[a] Corresponding author: Shanyong.wang@newcastle.edu.au

 DOI: 10.1051/e3sconf/20160915007

The cement grout used was in a water cement ratio (w/c) of 0.5. It was prepared by locally available Portland cement. The properties of the cement grout material are listed in Table 1. It has an initial setting time of 1.5-3 hours, and final setting time of 2.5-4 hours. The soundness is 1.0 mm and the specific gravity is 3.14.

Table 1. Fundamental properties of Portland cement

Properties	Values
Initial setting time	1.5-3 hours
Final setting time	2.5-4 hours
Soundness	1.0 mm
Fineness	330-410 m^2/kg
Specific gravity	3.14

2.2 Experimental Methods

2.2.1 Samples preparation

Sand at certain water content was firstly prepared by water spraying method from oven dried sand. A given quantity of the wet sand was then placed in the sample mould (for water retention test) or shear box (for shear test) and compacted to a desired void ratio of 0.80 (classified as loose sand). The sample with desired degree of saturation was thus obtained.

2.2.2 Water retention curve of the compacted sand

To obtain the water retention curve (WRC) of the compacted sand in this study, a tensiometer with measurement range of -100 to 100 kPa (±1.5% accuracy of the full range) was employed (see Figure 2a). Several samples at different water content (corresponding to different degree of saturation) were compacted in the testing moulds to the same void ratio of 0.80 (Figure 2b). After compaction, a special tube with the same diameter of the tensiometer probe (6 mm in diameter) was used to drill a hole for the tensiometer installation, the surface of the sample was covered to avoid the evaporation. The reading of the tensiometer was continually monitored, and the value at the stability was recorded as the soil suction.

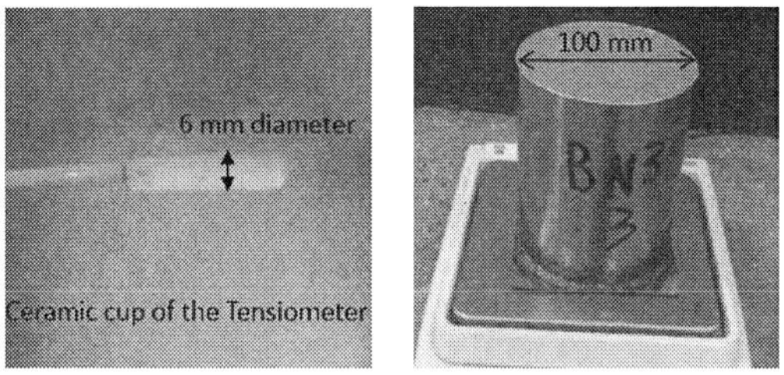

(a) Tensiometer probe (b) Soil mould

Figure 2 Instruments used for the WRC determination

2.2.3 Direct shear test at the interface

The interface shear stress between the unsaturated sand and cement grout was studied by using a conventional direct shear box apparatus shown in Figure 3. Both upper and bottom shear box is in dimension of Ø63 mm and 20 mm high.

Figure 3 Photo of the direct shear box apparatus used

For the preparation of the cement grout-sand interface, the sand sample at certain water content was first compacted in the interface sealed shear box (Figure 4a) to the void ratio of 0.80 Mg/m^3. Considering the producing of a cemented sand layer at the interface and the compaction of soil due to grouting pressure (Figure c) , the height of the soil was designed 1mm higher than the bottom shear box (see Figure 4a), as the shear plane will be most possible on the surface of this layer and sand. This value (1mm) was chosen based on several preparation tests.

The cement grout with w/c ratio of 0.5 was then obtained by mixing certain mass of water and cement powder. After mixed thoroughly using a blender, the grout was poured on the top surface of the compacted sand sample. A piston was the then put on the top to apply the vertical pressure by mechanical loading frame, which simulates the grouting pressure. The void between piston and the box wall was sealed by a special added o-ring to avoid the leakage of grout from the top. The same vertical pressure of 100 kPa was applied for all tests performed. Compaction, bleeding and penetration will be occurred at the interface of the grout and sand during this process, producing a cemented grout layer (Figure 4b). The cement grout was allowed to harden for 24 hours, during which the vertical pressure was kept constant.

Afterwards, the vertical pressure was released and the bolts used to fix the upper and bottom box was removed, a space was then created at the interface. Direct shear test was conducted under vertical pressure of 50 kPa at a shear speed of 1.00 mm/min. Corresponding strength value was measured automatically by the machine, allows the plotting of the displacement-shear strength curve. Totally eleven tests were performed on sand samples at different initial water content (degree of saturation) ranges from 2% to 22 %, corresponding to the degree of saturation of 6% to 73%.

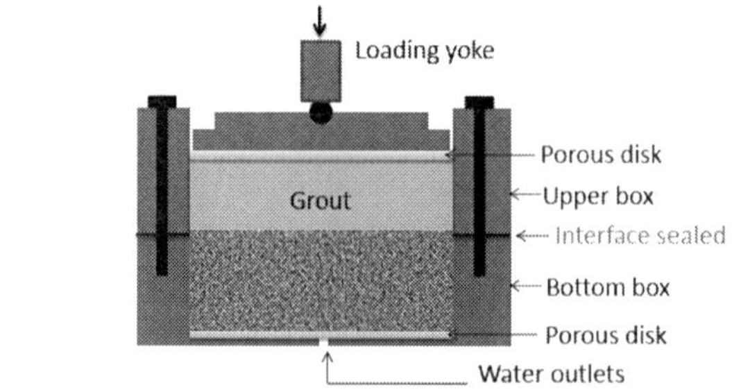

(a) Interface preparation process

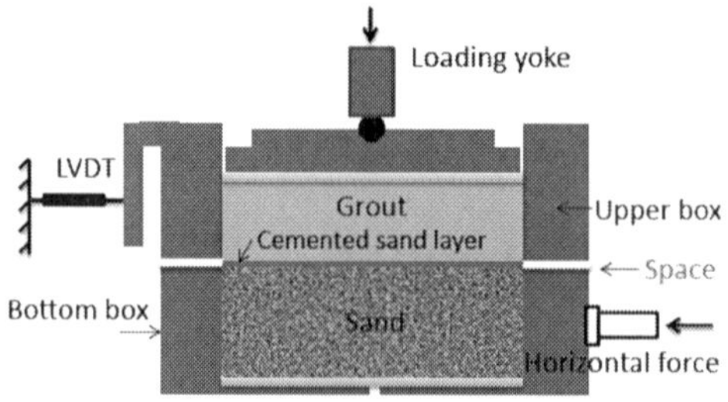

(b) Shear process

(c) Cemented sand layer

Figure 4 Schematic view of the experimental process

2.2.4 Microstructure observation of the interface

After the conduction of the interface shear test, the cemented layer of the interface obtained at different initial degree of saturation was observed by using a 200× magnification microscope. The penetration of cement grout into the sand voids could be visually investigated, which allows the explanation of the effect of degree of saturation on the interface shear stress.

3 Results and discussion

3.1. Water retention curve

The suction of the sand sample (void ratio of 0.80) at different water content (i.e. different degree of saturation) was measured by using a tensiometer and plotted with water content in Figure 5. An air entry value of 3.1 kPa could be estimated from the curve, which corresponds to water content of 26 %; while the residual water content was at around 2 %. It can be observed that from the water content of 2 %, which is located at the residual zone (Figure 6a), the water amount is very small (i.e. the water menisci is small) and resulting in a small cohesiveness between sand particles. The cohesion increases with the increase of water content (water menisci) and then entries into the transition zone (Figure 6b). The water content in the soil increases significantly with the decreasing suction, where the water menisci area in contact with the sand particles increases as saturation continues, thus leading to an increase in the cohesiveness. However, after a certain value of water content, where water phase are connected with air phase being isolated (Figure 6c), the surface tension force of the water menisci breaks, therefore the cohesiveness between sand particles became less and less until it was disappeared in the boundary effect stage (saturation state in Figure 6d). In the figure, the water contents at which cement-soil interface was prepared are also plotted, showing the positon of the samples on the water retention curve.

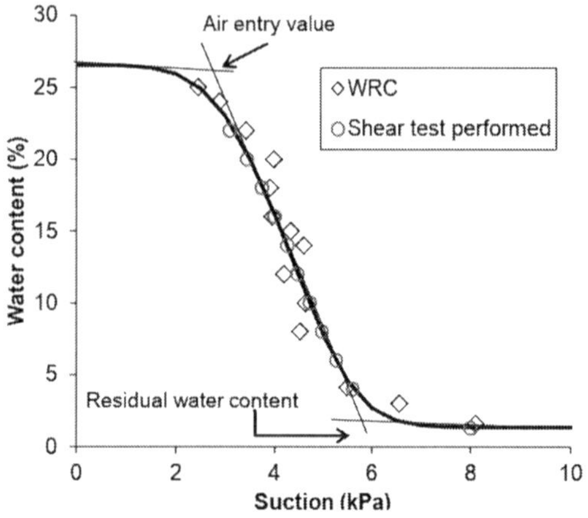

Figure 5 Water retention curve of the compacted sand

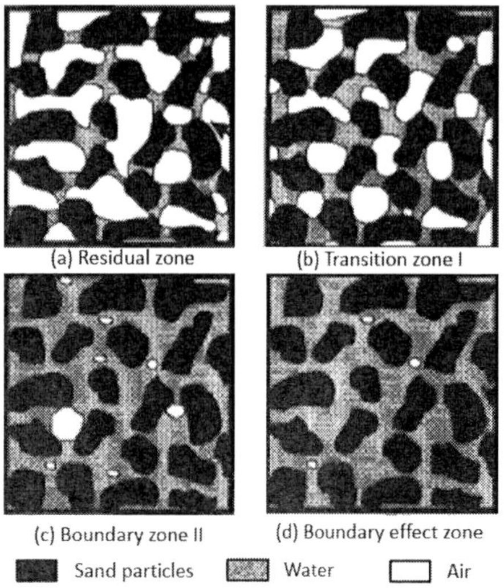

Figure 6 Water areas in different stages of the water retention curve (after Vanapalli et al. 1995) [9]

DOI: 10.1051/e3sconf/20160915007

3.2 Interface shear stress

The shear stress-horizontal displacement curves of the grout-sand interface prepared at different water content (degree of saturations) are showed in Figure 7. All curves show similar trend and shape: the interface shear stress gradually increases with shear displacement until it reaches the maximum shear strength, after which it decreases slightly or remains almost constant. Regarding the effect of water content, it can be seen that the curve for the sample at the lowest water content of 2% (6% of S_r) in this study locates at the top, indicating the largest shear stress at the interface; while the curves for initial water content of 4% and 22% followed subsequently. It seems that there is no obvious relationship between water content (degree of saturation) and interface shear stress, as the curves are not ordered with the increase of decrease of water content.

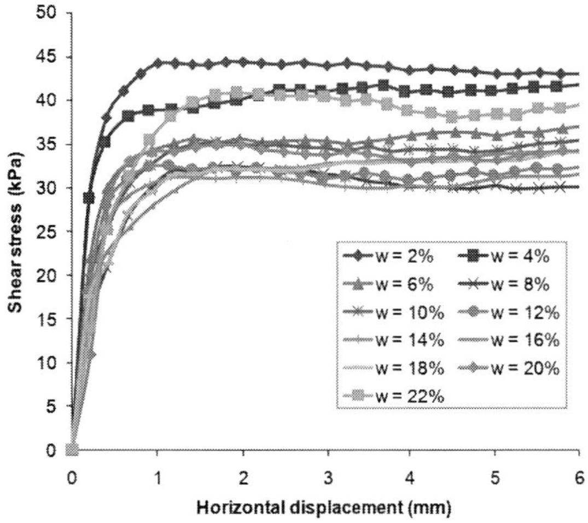

Figure 7 Shear stress Vs. horizontal displacement

To further analysis the increase of degree of saturation on the interface shear strength, the peak shear stress and residual stress were obtained from the curve and plotted versus the water content (degree of saturation) in Figure 8. It can be observed that the changes of interface shear strength with water content increase are not monotonic increase or decrease. Instead, with the water content increase from 2%, both the peak and residual stress decreases up to 6%, followed by an almost constant stage ranged from 6% to about 18%; afterwards, the shear stress increase with further water content increase. This variation trend is similar to the relationship between the injected grout volume and the degree of saturation (water content) for the same soil at the same void ratio [5]. These two phenomena are actually followed the same mechanism due to the preparation of the interface in this study was also intend to simulate the pressure grouting process. According to the previous study [6,7], the complicated mechanism of soil grout interaction during pressure injection involves the compaction of soil by grouting pressure as well as penetration and bleeding of the grout, which are affected by the mechanic behaviour of soil, i.e. bear capacity (compressibility) and shear stress (cohesiveness).

During injection, the pressurized grout may penetrate into the soil voids considerably, enhancing the bond strength and interface roughness between the grout part and surrounding soil. Therefore the shear stress of interface in Figure 8 is attributed to different bond strength and interface roughness obtained at different water content (degree of saturation). It can be seen from test positions in the water retention curve (Figure 6), due to the increase of sand cohesion consequently bearing capacity with increase of water content from 2% [8] (Figure 6a), less and less grout could be penetrated into the sand and the interface shear stress decreased. However, up to a certain water content until the surface tension force of the water film suspended between the pores breaks (Figure 6c), the subsequent increase of water content (decrease of suction) may induce a reduction in the cohesiveness, thus an increment of penetration leading to an increase of interface shear stress.

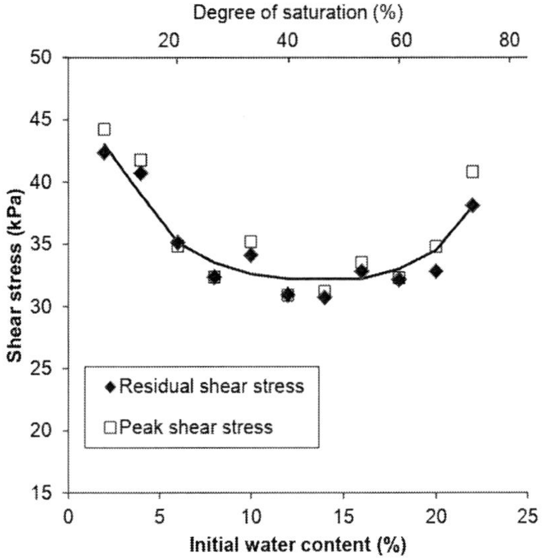

Figure 8 Shear stress Vs. initial water content (degree of saturation) of sand

3.3 Microstructure observation

The microstructures of the grout-sand interface obtained at different initial degree of saturation are presented in Figure 9. From the photo, it is clearly that the voids between sand particles at the degree of saturation of 6% (Figure 9a) are almost filled by the hardened grout; as the degree of saturation increase up to 45%, obvious less and less cement is filled between sand particles: at the degree of saturation of 20%, around half of voids is filled (Figure 9b), while for the 40% and 45% degree of saturation, there are just a few cement was attached between the contact surface of sand particle (Figure 9c and 9d). Thereafter, the subsequent increase of degree of saturation to 60% and 70% does not reduce the cement filled voids, instead, more and more voids between sand particles are filled by the hardened grout. The microstructure variation with initial degree of saturation is in agreement with the changes of interface shear stress (shown in Figure 8) as well as the corresponding explanation made on from water retention curve (Figure 6). The less the cohesiveness between sand particles, the more grout could be penetrated into the voids, and thus the larger the interface shear stress.

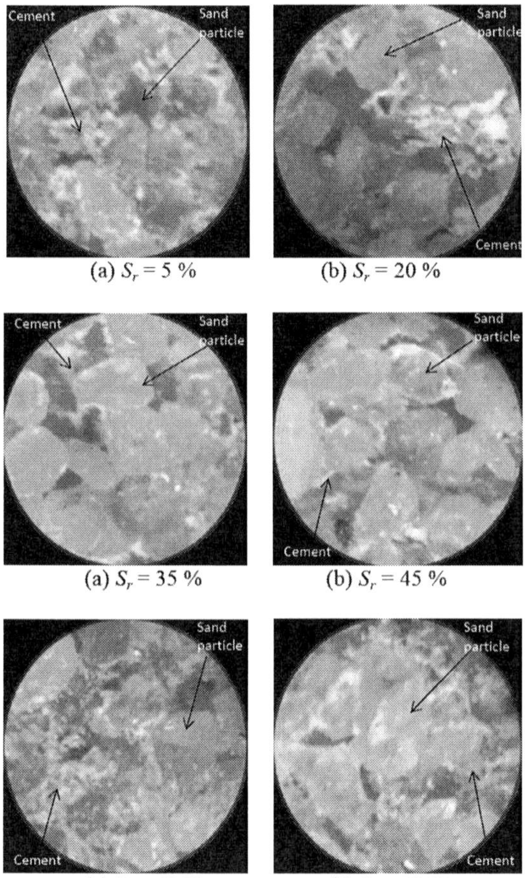

(a) S_r = 5 % (b) S_r = 20 %

(a) S_r = 35 % (b) S_r = 45 %

(a) S_r = 60 % (b) S_r = 70 %

Figure 9 Microscope photos of the interface

4 Conclusion

In order to investigate the effect of degree of saturation on the interface shear stress between the grout and soil in soil nailing, a series of direct shear tests were conducted on interface prepared at different soils degree of saturation/water content. Meanwhile, the soil water retention curve and interface microstructure were also investigated.

With the degree of saturation (water content) increase both the peak and residual shear stress at the interface decreases, followed by an almost constant stage; afterwards, the shear stress increase with further water content increase.

Interface shear stress changes are in agreement with the different bond strength and interface roughness obtained at different degree of saturation (water content): the less the cohesiveness between sand particles, the more grout is observed to ben penetrated into the voids, and thus the larger the interface shear stress.

References

1. F. Schlosser, Behaviour and design of soil nailing. Proceedings of theInternational Symposium on Recent Development in Ground Improvement Techniques, Bangkok: 399-413(1982)
2. A. Guilloux, and F. Schlosser, Soil nailing: practical applications,Proc. Symp. OnRecent Developments in Ground Improvement Techniques, Bangkok, 389-397 (1982)
3. M. A. Hossain, and J. H. Yin, Influence of Grouting Pressure on the Behavior of an Unsaturated Soil-Cement Interface, J. Geotech. Geoenviron. Eng. **138**:193-202 (2012)
4. R. Ajalloeian, H.S. Yu, and Allman, M.A., Physical and Mechanical Properties of Stockton Beach Sand, 7th Australia New Zealand Conference on Geomechanics: Geomechanics in a Changing World: Conference Proceedings. Barton, ACT: Institution of Engineers, Australia, 60-65 (1996)
5. Q. Wang. S.Y. Wang, S.W. Sloan, D.C. Sheng, R.Pakzad, Experimental investigation of pressure grouting in sand. Soils and Foundations, (Accepted for publication) (2015)
6. A. Bezuijen, M.P.M. Sanders, D. Hamer, A.F. Tol, Laboratory tests on compensation grouting, the influence of grout bleeding. Proc. WTC, Prague (2007)
7. F. Tuncdemir, and U. Ergun, Proceedings of the Institution of Civil Engineers Ground Improvement , , May 2009 Issue GI2, 93-101(2009)
8. T.G. Santhosh Kumar, M.A. Benny, A.Sridharan, T.J. Babu, Bearing Capacity Improvement of Loose Sandy Foundation Soils through Grouting, International Journal of Engineering Research and Applications (IJERA) Vol. **1**, Issue 3.1026-1033 (2011)
9. S. K. Vanapalli, D. G. Fredlund, D. E. Pufahl, A. W. Clifton, Model for the prediction of shear strength with respect to soil suction, Canadian Geotechnical Journal, **33**(3): 379-392 (1996)

Governing failure mode of unsaturated soil slopes under rainwater infiltration

Wengui Huang, Eng-Choon Leong[1,a] and Harianto Rahardjo[1]

[1]*School of Civil & Environmental Engineering, Nanyang Technological University, Singapore*

Abstract. Due to rainwater infiltration and loss of matric suction above the wetting font, both translational and rotational slips may occur in an unsaturated soil slope. It is useful to know which failure mode is more critical. The governing failure mode is likely to be affected by soil properties, slope geometry, depth of wetting front and contribution of matric suction to shear strength, which were investigated in this study. Specifically, upper bound limit analysis was adopted to develop stability charts based on rotational failure mechanism, and infinite slope model was used to develop stability charts based on translational failure mechanism. For a slope, the failure mechanism which gives the lower factor of safety is the governing failure mode. It was found that the failure mode of an unsaturated soil slope under rainfall is determined by parameter group $c'/\gamma H\tan\phi'$ for a given slope angle, depth of wetting front and contribution of matric suction to shear strength. With the increase in depth of wetting front and greater contribution of matric suction to shear strength, a slope is more prone to translational failure. Slopes with angle β = 45° are most vulnerable to translational failure, while gentler slopes (e.g. β = 15°) and steeper slopes (e.g. β = 75°) are more susceptible to rotational failure.

1 Introduction

For an unsaturated soil slope under rainwater infiltration, with the increase in depth of wetting front and the reduction of matric suction above the wetting front, both translational failure and rotational failure may occur, as shown in Figure 1. Most rainfall-induced landslides were reported to be shallow, with a depth of failure less than 3m above the groundwater table [1-2]. Deep-seated landslides after rainfall were also reported [3-4]. Therefore, there are generally two approaches to analyse rainfall-induced landslides. First, the slope is treated as "infinite" with the neglect of boundary effects and infinite slope analysis is performed to assess the slope stability [5-8]. Second, the slope is treated as "finite" with boundary conditions and extended limit equilibrium analysis is performed to assess the slope stability [9-11].

Infinite slope analysis can assess the stability based on translational failure mechanism, but it cannot assess the stability based on rotational failure mechanism. Limit equilibrium analysis can obtain the global factor of safety, but it may fail to search for the potential shallow slip surface [12]. Hence, if only infinite slope analysis or limit equilibrium analysis is performed, alternative failure mechanism may be missed.

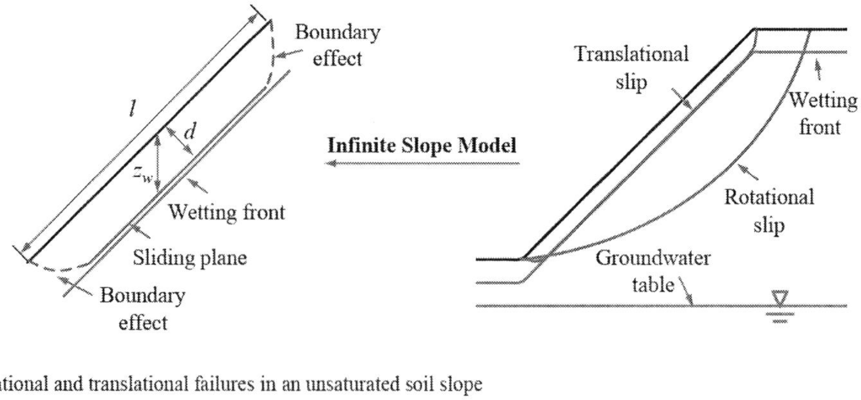

Figure 1. Rotational and translational failures in an unsaturated soil slope

[a] Corresponding author: CECLEONG@ntu.edu.sg

DOI: 10.1051/e3sconf/20160915008

The finite element analysis with a local factor of safety concept [12-13] can detect the potential shallow slip surface if it is more critical than the rotational slip surface. However, it is cumbersome to run a finite element analysis to determine whether a translational or rotational slip is more critical. Hence, the objective of this research is to investigate the controlling factors on the governing failure mode of unsaturated soil slopes under rainfall. The factors investigated include soil properties, slope geometry, depth of wetting front and contribution of matric suction to shear strength. The research will provide guidance in choosing the analysis method for rainfall-induced landslide analysis.

2 Methodology

In order to investigate the governing failure mode of unsaturated soil slopes under rainfall, the stability charts for unsaturated soil slopes under rainwater infiltration based on both translational failure mechanism and rotational failure mechanism were developed. For a slope with specific soil properties, the failure mechanism which gives the lower factor of safety is the governing failure mode of the slope. The stability of an unsaturated soil slope is directly related to the pore-water pressure profile before and after the rainwater infiltration, which will be introduced first. Afterwards, the methodology to develop stability charts based on both rotational failure mechanism and translational failure mechanism will be explained.

2.1 Pore-water pressure profiles

Pore-water pressure profile of an unsaturated soil slope is directly related to the location of the groundwater table and the surface flux condition [14]. Under no infiltration condition, for simplification the matric suction above groundwater table can be treated as hydrostatic relative to the groundwater table. Under rainwater infiltration, as shown in Figure 2, the matric suction can be partially lost (profile a), completely lost (profile b), or positive pore-water pressure may even build up above the wetting front (profile c) [15].

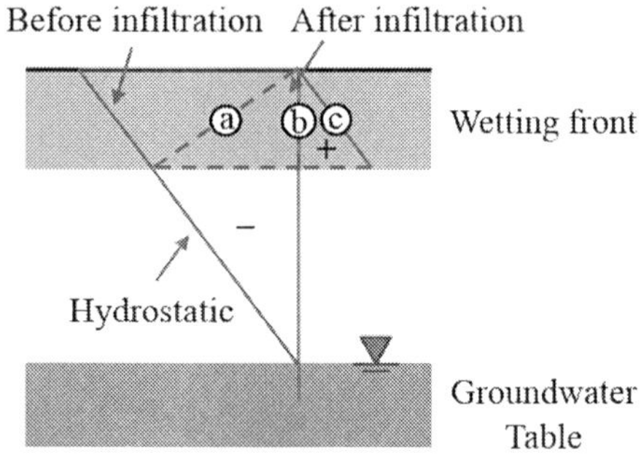

Figure 2. Pore-water pressure profiles before and after infiltration (modified from [15])

Research shows that profile a tends to occur in fine-grained soils, profile b tends to occur in coarse-grained soils [7] and profile c tends to occur in layered soils with permeable layer above wetting front and less permeable layer below wetting front [8]. In this research, profile c is not considered, as the research focuses on slopes made of homogeneous soils. The pore-water pressure profile b is adopted, which is a reasonable assumption for coarse-grained soils and a conservative assumption for fine-grained soils.

2.2 Stability chart based on rotational failure mechanism

Upper bound limit analysis has been widely used to develop stability charts for simple homogeneous soil slopes [16-17] and soil slopes under positive pore-water pressures [18-19]. Recently Huang et al. [20] incorporated negative pore-water pressure into the upper bound limit analysis and developed stability charts for unsaturated soil slopes under no infiltration condition. The research can be further extended by considering the depth of wetting front and stability charts for unsaturated soil slopes under infiltration condition can be developed subsequently.

According to the upper bound limit analysis, soil mass would fail if the rate of work done by soil weight and external forces exceeds the internal rate of energy dissipation. Hence, equating the external rate of work to the internal rate of energy dissipation gives an unsafe upper bound on the collapse or limit load [17].

For a simple homogeneous slope, the energy balance equation can be written as:

$$\dot{W}_\gamma = \dot{D}_{c'} \tag{1}$$

where $\dot{W}_\gamma$ is the rate of work done by soil weight, $\dot{D}_{c'}$ is the rate of internal energy dissipation.

If pore-water pressure is incorporated into the upper bound limit analysis and treated as external force, the energy balance equation can be expressed as (e.g., [21]):

$$\dot{W}_\gamma + \dot{W}_u = \dot{D}_{c'} \tag{2}$$

where $\dot{W}_u$ is the rate of work done by pore-water pressure u_w.

In the upper bound limit analysis, plastic shearing is assumed to follow associated flow rule, which requires that the velocity jump vector at slip surface is inclined to the slip surface at the effective angle of internal friction ϕ'. Hence, for a rotational failure, the slip surface must be a log-spiral curve [17, 21], as described by Equation (3) and shown in Figure 3, in which r_o is initial radius of the log-spiral curve, θ_o and θ_h are initial and final angles, $r(\theta)$ is the radius which is a function of θ. In Figure 3, H is the height of the slope, L is the length of the slope at the crest, α is the slope angle at the crest, β is the slope angle at the toe. If the slip surface passes below the toe (base failure), the inclination angle from the crest point (B) of the slope to the end point of the slip surface (C') is β', and $\beta' = \beta$ if the slip surface passes through the toe (toe failure).

$$r(\theta) = r_o \exp[(\theta - \theta_o)\tan\phi'] \quad (3)$$

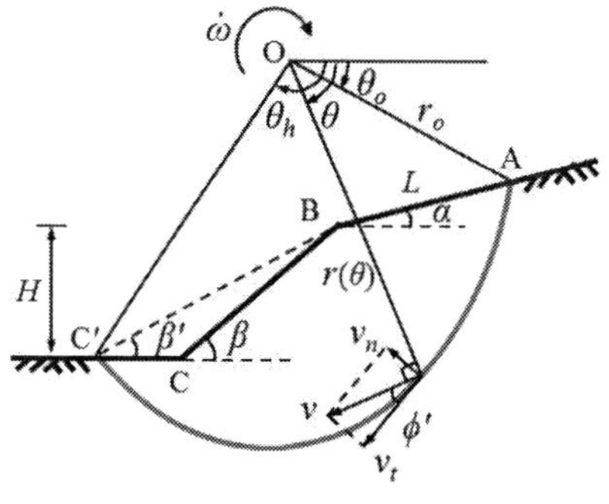

Figure 3. Log-spiral rotational failure

For rotational failure, $\dot{W}_\gamma$ can be calculated by a direct integration of the dot product of the unit weight vector $\boldsymbol{\gamma}$ and the velocity vector $\mathbf{v}$ over the region A - C' - C - B:

$$\dot{W}_\gamma = \int_V \boldsymbol{\gamma} \bullet \mathbf{v} \mathrm{d}V = \dot{\omega}\gamma r_o^3 (f_1 - f_2 - f_3 - f_4) \quad (4)$$

where $\dot{\omega}$ is the rate of rotation of the failed mass around the centre O, as shown in Figure 3. Coefficients f_1 to f_4 depend on the given values of α, β and ϕ', and variables θ_o, θ_h and β', and have been given by Chen [17].

$\dot{D}_{c'}$ can be calculated by the integration of the product of effective cohesion c' and tangential component of velocity v_t over the slip surface S:

$$\dot{D}_{c'} = \int_S c' v_t \mathrm{d}S = \frac{c'\dot{\omega} r_o^2}{2\tan\phi'} \{\exp[2(\theta_h - \theta_o)\tan\phi'] - 1\} \quad (5)$$

For a rigid rotation, the rate of work done by positive pore-water pressure can be calculated by [21]:

$$\dot{W}_u = \int u_w \tan\phi' \dot{\omega} r^2(\theta) \mathrm{d}\theta \quad (u_w \geq 0) \quad (6)$$

The rate of work done by negative pore-water pressure can be calculated by [20]:

$$\dot{W}_u = \int u_w \tan\phi^b \dot{\omega} r^2(\theta) \mathrm{d}\theta \quad (u_w < 0) \quad (7)$$

where ϕ^b is the rate of increase of shear strength due to matric suction.

Figure 4 shows the geometry of an unsaturated soil slope. The groundwater table is defined as a straight line [10, 11, 20], with a vertical distance H_{wt} between the toe of slope and groundwater table and an inclination angle i. The pore-water pressure u_w is assumed to be hydrostatic relative to the groundwater table before infiltration. After infiltration, the wetting front is assumed to be parallel to the ground surface with a vertical depth z_w, and $u_w = 0$ (profile b) for the soils above the wetting front.

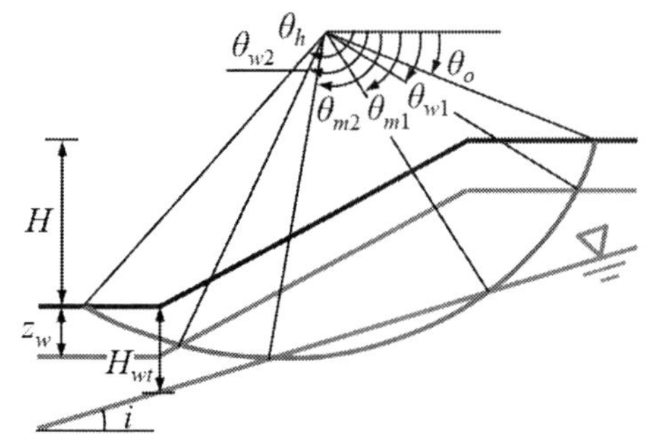

Figure 4. Geometry of an unsaturated soil slope under rainwater infiltration

The rate of work done by pore-water pressure u_w after infiltration can be calculated by:

$$\dot{W}_u = \int_{\theta_{w1}}^{\theta_{m1}} u_w(\theta)\tan\phi^b \dot{\omega} r^2(\theta)\mathrm{d}\theta + \int_{\theta_{m1}}^{\theta_{m2}} u_w(\theta)\tan\phi' \dot{\omega} r^2(\theta)\mathrm{d}\theta + \int_{\theta_{m2}}^{\theta_{w2}} u_w(\theta)\tan\phi^b \dot{\omega} r^2(\theta)\mathrm{d}\theta \quad (8)$$

where θ_{w1} and θ_{w2} are the angles correspond to the intersection points between wetting front and slip surface, θ_{m1} and θ_{m2} are the angles correspond to the intersection points between groundwater table and slip surface, as shown in Figure 4.

If ϕ^b is assumed to be a constant, $\dot{W}_u$ can finally be expressed as:

$$\dot{W}_u = \dot{\omega}\gamma r_o^3 f_{u_w} \quad (9)$$

where f_{u_w} depends on given values of α, β, i, γ_w/γ, H_{wt}/H, z_w/H, ϕ^b and ϕ', and variables θ_o, θ_h and β'. The detailed derivation of f_{u_w} can be found in Huang et al. [20].

Consequently, Equation (2) can be written as:

$$\frac{\gamma H}{c'} = \frac{H}{r_o} \frac{\exp[2(\theta_h - \theta_o)\tan\phi'] - 1}{2\tan\phi'(f_1 - f_2 - f_3 - f_4 + f_{u_w})} \quad (10)$$

In upper bound limit analysis, the soil is assumed to obey the Mohr-Coulomb failure criterion, and factor of safety F is defined as:

$$F = \frac{c'}{c_d} = \frac{\tan\phi'}{\tan\phi_d} \quad (11)$$

in which c_d and ϕ_d are the effective cohesion and effective internal friction angle, respectively, that are necessary to maintain energy balance in an admissible failure mechanism.

In order to present the stability charts as dimensionless forms, among the given conditions, i is normalized with β as i/β, and ϕ^b is normalized with ϕ' as ϕ^b/ϕ'. Furthermore, it is fixed that $\alpha = 0°$ and $\gamma_w/\gamma = 0.5$ for all the stability charts developed in this study. For a given slope angle β and ratios of H_{wt}/H, i/β, z_w/H, and ϕ^b/ϕ', the minimum value of $\gamma H/c'$ in Equation (10) can be found for a given value of ϕ' with θ_o, θ_h and β' being the variables. The obtained minimum $\gamma H/c'$ and corresponding ϕ' are denoted as $\gamma H/c_d$ and ϕ_d, respectively. Stability charts can be presented as $c_d/\gamma H$ being a function of ϕ_d (e.g. [22-24]). However, since F must be applied to both c' and $\tan\phi'$ [see Equation (11)], tedious iterations are required to obtain F from such charts [18, 24].

With the obtained $\gamma H/c_d$ and ϕ_d, Equation (10) can be rewritten as:

$$\frac{1}{\tan\phi_d} = \frac{2(\theta_h - \theta_o)}{\ln\left[1 + 2\frac{\gamma H \tan\phi_d}{c_d}\frac{r_o}{H}\left(f_1 - f_2 - f_3 - f_4 + f_{u_w}\right)\right]} \quad (12)$$

Parameter group $c'/\gamma H\tan\phi'$ is dimensionless and independent of F [25]:

$$\frac{c_d}{\gamma H \tan\phi_d} = \frac{c'/F}{\gamma H \tan\phi'/F} = \frac{c'}{\gamma H \tan\phi'} \quad (13)$$

According to Equations (11) and (13), Equation (12) can be rearranged to be:

$$\frac{F}{\tan\phi'} = \frac{2(\theta_h - \theta_o)}{\ln\left[1 + 2\frac{\gamma H \tan\phi'}{c'}\frac{r_o}{H}\left(f_1 - f_2 - f_3 - f_4 + f_{u_w}\right)\right]} \quad (14)$$

Stability charts can be developed by setting $F/\tan\phi'$ as a function of $c'/\gamma H\tan\phi'$ (e.g. [18-20]). The advantage of this format is that factor of safety of a slope can be obtained without iteration [18].

2.3 Stability chart based on translational failure mechanism

Factor of safety of a slope based on translational failure mechanism can be obtained by performing infinite slope analysis with the neglect of boundary effects. For a translational slip with pore-water pressure profile b (Figure 2) above wetting front and by using infinite slope model, F can be calculated as [15]:

$$F = \frac{c'}{\gamma z_w \sin\beta\cos\beta} + \frac{\tan\phi'}{\tan\beta} \quad (15)$$

Equation (15) can be rearranged to be:

$$\frac{F}{\tan\phi'} = \frac{c'}{\gamma H \tan\phi'}\frac{H}{z_w}\frac{1}{\sin\beta\cos\beta} + \frac{1}{\tan\beta} \quad (16)$$

Hence, for given z_w/H and β, stability chart based on translational failure mechanism can also be presented by setting $F/\tan\phi'$ as a function of $c'/\gamma H\tan\phi'$. However, it should be noted that infinite slope model would underestimate the factor of safety due to the neglect of boundary effects, and the accuracy of infinite slope analysis is directly related to the depth/length ratio of the translational slip [26-27].

3 Results and discussions

Figure 5 shows an example of the developed stability charts. The solid curve is developed based on translational failure mechanism with the condition that β = 45° and z_w/H = 0.1, the dash curve is developed based on rotational failure mechanism with the condition that β = 45°, z_w/H = 0.1, H_{wt}/H = 0.2, i = 0° and ϕ^b/ϕ' = 1/3. It is shown that there is an intersection point between the stability curves. The corresponding $c'/\gamma H\tan\phi'$ at the intersection point is defined as critical $c'/\gamma H\tan\phi'$. Below this critical value the translational failure mechanism would give a lower factor of safety and govern the failure mode. Above this critical value the rotational failure mechanism would give a lower factor of safety and govern the failure mode. For a slope, the higher the critical $c'/\gamma H\tan\phi'$, the more vulnerable the slope is to translational failure.

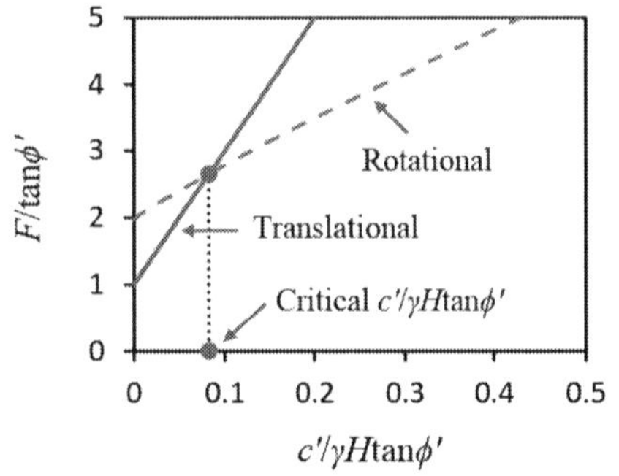

Figure 5. Stability charts for rotational failure and translational failure

A parametric study has been carried out to investigate the influence of slope angle (β), depth of wetting front (z_w/H), and contribution of matric suction (ϕ^b/ϕ', H_{wt}/H and i/β) on the critical $c'/\gamma H\tan\phi'$. The adopted parameters are shown in Table 1.

Table 1. Parametric study to the critical $c'/\gamma H\tan\phi'$

β =	15°, 30°, 45°, 60°, 75°
z_w/H =	0.05, 0.10, 0.15, 0.20
ϕ^b/ϕ' =	0, 1/3, 2/3, 1
H_{wt}/H =	0.2
i/β =	0

A typical plot indicating the influence of z_w/H on the critical $c'/\gamma H\tan\phi'$ is shown in Figure 6. It is found that the critical $c'/\gamma H\tan\phi'$ increases with the increase in z_w/H, and the curves are concave upward. The results reveal that with the increase in depth of wetting front, the slope becomes more prone to translational failure.

A typical plot representing the influence of ϕ^b/ϕ' on critical $c'/\gamma H\tan\phi'$ is shown in Figure 7. It is shown that the critical $c'/\gamma H\tan\phi'$ increases with the increase in ϕ^b/ϕ', but the curves are concave downward. The results reveal that due to the greater contribution of matric suction to shear strength, the slope becomes more susceptible to translational failure.

A typical plot showing the influence of β on critical $c'/\gamma H\tan\phi'$ is illustrated in Figure 8. It is shown that the critical $c'/\gamma H\tan\phi'$ obtains its peak value at β = 45°, and it decreases with both the increase and decrease of β. The results reveal that slopes with angle β = 45° are most vulnerable to translational failure, while gentler slopes (e.g. β = 15°) and steeper slopes (e.g. β = 75°) are more disposed to rotational failure.

However, it should be reminded that the stability chart for translational failure mechanism is developed based on the infinite slope model, which neglects boundary effects and the obtained factors of safety are underestimated. Hence, the critical $c'/\gamma H\tan\phi'$ shown from Figures 5 to 8 may be slightly overestimated.

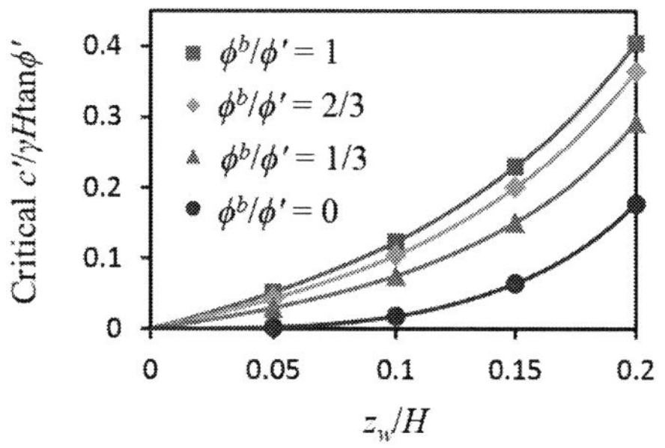

Figure 6. Influence of z_w/H on critical $c'/\gamma H\tan\phi'$ (β = 30°)

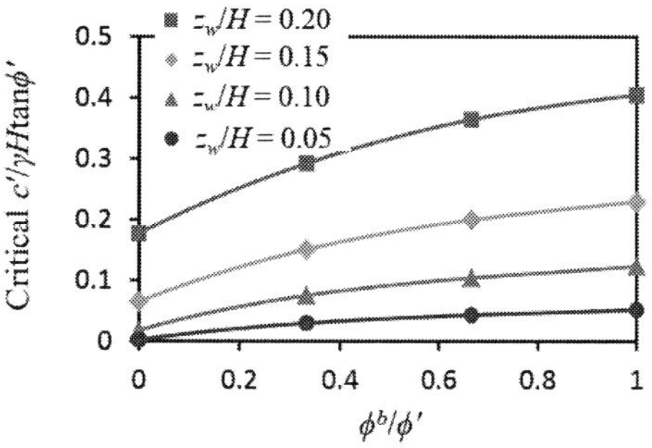

Figure 7. Influence of ϕ^b/ϕ' on critical $c'/\gamma H\tan\phi'$ (β = 30°)

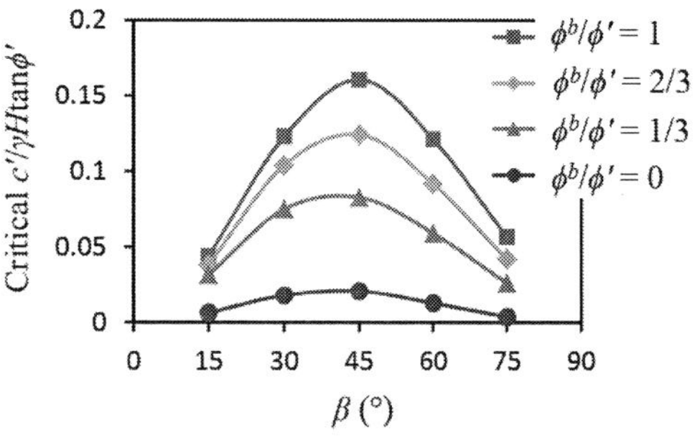

Figure 8. Influence of β on critical $c'/\gamma H\tan\phi'$ (z_w/H = 0.10)

4 Conclusions

For an unsaturated soil slope under rainwater infiltration, the increase in depth of wetting front and the reduction in matric suction above the wetting front may result in either translational or rotational failure mechanism. It is useful to know which failure mode is more critical for a specific slope. The governing failure mode is likely to be affected by soil properties, slope geometry, depth of wetting front and contribution of matric suction to shear strength, which were investigated in this study. Specifically, upper bound limit analysis was adopted to develop stability charts for unsaturated soil slopes under rainwater infiltration based on rotational failure mechanism, and infinite slope model was used to develop stability charts based on translational failure mechanism. The failure mechanism which gives the lower factor of safety is the governing failure mode.

The governing failure mode was found to be determined by the parameter group $c'/\gamma H\tan\phi'$. In addition, a critical $c'/\gamma H\tan\phi'$ exists below which translational failure governs and above which rotational failure governs. The influences of slope angle, depth of wetting front and contribution of matric suction to shear strength on the governing failure mode were preliminary investigated by a parametric study. It was found that with the increase in depth of wetting front and greater contribution of matric suction to shear strength, critical $c'/\gamma H\tan\phi'$ increases and slopes are more prone to translational failure under rainwater infiltration. Slopes with angle β = 45° have the highest critical $c'/\gamma H\tan\phi'$ and are most vulnerable to the translational failure, while gentler slopes (e.g. β = 15°) and steeper slopes (e.g. β = 75°) are more susceptible to the rotational failure. For the situation when translational failure has more possibility, infinite slope analysis is more preferable to assess the slope stability; otherwise limit equilibrium analysis is more desirable to assess the slope stability.

Acknowledgement

The first author gratefully acknowledges the research scholarship provided by Nanyang Technological University (Singapore).

References

1. F.C. Dai, C.F. Lee, S.J. Wang, Int. J. Remote Sens., **24**(23), 4817-4834 (2003)
2. K.A. Johnson, N. Sitar, Can. Geotech. J., **27**(6), 789-801 (1990)
3. D.M.S. Gerscovich, E.A. Vargas, T.M.P. De Campos, Eng. Geol., **88**(1), 23-40 (2006)
4. B.B. Huat, F.H. Ali, R.S.K. Rajoo, Am. J. Environ. Sci., **2**(4), 154-160 (2006)
5. D. Pradel, G. Raad, J. Geotech. Eng., **119**(2), 315-332 (1993)
6. B.D. Collins, D. Znidarcic, J. Geotech. Geoenviron. Eng., **130**(4), 362-372 (2004)
7. L.M. Lee, N. Gofar, H. Rahardjo, Eng. Geol., **108**(3), 272-285 (2009)
8. S.E. Cho, Eng. Geol., **105**(1), 32-43 (2009)
9. I. Tsaparas, H. Rahardjo, D.G. Toll, E.C. Leong, Comput. Geotech., **29**(1), 1-27 (2002)
10. H. Rahardjo, T.H. Ong, R.B. Rezaur, E.C. Leong, J. Geotech. Geoenviron. Eng., **133**(12), 1532-1543 (2007)
11. A. Rahimi, H. Rahardjo, E.C. Leong, J. Geotech. Geoenviron. Eng., **137**(5), 483–491 (2011)

DOI: 10.1051/e3sconf/20160915008

12. N. Lu, B. Şener-Kaya, A. Wayllace, J.W. Godt, Water Resour. Res., **48**(9) (2012)
13. E.E. Alonso, A. Gens, C.H. Delahaye, Hydrogeol. J., **11**(1), 174-192 (2003)
14. D.G. Fredlund, H. Rahardjo, *Soil mechanics for unsaturated soils*, John Wiley & Sons (1993)
15. H. Rahardjo, T.T. Lim, M.F. Chang, D.G. Fredlund, Can. Geotech. J., **32**(1), 60-77 (1995)
16. W.F. Chen, M.W. Giger, H.Y. Fang, Soils Found., **9**(4), 23-32 (1969)
17. W.F. Chen, *Limit analysis and soil plasticity*, Elsevier Science, Amsterdam, the Netherlands (1975)
18. R.L. Michalowski, J. Geotech. Geoenviron. Eng., **128**(4), 351-355 (2002)
19. R.L. Michalowski, S. Nadukuru, J. Geotech. Geoenviron. Eng., **139**(9), 1604-1610 (2013)
20. W. Huang, E.C. Leong, H. Rahardjo, J. Geotech. Geoenviron. Eng. (submitted)
21. R.L. Michalowski, Geotechnique, **45**(2), 283-293 (1995)
22. D.W. Taylor, J. Boston Soc. Civil Eng., **24**(3), 197-246 (1937)
23. A.W. Bishop, N. Morgenstern, Geotechnique, **10**(4), 129-153 (1960)
24. E. Spencer, Geotechnique, **17**(1), 11-26 (1967)
25. J.M. Bell, J. Soil Mech. and Found. Div., **92**(5), 51-65 (1966)
26. D.V. Griffiths, J. Huang, G.F. Dewolfe, Int. J. Numer. Anal. Met. Geomech., **35**(5), 569-585 (2011)
27. W. Huang, E. C. Leong, H. Rahardjo, *Unsaturated Soils: Research & Applications, Proceedings of the sixth Asia-Pacific conference on unsaturated soils, China*, 771-775 (2015)

Analysis of a rainfall-triggered landslide at rest and be thankful in Scotland

Brunella Balzano[1,a], Alessandro Tarantino[1], and Andrew Ridley[2]

[1] University of Strathclyde, Glasgow, United Kingdom
[2] Geotechnical Observations

Abstract. Weather and climate are key influences upon the triggering of landslides in Scotland. One of the primary factors influencing the landslide and debris flows occurrence is rainwater infiltration into an initially unsaturated slope. This leads to an increase in both degree of saturation and pore-water pressure and, consequently, to a decrease in the shear strength of the soil, which eventually triggers a landslide event. The A83 trunk road at the Rest and Be Thankful in Scotland is an area prone to risk from landslide and, because of climate change, this risk is going to become higher. An increased frequency of heavier rainfall has been observed in recent years and forecasts for the next decades are not more reassuring. In 2014, the biggest landslide event has occurred at the Rest and Be Thankful Pass. More than 2000 tonnes of landslip debris have been removed in efforts to reopen the route A83. Although the route is now open again, still many are the aspects related to the triggering of such event that have not been clarified yet. The aim of this paper is to address the problem of developing a simple numerical model to analyse rainfall-triggered shallow landslides.

1 Introduction

Rainfall affects the stability of natural and man-made slopes including road and railway embankments, cuttings, and slopes. Rainwater infiltration increases the degree of saturation and the pore-water pressure, which result in a decrease in shear strength along potential failure surfaces. A special class of problems is represented by rainfall-triggered shallow landslides as they often evolve into disastrous debris-flows. Shallow landslides are typically translational slope failures that involve up to a few meters of soil cover. They may occur wholly or partly in the unsaturated zone. A great percentage of landslide disasters worldwide have been caused by prolonged or intense rainfall.

Many areas in Scotland are affected by landslide risk due to abundant precipitation in the region. The Rest and be Thankful Pass is affected by shallow landslides evolving into debris flows occurring with return periods of only a few years. These events have caused significant disruption to the transport structures present in the area. The A83 trunk road at the Rest and Be Thankful is a very important road in Scotland and connects the Central Belt, via the A82 from Glasgow, with the south west of Scotland all the way down to Campbeltown Kintyre.

The most recent, and one of the biggest, event occurred in 2014. The road has been kept closed for several months and has been reopened only recently. Transport agencies are working in order to reduce the risk in that area.

[a] Corresponding author: brunella.balzano@strath.ac.uk

This paper addresses some aspects in the development of a hydro-mechanical to simulate the occurrence of hillslope debris-flows. In particular, it focuses on the problem of formulating a simple, yet convincing, hydraulic model.

2 Conceptual physically-based hydro-mechanical models

Figure 1 shows a 3D schematic view of a portion of the hillslope. As the hillsope is crossed by gullies (vertical wall channels), a typical transverse section includes a main hillslope, a side slope and a gully. The soil formation overlies a bedrock formation.

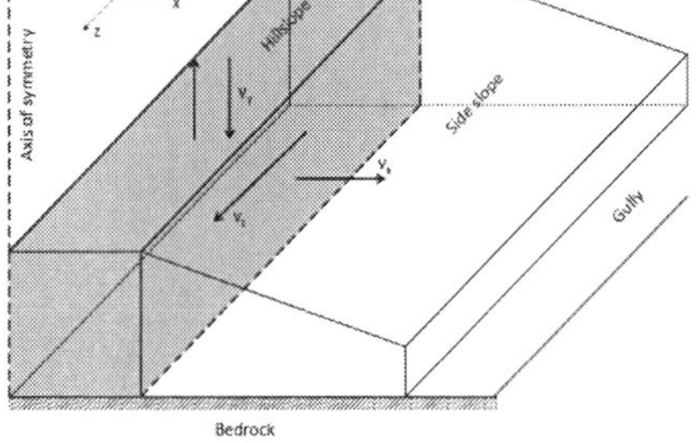

Figure 1. Schematic layout of the hillslope

 DOI: 10.1051/e3sconf/20160915009

Although the problem is 3-D, in terms of both hydraulic flow and collapse mechanisms, a full 3D hydro-mechanical model would be difficult to implement because of the computational burden and the uncertainties associated with the geotechnical characterisation of the slope (geometry and hydro-mechanical parameters). Simplifications in the hydraulic and mechanical models are therefore explored herein.

Concerning the mechanical model, instabilities may occur in the direction of the maximum slope (hillslope) and in the transverse direction (side slope). Considering that the thickness of the soli cover (less than 1m) is much lower that the length of the slope (a few hundredths of meters), an infinite slope model seems appropriate to interpret hillslope movements. This is the main mechanism targeted in this paper. As a result, failure is modelled as a translational movement in the y-z plane.

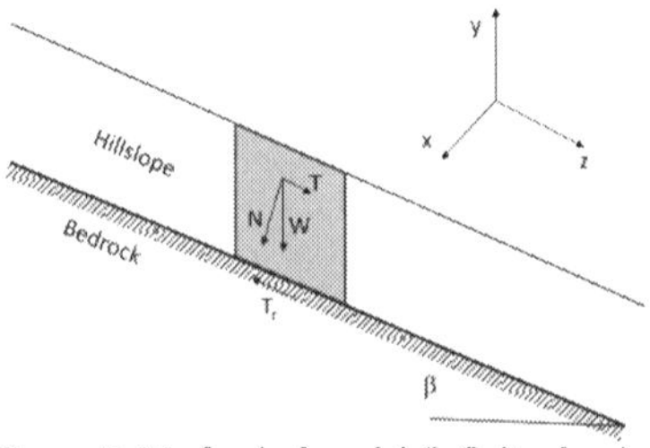

Figure 2. Mechanical model (infinite slope)

For the hydraulic model, the flow in *z*-direction can be neglected within the framework of an infinite slope model. Flows occur in *y*-direction (associated with rainfall and evapotranspiration at the ground surface) and *x*-direction (associated with lateral drainage into the gully) as shown in Figure 1.

As a result, two potential hydraulic models can be figured out. The first one would only envisage flow in *y*-direction and would therefore be a 1-D model. The second one would consider flow in the *xy*-plane and would therefore be a 2-D model.

The choice of the most appropriate hydraulic model model is strongly linked to the problem of the choice of the initial hydraulic condition at the onset of rainfall. The interconnection between these two aspects is not straightforward and is therefore discussed hereafter in more detail.

An initial condition is required in any time-dependent analysis. The initial condition at the onset of rainfall is difficult to predict and any arbitrary choice will have a significant influence on the numerical results (i.e. the slope may fail or not in the numerical simulation depending on the choice of the initial hydraulic condition).

A strategy to derive the hydraulic initial condition at the onset of current rainfall consists in analysing water flow over a sufficiently long antecedent period, say 10 years. In such an analysis, the initial condition at the beginning of the antecedent 10-year period can be set in an arbitrary way. However, it is assumed that the long hydrological history applied to the slope can cancel the 'memory' of the initial arbitrary hydraulic condition.

This concept is illustrated in Figure 3. If two arbitrary initial conditions are set at the beginning of the long antecedent period, the numerical simulation will tend to converge when the memory of the arbitrary initial condition is lost. To achieve the convergence shown by the solid lines in Figure 3, a hydrological balance should be accomplished over a very long period, in the sense that the simulation should return a degree of saturation fluctuating around an average level.

If the hydraulic model (including its boundary conditions) is not set properly, it may occur that the slope becomes rapidly either oversaturated or entirely dry (dashed lines in Figure 3). In other words, an essential requirement of the hydraulic model is that a hydrological balance can be reasonably accomplished.

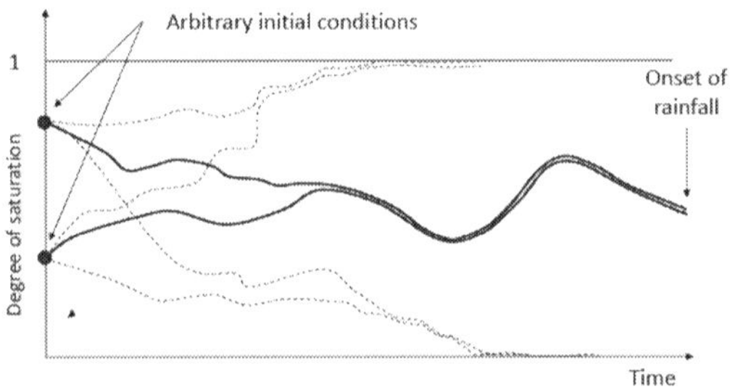

Figure 3. Analysis over long antecedent period to identify initial hydraulic condition at the onset of rainfall

The two hydraulic models that were tested are illustrated in Figure 4 and include a one-dimensional and a two-dimensional model.

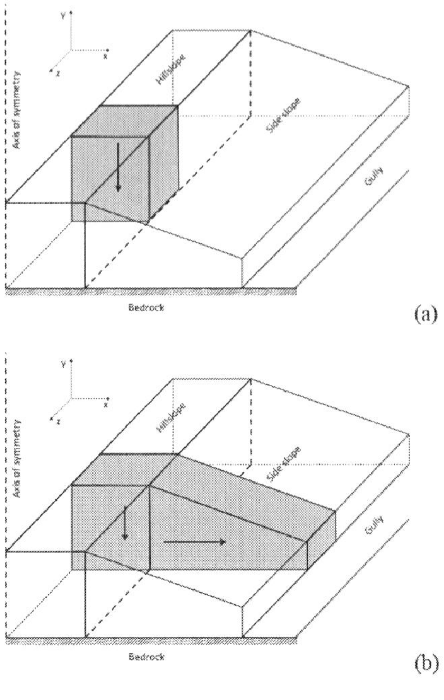

Figure 4. Candidate hydraulic models. (a) One-dimensional model (b) Two-dimensional model.

3 Numerical analyses

Flow of water within the slope under unsaturated and saturated conditions can be modelled by combining water mass balance and Darcy's law extended to unsaturated conditions. This gives the Richards equation in terms of hydraulic head.

$$C\frac{\partial h}{\partial t} = div[k \cdot grad(h)] \quad (1)$$

$$C = \gamma_w \frac{\partial \theta}{\partial u_w} \quad (2)$$

where Θ is the volumetric water content, t is the time, u_w is the pore-water pressure, k the unsaturated hydraulic conductivity, h is the hydraulic head, C is the water capacity and γ_w is the unit weight of water.

3.1. Soil properties

In order to perform a preliminary analysis to understand the phenomenon under study, the material used for the numerical modelling needs to be characterised hydraulically and mechanically.

To define the hydraulic behaviour of the soil the water retention curve and the hydraulic conductivity need to be estimated. The Vereecken model (Vereecken et al 1987, 1989) has been used for the estimation of the water retention curve of the material. The Vereecken model let us to evaluate the WRC and the permeability curve of a soil from the granulometric curve of the soil itself.

The formulations used are the following:

$$S_{r,e} = \frac{1}{(1+(\alpha H)^n)} \quad (3)$$

$$\theta = \theta_r + (\theta_s - \theta_r)S_e \quad (4)$$

$$\theta_r = 0.015 + 0.005 \cdot f_{clay} + 0.014 \cdot f_C \quad (5)$$

$$\theta_s = 0.81 - 0.283\rho_d + 0.00 f_{clay} \quad (6)$$

$$K_r = \frac{1}{(1+(bH)^\lambda)} \quad (7)$$

where $S_{r,e}$ is the effective degree of saturation, H is the hydraulic head expressed in cm, θ is the volumetric water content and θ_r and θ_s are the residual and saturated volumetric water content respectively; α, n and λ are the Vereecken coefficients. They have been estimated from the grain size distribution as follows:

$$\alpha = \exp\begin{pmatrix} -2.486 + 0.025 \cdot f_{sand} - 0.351 \cdot f_C + \\ -2.617 \cdot \rho_d - 0.023 \cdot f_{clay} \end{pmatrix} \quad (8)$$

$$n = \exp\begin{pmatrix} 0.053 - 0.019 \cdot f_{sand} + \\ -0.013 \cdot f_{clay} + 0.015 \cdot f_{sand}^2 \end{pmatrix} \quad (9)$$

$$b = \exp\begin{pmatrix} -2.64 - 0.019 \cdot f_{sand} + \\ 0.05 \cdot f_{clay} + 0.506\ln(k_{sat}) \end{pmatrix} \quad (10)$$

$$\lambda = \exp\left(1.186 - 0.194\ln(f_{clay}) - 0.0489\ln(f_{silt})\right) \quad (11)$$

where f_{sand}, f_{clay}, f_C, f_{silt} are the fractions of sand, clay, carbon and silt respectively (evaluated from the granulometric curve of the soils), k_{sat} is the saturated hydraulic conductivity of the soil and ρ_d is the dry density of the soil expressed in [g/cm^3].

For the purpose of this study, the parameters in Table 1 have been chosen to characterise the soil.

Table 1. Parameters to determine the soil water retention curve.

f_{sand}	f_{clay}	f_{silt}	f_C	k_{sat}	ρ_d
-	-	-	-	[m/s]	[g/cm^3]
20	35	37.5	7.5	10^{-7}	0.7

The water retention and the hydraulic conductivity curves are reported in Fig. 5 and Fig.6 respectively .

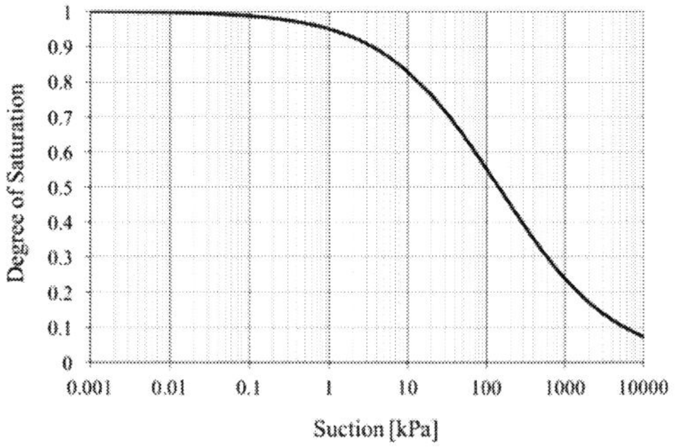

Figure 5. Water retention curve of the material.

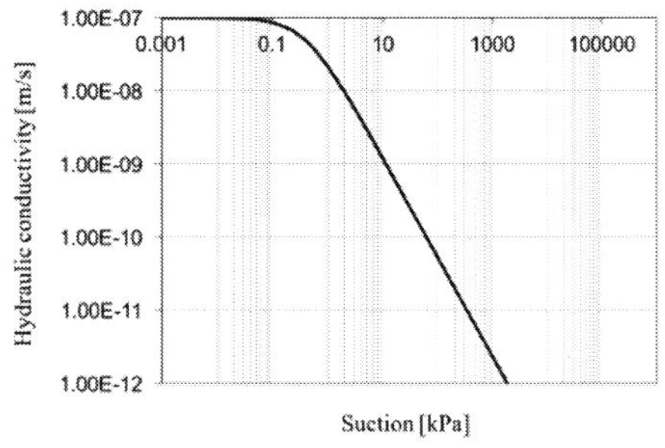

Figure 6. Hydraulic conductivity curve of the material.

3.2. Boundary conditions

To test the capability of the hydraulic models to accomplish a reasonable hydrological balance, two months have been analysed with two different rainfall histories. The fluxes applied are derived from the rain measured by a gauge installed directly on the site.

DOI: 10.1051/e3sconf/20160915009

The months analysed are January 2014 and July 2014 in order to analyse a relatively very wet period and a relatively dry period respectively.

A constant evaporation rate of 2mm/d has also been considered. It has not been directly applied as a boundary condition in the numerical analysis but it will be included in the final hydrological balance.

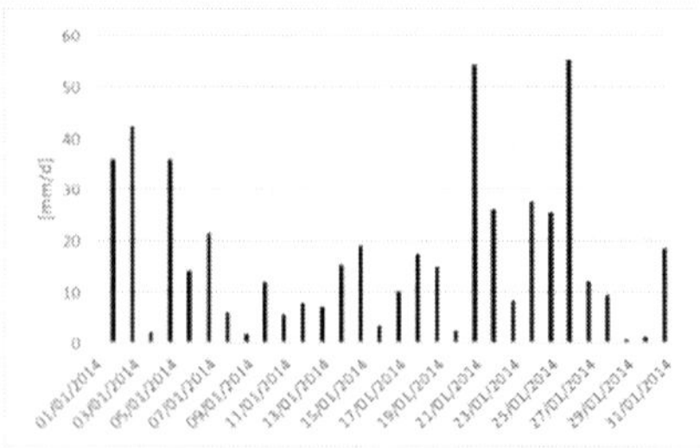

Figure 7. Rainfall history related to January 2014.

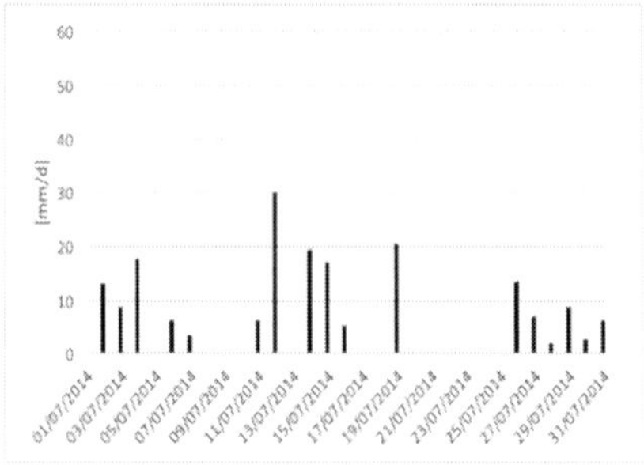

Figure 8. Rainfall history related to July 2014.

3.3. Geometry

The 1-D model

A slope thickness of 0.8m was considered. Since the slope dips to an angle of 30°, the thickness of the slope in direction perpendicular to the slope is equal to 0.8m·cos30°. According to Tarantino & Bosco (2000), the infiltration in the infinite slope can be analysed by considering 1-D column whose height is equal to the thickness in the perpendicular direction (Fig. 9).

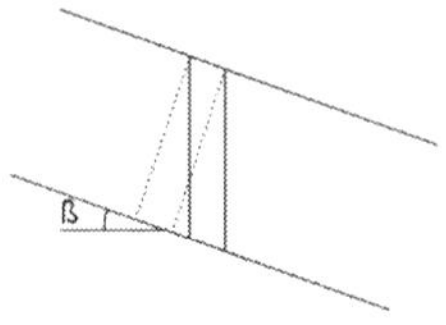

Figure 9. 1-D model geometry.

The boundary conditions have been assigned as follows:

- a vertical influx q_{in} at the top of the column (this boundary has also been set to be a seepage face review).
- impermeable on all the other boundaries.

The seepage face review is a boundary condition described as follows.

- If u_w<0 then $q= q_{in}$
- If u_w>0 at the time step, the boundary condition is reviewed and set to u_w=0

Basically, at the end of each iteration, the conditions along the specified potential seepage face are "reviewed" to check if the condition u_w<0 is met. Nodes with computed pressures greater than zero are not allowed, as positive pressure on the surface indicates ponding, which cannot happen along the sloping boundary. The water would run off, not pond. The specified boundary condition $q=q_{in}$ is therefore 'reviewed' to u_w=0. The difference between the externally applied influx q_{in} and the water that infiltrates under u_w=0 represents the runoff.

The 2-D model

In order to take into account and verify whether the contribution of the horizontal drainage to the hydrological balance is relevant, a 2-D model has been analysed. The transverse section of the infinite slope in the *y-z* plane has been considered for this numerical analysis.

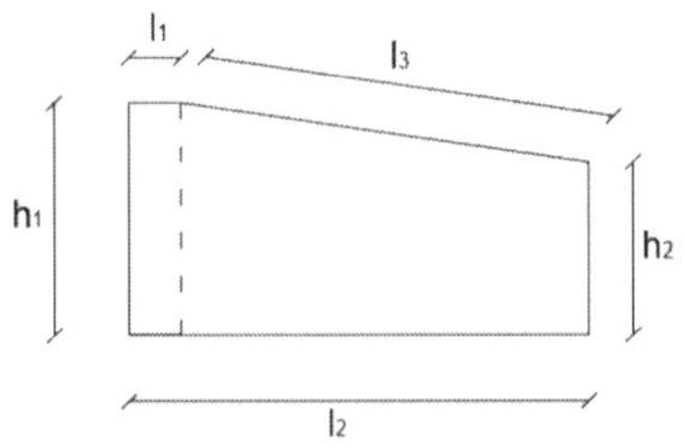

Figure 10. Geometry adopted for the 2D model.

Table 2. Parameters of the geometry of the 2D model.

h_1	h_2	l_1	l_2	l_3
[m]	[m]	[m]	[m]	[m]
0.8	0.6	0.2	1.77	0.57

The values of the geometrical parameters have been chosen in order to maintain an angle of 20° for the side slope.

The boundary conditions have been assigned as follows:

- boundary l_1 and l_3: a vertical influx at the boundary coupled with the seepage face review.
- boundary h_2: free flux (seepage face review only with q=0).
- boundary h_1:impermeable.
- boundary l_2: impermeable.

3.4. Results

The performance of the two hydraulic models has been assessed by computing a water mass balance of the control volume. The volumes entering the slope have been calculated by an integration of the unit fluxes on the relative boundaries. To achieve a reasonable hydrological balance, no or little accumulation of water into the slope should occur.

The 1-D model

For the 1-D model two different initial conditions have been considered:

- a hydrostatic pore water pressure profile with -10 kPa at the level of the bedrock.
- a hydrostatic pore water pressure profile with 0 kPa at the level of the bedrock

The results are shown in Figs. 11 and 12 for the 1-D model with initial u_w=-10 kPa at the bedrock and Figs. 13 and 14 for the 1-D model with initial u_w=0 kPa at the bedrock. Results are presented in terms applied rainfall, infiltration, and runoff (derived from the difference between rainfall and infiltration).

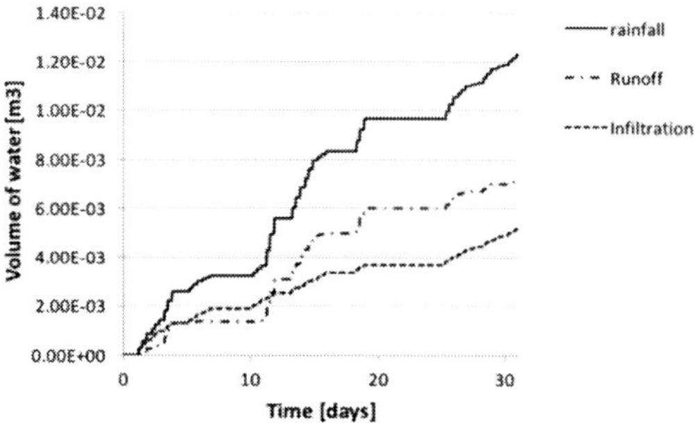

Figure 11. 1-D model July 2014, initial u_w=-10 kPa at the bedrock

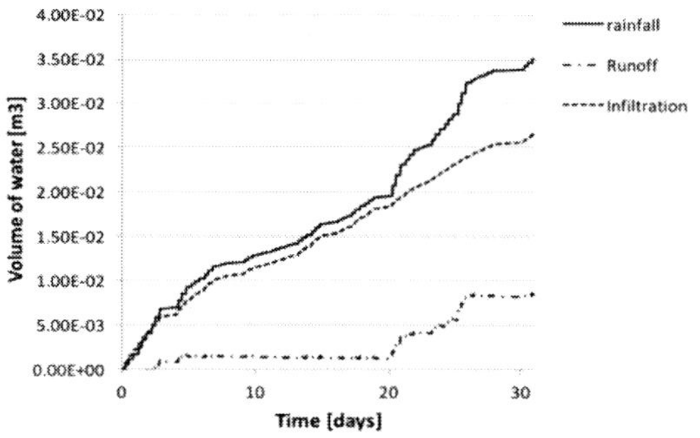

Figure 12. 1-D model January 2014, initial u_w=-10 kPa at the bedrock

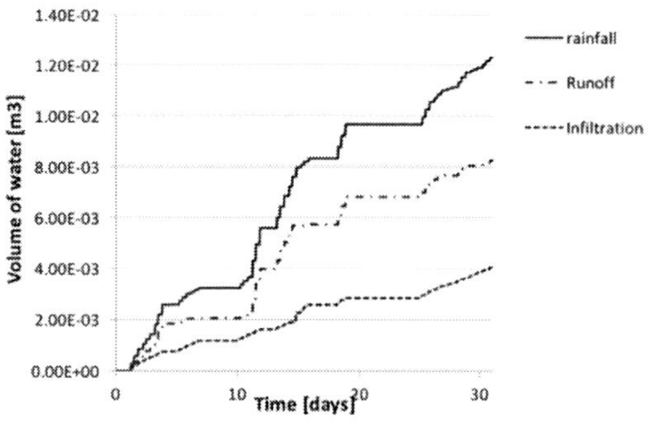

Figure 13. 1-D model July 2014, initial u_w=0 kPa at the bedrock.

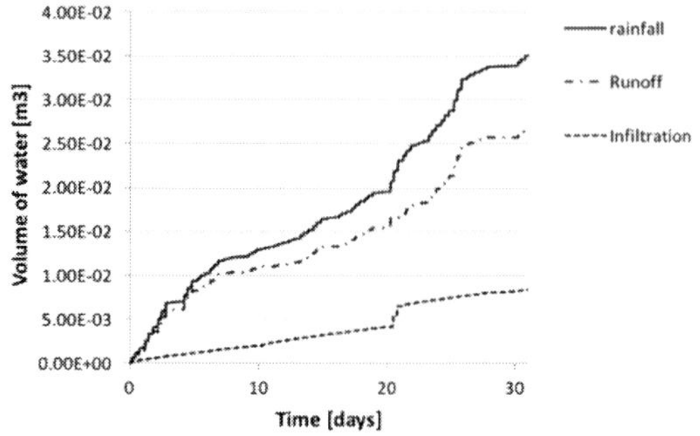

Figure 14. 1-D model January 2014, initial u_w=0 kPa at the bedrock

The cumulative amount of water ex-flux related to the evaporation was estimated by an integration of the unit flow rate, 2 mm/d, along the boundary where the evaporation rate is applied for entire duration of the analysis. For the 1D analysis, the cumulative volume of water related to the evaporation rate is 4.75e-003 m^3.

Figs. 15 to 18 show the hydrological balance for the four cases, months of January and July and initial pore-water pressure at the bedrock of u_w=-10 kPa or u_w=-0 kPa.

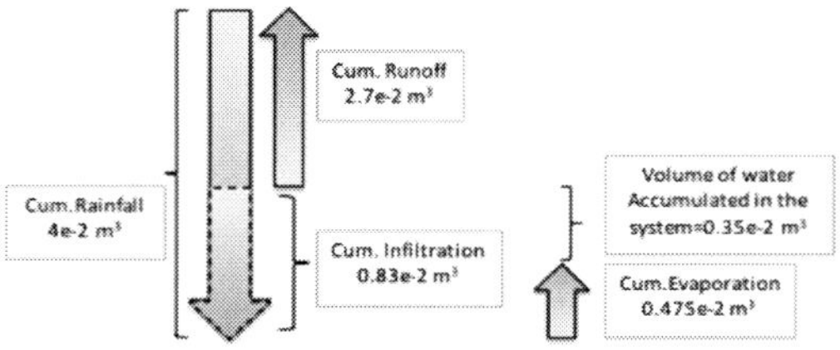

Figure 15. 1-D model January 2014, Hydrological balance. Initial u_w at the bedrock =0 kPa.

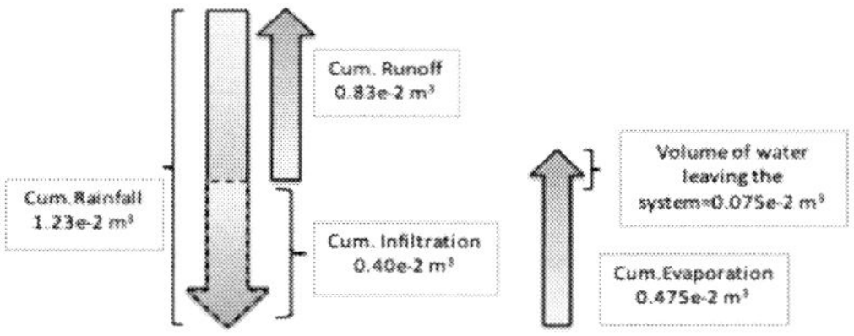

Figure 16. 1-D model July 2014, Hydrological balance. Initial u_w at the bedrock =0 kPa.

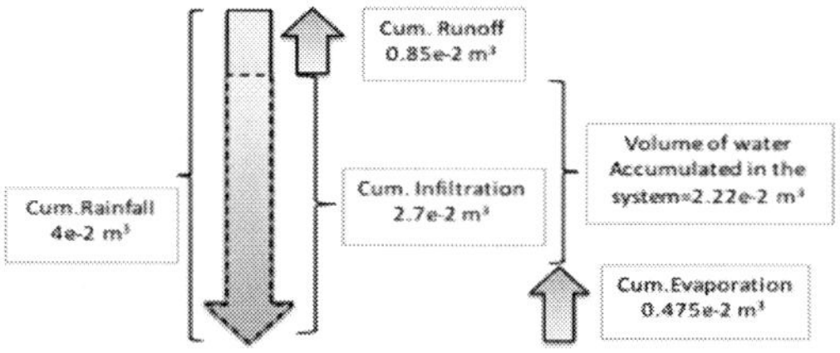

Figure 17. 1-D model January 2014, Hydrological balance. Initial u_w at the bedrock =-10 kPa.

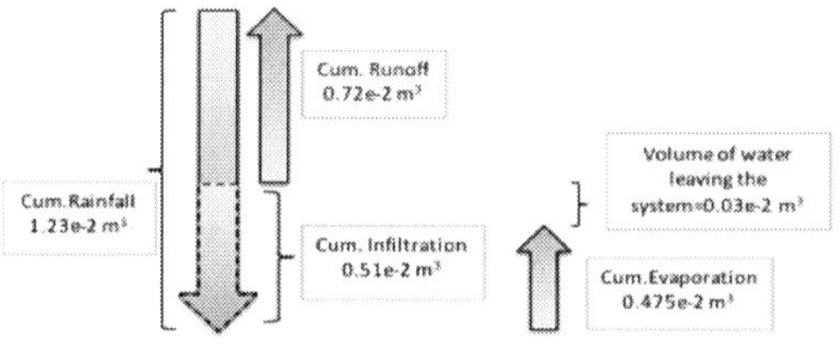

Figure 18. 1-D model July 2014, Hydrological balance. Initial u_w at the bedrock =-10 kPa.

In order for the hydrological balance to be satisfied, the volume of water accumulated into the column should be comparable with the volume outgoing via evaporation. In both cases the equilibrium seems to be almost satisfied just in July, although the balance is not exactly zero. On the other hand, in January there is a significant amount of water accumulating in the system. This causes the system to saturate very rapidly therefore leading to a failure after a few months after the beginning of the analysis.

The 2-D model

Figs 19 and 20 show the results of the analysis for the 2-D model in January and July. In this case, also the horizontal drainage takes part to the hydrological balance. The horizontal flow rate is the one related to the vertical dashed line as shown in Figure 10. For the 2-D analysis, the cumulative volume of water related to the evaporation rate is 1.24e-002 m^3.

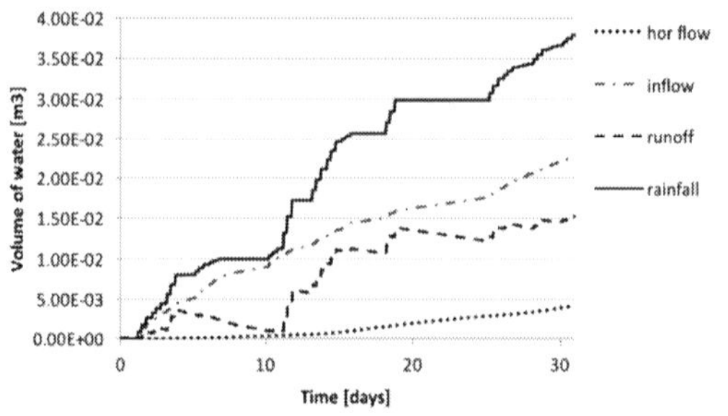

Figure 19. 2D model July initial wt=0 m.

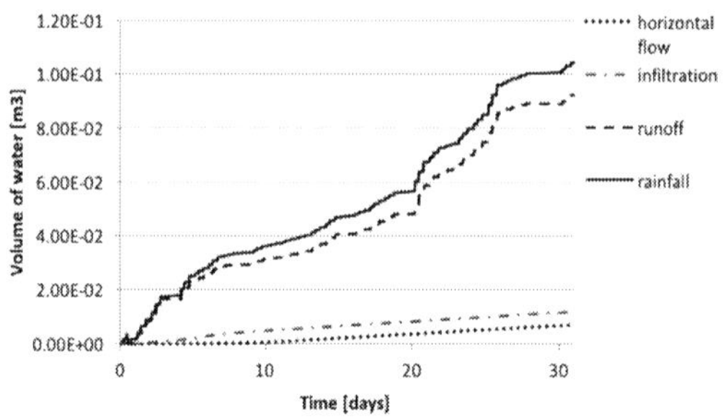

Figure 20. 2D model January initial wt=0 m.

As shown in Figs 21 and 22, the hydrological balance is reasonably achieved for the 2-D model. The horizontal drainage tends to avoid the accumulation of water in the system.

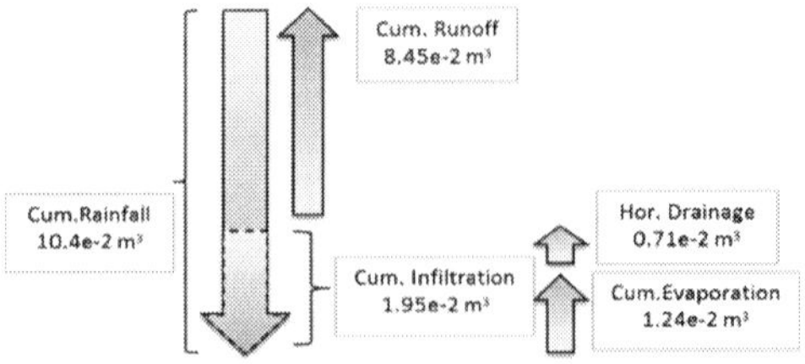

Figure 21. 2-D model January 2014, Hydrological balance. Initial u_w at the bedrock =0 kPa.

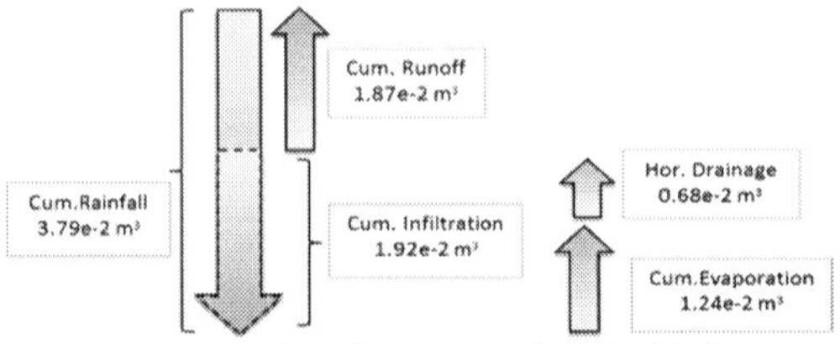

Figure 22. 2-D model July 2014, Hydrological balance. Initial u_w at the bedrock =0 kPa.

4 Conclusions

This paper has presented an approach to select an appropriate hydraulic model to analyse the stability of the Rest and Be Thankful slope. This approach could be extended to other problems of shallow landslides at the catchment scale.

Lateral flow towards the gully (horizontal drainage) seems to be critical in ensuring reasonable hydrological balance. As a result, a 2-D hydraulic model seems to be more appropriate to model to hydrological regime of the slope. On the other hand, a 1-D flux model appears to be not suitable as water tends to accumulate into the slope, i.e. runoff and evaporation alone do not compensate for the infiltrating rainwater.

The identification of a 2-D hydraulic model, to be coupled with a 1-D mechanical model (infinite slope), is the first step in the development of a full hydro-mechanical model for the slope.

The following step will be performing a long-term analysis in a period antecedent to the onset of critical rainfalls to be sufficiently long to cancel the effect of the arbitrary initial condition.

References

1. H. Vereecken, J. Feven, J. Maes, P. Darius, Estimating the soil moisture retention curve characteristic from texture, bulk density, and carbon conten, *Soil Science* 148: 389-403. (1989)
2. H. Vereecken, J. Maes, J. Feven, , Estimating the soil moisture retention curve characteristic from easily measured soil properties, *Soil Science* 149(1): 1-13. (1987)
3. A. Tarantino, G. Bosco, Role of soil suction in understanding the triggering mechanisms of flow slides associated with rainfall. In G.F. Wieczorek & N.D. Naeser (eds.). *Second International Conference on debris-flow hazards mitigation,* Taipei, Taiwan: 81-88.

DOI: 10.1051/e3sconf/20160915010

Numerical analysis of the stability of inhomogeneous slopes considering partially saturated conditions

Patrick P. Pichler[1,a] and Helmut F. Schweiger[1]

[1]*Graz University of Technology, Institute of Soil Mechanics and Foundation Engineering, Computational Geotechnics Group, 8010 Graz, Austria*

Abstract. It is well accepted that rainfall could play a significant role in instability of slopes. The main objective of the presented study is to quantify the influence of varying characteristics of water flow, its associated changes of pore-water pressures and shear strength on the stability of simplified, but inhomogeneous, slope geometries. The commonly used van Genuchten model was used to describe the Soil Water Characteristic Curve (SWCC) mathematically. In the context of this study, the influence of different hydraulic behaviour of soil layers, i.e. different SWCC, on the factor of safety (FoS) is evaluated by means of fully coupled flow-deformation analyses employing the finite element method. To quantify the slopes' factor of safety during rainfall events after specified times of infiltration or evaporation, the strength reduction method was applied. In addition to various combinations of soil layers, the influence of a water bearing high permeable soil layer between two less permeable soil layers, a situation which is often encountered in practice, on the factor of safety has been investigated.

1 Introduction

Numerous researchers discussed the most significant factors which control the instability of homogeneous soil slopes under rainfall infiltration [e.g. 1-2]. Basically, the initial FoS of a slope with given soil properties is controlled by its geometry, the depth of the groundwater table and the corresponding initial matric suction-profile in the unsaturated zone which is defined by the soil-water-characteristic curve (SWCC). The soil properties and the rainfall intensity have a greater impact on the factor of safety than the slope geometry and initial water table. The position of the initial groundwater level and the soil layers with different hydraulic behaviour defines the initial suction- and saturation-profile throughout the slope. The hydraulic characteristics of the soil such as saturation, which determines the relative permeability, and the intensity and duration of rainfall defines the actual infiltration rate into the slope. Additionally, the influence of different variations in geometry and impacts of various rainfall scenarios on the development of the FoS with time were evaluated.

The main objective of this study is to determine worst-case scenarios due to various combinations of different permeable soil layers under predefined hydraulic conditions. To estimate the FoS of the slope after predefined time periods of a rainfall event the well-known strength reduction technique was applied. The analyses were performed employing two-dimensional finite element models utilizing the code *PLAXIS 2D* [3].

2 Geometry and boundary conditions

The geometry of the slope and its boundary conditions are illustrated in figures 1 and 2. The height of the slope is 10 m and the slope angle α is 26.56° (horizontal to vertical is 2:1). The initial groundwater level was assumed to be inclined with the same angle as the boundary between the different soil layers. The left and the lower boundary were taken as closed flow boundaries, whereas water was able to flow through the upper and the right boundary of the model. The two dimensional finite element mesh consists of 1300 15-noded elements.

In a first step, the influence of three different locations of the boundary between the upper and lower soil layer (geometries 1, 2, 3 in figure 1) was evaluated. In geometry 1 (GEO 1), the layer boundary runs through the toe of the slope, in geometry 3 (GEO 3) the boundary lies at the same position as the initial groundwater level and geometry 2 (GEO 3) lies exactly between them. Figure 2 shows the geometry of the slope for the case, when a highly permeable sandy layer is situated between the upper and the lower soil layer. The groundwater table was located at the boundary between the middle and the lower soil layer. All the other model characteristics were kept the same as in the "2-layered model" (Figure 1).

The initial saturation - profiles through the slope are given by the groundwater level and the chosen SWCC. Figure 3 illustrates the initial saturation profiles for the configuration "sand-silt" and "silt-sand" for geometry 1 of the two-layered model.

[a] Corresponding author: patrick.pichler@tugraz.at

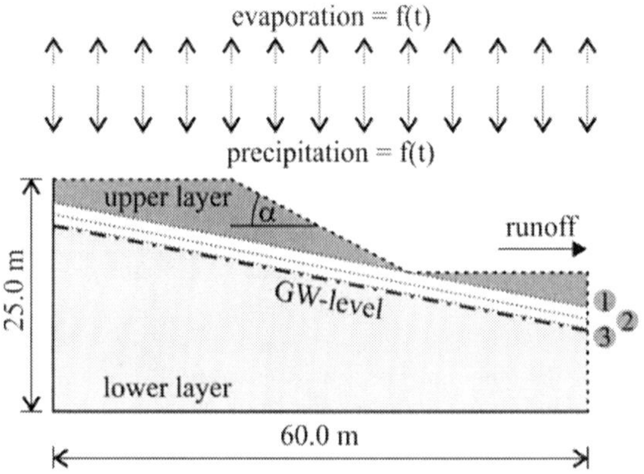

Figure 1. Geometries of the slope with two different soil layers

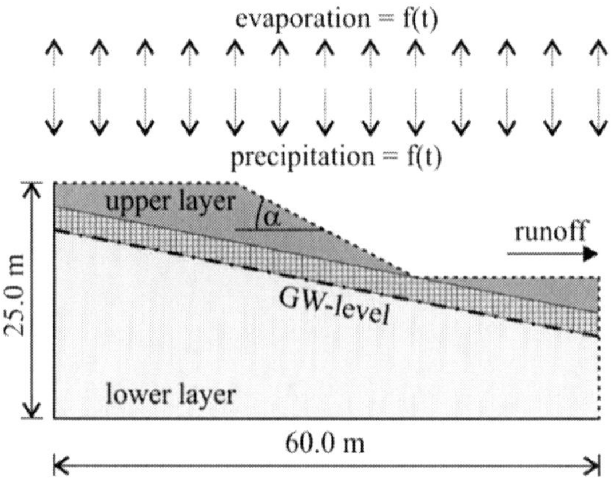

Figure 2. Geometry of the slope with three different soil layers

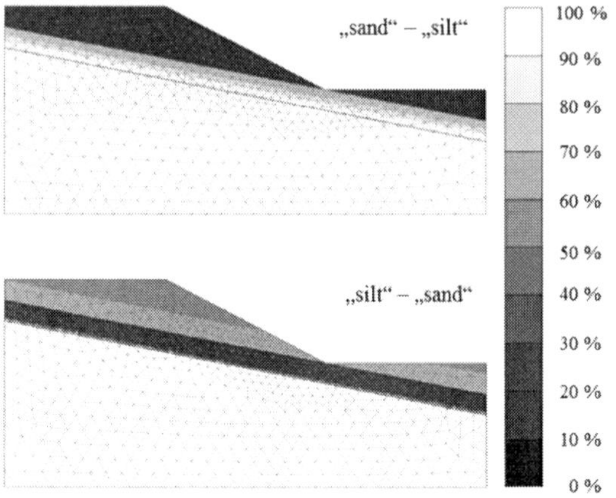

Figure 3. Initial saturation profile for the configurations "sand-silt" and "silt-sand" (GEO 1, two-layered model)

2.1 Soil properties

As the purpose of this study is to evaluate the influence of hydraulic properties and rainfall intensity on the FoS of the slope, the strength and stiffness parameters are assumed to be the same for all soil layers (see Table 1). Only the hydraulic properties differ for the various layers.

Table 1. Soil parameters for Mohr-Coloumb model

Description	Symbol	Value	Unit
Unit weight	γ	[kN/m³]	20
Elasticity modulus	E'	[kPa]	7500
Effective poissons' ratio	ν'	[-]	0.35
Effective cohesion	c'	[kPa]	10
Effective friction angle	φ'	[°]	20

In this study, two alternative hydraulic cases, one with a higher permeable upper layer and a less permeable lower layer and vice versa, have been investigated. The van Genuchten parameters used in this numerical analysis were taken from the USDA-Database, which is an international soil classification system [3]. The sand layer has a saturated hydraulic conductivity of $8.25*10^{-5}$ m/s, the silt layer $1*10^{-6}$ m/s, respectively. For the third soil layer (middle layer) also the SWCC for "sand" was taken, but the saturated permeability was increased to $1*10^{-3}$ m/s.

2.2 Hydrological changes

The influence of precipitation, evaporation and periods of no rainfall are analysed. Furthermore, various scenarios of applying the same amount of rain in different time periods were investigated. The rainfall intensities used in this study were adopted from the design rainfall for Graz (Austria) for 1 day with a return period of 20 years (around 120 mm /day) [4].

Different scenarios to apply the same amount of rainfall within 24 hours have been investigated:

- *Condition 1 (CC1):*
Constant precipitation during a predefined period with an infiltration discharge of 5 / 10 / 20 mm/hour for a duration of 24 hours.

- *Condition 2 (CC2):*
Alternating phases of precipitation and periods without precipitation (Figure 3).

Clarification: condition two means alternating periods with a duration of 3 hours with precipitation of 10 mm/hour and precipitation of 0 mm/hour (see Figure 4).

- *Condition 3 (CC3):*
Alternating phases of precipitation and evaporation (Figure 5).

Note: In order to quantify the differences between these climatic conditions, the sum of total applied rainfall (mm/day) was kept the same.

- q = constant with 5 mm/ hour for 24 h = 120 mm/ day
- q = constant with 10 mm/ hour for 24 h = 240 mm/ day
- q = constant with 20 mm/ hour for 24 h = 480 mm/ day

To keep the total amount of precipitation (Figure 4) or precipitation minus evaporation (Figure 5), independent of the hydraulic condition, equivalent daily discharges have been applied.

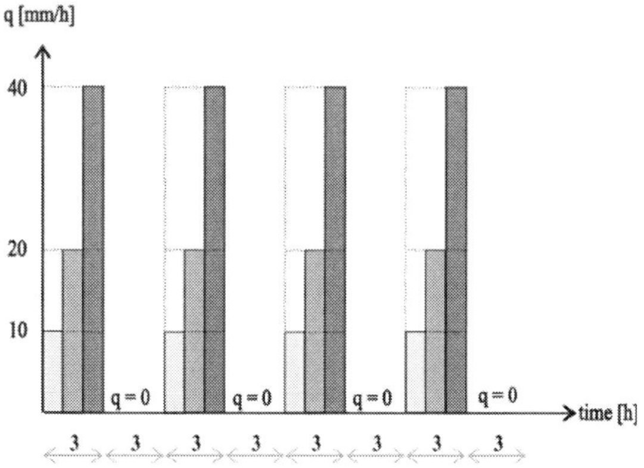

Figure 4. Condition 2 (CC2): Alternating phases of precipitation and periods without precipitation (q=0)

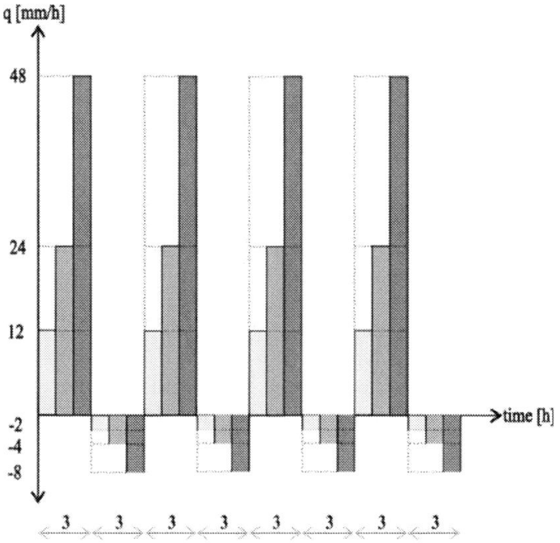

Figure 5. Condition 3 (CC3): Alternating phases of precipitation and evaporation

3 Soil Water Retention Curve

The hydraulic behaviour of unsaturated soils is described by the SWCC, which defines the soils capacity to keep water at different suction values. There are numerous models available to describe the SWCC mathematically, in this study the van Genuchten model was used [2].

$$S(\theta_p) = S_{res} + (S_{sat} - S_{res})\left[1 + \left(g_a|\theta_p|\right)^{g_n}\right]^{g_c} \quad (1)$$

$$\theta_p = \frac{p_w}{\rho_w g} \quad (2)$$

The van Genuchten function (Equation 1) is a three-parameter equation which relates the saturation to the pressure head θ_p. S_{res} is defined as the residual saturation which remains even at high suction heads, S_{sat} is the degree of saturation under saturated conditions, g_a, g_n and g_c are curve-fitting parameters.

Table 2. Soil parameters for Mohr-Coloumb model

Soil	k_{sat} [m/s]	g_a [1/m]	g_n [-]	g_l [-]
Sand	$8.3*10^{-5}$	14.5	2.68	0.5
Sand*	$1.0*10^{-3}$			
Silt	$1.0*10^{-6}$	1.6	1.37	0.5

Sand*: higher permeable sand layer in 3-layered model

Table 2 lists the hydraulic conductivity for saturated states and the van Genuchten curve-fitting parameters, which were used in the present study. For the "3-layered-model" (Figure 2) the permeability of the middle layer (Sand*) was increased to $1.0*10^{-3}$ m/s.

Figure 6 illustrates the SWCC for "sand" and "silt" (USDA, [3]).

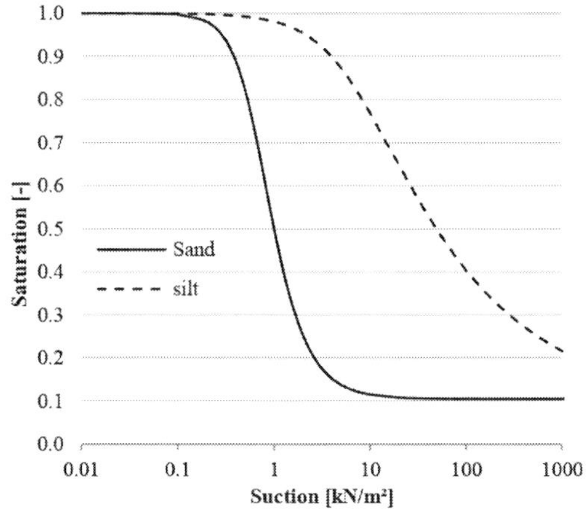

Figure 6. Soil Water Characteristic curves for sand and silt (USDA series)

4 Relative Permeability Curve

The coefficient of permeability in an unsaturated state depends on the saturation of the soil. The ratio of the permeability at a given saturation to the permeability in the saturated state is called relative permeability k_{rel} *(S)*. The hydraulic conductivity for unsaturated soil can be calculated by (Equation 3):

$$\underline{\underline{k}} = k_{rel}\,\underline{\underline{k}}^{sat} \quad (3)$$

In order to calculate the relative permeability k_{rel} *(S)* (Equation 5) in relation to Mualem -van Genuchten [5], the effective degree of saturation (S_e) is obtained as:

$$S_e = \frac{S - S_{res}}{S_{sat} - S_{res}} \quad (4)$$

$$k_{rel}(S) = (S_e)^{g_l}\left[1 - \left(1 - S_e^{\frac{g_n}{g_n - 1}}\right)^{\frac{g_n - 1}{g_n}}\right]^2 \quad (5)$$

Figure 7 shows the relative permeability depending on the degree of saturation for "sand" and "silt" (USDA), which were used in this study. Due to the initial depth of the groundwater level (see figures 1 & 2), the saturation of the upper soil layer is quite low. This, in combination with the chosen SWCC, led to very low values of unsaturated (relative) permeability.

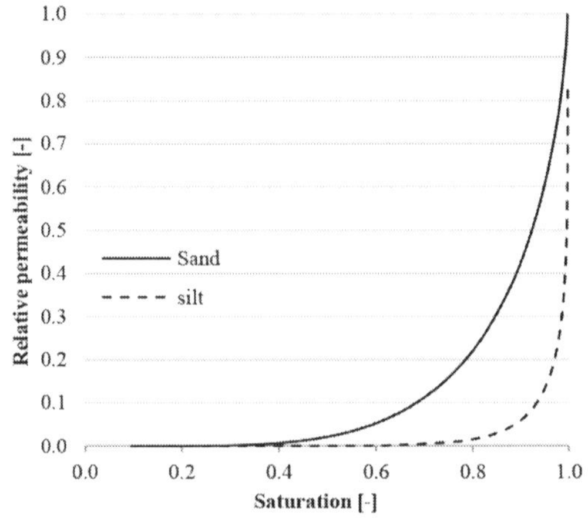

Figure 7. Relative permeability depending on saturation for sand and silt (USDA series)

In practice the permeability of a soil in situ is a parameter associated with a high degree of uncertainties. Due to that, a study with 2 orders higher permeability in the upper soil layer was done additionally (chapter 5.5).

5 Results

Table 3 gives an overview of the calculations which were done in this study. The influence of different geometries (GEO 1/2/3), permeability of the soil layers ("sand-silt" / "silt-sand"), hydraulic conditions (CC 1/2/3), differences between 2-layered and 3-layered models and the duration of applied rainfall events have been investigated.

To quantify the slopes' factor of safety during rainfall events after specified times of infiltration or evaporation, the well-known strength reduction method was applied (see Figure 8). The principle of the shear strength reduction technique is to reduce c' and $\tan\varphi'$ simultaneously in small increments until failure occurs in the finite element analysis. The factor of safety (FoS), with the shear strength parameters at failure c_r' and φ_r', can be written as:

$$FOS = \frac{\tan\varphi'}{\tan\varphi_r'} = \frac{c'}{c_r'} \tag{6}$$

Consequently, it is possible to evaluate the effects of variations in hydraulic conditions or geometry changes on the stability of the slope.

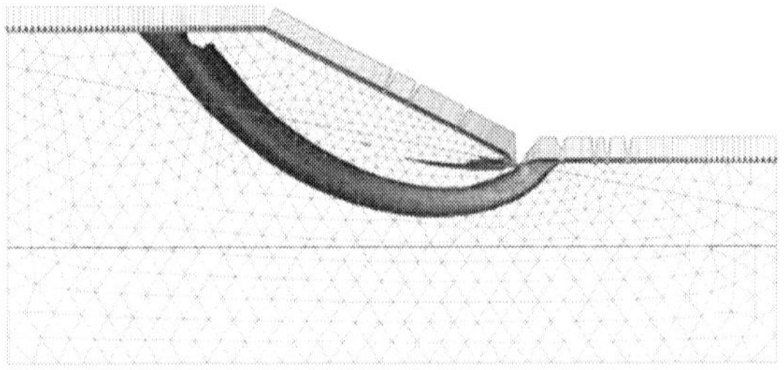

Figure 8. Incremental deviatoric strains after strength reduction method (5mm/hour, GEO1, CC1) after 24 hours

5.1 Influence of geometry changes of the soil layers

One major outcome of this study is that the configuration with a less permeable upper layer ("silt-sand") and climatic condition 1, independently of geometry (1/2/3), generally leads to an increase in the factor of safety. This is caused by the very low permeability of the upper soil layer.

If the applied precipitation rate is larger than the soils ability to infiltrate water, a predefined maximum head of 0.1 m builds up, which implicates an increase in the FoS (acts like a surcharge). Only if the predescribed maximum head (ψ_{max}) is reached, additional water is supposed to run off. This situation is certainly not realistic from a practical point of view, because runoff would probably be faster and inevitable cracks in the soil will allow water to infiltrate. However, it had been kept here in order to highlight the importance of correct boundary conditions.

In contrary, the results show a decreasing safety factor when the upper layer is more permeable ("sand-silt"). As the initial groundwater level is kept in the same location in all calculations, the initial factor of safety is given by the geometry of the model and the arrangement of the soil layers (geometry 1-3). Figure 9 shows the development of the FoS for different configurations of the geometry, whereas solely CC1 was applied. The differences between line types of the same kind lie within the applied total daily discharges (120/240/480 mm/day). It follows that the influence of geometry is more significant than rainfall events.

Table 3. Calculations Overview

Layer(s)			Geometry	Duration	Total q	Climatic condition CC
2-layered model			GEO 1			*CC1*
Upper layer		*Lower layer*				q=const
sand	k_s=8.25*10^{-5}	silt	GEO 2	1 day	120 mm	
sand	k_s=8.25*10^{-3}	silt				*CC2*
silt	k_s=1.0*10^{-6}	sand	GEO 3	3 days	240 mm	Precipitation / only flow
3-layered model						
Upper layer	*Middle layer*	*Lower layer*	1*		480 mm	*CC3*
sand	sand* (k_s=1*10^{-3})	silt				precipitation / evaporation

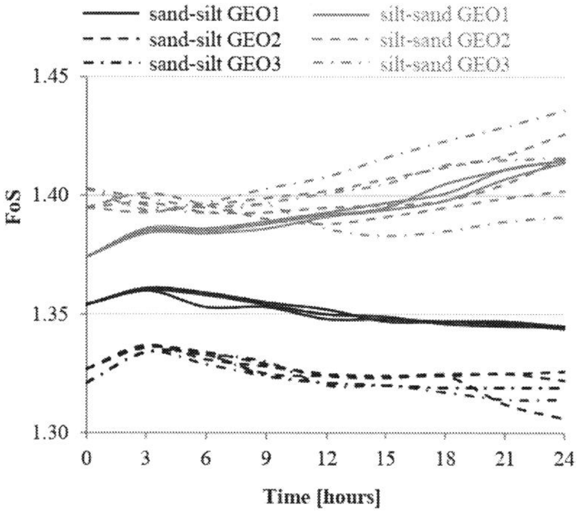

Figure 9. Development of the FoS of GEO 1/2/3 for 1 day and CC 1.

5.2 Influence of hydraulic conditions

Changes in the climatic conditions (CC 2/3) do not show any significant differences in the development of FoS with time compared to climatic condition 1 (CC1: constant infiltration rate over 24 hours) as described in chapter 4.1. CC2 and CC3 led to alternate periods with increasing and decreasing FoS, respectively.

In the present study, the constellation with a less permeable upper layer (silt) was no longer investigated. Due to the fact that changes in geometry (GEO 1/2/3) led to qualitatively comparable results, only GEO 1 is investigated in the following chapters.

5.3 Comparison of 2-layered and 3-layered geometries

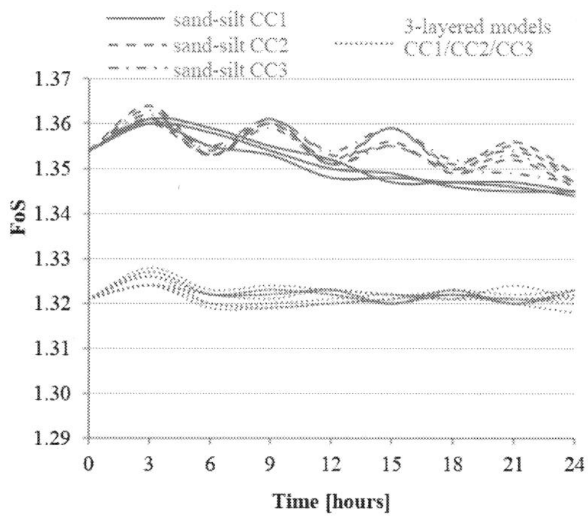

Figure 10. Development of the FoS of GEO 1 (2 layers) and the geometry with 3 layers for 1 day and CC 1/2/3.

The boundary conditions for the 3-layered soil model (figure 2) with a high-permeable soil layer between the upper and the lower layer showed negligibly small changes in the FoS with time for an observation time of 24 hours (see Figure 10).

5.4 Development of the FoS - period of three days

As the FoS slightly decreases when applying the total amount of rainfall within 24 hours, the influence of a three-day long lasting rainfall event (consequently with lower precipitation rates) was examined.

The results show a decrease of the FoS for the two-layered model with permeable upper layer (sand), whereas hydraulic condition 1 with a constant precipitation of 6.67 mm/hour for three days (equivalent to 20 mm/ hour for 1 day) induced the largest decrease in safety with time (see Figure 11).

The investigated three-layered soil model showed again only negligible changes of the FoS with time and was therefore no longer investigated.

DOI: 10.1051/e3sconf/20160915010

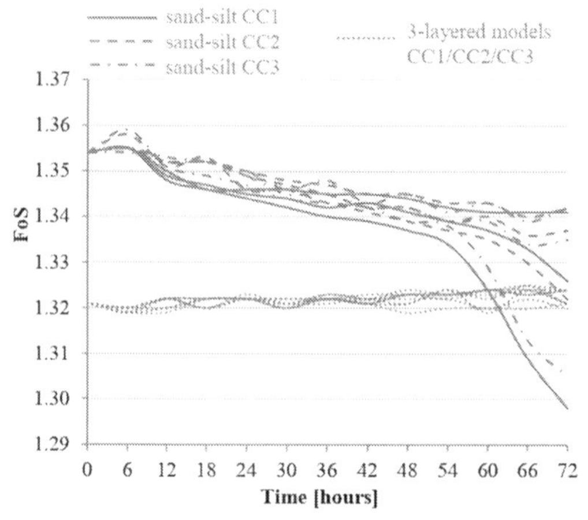

Figure 11. Development of the FoS of GEO 1 (2 layers) and the geometry with 3 layers for 3 days and CC 1/2/3.

5.5 Development of the FoS - period of three days and higher hydraulic permeability of the upper soil layer

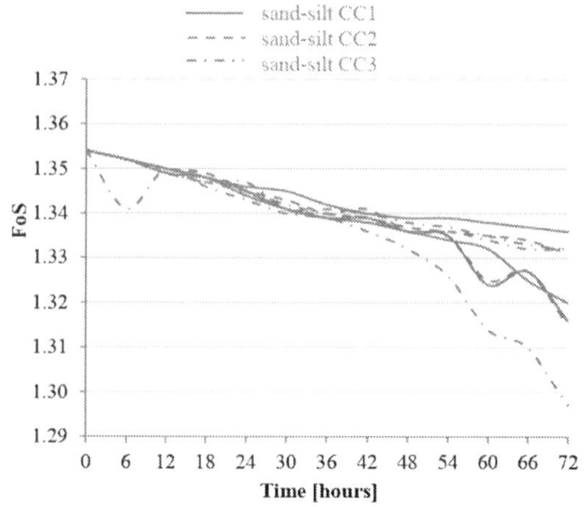

Figure 12. Development of the FoS of GEO 1 (2 layers) for 3 days, CC 1/2/3 and 2 orders higher hydraulic conductivity of the upper soil layer.

Figure 12 shows the development of the FoS of two-layered geometries under different climatic conditions (CC 1/2/3) and two orders higher hydraulic conductivity of the upper soil layer.

The main difference, compared to the results presented in chapter 4.4, is that there is, due to higher initial permeability, no increase in the factor of safety within the first period of precipitation (6 hours). The total decrease in the FoS after 72 hours is very small and comparable to those with lower initial permeability (compare figures 11 & 12).

6 Conclusion

In the present study, the changes in FoS of inhomogeneous soil slopes, which were subjected to different scenarios of rainfall, were investigated.

For the investigated boundary conditions, there are no significant differences in changes of the factor of safety between the geometries (GEO 1/2/3), which define the boundary between the two soil layers. The lower saturated permeability of fine-grained soils led to an increase in the FoS if they were taken as the upper soil layer, because a head, which acts like a surcharge, is built up. This is a situation which is unlikely to occur in practice but highlights the importance of careful selection of hydraulic boundary conditions.

Contrary to that, the factor of safety decreases if the upper soil layer has a higher saturated hydraulic conductivity, because water is able to infiltrate more easily.

There are only slight differences between the three conditions used in this study. The highest decrease in slopes' safety was generally reached under climatic condition 1, when a constant value of precipitation, without any periods of evaporation, was applied. The total sum of applied precipitation was always kept the same (120/240/480 mm/m²) and either applied within one or three days. Compared to a one day precipitation period with higher precipitation rates where only slight changes in the factor of safety occurred, the lower precipitation rate in the case of a 3 days period led to a decrease in the FoS with time.

The results also show, that under the given boundary conditions in this study, an increase of the upper soils' layer hydraulic conductivity does not change the FoS of the slope. Further studies using different SWCC are currently under way.

References

1. H. Rahardjo, T.H. Ong, R.B. Rezaur, E.C. Leong, J. geotech. Geoenviron. Eng, **12** (2007)
2. L.L. Zhang, J. Zhang, L.M. Zhang, W.H. Tang, P I Civil Eng-Geotec, **164**, Issue GE5, 299-316 (2011)
3. R.B.J. Brinkgreve, S. Kumarswamy, W.M. Swolfs, Plaxis 2D 2015 - Reference Manual, Plaxis bv, Delft, Netherlands (2015)
4. BMLFUW, Hydrografisches Jahrbuch in Österreich, **120,** Gitterpunkt 5214 (2012)
5. M.T. Van Genuchten, Soil Sci. Soc. Am. J., **44**, 892-898 (1980).

Characteristics of failure mass and safety factor during rainfall of an unsaturated slope

Thi Minh Hue Le[1,2,a], Domenico Gallipoli[3], Marcelo Sanchez[4], Simon Wheeler[5]

[1]*SAMCoT - Sustainable Arctic Marine and Coastal Technology, Centre for Research-based Innovation, Norwegian University of Science and Technology, Trondheim, Norway*
[2]*Klima2050 - Risk Reduction through Climate Adaptation of Building and Infrastructure, Centre for Research-based Innovation , SINTEF Building and Infrastructure, SINTEF, Trondheim, Norway*
[3]*Laboratoire SIAME, Université de Pau et des Pays de l'Adour, Anglet, France*
[4]*Zachry Department of Civil Engineering, Texas A&M University, College Station, United States*
[5]*School of Engineering, University of Glasgow, Glasgow, United Kingdom*

Abstract. Water is an important factor triggering destabilization of soil masses in slopes, particular in unsaturated condition. Rainfall and snow melting are some of the main sources of natural water which frequently induce unsaturated slope failure in arid/semi-arid region or in active layers (i.e. frequently frozen/thawed layers) in permafrost. If the slopes are in coastal environment or adjacent to a stream, the soil materials from the sliding masses might be carried away by waves and currents, which can further lead to soil erosion problem. Therefore, characterization of the failure mass and safety factor is critical both for evaluation of slope failure hazard but also for erosion problem. This paper presents a method to study unsaturated slope stability including characterizing the size of failure and the safety factor by analyzing a theoretical unsaturated homogeneous slope subjected to rainfall infiltration. The slope is modelled numerically by the finite element method. A shear strength reduction technique is modified for unsaturated soils and employed to evaluate the factor of safety and the extent of the sliding mass. The results show that the size of failure and the factor of safety vary over the rainfall event. Notably, the extent of the sliding mass for unsaturated slopes depends strongly on the infiltration depth. Large global sliding surface tends to be the dominant mechanism if the infiltration depth is very shallow or very deep. Conversely, local restricted sliding surface becomes more dominant if the infiltration is approximately at a certain critical value.

1 Introduction

Water is known to be one of the main factors contributing to triggering the destabilization of soil masses in unsaturated slopes. Some of the main sources of natural water are rainfall and snow melting. Rainfall has been known to induce slope failure in unsaturated soils in arid/semi-arid regions such as Hong Kong, Australia and the Middle East while snow melting can destabilize the unsaturated active layers in permafrost in Arctic regions such as Svalbard, Alaska and northern Russia. If the slopes are in coastal environments (e.g. in Svalbard or adjacent to a stream, the soil material from the sliding masses might be carried away by waves and currents, which can further lead to soil erosion problem. Therefore, characterization of the failure mass and the safety factor is an important task in order to evaluate the time and the volume of failure. This can subsequently be used for evaluating the hazard of a slope failure and for estimating the erosion rate, particular in the Arctic coastal areas.

This study limits to unsaturated slope failures caused by rainfall infiltration as the source of natural water. Even though catastrophic slope failures induced by rainfall infiltration is one of the major natural hazards facing many regions in the world, understanding of the triggering mechanisms still requires further research. This is particularly important and challenging for slopes in partially saturated soils because the wetting process can lead to a reduction in suction and, possibly, a rise in positive pore water pressure. Both these effects directly contribute to a reduction of the soil shear strength and, hence, destabilise the slope.

This study investigates an unsaturated homogeneous slope subjected to a rainfall event of constant intensity. The slope is modelled numerically by the finite element method (FEM) with application of a shear strength reduction technique (SRT) modified for unsaturated soils. The extent of the sliding mass and the factor of safety are evaluated at different times during and after the rainfall period. The development of the infiltrated depth is observed in parallel with the development of the slip surface in order to investigate the correlation between these factors.

[a] Corresponding author: thi.le@sintef.no

2 Method

2.1. Geometry and boundary condition

The slope model used in this study is detailed in Figure 1. The slope is 10 m height, having an angle of 2:1 (or 26.6°) and resting on a base of 20 m thickness. The initial stress distribution is in equilibrium with gravity (9.81 m^2/s) while the initial pore water pressure (p_w) distribution is in hydrostatic equilibrium with the initial water table position (5 m below the ground level at the toe of the slope). The hydrostatic condition results in a maximum initial suction (s) of approximately 150 kPa at the crest (AB in Fig. 1) which falls in the low end of the prevalent surface suction range of semiarid or arid environments such as in Australia [1].

A mesh of 1515 quadrilateral elements is employed. This mesh has been shown to produce a reliable estimate of the factor of safety (FoS) of the slope in saturated condition which remained essentially unchanged for finer discretisations [2]. The vast majority of elements are approximate squares or parallelograms (Fig. 1). Each square element, and the most of parallelogram elements, has an approximate area of 1 m^2. The uniformity of the mesh is important for the evaluation of the size of failure as discussed later. Approximately 1% of elements, in the centre of the mesh, are degenerated into smaller polygons (Fig. 1). The degenerated elements are located at considerable distances from the possible locations of slip surfaces and are, hence, unlikely to have any substantial impact on the evaluation of the size of the failure mass. The analysis was conducted using the finite element code Code_Bright [3].

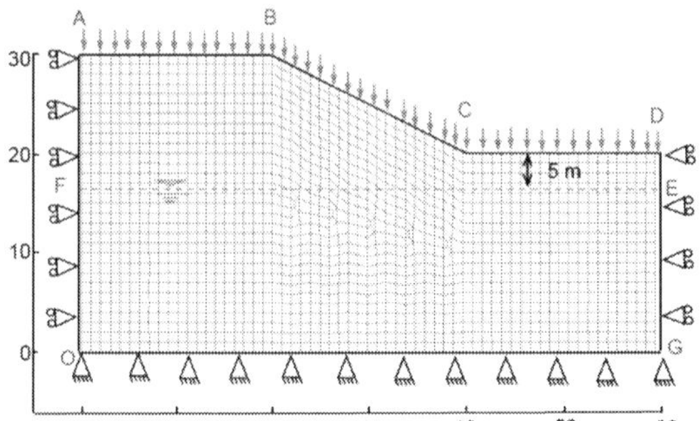

Figure 1: Slope dimensions, geometry and boundary conditions (scale in metre)

2.2 Hydraulic and mechanical models

The van Genuchten (1980) [4] water retention curve and van Genuchten and Nielsen (1985)[5] permeability function are selected because they are capable of representing the soil hydraulic behaviour in a rather simple and numerically stable way. They are also well-accepted in for modelling flows in unsaturated soils. These constitutive models are presented briefly in equation 1-5. More details about these relationships can be found in Le *et al.* (2010) ; Le (2011); Le *et al.*, (2012), Le *et al.*, (2013) and Le *et al.*, (2015) [2, 6-9]:

$$S_e = \frac{S - S_r}{S_s - S_r} \left(1 + \left(\frac{s}{s_e}\right)^{\frac{1}{1-m}}\right)^{-m} \quad (1)$$

$$s_e = s_{eo} \exp(\eta(\phi_o - \phi)) \quad (2)$$

$$k_s = k_{so} \frac{\phi^3}{(1-\phi)^2} \frac{(1-\phi_o)^2}{\phi_o^3} \quad (3)$$

$$k_r = \sqrt{S_e}(1 - (1 - S_e^{1/m})^m)^2 \quad (4)$$

$$\mathbf{q} = -k_s k_r \nabla(\frac{u_w}{\rho_w g} + z) \quad (5)$$

The soil water retention curve proposed by van Genuchten (1980) [4] is given by equation 1, which relates the effective degree of saturation (S_e) (calculated as a function of the current (S), the maximum (S_s), and the residual (S_r) degree of saturation) to suction s=-p_w through the air entry suction parameter s_e. In equation 2, the parameter s_e is in turn related to the porosity (ϕ) through parameter η which controls the rate at which s_e deviates from its reference value s_{eo} when ϕ deviates from its reference value ϕ_o [10, 11]. Similarly, equation 3 describes the dependency of the saturated permeability (k_s) on its reference value k_{so} and the difference between ϕ and its reference value ϕ_o, as proposed by Kozeny (1927) [12]. Equation 4 is the van Genuchten and Nielsen (1985) [5] permeability curve linking the relative permeability k_r to S_e through the parameter m, which can be geometrically interpreted as the curve gradient. The unsaturated permeability k_u is the product of the saturated and relative permeability (i.e. k_u = $k_s k_r$). Finally, the unsaturated flow **q** is calculated using the generalized Darcy's law (equation 5). The symbols u_w, ρ_w, g and z indicate the pore water pressure, the water density, the gravitational acceleration and the elevation coordinate, respectively.

The selected values of the required soil parameters are shown in Table 1. Rainfall intensity I_r is assumed to be constant at 43.2 mm/day. Constitutive parameters (e.g. m, k_{so}, s_{eo}) are selected to be at around the middle of the likely range for each parameter in order to avoid overly large or small input parameters and, hence, unrepresentative or unrealistic results. For example, an very low permeability (i.e. very small k_{so}) would result in the rainfall having no effect on slope stability because water would not have sufficient time to infiltrate into the unsaturated soil domain.

Table 1: Values of hydraulic parameters

Symbols	Units	Values	Symbols	Units	Values
m		0.2	ϕ_o		0.333
η		5	k_{so}	m/s	10^{-5}
S_s		1	s_{eo}	kPa	20
S_r		0.001	ρ_w	kg/m^3	1000

A linear elastic model with an extended Mohr-Coulomb (MC) failure criterion (equation 6) describes the mechanical behavior of the unsaturated soil [13]:

$$\tau = c' + \sigma \tan \phi' + s \tan \phi^b \qquad (6)$$

In equation 6, the shear stress at failure (τ) depends on the net normal stress (σ) and suction (s) through the friction angle (ϕ'), cohesion (c') and a parameter controlling the increase in shear strength with suction (ϕ^b). The component of strength contributed by suction (i.e. the 3rd term in equation 6) decreases with decreasing s and becomes zero for a fully saturated soil. The values of the strength parameters (c', ϕ' and ϕ^b) assumed in this study are typical of a clay and are based on the values reported by Bishop *et al.*, (1960) [14] for boulder clay and by Gan *et al.*, (1988) [15] for a compacted glacial till. The elastic parameter values, i.e. Young modulus (E) and Poisson ratio (v), are also typical of a clay and chosen within their respective typical ranges of variation [16]. The values of all mechanical parameters are summarized in Table 2.

Table 2: Values of mechanical parameters

Symbols	Units	Values	Symbols	Units	Values
E	MPa	100	ϕ'	°	20
v		0.3	c'	kPa	18
ρ_s	kg/m^3	2700	ϕ^b	°	18

3. Evaluating the FoS and the size of failure for unsaturated slopes

3.1. Shear strength reduction technique for unsaturated soils

The shear strength reduction technique (SRT), first suggested by Zienkiewicz *et al.* (1975) [17], is widely adopted to estimate the slope factor of safety (FoS) of saturated slopes analyzed by the FEM. The FoS is defined as the factor by which the strength parameters must be divided to bring a slope into the state of being "barely stable" [18]. Since the FEM allows tracking of the slope deformation up to and including early shear failure, this definition conveniently outlines a technique to calculate the FoS by gradually and systematically reducing cohesion, c' and the tangent of the friction angle, $\tan\phi'$ until "early failure" is achieved:

$$FoS = \frac{c'_{actual}}{c'_{fail}} = \frac{\tan \phi'_{actual}}{\tan \phi'_{fail}} \qquad (7)$$

In the unsaturated soil context, in addition to the conventional shear strength parameters c' and $\tan\phi'$, the Mohr-Coulomb criterion is also characterised by $\tan\phi^b$ which controls the expansion of the failure envelope with suction increase. The product of s and $\tan\phi^b$ has a similar role to the cohesion and is often combined with c' to create an "enhanced cohesion". Hence, it is reasonable to apply the strength reduction factor to $\tan\phi^b$ in the same manner as to the other two shear strength parameters. In other words, the FoS for unsaturated soil is defined as:

$$FoS = \frac{c'_{actual}}{c'_{fail}} = \frac{\tan \phi'_{actual}}{\tan \phi'_{fail}} = \frac{\tan \phi^b{}_{actual}}{\tan \phi^b{}_{fail}} \qquad (8)$$

During the SRT analysis, an algorithm is also executed to monitor and check the displacement history of all the nodes in the FE mesh. All the nodes that exhibit sudden increase in displacement during the SRT process, and subsequently abnormally large displacement at the end are recorded. The end of the analysis is the point at which the reduction factor could not increase further and is considered to be equal to the FoS. Details of different criteria which are used for this monitoring algorithm have been discussed in Le *et al.*, (2015) [6].

3.2. Simulation process

The simulation process involves three stages corresponding to (1) establishing the initial stresses, strains and p_w distributions under the action of gravity and the initial hydrostatic condition; (2) establishing stresses, strains and p_w distributions at times of interest over the rainfall period; and (3) estimating of the FoS and the size of failure at these times of interest via the SRT. The three stages are presented in detail below:

Stage 1: The p_w in equilibrium with the hydrostatic condition (i.e. stable horizontal water table at 5 m below the slope toe, EF in Fig. 1) is fixed at every node while the slope is assumed weightless. Subsequently, the unsaturated slope is gradually equilibrated with the standard gravity (9.81 m/s^2). Since this gravity loading process is entirely artificial (i.e. not based on an actual physical event) and has the only purpose of setting up a physically reasonable initial stresses/strains distributions, any deformations, and hence porosity changes, caused by this process are reset to zero after reaching equilibrium.

Stage 2: Rainfall is imposed on the crest, the inclined face and the base surface (ABCD) over a period of 10 days (Fig. 1) using the seepage boundary condition. After 10 days, the rainfall is terminated (i.e. removing the seepage boundary condition) but the simulation is continued for another 355 days to investigate recovery of slope stability.

The seepage boundary condition allows water to infiltrate the soil domain at a constant flux equal to I_r as long as the p_w at this boundary is less than zero. If the p_w becomes equal to or larger than zero, the seepage boundary condition automatically switches to a zero p_w boundary to avoid ponding of water on the infiltration boundary.

Boundaries OA, OG and GD (Fig. 1) are assumed to be impermeable which allows the infiltrated water to accumulate inside the soil domain and raises the water table during the rainfall period. During this stage, the p_w is no longer fixed at every node (as in Stage 1) to enable the natural response of the hydraulic regime to rainfall infiltration. The mechanical deformation induced by rainfall infiltration is fully coupled with the variation of p_w.

Stage 3: The SRT is employed to estimate the FoS and the size of failure at four times during (at 0, 0.5, 5 and 10 days) and four times after (at 15, 20, 100 and 365 days) the rainfall event. At each time of interest, the FoS is calculated in an independent SRT analysis in which the stresses and strains, obtained at the corresponding time from the simulation in Stage 2, are imposed as the initial stresses and strains conditions at each element. On the other hand, the p_w, also obtained from Stage 2, is fixed at every corresponding node throughout the SRT analysis. The restrain of p_w at a constant value is to ensure a fully drained process and hence prevent any p_w built up as the result of stress changes caused by the reduction of strength. This is desirable because the SRT process is purely a numerical technique to estimate the FoS and does not represent a physically based phenomenon.

3.3. Characterization of the sliding mass

In a conventional slope stability study, the FoS is usually the only variable used to evaluate risk. However, the hazard posed by a slope failure is normally judged, in reality, not only by this factor but also by the size of failure or the extent of the sliding mass. A large failure with an extensive sliding mass can cause serious damage while a small one might not cause any damage upon failure. Hence, a large slope failure is regarded from a practical standpoint as more dangerous than a small one even if the former might have a higher FoS than the latter. In addition, the estimation of the size of failure is very important to study soil erosion problem as it directly determines the amount of materials that might be transported way. In reality, the size of failure is often characterized by the volume of soil material sliding down. This study uses a two-dimensional model, hence the size of failure is characterized equivalently by the sliding area (A_s) and the maximum depth from the slope surface to the failure surface to (D_s).

In this study, the algorithm to monitor the displacement history at every node also allows the extent of the slope failure to be analysed. The value of A_s is equated to the total number of nodes which has displaced suddenly at some point during the SRT analysis and has displaced in total by a significant amount by the end of the SRT analysis. The reason is that each node lying on the sliding mass represents an area portion surrounding the node which is, in most cases, comprised of a quarter of each of the four elements sharing the node. Since the mesh consists mostly of square or parallelogram elements of similar areas (1 m^2), the portion representing by each node is also equal to 1 m^2. Some nodes locating close to or on the sliding mass boundaries represent portions of the sliding mass that are smaller than 1m^2. Therefore, the total number of nodes on the sliding mass slightly over-estimates the actual value of A_s. However, if A_s increases, the total number of nodes locating on the mass also increases and vice versa. The variation of the total number of nodes on the sliding mass therefore reflects the variation of A_s, and so it is considered satisfactory in this study to equate these two quantities. If the case studied was a practical slope, it is recommended that the estimation of A_s is improved either by using much finer mesh to reduce the discrepancy or by identifying nodes on or close to the boundaries and assigning smaller area portions to these nodes. The estimation of D_s involves calculating the distances from all the nodes on the sliding mass to the slope face in order to find the maximum value among those distances. Illustration of a sliding mass (after 100 days) imposed on the mesh is shown in Figure 2. The number of nodes within the shaded area is used to approximate A_s and D_s as discussed. The failure mode is also been categorized into 'global' when the failure surface extends the whole length of the slope face or 'local' when the failure surface limits to a portion of the slope face.

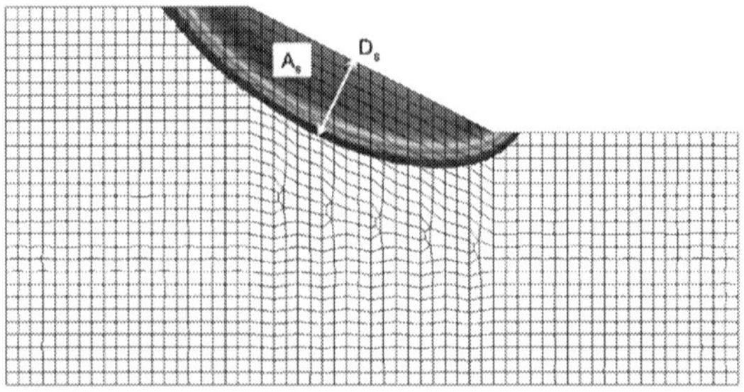

Figure 2: Estimation of slip surface area (A_s) and depth (D_s) after 100 days

4. Results and discussions

4.1. Qualitative analysis

The p_w contours and corresponding failure masses (indicated by displacement contours) during and after the period of rainfall are presented in Figures 3 and 4, respectively, for the homogenous unsaturated slope considered in this study. The values of FoS, A_s, D_s and the corresponding failure modes are tabulated in Table 3.

The use of the FEM allows natural and intuitive failure mechanisms to develop where the slip surfaces tend to pass through the weakest soil regions in the area of potential sliding (Figs. 3 and 4). It is important to note for unsaturated slopes that, unlike in dry or fully saturated slopes, the occurrence of the weakest region is significantly affected by the infiltration pattern which controls the reduction of suction and possibly the development of positive p_w close to the slope surface. Therefore, the soil area close to the slope face which has been wetted by the rainfall becomes significantly weaker than unsaturated soils in the area underneath the wetting front.

 DOI: 10.1051/e3sconf/20160915011

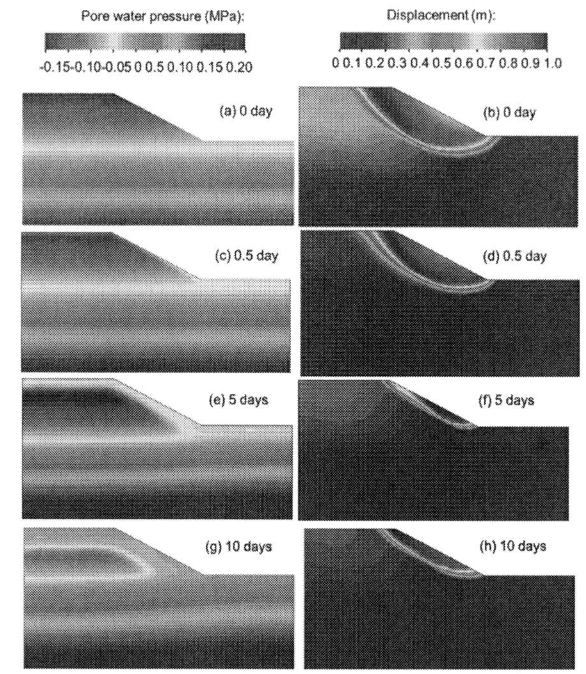

Figure 3: Evolution of pore water pressure contours (left) and slip surfaces at ultimate failure (right) during rainfall infiltration ($\mu(e)$ = 0.5, COV_e = 0.8, $\theta(e)$ = 8 m)

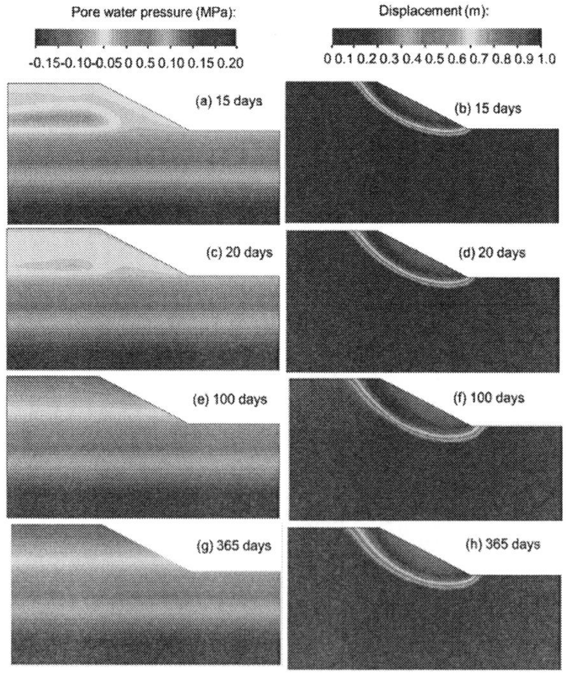

Figure 4: Evolution of pore water pressure contours (left) and slip surfaces at ultimate failure (right) after the rainfall ($\mu(e)$ = 0.5, COV_e = 0.8, $\theta(e)$ = 8 m)

Table 3: Values of studied variables at different times

Time (days)	FoS	A_s (m2)	D_s (m)	Failure mode
0	1.95	292	10.09	global
0.5	1.94	280	9.92	global
5	1.7	66	3.58	local
10	1.29	101	6.26	global
15	1.39	151	6.85	global
20	1.39	178	7.05	global
100	1.38	227	7.76	global
365	1.37	231	8.11	global

The results show that the FoS and the value of A_s vary considerably during the rainfall and during the post-infiltration period. At the beginning of the rainfall when the infiltration depth is insignificant or very shallow (i.e. 0 – 0.5 day), the size of failure tends to be large, however the FoS are quite high (see Fig. 3a-d and Table 3). As the rain infiltrates deeper into the slope face and the wetting front has migrated a certain distance from the surface (i.e. 0.5 – 5 days), the size of failure becomes smaller and the FoS drops at the same time (Fig. 3c-f and Table 3). Interestingly, as the rainfall continues and the wetting front migrates even deeper (i.e. 5 – 10 days), the FoS also continues to drop but the size of failure actually increases again. After the rainfall has stopped (i.e. after day 10), the size of failure keep increasing over time as the p_w re-adjusts to a new steady state during the post-infiltration period (Fig. 4 and Table 3). The FoS also increases after rainfall due to the recovery of suction in the soil region close to the slope face, then the FoS becomes almost stable from 15 days onward. Both the value of A_s and the FoS do not increase back to the initial levels (i.e. at time 0) after the rainfall stops due to the accumulation of water during the rainfall. This is the result of the impermeable condition assumed for the vertical and bottom boundaries (Fig. 1). Therefore, the new steady state at the end (day 365) has higher water table than the initial state.

Notably from Figure 3e, water infiltration into the slope face after 5 days seems to have reached a certain critical depth. At this depth, the wetted soil region is just sufficiently large to contain a shallow local failure mechanism (Fig. 3f). The values of A_s and D_s at 5 days are, in fact, the lowest and are significantly smaller than their corresponding values at earlier times or later times. The FoS reaches the lowest values at the end of the rainfall (10 days) which is caused by the combination of suction loss in the unsaturated zone and water table rise in the saturated zone. However, the size of failure at 10 days is larger than at 5 days (Fig 3. f and h) because the wetting front has migrated further which increases the size of the wetted region which contains the sliding mass.

Figures 5 and 6 compare the variation of the FoS, A_s and D_s over the rainfall period. The variation of failure depth is qualitatively similar to in the variation of failure area. The fact that the size of failure drops to the smallest value during the rainfall (5 days), but the FoS becomes the smallest at the end of the rainfall (10 days) suggests that the hazard evaluation of unsaturated slope failure should combine these two types of information (e.g. [9]).

Note that the results for the FoS in this study have been verified using the limit equilibrium method. This is performed using SLOPE/W (GEO-SLOPE International, Ltd, Alberta, Canada) to analyse the unsaturated slope with p_w distribution obtained from SEEP/W (GEO-SLOPE International, Ltd, Alberta, Canada). The results show good agreement between the FoS obtained by the FEM in this study and that obtained by GeoSLOPE commercial software.

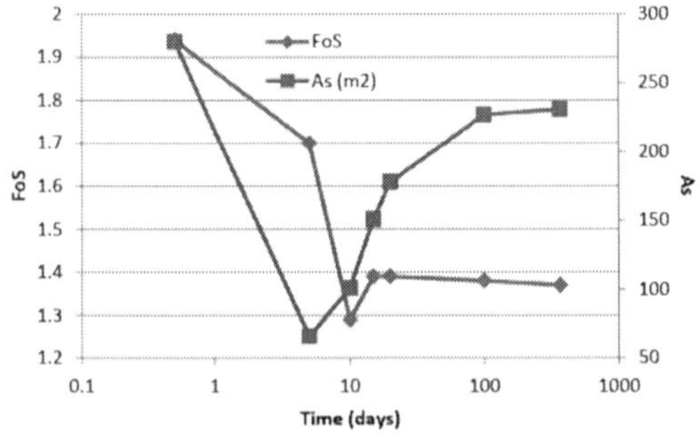

Figure 5: Variation of FoS and A_s with time

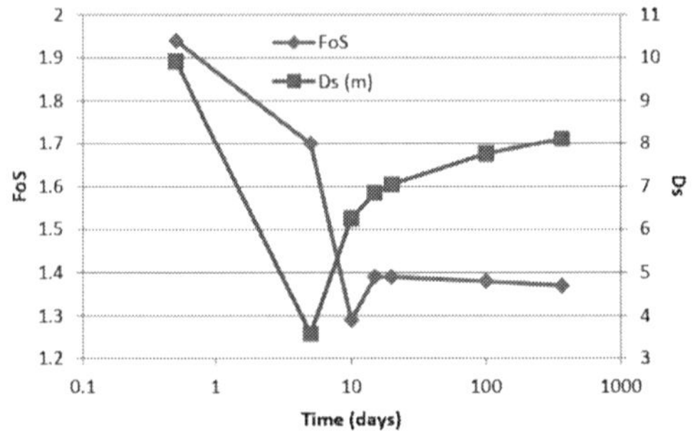

Figure 6: Variation of FoS and D_s with time

The localized failure mechanisms suggests that the method applied in this study has the potential to be use for investigating the retrogressive failure mechanisms triggered by water infiltration or similar factors (e.g. temperature rise in the cold region). If for example, the FoS corresponding to the localized failure at 5 days is smaller than 1 (e.g. the slope actually fails), then continuous rainfall can subsequently infiltrated into the exposed slip surface and cause subsequent failure once it has infiltrated deeper than the critical depth again. The process can continue if rainfall continues which lead to successive or retrogressive failure.

5. Conclusion

This paper has presented a method to study the failure mechanism of an unsaturated slope in which the failure has been characterized by both the factor of safety and its size in terms of area and depth. The conventional shear strength reduction technique was modified and adapted for unsaturated soils. The size of failure and factor of safety were evaluated by tracking the displacement history and all the nodes and recorded the nodes that are located on the sliding area.

The results show that the size of the failure and the factor of safety vary with time during the rainfall and also for a period after the rainfall has stopped. The size of failure reaches the smallest value at the transient state during the rainfall while the factor of safety drops to the lowest at the end of the rainfall. Notably, the extent of the sliding mass depends on the infiltration depth. Large global sliding surface tends to be the dominant mechanism if the infiltration depth is very shallow or very deep. Conversely, local restricted sliding surface becomes dominant if the infiltration is approximately a certain critical value which leads to the failure mass to confine to the wetted region.

Acknowledgements

This research has been supported by the centre for research-based innovation, Sustainable Arctic Marine and Coastal Technology (SAMCoT), and the centre for Risk Reduction through Climate Adaptation of Building and Infrastructure (Klima2050).

References

1. Cameron, D.A., et al., Influence of trees on expansive soils in southern Australia, in Expansive soils: recent advances in characterization and treatment, A.A. Al-Rawas and M.F.A. Goosen, Editors. 2006, Taylor & Francis: London, UK. p. 526.
2. Le, T.M.H., Stochastic Modelling of Slopes and Foundations on Heterogeneous Unsaturated Soils, in School of Engineering. 2011, The University of Glasgow: Glasgow, UK. p. 342.
3. Olivella, S., et al., Numerical formulation for a simulator (CODE-BRIGHT) for the coupled analysis of saline media. Engineering Computations, 1996. **13**(7): p. 87-112.
4. van Genuchten, M.T., A closed form equation for predicting the hydraulic conductivity of unsaturated soils. Soil Science Society of America Journal, 1980. **44**: p. 892-898.
5. van Genuchten, M.T. and D.R. Nielsen, On describing and predicting the hydraulic properties of unsaturated soils. Annales Geophysicae, 1985. **3**(5): p. 615-627.
6. Le, T.M.H., et al., Influences of spatial varying porosity on unsaturated flow through earth structures, in Fifth International Conference on Unsaturated Soils, E. Alonso and A. Gens, Editors. 2010, CRC Press/ Balkema: Barcelona, Spain. p. 1313-1319.
7. Le, T.M.H., et al., Stochastic analysis of unsaturated seepage through randomly heterogeneous earth embankments. International Journal for Numerical and Analytical Methods in Geomechanics, 2012. **36**(8): p. 1056-1076.

8. Le, T.M.H., et al., Rainfall-induced differential settlements of foundations on heterogeneous unsaturated soils. Geotechnique, 2013.
9. Le, T.M.H., et al., Stability and failure mass of unsaturated heterogeneous slopes. Canadian Geotechnical Journal, 2015. **52**(11): p. 1747-1761.
10. Rodríguez, R., et al., Experimental and numerical analysis of a mining waste desiccation. Canadian Geotechnical Journal, 2007. **44**: p. 644-658.
11. Zandarín, M.T., et al., The role of capillary water in the stability of tailing dams. Engineering Geology, 2009. **105**: p. 108-118.
12. Kozeny, J., Über kapillare Leitung des Wassers im Boden. Akad. Wiss. Wien, 1927. **136**(2a): p. 271-306.
13. Fredlund, D.G., N.R. Morgenstern, and R.A. Widger, The shear strength of an unsaturated soil. Canadian Geotechnical Journal, 1978. **15**(3): p. 313-321.
14. Bishop, A.W., et al., Factors controlling the strength of partly saturated cohesive soils, in Regional Conference on Shear Strength of Cohesive Soils. 1960: Boulder. p. 503-532.
15. Gan, J.K., D.G. Fredlund, and H. Rahardjo, Determination of the shear strength parameters of an unsaturated soil using the direct shear test. Canadian Geotechnical Journal, 1988. **25**(3): p. 500-510.
16. Zhu, T. Some Useful Numbers on the Engineering Properties of Materials -GEOL 615 Course note 2014 [cited 2014 23 May]; Available from: http://www.stanford.edu/~tyzhu/Documents/Some%20Useful%20Numbers.pdf.
17. Zienkiewicz, O.C., C. Humpheson, and R.W. Lewis, Associated and non-associated visco-plasticity and plasticity in soil mechanics. Géotechnique, 1975. **25**(4): p. 671-689.
18. Duncan, J.M., State of the art: Limit equilibrium and Finite-element analysis of slopes. Journal of Geotechnical and Geoenvironmental Engineering, 1996. **122**(7): p. 577-596.

Rainfall seasonality on landslide probability: a modelling approach

Estefanía Muñoz[1,a] and Andrés Ochoa[1]

[1]*Departament of Geosciences and Environment, National University of Colombia, Colombia*

Abstract. Precipitation is one of the main landslides triggering factors because the amount of infiltrated water into the ground causes changes in pore pressure due to the decrease in suction or the rising water table level. In this paper we estimated the return period of landslides from a rainfall stochastic behavior. Using rainfall intensity and duration probability density functions (pdf) and Philip's infiltration model, the pdf of water infiltrated volume into the soil was obtained. Then, assuming the soil reaches saturation, the pdf of depth wetting front (z_w) was obtained. Finally, with z_w and the Mohr-Coulomb rupture criteria, the pdf of the Factor of Safety (FS) was determined. Subsequently, landslide occurrence was analyzed under different hydroclimatic scenarios, such as dry and wet seasons and El Niño Southern Oscillation (ENSO) phases. The occurrence of shallow landslides depends more on intensity storm than duration storm, and they are more likely to happen during La Niña phase than during El Niño phase because both storm duration and storm intensity tend to increase.

1 Introduction

Most landslides occur on rainiest months [1], showing the close relation between the hydroclimatic conditions as a triggering factor and the occurrence of landslides.

Landslides triggered by rainfall, known as "soil slip", are characterized for having superficial failure surfaces (between 1 and 3 meters) and parallels to the slope face. This type of movements are due to changes on soil pore pressure when a wetting front product of infiltration advances [2].

The problem with rainfall as a triggering factor has been approached using physical and statistical models. Physical models are generally based on the combination of classical hydrological mechanisms (e.g. [3–7]). Statistical models have been used from different approaches, among which are the definition of critical thresholds that relate rainfall events intensity and duration with landslides occurrence (e.g. [8–12]), and the search for relations between landslide geometrical characteristics (perimeter, area and volume) and their frequency of occurrence for different climatological scenarios (e.g. [13–16]).

In this paper was analyzed the effect of annual and interannual seasonality (ENSO phenomenon) of precipitation on the stability of a hypothetical slope. From rainfall statistical behavior, the landslide occurrence probability on a slope was obtained, assuming that it fails when de factor of safety (FS) is less than 1. Through deterministic models of infiltration, wetting front depth and factor of safety, the rainfall probability density functions (pdf) was transformed into pdf of FS.

[a] Corresponding author: emunozh@unal.edu.co

2 Landslide probability

Starting from the historical behavior of the precipitation series and through infiltration models, it is possible to estimate infiltrated volume per unit area (e.g. [17,18]). There were used very simple models for rainfall, infiltration, wetting front depth and FS due to the mathematical difficulty to relate them analytically.

2.1 Rainfall model

Rainfall was modeled using a rectangular pulse Poisson model with exponential density distributions for average intensity i_r and total duration of the storms t_r, assumed as independent variables, as proposed by Eagleson [19].

To each point of the process are associated rectangular pulses with random durations and intensities. As shown in Figure 1, the intensity at each instant corresponds to the active rectangles overlapping at that time.

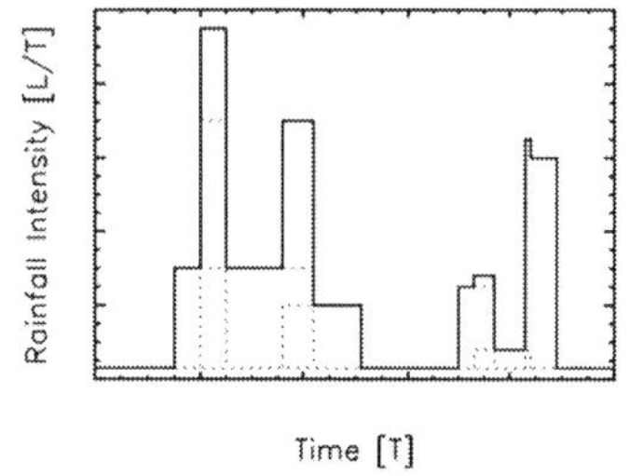

Figure 1. Rectangular pulse Poisson model.

Probability density functions of mean intensity and mean total duration of the storms are given by

$$f_{I_r}(i_r) = \beta e^{-\beta i_r},\ i_r \geq 0 \tag{1}$$

$$f_{T_r}(t_r) = \delta e^{-\delta t_r},\ t_r \geq 0 \tag{2}$$

where, β y δ are the inverse of the mean intensity and the total duration respectively.

2.2 Infiltration model

The model proposed by Philip [20] was used as an infiltration model, in which the infiltration capacity *f(t)* is defined according to saturated hydraulic conductivity k_s and soil sorptivity *S*. The potential infiltration capacity $f^*(t)$ and infiltrated volume $F^*(t)$ per unit area are

$$f^*(t) = k_s + \frac{1}{2} S t^{-\frac{1}{2}} \tag{3}$$

$$F^*(t) = k_s t + S t^{\frac{1}{2}} - k_s t_0 - S t_0^{\frac{1}{2}} + i t_0 \tag{4}$$

where, t_0 is time of ponding, that occurs when infiltration rate matches rainfall intensity, defined by

$$t_0 = \frac{S^2}{4(i - k_s)^2} \tag{5}$$

In order to calculate $f^*(t)$ and $F^*(t)$ it was stablished the relation between the storm mean intensity and soil saturated hydraulic conductivity. When the intensity is higher than the hydraulic conductivity it comes a time where the soil does not have enough capacity to infiltrate the water from the atmosphere (t_0), turning part of it on runoff water (Figure 2). When $i_r<k_s$ all the storm water infiltrates because the capacity of the soil to intake water is never exceeded (Figure 3).

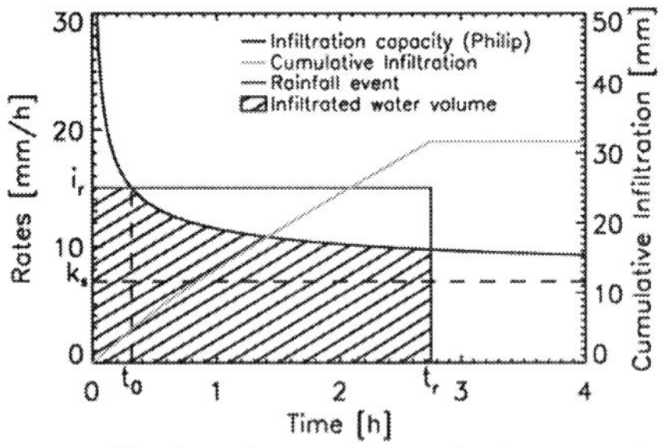

Figure 2. Infiltration when storm intensity is greater than or equal to saturated hydraulic conductivity.

In Figure 2 and Figure 3, blue rectangle represents the storm, the solid black line is the infiltration capacity given by Philip's model and the yellow line is the cumulative infiltration. The hatched area within the blue rectangle and below the infiltration capacity is the infiltrated water volume, and the unshaded area of the rectangle is the runoff volume *R(t)*, given by

$$R(t) = \begin{cases} 0, & i_r > k_s \wedge 0 < t < t_0 \\ \left[\dfrac{S}{2\sqrt{i_r - k_s}} - \sqrt{i_r - k_s t^{\frac{1}{2}}}\right]^2, & i_r > k_s \wedge t_0 < t < t_r \\ 0, & i_r < k_s \wedge \forall t \end{cases} \tag{6}$$

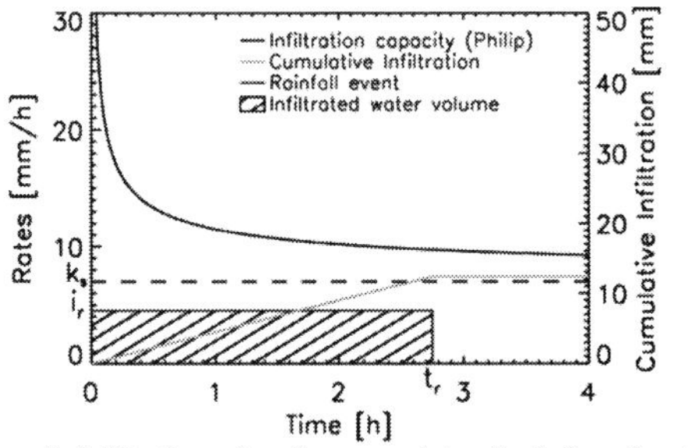

Figure 3. Infiltration when the storm intensity is less than the saturated hydraulic conductivity.

And the infiltrated volume per unit area *F(t)* by

$$F(t) = i_r t_r - R \tag{7}$$

2.3 Wetting front model

The depth of wetting front $z_w(t)$, that is to say, the maximum depth in which soil moisture varies with infiltrated volume per unit area, was calculated with modification of the equation proposed by Lumb [21]

$$z_w(t) = \frac{F(t)}{n(S_{wf} - S_{wi})} \tag{8}$$

where, *F(t)* is infiltrated water volume at time *t* of storm, *n* is soil porosity and S_{wf} and S_{wi} are final and initial degree of saturation respectively.

2.4 Slope stability model

With the Mohr-Coulomb rupture criteria and assuming an infinite slope, the relation to find the factor of safety FS was

$$FS(t) = \frac{c' + (\gamma_T - \gamma_w) z_w(t) \cos^2 \alpha \tan \phi'}{\gamma_T z_w(t) \sin \alpha \cos \alpha} \tag{9}$$

where, *c'* and ϕ' are soil cohesion and friction angle of soil respectively, γ_T is the soil unit weight, γ_w is the unit weight of water and α is the slope angle.

3 Rainfall seasonality effect on landslides

Mechanical and hydraulic parameters of hypothetical slope considered are presented on Table 1. The soil parameters correspond to a silt loam soil.

Table 1. Parameters of the slope and the soil that forms it.

Parameter	Value
S [mm h$^{1/2}$]	10
k_s [mm/h]	35
θ_{wi}	0.28
θ_{wf}	0.40
n	0.45
c [kPa]	10
ϕ [°]	25
α [°]	45
γ_T [kN/m^3]	17.5

The influence of the annual cycle on precipitation and the presence of macro climatic phenomenon like ENSO generate a great variability on rainfall, which implicates variability on soil infiltrated water content and consequently, on slope stability. The annual cycle indicates seasons with higher or lower precipitations along the year, while the presence of warm phase of ENSO (El Niño) leads to decreasing the average precipitation on different timescales and cold phase (La Niña) leads to increasing in most parts of Colombia [22].

For this paper the data analyzed correspond to Miguel de Aguinaga station, located in the city of Medellín, Colombia. This information has a temporary resolution of 15 minutes for an 11 year period (16/03/2001 to 24/05/2012).

The wet and dry seasons were defined from the annual cycle of the precipitation series, which presents rainy and dry periods, with a bimodal distribution. Periods between the months of April-May-June and September-October-November were defined as the wet season and December-January-February-March and July-August as the dry season, due to the fact that they have highest and lowest precipitations respectively.

Interannual variability due to ENSO phenomenon was estimated by dividing the historical precipitation series in three sets correspondent to the three ENSO phases: El Niño, Normal, and La Niña. The periods on which the phases occur are reported by the NOAA (National Oceanic and Atmospheric Administration) using the Oceanic Niño Index (ONI).

Table 2 and Table 3 show the inverse of the intensity and duration respectively for ENSO phases, climate seasons and combinations between them.

Table 4 shows the precipitation of the mean storm, calculated as the multiplication of the inverse of β and δ.

The wet season storms have longer durations but less intensities, while dry season storms are more intense but much shorter.

Table 2. Inverse of the mean intensity β [h/mm].

	El Niño	Normal	La Niña	General
Dry	0.44	0.41	0.36	0.40
Wet	0.41	0.43	0.36	0.41
General	0.42	0.43	0.36	**0.40**

Table 3. Inverse of the mean duration δ [h^{-1}].

	El Niño	Normal	La Niña	General
Dry	1.44	1.51	1.47	1.48
Wet	1.42	1.39	1.29	1.37
General	1.43	1.43	1.36	**1.41**

Table 4. Mean precipitation P [mm].

	El Niño	Normal	La Niña	General
Dry	1.60	1.62	1.89	1.70
Wet	1.72	1.66	2.14	1.79
General	1.68	1.65	2.03	**1.76**

Storms during La Niña phase are significantly more intense and lasting, and storms during El Niño phase, even though mean duration is the same as Normal phase, the mean intensity increases a little.

Storms in periods where dry season and La Niña phase concur are more intense, and the storms when dry season and El Niño phase concur are less intense, but rainfalls last more than those on the same season during La Niña and Normal phases.

During wet season, there are only high intensities when it concurs with La Niña phase, which also has the longest durations. When it concurs with El Niño and Normal phases, there is only an increase of duration.

During wet season, the average precipitation (P) is higher than during dry season. During cold phase of ENSO, P increases on both weather seasons, while on warm phase only increases when it concurs with wet season, because the storms tend to intensify.

The frequency curves of infiltrated water volume (F), wetting front depth (z_w) and factor of safety (FS) for different seasonal scenarios were calculated. Figure 4 shows the curves for dry, wet, and multi-annual seasons. The return period (Tr) for the three variables during wet and dry season is less than the average regime, as they are related to the probability that certain value of FS occurs and at the same time a dry or wet season occur.

The return period for FS equal to 1 is higher for wet season than for dry season, indicating that according to the models, the more intense rainfall events, although shorter, have the tendency to trigger more superficial landslides than those with longer duration and less intensity. This result was obtained under the assumption that the initial soil moisture is the same before each storm.

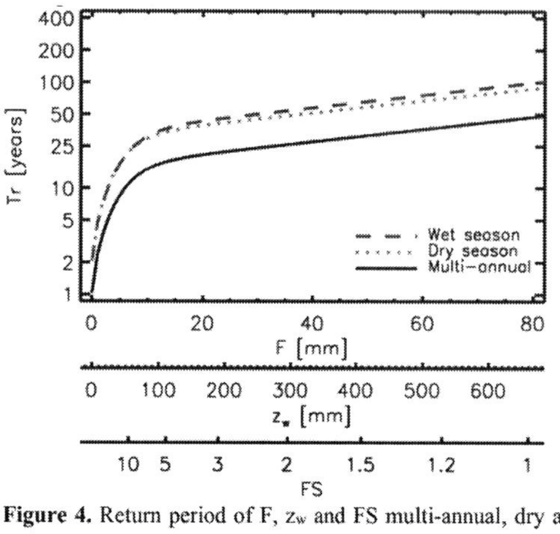

Figure 4. Return period of F, z_w and FS multi-annual, dry and wet seasons.

Figure 5 shows the return period of F, z_w and FS for the three phases of ENSO. Tr values on Normal phase are less than those of El Niño and La Niña, since most of the time it is on Normal phase. FS equal to 1, on average, occurs more frequently during La Niña than during El Niño, because during La Niña more rainfall events with longer durations and higher intensity occur.

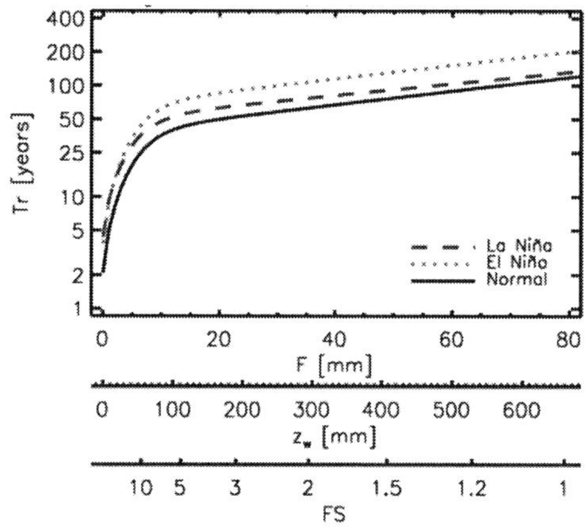

Figure 5. Return period of F, z_w and FS for the ENSO phases.

Figure 6 shows the return period for the seasons in which ENSO phases concur with the different rainfall seasons. During La Niña on wet and dry season, the return periods are very similar because the mean intensities of the rainfall events are very alike. The return period for El Niño during wet season is less than during wet season, since the intensities are higher. The same happens on the Normal periods.

The higher return period occur during El Niño for both rainfall seasons, and the lower for the Normal phase as they are the most probable situation.

4 Conclusions

The stability of a hypothetical slope has been analyzed under different climate scenarios, considering that probabilistic behavior of factor of safety is given by rainfall, while the other parameters involved are deterministic.

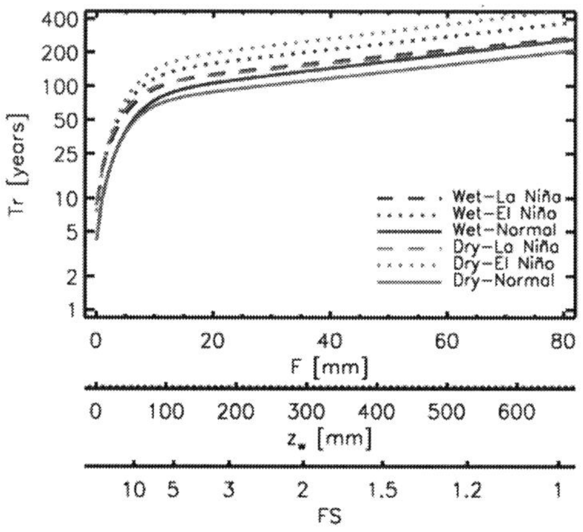

Figure 6. Return period of F, zw and FS for ENSO phases during wet and dry seasons.

The superficial landslides occurrence depends strongly on storm intensity more than on storm duration; therefore, it is more probable landslides occur during dry seasons, where storms tend to intensify.

During La Niña phase it is always more likely that landslides occur than during El Niño because storm durations and intensities tend to increase.

In the long term more landslides occur during Normal phase of ENSO than during El Niño and La Niña. This effect occurs because Normal phase is the most frequent.

These conclusions are valid only for slopes similar to the slope analyzed, and under climatic conditions such as the pluviometric station considered. Furthermore, the different models considered assume important simplifications. The Philip's infiltration model is a simplified solution to the Richard's equation in one dimension, so it does not consider lateral flows. Also, it assumes hydraulic conductivity and sorptivity coefficients constant in time and space. The wetting front model assumes a homogeneous and isotropic soil, and it reaches saturation to the depth z_w. In reality, moisture decreases progressively with depth. Additionally, the initial and final moisture must be known. The slope stability model assumes an infinite slope and a failure surface parallel to the ground surface, located at the same depth as z_w.

On the other hand, the simulation does not consider the effect of vegetation, heat water losses and the water table influence.

Acknowledgements

The authors thank COLCIENCIAS (Colombia) for their financial support.

References

1. É. Aristizábal, J. Gómez, E. Aristizábal, and J.

Gómez, Gestión Y Ambient. **10**, 17 (2007).
2. T. W. J. Van Asch, L. P. H. Van Beek, and T. A. Bogaard, in *1st Ital. Work. Landslides. Rainfall-Induced Landslides Mech. Monit. Tech. Nowcasting Model. Early-Warning Syst.*, edited by L. Picarelli, P. Tommasi, G. Urciuoli, and P. Versace (Naples, 2009).
3. L. Chen and M. H. Young, Water Resour. Res. **42**, 1 (2006).
4. S. E. Cho and S. R. Lee, Comput. Geotech. **28**, 185 (2001).
5. S. E. Cho, Eng. Geol. **105**, 32 (2009).
6. I. Egeli and H. F. Pulat, Sci. Iran. **18**, 1179 (2011).
7. S. Oh and N. Lu, Eng. Geol. **184**, 96 (2015).
8. P. Frattini, G. Crosta, and R. Sosio, Hydrol. Process. **23**, 1444 (2009).
9. T. V. Hromadka II, T. V. Hromadka III, and M. Phillips, Environ. Eng. Geosci. **XVI**, 343 (2010).
10. P. Tarolli, M. Borga, K.-T. Chang, and S.-H. Chiang, Geomorphology **133**, 199 (2011).
11. M. T. J. Terlien, Environ. Geol. **35**, 124 (1998).
12. M. N. Papa, V. Medina, F. Ciervo, and a. Bateman, Hydrol. Earth Syst. Sci. **17**, 4095 (2013).
13. B. D. Malamud, D. L. Turcotte, F. Guzzetti, and P. Reichenbach, Earth Surf. Process. Landforms **29**, 687 (2004).
14. J. F. Tiga, Análise Espacial E Estatística Dos Movimentos de Massa Deflagrados Pelas Chuvas Dos Dias 11 E 12 de Janeiro de 2011 Na Região Serrana Do Estado Do Rio de Janeiro, Brasil, Master, 2013.
15. M. Alvioli, F. Guzzetti, and M. Rossi, Geomorphology (2014).
16. F. C. Dai and C. F. Lee, Eng. Geol. **59**, 253 (2001).
17. P. S. Eagleson, Water Resour. Res. **14**, 741 (1978).
18. N. K. Goel, R. S. Kurothe, B. S. Mathur, and R. M. Vogel, J. Hydrol. **228**, 56 (2000).
19. P. S. Eagleson, Water Resour. Manag. **8**, 878 (1972).
20. J. R. Philip, Soil Sci. **83**, 345 (1957).
21. P. Lumb, Q. J. Eng. Geol. Hydrogeol. **8**, 31 (1975).
22. G. Poveda and O. Mesa, Meteorol. Colomb. **5**, 3 (2002).

 DOI: 10.1051/e3sconf/20160915013

Numerical simulation of rainfall tests on small-scale slope model

Fakher Hamrouni[1,a], Houcem Trabelsi[1] and Mehrez Jamei[1]

[1] National Engineering School of Tunis, Civil Engineering Department, Tunis El Manar University, Tunisia

Abstract. Rainfall-induced landslides are becoming ever more frequent in many regions in the world. In general, the slope failures occur during or immediately after the rainfall. In Tunisia, several regions are concerned by deep or shallow landslides induced by rainwater infiltration, particular in Beja (Tibar). However, due to the high cost of field tests, the quantification of rainwater effects still not adequately clarified. Different small scale slope models were used to study landslide as a result of change in saturation degree. In this paper, we present numerical simulation of rainfall tests on small scale model used clayey soil from a landslide site in Beja, Tibar. The aim of this paper is to study landslide in small scale clayey slope used the same characteristics and conditions of the experimental model realised by Rolando et al. (2004). For this simulation, we will use the numerical approach used to simulate rainfall tests already developed by Rolando et al. (2004). Numerical simulation is conducted using CODE-BRIGHT software.

1 Introduction

Numerous research studies have been conducted to investigate infiltration of rainwater into a slope and the effect of water infiltration on slope stability. The studies show that the increase of pore-water pressure (increase of water table level) is described as a mainly condition leading to slope failure. In addition, the increase of moisture content decreases the shearing resistance of soils. Indeed, until now there are difficulties to explain several cases of landslides induced by rainfall. In order to study the landslide as a result of change in saturation degree, and to clarify the rainwater infiltration effects, different small scale models slopes were used. The physical laboratory model with reduced dimensions has many advantages, like the identification of failure mechanisms, failure initiation and slope movements. In this part, we will present results of numerical simulation of rainfall test on small scale clayey slope model. Based on the assumption of effective stress, a simply modified Mohr-Coulomb model taking into account the suction has been used. The proposed model takes into account the reduction of cohesion under humidification.

2 Model properties

The numerical models chosen, like Mohr-Coulomb model used to simulate the failure plane in the small scale model, the Van Genuchten model for the retention curve, and Kozeny's model for the intrinsic permeability, help us to make a numerical simulation of a rainfall test on small scale already developed by Rolando et al (2004) that describe perfectly the experimental slope behaviour. The computed results obtained are similar to the experimental results. In the present paper, the objective is to study landslide in the clayey soil by numerical simulation of a rainfall test on reduced model using the same geometric characteristics, initial conditions and boundary conditions of the physical model used in the tests realised by Rolando et al (2004), only the soil used to construct the experimental slope is changed with the fact that the similitude conditions are not discussed in this paper (see Hamrouni et al., 2016).

Table 1. Geometric characteristics of the reduced model

Side	(mm)
Length	1970
Width	800
height	900

The soil properties as shown in Table 2, this soil provided from a landslide site in Beja, (Tibar) Tunisa.

[a] Corresponding author: fakher.hamrouni@gmail.com

 DOI: 10.1051/e3sconf/20160915013

Table 2. Properties of the clay soil

Parameter	Value
Permeability at saturation k (m/s)	10^{-7}
Intrinsic Permeability at saturation K(m^2)	10^{-14}
Specific gravity (gr/cm^3)	2.63
Effective friction angle (°)	10
Effective cohesion (KPa)	4
Plasticity index (%)	32
Liquid limit (%)	64

The experimental slope made of homogeneous layer of soil with thickness h = 30cm that was constructed over an impermeable base, the reduced slope is constructed under an initial water content w = 10%. Figure 1 shows the details of the reduced model.
We will retain the same conditions for the clayey slope.

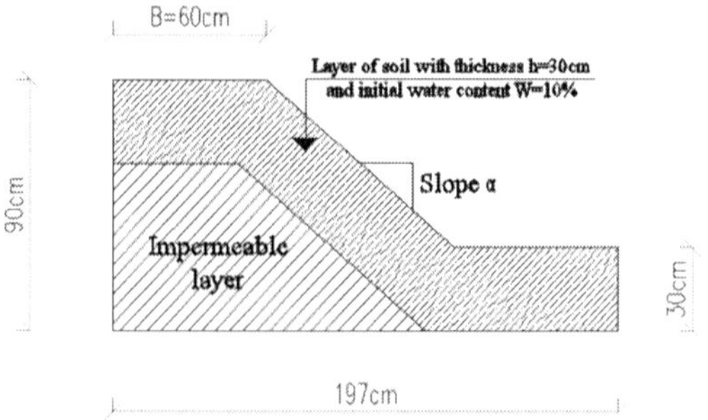

Figure 1. Details of the reduced model used

Hydraulic properties of the marly clay (water retention curve, permeability function, and intrinsic permeability) were investigated.

2.1 Retention curve

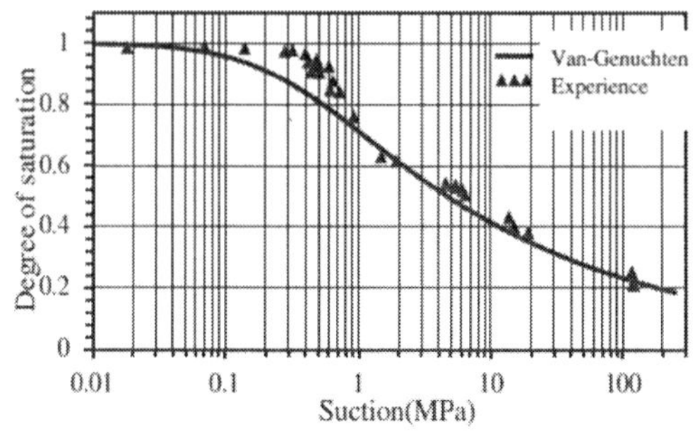

Figure 2. Water retention curve of the clay from Beja area

For the retention curve model, the Van-Genuchten model (1980) equation was calibrated on this experimental data as following:

$$S_e = \frac{S_l - S_{rl}}{S_{ls} - S_{rl}} = \left(1 + \left(\frac{u_g - u_w}{P}\right)^{\frac{1}{1-\lambda}}\right)^{-\lambda} \quad ; \lambda = 0.2;\ p = 0.1MPa$$

This equation contains the parameters λ and P. The first (λ) essentially controls the shape of the curve while the second (P) controls its height.

2.2 Intrinsic permeability

The permeability at saturation of this marly clay is $K = 10^{-7}$ m/s, the intrinsic permeability $k = 10^{-14}\ m^2$.

A simple function of the type of medium and its porosity is usually used like Kozeny's formula:

$$k = k_o \frac{\varphi^3}{(1-\varphi)^2} \frac{(1-\varphi_o)^2}{\varphi_o^3}, k_o = 10^{-14} m^2$$

Where φ_0 is the reference porosity and k_0 is the intrinsic permeability at the reference porosity.

2.3 Relative permeability

The function for calculating liquid relative permeability is a generalized power function of degree of saturation:

$$k_{rl} = AS_e^{\lambda}$$

This power function is independent of the retention curve assumed.

2.4 Mechanical model

In this case, we used the Mohr-Coulomb model to simulate the failure slope in the scale model. This model assumes that the state of stress of unsaturated soils is defined by two independent variables: the net stress and the suction as stress acting on the mechanical parameters, which are used to explain the unsaturated soil behaviour. In this model the relationship between cohesion and suction is simulated as a linear function. The proposed model is written in the context of the elastic-Viscoplasticity.

2.4.1 Elasticity

Linear elasticity with parameters E and ν, E is the Young modulus and ν is the poisson's ratio. For the clay soilt E = 44.52 MPa and ν = 0.3.

2.4.2 Viscoplasticity

The viscoplastic constitutive model:

$$\frac{d\varepsilon^p}{dt} = \Gamma \langle \Phi(F) \rangle \frac{\partial G}{\partial \sigma}$$

Where Γ is the viscosity, F is the yield function, G is the plastic potential (in this case F = G), Φ is the stress function and ε^p is the plastic deformation.
The total deformation ε is defined by the following expression:

$$\frac{d\varepsilon}{dt} = \frac{d\varepsilon^e}{dt} + \frac{d\varepsilon^p}{dt}$$

The expression of the stress function:

$$\Phi(F) = F^m$$

The yield function F is defined by the following expression:

$$F = q - \delta p' - c\beta$$

Where q is the deviatoric stress, c is the cohesion, m is a parameter which depends on the material, δ and β are defined by the following expression:

$$\delta = M = \frac{6\sin\phi'}{3-\sin\phi'} = 0.4 \qquad \beta = \frac{6\cos\phi'}{3-\sin\phi'} = 2.1$$

Where ϕ' is the friction angle, invariants used in the models are defined as:

$$p = \sigma_{oct} = \frac{1}{3}I_1 = \frac{1}{3}(\sigma_x + \sigma_y + \sigma_z)$$

$$q = \frac{3}{\sqrt{2}}\tau_{oct} = \frac{1}{\sqrt{2}}\sqrt{(\sigma_x-\sigma_y)^2 + (\sigma_y-\sigma_z)^2 + (\sigma_z-\sigma_x)^2 + 6(\tau_{xy}^2+\tau_{yz}^2+\tau_{zx}^2)}$$

3 Numerical simulation programs

A finite element (FE) analysis was carried to analyse the scale model tests. The numerical program is outlined in Table 3.

In the numerical modelling, the effect of temperature is not taken into account (isothermal conditions).

Table 3. Numerical program.

Case No.	Type of test	Material	Remarks
1	Rainfall	Silty sand	α = 40°,D_r=50%, W=10%, s=34.5 KPa , r = 42-72 mm/h, and β^* = 1
2	Rainfall	Clay	α=40°,D_r=50%, W=10%, s=51.402 MPa, r = 42-72 mm/h, and β^* = 0.7

β^* is the percentage of rain that infiltrates into the soil.

3.1 The obtained results

The numerical simulation of test 1 have been achieved, the idea was to simulate a rainfall test on small-scale model already developed by Rolando et al. 2004, in order to validate the proposed numerical models chosen. The numerical results are similar to the experimental results. The numerical model proposed perfectly describes the behavior of the experimental slope model.

The main experimental and numerical results of test1 (sandy slope),

- Test results showed that the slope failure was induced when the soil moisture content near the toe of the slope reached nearly full saturation.

- The failure process in terms of the "localized" plastic deformation, involved only the superficial portion of the slope (shallow landslide).

- The development of water table near the toe of the slope and the wetting front also moved upward, causing an increase in the volumetric water content near the surface soil at the slope toe.

The increase of the pore-water pressure due to the increase of water table level in the region near the toe of the slope is described as a mainly condition leading to slope failure in this case (for the sandy slope). The high permeability of this silty sand (K = 1.8 10^{-4} m/s) increase the rate of infiltration of the rainwater into the slope and increase the saturation degree in the slope which will lead to the decrease of the cohesion (apparent cohesion) and the decrease of the shear strength of the soil.

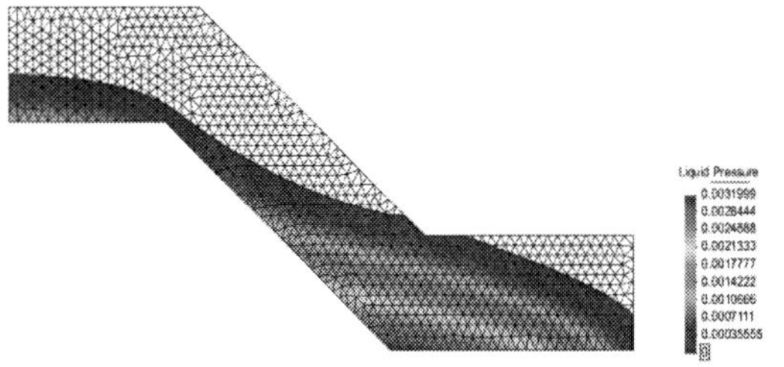

Figure 3. Numerical result of distribution of saturated zone at the time of failure (silty sand)

DOI: 10.1051/e3sconf/20160915013

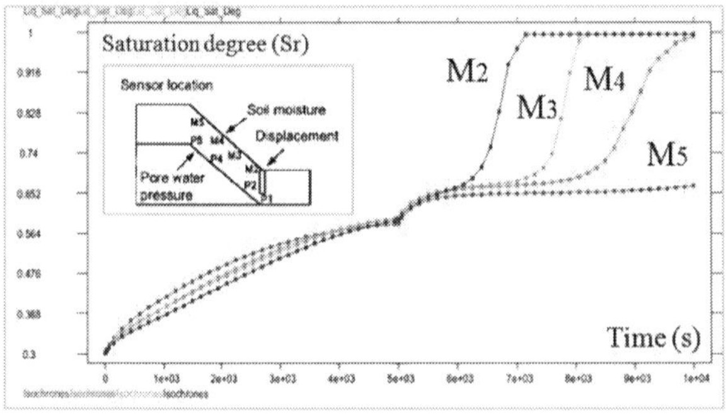

Figure 4. Time histories of saturation degree

In this paper, we will present numerical results of the rainfall test used the same model characteristics of the case 1 (reference case), the only difference is the material used to construct the slope (clay soil). This is in order to study the landslide in clayey soil, induced by rainfall. The main objective is to clarify the main parameters that cause landslide in this lower permeable soil.

3.2 Numerical simulation of test 2

The average size of the finite element mesh chosen for the simulation was 0.028 m and the number of triangular elements was 1839. The simulations are done according to the two following assumptions: small deformation and plane strain condition.

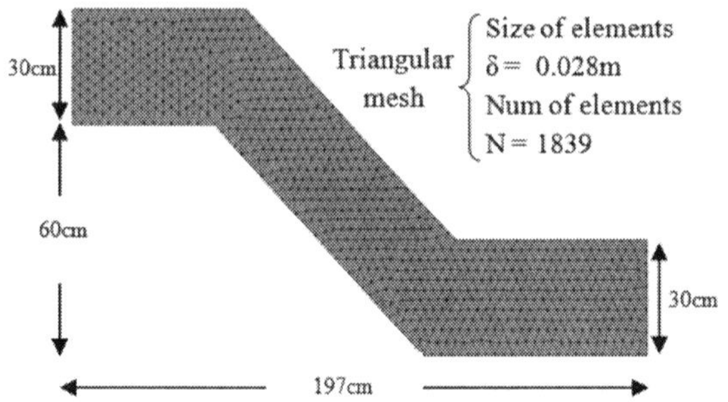

Figure 5. Mesh elements chosen for the simulation

4 Numerical results of rainfall test n°2

The boundary conditions and numerical modelling procedures adopted are identical to those used in the test1 (reference case). In this part, we will present results of numerical simulation of failure initiation in a reduced model of a clayey slope.

Numerical results: the time required for the slope failure in case 2 (clayey slope) was longer than in case 1 (silty sand) because of low permeability and the strong cohesion of the clay (c' = 4KPa).

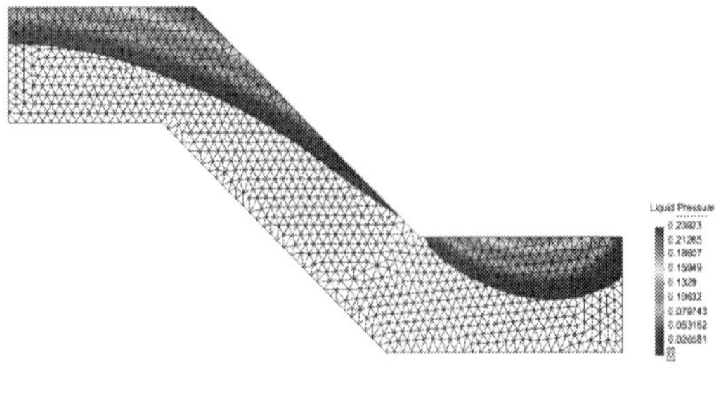

Figure 6. Numerical result of distribution of saturated zone at the time of failure (clay soil)

The low permeability of clay soil ($K = 10^{-14}$ m^2) does not allow rainwater to seep into the slope. The numerical results show that the water accumulates in the superficial layer with an increase of pore-water pressure in this region.

In this case, the decrease of cohesion and effective stress due to the increase of pore-water pressure ($u_w = 0.32$MPa) in the superficial layer is described as a mainly condition leading to slope failure (superficial portion).

Many slope failures occur during or after the rainfall in the Beja region (Tunisia), especially in a plastic clay soil. Stability problems are caused by soil saturation and the presence of cracks which are developed after along dry summer. The presence of cracks in the superficial layer (topsoil) increases the permeability of the clay soil.

In the aim to attain conditions near the real slope conditions, we simulate that the superficial layer (h = 10cm in reduced model) is a permeable layer (case of a cracked layer) by increasing the permeability.

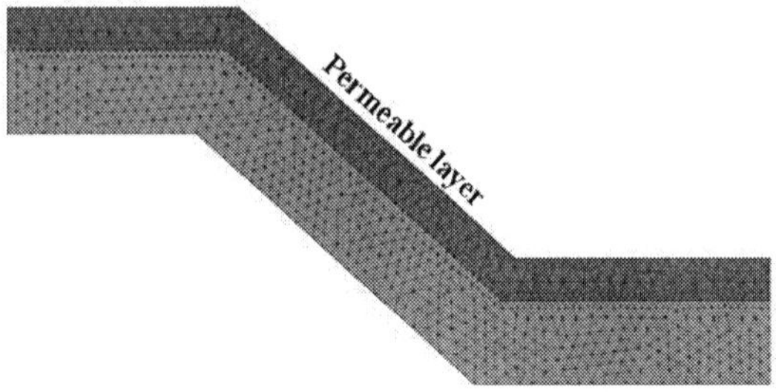

Figure 7. The chosen model for the simulation

In this part, we will study the effect of superficial layer (topsoil) on slope stability; the Table 4 shows the numerical program for this test.

Table 4. Numerical program.

Case No.	Type of test	Material	Remarks
3	Rainfall	Clay	α=40°, D_r=50%, W=10% et r = 42-72 mm/h, s=51.402 MPa β^*= 0.7, K_1 =10^{-11} and K_2=10^{-14}

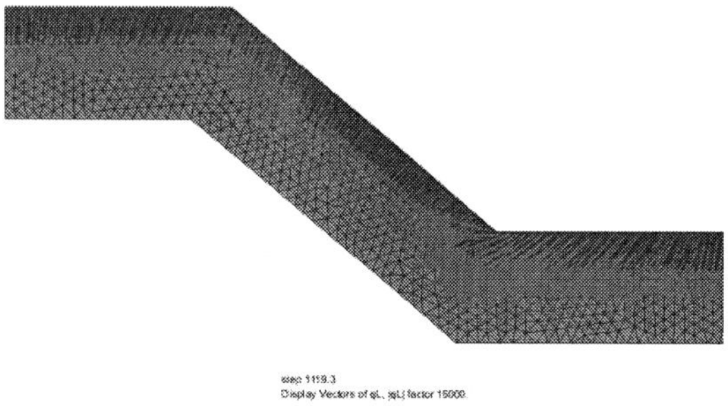

Figure 8. Water flow vectors at the start of rainfall

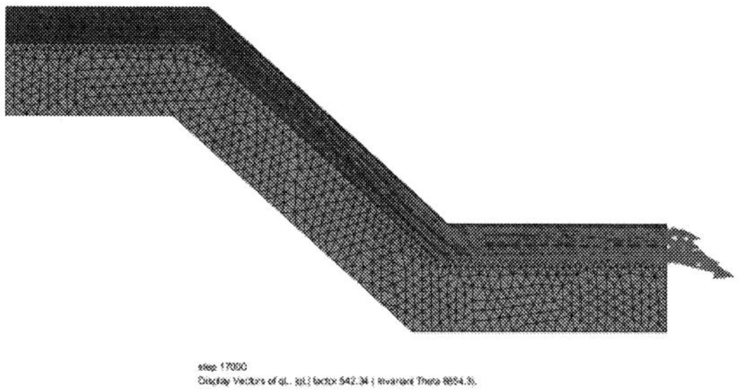

Figure 9. Water flow vectors at the end of rainfall

The clayey layer with low permeability represents an impermeable layer for this small scale slope model. The lowest layer requires a lot of time for it begins saturation. The above figure shows the direction of water flow.

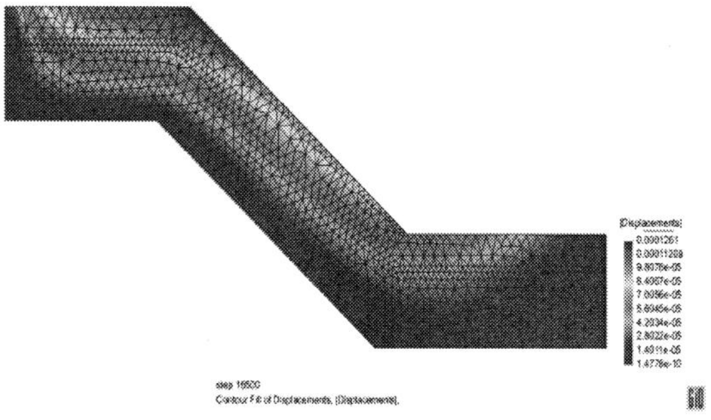

Figure 10. Numerical result of displacement

The failure process involved only the superficial portion of the slope as it's noted above.

5 Conclusion

Rainfall test on small scale slope model can help us to find an explanation of the phenomena of rainfall induced landslide especially for the clayey soil. Numerical simulation of the rainfall test on reduced model already developed by Rolando et al (2004), for the case of sandy slope allows us to validate the proposed numerical approach in the first part and to clarify the main parameters that cause landslide induced by rainwater infiltration in the both cases: the sandy slope (c'= 0 MPa and K = 1.8 10^{-4} m/s) and for a plastic clayey soil, in the second part. In this paper, we have not presented the numerical results of this test (test No 9 R.P.Orense 2004). The validation of the numerical approach was already conducted and published (see Hamrouni et al., 2015). The idea is here to use the numerical approach to simulate landslide in clayey slope using the same geometric characteristics of the reduced model. Some results obtained by numerical simulations of rainfall tests across a clayey soil in the scale model:

- In the case of homogenous sloping (test 2), landslide occur when a water table developed in the superficial portion of the slope and the pore-water pressure increases in this region. The increase of water pressure also the saturation degree in this part due to the low permeability of the clay (k = 10^{-14} m^2).

- In order to reach conditions near the real slope conditions, we will simulate that the superficial layer is a cracked layer by increasing the permeability (the value is shown according some laboratory œdometric tests on fissured clay, see Louati et al., 2015).

- Numerical result of test 3 shows that the development of cracks in the slope increases the permeability of soil and the rainwater infiltration into the slope. So it leads to increase of the moisture content and pore-water pressure in the slope and the decrease of shear strength of soil which leads to the slope failure

References

1. Rolando P.Orense, Suguru Shimoma, Kengo Maeda and Ikuo Towhata, "Instrumented Model Slope Failure due to Water Seepage, "Journal of Natural Disaster Science, vol. 26,pp. 15-26, 2004
2. Suguru Shimoma, Rolando P.Orense, Tsuyoshi Honda, Kengo Maeda and Ikuo Towhata, "
3. Trabelsi, H, Jamei, M, Zenzri, H, and Olivella, S, (2012). Crack patterns in clayey soils Experiments and modeling. International Journal for Numerical and Analytical methods in geomechanics, 36(11), 1410-1433.
4. Lyesse Laloui, Mathieu Nuth," On the use of the Generalized effective stress in the constitutive Modeling of unsaturated soils,"Computers and Geotechnics, vol. 36, pp. 20-23, 2009.
5. Chales W. W. Ng, Jian Liu, Rui Chen, and Jie Xu, " Physical and numerical modeling of an inclined three-layer (silty/gravelly sand/ clay) capillary barrier cover system under extreme rainfall, "waste Management, 2015.

6. M. Jamei, H. Guiras, and S. olivella," Analyse of slope movement initition induced by rainfall using the Elastoplastic Barcelone Basic Model," European journal of environnemental and civil Engineering. 2015.
7. Hong-Hu Zhu, Bin Shi, Jun-Fan Yan, Jie Zhang, and Jing Wang,"Investigation of the evolutionary process of a reinforced model slope using a fiber-optic monitoring network,"Engineering, Geology, vol. 186, pp.34-43, 2015.
8. F. Louati, H. Trabelsi, and M. Jamei, "Experimental and Numerical Investigation about Permeability of Fissured Clay", UNSAT, BATNA, 16-17 November 2015, pp. 161-166.
9. F. Hamrouni, H. Trabelsi, L. El Ghezal, and M. Jamei, "Numerical Simulation of Rainfall Test on Small-Scale model", UNSAT, BATNA, 16-17 November 2015, pp. 138-146.
10. F. Hamrouni, M. Jamei, H. Trabelsi, and L. Elghezal, "Numerical back analysis to optimize the laboratory scale model", Proposed for the 12th International Symposium on landslides, ISL Napoli, Italy, 12-19 June 2016.

DOI: 10.1051/e3sconf/20160916001

Simulation of water movement and its distribution in a soil column under a water source using pore - scale network modelling

Saeed Samadianfard[1,a], Amir Hossein Nazemi[1] and Ali Ashraf Sadraddini[1]

[1] Water Engineering Department, Faculty of Agriculture, University of Tabriz, Tabriz, Iran

Abstract. The relationship between capillary pressure and saturation has a remarkable value in investigating interactions of two immiscible fluids in porous media. Pore network models, as predictors of fluids movement in porous media, can play undeniable role in determining the mentioned relationship. In the current study, the results of numerical pore network modelling, which represents pore structure of a real porous medium with networks of pore bodies connected with pore throats, are used for computing the macroscopic relationship between capillary pressure and saturation. A notable volume of soil is influenced by water source and according to the results of previous studies, it seems practically impossible to simulate water movement in macro scale dimensions by use of pore scale models. Meanwhile a suitable solution was found in this study, by considering a thin vertical soil column under the water source which was divided to smaller volumes as cells, by horizontal crossings. Each cell was considered as a pore network unit, so the soil column was consisted of series of pore network units which were vertically jointed to each other. The moisture distribution and also wetting front movement in the column were determined by application of pore network model, using the dynamic update saturation solution method. The solution was conducted between each pair of consequent cells in an alternative manner within a time step since arrival time of the water from an upper cell to the lower one. Moreover, for evaluation of the model ability, soil moisture profiles in a sandy soil of an experimental tank under the water source were studied. Comparison of the simulation and observation data confirmed the high ability of the column pore network model in prediction of the moisture distribution and wetting front movement in a soil column.

1 Introduction

The vadose zone of soil is a fundamental segment of the hydrological cycle, straightforwardly influencing infiltration, storm runoff, evapotranspiration and aquifer recharge. Understanding the way of water movement in the vadose zone and the measurement of its physical characteristics are the key factors to solving a variety of engineering issues. Examples of such issues are: forecasting runoff from given precipitation events for erosion control; sediment transport and flood control; estimation of the water accessibility for plant growth and the amount of water recharge to the aquifer. Therefore, the investigation of soil-water movement has intrigued researchers from different disciplines, e.g. soil science, hydrology, agriculture and environmental sciences for years.

In recent decades, different methods have been presented for prediction of water movement in soils which can be classified in three groups including a) numerical and analytical methods, b) empirical models and c) pore scale models.

Numerical and analytical methods are typically based on Richards' equation. Numerical modeling of Brandt et al. [1] and Taghavi et al. [2] can be categorized in this group. In numerical methods, hydraulic characteristics of soil e.g. residual soil water content, saturated soil water content, saturated hydraulic conductivity, etc have to be measured experimentally before starting the iterative numerical procedure. In other words, performing time consuming experiments are prerequisite of numerical methods.

Another group of methods are empirical models, which have been applied by a lot of researchers using experimental data in specific soil types. Empirical models are derived from data observed either in field or in laboratory. The models of Ben-Asher [3], Schwartzman and Zur [4] and Amin and Ekhmaj [5] etc. fall in the category of empirical models.

The third group consists of network models like pore network models. In order to properly represent the soil, they need to have the same void ratio and grain size distribution as the soil. Therefore, macro scale constitutive parameters and coefficients can be obtained by pore network models using basic soil properties. In the mentioned pore network models, soil moisture profile can be simulated by defining the pore distributions of specified soil type. Pore network models have been used for studying solute transport in porous media and petroleum engineering. The movements of fluids such as

[a] Corresponding author: s.samadian@tabrizu.ac.ir

oil, water and gas in porous media have been the most interesting issues for researchers. Also, pore network models have been used for investigation of single and multi-phase flow processes [6,7], investigation of relationship between capillary pressure, saturation and interfacial area [8-11] and drainage and imbibition [12]. So, applying pore network models for investigating infiltration phenomenon from water source improved the insight to the water movement in pore structure of soil.

In spite of increasing use of new and innovative pore scale models such as pore network models, lattice boltzmann models and smoothed particle hydrodynamic approach in simulating fluid flows in porous media, these models are still less known in water and soil sciences. Although pore network models have innovative characteristics and researchers have used them for investigation of fluid flow in porous media, these models have been applied on pore scale and till now they have not been used for simulation of soil moisture profile in macro scale directly.

A pore network model for representing a porous medium such as soil usually has a volume of few cubic centimeters or less. The network with these dimensions has dozens of pore bodies and pore throats and solving fluid flow governing hydraulic equations in this pore network model demands few days to be fulfilled by new modern computers. So, practically it seems impossible to simulate fluid flow in soil in macro scale using one single pore network model. In the current research, a notable idea is presented for joining a series of pore networks to each other, and for this purpose as a case study, the vertical movement of water in a sandy soil under a water source is investigated.

2 Materials and Methods

2.1 Pore Network Model

2.1.1 Network topology

Pore network models can be classified in two groups: structured and unstructured. Up to now, most pore network models had a two-dimensional or three-dimensional lattice with uniform dividing. In such a network, pore bodies are placed at the lattice nodes, which are all similarly divided, and pore throats are lined up along the grid coordinates. The quantity of pore throats associated with a pore body is called coordination number.

2.1.2 Network geometry

Network geometry is identified with the geometrical state of pore bodies and cross sections of pore throats. Most researchers used pore bodies of cubic or spherical shapes; Joekar-Niasar and Hassanizadeh [13] introduced the pore bodies as truncated octahedrons and Ryazanov et al. [14] utilized n-cornered star-shaped pores in their researches.

2.1.3 Computational algorithms

Continuum-scale formulations of two-phase flow in porous media are typically solved for the pressure and saturation of wetting or nonwetting phases. Therefore, the accompanying system of equations for an inflexible porous medium and incompressible immiscible nonreactive fluids ought to be solved:

$$\varphi \frac{\partial S^{\alpha}}{\partial t} = -\nabla.\mathbf{V}^{\alpha}, \qquad \alpha = n, w \tag{1}$$

$$\mathbf{V}^{\alpha} = -\frac{1}{\mu^{\alpha}} k^{\alpha}\mathbf{K}\nabla P^{\alpha}, \qquad \alpha = n, w \tag{2}$$

$$S^{w} + S^{n} = 1 \tag{3}$$

$$P^{c} = P^{n} - P^{w} = f(S^{w}) \tag{4}$$

where, superscripts n and w denote nonwetting and wetting phases, respectively, φ is the porosity, μ is the viscosity of phase α, S is saturation of phase α, u is the velocity of phase α, K is the intrinsic permeability tensor, k is the relative permeability of phase α, P is the pressure of phase α and P^c is the capillary pressure.

In this study, we have used a pore network model which has been proposed by Joekar-Niasar et al. [11]. The pore network model has a three-dimensional regular lattice structure with constant coordination number of six. Pore bodies have cubic shape and pore throats have square cross sections. Figure 1 shows a schematic presentation of two pore bodies which are connected with a pore throat.

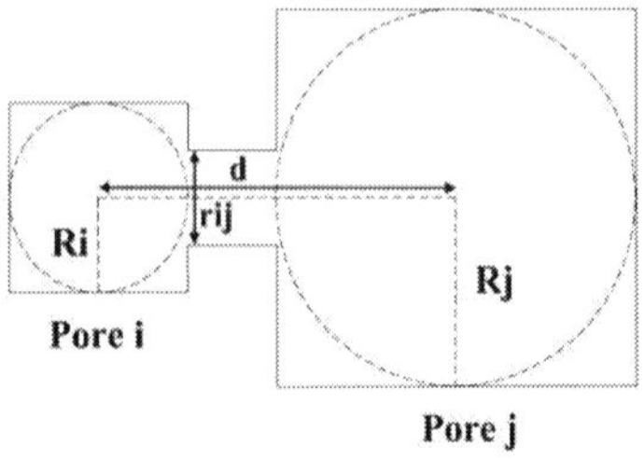

Figure 1. Schematic presentation of two pore bodies and the connected pore throats.

2.2 Numerical solution of Richards' equation using finite difference scheme

The Richards' equation is the most general method to compute soil moistures and hydrological fluxes, such as infiltration in porous media [15]. Consider two-and/or three-dimensional isothermal uniform Darcian flow of water in a variably saturated rigid porous medium and assume that the air phase plays an insignificant role in the liquid flow process. The governing flow equation for these conditions is given by the following modified form of the Richards' equation [16]:

$$C\frac{\partial h}{\partial t} = \frac{1}{r}\frac{\partial}{\partial r}\left(rK(h)\frac{\partial h}{\partial r}\right) + \frac{\partial}{\partial z}\left(K(h)\frac{\partial h}{\partial z}\right) - \frac{\partial K}{\partial z} - S(r,z,t,h) \tag{5}$$

where C is the specific capacity of water, h is the pressure head, r, z are radial and vertical directions, t is time, S is sink term and K denotes the unsaturated hydraulic conductivity function.

The general form of Richards' equation after discretization, linearization and simplification can be expressed as [17]:

$$C^n_{i,j}\frac{h^{n+1/2}_{i,j}-h^n_{i,j}}{\Delta z_i}=$$
$$\frac{K^{n+1/2}_{i+1/2,j}\left(h^{n+1/2}_{i+1,j}-h^{n+1/2}_{i,j}\right)-K^{n+1/2}_{i-1/2,j}\left(h^{n+1/2}_{i,j}-h^{n+1/2}_{i-1,j}\right)}{(\Delta r)^2}+ \quad (6)$$
$$\frac{K^n_{i,j+1/2}\left(h^n_{i,j+1}-h^n_{i,j}\right)-K^n_{i,j-1/2}\left(h^n_{i,j}-h^n_{i,j-1}\right)}{(\Delta z)^2}+$$
$$\frac{K^{n+1/2}_{i,j}}{i\Delta\Delta}\frac{\left(h^{n+1/2}_{i+1,j}-h^{n+1/2}_{i-1,j}\right)}{2\Delta\Delta}-\frac{\left(K^n_{i,j+1}-K^n_{i,j-1}\right)}{2\Delta\Delta}-S^n_{i,j}$$

In Eq. 6, superscript n refers to the current time step and superscript n+1/2 denotes the arithmetic mean of a parameter at time steps n and n+1. Furthermore, unsaturated hydraulic conductivity function (K) is defined by:

$$K=K_sS_e^{\,l}\left[1-\left(1-S_e^{1/m}\right)^m\right]^2 \quad (7)$$

where K_s is the saturated hydraulic conductivity, S_e is the effective saturation, m and l are shape parameters. For K, we take the geometrical mean as proposed by Vauclin et al. [18].

$$K_{i+1/2,j}=\sqrt{K_{i+1,j}\times K_{i,j}} \quad (8)$$

$$K_{i-1/2,j}=\sqrt{K_{i-1,j}\times K_{i,j}} \quad (9)$$

It should be noted that a third-type (Cauchy type) boundary condition is used to prescribe the water flux from water source in the soil surface and constant water content has been used as initial condition in numerical scheme [19].

2.3 Laboratory experiments

For evaluating the accuracy of the proposed model, experiments of water infiltration under surface water source were conducted on a sandy soil (90% sand, 5% silt and 5% clay). Laboratory experiments were carried out using a 120 cm × 120 cm × 120 cm transparent Plexiglas box (as shown in Fig. 2). Air dried sand with a mean particle size d_{50} = 0.4 mm was compacted at predetermined dry bulk density of 1.5 g.cm^{-3}. A polyethylene pipeline connected to water reservoir was laid on the soil surface which had a 16 mm outside diameter, a wall thickness of 2 mm and used for supplying water on top surface of the soil column. During operation, wetted soil dimensions were measured using high performance photography and analyzed by Digimizer Software. The observed soil moisture profile had a high degree of the horizontal symmetry and the flow in the centerline of this profile was considered as a pipe with only one dimensional vertical flow. So, the main aim of the experiments was to measure the depth of the wetted soil in the centerline in accordance with the location of the proposed column pore network model for simulation of vertical flow.

Figure 2. A sample of sandy soil moisture profile

2.4 Evaluation parameters

Several statistical criteria can be considered for evaluating the estimated depth of wetted soil. In this study the following criteria were used: correlation coefficient (R), mean absolute error (MAE), root mean squared error (RMSE) and index of agreement (IA).

$$R=\frac{\left(\sum_{i=1}^n x_iy_i-\frac{1}{n}\sum_{i=1}^n x_i\sum_{i=1}^n y_i\right)}{\left(\sum_{i=1}^n x_i^{\,2}-\frac{1}{n}\left(\sum_{i=1}^n x_i\right)^2\right)\left(\sum_{i=1}^n y_i^{\,2}-\frac{1}{n}\left(\sum_{i=1}^n y_i\right)^2\right)} \quad (10)$$

$$MAE=\frac{1}{n}\sum_{i=1}^n\left|x_i-y_i\right| \quad (11)$$

$$RMSE=\sqrt{\frac{1}{n}\sum_{i=1}^n(x_i-y_i)^2} \quad (12)$$

$$IA=1-\frac{\sum_{i=1}^n(x_i-y_i)^2}{\sum_{i=1}^n\left(\left|x_i-\frac{1}{n}\left(\sum_{i=1}^n y_i\right)\right|+\left|y_i-\frac{1}{n}\left(\sum_{i=1}^n y_i\right)\right|\right)^2} \quad (13)$$

where x_i is predicted distance from water source (cm), y_i is observed distance from water source (cm) and n denotes the number of values.

3 Results and discussion

3.1 Simulation of time dependent vertical water movement using column pore network model (CPNM)

A considerable volume of the soil is influenced by a water source and practically it seems impossible to simulate water movement in macro scale using pore scale models. So, a notable solution is proposed for simulating vertical flow in the current research and it is based on dividing a considered thin column in the center part of the volume using cells to smaller volumes, in which pore network model can be applied for each cell. In other words, the volume of the mentioned soil column consists of severed pore network models which connect to each other in vertical direction. The upper surface of each pore network model is considered as inflow boundary condition and the lower surface as drainage boundary condition.

In this research, the size distribution of pore bodies in a porous medium is computed using grain distribution of the experimental soil by considering a truncated log-normal distribution. Then, these pore bodies are arranged randomly in a porous medium. After that, the space between the centers of two adjacent pore bodies in a generated porous medium is considered twice the radial size of the maximum pore body (see Fig. 1). This established pore network size is selected in an iterative manner, so that to be as a representative volume of the studied soil. The selection procedure of this optimum-sized pore network is defined in the following section of the article. To facilitate the explanations in this discussion, the optimum size of pore network is named cell. For better imagination of the column pore network model, the connections of three typical cells, in which each cell has 4 pore bodies in each direction, are illustrated in Fig. 3.

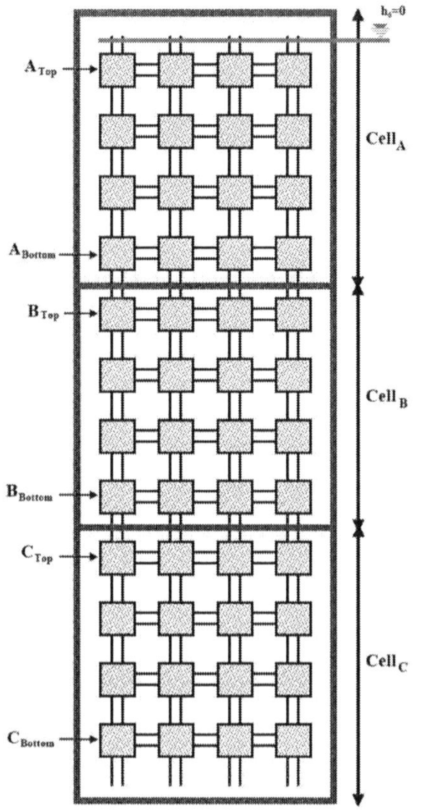

Figure 3. Connection of three typical cells in a vertical column, in which each cell has four pore bodies in each direction

3.2 Simulation of time-dependent soil moisture's vertical profile using pore network model coupled with numerical solution of Richards' equation (PNMRE)

It should be noted that the computation of PNMRE by finite difference scheme were done by an algorithm which has been developed in Wolfram Mathematica 10.0. To attain an optimum pore network size, arrangement of computations was carried out utilizing the same algorithm. For this reason, 6 different networks having 12, 16, 20, 24, 28 and 32 pore bodies in each bearing were considered. These simulations took 4 to 170 hours long with a high speed processing computer.

It was resulted that the capillary pressure - saturation curve changed with the network smaller than the network size of 28×28×28 pore bodies and afterward these changes were negligible for greater networks. Subsequently, the resulted curve from the selected network with 32 pore bodies in each direction was utilized for simulation of soil moisture profile in macro scale. In other words, the resulted pressure-saturation curve was used in solving the Richards' equation by finite difference scheme. The mentioned procedure started with initial pressures of all nodes of the network and the next pressures were computed by equation 6. Then, the correspondent saturations of all nodes were computed by the pressure-saturation curve resulted from pore network modeling. This procedure was continued until two pressure values of all nodes were less than a constant prescribed value.

Finally, by using the prepared algorithms, simulations of the pore network models were done numerically in two different scenarios including CPNM and PNMRE. In the CPNM, the simulations using developed algorithms took 15-20 days on the high speed processing computer. Consequently, distances of the wetting front from soil surface for different time durations from the beginning of the experiment were computed, which are presented in Fig. 4. Also, the corresponding statistical evaluations are given in Table 1.

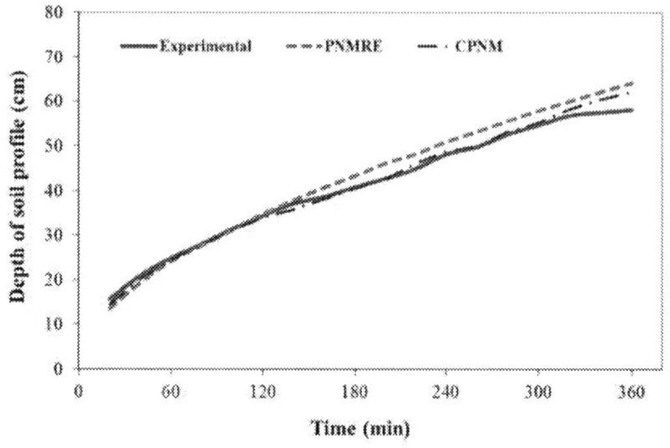

Figure 4. Observed and predicted distances of wetting front from soil surface for different time durations

DOI: 10.1051/e3sconf/20160916001

Table 1. Statistical results of PNMRE and CPNM models in experiments

Model	Statistical criteria			
	R	MAE	RMSE	IA
PNMRE	0.999	2.428	2.863	0.989
CPNM	0.998	0.939	1.362	0.997

As it is clear from Fig. 4 and Table 1, the CPNM by having less root mean squared error and higher index of agreement in comparison with PNMRE has been simulated the depth of soil moisture profile precisely.

Figure 5 shows the simulated water content distributions using PNMRE and CPNM models in different time durations from the beginning of the experiment. Each figure contains comparisons of the water contents simulated by the both mentioned methods in the vertical direction under the water source.

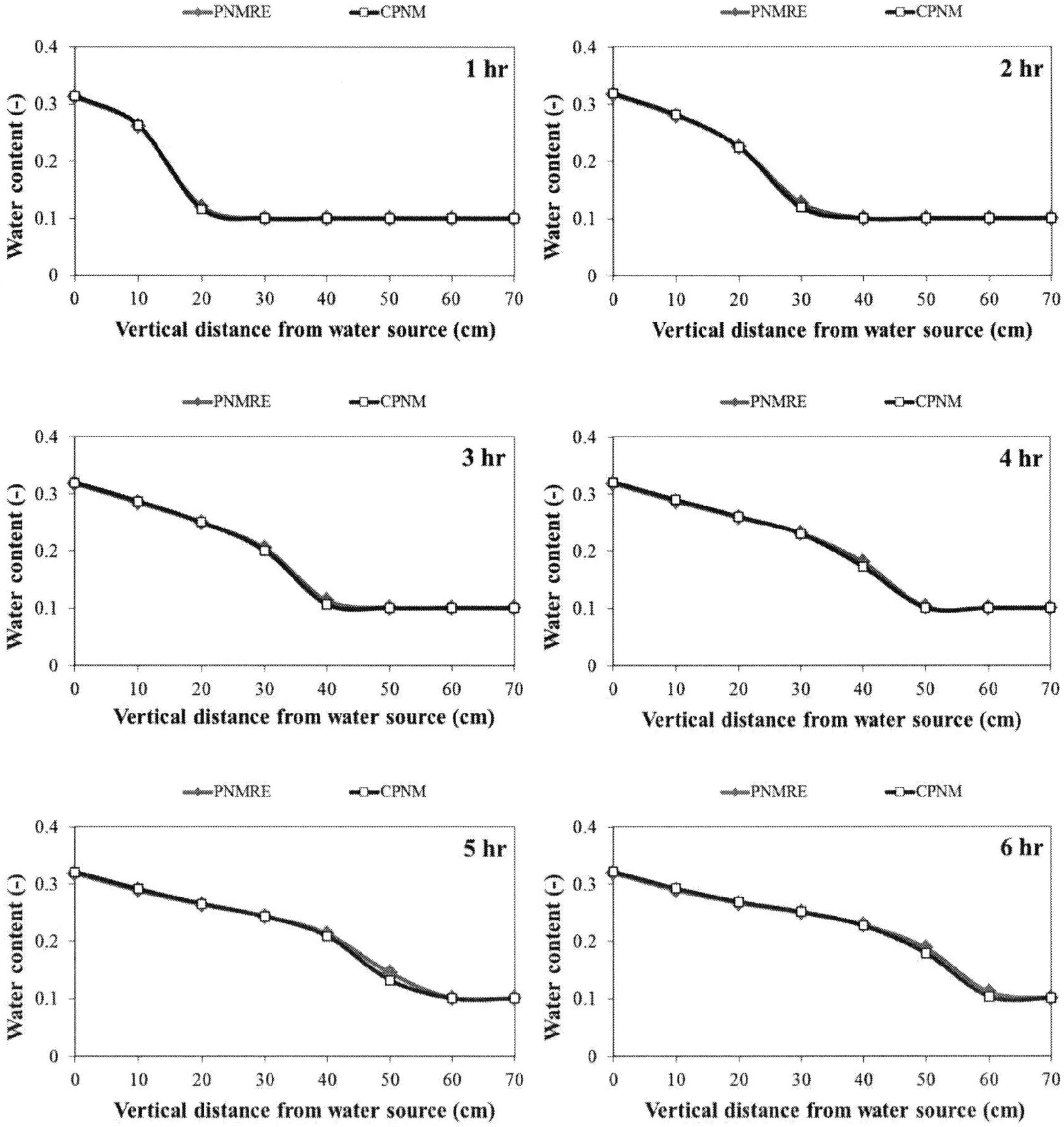

Figure 5. simulated water content distributions using PNMRE and CPNM models for different time durations

It is clear from the plots in Fig. 5 that, in general, the spatial distributions of the water content under the water source for the both models are in excellent agreement. But, since the curves have been plotted in the range of 70 cm in the distance and 0.3 in the water content, the differences (approximately 0.02 in the water content and few centimeters in the distance) between two simulated curves are not clearly visible. However, according to table 1, the CPNM gives precise predictions in comparison with the PNMRE.

4 Conclusion

In the current study, capability of the pore network model in simulating time-dependent moisture vertical movement under the water source has been investigated in two separate parts including the pore network model coupled with numerical solving of Richards' equation and column pore network model. For achieving the optimum size of pore network model, series of computations have been done using the developed algorithms and by considering grain size distribution of the experimental soil. As a result, the network with 32 pore bodies in each direction has been selected as the optimum network for simulation of soil moisture profile in macro scale. In column pore network modeling, the simulations using the developed algorithms took 15-20 days on the high speed processing computer. The results proved that the CPNM in comparison with PNMRE had been simulated the depth of soil moisture profile with higher precision. As a conclusion, it can be stated that the connecting pore network cells and generating column pore network have increased the precision of pore network modeling and proved the efficiency of the presented idea about connecting the pore network models.

References

1. Brandt, A., Breslker, E., Diner, N., Ben-Asher, J., Heller, J. and Goldberg, D. 1971. Infiltration from a trickle source: I. Mathematical models. Soil Science Society of American Journal. 35: 683-689.
2. Taghavi, S.A., Marino Miguel, A. and Rolston, D.E. 1984. Infiltration from trickle-irrigation source. Journal of Irrigation and Drainage Engineering-ASCE. 10: 331-341.
3. Ben-Asher, J. 1986. Infiltration and water extraction from trickle irrigation source: the effective hemisphere model. Soil Science Society of American Journal. 50 (4): 882-887.
4. Schwartzman, M. and Zur, B. 1986. Emitter spacing and geometry of wetted soil volume. Journal of Irrigation and Drainage Engineering. 112(3): 242-253.
5. Amin, M.S.M., Ekhmaj, A.I.M. 2006. DIPAC-drip irrigation water distribution pattern calculator. 7th International micro irrigation congress. September 10–16, Kuala Lumpur, Malaysia.
6. Rajaram, H., Ferrand, L.A. and Celia, M.A. 1997. Prediction of relative permeabilities for unconsolidated soils using pore-scale network models. Water Resources Research. 33: 43-52.
7. Fischer, U. and Celia, M.A. 1999. Prediction of relative and absolute permeabilities for gas and water from soil water retention curves using a pore-scale network model. Water Resources Research. 35: 1089-1100.
8. Reeves, P.C. and Celia, M.A. 1996. A functional relationship between capillary pressure, saturation, and interfacial area as revealed by a pore-scale network model. Water Resources Research. 32: 2345-2358.
9. Joekar-Niasar, V., Hassanizadeh, S.M., Pyrak-Nolte, L. and Berentsen, C. 2009. Simulating drainage and imbibition experiments in a high-porosity micro-model using an unstructured pore-network model. Water Resources Research. 45 (2): 1-15.
10. Joekar-Niasar, V., Prodanovic, M., Wildenschild, D. and Hassanizadeh, S.M. 2010a. Network model investigation of interfacial area, capillary pressure and saturation relationships in granular porous media. Water Resources Research. 46: 1-18.
11. Joekar-Niasar, V., Hassanizadeh, S.M. and Dahle, H.K. 2010b. Non-equilibrium effects in capillarity and interfacial area in two-phase flow: Dynamic pore-network modeling. Journal of Fluid Mechanics. 665: 38-71.
12. Hilfer, R. and Manwart, C. 2001. Permeability and conductivity for reconstruction models of porous media. Physical Review E. 64(021304): 1-4.
13. Joekar-Niasar, V. and Hassanizadeh, S.M. 2011. Effect of fluids properties on non-equilibrium capillarity effects: Dynamic pore-network modeling. International Journal of Multiphase Flow. 37(2): 198-214.
14. Ryazanov, A.V., Van Dijke, M.I.J. and Sorbie, K.S. 2009. Two-phase pore-network modelling: existence of oil layers during water invasion. Transport in Porous Media. 80(1): 79-99.
15. Richards, L.A. 1931. Capillary conduction of liquids in porous mediums. Physics. 1: 318-333.
16. Bear, J. 1988. Dynamics of Fluids in Porous Media, Dover Publications, New York.
17. Besharat, S., Nazemi, A.H. and Sadraddini, A.A. 2010. Parametric modeling of root length density and root water uptake in unsaturated soil. Turkish Journal of Agriculture and Forestry. 34, 439-449.
18. Vauclin, M., Haverkamp, R. and Vauchaud, G. 1989. Résolution de l'équation de l'infiltration de l'eau dans le sol : approches analytiques et numériques. Presses Universitaires de Grenoble, Grenoble, 183 pp.
19. Shan, Y. and Wang, Q. 2012. Simulation of salinity distribution in the overlap zone with double-point-source drip irrigation using HYDRUS-3D. Australian Journal of Crop Science. 6(2), 238-247.

On the hydrological properties of mountain soils, from measurement to the geotechnical implications

Stefano Barontini[1], Giovanna Grossi[1], Marco Peli[1], Roberto Ranzi[1] and Baldassare Bacchi[1]

[1]*Università degli Studi di Brescia, DICATAM, Via Branze, 43, I-25123, Brescia, Italy*

Abstract. Aiming at contributing with a hydrological perspective to the geotechnical investigation in mountain environments, particularly focusing on landslides triggered by perched water tables, we present some findings of a longlasting experimental and theoretical investigation of mountain—soils hydrology. After recalling a theoretical framework suitable to describe the hydrology of shallow, sloping and heterogeneous soils, we discuss some relevant difficulties concerning the measurement of the hydrological properties of heterogeneous and non mature soils, and we finally focus on the role played by the soil heterogeneity in the perched water tables onset.

1 Introduction

Mountain soils are often young and non mature as the gravity—driven mass movement is one of the major soil forming factors [1]. They are moreover seat of both freezing/thawing and wetting/drying cycles, and of intense biological activity, so that they are also heterogeneous due to the presence of cracks and biogenetic macropores in the uppermost layers [2]. On the other hand mountain soils play an important role in key hydrological processes, as runoff production, groundwater table recharge and evapotranspiration. Moreover in mountain environments, rainfall and the following redistribution of water content are the most common triggers of landslides worldwide [3] and rainfall itself is of crucial importance for the sediment production.

Focusing on landslides, diverse hydrological systems are expected to trigger diverse landslide typologies. According to [3], debris flow will be probably triggered by (surface) runoff; shallow soil slips, ranging from 1 to 2 m, might be caused by percolating water fronts or perched water tables generated by individual storm events; deeper landslides can be described only considering larger windows of meteorological inputs, vegetation and the role of the vadose zone. Within all these schemes, one has also to think about the role of preferential flow phenomena, which can greatly influence the triggering system. Moving from the understanding of the slope hydrology to the failure mechanisms, thus linking the soil hydrology with its geotechnical implications, is still a major challenge [4].

Aiming at contributing with a hydrological perspective to the geotechnical investigation of landslides triggering mechanisms, we present some results of a longlasting experimental and theoretical investigations of the hydrology of mountain soils. After recalling a theoretical framework suitable to describe the hydrology of shallow, sloping and heterogeneous soils (Section 2), we discuss some relevant difficulties concerning the measurement of the hydrological properties of heterogeneous and non mature soils (Section 3). Finally we focus on the role played by the soil heterogeneity in the perched water tables onset (Section 4).

2 Theoretical framework

In this work we will describe the soil water dynamics as a (deterministic) Darcian, isothermal and uncompressible flow in a continuum, undeformable and a priori unsaturated porous medium. These assumptions are very common for surface non—swelling soils and allow to describe many practical field conditions, as the soil can be also heterogeneous and anisotropic, although in a deterministic sense. With these assumptions the mass conservation principle, coupled with the Darcy law, takes one of the forms of the Richards equation:

$$\nabla \cdot (\mathbf{K} \nabla \Phi) = \frac{\partial \theta}{\partial t} . \quad (1)$$

In equation (1) $\mathbf{K}$ [L/T] is the soil—water conductivity tensor; Φ [L] is the generalized piezometric potential of the soil water; θ [L^3/L^3] is the volumetric soil water content. The piezometric potential $\Phi = h(\theta) + z$ is the summation of two components, that are the gravimetric potential z and the tensiometer—pressure potential h, which, in an unsaturated soil, is in turn a function of the water content θ by means of the soil—water retention function. In a heterogeneous and unsaturated soil, the conductivity tensor $\mathbf{K}$ depends both on the position $\mathbf{x}$ and on the soil—water content θ, and it is therefore related to the tensiometer—pressure potential h by means of the relative conductivity function $k(h)$, which will be introduced in the Section 2.1.

The Richards equation (1) can be solved once known the soil—water constitutive laws $\theta(h)$ and $k(h)$, in each point of the domain, and once fixed the initial and boundary conditions. In practical way and with a deterministic approach to the soil—water constitutive laws, the upper soil layers are either regarded to as subdivided into homogeneous and layered subdomains, or they are considered gradually varying with the depth in the transverse direction to the soil slope (x^* in Figure 1).

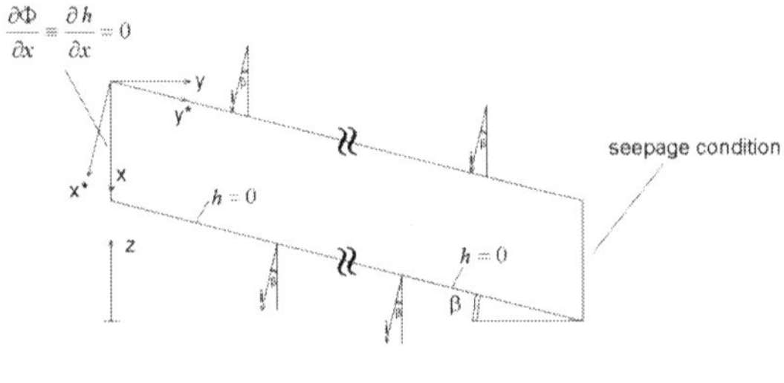

Figure 1. Sketch of a reference draining slope with indicated some typical boundary conditions for numerical simulation.

2.1 A form for the hydraulic conductivity tensor

An almost general form of the conductivity tensor **K** is given in the following equation (2):

$$\mathbf{K}(h,\mathbf{x})=K_{s,o}k(h)\mathrm{f}(x^*)\mathbf{k}^{(A)}, \quad (2)$$

in which $K_{s,o}$ is the hydraulic conductivity at soil saturation in the transverse direction x^* at the soil surface; $k(h)$ is the relative conductivity function; $\mathrm{f}(x^*)$ accounts for the soil heterogeneity and $\mathbf{k}^{(A)}$ is the anisotropy tensor. The product:

$$K_s(x^*)=K_{s,o}\mathrm{f}(x^*), \quad (3)$$

with $\mathrm{f}(x^*=0)=1$, describes the profile of the transverse conductivity at saturation $K_s(x^*)$. The tensor $\mathbf{k}^{(A)}$ accounts for how much lateral conductivities are greater than the transverse one, and in two dimensions takes the following form:

$$\mathbf{k}^{(A)}=\begin{bmatrix}1 & 0\\ 0 & r\end{bmatrix}, \quad (4)$$

in which r is the anisotropy coefficient. Equation (2) is written under the assumptions that (*a*) the soil is heterogeneous in the transverse direction x^* and it is homogeneous in the lateral direction y^*, (*b*) the principal directions of anisotropy are the transverse (x^*) and the lateral ones (y^*); (*c*) the heterogeneity does not affect the shape and the parameters of the relative conductivity function (and of the soil water retention function as well) but only the value of the hydraulic conductivity at saturation; (*d*) the anisotropy coefficient r is uniform also in the transverse direction. Assumptions (*a*) and (*b*) follow from the hypothesis that the transverse direction x^* is also the direction of the soil genetic development. Assumption (*c*) follows from the idea that the conductivity at saturation strongly depends on the soil macropores, while the constitutive laws $k(h)$ and $h(\theta)$ are sensitive to the micropores of the soil matrix.

As a consequence of genetic layering the soil hydraulic conductivity at saturation K_s decreases with depth in the transverse direction x^* across the upper soil horizons, being typically greater in A horizons, rich in macropores, and smaller in B ones [2]. On mountains, where strongly erosive processes act and mass movement is a key soil forming factor, the soil horizonation cannot fully develop and a gradual decrease of K_s typically occurs in the upper soil layers, as it will be showed in Section 3.1. Equation (2), with $\mathrm{f}(x^*)$ smooth and monotonically decreasing with x^*, is therefore effective in order to describe the K_s—field on mountain slopes.

2.2 Bottom boundary conditions and one dimensional formulation

A key issue which characterises the solution of equation (1), and therefore the prediction of the onset of a perched water table, is to define the boundary condition at the bottom of the investigated soil layer x_f^*. As a first approximation, we define here two reference soil profiles:

(*a*) Soil which lays on a massive rock or on a massive layer of parent material: the most suitable boundary condition is no flux in the transverse direction x^*, i.e.:

$$\left(\frac{\partial \Phi}{\partial x^*}\right)_{x_f^*}=0\,. \quad (5)$$

In this framework, if a perched water table onsets, the maximum value of the tensiometer—pressure potential will be predicted at the interface between the soil and the parent material;

(*b*) Soil which lays on fractured rock or on a greatly conductive layer of parent material which is not able to act any retention: the underlying layer behaves as a capillary barrier and suitable boundary conditions are a percolation—like condition, i.e.:

$$\left(\frac{\partial \Phi}{\partial x^*}\right)_{x_f^*}=-\cos\beta\,, \quad (6)$$

or a saturation one ($h\,(x_f^*)=0$). The first condition applies during transient flow, the latter during steady flux. In this case a perched water table can theoretically onset only if the soil is unhomogeneous and if the hydraulic conductivity at soil saturation monotonically decreases with x^* [5].

In case (*b*), as the percolation can take place, if one focuses on distributed processes (viz those related to rainfall or snow—melting infiltration, redistribution and evapotranspiration) and if the depth of the investigated soil layer is almost uniform along the slope (Figure 1), in the Richards equation (1) the condition of uniform flow:

$$\frac{\partial}{\partial y^*}(\cdot)=0 \, . \quad (7)$$

holds for each variable (with the exception of z). The equation (1) is therefore rewritten in a simplified one dimensional form:

$$\frac{\partial}{\partial x^*}\left[K_{s,o}k(h)f(x^*)\left(\frac{\partial h}{\partial x^*}-\cos\beta\right)\right]=\frac{\partial\theta}{\partial t} \, . \quad (8)$$

Due to the lateral effect of gravity, this simplification does not mean that the soil is seat of a purely transverse (one dimensional) flow, because a laterally—uniform flow q_{y^*}, given by:

$$q_{y^*}(x^*)=r\,K_{s,o}k(h)f(x^*)sen\beta \, . \quad (9)$$

takes place along the soil profile x^*.

3 Experimental evidences

3.1 Conductivity at saturation

Several methods are available both in the field and in the laboratory, and depending on the expected values, to determine the conductivity at saturation. Conductivities greater than 10^{-4} m/s are typically estimated by means of constant head permeameter or constant head infiltrometer; conductivities ranging from 10^{-8} m/s to 10^{-4} m/s are estimated also by means of falling head permeameter and falling head infiltrometer; conductivities smaller than 10^{-8} m/s are usually estimated by means of edometric compression. After a long lasting experimental practice, during which a great number of mountain soils was investigated in the Italian Alps (146 soils in the Toce river basin [6], 100 in the Oglio river basin [7], 127 in the Mella river basin [6]), we observed that falling—head methods were suitable for most of the cases.

In [6] and [7] we provided a detailed discussion about different approaches at interpreting the falling head permeameter and the single ring infiltrometer data. Here we will just recall two important aspects:

(1) The soil volume investigated by the single ring infiltrometer (~10 dm³) is much greater than that investigated by the falling head permeameter (~1 dm³), therefore the first method allows to better explore the soil discontinuities and macropores, and it is more effective for mountain young soils. Moreover the core disturbance is greater for the second method and the first method is therefore recommended when possible;

(2) The choice of reliable boundary conditions for the falling head permeameter is much easier than for the single ring infiltrometer. It is in fact often difficult to make a priori hypotheses about the shape of the flux and the values of the tensiometer—pressure potential at the bottom of the infiltrometer.

Single ring infiltrometer data can be interpreted by means of a number of methods, viz:

(1) Purely one—dimensional methods, based on the direct integration of Darcy's law across the soil core which is bounded by the infiltrometer;

(2) One—dimensional methods with scaling corrections, which account for the lateral water flow due to the non—saturation of the surrounding soil [8];

(3) Two—dimensional methods with axial symmetry [9];

(4) Inverse modelling (typically in two—dimensional axis—symmetric model) of the infiltration process.

Methods (2) and (3) account in some ways also for the surrounding soil, so that they better describe the infiltration process, but they require that the soil around and below the infiltrometer is uniform. In case the soil is not uniform, an inverse modelling approach is more realistic. By means of all these methods, reliable estimates are obtained if the suction below or around the infiltrometer is accounted for in some ways, otherwise conductivities are meaningfully overestimated.

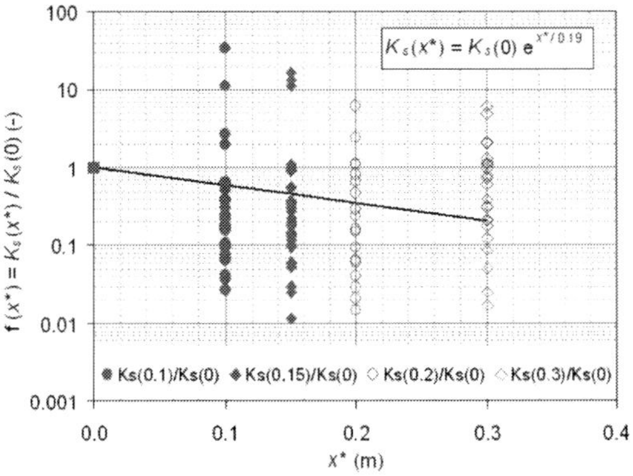

Figure 2. Experimental data about the soil heterogeneity function f(x^*) defined in eq. (2). Data from two subbasins of the Toce river basin, after [6], modified.

The first class of methods allows to investigate only the soil core inside the infiltrometer, therefore it is the most effective at obtaining conductivity profiles. In order to do so, after performing the first infiltration tests at the soil surface, the first soil layer is removed to perform other infiltration tests at fixed depth. In Figure 2 resulting values of the heterogeneity function f(x^*) measured in two subbasin of the Toce river basin are represented (after [6], modified). It is observed that in most of the profiles the conductivity values decrease with depth in the uppermost soil layers. According to a common hydrological practice [10] the heterogeneity function f(x^*) was interpreted by means of an exponential, and we estimated a length of decay L = 0.19 m and 0.12 m for the Toce river basin and for the Mella river basin, respectively.

3.2 Near—saturation conductivity

During the infiltration and the redistribution process in an unsaturated soil, a major role is played by the relative conductivity function $k(h)$. Typically the shape and the values of $k(h)$ are provided by predictive models [11]

DOI: 10.1051/e3sconf/20160916002

once known the soil—water retention function θ(h). Yet $k(h)$ is very sensitive to h, especially for nearly saturated soil (e.g. -30 cm < h < 0 cm), and it is therefore recommended to directly estimate it, at least for few values of h, by means of e.g. a tension infiltrometer [9]. This procedure reduces the uncertainties related to the use of the predictive models [12] and allows to estimate the contribution of the macropores to the hydraulic conductivity [13]. If we assume in fact that, in a capillary soil model, the soil—water content θ, in equilibrium at a certain tensiometer—pressure potential h, is the pore space volume characterised by local radii smaller than a certain radius $r(h)$, which is function of h by means of the Jourin—Borelli formula, then the macropores contribution to the conductivity is estimated comparing the conductivity determined by ponded infiltration with that determined at a certain tensiometer—pressure potential h. The Authors [13] observed a great decrease in the observed mean infiltration rate in a potential interval ranging from h = 0 cm to h = -3 cm, and they estimated that nearly 73% of the saturated flux occurred in the pores characterised by this potential interval.

3.3 Characteristic water contents and soil—water retention functions

The soil—water content dynamics are bounded by the characteristic water contents, that are the residual water content θ_r and the water content at saturation θ_s. These two values represent the minimum soil—water content for which the water phase is continuous, and the maximum soil—water content above which the gas phase is discontinuous, respectively. In traditional laboratory practice θ_r is estimated as the asymptote of the soil—water retention function and θ_s as the soil—water content after long imbibition, but they are far from being intrinsic soil properties. Their magnitudes, in fact, seasonally vary due to the soil swelling and to the soil biological activity, and they are moreover affected by the draining and wetting history. These quantities can be therefore regarded to as state variables, and a more reliable estimate is provided by analysing the extreme values reached by field—measured series of soil—water content. It is remarked that during transient analyses, especially of wetting processes and when a perched water table is expected to onset, a crucial role is played by the water content at saturation θ_s. As the relative conductivity $k(h)$ is very sensitive to h nearby saturation, then uncertainties on θ_s can sensitively affect the reliability of the description of a perched water table.

The soil—water retention relationships are usually determined in the laboratory by weighting the water content of soil samples at the equilibrium with certain tensiometer—pressure potentials. After that the experimental points are interpolated in order to fit analytical functions in the form θ(h) or $s(h)$, being s the effective soil saturation, defined by:

$$s=\frac{\theta-\theta_r}{\theta_s-\theta_r}. \tag{10}$$

Accounting for the uncertainties at determining the characteristic soil—water contents, and for the disturbance of the laboratory samples, it is suggested to substitute the volumetric soil—water content θ with the gravimetric one w, thus obtaining:

$$s=\frac{w-w_r}{w_s-w_r}, \tag{11}$$

being w_r and w_s the residual gravimetric water content and the gravimetric water content at saturation, respectively. Then the soil—water retention curve are interpolated in the form $s(h)$ and can be driven back to the field volumetric soil—water content by means of the field values of θ_r and θ_s. This procedure therefore uncouples the laboratory measurements from the field soil structure, and it is closely related to the idea that the retention is mainly due to the soil matrix and less to the soil macropores.

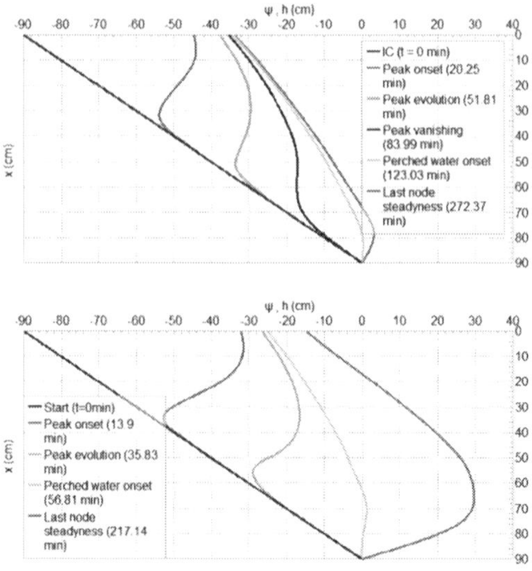

Figure 3. Soil—water content peaks and perched water tables onset in a horizontal, gradually layered and draining soil with decreasing conductivity at saturation with depth [14].

4 Geotechnical implications

If the hydraulic conductivity at saturation monotonically decreases with depth, subsurface peaks of water content can onset during long lasting infiltration at low infiltration rates. This pattern was proved for exponentially decreasing conductivity in a horizontal soil [5], and cannot be excluded in any case K_s monotonically decreases with depth. Afterwards this pattern was observed by means of a laboratory experiment during which artificial rainfall was poured over a multi—layered soil column characterised by decreasing K_s with depth [15]. During the same experiment it was observed that, if infiltration lasts after the water content peak formation, the soil saturation can be reached somewhere below the soil surface. Then a perched water table onsets and develops both upwards and downwards.

 DOI: 10.1051/e3sconf/20160916002

In a soil layer of finite thickness and inferiorly bounded by a capillary barrier, the perched water table may onset or it may not, depending on the infiltration rate and on the values of the conductivity at the bottom of the soil [14]. In Figure 3 two patterns of perched water onset are represented after numerical simulation of a soil with exponentially decreasing K_s with depth: in one case (Figure 3, above) a peak of tensiometer—pressure potential (and of water content as well) onsets and moves downward, then it vanishes and a perched water table onsets starting from the soil bottom; in the other case (Figue 3, below) the peak of tensiometer—pressure potential onsets and emphasizes until it reaches the soil saturation and originates a perched water table.

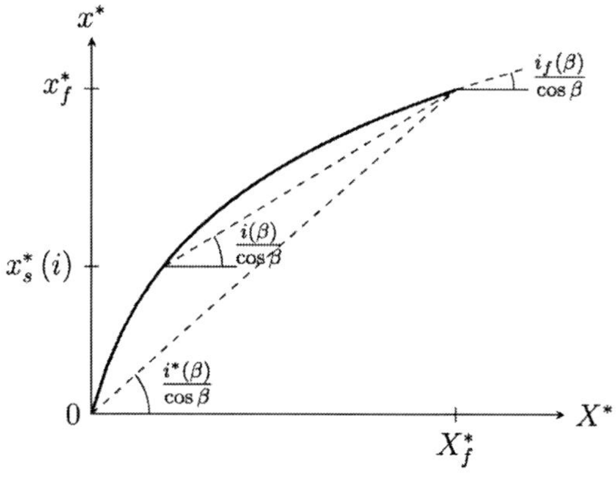

Figure 4. Infiltration thresholds and thickness of the perched water table as a function of $K_s(x^*)$ by means of the Zaslavsky transformation $X^*(x^*)$ (15) [16].

Infiltration thresholds and properties of the perched water tables (i.e. thickness of the perched water table, position and magnitude of the maximum tensiometer—pressure potential) were derived for horizontal soils [17] and then generalised for sloping soils with slope β [18]. Here we will recall that, in case of a sloping soil:

(*a*) The infiltration threshold i_f (β) at which a perched water table onsets is related to the transverse conductivity $K_s(x_f^*)$ at the soil bottom x_f^*, as given by:

$$i_f = K_s(x_f^*)\cos\beta\ ; \tag{12}$$

(b) The infiltration threshold $i^*(\beta)$ at which the soil reaches waterlogging is related to the equivalent transverse conductivity $K_{sx^*,eq}$ between the soil surface ($x^* = 0$) and the soil bottom (x_f^*) by means of :

$$i^* = K_{s,\mathrm{eq}}\cos\beta\ ; \tag{13}$$

(c) The depth of the maximum tensiometer—pressure potential $h_{\max}$ within a perched water table given by an infiltration rate i is at $x^*_{\max}$ where:

$$K_s(x^*_{\max}) = \frac{i}{\cos\beta}\ . \tag{14}$$

From equations (12), (13) and (14) emerges that the characterizing infiltration rates reduce with the decline from vertical axis cosβ because, if the slope increases also the gravitational—potential gradient is less effective at contributing to the transverse flow.

In Figure 4 the previously introduced results are effectively synthesised by means of the classical Zaslavsky transformation of variables [16]. The Zaslavsky transformation [19], defined as:

$$\mathrm{d}X^* = \frac{\mathrm{d}x^*}{K_s(x^*)}\ , \tag{15}$$

allows to symbolically map an unhomogeneous soil into a uniform one (with unitary conductivity), in which the saturated flow is characterised by the same gradients when it takes place along corresponding paths. The obtained map $x^*(X^*)$ is related to the properties of the perched water tables, which can take place in the investigated soil. Particularly the slope at X_f^* defines the threshold i_f (β) and the thickness of any perched water table given by an infiltration rate i (β) > i_f (β) is related to the chord starting in (X_f^* , x_f^*) and with slope i / cosβ.

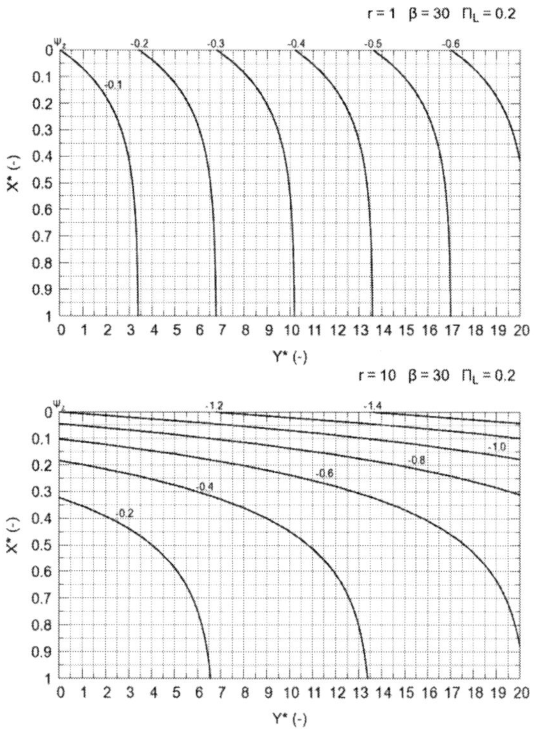

Figure 5. Lateral deflection of an infiltrated drop in a waterlogged sloping soil in presence of a perched water table. Only in these graphs X^* and Y^* refer to dimensionless variables, defined by $X^* = x^* / x_f^*$, and $Y^* = y^* / x_f^*$ [20].

As the soil conductivity profile is measured at certain depths and then interpolated on the basis of an analytical function $K_s(x^*)$, it is therefore of great importance the choice of the interpolating function, as different ones are characterised by different slopes and chords, thus affecting the prediction of the infiltration thresholds and of the properties of the perched water tables.

As it was evidenced by Zaslavsky and Sinai in the third of their 1981's papers [21], during an infiltration process in a uniform or sharply—layered soil layer, of undefined length, laying on a sloping capillary barrier and eventually anisotropic (Figure 1), an infiltrated drop crosses the soil layer and it is laterally deflected by the slope and by the effect of the anisotropy. The same pattern takes place in a heterogeneous soil, as it is represented in Figure 5 by means of the Lagrange stream function Ψ_{ζ^*} [20] which describes the flow field within a perched water table in a sloping soil at waterlogging, both in the case of isotropic (above, $r = 1$) and anisotropic (below, $r = 10$).

These results evidence that the soil layer is crossed by the drop in a finite lateral length (along the slope) and that, in a perched water table laying on a capillary barrier, the classical Dupuit—Forchheimer hypotheses for linear and gradually variable flow in the porous media do not apply, due to the meaningful transverse Darcian fluxes. On the other hand, as no hypotheses were introduced on the magnitude of the slope β, this framework to describe the flow within a perched water table is effective even in conditions of great slopes.

5 Conclusions

A theoretical framework to describe the relevant hydrological processes acting in a shallow, sloping and heterogeneous soil was presented (Section 2). The scheme evidences the important role played by the conductivity at saturation in the onset of perched water tables, which are among the most common shallow landslides triggering mechanisms. Some difficulties, some methodological aspects and some recommendations to measure the hydrological properties of soils in mountain environments were presented (Section 3). A focus was made on the importance of collecting information on the transverse heterogeneity of the hydraulic conductivity at soil saturation and on how to link the laboratory measured soil—water retention relationships with the field soil—water contents. Then the implications of the heterogeneity of the soil hydraulic conductivity at saturation on the properties of the perched water tables and on the characteristic infiltration thresholds were presented and discussed (Section 4).

References

[1] R. C. Graham. In: G. Certini, R. Scalenghe (eds), *Soils: basic concepts and future challenges*. pp. 151–163 (Cambridge University Press, 2006)

[2] M. Kirkby. In: R. Chorley (eds), *Water, Earth and Man*. pp. 215—227 (Taylor & Francis, 1969)

[3] T. W. J. van Asch, L. P. H. Van Beek, T. A. Bogaard. In: L. Picarelli, P. Tommasi, G. Urciuoli, P. Versace (eds), *Rainfall-induced landslides. Mechanisms monitoring techniques and nowcasting models for early warning systems. Proceedings of the 1st Italian Workshop on Landslides*. Vol. **1**. Napoli (2009)

[4] T. A. Bogaard, R. Greco. WIREs Water. (2015) doi:10.1002/wat2.1126

[5] S. Barontini, R. Ranzi, B. Bacchi. Wat. Resour. Res. **43**:W08411 (2007).

[6] S. Barontini, A. Clerici, R. Ranzi, B. Bacchi. In: *Climate and Hydrology in Mountain Areas*. pp. 101–121 (John Wiley & Sons, Ltd, 2005)

[7] S. Barontini, A. Clerici, R. Ranzi, B. Bacchi. In: *Hydrology in Mountain Regions: Observations, Processes and Dynamics*. pp. 112–118. IAHS Publ. **326** , ISBN 978-1-901502-89-3 (2009)

[8] L. Wu, L. Pan, J. Mitchell, B. Sanden. Soil Sci. Soc. Am. J. **63**: 788-792 (1999)

[9] W. D. Reynolds, D. E. Elrick. Soil Sci. Soc. Am. J. **55**: 633-639 (1991)

[10] K. J. Beven, M. J. Kirkby. Hydrol. Sci. Bull. **24**: 43-69 (1979)

[11] Y. Mualem. Wat. Resour. Res **12**: 513-522 (1976)

[12] T. Vogel, M. Cislerova. Transport in Porous Media **3**:1-15 (1988)

[13] K. W. Watson, R. J. Luxmoore. Soil Sci. Soc. Am. J. **50**:578-582 (1986)

[14] M. Peli, S. Barontini, T. A. Bogaard, B. Bacchi, R. Ranzi. In: C. Mancuso, C. Jommi and F. D'Onza (eds.), *Unsaturated Soils: Research and Applications, Vol. 2*. pp.167–174, (Springer, Heidelberg, 2012)

[15] S. Barontini, G. Belluardo, B. Bacchi, R. Ranzi. In: *Atti del XXXI Convegno di Idraulica e Costruzioni Idrauliche*, pp. 9 (Morlacchi, Perugia, 2008)

[16] S. Barontini, M. Falocchi, R. Ranzi. In: Lollino G., Giordan D., Crosta G.B., Corominas J., Azzam R., Wasowski J., Sciarra N. (Eds.) *Engineering Geology for Society and Territory. Volume 2: Landslide Processes*, pp: 2153–2156 (Springer International Publishing, 2015)

[17] S. Barontini, R. Ranzi. In: *Atti del XXXII Convegno Nazionale di Idraulica e Costruzioni Idrauliche*, 10 pp. (Walter Farina Editore, Palermo, 2010)

[18] S. Barontini, M. Peli, T.A. Bogaard, R. Ranzi. In: C. Margottini, P. Canuti, K Sassa (Eds.) *Landslide Science and Practice Volume 3*, pp: 143-149 (Springer Berlin Heidelberg, 2013)

[19] D. Zaslavsky. Soil Sci. **97**: 400-410 (1964)

[20] S. Barontini, M. Peli, R. Ranzi. In: *Proceedings of the Second IAHR European Division Congress*, 6 pp. (2012)

[21] D. Zaslavsky, G. Sinai. J. Hydr. Div. ASCE **107**: 37-52 (1981)

Moisture migration during loading and shearing of unsaturated sand

Bestun Shwan[a]

Department of civil and structural engineering, University of Sheffield, Mappin Building Street, Sheffield S1 3JD, UK.

Abstract. A series of drained unsaturated direct shear tests were conducted on a fine sand where suction was controlled by means of a hanging water column. A rise in the water level in the column was observed in conjunction with normal stress application for unsaturated samples. During shearing, the water level in the column also increased indicating continued water content loss from the sand. This is over and above any change in saturation that would arise due to volume change. Samples taken after the test for water content measurement confirmed the observed effect. Discussion of the phenomenon in the context of the soil water characteristic curve (SWCC) is presented and a simple hypothesis to explain this behaviour is proposed.

1 Introduction

The arrangement of water within the voids has a significant effect on the behaviour of unsaturated soils, [1] and [2]. Wheeler [3] defined two different forms of liquid water in unsaturated soil: bulk and meniscus water. Soil follows the bulk water form for any suction values less than the air entry value. The meniscus water form, however, is more pronounced when the applied suctions are greater than the air entry value, [2].

Understanding these two liquid forms during shearing and their effects on the behaviour of the unsaturated soil is significant. Romero and Vaunat [4] studied the soil water characteristic curve (SWCC) for deformable clays following which delimiting zones in the water retention curve were defined, which are controlled by water ratio (volume of water to volume of solids). Two zones: intra-aggregate water (adsorbed water) and inter-aggregate water were recognized. In the latter zone, air is occluded and the water ratio is highly sensitive to the mechanical effects.

The hydro-mechanical behaviour of unsaturated soils has been investigated in terms of the SWCC. Generally, the most recent studies of the SWCCs emphasised the cyclic drying and wetting behaviours. The cyclic behaviours were studied (for example, [5-6]) in terms of scanning path which can be defined as any intermediate point between the main drying and wetting curves, [6]. The effect of dilation during shearing on the drying behaviour and its consequence on the SWCC in a controlled suction test, however, is limited.

This paper presents, therefore, work in progress that investigates the drying behaviour during shearing at a low range of suctions for unsaturated sand in a context of the SWCC using a modified direct shear apparatus. The effect of the mechanical straining in a context of the SWCC on the basis of the experimental results performed in this study is also addressed qualitatively.

[a] bjnshwan@gmail.com

2 Experimental work

2.1 Soil properties and SWCC

A commercial fine sand, designated fraction D by the supplier- available from the David Ball Group-UK, was used. The physical properties for the sand are shown in Table 1.

Table 1. Physical properties for the test dry sand.

Gs	γ_{sat} kN/m^3	γ_{dry} kN/m^3	*e*	Particle range mm	Relative Density %
2.65	19.33	15.3	0.70	0.075-0.3	52

The SWCC was obtained using two different methods, the filter paper and hanging column technique (HCT) as shown in Fig. 1. For the HCT, a bespoke circular shear box of diameter 80 mm was used with additional features over standard laboratory shear boxes to be adaptable for the unsaturated experiments, see Fig. 2a. Suction was controlled by means of a hanging water column (burette) as shown in Fig. 2b. The SWCC was determined at null load condition. More details with regards to sample preparation and suction application is given by [7].

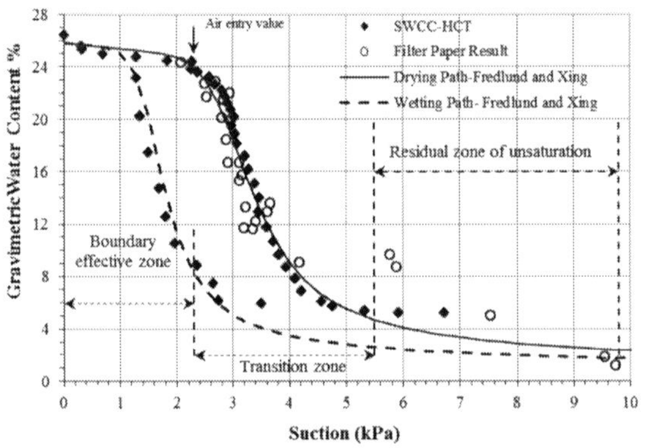

Figure 1. SWCC using the HCT and filter paper method.

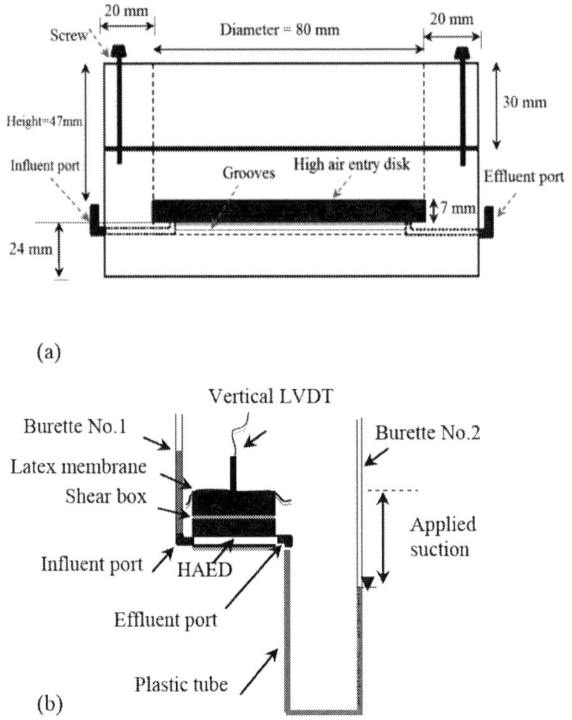

Figure 2. (a) Modified direct shear box (b) suction application using the HCT and filter paper method.

2.2 Moisture migration during loading and shearing

A series of drained direct shear tests were conducted on unsaturated sand at three normal stresses: 50, 100 and 200 kPa and at a range of nominal suctions: 2, 4 and 6 kPa where the movement of water was observed immediately after the normal stress application. The stated suction head measured at the soil surface was the distance from the water level in the burette to the soil surface. The applied suction was selected according to the SWCC for the sand used. Values of 2 and 6 kPa nominal suction were selected since they are close to the air entry suction and residual suction for the sand used, see Fig. 1. Finally, to investigate the behaviour of unsaturated sand at the transition zone (zone between air entry and residual suctions - see Fig. 1), a value of 4 kPa nominal suction was also utilised. Table 2 shows the nominal and suction values before and after the normal stress application for σ = 50 kPa for three repeat tests denoted as 1, 2 and 3 in Table 2. The water migration after normal stress application for σ = 100 and 200 kPa was identical with that obtained for σ = 50 kPa, however, only data for σ = 50 are presented. The idiom suction denotes the nominal suction value in the context of this paper and in the figure's legends.

During shearing, the water level in the burette also increased indicating continued water content loss from the sand.

Table 2. Nominal and suction values before and after application of the normal stress for σ = 50 kPa.

σ (kPa)	$s_{nominal}$ (kPa)	s_{before} (kPa)	s_{after} (kPa)
50-1	2	2.17	2.06
50-2	2	2.06	2.04
50-3	2	2.19	2.18
50-1	4	4.35	4.27
50-2	4	4.1	4.1
50-3	4	3.95	3.89
50-1	6	5.33	5.32
50-2	6	5.88	5.79
50-3	6	6.02	5.92

2.3 Determination of the Scanning path

The scanning path is a consequence of the suction change to a lower value. To determine the start of the scanning path on the main drying curve (point A in Fig. 2a), weight volume relationships were used. Prior to the test, the amount of water extracted from the sample by lowering the burette to apply a specific suction was collected in a container (water was expelled from an open-ended burette using a syringe) and measured to a scale of three decimal place accuracy and hence total weight of the sample was adjusted accordingly. Subsequently, the total unit weight of the sample (γ_t) at a known initial volume was calculated using the equation:

$$\gamma_t = \frac{W_t}{V_t} \tag{1}$$

where W_t is the total weight (kN) and V_t is the total volume (m^3). Degree of saturation (S_r) and water content (w) were calculated as follows:

$$S_r = \frac{\gamma_t (1+e) - G_S\,\gamma_w}{e\,\gamma_w} \tag{2}$$

$$w = \frac{S_r\,e}{G_s} \tag{3}$$

where e is the initial void ratio (= 0.7- see Table 1), G_S is the specific gravity and γ_w is the unit weight of water (kN/m^3). Suction head was measured as a distance from water level in the burette to the soil surface. By knowing the water content and suction value after

DOI: 10.1051/e3sconf/20160916003

application of the normal stress prior to the test, the start point of the scanning path on the main drying curve (point A in Fig. 2a) can be located.

At the end of the test, 5 samples were taken along height (47mm) of the sample to track the moisture content loss. The suction head for sample No. 1 (the samples were numbered from the top to bottom) was determined according to the last water level in the burette which was captured by a camera, installed in front of the burette during the test to take an image for the water level in the burette every 30 minutes. Suction head for sample No. 2 was less than the suction head of sample No. 1 by (47/5) mm. This calculation was replicated for other samples (i.e. samples No. 3, 4 and 5). By knowing both suction and water content, the end of the scanning path was determined. The measured water content was also compared with the calculated water content which was determined at different stages of the test using Eqs. 1, 2 and 3. Following this the calculated water content was paired with the inferred suction (obtained based on water level in the burette captured by the camera at different stages during the test) and plotted on the SWCC (see Fig. 2a-hollow triangle symbol). The calculation was based on adjusted height of the sample during shearing (compressibility and dilation behaviours) and unit weight. The height of the sample during shearing was obtained from a vertical LVDT reading. Then, void ratio was determined according to the adjusted total volume as follows:

$$V_v = V_t - V_s \quad (4)$$

$$e = \frac{V_v}{V_s} \quad (5)$$

where V_v is the volume of voids and V_s is the volume of solid.

3 Results

3.1 Drying behaviour during shearing at different low suctions and constant normal stress

Figures 3, 4 and 5 plot the gravimetric water content versus suction at constant σ = 50 kPa and different suctions of 2, 4 and 6 kPa showing the scanning paths. The start of the scanning path (point A in Fig. 3a) is plotted based on the adjusted values of the suction after the application of the normal stress (see Table 2). The beginning of the scanning path for a few cases was observed either above or below the drying path of SWCC-HCT (e.g. see Fig. 4a - circled). This was attributed to the effect of void ratio in which the sample initially was deemed to be looser (above the SWCC-HCT curve) or denser (below the SWCC-HCT curve). The denser sample provided a smaller void ratio and hence less water was retained between the voids.

In Figs. 3, 4 and 5, the most significant drying behaviour (water migration) occurred to the sample No. 1 (solid circle). The overall trend showed the drying behaviour at the different suction levels with a few cases of the wetting near the failure plane (e.g. see solid square-Fig. 3a).

The drying behaviour is attributed to (i) the effect of the normal stress application which caused water movement downwards and (ii) due to concentration of the suction in the shearing zone but not the entire thickness of the sample. This caused water migration into or out of the shearing zone. This may indicate why the wetting behaviour for most of the cases was observed near the failure plane (solid square in Fig. 3a - near the failure plane).

Concentration of the suction is due to the dilative behaviour in which dilation created bigger voids between the particles near the shear band for water to move into. The dilative behaviour was also observed for the unsaturated samples prepared at normal stresses of 100 and 200 kPa, detailed in [7].

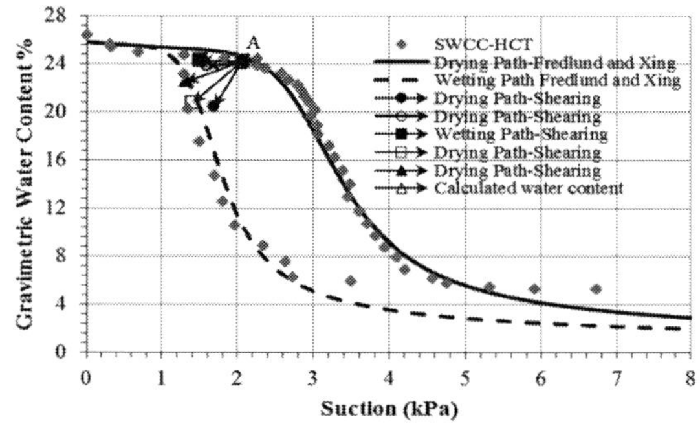

(a)

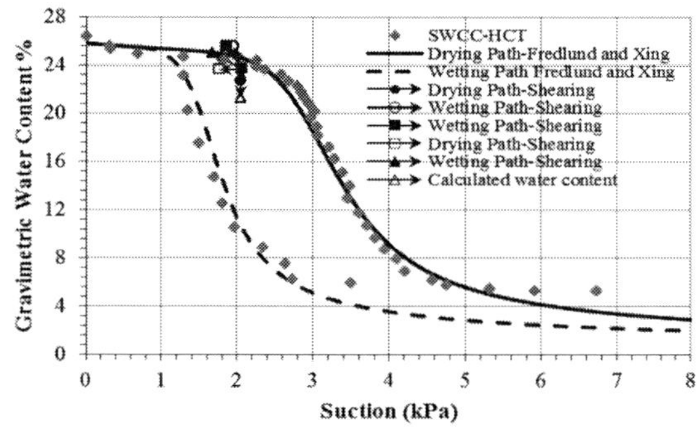

(b)

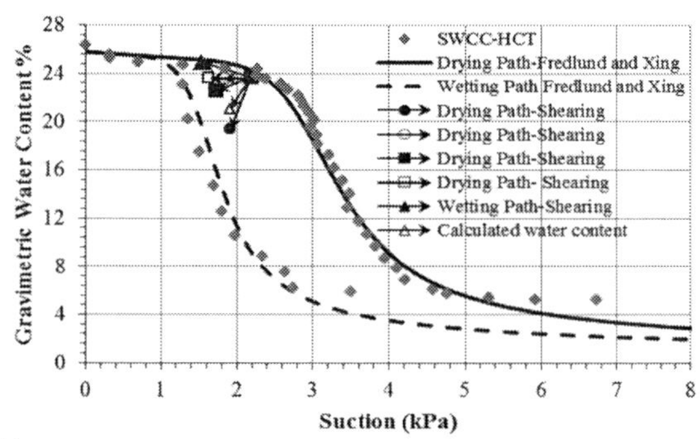

(c)

Figure 3. Scanning path for σ = 50 kPa and s = 2 kPa for (a) first test-1 (b) second test-2 (c) third test-3.

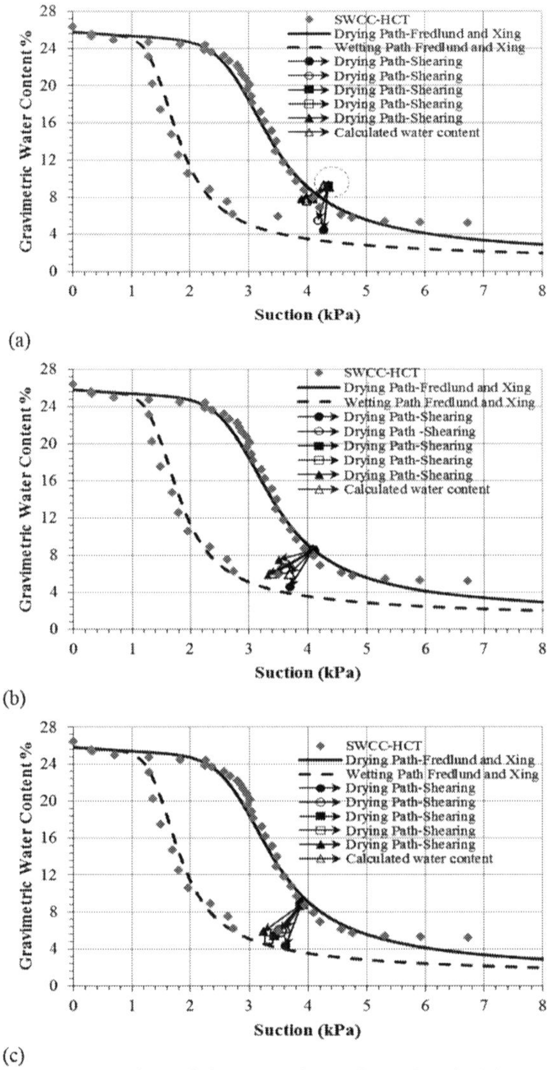

Figure 4. Scanning path for σ = 50 kPa and s = 4 kPa for (a) first test-1 (b) second test-2 (c) third test-3.

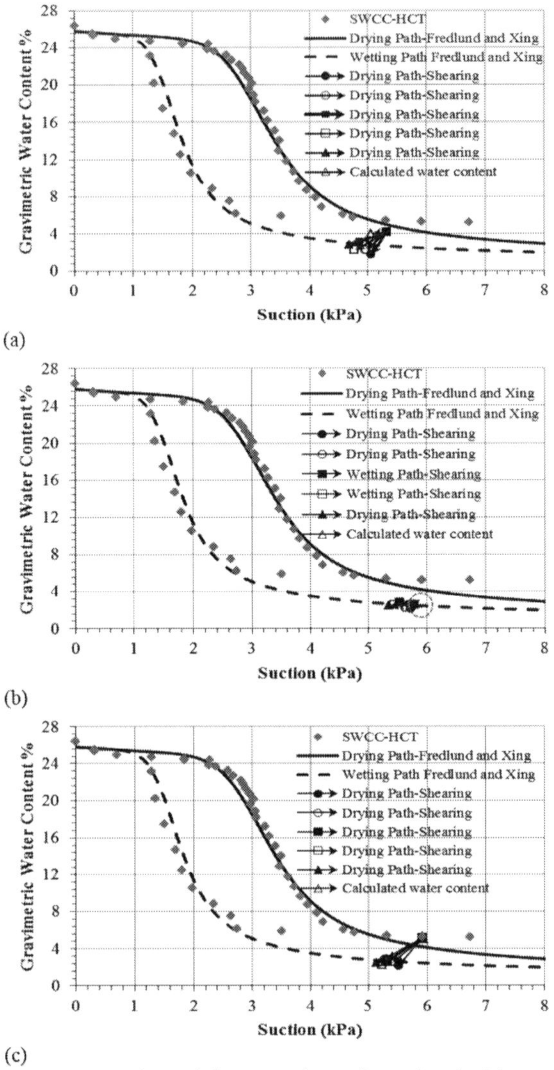

Figure 5. Scanning path for σ = 50 kPa and s = 6 kPa for (a) first test-1 (b) second test-2 (c) third test-3.

4 Discussion

The water level in the burette was seen to rise in conjunction with the normal stress application. During shearing, the water level in the burette also increased indicating continued water content loss from the samples. This was over and above any change in saturation that would arise due to volume change. Discussion of the phenomenon in the context of the SWCC is presented and a simple hypothesis to explain this behaviour is proposed.

The sample's paths on the SWCC in Figs. 3 to 5 typically showed drying behaviour with only a few wetting cases. The wetting path A to B in Fig. 6 is the expected behaviour which is attributed to an increase of degree of saturation during shearing (due to decrease in void ratio) and hence in suction. The wetting path is studied broadly in the field of unsaturated soil mechanics.

Although unexpected the drying path A to C in Fig. 6, on the other hand, corresponds to the observed increase in water level in the burette during shearing. As water is expelled from the sample towards the burette during shearing (S_r decreased), the distance from the water level in the burette to the sample surface decreased (s decreased). This means both S_r and s decreased when water is expelled from the sample. A similar situation occurs when water is imbibed by the sample during shearing (both S_r and s increased).

In general, drying behaviour was observed for most samples regardless of whether the samples follow path A to C or A to E. These paths are described in more detail below:

1. Scenario I- path A to C

The drying behaviour on the SWCC in Figs. 3 to 5 was analysed based on this scenario due to the assumed direct relationship between s and S_r.

2. Scenario II- path A to D

If suction is assumed to be constant during shearing (omitting the water level change in the burette), the sample will follow path A to D during drying. However, this scenario is unlikely to occur because suction will change with any change of degree of saturation.

3. Scenario III- path A to E

This scenario is the most likely case to be followed by the sample during drying. Here, the actual measurement of the suction, for example using tensiometers placed into the sample, is required.

The importance of studying the drying behaviour on the SWCC for the Scenario-III is highlighted in that such findings are not reported in the literature yet. The discussion in the next paragraphs is, therefore, devoted to the main affecting factors on the drying path in the context of the SWCC.

It is hypothesised that the drying behaviour is a normal stress and void ratio dependant. In terms of the former, water migration was observed both (i) immediately after the load application and (ii) during shearing. In terms of the latter, suction increased due to the dilative behaviour observed for the most unsaturated samples. The increase of suction caused a decrease in degree of saturation in which this decrease was reversible (dSr_e) and irreversible (dSr_p) change, where the subscripts e and p denote elastic and plastic components. The elastic (dSr_e) and plastic (dSr_p) changes of S_r are a combination of the mechanical process of straining of the soil skeleton under changes of applied stresses (see Fig. 7) and the hydraulic process of water inflow and outflow to individual voids (see Fig. 8).

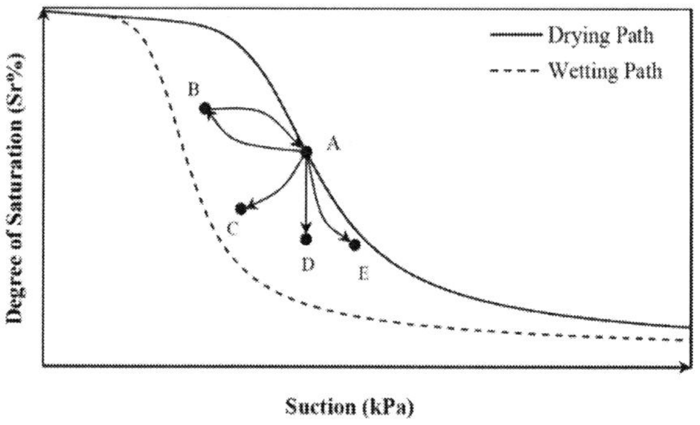

Figure 6. Typical SWCC showing scanning paths.

The elastic change (dSr_e) due to change of the normal stress is attributed to the soil particles and packet elastic deformations, [9]. The (dSr_e) due to the hydraulic process, however, is attributed to a change of radius of the meniscus, from position 2 to 3 in Fig. 8, and a contact angle.

The plastic change (dSr_p) due to the mechanical process, on the other hand, is because of slippage at inter-particle contacts which is controlled by the normal components of the contact forces (see Fig. 7b). The hydraulic effect on the (dSr_p) is due to further change of meniscus radius (position 3 to 4 in Fig. 8) in which air can break through into the voids.

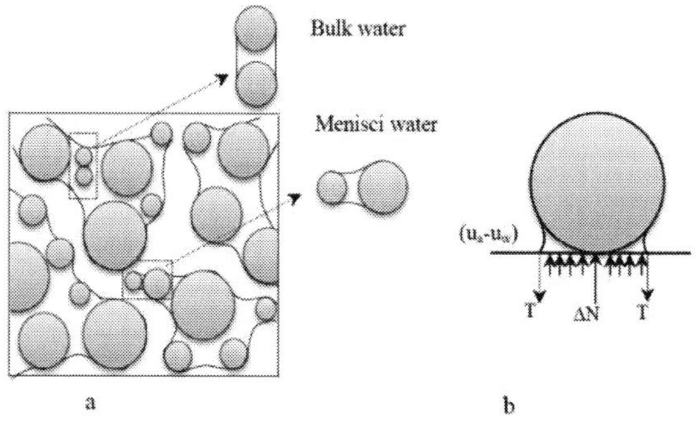

Figure 7. (a) Bulk and menisci water forms (b) Distribution of forces between two solid spheres, after [11].

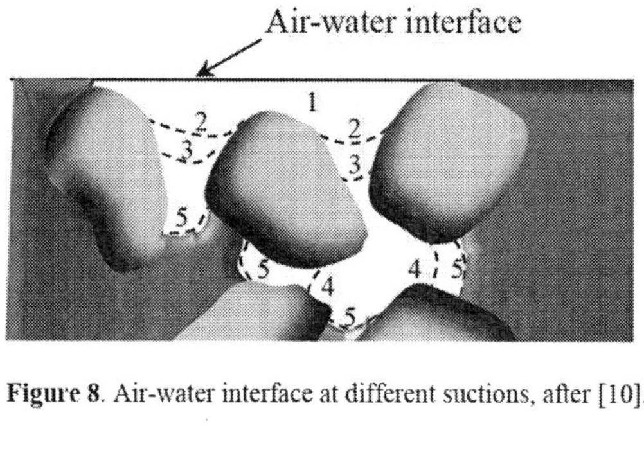

Figure 8. Air-water interface at different suctions, after [10].

5. Conclusions

1. A series of drained direct shear tests were conducted on unsaturated sand at a range of low suctions where the movement of water (drying) was observed immediately after the normal stress application and during shearing.
2. The drying behaviour was discussed in the context of the SWCC and a simple hypothesis was proposed which included the main factors affecting the water migration. It was hypothesised that the drying behaviour was attributed to the effects of the normal stress and dilative behaviour.

Acknowledgments

The author wishes to acknowledge the financial support which was provided by the Ministry of Higher Education-Kurdistan (Iraq) for this work.

References

1. S.J. Wheeler, D. Karube, *State of the art report: constitutive modeling*, in Proc. 1st Int. Conf. on Unsat. Soils, Paris pp. 1323-1356, (1995).
2. R. Vassallo, C. Mancuso, *Soil behavior in the small and the large strain range under controlled suction conditions*, in Int. Wksp on Unsat. Soils-Trento, Italy, pp. 10-12, (2000).
3. S.J. Wheeler, *Constitutive modelling of unsaturated soils*, in 4th Int. Conf. on Unsat. Soils, Arizona, USA, (2006).
4. E. Romero, J. Vaunat, *Retention curves of deformable clays*. In: Tarantino, A., Mancuso, C. (Eds.), Exp. Evid. and Theo. App. in Unsat. Soils, Balkema, Rotterdam, pp. 91-106, (2000).
5. D. Gallipoli, S.J. Wheeler, and M. Karstunen, *Modelling the variation of degree of saturation in a deformable unsaturated soil*. Ge'ot. J., **53**(1): pp 105-112, (2003).
6. M. Nuth, L. Laloui, *Advances in modelling hysteretic water retention curve in deformable soils*. Comp. and Geots. J., **35**(6), pp. 835-844, (2008).
7. B.J. Shwan, C. C. Smith, *Increase of the shear strength parameters for unsaturated sand*. Ge'ot. J., (2016), (to be published).
8. B.J. Shwan, C. C. Smith, *Effect of dilation on the degree of saturation at during shear*. Ge'ot. J. , 2016 (to be published).
9. S.J. Wheeler, R.S. Sharma, and M.S.R. Buisson, *Coupling of hydrulic hysteresis and stress-strain behaviour in unsaturated soils*. Ge'ot. J., **53**(1), pp 41-54, (2003).
10. R.A. Fisher, *On the capillary forces in an ideal soil*. Agric. Scie. J., **16**: pp 492-505, (1926).
11. E.C. Childs, *An introduction to the physical basis of soil water phenomena*. London/; Wiley-Interscience, (1969).

DOI: 10.1051/e3sconf/20160916004

Volumetric water content measurement probes in earth-dam construction

Michael Bardanis[1,a]

[1]*Edafos Engineering Consultants S.A., Laboratory, 2 Therapion st., 10444 Athens, Greece*

Abstract. Two frequency domain reflectometry (FDR) probes have been used. They were used on compacted soils both in the laboratory and in the field. Measurements in the laboratory were intended for calibration. The range of densities and types of materials where insertion of the probes can be achieved was investigated first. The effect of sporadic presence of coarser grains and density on these calibrations, once insertion could be achieved, were investigated second. Measurements on laboratory prepared samples with the same moisture content were different when the sample was kept in the mould from when it was extruded from it. Also both these measurements were different from that in a sample of the same density but significantly larger in diameter. It was found that measurements with these probes are affected by dilation exhibited by soil around the rods of the probes during insertion. Readings immediately after insertion of the sensors on samples extruded from their moulds were the ones closer to measured values. These readings combined with total volume and mass obtained from sand-cone tests during the construction of an earth-dam allowed fairly accurate estimation of the dry unit weight but not the gravimetric water content.

1 Introduction

Volumetric water content probes have been in use for years in the fields of soil science and lately unsaturated soil mechanics. In the field of soil science their use is limited practically to soft soils and initial density is not a problem for calibration, insertion of probes, and measurement duration. The ability of volumetric water content probes to measure field moisture content appears very appealing for earth-dam construction, where the need to control and measure field moisture and density are critical for appropriate construction. In these cases however, soils of various grain size distributions, densities and strengths may be encountered. The paper discusses the effect of these parameters on using this type of probes.

Volumetric water content measurement methods may be divided in three broad categories; nuclear methods, electromagnetic methods, and indirect measurement methods based on the direct or indirect measurement of suction and knowledge or some assumption on the soil-water characteristic curve. A detailed description of these methods is beyond the scope of the paper. The motivation behind the work presented in the paper was the investigation of the performance and accuracy of various small-size sensors currently available on the market with the intention to use them in the quality control of earthworks, more specifically earth dams. Various types of soils regarding grain-size distribution, plasticity and origin can be found in all types of earthworks. It is earth dams however that such a broad range of materials is found in the very same project, with all these materials requiring strict quality control, at a rate of testing and accuracy that will not inhibit construction rate or jeopardise safety. Perhaps the most common quality control test is the in-situ measurement of density and water content in order to investigate whether they satisfy the requirements of the design study or not. Nuclear methods have covered a lot of ground in this particular field, not without a cost however; financial, training, calibration, certification cost, not to mention an intrinsic –though nowadays unjustified- 'allergy' of personnel to methods described as 'nuclear'. More recently, electrical moisture/density gauges have become interesting alternatives, yet again with considerable cost considerations, although with an important advantage: probe size and strength that allows penetration even in coarser and denser materials often encountered in engineering projects. In an attempt to identify possible areas where cheaper sensors could be used with accuracy and efficiency, two types of commercial, off-the-shelf frequency domain reflectometry (FDR) sensors were used in various materials and ranges of densities and moisture contents. The experience gained is presented in this paper.

2 Penetration ability

Perhaps the most important requirement of whatever sensors used is the ability to penetrate the soil where a volumetric water content measurement is needed. This requirement was not investigated in a quantitative, but rather in a qualitative manner. Volumetric water content sensor ML-2x of Delta-T Devices is shown in Fig. 1a. It

[a] Corresponding author: mbardanis@edafos.gr, lab@edafos.gr

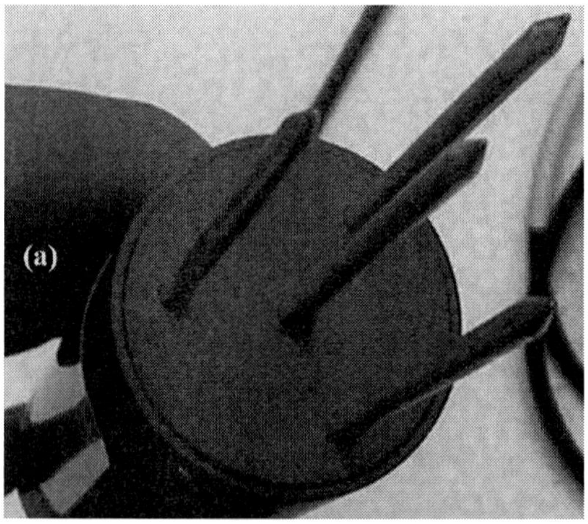

Figure 1. a) Empty plastic tube, cut to receive the MPS-2 sensor (green plasteline used to replaced the part of the tube.

was used in combination with the sand-cone density method for in-situ measurements of density and gravimetric water content as part of the quality control programme during the construction of the Ferekambos river dam on the island of Skiros, Greece. The shoulders of this dam are constructed of a compacted weathered serpentinite found locally in abundance (w_L=27%, I_p=8, G_s=2.61, w_{opt}=11%, $\gamma_{d,max}$=19.3 kN/m^3). The properties of this material have been thoroughly investigated [1, 2] and actual samples may be seen in Fig. 1b and c. Compacted samples of the material were prepared in the laboratory at the required dry unit weight (98% of the maximum dry unit weight obtained using the Proctor compaction test with standard energy) but dry, at and wet of optimum and the ML-2x sensor was used in each of the samples. The sensor was inserted relatively easily in the sample compacted wet of optimum, could not be inserted in the sample compacted at dry of optimum (Fig. 1b: insertion could not exceed more than 50% the length of the rods, by that length having already cracked the specimen) and only at the application of very large force was it inserted in the sample compacted at optimum moisture content, again however causing it to crack (Fig. 1c). Obviously therefore there is the need for a sensor of design and construction allowing the application of force sufficient for penetration of the rods in compacted samples, perhaps with mechanical means as well and not just by hand as this makes the use of the sensor not practical in most of the times.

Pre-drilling of pilot holes has been used with many sensors [indicatively 3]. This is an interesting approach which has been found to yield the same readings during comparisons between measurements on the same soil with the sensor inserted by applying force and by pushing the sensor in pre-drilled holes [3]. Also it has been found that for compressible soils, pre-drilling of pilot holes may in fact be the insertion method yielding more accurate measurements [4]. Still it must kept in mind that the motivation behind the findings presented in this paper is whether volumetric water content sensors could be used as a useful alternative during quality control of the construction of earthworks and especially earth-dams, enhancing speed of testing without compromising accuracy. It is the author's experience that the additional time and effort of pre-drilling pilot holes for the sensor rods essentially renders their use strenuous enough to inhibit the required rates of testing during quality control of compaction of geomaterials.

3 Effect of large-size grains

Another factor contributing directly to the penetration ability is the presence of large-size grains. It should be born in mind that even if there is interest only in the material of clay cores, as many specifications allow (for instance [5]) the fraction of large-size grains can be high (indicatively a material containing only 20% of plastic fines, and coarser grains smaller than 76mm can be used as material for the construction of a "clay" core provided it has plasticity index I_p>7, [5]). Obviously fractions of coarse grains of this magnitude can inhibit the insertion of the rods of volumetric water content sensors.

Except for this direct effect of large-size grains, their indirect effect on the volumetric water content of fine-grained soils with sporadic presence of coarse grains was investigated. The material used to construct the clay core of the aforementioned Ferekambos dam on Skiros island, Greece, was used. This is a a clayey silt with sand (w_L=44%, I_p=21, G_s=2.67, w_{opt}=23%, $\gamma_{d,max}$=16 kN/m^3). Compacted samples of the material in the range very wet of optimum (w=30%, 7% higher than optimum moisture content) were prepared and their initial volumetric water content was measured using the GS-3 frequency domain reflectometry (FDR) sensor by Decagon Devices Inc. Once initial values were obtained, compacted samples were cut vertical to their axis and fully saturated disks of various rocks and thicknesses were placed in the cut compacted samples. The gap around the periphery of the rock disks was covered with material of the same density and water content of the compacted samples and the volumetric water content was measured again, placing the sensor in a way that the rock disk was positioned in the middle of the distance between two of the three rods of the sensor. This sensor is shown in Fig. 2a, a disk of limestone placed in a cut compacted sample is shown in Fig. 2b, the sensor inserted for measurement in a sample containing a rock disk is shown in Fig. 2c, and a

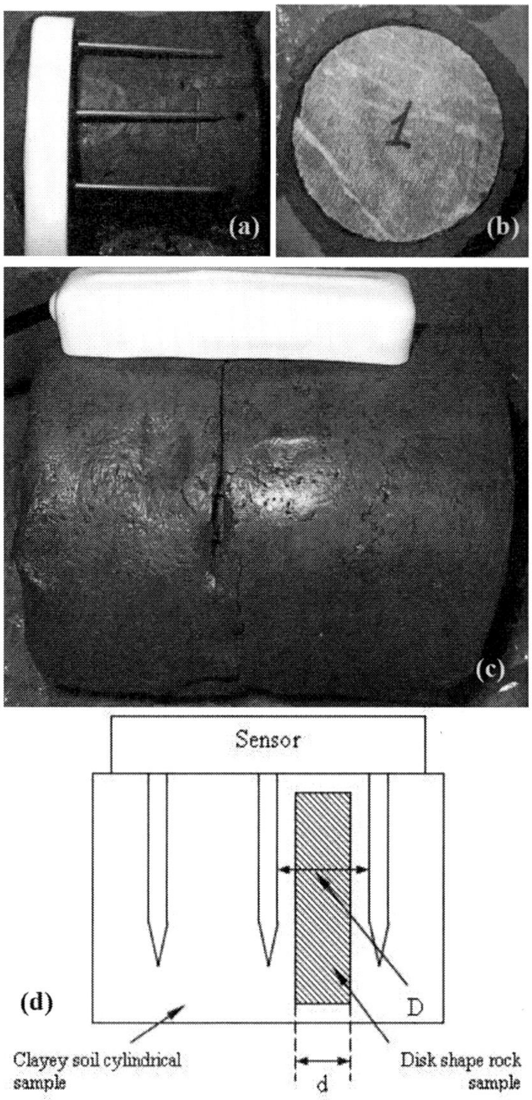

Figure 2. a) The GS-3 sensor before insertion, b) a disk of limestone inserted in the sample, c) insertion of the sensor in a compacted sample containing a rock disk, and d) schematic of the insertion of rock disks and definitions of d and D.

schematic of the insertion of rock disks in the compacted samples is shown in Fig. 2d. It is pointed out that measurements in the material of the samples prior to the insertion of the sensor were also performed on larger volume samples and no statistically significant differences were found in the measurements indicating that there are no boundary effects influencing measurements during this type of insertion of the sensor.

The results of this series of measurements are plotted in Fig. 3. Five different types of rock were used; sandstone, limestone, marble, granite and gneiss. Volumetric water content θ is plotted against the thickness of rock disks d in Fig. 3a. θ generally decreases with increasing rock disk thickness as expected due to smaller volumetric water content of the rock disks. It is notable that there seems to be a common trend

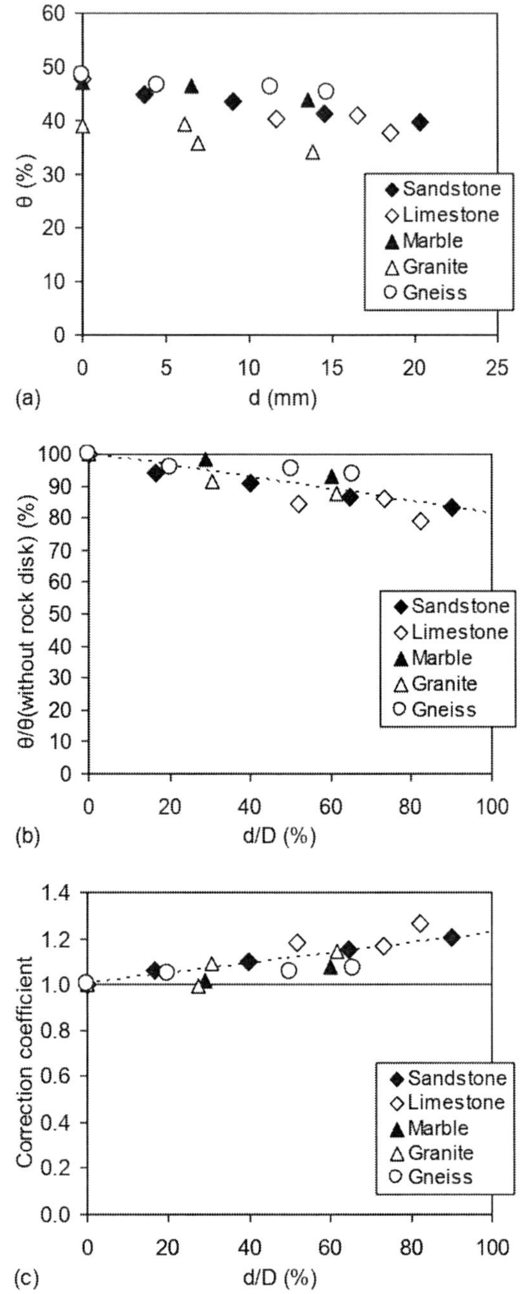

Figure 3. a) Volumetric water content θ vs rock disks' thickness d, b) volumetric water content over that of the initial sample vs the ratio of rock disks' thickness over rods' distance, D, and c) correction coefficient of volumetric water content of sample in the presence of rock disks for five different types of rock vs the ratio of rock disks' thickness over rods' distance, D.

irrespective of the rock type. Volumetric water content was then normalized by dividing it with its initial value measured on the compacted samples before insertion of the rock disks (θ_i) and the thickness of rock disks was normalized by dividing with the distance between the rods of the sensor D. Normalised results are plotted in

Fig. 3b. Again there seems to be a common trend irrespective of the rock type and θ/θ_i decreases with increasing d/D according to Equation 1 (obtained for all the results irrespective of the rock type). If one reverses θ/θ_i in order to obtain a correction coefficient with the d/D ratio, then the results are plotted in Fig. 3c, and Equation 2 yields the correction coefficient a_0 with ratio d/D used in Equation 3. Equations 1 and 2 may by no means be generalised in any respect. Still, this series of measurements and the formulation of Equations 1 and 2 can serve as a guide for investigating the effect of the sporadic presence of coarse grains in predominantly fine-grained materials and the way of correcting measurements with volumetric water content sensors.

$$\theta/\theta i = 100 - 0.187 \cdot d/D \qquad (1)$$

$$a_\theta = 1 + 0.0022 \cdot d/D \qquad (2)$$

$$\theta_{corrected} = a_\theta \cdot \theta_{measured} \qquad (3)$$

4 Indirect effect of density and effect of sample receivers

Except for the direct effect of density on penetration ability, the indirect effect due to the corresponding mechanical behaviour of soils was investigated. In an attempt to use the ML-2x sensor of Delta-T Devices in the quality control of the construction of the Ferekambos dam, the particular device was tested on compacted samples of the Skiros weathered Serpentinite used for the construction of the shoulders of the dam and part of the core. Compacted samples of various densities were prepared for calibration of measurements and these were carried out with the samples still in the moulds they were compacted in (Fig. 4a) and after they were extruded (Fig. 4b) with samples always covered with cling film to avoid drying. As these samples were relatively small (10cm high by 10cm in diameter) one much larger sample in diameter was prepared (Fig. 4c). In order to compare the effect of the presence of the mould with the previous measurements, another sample was prepared with a mould placed around the position where the sensor was measuring (Fig. 4d). Density of the samples was controlled by means of the number of strokes per layer for each of the three layers used to compact the samples. 8, 16, 25 and 50 strokes were used yielding dry unit weights between 16.7 and 17.8 kN/m^3. Measurements immediately after insertion of the sensors were recorded along with measurements after considerable time until readings were stable.

The measurements are presented in Fig. 5. In all cases, the sensor measured smaller values of volumetric water content when the sample had been extruded from the mould. The same was observed when the mould was placed in the larger sample, indicating that the metal of the moulds has the effect of increasing the measured value of volumetric water content. Smaller size of samples on the other hand increases the measured values as observed for a sample of the larger size sample that was trimmed and removed (denoted 'Air' in Fig. 5). The soil used in this series of measurements exhibits dilation

Figure 4. Samples of compacted Skiros weathered serpentinite with the ML-2x probe inserted for volumetric water content measurement a) still in the Proctor test compaction mould, b) after extrusion, c) of larger diameter, and d) of larger diameter but with the mould placed too.

during shear [2] for the densities the samples had. The insertion of sensor rods resembles shear and it is expected therefore to cause dilation of the soil in the vicinity of the sensor rods. Around the rods therefore there is a volume increase which results in attracting moisture from the soil lying further away. This causes the measured volumetric water content to increase with time as was observed (Fig. 5). This was not observed so clearly on the sample of the larger size sample that was trimmed and removed (denoted 'Air' in Fig. 5) as it must have been disturbed, and was not observed at all in the larger diameter sample. The latter cannot be clearly expected, yet this was the

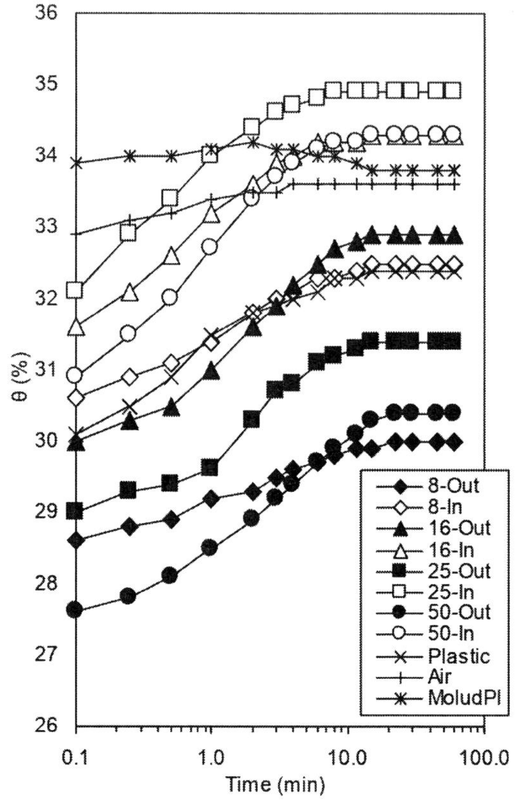

Figure 5. Volumetric water content measured with the ML-2x probe vs time after insertion of the probe in the samples of compacted Skiros weathered serpentinite. Number denotes the number of strokes, 'out' and 'in' that the sample was extruded from the mould or not respectively, 'plastic' the larger sample prepared, 'MouldPl' the larger sample with the mould placed too, and 'air' the sample trimmed from the larger sample without the mould.

most complicated sample to prepare of all samples and uncontrolled variations of densities and water content during preparation cannot be ruled out.

5 Calibration checks

The values of volumetric water content measured during this series of measurements were used in combination with the known volume of each sample and its total mass (exactly the same data available after performing a sand cone density test) to calculate the gravimetric water content and the dry unit weight of the samples (properties required during construction in order to decide whether to proceed with the next layer or not). Both measurements immediately after insertion and after stabilisation for samples inside and outside the moulds were used in these calculations and they were compared to the values obtained after actually measuring the gravimetric water content. These comparisons are shown in Fig. 6a for gravimetric water content and Fig. 6b for dry unit weight. As exhibited by these measurements, *volumetric water content readings immediately after insertion on samples extruded from their samples yield the values of gravimetric water content and dry unit weight closest to the actually measured values*. In the case of the dry unit weight deviation was only 1-3.5% of the measured value with the estimated value being in all cases smaller than the measured value by this magnitude which is on the safe side. For gravimetric water content estimated values were always larger than the measured values by 8-16% of the measured value. Perhaps the most important conclusion from this series of calibration checks for a material exhibiting dilation during insertion of the probe rods (as is usually the case in most compacted materials used in earth-dam construction), is the importance of using readings immediately after insertion as they yield estimations of gravimetric water content and dry unit weight (when combined with measurements of total volume and total mass, as in the case of the sand cone density test) closer to true values. Despite the disadvantage of small probes that are usually pushed in the ground by hand that there may be occasions of very dense and dry soils that they will not penetrate the soil; still, if they do penetrate, they have the advantage of

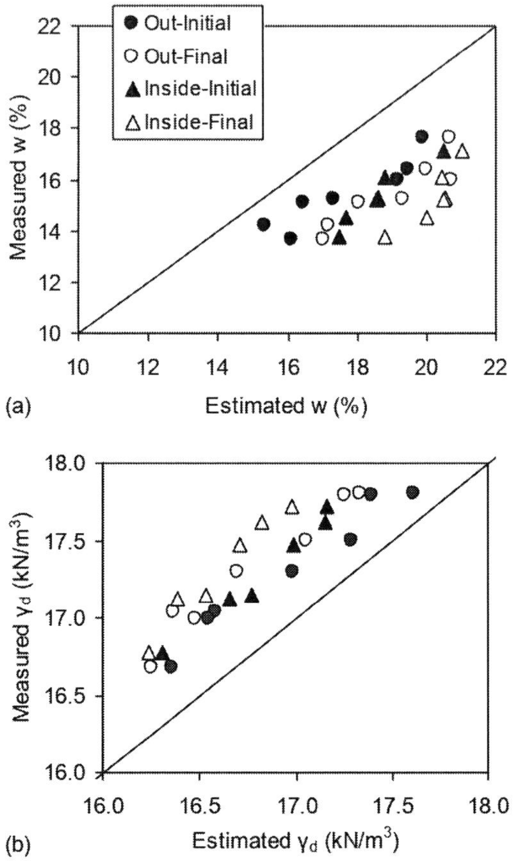

Figure 6. a) Measured vs estimated gravimetric water content, and b) measured vs estimated dry unit weight for samples of compacted Skiros weathered serpentinite from 'initial' and 'final' measurements performed with the ML-2x probe with the samples 'inside' the compaction moulds and 'out' of them.

yielding readings immediately compared to larger sensors with bigger rods that take more time to actually insert the rods in the ground and their readings will most likely be affected by dilation of soil around the rods.

6 An example of use

The ML-2x sensor was used in the quality control of the construction of the Ferekambos dam, especially during construction of the shoulders of the dam from Skiros weathered Serpentinite. It was used in combination with in-situ measurements of density by the sand-cone method and measurements of gravimetric water content by placement in the 110 °C oven. The idea behind its extensive use was that rather than waiting to obtain the gravimetric water content some time after retrieving the sample from the site and putting it into the oven, combine the measured value of volumetric water content on the soil of the in-situ test with the total volume and total mass known immediately after performing the sand cone test to obtain an estimation of gravimetric water content and dry unit weight. These estimations were not used during construction of the particular dam to take decisions on whether to proceed with construction of the next layer or not but rather in order to evaluate possible use of this method for future use. Comparison between measured and estimated values of gravimetric water content and dry unit weight are shown in Fig. 7a and 7b respectively. "1=1" lines are plotted for reference along with best-fit regression lines and for comparison the calibrations obtained on the same material from measurements of volumetric water content immediately after insertion of sensors in laboratory compacted samples extruded from their moulds. In both cases, points scatter practically equally around the "1=1" lines. Scatter was much larger for gravimetric water content (-45 to +70% of the measured values) and a best-fit line did not lie close to the "1=1" line. On the other hand scatter was very low for dry unit weight; only -4.6 to +5.3%, indeed a very low scatter allowing one to rely on this method for decisions regarding the dry unit weight of a compacted layer. It is notable how close the best-fit line lies to the "1=1" line.

7 Conclusions

Small-size volumetric water content sensors were used for various calibration tests on compacted soils used in the construction of earth-dams. Penetration ability of the sensors' rods is the first and foremost requirement. Sporadic presence of coarser grains decreases large values of volumetric water content and the need for an order of magnitude of the appropriate correction of measurements needs to be known if materials like that are encountered. Calibration checks revealed that the estimations of gravimetric water content and dry unit weight closest to their measured values are obtained by use of readings of volumetric water content taken immediately after insertion of the sensors in laboratory compacted samples extruded from their moulds. Comparison between measured and estimated gravimetric water content and dry unit weight on the basis of volumetric water content measured in-situ and combined with sand-cone density tests revealed that estimated values of dry unit weight are very close to the measured ones unlike gravimetric water content values.

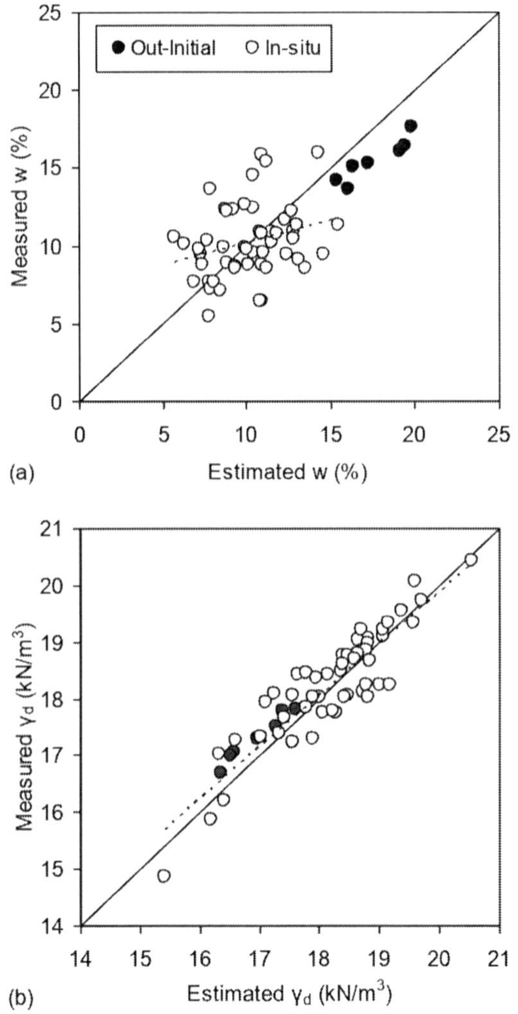

Figure 7. Measured vs estimated a) gravimetric water content, and b) dry unit weight from combination of volumetric water content measurements and sand-cone tests on compacted Skiros weathered Serpentinite during construction of the Ferekambos dam. Solid lines are "1=1" lines, dashed ones are best-fit.

References

1. M. E. Bardanis, S. Grifiza, proc. 5th Int. Conf. Unsaturated Soils, Barcelona, Spain, 2010, 1, 187-192.
2. M. E. Bardanis, S. Grifiza, proc. 6th Hel. Conf. on Geotech. and Geoenvironmental Engineering, Volos, 29 September - 1 October, 2010, Vol. 1, pp. 151-158.
3. G. C. Topp et al., Soil Sci. Soc. Am. J. 46 (1982) 678-684.
4. A. Rothe et al., Water Resour. Res., 3 (1997) 1585-1593.
5. Hellenic Technical Specification 1501-13-01-01-00, Impervious core of zoned earth and rockfill dams.

Evaluation of empirical methods for estimating potential evaporation values in northeast France

Kátia Vanessa Bicalho[1,a], Larissa Camporez Araujo[1], Yu-Jun Cui[2] and Bruno Teixeira Dantas[1]

[1]*Federal University of Espirito Santo, Civil Engineering Department, Vitória, ES, Brazil*
[2]*ENPC, Cermes, France*

Abstract. This study presents and discusses the field monitoring of an experimental embankment divided into two symmetrical sections constructed with two different fine-grained soils in the northeast of France. The field instrumentation included detailed monitoring of volumetric water content, matric suction and temperature within the embankment as well as measurements of meteorological data. The field measurements show that the used instrumentation worked well on monitoring the soil responses to the meteorological changes in the two soils. The study also evaluates different empirical methods for estimating potential evaporation (*ET*) values based on regional meteorological data. The results show a large variability observed on the *ET* values for the investigated site calculated by empirical methods or equations based on meteorological local date.

1 Introduction

It is accepted the importance of determining the changes in the soil properties, especially the changes in the soil water content and corresponding pore-water pressure, in the unsaturated region. The changes of soil water content in a drying soil sample are mainly dependent on the local environmental conditions, the initial water content and temperature values, the hydromechanical properties of the soil, and the boundary conditions at the soil-atmosphere interface.

Severe changes in weather conditions will lead to significant changes in pore-water pressure of the soil resulting in heave or shrinkage and causing damage to the structures. Soil pore-water pressures can be changed under two principal mechanisms: infiltration or evapotranspiration. The water balance calculation enables the geotechnical designer to estimate the amount of water entering, retained within and exiting the earth embankments or fills. Evaporation from soil, which represents an important term in the water balance equation, is a significantly complex mechanism and difficult to quantify ([1]), occurring in the form of combined liquid and vapor transport both at the depth and ground surface.

Under field conditions it is not possible to separate completely the evaporation from transpiration, so the actual evapotranspiration, or the total water loss, is used to describe the amount of evapotranspiration. When soil moisture is depleted, the actual evapotranspiration will be limited by monthly precipitation (lower limit) and potential evapotranspiration (upper limit). The potential evapotranspiration, *ET*, can be defined as the maximum evaporation rate in the case of water availability. In the months when the potential evapotranspiration is less than the rainfall, actual evapotranspiration is closer to the potential value ([2]).

This study presents some results of the field monitoring of an experimental embankment divided into two symmetrical sections constructed with two different fine-grained soils in the northeast of France. This paper focus on understanding of soil-atmosphere fluxes in the embankment system over time, therefore, it places emphasis on the soil suction, volumetric water content, temperature and site-specific meteorological measurements.

The paper evaluates five different empirical methods or equations for estimating *ET* values based on regional meteorological data: ([3]); ([4]); ([5]); ([6]); ([7]). Also, the study examines the use of simple water balance models based on precipitation and potential evaporation to assess soil suction and moisture conditions. It is assumed that the rainfall does not lead to ponding water on soil surface, i.e., the precipitation rate does not exceed the infiltration rate.

2 Materials and methods

2.1 Description of the site and soils

The embankment was constructed at Hericourt, in the Haute-Saone department (Franche-Comte region) in the northeast of France. It is exposed to a continental climate, with oceanic influences. The details of the embankment

[a] Corresponding author: kvbicalho@gmail.com

construction and the experimental program are presented in ([8]).

According to the unified soil classification system, the used two natural soils were classified as: CL, an inorganic clay with low plasticity, and CH, an inorganic clay with high plasticity. The French Soil Classification System classifies these soils into A2 and A4 groups, respectively. The clay-sized particles (≤ 0.002 mm) for the CL and CH soils are about 25-30% and 80-85% respectively. The used soils are quite common in occurrence in the region. Laboratory tests of the natural and treated soils were conducted to evaluate the appropriateness of the chosen treatment formulation and the batching.

The field instrumentation layout was symmetrical for the two sections of the embankment. The embankment consists of 17 layers made of the fill materials compacted to optimum water contents (Standard Proctor tests). The optimum water contents were determined by the intercept of the compaction curve with the degree of saturation line corresponding to 85%.

A layer of the slope, approximately at mid-slope, was selected for the investigations and analyses. The selected layer, located at about 1,8 m from the embankment base, is instrumented with sensors, for measuring suction and water content, located close to allow estimation of the *in situ* soil water retention curves. The field instrumentation program generated a large amount of data; therefore, this paper summarizes some of this data in graphical form from April 2011 to November 2011.

The embankment was also equipped with a site-specific meteorology station on the top surface to record the data every 30 minutes. The year of 2011 had a cumulative precipitation (rainfall) of 773 mm recorded by a French weather station at Luxeuil-les-Bains (located about 50 km from the site-specific meteorological station) compared with the average annual precipitation of 619 mm in France. It is possible to observe the different daily measurements between these two places due to the great distance between them.

In this paper, the meteorological data measured by the site-specific meteorological station is used in the investigations and analysis. The data used in this paper are: average monthly air temperature, *T*, relative humidity, *Rh*, solar radiation, *Rs*, and monthly per cent of daytime hours of the year *p* (see Table 1).

2.2 Description of the ET equations

Several methods or equations can be used to estimate ET from standard meteorological observations. The following five empirical equations were selected to estimate ET in this study: ([3]); ([4]); ([5]); ([6]); and ([7]). The equations are evaluated in this paper due to the advantage that the methods offer in calculating *ET* by using only standard meteorological. Often one chooses a model to estimate *ET* based on the available data to calculate the model. Moreover they are widely accepted methods under a variety of climatic regimes and conditions.

The *ET* equations are briefly summarized here and the cited references are suggested for a more detailed discussion. The *ET* was calculated for the investigated region in 2011 when the weather data were directly measured in the investigated slope.

The empirical ([3]) method is widely used, even though the method is not recommended for use in areas that are not climatically similar to the developed area, in the eastern region of USA, where sufficient moisture water was available to maintain active transpiration. Moreover one should be aware that the soil temperature fluctuates daily and yearly affected mainly by changes in air temperature and solar radiation ([8]). The method for monthly *ET* (mm/month) is:

$$ET'_p = 16\left(\frac{10T}{I}\right)^a \quad (1)$$

where *T* is the mean temperature for the month (in °C) for 0 T 26°C, *I* is the annual thermal index, *i.e.* the sum of monthly indices *I* [*i* = (*T*/5)1.514], *d* is a correction factor which depends on latitude and month. In this paper, *d* is defined equal to 1 ([9]), and,

a= 0.000000675 I 3- 0.0000771 I 2 + 0.0179 I + 0.492

The equation developed by ([4]) defines the daily ET as a function of the air temperature, *T* (°C), relative humidity, *Rh*, and solar radiation, *Rs* ($cal.cm^{-2}.d^{-1}$):

$$E_{Tur} = 0.013.\left(\frac{T}{T+15}\right).(R_s + 50).a_t \quad (2)$$

where a_t =1 if the average daily relative humidity for equal or higher than 50%, otherwise:

$$a_t = \left(1 + \frac{50 - Rh}{70}\right) \quad (3)$$

A combination of temperature and air relative humidity based equation (([5]) was used to estimate *ET*. The equation derived by ([5]) calculates the *ET* (mm/month) based on the mean monthly temperature, *T* (in °C), and air relative humidity, R_h (in %):

$$ET_{Rom} = 0.0018.(25 + T)^2.(100 - Rh) \quad (4)$$

The air relative humidity is the ratio of the absolute humidity to the saturation humidity for the air temperature. The saturation humidity is directly proportional to the air temperature and, the evaporation ceases when the air relative humidity approaches to 100%. Although ([9]) proposed an equation to calculated R_h, in this study the mean monthly relative humidity values measured at the instrumented area are adopted.

The equation proposed by ([6]) is well known in the western USA and has been used extensively elsewhere also ([6]):

$$ET_{BC1} = kp(0.46T + 0.83) \quad (5)$$

where T is mean temperature in °C, p is percentage of total daytime hours for the period used out of total daytime hours of the year (365 x 12), and k is a monthly consumptive use coefficient, depending on vegetation type, location and season. According to ([6]) for the growing season (May to October) k varies from 0.5 for orange tree to 1.2 for dense natural vegetation.

The equation suggested by ([7]) for estimating ET (mm/day) is:

$$ET_{Kha} = 0.34.p.T^{1.3} \quad (6)$$

Soil moisture at a location varies depending on the amount of precipitation (rain or snow) and evaporation. The atmospheric water balance (B), at a particular locality, can be written as:

$$B = P - ET \quad (7)$$

where P is the precipitation and ET is the potential evaporation. During the period of excess water (B positive), there is moisture available for ground-water recharge and runoff.

Table 1. The meteorological data at region from May to October in 2011.

Month (2011)	T (°C)	Rh (%)	P	Rs ($MJ.m^{-2}.dia^{-1}$)
May	13.17	75.1	0.27	3.48
June	17.18	77.9	0.32	3.46
July	16.56	82.8	0.32	3.29
August	18.30	85.8	0.31	3.16
September	15.22	89.3	0.24	2.44
October	9.01	93.0	0.16	1.47

3 Results and discussion

Storms with heavy rainfall exceeding 50 mm/hour were not recorded in 2011 in the investigated site. Some rainy periods of heavy rain (10-50mm/hour) were observed from May to July in 2011. A total of 34 rainfall events of moderate rain (2.5-10mm/hour) was recorded in July 2011, the month with the highest precipitation in 2011. A total of 13 rainfall events of moderate rain (2.5 -10 mm/hour) was recorded in June 2011, and only one moderate rainfall event was recorded in May 2011. Cumulative total rainfall in May, June and July were 37.2 mm/month, 85.8 mm/month, and 120.4 mm/month respectively.

([1]) stated that air temperature and relative humidity gradients are generally not constant with the height above the soil surface, but they are approximately constant at heights of 0.5-2m above the surface. The measurements show that the air temperatures are approximately constant at heights of 0.5m and 1.5m above the surface. A small difference of about 5-10% was observed in the air relative humidity recorded at 0.5m and 1.5m above the surface. The air close to the soil surface is warmer than it is higher up (at 0.5m and 1.5m above the soil surface) from April to August in 2011. The same trend is not observed in the months with lower temperatures (October and November 2011). Wind speeds between 0 and 5.5 m/s were recorded in 2011. The mean monthly wind speed was about 1.0 m/s for the entire monitoring period. Wind speed is important because stronger winds cause more evapotranspiration ([8]).

Figure 1 presents the computed ET values and the measured precipitation (rainfall) for the investigated site in 2011. It is a means of characterizing the local climate. Figure 1 also shows a comparison between the mean monthly potential evaporation values calculated by the selected ET equations. According to the ET methods, ([3]) and ([4]), a water deficit (B negative) was observed from May to October 2011. The ET methods ([6]) and ([7]) indicate a period of water surplus (B positive) from May to September 2011, Hericourt, France. In October 2011, B is negative (water deficit) according to the five adopted ET methods.

The results show a large variability observed on the ET values for the investigated site calculated by empirical methods or equations based on meteorological local date. The difference observed in the five methods can be attributed to variations of measured input data (i.e., air relative humidity) during 2011. The equations of ([3]) and ([6]) considers only temperature as input data while, depending on the season and region, other variables like wind speed, humidity and solar radiation can determine the magnitude of ET. ([5]) considered temperature and air relative humidity as input data. The mean monthly solar radiation and wind speed measured values in the region remained essentially unchanged during the evaluation period; therefore, it may be reasonable the assumption of no influence of the solar radiation and wind speed on the evaporation considered by the used methods for the site in 2011.

Even though, ([1]) showed that the ([3]) equation consistently underestimates the measured ET; ([10]) suggested that the ([3]) method overestimates ET where climate is relatively humid, while for arid and semiarid parts of China it produces an underestimation. ([9]) verified that for the region of their study, the constant 16 of the ([3]) method should be changed to 20. ([6]) suggest the values of k ranging from approximately 0.7 to 0.95. Therefore some calibrations or changes in the models parameters may be required to improve the accuracy of the proposed equations.

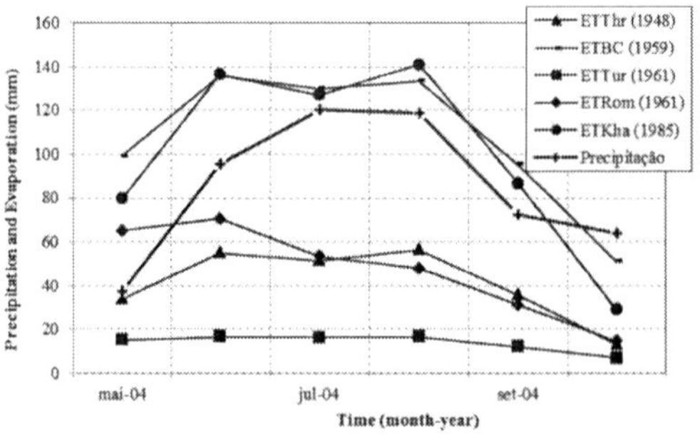

Figure 1. Precipitation (mm/month) and the ET (mm/month) values by the empirical equations from Thr = ([3]); Tur = ([4]); Rom = ([5]); BC = ([6]); Kha = ([7]) from May to October 2011, Hericourt, France.

Figure 2 shows the *in situ* soil suction, volumetric water content and rainfall measurements collected daily from April to November in 2011. Figure 2a shows the *in situ* soil suction measurements for the three sensors installed at the same layer along the face of the embankment constructed with the lime treated clay: CL +2%CaO. Suction is usually less than 200 kPa, and the pressure limit can be determined by the soil suction sensor. At each section of the slope the three Watermark soil suction sensors were placed at 0.25m, 0.5m, and 0.75m from the ground surface (mid-slope face). The measurements show similarities in the daily variations trend of suctions in the two treated fine-grained soils ([8]). The suction values are high near the ground surface, sensors placed at -0.25m of the slope face, during the period of water deficit (i.e., from April to June 2011) in the two investigated soils, and the fluctuation of the suction measurements generally decrease with depth.

The suction values observed in the high plasticity treated soil are generally lower than those at the correponding points in the lower plasticity treated soil, mainly due to the difference in the magnitude of the corresponding measured volumetric water content values: 39%-43% (lime treated CH) and 29-35% (lime treated CL). The maximum volumetric water content of a given soil volum is the saturated volumetric water content or the soil porosity. Considering the soil porosity value of about 53% for the lime treated CH and 43% for the lime treated CL, the degrees of saturation change from 74 to 81% in the investigated layer in the CH +4%CaO section and from 68 to 81% in the investigated layer in the CL +2%CaO section.

The water content of the soil surface (TDR 39, located between SUC08 and SUC11) gradually decreased as the water moved down into the soil or evaporated, and the water content of the deep soil (TDR 38, located at 0.875m from TDR39) remained almost constant in the lime treated CL soil. The same soil moisture response was not observed in the lime treated CH soil, in general, very small changes in the soil moisture were observed, even though the soil temperature values were different in the two slope locations. This can be explained by the differences in the permeability and water retention curves of the two soils ([11]). ([1]) reported that the rate of infiltration is affected by the permeability of the soil, its surface gradient and its water content or suction.

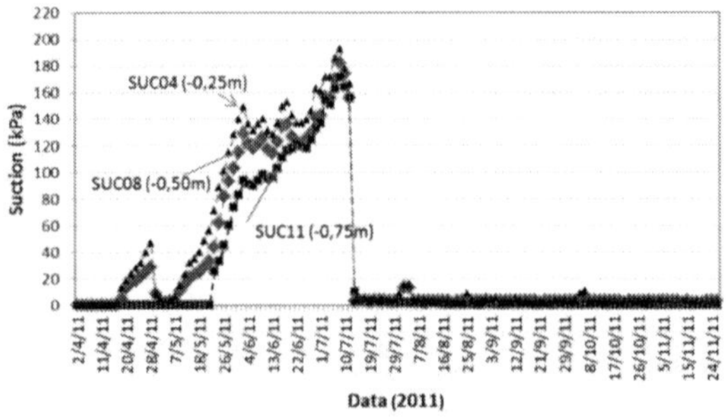

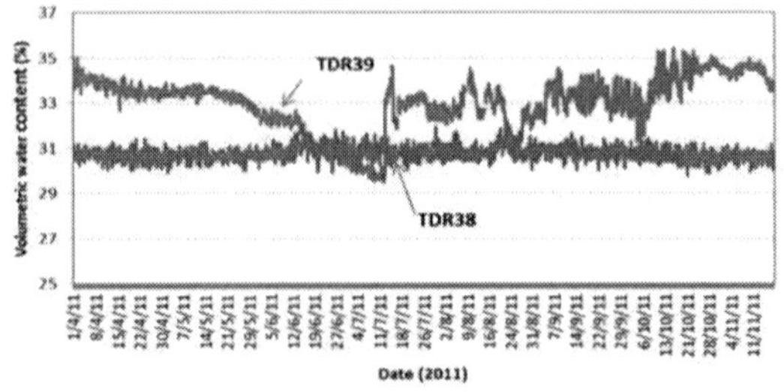

(b)

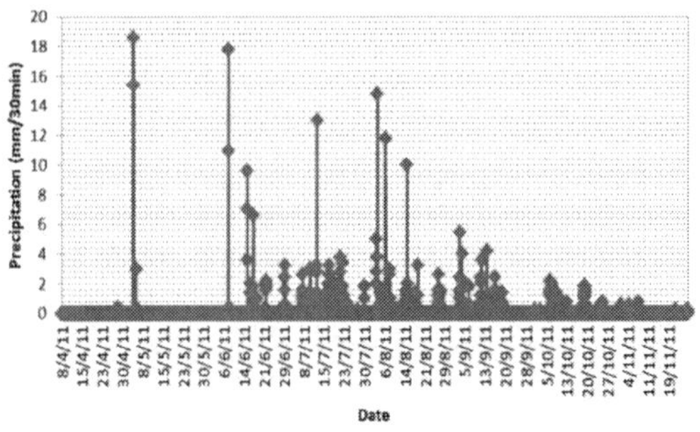

(c)

Figure 2. The daily measured (a) soil suctions, (b) volumetric water contents in different locations from slope face (mid-slope, CL+ 2%CaO section) , and (c) rainfall, from April to November, 2011.

4 Conclusions

In general, no significant response to the rainfall events was observed during the dry period in the treated soils. The rainfall events were recorded in 30 minutes period, and the results show that they have a significant effect on the suction changes at the start of the wet period (i.e., after long dry period) in the two soils.

The atmospheric water balance derived from monthly estimated potential evaporation based on standard meteorological observations may be used to estimate the period of water deficit observed in the daily suction and moisture measurements. However the *ET* obtained by the used methods involves some level of empiricism, and therefore some calibrations or changes in the models parameters may be required to improve the accuracy of the proposed equations.

References

1. E. Blight, Interactions between the atmosphere e the earth. *Geotech.*, 47, 4, 715-767 (1997).
2. C. W. Fetter, Applied Hydrogeology, 3rd ed. Macmillan College Publishing, Inc., New York, 616 p. (1994).
3. C. W. Thornthwaite, An approach toward a rational classification of climate. Geogr. Rev., 38: 55-94 (1948).
4. L. Turc, Estimation of irrigation water requirements, potential evapotranspiration: a simple climatic formula evolved up to date, Ann. Agron., 12, 13-49. (1961).
5. V. A. Romanenko, Computation of the autumn soil moisture using a universal relationship for a large area. Kiev: Ukrainian Hydro meteorological Research Institute, 3 p. (1961).
6. H. F. Blaney, W. D. Criddle, Determining consumptive use e irrigation water requirements. Tech. Bull., U. S. Dep. of Agric. 1275 (1962), 59 p.
7. N. S. Kharrufa (1985), Simplified equation for evapotranspiration in arid regions". Beiträge zur Hydrologie, Kirchzarten, Sonderheft 5.1, 39-47, (1985).
8. K. V. Bicalho, G. P. D. Vivacqua, Y-J Cui, M. Froumentin, D. Mercadier, A.M. Tang, Experimental Investigation of Soil-Atmosphere Interaction in an Instrumented Embankment constructed with two treted clays. Soils & Rocks, 38, 149 - 162 (2015).
9. C. –Y. Xu, V. P. Singh, A review on monthly water balance models for water resources investigations. *Water Resour. Manag* 12 (1998), 31-50.
10. D. Chen; G. Gao; C-Y Xu; J. Guo; G. Ren. Comparison of the Thornthwaite method and pan data with the standard Penman-Monteith estimates of reference evapotranspiration in China. Climate Research, 28:1, p. 123–132. (2005).
11. K. V. Bicalho, Modeling Water Flow in an Unsaturate Compacted Soil. PhD. Thesis, Department of Civil Engineering, University of Colorado, Boulder, 235 p. (1999).

DOI: 10.1051/e3sconf/20160916006

Formulation of a phase field model of multiphase flow in deformable porous media

Giulio Sciarra[1,a]

[1]*Department of Chemical Engineering Materials & Environment, University of Rome La Sapienza, via Eudossiana 18 I-00184, Rome Italy*

Abstract. In this contribution a new model for partially saturated porous media is presented, within the framework of gradient poromechanics, based on a phase field approach. While the standard retention curve is expected still to provide the intrinsic retention properties of the porous skeleton, depending on the porous texture, an enhanced description of the surface tension between the wetting and the non-wetting fluid, occupying the pore space, is stated considering a regularized phase field model based on an additional contribution to the overall free energy depending on the saturation gradient. This approach provides similar results as those of the model based on the concept of specific interfacial area. The dependence of the free energy on the gradient of the saturation also implies that Darcy law must be extended to become a fourth order partial differential equation.

1 Introduction

The constitutive characterization of partially saturated porous media became of interest at the middle of the last century when the scientific research started to face fundamental problems in geotechnics and petroleum engineering, concerning the response of partially imbibed soils, during drainage-imbibition cycles, or modeling the behavior of sedimentary reservoir rocks, when a multi-phase fluid flows through the porous space. Starting from the analysis of basic static problems, it became clear that the balance between capillary and driving forces, in particular gravitational forces, would have been the central subject of the modeling efforts. This pushed the research in the direction of finding out a relation between the curvature of the wetting/non-wetting fluid interface and the average content of the wetting fluid, say the retention curve for the macro-scale capillary pressure. For long time this last has been the only relation used to describe the hydraulic flow through partially saturated porous media, being also the pivot of the hydro-mechanical coupling with the constitutive law of the porous skeleton, see e.g. Alonso et al. [1].

In the same period Cahn & Hilliard [2] established the basic framework of modeling multi-phase flow in terms of space and time evolution of a phase field, continuously varying over thin interfaces. Surface tension is recovered, in this context, considering the integral, through the thickness of the layer, of the phase field gradient. This approach progressively attracted more and more interest, in particular within fluid mechanics, because of its advantages for numerical calculations; however limited contributions attempted at incorporating these ideas into modeling of unsaturated porous media [3].

In this paper a novel general approach is developed which aims at merging phase field modeling of multi-phase flow with unsaturated strain gradient poromechanics. While the standard retention curve is expected to still provide the intrinsic retention properties of the porous skeleton, depending on the porous texture, an enhanced description of surface tension between the wetting and the non-wetting fluid, occupying the pore space, is stated considering a regularized phase field model.

2 Thermodynamics

Thermodynamics of porous media has been summarized by Coussy [4] for two or more monophasic superimposed interacting continua, say the solid skeleton and the fluids saturating the porous space. In this case, the specific internal energies of the fluids are separately defined, whether they are in the liquid or in the gaseous phase; whilst the energy due to interfacial interactions between the fluids and among the solid and the fluids are incorporated into the macroscopic energy of the skeleton. Here a novel approach is adopted in order to account, within the framework of poromechanics, for the role of the interface between a liquid (water) and a gas (wet air), describing the mixture as a non-uniform diphasic fluid, which can reside in the liquid or in the gaseous phase. The fluid internal energy is given by:

$$\mathcal{E}_f = n\rho_f e_f(1/\rho_f, \mathrm{s}_f) + \kappa_f(\mathrm{f}_\rho), \qquad \mathrm{f}_\rho = \delta_{\alpha\beta}(n\rho_f)_{,\alpha}(n\rho_f)_{,\beta} \qquad (1)$$

where $\rho_f e_f$ is a double-well potential parametrized by the mass density, per unit volume of the mixture, ρ_f and the specific entropy s_f, n being the Eulerian porosity. The

[a] Corresponding author: giulio.sciarra@uniroma1.it

non-local energy κ_f penalizes the formation of interfaces and provides a regularization of the non-convex energy $\rho_f e_f$. The state equation of the fluid defines the thermodynamic pressure $\mathcal{P}$ and the chemical potential μ, in terms of the the mass density ρ_f:

$$\mathcal{P} = -\partial e_f/\partial(1/\rho_f), \quad \mu = \partial(\rho_f e_f)/\partial\rho_f \quad (2)$$

however, because of equation (1), these quantities are not sufficient to completely characterize the constitutive law of the non-uniform fluid, which, on the other hand, necessitates of specifying the so-called fluid hyper-stress vector, related to the derivative of κ_f with respect to $(n\rho_f)_{,\alpha}$. In the following incompressibility of the liquid phase is assumed which implies the variation of ρ_f to be univocally determined by the variation of the saturation degree S_r, $\rho_f = \rho_L S_r$, ρ_L being the intrinsic constant density of the liquid phase. The liquid (gaseous) phase therefore corresponds to $S_r=1$ ($S_r=0$).

From now on attention is focused on isothermal processes, a Duffing-like potential for the volumetric energy is assumed:

$$\Psi_f = \rho_f e_f = C(\gamma_{nw}/R)\, S_r^2(1-S_r)^2 \quad (3)$$

where γ_{nw} is the surface tension between the non-wetting and the wetting phase and R the average characteristic size of the channel through which the non-uniform fluid can pass. Let moreover the non-local term κ_f be quadratic in the gradient of nS_r. The first and the second principle of thermodynamics, together with the expression of the internal working relative to a strain gradient porous continuum [5], allow to state the constitutive prescriptions for the overall Piola–Kirchhoff stress and hyper-stress, S_{ij} and P_{ijk}, required to verify the overall balance of momentum $(S_{ij} - P_{ijk,k})_{,j} + b_i = 0$, as well as the generalized constitutive characterization of the capillary pressure $\mathcal{P}_c$ and the fluid hyper-stress vector γ_k. After cumbersome calculations, see [7], one gets

$$S_{ij} = S'_{ij} - (\mathcal{P} - S_r\mathcal{P}_c)\,\delta_{ij} - \gamma_k\,(2E_{ij,k} - \delta_{ij}\,(nS_r)_{,k}/(\phi_0 S_r)) \quad (4)$$

$$P_{ijk} = P'_{ijk} - \gamma_k\,\delta_{ij} \quad (5)$$

$$\phi\,\mathcal{P}_c = -\,\partial\Psi_s/\partial S_r + (\gamma_k/S_r)(J_{,k}/J), \quad \gamma_k/(\phi S_r) = -\,\partial\Psi_s/\partial(\phi S_r)_{,k} \quad (6)$$

ϕ (ϕ_0) being the Lagrangian (initial) porosity, and J the determinant of the Jacobian matrix, relative to the deformation of the overall porous continuum. $S'_{ij} = \partial\Psi_s/\partial E_{ij}$ and $P'_{ijk} = \partial\Psi_s/\partial E_{ij,k}$ are the generalized effective stresses, prescribed in terms of the partial derivatives of the skeleton free energy $\Psi_s = \Psi - \phi\,\Psi_f$ with respect to strain and strain gradient, Ψ being the free energy of the overall porous continuum. Equations (4)-(6) hold true within the assumption of small elastic strains.

Thermodynamics also implies the fluid dissipation to be positive which yields the generalized Darcy law:

$$-\,\mathcal{P}_{,k}/S_r + (\mathcal{P}_c - (\gamma_l/(nS_r))_{,l})_{,k} + b_k^f/(\phi S_r) = A_{kl}M_l/\rho_L \quad (7)$$

where M_l is the Lagrangian filtration vector and A_{kl} the inverse of permeability. Introducing the capillary energy U, so that $\partial\Psi_s/\partial S_r = \phi\,\partial U/\partial S_r$, implies equation (7) to be rephrased as follows:

$$-\,(\partial(\Psi_f + U)/S_r - (\partial\Psi_s/\partial(\phi S_r)_{,l})_{,l})_{,k} + b_k^f/(\phi S_r) = A_{kl}M_l/\rho_L. \quad (8)$$

The role of the capillary energy U is therefore that of modifying the double-well potential Ψ_f which prescribes the free energy of the fluid, in order to account for the wetting properties of the solid skeleton. The new free energy $\Psi_f + U$, which can be called effective energy of the pore-fluid, has not the same minima as Ψ_f, as they are shifted inward the interval (0,1) from below or from above, whether the solid skeleton is gas or liquid wet.

3 Characterization of the pore-fluid

Providing a constitutive characterization of the pore-fluid, say of the fluid within the pore network, is generally achieved, in unsaturated poromechanics, assuming the capillary pressure $\mathcal{P}_c$ to be a prescribed function of S_r. Here the liquid-gas mixture, which saturates the pore space, is regarded as a non-uniform fluid, the corresponding saturation ratio being used to characterize the state of the fluid at any current placement. No distinction is made explicit between the pressure of the liquid and the pressure of the gaseous phase, as well as between the corresponding chemical potentials. Moreover no explicit algebraic relation between the saturation degree and the capillary pressure or the effective chemical potential of the fluid μ^{eff}, defined in what follows, can a-priori be stated, without solving, at least at equilibrium, the generalized Richards equation:

$$d(\phi S_r)/dt - \{K_{sat}\,k(S_r)[\mu^{\text{eff}}_{,k} + b_k^f/(\phi S_r)]\}_{,k} = 0 \quad (9)$$

$$\mu^{\text{eff}} = \partial(\Psi_f + U)/S_r - (\partial\Psi_s/\partial(\phi S_r)_{,l})_{,l} \quad (10)$$

Here the permeability of the porous medium has been assumed isotropic: $(A^{-1})_{kl} = K_{sat}k(S_r)\delta_{ij}$, K_{sat} and $k(S_r)$ being the so-called saturated and relative permeabilities of the wetting phase, respectively. At the stationary state, equation (9) reduces to a fourth order partial differential equation in the space variable, which is definitely similar to the one prescribing the mass density distribution of a Cahn-Hilliard fluid at equilibrium. However a fundamental additional term is here accounted for, say the derivative of U with respect to S_r, which allows for describing the confining effect on the non-uniform fluid, due to the presence of the porous skeleton.

The role of Ψ_f and U in the characterization of the distribution of S_r, μ and $\mathcal{P}_c$ deserves a deeper discussion. Consider the chemical potential of the pure fluid, prescribed by equations (2)-(3), say

$$\mu = \partial\Psi_f/\partial S_r = \rho_L g h_f, \quad h_f = 2C\,(\gamma_{nw}/(\rho_L gR))\,S_r\,(1 - 3S_r + 2S_r^2) \quad (11)$$

and the derivative of the capillary energy, relative for

DOI: 10.1051/e3sconf/20160916006

instance to a sand, and given by the van Genuchten curve

$$\partial U/\partial S_r = -\rho_L g h_U, \quad h_U = (1/\alpha)\{[(S_r - S_r^{res})/(1-S_r^{res})]^{-1/m} - 1\}^{1/n} \tag{12}$$

S_r^{res} being the residual value of saturation; the two curves are depicted in Figure 1 (upper panel, gray dotted and gray dashed lines, respectively). Both the chemical potential and the derivative of the capillary energy are expressed in head units adopting the classical Leverett scaling, which prescribes the characteristic length R in terms of the intrinsic permeability of the soil κ and the initial porosity ϕ_0:$R = (\kappa/\phi_0)^{1/2}$ The profile of the negative chemical potential of the pore-fluid, say $-(\mu+\partial U/\partial S_r)$, plotted against S_r, is also drawn (solid line). It exhibits a non-monotonic behavior and two additional zeros (the spots) with respect to the one at S_r=1, which can always be found, when considering only the capillary pressure $\mathcal{P}_c$. This feature corresponds to the fact that the energy $\Psi_f + U$ maintains the double-well shape typical of Ψ_f, see Figure 1 (lower panel). However the two minima of $\Psi_f + U$ are no more isopotential and the one associated to the smallest value of S_r is shifted inwards the interval (0,1).

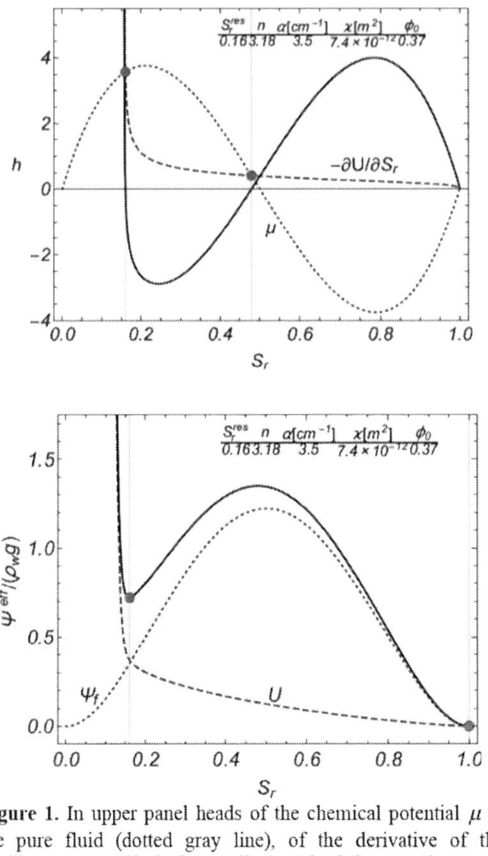

Figure 1. In upper panel heads of the chemical potential μ of the pure fluid (dotted gray line), of the derivative of the capillary energy (dashed gray line) and of the negative pore-fluid chemical potential $-(\mu+\partial U/\partial S_r)$ (solid black line) are depicted; in the lower panel the corresponding energies. The parameters which characterize the retention curve are those of a sand.

This non-monotonic behavior of the pore-fluid chemical potential resembles that one postulated by DiCarlo [7] in order to explain the formation of gravity fingers in soils. However it is worth to underline that the appropriate form of the potential $-\partial U/\partial S_r$ should be captured solving an inverse problem which stems from experimental evidence.

Notice that the effective generalized chemical potential μ^{eff} can be hydrostatic at equilibrium, and consistent with gravitational loading, even if the pore-fluid chemical potential $\mu + \partial U/\partial S_r$ is not.

4 Micro-scale interpretation of the generalized constitutive law of capillary pressure

Equation (6) explicitly improves the basic constitutive prescription of the macro-scale capillary pressure by a correction depending on the gradient of the Jacobian determinant J and the gradient of the water content nS_r, once the quadratic form κ_f for the non-local contribution to the energy of the fluid has been assumed. As a consequence when the hyper-stress acting on the non-uniform fluid vanishes, this correction vanishes as well; in other words the larger the gradient of the liquid content is, the wider the discrepancy between the standard and the enhanced constitutive prescriptions of $\mathcal{P}_c$ will be. The enhanced constitutive prescription of $\mathcal{P}_c$ is therefore expected to yield significant modifications of the capillary pressure, in the narrow subdomains of the current shape of the porous medium where significant gradients of the liquid content can be detected. This justifies an interpretation of this contribution in terms of the so-called specific interfacial area, see [8].

According with experimental data, reported among others in [9], and pore-network numerical simulation, see e.g. [10], local variations of the macroscopic capillary pressure are accompanied not only by changes in the saturation degree but also by changes in the average interfacial area a_{nw}, which accounts for the local cumulative measure of the interfaces between the non-wetting and the wetting phase (per unit volume of the Representative Volume Element, RVE). Even if changes in the interfacial area may not result into a (incoming/ outcoming) flow through the RVE, however, in this analysis, only such kind of variations of a_{nw} are considered. No account, on the other hand, is taken of those variations of the interfacial area which yield a pure remodeling of the internal structure of the RVE. This assumption will be clarified in the following when the homognization scheme adopted to upscale information from the micro to the macro scale is introduced and briefly discussed. Its rationale however resides in the identification of a relation between the interfacial area and the gradient of the saturation degree, to hold true only if the liquid particles are displaced to the boundary of the RVE.

To verify the validity of this hypothesis a heuristic micro-scale analysis is developed deforming a prototype reference configuration $\mathcal{B}_0$ of a RVE by a quasi-static loading path driven by the Lagrangian gradient of the macro-scale Jacobian determinant J; the liquid is squeezed out of the porous chamber along the same direction of the gradient of J. The saturation degree S_r is assumed not to be affected by the considered micro-scale deformation, which means that the ratio between the volume of the liquid and the volume of the pores within the RVE does not vary during the deformation process. This assumption allows to capture the correction to the constitutive prescription of $\mathcal{P}_c$ provided by the gradient term of the fluid energy only. In Figure 2 possible reference configurations of the RVE, parametrized by the degree of saturation are depicted.

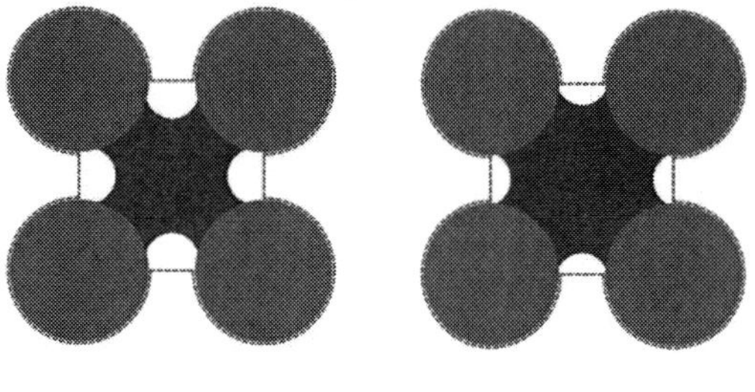

(a) (b)

Figure 2. Referential RVEs parametrized by the saturation degree: (a) S_r= 0.667298; (b) S_r= 0.834704; the dashed line indicates the boundary of the RVE. The distance between the beads, made dimensionless with respect to the characteristic size of the RVE, is L=0.25, whilst their radius is R=(1-L)/2.

Each RVE corresponds to the deformed current configuration $\mathcal{B}$ of a reference domain $\mathcal{B}_0$ constituted by four identical beads among which a suitable amount of liquid water (the wetting phase) is trapped by capillary forces. The amount of liquid (in volume) is prescribed by the value of the degree of saturation; on the other hand its spatial distribution is a function of the local wetting properties of the beads. Here, for the sake of simplicity, the same equilibrium Young contact angle is postulated at each loading step, moreover the interface between the non-wetting and the wetting fluid is kept constantly circular, during the evolution process. Tuning the parametrizing value of S_r provides a-posteriori the relation between the specific interfacial area and the saturation degree, and consequently allows to compare the results of the model with the benchmark experimental data, relative to drainage-imbibition cycles.

A suitable homogenization procedure, which stems from the averaging operations summarized in [11], can be developed in order to provide a motivation of the conjectured relation between the average interfacial area a_{nw} and the macroscopic capillary pressure. The average based homogenization scheme implies the corrective terms in equation (6) not to be zero only if the fluid trapped among the beads attains the boundary, during deformation. In this way the gradient of the liquid content does not fade away. Thus in order to prove a functional correlation between the gradient of ϕS_r and the interfacial area, all the micro-scale deformations which do not drive the liquid to the boundary of the RVE are not taken into account.

Considering the geometrical data reported in the caption of Figure 2, the corrective term of capillary pressure, introduced in equation (6), is explicitly calculated using the above mentioned homogenization techniques.

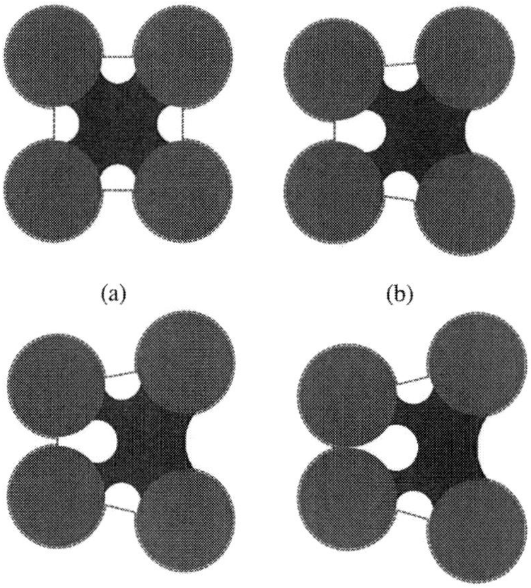

(a) (b)

(c) (d)

Figure 3. Admissible distributions of the liquid phase within different RVEs, obtained as the current configurations of the reference shape of panel (a). The micro-scale displacement of the beads is parametrized by the value of $J_{,1}$ via its dependence on the macro-scale second gradient of deformation. During the deformation the saturation ratio is kept constant: S_r= 0.667298.

At the same time the average interfacial area consistent with the envisaged liquid distribution, can be estimated for each value of $J_{,1}$. A contour plot of the constitutive law of $\mathcal{P}_c$, parametrized by the value of a_{nw}, can therefore be drawn, see Figure 5, which qualitatively resembles that one traced in Figure 11 of [10], for the values of the saturation degree which have been considered in the present analysis.

The considered deformation and flow regimes are however totally different with respect to those simulated in [10]. As a matter of facts they are driven by the gradient of the Jacobian J, at different values of S_r, whilst those ones of [10] are deduced simulating many scanning loops of drainage and imbibition. This is indeed an interesting point which corroborates the idea, introduced in [8], of resolving the hysteresis of the capillary pressure, between drainage and imbibition, introducing a suitable micro-structural parameter, say a_{nw}, which tunes the value of $\mathcal{P}_c$ for each value of S_r.

It is worth to underline that the considered micro-scale analysis stems from the assumption of non-negligible strain gradient, which in fact parametrizes the deformation process, even if only small strain of the solid skeleton are considered. If, on the other hand, this assumption were not valid, equation (6) whould abruptly simplify into the standard constitutive prescription of capillary pressure.

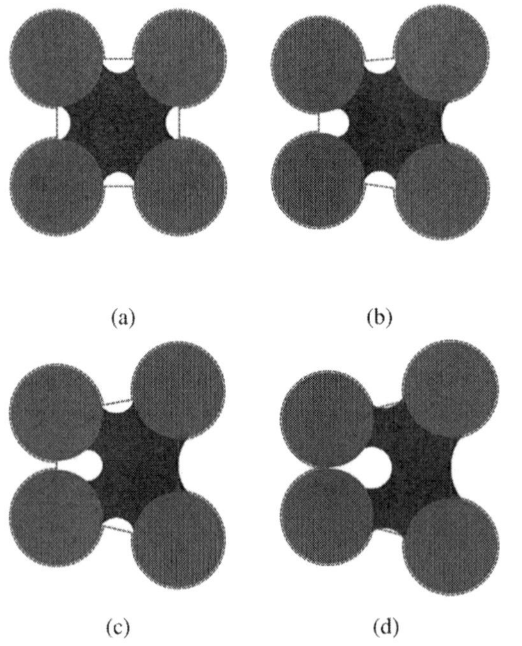

Figure 4. Admissible distributions of the liquid phase within different RVEs, obtained as the current configurations of the reference shape on panels (a). The micro-scale displacement of the beads is parametrized by the value of J_1 via its dependence on the macro-scale second gradient of deformation. During the deformation the saturation ratio is kept constant: S_r= 0.834704.

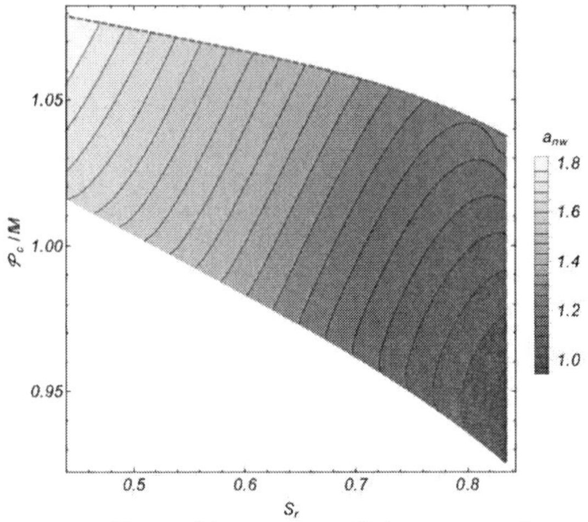

Figure 5. The capillary pressure $\mathcal{P}_c$ is represented as a function of the saturation ratio S_r, within the interval delimited by the considered maximum and minimum value of S_r and parametrized by the interfacial area a_{nw}.

5 Conclusions

In this paper gradient theory of poromechanics endowed with phase field modeling have been used to describe partial saturation. The spatial distribution of saturation, and strain, is regularized at phase coexistence, within the non-uniform fluid, and in the neighbors of possible heterogeneities in the porous skeleton. To do this the free energy of the overall porous medium has been regarded as a function not only of strain and saturation but also on their (Lagrangian) gradients.

As usual in gradient theories, the governing partial differential equations relative, in this case, to the solid skeleton and the non-uniform fluid are deduced using integration by parts twice, which implies the equations to be, in general, of the fourth order. An enhanced version of classical Richards' equation has been deduced from generalized Darcy's law, which in the considered regime of partial saturation does not depend only on the capillary pressure, but also on the so-called generalized chemical potential.

A novel constitutive characterization of the capillary pressure is established and a micro-scale interpretation is provided, which is consistent with that originally formulated in [8]. A comparison with experimental results also confirms that the contribution to capillary pressure due to the saturation gradient allows for recovering similar effects as those captured by the average interfacial area.

References

1. E.E. Alonso, A. Gens, A. Josa, Géotechnique **40**, 405–430 (1990)
2. J.W. Cahn. J.E. Hilliard, The Journal of Chemical Physics **28**, 258–267 (1958)
3. P. Papatzacos, S.M. Skjæveland, SPE Journal **9**, 47–56 (2004)
4. O. Coussy, *Poromechanics* (John Wiley & Sons Ltd, 2004)
5. G. Sciarra, F. dell'Isola, O. Coussy, International Journal of Solids and Structures, **44**, 6607–6629 (2007)
6. G. Sciarra, Phase field modeling of unsaturated porous media, submitted (2015)
7. D.A. DiCarlo, R. Juanes, T. LaForce, T.P. Witelski, Water Resources Research **44**, 2 (2008)
8. S.M. Hassanizadeh, W.G. Gray, Advances in Water Resources **16**, 53–67 (1993)
9. E. Dalla, M. Hilpert, C.T. Miller, Journal of Contaminant Hydrology **56**, 25–48 (2002)
10. V. Joekar-Niasar, M. Prodanović, D. Wildenschild, S.M. Hassanizadeh, Water Resources Research **46**, 6 (2010)
11. L. Dormieux, D. Kondo, F.J. Ulm Microporo-mechanics (John Wiley & Sons Ltd, 2006)

E3S Web of Conferences **9**, 16007 (2016) DOI: 10.1051/e3sconf/20160916007
E-UNSAT 2016

Estimation of the empirical model parameters of unsaturated soils

Salima Bouchemella[1,a] , Ibrahim Alimi Ichola[2] , Ahcène Séridi[3]

[1]*INFRARES Laboratory, Civil engineering Department, Univ-SoukAhras, Algeria*
LGCIE Laboratory, INSA- Lyon, France.
Civil engineering Department, Univ-Boumerdes, Algeria

Abstract. For each flow modelling in the unsaturated soils, it is necessary to determine the retention curve and the hydraulic conductivity curve of studied soils. Some empirical models use the same parameters to describe these two hydraulic properties. For this reason, the estimation of these parameters is achieved by adjusting the experimental points to the retention curve only, which is more easily measured as compared with the hydraulic conductivity curve. In this work, we show that the adjustment of the retention curve θ (h) is not generally sufficient to describe the hydraulic conductivity curve K (θ) and the spatio-temporal variation of the moisture in the soil θ (z). The models used in this study are van Genuchten- Mualem model (1980-1976) and Brooks and Corey model (1964), for two different soils; Gault clay and Givors silt.

1 Introduction

In practice, the retention curve θ (h) is easy to measure compared to the hydraulic conductivity curve K (θ). Therefore, some formulations, based on statistical pore-size distribution methods, have been proposed to predict the unsaturated hydraulic conductivity function K (θ) from knowledge the retention curve θ (h). Authors who have adopted this approach are numerous (Childs & Collis-Georges, Burdine, Marshall, Campbell, Mualem, Fredlund and Xing) [1-2-3-4-5-6]. The most widely models used are Burdine (1953) and Mualem (1976). Among a large number of the retention curve models proposed, only few can easily incorporate into these pore-size distribution models such as the functions proposed by Brooks and Corey, Brutsaert and van Genuchten [7-8-9]. Brooks and Corey used the Burdine model to predict K (θ), for against van Genuchten used Mualem model, usually noted van Genuchten-Mualem model.

For this reason, these empirical models estimate the hydrodynamic properties θ (h) and K(θ) using the same parameters, These parameters are obtained usually by fitting the experimental points of the retention curve θ (h) only, the hydraulic conductivity curve K (θ) is deduced after.

The aim of this work is to verify if the calculated hydraulic conductivity curve with these adjusted parameters can describe the measured one. And can also provide the spatio-temporal variation of the moisture in the soil θ (z). The models used in this study are the combined model of van Genuchten-Mualem (1980-1976). and Brooks and Corey model (1964). The choice of models is based on a comparative study conducted by Sillers [10] cited by Fredlund and Houston [11]. This choice depends also on the difference between the expressions of the models, their popularity and their use in the literature. This study was carried on two different soils: Gault clay and Givors silt.

2 hydraulic properties

The expressions of the water retention characteristics curve θ (h) and the hydraulic conductivity curve K (θ) of the used models in this work are defined as follows:

2.1 van Genuchten-Mualem model

The combined model of the hydraulic conductivity and retention curve van Genuchten-Mualem (1980-1976) is currently the most used model. Many authors have considered it as appropriate to a large range of soil, especially for fine soils [12-13]. This choice model takes also into consideration the strong nonlinearity of the hydrodynamic properties.

$$\theta_e = \frac{\theta - \theta_r}{\theta_s - \theta_r} = [1 + (\alpha h)^n]^{-m} \quad (1)$$

$$K(\theta) = K_s\left(\frac{\theta - \theta_r}{\theta_s - \theta_r}\right)^{0.5}\left[1 - \left[1 - \left(\frac{\theta - \theta_r}{\theta_s - \theta_r}\right)^{\frac{1}{m}}\right]^m\right]^2 \quad (2)$$

Where: θ_e= normalized volumetric water content; θ_s= volumetric water content at saturation [L^3/L^3]; θ_r= residual volumetric water content [L^3/L^3]; K_s= hydraulic conductivity at saturation [$L.T^{-1}$]. α = parameter related to the entry air pressure [L^{-1}]; n=a dimensionless coefficient

[a] Salima Bouchemella: sali.bouchemella@gmail.com

related to the pore size distribution with n≥1; m= Mualem coefficient defined by: m =1-1/n [5].

2.2 Brooks and Corey model

The simplicity of the expression of the Brooks and Corey model (1964) made that it is often used in numerical models to study unsaturated media; it is based on the assumption of the existence of the air entry pressure. Brooks and Corey used the Burdine model to predict hydraulic conductivity. θ (h) and K (θ) are written as follows:

$$\theta = \theta_s \text{ For } h \leq h_{ae} \quad (3a)$$

$$\theta_e(h) = \frac{\theta - \theta_r}{\theta_s - \theta_r} = \left(\frac{h_{ae}}{h}\right)^N \text{ For } h > h_{ae} \quad (3b)$$

$$K(\theta) = K_s\left(\frac{\theta - \theta_r}{\theta_s - \theta_r}\right)^{\frac{M}{N}} \quad (4)$$

Where: θ_e=normalized volumetric water content; θ_s= volumetric water content at saturation [L^3/L^3]; θ_r= residual volumetric water content [L^3/L^3]; K_s= hydraulic conductivity at saturation. [$L.T^{-1}$]; h_{ae}=the air entry pressure [L]; N = an empirical parameter often referred to as the pore size distribution index; M= a constant defined as M=2+3N [2].

2.3 Geotechnical characteristics of the tested soils

The tested soils in this study are the Gault clay and Givors silt. The geotechnical characteristics of these soils determined by Bentoumi [14-15] are presented in the table 1:

Table 1. Soil Geotechnical Characteristics

Properties	Gault clay	Givors silt
% element < 80μm	100	92
% element < 2μm	69	15
LL (%)	40	33
WL (%)	19	23
IP (%)	21	10
W_{opt} (%)	17.5	16.3
K_s(cm/mn)	1.5 10^{-6}	1.2 10^{-4}
θ_s (cm^3/cm^3)	0.365	0.355
θ_r (cm^3/cm^3)	0.125	0.025
(γ_d/γ_w)	1.77	1.73

3 Estimation of the parameters

3.1 Retention curve

Estimated parameter values for the studied soils are listed in tables 2 and 3, for van Genuchten-Mualem and Brooks and Corey empirical models respectively. The parameters values are obtained by fitting the models (equation 1 and 3) to the measured points of the retention curve θ (h) [14-16] using Curve Expert software _1.3. In general, the correlation coefficient will range from 0 to 1, with a correlation coefficient of 1 being the best. But in some peculiar circumstances, Curve Expert gives (r) greater that one, which an unrealistic values. This is indicative of a very poor data model.

From table 2 and 3, the correlation coefficient values (r) reflect the good accuracy of the retention model parameters in describing observed data.

Table 2. Values of van Genuchten-Mualem model parameters adjusted from θ (h)

Parameters	Gault clay	Givors silt
α (cm^{-1})	0.001975	0.002987
n	1.163	1.2845
m	0.149	0.221
r	0.971	0.9726

Table 3. Values of Brooks and Corey model parameters adjusted from θ (h)

Parameters	Gault clay	Givors silt
h_{ae} (cm)	190.54	77.446
N	0.0998	0.1498
M	2.299	2.449
r	0.978	0.989

3.2 Hydraulic conductivity curve

The calculated hydraulic conductivity curves obtained by the empirical model (equation 2 and 4), using the adjusted parameters listed in table 2 and 3 respectively for van Genuchten-Mualem model and Brooks & Corey model, are compared with the measured hydraulic conductivity curve determined by the instantaneous profiles method (Indirect measurement) [14-15]. The purpose of this comparison is to verify if the calculated k(θ), using equation 2 and 4, can describe the measured one, and if the estimation of the parameters by fitting the measured θ(h) only is sufficient. This comparison is done by determining the correlation coefficient, as is shown in tables 4 and 5 respectively for van Genuchten-Mualem model and Brooks & Corey model. These measured and calculated curves for the both models are shown in figure 1 for Gault clay, and in figure 2 for the Givors silt.

Table 4. Correlation coefficient value obtained by comparing the calculated and measured K(θ) in case of van Genuchten-Mualem model

Parameters	Gault clay	Givors silt
Measured K_s (cm/mn)	1.5 10^{-6}	1.2 10^{-4}
m	0.149	0.221
r	-	-

Table 5. Correlation coefficient value obtained by comparing the calculated and measured K (θ) in case of Brooks and Corey model

Parameters	Gault clay	Givors silt
Measured K_s (cm/mn)	1.5 10^{-6}	1.2 10^{-4}
N	0.0998	0.1498
M	2.299	2.449
r	-	-

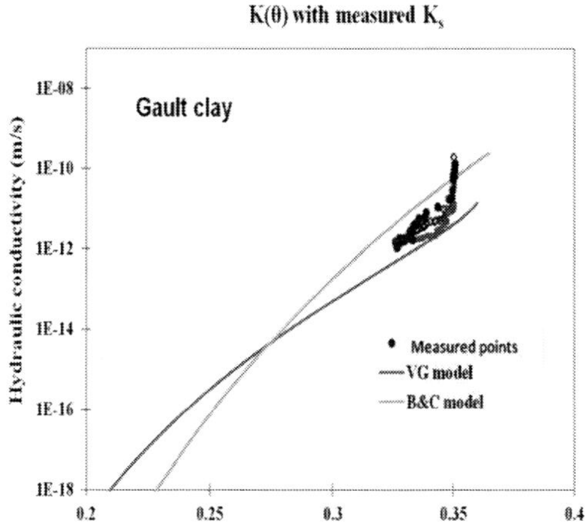

Figure 1. Hydraulic conductivity curves calculated with the measured value of K_s compared to the measured one for Gault clay ,

From tables 4 and 5 and from figure 1 and 2, it is found that the fitted parameters values of van Genuchten-Mualem and Brooks and Corey models from the retention curve cannot represent fairly the experimental hydraulic conductivity curve for the both soils. No (r) value was obtained in each case, which indicates no agreement between the calculated and measured curves of K (θ).

Because of the high difference observed, two adjustments are performed. The first one is to determine the parameters values of the empirical models by fitting the hydraulic conductivity curve. In the second one, a new value of K_s is estimated from measured hydraulic conductivity curve, by conserving the parameters adjusted from the water retention curve.

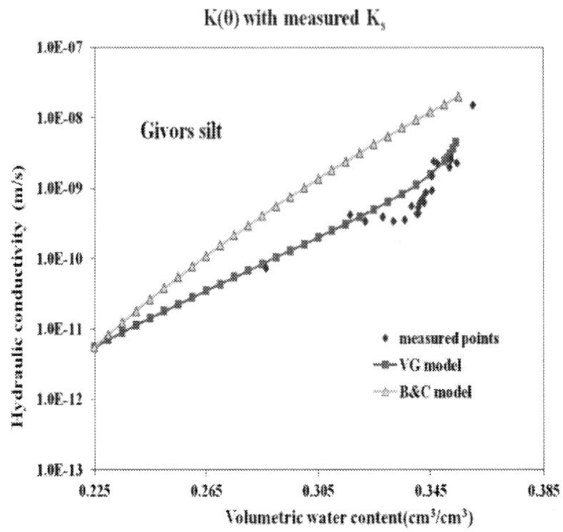

Figure 2. Hydraulic conductivity curves calculated with the measured value of K_s compared to the measured one for Givors silt.

3.2.1 Estimation of the model parameters by adjusting K (θ) keeping measured K_s.

The first adjustment of the hydraulic conductivity curve is to keep the value of K_s and to seek a new value of (n) for van Genuchten-Mualem model, and (N) for the Brooks & Corey model. Subsequently a comparison of the measured retention curve and that calculated with these new values is performed. The results of this adjustment are shown in tables 6 and 7.

Table 6. New values of van Genuchten-Mualem model parameters adjusted from K (θ) using measured K_s

Soils	n	m	r
Gault clay	0.456	0.313	0.363
Givors silt	1.358	0.264	0.668

Table 7. New values of Brooks and Corey model parameters adjusted from K (θ) using measured K_s

Soils	N	M	r
Gault clay	2.92	9.76	1.67
Givors silt	3.734	13.202	2.93

The correlation coefficients values obtained by fitting the measured hydraulic conductivity curves K (θ) are low. For van Genuchten-Mualem model they are of the order of 0.363 for Gault clay and of 0.668 for Givors silt. For Brooks & Corey model, the correlation coefficients values obtained for the two soils are greater than 1, this is indicative of a very poor data model. In addition, this adjustment does not determine the values of α and h_{ae}. So, the retention curve θ (h) of each model can't be defined. This leads us to not accept the found values.

 DOI: 10.1051/e3sconf/20160916007

3.2.2 Estimation of a new value of K_S keeping the adjusted parameter from θ (h).

In the second adjustment, and knowing that the permeability value K_s is obtained by the instantaneous profiles method (indirect measurement method), we try to find the best value of K_s which can give a good correlation of the two curves of K(θ) (measured and calculated), keeping the parameters values obtained by adjusting the retention curve θ(h) .

Table 8. Calculated values K_s for van Genuchten-Mualem model

Soils	Calculated K_s (cm/mn)	r
Gault clay	1.914 10^{-5}	0.630
Givors silt	7.98 10^{-5}	0.818

Table 9. Calculated values K_s for Brooks and Corey model

Soils	Calculated K_s (cm/mn)	r
Gault clay	1.92 10^{-6}	0.668
Givors silt	1.26 10^{-5}	0.770

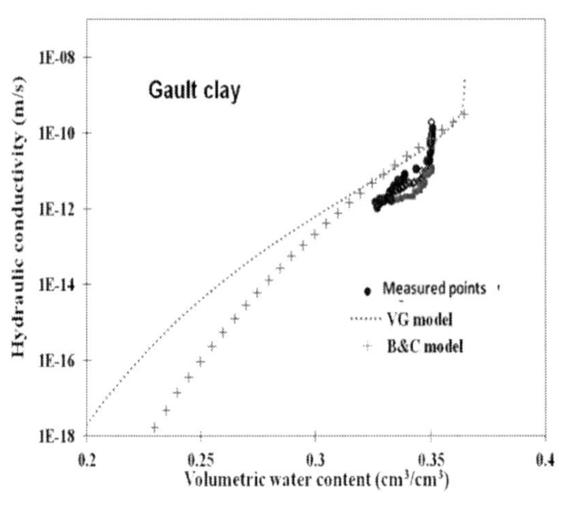

Figure 3. Hydraulic conductivity curves calculated with the new value of K_s for Gault clay.

Figure 3 and tables 8 and 9 show that for the Gault clay, the result of the second adjustment is not satisfactory; the correlation coefficients values are of the order of 0.630 for the van Genuchten-Mualem model and K_s is equal to 1.914 10^{-5} (cm/min), and of the order of 0.668 for Brooks & Corey model and K_s is equal to 1.92 10^{-6} (cm/min). The new estimated value of K_s is obtained with a low correlation coefficient because of few measured points used. Indeed the volumetric water content ranges from θ_i=0.325(cm³/cm³) to θ_s=0.365 (cm³/cm³), when θ_r= 0.125(cm³/cm³).

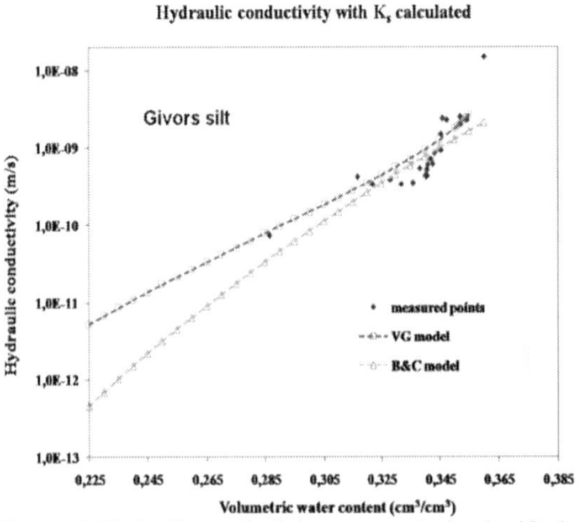

Figure 4. Hydraulic conductivity curves calculated with the new value of K_s for Givors silt.

But for the Givors silt a good correlation is obtained with the new estimated value of K_s (tables 8 and 9). They are of the order of 0.818 for the van Genuchten-Mualem model, and of 0.770 for the Brooks & Corey model. Figure 4 shows a clear improvement of the calculated curves comparing to the figure 2.

4 hydraulic profiles

Hydraulic profiles θ (z,t) present the spatio-temporal variation of the moisture in the soil. θ (z,t) are determined by the resolution of the Richards equation given by:

$$\frac{\partial \theta}{\partial t} = \frac{\partial}{\partial z}[K(\theta)\frac{\partial}{\partial z}(h+z)] \tag{5}$$

Where: θ: soil volumetric water content [L^3/L^3], t: time [T], K: hydraulic conductivity [L/T], h: the water pressure head [L], z: the depth [L].

In this study we use the numerical model developed by Bouchemella [16-17] based on resolution of capacitive form of Richards's equation, which is written as follows:

$$C(h)\frac{\partial h}{\partial t} = \frac{\partial}{\partial z}(K(h)\frac{\partial h}{\partial z} + K(h)) \tag{6}$$

Where: $C(h) = \partial\theta / \partial h$ is the specific soil water content capacity [L^{-1}], (h> 0 is a suction)

To solve equation (6), θ (h) and K (θ) are defined using the parameter values adjusted from measured θ (h) and by using also, the measured value of K_s for Gault clay and estimated value of K_s for Givors silt. In order to test the impact of the choice of the fitting method on describing the hydraulic profiles θ (z,t). In this section only the van Genuchten- Mualem model is used.

4.1 Gault clay

The tested problem is a vertical infiltration simulation conducted on 25 cm long soil column. The flow domain is a homogeneous Gault clay layer. van Genuchten-Mualem empirical model is used, with the parameters

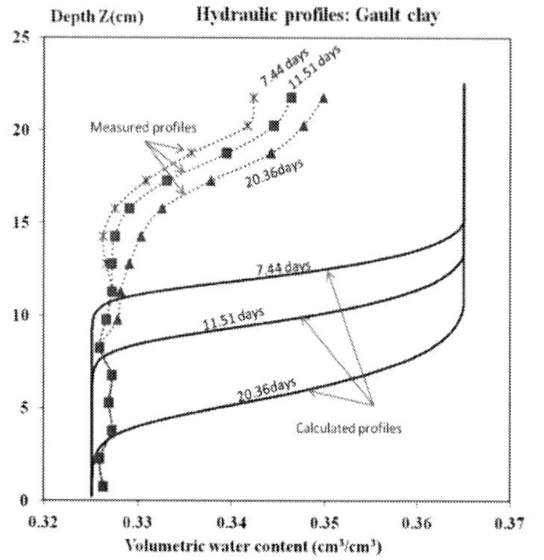

Figure 5. Hydraulic profiles of Gault clay

values listed in table 2 and the measured value of K_s. A water head pressure (h_0=-100 cm) is imposed at the top of the column combined with a zero flux at the bottom of the column. The soil was initially assumed in wetted a state with initial moisture content θ_i=0.325(cm³/cm³). The calculated hydraulic profiles are confronted to the measured ones under the same boundary and initial conditions by obtained by Bentoumi [14, 15] as shown on figure 5.

The infiltration test was carried out on wet initial state close to saturation θ_i=0.325 corresponding to the degree of saturation about S_r=93.39%, so the swelling potential is relatively low. And with no measurement of swelling soil during wetting path by Bentoumi [14, 15], the effect of change volume it is not takes account in this study.

From figure 5, we can observe that the calculated hydraulic profiles of Gault clay are in very ahead with respect to the measured ones. When wetting front Z_f at time 7.44 days is 10.5 cm obtained by the computed profile, it is equal to 4 cm from the measured one. So the infiltration estimated is faster than the measured one. We can deduce for the Gault clay that the parameters value of the hydraulic properties adjusted from the retention curve only, can't describe the hydraulic conductivity curve, and also the spatio-temporal variation of the moisture in the soil θ (z,t).

4.2 Givors silt

The simulation was carried out on a 25 cm long soil column. A zero water head pressure is imposed at the top of the column combined with a zero flux at the bottom of the column. The initial water content value is θ_i=0.215(cm³/cm³), the same as the one used in experimental tests. van Genuchten-Mualem empirical model is used, with the parameters values listed in table 2 and the calculated value of K_s (table 8). The results are shown in Figure 6.

Figure 6 shows that the computed profiles of Givors silt are close to the measured ones, especially at the time

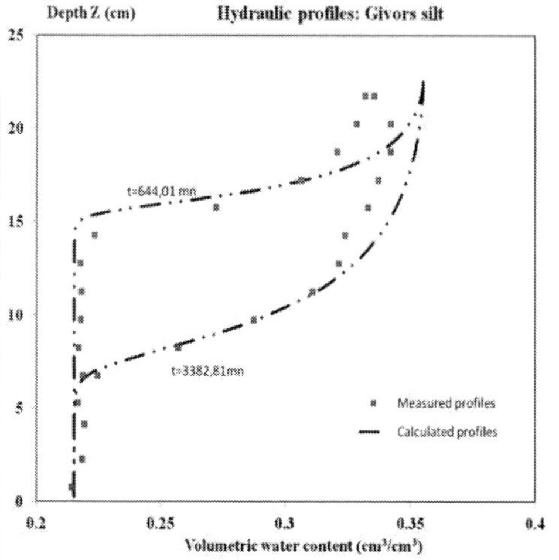

Figure 6. Hydraulic profiles of Givors silt

3382.81 mn. A slight difference of surface saturation is found. So we can deduce for the Givors silt, that the parameters value of the hydraulic properties adjusted from the retention curve can describe the hydraulic conductivity curve with a slight correction of the value of K_s. Therefore the spatio-temporal variation of the moisture in the soil θ (z,t).

5 Conclusions

In this study, we have shown that the parameters adjustment of the empirical models describing the hydraulic properties (retention curve and hydraulic conductivity), from the measured points of the retention curve, does not necessarily lead to well describe the curve hydraulic conductivity, and provide the progress of the moisture front presented by the water profile (case of the Gault clay).

We have also shown that some corrections made on the hydraulic conductivity at saturation (knowing that the value of the latter is vitiated by the errors) with keeping the adjusted parameters from the retention curve only, can lead to the good description of the hydraulic conductivity curve and the hydraulic profiles also (case of the Givors silt).

References

1. E. C. Childs, & G. N. Collis-George, *The permeability of porous materials*. Proc. Roy. Soc. London, Ser. A, **201**, pp. 392-405, (1950)
2. N.T. Burdine. *Relative Permeability Calculation from Size Distribution Data*. Pet. Trans. Am. Min. Metal. Pet. Eng., **198**, 71-78. (1953)
3. T.j. Marshall. *A relation between permeability and size distribution of pores*. J. Soil Sci., **9**, 1-8. (1958).
4. G.S. Campbell, *A simple model for determining unsaturated hydraulic conductivity from moisture retention data*. Soil Sci., **117**, pp. 3 11-3 14, (1974).

DOI: 10.1051/e3sconf/20160916007

5. Y. Mualem. *New model for predicting the hydraulic conductivity of unsaturated porous media*. Water Resour. Res., **12**, 513-296. (1976)
6. D.G. Fredlund, & A. Xing,. *Equations for the soil-water characteristic curve*. Can. Geotechn. Journal, **31(3)**, 521–532, (1994).
7. R. H. Brooks and A. T. Corey. *Hydraulic properties of porous media*. Hydrology Paper, **3**, Colorado state university, Fort Collins, CO. (1964).
8. W. Brutsaert, *Probability laws for pore size distributions*. Soil Science, **101(2),** 85-92, (1966)
9. M. TH. van Genuchten. *A closed form equation for predicting the hydraulic conductivity of unsaturated soils*. Soil SCi. Am. J., **44**, 892-898. (1980).
10. W. Sillers, *The mathematical representation of the soil-water characteristic curve*. M.Sc. thesis, University of Saskatchewan, Sask, Canada. (1996).
11. D. G. Fredlund , & S. L. Houston , *Protocol for the assessment of unsaturated soil properties in geotechnical engineering practice*. Canadian Geotechnical Journal, **46**, N°6,694-707(14) (2009).
12. D. Russo, *Determining soil hydraulic properties by parameter estimation: On the selection of a model for the hydraulic properties*, Water Resources Research, **24**, 453-459, (1988).
13. M. TH. van Genuchten, , F. J. Leij, & S. R. Yates, *The RETC Code for Quantifying the Hydraulic Functions of Unsaturated Soils*. Report N0. EPA/600/2-91/065, R. S. Kerr. Environ. Res. Laboratory, US Environmental Protection Agency, Ada, Oklahoma, p.85. (1991).
14. O. Bentoumi, and I. Alimi-Ichola. *Experimental determination of the hydraulic conductivity of an unsaturated soil in laboratory*. Bulletin of the International Association for Engineering Geology. Paris, **53**, 21-27(1996).
15. O. Bentoumi, *Transfer by infiltration in unsaturated compacted fine soils. Study of the diffusivity and conductivity*.Thesis, LGCIE Laboartory, Insa- Lyon, France (1995).
16. S. Bouchemella, A. Séridi, and I. Alimi-Ichola, *Numerical simulation of water flow in unsaturated soils: comparative study of different forms of Richards's equation*. Eur. J. Env. Civil Eng., **19**, Issue1,1-26(2015).DOI:10.1080/ 19648189.2014.926294.
17. S. Bouchemella. *Contribution to the numerical simulation of water flow in unsaturated porous media*. Thesis, Univ Guelma. Algeria (2015).

A New Mechanism of Canopy Effect in Unsaturated Freezing Soils

Jidong Teng[1,a], Sheng Zhang[1], Zuoyue He[1] and Daichao Sheng[1,2]

[1] *National Engineering Laboratory for High-Speed-Railway Construction, Central South University, China*
[2] *Centre of Excellence in Geotechnical Science and Engineering, the University of Newcastle, Australia*

Abstract. Canopy effect refers to the phenomenon where moisture accumulates underneath an impervious cover. Field observation reveals that canopy effect can take place in relatively dry soils where the groundwater table is deep and can lead to full saturation of the soil immediately underneath the impervious cover. On the other hand, numerical analysis based on existing theories of heat and mass transfer in unsaturated soils can only reproduce a minor amount of moisture accumulation due to an impervious cover, particularly when the groundwater table is relatively deep. In attempt to explain the observed canopy effect in field, this paper proposes a new mechanism of moisture accumulation in unsaturated freezing soils: vapour transfer in such a soil is accelerated by the process of vapour-ice desublimation. A new approach for modelling moisture and heat movements is proposed, in which the phase change of evaporation, condensation and de-sublimation of vapor flow are taken into account. The computed results show that the proposed model can indeed reproduce the unusual moisture accumulation observed in relatively dry soils. The results also demonstrate that soil freezing fed by vapour transfer can result in a water content close to full saturation. Since vapour transfer is seldom considered in geotechnical design, the canopy effect deserves more attention during construction and earth works in cold and arid regions.

1 Introduction

Significant frost heave damage was reported to airfield runways in an airport in north-western China, with runways broke-down or even shattered at some positions. It was found that a weak zone of subgrade soil formed along the centreline of the runway where the water content was observed to be close to full saturation (Li et al., 2014[1]). The airport is located in cold and arid north-western China. The typical annual precipitation in this region is as low as 328 mm, and the annual evaporation can be as high as 1438 mm. Moisture accumulation due to rainwater infiltration was considered unlikely as the pavement was much intact and impervious during the previous raining season (summer). The annual minimum and maximum temperature are around -20 °C and 35 °C, respectively. The groundwater table remains more than 20 m deep, eliminating the possibility of capillary action influence on the water content of the surface soil. The question is then: where is the water in the subgrade soil coming from?

According to Li et al (2014)[1], the pavement acts as a 'pot cover' and prevents evaporation, leading to moisture accumulation under the pavement. They used the terminology of 'pot cover effect', to reflect the fact that the moisture migration was driven by a temperature gradient in the ground and prevented by an impervious cover. However, as will be shown in the paper, this terminology does not represent what truly occurred under the airfield runway. The temperature gradient alone cannot be the only driving force for the moisture accumulation. There must exist some other mechanism that enables the moisture to accumulate, most likely in form of vapour transfer when the groundwater table is 20m deep. The biological terminology of canopy effect is borrowed in this paper, to reflect the dual mechanisms of vapour transfer. Canopy effect in plant biology refers to the phenomenon where crowns of individual trees or forest act as a special layer of absorption for energy from sunlight and water from tree roots. Significant water is consequently stored in the tree leaves for photosynthesis. The paper will provide a more detailed justification for the use of the terminology to refer to the large amount of moisture accumulation under impervious covers.

The objective of this work is to demonstrate and reproduce the unusual canopy effect observed in relatively cold and dry soils. Numerical simulation of coupled heat and mass transfer based on existing theories is first carried out, in attempt to study the effects of impervious covers on the moisture distribution in unsaturated soils. The limitations of existing theories are demonstrated through numerical examples. A new mechanism of vapour transfer in unsaturated freezing soils is then proposed, in which the phase changes of evaporation, condensation and desublimation during coupled heat and mass transfer are taken into account. The new model is then used to analyse the canopy effect observed in the airfield runways. Finally, some conclusions are drawn based on the numerical results and discussions.

[a] Corresponding author: jdteng@csu.edu.cn

 DOI: 10.1051/e3sconf/20160916008

2 Analysis of canopy effect using the existing theory

2.1 Canopy effect under impervious covers

When the ground surface is covered by an impervious cover, as shown in Figure 1, the moisture content profiles under the cover (profile A) and outside the cover (profile B) will be different. In addition, two possible moisture profiles may exist under the impervious cover: profile A_1 for a relatively high groundwater table where capillary action is the dominant mechanism of moisture transfer, and profile A_2 for a low groundwater table where vapour transfer is the dominant mechanism.

The occurrence of profile A_1 is straightforward, and indeed the only difference between profile A_1 and profile B is that the moisture content of the former is somewhat higher than the latter, due to the prevention of evaporation of the cover. The occurrence of profile A_2 is however less straightforward and is seldom mentioned in the literature. The case reported by Li et al (2014)[1] refers to profile A_2 in Figure 1. Eigenbrod and Kennepohl (1996)[2] also reported that significant moisture accumulation due to vapour transfer occurred underneath a road pavement and led to significant frost heave damage to the pavement. As profile A_2 represents a deep groundwater table and a dry foundation soil, issues associated with high moisture content are more likely to be ignored in geotechnical design. Therefore, the situation of profile A_2 represents a real threat to engineering practice, and is referred to as the canopy effect in the paper. The main objective here is to reveal the true physical mechanism of profile A_2 and to reproduce it via numerical modelling and laboratory experiments. This paper focuses on the numerical modelling, while the experimental verification will be discussed in a separate paper.

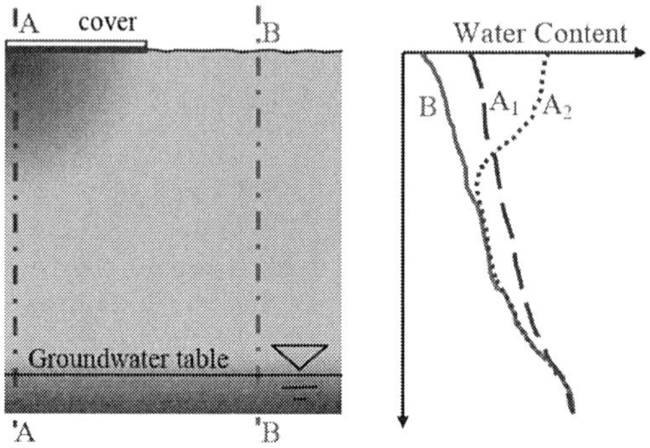

Figure 1. Canopy effect underneath an impervious cover.

2.2 Mathematical model

The theory of coupled moisture and heat transfer in unsaturated soils was pioneered by Philip and de Vries (1957)[3], who developed a mathematical model to describe the coupled movement of liquid water and water vapor under non-isothermal conditions (PDV model). Milly (1984)[4] adopted the matric head instead of the total volumetric water content in the PDV model. Nassar and Horton (1989)[5] further extended Milly's work by including osmotic effects on liquid and water vapor movement. Sakai et al. (2009)[6] presented a coupled model for water and vapor movement which considers condensation and evaporation effects under non-isothermal and low water content conditions.

According to the PDV model, the flow of liquid water and vapor in an unsaturated soil are driven either by a temperature gradient or by a matric head gradient. The governing equation can thus be expressed as:

$$\frac{\partial\theta}{\partial t}=\frac{\partial\theta_L}{\partial t}+\frac{\partial\theta_v}{\partial t}=\frac{\partial}{\partial z}\left[K_{Lh}\left(\frac{\partial h}{\partial z}+1\right)+K_{LT}\frac{\partial T}{\partial z}+K_{vh}\frac{\partial h}{\partial z}+K_{vT}\frac{\partial T}{\partial z}\right] \quad (1)$$

where θ is total water content ($m^3\ m^{-3}$); θ_L and θ_v are liquid water content and equivalent vapor content, respectively ($m^3\ m^{-3}$); h is water heat (m); T is temperature (K); z is the spatial coordinate positive upward (m); K_{Lh} ($m\ s^{-1}$) and K_{LT} ($m^2\ K^{-1}\ s^{-1}$) are the isothermal and thermal hydraulic conductivities for liquid water fluxes due to gradients in h and T, respectively; K_{vh} ($m\ s^{-1}$) and K_{vT} ($m^2\ K^{-1}\ s^{-1}$) are the isothermal and thermal vapor hydraulic conductivities, respectively (Teng et al., 2015[7]).

Considering vaporization and condensation processes in unsaturated soil, the governing equation of heat conservation is given as (de Veries, 1958[8]; Nassar and Horton, 1992[5]):

$$C_p\frac{\partial T}{\partial t}+L_w\rho_w\frac{\partial\theta_v}{\partial t}=\frac{\partial}{\partial z}\left[\lambda(\theta)\frac{\partial T}{\partial z}\right]-C_L\frac{\partial q_L T}{\partial z}-L_w\rho_w\frac{\partial q_v}{\partial z}-C_v\frac{\partial q_v T}{\partial z} \quad (2)$$

where C_p is defined as the sum of the volumetric heat capacity of the unsaturated soil ($MJ\ m^{-3}\ K^{-1}$) and is equal to the volumetric average of the solid (C_n= 1.92 $MJ\ m^{-3}\ K^{-1}$), liquid (C_L= 4.18 $MJ\ m^{-3}\ K^{-1}$), and air (C_v=6.3 $KJ\ m^{-3}\ K^{-1}$) phases (de Vries, 1958[8]); L_w is the latent heat of water vaporization ($J\ m^{-3}$); $\lambda(\theta)$ is the soil thermal conductivity ($W\ m^{-1}\ K^{-1}$); and q_L and q_v are the liquid water flux and vapor flux, respectively ($m\ s^{-1}$).

Neglecting hysteresis between drying and wetting curves, the van Genuchten model (van Genuchten, 1980[9]) is adopted to simulate the Soil Water Characteristic Curve (SWCC), as follows:

$$\Theta=\left[1+(-\alpha h)^n\right]^{-m} \quad (3)$$

where Θ is the effective saturation (dimensionless) and $\Theta=(\theta_L-\theta_r)/(\theta_s-\theta_r)$, θ_s and θ_r are the saturated and residual water content ($m^3\ m^{-3}$), respectively; α (m^{-1}), n (dimensionless) and m (=1-1/n) are empirical shape parameters. The isothermal unsaturated hydraulic conductivity of liquid water in both frozen and unfrozen soil is modeled following Mualem (1976) [10]:

$$K_{Lh}=K_s\Theta^l[1-(1-\Theta^{1/m})^m]^2 \quad (4)$$

DOI: 10.1051/e3sconf/20160916008

where K_s is the saturated hydraulic conductivity (m s^{-1}); l is a fitting parameter and was set to 0.5 by Mualem (1976)[10]. The definitions of thermal hydraulic conductivity of liquid water K_{LT}, and hydraulic conductivities of vapor due to temperature K_{vT}, and matric head K_{vh} refer to Saito et al. (2006) [11]:

$$K_{LT} = K_{Lh}(hG_{wT}\frac{1}{\gamma_0}\frac{d\gamma}{dT}) \quad (5)$$

$$K_{vh} = \frac{D}{\rho_w}(\rho_{vs}\frac{Mg}{RT}H_r) \quad (6)$$

$$K_{vT} = \frac{D}{\rho_w}\eta H_r(\frac{d\rho_{vs}}{dT}) \quad (7)$$

respectively. In Eq. (5), G_{wT} is the gain factor which amends the temperature dependence of the SWRC; γ is the surface tension of soil water (J m^{-2}); and γ_0 is the surface tension at 25 °C (= 71.89 g s^{-2}). In Eqs. (6) and (7); M is the molecular weight of water (= 0.018 kg mol^{-1}); g is the gravitational acceleration (m/s^2); R is the universal gas constant (= 8.341 J mol^{-1} K^{-1}); H_r is the relative humidity (dimensionless); η is an enhancement factor (dimensionless); ρ_{vs} is the saturated vapor density (kg m^{-3}); and D is the vapor diffusivity in soil (m^2 s^{-1}). The parameter H_r relates vapor density to matric head and temperature and is expressed is as follows:

$$H_r = \exp\left(\frac{hMg}{RT}\right) \quad (8)$$

Definitions of other parameters used in this model can be found in Teng et al. (2015) [7].

2.3 Numerical analysis for canopy effect

Equations (1) and (2), subjected to appropriate initial and boundary conditions, are solved numerically using the finite element method for spatial discretization and the finite difference method for temporal discretization. The solution procedure is facilitated by the COMSOL Multiphysics package (5.0). In order to reveal the mechanism of canopy effect, two cases are designed for computation as presented in Table 1. A one-dimensional soil column is chosen for the analysis, with a height of either 5 m or 20 m. The water table is located at the bottom of the soil column, while the top boundary either has an evaporating flux (10 cm/d) or is completely sealed (zero flux). The initial water content is set to 10% and is uniform over the depth. The initial temperature along the soil profile is 15 °C. The top and bottom boundary temperatures are constantly maintained at 0°C, and 15°C, respectively, for a period of 90 days.

Considering the engineering background of canopy effect as stated in Li et al. (2014) [1], this study chose the silty soil in Lanzhou area for analysis. The Soil Water Characteristic Curve of the silty soil is given in Figure 2, which was simulated by the van Genuchten model. Figure 3 presents the simulated unsaturated permeability by the Mualem model.

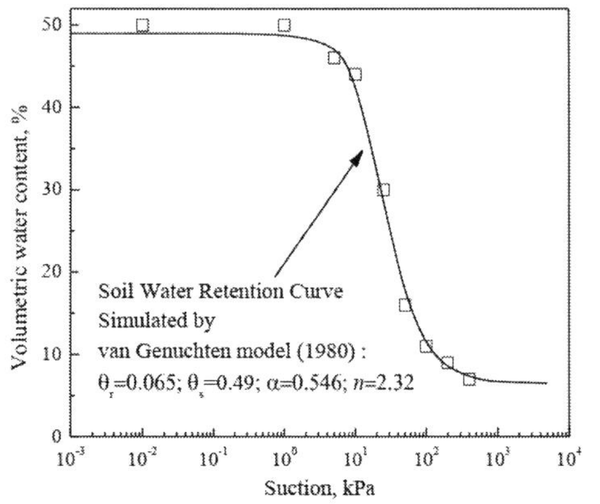

Figure 2. SWCC of silty soil at Lanzhou area, with symbols representing measured data, solid line the van Genuchten model.

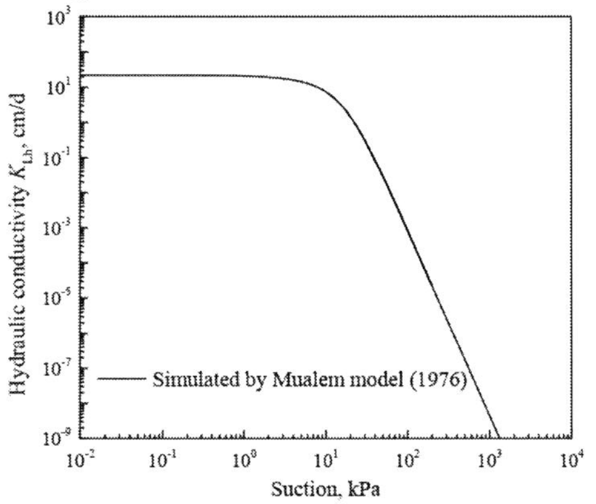

Figure 3. Unsaturated hydraulic conductivity of silty soil at Lanzhou area, simulated by the Mualem model.

The computed results of case (1) are shown in Figure 4. The figures numbered (a) and (b) denote the groundwater table located at 5 m and 20 m, respectively. The impervious cover affected the moisture content to a depth of 50 cm when the groundwater table is located at 5 m, and a depth of 100 m when the groundwater table is located at 20 m. The maximum difference between the covered and uncovered moisture profiles is about 4% of volumetric water content, which takes place at the ground surface. It is also interesting to note that the moisture content underneath the cover does not increase above the initial value of 10%, which is above the residual moisture content. The capillary rises in Figure 4(a) and 4(b) are around 4.5 m. The impervious cover seems to stop the evaporation at the ground surface, but fails to accumulate significant moisture. The canopy effect reported in Li et al. (2014)[1] is not apparent in this case.

Similar results shown in Figure 4 have also been obtained using commercial software such as Hydrus-1D[12]. Test runs using the COMSOL model and Hydrus-1D fail to reproduce a significant amount of moisture accumulation under impervious covers, which implies that existing models based on the PDV theory is incapable of generating the canopy effect of type A_2. A closer inspection of the PDV theory reveals its shortcomings. According to the PDV theory, liquid water flow and vapour flow is driven either by a matric suction gradient or by a temperature gradient. The vapour density is related to the matric suction and temperature through Eq. (8), while the matric suction is related to the moisture content via the SWCC (Eq. (3)). If evaporation is stopped by an impervious cover, or vapour condenses into liquid water, the liquid water content in the soil increases and the matric suction consequently decreases, which in turn increases the vapour density and reduces the moisture flow. Therefore, existing models based on the PDV theory predict an equilibrium state where moisture flow driven by temperature gradients in unsaturated soils is balanced by that driven by the suction gradient.

Such a case is very similar to a cup of hot water covered by a cold ceramic lid. Some vapour condensation will take place at the cold ceramic lid, but not much water will disappear from the cup, as the evaporation will stop when there is sufficient liquid water under the lid. However, the situation will change if the ceramic cover is replaced by a cover that absorbs water, for example, a cover made of dry wood. The water in the cup will transfer to the absorptive cover until it is fully saturated. The key question is then: what is the absorption mechanism for a moisture profile like A_2 shown in Figure 1? Or alternatively, under what mechanism can the moisture content continue to increase without increasing vapour density or decreasing matric suction?

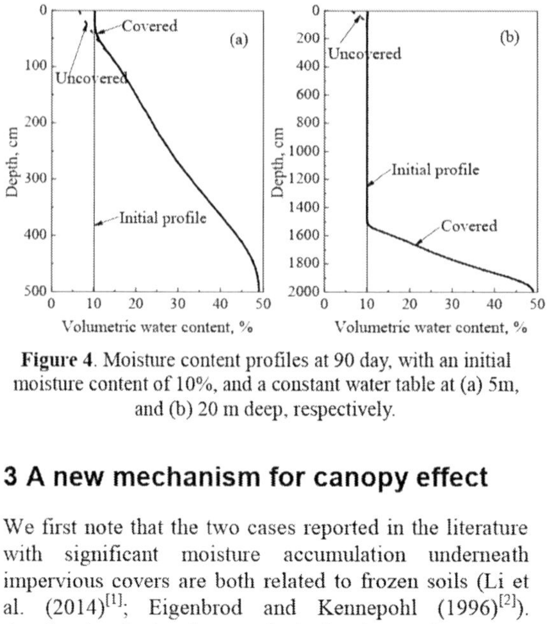

Figure 4. Moisture content profiles at 90 day, with an initial moisture content of 10%, and a constant water table at (a) 5m, and (b) 20 m deep, respectively.

3 A new mechanism for canopy effect

We first note that the two cases reported in the literature with significant moisture accumulation underneath impervious covers are both related to frozen soils (Li et al. (2014)[1]; Eigenbrod and Kennepohl (1996)[2]). Another inspirational example is that ice tends to grow continuously inside an old type freezer (the non-frost-free type) as long as warm and humid air enters the freezer. Realising these facts, we propose here a new mechanism of canopy effect as follows.

When the temperature in a soil drops below the frost point, vapour under an impervious cover will change into ice through desublimation. The formation of ice will further reduce air humility and increase the matric suction, which in turn accelerates the vapour transfer in the soil. In other words, the vapour-ice desublimation acts like a sink term in the moisture transfer equation, and releases energy (latent heat) in the heat transfer equation. This process could lead to significant accumulation of total water content (including ice) under the cover, and possibly a moisture profile of type A_2. The terminology 'canopy effect' is justified because of the absorption effect of the vapour-ice desublimation process.

A specific laboratory experimental program is designed to validate the above mechanism. The test results strongly support the freezing induced canopy effect. However, due to the page limitations, the experimental program and the test results will not be presented here. Instead, this paper focuses on the numerical simulation of the phenomenon.

3.1 Theory for vapour transfer in unsaturated freezing soil

Unsaturated soil mechanics so far mainly applies to soils with super-freezing temperatures. Thus, a model for coupled liquid water, vapour, and heat movements in unsaturated soils subjected to both sub- and super-freezing temperatures is not available in the literature. Such a model is proposed here, to reproduce the canopy effect observed in field.

In a sub-freezing condition, water exists in soil pores in three forms: liquid water, water vapour and ice. The endothermic and exothermic processes are accompanied with phase changes. Here, we extend the governing equations for coupled heat and mass transfer, i.e. Eq. (1) and (2), to include phase change. The mass conservation is expressed as,

$$\frac{\partial\theta}{\partial t}=\frac{\partial\theta_L}{\partial t}+\frac{\partial\theta_v}{\partial t}+\frac{\rho_i}{\rho_w}\frac{\partial\theta_i}{\partial t}= \frac{\partial}{\partial z}\left[K'_{Lh}\left(\frac{\partial h}{\partial z}+1\right)+K_{LT}\frac{\partial T}{\partial z}+K_{vh}\frac{\partial h}{\partial z}+K_{vT}\frac{\partial T}{\partial z}\right] \tag{9}$$

where θ_i is pore ice content (m^3 m^{-3}); K'_{Lh} is the permeability in unsaturated freezing soil (m/s) and differs from the permeability of unsaturated soil K_{Lh}. The expression of K'_{Lh} refers to Taylor and Luthin (1978)[13]:

$$K'_{Lh}=10^{-10\theta_i}K_{Lh} \tag{10}$$

The adjusting parameter $10^{-10\theta_i}$ denotes the obstruction of pore ice to vapour flow. When the soil temperature is over 0°C, θ_i becomes zero, and K'_{Lh} reverts into K_{Lh} as in Eq.(3).

The energy balance in an unsaturated freezing soil is described as follows,

$$C_p\frac{\partial T}{\partial t}-L_i\rho_i\frac{\partial\theta_i}{\partial t}+L_w\rho_w\frac{\partial\theta_v}{\partial t}= \frac{\partial}{\partial z}\left[\lambda'(\theta)\frac{\partial T}{\partial z}\right]-C_L\frac{\partial q_L T}{\partial z}-C_v\frac{\partial q_v T}{\partial z}-L_w\rho_w\frac{\partial q_v}{\partial z} \tag{11}$$

The four terms on the right-hand side of Eq. (11) are the soil heat flow by conduction, the convection of sensible heat due to flowing water, the transfer of latent heat by vapor diffusion, and the transfer of sensible heat by vapor diffusion, respectively. In Eq. (11), $\lambda'(\theta)$ is the heat conductivity of unsaturated freezing soil (W m^{-1} K^{-1}), and

DOI: 10.1051/e3sconf/20160916008

is obtained by replacing θ_L of $\lambda'(\theta)$ in Eq. (2) with $\theta_L+F\theta_i$ (Hansson et al., 2004)[14].

As for the movement of moisture and heat in the frozen zone, an additional constitutive relation is required to solve the governing equations. Here, the generalized Clapeyron equation is used to establish the relationship between the pressure of unfrozen liquid water and that of pore ice when these two phase are in equilibrium (Sheng et al., 2014)[15]:

$$\theta_L = (\theta_s - \theta_r)\left[1+(-\alpha\frac{L_f T}{gT_0})^n\right]^{-m} + \theta_r \qquad (12)$$

where T_0 is the freezing point of water (K); and T in this equation is in unit of °C.

3.2 Numerical reproduction of canopy effect

The proposed theory was numerically solved by using COMSOL Multiphysics (5.0), with vapour transfer and ice formation taken into account. In attempt to numerically reproduce the canopy effect using the proposed theory, we design a numerical example as follows. A soil column of 20 m in height is subjected to a temperature of -5 °C at its top and a temperature of 15 °C at its bottom. The top boundary of the soil is sealed for any water flux. The initial water content is 7%, and the computation period is 90 days.

Figure 5 shows the computed profiles of total water content ($\theta_L+\theta_v+\theta_i$), temperature, liquid water flux and vapour flux along the soil depth. Figure 5(a) depicts that the water content near the lower end is significantly raised from the initial value, with a capillary acting between 16-20 m at 90 days. The water content between 2.5 m and 16 m stays close to the initial level. For the soil within the top 2.5 m, the water content shows a significant increase, especially for the soil at top 1.0 m where a saturated water content of 42% is reached. Figure 6(b) gives the depths of the freezing front (0°C line), which is located at about 0.2 m, 0.5 m, 0.8 m respectively for 5 days, 30 days and 90 days. These positions coincide with the depths of maximum water content in Figure 5(a). Moreover, Figure 5(b) shows that the temperature profile below 8 m remains constant during the freezing period, which agrees with the real condition in Lanzhou area.

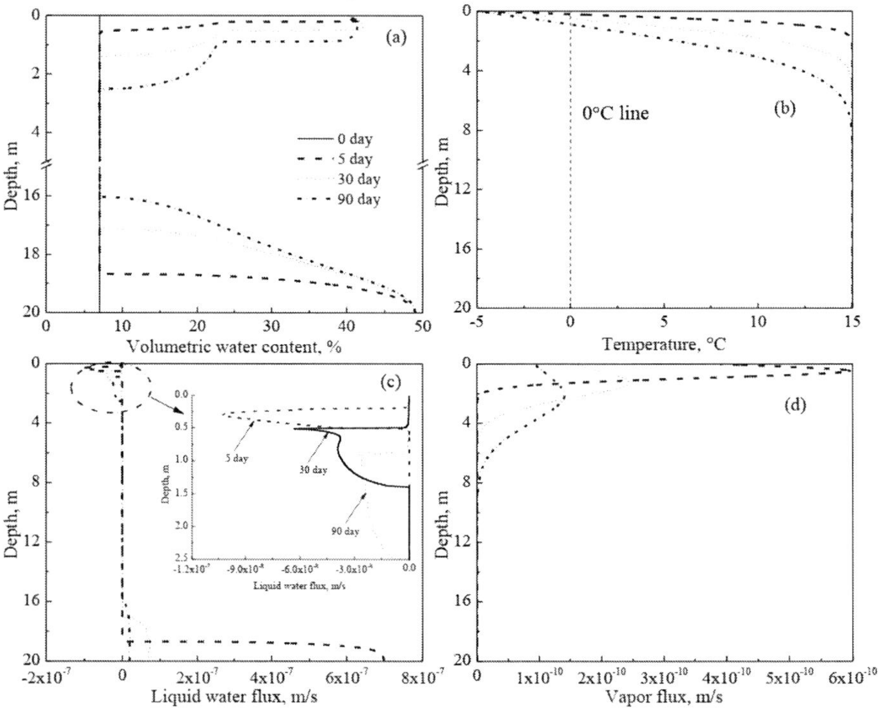

Figure 5. Computed water content profile (a), temperature profile (b), liquid water flux profile (c), and vapor flux profile (d).

In Figure 5(c), the positive value represents the upward flow of liquid water, and negative downward. The positive value of water flux between 16-20 m means that the liquid water in unsaturated soil migrates upwards under the capillary force, and the water content profile reaches an equilibrium state after a certain period. For the soil between 2.5-16 m, little water flux occurs there. The negative value of water flux in the top 2.5 m soil indicates a small downward liquid flow. The reason might be that the unfrozen water content near the surface is higher than that at a deeper position. It also implies that the increase of water content within the top region is not caused by the liquid water flow. The embedded chart in Figure 5(c) shows that the liquid water flux is close to zero within a top layer of soil (0-0.2 m at 5 days, 0-0.5 m at 30 days and 0-0.8 m at 90 days), which is due to the phase change of water into ice. Figure 5(d) presents the computed results of vapour flux, and shows a clear upward flux near the surface zone. It can be concluded that water accumulation beneath the impervious surface is caused by vapour transfer. The vapor flux reaches a peak value at a certain depth, which is likely due to the vapour condensation and ice formation that occupy the soil pores and impede vapor transfer.

It is interesting to note that moisture would not accumulate in an unsaturated freezing soil exposed to atmosphere. In this case, evaporation and sublimation at the ground surface prevents ice accumulation in the soil, a process analogous to the mechanism in a modern frost-free freezer. Inside such a freezer, air is circulated by means of one or more fans, and air circulation helps sublimate any ice or frost that may form in the freezer. Another example is that wet clothes hanging outside in cold winter will eventually dry up, even though some ice may initially form in the clothes. Numerical simulation of an unsaturated freezing soil with an open top surface is difficult, as the boundary condition of evaporation and sublimation is not easy to specify. However, field observation shows that no ice forms in very dry soils that are open to the atmosphere.

The canopy effect is reproduced in the numerical results in Figure 5, which shows that the canopy effect is an outcome of the interaction between vapour transfer in the unsaturated soil, the ice formation in the freezing soil and the impervious cover. The temperature gradient drives up some moisture flow towards to the top soil, but is not sufficient to saturate the soil, particularly when the groundwater table is low. Formation of ice in the soil further increases the matric suction and decreases the vapour density, which accelerates the upward vapour flow, leading to significant total moisture accumulation.

The model presented in this paper is one-dimensional only. In reality, vapour transfer in soil can take place in multiple directions. Humid air may enter the vapour-ice desublimation zone from horizontal directions, further enhancing the canopy effect.

4 Conclusions

The role of vapour flow is usually ignored when analysing water migration in unsaturated soils. Indeed, vapour flow is seldom considered in geotechnical design. However, there are circumstance where vapour transfer can lead to engineering problems and needs to be considered. One of such circumstances is the moisture accumulation caused by vapour transfer under impervious covers, and is referred to as the canopy effect. It was reported that significant moisture could accumulate in unsaturated freezing soils, which led to frost damages to airport runways and road pavements.

Existing models based on the Philip and de Vries theory of coupled heat and mass transfer in unsaturated soils fall short of capturing significant canopy effect.

A new approach of modelling the coupled movements of moisture and heat in unsaturated freezing soils is presented in this paper, in which phase changes of evaporation, condensation and de-sublimation are taken into account. The computed results demonstrate that soil freezing fed by vapour transfer can produce a water content close to full saturation for the soil under an impervious cover. Systematic experimental verification of the proposed numerical model is warranted.

Acknowledgements

This research was supported by National Basic Research Program of China (No. 2014CB047001) and National Natural Science Foundation of China (No. 51508578).

References

1. Q. Li, Y.P. Yao, L.M. Han, et al. Pot-Cover Effect of Soil. Industrial Construction, **44**, 2: 68-71 (in Chinese) (2014)
2. K.D. Eigenbrod, G.J.A. Kennepohl. Moisture accumulation and pore pressure at base of pavements. Transportation Research Record: Journal of the Transportation Research Board, **1546**(1):151-161 (1996)
3. J.R. Philip, D.A. de Vries. Moisture movement in porous materials under temperature gradient. Trans. Am. Geophys. Union, **38**: 222-232 (1957)
4. P.C.D. Milly. A simulation analysis of thermal effects on evaporation. Water Resource Research, **20**: 1087-1098 (1984)
5. I.N. Nassar, R. Horton. Simultaneous transfer of heat, water, and solute in porous media: I. Theoretical development. Soil Science Society of America Journal, **56**: 1350-1356 (1992)
6. M. Sakai, N. Toride, J. Šimůnek. Water and vapor movement with condensation and evaporation in a sandy column. Soil Science Society of America Journal, **73**, 3: 707-717 (2009)
7. J. Teng, D. Sheng, S. Liang, S. Zhang. A numerical model for heat and moisture transfer in unsaturated freezing soil. *Unsaturated soil mechanics from theory to practice*, CRC Press, 629-634 (2015)
8. D.A. de Vries. Simultaneous transfer of heat and moisture in porous media. Trans. Am. Geophys. Union, **39**: 909-916 (1958)
9. M.Th. van Genuchten. A closed-form equation for predicting hydraulic conductivity of unsaturated soils. Soil Science Society of America Journal, **44**: 892-898 (1980)
10. Y. Mualem. A new model for predicting the hydraulic conductivity of unsaturated porous media. Water Resources Research, **12**: 513-522 (1976)
11. H. Saito, J. Šimůnek, B.P. Mohanty. Numerical analysis of coupled water, vapor, and heat transport in the vadose zone. Vadose Zone Journal, **5**: 784-800 (2006)
12. Hydrus-1D, http://www.pc-progress.com/en/ Default. aspx?hydrus-1d
13. G. S. Taylor, J. N. Luthin. A model for coupled heat and moisture transfer during soil freezing. Canadian Geotechnical Journal, **15**, 4: 548-555 (1978)
14. K. Hansson, J. Šimůnek, M. Mizoguchi, L.C. Lundin, M.Th. van Genuchten. Water flow and heat transport in frozen soil: numerical solution and freeze-thaw applications. Vadose Zone Journal, **3**: 693-704 (2004)
15. D. Sheng, S. Zhang, F. Niu, G. Cheng. A potential new frost heave mechanism in high-speed railway embankments. Géotechnique, **64**, 2: 144-154 (2014)

DOI: 10.1051/e3sconf/20160916009

Modelling of imbibition process in an embankment scale model

Giada M. Rotisciani[1,a], Francesca Casini[2], Augusto Desideri[1] and Giulio Sciarra[3]

[1]*Sapienza Università di Roma, Dipartimento di Ingegneria Strutturale e Geotecnica, 00184 Roma, Italia*
[2]*Università degli Studi di Roma Tor Vergata, Dipartimento di Ingegneria Civile e Ingegneria Informatica, 00173, Roma, Italia*
[3]*Sapienza Università di Roma, Dipartimento Ingegneria Chimica Materiali Ambiente, 00184 Roma, Italia*

Abstract. This paper aims to investigate the hydro-mechanical behaviour of a loosely compacted embankment during an inundation event. This study is based on the results of a centrifuge test carried out on a small-scale embankment model made of an artificially compacted clay–sand mixture. The wetting-induced displacements are analyzed and interpreted by means of a constitutive model adapted to unsaturated conditions. The numerical predictions are presented in terms of time evolutions of settlements, as well as, spatial distributions of vertical displacements. These profiles are compared to those experimentally observed in order to validate the predictive capabilities of the model on a boundary value problem. Moreover, the stress paths followed by elementary soil elements located at different depths are analyzed to emphasize the stress and strain variations due to capillary rise.

1 Introduction

Soils used in the construction of roadway and railway embankments are usually in partial saturation regime. Once wetted, these materials can collapse causing extensive crest settlements. The triggering of the so-called volumetric collapse is controlled by several factors and, in particular, by the dry density and water content after compaction.

In this study, the centrifuge test conducted by [1] on an embankment scale model is numerically simulated. The back analyses of the experimental results aims to investigate the wetting-induced deformations and the evolution of the stress state due to capillary rise. The material used in the physical modelling is a sand-clay mixture compacted on the dry side of the optimum Proctor curve, at low compaction rate, to emphasize the settlements. The hydro-mechanical behaviour of the solid skeleton is described using the Modified Cam-Clay model extended to unsaturated conditions ([2]). The ability of this model to reproduce the response of elementary soil volumes has been verified over the years by several authors ([3], [4], [5]). The model has been validated also in boundary value problems such as a real dam by [6], and the back analyses of a shallow tunnel ([7]).

The numerical results presented here highlight the satisfactory capabilities of the Modified Cam Clay model in reproducing the effects induced by imbibition processes. The numerical results, validated against the experimental results, represent an important instrument to analyse the stress state evolution during an inundation event.

The paper is organized as follows. In the first part, the hydro-mechanical properties of the material, the sequence of the test and the experimental data are summarized. In the second part, the constitutive laws for solid and fluid phases are briefly described together with the main numerical results. In particular, the predicted settlement profiles are compared with the measurements in order to verify the predictive capabilities of the model. Finally, the stress paths resulting from the imbibition process and the volumetric behaviour are analysed with reference to two different soil elements.

2 Centrifuge test

The sequence of the test and the experimental data are briefly described in this section. More details on the centrifuge modelling can be found in [1].

The test is conducted at 100 g on an half embankment model scale of 21 cm in height made of 40% Speswhite kaolin and 60% Fontainebleau sand. The slope is inclined at 33° to the horizontal plane, according to the classical inclination for real road embankments. The final lengths of the base and the crest are 60 and 28.5 cm, respectively (Fig. 1).

The hydro-mechanical properties of the material, indicated in the following as SCM, are a combination of those of clayey and sandy soils. The standard Proctor curve is similar to that of fine grained soils and provides an optimum water content w_{opt} of 13.5% and a maximum dry density ρ_{MDD} of 1.87 kg/m^3.

[a] Corresponding author: giada.rotisciani@uniroma1.it

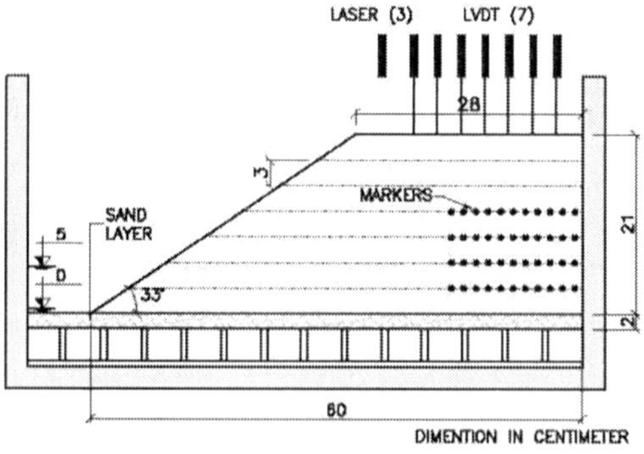

Figure 1. Cross section of the embankment model.

Similarly, the SCM behaviour along stress paths imposed by oedometric conditions does not qualitatively differ to that of clayey soils. The 1-D compression tests are performed by step at the target total vertical stress in saturated conditions and at constant water content for unsaturated samples. The water retention curve (WRC), depicted in Fig. 2 in the plane saturation degree S_r versus suction s is obtained using the osmotic method. The WRC exhibited by the material is typical of coarse grained soils, while the compressibility is similar to fine grained soils.

The model is built inside the strong-box in seven successive sub-layers, each approximately 3 cm thick. The SCM, with a water content of 10.5%, is compacted on the dry side of the Proctor curve to emphasize the displacement phenomena caused by wetting. A sand layer 2 cm thick is located on the bottom of the strong-box and is separated from the SCM layer by geotextile.

After deposition and compaction, a portion of SCM layer is removed to form a slope inclined at 33° to the horizontal plane. Upon ending the model preparation, the centrifuge acceleration is increased up to 100 ***g*** (Phase 1). Once stationary conditions are reached, the strong-box is hydraulically connected to a water tank and the water table is constantly maintained just above the sand-embankment interface level (Phase 2).

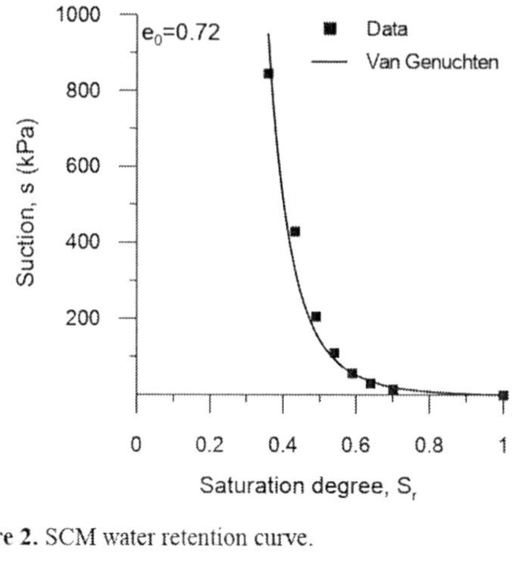

Figure 2. SCM water retention curve.

Finally, the water table inside the main compartment is raised up to 5 cm above the base (Phase 3).

The crest vertical displacements are recorded by three laser transducers, installed close to the crest slope, and seven regularly spaced LVDTs, installed in the central portion of the embankment top. The centrifuge is equipped with a digital video camera to monitor the soil movements during the test. The image analysis allows to determine the vertical displacements of the reference points located inside the embankment body (see Fig. 1).

Fig. 3 illustrates time evolution of the crest vertical displacements due to inundation. The zero values correspond to the settlements registered once ended the in-flight phase. The first imbibition event (Phase 2) induces settlements of about 5 mm, uniformly distributed on the embankment crest. The magnitude of the settlements increase in the Phase 3, when the water table progressively rises up to 5 mm above the base. The stationary values range from 12 to 14 mm, depending on the considered position. More specifically, the settlements are greater and develop more quickly near the slope than close to the axis of symmetry.

Fig. 5a illustrates the spatial distribution of the wetting-induced settlements, once ended the Phases 2 and 3. The vertical deformations mainly occur in the lower part of the embankment and the magnitude of the settlements appreciably increases as the water table moves upwards to 5 cm above the bottom. Similarly, volume changes mainly occur in the deep sub-layers and their magnitude increases as the embankment is inundated.

3 Numerical modelling

The centrifuge test has been numerically simulated using the commercial code Abaqus/Standard at the model scale. The plane strain simulation reproduces faithfully the geometry and the sequence of the test. In this section, after a brief description of the numerical model, the main numerical results are analyzed.

3.1 Numerical model

The soil layer is discretized by a finite element mesh 23 cm high and 60 cm wide with element size ranging from (0.5x0.5) to (1.0x0.5) cm^2. The adopted mesh is stable and accurate; it is mostly independent by element size. The nodes located along the vertical sides can move vertically while the nodes located along the base are constrained in both directions. Regarding hydraulic boundary conditions, fluid flow is prevented through all external surfaces.

The contact surface between the embankment and the sand base layer is modelled as an elastic-perfectly plastic interface with purely frictional Mohr Coulomb criterion.

The mechanical behaviour of the SCM is described using the Modified Cam-Clay model (MCCM) extended to unsaturated conditions following the approach proposed by [2].

This model is formulated within the framework of critical state soil mechanics in terms of Bishop effective stresses: $\sigma'_{ij} = \sigma_{ij} - u_a\delta_{ij} + \chi\,(u_a - u_w)\delta_{ij}$, where σ_{ij} is the total stress, χ coincides with the saturation degree, $(u_a\text{-}u_w)$ is the matrix suction and δ_{ij} is the Kronecker delta. MCCM is an elastic-plastic constitutive model, commonly used to reproduce the stress-strain response of fine grained soils. A detailed description of the mathematical formulation and the characteristics of this model can be found in [3, 5].

This model differs from that used for saturated soils in the hardening rule. In MCCM, the evolution of the yield surface is controlled by the increments of the plastic volumetric deformation and saturation degree. The introduction of the explicit dependence from S_r is needed to reproduce some specific aspects of the hydro-mechanical behaviour of unsaturated soils, such as the increase in the apparent preconsolidation stress with suction and the collapse upon wetting.

MCCM is completely defined by the six constitutive parameters summarized in Table 1 together with the hydraulic properties. More specifically, the constitutive parameters are: the slope of the unloading-reloading line URL in the compressibility plane (in semi-log scale) κ; the slope and the position of the saturated normal consolidation line NCL in the same plane, λ and N_0; Poisson's ratio ν; the slope of the critical state line in the stress invariant plane M and, finally, the parameter b that controls the relative position of the NCL associated to the current value of S_r with respect to the saturated NCL.

The model is calibrated against the results of oedometric (OED) and cone penetration tests (CPT). The first ones are conducted on saturated and unsaturated soil samples; the CPT are carried out on an embankment scale model similar to those used in the centrifuge test.

The fluid flow obeys the well-known Richards equations:

$$\nabla\left[\frac{k(S_r)}{\gamma_w}\nabla(s + \gamma_w z)\right] = \frac{\partial\theta}{\partial t} \quad (1)$$

where $k(S_r)$ represents the evolution law of the permeability with the saturation degree, γ_w is the specific weight of water and θ is the volumetric water content, defined as ratio between volume of water and total volume.

Table 1. SCM mechanical and hydraulic properties.

Mechanical parameters		Hydraulic parameters	
λ	0.085	α	0.35
κ	0.025	n	1.17
N_0	1.94	m	0.15
ν	0.25	k_{sat} (m/s)	$5\ 10^{-8}$
M	1.11		
b	7.4		

$k(S_r)$ is expressed as a power function of the degree of saturation: $k(S_r)=k_{sat}\ S_r^{\delta}$. The values for δ and the saturated permeability k_{sat} are deduced by back-analysing the time evolution of the crest vertical displacements (see Fig.4).

The WRC is fitted with the Van Genucthen expression; the parameters α, m and n are calibrated against the experimental data (see Fig. 3)

$$S_r = \frac{1}{[1+(\alpha s)^n]^m} \quad (2)$$

The numerical analysis accurately reproduces the sequence of the test and consists of four steps.

Under geostatic conditions, the total vertical stresses are completely negligible as the small dimensions of the embankment. The pore water pressures vary linearly with depth and assume an average value of 675 kPa, corresponding to a water content of 10.5%. The effective horizontal stresses are expressed as a function of the effective vertical stresses through the earth pressure coefficient K_0, placed equal to the unity.

To form the embankment, the mesh elements located above a surface inclined at 33° to horizontal plane are removed (Phase 0). The ***g*** level is increased by a factor of 100, coherently to the centrifuge test (Phase 1). Then, null pore water pressure is prescribed at the embankment base to simulate the rising of the water table (Phase 2). Finally, the inundation event is modelled by imposing a constant pore water pressure of 50 kPa at the sand-embankment interface level and applying a hydrostatic pore water pressure distribution on the slope, according to a water table located 5 cm above the base (Phase 3).

3.2 Numerical results

Time evolution of the predicted crest settlements is shown in Fig. 3 together with the data registered by the transducers. The numerical results refer to two nodes located at the top of the soil domain along the slope and the axis of symmetry. Globally, the model responses fit well the experimental curves.

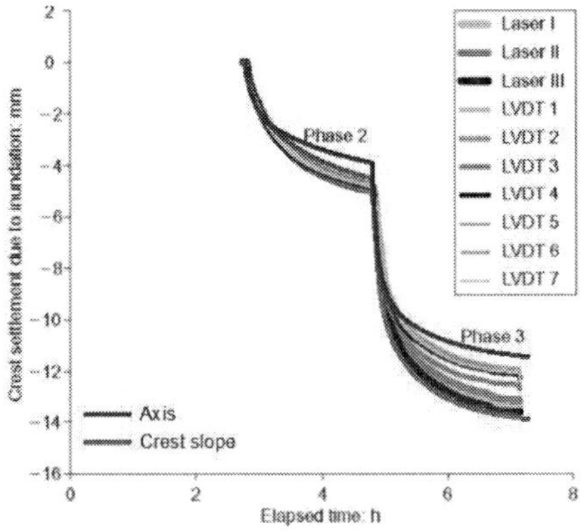

Figure 3. Time evolution of crest settlements.

 DOI: 10.1051/e3sconf/20160916009

In the Phase 2, the vertical displacements increase in magnitude up to values ranging from 4 to 5 mm. Differently to what is observed in the test, there is a small difference between the predicted settlements at the crest slope and the axis. In the Phase 3, the vertical displacements increase again in magnitude attaining values in the range of 11-14 mm. Similarly to the measurements, the predicted vertical displacements are more extensive and develop more quickly at the crest slope than at the axis. In this phase, the wetting-induced settlements are accompanied by significant lateral displacements, whose magnitude increases approaching the slope. As a consequence, the crest settlements near the axis, where no horizontal displacement is allowed, are smaller than those predicted at the slope, where lateral and vertical movements are combined.

The predicted settlements profiles are depicted in Fig. 4 together with the measurements. The numerical results refer to a vertical axis 0.5 cm far from the axis of symmetry. The satisfactory match between model predictions and experimental data confirms the ability of the Modified Cam-Clay model in reproducing wetting-induced deformations also in a boundary value problem.

The magnitude of the predicted vertical displacements appreciably increases as the water table rises up to 5 cm above the base (see Fig. 5). Despite the quantitative differences, the w profiles exhibit similar trends in the Phases 2 and 3. The settlements increase in the lower part of the embankment and keep an almost constant value in the upper part. Comparing the model predictions with the data, the major differences can be found in the upper part. Here, the MCCM predicts a soil response appreciably stiffer than that experimentally observed.

Upon validating the constitutive model, the hydro-mechanical response of the collapsible soil is analyzed in detail. The changes in the pore-water pressure u_w and degree of saturation, as well as, the stress paths and the deformations induced by capillary rise are examined with reference to two elements located at different depths along a vertical axis 5 cm far from the axis of symmetry (z=2.75 and 14.75 cm).

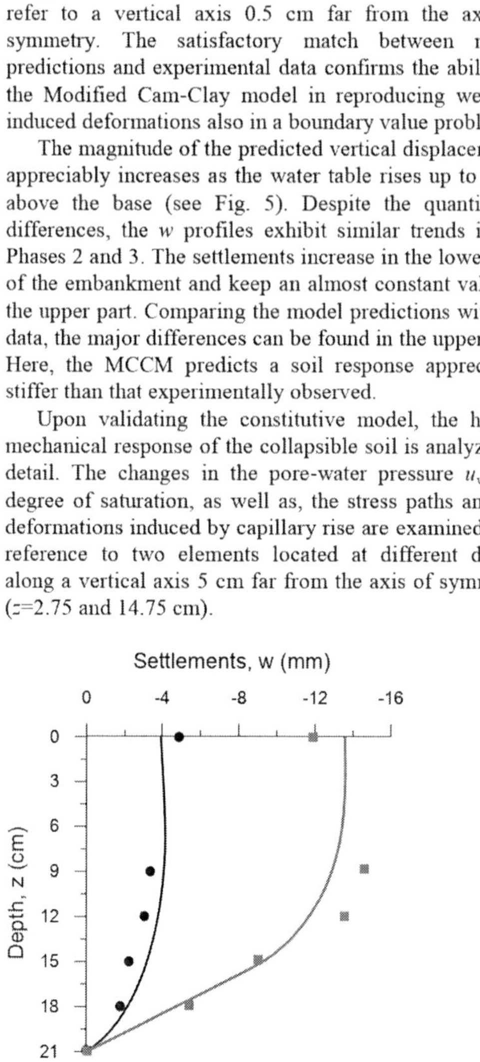

Figure 4. Spatial distribution of settlements.

As shown in Fig. 5, the application of the **g**-level entails a rotation of the u_w profile in a counter clockwise direction. This implies a reduction of the pore-water pressure of about 36 kPa in the shallow element and an increment of 84 kPa in the deep one. Despite the variations of u_w, there is no appreciable change in the S_r profile. This is mainly due to the shape of WRC and the initial conditions, in terms of saturation degree and suction level. In the in-flight phase, the point representative for the current conditions in both considered elements moves along the vertical branch of the WRC, where large variations of suction level are associated to small changes in S_r.

In the Phase 2, the pore-water pressures progressively increase and, once reached stationary conditions, exhibit a hydrostatic profile with a water table located at the bottom of the soil domain. Regardless of the considered element, u_w increases of about 520 kPa. Conversely, the variation in S_r depends on depth and is more pronounced in the deep element than in the shallow one. This result can be attributed to the shape of WRC and the differences in the suction level at the beginning of the Phase 2. Despite the increase of the pore-water pressure associated to the rising of the water level, the point representative for the shallow element lies again on the vertical branch of WRC. Conversely, the suction level in the deep element exceeds the maximum curvature point of the WRC and reaches the horizontal branch of the curve, where small variations of suction entail large changes in the degree of saturation.

In the Phase 3, as the water infiltrates through the base and the slope, the water table progressively rises within the embankment body and the pore-water pressures increase up to their hydrostatic stationary values. Consequently, the saturation degree increases and complete saturation is reached in soil elements located below the current position of the water table.

The above mentioned differences in the evolution of S_r significantly affect the hydro-mechanical responses of the two considered elements. The changes in the stress state and the corresponding volumetric deformations of the deep element are shown in Fig. 6 in terms of mean effective stress p', deviator q and specific volume v.

The initial stress state lies on the p'-axis inside the current yield surface (point A), as expected in highly over-consolidated soils (K_0=1).

The application of the centrifuge acceleration induces an increase of the gravity loading exerted on the solid skeleton and no appreciable change in the degree of saturation. Because of the oedometric boundary conditions, the stress state is forced to move along a stress path, whose slope is prescribed by the elastic coefficient $K_0^{el}=\nu/(1-\nu)$. Once reached the current yield surface, the stress path changes in slope and progressively tends towards the K_0 line (branch AB). The volumetric behaviour is initially controlled by the slope of the URL and, then, by NCL in oedometric conditions associated to the current saturation degree.

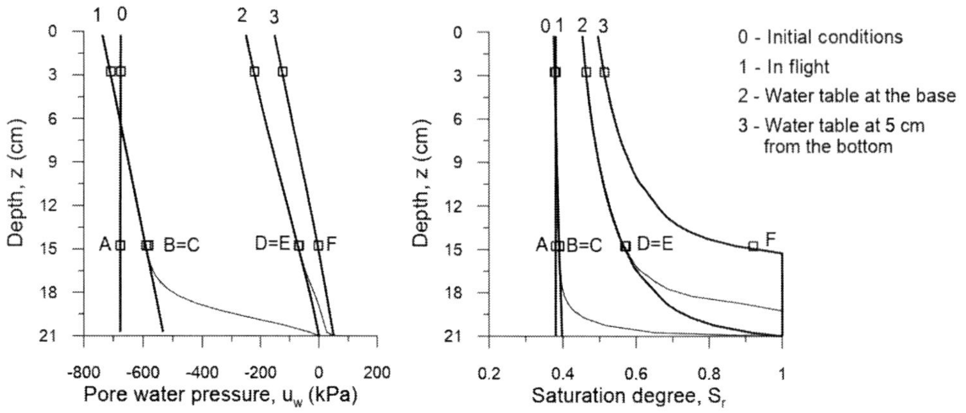

Figure 5. Time evolution of the pore water pressure and saturation degree profiles.

When null pore-water pressure is prescribed at the base, the mean effective stress decreases and the deviator slightly increases (branch BC). It is to notice that this increment of the stress ratio $\eta=q/p'$ occurs when the capillary front is still below the considered element (see Fig. 5). The propagation of the front is accompanied by a variation of the horizontal stress distribution due to the volumetric collapse of the deep elements. This variation involves also the soil region just above the current position of the front and justifies the reduction of the horizontal stress under a constant vertical load in the considered element. Once the front reaches the deep element, the stress ratio starts to decrease. Despite the stress release, the skinkage of the yield surface due to the increase of S_r keeps the stress state constantly in elastic-plastic regime (branch CD). As a consequence, the stress path tends towards the K_0-line and solid skeleton collapses. Significant volume changes occur only when S_r exceeds 0.5, which corresponds to the maximum curvature point of WRC. For S_r lower that this value, the positive volumetric deformations are well-balanced by the elastic swelling and the soil element deforms under a quasi constant volume. During the inundation event, the soil response is definitely similar to that observed in the Phase 2. As soon as the water starts to infiltrate in the embankment body through the bottom and the slope, the horizontal stress in the considered element decreases under a constant vertical stress. This implies an increment of the stress ratio and, consequently, a change in volume of the current yield surface (branch DE). When the front reaches the element, both stress invariants decrease, the elastic domain skinks and the volumetric collapse occurs (branch EF).

Consider now the responses of the shallow element.

At 1 ***g***, the mean effective stress is given by the suction level scaled by the saturation degree and, hence, there is no appreciable difference in the initial stress state between the deep and shallow elements. The application of the centrifuge acceleration induces an increment of the gravity loading exerted on the solid skeleton that is more extensive in the deep than in the shallow element. As a consequence, the evolution of the stress state in the deep element provokes the occurrence of plastic deformations while the shallow element behaves as an elastic material.

In the Phase 2, the stress state of the deep element constantly lies on the current yield surface. Conversely, the stress state of the shallow element moves again inside the elastic domain and its evolution is accompanied by positive variations in volume.

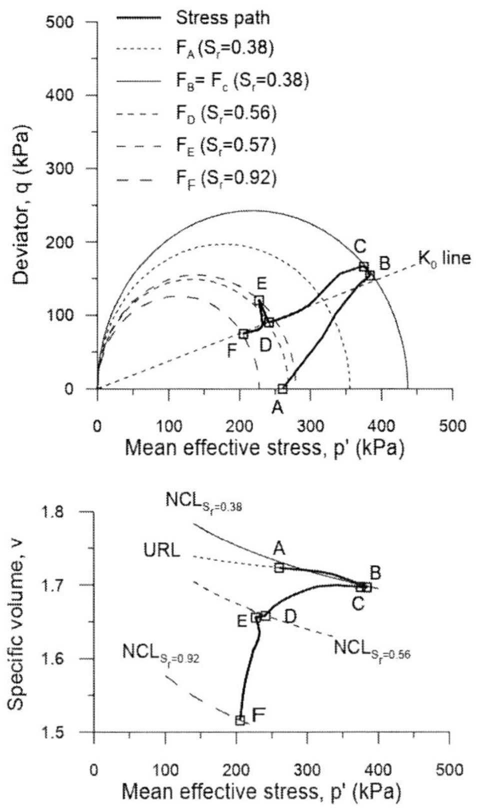

Figure 6. Stress paths and volumetric behaviour of deep elements (z=14.75 cm).

In the Phase 3, the ellipse captures the current stress state and irreversible deformations start to develop. However, differently to the deep element, the shallow element swells again because the elastic strains are greater than that associated to the volumetric collapse.

Once calibrated and validated, the numerical model has been adopted to investigate the influence of the base friction angle on the embankment response. A number of four additional numerical analyses has been performed modifying the contact friction angle between the sand layer and embankment φ_c in the range of 10-35°.

The results are depicted in Fig. 7, in terms of spatial distribution of the crest vertical and horizontal, u, displacements once reached the final stationary conditions. The comparison shows that there is no appreciable change in the embankment response when φ_c assumes values greater than 20°. Conversely, when φ_c is lower than 20°, the magnitude of the vertical and horizontal displacements depends on the base friction angle and, in particular, increases as φ_c decreases. However, it is to notice that significant variations of u and w occur only when φ_c becomes lower than 15°. These parametric analyses confirm the influence of the roughness of the contact surface on the performance of the embankment which it is non linear with the contact friction angle.

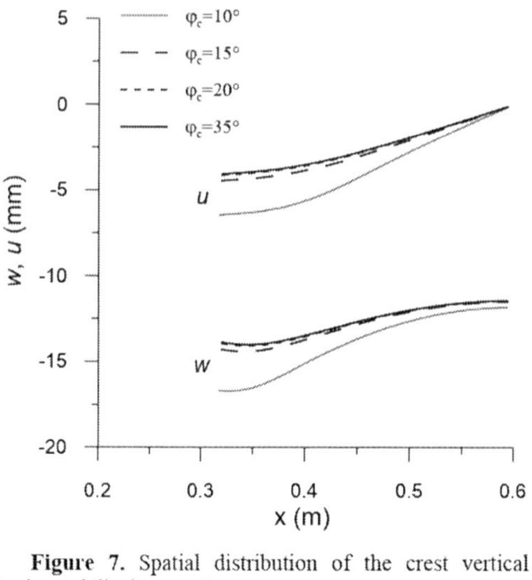

Figure 7. Spatial distribution of the crest vertical and horizontal displacements.

4 Conclusion

The results presented in this paper confirm the ability of the Modified Cam-Clay model to accurately predict wetting-induced deformations. The model responses, in terms of time evolution of crest settlements and spatial distribution of settlements and void ratio, well match the experimental data. Moreover, the numerical results underline the influence of the initial conditions, in terms of suction level and saturation degree, and the shape of the WRC on the occurrence of the volumetric collapse. The increase of S_r causes appreciable volumetric deformations only in the lower part of the embankment, where the suction level exceeds the maximum curvature point of the WRC. Therefore, to limit the risks associated to the triggering of the volumetric collapse, particular attention should be paid on the compaction procedure of the deep sub-layers. Here, the dry density should be maximized by keeping the water content close to the optimum value and/or increasing the compaction energy.

References

1. L. Thorel, V. Ferber, B. Caicedo, I.M. Khokhar, Géotechnique **61**, 5, 404-420 (2011).
2. C. Jommi, *Experimental Evidence and Theoretical Approaches in Unsaturated Soils* (A. Tarantino and C. Mancuso ed.), 139-153 (2000).
3. R. Tamagnini, Géotechnique **54**, 3, 223-228 (2004).
4. F. Casini, Can. Geot. J. **49**, 8, 954-960 (2012).
5. G. M. Rotisciani, G. Sciarra, F. Casini, A. Desideri, Int. J. Numer. Anal. Meth. Geomech., **39**, 1212–1234 (2014).
6. A. Burghignoli, A. Desideri, and R. Tamagnini, 3th Int. Conf. Unsat. Soils **1**, 221-225 (2002).
7. E. Soranzo, R. Tamagnini, W. Wu, Géotechnique **65**, 6, 454-467 (2015).

DOI: 10.1051/e3sconf/20160916010

Numerical study on the influence of entrapped air bubbles on the time-dependent pore pressure distribution in soils due to external changes in water level

Georg M. Ausweger[1,a] and Helmut F. Schweiger[1]

[1]*Graz University of Technology, Institute of Soil Mechanics and Foundation Engineering, Computational Geotechnics Group, 8010 Graz, Austria*

Abstract. In practical geotechnical engineering soils below the groundwater table are usually regarded as a two-phase medium, consisting of solids and water. The pore water is assumed to be incompressible. However, under certain conditions soils below the groundwater table may exhibit a liquid phase consisting of water and air. The air occurs in form of entrapped air bubbles and dissolved air. Such conditions are named quasi-saturated and the assumption of incompressibility is no longer justified. In addition the entrapped air bubbles influence the hydraulic conductivity of soils. These effects are usually neglected in standard problems of geotechnical engineering. However, sometimes it is required to include the pore fluid compressibility when modelling the hydraulic behaviour of soils in order to be able to explain certain phenomena observed in the field. This is for example true for fast fluctuating water levels in reservoirs. In order to study these phenomena, numerical investigations on the influence of the pore fluid compressibility on the pore water pressure changes in a soil layer beneath a reservoir with fast fluctuating water levels were performed. Preliminary results of this study are presented and it could be shown that numerical analysis and field data are in good agreement.

1 Introduction

In classical soil mechanics, soils below the ground water table are considered as a two-phase medium. All pores are filled with water and the pore water is assumed to be incompressible, at least for practical purposes. However, if a small amount of air is present in the pore water, the compressibility of the pore fluid and the hydraulic conductivity of the soil are influenced significantly. This air phase is discontinuous and occurs as small entrapped air bubbles and the air pressure in the air bubbles is no longer equal to the atmospheric pressure.

As soils below the ground water level with entrapped air bubbles are not fully saturated but the term unsaturated soils is in general used for zones in the subsoil with a continuous air phase and a suction higher than the air entry value, a new term "quasi-saturated soils" was introduced to describe soils with entrapped air bubbles in the pore water [1]

The term quasi-saturated soil is used for soils with positive pore water pressure and entrapped air bubbles as well as soils with a negative pore water pressure (suction) smaller than the air entry value. This contribution considers only states with a positive pore water pressure.

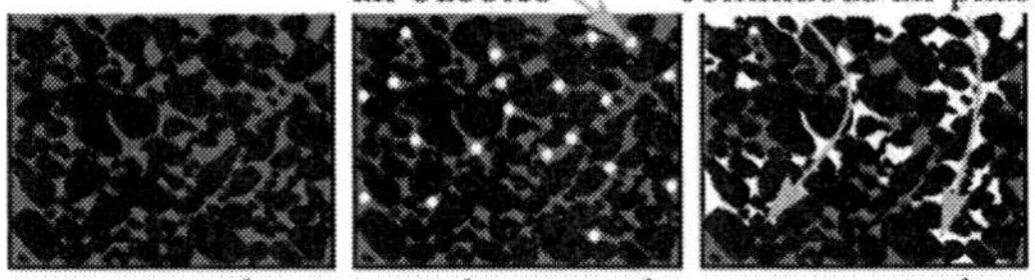

Figure 1. Comparison saturated – quasi-saturated – unsaturated soil. Adapted according to [2].

Observations in the field [3, 1] and laboratory tests [1, 4] show that soils below the ground water table are often in a quasi-saturated state. This has an influence on the behaviour of the pore fluid. The air bubbles in the soil can occur due to different reasons, for example fluctuating water tables [4]. Such fluctuating water tables are possible, amongst others, in storage basins of pumped storage power plants and in ship canals.

In this contribution the effects of air bubbles in the pore water are included in modelling the hydraulic behaviour of the soil as a two-phase medium. The aim is to explain, in a qualitative manner, the results of pore pressure measurements near a storage basin of a pumped storage power plant employing the finite element method. The focus is exclusively on matching the measured pore water pressures. In the first part of the paper the basics for the modelling of quasi-saturated soils are summarized for continuity. The second part presents the pore pressure

[a] Corresponding author: georg.ausweger@tugraz.at

measurement data and the comparison with results from finite element analyses.

2 Modelling of quasi-saturated soils

Soils in a quasi-saturated state consist of incompressible solids and a compressible pore fluid. The pore fluid is compressible due to entrapped air bubbles in the pore water. As the air bubbles reduce the hydraulic conductivity of soils, it is necessary to define a relative hydraulic conductivity k_{rel}, depending on the air content in the pore fluid. Furthermore, a water retention curve (WRC) has to be defined for quasi saturated soils.

The flow of pore fluid in a porous deformable material [5] can be described by

$$S\frac{\partial \varepsilon_{vol}}{\partial t} - n \cdot \left(\frac{S}{K_w} - \frac{\partial S}{\partial p_w}\right) \cdot \frac{\partial p_w}{\partial t} + \cdots$$

$$\cdots \nabla^T \left[\frac{k_{rel}}{\rho_w \cdot g} \mathbf{k}^{sat} \cdot \left(\nabla p_w + \rho_w \cdot \mathbf{g}\right)\right] = 0 \qquad (1)$$

where S is the degree of saturation, ε_{vol} are the volumetric strains, K_w is the bulk modulus of water and p_w is the pore water pressure. k_{rel} and $\mathbf{k}^{sat}$ represent the relative and saturated hydraulic conductivity respectively and ρ_w is the density of water.

In the following the influence of entrapped air bubbles on the relative hydraulic conductivity and the WRC, i.e. the relationship between degree of saturation S and pore water pressure p_w, is discussed.

2.1 Hydraulic conductivity of quasi-saturated soils

Laboratory tests on different soil types [1, 4, 6, 7] show a decrease in the hydraulic conductivity with increasing entrapped air content. Usually this decrease in hydraulic conductivity is described by a power law. An example for such a power law is the one according to Faybishenko [1].

$$\mathbf{k}^{qs} = \mathbf{k_0} + \left(\mathbf{k}^{sat} - \mathbf{k_0}\right) \cdot \left(1 - \frac{\omega}{\omega_{max}}\right)^n \qquad (2)$$

The quasi-saturated hydraulic conductivity $\mathbf{k}^{qs}$ is a function of the saturated hydraulic conductivity $\mathbf{k}^{sat}$, the minimum quasi-saturated hydraulic conductivity $\mathbf{k_0}$, which is the hydraulic conductivity at the transition from a quasi-saturated to an unsaturated state, i.e. at suction values slightly smaller than the air entry value, the volumetric content of entrapped air ω and the maximum entrapped air content ω_{max}. ω_{max} is the volumetric content of entrapped air, which corresponds to the minimum quasi-saturated hydraulic conductivity $\mathbf{k_0}$. n is a fitting parameter [4]. The relative hydraulic conductivity k_{rel} in Equation (1) is the ratio between the quasi-saturated hydraulic conductivity and the saturated hydraulic conductivity.

A comparison of different laboratory tests on the hydraulic conductivity of quasi-saturated soils [4] showed a significant difference in the relative hydraulic conductivity between coarse soils, like sand, and fine-grained soils – see Figure 2. The entrapped air saturation is the volumetric air content divided by the porosity.

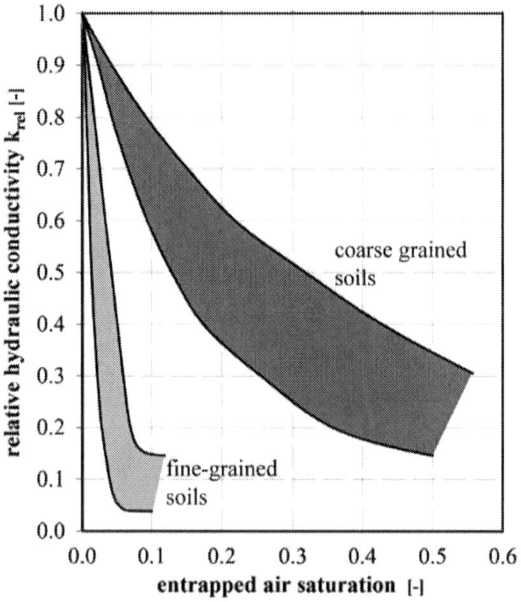

Figure 2. Relative hydraulic conductivity of quasi-saturated soils. Adapted according to [4].

2.2 Water retention curve of quasi-saturated soils

Water retention curves for quasi-saturated soils can be derived either from laboratory tests or on the basis of theoretical approaches. Recent literature provides different formulations of the WRC [6, 3, 2].

The WRC for a soil with positive pore water pressure p_w and entrapped air bubbles can be derived on the basis of Boyle's law and Henry's law for dissolving of free air into water.

$$S(p_w) = \frac{1}{1 - h + \dfrac{P_a}{p_w + P_a} \cdot \dfrac{S_{(p_w=0)} \cdot (h-1) + 1}{S_{(p_w=0)}}} \qquad (3)$$

The degree of saturation S depends on Henry's parameter h, the atmospheric pressure P_a, the pore water pressure p_w and a reference degree of saturation $S_{(pw=0)}$ at a pore water pressure $p_w = 0$. Equation (3) is equal to the WRC, which was derived by Boutonnier [2]. For the derivation of the WRC a pore air pressure p_a in the air bubbles equal to the surrounding pore water pressure p_w was assumed. That means the influence of the surface tension is neglected. Furthermore, the vapour pressure in the air bubble is neglected.

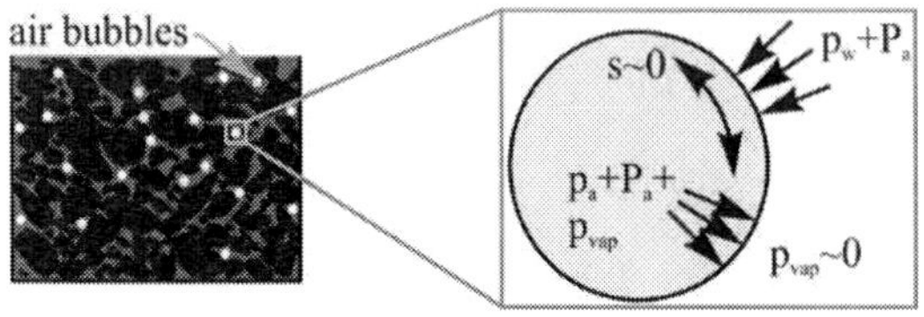

Figure 3. Assumptions for pressure in air bubble: surface tension s ~ 0 and vapour pressure p_{vap} ~ 0.

Henry's parameter for dissolving of free air in water can be assumed to h = 0.2. For fast loadings, this value should be set to zero, as the dissolving of free air in water needs some time under sustained pressure [2].

Figure 4 shows a comparison of different WRC from the literature for an arbitrary quasi-saturated soil with 10% entrapped air at a pore water pressure $p_w = 0$. It is emphasized that in this study the effect of air bubbles in the pore water is only investigated in the positive pore pressure range. A full WRC (with negative pore pressure range) is not used. Therefore Figure 4 does not represent a WRC for a particular soil type. According to Bicalho [6] the pore air pressure p_a is different from the pore water pressure p_w. For a better comparability $p_a = p_w$ was also assumed for the WRC according to Bicalho [6] in Figure 4.

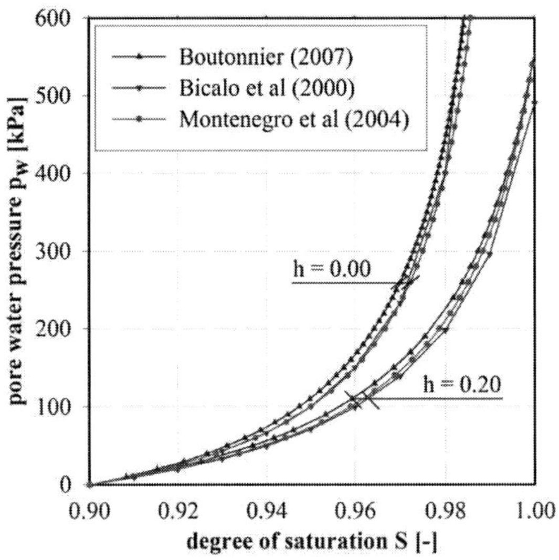

Figure 4. Comparison of water retention curves for $S_{(pw=0)}$ = *0,9 and h = 0.00 / 0.02.*

2.3 Compressibility of the pore fluid

The compressibility of the pore fluid results from the compressibility of air and the compressibility of water.

The isothermal compressibility of air C_{air} is inversely proportional to the absolute air pressure [8]

$$C_{air} = \frac{1}{P_a + p_a} \tag{4}$$

P_a is the atmospheric pressure and p_a the pore air pressure.

The compressibility of pure water C_{water} is about 4.5×10^{-7} [1/kPa].

The compressibility of the pore fluid, i.e. of the air-water mixture, can be derived based on the proportions of the constituents air and water [8].

$$C_{fluid}(p_w) = S(p_w) \cdot C_{water} + (1 - S(p_w) + h \cdot S(p_w)) \cdot C_{air}$$

$$C_{fluid}(p_w) = S(p_w) \cdot 4.5 \cdot 10^{-7} + \dots$$

$$\dots (1 - S(p_w) + h \cdot S(p_w)) \cdot \frac{1}{P_a + p_w} \tag{5}$$

As the degree of saturation and the compressibility of air C_{air} are a function of the pore water pressure p_w, also the compressibility of the pore fluid C_{fluid} is a function of the pore water pressure p_w. In case of h = 0.0, Equation (5) is consistent with the determination of the pore fluid compressibility according to Koning and Verruijt [8].

3 Comparison of finite element analysis and pore pressure measurements

Deformation measurements (inclinometer and geodetic measurements) identified movements of a slope adjacent to a storage basin of a pumped storage power plant. In order to identify the causes of the slope movements, additional pore pressure gauges were installed at the toe of the slope.

The results of the pore pressure measurements could not be explained on the basis of conventional soil mechanics. Thus, in a first finite element analysis with the software PLAXIS 2D 2015 [5] the pore fluid was no longer assumed to be incompressible. A simple 1D-finite element analysis was performed to check whether the compressibility of the pore fluid, due to air bubbles in the pore water, may be the reason for the at first unexpected measurement results. Of course the 1D-model cannot capture the influence of slope movements on the pore pressure development and its causes. However, this deficiency can be accepted at this stage of investigation as the primary objective of the finite element analysis was only a qualitative comparison with the measured trend of pore water pressure changes due to changes of water levels in the reservoir.

3.1 Site description

A slope with a slope angle of about 30° is situated next to a storage basin. Down to a depth of 25 m the slope consists mainly of weathered rock. At the toe of the slope, in the storage basin, the subsurface explorations identified a 20 m thick layer of silty sand, which is underlain by sandy clayey silt. Figure 5 shows a cross-section through the storage basin.

The water level in the storage basin changes two to four times a day, the differences in water levels being up to 8 m. The velocity of the water level change can be up to 25 m/day.

 DOI: 10.1051/e3sconf/20160916010

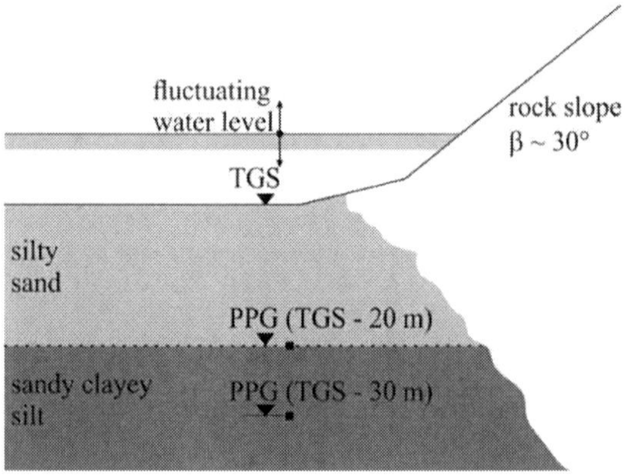

Figure 5. Cross-section through the storage basin.

3.2 Pore pressure measurements

The pore pressure gauges (PPG) were installed at the toe of the slope, in the storage basin, in 20 m and 30 m depth.

Figure 6 shows an example of typical pore pressure measurement results and the corresponding water level. This period was also used for the following comparative finite element analysis. The excess pore water pressure p_{excess} is defined as the difference between the measured pore water pressure p_w and the hydrostatic pore water pressure p_{hydro} with reference to the general ground water table. Pressure is positive.

$$p_{excess} = p_{water} - p_{hydro} \tag{6}$$

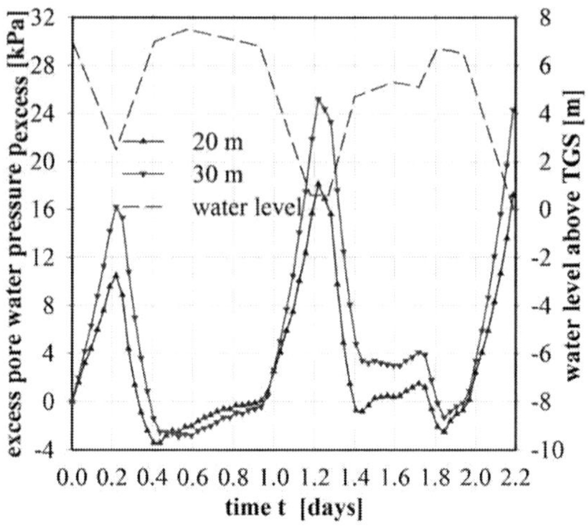

Figure 6. Typical pore pressure measurement data in 20 m and 30 m depth and corresponding water level.

In general Figure 6 shows an increasing excess pore pressure with a decreasing water level and vice versa. The value of excess pore pressure is higher the faster the water level changes. These observations might be a first indication for a compressible pore fluid as in case of quasi-saturated soils the value of the excess pore pressure is highly influenced by the ratio of loading time and hydraulic conductivity of the soil. Assuming an incompressible pore fluid, water level changes above the ground surface would not result in excess pore pressures. A further interesting detail can be observed during the period from 0.4 to 0.6 days. The recording of the water level shows a further increase in the water level but the velocity of the water level change is smaller than during the period from 0.2 to 0.4 days. These led to a further decrease of the excess pore water pressure in a depth of 30 m. However, the measurements in 20 m depth show a change in the course of the excess pore water pressure. The slower increasing water level led to an increase in excess pore pressure in 20 m depth. This means the behaviour is different in 20 m and 30 m depth. The change of the excess pore water pressure in 20 m depth could be an indication for a consolidation process. The decrease in excess pore water pressure due to the following slow increasing water level is smaller than the change of the excess pore pressure towards $p_{excess} = 0$ due to a reduction of the already existing excess pore pressure (consolidation). These observations led to the assumption of a compressible pore fluid for the finite element analysis.

3.3 Finite element model

For the qualitative comparison of the measurement results with the results of the finite element analysis a 1D-model, as shown in Figure 7, was used. The model is 30 m high. The upper 20 m consists of silty sand; the material for the lower 10 m is sandy clayey silt. As the permeability of the sandy clayey silt is very low the lower boundary was assumed impermeable. The boundary condition at the upper boundary is the changing water level (pore water pressure) according to the water level-time curve in Figure 7.

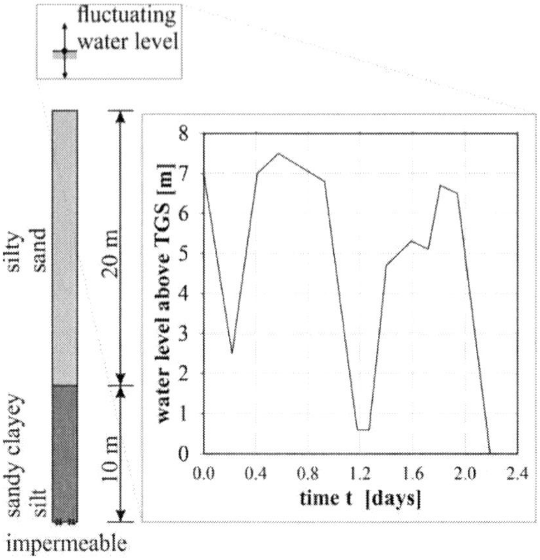

Figure 7. 1D-finite element model and water level-time curve as upper boundary condition.

The Hardening Soil Small-model (HSS-model) [5] was used as constitutive model for both soil types. The first calculations showed that the strains, due to a fluctuating water level, are very small. Thus, it was necessary to use the HSS-model, which incorporates the very stiff behaviour of soils at very small strains. The soil properties were determined on the basis of laboratory tests and from experience. They are summarised in Table 1. As the qualitative pore pressure distribution is the main issue at this stage, the same mechanical properties were used for both soil types; only the assigned hydraulic conductivities are different.

Table 1. Soil properties for the Hardening Soil Small-model.

Parameter	Unit	Silty sand	Sandy clayey silt
γ_{unsat}	kN/m³	18.00	
γ_{sat}	kN/m³	20.00	
$E_{50,ref}$	kPa	25 000	
$E_{oed,ref}$	kPa	20 000	
$E_{ur,ref}$	kPa	60 000	
ν'_{ur}	-	0.20	
p_{ref}	kPa	100	
m	-	0.80	
φ'	°	35.00	
c'	kPa	1.00	
ψ'	°	0.00	
k^{sat}	m/sec	10^{-6}	10^{-8}
G_0 / $\gamma_{0.7}$	kPa / -	62 500 / 20^{-4}	

3.4 Relative hydraulic conductivity and water retention curve

As the velocity of the water level changes were relatively fast, Henry's parameter was assumed to be zero. Thus, the water retention curve only depends on the reference degree of saturation $S_{(pw=0)}$. The calibration of this parameter was done on the basis of the measurement data from the first lowering in water level. Compared to the measurement data, a reference degree of saturation $S_{(pw=0)} = 0.987$ led to similar excess pore water pressures in the finite element analysis for the first water level lowering.

A parameter study showed that the influence of the relative hydraulic conductivity on the results is not very significant. This is due to the very small amount of air bubbles in the pore water ($S_{(pw=0)} = 0.987$). Furthermore, no data for the relative hydraulic conductivity was available. Thus relative hydraulic conductivity $k_{rel} = 1.00$ was assumed.

3.5 Results and comparison

Figure 8 shows a comparison of the measurement data with the results from the finite element analysis. For the comparison, the calculated change in the pore water pressure with time in 20 m and 30 m depth were analysed. Based on the calculated pore water pressure the excess pore water pressure was determined according to Equation (6).

The results of the finite element analysis are in a good agreement with the measurement data. For the period from 1.4 to 2.0 days the finite element analysis results show a slightly larger mismatch of analysis and measurements at a depth of 20 m. The reason for this could not be identified as yet. However, it has to be kept in mind that the measurements are real field data and thus some uncertainties in the data exist, e.g. due to inhomogeneity in the subsoil. Furthermore, one of the main sliding surfaces was detected in the depth of 20 m, thus this difference could be due to the movements of the adjacent slope, which cannot be considered within the 1D-model.

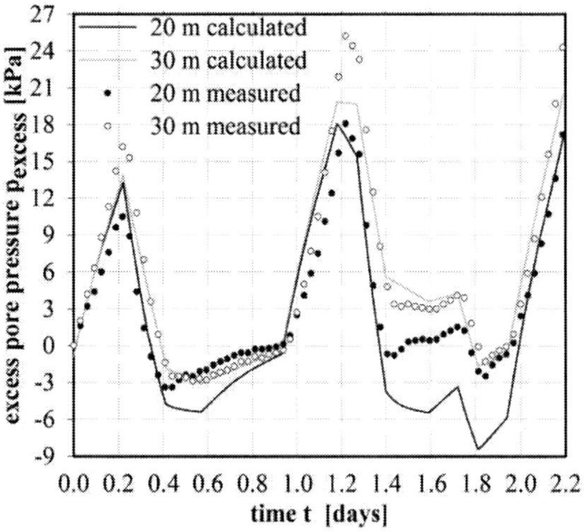

Figure 8. Comparison of measurement data with results from the finite element analysis.

In addition to the amount of air bubbles, the value of the excess pore water pressure depends mainly on the ratio between the velocity of the water level change and the hydraulic conductivity of the soil.

The change of the excess pore water pressures in a depth of 20 m during the period from 0.4 to 0.6 days, which is shown in the measurement data, cannot be modelled with the finite element model in very detail. However, the calculation results show a slower decrease of the excess pore water pressure in 20 m depth, compared to the calculation results in 30 m depth. This indicates a consolidation process during the further increase of the water level with a slower velocity as described before. The same behaviour can be observed during the period from 1.4 to 1.6 days.

4 Conclusion and Outlook

The presented example show that the presence of air bubbles in the pore water can have a significant influence on the development of excess pore pressures in soil layers below a reservoir with fluctuating water levels. The pore pressure measurements can be explained in a qualitative manner if the compressibility of the pore fluid, due to a small amount of air bubbles in the pore water, is considered in the finite element analysis. Eventually the excess pore water pressures could lead to a failure in the subsoil and therefore a realistic prediction of excess pore water pressures is essential in order to assess the risk of slope failure.

Further finite element analyses are currently performed including the real geometry of the slope in the vicinity of the storage basin. Finally, analyses considering full coupling of hydraulic and mechanical behaviour of the soil will be performed in order to obtain a reliable estimate of slope movements and possible failure scenarios.

References

1. B.A. Faybishenko, Water Resour. Res **31**, 2421 (1995)
2. L. Boutonnier, PhD Thesis (Institut National Polytechnique de Grenoble, 2007)
3. H. Montenegro, H.-J. Köhler, T. Holfelder, Unsaturated Soils: Numerical and Theoretical Approaches Vol. 2 (Springer, Berlin, 2003)
4. M. Marinas, J. W. Roy, J. E. Smith, Ground Water **51/1**, 41 (2013)
5. R.B.J. Brinkgreve, S. Kumarswamy, W.M. Swolfs, Manual Plaxis 2D 2015 (Plaxis bv, Delft, 2015)
6. K. V. Bicalho, D. Znidarcic, H.-Y. Ko, Geotech. SP. **99**, 517 (2000)
7. A. Sakaguchi, T. Nishimura, M. Kato, Vadose. Zone J. **4**, 139 (2005)
8. D. G. Fredlund, H. Rahardjo, Soil Mechanics for Unsaturated Soils (John Wiley & Sons, New York, 1993)

Modelling of coupled heat and moisture flows around a buried electrical cable

Hossein Eslami[1], Olivier Cuisinier[1,a], Farimah Masrouri[1]

[1]*Laboratoire d'Energétique et de Mécanique Théorique et Appliquée - UMR 7563 CNRS / Université de Lorraine – 2, Rue M. Roubault, TSA 70605, 54501, Vandœuvre-lès-Nancy, France*

Abstract. The admissible current within a buried electrical power cable is limited by the maximum allowed temperature of the cable (Joule effect). The thermal properties of the surrounding soil controls heat dissipation around the cable. The main focus of the study was to evaluate the coupled heat and moisture flow around such buried electrical cables. The heat dissipation of a buried power cable was simulated in the surrounding soil at unsteady conditions. The hydro-thermal coupling was modelled by taking into account the moisture flow of liquid water and vapour, and the heat flow in the soil by convection and advection. As the thermal vapour diffusion enhancement factor (η) appears to be a key parameter, the sensitivity study of the coupled heat and moisture flow in the ground regarding this parameter was performed. The variations of the degree of saturation and the temperature of the surrounding soil were studied over 180 days of heating. The results showed that the moisture flow was mainly caused by the vapour transport under temperature gradients. These results emphasized the significant effect of the hydrothermal characteristics of surrounding soil. The radius of influence of the power cable was also evaluated.

1 Introduction

Nowadays, new electrical power lines are generally buried to limit their impact on landscape. When a buried electrical power cable passes the current, the conductor heats and releases calories (Joule effect). The admissible current of buried electrical cables is limited by the maximum allowed temperature of the cable. Low thermal conductivity prevents heat dissipation in the ground, and thus the cable temperature increases. It is essential to ensure effective dissipation of this energy to avoid excessive heating of the cable that could lead to its rupture.

The study of the heat dissipated from the cable in the surrounding soil has been the subject of many works [1–10]. Previous works focused on the thermal behaviour of cable backfill materials, the ampacity of cables, the effect of the installation geometry including the dimensions of the trench, cable location and diameter, the thermal properties of the surrounding soils, the seasonal variation of temperature, etc. Nevertheless, in our knowledge the coupling between thermal and hydraulic phenomena was not considered. The moisture flow induced by thermal gradient and the effect of water content on the thermal conductivity have to be taken into account to evaluate the coupled heat and moisture flow around such buried cables.

Philip and De Vries [11] investigated the coupling between liquid water, water vapour and heat transport in soils. They accounted for the fluxes induced by both pressure head and soil temperature gradients. The water vapour flux under thermal gradients is an important part of water flow in soil. Philip and De Vries [11] also added a mechanistic enhancement factor (η) into the equation of thermally induced water vapour flow to consider the discrepancy between the predicted transfers using Fick's law and the measured water fluxes [12, 13]. Despite the importance of η in modelling of coupled heat and moisture flow, few works have been focused on evaluating it experimentally and quantitatively [14–17]. The available results showed that this parameter is not a constant and depends on water content and temperature variation. Other authors proposed empirical modifications for this parameter to improve its agreement with the observed experimental measurements [18, 19].

In this study, the heating of a buried electrical cable was simulated. The hydrothermal coupling was modelled taking into account the moisture flow in the form of liquid water and vapour, and the heat flow by convection and advection by the mean of PLAXIS®. The objective was to quantify the effect of the thermal vapour diffusion enhancement factor (mechanistic enhancement factor), η, on the thermal and hydraulic behaviour of the surrounding soil. The theory is first introduced. Then the numerical modelling will be described. The effects of η is then discussed in a last section.

[a] Corresponding author: Olivier.Cuisinier@ensg.univ-lorraine.fr

2 Theory

The moisture transfer in unsaturated soils takes place in two phases: liquid water and water vapour. The liquid water flow in unsaturated soils can be described using Darcy's law by replacing the saturated permeability with a relative permeability:

$$\underline{J}_w = \rho_w \underline{v}_w = \rho_w \frac{k_{rel}}{\mu} K\nabla(\rho_w g\, y + p_w) = \rho_w \frac{k_{rel}}{\mu} K(\rho_w \underline{g} + \nabla p_w) \quad (1)$$

Where J_w is the flux rate of mass of liquid water (kg/m^2/s), ρ_w is the density of water (kg/m^3), v_w is the flow rate of water (m/s), k_{rel} is the ratio of the permeability at a given saturation to the permeability in saturated state (-), μ is the dynamic viscosity of the water (N.s/m^2), K is the intrinsic permeability of the soil (m^2), g is the gravitational acceleration (m^2/s), y is the elevation (m) and p_w is the pore water pressure (N/m^2).

The diffusion of water vapour through the soil, based on Fick's law can be written [11]:

$$\underline{J}_v = -a\alpha D\nabla\rho_v \quad (2)$$

where J_v is the flux rate of mass of water vapour (kg/m^2/s), a is the air-filled porosity (m^3/m^3), α is the tortuosity (-), D is the diffusivity of water vapour in air, (m²/s) and ρ_v is the density of water vapour (kg/m^3).

The water vapour flow can be expressed as:

$$\underline{J}_v = -D_v\nabla\rho_v = -D_v\begin{pmatrix}\eta\left(\theta\frac{\partial\rho_{vS}}{\partial T} + \frac{\rho_v p_w}{\rho_w R T^2}\right)\nabla T \\ +\left(\frac{-\rho_v p_w}{\rho_w R T}\right)\nabla p_w\end{pmatrix} = -\eta D_v\left(\theta\frac{\partial\rho_{vS}}{\partial T} + \frac{\rho_v p_w}{\rho_w R T^2}\right)\nabla T - D_v\left(\frac{-\rho_v p_w}{\rho_w R T}\right)\nabla p_w \Rightarrow \underline{J}_v = -D_{Tv}\nabla T + D_{pv}\nabla p_w \quad (3)$$

Where ρ_{vS} is the saturated water vapour density (kg/m^3) and h_r is the relative humidity (-), T is the Temperature (K), R is the specific gas constant for water vapour, D_{pv} is the hydraulic vapour diffusion coefficient, D_{Tv} is the thermal vapour diffusion coefficient and η is the thermal vapour diffusion enhancement factor.

The water mass balance is given by this equation:

$$n\frac{\partial}{\partial t}(S\rho_w + (1-S)\rho_v) + (S\rho_w + (1-S)\rho_v)\left[\frac{\partial\varepsilon_v}{\partial t} + \frac{1-n}{\rho_s}\frac{\partial\rho_s}{\partial t}\right] = -\nabla.(\underline{J}_w + \underline{J}_v) \quad (4)$$

Where n is the porosity of porous medium, S is the degree of saturation, ρ_s is the soil grains density (kg/m^3) and ε_v is the volumetric strain. After expansion of the left-hand side of this equation by integrating the compressibility and volumetric thermal expansion of water and the volumetric thermal expansion of soil grains, the water mass balance can be expressed as [20]:

$$\left[n(\rho_w - \rho_v)\frac{\partial S}{\partial p_w} - nS\rho_w\beta_{wP0} - n(1-S)\frac{\rho_v}{\rho_w R_v T}\right]\frac{\partial p_w}{\partial t} + \left[n(\rho_w - \rho_v)\frac{\partial S}{\partial T} - nS\rho_w\beta_{wT0} - n(1-S)\left(\theta\frac{\partial\rho_{vS}}{\partial T} + \frac{\rho_v\rho_w}{\rho_w R_v T^2}\right)\right]\frac{\partial T}{\partial t} + [(S\rho_w + (1-S)\rho_v)(1-n)\beta_{sT}]\frac{\partial T}{\partial t} + (S\rho w + (1-S)\rho v)\frac{\partial\varepsilon v}{\partial t} + \nabla.(\underline{J}_w + \underline{J}_v) = 0 \quad (5)$$

Where β_{wP0}, β_{wT0} are the compressibility and volumetric thermal expansion of water and β_{sT} is the volumetric thermal expansion of soil grains.

The heat flow by conduction is governed by Fourier's law:

$$\underline{J}_c = -\lambda\nabla T \quad (6)$$

The thermal conductivity of the porous medium, λ, could be expressed as:

$$\lambda = (1-n)\lambda_s + nS\lambda_w + n(1-S)\lambda_v \quad (7)$$

Where λ_s, λ_w, λ_v, are the solid, the water and the vapour thermal conductivities. By calculating the thermal conductivity of the porous medium with the equation (7), the effect of the degree of saturation variation (e.g. soil drying) on this parameter is taken into account.

Heat flow by advection is given by:

$$\underline{J}_{Aw} = C_w T\underline{J}_w = C_w T\rho_w\left(\frac{k_{rel}}{\mu}\underline{\underline{\kappa}}^{\text{int}}(\nabla p_w + \rho_w\underline{g})\right) \quad (8)$$

Where C_w is the water specific heat capacity.

The energy balance equation for the porous medium can be written as follows:

$$\frac{\partial T}{\partial t}(nS\rho_w e_w + n(1-S)\rho_v e_v + (1-n)\rho_s e_s = -\nabla.(\underline{J}_{Aw} + \underline{J}_c) + Q_T \quad (9)$$

Where e_w, e_v and e_s are the internal energy in the water, vapour and solid phases, respectively and Q_T is the heat source term (i.e. heat generation rate per unit volume). The left-hand side of this equation could be replaced by $\rho C\frac{\partial T}{\partial t}$, where ρ$C$, the heat capacity of the porous medium is equal to:

$$\rho C = (1-n)\rho_s C_s + nS\rho_w C_w + n(1-S)\rho_v C_v \quad (10)$$

Where C_s, C_w and C_c are the solid, the water and the gas specific heat capacities.

Expanding the right-hand side of the equation (9) and adding a term to take into account the heat transfer at the surface in contact with the air, the heat balance equation can be written as:

$$\rho C\frac{\partial T}{\partial t} - \nabla.(\lambda\nabla T) + \rho_w C_w\left[\frac{k_{rel}}{\mu}\underline{\underline{\kappa}}^{\text{int}}(\nabla p_w + \rho_w\underline{g})\right].\nabla T + \rho_w C_w T\left[\nabla.\left(\frac{k_{rel}}{\mu}\underline{\underline{\kappa}}^{\text{int}}(\nabla p_w + \rho_w\underline{g})\right)\right] - Q_T - C_{az}(T - T_a) = 0 \quad (11)$$

Where T_a is the air temperature (K) and C_{as} (W/m^{2}K) is the convective heat transfer coefficient at the surface in contact with air.

3 Numerical modelling

The equations described is previous section have been implemented and released in the finite element code PLAXIS 2D [20]. Thermal module of this code offers a coupling between the thermal phenomena. This code was used to simulate the hydrothermal behaviour of a trench and the surrounding soil when the power cables were the heat sources. In this section, first the modelled trench is described. Then the different material characteristics are presented and the boundary conditions are explained.

3.1 Model description

To study the heating of the soil surrounding a buried electrical cable, it was decided to model the standard configuration for a typical buried electrical system. The trench is 1.50 m in depth and 0.60 m in width. Three-phase electric power cables are buried next to each other. The diameter of the cable is 86.6 mm. Concrete is installed around the cables to ensure mechanical stability of the cables after installation. Compacted soil is then installed above the concrete. Most of the time, the trenches are done under existing road, the top material of the trench was thus an asphaltic layer. Figure 1 shows the sketch of the trench in the surrounding soil, which was modelled in the framework of this study. Thus 4 materials, plus the cables, were defined:

- material 1: asphaltic concrete + base,
- material 2: sub-base,
- material 3: concrete,
- material 4: surrounding soil.

Thanks to the symmetry, the modelling was done only for the half of the trench relative to the Y axis. The mesh was automatically generated and the geometry was divided into 15-Node triangular elements with a refinement in the trench. Figure 1 shows the mesh that contains 1987 nodes and 16199 elements. The results of the preliminary studies showed that for the dimensions equal to 10 m, the scale effect is negligible.

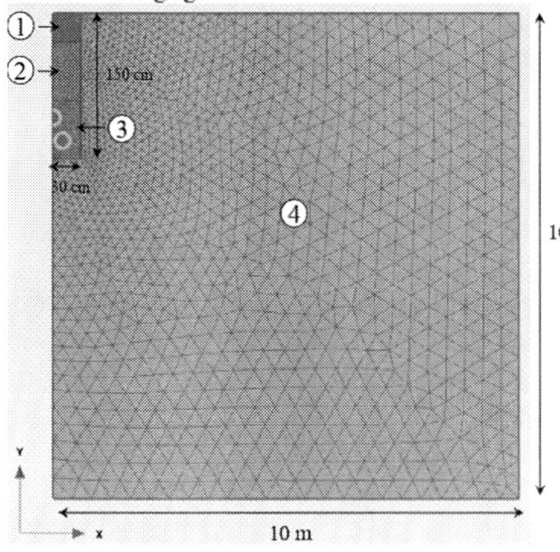

Figure 1. Geometry and meshing of the modelled trench.

3.2 Material properties

The thermal, hydraulic and hydrothermal properties of the surrounding soil are presented in Table 1. For the surrounding soil, a silt loam clay, the Van Genuchten model was used for the soil water retention curve.

Table 1. Thermal properties of the materials of the surrounding soil.

Property		Value
Thermal	C_s (kJ/t.K)	2000
	λ_s (W/m.K)	2.5
	ρ_s (kg/m3)	2700
Hydraulic	(D_v) (m2/day)	0.1033
	η	1 and 5
Hydraulic	$k_x = k_y$ (m/day)	0.1676
	S_{res} (-)	0.06203
	S_{sat} (-)	1.00
	g_n (-)	1.377
	g_a (-)	3.83
	g_l (-)	1.25

The diffusivity of water vapour (D_v) is obtained with:

$$D_v = a\alpha D \tag{12}$$

Where a is the air-filled porosity (m^3/m^3), α is the tortuosity (-) and D is the diffusivity of water vapour in air, (m²/s). The tortuosity (α) is described as a function of the volumetric water content (θ) [21] with:

$$\alpha = (n-\theta)^{2/3} \tag{13}$$

Therefore, for a soil with the porosity (n) = 0.3 and the air-filled porosity (a) = 0.15 and the volumetric water content (θ) = 0.15, the tortuosity (α) was equal to: $\alpha = (0.3-0.15)^{2/3} = 0.28$

The diffusivity of water vapour in air (D) is expressed as a function of temperature (Kelvin) [22] with:

$$D = 2.29\times10^{-5}\left(T/273.15\right)^{1.75} \tag{14}$$

Assuming that the soil temperature changes between 12 and 60 °C (i.e. the maximum allowed temperature of the cable) the mean temperature was 36 °C and the diffusivity of water vapour in air is:

$$D = 2.29\times10^{-5}\left(309.15/273.15\right)^{1.75} = 2.84\times10^{-5}\,m^2/s = 2.46\,m^2/day$$

The diffusivity of water vapour is:

$$D_v = a\alpha D = 0.15\times0.28\times2.46 = 0.1033\,m^2/jour\,.$$

Thermal vapour diffusion enhancement factor, η, varies between about 1 and 15 depending on the volumetric water content and temperature [14, 23–25].

 DOI: 10.1051/e3sconf/20160916011

This dependence was not considered explicitly by the model. A constant value of η was thus set for each simulation. To account for the dependency of η on water content, the value of η was assumed equal to 1 and 5 corresponding to dry and moderately humid soil initial conditions respectively. Using the same initial degree of saturation permitted to study explicitly the effect of η on the results.

The thermal properties of the different materials of the trench are presented in Table 2. The hydraulic properties of these materials were not taken into account in order to focus the study on the hydrothermal behaviour of the surrounding soil.

Table 2. Thermal properties of the materials of the trench.

Property	Material 1: asphaltic concrete + base	Material 2: sub-base	Material 3: concrete
C_s (kJ/t.K)	960	1500	900
λ_s (W/m.K)	1.95	1	1.2

3.3 Initial and boundary conditions

The initial conditions consisted of a homogeneous temperature of 12 °C (285.15 °K) and the water table depth at 4 m below the surface (Figure 2). The boundary conditions (hydraulic and thermal) were applied to each side of the model. Along the axe OY, the water flux was closed (Neumann boundary condition) and the temperature could change (Dirichlet boundary condition). For the right side boundary, the water and heat fluxes were open. For the bottom, the water flux was allowed and a constant temperature was imposed (285.15 °K). For the top boundary, the heat transfer at the surface in contact with the air was allowded by imposing the air temperature equal to 20 °C (293.15 °K) and defining C_{as} (W/m^{2}K), the convective heat transfer coefficient at the surface in contact with air [26] (convective boundary condition):

$$C_{as} = 7.2 + 3.8\upsilon \tag{15}$$

Where υ is the wind speed (m/s). The wind speed was considered equal to 20 km/h (= 5.56 m/s) as reference, C_{as} was:

$$C_{as} = 7.2 + 3.8 \times 5.56 = 28\ W / m^2 K$$

The evaporation in the interface of soil-atmosphere was simulated by a Neumann boundary condition with a flux equal to 0.5 mm/day. The heating was modelled by imposing a heat flux equal to 55 W/m^2. This flux was imposed in the interface between the cables and the concrete.

4 Sensitivity study for η values of the surrounding soil

The thermal vapour diffusion enhancement factor (η) varies with the variation of the water content. To quantify the effect of the variation of this parameter on the moisture and heat transfers, two models were compared. The initial conditions, boundary conditions and material characteristics were identical for the two models and the only difference was the η value of the surrounding soil. For one model η was equal to 1 (dry soil), and for the other, it was equal to 5 (moderately moist soil). The heating period was fixed to 180 days.
The degree of saturation (S) after 180 days of heating for the two values of η is presented in Figure 3. In order to focus on the soil behaviour in the vicinity of the trench, the profiles are shown for 5 m in each direction from the centre of the trench. We can see in this figure the variation of the profiles of the degree of saturation as a result of heating. The maximal drying was occurred at the interface soil-concrete and the final degree of saturation was 48.3% and 38.9% for η equal to 1 and 5 respectively. The global drying compared to the soil-atmosphere interface is related to the evaporation imposed at this boundary, letting moisture transfer to the outside of the system (0.5 mm/day). Approaching the water table (4 m of depth) the heating effect on the degree of saturation decreased.

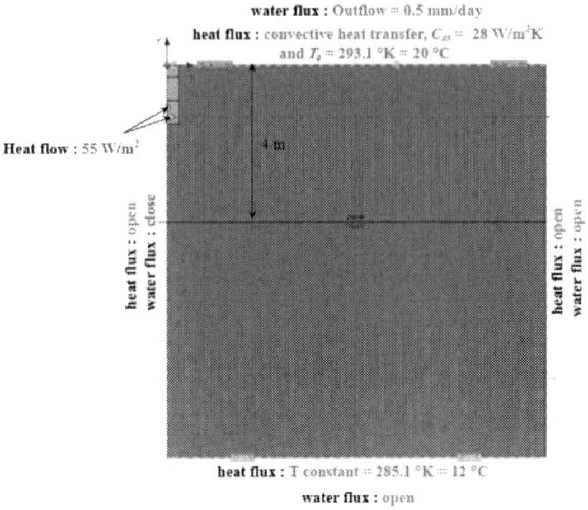

Figure 2. Model boundary conditions.

To study the radius of influence of the power cable, the degree of saturation (S) after 180 days of heating is drawn throughout the dotted red line in Figure 4. The initial degree of saturation was 80%. At the contact with the trench, S decreased to 48.9% for η = 1 while it decreased to 39.5% for η = 5. The larger decrease with η = 5 is related to the role of this parameter in vapour transfer: Increasing η accelerates the transfer of vapour resulting in a greater drying of the soil on one hand and a larger radius of influence on the other hand. The radius of influence is about 1.60 m for η = 1 but it is about 3 m for η = 5 (in addition to general drying caused by evaporation at the soil-atmosphere interface).

DOI: 10.1051/e3sconf/20160916011

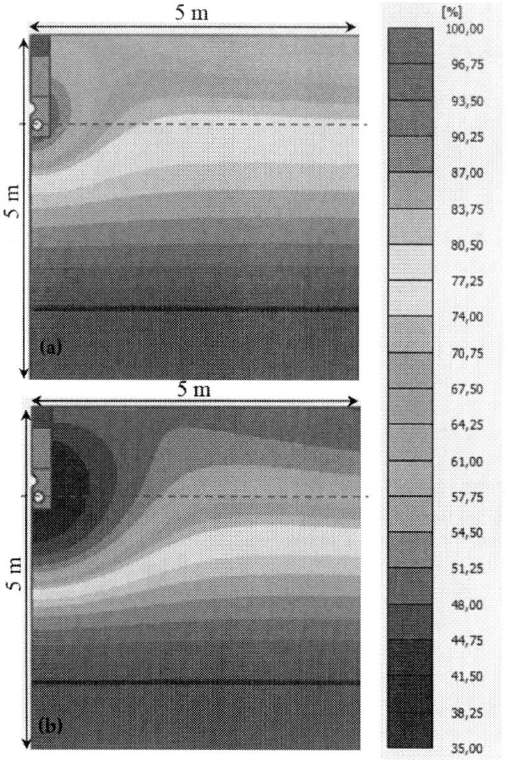

Figure 3. Degree of saturation after 180 days of heating: (a) $\eta = 1$; (b) $\eta = 5$.

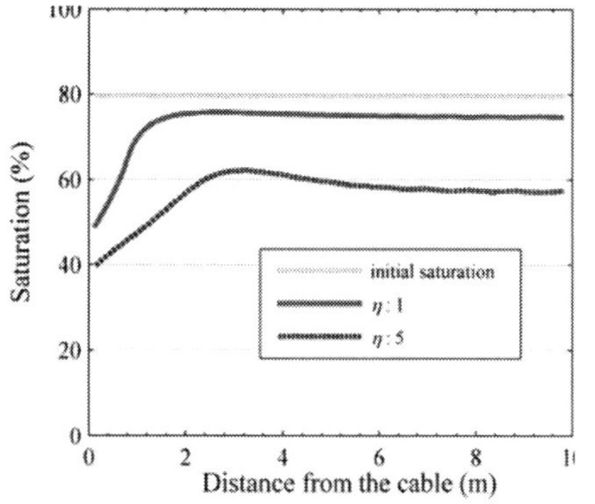

Figure 4. Degree of saturation after 180 days of heating throughout the dotted red line.

Figure 6 shows the temperature as a function of distance from the cable after 180 days. The heat propagation is almost the same. The maximal temperature was 49.5 °C and 48.3 °C for $\eta = 1$ and $\eta = 5$ respectively.

In order to study the kinetic of the temperature and degree of saturation variations, these parameters are drawn as a function of time for the point located at the right bottom of the trench (0.30,-1.50) (Figure 6). For temperature, the variation is identical in the first 25 days and then there is a small difference between two curves, reaching 1.3 °C of difference after 180 days with 31 °C for $\eta = 1$ and 29.7 for $\eta = 5$. For degree of saturation, when $\eta = 5$ there was an important decrease for the first 10 days and it was stabilized after 120 days. When $\eta = 1$, the decrease of S in the first days was less important than that of $\eta = 5$ but it continued to decrease with almost the same rate until 50 days. After 180 days it was not completely stabilized. The dependence of S variation to η is related to the diffusion of water vapour through the soil under thermal gradient (equation (3)). The thermal vapour diffusion coefficient, D_{Tv}, depends also on η and the increasing of η increases this parameters. The higher the value of D_{Tv}, the higher the vapour transfer induced by temperature gradient. Consequently the decrease of the degree of saturation is more important, and the soil will be drier in that case.

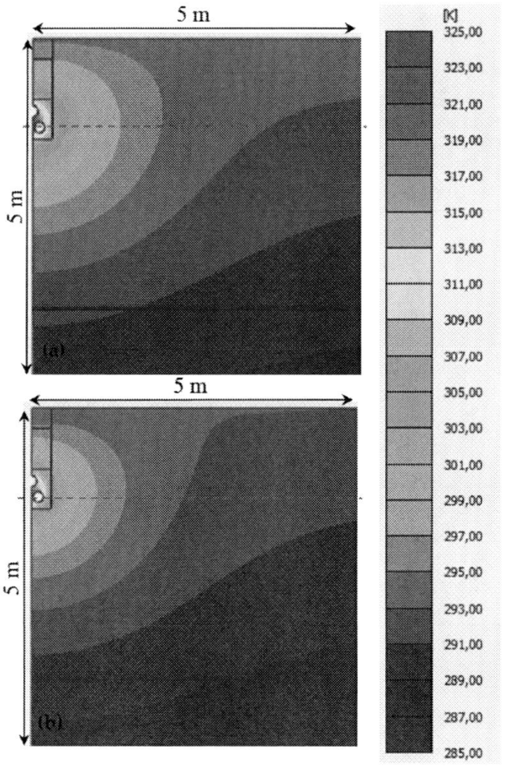

Figure 5. Temperature after 180 days of heating: (a) $\eta = 1$; (b) $\eta = 5$.

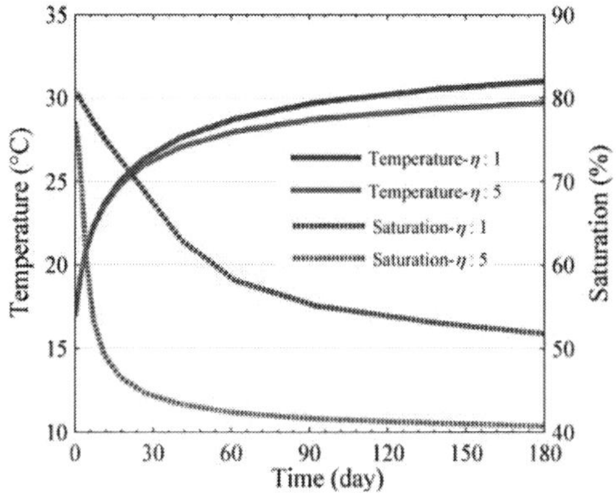

Figure 6. Temperature and saturation as a function of time for the point located at the right bottom of the trench.

5 Conclusion

The main objective of this work was to highlight the coupled heat and moisture flows around a buried electrical cable and quantify the effect of thermal vapour diffusion enhancement factor on this behaviour. The key parameters of this modelling were on one hand, the conventional parameters (the thermal conductivity, thermal capacity, retention curve and the permeability coefficient) and the diffusivity of water vapour (D_v) and on the other hand the thermal vapour diffusion enhancement factor (η). The increase of η resulted in a more significant drying and in a less important and slower temperature increase. This shows the high effect of transfer in vapour phase in the soil surrounding buried electrical cables. We observed also the dependence of the influence radius to the values of η. In this study, the energy used to evaporate water and the heat transported by vapour are not considered. Taking into account this part of energy balance by using the evaporation heat latent of water in the equations will increase the heat.

Acknowledgment

This study was part of the VALO CQFD research project funded by ADEME.

References

1. De Leon F, Anders GJ (2008) Effects of Backfilling on Cable Ampacity Analyzed With the Finite Element Method. IEEE Trans Power Deliv 23:537–543.
2. de Lieto Vollaro R, Fontana L, Vallati A (2011) Thermal analysis of underground electrical power cables buried in non-homogeneous soils. Appl Therm Eng 31:772–778.
3. Hanna MA, Chikhani AY, Salama MMA (1998) Thermal analysis of power cable systems in a trench in multi-layered soil. IEEE Trans Power Deliv 13:304–309.
4. Adams JI, Baljet AF (1968) The thermal behavior of cable backfill materials. IEEE Trans Power Appar Syst PAS-87:1149–1161.
5. El-Kady MA (1982) Optimization of Power Cable and Thermal Backfill Configurations. IEEE Trans Power Appar Syst PAS-101:4681–4688.
6. Mitchell JK, Abdel-hadi ON (1979) Temperature Distributions around Buried Cables. IEEE Trans Power App Syst 98:1158–1166.
7. Gela G, Dai JJ (1988) Calculation of thermal fields of underground cables using the boundary element method. IEEE Trans Power Deliv Vol 3, No 4, Oct 3:1341–1347.
8. Bernath A, Olfe DB, Ferguson BW (1984) Heat transfer measurements on uniqually loaded underground power cables with constant and cyclic currents. IEEE Trans Power Appar Syst PAS-103:2799–2806.
9. Al-Saud M s., El-Kady M a., Findlay RD (2006) Accurate Assessment of Thermal Field and Ampacity of Underground Power Cables. In: Can. Conf. Electr. Comput. Eng. pp 651–654.
10. Kellow MA (1981) A numerical procedure for the calculation of the temperature rise and ampacity of underground cables. IEEE Trans Power Appar Syst PAS-100:3322–3330.
11. Philip JR, De Vries D a. (1957) Moisture movements in porous materials under temperature gradients. Trans Am Geophys Union 38:222–232.
12. Gurr CG, Marshall TJ, Hutton JT (1952) Movement of water in soil due to a temperature gradient. Soil Sci 74:335–346.
13. Taylor SA, Cavazza L (1954) The movement of soil moisture in response to temperature gradients. Soil Sci Soc Am J 18:351–358.
14. Cass A, Campbell GS, Jones TL (1984) Enhancement of thermal water vapor diffusion in soil. Soil Sci Soc Am J 48:25–32.
15. Bonsu M (1997) Soil water management implications during the constant rate and the falling rate stages of soil evaporation. Agric Water Manag 33:87–97.
16. Hiraiwa Y, Kasubuchi T (2000) Temperature dependence of thermal conductivity of soil over a wide range of temperature (5–75 C). Eur J Soil Sci 211–218.
17. Wildenschild D, Roberts JJ (2001) Experimental tests of enhancement of vapor diffusion in Topopah Spring Tuff. J Porous Media 4:1–13.
18. Cary JW (1979) Soil Heat Transducers and Water Vapor Flow1. Soil Sci Soc Am J 43:835.
19. Jury WA, Letey J (1979) Water vapor movement in soil: reconciliation of theory and experiment. Soil Sci Soc Am J 43:823–827.
20. Haxaire A, Galavi V, Brinkgreve RBJ (2013) Definition and implementation of a fully coupled THM model for unsaturated soils. In: Kwasniewski M, Lydzba D (eds) Rock Mech. Resurces, Energy Environ. Taylor & Francis Group, Wroclaw, Poland, pp 507–512
21. Milly PCD (1984) A Simulation Analysis of Thermal Effects on Evaporation From Soil. Water Resour Res 20:1087.
22. Kimball B a., Jackson RD, Reginato RJ, et al. (1976) Comparison of Field-measured and Calculated Soil-heat Fluxes1. Soil Sci Soc Am J 40:18.
23. Lu S, Ren T, Yu Z, Horton R (2011) A method to estimate the water vapour enhancement factor in soil. Eur J Soil Sci 62:498–504.
24. Sakai M, Toride N, Šimůnek J (2009) Water and Vapor Movement with Condensation and Evaporation in a Sandy Column. Soil Sci Soc Am J 73:707.
25. Heitman JL, Horton R, Ren T, et al. (2008) A Test of Coupled Soil Heat and Water Transfer Prediction under Transient Boundary Temperatures. Soil Sci Soc Am J 72:1197. doi: 10.2136/sssaj2007.0234
26. Garzoli KV, Blackwell J (1987) An analysis of the nocturnal heat loss from a double skin plastic greenhouse. J Agric Eng Res 36:75–86.

DOI: 10.1051/e3sconf/20160917001

A finite strain elastoplastic constitutive model for unsaturated soils incorporating mechanisms of compaction and hydraulic collapse

Keita Nakamura[1] and Mamoru Kikumoto[1,a]

[1]*Yokohama National University, Tokiwadai 79-5, Hodogaya-ku, Yokohama 240-8501, Japan*

Abstract. Although the deformation of unsaturated soils has usually been described based on simple infinitesimal theory, simulation methods based on the rational framework of finite strain theory are attracting attention especially when solving geotechnical problems such as slope failure induced by heavy rain in which large a deformation is expected. The purpose of this study is to reformulate an existing constitutive model for unsaturated soils (Kikumoto et al., 2010) on the basis of finite strain theory. The proposed model is based on a critical state soil model, modified Cam-clay, implementing a hyperelastic model and a bilogarithmic ln$v$-ln$P$' (v, specific volume; P', effective mean Kirchhoff stress) relation for a finite strain. The model is incorporated with a soil water characteristic curve based on the van Genuchten model (1990) modified to be able to consider the effect of deformation of solid matrices. The key points of this model in describing the characteristics of unsaturated soils are as follows: (1) the movement of the normal consolidation line in ln$v$-ln$P$' resulted from the degree of saturation (Q, deviatoric Kirchhoff stress), and (2) the effect of specific volume on a water retention curve. Applicability of the model is shown through element simulations of compaction and successive soaking behavior.

1 Introduction

Although the stress-strain relationship of unsaturated soils has usually been described based on simple infinitesimal theory, application of the framework of finite strain theory to unsaturated soil mechanics is attracting attention (e.g. [1, 2]), especially when predicting geotechnical issues in which large deformation of the ground (such as a failure of slope or embankment owing to heavy rain or earthquake) is expected.

The constitutive framework of finite strain elastoplasticity has been developed [3–6] based on the multiplicative decomposition of the deformation gradient [7]. For geomaterials, several researchers [8–10] have proposed stress-update algorithms based on return mapping in principal space [3, 11, 12]. Borja and Tamagnini [8] developed the infinitesimal version of the modified Cam-Clay model for finite strain theory based on a multiplicative decomposition of the deformation gradient ($\mathbf{F} = \mathbf{F}^e \cdot \mathbf{F}^p$). In their formulation, a yield function is defined as a function of the mean Kirchhoff stress (P) and deviatoric Kirchhoff stress (Q) instead of mean Cauchy stress (p) and deviatoric Cauchy stress (q), respectively. In addition, for modeling the behavior of isotropic consolidation of soil, a bilogarithmic ln$v$-ln$P$ (v, specific volume) relation [13] is applied for some significant advantages [14, 15]. For unsaturated soil, Song and Borja [1, 2] developed a mathematical framework for coupled solid-deformation/fluid-diffusion in a finite strain range.

In this study, we aim to develop a method that can suitably predict the entire life of an embankment from its construction process to the failure stage owing to heavy rain, for instance. In order to achieve this, an infinitesimal constitutive model for unsaturated soils incorporating compaction and collapse mechanisms (Kikumoto et al. [16]) is extended to a finite-strain model that can capture the behavior of unsaturated soils at a large strain level. In the model, the shear strength of unsaturated soil is controlled by the movement of the normal consolidation line in the direction of the specific volume (v) axis with a variation in the degree of saturation (S_r), and a soil water characteristic curve incorporating the effect of changes in the void ratio (e) is adopted. In this paper, we present an outline of a constitutive model for unsaturated soils based on finite strain theory. The applicability of the model is finally discussed through element simulations of compaction and the soaking behavior of unsaturated soils.

2 Outline of a constitutive model for unsaturated soils

In this section, a model for unsaturated soils incorporating compaction and collapse mechanisms under a finite strain range is derived. This model consists of two parts, namely, an elastoplastic stress-strain relationship taking account of the effect of the degree of saturation, and an advanced soil water retention curve model considering effect of volumetric deformation.

[a] Corresponding author: kikumoto@ynu.ac.jp

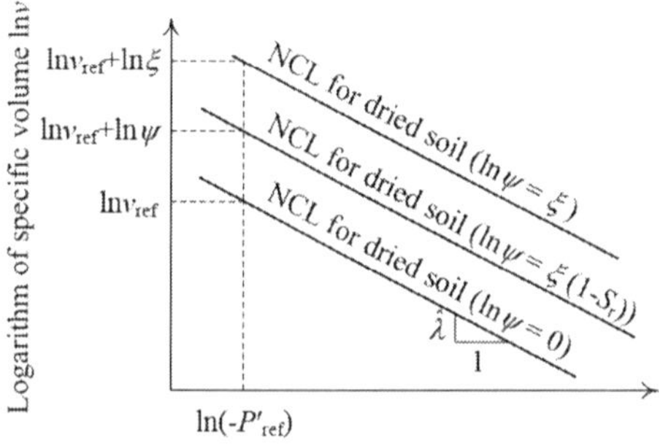

Figure 1. Assumed normal consolidation line for unsaturated soils: dependence of the position of NCL on variation of S_r.

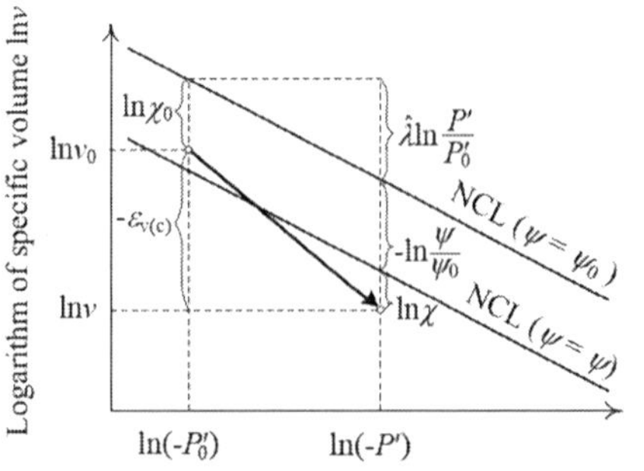

Figure 2. Volumetric behavior of unsaturated soil considering the effects of degree of saturation S_r and specific volume v.

2.1 Elastoplastic constitutive model

We first derive a stress-strain relationship for unsaturated soils based on the effective Kirchhoff stress:

$$\boldsymbol{\tau}' = \boldsymbol{\tau} + \{S_r \vartheta_w + (1 - S_r)\vartheta_a\}\mathbf{I} \quad (1)$$

where $\boldsymbol{\tau} = J\boldsymbol{\sigma}$ (J, Jacobian; $\boldsymbol{\sigma}$, Cauchy stress tensor) is the Kirchhoff stress tensor, S_r is degree of saturation, and $\vartheta = Ju_f$ (u_f, pore fluid pressure) is the Kirchhoff pore fluid pressure. $\mathbf{I}$ is the second-order identity tensor. Subscripts w and a denote water and air, respectively.

2.1.1 Hyperelastic model

In infinitesimal strain theory, a hypoelastic formulation is usually employed. This approach, however, causes an irrational dissipation of stored energy especially under conditions of cyclic loading [17]. On the other hand, the hyperelastic model guarantees the conservation of stored energy. Therefore, we apply a hyperelastic model [8] with pressure-dependent bulk and shear moduli to describe the elastic response, which was originally proposed by Houlsby [18] in an infinitesimal strain range and extended to finite strain theory by Borja and Tamagnini [8].

The hyperelastic potential function is defined as a function of the elastic volumetric strain ε_v^e and elastic deviator strain ε_s^e as

$$\mathrm{W}(\varepsilon_v^e, \varepsilon_s^e) = P'_{ref}\hat{\kappa}\exp\Omega + \frac{3}{2}\mu^e \varepsilon_s^{e2} \quad (2)$$

where Ω and μ^e are given as

$$\Omega = -\frac{\varepsilon_v^e - \varepsilon_{v,ref}^e}{\hat{\kappa}} \quad (3)$$

$$\mu^e = \mu_{ref} - \alpha P'_{ref}\exp\Omega \quad (4)$$

Here, $\varepsilon_{v,ref}^e$ is the elastic volumetric strain at a reference pressure P'_{ref}, and $\hat{\kappa}$ is the elastic compressibility index associated with the Kirchhoff pressure [8]. The elastic shear modulus μ^e contains a constant term μ_{ref} and variable term with a parameter α for pressure dependency. From Equation (2), the mean and deviatoric Kirchhoff stresses are given as follows:

$$P' = \frac{\partial \mathrm{W}(\varepsilon_v^e, \varepsilon_s^e)}{\partial \varepsilon_v^e} = P'_{ref}\exp\Omega\left[1 + \frac{3\alpha}{2\hat{\kappa}}\varepsilon_s^{e2}\right] \quad (5)$$

$$Q = \frac{\partial \mathrm{W}(\varepsilon_v^e, \varepsilon_s^e)}{\partial \varepsilon_s^e} = 3(\mu_0 - \alpha P'_{ref}\exp\Omega)\varepsilon_s^e \quad (6)$$

2.1.2 Yield function

Since unsaturated soils having a lower degree of saturation usually exhibit stiffer behavior, the normal consolidation line (NCL) defined in the $\ln P'$ - $\ln v$ plane is assumed to move in the direction of the specific volume v axis owing to the variation of degree of saturation S_r (Figure 1) in the proposed model.

$$\ln v_{NCL} = \ln v_{ref} - \hat{\lambda}\ln\frac{P'}{P'_{ref}} + \ln\psi \quad (7)$$

v_{ref} is the specific volume v on the normal consolidation line at $P' = P'_{ref}$. $\hat{\lambda}$ is the compression index representing the slope of the compression line in the $\ln P'$ - $\ln v$ relation, which is linked to the compression index $\tilde{\lambda}$ that defines the slope of the compression line in $\ln p'$ - $\ln v$ [8].

$$\hat{\lambda} = \frac{\tilde{\lambda}}{1 - \tilde{\lambda}} \quad (8)$$

$\ln\psi$ in Equation (7) is the separation between the NCL for current, the unsaturated state, and that for the saturated state in logarithmic scale. $\ln\psi$ works as a state variable considering the effect of the degree of saturation S_r on the strength of the unsaturated soils. $\ln\psi$ is assumed to increase as S_r decreases:

$$\ln\psi = \xi(1 - S_r) \quad (9)$$

ξ is a material parameter representing the vertical distance of the state boundary surface for dried and saturated samples in the compression plane. The volumetric movements of NCL and CSL are also considered in a similar way in the previous work [19].

In order to take the effect of density or the overconsolidation ratio into account in the proposed model, we also define a logarithmic volumetric distance $\ln\chi$ ($\chi \geq 1$) between the current specific volume v and specific volume v_{NCL} on the normal consolidation line at the current mean effective stress as a state variable

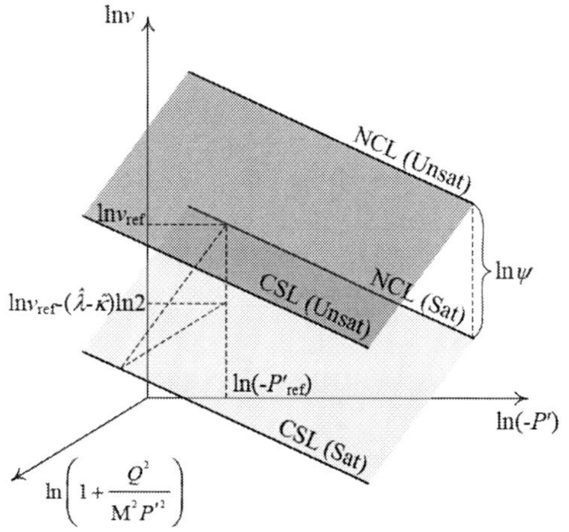

Figure 3. State boundary surface.

(Figure 2). Thus, the specific volume v can be written as follows:

$$\ln v = \ln v_{\mathrm{NCL}} - \ln\chi = \ln v_{\mathrm{ref}} - \hat{\lambda}\frac{P'}{P'_{\mathrm{ref}}} + \ln\psi - \ln\chi \quad (10)$$

The state variable χ is assumed to decrease with the development of the plastic deformation and finally converges to 1 as follows:

$$\dot{\chi} = -\dot{\gamma} a \ln\chi \quad (11)$$

a is a material parameter controlling the effect of the density or overconsolidation ratio, where $\dot{\gamma}$ is the increment of plastic multiplier.

From Equation (10), the volumetric strain for isotropic consolidation can be written as

$$\varepsilon_{\mathrm{v(c)}} = \ln\frac{v}{v_0} = -\hat{\lambda}\ln\frac{P'}{P'_0} + \ln\frac{\psi}{\psi_0} - \ln\frac{\chi}{\chi_0} \quad (12)$$

where subscript 0 denotes the values at initial state. The elastic component of the volumetric strain for pure isotropic loading is derived from Equation (5) as follows:

$$\varepsilon^{\mathrm{e}}_{\mathrm{v(c)}} = \varepsilon^{\mathrm{e}}_{\mathrm{v0}} - \hat{\kappa}\ln\frac{P'}{P'_0} \quad (13)$$

From Equations (12) and (13), we obtain the plastic volumetric strain for isotropic consolidation as Equation (14):

$$\begin{aligned}\varepsilon^{\mathrm{p}}_{\mathrm{v(c)}} &= \varepsilon_{\mathrm{v(c)}} - \varepsilon^{\mathrm{e}}_{\mathrm{v(c)}} \\ &= -(\hat{\lambda}-\hat{\kappa})\ln\frac{P'}{P'_0} + \ln\frac{\psi}{\psi_0} - \ln\frac{\chi}{\chi_0} - \varepsilon^{\mathrm{e}}_{\mathrm{v0}}\end{aligned} \quad (14)$$

Selecting an elliptic-shaped yield surface as the modified Cam-Clay model [20], the plastic volumetric strain owing to shearing, namely dilatancy, is defined as

$$\varepsilon^{\mathrm{p}}_{\mathrm{v(d)}} = -(\hat{\lambda}-\hat{\kappa})\ln\left(1+\frac{Q^2}{M^2P'^2}\right) \quad (15)$$

where M is a material constant and a critical state stress ratio, which controls the aspect ratio of the ellipsoidal yield surface. From Equations (14) and (15), the yield function can be finally written as

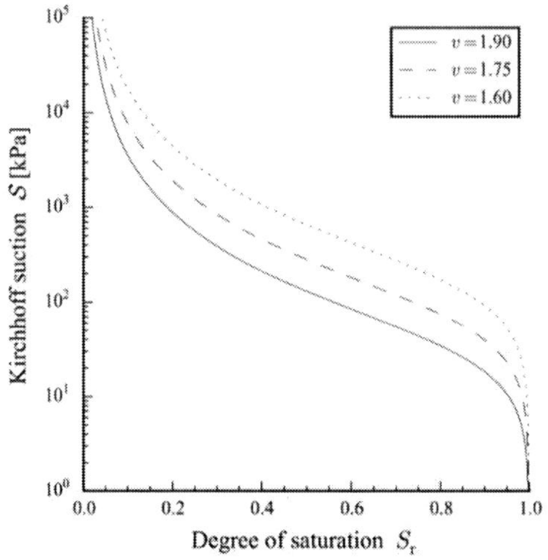

Figure 4. Soil water characteristic curve for three kinds of specific volumes.

Table 1. Material parameters for water retention curve

Maximum saturation $S_r^{\max}$	1.0
Minimum saturation $S_r^{\min}$	0.0
Fitting parameter α [1/kPa]	0.028
Fitting parameter n	1.5
Fitting parameter m	0.33
Reference specific volume v_{ref}	1.9
Effect of volumetric deformation ξ_{v}	9.4

$$\begin{aligned} f(P',Q,\psi,\chi,\varepsilon^{\mathrm{p}}_{\mathrm{v}}) &= (\hat{\lambda}-\hat{\kappa})\left[\ln\frac{P'}{P'_0} + \ln\left(1+\frac{Q^2}{M^2P'^2}\right)\right] \\ &\quad -\ln\frac{\psi}{\psi_0} + \ln\frac{\chi}{\chi_0} + \varepsilon^{\mathrm{e}}_{\mathrm{v0}} + \varepsilon^{\mathrm{p}}_{\mathrm{v}} = 0 \end{aligned} \quad (16)$$

where $\varepsilon^{\mathrm{p}}_{\mathrm{v}} = \varepsilon^{\mathrm{p}}_{\mathrm{v(c)}} + \varepsilon^{\mathrm{p}}_{\mathrm{v(d)}}$.

For the plastic flow rule, we employ a formulation of the exponential approximation [3, 12]:

$$\varepsilon^{\mathrm{e}}_{A,n+1} = \varepsilon^{\mathrm{e,tr}}_{A,n+1} - \Delta\gamma\frac{\partial f_{n+1}}{\partial\beta_{A,n+1}}\ ,\ A=1,2,3 \quad (17)$$

where $\varepsilon^{\mathrm{e,tr}}_{A,n+1} = \varepsilon_{A,n+1} - \varepsilon^{\mathrm{p}}_{A,n}$ is the trial elastic principal logarithmic strain, β_A is the principal Kirchhoff effective stress, and $\Delta\gamma = \dot{\gamma}\Delta t$ ($\Delta t = t_{n+1} - t_n$).

2.2 Water retention curve model

In this study, an extended version of a water retention curve model proposed by van Genuchten [21] is applied. This model is capable of considering the effect of volume change on retention characteristics. The model proposed by van Genuchten is first given as a function of the Kirchhoff suction as:

$$S_{\mathrm{r}} = S_{\mathrm{r}}^{\min} + (S_{\mathrm{r}}^{\max} - S_{\mathrm{r}}^{\min})\{1+(\alpha S)^n\}^{-m} \quad (18)$$

where $S_{\mathrm{r}}^{\min}$ and $S_{\mathrm{r}}^{\max}$ are the minimum and maximum degrees of saturation, respectively. α, n, m are material parameters, and $S = Js$ (s, suction) is the Kirchhoff suction. In order to consider the effect of the density of soils, we propose to replace S with a modified Kirchhoff suction S^* as Equations (19) and (20).

$$S_r = S_r^{\min} + (S_r^{\max} - S_r^{\min})\left\{1+(\alpha S^*)^n\right\}^{-m} \quad (19)$$

$$S^* = S\left(\frac{v}{v_{\mathrm{ref}}}\right)^{\xi_v} \quad (20)$$

Here, v_{ref} is a reference specific volume, and ξ_v is a material parameter for controlling the effect of the volumetric deformation of soil. Figure 4 shows the water retention curves for three kinds of specific volume (v = 1.90, 1.75, and 1.60). It is properly simulated by the model that soil with a higher density has a higher degree of saturation under the same value of suction, which is reported by experimental results [22].

2.3 Return mapping

Next, we derive the return mapping procedure based on the elastic principal logarithmic strains [23]. In the return mapping of proposed model, $\varepsilon^{e}_{A,n+1}$ (A = 1, 2, 3), χ_{n+1}, and $\Delta\gamma$ are unknown variables evaluated for given values $\varepsilon_{A,n+1}$ and S_{n+1}. Thus, the unknown variable vector $\mathbf{x}$ and residual vector $\mathbf{r}(\mathbf{x})$ take form, respectively, as

$$\mathbf{x} = \begin{bmatrix} \boldsymbol{\varepsilon}^{e}_{n+1} \\ \chi_{n+1} \\ \Delta\gamma \end{bmatrix}, \quad \mathbf{r}(\mathbf{x}) = \begin{bmatrix} \boldsymbol{\varepsilon}^{e}_{n+1} - \boldsymbol{\varepsilon}^{e,\mathrm{tr}}_{n+1} + \Delta\gamma \dfrac{\partial f_{n+1}}{\partial \boldsymbol{\beta}'_{n+1}} \\ \chi_{n+1} - \chi_n + \Delta\gamma a \ln \chi_{n+1} \\ f(P'_{n+1}, Q_{n+1}, \psi_{n+1}, \chi_{n+1}, \varepsilon^{p}_{v,n+1}) \end{bmatrix} \quad (21)$$

where $\boldsymbol{\beta}'$ is the principal effective Kirchhoff stress vector. The backward-Euler approximation is applied to Equation (11). ψ_{n+1} can be calculated from Equation (9) and (19). Equation (21) is nonlinear, so the solution $\mathbf{x}$ needs to be evaluated iteratively. To this end, we use Newton's method with a local Jacobian matrix $\partial\mathbf{r}(\mathbf{x})/\partial\mathbf{x}$ until $\|\mathbf{r}(\mathbf{x})\| <$ TOL is satisfied.

In the elastic regime (i.e., unloading), the above procedure is not implemented. This can be judged by the value of the yield function with trial values P'^{tr}_{n+1}, Q^{tr}_{n+1}, ψ_{n+1}, $\chi^{\mathrm{tr}}_{n+1}(=\chi_n)$, and $\varepsilon^{p,\mathrm{tr}}_{v,n+1}(=\varepsilon^{p}_{v,n})$ as follows:

$$f^{\mathrm{tr}}_{n+1} = f(P'^{\mathrm{tr}}_{n+1}, Q^{\mathrm{tr}}_{n+1}, \psi_{n+1}, \chi^{\mathrm{tr}}_{n+1}, \varepsilon^{p,\mathrm{tr}}_{v,n+1}) \quad (22)$$

The trial values except for χ^{tr}_{n+1} hold as updated values if $f^{\mathrm{tr}}_{n+1} < 0$ (elastic state), and χ_{n+1} is obtained by solving $f_{n+1} = 0$.

2.4 Algorithmic tangent moduli

An algorithmic tangent moduli that is consistent with a formulation based on the principal space is presented here. The use of the algorithmic tangent moduli is necessary to ensure the convergence of iterations.

A differentiation of the plastic flow rule (17) results in

Table 2. Material parameters for elastoplastic model

Elastic compressibility index $\hat{\kappa}$	0.012
Elastic shear modulus μ_0 [MPa]	20.0
Elastic parameter α	100.0
Elastoplastic compressibility index $\hat{\lambda}$	0.108
Critical stress ratio M	1.2
Reference pressure P'_{ref} [kPa]	-98.0
Reference specific volume v_{ref}	1.9
Effect of movement of NCL ξ	0.28
Effect of density and confining pressure a	0.03

Table 3. Initial state of soil for simulations

Suction s [kPa]	0.0
Elastic volumetric strain $\varepsilon_{v0}{}^{e}$	0.0
Mean Kirchhoff stress P'_0 [kPa]	-98.0
Specific volume v_0	1.85

$$d\boldsymbol{\varepsilon}^{e}_{n+1} = d\boldsymbol{\varepsilon}_{n+1} - \Delta\gamma\partial^2_{\boldsymbol{\beta}'\boldsymbol{\beta}'} f_{n+1} \cdot d\boldsymbol{\beta}'_{n+1} - \partial_{\boldsymbol{\beta}'} f_{n+1} d\Delta\gamma_{n+1}. \quad (23)$$

From Equations (2) and (23), the stress-strain relation with the Hessian matrix is derived as [4]

$$d\boldsymbol{\beta}'_{n+1} = \boldsymbol{\Xi}_{n+1} \cdot \left[d\boldsymbol{\varepsilon}_{n+1} - \partial_{\boldsymbol{\beta}'} f_{n+1} d\Delta\gamma_{n+1}\right] \quad (24)$$

$$\boldsymbol{\Xi}_{n+1} = \left[\mathbf{c}^{e\text{-}1}_{n+1} + \Delta\gamma\partial^2_{\boldsymbol{\beta}'\boldsymbol{\beta}'} f_{n+1}\right]^{-1} \quad (25)$$

where $\mathbf{c}^{e} = \partial^2_{\boldsymbol{\varepsilon}^e\boldsymbol{\varepsilon}^e} W(\varepsilon^{e}_{v}, \varepsilon^{e}_{s})$, and $\boldsymbol{\Xi}$ is the Hessian matrix. $d\chi_{n+1}$ is obtained by differentiating the backward-Euler approximation of Equation (11) as follows:

$$d\chi_{n+1} = -\frac{a\chi_{n+1}\ln\chi_{n+1}}{\chi_{n+1} + a\Delta\gamma} d\Delta\gamma \quad (26)$$

Inserting Equations (23), (24), and (26) into the consistency condition, we obtain

$$\begin{aligned} df_{n+1} &= \partial_{\boldsymbol{\beta}'} f_{n+1} \cdot d\boldsymbol{\beta}'_{n+1} + \partial_{\psi} f_{n+1} d\psi_{n+1} \\ &+ \partial_{\chi} f_{n+1} d\chi_{n+1} + \partial_{\varepsilon^p_v} f_{n+1} d\varepsilon^{p}_{v} = 0 \end{aligned} \quad (27)$$

Solving for $d\Delta\gamma$ gives

$$d\Delta\gamma_{n+1} = \left(\mathbf{N}_{n+1} \cdot \boldsymbol{\Xi}_{n+1} \cdot d\boldsymbol{\varepsilon}_{n+1} + \xi dS_r\right)/D_{n+1} \quad (28)$$

where

$$\mathbf{N}_{n+1} = \partial_{\boldsymbol{\beta}'} f_{n+1} + \boldsymbol{\delta} \cdot \Delta\gamma\partial^2_{\boldsymbol{\beta}'\boldsymbol{\beta}'} f_{n+1}, \quad (29)$$

$$D_{n+1} = \mathbf{N}_{n+1} \cdot \boldsymbol{\Xi}_{n+1} \cdot \partial_{\boldsymbol{\beta}'} f_{n+1} + \frac{a\ln\chi_{n+1}}{\chi_{n+1} + a\Delta\gamma} - \boldsymbol{\delta} \cdot \partial_{\boldsymbol{\beta}'} f_{n+1}. \quad (30)$$

$\boldsymbol{\delta} := (1,1,1)^{\mathrm{T}}$ is the vector in the principal space. The algorithmic tangent moduli $\partial\boldsymbol{\beta}'_{n+1}/\partial\boldsymbol{\varepsilon}_{n+1}$ and $\partial\boldsymbol{\beta}'_{n+1}/\partial S_{r,n+1}$ are obtained by inserting Equation (29) into Equation (24) as follows:

$$\frac{\partial\boldsymbol{\beta}'_{n+1}}{\partial\boldsymbol{\varepsilon}_{n+1}} = \boldsymbol{\Xi}_{n+1} - \left(\boldsymbol{\Xi}_{n+1} \cdot \partial_{\boldsymbol{\beta}'} f_{n+1} \otimes \mathbf{N}_{n+1} \cdot \boldsymbol{\Xi}_{n+1}\right)/D_{n+1} \quad (31)$$

$$\frac{\partial\boldsymbol{\beta}'_{n+1}}{\partial S_{r,n+1}} = \left(-\xi\boldsymbol{\Xi}_{n+1} \cdot \partial_{\boldsymbol{\beta}'} f_{n+1}\right)/D_{n+1} \quad (32)$$

These algorithmic tangent moduli correspond with conventional elastoplastic tangent moduli when $\Delta\gamma = 0$.

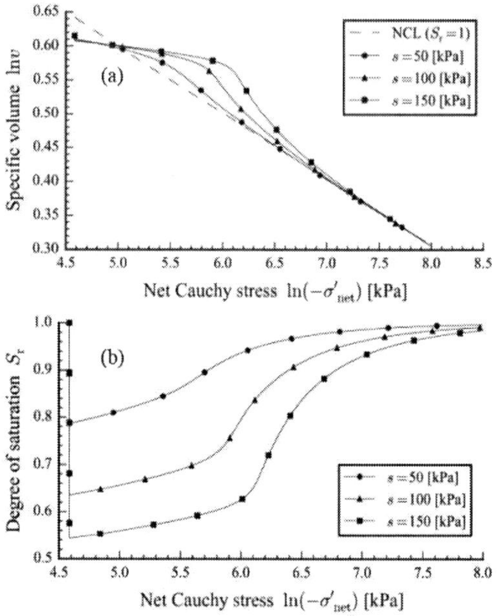

Figure 5. Simulations of isotropic compression of unsaturated soils under constant suction.

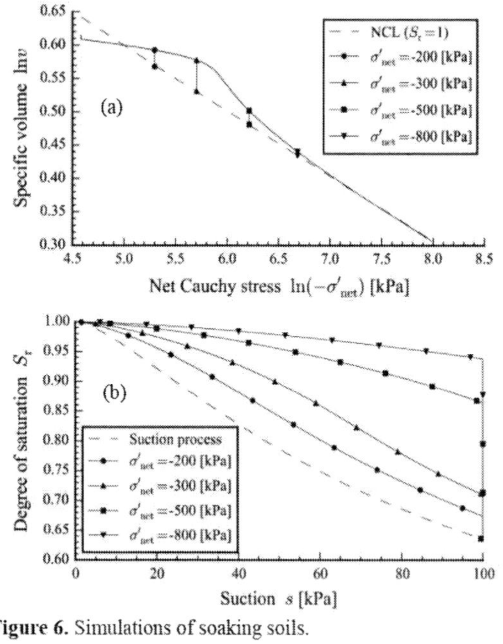

Figure 6. Simulations of soaking soils.

3 Simulations

Soaking and compaction behaviors of unsaturated soils are simulated by the proposed model, and the applicability of the model is discussed here. In the simulations presented here, the same sets of material parameters for the water retention curve and elastoplastic model summarized in Table 1 and 2 are adopted. The same set of initial parameters (such as density and the stress state) is also used in the simulations, which are summarized in Table 3. As shown in Table 3, the initial

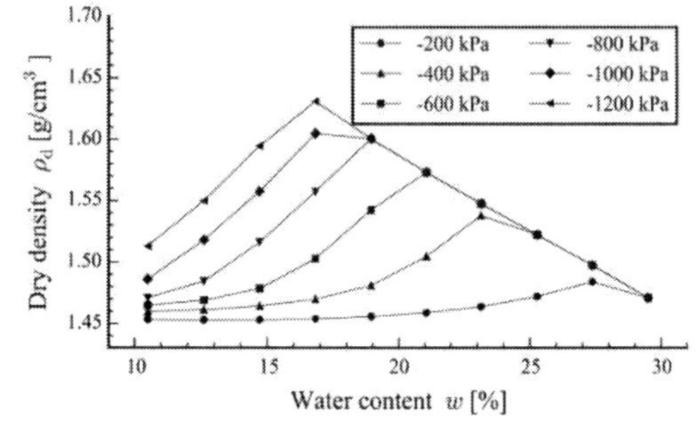

Figure 7. Compaction curves for several kinds of applied maximum stresses (final states of compaction simulations)

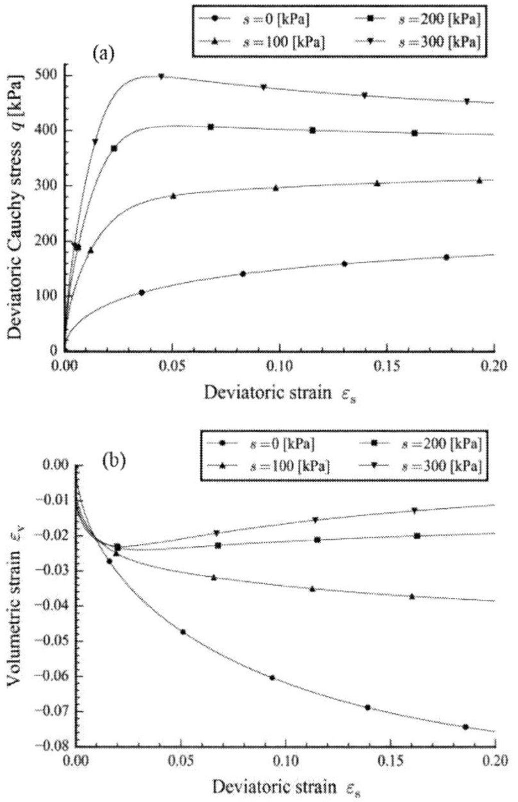

Figure 8. Simulations of triaxial test under constant suction and confining pressure.

state of the soil is assumed to be saturated.

The compression behavior of unsaturated soils having different values of degree of saturation is shown. Figure 5 shows the isotropic compression curves of unsaturated soil under constant suction. In this simulation, the suction was first increased to prescribed values (50, 100, and 150 kPa) under constant net stress ($\sigma'_{net} = -98\,\text{kPa}$) and three different values of the degree of saturation are obtained for each case. It is indicated that the proposed model can described the typical behaviors of unsaturated soils where: unsaturated soil is able to stay over the NCL for saturated soil in the beginning stage of compression; the degree of saturation increases owing to volumetric compression (even though suction is kept constant); there is rapid compression behavior to the NCL for saturated soil with an asymptotic increase in the degree of saturation to 100%. Such a tendency has been reported in past experimental works by authors such as Wheeler and Sivakumar [24] and Kayadelen [25].

We show a response of the proposed model from soaking. In Figure 6, the soil is soaked under constant net stress (-200, -300, -500, and -800 kPa) after the isotropic compression shown in Figure 5 (case: s = 100 [kPa]). In simulations, the soaking collapse behavior of soils can be seen by decreasing suction s to zero.

The compaction behavior of soils is simulated in Figure 7. Suction s is first increased under a constant net stress ($\sigma'_{\text{net}} = -98\text{kPa}$) until the prescribed water content w is achieved. Compaction behavior can be regarded as the exclusion of entrapped air without significant drainage of void water. Thus, in this simulation, the total Cauchy stress is increased from -98.0 kPa to a predetermined maximum value (-200, -400, -600, -800, -1000, and -1200 kPa) with constant water content, and returns to -98.0 kPa. Figure 7 shows that the proposed model can simulate the typical compaction behavior of soils.

We validate the proposed model in terms of the shear strength of unsaturated soils. In Figure 8, we simulate triaxial tests under constant suction (drained water and drained air). Suction was first increased to prescribed values (100, 200, and 300 kPa) before shearing. Shearing is then simulated under a constant confining pressure. Figure 7 shows that the proposed model can simulate the difference in the strength against shearing, i.e., unsaturated soil is stiffer than saturated soil.

4 Conclusion

A model for unsaturated soils that can describe compaction and collapse mechanisms under a finite strain range is presented here. The model is an extended version of an infinitesimal model proposed by Kikumoto et al. [16]. Essential concepts of the model in describing the behavior of unsaturated soils are as follows: shifting the NCL in the direction of the specific volume with a change in the degree of saturation, and a dependency of the degree of saturation on specific volume or density. The proposed model is formulated based on Kirchhoff stress invariants in order to predict the large deformation behavior of unsaturated soil based on finite strain theory. It is indicated through the simulations presented in this paper that the proposed model can describe the typical behavior of unsaturated soils such as soaking collapse phenomena, compaction of soils and shearing behavior.

References

1. X. Song, R.I. Borja, Vadose Zone J. **13**, 5 (2014)
2. X. Song, R.I. Borja, Int. J. Numer. Meth. Eng. **97**, 658-682 (2014)
3. J.C. Simo, *Handbook of Numerical Analysis*, 183–499 (1998)
4. J.C. Simo, T.J.R. Hughes, *Computational Inelasticity* 7 (1998)
5. F. Dunne, K. Runesson, *Introduction to Computational Plasticity* (2005)
6. E.A. de Souza Neto, D. Peric, D.R.J. Owen, *Computational Methods for Plasticity: Theory and Applications* (2008)
7. E.H. Lee, Elastic-plastic decomposition at finite strains, J. Appl. Mech. **36**, 1–6 (1969)
8. R. Borja, C. Tamagnini, Comput. Method. Appl. M. **155**, 73–95 (1998)
9. G. Meschke, W.N., Liu, Comput. Method. Appl. Mech. Eng. **173**, 167–187 (1999)
10. Y. Yamakawa, K. Hashiguchi, K.Ikeda, Int. J. Plasticity **26**, 634–658 (2010)
11. J.C. Simo, Comput. Method. Appl. Mech. Eng. **85**, 273–310 (1991)
12. J.C. Simo, Comput. Method. Appl. Mech. Eng. **99**, 61–112 (1992)
13. K. Hashiguchi, M. Ueno, Constitutive equations of soils, Proc., Ninth Int. Conf. Soil Mech. Found. Engrg., 73–82 (1977)
14. R. Butterfield, Geotechnique **29**, 469–480 (1979)
15. K. Hashiguchi, Int. J. Numer. Anal. Met. **19**, 367–376 (1995)
16. M. Kikumoto, H. Kyokawa, T. Nakai, H.M. Shahin, *Unsaturated Soils*, 849–855 (2010)
17. M. Zytynski, M.K. Randolph, R. Nova, C.P. Wroth, Int. J. Numer. Anal. Methods Geomech. **2**, 87–93 (1978)
18. G.T. Houlsby, Comput. Geotech **1**, 3–13 (1985)
19. D. Gallipoli, A. Gens, R. Sharma, J. Vaunat, Geotechnique **53**, 123–135.
20. K.H. Roscoe, J.B. Burland, *Engineering Plasticity*, 535–609 (1968)
21. M.T. van Genuchten, Soil Sci. Soc. Am. J. **44**, 892–898 (1980)
22. A. Tarantino, S. Tombolato, Geotechnique **55**, 307–317 (2005)
23. R.I. Borja, K.M. Sama, P.F. Sanz, Comput. Methods Appl. Mech. Eng. **192**, 1227–1258 (2003)
24. S.J. Wheeler, V. Sivakumar, Geotechnique 45, 35–53 (1995)
25. C. Kayadelen, Environ. Geol. (54), 325–334 (2008)

 DOI: 10.1051/e3sconf/20160917002

Development of a coupled thermo-hydro-mechanical double structure model for expansive soils

David Mašín[1]

[1]*Faculty of Science, Charles University in Prague, Czech Republic*

Abstract. In this paper, development of a thermo-hydro-mechanical model for expansive soils including double structure is described. The model is based on hypoplastic model by Mašín [6], in which the hydro-mechanical coupling is considered at each of the two structural levels. The model also includes separate effective stress definitions and water retention curves for the two levels of structure. In the proposed model, an approach by Mašín and Khalili [8] to include thermal effects into hypoplastic models is followed. This is combined with temperature-dependent water retention curve of macrostructure, temperature-induced deformation of microstructure and an enhanced double-structure coupling law. Good predictions of the model are demonstrated by comparing the model simulations with experimental data on MX80 bentonites taken over from literature.

1 Introduction

Thermo-hydro-mechanical modelling of the behaviour of expansive clays is important in a number of high-priority applications, such as design of nuclear waste repositories. Their behaviour is, however, remarkably complex, in particular due to their double-structure nature. Each of the structural levels respond differently to temperature change, suction change and mechanical action. In this paper, and advanced model is developed aiming to predict these complex phenomena in a unified manner.

2 Double structure model

The model described in this paper has been developed using double-structure framework, originaly proposed by Alonso et al. [1]. The model is based on the hydro-mechanical double structure hypoplastic model by Mašín [6]. This model is briefly described in Sec. 2.1.

2.1 Existing hydro-mechanical model

The double structure models are based on the assumption supported by various micro-mechanical studies that in expansive soils one can identify two levels of structure: So-called macrostructure, which is representing an assembly of silt-size aggregates of the clay particles, and so-called microstructure, which is representing the internal structure of these aggregates. A conceptual sketch of these two levels of structure is in Fig. 1.

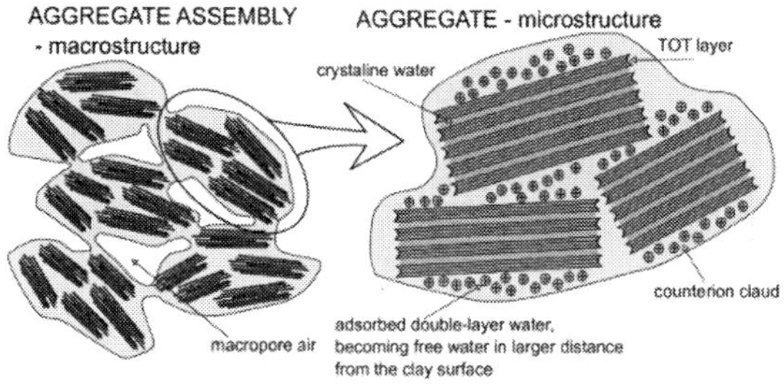

Figure 1. A conceptual sketch of two levels of structure considered in double-structure models (from [6]).

In the model by Mašín [6], separate models are considered for mechanical and hydraulic responses of microstructure and macrostructure. These responses are coupled at the given structural level, and additionally, the behaviour of the two structural levels is linked through the double-structure coupling function.

Mašín [6] based the mechanical model for the behaviour of macrostructure on the model by Mašín and Khalili [7]. Hydraulic response of macrostructure was based on the void-ratio dependent water retention model [4]. Microstructure has always been considered as fully saturated and its behaviour was governed by the Terzaghi effective stress principle: see Mašín and Khalili [9] for thorough discussion of this subject. The double structure coupling is controlled by a function of relative void ratio, which evolved from the original proposition by Alonso et al. [1].

[a] Corresponding author: masin@natur.cuni.cz

2.2 Proposed thermo-hydro-mechanical model

In the proposed model, thermal effects on the behaviour of macrostructure and microstructure have been incorporated. The model [6] has been enhanced in the following way:

- A new mechanical hypoplastic model for saturated soils (Mašín [5]) is adopted as a base mechanical model for macrostructure.
- The thermal behaviour of macrostructure is described by an approach developed by Mašín and Khalili [8]. The model assumes temperature-dependent normal compression lines of the form

$$\ln(1+e) = N(s,T) - \lambda^*(s,T)\ln\left(p^M/p_r\right) \tag{1}$$

where p^M is the effective mean stress of macrostructure, p_r is the reference stress of 1 kPa, s is suction and T is temperature (measured in Kelvins). $N(s,T)$ and $\lambda^*(s,T)$ are temperature- and suction-dependent positions and slopes of normal compression lines, respectively. They are defined as

$$N(s,T) = N + n_s\langle\ln\frac{s}{s_e}\rangle + n_T\ln\frac{T}{T_r} \tag{2}$$

$$\lambda^*(s,T) = \lambda^* + l_s\langle\ln\frac{s}{s_e}\rangle + l_T\ln\frac{T}{T_r} \tag{3}$$

where N, n_s, n_t, l_s, l_t are parameters and T_r is a reference temperature.

- The water retention model for macrostructure is based on the hysteretic model from [6], where the air-entry value of suction $s_{en}(T)$ is considered to be temperature-dependent. It is controlled by an equation

$$s_{en}(T) = s_{en}\left(\frac{a+bT}{a+bT_r}\right) \tag{4}$$

s_{en} the air-entry value of suction for macrostructure, which is void ratio dependent and it is calculated using approach from [4]. This model requires two parameters, namely reference air-entry value of suction s_{en0} for the reference macrostructural void ratio e_0^M. a and b in Eq. (4) are parameters. As pointed out by Grant and Salehzadeh [2], their values a=0.18 N/m and b=-0.00015 N/(mK) imply that the effects of temperature by water retention capacity are caused solely on its effect on surface tension. This has not been supported by experimental observation, however (Romero et al. [3]).

- Microstructure is considered to be fully saturated. Following the work by Mašín and Khalili [9], its mechanical response is considered to be governed by the Terzaghi effective stress principle with additional strains induced by temperature variation. The thermal deformation is considered to be fully reversible, governed by the coefficient α_s using

$$\dot{\boldsymbol{\epsilon}}^{mT} = \frac{\mathbf{1}}{3}\alpha_s\dot{T} \tag{5}$$

where $\dot{\boldsymbol{\epsilon}}^{mT}$ are thermal strains of microstructure and $\mathbf{1}$ is the second-order identity tensor.

- Integration of Eq. (5), together with the equation controlling volumetric response of microstructure due to the change of microstructural effective stress, can be used for initialisation of the microstructural void ratio e_m. The equation reads:

$$e^m = \exp\left[\kappa_m\ln\left(\frac{s_r}{p^m}\right) + \ln(1+e^m_{r0}) + \alpha_s(T-T_r)\right] - 1 \tag{6}$$

where p^m is the microstructural effective mean stress and e^m_{r0}, κ_m and s_r are parameters.

- In the original model, the double structure couling function f_m has been assumed as zero for aggregate shrinkage. In the present model, however, this assumption leads to underprediction of global shrinkage in cooling experiments. The experimental data (Fig. 4) indicate that the global shrinkage in cooling depends on suction. The following equation has been proposed for particle shrinkage which was found to leads to good representation of experimental data:

$$f_m = c_{sh}\left(\frac{s}{s_e}\right) \tag{7}$$

where c_{sh} is a parameter. f_m is bound within the range 0 to 1.

The complete model takes the following rate form:

$$\mathring{\boldsymbol{\sigma}}^M = f_s[\boldsymbol{\mathcal{L}} : (\dot{\boldsymbol{\epsilon}} - f_m\dot{\boldsymbol{\epsilon}}^m) + f_d\mathbf{N}\|\dot{\boldsymbol{\epsilon}} - f_m\dot{\boldsymbol{\epsilon}}^m\|] + f_u(\mathbf{H}_s + \mathbf{H}_T) \tag{8}$$

where $\boldsymbol{\mathcal{L}}$, $\mathbf{N}$, $\mathbf{H}_s$ and $\mathbf{H}_T$ are hypoplastic tensors, f_s, f_d and f_u are hypoplastic scalar factors, $\dot{\boldsymbol{\epsilon}}$ is the Euler stretching tensor, $\mathring{\boldsymbol{\sigma}}^M$ is the objective effective stress rate of macrostructure and $\dot{\boldsymbol{\epsilon}}^m$ is microstructural strain rate. Due to the space restrictions, it is not possible to describe Eq. (6) in detail, the interested readers can refer to [6, 7, 8] for thorough explanation of the model components.

It is to be pointed out that for the general case of a test with controled stretching, suction rate and temperature rate, and unknown rates of total stress and degree of saturation, solution of Eq. (6) is not straightforward: total stress rate appears both in the formulation of $\dot{\boldsymbol{\epsilon}}^m$ and in the formulation of $\mathring{\boldsymbol{\sigma}}^M$, thus on both the right- and left-hand side of Eq. (6). A numerical procedure has been developed to solve this equation, and it has been implemented into an in-house general purpose thermo-hydro-mechanical single element code. This implementation has been adopted in the evaluation of the model presented in Section 3.

2.3 Model parameters

In this section, physical meaning of model parameters which have not been mentioned in Sec. 2.2 is briefly described. These parameters are identical to the parameters of the basic hydro-mechanical model [6] and the readers are referred to the original publication for more details.

- The basic model requires, additionally to the parameters from Sec. 2.2, to specify parameters φ_c (critical state friction angle in a standard soil-mechanics meaning), ν (parameter controlling stiffness in shear) and κ^* (controls macrostructural volume strain in unloading).

- The parameter m is present at two places within the model formulation. First of all, it controls the factor f_u and thus the dependency of the wetting- and heating-induced compaction on the distance from the state boundary surface (the higher the value of m, the closer the state needs to be to the state boundary surface for the compaction to become significant). Second, the parameter m controls the double-structure coupling function and it thus affects the response to wetting-drying and heating-cooling cycles (see [6]).
- a_e is the ratio of air-entry and air-expulsion values of suction of the water retention model for macrostructure.

3 Model evaluation

3.1 Description of the material and experiments

The model has been evaluated with respect to experimental data on compacted bentonite by Tang and Cui [10] and Tang et al. [11]. They studied the behaviour of MX80 bentonite from Wyoming, USA, under non-isothermal conditions. Two experimental data sets have been adopted.

The first one has been published by Tang and Cui [10]. They studied water retention behaviour of a compacted bentonite in suction- and temperature-controlled isotropic cell. Prior to the test, the samples had the initial suction slightly lower than 145 MPa (140 MPa was assumed in the simulations) and the initial dry density was 16.5 kN/m^3. Subsequently, different values of total suction were applied using vapour equilibrium technique and water content of samples was measured until it has stabilised.

The second experimental data set has been published by Tang et al. [11]. The samples have been tested in suction- and temperature-controlled isotropic cell capable of application of high suctions using vapour equilibrium technique, high temperatures (up to 80°C reached in the experiments from [11]) and high mechanical isotropic stresses (up to 60 MPa applied in [11]). Prior to the testing, compacted specimens with an initial suction of 110 MPa and dry densities of approx. 17.5 kN/m^3 were machined to obtain the required dimensions (80 mm in diameter, 10-15 mm in height). Thereafter, suction was changed using vapour equilibrium technique to the desired values (9, 39 and 110 MPa in three experimental sets) while measuring the swelling deformation. Samples were then placed into the cell and loaded to the initial isotropic total stress of 0.1 MPa. This was the initial state for subsequent thermo-mechanical testing. For the detailed description of the tests the reader is referred to [11].

3.2 Description of the modelling procedure

In the simulations, complete thermo-hydro-mechanical histories of the samples have been followed. That is, the initial state for the given thermo-hydro-mechanical experimental stage has not been prescribed, but it has instead been simulated from the common initial state. All the water retention curve simulations were performed from the initial state of total suction s_t=140 MPa, e=0.64, T=25°C, a_{scan}=1 and zero total stress. The initial stage has been followed by a change of temperature to the desired value and subsequent suction variation under zero total stress. The initial void ratio was calculated from the initial dry densities using specific gravity of grains G_s=2.76, which was implied by the data in [11].

The samples tested in the suction- and temperature-controlled isotropic cell had all have the initial state of s=110 MPa, e=0.53, T=25°C, a_{scan}=0 and zero total stress. The initial stage has been followed by an (eventual) change of suction, increase of total stress to 0.1 MPa and increase of temperature. Subsequent testing followed the desired thermo-mechanical path.

3.3 Calibration of the model

Due to the limited number of available experiments, the model has been calibrated using the data to be predicted. Also, the experimental programme did not allow to calibrate all the model parameters. The additional parameters have been assumed using the previous experience. It is to be pointed out that the assumed parameters do not affect substantially the model predictions.

Parameters of the basic hypoplastic model λ^* and κ^* have been calibrated using isotropic compression experiments (Fig. 5). Parameters N, n_s, n_t, l_s, l_t and m were adjusted so the model properly predicted position of isotropic normal compression lines (Fig. 5) and also the heating-induced collapse/swelling strains (Fig. 4). Parameters φ_c and ν have been assumed.

Reference values s_r, e_0^M and T_r have been selected so they were within the range relevant for the present simulations. The corresponding e_{r0}^m has been adjusted for water retention curve predictions. The parameter κ_m controls both the swelling due to suction decrease (thermo-mechanical tests, Fig. 3) and water retention curves (Fig. 2). The value of κ_m has thus been selected to predict accurately the swelling tests, while it was observed that this value leads to overprediction of water content in water retention experiments. Note that as very large strains (up to 50%) were reached in swelling tests, the experimental data from [11] have been replotted in terms of natural strain for consistency with the modelling output.

The value of α_s is controlling the thermal-induced swelling, it has been calibrated using heating tests at the suction of 110 MPa (Fig. 4). The parameters controlling water retention curve of macrostructure have little influence on results at very high suctions, s_{e0} and a_e have thus been assumed and a and b have selected considering that the effects of temperature on water retention capacity are caused solely by its effect on surface tension.

The set of parameters adopted in all the simulations is given in Table 1.

Table 1. Parameters of the model adopted in all the simulations.

φ_c	λ^*	κ^*	N	ν	n_s	l_s
25	0.081	0.01	1.46	0.25	0.01	0.0045
n_t	l_t	m	α_s	κ_m	e_{r0}^m	c_{sh}
-0.07	0	10	0.00015 K^{-1}	0.2	0.1	0.002
s_{e0}	a	b	a_e	s_r	e_0^M	T_r
200 kPa	0.118	-0.000154	0.75	140 MPa	0.5	294 K

3.4 Model predictions

Model predictions are shown in Figures 2 to 5.

The predicted water retention curves are in Fig. 2. It is clear that, although the model predicts correctly the swelling due to wetting (Fig. 3), it is overpredicting water content at lower values of suction (Fig. 2). These two facts are contradictory, and they can be caused by the fact that the two experimental data sets originate from different experimental data sets (albeit from the same soil mechanics laboratory). The samples also had different initial conditions, there can thus be slight variations in the soil structure, which affects soil properties.

As the model predicts aggregate swelling with heating, it also predicts slightly higher retention capacity of heated soil: the experiments appear to show the contrary, but the difference is small.

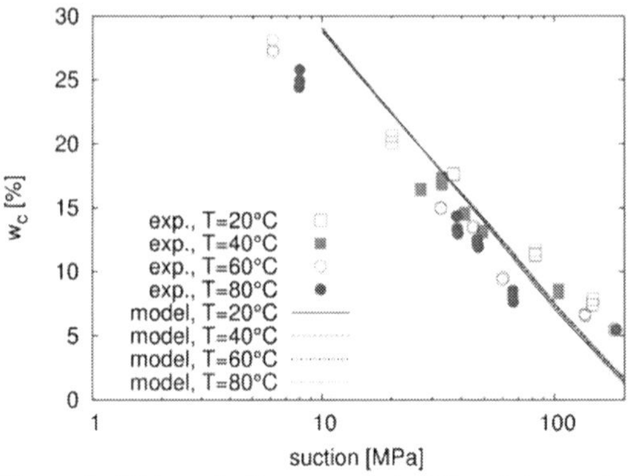

Figure 2. Water retention curves at different temperatures: experimental data [10] compared with model predictions.

Figure 3 shows swelling due to wetting at zero total stress and constant temperature 25°C as measured on samples which have later been placed into the suction- and temperature-controlled isotropic cell. Swelling is slightly underpredicted, which is a consequence of optimization of the model calibration so that also water retention curves are predicted reasonably. Recall that complete thermo-hydro-mechanical histories of the samples have been simulated, the state reached after the wetting stage thus represents the initial state for subsequent simulations.

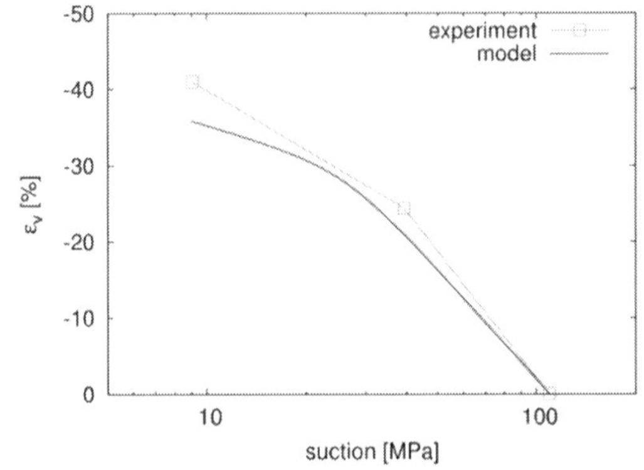

Figure 3. Swelling strains developed during wetting of the samples later tested in suction- and temperature-controlled isotropic cell. Experimental data [11] compared with model predictions.

Volume strains due to heating at various values of suction and mean total stress are shown in Figure 4. The model is accurately predicting the observed complex behaviour. In particular:

- At high suctions (110 MPa), the model is correctly predicting swelling, whose magnitude is controlled by the parameter α_s. The heating-induced swelling at high suctions is primarily reversible (Fig. 4b).
- At lower values of suction (9 MPa for total stress of 0.1 MPa and 39 MPa for total stress of 5 MPa), the model is predicting heating-induced compaction ("collapse"). This compaction is irreversible and it is controlled by the offset of normal compression lines at different temperatures (by the parameters l_T and n_T). Still, for stress 0.1 MPa and suction 39 MPa the state is well within the state boundary surface and heating-induced swelling is predicted in agreement with experimental data.
- Upon cooling, the model is predicting cooling-induced contraction. This contraction depends on suction, such that it is most pronounced at high suction of 110 MPa and least significant at lower values of suction (39 MPa and 9 MPa). These predictions are governed by the dependency of the double-structure coupling factor f_m on macrostructural degree of saturation (Eq. (7)).

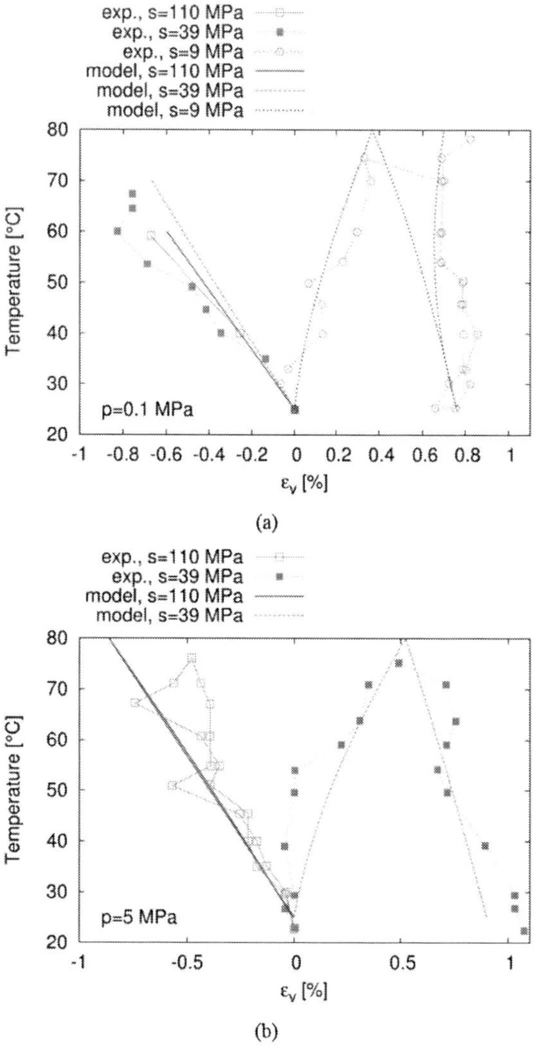

Figure 4. Volume change due to heating and cooling at total isotropic stresses of 0.1 MPa (a) and 5 MPa (b). Experimental data [11], compared with model predictions.

The isotropic compression at various values of suction and temperature is shown in Fig. 5. The model predicts reasonably the initial void ratio, implied by wetting-induced swelling during the preceding experimental stage. Considering the effect of temperature, the model predicts lower position of the isotropic normal compression lines at higher temperatures (the parameter n_t has a negative value). The difference appears to be insignificant in Fig. 5, but they are important to induce heating-induced compaction shown in Fig. 4.

Figure 5 shows that also the shape of the isotropic compression lines and the effect of suction on apparent preconsolidation pressure is predicted properly.

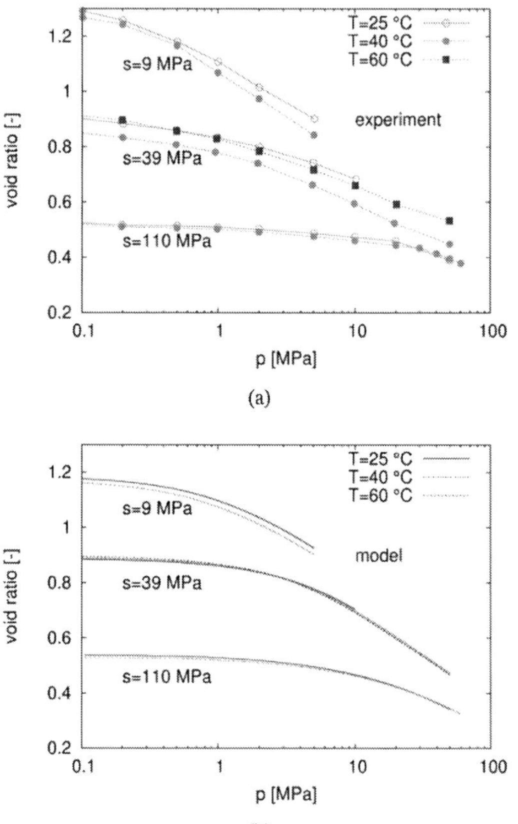

Figure 5. Isotropic compression tests at various suctions and temperatures. (a) experimental data from [11], (b) predictions.

4 Summary and conclusions

A new thermos-hydro-mechanical model for expansive soils based on double structure concept and hypoplasticity has been developed. In the paper, the most important properties of the model have been presented. It has been shown that the model provides correct predictions of the complex behaviour of MX80 bentonite under various thermo-hydro-mechanical paths.

In particular, the model properly predicts swelling or shrinkage in heating-cooling tests, depending on the current suction, total stress and void ratio. Also, global swelling of the samples due to wetting and the influence of suction and temperature on the shape if isotropic compression curves are well predicted. The model overpredicts the global water content at low values of suction, but this can potentially be caused by inconsistency in the two experimental data sets.

Acknowledgement

The author is grateful for the financial support by the research grant 15-05935S of the Czech Science Foundation.

 DOI: 10.1051/e3sconf/20160917002

References

1. E. E. Alonso, J. Vaunat, A. Gens, Engineering Geology **54**, 173-183 (1999)
2. S. Grant, A. Salehzadeh, Water Resources Research **32**, 261-270 (1996)
3. E. Romero, A. Gens, A. Lloret, Geotechnical and Geological Engineering **19**, 311-332 (2001)
4. D. Mašín, Int. J. Numer. Anal. Meth. Geomech. **34**, 73-90 (2010)
5. D. Mašín, Acta Geotechnica **8**, 481-49 (2013)
6. D. Mašín, Engineering Geology **165**, 73-88 (2013)
7. D. Mašín, N. Khalili, Int. J. Numer. Anal. Meth. Geomech. **32**, 1903-1926 (2008)
8. D. Mašín, N. Khalili, Int. J. Numer. Anal. Meth. Geomech. **36**, 17-35 (2012)
9. D. Mašín, N. Khalili, Canadian Geotechnical Journal (2015), doi: 10.1139/cgj-2014-0479
10. A. Tang, Y. Cui, Canadian Geotechnical Journal **42**, 287-296 (2007)
11. A. Tang, Y. Cui, Géotechnique **58**, 45-54 (2008)

Incorporating anisotropy in the Barcelona basic model

Muayad A. Al-Sharrad[1,a] and Domenico Gallipoli[2]

[1] Department of Civil Engineering, University of Anbar, Ramadi 31001, Iraq
[2] Université de Pau et des Pays de l'Adour, Anglet, France

Abstract. The paper presents an elasto-plastic model for unsaturated soils that extends the Barcelona basic model to introduce a dependency of mechanical behaviour on material anisotropy. The model builds upon a previous proposal by Al-Sharrad and Gallipoli [1] by linking material anisotropy to both the distortion and aspect ratio of the elliptical yield curve in the constant suction plane of deviator versus mean net stress. In order to take into account anisotropic behaviour, an additional constitutive equation is therefore proposed to uniquely relate the aspect ratio and the distortion of the yield ellipse. The new model is calibrated and validated against the experimental data by Al-Sharrad [2] via the simulation of triaxial tests conducted at various suctions on compacted kaolin samples. These simulations show that the new model predicts yield stresses and shear strains during anisotropic loading with greater accuracy compared to the model of Al-Sharrad and Gallipoli [1].

1 Introduction

Anisotropic soils exhibit properties that depend on the direction along which these properties are measured. The anisotropic mechanical behaviour of saturated soils has been studied since the 1970s (e.g. [3, 4, 5, 6]) and several elasto-plastic models, which incorporate directional dependency of stress-strain behaviour, have since then been proposed (e.g. [7, 8]). Similar studies have also been undertaken for unsaturated soils (e.g. [9, 10, 11]), though relatively fewer models have been proposed in this case (e.g. [12, 13, 14, 2, 1]). A review of existing elasto-plastic anisotropic models for both saturated and unsaturated soils can be found in Al-Sharrad [2].

In the present work, the Barcelona basic model (BBM) for unsaturated soils is modified to incorporate a dependency of yielding behaviour on material anisotropy. The new model is referred as ABBM3 because it builds upon previous proposals by Al-Sharrad and Gallipoli [1] and D'Onza et al. [13] which are referred as ABBM2 and ABBM1, respectively. The formulation of ABBM3 is similar to that of ABBM2 with one important difference consisting in the definition of the elliptical yield curve in the constant suction plane of deviator stress versus mean net stress. Unlike ABBM2, ABBM3 links material anisotropy not only to the distortion but also to the aspect ratio of the constant suction yield ellipse via an additional constitutive relationship between aspect ratio and distortion. By assuming an associated flow rule, ABBM3 provides improved predictions of shear strains and yield stresses compared to ABBM2.

2 Overview of ABBM2

The ABBM2 is formulated in terms of mean net stress p, deviator stress q and suction s. The model is capable of simulating the elasto-plastic anisotropic behaviour of unsaturated soils and reduces to BBM in the case of material isotropy. In particular, ABBM2 assumes that fabric anisotropy develops as a consequence of both plastic volumetric and plastic shear strains.

The yield curve takes the form of a distorted ellipse in the constant suction q:p plane (see Figure 1):

$$f = \left(q - \alpha(p + p_s(s))\right)^2 - \left(M^2 - \alpha^2\right)\left(p + p_s(s)\right)\left(p_m(s) - p\right) = 0 \quad (1)$$

where M is the slope of the critical state while $p_m(s)$ and α are the size and distortion of the yield ellipse, respectively (note that α, unlike $p_m(s)$, is independent of suction). Similar to BBM, ABBM2 assumes a critical state line in the constant suction q:p plane defined as:

$$q = \pm M\left[p + p_s(s)\right] \quad (2)$$

where the slope M of the critical state line controls also the aspect ratio of the distorted yield ellipse, which is equal to $\left(M^2 - \alpha^2\right)^{1/2}$. Likewise BBM, $p_s(s)$ defines the common intercept of both the yield ellipse and the critical state line with the negative p axis. This intercept varies

[a] Corresponding author: an_mission@yahoo.com

non-linearly with suction as proposed by Stropeit et al. [12]:

$$p_s(s) = a\left(1 - \exp\frac{-s}{a}\right) \quad (3)$$

where a is a model parameter.

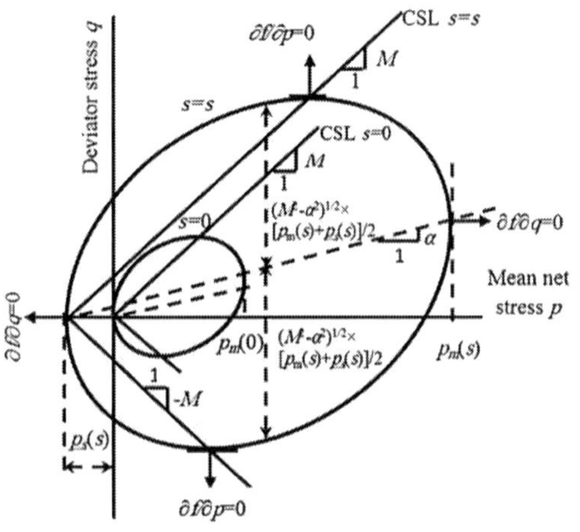

Figure 1. ABBM2 yield ellipses in constant suction q:p plane.

The model implements an associated plastic flow rule which, together with Equation 1, results in the following relationship between plastic volumetric strain ε_v^p and plastic shear strain ε_s^p:

$$\frac{d\varepsilon_s^p}{d\varepsilon_v^p} = \frac{2(\eta - \alpha)}{(M^2 - \eta^2)} \quad (4)$$

where η is the stress ratio defined as:

$$\eta = \frac{q}{p + p_s(s)} \quad (5)$$

Equation (4) predicts, at the intersection between the yield ellipse and the critical state line, a plastic flow vector that is always vertical (i.e. zero plastic volumetric strains), regardless of whether the soil is actually at critical state. This condition is however unnecessary because the flow vector only needs to be vertical if the soil is actually at critical state, and will therefore be relaxed in ABBM3.

In ABBM2, the volumetric hardening law is identical to that of BBM and relates the size of the yield ellipse $p_m(s)$ to plastic volumetric strains ε_v^p. A kinematic hardening law is also introduced to define the evolution of anisotropy by linking the changing distortion α of the yield ellipse to the increments of plastic volumetric and shear strains ε_v^p and ε_s^p:

$$d\alpha = \mu[f(\eta) - \alpha]\left[\left(d\varepsilon_v^p\right)^2 + \frac{9}{2}\left(d\varepsilon_s^p\right)^2\right]^{1/2} \quad (6)$$

where μ is a model parameter and $f(\eta)$ is a tri-linear relationship defined as:

$$f(\eta) = -cM \text{ for } \eta \le -M$$
$$f(\eta) = c\eta \text{ for } -M < \eta < M \quad (7)$$
$$f(\eta) = cM \text{ for } \eta \ge M$$

where c is a model parameter. According to Equations 5 and 6, if the soil is loaded with a stress ratio equal or greater than the critical one, i.e. $|\eta| \ge M$, the distortion of the yield ellipse will tend towards the following critical state value:

$$|\alpha_{cs}| = cM \quad (8)$$

Note that stress ratio η and distortion α are positive for triaxial compression and negative for triaxial extension.

3 Additional assumptions in ABBM3

Al-Sharrad el al. [11] showed that, for a given degree of anisotropy, experimental yield points are best fitted in a constant suction q:p plane by a distorted ellipse with an aspect ratio m generally greater than $\left(M^2 - \alpha^2\right)^{1/2}$, which is the aspect ratio predicted by ABBM2. A further improvement is obtained if different aspect ratios are used for the upper (compression) section and the lower (extension) section of the ellipse (see Figure 2), i.e. if $m = m_c$ for the section above the vertical tangent point ($\eta > \alpha$) and $m = m_e$ for the section below the vertical tangent point ($\eta < \alpha$). Therefore, unlike ABBM2, ABBM3 assumes constant suction yield ellipses with a different aspect ratio m:

$$f = (q - \alpha(p + p_s(s)))^2 - m^2(p + p_s(s))(p_m(s) - p) = 0 \quad (9)$$

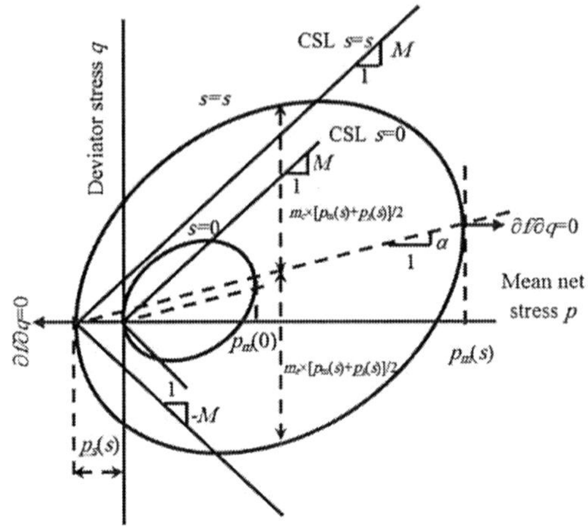

Figure 2. ABBM3 yield ellipses in constant suction q:p plane.

By assuming an associated plastic flow rule together with Equation 9, the following relationship between plastic strains is obtained in ABBM3:

$$\frac{d\varepsilon_s^p}{d\varepsilon_v^p} = \frac{2(\eta - \alpha)}{(m^2 + \alpha^2 - \eta^2)} \quad (10)$$

If critical state conditions are imposed in Equation 10, i.e. if $\frac{d\varepsilon_s^p}{d\varepsilon_v^p}=0$, $|\eta|=M$ and $|\alpha|$ $=|\alpha_{cs}|=cM$, the following aspect ratio at critical state is obtained:

$$m_{cs}=\left(M^2-\alpha_{cs}{}^2\right)^{1/2}=M\left(1-c^2\right)^{1/2} \quad (11)$$

The aspect ratio must be equal to m_{cs} when the soil attains critical state in order to ensure a vertical direction for the plastic flow vector as required by the definition of critical state. Due to the assumption of a different aspect ratio than in ABBM2, in ABBM3 the plastic flow vector is generally not vertical at the intersection between the yield ellipse and critical state line but becomes vertical only when the soil attains critical state.

4 Model performance

Most model parameters in ABBM3 are the same as in ABBM2, namely κ, κ_s, G, $\lambda(0)$, β, r, $N(0)$, p^c, a, M, μ and c. In particular, κ is the elastic swelling coefficient associated to stress, κ_s the elastic swelling coefficient associated to suction, G is the elastic shear modulus, $\lambda(0)$ is the slope of the saturated normal compression line, β is a parameter defining the change of the slope $\lambda(s)$ of constant suction normal compression lines with suction, r is a parameter defining the limit value of $\lambda(s)$ when suction tends to infinity and $N(0)$ is the specific volume at a reference pressure p^c on the saturated normal compression line. Parameters a, M, μ and c have been previously defined in Equations 2, 3, 6 and 7.

The values of parameters κ, κ_s, G, $\lambda(0)$, r, β, $N(0)$, p^c, a and M are independent of the particular form of the yield ellipse. Therefore, the same values presented in [1] for ABBM2 were also assumed for ABBM3. The values of parameters μ and c were instead specifically obtained in this work by following the iterative calibration procedure of [2]. This procedure involves performing model simulations with trial values of μ and c along predefined stress paths. The computed and experimental changes of α, caused by plastic straining along this stress path, are then compared and, in case of discrepancy, the values μ and c are refined until a satisfactory match is obtained. All the above parameter values are listed in Table 1.

For isotropically compacted samples, the initial value of the volumetric hardening parameter $p_m(0)$ was calculated as 22 kPa by following the procedure outlined in [15]. For anisotropically compacted samples, the initial value of the volumetric hardening parameter $p_m(0)$ was also calculated as 22 kPa according to the work of Al-Sharrad [2] by fitting the yield ellipse of Equation 9 to the yield data point at s=0. Similarly, the initial distortion α of the yield ellipse was calculated as 0 and 0.21 for isotropically and anisotropically compacted samples, respectively, based the work of Al-Sharrad [2].

These initial values of $p_m(0)$ and α, together with the parameter values in Table 1, were employed for simulating the behaviour of both isotropically and anisotropically compacted specimens during triaxial tests. However, due to space limitations, only few simulations of the behaviour of the isotropically compacted samples are shown later in this paper.

4.1 Additional parameters in ABBM3

In ABBM3 the aspect ratio of the yield ellipse is related to the current degree of anisotropy by means of an additional constitutive relationship between the aspect ratio and the distortion of the yield ellipse. In order to define this relationship, the experimental values of aspect ratios in compression and extension, i.e. m_c and m_e, are plotted in Figure 3 against the corresponding values of distortion α.

Equations 8 and 11, together with the parameter values in Table 1, calculate the critical state values of distortion and aspect ratio as $|\alpha_{cs}|$=0.30 and m_{cs}=0.67. Therefore, Figure 3 covers the entire range of distortion values from isotropic (α=0) up to critical state. Inspection of Figure 3 also indicates that the experimental values of m_c and m_e are practically constant over the measured range and equal to the isotropic aspect ratios m_{0c}=0.85 and m_{0e}=1.17, respectively.

The additional constitutive law must therefore predict an approximately constant value of aspect ratio over a relatively large range of distortions, followed by a sharp decrease towards m_{cs} when the distortion tends to $|\alpha_{cs}|$. The function chosen to provide this form of behaviour is:

$$m^2=\left[m_0^2-M^2\left(1-c^2\right)\right]\tanh\frac{F\left(cM-|\alpha|\right)}{|\alpha|}+M^2\left(1-c^2\right) \quad (12)$$

where m_0 is the isotropic aspect ratio at α=0 (either m_{0c} or m_{0e}) and F is an additional model parameter controlling the rate of variation of aspect ratio. The predictions by Equation 12 are plotted in Figure 3 for three different values of F. A value of F=20, or even greater, is required to fit the experimental data. However, preliminary simulations (not shown in the paper) indicated that large values of F result in the over-prediction of peak deviator stress during shearing. On the contrary, values of $F\leq5$ provide better predictions of shear strains especially if triaxial extension tests are simulated. Consequently, a value of F=5 was selected in the current work.

In summary, ABBM3 requires three extra parameters, namely m_{0c}, m_{0e} and F, in addition to the twelve of ABBM2 (Table 1).

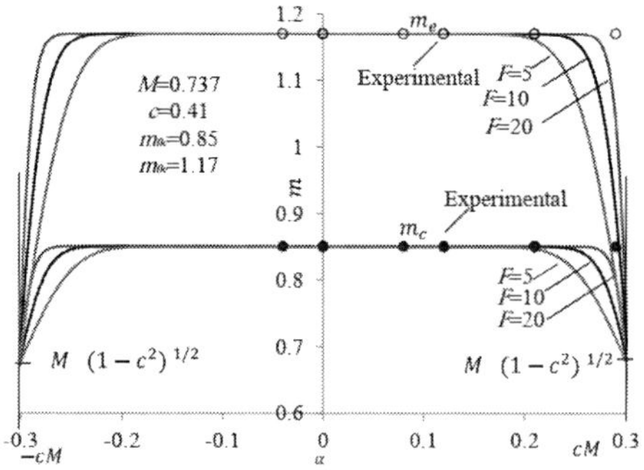

Figure 3. Variation of aspect ratio m with distortion α.

Table 1. Parameter values.

κ	0.012	M	0.737
κ_s	0.004	a	216.6 kPa
G	9960 MPa	μ	6.6
$\lambda(0)$	0.157	c	0.41
$N(0)$	1.80	m_{0c}	0.85
r	4.55	m_{0e}	1.17
β	0.0008 kPa^{-1}	F	5
p^c	696 kPa		

4.2 Predicted yield surface

Figure 4 shows the experimental yield stresses measured by Al-Sharrad [2] together with the yield ellipses predicted by ABBM3 and ABBM2 using the parameter values of Table 1. In each of the three plots in Figure 4, every yield point is obtained from a different test on ostensibly the same material. Specimens were initially equalized at s=300 kPa and then subjected to a load/unload cycle at a constant stress ratio of $\Delta q/\Delta p$=1.2 (Figure 4a), $\Delta q/\Delta p$=0 (Figure 4b) and $\Delta q/\Delta p$= -1 (Figure 4c). The load/unload cycle ranged from a small initial stress to p=200 kPa, which produces significant plastic deformations and hence a noticeable modification of material anisotropy (as confirmed by the differences between Figures 4a, 4b and 4c). Afterwards, specimens were sheared along different stress ratios to measure distinct points on the yield surface.

Inspection of Figure 4 indicates that ABBM3 provides, as expected, a better match to the experimental point than ABBM2. In particular, ABBM3 accurately predicts both the aspect ratio and the distortion of the yield ellipse, though it slightly under-predicts its size in Figures 4b and 4c. In contrast, ABBM2 significantly over-predicts the size of the yield ellipse in Figures 4a and 4c. Also, the aspect ratio predicted by ABBM2 is smaller compared to the experimental one.

4.3 Stress-strain behaviour

The performance of ABBM3 was investigated by simulating a number of triaxial tests conducted by Al-Sharrad [2] on specimens of isotropically compacted kaolin.

Figures 5 and 6 show the same set of tests in the v:ln$p$ and q:ε_s planes, respectively, where v is the specific volume and ε_s is the true shear strain. Three specimens were equalized at s=300 kPa, 100 kPa and 0 (saturated), respectively, and then isotropically loaded at constant suction to p=300 kPa before being sheared to failure at constant confining pressure. Figures 5a and 6a compare the predictions of ABBM3 and BBM while Figures 5b and 6b compare the predictions of ABBM3 and ABBM2. The predictions during isotropic loading by the three models are undistinguishable and agree well with the experimental data. During shearing, the experiments show a post-peak reduction of deviator stress caused by strain localization (observed in all tests). This experimental response is better captured by ABBM3 than BBM and ABBM2 (Figure 6). The better performance of ABBM3 is due to the assumed evolution of the yield ellipse with changing anisotropy in combination with an associated flow rule, which leads to the prediction of smaller shear strains compared to BBM and ABBM2. In many instances, ABBM3, unlike the other two models, is capable of predicting the observed peak of deviator stress during shearing as well as the subsequent reduction of deviator stress towards critical state. This happens when the stress point crosses the critical state line while the distortion and aspect ratio of the yield curve have not yet attained their respective critical state values calculated by Equations 8 and 11. Because of this, positive (i.e. compressive) plastic volumetric strains are still being predicted and the yield curve therefore keeps expanding which allows the stress state to cross the critical state line. Finally, as shearing progresses, the stress path converges back on the critical state line. At critical state, all models predict the same value of deviator stress q which is, in most cases, close to the experimental value.

Figures 7 and 8 show another set of tests in the v:ln$p$ and q:ε_s planes, respectively. In this case, three specimens were equalized at s=300 kPa, 100 kPa and 0 (saturated) and then sheared to failure with a stress ratio $\Delta q/\Delta p$=1 under constant suction. As before, Figures 7a and 8a compare the predictions of ABBM3 and BBM while Figures 7b and 8b compare the predictions of ABBM3 and ABBM2. Simulations in the v:ln$p$ plane (Figure 7) indicate that all models predict relatively well the yielding points marked by the sharp change in the slope of the compression curves. The slope of the post-yield compression line and the critical state values of specific volume are also reasonably captured by ABBM2 and ABBM3 but not BBM, which grossly over-predicts both of them.

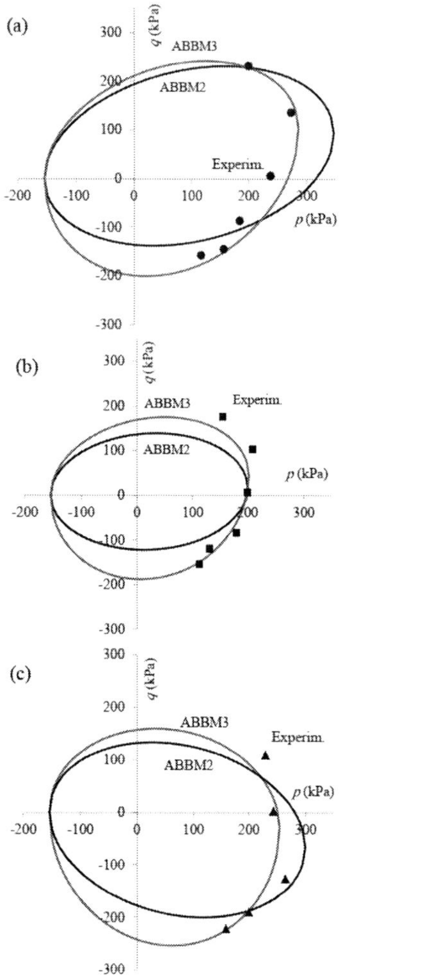

Figure 4. Experimental yield points and yield surfaces predicted by ABBM3 and ABBM2 for three soils with variable degrees of induced anisotropy.

5 Conclusions

An elasto-plastic anisotropic model for unsaturated soils, named ABBM3, has been presented. ABBM3 modifies the well-known BBM to incorporate material anisotropy and enhances a previous anisotropic model, named ABBM2, proposed by the same authors [1].

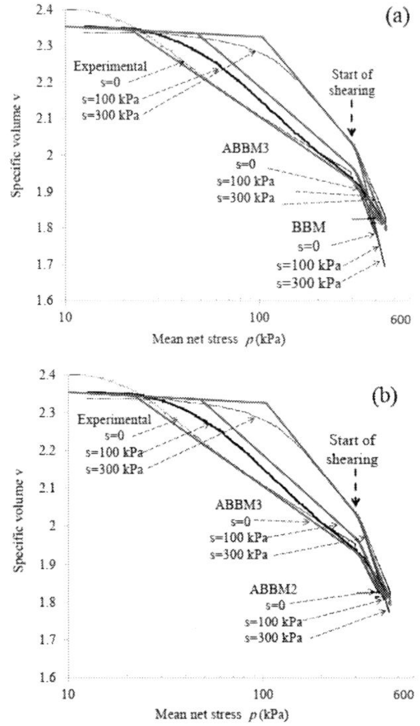

Figure 5. Experimental and predicted variation of v against ln$p$: (a) ABBM3 vs BBM and (b) ABBM3 and ABBM2.

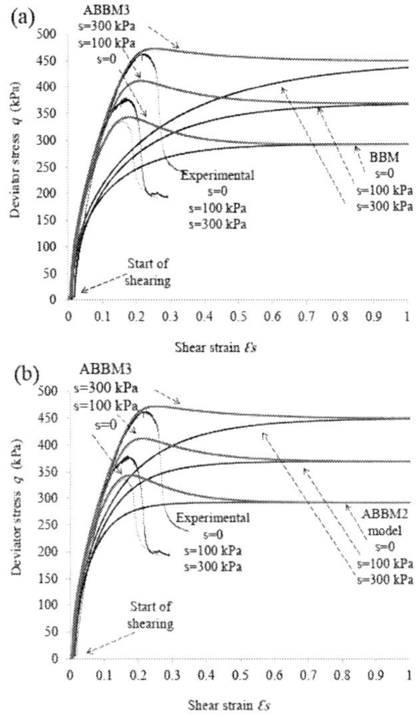

Figure 6. Experimental and predicted variation of q against ε_s: (a) ABBM3 vs BBM and (b) ABBM3 vs ABBM2.

DOI: 10.1051/e3sconf/20160917003

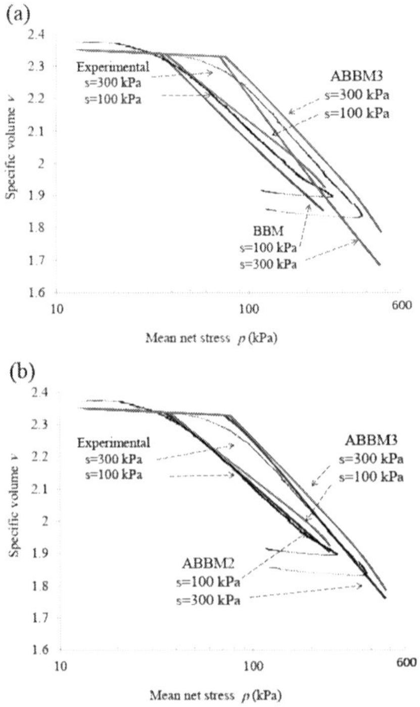

Figure 7. Experimental and predicted variation of v against ln$p$: (a) ABBM3 vs BBM and (b) ABBM3 and ABBM2.

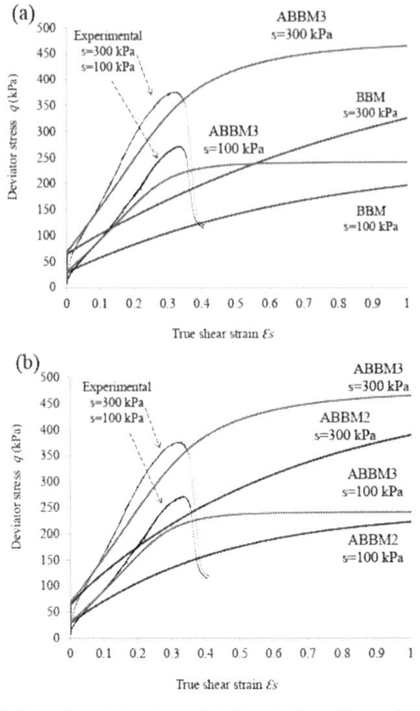

Figure 8. Experimental and predicted variation of q against ε_s: (a) ABBM3 vs BBM and (b) ABBM3 vs ABBM2.

In ABBM3, an improved form of elliptical yield surface with variable aspect ratio is proposed and an additional constitutive relationship is formulated between the aspect ratio and the distortion of the yield ellipse.

ABBM3 includes fifteen material parameters, ten of which are in common with BBM. The other five are necessary to define the evolution of anisotropy with plastic staining. This includes the change of both aspect ratio and distortion of the yield ellipse caused by plastic volumetric and shear deformations.

ABBM3 has been calibrated and validated against experimental data from Al-Sharrad [2]. Model simulations indicate that the incorporation of a variable aspect ratio in the mathematical formulation of the constant suction ellipse enhances the prediction of yield stresses and shear strains compared to BBM and ABBM2. Nevertheless, the number of model parameters becomes larger with increasing degree of model refinement, i.e. it increases from 10 in BBM, to 12 in ABBM2 and to 15 in ABBM3.

Finally, the calibration of model parameters controlling evolution of anisotropy requires a large number of unconventional triaxial tests, which suggests that further simplification is necessary to facilitate the application of the model to engineering practice.

References

1. M. A. Al-Sharrd & D. Gallipoli, *Proceedings of the 6th International Conference on Unsaturated Soils*, Sydney, Australia, 433-439, (2014).
2. M. A. Al-Sharrad, PhD thesis, University of Glasgow, UK, (2013).
3. J.H.A. Crooks & J. Graham, *Géotechnique*, **26**(2), 293–315, (1976).
4. J. Graham, M.L. Noonan & K. V. Lew, *Canadian Geotechnical Journal*, **20**(3), 502-516, (1983).
5. J. Graham, J.H.A. Crooks & S.L.K. Lau, *Géotechnique*, **38**(1), 125–134, (1988).
6. V. Sivakumar, I. G. Doran, J. Graham & A. Johnson, *Canadian Geotechnical Journal,* **38**(1), 125-137, (2001).
7. P.K. Banerjee & N.B. Yousif, *International Journal for Numerical and Analytical Methods in Geomechanics,* 10, 521-541, (1986).
8. S. J. Wheeler, A. Naatanen, M. Karstunen & M. Lojander, *Canadian Geotechnical Journal*, **40**(2), 403-418, (2003).
9. V. Sivakumar, R. Sivakumar, E.J. Murray, P. Mackinnon & J. Boyd, *Géotechnique,* **60**(8), 581–594, (2010).
10. V. Sivakumar, R. Sivakumar, J. Boyd & P. Mackinnon, *Géotechnique,* **60**(8), 595–609, (2010).
11. M. A. Al-Sharrad, S.J. Wheeler & D. Gallipoli, *Proceedings of the 2nd European Conference on Unsaturated Soils*, Napoli, Italy, **2**, 129-136, (2012).
12. K. Stropeit, S.J. Wheeler & Y.J. Cui, *Proceedings of 1st European Conference on Unsaturated Soils*, Durham, UK, 625-631, (2008).

13. F. D'Onza, D. Gallipoli & S.J. Wheeler, *Proceedings of the 5th International Conference on Unsaturated Soils,* Barcelona, Spain, **2**, 787-794, (2010).
14. G. Della Vecchia, C. Jommi & E. Romero, *International Journal for Numerical and Analytical Methods in Geomechanics*, 37 (5), 503-535, (2012).
15. D. Gallipoli, F. D'Onza, & S.J. Wheeler, *Canadian Geotechnical Journal*, **47**(11), 1175-1186, (2010).

E3S Web of Conferences 9, 17004 (2016)
DOI: 10.1051/e3sconf/20160917004
E-UNSAT 2016

Identification of potential strain heterogeneities during wetting-induced compaction

Constance Mihalache[1] and Giuseppe Buscarnera[1,a]

[1]Northwestern University, Evanston, IL 60208, United States

Abstract. Compaction upon wetting has been reported to occur in various types of unsaturated soils with damaging consequences for infrastructure in field applications. This deformation process is often referred to as "wetting-collapse", implying that it may be unstable in nature. Recent evidences also indicate that compaction localization represents a possible mode of deformation during wetting, even in the presence of oedometric loading paths. Despite this, little work has been done from a mechanistic perspective to assess the susceptibility of these processes to localization instability. Here we assess the potential for strain localization during wetting through controllability analyses defined in light of a second-order work expression for unsaturated soils. A hydro-mechanical constitutive model with suction-dependent hardening is used to simulate classic experimental data, and the controllability criteria are specialized to capture the potential for shear band formation for a range of band inclinations under water content-controlled and suction-controlled wetting paths. The effect of changes in material characteristics was evaluated, showing that the potential for strain localization upon water-injection increases with increasing values of suction-induced hardening, and that non-associativity may have an effect on both the potential for localization under rapid wetting, as well as on the range of band angles over which it may occur. Specifically, it is possible to distinguish two well-defined stress regions, one within which strain localization is first possible with horizontal band inclinations and another in which inclined localization zones tend to be more critical. Such results provide insight on the factors that may contribute to strain localization during wetting and find general applicability in the interpretation of the response of geo-structures subjected to intense hydrologic forcing.

1 Introduction

A classical problem of unsaturated soil mechanics is the study of compaction in soils subjected to wetting paths. In earthen systems, this compaction can be significant, causing substantial settlements in infrastructures such as dams and levees [1,2], as well as damage to structures [3] and transportation systems [4]. Due to its destructive nature, this phenomenon has been referred to as "wetting-induced collapse", which suggests that these events can be mechanically unstable. However, recent studies [5] inspecting the loss of wetting control in soils modelled with an elastoplastic constitutive law [6] have shown that both oedometric and isotropic saturation paths were controllable upon suction removal and water injection for the unsaturated soils considered in the study. To identify properties that may promote unstable wetting, the model parameters dictating the suction-induced hardening of the material were then increased to values higher than those indicated by the experiments. This analysis demonstrated the crucial effect of hydro-mechanical couplings on the onset of instabilities, suggesting that wetting paths may provoke unstable compaction only in soils characterized by significant water-sensitivity.

Such analyses, however, address only homogeneous modes of instability, and do not consider the potential for heterogeneous modes of deformation such as shear and/or compaction bands. Indeed, while oedometric paths imply homogenous uniaxial strains, recent experiments have identified zones of strain heterogeneity for wetting tests conducted in an oedometer [7]. Thus, it is necessary to consider the potential for strain localization also during this common class of saturation events.

While strain localization can be studied through a classic acoustic tensor approach [8,9], it is also possible to use controllability methods [10] to study such modes of instability under drained or undrained conditions [11]. In this paper the latter approach is used to assess strain heterogeneities during wetting on partially saturated soils, plotting controllability indices for wetting paths driven by fluid pressurization and/or water injection for a range of band angles. In addition, the role of various soil constants on the evolution of these indices has been inspected. In particular, it is shown that the predicted mode of instability is affected not only by the stress state, but also by suction-hardening and degree of non-associativity. In fact, depending on the considered parameters, different cases are identified where either horizontal (compactive) or inclined (shear) strain localization bands are favoured.

[a] Corresponding author: g-buscarnera@northwestern.edu

DOI: 10.1051/e3sconf/20160917004

2 Controllability criteria for shear and/or compaction strain localization

Similar to the analysis performed in [12], a material point is considered susceptible to shear localization at a certain inclination, θ (Figure 1), if it can no longer control an increment of shear stress acting along that inclination. Therefore, in addition to the axial and radial directions used in [5], we will also consider the role of shearing components perpendicular to the axial direction (Figure 1) in our controllability analysis. Consider a constitutive expression of the form,

$$\begin{Bmatrix} \dot{\sigma}_A^* \\ \dot{\sigma}_R^* \\ \dot{\tau} \\ n\dot{s} \end{Bmatrix} = \begin{bmatrix} D_{AA} & D_{AR} & D_{AS} & D_{AH} \\ D_{RA} & D_{RR} & D_{RS} & D_{RH} \\ D_{SA} & D_{SR} & D_{SS} & D_{SH} \\ D_{HA} & D_{HR} & D_{HS} & D_{HH} \end{bmatrix} \begin{Bmatrix} \dot{\varepsilon}_A \\ \dot{\varepsilon}_R \\ \dot{\gamma} \\ -\dot{S}_r \end{Bmatrix} \tag{1}$$

where τ is shear stress, γ is shear strain, n is porosity, s is suction, and the D_{ij} terms are partitions of the constitutive stiffness matrix, $\boldsymbol{D}$. Variables σ_A^* and σ_R^* are the axial and radial stresses work-conjugate to the axial and radial strains (ε_A and ε_R), as defined by the unsaturated second-order work expression [13], or specifically,

$$\dot{\sigma}^* = \dot{\sigma} - S_r \dot{u}_w - (1 - S_r)\dot{u}_a \tag{2}$$

where σ is the total stress, u_w the water pressure, u_a the air pressure, and a superimposed dot indicates the increment of the respective variables. For the sake of simplicity, we focus on the case in which only the degree of saturation affects the mechanical response in the plastic regime, but strains do not affect the retention response (i.e., one-way hydro-mechanical coupling, or $D_{HA} = D_{HR} = D_{HS} = 0$). The components of (1) can also be rearranged such that the constitutive expression is written in terms of net stresses (σ^{net}) and water ratio (e_w or the product of void ratio and degree of saturation).

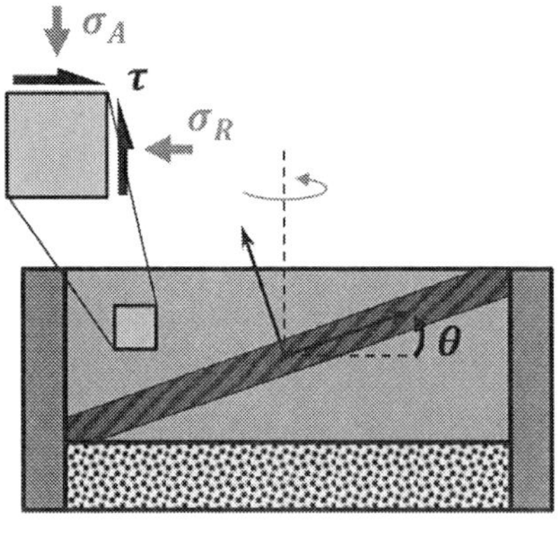

Figure 1. Schematic indicating the axis convention for the controllability analysis and the direction of shear localization

Using the loss of controllability approach [10], under the control of axial net stress, radial strain, shear stress and suction, the criterion for suction-controlled wetting becomes,

$$\det(X_{LOC}^S) = D_{HH}(D_{AA}D_{SS} - D_{AS}D_{SA}) = 0 \tag{3}$$

Under fulfillment of this criterion (during wetting under *fluid pressurization*) the material would not be able to control an increment of shear stress acting perpendicular to the axial direction. A similar criterion can also be obtained for wetting under *fluid injection* or water content control. In this case, the water ratio (e_w) is controlled in addition to the axial net stress, radial strain and shear stress, and the resulting criterion is,

$$\begin{aligned} &\det(X_{LOC}^{WC}) = D_{AA}^* D_{SS} - D_{AS} D_{SA}^* = 0 \\ &D_{AA}^* = D_{AA} - \left(D_{AH} - D_{HH}\frac{S_r}{n}\right)\frac{S_r}{n} \\ &D_{SA}^* = D_{SA} - D_{SH}\frac{S_r}{n} \end{aligned} \tag{4}$$

From the reasoning presented in [12], the formation of a deformation band at inclination, θ would require a stress perturbation parallel to the band direction. Thus, (3) and (4) would only be able to predict localization at a horizontal orientation (i.e., shear/compaction banding for the wetting-induced plastic processes considered in this work). To also consider the potential for localization under other band inclinations, the reference system must be rotated by an angle, θ so that the stress perturbation is oriented along the considered direction of potential strain localization (Figure 1). To this effect, a directional cosine rotation matrix for angle θ ($\boldsymbol{T_R}$) can be applied to the stiffness, $\boldsymbol{D}$ [12] such that,

$$\boldsymbol{D_R} = \boldsymbol{T_R} \boldsymbol{D} \tilde{\boldsymbol{T}}_{\boldsymbol{R}} \tag{5}$$

where the stiffness components in (3) and (4) can be replaced by their rotated counterparts. By computing $\boldsymbol{T_R}$ over a full range of localization angles (θ = −90° to 90°), the potential for heterogeneity due to shear localization at different rotations can be comprehensively examined. The results can also be compared to those for wetting under homogenous deformation ($\det X^S, \det X^{WC}$ in [5]).

Computation of these controllability indices requires the definition of stiffness, $\boldsymbol{D}$ via a constitutive model capable of reproducing wetting-induced compaction. In this work the elastoplastic constitutive model for unsaturated soils proposed by Buscarnera and Nova [6] is used. The model includes a hyperelastic law for the elasticity component, the ability to independently change the yield surface and plastic potential shape (i.e., for modelling non-associative materials), and a suction-dependent hardening law defined as,

$$\dot{p}_s = \frac{1}{B_p} p_s (\dot{\varepsilon}_v^p + \xi_s \dot{\varepsilon}_s^p) - r_{sw} p_s \dot{S}_r \tag{6}$$

where p_s is the yield surface size, and ε_v^p and ε_s^p are the volumetric and shear components of the plastic strain.

The parameter r_{sw} controls the contraction of the yield surface that would occur during saturation (i.e., when the suction increment is negative). Additional information on the model can be found in Buscarnera and Nova [6], and the specific parameters used in this study have been calibrated for the behaviour of a sandy clay [5]. The parameter values and descriptions are given in Table 1.

Table 1. Constitutive model parameters for sandy clay.

Description	Symbol	Value
Elastic Constants	$\hat{k}$	0.008
	G_0	13000 kPa
	p_r	1 kPa
Logarithmic Plastic Compressibility	B_p	0.045
Yield Surface Shape Parameters	a_f	0.63
	m_f	0.99
	M_c	1.23
	M_e	1.10
Suction-Hardening Parameter	r_{sw}	5.5
Dilatancy Parameter	ξ_s	0
SWCC Shape Parameters	a_{vg}	0.06
	m_{vg}	0.053
	n_{vg}	5.6

3 Potential for heterogeneity during oedometeric wetting

A drying, loading and wetting simulation under oedometric conditions was performed using the parameters in Table 1. The initially saturated material at low confinement was dried under constant net stress to 45% degree of saturation, then compressed under constant radial strain to a deviatoric stress around 210kPa, and then wet under oedometric conditions at constant axial net stress. The resulting stress path is shown in Figure 2, with a thin solid line indicating the final stress ratio reached during the simulation.

3.1 Material parameter calibration

The controllability indices for shear localization under suction-controlled wetting (3) and water content-controlled wetting (4) were monitored during the saturation path shown in Figure 2. The values for $\theta = 0°$ are plotted against the degree of saturation in Figure 3a, and the role of the angle of shear rotation is shown in Figures 3b and 3c at select points along the saturation path. The indices maintain positive values throughout the wetting simulation, implying that localized instabilities are not expected during this particular wetting-induced compaction process. This indicates that the material considered in this analysis is not prone to strain localization along the path imposed by oedometric wetting. As a result, its deformation response is expected to be homogeneous, thus corroborating earlier hypotheses of homogeneity of the wetting response used by Mihalache and Buscarnera [5] to perform their analyses.

Nevertheless, it is worth noting that the indices do change in value with rotation angle, suggesting that the inclination of shearing may affect the stability of the system under other conditions. The minimum values of both criteria also occur at relatively high values of degree of saturation, which may correspond with quasi-saturated conditions in which the air phase forms occluded bubbles within the pore fluid. For the sake of simplicity, hereafter we focus on strain localization analyses for degrees of saturation smaller than 75%, thus focusing only on states of saturation at which the two fluid phases can both be assumed to be continuously distributed across the sample.

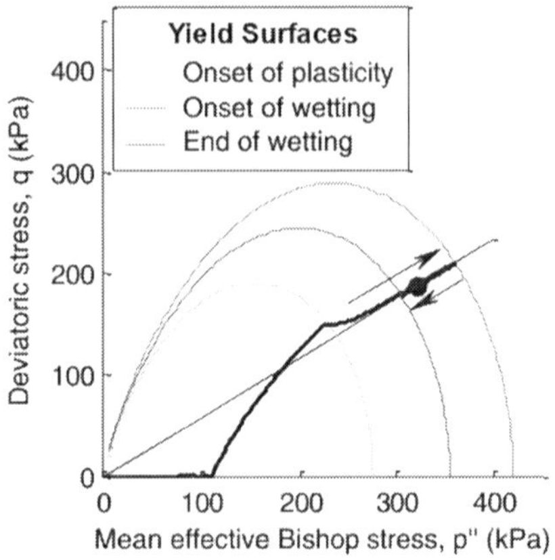

Figure 2. Oedometric stress path (thick line) with arrows indicating direction of loading and wetting. Yield surfaces for the onset of plasticity, onset of wetting and end of wetting are also shown. The point indicates the stress at which S_r=75%

3.2 Role of stress ratio during wetting

In addition to computing indices for the stress path simulated in Figure 2, it is also possible to observe how the indices vary at different stress ratios, for the same degree of saturation and along the same yield surface. Figure 4 shows the effect of considering stress ratios approximately 30% higher and 30% lower than the previous simulation. Though again all criteria remain positive throughout (so no instability is predicted at the given saturation state), the trends in the variation of the controllability indices with rotation angle do change. While for the initial simulation and for the simulation at a 30% lower stress ratio both criteria take minimum values at a horizontal band orientation, for the higher stress ratio the minimum values occur for non-zero band inclinations. The differences between the three situations suggest that the stress state at which wetting is imposed has a crucial role on the behaviour of both indices, similar to findings in the localization of rocks, where compaction bands and shear bands are expected at different stress ratios [14,15].

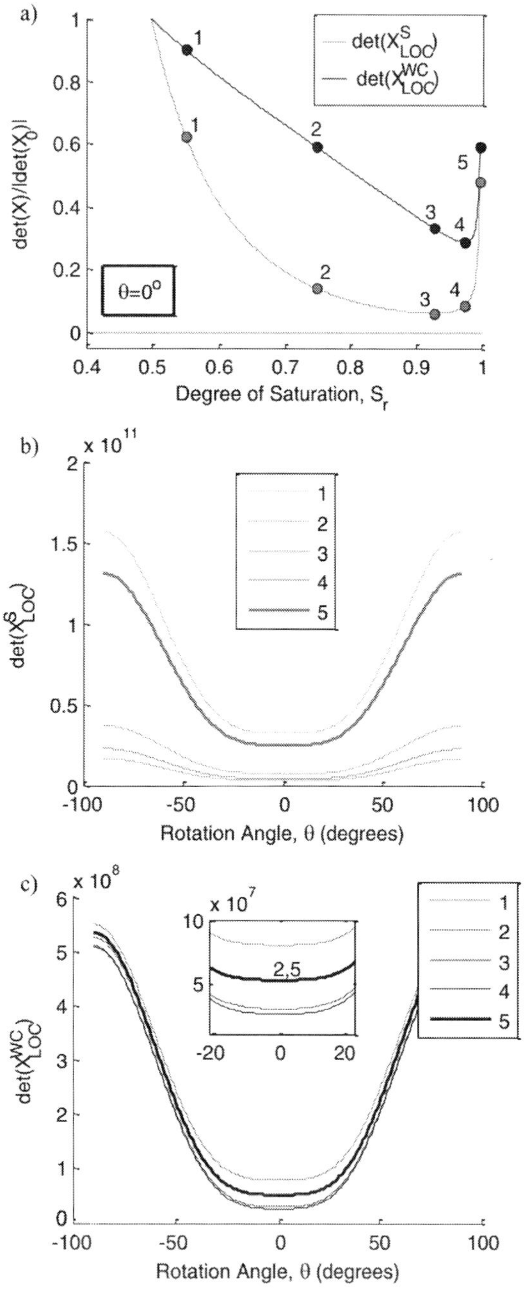

Figure 3. (a) Normalized controllability indices for suction controlled (light line) and water-content controlled wetting (dark line) for potential strain localization along horizontal bands, plotted against degree of saturation, (b) variation of loss of controllability index against shearing under suction control for a range of band angles, and (c) variation of loss of controllability index against shearing under water content control for a range of band angles.

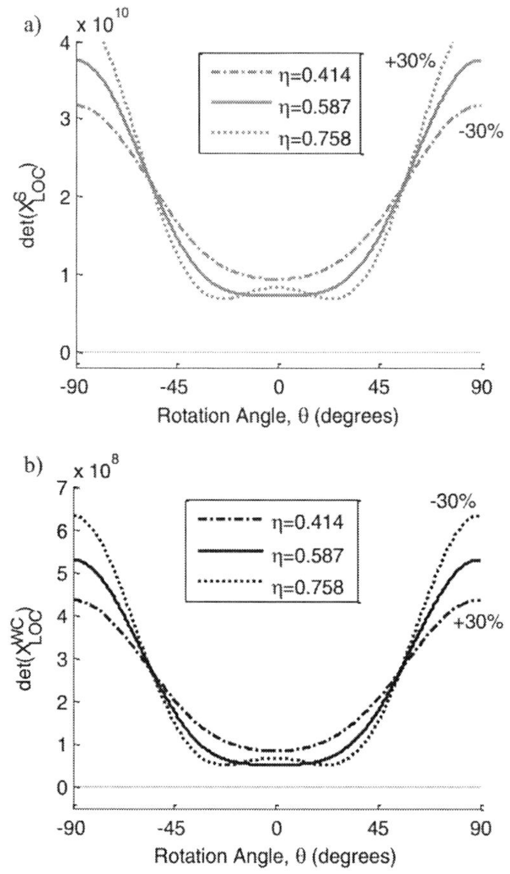

Figure 4. Controllability criteria for strain localization under suction-controlled (a) and water content-controlled wetting (b) plotted against band angle. Results are shown for stress ratios approximately corresponding with the oedometric wetting path at S_r=75% (solid lines), 30% higher stress ratio (dashed lines) and 30% lower stress ratio (dotted lines).

3.3 Influence of suction-induced hardening

In the previous work on homogenous modes of instability [5], a parametric analysis was performed to study the role of the suction-hardening parameter, r_{sw} on the stability upon wetting. It was found that increased values of this parameter made it possible for a loss of controllability to occur under water content control (injection).

The same type of analysis can also be performed to assess the potential of strain localization in materials with decreased suction-hardening during saturation. Figure 5 shows that indeed, higher values of r_{sw} reduce the magnitude of both indices, (3) and (4). In fact, for the case of water injection, the index reaches zero, with localization at $\theta = 0°$ being predicted as the first potential loss of controllability (i.e., compaction banding). With higher suction-hardening values, the range of localization angles increases for the considered degree of saturation, S_r=75%. Hypothetically, a shear perturbation under any of the inclinations within the range highlighted in Figure 5b would not be controllable during that point of the wetting path, for r_{sw}=11.

Figure 6 compares water content control indices at this higher value of suction-hardening for different band angles with the index for homogenous instability used by [5]. At this value of r_{sw}, the first strain localization index to vanish is associated with localized deformation along a horizontal band (i.e., $\theta = 0°$ or compaction banding). This condition is met when also the stability of wetting paths under homogeneous oedometric deformation is lost. This result suggests that prior to a loss of controllability, the sample remains homogenous, but once controllability against water-content control is lost, homogeneity can no longer be guaranteed.

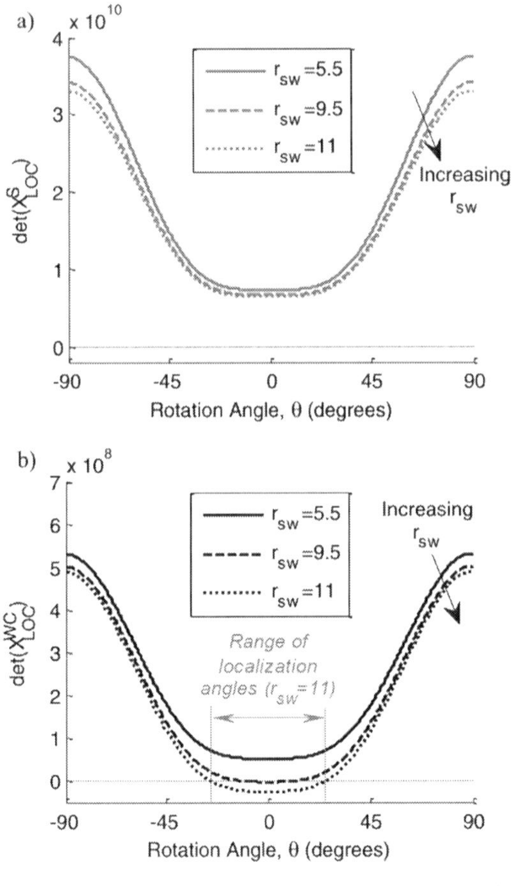

Figure 5. Controllability criteria for localization under suction-controlled wetting (a) and water content-controlled wetting (b) plotted against the angle of potential localization, for increasing suction-hardening.

3.4 Influence of non-associativity

Another material property that can influence the stability of soils is the degree of non-associativity. For instance, it is known that differences in the yield surface and plastic potential within an elastoplastic modelling context are needed to capture static liquefaction in sands [16,17]. Here the original wetting path simulation (r_{sw}=5.5) is repeated for different degrees of non-associativity. The stress ratio at the peak of the surface shape (M_c) is given different values for the yield surface (M_{cf}) and plastic potential (M_{cg}). Specifically, the latter, which is calibrated to match the critical state failure line, remains unchanged (i.e., $M_{cg} = M_c$), while M_{cf} is taken as a percentage of M_{cg}. The peak stress values in extension (M_{ef} and M_{eg}, respectively) are maintained at around 80% of those in compression, and the remaining shape parameters are unchanged.

Figure 7 shows the effect of the introduction of non-associativity. Not only does the shape of the localization profile change to favour strain localization along inclined band angles for increased non-associativity, but the water content control index (4) is actually reduced to zero or negative values at 75% degree of saturation. This indicates that non-associative materials can be prone to localization under water injection. In variance with the effect of increased values of suction-hardening (Figure 5b), controllability is not lost for stress perturbations at θ = 0°, but it is met for a range of band angles (Figure 7b). Although loss of suction control was not computed in any of the considered simulations, also index (3) was found to be sensitive to a change of shape parameters, approaching lower values upon an increase of non-associativity.

Again, these results can be compared to the criterion for homogenous compaction under water content control (Figure 8). The results suggest that if the test remains homogenous, it should always be controllable and that horizontal compaction bands are not expected. However, increased non-associativity makes loss of controllability possible for stress perturbations acting along certain band inclinations. Since under experimental conditions, small deviations from the idealization of homogeneous material points are possible (e.g., heterogeneity, boundary effects) local deformation zones not compliant with oedometric conditions can take place within wetted samples. In other words, a range of band angles similar to those in Figure 7 may become possible, and localized zones of shear and/or compaction may appear while the oedometric test would still remain externally controllable. This finding is in agreement with recent observation of localization during wetting [7] of a highly non-associative, loose sand [12].

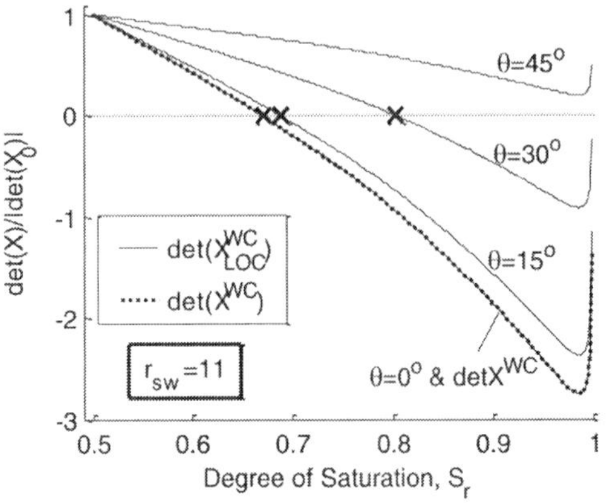

Figure 6. Comparison of the controllability indices computed for homogeneous (dotted line) and localized (solid line) wetting compaction for magnified values of suction-hardening (r_{sw}=11). The symbol X marks the point at which the criteria are violated.

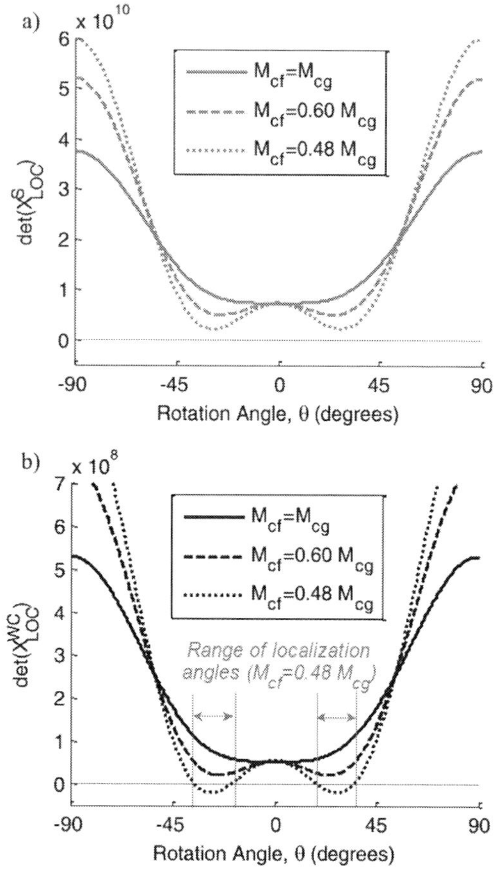

Figure 7. Effect of non-associativity on controllability criteria for (a) suction and (b) water content-controlled wetting at r_{sw}=5.5, plotted against the potential band angle.

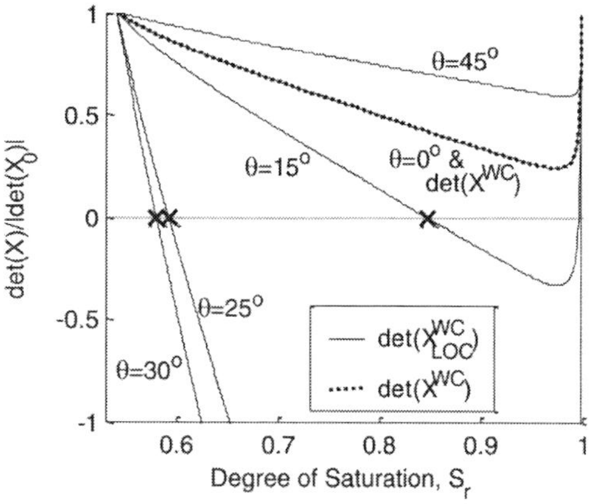

Figure 8. Comparison of water content control indices for homogeneous (dotted line) and localized (solid line) wetting compaction, for magnified non-associativity: $M_{cf} = 0.48M_{cg}$.

4 Conclusions

Conditions for loss of controllability against shearing were used to assess the potential for strain localization during wetting. Criteria were defined for saturation under suction removal and fluid volume injection. These criteria were computed for a range of band inclinations. While suction-controlled wetting has never found to cause any violations of the controllability criteria, different material properties did affect the potential for strain localization under water volume control. Variations in the stress ratio were found to change the range of band angles at which controllability indices reached a minimum, and increased values of the suction-hardening parameter were found to favour the formation of shear/compaction zones. For associated soils, such heterogeneities were found to be possible only at or after a loss of control for homogenous compaction. By contrast, it was found that an increase of the degree of non-associativity tended to promote shear and/or compaction bands even for degrees of saturation at which the homogenous compaction was still controllable. This suggests that small deviations from an ideal material point scheme (e.g., due to soil heterogeneity) may cause noticeable strain localization during wetting paths. These predictions are in agreement with recent experimental findings and can be used to gain a better understanding of the mechanics of wetting-induced deformations.

References

1. E. C. Lawton, R. J. Fragaszy, M. D. Hetherington, J. Geotech. Engrg. **118**, 9 (1992)
2. G. A. Miller, K. K. Muraleetharan, Y. Y. Lim, Transport. Res. Rec. **1755** (2001)
3. K. D. Walsh, W. N. Houston, S. L. Houston, J. Geotech. Engrg. **119**, 5 (1993)
4. S. L. Houston, J. Transp. Eng. **114**, 6 (1988)
5. C. Mihalache, G. Buscarnera, J. Geotech. Geoenviron. Eng. **141**, 2 (2014)
6. G. Buscarnera, R. Nova, Int. J. Numer. Anal. Meth. Geomech. **35** 179-200 (2011)
7. J. F. Bruchon, J.-M. Pereira, M. Vandamme, N. Lenoir, P. Delage, M. Bornert, Granul. Matter **15**, 6 (2013)
8. J. W. Rudnicki, J. R. Rice, J. Mech. Phys. Solids, **23**, 371-394 (1975)
9. R. I. Borja, Comput. Methods Appl. Mech. Engrg. **193**, 5301-5338 (2004)
10. S. Imposimato, R. Nova, Mechanics of Cohesive-Frictional Materials **3**, 65-87 (1998)
11. G. Buscarnera, C. di Prisco, Geotechnique Letters **1**, 85-90 (2011)
12. C. Mihalache, G. Buscarnera, Int. J. Numer. Anal. Meth. Geomech. **38**, 2 (2014)
13. G. Buscarnera, C. di Prisco, Int. J. Numer. Anal. Meth. Geomech. **36**, 36-49 (2012)
14. K. A. Issen, J. W. Rudnicki, J. Geophys. Res. **105**, B9 (2000)
15. F. Marinelli, G. Buscarnera, Int. J. Rock Mech. Min. **78**, 240-252 (2015)
16. G. Buscarnera, A.J. Whittle, Can. Geotech. J. **49**, 5, 499-511 (2012)
17. G. Buscarnera, A.J. Whittle, J. Geotech. Geoenviron. Eng. **139**, 3, 420-432 (2013)

Plasto-damage modelling for semi-brittle geomaterials

Ali Alizadeh [1] and Behrouz Gatmiri [1,2,a]

[1] *Civil Engineering Department, University of Tehran, Tehran, Iran*
[2] *ParisTech, Laboratoire Navier (École des Ponts) École des Ponts ParisTech (ENPC)*

Abstract. This paper presents an elastoplastic damage model for constitutive modelling of semi-brittle geomaterials showing two irreversible mechanisms. On one hand, the model deals with the plastic behaviour of a porous medium by a new variant of Barcelona Basic Model. On the other hand, the model combines the micromechanical definition of damage and phenomenological concepts in the framework of Continuum Damage Mechanics (CDM) for damage modelling. A second order tensorial damage variable is adopted for the model. Damaged effective stress variables are employed for formulation of elastoplastic behaviour laws and the plastic yield surface is a damage dependent one. The model has been validated by comparing the numerical results with experimental results of argillites.

1 Introduction

In the framework of radioactive waste disposal, argillaceous formations are under widespread consideration as host rock because of their favourable properties. Accordingly, constitutive modelling of clay formations as geological barrier is also an active topic for the researchers of this field.

Most of the currently existing geotechnical plastodamage models, mainly for the sake of simplicity, have made use of isotropic definition of damage (e.g. [1,2]). Contrary to this assumption, the experimental results on clay stones show an important plastic flow being coupled with an induced anisotropic damage [3,4]. As a result, an elastoplastic damage model might be an appropriate choice for constitutive modelling of semi-brittle geomaterials like argillites. In view of this, an anisotropic damage mechanism is formulated for the current study using a second order damage variable.

Although a number of plasto-damage models are also introduced in the literature, most of these models are based on the use of Bishop's like effective stress (e.g. [5,6]).

As pointed out by Sheng *et al.* [7], the complex stress variables (e.g. Bishop's effective stress) are dependent on material states and therefore are not straightforwardly controllable in conventional laboratory testing procedures. In the models which make use of the Bishop's effective stress with the parameter $\chi=S_r$, the inherent hydro-mechanical coupling is greatly dependent on the water retention relation used in the model. This entails utilizing sophisticated water retention models. Consequently the water retention properties of unsaturated medium should be specifically determined and it may necessitate the use of hysteretic water retention models in addition to mechanical formulation.

On the other hand the experimental results on the retention properties of damaged geomaterials are scarce in the literature. Therefore independent stress state variables including net stress and suction are adopted as stress variables of the proposed elastoplastic damage model.

This paper first presents the damage-related formulation of the model in the framework of Continuum Damage Mechanics (CDM). Afterwards, elastoplastic formulation is proposed based on Barcelona Basic Model concepts incorporating damage dependency of yield surface. At last, mechanical aspects of the model are validated by simulation of some triaxial tests and the results of the proposed model are compared with the experimental ones.

2 Formulation of the damage

2.1 Definition of an anisotropic damage variable

The damage variable is typically defined as a macroscopic measure of the microscopic degradation of a Representative Elementary Volume [8,9]. To date, different kinds of damage variables have been used in the context of Continuum Damage Mechanics to represent the density and orientation of micro-cracks. One or more scalar damage variables for isotropic damage modelling and vectorial damage variables, second and fourth rank damage tensors or even eighth order tensors [10] for anisotropic damage models. Induced damage in brittle geomaterials is strongly anisotropic in nature. Only second or higher even-order tensors can be adopted as damage variable to characterize the damage induced anisotropy [11]. As pointed out by [12], owing to the difficulties which would be encountered in postulating

[a] Corresponding author: behrouz.gatmiri@andra.fr

appropriate evolution laws for higher order tensors, the second-order tensor is preferred in the modelling of damage induced anisotropy. Having considered the above description, the second order variable $\underline{\underline{\Omega}}$ is adopted as the damage variable at the mesoscopic scale:

$$\underline{\underline{\Omega}} = \sum_{j=1}^{3} d_j \, \underline{n_j} \otimes \underline{n_j} \quad (1)$$

This definition is based on spectral decomposition. From the physical point of view the approximately parallel damages developed in the Representative Elementary Volume (REV) are considered as three orthogonal equivalent meso-cracks. As it can be illustrated for a two dimensional case, each bunch of micro-defects with approximately parallel direction is homogenized in an equivalent meso-scale crack. The equivalent meso-cracks are characterized by unit normal vector of crack plane, n_j, and a crack density parameter d_j.

2.2 Damaged stress variables

As the meso-cracks forms within the medium, the material starts losing effective load carrying surface. As a consequence, stress increases in the undamaged part of the material. Mechanically, the new stress field of the damaged medium can be obtained by removal of the defects and replacement of nominal stresses by effective stresses.

Fig. 1 (left) illustrates a cylindrical damaged Representative Elementary Volume (REV) submitted to a uniaxial tensile stress, σ. Fig. 1 (right) shows the undamaged counterpart of REV as a fictitious equivalent state in which the cross-sectional area of the medium is decreased due to the removal of the cracks and the nominal stress is replaced by the intensified stress, that is to say uniaxial effective damaged stress $\hat{\sigma} = \dfrac{\sigma}{1-d}$.

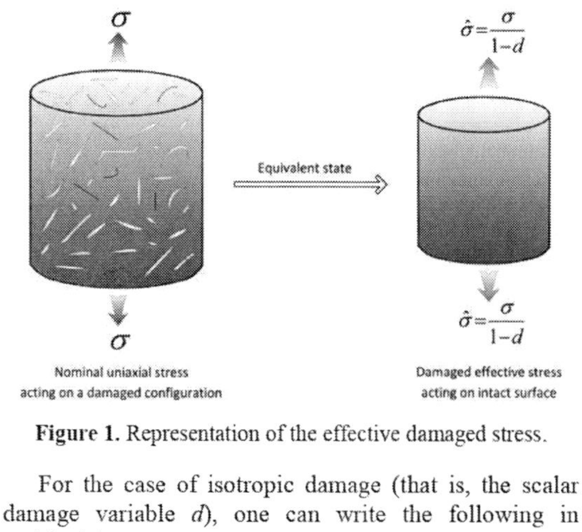

Figure 1. Representation of the effective damaged stress.

For the case of isotropic damage (that is, the scalar damage variable d), one can write the following in multiaxial stress space:

$$\underline{\underline{\hat{\sigma}}} = \frac{\underline{\underline{\sigma}}}{1-d} \quad (2)$$

However, in the case of the anisotropic damage, a fourth-order operator, $\underline{\underline{\underline{\underline{\mathbf{M}}}}}$ should be used for the definition of the damaged effective stress.

$$\underline{\underline{\hat{\boldsymbol{\sigma}}}} = \underline{\underline{\underline{\underline{\mathbf{M}}}}} : \underline{\underline{\boldsymbol{\sigma}}} \quad (3)$$

This is appropriate for a single phase damaged media, but for an unsaturated porous media both stress variables must be transformed to the effective state. Several mathematical forms have been proposed for the operator in the CDM context; however the operators which ensure symmetrisation of the damaged effective stress are preferred [13]. Therefore, the operator of Cordebois and Sidoroff [14] is extended to the unsaturated state variables for the definition of the damaged net stress, $\underline{\underline{\hat{\boldsymbol{\sigma}}''}}$, and the damaged suction, $\hat{s}$.

$$\begin{aligned} \underline{\underline{\hat{\boldsymbol{\sigma}}''}} &= (\underline{\underline{\boldsymbol{\delta}}} - \underline{\underline{\boldsymbol{\Omega}}})^{-1/2} . \underline{\underline{\boldsymbol{\sigma}''}} . (\underline{\underline{\boldsymbol{\delta}}} - \underline{\underline{\boldsymbol{\Omega}}})^{-1/2} \\ Tr\left(\hat{s}\,\underline{\underline{\boldsymbol{\delta}}}\right) &= Tr\left[(\underline{\underline{\boldsymbol{\delta}}} - \underline{\underline{\boldsymbol{\Omega}}})^{-1/2} . s\,\underline{\underline{\boldsymbol{\delta}}} . (\underline{\underline{\boldsymbol{\delta}}} - \underline{\underline{\boldsymbol{\Omega}}})^{-1/2}\right] \end{aligned} \quad (4)$$

where $\underline{\underline{\boldsymbol{\delta}}}$ is Kronecker delta. Having obtained the damaged stress variables (suction and net stress), the damaged medium behaviour can be formulated in the effective stress and actual strain space. This treatment can markedly simplify the computational implementation.

2.3 Damage criterion and damage evolution law

According to the experimental results reported in the literature, there is an obvious correlation between total tensile strains and the induced damage in brittle geomaterials. We can, therefore, assume that the damage growth is associated with the development of generalized tensile strain and the damage start developing within the medium as the strains reach a threshold which in itself is dependent on the current damaged state. Thus the Eq. 5 would be used as damaging yield function or damage criterion. It should be noted that this equation is inspired from [15].

$$F_d(\boldsymbol{\varepsilon}^+, \boldsymbol{\Omega}) = \sqrt{\frac{1}{2} g\,\boldsymbol{\varepsilon}^+ : g\,\boldsymbol{\varepsilon}^+} - C_0 - C_1(\boldsymbol{\delta} : \boldsymbol{\Omega}) \quad (5)$$

C_0, C_1 and are material parameters; wherein C_0 designates the initial damage threshold, and C_1 characterizes the degradation of the medium materials with the cumulated damage. This means that under the multiaxial stress states, the REV still continues damaging in tensile mode when the combination of major tensile strains attain damage driving threshold $[-C_0 - C_1(\boldsymbol{\delta}:\boldsymbol{\Omega})]$. In addition, g characterizes the resistance of the material to cracking of the medium and propagation of damage. It should be noted that the term $g\,\varepsilon^+$ is part of damage-conjugated stress or '*damage affinity*' that the product of it by damage is an energy release rate.

The sign convention of solid mechanics (compression negative) is adopted here. Accordingly, the positive part of total strain, ε^+, represents the extensional components of deformations of the medium. Thus the spectral decomposition of the total strains can be used as follows:

$$\boldsymbol{\varepsilon}^+ = \sum_{k=1}^{3} \varepsilon_k H(\varepsilon_k)\mathbf{W}_k \otimes \mathbf{W}_k \tag{6}$$

Where ε_k is k^{th} eigenvalue of the strain tensor which corresponds to k^{th} principal strain, $\mathbf{W}_k$ is eigenvector (principal vector) associated to ε_k and $H(\varepsilon_k)$ is the Heaviside function of the principal strain, $H(x) = 1$ for $x \geq 0$, $H(x) = 0$ for $x < 0$.

Assuming an associated flow rule (or normal dissipation scheme) for damage, one can write the damage evolution law as follows:

$$\dot{\boldsymbol{\Omega}} = \dot{\Lambda}_d \frac{\partial F_d(\boldsymbol{\varepsilon},\boldsymbol{\Omega})}{\partial g \boldsymbol{\varepsilon}^+} = \dot{\Lambda}_d \frac{\boldsymbol{\varepsilon}^+}{\sqrt{2tr(\boldsymbol{\varepsilon}^+ \boldsymbol{\varepsilon}^+)}} \tag{7}$$

This treatment can guarantee that, the evolution rate of damage variable and the tensor of tensile strains are developing in the same direction, as it was desirable from the beginning. The evolution of positive scalar damage multiplier, $\dot{\Lambda}_d$, can be determined by the consistency condition of the damage criterion $(\dot{F}_d = 0)$.

$$\dot{\Lambda}_d = \frac{\boldsymbol{\varepsilon}^+ : d\boldsymbol{\varepsilon}^+}{C_1 \boldsymbol{\delta} : \boldsymbol{\varepsilon}^+} \tag{8}$$

The loading–unloading conditions known as so-called Kuhn–Tucker conditions should be satisfied as follows:

$$\dot{\Lambda}_d \geq 0 \qquad F_d(\boldsymbol{\varepsilon},\boldsymbol{\Omega}) \leq 0 \qquad \dot{\Lambda}_d F_d(\boldsymbol{\varepsilon},\boldsymbol{\Omega}) = 0 \tag{9}$$

3 Characterization of elasto-plastic behaviour

3.1 Degradation of elastic properties

Degradation of elastic properties is one of main features of mechanical behaviour of argillites which has often been attributed to the induced damage by micro-cracks. Different forms of relations are introduced in the literature to reproduce such effect. Equation (10) is employed here to modify the bulk modulus and shear modulus of undamaged material undergoing damage propagation. Non-interacting micro-cracks and moderate damage assumption are made to obtain these relations [16].

$$B(\boldsymbol{\Omega}) = B_0(1 - \frac{1}{3}\alpha_1 tr\,\boldsymbol{\Omega}) \qquad G(\boldsymbol{\Omega}) = G_0(1 - \frac{1}{3}\alpha_2 tr\,\boldsymbol{\Omega}) \tag{10}$$

where α_1 and α_2 are material parameters characterizing the damage effects on the elastic properties of the geomaterial.

3.2 Formulation of plastic behaviour

Barcelona Basic Model (BBM) [17] and its variants are one of well known categories for constitutive modelling of clayey media. In this research, another elasto-plastic model based on the extension of BBM to damage effects is presented. The yield surface is modified so as to incorporate damage effects and new model is referred to as Damage Barcelona Basic Model (DBBM). It is assumed that the change of the yield function can be reproduced by variation of maximum apparent preconsolidation stress $\hat{p}_0$ and cohesion related variable $\hat{p}_s$. Thus one can express yield function as a function of damaged suction, accumulated volumetric thermoplastic strains ε_v^p as well as damage itself:

$$F_p = F_p(\hat{\sigma}_{ij}, \hat{s}, \varepsilon_v^p, \boldsymbol{\Omega}) = \hat{q}^2 - M^2(\hat{p} + \hat{p}_s)(\hat{p}_0 - \hat{p}) = 0 \tag{11}$$

In which:

$$\begin{aligned} \hat{p}_s &= \hat{p}_s(\hat{s},\boldsymbol{\Omega}) = k.\hat{s}.(1 + \frac{1}{3}\xi\, tr\,\boldsymbol{\Omega}) \\ \hat{p}_0 &= \hat{p}_0(\hat{s},\varepsilon_v^p,\boldsymbol{\Omega}) = p_{ref}\left(\frac{\hat{p}_c}{p_{ref}}\right)^{\frac{\lambda(0)-\kappa}{\lambda(\hat{s})-\kappa}} \\ \hat{p}_c &= p_{c0}.\exp(\gamma_H \varepsilon_v^p)(1 + \frac{1}{3}\xi\, tr\,\boldsymbol{\Omega}) \end{aligned} \tag{12}$$

$\hat{p}_c$ is the apparent preconsolidation stress for the saturated state and $\hat{p}_0$ is its counterpart for the unsaturated state. $\hat{p}_s$ represents the variation of cohesion with the corresponding suction or damage changes. k, ξ, γ_H are material parameters through which the above mentioned variables are defined. p_{ref} is the reference stress for the plastic yield surface.

It is evident that the yield function is written based on the damaged stress components $\hat{q}, \hat{p}$. In fact, the damaged effective stress variables play the role of coupling variables for the current elastoplastic damage model which can readily couple the damage influence with the elastic criterion and the plastic formulation. The term $(1 + \frac{1}{3}\xi\, tr\,\boldsymbol{\Omega})$ is multiplied by the undamaged variables p_s and p_c to reflect variation of plastic yield locus with damage.

Fig. 2 illustrates the yield loci of DBBM for the undamaged and damaged states. As can be seen, damage can shrink or swell the elastic domain. Adoption of positive value for ξ can enlarge the elastic domain while a negative value of ξ has an opposite effect resulting in shrinkage of the extent of the yield locus with the increase of damage intensity.

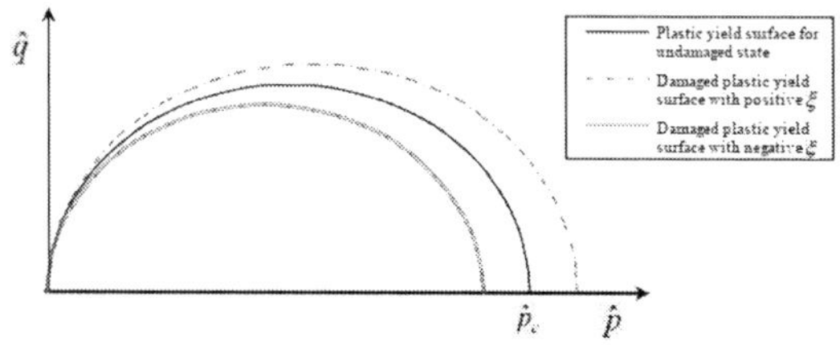

Figure 2. The yield surface of Elasto-plastic damage model (DBBM).

The slope of the normal compression line (NCL) or virgin compressibility index can be defined by following relation [17]:

$$\lambda(s) = \lambda(0)[(1-r)\exp(-\beta\hat{s}) + r] \quad (13)$$

Where r is a parameter controlling the maximum stiffness of the porous medium for an infinite suction, $r = \lambda(S \to \infty)/\lambda(0)$, and β is a parameter which characterizes the rate of increase of soil stiffness with suction.

With the assumption of small strains, the total strain rate can be decomposed into an elastic part and a plastic part as follows:

$$d\varepsilon = d\varepsilon^e + d\varepsilon^p \quad (14)$$

Plastic strain increments are deduced from an associated flow-rule (F_p=G_p):

$$d\varepsilon^p = d\Lambda_p \frac{\partial F_p}{\partial \boldsymbol{\sigma}} \quad (15)$$

In which the rate of plastic multiplier $d\Lambda_p$ is derived from the consistency condition given by:

$$\frac{\partial F_p}{\partial \boldsymbol{\sigma}} d\boldsymbol{\sigma} + \frac{\partial F_p}{\partial s} ds + \frac{\partial F_p}{\partial \boldsymbol{\Omega}} d\boldsymbol{\Omega} + \frac{\partial F_p}{\partial \varepsilon_v^p} d\varepsilon_v^p = 0 \quad (16)$$

This leads to:

$$d\Lambda_p = \frac{\left(\frac{\partial F_p}{\partial \sigma}\right)\mathbf{D}^e d\varepsilon + \left(\frac{\partial F_p}{\partial s} - \frac{\partial F_p}{\partial \boldsymbol{\sigma}}\mathbf{D}_{es}\right)ds - \left(\frac{\partial F_p}{\partial \boldsymbol{\Omega}}\right)d\boldsymbol{\Omega}}{H_1} \quad (17)$$

Substitution of the plastic multiplier into flow rule yields the following expression for incremental constitutive equation:

$$d\boldsymbol{\sigma} = (\mathbf{D}^e - \mathbf{D}^p)d\varepsilon - (\mathbf{D}_{es} + \mathbf{D}_{es}^p - \mathbf{D}_{ese}^p)ds - (\mathbf{D}_\Omega)d\boldsymbol{\Omega} \quad (18)$$

The terms of the equation are given in Appendix.

The proposed model has been implemented in Θ-STOCK Code which is developed by Gatmiri and his co-workers [18-20]. Nonlinear elastic, elastoplastic and elasto-damage constitutive laws as well as soil–atmosphere interaction model have already been incorporated in the code [21-25].

4 Numerical modelling and verification

To verify the adequacy of the model, triaxial testes have been modelled hereafter. The tests are conducted on a clay stone from the East region of France called 'Argilites de l'Est' [3,26,27]. The specimens were 40 mm in height and 20 mm in radius and the tests were carried out in isothermal state. In order to validate model for different loading paths, triaxial lateral extension as well as triaxial compression test with unloading – reloading cycles are simulated. Having considered the geometry and the loading condition of the tests, an axisymmetric configuration is adopted for the model.

Preliminary parameters for elastoplastic and damage behaviour were mainly taken from argillite parameters introduced in the literature (e.g. [25-27]). First, a triaxial lateral extension test was back-analyzed to adjust the preliminary parameters so as to obtain parameters that yielded the best fit to the experimental result. The Final model parameters obtained are summarized in Table 1 and the result of the back analysis is shown in Fig. 3.

Table 1. Parameters of the Model.

Elastic Parameters			
E_0 *(MPa)*	ν	κ	κ_s
11800	0.17	0.005	0.008
Plastic Parameters			
λ_0	r	β	k
0.028	0.06	0.0076	0.05
Plastic Parameters			
P_{c0} *(MPa)*	P_{ref} *(MPa)*	M	γ_H
16	1	2	49.32
Damage Parameters			
C_0 *(Pa)*	C_1 *(Pa)*	g	ζ
2.3E-4	5.2E-3	-1.414	0.05

Plastic compressibility modulus of argillite, γ_H, is not directly available in the literature. Hence the Eq. 19 is used to calculate it:

$$\gamma_H = \frac{1+e_0}{\lambda - \kappa} = 49.32 \quad (19)$$

For the parameters of degradation of elastic stiffness $\alpha_1 = 0$ and $\alpha_2 = 32(1-\upsilon_0)/15(2-\upsilon_0)$ are adopted. These values are based on theoretical estimations presented by Budiansky and O'Connell [16] which are also confirmed by Chiarelli's experimental results.

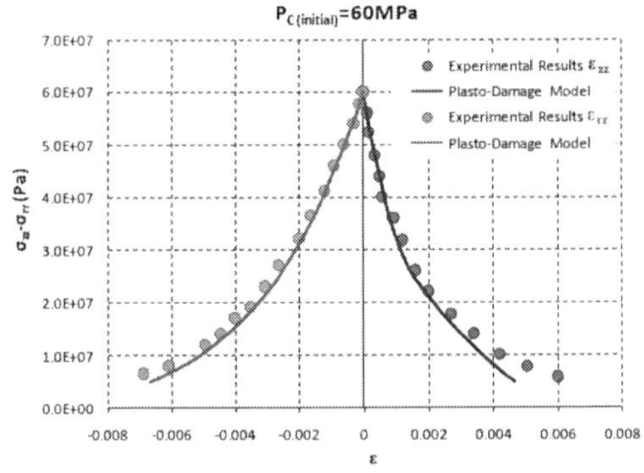

Figure 3. The results of back-analysis of triaxial lateral extension test.

To validate the proposed model in complex loading conditions, a triaxial compression tests with unloading-reloading cycles from Chiarelli's thesis is simulated. Comparison of numerical results and experimental data are illustrated in Fig. 4. As can be seen, the model provided a reasonable result for the test with unloading cycles and it could reproduce degradation of elastic properties properly.

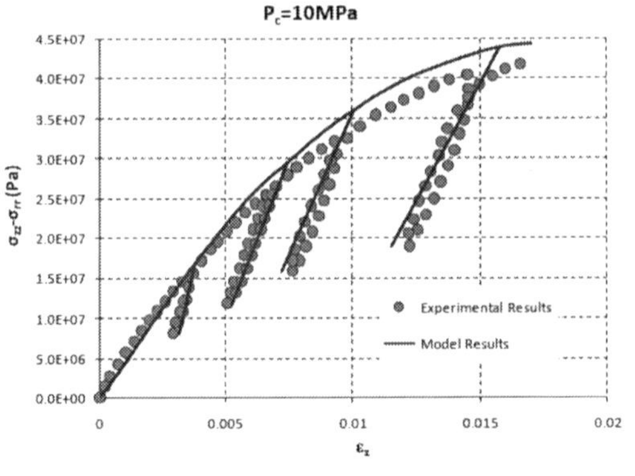

Figure 4. Comparison of model results and experimental data for unloading – reloading cycles.

Using the determined parameters, the authors have also simulated some triaxial compression tests with the proposed plastic damage model [28] and satisfactory results have been obtained although, for the sake of brevity of the text, these results are not included here.

5 Conclusions

This paper has dealt with the formulation of a poro-plasto-damage model for semi-brittle geomaterials. A second-order tensor is adopted as damage variable, characterizing crack density in principal directions. Damage is assumed to grow with tensile strains of medium and constitutive equations are formulated in terms of effective stress and actual strain. To this end, damaged effective terms of stress state variables including the net stress and suction are defined according to the Continuum Damage Mechanics principles.

A new elastoplastic formulation is presented based on Barcelona Basic Model (BBM) concepts in which damage dependency of the yield surface and the potential function have been taken into account by introducing new relationships. The proposed formulation is implemented in Θ-STOCK Code being a finite element code for modelling of coupled problems of multiphase media.

To investigate capability of the new model, it is verified against the experimental results from the literature. In order to validate model for different loading paths, triaxial lateral extension as well as triaxial compression tests with unloading – reloading cycles are simulated. Verification results indicated that the developed poro-plasto-damage model can provide an adequate framework for the modelling of the behaviour of the semi-brittle materials like argillite.

Appendix

Matrices of incremental constitutive equation are as follows:

$$\mathbf{D}^e = \frac{3B}{9B-E}\begin{bmatrix} 3B+E & 3B-E & 0 & 3B-E \\ 3B-E & 3B+E & 0 & 3B-E \\ 0 & 0 & E & 0 \\ 3B-E & 3B-E & 0 & 3B+E \end{bmatrix}$$

$$\mathbf{D}^p = \frac{\left(\left(\frac{\partial F_p}{\partial \boldsymbol{\sigma}}\right)^T \mathbf{D}^e\right)\left(\mathbf{D}^e \frac{\partial F_p}{\partial \boldsymbol{\sigma}}\right)}{H_1}$$

$$\mathbf{D}_s = \beta_s \mathbf{m}^T = \left(\frac{\kappa_s}{1+e_0}\frac{1}{s+p_{atm}}\right)\mathbf{m}^T$$

$$\mathbf{D}_{es} = \mathbf{D}^e \mathbf{D}_s^{-1}$$

$$\mathbf{D}_{es}^p = \frac{\frac{\partial F_p}{\partial s}\mathbf{D}^e \frac{\partial F_p}{\partial \boldsymbol{\sigma}}}{H_1}$$

$$\mathbf{D}_{ese}^p = \frac{\left(\frac{\partial F_p}{\partial \boldsymbol{\sigma}}\right)^T \mathbf{D}_{es}\left(\mathbf{D}^e \frac{\partial F_p}{\partial \boldsymbol{\sigma}}\right)}{H_1}$$

$$\mathbf{D}_{\Omega} = \frac{\left(\frac{\partial F_p}{\partial \boldsymbol{\Omega}}\mathbf{D}^e \frac{\partial F_p}{\partial \boldsymbol{\sigma}}\right)}{H_1}$$

$$H_1 = H - H_{cr} = \left(\frac{\partial F_p}{\partial \boldsymbol{\sigma}}\right)^T \mathbf{D}^e \frac{\partial F_p}{\partial \boldsymbol{\sigma}} - \frac{\partial F_p}{\partial \varepsilon_v^p}\frac{\partial F_p}{\partial p}$$

References

1. Shao, J.F., Jia, Y., Kondo, D., Chiarelli, A.S. A coupled elastoplastic damage model for semi-brittle materials and extension to unsaturated conditions. Mech. Mater. **38**: 218–232 (2006)
2. Zhou, H. Bian, H.B. Jia, Y., Shao, J.F. Elastoplastic damage modeling the mechanical behavior of rock-

like materials considering confining pressure dependency. Mech. Res. Commun. **53**: 1-8 (2013)
3. Chiarelli, A.S. Experimental investigation and constitutive modelling of coupled elastoplastic damage in hard argillites, Doctoral thesis (in French), University of Lille (2000)
4. Chiarelli, A.S., Shao, J.F., Hoteit, N. Modelling of elastoplastic damage behaviour of a claystone. Int. J. Plast. **19**: 23–45 (2003)
5. Shao, J.F., Ata, N., Ozanam, O. Study of desaturation and resaturation in brittle rock with anisotropic damage. Eng. Geol. **81**: 341–352 (2005)
6. Jia, Y., Song, X.C., Duveau, G., Su, K., Shao, J.F. Elastoplastic damage modelling of argillite in partially saturated condition and application. Phys. Chem. Earth. **32**: 656–666 (2007)
7. Sheng, D., Gens, A., Fredlund, D.G., Sloan, S.W. Unsaturated soils: From constitutive modelling to numerical algorithms. Comput. Geotech. **35**: 810-824 (2008)
8. Lemaitre, J., Chaboche, J.-L. Mechanics of Solid Materials. Cambridge University Press, London (1990)
9. Voyiadjis, G.Z., Deliktas, B. A coupled anisotropic damage model for the inelastic response of composite materials. Comput. Method. Appl. M. **183**: 159–199 (2000)
10. Chaboche, J.L. Continuum damage mechanics: part I and II. J. Appl. Mech. **55**: 59-72 (1988)
11. Kanatani, K. Distribution of directional data and fabric tensors. Int. J. Eng. Sci. **22**: 149–164 (1984)
12. Wu, J.-Y., Li, J. On the mathematical and thermodynamical descriptions of strain equivalence based anisotropic damage model. Mech. Mater. **40**: 377–400 (2008)
13. Lemaître, J., Desmorat, R. Engineering Damage Mechanics. Ductile, Creep, Fatigue and Brittle Failure. Springer: Berlin, Heidelberg (2005)
14. Cordebois, J.P., Sidoroff, F. Endomagement anisotrope en élasticité et plasticité. J. de Mécanique théorique et appliquée 45–60 (special issue) (1982)
15. Dragon, A., Halm, D. Modelisation de l'endommagement par meso-fissuration: comportement unilateral et anisotropie induite. CR Acad. Sci. Paris. **T322**(IIb), 275–282 (1996)
16. Budiansky, B., O_Connell, R. Elastic moduli of a cracked solid. Int. J. Solids Struct. **12**: 81–97 (1976)
17. Alonso, E.E., Gens, A., Josa, A. A constitutive model for partially saturated soils. Géotechnique. **40**(3): 405–430 (1990)
18. Gatmiri, B., Delage, P. A formulation of fully coupled thermal-hydraulic mechanical behavior of saturated porous media-numerical approach. Int. J. Numer. Anal. Meth Geomech **21**(3): 199–225 (1997)
19. Gatmiri, B., Seyedi, M., Delage, P., Fry, F. A new suction-based mathematical model for thermo-hydro-mechanical behavior of unsaturated porous media. Proceedings of the Sixth International Symposium on Numerical Models in Geomechanics, NUMOG VI, Quebec, Canada, 291–296 (1997)
20. Gatmiri, B., Jenab-Vossoughi, B., Delage, P. Validation of Θ-STOCK, finite element software for the analysis of thermo-hydro-mechanical behaviour of engineered clay barriers. In Proc NAFEMS WORLD CONGRESS 99 effective engineering analysis **1**: 645–56 (1999)
21. Arson, C., Gatmiri, B. Numerical study of a thermo–hydro-mechanical model for unsaturated porous media. Ann. Solid Struct. Mech. **1**: 59–78 (2010)
22. Arson, C., Gatmiri, B. Thermo-hydro-mechanical modeling of damage in unsaturated porous media: Theoretical framework and numerical study of the EDZ. Int. J. Numer. Anal. Meth. Geomech. **36**(3): 272–306 (2012)
23. Gatmiri, B., Hemmati, S., Arson, C., Amirzehni, E. A multiphase analysis for environmental impact assessement with θ-stock finite element program. J. Multiscale Modeling **2**: 23–68 (2010a)
24. Gatmiri, B., Maghoul, P., Duhamel, D. Two-dimensional transient thermo-hydro-mechanical fundamental solutions of multiphase porous media in frequency and time domains. Int. J. Solids Struct. **47**: 595-610 (2010b)
25. Gatmiri, B., Arson, C. Θ-stock, a powerful tool for thermohydromechanical behaviour and damage modeling of unsaturated porous media. Comput. Geotech. **35**(6): 890–915 (2008)
26. Chiarelli, A.S., Shao, J.F. Modélisation élastoplastique couplée à l'endommagement anisotrope induit pour des argilites. Revue Francaise de Génie Civil **6**(1): 115–130 (2002)
27. Homand, F., Chiarelli, A.S., Hoxha, D.D. Caractéristiques physiques et mécaniques du granite de la Vienne et de l'argilite de l'Est. Revue Française de Génie Civil **6**(1): 11–20 (2002)
28. Alizadeh, A. Modeling of Elastoplastic Damage in Unsaturated Porous Media. Doctoral thesis (in Persian), University of Tehran (2016)

An effective stress approach for hydro-mechanical coupling of unsaturated soils

Hiram Arroyo[1,a], Eduardo Rojas[1] and José Arroyo[2]

[1]*Faculty of Engineering, Universidad Autónoma de Querétaro, Centro Universitario, Cerro de las campanas, Qro., México*
[2]*Faculty of Civil Engineering, Universidad Michoacana de San Nicolás de Hidalgo, Centro Universitario, Avenida Francisco J. Mújica S/N, México*

Abstract. The simulation of the mechanical and hydraulic behaviour of soils is one of the most important tasks in soil mechanics. It is inaccurate to consider that the behaviour of saturated and unsaturated soils as if their governing laws were utterly different, this last condition is not sufficient to do so. However, unlike the laws governing the behaviour of saturated soils, those used to describe the behaviour of unsaturated soils lack the simplicity and predictability associated to the complexity of the phenomena that occur within these porous media. This is why it is important to establish a unified soil mechanics theory to reconcile saturated and unsaturated soil mechanics. In the present work, we describe a simple analytical equation to obtain effective stresses for any type of soil. The equation is coupled to an elastoplastic constitutive model which is capable to reproduce the stress-strain relationship of soil taking into account the hydro-dynamic coupling.

1 Introduction

Constitutive models for soils are intended to capture and predict the most important characteristics of their behavior.

On the matter of the Mechanics of Unsaturated Soils, the pioneering work by Alonso et al. [1] based, amongst others, on the work by Roscoe and Burland [2] are central pieces to understand the main trends on current scientific studies.

In this regard, the research stream related to constitutive modelling has paid attention on establishing a unified scheme for saturated and unsaturated behavior. Moreover, a constitutive model that couples both hydraulic and mechanical characteristics amongst a model is of paramount importance.

With this in mind, constitutive models research stream goes through two different ways. The first one concerns to the formulation of equations based on the assumption that the strains induced by suction s and the strains induced by mean net stresses p_{net} (or deviator stresses q) have no correlation. This is called, "independent state stress variables formulation".

The second one refers to the formulation of equations based on effective stress concept [3, 4]. These models are based on the "effective" stress concept by Terzaghi [5] for which it is stated that a single stress controls the hydraulic and mechanical behavior of soils. Governing relations formulated in terms of this type of stress are increasingly being used due to several advantages regarding the features they naturally capture for unsaturated soils [3, 4, 6].

Several constitutive models have been proposed to predict the mechanical behavior of unsaturated soils and more generally, the hydro-mechanical behavior of porous media [3, 6-9]. However, despite most of them are able to reproduce the most important features of these materials, complicated and elaborated laboratory testing procedures are required to determine the parameters contained within their relations.

In this research, an effective stress equation similar to that of Bishop [10] is analytically derived and helps to establish a complete elastoplastic constitutive framework for soils.

The main objective of this research is to propose highly coupled stress-strain-hydraulic relations for unsaturated soils to predict their mechanical behavior under cyclic loading and under any degree of saturation. Moreover, the input parameters for the constitutive model should be computed with simple soil mechanics test procedures, such as Soil-Water Characteristic Curve, saturated triaxial tests and soil classification.

2 Effective stress equation

[Rojas [11]] through an analytical analysis of a cross-section of a soil sample that contained saturated, dry and unsaturated fractions, derived Equation (1).

$$\sigma' = \sigma + \left(f^s + f^u S_r^u\right)s \quad (1)$$

[a] Corresponding author: hiramarroyo@gmail.com

A close-up to Equation (1) verifies that it is actually a Bishop-type equation with $\chi = f^s + f^u S_r^u$. Here, f^s and f^u represent the saturated and unsaturated fractions respectively; S_r^u stands for the degree of saturation of the unsaturated fraction. The mean effective stress will then be:

$$p' = \frac{\sigma_1' + 2\sigma_3'}{3} = p_{net} + \left(f^s + f^u S_r^u\right)s \quad (2)$$

Here, σ_1' and σ_3' are the effective stresses applied during a triaxial compression test on the principal directions 1 and 3. $p_{net} = p - u_a$ and $s = u_a - u_w$ stand for the mean net stress and suction, respectively.

3 The solid-porous model

The parameters contained within the parentheses in Equation (2) depend on the distribution of water inside the soil; the way in which their value is computed is described elsewhere [12, 13], for the sake of completeness we will roughly describe them herein.

A saturated fraction $f^s = V^s / V$ is the ratio of saturated volume of soil and the total volume of the soil V. A dry fraction $f^d = V^d / V$ and an unsaturated fraction $f^u = V^u / V$ are both defined in the same way. The concept of "unsaturated state" is better understood with the aid of Figure 1. Here, a solid particle is surrounded by four sites connected to four bonds, the total of them composing the BU entities. The hatched elements in Figure 1 form a Solid-Porous Unit SPU. The SPU will be part of the "saturated" fraction if all the BU entities are water-filled. If all of them are air-filled, the SPU will be part of the "dry" fraction. With this in mind, the SPU will be part of the "unsaturated" fraction if the BU entities of the SPU are a mixture of water-filled and air-filled entities.

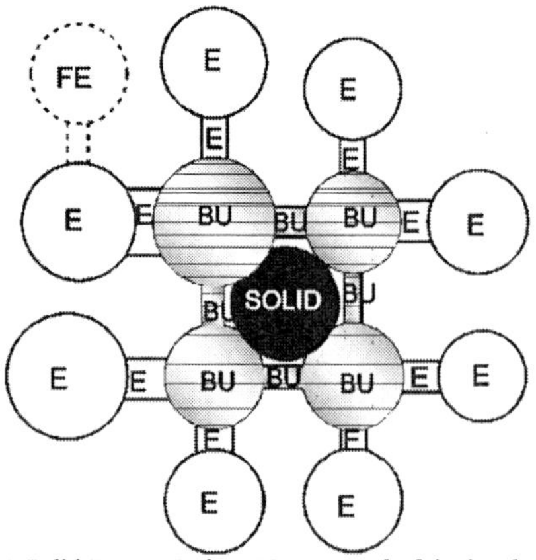

Figure 1 Solid-Porous Unit SPU composed of the hatched BU entities and the solid at the center.

In order to obtain the parameters contained within the parentheses in Equation (2), a Solid-Porous Model (SPM) is proposed. The SPM basically consist of a network of interconnected sites and bonds used to simulate wetting-drying processes. Consider a single site (circular entity) in Figure 1. If it is assumed that the whole volume of the soil is air-filled, suction must appertain a high value according to the Laplace's rule $s = 2Ts/R$. If s is increased in steps, with each Δs every site of the network is analyzed to see whether it complies with two conditions: 1) its radius must be smaller than the critical value $Rc = 2Ts/s$, and 2) it must already be connected to a bond that is already water-filled. If the two conditions are met, the site is considered saturated. The total volume of water V_w is obtained with every Δs by adding all the volumes of the saturated sites. With this procedure, it is possible to simulate a wetting curve and thus plot a Soil Water Characteristic Curve (SWCC) for the soil. Water retention curves are plots in Cartesian space where suction vs. degree of saturation is plotted. The drying process goes in the same way.

The frequency distribution of sites and bonds for the network is obtained taking as a basis its pore-size distribution PSD. The PSD can be obtained either by direct methods from which the intrusion of mercury is one of them, or by fitting the predicted SWCC to the experimental SWCC. This last method can be used to recover the PSD and has proven to be effective [14].

Figure 2 shows a representative part of a porous network composed of solids surrounded by sites. The network is currently at some particular stage of the wetting process as the water proceeds from the lower right corner site $P12$ through all the bonds and sites. This figure will help to understand and identify the meaning of the variables considered in the effective stress definition.

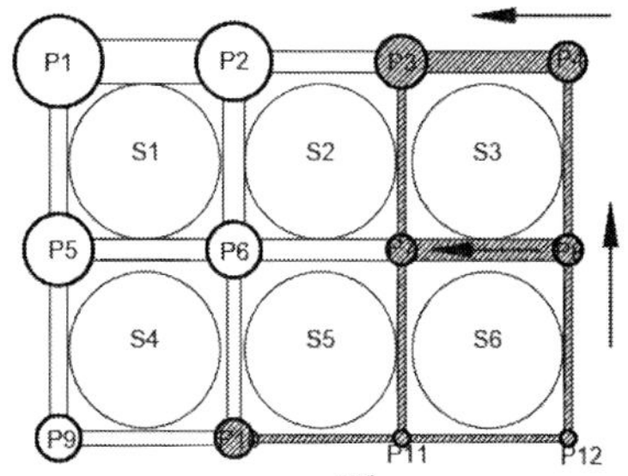

Figure 2 Sketch of the porous structure of a soil in wetting process conditions.

The void volume of the porous network consists of the volume of the sites P and the volume of bonds. Because the contribution of the bonds can be neglected [15], the total volume of the soil V will consist of the volume of solids $V_{SOL} = \sum_{i=1}^{6} S_i$ and the volume of sites $V_S = \sum_{i=1}^{12} P_i$. Accordingly to the previously established definition for the dry fraction, for the porous network in Figure 2, the dry fraction will be obtained by computing $f^d = \frac{V^d}{V} = \frac{V_{SOL}^d}{V} + \frac{V_S^d}{V} = \frac{S1}{V} + \frac{P1 + P2 + P5 + P6}{V}$. Note that, although $P9$ is air-filled, it is not included to compute the dry fraction since it is not a part of any dry or saturated solid.

For the case of the saturated fraction:

$$f^s = \frac{V^s}{V} = \frac{V^s_{SOL}}{V} + \frac{V^s_S}{V} = \frac{(S3+S6)}{V} + \frac{(P3+P4+P7+P8+P11+12)}{V}$$

Again, note that although $P10$ is water-filled, it was not included to compute the saturated fraction since it is not a part of any saturated nor any dry SPU.

The last parameter in equation (2) to be obtained is the degree of saturation of the unsaturated fraction. It is useful to apply the following relationship of void volume $V^u_S = V_S - V^s_S - V^d_S = P9 + P10$; meaning that, neglecting the bonds volume, the void volume is composed of the addition of the void volume of the unsaturated fraction, the void volume of the saturated fraction, and the void volume of the dry fraction, respectively.

Moreover, the volume of water contained within the unsaturated fraction can also be calculated from a simple difference: $V^u_w = V_w - V^s_S = P9$; meaning that the total volume of water contained within the porous structure is the addition of the water from the unsaturated fraction and the water from the saturated fraction.

Furthermore, the degree of saturation for the particular stage at which the soli-porous arrangement in Figure 2 is will be $S^u_r = V^u_w / V^u_S = P9/(P9+P10)$.

Finally, because the 100% of the soil volume is the addition of the dry, the saturated and the unsaturated fractions, the unsaturated fraction can be obtained with the difference $f^u = 1 - f^s - f^d$. Now, all the needed elements to compute the effective stress are identified.

Summing up, the previous elements enable obtaining, for every suction step Δs, the following hydraulic properties of the soil: the Water Retention Curves, the degree of saturation of the unsaturated fraction S^u_r, and the fractions f^s, f^d and f^u. All of them depend on the PSD of the material. In order to obtain an analytical equation for the PSD, we propose using a Normal Distribution. This way, parameters for the macroporosity (μ_{SM}, σ_{SM}) and for the microporosity (μ_{Sm}, σ_{Sm}) can be obtained by fitting a Normal Distribution to the Experimental porosimetry as shown in Figure 3.

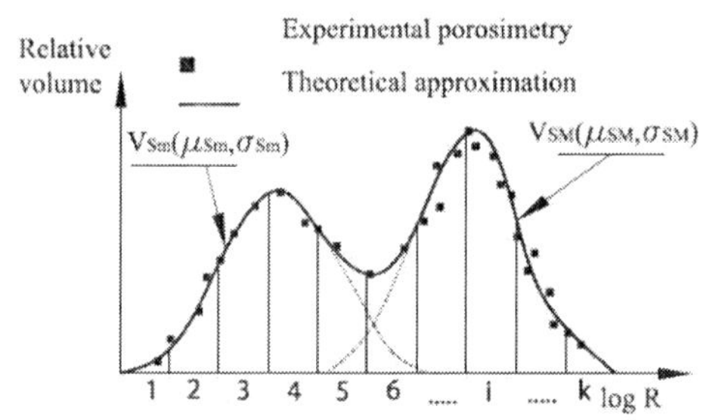

Figure 3 Experimental (black squares) and theoretical (solid line) pore-size distributions.

It has been shown experimentally that the mesopores are the ones that shrink when a soil sample is compressed. With this in mind, the following equation (3) allows obtaining the void ratio for every value that the parameters μ_{SM}, σ_{SM}, μ_{Sm} and σ_{Sm} take on. Because the basic assumption is that the volume strains are due to the shrinking of the macropores, it will be enough if the mean size of the Normal Distribution for macropores μ_{SM} is reduced.

$$e = e_M + e_m = \qquad (3)$$

$$fa \cdot \pi \sum_{i=1}^{k} \left\langle R_i^{\,2} \cdot f_i(\mu_{SM}, \sigma_{SM}) \cdot Fpv + R_i^{\,2} \cdot f_i(\mu_{Sm}, \sigma_{Sm}) \right\rangle = fa \cdot e_0$$

Note that displacing μ_{SM} towards lower values, reduce the void ratio of the material. This allows linking the mechanical behavior with the SWCC, which is one of the main issues in hydro-mechanical coupling in constitutive modelling. The SPM is completely described elsewhere [13].

4 Elastoplastic formulation

4.1 Yield locus in the $q-p'$ stress space

The mechanical part of the model is formulated under the Elastoplastic theory. The yielding space is composed of a yield locus expressed in terms of deviator stress $q = \sigma_1' - \sigma_3'$ and mean effective stress p' (see Figure 4) and a yield locus expressed in terms of the suction and the mean effective stress. An associated flow rule is assigned and an ellipse-shaped yield locus is used to identify the yielding in the stress space $q-p'$. The hardening parameter controlling the size of the yield function is p_0'. The flow rule is given by Equation (4) proposed Rojas et al. [16]:

$$d\varepsilon_v^p = (\lambda - \kappa)\frac{e}{1+e}\frac{dp_0'}{p_0'} \qquad (4)$$

Here, λ and κ are the compressibility indexes for normally consolidated and loading-unloading compression states.

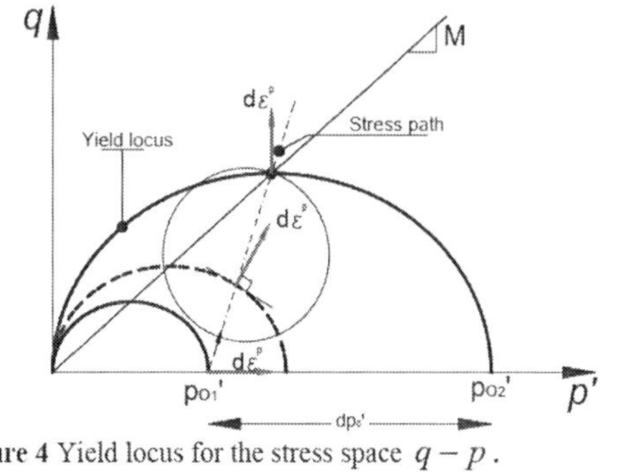

Figure 4 Yield locus for the stress space $q-p$.

The critical state theory is used herein. The plastic deformations are thus expressed with Equation (5).

$$\begin{bmatrix} d\varepsilon_v^p \\ d\varepsilon_d^p \end{bmatrix} = \qquad (5)$$

$$-\frac{1}{-M^2 p'\left(p_0'\frac{(1+e)}{e(\lambda-\kappa)}\right)}\begin{bmatrix} M^2(2p'-p_{0'}') & 2q \\ 2q & \frac{4q^2}{M^2(2p'-p_0')} \end{bmatrix}\begin{bmatrix} dp' \\ dq \end{bmatrix}$$

Here, parameter $M = 6\sin\varphi/(3-\sin\varphi)$, where φ is the friction angle of the soil; $d\varepsilon_v^p$ and $d\varepsilon_d^p$ are the plastic components of the isotropic and deviator strains respectively. The elastic part of the strains is given by

$$\begin{bmatrix} d\varepsilon_v^p \\ d\varepsilon_d^p \end{bmatrix} = \begin{bmatrix} \kappa & 0 \\ 0 & 1/3G \end{bmatrix}\begin{bmatrix} dp' \\ dq \end{bmatrix} \qquad (6)$$

Where G represents the shear modulus of the soil.

4.2 Yield locus in the $s-p'$ stress space

According to Equation (4), when suction increases from a normally consolidated stress state A to stress state B, p_0' changes as follows:

$$(dp_0')_{A-B} = \frac{1+e}{e}\frac{p'_A}{(\lambda-\kappa)}d\varepsilon_v^p = d(\chi s)_{A-B} \qquad (7)$$

Therefore, according to Equation (7) an increase in the effective preconsolidation stress p_0' is as much as $d(\chi s)$. Here, if suction goes from stress state A to B, the effective preconsolidation stress will increase up to $(p_0')_{A-B} = (p_0')_A + (dp_0')_{A-B} = (p_0')_A + 2d(\chi s)_{A-B}$. This means that the well-known Loading-Collapse Yield Surface (in reference to Alonso, Gens [1]) appertains the same shape as a drying path (see Figure 5).

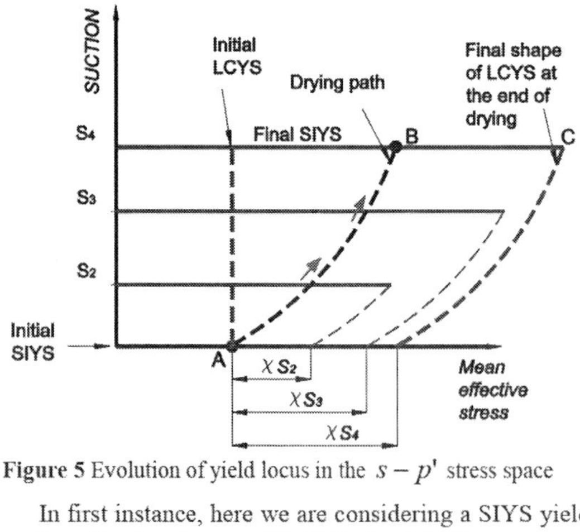

Figure 5 Evolution of yield locus in the $s-p'$ stress space

In first instance, here we are considering a SIYS yield surface (Suction Increase Yield Surface) in the same way as Alonso, Gens [1].

According to the outlined elastoplastic framework in Figure 5, at the end of an equalization stage, the soil must be subjected to an additional effective stress dp_{pnet} in order to reach elastoplastic deformation states. This preconsolidation behavior is experimentally experienced in many reported results. The main idea that need to be highlighted is that the model predicts the effective preconsolidation stress in unsaturated conditions.

5 Numerical-Experimental comparisons

The required parameters to predict the shear strength of soils with the proposed model will be the ones depicted in Table 1. It can be noticed that, the required parameters involve tests that can be obtained with classical geotechnical procedures.

Table 1. Input parameters for the proposed model

Parameter	Geotechnical test
Critical state slope M, Shearing Modulus G, preconsolidation stress $(p_0')_0$	Drained saturated triaxial test
Initial void ratio e	Soil classification
Pore-Size Distribution	Water Retention Curve
Compressibility indexes λ and κ	Isotropic compression test

The proposed model is validated herein considering the reported results of a set of triaxial compression tests equalized to $s = 300kPa$ by Futai and Almeida [17]. A set of drained triaxial compression tests where run with three different confining stresses: 100, 200 and 500. Table 2 contains the values of the required parameters to input to the model.

Table 2. Input parameters for the proposed model

Parameter	Parameter value
Critical state slope M	1.14
Preconsolidation stress in saturated conditions $(p_0')_0$	100 kPa
Shearing Modulus G	10 MPa
Initial void ratio e	1.18
Compressibility index λ	0.2
Compressibility index κ	0.05
Initial suction of the soil	300 kPa

For the hydraulic part of the model, **Table 3** contains the required parameters which arise from fitting the experimental SWCCs [17] to the ones that the model predict. The mean size and standard deviation for both the Solids 1 and the Solids 2 distribution were obtained by fitting the grain-size distribution of the soil to a pair of normal distribution.1

Table 3. Theoretical Pore-Size distribution of the soil.

	Mean size	Standard deviation
Macropores	10.0	2.5
Mesopores	0.0005	3.0
Macrobonds	8.0	4.0
Mesobonds	0.00001	13.0
Solids 1	6.0	3.5
Solids 2	0.0001	6.0

The procedure to fit the SWCCs to the model predictions is described elsewhere [13, 18].

Once the model parameters are all set up, it is possible to simulate the soil hardening due to suction increase. Initially, the sample is considered as though it where saturated and already in normally consolidated stages. In order to simulate an equalization process up to 300 kPa, a drying path must be simulated using the Solid-Porous Model. Initially, all the pores are considered in saturated conditions and suction is increased in steps. With each step, the porous network is all scanned to verify which pores are dry and which ones are saturated; the terms f^s, f^d, f^u, S_r^u and S_r are all computed with each step.

Once $s = 300kPa$ has been reached, the model predicts that the term $\chi = f^s + f^u S_r^u = 0.63$. Hence, according to Equation (7), the soil effective stress increases as much as $(dp')_{0\to300} = d(\chi s)_{0\to300} = (0.63)0.3 = 0.19$. Therefore, the mean effective preconsolidation stress at 300 kPa will thus be $(p_0')_0 + 2(dp_0')_{0\to300} = 0.1 + 2(0.19) = 0.48\ kPa$, and the mean net preconsolidation stress will thus be $(p_0')_0 + (dp_0')_{0\to300} = 0.1 + (0.19) = 0.29$.

Now, the samples subjected to confining net stresses of 100 and 200 kPa will be in preconsolidated states; a sample subjected to 500 kPa will be in normally consolidated states. The theoretical-experimental result comparisons for the triaxial compression state of these three samples are depicted Figures 4 to 6.

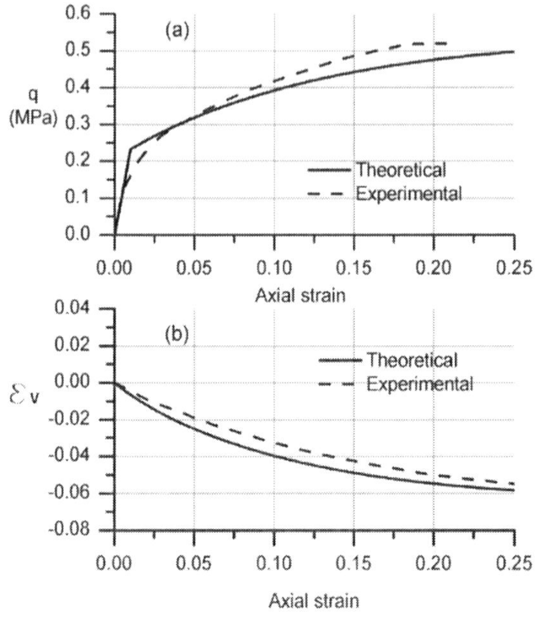

Figure 6 Theoretical-Experimental result comparisons of a soil subjected to a confining net stress p_{net} of 100 kPa and suction of 300 kPa.

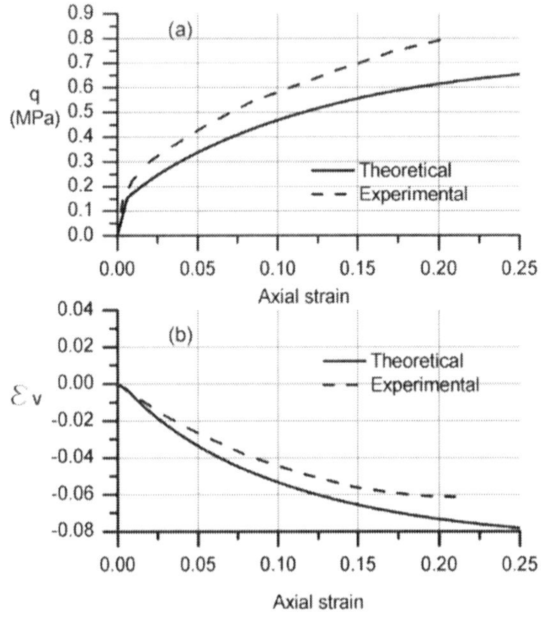

Figure 7 Theoretical-Experimental result comparisons of a soil subjected to a confining net stress p_{net} of 200 kPa and suction of 300 kPa.

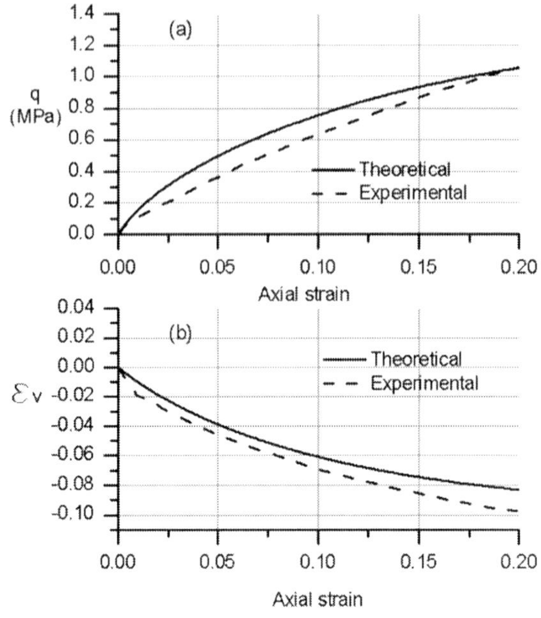

Figure 8 Theoretical-Experimental result comparisons of a soil subjected to a confining net stress p_{net} of 500 kPa and suction of 300 kPa.

Finally, the whole wetting-drying cycle for the expression $\chi = f^s + f^u S_r^u$ expressed in terms of the degree of saturation is plotted in Figure 9. From these results, it can be concluded, as reported by several cyclic suction-controlled triaxial tests by Khalili and

Zargarbashi [19], that despite the assumption $\chi = S_r$ provides a first approximation to Bishop's parameter, it is not accurate and will deliver incorrect results. This is far more important when dealing with cyclic loading such as when modeling wetting-drying cycles for slope stability analyses.

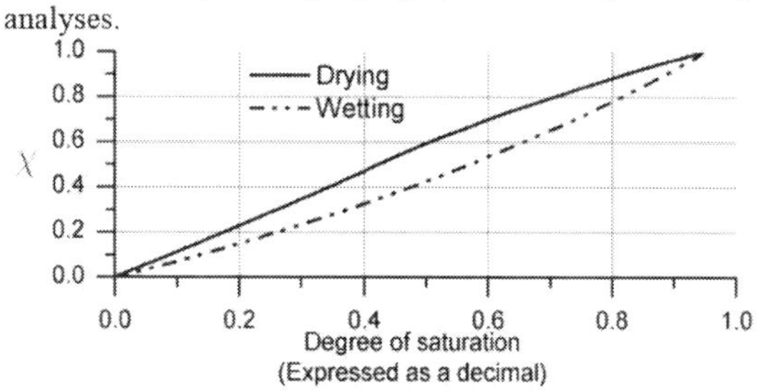

Figure 9 Effective stress parameter $\chi = f^s + f^u S_r^u$ for wetting-drying cycles.

6 Results and discussion

An elastoplastic model to predict the hardening of a soil under suction increments has been established. The Loading-Collapse yield function in the suction-mean effective stress is derived by considering that the hardening of the soil, equals the contribution to the effective stress due to suction increase. Because the effective stress equation depends on the SWCC through the porous model, the shape of the LC curve will depend on the shape of the SWCC.

The hydraulic and mechanical behaviors are highly coupled. A first approximation to a coupled formulation is achieved in a natural way by using effective stresses on the model formulations where the parameters to obtain such stresses depend on the SWCC. The second approximation to a constitutive model fully coupled model is establishing an equation for the SWCC that considers volume strains. Here, if the soil suffers any volume strain, the hydraulic consequences are that the SWCC displaces towards higher suctions.

The theoretical-experimental result comparisons show that the model is able of predicting the volume strains and axial strains of an unsaturated soil. The model parameters are determined with standard geotechnical tests such as drained triaxial compression tests and isotropic compression states, both in saturated conditions.

References

[1] E.E. Alonso, A. Gens, A. Josa, A constitutive model for partially saturated soils, Géotechnique 40 (1990) 405-430.

[2] K.H. Roscoe, J.B. Burland, On the generalized stress-strain behavior of 'wet' clay., in: (Eds.), Engineering Plasticity, Cambridge University Press, Cambridge 1968.

[3] M. Nuth, L. Laloui, Effective stress concept in unsaturated soils: Clarification and validation of a unified framework, Int. J. Numer. Anal. Met. 32 (2008) 771-801.

[4] L. Laloui, M. Nuth, On the use of the generalised effective stress in the constitutive modelling of unsaturated soils, Comput. Geotech. 36 (2009) 20-23.

[5] K. Terzaghi, Erdbaumechanik auf Bodenphysikalischer Grundlage, Franz Deuticke, Leipzig and Vienna, 1925.

[6] D. Sheng, Constitutive modelling of unsaturated soils: discussion of fundamental principles. , in: General Report presented at the 5th International Conference on Unsaturated Soils (6-8 Sept 2010, Barcelona), In: Unsaturated Soils, (EE Alonso & A Gens, eds), (2010).

[7] B. Loret, N. Khalili, A three-phase model for unsaturated soils, Int. J. Numer. Anal. Met. 24 (2000) 893-927.

[8] Q. Wang, D.E. Pufahl, D.G. Fredlund, A study of critical state on an unsaturated silty soil, Can. Geotech. J. 39 (2002) 213-218.

[9] Y.P. Yao, L. Niu, W.J. Cui, Unified hardening (UH) model for overconsolidated unsaturated soils, Can. Geotech. J. 51 (2014) 810-821.

[10] A.W. Bishop, The principle of effective stress, Teknisk Ukeblad 106 (1959) 859-863.

[11] E. Rojas, Equivalent Stress Equation for Unsaturated Soils. I: Equivalent Stress, International Journal of Geomechanics 8 (2008) 285-290.

[12] E. Rojas, J. Horta, T. López-Lara, J.B. Hernández, A probabilistic solid porous model to determine the shear strength of unsaturated soils, Probabilistic Eng. Mech. 26 (2011).

[13] H. Arroyo, E. Rojas, M.L. Pérez-Rea, J. Horta, J. Arroyo, A porous model to simulate the evolution of the soil–water characteristic curve with volumetric strains, C. R. Mecanique 343 (2015) 264-274.

[14] E. Rojas, M.L. Pérez-Rea, G. Gallegos, J. Leal, A porous model for the interpretation of mercury porosimetry tests, J. Porous Media 15 (2012) 517-530.

[15] J. Horta, E. Rojas, M.L. Pérez-Rea, T. López, J.B. Zaragoza, A random soild-porous model to simulate the retention curves of soils, Int. J. Numer. Anal. Met. 37 (2013).

[16] E. Rojas, M.L. Pérez-Rea, T. López-Lara, J.B. Hernández, J. Horta, Use of Effective Stresses to Model the Collapse upon, J. Geotech. Geoenviron. (2015).

[17] M.M. Futai, M.S.S. Almeida, An experimental investigation of the mechanical behaviour of an unsaturated gneiss residual soil,, Géotechnique 55 (2005) 201-213.

[18] H. Arroyo, E. Rojas, M.L. Pérez-Rea, J. Horta, J. Arroyo, Simulation of the shear strength for unsaturated soils, C. R. Mecanique 341 (2013) 727-742.

[19] N. Khalili, S. Zargarbashi, Influence of hydraulic hysteresis on effective stress in unsaturated soils, Géotechnique 60 (2010) 729-734.

Modelling of wetting tests for a natural pyroclastic soil

Mariagiovanna Moscariello[1], Sabatino Cuomo[1,a], Diego Manzanal[2], Vito Foresta and Manuel Pastor[2]

[1] *University of Salerno, Department of Civil Engineering Laboratory, of Geotechnics, Italy*

[2] *Universidad Politecnica de Madrid, Department of Applied Mathematics and Computer Science, Spain*

Abstract. The so-called wetting-induced collapse is one of the most common problems associated with unsaturated soils. This paper applies the Modified Pastor-Zienkiewicz model (MPZ) to analyse the wetting behaviour of undisturbed specimens of an unsaturated air-fall volcanic (pyroclastic) soil originated from the explosive activity of the Somma-Vesuvius volcano (Southern Italy). Both standard oedometric tests, suction-controlled oedometeric tests and suction-controlled isotropic tests are considered. The results of the constitutive modelling show a satisfactory capability of the MPZ to simulate the variations of soil void ratio upon wetting, with negligible differences among the measured and the computed values.

1. Introduction

The soils which exhibit a reduction in volume upon wetting at a constant total stress are called collapsible soils [1]. These soils generally experience three distinct stages of deformation during the wetting process: i) at high suction, a large decrease in suction generates a small reduction of soil volume; ii) at intermediate suction, soil wetting produces a significant volumetric reduction; iii) at low suction, further reductions of suction do not generate additional volume reduction. Collapsible soils are widespread in many parts of the world and they cause significant engineering problems, which range from a poor performance of foundations to the triggering of rainfall-induced landslides.

Soil collapse is influenced by several factors, i.e. the initial void ratio, the confining pressure, the initial matric suction. For this reason, a comprehensive investigation of the soil collapse behaviour should be performed through different types testing procedures, such as oedometer and triaxial tests [2].

The quantitative simulation of the soil collapse requires the use of advanced constitutive models, capable to deal with the hydro-mechanical behaviour of unsaturated soils. It is worth mentioning the Basic Barcelona Model [3] and the Wheeler-Sivakumar model [4], both based on the Classic Plasticity Theory [5]. The Generalized Plasticity-based model, proposed by Pastor et al. [6], represents an efficient alternative approach. The main advantage of Generalized Plasticity models is the ability to reproduce the stress strain behaviour of several types of soil with a good accuracy under both monotonic and cyclic loads, while not requiring the explicit definitions of yield and plastic potential surfaces [7]. Some modifications were later introduced to improve the capability of the original PZ model, by introducing a state parameter to incorporate the dependence of the flow rule and the plastic modulus on void ratio and confining pressure [8, 9, 10, 11].

In this paper, the Modified Pastor-Zienkiewicz constitutive model for unsaturated soils (MPZ) [11] is used to simulate the collapse behaviour of undisturbed specimens of a natural unsaturated air-fall volcanic (pyroclastic) soil of Southern Italy. The collapse is analyzed through three series of wetting tests and the predictive capability of the model is investigated.

2. Material

The paper deals with an unsaturated air-fall volcanic (pyroclastic) soil of Southern Italy, originated from the explosive activity of the Somma-Vesuvius volcano [12]. In particular, the pyroclastic deposits mantling the Pizzo d'Alvano massif have been extensively tested after the huge flowslides occurred in May 1998, which damaged four towns at the toe of the massif [12]. The laboratory tests selected for the calibration and validation of the constitutive MPZ model were carried out on undisturbed specimens, sampled in two tests areas, Sarno and Quindici, two small cities both damaged in 1998 [12].

The tested soil is classified as "Class A ashy soil" by Bilotta et al. [13], and is the class of the finest ashes present at this study area. The grain size distribution consists in 1.1 to 6.4% Gravel, 40.6 to 51.3% Sand, 40.9 to 53.61% Silt, 1.4 to 4.7% Clay [14]. The soils is characterized by very low dry unit weight (γ_d) and high void ratio (e). In particular, γ_d ranges from 6.93 to 8.65 kN/m^3, while e ranges from 2.595 to 1.982. In addition, the soil has a metastable behaviour in triaxial tests [14].

[a] Corresponding author: scuomo@unisa.it

DOI: 10.1051/e3sconf/20160917007

The Water Retention Curve (WRC), which relates the degree of saturation (S_r) to the matric suction, is available from previous experiments performed through both Volume-Extractor and Suction-Controlled-Oedometer, (Fig. 1) on specimens sampled at Sarno site and characterized by different void ratios [13]. In particular, the WRC is well described by the Van Genuchten model, with parameters α=0.19÷1.02, n=1.19÷1.79, and m=0.16÷0.44 [13].

The soil mechanical behavior was formerly investigated both in saturated and unsaturated conditions, as well as during the transition from unsaturated to saturated condition. In this paper, some literature tests will be used, whose main results are hereafter described, while specific details are provided in the references quoted below.

Soil collapsibility was investigated by Bilotta et al. [14] using different devices and testing procedures. In particular, suction-controlled wetting triaxial (ISO and Triaxial) and oedometric (SCED) tests were performed starting from an initial suction of 50 kPa. In addition, wetting tests were perfomed using the standard oedometer (ED) on specimens at natural water content corresponding to about 50 kPa. The stress paths performed were oedometric (k_0), isotropic (η=0) and triaxial (0.73<η<1), while the suction reduced in two manners: either gradually increasing the pore water pressure at the bottom of the specimens (test type: SCED, Triaxial and ISO) or suddenly flooding the specimen with distilled water (tests type: ED).

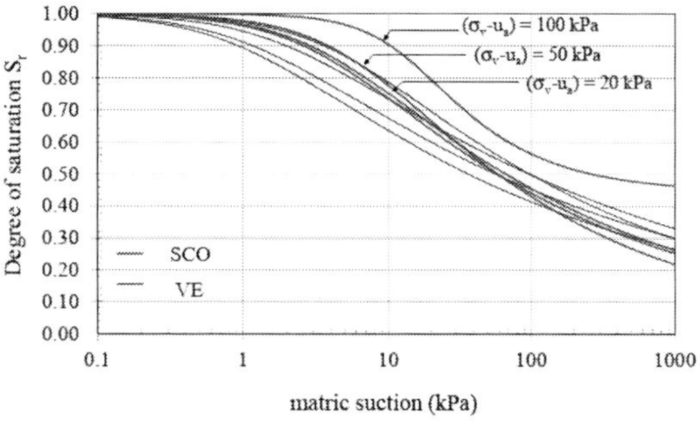

Figure 1. Water Retention Curve (WRC) of the tested material; SCED: Suction Controlled Oedometer and VE: Volume Extractor (modified from Bilotta et al. [13].

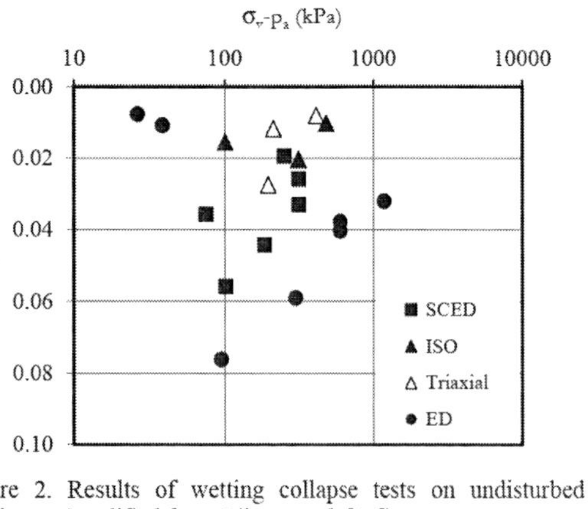

Figure 2. Results of wetting collapse tests on undisturbed specimens (modified from Bilotta et al. [14]).

Table 1. Oedometeric and isotropic tests performed on soil specimens (Sarno site) at initial suction (p_a-p_w) of 50 kPa, and at a fixed value of net vertical stress (σ_v–p_a), while the suction was decreased until zero.

Test type	Test	e_0	e_w	$(\sigma_v - p_a)$ (kPa)	ε_v
	ESA13_03	2.537	2.2295	95	0.068
SCED	ESA9_03	1.992	1.860	295	0.044
	ESA10_04	1.902	1.827	500	0.026
	USP6_06	2.217	2.167	100	0.016
ISO	USP1_07	1.889	1.830	311	0.020
	USP3_06	1.762	1.734	477	0.010
	ESL7_03	2.513	2.245	95	0.076
ED	ESL9_03	2.245	2.053	297	0.059
	ESL11_03	1.881	1.765	594	0.040

e_0 initial void ratio; e_w void ratio after wetting; $(\sigma_v - p_a)$ net vertical stress; ε_v volumetric strain after wetting.

Figure 2 reports the experimental results. The magnitude of soil collapse was strongly influenced by the initial void ratio, the stress level at the beginning of the wetting stage and the type of test performed, which influence in turn the stress path followed. This complex mechanical behaviour requires specific constitutive models to be quantitatively reproduced.

The paper focuses on a series of 9 tests performed on undisturbed specimens and carried out through standard oedometer, suction-controlled oedometer and suction-controlled triaxial apparatus. Three vertical net effective stresses were selected (100, 300 and 500 kPa), and the principal details of the tests are reported in Tab. 1.

3. Theoretical framework

The Generalized Plasticity Theory, introduced by Pastor and Zienkiewcz [6], is particularly suitable to describe the behaviour of either loose or dense granular soils, both in drained and undrained conditions, even along complex stress paths. The original PZ model was fully defined through: i) the elastic constitutive tensor; ii) three directions (the loading direction and the plastic flow directions); iii) two scalars (the plastic moduli). This original framework was extended to unsaturated sands subjected to large variations of both relative density (D_r) and confining pressure (p') [11]. Later, the MPZ model introduced a state parameter to reproduce the behaviour of a given sand with different densities or subjected to different confining pressures with a unique set of constitutive parameters [15, 16]. Particularly, the main modifications concerned: i) the dependency of dilatancy on soil relative density and confining pressure; ii) the yield surface associated with state parameter existing in an explicit or implicit form; iii) the dependence of isotropic plastic modulus on relative density variation.

The state parameter (ψ) represents the relationship between relative density, void ratio and confining pressure. In other words, the state parameter describes the relative position on the e-p' plane of the current state and the projection of the CSL. Therefore, the description of CSL is much important. The MPZ model considers the

formulation proposed by Li [17] , Eq. 1, where e_Γ is the void ratio at atmospheric pressure (p_{atm}), λ is the solpe of the CSL in e-$(p'/p_{atm})^\xi$ plane, e and p' are the void ratio and the confining pressure at critical state. The flow rule depends on the relative density and, therefore, on the state parameter [7]. The dilatancy equation (Eq. 2) used in the model is that proposed by Li and Dafalias [18] where M_g is the slope of the Critical State Line in the q-p' plane, η is the ratio between the mean effective stress p' and the deviatoric stress q, d_0 and m are material parameter of dilatancy law.
The parameter M_f (Eq. 3) is a function of the model constants h_1 and h_2, the parameter M_g and the ratio between the current void ratio e_0 and the void ratio at critical state e_{crit}.
The plastic modulus (H) is a function of the state parameter and the confining pressures (Eq. 4), where H_0 and β'_0 are constitutive parameters. H_{DM} is a function which incorporates the material memory into the plastic modulus equation.
Finally, the elastic shear modulus and bulk modulus are assumed depending on confining pressure and void ratio, using the relation proposed by Richart et al. [19].

$$e_{crit} = e_\Gamma + \lambda\left(\frac{p}{p_{atm}}\right)^{\xi_c} \quad (1)$$

$$d = \frac{d_0}{M_g}\left(M_g \exp^{m\psi} - \eta\right) \quad (2)$$

$$M_f = M_g\left[h_1 - h_2\left(\frac{e_0}{e_{crit}}\right)\right] \quad (3)$$

$$H = H'_0 \exp(-\beta'_0 \psi_q)\cdot\sqrt{p' \cdot p'_{atm}}\cdot H_{DM} \quad (4)$$

The Generalized Plasticity state parameter-based model for unsaturated soils is formulated in terms of effective stress tensor as proposed by Bishop [20]:

$$\sigma'_{ij} = \sigma_{ij} - p_a \cdot \delta_{ij} + S_{re}\cdot(p_a - p_w)\cdot\delta_{ij} \quad (5)$$

where σ_{ij} is the total stress tensor, p_a is the pore air pressure, p_w is the pore water pressure, p_a - p_w is the matrix suction (s), S_{re} is the relative degree of saturation and δ_{ij} is the Kronecker delta. In the triaxial plane, the Bishop equation is equal to:

$$p' = \bar{p} + \chi \cdot s \quad (6)$$

where $\bar{p}$ is the net stress and s is the suction.

In the model, the bonding parameter is introduced to take in account the effects of the suction due to capillary forces. The Critical State Line is defined starting from the equation proposed by Li and Wang [21], and is expressed by Eq. 8, where e_Γ, λ and ζ_c are the same model parameters of the saturated MPZ model, while ξ is the bonding parameter. ξ is the product of two factor: the saturation degree of air (1-S_r), which represents the number of water menisci per unit volume of solid fraction, and the function of suction f(s), which accounts for the effect of the suction on the stabilising inter-particle force exerted by a single meniscus. $g(\xi)$ is similar to a function proposed by Gallipoli et al. [22], which depends on the degree of saturation and suction (Eq. 8). Therefore, bounding parameter and the function g(ξ) define the dependence of CSL on matric suction.
The CSL in the $q - p'$ plane depends on the relative degree of saturation S_{re} (Eq. 10), which accounts for the residual degree of saturation (S_{r0}) related to the adsorbed water.
The Water Retention Curve (WRC) is considered in the constitutive model in order to simulate the hydro-mechanical behaviour of unsaturated soils. In the model, the WRC is described through a modified version of the equation proposed by Fredlund and Xing [23]. In the equation (Eq. 12), the suction is replaced by the normalized suction (Eq. 12) that accounts for the void ratio dependency. It entails:

$$e_c = e_\Gamma - \lambda\left(\frac{p'}{p_{atm}}\right)^{\zeta_c}(1+g(\xi))^{-\zeta_c} \quad (7)$$

$$g(\xi) = a[\exp(b\xi) - 1] \quad (8)$$

$$p'_{CSunsat} = p'_{CSsat}\cdot \exp[g(\xi)] \quad (9)$$

$$S_{re} = \frac{S_r - S_{r0}}{1 - S_{r0}} \quad (10)$$

$$S_r = S_{r0} + (1 - S_{r0})\left\{\ln\left[\exp(1) + \left(\frac{s^*}{a_w \cdot p_{atm}}\right)^n\right]\right\}^{-m} \quad (11)$$

$$s^* = e^{\Omega}\cdot s \quad (12)$$

where s^* is the normalized suction, defined as a function of the void ratio (e), the suction (s) and a parameter of the model (Ω).

Globally, it is necessary to evaluate some additional parameters to predict the unsaturated behaviour: 3 parameter account the bonding due to suction (a, b and c) and 5 or 9 parameter for accounting the water retention curve without and with hysteretic loop (S_{r0}, Ω, a_w, a_d, n_w, n_d, m_w, m_d, β_w). These parameters can be obtained from the results of saturated and unsaturated laboratory tests.

4. Constitutive modelling

4.1. Calibration

The MPZ model is calibrated using the experimental

results of isotropic and triaxial tests both on saturated and unsaturated soils, reported by Migliaro [24] and Lancellotta et al. [25]. Particularly, the constitutive parameters are respectively related to soil behaviour in: i) saturated conditions, such as for elasticity (G_0 and K_0), Critical State (M_g, e_Γ, λ and ζ_c), plastic flow (h_1, h_2, d_0 and m), and plastic modulus (H_0, β'_0, H_{v0}, and β_v), and ii) unsaturated conditions, such as bonding (a, b and c), and water retention curve (S_{r0}, Ω, a_w, a_d, n_w, n_d, m_w, m_d and β_w). The procedures adopted for the calibration of the constitutive parameters are explained.

The shear modulus G_{eso} is obtained by fitting the initial slope in the q-ε_q plane (ε_q is the distortional strain). The bulk modulus K_{evo} is adjusted by fitting the initial slope of p'-ε_v curves obtained through drained triaxial tests carried out at initial confining pressures ranged between 50 kPa and 100 kPa and initial void ratios between 2.007 and 2.385 [24].

The parameters related to Critical State (M_g, e_Γ, λ and ζ_c) are calibrated through triaxial tests performed on saturated soils and reported by Migliaro [24]. The tests show a hardening behaviour and do not reach any critical state, and it is difficult defining the CSL both in the q-p' and e-p' planes. Therefore, the CSL is drawn extrapolating the maximum deviatoric stress at a 40% ε_q. The M_g is calculated from the plot that relates the maximum (or critical) deviatoric stress to the corresponding mean effective stress of the drained triaxial tests. The M_g value is the maximum slope of the line that passed through the origin and better adjusted the experimental values.

The parameters e_Γ, λ and ζ_c are calibrated through the graph in the e-$(p'/p_{atm})^{\zeta c}$ plane that relates void ratio and the corresponding mean effective stress at critical state of drained triaxial tests [24]. ζ_c was chosen equal to 0.7 and then it is turned as proposed by Li and Wang [21].

The plastic flow parameters h_1 and h_2 are calibrated from the q-p' curves of undrained triaxial tests on loose saturated specimen carried out at initial mean effective pressure ranges between 50 kPa and 100 kPa; this is done on the basis that the M_f/M_g ratio approaches to 1.0 when the ratio e/e_{CS} reaches its lower limit.

The parameters related to dilatancy (m, d_0) are evaluated through the drained triaxial tests already adopted for the calibration of K_{evo}. The parameter m was determined from the eq. 2, at the phase transformation point (between contractive and dilative behavior), where the dilatancy is nil. The state parameter and the stress ratio, which appear in the eq. 2 must be previously found taking into account the experimental data available in [24] and [25].

The initial tangent slope of the volume strain-shear strain curve (d_0) is calibrated through drained triaxial compression, supposing that the elastic deformations were small and negligible, so the total strains were equal to the plastic strains. d_0 was calibrated through the method of the least square, fitting the experimental curve d_{exp}-ε_a and the model curve d_{mod}-ε_a.

The constant H_{v0}, H_0, and β'_v, β'_0 and r_0 – which measures the distance between the CSL and ISL (Isotropic State Line) in a saturated condition – were estimated by trial and error [26, 15, 16] in saturated condition. The six constant parameters (a_w, a_d, n_w, n_d, m_w, m_d) associated to the main drying and wetting curves are evaluate through the regression method proposed by Frendlund and Xing [23]. The WRC adopted for the constitutive modelling is labelled as SCO_3W[50] by Bilotta et al. [13]. This curve was obtained at net vertical stress of 50 kPa through the suction-controlled oedometer. The residual saturation degree (S_{r0}) can be evaluated from the S_r-s^* curve. The parameter Ω is calibrated to obtain a unique S_r-s^* curve for specimen with different void ratio.

The bounding parameters a and b are calibrated through the void ratio of unsaturated and saturated soils at critical state. The calibration is performed on triaxial tests on saturated and unsaturated specimens. The experimental data available [24, 25] show that it is not possible to outline a clear direct relationship between the existence and location of the CSL in unsaturated conditions and the soil suction. The experimental data of saturated condition instead are the same adopted to calibrate the critical state parameters. The values of void ratio at critical state are computed through the procedure proposed by Gallipoli et al. [22]. The slope and the intercept of the CSL in e-ln(p') plane are defined through experimental data on saturated specimens. Then, the net stresses, the saturation degrees and the suctions at critical state are obtained through the triaxial tests performed on unsaturated specimens. These experimental values are adopted to calculate the bounding parameter ξ and then g(ξ) at fixed values of p'. The values of g(ξ) are compared with the ratio e_s/e obtained from experimental data. The parameter c is calibrated through isotropic triaxial tests carried out at constant suction.

The parameters adopted for the constitutive modelling are partly taken from a previous calibration of MPZ made for saturated and unsaturated tests on pyroclastic soils [27, 28] and they are summarized in Tab. 2. The calibration of the MPZ model is carried out on three suction-controlled oedometric wetting tests performed at different net vertical stresses. The parameters are individuated through the procedures explained and then they are adjusted through a trial-and-error procedure. A unique set of parameters is capable to fit the volumetric strains experienced during the three tests (Fig. 3).

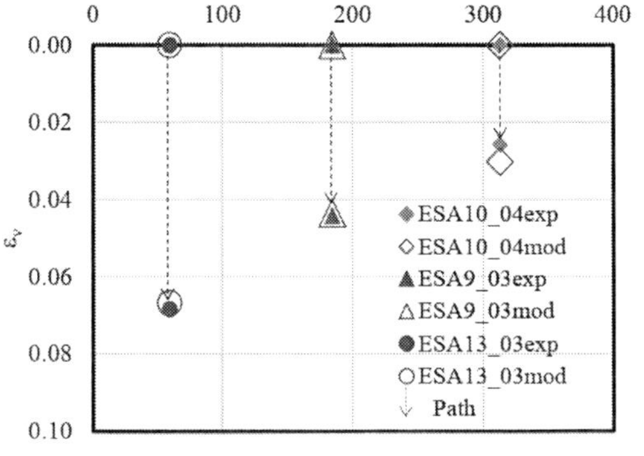

Figure 3. Experimental data (solid) and numerical results (empty) of the three suction-controlled oedometric wetting tests adopted for the calibration of MPZ model.

Table 2. Constitutive parameters calibrated for the MPZ model.

Elastic		Critical State				Bonding		
G_0	K_0	M_g	e_Γ	λ	ζ_c	a	b	c
1286	1746	1.55	2.3	0.27	0.8	1.29	0.16	0.7
Plastic modulus					Plastic flow			
H_0	β'_0	H_{v0}	β_v	r_0	h_1	h_2	d_0	m
30	1.1	20	2	1.44	1.0	0.35	0.8	2.5
Water Retention Curve								
S_{r0}	Ω	a_w	a_d	n_w	n_d	m_w	m_d	β_w
0.429	6.6	0.078	0.078	0.91	0.91	0.86	0.86	2

4.2. Validation and model performances

Isotropic and oedometric wetting tests were simulated (Fig. 4), referring to one single set of constitutive parameters reported in Tab. 2. Three different values of net vertical stress are chosen: low (100 kPa), intermediate (300 kPa) and high (500 kPa).

The tests at the same net vertical stress performed using different apparatus exhibit different volumetric strains. The isotropic wetting tests exhibit low values of volumetric strain, while the standard oedometric wetting tests experience high values of volumetric strain (solid markers in Fig. 4b). The reduction of void ratio during the wetting stage is also influenced by the stress path performed (Fig. 4a). The isotropic wetting tests experience the minimum variation of the void ratio. The manner of suction reduction influenced the magnitude of volumetric strain, while the variation of the void ratio is not influenced by the manner of the reduction of suction. The gradual reduction of the suction originated smaller volumetric strains than the sudden flooding. The net vertical stress influenced the volumetric strain, i.e. the specimens undergone at high net vertical stress exhibit the minimum variation of void ratio and the minimum volumetric strain.

The general agreement between the experimental results and the constitutive modelling is discussed with reference to ε_v-(p-p_a) and e-(p-p_a) planes. The best numerical simulations in the two planes are obtained for the standard oedomeric tests. In terms of void ratio, the differences among measured and computed values are evaluated in absolute and percentage values (Tab. 3).

Table 3. Comparison among experimental and numerical physical variables of the tests simulated. All the variables are computed at the collapse.

Test	e_{exp}	e_{mod}	$\lvert e_{mod}-e_{exp}\rvert$	$\lvert e_{mod}-e_{exp}\rvert / e_{exp}$
	-	-	-	(%)
ESA13_03	2.295	2.307	0.012	0.518
ESA9_03	1.860	1.865	0.005	0.242
ESA10_04	1.827	1.815	0.012	0.638
ESL7_03	2.245	2.249	0.004	0.189
ESL9_03	2.053	2.057	0.004	0.204
ESL11_03	1.765	1.766	0.001	0.394
USP6_06	2.167	2.158	0.009	0.395
USP1_07	1.831	1.849	0.018	0.985
USP3_06	1.734	1.738	0.005	0.264

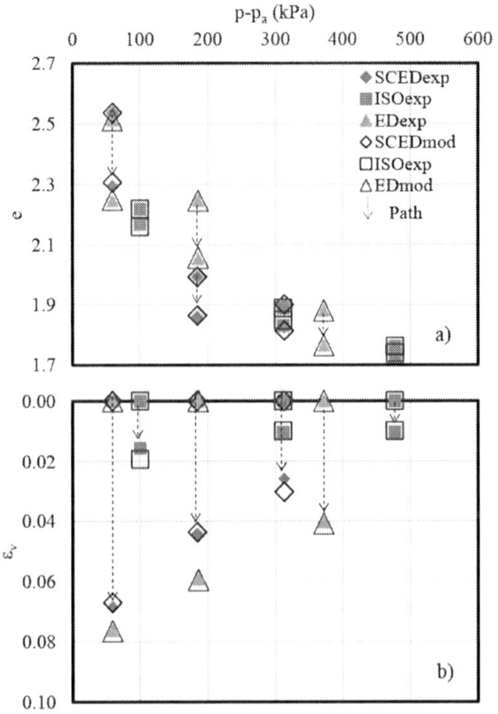

Figure 4. Comparison between experimental and numerical results in a) ε_v-(p-p_a) plane and b) e-(p-p_a) plane.

The minimum difference was equal to 0.039%, the maximum difference was 0.985% and the average difference was 0.386%. Therefore, the volumetric behaviour of the unsaturated pyroclastic soil in the wetting tests was excellently simulated by the MPZ model.

5. Conclusions

The paper dealt with the constitutive modelling of the mechanical behaviour of an unsaturated air-fall volcanic (pyroclastic) soil, originated by the explosive eruptions of the Vesuvius volcano in southern Italy. The wetting-induced collapse is analyzed through different type of tests: oedometric and isotropic tests. The capability of the constitutive model is evaluated with reference to the soil volumetric strain and the variation of void ratio.

It is highlighted that the MPZ model is capable to fit very well the void ratios and volumetric strains measured in the wetting tests. It is worth noting that this satisfactory simulation was performed by using a unique set of constitutive parameters. This peculiar feature of the MPZ model enhances the possibilities to reproduce realistic stress paths experienced by soil in boundary value problems.

Future researches should focus on the constitutive modelling of wetting tests in triaxial conditions along stress paths with non-zero deviatoric stresses.

References

1. Pereira, J. H., & Fredlund, D. G.. Volume change behavior of collapsible compacted gneiss soil. Journal of geotechnical and geoenvironmental engineering, 126(10), 907-916, (2000).
2. Vilar, O. M., and Davies G. I.. Collapse behavior analysis of a clayey sand using different testing procedures. Unsaturated Soils 2: 571-576, (2002).
3. Alonso, E.E., Gens, A., Josa, A. 1990. A constitutive model for partially saturated soils. Géotechnique, 40 No. 3: 405-430.
4. Wheeler, S. J., Sivakumar, V. An elasto-plastic critical state framework for unsaturated soil. Géotechnique, 45, No.1: 35-53, (1995).
5. Hill, R.. A variational principle of maximum plastic work in classical plasticity. The Quarterly Journal of Mechanics and Applied Mathematics, 1(1), 18-28, (1948).
6. Pastor, M., Zienkiewicz, O.C. and Chan, A.H.C. 1990. Generalized plasticity and the modelling of soil behaviour. International Journal for Numerical and Analytical Methods in Geomechanics, 14, 151-190, (1990).
7. Manzanal, D., Merodo, J.A., Pastor, M. 2011. Generalized plasticity state parameter-based model for saturated and un-saturated soils. Part I: Saturated state. International Jour-nal for Numerical and Analytical Methods in Geomechan-ics, 35, Issue 12, 1347–1362, (2011).
8. Bahda, F., Pastor, M., & Saitta, A.. A double hardening model based on generalized plasticity and state parameters for cyclic loading of sands. Numerical models in Geomechanics. The Netherlands, Rotterdam, 33-38 (1997).
9. Tonni, L., Gottardi, G., Simonini, P., Pastor, M., Mira, P.Use of Generalized Plasticity to describe the behaviour of a wide class of non-active natural soils. 3rd International Symposium on Deformation Characteristics of Geomaterials. Swets & Zeiglinger Lisse, (2003).
10. Ling, H. I., and Yang, S.. Unified sand model based on the critical state and generalized plasticity. Journal of Engineering Mechanics (2006).
11. Manzanal, D., Fernández Merodo, J. A., & Pastor, M.. Generalized plasticity theory revisited: new advances and applications. 17° European Young Geotechnical Engineers Conference. (2006).
12. Cascini, L., Cuomo, S. and Guida, D.. Typical source areas of May 1998 flow-like mass movements in the Campania region, Southern Italy. Engineering Geology, 96, 107-125 (2008).
13. Bilotta, E., Cascini, L., Foresta, V., and Sorbino, G.. Geotechnical characterization of pyroclastic soils involved in huge flowslides. Geotechnical and Geological Engineering, 23: 365-402, (2005).
14. Bilotta, E., Foresta, V., Migliaro, G.. The influence of suction on stiffness, viscosity and collapse of some volcanic ashy soils. In Unsaturated Soils: Advances in Geo-Engineering, Toll et al. eds, (2008).
15. Manzanal, D.. Modelo constitutivo basado en la teorìa de la Plasticidad Generalizada con la incorporaciòn de paràmetros de estado para arenas saturadas y no saturadas, PhD thesis, Universidad Politecnica de Madrid, Spain, (2008).
16. Manzanal, D., Pastor, M., Merodo, J.A.F.. Generalized plasticity state parameter-based model for saturated and unsaturated soils. Part II: Unsaturated soil modeling. International Journal for Numerical and Analytical Methods in Geomechanics, 35 (18), 1899-1917, (2011).
17. Li, X. S. Modeling of dilative shear failure. Journal of geotechnical and geoenvironmental engineering 123.7: 609-616, (1997).
18. Li X.S., Dafalias Y.F.. Dilatancy for cohesionless soils. Géotechnique, 50, No.4, 449 -460, (2000).
19. Richart, F. E., Hall, J. R., Woods, R. D.. Vibrations of soils and foundations, Prentice-Hall (1970).
20. Bishop AW. The principle of effective stress. Tek Ukeblas; 39:859–863 (1959).
21. Li, X.S., Wang, Y.. Linear representation of steady-state line for sand. Journal of Geotechnical and Geoenviromental Engineering, 124, No. 12, 1215-1217, (1998).
22. Gallipoli, D., Gens, A., Sharma, R., & Vaunat, J.. An elasto-plastic model for unsaturated soil incorporating the effects of suction and degree of saturation on mechanical behaviour. Géotechnique., 53(1), 123-136, (2003).
23. Fredlund, D. G., & Xing, A.. Equations for the soil-water characteristic curve. Canadian geotechnical journal, 31(4), 521-532, (1994).
24. Migliaro, G.. Il legame costitutivo dei terreni piroclastici per la modellazione di scavi in ambiente urbanizzato ed in-fluenza della parziale saturazione. PhD Thesis, Università degli studi di Salerno (Italy), 305 pages, (2008).
25. Lancellotta, R., Di Prisco, C., Costanzo, D., Foti, S., Sorbino, G., Buscarnera, G., Cosentini, R.M., Foresta, V. Caratterizzazione e modellazione geotecnica. In: Criteri di zona-zione della suscettibilità e della pericolosità da frane innescate da eventi estremi (piogge e sisma). Leonardo Ca-scini. Composervice srl, Padova, 266-319, (2012).
26. Tonni, L., Gottardi, G., Simonini, P., Pastor, M., & Mira, P.. Use of Generalized Plasticity to describe the behaviour of a wide class of non-active natural soils. In 3rd International Symposium on Deformation Characteristics of Geomaterials pp. 1145-1153, (2003).
27. Cuomo, S., Manzanal D. Moscariello, M., Pastor M., Foresta V.. Application of a generalized plasticity constitutive model to a saturated pyroclastic soil of Southern Italy. Advances in Soil Mechanics and Geotechnical Engineering. Eds. D. Manzanal & A. Sfriso. Buenos Aires, IOS Press (2015).
28. Cuomo, S., Moscariello, M., Foresta V., Manzanal D., Pastor M.. Experimental investigation and constitutive modelling for an unsaturated pyroclastic soil Proceedings of Workshop on Volcanic Rocks and Soils, Ischia (Italy), 24-25 September 2016.

Fully undrained cyclic loading simulation on unsaturated soils using an elastoplastic model for unsaturated soils

Veerayut Komolvilas[1] and Mamoru Kikumoto[1,a]

[1]*Yokohama National University, Tokiwadai 79-5, Hodogaya, Yokohama, Kanagawa 240-8501, Japan*

Abstract. Several researchers have reported that Bishop's mean effective stress decreases in unsaturated soils under fully undrained cyclic loading conditions, and unsaturated soils are finally liquefied in a similar manner as saturated soils. This paper presents a series of simulations of such fully undrained cyclic loading on unsaturated soils using an elastoplastic model of the unsaturated soil. This model is formulated using the Bishop's effective stress tensor incorporating the following concepts: the volumetric movement of the state boundary surface containing the normal consolidation line and the critical state line due to the variation in the degree of saturation, a soil water characteristic curve model considering the effect of specific volume and hysteresis, the subloading surface model, and Boyle's law. Comparisons between the simulation results and the experimental ones show that the model agreed well with the unsaturated soil behavior under cyclic loading. Finally, the typical cyclic behavior of unsaturated soils under fully undrained conditions, such as the mechanism of liquefaction of unsaturated soils, the compression behavior, and an increase in the degree of saturation, are described through the proposed simulation results.

1 Introduction

Soils are often subjected to a cyclic loading under unsaturated conditions in actual fields such as the deformation of embankments and reclaimed grounds during an earthquake. In Japan, the Sanriku-Minami earthquake triggered a landslide in the town of Tsukidate on May 26, 2003. An artificial fill in this disaster area classified as a volcanic sandy soil lost its effective stress under cyclic loading although the degree of saturation is about 70% [1]. In 2011, the landfills along the northeastern shorelines of the Tokyo Bay liquefied because of the Tohoku earthquake, which caused soil subsidence around an area of 42 km^2 [2]. Until now, questions have been raised about the liquefaction potential of unsaturated soils.

Recently, several researchers conducted cyclic shear tests on unsaturated soils to investigate the cyclic shear behavior of the unsaturated soil. Ishihara et al. [3] studied the effects of relative density and the degree of saturation on the undrained behavior of near-saturated sand through multiple series of monotonic and cyclic triaxial tests. Altun and Goktepe [4] conducted a torsional shear test on unsaturated silty clay to explore the small and large strain behavior of unsaturated soils. Unno et al. [1, 5] conducted a series of strain-controlled cyclic triaxial tests on the unsaturated sand under fully undrained conditions, namely, unexhausted air and undrained water, to study the general liquefaction state of unsaturated soils. Okamura et al. [6, 7] observed the influence of air and suction pressures on the liquefaction resistance of unsaturated soils through a series of cyclic triaxial tests on a fine clean sand and non-plastic silt under fully undrained conditions. Tsukamoto et al. [8] conducted a series of undrained stress-controlled cyclic triaxial tests on the unsaturated sand in order to examine the changes in the cyclic resistance of silty sand with different grain compositions.

All the studies reviewed here support the hypothesis that the mean value of the Bishop's effective stress of the unsaturated soil gradually decreases under fully undrained cyclic loading conditions and the soil is finally liquefied in a similar manner as the saturated soil. The decrease in the effective stress under fully undrained cyclic loading is caused by an increase in pore-air and pore-water pressures, and the liquefaction of the unsaturated soil will occur when both air and water pressures reach the initial value of the total confining pressure [5]. It can be observed that the liquefaction resistance of the unsaturated soil depends on the initial confining pressure; the higher the initial confining pressure, the higher the liquefaction resistance will be [5, 6]. The degree of saturation, which has a significant effect on the cyclic behavior of the unsaturated soil, has also been recognized. The cyclic shear strength of the unsaturated soil increases with the decreasing degree of saturation [1, 3-8]. Some results showed that the cyclic stress ratio almost doubled as the degree of saturation decreased from 100% to 90% [4, 6]. However, the effect of the degree of saturation on cyclic shear strength will reduce if a low confining pressure is exerted on the unsaturated soil [6]. The compressibility of the

[a] Corresponding author: kikumoto@ynu.ac.jp

unsaturated soil, which depends on the density and soil particle structure, is another important factor that enhances liquefaction. The unsaturated soil with a low relative density (loose) or a highly compressible soil structure may easily lose its effective stress under cyclic loading [3, 5]. Moreover, the difference in the development of pore-air and pore-water pressures, which causes the liquefaction, varies with the volume change characteristics dependent on the grain size compositions [8]. A change in the volume of the unsaturated soil during undrained cyclic shearing is directly influenced by pore air. According to the hypothesis that the individual soil particles and water are incompressible in comparison with the soil skeleton, a change in the volume of the soil skeleton is assumed to be equal to that of pore air under fully undrained cyclic loading [5]. The pore air absorbs the generated excess pore pressures resulted in a reduction in the pore-air volume [6, 7]. A reduction in the volume of unsaturated soils during fully undrained cyclic loading causes an increase in the degree of saturation [1, 5]. Suction is a mechanism that affects the cyclic behavior of unsaturated soils [7]. It is usually indicated that suction increases the stiffness of the soil. During cyclic loading, suction decreases because the development of air pressure is less than that of water pressure [1, 5-7]. It should be noted that liquefaction does not occur when suction becomes zero [5].

The main purpose of this paper is to present a series of simulations of fully undrained cyclic loading on unsaturated soils using an elastoplastic constitutive model for unsaturated soils [9]. This model is formulated using the Bishop's effective stress tensor incorporating the following concepts: the volumetric movement of the state boundary surface containing the critical state line due to the variation in the degree of saturation, a soil water characteristic curve model considering the effects of specific volume and hydraulic hysteresis, the subloading surface model, and Boyle's law. Comparisons between the simulation and experimental results show that the model agreed well with the unsaturated soil behavior under cyclic loading. Finally, the typical cyclic behavior of unsaturated soils under fully undrained conditions, such as the mechanism of liquefaction of unsaturated soils, the compression behavior, and an increase in the degree of saturation, are described through the proposed simulation results.

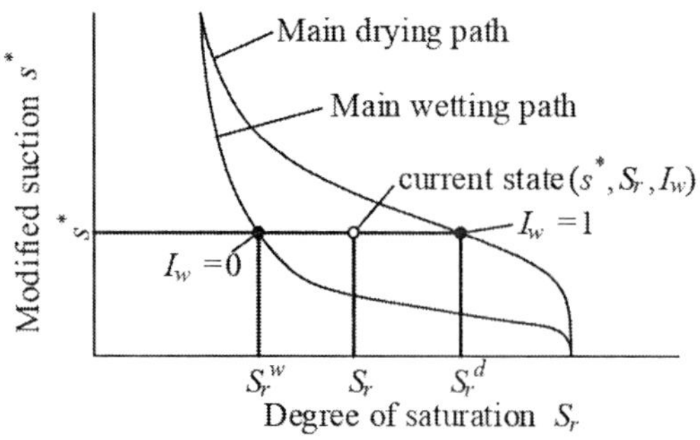

Figure 1. Modeling of water retention curve [9].

2 Basic concepts

This section describes the basic concepts applied to formulate a model for unsaturated soils [9], which is used to predict the cyclic behavior of unsaturated soils under fully undrained conditions.

2.1 The Bishop's effective stress

In this paper, the effective stress of unsaturated soils is defined based on Bishop's effective stress [10] as shown in equation (1).

$$\boldsymbol{\sigma}'' = \boldsymbol{\sigma} - u_a \mathbf{I} + \chi(u_a - u_w)\mathbf{I} = \boldsymbol{\sigma}^{net} + S_r s \mathbf{I} \qquad (1)$$

where $\boldsymbol{\sigma}$, $\boldsymbol{\sigma}^{net}$, u_a, u_w, S_r, and s represent Cauchy total stress tensor, Cauchy net stress tensor, air pressure, water pressure, degree of saturation, and suction, respectively. χ is a variable given as a function of S_r and assumed to be equal to S_r in this study for simplicity.

2.2 Soil water retention curve model considering the effects of density and hydraulic hysteresis

As described earlier, the degree of saturation, suction, and a volume change affect the cyclic behavior of the unsaturated soil. Therefore, a proper water retention curve model is necessary to formulate an elastoplastic constitutive model for the unsaturated soil. This study uses a modified water retention curve considering the effects of density and hydraulic hysteresis [9].

To consider the effects of density, the van Genuchten model [11] is modified as

$$S_r = S_{r_{min}} + \left(S_{r_{max}} - S_{r_{min}}\right)\left[1 + (\alpha s^*)^n\right]^{-m} \qquad (2)$$

where

$$s^* = s\left(\frac{e}{e_{ref}}\right)^{\xi_e} \qquad (3)$$

where $S_{r_{min}}$, $S_{r_{max}}$, α, n, and m are material parameters, s^* is the modified suction, e is the void ratio, e_{ref} is the reference void ratio of saturated, normally consolidated soil under atmospheric pressure, and ξ_e is the parameter controlling the effect of density.

A water retention curve is generally described as the relationship between suction and the degree of saturation dependent on the drying and wetting histories. We define the drying and wetting curves as

$$S_r^A = S_{r_{min}} + \left(S_{r_{max}} - S_{r_{min}}\right)\left[1 + (\alpha^A s^*)^n\right]^{-m}, A = \mathrm{d}, \mathrm{w} \qquad (4)$$

where d and w denote the main drying and wetting curves, respectively. Figure 1 illustrates the differences in the main drying and wetting paths which are the highest and lowest boundaries of the degree of saturation.

DOI: 10.1051/e3sconf/20160917008

The current degree of saturation is the locus of a point between the two main curves represented by I_w. I_w is defined as the ratio of interior division of the current state between two reference states on the main curves expressed by equation (5).

$$I_w = \left(S_r - S_r^w\right)\left(S_r^d - S_r^w\right)^{-1} \tag{5}$$

Substituting equation (4) in (5) and solving for S_r, we obtain

$$\begin{aligned} S_r = I_w\left\{S_{r_{min}} + \left(S_{r_{max}} - S_{r_{min}}\right)\left[1 + \left(\alpha^d s^*\right)^n\right]^{-m}\right\} + \\ \left(1 - I_w\right)\left\{S_{r_{min}} + \left(S_{r_{max}} - S_{r_{min}}\right)\left[1 + \left(\alpha^w s^*\right)^n\right]^{-m}\right\} \end{aligned} \tag{6}$$

We assume that state variable I_w increases monotonically from 0 to 1 with decreasing S_r, and vice versa. Thus, an evolution law for I_w can be written as

$$dI_w = \begin{cases} -\xi_w\left(1 - I_w\right)^3 dS_r & \text{when } dS_r \le 0 \\ -\xi_w I_w{}^3 dS_r & \text{when } dS_r \ge 0 \end{cases} \tag{7}$$

where ξ_w is the material constant controlling the effect of suction histories.

2.3 Elastoplastic stress-strain relationship for unsaturated soil

This section presents an elastoplastic constitutive model for the unsaturated soil [9] formulated by the subloading surface model and the volumetric movement of the state boundary surface containing the critical state line due to the variation in the degree of saturation.

The state boundary surface of the saturated soil, which is the loosest state, can be expressed by

$$v_{NC} = N - \lambda \ln \frac{p'}{p_a} \tag{8}$$

where v_{NC} is the specific volume of the state boundary surface, N is the reference specific volume of saturated, normally consolidated soil under atmospheric pressure, λ is the compression index, p' is the mean effective stress, and p_a is the atmospheric pressure.

As the unsaturated soil shows a relatively high stiffness and retains a larger specific volume than the saturated soil, the state boundary surface is assumed to shift upward (or downward) with the variation in S_r in the direction of the specific volume axis. The effect of S_r on the volumetric movement of the state boundary surface can be represented by

$$v_{NC} = N - \lambda \ln \frac{p'}{p_a} + \psi \tag{9}$$

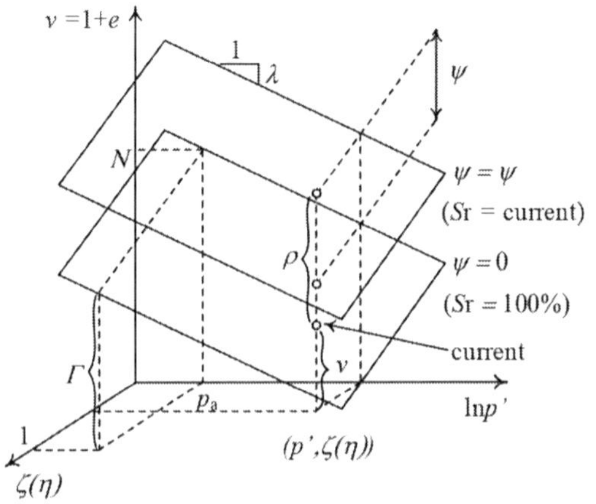

Figure 2. Modeling of the volumetric behavior of the unsaturated soil considering the effects of state variables ρ and ψ.

The concept of the state variable ψ (≥ 0) is shown in Figure 2. The variable ψ can be expressed as a function of S_r.

$$\psi = \xi\left(1 - S_r\right) \tag{10}$$

where ξ is the material parameter representing the vertical distance of the state boundary surface for dried and saturated samples in the compression plane.

To consider the behavior of an overconsolidated soil, the subloading surface concept is applied to our model. According to the subloading surface concept [12], a soil exhibits elastoplastic deformation even in an overconsolidated state and then gradually approaches the normal consolidation plane with the increase in the stress level. An arbitrary specific volume can be represented by

$$v = v_{NC} - \rho = N - \lambda \ln \frac{p'}{p_a} + \psi - \rho \tag{11}$$

The concept of state variable ρ (≥ 0) is shown in Figure 2. State variable ρ is the difference between the specific volume of the current state and that of the state on the state boundary surface under the same stress $(p, \zeta(\eta))$. As ρ may decrease with the development of plastic deformation and finally converge to zero, an evolution law of ρ can be represented by

$$\frac{d\rho}{v_0} = -a\rho|\rho|\left\|d\varepsilon^P\right\| \tag{12}$$

where a is a parameter controlling the effect of density.

A monotonic increasing function $\zeta(\eta)$ based on the modified Cam-clay [13] is then applied. We obtain

$$v = N - \lambda \ln \frac{p'}{p_a} + (\Gamma - N)\zeta(\eta) + \psi - \rho \tag{13}$$

DOI: 10.1051/e3sconf/20160917008

Table 1. Parameters for stress-strain characteristics.

λ	0.105	Compression index
κ	0.025	Swelling index
M	1.3	Stress ratio at critical state
υ	0.3	Poisson ratio
N	1.80	Reference specific volume
a	100.0	Effect of density and confining pressure
ξ	0.80	Effect of S_r on the position of state boundary surface

where Γ is the reference specific volume of the saturated soil under atmospheric pressure at a critical state, and $\zeta(\eta)$ is defined as:

$$\zeta(\eta)=\frac{\ln\left[1+\left(\frac{\eta}{M}\right)^2\right]}{\ln 2} \tag{14}$$

where η is the stress ratio, which is equal to M at the critical state.

By substituting v_0, ψ_0, ρ_0, p'_0, and $q=0$ as the initial state in equation (13), we obtain

$$v_0=\mathrm{N}-\lambda\ln\frac{p'_0}{p_a}+\psi_0-\rho_0 \tag{15}$$

Total volumetric strain is given by

$$\varepsilon_V=\frac{-\Delta v}{v_0}=\frac{v_0-v}{v_0} \tag{16}$$

By substituting equations (13) and (15) in equation (16), we obtain

$$\varepsilon_V=\frac{1}{v_0}\left[\lambda\ln\frac{p'}{p'_0}+(\mathrm{N}-\Gamma)\zeta(\eta)-(\psi-\psi_0)+(\rho-\rho_0)\right] \tag{17}$$

Elastic volumetric strain can be obtained from a usual elastic relationship as

$$\varepsilon_V^e=\frac{\kappa}{v_0}\lambda\ln\frac{p'}{p'_0} \tag{18}$$

where κ is the swelling index.

Plastic volumetric strain can be determined by subtracting the elastic volumetric strain from the total volumetric strain, which can be expressed as

$$\varepsilon_V^p=\frac{\lambda-\kappa}{v_0}\ln\frac{p'}{p'_0}+\frac{\mathrm{N}-\Gamma}{v_0}\zeta(\eta)-\frac{\psi-\psi_0}{v_0}+\frac{\rho-\rho_0}{v_0} \tag{19}$$

From equation (19), yield function can be written as

$$f=\frac{\lambda-\kappa}{v_0}\ln\frac{p'}{p'_0}+\frac{\mathrm{N}-\Gamma}{v_0}\zeta(\eta)-\frac{\psi-\psi_0}{v_0}+\frac{\rho-\rho_0}{v_0}-\varepsilon_V^p \tag{20}$$

Table 2. Parameters for water retention curve.

$S_{r_{max}}$	1.0	Parameters for Van Genuchten model
$S_{r_{min}}$	0.20	
α_d (1/kPa)	0.04	
α_w (1/kPa)	2.00	
n	1.67	
m	0.40	
ξ_w	34.0	Influence of suction histories
ξ_e	2.5	Influence of void ratio
e_{ref}	0.80	Reference void ratio

An associated flow is assumed in the model. As the soil exhibits an unlimited distortional strain at the critical state with no change in stress or volume, the derivatives of the yield function with respect to p' becomes zero. $(\mathrm{N}-\Gamma)$ is thus equal to $(\lambda-\kappa)\ln 2$ in case equation (14) is applied. Equation (20) can be finally rearranged to

$$f=\frac{\lambda-\kappa}{v_0}\left\{\ln\frac{p'}{p'_0}+\ln\left[1+\left(\frac{\eta}{M}\right)^2\right]\right\}-\frac{\psi-\psi_0}{v_0}+\frac{\rho-\rho_0}{v_0}-\varepsilon_V^p \tag{21}$$

The fully undrained condition, namely, unexhausted air and undrained water, is the condition where air and water are unable to drain out of the soil. In other words, the mass of water and air are constant.

In order to simulate the unexhausted air condition, we assume that air is an ideal gas and the temperature is constant. Therefore, Boyle's law, which states that the pressure of a given mass of an ideal gas is inversely proportional to its volume at a constant temperature, can be used as shown in equation (22).

$$u_a V_a=\text{constant} \tag{22}$$

where V_a is the volume of air.

A classic equation for solving problems involving three-phrase relationships (solid, water, and air) can be used to satisfy the undrained water condition as

$$w=\frac{eS_r}{G_s}=\text{constant} \tag{23}$$

where w, e, and G_s are the water content, void ratio, and specific gravity of the soil, respectively.

Finally, an elastoplastic model for unsaturated soils [9] can be formulated by applying these basic concepts in order to predict the cyclic behavior of unsaturated soils under fully undrained conditions.

Table 3. Initial state of cyclic shearing simulation

Case No.	c-2	c-3
Air pressure (kPa)	6.0	14.8
Water pressure (kPa)	0.0	0.0
Suction (kPa)	6.0	14.8
Net stress (kPa)	18.5	19.9
Mean effective stress (kPa)	23.2	30.8

Note mean effective stress is calculated based on the Bishop's effective stress equation, and the pressure is gauge pressure, which excludes atmospheric pressure.

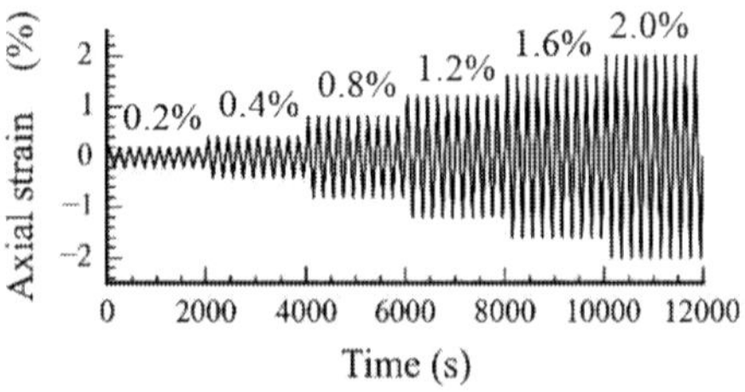

Figure 3. Time history of axial strain during a cyclic shearing process.

3 Simulations

A series of simulations of cyclic triaxial tests on unsaturated soils under fully undrained conditions are performed here. The analysis had been carried out using parameters for Tsukidate volcanic sand (non-plastic sand), which has a specific gravity of 2.478, as shown in Tables 1 and 2.

Cyclic triaxial tests, conducted by Unno et al. [14], have been performed on two types of unsaturated samples (initial degree of saturation = 78.9% and 73.5% with the same initial void ratio of 0.93). In the simulation, the initial state of cyclic shearing simulation is first set as shown in Table 3. Cyclic axial strain with a loading frequency of 0.005 Hz as shown in Figure 3 is then applied to the specimens under the unexhausted air and undrained water conditions at a constant confining pressure. Finally, the simulation results, i.e., the time histories of suction, mean effective stress, air pressure, water pressure, and void ratio, have been obtained as shown in Figures 4 and 5.

Figures 4 and 5 show the comparison between the experimental results for cases c-2 and c-3 conducted by Unno et al. [14] and their corresponding calculations.

According to Figures 4 and 5, the proposed model can precisely describe the cyclic behavior of unsaturated soils under fully undrained cyclic loading conditions. The unsaturated soil lost its effective stress because of the development of pore-air and pore-water pressures and a decrease in suction because the development of air pressure is less than that of water pressure during cyclic shear.

The proposed model can illustrate the fact that liquefaction does not occur when suction becomes zero. Based on the Bishop's effective stress equation, liquefaction of the unsaturated soil will occur when the suction and the net stress become zero. This situation means that air pressure, water pressure, and total confining pressure must be equal.

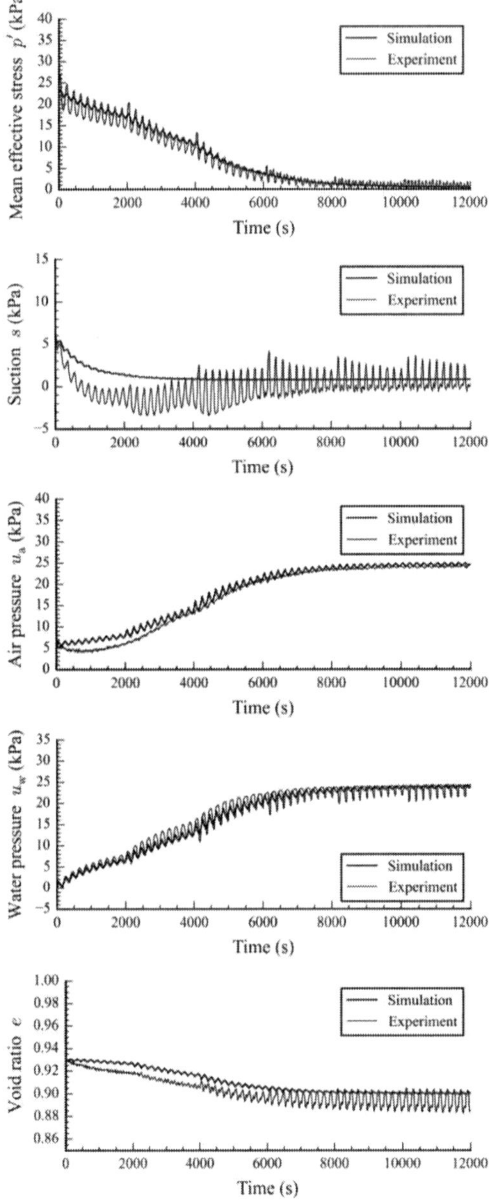

Figure 4. Comparison between the simulation results and the experimental results of case c-2 (S_r = 78.9%, s = 6.0 kPa).

Moreover, the proposed model incorporating Boyle's law can capture the compression behavior of unsaturated soils under fully undrained conditions. As air pressure increases during cyclic loading, air volume will automatically decrease following Boyle's law. The magnitude of the decrease in the void ratio was also predicted accurately.

The proposed model for water retention curve can predict the increase in the degree of saturation due to volumetric contraction as shown in Figure 6.

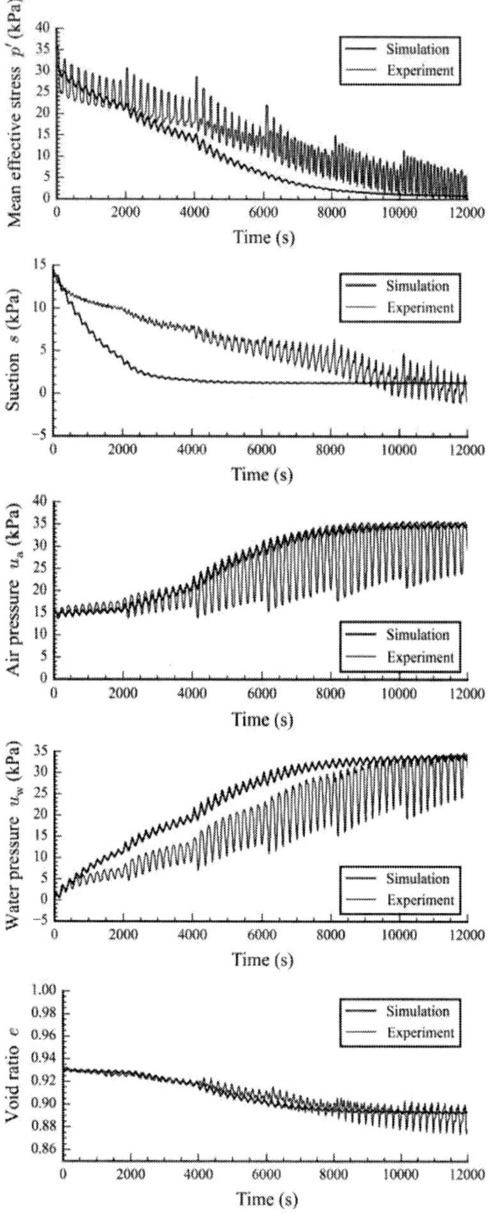

Figure 5. Comparison between the simulation results and the experimental results of case c-3 (S_r = 73.5%, s = 14.8 kPa).

4 Conclusion

An elastoplastic constitutive model for the unsaturated soil, which is able to predict the cyclic behavior under fully undrained conditions, has been developed.

This model is formulated using the Bishop's effective stress tensor incorporating the following concepts: the volumetric movement of the state boundary surface containing the critical state line due to the variation in the degree of saturation, the soil water characteristic curve model considering the effects of specific volume and hydraulic hysteresis, the subloading surface model, and Boyle's laws.

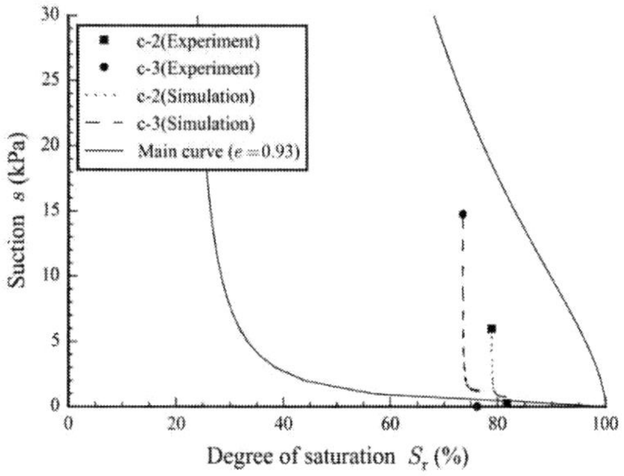

Figure 6. Water retention curve during fully undrained cyclic triaxial test.

The validity of the proposed model is checked through a series of cyclic triaxial tests on unsaturated soils under fully undrained conditions. It is indicated through the simulations that the proposed model properly describes the fully undrained cyclic behavior of unsaturated soils such as liquefaction phenomena, compression behavior, and an increase in the degree of saturation.

References

1. T. Unno, M. Kazama, N. Sento, R. Uzuoka, *The 4th International Conference on Unsaturated Soils*, 1133-1144 (2006)
2. K. Konagai, T. Kiyota, S. Suyama, T. Asakura, K. Shibuya, C. Eto, Soil Dyn. Earthq. Eng., **53**, 240-253 (2013)
3. K. Ishihara, Y. Tsukamoto, K. Kamada, *Cyclic Behavior of Soils and Liquefaction Phenomena*, 27-39 (2004)
4. S. Altun, A.B. Goktepe, *The 4th International Conference on Unsaturated Soils*, 497-507 (2006)
5. T. Unno, M. Kazama, R. Uzuoka, N. Sento, Soils. Found., **48**, 87-99 (2008)
6. M. Okamura, Y. Soga, Soils. Found., **46**, 695-700 (2006)
7. M. Okamura, K. Noguchi, Soils. Found., **49**, 221-229 (2009)
8. Y. Tsukamoto, S. Kawabe, J. Matsumoto, S. Hagiwara, Soils. Found., **54**, 1094-1103 (2014)
9. M. Kikumoto, H. Kyokawa, T. Nakai, H.M. Shahin, *The 5th International Conference on Unsaturated Soils*, 849-855 (2010)
10. A.W. Bishop, Teknisk Ukeblad, **39**, 859-863 (1959)
11. M. Th. van Genuchten, Soil Sci. Soc. Am. J., **44**, 892-898 (1980)
12. K. Hashigushi, M. Ueno, *The 9th International Conference on Soil Mechanics and Foundation Engineering*, 73-82 (1977)
13. K. Roscoe, J.B. Burland, *Engineering Plasticity*, 533-609 (1968)
14. T. Unno, R. Uzuoka, N. Sento, M. Kazama, J.S.C.E., **69**, 386-403 (2013)

Essential features of compacted silty sand behavior via suction-controlled triaxial testing

Ujwalkumar D. Patil[1,a], Laureano R. Hoyos[2], and Anand J. Puppala[3]

[1]*Faculty Research Associate, University of Texas at Arlington, Arlington, Texas 76019, USA.*
[2]*Professor, University of Texas at Arlington, Arlington, Texas 76019, USA.*
[3]*Professor, University of Texas at Arlington, Arlington, Texas 76019, USA.*

Abstract. The main intent of this work was to gain further insight into some of the limitations (and potential) of the Barcelona Basic Model (BBM) in predicting the stress-strain response of compacted unsaturated soils that are prone to post-peak strain-induced softening during suction-controlled shearing. To this aim, a comprehensive series of suction-controlled tests were conducted on statically compacted specimens of silty sand using a recently implemented, servo/suction-controlled, double-walled triaxial apparatus. The brittle and dilatant nature of the test soil, as expected, did not allow for good agreement between the experimental and BBM predicted stress-strain responses of compacted silty sand, since post-peak softening was not adequately captured. BBM simulations, however, held reasonably good during the early shearing stage (up to about 1-2% total shear strain), and at higher values of shear strain, that is, close to critical state condition.

1 Introduction

Despite several elasto-plastic constitutive models being proposed, the critical-based framework originally proposed by [1] remains to date; the most widely cited and used model for unsaturated soils. Soil modelers face a challenge to accommodate all possible suction ranges of unsaturated soil responses. These include post-peak softening, accompanied by stress-induced dilatancy that is exhibited typically by dense soils. Other challenges include strain-hardening type responses, accompanied by volumetric compressive behavior, which is typical of loose soils, swelling and collapse under wetting, as well as shrinkage and cracking while drying.

The verification of proposed constitutive relations must be conducted for a wide range of soils in order to ensure uniqueness and subsequent confidence on the part of researchers and practicing geotechnical engineers [4]. In the last two decades, the model has been the subject of increasingly sophisticated enhancements, mainly from a theoretical standpoint (e.g., [8], [14], [3], [2], [13], [5], [15], [11], and [6]). However, as previously mentioned, conclusive evidence of its validation for heavily consolidated materials which are distinctly brittle and dilatant in nature, particularly their post-peak softening behavior, is still lacking, mainly due to a large scarcity of experimental data of this kind. In this study, constitutive parameters postulated by the BBM framework are experimentally calibrated from the series of suction-controlled triaxial tests and then used for predicting compacted silty sand response at matric suction states from 50 to 750 kPa.

2 Test material

The test soil comprising of sand (55%), silt (37%), and clay (8%), is classified as silty sand (SM) according to the Unified Soil Classification system (USCS). The standard proctor test resulted in maximum dry unit weight of 1.87 g/cm^3 and optimum moisture content of 12.2%.

3 Specimen preparation and suction-controlled triaxial system

Each specimen was statically compacted in nine equal layers at moisture content of 14.2% (+2.2% wet of optimum) and target dry density of 1.8 g/cm^3. Each layer was compressed statically in a mold using a 50 KN load frame at a constant rate of 1 mm/min to a total vertical stress of 1600 kPa to produce uniformly compacted specimens of 2.8 inch diameter and 5.6 inch height.

A photograph of automatic new double-walled triaxial equipment with frictionless automatic volume change device and flushing device assembly is illustrated in Fig. 1. The efficacy of this device to duplicate tests and its functioning has been detailed by [9] and [10]. Axis-translation technique as proposed by [7] is used to impose matric suction.

[a] Corresponding author: ujwalkumar.patil@mavs.uta.edu

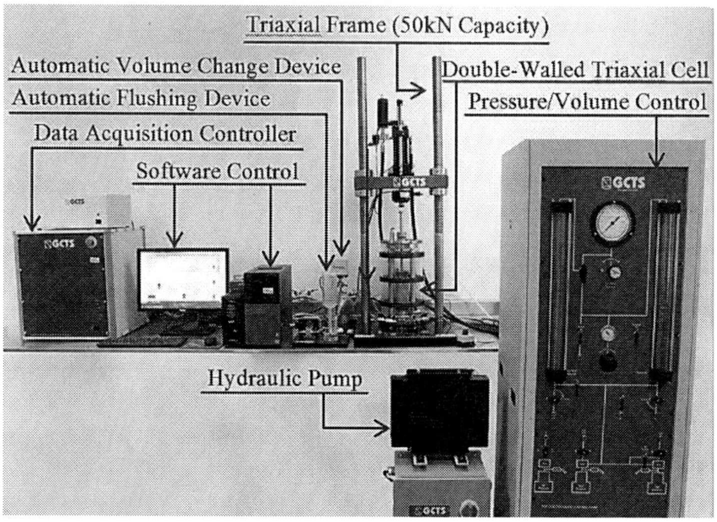

Figure 1. Servo/suction-controlled, double-walled triaxial apparatus.

4 Calibration of BBM parameters

[1] postulated a unified, critical state based, constitutive framework for unsaturated soils by extending the modified Cam Clay model [12], from saturated to unsaturated form, using suction as an independent stress variable, while introducing the concept of the loading-collapse (LC) yield surface.

4.1 Calibration of BBM parameters under isotropic loading

BBM postulated yield function represents an ellipse-shape yield surface in p:q plane, beyond which plastic compression occurs on account of increased stress or decreased suction. The model assumes that only elastic strains will be accumulated when the soil state lies inside a state boundary hypersurface, as given by the following:

$$f_1 = f(p, q, s) = q^2 - M^2(p + p_s)[p_o(s) - p] = 0 \quad (1)$$

where, M = slope of the critical state line in p:q plane, and

$$p_s = ks \quad (2)$$

where, k = slope of the apparent tensile strength locus in p:s plane. The model assumes a monotonic decrease of the volumetric stiffness parameter λ(s) with increasing suction, expressed as follows:

$$\lambda(s) = \lambda(0)\left[(1 - r)e^{(-\beta s)} + r\right] \quad (3)$$

where, $r = \frac{\lambda(s\to\infty)}{\lambda(0)}$, and

β = parameter controlling the rate of increase of soil stiffness with matric suction. The yield surface at constant suction may be described by an ellipse marked by an isotropic consolidation stress, $p_o(s)$, that lies on the loading-collapse (LC) yield curve, which in turn takes the following form in p:s plane:

$$\frac{p_o(s)}{p^c} = \left[\frac{p_o(0)}{p^c}\right]^{\frac{\lambda(0)-\kappa}{\lambda(s)-\kappa}} \quad (4)$$

where, $p_o(s)$ = yield stress that varies with matric suction; $p_o(0)$ = preconsolidation stress under saturated condition; p^c = reference preconsolidation stress at which the LC locus becomes a straight line, i.e., $p_o(s) = p^c$; λ(0) = compressibility coefficient for saturated state along virgin loading; λ(s) = soil stiffness parameter that varies with matric suction; κ = elastic rebound index with respect to net mean stress. A non-associative flow rule (Eqn. 5) was adopted for the direction of plastic shear strain increments $d\varepsilon^p{}_q$ and is given as:

$$d\varepsilon_q^p = \frac{2\alpha q_C}{M^2[2p_C + p_s - p_o^C(s)]} d\varepsilon_{vp}^p \quad (5)$$

where the constant α is defined as follows:

$$\alpha = \frac{M(M-9)(M-3)}{9(6-M)}\left[\frac{1}{1-\frac{\kappa}{\lambda(0)}}\right] \quad (6)$$

The least square method was used to solve the overdetermined system in Equation (3), allowing for the determination of constitutive parameters λ(0), r, and β. Using experimental values, λ(50) = 0.01381, λ(250) = 0.012146, λ(500) = 0.005097, and λ(750) = 0.0061339,

 DOI: 10.1051/e3sconf/20160917009

best-fit values for these three unknowns were found to be $\lambda(0)$ = 0.02, r = 0.26, and β = 2.0 MPa^{-1}, with a coefficient of determination, R^2 = 0.85. Figures 2 show the experimentally calibrated values of the stiffness parameter $\lambda(s)$, along with its parametric predictions, using Eq. (3), for different values of parameters β. Reasonably good agreement is observed between experimental and predicted values of volumetric stiffness $\lambda(s)$ for this partial set of calibrated parameters.

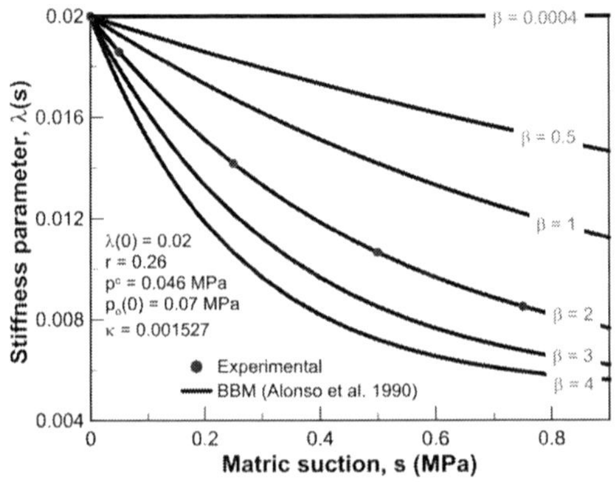

Figure 2. Experimental stiffness parameter, $\lambda(s)$ for test soil and predicted curves for various values of β using Equation (3).

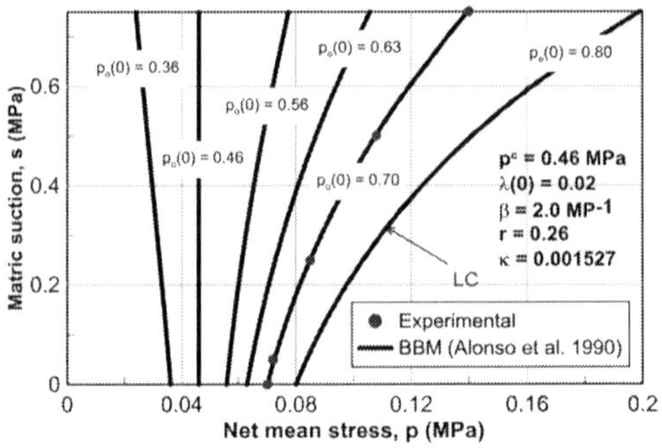

Figure 3. Experimental yield stress values on best-fit LC curve and typical curves predicted for different values of $p_o(0)$.

The same method is used to solve the overdetermined system in Eq. (4), thus allowing for the assessment of parameters p^c and $p_o(0)$. Using experimental values, $p_o(50)$ = 72 kPa, $p_o(250)$ = 83 kPa, $p_o(500)$ = 120 kPa, and $p_o(750)$ = 140 kPa, best-fit values for these two unknowns were found to be p^c = 46 kPa, and $p_o(0)$ = 70 kPa, with a coefficient of determination, R^2 = 0.99.

The experimental data points from compacted silty sand used in this research show a very close agreement with the LC curve predicted for $p_o(0)$ = 70 kPa. The figure also illustrates the reference preconsolidation stress, for which the LC curve becomes a perfectly straight line, calibrated to a value equal to 46 kPa.

4.2 Calibration of BBM parameters under axisymmetric shearing

Figure 4 shows the best-fit critical state lines (CSLs) obtained from the series of suction-controlled CTC tests, as plotted in p:q:s space, for matric suction values, s = 0, 50, 250, 500, and 750 kPa. It can be observed that the slope M(s) of the CSLs remains virtually constant regardless of the level of induced matric suction, in close agreement with the original BBM formulation. This, therefore, establishes a unique relationship between deviatoric and net mean stresses for suction-controlled CD triaxial tests on compacted silty sand.

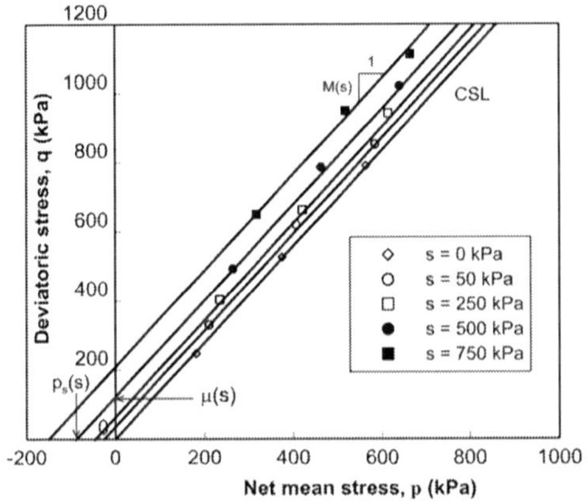

Figure 4. Critical state lines in p-q plane generated from suction-controlled CD triaxial tests via axis-translation.

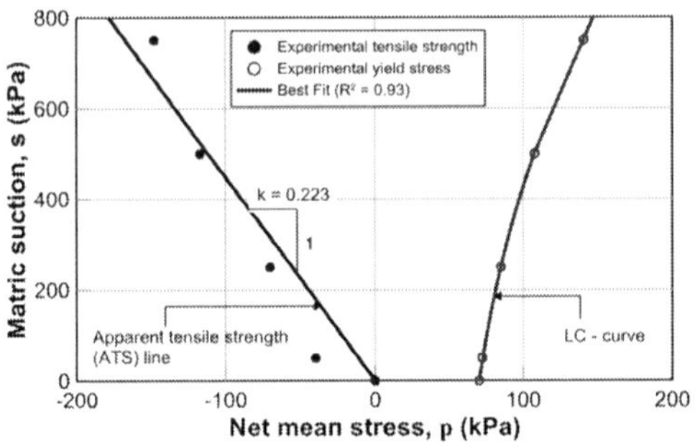

Figure 5. Apparent tensile strength (ATS) and loading-collapse (LC) loci experimentally assessed from suction-controlled isotropic loading.

It can be readily noticed from Figure 4 that the slope of the critical state lines remains virtually constant despite the introduction of suction. Thus, it establishes a unique relationship between deviator and net mean stresses for suction-controlled CD triaxial tests. However, a noticeable upward shift in the position of critical state line is observed with the introduction of matric suction, as compared to its saturated counterpart. The intersection of the CSLs with the q-axis gives the magnitude of the apparent cohesion, $\mu(s)$; while the intersection of the CSLs with the negative p-axis gives the magnitude of the additional tensile strength, $p_s(s)$, offered by the soil with increasing matric suction.

Additional constitutive parameters postulated by the BBM framework, including the slope of the apparent tensile strength (ATS) locus in p:s plane, and the slope of the CSLs, as defined by Eqs. (1) and (2), can now be experimentally calibrated. Figure 5 shows the LC yield curve assessed for compacted silty sand from suction-controlled isotropic consolidation tests, along with the ATS locus assessed in p:s plane from the best-fitted CSLs obtained from suction-controlled CTC tests. Within the range of stresses applied in this work, i.e., matric suction, s = 50 to 750 kPa, and initial net mean stress, $p = (\sigma_3 - u_a)$ = 100 to 300 kPa, a three-dimensional plain surface plane, as given by Eq. (7), can be used to best-fit all of the experimental values of deviatoric stress at critical state:

$$q = Mp + Mks = 1.42\,p + 0.316\,s \tag{7}$$

The plane yields best-fit values, M = 1.42, for the slope of CSLs and, k = 0.223, for the rate of increase in ATS with suction. Predicted values of deviator stress are reasonably close to those experimentally obtained from suction-controlled CTC tests, yielding a coefficient of determination, $R^2 = 0.93$, as shown in Figure 5.

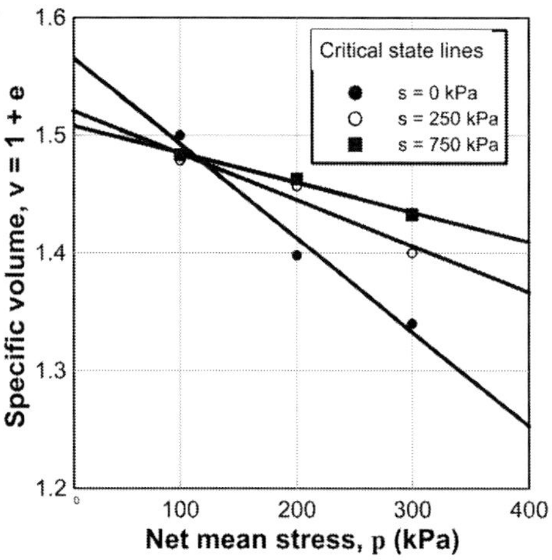

Figure 6. Critical state lines at failure in *p-v* plane.

As a consequence of the non-associative flow rule postulated by the BBM through Equations (5) and (6), no plastic shear strain is expected to occur during initial shearing when the soil is under lightly overconsolidated condition. Hence, the elastic shear modulus can be estimated from the initial portion of the deviator stress vs. total shear strain response, which yields an average value, G = 25 MPa, for the compacted silty sand tested in this work.

Post-consolidation states under isotropic loading are assumed to follow a normal consolidation line (NCL) in v:p plane. However, as the soil gets sheared, the soil state moves from its initial position on the NCL, eventually reaching a CSL. All the unsaturated specimens showed initial contraction followed by significant dilation, modulated by the level of induced matric suction, just as the volume change paths in p:v plane, where the slope of the CSL is observed to decrease with increasing suction, as postulated by the BBM.

Figure 6 shows the experimentally assessed CSLs in p:v plane for matric suction values, s = 0, 250, and 750 kPa.

4.3 BBM predictions of soil response

The type of test performed in this research followed the conventional triaxial compression (CTC) stress path. The soil specimen was subjected to constant matric suction and to constant confining net mean pressure, p_{ini}. The specimen failed by monotonic increase of the net axial pressure, σ_1, while keeping the isotropic confining pressure, $(\sigma_3 - u_a)$, constant. Therefore, the stress path followed by the stress state in p-q space had a slope of 1H:3V. Explicit integration of key constitutive relationships has been used in this work to simulate soil response through BBM predictions of a drained (constant-s) conventional triaxial compression (CTC) test.

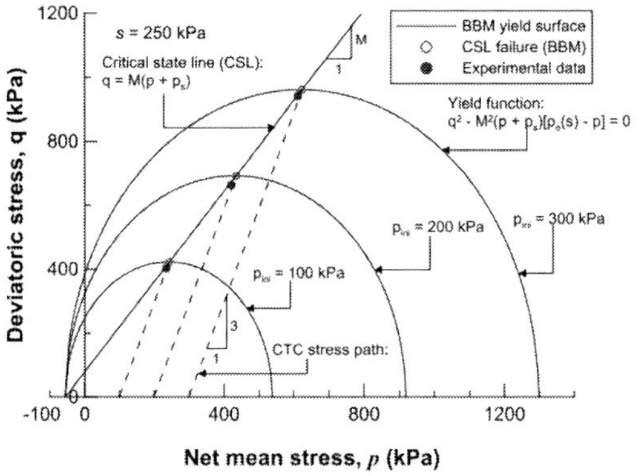

Figure 7. BBM prediction of yield surface for drained CTC tests conducted at constant matric suction, s = 250 kPa, and initial net mean stresses, p_{ini} = 100, 200, and 300 kPa.

For hardening materials, continued shearing along a suction-controlled CTC stress path (Fig. 7) is expected to cause the elliptical yield surface in p:q plane to move outward (or increase in size) from the current point, while it would move inward (or shrink in size) for softening materials. The relative position of the current stress state with respect to the CSL governs whether the material sustains plastic dilatancy or contractancy.

Figures 8 and 9 show comparisons between BBM predicted and experimental deviator stress vs. shear strain response of compacted silty sand from fully drained (constant suction) CTC tests conducted at two different values of matric suction, s = 250 kPa and 500 kPa, and for initial value of net mean stress, $p = (\sigma_3 - u_a)$ = 100 kPa and 300 kPa, respectively. In general, no close agreement is observed between the experimental and predicted soil responses, primarily given the largely brittle and dilatant nature of the test soil before it reaches critical state.

The original BBM framework, however, does not contemplate a stress state that fails to overcome the CSL during plastic loading, since the peak stress is always assumed to be reached at critical state. Therefore, it is not expected to be suitable for reproducing the transition from initial contractancy to dilatancy, and hence the post-peak strain-softening commonly observed in dense or overconsolidated geomaterials (Figs. 8-9).

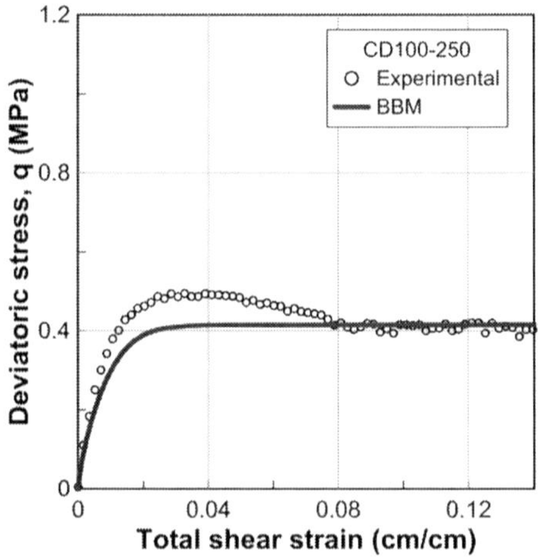

Figure 8. Experimental and BBM predicted response of silty sand from triaxial test CD100-250.

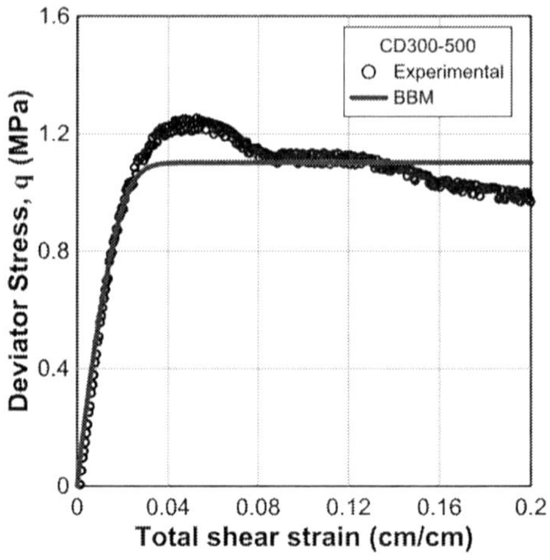

Figure 9. Experimental and BBM predicted response of silty sand from triaxial test CD300-500.

In effect, although predictions of deviatoric stress at critical state (end values) proved to be reasonably close to those experimentally observed (Fig. 8 and 9), BBM predictions considerably deviate from the stress-strain response of compacted silty sand, particularly at higher matric suctions and net confining pressures, which are precisely the stress state for which the test soil exhibits largest post-peak softening, accompanied by significant dilation.

From a qualitative standpoint, however, reasonably good predictions are generally observed for most of the initial shearing stage, up to about 1-2% shear strain, and at considerably large values of shear strain, i.e., critical state. As previously stated, these results are valuable to unsaturated soil constitutive modelers in validating more suitable frameworks for compacted or overconsolidated intermediate geomaterials when subjected to suction-controlled monotonic shearing.

5 Concluding remarks

BBM predictions of continuous stress-strain response of compacted silty sand were made by using the experimentally calibrated parameters, and compared with experimental results. Numerical predictions were implemented by using the explicit integration technique.

Given the evidently "brittle" and "dilatant" nature of the test soil, no good agreement was observed between the experimental and BBM predicted stress-strain responses. The post-peak softening of the test soil was not adequately captured. BBM simulations, however, hold reasonably good mostly during the early shearing stage, up to about 1-2% shear strain, and at higher values of shear strain (i.e. at critical state).

Acknowledgements

The experimental work described in this paper is part of an ongoing research project funded by National Science Foundation (NSF-MRI award ID # 1039956). This support is gratefully acknowledged. Any findings, conclusions, or recommendations expressed in this material are those of the authors and do not necessarily reflect the views of the National Science Foundation.

References

1. Alonso, E.E., Gens, A., and Josa, A. "A constitutive model for partially saturated soils." Géotechnique, **40** (3), 405-530, (1990).
2. Bolzon, G., Schrefler, B. A. and Zienkiewicz, O. C. "Elastoplastic soil constitutive laws generalized to partially saturated states." Géotechnique, **46** (2), 279-289, (1996).
3. Cui, Y.J., and Delage, P. "Yielding and plastic behavior of an unsaturated compacted silt." Géotechnique, **46**(2), 291-311, (1996).
4. Fredlund, D. G., and Rahardjo, H. "*Soil mechanics for unsaturated soils.*" New York, Wiley, (2012).
5. Georgiadis, K., Potts, D. M., and Zdravkovic, L. "Three-dimensional constitutive model for partially and fully saturated soils." International Journal of Geomechanics, **5** (3), 244-255, (2005).
6. Hoyos, L. R., Diego D. Perez-Ruiz and Puppala, A.J. "Modeling unsaturated soil response under suction-controlled true triaxial stress paths." International Journal of Geomechanics, *ASCE*. **12** (3), 292-308, (2012).
7. Hilf, J. W. "An investigation of pore water pressures in compacted cohesive soils." *U.S. Dept. of the interior, Bureau of Reclamation*, Tech. Mem. 654, Denver, Col., U.S.A (1956).

8. Josa, A., Balmaceda, A., Gens, A., and Alonso, E. E. "An elastoplastic model for partially saturated soils exhibiting a maximum of collapse." Proceedings, *3rd International Conference on Computational Plasticity*, Barcelona, **1**, 815-826 (1992).
9. Patil, U. D., Puppala, A. J., and Hoyos, L. R. (2014). "Strength/volume change behavior of compacted silty sand using a novel double-walled suction-controlled triaxial system." *Unsaturated Soils: Research and Applications – Proceedings of the 6th International conference on unsaturated soils*, UNSAT 2014, **2**, 1727-1733 (2014).
10. Patil, U. D. "Response of unsaturated silty sand over a wider range of suction states using a novel double-walled triaxial testing system." Ph.D. dissertation, Univ. of Texas at Arlington, Arlington, TX, USA (2014).
11. Russel, A. R. and Khalili, N. "A unified bounding surface plasticity model for unsaturated." International Journal of Numerical and analytical methods in Geomechanics, **30** (3), 181-212 (2006).
12. Schofield, A., and Wroth, C. P. "*Critical state soil mechanics*." McGraw-Hill, London (1968).
13. Vaunat, J, Romero, E., and Jommi, C. "An elasto-plastic hydro-mechanical model for unsaturated soils." Experimental evidence and theoretic approaches in unsaturated soils (Eds. Tarantino, A. and Mancuso, C), 121-138, Rotterdam: Balkema, (2000).
14. Wheeler, S. J., and Sivakumar, V. "An elasto plastic critical state framework for unsaturated soil." Géotechnique, **45** (1), 35-53 (1995).
15. Yu, H. S. "*Plasticity and geotechnics*." Springer, New York, (2006).

 DOI: 10.1051/e3sconf/20160917010

A Constitutive Model for Unsaturated soils based on a Compressibility Framework dependent on Suction and Degree of Saturation

Panagiotis Sitarenios[1,a] and Michael Kavvadas[1]

[1]National Technical University of Athens, School of Civil Engineering, Geotechnical Department, Athens, Greece

Abstract. The Modified Cam Clay model is extended to account for the behaviour of unsaturated soils using Bishop's stress. To describe the Loading – Collapse behaviour, the model incorporates a compressibility framework with suction and degree of saturation dependent compression lines. For simplicity, the present paper describes the model in the triaxial stress space with characteristic simulations of constant suction compression and triaxial tests, as well as wetting tests. The model reproduces an evolving post yield compressibility under constant suction compression, and thus, can adequately describe a maximum of collapse.

1 Introduction

This paper presents a simple constitutive model for the numerical analyses of unsaturated soils. The proposed model is an extension of the Modified Cam Clay (MCC), in the unsaturated regime. It incorporates a Loading – Collapse (LC) Surface, derived on the bases of a compressibility framework proposed by the authors [1].

The utilized framework includes compression lines that are both dependent on suction (s) and on the effective (macro-structural) degree of saturation (S_r^e). This double dependence allows for a comprehensive description of an evolving compressibility during constant suction tests.

2 Compressibility framework

Sitarenios and Kavvadas [1] extended an intrinsic compressibility framework for saturated soils to account for the effects of partial saturation. The framework utilizes Bishop's average skeleton stress, assuming $\chi = S_r^e$, while in general it is in favour of an increased yield stress and a decreased post yield compressibility with increasing suction (s).

It comprises an enhancement of the Barcelona Basic Model (BBM) to account for the slope of the unsaturated compression lines in the Bishop's stress domain, based on a relatively simple idea to scale the effect of suction by incorporating the effective degree of saturation:

$$\lambda(s,S_r^e) = \lambda - \left(\lambda - \lambda(s)\right)\left(1-S_r^e\right)^{\gamma} \tag{1}$$

The term $(1\text{-}S_r^e)^{\gamma}$ in fact implies that an increase in degree of saturation acts as a structure degradation mechanism which progressively increases the post yield compressibility of the unsaturated material. Material constant γ is added (it was not included in [1]), to scale the effect of degree of saturation, allowing for increased versatility. For the $\lambda(s)$, the well-known BBM equation is used:

$$\lambda(s) = \lambda\left[\left(1-r\right)e^{-\beta s} + r\right] \tag{2}$$

and after some algebra we end up with:

$$\lambda(s,S_r^e) = \lambda\left[1-\left(1-r\right)\left(1-S_r^e\right)^{\gamma}\left(1-e^{-\beta s}\right)\right] \tag{3}$$

In equation 3, λ is the slope of the compression lines under saturated conditions and r, β, γ are material constants. Parameters r, β originate from the BBM, nevertheless they do not fully retain their original role.

The proposed framework (see fig. 1) suggests that unsaturated soil states, lay on a compression curve which originates from the saturated compression line on a reference pressure P_c, while its slope is given by eq. 3. Thus they plot to the right of the saturated compression line in the metastable domain, a common assumption, reflecting the beneficial effect of inter-particle forces (induced by partial saturation) on soil stiffness. It additionally accounts for the increased yield stress that unsaturated soils exhibit when compressed. The latter, in the present framework can be calculated using a similar to BBM concept to obtain:

$$P_0(s,S_r^e) = P_c\left(\frac{P_0^*}{P_c}\right)^{\frac{\lambda-\kappa}{\lambda(s,S_r^e)-\kappa}} \tag{4}$$

Equation 4 gives the evolution of the apparent preconsolidation pressure with partial saturation for a material with a saturated preconsolidation pressure equal to P_0*.

[a] Corresponding author: pxsitarenios@yahoo.gr

DOI: 10.1051/e3sconf/20160917010

The similarity of eq. 4 with its BBM counterpart is apparent, nevertheless in the proposed framework its additional dependence on the effective degree of saturation, ensures that saturated states will always lay on the compression line of the saturated material, irrespectively of the level of the applied suction. This is a prerequisite for frameworks that utilize Bishop's stress as the latter coincides with Terzaghi's effective stress upon saturation. Frameworks solely dependent on suction cannot capture such a behaviour unless some kind of discontinuity, associated with the air entry value of the material, is assumed.

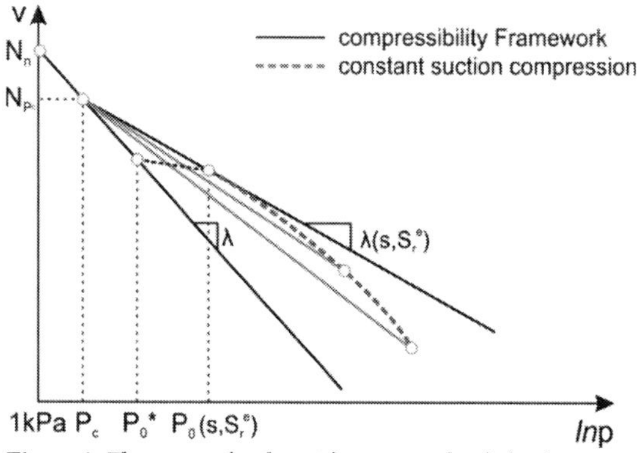

Figure 1. The assumed volumetric compression behaviour.

Moreover, the dependence on degree of saturation, can reproduce a constantly evolving compressibility for soils compressed under constant suction, as a natural outcome of the soil's increased capacity for water retention with decreasing void ratio. Such a behaviour is schematically depicted by the grey dotted line in figure 1.

In the last few years, many researches proposed that the slope (i.e., [2]) and/or the position (i.e., [3]) of the unsaturated compression lines in the Bishop's stress domain can be solely dependent on degree of saturation with suction only implicitly influencing the behaviour through the Water Retention Curve and Bishop stress. Such frameworks address the issue of unique compression lines for saturated states, and can also represent a constantly evolving compressibility under a constant given suction level.

Nevertheless, we believe that a double dependence is more representative of the fundamental mechanisms describing the shift of the compression lines in the metastable domain. As was mentioned before, this "structure" inducing mechanism is attributed to the bonding effect of the forming inter-particle forces. As explained in [1], similar to Gallipoli's proposal [4] we believe that the intensity of this bonding effect depends both on suction and on the quantity of water in the soil's pores.

The latter can be better understood in a relatively coarse-grained soil, where capillary phenomena prevail. In such a case, inter-particle forces are mainly attributed to the forming water menisci. The strength of the forming menisci depends mainly on suction, while how many menisci are present in a given soil mass depends on degree of saturation. As a soil progressively saturates under a constant suction level, some of the menisci are destroyed and thus the intensity of the inter-particle forces drops.

On the other hand, examining the theoretical case, where suction changes under a given degree of saturation, the strength of "constant" menisci increases or decreases, following suction changes. This example, although severely simplifying and fictitious to some extent, is illustrative of the assumed behaviour.

3 Constitutive equations

The aforementioned framework is incorporated in the MCC model, to describe the LC curve. The model uses Bishop's average skeleton stress as its First Constitutive Variable (FCV), while also suction (s) and effective degree of saturation (S_r^e) are incorporated as extra constitutive variables, the three of them comprising the ensemble of the model's external variables. No further modifications are implemented, in fact subscribing to the idea that the shear strength behaviour as well as the behaviour in the elastic domain are fully interpretable in terms of Bishop's average skeleton stress, when the macro-structure's degree of saturation is used.

At this point we shall mention that although degree of saturation is assumed an external variable, in fact it is not an independent variable, as it is an outcome of the soil's Water Retention Curve (WRC).

Regarding the set of internal variables, alike MCC, the proposed model utilizes the specific volume (v), needed for the porous elastic behaviour and for the hardening rule, while the saturated preconsolidation pressure P_0^* is used as the hardening variable.

In the following lines, the model formulation is presented in the triaxial effective stress space ($p = \bar{p} + s \cdot S_r^e$, q) with the primes dropped for simplicity, as net stress ($\bar{p}$) does not appear in the formulation, while dot over a symbol denotes an infinitesimal increment of the corresponding quantity.

3.1. Yield function

The yield surface is described by the following yield function:

$$f\left(p,q,a,S_r^e,P_0^*\right) = q^2 - M^2 p\left(P_0(s,S_r^e) - p\right) = 0 \quad (5)$$

in combination with equations (4) and (3). M is the slope of the Critical State Line (CSL) in the p-q. The behaviour inside the Yield Surface is assumed elastic, while plastic states always lay on the Yield Surface. Figure 2 presents the yield surface in the p-q-s space.

3.2 Elasticity

The standard porous elasticity is utilized to describe the behaviour inside the elastic domain. The corresponding elastic increment of Bishop's mean stress $\dot{p}$ and of the deviatoric stress $\dot{q}$ are calculated through:

$$\dot{p} = K \cdot \dot{\varepsilon}_{vol} \quad (6)$$

$$\dot{q} = \frac{1}{3G} \cdot \dot{\varepsilon}_q \quad (7)$$

where K, G are the Bulk and Shear Modulus, respectively. K depends on void ratio and on the level of the applied stress through:

$$K = \frac{v}{\kappa} p \quad (8)$$

where κ is the slope of the swelling lines.

A pressure dependant Shear Modulus is also assumed depending on; a) the value of the Bulk Modulus and b) the Poisson's ratio value (ν):

$$G = \frac{1}{2}\left(\frac{3(1-2\nu)}{1+\nu}\right) K \quad (9)$$

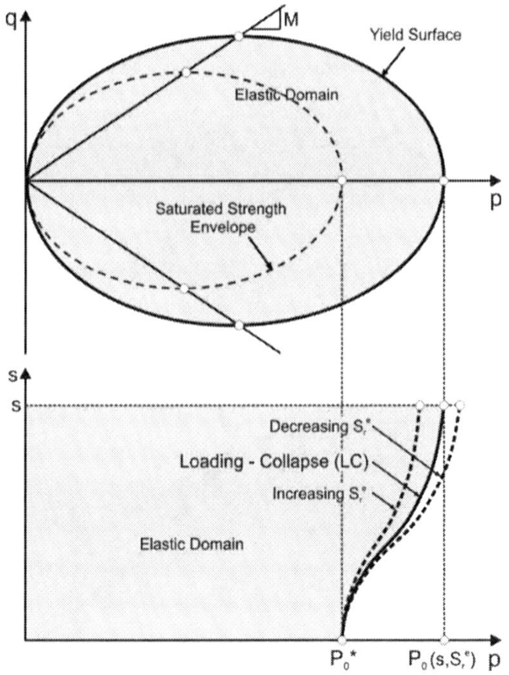

Figure 2. The model's yield surface in p-q-s space.

3.3 Flow rule

The flow rule describes the onset of plastic strains during plastic loading. The MCC assumes an associated flow rule, where the plastic potential function (g) coincides with the yield function ($g \equiv f$) and thus, plastic strain increments are described by the following expressions:

$$\dot{\varepsilon}_{vol}^{p} = \dot{\Lambda} \cdot \frac{\partial f}{\partial p} \quad (10)$$

$$\dot{\varepsilon}_{q}^{p} = \dot{\Lambda} \cdot \frac{\partial f}{\partial q} \quad (11)$$

$\dot{\Lambda}$ is the so called plastic multiplier and is defined as:

$$\dot{\Lambda} = \frac{1}{H}\left(\frac{\partial f}{\partial p}\dot{p} + \frac{\partial f}{\partial q}\dot{q} + \frac{\partial f}{\partial s}\dot{s} + \frac{\partial f}{\partial S_r^e}\dot{S}_r^e\right) \quad (12)$$

where H is the Plastic Modulus, derived on existing assumptions, through the consistency condition.

Having defined the evolution of plastic strain increments, by further utilizing the basic kinematic decomposition of total strains to an elastic and a plastic part ($\varepsilon = \varepsilon^e + \varepsilon^p$), we may write that:

$$\dot{p} = K \cdot \dot{\varepsilon}_{vol}^{e} = K \cdot (\dot{\varepsilon}_{vol} - \dot{\varepsilon}_{vol}^{p}) = K \cdot (\dot{\varepsilon}_{vol} - \dot{\Lambda} \cdot \frac{\partial f}{\partial p}) \quad (13)$$

$$\dot{q} = 3G \cdot \dot{\varepsilon}_{q}^{e} = 3G \cdot (\dot{\varepsilon}_{q} - \dot{\varepsilon}_{q}^{p}) = 3G \cdot (\dot{\varepsilon}_{q} - \dot{\Lambda} \cdot \frac{\partial f}{\partial q}) \quad (14)$$

In eq.12 we may substitute $\dot{p}$ and $\dot{q}$ with eq.13 and eq.14 respectively, to finally compute $\dot{\Lambda}$ in terms of the total strain increments, suction and effective degree of saturation increments.

$$\dot{\Lambda} = \frac{K\frac{\partial f}{\partial p}\dot{\varepsilon}_{vol} + 3G\frac{\partial f}{\partial q}\dot{\varepsilon}_q + \frac{\partial f}{\partial s}\dot{s} + \frac{\partial f}{\partial S_r^e}\dot{S}_r^e}{H + K\left(\frac{\partial f}{\partial p}\right)^2 + 3G\left(\frac{\partial f}{\partial q}\right)^2} \quad (15)$$

3.4 Hardening rule

As already mentioned, in the proposed model, the saturated preconsolidation pressure is selected as the hardening variable. The MCC volumetric, hardening rule is used to describe its evolution and thus, we may write:

$$\dot{P}_0^* = \frac{v}{\lambda - \kappa} P_0^* \dot{\Lambda} \frac{\partial f}{\partial p} \quad (16)$$

where λ the MCC compressibility parameter describing the slope of virgin compression lines under saturated conditions.

3.5 Consistency condition – Plastic modulus

Finally, the consistency condition is employed to calculate the Plastic Modulus H. Consistency condition ensures that plastic states remain on the Yield Surface and is mathematically described by:

$$\dot{f} = \frac{\partial f}{\partial p}\dot{p} + \frac{\partial f}{\partial q}\dot{q} + \frac{\partial f}{\partial s}\dot{s} + \frac{\partial f}{\partial S_r^e}\dot{S}_r^e + \frac{\partial f}{\partial P_0^*}\dot{P}_0^* = 0 \quad (17)$$

Using eq.12 in eq.15 we can write that:

$$\dot{\Lambda} H + \frac{\partial f}{\partial P_0^*}\dot{P}_0^* = 0 \quad (18)$$

While further substituting $\dot{P}_0^*$ with eq.16, Plastic Modulus can be finally expressed as:

$$H = -\frac{\partial f}{\partial P_0^*}\frac{v}{\lambda - \kappa} P_0^* \frac{\partial f}{\partial p} \quad (19)$$

The derivatives of the yield surface with respect to the external variables and the hardening variable, necessary in the previous calculations, are given in Appendix I.

4 Model predictions

This section evaluates the model's predictions through a series of simple material point simulations, based on a fictitious set of parameters.

To do so, the model is implemented in a single point algorithm, while an explicit integration scheme is used to solve the constitutive equations. The algorithm is capable of imposing common laboratory stress paths.

Regarding simulations under unsaturated conditions, suction is an input to the algorithm while degree of saturation is calculated through the Gallipoli's WRM [5].

$$S_r = \frac{1}{\left[1+\left(\left(\phi(v-1)^{\psi}\right)s\right)^n\right]^m} \tag{20}$$

The selected WRM can reproduce an evolving degree of saturation even during compression under constant suction due to its void ratio dependence. Equation 21 is used to calculate the increment of degree of saturation with suction and volumetric strains.

$$\dot{S}_r = \frac{\partial S_r}{\partial s}\dot{s} + \frac{\partial S_r}{\partial v}(-v)\dot{\varepsilon}_{vol} \tag{21}$$

Finally, to scale degree of saturation for macro-structure the Alonso et. al. [6] power law is used. The effective degree of saturation and its corresponding increment, are calculated according to:

$$S_r^e = (S_r)^a \tag{22}$$

$$\dot{S}_r^e = \frac{\partial S_r^e}{\partial S_r}\dot{S}_r^e = a(S_r)^{a-1}\dot{S}_r \tag{23}$$

Table 1 presents the material parameters selected for the analysis. The selected parameters, although fictitious, are coherent, and representative of the typical range of values for a silt. Four (4) extra parameters with respect to the MCC model are needed for the mechanical behaviour. Of course additional variables are required for the WRM and to scale degree of saturation for macro-structure.

Table 1. Material parameters used in the analysis.

Mechanical MCC		**Mechanical (UNSAT)**	
M	1.15	r	0.8
λ	0.07	β	100 kPa^{-1}
κ	0.01	γ	0.8
v	0.333	P_c	1 kPa
Water Retention Model (WRM)			
ϕ	0.011	m	0.567
ψ	4.0	n	1.005
Alonso et. al. power law (2005)			
a	2.5		

Figure 3 plots the WRCs assumed for different specific volume values as well as the corresponding evolution of the $s \cdot S_r^e$ term.

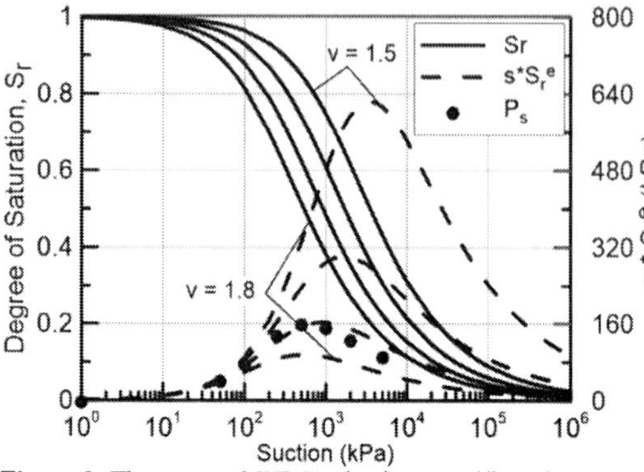

Figure 3. The assumed WRC's for four specific volume values (v=1.5, 1.6, 1.7, 1.8) and the corresponding $s \cdot S_r^e$ term.

4.1 Constant suction isotropic compression

Figure 4 presents the results of isotropic, constant suction compression tests under seven (7) different suction levels, ranging from $s = 50\,\text{kPa}$ to $s = 5\,\text{MPa}$, thus covering an extended portion of the WRC. Initially over-consolidated soil samples with $P_0^* = 100\,\text{kPa}$ are dried under constant net mean stress $\bar{p} = 50\,\text{kPa}$ by raising suction to the desired level. To calculate the initial specific volume, $N_{iso} = 2.10$ was assumed. Isotopic compression up to a men net stress of $\bar{p} = 5$ MPa is finally imposed.

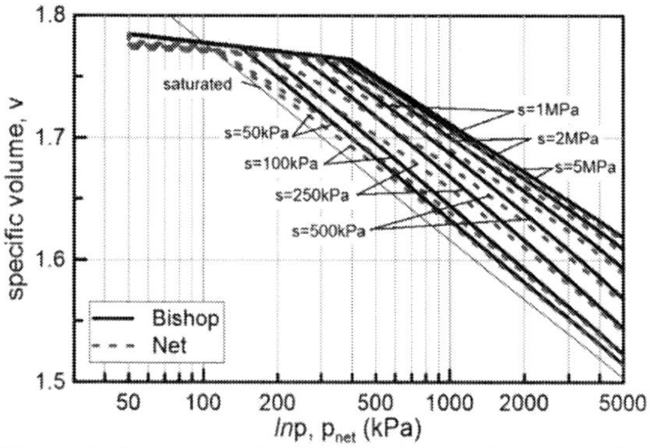

Figure 4. Constant suction isotropic compression tests under various suction levels. Behaviour on the $v - \ln p, \bar{p}$ planes.

We may observe that the proposed framework can successfully describe a non-linear increase in the yield stress with increasing suction, as well as a constantly increasing post yield compressibility for compression under constant suction.

In the elastic domain, as expected, the results coincide when plotted in terms of Bishop stress. Nevertheless, in terms of net stress, there is a clear decrease in the specific volume under constant net stress, associated with the drying stage, while during the subsequent compression under constant suction different swelling lines apply to different suction levels, while their slope evolves with net stress, as has been explained in Jommi [7].

Regarding the post yield compressibility, we may observe that the higher the suction the higher the specific volume that the soil can retain under any given net stress level, reflecting the beneficial effects of partial saturation in stiffening the response. Moreover, the constantly evolving compressibility reproduces convex net stress compression lines, indicating a decreasing initial post yield compressibility with suction, which constantly increases with compression, while at some point becomes even greater than that of the same material under saturated conditions. The reproduced compression lines are very reminiscent of frameworks usually underlying constitutive models which try to reproduce a maximum of collapse, while in the proposed model it is a natural outcome of the dependence on both suction and degree of saturation.

4.2 Constant suction triaxial compression

The isotropic compression simulations presented in the previous section, are repeated, and for each suction level imposed. Compression is now stopped at three different mean net stress levels ($\bar{p}$ = 500, 1000 and 1500kPa), and the soil elements are subjected to triaxial compression under drained, constant suction conditions until critical state is reached.

All elements are sheared from normally consolidated condition (wet side), as the selected net stress levels are higher than the apparent preconsolidation pressure developed during drying. All elements exhibit a strain-hardening response and a contractant volumetric behaviour, while ultimate conditions dependent on both suction and net stress.

Figure 5 presents the ultimate stress points reached in terms of net stress. We may observe that as expected, different CSLs apply for different suction levels in the net stress domain. The tensorial translation P_s corresponding to each line has been included in figure 2, to compare with the $s \cdot S_r^e$.

Figure 6 plots the normalized ratio ($q_{u,s} / q_{u,s=0}$) along with M(s), representing the strength increase with suction, with respect to the shear strength of the same material under the same confining stress when saturated under zero suction.

As expected, in general it follows the $s \cdot S_r^e$ term, nevertheless due the fact that the points corresponding to the same suction level, do not correspond to a common S_r^e value the obtained P_s values represent a weighted behaviour.

Another effect of the evolving S_r is that the CSLs are not parallel to the saturated CSL but their slope depends on suction. Figure 6 plots the evolution of M with suction. We may observe that it slightly increases with suction until s=1000kPa and then practically stabilizes. We shall mention that various researches have experimentally defined such suction dependent CSLs with Wheeler and Sivakumar [8] being amongst the first.

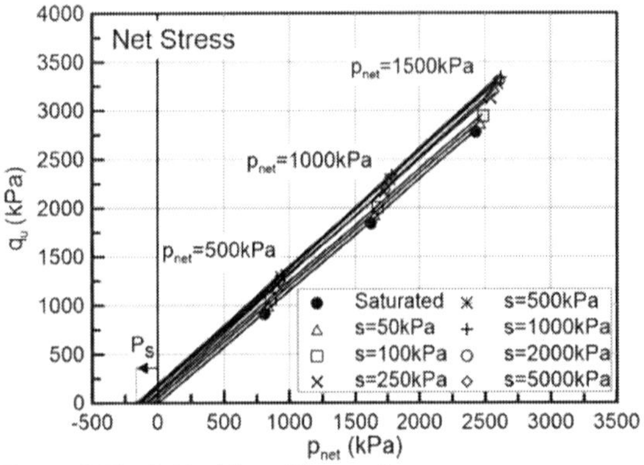

Figure 5. The Critical State Lines on the p_{net}-q.

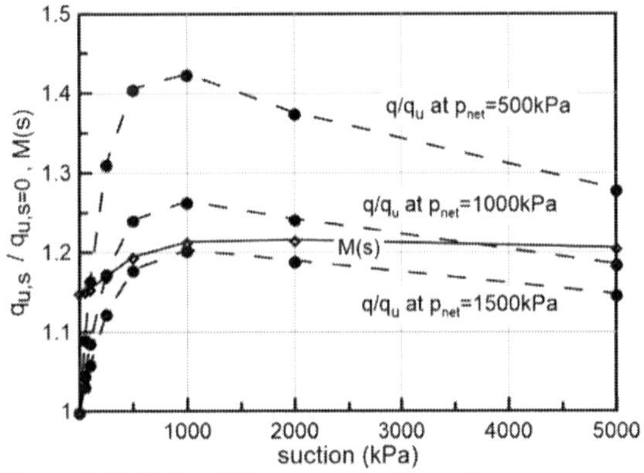

Figure 6. Evolution of shear strength with suction.

The model can represent a non-linear increase in the shear strength with suction, similar to what is usually observed in experimental results. A maximum is reached for s=1MPa (for the selected parameters), while shear strength drops for higher suction values. Moreover, the results indicate that as net stress increases, the effect of partial saturation in strengthening the soil behaviour decreases.

Qualitatively, the evolution of shear strength follows the evolution of the $s \cdot S_r^e$, a common outcome of Bishop's stress incorporation in effective stress models, and also very helpful in terms of calibration. Scaling the degree of saturation for macrostructure though is very essential, for the $s \cdot S_r^e$ term not to obtain extreme values.

4.3 Wetting tests – volumetric collapse

Finally, once again based on the isotropic compression tests previously reported, wetting tests are simulated under different suction and net stress levels. Wetting tests are simulated by gradually reducing suction to zero, while maintaining the net mean stress constant.

Figure 7 presents indicative results for simulations regarding s=100kPa under seven different constant confining pressures given in the graph. We may observe that swelling behaviour is reproduced inside the elastic region while for stress levels higher than the preconsolidation pressure, the stress state lays on the LC curve and thus, collapse strains are reproduced.

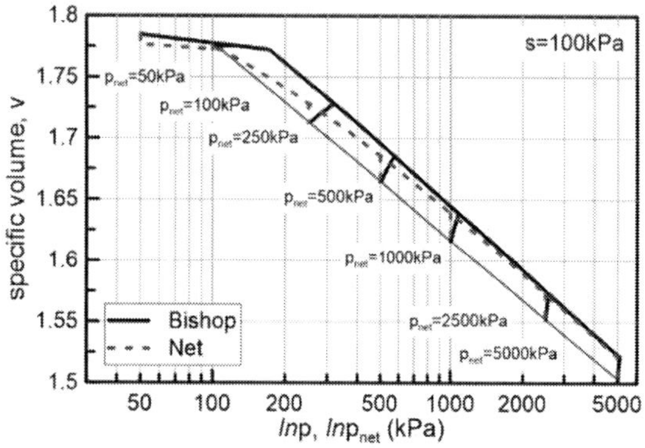

Figure 7. Wetting tests at various stress levels for $s = 100kPa$.

Figure 8 summarizes the obtained volumetric response. Only the volumetric strains associated with the wetting phase are plotted. The results reveal that, the proposed model can reproduce typical collapse behaviour, usually observed in experimental results (i.e., [9]).

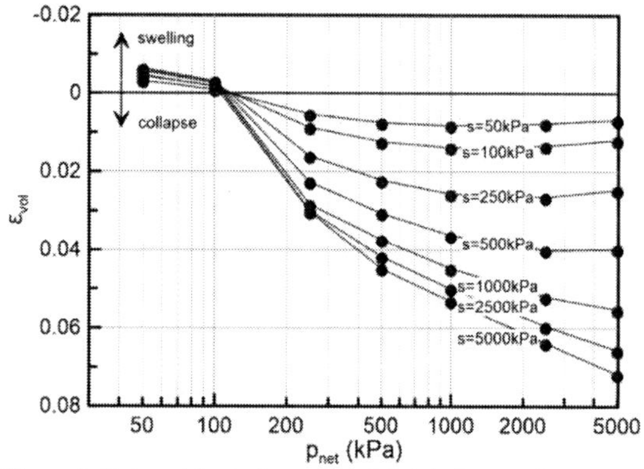

Figure 8. Volumetric strains accumulated during wetting tests.

In more details, swelling or collapse behaviour is reproduced depending on the level of the applied net stress. Moreover, the calculated strains increase with increasing suction under any given net stress level, while when examined under a constant suction level, collapse strains increase with increasing net stress, until a maximum is reached, followed by a reduction in the amount of collapse.

For the parameters selected, a maximum of collapse appears at $\bar{p} = 1\,\text{MPa}$, for the elements with an initial suction up to $s = 250\,\text{kPa}$, while for higher initial suction no maximum appears up to the net stress tested. Compression to higher net stress levels is required with increasing suction, due to the increased soil stiffness.

5 Conclusions

The mathematical formulation and typical predictions of a constitutive model for unsaturated soils were presented. The model is based on the MCC with the addition of a relatively simple framework to account for structure induced by partial saturation.

The simulation results indicate that at least in a qualitative way the model can represent typical behaviour of unsaturated soils, like increasing yield stress and decreasing post yield stiffness with increasing suction, as well as non-linear evolution of the shear strength with suction. Moreover, as far as a void ratio dependent WRM is used, it can also reproduce a maximum of collapse.

Further evaluation of the proposed framework against actual experimental data is still necessary.

Appendix I

$$\frac{\partial f}{\partial p} = -M^2 p\left(P_0\left(s, S_r^e\right) - p\right)$$

$$\frac{\partial f}{\partial q} = 2q$$

$$\frac{\partial f}{\partial s} = \frac{\partial f}{\partial P_0(s,S_r^e)} \cdot \frac{\partial P_0(s,S_r^e)}{\partial \lambda(s,S_r^e)} \cdot \frac{\partial \lambda(s,S_r^e)}{\partial s}$$

$$\frac{\partial f}{\partial S_r^e} = \frac{\partial f}{\partial P_0(s,S_r^e)} \cdot \frac{\partial P_0(s,S_r^e)}{\partial \lambda(s,S_r^e)} \cdot \frac{\partial \lambda(s,S_r^e)}{\partial S_r^e}$$

$$\frac{\partial f}{\partial P_0^*} = \frac{\partial f}{\partial P_0(s,S_r^e)} \cdot \frac{\partial P_0(s,S_r^e)}{\partial P_0^*}$$

where:

$$\frac{\partial f}{\partial P_0(s,S_r^e)} = -M^2 p$$

$$\frac{\partial P_0(s,S_r^e)}{\partial \lambda(s,S_r^e)} = P_0(s,S_r^e)\ln\left(\frac{P_0}{P_c}\right)\frac{\kappa - \lambda}{\left(\lambda(s,S_r^e) - \kappa\right)^2}$$

$$\frac{\partial \lambda(s,S_r^e)}{\partial s} = -\lambda\beta(1-r)\left(1-S_r^e\right)^{\gamma} e^{-\beta s}$$

$$\frac{\partial \lambda(s,S_r^e)}{\partial S_r^e} = \lambda\gamma(1-r)\left(1-e^{-\beta s}\right)\left(1-S_r^e\right)^{\gamma-1}$$

$$\frac{\partial P_0(s,S_r^e)}{\partial P_0^*} = \frac{P_0(s,S_r^e)}{P_0^*} \cdot \frac{\lambda - \kappa}{\lambda(s,S_r^e) - \kappa}$$

References

1. P. Sitarenios, M. Kavvadas, Unsaturated Soils: Research & Applications **1**, 449 (2014)
2. A.N. Zhou, D. Sheng, S.W. Sloan, A. Gens, Computers and Geotechnics. **43**, 178 (2012)
3. F. Casini, Canadian Geotechnical Journal. 49, 954 (2012)
4. D. Gallipoli, A. Gens, R. Sharma, J. Vaunat, Géotechnique. **53**, 123 (2003)
5. D. Gallipoli, S. Wheeler, M. Karstunen, Géotechnique. **53**, 105 (2003)
6. E.E. Alonso, J.M. Pereira, J. Vaunat, S. Olivella, Géotechnique. **60**, 913 (2010)
7. C. Jommi, Experimental evidence and theoretical approaches in unsaturated soils. 139 (2000)
8. S.J. Wheeler, V. Sivakumar, Géotechnique **45**, 35 (1995)
9. E.C. Lawton, R.J. Fragaszy, J.H. Hardcastle, Journal of Geotechnical Engineering **115**, 1252 (1989).

Prediction of wetting-induced volume change behaviour of compacted unsaturated soils in the context of the MPK framework

Arunodi Abeyrathne[1,a], Jayantha Kodikara[1] and Ha Bui[1]

[1]*Monash University, Melbourne, Australia*

Abstract. Compacted unsaturated soils show distinct behavioural patterns when compared with their saturated counterparts, and these behavioural patterns become more and more complex when soil is subjected to wetting, such as swelling or collapse and loss of shear strength. Ignoring these possible complex behaviours of unsaturated soils coupled with varying climatic conditions can lead to inaccurate assessment of the performance of geotechnical structures. Therefore, the accurate prediction of the wetting-induced volume change behaviour of compacted unsaturated soils is a key step in understanding and modelling unsaturated soil behaviour. Of the different constitutive models introduced to date which can define the wetting-induced volume change behaviour of compacted unsaturated soils reasonably satisfactorily, the MPK framework for the volumetric constitutive behaviour of compacted unsaturated soils proposed by Kodikara (2012) has shown favourable results with its direct relationship to moisture variation. This approach provides a practical approach, where reliance is placed on the variations in the moisture or degree of saturation as input. From this perspective, this study evaluates the volumetric behaviour of compacted unsaturated soils due to various wetting state paths, and the results are compared with the predictions of the MPK framework. The results show that the MPK framework is capable of predicting wetting-induced volume change behaviour, including the wetting-induced yielding of compacted unsaturated soils.

1 Introduction

The volume change behaviour of unsaturated soils has been a topic of debate among researchers for many years. This is due to the co-existence of air and water in void spaces, which complicates the mechanical behaviour of unsaturated soils in contrast to the principles of saturated soil mechanics. Over the last few decades considerable progress has been made into understanding and modelling unsaturated soil behaviour and there are now a number of models that can define unsaturated soil behaviour reasonably satisfactorily. (Alonso et al. [1], Wheeler and Sivakumar [2], Gallipoli et al. [3], Sun et al. [4]) Among them, the models capable of explaining compacted unsaturated soil behaviour are of significant practical concern, as the occurrence of unsaturated soils as a result of compaction is common in engineering practice in the form of embankments, earth dams and backfilling for other geo-structures.

Although compacted unsaturated soils characteristically have high strength and low compressibility, they may undergo large deformations and loss of shear strength as a result of being subjected to varying environmental loading conditions. One example is the collapse settlement of a compacted sandy clay fill under its own weight several years after construction as a result of being subjected to strong increase in surface infiltration, as reported by Brandon et al. [5]. This is due to the potential for swelling or collapse upon wetting depending on the confining stress, which is one of the key volumetric behavioural patterns of compacted unsaturated soils. Therefore, the accurate prediction of the wetting-induced volume change behaviour of compacted unsaturated soils is an important step in understanding and modelling compacted unsaturated soil behaviour.

The common practice in modelling the volumetric behaviour of unsaturated soils is in terms of two independent state variables: mean net stress and matric suction, and wetting-induced volume change behaviour is associated with the variation of suction. The very first model was proposed by Alonso et al. [1] in the net stress-suction space, known as the Barcelona Basic Model (BBM), which was then followed by many other researchers, including Wheeler and Sivakumar [2], who utilised the same stress space to model the mechanical behaviour of compacted unsaturated soils. Some other researchers have tried to deviate from the traditional choice of state variables, employing alternative pairs such as effective stress-suction, and the net stress-moisture ratio. One such model is the MPK framework introduced by Kodikara [6] to model compacted unsaturated soil behaviour in the net stress-moisture ratio space.

The MPK framework for the volumetric behaviour of compacted unsaturated soils utilises the net stress and moisture ratio, which is the product of gravimetric

[a] Corresponding author: Arunodi.Abeyrathne@monash.edu

moisture content and the specific gravity (giving the ratio of volume of water divided by volume of solids). This model has shown favourable results to date, as reported by Islam [7], Islam and Kodikara [8] and Kodikara et al. [9]. More importantly, with its direct relationship to moisture variation, wetting-induced volume change behaviour, including wetting-induced yielding, can be defined in a relatively straightforward manner in the context of this framework. In this perspective, the present paper examines the application of the MPK framework to the interpretation of the wetting- induced volume change behaviour of compacted unsaturated soils.

2 MPK Framework

The basic concept of the MPK framework is that there exists a state boundary surface in void ratio (e), moisture ratio (e_w) and net stress (p) space, referred to as the Loading Wetting State Boundary Surface (LWSBS). The LWSBS represents the loosest states that compacted unsaturated soil can obtain from loading and (or) wetting paths and it can be produced by a family of compaction curves at different compaction stresses. The LWSBS as presented by Kodikara [6] is shown in Figure 1.

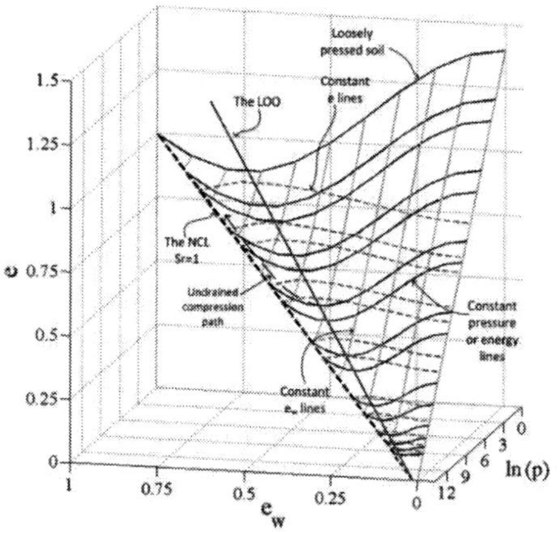

Figure 1. The LWSBS in $e - e_w - ln(p)$ (adapted from Kodikara,[6])

One of the striking features of the LWSBS as shown in Figure 1 is the Line of Optimums (LOO), marked by the line joining the optima of each compaction curve, which defines the boundary for the continuous air phase of compacted unsaturated soils. According to the LWSBS concept, on the dry side of LOO, the air phase is postulated to be continuous with air free to drain during loading and wetting, while on the wet side of LOO air is trapped within the water menisci. Compaction curves represent the constant net stress contours of the LWSBS. However, when constructing the LWSBS for the wet side of LOO, curves corresponding to drained paths have been used as the compaction curves symbolize the (water) undrained compression paths where achievement of full saturation is difficult. The normally-consolidated line (NCL) for saturated soils (from the loosely compacted state) marks the intersection of the LWSBS with the saturation plane, where constant net stress contours are generally pinned.

In the MPK framework this state boundary surface is used to define the loosest states of compacted soil governing various volume change behavioural patterns of compacted unsaturated soils under loading and/or wetting, including yielding due to loading at constant moisture content, collapse and/or swelling due to wetting at constant operational stress, dependence of collapse potential on the operational stress, and change of yield pressure during wetting. It could be considered analogous to the extension of the saturated normally-consolidated line into unsaturated space. For example, a loading of a loose soil at constant moisture content state path should follow the constant moisture ratio contour on the LWSBS from a nominal stress to a certain net compressive stress. If that soil is wetted at that compressive stress, the soil should collapse following the current net stress contour of the LWSBS to the final moisture content.

3 Experimental Work

3.1 Materials and methods

Commercially-available kaolin, known by the trade name of Eckalite 1, was used throughout the experimental program. Kaolin is white-coloured lightly reactive clay which has been widely used in unsaturated soil modelling for many years. The kaolin used in the present research program had a clay content of 75% with liquid and plastic limits of 60.5% and 27.9% respectively.

A number of laboratory experiments were carried out on kaolin soil under isotropic compression using the GCTS unsaturated soils tri-axial test system to develop the LWSBS and to validate it under independent state paths. The GCTS unsaturated soils tri-axial test system is an electro-pneumatic system with closed-loop digital servo control capable of performing fully automated tri-axial tests on saturated or unsaturated test specimens. The LWSBS was generated through constant moisture content isotropic compression tests in tri-axial test system performed on samples prepared as loosely as possible at different moisture contents. Validation tests were performed independently employing various loading/unloading, wetting or combination state paths with non-decreasing degrees of saturation.

Two validation experimental results, a loading/wetting state path validation test and a loading/unloading/wetting/loading state path validation test are presented here, giving emphasis to the wetting state paths. The loading/wetting state path test was performed by initially mixing dry powder kaolin with water to a moisture content of 19.2% (e_w = 0.51), compacting in five layers at a vertical compaction stress of 50 kPa to prepare the test specimen, then isotropically compressing the test specimen to a maximum of net stress of 600 kPa in tri-axial test system at constant moisture content under the loading stage and finally wetting to a moisture content of 28.3% (e_w = 0.75) at constant net stress of 600 kPa under the wetting stage. For the loading/unloading/wetting/loading state path test the procedure was similar until the first loading stage to a net stress of 700 kPa. Thereafter, the soil was unloaded to a net stress of 400 kPa at constant moisture content of 17.1% ($e_w = 0.45$) before wetting to a moisture content of 33.6% ($e_w = 0.88$) at constant net stress, followed by a second loading stage.

3.2 Results and discussion

Figures 2(a) and 3(a) show the results of the two validation experimental tests described above in $e - e_w$ plane while Figures 2(b) and 3(b) show their three-dimensional representations in $e - e_w - p$ space respectively.

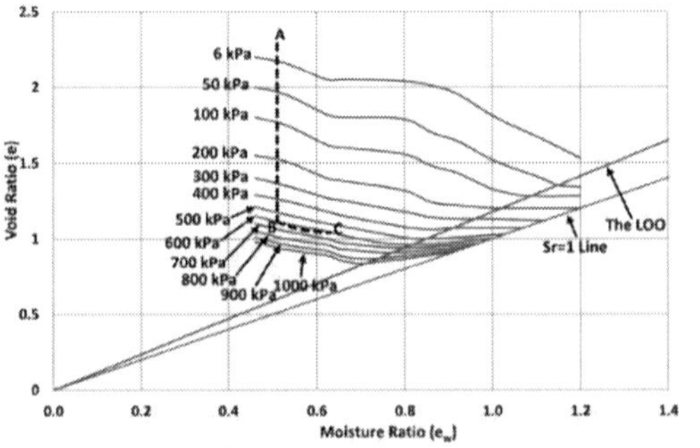

(a) In $e - e_w$ plane

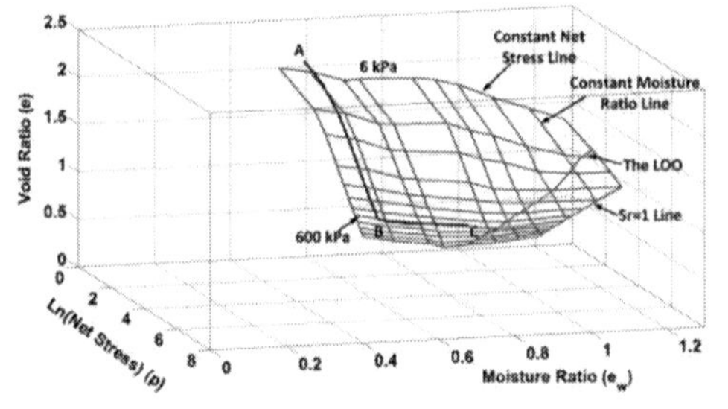

(b) In $e - e_w - lnp$ space

Figure 2. Loading – wetting state path test

As shown in Figures 2(a) and 3(a), the void ratio decreased during isotropic loading at the respective constant moisture content and during unloading the void ratio increased. During wetting the void ratio continually decreased in Figure 2(a) where as in Figure 3(a) the void ratio slightly increased until point D and thereafter the void ratio decreased.

According to the MPK framework, the loading/wetting state path test should follow the LWSBS, and it is clear in Figure 2 that the soil compressed following the constant moisture ratio contour of the LWSBS during loading (the AB line in Figure 2) and the constant net stress contour of the LWSBS during wetting (the BC line in Figure 2). Similarly, as explained in Abeyrathne et al. [10], it is evident in Figure 3 that for the loading/unloading/wetting/loading test the soil also followed the LWSBS, thus confirming the state boundary concept of the MPK framework. Since soil has moved into the unloading space of LWSBS during unloading, the soil swelled until it reached the LWSBS at D, indicating a volume increase during subsequent wetting. It is important to note here that wetting-induced volume change behaviour can take two forms: either swelling or collapse, depending on the position of soil with respect to the LWSBS, as is evident in Figures 2 and 3. If the soil has already yielded on the LWSBS, the soil will continue to collapse during wetting (the BC line in Figure 2). However, if the soil has previously yielded and has moved into the unloading space inside the LWSBS during unloading, the soil could swell until it intercepts the corresponding net stress contour of the LWSBS (the CD line in Figure 3). Wetting beyond that would lead to collapse following the LWSBS.

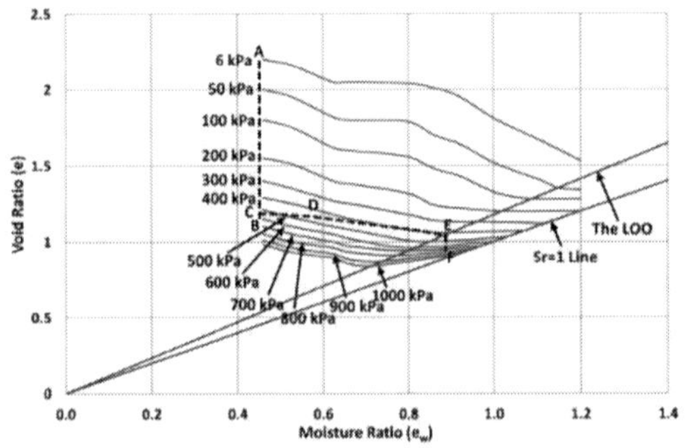

(a) In $e - e_w$ plane

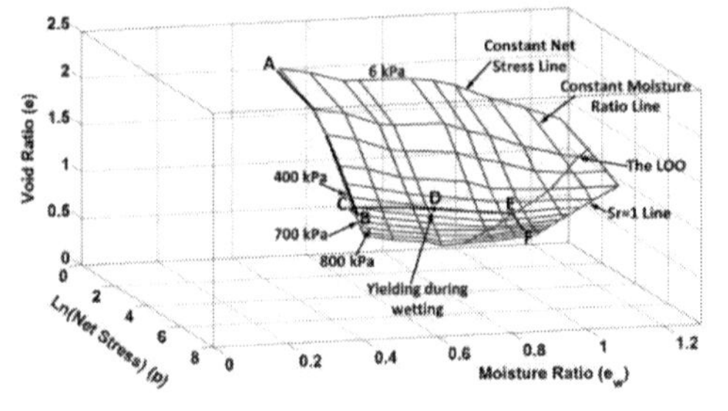

(b) In $e - e_w - lnp$ space

Figure 3. Loading – unloading – wetting – loading state path test

Another important feature of the LWSBS concept is its ability to explain yielding of compacted unsaturated soils during loading and/or wetting. The traditional approach to defining yielding of unsaturated soils is in the form of yield curves, where the loading-collapse (LC) yield curve is specifically used in volumetric space. The LWSBS of the MPK framework acts as a yield surface for compacted unsaturated soils in volumetric space and the soil will start yielding and generate plastic volumetric strains once it reaches the LWSBS. Therefore, the yielding of compacted unsaturated soils during wetting state paths can be easily captured by the LWSBS. For example, the wetting path CE in Figure 3 will yield at D when it reaches the corresponding net stress contour of LWSBS during swelling.

If the LC curve for the MPK framework is plotted in the moisture ratio-net stress space, which can be approximated to the projection of LWSBS into that stress space, the two yield points B and D would lie on two different LC curves, indicating the yielding of compacted unsaturated soils during wetting from an unloaded position, as shown in Figure 4. P_B and P_D are the yield stresses of 700 kPa and 400 kPa respectively. This is in contrast to one of the key assumptions of the LC curve model in the BBM that any soil behaviour inside the LC curve is elastic. This has been pointed out by some other researchers, including

Zakaria [11]. The path DE is also marked in Figure 4, where subsequent collapse deformation has increased the yield point again at the LOO, as shown.

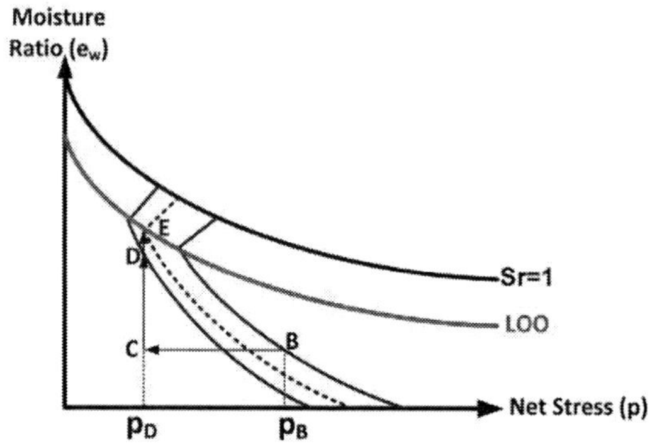

Figure 4. Loading-collapse yield curves for CE wetting path of Figure 3

4 Conclusion

The applicability of the LWSBS concept of the MPK framework in predicting the wetting-induced volume change behaviour of compacted unsaturated soils was evaluated using loading/unloading, wetting or combination state paths with non-decreasing degrees of saturation. The results prove that wetting-induced volume change behaviour, including wetting-induced yielding, is well captured by the LWSBS of the MPK framework.

The major advantage of the model is its reliance on relatively simple constant moisture content compaction testing in constructing the LWSBS (or the compaction surface). Furthermore, its direct relationship with moisture content facilitates the straightforward prediction of wetting-induced volume change behaviour, in contrast to suction-based models where change in suction due to wetting is measured.

On the basis of current progress in modelling compacted unsaturated soil behaviour in the context of the MPK framework, the framework may be used to answer a range of state paths applicable to field applications.

References

1. E. E. Alonso, A. Gens, A. Josa, Geotechnique, *A constitutive model for partially saturated soils*, **40**, 405-430 (1990)
2. S. J. Wheeler, V. Sivakumar, Geotechnique, *An elasto-plastic critical state framework for unsaturated soil*, **45**, 35-53 (1995)
3. T. Brandon, J. Duncan, W. Gardner, J. Geotech. Eng., *Hydro-compression Settlement of Deep Fills,* **116**, 1536-1548 (1990)
4. D. Gallipoli, A. Gens, R. Sharma, J. Vaunat, Geotechnique, *An elasto-plastic model for unsaturated soil incorporating the effects of suction and degree of saturation on mechanical behaviour*, **53**, 123-135(2003)
5. D. A. Sun, H. B. Cui, H. Matsuoka, D. C. Sheng, Soils Found, *A three-dimensional elastoplastic model for unsaturated compacted soils with hydraulic hysteresis*, **47**, 253-264 (2007)
6. J. Kodikara, Can. Geotech. J., *New framework for volumetric constitutive behaviour of compacted unsaturated soils*, **49**, 1227-1243 (2012)
7. T. Islam, PhD thesis, *A study of volumetric behaviour of compacted clayey soils in void ratio, moisture ratio and net stress space*, Monash University, Melbourne (2015)
8. T. Islam, J. Kodikara, Can. Geotech. J., *Interpretation of the loading/wetting behaviour of compacted soils within the MPK framework: Part I Static compaction*, (2015)
9. J. Kodikara, T. Islam, P. Rajeev, Can. Geotech. J., *Interpretation of the loading/wetting behaviour of compacted soils within the MPK framework: Part II Dynamic compaction*, (2015)
10. A. Abeyrathne, T. Islam, J. Kodikara, H. Bui, Proceedings of 6th Asia-Pacific conference on Unsaturated Soils, *Application of MPK framework to interpret volume change behaviour of compacted unsaturated soils*, 749-752 (2015)
11. I. Zakaria, PhD thesis *Yielding of unsaturated soil*, University of Sheffield, Sheffield (1994)

 DOI: 10.1051/e3sconf/20160918001

Changes in the microstructure of compacted bentonite caused by heating and hydration

M.V. Villar[a], V. Gutiérrez-Rodrigo, R.J. Iglesias, R. Campos and L. Gutiérrez-Nebot

CIEMAT. Avd. Complutense 40, 28040 Madrid, Spain

Abstract. Two twin 40-cm long columns of compacted FEBEX bentonite were tested in Teflon cells; water was supplied through the top surface of the columns and in one of them a heater was placed at the base and set to 100°C. The purpose of these tests was to simulate the behaviour of an engineered barrier in a radioactive waste repository and investigate the effect of the thermal gradient on saturation. In particular, changes in the pore size distribution and interlayer size have been investigated in this work. The thermal gradient had a strong influence on the water intake and distribution. Water content and dry density gradients persisted in the two columns after 12 years of testing. These changes gave place to the modification of the bentonite microstructure, overall increasing the microstructural void ratio and the proportion of adsorbed, interlayer water.

1 Introduction

Geological disposal is the preferred, internationally accepted option for wastes with high levels of radioactivity or significant contents of long-lived radionuclides. The safety of repositories for radioactive waste is based on the multiple barrier concept, whereby both engineered and natural barriers between the waste and the surface act in concert to contain the wastes [1]. Deep geological repository concepts for nuclear waste disposal may include a clay barrier around the waste containers, which is the main retarding/retention component once the failure of the canister and the oxidation/dissolution of the combustible matrix are produced. Bentonites or bentonite-based materials have been usually proposed to construct this buffer around the waste container, because of their high retention capacity, low permeability and high swelling ability. The heat released by the waste will induce a thermal gradient through the bentonite barrier while groundwater will tend to flow into it. As a consequence, coupled thermal, hydraulic, mechanical and geochemical processes will take place during the transient period of the repository live. The understanding of these processes is key for the prediction of the repository evolution.

For example, it still remains unclear whether the high temperatures around the canister would hinder the full saturation of the inner part of the barrier or just delay it, but this seems to closely depend on the actual temperatures reached in the barrier and on its thickness. In most repository concepts it is expected that the temperature in the buffer be less than 100°C. At the Buffer Container Experiment performed at Lac du Bonnet (Canada), the areas of the buffer –whose thickness was 25 cm– adjacent to the heater had water contents below the initial one after 2.5 years of heating [2]. In the FEBEX in situ test, the bentonite closer to the heater had water contents below the initial ones after five years of heating (the surface temperature of the heater was of 100°C), although they were recovering after the intense initial drying (Figure 1). On the contrary, in the same period of time, the sensors located at the same distance from the gallery wall, but in an area not affected by the thermal gradient, recorded much higher relative humidity. The thickness of the bentonite barrier in this case was of 65 cm [4].

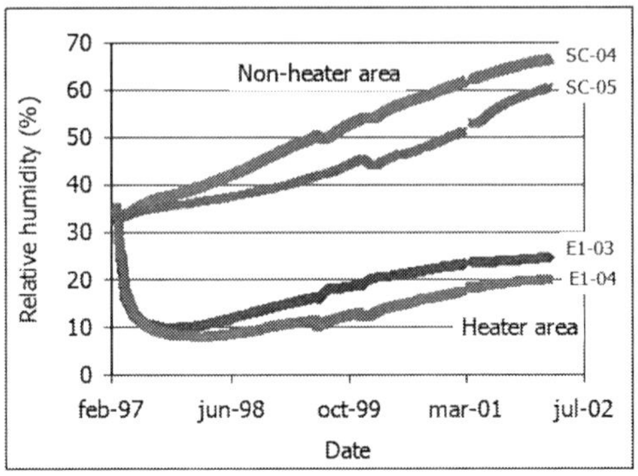

Figure 1. Evolution of the relative humidity of the bentonite recorded by two capacitive sensors located at 3.5 cm from the heater and two sensors located at approximately the same distance from the gallery wall (54 cm) in a section not affected by the heater of the FEBEX in situ test (AITEMIN data base, [3]).

[a] Corresponding author: mv.villar@ciemat.es

Conducting large-scale tests as those mentioned above is complicated and extremely costly. However, laboratory tests of different scales are very useful to identify and quantify processes in shorter periods of time. With the aim of checking the effect of the thermal gradient on water intake and redistribution, two tests were performed in cylindrical cells in which compacted bentonite was subjected to hydration through the upper surface and, in one of the cells, to heating through the base. These are called thermo-hydraulic (TH) tests and the modifications underwent by the bentonite after 12 years of testing are described below.

2 Materials and methods

2.1 Material

The FEBEX bentonite was extracted from the Cortijo de Archidona deposit (Almería, Spain) and at the factory it was disaggregated and gently ground, dried at 60°C and sieved by 5 mm, which is the maximum grain size of the material [4]. The montmorillonite content of the FEBEX bentonite is above 90 wt.%. The smectitic phases are actually made up of a smectite-illite mixed layer, with 10-15% of illite layers. Besides, the bentonite contains variable quantities of quartz, plagioclase, K-felspar, calcite and opal-CT (cristobalite-trydimite). The cation exchange capacity is 102±4 meq/100g, the main exchangeable cations being calcium (35±2 meq/100g), magnesium (31±3 meq/100g) and sodium (27±1 meq/100g).

The density of the solid particles is 2.70±0.04 g/cm^3, the hygroscopic water content 13.7±1.3 % and 67±3 percent of particles are smaller than 2 μm.

The saturated hydraulic conductivity of compacted bentonite samples is exponentially related to their dry density. For a dry density of 1.6 g/cm^3 the saturated permeability of the bentonite is about $5 \cdot 10^{-14}$ m/s at room temperature. The swelling pressure is also exponentially related to the bentonite dry density, and when the bentonite compacted at dry density 1.6 g/cm^3 is saturated with deionised water at room temperature, the swelling pressure has a value of about 6 MPa.

2.2 Thermo-hydraulic tests

The TH tests were performed in cylindrical cells whose internal diameter was 7 cm and inner length 40 cm (Figure 2). They were made of Teflon to prevent as much as possible lateral heat conduction, and externally covered with steel semi-cylindrical pieces to avoid the deformation of the cell by bentonite swelling.

Five blocks of FEBEX clay compacted with its hygroscopic water content (around 13%) at an initial nominal dry density of 1.65 g/cm^3 were stacked inside each cell. Three of the blocks had a length of 10 cm, whereas the two placed at the ends of the cells had a length of 5 cm. An average compaction pressure of 30 MPa was applied to manufacture the blocks. The bottom part of the cells was a plane stainless steel heater. In one of the tests (GT40) the clay was heated through the bottom surface at a temperature of 100°C, which is the temperature expected on the surface of the waste container in the Spanish concept [5]. The other test (I40) was carried out at isothermal conditions (laboratory temperature). The upper closing of the cells was made by means of a stainless steel plug. Inside this plug there was a deposit in which water circulated at room temperature (20-30°C). A commercial granitic water of salinity 0.02% was injected through the upper plug of the cell at a pressure of 1.2 MPa. Its chemical composition simulated the water that saturates the barrier in a repository excavated in granitic rock.

The cells were instrumented with capacitive-type sensors placed inside the clay at three different levels separated 10 cm.

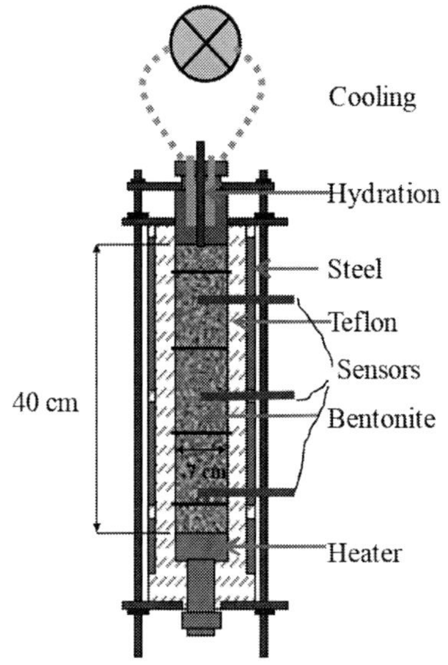

Figure 2. Schematic design of the thermo-hydraulic cells

2.3 Postmortem test

After 12 years of TH treatment the cells were dismantled and the clay columns extracted. Then the bentonite columns were sawed in 20 cylindrical sections of 2 cm in thickness used for the different analyses.

The gravimetric water content of each subsample was determined by oven drying at 110°C for 48 hours. To calculate the dry density, the volume of the specimens was determined by immersing them in a recipient containing mercury and by weighing the mercury displaced.

The pore size distribution of each subsample was determined by mercury intrusion porosimetry (MIP). Samples smaller than 3 cm^3 were lyophilised to eliminate the water in the pores. The porosimeter used was a Micromeritics AutoPore Series IV 9500, allowing the exploration of pore diameter sizes between 0.006 and 600 μm.

Other subsamples were preserved in paraffined foil and within less than an hour, the X-ray profile of a plane surface of them was registered at laboratory temperature after removing the foil and without any further treatment.

DOI: 10.1051/e3sconf/20160918001

An anticatode of Cu (CuKα) radiation was used with a Philips model X'Pert-MPD diffractometer at 40 mA, 45 kV operating condition (see [6] for more details). The *d*(001) reflection or basal spacing gives the distance along the crystallographic c-axis between clay lamellae, and for a given material depends on the degree of hydration of the interlayer.

3 Results

Online results provided by the sensors during operation were presented in [7] and [8]. After 12 years of testing the relative humidity recorded by the three sensors in column I40 was about 90%. At that time, test GT40 showed a state similar to I40 in the upper 20 cm, but the sensor located at 10 cm from the heater had recorded a strong initial drying that was not recovered after 12 years of hydration. Thus, the final average relative humidity of the bentonite column I40 was higher, because the hot zones of test GT40 remained desiccated for the whole duration of the test. In this column the temperatures were practically constant during the entire test: 30±3°C at 30 cm from the heater, 36±3°C at 20 cm and 51±3°C at 10 cm (Figure 3). The latter value indicates that the thermal gradient was very steep close to the heater.

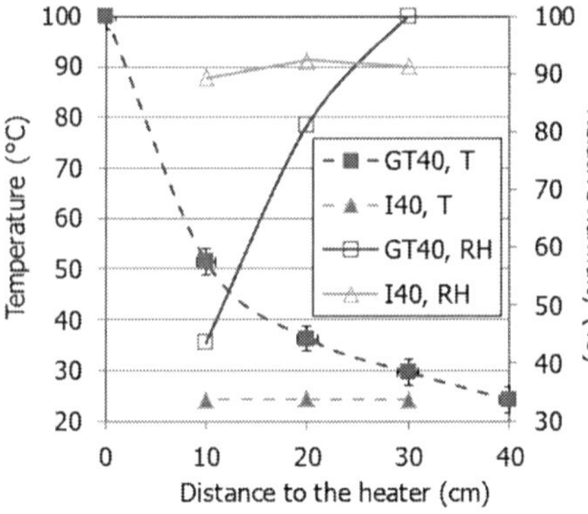

Figure 3. Steady temperatures (T) inside the bentonite during operation and final relative humidity (RH)

Upon dismantling it was observed that the five blocks that formed column I40 were sealed, whereas only the three upper ones were sealed in column GT40. It was also checked that the o-rings at the base of the latter had seriously degraded, which let evaporation take place through the bottom part of the cell for an unknown period of time during operation.

3.1 Final physical state

The gravimetric water content measured at the end of the tests ranged from 35% at the top in both columns to 23% at the bottom in the isothermal test (test I40) and to 3% near the heater in the column subjected to thermal gradient (test GT40). The dry densities had changed according to the swelling caused by hydration, being lower towards the hydration surface (Figure 4). The high swelling pressure developed by the bentonite upon saturation deformed the Teflon cells in spite of the external steel reinforcement, which allowed an overall increase of the columns' volume, causing a decrease of the bentonite dry density from 1.65 g/cm^3 to 1.46 g/cm^3 in test I40 and to 1.55 g/cm^3 in test GT40. The final average water content was higher in test I40 (27.7% *vs.* 20.4%) but the water content and dry density gradients were steeper in test GT40. The final degree of saturation was about 100% in the 5 and 10 cm closest to the hydration surface in column GT40 and I40, respectively and lower in the rest of the columns [9].

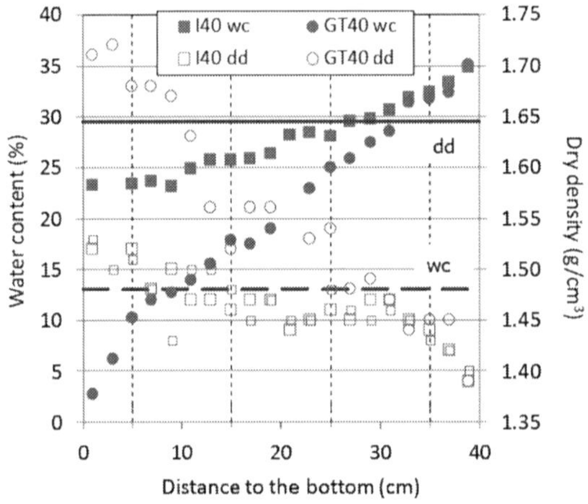

Figure 4. Final water content (wc) and dry density (dd) along the columns (the horizontal thick lines indicate the initial values and the vertical dotted lines the separation between blocks inside the columns)

3.2 Pore size distribution

The pore size distribution of samples taken along the columns was analysed by mercury intrusion porosimetry. Due to the limitations of the method, only part of the macropores (pores of sizes between $6 \cdot 10^5$ and 50 nm) and part of the mesopores (pores between 50 and 6 nm) were explored. It is not expected to find pores larger than $6 \cdot 10^5$ nm in compacted clay materials, but the number of pores smaller than 6 nm can be very relevant. To overcome this undervaluation of porosity, an estimation of the percentage of pores not intruded by mercury can be made by comparing the actual void ratio of the samples (computed from their dry density and density of solid particles) and the apparent void ratio calculated from mercury intrusion. Thus, the pore size distribution curves obtained by MIP were corrected to take into account the percentage of pores not intruded. For the sake of clarity only a few curves of each test are shown in Figure 5 and Figure 6. Both Figures include the pore size distribution curve for a sample compacted to 1.6 g/cm^3 with hygroscopic water content, which would represent the initial conditions of the bentonite blocks.

Two pore families appeared systematically in all the samples, one in the size range of macropores and another one in the size range of mesopores. This is the usual pore size distribution pattern obtained by MIP in compacted

FEBEX bentonite, irrespective of the water content or dry density. The percentage of pores intruded was on average of 50%, which is slightly below the percentage of pores intruded in the original material. Additionally, more than 60% of the pores of all the samples had a diameter smaller than 50 nm after testing, which meant a slight increase in this pore size range with respect to the original material. Both observations point to an increase in the number of the smallest pores after testing.

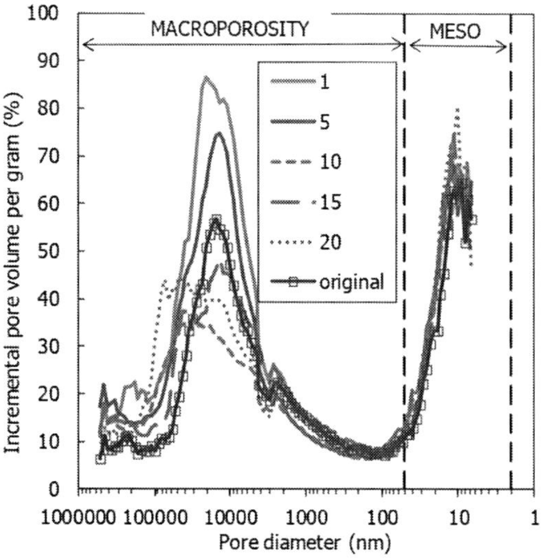

Figure 5. Pore size distribution of samples from test I40 (1: 39 cm from bottom, 5: 35 cm, 10: 21 cm, 15: 11 cm, 20: 1 cm) and of a sample compacted at 1.6 g/cm^3 with *w*=14% (original)

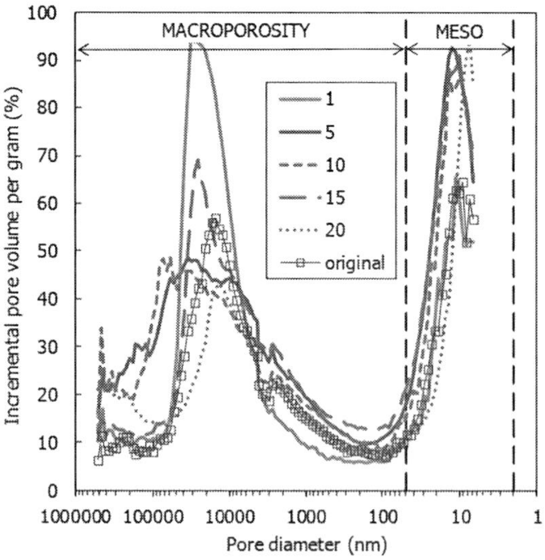

Figure 6. Pore size distribution of samples from test GT40 (1: 39 cm from bottom, 5: 35 cm, 10: 23 cm, 15: 11 cm, 20: 1 cm) and of a sample compacted at 1.6 g/cm^3 with *w*=14% (original)

Assuming that the percentage of pores not intruded by mercury corresponded entirely to pores of size smaller than 6 nm (the limit of the equipment) and assimilating this percentage to the micropore size (which is not totally right, since micropores are those smaller than 2 nm), an estimation of the percentage of micropores could be inferred from the percentage of pores non-intruded. The percentage of each pore size could thus be recalculated. Following the terminology proposed in some constitutive models [10], the pores of sizes <50 nm (i.e. micro and mesopores) would constitute the microstructure, where physico-chemical and other phenomena take place at particle level, whereas larger pores would represent the capillary-controlled macrostructure. Taking into account the measured void ratio of the samples and the recalculated percentage of each pore size, the micro and macrostructure void ratios were computed and are plotted in Figure 7. Hydration brought an increase in void ratio with respect to the original values, due to the swelling allowed by the cell deformation (see section 3.1). In column I40 there was an overall significant increase in microstructural void ratio with respect to the original material, whereas the macro void ratio increased only in the most hydrated part of the column. The macro void ratio evolution was similar in test GT40, as was the increase in microstructural void ratio in the most hydrated upper part of the column. However, at approximately 15 cm from the hydration surface, the microstructural porosity started to decrease sharply towards the heater. In fact, the final micro void ratio in the 10 cm closest to the heater in test GT40 was below the initial one. This was the part of the column where the dry density increased above the initial value due to water evaporation and shrinkage (Figure 4).

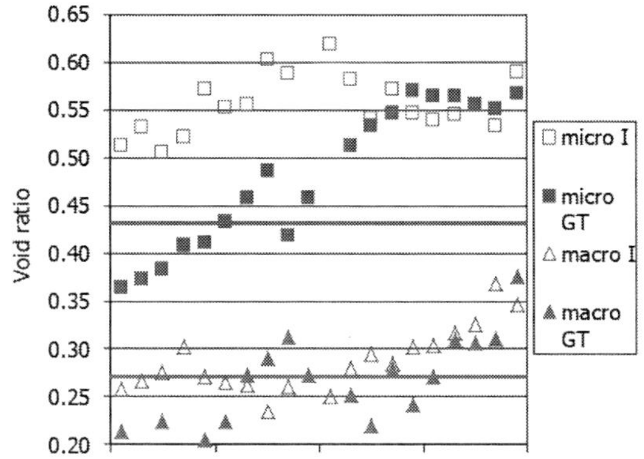

Figure 7. Microstructural and macrostructural void ratio along the two columns at the end of the tests (the horizontal thick lines indicate the initial values)

3.3 Interlayer distance

The basal spacing of the smectite was determined by X-ray diffraction in samples preserved to keep the same density and water content as upon dismantling. This value gives a measure of the interlayer distance between

 DOI: 10.1051/e3sconf/20160918001

smectite particles, a minimum value of about 1 nm corresponding to a collapsed interlayer with no water in it. It is considered that smectites display basal spacings of about 1.25, 1.55 and 1.85 nm for the homogeneous 1, 2 and 3 water layers hydration states, respectively.

The values measured were similar all along column I40 with an average value of 1.64 nm, whereas in test GT40 changed from 1.83 to 1.26 nm from the top to the bottom of the heater. This means that the interlayer of the bentonite would have between 2- and 3-layer hydrates in all the samples of test I40 and from almost 3-layer hydrates near the hydration surface to 1-layer hydrate close to the heater in test GT40 (Figure 8). Under hygroscopic water content, which was the condition of the bentonite at the beginning of the tests, the basal spacing of the FEBEX bentonite is about 1.48 nm. Hence, the interlayer distance increased during the tests all along column I40 and in the upper 25 cm of column GT40, i.e. where the water content was higher than the initial one (Figure 4).

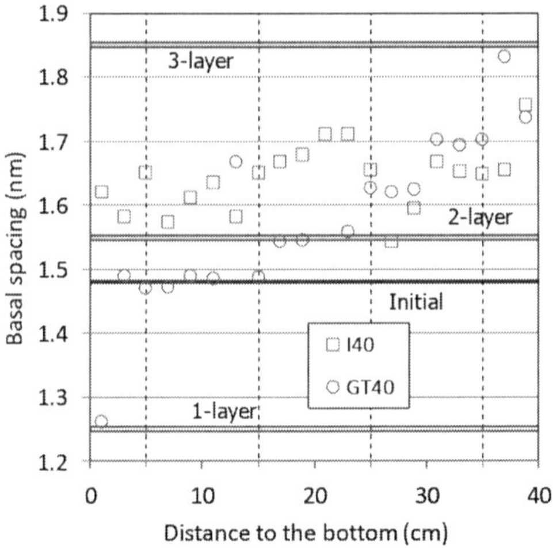

Figure 8. Basal spacings measured by XRD in unaltered samples at the end of the TH tests and in the initial sample (horizontal line). The horizontal bands indicate the range of spacings corresponding to the different hydration states

In fact, the basal spacing can be related to the water content almost lineally, although with a large dispersion for the higher water contents, as can be seen in Figure 9, where the results have been plotted again expressed in Å (1 nm = 10 Å). The basal spacings measured in compacted FEBEX bentonite saturated with water in the liquid or in the vapour phase and under free or confined volume conditions (taken from [6] are also plotted in the same figure). The dispersion of these data was attributed to the different dry density and hydration time of the samples, which was in all cases much shorter than the duration of tests I40 and GT40. It can be observed that the values for the two TH tests follow the same trend, which shows that the interlayer water retention capacity did not decrease over time, on the contrary, the values obtained after 12 years of TH treatment tended to be higher for the same water content.

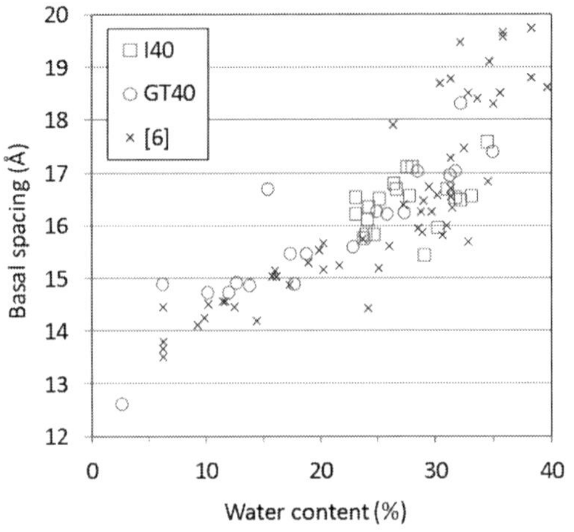

Figure 9. Basal spacing of FEBEX samples compacted to different dry densities and saturated in different ways [6] and those measured at the end of tests I40 and GT40

4 Concluding remarks

Two thermo-hydraulic tests in cells performed with compacted FEBEX bentonite run for almost 12 years. In one of them the column of bentonite was hydrated under thermal gradient (GT40) and in the other one at isothermal conditions (I40). The tests provided online information about the temperature and relative humidity inside the bentonite [7, 8]. Finally they were dismantled and the physical state of the bentonite was analysed.

There were water content and dry density gradients along both bentonite columns, sharper for cell GT40. The water content had increased with respect to the initial one all along column I40, whereas in the 15 cm closest to the heater in test GT40 the water content was lower than the initial and the dry density higher, due to evaporation and shrinkage. Despite the fact that the macrostructural void ratio increased with respect to the original value in the most hydrated part of the two columns, the final micro/macro void ratio relation was higher than the initial all along the two columns. However, the microstructural void ratio decreased towards the heater in column GT40.

The interlayer distances measured were correlated with the water content, increasing with it as expected. However, for a given water content, the basal spacings measured tended to be higher than those in samples that had the same water contents for much shorter periods of time (reported in previous investigations). This could be a consequence of the water redistribution over time, which brought an increase in the number of layer hydrates with time probably correlated with the increase in microporosity observed in the MIP tests. This implies that the maturation of the barrier would lead to a decrease in free water availability, since most water would eventually enter the interlayer porosity.

The fact that the Teflon cells could deform during the tests due to the high swelling of the bentonite upon saturation, giving place to an overall decrease of the columns density, and that evaporation took place through the bottom of column GT40 are experimental artefacts that must be taken into account when extrapolating these results to the real barrier case, in which none of these processes will take place. As well, the temperatures in column GT40 were lower than those expected in a real repository, due to the different geometry and heat dissipation allowed by the experimental setup [8].

These tests have highlighted the influence of the thermal gradient on the hydration kinetics, the persistence of the water content and dry density gradients and the slowness of the saturation process. As well, the changes in water content and density occurred gave place to the modification of the bentonite microstructure, on the whole increasing the microstructural void ratio and the proportion of adsorbed, interlayer water.

References

1. N. Chapman, C. McCombie, *Waste Management Series, 3* (Elsevier, Amsterdam, 2003)
2. D. Dixon, N. Chandler, J. Graham, M.N. Gray, Can. Geotech. J. **39**, 503-518 (2002)
3. M.V. Villar, *Informes Técnicos CIEMAT 1044* (CIEMAT, Madrid, 2004)
4. ENRESA, *Publicación Técnica ENRESA 05-0/2006* (ENRESA, Madrid, 2006)
5. ENRESA, *Publicación Técnica ENRESA 11/95* (ENRESA, Madrid, 1995)
6. M.V. Villar, R. Gómez-Espina, L. Gutiérrez-Nebot, Applied Clay Science **65-66**, 95-105 (2012a)
7. M.V. Villar, R. Gómez-Espina, *Informes Técnicos CIEMAT 1178* (CIEMAT, Madrid, 2009)
8. M.V. Villar, P.L. Martín, I. Bárcena, J.L. García-Siñeriz, R. Gómez-Espina, A. Lloret, Eng. Geology **149-150**, 57-69 (2012b)
9. M.V. Villar, R. Gómez-Espina, P.L. Martín, F.J. Romero, R.J. Iglesias, *XXIII Reunión SEA 2014, Libro de resúmenes*, 79-80 (2014)
10. A. Gens, E. Alonso, Canadian Geotech. J. **29**, 1013-1032 (1992)

Investigation on creep behavior of geo-materials with suction control technique

Tomoyoshi NISHIMURA[1]

[1]*Department of Civil Engineering, Ashikaga Institute of Technology, Tochigi, Japan*

Abstract. The compacted bentonite which has typical couple problem associated to thermal - hydration - mechanical – chemical (THMC) consist of one component of engineered barrier. Recently, the couple THMC formulation modelling suggested by some researchers can be predicted basically phenomena for engineered barrier that approach to correct evaluate satisfied facilities. The compacted bentonite is essentially unsaturated condition, some behaviors for bentonite has similar or close with generally expansive unsaturated soils. Therefore, hydrations have given significant influence on deformation of compacted bentonite such as swelling. There are many researches for swelling behavior of compacted bentonite within soaking. Extended theoretical or experimental investigations for unsaturated soil mechanics are possible to describe the strength-deformation behavior of compacted bentonite with suction controlling principle. A new method of determining the failure phase such as great axis deformation and destructions like strip of surface in the laboratory is described and the creep behavior of compacted bentonite is considered under maintain of high relative humidity environment. The creep deformation measured using improved cyclic relative humidity control apparatus in terms of specific suction control technique.

1 Introduction

Atomic power plant supply enough electrical power to industrial activity or human life that must be administrate under strict surveillance. Simultaneously, the radioactive waste disposal have been produced from atomic plant that the engineers should establish the radioactive waste management. The hydro-thermal-mechanical properties of bentonite have been investigated the past decades in the area such as geotechnical engineering and geo-environmental engineering. To understanding its properties is useful to evaluate safety and verification in high level radioactive waste disposal system for terms significant long-time period. Geotechnical engineers have required to deal with accurate performance assessment of high level radioactive waste disposal, is related to effect of deformation due to placement heavy waste on the compacted bentonite as barrier system. In drained condition or undrained condition, volume changes response of clayey rock is known to sensitive to temperature elevation or decreasing of effective stress soil particle together with respect to overconsolidation ratio. Recently, further factors such as gas migration, gas pressure, thermal heat and unsaturation-saturation couple properties have been focused on, and mathematical models imposed these parameters are processed.

Generation of hydration efforts give that the resulting decrement of effective stress in macro-micro structures lead to negative consequence in terms of mechanical stability around barrier systems. The even lower permeability of compacted bentonite, the increasing soil moisture occur under high relative humidity conditions. This study investigated the creep test of compacted bentonite consisted of barrier system of high level radioactive waste disposal. In essentially, the creep behaviour which soil occurrence deformation with long time consuming is applied at constant effective stress (L. Bjerrum (1967), A. Singh, JK. Mitchell (1968), J-H. Yin, J. Graham (1999)). The creep behaviour of soils that were confirmed by experimental works described much significant time-dependency (L. Bjerrum (1967), J. Graham, et al. (1983)). Their materials were classified as clay, practice problems of which was typical consolidation phenomena. A several constitute models were established based on oedometer tests (G. Mesri, PM. Godlewski (1977), J-H. Yin, J. Graham (1999)). The oedometer test results indicated time dependent stress- strain behaviour of clay which contributed to propose accurate constitutive model. The empirical simulation models had advantage to explain time-characteristic comparison to theoretical models. Almost of previous literatures considered saturated soils. There were quite a few attempt for unsaturated soils (G. Mesri, et al. (1981), V. De Gennaro et al. (2003), V. Schwarz, et al. (2006), LA. Oldecop, EE. Alonso (2007), Nawir, F. Tatsuoka (2011)). To understand, however, is interesting and useful into engineering practices. Recently, creep behaviour of unsaturated clay which was evaluated assessing suction control by X.L. Lai, et al. (2014) obtained from conventional unsaturated triaxial test.

tomo@ashitech.ac.jp

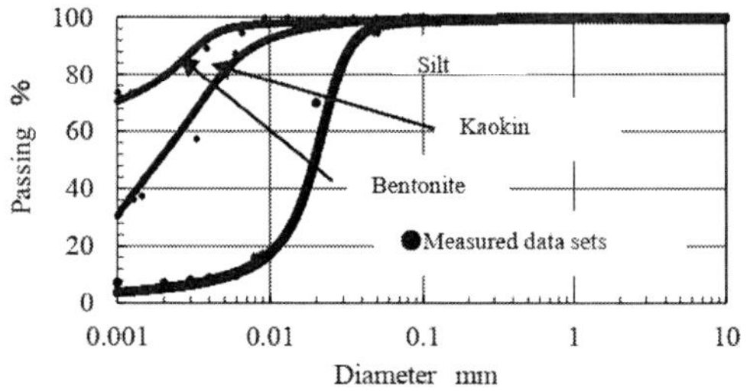

Figure 1. Grain size distribution for materials used in this test.

A series of creep test with high suction control by circulation relative humidity apparatus were performed, which used vapour pressure technique. All of creep tests remained at constant vertical stress under no lateral confining pressure. These test results presented in terms of vertical deformation-time in the constant relative humidity. Then, effect of initial water content, stress level and suction on deformation were described during long-time, and strain-rate was calculated. Not only compacted bentonite, creep behavior of kaolin and silt were measured to compare with compacted bentonite which were not expansive properties.

2 Materials and testing procedure

2.1 Materials

The bentonite, kaolin and silt were prepared as soil materials for this testing that the grain size distribution were shown in Fig.1. Fine contents of each material was were evaluated as following; 95.6%, 80.8% and 9.2%. It was clear that bentonite had most large content comparison to kaolin and silt. Also, the bentonite was made in Kunigeru Co. Ltd which was Sodium type. To obtain accurate measurement results that soil specimen maintained uniformity by static compaction.

2.2 Testing procedure

2.2.1 SWCC test

Three difference soil materials were prepared in powder condition without compaction performance. Initial water contents had a range from 5.7 % to 6.2 %, and the weight of sample was from 6 g to 14 g. Highly suctions more than 2.8 MPa that can be controlled using salt solutions. The conformed suction had a range from 2.8 MPa to 296 MPa that corresponded to a range from 98 % to 11 % in relative humidity. Seven difference salt solutions were used to control suction. The powder bentonite in air-dried condition and salt solutions were put into a glass container. For suction equilibrium, the change of soil moisture is quite slowly reason why hydration activity was performed as vapour migration. T. Nishimura (2015) described suction equilibrate of compacted bentonite which SWCC test required at least one month. Also, T. Nishimura (2015) investigated the influence of thermal effort on SWCCs for various soil materials. The changes of water

Table 1. Summary of creep test.

Specimen	Water content at compaction %	Dry density g/cm^3	Applied RH %	Vertical stress kPa
Silt	10.00	1.387	98	9.9
Kaolin	10.00	1.600	98	9.9
Bentonite	6.80	1.600	98	9.9
Bentonite	9.70	1.600	98	9.9
Bentonite	17.20	1.600	98	9.9
Bentonite	21.20	1.600	98	11.9, 20.9, 119.8, 474.4

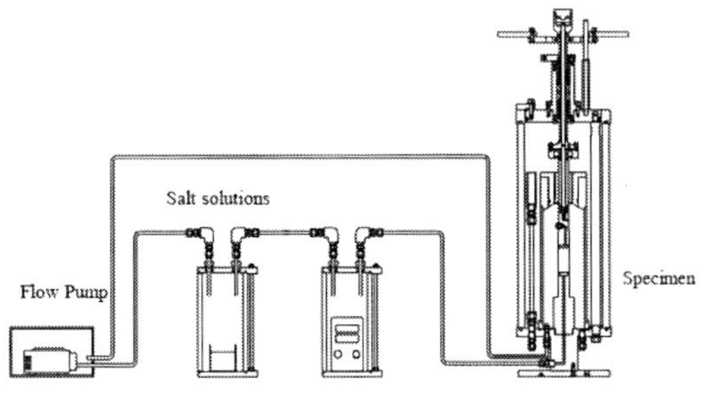

Figure 2. Modified creep test apparatus corresponding to RH control.

content with days was a likely hyperbolic curve, which was slightly a little. Therefore, employment of one confirmed suction was maintained a period of more one month. Measurement of weight of specimen realized the drainage or adsorption of soil moisture. Hyperbolic curve can useful to refer for decision of suction equilibrate in this SWCC test.

2.2.2 Unconfined compression test

All specimens with a dry density of 1.60 g/cm^3, which had a height of 10 cm, a diameter of 5 cm in size. Specimens were put on the pedestal, the cap was placed the top face of specimen. The triaxial cell was fixed to basement. The axial displacement was measured using a digital transfer sensor having a capacity of 50 mm. At once set up the specimen, axial displacement sensor and loading sensor were initialized, taking zero values. The axial strain ratio of 1.0 % per min was applied to specimen without confining pressure, which this test procedure applied correspondingly to the following JIS A 1216 code. The end of test was urged when the deviator stress described quite large reduction comparison with maximum deviator stress or occurred critical damage by shearing.

2.2.3 Creep test

For preparation of specimen on creep test, firstly bentonite and kaolin was no air-dried, and the silt was air-dried. Then, the dis-air water were sprayed to all soil materials and the required water contents had a range from 6.80 % to 21.2%. The soil regulated the water content was put into sealed plastic bag at least three days. Statically compaction with one layer which was performed in the stiffness steel mould had to avoid the influence of non-uniformity in the specimen. Finally, the specimens had a diameter of 50 mm, and a height of 100 mm which were unsaturated conditions in various suctions. Table 1 summarized basic properties of specimens.

At once specimen was mounted on the pedestal in the triaxial chamber of modified creep test apparatus (Fig. 2). The creep test apparatus was employed a conventional cyclic relative humidity control system, which can applied required RH using vapour pressure technique. This apparatus consist of triaxial chamber, air cyclic flow pump, the chamber including salt solution and dial gage sensor (maximum capacity was 100 mm). A dynamic activity of the conventional pump had at least 10kPa in pressure, which maintained steady air flow. The suctions corresponding to its relative humidity were imposed completely to the specimen in the triaxial chamber that the specimen occurred hydration pressurization following the vapour pressure technique. The air in the system remained circulation by performance of conventional pump. It was true that continue hydration pressurization diffusion from surface to inside.

An axial force up to 2 kg corresponding to axial stress up to 9.9 kPa, was applied in the axial direction using a cylinder. Axial displacement can be monitored on PC display using digital transducer equipment and RS232C electrical code. The critical programs were conducted out in this creep test, all test series were shown in Table 1. The laboratory room temperature was remained at 20 degrees.

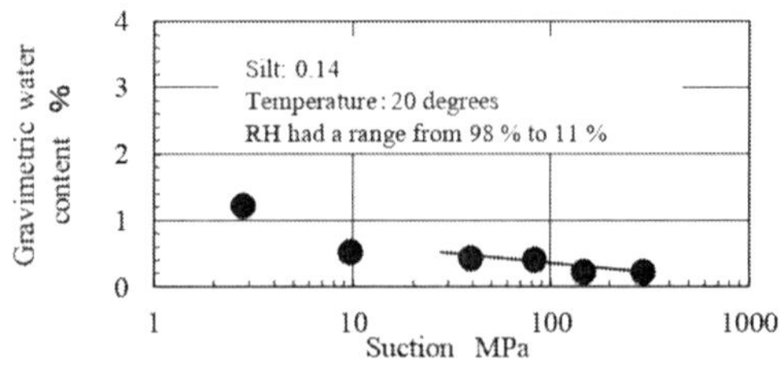

Figure 3. Soil-water characteristic curve for silt.

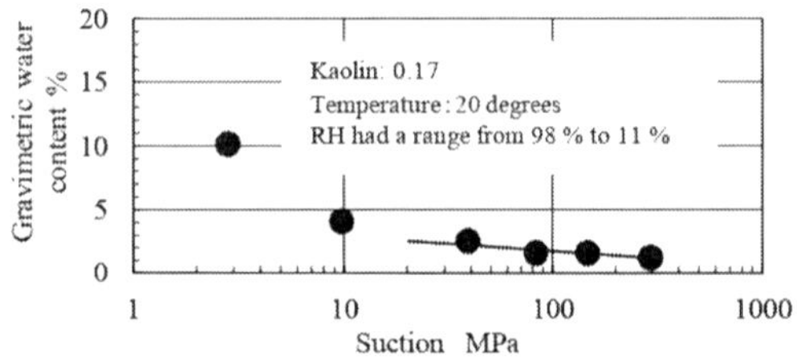

Figure 4. Soil-water characteristic curve for kaolin.

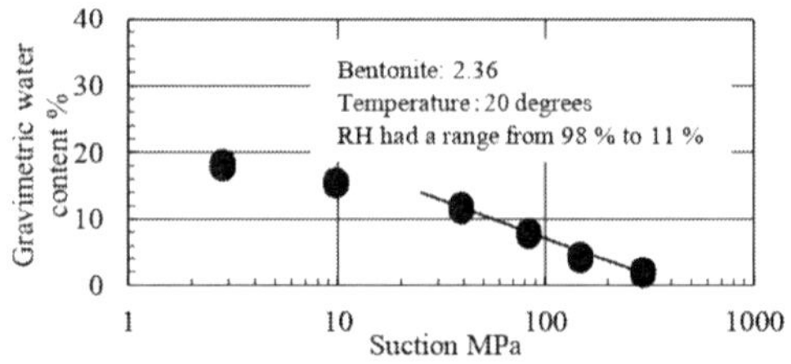

Figure 5. Soil-water characteristic curve for bentonite.

3 Test results

3.1 Soil-water characteristic curves

High suctions were applied to three different materials which each fine component had a significant difference. It was clear that grain size distribution curve (Fig. 1) indicated difference of fine contents among them. Each soil-water characteristic curve was shown in Figs.3 to 5. Firstly, each water content of silt and kaolin was slightly a little at suction of 2.8 MPa (corresponding to RH 98 %), which was less than 10.0 %. Retention of water content for bentonite remained relatively much, it was 18 % at least. After suction of 2.8 MPa, water content indicated large reduction till suction was 9.8 MPa with exception of bentonite. Subsequently, ratio of water content reduction to suction was rather slight, and the incline was remained to maximum suction. In case of bentonite, water content leaved more than 10 % till suction 39 MPa, which retention activity was considerably difference comparison to other two materials. Three incline straight lines on logarithmic scale were determined on conventional diagram, which each inclination value was shown in figures. The inclination value was recognized as one of distinguishing feature parameters, and included into expression for prediction of soil-water characteristic curves with whole suction ranges (E.C. Leong, H. Rahardjo (1997)). The inclination for bentonite had more than ten times to other two materials.

3.2 Unconfined compression strength

Results of unconfined tests at different water contents are described as stress-strain curve. Here, saturated condition was prepared which unsaturated, compacted bentonite was swelled at constant volume for a long time before unconfined compression test. The different suctions were employed corresponding to water content as shown in SWCC (Fig. 5). Suction of bentonite having water content of 9.70 % can be evaluated around 75 MPa. When water content were 17.20 % and 21.20 %, suctions were determined as following; 8.0MPa and 2.8 MPa.

The stress-strain curves were shown in Fig. 6 that all curves clearly had the influence of suction or water content. Deviator stress increased rapidly at beginning of compression for water content 9.70 %, which typical stress-strain behavior was observed as like to rather stiffness material. At once, specimen approached to maximum deviator stress, which unconfined compression strength can be determined till axial strain was 2.0 %. Subsequently, deviator stress had large reduction. The behavior which specimen having water content of 17 % was almost similar to water content of 9.70 %. The unconfined compression strength, however, clearly decreased that its reduction corresponded to 743.2 kPa.

According to decrease suction such as increasing water content, shear resistance described further. As suction was zero, saturated bentonite evidenced lowest

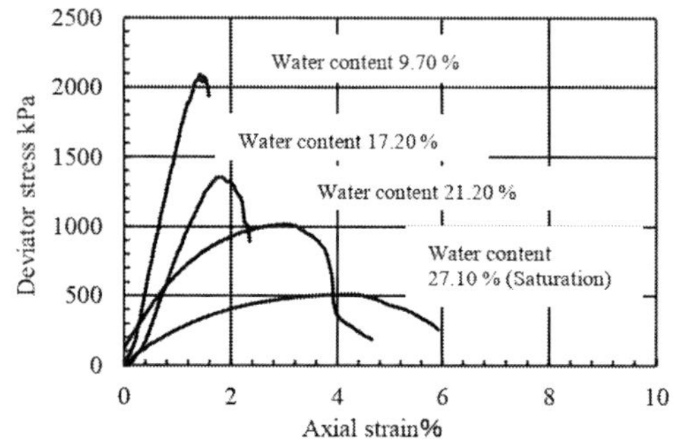

Figure 6. Stress-strain curve with various water contents for bentonite.

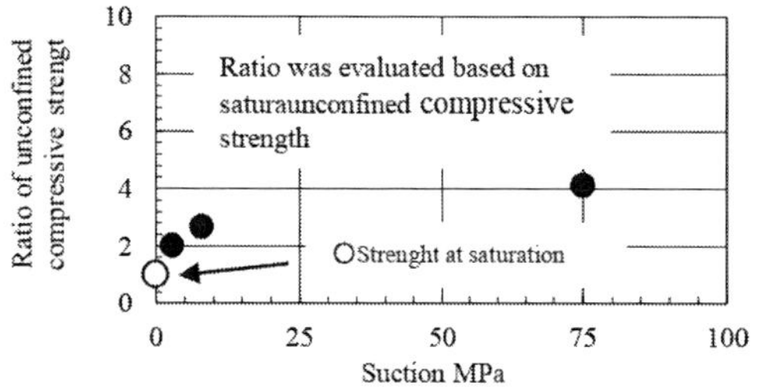

Figure 7. Influence of hydration effort on unconfined compressive strength.

strength. Also, stress-strain curve of saturated bentonite was quite difference with unsaturated bentonite. During axial strain processed in a range of 1.75 %, the deviator stresses of saturated bentonite were remained constant

Relationship between suction of bentonite and increment of unconfined compression strength was shown in Fig. 7. Hence, increment meant the difference of unconfined compression strength among each value of unsaturated bentonites and a saturated bentonite. The increment for the strength associated closely with suction, which the feature was seemed to be a hyperbolic curve. Increment of suction (i.e., contrast with decreasing of water content) induced a decreasing of shear resistance.

3.3 Creep phenomena

3.3.1 Silt and kaolin

Measured axial strain and elapsed time was presented for silt and kaolin in Fig. 8. There was a difference between silt and kaolin in axial strain and time relationship. At end of creep test, elapsed time approached to 110 days at least. Silt indicated shrinkage strain (i.e., compression) corresponding to deviator stress of 9.9 kPa at once, which the strain of 0.05 % was measured. The axial strain behavior seemed to be steady when elapsed time was over 10 days. After that, the axial strain was 0.08 % whole creep test. An axial strain of silt was measured without cracks and damages on lateral surface.

Kaolin occurred expansion in axial direction that was quite contrast creep behavior comparison to silt. The axial strain of 0.09 % was calculated at beginning of creep test

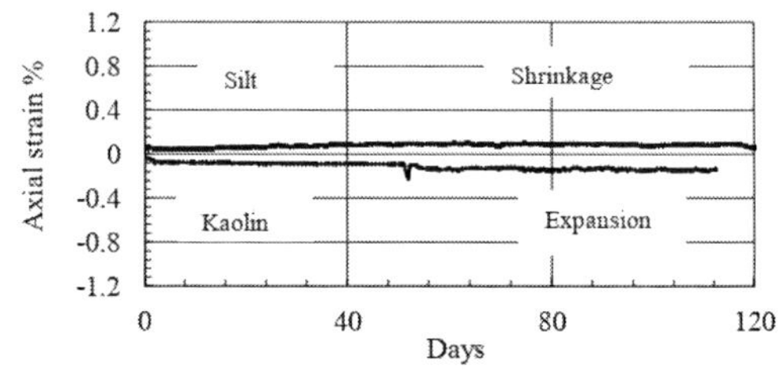

Figure 8. Change of axial strain with time for silt and kaolin.

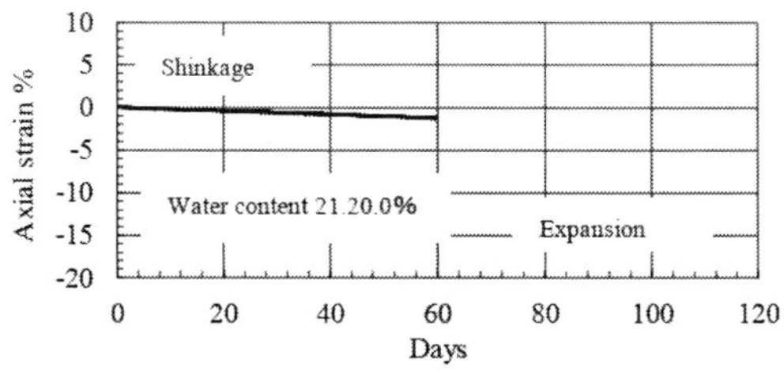

Figure 9. Creep behaviour for bentonite with water content of 21.20 %.

with a deviator stress of 9.9 kPa. With time, accumulation of expansion strain due to hydration was slightly. Measured creep strain was less than 0.15 % at end of test. There were no disturbance damages or cracks on lateral surface of specimen that was similar to phenomena for silt.

3.3.2 Bentonite at vertical stress of 9.9 kPa

Results of creep tests of bentonite compacted with various water content, all of which were presented Figs. 9 to 12. Here, water content mentioned in figures meant that initial specimen had before creep test. All of desired bentonite specimens described expansion behavior obviously in vertical direction. Hydration due to vapour pressure activity induced to encourage expansion to unsaturated compacted bentonite, in which was defined the influence of water content on magnitude of expansion.

In case of water content of 21.20 % (Fig. 9), considerable small expansive strain was measured at constant vertical stress of 9.9 kPa. The expansive strain was accumulated with slightly increment till end of test, and indicated the 1.24 %. The relationship between strain and elapsed days was assumed as straight line. More days was required to determine the equilibrium strain. In case of 17.20 %, expansive strain behavior was similar to the water content of 21.20 %, which the growing of strain increased slowly. Beyond elapsed times of 100 days, axial strain seemed to be equilibrium. Measured axial strain was large comparison with water content of 21.20 %. Both water content of 9.70% and 6.80 % described expansion-time curves which were quite difference comparison with above mentioned in Figs. 9 and 10. At beginning of hydration performance, the expansion strain had full growth, and approached to 5 % until elapsed time was five days. The strain rates obtained from Figs. 11 and 12 as shown in Figs. 13 and 14. The strain rates increased rapidly at beginning of hydration, which

 DOI: 10.1051/e3sconf/20160918002

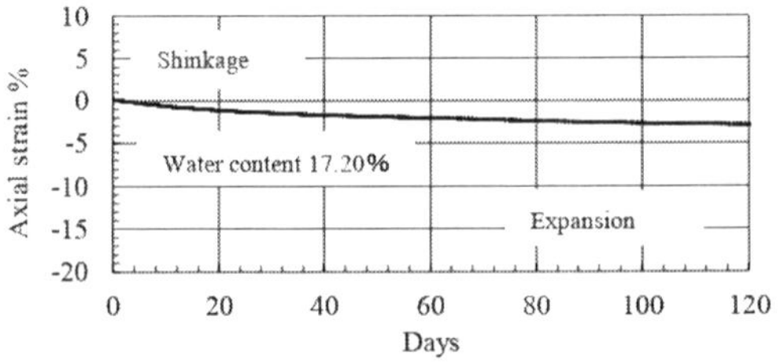

Figure 10. Creep behaviour for bentonite with water content of 17.20 %.

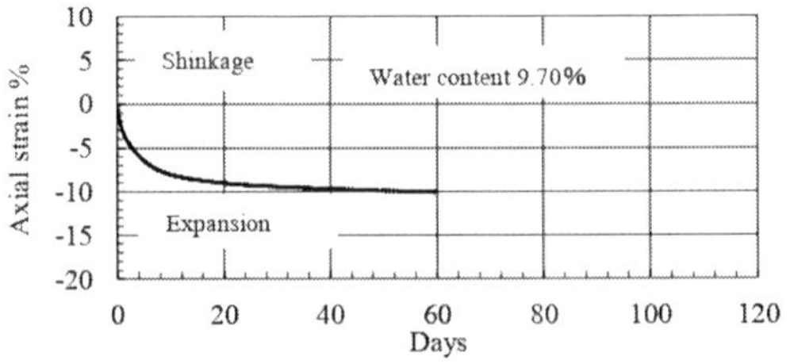

Figure 11. Creep behaviour for bentonite with water content of 9.70 %.

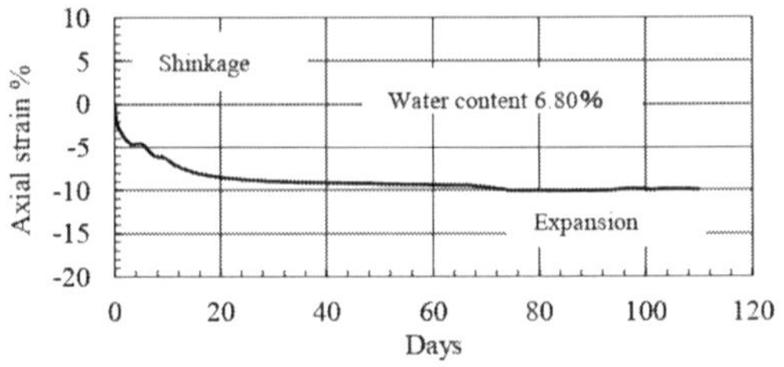

Figure 12. Creep behaviour for bentonite with water content of 6.80 %.

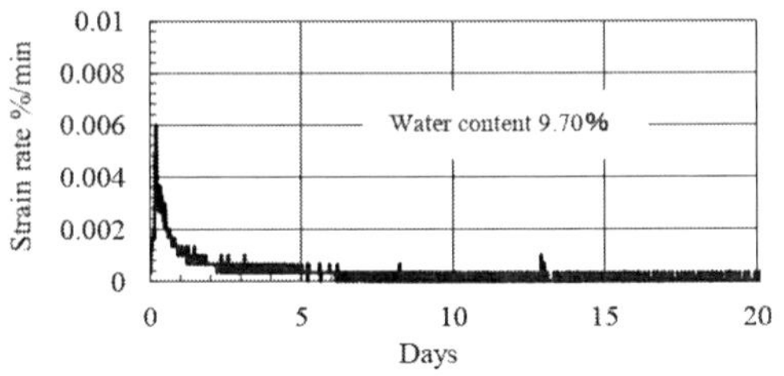

Figure 13. Strain rate in creep behaviour for bentonite with water content of 9.70 %.

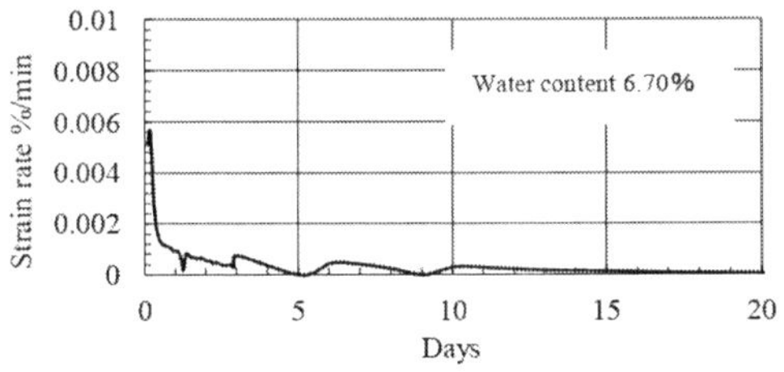

Figure 14 Strain rate in creep behaviour for bentonite with water content of 6.70 %.

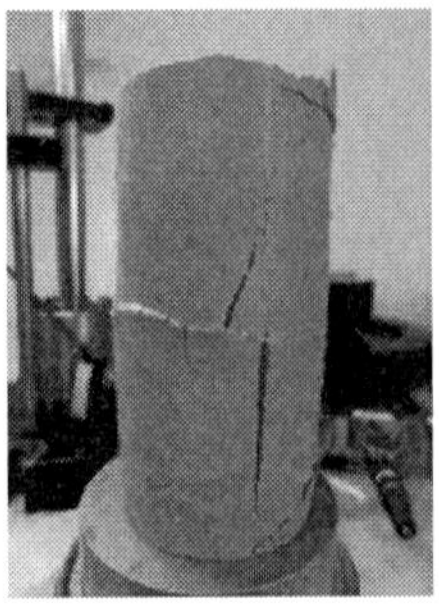

Figure 15. Creep failure of bentonite with water content of 6.80 % at vertical stress of 9.9 kPa.

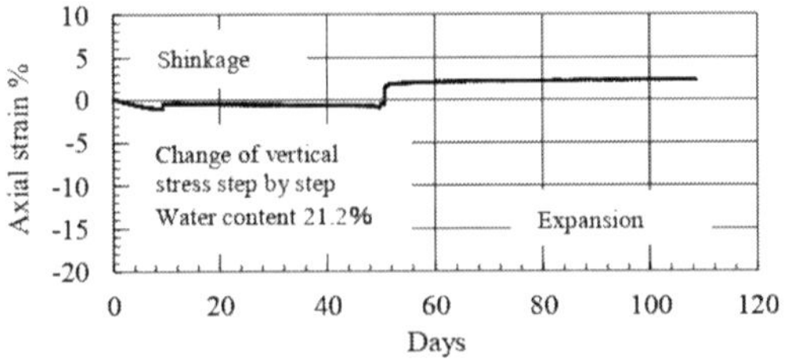

Figure 16. Creep behaviour for bentonite with water content of 21.20 % under different vertical stresses.

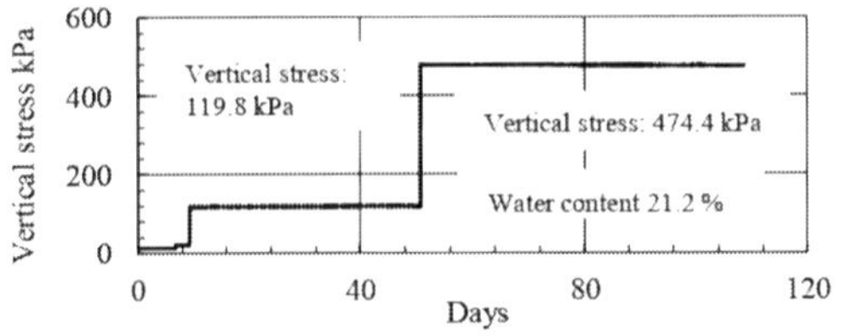

Figure 17. Different vertical stresses with elapsed time.

approach to 0.006 % per minute. At once, their strain rates were decreased that strain rate-time curves had a sharp point (i.e. maximum value). Beyond one day, the strain rates were less than 0.002 % per minute, which quite slow reduction was maintained till up to end of test. In addition, specimen with water content of 6.80 % produced so many cracks at lateral surface (Fig. 15).

3.3.3 Bentonite under various vertical stress

Fig. 16 showed changing of strain with increasing of vertical stress. The given vertical stresses which had a range of 11.9 kPa to 474.4 kPa remained at least 100 days (Fig. 17). The largest vertical stress was 474.4 kPa, which correspond to approximately 47 % to measured unconfined compression strength above mentioned in Fig. 6 (unconfined compression test). Firstly, the specimen indicated slightly little value at vertical stress of 11.9 kPa. After elapsed time was over 10 days, the vertical stress was up to 119.8 kPa. The expansion strain was accumulated slightly increment that described 0.94 %. Significant increment of vertical stress caused shrinkage deformation, which loaded from 119.8 kPa to 474.4 kPa. At once, the specimen occurred axial strain of 0.8 % in shrinkage,

DOI: 10.1051/e3sconf/20160918002

which maintained to reduce its height. Measured axial strain was equilibrium to 2.3 % in shrinkage at end of creep test.

4 Conclusions

This study conducted out three difference testing (i.e., SWCC test, unconfined compression test and creep test). Particularly, the creep testing apparatus was modified, which applying of hydration was maintained at constant RH environment with air circulation. The creep behaviour of compacted bentonite was summarized which provided the influence of initial water content at compaction.

(1) The inclination at high suction ranges was defined as ratio of water content reduction to suction. The inclination value of bentonite was more than ten times to silt or kaolin.

(2) The increasing of unconfined compressive strength closely associated with suction (or water content). Also, the stress-strain behaviour was influenced by similar effort.

(3) The bentonite described interesting phenomena regard to creep behaviour, which large axial strain in expansion was proved at relatively low water content. After long equilibrium time, so many cracks were occurred at lateral surface.

Acknowledgement

Ashikaga Institute of Technology supported to this study, a precious grant as research fund was effectively to process well of experimental works. Author expresses gratitude and appreciation.

References

1. A. Singh, JK. Mitchell, General stress-strain-time function for soils, *Journal of Soil Mechanics and Found Engineering Division, ASCE*, 94(1), 21-46 (1968).
2. E.C. Leong, H. Rahardjo, Review of soil-water characteristic curve equations, *International of Geotechnical and Geoenvironmental Engineering*, 123(12), 1106-1117 (1997).
3. G. Mesri, PM. Godlewski, Time and stress compressibility interrelationship. *Journal of Geotechnical Engineering*, 103(5), 417-430 (1977).
4. G. Mesri, E. Rebres-Cordero, DR. Shields, A. Castro, Shear stress-strain-time behaviour of clays. *Géotechnique*, 31(4), 537-552 (1981).
5. J. Graham, JHA. Crooks, AL. Bell, Time effects on the stress–strain behaviour of natural soft clays, *Geotechnique*, 33(3), 327-340 (1983).
6. J-H. Yin, J. Graham, Elastic viscoplastic modelling of the time-dependent stress-strain behaviour of soils, *Canadian Geotechnical Journal*, 36(4), 736-745 (1999).
7. J-H. Yin, J. Graham, Elastic viscoplastic modelling of the time-dependent stress-strain behaviour of soils, *Canadian Geotechnical Journal*, 36(4), 736-745 (1999).
8. JL. Deng, H. Nawir, F. Tatsuoka, Effects of viscous property and wetting on 1-D compression of clay and model simulation, *Soils and Foundations*, 51(5), 897-913 (2011).
9. L. Bjerrum, Engineering geology of Norwegian normally-consolidated marine clays as related to the settlements of buildings, *Géotechnique*, 17(2), 83-119 (1967).
10. LA. Oldecop, EE. Alonso, Theoretical investigation of the time-dependent behaviour of rockfill, *Géotechnique*, 57(3), 289-301 (2007).
11. T. Nishimura, Effect of heating to soil-water characteristic curve of compacted bentonite, Proceedings of the Sixth Asia-Pacific Conference on Unsaturated Soils, Guilin, China October 2015, 419-424 (2015).
12. V. De Gennaro, P. Delage, YJ. Cui, CH. Schroeder, F. Collin, Time-dependent behaviour of oil reservoir chalk: a multiphase approach, *Soils and Foundations*, 43(4), 131-147 (2003).
13. V. Schwarz, A. Becker, C. Vrettos, An initial study on the viscous behaviour of a partially saturated kaolinite clay based on triaxial tests, *In Proceedings of the Fourth International Conference on Unsaturated Soils*, Carefree, Ariz. ASCE. 1811–1820 (2006).
14. YT. Kim, S. Leroueil, Modeling the viscoplastic behaviour of clays during consolidation: application to Berthierville clay in both laboratory and field conditions, *Canadian Geotechnical Journal*, 38(3), 484-497 (2001).
15. X.L. Lai, S.M. Wang,W.M. Ye, Y.Y. Cui, Experimental investigation on the creep behavior of an unsaturated clay, Canadian Geotechnical Journal, 51(6), 621-628 (2014).

Investigation on hydraulic properties of compacted GMZ bentonite used as buffer/backfill material

W. M. Ye [1,2,a], Q. Xin[1], B. Chen [1], Y.G. Chen [1], Y. J. Cui[3]

[1]*Key Laboratory of Geotechnical and Underground Engineering of Ministry of Education, Tongji University, Shanghai 200092, China*
[2]*United Research Center for Urban Environment and Sustainable Development, the Ministry of Education, Shanghai 200092, China*
[3]*Ecole des Ponts ParisTech, UR Navier/CERMES, 77455, France*

Abstract. During the past decades, GMZ bentonite has been widely investigated for its use as buffer/backfill materials in China. Based on a comprehensive review of the former studies, achievements on experimental and theoretic works on the hydraulic aspects of compacted GMZ bentonite with consideration of temperature effects are presented in this paper. Water retention property of compacted GMZ bentonite depends on constraint conditions. Temperature effects on water-retention depend on constraint conditions and suction. The hysteresis behaviour is not obvious. Based on the test results, a revised water retention model was developed for considering the temperature effect. The saturated hydraulic conductivity of the densely compacted GMZ bentonite increases as dry density and temperature increases. A revised model, which considers temperature influence on water viscosity and the effective flow cross-sectional area of porous channels, for prediction of saturated hydraulic conductivity have been developed and verified. The unsaturated hydraulic conductivity of confined densely compacted GMZ bentonite samples decreases first and then increases with suction decrease from an initial value of 80 MPa to zero. With consideration of temperature effects and microstructure changes, a revised model for prediction of unsaturated hydraulic conductivity of compacted GMZ01 bentonite was proposed.

1 Introduction

Bentonite has generally been chosen as buffer and backfill material for construction of engineered barrier in deep geological repository because of its high swelling capacity, low permeability, micro-porous structure and good sorption properties. This clay-based engineering barrier has multiple purposes of providing mechanical stability for the waste canister (by absorbing stresses and deformations); serving as a buffer around it; sealing discontinuities in the emplacement boreholes and drifts; and delaying the water infiltration from the host rock. During the long-term operation of the repository, the engineered clay barrier and adjacent host rock will be subjected to the heating effect of the nuclear waste, to various associated hydraulic and mechanical phenomena that interact in a complex way. Therefore, the hydraulic property of the compacted bentonite is one of the key properties for the design of such a disposal system (Dixon et al., 1987; Nachabe, 1995; Liu and Wen, 2003; Ye et al., 2013; 2014a,b; Villar, 2000; 2002; Komine, 2004). Furthermore, compacted bentonite is initially unsaturated, and will therefore be subjected to hydration from the surrounding rock, triggering further coupled thermo-hydro-mechanical (THM) phenomena.

Chinese high-level radioactive waste repository program was launched in the middle of 1980s (Wang et al., 2006). It was suggested that a high-level radioactive waste repository will be built in the middle of the 21st Century. Beishan in Gangsu Province has been selected as the potential site for construction of the first repository in China. Gaomiaozi (GMZ) bentonite has been considered as a possible material for the construction of engineered barrier in the recent Chinese program of radioactive waste disposal (Liu et al., 2001; Ye et al. 2009a).

In the past decades, investigations on mineralogy and chemical composition, mechanical properties, hydraulic behavior, swelling behavior, thermal conductivity, microstructure and volume change behavior were performed on GMZ bentonite (Liu and Wen, 2003; Wang et al., 2006; Wen, 2006; Chen et al., 2006; Niu, 2008; Ye et al., 2009a,b; 2010; 2012; Sun et al., 2015). Related results have been reported.

In this paper, based on a comprehensive review of the former studies, the achievements on experimental and theoretic works on hydraulic properties of GMZ bentonite are presented.

2 Basic properties of GMZ bentonite

Gaomiaozi (GMZ) bentonite originates from Inner Mongolia, China, 300 km northwest from Beijing. There are 160 million tons bentonite (with 120 million tons Na-

[a] Corresponding author: ye_tju@tongji.edu.cn

Bentonite) reserves in the deposit, which occupies a mining area about 72 km^2. The deposit was formed in the later Jurassic period (Liu et al., 2001). GMZ Bentonite is bedded, with a soapy texture and waxy appearance.
The Cation Exchange Capacity (CEC) of GMZ bentonite is showed in Table1 (Ye et al., 2010).

Table 1. Cation exchange capacity of some bentonites (Ye et al., 2010)

Sample	CEC (meq/100 g)	Exchangeable cation (meq/100 g)			
		E(K+)	E(Na^+)	E($1/2Ca^{2+}$)	E($1/2Mg^{2+}$)
GMZ	77.30	2.51	43.36	29.14	12.33

Some basic physical properties of GMZ bentonite are presented in Table 2.

Table 2. Physical properties of GMZ01 bentonite (Ye et al., 2014)

Property	Description
Specific gravity of soil grain	2.66
pH	8.68–9.86
Liquid limit (%)	276
Plastic limit (%)	37
Total specific surface area/ ($m^2 \cdot g^{-1}$)	570
Main minerals	Montmorillonite(75.4%), quartz (11.7%), feldspar (4.3%), cristobalite (7.3%)

Results in Tab. 1 and Tab. 2 show that GMZ bentonite has high montmorillonite content, which gives it a high Cations Exchange Capacity (CEC = 77.30 meq/100g), a large plasticity index (Ip = 239), and a large specific surface area (S = 570 m^2/g).

3. Water retention properties

Using the vapour phase technique and the osmotic technique, the WRCs of both the confined and unconfined GMZ bentonite specimens were measured at room temperature (20°C) by Chen et al. (2006) (Fig. 1). The result in Fig. 1 shows that, starting from the initial suction of 4.2 MPa, the difference between different confining conditions becomes less significant as the suction increases. However, for suctions lower than 4 MPa, the confined specimen gives much lower water content: at the lowest suction (0.013 MPa) measured the difference in water content is as large as 140%.

The water retention curves obtained under temperatures following the wetting path for the unconfined GMZ01 bentonite specimens with a dry density of 1.70 g/cm^3 are presented in Fig. 2. It is observed that temperature effect on the water retention capacity is insignificant.

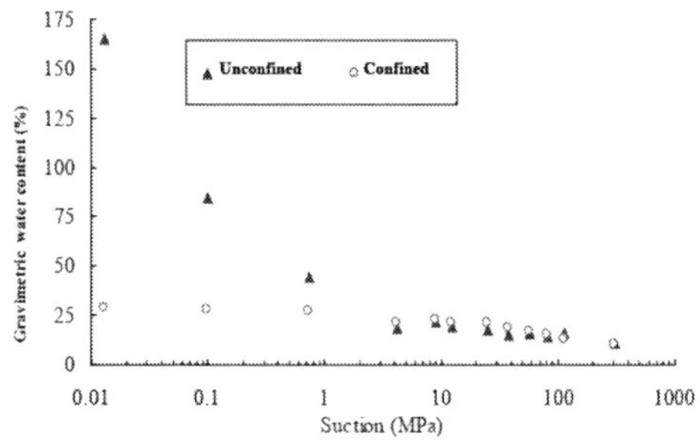

Fig. 1. WRCs of the GMZ bentonite with dry density of 1.7 g/cm^3 under confined and unconfined conditions at room temperature (Chen et al., 2006)

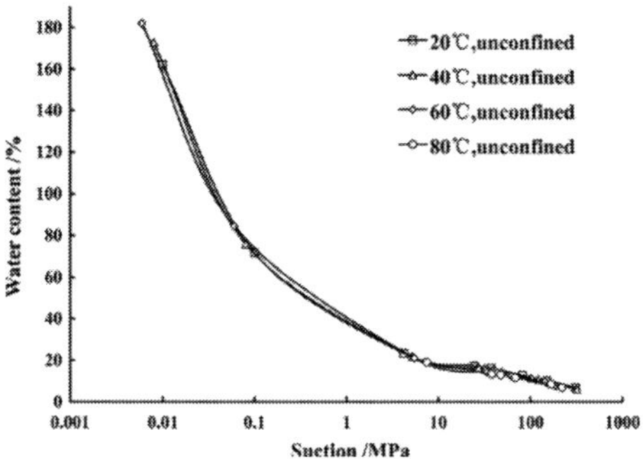

Fig. 2 SWRCs of compacted GMZ bentonite under unconfined conditions (20, 40, 60 and 80 °C) (Wan et al., 2015)

For measuring the SWRCs at temperatures, salt solutions and their corresponding suctions at temperatures in Table 2 (Tang and Cui, 2005) were employed. The temperature effect on the suction imposed by PEG solutions is 1%/°C (Tang et al., 2010) was accepted for application of osmotic technique.

The water retention curves of confined compacted GMZ bentonite following wetting path at different temperatures are presented in Fig. 3.

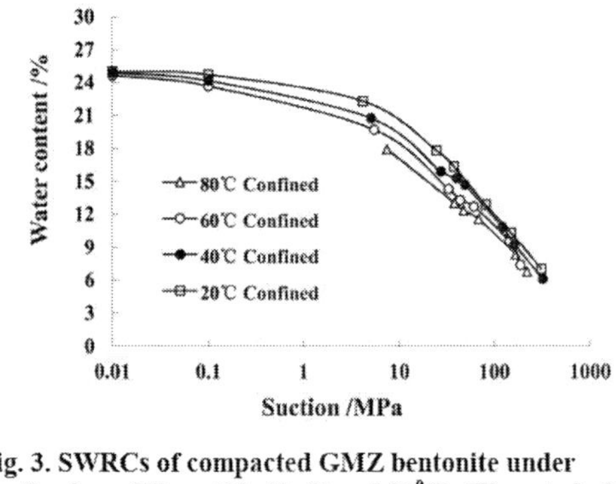

Fig. 3. SWRCs of compacted GMZ bentonite under confined conditions (20, 40, 60 and 80^{0}C) (Wan et al., 2015)

Curves in Fig. 3 show that, for confined specimens, the water retention capacity decreases as temperature rises. Moreover, this temperature influence strongly depends on suction. When the suction is close to zero (0.01 MPa), the temperature effect is hardly observed. A possible explanation of this observation is that the lateral swelling stress applied to the compacted bentonite sample increases with suction decrease under confined conditions. Meanwhile, decreasing suction could not make more water moving into large pores, as most of the pores have already been filled with water.

Comparison between the water retention curves shown in Figs. 2 and 3 reveals that temperature effects on water retention capacity strongly depends on the constraint conditions. This conclusion agrees well with the observations reported by Villar and Lloret (2004).

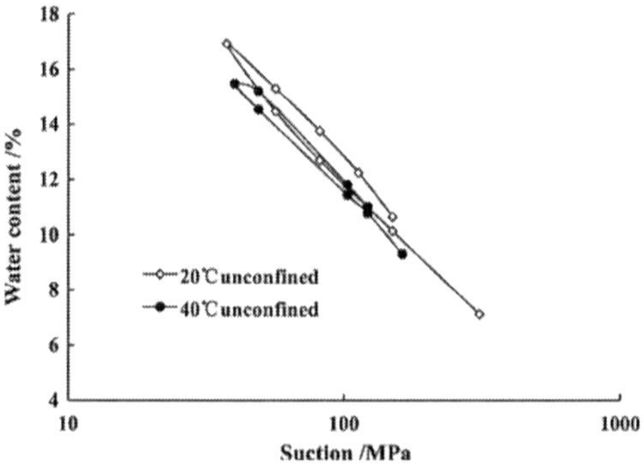

Fig. 4. Hysteresis behavior of unconfined GMZ bentonite at 20 and 40 °C temperatures (Wan et al., 2015)

The hysteresis curves of unconfined compacted GMZ bentonite following wetting / drying cycles at 20 and 40 °C are shown in Fig. 4. It is observed that the hysteresis behavior becomes less obvious with temperature increase, in agreement with the observation reported by Ye et al. (2009b). This indicates that, as temperature increases, the water retention capacity and the hysteresis decrease regardless of the confining conditions. Fig. 4 also shows that the influence of temperature on the hysteresis behavior depends on the suction level: the higher the temperature, the less obvious the hysteresis behavior.

For modelling the soil water retention behavior of compacted GMZ bentonite with consideration of temperature effects, based on a comprehensive review of literatures (Brooks and Corey 1964; van Genuchten 1980; Fredlund and Xing 1994; Leij et al. 1997; Sánchez, 2004; Jacinto et al., 2009), Wan et al. (2015) established the following equation to describe the water retention curves of the densely compacted GMZ bentonite (1.7 Mg/m^3) with consideration of temperature effects:

$$w=\left(1-\frac{\ln(1+\frac{\psi}{\psi_r})}{1+\frac{1000}{\psi_r}}\right)\frac{w_{sat}^0+\zeta(T-T_0)}{\left\{\ln[e+(\psi/(\eta Ln(T-273)+\varepsilon))^n]\right\}^m} \quad (1)$$

Where $w^0_{sat}(T)$= 0.25; ζ=0.00018; T_0=273.4. By fitting the SWRC measured at 20°C using equation (1), for the confined compacted GMZ01 bentonite with a dry density of 1.7 Mg/m^3, parameters m and n are determined, equal to 0.5864 and 0.8086, respectively.

With equation (1), the SWRCs of compacted GMZ01 bentonite at 60°C and 80°C were calculated and compared to the measured ones. Comparison demonstrates that the revised model can satisfactorily describe the soil water retention properties of the confined compacted GMZ01 bentonite changing with temperature.

4. Hydraulic conductivity

4.1 Saturated hydraulic conductivity

With self-made apparatus, Ye et al. (2013) measured the saturated hydraulic conductivity of densely (1.70g/cm^3) compacted GMZ bentonite at temperatures (Fig. 5).

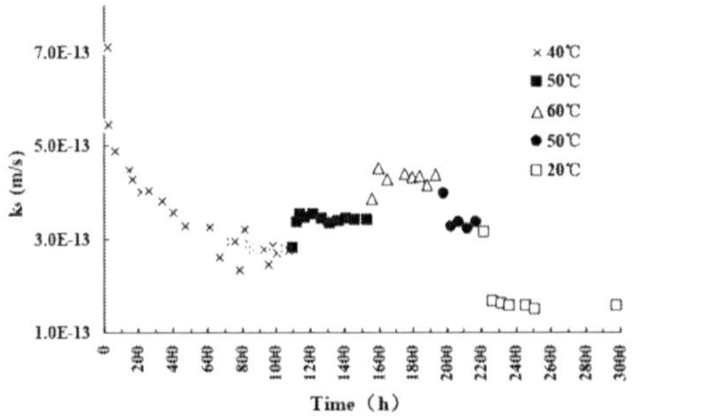

Fig. 5. Evolution of saturated hydraulic conductivity of the compacted GMZ bentonite with time at temperatures (Ye et al., 2013)

According to the test results, the following equation was proposed for prediction of the saturated hydraulic conductivity of the compacted GMZ bentonite considering temperature influences on both water viscosity and effective flow cross area (apparent porosity) in its porous channels (Ye et al., 2013).

$$k_s=\frac{k_{in}(T)\rho_w g}{\eta(T)}=\frac{\exp(-45.703+0.0054\times(T-273))\rho_w g}{0.0002601+0.001517\exp[-0.034688\times(T-273)]} \quad (2)$$

Where k_s is saturated hydraulic conductivity, k_{in} intrinsic hydraulic conductivity, η water viscosity, T is temperature, ρ_w is water density and g is acceleration of gravity.

Calculation shows that the proposed model (eq. (2)) can well describe the temperature influences on the saturated hydraulic conductivity of the compacted GMZ bentonite.

It should be noted that the values measured during the first 600 h in Fig. 5 probably correspond to a period when the sample was still unsaturated.

4.2 Unsaturated hydraulic conductivity

Using a self-designed cell, Ye et al. (2009a) measured the unsaturated hydraulic conductivity of GMZ bentonite with a dry density of 1.7 g/cm^3 under confined conditions at ambient temperature. With the instantaneous profile method, the unsaturated hydraulic conductivities measured at different locations along the height of the specimen evolving with suction are presented in Fig. 6. Results in Fig. 6 indicate that, when suction is reduced from the starting point about 80 MPa to zero, the measured unsaturated hydraulic conductivity firstly decreases from 2E-14 m/s at 80 MPa suction, then increases from 7E-15 m/s at 70 MPa suction, and 10E-13 m/s at zero suction. The decrease can be attributed to the large pore clogging due to the possible exfoliation process (Ye et al., 2009a).

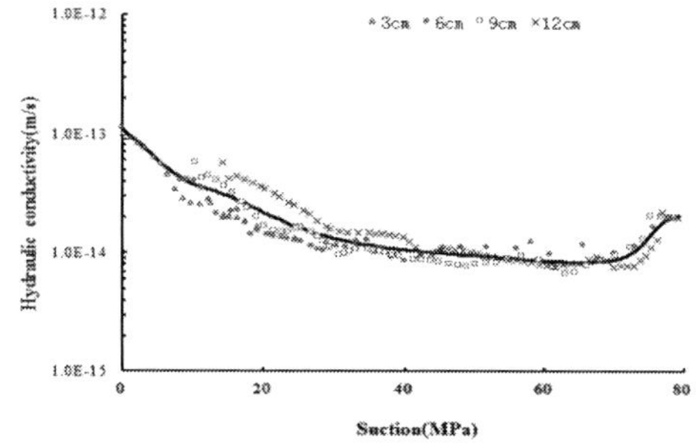

Fig. 6 Evolutions of hydraulic conductivity of the GMZ bentonite at different heights of the specimen with suction under confined conditions (Ye et al., 2009a)

Niu (2008) tested the unsaturated hydraulic conductivity of the densely compacted GMZ bentonite under unconfined conditions at room temperature. Results (Fig. 7) show that the unsaturated hydraulic conductivity is in a larger range of 1.0E-12 and 1.0E-15 m/s and it develops gently with suction except for fluctuations both at the initial and the final stages of the experiment.

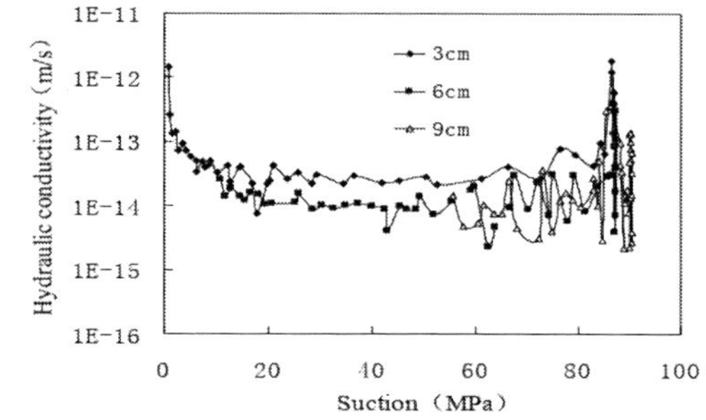

Fig. 7 Evolution of hydraulic conductivity of the GMZ bentonite with suction under unconfined conditions (Niu, 2008)

Comparison of results in Figs. 6 and 7 reveals that, for a same suction, the unsaturated hydraulic conductivity of compacted GMZ bentonite under unconfined conditions is higher than that of under confined conditions. This is possibly induced by the difference in the mechanism of microstructural changes during hydration under different confining conditions.

With self-made apparatus, Ye et al. (2009b; 2012) measured the unsaturated hydraulic conductivity of the confined GMZ specimens at 20°C, 40°C and 60°C, which were compared in Fig. 8. Results show that under confined conditions, the unsaturated hydraulic conductivity of the highly compacted GMZ bentonite increases with temperature rise. This temperature effect can be explained by changes of water viscosity (Duley and Domingo, 1943; Hopmans and Dane, 1986; Constantz and Murphy, 1991), water density, and to some extent the intrinsic hydraulic conductivity (Villar et al., 2005). Among which, the water viscosity has been found to be the most significant factor (Delage et al., 2000).

Moreover, the rate of change also decreases as temperature increases. The temperature effect becomes more significant at higher suctions (above 20 MPa). In the range of lower suctions (less than 20 MPa), it is observed that the lower the suction, the less the temperature effect. The possible explanation is that for lower suctions the moisture absorbed by the bentonite is mainly associated with microstructure changes and the temperature effect on the microstructure is not significant.

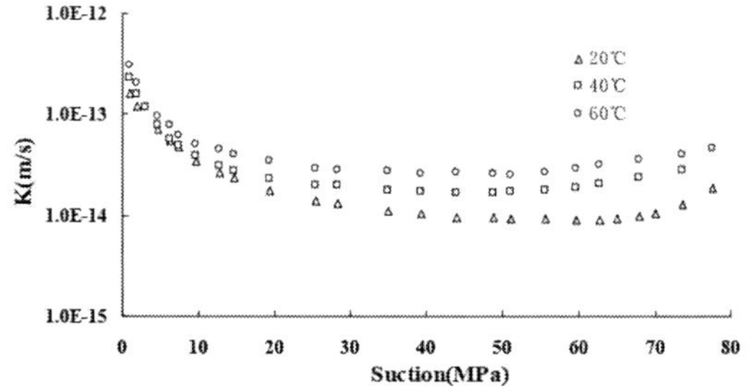

Fig. 8 Evolution of unsaturated hydraulic conductivity with suction for the confined GMZ bentonite at temperatures (Ye et al., 2012)

According to the test results, a revised fractal model for prediction of unsaturated hydraulic conductivity of compacted bentonite with consideration of temperature effects and microstructure changes was developed by Ye et al. (2014b).

$$k_w = k_s(T)k_r(s,T) = k_s(T) \times A \times B\left(\frac{s(T)}{a(T)}\right)^{-\lambda(n+1+2/\lambda)} \quad (3)$$

Where, $k_s(T)$ is saturated hydraulic conductivity of compacted GMZ bentonite with consideration of temperature effects, which can be expressed by equation (2). Parameter A is the influencing factor of microstructure changes, B is the influence coefficient of the effective degree of saturation on the variation of void ratio. For the compacted GMZ01 bentonite specimen tested: B = 0.515. Based on the measured unsaturated hydraulic conductivity of densely compacted GMZ01 bentonite with a dry density of 1.7 Mg/m^3 under confined conditions at temperatures 40° C and 60° C in Fig. 8, parameters A and n were fitted: n=-5.371 and A=0.51. With MIP observations, the value of the fractal dimension D = 2.712 and the fractal dimension coefficient λ = 0.288.

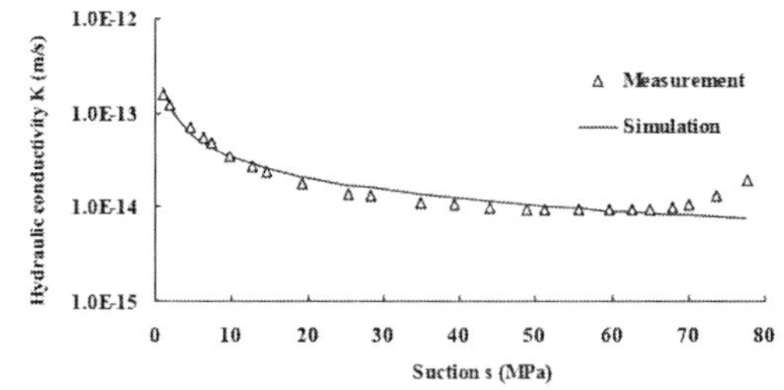

Fig. 9 Comparison between the prediction and measurement at 20°C temperature (Ye et al., 2014)

Verification result (Fig. 9) indicates that equation (3) can give good prediction of the unsaturated hydraulic conductivity of densely compacted GMZ01 bentonite under confined conditions in a suction range of 0-70MPa. However, deviation occurs in higher suctions (>70MPa). This deviation may result from the inter-aggregate pores clogging (Ye et al., 2009b), which probably induced by possible exfoliation process that occurs around the macro-pores during aggregate hydration (Pusch et al. 2012; Pusch and Yong 2003; Pusch et al. 2010).

5 Conclusions

Based on a review of the former studies, achievements on experimental and theoretic results obtained on the hydraulic behaviour of compacted GMZ bentonite are presented in this paper.

The influences of constraint conditions on water retention properties of GMZ bentonite depends on the range of suction. Temperature effects depends on constraint conditions and suction. Under unconfined conditions, the effect of temperature at high suctions is different from that at low suctions. The hysteresis behaviour of unconfined compacted bentonite is not obvious on the whole, even it was found reduced with temperature increase. A revised water retention model was developed for considering the temperature effect. Comparison between the calculated results and the experimental data shows the relevance of the model developed in describing the water retention properties of compacted GMZ bentonite at different temperatures under confined conditions.

The saturated hydraulic conductivity of the GMZ bentonite decreases as dry density increases and increases as temperature increases. With consideration of both the temperature influence on water viscosity and the effective flow cross-sectional area of porous channels, the proposed model can satisfactorily account for the temperature effects.

The unsaturated hydraulic conductivity of GMZ bentonite firstly decreases and then increases with suction decrease from an initial value of 80 MPa to zero. This phenomenon could be attributed to the large pore clogging due to exfoliation process. The unsaturated hydraulic conductivity of compacted GMZ bentonite under unconfined conditions is higher than that under confined conditions. Under confined conditions, the unsaturated hydraulic conductivity of the highly compacted GMZ bentonite increases with temperature rise. Moreover, the changing rate depends on suction.

Acknowledgments

The authors are grateful to the National Natural Science Foundation of China (Projects No. 41527801, 41272287), the China Atomic Energy Authority (Project [2011]1051) for the financial supports.

References

1. Brooks, R.H., Corey, A.T., 1964. Hydraulic properties of porous media Colorado State Univ. Hydrol. Paper, no.3.
2. Chen, B., Qian, L.X., Ye, W.M., Cui, Y.J., Wang, J. 2006. Soil-water characteristic curves of Gaomiaozi Bentonite. Chinese Journal of Rock Mechanics and Engineering. 21:1054-1058.
3. Constantz, J., Murphy, F. 1991. The temperature dependence of ponded infiltration under isothermal conditions. J Hydrol 122, 119-128.
4. Cui, Y.J., Tang, A.M., Loiseau, C., Delage, P. 2008. Determining the unsaturated hydraulic conductivity of a compacted sand-bentonite mixture under constant-volume and free-swell conditions. Physics and Chemistry of the Earth 33, 462-471.
5. Delage, P., Sultan, N., Cui, Y.J. 2000. On the thermal consolidation of Boom clay. Can Geotech J 37 (2), 343-354.
6. Dixon, D.A., Cheung, S.C.H., Gray, M.N., Davidson, B.C. 1987. The hydraulic conductivity of dense clay soils. Proceedings of the 40th Canadian Geotechnical Conference, Regina, Saskachewan – Canada, pp. 389-396.
7. Duley, F.L., Domingo, C.E. 1943. Effect of wáter temperatura on rate of infiltration. Soil Sci Soc proc 31, 129-131.
8. Fredlund, D.G., Xing, A. 1994. Equations for the soil-water characteristic curve. Canadian Geotechnical Journal, Volume 31, 521-532.
9. Hopmans, J., Dane, J. 1986. Temperature dependence of soil hydraulic properties. Soil Sci Soc Am J 50, 4-9.
10. Jacinto, A.C., Villar, M.V., Gómez-Espinab R., Ledesma A. 2009. Adaptation of van Genuchten expression to the effects of temperature and density for compacted bentonites. Applied Clay Science 42, 575–582.
11. Komine, H. 2004. Simplified evaluation on hydraulic conductivities of sand-bentonite mixture backfill. Applied Clay Science 26 (1-4), 13-19.
12. Leij, F.J., Russell, W.B., Lesch, S.M. 1997. Closed-form expressions for water retention and conductivity data. Ground Water 35, 848–858.
13. Liu, Y.M., Xu, G.Q., Liu, S.F. 2001. Study on the Basic Property of Gaomiaozi Bentonite, Inner Mongolia. China Nuclear Industry Audio and Visual Publishing House, Beijing, 1-20 (in Chinese).
14. Liu, Y.M., Wen, Z.J. 2003. An investigation of the physical properties of clayey materials used in nuclear waste disposal at great depth. Mineral rocks 23 (4), 42-45 (in Chinese).
15. Nachabe, H.M. 1995. Estimating hydraulic conductivity for models of Soils with Macro-pores. Journal of Irrigation and Drainage Engineering 121 (1), 95-102.

16. Niu, W.J. 2008. Study on unsaturated permeability of densely compacted bentonite under free swelling conditions. Ph.D-thesis, Tongji University, Shanghai (in Chinese).
17. Pusch R., Prikryl R., Weishauptová Z, Liu X. D. , Knutsson S. 2012. Role of clay micro-structure in expandable buffer clay. Journal of Purity, Utility Reaction and Environment Vol.1 No.6, 297-322.
18. Pusch R., Yong R. 2003. Water saturation and retention of hydrophilic clay buffer—micro structural aspects. Applied Clay Sci 23, 61– 68
19. Pusch, R., Yong, R.N., Nakano, M., 2010. Stiffening of smectite buffer clay by hydrothermal effects. Engineering Geology, Vol.116. pp. 21-31
20. Sun, W.J., Wei Z.F., Sun, D.A., Liu, S.Q., Fatahi, B., Wang, X.Q., 2015. Evaluation of the swelling characteristics of bentonite-sand mixtures. Engineering Geology, 199. 1-11.
21. Sánchez, M. 2004. Thermo-hydro-mechanical coupled analysis in low permeability media. The Polytechnic University of Catalunya, Barcelona.
22. Tang A.M.,Cui Y.J. 2005. Controlling suction by the vapour equilibrium technique at different temperatures and its application in determining the water retention properties of MX80 clay. Can. Geotech. J. 42:1-10.
23. Villar, M.V. 2000. Caracterización termo-hydro-mecánica de una bentonita de Cabo de Gata: Ph.D. Thesis. Universidad Complutense de Madrid. Madrid (in Spanish).
24. Villar, M.V. 2002. Thermo-hydro-mechanical characterization of a bentonite from Cabo de Gata. A study applied to the use of bentonite as sealing material in high level radioactive waste repositories. Publicación Técnica ENRESA, Madrid, Spain.
25. Villar M.V., Lloret A. 2004. Influence of temperature on the hydromechanical behaviour of a compacted bentonite. Appl Clay Sci 26: 337–350
26. Villar, M.V.; Romero, E. & Lloret, A. 2005. Thermo-mechanical and geochemical effects on the permeability of high-density clays. In: ALONSO, E.E. & LEDESMA, A. (eds.): Advances in understanding engineered clay barriers. A.A. Balkema Publishers, Leiden. 177-191. ISBN 04-1536-544-9.
27. Van Genuchten, M. Th. 1980. A closed-form equation for predicting the hydraulic conductivity of unsaturated soils. Soil Sci. Soc. Am. 44, 892–898.
28. Wan M, Ye WM*, Chen YG, Cui YJ, Wang J. 2015. Influence of temperature on the water retention properties of compacted GMZ01 bentonite. Environmental Earth Sciences. 73(8): 4053-4061.
29. Wang, J., Sui, R., Chen, W., Guo, Y.H., Jin, Y.X., Wen, Z.J., Liu, Y.M. 2006. Deep geological disposal of high-level radioactive wastes in China. Chinese Journal of Rock Mechanics and Engineering, 25 (4), 649-658.
30. Wen, Z.J. 2006. Physical property of China's buffer material for high level radioactive waste repositories. Chinese Journal of Rock Mechanics and Engineering 25, 794-800 (in Chinese).
31. Ye, W.M., Cui, Y.J., Qian, L.X., Chen, B. 2009a. An experimental study of the water transfer through confined compacted GMZ bentonite. Engineering Geology, Volume 108, 169-176.
32. Ye, W.M., Wan, M., Chen, B., Chen, Y.G., Cui, Y.J., Wang, J. 2009b. Effect of temperature on soil-water characteristics and hysteresis of compacted Gaomiaozi bentonite. J of Central South Univ of Tech 16 (5), 821-826.
33. Ye, W.M., Chen, Y.G., Chen, B., Wang, Q., Wang, J. 2010. Advances on the knowledge of the buffer/backfill properties of heavily-compacted GMZ bentonite. Engineering Geology, Volume 116, Issues 1-2, Pages 12-20.
34. Ye, W.M., Wan, M., Chen, B., Chen, Y.G., Cui, Y.J., Wang, J. 2012. Temperature effects on the unsaturated permeability of the densely compacted GMZ01 bentonite under confined conditions. Engineering Geology, Vol 126, 1-7
35. Ye, W.M., Wan, M., Chen, B., Chen, Y.G., Cui, Y.J. 2013. Temperature effects on the swelling pressure and saturated hydraulic conductivity of the compacted GMZ01 bentonite. Environmental Earth Science, Vol 68, 281-288.
36. Ye WM, N.C. Borrell, J.Y. Zhu, B. Chen, Y.G. Chen. 2014a. Advances on the investigation of the hydraulic behaviour of compacted GMZ bentonite. Engineering Geology 169: 41–49
37. Ye, W.M., Wan, M., Chen, B., Chen, Y.G., Cui, Y.J., Wang, J. 2014b. An Unsaturated Hydraulic Conductivity Model for Compacted GMZ01 Bentonite with Consideration of Temperature. Environmental Earth Sciences. 71(4): 1937-1944

Stress release and suction generation in the Callovo-Oxfordian claystone

Pierre Delage[1,a], Hamza Menaceur[1,2], Anh Minh Tang[1] and Jean Talandier[3]

[1] *Ecole des Ponts ParisTech, Navier/CERMES, Marne la Vallée, France*
[2] *Now in LEMTA, Université de Lorraine, Nancy, France*
[3] *Andra, Châtenay-Malabry, France*

Abstract. Stress release effects in a specimen of claystone extracted at great depth (465 m) in the layer of the Callovo-Oxfordian claystone in Eastern France is examined by mimicking the stress change during extraction in the oedometer. To do so, a specimen first submitted to hydration with swelling impeded is afterwards unloaded at constant water content. Comparable mercury intrusion pore size distribution curves are obtained before and after stress release, showing little change in the configuration of inter-platelets pores within the clay matrix of the claystone. The swelling observed at constant water content is then related, to a water transfer from inter-platelets pores to intra-platelets pores, resulting in crystalline swelling within the smectite fraction of the inter-layer illite-smectite minerals.

1 Introduction

The Callovo-Oxfordian (COx) claystone is considered as a possible host rock for deep geological disposal of radioactive waste in France. Many investigations on its hydromechanical behaviour are carried in the Underground Research Laboratory (URL) managed by Andra (the French Agency for radioactive waste disposal) close to the village of Bure (Eastern France). The URL has been excavated at a depth of 490 m in a 150 m thick layer COx claystone.

Prior to be tested in surface laboratory, claystone specimens have to be cored and extracted from the claystone mass. It is important to figure out how far the specimens tested in the laboratory are representative of the in-situ intact state, prior to extraction. The effect of stress release is quite significant at depths as high as 490 m. Given that COx claystones specimens provided to surface laboratory are generally unsaturated, particular attention has been paid to ensure fully saturation under stress conditions close to in-situ and full drainage conditions during compression tests. In low permeability rocks (the saturated permeability is in the order of 10^{-13} m/s in the COx claystone), saturation can be quite long and pore pressure homogeneity during testing can be problematic. In this regards, specific devices with reduced length of drainage have been developed (Monfared et al., 1, Hu et al., 2).

Desaturation of specimens of the COx claystone also occur in-situ in the gallery walls due to ventilation. This will not be considered here and only the effects of stress release will be considered. It is well known that saturated clay specimens carefully extracted at standard geotechnical depths (various tens of meters) exhibit suction (Skempton and Sowa, 3, Bishop et al., 4, Doran et al., 5, Delage et al., 6). In the case of the "perfect sampling" of an isotropic elastic sample and provided the sample remain fully saturated once extracted, this suction is supposed to be equal to the mean effective stress the specimen was submitted to prior to extraction (4). A necessary condition for the sample to remain saturated is that its air entry value should not exceed the mean effective in-situ stress. In the case of the Boom clay, suction effects have been investigated by (6) who found that specimens extracted at a depth of 225 m in the URL of Mol (Belgium) exhibited a suction value between 2 and 3 MPa, close to the mean in-situ effective stress (2.12 MPa). This work also confirmed that swelling clays to be tested in the triaxial apparatus should not be put in contact in water without previously being submitted to a mean effective stress as close as possible to that existing in-situ to avoid swelling and damage that would degrade the mechanical response and reduce the volumetric yield stress.

Compared to clays, claystones have been consolidated and strengthened during very long diagenesis periods during much longer geologic history, resulting in significantly lower porosity, strong diagenesis bonds and higher mechanical resistance. Following the previous investigation in the Boom clay, suction effects were investigated in the Callovo-Oxfordian (COx) claystone by testing specimens extracted at a depth of 479 m in the Bure URL in Eastern France. Investigation of the effects of desaturation in claystones (Pham et al. ,7, Jougnot et al., 8, and Wan et al., 9) is of some interest due to the consequences of the ventilation of the galleries along the gallery walls.

[a] Corresponding author: delage@cermes.enpc.fr

DOI: 10.1051/e3sconf/20160918004

In this paper, one examines what happens during stress release at constant water content on a specimen of COx claystone. To do so, a COx specimen was submitted to hydration at constant volume in the oedometer and stress release was imposed at constant water content. The change in microstructure were investigated by using mercury intrusion porosimetry.

2 Experimental methods

Tests were carried out in a room at controlled temperature and relative humidity (20°C and 80% respectively). Following Mohajerani et al. (10), the saturation of the sample was conducted on a high stress double lever arm oedometer (Marcial et al., 11) by carefully limiting volume changes during water infiltration. To do so, the vertical displacement gauge was manually followed and the specimen was progressively loaded to compensate any swelling with a limit fixed at 2 μm.

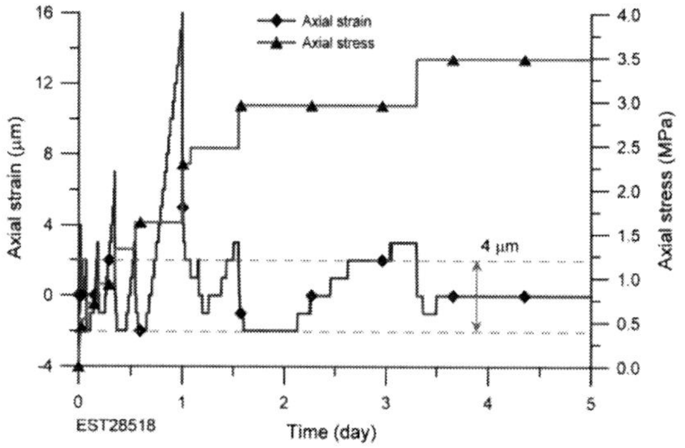

Figure 1. Resaturation with limited swelling

The detail of the procedure is illustrated in **Figure 1** that shows that the maximum stress applied to neutralize the swelling was equal to 3.48 MPa for a sample with an initial degree of saturation (S_r) of 77% and a suction of 34 MPa (WP4 measurement). The relatively high displacement of 16 □m observed at the end of first day is due to imperfect control during night time. To investigate suction effects on such a specimen resaturated with limited volume change, the specimen was afterwards quickly unloaded at constant water content. To do so, the porous stones were first desaturated by air flushing.

Pore-size distribution measurements (Mercury Porosimeter micromeritics-AutoPore IV 9500) were conducted on freeze-dried specimens to investigate microstructure changes. Specimens were previously frozen at - 210°C by using slush nitrogen (obtained by previously putting liquid nitrogen under vacuum, Delage et al. 2006, 12). The range of pores investigated was between 363.6 μm (minimum pressure of 3.4 kPa) to 5.5 nm (maximum pressure of 227.5 MPa).

3 The Callovo-Oxfordian claystone

The COx claystone is a sedimentary rock deposited 155 millions years ago between the Callovian and the Oxfordian ages in the western area of the Parisian sedimentary basin on top of a layer of Dogger limestone. The claystone was afterwards covered by an Oxfordian limestone layer. It has been since that time submitted to various tectonic effects including some horizontal stresses resulting from the Alpine orogenese. The COx claystone is composed of a clay matrix containing some detritic grains of quartz and calcite. At the depth of the Bure URL (490 m), the average mineralogical composition of COx claystone is as follows (Gaucher et al., 13): 45-50% clay fraction (mainly interstratified minerals of illite–smectites with a fraction between 50 and 70% of smectites, 20% carbonate, 22% quartz and 9% other minerals (feldspars, pyrite, dolomite and siderite). The total porosity varies in the COx layer between 14% in carbonated levels and 19.5% in the more argillaceous levels (Yven et al., 14).

The stress state measured in the 490 m deep Bure URL is as follows (Wileveau et al., 15):

- $\sigma_v = 12.7$ MPa
- $\sigma_h = 12.4$ MPa
- $\sigma_H = 12.7 - 14.8$ MPa
- $u = 4.9$ MPa

The vertical stress σ_v is close to the minor horizontal stress □h and smaller than the major horizontal stress □H. For this reason and some others (Mohajerani et al., 10), usual concepts of preconsolidation valid in clays do not apply to the COx claystone. Based in the above values, the equivalent effective Terzaghi mean stress:

$p'=1/3\ (\sigma'_v + 2\ \sigma'_h)$

is between 7.7 and 8.4 MPa (with $\sigma'_i = \sigma_i - u$).

Table 1 provides a comparison between the geotechnical properties of the Boom clay (at a depth of 225 m) and that of the COx clay (at a depth of 479 m with maximal clay fraction of 48-50%, Gaucher et al., 13).

Table 1. Geotechnical characteristics of the Boom clay and COx claystone

	Boom clay	Callovo-Oxfordian claystone
Geology	Oligocene (Rupelian)	Jurassic (Callovo-Oxfordian)
Age	30 millions years	160 millions years
Depth	225 m	479 m
Water content w (%)	20-30	< 6.5
Porosity n (%)	35-40	14-19
Clay fraction (%)	40-70	48-50
Young's modulus (GPa)	0.2-0.4	3.6-8.5
Initial specimen suction	2.8 MPa[1]	34 MPa[2]

[1] *Measured by the filter paper method*
[2] *Measured by the Decagon dew-point tensiometer*

The differences in water content (w), Young's modulus and unconfined shear strength illustrate the

DOI: 10.1051/e3sconf/20160918004

significant difference between a stiff clay (Boom clay) and a claystone (COx).

Further understanding about the microstructure, mineralogy and porosity of the COx claystone was gained by Yven et al. (14) by using different methods (including scanning electron microscope, autoradiography, mercury intrusion porosimetry, oil, helium and nitrogen adsorption), resulting in the conceptual model presented in Figure 4. The model shows how individual calcite or quartz grains are contained into a clay matrix with interconnected porosity.

Figure 2 presents a typical pore size distribution curve of the COx claystone obtained by mercury intrusion porosimetry (MIP) on a freeze-dried specimen, showing a well defined single pore population (with an average diameter of 32 nm) that is typical of the clay matrix.

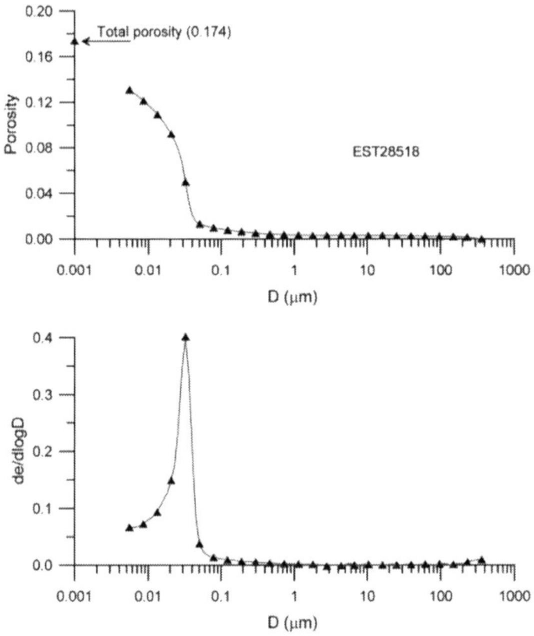

Figure 2. Pore size distribution curve initial state

The intruded porosity represents 75% of the total porosity (n = 17.4% and also represented in the graph). This data is in good agreement with that of Yven et al., (14) who estimated the porosity of small dimension (3 - 10) to 33.5 % of the total porosity in accordance with the 34% of porosity smaller than 10 nm found here.

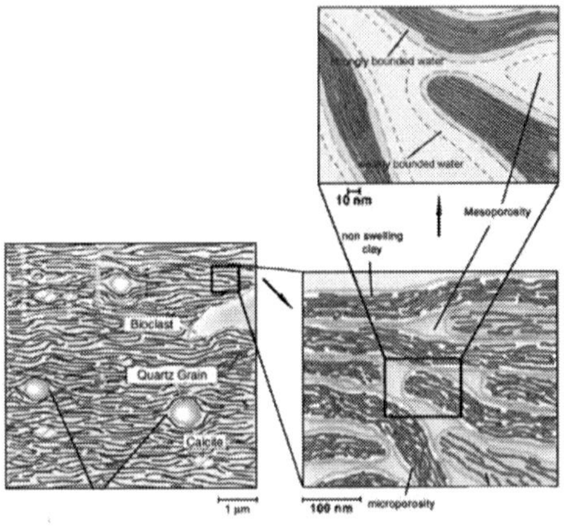

Figure 3. Microstructure model (after Yven et al., 14)

The remaining porosity is located within the clay matrix in smaller pores that cannot be investigated by MIP. This is a well known feature in clay materials containing smectites, due to very thin intra-platelets (or intra-crystalline) porosity (see Delage et al., 12 in compacted smectites).

The interpretation of the single pore population observed in Figure 2 can be made by assimilating the clay matrix to an assembly of bricks made up of platelets, as can be seen in Figure 3. The mean pore radius hence provides an estimate of the average platelets thickness. With an inter-basal spacing of 9.6 Å (0.96 nm) typical of dry smectites and illites and an average platelet thickness of 20 nm corresponding to the average entrance diameter determined from the PSD curve of the dried specimen, an average number of 21 layers by platelets can be roughly estimated.

4 Results

The stress-strain path corresponding to the instantaneous constant water unloading phase from the 3.48 MPa vertical stress is represented in Figure 4.

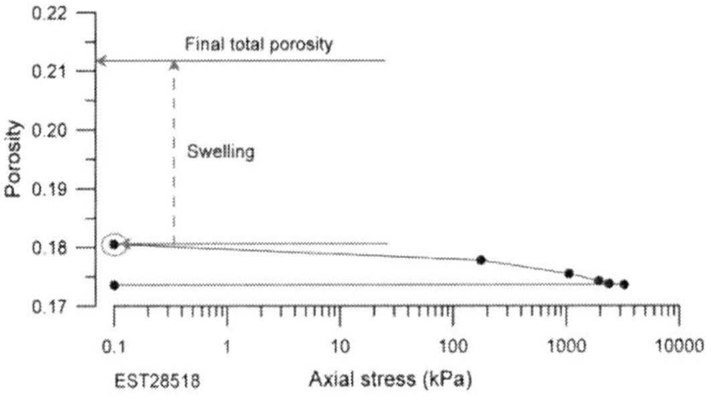

Figure 4. Unloading phase at constant water content

Unloading was performed stepwise at intermediate stresses of 0.8 MPa (in the semi-log graph, complete unloading is arbitrarily represented at a stress of 0.1 kPa). The unloading process lasted 5 minutes and exhibited an increase in porosity from 17.4 to 18% calculated from the swelling observed in Figure 4. The sample was afterwards extracted from the oedometer cell allowing for (slight) radial and axial expansion as shown in Table 2.

Table 2. Characteristics of the tested specimen

	Before test	After test
Height h (mm)	8.66	9.19
Diameter d (mm)	37.92	38.06
Mass m (g)	23.75	24.4
Water content w (%)	6.3	8.4
Void ratio e	0.21	0.27
Porosity n	17.4	21.2
Degree of saturation S_r	77	81
Suction s (MPa)	34	21

Careful calliper measurements carried out 15 minutes after unloading sample provided a porosity value of 20.9 %. Obviously, significant axial swelling occurred at constant water content once the sample was extracted from the ring. Note that hydrostatic weighing provided a porosity value of 21.2 % quite close to that obtained by calliper measurements, also represented in Figure 4. The specimen was afterwards isolated from evaporation by wrapping in a plastic film and rapidly frozen to keep constant water content.

One can observe that, in spite of having a constant water content (w = 8.4%), the swelling observed (final porosity n = 21.2% compared to 17.4%) resulted in a simultaneous desaturation of the specimen with a final value of the degree of saturation S_r = 81% corresponding to a 21 MPa suction (WP4 measurement). This means that the air-entry value of the sample has been exceeded. Indeed, this value, estimated between 7 and 9 MPa according to gas penetration tests (De La Vaissière and Talandier, 16) is significantly smaller than the final value of 21 MPa.

The pore size distribution (PSD) curve of the swollen sample is presented in Figure 5. Also reported in the graph is the total porosity (21.2%).

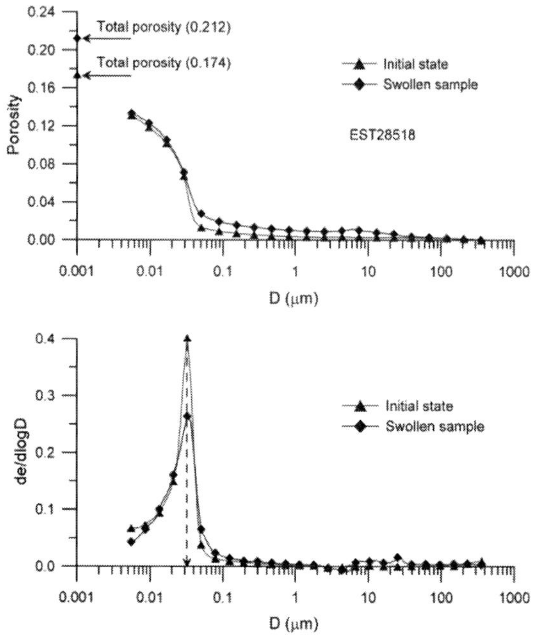

Figure 5. PSD curve of the swollen specimen

Surprisingly, the shape of the PSD of the swollen unsaturated specimen at constant water content is quite comparable to the initial one presented in Figure 2. It exhibits a single pore population with a mean radius of 32 nm, comparable to that of the intact sample (although slightly less well classified as indicated by the density function curve). No large pore population representing possible cracks due to hydraulic damage and swelling (Yang et al., 17, Wan et al., 9) is observed. Also, visual examination of the sample didn't evidence any visible cracks with thickness larger than the maximum diameter identified by mercury intrusion (363.6 μm).

5 Discussion

The global specimen expansion at constant water content is then only possible within the intra-platelet infra-porosity (too small to be intruded by mercury) with a transfer of water from the larger inter-platelets pores (average 32 nm) to the smaller intra-platelet pores that results in crystalline swelling. Here, swelling occurs within the inter-stratified illite smectite clay platelets.

As any clay material, the pore water of the COx clay includes both capillary and hydration water (e.g. Revil and Lu, 18). By only considering capillary effects, the suction s exerted when desaturating 32 nm diameter pores is given by the Young-Laplace equation:

$$s = \frac{2\sigma_{w/a}\cos\theta}{r} \quad (1)$$

in which $\sigma_{w/a}$ is the water/air interfacial tension equal to 72.8×10^{-3} N/m and θ the solid-liquid contact angle with $\cos\theta = 1$ for water. A pressure value of 9.1 MPa is obtained, slightly higher than the air-entry value of the

COx claystone, confirming the predominance of capillary effects in the clay water interaction in these inter-platelets pores schematically represented in Figure 3.

Comparing this value with the final suction of 21 MPa finally measured in the sample indicates that inter-platelets pores have been desaturated. The high suction value is hence mainly due to strong clay-water interaction occurring inside the platelets, confirming the intra-crystalline nature of the swelling.

The question about the rate of water transmission between the large inter-platelets pores and the thin intra-platelets ones arises given the very low permeability and slow water transfer suspected. This internal water transfer, comparable in nature to long term internal transfers observed in compacted bentonite at constant water content (Delage et al., 12) has been completed in around 30 minutes after unloading. Actually, the slowing effects due to the low permeability (around 10^{-20} m^2 in the COx claystone) are compensated by the small distance that water molecules have to travel from the inter-platelets pore to the intra-platelets ones.

6 Conclusion

Suction effects were investigated in a specimen of the Callovo-Oxfordian claystone extracted at great depth (479 m) in the underground research laboratory of ANDRA in Bure (France). The specimen was first saturated with limited volume change by progressively increasing vertical stress up to 3.48 MPa. The specimen was afterwards quickly unloaded at constant water content, exhibiting significant swelling and desaturation, 30 minutes after extraction from the oedometer ring with a suction significantly larger than the air entry value. The pore size distribution of the swollen sample was comparable to that of the initial one with no appearance of large pores, indicating that swelling was due to a water transfer within the clay matrix from inter-platelets pores to intra-platelet pores, in a pore range that is too small to be detected by mercury intrusion. This phenomenon may occur during gallery excavation in areas of the excavation damaged zone submitted to instantaneous stress release, as a first stage of desaturation, prior to the subsequent evaporation due to the ventilation of the galleries.

Acknowledgments

The authors are grateful to ANDRA (French Agency for Radioactive Waste Management) for funding this research.

References

[1] Monfared M, Delage P, Sulem J, Mohajerani M, Tang AM, De Laure E (2011). A new hollow cylinder triaxial cell to study the behaviour of geomaterials with low permeability, Int. J. Rock Mech. Min. Sc. 48 (4): 637-649.

[2] Hu DW, Zhang F, Shao JF (2013). Experimental study of poromechanical behavior of saturated claystone under triaxial compression. Acta Geotech. 9: 207-214.

[3] Skempton, A. W. and Sowa, V. A. 1963; The behaviour of saturated clays during sampling and testing. Géotechnique 13, No. 4, 269-290.

[4] Bishop A.W., Kumapley N.K. and El Ruwayih A. 1974; The influence of pore water tension on the strength of a clay. Proc. Royal Soc. London vol. 1286, 511-554.

[5] Doran, I.G, Sivakumar, V, Graham, J. and Johnson, A. 2000; Estimation of in-situ stresses using anisotropic elasticity and suction measurements. Géotechnique 50, No. 2, 189-196.

[6] Delage P., Le T.T., Tang A.M., Cui Y.J., Li X.L. 2007. Suction effects in deep Boom clay block samples. Géotechnique 57 (1), 239-244.

[7] Pham QT, Vales F, Malinsky L, Nguyen Minh D, Gharbi H (2007). Effects of desaturation-resaturation on mudstone. Phys. Chem. Earth 32: 646–655.

[8] Jougnot D., A. Revil, N. Lu, and A. Wayllace. 2010. Transport properties of the Calloxo-Oxfordian clay-rock under partially saturated conditions, Water Res. Research, 46, W08514.

[9] Wan M, Delage P, Tang AM, Talandier J (2013). Water retention properties of the Callovo-Oxfordian claystone. Int. J. Rock Mech. Min. Sc. 64: 96–104

[10] Mohajerani M., Delage P., Monfared M. et al. 2011. Oedometric compression and swelling behavior of the Callovo-Oxfordian argillite. Int. J. Rock Mech. Min. Sc. 48(4): 606–615.

[11] Marcial D., Delage P., Cui Y.J. 2002; On the high stress compression of bentonites. Can. Geotech. J. 39(4): 812-820.

[12] Delage P., Marcial D., Cui Y.J., Ruiz X. 2006; Ageing effects in a compacted bentonite: a microstructure approach. Géotechnique; 56(5): 291–304.

[13] Gaucher G, Robelin C, Matray JM et al. (2004). ANDRA underground research laboratory: interpretation of the mineralogical and geochemical data acquired in the Callovian-Oxfordian formation by investigative drilling. Phys. Chem. Earth 29: 55-77

[14] Yven B, Sammartino S, Geroud Y et al. (2007). Mineralogy texture and porosity of Callovo-Oxfordian claystones of the Meuse/Haute-Marne region (eastern Paris Basin), Mém. Soc. Géol. France, 178: 73-90

[15] Wileveau Y., Cornet F.H., Desroches J., Blumling P. 2007; Complete in situ stress determination in an argillite sedimentary formation. Phys. Chem. Earth 32(8–14): 866–878.

[16] De La Vaissière R., Talandier J. 2012; Gas injection tests in the Meuse/Haute Marne underground research laboratory. Proc. Transfert 2012 Conf., Ecole Centrale Lille Eds, 360–368.

[17] Yang D., Chanchole S., Valli P., Chen L. 2012; Study of the Anisotropic Properties of Argillite Under Moisture and Mechanical Loads. Rock Mech. Rock Eng. 46(2): 247–257.

[18] Revil A. and N. Lu, Unified water sorption and desorption isotherms for clayey porous materials, Water Res. Research, 49(9), 5685-5699.

DOI: 10.1051/e3sconf/20160918005

Hydro-mechanical modelling of MX-80 bentonite: one dimensional study

Ayman A. Abed[1], Mika Laitinen[2], Joni Lämsä[3], Tero Harjupatana[3], Wojciech T. Sołowski[1,a] and Markku Kataja[3]

[1]*Aalto University, Department of Civil and Environmental Engineering, P.O.Box 12100, FI-00076 Aalto, Finland*
[2]*Numerola Oy, P.O.Box 126, 40101 Jyväskylä, Finland*
[3]*University of Jyväskylä, Department of Physics, P.O.Box35 (YFL), FI-40014Jyväskylä, Finland*

Abstract. As a first step towards modelling the coupled Thermo-Hydro-Mechanical-Chemical (THMC) behaviour of bentonite, the Barcelona Basic Model (BBM) has been implemented into Numerrinfinite element code. This model has beenfully coupled with the single phase flow equation for unsaturated soils which models liquid water transport. Suction obtained from solving the flow equation is used as an input for the BBM model and the volumetric deformations from the mechanical analysis are used to update the pore water pressure field. As an alternative, BBM is used alongside the Kröhn's model which assumes that bentonite re-saturation is mainly driven by water vapour diffusion. The paper simulates one dimensional infiltration test on MX-80bentonite with both implemented modelsfor water transport and compares the results withthose from a laboratory experiments based on X-ray tomography.The numerical results of the simulations are similar despite taking into account different physical phenomena.

1 Introduction

The paper initially describes implementation of the well-known Barcelona Basic Model (BBM)[1]into Numerrin numerical solver [2]. The implementation of BBM bases on standard explicit sub-stepping scheme with error control for stress integration[3-4]. In addition in the subsequent finite element simulations two hydraulic models are tested:the single phase flow equation[5] and the vapour diffusion model (Kröhn's model)[6]. The single phase flow equation is fully coupled withBBM.TheKröhn's model,on the other hand,does not support coupling directly and is used alongside BBM. The paper gives a brief description of the mechanical and hydraulic models before using them in the simulation of a 1D wettingexperimentof MX-80 bentonite under constant volume conditions.

2 Barcelona Basic Model

The Barcelona Basic Modelbases on the Modified Cam Clay model[7] and can be viewed as its extensionwhich accounts for unsaturated soil behaviour.BBM uses net stress$\sigma = \sigma^{tot}$-u_a and suction $s = u_w - u_a$ as independent variables, where σ^{tot}, u_w and u_a are the total stress, pore water pressure and pore air pressure respectively.In this paper u_a is taken as zero.

2.1 BBM in elasticity

The rate of total strain$\dot{\boldsymbol{\varepsilon}}$ in BBM is decomposed into elastic $\dot{\boldsymbol{\varepsilon}}_e$ and plastic part $\dot{\boldsymbol{\varepsilon}}_p$:

$$\dot{\boldsymbol{\varepsilon}} = \dot{\boldsymbol{\varepsilon}}_e + \dot{\boldsymbol{\varepsilon}}_p \tag{1}$$

During elasticity, the plastic part of the total strain ratevanishes $\dot{\boldsymbol{\varepsilon}}_p = 0$. The rate of elastic strain$\dot{\boldsymbol{\varepsilon}}_e$is decomposed further intothe elastic strain rate due to net stress$\dot{\boldsymbol{\varepsilon}}_e^\sigma$ and the elastic strain rate related to suction $\dot{\varepsilon}_e^s$:

$$\dot{\boldsymbol{\varepsilon}}_e = \dot{\boldsymbol{\varepsilon}}_e^\sigma + \mathbf{m}^T \dot{\varepsilon}_e^s \tag{2}$$

with$\mathbf{m}$ = {1,1,1,0,0,0}. Thus, the elastic net stress rate can be calculated as:

$$\dot{\boldsymbol{\sigma}} = \mathbf{D}_e \dot{\boldsymbol{\varepsilon}}_e^\sigma = \mathbf{D}_e \left(\dot{\boldsymbol{\varepsilon}} - \mathbf{m}^T \dot{\varepsilon}_s^e\right) \tag{3}$$

where$\mathbf{D}_e$ is the elastic stiffness matrix :

$$\mathrm{D}^e = \begin{bmatrix} K+\frac{4}{3}G & K-\frac{2}{3}G & K-\frac{2}{3}G & 0 & 0 & 0 \\ & K+\frac{4}{3}G & K-\frac{2}{3}G & 0 & 0 & 0 \\ & & K+\frac{4}{3}G & 0 & 0 & 0 \\ & & & G & 0 & 0 \\ & sym & & & G & 0 \\ & & & & & G \end{bmatrix} \tag{4}$$

with $K = \dfrac{(1+e)\cdot p}{\kappa}; G = \dfrac{3K(1-2\nu)}{2(1+\nu)}$

in the above formulation$K$ and G are the bulk and the shear modulus respectively, e stands for thesoil void ratio, $\kappa$for the soil swelling index, $\nu$is the Poisson's ratio and$p = tr(\sigma)/3$is the isotropic net pressure.The strain rate relatedto suction changeis:

$$\dot{\varepsilon}_e^s = \frac{\kappa_s \dot{s}}{3(1+e)(s+p_{atm})} \tag{5}$$

whereκ_s is the soil swelling index with respect to suction variation and p_{atm} is the atmospheric pressure.

[a]Corresponding author: wojciech.solowski@aalto.fi

2.2 BBM in plasticity

Figure 1 shows a 3D representation of the yield surface ofBBM which equation is:

$$F=q^2-M^2(p+ks)(p_o-p)=0 \quad (6)$$

The symbol q stands for the deviatoric stress:

$$q=\frac{1}{\sqrt{2}}\sqrt{(\sigma_1-\sigma_2)^2+(\sigma_2-\sigma_3)^2+(\sigma_3-\sigma_1)^2} \quad (7)$$

whereσ_1, σ_2 and σ_3 are the principal stresses. The trace of the yield surface in the plane $q=0$ is the *Loading Collapse* (LC) curve.

The plastic strain direction is determined using a plastic potential function:

$$G=\alpha q^2-M^2(p+ks)(p_o-p)=0 \quad (8)$$

The factor α which allows forthe recovery oftheJaky'sapproximationof coefficient of at rest pressure K_0in 1D compression isdefined as:

$$\alpha=\frac{M(M-9)(M-3)}{9(6-M)}\frac{\lambda}{\lambda-\kappa} \quad (9)$$

whereλ is the slope of normal consolidation line.

BBM assumes that the soil shear strength increases linearly with product of suction and model constant k.The soil preconsolidation pressure p_o is also considered to be suction dependent through the equation:

$$p_o=p^c\left(\frac{p_o^*}{p^c}\right)^{\frac{\lambda-\kappa}{\lambda_s-\kappa}} \quad (10)$$

with

$$\lambda_s=\lambda\left[(1-r)e^{-\beta s}+r\right] \quad (11)$$

The preconsolidation pressure at full saturation is indicated by p_o^*. Equations(10) and (11) introduce p^c, β and r as new BBM parameters. These parameters are used to define the effect of suction on the preconsolidation pressure and the post-yielding stiffness.

During plasticity the net stress rate is calculated as:

$$\dot{\boldsymbol{\sigma}}=\mathbf{D}_e\dot{\varepsilon}_e^{\sigma}=\mathbf{D}_e\left(\dot{\boldsymbol{\varepsilon}}-\mathbf{m}^T\dot{\varepsilon}_s^{e}-\dot{\boldsymbol{\varepsilon}}_p\right) \quad (12)$$

where the plastic strain rate $\dot{\boldsymbol{\varepsilon}}_p$ is determined by the flow rule:

$$\dot{\boldsymbol{\varepsilon}}_p=\dot{\Lambda}\frac{\partial G}{\partial \boldsymbol{\sigma}} \quad (13)$$

The plastic multiplier $\dot{\Lambda}$ can be derived based on plasticity theory[4]:

$$\dot{\Lambda}=\frac{\mathbf{a}^T\mathbf{D}_e\dot{\varepsilon}+\left[\mathbf{c}-\mathbf{a}^T\mathbf{D}_e\mathbf{b}\right]\dot{s}}{\mathbf{a}^T\mathbf{D}_e\mathbf{g}-d} \quad (14)$$

, where:

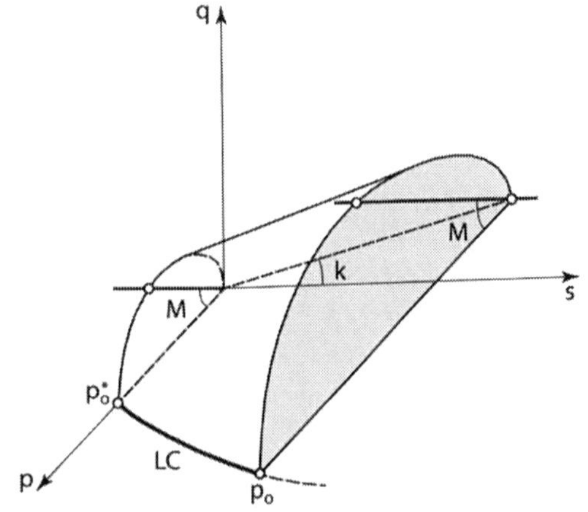

Figure 1.3D representation of the yield surface in BBM.

$$\mathbf{a}=\frac{\partial F}{\partial \boldsymbol{\sigma}};\mathbf{b}=\mathbf{m}\frac{\kappa_s\dot{s}}{3(1+e)(s+p_{atm})};$$
$$\mathbf{c}=\frac{\partial F}{\partial s};d=\frac{\partial F}{\partial \varepsilon_p^v}\mathbf{m}^T\frac{\partial G}{\partial \sigma};\mathbf{g}=\frac{\partial G}{\partial \boldsymbol{\sigma}} \quad (15)$$

Substituting equations (5) and (13) into equation(12) yields the net stress rate:

$$\dot{\boldsymbol{\sigma}}=\mathbf{D}_e\dot{\boldsymbol{\varepsilon}}-\mathbf{D}_e\mathbf{m}^T\frac{\kappa_s\dot{s}}{3(1+e)(s+p_{atm})}-\mathbf{D}_e\dot{\Lambda}\frac{\partial G}{\partial \boldsymbol{\sigma}} \quad (16)$$

Finally, the preconsolidation pressure is updated as:

$$p_o^*=p_{oi}^*e^{\frac{(1+e)\dot{\varepsilon}_p^v}{\lambda-\kappa}} \quad (17)$$

where p_{oi}^* is the initial preconsolidation pressure and $\dot{\varepsilon}_p^v=tr(\dot{\boldsymbol{\varepsilon}}_p)$.

2.3 BBM implementation into Numerrin code

BBM is implemented into Numerrinand used inthe Finite Element Method simulations. Stress is integrated with a standard explicit sub-stepping scheme with error control [3-4].In such a scheme if plasticity is detected then an automatic sub-stepping of the strain increment is initiated. The final number of sub-steps is dependent on the required accuracy by the user.

In order to validate the implementation, the data provided in literature [1] has been reproduced. For the sake of brevity only two numerical tests are shown:(a) wetting of unsaturated soil under isotropic condition and(b) shearing under constant confining pressure but at different suction levels. Both testsused model parametersgiven in Table 1.

Table 1. BBM parameters as used in the verification tests, after [1].

κ [-]	λ [-]	G [kPa]	κ_s [-]	M [-]	r [-]	β [kPa^{-1}]	p^c [kPa]	p_o* [kPa]	k [-]
0.02	0.2	10000.0	0.008	1.0	0.75	0.0125	100.0	200.0	0.6

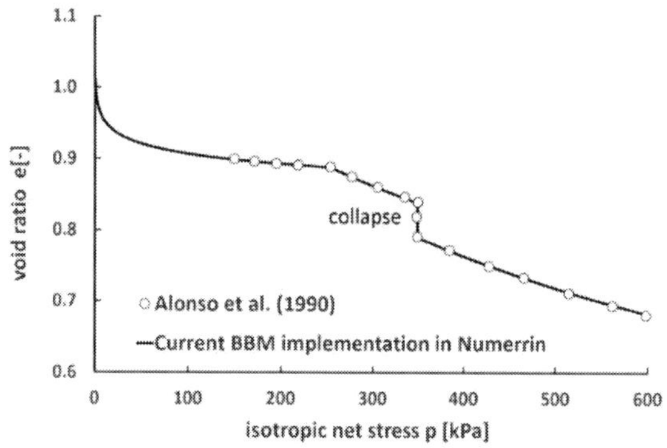

Figure 2. Numerical results versus reference data: isotropic loading followed by wetting.

2.3.1 Isotropic loading followed by wetting

A soil sample with an initial suction of 200 kPa is isotropically loaded up to a confining pressure of 350 kPa. At this confining pressure the soil is wetted till full saturation, which leads to soil collapse. After the full wetting the soil is isotropically loaded in fully saturated conditions up to 600 kPa. Figure 2shows impeccable match between the numerical results and the reference data.

2.3.2 Shear at different suction levels

The model implementation is furthertested by replicating the shear paths (Figure 3). The soil is sheared under constant confining pressures at three different suction levels (s=100, 200 and 300 kPa). The numerical results and the reference data match perfectly.

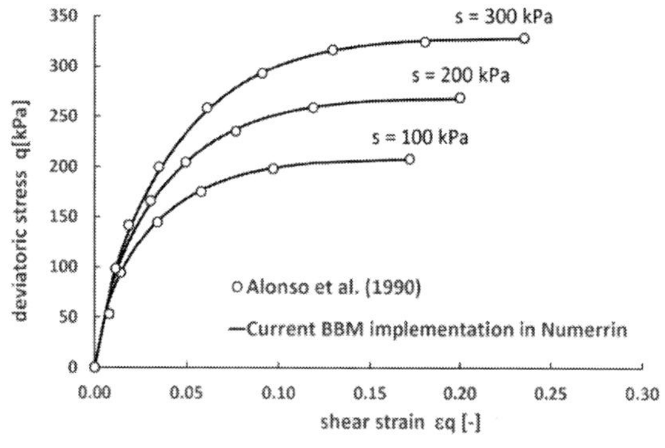

Figure 3. Numerical results versus reference data: shear test.

3 Single phase flow formulation

For the unsaturated water flow, the water mass conservation equation can be written as:

$$-\frac{\partial}{\partial x}(n\rho_w \mathbf{v}) = \frac{\partial}{\partial t}(n\rho_w S_r) \tag{18}$$

where$n$, ρ_w and S_r stand for soil porosity, water density and degree of saturation respectively. The *average water velocity*$\mathbf{v} = \mathbf{q} / n$where $\mathbf{q}$is the *specific discharge*given by Darcy's law:

$$\mathbf{q} = -\mathbf{k}(h)\left[\frac{\partial h}{\partial \mathbf{x}}\right] \tag{19}$$

where$k(h)$ is the suction dependent hydraulic conductivity and h is the total head being the sum of the pressure head ψ and the geodetic head z.Equation (18) shows that the net water flux expressed by the left hand side should be equal to the water mass stored or expelled by the soil expressed by the right hand side. On expandingequation (18)it becomes:

$$\frac{\partial \mathbf{v}}{\partial x} + n\frac{\partial S_r}{\partial \psi}\frac{\partial \psi}{\partial t} + nS_r\gamma_w\beta_w\frac{\partial \psi}{\partial t} - (1-n)S_r\frac{\partial \varepsilon_v}{\partial t} = 0 \tag{20}$$

The flow equation(20)includes mechanical couplingby acknowledging the effects of volumetric deformations ε_v on the water flow field. It also takes into account the effect of water compressibility through the parameterβ_w.

3.1 Implementation into Numerrin code

The Finite Element discretization of equation (20)over domain Ωyields:

$$\begin{aligned}&\int_\Omega \mathbf{N}^T nS_r\gamma_w\beta_w\mathbf{N}\frac{d\boldsymbol{\psi}}{dt}d\Omega + \int_\Omega \mathbf{N}^T nC\mathbf{N}\frac{d\boldsymbol{\psi}}{dt}d\Omega\\ &-\int_\Omega \mathbf{N}^T(1-n)S_r\mathbf{m}^T\nabla\mathbf{N}\frac{d\mathbf{u}}{dt}d\Omega + \int_\Omega (\nabla\mathbf{N})^T\mathbf{k}\nabla\mathbf{N}\boldsymbol{\psi}d\Omega\\ &+\int_\Omega (\nabla\mathbf{N})^T\mathbf{k}\,d\Omega = 0\end{aligned} \tag{21}$$

where$\mathbf{N}$ and $\mathbf{u}$ are the element shape functions and the nodal displacement respectively, and∇stands for gradient. The *soil specific moisture capacity* $C$is:

$$C = n\frac{\partial S_r}{\partial \psi} \tag{22}$$

By solving Equation (21) with suitable boundary conditions, one will be able to determine the suction variation overtime. These values of suction are passed to BBM for the deformation analysis. The resultingdisplacement fieldis passed back to the flow equation with the required volumetric deformations to update the flow field and to progress into the next time step in solving the coupled hydro-mechanical system.
The basic input data needed to solve the unsaturated flow equation are the soil water retention curve (SWRC) and the soil permeability curve. These curves identify how the soil water content and permeability are changingwith suction variation. In most case some points of the SWRC are experimentally determined and then fitted to one of the common SWRC models.For example van Genuchten[8] or Brooks-Corey [9]modelscan be used for fitting the soil water retention data. The permeability curve is more difficult to establish and is usually based on the SWRC data.

DOI: 10.1051/e3sconf/20160918005

4 Extended vapour diffusion model

The extended vapour diffusion model or Kröhn's model [6] is based on a phenomenological description of bentonite re-saturation. The model considers that the saturation process is driven by the vapour diffusion in bentonite pores and by the inter-lamellar water diffusion. The fundamental balance equation for this model is:

$$\frac{\partial(n\rho_v)}{\partial t}-\frac{\partial}{\partial \mathbf{x}}\left(nD_v\frac{\partial \rho_v}{\partial \mathbf{x}}\right)+\rho_d\frac{\partial w}{\partial t}-\rho_d\frac{\partial}{\partial \mathbf{x}}\left(D_w\frac{\partial w}{\partial \mathbf{x}}\right)=0 \quad (23)$$

whereρ_v and ρ_d arethe water vapour density and the bentonite dry density respectively. Diffusion is controlled by the vapour diffusion coefficient D_v and the inter-lamellar water diffusion coefficient D_w.

The extended vapour diffusion model implementation has been used alongside BBM. The water content *w*has been converted to suction using the SWRC which affected the deformations. However, those mechanical deformationshaveno effect on the water transport as indicated inequation (23).

5 Simulation of 1D wetting test of MX-80 bentonite

5.1 Experimental procedure

One-dimensional wetting and swelling of a compacted cylindrical MX-80-bentonite sample confined in a constant volume were measured using a non-invasive method based on X-ray micro-tomography[10]. The experiments were carried out using a table-top X-ray tomographic scanner (SkyScan 1172) and the sample holder shown in Figure 4. The method is based on comparison of X-ray tomographic images of the sample in the original unwetted state and in the wetted and deformed state. Full details of the experimental procedure and data analysis can be found in[11].

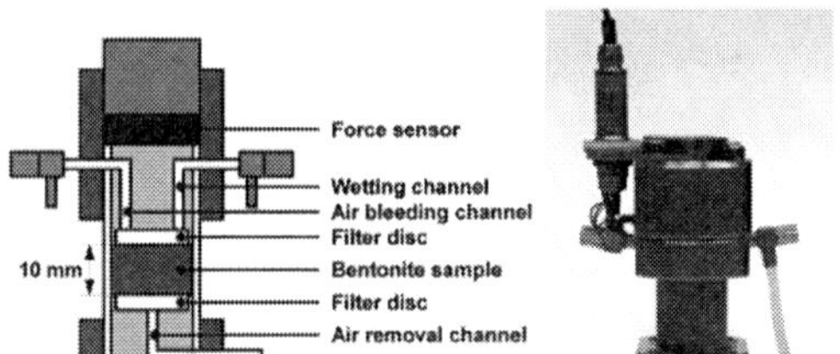

Figure 4.Schematic illustration (a) and design (b) of the sample holder used in X-ray tomographic experiments of 1D axial wettingof MX-80 bentonite [10].

The initial dry density and water content of the bentonite sample were ρ_d = 1600 kg/m^3 and w = 12.1%, respectively. The duration of the measurement was 7.2 days. With the used sample holder geometry the measurement yields time evolution of the axial distribution of dry density, water content and swelling pressure during the wetting process.

5.2 Finite Element model

Figure 5a shows the finite element model used to simulate the wetting test. The model consists of 80 rectangular 4-noded elements with 4 integration points per element. To simulate a constant volume conditions the model is constrained in the directions normal to its boundaries as it clear in Figure 5b. The hydraulic boundaries along both sides of the sample are closed while keeping the top boundary open for discharge. The bottom boundary is kept under fixed zero pressure head to simulate continuous wetting, see Figure 5c.

The simulationmonitors the variation of pressure head along the sample height in time as well as the development of the swelling pressure at a control point at the top of the sample (seeFigure 5a).

The experiment is simulated twice:the first timethe BBM is coupled with the single phase flow equation(20), the other time with the Kröhn's model(23).

5.3 Models parameters

5.3.1 Mechanical parameters

The BBM mechanical parameters have been calibrated using the experimental data[10].The calibration processwas performed in such way that parameters are kept in the ranges of typical values used for MX-80 bentonite [12-13].

Table 2 lists the parameters as used in the analyses. During the calibration process it was noticed that the values of the preconsolidation pressure p_o^* and theswelling index with respect to suction κ_s have the biggest impact on fitting the swelling pressure data.

In fact, the preconsolidation pressure can be estimated depending on the pressing pressure during the preparation of the bentonite sample. However such information is not available for the current study, therefore the preconsolidation pressure was assumed and calibrated in light of data available in literature [14-15].

5.3.2 Hydraulic parameters

The soil water retention curve was estimated using the data from literature [16-17] as well as measurements [18] for the tested bentonite.

Table 2. Calibrated BBM parameters for MX-80 bentonite.

κ [-]	λ [-]	ν [-]	κ_s [-]	M [-]	r [-]	β [kPa^{-1}]	p^c [kPa]	p_o* [kPa]	k [-]
0.05	0.15	0.3	0.25	1.07	0.8	2.0E-05	100.0	2200.0	0.001

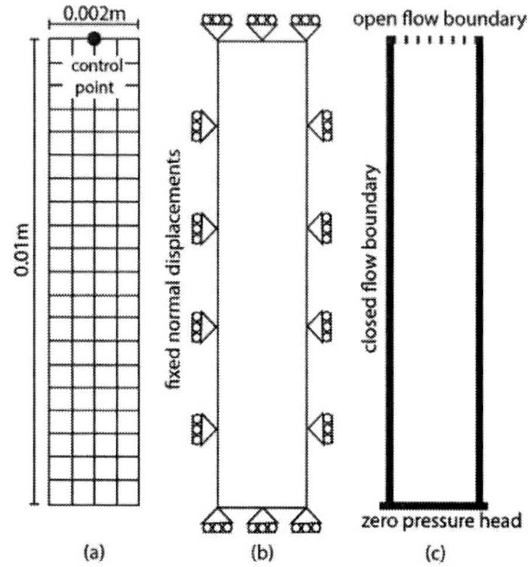

Figure 5.Finite element model for the constant volume test on MX-80 bentonite.

The approximation of the datawith van Genuchtenmodel is shown in Figure 6. As the initial water content is 12.1% the SWRC gives a corresponding initial suction estimate of about 70 MPa. The saturated hydraulic conductivity was assumed ask_{sat} = 4.0E-14 m/s.Similarly to[13] the soil relative permeability curve is takenas:

$$k = k_{sat}\left(\frac{S_r - S_{res}}{S_{sat} - S_{res}}\right)^3 \qquad (24)$$

whereS_{sat} and S_{res} are the degree of saturation at full saturation (usually S_{sat} = 1) and at the residual state respectively, calibration gave a value of S_{res} = 0.1.

ForKröhn's model the calibration yieldeda value ofD_v = 10.0E-6 m^2/s for the vapour diffusion coefficient and D_w = 4.0E-10 m^2/s for the inter-lamellar water diffusion coefficient.

5.4 Simulation results

5.4.1 Using coupled BBM

Figure 7 shows that the coupled BBM results fit well the experimental measurements for the variation of water pressure head in time. However, the prediction for the evolution of swelling pressure at the control point is less accurate, even though the final swelling pressure value is matched (Figure 10).In the simulation,stresses follow a nonlinear elastic path upon wetting till the suction reaches a value of approximately 10 MPa. Below that suction value plastic deformation occurs and the stress path follows the LC yield curve.

5.4.2 Using decoupled BBM-Kröhn's model

Figure 8shows that the decoupled BBM-Kröhn's model can reproduce the variation of suction during bentonite re-saturation process. The final value of swelling pressure is well captured as shown in Figure 10.

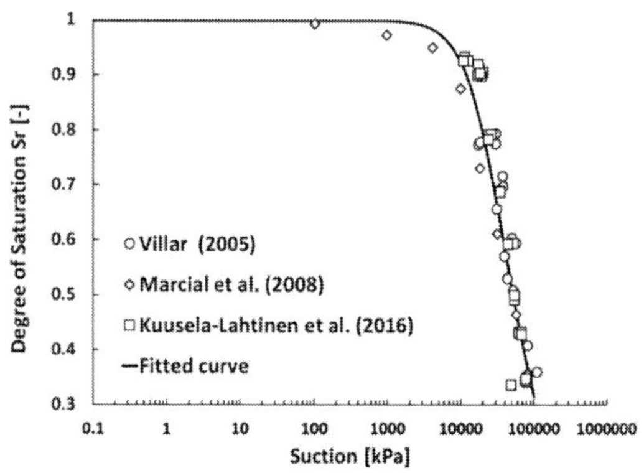

Figure 6.Fitted SWRC for MX-80 bentonite.

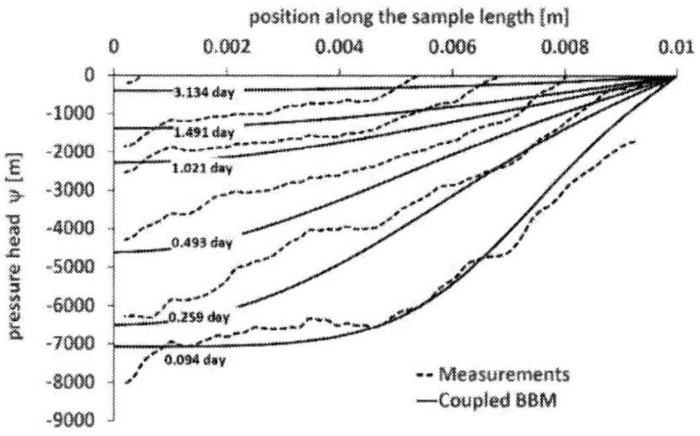

Figure 7.Pressure head in time: coupled BBM versus measurements.

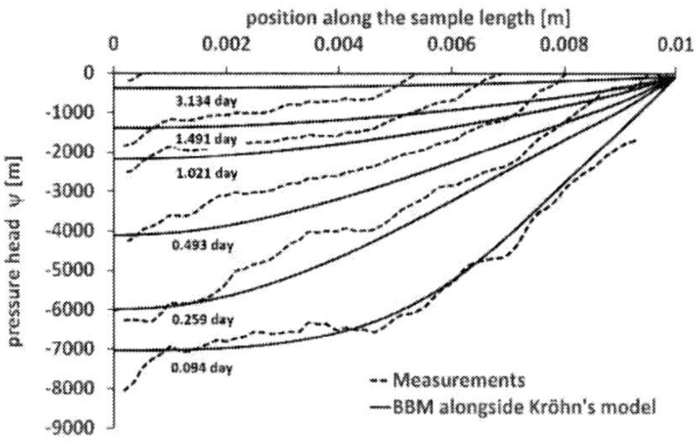

Figure 8.Pressure head in time: decoupled BBM-Kröhn's versus measurements.

6 Discussion

The paper presented finite element simulations of a simple wetting experiment.In the calculations BBM model coupled with two different formulations for water transport was used. Both obtained predictionsare very similar no matter which formulation for flow is chosen (Figure 9 and Figure 10). That is puzzling as the single phase flow equation and the extended vapour diffusion equation are based on quitedifferent physical processes. In fact,bentonite re-saturation is most likely happening due to both water transport and vapour diffusion.

DOI: 10.1051/e3sconf/20160918005

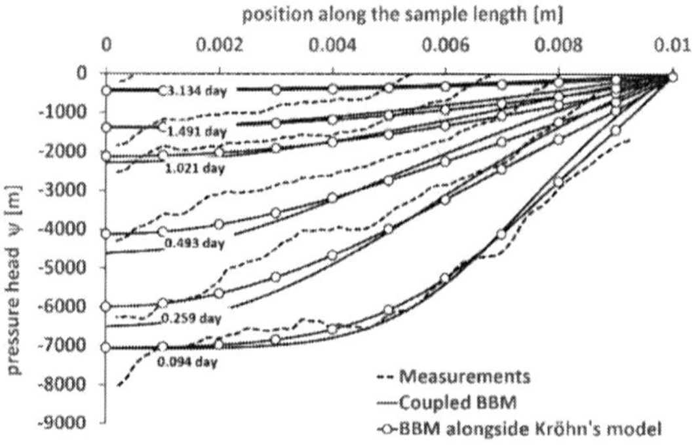

Figure 9.Coupled BBM and decoupled BBM-Kröhn's model predictions for pressure head variation in time.

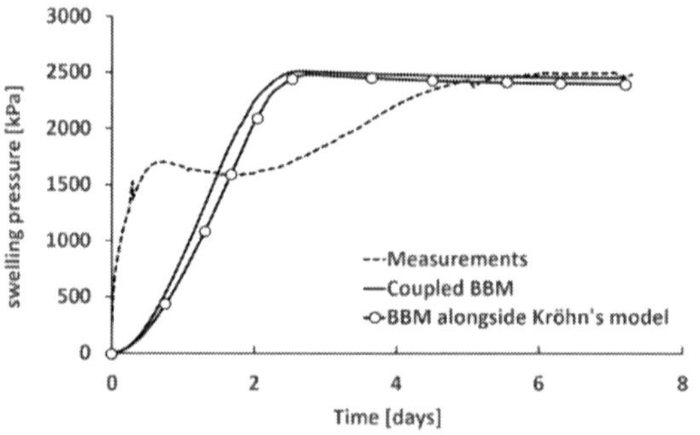

Figure 10.Coupled BBM and decoupled BBM-Kröhn's model predictions for swelling pressure in time.

However, based on this short study it seems that simply adding these two effects togethermaynot lead to improvement in the simulation results.

Deeper links to the physicalprocesses behind wetting of bentonite must be included in the future, which may allow for some judgment on the weight of each transport mode during re-saturation.

Furthermore, the mechanical model needs to be improvedin order to better capture the swelling pressure development in time. This might be achieved by incorporating the role of the bentonite micro-structural level intothe constitutive modelling framework.

Acknowledgement

The authors would like to gratefully acknowledge that the presented research has been funded by KYT2018 Finnish Research Programme on Nuclear Waste Management via THEBES project.

References

1. E.E. Alonso, A. Gens, A. Josa, A constitutive model for partially saturated soils, Géotechnique. 40 (1990) 405-430.
2. M. Laitinen, Numerrin 4.0 Manual, (2013).
3. S.W. Sloan, A.J. Abbo, D. Sheng, Refined explicit integration of elastoplastic models with automatic error control, Eng. Comput. 18 (2001) 121-194.
4. W.T. Sołowski, D. Gallipoli, Explicit stress integration with error control for the Barcelona Basic Model: Part I: Algorithms formulations, Comput. Geotech. 37 (2010) 59-67.
5. L.A. Richards, Capillary conduction of liquids through porous mediums, J. Appl. Phys. 1 (1931) 318-333.
6. K. Kröhn, Simulating non-isothermal water uptake of compacted bentonite without coupling of hydraulics to mechanics, Appl. Clay. Sci. 47 (2010) 28-35.
7. A. Schofield, P. Wroth, Critical state soil mechanics, (1968).
8. M.T. Van Genuchten, A closed-form equation for predicting the hydraulic conductivity of unsaturated soils, Soil Sci. Soc. Am. J. 44 (1980) 892-898.
9. R.H. Brooks, A. Corey, Properties of porous media affecting fluid flow, Journal of the Irrigation and Drainage Division. 92 (1966) 61-90.
10. J. Lämsä, Measuring Water Transport and Deformations in MX-80 -Bentonite with X-Ray Micro Tomography, MSc thesis (in Finnish) ed., University of Jyväskylä, Jyväskylä, 2015.
11. T. Harjupatana, J. Alaraudanjoki, M. Kataja, X-ray tomographic method for measuring three-dimensional deformation and water content distribution in swelling clays, Appl. Clay. Sci. 114 (2015) 386-394.
12. E. Toprak, N. Mokni, S. Olivella, X. Pintado, Thermo-hydro-mechanical modelling of buffer, synthesis report. (2013).
13. M. Villar, M. Sánchez, A. Gens, Behaviour of a bentonite barrier in the laboratory: Experimental results up to 8years and numerical simulation, Physics and Chemistry of the Earth, Parts A/B/C. 33 (2008) S476-S485.
14. L. Nguyen-Tuan, M. Datcheva, M. Khan, T. Schanz, Numerical Investigation and Back Analysis for High Swelling Pressure in Constant Volume Test, (2011).
15. O. Kristensson, M. Åkesson, Mechanical modeling of MX-80–Quick tools for BBM parameter analysis, Physics and Chemistry of the Earth, Parts A/B/C. 33 (2008) S508-S515.
16. M. Villar, MX-80 bentonite, thermo-hydro-mechanical characterisation performed at CIEMAT in the context of the prototype project, Informes Técnicos Ciemat. 1053 (2005) 39.
17. D. Marcial, P. Delage, Y.J. Cui, Hydromechanical couplings in confined MX80 bentonite during hydration, (2008).
18. A. Kuusela-Lahtinen, G. Sinnathamby, J. Mendez, W. Sołowski, D. Gallipoli, X. Pintado, L. Korkiala-Tanttu, Estimation of water retention behaviour of bentonite partially saturated with saline solution, (2016).

 DOI: 10.1051/e3sconf/20160918006

Study on GMZ bentonite-sand mixture by undrained triaxial tests

Wen-jing Sun[1,2,a], Chao Chen[1], Shi-qing Liu[1], De-an Sun[1], Xue-hai Liang[1], Yun-Zhi Tan[3] and Behzad Fatahi[4]

[1]*Civil Engineering Department, Shanghai University, Shanghai, 200444, China*
[2]*State Key Laboratory of Geomechanics and Geotechnical Engineering, Institute of Rock and Soil Mechanics, Chinese Academy of Sciences, Wuhan 430071, China*
[3]*College of civil engineering&architecture, China Three Gorges University, Yichang, 443002,China*
[4]*Centre for Built Infrastructure Research, School of Civil and Environmental Engineering, Faculty of Engineering and IT, University of Technology, Sydney, 15 Broadway, Ultimo, NSW2007, Australia*

Abstract. It is particularly necessary to study the deformation, strength and the changes of pore water pressure of bentonite-based buffer/backfill materials under the undrained condition. A series of isotropic compression tests and triaxial shear tests under undrained conditions were conducted on the compacted saturated/unsaturated GMZ bentonite-sand mixtures with dry mass ratio of bentonite/sand of 30:70. During the tests, the images of the sample were collected by photographic equipment and subsequently were cropped, binarized and centroids marked by image processing technique. Based on identification of the variation of the position of marked centroids, the deformation of the sample can be determined automatically in real-time. Finally, the hydro-mechanical behaviour of saturated and unsaturated bentonite-sand mixtures under the undrained condition can be obtained. From results of triaxial shear tests on unsaturated samples under constant water content, inflated volumetric deformation transforms to contractive volumetric deformation due to the increase of the confining pressure and lateral expansion deformation are observed due to the increase in the shearing stress. Moreover, the net mean stress affects the initial stiffness, undrained shear strength and deformation of the sample during the undrained shear tests.

1 Introduction

Currently, it is widely accepted that the deep geological disposal is the feasible solution for the disposal of high-level radioactive wastes [1]. The project adopts the multiple engineering barrier system, and respectively from the inside to the outside are waste storage containers, engineered barrier based on the bentonite and the surrounding rock geological barrier. Addition of a certain proportion of the quartz sand to bentonite can efficiently enhance the overall thermal conductivity, the mechanical strength and long-term stability ([2], [3], [4]).

Because of bentonite's low hydraulic conductivity, the unsaturated bentonite-sand mixture can be considered in an undrained condition during construction and in-service stages of deep geological repository. During undrained shearing process, the negative pore water pressure is produced in soil. With the passage of time, the pressure will dissipate gradually, and the strength of bentonite-sand mixture will correspondingly decreases, which brings the potential hazards to deep geological disposal project. Therefore, it is particularly necessary to study the hydro-mechanical behaviour on bentonite-sand mixtures under the undrained condition.

In recent years, several researchers have conducted some research about hydro-mechanical behaviour of unsaturated soils under constant water content using triaxial test. Rahardjo et al. studied the strength characteristics of compacted residual soils [5]. Thu et al. investigated the variation of shear strength and pore pressure of compacted kaolin soil [6]. Chen et al. studied the suction and mechanical properties of unsaturated undisturbed loess at constant moisture content triaxial condition [7]. However, the triaxial test results on unsaturated bentonite-based materials under constant water content are relatively small which might be due to the availability of testing apparatus and long duration of triaxial tests on unsaturated bentonite-based materials.

This paper aims to study the hydro-mechanical properties of bentonite-sand mixture used in the deep geological disposal project at the undrained condition through triaxial isotropic compression and triaxial shear tests. During the tests, the images of the sample were collected by photographic equipment and subsequently the deformation of the sample could be determined automatically in real-time by image processing technique. Through the undrained triaxial tests, the variation of deformation, strength and pore water pressure of saturated and unsaturated bentonite-sand mixtures could be obtained under undrained condition. The tests provide

[a] Corresponding author: wjsun@shu.edu.cn

parameters and data to the model which is capable of predicting the coupled hydro-mechanical behaviour of unsaturated expansive soils under both drained and undrained conditions.

2 Testing program

2.1 Testing materials

According to the performance analysis, GMZ sodium bentonite produced in the Inner Mongolia autonomous region is considered for using as a matrix material of engineered barrier in deep geological disposal system in China because of its high swelling property, low hydraulic conductivity, good self-sealing capacity, and excellent radionuclide sorption property [8].

In the tests conducted in this study, GMZ07 Na-bentonite was used, and Fujian standard sand was also included. The Fujian standard sand was firstly seived to reserve the sand particles with the size between 0.074 mm and 0.5 mm. The main physical parameters of GMZ07 bentonite and seived Fujian standard sand are listed in Table 1.The samples, 50 mm in diameter and 100 mm in height, were prepared by compaction in a mold at certain initial water content and dry density, where the ratio of bentonite/sand in terms of mass was 30:70.The deionized water was used during the tests to minimize the influence of ion exchange on the hydro-mechanical behaviour because of bentonite's very high cation exchange capacity.

Table1. Properties of GMZ 07-Na bentonite and Fujian standard sand

GMZ07-Na	Index	Fujian Sand	Index
G_s	2.76	G_s	2.65
w_L(%)	125	D_{50}(mm)	0.34
w_P(%)	30	C_u	1.97
I_p	95		

G_s: Specific gravitity; w_L: Liquid limit; w_P: Plastic limit; I_P: Plastic index; D_{50}: Average diameter; C_u: Uniformity coefficient.

2.2 Test apparatus & deformation measurement method

The unsaturated/saturated triaxial test apparatus adopting digital image measurement method is shown in Figure 1. The apparatus was modified on the basis of a triaxial test apparatus for saturated soils, adding an air pressure controller, and replacing the original pedestal of the cell with a new one with a high air entry value ceramic disk. In order to measure the volume change of sample during consolidation and shearing precisely, a digital image measurement system, designed and developed at the State Key Laboratory of Structural Analysis for Industrial Equipment at Dalian University of Technology, based on the sub-pixel accurate corner locator was used to measure the strain of sample ([9],[10]). The system consisted of a complementary metal-oxide-semiconductor (CMOS) digital photographic camera (1280×1024 pixels) containing the USB2.0 data interface, 16 mm prime lens, and the image processing software. The digital camera model was DH-HV03 manufactured by China Daheng (Group) Co., Ltd. with advantages of high accuracy, high resolution, and low noise characteristics. In order to facilitate the acquisition and analysis of images, the inner-wall of triaxial cell was painted black with LED lights placed at the top and bottom. A black rubber membrane was used to insulate soil from water and the white square marks were printed on its surface.

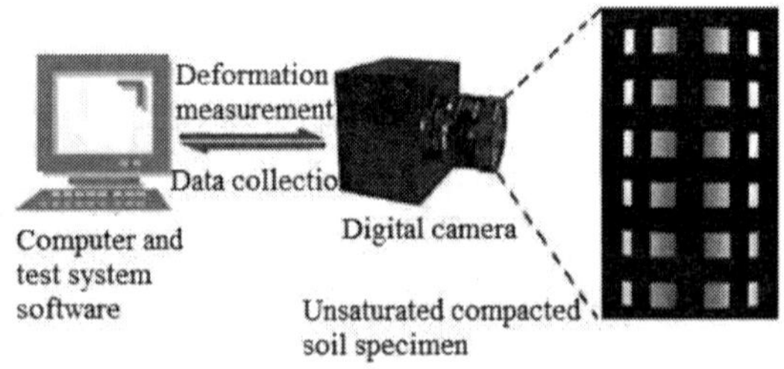

(a) Digital image measurement system [9]

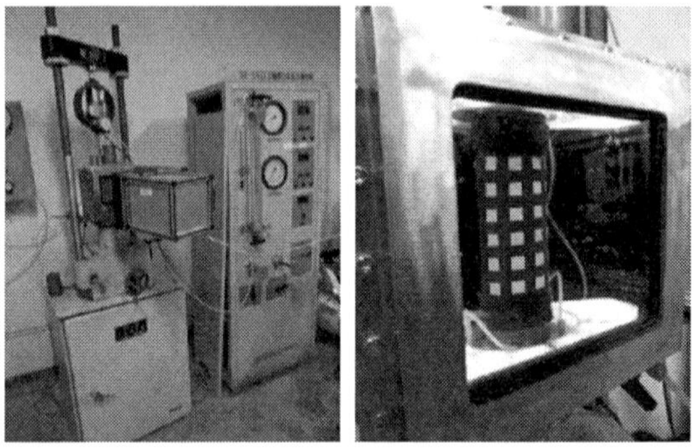

(b) Modified triaxial apparatus and cell

Fig.1 Triaxial test apparatus of unsaturated/saturated soil

Advantages of this digital image measurement technology include noncontact measurements, minimal disturbance to samples, and high accuracy [9]. When installing the sample, the sample wrapped with membrane must be kept upright. Moreover, those four columns and middle six lines as a total of 24 white square marks printed on the surface of membrane should be located in the central picture area, as shown in Figure 1. During the process of consolidation and shearing, the images of the unsaturated/saturated samples were collected by photographic equipment. Subsequently, the images were cropped, binarized and centroids marked by image processing technique, as shown in Figure 2. The marked centroids as well as coordinates were expressed as $c_{ij}(x_{ij}, y_{ij})$, where i =1, 2, 3, 4, 5, 6; j =1, 2, 3, 4. The deformation of the sample could be determined based on identification of the variation of the position of marked centroids.

DOI: 10.1051/e3sconf/20160918006

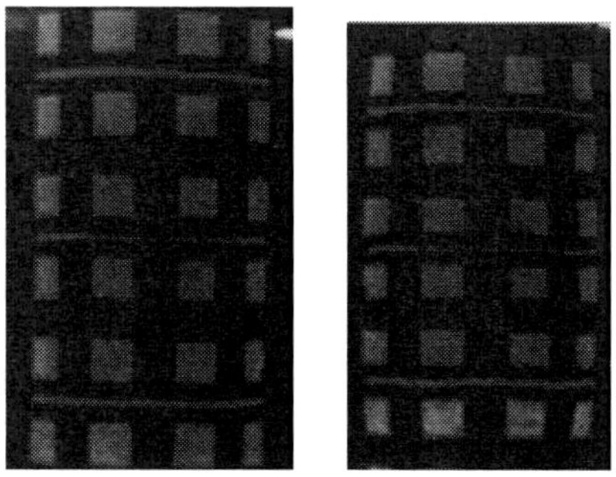

Initial Time t
(a) Images after cropped

Initial Time t
(b) Images after binarized

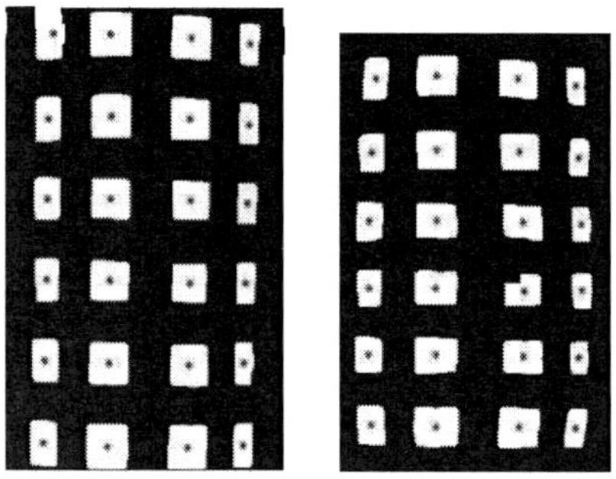

Initial Time t
(c) Images after centroids marked

Fig.2 Image processing

Parts of the collected images were selected and the pixel coordinates of marked points on the images were expressed to determine the centroids' distribution. Then the axial and radial strains of the sample could be calculated according to the pixel coordinates of centroids at each moment. The collected pictures reflected the projection of three-dimensional entity. After the transformation of pixel coordinates, the radial strain of samples could be calculated using the similar triangle principle, as shown in Figure 3. In Figure 3, r is the initial radius of the sample, R is the radius of the sample at time t, a is the initial radial projection length, b is the radial projection length at time t, x_{i1} and x_{i4} are the horizontal data of centroids at the first and the fourth columns. The transformation process of radial strain from initial to time t can be expressed by Eq. (1).

$$\varepsilon_r = (R-r)/r = (b-a)/a \tag{1}$$

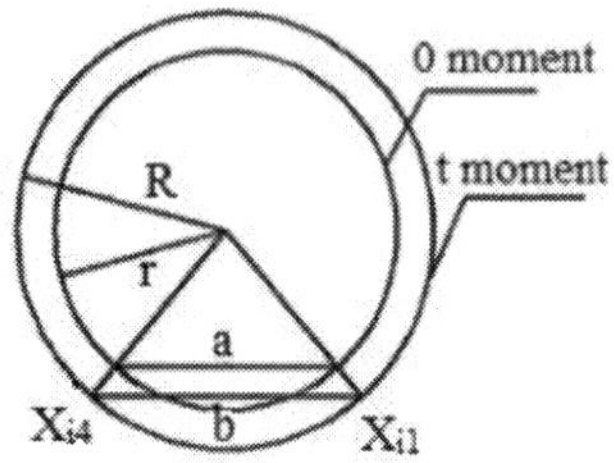

Fig.3 Sketch of change in samples cross section

Figure 4 illustrates the distribution of the centroid location during the triaxial shearing test. The strain calculation process is as follows [9]:

The axial average strain ε_a^t at time t is:

$$\varepsilon_a^t = \frac{1}{m}\sum_{j=1}^{m}\frac{(y_{6j}^0 - y_{1j}^0)-(y_{6j}^t - y_{1j}^t)}{(y_{6j}^0 - y_{1j}^0)} \tag{2}$$

where m is the numbers of columns of marked points in the axial direction, m=4, y_{6j} and y_{1j} are the vertical coordinate values of centroids at the sixth line and the first line.

The radial average strain ε_r^t at time t is:

$$\varepsilon_r^t = \frac{1}{n}\sum_{i=1}^{n}\frac{(x_{i4}^0 - x_{i1}^0)-(x_{i4}^t - x_{i1}^t)}{(x_{i4}^0 - x_{i1}^0)} \tag{3}$$

where n is the numbers of rows of marked points in the radial direction, n=6, x_{i4} and x_{i1} are the horizontal data of centroids at fourth column and the first column.

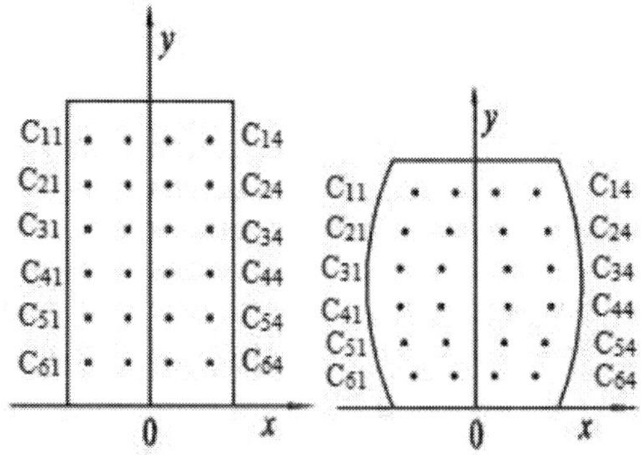

(a) Initial (b) Time t

Fig.4 Sketch of deformation and distribution of centroid location during triaxial shearing test.

 DOI: 10.1051/e3sconf/20160918006

To verify the accuracy of the obtained strains using image processing technology, an auxiliary dial indicator was also used in the process of shearing. Figure 5 shows the comparisons of the axial strain measured by the image processing and by the dial indicator. It can be seen that the results are in good agreement, which testified the feasibility of applying digital image measurement technology for the deformation measurement in the triaxial tests.

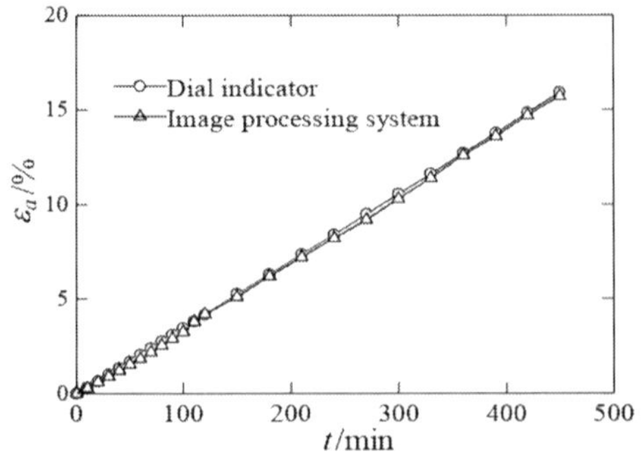

Fig.5 Comparisons of axial strain results measured by dial indicator and image processing

3 Testing result and analysis

The six samples, No.1-No.6, were firstly compressed isotropically under undrained condition. After compression test, undrained triaxial shear tests were carried out under constant net confining pressure with the axial shear rate of 0.033mm/min. The state parameters on samples No.1-No.6 in the tests are listed in Table 2.

No.1-No.4 samples were in the unsaturated state with initial water content of 17.6% and the initial suction of 150kPa measured by the filter paper method [11]. The undrained shear tests under the net confining pressure of50, 100, 200 and 400kPa were conducted respectively. The pore air pressures applied on No.1-No.3 samples were kept constant at 300 kPa, however, the air pressure was controlled at 150kPa for No.4 sample because the maximum pressure burdened by the cell was less than 600kPa. From Table 2, it can be seen that the void ratio decreases after isotropic compression, and the suction decreases due to the increasing of degree of saturation and the decreasing of void ratio. Moreover, the suction reduction increases with the net confining pressure.

Figure 6 shows the stress-strain behaviour of four unsaturated samples (No.1-No.4) in the undrained triaxial shear tests under the net confining pressure of 50kPa, 100 kPa, 200 kPa and 400kPa respectively. It can be observed that the q-ε_a curves of samples No.2-No.4 under the net confining pressure of 100 kPa, 200 kPa and 400kPa show the hardening behaviour, while the deviatoric stress stabilizes after 10% axial strain for No.1 sample under the net confining stress of 50kPa. The strength, hardening degree and the initial stiffness of samples increase with the net confining pressure. This is due to the fact that the structure of unsaturated soil sample gradually degrades with the increase of net confining pressure and the decrease of the suction. In the process of shearing, all samples produce lateral expansion deformation that increases with the net confining pressure. Moreover, from the ε_a-ε_v curve, it is evident that shear dilatancy occurs under the net confining pressure of 50 kPa, and the volumetric change converts from shear dilatancy to shear shrinkage when the net confining pressure increases to 100 kPa, 200 kPa and 400 kPa, simultaneously the volume contraction increases with the net confining pressure.

Table2 .State properties of samples in undrained triaxial tests

No.	Initial state			After compression			After shearing			u_a	σ_{3n}
	w_0 /%	s_0 /kPa	e_0	s_c /kPa	e_c	S_{rc}/ %	s_s /kPa	e_s	S_{rs}/ %	/kPa	/kPa
1	17.7	150	0.80	132	0.80	60.30	84	0.83	56.85	300	50
2	17.5	150	0.80	127	0.79	59.97	68	0.76	61.46	300	100
3	17.7	150	0.79	122	0.77	62.67	56	0.72	66.28	300	200
4	17.7	150	0.79	90	0.75	63.14	56	0.66	71.64	150	400
5	29.6	0	0.85	0	0.84	94.51	0	0.82	96.81	0	100
6	29.7	0	0.85	0	0.83	95.83	0	0.82	97.14	0	200

Note: w_0, e_0, s_0, S_{r0}—initial water content, void ratio, suction, degree of saturation; e_c, s_c, S_{rc}—void ratio, suction, degree of saturation after compression; e_s, s_s, S_{rs}—void ratio, suction, degree of saturation after shearing; σ_3—total confining pressure; u_a—pore air pressure; σ_{3n}—net confining pressure.

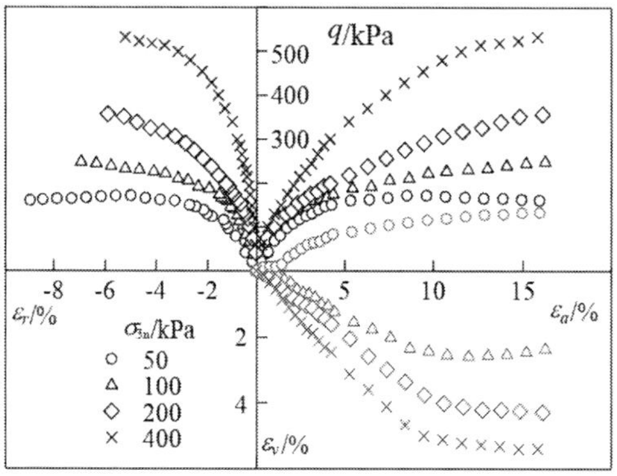

Fig.6 Stress-strain relations of unsaturated samples in constant water content triaxial shear tests

The void ratio and suction decrease after the isotropic compression for four unsaturated samples No.1-No.4 with almost the same initial state, as shown in Table 2. During the undrained shear test, the hydro-mechanical properties of the unsaturated samples also change, as shown in Figure 7. Figure 7(a) shows the variation of the suction and degree of saturation while Figure 7(b) shows the relationship between deformation and average net stress. It can be observed that the void ratio and suction decrease and the degree of saturation increases with the shear stress under the net confining pressure of 100 kPa, 200kPa and 400 kPa. The dilatancy deformation occurs and the degree of saturation decreases for sample No.1 under the net confining pressure of 50 kPa. Moreover the suction decreased and stabilized after 10% axial strain.

In order to determine the shear strength of saturated soil samples and compare with the results of undrained triaxial test of unsaturated soil samples, the undrained triaxial tests at the confining pressure of 100 kPa and 200kPa were carried out on the two saturated samples No. 5 and No.6 respectively. The state parameters of samples No.5-No.6 in the tests are summarized in Table 2.

Figure 8 shows the Mohr circles and shear strength envelopes obtained from undrained triaxial shear tests. Considering the variation of pore water pressure of samples during the undrained shear tests, the average skeleton stress is adapted in horizontal axis. The average skeleton stress is defined as:

$$\sigma' = \sigma - u_a + S_r s \quad (4)$$

where σ is the total stress, u_a is the pore air pressure, $\sigma - u_a$ is the net stress, S_r is the degree of saturation, and s is the suction. The unconsolidated undrained triaxial tests on saturated samples were conducted, so the shear failure envelope was a horizontal line, and the undrained shear strength τ_f of saturated sample was 34.5kPa. The internal friction angle φ of unsaturated GMZ Na bentonite - sand mixture with a sand content of 70% was 21° according to the shear failure envelope of the unsaturated samples.

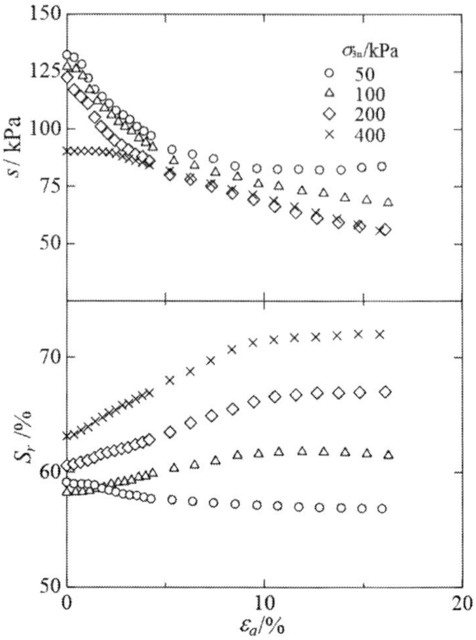

(a) s–ε_a & S_r–ε_a

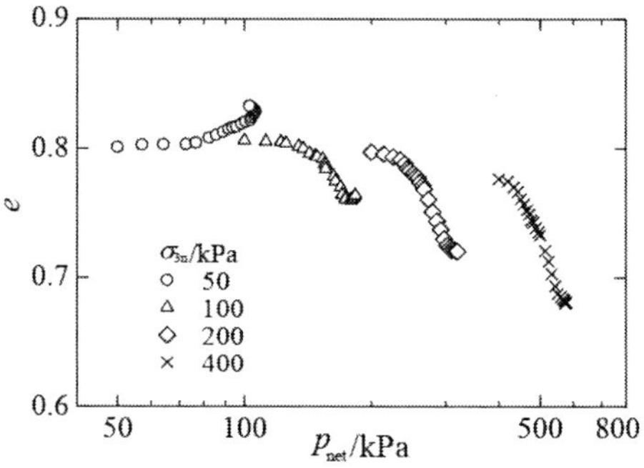

(b) e - p_{net}

Fig.7 Hydro-mechanical behaviour in undrained triaxial shear tests

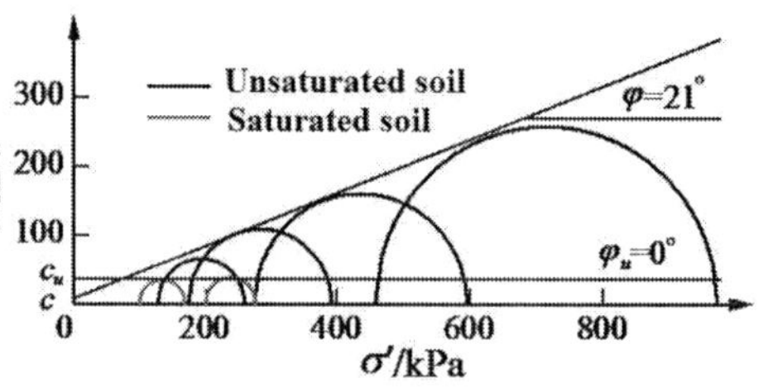

Fig.8 Mohr's stress circles and strength envelopes for unsaturated and saturated samples

Figure 9 depicts the stress-strain curve of unsaturated and saturated samples under the same net confining pressure (σ_{3n}=100kPa). It can be concluded that, the shear strength of unsaturated sample is higher than that of saturated sample under the same net confining pressure. This is because the air in the unsaturated sample is compressed or partially dissolved in water during shearing, which subsequently increases the shear strength of unsaturated soil sample.

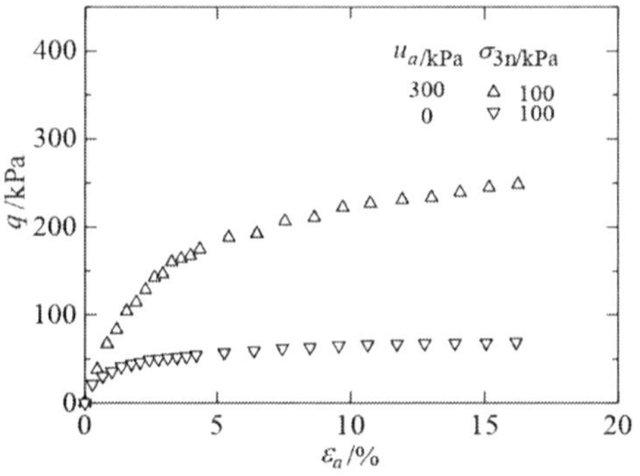

Fig.9 Stress-strain curves of saturated and unsaturated sample at the same net confining pressure

4 Conclusions

A series of undrained isotropic compression tests and undrained triaxial shear tests were conducted on GMZ bentonite-sand mixture with 70% sand content, and the strength, deformation and hydro-mechanical characteristics were measured and the following conclusions could be drawn:

1) After the undrained compression tests on unsaturated soil samples, the void ratio and suction decrease, while the rate of change increases with the net confining pressure.
2) In the undrained triaxial shear test, the undrained shear strength increases with the net confining pressure, and the shear strength of unsaturated sample is higher than that of saturated sample under the same net confining pressure.
3) Lateral expansion can be observed during the undrained shear tests, and the inflated volumetric deformation converts to the contractive one with the increase of the net confining pressure.

Acknowledgements

The authors appreciate the financial support of the National Sciences Foundation of China (Grant No. 41102163, 41572284) and the Open Research Fund of State Key Laboratory of Geomechanics and Geotechnical Engineering, Institute of Rock and Soil Mechanics, Chinese Academy of Sciences (Grant No. Z013008)

References

1. KASAM, Nuclear waste, state-of-the-art reports. (2001)
2. Chapuis.R.P, Sand-bentonite liners. Can Geotech J. **27** (1): 47-57(1990)
3. Kenney.T, Vanveen.W, Swallow.M, Sungaila.M. Can Geotech J. **29**(3): 364-374.(1992)
4. Chen Y.G, Ye W.M,Wang Q, Chen B. Eng Geol. **18** (3): 357-362.(2010)
5. Rahardjo H, Heng O B, Leong E C. Can Geotech J. **41**(3): 421–436.(2004)
6. Thu T M, Rahardjo H, Leong E C. J Geotech Geoenviron.**132**(3): 411–419.(2006)
7. Chen C.L, Zhang D.F, Dong Y.Z, Chen H, Yu D. B, Xue J.X. J Geotech Geoenviron.**36**(7): 1195-1202. (in Chinese, 2014)
8. Wen Z.J. Acta Geologica Sinica (English Edition). **82**(5): 1050-1055.(2008)
9. Shao L.T, Wang Z.P, Han G.C, et al. Chinese Journal of Geotechnical Engineering. **23**(3): 337-341. (in Chinese, 2001)
10. Sang Y, Shao L.T, Guo X.X, Huang C. Instrum Sci Technol. 43:429－445, (2015)
11. Sun D.A, Meng D.L, Sun W.J, et al. Rock Soil Mech. **32**(4): 973-978 (in Chinese, 2010)

Investigation of the influence of animal burrowing on the failure of the levee of San Matteo along the Secchia river

Maria Luisa Taccari[1,2,a] and Raymond van der Meij[2]

[1]Delft University of Technology, Civil Engineering, 2628 CD Delft, Netherlands
[2]Deltares, 2629 HD Delft, Netherlands

Abstract. Animal burrowing can greatly influence the water pressures in a flood protection embankment and thereby be a cause of breaching of flood defences. However, little guidance and literature is available on this subject. This paper investigates the contribution of badgers, porcupines and foxes to the failure of the levee of San Matteo (Modena, Italy) on 19th January 2014. The proposed method evaluates their influence on the water pressures in the embankment during rainfall and a high water tide. The influence of the burrowing is assessed through a transient FEM flow analysis. Starting from the documented entrances situated in the vadose zone, different scenarios for the internal distribution of tunnels and chambers are proposed. The most likely representative network for the loss of stability of the dike is assessed.

1 Introduction

On the 19th January 2014, around 6am, a high water level in the river Secchia created a breach along a dike in San Matteo (Modena, Italy). The water volume flowing out of the breach flooded the surrounding countryside and the villages of Bastiglia and Bomporto. Around a thousand of people had to be evacuated and one man died during the relief operations.

Figure 1. Breach on the right side of Secchia river. (R. Ferrari, Protezione Civile di Modena, 19-01-2014)

During the months after the failure, a scientific committee composed of six members of five Italian universities investigated the causes leading to the event. In July 2014, the Investigative Evaluation Report *"Relazione tecnico-scientifica sulle cause del collasso dell'argine del fiume Secchia avvenuto il giorno 19 gennaio 2014 presso la frazione San Matteo"* (Technical-scientific rapport about the causes leading to the collapse of the levee of Secchia river, which took place on 19th January 2014) was published [1]. The report presents hypotheses on failure mechanisms, indicating that internal erosion and macro-instability are possible causes of the breach. The presence of animal burrows could have contributed to the formation of these phenomena.

The current paper investigates the influence of animal burrowing on the pore water pressures to predict the failure mechanism for the Secchia river. When the outside water level increases, the phreatic line changes in shape and with a rate depending on the properties of the soil body. The heterogeneity introduced by the animal burrows influences this trend and the pore pressure distribution. The groundwater flow inside the dike body gives an indication of possible failure mechanisms, which can be derived for several scenarios of animal burrows. Finally, the paper concludes over the likely holes distribution and failure mechanism leading to the breach.

2 Initial and boundary conditions

2.1 Soil properties and stratigraphy

In order to define the stratigraphy and soil properties, field and laboratory tests were performed in three sections near to the location of the breach. Core samples have also been taken in these three locations and laboratory tests have been performed [1].

The dike body, defined as Unit A and whose top is placed at 37.66 MAMSL (Figure 2), consists of silt with

[a] Maria Luisa Taccari: MariaLuisa.Taccari@deltares.nl

sand and sandy silt. Underneath the levee body, Unit B consists of sandy silt, locally with clay, and it is formed with sub-horizontal layers; it extends from 31.55 MAMSL till 25.16 MAMSL. Finally Unit C is mostly composed by clay and, locally, sand and silt. The phreatic level is placed at 29 MAMSL and the river is situated on the left side.

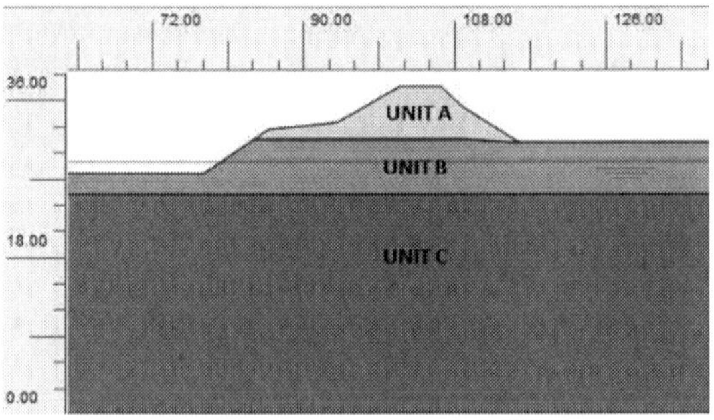

Figure 2. Profile of the dike.

The soil is classified according to the USDA classification system. The results of the classification proofs are made public available [1] and used in the current study. The parameters for the strength and deformability of the soil have been obtained through interpretation of the laboratory tests and the field tests (Table 1) [1].

Table 1. Soil properties of the units, used in the analysis.

Unit	A- Dike Body	B	C
Soil type	Silty Sand	Silty Sand	Clay
γ_{sat} (kN/m3)	21.7	18.8	18.9
γ_{unat}(kN/m3)	19.5	16.0	16.0
E' (kPa)	78.5E3	48.7E3	54.2E3
m (-)	0.2-0.3	0.2-0.3	0.4-0.5
c (kPa)	7	7	-
φ' (°)	32	28.8	-
su (kN/m^2)	-	-	55
k_x (m/day)	1	0.162	10E-3
K_y (m/day)	1	0.162	10E-4
e (-)	0.8	0.8	1

The saturated hydraulic conductivity of the soil of the dike is made available in the Investigative Evaluation Report [1]. The permeability of the other two layers is estimated from CPT through correlations available in literature. Since the soil of the dike body is artificially placed, the same value is assumed for vertical and horizontal permeability. Instead, the deeper unit is formed by stratified layers of silt and sand, thus the horizontal permeability largely differs from the vertical.

The flow parameters for the soil of the dike body are described by the Soil Water Retention Curve and the Hydraulic Conductivity Function (Figure 3).

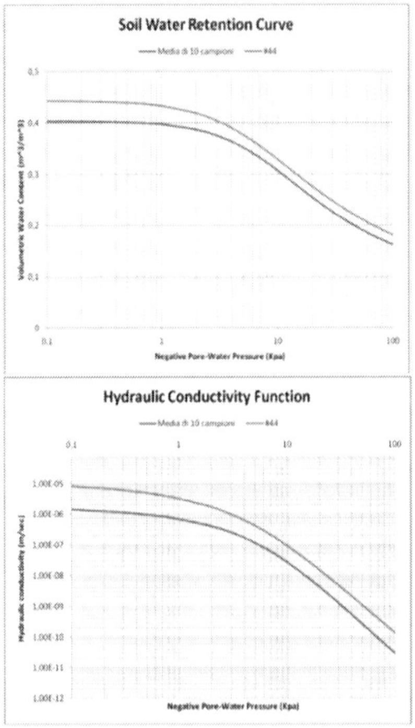

Figure 3 Soil Water Retention Curve and Hydraulic Conductivity Function, obtained by averaging 10 samples (in red) and for the sample considered the most representative (in blue). [1]

The curves are constructed with the model of Van Genuchten-Mualem and the used parameters are summarized in Table 2.

Table 2. Average values of input parameters for the Van Genuchten-Mualem equations.

Input parameter	**Average value**
a (cm^{-1})	0.0154
n (-)	1.4938
m (-)	0.3306
Θ_r (-)	0.0785
Θ_n (-)	0.4037
K_s (cm h^{-1})	0.6771

2.2 Hydraulic boundary conditions

During the night of the breach, the values of river water levels and rainfall leading to the failure were not exceptional, compared to other events during the previous year. Few months before the breach (05/04/14, [2]) the water level had a peak just 30cm below the maximum water level at 6:00 at the moment of the breach, while, during the whole year 2014, higher values for rainfall were also recorded.

2.2.1 Rainfall

The rain was light during all the day before the failure until 8pm on 18/01/2014, when its intensity increased presenting a peak of heavy rain during the night of the 19th January. Around 2am the intensity decreased and moderate rain was present at 6am when the failure took place [1].

2.2.2 Changing water level

The model in the Investigative Evaluation Report with simulation time between 15-01-2014 and 27-01-2014 presents a minimum water level of 29 MAMSL on 17th January 2014. The water level rose during the afternoon of the 17th January 2014 until reaching its peak after the moment of the breach. At 6.00am on 19/01/2014, when failure took place, the water level in the river was equal to 36 MAMSL, that is 1.50meter below the levee crest (Figure 4). There was no significant wave run-up, therefore failure due to overtopping can be excluded

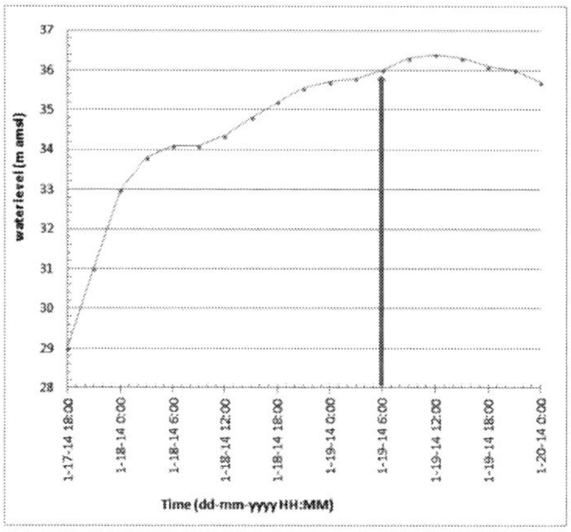

Figure 4. Water level in the river between 17-01-2014 at 18:00 and 20-01-2014 at 0:00 in the section of the breach [1]

The breach occurred when the water in the river was not at its maximum level yet and the peak that came after didn't cause other breaches along the levee. This observation confirms the hypothesis of a localized weak spot along the dike where failure occurred.

3 Animal burrows

3.1 The environment

The levee of San Matteo offers an ideal environment for the diffusion of badgers, porcupines and foxes. Even if little grass is present along the slopes, the vegetation at the inner side of the dike and the isolation of the levee from human presence attract the animals. Few houses are present next to the embankment and the road is around 100 meters far from it.

The height of the levee is around 7-8 meters, ensuring dry places where to build setts, which are usually not reached by the water level in the river. The soil consisting on silty sand is also an ideal soil where to dig on.

3.2 The entrances from the Italian report

The presence of four burrow entrances is indicated in the Investigative Evaluation Report, where their location is calculated from an aerial photo from 2012 [1]. Only one entrance is located along the outer slope (Figure 5) and the entrances are located almost at the top of the levee, since badgers, porcupines and foxes dig in dry places. Only in rare occasions, the water level in the river covers the entrance placed along the outer slope. During the night of the 19th January 2014, the water level was around 30cm below it.

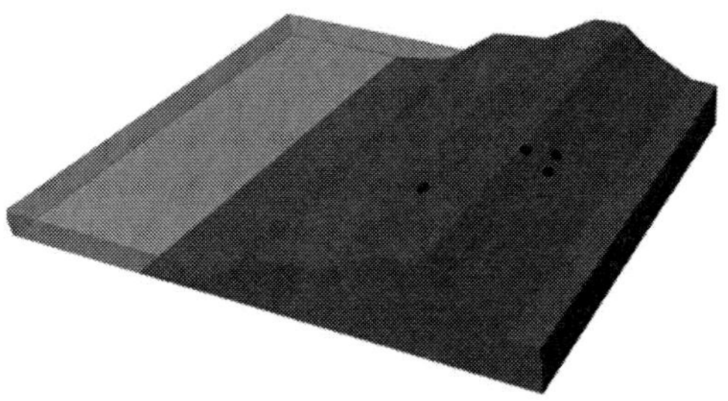

Figure 5. Position of burrow entrances along the dike slopes

3.3 Likely geometry of burrows

Setts, which are the underground system excavated by badgers, usually have a mean number of entrances between 7 and 12. The width of the sett is often very large, from 50 to 150m [3].

The system of the levee of San Matteo under analysis has 4 entrances and the maximum distance between them is around 10 meters. Thus, it is reasonable to assume that the entrances belong to the same main sett: the three entrances along the inner slope are most likely linked to the one in the outer slope.

The burrows of badgers are tunnels with semi-circular (arched) or squashed elliptical form, wider than the high [4]. The tunnels have a typical height of 0.20 m and are 0.30-0.35 m wide [5]. The nesting chambers have squashed spherical shape with arched roof and their horizontal section is roughly circular [6]. The dimensions of the chamber can be wider if more badgers occupy it.

The geometry of the subsurface network under analysis is not a-priori known but good assumptions can be made based on the available observations. The tunnels, modelled as cylinders with diameter equal to 0.2m [5], start from the known entrances and go down for 1 meter with slope of 45°; they continue then horizontally until reaching the centre of the levee. There the tunnels are linked together forming a whole system.

3.4 Possible influence of burrows on water pressure

The phreatic line inside the dike body changes with the hydraulic boundary conditions depending on the properties of the soil body. In presence of animal burrows, the phreatic line can rise and thereby decrease the effective stress and strength and posing a risk to the stability of the dike.

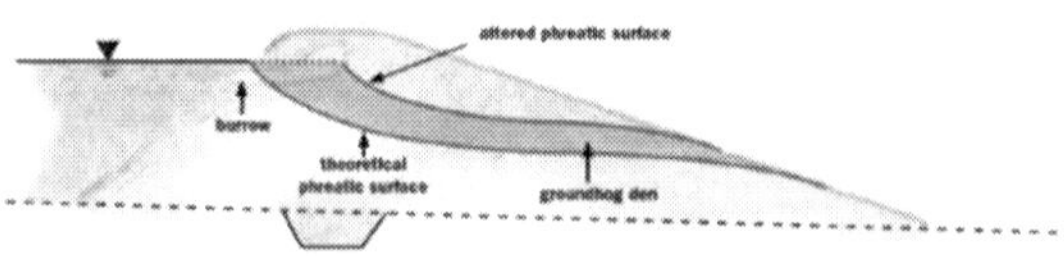

Figure 6. Alteration of phreatic line due to animal burrow [7].

If the phreatic line exits above the inner toe of the dike and the seepage forces are high enough, the probability of seepage and micro-instability increases.

The presence of animal burrows can also facilitate "internal erosion as evolution of defects", such as piping is defined when cracks and microfissures are present in the soil matrix [8]. As water flows into the cavities, the seepage forces act along the surface, eroding them and removing the subsurface soil.

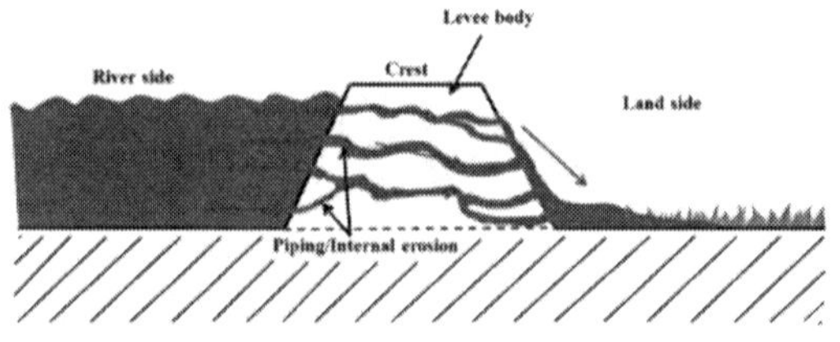

Figure 7. Internal erosion in presence of animal burrows [9].

4 Possible failure mechanisms

4.1 Macro-Stability

4.1.1 Analyses

Groundwater flow analyses are performed with the software PLAXIS 2D, which is a finite element computer program used to perform deformation and stability analyses for several geotechnical applications. The program is chosen to investigate the case study in order to analyse the groundwater flow while implementing transient boundary conditions. Once the initial conditions are implemented, a first steady state calculation is performed with the water level in the river posed equal to its maximum value. Transient groundwater flow analyses are instead carried out for a simulation period of 18 days. Mohr-Coulomb model is used for the stability analysis during the phi/c reduction phase. The input parameters for the unsaturated soil described by the Van Genuchten-Mualem model and the other input values used for the analyses have been introduced in paragraph 2.1. In the Investigative Evaluation Report an initial negative pore water pressure distribution (suction) is assumed inside the dike body, which is based on data available in literature for similar levees during winter conditions: it presents a minimum value of -40 kPa in the centre of the dike body while the water pressure is equal to -10 kPa at the top of the levee [1]. Due to the unrealistic low value of permeability associated with the high suction at the top of the levee that the hydrostatic distribution of pore water pressures would give in the unsaturated zone, this assumption is accepted also for the current study.

First, the analysis is carried out without considering the presence of animal burrows (case a). The burrows are then implemented as soil tunnels with unique values of permeability and porosity in their all domain, according to the dual porosity and dual permeability model [10]. In order to simulate the void of the burrow and the ease of the water to flow into the cavity, high values of porosity and permeability are implemented. In particular, the porosity is equal to n=0.999 and the permeability is assumed 100 times higher than the one of the surrounding soil. The analysis is carried out for a cross-section along the entrance at the outer slope (case b) and for one entrance at the inner slope (case c). Finally, the last scenario considers the case in which a more developed system was present when the breach occurred. As a matter of fact, the entrances described in the Investigative Evaluation Report are derived from data from 2012. The geometric configuration of the system is obtained considering the average growing rate of badger setts [6] and geometrical assumptions. The assumed sett present in 2014 consists on a system with one new entrance, placed along the outer slope at 35.7 MAMSL. It is connected to the burrows at the inner slope, thus forming a whole system (case d).

4.1.2 Results

In all the analysed cases, the phreatic line of the steady state solution crosses the levee body and it exits in a point placed 1.5 meter above the toe of the dike. The transient flow analysis presents lower pore water pressures that the steady state solution, while, at the inner side, the water level exits at the dike toe due to the influence of rain. The burrow situated at the outer slope below the maximum water level (case d), has the main influence to the pore pressures for both steady state and transient flow analyses (Figure 8).

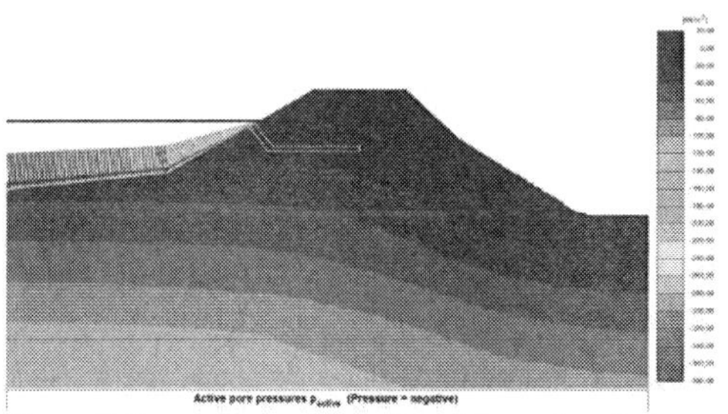

Figure 8. Pore pressure distribution at the end analysis, case d. The phreatic line for the transient flow analysis is in red, while the phreatic line of the steady state solution is the blue line. The contours of the animal burrow are in green.

The safety factors obtained with the φ-c reduction method are summarized in Table 3. If the animal burrows are not considered in the analysis (case a), the safety factor is equal to SF=1.333 for the steady state calculation and SF= 1.812 for transient flow analysis. Steady state solution gives lower results than transient flow because of the higher pore water pressures and so the lower effective stresses. Then, the animal burrows are implemented in the analysis. The results of the 2d calculations for cross sections along the burrow entrances show that the burrows recorded in the Investigative Evaluation Report and located at the outer and inner slope have relatively small influence to the results (cases b and c). Instead, the burrow assumed in the scenario for 2014 and situated along the outer slope below maximum water level (case d), decreases the safety factor to 1.173 for steady state solution and to 1.412 for the transient flow: this burrow gives the biggest impact to the macro-stability of the inner slope.

Table 3. Safety Factors for the analysis performed in Plaxis 2d

Case	SF – 2D steady state	SF – 2D transient flow
a)No burrow	1.333	1.812
b)Burrow at the outer slope	1.258	1.733
c)Burrow at the inner slope	1.390	1.786
d)Burrow in 2014	1.173	1.412

The dike proves to be safe to macro-instability: the safety factor is largely above one for all the performed analyses (2d analysis, steady state and transient flow, with and without animal burrowing).

4.2 Micro-Stability

If the phreatic line exits at a point above the toe of the inner slope and seepage flow develops, micro-instability can occur [11]. Failure starts at the exit of the phreatic line at the inner slope when the weight above and the forces created by the flow exceed the resisting shear resistance. Particles at the exit point are detached and carried away by the seepage flow and the process gradually retrocedes towards the upstream side of the levee.

In all the analysed cases, the phreatic level for the steady state solution exits in a point placed 1.5 meter above the toe of the dike. However, it is the transient flow solution which simulates the conditions when failure took place, rather than the steady state analysis. The solutions given by the transient calculations in all the analysed cases present the water level crossing the dike body, reaching the layer underneath the embankment and finally exiting at the inner toe of the dike (Figure 8). The intense rain locally hits the surface, increasing the pore water pressures and so raising the water table at the inner side. However, no real seepage point, which exercises seepage forces against the inner slope, is present and can cause micro-instability.

In conclusion, the water table at the end of the transient flow analysis never exits at the inner toe of the dike; it remains below it. No seepage point and forces against the inner slope are present, excluding so micro-instability as a possible failure mechanism for the levee of San Matteo.

4.3 Internal erosion

The failure of the Truckee Canal Embankment (2008) due to a large network of animal burrowing was the incentive for performing Hole Erosion Test (HET) by the U.S. Department of the Interior (Denver, Colorado). Bezzazi et al. [12] have presented a simplified analytical model to the Hole Erosion Test which estimates the evolution of the inner tube radius as a function of time. This paper applies the approach derived by the Hole Erosion Test for the levee of San Matteo in the conditions that led to the breach. When the water level in the river covers the lower burrow entrance at the outer slope and the burrow is connected to the inner slope, a pressure difference develops between the outer and inner side. If then this pressure difference overcomes the critical shear stress, internal erosion starts and causes the hole to increase.

The scenario used for the analysis of internal erosion corresponds to the network which has been assumed to be present in 2014. A burrow entrance is placed at 35.70 MAMSL, under the maximum water level that led to the breach. This burrow is connected to the lower burrow placed at the inner slope through a central tunnel which longitudinally crosses the levee. The total length L of the cavities connecting them is approximately 20 meters. The initial radius of the burrow R(0) is equal to 0.2 meter, the soil density ρ_d is obtained by the available data and the critical shear stress τ_s, for which erosion starts, is set equal to 50 Pa , that is a value found in the literature for similar soil [8]. The soil of the levee is diggable and erodible as proved by the animal burrowing activity, but it also maintains some erosion resistance while standing after the same burrowing activity. As suggested in the literature for similar soils [13], the surface erosion coefficient C_{er} , which varies among several orders of magnitude and indicates the ease of the material to erode, assumed equal to 1.58E-04 s/m.

Table 4. Input parameters for the analysis.

R(0) (m)	L (m)	ρ_d (kg/m^3)	τ_s (Pa)	C_{er} (s m^{-1})
0.2	20	1989	50	1.58E-04

The pressure loss is equal to 0 until the water level is below the entrance of the tunnel. At 0:00 on 19-01-2014 the water in the river is at the same level of the entrance of the burrow at the outer slope; the water level then increases and water enters into the cavity. At 0:00 on 19-01-2014, the water overcomes the entrance of the burrow and erosion starts (Figure 9). The phenomenon develops in time: at 3:00 the radius is equal to 0.60m and it reaches dimension of 1 meter at 04:30.

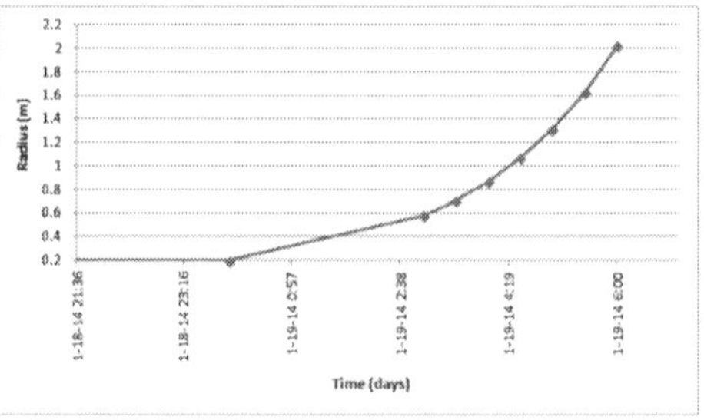

Figure 9. Development of Radius with Time due to Internal Erosion, using the approach derived by the Hole Erosion Test.

5 Conclusion

The current paper has analysed the failure of the levee of San Matteo (Italy, 19th January 2014) previously investigated through the Investigative Evaluation Report (*"Relazione tecnico-scientifica sulle cause del collasso dell'argine del fiume Secchia avvenuto il giorno 19 gennaio 2014 presso la frazione San Matteo"*) which revealed the possible influence of animal burrowing in the creation of the breach. The study gains a better insight on the influence of animal burrows on groundwater flow and stability analysis.

In the Investigative Evaluation Report, entrances of animal burrows along the section that failed on 19th January 2014 are illustrated. They belong to European Badgers, Foxes and Crested Porcupines. The entrances, located along the inner and outer slope, have maximum distance of 10 meters: it is likely that they form a unique connected system with the function of a main sett of reduced dimension. Moreover, the Investigative Evaluation Report presents the burrow entrances deduced by aerial photos of the levee in 2010 and 2012; when the breach took place, in 2014, the system was likely more developed with a new entrance located below the maximum water level at the outer slope and connected to the main sett.

The analyses performed in the current study consider the possibility of Instability of the Inner Slope, Micro-Instability and Internal Erosion to occur, with and without animal burrowing. The input parameters are derived by field and laboratory tests and made available in the Investigative Evaluation Report. Insufficient tests results are published to statistically determine a reliable lower bound of the strength parameters.

First, in the performed analyses the dike shows to be safe to macro-instability: the safety factor is largely above one, with and without animal burrowing. Second, in all the performed transient groundwater flow analyses, with water level changing in the river and rainfall which happened before the event, the water level presents no seepage point exiting above the inner toe. Also when implementing the animal burrows, the water flow doesn't exercise forces against the inner slope which could start eroding it. For this reason, we concluded that micro-instability unlikely occurred. As a matter of fact, the analyses performed with the available information give such high safety factors, that the uncertainty in the strength is not likely to cause failure due to micro- or macro-instability.

Finally, we concluded that internal erosion was the likely mechanism leading to failure. No tests are performed to determine the parameters for the Hole Erosion Test. However, this is the only mechanism that leads to failure with realistic soil parameters. The connected underground system with burrows running from the outer to the inner slope and the entrance of a burrow located below maximum water level gave the conditions for water to fill the burrow and erode the tunnels in the whole connected network. In view of the performed analyses, we found that hole erosion after animal borrowing is a most suitable explanation for the collapse of the San Matteo levee.

References

1. L. D'Alpaos et al., *Relazione tecnico-scientifica sulle cause del collasso dell'argine del fiume Secchia avvenuto il giorno 19 gennaio 2014 presso la frazione San Matteo* (2014)
2. http://www.arpa.emr.it/ (Accessed : 20th October 2015)
3. A. W. Byrne et al., Royal Irish Academy, *The ecology of the European Badger (Meles Meles) In Ireland - a review*, **112** (2012)
4. V. G. Heptner, A. A. Sludskii, *Mammals of the Soviet Union*, **2** (1988)
5. D. Nichol et al., Q. J. Eng. Geol. Hydroge, *Application of ground-penetrating radar to investigate the effect of badger setts on slope stability at St Asaph Bypass, North Wales*, **36**, 143-153 (2003)
6. T. J. Roper, *Badger* (2010)
7. Deltares, *Diagnosis of urban flood embankment performance* (2012)
8. S. Bonelli, *Interpreting Hole Erosion Tests with a constant flow rate* (2006).
9. H. K. Chlaib et al., J. Appl. Geophys., *Using ground penetrating radar in levee assessment to detect small scale animal burrows*, **103**, 121-131 (2014)
10. L. Stadler et al., Transp Porous Med, *Modeling Macroporous Soils with a Two-Phase Dual-Permeability Model*, **95**, 585-601 (2012)
11. S. Vorogushyn et al., Nat. Hazards Earth Syst. Sci., *Development of dike fragility curves for piping and micro-instability breach mechanisms*, **9**, 1383-1401 (2009)
12. M. Bezzazi et al., Am. J. Eng. Appl. Sci., *A Simplified Analytical Modeling of the Hole Erosion Test*, **4**, 765-768 (2010)
13. E. Erdogan et T. Wahl, *Results of Laboratory Physical Properties and Hole Erosion Tests, Truckee Canal Embankment Breach, Newlands Project, Nevada.* (2008)

On the role of partially saturated soil strength in the stability analysis of a river embankment under steady-state and transient seepage conditions

Guido Gottardi[1], Carmine Gerardo Gragnano[1,a]

[1]*Department of Civil, Chemical, Environmental and Materials Engineering, DICAM, University of Bologna, Italy*

Abstract. River flood risk is considered being one of the most costly hazards in Europe and under a further major potential impact of climate change, in combination with land-use changes and water management practices, flood risk is expected to increase for many river basins. In engineering practice, the design of river embankments is usually performed using simplified approaches, considering steady-state flow conditions induced by the retained water and almost neglecting issues related to partially saturated soils, thus leading to potentially heavily over conservative results of stability analyses. To provide a realistic assessment of river bank stability conditions and to get a more accurate prediction of flood risk it is necessary to consider river bank soil behaviour at different saturation degrees in connection with transient seepage flow. A numerical study on stability conditions of a specific river embankment focusing on the partially saturated soil strength contribution is presented herein. Seepage and stability analyses have been carried out using the information collected on river Secchia flooding case study, occurred north of the city of Modena (Italy) in January 2014. Limit equilibrium method has been adopted for assessing the overall stability in steady-state and transient flow conditions. Useful indications for accounting unsaturated soil strength in similar circumstances are finally provided in the paper.

1 Introduction

Riverbank stability analysis represents a key problem in engineering practice and a topical geotechnical research issue. Results of the studies related to this subject can provide useful suggestions both for risk assessment problems and for the definition of design approach criteria. However, these two aspects can substantially differ in relation to their main assumptions on which they are based. Typically, design approach analyses are conducted under steady state flow regime, assuming simplified initial conditions and mostly disregarding issues related to soil partial saturation, thus often leading to deliberately strongly over-conservative results in terms of actual probability of failure. On the other hand, risk susceptibility analysis should be founded on more realistic hypotheses, considering transient seepage through the embankment induced by time dependent boundary conditions, assuming pertinent initial conditions, and taking into account unsaturated and heterogeneous soils, thus influencing seepage characteristics and relevant strength parameters.,. A significant influence in riverbank stability can in fact be certainly ascribed to partially saturated soil strength, which in turn directly depends on soil hydraulic parameters and effective saturation degree, providing an additional, but variable, term able to substantially modify the bank probability of failure with time.

In order to assess the importance of taking into account soil river bank as partially saturated, a probabilistic study has been carried out using evidences, observations and data collected on a specific flooding case study, referring to bank's total sudden collapse occurred on January 19, 2014 on river Secchia, north of the town of Modena (Italy). In that case, decrease in soil suction values due to persistent high water levels along the river together with the widespread presence of animal burrows, clearly detected near the failed section and possibly existing in it before collapse, are supposed to have triggered an overall instability mechanism [1]; consequently, overtopping and erosion rapidly contributed to create a large breach in the river bank (Figure 1).

Aim of this study is to quantify the incidence of unsaturated soil strength on assessing the probability of bank instability under steady-state and transient seepage conditions and to stress its relevant influence on future risk assessment and design approach. Therefore, it is not meant to discuss further herein of the actual mechanisms that led Secchia's river bank to sudden total collapse and to the subsequent severe flooding of the surrounding countryside which caused one casualty, severe property losses and € 221 million euros of total estimated damages.

[a] Corresponding author: carmine.gragnano2@unibo.it

Figure 1. The breach area after 19/01/2014 flooding event.

2 The case study

The flooding area is located in the southern part of the Po valley (Italy), resting on a deep alluvial Pleistocene formation, locally at least 400 m thick. The morphology is resulted from the depositional and erosional activity of the Secchia and Panaro rivers, both Po's tributaries, which drain large areas of the Emilian Apennines. The geotechnical properties of the river bank's soil and subsoil were estimated by carrying out extensive site and laboratory investigations.

Site investigation involved the execution of cone penetration tests including pore-pressure measurements (CPTUs) running up to a 14-26 m depth, distributed on three different cross sections considered representative of the collapsed bank and located near the breach: section 1 was located on the right river side, 900 m upstream the breach area, section 2 was selected on the left river side, in front of the breach area and section 3 was taken again on the right river side, but immediately downstream the breach area. For each cross section, three CPTU tests were executed: on the river side, on the river bank top and on the landward side.

CPTUs performed on top were equipped with a seismic module for measuring the shear wave velocity propagation (SCPTUs). During their execution, pore pressure dissipation tests were also performed.

For the CPTUs result interpretation, semi-empirical correlations proposed by Robertson [2] were used.

In particular, soil classification was derived by an iterative procedure, based on the value of the Soil Behavior Type Index (Icn), estimated through the normalized cone resistance (Qt) and the normalized friction ratio (Fr).

Figure 2 plots the soil stratigraphic profile consistently defined by the results of in situ tests. Three different soil types were identified: firstly Unit R, composing the river bank soil, consisting of a complex alternation of silt with sand and sandy silt, with frequent thin sandy layers; the second, Unit B, composing the river bank foundation, consisting of 6.5 m average thickness sandy silt layers, locally clayey, primarily horizontally stratified; finally, Unit C, composing the subsoil, consisting of a primarily clayey layer with local sandy-silt sublayers, up to the maximum investigated depth.

Remoulded and undisturbed samples were taken from each section to determine relevant geotechnical properties by a comprehensive set of standard physical and mechanical tests. Fine grained soil percentage in Unit R turned out to be between 5 and 10 %, characterized by a PI < 10% and by a natural water content lower than the PL up to a 4-5 m depth from the river bank top. Soils sampled from Unit B are lithologically similar to the ones collected from Unit R; their fine grained fraction is again characterized by low plasticity, but with a natural water content typically greater than LL. The deeper clays, identified as the Unit C, are characterized by high PI values. Besides, local carbonate concretions were found at all depths, even in significant percentage. Unit R and Unit B soil strength parameters, distinctly for critical and peak conditions, were determined from in situ (CPTU and SCPTU) and laboratory tests (TxCID, TxCIU, DS). Unit C geotechnical parameters were estimated only by well-established correlations from CPTU and SCPTU results [3]. All estimated strength values are listed in Table 1.

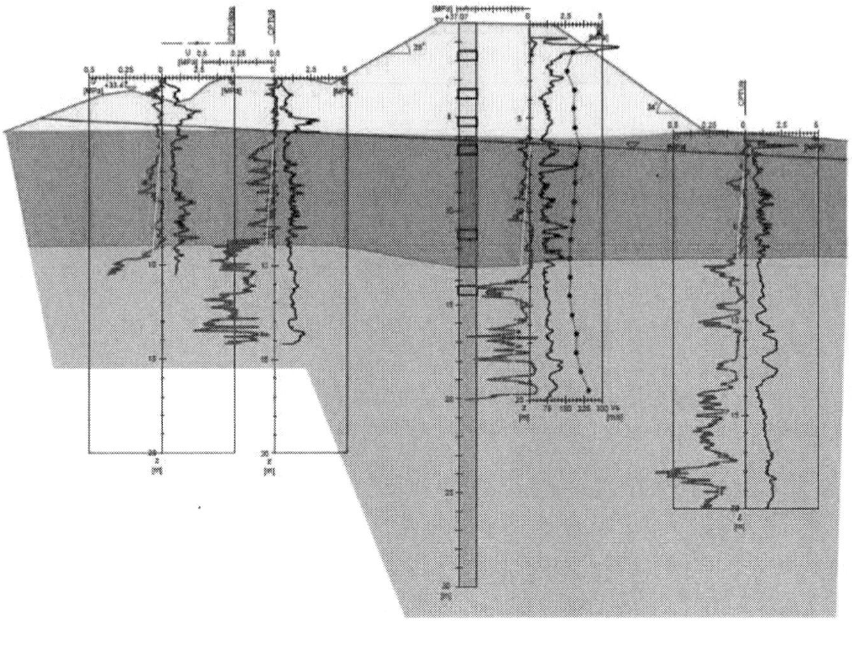

Figure 2. Stratigraphic profile and CPTUs results, Section 3.

Table 1. Estimated strength soil properties.

Soil Type	φ' (°)		c' (kPa)
	Mean value	Standard deviation	Estimated value
Unit R	32.0	1.94	0.0
Unit B	28.8	3.20	0.0
Unit C	24.9	2.40	0.0

Unit R soil retention and hydraulic properties have been carefully studied in laboratory through a series of nine evaporation tests conducted on undisturbed samples taken 600 m downstream the breach area at depths between 0.65 and 1.7 m. The idea of using evaporation data to determine unsaturated soil hydraulic properties was introduced by Wind [4] and then slightly modified by several authors [5, 6]. The evaporation tests of this study were performed according to the procedure proposed by Romano and Santini [7] for evaluating a parameter optimization method to determine the unsaturated hydraulic properties of different soil types. A widely used hydraulic model, the van Genuchten – Mualem model [8] has been then adopted to fit the experimental data; this model comprises a non-hysteretic relationship (soil water retention curve) to estimate the effective degree of saturation, Se, from soil suction, s (difference between pore air pressure, u_a, and pore water pressure, u_w):

$$S_e(h) = \frac{\theta-\theta_r}{\theta_s-\theta_r} = \left[\frac{1}{1+(\alpha_{VG}\cdot s)^{n_{VG}}}\right]^{m_{VG}} \quad (1)$$

θ, θ_s θ_r are the current, maximum and residual values of soil water content respectively, α_{VG} and n_{VG} are model parameters dictating the inflection point and the shape of the retention curve respectively. Notice that n_{VG} and m_{VG} are dependent parameters, being:

$$m_{VG} = 1 - \frac{1}{n_{VG}} \quad (2)$$

The van Genuchten – Mualem model is employed to predict the variation of relative permeability k_r with the effective degree of saturation, according to the following hydraulic conductivity function:

$$k_r(h) = S_e^{0.5}\left[1-\left(1-S_e^{1/m_{VG}}\right)^{m_{VG}}\right]^2 \quad (3)$$

The hydraulic permeability, $k(S_e)$, is then expressed as the product of the saturated permeability, k_0, estimated from in situ and laboratory test results, and relative permeability, k_r. By adopting such constitutive model and parameters, nine soil-water retention curves (SWRC) and hydraulic conductivity functions (HCF) were thus obtained.

For Unit B and Unit C, an estimation of hydraulic conductivity at saturation was made. Average values of all soil hydraulic properties mentioned above are listed in Table 2.

Table 2. Average values of estimated hydraulic soil properties.

Soil type	θ_r (m^3/m^3)	θ_s (m^3/m^3)	α_{VG} (kPa^{-1})	n_{VG} (-)	k_0 (log(m/s))
Unit R	0.079	0.395	0.164	1.328	-5.804
Unit B	-	-	-	-	-5.725
Unit C	-	-	-	-	-8.886

3 Flow modelling and bank stability

Seepage characteristics and stability conditions of the collapsed Secchia's river bank have been evaluated using a 2D numerical model based on actual geometrical, stratigraphical, geotechnical and hydrological data summarised in the previous section. Limit equilibrium analyses were performed both in steady-state and in transient flow conditions, as induced by river water levels and rainfall events registered since one month before the river bank failure and using the hydraulic parameters as deterministic variables, thus allowing only the soil shear strength parameter to account for the inherent soil variability.

3.1. Saturated and unsaturated flow model

Saturated and unsaturated flow was modelled using the software SEEP/W [9]. The code can perform a two-dimensional, finite element seepage analysis using the equations of motion and mass conservation. Partial differential equations governing steady-state free surface flow in saturated soil conditions is given below:

$$-k_0\nabla\left(\frac{u_w}{\gamma_w}+z\right) = \mathbf{0} \quad (4)$$

where k_0 is the soil hydraulic permeability at saturation, u_w is the pore water pressure and z is the vertical elevation coordinate. In the case of unsaturated soils, the hydraulic conductivity is a function of water content and pore water pressure. Relationships between volumetric water content and pore water pressure and between hydraulic conductivity and pore water pressure are defined by soil-water retention curves (SWRC) and hydraulic conductivity functions (HCF), respectively. In transient seepage analysis, the water flow per unit area, **q**, is calculated using generalized Darcy's law:

$$-k_r k_0\nabla\left(\frac{u_w}{\gamma_w}+z\right) = q \quad (5)$$

Notice that only flow of water in the liquid phase has been considered in this work. Pore water pressure distributions were computed using adequate boundary conditions, described below, and assuming hydraulic parameters as deterministic variables, equal to their average values. Seepage analysis results, in terms of pore water pressure (PWP) distribution, are showed in Figure 3. In particular, are showed PWP distributions evaluated performing steady state and transient analysis using pertinent initial and boundary conditions, hereafter described; the PWP increment between two adjacent isolines is 10 kPa. Different suction distribution estimated in the riverbank soil significantly influence stability analyses.

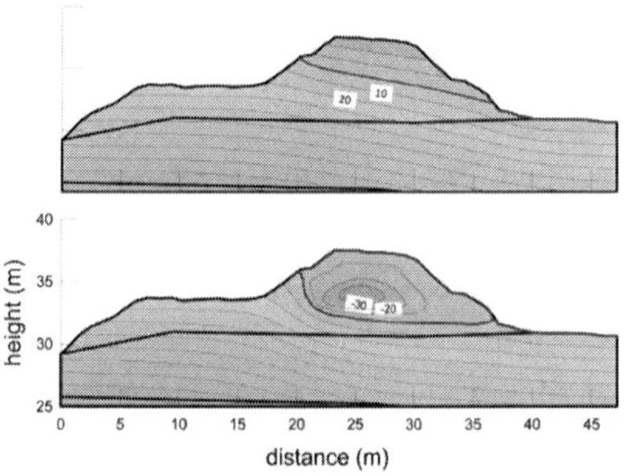

Figure 3. Results of steady state (top) and transient (bottom) seepage analysis in terms of PWP distribution.

3.2 Initial and boundary conditions

Numerical flow regime determination was conducted using specific initial and boundary conditions for steady-state and transient seepage analyses. Steady-state seepage analysis was conducted assuming a 35.9 m above msl retained water level on the river side, which is the maximum water level recorded during the specific flooding event (19 January 2014), water table at ground level on the landward side and soil permeabilities constant and equal to their saturated values. For a more accurate study of the embankment stability evaluation for that specific flooding event, flow characteristics have been also determined in transient conditions induced by the actual sequence of retained water levels and rainfall infiltrations, starting from realistic initial conditions. To this purpose, initial pore water pressure distribution was assigned starting from a water level at 30 m above msl, suction linearly decreasing to about -40 kPa in the middle of the bank, then increasing back up to zero value at the surface top level. The aim was to simulate a typical suction distribution in a wet period, characterized by rainfalls that tend to reduce the soil suction near the top surface by heavy water infiltration. This hypothesis has been considered admissible [10, 11, 12] also in absence of specific site measurements. It is worth noticing here how the actual initial distribution of pore water pressures is of vital importance, as it strongly affects the final results. Hydrostatic pore water pressure regime was considered beneath the water level. Retained water level and rain infiltration were then introduced as time-dependent boundary conditions in the transient seepage analysis, assigning linearly variable total hydraulic head values on the river side nodes and unit rate of flow to boundary surfaces respectively. Input values have been analyzed using data recorded during river water level and rainfall events starting 25 December 2014 [1]. In Figure 4 recorded water levels and rainfalls are plotted together with the interpolating linear function used in the numerical model as total hydraulic head boundary condition.

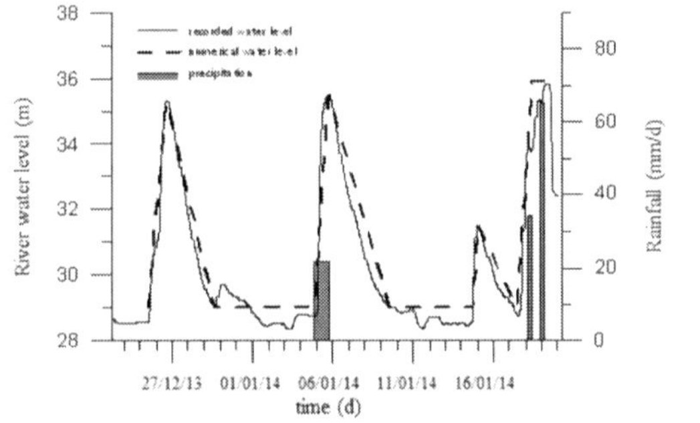

Figure 4. Flood hydrograph, rainfall hyetograph and numerical total hydraulic head boundary condition recorded and modelled from 25 December 2013 to 19 January 2014.

3.3 Riverbank stability analysis

Positive and negative pore water pressure distributions thus obtained were used as input data for the stability analyses performed by limit equilibrium with the Morgenstern-Price method [13], both under steady-state and transient flow conditions. Limit equilibrium analyses were performed for two critical surfaces, along landward and riverside slopes, identified in preliminary analyses and aimed at reproducing the possible onset of the actual river bank collapse (Figure 5).

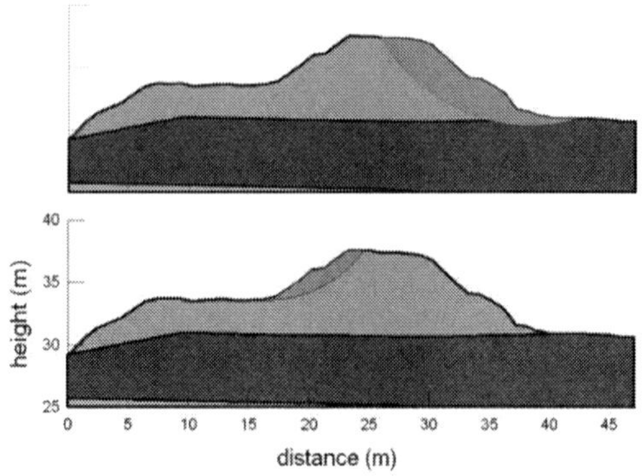

Figure 5. Landward (top) and riverside (bottom) critical surfaces identified for the stability analyses.

Vanapalli [14] failure criterion was used in limit equilibrium analysis for accounting unsaturated soil strength contribution, expressed in the form:

$$\tau = c' + \sigma_n \tan\varphi' + (u_a - u_w)\, S_e \tan\varphi' \qquad (6)$$

where τ = shear strength, σ_n = total normal stress, u_a = pore air pressure, u_w = pore water pressure, c' = effective cohesion, φ' = friction angle in terms of effective stress, S_e = effective degree of saturation. The shear strength contribution due to the matric suction ($u_a - u_w$) is equal to $\tan\varphi'$ up to the air-entry value (AEV) of the soil. In other words, the conventional equation for estimating the shear strength of saturated soils can be used up to the AEV for unsaturated soils. This equation suggests that there is a one to one relationship between the degree of saturation, S, and the area of water contact along the shear plane in the soil. A previous formulation had been provided by Fredlund [15], expressed as:

$$\tau = c' + \sigma_n \tan\varphi' + (u_a - u_w) \tan\varphi^b \quad (7)$$

where φ^b = angle expressing the rate of increase in strength relative to the matric suction ($u_a - u_w$). As often observed [16, 17], φ^b is a parameter strongly dependent on matric suction, varying up to a maximum value of φ' when the soil approaches saturation. Using a constant value of φ^b for accounting unsaturated soil strength contribution is certainly an approximation and could lead to errors in stability evaluation process when the matric range and the φ^b variability is not well defined. In this study, useful indication about φ^b value is provided for similar cases.

4 Probabilistic stability analyses

The probabilistic stability analysis of the river bank was performed with reference to both landward and river side slopes. Generally, in slope stability analyses only the soil shear strength parameters φ' and c' are considered to provide the probability of failure, assuming fully saturated conditions below the water table. However, in this work the role of partially saturated soil strength was also considered to assess the riverbank stability, in order to stress the importance of the soil hydraulic and retention parameters. Two cases were considered:
- Case 1, effective shear strength angle (φ') as random variable and soil hydraulic permeabilities (k_0) constant and equal to saturated values, not taking into account unsaturated soil state;
- Case 2, effective shear strength angle (φ') as random variable and soil hydraulic parameters (θ_r, θ_{sat}, α_{VG}, n_{VG}, k_0) as deterministic variables, equal to respective mean values (see Table 2).

The strength parameters of the soil beneath the embankment were assumed as deterministic values. Analyses of Case 1 have been performed only in steady-state conditions; for Case 2, steady-state (Case 2a) and transient (Case 2b) seepage analyses have been carried out, using the above described initial and boundary conditions to obtain the resulting pore water pressure distributions. A number of 100 000 trials by Monte Carlo method were performed for estimating mean values and standard deviations of FS distribution together with the reliability index of each considered case and failure surface. The effective shear strength angle φ' of the soil bank was therefore the only random variable, with normal distribution defined by a mean value of 32° and a standard deviation of 1.8°, with a covariance (CoV.) of around 6%. Analysis results, summarized in Table 3, are presented in terms of mean values and standard deviations of Factor of Safety (FS), together with the reliability index and the probability of failure, for the landward and river side bank slope, respectively. In Case 1, probabilistic stability analyses in steady state seepage conditions provide very low reliability index values (high probability of failure), highlighting a possibly serious critical condition for the riverbank. Instead, high reliability index values (probability of failure quite close to zero, but not null) are provided for the river side critical surface. Even if not representative for the specific flooding event, these results are, however, meaningful to stress the limited riverbank safety state in case of persistent high water levels in conjunction with harsh environmental conditions, as extreme rainfall, that could lead the bank soil close to saturation. In Case 2a, stability analyses provide much lower values of probability of failure for both critical surfaces, reducing of at least one order of magnitude with respect to Case 1; mean values and standard deviations of factor of safety are increased, as well as its CoV; this could be ascribed to the fact that in Case 2, φ' is twice used in the failure criterion, leading to a major dispersion of the results in terms of FS with respect to Case 1. Probabilistic stability analyses in transient seepage conditions (Case 2b) provide very high values of reliability index (very low probabilities of failure), for both landward and river side slope, not showing real chances of a riverbank instability. This result would suggest that the considered conditions are not able alone to account for the collapse occurred; in fact other issues, as hydraulic parameter variability and possible presence of animal burrows have been called upon to have triggered the instability mechanism leading to final collapse. Finally, a set of probabilistic analyses was carried out, for Case 2, both in steady-state and transient seepage conditions, in order to obtain the equivalent value of φ^b able to provide similar results in terms of mean FS; this was obtained using a deterministic value of φ^b equal to 23.5°. These analysis provided, thus, lower values for standard deviations and probabilities of failure, since φ' variability is not taken into account for unsaturated soil strength contribution. Equivalent values for φ^b are specific for a considered suction range and, therefore, could vary depending on the assumed critical surface.

Table 3. Results of MP limit equilibrium analyses using M-C method in steady state and transient seepage conditions.

Case 1		Landward	Riverside
Mean value	μFS	1.031	1.418
Standard deviation	σFS	0.059	0.108
Reliability index	β	0.528	3.871
Probability of failure	PF (%)	30.441	8.10×10^{-2}

Case 2a		Landward	Riverside
Mean value	μFS	1.145	1.735
Standard deviation	σFS	0.069	0.132
Reliability index	β	2.100	5.568
Probability of failure	PF (%)	1.302	0.000

Case 2b		Landward	Riverside
Mean value	μFS	1.388	2.088
Standard deviation	σFS	0.084	0.159
Reliability index	β	4.588	6.837
Probability of failure	PF (%)	0.000	0.000

5 Conclusions

In this paper, a probabilistic study of riverbank stability based on an accurate geotechnical and hydrological characterization, has been presented. Through the use of different flow model types and hypotheses on boundary and initial conditions, the importance of considering partially saturated soil strength contribution in assessing the probability of failure for riverbanks has been underlined. On the basis of results conducted with different failure criteria, an accurate estimate of an equivalent φ_b has been provided and could represent a useful indication for similar applications, in terms of specific soil types and matric suction ranges.

Accounting for unsaturated soil strength in steady-state conditions leads to less conservative and more realistic results, but it also requires to make use of many further information, as hydraulic properties of soil and initial and boundary conditions for a significantly long enough period. In addition, hydraulic parameters have been largely demonstrated to be strongly variables, even for mechanically and physically rather homogenous soils and therefore this issue should be also taken into account in order to achieve values of probability of failure consistent with the actual risk and, above all, to examine their evolution in conjunction with boundary condition evolution. Therefore hydraulic and retention parameters of riverbank soils, influencing seepage and strength characteristics, should be also treated as random variables; a specific uncertainty propagation analysis could thus lead to higher and more realistic values of probability of failure both under steady-state and transient seepage conditions.

References

1. L. D'Alpaos, A. Brath, V. Fioravante, G. Gottardi, P. Mignosa, and S. Orlandini, Available from http://Ambiente.Regione.Emilia-Romagna.it/geologia/notizie/notizie-2014/fiume-Secchia (2014)
2. P.K. Robertson. Can. Geotech. J., **46**(11): 1337-1355. (2009)
3. P.W. Mayne, and R.G. Campanella. 16th International Conference on Soil Mechanics and Geotechnical Engineering, Osaka, Japan, 12-16 September, 2005. pp. 721-724. (2005)
4. G.P. Wind. *Water in Unsaturated Zone*, Wageningen, Netherlands, June 1966. pp. 181-191. (1968)
5. S. Tamari, L. Bruckler, J. Halbertsma, J. Chadoeuf. Soil Science America J., **57**(3): 642-651. (1993)
6. O. Wendroth, W. Ehlers, J.W. Hopmans, H. Kage, J. Halbertsma, J.H.M. Wosten. Soil Science of America J., **57**(6): 1436-1443. (1993)
7. N. Romano and A. Santini. Water resources research, **35**(11): 3343-3359. (1999)
8. M. Th. van Genuchten. Soil Science of America J, **4**(5): 892–898. (1980)
9. Geo-Slope International Ltd., Calgary, Canada. (2008)
10. N. Casagli, M. Rinaldi, A. Gargini, A. Curini. Earth Surf. Proc. and Landforms, **24**: 253-277. (1999)
11. M. Rinaldi, N. Casagli, S. Dapporto, A. Gargini Earth Surf. Proc. and Landforms, **29**:237-254. (2004)
12. G. Calabresi, F. Coleselli, D. Danese, G.P. Giani, C. Mancuso, L. Montrasio, A. Nocilla, L. Pagano, E. Reali, A. Sciotti. Canad. Geotech. J. **50**(9): 947-960. (2013)
13. N. R. Morgenstern, and V.E. Price. Geotechnique, **15**: 70-93. (1965)
14. S.K. Vanapalli, D.G. Fredlund, D.E. Pufahl and A.W. Clifton. Canad. Geotech. J., **33**(3): 379-392. (1996)
15. D.G. Fredlund, N.R. Morgenstern and R.A. Widger. Canad. Geotech. J., **15**(3): 313 – 321. (1978)
16. D.G. Fredlund, H. Rahardjo. *John Wiley & Sons*. (1993)
17. N. Lu and W.J. Likos. *John Wiley & Sons*. (2004)

Swelling and collapse of compacted soils to be used as earth dam cores

Michael Bardanis[1,a] and Sofia Grifiza[1]

[1]*Edafos Engineering Consultants S.A., Laboratory, 2 Therapion st., 10444 Athens, Greece*

Abstract. The mechanical behaviour of fine-grained soil materials to be used as impermeable cores for earth dams has been extensively studied by numerous researchers. The required properties of these materials have also been very well described by standards and specifications. Yet, more than often it is required to assess their swelling/collapse potential, especially at various vertical stresses, as a means to estimate their volume changes upon inundation which is going to be caused by filling the dam reservoir. In the paper, experimental results of five different soils are presented. The soils tested ranged from non-plastic silty sands with clay to medium plasticity clayey silts that were compacted in conditions dry, at and wet of optimum moisture content as derived from standard compaction energy Proctor tests, then subjected to one-dimensional loading conditions and then inundated. The vertical stress was up to 7.6 MPa. The experimental results are categorized according to initial moisture content relative to Proctor optimum and indicate expected magnitudes of strains due to inundation for various grain-size distributions and plasticities.

1 Introduction

Earth and rockfill dams have become increasingly more attractive relative to other types of dams over the past decades. More than often, clayey materials for the construction of the clay core in these types of dams may be hard to obtain in the necessary quantities at close proximity to the dam under construction, pose various concerns regarding their mechanical behaviour and constitute the subject of extensive experimental programmes before construction and quality control during construction. The required properties of these materials are generally very well described by specifications. These describe in detail the tests needed during geotechnical investigation and quality control and the standards that should be followed [1]. Yet, in addition to usual classification and mechanical properties tests, it is often required to assess the swelling/collapse potential of these soils, especially at various vertical stresses, as a means to estimate their volume changes upon inundation which is going to be caused by filling the dam. Given the fact that earth dams can be very high, the range of vertical stress that needs to be applied can be quite large.

Experimental results from five different soils examined for use in earth dam clay cores are presented in the paper along with results from a sedimentary clayey soil for comparison. The soils tested, ranged from non-plastic silty sands with clay, to medium plasticity clayey silts and were compacted in conditions dry, at and wet of optimum moisture content as derived from standard compaction energy Proctor tests. They were then subjected to various vertical stresses under one-dimensional loading conditions and then inundated. The vertical stress was up to 7.6 MPa. The experimental results are categorized according to initial moisture content relative to Proctor test optimum moisture content and indicate expected magnitudes of strains due to inundation for various grain-size distributions, plasticity indices and geological origins.

2 The soils tested

The soils tested by this stage of the ongoing research are Livadero clay, Skiros clay, Skiros weathered Serpentinite, and two Platy river non-plastic sandy silt to silty sand. For comparison, results from Maroussi clay, a sedimentary clayey silt with sand found in abundance in the north of Athens are also presented. Livadero clay is a sandy silt with clay found in the region of Drama close to the city of Livadero in northern Greece and has been examined for use in the construction of the clay core of an earth dam. Skiros clay is a clayey silt with sand used for the construction of the Ferakambos river dam on the island of Skiros in the Aegean Sea. In the same dam a locally found weathered Serpentinite found in abundance has been used for the construction of the shoulders of the same dam. Given its low permeability and that the clay found in the vicinity eventually was not found in the necessary quantities, the same material (in its more plastic and fine-grained phase) has been used for the construction of the top 5m of the clay core (the dam is 27m high). After all a preliminary study for the dam (before finalisation of the design study to the design that was adopted for construction) had proposed the construction of a uniform dam made totally with weathered Serpentinite and a major central drain. Details of the mechanical behaviour of the Skiros weathered Serpentini-

[a] Corresponding author: mbardanis@edafos.gr, lab@edafos.gr

Table 1. Index properties of the soils tested.

Soil	Clay (%)	Silt (%)	w_L (%)	I_P	G_s	USCS
Maroussi clay (MC)	40.1	41.9	47	28	2.66	CL
Livadero clay (LC)	22.0	45.8	37	20	2.70	CL
Skiros clay (SC)	35.3	52.4	44	21	2.67	CL
Skiros weathered Serpentinite (SWS)	2.1	48.5	27	8	2.61	CL
Platy River sandy silt (M1)	5.4	51.2	N.P.		2.75	ML
Platy River silty sand (M4B)	4.5	35.6	N.P.		2.68	SM

Table 2. Index properties of the soils tested.

Soil	$\gamma_{d,max}$ (kN/m^3)	w_{opt} (%)	w_{dry} (%)	w_{wet} (%)
Maroussi clay (MC)	17.1	18.5	15.0	21.5
Livadero clay (LC)	17.4	17	15	19
Skiros clay (SC)	16.0	23	16	26
Skiros weathered Serpentinite (SWS)	19.3	11	8.5	13.5
Platy River sandy silt (M1)	18.6	14	11	17
Platy River silty sand (M4B)	19.3	12	10	14

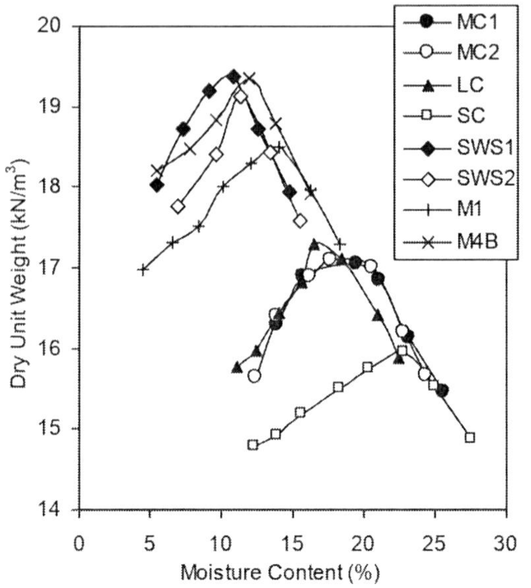

Figure 1. Dry unit weight-moisture content curves from standard compaction energy Proctor tests for the soils tested (two tests shown for Maroussi clay and Skiros weathered Serpentinite, average values from these reported in Table 2).

te have already been presented [2, 3]. Another dam for irrigation purposes on the Platy river in the Rethimno region of southern Crete has been examined. The 55m high dam will be an earth dam with clay core built from clays found in the reservoir basin and shoulders built from coarse-grained alluvial deposits found also in the reservoir basin. As part of the geotechnical investigation and in search of clay deposits for use in the construction of the clay core, two non-plastic materials found in abundance in the reservoir basin were also examined. One was a sandy silt with clay (M1) and the other a silty sand with clay (M4B). Index properties of the soils tested are summarised in Table 1.

Standard energy Proctor compaction tests have been performed on all the soils tested. The compaction characteristics of the soils are presented in Table 2 and their dry unit weight-moisture content compaction curves are presented in Fig. 1. Depending on the shape of the compaction curves in Fig. 1, moisture contents for preparation of samples in conditions dry and wet of optimum moisture content were decided so as samples with dry unit weight 95% of the maximum dry unit weight could be prepared. These moisture content values are also shown in Table 2.

3 Experimental method

Two methods were employed for samples' preparation. One was dynamic compaction in the Proctor test mould using an automatic compactor and then trimming the specimens to be compressed and inundated from these larger samples, and the other dynamic compaction by hand in the oedometer cells by controlling dry soil mass, water mass and the height of the sample by gradual compaction until the required dry unit weight is achieved. Maroussi clay and Skiros weathered Serpentinite samples were prepared using the former method, and samples of all the other soils using the latter.

A review of available methods for preparation of compacted samples for swelling/collapse tests and their comparison is beyond the scope of this paper; still the two methods used constitute widely used and well accepted methods for preparation of compacted samples. If one had to point to a difference standing out is that the preparation of large size samples in the Proctor test moulds allows the measurement of suction of each sample before trimming the specimens to be tested. Therefore, although conventional oedometers may be used for the swelling/collapse tests, still the magnitude of suction in each moisture condition relative to Proctor test optimum moisture content is known and can be associated with the observed strains due to inundation other than just the moisture content. As already reported [3] the Skiros weathered Serpentinite exhibited matric suction of 12 kPa and total suction of 240 kPa for samples compacted wet of optimum, 30 kPa and 240 kPa respectively for samples compacted at optimum, and finally 70 kPa and 380 kPa respectively for samples compacted dry of optimum moisture content. Matric suction in this particular case was measured using a Soilmoisture Equipment Corp. 2100F laboratory tensio-

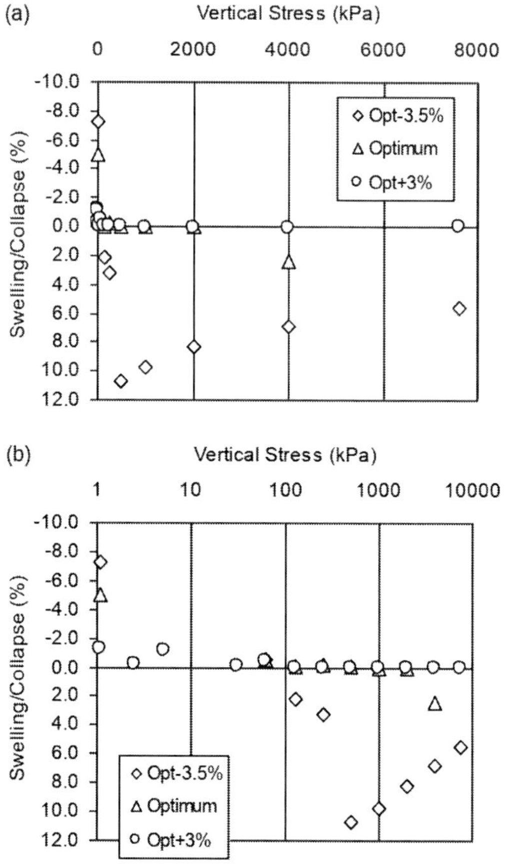

Figure 2. Strain due to inundation versus vertical stress for Maroussi clay at each of the three moisture contents tested for vertical stress in a) linear scale, and b) logarithmic scale.

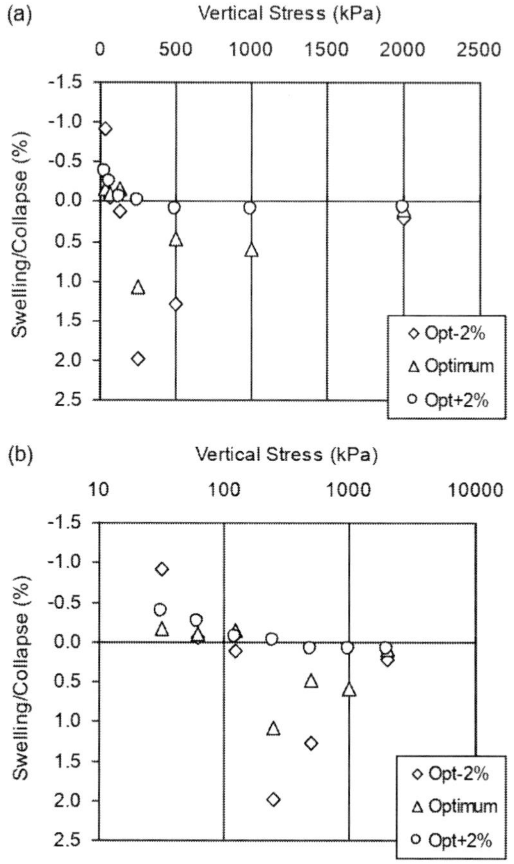

Figure 3. Strain due to inundation versus vertical stress for Livadero clay at each of the three moisture contents tested for vertical stress in a) linear scale, and b) logarithmic scale.

meter and total suction using a Decagon Devices, Inc. chilled mirror hygrometer. Matric suction was measured by installing the tensiometer in the large size sample before trimming the specimen for the swelling/collapse test and total suction on trimmings from locations immediately above and below the trimmed samples. Similar measurements on compacted samples of Maroussi clay [4] yielded 60 and 130 kPa for samples compacted wet of optimum, 80 and 330 kPa for samples compacted at optimum, and a total suction of 2500 kPa for samples compacted dry of optimum. In this last case the tensiometer could not measure the matric suction of the samples. These measurements of suction and the insight they offer in the interpretation of observed swelling/collapse upon inundation come at the cost of larger inhomogeneity of specimens finally trimmed for actual swelling/collapse testing, and at a cost and effort that cannot be routinely applied in geotechnical practice.

4 Swelling/collapse strains due to inundation

The swelling/collapse strains upon inundation for each of the soils tested are plotted against vertical stress in Fig. 2 to 7. Diagrams denoted (a) present results with the vertical stress axis in linear scale, and diagrams denoted (b) present results with the vertical strain axis in logarithmic scale. Curves for uniform moisture conditions are presented in each figure, i.e. curves for samples compacted dry of optimum, at optimum, and wet of optimum moisture content. Swelling/collapse strains upon inundation for Maroussi clay are plotted in Figure 2. Tests were performed both at very low vertical stress (just 1 kPa) and at very high vertical stress (7.6 MPa). The strains observed are really impressive ranging from 7.3% swelling for 1 kPa vertical stress (practically free swelling) to 10.7% collapse for 500 kPa of vertical stress which is also the maximum collapse strain observed with values for higher stresses decreasing linearly with the logarithm of increasing vertical stress. Another important observation is that these very high values of strain due to inundation correspond to a soil and a moisture condition that yields a value of swelling pressure of just 75 to 80 kPa (defined as the intersection of the strain-log stress line with the zero strain axis); a value that would not cause serious concern in everyday geotechnical practice although experimental results indicate that strains due to inundation above and below this value are remarkably

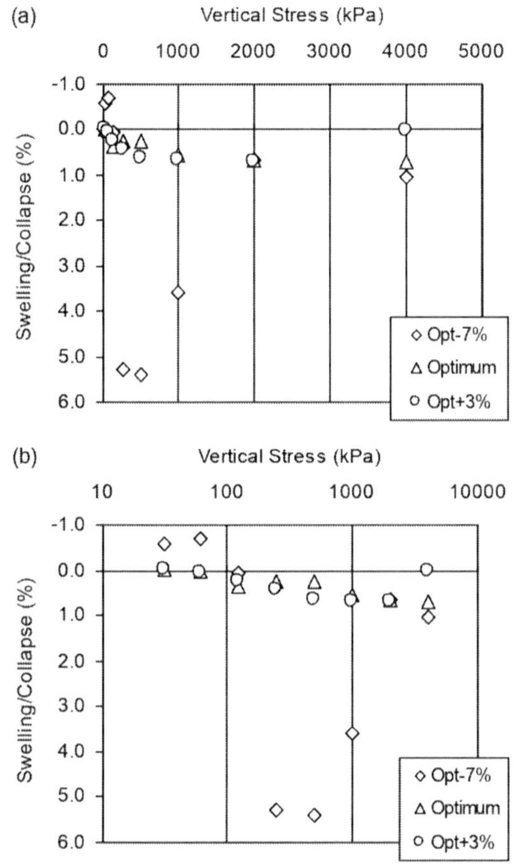

Figure 4. Strain due to inundation versus vertical stress for Skiros clay at each of the three moisture contents tested for vertical stress in a) linear scale, and b) logarithmic scale.

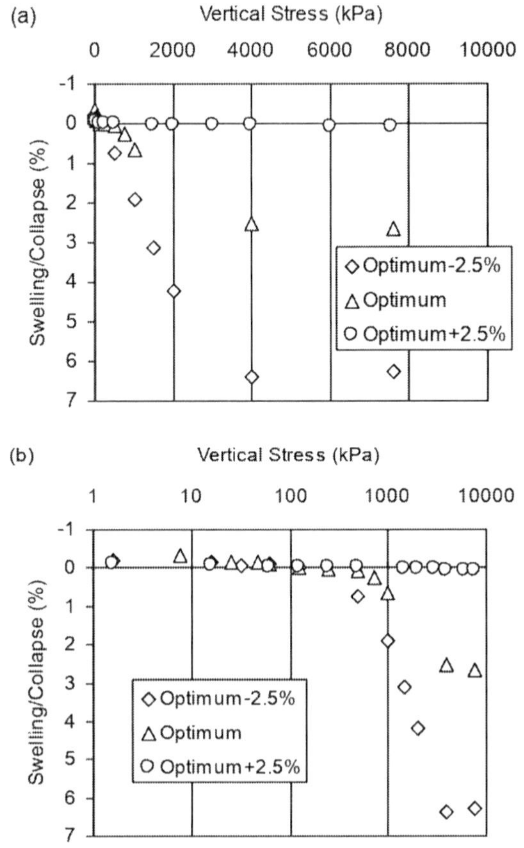

Figure 5. Strain due to inundation versus vertical stress for Skiros weathered Serpentinite at each of the three moisture contents tested for vertical stress in a) linear scale, and b) logarithmic scale.

high to the point of being called alarming. To add to this point, strains due to inundation for this soil compacted at optimum moisture content indicate a value of swelling pressure in the order of 700 kPa but strains below that value as low as 60 kPa and above that value as high as 2000 kPa are really small (between -0.5% and +0.02%) and only at very low stresses and very high stresses exhibit very high values (-5% for 1 kPa and +4% for 7.6 MPa). In this case therefore, a truly alarming value of swelling pressure (700 kPa) is not associated with similarly significant strains due to inundation in the range of usually applied vertical stresses in engineering works. Finally, strains due to inundation for this soil compacted wet of optimum moisture content indicate much smaller strain values especially for collapse and also for swelling, although the variation there is much higher (yet values for vey low values of vertical stress may reach even 1.2% swelling). Swelling pressure in this case is again between 70 and 80 kPa. The behaviour of this material upon inundation justified the investigation of its mineralogy. The XRD method was employed indicating a total of 65% clayey minerals with the rest being equal proportions of calcitic and silica minerals. Clayey minerals included 15% illite, 11% montmorillonite, 10% kaolinite, 10% Muscovite, 8% Halloysite, 5% Chlorite and 4% Serpentinite. A mineralogy like that is not commonly found, yet the observations on the behaviour of Maroussi clay upon inundation shed considerable light into the relation (literally the lack of it) between the magnitude of swelling pressure and the magnitude of strains due to inundation.

Similar observations can be made for Livadero clay (Fig. 3), although not as pronounced as with Maroussi clay. Dry of optimum the soil exhibited the highest swelling strains at low stresses and the highest collapse strains at higher stresses, again with the maximum exhibited at 250 kPa with considerable decrease with stress after that value. Swelling pressure was 65-70 kPa. At optimum the soil exhibited a similar behaviour with smaller strains in both ranges, an increase of swelling pressure to 150 kPa and similar decrease of collapse strains with vertical stress after a maximum exhibited at 250 kPa. Finally, for wet of optimum, the strains observed were the smallest (maintaining considerable values for swelling at low stresses,) with the interesting observation that swelling pressure climbed to 270 kPa.

Slightly different observations can be made for

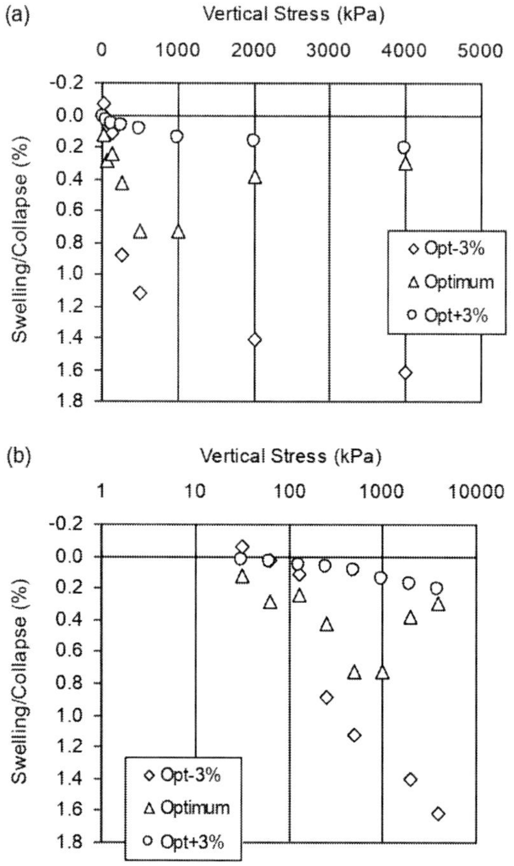

Figure 6. Strain due to inundation versus vertical stress for Platy River Sandy Silt (M1) at each of the three moisture contents tested for vertical stress in a) linear scale, and b) logarithmic scale.

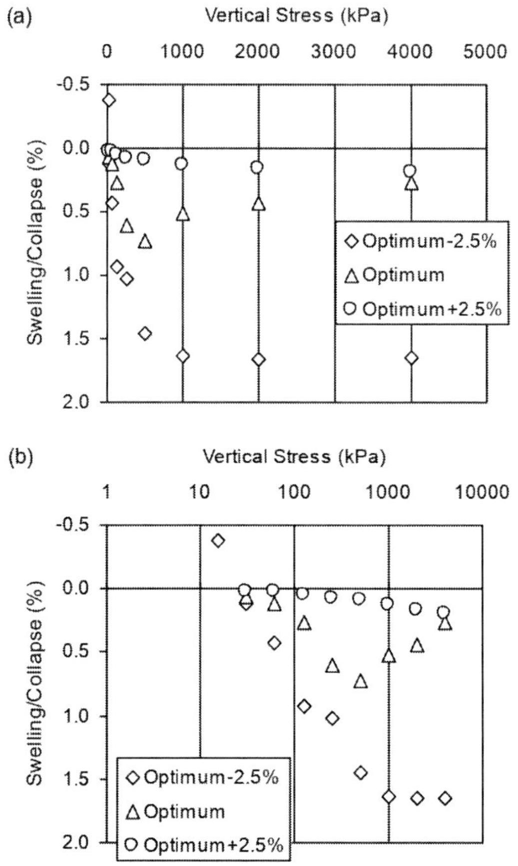

Figure 7. Strain due to inundation versus vertical stress for Platy River Silty Sand (M4B) at each of the three moisture contents tested for vertical stress in a) linear scale, and b) logarithmic scale.

Skiros clay (Fig. 4). Dry of optimum the soil exhibited the highest swelling strains at low stresses and the highest collapse strains at higher stresses, with the maximum exhibited at 500 kPa with considerable decrease with stress after that value. Swelling pressure was 100-110 kPa. At optimum the soil exhibited a continuous increase of collapse strain up to the maximum of 4000 kPa of vertical stress. Finally, for wet of optimum, the strains observed were the smallest (maintaining considerable values for collapse at high stresses) with the interesting observation that collapse strains started decreasing after 2000 kPa.

Skiros weathered Serpentinite was the only one of the soils tested that did not exhibit a maximum of collapse strain at any moisture conditions (Fig. 5). For moisture content 2.5% dry of optimum, from 250 kPa of vertical stress the collapse strain started to increase rapidly up to 4000 kPa where a tendency for stabilization of the collapse strain with vertical stress was observed. Maximum collapse strain reached 6.4%. Zero deformation upon wetting was observed at 125 kPa. For lower stresses swelling was observed but its maximum value was very small (-0.2%). For optimum moisture content a similar behaviour was observed although the increase in collapse strain is less rapid and the maximum collapse strain smaller (it reached a maximum value of 2.7%). The rapid increase started between 500 and 750 kPa and lasted up to 4000 kPa where stabilization of the collapse strain occurred again. Zero deformation upon wetting was observed at 225 kPa. For lower stresses, swelling was observed but its maximum value was only -0.34%. Finally for wet of optimum, collapse strain observed even at vertical stress as high as 7600 kPa was practically negligible (max. 0.07%). In this particular case of so low collapse strains, it may be argued that there was essentially no strain due to wetting as the deformation observed after wetting for 24hrs was only secondary compression deformation of the sample. Zero deformation upon wetting was observed at 500 kPa (a value already reported and discussed [2]). For lower stresses swelling was observed but its maximum value was only -0.09% for the minimum applied stress.

The Platy river sandy silt (M1, Fig. 6) and silty sand (M4B, Fig. 7), although non-plastic, they did exhibit for conditions dry of optimum a swelling pressure of 40 and 25 kPa respectively. In conditions dry of optimum, M1

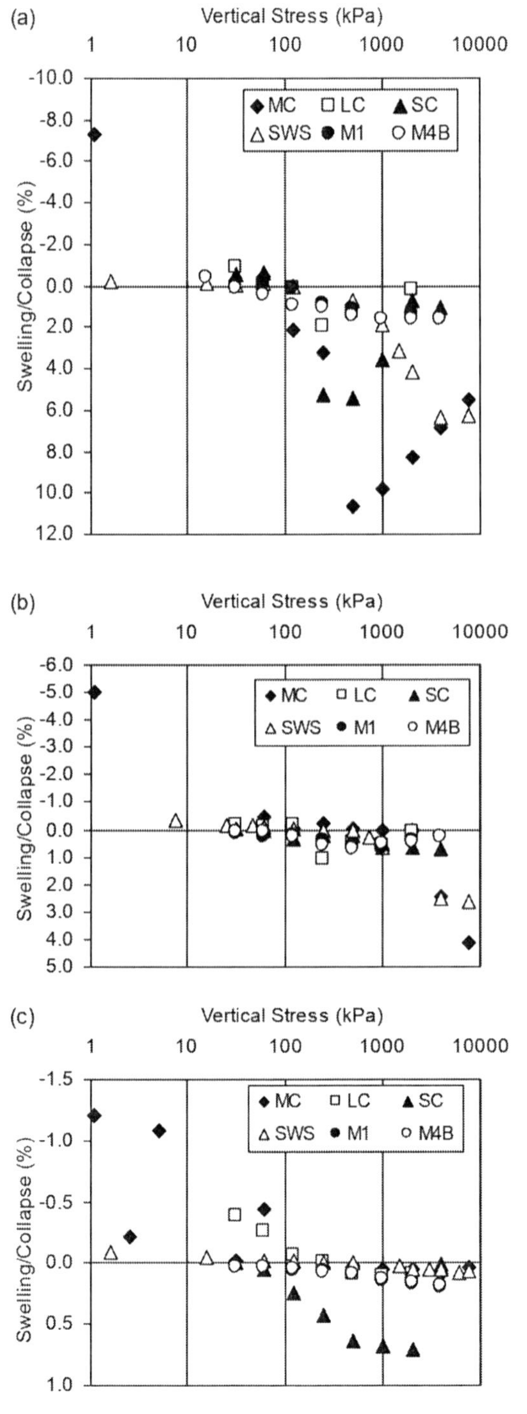

Figure 8. Strain due to wetting versus vertical stress for all soils tested at a) dry of optimum, b) optimum, and c) wet of optimum moisture content.

exhibited a rapid increase of collapse strains between 125 and 250 kPa and no signs of stabilization of collapse after that stress, while M4B did not exhibit a rapid increase and collapse stabilized after 1000 kPa. In conditions at optimum, both soils exhibited a maximum of collapse strain of 0.7% at 500 kPa and a subsequent decrease of collapse strain values. Finally, for conditions wet of optimum, both soils exhibited no swelling with the collapse strains being the smallest and with a linear increase in their values with the logarithm of the vertical stress.

Finally in Fig. 8 swelling/collapse strain is plotted against vertical stress (in logarithmic scale) for all soils at the same moisture conditions; dry (Fig. 8a), at optimum (Fig. 8b) and wet (Fig. 8c). All soils in conditions dry of optimum exhibit a maximum of collapse strain or stabilisation. Something similar cannot be observed for conditions at optimum moisture content, and finally for conditions wet of optimum usually a linear increase of collapse strains is observed although stabilisation has been observed too (Skiros clay). As found by other researchers too [5], both swelling and collapse may be reduced practically to extinction by compacting the soil wet of optimum.

5 Conclusions

Swelling/collapse of a total of six soils was investigated experimentally. Five of these soils have been used for or are intended (or have been examined) for the construction of clay cores of actual dams. All soils in conditions dry of optimum exhibit a maximum of collapse strain or stabilisation of collapse. Something similar cannot be observed for conditions at optimum moisture content, and finally for conditions wet of optimum usually a linear increase of collapse strains is observed although stabilisation has been observed too. There seems to be no direct relation between the magnitude of swelling pressure and the magnitude of strains due to inundation whether that is swelling strains or collapse strains.

Acknowledgements

Owner of the Ferekambos Dam on Skiros island and the Platy river dam in Crete is the Agricultural Development and Foods Ministry of Greece. Prime contractor for the Ferekambos dam was "Elliniki Hydrokataskevi S.A.". Owner of the Milorema dam in Livadero region is the Eastern Macedonia-Thrace prefecture of Greece. The authors wish to acknowledge the help received from the Edafos S.A. Lab technicians, Mr C. Vagenas and Mr M. Tsoukaladakis.

References

1. Hellenic Technical Specification 1501-13-01-01-00, Impervious core of zoned earth and rockfill dams.
2. M. E. Bardanis, S. Grifiza, proc. 5th Int. Conf. Unsaturated Soils, Barcelona, Spain, 2010, Vol. 1, pp. 187-192.
3. M. E. Bardanis, S. Grifiza, proc. 6th Hel. Conf. on Geotech. and Geoenvironmental Engineering, Volos, 29 September - 1 October, 2010, Vol. 1, pp. 151-158.
4. M. E. Bardanis, 2016, PhD Thesis.
5. E. C. Lawton, R. J. Fragaszy, J. H. Hardcastle, J. Geotech. Engng, 115, 9, 1989, pp. 1252-1267.

Distribution of pore water pressure in an earthen dam considering unsaturated-saturated seepage analysis

Kumar Venkatesh[1a], Siva Ram Karumanchi[2]

[1]Assistant Professor, Civil Engineering Department, MNNIT, Allahabad, India
[2]M.Tech Scholar, Civil Engineering Department, MNNIT, Allahabad, India

Abstract. The variation of pore water pressure in earthen dams plays an important role in maintaining its stability. The pore water pressure within the dam are altered by the external loading conditions like rapid drawdown of reservoir water, earthquake loading and raise of water table caused by infiltration of rainfall. The seepage through an earthen dam involves saturated and unsaturated flows but to avoid complexity in solving the non-linear partial differential equations, the flow in unsaturated zone is neglected and seepage analysis is carried by constructing the flow net in which the pore water pressures beyond the free surface is taken as zero. In actual conditions negative pore water pressure develops beyond the free surface due to the capillarity which leads development to the matrix suction of the soil. In this paper a comparative study on distribution of pore pressure in a zoned earthen dam under steady state and transient conditions had been carried out considering unsaturated-saturated seepage theory. To solve the non-linear partial differential equations, finite element method has been adopted in the present study. The earthen dam has been modeled in different stages. At each stage a new parameter was added and parametric analysis was carried out. The results indicate that negative pore water pressure developed at the downstream side and the pore pressures at the mid-levels of the core are high. This specifies that, soils with low permeability have higher pore pressure. The pore pressures appeared to be higher in upstream side during rapid drawdown compared to steady state.

1 Introduction

The seepage analysis is mainly in the interest of slope stability of the earth retaining structures, hydraulic structures etc. from geotechnical point of view. But in general there are seepage related problems that may be ignored like piping failure in earthen dams, capillary siphoning, infiltration of rainfall etc. These factors are of certain interest for the detailed stability analysis of an earthen dam and its maintenance.

Since Casagrande's (1937) classic paper "Seepage through dams," seepage problems in geotechnical engineering have been generally solved by sketching flow nets. This method was based on the assumption that water flows only in the saturated zone. This method of solution is practical for simple steady state problems where the boundary of the flow region is clearly defined and the soil conditions are not too complex. However, many seepage problems of practical interest are complex, and a flow-net solution is not feasible [1].

Over the years, the calculation of seepage had been simplified with application of numerical methods like finite element method, finite difference method etc. The first step in the determination of automatic flow net by finite element method was given by Fred T. Tracy and N. Radhakrishnan, (1989). They used Cauchy-Riemann equations to determine the correct boundary conditions. In saturated condition the phreatic line was assumed to be upper boundary for confined flow conditions [1].

The results obtained from these assumptions are approximate and analysis was also done based on several assumptions ignoring some criteria. To overcome these defects, a unsaturated condition was also include in determintaion of seepage where the upper boundary varies with the degree of saturation which inturn depends on the capillarity of the soil. In the capillary zone the pore water pressures are taken as negative with respect to the atmosperic pressure and the present magnitude of matrix suction have found to be crucial in determining the properties of the unsaturated soils which are required for the seepage analysis [2]. The behavior of flow through soil in the analysis of saturated-unsaturated condition is based on the soil-water charateristic curve (SWCC) which is defined as the relation between degree of saturation and matrix suction [4]. The matrix suction is usually obtained by drying or wetting of a soil sample under a constant stress while monitoring the water released from or absorbed by the specimen and the total volume change of the specimen (centrifuge method).

[a] Kumar Venkatesh: venkatesh@mnnit.ac.in

2 Formulation

2.1 Steady state analysis

The steady state seepage analysis depends only on the permeability of soil materials. In steady-state conditions, equation for laminar, two-dimensional flow in homogeneous, isotropic, porous media was developed by Terzaghi (1943) by using Laplace equation, rigorous developments can be found in Bear (1972), Cedergren (1977), and Harr (1962).

The governing differential equation for 3-dimensional seepage analysis is

$$\frac{\partial}{\partial x}\left(k_x \frac{\partial h}{\partial x}\right) + \frac{\partial}{\partial y}\left(k_y \frac{\partial h}{\partial y}\right) + \frac{\partial}{\partial z}\left(k_z \frac{\partial h}{\partial z}\right) = 0 \qquad (1)$$

Where,
h = head in m
k_x, k_y, k_z = permeability in x, y, z directions

2.2 Transient seepage analysis

The transient state condition is a variable of time and degree of saturation of the soil. So an initial condition needs to be described aside from boundary conditions. In this paper, the standard H-based Richards's equation is used to determine the seepage and is expressed as follows

$$\frac{\partial}{\partial x}\left(k_x \frac{\partial h}{\partial x}\right) + \frac{\partial}{\partial y}\left(k_y \frac{\partial h}{\partial y}\right) + \frac{\partial}{\partial z}\left(k_z \left(\frac{\partial h}{\partial z}\right)\right) = \gamma_w m_w^2 \frac{\partial h}{\partial t} \qquad (2)$$

Where γ_w, m_w^2 are unit weight of water, coefficient of water storage constant with the slope of SWCC. Figure 1 show suction vs volumetric water content for different soil.

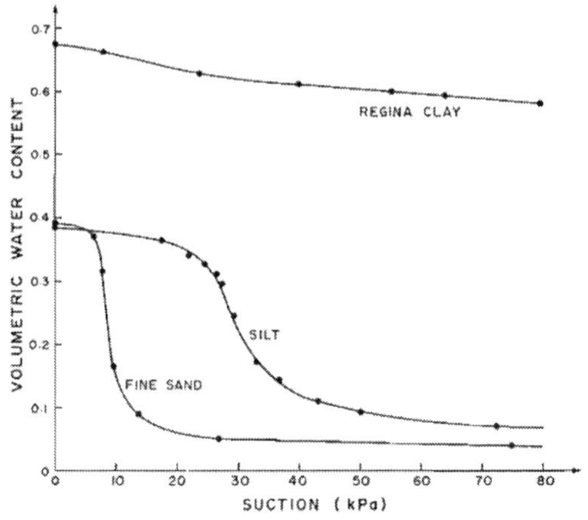

Figure1. SWCC for fine sand, silt and regina clay

3 Numerical modelling

The analysis of flow through an earthen dam by conventional and analytical methods is tedious and the main complexity is involved in transient flow conditions, interaction between sheet pile and soil. Though, finite element analysis is efficient in analyzing the flow through an earthen dam.

In this paper, the numerical model of an earthen dam was developed and parametric study on pore water pressure distribution in the earthen dam was studied for different condition. Three dimensional 4 noded tetrahedron elements were used for the finite element modeling.

3.1 Geometry of the model

The selection of geometry of an earthen dam is based on various literature review and suitable recommendations as shown in Figure 2. 3D meshing of earthen dam with sheet pile and horizontal drains are shown in Fgure 3.The height of dam is fixed to 30m. The inclination of upstream and downstream slopes is 3 (horizontal):1 (vertical) and 2.5 (horizontal):1 (vertical). The top width is restricted to 10m length and base width is 175m in length. The thickness of the dam is taken as 50m. In the case of pervious foundations, the depth of foundation is taken as 40m. Table 1 represents the different soil properties used in components of earthen dam where as Table 2 shows the interface properties between soil and sheet pile.

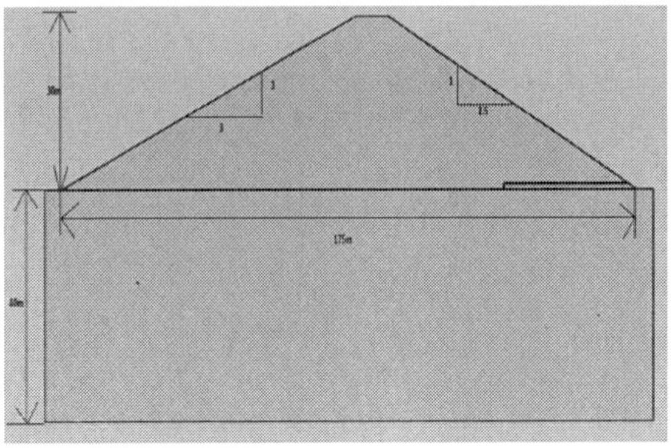

Figure 2. Geometry of a homogeneous earthen dam with blanket drain

DOI: 10.1051/e3sconf/20160919004

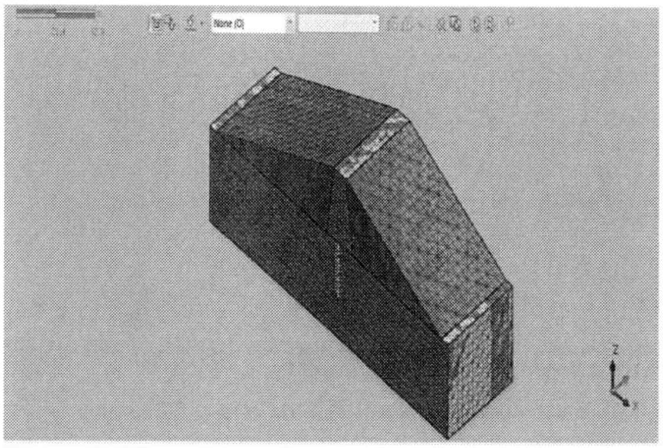

Figure 3. 3D meshing of a zoned earth dam with sheet pile and horizontal drain

Table 1. The properties of different types of soils used in components of earthen dam

Components of earthen dam	E	γ	γ_{sat}	e_0	$K_{x,y,z}$	C	φ
Core	40000	21	22	0.5	1.00E-08	30	35.6
Filter	50000	20	23	0.5	0.0001	0	33
Dam shell	52000	17	22	0.5	1.00E-06	0	33
Drain	50000	19	21	0.5	2.78E-06	0	39
Foundation	2.00E+06	23	26	0.5	1.00E-08	50	39

Where,
E = Elastic modulus, kN/m^2
γ = Unit weight of soil, kN/m^3
γ_{sat} = Saturated unit weight of soil, kN/m^3
e_0 = Initial void ratio
$K_{x,y,z}$ = Coefficient of permeability, m/sec
C = Cohesion, kN/m^2
φ = Angle of internal friction in degree

Table 2. The interface properties between soil and sheet pile

Name	K_n (kN/m^3)	K_t (kN/m^3)	C (kN/m^2)	φ
Interface Material (Wizard)	5.50E+07	5.00E+06	26	36^0

where,
K_n = Coefficient of rigidity (normal elastic constant)
K_t = Coefficient of rigidity (tangential elastic constant)

The boundary condition for steady state seepage is as follows:

a. Constant head of 26m at upstream slope,

b. Constant head at downstream slope has been kept as zero seepage line and

c. Discharge below the base is zero for confined flow.

The boundary condition for transient seepage is as follows

a. The upstream slope is subjected to varying water level with respect to time.
b. Drawdown rate of 2m/day has been considered in the present study, which is shown in Table 3.

Table 3 Drawdown rate with respect to time for five days

Time (Hrs.)	Water level (2m/day) (m)
24	24
48	22
72	20
96	18
120	16

4 Results and Discussion

4.1 Effect of horizontal drain on pore water pressure

It has been observed from Figure 4 that increase in length of the drain reduces the pore water pressure (PWP in kN/m^2) at the downstream side of the dam. But at the upstream side, the pore water change is negligible and it coincides with the values of pore water pressure of earth dam without drain. Figure 5 shows the contours of pore water pressure with horizontal drain.

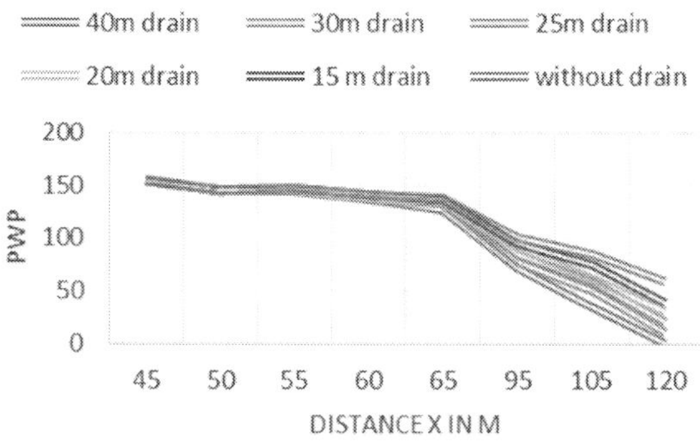

Figure 4. Pore water pressure (kN/m^2) variation in earthen dam with different drain length

DOI: 10.1051/e3sconf/20160919004

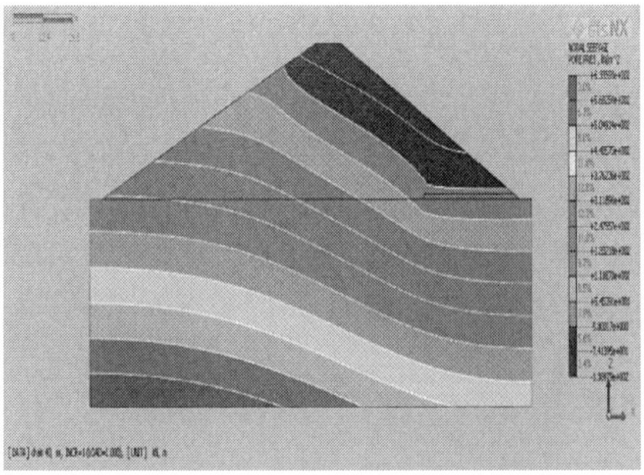

Figure 5. Pore water pressure variation of an earth dam with horizontal drain

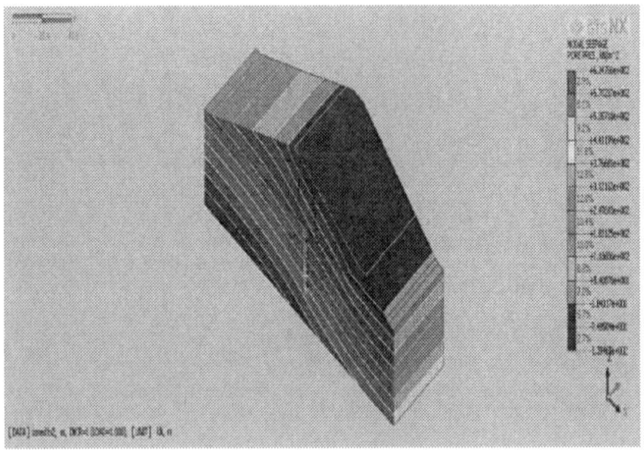

Figure 7. 3D pore water pressure variation of an earth dam with sheet pile

4.2 Effect of sheet pile length on pore water pressure

Effect of sheet pile length at B/8 distance from downstream had been studied. Figure 4 depict that pore water pressure is increasing with the increase in length of sheet pile up to the location of sheet pile whereas at downstream side of sheet pile, the pore water pressures is decreasing with the increase in length of the sheet pile. 3D model with pore water pressure is depicted in Figure 7.

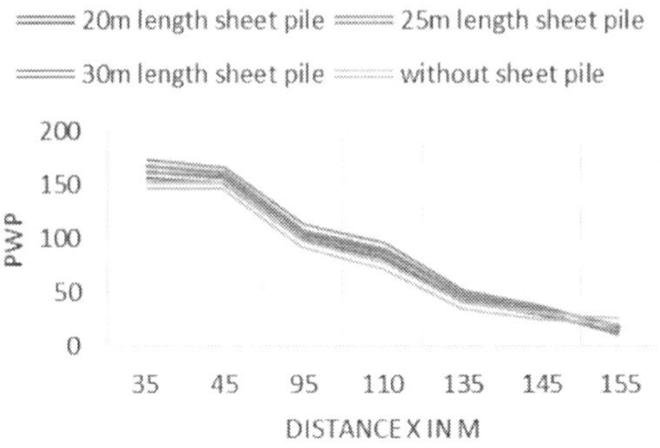

Figure 6. Variation of pore water pressure in an earthen dam for different lengths of sheet pile at a position B/8 distance from d/s

4.3 Effect of sheet pile location on pore water pressure

Now effect of sheet pile location for 30m length of sheet pile as well as without sheet pile has been analyzed. The location of sheet pile shifts from downstream to upstream. It has been observed from Figures 8 and 9 that the pore water pressure increases at the upstream side of the dam. Pore water pressure reduces at the downstream side in comparison to upstream side due to ease of drainage.

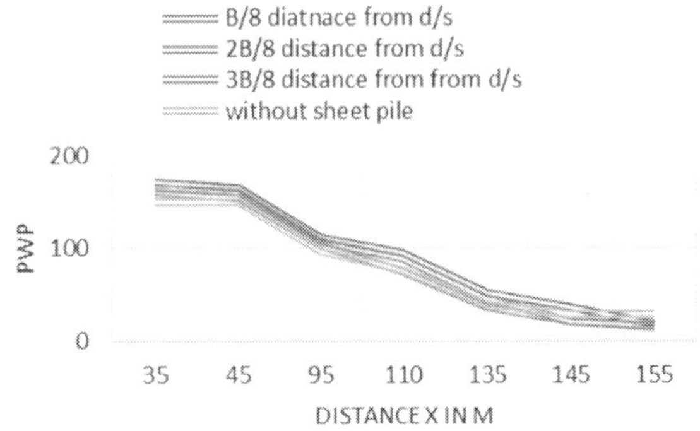

Figure 8. Variation of pore water pressure in earthen dam with different locations in d/s with 30m length sheet pile

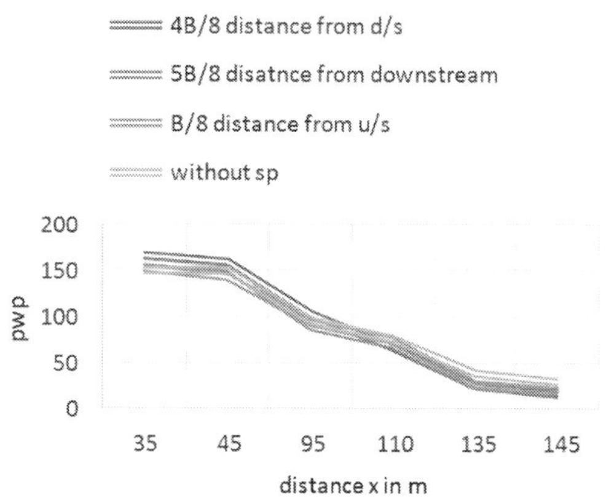

Figure 9. Variation of pore water pressure in earthen dam with different locations in d/s with 30m length sheet pile

4.4 Effect of drawdown rate on pore water pressure

Drawdown rate of 2m/day up to 5days has been considered for analysis. It has also been compare with the steady state seepage analysis as shown in Figure 10. On 5th day the pore water pressure are higher compared to the steady state in a homogeneous earthen dam.

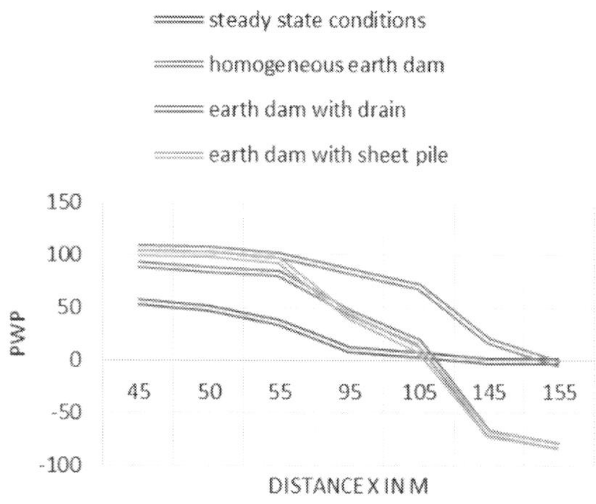

Figure 10. Variation of pore water pressure in an earthen dam under transient conditions (day 5)

The pore water pressure remains high at upstream side whereas pore water pressure gradually decreases at the downstream side. In comparison to sheet pile horizontal drain have relatively less pore water pressure at upstream side. The negative pore water pressure is predominant in the earthen dam with sheet pile and with drain. Though these are negligible in the case of homogeneous earthen dam.

5 Conclusions

The three dimensional analysis to determine the distribution of seepage characteristics in an earth dam is complex. Attempt has been made in this work to obtain these distributions by means of three dimensional finite element simulations. These simulations and output yields some conclusions which are summarized in the following section.

a) The pore water pressure during steady state and transient state had been decreased phenomenally at the upstream face due to the presence of horizontal drain at the downstream side.

b) The pore water pressure values decreased, as the location of sheet pile shifted towards upstream side of the dam.

c) The pore water pressure during drawdown conditions remained higher at the upstream side of the dam and also at the central core.

d) During drawdown conditions, an analysis should be done to decrease the pore water pressure the installation of horizontal blanket drains at upstream side of the dam.

References

1. L. Lam, D. G. Fredlund, and S. L. Barbour. "Transient seepage model for saturated-unsaturated soil systems: a geotechnical engineering approach". Can. Geotech. J. 24. 565-580 (1987)
2. C. W. W. Ng and Shi. "A numerical investigation of stability of unsaturated soil slopes subjected to transient seepage". Computer geotechnics, vol 22, No. 1, pp. 1-28,1998.
3. An-Nan Zhou, Daichao Sheng, Jie Li. " Modelling water retention and volume changes behaviours of unsaturated soils in non-isothermal conditions. Computer and geotechnics 55(2014) 1-13
4. C.Yang, D. Sheng and J.P. Carter. "Hysteretic seepage analysis in unsaturated soil covers". Unsaturated soil theory and practice 2011, ISBN 978-616-7522-77-7.
5. Kenneth Gavin, Jianfeng Xue."A simple method to analsze infilteration into unsaturated soil slopes". Computer and geotechnics 35 (2008) 223-230.
6. Chao Yang, Daichao Sheng, John P. Carter. Effect of hydraulic hysteresis on seepage analysis for unsaturated soil". Computer and geotechnics 41 (2012) 36-56.

Reliability analysis of flood embankments taking into account a stochastic distribution of hydraulic loading

Alessia Amabile[1,a], Manoel P. Cordão-Neto[2], Fabio De Polo[3], and Alessandro Tarantino[1]

[1]*University of Strathclyde, Civil and Environmental Engineering Department, Glasgow, UK*
[2]*Universidade de Brasilia, Brasilia, Brazil*
[3]*Hydraulic Department, Autonomous Province of Bolzano, Bolzano, Italy*

Abstract. Flooding is a worldwide phenomenon. Over the last few decades the world has experienced a rising number of devastating flood events and the trend in such natural disasters is increasing. Furthermore, escalations in both the probability and magnitude of flood hazards are expected as a result of climate change. Flood defence embankments are one of the major flood defence measures and reliability assessment for these structures is therefore a very important process. Routine hydro-mechanical models for the stability of flood embankments are based on the assumptions of steady-state through-flow and zero pore-pressures above the phreatic surface, i.e. negative capillary pressure (suction) is ignored. Despite common belief, these assumptions may not always lead to conservative design. In addition, hydraulic loading is stochastic in nature and flood embankment stability should therefore be assessed in probabilistic terms. This cannot be accommodated by steady-state flow models. The paper presents an approach for reliability analysis of flood embankment taking into account the transient water through-flow. The factor of safety of the embankment is assessed in probabilistic terms based on a stochastic distribution for the hydraulic loading. Two different probabilistic approaches are tested to compare and validate the results.

1 Introduction

Current methods for assessing the stability of flood embankments are mostly deterministic. This means they handle hydro-mechanical loading as if it was precisely known; this is not the case when dealing with natural hazards such as floods, which are stochastic phenomena. The natural variability of soil properties or the scarce amounts of field and laboratory experimental data also add uncertainty to the characterisation of mechanical and hydraulic properties of materials.

Probabilistic approaches enable randomness and uncertainty in the loading and soil hydro-mechanical properties to be taken into account by defining them as random variables described by a probability distribution.

This paper focuses on the uncertainty of the hydraulic loading in the stability analysis of flood embankments. The hydraulic loading is represented by a hydrograph, i.e. the change of river level over time. The uncertainty in the hydraulic loading has been taken into account by:

- identifying a number of key variables to characterise the hydrograph;
- identifying the probability distribution function of each key variable;
- assessing the relative influence of each key variable on the result of the stability analysis.

The output of a deterministic stability analysis is a single value of factor of safety; if this value is greater than unity the embankment is considered to be stable. On the other hand, the output of a probabilistic analysis is the probability distribution of the factor of safety. Its mean and standard deviation can be used to calculate the probability of failure of the embankment, i.e. the probability that the factor of safety is lower than unity.

Two different probabilistic approaches, i.e. FOSM and Monte Carlo method, have been applied to assess the stability of an ideal embankment. An embankment in the North of Italy on the Adige River has been chosen as a reference.

2 Probabilistic methods

2.1 First Order Second Moment method

The First Order Second Moment (FOSM) method is based on a Taylor's series expansion of the performance function $Y=g[X_1, X_2, \dots, X_n]$ around its mean value, where $X_1,\dots,X_n$ are the input random variables. Only first order terms of the series are taken into account and only the first and second moment, i.e. mean and standard deviation σ, are considered, hence the name First Order Second Moment method. If the input variables are not correlated, the mean $E[Y]$ and the variance $\sigma^2[Y]$ of the performance function are given by Eq. (1) and Eq. (2) respectively:

$$E[Y] \cong g(E[X_1], E[X_2], \dots, E[X_n]) \tag{1}$$

[a] Corresponding author: alessia.amabile@strath.ac.uk

$$\sigma^2[Y] \cong \sum_1^n \left(\frac{\partial g}{\partial X_i}\right)^2 \sigma^2[X_i] \qquad (2)$$

If a probability distribution function is assumed for the performance function *Y*, its mean and standard deviation can be used to calculate the probability for any of its values. If input variables are not independent, correlation has to be taken into account for the calculation of the variance $\sigma^2[Y]$. The assumption of the probability distribution function for the performance function is one of the limitations of the FOSM method.

The analytical calculation of the partial derivatives in Eq. (2) may be complex for many geotechnical problems; they can be estimated numerically by computing the performance function at two different points. A common practice [1] is to select two points over a range of plus and minus one standard deviation with respect to the mean value of the random variable X_i, in order to capture the nonlinear behaviour of the function in a range of likely values.

One of the major advantages of FOSM method is that the terms $\left(\frac{\partial g}{\partial X_i}\right)^2 \sigma^2[X_i]$ provide an immediate quantitative assessment of the influence of the variability of each input variable on the variance of the performance function.

2.2 Monte Carlo method

Monte Carlo method is very commonly used to obtain estimates of the solution of complex mathematical problems. It is a useful tool when the physical laws governing a certain process are known, but the problem cannot be solved analytically. This is frequently the case in geotechnical engineering, where closed-form solutions are not available for many common problems [2].

The Monte Carlo method is based on the repeated sampling of random numbers [3]. Each random number is associated to a value of the input variable X_i via its probability distribution. The performance function is evaluated at different points, corresponding to different sets of random values of the input variables X_i. This process is repeated many times in order to cover all the possible range of outcomes. All the resulting values of the performance function are then aggregated to obtain an estimate of its probability distribution, from which the mean and standard deviation can be calculated.

The number of points at which the performance function *Y* needs to be evaluated to have a good estimate of the probability distribution, i.e. the number of simulations to run, depends on the number of input variables. If the number of input variables increases the number of required simulations increases accordingly: the process becomes more and more time-consuming and computationally heavy.

3 Hydrological loading

Hydrological data used in this paper are referred to measurements taken in Bronzolo (BZ) on the Adige River, in the North of Italy, where a hydrometer is located. In this section river level variations have been consistently recorded for the past 30 years, thus allowing for the development of a probabilistic model for the hydraulic loading.

The dataset consists of:

- 82 records of annual maximum river level h_{peak} from 1924 to 2005;
- 30 recorded hydrographs of all the flood events occurred between 1977 and 2014.

3.1 The hydrograph

In common approaches for embankment stability analysis, a steady-state flow is considered and the hydraulic loading is described by a single parameter, i.e. the maximum river level h_{peak}. In transient-state analysis, on the other hand, a time-dependent hydraulic loading has to be considered. This means that the flood event is not defined by the single value h_{peak}, but by a function describing the evolution in time of the river level, i.e. the hydrograph. In this case not only the peak, but also factors like duration or parameters controlling the shape of the hydrograph have to be taken into account.

Between 1977 and 2014, 30 flood events have been recorded by the hydrometer in Bronzolo. From the analysis of the recorded flood hydrographs an attempt has been made to develop a mathematical function able to describe a "typical" hydrograph in the section under consideration. The advantage of this approach is that a function makes the model flexible and easily adaptable to hydrographs recorded at different locations; moreover it can be conveniently reproduced and implemented in any software.

Four key variables have been selected and identified to describe a flood event:

- h_{peak}: maximum river level reached during the flood event;
- h_{start}/h_{peak}: ratio between the initial river level h_{start}, at the beginning of the flood event, and the maximum river level;
- t_{rise}: time elapsed between the beginning of the flood event (h_{start}) and the moment corresponding to the maximum river level (h_{peak});
- α: shape parameter controlling the falling limb of the hydrograph.

A representation of the hydrograph and the key variables is shown in Fig. (1).

The rising limb of the hydrograph, before the peak, can be well described by a linear function; the parameters governing this part are t_{rise}, h_{start} and h_{peak}, as shown in Eq. (3). For the falling limb of the hydrograph ($t>t_{rise}$) among different functions the power decay has shown the best fitting to the recorded hydrographs. The shape of the decreasing part of the hydrographs therefore depends not only on the shape parameter α but also on t_{rise}, as shown in Eq. (4).

$$h = h_{start} + \frac{h_{peak}-h_{start}}{t_{rise}} t \quad (t \le t_{rise}) \qquad (3)$$

$$h = h_{peak} * \left(\frac{t}{t_{rise}}\right)^{(-\alpha)} \quad (t > t_{rise}) \qquad (4)$$

 DOI: 10.1051/e3sconf/20160919005

For each recorded hydrograph the best-fitting values of t_{rise}, h_{start} and α have been found by minimising the least square error between the recorded hydrograph and the one described by functions in Eq. (3) and Eq. (4).

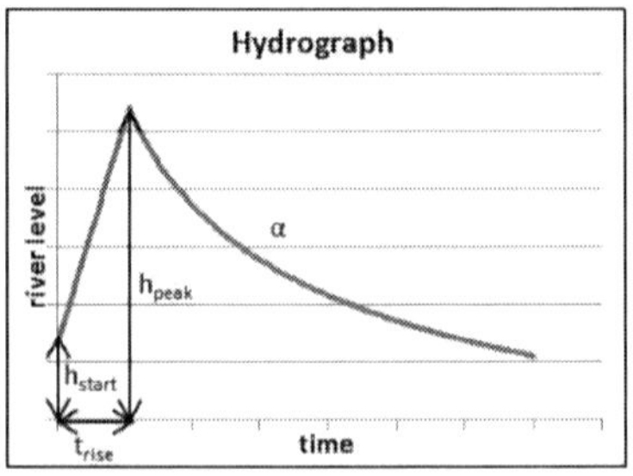

Figure 1. Representation of the hydrograph.

3.2 Probabilistic modelling of floods

After identifying a function to define the hydrograph the probabilistic model for the flood has been developed by identifying the probability distribution of each key variable, under the assumption that they are independent.

The probabilistic distribution of the peak river level of the hydrograph h_{peak} has been assessed in the conventional way, i.e. deriving the cumulative probability distribution function from the annual maximum series. The empirical cumulative distribution function has been fitted with a Gumbel distribution, which is commonly employed in flood frequency analysis.

The empirical cumulative distribution functions for all the other parameters have been derived from the analysis of past flood events, not necessarily associated with the maximum annual river level. The empirical distribution functions have been fitted with a normal distribution. Results are reported in Fig. (2) for h_{peak} and Table (1) for h_{start}/h_{peak}, t_{rise} and α.

Table 1. Mean and standard deviation of normal probability distribution function for the variables h_{start}/h_{peak}, t_{rise} and α.

Variable	μ-Normal	σ-Normal
h_{start}/h_{peak}	0.613	0.094
t_{rise} [d]	1.077	0.371
α	0.278	0.095

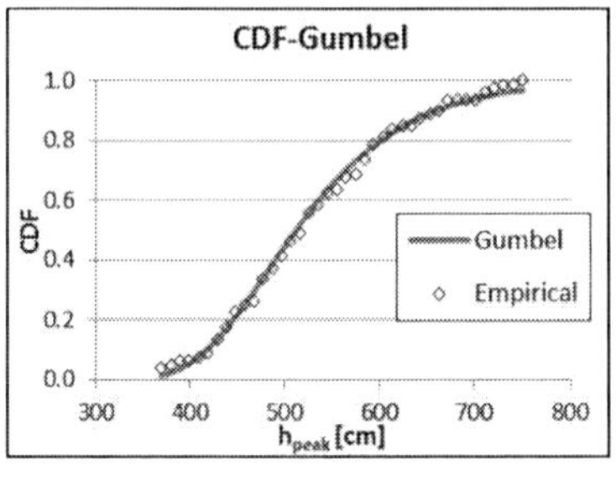

Figure 2. Empirical and Gumbel cumulative probability distribution function for the variable h_{peak}.

4 The hydro-mechanical model

4.1 Geometry and materials

An embankment in the North of Italy on the Adige River has been chosen as a reference. The embankment is about 8m high from the riverbed and about 5.5 m high from the ground surface on the landside. The simplified geometry and soil profile of the cross-section is shown in Fig. (3).

The original soil profile has been simplified into two homogeneous layers, the first one F including the foundation and the second one E including the embankment and the shallow layer of agricultural soil on the landside.

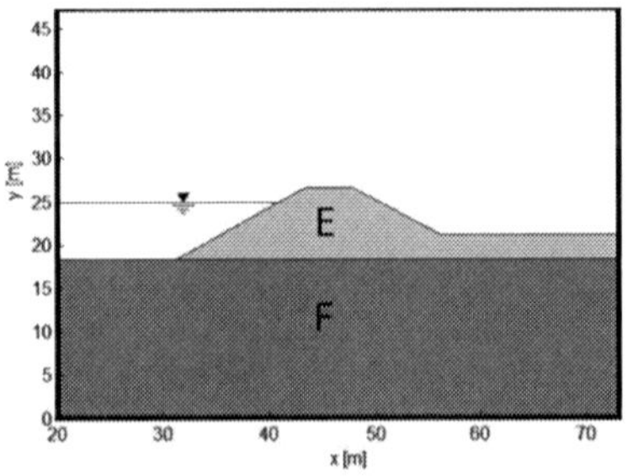

Figure 3. Geometry and soil profile of the embankment.

Van Genuchten model has been used to characterise the hydraulic behaviour of materials under unsaturated condition [4]. The equations for the effective saturation, the volumetric water content, and the relative hydraulic conductivity are defined in Eq. (5), Eq. (6) and Eq. (7) respectively:

$$S_e = \frac{1}{(1+(\alpha_{VG} s)^n)^m} \tag{5}$$

$$\theta = \theta_{res} + (\theta_{sat} - \theta_{res}) S_e \tag{6}$$

$$k_{rel} = S_e^{-\lambda}\left(1-\left(1-S_e^{\frac{1}{m}}\right)^m\right)^2 \tag{7}$$

where

$$\mathrm{m} = 1 - \frac{1}{\mathrm{n}} \tag{8}$$

s is the suction and θ_{res} and θ_{sat} are the residual volumetric water content and the saturated volumetric water content respectively.

Plots of water retention and relative hydraulic conductivity curves are shown in Fig. (4). The material in the foundation layer is always saturated; for this reason the value of the coefficient λ has been chosen in order to have a drop of just one order of magnitude in the hydraulic conductivity. A small drop in the hydraulic conductivity makes the computation numerically easier without affecting the accuracy of results since this range of suction is never explored in the simulation.

For each material, the values of θ_{res} and θ_{sat}, α, n and λ are reported in Table (2), together with the mechanical properties and the values of saturated hydraulic conductivity.

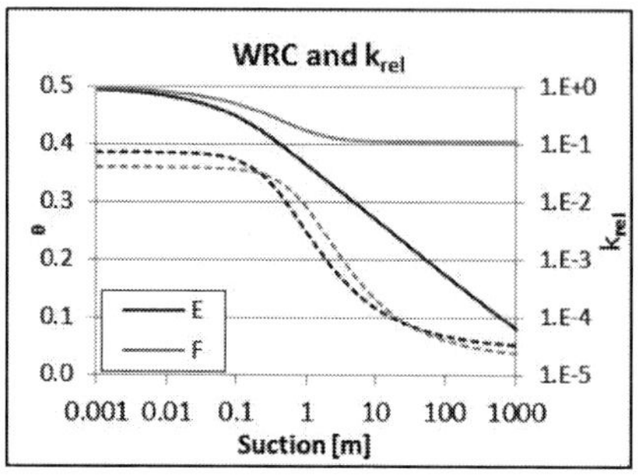

Figure 4. Water retention curve (continuous lines) and relative hydraulic conductivity curve (dashed lines) for the foundation layer (F) and the embankment/shallow layer on landside (E).

Table 2. Material properties for foundation layer (F) and embankment and shallow layer on landside (E).

Material properties		F	E
Friction angle	φ' [°]	33	30
Cohesion	c' [kPa]	0	0
Unit weight	γ [kN/m³]	20	20
Sat. hydraulic conductivity	k_H [m/s]	0.001	0.0005
Res. water content	θ_{res}	0.029	0.045
Sat. water content	θ_{sat}	0.360	0.386
Parameters for Van Genuchten hydraulic functions	α_{VG} [m⁻¹]	1.0	2.5
	n_{VG}	1.50	1.48
	λ_{VG}	6	4.2

4.2 The seepage analysis

A transient-state seepage analysis has been performed in order to evaluate the distribution of pore water pressure in the embankment during the flood event. The boundary condition on the riverside is given by the time-dependent hydraulic head represented by the hydrograph; the boundary condition on the slope of the embankment and the ground surface is a seepage face.

The initial condition in the transient-state seepage analysis has been obtained from a steady-state analysis associated with a river level equal to h_{start}, i.e. the initial river level for the flood event. For the steady-state analysis on the far end vertical boundary on the landside, 80 m away from the toe of the embankment, the ground water table has been assumed to be located at the interface between the two layers.

4.3 The stability analysis

For the stability analysis Bishop's simplified method has been used. This is one of the so called methods of slices, which discretise the soil mass by dividing it in slices and are commonly used when analysing stability of slopes in 2D. The soil is assumed to have rigid-perfectly plastic behaviour, i.e. that the soil does not show any deformation before failure; for this reason the solution of the problem is based only on equilibrium equations.

Bishop's simplified method is based on the assumptions of circular slip surface and zero inter-slice shear forces; it satisfies vertical force equilibrium for each slice and overall moment equilibrium about the centre of the circular slip surface. In order to solve the problem an iterative procedure is required.

5 Application and results

5.1 FOSM

In the application of the FOSM method to the stability analysis of the embankment taking into account the uncertainty in the hydrological loading, the input variable $X_1,\dots,X_n$ are represented by the four key variables identified to describe the hydrograph (h_{peak}, t_{rise}, h_{start}/h_{peak} and α) and the performance function $Y=g[X_1, X_2,\dots, X_n]$ is represented by the factor of safety FS.

The mean and standard deviation of the factor of safety can be calculated as:

$$E[FS] \cong g\left(E\left[h_{peak}\right], E\left[h_{2start}\right], E\left[t_{rise}\right], E[\alpha]\right) \quad (9)$$

$$\sigma^2[FS] \cong \left(\frac{\partial g}{\partial h_{peak}}\right)^2 \sigma^2\left[h_{peak}\right] + \left(\frac{\partial g}{\partial h_{start}}\right)^2 \sigma^2\left[h_{start}\right] + \left(\frac{\partial g}{\partial t_{rise}}\right)^2 \sigma^2\left[t_{rise}\right] + \left(\frac{\partial g}{\partial \alpha}\right)^2 \sigma^2[\alpha] \quad (10)$$

The number of simulations required to apply the FOSM method is $1+2n$, where n is the number of input variables. In order to calculate the mean value of the factor of safety, the mean value of each input variable is considered (*1* simulation). In order to calculate the variance of the factor of safety, for each input variables the factor of safety is calculated in two points, correspondent to plus and minus one standard deviation with respect to the mean value (*2n* simulations). By keeping all variables except one equal to their mean values nine hydrographs have been produced by combining these values, as shown in Table (3) and Fig. (5).

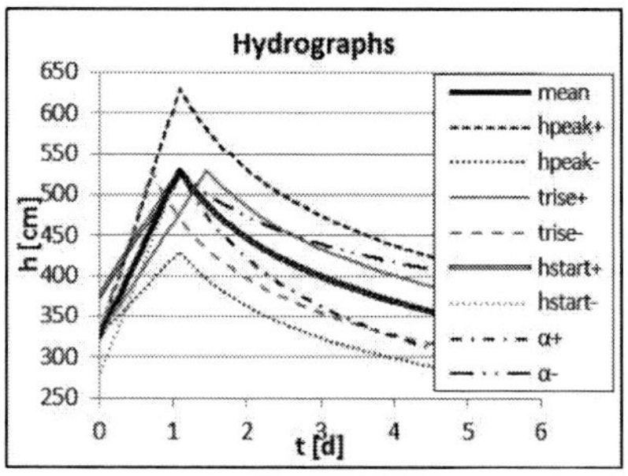

Figure 5. Hydrographs obtained from combinations of values of input variables for the application of FOSM method.

A normal probability distribution function is assumed for the factor of safety; the need to assume the distribution function for the performance function is one of the limitations of FOSM method. Another limitation in the application to stability analysis is the need to take into account a single slip surface. The critical slip surface

DOI: 10.1051/e3sconf/20160919005

obtained at the peak from the analysis with the mean values of all the input variables is considered.

Seepage analyses have been performed with the software SEEP/W and stability analyses have been performed with the software SLOPE/W. The mean value and standard deviation of the factor of safety have been calculated at different times and results are shown in Fig. (6). The minimum value of the mean of the factor of safety corresponds to the peak of the hydrograph.

Table 3. Combinations of values of input variables for the application of FOSM method.

	h_{peak} [cm]	t_{rise} [d]	h_{start} [cm]	α
mean	530	1.083	325	0.278
h_{peak}+	*630*	1.083	325	0.278
h_{peak}-	*430*	1.083	325	0.278
t_{rise}+	530	*1.458*	325	0.278
t_{rise}-	530	*0.708*	325	0.278
h_{start}+	530	1.083	*375*	0.278
h_{start}-	530	1.083	*275*	0.278
α+	530	1.083	325	*0.373*
α-	530	1.083	325	*0.184*

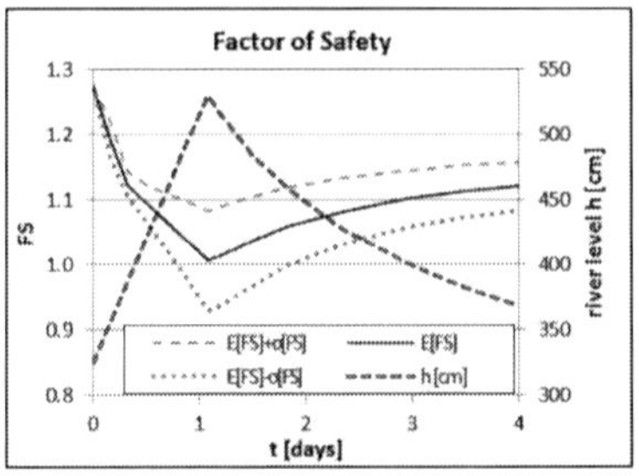

Figure 6. Factor of safety versus time and hydrograph.

The relative contribution of each variable to the variance of the factor of safety can be evaluated as:

$$w[X_i] = \frac{\left(\frac{\partial g}{\partial X_i}\right)^2 \sigma^2[X_i]}{\sigma^2[Y]} \quad (11)$$

Results are shown in Fig. (7). The sum of all the terms $w[X_i]$ is equal to 1 and the contribution of each term is represented by its area. For every instant, the contribution of each term $w[X_i]$ is represented by the distance between the corresponding area's top and bottom boundaries. As expected, the contribution of h_{start} is 100% at the initial time and becomes zero after the peak; on the other hand, α has a contribution, although not very relevant, only after the peak. At the peak (*1.083 days*), when the mean of the factor of safety reaches its minimum, more than 80% of the contribution to the variance of the factor of safety is given by the input variable h_{peak}. For this reason when applying the Monte Carlo method only the uncertainty in h_{peak} is taken into account.

5.2 Monte Carlo

For the application of Monte Carlo method to stability analysis, the values of the input variables are randomly sampled from their cumulative probability distribution function in order to generate random hydrographs. Every random hydrograph thus generated represents the boundary condition for one seepage analysis; the distribution of pore water pressure derived from each seepage analysis is used to perform one stability analysis and calculate one value of the factor of safety. By repeating this process a number of times, a large number of values of the factor of safety are obtained and therefore its empirical cumulative distribution function can be derived. The empirical distribution function is fitted with a normal distribution in order to compare results with the ones obtained with FOSM method.

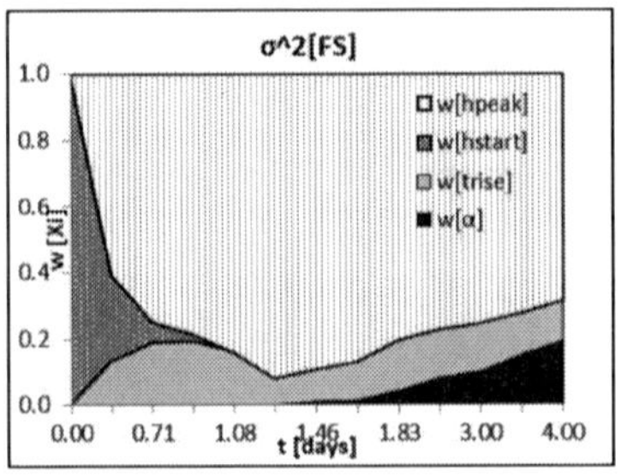

Figure 7. Relative contribution of each random variable to the variance of the Factor of Safety with respect to time.

The first step in the application of Monte Carlo method is therefore the generation of a set of random numbers between 0 and 1; to every random number a value of the input variable h_{peak} can be associated via its cumulative probability distribution function shown in Fig. (2). The variability of h_{start}, t_{rise} and α is neglected and these variables are kept constant and equal to their mean values for the definition of all the hydrographs, as explained in section 5.1.

The seepage analyses have been run with the software Comsol Multiphysics (*Subsurface flow and porous media package*). In each seepage analysis the distribution of pore water pressure on the critical slip surface has been evaluated and used as input for the stability analysis.

A routine in *Matlab* has been written to perform the following tasks:

1. Generate one random number r_i between 0 and 1;
2. Associate r_i to a value of h_{peak} through its Gumbel cumulative probability distribution function;
3. Generate the hydrograph and apply it as boundary condition for the seepage analysis in Comsol Multiphysics;
4. Run seepage analysis in Comsol Multiphysics;
5. Extract the distribution of pore water pressure on the critical slip surface;
6. Assess the factor of safety with Bishop's simplified method;
7. Repeat steps from 1 to 6 for 400 times.

The resulting 400 values of factor of safety at the peak are plotted in Fig. (8), while the evolution in time of

DOI: 10.1051/e3sconf/20160919005

its mean value and standard deviation is plotted in Fig. (9). Comparison with FOSM results show a very good match for both plots.

The empirical cumulative probability distribution function of the factor of safety is plotted in Fig. (10). It has been fitted with a normal distribution with maximum likelihood estimation; the mean and standard deviation are reported in Table (4). They match satisfactorily the results obtained with FOSM.

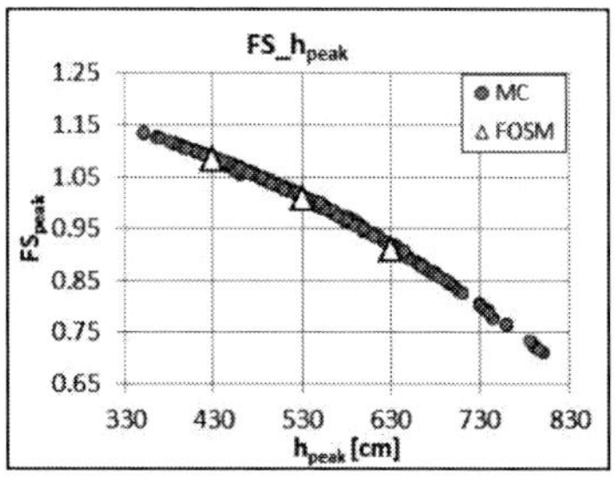

Figure 8. Factor of Safety versus river level at peak.

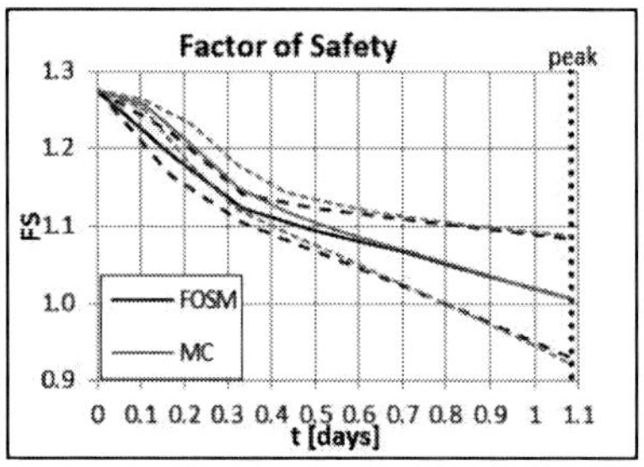

Figure 9. Factor of safety versus time obtained with FOSM and Monte Carlo methods.

Table 4. Mean and standard deviation of the Factor of Safety estimated with FOSM and Monte Carlo methods.

FOSM	E[FS]	1.007
	σ[FS]	0.077
MC	E[FS]	1.006
	σ[FS]	0.083

6 Conclusions

For the stability analysis of flood embankments it is important to take into account uncertainties in the hydraulic loading, which is represented by natural phenomena such as flood events. Results in this paper show that the variability of the maximum river level h_{peak} has a significant effect on the variance of the factor of safety of the embankment and it cannot be neglected.

Probabilistic methods are a suitable approach in this case. FOSM method offers a very simple method; it has some limitations because an assumption has to be made about the probability distribution function of the factor of safety. Moreover it is not a suitable tool when the performance function has a highly non-linear behaviour.

Comparison with results obtained with Monte Carlo method show that the assumption of a normal distribution for the factor of safety is appropriate and that the factor of safety varies almost linearly with the maximum river level h_{peak}. The match between mean and standard deviation of the factor of safety obtained with FOSM and Monte Carlo method is very satisfactory. The number of simulations required by FOSM to obtain the same accuracy is more than 40 times less than the one required by Monte Carlo method, that can become quite heavy and time-consuming when more than one input random variable needs to be taken into account. Results show that FOSM method is a suitable and affordable approach in terms of time and number of simulations required to achieve sufficiently accurate results.

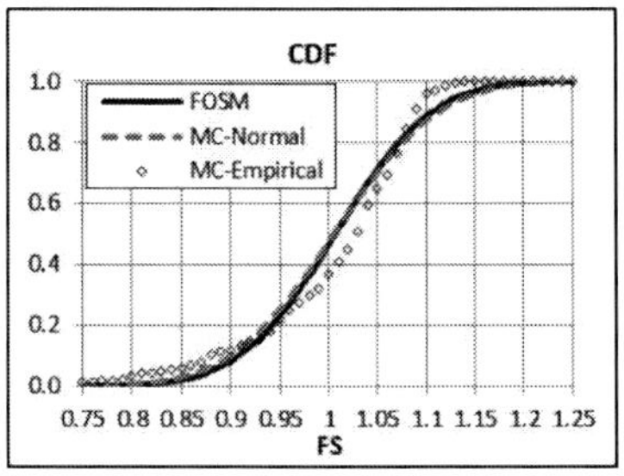

Figure 10. Cumulative distribution function of the Factor of Safety from FOSM and Monte Carlo methods.

Acknowledgements

The authors wish to acknowledge the support of the European Commission via the Marie Curie IRSES project GREAT 'Geotechnical and geological Responses to climate change: Exchanging Approaches and Technologies on a world-wide scale' (FP7-PEOPLE-2013-IRSES-612665). The authors also wish to acknowledge the support of the Autonomous Province of Bolzano, Italy.

References

1. I. F. Otálvaro, M. P. Cordão-Neto, *Probabilistic analyses of slope stability under infiltration conditions. Advances in Unsaturated Soils* (Caicedo et al. (eds), Taylor & Francis Group, 2013).
2. G. A. Fenton, *Simulation. ALERT Doctoral School 2014 – Stochastic analysis and inverse modelling* (M. A Hicks, C. Jommi (eds), 2014).
3. M. Marseguerra, E. Zio, *Basics of the Monte Carlo Method with Application to System Reliability* (LiLoLe-Verlag, 2002).
4. M. Th. Van Genuchten, *A closed-form equation for predicting the hydraulic conductivity of unsaturated soils,* Soil Science Society American Journal 44(5): 892-898 (1980).

Conceptual considerations of evaluate internal erosion phenomenon via no-erosion filter test and continuing erosion filter test

Johnatan Ramos-Rivera[1], Oscar Echeverri-Ramirez[2]

[1]*Geotechnical Engineer, National University of Colombia, Geotechnical Research Group, School of Mines*
[2]*Associate Professor, National University of Colombia, Geotechnical Research Group, School of Mines*

Abstract. Some widely-graded soils may exhibit, under the influence of steady seepage flow, a behaviour in which grains of the finer fraction migrate through the interstices of the matrix formed by the coarser fraction. The migrating fines may accumulate at a downstream location within the soil. Alternatively, and where there is no capacity for retention at the downstream or exit boundary, the behaviour may lead to a washing out and consequent loss of the finer fraction. The phenomenon of erosion is termed internal instability, and the soils are considered internally unstable. Taking into consideration (i) the specimen reconstitution by method of compaction, (ii) the application of a vertical stress to the specimen, and (iii) the use of multi-stage seepage flow with head-control, to measure the origin of a conduit through the coarser fraction, some test devices were conducted by different authors to evaluate this phenomenon, the purpose of this paper is to present some considerations and key aspects about internal erosion in dams and filter compatibility with core material (specimen reconstitution, test procedure, consolidation, seepage flow, test program and its relevance to the reality). The main reason to present this investigation is due to the absence of any specified regulatory or standard test method. Given the importance of filter compatibility of the zoned earth core dam and filter materials, as well the grading stability of each zone in the presence of seepage flow, additional consideration will be given to performing Continuing Erosion Filter (CEF) tests on the core-filter interface, using the laboratory permeameter device.

1 General framework of internal erosion

Soil erosion is the cause of failure of the majority of dikes and dams whether through internal erosion, wave overtopping or overflow. However, for internal erosion, the definitions of the ICOLD European Working Group on Internal Erosion of Embankment Dams will be followed, starting that internal erosion occurs when soil particles within an embankment dam or its foundation, are carried downstream by seepage flow. Internal erosion can initiate by concentrated leak erosion, backward erosion, suffusion and soil contact erosion.

1.1 Context, overview and general description

The good performance of embankment dams with filters designed in accordance with modern design criteria have proven that these filters are capable of reliably sealing concentrated leaks without significant erosion (Sherard and Dunnigan 1989, Peek 1990). However, many existing dams have filters that do not satisfy these criteria, being too coarse by design or having segregated during construction.

Modern design criteria are based on laboratory tests that simulate the seepage forces acting on the sample of the core and filter material. One of the most recently adapted criteria is via no-erosion filter (NEF) test, which allows soil erosion detachment and migration under a static or multi-stage hydraulic gradient. A variation of NEF Test is a called CEF (Continue erosion) test.

We can distinguish four successive phases of internal erosion: (a) initiation when one of the phenomena of detachment of particles occurs, (b) continuation when erosion process can be (or not) stopped by filtering, (c) progression when internal erosion comes to a pipe through the structure or increase pore pressure in the downstream part, (d) failure and breach (N. Benahmed, 2012) resulting in uncontrolled release of water in the plain (Figure 1). Ultimately, potential scenarios of failure can be constructed by combination of the basic mechanisms of internal erosion together with others elementary processes (settlement, uplift, sinkhole, clogging…).

Depending on the ratio of particle and pore sizes, the erosion will either:

[a] Corresponding author: jramosr@unal.edu.co

- Not continue (i.e., no erosion); or
- Stop after only minor erosion (i.e., some erosion); or
- Stop only after a significant amount of erosion (i.e., excessive erosion); or
- Continue (i.e., continuing erosion)

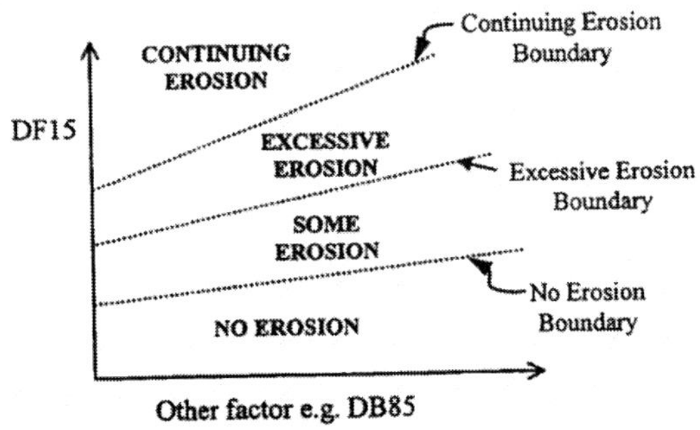

Figure 1 Conceptual filter erosion boundaries (Foster, 1999), Foster & Fell (2001)

The proposals from national and international agencies and laboratory tests, completed or in progress, e.g the ICOLD bulletin on internal erosion (International Commission On Large Dams). Before these studies, specific recommendations about internal erosion risk were extremely limited, only the use of filters is recommended, without any quantitative criteria or description (or even definition) of the different mechanisms. The aim here is to unify these recent approaches in the same general framework to present a summary review of internal erosion for levee managers, engineers and technicians working on the safety of hydraulic structures and (urban) flood defenses.

1.2 Empirical methods to evaluate internal erosion

Methods used to assess the material susceptibility of a soil gradation to seepage-induced internal instability are empirical in origin. Taken collectively, the state-of-practice and the state-of-art comprise a total of six methods (Kezdi, 1979; Sherard, 1979; Kenney and Lau, 1985, 1986; Burenkova, 1993; Li and Fannin, 2008; Wan and Fell, 2008). The Kenney and Lau (CDA, 2007) and Sherard (USBR, 2011) methods are considered representative of the state-of-practice; the other four methods represent the state-of-art.

The six methods may be clustered in three groups, based on similarity of the underlying concepts: (i) Kezdi (1979) and Sherard (1979); (ii) Kenney and Lau (1985, 1986), and its adaptation by Li and Fannin (2008); and (iii) Burenkova (1993), and its adaptation by Wan and Fell (2008).

The method of Sherard (1979) was proposed to evaluate the potential for instability in glacial tills that are commonly used as core material in a zoned embankment dam. It does not appear to have been developed from the findings of laboratory testing of soil. Rather, the approach entails splitting the gradation into two fractions at any delimiting grain size value "exceeding about 0.1 or 0.2 millimeters" (no maximum value was given), and evaluating the grain size distribution of the coarser and finer fractions with respect to the Terzaghi (1939) empirical filter retention criterion, expressed as $(D15'/d85')\max \leq 4$ to 5.

The method of Kezdi (1979) similarly invokes a split-gradation analysis, with a more conservative limiting criterion of $(D15'/d85')\ \max \leq 4$. In contrast to the Sherard method, no recommendation was made on the value of delimiting grain size. Neither was an explicit recommendation provided on the soil types to which the method should be applied, although it may inferred from example applications in the original publication that it is suitable for sands and clayey silts (see Kezdi, 1979,).

The method of Kenney and Lau (1985) was proposed to evaluate the potential for instability in coarse-grained soil. It was conceived from interpretation of laboratory permeameter tests on gravel and sand. The method requires calculation of a stability index H/F along a prescribed length of the gradation curve. For each grain size D, and corresponding percentage of mass passing F, a value H is defined as the percentage of mass passing between D and 4D, which establishes the H/F ratio. The method is applied along the finer end of the gradation curve to maximum value of F = 20 % for soil with a coarse fraction that is widely-graded (WG), and F = 30% for soil with a coarse fraction that is narrowly graded (NG). Kenney and Lau (1986) suggest a limit value to stability of $(H/F)\ \min \geq 1$.

The method of Burenkova (1993) was proposed to evaluate the potential for instability in soil, from interpretation of laboratory permeameter tests on gravel and sand with unreported fines content. The method requires calculation of two shape parameters, d90/d15 and d90/d60, which are compared to an upper and lower boundary that enclose a region within which internally stable gradations are found to plot. The lower boundary is defined by $d90/d60 = 0.76 \log (d90/d15) + 1$; the upper boundary is defined by $d90/d60 = 1.86 \log (d90/d15) + 1$.

The method of Li and Fannin (2008) was proposed in order to address some of the apparent conservatism of the Kenney and Lau method, by combining the limit value of Kenney and Lau with that of Kezdi to yield a hybrid threshold to material susceptibility. It was developed with reference to a database of laboratory permeameter tests on gravel and sand. A gradation with $(H/F)\ \min < 1$ for $F \leq 15\%$, or $H < 15\ \%$ for $15\ \% \leq F$, is designated potentially unstable.

The method of Wan and Fell (2008) was proposed from interpretation of laboratory permeameter tests on gravel-sand-silt-clay mixtures, with a fines content of up to 45 % that exhibited zero to low plasticity. It is explicitly declared that the method "is not able to identify the internal instability of gap-graded soils". The method similarly requires calculation of two shape parameters, d20/d5 and d90/d60, which are compared to boundaries that enclose regions within which stable, unstable and "transition" gradations are found to plot.

2 Multi-Scale test to assess internal erosion

2.1 No-erosion filter test

The No Erosion Filter (NEF) test was proposed by Sherard & Dunnigan (1985), for testing cohesive base soils. The typical apparatus is shown in Figure 2. In this test a sample of base soil is compacted on top of the filter. A pinhole is pushed through the base soil to simulate a concentrated leak through a crack. Water is forced through the pinhole under high pressure, which is sufficient to erode the base soil. The test is considered successful if no visible erosion occurs before the filter interface seals, and unsuccessful if some erosion is necessary to reach steady filtration (Delgado, 2000).

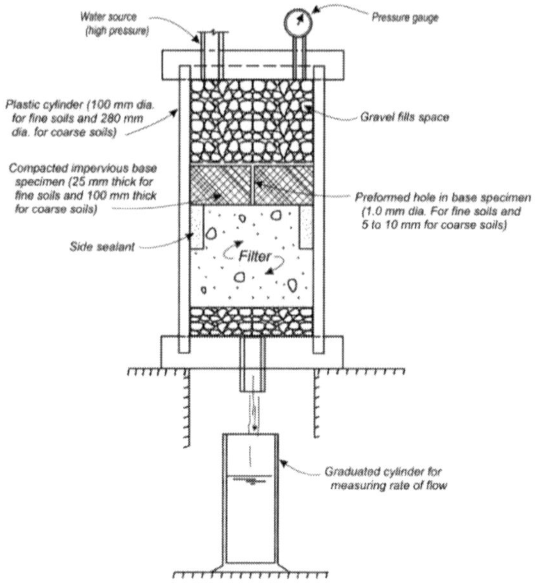

Figure 2 Typical NEF Apparatus (Sherard and Dunnigan (1985)

As Kenney and Lau (1985) observe, "materials that exhibit unstable gradings show their potential for instability, but whether or not they would behave as unstable materials in practice would depend on the conditions of particle transport to which they were subjected". Laboratory permeameter tests provide a means to establish whether the onset of instability may be triggered by seepage flow.

Permeameter testing is most commonly undertaken using a rigid-wall device, which simplify the process of specimen reconstitution, instrumentation, operation and maintenance. The design details of each permeameter test device reported in the literature vary with the objectives of the test program, and it is reasonable to make a distinction between (i) project-specific permeameter tests in support of materials characterization for engineering design, and (ii) research-specific testing in support of fundamental advances in engineering science.

2.2 Continue-erosion filter test

Foster & Fell (1999b, 2001) developed a method for assessing filters in existing dams that do not satisfy modern design criteria. This involves assessing the particle size distribution of the filters compared to the core that filter is protecting, to determine whether the filters are sufficiently fine to satisfy no excessive or continuing erosion criteria (See Figure 3). The method is based on the analysis of laboratory tests and the characteristics of dam that have experienced piping, clogging and sinkholes incidents.

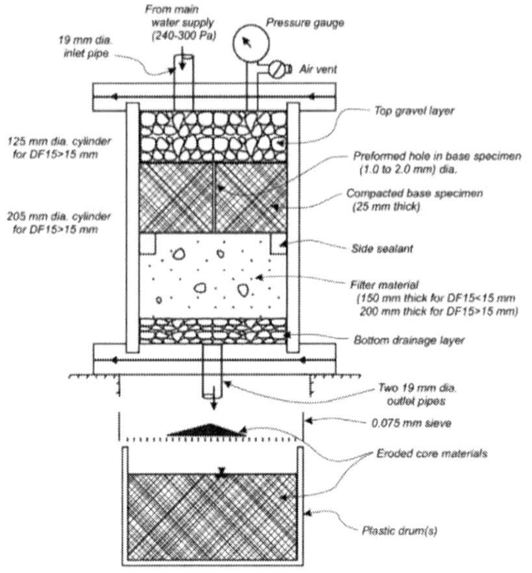

Figure 3 CEF Test Configuration (Foster and Fell 1999, 2001)

Filter zones and adjacent materials which are coarser than required by modern design methods based on particle size will often be quite effective in controlling internal erosion (Foster and Fell 1999, 2001). Downstream rockfill and sand/gravel zones which were not designed as filters may provide some protection against continuation of internal erosion.

2.2 General attributes of permeameter test devices

The internal diameter (D) of rigid-wall devices has varied to a maximum diameter of 275 mm or greater (see for example, Kenney and Lau, 1985; Moffat, 2005; Li,

2008; Wan and Fell, 2008; Ahlinhan and Achmus, 2010; and, Sail et al., 2011). A ratio D/D100 ≤ 8 to 12 is typically respected in matching the grain size distribution of a test specimen to permeameter device, to achieve general accordance with ASTM D-2434. Typically, the specimen is reconstituted to length (L) that yields a ratio L/D ≈ 1, although testing has been conducted in the approximate range 0.5 ≤ L/D ≤ 2.

A rigid-wall cell of transparent material permits inspection of the specimen during the permeameter test, and such visual observations provide valuable insight to the response of the specimen.

Two methods of specimen reconstitution are used in testing: (i) compaction of the soil in a moist (unsaturated, hence three-phase) condition, after which effort may be made to fully saturate the soil; (ii) deposition of the soil in a saturated (two-phase) condition.

Axial loading is controlled in some devices, however a stress inhomogeneity may arise at large axial loads due to the effects of side-wall friction in the permeameter cell, especially in specimens with greater L/D ratio.

Seepage flow is imposed in a vertical direction, typically in a multi-stage sequence of increasing hydraulic gradient by means of head-control. Downward seepage flow may facilitate easy collection of eroded soil, however integrity of the specimen may be compromised by loss of soil through the basal (outflow) boundary (a wire mesh-screen, else granular layer) during reconstitution of the specimen. In contrast, upward seepage flow compromises the collection of eroded soil, but allows for selection of a basal (inflow) boundary through which little or no soil loss occurs during specimen reconstitution

2.3 Detailed observations on the configuration of the permeameter test device

As noted above, the adverse influence of sidewall friction in a rigid-wall permeameter is greater with increasing length/diameter (L/D) ratio of the test specimen. It is recommended that testing be performed at a maximum ratio of L/D ≈ 1.

Downward seepage flow may facilitate easy collection of eroded soil from the specimen, and best-approximate flow conditions in the field; however the integrity of the specimen may be compromised by loss of soil particles through the basal (outflow) boundary (an assembly of mesh-screens) during reconstitution of the specimen. In contrast, although upward seepage flow compromises the collection of soil particles eroded from the specimen, it enables the use of a basal (inflow) boundary of relatively small openings through which little or no soil loss need occur during specimen reconstitution. Careful consideration should be given to these factors when selecting the flow direction for permeameter testing: the final decision should be informed by experienced gained in the reconstitution of materials from the Ituango site, and with upward seepage flow being taken as the default (conservative) direction.

It is important, in any systematic study of soil behaviour, to ensure reproducibility of the test specimen throughout the program of testing. It is important, when reconstituting test specimens by means of compaction, that a routine be established in which the blowing effort per unit volume of soil is identical in all specimens. The decision to test specimens in a relatively loose or, in contrast a relatively dense state, should be informed by project-considerations.

While axial loading may be applied to the top surface of many test specimens in the program of laboratory testing, it may prove instructive, should time permit, to test a specimen under the conditions of zero top stress and upward seepage flow, in a manner that reproduces the classic method of Skempton and Brogan (1994) that has also been used by Li (2008).

A multi-stage seepage flow test on a relatively permeable soil can, at a relatively large hydraulic gradient, result in significant volumetric flow through the test specimen. Careful consideration should be given to the sizing of the inflow and outflow ports on the permeameter, to ensure that they do not impose an undesirable restriction on flow rate through the test device.

The program of laboratory testing involves multi-stage permeameter (MSP) tests on the zoned core material, and on the zoned filter material, of the earth-fill dam. The MSP tests are intended to investigate the potential for seepage-induced internal instability in these two materials.

3 Conclusions and remarks

It is not recommended to use the method of Kenney and Lau, or its adaptation by Li and Fannin in core materials, because the respective databases of these two empirical methods do not include any gradations that are comparable with the core Dam material: the two methods are proposed for assessment of gravel and sand mixtures only.

Materials that are susceptible to seepage-induced internal instability are generally recognized as also being susceptible to the action of segregation during excavation, transport and placement in construction. The likely influence of any segregation should therefore be considered in the selection of material gradations for assessment using these methods.

References

1. Ahlinhan, M.F. and Achmus, M. (2010). Experimental investigation of critical hydraulic gradients for unstable soils. International

Conference on Scour and Erosion (ICSE-5), San Francisco, USA, 7-10 Nov. 2010: 599-608.
2. ASTM D-2434 Standard test method for permeability of granular soils (constant head), in the Annual Book of ASTM Standards, Vol.04.09, ASTM Philadelphia.
3. Burenkova, V. V. (1993). Assessment of suffosion in noncohesive and graded soils. 1st International Conference on Filters in Geotechnical and Hydraulic Engineering: 357–360.
4. CDA (2007). Canadian Dam Association: Dam Safety Guidelines.
5. Garner, S.J. and Fannin, R.J. (2010). Understanding internal erosion: a decade of research following a sinkhole incident. International Journal of Hydropower and Dams, 17(3): 93-98.
6. ICOLD (2013). 22 January 2013 pre-print of the International Commission on Large Dams Bulletin 164: Internal erosion of existing dams, levees and dikes and their foundations, Vol. 1: Internal erosion processes and engineering assessment, 151p.
7. Kenney, T.C. and Lau, D. (1985). Internal stability of granular filters. Canadian Geotechnical Journal, 22: 215-225.
8. Kenney T.C. and Lau, D. (1986). Internal stability of granular filters: Reply. Canadian Geotechnical Journal, 23: 420-423.
9. Kezdi, A. (1979). Soil physics – selected topics. Elsevier Scientific Publishing Company, Amsterdam, 160p.
10. Li, M. 2008. Seepage induced instability in widely graded soils. Ph.D. thesis, The University of British Columbia, Vancouver, Canada, 300p.
11. Li, M., and Fannin, R.J. (2008). Comparison of two criteria for internal stability of granular soil. Canadian Geotechnical Journal 45: 1303–1309.
12. Moffat, R. (2005) Experiments on the internal stability of widely graded cohesionless soils. Ph.D. thesis, The University of British Columbia, Vancouver, Canada, 279p.
13. Moffat, R. and Fannin, R.J. (2006). A large permeameter for study of internal stability in cohesionless soils. ASTM Geotechnical Testing Journal, 29: 273-279.
14. Ronnqvist, H. (2010). Predicting surfacing internal erosion in moraine core dams. Licentiate thesis, Royal Institute of Technology, Stockholm, Sweden, 197p.
15. Sail, Y., Marot, D. Sibille, L. and Alexis, A. (2011). Suffusion tests on cohesionless granular matter. European Journal of Environmental and Civil Engineering, 15:799-817.
16. Sherard, J.L. (1979). Sinkholes in dams of coarse, broadly graded soils. Trans. 13th Int. Cong. on Large Dams, New Dehli, Vol. 2: 25-34.
17. Terzaghi, K. (1939). Soil mechanics: a new chapter in engineering science. J. Instn. Civ. Engrs., 12: 106-141.
18. USBR (2011). US Bureau of Reclamation, Design Standards No. 13, Embankment dams: Chapter 5, Protective Filters. 254p.
19. Wan, C.F., and Fell, R. (2008). Assessing the potential of internal instability and suffusion in embankment dams and their foundations. Journal of Geotechnical and Geoenvironmental Engineering 134:401-407.

Study of chemical-mineralogical properties of modified soils with polymers addition

Jonny Patricio[1a], John Kennedy Rodrigues[2] and Lêda Christiane Lucena[2]

[1]*UFPE, Civil Engineering Postgraduate Department, 1235 Professor Morais do Rego, Recife, Brasil*
[2]*UFCG, Civil Engineering Department, 882 Aprígio Veloso, Campina Grande, Brasil*

Abstract. On highways, the soil is considered a supported material and compound pavements layers. For this, they must have such characteristics that confer stability and mechanical resistance to traffic internal forces during the pavement life. When soils do not have required characteristics by the project can be used stabilization techniques that make the natural soil adequately to roads requirement. Based on this assumption, this study aimed to evaluate the efficacy of polymer association in soil stabilization for use in roads pavements. Were evaluated chemical and mineralogical properties on two (2) different soils with sample of pure soil and with the addition of the polymer association. Based on the obtained results, polymer association changes was observed on X-ray fluorescent spectrometry (XRF); X-ray diffraction (XRD); scanning electron microscopy (SEM) and Methylene blue. In general, the polymeric association studied in this research was effective in chemical and mineralogical analyzes for use on stabilized soils, making this technique efficient for use in layers of road pavements. Key-words: polymers, stabilization, pavements.

1 Introduction

The Brazilian road sector is particularly important because it is the main modal used in loads transportation. During the decades of 1990 and 2000, road transport accounted for over 60% of the total transported in the country. This reflects a process that spanned several decades in which predominated the fast growth of road segment on all the other modalities (IPEA, 2010).

On highways, the soil is considered a supported material and compound pavements layers. For this, they must have such characteristics that confer stability and mechanical resistance to traffic internal forces during the pavement life.

When soils do not have required characteristics by the project two procedures can be used. Firstly, withdrawal of the original material followed by the replacement with desired geotechnical characteristics. This procedure can become impractical in the case of roads that require a large volume of replacement or large transport distances. The second procedure is to use stabilization techniques that make the natural soil adequately to roads requirement.

Chemical treatment of pavement base, subbase, and subgrade materials is undertaken to improve workability during compaction, to create a firm working surface for paving equipment, to increase the strength and stiffness of a foundation layer, to reduce potential shrink and swell due to moisture changes or frost action, or both, or to control dust on unpaved roads (Rauch et al, 2002).

Tingle et al (2007) aims to understanding the chemical and physical bonding mechanisms associated with selected nontraditional stabilizers. It was observed that polymer stabilizers coat soil particles and form strong physical bonds. Polar components present in the polymer may adsorb strongly onto soil particle surfaces, promoting adhesion.

Based on this assumption, this study proposes to study chemical-mineralogical properties of modified soils with polymers addition seeing that, this technique improve best mechanical behavior of the soil and reduces deterioration of roads layers.

2 Materials

For the present study was used the association of followed three polymers: Polymer A, Polymer B and Polymer C. The functions of those polymers are:

- Polymer A is water based heavy duty formulation of acrylic polymer specifically for road building or creating a hardened surface suitable for all types of traffic. When mixed into soil and compacted yields an extremely strong three dimensional substrate that can easily withstand traffic loads;
- Polymer B is a combination of two complementary components developed specifically to fill the capillary network with a silicone resin thus rendering the soil water repellent. So this is product is a water-proofing agent employed to reduce the water sensitivity of soil particles, especially in soils containing significant levels of clays or

[a] Corresponding author: jonny_dantas@hotmail.com

silts. It functions to hydrophobize the soil surface and resist water intrusion;

- Polymer C was formerly thought to function as a catalyst to enhance the drying process and impart early strength development.

According to the supplier of the products, a good formulation of polymer association was 6% of Polymer A, 2% of Polymer B and 5% of Polymers C (percent by weigh based on soil). This research evaluated lower content but following the same percentages between the three products.

Table 1 describes contents of polymer association utilized in this research.

Table 1. Contents of polymer association.

Polymer A (%)	Polymer B (%)	Polymer C (%)	Total contents of polymer association (%)
3	0,9	2,6	6,5

It was studied two (2) different soils with the followed TRB classification: Soil 1 (A-2-6) and Soil 2 (A-2-4). Table 2 presents a summary of the geotechnical properties of these soils.

Table 2. Physical properties of research soils.

Properties	Soil 1	Soil 2	Standard Method
Passing 4,8 mm sieve (%)	96,34	99,98	
Passing 0,42 mm sieve (%)	37,29	74,79	ABNT- NBR 7181/1984
Passing 0,074 mm sieve (%)	11,12	12,98	
Liquid limit (%)	34,22	NL	ABNT- NBR 6459/1984
Plastic limit (%)	NP	NP	ABNT- NBR 7180/1984
Specific Gravity	2,655	2,703	ABNT- NBR 6508/1984

3 Testing Program

The study of physical-chemical-mineralogical properties of polymers addition on soil stabilization was analyzed by the results of (i) X-ray fluorescent spectrometry (XRF); (ii) X-ray diffraction (XRD); (iii) scanning electron microscopy (SEM) and (iv) Methylene blue.

3.1. X-ray fluorescent spectrometry

For determination of chemical composition of studied soils, it was used X-ray fluorescent spectrometry. This technique is based on the principle that the absorption of X-rays by the material causes internal ionization of atoms, generating a radiation characteristic, known as "fluorescence". This analysis allowed observes the presence of oxides such as SiO2, Al2O3, Fe2O3, CaO, MgO, K2O, Ti2O, among others. For X-ray fluorescent spectrometry tests was used the content of 6,4% of polymer association on stabilizers samples.

3.2 X-ray diffraction

For mineralogical characterization of studied soils, it was used X-Ray diffraction. It was utilized Cukα radiation, 40kV voltage, 30 mA current and scan of $2° < 2\theta < 30°$ and $\lambda = 1{,}54$Å. For X-Ray diffraction was used the content of 6,4% of polymer association on stabilizers samples.

3.3 Scanning electron microscopy

Electron Microscopy method is the generation of data that allows the observation and characterization of organic materials and inorganic heterogeneous on a micrometer scale. The SEM provides a magnified image of the surface of the material that is very similar to that expected if one could actually "look" this surface. Not only is the material topography of information produced as well as, information on their composition. For scanning electron microscopy was used the content of 6,4% of polymer association on stabilizers samples.

3.4 Methylene blue

The methylene blue adsorption test is an alternative very efficient, fast and inexpensive to determine the expanding activity plastic soils.

The mechanism of adsorption of methylene blue by the particles is the ion exchange between the cations existing on the surface of these particles, such as calcium, sodium, magnesium and potassium cations, and cations resulting from the dissociation of methylene blue molecule aqueous solution. As a result of this cation exchange is formed around the particle a monomolecular methylene blue layer. For methylene blue adsorption test was used the content of 6,4% of polymer association on stabilizers samples.

DOI: 10.1051/e3sconf/20160920001

4 Results

4.1. X-ray fluorescent spectrometry

Table 3 shows the chemical composition of studied soil without stabilizers and with the addition of polymer association.

Table 3. X-ray fluorescent spectrometry.

Properties	Soil 1		Soil 2	
	Pure	Polymer Add	Pure	Polymer Add
FL*	12,25	16,4	6,82	19,5
SiO_2	49,04	39,60	57,70	35,03
Al_2O_3	31,67	32,55	26,56	32,54
Fe_2O_3	1,28	8,0	5,71	4,50
TiO_2	-	0,96	1,23	1,20
K_2O	4,59	-	-	-
MgO	0,89	-	-	-
CaO	-	1,91	-	4,70
Others	0,28	0,58	1,98	2,53

***FL = Fire lost**

According with the results, Soil 1 chemical composition was basically formed by SiO2 (49%), Al2O3 (32%) and K2O (5%). Soil 2 was basically formed by SiO2 (58%), Al2O3 (27%) and Fe2O3 (6%).

The results indicate that the addition of the polymeric association caused an elevation of the fire lost, possibly because it is an organic material. Could be verified the presence of CaO in all soil samples with polymer addition, justified by the presence of inorganic calcium in the chemical composition of Polymer C.

4.2 X-ray diffraction

Figure 1 shows the diffractogram of studied soil without stabilizers (Soil 1-Figure 1.a and Soil 2 – Figure 1.b) and with the addition of polymer association (Soil 1- Figure 1.c and Soil 2 – Figure 1.d).

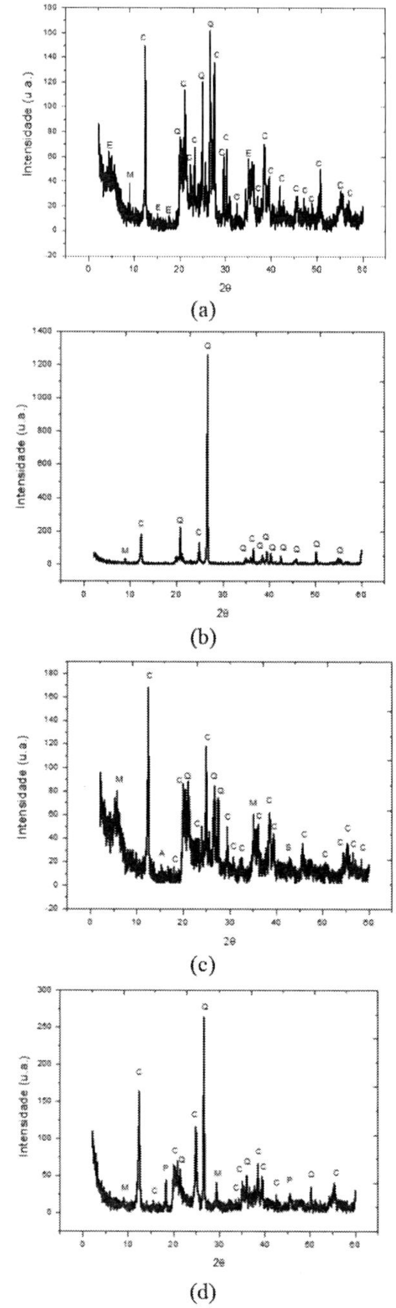

Figure 1. Diffractogram for studied soils.

Legend: A = anorthite/ C = kaolinite/ E= smectite/ M= mica/ Q= quartz/ S= sanidine/P= portlandite.

According to the diffractogram, Soil 1 showed the mineralogical phases: Anorthite, kaolinite, quartz, mica and sanidine. Soil 2 showed the mineralogical phases:mica, portlandite, quartz and kaolinite.

4.3 Scanning electron microscopy

Figure 2 shows the SEM of studied soil without stabilizers (Soil 1- Figure 2.a and Soil 2 – Figure 2.b) and with the addition of polymer association (Soil 1- Figure 2.c and Soil 2 – Figure 2.d).

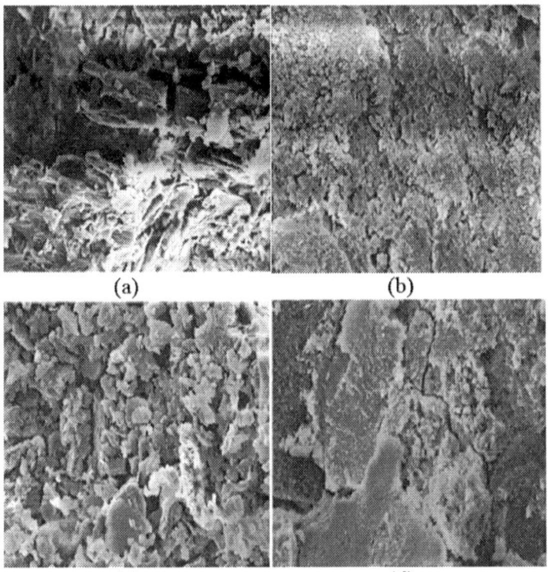

Figure 2. Scanning electron microscopy of studied soils.

According to Figure 2.a, Soil 1 shows a rough morphological structure and heterogeneous with the presence of pores, and dispersed particles, indicating that there was not a good interaction between the particles constituent material. Soil 2 shows a rough morphological structure with the densified regions and the presence of pores, characteristics of soils with lower average particle diameter

According to Figure 2.c, it was observed that the Soil 1 shows heterogeneous morphologic structure with the presence of a few pores, hexagonal crystals and needle-shaped. Soil 2 (Figure 2.d) shows very densified structure with few pores and some particles dispersed in the matrix, which could be quartz.

4.4 Methylene blue

Table 4 shows the methelyne blue adsorption of studied soil without stabilizers and with the addition of polymer association.

Table 4. Methelyne blue absorption os studied soil.

Methelyne blue absorption (meq/100 g dry clay)	Pure	Polymer Add
Soil 1	28	20
Soil 2	28	20

The analysis of the results reveals that studied soils showed low cation exchange index. These values are justified by the presence of iron and aluminum oxides which have low capacity of cation change (20-50 meq/ 100g dry clay), and the presence of kaolinite which contributes significantly to the reduction of this index.

It was observed that after addition of the polymer association occurred reducing the adsorption of methylene blue at Soil 1, indicating that under these conditions these soils have a smaller cation exchange. This occurrence is explained by the crystallization process of the grains of clay minerals present in the soil. Particles soils were involved by polymers generating a barrier that prevents the exchange cations.

This reduction may indicate a significant decrease in the activity of clay minerals causing a decrease in the plasticity of soils.

Conclusion

On the basis of the study presented above, the following conclusions are drawn:

The results indicate that the addition of the polymeric association caused an elevation of the fire lost, possibly because it is an organic material. Could be verified the presence of CaO in all soil samples with polymer addition, justified by the presence of inorganic calcium in the chemical composition of Polymer C.

Evaluating scanning electron microscopy results, can be observed that the presence of polymeric association form physical bound, causing union between the particles of soil.

Besides, it was observed the addition of polymeric association filled the capillary network with a silicone resin.

It was observed that after addition of the polymer association occurred reducing the adsorption of methylene blue indicating that under these conditions these soils have a smaller cation exchange.

References

1. IPEA, Brasil em desenvolvimento: Estado, planejamento e politicas públicas. Instituto de Pesquisa Econômica Aplicada, Brasília, 2010;
2. Rauch, A. F., Harmon, J. S., Katz, L. E., Liljestrand, H. M. Measured effects of liquid soil stabilizers on engineering properties of clay. Transp. Res. Rec., 1787, 33–41, 2002.
3. Tingle, J. S., Newman, J. K., Larson, S. L., Weiss, C. A., and RUSHING J. F. Stabilization Mechanisms of Nontraditional Additives. Journal of the Transportation Research Board, No. 1989, Vol. 2, Transportation Research Board of the National Academies, Washington, D.C., 2007.

Modeling of unsaturated granular materials in flexible pavements

Fan Gu[1,a], Xue Luo[1], Yuqing Zhang[2], Robert Lytton[1] and Hakan Sahin[1]

[1]*Texas A&M Transportation Institute, Texas A&M University, 77843 College Station, Texas, USA*
[2]*School of Engineering and Applied Science, Aston University, B4 7ET, Birmingham, UK*

Abstract. The unsaturated granular material (UGM) is found to exhibit the moisture-sensitive and stress-dependent nonlinear cross-anisotropic behaviour in flexible pavements. This paper aims at developing a finite element (FE) model for pavement structure, which takes into account this behaviour of UGM. First, the Lytton model is employed to characterize the moisture-sensitive and stress-dependent behaviour of UGM, which incorporated a matric suction term to the existing stress-dependent constitutive model. The Lytton model is validated by the laboratory resilient modulus tests on the selected UGMs at different moisture contents. Second, the nonlinear cross-anisotropic constitutive equation of UGM is derived from the generalized Hooke's Law. The coefficients of the constitutive model are determined by the rapid triaxial test. Third, a User-Defined Material (UMAT) subroutine is developed to characterize this constitutive behaviour in the FE software ABAQUS. The UMAT subroutine adopts the secant stiffness approach with multiple damping factors. The UMAT subroutine is then implemented in the FE model of flexible pavement structures. The FE simulation results indicate the nonlinear cross-anisotropic model predicts greater pavement responses than the isotropic model. When the UGM is suction sensitive, it is found that the moisture content of UGM significantly affects the moduli distribution of base layer and the critical strains (i.e., tensile strain at the bottom of asphalt concrete, and compressive strains in base and subgrade layers) of pavement structures.

1 Introduction

Unsaturated granular materials (UGMs) are often used as base layers for flexible pavements. An unbound granular base provides the foundational support to the pavement structure, and dissipates the stresses induced by traffic loading to the underlying subgrade. Understanding the constitutive behaviour of UGM is crucial to the accurate performance prediction of the pavement structures. In the conventional pavement design, the granular base is assumed to be linear elastic. By using this assumption, the linear isotropic model predicted an unexpected tensile stress at the bottom of the base layer, which conflicts with the fact that the UGM cannot transfer the tensile stress among the aggregate particles. A number of recent studies have revealed that the UGM exhibits the nonlinear cross-anisotropic behaviour, which means the resilient modulus of the granular base is stress-dependent, and its horizontal modulus is smaller than the vertical modulus [1-2]. It was found that modeling the UGM as a cross-anisotropic material can significantly reduce or eliminate the tensile stresses in base layer [3-4]. Field studies further concluded that the nonlinear cross-anisotropic model provided better agreement with the field measurements [5-6]. Hence, modeling UGM as nonlinear cross-anisotropic material should be taken into account for the pavement design and analysis. However, a review of these existing studies showed that the moisture condition of the granular base was often assumed at the optimum, or the same condition that the granular material was tested in the laboratory. This assumption ignores the fact that the moisture condition of the UGM is significantly affected by the weather, the groundwater table level, the drainage condition, and the surface properties in the field.

The moisture content affects the constitutive behaviour of UGM, which further influences the performance of pavement structures in the field. It is reported that the resilient modulus of UGM is moisture-sensitive, i.e., the modulus decreases with the growing saturation level [7]. Salour and Erlingsson [8] investigated the pavement response to variations of moisture content of base layers using falling weight Deflectometer tests. They concluded that increasing the water content of UGM significantly reduces the back-calculated modulus of base layer. These studies therefore suggest that the moisture-sensitive behaviour of UGM should be taken into account for modeling the pavement structures.

To address the aforementioned problems, this study aims at proposing a new constitutive model for UGM considering both nonlinear cross-anisotropic behavior and moisture-sensitive characteristics, and incorporating the proposed constitutive model into the finite element model of the base layer to quantify the influence of moisture content on the pavement performance. More specifically, the saturation factor and the matric suction of the UGM will be applied to the proposed constitutive

[a] Corresponding author: tracygufan@tamu.edu

model to reflect the moisture dependency. A new user-defined material subroutine (UMAT) will be developed to characterize the moisture-sensitive and stress-dependent nonlinear cross-anisotropic behavior of base material in the software ABAQUS.

The paper is organized as follows. The next section presents the proposed constitutive model to capture the moisture-sensitive characteristic of UGM. The following section develops a UMAT to define the UGM in the finite element model. The developed UMAT is also verified by the designed numerical experiment. After that, the finite element models for flexible pavement structures are setup, which consider asphalt concrete as a viscoelastic material, granular base as a moisture-sensitive and stress-dependent nonlinear cross-anisotropic material, and subgrade as an elastic material. The importance of nonlinear cross-anisotropic model for predicting the pavement response, the effect of moisture content of unbound base material on the pavement response, are investigated in this section. The final section summarizes the findings of this paper.

2 Constitutive Model for Unsaturated Granular Materials

The generalized Hooke's law is used to define the cross-anisotropic behaviour of UGM for an axisymmetric problem, which is shown in Equation 1 [3].

$$\begin{bmatrix} \frac{1}{E_x} & -\frac{\nu_{xy}}{E_x} & -\frac{\nu_{xx}}{E_x} & 0 \\ -\frac{\nu_{xy}}{E_x} & \frac{1}{E_y} & -\frac{\nu_{xy}}{E_x} & 0 \\ -\frac{\nu_{xx}}{E_x} & -\frac{\nu_{xy}}{E_x} & \frac{1}{E_x} & 0 \\ 0 & 0 & 0 & \frac{1}{G_{xy}} \end{bmatrix} \begin{bmatrix} \sigma_x \\ \sigma_y \\ \sigma_x \\ \tau_{xy} \end{bmatrix} = \begin{bmatrix} \varepsilon_x \\ \varepsilon_y \\ \varepsilon_x \\ \gamma_{xy} \end{bmatrix} \qquad (1)$$

where E_x is the horizontal modulus; E_y is the vertical modulus; G_{xy} is the shear modulus; ν_{xy} is the Poisson's ratio to characterize the effect of vertical strain on horizontal strain; ν_{xx} is the Poisson's ratio to characterize the effect of horizontal strain on horizontal strain.

In ABAQUS, this constitutive model needs to be rewritten as a strain-stress relationship. Converted from Equation 1, the strain-stress relationship for the cross-anisotropic material can be expressed as,

$$\begin{bmatrix} \sigma_x \\ \sigma_y \\ \sigma_x \\ \tau_{xy} \end{bmatrix} = \frac{E_y}{\alpha\beta} \begin{bmatrix} n\left(1-n\nu_{yx}^2\right) & n\nu_{yx}\alpha & n\left(\nu_{xx}+n\nu_{yx}^2\right) & 0 \\ n\nu_{yx}\alpha & 1-\nu_{xx}^2 & n\nu_{yx}\alpha & 0 \\ n\left(\nu_{xx}+n\nu_{yx}^2\right) & n\nu_{yx}\alpha & n\left(1-n\nu_{yx}^2\right) & 0 \\ 0 & 0 & 0 & m\alpha\beta \end{bmatrix} \begin{bmatrix} \varepsilon_x \\ \varepsilon_y \\ \varepsilon_x \\ \frac{1}{2}\gamma_{xy} \end{bmatrix} \qquad (2)$$

where $n = \frac{E_x}{E_y}$; $m = \frac{G_{xy}}{E_y}$; $\alpha = 1 + \nu_{xx}$; $\beta = 1 - \nu_{xx} - 2n\nu_{yx}^2$.

This strain-stress relationship is used to compute the incremental stress for a given incremental strain in ABAQUS. In Equation 2, the vertical modulus E_y is dependent on both the stress state and the moisture content. In order to characterize this behavior, the Lytton model is used as shown in Equation 3, which incorporates a matric suction term into the generalized resilient modulus model [9-10].

$$E_y = k_1 P_a \left(\frac{I_1 - 3\theta f h_m}{P_a}\right)^{k_2} \left(\frac{\tau_{oct}}{P_a}\right)^{k_3} \qquad (3)$$

where I_1 = the first invariant of the stress tensor; P_a = the atmospheric pressure; θ = the volumetric water content; h_m = the matric suction in the aggregate matrix; f = the saturation factor, $1 \le f \le \frac{1}{\theta}$; τ_{oct} = the octahedral shear stress; and k_1, k_2 and k_3 are regression coefficients.

In order to validate the Lytton model, the repeated load triaxial tests are conducted on the 3 selected materials at 3 different moisture contents (i.e., optimum moisture condition, and 1.5 percent below and above the optimum moisture condition). Figure 1 shows the comparison between the predicted moduli using Equation 3 and the measured moduli from the triaxial tests. The model prediction provides a good agreement with the test measurements. This indicates that the constitutive model proposed in Equation 3 is able to reflect the moisture-sensitive and stress-dependent behavior of UGM.

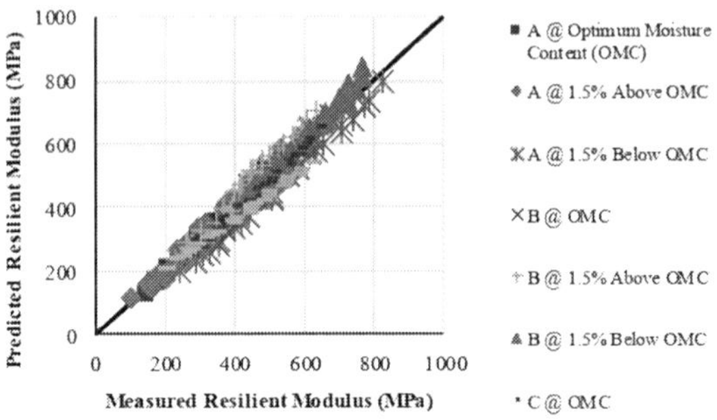

Figure 1. Comparison of predicted and measured resilient moduli for selected materials

3 Development of a User-Defined Material Subroutine for Unsaturated Granular Material

Several studies have been carried out to program different UMAT subroutines to define the stress-dependent behavior of UGM. One UMAT subroutine was developed for the UGM based on the tangent stiffness method [11]. The nonlinear stress-dependent resilient modulus model was formulated as a function of the strain state. Kim et al. [12] adopted a direct secant stiffness approach to determine the nonlinear resilient modulus solution in each iteration. This nonlinear solution technique is less complicated than the tangent stiffness approach and Newton-Raphson approach, but it is accurate enough to provide convergence of the iterations. Based on the secant stiffness approach, Wang and Al-Qadi [6] programmed a UMAT subroutine for the UGM by incorporating an anisotropic constitutive model. Using this nonlinear cross-anisotropic UMAT subroutine, they successfully analyzed the response of a 3-Dimensional pavement model under the moving vehicular loading. In this study, a similar UMAT subroutine is also programmed to define the moisture-sensitive and stress-dependent behavior by using the secant stiffness technique with the damping factor λ. The trial vertical modulus is computed by Equation 4 in each iteration.

$$E_y^i = (1-\lambda) E_y^{i-1} + \lambda E_{ycomputed}^i \quad (4)$$

where E_y^i is the vertical modulus output from the ith iteration; E_y^{i-1} is the vertical modulus output from the (i-1)[th] iteration; λ is the damping factor (e.g., initial λ is 0.95); $E_{ycomputed}^i$ is the vertical modulus computed from Equation 3 at the ith iteration [12]. The convergence criteria used in this study are shown in Equations 5 and 6.

$$Error_i = \frac{\left|E_y^i - E_y^{i-1}\right|}{\left|E_y^i\right|} \le 2\% \quad (5)$$

$$Error_c = \frac{\sum_{i=1}^{n} \left(E_y^i - E_y^{i-1}\right)^2}{\sum_{i=1}^{n} \left(E_y^i\right)^2} \le 0.5\% \quad (6)$$

where $Error_i$ is the individual error for each node; $Error_c$ is the cumulative error for the entire model; n is the number of nodes in the model. The moisture-sensitive and stress-dependent cross-anisotropic constitutive models, as shown in Equations 2 and 3, are coded into the UMAT. The Mohr-Coulomb failure theory is also applied to adjust the initially computed horizontal stresses so that the yield stress of the material will not be exceeded. This method originally proposed by ILLI-PAVE and KENLAYER is incorporated in the development of the UMAT for nonlinear cross-anisotropic unbound aggregates in this study. Figure 2 is the flowchart of the developed UMAT subroutine.

4 Finite Element Modeling of Flexible Pavement Structures

As shown in Figure 3, the axisymmetric pavement structures analysed in this section consist of a 15-cm hot mix asphalt (HMA) layer, a 25-cm granular base and 1.4-meter subgrade. The pavement structures are subjected to a half-sine impact load with a loading amplitude of 40.03 kN and a pulse duration of 0.1 second. The load is assumed as a uniform pressure over a 0.15m radius of circular area at the left edge of the axisymmetric pavement structures. The meshed finite element model is constructed according to the pavement structures in Figure 3. Fine mesh is used in the loading area. The 8-node biquadratic axisymmetric elements with reduced integration are used in the whole finite domain. The interfaces between the HMA layer, granular base and subgrade are assumed to be fully bonded.

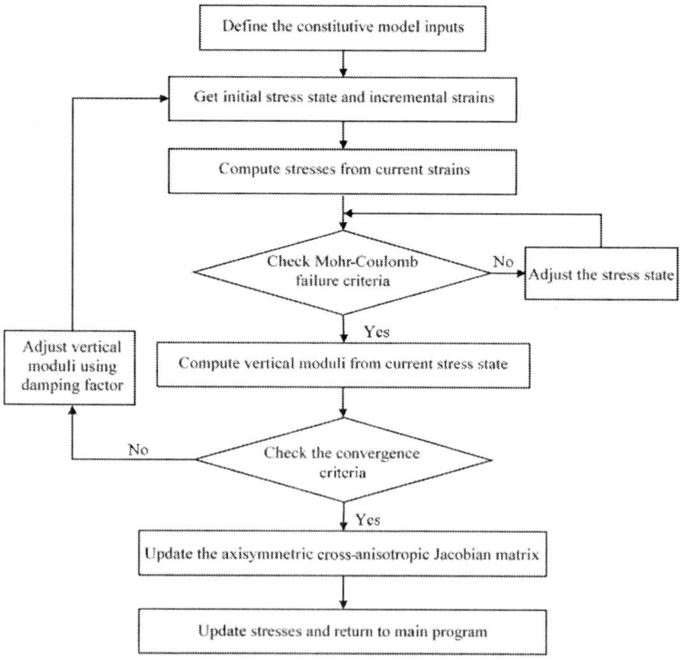

Figure 2. Flowchart of developed UMAT subroutine

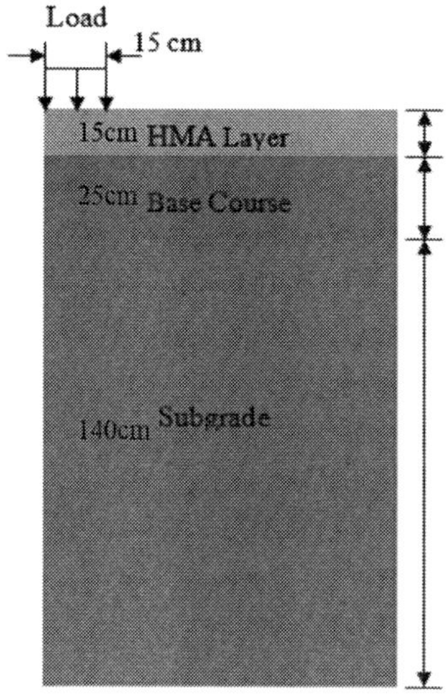

Figure 3. Schematic plot of pavement structures

HMA is considered as a viscoelastic material in the numerical analysis. In ABAQUS, the Prony-Series models are used to characterize the time-dependent behaviour of HMA, which are shown in Equations 7 and 8.

$$G(t) = G_0\left(1 - \sum_{i=1}^{n} G_i\left(1 - e^{-t/\tau_i}\right)\right) \quad (7)$$

$$K(t) = K_0\left(1 - \sum_{i=1}^{n} K_i\left(1 - e^{-t/\tau_i}\right)\right) \quad (8)$$

where $G(t)$ and $K(t)$ are the relaxation shear modulus and bulk modulus, respectively; G_0 and K_0 are the corresponding instantaneous shear modulus and bulk modulus ; G_i, K_i and τ_i are the input coefficients. The coefficients of the Prony-Series model are determined by fitting the dynamic modulus test results. Table 1a lists the coefficients of the Prony-Series model for the HMA. A constant Poisson's ratio is assumed during the analysis. The nonlinear cross-anisotropic properties of UGM are presented in Table 1b. Figure 4 is the soil-water characteristic curve to characterize the moisture-sensitivity of UGM. As shown in Table 1c, subgrade is simplified as a linear-elastic material with constant Poisson's ratio.

Table 1. Determined Model Inputs for Paving Materials

a. Determined Prony-Series Model Coefficients for HMA

Series Number	Prony-Series Coefficients		
i	Gi	Ki	τi
1	0.362	0.362	4.09E-06
2	0.363	0.363	2.56E-04
3	0.177	0.177	7.71E-03
4	0.074	0.074	2.10E-01
5	0.017	0.017	3.88E+00
6	0.006	0.006	6.53E+01
Note: Instantaneous modulus=18,130MPa, Poisson's ratio=0.35			

b. Nonlinear Cross-Anisotropic Properties for UGM

Input Variables	k_1	k_2	k_3	n	m	ν_{xy}	ν_{xx}
Values	1281	0.81	-0.08	0.45	0.35	0.17	0.43

c. Linear-Elastic Material Properties for Subgrade

Input Variables	E (MPa)	ν
Values	56	0.4

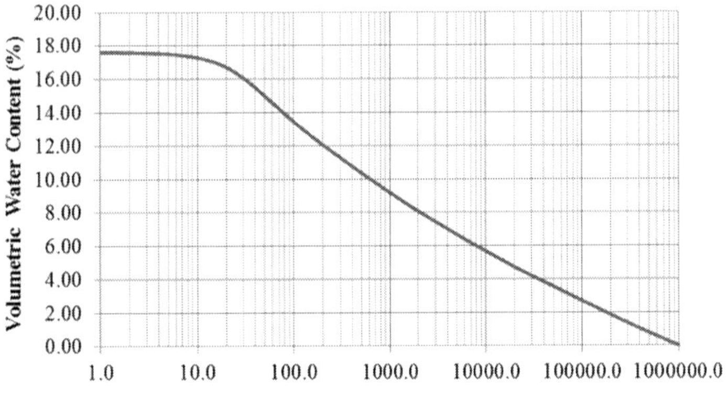

Figure 4. Soil-Water Characteristic Curve for UGM

5 Results and Discussion

5.1 Effect of Anisotropic Model on Pavement Responses

The comparison of pavement responses predicted by the anisotropic model and the isotropic model are shown in Figures 5a, 5b and 5c. As mentioned previously, the primary difference between the anisotropic model and the isotropic model is that the horizontal moduli of UGM are smaller than the vertical moduli in the anisotropic model, while the horizontal and vertical moduli of UGM are assumed to be equal in the isotropic model. It is seen from Figure 5a that the anisotropic model predicts approximately 10% higher tensile strain at the bottom of HMA layer than the isotropic model. This indicates that the pavement fatigue life estimated by the anisotropic model is shorter than the fatigue life predicted by the isotropic model. Figures 5b and 5c show that the anisotropic model also predicts 8% higher average compressive strain in base and 12% higher compressive strain at the top of subgrade when compared to the isotropic model. This demonstrates that using the anisotropic model for pavement design will predict a higher rutting depth. Hence, it is concluded that using anisotropic model to predict the pavement performance provides more conservative results for pavement ME design.

 DOI: 10.1051/e3sconf/20160920002

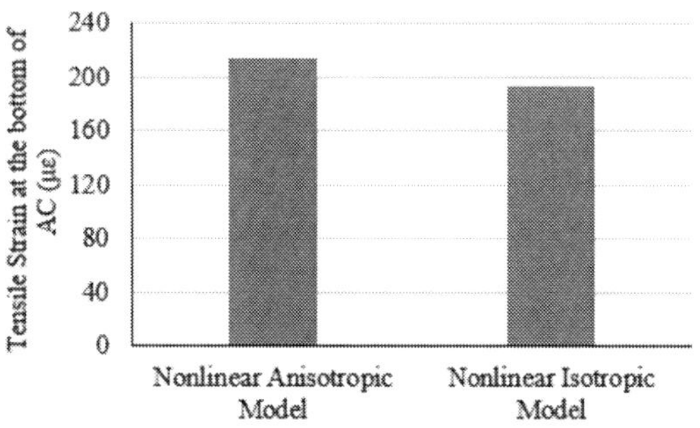

a. Tensile Strain at the Bottom of HMA Layer

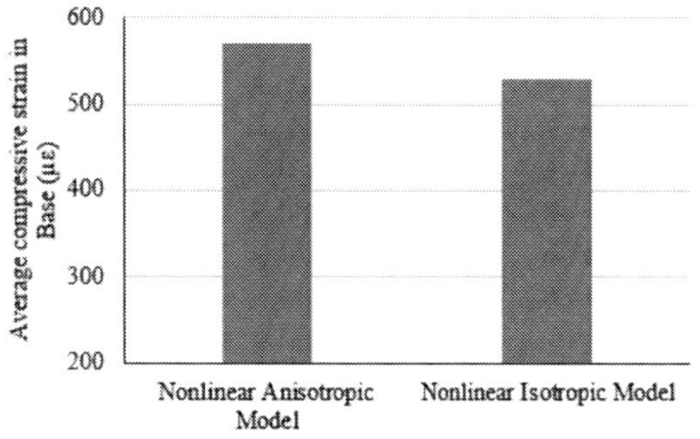

b. Average Compressive Strain in Base Layer

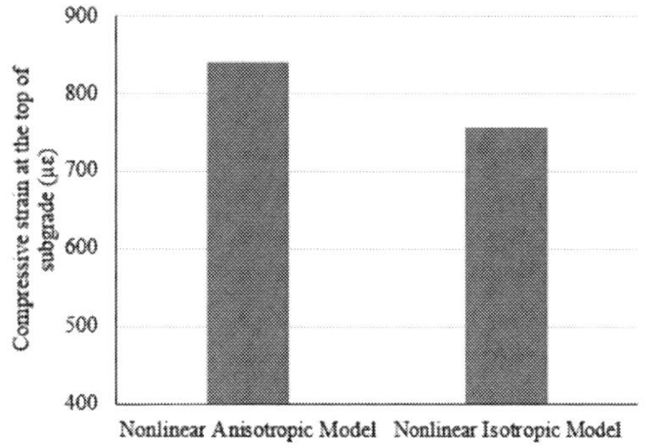

c. Compressive Strain at the Top of Subgrade

Figure 5. Effect of Anisotropic Model on Pavement Responses

5.2 Effect of Moisture Content of Unbound Base Layer on Pavement Responses

To investigate the effect of moisture content of UGM on pavement responses, three moisture conditions are simulated in the numerical model, which include the low moisture condition (i.e., the degree of saturation is 0.7), the optimum moisture condition (i.e., the degree of saturation is 0.85), and the saturated condition (i.e., the degree of saturation is 1.0). Figure 6 compares the vertical moduli (i.e., SDV1) distribution in base course at different moisture conditions. It is seen that the moduli of UGM in the vicinity of load area are significantly larger than those far away the load area. This is because the UGM is modelled as a stress-hardening material. It is also shown that the modulus of UGM decreases from the top to the bottom of base layer, which indicates the stress state varies in the base layer. One of objectives in this study is to model the moisture-sensitive behaviour of UGM. The comparison among Figures 6a, 6b and 6c demonstrates that the model-predicted moduli of UGM is sensitive to the moisture condition. It is seen that the increasing of moisture content of unbound base significantly reduces the moduli of UGM. A further observation shows that the moduli of UGM at the low moisture condition are nearly twice as large as those of base material at the saturated condition.

This variation thereby results in the change of pavement responses, such as the surface deflection, the tensile strain at the bottom of HMA layer, and the compressive strains in base course and subgrade. It is seen from Figure 7 that the model-predicted surface deflections, tensile strain at the bottom of HMA layer and compressive strain in base course are significantly sensitive to the moisture condition in base course, while the model-predicted compressive strain at the top of subgrade is merely slightly affected by the moisture variation in base course. It is obvious that increasing the moisture content of base course results in larger surface deflections, higher tensile strain at the bottom of HMA layer, and higher compressive strains in base and subgrade. This indicates that the developed finite element model can properly reflect the influence of moisture content of UGM on pavement responses.

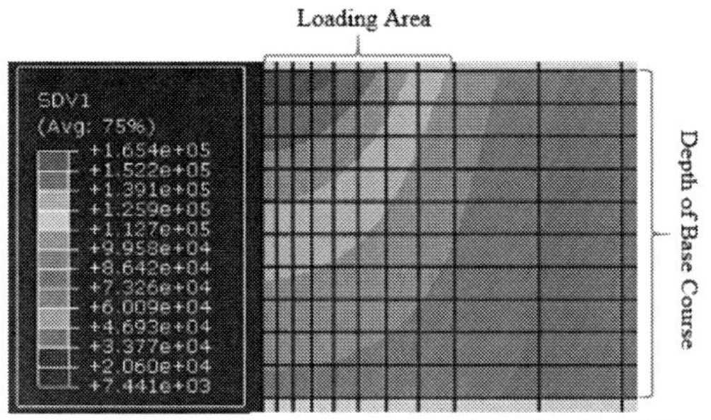

a. UGM at High Moisture Condition (Unit: kPa)

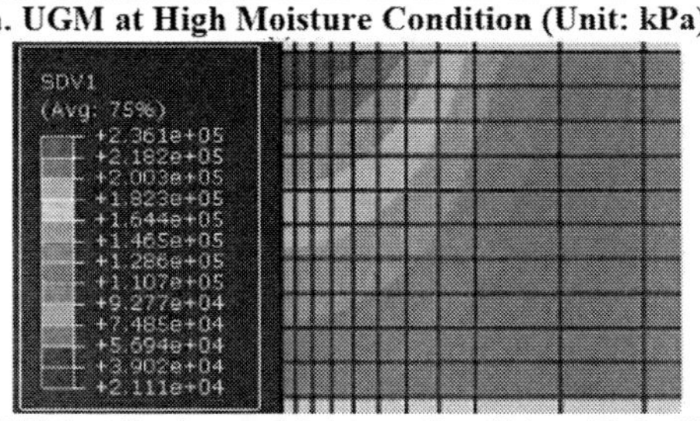

b. UGM at Optimum Moisture Condition (Unit: kPa)

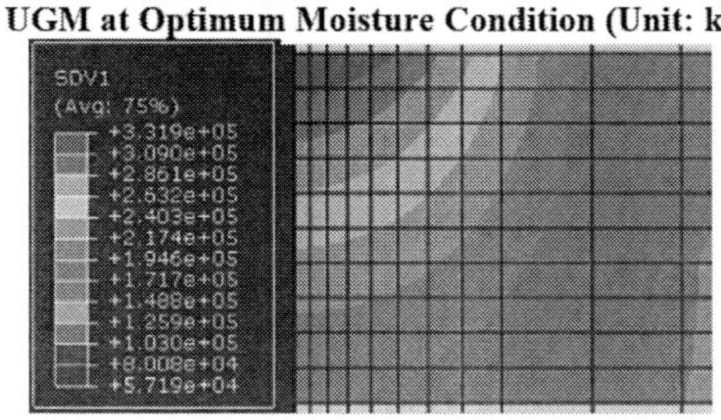

c. UGM at Low Moisture Condition (Unit: kPa)

Figure 6. Vertical Moduli Distribution in Base Layer

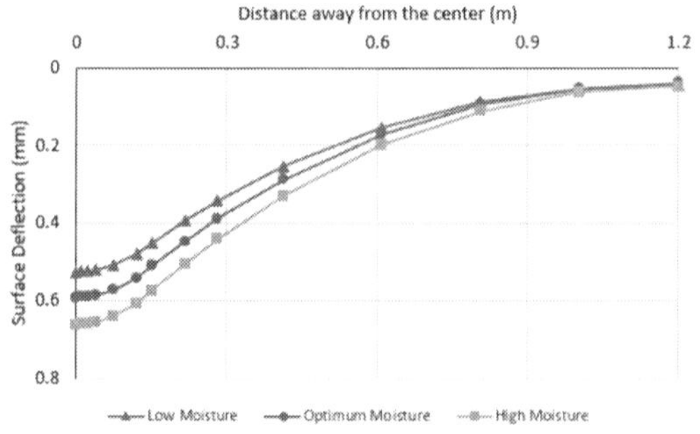

a. Surface Deflections of Flexible Pavement

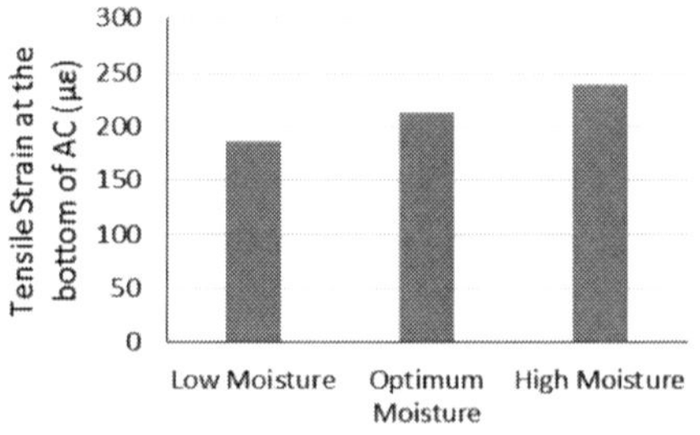

b. Tensile Strain at the Bottom of HMA layer

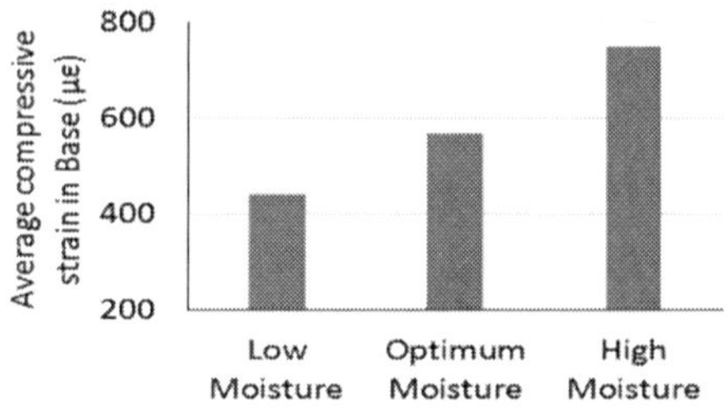

c. Average Compressive Strain in Base Layer

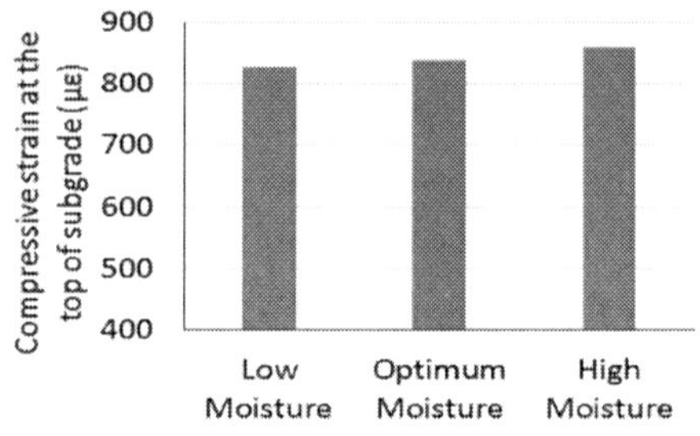

d. Compressive Strain at the Top of Subgrade

Figure 7. Effect of Moisture Content of UGM on Pavement Responses

6 Conclusions

The Lytton model is used to characterize the stress-dependence and moisture-dependence of resilient modulus of UGM. The degree of saturation and the matric suction are incorporated to discriminate the effect of the moisture variations. The moisture dependence of the Lytton model is validated by comparing the model-predicted resilient moduli of UGM at different moisture contents to those measured from the laboratory tests. It is demonstrated that the matric suction of the UGM is a key element to reflect the moisture dependence of the resilient modulus. The FE approach is then employed to model the stress-dependent and moisture-sensitive cross-anisotropic behaviour of UGM. The secant stiffness method with the multiple damping factors are efficient to program the UMAT subroutine. The UMAT subroutine is then implemented in the FE model of flexible pavement structures. The FE simulation results indicate the nonlinear cross-anisotropic model predicts greater pavement responses than the isotropic model. When the UGM is suction sensitive, it is found that the moisture content of UGM significantly affects the moduli distribution of base layer and the critical strains (i.e., tensile strain at the bottom of asphalt concrete, and compressive strains in base and subgrade layers) of pavement structures.

References

[1] A. Adu-Osei, D. Little, and R. Lytton. *Cross-anisotropic characterization of unbound granular materials*. Transp. Res. Rec., **1789**, 82-91 (2001)

[2] E. Tutumluer, and M. Thompson. *Anisotropic modeling of granular bases in flexible pavements*. Transp. Res. Rec., **1577**, 18-26 (1997)

[3] J. Oh, R. Lytton, and E. Fernando. *Modeling of pavement response using nonlinear cross-anisotropy approach*. J. Transp. Eng., **132**, 6, 458-486 (2006)

[4] I. Al-Qadi, H. Wang, and E. Tutumluer. *Dynamic analysis of thin asphalt pavements by using cross-anisotropic stress-dependent properties for granular layer*. Transp. Res. Rec., **2154**, 156-163 (2010)

[5] E. Tutumluer, D. Little, and S. Kim. *Validated model for predicting field performance of aggregate base courses*. Transp. Res. Rec., **1837**, 41-49 (2003)

[6] H. Wang, and I. Al-Qadi. *Importance of nonlinear anisotropic modeling of granular base for predicting maximum viscoelastic pavement responses under moving vehicular loading*. J. Eng. Mech., **139**, 1, 29-38 (2013)

[7] F. Lekarp., U. Isacsson, and A. Dawson. *State of art: Resilient response of unbound aggregates*. J. Transp. Eng., **126**, 1, 66-75 (2000)

[8] F. Salour, and S. Erlingsson. *Investigation of a pavement structural behaviour during spring thaw using falling weight deflectometer*. Road Mater. Pavement, **14**, 1, 141-158 (2013)

[9] R. Lytton. *Foundations and pavements on unsaturated soils*. Proceedings, First International Conference on Unsaturated Soils, ISSMFE, Paris, 1201-1220 (1995)

[10] F. Gu, H. Sahin, X. Luo, R. Luo, and R. Lytton *Estimation of resilient modulus of unbound aggregates using performance-related base course properties*. J. Mater. Civ. Eng., **27**, 6, 04014188 (2015)

[11] K. Hjelmstad, and E. Taciroglu. *Analysis and implementation of resilient modulus models for granular solids*. J. Eng. Mech., **126**, 8, 821-830 (2000)

[12] M. Kim, E. Tutumluer, and J. Kwon. *Nonlinear pavement foundation modeling for three-dimensional finite element analysis of flexible pavements*. IJOG, **9**, 5, 195-208 (2009)

Experiences from in-situ monitoring of pavement under weather conditions change

Barbara Likar[1,a], Stanislav Lenart[1], Karmen Fifer Bizjak[1] and Anh Minh Tang[2]

[1] *Slovenian National Building and Civil Engineering Institute, Dimiceva ulica 12, 1000 Ljubljana, Slovenia*
[2] *Ecole des Ponts ParisTech, U.R. Navier/CERMES, 6-8 av. Blaise Pascal, Cité Descartes, Champs-sur-Marne, 77455 Marne-la-Vallee, France*

Abstract. Very strong winters with temperatures under 0°C and hot summers with temperatures more than 30°C are observed in the South East part of Slovenia. Those big differences in temperature during the year and especially temperatures below freezing point have strong influence on asphalt layer and sub base of road pavement. The freeze/thaw cycles lead to formation of ice lenses in base course causing cracks in asphalt layers and degrade the pavement usually in a few years. For this reason one section of the national road in the South East part of Slovenia was rebuilt with cold in-situ pavement retreatment. A test field with inbuilt sensors for measuring water content, temperature and deformation in various depths and locations was constructed during the remediation works to study the mechanism of freeze-thaw degradation of pavements. The main goal of the test field is to determine water content in sub base, freezing depth, temperature distribution and deformations, which lead to cracks in asphalt layer after the remediation work in the road construction.

1 Introduction

The main purpose of constructing test field for monitoring water content, temperatures and deformations in retreatment pavement was to observe changes in freezing depth according to cold in-situ retreatment layer with foamed bitumen and cement.

In Slovenia, normal pavement construction for National road in the area with potential presence of freezing consists of two layers above the undisturbed ground soil: (i) the sub base layer in thickness of 0.50 m with of freezing resistant gravel 0/90 mm; (ii) and the base layer in thickness of 0.20 m with of gravel 0/32 mm with prescribed granulation. Asphalt layer, of a thickness depending on the traffic load, is above the base layer. Many roads are damaged according to fouled base layer. With an aim to find cost-effective and durable remediation technology, which retains the existing structure of pavement and adds only a new asphalt layer, cold in-situ retreatment with foamed bitumen and cement was adapted.

Many authors present the effects of climate on soil strength as well as pavement dependence on climate, soil type, drainage and depth of ground water table ([1, 3, 7]). Most investigations can be divided into two main groups: determining moisture and temperature effect on pavement strength or determining freeze-thaw cycle effect on pavement strength. Moisture content and temperature variations change the deformation properties of unbound material ([7]), which leads to variations - mainly continuous annual decrease of permanent deformation and settlements. Also soil temperature as an important parameter for prediction of freezing and thawing period, is affected by weather conditions, its properties, water content, loading and surface conditions ([12, 13, 14]). Pavement stiffness rapidly decreases during spring thaw, which causes large settlements ([14]), while during freezing no changes in settlement of road surface are observed ([14]). However spring thawing and snow melting can increase water content in base or sub base layer to saturated state, which also leads to reduction of pavement strength, decreasing dry density and increasing the porosity ([2, 8, 10]).

The effects of freezing and thawing on sub base and recycled layer of pavement were investigated in the presented paper. Monitoring of water content, temperature distribution and strains in sub base and recycled layer of the reconstructed pavement of national road in Slovenia is presented hereafter.

2 Description of the test field

The test field is located in the South East part of Slovenia, on the national road Črmošnjice – Črnomelj: R1-216/1178 at kilometre 4+490 right, in the town

[a] Corresponding author: barbara.likar@zag.si

Vrčice. The test field was constructed in November 2011 during the remediation work in the road construction. The remediation was done by in-situ cold retreatment with foamed bitumen and cement in depth of 30 cm. During the remediation works, soil was sampled in the locations of sensor installation for the purpose of field characterisation. Figure 1 shows the soil profile in the studied section with all installed sensors. The thickness of the recycled layer is about 0.30 m. The old sub base layer consists of silty gravel in thickness of 0.48 m and gravel with clay in thickness of 0.10 m. The sub grade consists of dolomite gravel, mainly clayish. No ground water was found during the excavation. The pavement construction is buried under the natural ground surface.
The field instrumentation was done based on the field characterisation with sensors for temperature, water content and horizontal strains. Sensors were installed in recycled layer and in old sub base layer in two horizontal positions: near the wheel tracks and near the middle of a pavement. Seven temperature sensors (T1 to T7) were installed in the pavement structure while two (T8 and T9) were installed near the data logger (Figure 1). The temperature sensors are DS 1825 type. Water content is measured with 5 TDR probes (TDR1 to TDR5), type TRIME-PICO 32. TDR probes allow also temperature measurements. Four strain gauges were installed to measure horizontal strain, two at the bottom of asphalt layer (S3 and S4) and two (S1 to S2) in the middle of recycled layer. As strain gauge can be easily destroyed during excavation works, they were protected with bitumen tape. These sensors were developed within the EU FP6 research project SPENS ([9]). All sensors are connected with data logger with saving interval of four hours.

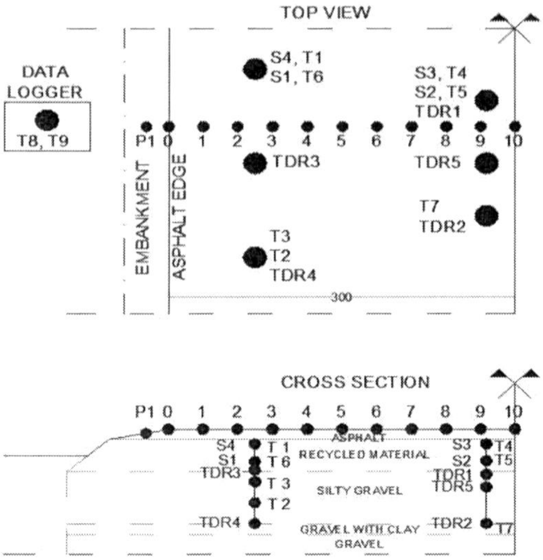

Figure 1. Sensors position in test field

After completion of asphalt protection layer geodetic measurement of vertical displacement are performed annually in 12 points (P1, 0, 1 to 10). All sensors and measuring points are presented in Figure 1.

Slovenian agency for environment (ARSO) has two weather stations (Semič and Černomelj) near the test field to monitor temperature, precipitation and snow height. Those data were used for analysis of monitoring results. They are available on the web page http://meteo.arso.gov.si.

2.1 Material of pavement

During the excavation works, soil samples of recycled material and old sub base layer were obtained for laboratory investigation. Results of dry density on test field, specific gravity, water permeability and sieving results are presented in Table 1 and Figure 2.

The remediation was done with cold in-situ retreatment by 3 % of foamed asphalt and 1 % of cement, which were mixed with old base layer (64 %) and old asphalt layer (32 %). Old base layer was fouled gravel 0/45 mm.

2.2 Installation of sensors

Dry densities of materials were measured by sand method on compacted layers. Figure 3 shows installation of TDR probe.

Table 1. Material characteristics

		Recycled material	Old sub base
Density	ρ_d [t/m³]	2.195	2.272
Particle density	ρ_s [t/m³]	2.61	2.72
Water permeability	k [m/s]	$1.8*10^{-4}$	$4.6*10^{-5}$
Particle size distribution	Cu [-]	95.6	174.2
	Cc [-]	10.6	13.6
	< 0.063mm [%]	10.0	10.3

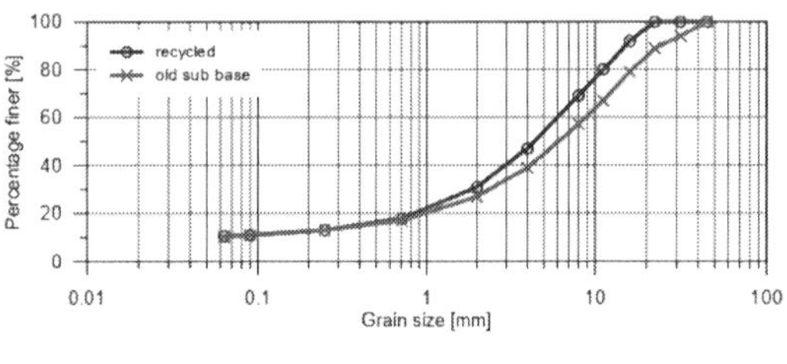

Figure 2. Grain size distribution

On Figure 4 installation of strain gauges is also presented. Long strain gauge is glued on bituminous tape, which is stabilised with two cardboard rectangles on each field, like letter H. The cardboard rectangles enable strain gauges good contact with bedding material. Manufactured and calibrated strain gauge was then laid on the bedding layer surface on thin bitumen spot at the field. The bitumen spot is placed just in small area around strain gauge. If strain gauge is installed in the soil, then sensor is covered with excavated soil and compacted. In other hand, when strain gauge sensor is installed under the asphalt layer, it is protected with asphalt on upper field and then asphaltation with machine can be done. ([7])

All sensors were checked and calibrated in laboratory before installation work. Calibration of TDR probes included the effects of temperature, soil type and soil unit mass upon the measured values ([16, 18, 19, 20, 21, 22, 23]). The influence of temperature upon the measurement with strain gauges was also checked in laboratory and included into calibration. To minimize temperature effect dummy sensors were installed next to the test section.

Figure 3. Installation of TDR probe.

Figure 4. Installation of strain gauge.

3 Experimental results

In the Figure 5 and Figure 6 all monitored data are presented. While many sensors of temperature were installed, all are presented but only minimum and maximum temperature recorded on weather station are showed in legend. Temperatures in the pavement are between daily minimum and maximum air temperature, only two sensors during summer times recorded higher temperature. Those two sensors T8 and T9 are installed in data logger covered with iron plate.

The frost can be detected like suddenly fall in volumetric water content on the diagram of water content. This happened in December 2011 for all probes and in January 2015 only in recycled layer. During the whole monitoring period volumetric water content in all layers grow to the predicted value of saturation state after longer rainfall period.

Horizontal strains that were measured in recycled layer by strain gauges are low. However some findings can be observed. Main changes are recorded under the wheel tracks during winter times. After five years of strains monitoring at the bottom of asphalt layer (S4) the edge zone showed compression behaviour, while in recycled layer in the edge zone (S1) strains have increased slowly to extension of 0.5 ‰.

Horizontal strains in recycled layer in the middle zone (S2) have shown almost no change, while S3, which has been installed at the bottom of asphalt layer has failed.

Geodetic measurements of vertical displacement at the surface show that whole pavement settled down. While embankment and edge zone (width of 1 m) of road surface settles for 10 to 18 mm, the centre zone settles no more than 6 mm.

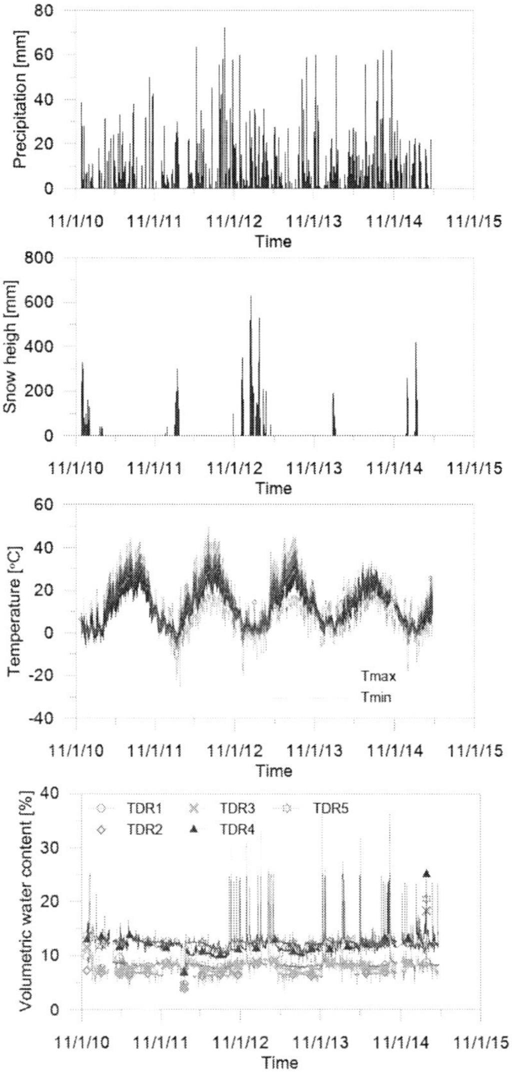

Figure 5. Precipitation, snow height, temperature and volumetric water content against time.

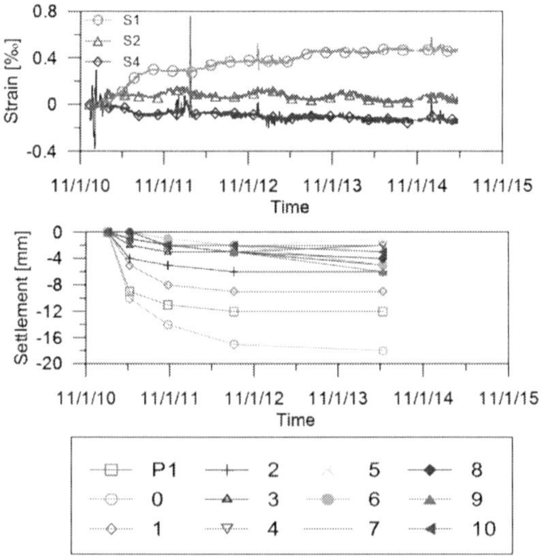

Figure 6. Strains and settlement against time.

During winter, when also underground temperatures fall under zero, soil start freezing (Figure 7). It is observed for recycled layer, that freezing starts when temperature in pavement is at least few days under zero. At the freezing point the TDR probes show value between 4 to 7 %. Another effect can be observed during the thawing which starts when temperature reach around -2 °C. Water content on the edge grows more than water content in middle part of pavement. Values of volumetric water content in recycled layer and in the old sub base layer in the middle zone of pavement are higher after thawing than before freezing. Described mechanism is known as thaw weakening, while during the freezing time water from deeper depths drawn up to the shallow frost zone. After spring thawing, water content in the layer increases in some cases to saturated value ([2, 8]). Sensor TDR3 expresses described mechanism. TDR2 sensor shows some problems during this winter. August and September 2014 had many rainy days with high values of precipitation. When daily precipitation exceeds 20 mm, the old sub base layer on the edge part of pavement reached predicted full saturation value of volumetric water content (Figure 8). While the TDR probes were installed in finer material than the layer materials are and the real values of dry density around the probes are not know, it is not possible to speak about saturation ([15]). Values of volumetric water content for old sub base layer never exceed 26 %, which lead to prediction that this value correspond to full saturation state. Same prediction could be done for recycled layer, but value of saturation volumetric water content is 36 %. After few days of daily precipitation higher than 20 mm in all measuring points increase of volumetric water content is observed. In old sub base layer predicted full saturation is reached.

The same is observed for recycled layer on the edge part of the pavement; while in the middle part of the pavement the increase of volumetric water content for about 20 % is observed. Recycled layer is protected with asphalt layer and water from precipitation flow from middle part to the edge part, where drainage is not regulated. Thus, volumetric water content is higher in the edge part of the pavement compared to the rest of the pavement. Similarly, also other authors observe that water content in unbound layers is highly dependent on permeability properties of asphalt layer ([10]) as well as permeability properties of cold in-situ retreatment with foamed bitumen layer ([4]).

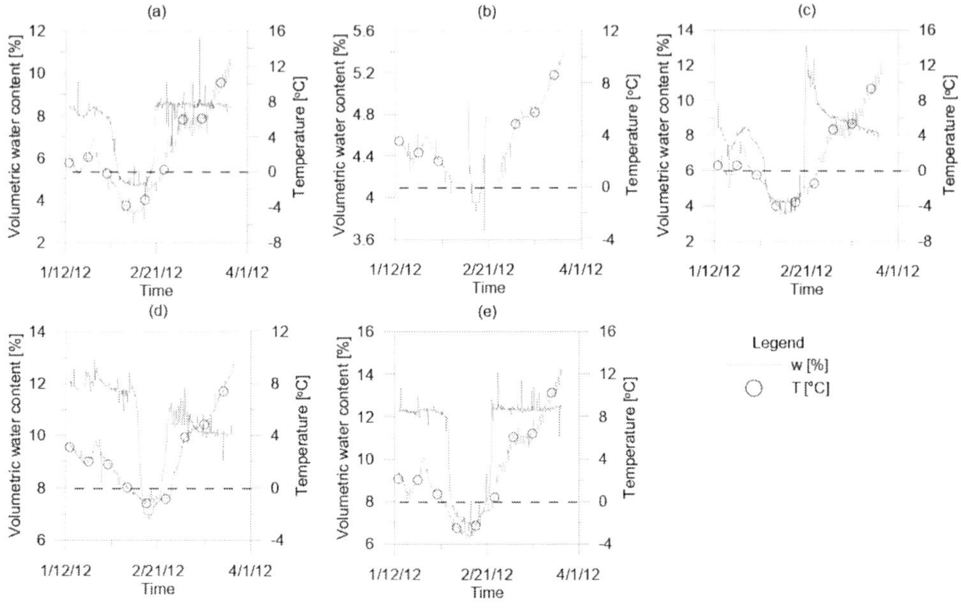

Figure 7. Freezing period for each TDR probe: (a) TDR1, (b) TDR2, (c) TDR3, (d) TDR4 and (e) TDR5.

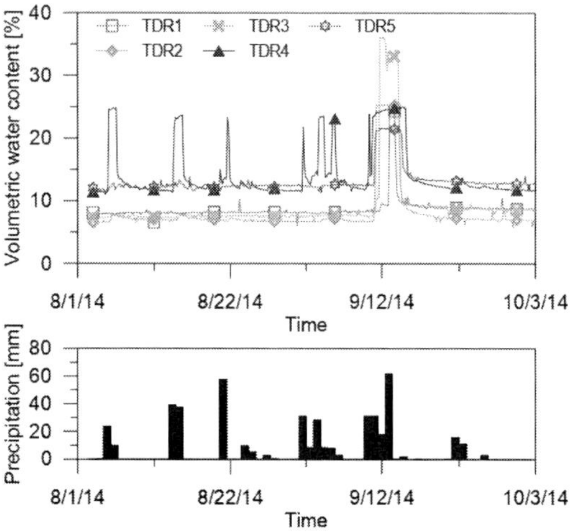

Figure 8. Relation between volumetric water content and precipitation (August and September 2014).

The same water flow is observed during melting of snow (Figure 9). Water from melting of snow is firstly recorded in edge zone of old sub base layer. After few days water start to penetrate in the middle zone of old sub base layer and also in the edge zone in recycled layer. The TDR probe in the middle zone of recycled layer does not record any water change due to melting of snow, and thus gives an assumption that protection with asphalt layer is enough impermeable.

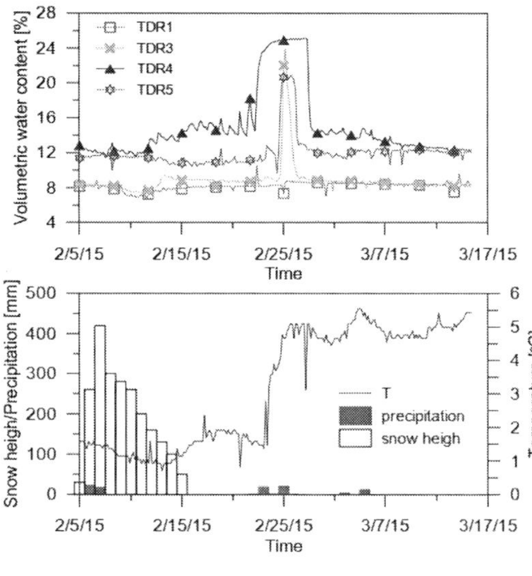

Figure 9. Relation between volumetric water content and melting of snow (February 2015).

Figure 10 and Figure 11 represent relation between deformation and temperature measured by strain gauges. In February 2012 horizontal strain increase after temperature exceeds 0 °C. At this time freezing was also observed. Thawing corresponds to extension at the bottom of asphalt layer in the edge zone. During thawing time water content in soil increase and this leads to extension in recycled layer. Different behaviour is recorded in December 2012 under asphalt layer and in the recycled layer, while in the middle zone only yearly changes are recorded. During winter times strains increased, while during summers returns almost to the start value (Figure 6). Very small changes can be influenced by temperature response of strain gauges. Other sensors show increasing of strain when temperatures decrease less than zero and when temperatures start increasing strains decrease to the values 0.02 ‰ lower than before freezing. During freezing time extension in recycled layer is present and during thawing compression is recorded.

This behaviour corresponds to ice lenses formation. While temperature in top falls under zero, water from under layer starts penetrating to surface. Bituminous protection type of strain gauge prevents water to penetrate higher and in the case of temperature lower than 0 °C, ice lenses start creating under strain gauge which leads to extension. During all five years of monitoring a small compression is observed under asphalt layer, which is in contrast with [5] observations. In the case of explaining deformation under the wheel tracks, extension at the bottom of asphalt layer is presented. In this way, freeze-thaw cycles dominate on strain response.

Measurements of vertical displacements show that edge zone of pavement settle more than the middle zone. Traffic load on new or rebuilt pavement can causes higher settlements of road surface ([5, 14]) as well as frequent and high volumetric water content changes ([2, 8, 10]). While settlements have been measured only few times out of the freezing period and not continuously during the whole monitoring period, results do not capture possible heaving due to freezing. Furthermore, it is difficult to decide about the reason for settlements. On the authors opinion, measured settlements are caused mostly by the traffic loading.

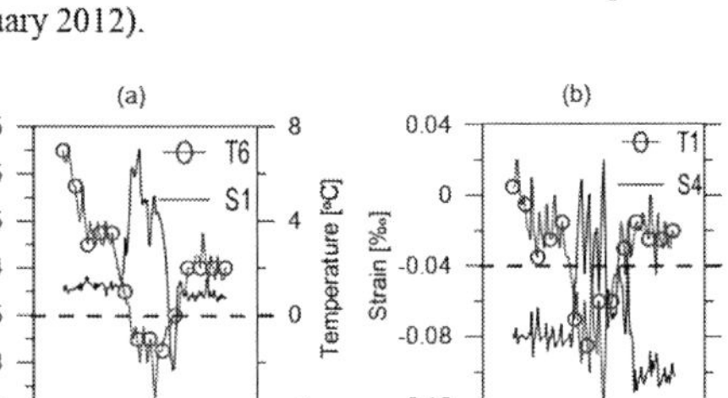

Figure 10. Relation between deformation and temperature (February 2012).

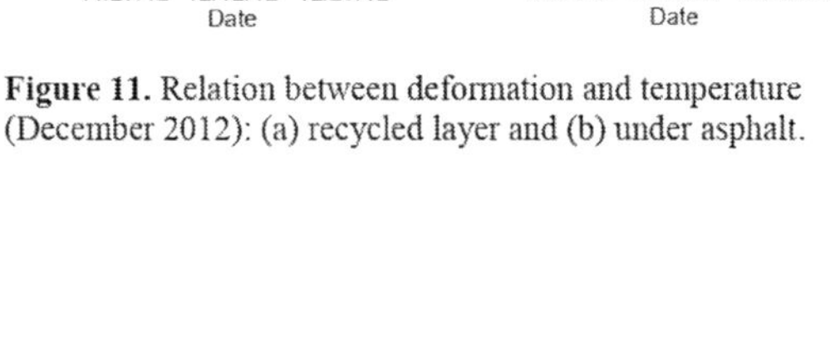

Figure 11. Relation between deformation and temperature (December 2012): (a) recycled layer and (b) under asphalt.

4 Conclusions

This paper describes hydraulic and deformation observation in pavement at National road in Slovenia during five years. The monitoring system improves knowledge about water transition in pavement and freezing depth. Presented results in this work also enable to compare behaviour of in-situ cold retreatment layer with fouled old sub base layer.

The remediation work of old pavement was done with cold in-situ pavement retreatment using foamed bitumen and cement. While this is not a common solution in Slovenia it was expected, that recycled layer would reduce the freezing depth. With recorded data it is observed that freezing depth is still 80 cm, when temperatures in pavement are at least 5 days under zero. Freezing process leads to extension horizontal strain at the bottom of asphalt layer and also in recycled layer, while thawing cause compression. Horizontal deformation at the bottom of asphalt layer after five years becomes lightly compression and in recycled layer extension in edge zone. This behaviour corresponds to geodetic measurements of surface settlements, which cause higher settlement in edge zone than in the middle zone. At least no cracks are observed on asphalt surface after five years which corresponds to small recorded deformations in recycled layer.

High values of water content are reached in the whole pavement. The main reason is no drainage system. The water from precipitation and melting of snow penetrates from edges to the middle part of pavement, firstly in old sub base layer. In case of few days high precipitation old sub base layer becomes saturated and water start penetrating to recycled layer, while no drainage is in pavement.

Acknowledgements

The preparation of this work was assisted by author's participation within the COST Action TU1202 and a short research visit of a first author at ENCP. Authors are very grateful for this opportunity.

References

1. N. Bandara, T.H. Binoy, H.S. Aboujrad, *Cold Regions Engineering 2015*, 135 (2015)
2. P.J. Becker, D.J. White, P.K.R. Vennapusa, M.J. Dunn, *Transportation Research Board* (2014)
3. T. Cackler, D.J. White, R.L. Handy, P.K.R. Vennapusa, *CEER Research Project* (2014)
4. M. Iwanski, P. Buczynski, G. Mazurek, *Sustainability, Eco-efficiency and Conservation in Transportation Infrastructure Asset Management – Losa & Papagiannakis* (2014)
5. D. Jones, R. Wu, S. Louw, TRR Journal **2462**, 126 (2014)
6. W. Li, M. Pour-Ghaz, J. Castro, J. Weiss, J. Mater. Civ. Eng. **24**, 299 (2012)
7. O. Maadani, A.O. Abd El Halim, N. Mostafa, J. Cold Reg. Eng. (to be published, 2014)
8. E. Simonsen, U. Isacsson, Cold Regions Science and Technology **29**, 135 (1999)
9. M. Tušar, M. Ravnikar Turk, L.G. Wiman, D. Kokot, S. Lenart, E. Kemperle, E., *Brussels: European commission, DG research,* (2011)
10. Z. Wen, M. Zhang, W. Ma, Q. Wu, F. Niu, Q. Yu, Z. Fan, Z. Sun, European Journal of Environmental and Civil Engineering **19**, 387 (2014)
11. D.J. White, P.K.R. Vennapusa, P. Becker, C. White, *CEER Research Project* (2013)
12. H. Xu, J.D. Spitler, Renewable Energy 72, 1 (2014)
13. X.Z. Xu, J.C. Wang, Y.S. Deng, (2010). Thermo-physic of Frozen Soil. Science Press, Beijing 113 (in Chinese)
14. F. Yu, J. Qi, X. Yao, Y. Liu, Engineering Geology **160**, 44 (2013)
15. A.M. Tang, A.N. Ta, Y.J. Cui, J. Thiriat, GTJ **32**, 1 (2009)
16. G.C. Topp, J.L Davis, A.P. Annan, Water Resources Research **16**, 574 (1980)
17. W.A. Take, D.N. Arnepalli, R.W.I. Brachman, R.K. Rowe, *OttawaGeo 2007,* 1865 (2007)
18. T.V. Duong, A.M. Tang, Y.J. Cui, V.N. Trinh, N. Calon, Geotech. Test. J. **36**, 55 (2013)
19. T.V. Duong, Y.J. Cui, A.M. Tang, J.C. Dupla, N. Calon, Can. Geotech. J. **51**, 735 (2014)
20. O.H. Jacobsen, P. Schjonning, Journal of Hydrology **151**, 147 (1993)
21. J. Stolte, J.M. Halbertsina, M. Veerman, G.J. Wösten, J.H.M. Freijer, J.I. Bouten, W. Dirksen, C.Van Dam, J.C. Van den Berg, Soil Science Society of Americal journal **58**, 1596 (1994)
22. Y. Gong, Q. Cao, Z. Sun, Hydrological Processes **17**, 3601 (2003)
23. J.M. Schneider, D. Fratta, Can. Geotech. J. **46**, 753 (2009)
24. TSC 06.5012:2003, Projektiranje klimatski in hidrološki pogoji

E3S Web of Conferences 9, 20004 (2016) DOI: 10.1051/e3sconf/20160920004
E-UNSAT 2016

A site assessment of pavement cracking in a drought environment: a case history

James B. Nevels, Jr.[1,a], Christopher R. Clarke[2], Lizhou Chen[3], and Rifat Bulut[4]

[1]*Geotechnical Engineering Consultant, Norman, Oklahoma, USA*
[2]*Geotechnical Engineer, Oklahoma Department of Transportation, Oklahoma City, Oklahoma, USA*
[3]*Graduate Student, Oklahoma State University, Stillwater, Oklahoma, USA*
[4]*Associate Professor, Oklahoma State University, Stillwater, Oklahoma, USA*

Abstract. This paper builds upon the findings of a site investigation into the causes of longitudinal cracks that developed soon after the re-construction of Himes and Johnson Streets in the City of Norman, Oklahoma between May to September 2012. The re–construction of the city street pavement and base course were completed during a protracted drought period, and shortly after completion of the pavement section longitudinal cracks occurred predominantly along the curb line. The case history concluded that the drought conditions resulted in significant subgrade soil shrinkage. The shrinkage caused changes in soil suction with depth resulting in tensile stresses in the asphalt pavement and stabilized subbase that lead to the pavement cracking. A simple practical moisture/suction variation model with depth (Hamberg) was used in conjunction with the soil water characteristic curve (SWCC) to make predictions of the soil suction changes that are resulting in the pavement cracking. The paper documents the comparison of the Hamberg model used in the case history with the Mitchell and finite element (FEM) models. The purpose of the comparison was to see if improved accuracy of the Mitchell model and a finite element model resulted in significant changes in the prediction of soil suction.

1 Introduction

In a recent case history involving longitudinal cracks that developed in two re–constructed city streets within weeks following the completion of the paving and opening up to traffic, a forensic study was performed to explain the causes for the cracking. The question was asked by the City of Norman of the project geotechnical engineer, Burgess Engineering and Testing, Inc. (BETI), to explain why the newly constructed pavements developed predominantly longitudinal cracks within weeks following completion of the paving. A compounding factor in this forensic study was the fact that the (BETI) geotechnical engineer requested that the explanation of the longitudinal cracking be done with minimum of testing and cost. This request necessitated the use of estimated soil properties. Documentation of the case history is presented by Clarke, Nevels, and Cabrera [1].

2 Pavement typical section

The street construction required the removal of the old 150-mm concrete slab pavement installed in the early 1950's. The new pavement section consisted of 150-mm asphalt pavement placed in two lifts (bottom 100-mm layer and a top 50-mm layer) underlain by a 150-mm thick hydrated lime modified subgrade base course.

During and following the pavement re-construction in the spring and summer of 2012, the asphalt separated from both integral mountable concrete curbs throughout the lengths of the pavement re–construction for both Himes and Johnston Streets. The crack widths were observed to range from a 3 to 13 mm. Additionally, longitudinal cracks were also seen to develop under the canopies of the larger trees and in a few non–tree areas.

At the time of the re-construction of Himes and Johnson Streets, the City of Norman as well as much of Oklahoma experienced a very dry year. A snap-shot of the climatic conditions for approximately one year period from the time of the project letting forward based on the Standard Precipitation Index (SPI) ranged from a moderate to an extreme drought.

3 Subgrade characterization

According to the current US Department of Agriculture (USDA) Natural Resources Conservation Service (NRCS) [2], the predominantly mapped soil series underlying the lineal extents of both Himes and Johnson Streets is the Kirkland soil series, 0 to 3 percent slopes. The soil profile presented in Figure 1 was used in the subgrade analysis to characterize the Kirkland soil profile with depth. An estimate of the shrinkage limit with depth was made from Arthur Casagrande equation, Shrinkage

[a] Corresponding author: jnevels1@cox.net

Limit (SL) = 20 +/– Δpi, where Δpi is the distance above or below the A–line in the soil plasticity chart [3]. The dry unit weight (γ_d) was estimated from the tabulated primary characterization data found in the Web Soil Survey for selected Kirkland soil series pedons in Cleveland County wide basis [4]. The soil properties in the (NRCS) data set that matched the depths of the hand auger boring shown in Figure 1 was selected for the subgrade soil analysis. The total unit weight (γ_t) and shrinkage limit (SL) estimates are shown in Table 1. The specific gravity of the Kirkland soil series soil with depth was estimated as 2.70.

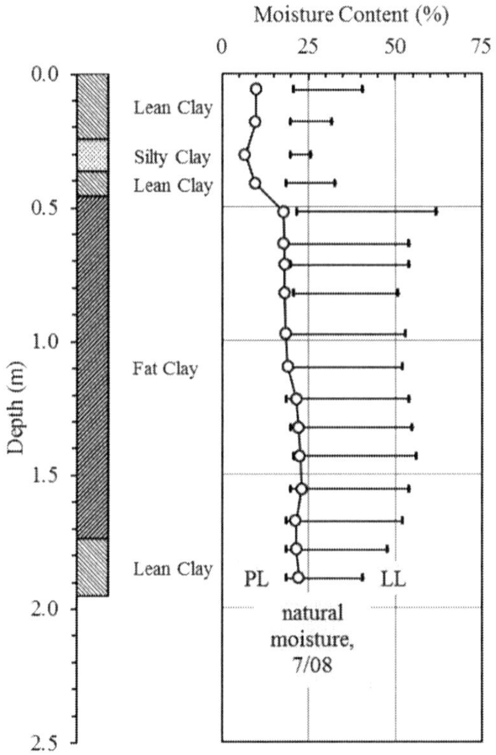

Figure 1. Kirkland series profile giving gravimetric moisture content versus depth.

The in situ moisture contents used in the analysis were taken from an average of the four auger borings in the BETI report. These auger borings were augured to a depth of 0.91 m. The four gravimetric moisture content (MC) measurements were all made in the subgrade soils at 0.30- and 0.76-m depths on Himes and Johnson Streets, and these moisture contents were at or exceeded the plastic limit. The average moisture content at the 0.30- and 0.76-m depths respectively were used the wet side points in the analysis of the Kirkland soil profile in Figure 2.

4 Hamberg moisture variation model

The problem encountered in the case history is one where the effect of drought drying conditions can be evaluated on an exposed subgrade where the initial moisture content is at approximately the plastic limit near the ground surface (following pavement removal) and approaches the shrinkage limit an extremely dry soil profile prior to paving.

Table 1. Kirkland soil series shrinkage limit estimate.

Horizon	Depth (m)	γ_d (kg/m^3)	MC (%)	γ_t (kg/m^3)	SL (%)
Ap1	0.00-0.12	1710	9.99	1880	*
Ap2	0.12-0.27	1640	9.77	1800	*
Bt1	0.27-0.37	1690	6.73	1800	*
Bt2	0.37-0.51	1860	9.68	2040	13
Bt3	0.51-0.67	1800	18.07	2130	10
Bt4	0.67-0.85	1940	18.13	2290	11
Btk1	0.85-1.17	1660	18.78	1970	12
Btk2	1.17-1.44	1880	22.08	2300	11
Btk3	1.44-1.74	1800	23.06	2220	11
Btk4	1.74-2.04	1930	22.37	2360	13
Btk5	2.04-2.38	1760	24.75	2200	14

* Profile was excavated to 0.4 m. In Table 1, MC is the gravimetric water content and SL is the shrinkage limit.

Estimates of idealized of initial and final moisture contents/suction in the soil profile are required in order to predict the effect of the drought climatic conditions had in causing the pavement cracking. The Hamberg model [5] which is based on the assumptions that the initial moisture content near the ground surface does not fall much below the shrinkage limit and the maximum moisture content under a paved surface does not significantly exceed the plastic limit was utilized in the analysis. Choosing the plastic limit and the shrinkage limit as the initial and final moisture contents, respectively allows for a straight line assumption between the ground surface and the active depth, see Figure 2. The average moisture contents at the 0.30- and 0.76-m depths from the BETI report respectively were used the wet side points in the analysis of the Kirkland soil profile in Figure 2.

5 Suction comparison using different methods

It is well established in the literature that suction change is a primary reason leading to the development of longitudinal cracks in pavements. Based on the measured moisture contents at different depths, the suction can be obtained using different methods. In this section, the suction profile from soil water characteristic curve (SWCC) will be compared to the suction profile from Mitchell model and FEM analysis respectively.

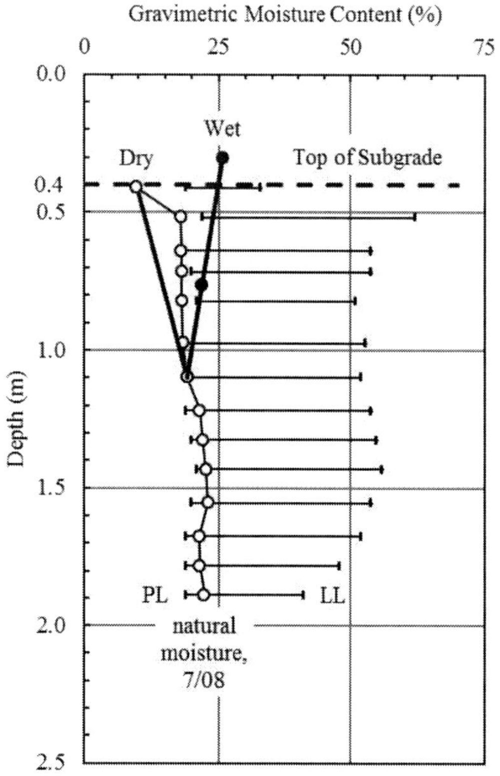

Figure 2. Kirkland series gravimetric moisture content profile.

5.1 Suction profile from SWCC

Zapata [6] proposed that the fitting parameters of Fredlund and Xing model [7] were closely related to the plasticity index and the percentage passing No. 200 sieve for plastic soils. The following equations were put forward to calculate those fitting parameters:

$$a_f = 0.00364(wPI)^{3.35} + 4(wPI) + 11 \quad (1)$$

$$m_f = 0.0514(wPI)^{0.465} + 0.5 \quad (2)$$

$$n_f = m_f[-2.313(wPI)^{0.14} + 5] \quad (3)$$

$$h_r = a_f[32.44e^{0.0186(wPI)}] \quad (4)$$

where e is Euler's number (2.71828), wPI is the product of plasticity index and decimal passing No. 200.

For Kirkland soils, the average product of plasticity index and percentages passing No. 200 sieve is 33.47. Substituting those values into Equation (1), (2), (3) and (4), the four fitting parameters obtained are: $a_f = 611$, $m_f = 0.76$, $n_f = 0.93$, and $h_r = 36945$. The saturated gravimetric moisture content is 0.25. Equation (5) is obtained by substituting the data above into the model of Fredlund and Xing [7]. Figure 3 is the soil water characteristic curve (SWCC) generated using Equation (5):

$$MC = \left\{1 - \frac{\ln\left(1+\frac{\psi}{h_r}\right)}{\ln\left[1+\left(\frac{10^6}{h_r}\right)\right]}\right\}\left\{\frac{MC_s}{\left[\ln\left(e+\left(\frac{\psi}{a_f}\right)^{n_f}\right)\right]^{m_f}}\right\} \quad (5)$$

where, MC is the gravimetric water content, MC_s is the saturated gravimetric water content, ψ is the suction, and a_f, m_f, n_f, and h_r are the curve-fitting parameters.

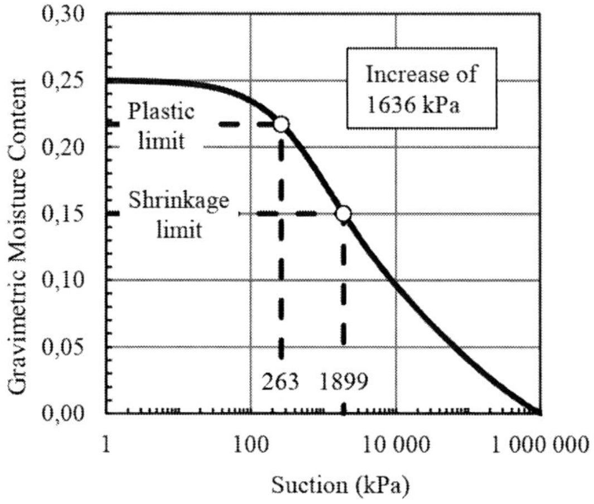

Figure 3. Kirkland series soil water characteristic curve (SWCC).

In this analysis, the equilibrium moisture content is equal to plastic limit (i.e., 18.13%), the moisture content at the dry side and wet side are 15.00% (i.e., shrinkage limit) and 21.68%, respectively. Substituting those values into Equation (5), the suction is obtained, and it is 812.8 kPa at equilibrium state, 1905.5 kPa at the dry side, and 263.0 kPa at the wet side. It is noted that the SWCC is generated using the boundary gravimetric water contents predicted by the Hamberg model, which is utilized in comparing the results with the Mitchell and FEM models as described below.

5.2 Suction profile from Mitchell model

Another widely used model for predicting suction profile is Mitchell's model [8]. The governing differential equation is given by Equation (6), and described in detail in Lytton et al. [9]. Mitchell solved the partial differential Equation (6) when the soil is subjected to a constant suction at the surface of subgrade, and the solution for these boundary conditions is provided in Equation (7):

$$\frac{\partial u}{\partial t} = \alpha \frac{\partial^2 u}{\partial x^2} \quad (6)$$

$$u = u_0 + (u_f - u_0)\left(1 - \operatorname{erf}\frac{x}{2\sqrt{\alpha t}}\right) \quad (7)$$

where, u is the suction, t is the time, x is the depth, and α is the diffusion coefficient, u_0 is the initial equilibrium suction, u_f is the final surface suction, and the term "erf" is the error function.

In order to make a reasonable comparison between the Mitchell and Hamberg (based on SWCC), it is assumed that their initial suctions and final surface suctions are the same. Also, the drying time (time from when the subgrade was exposed to the atmosphere until the subgrade was paved) is 3 months, and the diffusion coefficient is 1.0×10^{-1} mm/s, which is a typical value of diffusion coefficient of soils in Oklahoma. The diffusion coefficient is a back-calculation in the model to fit active depth and the moisture content limits at the surface.

5.3 Suction profile from finite element analysis

Besides the Hamberg and Mitchell models, a numerical method (FEM) is also utilized to study the change of suction with depth. Due to the similarity between suction diffusion and heat diffusion, the "Heat Transfer Model" in Abaqus software is utilized for analyzing the distribution of suction in subgrade soils.

In this analysis, the final surface suction, the equilibrium suction, the drying time and the diffusion coefficients are the same as those used in the Mitchell and Hamberg models. The boundaries of the FEM model are shown in Figure 4(a). The CD side is the surface of subgrade soil which is exposed to constant suctions. AB side is the bottom of subgrade, and regarded as no suction change, and BC side and AD side are the vertical boundaries of the subgrade soil. Figure 4(b) is the screenshot of dry suction profile obtained from Abaqus. The red color means the highest suction 1905.5 kPa, and the blue color represent the equilibrium suction 812.30 kPa, and the other colors are for transitional suctions. The suction profiles predicted by the three models are presented in Figure 5.

5.4 Comparisons

Figure 5 is generated using the suctions from SWCC (using the Hamberg model), Mitchell model and finite element analysis. It can be seen that the shapes of the suction profiles predicted by the Mitchell and FEM models are similar, as compared to the Hamberg model.

Suction decreases with depth on the dry side, and increases with depth on the wet side. Also, all curves intersect at 1.0-m depth, which is the depth of the moisture active zone. Meanwhile, it is found that the suction envelope predicted by the Hamberg (SWCC) is significantly larger than the suction envelopes predicted by Mitchell and FEM models, especially on the dry side as compared to the wet side of the suction envelopes.

Although the governing differential equation for moisture flow for both the Mitchell and FEM models is the same, there are some differences between them, especially at shallow depths. Even though the Hamberg model is very simple and practical to use, one should be cautious in terms of its validity for realistic prediction of suction profiles. On the other hand, the Hamberg model can be accurate enough to generate the suction profile on the wet side for practical purposes.

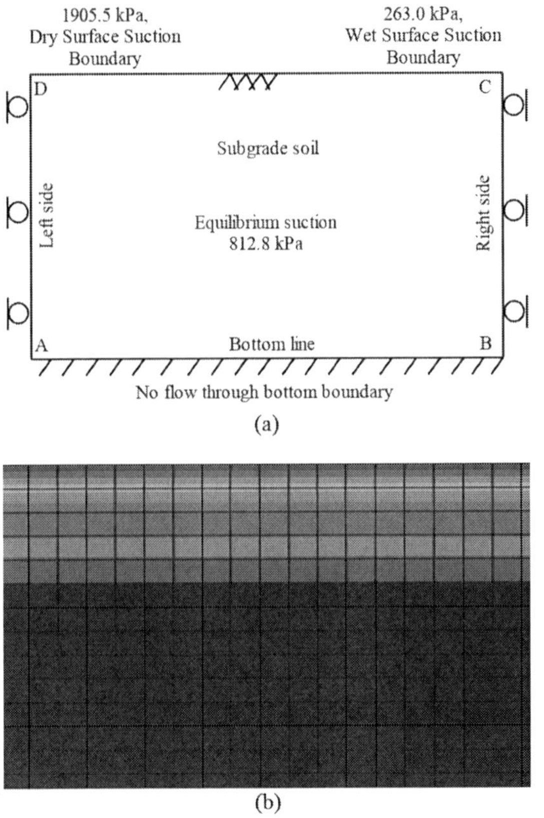

Figure 4. (a) Geometrical model of FEM; (b) Distribution of suction with depth at dry side.

6 Conclusions

The SWCC provides an expedient way of relating moisture/soil suction with depth for a soil profile. The data presented is in terms of soil suction for the case history where a pavement was removed with the underlying subgrade soil moisture at or exceeding the plastic limit and allowed to dry out to approximately the shrinkage limit before the pavement was laid. Based on the data presented in the three models in Figure 5, the following conclusions can be made:

- The Mitchell and FEM analysis predict similar estimates of the soil suction with depth. The Mitchell and FEM models are very close on the wet side but diverge slightly more with depth on the dry side.
- Integrating the area between the wet and dry curves for each model in Figure 5, the areas between the subgrade surface and the active depth are as follows: FEM = 414 kPa·m; Mitchell = 482 kPa·m; and Hamberg = 827 kPa·m. On a ratio basis and assuming that the FEM provides the best estimate with depth at 1.0, the Mitchell has a ratio of 1.16, and the Hamberg model has a ratio of 2.00.

- The variation in the model curves is consistent with published field data in the dry side versus wet side moisture content/suction profiles with depth.
- Regarding the prediction of moisture content/suction with depth, the Hamberg model is a good first indicator but very conservative approach considering especially the dry side.
- The Hamberg model is appropriate only for preliminary analysis.

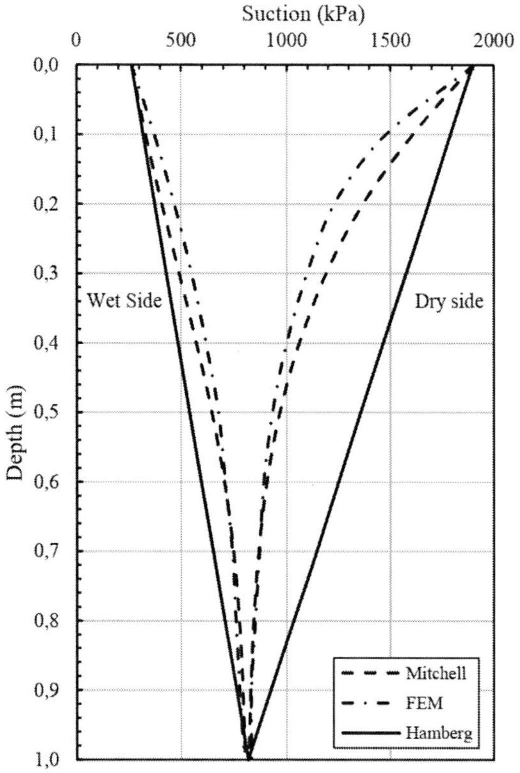

Figure 5. Comparison of suction profiles from different models.

References

1. C.R. Clarke, J.B. Nevels, Jr., A.S. Cabrera, "Himes and Johnson Street Subgrade Soil Analysis", *UNSAT2014*, Sydney, Australia (2014)
2. Natural Resources Conservation Service (NRCS), Web Soil Survey 3.1, Washington, DC: US Department of Agriculture (2014)
3. R.D. Holtz, W.D. Kovacs, *An Introduction to Geotechnical Engineering* (Prentice–Hall, Inc., Englewood Cliff, NJ, 1981)
4. Soil Conservation Service (SCS), Soil Survey of Cleveland County, Oklahoma. Washington DC: US Department of Agriculture (1987)
5. J.D. Nelson, D.J. Miller, *Expansive Soils: Problems and Practice* (John Wiley & Sons, Inc., New York, 1992)
6. C.E. Zapata, "Uncertainty in soil-water-characteristic curve and impacts on unsaturated shear strength predictions." Ph.D. Dissertation, Arizona State University (1999)
7. D.G. Fredlund, A. Xing, "Equations for the Soil–Water Characteristic Curve", Canadian Geotechnical Journal, **31**(3): pp. 521–532 (1994)
8. P.W. Mitchell, "The structural analysis of footings on expansive soil." *Research Report No.1*, Kenneth W.G Smith and Associates Pty. Ltd. Newton, South Australia (1979)
9. R.L. Lytton, C.P. Aubeny, R. Bulut, "Design procedure for pavements on expansive soils." *TxDOT Report No. 0-4518-1*, Texas Transportation Institute, College Station (2005)

Water balance of an earth fill built of organic clay

Emanuel Birle[1a], Dirk Heyer[1]

[1]*TU München, Zentrum Geotechnik, 82284 München, Germany*

Abstract. The paper presents investigations on the water balance of an earth fill built of organic clay in humid climate. As the organic soil used for the fill contains geogenetically elevated concentrations of arsenic, particular attention is paid on the seepage flow through the fill. The test fill is 5 m high, 30 m long and 25 m wide. The fill consists of the organic clay compacted at water contents wet and dry of Proctor Optimum covered by a drainage mat and a 60 cm thick top layer. For the determination of the water balance extensive measuring systems were installed. The seepage at the bottom measured so far was less than 2 % of the precipitation. The interflow in the drainage mat above the compacted organic clay was of similar magnitude. The estimated evapotranspiration reached approx. 84 % of the precipitation. According to these measurements the percolation is much lower than the percolation of many landfill covers in humid climates.

1 Introduction

In the area of the "Erdinger Moos", where the Munich Airport is located, the top soil consists of clays with organic content. These organic soils contain geogenetically elevated concentrations of arsenic. The organic top soils borrowed in the course of building activities are to be reused for noise and view protection embankments. Laboratory investigations were performed to investigate the compaction properties as well as the mechanical and hydraulic behaviour of the organic soils under compacted conditions. The results show that these organic soils can principally be used for such embankments. As the permeability of the soil is low, additional sealing layers are not necessary. For the detailed investigation of the hydro-mechanical behaviour and for the determination of seepage quantities a large scale, 5 m high, 30 m long and 25 m wide fill was constructed. For the measurement of the hydrology of the fill including seepage flow through the compacted organic soil extensive measuring systems were installed. The fill was constructed in autumn 2008. First results of the water balance of the fill were published at the fifth international conference on unsaturated soils 2010 in Barcelona [1] and at the GeoCongress 2012 [2]. Since then further measurements were taken and valuable information about the water balance could be gained.

In the following the design of the fill and the measuring instrumentation is briefly described. Thereafter, the results of the measurements are presented and the water flow within the fill is analysed.

2 Soil properties

According to DIN 18196, the soil used for the field test is a clay with organic content (OH), with a liquid limit of 124 % and a plastic limit of 44 %. The organic content determined by the ignition method lay between 25 % and 30 %. For the determination of the compaction behaviour a Standard Proctor test was carried out. Other than in mineral soils no distinct Proctor Optimum could be determined. Furthermore, a dry density of approximately 0.8 g/cm^3 and a water content ranging between 55 % und 75 % was noted. The coefficient of permeability depends strongly on the compaction water content. For a water content of 55 % a coefficient of permeability of $k = 5 \cdot 10^{-8}$ m/s and for a higher water content of 75 % a lower coefficient of $k = 5 \cdot 10^{-9}$ m/s was determined. The natural water content in the field varies, depending on climatic conditions, between 60 % and 85 %.

The soil water retention curve (SWRC) was determined for compacted samples with an initial water content of 75 % and a dry density of 0.8 g/cm^3. The low suction range up to 100 kPa was investigated by means of tensiometers (matric suction). For the high suction range a chilled mirror dew-point hygrometer was used (total suction). Figure 1 shows the obtained drying and wetting curves. For the determination of the drying path, the samples were not previously saturated, the drying was rather started from the initial state after compaction. Starting from the initial state after compaction also a wetting path was determined. The main wetting path of the SWRC was investigated with samples previously dried from the initial state after compaction to a hygroscopic water content of approximately 3 %. To

[a] Corresponding author: e.birle@tum.de

reduce the time required for the determination of the SWRC, several samples were prepared and dried to the desired water contents. Thereafter the matric suction was measured by tensiometers and the exact water content was determined by oven drying.

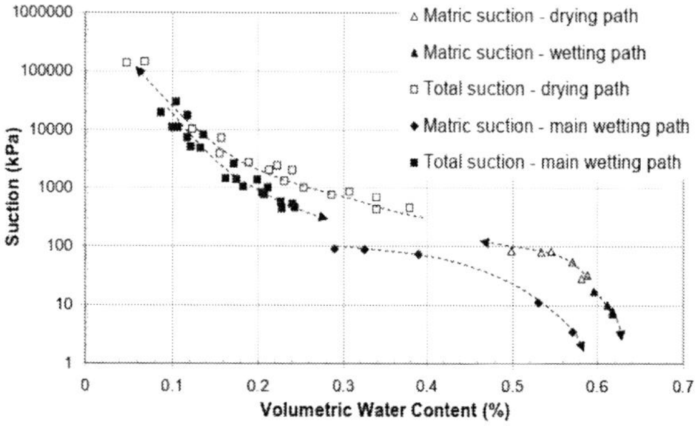

Figure 1. Soil water retention curve

3 Design of the test fill

The design of the fill is shown in Figure 2. The compacted organic soil is covered by a drainage mat and a 0.6 m thick top layer of organic clay. The fill is 5 m high, 30 m long and 25 m wide. As the hydraulic behaviour depends strongly on the compaction water content, it was decided to divide the fill in two parts. On the west side the soil was compacted with an initial water content of approximately 55 % and on the east side with an initial water content of approx. 75 %. The compaction was performed by means of a heavy polygonal drum roller compactor (25 tons). With it a dry density of the dry and wet compacted material between 0.75 g/cm^3 and 0.85 g/cm^3 was achieved. After compacting the organic material the drainage mat was installed. For the non-compacted top layer, a clay with organic contents of similar properties was used.

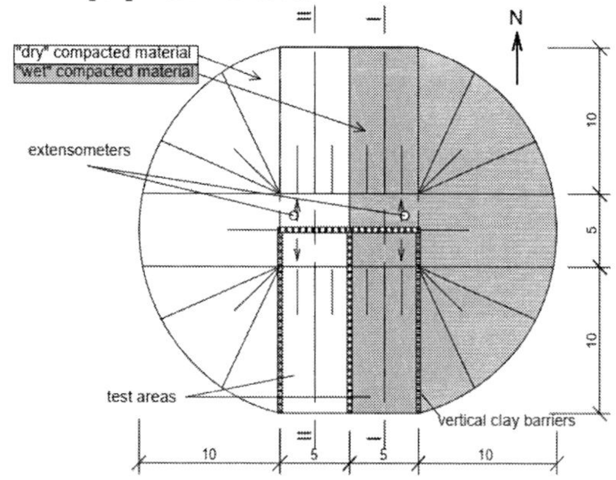

Figure 2. Plan View of the Test Embankment

On the south side of the fill, two areas of 5 m width and 12.5 m length were separated from the remaining part of the fill by vertical clay barriers of very low permeability (see figure 1). In these areas, two-dimensional flow conditions can be assumed. For the analysis of the seepage extensive measuring systems were installed in these test areas. The leachate at the bottom of the embankment is collected by a drainage mat and an underlying geomembrane. A tube connected to the geomembrane discharges the water to a gauge well situated on the south side of the fill, where the leachate discharge is measured by electronic balances. The interflow in the drainage mat covering the compacted organic soil is collected at the bottom of the test areas and measured in the gauge in a like manner. The flow of water in the compacted organic soil is monitored by means of tensiometers and TDR-probes. The position of these sensors after compaction is shown in Figure 3.

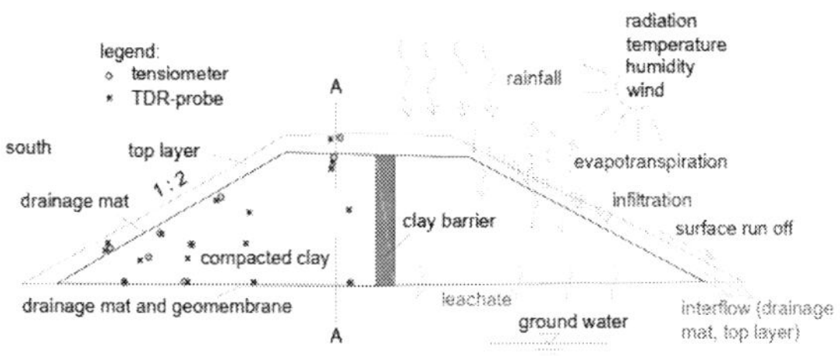

Figure 3. Cross section of the fill with the final arrangement of the tensiometers and TDR-probes (left section) and the description of the water balance (right section)

For the determination of the volumetric water content by means of the signals of the TDR-probes, a calibration of the TDR-probes had been performed.

For the simulation of the water balance of the fill the climatic conditions have to be known. The climate is measured by a meteorological station close to the fill. It consists of several sensors measuring the air temperature and humidity, the effective radiation, the wind velocity and direction as well as the amount of precipitation.

The vertical deformations of the fill are measured by means of four extensometers, installed in two boreholes; one in the wet and one in the dry compacted material (see Figure 2). The water balance of the fill is shown schematically in Figure 3.

4 Results and analysis

4.1 Precipitation

The precipitation recorded at the fill between November 2008 and June 2014 is shown in Figure 4. The precipitation pattern is typical for the south of Germany showing a maximum in the summer months. This is caused by the accumulating effect of the Alps that is particularly significant during the summer months. In total a precipitation of 4308 mm was measured, which leads to an average annual precipitation of 760 mm. Especially the years 2013 and 2014 were slightly drier and the years 2010 and 2012 significantly wetter than the average.

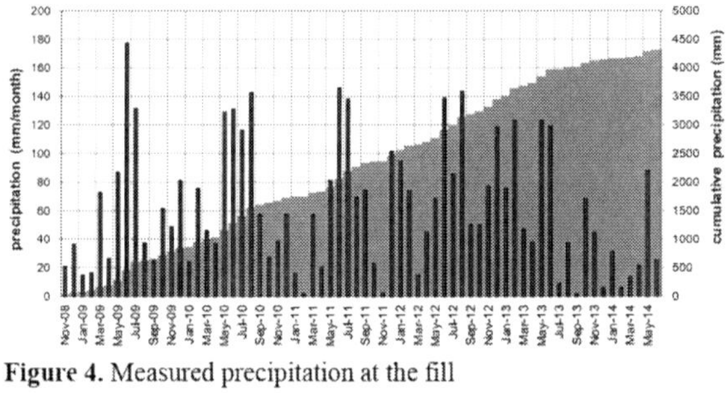

Figure 4. Measured precipitation at the fill

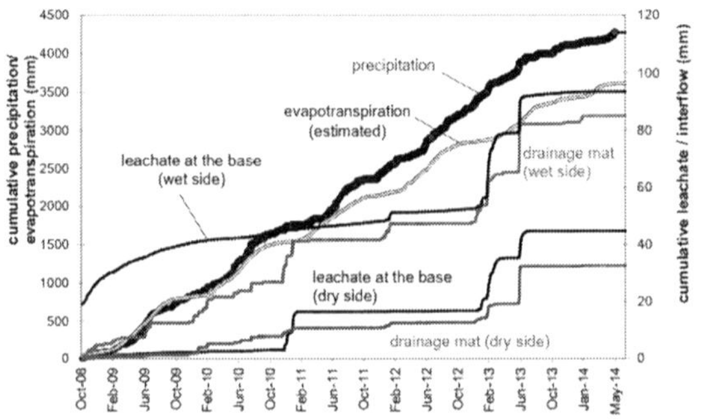

Figure 5. Cumulative water volumes of precipitation, evapotranspiration, drainage interflow and leachate for the wet and dry compacted clay

4.2 Seepage and interflow

Figure 5 shows the accumulated water volumes of precipitation, evapotranspiration, interflow in the drainage mat and leachate at the bottom of the wet and dry compacted material. For the wet side 93 mm were collected at the base of the test area in the entire observation period and 85 mm from the drainage mat placed above the compacted organic clay. In comparison to the total precipitation of 4308 mm these are very small fluxes (approx. 2 %). For the dry compacted clay even smaller fluxes for the leachate at the base (45 mm) as well as for the drainage mat (33 mm) were recorded. This is caused by the low water content of the "dry" compacted material for which reason main part of the seepage water is retained by it. The small cumulative interflows collected from the drainage mats of the wet and dry side show that the top soil has a high water retention capacity and that evapotranspiration plays a significant role.

In detail the amount of leachate collected at the base of the wet and dry compacted soil is given in Figure 6. For the wet side comparatively large leachate arose directly after compaction in September 2008. Taking the measured pore water pressures into consideration, a consolidation process was initiated by the compaction of the wet clay. In some areas the water content of the wet compacted material was too high, so that excess pore water pressures were generated by the heavy compaction equipment. The consolidation phase is assumed to be widely completed at the end of April 2009. After April 2009 the leachate flow reduced further and took on an approximately constant value in January 2010. Thenceforth a mean leachate flux of approx. 0.3 mm/month was measured at the base of the wet side except of the periods between November 2010 to February 2011, December 2011 to February 2012 and January 2013 to June 2013. High seepage of approx. 37 mm was measured especially between January 2013 and June 2013. This was caused by very high precipitation within these 6 months. Due to the comparatively dry climate conditions since August 2013 no significantly seepage was measured.

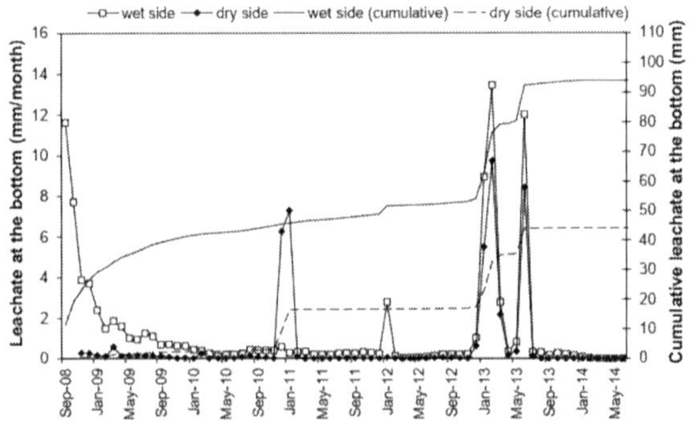

Figure 6. Leachate at the Base of the Embankment

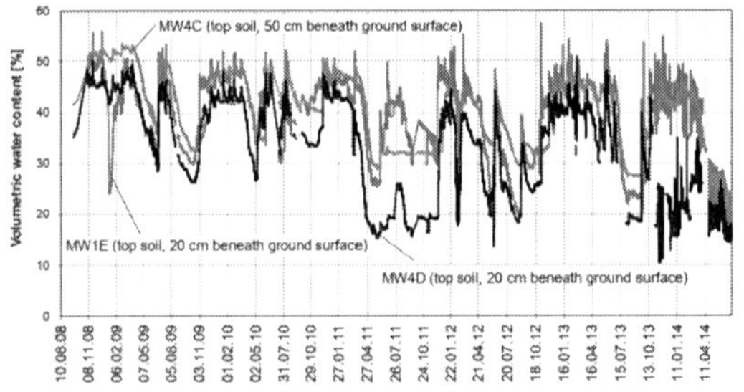

Figure 7. Volumetric Water Content in the Top Soil

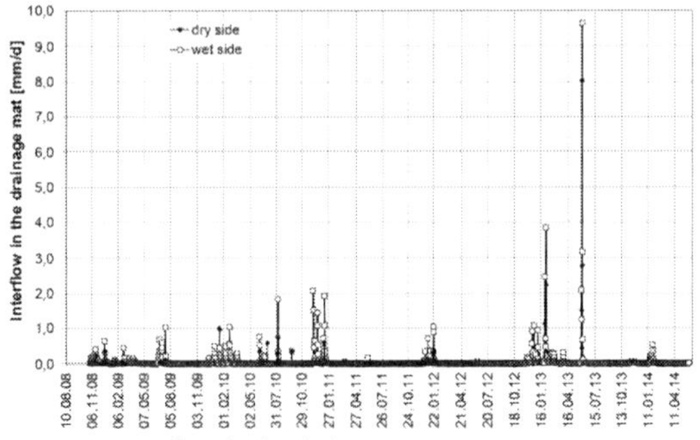

Figure 8. Interflow in the drainage mat

From the measurements of the TDR-probes located in the top soil it can be seen (Figure 7), that the interflow in the drainage mat (Figure 8) correlates with the time dependent variation of the water content in the top layer. An interflow in the drainage mat occurs only when the top soil is nearly saturated. This happens usually in winter, when no evapotranspiration takes place or in summer after a wet period with high precipitation (e.g. in

 DOI: 10.1051/e3sconf/20160920005

summer 2013). As the wet compacted clay is almost saturated after compaction, less water is absorbed by the soil and a higher interflow in the drainage mat is noticeable in comparison to the dry compacted soil. However, similar to the dry side, an interflow in the drainage mat takes place only occasionally, when the top soil reaches water contents above the water content at field capacity and thus infiltrating water can not be retained.

4.4 Evapotranspiration

The potential evapotranspiration was determined by the Penman-Monteith equation as described by Allen et al. (1998) [3]. For the determination of the aerodynamic resistance a crop height of the grass vegetation of 15 cm was assumed. The leaf area index was estimated according to [3] at 3.6, with which a surface resistance of 56 s/m was determined. Due to the inclination of the test fill and the resulting different net solar radiation on the inclined slopes, the potential evapotranspiration determined by Penman-Monteith had to be corrected. According to Golf [4] the evapotranspiration on the south facing slopes was increased by a factor of 2 for the winter half-year and a factor of 1.25 for the summer half-year. The potential evapotranspiration ETp describes the maximum evapotranspiration that is possible for saturated soils. The actual evapotranspiration depends on the water content of the top soil. For the determination of the actual evapotranspiration ETa the following equation of Disse [5] was used:

$$ET_a = \frac{1-e^{-r\cdot\frac{\theta-\theta_{PWP}}{\theta_{FK}-\theta_{PWP}}}}{1-2\cdot e^{-r}+e^{-r\cdot\frac{\theta-\theta_{PWP}}{\theta_{FK}-\theta_{PWP}}}}\cdot ET_p \qquad (1)$$

By means of this equation the actual evapotranspiration ETa can be determined from the potential evapotranspiration ETp considering the actual water content of the top soil θ, the water content at field capacity θ_{FK} and the water content at the permanent wilting point θ_{PWP}. The parameter r depends on the vegetation and was taken to be r = 4 according to [5]. The volumetric water content at the wilting point was determined from the soil-water retention curve of the top soil as the water content at a matric suction of 1500 kPa. For the determination of the volumetric water content at field capacity the volumetric water contents measured at the beginning of the growing seasons (see Figure 8) were analyzed. By doing so a value of 0.45 was taken. The actual volumetric water content θ was taken from field measurements. As the root depth of the grass vegetation was assumed to be larger than 20 cm only the measurements of the sensor MW4C located approx. 50 cm beneath the ground surface (see Figure 8) were used, allowing the determination of the actual evapotranspiration given in figure 9. According to the climatic conditions in the summer half-years a significant evapotranspiration take place while in the winter half-years the evapotranspiration is comparatively small. For the entire observation period between November 2008 and June 2014 an evapotranspiration of 3605 mm was determined (see figure 9). This means, that approx. 84 % of the precipitation removed from the ground surface and the top soil by evaporation and transpiration. This explains the small seepage fluxes recorded at the drainage mat of the wet side.

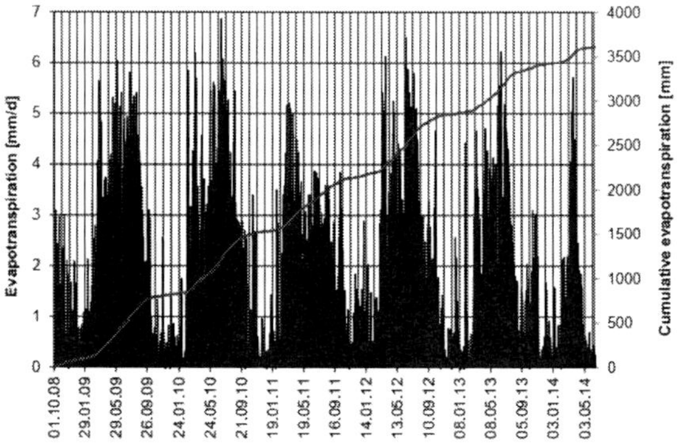

Figure 9. Estimated Evapotranspiration

4.5 Storage

For the wet side it can be assumed that for the period between 2008 and 2014 no relevant storage has occurred as the material was almost saturated after compaction. This is confirmed by the measurements of the water content sensor at the base of the fill which show no significant changes over the considered time period (see Figure 10).

In contrast, the dry compacted organic clay retains a significant part of the infiltrating water and thus the seepage at the base of the dry side and the interflow from the drainage mat of the dry side is only half of that collected from the wet side. From Figure 11 it can be seen how the water content of the dry compacted material at the base of the fill increased with time. The initial volumetric water content was between 40 % and 50 %. In summer 2014 the water content has reached values between 48 % and 60 %.

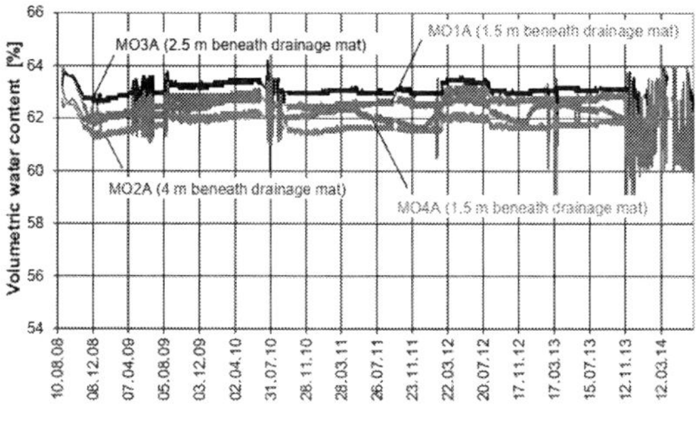

Figure 10. Volumetric water contents at the base of the wet side

DOI: 10.1051/e3sconf/20160920005

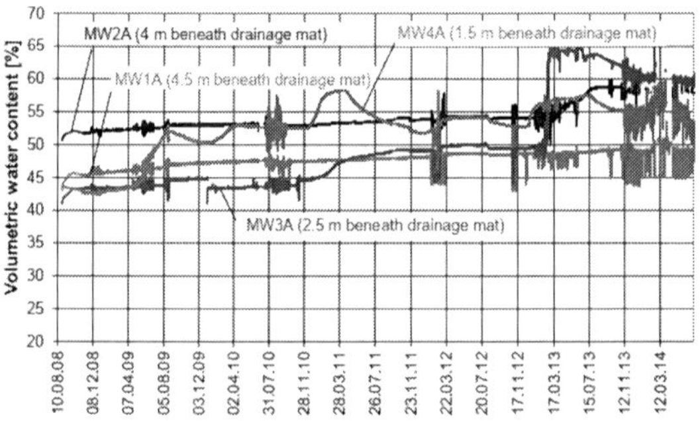

Figure 11. Volumetric water contents at the base of the dry side

4.5 Surface run off

According to the water balance equation the precipitation (N) must be equal to the sum of evapotranspiration (ET), surface run off (SR), seepage flow (seepage at the base and interflow - S) and storage (W):

$$N = ET + SR + S + W \tag{2}$$

As described in chapter 4.4 for the wet compacted organic clay the storage can be neglected. Thus the surface run off, which is not measured directly, can be calculated from the precipitation, the evapotranspiration and the seepage flow as follows:

$$SR = N - ET - S \tag{3}$$

According to the cumulative fluxes shown in Figure 5 the accumulated surface run off between 2008 and 2014 is 503 mm. As the same material was used for the top layer of the wet and the dry side, for the dry side a similar surface run off is expected.

5 Conclusions

For the detailed investigation of seepage quantities through a slightly contaminated organic clay a 5 m high, 30 m long and 25 m wide test fill was constructed. In view of the low permeability of the soil after compaction it was covered only by a drainage mat and a top soil without an additional sealing liner. As the hydraulic behavior depends strongly on the compaction water content, the test fill was divided in two parts. In the west side the soil was compacted at an initial water content dry of Proctor Optimum and in the east side at a water content wet of Proctor Optimum. For the measurement of the water balance of the fill including seepage flow through the compacted organic soil extensive measuring devices were installed. From the measurements the following results were obtained:

The seepage collected at the base of the wet compacted clay in the period between November 2008 and June 2014 amounts to less than 2 % of the precipitation. Also the interflow in the drainage mat above the wet compacted clay is small (2 % of precipitation).

The seepage collected at the base of the dry compacted clay between November 2008 and June 2014 is 1 % of precipitation and so even smaller than that of the wet compacted clay. Also the interflow in the drainage mat above the dry compacted clay is small as significant part of the seepage water is retained by the dry compacted clay. This means that a saturation process takes place and that the seepage at the base of the dry compacted clay could increase once the retention capacity of the soil is reached.

Evapotranspiration plays an important role for the water balance of the test fill. Due to the high retention capacity of the soil and the underlying drainage mat which acts as a capillary break, the evapotranspiration reaches approx. 84 % of the precipitation.

In comparison to measurements obtained for landfill covers in humid climates ([6], [7], [8]) the percolation of the top layer is very low and the evapotranspiration is very high. This is caused by the high organic content of the top layer, that results in a high water retention capacity, the high inclination of the test fill slope (26,6°) and the low permeability of the compacted organic clay.

References

1. E. Birle, M. Boso, D. Heyer. 5th Int. Conf. on Unsat. Soils. Vol. II, pp. 1299-1304 (2004)
2. E. Birle, D. Heyer: GeoCongress 2012, Oakland, (2012)
3. R. G. Allen, L. S. Pereira, D. Raes, M. Smith. Computing crop water requirements - FAO irrigation and drainage paper 56, (1998) http://www.fao.org/docrep/X0490E/x0490e00.htm
4. W. Golf. Ermittlung der Wasserressourcen im Mittelgebirge, Wasserwirtschaft und Wassertechnik 31, pp. 93-95 (1981), cited in DVWK Merkblatt 238/1996: Ermittlung der Verdunstung von Land- und Wasserflächen [Determination of Evaporation on Land- and Water Surfaces]
5. M. Disse M. (1999). Physics and Chemistry of the Earth, Vol. 24, No. 4, pp. 325-330, Elsevier Science Ltd. (1999)
6. W. H. Albright, C. H. Benson, G. W. Gee, A. C. Roesler, T. Abichou, P. Apiwantragoon, B. F. Lyles, S. A. Rock. J. Environ. Qual. Vol. 33, pp. 2317–2332 (2004)
7. W. U. Henken-Mellies. Höxteraner Berichte zu angewandten Umweltwissenschaften; Vol. 03, pp. 117-126 (2002)
8. S. Melchior, K. Berger, B. Vielhaber, G. Miehlich. Höxteraner Berichte zu angewandten Umweltwissenschaften; Vol. 03, pp. 57-74 (2002)

Earthwork laboratory tests and suction – experimental and numerical approach

Dino Mahmutovic[1,2,a], Soanarivo Rinah Andriantrehina[3], Luc Boutonnier[1] , Jacques Monnet[4], Saïd Taibi[3], Thierry Dubreucq[5]

[1] *EGIS-Géotechnique, 3, rue Docteur Schweitzer, 38180, Seyssins, France*
[2] *UNIV. GRENOBLE ALPES, 3SR, F-39800 Grenoble, France*
[3] *LOMC, CNRS UMR 6294, Université du Havre, BP 540, 75058 Le Havre, France*
[4] *GAIATECH, 22 rue Antoine Chollier, 38170, Seyssinet, France ;*
[5] *IFFSTAR, Route de Bouaye CS4, 44344 Bouguenais cedex*

Abstract. The research presented in this publication is funded by the French National Project « Terredurable », which is dedicated to the study of soils in quasi-saturated conditions (close to saturation). Stability and deformation of earth structures are reanalyzed with a quasi-saturated hydromechanical model. In this paper, IPI tests are modeled with a two-dimensional explicit finite difference program (Flac 2D). The model presented considers mechanical soil theory to approach earthwork tests by integrating a compressible bulk modulus. Results of the numerical approach show how it is possible to link parameters of near saturation soil mechanic (saturation, suction) to earthwork strength of soil parameters (IPI value).

This communication is devoted to the simulation of IPI tests. In the first part, details about Proctor test and IPI tests will be given, with main experimental results. Then oedometric tests, suction measurement, triaxial tests, mercury porosimetry and grain density (γ_s) measurements have been performed on Proctor samples in order to feed our model to simulate soils near saturation. Finally simulations are carried out using FLAC to reproduce IPI settlement curves.
This paper aims at a better comprehension of IPI tests and the hydro-mechanical behavior related to those tests.

1 Description of laboratory tests

IPI laboratory tests (standard NF-EN-13286-47) are carried out in 8 samples from the same compacted soil with different water contents. Results of IPI test are represented in Figure 1. For a better comparison with the numerical results, the complete loading curve of the force applied on the soil with respect to the settlement is drawn. The elastic and the plastic part of the curve can so be appreciated.
The main characteristics of the studied soil are presented in Table 1 below.

Table 1. Main geotechnical characteristics of the studied soil, which come from A304 site (marl soil)

Type of soil	GTR class	% $CaCO_3$	w_L	IP	% 80 µm
marle	A4	18,5	80	49	94

[a] Corresponding author: dino.mahmutovic@egis.fr

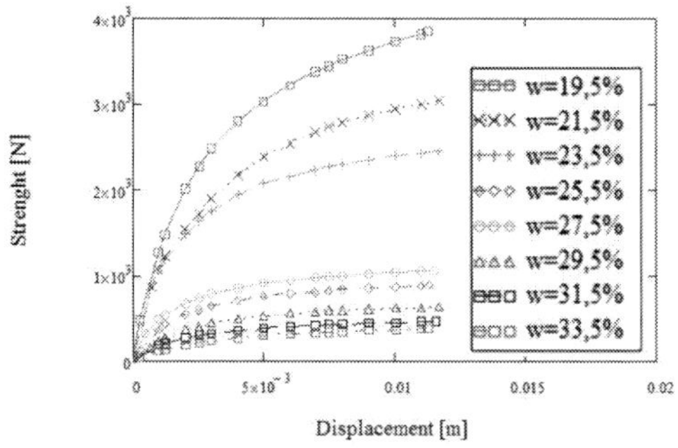

Figure 1. IPI experimental curves for different values of water content

In Figure 1, decrease of water content coincides with increase of reaction force. These obtained curves can be categorized into two groups: the first group with water content under 25.5% and the second group over this value. For the first group of curves, a high increase of the ground reaction force is observed for a small decrease of the water content. For the second group, the influence of water content on the reaction force is smaller.

This tendency can be also observed from other parameters, as suction. Proctor tests with suction measurements (measured with a filter paper method) are used. The initial suction and densities determined by Proctor tests (as shown in Figure 2) are then used for numerical simulation:

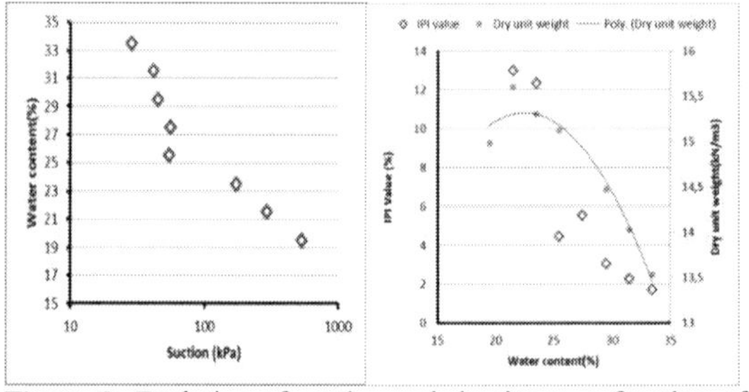

Figure 2. Evolution of suction and density as a function of water content

It can be found in Figure 2 that the suction values increase when the water content values decrease. These results show a change of slope between 23% and 25% of water content, which correspond the optimum Proctor water content. A same value can be also identified in the Figure 1. This value also represents the value of air entry (Boutonnier 2007).

Depending on if the water content locates at the dry or wet side (the air entry) of the Proctor curve, two different hydro-mechanical behaviours can be observed (Cui 1993).

In this paper only the wet side will be considered, since it concerns most earthworks, as illustrated in Figure 4. A theoretical model which is able to simulate the IPI tests will be introduced in the later section.

2 Description of laboratory tests

2.1 Determination of the quasi-saturated parameters

In this model, as shown in Figures 3 and 4, the soil can be divided into 4 domains of saturation (D1 to D4 – unsaturated to saturated) (Boutonnier 2007).

A particular attention is paid to the transition zone from saturation to desaturation (domains D2 and D3)) which are the domain concerned by the IPI tests and earthworks (at optimum Proctor and on the wet side).

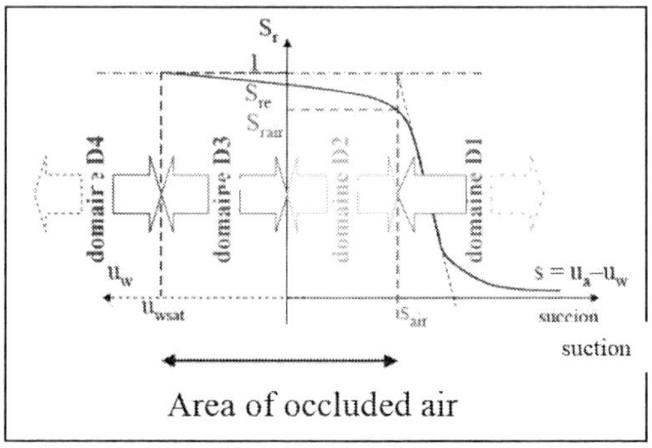

Figure 3. Division of retention curve in 4 domains of saturation

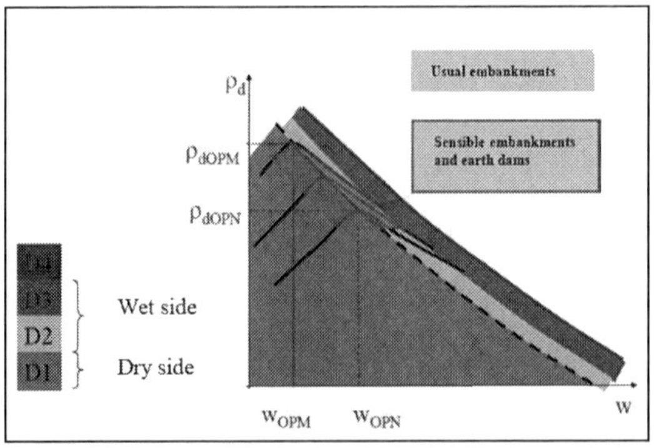

Figure 4. Division of Proctor curve in 4 domains of saturation

In the quasi-saturated areas, the model is based on the compressibility coefficient of the interstitial fluid c_f (Equation for domain D2 (1) and Equation for domain D3 (2)) (Figures 3 and 4) which varies with the degree of water saturation in the soil.

$$c_f = \frac{1}{S_r} \cdot \frac{1}{s_{air}} \left(S_{re} - S_{rair} \right) + c_w \tag{1}$$

$$c_f = \frac{1 - S_r + h.S_r}{u_w + r_{bm} + P_a - u_{wg}} + c_w \tag{2}$$

This model contains 5 parameters: the degree of saturation for a null suction S_{re}, the degree of saturation at air entry S_{rair}, the air entry suction s_{air}, the mean radius of air bubbles r_{bm}, and the Henry's constant h.

The value of air entry suction coincides with the value of suction at the optimum Proctor (200 kPa). With a measured value of γ_s of 27,5 kN/m^3, a dry unit weight of 15,3 kN/m^3, an optimum water content of 23%, the value of saturation degree corresponding to the air entry (equation (3)) is 84,6 %.

$$S_{rair} = \frac{\gamma_s}{\gamma_w} \frac{\gamma_{dopt}\, w_{opt}}{\gamma_s - \gamma_{dopt}} \tag{3}$$

The degree of saturation for a null suction can be calculated for each sample (Table 2) if the suction and the degree of saturation have been tested at the beginning after the IPI test (equation (4)).

$$S_r = S_{re} - u_w \frac{(S_{re} - S_{rair})}{u_{wair}} \tag{4}$$

Table 2. Value of S_{re} for different water content.

Water content	S_{re}	$S_{re\ corrected}$
23,5%	0,885	0,885
25,5%	0,919	0,920
27,5%	1,000	0,945
29,5%	0,968	0,965
31,5%	0,965	0,97
33,5%	0,940	0,975

DOI: 10.1051/e3sconf/20160920006

Two values of S_{re} have been corrected in order to obtain a smooth curve. The mean radius of air bubbles obtained with Mercury porosimetry tests is close to 10^{-5} m.

2.2 Determination of the constitutive law parameters

The quasi saturated parameters and the constitutive parameters have to be determined for the numerical simulation using modified Cam-Clay model. The parameters are the slope of the plastic loading λ (Equation (5)), the slope of the elastic loading κ (Equation (6)), the critical state line M , the void ratio for the reference mean pressure of 1 kPa N_0 (Equation (7)) and the mean pressure of preconsolidation p'_c.
A consolidated and drained oedometric test allows the determination of λ, κ, and N_0 parameters, as illustrated below (figure 5).

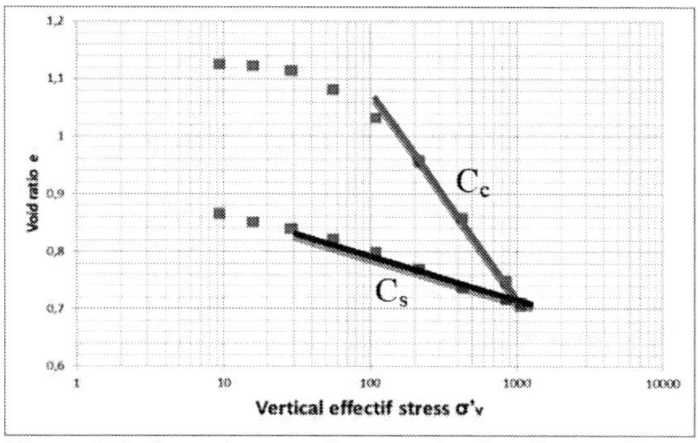

Figure 5. Oedometric curves for a sample taken from a soil compacted at the optimum Proctor and saturated before the test

$$\lambda = \frac{C_c}{\ln 10} \quad (5)$$

$$\kappa = \frac{C_s}{\ln 10} \quad (6)$$

$$N_0 = e + \lambda \, ln \frac{p'}{p'_{ref1kpa}} \quad (7)$$

Three consolidated and drained triaxial tests are carried out (samples taken from a soil compacted at the optimum Proctor), under different consolidation pressures: 100 kPa, 200 kPa and 500 kPa. The experimental results are represented below (Figure 6):

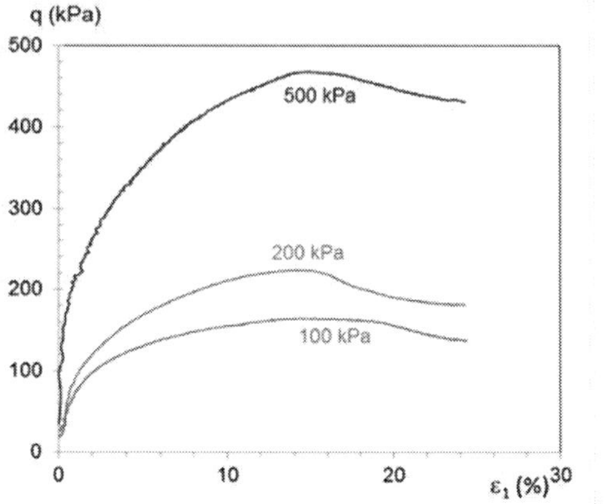

Figure 6. Shearing results of the triaxial C_{u+u} test on a sample taken from the soil compacted at the optimum Proctor

Figure 7 shows that the levels of pressure 100 kPa and 200 kPa are overconsolidated, whereas the 500 kPa level is normally consolidated. The normally consolidated parameters will be used in the simulations with a value of M equal to 0.71.

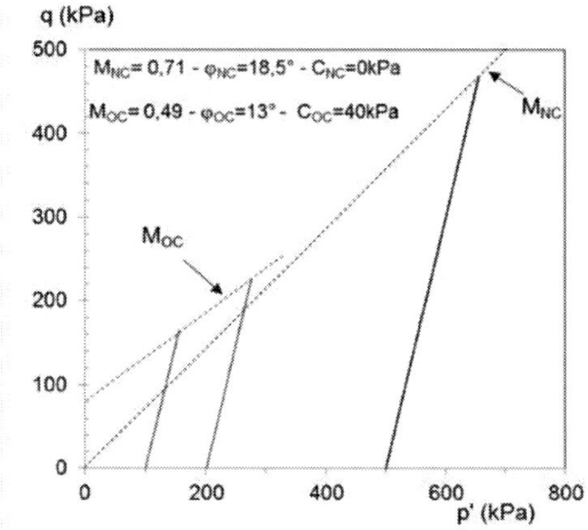

Figure 7. Limit shear strength for different lateral stress

The preconsolidation stress reflects the history of the ground in terms of stresses. Each Proctor test, compacted with a different water content, results from a different preconsolidation stress. These differences of values explain the variations observed in the IPI tests. The significance of the parameter p'_c will be discussed with the future simulations.
In a practical way, p'_c is the interception of the κ-slope and the λ-slope lines (Figure 8). If the initial state of stress of the soil at a given point in time is known as well as its void ratio, p'_c can be calculated using the virgin curve. This is where the Proctor tests and the suction measurements can step in.

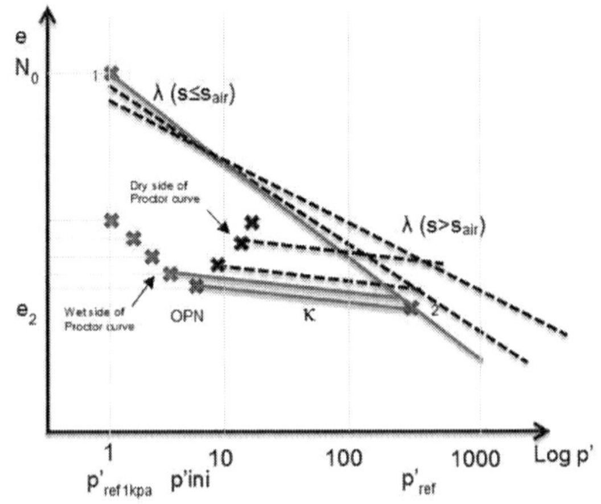

Figure 8. Evolution of the void ratio along the Proctor compaction.

γ_d values determined through the Protor tests allow to calculate the initial void ratio before the IPI test, and the suction measurements allow to identify the initial vertical effective stress value (and the initial mean effective stress value afterwards). The value of the preconsolidated effective stress is given by the following formula (Eq. 8):

$$\frac{p'_c}{p'_{ref1kpa}} = e^{\frac{e_s - e_1 - \kappa \ln\frac{p'_{ref1kpa}}{p'_{ini}}}{(\kappa-\lambda)}} \qquad (8)$$

The value of the mean effective stress p'_{ini} can be deduced from the vertical effective stress and the coefficient K_0 (Equation (9)) with the following equation:

$$p'_{ini} = \frac{\sigma'_{vini} + K_0 . \sigma'_{vini}}{3} \qquad (9)$$

2.3 Summary of parameters deduced from laboratory tests

Parameters derived from laboratory tests are presented in the Table 3 and Table 4 below.

Table 3. Value of constitutive parameters for different values of water content

W_{nat}	λ	κ	p'c [kPa]	M	N_0
25,5 %	0,153	0,04	1360	0,7	1,73
27,5 %			975		
29,5 %			655		
31,5 %			369		
33,5 %			184		

Table 4. Value of hydro-mechanical parameters for different values of water content

W_{nat}	Suction s [kPa]	Dry density	Void ratio e	Gs [kN/m³]	K_0
25,5 %	56	1,51	0,75	27,5	1,75
27,5 %	54	1,53	0,73		1,55
29,5 %	45	1,45	0,83		1,45
31,5 %	42	1,40	0,89		1,30
33,5 %	29	1,35	0,96		1,20

K0 parameters are determined using Schmid (1966) relation (Equation (10)).

$$K_0 = (1 - \sin\varphi)\left(\frac{\sigma'_{vmax}}{\sigma'_v}\right)^{0,5} \qquad (10)$$

3 Description of the numerical simulation

Flac software is used to implement the quasi saturated theory and simulate the IPI tests. Flac is based on a finite differences calculation code.

3.1 Geometry and boundaries condition

An axisymmetric calculation is used because of the axial symmetry of the IPI test. The geometry of the numerical model is drawn with the dimensions of 12 cm in height and 7,5 cm in radius which corresponds to the size of the CBR mould .

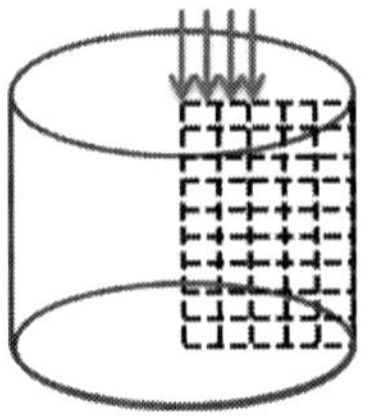

Figure 9. The illustration of the numerical simulation model

No displacement is allowed in x-y directions at the base of the model and in x direction at the lateral boundary of the model. Large strain and a re-mesh of the grid are used during the simulation

3.2 Stress fields initialisation

The IPI test is performed after the Proctor test, thus after the compaction of the soil. The effect of this compaction has to be considered in the initialization of the stress field.
In order to implement this stress field, a silo type calculation was initially used considering the friction of the metallic mold with the soil. This calculation has been then simplified using a null friction as lubrication was used on the mold before compaction.

Finally only weight of the soil and K_0 coefficient are considered.
Suction is considered to be homogeneous inside each sample of soil (with different water content).The hypothesis is true only at the beginning of the IPI test (before the penetration of the piston).

3.3 Use of a compressibility interstitial fluid

The Equations (1) and (2) allow the use of a two phases model (soil and interstitial fluid) in our code. The value of the compressibility coefficient changes during the test with the value of the degree of saturation of water. The Terzaghi effective stress principle is used because the value of suction is under the air entry value.

4 Results of the simulation and comparison with experimental results

A numerical modeling is conducted for each Proctor test on the wet side of the optimum. The experimental Proctor values are then compared to those obtained by the proposed theoretical model, as shown in the Figures 10.
An accurate agreement is observed between experimental and theoretical models, which validates the proposed model. The same constitutive parameters have been used for all the simulations (with different water contents) .

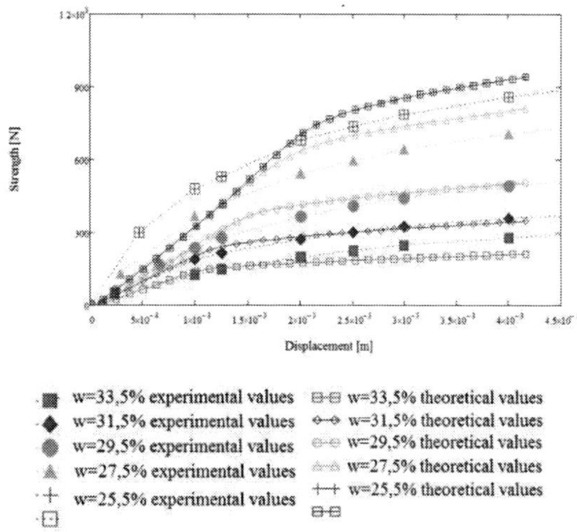

Figure 10. Comparison between experimental and numerical results of IPI test on the wet side of the Proctor curve.

The small discrepancies between numerical and experimental results may result from experimental uncertainties on the measure of suction and dry density of the soil. Influence of different parameters has been tested, including the suction, void ratio, compressibility coefficient.
Results of this parametric analysis are plotted in the Figures 11, 12 and 13.

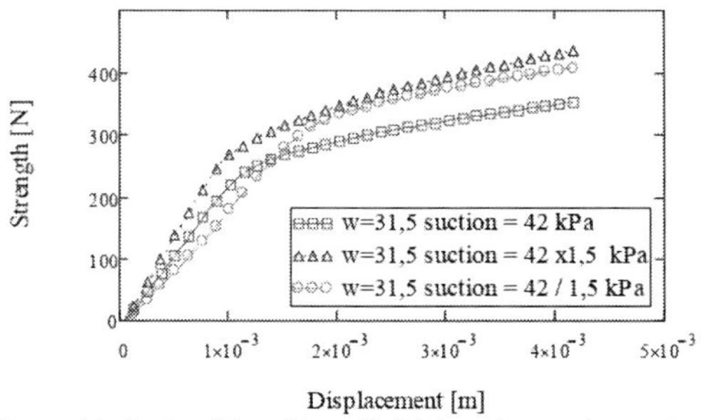

Figure 11. Study of the effect of initial suction on the result of reaction strength of soil for one example of IPI test (W_{nat} = 31,5%)

Figure 11 shows that the suction has a large influence in the initial slope of the curve and in the final value.

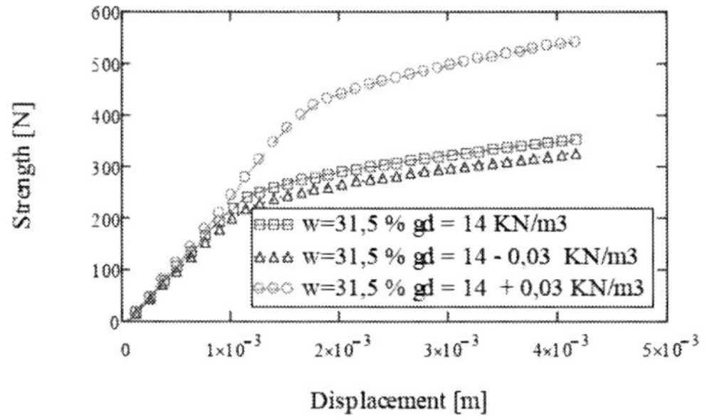

Figure 12 Study of the effect of initial void ratio on the result of reaction strength of soil for one example of IPI test (W_{nat} = 31,5 %)

Figure 12 shows that the lower is the void ratio (and the bigger is the dry unit weight), the bigger is reaction force of the soil. Figure 8 shows how the variation of the void ratio influences the value of preconsolidation pressure p'_c.

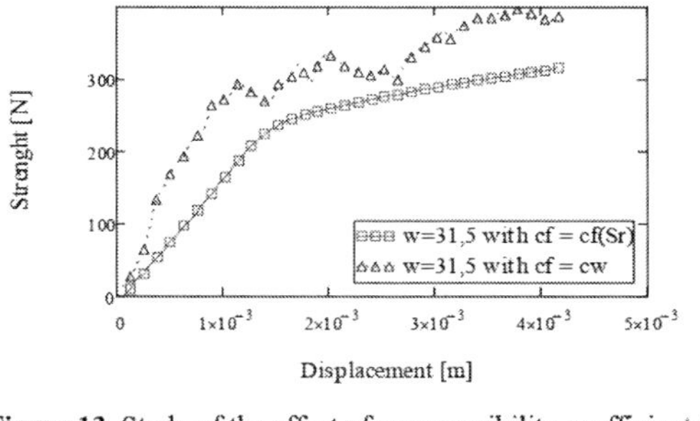

Figure 13. Study of the effect of compressibility coefficient (c_f) for air/water phases on the result of reaction strength of soil for one example of IPI test (W_{nat} = 31,5 %)

Figure 13 attest that the reaction force of the soil is strongly dependent of the value of the compressibility coefficient of interstitial fluid.

Conclusion

IPI tests on marl have been presented. These tests are completed by, suction measurements, oedometric and triaxial tests. These different tests give us a better understanding of hydro-mechanical response of earthworks tests. A model of compressibility of the interstitial fluid has been validated on the wet side of the Proctor curve, as well as an elasto-plastic constitutive law of Cam-Clay type.

Based on the newly proposed model, it is found the reaction force of the soil during IPI test (on the wet side) is governed by 3 mains parameters: suction, preconsolidated effective stress, and compressibility coefficient (c_f) for the air/water phases.

The next step will be numerical simulation of IPI test on the dry side of the Proctor curve. This work will take into account an extension of effective stress principle which depends on suction and the χ coefficient (Khalili et Khabbaz, 1998).

Acknowledgment

The authors would like to thank the French ANR project TerreDurable (ANR 2011 VILD 004 01 convention, programme Bâtiments et villes durables), which permit the study of soils near to saturation and the provision of their laboratory test.

References

1. Afnor 2005, norme NF-EN-13286-47, « Méthodes d'essai pour la détermination de l'indice portant Californien (CBR), de l'indice de portance immédiat (IPI) et du gonflement ».
2. Andrianatrehina R., Mahmutovic, D., Boutonnier, L. taibi, S. Fleureau, J-M., Fry, J-J., Monnet, J. (2015). Observation of craking evolution during free dessication – experimental and numerical approaches. Symposium international SEC 2015. Marne-la-vallée.
3. Boutonnier, L., Mahmutovic, D., Fry, J-J. (2015). Génération de pression interstitielle dans les fondations argileuses des barrages en remblai. Colloque CBR 2015.
4. Boutonnier, L (2007 Comportement hydromécanique des sols fins proches de la saturation : cas des ouvrages en terre : coefficient B, déformations instantanées et différées, retrait / gonflement. thèse INP Grenoble.
5. Khalili, N., Khabbaz, M.H, (1998). A unique relationship for the shear strength of unsaturated soils. Geotechnique vol 48, n°5, p.681-687.
6. Mahmutovic, D., Monnet, J., Loret, B., Ple, O., Pelizzari, B., Boutonnier L., Lai B. (2014). A model for weakly unsaturated soils taking into account the three phases (air, water and soil). Unsat 2014. Sydney.
7. Schmidt B., (1966), Discussion of "Earth pressures at rest related to stress history", Canadian Geotechnical Journal, vol. 3,4, pp.239-242.

 DOI: 10.1051/e3sconf/20160920007

Behavior of railroads on shrink-swell soils

Dong Wang[1], Marcelo Sánchez[1,a] Jean-Louis Briaud[1]

[1]Zachry Department of Civil Engineering, Texas A&M University, College Station, USA

Abstract. Shrink-swell soils are characterized by significant volume expansions and contractions during wetting and drying processes, respectively. Railroads on shrink-swell soils are subjected to huge solicitations associated with the volume change of the soil. The highly permeable ballast layer exacerbates soil volume changes due to moisture variation, which generally lead to unacceptable uneven settlement. The mechanism behind the observed railroads problems on expansive soils, and the associated (possible) remedial solutions, are still unclear. A site in Texas has been chosen recently to study in detail the aforementioned problems. In this paper, the typical geomechanical problems associated with railroads on shrink-swell soils in Texas are discussed. We also presented the test site information and preliminary lab tests to obtain water retention curve of the site soils.

1 Introduction

Railroad embankments and subgrades are subjected not only to cyclic train loadings (which has been continuously increasing due to freight requirement), but also to weather variations. Texas, and many other regions in the world with abundant shrink-swell soil has been observing continuous settlement problems. The settlement problems in Texas become especially severe in harsh weather condition such as during the drought summer of 2011 with an average of only 37.6cm of rain over the year. On the other hand, in October 2015, when a storm with 50cm daily rainfall hit Texas, train derailed passing by a soaked embankment in the flood region.

Existing railroads in the United States are facing challenging geotechnical problems, such as i) slurry pumping (e.g.[1]); ii) subgrade subsidence (e.g.[2]); iii) progressive shear failure (e.g.[3]); iv) ballast pocket (e.g.[4]); v) massive shear failure (e.g.[5]); vi) subgrade cavities and collapse (e.g.[6]); vii) issues associated with expansive soil behaviour (e.g.[7]); and viii) problems related to frozen soils (e.g.[8]). Typical problems observed in Texas railroads are types (i), (ii), (iii), (iv), and (vii), as described above. To advance our understanding on the mechanism of the aforementioned subgrade problems, we selected a test railroad site in Texas to study in detail the interactions between railroad, natural soils, and weather variations. This work is in progress. In this paper we present some preliminary site investigation alongside a literature review and preliminary lab test results on water retention curve.

2 Typical railroad geotechnical problems in Texas

As the state with highest expansive soil coverage in the United States, Texas encountered railroads settlement problems not only related to subgrades, but also embankments, because local shrink-swell soils have been historically used as fill materials.

The problems of expansive soils are aggravated in areas where moisture changes repeatedly. Significant volume changes observed in these soils have very negative impact on the long term performance of railroads. Uneven longitudinal and transversal settlements are very characteristics in railways constructed in swell-shrink soils.

Different factors can be associated with the development of uneven railroad settlements in shrink-swell soils. For example, non-homogeneous distributions of soils properties may induce different volume changes in different zones, leading to uneven settlements. Non-homogeneous distribution of moisture in the ground can also induce uneven settlements. For instance, the presence of trees close to the railroad may lead to important changes in the distribution of water content. Relatively high matric suction can develop in the root zone, which alter the distribution of matric suction in the surrounding soil with the consequence of the associated soil shrinkage. A non-homogeneous distribution of water content can also be related to other factors, as for example local topographic conditions. Sometimes railroad acts as an artificial barrier for the movement of the run-off water. Despite the proper design of culvert and drains, it is quite common to observe after intense rains ponding water at one side of the embankment. Figure 1.a) represents this problem schematically showing a railroad embankment crossing a terrain with a natural gradient. Under this scenario, the water accumulated on one side of the embankment will

[a] Corresponding author: msanchez@civil.tamu.edu (M. Sánchez)

infiltrate, wetting the natural soils and embankment leading to the expansion of the swelling clays. The other side of the embankment will be exposed to dry air and the soils will shrink. The final outcome is uneven transversal settlements of the railroad (Figure 1.b).

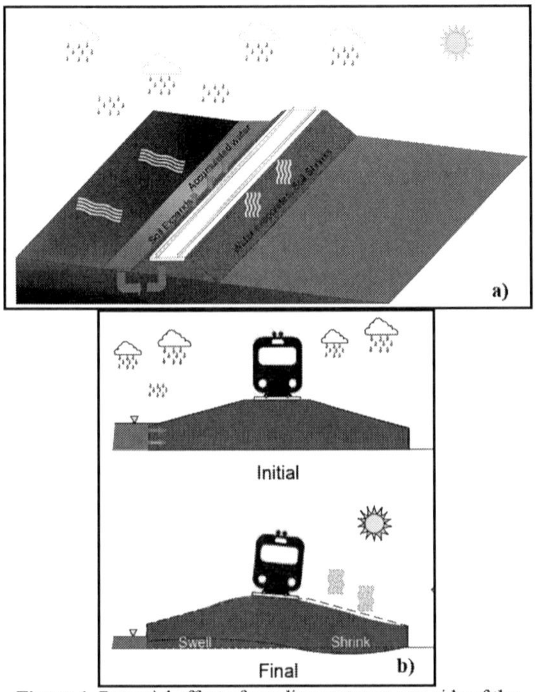

Figure 1. Potential effect of ponding water on one side of the railroad, a) 3D schematic representation, b) uneven settlements induced by wetting and drying of shrink-swell soils.

Another problem associated with expansive clays is the formation of cracks during drying. Cracks affect both mechanical and hydraulic properties of soils. Drying cracks extend the 'active zone' of expansive clays because the movement of water is favoured by the presence of vertical cracks (e.g.[9]). The presence of drying cracks in rail embankments may trigger stability problems, associated e.g. with the internal erosion induced by the water flowing through cracks ay higher velocities.

The presence of 'ballast pocket' is also very common in Texas railroads. As ballast is pushed into subgrade due to the train load, the underlying subgrade will form a pocket-like structure in which ballast is held inside. Ballast pocket usually occurs in sections with soft clays. The formation of ballast pocket is triggered because ties tend to form water-trapping ponds (e.g. with water from rains). This water gradually infiltrates through the ballast and saturates the underlying soft subgrade soils. The train dynamic loads reduce the effective stresses of the subgrade and the soils deform. Discrete cavities are then formed under each tie. Once these cavities are formed, rainfall and ground water will be conducted into deeper layers in the subgrade. In this way, a vicious circle continues, leading to the formation of large cavities when discrete ballast pockets connect to each other. Figure 2 shows a typical ballast pocket.

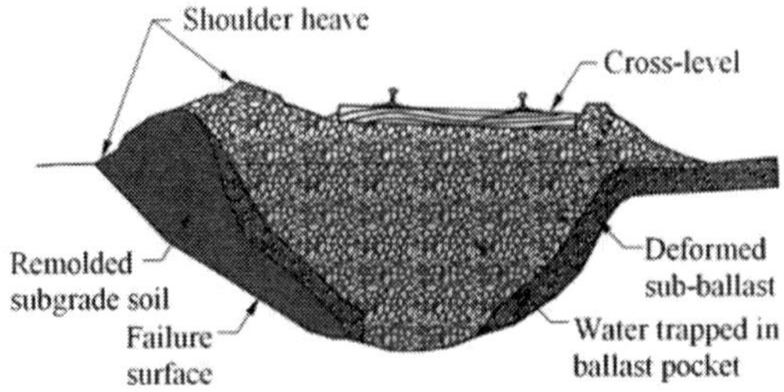

Figure 2. A sketch of railroad ballast pocket [10]

Progressive shear failure is another mechanism that can be responsible for uneven settlements in railroads in clayey subgrades (e.g.[3]). This failure mode will most likely develop in the top part of the subgrade, especially when a soft upper layer overlaid a rigid one. The trains load causes a higher stress concentration on the top layer. Overstressed clays are progressively squeezed sideways and upwards [3]. In the United Kingdom, this failure is known as 'Cess Heave'[11].

'Slurry pumping' is another phenomenon that has been observed in Texas. This phenomenon occurs in sections where the subgrade soil is hard, fine grained and poorly drained. It develops in stages, starting with water accumulation, continuing with local slurry pumping, and finishing with large-range slurry pumping. Generally the slurry pumped from the subgrade infiltrates into the ballast layer and coagulates with the ballast and ends up with a bonded crust layer. This causes uneven strength in the ballast. Variation of moisture in the slurry, once it migrates to the ballast layer, may also occur. Furthermore, infiltrated water (e.g. through small cracks and fissures in the crust layer) is subjected to huge pressure induced by the relatively closed space and dynamic train loading. This high pressure assists the pumping of slurry and another cycle starts. Ties lying on this kind of ballast are under uneven loading and will be prone to damage.

3 Test site

We found few publications reporting remedial solutions for railroads on shrink-swell soils. To advance our understanding in this area, a railroad test-site on shrink-swell soils has been selected to investigate in detail possible remedial solutions.

The test site is a single main track with an annual tonnage of 58 MGT. The issue of slow orders is very frequent. Particularly dramatic was the very dry summer of 2011, a large number of slow orders was issued. Based on this fact we hypothesize the shrinkage of local soils upon drying is the main problem.

3.1 Site location

Figure 3 presents a map with the distribution of swelling clays around the area of Texas. It can be observed that the soil units with the most high expansive potential clays, (i.e. unit in red), and soils that contain abundant content of swelling potential clays (i.e. unit in blue) dominate the soils in Texas. The location of the selected test site falls in the red zone.

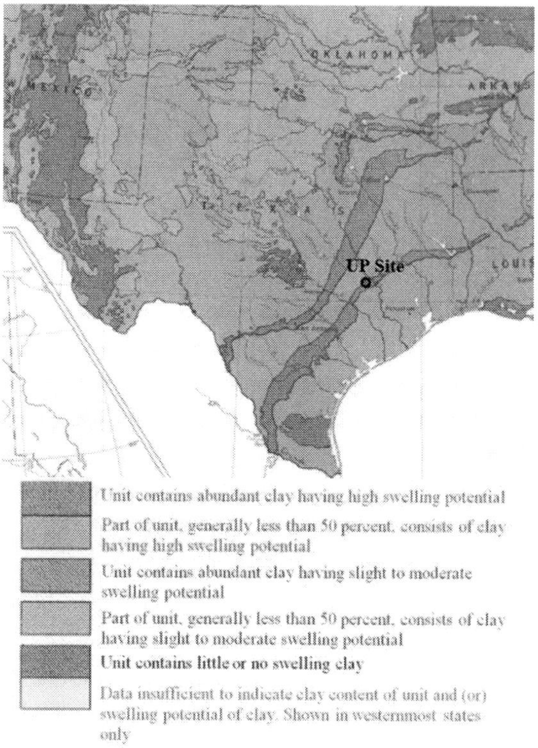

Figure 3. Test site location on the swelling clays map (map from [12])

As explained above, expansive soils experience large volume changes upon moisture changes. They also develop permanent deformations during wet and dry cycles. Therefore, it is of interest to look at the interaction between soil and atmosphere. The Thornthwaite Moisture Index (TMI,[13]) is a very good indicator of the hydraulic balance between soil and atmosphere. It relates the supply of water (precipitation) to the soil to the potential evaporation. In this context, a moist climate has a positive TMI and a dry one has a negative TMI. Figure 4 shows the TMI map for Texas. It is noted that the test site is at a TMI=0 region. This implies a perfect balance between precipitation and evaporation at this location.

Figure 5 shows the monthly average precipitation from year 1971 to 2000 at the site. Two dry seasons (DS) and two wet seasons (WS) can be noted each year. This is a very unfavourable weather condition for swell-shrink soils, because the soil is exposed to two balanced (according to the TMI) wet-dry cycles.

The field investigation focuses on two main problematic sections: a cut and a fill section. A reference section with similar properties to the previous ones, but with no major problems reported in the past was chosen as a control group. The proposed field study encompasses two main components: long term monitoring, and site investigation. More details can be found elsewhere [14].

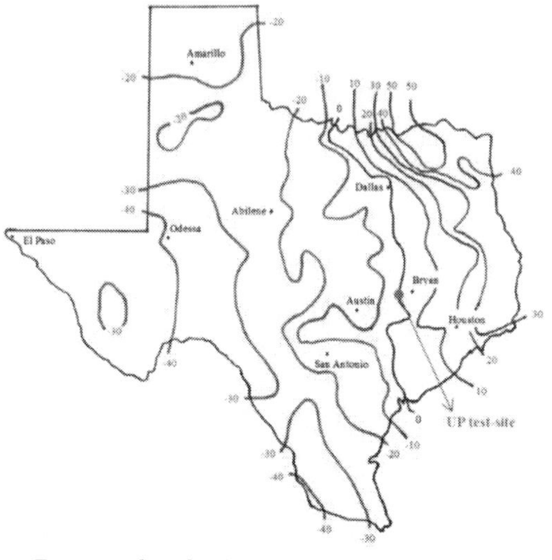

Figure 4. Thornthwaite Moisture Index for Texas [13]

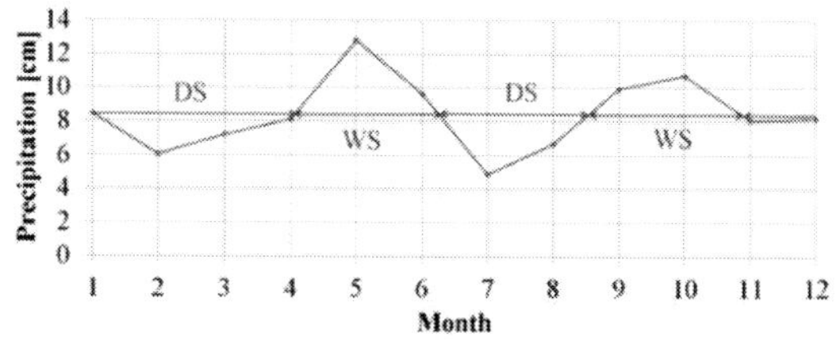

Figure 5. Monthly average precipitation (U.S. Climate Normals 1971-2000) in the area under study

3.2 Proposed monitoring

There are two main aspects that we want to monitor: i) soil movements, and ii) changes in soil moisture content at different depths. The study of soil movements is essential for diagnosis and understanding of the mechanisms behind the observed settlements. It is also crucial to tack the variation of the moisture content, because expansive soils respond with large volume changes upon moisture variations. For a continuous and unattended monitoring of the transversal movements at the cut and fill sections we have installed two track twist sensors [15] (one at the fill section, the other at the cut section) at site. The monitoring of the movements will be completed with inclinometer readings at the three selected sections. To measure the absolute embankment settlements, a benchmark resting on the bedrock was installed near the fill section. As for the monitoring of the variation of soil moisture, we have selected the EnviroSCAN multisensors to measure the moisture content profiles at the cut and fill sections in a continuous and automatic manner. This sensor is based on frequency domain reflectometry to measure soil volumetric water content. It can monitor moisture content from shallow depths (i.e. 0 to 10 [cm]) up to very deep positions (around 40 [m]), and a maximum of 16 sensors can be included in one probe. A data acquisition system is installed at the site for automatic data collection of the two twist sensors and two moisture probes every 6 hours.

 DOI: 10.1051/e3sconf/20160920007

3.3 Site investigation

The site investigation had the following multiple objectives: i) to gather soil samples; ii) to perform the in-situ tests, iii) to install the twist sensors, water content probes and data acquisition system; iv) to install inclinometer casings; v) to install the monitoring well; and vi) to construct the surveying benchmark. Figure 6 presents a detailed description for the fill section. The section profile, soil stratigraphy, instrumentation layout, and SPT blow counts are included in Figure 6. The cut section has a similar instrumentation layout. According to the USCS Classification System, the soil types in both sections are mainly high plasticity clays.

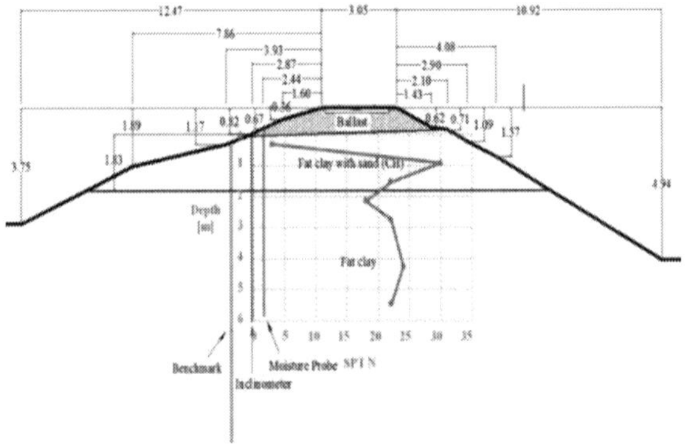

Figure 6. UP test site profile and instrumentation layout at the fill section

4. Soil water retention curve test

Water retention curve is the relationship between suction and water content (or degree of saturation) in soils. There are many techniques to measure suction in soils, such as tensiometer, Fredlund thermal conductivity sensor, Decagon WP4C, capacitive hygrometer, and filter paper. In this paper, filter paper and WP4 device are used to measure soil suction values. The filter paper test is an easy and cheap lab test which can readily be carried out in laboratories; however it requires sometime to obtain the results. The Decagon WP4 device, on the other hand, provide soil suction values in minutes. Both techniques are applied on the site soil and the results are compared.

4.1 Filter paper test

Reconstituted samples were statically compacted to the same dry density as observed in the site in order to capture the natural condition characteristics. The natural samples were firstly air dried. Then, they were grinded and sieved. All soil particles passed No. 40 sieve. Reconstituted soil samples were made according to the weight percentage retained on different size sieves.

A known volume (i.e. 155.68cm^3) split container was used for sample compaction. Based on the known volume and known dry density the necessary dry soil mass was calculated. The hydroscopic water content was considered when calculating the amount of water needed. The target degree of saturation and its corresponding water content needed in a soil sample with dry density=1.465g/cm^3 is shown in Table 1. A larger number of samples were prepared around the saturated condition to detect the air entry value. Over saturated samples were prepared in order to ensure saturated condition. Degree of saturation were recalculated after the test by measuring the final sample water content.

Table 1. Target water content in sample for different degrees of saturation

Target degree of saturation	Target water content	water mass needed in dry soil(g)	water mass needed in air dry soil powders (g)
1.1	0.335980895	76.60613	63.3218
1.05	0.320709036	73.12403	59.8397
1	0.305437177	69.64194	56.35761
0.95	0.290165318	66.15984	52.87551
0.9	0.274893459	62.67774	49.39341
0.7	0.213806	48.74936	35.46503
0.5	0.152719	34.82097	21.53664
0.3	0.091631	20.89258	20.89258

After mixing the soil powders with required amount of water, the wetted soil powders were sealed in plastic bags and stored in the moisture room for 24 hours for moisture redistribution in the sample. Then the samples were compacted to target volume (155.68cm^3) so that all samples have the same dry density.

The ASTM D5298-10 was followed to carry out the filter paper tests. The equilibrium time for this test was around 2 weeks. In order to compare the effect of the dry density on water retention, another set of filter paper test on the same soil sample with a different dry density=1.221g/cm^3 was also carried out.

4.2 WP4 test

The WP4 device is a fast instrument based on the chilled-mirror dewpoint technique to measure the water potential. The sample was sealed in a small space with the ceiling of a mirror and a means of detecting condensation on the mirror. At equilibrium, the water potential of the air in the chamber is the same as the water potential of the sample. [16]

After finishing the filter paper test, a small portion of the sample was taken and placed in the WP4 device for water potential test. The test results for both the filter paper test and WP4 device are presented in the following Section

4.3 Water retention curve results

The water retention curve results tested by filter paper method and the WP4 method are shown in Figure 7.

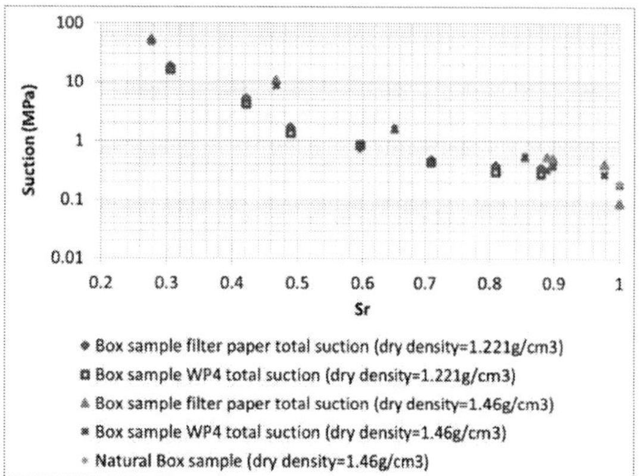

Figure 7. Water retention curves using filter paper method and WP4 method on different dry density soils.

It can be seen from Figure 7 that the filter paper method and the WP4 method gives quite similar results. The air entry value for the site soil is around 500kPa. It can also be observed that the reconstituted soil sample gives the same suction value as the natural sample. As expected, the lower dry density soil has less capacity to retain water in the soil skeleton under the same suction condition than higher dry density soils.

5 Summary and conclusion

In this paper we focused the attention on typical geotechnical problems observed in Texas railroads. Severe damages in railroads on shrink-swell soils (generally associated with uneven settlements) are induced by moisture changes of the subgrade and embankments, both composed of expansive clay minerals. In this paper we discussed in detail the different problems observed in these types of railroads. It is not obvious that remedial solutions for expansive soils typically used in building constructions and roads can be directly applied to railroads. Because of that, a test site in Texas was recently selected to study the performance of railroads on expansive soils. The main aspects of the test site, including the monitoring instrumentation and site investigation are discussed. From the site investigation, it was confirmed that the dominant soils at the test site correspond to high plasticity expansive clays. As part of the lab investigation. filter paper and WP4 test were carried out for the in-situ soils. Test results using these two different techniques show good agreement. This is a work in progress, following stages are associated with continuous monitoring of settlements and moisture changes, suction-controlled lab tests such as consolidation test, the proposal of possible remedial solutions and also the numerical modelling.

References

1. T.V. Duong, Y.J. Cui, A.M. Tang, J.C. Dupla, J. Canou, N. Calon, A. Robinet. Engineering Geology 171, 45-58 (2014)
2. Y. Luo, S.S. Peng. Proc. 13th international conference, ground control in mining, Morgantown, WV, 308–319 (1994)
3. E.T. Selig, J.M. Waters. Track Geotechnology and Substructure Management, Thomas Telford (1994)
4. D.Q. Li, E.T. Selig. J. Geotech. and Geoenviron. Eng. 124(4) 316-322 (1998)
5. R.J. Chandler, M. Pachakis, J. Mercer, J. Wrightman. Quarterly Journal of Engineering Geology and Hydrogeology 6, 405-422 (1973)
6. R.L. Zhang, Q.C. Wang, J. Zhang. Subgrade Engineering 4, 58-61 (2010)
7. S.W. Liao. Expansive soil and railway engineering, Chinese Railway Publishing Press, Beijing (1984)
8. W. Ma, G. Cheng, Q. Wu. Cold Regions Science and Technology 59, 3-11 (2009)
9. M. Flury, H. Fluhler, W.A. Jury, J. Leuenberger. Water resources research 30 (7), 1945-1954 (1994)
10. Union Pacific. Union Pacific railroad engineering strandards: Trench drains (2011)
11. M.P.N. Burrow, D. Bowness, G.S. Ghataora. Journal of Rail and Rapid Transit 221(1),1-12 (2007)
12. W.W. Olilve, A.F. Chleborad, C.W. Frahme, J. Schlocker, R.R. Schneider, R.L. Schuster. Swelling clays map of the conterminous United States. Dept. Interior US Geological Survey (1989)
13. W.K. Wary. PhD dissertation, Texas A&M University, College Station (1978)
14. M. Sanchez, D. Wang, J.L. Briaud, C. Douglas. Transportation Geotechnics 1, 257-275 (2014)
15. DGSI. EL Track Monitoring System (2008)
16. Decagon Devices, Inc. Wp4 Operator's Manual Version 2.1

AUTHOR INDEX

AUTHOR INDEX

AUTHOR INDEX

AUTHOR INDEX

AUTHOR INDEX

EDP Sciences – Web of Conferences
17, Avenue du Hoggar
Parc d'Activité de Courtabouf
BP 112
F-91944 Les Ulis Cedex A
France

ISBN 978-1-5108-3018-9

CURRAN ASSOCIATES INC.
proceedings.com